# BURGER'S
# MEDICINAL CHEMISTRY
# AND
# DRUG DISCOVERY

# BURGER'S MEDICINAL CHEMISTRY AND DRUG DISCOVERY

# BURGER'S MEDICINAL CHEMISTRY AND DRUG DISCOVERY

## Sixth Edition

## Volume 5: Chemotherapeutic Agents

Edited by

**Donald J. Abraham**
Department of Medicinal Chemistry
School of Pharmacy
Virginia Commonwealth University
Richmond, Virginia

*Burger's Medicinal Chemistry and Drug Discovery*
is available Online in full color at
www.mrw.interscience.wiley.com/bmcdd.

**WILEY-
INTERSCIENCE**

## A John Wiley and Sons, Inc., Publication

***Cover Description*** The molecule on the cover is hemoglobin with the allosteric effector RSR 13 attached. Three groups initiated structure-based drug design in the middle to late 1970s. Two of the groups, Peter Goodford's at the Burroughs Wellcome Laboratories in London and the editors' at the School of Pharmacy at the University of Pittsburgh, worked on hemoglobin while David Matthews at Agouron Pharmaceuticals worked on dihydrofolate reductase. Max Perutz and his coworkers' solution of the phase problem produced the first three-dimensional structure of a protein whose coordinates were of interest for drug design. The Editor worked with Max Perutz from 1980 to 1988, attempting to design antisickling agents. One of the active antisickling molecules, clofibric acid, would not lead to a sickle cell drug but to an allosteric effector RSR 13 designed and synthesized at Virginia Commonwealth University, which has been studied clinically for treatment of metastatic brain cancer. Max Perutz, whose work would provide the underpinnings for structure-based drug design, passed away in February 2002. His spirit and love for science that transferred to his students, postdoctoral fellows, visiting scientists, and colleagues worldwide, like Professor Burger's, is the leaven and inspiration for new discoveries.

***Library of Congress Cataloging-in-Publication Data:***

Burger's medicinal chemistry and drug discovery.—6th ed., Volume 5: chemotherapeutic agents/ Donald J. Abraham, editor

ISBN 0-471-37031-2 (v. 5: acid-free paper)

Printed in the United States of America.

10  9  8  7  6  5  4  3  2  1

*To Alfred Burger and Max Perutz for their mentorship
and life-long passion for science.*

# BURGER MEMORIAL EDITION

The Sixth Edition of Burger's Medicinal Chemistry and Drug Discovery is being designated as a Memorial Edition. Professor Alfred Burger was born in Vienna, Austria on September 6, 1905 and died on December 30, 2000. Dr. Burger received his Ph.D. from the University of Vienna in 1928 and joined the Drug Addiction Laboratory in the Department of Chemistry at the University of Virginia in 1929. During his early years at UVA, he synthesized fragments of the morphine molecule in an attempt to find the analgesic pharmacophore. He joined the UVA chemistry faculty in 1938 and served the department until his retirement in 1970. The chemistry department at UVA became the major academic training ground for medicinal chemists because of Professor Burger.

Dr. Burger's research focused on analgesics, antidepressants, and chemotherapeutic agents. He is one of the few academicians to have a drug, designed and synthesized in his laboratories, brought to market [Parnate, which is the brand name for tranylcypromine, a monoamine oxidase (MAO) inhibitor]. Dr. Burger was a visiting Professor at the University of Hawaii and lectured throughout the world. He founded the *Journal of Medicinal Chemistry, Medicinal Chemistry Research,* and published the first major reference work *"Medicinal Chemistry"* in two volumes in 1951. His last published work, a book, was written at age 90 (*Understanding Medications: What the Label Doesn't Tell You,* June 1995). Dr. Burger received the Louis Pasteur Medal of the Pasteur Institute and the American Chemical Society Smissman Award. Dr. Burger played the violin and loved classical music. He was married for 65 years to Frances Page Burger, a genteel Virginia lady who always had a smile and an open house for the Professor's graduate students and postdoctoral fellows.

# PREFACE

The Editors, Editorial Board Members, and John Wiley and Sons have worked for three and a half years to update the fifth edition of Burger's Medicinal Chemistry and Drug Discovery. The sixth edition has several new and unique features. For the first time, there will be an online version of this major reference work. The online version will permit updating and easy access. For the first time, all volumes are structured entirely according to content and published simultaneously. Our intention was to provide a spectrum of fields that would provide new or experienced medicinal chemists, biologists, pharmacologists and molecular biologists entry to their subjects of interest as well as provide a current and global perspective of drug design, and drug development.

Our hope was to make this edition of Burger the most comprehensive and useful published to date. To accomplish this goal, we expanded the content from 69 chapters (5 volumes) by approximately 50% (to over 100 chapters in 6 volumes). We are greatly in debt to the authors and editorial board members participating in this revision of the major reference work in our field. Several new subject areas have emerged since the fifth edition appeared. Proteomics, genomics, bioinformatics, combinatorial chemistry, high-throughput screening, blood substitutes, allosteric effectors as potential drugs, COX inhibitors, the statins, and high-throughput pharmacology are only a few. In addition to the new areas, we have filled in gaps in the fifth edition by including topics that were not covered. In the sixth edition, we devote an entire subsection of Volume 4 to cancer research; we have also reviewed the major published Medicinal Chemistry and Pharmacology texts to ensure that we did not omit any major therapeutic classes of drugs. An editorial board was constituted for the first time to also review and suggest topics for inclusion. Their help was greatly appreciated. The newest innovation in this series will be the publication of an academic, "textbook-like" version titled, "Burger's Fundamentals of Medicinal Chemistry." The academic text is to be published about a year after this reference work appears. It will also appear with soft cover. Appropriate and key information will be extracted from the major reference.

There are numerous colleagues, friends, and associates to thank for their assistance. First and foremost is Assistant Editor Dr. John Andrako, Professor emeritus, Virginia Commonwealth University, School of Pharmacy. John and I met almost every Tuesday for over three years to map out and execute the game plan for the sixth edition. His contribution to the sixth edition cannot be understated. Ms. Susanne Steitz, Editorial Program Coordinator at Wiley, tirelessly and meticulously kept us on schedule. Her contribution was also key in helping encourage authors to return manuscripts and revisions so we could publish the entire set at once. I would also like to especially thank colleagues who attended the QSAR Gordon Conference in 1999 for very helpful suggestions, especially Roy Vaz, John Mason, Yvonne Martin, John Block, and Hugo

Kubinyi. The editors are greatly indebted to Professor Peter Ruenitz for preparing a template chapter as a guide for all authors. My secretary, Michelle Craighead, deserves special thanks for helping contact authors and reading the several thousand e-mails generated during the project. I also thank the computer center at Virginia Commonwealth University for suspending rules on storage and e-mail so that we might safely store all the versions of the author's manuscripts where they could be backed up daily. Last and not least, I want to thank each and every author, some of whom tackled two chapters. Their contributions have provided our field with a sound foundation of information to build for the future. We thank the many reviewers of manuscripts whose critiques have greatly enhanced the presentation and content for the sixth edition. Special thanks to Professors Richard Glennon, William Soine, Richard Westkaemper, Umesh Desai, Glen Kellogg, Brad Windle, Lemont Kier, Malgorzata Dukat, Martin Safo, Jason Rife, Kevin Reynolds, and John Andrako in our Department of Medicinal Chemistry, School of Pharmacy, Virginia Commonwealth University for suggestions and special assistance in reviewing manuscripts and text. Graduate student Derek Cashman took able charge of our web site, *http://www.burgersmedchem.com*, another first for this reference work. I would especially like to thank my dean, Victor Yanchick, and Virginia Commonwealth University for their support and encouragement. Finally, I thank my wife Nancy who understood the magnitude of this project and provided insight on how to set up our home office as well as provide John Andrako and me lunchtime menus where we often dreamed of getting chapters completed in all areas we selected. To everyone involved, many, many thanks.

DONALD J. ABRAHAM
Midlothian, Virginia

**Dr. Alfred Burger**

Photograph of Professor Burger followed by his comments to the American Chemical Society 26th Medicinal Chemistry Symposium on June 14, 1998. This was his last public appearance at a meeting of medicinal chemists. As general chair of the 1998 ACS Medicinal Chemistry Symposium, the editor invited Professor Burger to open the meeting. He was concerned that the young chemists would not know who he was and he might have an attack due to his battle with Parkinson's disease. These fears never were realized and his comments to the more than five hundred attendees drew a sustained standing ovation. The Professor was 93, and it was Mrs. Burger's 91st birthday.

# Opening Remarks

# ACS 26<sup>th</sup> Medicinal Chemistry Symposium

June 14, 1998
Alfred Burger
University of Virginia

It has been 46 years since the third Medicinal Chemistry Symposium met at the University of Virginia in Charlottesville in 1952. Today, the Virginia Commonwealth University welcomes you and joins all of you in looking forward to an exciting program.

So many aspects of medicinal chemistry have changed in that half century that most of the new data to be presented this week would have been unexpected and unbelievable had they been mentioned in 1952. The upsurge in biochemical understandings of drug transport and drug action has made rational drug design a reality in many therapeutic areas and has made medicinal chemistry an independent science. We have our own journal, the best in the world, whose articles comprise all the innovations of medicinal researches. And if you look at the announcements of job opportunities in the pharmaceutical industry as they appear in *Chemical & Engineering News,* you will find in every issue more openings in medicinal chemistry than in other fields of chemistry. Thus, we can feel the excitement of being part of this medicinal tidal wave, which has also been fed by the expansion of the needed research training provided by increasing numbers of universities.

The ultimate beneficiary of scientific advances in discovering new and better therapeutic agents and understanding their modes of action is the patient. Physicians now can safely look forward to new methods of treatment of hitherto untreatable conditions. To the medicinal scientist all this has increased the pride of belonging to a profession which can offer predictable intellectual rewards. Our symposium will be an integral part of these developments.

# CONTENTS

1 MOLECULAR BIOLOGY OF
CANCER, 1

Jesse D. Martinez
Michele Taylor Parker
Kimberly E. Fultz
Natalia A. Ignatenko
Eugene W. Gerner
*Departments of Radiation Oncology/*
*Cancer Biology Section*
*Molecular and Cellular Biology*
*Biochemistry and Molecular*
*Biophysics*
*Cancer Biology Graduate Program*
*The University of Arizona*
*Tuscon, Arizona*

2 SYNTHETIC DNA-TARGETED
CHEMOTHERAPEUTIC AGENTS
AND RELATED TUMOR-
ACTIVATED PRODRUGS, 51

William A. Denny
*Auckland Cancer Society Research*
*Centre*
*Faculty of Medical and Health*
*Sciences*
*The University of Auckland*
*Auckland, New Zealand*

3 ANTITUMOR NATURAL
PRODUCTS, 107

Lester A. Mitscher
Apurba Dutta
*Department of Medicinal Chemistry*
*Kansas University*
*Lawrence, Kansas*

4 RADIOSENSITIZERS AND
RADIOPROTECTIVE AGENTS, 151

Edward A. Bump
*DRAXIMAGE Inc.*
*Kirkland, Quebec, Canada*

Stephen J. Hoffman
*Allos Therapeutics, Inc.*
*Westminster, Colorado*

William O. Foye
*Massachusetts College of Pharmacy*
*and Health Sciences*
*Boston, Massachusetts*

5 SYNTHETIC ANTIANGIOGENIC
AGENTS, 215

Orest W. Blaschuk
*Mc Gill University, Royal Victoria*
*Hospital*
*Montreal, Quebec, Canada*

J. Matthew Symonds
*Adherex Technologies Inc.*
*Ottawa, Ontario, Canada*

6 FUTURE STRATEGIES IN
IMMUNOTHERAPY, 223

Douglas F. Lake
Sara O. Dionne
*University of Arizona Cancer Center*
*Tucson, Arizona*

7 SELECTIVE TOXICITY, 249

John H. Block
*College of Pharmacy*
*Oregon State University*
*Corvallis, Oregon*

8 DRUG RESISTANCE IN CANCER
CHEMOTHERAPY, 281

Amelia M. Wall
*Division of Clinical Pharmacology*
*and Therapeutics*
*Children's Hospital of Philadelphia*
*Cancer Pharmacology Institute*
*University of Pennsylvania*
*Philadelphia, Pennsylvania*

9 ANTIVIRAL AGENTS, DNA, 293

Tim Middleton
Todd Rockway
*Abbott Laboratories*
*Abbott Park, Illinois*

10 ANTIVIRAL AGENTS, RNA
VIRUSES (OTHER THAN HIV),
AND ORTHOPOXVIRUSES, 359

Christopher Tseng
Catherine Laughlin
*NIAID*
*National Institutes of Health*
*Bethesda, Maryland*

11 RATIONALE OF DESIGN OF
ANTI-HIV DRUGS, 457

Ahmed S. Mehanna
*Massachusetts College of Pharmacy*
*and Health Sciences*
*Department of Pharmaceutical*
*Sciences*
*School of Pharmacy*
*Boston, Massachusetts*

12 ORGAN TRANSPLANT DRUGS,
485

Bijoy Kundu
*Medicinal Chemistry Division*
*Central Drug Research Institute*
*Lucknow, India*

13 SYNTHETIC ANTIBACTERIAL
AGENTS, 537

Nitya Anand
*B-62, Nirala Nagar,*
*Lucknow, India*

William A. Remers
*University of Arizona*
*College of Pharmacy*
*Tucson, Arizona*

14 β-LACTAM ANTIBIOTICS, 607

Daniele Andreotti
Stefano Biondi
Enza Di Modugno
*GlaxoSmithKline Research Center*
*Verona, Italy*

15 TETRACYCLINE,
AMINOGLYCOSIDE, MACROLIDE,
AND MISCELLANEOUS
ANTIBIOTICS, 737

Gerard D. Wright
*Antimicrobial Research Centre*
*Department of Biochemistry*
*McMaster University*
*Hamilton, Ontario, Canada*

Daniel T. W. Chu
*Chiron Corporation*
*Emeryville, California*

16  ANTIMYCOBACTERIAL
    AGENTS, 807

    Piero Sensi
    *University of Milan*
    *Milan, Italy*

    Giuliana Gialdroni Grassi
    *University of Pavia*
    *Pavia, Italy*

17  ANTIFUNGAL AGENTS, 881

    William J. Watkins
    Thomas E. Renau
    *Essential Therapeutics*
    *Mountain View, California*

18  ANTIMALARIAL AGENTS, 919

    Dee Ann Casteel
    *Department of Chemistry*
    *Bucknell University*
    *Lewisburg, Pennsylvania*

19  ANTIPROTOZOAL AGENTS,
    1033

    David S. Fries
    *TJ Long School of Pharmacy*
    *University of the Pacific*
    *Stockton, California*

    Alan H. Fairlamb
    *Division of Biological Chemistry*
    *and Molecular Microbiology*
    *The Wellcome Trust Biocentre*
    *University of Dundee*
    *Dundee, United Kingdom*

20  ANTHELMINTICS, 1089

    Robert K. Griffith
    *West Virginia University*
    *School of Pharmacy*
    *Morgantown, West Virginia*

    INDEX, 1097

# BURGER'S
# MEDICINAL CHEMISTRY
# AND
# DRUG DISCOVERY

# CHAPTER ONE

# Molecular Biology of Cancer

Jesse D. Martinez
Michele Taylor Parker
Kimberly E. Fultz
Natalia A. Ignatenko
Eugene W. Gerner
Departments of Radiation Oncology/Cancer Biology Section
Molecular and Cellular Biology
Biochemistry and Molecular Biophysics
Cancer Biology Graduate Program
The University of Arizona
Tuscon, Arizona

## Contents

1 Introduction, 2
2 Tumorigenesis, 2
    2.1 Normal-Precancer-Cancer Sequence, 2
    2.2 Carcinogenesis, 3
    2.3 Genetic Variability and Other Modifiers of
        Tumorigenesis, 5
        2.3.1 Genetic Variability Affecting Cancer, 5
        2.3.2 Genetic Variability in
            c-myc–Dependent Expression of
            Ornithine Decarboxylase, 7
    2.4 Epigenetic Changes, 7
3 Molecular Basis of Cancer Phenotypes, 10
    3.1 Immortality, 10
    3.2 Decreased Dependence on Growth Factors to
        Support Proliferation, 11
    3.3 Loss of Anchorage-Dependent Growth and
        Altered Cell Adhesion, 12
    3.4 Cell Cycle and Loss of Cell Cycle Control, 14
    3.5 Apoptosis and Reduced Sensitivity to
        Apoptosis, 16
    3.6 Increased Genetic Instability, 19
    3.7 Angiogenesis, 20
4 Cancer-Related Genes, 21
    4.1 Oncogenes, 21
        4.1.1 Growth Factors and Growth Factor
            Receptors, 21
        4.1.2 G Proteins, 23
        4.1.3 Serine/Threonine Kinases, 24
        4.1.4 Nonreceptor Tyrosine Kinases, 24
        4.1.5 Transcription Factors as Oncogenes,
            25
        4.1.6 Cytoplasmic Proteins, 26

*Burger's Medicinal Chemistry and Drug Discovery*
Sixth Edition, Volume 5: Chemotherapeutic Agents
Edited by Donald J. Abraham
ISBN 0-471-37031-2    © 2003 John Wiley & Sons, Inc.

    4.2 Tumor Suppressor Genes, 26
        4.2.1 Retinoblastoma, 27
        4.2.2 p53, 27
        4.2.3 Adenomatous Polyposis Coli, 29
        4.2.4 Phosphatase and Tensin Homologue,
            30
        4.2.5 Transforming Growth Factor-$\beta$, 30
        4.2.6 Heritable Cancer Syndromes, 32
5 Interventions, 32
    5.1 Prevention Strategies, 32
    5.2 Targets, 33
        5.2.1 Biochemical Targets, 33
        5.2.2 Cyclooxygenase-2 and Cancer, 33
        5.2.3 Other Targets, 35
    5.3 Therapy, 35
        5.3.1 Importance of Studying Gene
            Expression, 35
        5.3.2 cDNA Microarray Technology, 35
        5.3.3 Discoveries from cDNA Microarray
            Data, 37

    5.3.4 Limitations of Microarray
        Technologies, 37
    5.4 Modifying Cell Adhesion, 37
        5.4.1 MMP Inhibitors, 37
        5.4.2 Anticoagulants, 38
        5.4.3 Inhibitors of Angiogenesis, 38
    5.5 Prospects for Gene Therapy of Cancer, 39
        5.5.1 Gene Delivery Systems, 39
            5.5.1.1 Viral Vectors, 40
            5.5.1.2 Non-Viral Gene Delivery
                Systems, 42
    5.6 Gene Therapy Approaches, 43
        5.6.1 Immunomodulation, 43
        5.6.2 Suicidal Gene Approach, 44
        5.6.3 Targeting Loss of Tumor Suppressor
            Function and Oncogene
            Overexpression, 44
        5.6.4 Angiogenesis Control, 45
        5.6.5 Matrix Metalloproteinase, 45
6 Acknowledgments, 46

# 1   INTRODUCTION

Cancer is a major human health problem worldwide and is the second leading cause of death in the United States (1). Over the past 30 years, significant progress has been achieved in understanding the molecular basis of cancer. The accumulation of this basic knowledge has established that cancer is a variety of distinct diseases and that defective genes cause these diseases. Further, gene defects are diverse in nature and can involve either loss or gain of gene functions. A number of inherited syndromes associated with increased risk of cancer have been identified.

This chapter will review our current understanding of the mechanisms of cancer development, or carcinogenesis, and the genetic basis of cancer. The roles of gene defects in both germline and somatic cells will be discussed as they relate to genetic and sporadic forms of cancer. Specific examples of oncogenes, or cancer-causing genes, and tumor suppressor genes will be presented, along with descriptions of the relevant pathways that signal normal and cancer phenotypes.

While cancer is clearly associated with an increase in cell number, alterations in mechanisms regulating new cell birth, or cell proliferation, are only one facet of the mechanisms of cancer. Decreased rates of cell death, or ap-

optosis, are now known to contribute to certain types of cancer. Cancer is distinctive from other tumor-forming processes because of its ability to invade surrounding tissues. This chapter will address mechanisms regulating the important cancer phenotypes of altered cell proliferation, apoptosis, and invasiveness.

Recently, it has become possible to exploit this basic information to develop mechanism-based strategies for cancer prevention and treatment. The success of both public and private efforts to sequence genomes, including human and other organisms, has contributed to this effort. Several examples of mechanism-based anti-cancer strategies will be discussed. Finally, potential strategies for gene therapy of cancer will also be addressed.

# 2   TUMORIGENESIS

## 2.1   Normal-Precancer-Cancer Sequence

Insight into tumor development first came from epidemiological studies that examined the relationship between age and cancer incidence that showed that cancer incidence increases with roughly the fifth power of elapsed age (2). Hence, it was predicted that at least five rate-limiting steps must be overcome before a clinically observable tumor could arise. It is now known that these rate-limiting steps are genetic mutations that dysregulate the activities of genes that control cell growth, regulate sensitivity to programmed cell death, and maintain genetic stability. Hence, tumorigenesis is a multistep process.

Although the processes that occur during tumorigenesis are only incompletely understood, it is clear that the successive accumulation of mutations in key genes is the force that drives tumorigenesis. Each successive mutation is thought to provide the developing tumor cell with important growth advantages that allow cell clones to outgrow their more normal neighboring cells. Hence, tumor development can be thought of as Darwinian evolution on a microscopic scale with each successive generation of tumor cell more adapted to overcoming the social rules that regulate the growth of normal cells. This is called clonal evolution (3).

Given that tumorigenesis is the result of mutations in a select set of genes, much effort by cancer biologists has been focused on identifying these genes and understanding how they function to alter cell growth. Early efforts in this area were lead by virologists studying retrovirus-induced tumors in animal models. These studies led to cloning of the first oncogenes and the realization that oncogenes, indeed all cancer-related genes, are aberrant forms of genes that have important functions in regulating normal cell growth (4). In subsequent studies, these newly identified oncogenes were introduced into normal cells in an effort to reproduce tumorigenesis *in vitro*. Importantly, it was found that no single oncogene could confer all of the physiological traits of a transformed cell to a normal cell. Rather this required that at least two oncogenes acting cooperatively to give rise to cells with the fully transformed phenotype (5). This observation provides important insights into tumorigenesis. First, the multistep nature of tumorigenesis can be rationalized as mutations in different genes with each event providing a selective growth advantage. Second, oncogene cooperativity is likely to be cause by the requirement for dysregulation of cell growth at multiple levels.

Fearon and Vogelstein (6) have proposed a linear progression model (Fig. 1.1) to describe tumorigenesis using colon carcinogenesis in

humans as the paradigm. They suggest that malignant colorectal tumors (carcinomas) evolve from preexisting benign tumors (adenomas) in a stepwise fashion with benign, less aggressive lesions giving rise to more lethal neoplasms. In their model, both genetic [e.g., adenomatous polyposis coli (APC) mutations] and epigenetic changes (e.g., DNA methylation affecting gene expression) accumulate over time, and it is the progressive accumulation of these changes that occur in a preferred, but not invariable, order that are associated with the evolution of colonic neoplasms. Other important features of this model are that at least four to five mutations are required for the formation of a malignant tumor, in agreement with the epidemiological data, with fewer changes giving rise to intermediate benign lesions, that tumors arise through the mutational activation of oncogenes and inactivation of tumor suppressor genes, and that it is the sum total of the effect of these mutations on tumor cell physiology that is important rather than the order in which they occur.

An important implication of the multistep model of tumorigenesis is that lethal neoplasms are preceded by less aggressive intermediate steps with predictable genetic alterations. This suggests that if the genetic defects which occur early in the process can be identified, a strategy that interferes with their function might prevent development of more advanced tumors. Moreover, preventive screening methods that can detect cells with the early genetic mutations may help to identify these lesions in their earliest and most curable stages. Consequently, identification of the genes that are mutated in cancers and elucidation of their mechanism of action is important not only to explain the characteristic phenotypes exhibited by tumor cells, but also to provide targets for development of therapeutic agents.

## 2.2 Carcinogenesis

Carcinogenesis is the process that leads to genetic mutations induced by physical or chemical agents. Conceptually, this process can be divided into three distinct stages: initiation, promotion, and progression (7). Initiation involves an irreversible genetic change, usually a mutation in a single gene. Promotion is gen-

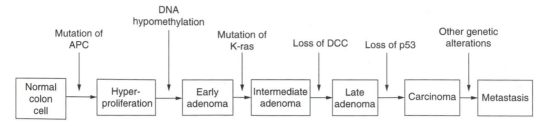

**Figure 1.1.** Adenoma-carcinoma sequence. Fearon and Vogelstein (6) proposed this classic model for the multistage progression of colorectal cancer. A mutation in the APC tumor suppressor gene is generally considered to be the initiation event. This is followed by the sequential accumulation of other epigenetic and genetic changes that eventually result in the progression from a normal cell to a metastatic tumor.

erally associated with increased proliferation of initiated cells, which increases the population of initiated cells. Progression is the accumulation of more genetic mutations that lead to the acquisition of the malignant or invasive phenotype.

In the best-characterized model of chemical carcinogenesis, the mouse skin model, initiation is an irreversible event that occurs when a genotoxic chemical, or its reactive metabolite, causes a DNA mutation in a critical growth controlling gene such as Ha-ras (8). Outwardly, initiated cells seem normal. However, they remain susceptible to promotion and further neoplastic development indefinitely. DNA mutations that occur in initiated cells can confer growth advantages, which allow them to evolve and/or grow faster bypassing normal cellular growth controls. The different types of mutations that can occur include point mutations, deletions, insertions, chromosomal translocations, and amplifications. Three important steps involved in initiation are carcinogen metabolism, DNA repair, and cell proliferation. Many chemical agents must be metabolically activated before they become carcinogenic. Most carcinogens, or their active metabolites, are strong electrophiles and bind to DNA to form adducts that must be removed by DNA repair mechanisms (9). Hence, DNA repair is essential to reverse adduct formation and to prevent DNA damage. Failure to repair chemical adducts, followed by cell proliferation, results in permanent alterations or mutation(s) in the genome that can lead to oncogene activation or inactivation of tumor suppressor genes.

Promotion is a reversible process in which chemical agents stimulate proliferation of initiated cells. Typically, promoting agents are nongenotoxic, that is they are unable to form DNA adducts or cause DNA damage but are able to stimulate cell proliferation. Hence, exposure to tumor promoting agents results in rapid growth of the initiated cells and the eventual formation of non-invasive tumors. In the mouse skin tumorigenesis model, application of a single dose of an initiating agent does not usually result in tumor formation. However, when the initiation step is followed by repeated applications of a tumor promoting agent, such as 12-*O*-tetradecanoyl-phorbol-13-acetate (TPA), numerous skin tumors arise and eventually result in invasive carcinomas. Consequently, tumor promoters are thought to function by fostering clonal selection of cells with a more malignant phenotype. Importantly, tumor formation is dependent on repeated exposure to the tumor promoter. Halting application of the tumor promoter prevents or reduces the frequency with which tumors form. The sequence of exposure is important because tumors do not develop in the absence of an initiating agent even if the tumor promoting agent is applied repeatedly. Therefore, the genetic mutation caused by the initiating agent is essential for further neoplastic development under the influence of the promoting agent.

Progression refers to the process of acquiring additional mutations that lead to malignancy and metastasis. Many initiating agents can also lead to tumor progression, strong support for the notion that further mutations are

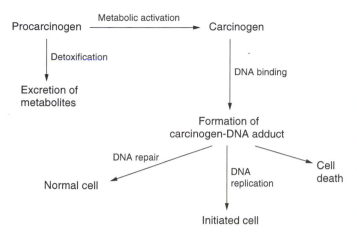

Procarcinogen — Metabolic activation → Carcinogen

Detoxification ↓

Excretion of metabolites

DNA binding ↓

Formation of carcinogen-DNA adduct

DNA repair ↙        ↓ DNA replication        ↘ Cell death

Normal cell

Initiated cell

**Figure 1.2.** Possible outcomes of carcinogen metabolic activation. Once a carcinogen is metabolically activated it can bind to DNA and form carcinogen-DNA adducts. These adducts will ultimately lead to mutations if they are not repaired. If DNA repair does not occur, the cell will either undergo apoptosis or the DNA will be replicated, resulting in an initiated cell.

needed for cells to acquire the phenotypic characteristics of malignant tumor cells. Some of these agents include benzo(a)pyrene, $\beta$-napthylamine, 2-acetylaminofluorene, aflatoxin $B_1$, dimethylnitrosamine, 2-amino-3-methylimidazo(4,5-$f$)quinoline (IQ), benzidine, vinyl chloride, and 4-(methylnitrosamino)-1-(3-pyridyl)-1-butanone (NNK) (10). These chemicals are converted into positively charged metabolites that bind to negatively charged groups on molecules like proteins and nucleic acids. This results in the formation of DNA adducts which, if not repaired, lead to mutations (9) (Fig. 1.2). The result of these mutations enables the tumors to grow, invade surrounding tissue, and metastasize.

Damage to DNA and the genetic mutations that can result from them are a central theme in carcinogenesis. Hence, the environmental factors that cause DNA damage are of great interest. Environmental agents that can cause DNA damage include ionizing radiation, ultraviolet (UV) light, and chemical agents (11). Some of the DNA lesions that can result include single-strand breaks, double-strand breaks, base alterations, cross-links, insertion of incorrect bases, and addition/deletion of DNA sequences. Cells have evolved several different repair mechanisms that can reverse the lesions caused by these agents, which has been extensively reviewed elsewhere (12).

The metabolic processing of environmental carcinogens is also of key importance because this can determine the extent and duration to which an organism is exposed to a carcinogen. Phase I and phase II metabolizing enzymes play important roles in the metabolic activation and detoxification of carcinogenic agents. The phase I enzymes include monooxygenases, dehydrogenases, esterases, reductases, and oxidases. These enzymes introduce functional groups on the substrate. The most important superfamily of the phase I enzymes are the cytochrome P450 monooxygenases, which metabolize polyaromatic hydrocarbons, aromatic amines, heterocyclic amines, and nitrosamines. Phase II metabolizing enzymes are important for the detoxification and excretion of carcinogens. Some examples include epoxide hydrase, glutathione-$S$-transferase, and uridine 5'-diphosphate (UDP) glucuronide transferase. There are also some direct acting carcinogens that do not require metabolic activation. These include nitrogen mustard, dimethylcarbamyl chloride, and $\beta$-propiolactone.

## 2.3 Genetic Variability and Other Modifiers of Tumorigenesis

**2.3.1 Genetic Variability Affecting Cancer.** Different types of cancers, as well as their severity, seem to correlate with the type of mutation acquired by a specific gene. Mutation "hot spots" are regions of genes that are frequently mutated compared with other regions within that gene. For example, observations that the majority of colon adenomas are associated with alterations in the adenomatous polyposis coli (APC) have been based on immunohistochemical analysis of $\beta$-catenin localization and formation of less than full

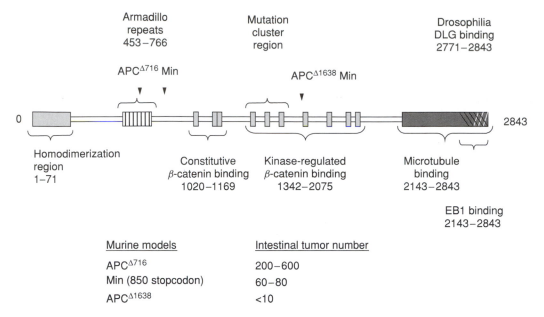

**Figure 1.3.** Diagram of APC protein regions, relating risk of intestinal carcinogenesis to length of APC peptide translated. APC contains 2833 amino acids. Mutation hot spot regions are found in areas between amino acids 1500–2000. Three genetically altered mouse models of APC-dependent intestinal carcinogenesis have been developed. Min mice have a stop codon mutation in codon 850 of the murine APC homolog. Two transgenic mice, APC$^{\Delta716}$ and APC$^{\Delta1635}$, also have been developed. Intestinal tumor number in these models is inversely related to size of the APC peptide translated.

length APC protein production after *in vitro* translation of colonic mucosal tissue RNA. These studies have not documented specific gene mutations in APC. This is important, because it is known from animal studies that the location of APC mutations can have a dramatic effect on the degree of intestinal carcinogenesis. Thus, it is possible that colon adenoma size, and subsequent risk of colon cancer could be dictated by location of specific mutations in APC (Fig. 1.3).

As suggested by the model depicted in Fig. 1.3, high risk might be associated with mutations causing stop codons in the amino terminal end of the protein. Low risk might be associated with mutations resulting in peptides of greater length. Current research is testing the hypothesis that specific genetic alterations in APC alone may be sufficient as a prognostic factor for risk of adenoma recurrence and subsequently, colon cancer development.

One type of genetic alteration that is gaining increasing attention is the single nucleotide polymorphism (SNP). This polymorphism results from a single base mutation that leads to the substitution of one base for another. SNPs occur quite frequently (about every 0.3–1 kb within the genome) and can be identified by several different techniques. A common method for the analysis of SNPs is based on the knowledge that single-base changes have the capability of destroying or creating a restriction enzyme site within a specific region of DNA. Digestion of a piece of DNA, containing the site in question, with the appropriate enzyme can distinguish between variants based on the resulting fragment sizes. This type of analysis is commonly referred to as restriction fragment length polymorphism (RFLP).

The importance of analyzing SNPs rests on the premise that individuals with a nucleotide at a specific position may display a normal phenotype, whereas individuals with a different nucleotide at this same position may exhibit increased predisposition for a certain disease or phenotype. Therefore, many studies are being conducted to determine the fre-

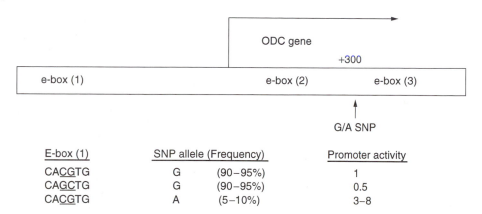

| E-box (1) | SNP allele (Frequency) | | Promoter activity |
|---|---|---|---|
| CA<u>C</u>GTG | G | (90–95%) | 1 |
| CA<u>GC</u>TG | G | (90–95%) | 0.5 |
| CA<u>C</u>GTG | A | (5–10%) | 3–8 |

**Figure 1.4.** Influence of specific genetic changes on ODC promoter activity. These data were derived from transient transfection experiments in human colon tumor–derived HT29 cells. The arrow in this figure 1.4 shows the SNP. The SNP occurs between two E-boxes that are located 3′ of the transcription start site. The effects of this genetic change are taken from Guo et al. (56). It is important to point out that the constructs used to assess the promoter activity of the polymorphic region containing the SNP and E-boxes 2 and 3 contained some of the 5′ promoter region, but not E-box 1 (56). The constructs used to assess the role of E-box 1 in HT-29 contained the major, c-myc unresponsive allele between E-boxes 2 and 3.

quency of specific SNPs in the general population and to use these findings to explain phenotypic variation.

For example, a recent study found an association between a polymorphism leading to an amino acid substitution (aspartate to valine) in codon 1822 of the APC gene and a reduced risk for cancer in people eating a low-fat diet (13). The variant valine had an allele frequency of 22.8% in a primarily Caucasian control population. This non-truncating mutation has not yet been shown to have functional significance. If functional, such a polymorphism could cooperate with single allele truncating mutations that occur with high frequency in sporadic colon adenomas (14), to increase colon cancer risk. This polymorphism is especially interesting, because dietary factors, specifically fat consumption, may contribute to risk in only specific genetic subsets.

### 2.3.2 Genetic Variability in c-myc–Dependent Expression of Ornithine Decarboxylase.

The proliferation-associated polyamines are essential for cell growth but may contribute to carcinogenesis when in excess. Various studies have shown that inhibition of polyamine synthesis impedes carcinogenesis. Ornithine decarboxylase (ODC), the first enzyme in poly-

amine synthesis, may play a key role in tumor development. Therefore, elucidation of the mechanisms by which ODC is regulated is essential. The literature indicates that ODC is a downstream mediator of APC and suggests that ODC may be an APC modifier gene. Thus, polymorphisms in the ODC promoter affecting c-myc–dependent ODC transcription could be a mechanism of genetic variability of APC-dependent carcinogenesis.

O'Brien and colleagues (15) have measured the incidence in several human subgroups of a SNP in a region of the ODC promoter, 3′ of the transcription start site, that is flanked by two E-boxes (CACGTG) (Fig. 1.4). The E-box is a DNA sequence where specific transcription factors bind. The two resulting alleles are identified by a polymorphic *Pst*I RFLP. The minor allele (A at position +317) is homozygous in 6–10% of individuals, whereas the major allele (G at position +317) is homozygous or heterozygous in 90–94% of these groups. They have also measured functionality of the polymorphisms. When ODC promoter-reporter constructs are expressed in rodent cells, the minor allele confers 3–8 times the promoter activity compared with the major allele. Further, expression of the minor allele is enhanced by c-myc expression to a greater extent than the major allele.

## 2.4  Epigenetic Changes

Gene function can be disrupted either through genetic alterations, which directly mutate or delete genes, or epigenetic alterations, which alter the state of gene expression. Epigenetic mechanisms regulating gene expression include signal transduction pathways, DNA methylation, and chromatin remodeling. Methylation of DNA is a biochemical addition of a methyl group at position 5 of the pyrimidine ring of cytosine in the sequence CG. This modification occurs in two ways: (1) from a preexisting pattern on the coding strand or (2) by *de novo* addition of a methyl group to fully unmethylated DNA. Cleavage of DNA with the restriction endonuclease *Hpa*II, which cannot cut the central C in the sequence CCGG if it is methylated, allows detection of methylated sites in DNA. Small regions of DNA with methylated cytosine, called "CpG islands," have been found in the 5′-promoter region of about one-half of all human genes (including most housekeeping genes).

There are three DNA methyltransferases (Dnmt), Dnmt1, Dnmt3a, and Dnmt3b, that have been identified in mammalian cells (16). The most abundant and ubiquitous enzyme, Dnmt1, shows high affinity for hemimethylated DNA, suggesting a role of Dnmt1 in the inheritance of preexisting patterns of DNA methylation after each round of DNA replication. The other two enzymes, Dnmt3a and Dnmt3b, are tissue specific and have been shown to be involved in *de novo* methylation. *De novo* CpG island methylation, however, is not a feature of proliferating cells, and can be considered a pathologic event in neoplasia.

Over the years, a number of different methyl-CpG binding proteins, such as methyl-CpG-binding domain-containing proteins (MBD1-4) were identified (17) that compete with transcription factors and prevent them from binding to promoter sequences. These methyl-CpG binding factors can also recruit histone deacetylases (HDACs), resulting in condensation of local chromatin structure (Fig. 1.5). This makes the methylated DNA less accessible to transcription factors and results in gene silencing.

Gene expression is inhibited by DNA methylation. DNA methylation patterns dramatically change at different stages of cell development and differentiation and correlate with changes in gene expression (18). Demethylation releases gene expression in the first days of embryogenesis. Later, *de novo* methylation establishes adult patterns of gene methylation. In differentiated cells, methylation status is retained by the activity of the Dnmt1 enzyme. In normal tissues, DNA methylation is associated with gene silencing, chromosome X inactivation (19), and imprinting (20). Because the most normal methylation takes place within highly repeated transposable elements, it has been proposed that such methylation plays a role in genome defense by suppressing potentially harmful effects of expression at these sites.

Neoplastic cells are characterized by simultaneous global DNA hypomethylation, localized hypermethylation that involves CpG islands and increased HDAC activity (21). Hypomethylation has been linked to chromosomal instability *in vitro* and it seems to have the same effect in carcinogenesis (22). 5-Methylcytosine is a relatively unstable base because its spontaneous deamination leads to the formation of uracil. Such changes can also contribute to the appearance of germline mutations in inherited disease and somatic mutations in neoplasia. Aberrant CpG island hypermethylation in normally unmethylated regions around gene transcription start sites, which results in transcriptional silencing of genes, suggests that it plays an important role as an alternate mechanism by which tumor suppressor genes are inactivated in cancer (21). Hypermethylated genes identified in human cancers include the tumor suppressor genes that cause familial forms of human cancer when mutated in the germline, as well as genes that are not fully documented tumor suppressors (Table 1.1). Some of these genes, such as APC, the breast cancer gene BRCA-1, E-cadherin, mismatch repair gene hMLH1, and the Von Hippel-Lindau gene can exhibit this change in non-familial cancers.

Recent studies indicate that promoter hypermethylation is often an early event in tumor progression. It has been shown in the colon that genes that have increased hyper-

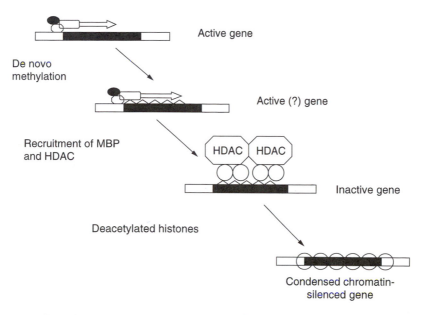

**Figure 1.5.** Effect of methylation and histone deacetylation on gene expression. When a gene is active, the promoter region is occupied by transcription factors that direct production of messenger RNA. *De novo* methylation has minimal effects on gene expression. However, methylated DNA attracts methyl-binding proteins (MBP). These methyl-binding proteins in turn attract a protein complex that contains histone deacetylase (HDAC). This results in inhibition of messenger RNA synthesis, and no functional protein can be made from the gene. Through the action of MBP and HDAC, the DNA structure changes to a compact, "condensed chromatin" configuration, which results in permanent inhibition of messenger RNA and protein synthesis (silencing).

methylation in the promoter region in normal tissue as a function of aging are the same as genes with the highest rate of promoter hypermethylation in tumors (9). Interestingly, this group of genes does not include classic tumor suppressor genes. Some genes, such as the estrogen receptor where age-related hypermethylation in the colon was first discovered, may be important for the modulation of cell growth and differentiation in the colonic mucosa.

Promoter hypermethylation of genes, which are normally unmethylated at all ages, has also been found early in tumorigenesis. These epigenetic alterations can produce the early loss of cell cycle control, altered regulation of gene transcription factors, disruption of cell-cell interactions, and multiple types of genetic instability, which are all characteristic of neoplasia. For example, hypermethylation of the APC gene has recently been reported for a subset of colon cancers (23). Hypermethylation of hMLH1, which is associated with microsatellite instability in colon, endometrial,

and gastric neoplasia, has been seen in early stages of cancer progression (24). Finally, hypermethylation of the E-cadherin promoter frequently occurs in early stages of breast cancer and can trigger invasion (25).

Loss of gene function through epigenetic changes differs from genetic changes in terms of its consequences for tumor biology. First, gene function loss caused by aberrant promoter methylation may manifest in a more subtle, selective advantage than gene mutations during tumor progression. Second, although promoter hypermethylation causing gene silencing is usually stable in cancer cells, this change, unlike mutation, is potentially reversible. It has become evident that not only the mutagens, but various factors influencing cell metabolism, particularly methylation, lie at the origin of carcinogenesis.

Silencing of gene expression by methylation may be modulated by biochemical or biological manipulation. It has been shown that pharmacological inhibition of methyltrans-

**Table 1.1  Hypermethylated Genes in Cancer**

| Gene | Function | Type of Tumor |
|---|---|---|
| **Familial Cancers** | | |
| APC | Signal transduction | Colon cancer |
| BRCA1 | DNA repair | Breast cancer |
| E-cadherin | Adhesion and metastasis | Multiple cancers |
| hMLH1 | DNA mismatch repair | Colon, gastric, and endometrial cancer |
| p16/CDKN2A | Cell cycle regulation | Multiple cancers |
| RB1 | Cell cycle regulation | Retinoblastoma |
| VHL | Cytoskeletal organization, angiogenesis inhibition | Renal-cell cancer |
| **Other Cancers** | | |
| Androgen receptor | Growth and differentiation | Prostate cancer |
| c-ABL | Tyrosine kinase | Chronic myelogenous leukemia |
| Endothelin receptor B | Growth and differentiation | Prostate cancer |
| Estrogen receptor $\alpha$ | Transcription | Multiple cancers |
| FHIT | Detoxification | Esophageal cancer |
| GST-$\pi$ | Drug transport | Prostate cancer |
| MDR1 | Drug transport | Acute leukemias |
| O6-MGMT | DNA repair | Multiple cancers |
| p14/ARF | Cell cycle regulation | Colon cancer |
| p15/CDKN2B | Cell cycle regulation | Malignant hematologic disease |
| Progesterone receptor | Growth and differentiation | Breast cancer |
| Retinoic acid receptor $\beta$ | Growth and differentiation | Colon and breast cancer |
| THBS1 | Angiogenesis inhibition | Colon cancer, glioblastoma multiforme |
| TIMP3 | Metastasis | Multiple cancers |

ferases resulted in reactivation of gene expression *in vitro* (26) and prevented tumor growth in animal models (27). These studies generated interest in the clinical uses of hypomethylating agents in humans.

## 3  MOLECULAR BASIS OF CANCER PHENOTYPES

Cancer is a multistep process that requires the accumulation of multiple genetic mutations in a single cell that bestow features characteristic of a neoplastic cell. Typically, tumor cells differ from normal cells in that they exhibit uncontrolled growth. Because features that distinguish tumor from normal cells may be key to understanding neoplastic cell behavior and may ultimately lead to therapies that can target tumor cells, considerable effort has been directed at identifying the phenotypic characteristics of *in vitro*–transformed cells and of tumor cells derived from natural sources. This work has resulted in a list of

properties that are characteristic of tumor cells and that are now known to be the basis for the behaviors exhibited by neoplastic cells. Some of the features that will be discussed in detail include immortality, decreased dependence on growth factors to support proliferation, loss of anchorage-dependent growth, loss of cell cycle control, reduced sensitivity to apoptotic cell death, and increased genetic instability. Other morphological and biochemical characteristics used to identify the transformed phenotype are cytological changes, altered enzyme production, and the ability to produce tumors in experimental animals (28).

### 3.1  Immortality

Normal diploid fibroblasts have a limited capacity to grow and divide both *in vivo* and *in vitro*. Even if provided with optimal growth conditions, *in vitro* normal cells will cease dividing after 50–60 population doublings and then senesce and die. In contrast, malignant cells that have become established in culture proliferate indefinitely and are said to be immortalized. The barrier that restricts the life span of normal cells is known as the Hayflick limit and was first described in experiments that attempted immortalization of rodent cells (29). Normal embryo-derived rodent cells, when cultured *in vitro*, initially divide rapidly. Eventually, however, these cultures undergo a crisis phase during which many of the cells senesce and die. After extended maintenance, however, proliferation in the cultures increases and cells that can divide indefinitely emerge. The molecular changes that take place during crisis have revealed at least two important restrictions that must be overcome for cells to become immortalized and both of these changes occur in natural tumor cells.

One barrier to cellular immortalization is the inability of the DNA replication machinery to efficiently replicate the linear ends of DNA at the 5′ ends, which leads to the shortening of the chromosome. In bacteria, the end-replication problem is solved with a circular chromosome. In human cells, the ends of chromosomes are capped with 5–15 kb of repetitive DNA sequences known as telomeres. Telomeres serve as a safety cap of noncoding DNA that is lost during normal cell division without consequence to normal function of the cell. However, because telomere length is shortened with each round of cell division, indefinite proliferation is impossible because eventually the inability to replicate chromosomal ends nibbles into DNA containing vital genes.

Telomeres seem to be lengthened during gametogenesis as a consequence of the activity of an enzyme called telomerase. Telomerase activity has been detected in normal ovarian epithelial tissue. More importantly, telomerase activity is elevated in the tumor tissue but not the normal tissue from the same patient. This implies that one mechanism by which tumor cells overcome the shortening telomere problem and acquire the capacity to proliferate indefinitely is through abnormal up-regulation of telomerase activity. The finding that telomerase activity is found almost exclusively in tumor cells is significant because it suggests that this enzyme may be a useful therapeutic target (30). Therapies aimed at suppressing telomerase would eliminate a feature essential for tumor cell survival and would be selective.

A second feature of immortalization is loss of growth control by elimination of tumor suppressor activity. Recent evidence suggests that inactivating mutations in both the Rb and p53 tumor suppressor genes occurs during crisis. Both of these genes are discussed in more detail later in this chapter and both function to inhibit cell proliferation by regulating cell cycle progression. Consequently, loss of tumor suppressor function also appears to be a critical event in immortalization.

## 3.2 Decreased Dependence on Growth Factors to Support Proliferation

Cells grown in culture require media supplemented with various growth factors to continue proliferating. In normal human tissues, growth factors are generally produced extracellularly at distant sites and then are either carried through the bloodstream or diffuse to their nearby target cells. The former mode of growth factor stimulation is termed endocrine stimulation, and the latter mode, paracrine stimulation. However, tumor cells often produce their own growth factors that bind to and stimulate the activity of receptors that are also present on the same tumor cells that are producing the growth factor. This results in a continuous self-generated proliferative signal known as autocrine stimulation that drives proliferation of the tumor cell continuously even in the absence of any exogenous proliferative signal. Autocrine stimulation is manifested as a reduced requirement for serum because serum is the source of many of the growth factors in the media used to propagate cells *in vitro*.

Because of the prominent role that growth factors and their cognate receptors play in tumor cell proliferation, they have also become favorite therapeutic targets. For example, the epidermal growth factor receptor (EGFR) is known to play a major role in the progression of most human epithelial tumors, and its overexpression is associated with poor prognosis. As a consequence, different approaches have been developed to block EGFR activation function in cancer cells, including anti-EGFR blocking monoclonal antibodies (MAb), epidermal growth factor (EGF) fused to toxins, and small molecules that inhibit the receptor's tyrosine kinase activity (RTK). Of these, an orally active anilinoquinazoline, ZD1839 ("Iressa") shows the most promise as an antitumor agent by potentiating the antitumor activity of conventional chemotherapy (31).

### 3.3 Loss of Anchorage-Dependent Growth and Altered Cell Adhesion

Most normal mammalian cells do not grow, but instead undergo cell death if they become detached from a solid substrate. Tumor cells, however, frequently can grow in suspension or in a semisolid agar gel. The significance of the loss of this anchorage-dependent growth of cancer cells relates to the ability of the parent tumor cells to leave the primary tumor site and become established elsewhere in the body. The ability of cancer cells to invade and metastasize foreign tissues represents the final and most difficult-to-treat stage of tumor development, and it is this change that accompanies the conversion of a benign tumor to a life-threatening cancer.

Metastasis is a complex process that requires the acquisition of several new characteristics for tumor cells to successfully colonize distant sites in the body. Epithelial cells normally grow attached to a basement membrane that forms a boundary between the epithelial cell layers and the underlying supporting stroma separating the two tissues. This basement membrane consists of a complex array of extracellular matrix proteins including type IV collagen, proteoglycans, laminin, and fibronectin, which normally acts as a barrier to epithelial cells. A common feature of tumor cells with metastatic potential is the capacity to penetrate the basement membrane by proteolysis, to survive in the absence of attachment to this substrate, and to colonize and grow in a tissue that may be foreign relative to the original tissue of origin.

Consequently, metastasis is a multistep process that begins with detachment of tumor cells from the primary tumor and penetration through the basement membrane by degradation of the extracellular matrix (ECM) proteins. This capacity to proteolytically degrade basement membrane proteins is driven, in part, by the expression of matrix metalloproteinases. Matrix metalloproteinases, or MMPs, are a family of enzymes that are either secreted (MMPs 1–13, 18–20) or anchored in the cell membrane (MMPs 14–17) (Table 1.2). Regulation of MMPs occurs at several levels: transcription, proteolytic activation of the zymogen, and inhibition of the active enzyme (32). MMPs are typically absent in normal adult cells, but a variety of stimuli, such as cytokines, growth factors, and alterations in cell-cell and cell-ECM interactions, can induce their expression. The expression of MMPs in tumors is frequently localized to stromal cells surrounding malignant tumor cells. Most of the MMPs are secreted in their inactive (zymogen) form and require proteolytic cleavage to be activated. In some cases, MMPs have been shown to undergo mutual and/or autoactivation *in vitro* (33).

Several lines of evidence implicate MMPs in tumor progression and metastasis. First, MMPs are overexpressed in tumors from a variety of tissues and the expression of one, matrilysin, is clearly elevated in invasive prostate cancer epithelium (34–36). Second, reduction of tissue inhibitor of matrix metalloproteinases-1 (TIMP-1) expression in mouse fibroblasts (Swiss 3T3), using antisense RNA technology, increased the incidence of metastatic tumors in immunocompromised mice. Similarly, overexpression of the various MMPs has provided direct evidence for their role in metastasis. Importantly, synthetic MMP inhibitors have also been produced and they lead to a reduction in metastasis in several experimental models of melanoma, colorectal carcinoma, and mammary carcinoma, suggesting a mechanism by which the invasive potential of tumors may be reduced (37).

Once tumor cells escape through the basement membrane, they can metastasize through two major routes, the blood and lymphatic vessels. Tumors originating in different parts of the body have characteristic patterns of invasion. Some tumors, such as those of the head and neck, spread initially to regional lymph nodes. Others, such as breast tumors, have the ability to spread to distant sites relatively early. The site of the primary tumor generally dictates whether the invasion will occur through the lymphatic or blood vessel system. The cells that escape into the vasculature must evade host immune defense mechanisms to be successfully transported to regional or distal locations. Tumor cells then

**Table 1.2  MMPs**

| MMP | Common Name | Substrates | Cell Surface |
|---|---|---|---|
| 1 | collagenase-1, interstitial collagenase | collagen I, II, III, VII, X, IGFBP | yes |
| 2 | gelatinase A | gelatin, collagen I, IV, V, X, laminin, IGFBP, latent TGF-$\beta$ | yes |
| 3 | stromelysin-1 | collagen III, IV, V, IX, X, gelatin, E-cadherin, IGFBP, fibronectin, elastin, laminin proteoglycans, perlecan, HB-EGF, proMMP-13 | unknown |
| 7 | matrilysin | laminin, fibronectin, gelatin, collagen IV, proteoglycans FasL, proMMP-1, HB-EGF | yes |
| 8 | collagenase-2, neutrophil collagenase | collagen I, II, III, VII, X | unknown |
| 9 | gelatinase B | collagen I, IV, V, X, gelatin, IGFBP, latent TGF-b | yes |
| 10 | stromelysin-2 | collagen III, IV, IX, X, gelatin, laminin, proteoglycans, proMMP-1, proMMP-13 | unknown |
| 11 | stromelysin-3 | IGFBP, a-1-antiprotease | unknown |
| 12 | metalloelastase | elastin, proMMP-13 | unknown |
| 13 | collagenase-3 | collagen I, II, III, IV, VII, X, XIV, fibronectin, proMMP-9, tenascin, aggrecan | unknown |
| 14 | MT1-MMP | gelatin, collagen I, fibrin, proteoglycans, laminin, fibronectin, proMMP-2 | yes |
| 15 | MT2-MMP | laminin, fibronectin, proMMP-2, proMMP-13, tenascin | yes |
| 16 | MT3-MMP | gelatin, collagen III, fibronectin, proMMP-2 | yes |
| 17 | MT4-MMP | unknown | yes |
| 18/19 | RASI-1 | unknown | unknown |
| 20 | Enamelysin | amelogenin | unknown |

exit blood vessels and escape into the host tissue by again compromising a basement membrane, this time the basement membrane of the blood vessel endothelium. Projections called invadopodia, which contain various proteases and adhesive molecules, adhere to the basement membrane, and this involves membrane components such as laminin, fibronectin, type IV collagen, and proteoglycans. The tumor cells then produce various proteolytic enzymes, including MMPs, which degrade the basement membrane and allow invasion of the host tissue. This process is referred to as extravasation.

The interaction between cells and extracellular matrix proteins occurs through cell-surface receptors, the best characterized of which is the fibronectin receptor that binds fibronectin. Other receptors bind collagen and laminin. Collectively these receptors are called integrins, and their interaction with matrix components conveys regulatory signals to the cell (38). They are heterodimeric molecules consisting of one of several alpha and beta subunits that may combine in any number of permutations to generate a receptor with distinct substrate preferences. Changes in the expression of integrin subunits is associated with invasive and metastatic cells facilitating invasion by shifting the cadre of integrins to integrins that preferentially bind the degraded subunits of extracellular matrix proteins produced by MMPs. Hence, integrin expression has served as a marker for the invasive phenotype and may be a logical target for novel therapies that interfere with the progress of advanced tumors.

In addition to their role in invasion, the evidence also indicates that MMPs may play a role in tumor initiation and in tumorigenicity. Expression of MMP-3 in normal mammary epithelial cells led to the formation of invasive tumors (39). A proposed mechanism for this initiation involves the ability of MMP-3 to

cleave E-cadherin. E-cadherin is a protein involved in cell-cell adhesion together with
other proteins such as $\beta$-catenin and $\alpha$-cati-
nin. Loss of E-cadherin function is known to
lead to tumorigenicity and invasiveness as a
result of loss of cellular adhesion. Interestingly, inhibition of MMP-7 and MMP-11, using antisense approaches, did not affect invasiveness or metastatic potential *in vitro*.
However, tumorigenicity was altered (40).
Matrilysin, MMP-7 messenger RNA (mRNA),
are present in benign tumors and malignant
tumor cells of the colon. The relative level of
matrilysin expression correlates with the
stage of tumor progression.

### 3.4   Cell Cycle and Loss of Cell Cycle Control

Proliferation is a complex process consisting
of multiple subroutines that collectively bring
about cell division. At the heart of proliferation is the cell cycle, which consists of many
processes that must be completed in a timely
and sequence specific manner. Accordingly,
regulation of cell cycle events is a multifaceted
affair and consists of a series of checks and
balances that monitor nutritional status, cell
size, presence or absence of growth factors,
and integrity of the genome. These cell cycle
regulatory pathways and the signal transduction pathways that communicate with them
are populated with oncogenes and tumor suppressor genes.

Cell division is divided into four phases: G1,
S, G2, and M (Fig. 1.6). The entire process is
punctuated by two spectacular events, the
replication of DNA during S phase and chromosome segregation during mitosis or M
phase. Of the four cell cycle phases, three can
be assigned to replicating cells and only the G1
phase, and a related quiesent phase, G0, are
nonreplicative in nature. Normal cycling cells
that cease to proliferate enter the resting
phase, or G1, and their exit into the replicative
phases is strongly dependent on the presence
of growth factors and nutrients. However,
once the cells enter the replicative phase of the
cell cycle, they become irrevocably committed
to completing cell division. Hence, the conditions that lead to exit from G1 and entry into S
are tightly regulated and are frequently mis-
regulated in neoplastic cells that exhibit un-

controlled proliferation. Studies first conducted by Arthur Pardee revealed the
existence of a point in G1 that restricted the
passage of cells into S phase, and this was postulated to be controlled by a labile protein factor (41). Passage across this restriction point,
or R point, is now known to be sensitive to
growth factor stimulation.

Movement through the cell cycle is controlled by two classes of cell cycle proteins,
cyclins and cyclin dependent kinases (CDKs),
which physically associate to form a protein
kinase that drives the cell cycle forward (42).
At least 8 cyclins and 12 CDKs have been identified in mammalian cells. The name "cyclin"
derives from the characteristic rise and fall in
abundance of cyclin B as cells progress
through the cell cycle. The accumulation of
cyclin proteins occurs through cell cycle-de-
pendent induction of gene transcription, but
elimination of cyclins occurs by carefully regulated degradation that is enabled through
protein sequence tags known as destruction
boxes and PEST sequences. Although not all
of the cyclin types exhibit this oscillation in
protein quantity, those cyclins that play key
roles in progression through the cell cycle (cyclins E, A, and B) are most abundant during
discrete phases of the cell cycle. Cyclin D1 is
synthesized during G1 just before the restriction point and plays an important role in regulation of the R point. Cyclin E is most abundant during late G1 and early S and is
essential for exit from G1 and progression into
S phase. Elevated levels of these two G1 cyclins can result in uncontrolled proliferation.
Indeed, both cyclin D1 and cyclin E are overexpressed in some tumor types, suggesting
that the cyclins and other components of the
cell cycle may be useful therapeutic targets
(43).

The second component of the enzyme complex is CDK that, as the name implies, requires an associated cyclin to become active.
At least 12 of the protein kinases have been
isolated from humans, *Xenopus*, and *Drosophila*, and are numbered according to a standardized nomenclature beginning with CDK1,
which for historical reasons, is most frequently referred to as cell division cycle 2
(cdc2). Unlike the cyclins, abundance of the
CDK proteins remains relatively constant

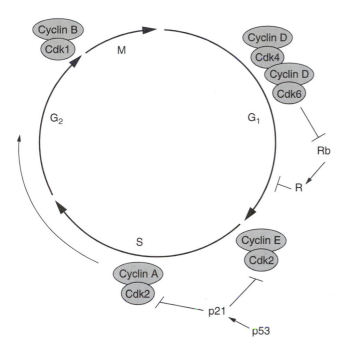

**Figure 1.6.** Model of the cell cycle and the cyclin/cdk complexes that are required at each cell cycle phase. CyclinD/cdk4-6 complexes suppress Rb function by phosphorylating the protein allowing transition across the restriction R-point. P53 suppresses cell cycle progression by stimulating the expression of the cyclin dependent kinase inhibitor p21, which binds with and inactivates a variety of cyclin/cdk complexes.

throughout the cell cycle. Instead, their activity changes during different phases of the cell cycle in accordance with whether or not an activating cyclin is present and whether or not the kinase itself is appropriately phosphorylated. Both cyclins and CDKs are highly conserved from yeast to man and function similarly, suggesting that the cell cycle is controlled by a universal cell cycle engine that operates through the action of evolutionarily conserved proteins. Hence, drug discovery studies aimed at identifying agents that regulate the cell cycle may be performed in model organisms, such as yeast, *C. elegans*, and *Drosophila* with some assurance that the targeted mechanisms will also be relevant to humans.

It is now clear that specific cyclin/cdk complexes are required during specific stages of the cell cycle. Cyclin D1/cdk4,6 activity is essential for crossing the restriction point and pushing cells into replication. A major substrate of the cyclin D1/cdk4,6 complex is the retinoblastoma (Rb) tumor suppressor protein, which when phosphorylated by this kinase complex, is inactivated. This frees the cell from the restrictions on cell proliferation imposed by the Rb protein. It is this event that is believed to be decisive in the stimulation of resting cells to undergo proliferation. Cyclin

E/cdk2 plays a role later in the cell cycle for proliferating cells by pushing them from G1 into S phase. Cyclin E is overexpressed in some breast cancers where it may enhance the proliferative capacity of tumor cells. Cyclin A/cdk2 sustains DNA replication and is therefore required during S phase. Cyclin B/cdc2 is required by cells entering mitosis up through metaphase. At the end of metaphase, cyclin B is degraded, and cdc2 becomes inactivated, allowing mitotic cells to progress into anaphase and to complete mitosis. Sustaining the activity of cyclin B/cdc2 causes cells to arrest in metaphase. Hence, it is the collective result brought about by the activation and deactivation of cyclin/cdk complexes that pushes proliferating cells through the cell cycle.

Superimposed on the functions of the cell cycle engine is a complex network of both positive and negative regulatory pathways. Important negative regulators are the cyclin dependent kinase inhibitors or CKIs. There are two families of CKIs, the Cip/Kip family and the INK4 family (44). The Cip/Kip family consists of three members, p21/Cip1/waf1/Sdi1, p21/Kip1, and p57/Kip2. All of the proteins in this family have broad specificity and can bind to and inactivate most of the cyclin/cdk complexes that are essential for progression through the cell cycle. p21$^{waf1}$, the first discovered and best characterized member of the Cip/Kip family, is stimulated by the p53 tumor suppressor protein in response to DNA damage and halts cell cycle progression to allow for DNA repair (45). The INK4 family of CKIs contains four member proteins, p16/INK4a, p15/INK4b, p18/INK4c, and p19/INK4d. Unlike the Cip/Kip family, the INK4 proteins have restricted binding and associate exclusively with cdk4/6. Consequently, their principal function is to regulate cyclin D1/cdk4/6 activity, and therefore, the phosphorylation status of the Rb tumor suppressor. p16/INK4a is itself a tumor suppressor that is frequently mutated in melanoma (46). Indeed, at least one component of the p16/cyclin D1/Rb pathways is either mutated or deregulated in some fashion in over 90% of lung cancers, emphasizing the importance of this pathway in regulating tumor cell proliferation.

Transit through the cell cycle is regulated by two types of controls. In the first type, the cumulative exposure to specific signals, such as growth factors, is assessed and if the sum of these signals satisfies the conditions required by the R point, proliferation ensues. In the second, feedback controls or checkpoints monitor whether the genome is intact and whether previous cell cycle steps have been completed. At least five cell cycle checkpoints have been identified, two that monitor integrity of the DNA and halt cell cycle progression in either G1 or G2, one that ensures DNA synthesis has been completed before mitosis begins, one that monitors completion of mitosis before allowing another round of DNA synthesis, and one that monitors chromosome alignment on the equatorial plate before initiation of anaphase. Of these, the two checkpoints that monitor integrity of DNA have been the most extensively studied, and as might be expected, these checkpoints and the genes that enforce them are critically important for the response that cells mount to genotoxic stresses. Abrogation of checkpoints leads to genomic instability and an increased mutation frequency (47).

Progress in elucidating the mechanisms of checkpoint function reveals that a number of checkpoint genes are frequently mutated in human cancers. For example, the p53 tumor suppressor functions as a cell cycle checkpoint

that halts cell cycle progression in G1 by inducing the expression of the p21$^{waf1}$ gene in the presence of damaged DNA (45). The p53 gene is frequently mutated in human cancers and consequently, most tumor cells lack the DNA damage-induced p53-dependent G1 checkpoint, increasing the likelihood that mutations will be propagated in these cells. Because p53 also promotes apoptosis, the lack of p53 in these cells also makes them more resistant to the DNA damage-induced apoptosis. Because most chemotherapeutic agents kill cells through DNA damage-induced apoptosis, tumor cells with mutant p53 are also more resistant to conventional therapies (48).

### 3.5 Apoptosis and Reduced Sensitivity to Apoptosis

Apoptosis is a genetically controlled form of cell death that is essential for tissue remodeling during embryogenesis and for maintenance of the homeostatic balance of cell numbers later in adult life. The importance of apoptosis to human disease comes from the realization that disruption of the apoptotic process is thought to play a role in diverse human diseases ranging from malignancy to neurodegenerative disorders. Because apoptosis is a genetically controlled process, much effort has been spent on identifying these genetic components to better understand the apoptotic process as well as to identify potential therapeutic targets that might be manipulated in disease conditions where disruption of apoptosis occurs.

Although multiple forms of cell death have been described, apoptosis is characterized by morphological changes including cell shrinkage, membrane blebbing, chromatin condensation and nuclear fragmentation, loss of microvilli, and extensive degradation of chromosomal DNA. In general, the apoptotic program can be subdivided into three phases: the initiation phase, the decision/effector phase, and the degradation/execution phase (Fig. 1.7). In the initiation phase, signal transduction pathways that are responsive to external stimuli, such as death receptor ligands, or to internal conditions, such as that produced by DNA damage, are activated. During the ensuing decision/effector phase, changes in the mitochondrial membrane occur that result in

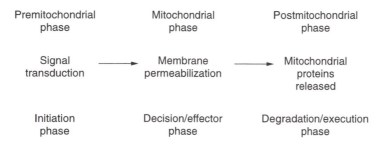

**Figure 1.7.** Mitochondria-mediated apoptosis. Mitochondria-mediated apoptosis is divided into three phases. Mitochondrial stress stimulates signal transduction and constitutes the initiation phase. During the second phase, changes in the structure of the mitochondrial membrane make it permeable to large proteins, allowing the release of cytochrome c and induction of the third and final phase, during which degradation of cellular proteins occurs.

disruption of the mitochondrial membrane potential and ultimately loss of mitochondrial membrane integrity. A key event in the decision/effector phase is the release of cytochrome c into the cytoplasm and activation of proteases and nucleases that signal the onset of the final degradation/execution phase. An important concept in understanding apoptosis is that the mitochrondrion is a key target of apoptotic stimuli and disruption of mitochondrial function is central to subsequent events that lead to degradation of vital cellular components.

Of the signal transduction pathways that initiate apoptosis, the best understood at the molecular level involves the death receptors including Fas/cluster of differentiation 95 (CD95), tumor necrosis factor receptor 1 (TNFR1), and death receptors 3, 4, and 5 (DR 3,4,5) (Fig. 1.8). All death receptors share an amino acid sequence known as the death domain (DD) that functions as a binding site for a specific set of death signaling proteins. Stimulation of these transmembrane receptors can be induced by interaction with its cognate ligand or by binding to an agonistic antibody, which results in receptor trimerization and recruitment of intracellular death molecules and stimulation of downstream signaling events. Here death receptors are classified as either CD95-like (Fas/CD95, DR4, and DR5) or TNFR1-like (TNF-R1, DR3, and DR6) based on the downstream signaling events that are induced as a consequence of receptor activation.

Activation of Fas/CD95 leads to clustering and recruitment of Fas-associated death domain (FADD; sometimes called Mort1) to the Fas/CD95 intracellular DD (49). FADD contains a C-terminal DD that enables it to interact with trimerized Fas receptor as well as an N-terminal death effector domain (DED), which can associate with the prodomain of the serine protease, caspase-8. This complex is referred to as the death-inducing signaling complex (DISC). As more procaspase-8 is recruited to this complex, caspase-8 undergoes transcatalytic cleavage to generate active protease. Activation of TNFR1-like death receptors results in similar events except that the first protein to be recruited to the activated receptor is the TNFR-associated death domain (TRADD) adaptor protein that subsequently recruits FADD and procaspase-8. Signaling through the TNFR1-like receptors is more complex and includes recruitment of other factors that do not interact with Fas/CD95. For example, TRADD also couples with the receptor interacting protein (RIP), which links stimulation of TNFR1 to signal transduction mechanisms, leading to activation of nuclear factor-kappa B (NF-κB). Because RIP does not interact with Fas/CD95, this class of receptors does not activate NF-kappa B.

The critical downstream effectors of death receptor activation are the caspases, and these are considered the engine of apoptotic cell death (50). Caspases are a family of cysteine proteases with at least 14 members. They are synthesized in the cells as inactive enzymes that must be processed by proteolytic cleavage at aspartic acid residues. These cleavage sites are between the N-terminal prodomain, the

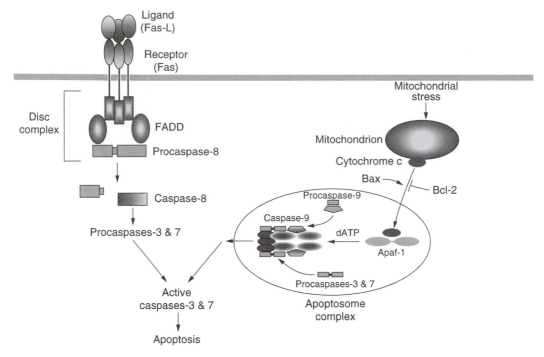

**Figure 1.8.** Apoptosis—receptor-mediated and mitochondrial apoptosis cascades. Trimerization of the Fas receptor initiates recruitment of the death domain-containing adaptor protein FADD, which binds to procaspase-8 promoting trans-catalytic cleavage of prodomain. Caspase-8 initiates the caspase cascade by acting on downstream effector caspases 3 and 7. In mitochondria-mediated apoptosis cytochrome c, release is a key event in apoptosis and is stimulated by Bax and suppressed by Bcl-2. The released cytochrome c binds with Apaf-1 and in conjunction with dATP induces a conformational change in Apaf-1 that permits oligomerization into a $\sim$700-kDa complex, which is called the apoptosome complex and is capable of recruiting caspases-9, -3, and -7.

large P20, and small P10 domains. The activated proteases cleave other proteins by recognizing an aspartic acid residue at the cleavage site and are consistent with an auto- or trans-cleavage processing mechanism for activation when recruited to activated death receptors.

Importantly, biochemical studies support the notion of a caspase hierarchy that consists of initators and effectors that are activated in a cascade fashion. Initiator caspases such as caspase-8 and -9 are activated directly by apoptotic stimuli and function, in part, by activating effector caspases such as caspase-3, -6, and -7 by proteolytic cleavage. It is the effector caspases that result in highly specific cleavage of various cellular proteins and the biochemical and morphological degradation associated with apoptosis.

In contrast to death receptor-mediated apoptosis that functions through a well-defined

pathway, mediators of stress-induced apoptosis such as growth factors, cytokines, and DNA damage activate diverse signaling pathways that converge on the mitochondrial membrane (51). Many proapoptotic agents have been shown to disrupt the mitochondrial membrane potential ($\Delta\Psi_m$), leading to an increase in membrane permeability and release of cytochrome c into the cytosol. Cytochrome c release is a common occurrence in apoptosis and is thought to be mediated by opening of the permeability transmembrane pore complex (PTPC), a large multiprotein complex that consists of at least 50 different proteins. The cytosolic cytochrome c interacts with apoptosis activating factor-1 (Apaf-1), dATP/ATP, and procaspase-9 to form a complex known as the apoptosome. Cytochrome c and dATP/ATP stimulate Apaf-1 self-oligomerization and trans-catalytic activation of procaspase-9 to the active enzyme. Active caspase-9 activates effector caspases-3 and -7 and leads to the cellular protein degradation characteristic of apoptosis.

As release of cytochrome c can have dire consequences for viability of the cell, its release is tightly regulated. Indeed, a whole family of proteins, of which B-cell lymphoma-2 (Bcl-2) is the founding member, that share homology in regions called the Bcl-2 homology domains are dedicated to regulation of cytochrome c release from the mitochondria (52). Both positive regulators (Bax, Bak, Bik, and Bid) that promote apoptosis and negative regulators (Bcl-2 and Bcl-$x_L$), which suppress apoptosis, act by regulating permeability of the mitochondrial membrane to cytochrome c. Bcl-2 family members have been found in both the cytosol and associated with membranes. Bax is normally found in the cytosol, but subcellular localization changes during apoptosis. Bax has been shown to insert into the mitochondrial membrane where, because of its structure that is similar to other pore-forming proteins, it is thought to promote release of cytochrome c. Bcl-2 functions by inhibiting insertion of Bax into the mitochondrial membrane. Hence, a key factor that determines whether a cell will undergo apoptosis is the ratio of proapoptotic to antiapoptotic Bcl-2 family proteins.

Because apoptosis serves to eliminate cells with a high neoplastic potential, cancer cells have evolved to evade apoptosis primarily through two mechanisms. In the first of these, Bcl-2, which suppresses apoptosis, is overexpressed. The Bcl-2 oncogene was first identified as a break point in chromosomal translocations that frequently occurred in B-cell–derived human tumors. Characterization of the rearrangements revealed that the Bcl-2 gene is overexpressed by virtue of being placed adjacent to the powerful IgH promoter. Cloning of the Bcl-2 gene and overexpression in cells of B-cell lineage reduced the sensitivity of these cells to apoptosis and allowed them to survive under conditions that ordinarily caused normal cells to die.

The second mechanism that provides cancer cells with resistance to apoptosis is the suppression of the Fas receptor. As with other receptors, mutations can occur in either the ligand binding domain or in the intracellular domain interfering with activation of the death signaling pathway. More recently a novel mechanism for suppressing Fas-receptor activation has been identified in which cancer cells synthesize decoy receptors to which ligands can bind but are unable to induce apoptosis (53).

## 3.6 Increased Genetic Instability

A hallmark of tumor cells is genetic instability that is manifested at the chromosomal level as either aneuploidy (the gain or loss of one or more specific chromosomes) or polyloidy (the accumulation of an entire extra set of chromosomes). Acquisition of extra chromosomes is one mechanism by which extra copies of a growth promoting gene can be acquired by cancer cells, providing them with a selective growth advantage. Structural abnormalities are also common in advanced tumors that lead to various types of chromosomal rearrangements. Translocations and random insertion of genetic material into one chromosome from another can place genes that are not normally located adjacent to one another in close proximity usually leading to abnormal gene expression. Some of these rearrangements are routinely observed in some cancers such as in Burkitt's lymphoma where rearrangements involving chromosome 8 and 14 lead to abnormal expression of the c-myc protooncogene as a consequence of being placed adjacent to the immunoglobulin heavy chain promoter.

In chronic myelogenous leukemia (CML), an abnormal chromosome known as the Philadelphia chromosome results from a translocation involving chromosomes 9 and 22. The genes for two unrelated proteins, c-Abl and Bcr, a tyrosine kinase, and a GTPase activating protein (GAP), are spliced together, forming a chimeric protein that results in a powerful and constitutively active kinase that drives proliferation of the cells in which it is expressed.

Other forms of genetic instability include gene amplification. Under normal conditions, all DNA within the cell is replicated uniformly and only once per cell cycle. However, in cancer cells some regions of a chromosome can undergo multiple rounds of replication such that multiple copies of a growth-promoting gene(s) is obtained. These can result in chromosomes with regions of DNA that stain uniformly during karyotype analysis of a tumor cell or in the production of extrachromosomal DNA-containing bodies known as double minute chromosomes. A typical example of this type of amplification targets the N-myc gene, which is amplified in $\sim$30% of advanced neuroblastomas (54).

More subtle changes at the sequence level affecting growth-controlling genes is also common in human tumors. Mutations can occur as a consequence of either defects in DNA repair or decreased fidelity during DNA replication. The components of these pathways are critical for maintenance of genome integrity and inherited mutations in the genes of DNA repair proteins and proteins that repair misreplicated DNA explains some inherited cancer-prone syndromes (55).

### 3.7 Angiogenesis

Without the production of new blood vessels, tumor growth is limited to a volume of a few cubic millimeters by the distance that oxygen and other nutrients can diffuse through tissues. As tumor size increases, intratumoral $O_2$ levels fall and the center of the mass becomes hypoxic, leading to up-regulation of the hypoxia inducible factor (HIF1). HIF1 is a heterodimeric transcription factor composed of a constitutively expressed HIF-1 beta subunit and an $O_2$ regulatable HIF-1 alpha subunit (56). Under normoxic conditions, levels of HIF1 are kept low through the actions of the VHL tumor suppressor protein, which functions as a ubiquitin ligase that promotes degradation through a proteosome mediated pathway (57). An important transcriptional target of HIF1 is the VEGF growth factor, which in conjunction with other cytokines, induces neovascularization of tumors and allows them to grow beyond the size limitation imposed by oxygen diffusion. This increased production of proangiogenic factors and reduction of anti-angiogenic factors is known as the "angiogenic switch" and is a significant milestone in tumorigenesis that leads to the development of more lethal tumors.

Angiogenesis is the sprouting of capillaries from preexisting vessels during embryonic development and is almost absent in adult tissues with the exception of transient angiogenesis during the female reproductive cycle and wound healing, and the soluble factor that plays a critical role in promoting angiogenesis is vascular endothelial growth factor (VEGF) (58). VEGF was first implicated in angiogenesis when it was identified as a factor secreted by tumor cells, which caused normal blood vessels to become hyperpermeable (59). The following evidence supports a role for VEGF in tumor angiogenesis.

1. VEGF is present in almost every type of human tumor. It is especially high in concentration around tumor blood vessels and in hypoxic regions of the tumor.
2. VEGF receptors are found in blood vessels within or near tumors.
3. Monoclonal neutralizing antibodies for VEGF can suppress the growth of VEGF-expressing solid tumors in mice. These lack any effect in cell culture where angiogenesis is not needed.

Ferrara and Henzel (60) identified VEGF as a growth factor capable of inducing proliferation of endothelial cells but not fibroblasts or epithelial cells. Inhibition of one of the identified VEGF receptors, FLK1, inhibits the growth of a variety of solid tumors (61). Similarly, the injection of an antibody to VEGF strongly suppresses the growth of solid tumors of the subcutaneously implanted human fibrosarcoma cell line HT-1080 (62).

There are several forms of VEGF that seem to have different functions in angiogenesis. These isoforms are VEGF, VEGF-B, VEFG-C, and VEGF-D. VEGF-B is found in a variety of normal organs, particularly the heart and skeletal muscle. It can form heterodimers with VEGF and can affect the availability of VEGF for receptor binding (63). VEGF-D seems to be regulated by c-fos and is strongly expressed in the fetal lung (64). However, in the adult it is mainly expressed in skeletal muscle, heart, lung, and intestine. VEGF-D is also able to stimulate endothelial cell proliferation (65).

VEGF-C is about 30% homologous to VEGF. Unlike both VEGF and VEGF-B, VEGF-C does not bind to heparin. It is able to increase vascular permeability and stimulate the migration and proliferation of endothelial cells, although at a significantly higher concentration than VEGF. VEGF-C is expressed during embryonal development where lymphatics sprout from venous vessels (66). It is also present in adult tissues and may play a role in lymphatic endothelial differentiation. Flt-4, the receptor for VEGF-C, is expressed in angioblasts, veins, and lymphatics during embryogenesis, but it is mostly restricted to the lymphatic endothelium in adult tissues. Because of these expression patterns, VEGF-C and Flt-4 may be involved in lymphangiogenesis. This is the process of lymphatic generation. Lymphatic vasculature is very important because of its involvement in lymphatic drainage, immune function, inflammation, and tumor metastasis.

Other cytokines and growth factors also play an important role in promoting angiogenesis. Some of these act directly on endothelial cells, whereas others stimulate adjacent inflammatory cells. Some can cause migration but not division of endothelial cells such as angiotropin, macrophage-derived factor, and TNF$\alpha$, or stimulate proliferation such as EGF, acidic and basic fibroblast growth factors (aFGF, bFGF), transforming growth factor $\beta$ (TGF$\beta$), and VEGF (67). Tumors secrete these factors, which stimulate endothelial migration, proliferation, proteolytic activity, and capillary morphogenesis (68).

Several angiogenic factors have been identified that can be secreted from tumors. Many of these are growth factors that are described as heparin-binding growth factors. Specifically, these include VEGF, FGFs, TGF-$\beta$, and the hepatocyte growth factor (HGF). The binding of these factors to heparin sulphate proteoglycans (HSPG) may be a mechanism for bringing the growth factors to the cell surface and presenting them to their appropriate receptors in the proper conformation. This facilitates the interaction between the growth factors and receptors. Studies have shown that tumor growth is adversely affected by agents that block angiogenesis (69) but is stimulated by factors that enhance angiogenesis (70).

Angiogenesis may be useful as a prognostic indicator. Tumor sections can be stained immunohistochemically for angiogenic determinants, such as VEGF, to determine the density of vasculature within the tumor, and there is a strong correlation between high vessel density and poor prognosis (71). This correlation implies a relationship between angiogenesis and metastasis.

## 4  CANCER-RELATED GENES

### 4.1  Oncogenes

Oncogenes are derived from normal host genes, also called protooncogenes, that become dysregulated as a consequence of mutation. Oncogenes contribute to the transformation process by driving cell proliferation or reducing sensitivity to cell death. Historically, oncogenes were identified in four major ways: chromosomal translocation, gene amplification, RNA tumor viruses, and gene transfer experiments. Gene transfer experiments consist of transfecting DNA isolated from tumor cells into normal rodent cells (usually NIH-3T3 cells) and observing any morphological changes. These morphological changes became the hallmarks for cell transformation, the process of becoming tumorigenic. As previously discussed, the characteristics of transformed cells are as follows: (1) the ability to form foci instead of a monolayer in tissue culture; (2) the ability to grow without adherence to a matrix, or "anchorage-independent growth"; and (3) the ability to form tumors when injected into immunologically compromised animals.

There are seven classes of oncogenes, classified by their location in the cell and their biochemical activity (Table 1.3). All of these oncogenes have different properties that can lead to cancer. The classes of oncogenes are growth factors, growth factor receptors, membrane-associated guanine nucleotide-binding proteins, serine-threonine protein kinases, cytoplasmic tyrosine kinases, nuclear proteins, and cytoplasmic proteins that affect cell survival.

### 4.1.1  Growth Factors and Growth Factor Receptors.
Cell growth and proliferation are subject to regulation by external signals that are typically transmitted to the cell in the

**Table 1.3  Oncogenes**

| Oncogenes | Protein Function | Neoplasm(s) |
|---|---|---|
| **Growth Factors** | | |
| sis | Platelet-derived growth factor | fibrosarcoma |
| int-2 | Fibroblast growth factor | breast |
| trk | Nerve growth factor | neuroblastoma |
| **Growth Factor Receptors** | | |
| erb-B1 | Epidermal growth factor receptor | squamous cell carcinoma |
| erb-B2/HER2/neu | Heregulin | breast carcinoma |
| fms | Hematopoietic colony stimulating factor | sarcoma |
| ros | Insulin receptor | astrocytoma |
| **Tyrosine kinases** | | |
| bcr-abl | Tyrosine kinase | chronic myelogenous leukemia |
| src | Tyrosine kinase | colon |
| lck | Tyrosine kinase | colon |
| **Serine-Threonine protein kinases** | | |
| raf | Serine-threonine kinase | sarcoma |
| mos | Serine-threonine kinase | sarcoma |
| **Guanine nucleotide binding proteins** | | |
| H-ras | GTPase | melanoma; lung, pancreas |
| K-ras | GTPase | leukemias; colon, lung, pancreas |
| N-ras | GTPase | carcinoma of the genitourinary tract and thyroid; melanoma |
| **Cytoplasmic proteins** | | |
| bcl-2 | Anti-apoptotic protein | non-Hodgkin's B-cell lymphoma |
| **Nuclear proteins** | | |
| myc | Transcription factor | Burkitt's lymphoma |
| jun | Transcription factor (AP-1) | osteosarcoma |
| fos | Transcription factor (AP-1) | sarcoma |

form of growth factors that bind to and activate specific growth factor receptors. Predictably, one class of oncogenes consists of growth factors that can stimulate tumor cell growth. In normal cells and tissues, growth factors are produced by one cell type that then act on another cell type. This is termed paracrine stimulation. However, many cancer cells secrete their own growth factors as well as express the cognate receptors that are stimulated by those factors. Because of this autocrine stimulation, cancer cells are less dependent on external sources of growth factors for proliferation and their growth is unregulated. Examples of oncogenic growth factors include v-sis, which is the viral homolog of the platelet-derived growth factor (PDGF) gene. PDGF stimulates

the proliferation of cells derived from connective tissue such as fibroblasts, smooth muscle cells, and glial cells. Thus, tumors caused by excess stimulation by v-sis include fibrosarcomas and gliomas.

The receptors that interact with growth factors are also another large family of oncogenes. Growth factor receptors are composed of three domains: an extracellular domain that contains the ligand binding domain that interacts with the appropriate growth factor, a hydrophobic transmembrane domain, and a cytoplasmic domain that typically contains a kinase domain that can phosphorylate tyrosine residues in other proteins. Hence, these receptors are frequently referred to as receptor tyrosine kinases (RTK). It is this kinase

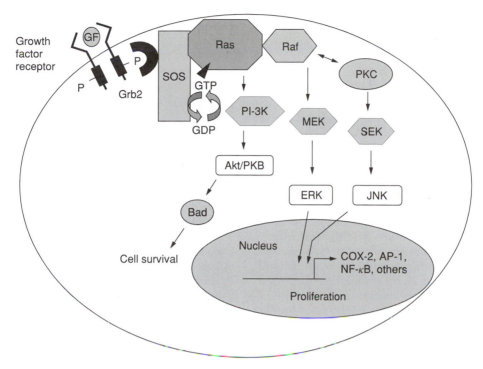

**Figure 1.9.** Ras signaling pathway. Growth factor (GF) binds to its receptor and initiates dimerization and autophosphorylation. Grb2 interacts with SOS, which activates ras by promoting the GTP-bound form. Ras recruits Raf to the plasma membrane and initiates the Raf/MAPK signaling cascade. Protein kinase C also stimulates this pathway as well as another cascade of stress-activated kinases (SEK/JNK). Both of these signaling pathways promote cell proliferation by stimulating the transcription of genes like cyclooxygenase-2, activator protein-1, and nuclear factor-κB. Ras also signals phosphoinositol-3-kinase and Akt/protein kinase B for cell survival.

activity that is essential to the intracellular signaling that is stimulated by an activated receptor and in all oncogenic receptors mutations that lead to constitutive intracellular signaling promote unregulated cellular proliferation. RTKs can become oncogenically activated by mutations in each of the protein domains. Genetic mutations that result in the production of an epidermal growth factor receptor (EGFR) lacking the extracellular ligand binding domain leads to constitutive signaling. This oncogenic EGFR is known as erb-B1 (Fig. 1.9).

Normally, EGF binds to the extracellular portion of the EGFR and causes dimerization of the intracellular part of the receptor and association with adaptor proteins, Son of Sevenless (SOS), and growth factor receptor binding protein 2 (Grb 2). These proteins interact through src-homology (SH) do-

mains SH2 and SH3, respectively. Through an unknown mechanism, the SOS-Grb 2 complex activates the oncogene ras. Ras induces an intracellular cascade of kinases to promote proliferation. These signaling cascades become constitutive when the extracellular portion of the EGFR becomes truncated, as in the case of erb-B1. Oncogenic activation of a related RTK, erb-B2, occurs as a consequence of a single point mutation that falls within the transmembrane region of this receptor (72). This mutated receptor is frequently found in breast cancers. Finally, mutations in the cytoplasmic kinase domain can also cause constitutive activity leading to constitutive signaling.

**4.1.2 G Proteins.** In many cases, signaling that is initiated by growth factors activating their receptors passes next to membrane associated guanine nucleotide-binding proteins, which when activated by mutation, constitute another class of oncogenes. The prototypical member of this family of oncogenes is the ras oncogene. There are three ras genes in this family of oncogenes, which include H-ras, K-ras, and N-ras. These genes differ in their expression patterns in different tissues. All have been found to have point mutations in human cancers including liver, colon, skin, pancreatic, and lung cancers, which lead to constitutive signaling of genes involved in proliferation, cell survival, and remodeling of the actin cytoskeleton. Ras is a small molecular weight protein that is post-translationally modified by attachment of a farnasyl fatty acid moiety to the C-terminus. Because this post-translational modification is essential for activity of the ras oncogenes, this process has become a target for drug development aimed at interfering with ras activity (73).

Ras binds both guanosine 5'-triphosphate (GTP) and guanosine 5'-diphosphate (GDP) reversibly but is only in the activated state and capable of signaling when bound to GTP. The activated, GTP-bound form of ras signals a variety of mitogen-induced and stress-induced pathways, leading to transcription of genes necessary for cell growth and proliferation (74). Mitogens such as growth factors can activate ras through the epidermal growth factor receptor, and stress factors affecting ras include ultraviolet light, heat, and genotoxins. Guanine nucleotide exchange factors (GEFs) foster ras activation by promoting the exchange of GDP for GTP. In contrast, GTPase activating proteins (GAPs) suppress ras activity by promoting GTP hydrolysis by ras, resulting in the GDP-bound inactive form of ras (75). Importantly, because GAPs function to suppress cell proliferation, they can be thought of as tumor suppressors. Indeed, the neurofibromatosis gene, NF-1, is a GAP that acts as a tumor suppressor gene and can be inherited in a mutated and nonfunctional form giving rise to the Von Recklinghausen neurofibromatosis or neurofibromatosis type 1 cancer syndrome (76).

**4.1.3 Serine/Threonine Kinases.** Once activated, ras then transmits the growth signal to a third class of signaling molecules that is comprised of the serine/thereonine kinases. The best studied of these serine-threonine protein kinases is the raf oncogene, which is activated when it is recruited to the plasma membrane by ras (77). Raf then initiates a cascade of mitogen-induced protein kinases (MAPKs), which culminate in the nucleus with the activation of genes containing Elk-1 transcription factor binding sites. Raf can also directly activate protein kinase C, which signals another set of kinases that phosphorylate the c-jun transcription factor.

Another ras effector gene is phosphoinositol 3-kinase (PI-3K), which initiates a signaling pathway for cell survival (78). PI-3K phosphorylates phosphatidalinositol (3,4,5)-triphosphate (PtdIns-3,4,5-P3), an important intracellular second messenger, thus aiding in the transmission of signals for proliferation to the nucleus. PI-3K consists of a catalytic subunit, p110, and a regulatory subunit, p85, and there are five isoforms of each subunit. PI-3K phosphorylates protein kinase B (Akt/PKB) on serine and threonine residues, which in turn modulate cellular processes like glycolysis and translation initiation and elongation. Akt/PKB also phosphorylates Bad, a pro-apoptotic protein. When Bad is phosphorylated, it is sequestered by the 14-3-3 protein, rendering it incapable of binding to the anti-apoptotic protein, bcl-2, and thus, results in apoptosis. Akt's phosphorylation of Bad serves to inhibit apoptosis and promote cell survival. This has deleterious effects for the organism because tumor cells are not permitted to undergo apoptosis and will survive and divide.

PI-3K has been linked to the development of colon cancer by a study showing that genetic inactivation of the p110gamma catalytic subunit of PI-3K leads to the development of invasive colorectal adenocarcinomas in mice (79). This pathway is not completely separate from the Raf/MAPK pathway, because Akt has been found to inhibit Raf activity. In fact, none of the aforementioned ras-mediated pathways operate completely independently; there are multiple examples of crosstalk between these signaling pathways.

**4.1.4 Nonreceptor Tyrosine Kinases.** In addition to growth factor receptors, other nonreceptor kinases target protein tyrosines for phosphorylation and can become activated as oncogenes. Indeed, one of the first oncogenes to be discovered, src, is the best characterized member of a family of proteins that have oncogenic potential. The src family of proteins are post-translationally modified by attachment of a myristate moiety to the N-terminus, which enables association with the plasma membrane. The members of the src family of proteins exhibit 75% homology at the amino acid level with the greatest degree of similarity found in three regions that have been labeled src homology domains 1, 2, and 3 (e.g., SH1, SH2, and SH3). The SH1 domain encompases the domain that contains kinase activity. The SH2 and SH3 domains are located adjacent to and N-terminal to the kinase domain and function to promote protein/protein interactions. The SH2 domain binds with phosphorylated tyorsines, whereas the SH3 domain has affinity for the proline rich regions of proteins. Importantly, SH2 and SH3 domains are found in a large number of other proteins that are involved in intracellular signaling and that have oncogenic potential, and the structure of these domains are strongly conserved. Because SH2 and SH3 domains serve to potentiate signal transduction, they have also become targets for drug discovery programs aimed at disrupting the constitutive signaling generated by oncogenic activity (80).

A second oncogenic protein tyrosine kinase of considerable clinical importance is the Bcr-Abl oncogene. The Bcr-Abl protein is a chimeric fusion protein formed by a reciprocal translocation involving chromosomes 9 and 22. This chromosomal rearrangement is diagnostic for the hematopoietic malignancy, chronic myelogenous leukemia (CML), and the rearranged chromosome is known as the Philadelphia chromosome (81). The c-Abl gene maps to chromosome 9 and is a tyrosine kinase, whereas the BCR gene is now known to be GTPase-activating protein (GAP), which when fused to Abl results in an unregulated tyrosine kinase that functions to promote cellular proliferation (82). The bcr-abl protein interacts with SH2 domains on Grb 2 and relocates to the cytoskeleton and initiates ras signaling, a primary mode of tumorigenic potential. Bcr-abl reduces growth factor dependence, alters adhesion properties, and enhances viability of CML cells. Consequently, the kinase activity of Bcr-Abl is a primary factor in stimulating the proliferation of CML cells, and therefore, has become the target for drug therapies aimed at combating this cancer. Indeed, the drug STI571 has been spectacularly successful in the clinic at causing remission of this disease (83).

### 4.1.5 Transcription Factors as Oncogenes.

Another class of oncogenes are those that encode nuclear proteins, or transcription factors. Two examples of this class of oncogenes are AP-1 and c-myc. Activator protein-1 (AP-1) consists of Fos family members (c-fos, fos B, Fra 1, and Fra 2) and Jun family members (c-jun, jun B, and jun D), which can dimerize through a lucine rich protein/protein interaction domain known as the leucine zipper (84). Fos-jun heterodimers are the most active, jun-jun homodimers are weakly active, and fos-fos homodimers form only in extremely rare circumstances. These dimers bind to AP-1 DNA binding sites, which are also called the tumor promoter TPA-responsive element (TRE) or glucocorticoid response element (GRE). AP-1 can be activated by ionizing and ultraviolet irradiation, DNA damage, cytokines, and oxidative and cellular stresses (85).

AP-1 has several functions in the cell, including the promotion of cell proliferation and metastasis. AP-1 is a nuclear target for growth factor-induced signaling such as the aforementioned EGFR-mediated kinase cascade. AP-1–regulated genes include genes necessary for metastasis, and invasion like the MMPs matrilysin and stromelysin, as well as collagenase two proteins that aid in cell migration through connective tissue.

Deregulation of c-myc often occurs either by gene rearrangement or amplification in human cancers. Here again the hematologic cancers are instructive. In Burkitt's lymphoma, a frequent reciprocal translocation between chromosomes 8 and 14 leads to juxtapositioning of the myc gene adjacent to the Ig heavy chain promoter/enhancer complex, causing uncontrolled expression and production of the myc protein (86). Translocations between chromosomes 2 and 8 and between 8 and 22 also occur and involve other immunoglobulin producing gene complexes. In all cases the overproduction of myc results in uncontrolled cell proliferation.

Myc overexpression also occurs in solid tumors, but is usually the result of gene amplification (87). The oncogenic potential of c-myc has been studied most widely as it pertains to the development of colon cancer. Both c-myc RNA and protein are overexpressed at the early and late stages of colorectal tumorigenesis. The cause for this overexpression is still unknown, but a strong possibility may be that it is regulated by the APC pathway. The APC tumor suppressor gene is mutated in approximately 90% of colorectal tumors, both sporadic and inherited forms. APC will be discussed in detail in the "tumor suppressor" section of this chapter.

He et al. (88) found that when APC expression was induced in stably transfected APC$^{-/-}$ colon cancer cells (using an inducible metallothionine promoter linked to the APC gene), they observed a time-dependent decrease in the RNA and protein levels of c-myc. This suggested that c-myc may be regulated by APC through the $\beta$-catenin/T-cell factor-4 (Tcf-4) transcription complex. They also showed that constitutive expression of mutant $\beta$-catenin (mutated so that it is insensitive to APC) in embryonic kidney cells resulted in a significant increase of c-myc expression. Analysis of the c-myc gene revealed two possible Tcf-4 transcription factor binding sites. Mobility shift assays demonstrated that Tcf-4 binds to both of the potential binding sites, leading to c-myc gene expression. Expression of dominant-negative Tcf-4 in HCT116 (mutant $\beta$-catenin) or SW480 (mutant APC) reduced endogenous levels of c-myc (88).

The c-myc protein binds to DNA through its basic, helix-loop-helix/leucine zipper domain. Many target genes of c-myc have been identified that are involved in cell growth and proliferation. Some of these genes include ODC, cell cycle genes cyclins A, E, and D1, as well as cdc2, cdc25, eukaryotic initiation factor 4E (eIF4E), heat shock protein 70 (hsp70), and dihydrofolate reductase. Overexpression of c-myc may therefore affect the transcription of these genes, thus promoting hyperproliferation and tumorigenesis.

C-myc is also found to be amplified in promyelocytic leukemia and small cell lung cancer. The c-myc protein requires dimerization with Max to initiate transcription, and Max homodimers serve as an antagonist of transcription. The formation of Mad-Max dimers also suppresses transcription. It is also interesting to note that the full oncogenic potential of c-myc relies on cooperation with other oncogenes like ras.

**4.1.6 Cytoplasmic Proteins.** Bcl-2 is an example of a cytoplasmic oncogene that has anti-apoptotic potential. Increased production of bcl-2 protein is seen in a variety of tumor types and is associated with poor prognosis in carcinomas of the colon and prostate. The function of bcl-2 is explained in detail in the "apoptosis" section of this chapter.

## 4.2 Tumor Suppressor Genes

In contrast to oncogenes, tumor suppressor genes can directly or indirectly inhibit cell growth. Those that directly inhibit cell growth or promote cell death are known as "gatekeepers" and their activity is rate limiting for tumor cell proliferation. Hence, both copies of gatekeeper tumor suppressors must be functionally eliminated for tumors to develop. This characteristic requirement is a hallmark of tumor suppressor genes. Mutations that inactivate one allele of a gatekeeper gene can be inherited through the germline, which in conjunction with somatic mutation of the remaining allele, leads to cancer predisposition syndromes. For example, mutations of the APC gene lead to colon tumors. Somatic mutations that inactivate both gatekeeper alleles occur in sporadic tumors.

Those tumor suppressor genes that do not directly suppress proliferation, but function to promote genetic stability are known as "caretakers." Caretakers function in DNA repair pathways and elimination of caretakers results in increased mutation rates. Because numerous mutations are required for the full development of a tumor, elimination of caretaker tumor suppressors can greatly accelerate tumor progression. As with gatekeepers, mutations can be inherited through the germline and can give rise to cancer predisposition syndromes. An example of a caretaker gene is

**Table 1.4   Tumor Suppressor Genes**

| TS Gene | Protein Function | Neoplasm(s) |
|---|---|---|
| APC | cell adhesion | colon |
| BRCA 1 | transcription factor | breast and ovary |
| BRCA 2 | DNA repair | breast and ovary |
| CDK4 | cyclin D kinase | melanoma |
| hMLH1 | DNA mismatch repair | HNPCC[a] |
| hMSH2 | DNA mismatch repair | HNPCC |
| hPMS1 | DNA mismatch repair | HNPCC |
| hPMS2 | DNA mismatch repair | HNPCC |
| MEN1[b] | Ret receptor | thyroid |
| NF1 | GTPase | neuroblastoma |
| p53 | transcription factor | colon, lung, breast |
| Rb | cell cycle checkpoint | retinoblastoma |
| WT-1 | transcription factor | childhood kidney |

[a]Hereditary non-polyposis colon cancer.
[b]Multiple endocrine neoplasia.

MSH2, which functions in the mismatch DNA repair system, and inherited mutations in this gene gives rise to the hereditary nonpolyposis colorectal cancer (HNPCC) syndrome (Table 1.4).

### 4.2.1 Retinoblastoma.

Retinoblastoma (Rb) is a childhood disease. There are both hereditary and nonhereditary forms of the disease. Approximately 60% of patients develop the nonhereditary form and present with unilateral tumor development (one eye is affected). About 40% of Rb patients have a germline mutation that predisposes them to the disease. Of these patients, 80% of the cases are bilateral, 15% are unilateral, and about 5% are asymptomatic carriers of the mutation. It is an autosomal dominant trait and is caused by mutations in the Rb gene on chromosome 13. Abnormalities of the Rb gene have also been seen in breast, lung, and bladder cancers.

Retinoblastoma arises when both of the Rb alleles are inactivated. In the inherited form, one parental chromosome carries a defect (most often a deletion) at the Rb locus. A second somatic mutation must occur in retinal cells to cause the loss of the other (normal) Rb allele. In sporadic cases, both of the parental chromosomes are normal and both Rb alleles are lost as a result of individual somatic mutations. Approximately one-half of all retinoblastoma cases show a deletion at the Rb locus. The locus is very large, >150 kb, and there-

fore may be more susceptible to mutations because it is such a large target.

Rb was the first human tumor suppressor gene identified, and the loss of RB protein function leads to malignancy. The RB protein is localized in the nucleus where it is either phosphorylated or unphosphorylated (Fig. 1.10). When unphosphorylated, RB binds to the E2F transcription factor and prevents transcriptional activation of E2F target genes. This normally occurs during the M and early G1 phases of the cell cycle. During late G1, S, and G2 phases, RB is phosphorylated. When phosphorylated, RB can no longer bind to E2F. This release from inhibition allows E2F to activate transcription of S-phase genes and the cell cycle progresses. When loss of RB function occurs because of various mutations in the Rb gene, the cell cycle becomes deregulated, and uncontrolled cell division results. This is because RB can no longer bind to and inhibit E2F. Therefore, the transcription factor can constitutively activate its target genes. This ultimately leads to tumor development (89).

### 4.2.2 p53.

The p53 tumor suppressor is activated in response to a wide variety of cellular stresses including DNA damage, ribonucleotide depletion, redox modulation, hypoxia, changes in cell adhesion, and the stresses created by activated oncogenes. The p53 protein functions as a transcription factor that, when

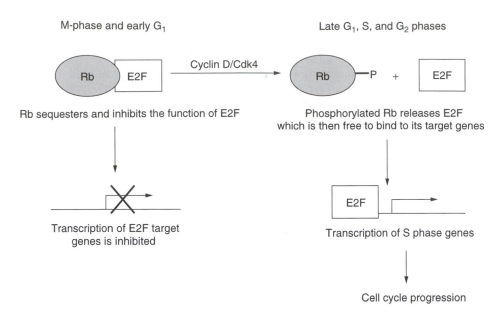

M-phase and early $G_1$              Late $G_1$, S, and $G_2$ phases

Cyclin D/Cdk4

Rb sequesters and inhibits the function of E2F

Phosphorylated Rb releases E2F
which is then free to bind to its target genes

Transcription of E2F target
genes is inhibited

Transcription of S phase genes

Cell cycle progression

**Figure 1.10.** Cell cycle control by the retinoblastoma (Rb) tumor suppressor protein. Unphosphorylated Rb negatively regulates progression into the S phase of the cell cycle by binding to the E2F transcription factor. In this complex, E2F is prevented from activating transcription of its target genes. During late G1, Rb is phosphorylated by the cyclin D/Cdk4 complex and can no longer sequester the E2F transcription factor. E2F then binds to its target S-phase genes, promoting their transcription and allowing the cell cycle to progress.

activated, stimulates the expression of a variety of effectors that bring about growth arrest, promote DNA repair, and stimulate cell death by apoptosis. Collectively these activities act to maintain genomic stability. Elimination of p53 function leads to increased rates of mutation and resistance to apoptosis. Thus, p53 sits at the crux of several biochemical pathways that are disrupted during tumorigenesis. Consequently, mutations in p53 are the most frequent genetic change encountered in human cancers.

p53 activity can be eliminated by at least three mechanisms. The most common event that leads to a nonfunctioning protein is mutation of the p53 gene, which occurs in about 50% of all sporadic human tumors. As with other tumor suppressors, mutations can occur in somatic tissues or can be inherited through the germline. Inherited p53 mutations give rise to the Li-Fraumeni syndrome in which affected individuals develop bone or soft-tissue sarcomas at an early age. In addition, nonmutational inactivation of p53 can occur in the presence of viral transforming antigens. For example, the simian virus 40 (SV40) large T antigen binds with p53 and forms an inactive complex, whereas the papilloma virus E6 protein eliminates p53 by causing premature degradation of the protein through the 26S proteosome. Clearly, the interaction between these transforming antigens and p53 is critical because viral antigens that are incapable of doing so lose their transforming ability. The third mechanism by which p53 activity can be eliminated is by cytoplasmic sequestration. p53 that is unable to enter the nucleus cannot induce the expression of downstream effector genes that are necessary for mounting the cellular response to genotoxic stress.

Activation of p53 by ionizing radiation (IR) and other DNA damaging agents involves a complex set of interdependent post-translational modifications that control protein/protein associations, protein turnover, and subcellular localization. Under normal conditions, levels of p53 are kept minimal by ubiquitination and proteosome-mediated degradation that contributes to the short half-life (3–20 min) of the protein. A key player in maintenance of low p53 levels is mdm2. Mdm2 performs this function by interacting with p53 at its $N$-terminus and targets p53 for proteosome-mediated degradation. Exposure to IR results in a series of, as yet incompletely understood, phosphorylation events in p53's $N$-terminus, which inhibits Mdm2 binding and results in increased intracellular p53 levels. Mdm2 and p53 function in a feedback loop where activated p53 stimulates the expression of Mdm2, which in turn reduces the duration of up-regulated p53 activity. Overexpression of Mdm2 suppresses p53 by preventing its accumulation in response to DNA damage. Consequently, Mdm2 can function as an oncogene that acts in much the same way as the papilloma virus E6 protein. In fact, Mdm2 is overexpressed in some tumors such as osteosarcomas.

The p53 protein can be divided into three structural domains that are essential for tumor suppressor function. The $N$-terminus consists of a transactivation domain that interacts with various basal transcription factors and cellular and viral proteins that modify its function. The central domain contains the sequence specific DNA binding activity. Most mutations in the p53 gene fall within this domain that disrupts the structure of this region and eliminates DNA binding activity. The importance of DNA binding is emphasized by the fact that mutations accumulate preferentially in several amino acids that are involved in directly contacting DNA. The C-terminus has been assigned several activities including non-specific DNA binding activity, acting as a binding site for other p53 molecules, and formation of p53 tetramers, and functioning as a pseudosubstrate domain that occludes the central DNA binding domain.

Because of the frequency with which p53 is mutated in human tumors, much attention has been directed at developing methods that compensate for the loss of wild-type function or can reactivate wild-type p53 activity in mutant proteins. For example, strategies aimed at manipulating the conformation of mutant proteins have led to the discovery that peptides that bind the C-terminus can reactivate wild-type function in some mutant proteins. Strategies that take advantage of the vast knowledge of virus biology and p53 function

have lead to the construction of viral vectors that can introduce a wild-type p53 into tumor cells. One clever approach takes advantage of the fact that adenoviruses with a defective E1B 55K protein cannot replicate in normal human cells. For adenoviruses to replicate in cells, they must suppress p53 activity, which functions to limit the uncontrolled DNA replication that is required for production of virus genomes. However, adenoviruses with a defective E1B 55K gene can replicate in tumor cells because they lack a functional p53. Thus, these viruses kill tumor cells specifically and leave normal cells untouched (90).

**4.2.3 Adenomatous Polyposis Coli.** The tumor suppressor gene, APC, is mutated in almost 90% of human colon cancers and 30% of melanoma skin cancers. The inherited loss of APC tumor suppressor function results in familial adenomatous polyposis (FAP). FAP patients develop hundreds to thousands of colon polyps by their second or third decade of life. By age 40, one or two of these polyps usually develops into a malignant carcinoma, and thus, many of these patients choose to have a colectomy to prevent carcinoma formation. Mutations in APC occur in the majority of sporadic colon cancers too.

APC mutation is an early event in colon carcinogenesis, and is therefore, considered to be the initiating event. Loss of this tumor suppressor gene results in constitutive activity of the oncogene, c-myc, through an intricate collection of protein-protein interactions. Briefly, APC interacts with other cellular proteins, including the oncogene $\beta$-catenin (Fig. 1.11). Axin, an inhibitor of Wnt signaling, forms a complex with glycogen synthase kinase 3$\beta$ (GSK3$\beta$), $\beta$-catenin, and APC and stimulates the phosphorylation of $\beta$-catenin by GSK3$\beta$, thus causing down-regulation of gene expression mediated by $\beta$-catenin/Tcf complexes (91). Dissociation of the axin, GSK3$\beta$, $\beta$-catenin, and APC complex by Wnt family members leads to stabilization of $\beta$-catenin and activation of Tcf-mediated transcription. Deletion of APC alleles, or mutations causing truncations in APC that influence its interaction with $\beta$-catenin, also leads to stabilization of $\beta$-catenin and activation of Tcf/lymphoid enhancing factor (Lef)–dependent gene expression. At least one member of the Tcf/Lef family of transcriptional activators has been identified in human colon mucosal tissues. This member is termed hTcf-4. Several target genes for Tcf/Lef have been identified, including the c-myc oncogene. Overexpression of wild-type APC cDNA in human colon tumor-derived HT29 cells, which lack a normal APC allele, causes down-regulation of c-myc transcription. Upregulation of β-catenin in cells expressing normal APC alleles causes increased c-myc expression. Thus, wild-type APC serves to suppress c-myc expression. Either normal regulation by Wnt signaling, or mutation/deletion of APC, activates c-myc expression. In many colon cancers, the APC gene is not necessarily mutated, but the mutation in the pathway is found in β-catenin, which yields the same constitutive signaling from the pathway.

APC regulates the rates of proliferation and apoptosis by several different mechanisms. Wild-type APC is important for cytoskeletal integrity, cellular adhesion, and Wnt signaling. APC plays a role in the G1/S transition of the cell cycle by modulating expression levels of c-myc and cyclin D1. Wild-type, full length APC is also important in maintaining intestinal cell migration up the crypt and inducing apoptosis.

### 4.2.4 Phosphatase and Tensin Homologue.

The phosphatase and tensin homologue (PTEN) or mutated in multiple advanced cancers (MMAC) tumor suppressor gene was first identified in the most aggressive form of brain cancer, glioblastoma multiform. PTEN also is mutated in a significant fraction of endometrial carcinomas, prostate carcinomas, and melanomas. PTEN's primary functions as a tumor suppressor gene are the induction of cell cycle arrest and apoptosis (92). PTEN is a dual-specificity phosphatase, meaning that it can dephosphorylate proteins on serine, threonine, and tyrosine residues. It specifically dephosphorylates PtdIns-3,4,5-P3, antagonizing the function of PI-3K. PTEN, therefore, acts as a negative regulator of Akt activation. Because Akt can suppress apoptosis by the phosphorylation of the pro-apoptotic protein Bad, PTEN can induce apoptosis of mutated or stressed cells to prevent tumor formation.

In addition to modulating apoptosis, PTEN plays a role in angiogenesis. PTEN suppresses the PI-3K-mediated induction of blood vessel growth factors like VEGF. EGF and ras act to induce genes regulated by the hypoxia-induced factor (HIF-1), which is blocked by PTEN activity. PTEN also inhibits cell migration and formation of focal adhesions when overexpressed in glioblastoma cell lines, suggesting that it helps to inhibit metastasis as well (93).

PTEN also inhibits signaling from the insulin growth factor receptor (IGF-R). Insulin receptor substrates-1/2 (IRS-1/2) are docking proteins that are recruited by the insulin receptor and in turn, recruit PI-3K for signal transduction. The tumor suppressor function of PTEN helps to prevent aberrant signaling when insulin binds to its cell surface receptor.

### 4.2.5 Transforming Growth Factor-β.

Transforming growth factor-β (TGF-β) is growth stimulatory in endothelial cells but growth inhibitory for epithelial cells, rendering it a tumor suppressor gene in epithelial-derived cancers. The TGF-β family of growth factors binds to two unique receptors, TGF-β type I and type II. Tumor cells lose their response to the growth factor and mutations in the receptors also contribute to carcinogenesis. Ligand binding to the TGF-β receptors causes intracellular signaling of other tumor suppressor genes, the Smad proteins. Smads help to initiate TGF-β-mediated gene transcription.

TGF-β1 normally inhibits growth of human colonic cells, but in the process of becoming tumorigenic, these cells obtain a decreased response to the growth inhibitory actions of TGF-β. TGF-β1 also serves as an inhibitor of immune surveillance (94). TGF-β1 indirectly suppresses the function of the immune system by inhibiting the production of TNF-α and by inhibiting the expression of class II major histocompatibility complex (MHC) molecules. TGF-β1 also promotes tumor progression by modulating processes necessary for metastasis such as degradation of the extracellular matrix, tumor cell invasion and VEGF-mediated angiogenesis.

The TGF-β receptor type II (TβRII) is mutated in association with microsatellite instability in most colorectal carcinomas (95). As

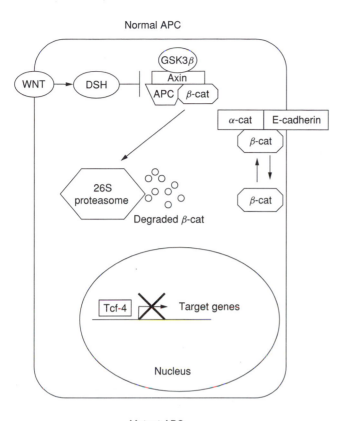

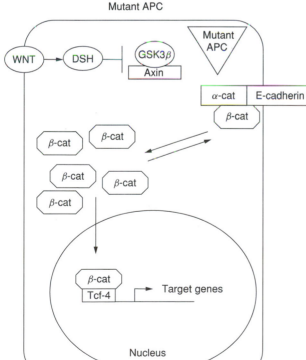

**Figure 1.11.** The APC signaling pathway. In a normal cell, APC forms a complex with axin, GSK-3$\beta$, and $\beta$-catenin. This promotes proteosomal degradation of $\beta$-catenin and prevents transcription of $\beta$-catenin/Tcf4 target genes. When APC is mutated, the multi-protein complex cannot form and $\beta$-catenin is not degraded. Instead, $\beta$-catenin is translocated to the nucleus where it binds with Tcf4 to activate transcription of various target genes. Some of the known target genes, like c-myc and cyclin D1, play important roles in cell proliferation.

many as 25% of colon cancers have missense mutations in the kinase domain of this receptor. A missense mutation in the kinase domain of the T$\beta$RI has also been identified in metastatic breast cancer. It was also found that the expression of the TGF-$\beta$2 receptor is suppressed in metastatic oral squamous cell carcinomas compared with the primary tumor.

**4.2.6 Heritable Cancer Syndromes.** There are several known inheritable DNA repair-deficiency diseases. Four of these are autosomal recessive diseases and include Xeroderma pigmentosum (XP), ataxia telangiectasia (AT), Fanconi's anemia (FA), and Bloom's syndrome (BS). XP patients are very sensitive to UV light and have increased predisposition to skin cancer (approximately 1000-fold) (96). AT patients exhibit a high incidence of lymphomas, and the incidence of lymphoma development is also increased for both FA and BS patients.

HNPCC arises due to a defect in mismatch repair (MMR). The incidence of HNPCC is often quoted as 1–10% of all colorectal cancers (97). It is an autosomal dominant disease and results in early onset of colorectal adenocarcinoma. Many of these tumors demonstrate microsatellite instability and are termed replication error positive (RER+). Endometrial and ovarian cancers are the second and third most common cancers in families with the HNPCC gene defect.

The most common mutations in HNPCC are in the mismatch repair genes, MSH2 and MLH1 (>80%) (98). The mismatch repair system normally corrects errors of 1–5 base pairs made during replication. Therefore, defects in this system result in many errors and create microsatellite instability. A suggested model for HNPCC development starts with a mutation in the MMR genes followed by another mutation in a gene such as APC. These two events lead to cellular hyperproliferation. Next, a mutation occurs leading to the inactivation of the wild-type allele of the MMR gene. Because of this MMR defect, mutations in other genes involved in tumor progression, such as deleted in colon cancer (DCC), p53, and K-ras, occur.

A variety of genes are responsible for the different inherited forms of GI cancers. For example, individuals with FAP, bearing germline mutations/deletions in the APC tumor suppressor gene, account for only a small fraction of colon cancers in the United States (<1%). However, the majority of sporadic colon adenomas have also been found to contain single allele alterations in APC and exhibit altered signaling of $\beta$-catenin, a protein negatively regulated by APC. Altered $\beta$-catenin signaling is inferred from immunohistochemical studies demonstrating that $\beta$-catenin is translocated to the nucleus in the majority of epithelial cells in adenomas, whereas $\beta$-catenin is generally seen associated with the cell membrane in normal colonic epithelia. These data suggest that the process of adenoma development selects for alterations in APC.

# 5  INTERVENTIONS

## 5.1  Prevention Strategies

Numerous investigators are taking advantage of our current knowledge of the mechanisms of carcinogenesis in human epithelial tissues to develop strategies for disrupting this process and thereby preventing cancer. As discussed earlier in this chapter, carcinogenesis proceeds by a multistep process, in which normal epithelial tissues acquire aberrant growth properties. These neoplastic cells progress to become invasive cancer. Historically, cancer therapy has addressed only the last phase of this process. Prevention strategies are now focusing on pre-invasive, yet neoplastic lesions.

Prevention strategies generally influence one or more of five processes in carcinogenesis (99). One strategy has been to inhibit carcinogen-induced initiation events, which lead to DNA damage. An important caveat to this strategy is that the intervention must be present at the time of carcinogen exposure to be effective. Once irreversible DNA damage has occurred, this type of strategy is ineffective in preventing cancer development.

Another strategy has been to inhibit initiated cell proliferation associated with the promotion stage of carcinogenesis. An advantage to this type of strategy is that interventions affecting promotion are effective after initiating events have occurred. Because humans are exposed to carcinogenic agents (e.g., chemicals in tobacco smoke, automobile exhaust) throughout their lifetimes, cancer preventive agents that work after initiating events have occurred are desirable. Two strategies of decreasing cell proliferation are induction of apoptosis, or cell death, and differentiation, which may or may not be associated with apoptosis. Induction of either differentiation or apoptosis will stabilize or decrease, respectively, overall cell number in a tissue.

A final strategy for preventing cancer is to inhibit development of the invasive phenotype in benign, or non-invasive, precancers that occur during the process of epithelial carcinogenesis.

Investigators are beginning to address the possibility that the efficacy of cancer prevention strategies may depend on both genetic and environmental risk factors affecting specific individuals. Mutations/deletion of the APC tumor suppressor gene, discussed earlier, causes intestinal tumor formation in both rodents and humans. Increasing levels of dietary fat increases intestinal tumor number in rodent models (100). However, mice with a defective APC gene develop tumors even on low-fat diets. Thus, dietary modifications may reduce carcinogenesis in individuals without, but may be ineffective in individuals with, certain genetic risk factors for specific cancers. Recently, several large randomized studies conducted in the United States have failed to detect any protective effect of dietary fiber increase or dietary fat decrease on colon polyp recurrence (101).

## 5.2 Targets

Targets for cancer prevention strategies can be either biochemical species produced by the action of a physical or chemical carcinogen or an enzyme/protein aberrantly expressed as a consequence of a genetic or environmental risk factor (the latter would include exposure to environmental carcinogens). In developing mechanism-based prevention or treatment strategies based on specific "targets," it is crucial to establish that the "target" is present in the target tissue (or cells influencing target tissue behaviors), causatively involved in the disease process in question and modulated by the intervention.

**5.2.1 Biochemical Targets.** One example of a biochemical target produced by carcinogens is reactive oxygen species (ROS). Ionizing radiation is a complete carcinogen and produces much of its DNA damage through ROS (102). Several strategies for preventing ROS-induced cell damage have been developed. The aminothiol, amifostine, inhibits radiation-induced DNA damage to a large degree by scavenging free radicals produced by ionizing radiation. Amifostine and its derivatives suppress ionizing radiation-induced transformation and carcinogenesis. Antioxidants, including protein and non-protein sulfhydrals and certain vitamins, are effective modulators of ROS produced by physical and chemical carcinogens (103). Antioxidants are effective in inhibiting carcinogenesis in some experimental models, but their roles in human cancer prevention remains unclear. At least some agents with antioxidant activity may increase carcinogenesis in some tissues. Heavy smokers receiving combinations of beta-carotene and vitamin A had excess lung cancer incidence and mortality, compared with control groups not receiving this intervention (104).

Other examples of biochemical targets are the dihydroxy bile acids, which are tumor promoters of colon cancer (105). Both genetic and dietary factors are known to influence intestinal luminal levels of these steroid-like molecules, whose levels are associated with colon cancer risk. Calcium reduces intestinal luminal bile acid levels by several possible mechanisms, and dietary calcium supplementation is associated with a small (∼25%), but statistically significant, reduction in colon polyp recurrence (106). This result requires cautious evaluation, however, as similar levels of calcium supplementation have been associated with increased risk of prostate cancer (107). This example and the result of the beta-carotene study mentioned above underscore the tissue-specific differences in carcinogenesis and the difficulties of applying common dietary components (e.g., calcium, antioxidants) in cancer prevention strategies in humans.

**5.2.2 Cyclooxygenase-2 and Cancer.** Cyclooxygenase (COX) enzymes catalyze prostaglandins from arachidonic acid. Prostaglandins play a role in biological processes

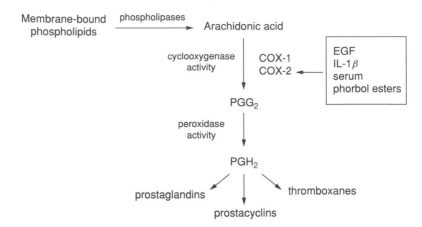

**Figure 1.12.** Cyclooxygenases catalyze prostaglandins from arachidonic acid. COX-2 is inducible by a variety of stimuli including growth factors, cytokines, and tumor promoters. PGH2 forms three classes of eicosanoids: prostaglandins, prostacyclins, and thromboxanes. Adapted from C. S. Williams and R. N. DuBois, *Am. J. Physiol.,* **270,** G393 (1996).

including blood clotting, ovulation, bone metabolism, nerve growth and development, and immune responses (108). There are two COX isoforms, COX-1 and COX-2. COX-1 is constitutively expressed in most cell types and is necessary for homeostasis of colonic epithelium and platelet aggregation. COX-2, on the other hand, is inducible by a variety of stimuli including growth factors, stress conditions, and cytokines (Fig. 1.12).

Several studies have implicated COX-2 in carcinogenesis. COX-2 protein levels, and therefore, prostaglandin production, are upregulated in many tumor types, including pancreatic, gastric, breast, skin, and colon cancers. Several lines of evidence suggest that overexpression of COX-2 plays an important role in colonic polyp formation and cancer progression. COX-2 modulates metastatic potential by inducing MMPs, which can be directly inhibited by COX-2 inhibitors. In addition, cells overexpressing COX-2 secrete increased levels of angiogenic factors like VEGF and bFGF. COX-2 not only aids in invasion but also inhibits apoptosis by up-regulating Bcl-2.

COX-2 has come under intensive study as a target for colon cancer prevention. Multiple studies have illustrated that COX-2 selective inhibitors suppress tumorigenesis in multiple intestinal neoplasia (Min) mice. COX-2 inhibitors also inhibit tumor cell growth in immunocompromised mice (109). The same phenomena has been illustrated in human chemoprevention trials. Recent studies have linked prolonged use of nonsteroidal anti-inflammatory drugs (NSAIDs) to decreased colon cancer risk and mortality. NSAIDs inhibit the cyclooxygenase enzymes, and new COX-2 selective agents are gaining popularity in the treatment of inflammation. NSAIDs that inhibit both COX-1 and COX-2 have been associated with reduced cancer risk in several large epidemiology studies. Whether inhibition of COX-1 and/or COX-2 is the optimal strategy for reducing risks of certain cancers is unknown.

Because COX-2 is induced in certain neoplastic tissues, the molecular regulation of its expression is being studied in a variety of experimental models. Human and rodent cell lines expressing various levels of COX-2 are being studied for genetic modifications that lead to the dysregulation of COX-2. COX-2 regulation occurs both transcriptionally and translationally, and this regulation differs depending on the species studied and the mutational status of the cell lines.

Signaling pathways leading to modulation of COX-2 expression are also being investigated. Both oncogenes and tumor suppressor genes have been shown to modulate COX-2 in cell model systems. The activation of the H-ras and K-ras oncogenes leads to induction of COX-2 expression in colon cancer cells. This induction is mediated by the stabilization of COX-2 mRNA. Wild-type, full-length APC suppresses COX-2 expression, suggesting that normal activity of this tumor suppressor gene may prevent cancer by inhibiting expression of cancer-promoting genes like COX-2. APC down-regulates COX-2 protein without affecting COX-2 mRNA levels. Thus, both ras and APC regulate COX-2 expression by post-transcriptional mechanisms. TGF-$\beta$1 is another tumor suppressor gene that influences expression of COX-2. TGF-$\beta$1-mediated transformation of rodent intestinal epithelial cells causes a significant induction of COX-2 protein expression. TGF-$\beta$1 synergistically enhances ras-induced COX-2 expression by stabilizing COX-2 mRNA. COX-2 expression is also influenced by the PI-3K pathway. Pharmacological inhibition of PI-3K or downstream PKB/Akt, as well as dominant-negative forms of Akt dramatically reduce COX-2 protein levels.

**5.2.3 Other Targets.** Technologies such as DNA microarrays are identifying genes that are aberrantly up-regulated in human intra-epithelial neoplasia (IEN). As discussed earlier, ODC, the first enzyme in polyamine synthesis, is up-regulated in a variety of IEN as a consequence of specific genetic alterations. Difluoromethylornithine (DFMO), an enzyme activated irreversible inhibitor of ODC, is a potent suppressor of several experimental models of epithelial carcinogenesis and is being evaluated in human cancer prevention trials (110). Pathways signaling cell behaviors are also activated in specific cancers. A number of agents, including NSAIDs and components of green and black teas, have been shown to inhibit certain signaling pathways in cell-type and tissue-specific manners.

## 5.3 Therapy

**5.3.1 Importance of Studying Gene Expression.** Cancer, among other diseases, is caused by the deregulation of gene expression. Some genes are overexpressed, producing abundant supplies of their gene products, whereas other crucial genes are suppressed or even deleted. The expression levels of genes associated with cancer influence processes such as cell proliferation, apoptosis, and invasion. Genes in-

volved in growth, for example, are often over-expressed in tumor tissues compared with normal adjacent tissue from the same organ. It is imperative to elucidate which genes are overexpressed or down-regulated in tumors because these genes represent critical therapeutic targets.

Researchers today generally concentrate on a few particular genes and study their regulation, expression, and downstream signaling using conventional molecular biology tools. With the onslaught of new genome data, and the development of the GeneChip, scientists are now able to study the expression levels of numerous genes simultaneously. The ability to analyze global profiles of gene expression in normal tissue compared with tumor tissue can help reveal how gene expression affects the overall process of carcinogenesis.

**5.3.2 cDNA Microarray Technology.** cDNA microarray technology is based on the simple concept of DNA base pairing. cDNA from tumor samples hybridize with the complementary DNA sequences on the chip. The DNA sequences are the target genes that will be studied for expression levels in particular tissues. These sequences, or probes, can be in the form of known oligos, DNA encoding the full-length gene, open reading frames (ORFs), or sometimes even the entire genome of an organism like *Saccharomyces cerevisiae*. Genes can be chosen by their proximity to each other on a chromosome or their similar functions. cDNA probes are then spotted onto a glass slide or computer chip (GeneChip), using a variety of different robotic techniques. A typical microarray slide will contain approximately 5000 genes.

cDNA microarray is particularly useful to the field of cancer biology because it allows scientists to study changes in gene expression caused when a normal tissue becomes neoplastic. In addition, normal tissue can be compared with preneoplastic lesions as well as metastatic cancer, to fully examine the entire tumorigenic process. The mRNA is extracted from cell lines or tissue and is reverse transcribed into the more stable form of cDNA. The cDNA is then labeled with reporters containing two colored dyes, rhodamine red, Cy3,

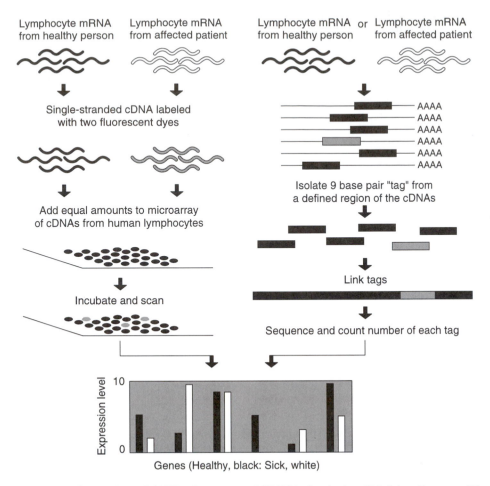

**Figure 1.13.** Comparison of cDNA microarray and SAGE technologies. At left is a diagram of the microarray assay for gene expression; the SAGE technique is illustrated at right. Here, the procedures assess how gene expression differs in lymphocytes from a healthy person and those from a person fighting off an infection. Reprinted with permission from K. Sutliff, *Science*, **270**, 368 (1995). American Association for the Advancement of Science.

and fluorescien green, Cy5. The cDNA is then hybridized to the DNA on the microarray slide. The slides are exposed to a laser beam, causing the dyes to give off their respective emissions and the relative expression levels of that gene are read and processed.

A similar technique to cDNA microarray that allows for multigene expression analysis is serial analysis of gene expression (SAGE) (Fig. 1.13). SAGE is based on the principle that a 9–10 nucleotide sequence contains sufficient information to identify a gene. These short nucleotide sequences are amplified by polymerase chain reaction (PCR) and then

30–50 of these SAGE "tags" are linked together as a single DNA molecule. These long DNA molecules are sequenced and the number of times that a single "tag" appears correlates to that gene's expression level. Proof of concept for this technique was illustrated in a study of gene expression in pancreatic cells. The most abundant "tags" found were those that encoded highly expressed pancreatic enzymes like trypsinogen 2. cDNA microarray methodology has also been validated by studies showing that expression data for tumor cell lines grown in tissue culture conditions can be classified according to their tissue of origin.

### 5.3.3 Discoveries from cDNA Microarray Data.
The contribution of microarray technology is influential in both the basic understanding of cancer pathology as well as in drug discovery and development. These studies reveal genes that may prove to be important diagnostic or prognostic markers of disease. They also can be used to predict adverse reactions to chemotherapies if mRNA from drug-treated cells is hybridized to panels of genes related to liver toxicity or the immune response.

Microarray technology also corroborates many *in vitro* cell studies that are criticized for ignoring the important role of other cell types in the tumor microenvironment. This technology can aid in distinguishing between cell type–specific or tumor-specific gene expression. For example, SAGE analysis of colon tumors and colon cancer cell lines showed 72% of the transcripts expressed at reduced levels in colon tumors were also expressed at reduced levels in the cell lines. One interesting finding from this study was that two commonly mutated oncogenes, c-fos and c-erbB3, were found to be expressed at higher levels in normal colonic epithelium than in colonic tumors; this contradicts reports that these oncogenes are up-regulated in transformed cells compared with normal cells. Again, microarray analysis is helping to merge cell biology studies with whole tumor biology.

Activation of the c-myc oncogene is a common genetic alteration occurring in many cancers. A cDNA microarray study found that c-myc activation leads to down-regulation of genes encoding extracellular matrix proteins, and thus, may play a role in regulating cell adhesion and structure. C-myc has also been associated with cell proliferation, which was illustrated by up-regulation of the genes eIF-5A and ODC. Another study of colon tumors revealed that only 1.8% of the 6000 transcripts studied were differentially expressed in normal tissues and tumors (111). Studies such as these suggest the critical importance of these differentially regulated genes in the cancer phenotype.

In addition to oncogene activation, the effects of tumor suppressor genes have been investigated through microarray technology. Over 30 novel transcripts were identified as regulated by p53 induction (112). Such a great number of genes simultaneously linked to p53 expression would not have been possible without SAGE technology. However, only 8% of these new genes were induced in normal cells compared with p53 knockout cells, suggesting that most of these p53-dependent genes are also dependent on other transcription factors. This is just one example of how microarray technology may be able to look at crosstalk in signaling pathways.

### 5.3.4 Limitations of Microarray Technologies.
Although cDNA microarray and SAGE technologies are quickly identifying new genes involved in tumorigenesis, there are significant limitations to these strategies. First, the expression pattern of a gene only provides indirect information about its function; a new gene may be classified as necessary for a certain biological process, but its exact role in that process cannot be determined. Second, mRNA levels do not always correlate with protein levels, and even protein expression may not translate into a physiological effect. Third, the up-regulation or suppression of a gene may be either the cause or the effect of a disease state and microarray technology does not distinguish between the two possibilities.

Both cDNA microarray and SAGE analyses require verification of changes in gene expression by Northern blots. Modest changes in gene expression are often overlooked when data is reported in terms of fourfold or greater changes. Because the ability to detect differences in gene expression is dependent on the magnitude of variance, a small induction or suppression of a gene may be discarded as inconsequential when it may actually be critical for downstream signaling of other genes.

## 5.4 Modifying Cell Adhesion

### 5.4.1 MMP Inhibitors.
Several MMP inhibitors are currently being developed for cancer treatment. If MMPs do play an integral role in malignant progression, then pharmacological inhibition of MMPs could inhibit tumor invasiveness. The inhibition of MMP function is currently the focus of most antimetastatic efforts. MMP inhibitors fall into three categories: (*1*) collagen peptidomimetics and non-

peptidomimetics, (2) tetracycline derivatives, and (3) bisphosphonates. The peptidomimetic MMP inhibitors have a structure that mimics that of collagen at the site where the MMP binds to it. Batimastat, a peptidomimetic inhibitor, was the first MMP inhibitor to be evaluated in cancer patients and is not orally available. Matimastat is orally available and is currently in phase II and III clinical trials (113). When bound to the MMP, these inhibitors chelate the zinc atom in the enzyme's active site. There are several nonpeptidomimetic inhibitors that are also in various phases of clinical trials. These are more specific than their peptidic counterparts and have exhibited antitumor activity in preclinical studies (113).

Tetracycline derivatives inhibit both the activity of the MMPs and their production. They can inhibit MMP-1, -3, and -13 (the collagenases) and MMP-2 and -9 (the gelatinases) by several different mechanisms. These mechanisms include (1) blocking MMP activity by chelation of zinc at the enzyme active site, (2) inhibiting the proteolytic activation of the pro-MMP, (3) decreasing the expression of the MMPs, and (4) preventing proteolytic and oxidative degradation of the MMPs.

The mechanism of action of the bisphosphonates has not been elucidated, but they have been used extensively for disorders in calcium homeostasis and recently in breast cancer and multiple myeloma patients to prevent bone metastases (114). Clodronate, a bisphosphonate, inhibited expression of MT1-MMP RNA and protein in a fibrosarcoma cell line and effectively reduced the invasion of melanoma and fibrosarcoma cell lines through artificial basement membranes (115).

### 5.4.2 Anticoagulants.

One theory surrounding the invasion process is that blood-clotting components may play a role in metastasis by either trapping the tumor cells in capillaries or by facilitating their adherence to capillary walls. Large numbers of tumor cells are released into the bloodstream during the metastatic process, and they must be able to survive the wide range of host defense mechanisms. Tumor cells have been shown to interact with platelets, lymphocytes, and leukocytes, and this may serve to promote metastasis. Studies have been done that inhibit tumor cell-platelet interactions, and these have resulted in a decreased probability of metastasis formation. It has also been shown that fibrin is always located in and around cancerous lesions, which may indicate that the cells use the fibrin structure as a support on which to attach themselves and grow. It may also serve as protection against host inflammatory cells so that the tumor is not destroyed.

Treating hepatic metastases of a human pancreatic cancer in a nude (lacking a thymus) mouse with prostacyclin, a potent inhibitor of platelet aggregation, led to a significant reduction in the mean surface area of the liver covered with tumor compared with the untreated control group (116). Many other groups have reported a reduction in metastatic potential with treatment of prostacyclin and prostacyclin-analogues, such as iloprost and cicaprost. There are currently over 50 different clinical trials in varying phases underway to determine the efficacy of these anticoagulant therapies. Most of these trials are in combination with other conventional anti-cancer regimens. So far, the experimental evidence indicates that anticoagulants or inhibitors of platelet aggregation are useful in the prevention of metastases.

### 5.4.3 Inhibitors of Angiogenesis.

The growth and expansion of tumors and their metastases are dependent on angiogenesis, or new blood vessel formation. Angiogenesis is regulated by a complex of stimulators and inhibitors (Fig. 1.14). The balance between the positive and negative regulators of angiogenesis inside a tumor environment is important for the homeostasis of microvessels. Tumor cells can secrete proangiogenic paracrine factors, which stimulate endothelial cells to form new blood vessels. The use of angiogenesis inhibitors may be a potential mode of therapy and is still in early clinical trials. This type of therapy would be a way of controlling the disease rather than eliminating it. Whereas toxicity may not be a major problem, adverse effects may be expected in fertility and wound healing.

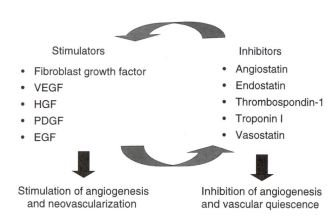

| Stimulators | Inhibitors |
|---|---|
| • Fibroblast growth factor | • Angiostatin |
| • VEGF | • Endostatin |
| • HGF | • Thrombospondin-1 |
| • PDGF | • Troponin I |
| • EGF | • Vasostatin |

Stimulation of angiogenesis and neovascularization

Inhibition of angiogenesis and vascular quiescence

**Figure 1.14.** Stimulators and inhibitors of angiogenesis. Under physiological conditions, the balance of factors that affect angiogenesis is precisely regulated. However, under pathophysiological conditions, normal angiogenesis is disturbed because of the continued production of stimulators.

## 5.5 Prospects for Gene Therapy of Cancer

Gene therapy is the transfer of genetic material into cells for therapeutic purpose. Gene transfer technology has become available after extensive study of molecular mechanisms of many diseases and improvement of techniques for manipulating genetic materials in the laboratory. Concepts for genetic therapy of cancer were developed based on knowledge that neoplasia is a molecular disorder resulting from loss of expression of recessive tumor suppressor genes and activation of dominant oncogenes.

Cancer gene therapy is aimed at correcting genetic mutations found in malignant cells or delivering biologically active material against cancer cells. One approach used in gene therapy of cancer is gene replacement/correction to restore the function of a defective homologous gene or to down-regulate oncogenic expression in somatic cells. Another approach is immune modulation by introduction of therapeutic genes, such as cytokines, into the target cells to treat cancer by stimulating an immune response against the tumor. Molecular therapy by activating prodrugs (e.g., ganciclovir, 5-fluorocytosine) within tumor cells and suicide gene therapy approaches have already been successful in early clinical trials. The high performance of these approaches fully depends on the efficacy and specificity of therapeutic gene expressing and delivery systems.

**5.5.1 Gene Delivery Systems.** The exogenous genetic material (the transgene) is usually introduced into tumor cells by a vector. A vector, or plasmid, is a circular DNA sequence that is designed to replicate inserted foreign DNA for the purpose of producing more protein product. Plasmids designed for gene therapy applications usually contain the gene of interest and regulatory elements that enhance the gene's expression. The ideal vector for gene therapy is one that would be safe, have high transfection efficiency, and be easy to manipulate and produce in large quantities. It would be efficient at delivering genetic material and selectively transducing cells within a tumor mass. The vector would be immunogenic for the recipient and would express the gene in a regulated fashion and at high levels as long as required.

There are two main approaches for the insertion of gene expressing systems into cells. In the *ex vivo* technique, cells affected by the disease are transfected with a therapeutic gene *in vitro* for the expression of exogenous genetic material. After viral propagation, replication is rendered incompetent and these cells can be transplanted into the recipient. In the *in vivo* technique, vectors are inserted directly into target tissue by systemic injections of the gene expressing system.

The simplest delivery system is a plasmid by itself, or so-called naked DNA. Direct injections of DNA have been successfully used to transfect tissues with low levels of nuclease activity in muscle tissue (117), liver (118), and experimental melanoma (119). Systemic injection of naked DNA is, in general, much less efficient because serum nucleases degrade plasmid DNA in the blood within minutes (120).

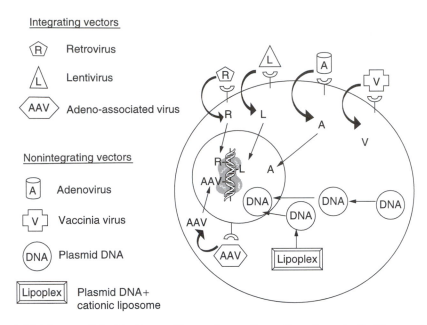

**Figure 1.15.** Virus particles bind to specific receptors on the surface of target cells. These vectors are internalized and their genome enters the cells. In the case of retroviruses, the single-stranded RNA genome is converted into double-stranded DNA by the reverse transcriptase enzyme encoded by the virus. The double-stranded DNA is taken up by the nucleus and integrated within the host genome as a provirus. The integration is random for retroviruses. Lentiviruses have a similar life cycle. Adenovirus binds to specific receptors on the surface of susceptible cells and are then absorbed and internalized by receptor-mediated endocytosis. The viral genome enters the cytoplasm of the cell and the double-stranded DNA genome is taken up by the nucleus. Vaccinia virus replicates in the cytoplasm of cells. DNA delivered by lipoplex and other nonviral systems enters cells through electrostatic interactions (endocytosis, phagocytosis, pinocytosis, and direct fusion with cell membrane). DNA is released before entry into the nucleus, where it stays as an episome.

To protect DNA on systemic application, it is usually complexed with viruses or with cationic lipids, polymers, or peptides. The resulting complex protects the DNA from the attack of nucleases and potentially improves transfection efficiency and specificity on multiple levels through interaction of DNA complexes with the various biological barriers.

The choice of viral or non-viral (synthetic) delivery strategy depends on localization and type of affected tissue, as well as on therapeutic approach. Viral vectors use the ability of viruses to overcome the cellular barriers and introduce genetic material either through the integration of the vector into the host genome (retroviruses, lentiviruses, adeno-associated viruses) or by episomal delivery (adenoviruses) followed by stable gene expression (Fig. 1.15).

*5.5.1.1 Viral Vectors.* Retroviral vectors have been used for *ex vivo* gene delivery and are the most useful vectors for stably integrating foreign DNA into target cells. Retroviruses are enveloped viruses that contain 7- to 12-kb RNA genomes. After the virus enters the cells through specific cell surface receptors, its genome is reverse transcribed into double-stranded DNA and subsequently integrated into the host chromosome in the form of a provirus. The provirus replicates along with the host chromosome and is transmitted to all of the host cell progeny. Because the retrovirus genome is relatively small and well characterized, it was possible to engineer a vector encoding only the transgene without replication competent viruses (RCV) or virus structural genes.

The most widely used retrovirus vectors are based on murine leukemia viruses (MLV). The lack of specificity of these vectors is a major obstacle for appropriate and controlled expression of foreign genes. Retroviruses are not efficient for direct *in vivo* injection because of inactivation by the host immune system (121). To circumvent this, *cis*-acting viral sequences, such as long terminal repeats (LTRs), transfer RNA (tRNA) primer binding sites, and polypurine tracts, have been used for developing packaging systems of retrovirus vectors. Many recombinant retrovirus vectors are designed to express two genes, one of which is often a selectable marker. New strategies for expression, such as splicing, transcription from heterologous promoters, and translation directed by an internal ribosome entry signal (IRES), have been used for expression of the second gene. Attempts have also been made to achieve efficient gene delivery by targeting retroviral integration through modifying protein sequences in the viral envelope (122). These modifications include various targeting ligands, particularly ligands for the human EGFR, erythropoetin receptor, and single chain antibody fragments against the low-density lipoprotein (LDL) receptor.

Examples of lentiviruses are the human immunodeficiency virus (HIV-1), the equine infectious anemia virus, and the feline immunodeficiency virus (123). Although lentiviruses also have an RNA genome, their advantage compared with other retroviruses is the ability to infect and stably integrate into non-dividing cells. To create a safe gene transfer vector based on the HIV-1 genome, the genome was altered and mutated to produce replication-defective particles. Several studies, both *in vitro* and *in vivo*, have shown successful gene transfer, including transduction of non-dividing hematopoetic cells at high efficiencies (up to 90%) and stable gene expression in several target tissues of interest such as liver (8 weeks) and muscle and brain (6 months) with no detectable immune response (124). At the same time, the safety concern still remains for *in vivo* applications of this vector.

Vaccinia virus is a member of the Poxviridae family, which possesses a complex DNA genome encoding more than 200 proteins. The advantages of using vaccinia viruses for gene transfer include their ability to accommodate large or multiple gene inserts, to infect cells during different stages of the cell cycle, and their unique feature to replicate in the cytoplasm. Recombinant vaccinia vectors can be constructed using homologous recombination after transfection of vaccinia virus-infected cells with plasmid DNA constructs. This vector has been used in clinical trials to deliver genes encoding tumor antigens such as melanoma antigen (MAGE-1), carcinoembryonic antigen (CEA), prostate-specific antigen (PSA), interleukins (e.g., IL-1$\beta$, IL-12), and costimulatory molecule B7 (125).

In recent years, there has been a great interest in the use of adenoviral vectors for cancer gene therapy. The main reasons for this are the ease in construction of adenoviruses in the laboratory and their ability to grow to high titers, infect a variety of cell types, and produce the heterologous protein of interest in dividing and non-dividing cells. Adenoviruses are also characterized by efficient receptor-mediated endocytosis, mediated by its fiber protein, and on infection of cancer cells, they exhibit high levels of transgene expression (126). They often are used to transfer genes of large sizes because of their high packaging capacity (up to 36 kb). Adenoviral vectors do not integrate into the host chromosomes, and therefore, they are degraded by the host. This results in a short-term expression of the transduced gene, which, nevertheless, could be sufficient to achieve the cancer gene therapy efficacy. Adenoviruses are widely used for direct *in vivo* injections. Adenoviruses are DNA-containing, non-enveloped viruses.

The two most commonly used adenoviruses for recombinant vectors are Ad2 and Ad5, mainly because their genomes have been best characterized and because these viruses have never been shown to induce tumors. Adenoviruses, like other viral vectors, lack cell and tissue specificity. To improve targeted gene delivery, attempts have been made to couple ligands or antibodies to the adenovirus capsid proteins (127). Specificity of the transgene expression also can be introduced by using tissue specific antigens, such as CEA for the treatment of pancreatic and colon cancers, mucin (MUC-1) promoters for breast cancer cells, al-

**Table 1.5  Cell-Type Specific Promoters for Targeted Gene Expression**

| Promoter | Target Cell/Tissue | Therapeutic Gene |
|---|---|---|
| PEPCK promoter | Hepatocytes | Neomycin phosphotransferase, Growth hormone |
| AFP promoter | Hepatocellular carcinoma | HSV-tk, VZV-tk |
| MMTV-LTR | Mammary carcinoma | TNFα |
| WAP promoter | Mammary carcinoma | Recombinant protein C |
| β-casein | Mammary carcinoma | In development |
| CEA promoter | Colon and lung carcinoma | HSV-tk, CD |
| SLPI promoter | Carcinomas | HSV-tk, CD |
| Tyrosinase promoter | Melanomas | HSV-tk, IL-2 |
| c-erbB2 promoter | Breast, pancreatic, gastric carcinomas | CD, HSV-tk |
| Myc-max-responsive element | Lung | HSV-tk |
| *Therapy-Inducible Tissues* | | |
| Egr-1 promoter | Irradiated tumors | TNFα |
| Grp78 promoter | Anoxic, acidic tumor tissue | Neomycin phosphotransferase |
| MDR1 promoter | Chemotherapy-treated tumors | TNFα |
| HSP70 | Hyperthermy-treated tumors | IL-2 |

pha-fetoprotein promoters for hepatocellular carcinoma, and the tyrosinase promoter for melanoma (126). *In vivo* administration of adenoviral vector has been extensively used in preclinical and clinical cancer therapy (128).

There are many regulatory elements controlling cell type–specific gene expression and inducible sequences within promoters that have been used in construction of viral vectors for cancer therapy. Vector systems that include cell type–specific promoters or elements responding to regulatory signals represent a way for a safe, selective, and controlled expression of therapeutic genes that could increase efficacy and stability of gene expression (Table 1.5).

Vectors based on adenoassociated viruses (AAV) also have been successfully used to transfer genes. AAV is a small, single-stranded DNA virus that requires a helper virus for infection, usually an adenovirus or herpesvirus. AAV vectors can be used for the delivery of antisense genes, "suicide" gene therapy, and recently, for the delivery of anti-angiogenic factors. Recent studies in the area of vector design have been focused on conditional expression that can be induced by antibiotics (129), heat shock (130), or other small molecules (131).

*5.5.1.2 Non-Viral Gene Delivery Systems.* Non-viral gene delivery systems are based on non-covalent bonds between cationic carrier molecules (e.g., lipids or polymers) and the negatively charged plasmid DNA. Complexes of DNA with three main groups of materials, i.e., cationic lipids (lipoplex) such as CTAB and DMRI, polymers (polyplex) such as poly-L-lysine and polyathylenimine, or peptides have been evaluated as synthetic gene delivery systems (132). The formation of these complexes, which is generally based on electrostatic interactions with the plasmid DNA, is difficult to control as they depend on both the stoichiometry of DNA and complexing agent and on kinetic parameters (e.g., speed of mixing and volumes). It has been shown that DNA is efficiently condensed and protected from nucleases at higher lipid:DNA ratios, proving that the positive charges of the complexes are important for the interaction with cells *in vitro* and *in vivo*. Although the resulting particles are stable, they have a high tendency to interact non-specifically with biological surfaces and molecules.

Lipoplexes are actively used in clinical trials for *in vivo* and *ex vivo* delivery of genes encoding cytokines, immunostimulatory molecules, and adenoviral genes (133). *In vivo*, these interactions may compromise the tissue-specific delivery of the complexes, creating uneven biodistribution and transgene expression in the body, particularly, in lungs. To overcome this problem, the complexes can be injected either into the vasculature or directly into the affected organ (134).

The combinatorial gene delivery approach uses the whole virus, either replication deficient or inactivated, or only essential viral components, together with the non-viral system. Systems, based on adenovirus ("adenofection"), or viral proteins that are required to trigger efficient endosomal escape, and polyplex and lipoplex non-viral systems have shown improvement in transfection efficiency and resistance to endosomal degradation (135).

## 5.6  Gene Therapy Approaches

**5.6.1 Immunomodulation.** This approach employs the patient's physiological immune response cascade to amplify therapeutic effects (136). Most patients with cancer lack an effective immune response to their tumors. This could be caused by defects in antigen presentation, stimulation, or differentiation of activated T cells into functional effector cells. Antitumor immunity response requires participation of different immune cells, including helper effector T-cells (Th), cytotoxic T-lymphocytes (CTLs), and natural killer (NK) cells. Activation of $CD4^+$ and $CD8^+$ T-cells requires at least two major signals. The first signal is triggered by binding of complexes of T-cell receptor (TCR) and specific antigenic peptide with MHC-class II or I molecules, respectively. The second signal for $CD4^+$ T-cells is provided by engagement of CD28 on the T-cell surface by members of the B7 family of costimulatory molecules on the surface of professional antigen-presenting cells. The nature of second signal for $CD8^+$ T-cells has not been completely understood but requires the presence of helper $CD4^+$ T-cells. Following activation and clonal expansion, activated $CD4^+$ T-cells differentiate into helper effector cells of either the Th1 or Th2 phenotype. Th1 cells produce cytokines, such as IL-2, interferon-$\gamma$, and TNF, that stimulate monocytes and NK cells and promote the differentiation of activated $CD8^+$ T-cells into CTLs.

The growing understanding of the biological basis of antigen-specific cellular recognition and experimental studies of an antitumor effect mediated through the cellular immune system helped to develop various immuno-modulation strategies. Modulation of immune response can be achieved through stimulation and modification of immune effector cells, enabling them to recognize and reject cells that carry a tumor antigen. Additionally, tumor cells can be genetically modified to increase immunogenicity and trigger an immune response.

Cytokine levels are relatively low in cancer patients. To correct for this deficiency, cytokines can be introduced as recombinant molecules, and this is advantageous in controlling their blood concentration and biological activity. Because cytokines are relatively unstable *in vivo*, cancer patients have to receive a large amount of the recombinant protein to maintain the required blood concentration for biological activity. Administration of the protein is often toxic to the patients. Another therapeutic approach is the introduction of genes encoding various cytokines, costimulatory molecules, allogenic antigens, and tumor-associated antigens into tumors (137). Previous preclinical studies have shown that cytokines that facilitate Th1 cell-mediated immune reactions but not Th2 cell-mediated reactions, when produced in tumors, are effective for antitumor responses. In addition, cytokines or costimulatory molecules delivered to tumor cells may enhance the transfer of tumor antigens to antigen-presenting cells. The most potent known antigen-presenting cells for actively stimulating specific cellular immune responses are dendritic cells. *Ex vivo* gene delivery to cultured dendritic cells or direct *in vivo* gene delivery to antigen-presenting cells can be more efficient in stimulating cellular antitumor immunity (138).

Several technical problems of expressing sufficient amounts of immunostimulatory proteins in appropriate target cells remain unsolved, but the potential of immune modulation gene therapy is high. Immunotherapy trials also contribute to the present knowledge of

**Figure 1.16.** Mechanisms of thymidine kinase (TK) ganciclovir (GCV)-induced apoptosis. TK phosphorylates the nontoxic prodrug GCV to GCV-triphosphate (GCV-PPP), which causes chain termination and single-strand breaks on incorporation into DNA. TK/GCV induces p53 accumulation, which can cause translocation of preformed death receptor CD95 from the Golgi apparatus to the cell surface without inducing *de novo* synthesis of CD95. The signaling complex then is formed by CD95, the adapter molecule Fas-associated death domain (FADD) protein, and the initiator caspase-8, which leads to cleavage of caspases causing apoptosis. TK/GCV also leads to mitochondria damage, including loss of mitochondrial membrane potential and the release of cytochrome c inducing caspase activation and nuclear fragmentation.

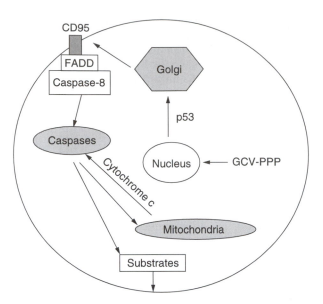

how antitumor responses can be effectively produced in cancer patients.

**5.6.2 Suicidal Gene Approach.** Elimination of cancer cells can be accomplished by the introduction of vectors that specifically express death promoting genes in tumor cells. One method, called suicide gene therapy, involves the expression of a gene encoding an enzyme, normally not present in human cells, that converts a systemically delivered nontoxic prodrug into a toxic agent. The toxin should kill the cancer cells expressing the gene as well as the surrounding cells not expressing the gene (bystander effect).

The herpes simplex virus thymidine kinase type 1 (HSV-tk) gene was initially used for long-term replacement gene therapy because it is about 1000-fold more efficient than mammalian thymidine kinase at phosphorylating the nontoxic prodrug ganciclovir (GCV) into its toxic metabolite ganciclovir triphosphate. The efficacy of HSV-tk transduction of tumors followed by ganciclovir therapy has been confirmed by systemic administration of ganciclovir after intratumoral injection of fibroblasts transduced with an HSV-tk retroviral vector in several preclinical models (139). The molecular mechanism of HSV-tk therapy is based on induction of apoptosis in target cells through accumulation of p53 protein (Fig. 1.16). Clinical trials of HSV-tk suicide gene therapy,

where ganciclovir was given after the retroviral or adenoviral introduction of HSV-tk gene, have been conducted in patients with brain tumors, melanoma, or mesothelioma (140–142) (Fig. 1.16).

Another suicide gene under active investigation for cancer therapy is the cytosine deaminase (CD) gene. CD converts the nontoxic fluoropyrimidine 5-fluorocytosine to 5-fluorouracil. Transduction of the CD renders tumor cells sensitive to 5-fluorocytosine *in vitro* and *in vivo*. The CD/5-fluorocytosine system has been used in a clinical trial, where adenovirus expressing the CD gene was injected intratumorally into hepatic metastases from colorectal cancer (143). As with HSV-tk gene transfer, evidence exists that cytosine-deaminase gene transfer into tumor cells promotes antitumor immune responses. The malignancies targeted with suicide gene therapy in the field of pediatric oncology are brain tumors, neuroblastoma, and acute lymphoblastic leukemia (144).

**5.6.3 Targeting Loss of Tumor Suppressor Function and Oncogene Overexpression.** Several tumor suppressor genes, including p53, Rb, and APC, have been identified by their association with hereditary cancers. Many sporadic tumors harbor inactivating or recessive mutations in one or more tumor suppressor genes. Gene transfer techniques can be applied to introduce wild-type copies of tumor suppressor genes into malignant cells, thus potentially reversing the neoplastic phenotype. The p53 tumor suppressor gene has been of interest because p53 mutation occurs commonly in a variety of human cancers, including breast, lung, colon, prostate, bladder, and cervix. The use of adenoviral vectors to deliver the p53 transgene to human tumors is now under evaluation in several clinical trials (145). The overexpression of Fas ligand caused by adenovirus-mediated wild-type p53 gene transfer induces neutrophil infiltration into human colorectal tumors, which may play a critical role in the bystander effect of p53 gene therapy (146).

Besides the p53 gene, other tumor suppressor genes that regulate the cell cycle have been used in cancer gene therapy. Among them are Rb, BRCA1, PTEN, p16, E2F, and fragile histidine triad (FHIT) genes. Clinical trials with BRCA1 and Rb have been initiated (147).

Protooncogenes, in contrast to tumor suppressor genes, gain dominant mutation resulting in excessive expression of their protein products, which lead to development of the malignant phenotype. Three members of the Ras family of oncogenes (H-ras, K-ras, and N-ras) are among the most commonly activated oncogenes in human cancers. Several strategies have been designed to combat K-ras mutations, including antisense nucleotide, ribozymes (148–150), and intracellular single-chain antibodies (151). cDNA encoding antisense RNA can be delivered using the viral vector system approach. *In vivo* gene therapy with K-ras, c-fos, and c-myc antisense nucleotides is currently being applied in clinical trials.

**5.6.4 Angiogenesis Control.** Gene therapy offers a new strategy for the delivery of angiogenesis inhibitors. By engineering and delivering vectors that carry the coding sequence for an antiangiogenic protein, it is possible to produce high levels of antiangiogenic factors in the tumor location or to systemically prevent the growth of distant metastasis. Several angiogenic inhibitors, such as angiostatin (152), endostatin (153), plasminogen activator inhibitor type 1 (154), and truncated VEGF receptor (155), have been tested using this approach. These studies have demonstrated that retroviral and adenoviral vectors could be used to inhibit endothelial cell growth *in vitro* and angiogenesis *in vivo*. The inhibition of tumor-associated angiogenesis results in increased apoptotic tumor cell death, leading to inhibition of tumor growth.

**5.6.5 Matrix Metalloproteinase.** As mentioned earlier in the chapter, MMPs are capable of proteolytic degradation of stromal ECM, which is essential in cancer cell migration and invasion, as well as in tumor-induced angiogenesis. The activity of MMPs *in vivo* is inhibited by TIMPs, small secreted proteins with molecular weight of between 20 and 30 kDa. TIMPs inhibit MMPs by binding to both the latent and active forms of MMPs. The following properties of TIMPs such as secretion, diffusion (TIMP-1, -2 and -4), induction of apoptosis (TIMP-3), and inhibition of multiple MMPs make them very attractive tools for gene therapy application.

Inhibition of cancer cell invasion after overexpression of TIMPs using different gene delivery vectors has been shown *in vitro* in gastric cancer cells and mammary carcinoma cells (156, 157). Overexpression *in vitro* of TIMP-2, which was delivered by a recombinant adenovirus (AdTIMP-2), inhibited the invasion of both tumor and endothelial cells in three murine models without affecting cell proliferation (158). Its *in vivo* efficiency has been evaluated in the LLC murine lung cancer model, the colon cancer C51 model, as well as in MDA-MB231 human breast cancer in athymic mice. Preinfection of tumor cells by AdTIMP-2 resulted in an inhibition of tumor establishment in more than 50% of mice in LLC and C51 models and in 100% of mice in the MDA-MB231 model. A single local injection of AdTIMP-2 into preestablished tumors of these three tumor types reduced tumor growth rates by 60–80%, and the tumor-associated angiogenesis index by 25–75%. Lung metastasis of LLC tumors was inhibited by >90%. In addition, AdTIMP-2-treated mice showed a significantly prolonged survival in all the cancer models tested. These data demonstrate the potential of adenovirus-mediated TIMP-2 therapy in cancer treatment.

## 6  ACKNOWLEDGMENTS

We would like to acknowledge the expert and dedicated assistance of Kim Nicolini without whom this chapter would never have been completed. Her perseverance and insistence that we keep pushing forward to the end are much appreciated.

## REFERENCES

1. M. H. Myers and L. A. Ries, *Can. Cancer J. Clin.*, **39**, 21–32 (1989).
2. R. Peto, *IARC Sci. Publ.*, **39**, 27–28 (1982).
3. P. C. Nowell, *Science*, **194**, 23–28 (1976).
4. D. Stehelin, H. E. Varmus, J. M. Bishop, and P. K. Vogt, *Nature*, **260**, 170–173 (1976).
5. H. Land, L. F. Parada, and R. A. Weinberg, *Nature*, **304**, 596–602 (1983).
6. E. R. Fearon and B. Vogelstein, *Cell*, **61**, 759–767 (1990).
7. H. Hennings, A. B. Glick, D. A. Greenhalgh, D. L. Morgan, J. E. Strickland, T. Tennenbaum, and S. H. Yuspa, *Proc. Soc. Exp. Biol. Med.*, **202**, 1–8 (1993).
8. M. A. Nelson, B. W. Futscher, T. Kinsella, J. Wymer, and G. T. Bowden, *Proc. Natl. Acad. Sci. USA*, **89**, 6398–6402 (1992).
9. T. Minamoto, M. Mai, and Z. Ronai, *Carcinogenesis*, **20**, 519–527 (1999).
10. I. Tannock and R. P. Hill, *The Basic Science of Oncology*, Pergamon Press, New York, 1987.
11. J. S. Bertram, *Mol. Aspects Med.*, **21**, 167–223 (2000).
12. E. C. Friedberg, G. C. Walker, and W. Siede, *DNA Repair and Mutagenesis*, American Society for Microbiology Press, Washington, DC, 1995.
13. M. L. Slattery, W. Samowitz, L. Ballard, D. Schaffer, M. Leppert, and J. D. Potter, *Cancer Res.*, **61**, 1000–1004 (2001).
14. M. Iwamoto, D. J. Ahnen, W. A. Franklin, and T. H. Maltzman, *Carcinogenesis*, **21**, 1935–1940 (2000).
15. Y. Guo, R. B. Harris, D. Rosson, D. Boorman, and T. G. O'Brien, *Cancer Res.*, **60**, 6314–6317 (2000).
16. T. Bestor, A. Laudano, R. Mattaliano, and V. Ingram, *J. Mol. Biol.*, **203**, 971–983 (1988).
17. A. P. Bird and A. P. Wolffe, *Cell*, **99**, 451–454 (1999).
18. W. Doerfler, *Annu. Rev. Biochem.*, **52**, 93–124 (1983).
19. T. Goto and M. Monk, *Microbiol. Mol. Biol. Rev.*, **62**, 362–378 (1998).
20. D. P. Barlow, *Science*, **270**, 1610–1613 (1995).
21. S. B. Baylin, M. Esteller, M. R. Rountree, K. E. Bachman, K. Schuebel, and J. G. Herman, *Hum. Mol. Genet.*, **10**, 687–692 (2001).
22. R. Z. Chen, U. Pettersson, C. Beard, L. Jackson-Grusby, and R. Jaenisch, *Nature*, **395**, 89–93 (1998).
23. M. O. Hiltunen, L. Alhonen, J. Koistinaho, S. Myohanen, M. Paakkonen, S. Marin, V. M. Kosma, and J. Janne, *Int. J. Cancer*, **70**, 644–648 (1997).
24. M. F. Kane, M. Loda, G. M. Gaida, J. Lipman, R. Mishra, H. Goldman, J. M. Jessup, and R. Kolodner, *Cancer Res.*, **57**, 808–811 (1997).
25. S. J. Nass, J. G. Herman, E. Gabrielson, P. W. Iversen, F. F. Parl, N. E. Davidson, and J. R. Graff, *Cancer Res.*, **60**, 4346–4348 (2000).
26. C. M. Bender, J. M. Zingg, and P. A. Jones, *Pharm. Res.*, **15**, 175–187 (1998).
27. L. E. Lantry, Z. Zhang, K. A. Crist, Y. Wang, G. J. Kelloff, R. A. Lubet, and M. You, *Carcinogenesis*, **20**, 343–346 (1999).
28. R. W. Ruddon,. *Cancer Biology*, 3rd ed., Oxford University Press, New York, (1995).
29. J. C. Houck, V. K. Sharma, and L. Hayflick, *Proc. Soc. Exp. Biol. Med.*, **137**, 331–333 (1971).
30. D. J. Bearss, L. H. Hurley, and D. D. Von Hoff, *Oncogene*, **19**, 6632–6641 (2000).
31. F. Ciardiello, *Drugs*, **60**, 25–32 (2000).
32. I. Stamenkovic, *Sem. Cancer Biol.*, **10**, 415–433 (2000).
33. H. Nagase, *Biol. Chem.*, **378**, 151–160 (1997).
34. A. Lochter and M. J. Bissell, *Apmis.*, **107**, 128–136 (1999).
35. B. Davidson, I. Goldberg, P. Liokumovich, J. Kopolovic, W. H. Gotlieb, L. Lerner-Geva, I. Reder, G. Ben-Baruch, and R. Reich, *Int. J. Gynecol. Pathol.*, **17**, 295–301 (1998).
36. J. D. Knox, J. Wolf, K. McDaniel, V. Clark, M. Loriot, G. T. Bowden, and R. B. Nagle, *Mol. Carcinog.*, **15**, 57–63 (1996).
37. R. G. Chirivi, A. Garofalo, M. J. Crimmin, L. J. Bawden, A. Stoppacciaro, P. D. Brown, and R. Giavazzi, *Int. J. Cancer*, **58**, 460–464 (1994).
38. A. E. Aplin, A. Howe, S. K. Alahari, and R. L. Juliano, *Pharmacol Rev.*, **50**, 197–263 (1998).
39. Z. Dong, R. Kumar, X. Yang, and I. J. Fidler, *Cell*, **88**, 801–810 (1997).
40. J. P. Witty, S. McDonnell, K. J. Newell, P. Cannon, M. Navre, R. J. Tressler, and L. M. Matrisian, *Cancer Res.*, **54**, 4805–4812 (1994).

41. A. B. Pardee, *Proc. Natl. Acad. Sci. USA*, **71**, 1286–1290 (1974).

42. C. Hutchison and D. M. Glover, *Cell Cycle Control*, IRL Press, Oxford, 1995.

43. B. T. Zafonte, J. Hulit, D. F. Amanatullah, C. Albanese, C. Wang, E. Rosen, A. Reutens, J. A. Sparano, M. P. Lisanti, and R. G. Pestell, *Front Biosci.*, **5**, D938–D961 (2000).

44. J. W. Harper, *Cancer Surv.*, **29**, 91–107 (1997).

45. W. S. el-Deiry, J. W. Harper, P. M. O'Connor, V. E. Velculescu, C. E. Canman, J. Jackman, J. A. Pietenpol, M. Burrell, D. E. Hill, and Y. Wang, *Cancer Res.*, **54**, 1169–1174 (1994).

46. L. Liu, N. J. Lassam, J. M. Slingerland, D. Bailey, D. Cole, R. Jenkins, and D. Hogg, *Oncogene*, **11**, 405–412 (1995).

47. T. Weinert, *Cancer Surv.*, **29**, 109–132 (1997).

48. S. W. Lowe, H. E. Ruley, T. Jacks, and D. E. Housman, *Cell*, **74**, 957–967 (1993).

49. S. N. Farrow, *Biochem. Soc. Trans.*, **27**, 812–814 (1999).

50. J. Wang and M. J. Lenardo, *J. Cell Sci.*, **113**, 753–757 (2000).

51. M. Loeffler and G. Kroemer, *Exp. Cell Res.*, **256**, 19–26 (2000).

52. A. Gross, J. M. McDonnell, and S. J. Korsmeyer, *Genes Dev.*, **13**, 1899–1911 (1999).

53. A. Ashkenazi and V. M. Dixit, *Curr. Opin. Cell Biol.*, **11**, 255–260 (1999).

54. R. C. Seeger, G. M. Brodeur, H. Sather, A. Dalton, S. E. Siegel, K. Y. Wong, and D. Hammond, *N. Engl. J. Med.*, **313**, 1111–1116 (1985).

55. E. C. Friedberg, *Cancer J. Sci. Am.*, **5**, 257–263 (1999).

56. G. L. Semenza, *J. Appl. Physiol.*, **88**, 1474–1480 (2000).

57. J. R. Gnarra, S. Zhou, M. J. Merrill, J. R. Wagner, A. Krumm, E. Papavassiliou, E. H. Oldfield, R. D. Klausner, and W. M. Linehan, *Proc. Natl. Acad. Sci. USA*, **93**, 10589–10594 (1996).

58. I. J. Fidler, and L. M. Ellis, *Cell*, **79**, 185–188 (1994).

59. D. R. Senger, S. J. Galli, A. M. Dvorak, C. A. Perruzzi, V. S. Harvey, and H. F. Dvorak, *Science*, **219**, 983–985 (1983).

60. N. Ferrara, and W. J. Henzel, *Biochem. Biophys. Res. Commun.*, **161**, 851–858 (1989).

61. B. Millauer, M. P. Longhi, K. H. Plate, L. K. Shawver, W. Risau, A. Ullrich, and L. M. Strawn, *Cancer Res.*, **56**, 1615–1620 (1996).

62. M. Asano, A. Yukita, T. Matsumoto, S. Kondo, and H. Suzuki, *Cancer Res.*, **55**, 5296–5301 (1995).

63. B. Olofsson, K. Pajusola, A. Kaipainen, G. von Euler, V. Joukov, O. Saksela, A. Orpana, R. F. Pettersson, K. Alitalo and U. Eriksson, *Proc Natl Acad Sci U S A*, **93**, 2576–81 (1996).

64. M. Orlandini, L. Marconcini, R. Ferruzzi and S. Oliviero., *Proc. Natl. Acad. Sci. U S A*, **93**, 11675–80 (1996).

65. M. G. Achen, M. Jeltsch, E. Kukk, T. Makinen, A. Vitali, A. F. Wilks, K. Alitalo and S. A. Stacker, *Proc. Natl. Acad. Sci. U S A*, **95**, 548–53 (1998).

66. E. Kukk, A. Lymboussaki, S. Taira, A. Kaipainen, M. Jeltsch, V. Joukov and K. Alitalo, *Development*, **122**, 3829–37 (1996).

67. S. Tessler, P. Rockwell, D. Hicklin, T. Cohen, B. Z. Levi, L. Witte, I. R. Lemischka, and G. Neufeld, *J. Biol. Chem.*, **269**, 12456–12461 (1994).

68. W. Risau, *Prog. Growth Factor Res.*, **2**, 71–79 (1990).

69. K. J. Kim, B. Li, J. Winer, M. Armanini, N. Gillett, H. S. Phillips, and N. Ferrara, *Nature*, **362**, 841–844 (1993).

70. N. Ueki, M. Nakazato, T. Ohkawa, T. Ikeda, Y. Amuro, T. Hada, and K. Higashino, *Biochim. Biophys. Acta.*, **1137**, 189–196 (1992).

71. N. Weidner, J. P. Semple, W. R. Welch, and J. Folkman, *N. Engl. J. Med.*, **324**, 1–8 (1991).

72. C. I. Bargmann, M. C. Hung, and R. A. Weinberg, *Cell*, **45**, 649–657 (1986).

73. M. Crul, G. J. de Klerk, J. H. Beijnen, and J. H. Schellens, *Anticancer Drugs*, **12**, 163–184 (2001).

74. A. B. Vojtek and C. J. Der, *J. Biol. Chem.*, **273**, 19925–19928 (1998).

75. H. R. Bourne, D. A. Sanders, and F. McCormick, *Nature*, **349**, 117–127 (1991).

76. M. R. Wallace, D. A. Marchuk, L. B. Andersen, R. Letcher, H. M. Odeh, A. M. Saulino, J. W. Fountain, A. Brereton, J. Nicholson, and A. L. Mitchell, *Science*, **249**, 181–186 (1990).

77. D. Stokoe, S. G. Macdonald, K. Cadwallader, M. Symons, and J. F. Hancock, *Science*, **264**, 1463–1467 (1994).

78. P. Rodriguez-Viciana, P. H. Warne, R. Dhand, B. Vanhaesebroeck, I. Gout, M. J. Fry, M. D. Waterfield, and J. Downward, *Nature*, **370**, 527–532 (1994).

79. T. Sasaki, J. Irie-Sasaki, Y. Horie, K. Bachmaier, J. E. Fata, M. Li, A. Suzuki, D. Bouchard, A. Ho, M. Redston, S. Gallinger, R. Khokha, T. W. Mak, P. T. Hawkins, L. Stephens, S. W. Scherer, M. Tsao, and J. M. Penninger, *Nature*, **406**, 897–902 (2000).

80. M. Vidal, V. Gigoux, and C. Garbay, *Crit. Rev. Oncol. Hematol.*, **40**, 175–186 (2001).

81. G. Q. Daley and Y. Ben-Neriah, *Adv. Cancer Res.*, **57**, 151–184 (1991).

82. D. Diekmann, S. Brill, M. D. Garrett, N. Totty, J. Hsuan, C. Monfries, C. Hall, L. Lim, and A. Hall, *Nature*, **351**, 400–402 (1991).

83. M. J. Mauro, M. O'Dwyer, M. C. Heinrich, and B. J. Druker, *J. Clin. Oncol.*, **20**, 325–334 (2002).

84. T. Hai, and T. Curran, *Proc. Natl. Acad. Sci. USA*, **88**, 3720–3724 (1991).

85. M. Karin, Z. Liu, and E. Zandi, *Curr. Opin. Cell. Biol.*, **9**, 240–246 (1997).

86. K. F. Mitchell, J. Battey, G. F. Hollis, C. Moulding, R. Taub, and P. Leder, *J. Cell. Physiol.*, Suppl 3, 171–177 (1984).

87. P. J. Koskinen and K. Alitalo, *Sem. Cancer Biol.*, **4**, 3–12 (1993).

88. T. C. He, A. B. Sparks, C. Rago, H. Hermeking, L. Zawel, L. T. da Costa, P. J. Morin, B. Vogelstein, and K. W. Kinzler, *Science*, **281**, 1509–1512 (1998).

89. L. Zheng and W. H. Lee, *Exp. Cell Res.*, **264**, 2–18 (2001).

90. J. R. Bischoff, D. H. Kirn, A. Williams, C. Heise, S. Horn, M. Muna, L. Ng, J. A. Nye, A. Sampson-Johannes, A. Fattaey, and F. McCormick, *Science*, **274**, 373–376 (1996).

91. S. Ikeda, S. Kishida, H. Yamamoto, H. Murai, S. Koyama, and A. Kikuchi, *EMBO J.*, **17**, 1371–1384 (1998).

92. L. Simpson and R. Parsons, *Exp. Cell Res.*, **264**, 29–41 (2001).

93. M. Tamura, J. Gu, K. Matsumoto, S. Aota, R. Parsons, and K. M. Yamada, *Science*, **280**, 1614–1617 (1998).

94. G. Torre-Amione, R. D. Beauchamp, H. Koeppen, B. H. Park, H. Schreiber, H. L. Moses, and D. A. Rowley, *Proc. Natl. Acad. Sci. USA*, **87**, 1486–1490 (1990).

95. S. Markowitz, J. Wang, L. Myeroff, R. Parsons, L. Sun, J. Lutterbaugh, R. S. Fan, E. Zborowska, K. W. Kinzler, and B. Vogelstein, *Science*, **268**, 1336–1338 (1995).

96. J. E. Cleaver, *J. Dermatol. Sci.*, **23**, 1–11 (2000).

97. H. T. Lynch, T. C. Smyrk, P. Watson, S. J. Lanspa, J. F. Lynch, P. M. Lynch, R. J. Cavalieri, and C. R. Boland, *Gastroenterology*, **104**, 1535–1549 (1993).

98. P. Peltomaki and H. F. Vasen, *Gastroenterology*, **113**, 1146–1158 (1997).

99. P. Greenwald, *Sci. Am.*, **275**, 96–99 (1996).

100. H. S. Wasan, M. Novelli, J. Bee, and W. F. Bodmer, *Proc. Natl. Acad. Sci. USA*, **94**, 3308–3313 (1997).

101. D. S. Alberts, M. E. Martinez, D. J. Roe, J. M. Guillen-Rodriguez, J. R. Marshall, J. B. van Leeuwen, M. E. Reid, C. Ritenbaugh, P. A. Vargas, A. B. Bhattacharyya, D. L. Earnest, and R. E. Sampliner, *N. Engl. J. Med.*, **342**, 1156–1162 (2000).

102. S. A. Leadon, *Sem. Radiat. Oncol.*, **6**, 295–305 (1996).

103. M. Abdulla, and P. Gruber, *Biofactors*, **12**, 45–51 (2000).

104. G. S. Omenn, G. E. Goodman, M. D. Thornquist, J. Balmes, M. R. Cullen, A. Glass, J. P. Keogh, F. L. Meyskens Jr., B. Valanis, J. H. Williams Jr., S. Barnhart, M. G. Cherniack, C. A. Brodkin, and S. Hammar, *J. Natl. Cancer Inst.*, **88**, 1550–1559 (1996).

105. B. Reddy, A. Engle, S. Katsifis, B. Simi, H. P. Bartram, P. Perrino, and C. Mahan, *Cancer Res.*, **49**, 4629–4635 (1989).

106. J. A. Baron, M. Beach, J. S. Mandel, R. U. van Stolk, R. W. Haile, R. S. Sandler, R. Rothstein, R. W. Summers, D. C. Snover, G. J. Beck, J. H. Bond, and E. R. Greenberg, *N. Engl. J. Med.*, **340**, 101–107 (1999).

107. E. Giovannucci, E. B. Rimm, A. Wolk, A. Ascherio, M. J. Stampfer, G. A. Colditz, and W. C. Willett, *Cancer Res.*, **58**, 442–447 (1998).

108. R. N. Dubois, S. B. Abramson, L. Crofford, R. A. Gupta, L. S. Simon, L. B. Van De Putte, and P. E. Lipsky, *Faseb J.*, **12**, 1063–1073 (1998).

109. A. T. Koki, K. M. Leahy, and J. L. Masferrer, *Expert Opin. Investig. Drugs*, **8**, 1623–1638 (1999).

110. F. L. Meyskens Jr. and E. W. Gerner, *Clin. Cancer Res.*, **5**, 945–951 (1999).

111. D. A. Notterman, U. Alon, A. J. Sierk, and A. J. Levine, *Cancer Res.*, **61**, 3124–3130 (2001).

112. J. Yu, L. Zhang, P. M. Hwang, C. Rago, K. W. Kinzler, and B. Vogelstein, *Proc. Natl. Acad. Sci. USA*, **96**, 14517–14522 (1999).

113. M. Hidalgo and S. G. Eckhardt, *J. Natl. Cancer Inst.*, **93**, 178–193 (2001).

114. P. D. Delmas, *N. Engl. J. Med.*, **335**, 1836–1837 (1996).

115. O. Teronen, P. Heikkila, Y. T. Konttinen, M. Laitinen, T. Salo, R. Hanemaaijer, A. Teronen, P. Maisi, and T. Sorsa, *Ann. NY Acad. Sci.*, **878**, 453–465 (1999).

116. M. A. Schwalke, G. N. Tzanakakis, and M. P. Vezeridis, *J. Surg. Res.*, **49**, 164–167 (1990).

117. M. E. Barry, D. Pinto-Gonzalez, F. M. Orson, G. J. McKenzie, G. R. Petry, and M. A. Barry, *Hum. Gene Therap.*, **10**, 2461–2480 (1999).

118. M. A. Hickman, R. W. Malone, K. Lehmann-Bruinsma, T. R. Sih, D. Knoell, F. C. Szoka, R. Walzem, D. M. Carlson, and J. S. Powell, *Hum. Gene Therap.*, **5**, 1477–1483 (1994).

119. J. P. Yang, and L. Huang, *Gene Therap.*, **3**, 542–548 (1996).

120. R. Niven, R. Pearlman, T. Wedeking, J. Mackeigan, P. Noker, L. Simpson-Herren, and J. G. Smith, *J. Pharm. Sci.*, **87**, 1292–1299 (1998).

121. R. M. Bartholomew, A. F. Esser, and H. J. Muller-Eberhard, *J. Exp. Med.*, **147**, 844–853 (1978).

122. F. Bushman, *Science*, **267**, 1443–1444 (1995).

123. S. V. Joang, E. B. Stephens, and O. Narayan in B. N. Fields, D. M. Knipe and P. M. Howley, Eds., *Virology*, Lippincott-Raven, Philadelphia, PA, 1996, 1997–1996.

124. U. Blomer, L. Naldini, T. Kafri, D. Trono, I. M. Verma, and F. H. Gage, *J. Virol.*, **71**, 6641–6649 (1997).

125. J. C. Cusack Jr. and K. K. Tanabe, *Surg. Oncol. Clin. North Am.*, **7**, 421–469 (1998).

126. P. Seth, *Gene Therapy for the Treatment of Cancer*, Vol. **44–77**, Cambridge University Press, Cambridge, 1998.

127. V. N. Krasnykh, G. V. Mikheeva, J. T. Douglas, and D. T. Curiel, *J. Virol.*, **70**, 6839–6846 (1996).

128. J. A. Roth and R. J. Cristiano, *J. Natl. Cancer Inst.*, **89**, 21–39 (1997).

129. A. Iida, S. T. Chen, T. Friedmann, and J. K. Yee, *J. Virol.*, **70**, 6054–6059 (1996).

130. E. W. Gerner, E. M. Hersh, M. Pennington, T. C. Tsang, D. Harris, F. Vasanwala, and J. Brailey, *Int. J. Hyperthermia*, **16**, 171–181 (2000).

131. T. Clackson, *Curr. Opin. Chem. Biol.*, **1**, 210–218 (1997).

132. P. L. Felgner and G. M. Ringold, *Nature*, **337**, 387–388 (1989).

133. N. S. Templeton, D. D. Lasic, P. M. Frederik, H. H. Strey, D. D. Roberts, and G. N. Pavlakis, *Nature Biotechnol.*, **15**, 647–652 (1997).

134. A. Kikuchi, Y. Aoki, S. Sugaya, T. Serikawa, K. Takakuwa, K. Tanaka, N. Suzuki, and H. Kikuchi, *Hum. Gene Therap.*, **10**, 947–955 (1999).

135. C. Meunier-Durmort, R. Picart, T. Ragot, M. Perricaudet, B. Hainque, and C. Forest, *Biochim. Biophys. Acta*, **1330**, 8–16 (1997).

136. T. Tuting, W. J. Storkus, and M. T. Lotze, *J. Mol. Med.*, **75**, 478–491 (1997).

137. M. Lindauer, T. Stanislawski, A. Haussler, E. Antunes, A. Cellary, C. Huber, and M. Theobald, *J. Mol. Med.*, **76**, 32–47 (1998).

138. J. Banchereau, F. Briere, C. Caux, J. Davoust, S. Lebecque, Y. J. Liu, B. Pulendran, and K. Palucka, *Annu. Rev. Immunol.*, **18**, 767–811 (2000).

139. B. M. Davis, O. N. Koc, K. Lee, and S. L. Gerson, *Curr. Opin. Oncol.*, **8**, 499–508 (1996).

140. Z. Ram, *Isr. Med. Assoc. J.*, **1**, 188–193 (1999).

141. D. Klatzmann, P. Cherin, G. Bensimon, O. Boyer, A. Coutellier, F. Charlotte, C. Boccaccio, J. L. Salzmann, and S. Herson, *Hum. Gene Therap.*, **9**, 2585–2594 (1998).

142. D. H. Sterman, K. Molnar-Kimber, T. Iyengar, M. Chang, M. Lanuti, K. M. Amin, B. K. Pierce, E. Kang, J. Treat, A. Recio, L. Litzky, J. M. Wilson, L. R. Kaiser, and S. M. Albelda, *Cancer Gene Therap.*, **7**, 1511–1518 (2000).

143. R. G. Crystal, E. Hirschowitz, M. Lieberman, J. Daly, E. Kazam, C. Henschke, D. Yankelevitz, N. Kemeny, R. Silverstein, A. Ohwada, T. Russi, A. Mastrangeli, A. Sanders, J. Cooke, and B. G. Harvey, *Hum. Gene Therap.*, **8**, 985–1001 (1997).

144. C. Beltinger, W. Uckert, and K. M. Debatin, *J. Mol. Med.*, **78**, 598–612 (2001).

145. J. Nemunaitis, S. G. Swisher, T. Timmons, D. Connors, M. Mack, L. Doerksen, D. Weill, J. Wait, D. D. Lawrence, B. L. Kemp, F. Fossella, B. S. Glisson, W. K. Hong, F. R. Khuri, J. M. Kurie, J. J. Lee, J. S. Lee, D. M. Nguyen, J. C. Nesbitt, R. Perez-Soler, K. M. Pisters, J. B. Putnam, W. R. Richli, D. M. Shin, and G. L. Walsh, *J. Clin. Oncol.*, **18**, 609–622 (2000).

146. T. Waku, T. Fujiwara, J. Shao, T. Itoshima, T. Murakami, M. Kataoka, S. Gomi, J. A. Roth, and N. Tanaka, *J. Immunol.*, **165**, 5884–5890 (2000).

147. D. L. Tait, P. S. Obermiller, A. R. Hatmaker, S. Redlin-Frazier, and J. T. Holt, *Clin. Cancer Res.*, **5**, 1708–1714 (1999).

148. K. R. Birikh, P. A. Heaton, and F. Eckstein, *Eur. J. Biochem.*, **245**, 1–16 (1997).

149. Y. A. Zhang, J. Nemunaitis, and A. W. Tong, *Mol. Biotechnol.*, **15**, 39–49 (2000).

150. H. J. Andreyev, P. J. Ross, D. Cunningham, and P. A. Clarke, *Gut*, **48**, 230–237 (2001).

151. W. A. Marasco, *Gene Therap.*, **4**, 11–15 (1997).

152. F. Griscelli, H. Li, A. Bennaceur-Griscelli, J. Soria, P. Opolon, C. Soria, M. Perricaudet, P. Yeh, and H. Lu, *Proc. Natl. Acad. Sci. USA*, **95**, 6367–6372 (1998).

153. P. Blezinger, J. Wang, M. Gondo, A. Quezada, D. Mehrens, M. French, A. Singhal, S. Sullivan, A. Rolland, R. Ralston, and W. Min, *Nature Biotechnol.*, **17**, 343–348 (1999).

154. D. Ma, R. D. Gerard, X. Y. Li, H. Alizadeh, and J. Y. Niederkorn, *Blood*, **90**, 2738–2746 (1997).

155. H. L. Kong, D. Hecht, W. Song, I. Kovesdi, N. R. Hackett, A. Yayon, and R. G. Crystal, *Hum. Gene Therap.*, **9**, 823–833 (1998).

156. M. Watanabe, Y. Takahashi, T. Ohta, M. Mai, T. Sasaki, and M. Seiki, *Cancer*, **77**, 1676–1680 (1996).

157. D. F. Alonso, G. Skilton, M. S. De Lorenzo, A. M. Scursoni, H. Yoshiji, and D. E. Gomez, *Oncol. Rep.*, **5**, 1083–1087 (1998).

158. H. Li, F. Lindenmeyer, C. Grenet, P. Opolon, S. Menashi, C. Soria, P. Yeh, M. Perricaudet, and H. Lu, *Hum. Gene Therap.*, **12**, 515–526 (2001).

# CHAPTER TWO

# Synthetic DNA-Targeted Chemotherapeutic Agents and Related Tumor-Activated Prodrugs

WILLIAM A. DENNY
Auckland Cancer Society Research Centre
Faculty of Medical and Health Sciences
The University of Auckland
Auckland, New Zealand

## Contents

1 General Introduction, 52
2 Alkylating Agents, 53
   2.1 Introduction, 53
   2.2 Clinical Examples of Alkylating Agents, 53
   2.3 Mustards, 53
      2.3.1 History, 53
      2.3.2 Mechanism and SAR, 53
      2.3.3 Biological Activity and Side Effects, 56
      2.3.4 Recent Developments: Minor Groove
         Targeting, 57
   2.4 Platinum Complexes, 59
      2.4.1 History, 59
      2.4.2 Mechanism and SAR, 60
      2.4.3 Biological Activity and Side Effects, 61
      2.4.4 Recent Developments: Increased
         Interstrand Crosslinking, 61
   2.5 Cyclopropylindoles, 61
      2.5.1 History, 61
      2.5.2 Mechanism and SAR, 62
      2.5.3 Biological Activity and Side Effects, 64
      2.5.4 Recent Developments, 64
   2.6 Nitrosoureas, 64
      2.6.1 History, 64
      2.6.2 Mechanism and SAR, 65
      2.6.3 Biological Activity and Side Effects, 65
   2.7 Triazenes, 65
      2.7.1 History, 65
      2.7.2 Mechanism and SAR, 65
      2.7.3 Biological Activity and Side Effects, 66

*Burger's Medicinal Chemistry and Drug Discovery*
Sixth Edition, Volume 5: Chemotherapeutic Agents
Edited by Donald J. Abraham
ISBN 0-471-37031-2   © 2003 John Wiley & Sons, Inc.

3 Synthetic DNA-Intercalating Topoisomerase
   Inhibitors, 67
   3.1 Introduction, 67
   3.2 Clinical Use of Agents, 68
   3.3 Topo II Inhibitors, 68
      3.3.1 History, 68
      3.3.2 Mechanism of Action and SAR, 69
      3.3.3 Biological Activity and Side Effects, 71
      3.3.4 Recent Developments: Compounds
         with Lower Cardiotoxicity, 71
   3.4 Dual Topo I/II Inhibitors, 72
      3.4.1 History, 72
      3.4.2 Mechanism and SAR, 74
      3.4.3 Biological Activity and Side Effects, 74
      3.4.4 Recent Developments: bis Analogs as
         Dual Topo I/II Inhibitors, 74
4 Antimetabolites, 75
   4.1 Introduction, 75
   4.2 Clinical Use of Agents, 76
   4.3 Antifolates, 76
      4.3.1 History, 76
      4.3.2 Mechanism and SAR, 78
      4.3.3 Biological Activity and Side Effects, 78
      4.3.4 Recent Developments: Lipophilic
         Antifolates, 79
   4.4 Pyrimidine Analogs, 79
      4.4.1 History, 79
      4.4.2 Mechanism and SAR, 80
      4.4.3 Biological Activity and Side Effects, 81
   4.5 Purine Analogs, 81
      4.5.1 History, 81
      4.5.2 Mechanism and SAR, 81
      4.5.3 Biological Activity and Side Effects, 82
5 Tumor-Activated Prodrugs, 82

5.1 Introduction, 82
5.2 Clinical Use of Tumor-Activated Prodrugs,
   83
5.3 Hypoxia-Activated Prodrugs (Bioreductives),
   83
   5.3.1 History, 83
   5.3.2 Mechanism and SAR, 85
   5.3.3 Biological Activity and Side Effects, 86
5.4 Prodrugs for ADEPT, 87
   5.4.1 History, 87
   5.4.2 Mechanism and SAR, 87
      5.4.2.1 Prodrugs for Phosphatase
         Enzymes, 87
      5.4.2.2 Prodrugs for Peptidase
         Enzymes, 87
      5.4.2.3 Prodrugs for $\beta$-Lactamase
         Enzymes, 88
      5.4.2.4 Prodrugs for Glucuronidase
         Enzymes, 89
5.5 Prodrugs for GDEPT, 90
   5.5.1 History, 90
   5.5.2 Mechanism and SAR, 90
      5.5.2.1 Prodrugs for Kinase Enzymes,
         90
      5.5.2.2 Prodrugs for Cytosine
         Deaminase, 90
      5.5.2.3 Prodrugs for Oxidative
         Enzymes, 91
      5.5.2.4 Prodrugs for Reductase
         Enzymes, 91
5.6 Antibody-Toxin Conjugates, 92
   5.6.1 History, 92
   5.6.2 Mechanism and SAR, 92

## 1 GENERAL INTRODUCTION

Synthetic drugs have always played an important role in cancer therapy. In fact, systemic chemotherapy for cancer began in the 1940s and 1950s with the nitrogen mustards developed from war gases (1) and with antimetabolites developed from early knowledge about DNA metabolism (2). Large-scale random screening programs over the next 25 years (mainly by the U.S. National Cancer Institute) (3) seeking cytotoxic agents resulted in the identification of a number of cytotoxic natural products that target DNA. Many of these (e.g., anthracyclines, epipodophylloxins, and vinca alkaloids) became very useful drugs that are still widely used today. Most of the natural products were so complex that neither they

nor close analogs could be economically produced by synthesis, limiting the role of synthetic chemistry in optimizing their potencies or pharmacokinetic properties. However, the discovery of their activity and mechanism of action sparked much work on simpler synthetic analogs. One result was the development of the large class of synthetic topoisomerase inhibitors that are now an important group of drugs. More recently, the increasing power of organic synthesis has greatly improved chances that quite complex natural product leads can be synthesized economically, and therefore that close analogs can be made to try and optimize physicochemical properties; recent examples are cyclopropylindolines and epothilones. However, the primary focus in this chapter are synthetic compounds that have not been derived from a natural product lead. Finally, our increasing understanding of tumor physiology and genetics has allowed the development of a new class of synthetic agents, tumor-activated prodrugs. These attempt to exploit tumor-specific phenomena, such as unique antigen expression, low pH, and hypoxia, to activate prodrugs of the more classical cytotoxins only in tumors, thus increasing their therapeutic range.

## 2   ALKYLATING AGENTS

### 2.1   Introduction

Compounds that alkylate DNA have long been of interest as anticancer drugs. Alkylating agents can be strictly defined as electrophiles that can replace a hydrogen atom by an alkyl group under physiological conditions, but the term is usually more broadly interpreted to include any compound that can replace hydrogen under these conditions, including metal complexes forming coordinate bonds. Many different types of chemical are able to alkylate DNA, and several are used as anticancer drugs, but the most important classes of such agents in clinical use are the nitrogen mustards and the platinum complexes, the nitrosoureas, and the triazene-based DNA-methylating agents. The DNA minor groove-alkylating cyclopropylindoles are also a fascinating group of compounds that may not yet have found their correct niche in cancer therapy. Other important classes of DNA alkylating agents, the pyrrolobenzodiazepines and the mitosenes, are covered in a different chapter, although the bioreductive properties of the mitosenes are mentioned in Section 5.3.

### 2.2   Clinical Examples of Alkylating Agents

The most commonly used mustards and platinum complexes are listed in Table 2.1, along with other recent DNA-alkylating agents that have received clinical trial. These compounds are invariably used in combination with other agents in multidrug therapy regimens.

### 2.3   Mustards

**2.3.1  History.** As noted above, the mustards were among the very earliest class of

anticancer agents developed, and they have been extensively reviewed. Mechlorethamine (**1**) (4) was the first systemic agent approved for use in cancer therapy in 1949. Chlorambucil (**2**) (5) was approved in 1957, melphalan (**3**) (6) in 1964, cyclophosphamide (**4**) (7) in 1959, and ifosfamide (**5**) (8) in 1988. The phosphoramide mustard cyclophosphamide (**4**) is currently the most widely used mustard, while chlorambucil (**2**) and melphalan (**3**) are still in use as components of many combination chemotherapy regimens.

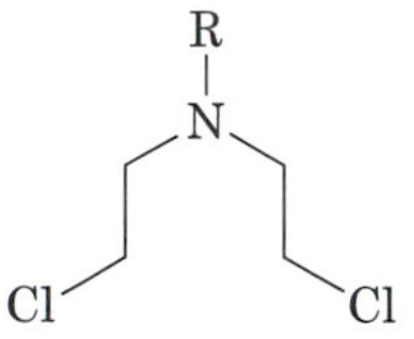

(**1**) R = $CH_3$
(**2**) R = $(CH_2)_3CO_2H$
(**3**) R = $CH(NH_2)CO_2H$

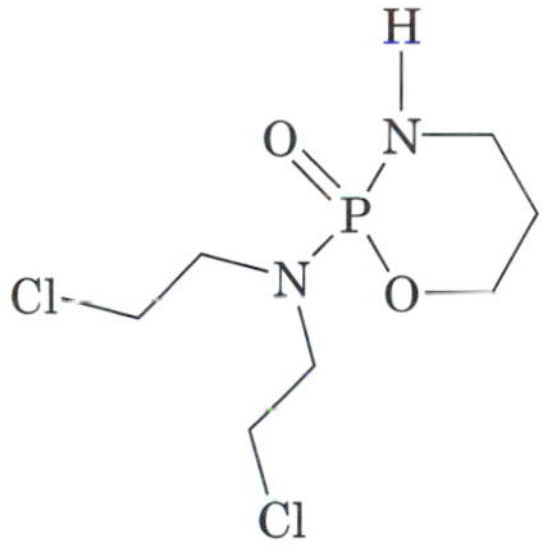

(**4**)

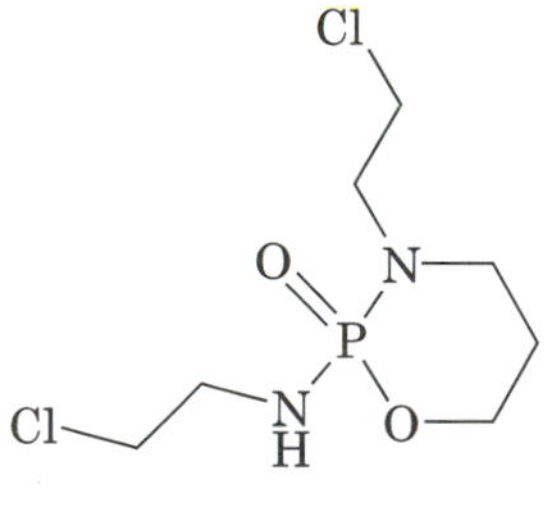

(**5**)

**2.3.2  Mechanism and SAR.** The biologically important initial lesion formed by mustards in cells is interstrand cross-links between different DNA bases (9), although there is also evidence that they cause termination of

**Table 2.1   Alkylating Agents Used in Cancer Chemotherapy**

| Generic Name (Structure) | Trade Name | Originator | Chemical Class |
|---|---|---|---|
| Mustards | | | |
| mechlorethamine (**1**) | Mustargen | Merck | aliphatic mustard |
| chlorambucil (**2**) | Leukeran | | aromatic mustard |
| melphalan (**3**) | Alkeran | | aromatic mustard |
| cyclophosphamide (**4**) | Cytoxan | Bristol-Myers | phosphoramide mustard |
| ifosfamide (**5**) | Ifex | Bristol-Myers | phosphoramide mustard |
| Platinum complexes | | | |
| cisplatin (**14**) | Cisplatin | Bristol-Myers | platinum complex |
| carboplatin (**15**) | Paraplatin | Bristol-Myers | platinum complex |
| tetraplatin (**16**) | Ormaplatin | | platinum complex |
| oxaliplatin (**17**) | JM-83 | Sanofi | platinum complex |
| ZD-0473 (**18**) | | AstraZeneca | platinum complex |
| satraplatin (**19**) | JM-216 | Johnson-Matthey | platinum complex |
| BBR 3464 (**20**) | | Boehringer | triplatinum complex |
| Cyclopropylindoles | | | |
| Adozelesin (**22**) | | Upjohn | cyclopropylindole |
| Carzelesin (**23**) | | Upjohn | cyclopropylindole |
| KW 2189 (**24**) | | Kyowa Hakko | cyclopropylindole |
| Nitrosoureas | | | |
| CCNU (**29**) | Lomustine | Bristol-Myers | nitrosourea |
| BCNU (**30**) | Carmustine | Bristol-Myers | nitrosourea |
| Streptozotocin (**31**) | Zanosar | Upjohn | nitrosourea |
| Methylating agents | | | |
| dacarbazine (**29**) | DTIC | | triazene |
| mitozolomide (**30**) | azolastone | | triazene |
| temozolomide (**31**) | Temodar | Shering-Plough | triazene |

transcription (10). The overall process of alkylation is a two-step sequence involved formation of a cyclic cationic intermediate, followed by nucleophilic attack on that intermediate by DNA (Fig. 2.1a). Mustards can be divided into two broad classes, depending on the mechanism of the rate-determining step in this process. The less basic compounds aromatic mustards have formation of the solvated cyclic carbocation (which is in equilibrium with the aziridinium cation) as the rate-determining step, following first-order kinetics (11) (Fig. 2.1b). Nucleophilic attack on this is then rapid, so that the cyclic form does not accumulate, and the overall reaction is first-order ($S_N1$), with the rate depending only on the concentration of the mustard (Equation 2.1).

$$R—X \rightarrow R^+ \rightarrow R—DNA \quad (2.1)$$

For the more basic aliphatic mustards, the first step (formation of the aziridinium cation), is rapid, and the rate-determining step is a second-order nucleophilic substitution on this by DNA (11). In these cases, the aziridinium cation can be detected as an intermediate, and the overall reaction is second-order ($S_N2$), with the rate depending on the concentrations of both the mustard and the DNA (Equation 2.2).

$$DNA + R—X \rightarrow DNA \ldots ; R \ldots H \rightarrow R—DNA + X \quad (2.2)$$

This kinetic classification is only broad, but it is useful as a rough predictor of the spectrum of adducts formed. Generally, $S_N1$-type compounds are expected to be less discriminating in their pattern of alkylation (reaction at N, P, and O sites on DNA), whereas most $S_N2$ type compounds tend to alkylate only at N sites on the DNA bases (12).

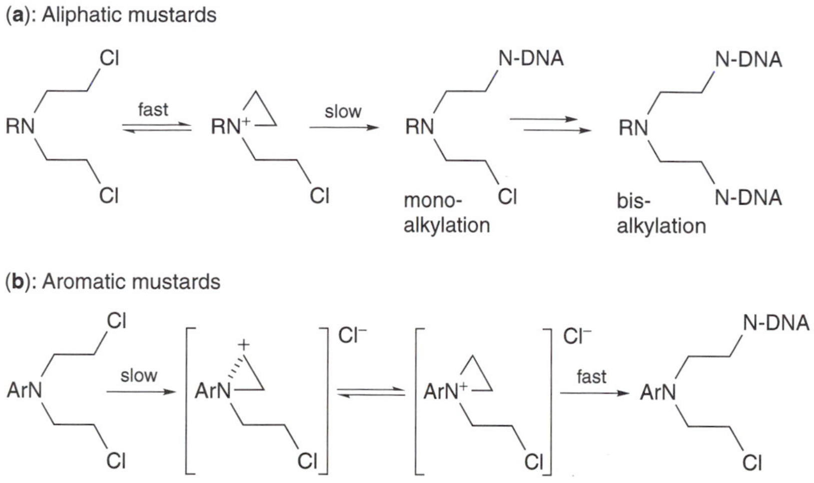

(**a**): Aliphatic mustards

(**b**): Aromatic mustards

**Figure 2.1.** Mechanism of alkylation by mustards.

The primary site of DNA alkylation by mustards is at the N7 position of guanine, particularly at guanines in contiguous runs of guanines (13), which have the lowest molecular electrostatic potentials (14). However, the level of selectivity of the initial attack by mustards (to form monoadducts) is quite low, with evidence (15) that most guanines are attacked. Studies with alkyl mustards have also shown significant levels of alkylation at the N3 position of adenine (16, 17). However, the sequence selectivity of cross-link formation by mustards is necessarily higher, because of the requirement to have two suitable sites juxtaposed. Early work (18) on the interaction of mechlorethamine (**1**) with DNA resulted in isolation of the 7-linked bis-guanine adduct, and it has been widely assumed that the cross-links were between adjacent guanines (i.e., at 5'-GC or 5'-CG sites). However, later work (19) showed that the preferred cross-links are between non-adjacent guanines (i.e., at 5'-GNC sites).

Cyclophosphamide (**4**) is a non-specific prodrug of the active metabolite phosphoramide mustard, requiring enzymic activation by cellular mixed function oxidases (primarily in the liver) to 4-hydroxycyclophosphamide, which is in equilibrium with the open-chain aldophosphamide. Spontaneous elimination of this then gives acrolein and phosphoramide mustard (Fig. 2.2). The isomeric ifosfamide (**5**) is activated more slowly, but in a broadly similar fashion to give the analogous isophosphoramide mustard (20). A significant difference in the metabolism between the two isomers is a higher level of dechloroethylation with ifosfamide, which may account for its greater neurotoxicity (21).

The rates of the various reactions of aromatic nitrogen mustards (hydrolysis, alkylation of DNA) can be correlated closely with the basicity of the nitrogen, that in turn, can be systematically altered by ring substituents. The rates of hydrolysis ($K_H$) of a series of substituted aromatic nitrogen mustards in aqueous acetone can be described (22) by Equation 2.3, where $\sigma$ is the Hammett electronic parameter.

$$\log K_H = -1.84\sigma - 4.02 \qquad (2.3)$$

The negative slope is evidence for an $S_N1$ mechanism, indicating that electron-releasing substituents (negative $\sigma$ values) increase the rate of hydrolysis by accelerating formation of the carbocation. The same broad correlations hold for how well the compounds alkylate DNA, with a similar equation (Equation 2.4) describing the rates of alkylation ($K$) of 4-(4-nitrobenzyl)pyridine (a nucleophile similar to DNA nucleophilic sites) by substituted aromatic nitrogen mustards (22), where $\sigma^-$ is an electronic parameter closely related to $\sigma$.

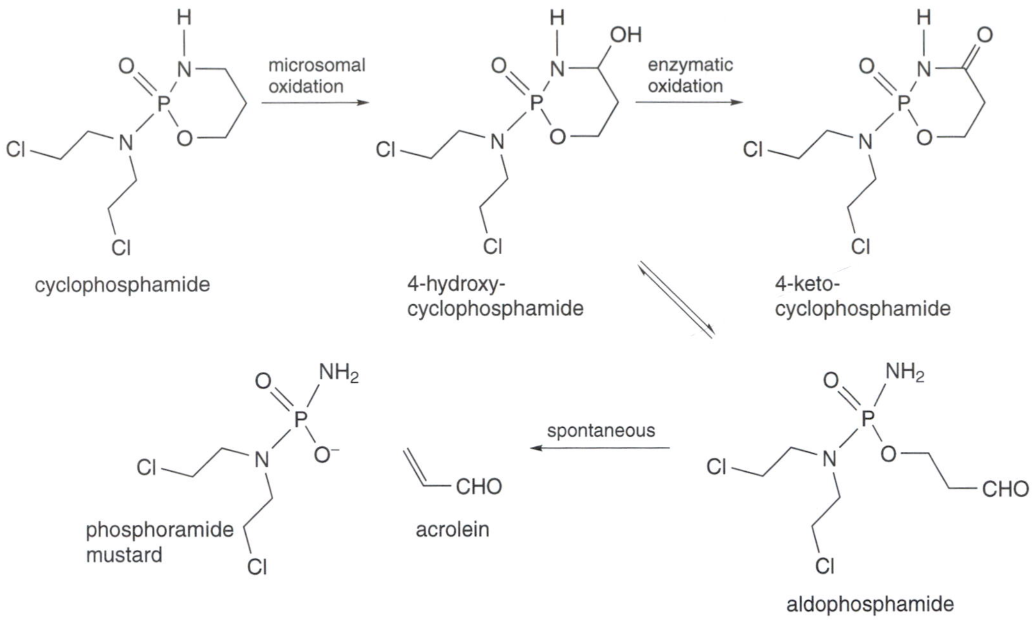

**Figure 2.2.** Metabolism of cyclophosphamide.

$$\log K = -1.92\sigma^- - 1.17 \qquad (2.4)$$

The cytotoxicities of the above compounds ($1/IC_{50}$ values) also correlate well with substituent $\sigma$ values, with the more reactive compounds (bearing electron-donating substituents) being the more cytotoxic, as in Equation 2.5 (23).

$$\log(1/IC_{50}) = -2.46\sigma + 0.53 \qquad (2.5)$$

The cytotoxicity of aromatic mustards can thus be predictably varied over a very wide range by controlling the basicity of the mustard nitrogen through ring substitution or other means.

**2.3.3 Biological Activity and Side Effects.**
The (necessarily) high chemical reactivity of mustards leads to rapid loss of drug by interaction with other cellular nucleophiles, particularly proteins and low molecular weight thiols. This results in the development of cellular reistance by increases in the levels of low molecular weight thiols (particularly glutathione) (24, 25). Of equal importance for efficacy, much of the drug can reach the DNA with only

one alkylating moiety intact, leading to monoalkylation events which are considered to be genotoxic rather than cytotoxic (26). The fact that cross-linking is a two-step process adds to the proportion of (genotoxic) monoalkylation events, because the second step is very dependent on spatial availability of a second nucleophilic DNA site. Mustards have no intrinsic biochemical or pharmacological selectivity for cancer cells, and they act as classical antiproliferative drugs, whose therapeutic effects are primarily cytokinetic. They target rapidly dividing cells rather than cancer cells, and this, together with their generally systemic distribution, causes killing of rapidly dividing normal cell populations in the bone marrow and gut, usually resulting in myelosupression, the dose-limiting side effect. Because of their genotoxicity, there is a risk of the development of second cancers from their mutagenic effects (27). The most frequent alkylator-induced malignancy is acute leukemia, usually occurring a long period (3–7 years) after treatment. These usually demonstrate deletions of chromosome 13 and loss of parts or all of chromosomes 5 or 7 (loss of the coding regions for tumor-suppressor genes). The induced tumors are typically myelodysplasias (28). In one study (29), 6% of all myeloid leukemias were therapy-related, with mustards, nitrosoureas, and procarbazine producing the greatest levels of induction.

### 2.3.4 Recent Developments: Minor Groove Targeting.

Many of the limitations noted above could in principle be ameliorated by targeting the mustard moiety more specifically to the DNA-affinic carrier molecule. This has resulted in much work where mustards have been attached to DNA-affinic compounds (30–32). This could mean less chance of losing active drug by reaction with other cell components, rendering less effective the development of cellular resistance by elevation of thiol levels. A higher proportion of bifunctional alkylating agent delivered intact to the DNA would also contribute to a higher proportion of cross-links over to monoalkylation events. The use of a carrier with sequence-specific reversible binding ability should also result in greater specificity of alkylation, both sequence-specifically (at the favored reversible binding site of the carrier) and regio-specifically (at particular atoms on the DNA bases).

Attachment to DNA-intercalating carriers goes back to the work of Creech et al. (33), who originally suggested that the attachment to acridine carriers might serve to target the reactive center to DNA. They showed that such "targeted mustards" such as (6) were more potent than the corresponding untargeted moiety against ascitic tumors *in vivo*, but these proved to be exceptionally potent frameshift mutagens in bacteria, and this property has tended to dominate the perception of these compounds. Later work showed that such targeting by an intercalator could also drastically modify the pattern of DNA alkylation by the mustard. Thus, whereas untargeted mustards react largely at the N7 of guanines in runs of guanines, quinacrine mustard (7) also alkylates at guanines in 5'-GT sites (13). Isolation and identification of DNA adducts showed that whereas the acridine-linked mustard (8) formed primarily guanine N7 adducts, the similar analog (9) formed exclusively adenine N1 adducts (34), showing the extent of which

DNA targeting by attachment to carrier molecules can alter the usual pattern of DNA alkylation by mustards.

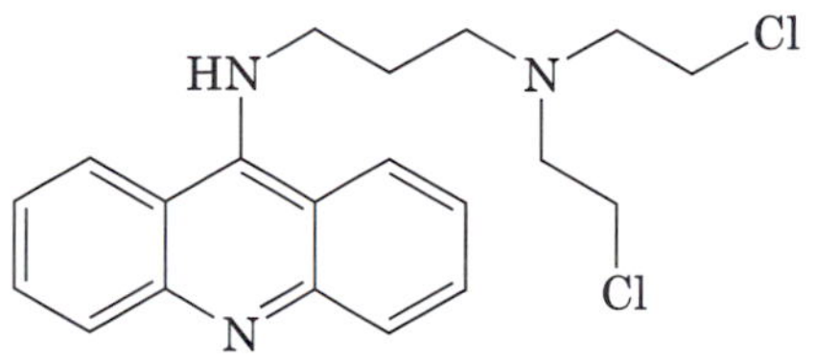

(6)

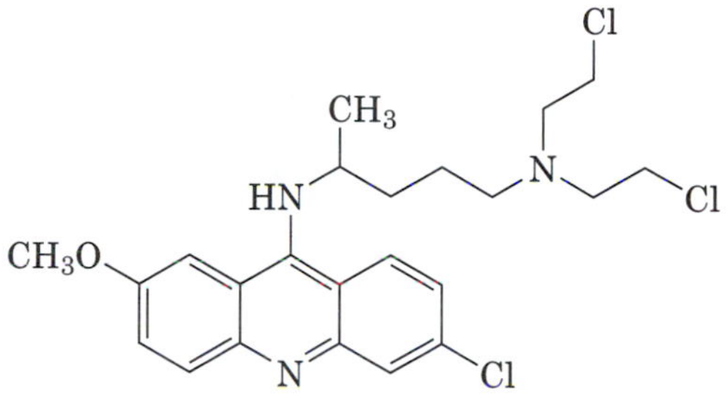

(7)

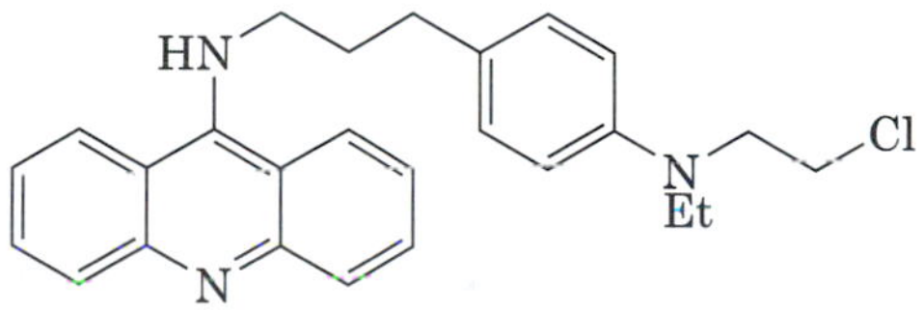

(8)

However, most DNA-targeting of mustards has been done using minor groove-binding carriers. These ligands offer much larger binding site sizes (up to 5–6 base pairs) than intercalators, together with a highly defined binding orientation. Whereas several other minor groove-binding carriers have been used (35–37), most work has employed polypyrrole and related ligands. These compounds have been well documented as reversible AT-specific minor groove binders (38), and early work using a variety of alkylating units (e.g., bromoacetyl) showed highly specific alkylation at adenines in runs of adenines (39). The benzoic acid mustard derivative tallimustine (10; FCE 24517) was selected for further development on the basis of its broad-spectrum solid tumor activity (40). Despite possessing a difunctional

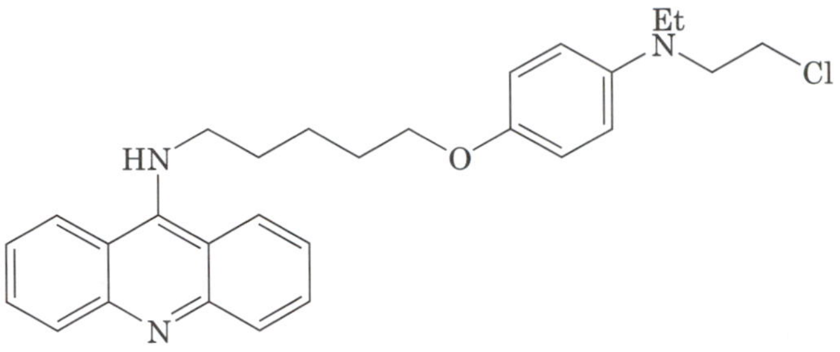

**(9)**

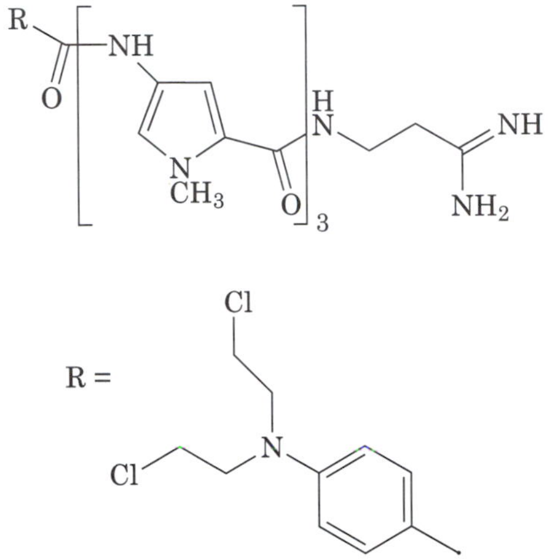

alkylator, this compound monoalkylates DNA at the N3 of adenine in the minor groove, almost exclusively at the sequence 5'-TTTTGA (41), with a single base modification in the hexamer completely abolishing alkylation (42). The number of pyrroleamide units also affected the pattern of DNA alkylation, with a monopyrrole analog showing mainly guanine-N7 alkylation similar to that of the untargeted mustard, but with additional adenine-N3 lesions (43). Di- and tripyrrole conjugates alkylated only in AT tracts, with increasing specificity for alkylation at the 3'-terminal units in two 5'-TTTTGG and 5'-TTTTGA sequences (guanine N3 and adenine N3 lesions, respectively).

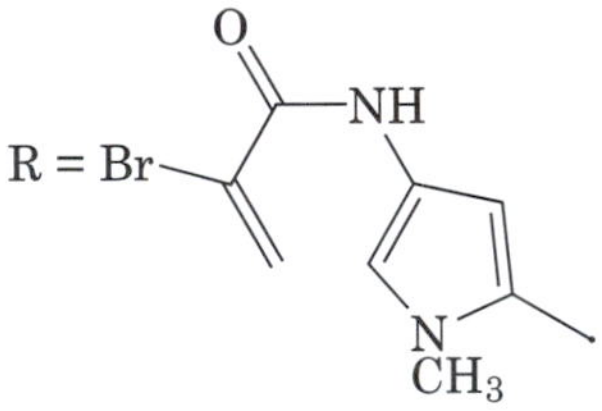

Tallimustine was developed for clinical trial (44) and shows biological effects somewhat different to those of mustards like melphalan. It induces blockage of the cell cycle in G2 but without the delay through S-phase normally seen with untargeted mustards, suggesting a different mechanism of cytotoxicity through monoadduct formation (45). As a highly sequence-specific alkylator, it selectively blocks the binding of transcription binding protein and complexes to their AT-rich cognate sequences (46). Clinical trials of tallimustine (47, 48) reported severe myelosuppression as the dose-limiting toxicity. Recent work with halogenoacrylic derivatives (e.g., **11**) show these may work differently, possibly through Michael-type reactions (49), with much better cytotoxicity/myelotoxicity indices (50).

Perhaps the ultimate in targeting mustards to specific DNA sites has been achieved by Dervan and co-workers, who have developed the "hairpin polyamide" concept where

**(10)**

**(11)**

poly(pyrrole/imidazole) compounds bind in a side-by-side manner in the minor groove (51). These compounds can bind tightly and selectively to individual designated sequences of up to 12 bp long (52). As an example, polyamide **12**, with an incipient mustard side chain attached, binds to its designated sites 5'-AGCT-GCT and 5'-TGCAGCA with equilibrium association constants $K$ of 1.6 and 1.3 $\times$ $10^{10}$ $M^{-1}$, respectively, and >100-fold less strongly to double mismatch sites (53). The corresponding mustard (**13**) alkylated at adenine N3 sited in target 5′-(A/T)GC(A/T)GC(A/T) sequences on a 241-bp HIV-1 promotor sequence in high yield and about 20-fold selectively over double mismatch sites (53).

The first has been to seek compounds with lower neurotoxicity than cisplatin. Whereas better clinical management has improved things, one of the main drivers of analog development has been agents with less neurotoxicity. Carboplatin (**15**) has carboxylate instead

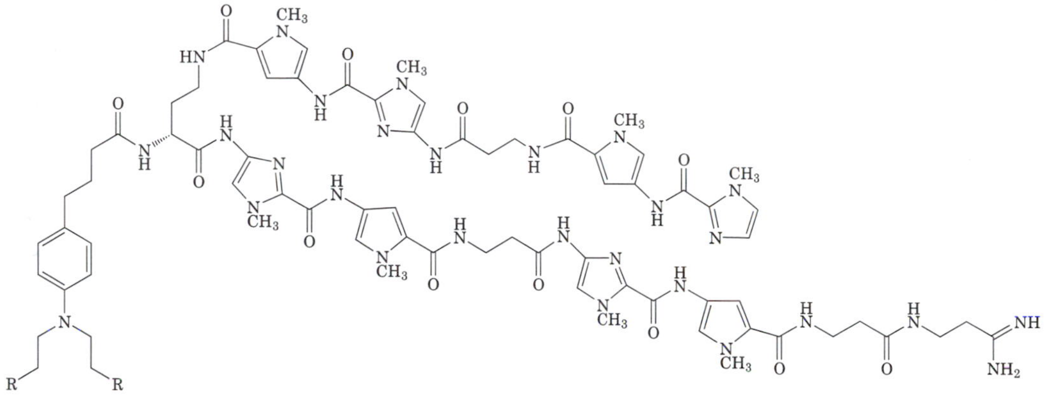

(**12**) R = OH
(**13**) R-Cl

## 2.4 Platinum Complexes

**2.4.1 History.** The complex *cis*-diamminodichloroplatinum (II) (cisplatin; **14**) was first described in 1845, but it was not until 1969 that it was reported to have antitumor activity. These studies were sparked by experiments by Rosenberg on the effects of electric fields on bacteria, when the peculiar effects seen with *E. coli* cells were shown to be caused by the electrochemical synthesis of cisplatin from the ammonium chloride electrolyte and the platinum electrodes (54). Clinical trials began in 1972, and after slow progress because of high toxicity, cisplatin became one of the most widely used anticancer drugs; it is the main reason for the spectacular successes in drug treatment of testicular and ovarian cancer. Thousands of analogs of cisplatin have been made and evaluated, with two major driving forces.

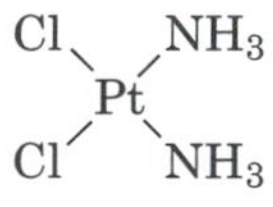

of chloride-leaving groups. These hydrolyze much less rapidly, resulting in lower nephro- and neurotoxicity (the dose-limiting toxicity of carboplatin is myelosuppression), while retaining the broad spectrum of activity of cisplatin (55).

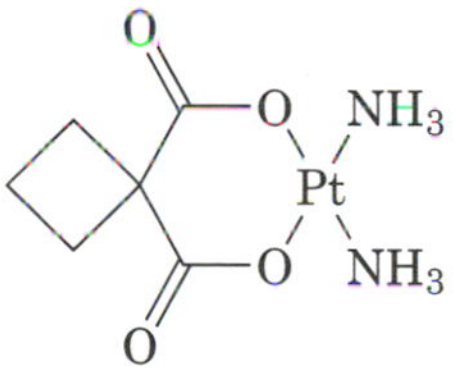

(**15**)

The second impetus to analog development has been to seek agents active in cell lines that become resistant to cisplatin. One mechanism of resistance to cisplatin is an increased ability to repair the DNA adducts formed (56), and analogs such tetraplatin (**16**; ormaplatin) and oxaliplatin (**17**), with *trans*-1,2-diaminocyclohexane (DACH) ligands, were shown to be more effective against such resistant cell lines (57). These compounds proved to be neurotoxic and tetraplatin was difficult to formulate, but oxaliplatin has shown promise (58),

(**14**)

especially in colorectal cancer, where it is synergistic with 5-fluorouracil (59). A second significant mechanism of resistance is elevation of thiol levels (primarily glutathione) in cells (60). The drug ZD-0473 (**18**; JM-473) is more resistant than cisplatin to thiols, possibly because of steric hindrance by the pyridine ligand (61) and is in phase II clinical trials as an IV formulation; an oral formulation is also in development (62).

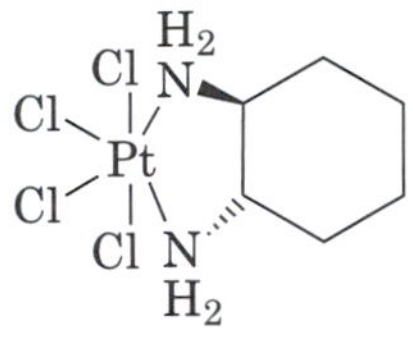

(16)

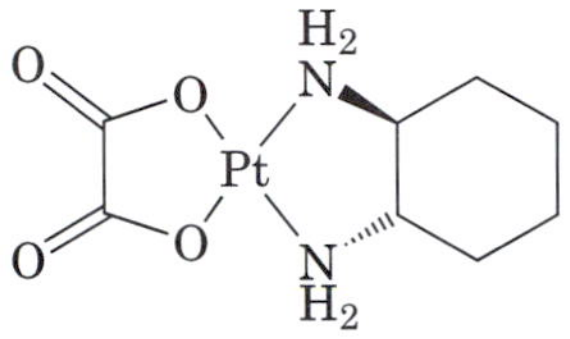

(17)

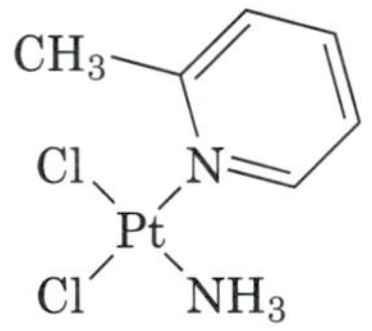

(18)

Satraplatin (**19**; JM216) is also being developed as an orally available platinum agent (63). It has potent *in vitro* cytotoxicity against a variety of tumor cell lines and also had oral antitumor activity against a variety of murine and human subcutaneous tumor models *in vivo*, broadly comparable with the level of activity obtainable with parenterally administered cisplatin (64). Satraplatin has shown activity in phase I trials in lung cancer, with no neurotoxicity or nephrotoxicity (63), and responses have also been seen in small cell lung cancer and hormone refractory prostate cancer.

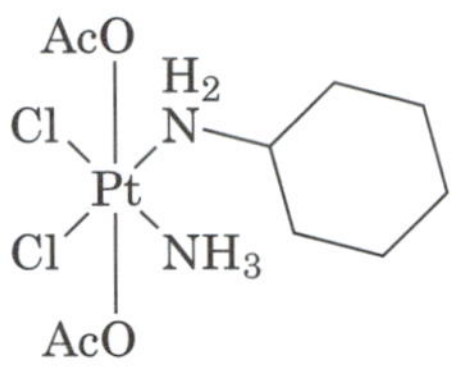

(19)

**2.4.2 Mechanism and SAR.** As with the nitrogen mustards, the mechanism of action of the platinum complexes involves formation of DNA cross-links. In the platinum complexes, the chloride or carboxylato ligands are the leaving groups, with the amine ligands being substitutionally inert and serving to modulate other properties. The bonds formed and broken in this case are coordinate metal-ligand bonds that are not permanent but have characteristic half-lives (although these may be very long), making the chemistry quite different to that of the mustards. Thus the Pt-Cl bonds in cisplatin (**14**) are more stable in the relatively high chloride conditions in plasma than they are in the lower chloride conditions inside cells, where the reaction with water to form aquo species is more facile (65). The cationically charged aquo species have higher affinity for DNA, and react primarily at guanine N7 sites in the major groove to form long-lived ammine complexes (Fig. 2.3).

Cisplatin reacts with DNA to form a number of different adducts. However, by far the most common are intrastrand guanine N7-guanine N7 adducts between adjacent guanines on the same strand (ca. 65%), followed by similar intrastrand guanine N7-adenine N7 adducts (ca. 25%), with DNA-protein cross-links and monofunctional adducts making up less than 10% and DNA interstrand adducts less than 1% of the total adducts. A major difference between mustards and platinum complexes is that whereas hydrolysis in the former case is a deactivating event, leading to loss of bifunctionality (and thus cross-linking ability), with platinum complexes, formation of the aquo species is a necessary activating process. Thus there is a much higher proportion of cross-links to monoadducts formed with platinum complexes than with mustards. The use of [$^1$H, $^{15}$N] heteronuclear single quantum coherence (HSQC) 2D

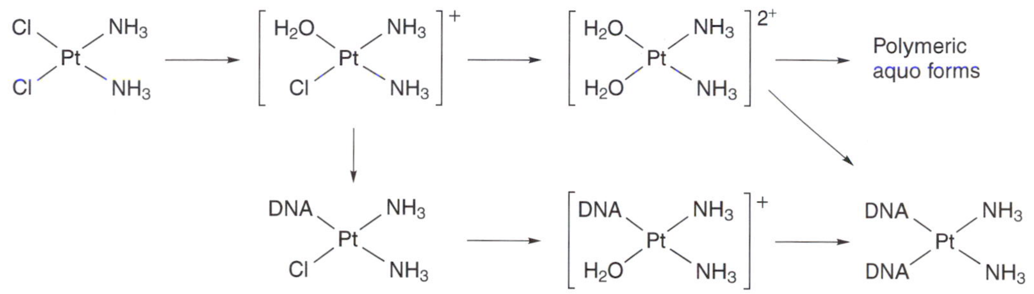

**Figure 2.3.** Reaction of platinum species with DNA.

NMR has recently allowed a better understanding of the kinetics of the multiple processes involved in the reaction of platinum drugs with DNA (66).

**2.4.3 Biological Activity and Side Effects.** While cisplatin is an extremely useful drug, it has many side effects. In addition to the myelosuppressive activity typical of a DNA alkylating agent, it also showed severe renal and neural toxicity. The analog development work described above has been aimed primarily to overcoming some of the these side effects. Thus carboplatin is less nephro- and neurotoxic; tetraplatin, oxaliplatin, and ZD-0473 are more effective against various types of resistance mechanisms; and satraplatin is orally effective. However, none of these compounds are overall more effective than cisplatin, and they do not show major differences in their interaction with DNA.

**2.4.4 Recent Developments: Increased Interstrand Crosslinking.** The tetracationic triplatinum complex BBR 3464 (**20**) seems to represent a new structural class of DNA-modifying anticancer agents (67). It reacts with DNA faster than does cisplatin, suggesting rapid cellular uptake and nuclear access (68), to give a different profile of adducts than cisplatin, with about 20% being interstrand GG cross-links. DNA modified by BBR3464 cross-reacted with antibodies raised to transplatin-adducted DNA but not to antibodies raised to cisplatin-adducted DNA (67). BBR 3464 was 30-fold more cytotoxic than cisplatin in L1210 cells and showed no cross-resistance in sublines resistant to cisplatin because of impaired accumulation and lower DNA binding (69).

Consistent with this, it was also highly active in a panel of cisplatin-resistant xenografts, giving longer growth delays (70). Unlike cisplatin, BBR 3464 was able to induce the p53/p21 pathway to a similar extent in both cisplatin-sensitive and -resistant cells (71), and had a quite different sensitivity profile to cisplatin in the U.S. National Cancer Institute's 60-cell-line screening panel (70). In a phase I trial using a single-dose schedule, no significant neural or renal toxicity was observed; the side-limiting toxicity at 0.17 mg/m$^2$ was short-lasting neutropenia (72).

**2.5  Cyclopropylindoles**

**2.5.1 History.** Interest in DNA minorgroove alkylating agents was stimulated by the discovery (73) of the natural product CC-1065 (**21**) from *Streptomyces zelensis* (74), which showed extraordinary potency in a number of animal tumor models (75) but with concomitant fatal delayed hepatotoxicity at therapeutic doses (76). An extensive synthesis program at Upjohn prepared a large number of analogs in an attempt to understand structure-activity relationships for the class (77) and succeeded in developing the structurally simpler agents adozelesin (**22**) and the open-chain form carzelesin (**23**), which did not show the delayed hepatotoxicity of CC-1065 (78, 79). The related semi-synthetic duocarmycin analog KW 2189 (**24**) is a carbamate prodrug, releasing the active moiety DU-86 (**25**) by esterase hydrolysis (80). KW 2189 has been prepared on a large scale by a three-step synthesis with an overall 55% yield from natural duocarmycin B2 (81). Although it is less potent than duocarmycin in cell culture assays,

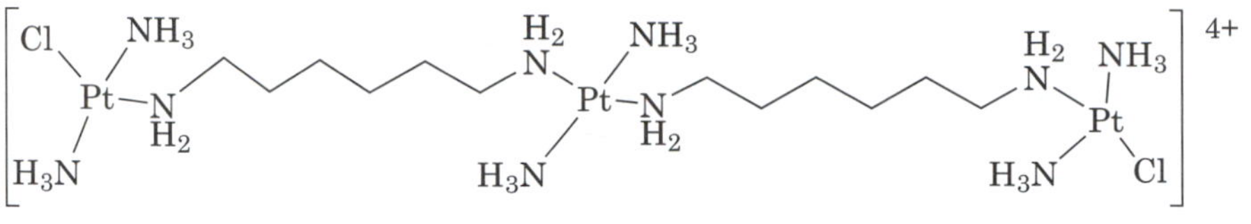

(20)

it has high activity in a wide range of human solid tumor xenografts in mice and lacks the delayed lethal toxicity seen with some other cyclopropylindoles (82).

**2.5.2 Mechanism and SAR.** These compounds bind initially reversibly in the minor groove of DNA with minimal structural distortion, and subsequently, alkylate specifically at the N3 position of adenine (83). This provides further evidence that targeting alkylating functionality to the DNA minor groove can provide compounds of very high cytotoxic potency. Whereas the lead compound is a natural

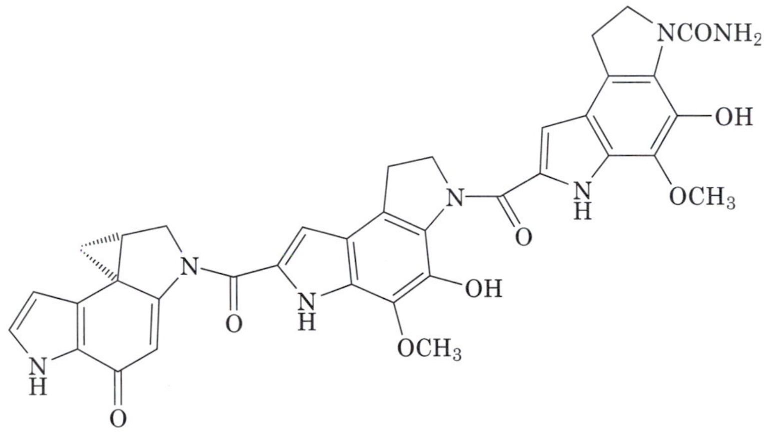

(21)

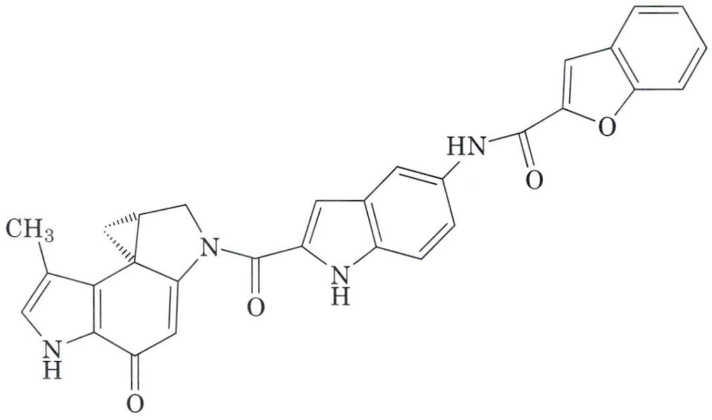

(22)

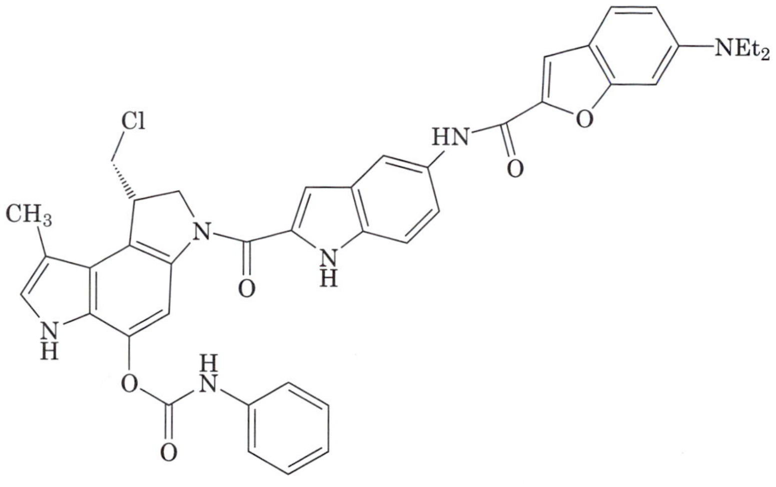

(23)

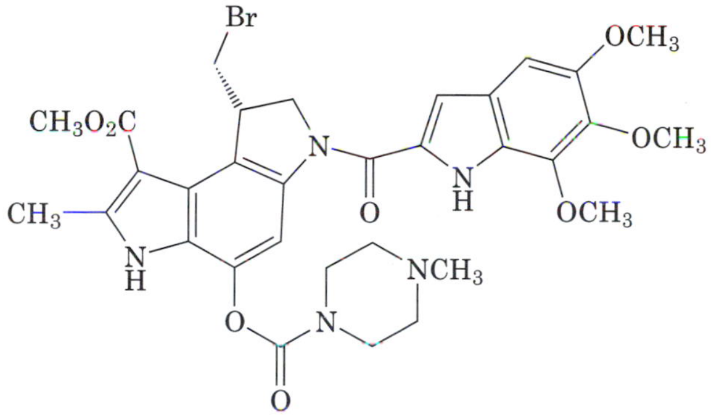

(24)

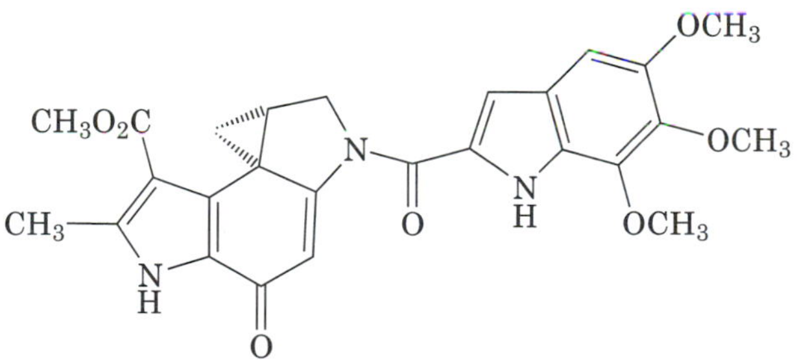

(25)

product, it has sparked a vast amount of synthetic chemistry, and the analogs developed for clinical studies are synthetic. It has been proposed (84) that binding of these com- pounds in the minor groove of DNA requires a propellor twist of the cyclopropyldienone and indole subunits around the amide bond, and that this interrupts the vinylogous amide sta-

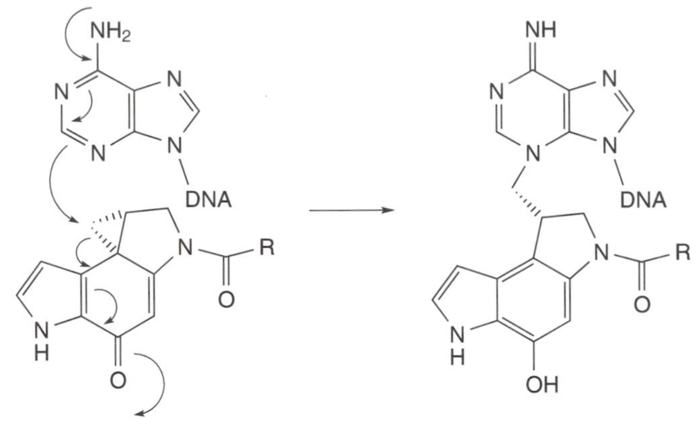

**Figure 2.4.** Alkylation of DNA by cyclopropylindoles.

bilization of the cyclopropyldienone, activating the conjugated cyclopropane electrophile (Fig. 2.4). Changes in the DNA binding side chain have only minor effects on the sequence selectivity of alkylation; both adozelesin (**22**) and carzelesin (**23**) alkylate DNA at the consensus sequences 5'-(A/T)(A/T)A and 5'-(A/T)(G/C)(A/T)A are broadly similar (85) to the consensus sequence 5'-(A/T)(A/T)A for CC-1065. A series of analogs of KW 2189 with water-solubilizing cinnamate side-chains were reported to have potent *in vivo* antitumor activity and low peripheral blood toxicity compared with the trimethoxyindole congeners (86), and more potent ring A dialkylaminoalkyl derivatives have also been reported (87).

**2.5.3 Biological Activity and Side Effects.** Many of the synthetic compounds developed from the original natural product lead were also extremely potent and showed broad-spectrum activity in human tumor colony–forming assays (88), and both adozelesin (**22**) and carzelesin (**23**) proceeded to clinical trial. However, adozelsin had only marginal efficacy in a phase II trial of untreated metastatic breast carcinoma (89). Similarly, carzelesin showed no activity in a phase II trial in patients with a variety of advanced solid tumors (90). A phase I trial of KW 2189 (**24**) established the maximum tolerated dose at 0.04 mg/m$^2$/d when given daily for 5 days, with leukopenia, neutropenia, and thrombocytopenia as dose-limiting toxicities (91). A phase II pilot study in metastatic renal cell carcinoma showed a good safety profile but no activity (92).

**2.5.4 Recent Developments.** Amino analogs (e.g., **26**) of the corresponding phenolic open-chain forms (e.g., **27**) were reported to have comparable cytotoxicity (93) and similar patterns of DNA interaction, alkylating preferentially at 5'-A(A/T)AN sequences (94). These have been proposed as effectors for tumor-activated prodrugs (see Section 5).

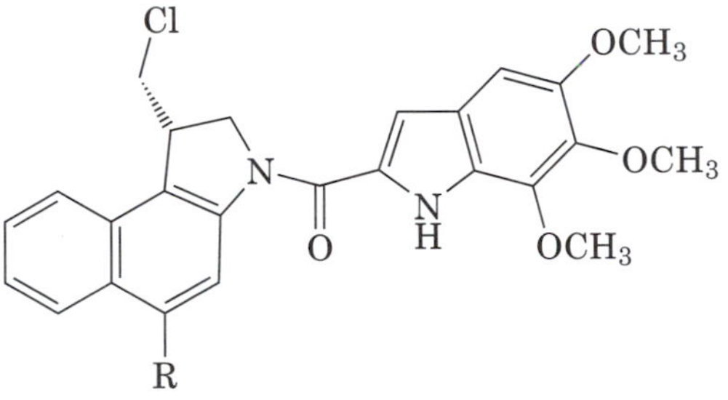

(**26**) R = NH$_2$
(**27**) R = OH

**2.6 Nitrosoureas**

**2.6.1 History.** This class of compounds has a long history, and extensive reviews exist on all aspects of their chemistry and biology (95). The initial impetus for their development came from screening at the U.S. National Cancer Institute, where 1-methyl-3-nitro-1-nitrosourea (**28**) showed some activity in the *in vivo* leukemia screen (96). Development of this lead resulted in the urea-based clinical agents BCNU (**29**; carmustine) and CCNU (**30**; lomustine). These reactive compounds have very short half-lives (a few minutes) (97), but their very lipophilic nature suggested they might cross the blood-brain barrier and be useful in brain tumors (98). The more hydrophilic streptozotocin (**31**) is a natural product isolated from *Streptomyces* species. It was evaluated initially as an antibacterial agent but proved to be too toxic (99).

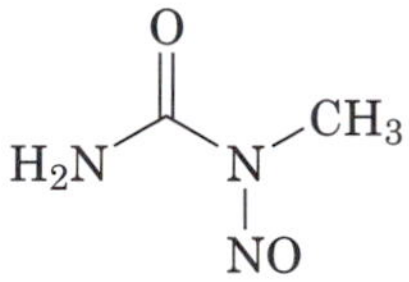

(28)

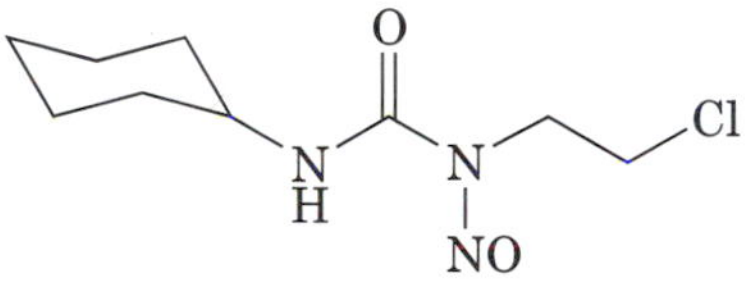

(29)

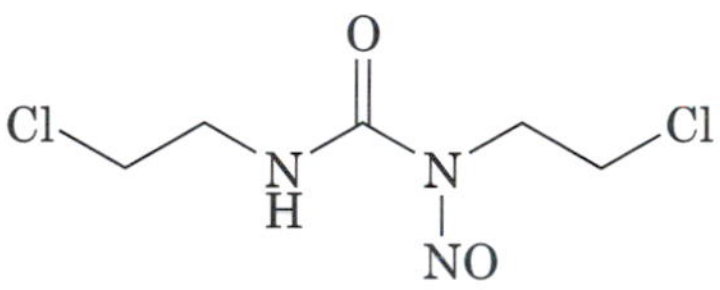

(30)

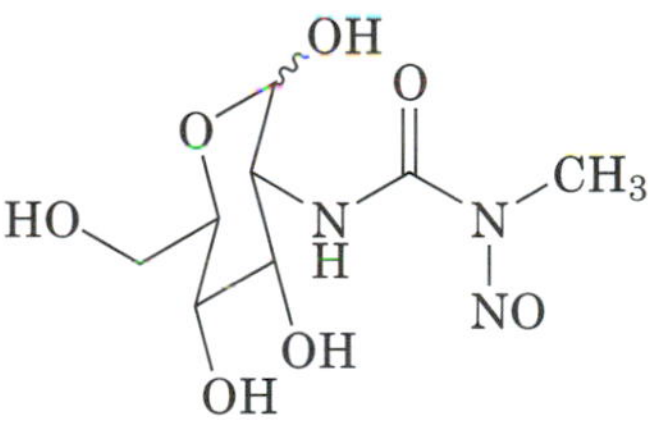

(31)

**2.6.2 Mechanism and SAR.** The mechanism of the nitrosoureas is complex. They possess both alkylating and carbamoylating activities (100). Decomposition occurs spontaneously in aqueous media by cleavage of the N—CO bond to give a diazoacetate (alkylating agent) and an isocyanic acid (carbamoylating agent) (101) (Fig. 2.5).

**2.6.3 Biological Activity and Side Effects.** Streptozotocin has been used as a component of multidrug protocols for Hodgkin's disease (102), and for pancreatic (103) and colorectal (104) carcinomas, where some responses were seen, but the drug is not now widely used. A recent report (105) notes activity of a streptozocin/o,p'-DDD combination in adrenocortical cancer.

## 2.7 Triazenes

**2.7.1 History.** Dacarbazine (**32**; DTIC) came from studies by Shealy and co-workers, and has been well reviewed (106, 107). Mitozolomide (**33**) was developed by Stevens and co-workers as a potential prodrug of linear triazenes such as [5-(3-3-dimethyl-l-triazenyl)imidazole-4-carboxamide] (DTIC) (108). The same workers later followed up with the development of the related temozolomide (**34**), which lacks the 2-chloroethyl group (109).

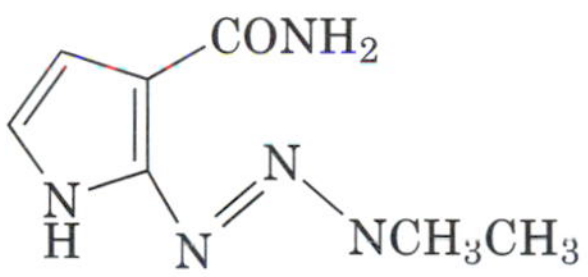

(32)

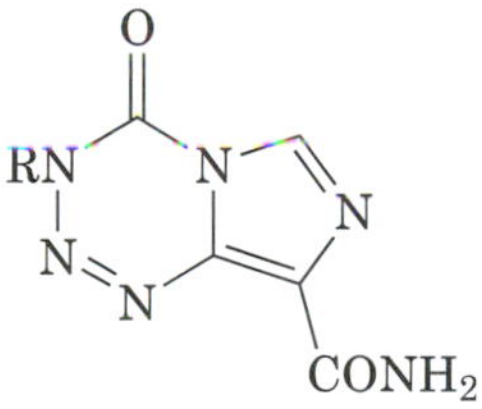

(33) R = CH$_2$CH$_2$Cl
(34) R = CH$_3$

**2.7.2 Mechanism and SAR.** The cyclic triazenes undergo base-catalyzed ring opening, followed by spontaneous decarboxylation. Thus, temozolomide (**34**) forms the open-chain triazene [5(3-methyl-l-triazenyl)imidazole-4-carboxamide] (MTIC), which then fragments to a methyldiazonium species, the DNA-methylating agent (Fig. 2.6) (110). Temozolomide alkylates DNA primarily at the

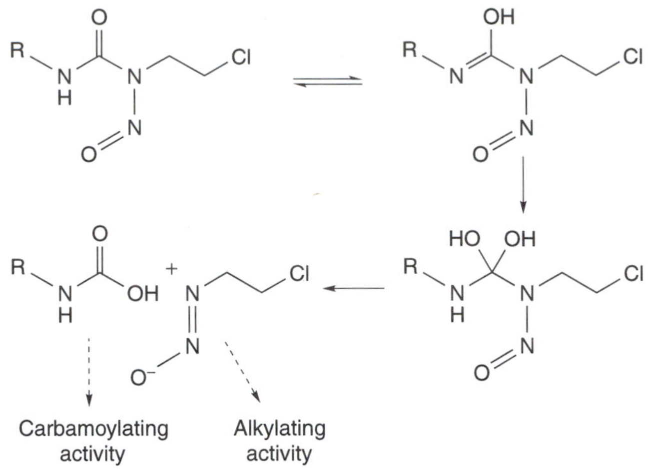

**Figure 2.5.** Mechanism of action of nitrosoureas.

Carbamoylating activity          Alkylating activity

N7 of guanine, but O6 guanine alkylation also occurs. The rate of conversion to the alkylating species is not influenced by the presence of DNA, suggesting no or very weak binding of the prodrug (110). As with mitozolomide, cytotoxicity correlates with the alkylation of the O6-position of guanine (111). L1210 cells treated with mitozolomide form DNA interstrand cross-links, presumably through the 2-chloroethyldiazo metabolite, suggesting this is a major mechanism of cytotoxicity (112). Mitozolomide preferentially alkylates DNA at guanines in runs of guanines, forming 7-hydroxyethyl and 7-chloroethyl adducts (113).

**2.7.3 Biological Activity and Side Effects.** Dacarbazine has been widely used for many years, and in particular, has been the corner-stone of drug therapy for malignant melanoma (114, 115). It is metabolized by *N*-hydroxylation, followed by *N*-demethylation, to give a monomethyltriazene that then methylates DNA (116). No surprisingly, dacarbazine is strongly carcinogenic in animal models (117), suggesting it may also be a human carcinogen.

Mitozolomide proved curative against a broad range of murine tumor models *in vivo* (118) and showed very pronounced antitumor effects in a range of human tumor xenografts (119). Cell lines with constitutive levels of O6-methylguanine-DNA methyl-transferase (Mer+ phenotype) were less sensitive to the cytotoxic effects of mitozolomide, consistent with alkylation of the O6-position of guanine being the cytotoxic event (111). In 1998, mito-

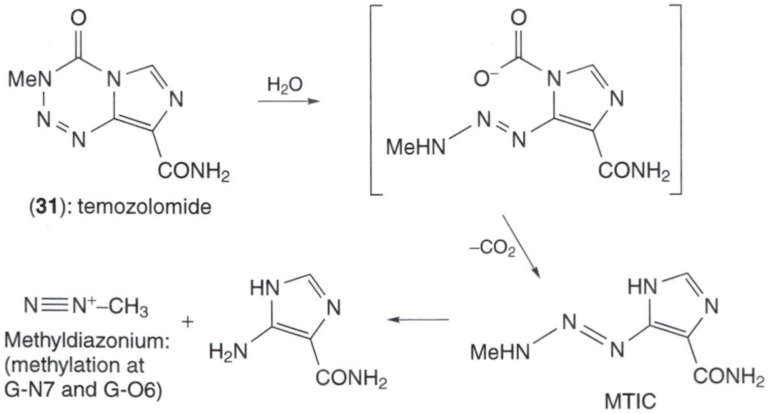

**Figure 2.6.** Mechanism of activation of temozolomide.

zolomide entered phase I clinical trials (120), but despite demonstrable activity in small-cell lung cancer (SCLC) and melanoma (121), unpredictable myelosuppression precluded further development (121, 122). However, recent work in the successful transduction of human hematopoietic progenitor cells with variants of this enzyme has led to suggestions that this could be used clinically to protect against myelosuppression, to allow safer use of agents like mitozolomide and temozolomide in conjunction with O6-benzylguanine (123, 124).

Temozolomide also demonstrated good *in vivo* activity against a variety of mouse tumor models, including the TLX5 lymphoma (125), and excellent antitumor activity, including cures, on oral administration to athymic mice bearing both subcutaneous and intracerebral human brain tumor xenografts (126). Many later studies confirmed the good activity in brain tumor models, and this, together with the lesser myelosupression seen in toxicology screens, led to phase I trials (127). Trials in radiotherapy-resistant astrocytomas confirmed animal data suggesting that temozolomide efficiently passes the blood-brain barrier (128). Recent reviews show that oral temozolomide has almost 100% bioavailability, acceptable non-cumulative myelosuppression, and is clinically useful in the treatment of gliomas (129, 130) and brain metastases in advanced melanoma (131).

## 3  SYNTHETIC DNA-INTERCALATING TOPOISOMERASE INHIBITORS

### 3.1  Introduction

Intercalation as a mode of the reversible binding of ligands to DNA was first described by Lerman (132) for the acridine proflavine (**35**). Intercalation involves insertion of the chromophore between the base pairs, and is now understood to be the major DNA binding mode of virtually any flat polyaromatic ligand of sufficiently large surface area and suitable steric properties. Intercalative binding is driven primarily by stacking (charge-transfer and dipole-induced dipole) and electrostatic interactions, with entropy (dislodgement of ordered water around the DNA) of lesser and variable importance (133). A great deal of work has

been done delineating the ligand structural properties that favor intercalation, the geometry, kinetics, and DNA sequence-selectivity of the binding process, and the effect of such binding on the structure of the DNA substrate (134). A very large number of compounds have been shown to be DNA intercalating agents, and many of these show cytotoxic activity. In the early 1980s, it was shown that these cytotoxic effects were primarily caused by the compounds forming ternary complexes with DNA and the enzyme topoisomerase II, altering the position of equilibrium and trapping a reaction intermediate termed the "cleavable complex" (135, 136). The DNA intercalators are now recognized collectively as a major class of topoisomerase poisons.

(**35**)

The topoisomerases are enzymes that regulate DNA topology by successive cleavage-religation reactions and are a major target for anticancer drugs. Topoisomerase (topo) II is a homodimeric protein, associated with the mitotic chromosome scaffold. It initially binds to DNA reversibly and then executes a series of concerted strand-breaking and religation processes to relieve torsional stresses generated during DNA replication (137). Whereas many types of agents interfere to some extent with the normal function of topo II, DNA-intercalating agents in particular have the ability to cause lethal DNA double-strand breaks. The observation of these breaks, characterized by protein covalently attached to the 5′-ends, first led to suggestions (135) that a topoisomerase enzyme was involved. It is now clear that the primary mode of cytotoxicity of most DNA intercalating agents involves inhibition of the religation step of the action of the enzyme mammalian DNA topo II (138, 139). The topo II enzyme has major isozymes coded by two separate genes (140). The II$\alpha$ isozyme (170 kDa) maps (141, 142) to chromosome 17, is the regulated during the cell cycle, and is the target of virtually all of the DNA intercalators.

**Table 2.2   Synthetic Topoisomerase (topo) Inhibitors Used in Cancer Chemotherapy**

| Generic Name (Structure) | Trade Name | Originator | Chemical Class |
|---|---|---|---|
| Topo II inhibitors | | | |
| amsacrine (**39**) | Amsidyl | Warner-Lambert | 9-anilinoacridine |
| asulacrine (**40**) | | Sparta | 9-anilinoacridine |
| mitoxantrone (**41**) | Novantrone | Wyeth | anthracenedione |
| BBR 2778 (**43**) | | Boehringer | |
| losoxantrone (**45**) | | Warner-Lambert | |
| piroxantrone (**46**) | | Warner-Lambert | |
| Dual topo I/II inhibitors | | | |
| DACA (**51**) | XR-5000 | Xenova | Acridine |
| intoplicine (**52**) | | Ilex | Pyridoindole |
| TAS 103 (**53**) | | Taisho | Indenoquinolone |
| DMP 840 (**58**) | | Knoll | Bis(naphthalimide) |

The II$\beta$ isozyme (180 kDa) maps (143) to chromosome 3 and becomes the predominant isozyme in both non-cycling cells and in cells resistant to "classical" topo II agents (138, 143).

## 3.2   Clinical Use of Agents

The most commonly used synthetic topoisomerase inhibitors are listed in Table 2.2, together with some interesting new agents in early clinical development. A large number of synthetic DNA-intercalating agents have been developed, representing a broad range of chemistries. Many of these have been evaluated in early clinical trials, but relatively few have shown useful activity. Apart from synthetic analogs of natural products such as the anthracyclines (which will not be covered here), the major subclasses of clinically useful synthetic DNA intercalators are the acridines and the anthracenediones. In contrast, the most important topo I inhibitors are derived from the natural product camptothecin and will also not be covered here.

## 3.3   Topo II Inhibitors

**3.3.1   History.**  The two main classes of synthetic topo II inhibitors are the 9-anilinoacridines and the anthracenediones. The 9-anilinoacridine amsacrine evolved from work carried out by Cain and associates on anti-leukemic quinolinium-type agents (144), which they suggested intercalated following initial minor groove binding of the remainder of the molecule (145). A series of acridinium analogs

(e.g., **36**) were more active (146), and later work showed that these compounds, with the larger chromophore, did intercalate DNA (147). The acridinium series proved unique in that unquarternized derivatives (e.g., **37**) were also active, and a series of progressively simpler analogs led to the methanesulfonamide (**38**) (148), which had superior water-solubility, stability, and biological activity. Further work, on the basis of a theory that high electron density at the 6'-position was favored (149), showed that a 3'-methoxy group greatly increased potency, resulting in amsacrine (**39**; m-AMSA). Following detailed animal testing, the U.S. National Cancer Institute initiated clinical trials in 1974 (150). Encouraging results in both leukemias and lymphomas (150), with an apparent lack of cross-resistance to doxorubicin (151), resulted in amsacrine becoming the first synthetic DNA intercalator to show clinical efficacy (152).

The success of amsacrine led to a search for analogs with a broader spectrum of action. Because the high p$K$a (8.02) of amsacrine was thought to play a part in limiting its distribution, analogs with a lower p$K$a that still retained high DNA binding and had improved aqueous solubility were sought. QSAR studies (153) suggested that the anilino side chain was close to optimal, and focused attention on the 4- and 5-positions as being the most suitable for modification. Carboxamide substituents were seen as favourable for lowering p$K$a, and several 4- and 3-carboxamides, including the 4-methyl-5-methylcarboxamide (**40**; asula-

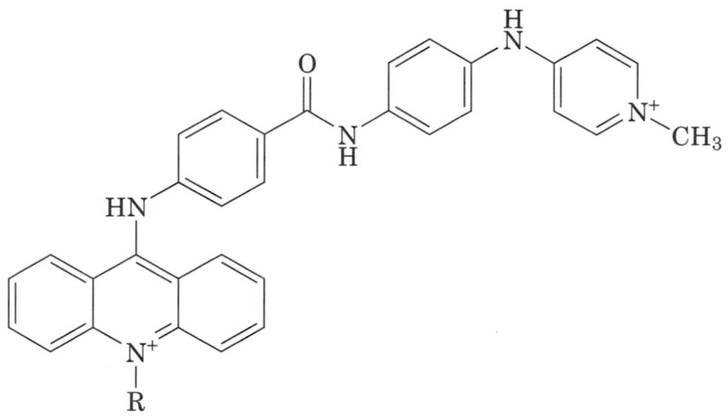

**(36)** R = CH$_3$
**(37)** R = H

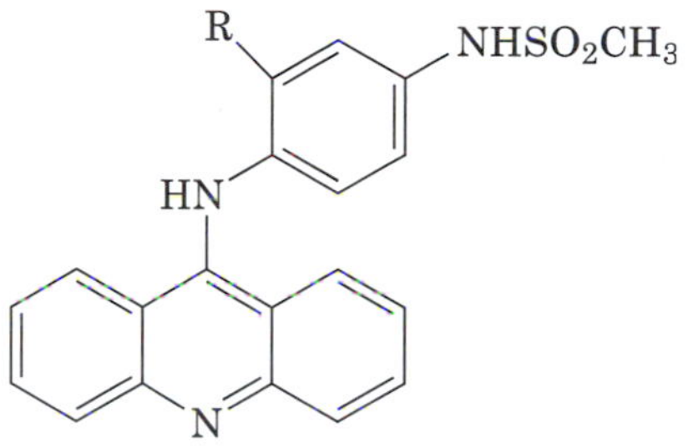

**(38)** R = H
**(39)** R = OCH$_3$

crine) were studied in detail (154). This proved the most active of a series evaluated for oral activity (155) was the best against both a human solid tumor cell-line panel (156) and a wide range of murine solid tumors *in vivo* (157), and was selected for clinical trial.

Mitoxantrone (**41**) was discovered through screening of industrial dye compounds (158).

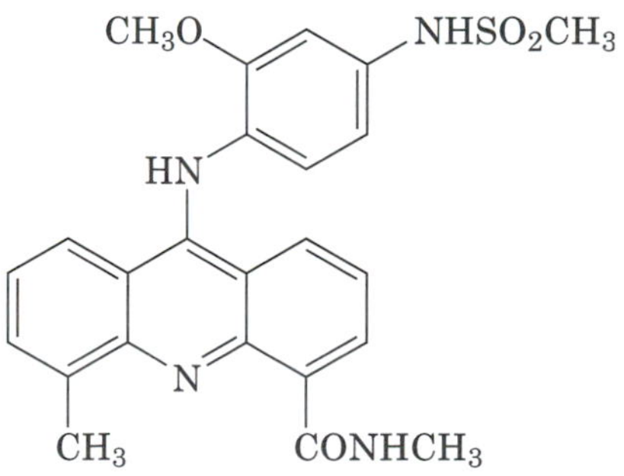

**(40)**

Both it and the des-hydroxy analog amet-antrone (**42**) showed broad-spectrum activity in animal tumor models (159), and mitox-antrone has become probably the most widely used synthetic DNA intercalating agent.

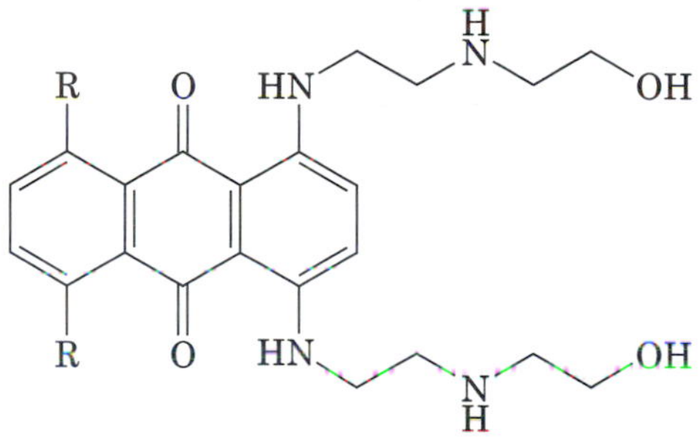

**(41)** R = OH
**(42)** R = H

**3.3.2 Mechanism of Action and SAR.** Drugs aimed at topoisomerases can work in one or both of two ways; by inhibiting the ability of the enzyme to relax DNA by preventing its initial cleavage function or by preventing religation of the transient "cleavable complex" (stabilization of the cleavable complex), resulting in enhanced strand breaks. The second is the more cytotoxic process and is the mechanism by which the majority of topo inhibitors (or, more accurately, topo poisons) work. Much of the early structure-activity relationship (SAR) work on the DNA intercalator class of drugs focused around their interaction with DNA, delineating the requirements for successful intercalation and tight binding (134). Early SAR studies for several classes of intercalators showed positive relationships between cytotoxic potency and strength of DNA binding (160, 161) and long residence times for the intercalators at individual DNA sites (162). However, the discovery that the primary mode of cytotoxicity of these compounds was inhibition of topo II through formation of a ternary drug/DNA/protein complex (138, 139) made it clear that drug design through modeling of DNA binding properties alone could be misleading.

Amsacrine binds to DNA by reversible, enthalpy-driven (163) intercalation of the acridine chromophore, with an association constant of $1.8 \times 10^5 \, M^{-1}$ for calf thymus DNA in $0.01 \, M$ salt (164). By analogy with the crystal structure determined for 9-aminoacridine binding to a dinucleotide (165), amsacrine was postulated to bind with the anilino ring lodged in the minor groove, with the 1'-substituent pointing tangentially away from the helix, and the possibility of it thus interacting with another (protein) macromolecule to form a ternary complex was noted (164). This conformation was also supported by energy calculations (133, 166). In the ternary cleavable complexes, DNA intercalation of the acridine occurs, and the anilino side-chain seems to specifically interact with the enzyme as well (167, 168).

Amsacrine causes comparable levels of cell killing in yeast transfected with either human topo II$\alpha$ or II$\beta$, whereas etoposide, doxorubicin, and mitoxantrone produced higher degrees of cell killing with topo II$\alpha$ (169). However, the sensitivity of a panel of human breast cancer cell lines to amsacrine was shown to correlate better with the level of expression of the topo II$\alpha$ protein (although not with the level of topo II$\alpha$ mRNA) (170), suggesting that the former is the most important mechanism of resistance to these topo II inhibitors. A human SCLC line (GLC4) with acquired resistance to amsacrine did not overexpress P-glycoprotein, and it had an 82% decrease in topoII$\beta$ protein but no change in topoII$\alpha$ protein level (171). A classification of antitumor drugs by their topo II-induced DNA cleavage activity and sequence preference placed amsacrine in the class that enhanced the stabilization of cleavable complexes at a single major

site, acting upstream of the DNA cleavage step through enhancement of cleavage (172). Amsacrine seems unique among topo II poisons in that its ability to trap both topo II$\alpha$– and topo II$\beta$–induced lesions is only modestly reduced in ATP-depleted cells; it is suggested that amsacrine produces mainly prestrand passage DNA lesions, whereas other topo II poisons only stabilize poststrand passage DNA lesions in intact cells (173). Studies with amsacrine and other topo II poisons in HeLa (174) and AHH-1 human lymphoblastoid (175) cells suggest that these compounds can induce cell death by apoptosis. However, whereas amsacrine induced apoptosis in wild-type SCLC cells, it did not do so in an amsacrine/camptothecin-resistant subline, and no significant difference in the expression of several genes (c-myc, bc12, c-jun, p53) involved in the apoptotic process was seen in either the parental and resistant cells after drug treatment. These data suggest that modulation in the apoptotic pathway could be an additional mechanism of resistance to amsacrine and other topo II agents (176).

Mitoxantrone also binds reversibly to DNA by intercalation (177), with an unwinding angle of 23°, probably with the chromophore inserted perpendicular to the base-pair axis with the side-chains lying in the major groove (178), although this has not been rigorously proven. Footprinting studies show the preferred intercalation site for mitoxantrone to be 5'-(A/T)CG or 5'-(A/T)CA sites (179). Mitoxantrone and the related ametantrone (**42**) bind tightly and about equally well to DNA with association constants of about $5 \times 10^6$ $M^{-1}$ at physiological salt concentrations (180), but mitoxantrone has about fourfold slower dissociation kinetics (177, 181). The higher cytotoxicity of mitoxantrone compared with ametantrone correlated with its higher capacity to induce topo II-mediated cleavable complexes, suggested because of greater stability of the ternary complex (182). However, whereas mitoxantrone showed a similar capacity to amsacrine at inducing cleavable complexes, it is considerably more potent and able to induce much more long-lived blocks at the G2 stage of the cell cycle (183). It induces DNA fragmentation and activates caspases, demonstrating that the ultimate cytotoxic effect is induction of apoptosis (184). Mitoxantrone is readily oxidized (for example by human myeloperoxidase) to metabolites that covalently bind to DNA (185). It also, like the anthracyclines, forms formaldehyde-induced adducts that function as virtual interstrand cross-links (186). Both of these properties may be relevant to its biological activity. However, as with many other intercalators, mitoxantrone's cytotoxicity is caused largely by inhibition of topo II.

### 3.3.3 Biological Activity and Side Effects.
The clinical use of amsacrine is mainly in acute myeloid leukemia. Amsacrine/etoposide therapy with or without azacitidine in relapsed childhood acute myeloid leukemia was effective (34% complete responses), with azacytidine not improving response rates (187). Recent successful use in various adult leukemias has also been reported (188–190). Amsacrine has generally not been successful in the treatment of solid tumors, except for some responses in head-and-neck cancer, where high-dose amsacrine was a toxic but very effective drug for first-line treatment (191), with a response rate of 65%. Whereas much less cardiotoxic than the anthracyclines, pre-exposure to amsacrine is a risk factor for cardiotoxicity after anthracycline treatment for childhood cancer (192).

Asulacrine showed a similar mechanism of action to amsacrine, generating DNA protein cross-links and DNA breaks through inhibition of topo II (193, 194). In initial phase II trials, some drug-induced remissions were seen in non-small-cell lung cancer and breast cancer but not in colorectal and gastric cancer (195, 196). A pilot study of a trial of oral administration has been reported (197), but there are no reports of asulacrine being used clinically in combination therapy.

Mitoxantrone is used in first-line therapy for acute myelocytic leukemia (AML) (198), and along with cytosine arabinoside, is suggested as salvage therapy in AML and chronic myelocytic leukemia (CML) (199). In combination with a steroid, it is the drug of choice for palliative treatment of hormone-resistant prostate cancer (200). It is also an effective treatment for secondary progressive multiple sclerosis, but the duration of treatment is limited by cumulative cardiotoxicity (201). Again, whereas less cardiotoxic than the anthracyclines, mitoxantrone has been shown to have cumulative cardiotoxic effects (202). Mitoxantrone is genotoxic in the *in vitro* micronucleus test and in mutation assays (203) and has been reported to induce secondary cancers after use in the treatment of breast cancer (204). Resistance to mitoxantrone can develop in a number of ways: by lower expression of topo II (205), by expression of a topo II with altered DNA cleavage activity (206), by decreased drug uptake even in the absence of elevated levels of P-glycoprotein (207), and by inherent resistance to the induction of apoptosis (208). Many cells develop multifactorial resistance to mitoxantrone (209).

Topoisomerase inhibitors are also known to be tumorigenic, related to the formation of multiple DNA strand breaks. A frequent chromosomal translocation is at 11q23, where the myeloid-lymphoid leukemia (MLL) gene is located (28), but other translocations are also seen. The onset of induction of AML is shorter than with alkylating agents, with the average around 2 years and an incidence of 2–12% (210). Anthracyclines, mitoxantrone, and epipodophyllotoxins have all been shown to induce AML (211).

### 3.3.4 Recent Developments: Compounds with Lower Cardiotoxicity.
The development of analogs of mitoxantrone has been driven largely by the requirement for lower cardiotoxicity. Two broad classes of analogs can be distinguished, and much work has been done on both. The first are close analogs of mitoxantrone, where the tricyclic chromophore has been maintained and variations occur in the side-chains or the chromophore atoms. In a study of aza analogs, Krapcho and co-workers found that the positioning of the aza group was critical, with the 2-aza derivative (**43**; BBR 2778) being the most potent (212). This bound less tightly to DNA but induced topo II–mediated DNA cleavage (213). Preclinical studies showed (**43**) has a better therapeutic index and lower cardiotoxicity than mitoxantrone (214), and a phase I trial has been reported (215).

The second broad class are tetracyclic compounds, primarily the imidazoacridinones and

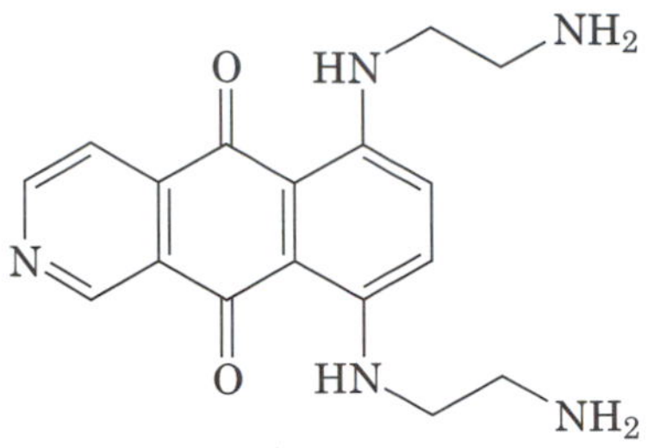

(43)

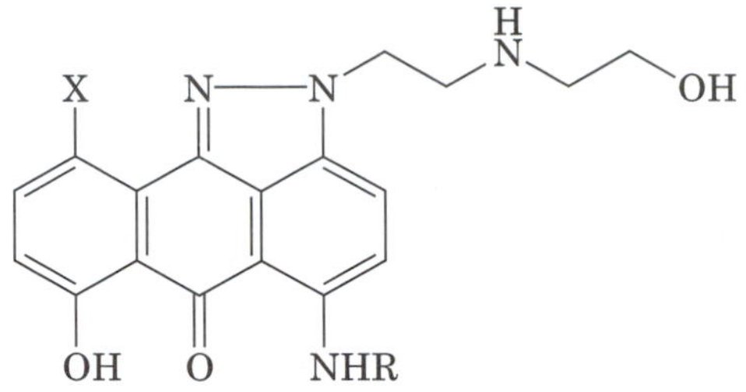

(44) X = OH, R = (CH$_2$)$_2$NHCH$_3$
(45) X = H, R = (CH$_2$)$_2$NH(CH$_2$)$_2$OH
(46) X = OH, R = (CH$_2$)$_3$NH$_2$

the anthrapyrazoles. Showalter and colleagues at Parke-Davis, in search of less cardiotoxic agents, developed these initially. They laid down the basic SAR, showing that activity was maximal with alkylamino side-chains at the N-2 and C-5 positions with two to three carbon spacers between proximal and distal nitrogens and showed they induced less oxygen consumption than doxorubicin in the rat liver microsomal system (216). These compounds bind very tightly to DNA by intercalation, with association constants around 2 × 10$^8$ $M^{-1}$ (217). They were highly active in murine leukemias and a range of human tumor xenografts (218), and three (44–46) were selected for preclinical evaluation (219). This early judgment was vindicated by the fact that all three of these later went forward to clinical trial. Teloxantrone (44) did receive a clinical trial (220) but has not been further reported on. However, losoxantrone (45) and piroxantrone (46) have been more widely studied. Piroxantrone (46) showed some responses in phase I trials (221), but this was not borne out in phase II trials (222, 223). Losoxantrone (45) showed classical topo II inhibition (224), and a number of phase I trials were conducted in which the dose limiting toxicity was leucopenia; some non-cumulative cardiotoxicity was also seen (225). The major metabolites detected in humans resulted from oxidation of the hydroxymethylene side-chains to either mono- or dicarboxylic acid derivatives (226, 227). A phase II trial in hormone-refractory metastatic prostate cancer showed improvement of clinical symptoms in one-third of patients (228), and the drug is reported to be currently in phase III development (227).

Structure-activity studies on imidazo-acridinones (229) identified (47) (C-1311) as a

potential anticancer drug that intercalates DNA (230), inhibits the catalytic activity of topo II, and has broad-spectrum solid tumor activity (231). It is reported to be in a phase I clinical trial (232).

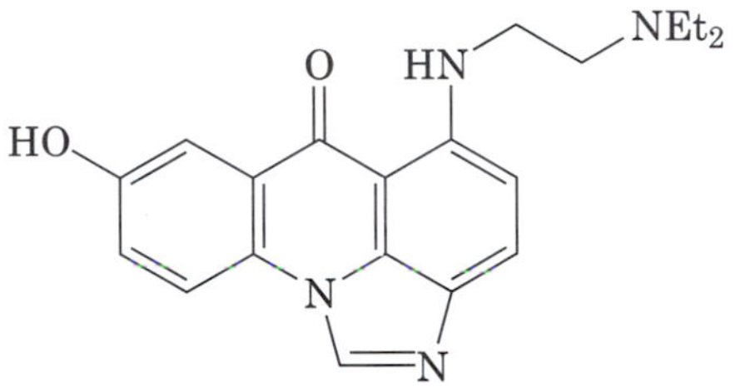

(47)

### 3.4 Dual Topo I/II Inhibitors

**3.4.1 History.** The topo I and topo II enzymes are expressed at different absolute levels in different cell types. Topo II levels are reported to be high in many breast and ovarian lines (233), whereas topo I levels are reported to be high in many colon cancer lines (234); the good clinical activity of camptothecin analogs against colon tumors has been suggested because in part of this high level of topo I expression (235). The time-course of expression of topo I and topo II also differs markedly, with topo II levels at their highest during S-phase, whereas levels of topo I remain relatively constant through the cell cycle (236). Because expression of either enzyme seems to be sufficient to support cell division, the development of resistance to topo I inhibitors is often accompanied by a concomitant rise in the level of topo II and vice versa (237, 238).

Thus, one of the recent interests in topo inhibitors has been in agents capable of simultaneous inhibition of both enzymes, although relatively few compounds have been reported as dual topo I/II inhibitors (239). The anthraquinone saintopin (48) is a potent poison of both topo I and topo II (240) but has not been developed as a drug. The quaternary alkaloid nitidine (49) is reported (241) to be a dual poison, although more active against topo I. The related quaternary salt NK 109 (50) is described as a topo II poison, but etoposide-resistant lines with reduced topo II levels are still sensitive (242), suggesting a dual activity. Most work has been focused on the DNA intercalators DACA (51; XR-5000), intoplicine (52), and TAS 103 (53).

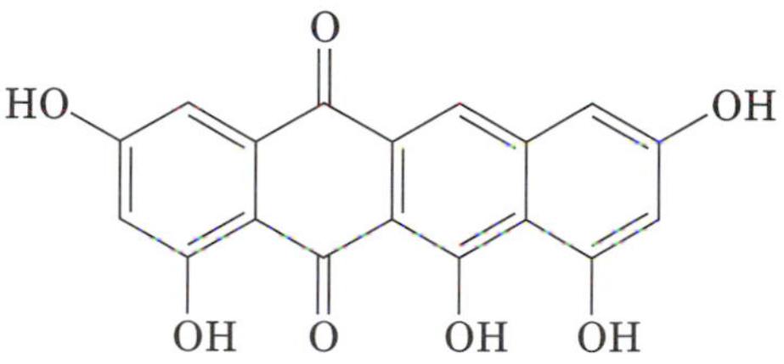

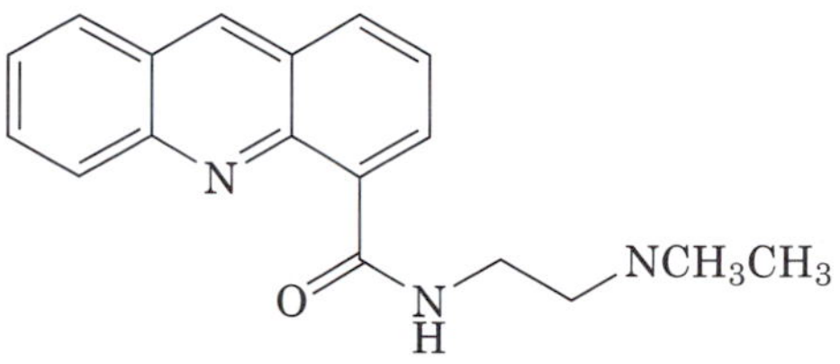

(51)

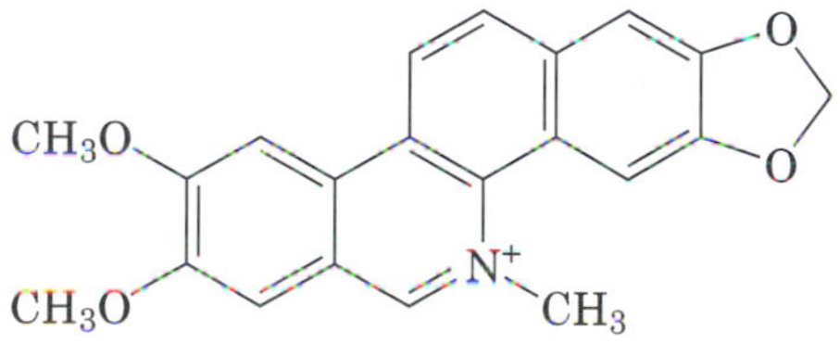

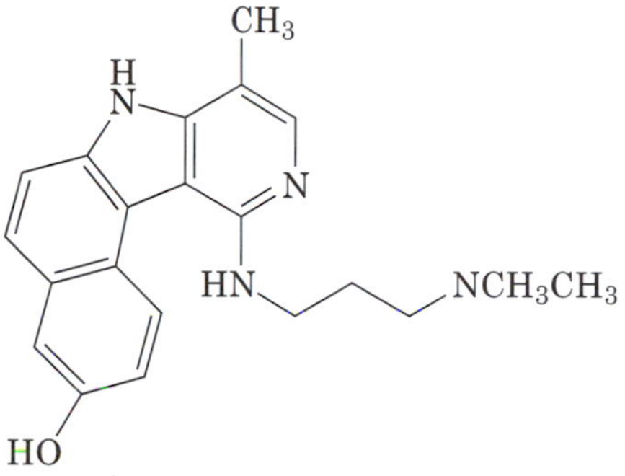

(52)

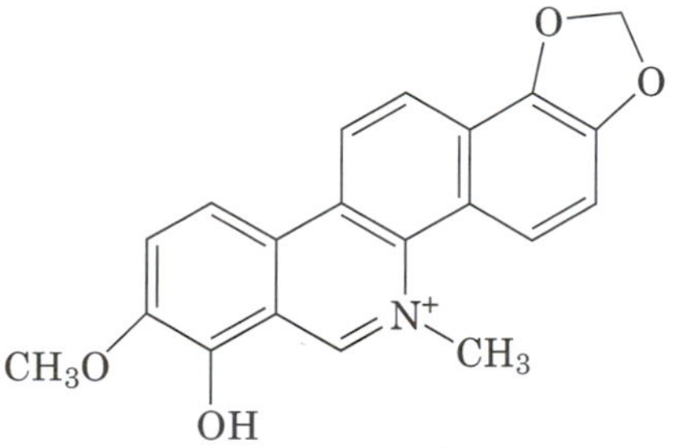

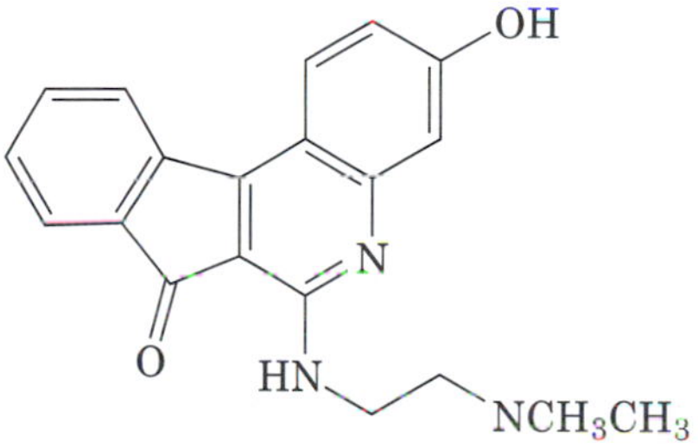

(48)

(49)

(53)

(50)

The 9-aminoacridine-4-carboxamides were first reported in 1984 as a new class of DNA-intercalating agents (243) with well-defined structure-activity relationships for both chromophore and side-chain (244). The derived acridine-4-carboxamide analog (51) (DACA) also binds to DNA by intercalation and induces DNA cleavage in the presence of either topo I or topo II enzymes, being unaffected by either P-glycoprotein–mediated multidrug resistance or "atypical" multidrug resistant caused by low topo II activity (245). DACA showed remarkable activity against multidrug resistant cells (246) and *in vivo* activity against the Lewis lung carcinoma (247), leading to clinical evaluation.

The DNA-intercalating (248) pyridoindole

intoplicine (**52**) is reported to also be a dual topo I/II poison (248, 249). Analogs of intoplicine that were only topo I or topo II poisons were less cytotoxic (248), suggesting the possible use of a dual poisoning ability. Intoplicine showed activity in a variety of human tumor explants in a soft agar cloning assay (250) and in transplantable mouse tumors *in vivo* (251). Phase I trials of intoplicine have been conducted (252, 253), but phase II trials have not been reported.

The indenoquinolone TAS-103 (**53**) is also reported to be a DNA-intercalating agent (254) and to enhance both topo I– and topo II–mediated DNA cleavage in treated cells (255), but it is now considered that topo II is the primary cellular target (256). TAS-103 showed broad-spectrum activity against a number of cell lines, with no cross-resistance in cells with lower topo I expression and only slight cross-resistance in those where topo II was down-regulated (257). A phase I clinical trial of TAS 103 recommended a dose of 130–160 mg/m$^2$ for phase II trials (258), but these have not yet been reported.

**3.4.2 Mechanism and SAR.** There seems to be no clear structural features predisposing to dual topo I/II activity. Raman and CD studies of intoplicine analogs suggest that the dual poisoning abilities of intoplicine are a result of its ability to simultaneously form two types of DNA complexes: a "deep intercalation mode" responsible for topo I–mediated cleavage and an "outside binding mode" responsible for topo II–mediated cleavage (259).

**3.4.3 Biological Activity and Side Effects.** The primary route of metabolism of DACA is oxidation at C-9 by aldehyde oxidase to give the acridone (**54**), although oxidative demethylation of the side-chain dimethylamino group has also been observed (260). Pharmacological studies showed high binding to human $\alpha$1-acid glycoprotein, followed by albumin (261). In phase I clinical trials, the major urinary metabolite was the *N*-oxide (**54**), whereas the major plasma metabolites (262) were (**54**) and (**55**). The maximum tolerated dose in initial phase I trials was 750 mg/m$^2$ using a 3-h infusion, with the dose-limiting toxicity being arm pain of unknown cause at the infusion site

(263) (avoidable using a 5-day infusion). Phase I trial reports for intoplicine noted hepatotoxicity rather than myelosuppression as the major dose-limiting toxicity (264).

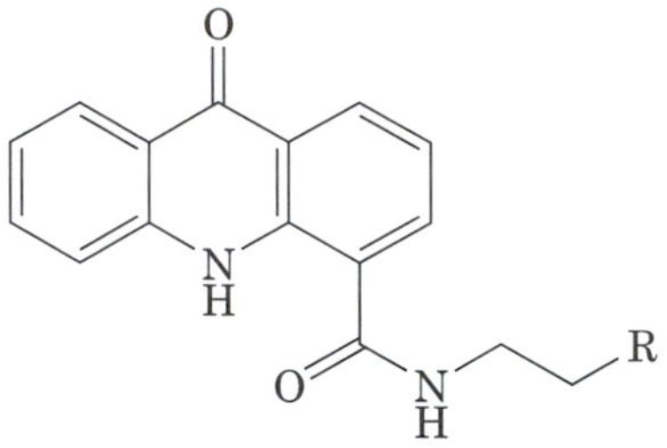

(**54**) R = N(O)CH$_3$CH$_3$
(**55**) R = NCH$_3$CH$_3$

**3.4.4 Recent Developments: bis Analogs as Dual Topo I/II Inhibitors.** Because of the early SAR suggesting a positive correlation between cytotoxic potency and the strength of DNA binding and because bis-intercalation would theoretically greatly increase DNA binding, many dimeric compounds designed as bis-intercalators were evaluated as anticancer drugs (134, 265). However, the biological activities of these compounds were generally disappointing. The bis(acridine) (**56**) was considered for clinical trial (266) but had significant

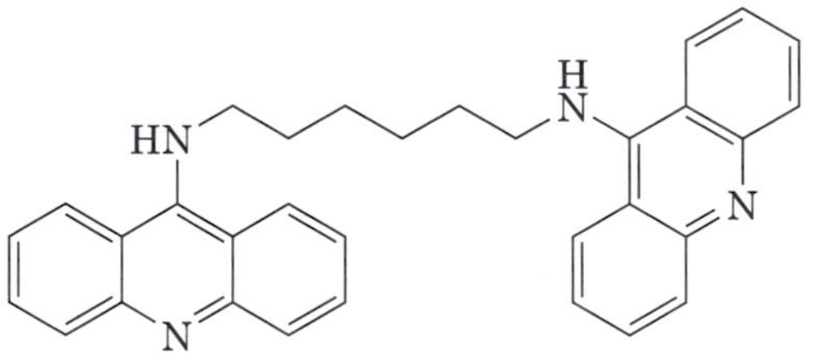

(**56**)

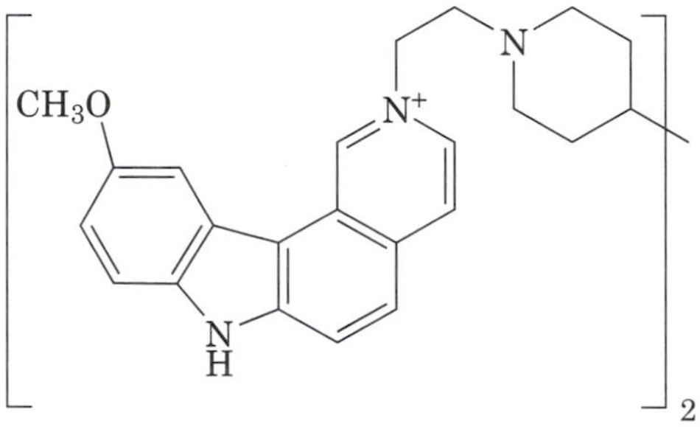

(**57**)

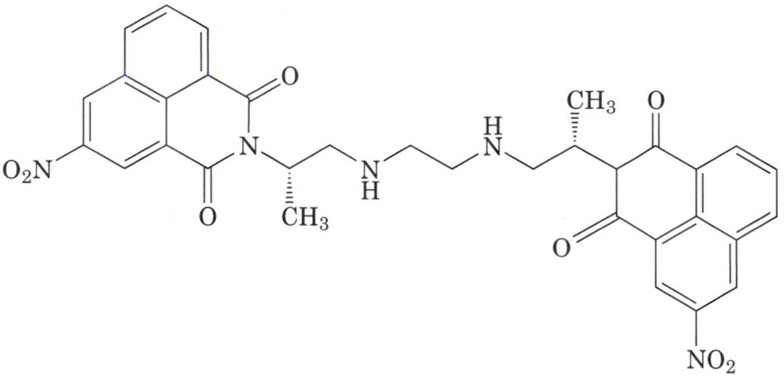

(58)

CNS toxicity, and the bis(ellipticine) analog ditercalinium (57) had unacceptable mitochondrial toxicity (267).

More recently, several series of dimers of more lipophilic chromophores have shown potent and broad-spectrum activity against a variety of human solid tumor cell lines, both in culture and as xenografts in nude mice. The bis(naphthalimide) analog DMP 840 (58) was curative in a variety of human solid tumor xenografts in nude mice (268). A series of bis(imidazoacridinones) (e.g., 59; WMC-26) showed highly selective cytotoxicity towards human colon carcinoma cells both in culture and in xenografts, although it seems that it is not a bis-intercalating agent (269). Several series of bis analogs of tri- and polycyclic carboxamides, including acridines (270) (e.g., 60), phenazines (271) (e.g., 61), and indenoquinolines (272) (e.g., 62), are also potent cytotoxic agents and dual topo I/II inhibitors. SAR studies of these compounds (270–273) show that both chromophore substitution and linker chain variations can significantly affect potency. The dicationic bis(phenazine) (63; XR5944) is of particular interest, with subnanomolar potency in a range of human cell lines (274) and active in multidrug-resistant cell lines *in vitro* and *in vivo* (275).

## 4   ANTIMETABOLITES

### 4.1   Introduction

The class of compounds known broadly as antimetabolites interfere in varying ways with

(59)

the synthesis of DNA. Along with the alkylating agents, antimetabolites such as methotrexate (65), 5-fluorouracil (73), cytosine arabinoside (74), and 6-mercaptopurine (76)

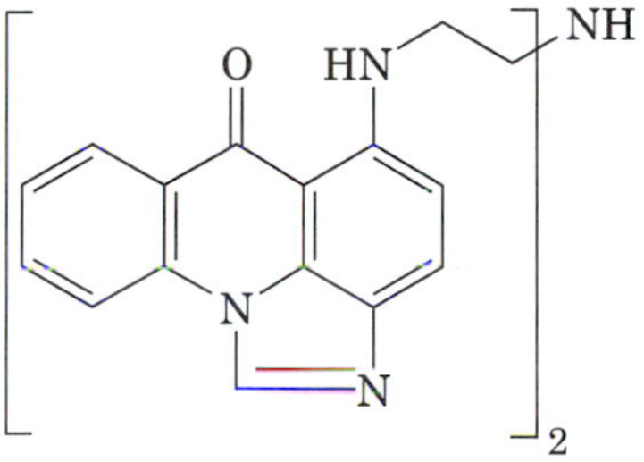

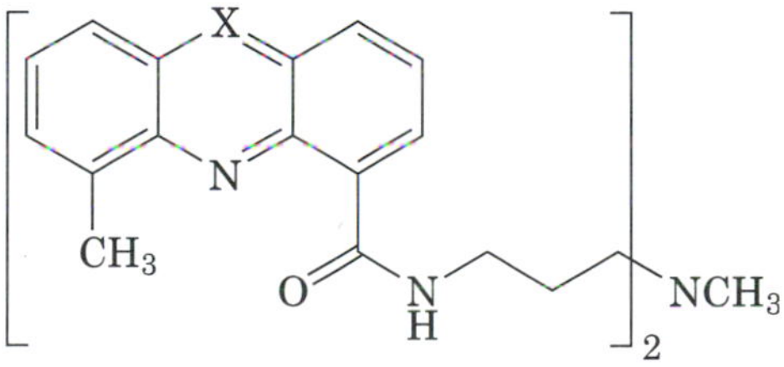

(60) X = CH
(61) X = N

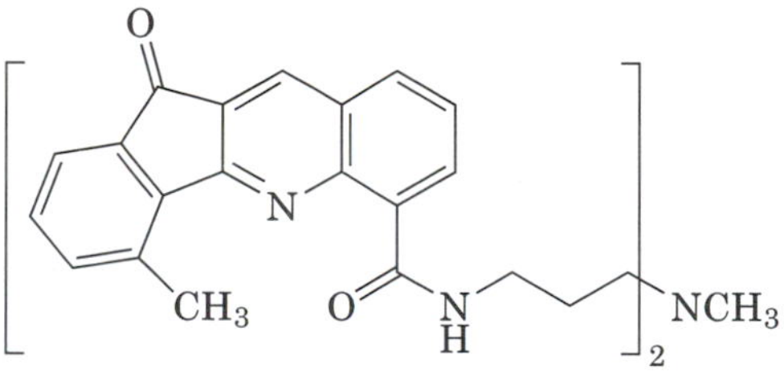

(62)

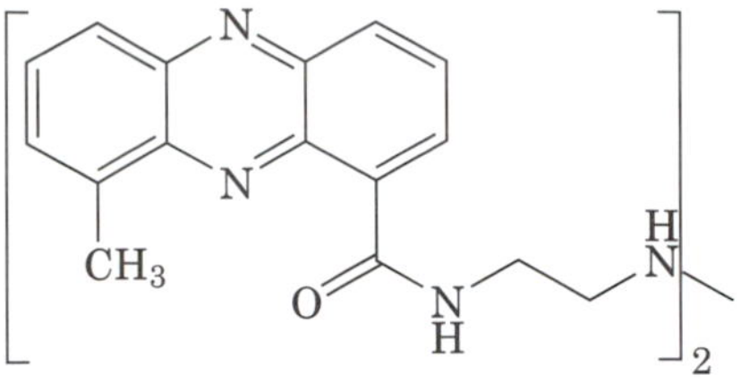

(63)

were some of the earliest drugs used in cancer chemotherapy.

## 4.2 Clinical Use of Agents

The most commonly used antimetabolites are listed in Table 2.3. These compounds are invariably used in combination with other agents in multidrug therapy regimens.

## 4.3 Antifolates

**4.3.1 History.** Antifolates interfere at various points in the process (folic acid metabolism) that provides the one-carbon unit required to convert deoxyuridine monophosphate to thymidylic acid for synthesis of the pyrimidines (Fig. 2.7). They are also key intermediates in the glycinanide ribonucleotide (GAR)-formyltransferase- and aminoimidazole carboxamide ribonucleotide (AICR)-

formyltransferase-mediated construction of the purines (276). The first antifolate used clinically was aminopterin (**64**) and was rapidly followed by methotrexate (**65**), which was registered for clinical use in 1953. These "classical" (glutamate-containing) antifolates bind tightly to the enzyme dihydrofolate reductase (DHFR; Fig. 2.7). Methotrexate has been very widely used and has been extensively reviewed (277, 278). A more recent classical antifolate is the 10-ethyl analog edatrexate (**66**). This was developed following observations that 1-alkyl analogs showed better relative uptake into tumor tissue, and edatrexate shows enhanced uptake, retention, and polyglutamate formation in tumor cells (279). Whereas edatrexate binds to DHFR similarly to methotrexate, it showed better activity in animal tumor models (280), including models resistant to methotrexate (281). Resistance to methotrexate arises in several ways, the most important of which are elevation of DHFR levels and lowering of both folate transport and polyglutamylation activities (282).

The enzyme thymidylate synthase (TS) is also intimately involved in folate metabolism, catalyzing the reductive methylation of deoxyuridine monophosphate (dUMP) to thymidylate (dTMP), a reaction in which $N^5,N^{10}$-methylenetetrahydrofolate is a cofactor (Fig. 2.7). Whereas the pyrimidine-binding site on

**Table 2.3 Antimetabolites Used in Cancer Chemotherapy**

| Generic Name (Structure) | Trade Name | Originator | Chemical Class |
|---|---|---|---|
| Folic acid analogs | | | |
|   methotrexate (**65**) | | | folate analog |
|   edatrexate (**66**) | | | folate analog |
|   raltitrexed (**68**) | Tomudex | Lilly | |
|   permetrexed (**69**) | | | |
|   trimetrexate (**70**) | NeuTrexin | US BioScience | |
|   piritrexim (**71**) | | Burroughs-Wellcome | |
|   nolatrexted (**72**) | Thymitaq | Zarix | |
| Pyrimidine analogs | | | |
|   5-fluorouracil (**73**) | Adrucil | Roche | pyrimidine |
|   cytosine arabinoside (**74**) | Cytosar | Pharmacia & Upjohn | pyrimidine |
|   gemcitabine (**75**) | Gemzar | Lilly | pyrimidine |
| Purine analogs | | | |
|   6-mercaptopurine (**76**) | Purinethol | Burroughs-Wellcome | |
|   6-thioguanine (**77**) | Lanvis | Glaxo-Wellcome | purine |
|   fludarabine (**78**) | Fludara | Berlex Laboratories | purine |
|   2′-deoxycoformycin (**79**) | Pentostatin | Supergen Inc | purine analog |
|   2-chloro-2′deoxyadenosine (**80**) | Cladribine | Bedford Laboratories | purine |

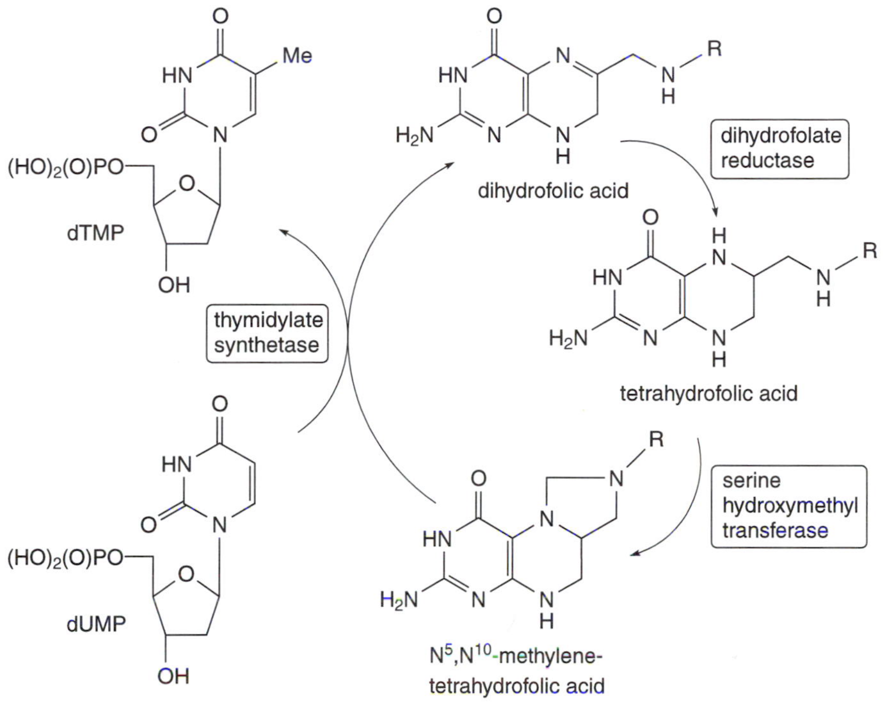

**Figure 2.7.** Folate biosynthesis.

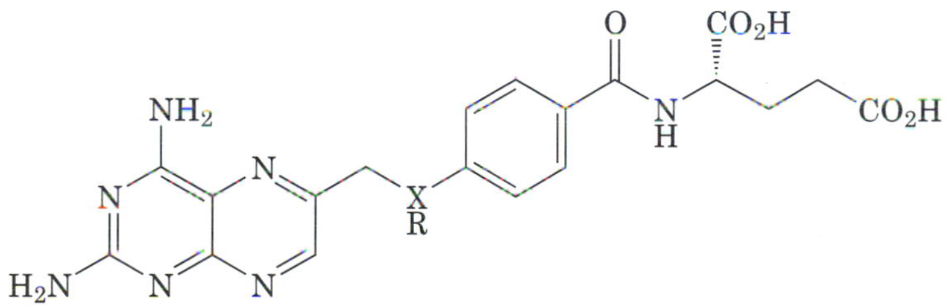

(**64**) R = H, X = N
(**65**) R = CH$_3$, X = N
(**66**) R = Et, X = CH

the Ts enzyme has been a major target for anticancer drugs such as 5-fluorouracil (see Section 4.4), it also has a folate-binding site that has been a target for drug development. Methotrexate itself binds weakly to this site, and can exercise cytotoxicity through TS inhibition in cells that highly overexpress DHFR (283). The design of highly specific inhibitors of the folate binding site of TS led initially to the quinazoline derivative CB 3717 (**67**) (284, 285). This proved to be a tight-binding inhibi-

tor of TS ($K_i$ 4.5 n$M$), with 10-fold selectivity over DHFR, with the ability to undergo polyglutamylation in cells to metabolites that are more potent and more selective for TS over DHFR (286). CB 3717 showed some activity in a number of phase I/II clinical trials, but severe nephrotoxicity, caused probably by precipitation of drug in the kidneys (287), led to its withdrawal (288).

Raltitrexed (**68**; tomudex) is another "classical" folic acid derivative that exerts its ther-

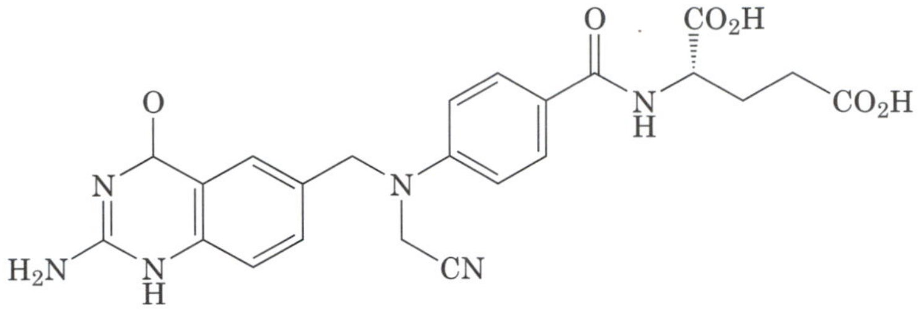

(67)

apeutic effect through by inhibition of the folate site of TS (289). It is polyglutamylated in cells into metabolites that are more potent inhibitors of TS than the parent drug and are retained in cells. Raltitrexed showed activity in a number of tumor types in phase I/II trials, but a major use may be in colon cancer. Here it shows activity similar to 5-fluorouracil (response rates of 14–19%) but with lesser toxicity (290), although the results of a recent phase II/III trial question this (291).

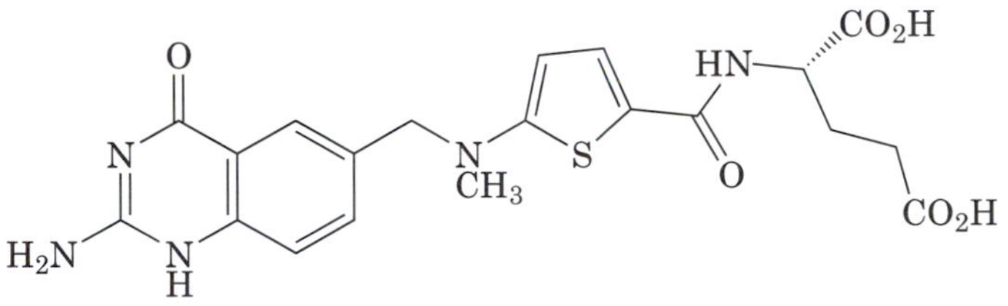

**4.3.2 Mechanism and SAR.** Methotrexate (**65**), introduced in 1953, and the related aminopterin (**64**) bind to DHFR, preventing transfer of the one-carbon unit from dihydrofolic acid to methylenetetrahydrofolic acid and ultimately to thymidine (Fig. 2.7). Methotrexate is taken into cells by the folate transporter and converted in cells to active polyglutamate metabolites by folylpolyglutamate synthase (297); this also has the effect of trapping the drug in cells (279). A large amount of work has

(68)

Permetrexed (**69**; MTA) also has TS as a major target, with DHFR and glycinamide ribonucleotide formyltransferase (GARFT) being important secondary sites of action (292). Permetrexed is an excellent substrate for FPGS, and it and its polyglutamylated metabolites are potent inhibitors for all of the above enzymes (293). Permetrexed performed well in the human tumor-cloning assay against colorectal (32% of cell lines inhibited) and non-small-cell lung cancer (25% of cell lines inhibited) (294). It showed broad antitumor activity in phase II trials with breast, colon, pancreatic, bladder, head-and-neck, and cervical carcinomas, and non-small-cell lung cancer, both as a single agent and in combination with agents such as gemcitabine and cisplatin, and it is in phase III evaluation (295, 296).

been done to delineate the SAR for 2,4-diaminopteridines binding to DHFR, but no clinical successor to methotrexate as a DHFR inhibitor has yet been found among the "classical" antifolates (287), although edatrexate (**66**) is still in development. Because there is also a folate site on TS, these compounds have some level of binding to this as well. CB 3717 and raltitrexed were designed specifically to target TS rather than DHFR, whereas permetrexed is closer to a general folate pathway inhibitor.

**4.3.3 Biological Activity and Side Effects.** Methotrexate has broad-spectrum clinical activity and is still the most widely used antifolate, despite high myelosupressive activity and frequent development of resistance by various mechanisms. The newer antifolates have broadly similar toxicity profiles.

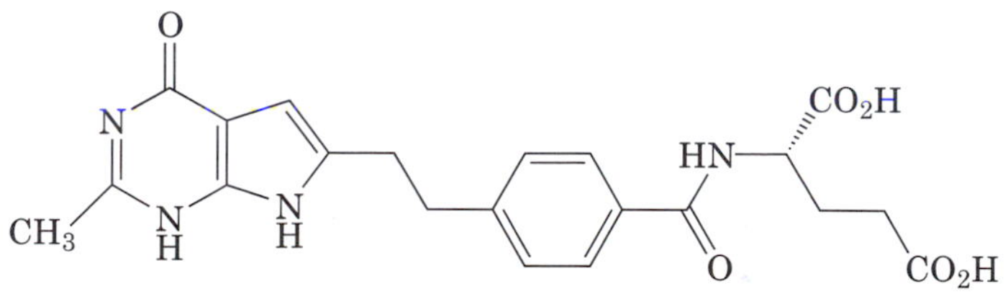

(69)

**4.3.4 Recent Developments: Lipophilic Antifolates.** These compounds were designed to circumvent resistance to methotrexate that arises by reduced folate uptake or reduced polyglutamylation. They are relatively lipophilic compounds, lacking a glutamate residue, that get into cells by passive diffusion. The first examples to receive clinical evaluation were trimetrexate (**70**) and piritrexim (**71**). Trimetrexate was superior to methotrexate in animal models, with activity in methotrexate-resistant lines (298) but (unlike methotrexate) is susceptible to P-glycoprotein–mediated multidrug resistance (299). Trimetrexate (**70**) has had extensive clinical trials and has shown activity in a number of tumors, including breast, non-small-cell lung, head-and-neck, and prostate (300), and particularly in colon cancer in conjunction with 5-fluorouracil/leucovorin (301). Piritrexim (**71**) was chosen for development from a range of lipid-soluble diaminoheterocyclic compounds on the basis of potent DHFR inhibition and minimal effects on histamine metabolism (302). Piritrexim is about 75% bioavailable when given orally (303), and in phase II trials showed some activity using oral dosing in bladder cancer (304). It is also more effective than methotrexate in severe psoriasis, because its lack of polyglutamylated metabolites makes it less hepatotoxic in long-term dosing (305).

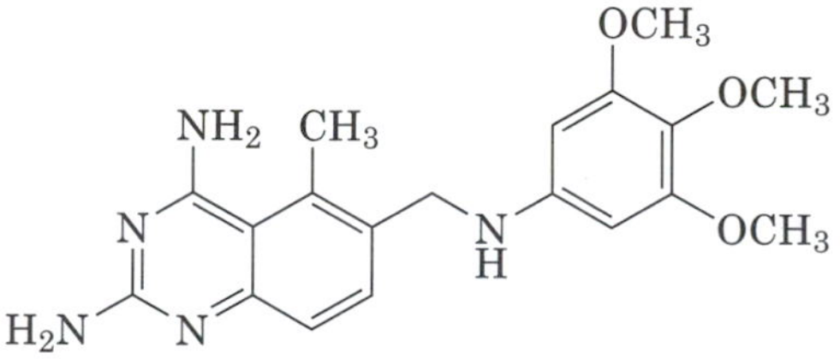

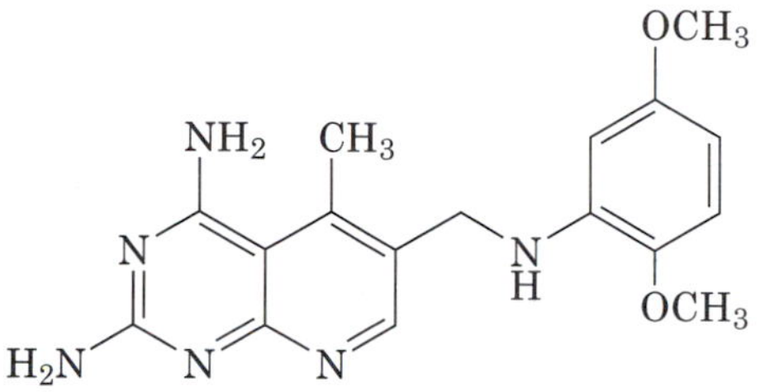

(71)

Nolatrexed (**72**; thymitaq) is a lipophilic folate analog designed as a TS inhibitor, using structure-based methods to maximize binding at the folate site (306). It is a potent ($K_i$ 11 n$M$), non-competitive inhibitor of human TS, with modest growth-inhibitory effects (IC$_{50}$s 0.4–7 $\mu M$) against a wide variety of murine and human cell lines. Nolatrexed does not enter cells by the reduced folate carrier, is not polyglutamylated, and does not inhibit DHFR. The activity of the drug is abrogated by thymidine (but not hypoxanthine), and TS overexpressing cells are strongly resistant, demonstrating that the primary target is TS (307). Oral bioavailability in rats was 30–50%, and oral nolatrexed showed curative activity against both IP- and IM-implanted thymidine kinase-deficient murine L5178Y/TK-lymphomas (306). Combinations of nolatrexed and cisplatin showed synergistic activity in both 5-FU- and cisplatin-resistant ovariant and colon cancer cells (306). Modest effects were seen in phase II trials of nolatrexed in advanced hepatocellular carcinoma (307), and phase I combination studies with paclitaxel are ongoing (310).

## 4.4 Pyrimidine Analogs

**4.4.1 History.** The pyrimidine analogs 5-fluorouracil (**73**) and cytosine arabinoside

(70)

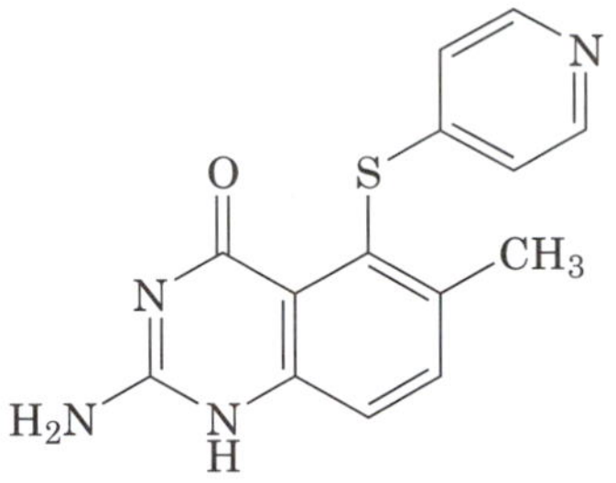

(72)

(75)

**74**; ara-C, cytosar) were developed from a knowledge of DNA metabolism (2) and were registered for clinical use in 1962 and 1969, respectively. A huge amount of work has gone into developing further analogs, and this has recently begun to pay off with the more recent introduction of gemcitabine (**75**; gemzar).

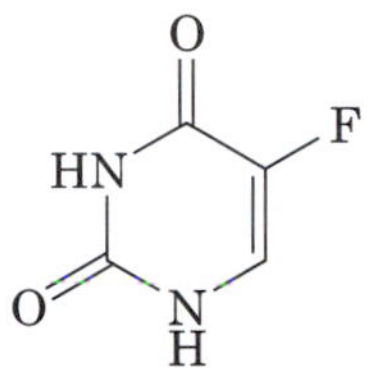

(73)

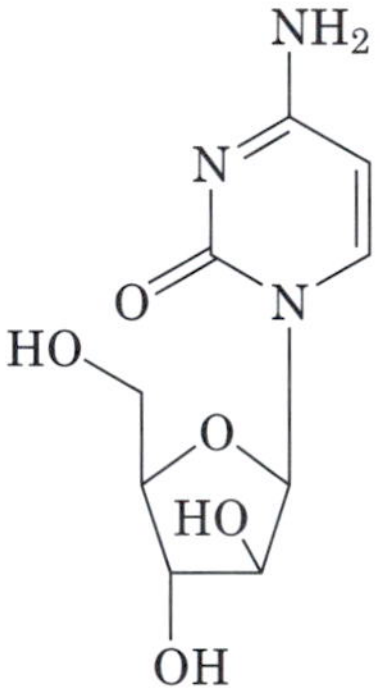

(74)

**4.4.2 Mechanism and SAR.** The mechanisms by which 5-fluorouracil (**73**) exerts its cytotoxicity have been extensively reported (311, 312). It is converted in cells to the monophosphate 5-FdUMP, which binds initially reversibly at the dUMP site of the enzyme thymidylate synthetase (Fig. 2.7). This is followed by Michael-type attack of an SH group on the

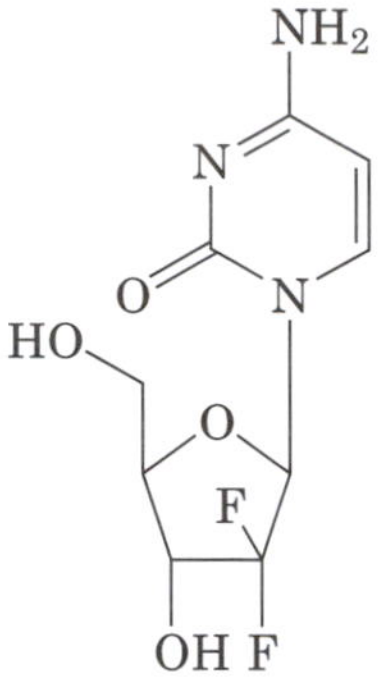

enzyme to given an enolate-type intermediate, which reacts at the methylene moiety of $N^5,N^{10}$-methylenetetrahydrofolate to form a covalent drug-enzyme-cofactor ternary complex (Fig. 2.8). Because the fluorine cannot be displaced, as with the natural (non-fluorinated) substrates, this results in permanent poisoning of the enzyme with 1:1 stoichiometry. 5-Fluorouracil is also converted into the triphosphate 5-FdUTP, which is incorporated into both RNA and DNA.

The mechanism of action of cytosine arabinoside (**74**) has been well reviewed (313). It acts primarily as a chain terminator during the elongation phase of DNA synthesis, incorporating into the growing chain and preventing the action of DNA polymerases (314). Gemcitabine also acts primarily as a chain terminator, but has additional effects, through rapid phosphorylation by deoxycytidine kinase to di- and tri-phosphate metabolites. The diphosphate inhibits ribonucleotide reductase (RR), the enzyme responsible for producing the deoxynucleotides required for DNA synthesis and repair, and the subsequent depletion of cellular deoxynucleotides favors gemcitabine triphosphate incorporation into DNA over the normal dCTP, in a "self-potentiating" mechanism (315). Incorporation of gemcitabine into the elongating DNA strand results in the halting of DNA polymerases after the addition of one more additional deoxynucleotide, in a "masked chain termination" event that seems to lock the drug into DNA, preventing proof-reading exonucleases from removing it. Gemcitabine is synergistic with cisplatin because of the triphosphate preventing chain

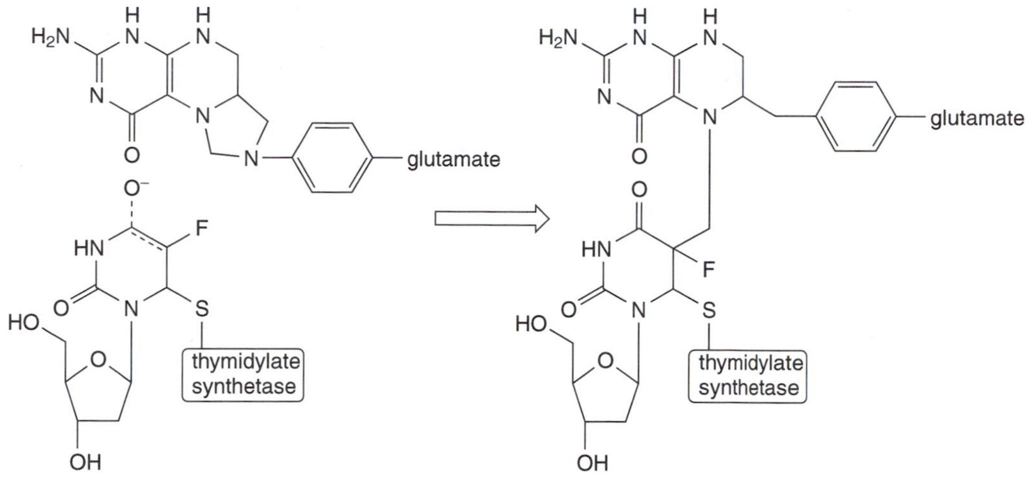

**Figure 2.8.** Mechanism of 5-fluorouracil inhibition of thymidylate synthetase.

elongation during the DNA resynthesis process after nucleotide excision repair of the lesions (316). The mechanism of RR inhibition by gemcitabine has been studied in *E. coli* and seems to be different to that of other 2′-substituted nucleotide inhibitors, involving inactivation of the R1 subunit (317), and overexpression of RR is a resistance mechanism for gemcitabine (318). It is also an effective radiation sensitizer, probably through depletion of dATP pools in cells (319) and can increase cellular apoptosis in irradiated cells (320).

**4.4.3 Biological Activity and Side Effects.** Both 5-fluorouracil and cytosine arabinoside remain widely used in combination cancer chemotherapy. 5-Fluorouracil is one of the most effective drugs against colon cancer (311). Cytosine arabinoside is effective in leukemias and lymphomas but has a very short half-life (ca. 12 min in man), because of catabolism by cytidine deaminase (321), and various non-specific prodrug forms are used (322). Gemcitabine was shown in phase I trials to be active in a number of cancers, especially in non-small-cell lung cancer, where it showed >20% responses as a single agent and up to 54% in combination with cisplatin (323). In phase II trials, it has proved active in a wide range of tumors, including non-small-cell lung cancer (>60% responses in combination with cisplatin) (324), urothelial (22–28% responses as monotherapy, 42–66% in combination with

cisplatin) (325), advanced breast cancer (25.0% responses as monotherapy) (326), and metastatic bladder cancer 42–66% responses in combination with cisplatin) (327). The main adverse effects were hematological but were generally mild. A number of large phase III trials are in progress.

## 4.5 Purine Analogs

**4.5.1 History.** The purine analogs 6-mercaptopurine (**76**) and 6-thioguanine (**77**) were among the first anticancer drugs to be used, registered in 1953 and 1966, respectively. Later, the purine nucleoside analogs fludarabine (**78**) and pentostatin (**79**; 2′-deoxycoformycin) were registered in 1991, and cladribine (**80**; 2-chloro-2′-deoxyadenosine) was registered in 1992.

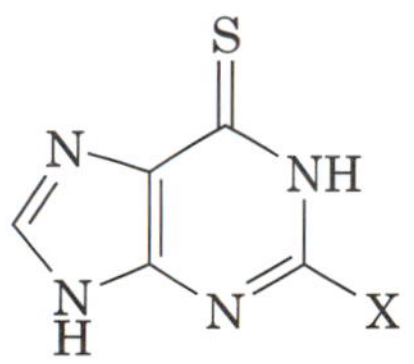

(**76**) X = NH$_2$
(**77**) X = N

**4.5.2 Mechanism and SAR.** Cytosine arabinoside, fludarabine, and cladribine are taken into cells through a specific nucleoside trans-

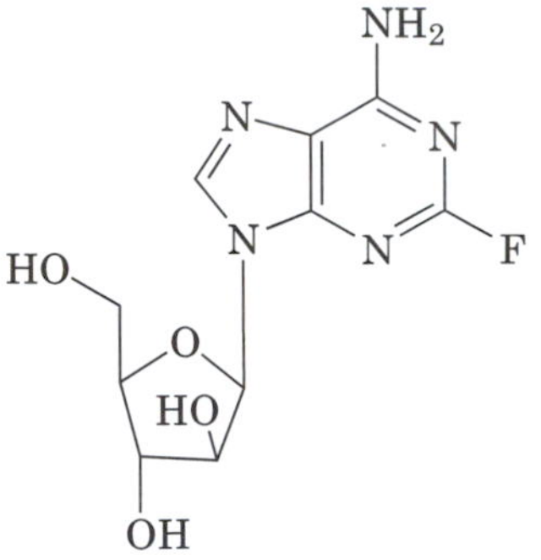

(78)

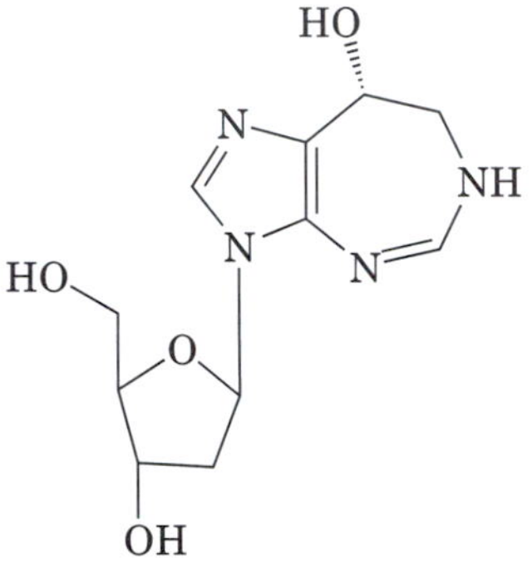

(79)

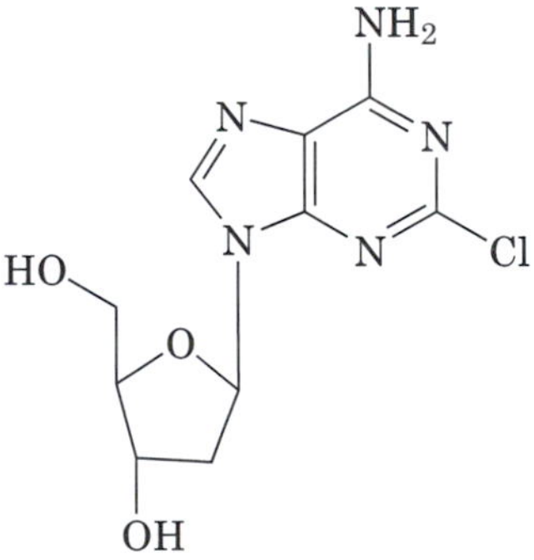

(80)

porter protein and are phosphorylated to the mono-, di-, and triphosphates, with the first phosphorylation mainly by deoxycytidine kinase (328). The active triphosphate derivatives are incorporated into DNA, blocking polymerase function and thus DNA synthesis. Cladribine is resistant to degradation by adenosine deaminase (329) and induces apoptosis

in leukemia cell lines through the Fas/Fas ligand pathway (330). It also interrupts deoxyadenosine metabolism, blocking both phosphorylation and deamination (329). Pentostatin is also converted to the triphosphate and incorporated into DNA, where it blocks polymerase function (331) but is also an extremely potent inhibitor of adenosine deaminase ($K_i$ $2.5 \times 10^{12}$ $M^{-1}$) (332).

**4.5.3 Biological Activity and Side Effects.** These three adenosine analogs, which are cytotoxic to both dividing and resting lymphocytes, have revolutionized the treatment of indolent lymphoid malignancies such as chronic lymphocytic leukemia, non-Hodgkin's lymphoma, cutaneous T cell lymphoma, and hairy cell leukemia. Both fludarabine and cladribine showed similar good response rates, but were cross-resistant, in refractory non-Hodgkin's lymphoma (333). Cladribine is active in hairly cell leukemia (>80% complete responses) (334), non-Hodgkin's lymphoma (89% responses) (335), refractory chronic lymphocytic leukemia (44% responses) (336), untreated chronic lymphocytic leukemia (85% response rate) (337), and cutaneous T-cell lymphomas (28% responses) (338), but it showed little activity in solid tumors.

## 5 TUMOR-ACTIVATED PRODRUGS

### 5.1 Introduction

As noted above, the majority of clinically used anticancer drugs are systemic anti-proliferative agents (cytotoxins). These kill cells by a variety of mechanisms primarily by attacking their DNA at some level (synthesis, replication, or processing). However, a large part of their selectivity for cancer cells is based on cytokinetics, in that they (to varying extents) are preferentially toxic to cycling cells. For this reason, their therapeutic efficacy is limited by the damage they also cause to proliferating normal cells such as those in the bone marrow and gut epithelia. This is especially true in the treatment of solid tumors, where cell doubling times may be very long. Whereas efforts to physically target cytotoxins to tumor tissue has not been very successful, the development of relatively nontoxic prodrug forms

**Table 2.4    Tumor-Activated Prodrugs in Clinical Trial for Cancer Chemotherapy**

| Generic Name (Structure) | Trade Name | Originator | Chemical Class |
|---|---|---|---|
| Hypoxia-activated prodrugs | | | |
|   tirapazamine (**82**) | Tirazone | Sanofi | benzotriazine-di-*N*-oxide |
|   AQ4N (**84**) | | British Technology Group | aliphatic *N*-oxide |
|   porfiromycin (**86**) | | Vion | aziridinylquinone |
| ADEPT prodrugs | | | |
|   ZD 2767P (**93**) | | AstraZeneca | aromatic mustard |
| GDEPT prodrugs | | | |
|   ganciclovir (**99**) | Cytovene | Hoffmann LaRoche | |
|   CB 1954 (**102**) | | Cobra Therapeutics | dinitrophenylaziridine |
| Antibody-toxin conjugates | | | |
|   SGN-15 (**104**) | | Seattle Genetics | antibody/doxorubicin |
|   Gemtuzumab ozogamycin | | | |
|     (**105**) | Mylotarg | Wyeth-Ayerst | antibody-enediyne |
|   SB 408075 (**106**) | | Immunogen | antibody/maytansinoid |
|   KM231-DU257 (**107**) | | Kyowa Hakko | antibody/cyclopropylindole |

of cytotoxins, which can be selectively activated in tumor tissue, is beginning to achieve some success, and in the future, may become a major strategy.

Prodrugs can be defined broadly as agents that are transformed after administration, either by metabolism or by spontaneous chemical breakdown, to form a pharmacologically active species. Strictly speaking, agents such as cyclophosphamide (**4**) are prodrugs, but these undergo non-specific activation in all tissues. Of more interest are tumor-activated prodrugs that exploit various aspects of tumor physiology and other techniques to become selectively activated in tumor tissue to toxic species. The multiple criteria required of a tumor-activated prodrug has meant that these compounds, whereas sometimes using natural products such as doxorubicin as the toxins, are primarily synthetic agents. Most tumor-activated prodrugs fall under one of four categories: hypoxia-selective prodrugs (bioreductives), prodrugs for antibody-directed enzyme-prodrug therapy (ADEPT prodrugs), prodrugs for gene-directed enzyme-prodrug therapy (GDEPT prodrugs), and antibody-toxin conjugates (armed antibodies).

## 5.2    Clinical Use of Tumor-Activated Prodrugs

Because interest in tumor-activated prodrugs is relatively recent, only the hypoxia-selective agent tirapazamine has had extensive clinical use, and even this is still in development, although it looks likely to become the first clinically useful hypoxia-selective drug (339). The limited clinical experience with these various drugs is discussed below, in each subclass (Table 2.4).

## 5.3    Hypoxia-Activated Prodrugs (Bioreductives)

**5.3.1  History.**  The imperfect neovascularization that develops in growing solid tumors results in limited and inefficient blood vessel networks and restricted and often chaotic blood flow (340). This generates chronic or diffusion hypoxia, where cells sufficiently distant from the nearest blood capillary are hypoxic for long periods, caused by the steep diffusion gradient of oxygen in tissue. The high and variable interstitial pressures caused by the growing tumor (341) can also result in transient or perfusion hypoxia, resulting from the temporary shut down of blood vessels placing sections of tissue under hypoxia for shorter periods (342). Because severe hypoxia is a common and unique property of cells in solid tumors, it is thus an important potential mechanism for the tumor-specific activation of prodrugs. This concept grew initially out of the development of radiosensitizers, drugs designed to take the place of oxygen in hypoxic tissue by oxidatively "fixing" the initial DNA radicals formed by ionizing radiation to gener-

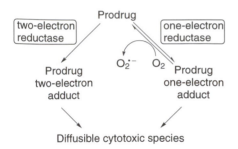

**Figure 2.9.** Hypoxia-activated prodrugs.

(82)

ate cytotoxic strand breaks (343). Such compounds tended to be easily reduced electron-deficient species such as misonidazole (**81**). In addition to their radiosensitizing properties as "oxygen-mimetics," many of these compounds were also found to have modestly higher levels (ca. 10-fold) of cytotoxicity in hypoxic compare with oxygenated cells in culture (344). THe mechanism of such hypoxia-selective cytotoxicity is the ability of the prodrug to the metabolized by reductive enzymes such as cytochrome P450 reductase and xanthine oxidase (345) to a transient one-electron intermediate. In normal oxygenated tissue, this is efficiently back-oxidized by molecular oxygen to the parent compound, but in hypoxic cells, it is further metabolized or spontaneously breaks down to more cytotoxic species (346) (Fig. 2.9).

(81)

The most well-studied hypoxia-activated prodrug is the synthetic agent tirapazamine (**82**; 3-aminobenzotriazone-1,4-di-N-oxide). This drug was originally evaluated as an antimicrobial agent (347) but was discovered to have hypoxia-selective cytotoxicity in a screening program. It is now in widespread phase III studies (348) and may become the first clinically useful hypoxia-selective drug (339).

Aliphatic N-oxides of DNA-binding agents have also been explored as synthetic hypoxia-selective prodrugs. This was first demonstrated by the drug nitracrine-N-oxide (**83**). This is much more hypoxia-selective (>1000-fold in cell culture) (349) than the free amine itself, which also shows moderate hypoxic selectivity through reductive activation of the nitro group (350, 351). The most advanced drug of this type in development is the bis-N-oxide AQ4N (**84**) (352), which is due to begin phase I clinical trials shortly.

(83)

(84)

The natural product and widely used clinical agent mitomycin C (**85**) shows modest hypoxic selectivity (353), but this is not the main basis of its usefulness. However, its analog porfiromycin (**86**) does show greater selectivity (354), and it has been developed primarily as a hypoxia-activated prodrug, to the extent of phase I clinical trials for head- and neck-cancer in combination with radiotherapy (355).

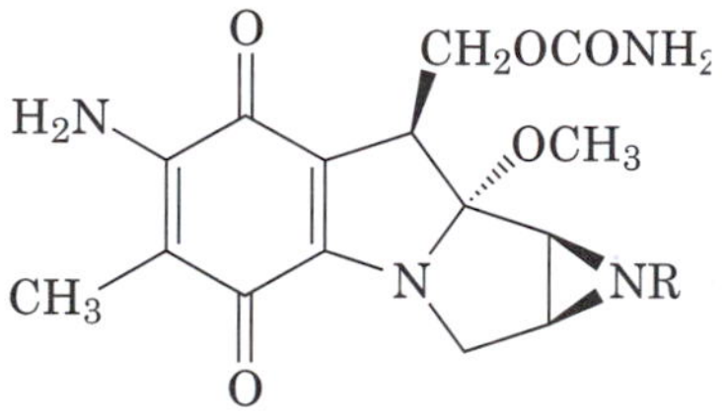

(**85**) R = H
(**86**) R = CH$_3$

**5.3.2 Mechanism and SAR.** The classes of hypoxia-activated prodrugs discussed above work by a variety of different mechanisms. Tirapazamine (**82**) was found to undergo enzymic one-electron reduction to a transient oxidizing nitroxyl (356) or carbon-centered radical and ultimately to the two-electron mono-*N*-oxide reduction product (Fig. 2.10). The transient radicals were shown to cause breaks at the C-4′ ribose site of DNA, followed by the oxidation of these by oxygen or other oxidants (including tirapazamine itself) (357), through formation of a covalent adduct at the *N*-oxide oxygen. The main reducing enzymes responsible for the hypoxia-selective cytotoxic metabolism of tirapazamine are cytochrome P450 and cytochrome P450 reductase (345),

although it is also reduced under hypoxia by aldehyde oxidase, xanthine oxidase (358), and nitric oxide synthase (359). A critical feature is that tirapazamine, although only forming a monofunctional radical, generates a high proportion of double-strand DNA breaks. This is suggested to be caused by high local radical concentrations generated by an undefined intranuclear reductase associated with DNA (339).

The aliphatic tertiary amine *N*-oxides of the bis-bioreductive prodrug AQ4N (**84**) are also reduced (to the free amines) largely by the CYP3A isozyme of nicotinamide adenine dinucleotide phosphate (NADPH):cytochrome C (P-450) reductase (360). Although this in not a one-electron process, it is still oxygen-inhibited, with a direct competition between oxygen and the drug at the enzyme site. Regeneration of the cationic side-chains of (**84**) allows tight binding to DNA and an ability to function as a topo II poison, similarly to the closely related drug mitoxantrone (**41**) (361). AQ4N is not significantly active as a single agent in most murine solid tumors *in vivo*, but it potentiates the effects of radiation therapy (which kills the oxygenated tumor cells) in a dose-dependent manner (362). Increased efficacy was also seen with combinations of AQ4N and cyclophosphamide in murine tumor models (363). AQ4N is due to begin phase I clinical trials shortly. This approach seems quite general, with compounds like DACA *N*-oxide (**87**) also showing significant hypoxic selectivity in cell culture (364). Nitracrine *N*-oxide (**83**) is an

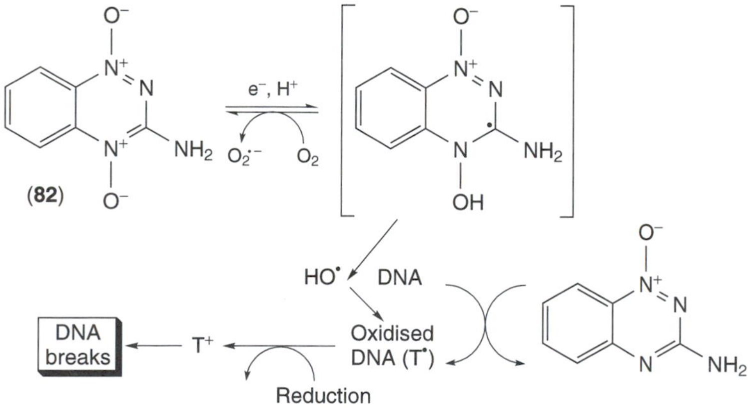

**Figure 2.10.** Metabolism of tirapazamine.

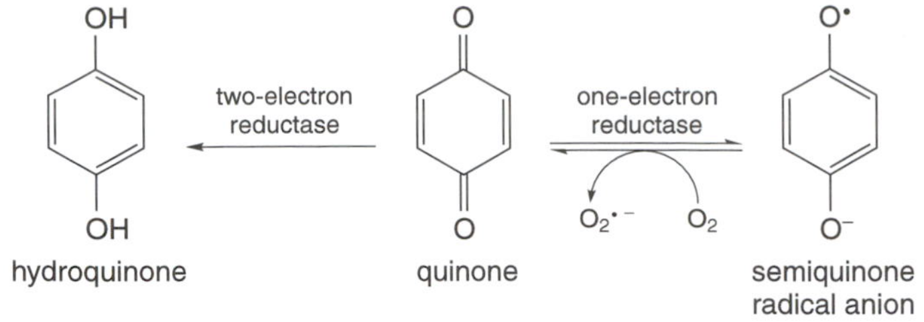

**Figure 2.11.** Bioreductive metabolism of quinones.

interesting example of another bis-bioreductive prodrug with two different reductive centers (nitro and N-oxide). Both centers need to be reduced for full activation, with the N-oxide demasking needing to occur before nitro reduction (365). Whereas (**83**) has exceptional hypoxic selectivity (>1000-fold) in cell culture (349), it shows little activity *in vivo* against the hypoxic subfraction of cells in KHT tumors (366) because of rapid metabolism. Attempted modulation of this by either lowering the reduction potential of the nitro group (366) or changing the steric environment of the N-oxide (367) was not useful.

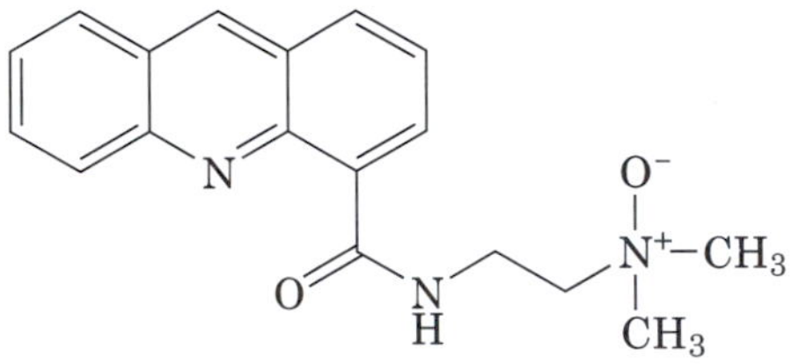

(**87**)

Mitomycin C (**85**) and porfiromycin (**86**) are quinones that readily undergo one-electron reduction, primarily by NADPH:cytochrome C (P-450) reductase (368), to the corresponding semiquinone radical anion that is capable of back-oxidation by molecular oxygen (Fig. 2.11). Following this, mitomycin C undergoes a well-documented fragmentation to DNA cross-linking agents that form guanine–guanine crosslinks in the major groove (369). One potential drawback to quinone-based compounds as hypoxia-activated prodrugs is that they are also often good substrates for two-electron reductases, particularly DT diaphorase (DTD; NQO1; NAD(P)H:quinone-acceptor oxidoreductase) (370).

**5.3.3 Biological Activity and Side Effects.** Tirapazamine shows high selective toxicity (100- to 200-fold) toward hypoxic cells in culture, but its diffusion through tissue is limited by its ready metabolism to the (non-diffusible) radical species (371). Tirapazamine has an ability to kill cells over a much wider range of oxygen concentrations (as high as 2% $O_2$) (372) than most other hypoxia-selective cytotoxins, so that its activation is not restricted to completely anoxic tissues (373). In animal studies, tirapazamine enhanced the effect of both single-dose (374) and fractionated (375) radiation. Combinations of tirapazamine with both cisplatin (376), cyclophosphamide (377), and other cytotoxic agents, including etoposide, bleomycin, and paclitaxel (378, 379) showed additive or greater than additive effects on both tumor cell killing and tumor growth delay. Combinations with the blood flow inhibitor 5,6-dimethyl-xanthenone-y-acetic acid (DMXAA) showed marked increases in activity in a variety of tumor models (380). Tirapazamine has had extensive clinical trials in head-and-neck cancer in conjunction with radiation (to kill oxygenated cells) (381) with encouraging results (382). Combinations with cisplatin (**7**) are also promising, the tirapazamine enhancing its effects, probably by delaying the repair of cisplatin-induced DNA cross-links in hypoxic cells (376). This has resulted in superior response rates compared with cisplatin alone in cervical cancer (383), mesothelioma (384), malignant melanoma (385), and particularly non-small-cell lung cancer (386). Clinical toxicities of tirapazamine include ototoxicity and muscle cramping (387). A laboratory study showed that tirapazamine caused time- and dose-dependent retinal damage in mice (388), but this does not seem to be a clinical issue.

Related quinoxalinecarbonitrile-1,4-di-*N*-oxides (e.g., **88**), where the 2-nitrogen in the benzotriazine unit of tirapazamine is replaced with a C—CN unit, are also potent and highly selective hypoxia-selective drugs (389). Structure-activity studies with these compounds show that hypoxic selectivity is retained when H or NHR replaces the 3-amino group.

(88)

## 5.4  Prodrugs for ADEPT

**5.4.1 History.** Antibody-directed enzyme-prodrug therapy (ADEPT) is an adaption of the earlier concept (390, 391) of immunotoxins. The difference is that instead of the toxin being attached to the antibody for localization on tumors, an enzyme (usually non-human) is attached and thus localized instead (392, 393) (Fig. 2.12). A prodrug that can be activated efficiently and selectively by the enzyme is then administered and is catalytically activated by the localized enzyme only in the vicinity of the tumor cells. The advantage of using non-human enzymes is the enhanced ability to find prodrugs that can be selectively activated. ADEPT shares with the original immunotoxin concept the problems of limited access of the (large) antibody-enzyme conjugate to tumors and the usually heterogeneous expression of the target antigen on tumor cells. However, provided the released cytotoxin has the appropriate properties (high potency and an efficient bystander effect) it can ameliorate these problems by diffusing from the cells where it is generated to enter and kill surrounding tumor cells that may not possess prodrug activating ability. A further increase in efficacy can be achieved if the prodrug is designed to be excluded from cells until it is activated (394).

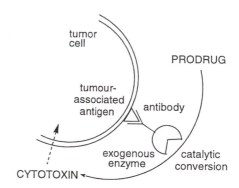

**Figure 2.12.** Antibody-directed enzyme-prodrug therapy (ADEPT).

**5.4.2 Mechanism and SAR.** The specific mechanism of action depends on the type of enzyme used to activate the prodrug. Particular requirements of the prodrug include being a selective and efficient substrate for the enzyme used. General requirements are an ability to be excluded from cells (usually achieved by high hydrophilicity and/or possession of a negative charge) until activation and the capability to then release a potent and diffusible toxin with a substantial bystander effect.

*5.4.2.1 Prodrugs for Phosphatase Enzymes.* Phosphates have been employed as ADEPT prodrugs because both aromatic (e.g., **89**; etoposide phosphate) (395) and aliphatic (e.g., **90**; mitomycin phosphate) (396) examples are efficiently cleaved by alkaline phosphatases and are substantially cell-excluded. However, it proved difficult to achieve selectivity because there is an abundance of such phosphatase enzymes in human serum and other tissues, and phosphates are primarily now used directly as non-specific prodrugs; the antivascular agent **(91)** (combretastatin phosphate) is an example (397).

*5.4.2.2 Prodrugs for Peptidase Enzymes.* Glutamate-type prodrugs of mustards (e.g., **92, 93**) are effectively excluded from cells by the diacid side-chain and can be cleaved by the *Pseudomonas*-derived enzyme carboxypeptidase G2 (398). Cleavage of the amide or carbamate releases more lipophilic agent that is also activated by electron release through the aromatic ring to the mustard. The amide **(92)** (CMDA) was the first ADEPT prodrug evaluated clinically (399), and the carbamate pro-

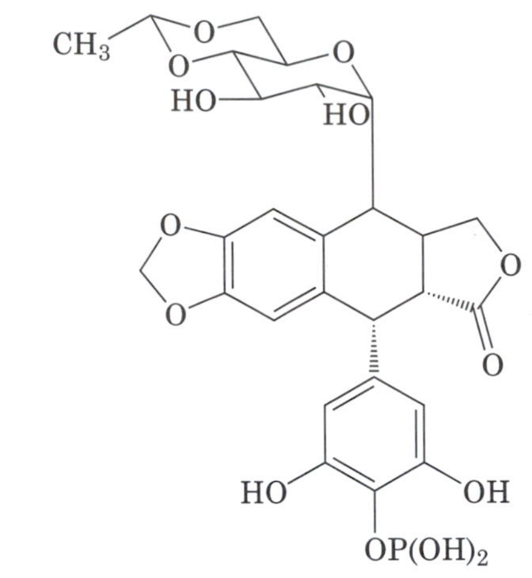

(89)

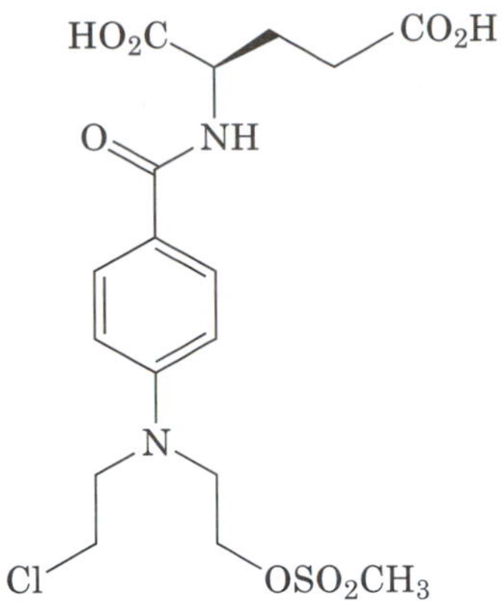

(92)

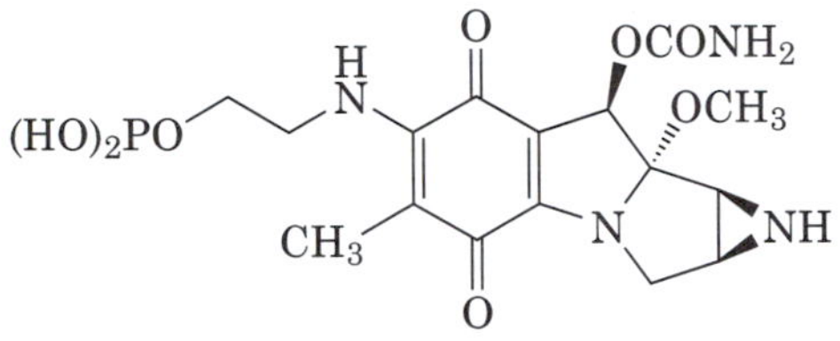

(90)

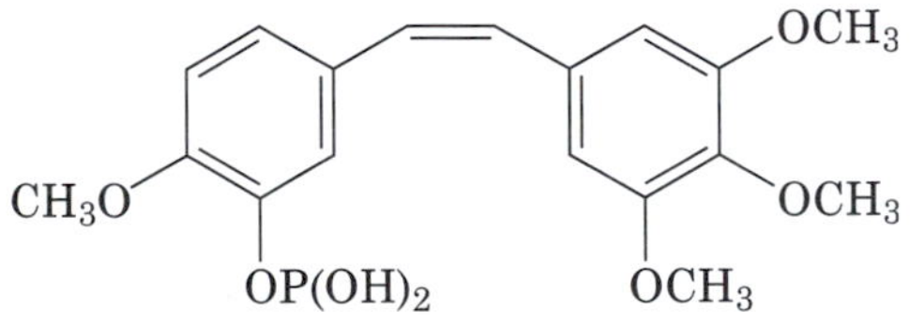

(91)

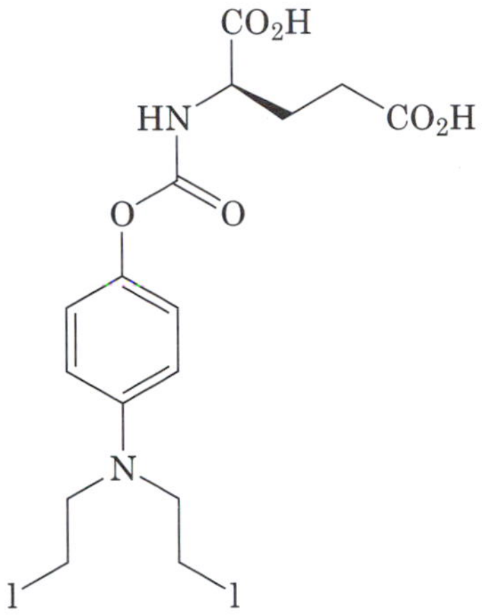

(93)

drug (**93**) (ZD 2767), releasing a more cyto-toxic phenol iodomustard, is currently in phase I clinical trial (400).

*5.4.2.3 Prodrugs for β-Lactamase Enzymes.* These enzymes from *Enterobacter* species can selectively hydrolyze the four-membered β-lactam ring of penicillins and cephalospo-rins and have been used in a variety of pro-drugs (401). Hydrolysis is followed by sponta-neous fragmentation of the carbamate side-chain and the release of a variety of toxic amines (Fig. 2.13). Carboxy and sulfoxide groups on the cephem nucleus assist with cell exclusion. Several nitrogen mustard prodrugs

for β-lactamase have been evaluated. The cephem analog (**94**) affected cures in mice bearing xenografts of human melanoma cells if given subsequent to treatment with 96.5/bL, a mAb/β-lactamase conjugate that binds to specific surface antigens on these cells (402). A cephem derivative of doxorubicin (**95**) showed higher intratumoral levels of doxorubicin af-ter treatment with the conjugate than with doxorubicin alone (403). However, the differ-ential cytotoxicities between drug and pro-drug in this approach are only moderate, and no such prodrugs have yet proceeded to clini-cal trial.

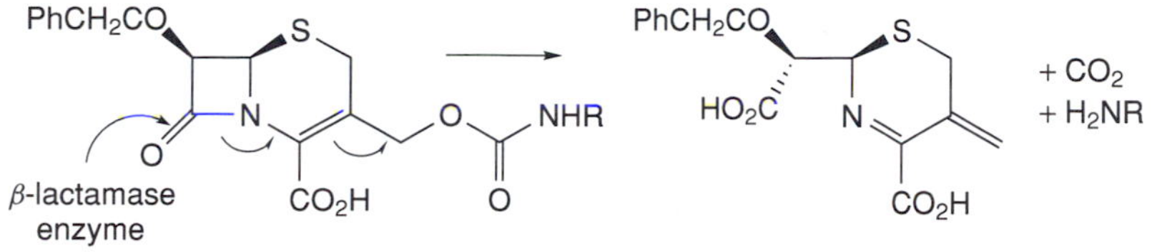

**Figure 2.13.** Fragmentation of $\beta$-lactamase prodrugs.

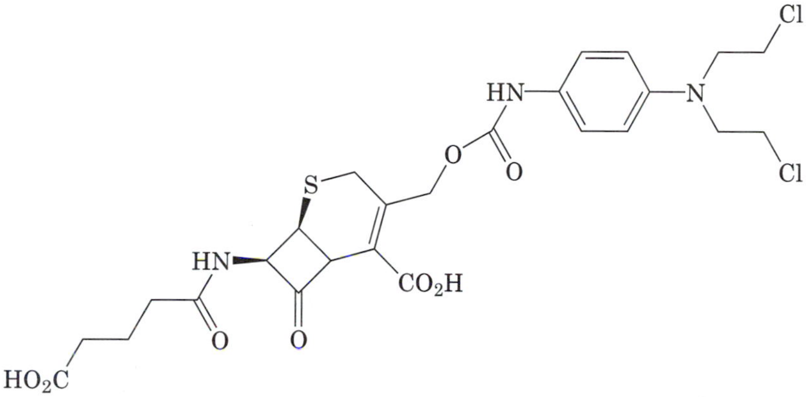

(94)

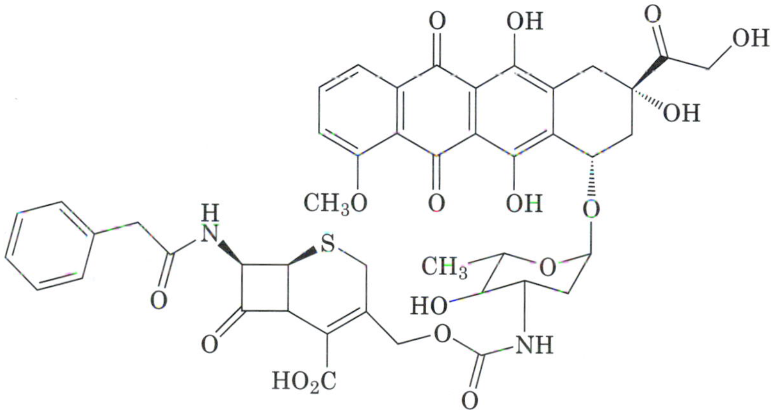

(95)

**5.4.2.4 Prodrugs for Glucuronidase Enzymes.** Because serum levels of $\beta$-glucuronidase (GUS) are very low (they are largely confined to lysozymes in cells), it is possible to use the human version as an activating enzyme in ADEPT, avoiding potential immunogenicity problems (404). Work with this enzyme has tended to focus on anthracycline effectors (405). The epirubicin $O$-glucuronide prodrug (96) was 100- to 1000-fold less cytotoxic than epirubicin itself *in vitro* (406), but pretreatment of antigen-positive cells with an 323/A3-GUS-*E. coli* immunoconjugate gave equivalent cytotoxicity to that of the free drug. The doxorubicin prodrug (97) used an immolative spacer unit (405), and although only 10-fold less cytotoxic than free doxorubicin, was a better substrate for the enzyme (407). The doxorubicin prodrug DOX-GA3 (**98**) was 12-fold less toxic than doxorubicin in cells of the human ovarian cancer cell line FMa and was somewhat superior to doxorubicin against FMa xenografts in mice in conjunction with 323-A3/human β-glucuronidase conjugate (408). However, the use of such tightly DNA binding, cell cycle–specific (topo II inhibitor) effectors is not yet known.

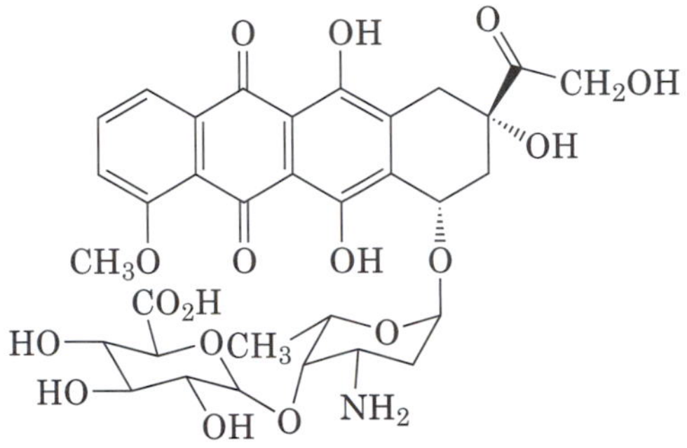

(96)

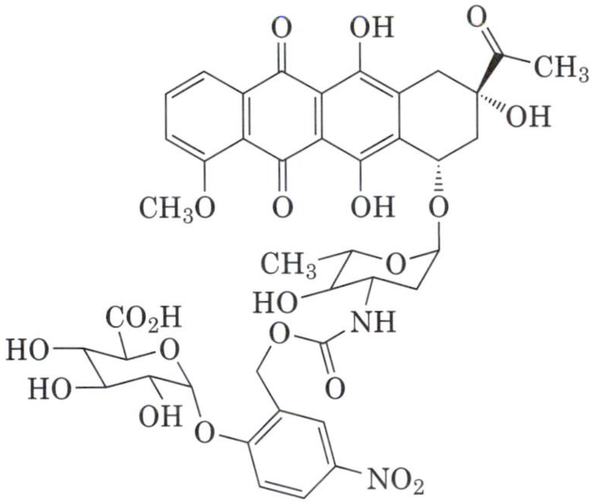

(97)

## 5.5 Prodrugs for GDEPT

**5.5.1 History.** The ADEPT approach is generally limited to the use of enzymes that do not require energy-producing cofactors, and it also has the likelihood of generating immune responses to the foreign proteins used. In gene-directed enzyme prodrug therapy (GDEPT), the enzyme is targeted to tumor cells by integrating the gene that produces it

into the genome of the tumor cells, followed by administration of the prodrug. A small but growing proportion of the large number of "gene therapy" trials now in progress is for gene-directed enzyme-prodrug therapy, or "suicide gene therapy," although problems with systemic gene delivery remain (409). This concept theoretically retains the advantages of ADEPT in terms of selective and sufficient access of the activated drug to tumor cells and expands the class of available enzymes to those that require endogenous cofactors. However one approach to design in selectivity between prodrug and toxin is lost compared with ADEPT, because the prodrugs must be able to enter cells freely.

**5.5.2 Mechanism and SAR.** As with ADEPT prodrugs, the specific mechanism of action depends on the type of enzyme used to activate the prodrug. GDEPT offers a wider choice of enzymes, because those with cofactors not readily available outside cells can also be used. The protocol is also generally less immunogenic than ADEPT.

*5.5.2.1 Prodrugs for Kinase Enzymes.* The most widely used prodrug in GDEPT protocols is the antifungal agent ganciclovir (**99**), which is activated by the thymidine kinase enzyme from Herpes simplex virus, converting it into the monophosphate (**100**). This can then be converted by cellular enzymes into the toxic triphosphate, which acts as an antimetabolite. This combination has been evaluated in numerous clinical trials, primarily in gliomas by intratumoral injection (410). A limitation of the approach is the poor bystander properties of the active drug, which cannot enter cells by passive diffusion, but instead uses gap junction connections (411) that are not well developed in many types of tumors (412).

*5.5.2.2 Prodrugs for Cytosine Deaminase.* The yeast enzyme cytosine deaminase (413) has also been widely studied as a GDEPT system in conjunction with 5-fluorocytosine (**101**), which it converts to the thymidylate synthetase inhibitor 5-fluorouracil (**73**). This has good diffusion properties and shows better bystander effects (414). Experimental studies have focused mainly on colon cancer models for the use of this combination, because clinically 5-fluorouracil is one of the most effective

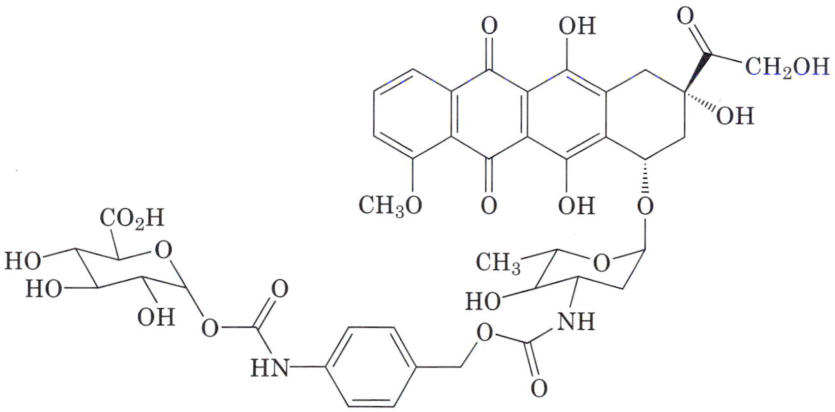

**(98)**

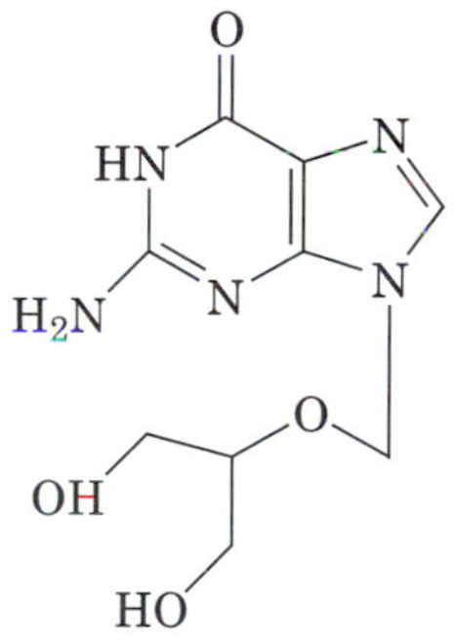

**(99)**

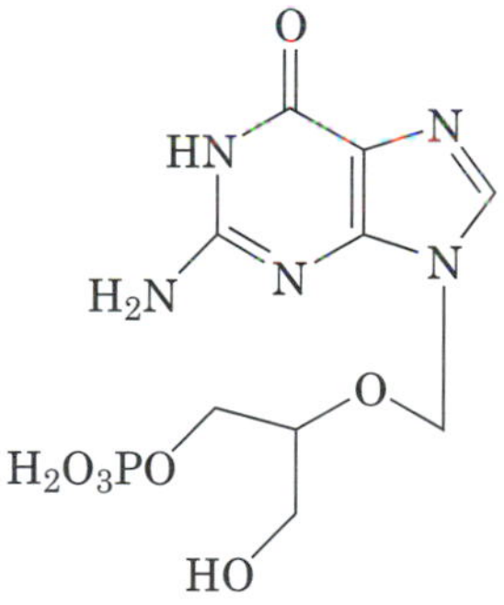

**(100)**

**(101)**

drugs for colon cancer. Possible drawbacks include the relatively low potency of **73**, coupled with its pronounced cell cycle selectivity (415), and no clinical trials of the protocol have yet been reported.

**5.5.2.3 Prodrugs for Oxidative Enzymes.** The cytochrome P450 enzymes that non-specifically activate the clinical agent cyclophosphamide (**4**) in the liver to the active species phosphoramide mustard (see Section 2.3.2 and Fig. 2.2) have also been employed in a GDEPT protocol with cyclophosphamide. Treatment of sc 9L gliosarcoma tumors transduced with various isozymes, especially CYP2B6 or CYP2C18-Met, with (**4**) gave large enhancements over the normal liver P450-dependent antitumor effect seen with control 9L tumors (growth delays of 25–50 days compared with 5–6 days), with no apparent increase in host toxicity (416).

**5.5.2.4 Prodrugs for Reductase Enzymes.** The dinitrophenylaziridine (**102102**; CB 1954) is activated by the aerobic nitroreductase (NTR) from *E. coli* (417), in conjunction with NADH or NADPH, to a mixture of hydroxylamines (Fig. 2.14). The 4-hydroxylamine (**103**) is then further metabolized by cellular enzymes to DNA cross-linking species. CB 1954 shows high selectivity (up to 1000-fold) for a variety of cell lines transduced

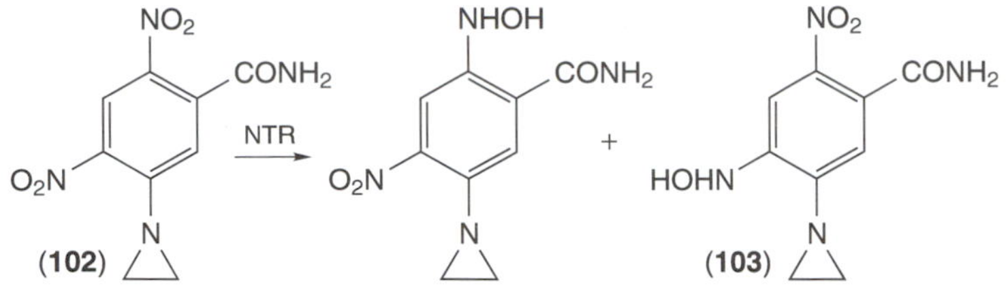

**Figure 2.14.** Metabolism of CB 1954 by *E. coli* NTR.

with the enzyme over the corresponding wild-type cell lines (418). It is now in clinical trial in conjunction with NTR, using a GDEPT protocol (419).

### 5.6 Antibody-Toxin Conjugates

**5.6.1 History.** This is a direct development of the "immunotoxin" approach, again exploiting the fact that many types of tumor cells present characteristic tumor-associated antigens on their surface (391). Despite much work, this approach has not been particularly successful until recently, with a combination of the availability of more resurgence of interest. The hypothesis is that conjugation of toxic drugs to the antibodies deactivate the drug (by limiting diffusional access to cells) without changing the selectivity of binding of the antibody. This allows it to locate on (antigen-bearing) tumor cells, internalize, and release the toxin (often through an acid-labile linker) when it is taken up into acidic endosomes (Fig. 2.15).

**5.6.2 Mechanism and SAR.** A wide variety of antibodies, linkers, and toxins are currently being explored in this approach. Doxorubicin continues to be widely used as a toxin because

it is so well characterized, although it is not exceptionally potent. The most advanced doxorubicin-containing conjugate is SGN-15 (**104**), in which an average of eight molecules of doxorubicin are linked through an acid-labile hydrazone link, through the C-14 carbonyl, to the chimeric mAb BR96, which binds to a modified Le$^y$ antigen on tumor cells. The major route of breakdown of (**104**) *in vitro* has been shown to be acid-catalyzed hydrazone hydrolysis, as designed (420). SGN-15 induced cures of established subcutaneous human colon carcinomas in athymic mice and rats (421), where free doxorubicin at its maximum-tolerated doses were ineffective. A recent phase I clinical trial of SGN-15 in patients with metastatic colon and breast cancers expressing the Le$^y$ antigen determined the optimal dose to be 700 mg/m$^2$ (equivalent to 19 mg/m$^2$ of doxorubicin), with only mild toxicity (422).

Conjugates of the extremely potent calicheamicin-type DNA cleaving agents have been under development for some time (423, 424). The conjugate (**105**) (gemtuzumab ozogamicin; mylotarg) was the first antibody armed with a small-molecule cytotoxin to reach clinical trial. Mylotarg has an average of four to five calicheamicin molecules linked through an acid-labile hydrazone linker, through a sterically-hindered disulfide, to a humanized hP67.6 IG1-based antibody that recognizes the CD33 antigen on normal and leukemic myeloid progenitor cells (425). Cleavage of the linker in the low pH endosomic environment in cells is followed by intramolecular cyclization to generate the transient benzenoid diradical that results in DNA double-strand cleavage (426). In phase II studies in relapsed AML patients, an overall 30% response rate was seen (427), with delayed hepatotoxicity as a possible side effect (428). Conjugate (**105**) has also been reported to be active clinically in brc/abl-positive CML (429).

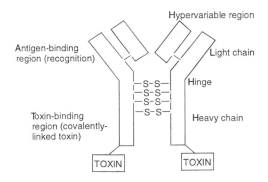

**Figure 2.15.** Schematic of toxin-armed antibody.

(104)

(105)

(106)

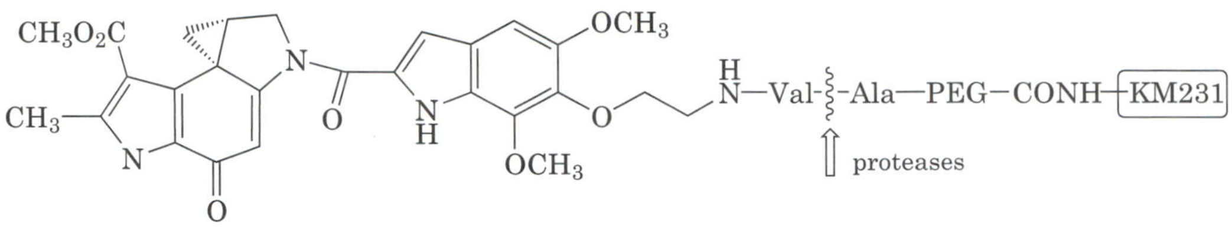

(107)

The conjugate SB 408075 (**106**) employs the very potent synthetic maytansinoid-type tubulin inhibitor DM1 (430), with an average of four molecules of the toxin attached by a disulfide link to hC242, an antibody to a mucin-like glycoprotein on colorectal cancer cells. This conjugate affected cures in mice bearing large COLO205 human colon tumor xenografts (431) and is reported to be in phase II clinical trials.

Finally, members of the class of very cytotoxic DNA minor groove alkylators exemplified by the natural products CC-1065 and duocarmycin (83) have also been used to arm antibodies. As discussed in Section 2.5, these DNA minor groove alkylators were also evaluated clinically in their own right, but proved too toxic. Conjugate (**107**) (KM231-DU257) contains an average of two molecules of the duocarmycin analog DU257, linked through a PEGylated dipeptide (HO$_2$C-Val-Ala-NH$_2$) to an M231 antibody that targets the sLe$^a$ antigen (432). The PEGylated linker prolongs plasma half-life, and the Val-Ala link is cleaved by tumor proteases to primarily release the DU257-Val conjugate, which has similar potency to DU257 itself.

## REFERENCES

1. L. S. Goodman, M. M. Wintrobe, W. Damesheck, M. J. Goodman, A. Gilman, and M. T. McLennan, *J. Am. Med. Assoc.*, **132**, 126 (1946).

2. G. B. Elion, *Bioscience Rep.*, **9**, 509 (1989).

3. R. K-Y. Zee-Cheng and C. C. Cheng, *Meth. Find. Exp. Clin. Pharmacol.*, **10**, 67 (1988).

4. P. Brookes, *Mutat. Res.*, **233**, 3 (1990).

5. A. Begleiter, M. Mowat, L. G. Israels, and J. B. Johnston, *Leuk. Lymphoma*, **23**, 187 (1996).

6. R. L. Furner and R. K. Brown, *Cancer Treat. Repts.*, **64**, 559 (1980).

7. O. M. Colvin, *Curr. Pharm. Des.*, **5**, 555 (1999).

8. A. V. Boddy and S. M. Yule, *Clin. Pharm.*, **38**, 291 (2000).

9. J. Hansson, R. Lewensohn, U. Ringborg, and B. Nilsson, *Cancer Res.*, **47**, 2631 (1987).

10. R. O. Peiper, B. W. Futscher, and L. C. Erickson, *Carcinogenesis*, **10**, 1307 (1989).

11. I. Niculescu-Duvaz, I. Baracu, and A. T. Balaban, In: D. E. V. Wilman, Ed., *Chemistry of Antitumour Agents*, Blackie, London, 1989, 63.

12. D. E. V. Wilman and T. A. Connors, In: S. Neidle and M. J. Waring, Eds., *Molecular Aspects of Anticancer Drug Action*, MacMillan, London, 1983, 233.

13. K. W. Kohn, J. A. Hartley, and W. B. Mattes, *Nucleic Acids Res.*, **15**, 10531 (1987).

14. A. Pullman and B. Pullman, *Quart. Rev. Biophys.*, **14**, 289 (1981).

15. A. S. Prakash, W. A. Denny, T. A. Gourdie, K. K. Valu, P. D. Woodgate, and L. P. G. Wakelin, *Biochemistry*, **29**, 9799 (1990).

16. M. R. Osborne, D. E. V. Wilman, and P. D. Lawley, *Chem. Res. Tox.*, **8**, 316 (1995).

17. M. R. Osborne and P. D. Lawley, *Chem.-Biol. Int.*, **89**, 49 (1993).

18. P. D. Lawley and P. Brookes, *J. Mol. Biol.*, **25**, 143 (1967).

19. J. T. Millard, S. Raucher, and P. B. Hopkins, *J. Am. Chem. Soc.*, **112**, 2459 (1990).

20. N. E. Sladek, *Pharmacol. Ther.*, **37**, 301 (1988).

21. M. P. Goren, R. K. Wright, and C. B. Pratt, *Lancet*, 1219 (1986).

22. A. Panthananickal, C. Hansch, and A. Leo, *J. Med. Chem.*, **21**, 16 (1978).

23. B. D. Palmer, W. R. Wilson, S. M. Pullen, and W. A. Denny, *J. Med. Chem.*, **33**, 112 (1990).

24. A. L. Wang and K. D. Tew, *Cancer Treatment Repts.*, **69**, 677 (1985).

25. K. Suzukake, B. P. Vistica, and D. T. Vistica, *Biochem. Pharmacol.*, **32**, 165 (1983).

26. M. Brendel and A. Ruhland, *Mutation Res.*, **133**, 51 (1984).

27. R. G. Palmer and A. M. Denman, *Cancer Treat. Rev.*, **2**, 121 (1984).

28. G. Leone, M. T. Voso, S. Sica, and R. Morosetti, *Leukemia & Lymphoma*, **41**, 255 (2001).

29. G. Leone, L. Mele, A. Pulsoni, F. Equitani, and L. Pagano, *Haematologica*, **84**, 937 (1999).

30. W. A. Denny, *Curr. Med. Chem.*, **8**, 533 (2001).

31. T. A. Gourdie, K. K. Valu, G. L. Gravatt, T. J. Boritzki, B. C. Baguley, W. R. Wilson, P. D. Woodgate, and W. A. Denny, *J. Med. Chem.*, **33**, 1177 (1990).

32. W. B. Mattes, J. A. Hartley, and K. W. Kohn, *Nucleic Acids Res.*, **14**, 2971 (1986).

33. H. J. Creech, R. K. Preston, R. M. Peck, A. S. O'Connell, and B. N. Ames, *J. Med. Chem.*, **15**, 739 (1972).

34. T. J. Boritzki, B. D. Palmer, J. M. Coddington, and W. A. Denny, *Chem. Res. Tox.*, **7**, 41 (1994).

35. M. Lee, A. L. Rhodes, M. D. Wyatt, M. D'Incalci, S. Forrow, and J. A. Hartley, *J. Med. Chem.*, **36**, 863 (1993).

36. G. L. Gravatt, B. C. Baguley, W. R. Wilson, and W. A. Denny, *J. Med. Chem.*, **37**, 4338 (1994).

37. G. J. Atwell, B. M. Yaghi, P. R. Turner, M. Boyd, C. J. O'Connor, L. R. Ferguson, B. C. Baguley, and W. A. Denny, *Bioorg. Med. Chem.*, **6**, 679 (1995).

38. J. G. Pelton and D. E. Wemmer, *J. Am. Chem. Soc.*, **112**, 1393 (1990).

39. B. F. Baker and P. B. Dervan, *J. Am. Chem. Soc.*, **107**, 8266 (1985).

40. F. M. Arcamone, F. Animati, B. Barbieri, E. Configliacchi, R. D'Alessio, C. Geroni, F. C. Giuliani, E. Lazzari, M. Menozzi, N. Mongelli, S. Penco, and M. A. Verini, *J. Med. Chem.*, **32**, 774 (1989).

41. M. Broggini, E. Erba, M. Ponti, D. Ballinari, C. Geroni, F. Spreafico, and M. D'Incalci, *Cancer Res.*, **51**, 199 (1991).

42. M. Broggini, H. M. Coley, N. Mongelli, E. Pesenti, M. D. Wyatt, J. A. Hartley, and M. D'Incalci, *Nucleic Acids Res.*, **23**, 81 (1995).

43. M. D. Wyatt, M. Lee, B. J. Garbiras, R. L. Souhami, and J. A. Hartley, *Biochemistry*, **34**, 13034 (1995).

44. P. Cozzi and N. Mongelli, *Curr. Pharm. Des.*, **4**, 181 (1998).

45. E. Erba, E. Mascellani, A. Pifferi, and M. D'Incalci, *Int. J. Cancer*, **62**, 170 (1995).

46. M. Bellorini, V. Moncollin, M. D'Incalci, N. Mongelli, and R. Mantovani, *Nucleic Acids Res.*, **23**, 1657 (1995).

47. G. R. Weiss, I. Poggesi, M. Rocchetti, D. Demaria, T. Mooneyham, D. Reilly, L. V. Vitek, F. Whaley, E. Patricia, D. D. von Hoff, and P. A. O'Dwyer, *Clin. Cancer Res.*, **4**, 53 (1998).

48. C. J. A. Punt, Y. Humblet, E. Roca, L. Y. Dirix, R. Wainstein, A. Polli, and I. Corradino, *Brit. J. Cancer*, **73**, 803 (1996).

49. P. Cozzi, I. Beria, M. Caldarelli, L. Capolongo, C. Geroni, and N. Mongelli, *Bioorg. Med. Chem. Lett.*, **10**, 1269 (2000).

50. P. Cozzi, *Farmaco*, **55**, 168 (2000).

51. J. W. Trauger, E. E. Baird, and P. B. Dervan, *Nature*, **382**, 559 (1996).

52. J. W. Trauger, E. E. Baird, and P. B. Dervan, *J. Am. Chem. Soc.*, **120**, 3534 (1998).

53. N. R. Wurtz and P. B. Dervan, *Chem. Biol.*, **7**, 153 (2000).

54. B. Rosenberg, L. Van Camp, E. B. Grimley, and A. J. Thomson, *J. Biol. Chem.*, **242**, 1347 (1967).

55. R. B. Weiss and M. C. Christian, *Drugs*, **46**, 360 (1993).

56. K. J. Scanlon, M. Kashani-Sabet, T. Tone, and T. Funato, *Pharmacol. Ther.*, **52**, 385 (1991).

57. J. H. Burchenal, K. Kalaher, T. O'Toole, and J. Chisholm, *Cancer Res.*, **37**, 3455 (1977).

58. P. J. O'Dwyer, J. P. Stevenson, and S. W. Johnson, In: L. R. Kelland and N. P. Farrell, Eds., *Platinum-Based Drugs in Cancer Therapy*, Humana Press, Totowa, NJ, 2000, 231.

59. R. J. Pelley, *Curr. Oncol. Rep.*, **3**, 147 (2001).

60. B. C. Behrens, T. C. Hamilton, H. Masuda, K. R. Grotzinger, J. Whang-Peng, K. G. Louie, T. Knutsen, W. M. McKoy, R. C. Young, and R. F. Ozols, *Cancer Res.*, **47**, 414 (1987).

61. L. R. Kelland, S. Y. Sharp, C. F. O'Neill, I. F. Raynaud, P. J. Beale, and I. R. Judson, *J. Inorg. Biochem.*, **77**, 111 (1999).

62. M. P. Hay, *Curr. Opin. Invest. Drugs*, **1**, 263 (2000).

63. L. R. Kelland, *Exp. Opin. Invest. Drugs*, **9**, 1373 (2000).

64. E. M. Bengtson and J. R. Rigas, *Drugs*, **58**(Suppl 3), 57 (1999).

65. S. J. Berners-Price and T. G. Appleton, *Platinum-Based Drugs Cancer Therap.*, **3**, 3–20 (2000).

66. M. S. Davies, S. J. Berners-Price, and T. W. Hambley, *J. Inorg. Biochem.*, **79**, 167 (2000).

67. V. Brabec, J. Kasparkova, O. Vrana, O. Novakova, J. W. Cox, Y. Qu, and N. Farrell, *Biochemistry*, **38**, 6781 (1999).

68. J. D. Roberts, J. Peroutka, and N. Farrell, *J. Inorg. Biochem.*, **77**, 51 (1999).

69. P. Di Blasi, A. Bernareggi, G. Beggiolin, L. Piazzoni, E. Menta, and M. L. Formento, *Anticancer Res.*, **18**, 3113 (1998).

70. C. Manzotti, G. Pratesi, E. Menta, R. Di Domenico, E. Cavalletti, H. H. Fiebig, L. R. Kelland, N. Farrell, D. Polizzi, R. Supino, G. Pezzoni, and F. Zunino, *Clin. Cancer Res.*, **6**, 2626 (2000).

71. T. Servidei, C. Ferlini, A. Riccardi, D. Meco, G. Scambia, G. Segni, C. Manzotti, and R. Riccardi, *Eur. J. Cancer*, **37**, 930 (2001).

72. C. Sessa, G. Capri, L. Gianni, F. Peccatori, G. Grasselli, J. Bauer, M. Zucchetti, L. Vigano, A. Gatti, C. Minoia, P. Liati, S. Van den Bosch, A. Bernareggi, G. Camboni, and S. Marsoni, *Ann. Oncol.*, **11**, 977 (2000).

73. L. H. Hurley, C-S. Lee, J. P. McGovren, M. A. Warpehoski, M. A. Mitchell, R. C. Kelly, and P. A. Aristoff, *Biochemistry*, **27**, 3886 (1988).

74. L. J. Hanka, A. Dietz, S. A. Gerpheide, S. L. Kuentzel, and D. G. Martin, *J. Antibiot.*, **31**, 1211 (1978).

75. D. G. Martin, C. Biles, S. A. Gerpheide, L. J. Hanka, W. C. Krueger, J. P. McGovren, S. A. Mizak, G. L. Neil, J. C. Stewart, and J. Visser, *J. Antibiot.*, **34**, 1119 (1981).

76. J. P. McGovren, G. L. Clarke, E. A. Pratt, and T. F. DeKoning, *J. Antibiot.*, **37**, 63 (1984).

77. M. A. Warpehoski, *Drugs Future*, **16**, 131 (1991).

78. L. H. Li, R. C. Kelly, M. A. Warpehoski, J. P. McGovren, I. Gebhard, and T. F. DeKoning, *Inv. New Drugs*, **9**, 137 (1991).

79. L. H. Li, T. F. DeKoning, R. C. Kelly, W. C. Krueger, J. P. McGovren, G. E. Padbury, G. L. Petzold, T. L. Wallace, R. J. Ouding, M. D. Prairie, and I. Gebhard, *Cancer Res.*, **52**, 4904 (1992).

80. H. Ogasawara, K. Nishio, Y. Takeda, T. Ohmori, N. Kubota, Y. Funayama, T. Ohira, Y. Kuraishi, Y. Isogai, and N. Saijo, *Jpn. J. Cancer Res.*, **85**, 418 (1994).

81. M. Kinugawa, S. Nakamura, A. Sakaguchi, Y. Masuda, H. Saito, T. Ogasa, and M. Kasai, *Org. Process Res. Dev.*, **2**, 344 (1998).

82. E. Kobayashi, A. Okamoto, M. Asada, M. Okabe, S. Nagamura, A. Asai, H. Saito, K. Gomi, and T. Hirata, *Cancer Res.*, **54**, 2404 (1994).

83. D. L. Boger and D. S. Johnson, *Angew. Chem., Int. Ed. Engl.*, **35**, 1438 (1996).

84. D. L. Boger, B. Bollinger, D. L. Hertzog, D. S. Johnson, H. Cai, P. Mesini, R. M. Garbaccio, Q. Jin, and P. A. Kitos, *J. Amer. Chem. Soc.*, **119**, 4987 (1997).

85. J. H. Yoon and C. S. Lee, *Arch. Pharm. Res.*, **21**, 385 (1998).

86. N. Amishiro, S. Nagamura, E. Kobayashi, K. Gomi, and H. Sato, *J. Med. Chem.*, **42**, 669 (1999).

87. N. Amishiro, A. Okamoto, C. Murakata, T. Tamaoki, M. Okabe, and H. Saito, *J. Med. Chem.*, **42**, 2946 (1999).

88. M. Hidalgo, E. Izbicka, C. Cerna, L. Gomez, E. K. Rowinsky, S. D. Weitman, and D. D. von Hoff, *Anti-Cancer Drugs*, **10**, 295 (1999).

89. M. Cristofanilli, W. J. Bryan, L. L. Miller, A. Y. C. Chang, W. J. Gradishar, D. W. Kufe, and G. N. Hortobagyi, *Anti-Cancer Drugs*, **9**, 779 (1998).

90. N. Pavlidis, S. Aamdal, A. Awada, H. Calvert, P. Fumoleau, R. Sorio, C. Punt, J. Verweij, A. van Oosterom, R. Morant, J. Wanders, and A. R. Hanauske, *Cancer Chemother. Pharmacol.*, **46**, 167 (2000).

91. S. R. Alberts, C. Erlichman, J. M. Reid, J. A. Sloan, M. M. Ames, R. L. Richardson, and R. M. Goldberg, *Clin. Cancer Res.*, **4**, 2111 (1998).

92. E. J. Small, R. Figlin, D. Petrylak, D. J. Vaughn, O. Sartor, I. Horak, R. Pincus, A. Kremer, and C. Bowden, *Inv. New Drugs*, **18**, 193 (2000).

93. G. J. Atwell, M. Tercel, M. Boyd, W. R. Wilson, and W. A. Denny, *J. Org. Chem.*, **63**, 9414 (1998).

94. M. A. Gieseg, J. Matejovic, and W. A. Denny, *Anti-Cancer Drug Design*, **14**, 77 (1999).

95. J. A. Montgomery and T. P. Johnson, In: D. E. V. Wilman, Ed., *Chemistry of Antitumour Agents*, Blackie, London, 1989, 131.

96. H. E. Skipper, F. M. Schabel, M. W. Trader, and J. R. Thomson, *Cancer Res.*, **21**, 1154 (1961).

97. V. T. DeVita, C. Denham, J. D. Davidson, and V. T. Olivero, *Clin. Pharm. Ther.*, **8**, 566 (1967).

98. S. E. Salmon, *Cancer Treatment Rep.*, **60**, 789 (1976).

99. R. B. Weiss, *Cancer Treat. Rep.*, **66**, 427 (1982).

100. J. A. Mongomery, R. James, G. S. McCaleb, and T. P. Johnson, *J. Med. Chem.*, **10**, 568 (1975).

101. N. Buckley, *J. Org. Chem.*, **52**, 484 (1987).

102. P. S. Schein, M. J. O'Connell, J. Blom, S. Hubbard, I. T. Magrath, P. A. Bergevin, P. H. Wiernik, J. L. Zeigler, and V. T. DeVita, *Cancer*, **34**, 933 (1974).

103. R. M. Bukowski, R. T. Abderhalden, J. S. Hewlett, J. K. Weick, and C. W. Groppe, *Cancer Clin. Trials*, **3**, 321 (1980).

104. N. Kemeny, A. Yagoda, D. Braun, and R. Golbey, *Cancer*, **45**, 876 (1980).

105. T. S. Khan, H. Imam, C. Juhlin, B. Skogseid, S. Grondal, S. Tibblin, E. Wilander, and K. Oberg, *Ann. Oncol.*, **11**, 1281 (2000).

106. Y. F. Shealy, C. A. Krauth, and J. A. Montgomery, *J. Org. Chem.*, **27**, 2150 (1962).

107. Y. F. Shealy, J. A. Montgomery, and W. R. Laster, *Biochem. Pharmacol.*, **11**, 674 (1962).

108. M. F. G. Stevens, J. A. Hickman, R. Stone, N. W. Gibson, G. U. Baig, E. Lunt, and C. G. Newton, *J. Med. Chem.*, **27**, 196 (1984).

109. E. S. Newlands, M. F. G. Stevens, S. R. Wedge, R. T. Wheelhouse, and C. Brock, *Cancer Treat. Rev.*, **23**, 35 (1997).

110. A. S. Clark, B. Deans, M. F. G. Stevens, M. J. Tisdale, R. T. Wheelhouse, B. J. Denny, and J. A. Hartley, *J. Med. Chem.*, **38**, 1493 (1995).

111. M. J. Tisdale, *Biochem. Pharmacol.*, **36**, 457 (1987).

112. N. W. Gibson, L. C. Erickson, and J. A. Hickman, *Cancer Res.*, **44**, 1767 (1984).

113. J. A. Hartley, N. W. Gibson, K. W. Kohn, and W. B. Mattes, *Cancer Res.*, **46**, 1943 (1986).

114. G. L. Cohen and C. I. Falkson, *Drugs*, **55**, 791 (1998).

115. L. Dreiling, S. Hoffman, and W. A. Robinson, *Adv. Intern. Med.*, **41**, 553 (1996).

116. T. A. Connors, P. M. Goddard, K. Merai, W. C. J. Ross, and D. E. V. Wilman, *Biochem. Pharmacol.*, **25**, 241 (1976).

117. L. Meer, R. C. Janzer, P. Kleihues, and G. F. Kolar, *Biochem. Pharmacol.*, **35**, 3243 (1986).

118. J. A. Hickman, M. F. G. Stevens, N. W. Gibson, S. P. Langdon, C. Fizames, F. Lavelle, G. Atassi, E. Lunt, and R. M. Tilson, *Cancer Res.*, **45**, 3008 (1985).

119. O. Fodstad, S. Aamdal, A. Pihl, and M. R. Boyd, *Cancer Res.*, **45**, 1778 (1985).

120. E. S. Newlands, G. Blackledge, J. A. Slack, C. Goddard, C. J. Brindley, L. Holden, and M. F. Stevens, *Cancer Treat. Rep.*, **69**, 801 (1985).

121. M. Harding, V. Docherty, R. Mackie, A. Dorward, and S. Kaye, *Eur. J. Cancer Clin. Oncol.*, **25**, 785 (1989).

122. R. Somers, A. Santoro, J. Verweij, P. Lucas, J. Rouesse, T. Kok, A. Casali, C. Seynaeve, and D. Thomas, *Eur. J. Cancer*, **28A**, 855 (1992).

123. I. Hickson, L. J. Fairbairn, N. Chinnasamy, L. S. Lashford, N. Thatcher, G. P. Margison, T. M. Dexter, and J. A. Rafferty, *Gene Therap.*, **5**, 835 (1998).

124. N. Chinnasamy, J. A. Rafferty, I. Hickson, L. S. Lashford, S. J. Longhurst, N. Thatcher, G. P. Margison, T. M. Dexter, and L. J. Fairbairn, *Gene Therap.*, **5**, 842 (1998).

125. M. F. Stevens, J. A. Hickman, S. P. Langdon, D. Chubb, L. Vickers, R. Stone, G. Baig, C. Goddard, N. W. Gibson, and J. A. Slack, *Cancer Res.*, **47**, 5846 (1987).

126. J. Plowman, W. R. Waud, A. D. Koutsoukos, L. V. Rubinstein, T. D. Moore, and M. R. Grever, *Cancer Res.*, **54**, 3793 (1994).

127. E. S. Newlands, G. R. Blackledge, J. A. Slack, G. J. Rustin, D. B. Smith, N. S. Stuart, C. P. Quarterman, R. Hoffman, M. F. Stevens, and M. H. Brampton, *Br. J. Cancer.*, **65**, 287 (1992).

128. S. M. O'Reilly, E. S. Newlands, M. G. Glaser, M. Brampton, J. M. Rice-Edwards, R. D. Illingworth, P. G. Richards, C. Kennard, I. R. Colquhoun, and P. Lewis, *Eur. J. Cancer*, **29A**, 940 (1993).

129. H. S. Friedman, T. Kerby, and H. Calvert, *Clin. Cancer Res.*, **6**, 2585 (2000).

130. M. D. Prados, *Sem. Oncol.*, **27**(Suppl 6), 41 (2000).

131. S. S. Agarwala and J. M. Kirkwood, *Oncologist*, **5**, 144 (2000).

132. L. S. Lerman, *J. Mol. Biol.*, **3**, 18 (1961).

133. R. M. Wadkins and D. E. Graves, *Biochemistry*, **30**, 4277 (1991).

134. W. A. Denny, *Anti-Cancer Drug Design*, **4**, 241 (1989).

135. L. A. Zwelling, S. Michaels, L. C. Erickson, R. S. Ungerleider, M. Nichols, and K. W. Kohn, *Biochemistry*, **20**, 6553 (1981).

136. E. M. Nelson, K. M. Tewey, and L. F. Liu, *Proc. Natl. Acad. Sci. USA*, **181**, 1361 (1984).

137. P. Deisabella, G. Capranico, and F. Zunino, *Life Sci.*, **48**, 2195 (1991).

138. L. F. Liu, *Ann. Rev. Biochem.*, **58**, 351 (1989).

139. M. J. Robinson and N. Osheroff, *Biochemistry*, **30**, 1807 (1991).

140. T. D. Y. Chung, F. H. Drake, S. R. Tan, M. Per, S. T. Crooke, and C. K. Mirabelli, *Proc. Natl. Acad. Sci. USA*, **86**, 9431 (1989).

141. M. Tsai-Pflugfelder, L. F. Liu, A. A. Lui, K. M. Tewey, J. Whang-Peng, T. Knutsen, K. Huebner, C. M. Croce, and J. C. Wang, *Proc. Natl. Acad. Sci. USA*, **85**, 7177 (1988).

142. K. B. Tan, T. F. Dorman, K. M. Falls, T. D. Y. Chung, C. K. Mirabelli, S. T. Crooke, and J. Mao, *Cancer Res.*, **52**, 231 (1992).

143. F. H. Drake, G. A. Hofmann, H. F. Bartus, M. R. Mattern, S. T. Crooke, and C. K. Mirabelli, *Biochemistry*, **28**, 8154 (1989).

144. W. A. Denny, G. J. Atwell, B. C. Baguley, and B. F. Cain, *J. Med. Chem.*, **22**, 134 (1979).

145. B. F. Cain, G. J. Atwell, and R. N. Seelye, *J. Med. Chem.*, **12**, 199 (1969).

146. B. F. Cain, G. J. Atwell, and R. N. Seelye, *J. Med. Chem.*, **14**, 311 (1971).

147. A. W. Braithwaite and B. C. Baguley, *Biochemistry*, **19**, 1101 (1980).

148. G. J. Atwell, B. F. Cain, and R. N. Seelye, *J. Med. Chem.*, **15**, 611 (1972).

149. B. F. Cain, G. J. Atwell, and W. A. Denny, *J. Med. Chem.*, **18**, 1110 (1975).

150. W. R. Grove, C. L. Fortner, and P. H. Wiernik, *Clinical Pharmacy*, **1**, 320 (1982).

151. H. J. Lawrence, C. A. Ries, B. D. Reynolds, J. P. Lewis, M. M. Koretz, and F. M. Torti, *Cancer Treatment Rep.*, **66**, 1475 (1982).

152. W. A. Denny, In: D. Lednicer and J. Bindra, Eds., *Chronicles of Drug Discovery*, Vol. **3**, ACS Publications, Washington, 1993, 381.

153. W. A. Denny, G. J. Atwell, B. F. Cain, A. Leo, A. Panthananickal, and C. Hansch, *J. Med. Chem.*, **25**, 276 (1982).

154. B. F. Cain, G. J. Atwell, and W. A. Denny, *J. Med. Chem.*, **20**, 987 (1977).

155. W. A. Denny, G. J. Atwell, and B. C. Baguley, *J. Med. Chem.*, **27**, 363 (1984).

156. B. C. Baguley, W. A. Denny, G. J. Atwell, G. F. Finlay, G. W. Rewcastle, S. J. Twigden, and W. R. Wilson, *Cancer Res.*, **44** (1984).

157. W. R. Leopold, T. H. Corbett, D. P. Griswold, J. Plowman, and B. C. Baguley, *J. Natl. Cancer Inst.*, **79**, 343 (1987).

158. R. K. Y. Zee-Cheng and C. C. Cheng, *J. Med. Chem.*, **21**, 291 (1978).

159. R. K. Y. Zee-Cheng and C. C. Cheng, *Drugs Future*, **8**, 229 (1983).

160. B. C. Baguley, W. A. Denny, G. J. Atwell, and B. F. Cain, *J. Med. Chem.*, **24**, 520 (1981).

161. L. Valentini, V. Nicolella, E. Vannini, M. Menuzzi, S. Penco, and F. M. Arcamone, *Farmaco*, **40**, 376 (1985).

162. J. Feigon, W. A. Denny, W. Leupin, and D. R. Kearns, *J. Med. Chem.*, **23**, 450 (1984).

163. R. M. Wadkins and D. E. Graves, *Nucleic Acids Res.*, **17**, 9933 (1989).

164. W. R. Wilson, B. C. Baguley, L. P. G. Wakelin, and M. J. Waring, *Mol. Pharmacol.*, **20**, 404 (1981).

165. T. D. Sakore, B. S. Reddy, and H. M. Sobell, *J. Mol. Biol.*, **135**, 763 (1979).

166. K. X. Chen, N. Gresh, and B. Pullman, *Nucleic Acids Res.*, **16**, 3061 (1988).

167. I. Chourpa, H. Morjani, J-F. Riou, and M. Manfait, *FEBS Lett.*, **397**, 61 (1996).

168. G. J. Finlay, G. J. Atwell, and B. C. Baguley, *Oncol. Res.*, **11**, 249 (1999).

169. E. L. Meczes, K. I. Marsh, M. L. Fisher, P. M. Rogers, and C. A. Austin, *Cancer Chemotherap. Pharmacol.*, **39**, 367 (1997).

170. S. Houlbrook, C. M. Addison, S. L. Davies, J. Carmichael, I. J. Stratford, A. L. Harris, and I. D. Hickson, *Br. J. Cancer*, **72**, 1454 (1995).

171. S. Withoff, E. G. E. de Vries, W. N. Keith, E. F. Nienhuis, W. T. A. van der Graaf, D. R. A. Uges, and N. H. Mulder, *Br. J. Cancer*, **74**, 1869 (1996).

172. B. Van Hille, D. Perrin, and B. T. Hill, *Anti-Cancer Drugs*, **10**, 551 (1999).

173. M. Sorensen, M. Sehested, and P. B. Jensen, *Mol. Pharmacol.*, **55**, 424 (1999).

174. C. Negri, R. Bernardi, M. Donzelli, and A. I. Scovassi, *Biochimie*, **77**, 893 (1995).

175. S. M. Morris, O. E. Domon, L. J. McGarrity, J. J. Chen, and D. A. Casciano, *Mutat. Res.*, **329**, 79 (1995).

176. S. Prost and G. Riou, *Cell. Pharmacol.*, **2**, 103 (1995).

177. W. A. Denny and L. P. G. Wakelin, *Anti-Cancer Drug Design*, **5**, 189 (1990).

178. J. W. Lown and C. J. Hanstock, *Biomol. Struct. Dyn.*, **2**, 1097 (1985).

179. C. Bailly, S. Routier, J. L. Bernier, and M. J. Waring, *FEBS Lett.*, **379**, 269 (1996).

180. J. Kapuscinski and Z. Darzynkiewicz, *Biochem. Pharmacol.*, **34**, 4203 (1985).

181. C. R. Krishnamoorthy, S. F. Yen, J. C. Smith, J. W. Lown, and W. D. Wilson, *Biochemistry*, **25**, 5933 (1986).

182. P. De Isabella, G. Capranico, M. Palumbo, C. Sissi, A. P. Krapcho, and F. Zunino, *Mol. Pharmacol.*, **43**, 715 (1993).

183. M. E. Fox and P. J. Smith, *Cancer Res.*, **50**, 5813 (1990).

184. B. Bellosillo, D. Colomer, G. Pons, and J. Gil, *Br. J. Haematol.*, **100**, 142 (1998).

185. C. Panousis, A. J. Kettle, and D. R. Phillips, *Anti-Cancer Drug Design*, **10**, 593 (1995).

186. B. S. Belinda, C. Cullinane, and D. R. Phillips, *Nucleic Acids Res.*, **27**, 2918 (1999).

187. C. P. Steuber, J. Krischer, T. Holbrook, B. Camitta, V. Land, C. Sexauer, D. Mahoney, and H. Weinstein, *J. Clin. Oncol.*, **14**, 1521 (1996).

188. E. Estey, P. F. Thall, S. Pierce, H. Kantarjian, and M. Keating, *J. Clin. Oncol.*, **15**, 483 (1997).

189. A. W. Dekker, M. B. Van't Veer, W. Sizoo, H. L. Haak, J. Van Der Lelie, G. Ossenkoppele, P. C. Huijgens, H. C. Schuouten, and P. Sonneveld, *J. Clin. Oncol.*, **15**, 476 (1997).

190. L. Mollgard, U. Tidefelt, B. Sundman-Engberg, C. Lofgren, S. Lehman, and C. Paul, *Therap. Drug Monit.*, **20**, 640 (1998).

191. S. Jelic, Z. Nikolic-Tomasevic, V. Kovcin, N. Milanovic, Z. Tomasevic, V. Jovanovic, and M. Vlajic, *J. Chemother. (Firenze)*, **9**, 364 (1997).

192. J. P. Krischer, S. Epstein, D. D. Cuthbertson, A. M. Goorin, M. L. Epstein, and S. E. Lipshultz, *J. Clin. Oncol.*, **15**, 1544 (1997).

193. J. M. Covey, K. W. Kohn, D. Kerrigan, E. J. Tilchen, and Y. Pommier, *Cancer Res.*, **48**, 860 (1988).

194. E. Schneider, S. J. Darkin, P. A. Lawson, L-M. Ching, R. K. Ralph, and B. C. Baguley, *Eur. J. Cancer Clin. Oncol.*, **24**, 1783 (1988).

195. J. R. Hardy, V. J. Harvey, J. W. Paxton, P. Evans, S. Smith, W. Grove, A. J. Grillo-Lopez, and B. C. Baguley, *Cancer Res.*, **48**, 6593 (1988).

196. V. J. Harvey, J. R. Hardy, S. Smith, W. Grove, and B. C. Baguley, *Eur. J. Cancer*, **27**, 1617 (1991).

197. D. Fyfe, F. Reynaud, J. White, K. Clayton, L. Mason, C. Gardner, P. J. Woll, I. Judson, and J. Carmichael, *Br. J. Cancer*, **78**(Suppl 2), 27 (1998).

198. C. A. Koller, H. M. Kantarjian, E. J. Feldman, S. O'Brien, M. B. Rios, E. Estey, and M. Keating, *Cancer*, **86**, 2246 (1999).

199. X. Thomas and E. Archimbaud, *Hematol. Cell Therap.*, **39**, 163 (1997).

200. W. K. Oh, *Cancer*, **88**(Suppl), 3015 (2000).

201. K. K. Jain, *Exp. Opin. Invest. Drugs*, **9**, 1139 (2000).

202. C. J. Dunn and K. L. Goa, *Drugs Aging*, **9**, 122 (1996).

203. G. Boos and H. Stopper, *Toxicol. Lett.*, **116**, 7 (2000).

204. P. M. Carli, C. Sgro, N. Parchin-Geneste, N. Isambert, F. Mugneret, F. Girodon, and M. Maynadie, *Leukemia*, **14**, 1014 (2000).

205. D. Nielsen, J. Eriksen, C. Maare, T. Litman, E. Kjaersgaard, T. Plesner, E. Friche, and T. Skovsgaard, *Biochem. Pharmacol.*, **60**, 363 (2000).

206. D. M. Sullivan, P. W. Feldhoff, R. B. Lock, N. B. Smith, and W. M. Pierce, *Int. J. Oncol.*, **7**, 1383 (1995).

207. U. Kellner, L. Hutchinson, A. Seidel, H. Lage, D. Hermann, M. K. Danks, M. Dietel, and S. H. Kaufmann, *Int. J. Cancer*, **71**, 817 (1997).

208. J. D. Bailly, A. Skladanowski, A. Bettaieb, V. Mansat, A. K. Larsen, and G. Laurent, *Leukemia*, **11**, 1523 (1997).

209. L. A. Hazlehurst, N. E. Foley, M. C. Gleason-Guzman, M. P. Hacker, A. E. Cress, L. W. Greenberger, M. C. De Jong, and W. S. Dalton, *Cancer Res.*, **59**, 1021 (1999).

210. C. A. Felix, *Biochimica et Biophysica Acta*, **1400**, 233 (1998).

211. K. Seiter, E. J. Feldman, C. Sreekantaiah, M. Pozzuoli, J. Weisberger, D. Liu, C. Papageorgio, M. Weiss, R. Kancherla, and T. Ahmed, *Leukemia*, **15**, 963 (2001).

212. A. P. Krapcho, M. E. Petry, Z. Getahun, J. J. Landi, J. Stallman, J. F. Polsenberg, C. E. Gallagher, M. J. Maresch, M. P. Hacker, F. C. Giuliani, G. Beggiolin, G. Pezzioni, E. Menta, C. Manzotti, A. Oliva, S. Spinelli, and S. Tognnella, *J. Med. Chem.*, **37**, 828 (1994).

213. P. De Isabella, M. Palumbo, C. Sissi, G. Capranico, N. Carenini, E. Menta, A. Oliva, S. Spinelli, A. P. Krapcho, F. C. Giuliani, and F. Zunino, *Mol. Pharmacol.*, **48**, 30 (1995).

214. E. B. Skibo, *Curr. Opin. Oncol. Endocr. Metab. Invest. Drugs*, **2**, 332 (2000).

215. L. K. Dawson, D. I. Jodrell, A. Bowman, R. Rye, B. Byrne, A. Bernareggi, G. Camboni, and J. F. Smyth, *Eur. J. Cancer*, **36**, 2353 (2000).

216. H. D. Showalter, J. L. Johnson, J. M. Hoftiezer, W. R. Turner, L. M. Werbe, W. R. Leopold, J. L. Shillis, R. C. Jackson, and E. F. Elslager, *J. Med. Chem.*, **30**, 121 (1987).

217. J. A. Hartley, K. Reszka, E. T. Zuo, W. D. Wilson, A. Morgan, and J. W. Lown, *Mol. Pharmacol.*, **33**, 265 (1988).

218. W. R. Leopold, J. M. Nelson, J. Plowman, and R. C. Jackson, *Cancer Res.*, **45**, 5532 (1985).

219. H. D. Showalter, D. W. Fry, W. R. Leopold, J. W. Lown, J. A. Plambeck, and K. Reszka, *Anti-Cancer Drug Design*, **1**, 73 (1986).

220. K. Belanger, J. Jolivet, J. Maroun, D. Stewart, A. Grillo-Lopez, L. Whitfield, N. Wainman, and E. Eisenhauer, *Inv. New Drugs*, **11**, 301 (1993).

221. M. M. Ames, C. L. Loprinzi, J. M. Collins, C. van Haelst-Pisani, R. L. Richardson, J. Rubin, and C. G. Moertel, *Cancer Res.*, **50**, 3905 (1990).

222. K. S. Albain, P. Y. Liu, A. Hantel, E. A. Poplin, R. V. O'Toole, J. L. Wade, A. M. Maddox, and D. S. Alberts, *Gyn. Oncol.*, **57**, 407 (1995).

223. M. M. Zalupski, J. Benedetti, S. P. Balcerzak, L. F. Hutchins, R. J. Belt, A. Hantel, and J. W. Goodwin, *Inv. New Drugs*, **11**, 337 (1993).

224. F. Leteurtre, G. Kohlhagen, K. D. Paull, and Y. Pommier, *J. Natl. Cancer Inst.*, **86**, 1239 (1994).

225. S. G. Diab, S. D. Baker, A. Joshi, H. A. Burris, P. W. Cobb, M. A. Villalona-Calero, S. G. Eckhardt, G. R. Weiss, G. I. Rodriguez, R. Drengler, M. Kraynak, L. Hammond, M. Finizio, D. D. Von Hoff, and E. K. Rowinsky, *Clin. Cancer Res.*, **5**, 299 (1999).

226. J. Blanz, U. Renner, K. Schmeer, G. Ehninger, and K. P. Zeller, *Drug Met. Disp.*, **21**, 955 (1993).

227. A. S. Joshi, H. J. Pieniaszek, E. E. Vokes, N. J. Vogelzang, A. F. Davidson, L. E. Richards, M. F. Chai, M. Finizio, and M. J. Ratain, *Drug Met. Disp.*, **29**, 96 (2001).

228. S. D. Huan, R. B. Natale, D. J. Stewart, G. P. Sartiano, P. J. Stella, J. D. Roberts, A. L. Symes, and M. Finizio, *Clin. Cancer Res.*, **6**, 1333 (2000).

229. W. M. Cholody, S. Martelli, J. Paradziej-Lukowicz, and J. Konopa, *J. Med. Chem.*, **33**, 49 (1990).

230. A. M. Burger, T. C. Jenkins, J. A. Double, and M. C. Bibby, *Br. J. Cancer*, **81**, 367 (1999).

231. A. Skladanowski, S. Y. Plisov, J. Konopa, and A. K. Larsen, *Mol. Pharmacol.*, **49**, 772 (1996).

232. C. R. Calabrese, P. M. Loadman, L. S. E. Lim, M. C. Bibby, J. A. Double, J. E. Brown, and J. H. Lamb, *Drug Metab. Disp.*, **27**, 240 (1999).

233. J. A. Holden, D. H. Rolfson, and C. T. Wittwer, *Biochemistry*, **29**, 2127 (1990).

234. I. Husain, J. L. Mohler, H. F. Seigler, and J. M. Easterman, *Cancer Res.*, **54**, 539 (1994).

235. B. Giovanella, J. S. Stehlin, M. E. Wall, M. C. Wani, A. W. Nicholas, L. F. Liu, R. Silber, and M. Potmesil, *Science*, **246**, 1946 (1989).

236. M. M. S. Heck, W. N. Hittelman, and W. C. Earnshaw, *Proc. Natl. Acad. Sci. USA*, **85**, 1086 (1988).

237. K. B. Tan, M. R. Mattern, W. K. Eng, F. L. McCabe, and R. K. Johnson, *J. Natl. Cancer Inst.*, **81**, 1732 (1989).

238. C. M. Whitacre, E. Zborowska, N. H. Gordon, W. MacKay, and N. A. Berger, *Cancer Res.*, **57**, 1425 (1997).

239. W. A. Denny, *Exp. Opin. Inv. Drugs*, **6**, 1845 (1997).

240. Y. Yamashita, S. Kawada, N. Fuji, and H. Nakano, *Biochemistry*, **30**, 5838 (1991).

241. L. K. Wang, R. K. Johnson, and S. M. Hecht, *Chem. Res. Tox.*, **6**, 813 (1993).

242. F. Kanzawa, K. Nishio, T. Ishida, M. Fukuda, H. Kurokawa, H. Fukumoto, Y. Nomoto, K. Fukuoka, K. Bojanowski, and N. Saijo, *Br. J. Cancer*, **76**, 571 (1997).

243. G. J. Atwell, B. F. Cain, B. C. Baguley, G. J. Finlay, and W. A. Denny, *J. Med. Chem.*, **27**, 1481 (1984).

244. W. A. Denny, G. J. Atwell, G. W. Rewcastle, and B. C. Baguley, *J. Med. Chem.*, **30**, 658 (1987).

245. G. J. Finlay, J. F. Riou, and B. C. Baguley, *Eur. J. Cancer*, **32A**, 708 (1996).

246. G. J. Finlay, E. S. Marshall, J. H. L. Matthews, K. D. Paull, and B. C. Baguley, *Cancer Chemother. Pharmacol.*, **31**, 401 (1993).

247. G. J. Finlay and B. C. Baguley, *Eur. J. Cancer Clin. Oncol.*, **25**, 271 (1989).

248. J. F. Riou, P. Fosse, C. H. Nguyen, A. K. Larsen, M. C. Bissery, L. Grondard, J. M. Saucier, E. Bisagni, and F. Lavelle, *Cancer Res.*, **53**, 5987 (1993).

249. B. Poddevin, J. F. Riou, F. Lavelle, and Y. Pommier, *Mol. Pharmacol.*, **44**, 767 (1993).

250. J. R. Eckhardt, H. A. Burris, J. G. Kuhn, M. C. Bissery, M. Klink-Alaki, G. M. Clark, and D. D. von Hoff, *J. Natl. Cancer Inst.*, **86**, 30 (1994).

251. M. C. Bissery, C. H. Nguyèn, E. Bisagni, P. Vrignaud, and F. Lavelle, *Inv. New Drugs*, **11**, 263 (1993).

252. R. van Gijn, W. W. ten Bokkel Huinink, S. Rodenhuis, J. B. Vermorken, O. van Tellingen, H. Rosing, L. J. van Warmerdam, and J. H. Beijnen, *Anti-Cancer Drugs*, **10**, 17 (1999).

253. R. A. Newman, J. Kim, B. M. Newman, R. Bruno, M. Bayssas, M. Klink-Alaki, and R. Pazdur, *Anti-Cancer Drugs*, **10**, 889 (1999).

254. J. M. Fortune, L. Velea, D. E. Graves, T. Utsugi, Y. Yamada, and N. Osheroff, *Biochemistry*, **38**, 15580 (1999).

255. T. Utsugi, K. Aoyagi, T. Asao, S. Okazaki, Y. Aoyagi, M. Sano, K. Wierzba, and Y. Yamada, *Jpn. J. Cancer Res.*, **88**, 992 (1997).

256. J. A. Byl, J. M. Fortune, D. A. Burden, J. L. Nitiss, T. Utsugi, Y. Yamada, and N. Osheroff, *Biochemistry*, **38**, 15573 (1999).

257. Y. Aoyagi, T. Kobunai, T. Utsugi, T. Oh-hara, and Y. Yamada, *Jpn. J. Cancer Res.*, **90**, 578 (1999).

258. R. B. Ewesuedo, L. Iyer, S. Das, A. Koenig, S. Mani, N. J. Vogelzang, R. L. Schilsky, W. Brenckman, and M. J. Ratain, *J. Clin. Oncol.*, **19**, 2084 (2001).

259. I. Nabiev, I. Chourpa, J. F. Riou, C. H. Nguyen, F. Lavelle, and M. Manfait, *Biochemistry*, **33**, 9013 (1994).

260. I. G. C. Robertson, B. D. Palmer, M. Officer, D. J. Siegers, J. W. Paxton, and G. J. Shaw, *Biochem. Pharmacol.*, **42**, 1879 (1991).

261. S. M. H. Evans, I. G. C. Robertson, and J. W. Paxton, *J. Pharm. Pharmacol.*, **46**, 63 (1994).

262. P. C. Schofield, I. G. C. Robertson, J. W. Paxton, M. R. McCrystal, B. D. Evans, P. Kestell, and B. C. Baguley, *Cancer Chemotherap. Pharmacol.*, **44**, 51 (1999).

263. M. R. McCrystal, B. D. Evans, V. J. Harvey, P. L. Thompson, D. J. Porter, and B. C. Baguley, *Cancer Chemotherap. Pharmacol.*, **44**, 39 (1999).

264. D. Abigerges, J. P. Armand, G. G. Chabot, R. Bruno, M. C. Bissery, M. Bayssas, M. Klink-Alaki, M. Clavel, and G. Catimel, *Anti-Cancer Drugs*, **7**, 166 (1996).

265. L. P. G. Wakelin, *Med. Res. Rev.*, **6**, 276 (1986).

266. A. Goldin, J. M. Vendetti, J. S. McDonald, F. M. Muggia, J. E. Henney, and V. T. DeVita, *Eur. J. Cancer*, **17**, 129 (1981).

267. E. Segal-Bendirjian, D. Coulaud, B. P. Roques, and J. B. Le Pecq, *Cancer Res.*, **48**, 4982 (1988).

268. R. J. McRipley, P. E. Burns-Horwitz, P. M. Czerniak, R. J. Diamond, M. A. Diamond, J. L. D. Miller, F. J. Page, D. L. Dexter, and S. F. Chen, *Cancer Res.*, **54**, 159 (1994).

269. W. M. Cholody, L. Hernandez, L. Hassner, D. A. Scuderio, D. B. Djurickovic, and C. J. Michejda, *J. Med. Chem.*, **38**, 3043 (1995).

270. S. A. Gamage, J. A. Spicer, G. J. Atwell, G. J. Finlay, B. C. Baguley, and W. A. Denny, *J. Med. Chem.*, **42**, 2383 (1999).

271. J. A. Spicer, S. A. Gamage, G. W. Rewcastle, G. J. Finlay, D. J. A. Bridewell, B. C. Baguley, and W. A. Denny, *J. Med. Chem.*, **43**, 1350 (2000).

272. L. W. Deady, J. Desneves, A. J. Kaye, M. Thompson, M. Finlay, B. C. Baguley, and W. A. Denny, *Bioorg. Med. Chem.*, **7**, 2801 (1999).

273. M. F. Brana, J. M. Castellano, M. Moran, M. J. Perez de Vega, D. Perron, D. Conlon, P. F. Bousquet, C. A. Romerdahl, and S. P. Robinson, *Anti-Cancer Drug Design*, **11**, 297 (1996).

274. S. A. Gamage, J. A. Spicer, G. J. Finlay, A. J. Stewart, P. Charlton, B. C. Baguley, and W. A. Denny, *J. Med. Chem.*, **44**, 1407 (2001).

275. A. J. Stewart, P. Mistry, W. Dangerfield, D. Bootle, M. Baker, B. Kofler, S. Okiji, B. C. Baguley, W. A. Denny, and P. Charlton, *Anticancer Drugs*, **12**, 359 (2001).

276. A. G. Nair, In: D. E. V. Wilman, Ed., *Chemistry of Antitumour Agents*, Blackie, London, 1989, 202.

277. W. A. Bleyer, *Cancer*, **41**, 36 (1978).

278. F. M. Huennekens, *Adv. Enz. Reg.*, **34**, 397 (1994).

279. F. M. Sirotnak, J. I. DeGraw, D. M. Moccio, L. L. Samuels, and L. J. Goutas, *Cancer Chemotherap. Pharmacol.*, **12**, 18 (1984).

280. F. M. Sirotnak, J. I. DeGraw, F. A. Schmid, L. J. Goutas, and D. M. Moccio, *Cancer Chemotherap. Pharmacol.*, **12**, 26 (1984).

281. D. H. Brown, B. J. Braakhuis, G. A. van Dongen, M. van Walsum, M. Bagnay, and G. B. Snow, *Anticancer Res.*, **9**, 1549 (1989).

282. B. F. van der Laan, G. Jansen, I. Kathmann, J. H. Schornagel, and G. J. Hordijk, *Eur. J. Cancer*, **27**, 1274 (1991).

283. R. C. Jackson and D. Niethammer, *Eur. J. Cancer*, **13**, 567 (1977).

284. T. R. Jones, M. J. Smithers, M. A. Taylor, A. L. Jackman, A. H. Calvert, S. J. Harland, and K. R. Harrap, *J. Med. Chem.*, **29**, 468 (1986).

285. R. C. Jackson, A. L. Jackman, and A. H. Calvert, *Biochem. Pharmacol.*, **32**, 3783 (1983).

286. E. Sikora, A. L. Jackson, D. R. Newell, and A. H. Calvert, *Biochem. Pharmacol.*, **37**, 4047 (1988).

287. E. M. Berman and L. M. Werbel, *J. Med. Chem.*, **34**, 479 (1991).

288. A. L. Jackman and A. H. Calvert, *Ann. Oncol.*, **6**, 871 (1995).

289. I. R. Judson, *Anti-Cancer Drugs*, **8**(Suppl 2), 5 (1997).

290. N. S. Gunasekara and D. Faulds, *Drugs*, **55**, 423 (1998).

291. P. Comella, F. De Vita, S. Mancarella, L. De Lucia, M. Biglietto, R. Casaretti, A. Farris, G. P. Ianniello, V. LoRusso, A. Avallone, G. Carteni, S. S. Leo, G. Catalano, M. De Lena, and G. Comella, *Ann. Oncol.*, **11**, 1323 (2000).

292. R. M. Schultz, V. F. Patel, J. F. Worzalla, and C. Shih, *Anticancer Res.*, **19**, 437 (1999).

293. L. G. Mendelsohn, C. Shih, V. J. Chen, L. L. Habeck, S. B. Gates, and K. A. Shackelford, *Semin. Oncol.*, **26**(Suppl 6), 42 (1999).

294. C. D. Britten, E. Izbicka, S. Hilsenbeck, R. Lawrence, K. Davidson, C. Cerna, L. Gomez,

E. K. Rowinsky, S. Weitman, and D. D. Von Hoff, *Cancer Chemotherap. Pharmacol.*, **44**, 105 (1999).

295. P. J. O'Dwyer, K. Nelson, and D. E. Thornton, *Semin. Oncol.*, **26**(Suppl 6), 99 (1999).

296. A. A. Adjei, *Ann. Oncol.*, **11**, 1335 (2000).

297. P. Kumar, L. Kisliuk, Y. Guamont, M. G. Nair, C. M. Baugh, and B. T. Kaufmann, *Cancer Res.*, **46**, 5020 (1986).

298. P. J. O'Dwyer, D. D. Shoemaker, J. Plowman, J. Cradock, A. Grillo-Lopez, and B. Leyland-Jones, *Inv. New Drugs*, **3**, 71 (1985).

299. W. D. Klohs, R. W. Steinkampf, J. A. Besserer, and D. W. Fry, *Cancer Lett.*, **31**, 256 (1986).

300. D. G. Haller, *Semin. Oncol.*, **24**(Suppl 18), 71 (1997).

301. C. D. Blanke, M. Messenger, and S. C. Taplin, *Semin. Oncol.*, **24**(Suppl 18), 57 (1997).

302. D. S. Duch, M. P. Edelstein, S. W. Bowers, and C. A. Nichol, *Cancer Res.*, **42**, 3987 (1982).

303. R. de Wit, S. B. Kaye, J. T. Roberts, G. Stoter, J. Scott, and J. Verweij, *Br. J. Cancer*, **67**, 388 (1993).

304. M. Khorsand, J. Lange, L. Feun, N. J. Clendeninn, M. Collier, and G. Wilding, *Inv. New Drugs*, **15**, 157 (1997).

305. W. Perkins, R. E. Williams, J. P. Vestey, M. J. Tidman, A. M. Layton, W. J. Cunliffe, E. M. Saihan, M. R. Klaber, V. K. Manna, and H. Baker, *Br. J. Dermatol.*, **129**, 584 (1993).

306. S. Webber, C. Bartlett, T. J. Boritzki, J. A. Hilliard, E. F. Howland, A. L. Johnston, M. Kosa, S. A. Margosiak, C. A. Morse, and B. V. Shetty, *Cancer Chemotherap. Pharmacol.*, **37**, 509 (1996).

307. M. S. Rhee, S. Webber, and J. Galivan, *Cell. Pharmacol.*, **2**, 97 (1995).

308. E. Raymond, S. Djelloul, C. Buquet-Fagot, J. Mester, and C. Gespach, *Anti-Cancer Drugs*, **7**, 752 (1996).

309. K. Stuart, J. Tessitore, J. Rudy, N. Clendennin, and A. Johnston, *Cancer*, **86**, 410 (1999).

310. A. N. Hughes, M. J. Griffin, D. R. Newell, A. H. Calvert, A. Johnston, B. Kerr, C. Lee, B. Liang, and A. V. Boddy, *Br. J. Cancer*, **82**, 1519 (2000).

311. W. B. Parker and Y. C. Cheng, *Pharmacol. Therap.*, **48**, 381 (1990).

312. A. D. Broom, *J. Med. Chem.*, **32**, 2 (1989).

313. M. G. Pavallicini, *Pharmacol. Therap.*, **25**, 207 (1984).

314. S. S. Cohen, *Med. Biol.*, **54**, 299 (1976).

315. W. Plunkett, P. Huang, and V. Gandhi, *Anti-Cancer Drugs*, **6**(Suppl 6), 7 (1995).

316. L. Y. Yang, L. Li, H. Jiang, Y. Shen, and W. Plunkett, *Clin. Cancer Res.*, **6**, 773 (2000).

317. W. A. van der Donk, G. Yu, I. Perez, R. J. Sanchez, J. Stubbe, V. Samano, and M. J. Robins, *Biochemistry*, **37**, 6419 (1998).

318. Y. G. Goan, B. Zhou, E. Hu, S. Mi, and Y. Yen, *Cancer Res.*, **59**, 4204 (1999).

319. T. S. Lawrence, E. Y. Chang, T. M. Hahn, L. W. Hertel, and D. S. Shewach, *Int. J. Radiat. Oncol. Biol. Phys.*, **34**, 867 (1996).

320. T. S. Lawrence, M. A. Davis, A. Hough, and A. Rehemtulla, *Clin. Cancer Res.*, **7**, 314 (2001).

321. D. H. W. Ho and E. Frei, *Clin. Pharmacol. Therap.*, **12**, 944 (1971).

322. M. MacCoss and M. J. Robins, In: D. E. V. Wilman, Ed. *Chemistry of Antitumour Agents*, Blackie, London, 1989, 261.

323. G. Cartei, C. Sacco, A. Sibau, N. Pella, A. Iop, and G. Tabaro, *Ann. Oncol.*, **10**(Suppl 5), 57 (1999).

324. G. V. Scagliotti, *Oncology*, **14**(Suppl 4), 15 (2000).

325. S. Culine, *Anti-Cancer Drugs*, **11**(Suppl 1), 9 (2000).

326. J. Carmichael, K. Possinger, P. Phillip, M. Beykirch, H. Kerr, J. Walling, and A. L. Harris, *J. Clin. Oncol.*, **13**, 2731 (1995).

327. H. von der Maase, *Eur. J. Cancer*, **36**(Suppl 2), 13 (2000).

328. M. Sasvari-Szekely, T. Spasokoukotskaja, M. Szoke, Z. Csapo, A. Turi, I. Szanto, S. Eriksson, and M. Staub, *Biochem. Pharmacol.*, **56**, 1175 (1998).

329. K. Fabianowska-Majewska, K. Tybor, J. Duley, and A. Simmonds, *Biochem. Pharmacol.*, **50**, 1379 (1995).

330. Y. Nomura, O. Inanami, K. Takahashi, A. Matsuda, and M. Kuwabara, *Leukemia*, **14**, 299 (2000).

331. M. F. E. Siaw and M. S. Coleman, *J. Biol. Chem.*, **259**, 9426 (1984).

332. R. P. Agarwal, *Pharmacol. Therap.*, **17**, 399 (1982).

333. C. Tondini, M. Balzarotti, I. Rampinelli, P. Valagussa, M. Luoni, A. De Paoli, A. Santoro, and G. Bonadonna, *Ann. Oncol.*, **11**, 231 (2000).

334. B. D. Cheson, J. M. Sorensen, D. A. Vena, M. J. Montello, J. A. Barrett, E. Damasio, M. Tallman, L. Annino, J. Connors, B. Coiffier, and F. Lauria, *J. Clin. Oncol.*, **16**, 3007 (1998).

335. A. Saven, S. Emanuele, M. Kosty, J. Koziol, D. Ellison, and L. D. Piro, *Blood*, **86**, 1710 (1995).

336. A. Saven and L. D. Piro, *Ann. Int. Med.*, **120**, 784 (1994).

337. A. Saven, R. H. Lemon, M. Kosty, E. Beutler, and I. D. Piro, *J. Clin. Oncol.*, **13**, 570 (1995).

338. T. M. Kuzel, A. Hurria, E. Samuelson, M. S. Tallman, H. H. Roenigk, A. W. Rademaker, and S. T. Rosen, *Blood*, **87**, 906 (1996).

339. J. M. Brown, *Cancer Res.*, **59**, 5863 (1999).

340. C. N. Coleman, *J. Natl. Cancer Inst.*, **80**, 310 (1988).

341. J. W. Baish, P. A. Netti, and R. K. Jain, *Microvascular Res.*, **53**, 128 (1997).

342. J. M. Brown and A. J. Giaccia, *Cancer Res.*, **58**, 1405 (1998).

343. V. L. Narayanan and W. W. Lee, *Adv. Pharmacol. Chemotherap.*, **19**, 155 (1982).

344. J. M. Brown, *Int. J. Radiat. Oncol. Biol. Phys.*, **8**, 675 (1982).

345. A. V. Patterson, M. P. Saunders, E. C. Chinje, L. H. Patterson, and I. J. Stratford, *Anti-Cancer Drug Design*, **13**, 541 (1998).

346. W. A. Denny, W. R. Wilson, and M. P. Hay, *Br. J. Cancer*, **74**(Suppl 27), 32 (1996).

347. F. Seng, K. Ley, B. Hamburger, and F. Bechlars, *Ger. Pat.* DE 2255825 (1974); *Chem. Abstr.*, **81**, 105578 (1974).

348. W. A. Denny and W. R. Wilson, *Exp. Opin. Inv. Drugs*, **9**, 2889 (2000).

349. W. R. Wilson, P. van Zijl, and W. A. Denny, *Int. J. Radiat. Oncol. Biol. Phys.*, **22**, 693 (1992).

350. W. R. Wilson, W. A. Denny, S. J. Twigden, B. C. Baguley, and J. C. Probert, *Br. J. Cancer*, **49**, 215 (1984).

351. W. R. Wilson, W. A. Denny, G. M. Stewart, A. Fenn, and J. C. Probert, *Int. J. Radiat. Oncol. Biol. Phys.*, **12**, 1235 (1986).

352. P. J. Smith, N. J. Blunt, R. Desnoyers, Y. Giles, and L. H. Patterson, *Cancer Chemotherap. Pharmacol.*, **39**, 455 (1997).

353. S. Rockwell, A. C. Sartorelli, M. Tomasz, and K. A. Kennedy, *Cancer Met. Rev.*, **12**, 165 (1993).

354. M. F. Belcourt, W. F. Hodnick, S. Rockwell, and A. C. Sartorelli, *Adv. Enz. Reg.*, **38**, 111 (1998).

355. B. G. Haffty, Y. H. Son, L. D. Wilson, R. Papac, D. Fischer, S. Rockwell, A. C. Sartorelli, D. Ross, C. T. Sasaki, and J. J. Fischer, *Radiat. Oncol. Invest.*, **5**, 235 (1997).

356. B. Ganley, G. Chowdhury, J. Bhancali, J. R. Daniels, and K. S. Gates, *Bioorg. Med. Chem.*, **9**, 2395–2401 (2001).

357. J. T. Hwang, M. M. Greenberg, T. Fuchs, and K. S. Gates, *Biochemistry*, **38**, 14248 (1999).

358. M. L. Walton and P. Workman, *Biochem. Pharmacol.*, **39**, 1735 (1990).

359. A. P. Garner, M. J. Paine, L. Rodriguez-Crespo, E. C. Chinje, D. E. Ortiz, P. Montellano, I. J. Stratford, D. G. Tew, and C. R. Wolf, *Cancer Res.*, **59**, 1929 (1999).

360. S. M. Raleigh, E. Wanogho, M. D. Burke, S. R. McKeown, and L. H. Patterson, *Int. J. Radiat. Oncol. Biol. Phys.*, **42**, 763 (1988).

361. L. H. Patterson, *Cancer Met. Rev.*, **12**, 119 (1993).

362. L. H. Patterson, S. R. McKeown, K. Ruparelia, J. A. Double, M. C. Bibby, S. Cole, and I. J. Stratford, *Br. J. Cancer*, **82**, 1984 (2000).

363. O. P. Friery, R. Gallagher, M. M. Murray, C. M. Hughes, E. S. Galligan, I. A. McIntyre, L. H. Patterson, D. G. Hirst, and S. R. McKeown, *Br. J. Cancer*, **82**, 1469 (2000).

364. W. R. Wilson, W. A. Denny, S. M. Pullen, K. M. Thompson, A. E. Li, L. H. Patterson, and H. H. Lee, *Br. J. Cancer*, **74**(Suppl 27), 43 (1996).

365. B. G. Siim, K. O. Hicks, S. M. Pullen, P. L. van Zijl, W. A. Denny, and W. R. Wilson, *Biochem. Pharmacol.*, **60**, 969 (2000).

366. H. H. Lee, W. R. Wilson, D. M. Ferry, P. van Zijl, S. M. Pullen, and W. A. Denny, *J. Med. Chem.*, **39**, 2508 (1996).

367. H. H. Lee, W. R. Wilson, and W. A. Denny, *Anti-Cancer Drug Design*, **14**, 487 (1999).

368. M. F. Belcourt, W. F. Hodnick, S. Rockwell, and A. C. Sartorelli, *Proc. Natl. Acad. Sci. USA*, **93**, 456 (1996).

369. M. Tomasz and Y. Palom, *Pharmacol. Therap.*, **76**, 73 (1997).

370. H. D. Beall and S. L. Winski, *Front. Biosci.*, **5**, D639 (2000).

371. K. O. Hicks, Y. Fleming, B. G. Siim, C. J. Koch, and W. R. Wilson, *Int. J. Radiat. Oncol. Biol. Phys.*, **42**, 641 (1998).

372. E. Lartigau and M. Guichard, *Int. J. Radiat. Biol.*, **67**, 211 (1995).

373. C. J. Koch, *Cancer Res.*, **53**, 3992 (1993).

374. E. M. Zeeman, V. K. Hirst, M. J. Lemmon, and J. M. Brown, *Radiother. Oncol.*, **12**, 209 (1988).

375. M. J. Dorie, D. Menke, and J. M. Brown, *Int. J. Radiat. Oncol. Biol. Phys.*, **28**, 145 (1994).

376. M. S. Kovacs, D. J. Hocking, J. W. Evans, B. G. Siim, B. G. Wouters, and J. M. Brown, *Br. J. Cancer*, **80**, 1245 (1999).

377. V. K. Langmuir, J. A. Rooker, M. Osen, H. L. Mendonca, and K. R. Laderoute, *Cancer Res.*, **54**, 2845 (1994).

378. E. Lartigau and M. Guichard, *Br. J. Cancer*, **73**, 1480 (1996).

379. M. J. Dorie and J. M. Brown, *Cancer Chemotherap. Pharmacol.*, **39**, 361 (1997).

380. S. Cliffe, M. L. Taylor, M. Rutland, B. C. Baguley, R. P. Hill, and W. R. Wilson, *Int. J. Radiat. Oncol. Biol. Phys.*, **29**, 373 (1994).

381. J. Del Rowe, C. Scott, M. Werner-Wasik, J. P. Bahary, W. J. Curran, R. C. Urtasun, and B. Fisher, *J. Clin. Oncol.*, **18**, 1254 (2000).

382. D. J. Lee, A. Trotti, S. Spencer, R. Rostock, C. Fisher, R. von Roemeling, E. Harvey, and E. Groves, *Int. J. Radiat. Oncol. Biol. Phys.*, **42**, 811 (1998).

383. C. Aghajanian, C. Brown, C. O'Flaherty, A. Fleischauer, L. Curtin, R. Roemeling, and D. R. Spriggs, *Gyn. Oncol.*, **67**, 127 (1997).

384. A. Y. Bedikian, S. S. Legha, O. Eton, A. C. Buzaid, N. Papadopoulos, S. Coates, T. Simmons, J. Neefe, and R. von Roemeling, *Ann. Oncol.*, **8**, 363 (1997).

385. A. Y. Bedikian, S. S. Legha, O. Eton, A. C. Buzaid, N. Papadopoulos, C. Plager, S. McIntyre, and J. Viallet, *Anti-Cancer Drugs*, **10**, 735 (1999).

386. J. Von Pawel, R. von Roemeling, U. Gatzemeier, M. Boyer, I. O. Elisson, P. Clark, D. Talbot, A. Rey, T. W. Butler, V. Hirsh, I. Olver, B. Bergman, J. Ayoub, G. Richardson, D. Dunlop, A. Arcenas, R. Vescio, J. Viallet, and J. Treat, *J. Clin. Oncol.*, **18**, 1351 (2000).

387. V. A. Miller, K. K. Ng, S. C. Grant, H. Kindler, B. Pizzo, R. T. Heelan, R. von Roemeling, and M. G. Kris, *Ann. Oncol.*, **8**, 1269 (1997).

388. A. E. Lee and W. R. Wilson, *Toxicol. Appl. Pharmacol.*, **163**, 50 (2000).

389. A. Monge, F. J. Martinez-Crespo, A. Lopez de Cerain, J. A. Palop, S. Narro, V. Senador, A. Marin, Y. Sainz, M. Gonzalez, and E. J. Hamilton, *J. Med. Chem.*, **38**, 4488 (1995).

390. M. C. Green, J. L. Murray, and G. N. Hortobagyi, *Cancer Treat. Rev.*, **26**, 269 (2000).

391. G. M. Dubowchik and M. A. Walker, *Pharm. Therap.*, **83**, 67 (1999).

392. M. P. Deonarain and A. A. Epenetos, *Br. J. Cancer*, **70**, 786 (1994).

393. I. Niculescu-Duvaz and C. J. Springer, *Adv. Drug Del. Rev.*, **26**, 151 (1997).

394. M. P. Hay and W. A. Denny, *Drugs Future*, **12**, 917 (1996).

395. P. Senter, M. Saulnier, G. Schreiber, D. Hirschberg, J. Brown, L. Hellstrom, and K. Hellstrom, *Proc. Natl. Acad. Sci. USA*, **85**, 4842 (1988).

396. P. D. Senter, G. J. Schreiber, D. L. Hirschberg, S. A. Ashe, C. E. Hellstrom, and I. Hellstrom, *Cancer Res.*, **49**, 5789 (1989).

397. J. Griggs, R. Hesketh, G. A. Smith, K. M. Brindle, J. C. Metcalfe, G. A. Thomas, and E. D. Williams, *Br. J. Cancer*, **84**, 832 (2001).

398. I. Niculescu-Duvaz, F. Friedlos, D. Niculescu-Duvaz, L. Davies, and C. J. Springer, *Anti-Cancer Drug Design*, **14**, 517 (1999).

399. K. D. Bagshawe, S. K. Sharma, C. J. Springer, and G. Rogers, *Anal. Oncol.*, **5**, 879 (1994).

400. N. R. Monks, J. A. Calvete, N. J. Curtin, D. C. Blakey, S. J. East, and D. R. Newell, *Br. J. Cancer*, **83**, 267 (2000).

401. T. P. Smyth, M. E. O'Donnell, M. J. O'Connor, and J. O. St. Ledger, *Tetrahedron*, **56**, 5699 (2000).

402. D. E. Kerr, G. J. Schreiber, V. M. Vrudhula, H. P. Svensson, I. Hellstrom, K. E. Hellstrom, and P. D. Senter, *Cancer Res.*, **55**, 3558 (1995).

403. H. P. Svensson, V. M. Vrudhula, J. E. Emswiler, J. F. MacMaster, W. L. Cosand, P. D. Senter, and P. M. Wallace, *Cancer Res.*, **55**, 2357 (1995).

404. S. M. Wang, J. W. Chern, M. Y. Yeh, J. C. Ng, E. Tung, and S. R. Roffler, *Cancer Res.*, **52**, 4484 (1992).

405. R. G. G. Leenders, E. W. P. Damen, E. J. A. Bijsterveld, P. H. J. Houba, I. H. van der Meulen-Muilman, E. Boven, and H. J. Haisma, *Bioorg. Med. Chem.*, **7**, 1597 (1999).

406. H. J. Haisma, E. Boven, M. van Muijen, L. De Jong, W. J. Van der Vigh, and H. M. Pinedo, *Br. J. Cancer*, **66**, 474 (1992).

407. H. J. Haisma, M. Van Muijen, H. M. Pinedo, and E. Boven, *Cell Biophys.*, **24/25**, 185 (1994).

408. P. H. J. Houba, E. Boven, I. H. Van Der Meulen-Muileman, R. G. G. Leenders, J. W. Scheeren, H. M. Pinedo, and H. J. Haisma, *Int. J. Cancer*, **91**, 550 (2001).

409. W. F. Anderson, *Nature Med.*, **6**, 862 (2000).

410. O. Wildner, *Ann. Med.*, **31**, 421 (1999).

411. H. Ishii-Morita, R. Agbaria, C. A. Mullen, H. Hirano, D. A. Koeplin, Z. Ram, E. H. Oldfield, D. G. Johns, and R. M. Blaese, *Gene Therap.*, **4**, 244 (1997).

412. M. Mesnil and H. Yamasaki, *Cancer Res.*, **60**, 3989 (2000).

413. E. Kievit, E. Bershad, E. Ng, P. Sethna, I. Dev, T. S. Lawrence, and A. Rehemtulla, *Cancer Res.*, **59**, 1417 (1999).

414. B. Huber, E. Austin, C. Richards, S. Davis, and S. Good, *Proc. Natl. Acad. Sci. USA*, **91**, 8302 (1994).

415. C. G. Leichman, *Oncology*, **13**, 26 (1999).

416. Y. Jounaidi, J. E. Hecht, and D. J. Waxman, *Cancer Res.*, **58**, 4391 (1998).

417. G. M. Anlezark, R. G. Melton, R. F. Sherwood, B. Coles, F. Friedlos, and R. J. Knox, *Biochem. Pharmacol.*, **44**, 2289 (1992).

418. J. I. Grove, P. F. Searle, S. J. Weedon, N. K. Green, L. A. McNeish, and D. J. Kerr, *Anti-Cancer Drug Design*, **14**, 461 (1999).

419. C. Palmer, G. Chung-Faye, R. Barton, D. Ferry, J. Baddeley, D. Anderson, L. Seymour, and D. J. Kerr, *Br. J. Cancer*, **83**(Suppl 1), 71 (2000).

420. N. P. Barbour, M. Paborji, T. C. Alexander, W. P. Coppola, and J. B. Bogardus, *Pharm. Res.*, **12**, 215 (1995).

421. H. O. Sjogren, M. Isaksson, D. Willner, I. Hellstrom, K. E. Hellstrom, and P. A. Trail, *Cancer Res.*, **57**, 4530 (1997).

422. M. N. Saleh, S. Sugarman, J. Murray, J. B. Ostroff, D. Healey, D. Jones, C. R. Daniel, D. LeBherz, H. Brewer, N. Onetto, and A. F. LoBuglio, *J. Clin. Oncol.*, **18**, 2282 (2000).

423. L. M. Hinman, P. R. Hamann, R. Wallace, A. T. Menendez, F. E. Durr, and J. Upeslacis, *Cancer Res.*, **53**, 3336 (1993).

424. K. Knoll, W. Wrasidlo, J. E. Scherberich, G. Gaedicke, and P. Fischer, *Cancer Res.*, **60**, 6089 (2000).

425. M. M. Siegel, K. Tabei, A. Kunz, I. J. Hollander, P. R. Hamann, D. H. Bell, S. Berkenkamp, and F. Hillenkamp, *Anal. Chem.*, **69**, 2716 (1997).

426. J. S. Thorson, E. L. Sievers, J. Ahlert, E. Shepard, R. E. Whitwam, K. C. Onwueme, and M. Ruppen, *Curr. Pharm. Des.*, **6**, 1841 (2000).

427. E. L. Sievers, R. A. Larson, E. Estey, B. Lowenberg, H. Dombret, C. Karanes, M. Theobald, J. M. Bennett, M. L. Sherman, M. S. Berger, C. B. Eten, M. R. Loken, J. J. van Dongen, I. D. Bernstein, and F. R. Appelbaum, *J. Clin. Oncol.*, **19**, 3244 (2001).

428. P. F. Bross, J. Beitz, G. Chen, X. O. Chen, E. Duffy, L. Kieffer, S. Roy, R. Sridhara, A. Rahman, G. Williams, and R. Pazdur, *Clin. Cancer Res.*, **7**, 1490 (2001).

429. M. P. De Vetten, J. H. Jansen, B. A. van der Reijden, M. S. Berger, J. M. Zijlmans, and B. Lowenberg, *Br. J. Haematol.*, **111**, 277 (2000).

430. C. Liu and R. V. J. Chari, *Expert Opin. Invest. Drugs*, **6**, 169 (1997).

431. C. Liu, B. M. Tadayoni, L. A. Bourret, K. M. Mattocks, S. M. Derr, W. C. Widdison, N. L. Kedersha, P. D. Ariniello, V. Goldmacher, J. M. Lambert, W. A. Blattler, and R. V. Chari, *Proc. Natl. Acad. Sci. USA*, **93**, 8618 (1996).

432. T. Suzawa, S. Nagamura, H. Saito, S. Ohta, N. Hanai, and M. Yamasaki, *Bioorg. Med. Chem.*, **8**, 2175 (2000).

# CHAPTER THREE

# Antitumor Natural Products

LESTER A. MITSCHER
APURBA DUTTA
Department of Medicinal Chemistry
Kansas University
Lawrence, Kansas

## Contents

1 Introduction, 109
2 Drugs Attacking DNA, 111
   2.1 Dactinomycin (Cosmegen), 111
      2.1.1 Introduction, 111
      2.1.2 Medicinal Uses, 111
      2.1.3 Contraindications and Side Effects, 111
      2.1.4 Pharmacokinetic Features, 111
      2.1.5 Medicinal Chemical Transformations, 112
      2.1.6 Molecular Mode of Action, 113
      2.1.7 Biosynthesis, 113
   2.2 Bleomycin (Blenoxane), 115
      2.2.1 Introduction, 115
      2.2.2 Medicinal Uses, 115
      2.2.3 Contraindications and Side Effects, 115
      2.2.4 Pharmacokinetic Features, 116
      2.2.5 Medicinal Chemistry, 116
      2.2.6 Biosynthesis, 118
      2.2.7 Molecular Mode of Action and Resistance, 118
      2.2.8 Recent Developments and Things to Come, 120
   2.3 Mitomycin (Mutamycin), 120
      2.3.1 Introduction, 120
      2.3.2 Clinical Use, 120
      2.3.3 Contraindications and Side Effects, 120
      2.3.4 Pharmacokinetics, 121
      2.3.5 Medicinal Chemistry, 121
      2.3.6 Molecular Mode of Action and Resistance, 121
      2.3.7 Medicinal Chemistry, 122
   2.4 Plicamycin (Formerly Mithramycin; Mithracin), 122
      2.4.1 Introduction, 122
      2.4.2 Clinical Uses, 123
      2.4.3 Contraindications and Side Effects, 123
      2.4.4 Pharmacokinetics, 123
      2.4.5 Mode of Action and Resistance, 123

*Burger's Medicinal Chemistry and Drug Discovery*
Sixth Edition, Volume 5: Chemotherapeutic Agents
Edited by Donald J. Abraham
ISBN 0-471-37031-2 © 2003 John Wiley & Sons, Inc.

2.4.6 Medicinal Chemistry, 124
2.4.7 Biosynthesis, 124
3 Drugs Inhibiting Enzymes That Process DNA, 124
   3.1 Anthracyclines, 124
      3.1.1 Daunorubicin (Daunomycin; Cerubidine, Rubidomycin; 18), 126
         3.1.1.1 Therapeutic Uses, 126
         3.1.1.2 Side Effects and Contraindications, 126
         3.1.1.3 Pharmacokinetics, 126
         3.1.1.4 Mechanism of Action and Resistance, 127
      3.1.2 Doxorubicin, 128
         3.1.2.1 Therapeutic Uses, 128
         3.1.2.2 Side Effects and Contraindications, 128
         3.1.2.3 Pharmacokinetics, 128
         3.1.2.4 Molecular Mode of Action and Resistance, 128
      3.1.3 Epirubicin, 128
      3.1.4 Valrubicin, 128
         3.1.4.1 Biosynthesis, 129
         3.1.4.2 Medicinal Chemistry, 129
         3.1.4.3 Biosynthesis, 129
         3.1.4.4 Recent Developments and Things to Come, 130
   3.2 Camptothecins, 130
      3.2.1 Irinotecan (CPT-11), 131
         3.2.1.1 Clinical Uses, 131
      3.2.2 Topotecan, 131
         3.2.2.1 Clinical Uses, 131
         3.2.2.2 Contraindications and Side Effects, 132
         3.2.2.3 Pharmacokinetic Features, 132
         3.2.2.4 Molecular Mode of Action and Resistance, 132
         3.2.2.5 Medicinal Chemistry, 133
         3.2.2.6 Quantitative Structure-Activity Relationships (QSARs), 134
         3.2.2.7 Recent Developments and Things to Come, 134
   3.3 Isopodophyllotoxins, 134
      3.3.1 Etoposide, 134
         3.3.1.1 Therapeutic Uses, 134
         3.3.1.2 Side Effects and Contraindications, 135
         3.3.1.3 Pharmacokinetics, 135
         3.3.1.4 Mode of Action and Resistance, 135
         3.3.1.5 Medicinal Chemistry (290, 291), 135

3.3.2 Teniposide, 135
   3.3.2.1 Therapeutic Uses, 135
   3.3.2.2 Pharmacokinetics, 136
   3.3.2.3 Mode of Action and Resistance, 136
   3.3.2.4 Structure-Activity Relationships, 136
4 Drugs Interfering with Tubulin Polymerization/ Depolymerization, 136
   4.1 Taxus Diterpenes, 136
      4.1.1 Paclitaxel/Taxol, 136
         4.1.1.1 Clinical Uses, 137
         4.1.1.2 Side Effects and Contraindications, 137
         4.1.1.3 Pharmacokinetics, 138
         4.1.1.4 Molecular Mode of Action and Resistance, 138
      4.1.2 Docetaxel/Taxotere, 138
         4.1.2.1 Clinical Applications, 138
         4.1.2.2 Side Effects and Contraindications, 138
         4.1.2.3 Pharmacokinetics, 138
         4.1.2.4 Mode of Action and Resistance, 138
         4.1.2.5 Chemical Transformations, 138
         4.1.2.6 Biosynthesis, 139
         4.1.2.7 Things to Come, 139
   4.2 Dimeric Vinca Alkaloids, 139
      4.2.1 Vinblastine (Velban), 140
         4.2.1.1 Medicinal Uses, 140
         4.2.1.2 Side Effects and Contraindications, 141
         4.2.1.3 Pharmacokinetic Features, 141
         4.2.1.4 Medicinal Chemical Transformations, 141
         4.2.1.5 Molecular Mode of Action and Resistance, 141
      4.2.2 Vincristine (Oncovin, Vincasar PFS), 142
         4.2.2.1 Medical Uses, 142
         4.2.2.2 Pharmacokinetic Features, 142
         4.2.2.3 Side Effects and Contraindications, 142
         4.2.2.4 Resistance, 142
      4.2.3 Vinorelbine (Navelbine), 142
         4.2.3.1 Medicinal Uses, 142
         4.2.3.2 Pharmacokinetic Features, 142
         4.2.3.3 Side Effects and Contraindications, 142
         4.2.3.4 Things to Come, 142

# 1  INTRODUCTION

Cancer is now believed to be the number one cause of premature death in industrialized nations. The market for anticancer agents was estimated at about US$10 billion in 1997 and continues to escalate. Because of the need and the value of these drugs, many laboratories are intensively investigating the chemistry and biology of novel anticancer agents. Major advances have been made in understanding the nature and vulnerability of cancerous cells, resulting in development of novel screens and approaches. For the present, however, cytotoxic agents, many of natural origin, are the mainstays of anticancer chemotherapy.

A wide array of complex terrestrial and marine natural products possesses antitumor activity (1–5). A few of these were in folkloric use in fairly ancient times, whereas many have been discovered very recently as the result of directed screening programs. In earlier times screening was principally carried out against P388 and L1210 (murine leukemia models), but now there is greater emphasis on slower growing solid tumors. Antitumor natural products possess some of the most intricate structures of any compounds finding medicinal use today, and most are so toxic that each patient must be carefully titrated with them. Even with this care, patients still find the side effects attendant on their use hard to bear.

The question of why these substances occur in nature is endlessly debated. Many believe that they are defensive secretions that allow the organisms that produce them to survive in a hostile world. Others believe that they represent growth regulators that allow organized and controlled growth of cells and that they are not particularly toxic in the quantities normally found in the producing cells. It is not easy to resolve such arguments, but the point remains beyond dispute that such compounds are widespread, are easily detected, and that individual plants or animals have evolved widely disparate structural solutions to whatever needs these compounds actually fulfill.

These agents are collectively the most complex nonpolymeric organic medicinal agents in present use. At the time of their discovery, elucidation of their chemical structures frequently pushed the limits of chemical science. Unraveling their molecular modes of action in many cases revealed previously unsuspected complexities in cellular growth regulation and biochemistry. Successful synthesis of these and related compounds has greatly enriched our synthetic capabilities and a number of these syntheses have become classics of the art. Learning how to administer them safely to patients required the highest level of clinical expertise.

It is also interesting to contemplate their structural diversity from a biosynthetic standpoint. Starting with fairly ordinary monomeric units, complex enzymic pathways ultimately produced these cytostatic/cytotoxic agents without at the same time poisoning the microorganism or plant producing them. It is not credible to suppose that the organisms produced these substances as a gift to humanity. Each of these products represents such a finely crafted idiosyncratic design that one wonders why so many different organisms came up with such different solutions, given that the starting materials are basically similar. One might have guessed that fewer general solutions would have developed over biological time if one wishes to believe that they serve a role in regulating the growth of the producing organisms. In any event, mankind is fortunate that their activity spectrum is broad enough for us to use.

The toxicity of these agents is not particularly surprising, in that the screens employed in their discovery have historically depended on lethality to cells as an endpoint (6–8). It has historically been considered that rapidly growing cells, including cancer cells in particular, have a greater appetite for nutrients than more quiescent cells and so are selectively intoxicated on a kinetic basis. Thus the safety margin toward untransformed cells is not great. Furthermore, comparatively slow growing tumors are particularly hard to treat with such agents. Host cells that have a high growth fraction are also killed. Thus the usual constellation of side affects [allopecia, gastrointestinal (GI) ulceration, fertility impairment, immune suppression, blood dyscrasias, etc.] is relatively unavoidable. Very recently, synthetic agents able to interfere with aberrant cytokine-mediated growth signals

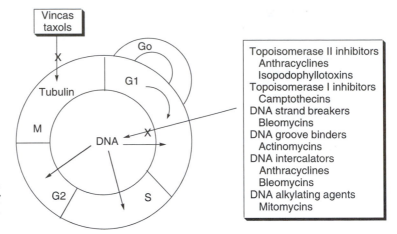

**Figure 3.1.** Synopsis of molecular modes of action of various prominent antitumor natural products.

have begun to appear on the market. Gleevec is the first commercial success embodying this approach and its antitumor application is comparatively nontoxic. It seems likely that natural products can be found sharing these characteristics. If so, a new era of natural product chemotherapeutic agents with minimal toxicity to normal cells will dawn.

The results of one of the principal screening methods in present use is collected in the National Cancer Institute database that was established in 1990 (6). This is based on comparative potency against 60 different human cancer cell lines grown in tissue culture. More than 70,000 compounds have been put through this screen and the data for each are presented in graphical form. From this, insights into mechanism of action and mode of resistance can be drawn (9). Many other tests are in present use, including screens for signal transduction inhibitors, antiangiogenesis, cell-cycle inhibition, exploitation of functional genomics, immunotherapeutics, vaccines, and chemoprevention. Much inventive biology is coming forward and exciting days appear to lie ahead.

The natural agents presently in use can be conveniently classified according to their molecular modes of action as follows:

1. Drugs attacking DNA
   - Dactinomycin
   - Bleomycin
   - Mitomycin
   - Plicamycin (mithramycin)

2. Drugs inhibiting enzymes that process DNA

   - Anthracyclines (daunorubicin, doxorubicin, epirubicin, idarubicin, valrubicin)
   - Camptothecins (topotecan, irinotecan)
   - Isopodophyllotoxins (etoposide, tenipocide)

3. Drugs interfering with tubulin polymerization/depolymerization

   - Taxus diterpenes (docetaxel, paclitaxel/taxol)
   - Vinca dimeric alkaloids (vinblastine, vincristine, vinorelbine)

Figure 3.1 illustrates in summary form the various points of attack of prominent natural antitumor agents on growing cells. One notes that DNA or tubulin in one way or another (either by direct attack or by interference with enzymes processing these important cellular macromolecules) is the primary target of all of these agents and that most phases of the cell cycle are involved, especially when mixtures ("cocktails") are employed.

Added biological detail of the properties and applications of these naturally occurring natural products can be found in *The AHFS Drug Information 2001* book (10) and in *Goodman and Gilman's Pharmacological Basis of Therapeutics* (11).

## 2  DRUGS ATTACKING DNA

### 2.1  Dactinomycin (Cosmegen)

**2.1.1  Introduction.** The actinomycins are a family of yellow-red peptide-containing antitumor antibiotics produced by fermentation of various *Streptomyces* and *Micromonospora* species. The first members of the actinomycin family were discovered in the early 1940s in the hopes of finding nontoxic antibacterial antibiotics in fermentations of soil microorganisms (12), although in the actinomycin case this ambition was dashed by their high toxicity. Somewhat later (since about 1958) this was compensated for by the discovery that the toxicity to rapidly growing cells could be useful in cancer chemotherapy. One should note, however, that later discoveries demonstrated that the potencies against microbes and against tumors do not parallel well. At the present time about seven different complexes of actinomycins have been identified, each differing from the others primarily by the various amino acids constituting the two cyclic depsipentapeptide side chains pendant from the common phenoxazinone chromophore [called actinocin (**1**)]. When the two cyclic pep-

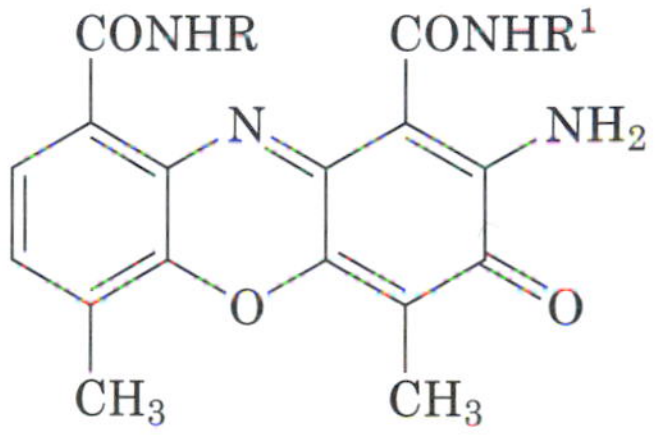

(1)

tide side chains are identical, these agents are referred to as isoactinomycins (R = R′). When they are different from each other, they are known as anisoactinomycins (R ≠ R′). Of the 20 or so natural actinomycins and a much larger number of synthetic and biosynthetic analogs, actinomycin D [(**2**), from which the generic name dactinomycin is derived] is the most prominent medicinally. A useful trivial nomenclatural system has also grown up. In this system, dactinomycin is referred to as Val-2-AM and other analogs are named by the

position and identity of the amino acids that are exchanged. Actinomycin C (cactinomycin-3) is thus known as Ile-2-AM.

**2.1.2  Medicinal Uses.** As noted in the summarizing table, dactinomycin is used medicinally by intravenous (i.v.) injection for the treatment of Wilm's tumor, rhabdomyosarcoma, metastatic and nonmetastatic choriocarcinoma, nonseminomatous testicular carcinoma, Ewing's sarcoma, nonmetastatic Ewing's sarcoma, and sarcoma botryoides. The usual dose is 10–15 $\mu$g/kg i.v. for 5 days. If no serious symptoms develop from this, additional treatments are given at 2- to 4-week intervals. Other treatment schedules have also been used. The drug is often combined with vincristine and cyclophosphamide in a cocktail to enhance the cure rate (13).

**2.1.3  Contraindications and Side Effects.** Dactinomycin is contraindicated in the presence of chicken pox or herpes zoster, wherein administration may result in severe exacerbation, occasionally including death. The drug is extremely corrosive in soft tissues, so extravasation can lead to severe tissue damage (14). To avoid this the drug is usually injected into infusion tubing rather than being injected directly into veins. When combined with radiation therapy, exaggerated skin reactions can occur as can an increase in GI toxicity and bone marrow problems. Secondary tumors can be observed in some cases that can be attributed to the drug. Dactinomycin is carcinogenic and mutagenic in animal studies and malformations in animal fetuses have also been observed. Nausea and vomiting are common along with renal, hepatic, and bone marrow function abnormalities. The usual alopecia, skin eruptions, GI ulcerations, proctitis, anemia, and other blood dyscrasias, esophagitis, anorexia, malaise, fatigue, and fever, for example, are also observed. Clearly this is a very toxic drug.

**2.1.4  Pharmacokinetic Features.** Dactinomycin is not very available after oral administration, so it is primarily administered by injection. About 2 h after i.v. administration very little circulating dactinomycin can be detected in blood. It is primarily excreted in the

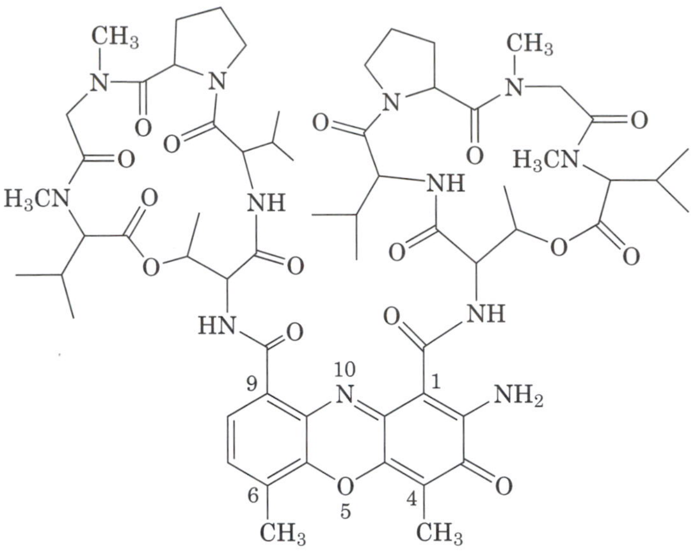

(2)

bile and the urine. It does not pass the blood-brain barrier. Dactinomycin is only slightly metabolized. Despite these factors it has a half-life of about 36 h. Persistence is largely accounted for by tight binding of the drug to DNA in nucleated cells (15–17).

**2.1.5  Medicinal Chemical Transformations.** Total synthesis of dactinomycin has been accomplished but this has not proved as yet to be of practical value (see Fig. 3.2). Analogs can be assembled from appropriately substituted benzenoid analogs. The overall strategy commonly involves construction of the external aromatic rings, attachment of the depsipeptide side-chain precursors, oxidative generation of the actinocin ring system, and functional group transformations to complete the synthesis (18–26).

Semisynthetic side-chain analogs of the actinomycins are prepared by removal of the depsipentapeptide side chains and their replacement by synthetic moieties. Analogs with altered peptide side chains are also prepared by directed biosynthetic manipulation of the fermentations. The synthetic replacement has been done in a combinatorial mode as well (27). Replacement of the normal side chains by simple amines leads to inactive products. Most of the other side-chain variations have led to compounds with reduced *in vivo* potency. None of those few analogs where this is not true has been commercialized.

Some chemical alterations in the chromophoric phenoxazinone moiety have also been accomplished. After considerable work it has emerged that the C-2 and the C-7 positions can be substituted with retention of signifi-

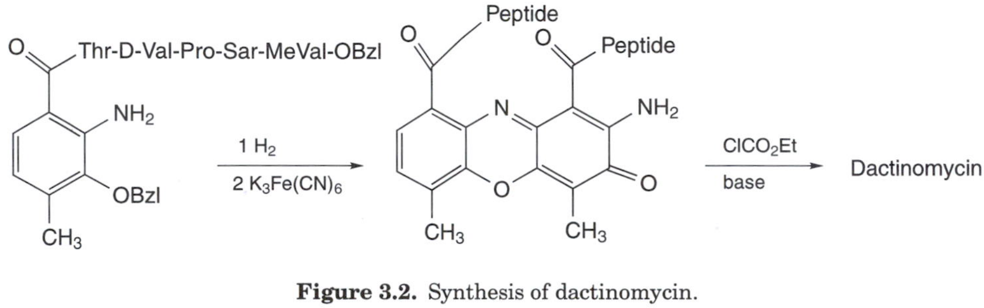

**Figure 3.2.** Synthesis of dactinomycin.

cant activity. Among some of the useful reactions leading to testable analogs are a series of addition/elimination reactions, starting with careful alkali hydrolysis to produce the C-2 OH analog. This can be converted by thionyl chloride treatment to the C-2 Cl analog. This in turn can be reacted with a variety of amines to produce alkylated C-2 amino substances (28). Catalytic reduction of the 2-Cl analog results in the protio analog, which is inactive. The C-2 chloro analog can be halogenated with chlorine or bromine to produce the C-2 chloro-C7 chloro or bromo analogs. These in turn can be solvolyzed to the C-2 amino-C7 halo analogs (29–32). Nitration and hydroxylation at the C-7 position can be accomplished but require prior protection with pyruvate. After nitration or oxidation to the quinone imine- and reduction, careful alkaline hydrolysis of the blocking pyruvate moiety leads to the desired analogs (28, 29, 33). The C-7 OH analog can be converted to the allyl ether and this can be epoxidized to produce an analog that not only can intercalate by virtue of its aromatic rings but can also alkylate DNA. The nitrogen analog of the epoxide (aziridinylmethylene) can be prepared by a somewhat different route. Hydrogenation of this last opens the aziridine ring to produce the primary amine (34–36). (See Fig. 3.3.)

The central chromophoric ring can also be modified to, for example, the phenazine (37, 38) analogs and to oxazinone and oxazole ring analogs (39, 40). These products have not become important (41).

In sum, these studies demonstrate that the side-chains are important determinants of activity as is the basic chromophoric three-ring system. Peripheral adornments are tolerated but not superior (42–48). Considering the putative molecular mode of action described below, this definition of the pharmacophore is not surprising. This definition of the pharmacophore is schematically represented in Fig. 3.4, where the pharmacologically successful transformations that take place are represented by the boxes.

**2.1.6 Molecular Mode of Action.** The flat three-ring fused aromatic portion of dactinomycin intercalates into double-helical DNA between the stacked bases (preferring guanine-cytosine pairs), whereas the attached cyclic peptide side chains of the drug bind into the minor grooves, thus further anchoring the complex (49–58). These combined interactions produce a tight and long-lasting binding. This model is supported by extensive X-ray studies with model nucleotides. As with other intercalating drugs, this interaction stretches the DNA and interferes with DNA transcription into RNA by RNA polymerase. The interference with the functioning of DNA-dependent RNA polymerase by dactinomycin is much stronger than the interference with DNA polymerases themselves. The consequences of intercalation are believed to be responsible for the antitumor action and most of the toxicity of dactinomycin. Some strand breaks are also reported. These broken products are believed to result from redox reactions of the quinonelike central chromophoric ring (57). Although relatively non-cell-cycle specific, dactinomycin's action is particularly prominent in the G-1 phase. The cytotoxic action of dactinomycin on rapidly proliferating cells is pronounced, resulting not only in antitumor activity but also in severe toxicities to certain host organs. Figure 3.5 illustrates the intercalation and minor-groove binding of dactinomycins.

Resistance to dactinomycin is primarily attributable to drug export through overexpression of P-glycoprotein and to alterations in tumor cell differentiation mechanisms (58–63).

**2.1.7 Biosynthesis.** The actinomycins are biosynthesized starting with tryptamine (see Fig. 3.6). This passes through kynurenine to 3-hydroxyanthranilic acid then to 4-methyl-3-hydroxyanthranilic acid. To this last the peptide side chains are added. Oxidative dimerization then results in completion of the phenoxazinone ring chromophore. This process is rather similar to that used in total chemical synthesis of dactinomycin. The unusual amino acids in the side chains provide strong evidence for very significant post-translational modifications. The various D-amino acids are converted from the L-stereoisomers and, in the case of dactinomycin, sarcosine is N-methylated (64). By varying the amino acid composition of the medium, a variety of actinomycin analogs can be made by directed fermentation (65, 66).

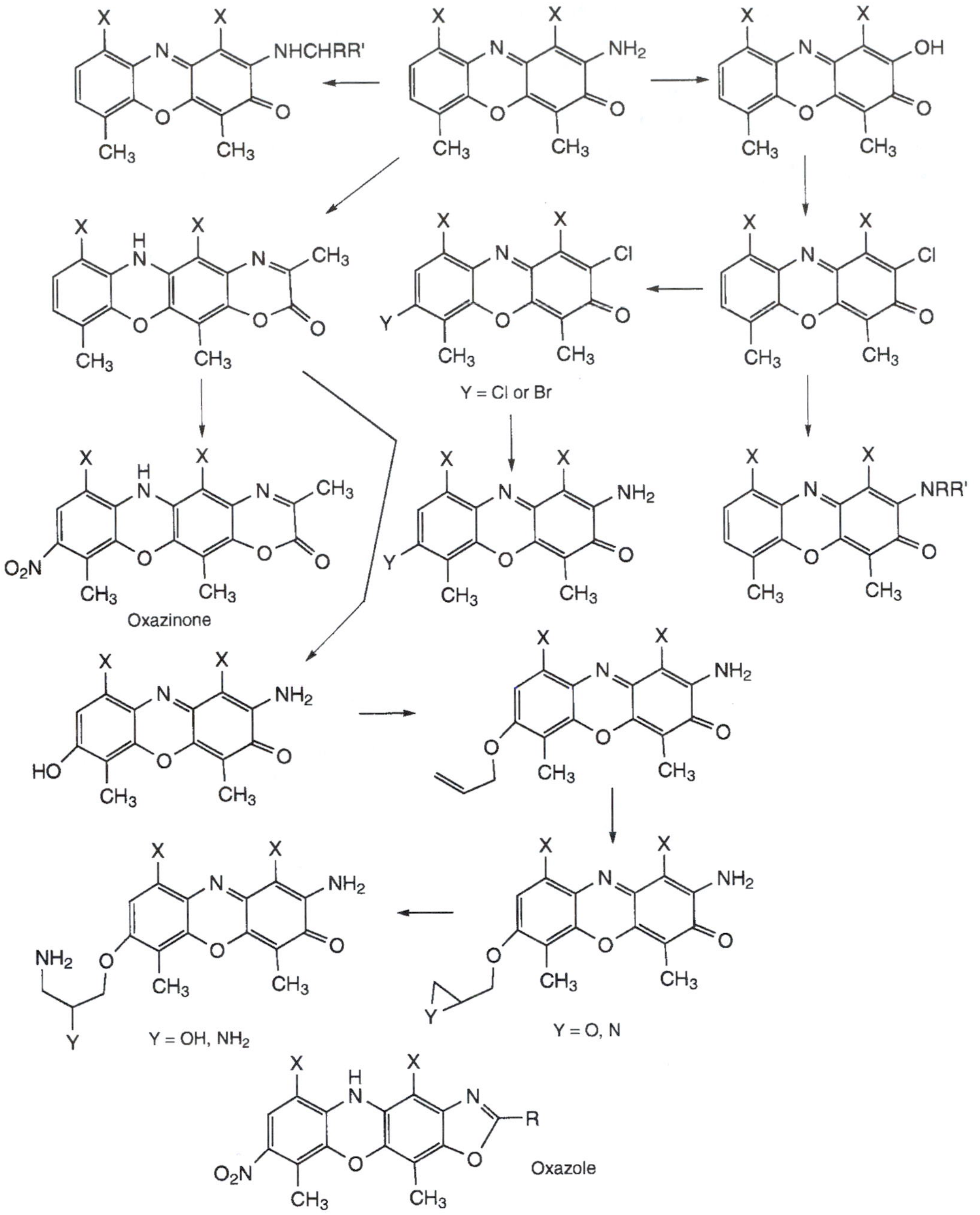

**Figure 3.3.** Synthesis of dactinomycin analogs.

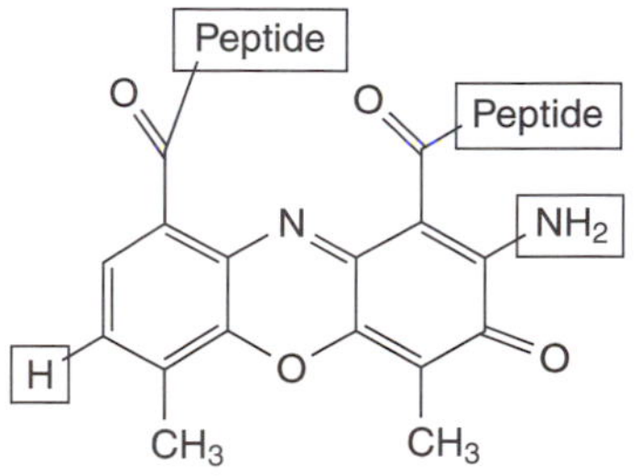

The "boxed" functional groups can be changed with retention of significant biological activity. Not all such changes, however, are successful.

**Figure 3.4.** Synopsis of pharmacologically successful transformations of actinomycins.

The chemistry of actinomycins has been the subject of a number of detailed reviews (67–71).

## 2.2  Bleomycin (Blenoxane)

**2.2.1  Introduction.** Bleomycin sulfate is a mixture of cytotoxic water-soluble basic glycopeptide antibiotics isolated by the Umezawa group from fermentation broths of *Streptomyces verticillus*. The commercial form consists of cuprous chelates primarily of bleomycins A-2 (**3**) and B-2 (**4**). Subsequently, many analogs have been isolated by various groups and been given various names. Among these are the pepleomycins (**5**), phleomycins, (**11**) cleomycins, (**12**) tallysomycins, (**13**), and zorbamycins (**14**).

**2.2.2  Medicinal Uses.** Bleomycin is used intramuscularly (i.m.), subcutaneously (s.c.),

i.v., or intrapleurally, often in combination with other antibiotics, for the clinical treatment of squamous cell carcinomas, Hodgkin's disease, testicular and ovarian carcinoma, and malignant pleural effusion. It is also instilled into the bladder for bladder cancer so that less generalized side effects are obtained. It is often coadministered with a variety of other antitumor agents to enhance its antitumor efficacy. One advantage that bleomycin has in such combinations is that it possesses little bone marrow toxicity and is not very immune suppressant, so it is compatible therapeutically with other agents (72–75).

**2.2.3  Contraindications and Side Effects.** Bleomycin is contraindicated when idiosyncratic or hypersensitive reactions are observed. Immediate or delayed reactions resembling anaphylaxis occur in about 1% of lymphoma patients. Because of the possibility of anaphylaxis, it is wise to treat lymphoma patients with 2 units or less for the first two doses. If no acute reaction occurs, then the normal administration schedule can be followed.

The most severe toxicity of bleomycin is pulmonary fibrosis and is more common with higher doses. This toxicity is observed in about 10% of patients and is difficult to anticipate, hard to detect in its early stages, and in about 10% of those affected it progresses to fatal lung compromise (76–78). Renal damage occurs occasionally and further decreases the rate of excretion of the drug. In rats, bleomycin has been observed to be tumorigenic. In pregnant females, fetal damage can result.

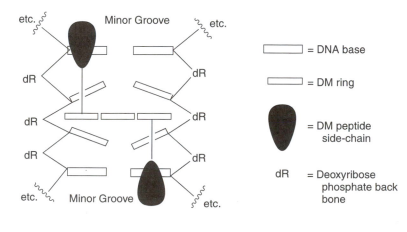

= DNA base

= DM ring

= DM peptide side-chain

dR  = Deoxyribose phosphate back bone

**Figure 3.5.** Cartoon of intercalation and minor groove binding of dactinomycins.

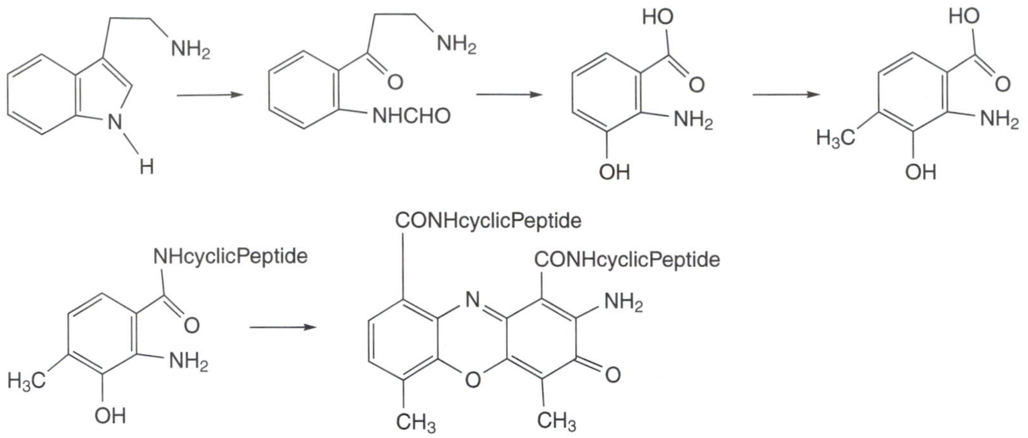

**Figure 3.6.** Biosynthesis of dactinomycin.

Skin and mucous membrane damage, hair loss, rash, and itching, for example, are not uncommon and may require discontinuation of the drug. In addition, the common constellation of fever, nausea, chills, vomiting, anorexia, weight loss, pain at the tumor site, and phlebitis are seen.

Coadministration with digoxin and phenytoin may lead to a decrease in blood levels.

The side effects of bleomycin generally do not reinforce the toxicities of other antitumor agents, so it is often used in anticancer cocktails.

**2.2.4 Pharmacokinetic Features.** When injected i.v., bleomycin is rapidly distributed and has a half-life of 10–20 min. Intramuscular injections peak in 30–60 min, although the peak levels are less than about one-third those obtained i.v. The overall half-life of bleomycin is about 3 h. Skin and lungs accumulate particularly high concentrations of the drug, in part because these are apparently the only tissues that do not rapidly deactivate it by enzymatic hydrolysis. It does not cross the blood-brain barrier efficiently because of its size and polarity. About 60–70% of the administered dose is recoverable as active bleomycin in the urine. Excretion is progressively delayed when the kidneys are damaged, so the doses are reduced by reference to creatinine levels (79).

**2.2.5 Medicinal Chemistry.** The essential central core of bleomycin provides a chelating environment for transition metals, especially

Cu(I) and Fe(II) (3). The branched glycopeptide side chain is less essential for activity and appears to serve in facilitating passage across cell membranes and to assist in oxygen binding. Removal of the sugars and the oxygen to which they are attached produces molecules that are fully active but distinct from bleomycin itself. The dipeptide unit is a linker arm but contributes key hydrogen bonding and perhaps other binding interactions that intensify activity and produce degrees of base specificity to the cleavages. The bithiazole unit and its pendant terminal cation are important in DNA targeting of the drug. These contributions were uncovered by the chemical synthesis of analogs that could not readily have been prepared by degradation of bleomycin itself or by directed biosynthesis.

Partial chemical synthesis, with or without the aid of enzymes, has also produced a variety of analogs through modifications of this peripheral side-chain array (80–91). Bleomycin is a conglomerate molecule built up from a collection of unusual subunits. Most of these were prepared independently by synthesis, in preparation for ultimate assembly into bleomycin itself or its analogs. The terminal bithiazole and its pendant amides are the portion of the molecule that binds to DNA. For the purpose of making analogs, the charged dimethylsulfonium group is monodemethylated through the agency of heat. The resulting compound is then cleaved to bleomycinic acid (**6**) by use of cyanogen bromide followed by mild alkaline treatment. Some soil

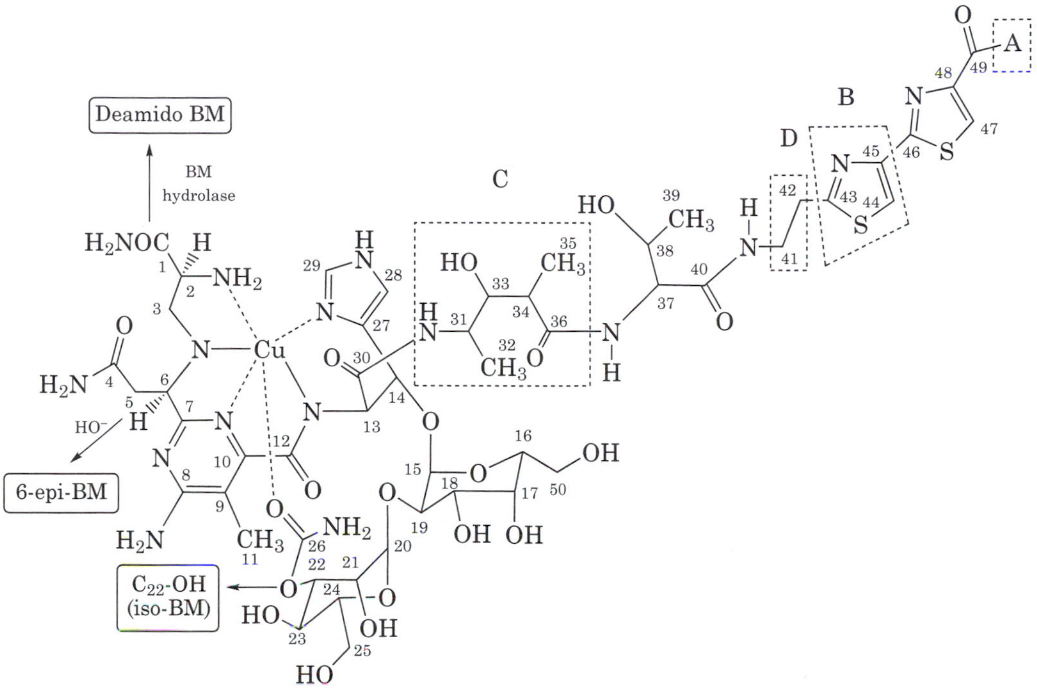

(3) A = ——NH(CH$_2$)$_3\overset{+}{S}$(CH$_3$)$_2$X$^-$
(4) A = ——NH(CH$_2$)$_4$NHC(NH)NH$_2$
(5) A = ——NH(CH$_2$)$_3$NH(CH$_2$)$_2$——2-pyridyl
(6) A = ——OH
(7) A = ——NH(CH$_2$)$_3$S(O)CH$_3$
(8) A = ——NH(CH$_2$)$_3$NH(CH$_2$)$_4$NH$_2$
(9) A = ——NH(CH$_2$)$_3$NH(CH$_2$)$_4$NH(CH$_2$)$_3$NH$_2$
(10) A = ——NH$_2$
(11) A = as in (3); B = (44, 45) Dihydrothiazole

(12) A = as in (3); C = {—NH\overset{OH}{\bigtriangleup}}

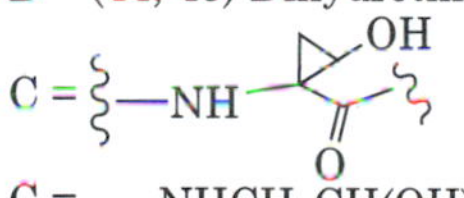

(13) A = as in (3); C = ——NHCH$_2$CH(OH)CH(CH$_3$)CO—— ;
     D = ——CH(β-O-4-amino-4,6-dideoxy-L-tallose)——CH(OH)——
(14) C$_{50}$ = CH$_3$; A = as in (3); B = as in (11);
     C = ——NHCH$_2$(CH$_2$OH)CH(OH)CH(CH$_3$)CO—— ;

microorganisms possess acylagmatine amidohydrolase capable of converting bleomycin to bleomycinic acid. Bleomycinic acid is then converted to the desired amides by use of water-soluble carbodimide chemistry. Whereas the chemical method is capable of producing greater structural variation, in practice the semisynthetic method has proved more convenient.

Although bleomycin and its analogs have also been totally synthesized in various laboratories, the processes are too complex to be of commercial value (92–95).

The phleomycins (11) are related in that one of the thiazole rings has been reduced to its C-44,45-dihydro analog. The phleomycins have substantial antitumor activity but are too nephrotoxic for clinical use. The cliomycins (12), tallysomycins (13), zorbamycins (14), zorbonamycins, platomycins, and victomycins are also structurally related to the bleomycins. None of these various alternative substances has displaced the bleomycin complex from the market, even though many possess significant antitumor properties. The

$$BM-Fe^{++} \xrightarrow[O_2]{H^+} BM-Fe^{+++} + H-O-O^{\bullet} \xrightarrow[H^+]{BM-Fe^{++}} H-O-O-H + BM-Fe^{+++}$$

$$H-O-O-H \xrightarrow[H^+]{BM-Fe^{++}} H-O-H + H-O^{\bullet} + BM-Fe^{+++}$$

**Figure 3.7.** Generation of reactive oxygen species by transition metal chelates of bleomycins.

specific potencies and toxicities vary widely with structural variations.

In the presence of a mild base, metal-free bleomycin isomerizes to isobleomycin through an O-to-O acyl migration of the carbamoyl moiety from position 22 to 23 of the mannosyl group. Copper (II) bleomycin, under the same conditions, slowly isomerizes at its masked aspartamine moiety attached to the pyrimidine substituent (at C-6). This isomer is substantially less active than bleomycin itself.

Bleomycin chelates with various transition metals, the most relevant of which are iron (II) and copper (I), to form the corresponding complexes. The iron complex binds oxygen and becomes oxidized, producing the hydroxyl radical and the hydroperoxyl radical. This is schematically illustrated in Fig. 3.7. The bithiazole moiety intercalates into DNA and the complex is stabilized by electrostatic attractions between the sulfonium or ammonium side chains with the phosphate backbone of DNA. This fixes the drug at DNA, whereupon the reactive oxygen species generated by its transition-metal complex breaks the DNA molecule at the sugar backbone, thus releasing purine and pyrimidine bases. This important reaction is illustrated in Fig. 3.8. Specific details of this complex interaction are still emerging.

Given that the biological action of bleomycin depends collectively on its ability to intercalate, to stabilize the intercalation complex by electrostatic forces, and to complex transition metals capable of generating oxygen radicals, the pharmacophore is distributed through the molecule. Acceptable variations involve substitution of various groups onto bleomycinic acid and a variety of other comparatively trivial changes such as partial reduction of the thiazole moieties and alterations of the amino acids near the bleomycinic acid carboxyl group.

Recently, efforts have been directed to the synthesis of various macromolecular conjugates of bleomycin, in an attempt to produce tissue selectivity and, perhaps, reduce lung toxicity. Some of these agents retain very significant nucleic acid clastogenicity *in vitro* (96).

**2.2.6 Biosynthesis.** Many analogs of bleomycin have been prepared by directed biosynthesis through appropriate media supplementation (97–100). Approximately 10 naturally occurring bleomycins have been reported (3–4, 7–10, etc.). These differ from one another by possessing a variety of different diamino analogs in place of the sulfoniumamino side chain attached to C-49 of bleomycinic acid (6). In addition, directed biosynthetic methods involving media supplementation with suitable precursors have produced approximately 21 others, which also consist of a variety of diamino analogs in which the C-49 moiety has been replaced. Thus the biosynthesis of bleomycinic acid is relatively tightly controlled, although the amide synthase that puts on the various side chains is not very specific in its substrate tastes.

**2.2.7 Molecular Mode of Action and Resistance.** The precise molecular mode of action of bleomycin is incompletely understood because it has numerous actions in test systems. The bleomycins are known to bind preferentially to the minor groove of DNA, although the specific details of this host-guest interaction are still elusive. The cytotoxicity of the bleomycins is enhanced when a DNA-binding region is present and the specific nature of the DNA-binding moiety can convey sequence specificity. The nucleic acid-cleaving capacity is metal ion and oxygen dependent and it is believed that the complexes generate reactive oxygen species that are responsible for the single- and double-strand nucleic acid cleavages observed (see Fig. 3.8). This DNA destruction is generally believed to account for its cytotoxicity

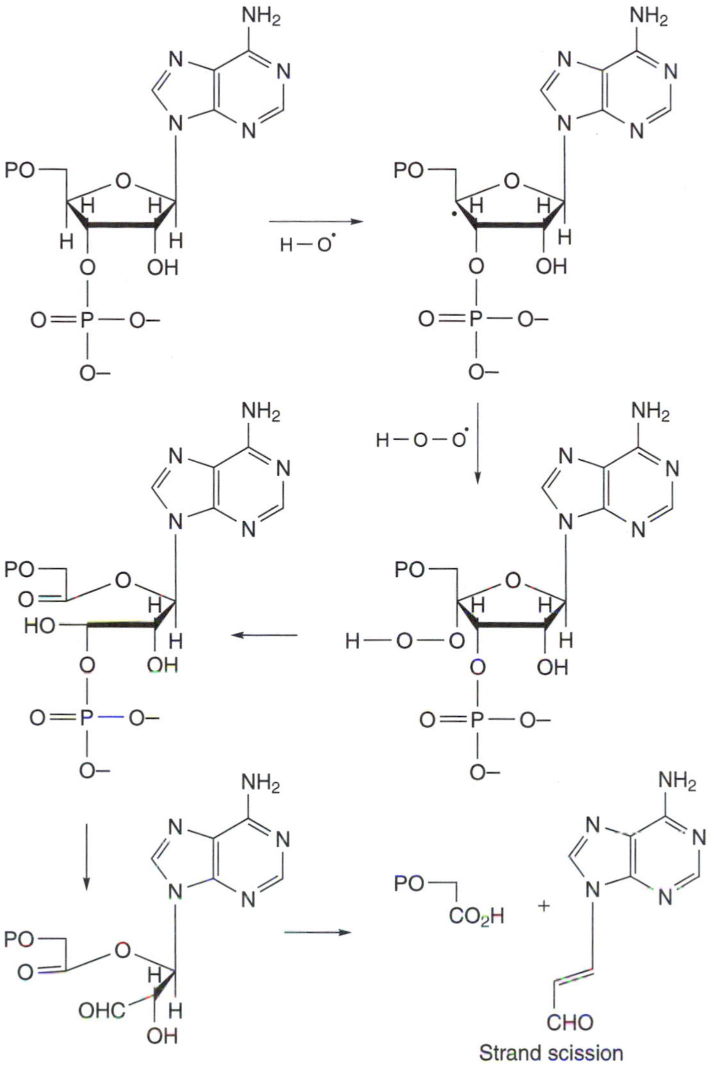

**Figure 3.8.** DNA backbone cleavage catalyzed by bleomycins.

(101–104). Interestingly, in the absence of DNA, bleomycin is also capable of destroying itself instead, presumably through the action of the same reactive oxygen species (105). A number of artificial analogs have been prepared to explore the contribution of various molecular features of these drugs and to exploit these features. Some of these products include agents that are inert by themselves but that enhance the cytotoxicity of bleomycin fragments when attached craftily to them. These agents usually contain aromatic moieties and have the capacity to have a cationic moiety as well. Bleomycin is known to generate oxygen-based free radicals when chelated to certain metal ions, notably ferrous iron and copper. When chelated to ferric iron, a reducing agent adds an electron to convert the complex to ferrous iron. This, in turn, transfers an electron to oxygen, producing either the superoxide radical or the hydroxide radical (see Fig. 3.7). These radicals attack ribosyl moieties in DNA and RNA, leading to nucleic acid fragmentation and subsequent interference with their biosynthesis. This action is believed to be primary in the cytotoxic action of bleomycin. Bleomycin's action is cell cycle specific, causing major damage in the G-2 and less in the M phase.

Resistance to bleomycin occurs primarily through the action of bleomycin hydrolase, which attacks metal-free bleomycin at the C-4 carboxamide moiety to produce deamidobleomycin (106). This last produces radicals at a much lower frequency than that of bleomycin itself. This causes a much lower cleavage of DNA and removes the majority of the antitumor action of bleomycin. In support of this idea, resistant cells usually possess a higher concentration of bleomycin hydrolase than do sensitive cells. The hydrolase is present in normal tissues, particularly in the liver. Interestingly, recent evidence implicates this enzyme in the formation of amyloid precursor protein characteristic of Alzheimer's disease (107). Other experts implicate enhanced DNA repair capacity or decreased cellular uptake as contributory to resistance.

### 2.2.8 Recent Developments and Things to Come.
Considering bleomycin's particular ability to destroy DNA and RNA molecules, there is comparatively little likelihood that molecular manipulation of bleomycin will soon produce a nontoxic version of the drug.

The chemical properties of the bleomycins have been reviewed recently (104, 108–113).

## 2.3 Mitomycin (Mutamycin)

### 2.3.1 Introduction.
Mitomycin C (**15**) was discovered initially at the Kitasato Institute (114) and at the Kyowa Hakko Kogyo labora-

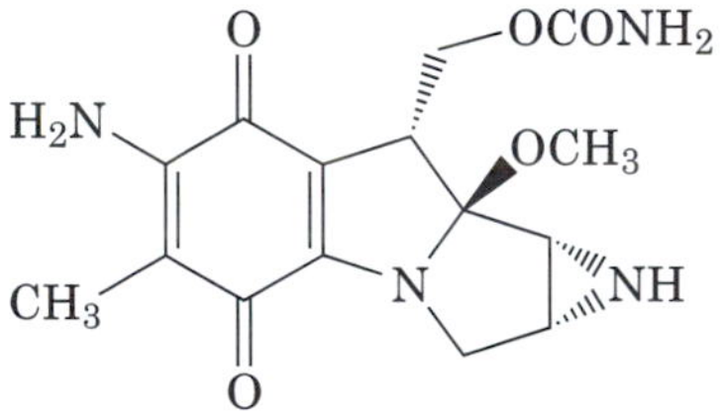

(**15**)

tories in Japan, as a metabolite of *Streptomyces caespitosus* (115), and elsewhere (116). A number of analogs have been discovered at several other places. These drugs are a group of blue aziridine-containing quinones, of which mitomycin C is the most important from a clinical perspective. Mitomycin A and

porphiromycin also belong to this group but have not been marketed. Mitomycins apparently were the first of the useful bioreductively activated DNA alkylating agents to be discovered. Literally thousands of alkylating agents, notably the $\alpha,\beta$-unsaturated sesquiterpene lactones of the Compositae, have been found in nature, and an enormous effort has been expended in their synthesis and evaluation without notable success. The contrasting success of the mitomycins seems to derive from the finding that they are relatively inert until bioreductively activated, so they show greater biological selectivity compared with that of many other naturally occurring alkylating agents.

### 2.3.2 Clinical Use.
Mitomycin is administered i.v. in combinations of antitumor agents for treatment of disseminated adenocarcinoma of the stomach, colon, or pancreas, or for treatment of other tumors where other drugs have failed (117–120).

### 2.3.3 Contraindications and Side Effects.
It is contraindicated in cases of hypersensitivity or idiosyncratic responses to the drug or where there are preexisting blood dyscrasias. The drug can cause a serious cumulative bone marrow suppression, notably thrombocytopenia and leukopenia (121, 122), that can contribute to the development of overwhelming infectious disease. This requires reducing dosages. Irreversible renal failure as a consequence of hemolytic uremic syndrome is also possible (121). Occasionally adult respiratory distress syndrome has also been seen. When extravasation is seen during administration, cellulitis, ulceration, and sloughing of tissue may be the consequence (123, 124). The drug is known to be tumorigenic in rodents. Its safety in pregnancy is unclear and teratogenicity is seen in rodent studies. Other side effects include fever, anorexia, nausea, vomiting, headache, blurred vision, confusion, drowsiness, syncope, fatigue, edema, thrombophlebitis, hematemesis, diarrhea, and pain. It is not clear that all of these are related to the use of mitomycin or whether they are at least partly the consequence of other agents in antitumor cocktails.

**2.3.4    Pharmacokinetics.** Mitomycin is poorly absorbed orally and is rapidly cleared when injected i.v., with a serum half-life of about 30–90 min after a bolus dose of 30 mg. Metabolism takes place primarily in the liver and is saturable. As a consequence of the saturability, the amount of free drug in the urine increases with increasing doses. Only about 10% of an average administered dose is excreted unchanged in the urine and the bile because extensive metabolism takes place. The drug is distributed widely in the tissues, with the exception of the brain, where very little penetrates (125–128).

Because mitomycin C is activated as an antitumor agent by reduction, significant effort has been expended on trying to decide whether DT-diaphorase activity correlates well with antitumor activity *in vivo*. This is as yet imperfectly resolved but the correlation appears to be poor. Other studies suggest that NADPH:cytochrome P450 reductase (a quinone reductase) contributes strongly under some circumstances.

Inactivation and activation occur by metabolism and/or by conjugation, and a number of metabolites, principally 2,7-diaminomitosene, have been identified (129–131). The ratio between inactivation and activation is partially a function of whether DNA intercepts the reduced species before it is quenched by some other molecular species.

**2.3.5    Medicinal Chemistry.** Much exploration of the chemistry of the mitomycins has been carried out accompanied by excellent reviews in the literature (132–134). Total chemical syntheses of mitomycins A and C have been achieved, but these are not practical for production purposes (135–137). More than a thousand analogs have been prepared by semisynthesis but none of these agents has succeeded in replacing mitomycin C itself. Generally, it has been found that mitomycin C analogs are less toxic than mitomycin A derivatives. Most modifications have been achieved at the N-1a, C-7, C-6, and C-10 positions. The C-7 position is particularly conveniently altered through addition/elimination sequences, and some of these agents have received extensive evaluation. It is noted that the C-6 and C-7 positions play only an indirect role in the activation of the ring system, so

substitutions there might be regarded as primarily significant in altering the pharmacokinetic properties of the mitomycins. It has been found quite recently, however, that the participation of the C-7 substituent in activation by thiols differs significantly when C-7 bears a methoxyl group (the mitomycin A series) compared to the activation when C-7 bears an amino group (the mitomycin C series). Indeed, thiols activate the methoxy analogs but not the amino analogs. Mechanistically, both series arrive at the same bisalkylating species *in vivo* but through different routes. This may help rationalize why mitomycin A is both more potent and more cardiotoxic than mitomycin C (138). The results of a comparison of physicochemical properties and biological activity of the mitomycins led to the conclusion that potency correlates with uptake, as influenced primarily by log $P$, and also with the redox potential (E1/2) (139).

The metabolism of mitomycin C *in vivo* primarily leads through reduction and loss of methanol to a dihydromitosene end product. Interception by DNA, on the other hand, leads to alkylation of the latter instead (138, 139).

**2.3.6   Molecular Mode of Action and Resistance.** Mitomycin C undergoes enzymatic reductive activation to produce reactive species capable of bisalkylation and crosslinking of DNA, resulting in inhibition of DNA biosynthesis (140–142). This effect is particularly prominent at guanine-cytosine pairs. The reductive activation of mitomycin C makes it particularly useful in anaerobic portions of tumor masses that have a generally reducing environment. Mitomycin is also capable of causing single-strand breaks in DNA molecules.

The apparent chemical mechanism by which mitomycin is reductively alkylated to a bisalkylating agent is illustrated in Fig. 3.9. The process is initiated by a quinone reduction followed by elimination of methanol, opening of the aziridine ring, conjugate addition of DNA, ejection of the carbamate function, and further addition of DNA.

The bisalkylation of DNA can be either intrastrand or interstrand, as illustrated in Fig. 3.10.

Resistance is attributed to failure of reduction (143), to premature reoxidation (143,

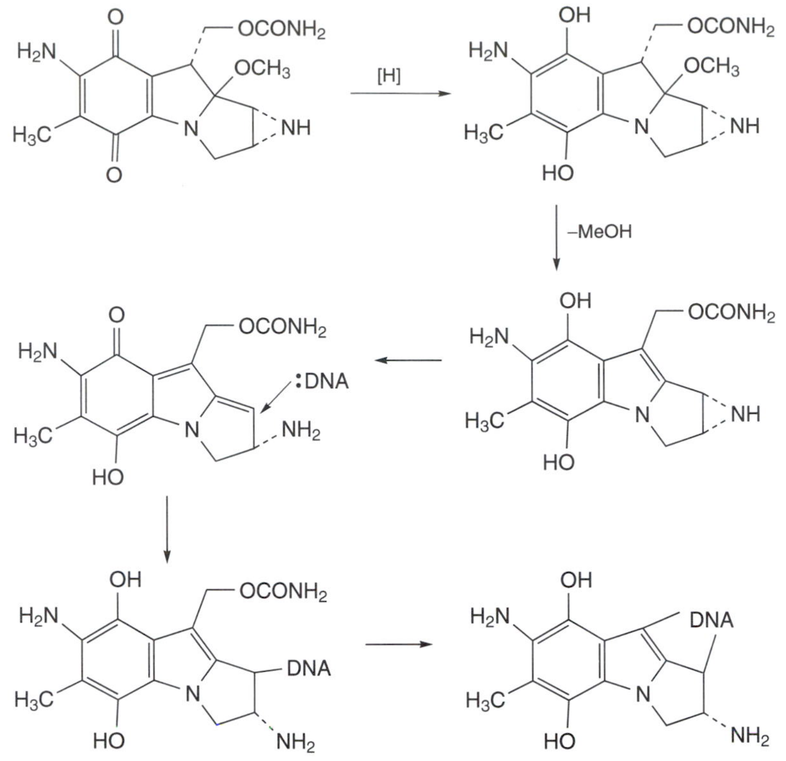

**Figure 3.9.** Reductive activation and bisalkylation of DNA by mitomycin C.

144), binding to a drug-intercepting protein that also has oxidase activity (145), and to P-glycoprotein–mediated efflux from cancer cells (146, 147).

**2.3.7 Medicinal Chemistry.** The pharmacologically successful chemical transformations of mitomycin are schematically summarized in Fig. 3.11.

The chemistry and pharmacological actions of the mitomycins have been reviewed (132–134, 148).

## 2.4 Plicamycin (Formerly Mithramycin; Mithracin)

**2.4.1 Introduction.** Plicamycin (**16**), produced by fermentation of *Streptomyces plicatus* and *S. argillaceus*, was isolated in 1953 (149). It is a member of the aureolic acid family of glycosylated polyketides, which also includes chromomycins, chromocyclomycins, olivomycins, and UCH9. It was subsequently

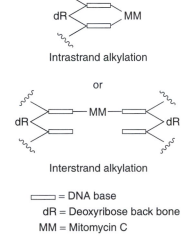

Intrastrand alkylation

or

Interstrand alkylation

▭ = DNA base
dR = Deoxyribose back bone
MM = Mitomycin C

**Figure 3.10.** Interstrand and intrastrand alkylation of DNA by bioreductively activated mitomycin C.

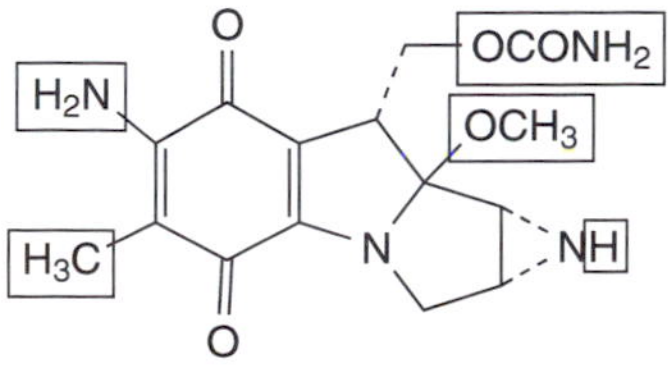

The "boxed" functional groups can be changed with retention of significant biological activity. Not all such changes, however, are successful.

**Figure 3.11.** Pharmacologically successful modifications of mitomycin C.

found to be identical to mithramycin, a fermentation product of *S. argillaceus* and *S. tanashiensis*.

**2.4.2 Clinical Uses.** Plicamycin is highly toxic but is nevertheless administered i.v. for treatment of testicular tumors (150–153). In lower doses it is used for treatment of hypercalcemia and hypercalciuria associated with advanced cancer, particularly involving Paget's bone disease (154–157).

**2.4.3 Contraindications and Side Effects.** Severe thrombocytopenia, hemorrhagic tendency, and death can be encountered with the use of plicamycin (158, 159). Renal impairment, mutagenicity, and interference with fertility are also known to occur with the use of plicamycin. Anorexia, nausea, vomiting, diarrhea and stomatitis, fever, drowsiness, weakness, lethargy, malaise, headache, depression phlebitis, facial flushing, skin rash, hepatotoxicity, and electrolyte disturbances (decrease in serum calcium, potassium, and phosphate levels) are also encountered.

Plicamycin is contraindicated with coagulation disorders, thrombocytopenia, thrombocytopathy, impairment of bone marrow function, and in pregnancy.

The toxic reactions of plicamycin are much less severe and frequent in the lower dosages employed to lower calcium ion levels.

**2.4.4 Pharmacokinetics.** Plicamycin is given i.v., whereupon a complex excretion pattern ensues, with a half-life of approximately 11 h having been reported (160).

**2.4.5 Mode of Action and Resistance.** The exact mechanism of action of plicamycin is elusive but it is known to intercalate into DNA, favoring G-C base pairs, resulting in the inhibition of enzymes that process DNA (161–163). Plicamycin also interferes in the biosynthesis of RNA (163). The effect of plicamycin is enhanced in the presence of divalent metal ions such as magnesium (II). Its hypocalcemic action is unrelated to this but is rather mediated by interference with the function of vitamin D in some unclear manner (164). Plicamycin also acts on osteoclasts and blocks the action of parathyroid hormone (165, 166).

Resistance to plicamycin involves efflux

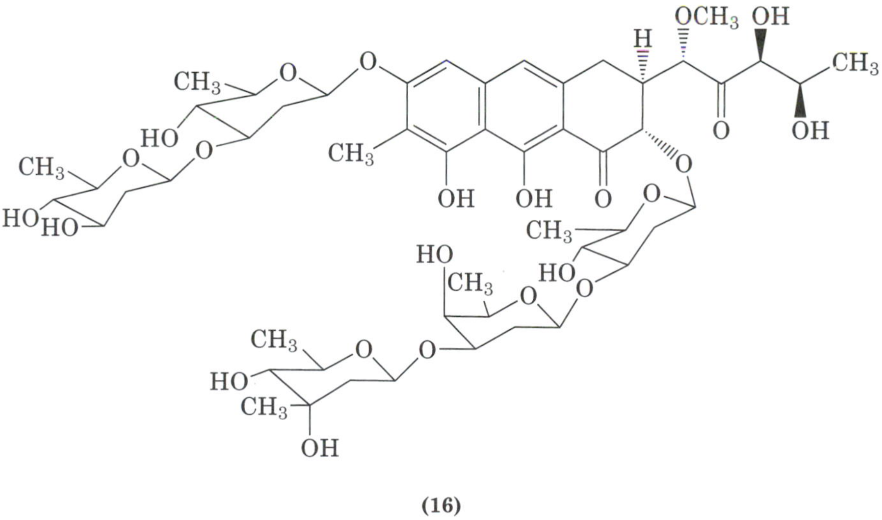

(16)

through the action of P-glycoprotein (167), although recent publications suggest that plicamycin has the capacity to suppress *MDR 1* gene expression *in vitro*, thereby modulating multidrug resistance (168).

**2.4.6 Medicinal Chemistry.** The chemistry of plicamycin and its analogs has been reviewed (169). For a long time there was considerable confusion about the precise chemical structure of plicamycin (mostly with respect to the number and arrangement of the sugars) but this has now apparently been resolved by careful NMR studies (170).

The sugars must be present in plicamycin for successful DNA binding and magnesium ion also promotes the interaction.

**2.4.7 Biosynthesis.** Biosynthesis of the aureolic acid group of antitumor antibiotics begins with condensation of 10 acetyl units to produce a formal polyketide that, on condensation, produces a tetracyclic intermediate whose structure and that of the subsequent intermediates is reminiscent of those involved in tetracycline biosynthesis (171). After the formation of premithramycinone, a rather complex sequence of reactions ensues, as illustrated in Fig. 3.12. A sequence of methylations and glycosylations lead to premithramycin A3. Of particular interest in the remaining sequence is an oxidative ring scission and decarboxylation, which leads to the final tricyclic ring system. This is followed by oxidation level adjustment, producing plicamycin itself, or to one of the other members of this class, depending on the specifics of the biosynthetic intermediates (172, 173). Omission of the key C-7 methylation step leads, for example, through a parallel pathway to the formation of 7-demethylmithramycin (174).

# 3 DRUGS INHIBITING ENZYMES THAT PROCESS DNA

## 3.1 Anthracyclines

The anthracyclines are an important class of streptomycete-derived tetracyclic glycosidic and intercalating red quinone-based drugs. None of the first generation of this widespread

class of natural products became clinically prominent. The structures of some of these chemically interesting compounds, generally named as rhodomycins, including pyrromycin, musettamycin, and marcellomycin (whose names will please opera buffs), are given in Fig. 3.13. Those anthracyclines of clinical value were discovered initially in the Pharmitalia Laboratories in Italy and subsequently in a number of other places (175, 176). The first of the clinically useful group was the *Streptomyces peucetius* metabolite, daunorubicin (**18**). This was followed by its hydroxylated analog doxorubicin (**17**), a metabolite of *S. peuceteus* var. *caesius*. Many synthetic anthracyclines resulted from intense study in many laboratories. These synthetic methods led to a number of marketed products, including daunomycin's desmethoxy analog idarubicin (**20**) and doxorubicin's diastereomer epirubicin (**19**) (177, 178), and the bisacylated product of doxorubicin, valrubicin (**21**). Daunomycin and idarubicin are primarily used for the treatment of acute leukemia, and epirubicin is used for solid tumors, but doxorubicin is used for a much wider range of cancers.

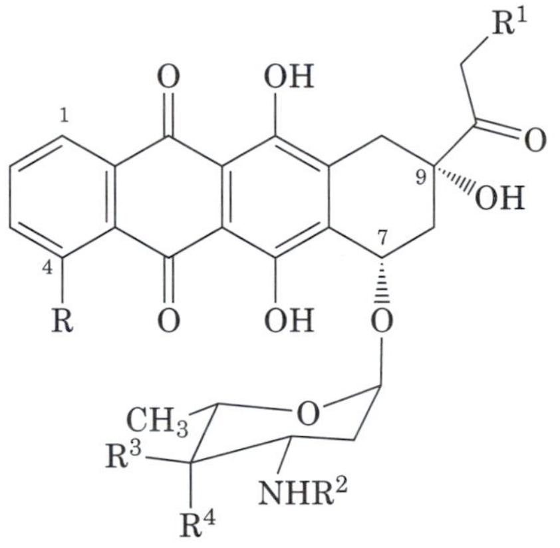

| Compd. | R | $R^1$ | $R^2$ | $R^3$ | $R^4$ |
|--------|------|---------|---------|-----|-----|
| (17) | $OCH_3$ | OH | H | H | OH |
| (18) | $OCH_3$ | H | H | H | OH |
| (19) | $OCH_3$ | OH | H | OH | H |
| (20) | H | H | H | H | OH |
| (21) | $OCH_3$ | OCOBu | $COCF_3$ | H | OH |

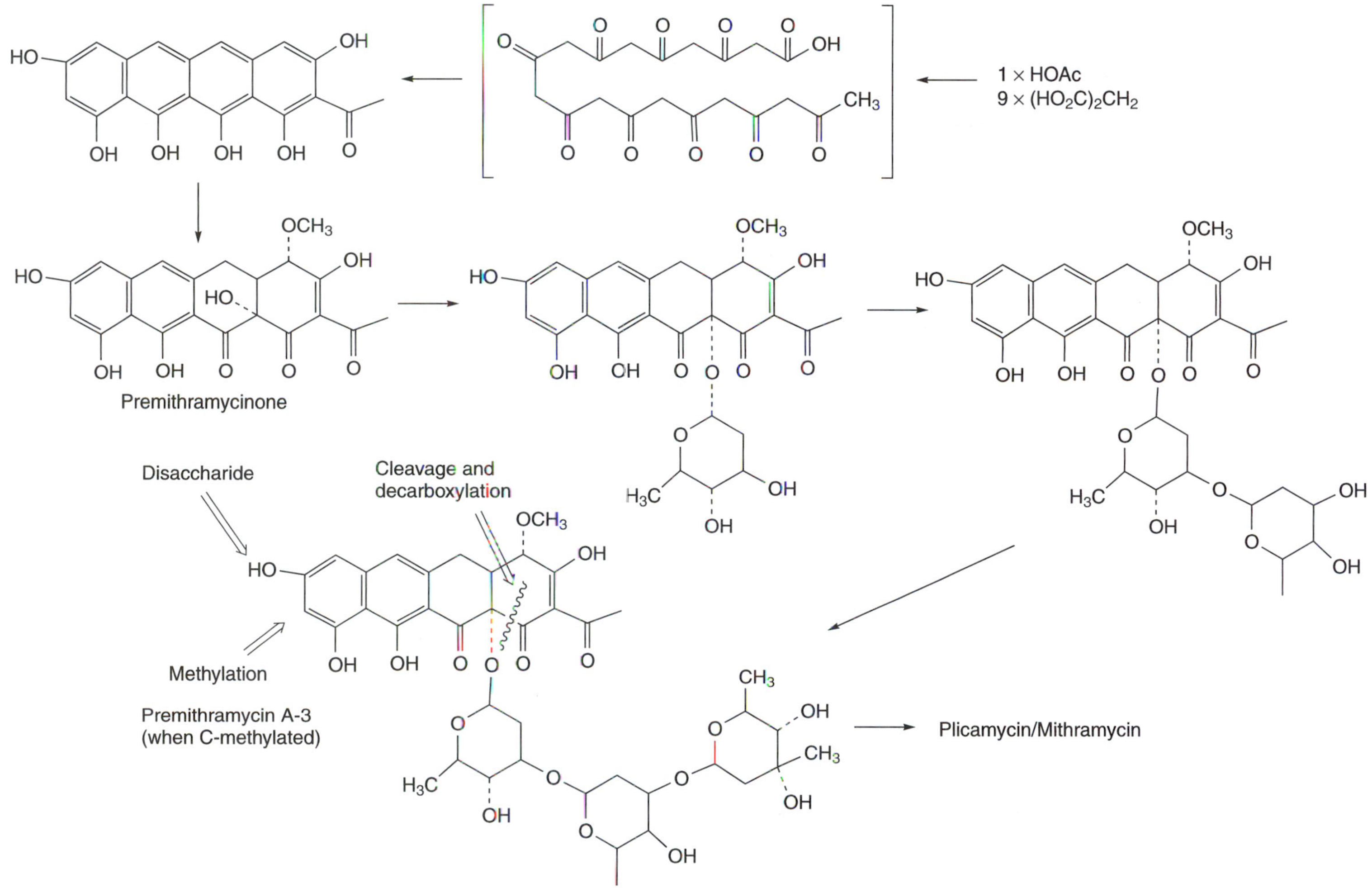

**Figure 3.12.** Biosynthesis of plicamycin.

125

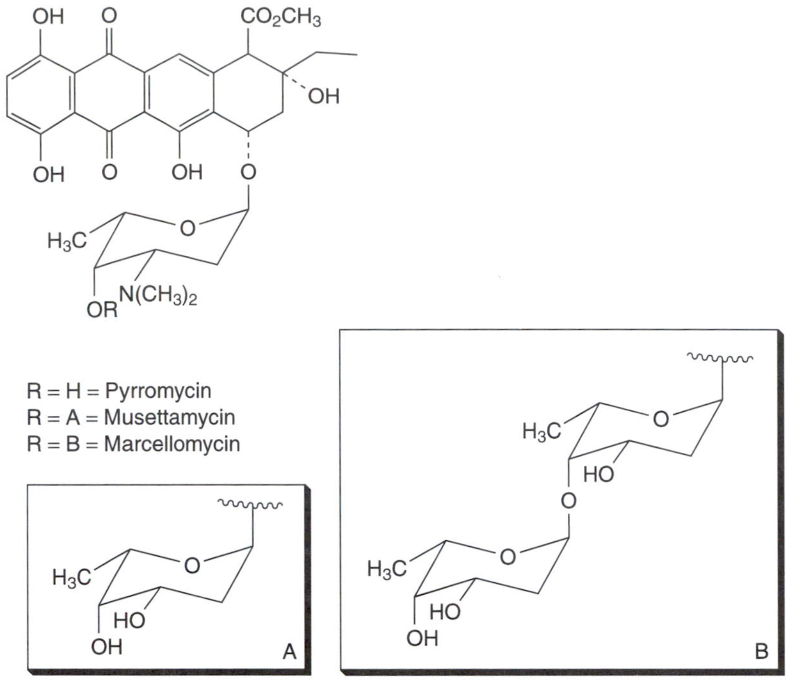

**Figure 3.13.** Structures of some unmarketed anthracyclines.

### 3.1.1 Daunorubicin (Daunomycin; Cerubidine, Rubidomycin; 18)

*3.1.1.1 Therapeutic Uses.* Daunorubicin is used in combination with other agents by i.v. infusion for treatment of acute myelogenous and lymphocytic leukemias (179–181).

*3.1.1.2 Side Effects and Contraindications.* It is not generally used i.m. or s.c. because of the severe tissue damage that may accompany extravasation (182). It is contraindicated when hypersensitivity reactions are present. Among the side effects that are encountered are severe cumulative myocardial toxicity that can include acute congestive heart failure after cumulative doses above 400–550 mg/m$^2$ of body surface in adults and less in infants (183), severe myelosuppression (hemorrhage, superinfections), bone marrow suppression, secondary leukemia, renal/hepatic failure, carcinogenesis, mutagenesis, teratogenicity, and fertility impairment. The cardiomyopathy is characteristic of the anthracycline class and can occur long after therapy is concluded (184, 185). The highly colored nature of the drug can lead to urine discoloration that alarms the patient because of drug excretion. In addition,

alopecia, rash, contact dermatitis, urticaria, nausea, vomiting, mucositis, diarrhea, abdominal pain, fever, chills, and (occasionally) anaphylaxis are observable. When given along with cyclophosphamide, its cardiotoxicity is enhanced and enhanced toxicity is seen when given concurrently with methotrexate.

*3.1.1.3 Pharmacokinetics.* On i.v. administration the drug is rapidly distributed into tissues but does not enter the central nervous system. Rapid liver reduction to daunomycinol is seen followed by hydrolytic or reductive loss of the sugar along with the oxygen atom with which it is attached to the ring system. These two reactions also can take place before reduction. Demethylation of the *O*-methyl ether moiety also occurs followed by sulfation or glucuronidation of the resulting phenolic OH. These and other transformation products have lesser bioactivity (186). Patients with decreased liver function should receive smaller doses because they are not able to detoxify the drug effectively. The half-life is about 8.5 h and about 25% of the active drug is found in the urine along with about 40% in the bile (187). Liposomally encased daunorubicin ci-

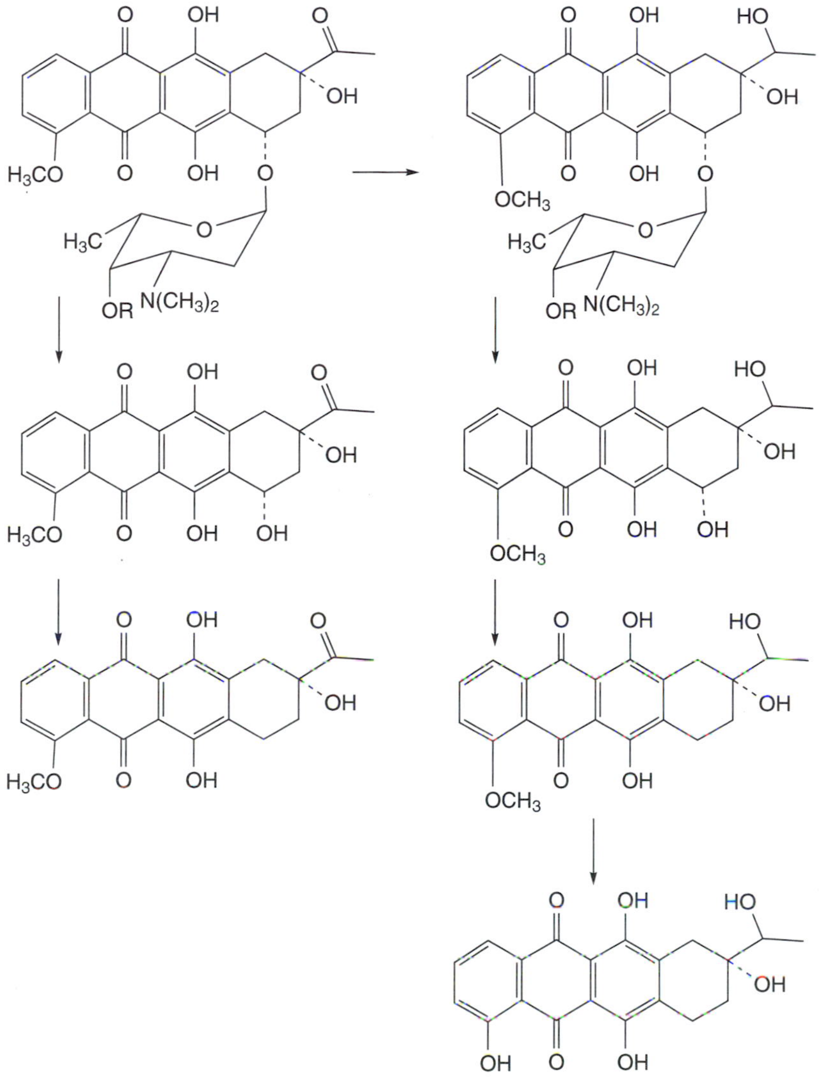

**Figure 3.14.** Metabolism of daunorubicin.

trate shows greater selectivity for solid tumors and is translocated in the lymph (180). Nuclear binding of anthracyclines is sufficiently strong to complicate the excretion pattern and to determine the tissue distribution of these agents (188). Different tissues bind doxorubicin in direct proportion to their DNA content. The metabolism of daunorubicin is illustrated in Fig. 3.14.

**3.1.1.4 Mechanism of Action and Resistance.** The mode of action of daunorubicin and the other clinically useful anthracyclines is multiple. Authorities differ with respect to which is the most significant but most attribute this to inhibition of the mammalian topoisomerase II, essential for shaping DNA, so that it can function and be processed (189). The drug also intercalates into DNA, inhibits DNA and RNA polymerases, and also causes free-radical single- and double-strand damage to DNA (190). These drugs are, therefore, also mutagenic and carcinogenic. Free-radical (reactive oxygen species) generation is promoted by the interaction of these drugs with P450

(191) and with iron, which they chelate (192). The reactive oxygen species that they can generate also cause severe damage to membranes and this may contribute not only to their antitumor efficacy but also to the cardiomyopathy that they cause (193).

Resistance to daunorubicin and the other anthracyclines is attributed to efflux mediated by P-glycoprotein, whose expression is amplified in response to their use (187, 194–196). A number of other mechanisms have been advanced as contributory such as use of other export mechanisms, increased endogenous antioxidant mechanisms, and decreased action of mammalian topoisomerase II (197).

### 3.1.2 Doxorubicin.
Doxorubicin [adriamycin, rubex; (**17**)] is a hydroxylated analog of daunorubicin but finds much wider anticancer use.

#### 3.1.2.1 Therapeutic Uses.
Doxorubicin is given i.v. by rapid infusion for the treatment of disseminated neoplastic conditions such as acute lymphoblastic leukemia, acute myeloblastic leukemia, Wilms' tumor, neuroblastoma, soft tissue and bone sarcomas, breast carcinoma, ovarian carcinoma, transitional cell bladder carcinoma, thyroid carcinoma, Hodgkin's and non-Hodgkin's lymphomas, bronchogenic carcinoma, and gastric carcinoma.

#### 3.1.2.2 Side Effects and Contraindications.
Doxorubicin is contraindicated in patients with preexisting severe myelosuppression consequent either to other antitumor treatments or to radiotherapy. It is also contraindicated when hypersensitivity to anthracyclines is present or when significant previous doses of other anthracyclines have been administered, given that their doses coaccumulate toward congestive heart failure.

Side effects are generally similar to those seen with daunorubicin (which see), with particular reference to cumulative drug-related congestive heart failure, extravasation problems, myelosuppression, and hepatic damage.

#### 3.1.2.3 Pharmacokinetics.
As with daunorubicin, the tissue distribution of doxorubicin is strongly influenced by the cellular content of DNA in various parts of the body (188). Metabolites of doxorubicin are its aglycone, its deoxyaglycone, doxorubicinol and its deoxyaglycone, and demethyldeoxyadriamycinol aglycone as its 4-*O*-β-glucuronide and *O*-sulfides. Thus carbonyl reduction is the main metabolic reaction and this is followed by various hydrolytic and reductive losses of the sugar, O-demethylation, and various conjugative reactions (198). These reactions quite parallel the findings with those of doxorubicin.

#### 3.1.2.4 Molecular Mode of Action and Resistance.
The manifold cytotoxic actions of doxorubicin on cells are qualitatively the same as those of daunorubicin. Likewise, the resistance mechanisms, especially those involving P-glycoprotein expulsion, are closely similar. Interestingly, expulsion is significantly lessened by liposome encapsulation (199).

### 3.1.3 Epirubicin.
Epirubicin (Ellence, (**19**) is a C-4′-diastereoisomer of doxorubicin given by i.v. infusion as an adjunct to the use of other agents for the treatment of breast cancer, when axillary node tumor involvement is seen after breast removal surgery (200). The toxicities of epirubicin are analogous to those described above for daunorubicin and doxorubicin (which see). Particular note should be paid to drug-related cumulative congestive heart failure, extravasation problems, myelosuppression, and hepatic damage.

### 3.1.4 Valrubicin.
Valrubicin (**21**) and idarubicin (**20**) are also anthracyclines that have seen significant clinical use (201). Idarubicin differs from doxorubicin in lacking the methoxy group in the chromophore and has an epimeric hydroxyl group in the sugar (202). This molecule is comparatively lipophilic, resulting in increased cellular uptake (the cellular concentrations exceed 100 times those achieved in plasma) (203) and strong serum protein binding (204). Extensive extrahepatic metabolism to the 13-dihydro analog occurs (205).

Valrubicin is the valeric ester trifluoroacetic amide of doxorubicin (206–208). It is instilled into the bladder through a urethral catheter after bladder drainage and is voided after 2 h (209). It is highly toxic on contact with tissues but its means of administration limits systemic exposure. Its local adverse reactions are usually comparatively mild and resolve in about 24 h. Evidence indicates that the valeryl ester moiety is removed enzymically *in vivo* before exerting its cytotoxic effect (210, 211).

**3.1.4.1 Biosynthesis.** The anthracyclines are polyketides, as can be readily discerned from their structures. Doxorubicin is produced from daunorubicin by a late hydroxylation step that is genetically unstable. As a consequence, it is apparently produced commercially by an efficient chemical transformation instead of by fermentation (212).

**3.1.4.2 Medicinal Chemistry.** Many hundreds of analogs have been prepared either by chemical transformation of the natural products themselves or by total synthesis. As a result a reasonable understanding of their structure-activity relationships is at hand (213–218).

The impressive anticancer activity and clinical potential of the anthracyclines resulted in intensive research toward total synthesis and structural modification studies of these compounds (175, 176, 219, 220). From a structure-activity relationship (SAR) viewpoint, the anthracycline structural core can be divided into three major components: (*1*) Ring D, the alicyclic moiety bearing the two-carbon side-chain group and the tertiary hydroxy group at C-9, concomitantly having a chiral secondary hydroxy group at C-7, which in turn is connected to the aminosugar unit; (*2*) the aminosugar residue, attached to the C-7 hydroxy group through an α-glycosidic linkage; (*3*) the anthraquinone chromophore, consisting of a quinone and a hydroquinone moiety on adjacent rings. The C-13 and C-14 positions of the various anthracyclines are obvious functional sites for derivatization. Thus, the 13-keto functionality has been subjected to reduction, deoxygenation, hydrazide formation, and so forth, without adversely affecting the bioactivity. Similarly, incorporation of various ester and ether functionalities at C-14, through initial halide formation and subsequent displacement of the halogen with nucleophiles, was found to be a useful approach in modulating the activity of the parent anthracyclines. However, homologation of the C-9 alkyl chain or introduction of amine functionalities at C-14 is detrimental to activity. Additionally, formation of 9,10-anhydro or the 9-deoxy analogs results in decreased activity. Interest-

ingly, the natural stereochemical configurations at C-7 and C-9 were found to be an important contributor to bioactivity, wherein it has been proposed that H-bonding between the two *cis*-oxygen functionalities at these positions stabilizes the preferred half-chair conformation of the D-ring.

The amino sugar residue of the various anthracyclines is an essential requirement for bioactivity. Among the various SAR studies involving the carbohydrate core, it has been seen that attachment of this moiety to the anthracycline nucleus through an α-anomeric bond is necessary for optimum activity. Conversion of the C-3′ amine group to the corresponding dimethylamino or morpholino functionalities confers improved activity; however, acylation of the amine (the exception being trifluoroacetyl) or its replacement with a hydroxy group results in loss of activity. Interestingly, conversion of the C-4′ hydroxy group to its corresponding methyl ether, C-4′ epimerization, or deoxygenation has a negligible effect on bioactivity. In more recent studies, novel disaccharide analogs of doxorubicin and idarubicin have been found to exhibit impressive antitumor activity (221).

The anthraquinone chromophore is an important structural feature of the anthracyclines. The various oxygenated functionalities present in this fragment have been the focus of considerable synthetic activity in search of analogs with improved activity. Thus, the phenolic hydroxy groups present in this core were found to undergo ready acylation and alkylation under standard reaction conditions. It has been shown that, *O*-methylation of the C-6 or C-11 phenolic groups results in analogs with markedly reduced activity, whereas C-4 modifications such as demethylation and deoxygenation do not affect bioactivity. Interestingly, a serendipitous transformation of the C-5 carbonyl to the corresponding imino functionality resulted in an analog that retained activity and was found to be significantly less cardiotoxic than the parent compound.

**3.1.4.3 Biosynthesis.** The proposed biogenesis of the anthracyclines invokes the involvement of a polyketide synthon. In studies involving various blocked mutants of anthracycline-producing *Streptomyces* and utilization of $^{14}$C-labeled acetate and propionate pre-

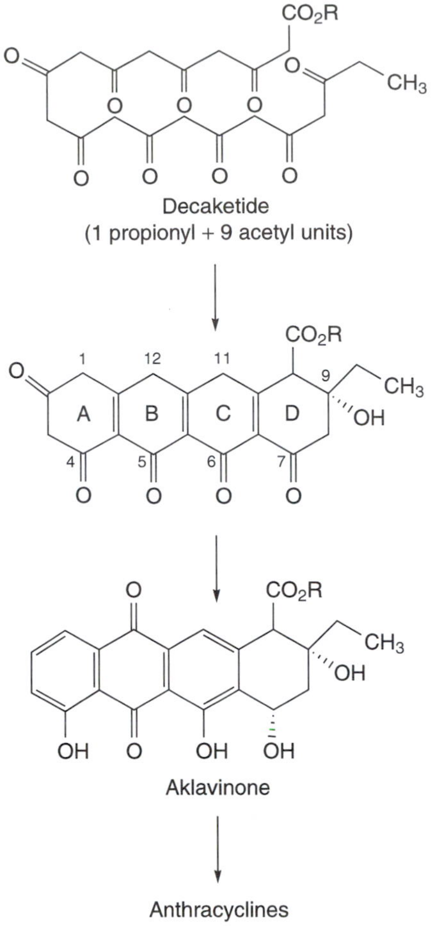

**Figure 3.15.** Proposed biosynthetic pathway leading to anthracyclines.

cursors, it has been shown that there are two biosynthetic pathways responsible for the formation of the polyketide fragment. Daunomycinone, pyrromycinone, and related aglycones are derived from a polyketide synthon having one propionate and nine acetate units, whereas deviant members such as steffimycinone and nogalanol are obtained from a 10-acetate polyketide unit. Thus, a "head-to-tail" condensation of the decaketide chain forms the parent tetracyclic core and the C-9 quaternary center of the anthracyclines. A sequence of biotransformations involving C-2 and C-7 carbonyl reduction, dehydration (C-2/C-3), enolization/aromatization, and B-ring oxidation leads to aklavinone. Further oxidation,

decarboxylation (for some class of compounds), and glycosidation finally result in the corresponding bioactive glycosides. See Fig. 3.15 for a schematic illustration of the proposed biosynthetic pathway leading to anthracyclines.

***3.1.4.4 Recent Developments and Things to Come.*** Reviews of this topic are available (203, 210, 222).

### 3.2 Camptothecins

Camptothecin (**22**) was discovered almost at the same time (1966) as was taxol and by the same research group (223). It is present in the extractives of the Chinese tree *Camptotheca acuminata* (growing in California) and has subsequently been found to be abundant in the extractives of *Mappia foetida*, a weed that grows prolifically in the Western Ghats of India. Despite its early promise in laboratory and rodent studies, it was disappointing in clinical studies because of severe toxicity and so it has not found clinical use by itself, but serves as the inspiration for the preparation of its clinical descendants prepared both by partial and total chemical synthesis methods. Camptothecin itself is very insoluble. This made early evaluation difficult. Tests were performed on its sodium salt (prepared by hydrolysis of the lactone ring) but clinical trials of this salt had to be discontinued because of severe, unpredictable hemorrhagic cystitis, even though some patients with gastric and colon cancers were responding to the drug. A quiet period followed. Much later came a re-

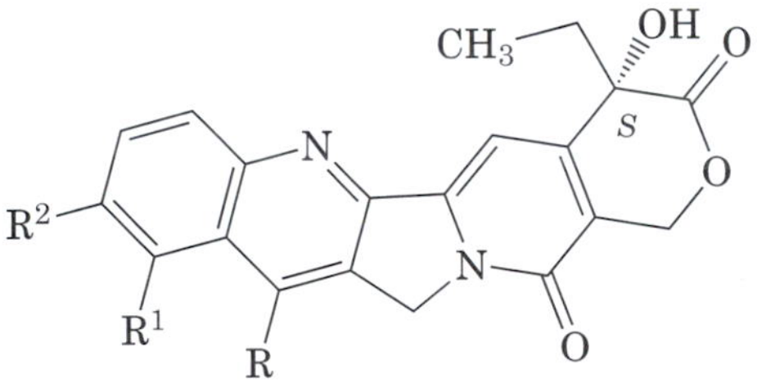

| Compd. | R | $R^1$ | $R^2$ |
|---|---|---|---|
| (**22**) | H | H | H |
| (**23**) | H | $CH_2N(CH_3)_2$ | OH |
| (**24**) | Et | H | 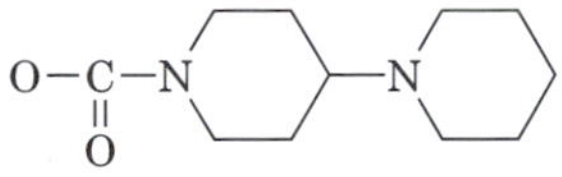 $O-C(=O)-N$(piperidine)$-N$(piperidine) |

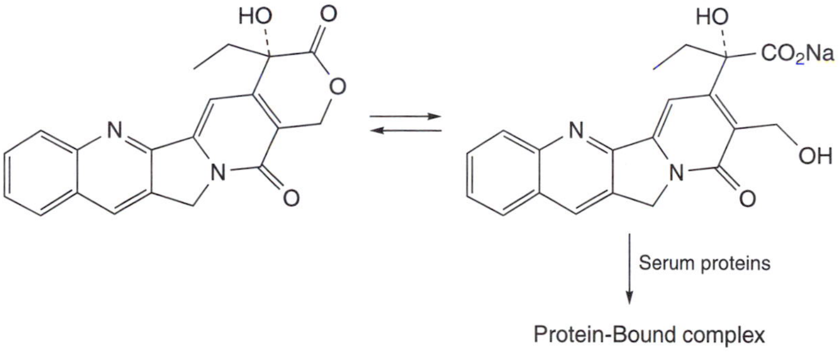

**Figure 3.16.** Hydrolysis of camptothecin analogs.

surgence of interest because of the discovery that the drug works by inhibiting nuclear mammalian topoisomerase I, a novel mechanism of action among contemporary antitumor agents (224). Topoisomerase I is a ubiquitous enzyme essential for changing the twisting number of DNA molecules (relaxing supercoils) so that they can be transcribed and repaired. The levels of topoisomerase I are often raised in tumor cells. Topoisomerase I exerts its action by making transient single-strand breaks in duplex DNA, rotating the molecule, and resealing again. Camptothecin and its analogs form a ternary complex with the cut DNA and topoisomerase I, which prevents progression or regression. The cut DNA is unavailable to the cell, so that it is stranded in the S-phase of the cell cycle and the DNA is degraded, thus leading to cell death.

Camptothecin itself is very water insoluble, thus impeding its use by injection. Furthermore, it is quite unstable in the body because of ease of hydrolysis of the lactone ring under physiological conditions, to produce the highly toxic acid analog (Fig. 3.16). The ring-opened form is also highly serum protein bound, helping to account for its comparatively poor activity *in vivo*. This high level of binding also displaces the equilibrium further in the direction of the undesirable ring-opened acid form. These factors apparently are less limiting in mice, producing a significant species difference in behavior. This raised the level of disappointment when, despite favorable animal studies, the drug performed poorly in the clinic. Many analogs were subsequently prepared by total synthesis and by conversions of camptothecin itself. The more promising of these newer analogs are much more soluble in water and less serum protein bound, helping them to overcome some of the defects of camptothecin itself.

*Metabolism.* Hydrolysis to the less-active and toxic ring-opened lactone occurs readily *in vivo* under physiological conditions (Fig. 3.16). Further, the lactone binds to serum proteins approximately 200 times more than does camptothecin itself. By mass action, this shifts the equilibrium toward ring opening. The lactone-opened analogs are significantly more water soluble than the lactone forms but are generally rather less active.

### 3.2.1 Irinotecan (CPT-11)

*3.2.1.1 Clinical Uses.* Irinotecan (**24**) is an analog hydroxylated in the quinoline ring and further converted to an amine-bearing prodrug carbamate linker. It is given by i.v. infusion, often in combination with 5-FU and leucovorin (which combination is particularly toxic) for the treatment of metastatic carcinoma of the colon or rectum (225). Irinotecan and its metabolites are much less serum protein bound than topotecan and have a somewhat longer half-life in serum. Irinotecan, however, is poorly orally bioavailable and is also subject to a significant first-pass metabolism.

### 3.2.2 Topotecan

*3.2.2.1 Clinical Uses.* Topotecan (**23**) is used for ovarian (226, 227) and small-cell lung cancers (228–235). Topotecan is rapidly me-

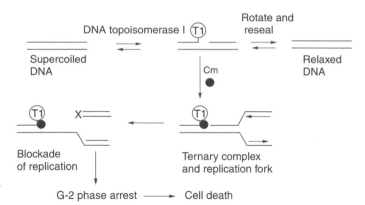

**Figure 3.17.** Schematic of camptothecin molecular mode of action.

tabolized by hydrolysis and the majority of the drug (75–80%) is hydrolyzed in the plasma, with a half-life of a couple of hours. *N*-De-methyltopotecan and its glucuronide are found to a lesser extent.

*3.2.2.2 Contraindications and Side Effects.* Extravasation of the camptothecin-derived drugs leads to tissue damage, and the drugs are strongly emetic and neutropenia is common. Hypersensitivity to irinotecan is observed and is a contraindication. Diarrhea occurs by various mechanisms. Prior exposure to pelvic/abdominal irradiation enhances the risk of severe myelosuppression and deaths have been observed attributed to consequent infections (236). Orthostatic hypertension, vasodilation, insomnia, dizziness, alopecia, rash, anorexia, constipation, dyspepsia, anemia, weight loss, dehydration, colitis/ileus, renal function, and fertility impairment are also seen with irinotecan but are generally considered to be mild (237).

*3.2.2.3 Pharmacokinetic Features.* The pharmacokinetic features of topotecan are very complex. The drug is subject to alteration by esterases and the products are variously glucuronidated as well as oxidized by CYP 3A4, so is subject to a number of possible drug-drug interactions (238). After oral administration, about 30–40% of the drug is bioavailable (239–241). After i.v. dosage of prodrug irinotecan, rapid metabolic conversion by hydrolysis of the carbamoyl moiety to an active phenolic metabolite (SN-38) occurs as a result of the action of liver carboxylesterase; this is followed by glucuronidation to a metabolite that is much less potent. Metabolite SN-38 is about

1000-fold more active than irinotecan itself and accounts for the bulk of the antitumor activity of the drug. Fortunately, SN-38 has much less affinity toward serum proteins and this shifts the equilibrium toward retention of the active lactone form. Irinotecan is also converted in part to a metabolite in which the piperazine ring is oxidatively opened to produce an acid analog (presumably through its lactam) (242). About 11–20% of active irinotecan is excreted in the urine but the majority of the drug and its metabolites are excreted in the bile. There appears to be a significant patient-to-patient variation in ability to metabolize irinotecan (236).

*3.2.2.4 Molecular Mode of Action and Resistance.* The camptothecins are inhibitors of the action of mammalian topoisomerase I. The normal function of this essential enzyme is to produce temporary single-strand breaks by which the topography of DNA can be altered, so that the molecule can be processed. In the presence of camptothecin and its analogs a ternary complex forms (camptothecin analogs + DNA + enzyme) that results in single-strand breaks that cannot be resealed and this leads to defective DNA. In particular, when the replicating fork of DNA reaches the cleavable complex generated by camptothecin derivatives, irreversible strand breaks result, causing a failure in DNA processing, thus causing the cells to die. The camptothecins are thus S-phase poisons (Fig. 3.17). The specific molecular details are still obscure, however (237).

Resistance to the camptothecins is believed to result in part from excretion mediated by

P-glycoprotein and MRP-3 (multidrug resistance-associated protein) mechanisms (237, 243), although a number of other biological effects are seen in *in vitro* studies (244–247). Their meaning in clinical cases is as yet unclear. In some resistant cells, reduced levels of hydrolases capable of cleaving irinotecan to SN-38 seem to contribute. Another mode of resistance involves decreases in content and potency of topoisomerase I (237, 248, 249).

*3.2.2.5 Medicinal Chemistry.* Several papers and reviews on the various total synthesis and analog studies of camptothecin and related molecules have been published (250–254).

The objectives of much of this work are clear. Drawbacks of camptothecin that have to be overcome are its poor water solubility, ease of hydrolytic lactone opening to the undesirable acid form, high serum protein binding, and the reversibility of its drug-target interaction. The solubility problem has been approached, interestingly, in quite opposite directions. Some groups have sought to increase water solubility and others to make the molecules even more lipophilic. Each approach has worked significantly.

After considerable effort it was discovered that placing substituents at the C-9 and C-11 positions considerably decreased serum protein binding, even with the lactone-ring opened analogs, and yet this did not interfere with antitumor activity. Among the analogs that have received clinical examination but have not (yet) been marketed are lurotecan (**26**), 9-nitro and 9-aminocamptothecin (255), and DX-8951f (256). A number of other analogs stand out from the many that have been made. Among these are the hexacyclic 1,4-ox-

(**26**)

azines (257), Ring E homocamptothecins, 7-cyanocamptothecins (258), and the silatecans (**25**) (259). The latter are structurally unusual, in that very few candidate drugs contain silicon atoms. Furthermore, the best analogs are quite lipid soluble and, despite this, display superior stability in human blood and decreased albumin binding combined with significant potency. A summary of camptothecin SARs is illustrated in Fig. 3.18.

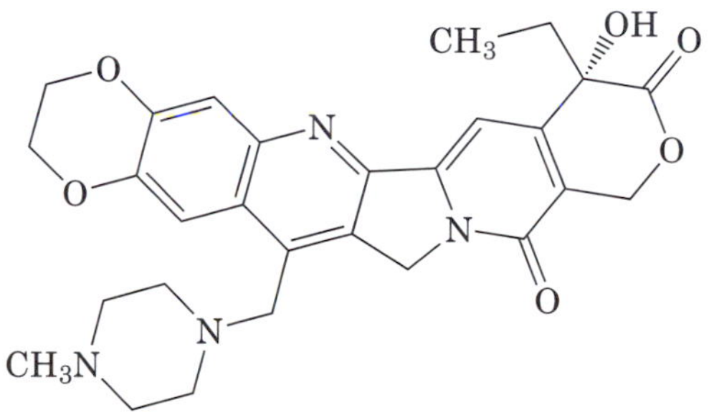

Much effort has been expended also to enhance the water solubility of the camptothecins by inventive formulations. A number of prodrugs have also been made in attempts to enhance stability and water solubility. Among these are the C-16 esters, such as the butyrates and propionates, some sugar-containing molecules, and the C-11 carbamates, of which irinotecan is the most successful to date.

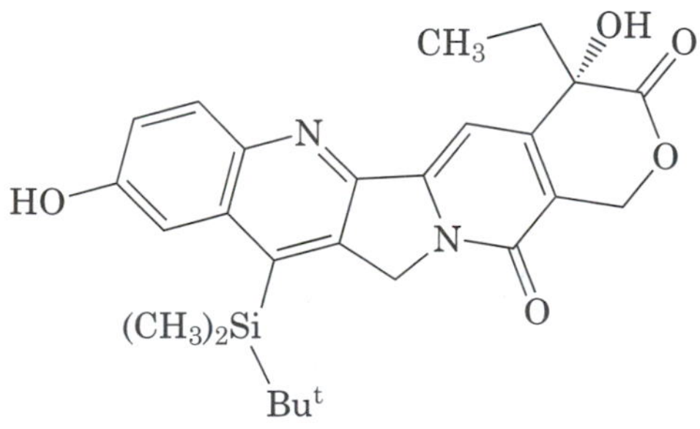

Topotecan is likewise hydroxylated in the quinoline ring but with a dimethylamino-

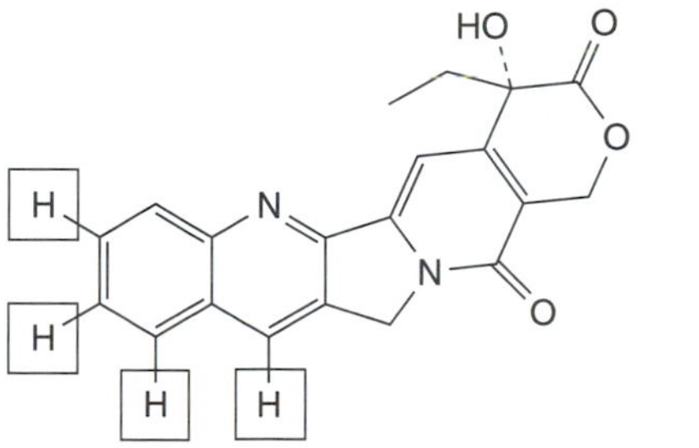

The "boxed" functional groups can be changed with retention of significant biological activity. Not all such changes, however, are successful

**Figure 3.18.** Summary of camptothecin structure-activity relationships.

(**25**)

methylene moiety adjacent. It is administered i.v. for the treatment of ovarian and small-cell lung cancer. As with the other camptothecins, topotecan undergoes a reversible pH-dependent hydrolysis of its lactone moiety. It is the lactone form that is pharmacologically active. The drug has a complex excretion pattern, with a terminal half-life of about 2–3 h, and about 30% of the drug appears in the urine. Kidney damage decreases the excretion of the drug. Binding of topotecan to serum proteins is about 35%. The clinical side effects of topotecan are similar to those of irinotecan.

The chiral center of the camptothecins is $S$; the $R$-enantiomers are much less (10- to 100-fold) potent.

*3.2.2.6   Quantitative Structure-Activity Relationships (QSARs).* Many synthetic camptothecin analogs have been prepared in attempts to stabilize the active lactone form and to enhance water solubility. A QSAR correlation has been published based on the NCI database information for 167 camptothecin analogs. The key functions that emerged from this are the presence and comparative positions of the E-ring hydroxyl and lactone carbonyl and the D-ring carbonyl (260).

*3.2.2.7   Recent Developments and Things to Come.* Topoisomerase I inhibition is a popular area of contemporary research and a number of analogs are in various stages of preclinical and clinical workup. It seems likely that the immediate future will see the emergence of additional agents in this class (261–264).

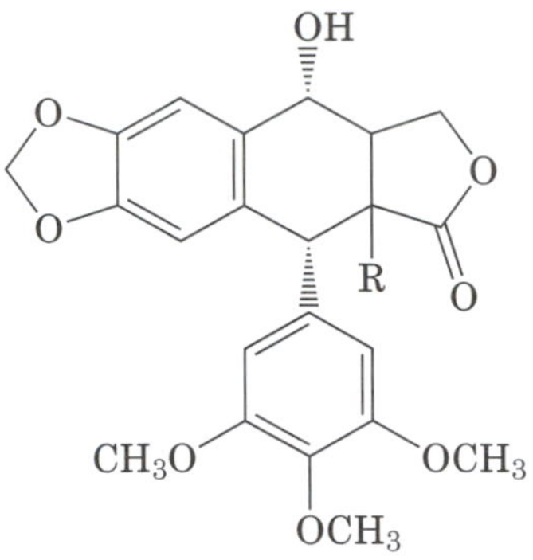

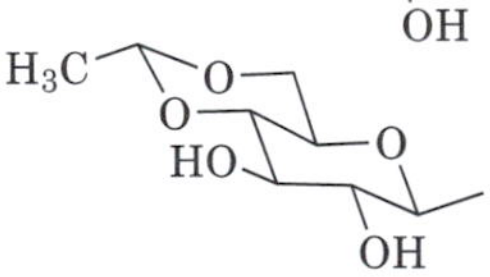

### 3.3   Isopodophyllotoxins

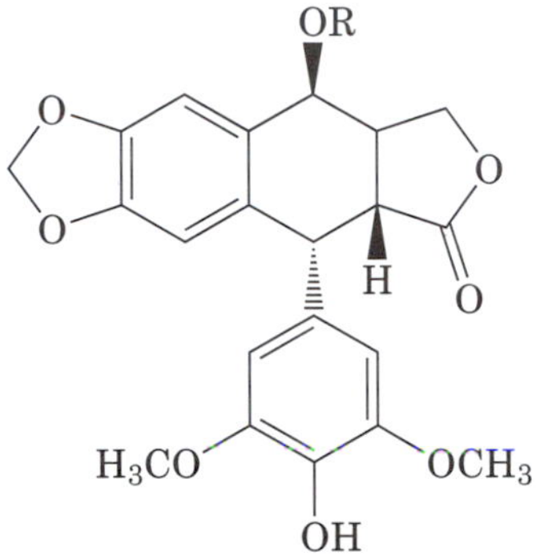

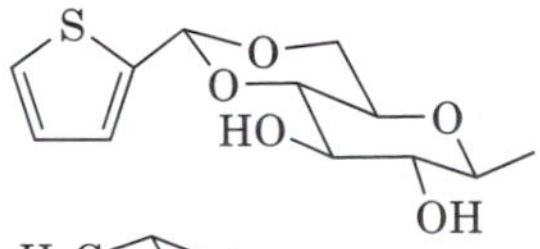

The lignan podophyllotoxin (**27**) is an ancient folk remedy (classically used for treatment of gout) found in the May apple, *Podophyllum peltatum* (265, 266). Interestingly podophyllotoxin binds to tubulin at a site distinct from that occupied by taxol and the vinca bases, although its molecular mode of action does not involve this in any obvious way, and modern clinical interest lies in its isomers instead. The isopodophyllotoxins are semisynthetic analogs resulting from acid-catalyzed reaction with suitably protected sugars followed by additional transformations. This results in attachment of the sugars to the ring system, with opposite stereochemistry to podophyllotoxin itself. Etoposide

(**27**) R = β-H
(**30**) R = α-H

(**28**) R =

(**29**) R =

(**28**) and teniposide (**29**) are the most prominent analogs so produced and these possess a different mode of action than that of podophyllotoxin (267)! Another diastereoisomer, picropodophyllotoxin (**30**), is produced by epimerization of podophyllotoxin at the lactone ring but it has not led to interesting analogs.

### 3.3.1   Etoposide

*3.3.1.1   Therapeutic Uses.* Etoposide is injected for the treatment of refractory testicu-

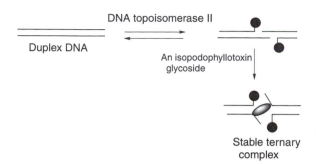

**Figure 3.19.** Illustration of the formation of a ternary complex between DNA, DNA topoisomerase II, and an isopodophyllotoxin glycoside.

lar tumors (268), small-cell lung cancer, and other less thoroughly established tumor regimes (269–271).

***3.3.1.2 Side Effects and Contraindications.*** Hypersensitivity to etoposide or to the cremophor EL vehicle are contraindications (272, 273). Myelosuppression, alopecia, nausea, vomiting, anorexia, diarrhea, hepatic damage, leukopenia, and thrombocytopenia are among other side effects. In a small number of patients treated with etoposide, a therapy-related leukemia results (274, 275).

***3.3.1.3 Pharmacokinetics.*** The drug is given by slow i.v. infusion and also can be given orally. About half of the administered dose is bioavailable and follows a biphasic elimination kinetic profile after infusion (276). Oxidative de-*O*-methylation is seen as a result of the action of human cytochrome P450 3A4 (277). Hydrolysis of the lactone is seen and conjugates are excreted in the urine (278). About half of the administered dose is excreted unchanged in the urine. Etoposide binds 76–96% to serum proteins and is displaced therefrom by bilirubin, so liver damage can require reduction in dosage (279, 280). The drug does not effectively pass the blood-brain barrier (281). The drug distributes best into small bowel, prostate, thyroid, bladder, spleen, and testicle but does not stay in the body for extended times after cessation of treatment (282).

***3.3.1.4 Mode of Action and Resistance.*** Etoposide administration causes DNA single- and double-strand breaks and DNA-protein links. This effect appears to be based on inhibition of topoisomerase II (283, 284). Furthermore, it is not an intercalator nor does it bind directly to DNA in the absence of the enzyme. Its action is most prominent in the late S or

early G-2 cell-cycle phases; thus cells do not enter the M-phase. The details of the interaction between topoisomerase II, DNA, and the isopodophyllotoxins are still emerging. Human topoisomerase II is a homodimeric enzyme responsible for manipulating DNA supercoiling, chromosomal condensation/decondensation, and unlinking of intertwined daughter chromosomes. These steps require energy gained by hydrolysis of ATP. Etoposide (and teniposide) act by stabilizing the covalent topoisomerase II-DNA intermediate and this stabilized ternary complex containing enzyme-cleaved DNA acts as a cellular poison. Figure 3.19 illustrates the formation of a ternary complex between DNA, DNA topoisomerase II, and an epipodophyllotoxin glycoside. During one topoisomerase II catalytic cycle, two ATP atoms are hydrolyzed. Etoposide and teniposide inhibit release of the ADP resulting from hydrolysis of the first ATP in a manner yet to be determined precisely, although the net result is that the ATPase activity of the enzyme is inhibited and resealing is prevented (285). Resistance takes the form of P-glycoprotein-related efflux (286), decreased expression and biosynthesis of topoisomerase II (287), or mutations in human topoisomerase IIα (288) or *p53* tumor-suppressor gene (289).

***3.3.1.5 Medicinal Chemistry (290, 291).*** Although etoposide is widely used, it is inconveniently water insoluble. A water-soluble prodrug, etopophos, has been introduced. This agent is rapidly and extensively converted back to etoposide after injection (267, 292).

### 3.3.2 Teniposide

***3.3.2.1 Therapeutic Uses.*** Teniposide is injected for the treatment of acute nonlymphocytic leukemia, Hodgkin's disease and other

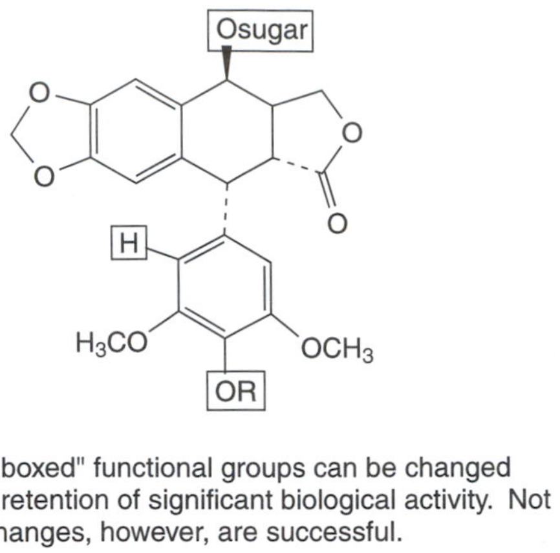

Osugar
H
H3CO
OCH3
OR
O
O
O
O
The "boxed" functional groups can be changed with retention of significant biological activity. Not all changes, however, are successful.

**Figure 3.20.** Summary of isopodophyllotoxin glycoside structure-activity relationships.

lymphomas, Kaposi's sarcoma, neuroblastoma, and other less thoroughly validated tumor situations. Unfortunately, in a number of children treated for leukemia, later development of a teniposide-generated leukemia may occur (267).

***3.3.2.2 Pharmacokinetics.*** The drug is given by slow i.v. infusion and also can be given orally. About half of the administered dose is bioavailable and follows a biphasic elimination kinetic profile after infusion. The drug does not efficiently pass the blood-brain barrier. Its tissue distribution and persistence are similar to those of etoposide (282). Hydrolysis of the lactone is seen and conjugates are excreted in the urine after oxidative demethylation, more so than with etoposide (277).

***3.3.2.3 Mode of Action and Resistance.*** Teniposide administration causes DNA single- and double-strand breaks and DNA-protein links. This effect appears to be based on inhibition of topoisomerase II because the drug is not an intercalator nor does it bind to DNA. Its action is most prominent in the late S or early G-2 cell-cycle phases; thus cells do not enter the M-phase. Resistance takes the form of P-glycoprotein–related efflux, decreased biosynthesis of topoisomerase II, or mutations in *p53* tumor-suppressor gene (293).

***3.3.2.4 Structure-Activity Relationships.*** Figure 3.20 illustrates the comparatively limited information relating to isopodophyllotoxin glycosides.

Detailed reviews of the properties of the isopodophyllotoxins are available (294–297).

## 4 DRUGS INTERFERING WITH TUBULIN POLYMERIZATION/DEPOLYMERIZATION

Microtubules provide a sort of cytoskeleton for cells so that they can maintain their shapes. They also form a sort of "rails," along which the chromosomes move during mitosis. These microtubules are constructed by the controlled polymerization of monomeric tubulin proteins of which there are two types, $\alpha$ and $\beta$. Figure 3.21 illustrates this process. The dimeric vinca alkaloids interfere with polymerization, thus preventing cell division by preventing the formation of new microtubules. The taxus alkaloids, on the other hand, promote the polymerization into new microtubules but stabilize these and prevent their remodeling. This prevents cell growth and repair. These mechanisms are compatible with the modes of action of other antitumor agents, thereby allowing for synergy when combined with these substances in cocktails.

### 4.1 Taxus Diterpenes

**4.1.1 Paclitaxel/Taxol.** Taxol (**31**), a diterpene ester, was discovered initially as a minor component in the bark of the Pacific yew at approximately the same time as camptothecin was found by the same group in quite another source

α tubulin
β tubulin
Tubulin
dimers
Slows
dynamics
Depolymerizes
Microtubule

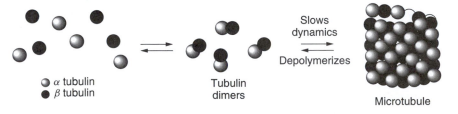

**Figure 3.21.** Tubulin polymerization to microtubules and their disassembly.

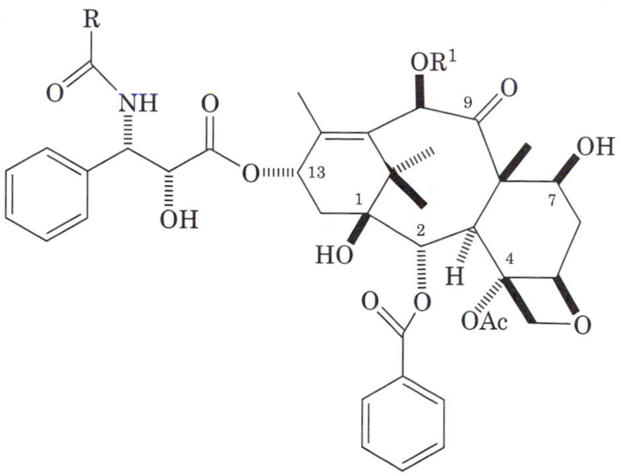

(31) R = Ph; R$^1$ = COCH$_3$
(32) R = $^t$BuO; R$^1$ = H

(223, 298). It took a great many years for taxol to come to the clinic because its initial performance in tumor-bearing mice was comparatively unimpressive. The progress to market was accelerated materially by the discovery of a novel (at the time) molecular mode of action. Taxol stimulates the formation of microtubules from tubulin and stabilizes this polymer, which stops cells from dividing (299–301).

After an enormous effort, semisynthesis from 10-deacetylbaccatin III (33), itself available in quantity from the much more renewable needles of various abundant yew species, proved economical and also allowed the synthesis of many analogs, of which docetaxel

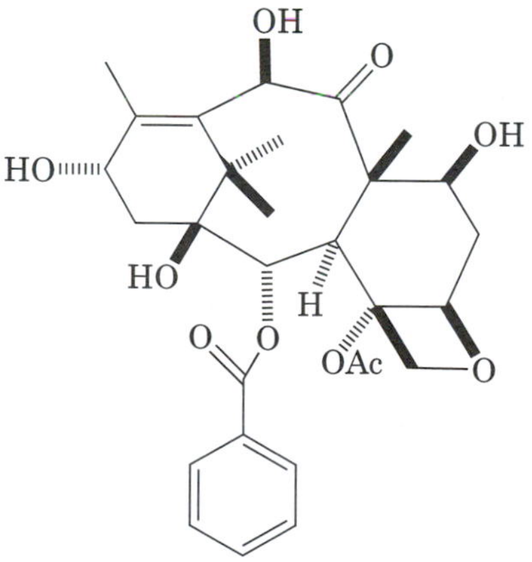

(33)

(32) is the most prominent (302). Many partial and total syntheses of taxol have been reported but none of these has as yet proved to be practical. Despite a long tradition allowing the discoverer to name an important compound, taxol was renamed paclitaxel for commercial purposes by the CRADA winner, Bristol-Myers Squibb.

**4.1.1.1 Clinical Uses.** Taxol is widely used in combinations for the therapy of refractory ovarian, breast, lung, esophageal, bladder, and head and neck cancers.

**4.1.1.2 Side Effects and Contraindications.** Taxol is a very toxic drug and it must be used with care. Bone marrow suppression (neutropenia) is a major dose-limiting side effect (303). A few patients develop severe cardiac conduction abnormalities (304). Patients with liver abnormalities may be especially sensitive to taxol. Fertility impairment and mutagenesis is seen in experimental animals, so taxol should only be given to pregnant patients with special care. Hypersensitivity is not uncommon and is often associated with the solvent in which this especially water insoluble drug must be administered (cremophor EL, a polyethoxylated castor oil) (305). Peripheral neuropathy occurs frequently and requires reduced dosage. In contrast to many other antitumor agents, extravasation, although causing discomfort and local pathologies, does not generally lead to severe necrosis. Gastrointestinal distress (diarrhea, fever, anemia, mucositis, nausea, and vomiting), alopecia, edema, and opportunistic infections are also reported by many patients. Paclitaxel is metabolized by the P450 system, so coadministration of drugs requiring processing by CYP2C8 and CYP3A4 requires caution. This statement is also true of docetaxol (306).

*4.1.1.3  Pharmacokinetics.* The drug is generally given by long-time (3 or 24 h) infusion at 3-week intervals or short-time (1 h) infusions at weekly intervals and is heavily protein bound (90–98%). Attempts to infuse the drug over very long times (96 h) have been made but involve significant practical limitations. The drug is excreted after a biphasic mode, with an initial rapid serum decline as the drug is distributed to the tissues and the overflow excreted. Return from peripheral tissues is slow and accounts for the second part of the excretion curve. The excretion half-life is fairly long ($\sim$ 13–55 h) (307–309). Extensive clearance other than by urine takes place, given that only 1–13% of the drug is found in the urine. Metabolism is primarily oxidative, with the main metabolite being the 6$\alpha$-hydroxy analog and lesser amounts of the 3$'$-parahydroxybenzamide and the 6$\alpha$-hydroxy, 3$'$-parahydroxybenzamide analogs being detected (310–312).

*4.1.1.4  Molecular Mode of Action and Resistance.* Taxol binds to the $\beta$-tubulin component and stimulates the formation of microtubules. These, however, do not break down, so the cell is unable to repair and to undergo mitosis (301). Resistant cells in culture are often seen to produce P-glycoprotein to excrete the drug (313), and also to have mutations in the $\beta$-tubulin component (314, 315). Whether this is responsible for clinical resistance is still being studied. Overexpression of the *ErbB2* gene occurs fairly often in breast tumors and this leads to overproduction of a transmembrane growth factor receptor belonging to the ErbB receptor tyrosine kinase subfamily. Cells with this characteristic have reduced responsiveness to taxol (316). Other growth factor anomalies involving, for example, EGFRviii and HER-2 are also seen in some cell lines (317).

**4.1.2  Docetaxel/Taxotere.** Docetaxel (**32**) is a semisynthetic analog of taxol prepared by a variety of chemical means, starting with the more abundant 10-deacetylbaccatin III (**33**) (302). It has found a significant place in anticancer chemotherapy but is still a significantly toxic drug that must be used with care.

*4.1.2.1  Clinical Applications.* Docetaxel is administered by i.v. infusion for the treatment of breast cancer, non-small-cell lung cancer and a variety of other less well established antitumor indications (318).

*4.1.2.2  Side Effects and Contraindications.* Many of the adverse effects of docetaxel are similar to those of taxol itself. The drug, however, is administered in polysorbate 80 rather than cremophor, so allergy is more commonly to the drug itself and can be severe. Poor liver function greatly enhances patient sensitivity to docetaxel. Severe fluid retention can also be observed. Patients are often administered corticoids before being exposed to docetaxel to assist in their tolerance of the drug. Myelotoxicity is potentially severe, so blood cell counts should be monitored. The toxicity of docetaxel is exaggerated when liver disease is present (319).

*4.1.2.3  Pharmacokinetics.* In contrast to paclitaxel, docetaxel has linear pharmacokinetics at the doses used in the clinic. As with paclitaxel, metabolism takes place in the liver through cytochrome P450 enzymic oxidation and the metabolites are excreted primarily in the bile. The involvement of P450 3A4 and 3A5 requires care in coadministering drugs that are also metabolized by these common enzymes (320). The metabolites are generally less toxic and less potent than docetaxel itself (321).

*4.1.2.4  Mode of Action and Resistance.* See paclitaxel.

*4.1.2.5  Chemical Transformations.* Although several total syntheses of taxol have been achieved during the last few years, low overall yields and high costs preclude them from being of commercial importance. Fortunately, isolation of the two taxol biosynthetic precursors, baccatin III and 10-deacetyl baccatin III, initially from the regenerable needles of the yew species *T. baccata* and subsequent development of highly efficient semisynthesis of taxol and taxotere from the above precursors have apparently solved the present supply problem of these precious drugs. Moreover, the semisynthetic routes have also provided means to carry out extensive SAR studies and

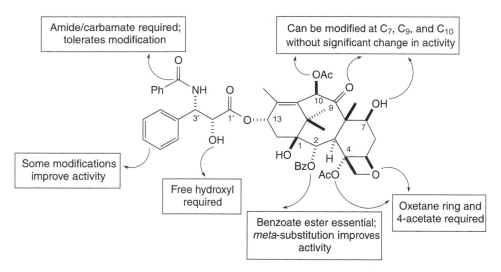

**Figure 3.22.** Structure-activity relationships of the taxol series.

consequent access to a large number of taxol analogs. The SAR studies on taxol have demonstrated that the C-3 phenylisoserinate side chain is an essential component for bioactivity, wherein limited modifications can be carried out at the 3'-phenyl and the 3'-$N$-benzoyl sites toward attenuating activity. Similarly, the presence of an intact oxetane ring in the diterpenoid core appears to be essential for bioactivity. Additionally, although the oxygen-bearing functionalities at C-7, C-9, and C-10 allow various modifications, an acetoxy at C-4 and an aroyloxy group at C-2 are indispensable for optimum activity. Interestingly, A-ring-contracted taxol analogs (A-nortaxol) were found to retain tubulin assembly activity, albeit with significantly diminished cytotoxicity (322–325).

The SARs of the taxol series are summarized in Fig. 3.22.

*4.1.2.6 Biosynthesis.* Taxol is one of the structurally more complex members of the diterpene family, characterized by the presence of the unusual taxane ring system. The initial steps in taxol biosynthesis involve the cyclization of geranylgeranyl diphosphate to taxa-4(5),11(12)diene, forming the taxane core structure. Subsequent cytochrome P450-mediated hydroxylation at C-5 of the olefin is followed by several other cytochrome P450-dependent oxygenations at C-1, C-2, C-4, C-7, C-9, C-10, and C-13 (the precise order of these

regiospecific oxidations, however, is not yet known) and CoA-dependent acylations of the taxane core, en route to taxol (326). Biosynthetically, the $N$-benzoyl phenylisoserine side chain has been shown to originate from phenylalanine and its further elaboration involves a late-stage esterification at C-13 of an advanced baccatin III intermediate (327).

The biosynthetic pathway between geranylgeranyl diphosphate and taxol remains to be fully elucidated but apparently passes through taxa-4(5),11(12)-diene and taxa-4(20),11(12)-diene-5α-ol, as shown in Fig. 3.23.

*4.1.2.7 Things to Come.* Recent interest has developed about the properties of the epothilones. These apparently bind to tubulin at approximately the taxol site but resistance by P-glycoprotein expulsion is apparently not significant with these fermentation products and they are active against a number of taxol-resistant cell lines. Elutherobin is another natural product binding to the taxane-binding site but this agent is cross-resistant with taxol (328). The clinical future of these agents is as yet uncertain and they have inspired much synthetic and biochemical attention.

## 4.2 Dimeric Vinca Alkaloids

The dimeric indole-indoline alkaloids were initially isolated from the Madagascan periwinkle, *Catharanthus rosea* (formerly named *Vinca rosea*). The plant was originally investi-

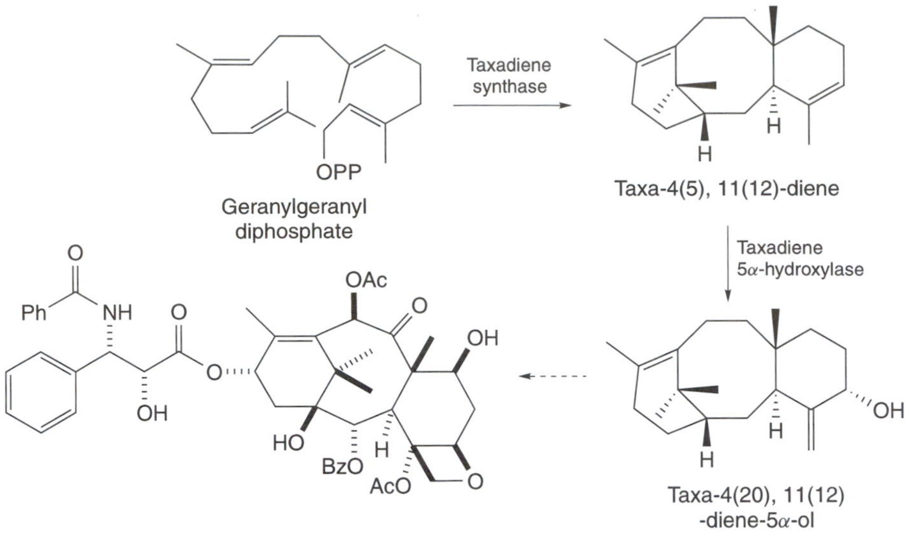

**Figure 3.23.** Biosynthesis of taxol.

gated as a follow-up to folkloric reports of hypoglycemic activity, and it was hoped to be of value in treating diabetes mellitus. This did not prove to be true but, during the investigation of extracts, certain fractions produced granulocytopenia and bone marrow suppression in animals. The active alkaloids were isolated from a matrix of indole alkaloids and were found to be active antileukemic agents against P-1534 cells. Development for human use followed after extensive experimentation. Four of these unsymmetrical dimeric alkaloids ultimately found use as antitumor agents. The best agents contain C-linked vindoline and 16β-carbomethoxy velbanamine units. Apparently minor structural differences between the alkaloids led to major differences in potency and utility (329). Because of their relative scarcity and medicinal value, these dimers have been attractive synthetic targets and a rich synthetic and biosynthetic literature has grown up around them. Inspection of their structures readily leads to the inference that they are the products of unsymmetrical free-radical coupling. After much work, two groups, those of Potier in France (330, 331) and of Kutney in Canada (332), succeeded in stereoselective dimerization. Treatment of the abundant alkaloid catharanthine as its N-oxide with trifluoroacetic anhydride leads to a fragmentation into an enamine that can be intercepted by vindoline, another comparatively abundant alkaloid, and the product reduced by sodium borohydride. Under low temperature conditions the condensation is stereospecific in the desired manner. This is believed to reflect a concerted interaction. When the reaction is run at higher temperatures, a mixture of diastereomers is produced instead. This is believed to be the result of a stepwise condensation. Variation of this chemistry leads to the formation of useful synthetic analogs and interconversions into natural analogs as well. Figure 3.24 illustrates the partial chemical synthesis of vinca dimers with the natural stereochemistry.

It is interesting to note that dolastatin 10, a marine natural product with exceptional antitumor properties, also binds near to the vinca alkaloid binding domain and inhibits tubulin polymerization (333, 334).

### 4.2.1 Vinblastine (Velban)

*4.2.1.1 Medicinal Uses.* Vinblastine sulfate (**34**) is given i.v. with great care, to avoid damaging extravasation (123), for the treatment of metastatic testicular tumors (usually in combination with bleomycin and cisplatin). Various lymphomas also may respond. It has only limited neurotoxicity, thus enhancing its utility.

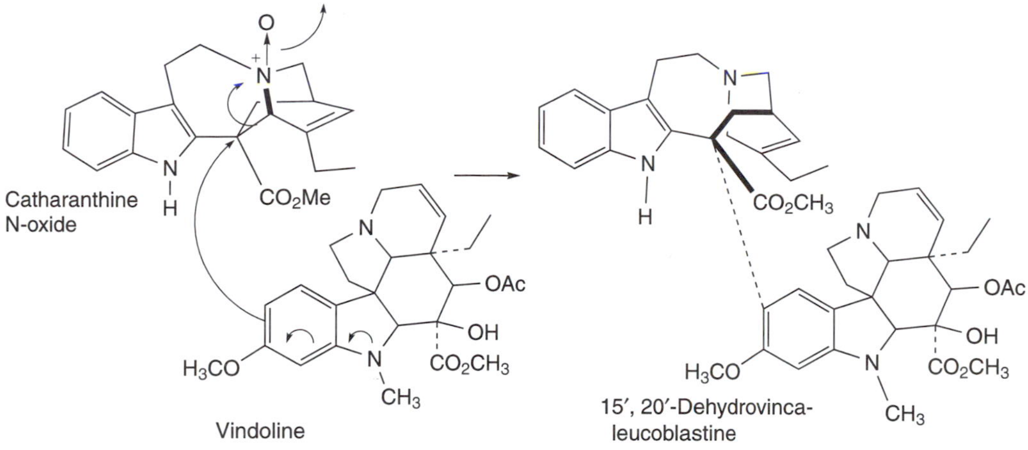

**Figure 3.24.** Partial chemical synthesis of vinca dimers with the natural stereochemistry.

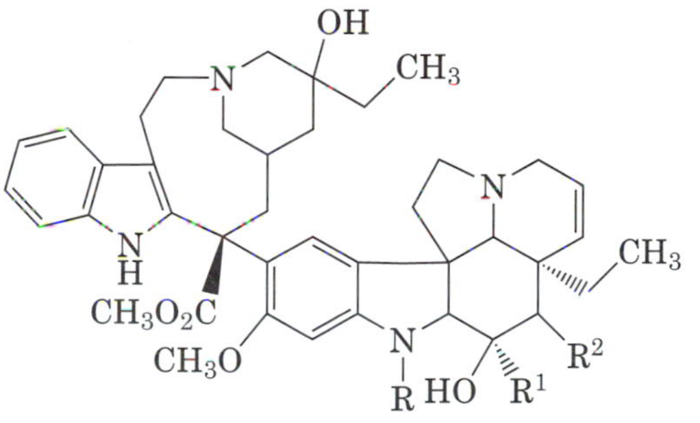

| Compd. | R | R$^1$ | R$^2$ |
|--------|-----|--------|--------|
| (34) | CH$_3$ | CO$_2$CH$_3$ | OCOCH$_3$ |
| (35) | CHO | CO$_2$CH$_3$ | OCOCH$_3$ |
| (36) | CH$_3$ | CONH$_2$ | OH |

**4.2.1.2 Side Effects and Contraindications.** Vinblastine causes severe tissue necrosis upon extravasation. Mild neurotoxicity and myelosuppression occur and these effects should be monitored to prevent significant toxicity to the patient. The other side effects of vinblastine are common to antitumor agents (alopecia, ulceration, nausea, etc.).

**4.2.1.3 Pharmacokinetic Features.** Vinblastine is extensively metabolized in the liver and the metabolites are excreted as conjugates in the bile. About 15% of the drug is found unchanged in the urine (335, 336). Oxidative degradation of the catharanthus alkaloids occurs in part catalyzed by the action of myeloperoxidase.

Cleavage occurs between C-20' and C-21' and is structurally facilitated by the presence of a C-20' hydroxyl moiety (337). Peroxidase and ceruloplasmin also catalyze oxidative transformations of vinblastine (338, 339).

**4.2.1.4 Medicinal Chemical Transformations.** Hydrolysis of the acetyl group at C-4 of vinblastine abolishes its antileukemic activity. Furthermore, acetylation of the free hydroxyl groups also inactivates the molecule. The dimeric structure is required as is the stereochemistry of the point of attachment. Hydrogenation of the olefinic linkage and reduction to the carbinol also greatly diminish potency. Thus the antileukemic activity is substantially dependent on the specific structural groups present in the molecule.

**4.2.1.5 Molecular Mode of Action and Resistance.** Vinblastine blocks cells in the M-phase. It binds to the β-subunit of tubulin in its dimer in a one-to-one complex, thus preventing its polymerization into microtubules. The binding site is near to, but different from, that of colchicines but similar to that of maytansine and rhizoxin (although the consequences of binding of the latter are different from those of vinca binding). Nontubulin oligomers form from the component parts as a consequence, and preformed tubulin depolymerizes and the complex with vinblastine crystallizes (340). Failure to produce functional microtubules prevents proper chromosome formation and thus prevents cell division. The blocked cells then die (become apoptotic). Other cellular processes dependent on microtubules are also interfered with, although the blockade of chromosome formation is regarded as central to their action (341). Resistance mainly takes the form of elaboration of P-glycoproteins that export vinblastine, and this cross-resistance is broad enough to include the other vinca alkaloids and other antitumor agents as well (342). Resistance is also attributed to alterations in the tubulin subunits (343).

### 4.2.2 Vincristine (Oncovin, Vincasar PFS)

*4.2.2.1 Medical Uses.* Vincristine (**35**) is a common component of antitumor cocktails used in treating acute lymphoblastic leukemia and solid tumors of youngsters and in adult lymphoma. It is commonly used with corticosteroids. Study of its use in the form of liposomes has also been carried out (344). Its use produces limited myelosuppression, so it is an attractive component in cocktails. The reduced myelotoxicity may be attributable to oxidative degradation of the drug by myeloperoxidase, a heme-centered peroxidase enzyme present in acute myeloblastic leukemia but not in acute lymphoblastic leukemia (345, 346).

*4.2.2.2 Pharmacokinetic Features.* Vincristine is extensively metabolized in the liver and the metabolites are excreted as conjugates in the bile. About 15% of the drug is found unchanged in the urine.

*4.2.2.3 Side Effects and Contraindications.* Vincristine causes severe tissue necrosis upon extravasation (123). Neurotoxicity is a significant potential problem with vincristine and is often treated in part by reducing the dose of the drug (347). Myelosuppression also occurs but to a lesser extent and this effect should be monitored to prevent significant toxicity to the patient. Gout can occur with vincristine administration and can be controlled by use of allopurinol. The other side effects of vinblastine are common to antitumor agents (alopecia, ulceration, nausea, diarrhea, etc.).

*4.2.2.4 Resistance.* Resistance to vincristine is mediated in part by export resulting from the multidrug resistance protein and, interestingly, is characterized by cotransport with reduced glutathione (348).

### 4.2.3 Vinorelbine (Navelbine)

*4.2.3.1 Medicinal Uses.* Vinorelbine (**37**) is used against non-small-cell lung cancer and

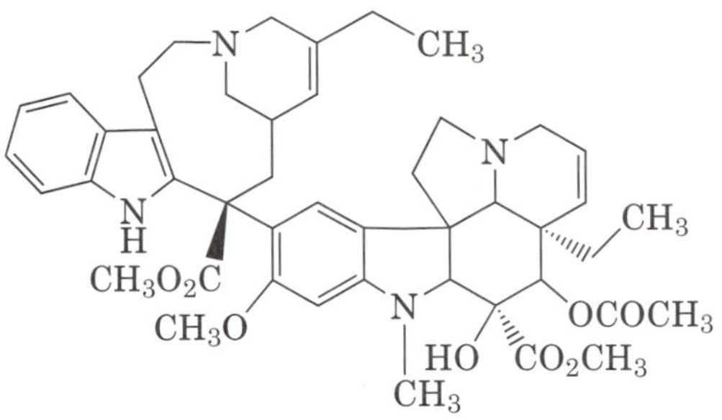

(37)

against breast cancer (349–353). It appears to be intermediate in its neurotoxicity and myelosuppression compared to that of the other vinca antitumor agents (354).

*4.2.3.2 Pharmacokinetic Features.* Vinorelbine is extensively metabolized in the liver and the metabolites are excreted as conjugates in the bile. About 15% of the drug is found unchanged in the urine.

*4.2.3.3 Side Effects and Contraindications.* Vinorelbine causes severe tissue necrosis upon extravasation as well as phlebitis (355). Prior i.v. administration of cimetidine partially avoids this. Mild neurotoxicity and myelosuppression occur and these effects should be monitored to prevent significant toxicity to the patient. Its most notable toxic side effect appears to be granulocytopenia. The other side effects of vinorelbine are common to antitumor agents (alopecia, ulceration, nausea, etc.).

*4.2.3.4 Things to Come.* Vindisine (**36**) is an analog prepared from vinblastine (**34**). Its antitumor spectrum, however, is more closely similar to that of vincristine. Clinical studies show activity against acute leukemia; lung cancer; breast carcinoma; squamous cell carcinoma of the esophagous, head, and neck; Hodgkin's disease; and non-Hodgkin's lymphomas. Its toxicities include myelosuppression and neurotoxicity. Despite these promising findings, it has yet to be introduced into the clinic (356).

Vinflunine (**38**) is a dimeric alkaloid, containing two gem-fluorine atoms, prepared by a mechanistically interesting process using super acidic reactants on vinorelbine. This com-

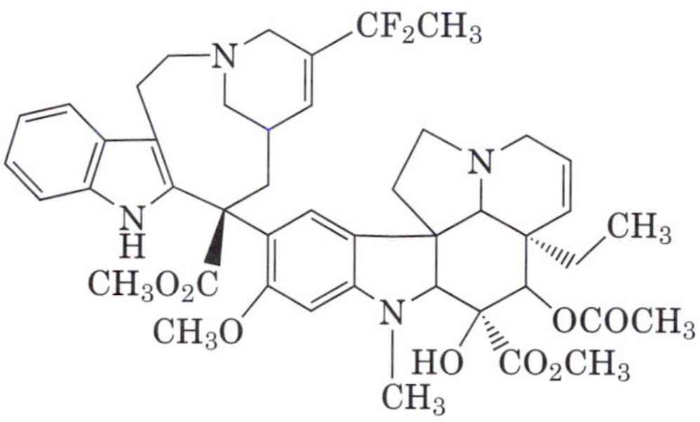

**(38)**

pound has improved antitumor potency in a variety of model tumor systems, shows less drug resistance (357), and has entered clinical trials (358, 359).

Some additional reviews of this topic are available (328, 360–362).

## REFERENCES

1. A. B. daRocha, R. M. Lopes, and G. Schwartsmann, *Curr. Opin. Pharmacol.*, **1**, 364 (2001).

2. A. K. Mukherjee, S. Basu, N. Sarkar, and A. C. Ghosh, *Curr. Med. Chem.*, **8**, 1467 (2001).

3. J. M. Pezzuto, *Biochem. Pharmacol.*, **53**, 121 (1997).

4. G. R. Pettit, G. M. Cragg, and C. L. Herald, Biosynthetic Products for Cancer Chemotherapy, **Vols. 1–4**, Elsevier, New York, 1984, 1985, 1989, and earlier.

5. G. R. Pettit, F. H. Pierson, and C. L. Herald, *Anticancer Drugs from Animals, Plants, and Microorganisms*, Wiley Interscience, New York, 1994.

6. G. M. Cragg, *Med. Res. Rev.*, **18**, 315 (1998).

7. NIH Workshop: Bioassays for Discovery of Antitumor and Antiviral Agents from Natural Sources, 18–19 October 1988.

8. F. A. Valeriote, T. H. Corbett, and L. H. Baker, Eds., *Cytotoxic Anticancer Drugs: Models and Concepts for Drug Discovery and Development*, Kluwer Academic, Dordrecht, The Netherlands, 1992.

9. G. M. Cragg and D. J. Newman, *Cancer Invest.*, **17**, 153 (1999).

10. G. V. McEvoy, Ed., *AHFS Drug Information 2001*, American Society of Health System Pharmacists, Bethesda, MD, 2001.

11. J. G. Hardman, L. E. Limbard, and A. G. Gilman, Eds., *Goodman and Gilman's Pharmaco-logical Basis of Therapeutics*, 10th ed., McGraw-Hill, New York, 2001.

12. S. A. Waksman and H. B. Woodruff, *Proc. Soc. Exp. Biol. Med.*, **45**, 609 (1940).

13. P. Calabrisi and B. A. Chabner in J. G. Hardman, L. E. Limbard, and A. G. Goodman, Eds., *Goodman and Gilman's Pharmacological Basis of Therapeutics*, 10th ed., McGraw-Hill, New York, 2001, p. 1425.

14. M. J. Coppes, K. Jorgenson, and J. P. Arlette, *Med. Pediatr. Oncol.*, **31**, 128 (1997).

15. R. L. Juliano and D. Stamp, *Biochem. Pharmacol.*, **27**, 21 (1978).

16. R. S. Benjamin, S. W. Hall, M. A. Burgess, et al., *Cancer Treat. Rep.*, **60**, 289 (1976).

17. M. H. Tattersall, J. E. Sodergren, S. K. Dngupta, et al., *Clin. Pharmacol. Ther.*, **17**, 701 (1975).

18. K. Nakajima, T. Tanaka, M. Neya, and K. Okawa, *Pept. Chem.*, **19**, 143 (1982).

19. K. Okawa, K. Nakajima, and T. Tanaka, *J. Heterocycl. Chem.*, **17**, 1815 (1980).

20. K. Okawa, K. Nakajima, and T. Tanaka, *Pept. Chem.*, **15**, 131 (1977).

21. T. Tanaka, K. Nakajima, K. Okawa, *Bull. Chem. Soc. Jpn.*, **53**, 1352 (1980).

22. G. P. Vlasov, V. N. Lashkov, S. V. Kulikov, and O. F. Ginzburg, *Zh. Org. Khim.*, **14**, 1961 (1978).

23. J. Meienhofer, *J. Am. Chem. Soc.*, **92**, 3771 (1970).

24. J. Meienhofer, *Experientia*, **24**, 776 (1968).

25. H. Brockmann and H. Lackner, *Chem. Ber.*, **101**, 1312 (1968).

26. H. Brockmann and H. Lackner, *Chem. Ber.*, **101**, 2231 (1968).

27. G. Tong and J. Nielsen, *Bioorg. Med. Chem.*, **4**, 693 (1996).

28. H. Brockmann, G. Pampus, and R. Mecke, *Chem. Ber.*, **92**, 3082 (1959).

29. H. Brockmann, J. Ammann, and W. Mueller, *Tetrahedron Lett.*,3595 (1966).

30. J. Meienhofer, *Cancer Chemother. Rep.*, **58**, 21 (1974).

31. S. K. Sengupta, J. E. Anderson, K. Yury, et al., *J. Med. Chem.*, **24**, 1052 (1981).

32. M. S. Madhavarao, M. Chaykovsky, and S. K. Sengupta, *J. Med. Chem.*, **21**, 958 (1978).

33. H. Brockmann, W. Mueller, and H. Peterssen-Borstel, *Tetrahedron Lett.*,3531 (1966).

34. S. K. Sengupta, D. P. Rosenbaum, R. K. Sehgal, et al., *J. Med. Chem.*, **31**, 1540 (1988).

35. R. K. Sehgal, B. Almassian, D. P. Rosenbaum, et al., *J. Med. Chem.*, **30**, 1626 (1987).

36. S. K. Sengupta, J. E. Anderson, and C. Kelley, *J. Med. Chem.*, **25**, 1214 (1982).

37. C. W. Mosher, D. Y. Lee, R. M. Enanoza, et al., *J. Med. Chem.*, **22**, 1051 (1979).

38. S. K. Sengupta, D. H. Trites, M. S. Madhavarao, and W. R. Beltz, *J. Med. Chem.*, **22**, 297 (1979).

39. S. K. Sengupta, S. K. Tinter, and E. J. Modest, *J. Heterocycl. Chem.*, **15**, 129 (1978).

40. S. K. Sengupta, M. S. Madhavarao, C. Kelly, and J. Blondin, *J. Med. Chem.*, **26**, 1631 (1983).

41. S. K. Sengupta, C. Kelly, and R. K. Sehgal, *J. Med. Chem.*, **28**, 620 (1985).

42. A. K. A. Chowdhury, J. Brown, and R. B. Longmore, *J. Med. Chem.*, **21**, 607 (1978).

43. S. Moore, M. Kondo, M. Copeland, et al., *J. Med. Chem.*, **18**, 1098 (1975).

44. R. K. Sehgal, B. Almassian, D. P. Rosenbaum, et al., *J. Med. Chem.*, **31**, 790 (1988).

45. F. Seela, *J. Med. Chem.*, **15**, 684 (1972).

46. H. Brockmann and F. Seela, *Chem. Ber.*, **104**, 2751 (1971).

47. A. Lifferth, I. Bahner, H. Lackner, and M. Schafer, *Z. Naturforsch. B: Chem. Sci.*, **54**, 681 (1999).

48. E. N. Glibin, N. G. Plekhanova, D. V. Ovchinnikov, and Z. I. Korshunova, *Zh. Org. Khim.*, **32**, 406 (1996).

49. S. Kamitori and F. Takusagawa, *J. Mol. Biol.*, **225**, 445 (1992).

50. X. Liu, H. Chen, and D. J. Patel, *J. Biomol. NMR*, **1**, 323 (1991).

51. R. M. Wadkins, E. A. Jares-Erijman, R. Klement, et al., *J. Mol. Biol.*, **262**, 53 (1996).

52. H. Chen, X. Liu, and D. J. Patel, *J. Mol. Biol.*, **258**, 457 (1996).

53. R. L. Rill, G. A. Marsch, and D. E. Graves, *J. Biomol. Struct. Dyn.*, **7**, 591 (1989).

54. F. M. Chen, *Biochemistry*, **27**, 6393 (1988).

55. R. L. Jones, E. V. Scott, G. Zon, et al., *Biochemistry*, **27**, 6021 (1988).

56. H. M. Sobell, *Mol. Biol.*, **13**, 153 (1973).

57. R. K. Sehgal, S. K. Sengupta, D. J. Waxman, and A. I. Tauber, *Anticancer Drug Des.*, **1**, 13 (1985).

58. T. Knutsen, L. A. Mickley, T. Ried, et al., *Genes Chromosomes Cancer*, **23**, 44 (1998).

59. J. Prados, C. Melguizo, A. Fernandez, et al., *J. Pathol.*, **180**, 85 (1996).

60. J. Prados, C. Melguizo, A. Fernandez, et al., *Cell. Mol. Biol.*, **40**, 137 (1994).

61. S. E. Devine, V. Ling, and P. W. Melera, *Proc. Natl. Acad. Sci. USA*, **89**, 4564 (1992).

62. C. D. Selassie, C. Hansch, and T. A. Khwaja, *J. Med. Chem.*, **33**, 1914 (1990).

63. M. Inaba and R. K. Johnson, *Cancer Res.*, **37**, 4629 (1977).

64. G. H. Jones, *Antimicrob. Agents Chemother.*, **44**, 1322 (2000).

65. S. T. Crooke in B. A. Chabner and H. M. Pinedo, Eds., *The Cancer Pharmacology Annual*, Excerpta Medica, Amsterdam, 1983, p. 69.

66. E. Katz, H. A. Lloyd, and A. B. Mauger, *J. Antibiot. (Tokyo)*, **43**, 731 (1990).

67. S. A. Waksman, Ed., *Actinomycin: Nature, Formation, and Activities*, Wiley-Interscience, New York, 1968.

68. H. Brockmann, *Progress in the Chemistry of Organic Natural Products*, Vol. **18**, Springer, New York, 1960, p. 1.

69. J. Meienhofer and E. Atherton in D. Perlman, Ed., *Structure-Activity Relationships among the Semisynthetic Antibiotics*, Academic Press, New York, 1977, p. 427.

70. A. B. Mauger in P. G. Sames, Ed., *Topics in Antibiotic Chemistry*, Ellis Horwood, Chichester, UK, 1980, p. 224.

71. A. B. Mauger in S. V. Wilman, Ed., *The Chemistry of Antitumor Agents*, Blackie, Glasgow/London, 1990, pp. 403–409.

72. M. A. Friedman, *Recent Results Cancer Res.*, **63**, 152 (1978).

73. W. E. Evans, G. C. Yee, W. R. Coom, C. B. Pratt, and A. A. Green, *Drug Intell. Clin. Pharm.*, **16**, 448 (1982).

74. Y. Muraoka and T. Takita, *Cancer Chemother. Biol. Response Modif.*, **10**, 40 (1988).

75. J. S. Lazo, *Cancer Chemother. Biol. Response Modif.*, **18**, 39 (1999).

76. K. Jules-Elysee and D. A. White, *Clin. Chest Med.*, **11**, 1 (1990).

77. R. A. Chisholm, A. K. Dixon, M. V. Williams, and R. T. Oliver, *Cancer Chemother. Pharmacol.*, **30**, 158 (1992).

78. J. Hay, S. Shahzeidi, and G. Laurent, *Arch. Toxicol.*, **65**, 81 (1991).

79. R. T. Dorr, *Semin. Oncol.*, **19**, 3 (1992).

80. M. J. Rishel and S. M. Hecht, *Org. Lett.*, **3**, 2867 (2001).

81. J. A. Highfield, L. K. Mehta, J. Parrick, and P. Wardman, *Bioorg. Med. Chem.*, **8**, 1065 (2000).

82. L. Huang, J. C. Quada, and J. W. Lown, *Heterocycl. Commun.*, **1**, 335 (1995).

83. M. D. Levin, K. Subrahamanian, H. Katz, et al., *J. Am. Chem. Soc.*, **102**, 1452 (1980).

84. T. Tsuchiva, T. Miyake, S. Kageyama, et al., *Tetrahedron Lett.*, **22**, 1413 (1981).

85. T. Ohgi and S. M. Hecht, *J. Org. Chem.*, **46**, 3761 (1981).

86. C. Leitheiser, M. J. Rishel, X. Wu, and S. M. Hecht, *Org. Lett.*, **2**, 3397 (2000).

87. D. L. Boger, S. Teramoto, and H. Cai, *Bioorg. Med. Chem.*, **5**, 1577 (1997).

88. D. L. Boger, S. Teramoto, and H. Cai, *Bioorg. Med. Chem.*, **4**, 179 (1996).

89. D. L. Boger, S. L. Colletti, S. Teramoto, et al., *Bioorg. Med. Chem.*, **3**, 1281 (1995).

90. W. J. Vloon, C. Kruk, U. K. Pandit, et al., *J. Med. Chem.*, **30**, 20 (1987).

91. T. Takita, A. Fujii, T. Fukuoka, and H. Umezawa, *J. Antibiot. (Tokyo)*, **26**, 252 (1973).

92. S. Saito, Y. Umezawa, T. Yoshioka, et al., *J. Antibiot. (Tokyo)*, **36**, 92 (1983).

93. T. Takita, Y. Umezawa, S. Saito, et al., *Proc. Am. Pept. Symp.*, **7**, 29 (1981).

94. Y. Aoyagi, K. Katano, H. Suguna, et al., *J. Am. Chem. Soc.*, **104**, 5537 (1982).

95. S. Saito, Y. Umezawa, T. Yoshioka, et al., *J. Antibiot. (Tokyo)*, **36**, 95 (1983).

96. A. K. Choudhury, Z. F. Tao, and S. M. Hecht, *Org. Lett.*, **3**, 1291 (2001).

97. B. Shen, L. Du, C. Sanchez, et al., *Bioorg. Chem.*, **27**, 155 (1999).

98. T. Takita and Y. Muraoka, in H. Kleinkauf and H. Von Dohren, Eds., *Biochemistry of Peptide Antibiotics*, de Gruyter, Berlin, 1990, pp. 289–309.

99. T. Nakatani, A. Fujii, H. Naganawa, et al., *J. Antibiot. (Tokyo)*, **33**, 717 (1980).

100. A. Fujii, T. Takita, N. Shimada, and H. Umezawa, *J. Antibiot. (Tokyo)*, **27**, 73 (1974).

101. S. M. Hecht, *Fed. Proc.*, **45**, 2784 (1986).

102. H. Umezawa, *Lloydia*, **40**, 67 (1977).

103. P. R. Twentyman, *Pharmacol. Ther.*, **23**, 417 (1983).

104. S. M. Hecht, *J. Nat. Prod.*, **63**, 158 (2000).

105. M. Nakamura and J. Peisach, *J. Antibiot. (Tokyo)*, **41**, 638 (1988).

106. S. M. Sebti and J. S. Lazo, *Pharmacol. Ther.*, **38**, 321 (1988).

107. I. M. Lefterov, R. P. Koldamova, and J. S. Lazo, *FASEB J.*, **14**, 1837 (2000).

108. A. Caputo, Ed., *Biological Basis for the Clinical Effect of Bleomycin*, S. Karger AG, Basel, 1976.

109. B. I. Sikic, M. Rozensweig, and S. Carter, Eds., *Bleomycin Chemotherapy*, Academic Press, San Diego, CA, 1985.

110. S. K. Carter, S. T. Crooke, and H. Umezawa, Eds., *Bleomycin: Current Status and New Developments*, Academic Press, San Diego, CA, 1978.

111. B. Meunier, Ed., *Metal-oxo and Metal-peroxo Species in Catalytic Oxidations*, Springer-Verlag, New York, 2000.

112. H. Umezawa, *Pure Appl. Chem.*, **28**, 665 (1971).

113. H. Umezawa, *Rev. Infect. Dis.*, **9**, 147 (1987).

114. T. Hata, Y. Sano, R. Sugawara, et al., *J. Antibiot. (Tokyo)*, **9**, 141 (1956).

115. S. Wakaki, H. Marumo, K. Tomioka, et al., *Antibiot. Chemother.*, **8**, 228 (1959).

116. C. DeBoer, A. Dietz, N. E. Lummus, and G. M. Savage, *Antimicrob. Agents Chemother. Ann.*, 17 (1961).

117. J. P. Sculier, L. Ghisdal, T. Berghmans, et al., *Br. J. Cancer*, **84**, 1150 (2001).

118. W. T. Bradner, *Cancer Treat. Rev.*, **27**, 35 (2001).

119. J. Verweij, A. Sparreboom, and K. Nooter, *Cancer Chemother. Biol. Response Modif.*, **18**, 46 (1999).

120. J. Verweij, *Cancer Chemother. Biol. Response Modif.*, **17**, 46 (1997).

121. P. J. Medina, J. M. Sipols, and J. N. George, *Curr. Opin. Hematol.*, **8**, 286 (2001).

122. Y. Nishiyama, Y. Komaba, H. Kitamura, and Y. Katayama, *Intern. Med.*, **40**, 237 (2001).

123. E. Kassner, *J. Pediatr. Oncol. Nurs.*, **17**, 135 (2000).

124. J. S. Patel and M. Krusa, *Pharmacotherapy*, **19**, 1002 (1999).

125. V. J. Spanswick, J. Cummings, A. A. Ritchie, and J. F. Smyth, *Biochem. Pharmacol.*, **56**, 1497 (1998).

126. R. M. Phillips, A. M. Burger, P. M. Loadman, et al., *Cancer Res.*, **60**, 6384 (2000).

127. V. J. Spanswick, J. Cummings, and J. F. Smyth, *Gen. Pharmacol.*, **31**, 539 (1998).

128. M. Theisohn, R. Fischbach, R. Joseph, et al., *Int. J. Clin. Pharmacol. Ther.*, **35**, 72 (1997).

129. G. S. Kumar, R. Lipman, J. Cummings, and M. Tomasz, *Biochemistry*, **36**, 14128 (1997).

130. M. Tomasz and R. Lipman, *Biochemistry*, **20**, 5056 (1981).

131. B. S. Iyengar, R. T. Dorr, N. G. Shipp, and W. A. Remers, *J. Med. Chem.*, **33**, 253 (1990).

132. W. A. Remers and B. S. Iyengar in G. Lukacs and M. Ohno, Eds., *Recent Progress in the Chemical Synthesis of Antibiotics*, Springer-Verlag, Berlin, 1990.

133. W. A. Remers and R. T. Dorr in S. W. Pelletier, Ed., *Alkaloids: Chemical and Biological Properties*, John Wiley & Sons, New York, 1988.

134. R. W. Frank and M. Tomasz in E. Wilman, Ed., *The Chemistry of Antitumor Agents*, Blackie, Glasgow/London, 1989.

135. F. Nakatsubo, A. J. Cocuzza, D. E. Keeley, and Y. Kishi, *J. Am. Chem. Soc.*, **99**, 4835 (1977).

136. T. Fukuyama and I. Yang, *J. Am. Chem. Soc.*, **109**, 7881 (1987).

137. T. Fukuyama and I. Yang, *J. Am. Chem. Soc.*, **111**, 8303 (1989).

138. M. M. Paz, A. Das, Y. Palom, Q. Y. He, and M. Tomasz, *J. Med. Chem.*, **44**, 2834 (2001).

139. K. R. Kunz, B. S. Iyengar, R. T. Dorr, et al., *J. Med. Chem.*, **34**, 2286 (1991).

140. M. Tomasz and Y. Palom, *Pharmacol. Ther.*, **76**, 73 (1997).

141. M. J. Boyer, *Oncol. Res.*, **9**, 391 (1997).

142. R. H. Hargreaves, J. A. Hartley, and J. Butler, *Front. Biosci.*, **5**, E172 (2000).

143. R. P. Baumann, W. F. Hodnick, H. A. Seow, et al., *Cancer Res.*, **61**, 7770 (2001).

144. J. Cummings, *Drug Resist. Update*, **3**, 143 (2000).

145. M. He, P. J. Sheldon, and D. H. Sherman, *Proc. Natl. Acad. Sci. USA*, **98**, 926 (2001).

146. M. F. Belcourt, P. G. Penketh, W. F. Hodnick, et al., *Proc. Natl. Acad. Sci. USA*, **96**, 10489 (1999).

147. R. T. Dorr, *Semin. Oncol.*, **32**, 1584 (1988).

148. S. K. Carter and S. T. Cook, Eds., *Mitomycin C: Current Status and New Developments*. Academic Press, San Diego, CA, 1980.

149. W. E. Grundy, A. W. Goldstein, C. J. Rickher, et al., *Antibiot. Chemother.*, **3**, 1215 (1953).

150. B. J. Kennedy and J. L. Torkelson, *Med. Pediatr. Oncol.*, **24**, 327 (1995).

151. G. J. Hill 2nd, N. Sedransk, D. Rochlin, et al., *Cancer*, **30**, 900 (1972).

152. B. J. Kennedy, *J. Urol.*, **107**, 429 (1972).

153. N. W. Ream, C. P. Perlia, J. Wolter, and S. G. Taylor 3rd, *JAMA*, **204**, 1030 (1968).

154. N. Zojer, A. V. Keck, and M. Pecherstorfer, *Drug Saf.*, **21**, 389 (1999).

155. S. H. Ralston, *Cancer Surv.*, **21**, 179 (1994).

156. J. P. Bilezikian, *N. Engl. J. Med.*, **326**, 1196 (1992).

157. G. Harrington, K. B. Olson, J. Horton, et al., *N. Engl. J. Med.*, **283**, 1172 (1970).

158. B. Singh and R. S. Gupta, *Cancer Res.*, **45**, 2813 (1985).

159. D. J. Ahr, S. J. Scialla, and D. B. Kimbali, Jr., *Cancer*, **41**, 448 (1978).

160. K. Fang, C. A. Koller, N. Brown, et al., *Ther. Drug Monit.*, **14**, 255 (1992).

161. S. Chakrabarti, M. A. Mir, and D. Dasgupta, *Biopolymers*, **62**, 131 (2001).

162. D. Dasgupta, B. K. Shashiprabha, and S. K. Podder, *Indian J. Biochem. Biophys.*, **16**, 18 (1979).

163. O. A. Kiseleva, N. G. Volkova, E. A. Stukacheva, et al., *Mol. Biol.*, **7**, 741 (1974).

164. S. J. Wimalawansa, *Semin. Arthritis Rheum.*, **23**, 267 (1994).

165. E. P. Cortes, J. F. Holland, R. Moskowitz, and E. Depoli, *Cancer Res.*, **32**, 74 (1972).

166. C. Minkin, *Calcif. Tissue Res.*, **13**, 249 (1973).

167. G. Carulli, M. Petrini, A. Marinia, et al., *Haematologica*, **75**, 516 (1990).

168. M. Tagashira, T. Kitagawa, S. Isonishi, et al., *Biol. Pharm. Bull.*, **23**, 926 (2000).

169. H. Umezawa in V. T. DiVita Jr., and H. Busch, Eds., Methods in Cancer Research, Vol. **XVI**, Part A, Academic Press, New York, 1979, pp. 43–72.

170. S. E. Wohlert, E. Kuenzel, R. Machinek, et al., *J. Nat. Prod.*, **62**, 119 (1999).

171. F. Lombo, E. Kunzel, L. Prado, et al., *Angew. Chem. Int. Ed. Engl.* **39**, 796 (2000).

172. M. J. Lozano, L. L. Remseng, L. M. Quiros, et al., *J. Biol. Chem.*, **275**, 3065 (2000).

173. G. Blanco, H. Fu, C. Mendez, et al., *Chem. Biol.*, **3**, 193 (1996).

174. L. Prado, E. Fernandez, U. Weissbach, et al., *Chem. Biol.*, **6**, 19 (1999).

175. J. W. Lown, *Chem. Soc. Rev.*, **22**, 165 (1993).

176. F. Arcamone, Doxorubicin Anticancer Antibiotics (Medicinal Chemistry Monograph), Vol. **17**, Academic Press, New York, 1981.

177. R. Silber, L. F. Liu, M. Israel, et al., *NCI Monogr.*, 111 (1987).

178. D. Horton, W. Priebe, and O. J. Varela, *Antibiotics*, **37**, 1635 (1984).

179. R. C. Young, R. F. Ozols, and C. E. Myers, *N. Engl. J. Med.*, **305**, 139 (1981).

180. J. Cortes, E. Estrey, S. O'Brien, et al., *Cancer*, **92**, 7 (2001).

181. B. N. Hortobagyi, *Drugs*, **54S4**, 1 (1997).

182. S. W. Langer, M. Sehested, and P. B. Jensen, *Clin. Cancer Res.*, **6**, 3680 (2000).

183. P. K. Singal, T. Li, D. Kumar, et al., *Mol. Cell. Biochem.*, **207**, 77 (2000).

184. D. E. Semenov, E. L. Lushnikova, and L. M. Nepomnyashchikh, *Bull. Exp. Biol. Med.*, **131**, 505 (2001).

185. D. L. Keefe, *Semin. Oncol.*, **28S12**, 2 (2001).

186. S. Takanashi and N. R. Bachur, *J. Pharmacol. Exp. Ther.*, **195**, 41 (1975).

187. P. Galettis, J. Boutagy, and D. D. Ma, *Br. J. Cancer*, **70**, 324 (1994).

188. T. Terasaki, T. Iga, Y. Sugiyama, et al., *J. Pharmacobiodyn.*, **7**, 269 (1984).

189. W. T. Beck and M. K. Danks, *Semin. Cancer Biol.*, **2**, 235 (1991).

190. K. Kiyomiya, S. Matsuo, and M. Kurebe, *Cancer Chemother. Pharmacol.*, **47**, 51 (2001).

191. J. Serrano, C. M. Palmeira, D. W. Kuchi, and K. B. Wallace, *Biochim. Biophys. Acta*, **1411**, 201 (1999).

192. S. Rajagopalan, P. M. Politi, B. K. Sinha, and C. E. Myers, *Cancer Res.*, **48**, 4766 (1988).

193. T. R. Tritton, S. A. Murphree, and A. C. Sartorelli, *Biochem. Biophys. Res. Commun.*, **84**, 802 (1978).

194. J. A. Endicott and V. Ling, *Annu. Rev. Biochem.*, **58**, 137 (1989).

195. P. R. Wielinga, H. V. Westerhoff, and J. Lankelma, *Eur. J. Biochem.*, **267**, 649 (2000).

196. M. Michieli, D. Damiani, A. Ermacora, et al., *Br. J. Haematol.*, **104**, 328 (1999).

197. W. Ax, M. Soldan, L. Koch, and E. Maser, *Biochem. Pharmacol.*, **59**, 293 (2000).

198. S. Takanashi and N. R. Bachur, *Drug Metab. Dispos.*, **4**, 79 (1976).

199. G. D. Kruh and L. J. Goldstein, *Curr. Opin. Oncol.*, **5**, 1029 (1993).

200. G. L. Plosker and D. Faulds, *Drugs*, **45**, 788 (1993).

201. M. Grandi, G. Pezzoni, D. Ballinari, et al., *Cancer Treat. Rev.*, **17**, 133 (1990).

202. W. Cabri, *Chim. Ind. (Milan)*, **75**, 314 (1993).

203. F. Arcamone, F. Animati, M. Bigioni, et al., *Biochem. Pharmacol.*, **57**, 1133 (1999).

204. G. Toffoli, R. Sorio, P. Aita, et al., *Clin. Cancer Res.*, **6**, 2279 (2000).

205. D. D. Ross, L. A. Doyle, W. Yang, et al., *Biochem. Pharmacol.*, **50**, 1673 (1995).

206. M. Israel, P. G. Potti, and R. Seshadri, *J. Med. Chem.*, **28**, 1223 (1985).

207. C. Ma, Y.-G. Wang, J.-F. Shi, et al., *Synth. Commun.*, **29**, 3581 (1999).

208. S. V. Onrust and H. M. Lamb, *Drugs Aging*, **15**, 69 (1999).

209. D. D. Kuznetsov, N. F. Alsikafi, R. C. O'Connor, and G. D. Steinberg, *Expert Opin. Pharmacother.*, **2**, 1009 (2001).

210. F. Arcamone, F. Animati, G. Capranico, et al., *Pharmacol. Ther.*, **76**, 117 (1997).

211. R. Abbruzzi, M. Rizzardini, A. Benigni, et al., *Cancer Treat. Rep.*, **64**, 873 (1980).

212. F. Arcamone and G. Cassinellit, *Curr. Med. Chem.*, **5**, 391 (1998).

213. D. Farquhar, A. Cherif, E. Bakina, and J. A. Nelson, *J. Med. Chem.*, **41**, 965 (1998).

214. M. Binashi, M. Bigioni, A. Cipollone, et al., *Curr. Med. Chem.*, **1**, 113 (2001).

215. A. Suarato, F. Angelucci, and C. Geroni, *Curr. Pharm. Des.*, **5**, 217 (1999).

216. D. W. Cameron, G. I. Feutrill, and P. G. Griffiths, *Aust. J. Chem.*, **53**, 25 (2000).

217. J. Nafziger, G. Averland, E. Bertounesque, et al., *J. Antibiot. (Tokyo)*, **48**, 1185 (1995).

218. S. Penco, *Chim. Ind. (Milan)*, **75**, 369 (1993).

219. T. Oki in H. S. El-Khadem, Ed., *Anthracycline Antibiotics*, Academic Press, New York, 1982, p. 75.

220. W. Priebe, Ed., *Anthracycline Antibiotics: New Analogues, Methods of Delivery, and Mechanisms of Action.*, ACS Symposium Series **574**, American Chemical Society, Washington, DC, 1995.

221. F. Zunino, G. Pratesi, and P. Perego, *Biochem. Pharmacol.*, **61**, 933 (2001).

222. S. T. Crooke and S. D. Reich, Eds., *Anthracyclines: Current Status and New Developments*, Academic Press, San Diego, CA, 1980.

223. M. E. Wall, *Med. Res. Rev.*, **18**, 299 (1998).

224. Y. H. Hsiang, R. Hertzberg, S. Hecht, and L. F. Liu, *J. Biol. Chem.*, **260**, 14873 (1985).

225. N. Saijo, *Ann. N. Y. Acad. Sci.*, **922**, 92 (2000).

226. R. F. Ozols, *Semin. Oncol.*, **6S18**, 34 (1999).

227. A. B. Sandler, *Oncology*, **1S1**, 11 (2001).

228. A. A. Adeji, *Curr. Opin. Pulm. Med.*, **6**, 384 (2000).

229. C. Forbes, L. Shirran, A. M. Bagnall, S. Duffy, et al., *Health Technol. Assess.*, **5**, 1 (2001).

230. C. F. Stewart, *Cancer Chemother. Biol. Response Modif.*, **19**, 85 (2001).

231. C. H. Huang and J. Treat, *Oncology*, **61S1**, 14 (2001).

232. J. H. Schiller, *Oncology*, **61S1**, 1 (2001).

233. G. H. Eltabbakh and C. S. Awtrey, *Expert Opin Pharmacother.*, **2**, 109 (2001).

234. D. F. Kehrer, O. Soepenberg, W. J. Loos, et al., *Anticancer Drugs*, **12**, 89, (2001).

235. W. Schuette, *Lung Cancer*, **33S1**, S107 (2001).

236. B. Arun and E. P. Frenkel, *Expert Opin. Pharmacother.*, **2**, 491 (2001).

237. P. Rivory and J. Robert, *Bull. Cancer*, **82**, 265 (1995).

238. R. H. Mathijssen, R. J. van Alphen, J. Verweij, et al., *Clin. Cancer Res.*, 2182 (2001).

239. L. Iyer and M. J. Ratain, *Cancer Chemother. Pharmacol.*, **42S**, S31 (1998).

240. M. Boucaud, F. Pinguet, S. Poujol, et al., *Eur. J. Cancer*, **37**, 2357 (2001).

241. C. Kollmannsberger, K. Mross, A. Jakob, et al., *Oncology*, **56**, 1 (1999).

242. J.-D. Bourzat, M. Vuilhorgne, L. P. Rivory, et al., *Tetrahedron Lett.*, **37**, 6327 (1996).

243. A. Fedier, V. A. Schwarz, H. Walt, et al., *Int. J. Cancer*, **93**, 571 (2001).

244. O. Popanda, C. Flohr, J. C. Dai, et al., *Mol. Carcinog.*, 171 (2001).

245. A. Fedier, V. A. Schwarz, H. Walt, et al., *Int. J. Cancer*, **93**, 571 (2001).

246. Y. Urasaki, G. S. Laco, P. Pourquier, et al., *Cancer Res.*, **61**, 1964 (2001).

247. J. H. Schellens, M. Maliepaard, R. J. Scheper, et al., *Ann. N. Y. Acad. Sci.*, **922**, 188 (2000).

248. A. Saleem, T. K. Edwards, Z. Rasheed, and E. H. Rubin, *Ann. N. Y. Acad. Sci.*, **922**, 46 (2000).

249. S. D. Desai, T. K. Li, A. Rodriguez-Bauman, et al., *Cancer Res.*, **61**, 5926 (2001).

250. J. M. D. Fortunak, J. Kitteringham, A. R. Mastrocola, et al., *Tetrahedron Lett.*, **37**, 5683 (1996).

251. D. P. Curran, S.-B. Ko, and H. Josien, *Angew. Chem. Int. Ed. Engl.*, 2683 (1996).

252. S. Sawada, T. Yokokura, and T. Miyasaka, *Curr. Pharm. Des.*, **1**, 113 (1995).

253. S. Sawada and T. Yokokura, *Ann. N. Y. Acad. Sci.*, **803**, 13 (1996).

254. K. E. Henegar, S. W. Ashford, T. A. Baughman, et al., *J. Org. Chem.*, **62**, 6588 (1997).

255. C. H. Takimoto and R. Thomas, *Ann. N. Y. Acad. Sci.*, **922**, 224 (2000).

256. M. Ishii, M. Iwahana, I. Mitsui, et al., *Anticancer Drugs*, **11**, 353 (2000).

257. D. K. Kim, D. H. Ryu, J. Y. Lee, et al., *J. Med. Chem.*, **44**, 1594 (2001).

258. S. Dellavalle, T. Delsoldato, A. Ferrari, et al., *J. Med. Chem.*, **43**, 3963 (2000).

259. D. Bom, D. P. Curran, S. Kruszewski, et al., *J. Med. Chem.*, **43**, 3970 (2000).

260. Y. Fan, L. M. Shi, K. W. Kohn, et al., *J. Med. Chem.*, **44**, 3254 (2001).

261. O. Lavergne and D. C. Bigg, *Bull. Cancer*, Spec. No. 51 (1998).

262. E. Boven, A. H. Van Hattam, I. Hoogsteen, et al., *Ann. N. Y. Acad. Sci.*, **922**, 175 (2001).

263. D. Demarquay, M. Huchet, H. Coulomb, et al., *Anticancer Drugs*, **12**, 9 (2001).

264. A. K. Larsen, C. Gilbert, G. Chyzak, et al., *Cancer Res.*, **61**, 2961 (2001).

265. C. Canel, R. M. Moraes, F. E. Dayan, and D. Ferreira, *Phytochemistry*, **54**, 115 (2000).

266. T. F. Imbert, *Biochemie*, **80**, 207 (1998).

267. K. R. Hande, *Eur. J. Cancer*, **34**, 1514 (1998).

268. R. T. Oliver, *Curr. Opin. Oncol.*, **13**, 191 (2001).

269. W. Schuette, *Lung Cancer*, **33S1**, S99 (2001).

270. D. Mavroudis, E. Papadakis, M. Veslemes, et al., *Ann. Oncol.*, **12**, 463 (2001).

271. K. R. Hande, *Biochem. Biophys. Acta*, **1400**, 173 (1998).

272. R. M. Hoetelmans, J. H. Schornagel, W. W. ten Bokkel Huinink, and J. H. Beijnen, *Ann. Pharmacother.*, **30**, 367 (1996).

273. B. J. Bernstein and M. B. Troner, *Pharmacotherapy*, **19**, 989 (1999).

274. C. A. Felix, *Biochim. Biophys. Acta*, **1400**, 233 (1998).

275. J. E. Rubnitz and A. T. Look, *J. Pediatr. Hematol. Oncol.*, **20**, 1 (1998).

276. H. L. McLeod and W. E. Evans, *Cancer Surv.*, **17**, 253 (1993).

277. M. V. Relling, J. Nemec, E. G. Schuetz, et al., *Mol. Pharmacol.*, **45**, 352 (1994).

278. R. Kitamura, T. Bandoh, M. Tsuda, and T. Satoh, *J. Chromatogr. B Biomed. Sci. Appl.*, **690**, 283 (1997).

279. L. Nguyen, E. Chatelut, C. Chevreau, et al., *Cancer Chemother. Pharmacol.*, **41**, 125 (1998).

280. B. Liu, H. M. Earl, C. J. Poole, et al., *Cancer Chemother. Pharmacol.*, **36**, 506 (1995).

281. C. L. Chen, J. Rawwas, A. Sorrell, et al., *Leuk. Lymphoma*, **42**, 317 (2001).

282. D. J. Stewart, D. Grewaal, M. D. Redmond, et al., *Cancer Chemother. Pharmacol.*, **32**, 368 (1993).

283. J. A. Holden, *Ann. Clin. Lab. Sci.*, **27**, 402 (1997).

284. J. A. Byl, S. D. Cline, T. Utsugi, et al., *Biochemistry*, **40**, 712 (2001).

285. S. K. Morris and J. E. Lindsley, *J. Biol. Chem.*, **274**, 30690 (1999).

286. W. L. Chiou, S. M. Chung, T. C. Wu, and C. Ma, *Int. J. Clin. Pharmacol. Ther.*, **39**, 93 (2001).

287. M. Koshiyama, H. Fujii, M. Kinezaki, and M. Yoshida, *Anticancer Res.*, **21**, 905 (2001).

288. Y. Mao, C. Yu, T. S. Hsieh, et al., *Biochemistry*, **38**, 10793 (1999).

289. Y. Matsumoto, H. Takano, K. Kunishio, et al., *Jpn. J. Cancer Res.*, **92**, 1133 (2001).

290. O. Middel, H. J. Woerdenbag, W. van Uden, et al., *J. Med. Chem.*, **38**, 2112 (1995).

291. H. F. Stahelin and A. von Wartburg, *Cancer Res.*, **51**, 5 (1991).

292. L. Schacter, *Semin. Oncol.*, **6S13**, 1 (1996).

293. N. Jain, Y. M. Lam, J. Pym, and B. G. Campling, *Cancer*, **77**, 1797 (1996).

294. B. F. Issell, F. M. Muggia, and S. K. Carter, *Etoposide (VP-16): Current Status and New Developments*, Academic Press, San Diego, CA, 1984.

295. B. Botta, G. Delle Monache, D. Misiti, et al., *Curr. Med. Chem.*, **8**, 1363 (2001).

296. U. Kellner, P. Rudolph, and R. Parwaresch, *Onkologie*, **23**, 424 (2000).

297. M. Gordaliza, J. M. Miguel del Corral, M. Angeles Castro, et al., *Farmaco*, **56**, 297 (2001).

298. D. G. I. Kingston, *Chem. Commun.*, 867 (2001).

299. P. B. Schiff, J. Fant, and S. B. Horwitz, *Nature*, **277**, 665 (1979).

300. J. J. Manfredi and S. B. Horwitz, *Pharmacol. Ther.*, **25**, 83 (1984).

301. S. B. Horwitz, *Ann. Oncol.*, **5S6**, S3 (1994).

302. P. Potier, *C. R. Acad. Agric. Fr.*, **86**, 179 (2000).

303. C. H. Wu, C. H. Yang, J. N. Lee, et al., *Int. J. Gynecol. Cancer*, **11**, 295 (2001).

304. G. Minotti, A. Saponiero, S. Licata, et al., *Clin. Cancer Res.*, **7**, 1511 (2001).

305. H. Gelderblom, J. Verweij, K. Nooter, and A. Sparreboom, *Eur. J. Cancer*, **37**, 1590 (2001).

306. K. T. Kivisto, H. K. Kroemer, and M. Eichelbaum, *Br. J. Clin. Pharmacol.*, **40**, 523 (1995).

307. M. O. Karlsson, V. Molnar, A. Freijs, et al., *Drug Metab. Dispos.*, **27**, 1220 (1999).

308. H. Maier-Lenz, B. Hauns, B. Haering, et al., *Semin. Oncol.*, **24S6**, S19-16 (1997).

309. C. M. Kearns, L. Gianni, and M. J. Egorin, *Semin. Oncol.*, **22S6**, S16-23 (1995).

310. B. Monsarrat, I. Royer, M. Wright, and T. Cresteil, *Bull. Cancer*, **84**, 125 (1997).

311. A. Rahman, K. R. Korzekwa, J. Grogan, et al., *Cancer Res.*, **54**, 5543 (1994).

312. A. Sparreboom, O. Van Tellingen, E. J. Scherrenburg, et al., *Drug Metab. Dispos.*, **24**, 655 (1996).

313. S. H. Jang, M. G. Wientjes, and J. L.-S. Au, *J. Pharmacol. Exp. Ther.*, **298**, 1236 (2001).

314. P. Giannakakou, D. L. Sackett, Y.-K. Ksang, et al., *J. Biol. Chem.*, **272**, 17118 (1997).

315. M. Kavallaris, D. Y. Kuo, C. A. Burkhart, et al., *J. Clin. Invest.*, **100**, 1282 (1997).

316. D. Yu, *Semin. Oncol.*, **28–5S16**, S12 (2001).

317. R. B. Montgomery, J. Guzman, D. M. O'Rourke, and W. L. Stahl, *J. Biol. Chem.*, **275**, 17358 (2000).

318. T. J. Lynch, Jr., *Semin. Oncol.*, **3S9**, 5 (2001).

319. R. Bruno, N. Vivier, C. Veyrat-Follet, et al., *Invest. New Drugs*, **19**, 163 (2001).

320. M. Shou, M. Martinet, K. R. Korzekwa, et al., *Pharmacogenetics*, **8**, 391 (1998).

321. U. Vaishampayan, R. E. Parchment, B. R. Jasti, and M. Hussain, *Urology*, **54S6A**, 22 (1999).

322. G. I. Georg, T. C. Boge, Z. S. Cheruvallath, et al. in M. Sufness, Ed., *Taxol: Science and Applications*, CRC Press, Boca Raton, FL, 1995, p. 317.

323. E. Ter Hart, *Expert Opin. Ther. Patents*, **8**, 57 (1998).

324. S. Lin and I. Ojima, *Expert Opin. Ther. Patents*, **10**, 1 (2000).

325. F. Gueritte, *Curr. Pharm. Des.*, **7**, 1229 (2001).

326. S. Jennewein, C. D. Rithner, R. M. Williams, and R. B. Croteau, *Proc. Natl. Acad. Sci. USA*, **98**, 13595 (2001).

327. H. G. Floss and U. Mocek in M. Sufness, Ed., *Taxol: Science and Applications*, CRC Press, Boca Raton, FL, 1995, p. 191.

328. Q. Li, H. L. Sham, and S. H. Rosenberg, *Annu. Rep. Med. Chem.*, **34**, 139 (1999).

329. J. Fahy, *Curr. Pharm. Des.*, **7**, 1181 (2001).

330. P. Potier, N. Langlois, Y. Langlois, and F. Gueritte, *Chem. Commun.*, 670 (1975).

331. P. Mangeney, R. Z. Andriamialisoa, N. Langlois, et al., *J. Am. Chem. Soc.*, **101**, 2243 (1979).

332. J. P. Kutney, T. Hibino, E. Jahngen, et al., *Helv. Chim. Acta*, **59**, 2858 (1976).

333. R. L. Bai, G. R. Pettit, and E. Hamel, *Biochem. Pharmacol.*, **39**, 1941 (1990).

334. R. L. Bai, G. R. Pettit, and E. Hamel, *J. Biol. Chem.*, **265**, 17141 (1990).

335. X. J. Zhou and R. Rahmani, *Drugs*, **44S4**, 1 (1992).

336. O. Van Tellingen, J. H. Beijnen, W. J. Nooijen, and A. Bult, *Cancer Chemother. Pharmacol.*, **32**, 286 (1993).

337. D. Schlaifer, M. R. Cooper, M. Attal, et al., *Leukemia*, **8**, 668 (1994).

338. S. A. Elmarakby, M. W. Duffel, and J. P. Rosazza, *J. Med. Chem.*, **32**, 2158 (1989).

339. S. A. Elmarakby, M. W. Duffel, A. Goswami, et al., *J. Med. Chem.*, **32**, 674 (1989).

340. D. L. Sackett, *Biochemistry*, **34**, 7010 (1995).

341. V. Prakash and S. N. Timasheff, *Biochemistry*, **24**, 5004 (1985).

342. K. E. Luker, C. M. Pica, R. D. Schreiber, and D. Piwnica-Worms, *Cancer Res.*, **61**, 6540 (2001).

343. M. Kavallaris, A. S. Tait, B. J. Walsh, et al., *Cancer Res.*, **61**, 5803 (2001).

344. R. Krishna, M. S. Webb, G. St. Onge, and L. D. Mayer, *J. Pharmacol. Exp. Ther.*, **298**, 1206 (2001).

345. D. Schlaifer, M. R. Cooper, M. Attal, et al., *Blood*, **81**, 482 (1993).

346. D. Schlaifer, E. Duchayne, C. Demur, et al., *Leuk. Lymphoma*, 441 (1996).

347. R. Naumann, J. Mohm, U. Reuner, and F. R. Kroschinsky, *Br. J. Haematol.*, **115**, 323 (2001).

348. D. W. Loe, R. G. Deeley, and S. P. Cole, *Cancer Res.*, **58**, 5130 (1998).

349. J. Montalar, S. Morales, I. Maestu, et al., *Lung Cancer*, **34**, 305 (2001).

350. M. Namer, P. Soler-Michel, F. Turpin, et al., *Eur. J. Cancer*, **37**, 1132 (2001).

351. R. Lilenbaum, *Oncologist*, **6S1**, 16 (2001).

352. G. Masters, *Oncologist*, **6S1**, 12 (2001).

353. H. J. Burstein, C. A. Bunnell, and E. P. Winer, *Semin. Oncol.*, **28**, 344 (2001).

354. K. Sonoda, T. Ohshiro, M. Ohta, et al., *Gan To Kagaku Ryoho*, **28**, 1397 (2001).

355. M. Vassilomanolakis, G. Koumakis, V. Barbounis, et al., *Support Care Cancer*, **9**, 108 (2001).

356. R. J. Cersosimo, R. Bromer, J. T. Licciardello, and W. K. Hong, *Pharmacotherapy*, **3**, 259 (1983).

357. C. Etievant, A. Kruczynski, J. M. Barret, et al., *Cancer Chemother. Pharmacol.*, **48**, 62 (2001).

358. B. T. Hill, *Curr. Pharm. Des.*, **7**, 1199 (2001).

359. A. Kruczynski and B. T. Hill, *Crit. Rev. Oncol. Hematol.*, **40**, 159 (2001).

360. W. I. Taylor, *The Catharanthus Alkaloids: Botany, Chemistry, Pharmacology, and Clinical Use*, Marcel Dekker, New York, 1975.

361. E. K. Rowinsky and R. C. Donehower in V. T. DeVita, Jr., S. Hellman, and S. A. Rosenberg, Eds., *Cancer: Principle and Practice of Oncology*, 5th ed., Lippencott-Raven, Philadelphia, 1997, pp. 467–483.

362. Q. Shi, K. Chen, S. L. Morris-Natchke, and K. H. Lee, *Curr. Pharm. Des.*, **4**, 219 (1998).

## SOME ADDITIONAL REFERENCES NOT SPECIFICALLY QUOTED PREVIOUSLY

W. A. Remers, Ed., *Antineoplastic Agents*, Wiley-Interscience, New York, 1984.

M. Suffness, Ed., *Taxol: Science and Applications*, CRC Press, Boca Raton, FL, 1995.

M. Potmesil and H. Pinedo, Eds., *Camptothecins: New Anticancer Agents*, CRC Press, Boca Raton, FL, 1995.

D. G. I. Kingston, A. A. Molinero, and J. M. Rimoldi, *Prog. Chem. Org. Nat. Prod.*, **61**, 1 (1993).

D. G. I. Kingston, P. G. Jagtap, H. Yuan, and L. Samala, *Prog. Chem. Org. Nat. Prod.*, **84**, 53 (2002).

# CHAPTER FOUR

# Radiosensitizers and Radioprotective Agents

EDWARD A. BUMP
DRAXIMAGE Inc.
Kirkland, Quebec, Canada

STEPHEN J. HOFFMAN
Allos Therapeutics, Inc.
Westminster, Colorado

WILLIAM O. FOYE
Massachusetts College of Pharmacy and Health Sciences
Boston, Massachusetts

## Contents

1 Protective Agents against Ionizing Radiation, 152
  1.1 Introduction, 152
  1.2 Radiation Damage, 153
  1.3 Antiradiation Testing, 154
  1.4 Protective Compounds, 154
    1.4.1 Thiols and Thiol Derivatives, 155
    1.4.2 Other Sulfur-Containing Compounds, 160
    1.4.3 Metabolic Inhibitors, 162
    1.4.4 Agents Involving Metal Ions, 163
    1.4.5 Hydroxyl-Containing Compounds, 163
    1.4.6 Heterocyclic Compounds, 163
    1.4.7 Physiologically Active Substances, 165
    1.4.8 Metabolites and Naturally Occurring Compounds, 166
    1.4.9 Polymeric Substances, 166
    1.4.10 Miscellaneous Substances, 167
  1.5 Mechanisms of Protective Action, 167
    1.5.1 Protection by Anoxia or Hypoxia, 167
    1.5.2 Inhibition of Free-Radical Processes, 168
    1.5.3 Mixed Disulfide Hypothesis, 168
    1.5.4 Biochemical Shock, 169
    1.5.5 Control of DNA Breakdown, 169
    1.5.6 Metabolic Effects, 170
    1.5.7 Use of Radioprotective Agents in Radiotherapy and Chemotherapy of Cancers, 172
2 Radiosensitizers, 173
  2.1 Introduction, 173

*Burger's Medicinal Chemistry and Drug Discovery*
Sixth Edition, Volume 5: Chemotherapeutic Agents
Edited by Donald J. Abraham
ISBN 0-471-37031-2   © 2003 John Wiley & Sons, Inc.

2.2 Radiosensitization by Alteration of Energy
Absorption, 173
2.2.1 Boron Neutron Capture Therapy
(BNCT), 173
2.2.2 k-Edge Absorption and Photoactivation
of Elements of High Atomic Number,
174
2.2.3 Photodynamic Therapy, 174
2.3 Alteration of the Primary
Radiolytic Products, 175
2.4 Radiosensitization by Reaction with DNA
Radicals, 176
2.5 Additional Applications for Electron-Affinic
Drugs in Cancer Therapy, 180
2.5.1 Binding of Nitroimidazoles to Hypoxic
Cells: Use in Detection of Hypoxia, 180
2.5.2 Additional Sensitization by Hypoxic
Metabolism of Nitroimidazoles, 180
2.5.3 Bioreductive Drugs, 181
2.6 Radiosensitization by Alteration of Oxygen
Delivery, 183
2.6.1 RSR13, 184
2.6.1.1 Mechanism of Action, 185
2.6.1.2 Tumor Hypoxia, 186
2.6.1.3 Radiation Therapy
Sensitization with RSR13, 186

2.6.1.4 Clinical Trials with RSR13, 188
2.6.1.5 Summary of RSR13 Results, 190
2.7 Radiosensitization by Depletion
of Endogenous Protectors, 190
2.8 Radiosensitization by Inhibition
of DNA Repair, 191
2.8.1 PLD Repair Inhibitors, 191
2.8.2 Radiosensitization by Reaction with
Protein Sulfhydryls, 192
2.9 Radiosensitization by Perturbation
of Cellular Metabolism, 193
2.9.1 Perturbation of Energy Metabolism,
193
2.9.2 Abrogation of $G_2$ Delay, 194
2.9.3 Radiosensitization by Growth Factors
and Cytokines, 194
2.9.4 Halogenated Pyrimidines, 194
2.10 Radiosensitizers for which the Mechanism
of Sensitization has not been Established,
196
2.10.1 Metal Ion Complexes, 196
2.10.2 Thiols and Miscellaneous Compounds,
196
2.10.3 Bacterial Sensitizers, 197
3 Summary and Prospects for Future Development
of Clinical Radiation Modifiers, 197

# 1 PROTECTIVE AGENTS AGAINST IONIZING RADIATION

## 1.1 Introduction

The protective action of certain substances against the damaging effects of ionizing radiation was first noted in 1942 (but not published until 1949) by Dale, Gray, and Meredith. A decrease in the inactivation of two enzymes by X-rays was observed upon addition of several substances, including colloidal sulfur and thiourea, to aqueous preparations of the enzymes (1). Radioprotective effects for a bacteriophage were observed by Latarjet and Ephrati in 1948, using cysteine, cystine, gluta-thione, thioglycolic acid, and tryptophan (2). Radioprotection of mice against X-rays was achieved shortly thereafter in three different laboratories, in Belgium, the United States, and Britain, by use of cyanide (3), cysteine (4), and thiourea (5), respectively. These protective effects were attributed at the time to inhibition of, or reaction with, cellular enzymes. The importance of sulfur-containing molecules for radioprotection was thus demon-

strated from the very earliest experiments with living systems, although the reasons for selection of sulfur compounds were not clear.

The importance of the mercapto (or thiol) function was demonstrated in 1951 by Bacq et al. (6), a Belgian physiologist, who removed the carboxyl group of cysteine and obtained 2-mercaptoethylamine (MEA, or cysteamine; $NH_2CH_2CH_2SH$), which proved to be a much stronger protective agent in mice than any previously tested. The presence of the amino group was also considered essential for good radioprotection, and most of the mercaptans and other sulfur-containing molecules, later synthesized, also contained an amino or other basic function. MEA and its derivatives, particularly those having greater lipophilic character, are still regarded as the most potent of the whole-body radioprotective agents.

Since 1952 other types of structures with radioprotective activity have been found, including a number of physiologically important agents, notably serotonin, but none has yet exceeded the amino alkyl mercaptans in effectiveness on a molar basis. Various explanations have been put forward for the protective activity of the thiols, but it was not until more knowledge was available regarding the radicals generated by ionizing radiation, and their effect on DNA, that present concepts became established.

Research attempts to explain the action of chemical radiation protectors have involved the use not only of mammals but also of plants, bacteria, distinct types of nonmammalian cells, and even some synthetic plastics affected by ionizing radiation. This discussion attempts to categorize the various types of chemical structures that afford some protection against the deleterious effects of ionizing radiation in mammals and to describe the most widely held mechanistic theories of radioprotection.

## 1.2 Radiation Damage

Biologic effects of radiation have been reviewed in detail by a number of authors, including von Sonntag (7), Pizzarello and Witcofski (8), Casarett (9), Okada (10), and Dertinger and Jung (11). Ionizing radiation can have three types of biological effect: perturbation of cellular regulation, mutation, or cell death. Most of the research on radioprotection has focused on cell death and consequences at the level of tissue injury and death of the organism. In mammals, death can result from damage to the blood-forming organs, gastrointestinal system, or central nervous system, depending on radiation dose. Hematopoietic death from bleeding, infection, or anemia is the endpoint that was used in most of the early studies on radioprotection, and follows 7–30 days after exposure to a potentially lethal single dose (~400 rads) in mice. These whole-body effects of relatively high doses of ionizing radiation are most easily explained in terms of depletion of stem cells by cell killing that in turn is most directly explained by DNA damage that occurs at the time of irradiation. However, the mechanisms of some other radiation effects, such as carcinogenesis, teratogenesis, and delayed vascular injury, could include altered cellular regulation and secondary DNA damage. The full spectrum of biologically relevant mechanisms of radiation injury has not been fully elucidated. Many mechanisms of molecular damage that have

been elucidated may not be relevant to biological damage. It is therefore difficult to classify radioprotectors according to mechanism. They are organized in this discussion mostly according to chemical features, although it should be kept in mind that structurally similar compounds could be acting by completely different mechanisms, and even a single compound may protect by different mechanisms depending on the model system used for the study.

Absorption of radiation energy by biological molecules has been considered to be either direct or indirect (12–14), although they both can result in the same kind of damage to a target molecule. Direct action involves absorption of radiation energy by a target molecule, such as DNA. The absorbed energy is sufficient to cause the ejection of an electron from an atom of the target molecule (hence the term *ionizing* radiation), leaving the target molecule with an unpaired electron: that is, converting it into a free radical. Indirect action involves the absorption of radiation energy by a molecule (such as $H_2O$) other than the target molecule, and subsequent transfer of the energy to the target molecules by reaction of radiolytically produced nontarget free radical with the target molecule. In either case, the result is a target molecule free radical. Subsequent reactions of the target molecule free radical can result in permanent chemical alteration, leading to a biological consequence. Reaction of the target molecule free radical with a hydrogen donor, such as a thiol, can restore the lost electron, thus restoring the target molecule (13). Repair usually refers to an enzymatic process, whereas the term *restoration* is the preferred usage to describe this type of chemical radioprotection.

Another mechanism of radioprotection is to scavenge the nontarget free radicals produced by radiation before they can react with the target molecule. The most important diffusible free radical involved in the indirect effect is the hydroxyl radical, formed by radiolysis of water. Hydroxyl-radical scavenging can be the most effective mechanism of radioprotection of target molecules in dilute solution. However, radioprotection of mammalian cells by hydroxyl-radical scavenging is difficult to achieve because these highly reactive molecules will react with cellular constituents very rapidly at the high solute concentrations that exist in cells, and a very high concentration of hydroxyl-radical scavenger is required to intercept the hydroxyl radicals before this happens (7).

## 1.3   Antiradiation Testing

Much of the early testing of radioprotective agents employed either X-rays or γ-rays from an external source. Animals for initial *in vivo* testing most often have been mice or rats; guinea pigs have been used less frequently. Large animal radioprotector testing, with dogs or monkeys, has been limited to the more effective compounds, as determined from screening with mice or rats. Further information on this testing, through the use of 30-day survival [lethal dose for 50% survival at 30 days $(LD_{50/30})$] as a criterion for protection, may be found in texts devoted to radiobiology (15, 16) from that era, the 1960s.

Various physiological effects may be observed, depending on the dose and type of radiation, as well as on the type of animal used. In theory, the appearance of any observable symptom of radiation damage may be used as the basis of a testing procedure but, historically, lethality has generally been the criterion for protection. Sufficient numbers of animals must be employed for statistical significance, and in the case of mice irradiated with a lethal dose of X- or γ-rays, the endpoint for survival is generally taken to be survival for 30 days after irradiation. Testing results are expressed most commonly as the percentage survival for the observation period compared to the survival of control animals. Another method of expression of test data is in terms of the dose reduction factor (DRF, which is the ratio of the radiation dose causing an effect such as $LD_{50}$ in the treated animals or cells to the dose causing the same effect in the unprotected animals or cells). Recently, radioprotection studies have been most frequently carried out in cell culture (17). Particular mechanisms of protection may be more effective with regard to certain endpoints (17). For example, protection against mutagenesis has been observed at lower concentrations of amino thiols than are required for protection against cell killing.

The dose of protective agent employed is usually the maximum tolerated dose (MTD), that is, the dose causing no deleterious effects. In a drug-screening program, candidate compounds are usually tested at their MTD level by use of a radiation dose that is lethal to all control animals in 30 days. The time interval between administration of the drug and irradiation of the animals is usually 15–30 min for intraperitoneal dosing and 30–60 min for oral dosing. Drugs believed to act by inducing hypoxia or other metabolic changes usually must be administered several hours before irradiation. The rate of irradiation in screening programs has commonly been 50–250 rads/min. At lower dose rates, the time for maximum effectiveness of the radioprotector can be exceeded before the total radiation dose is administered. In addition, repair processes could become significant before the irradiation is complete. Chronic radiation studies have been carried out with repeated administration of protector, but results have been less decisive (18).

Several concepts should be kept in mind with regard to testing results of radiation protectors (1). Because many compounds have been tested at the maximum tolerated dose, the best protectors are not necessarily the most active on a molar basis, but rather the most effective at a given level of toxicity (2). Although many of these compounds are as active as the parent compound, some are activated metabolically, or may act indirectly by inducing an endogenous protective mechanism.

Other testing procedures used to a lesser extent include the inhibition of bacterial or plant growth and the prevention of depolymerization of polymethacrylate or polystyrene (19) or of DNA (20). Other test procedures for evaluation of radioprotectors have included the plaque-forming ability of coliphage T (21), effect on Eh potential (22), inhibition of the formation of peroxides of unsaturated lipids or 13-carotene (22), inhibition of chemiluminescence of γ-irradiated mouse tissue homogenates (23), and use of spleen colony counts (24).

## 1.4   Protective Compounds

The more extensively investigated compounds have been discussed in books by Thomson (16); Bacq (25, 27); Balubukha (26); Nygaard and Simic (27); Livesey, Reed, and Adamson (28); and Bump and Malaker (17). A catalog of compounds tested for radiation protection up to 1963 was compiled by Huber and Spode (29). Extensive reviews on protective agents since 1963 have been written by Melching and Streffer (30), Overman and Jackson (31), Romantsev (32), Foye (33), Klayman and Copeland (34), Yashunskii and Kovtun (35), and Bump and Brown (36). Reports of two international symposia on radioprotective and radiosensitizing agents have been published by Paoletti and Vertua (37) and by Moroson and Quintiliani (38). A series of symposia on radiation modifiers has also been held (39). Chapters on radioprotective agents have appeared in *Annual Reports in Medicinal Chemistry* in 1966 and 1967 (40) and in 1968 and 1970 (41), and in *Military Radiobiology in 1987* (42).

In the following discussion of structure-activity relationships, results on radioprotection of mice are compared unless otherwise stated. Relevant details concerning radiation dose, compound dose, route of administration, or strain of test animal, variations of which can alter results significantly, may be found in the original references.

**1.4.1 Thiols and Thiol Derivatives.** 2-Mercaptoethylamine (MEA, cysteamine) and its derivatives have constituted the most effective class of radioprotective compounds. Since the initial discoveries of the protective action in mice of cysteine by Patt et al. (4) and its decarboxylated derivative, MEA, by Bacq et al. (6), hundreds of derivatives and analogs of the mercaptoethylamine structure have been synthesized and tested for radioprotective activity. In the United States, the Walter Reed Army Institute of Research (WRAIR) funded a large synthetic program and developed a screening procedure for compounds mainly of this type, during the period from 1959 to 1986. A compilation of the compounds tested in this program was made by T. R. Sweeney of the WRAIR in 1979. Many European countries also supported research programs on the development of radioprotective compounds and have been joined more recently by China and India. Other types of agents have been found

with protective activities, but sulfur-containing molecules have been by far the most numerous.

Several structural requirements for activity in this series have become established. The presence of a free thiol group, or a thiol derivative that can be converted to a free thiol *in vivo*, is essential for activity. The presence of a basic function [amino, amidino, or guanidino (43)] (**1**) located two or three carbon atoms

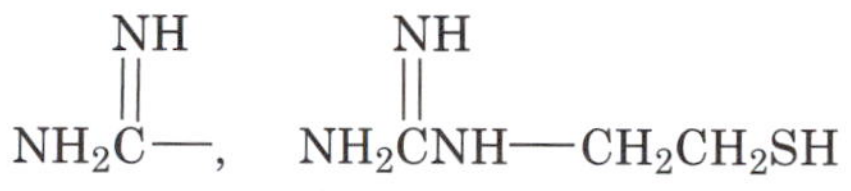

$$\overset{NH}{\underset{\|}{NH_2C}}—, \quad \overset{NH}{\underset{\|}{NH_2CNH}}—CH_2CH_2SH$$

(**1**)

distant from the thiol group is favorable for the best activity. Activity for these basic thiols drops off drastically with more than a three-carbon distance (44). The benefit arising from the basic group has not yet been explained, however. Several acyl thiol derivatives (**2**),

$$^+NH_3CH_2CH_2SX \quad X = SO_3^-, PO_3H^-, CS_2^-$$

(**2**)

such as the thiosulfuric acid (45), phosphorothioic acid (46), and trithiocarbonic acid (47), most likely liberate free thiol in the animal.

Some radioprotectors may act by releasing endogenous nonprotein thiols normally bound by disulfide linkages with serum or interstitial proteins. An increase in tissue nonprotein thiol levels results after administration of the thiosulfate and phosphorothioate of MEA (48).

Alkylation of the nitrogen of MEA causes loss of activity in some cases, but has also resulted in some of the most potent of the MEA derivatives, of which WR2721 (**3**) is the best

$$^+NH_3CH_2CH_2CH_2NHCH_2CH_2SPO_3H^-$$

(**3**)

known. The *N*-β-phenethyl and *N*-β-thienylethyl derivatives have good activity (49). Dialkylation of the nitrogen of MEA usually results in some loss of activity. Whereas the *N,N*-dimethyl and *N,N*-diethyl derivatives retain much of the activity of the MEA, the *N,N*-dipropyl and *N,N*-diisobutyl derivatives retain a little activity, and the di-*n*-butyl derivative is inactive (44). *N,N'*-Polymethylene bridging of the MEA structure provides compounds (**4**)

$$XS(CH_2)_2NH(CH_2)_nNH(CH_2)_2SX$$

(**4**)

that are active, where X is $PO_3H_2$ and *n* is 3 or 4, but inactive where X is $SO_3H$ (50).

Alkylation of the carbon atoms of the MEA structure has given varied results. Active compounds have been found among C-monoalkyl derivatives, with 2-aminopropan-1-thiol having moderate activity and 1-aminopropan-2-thiol having good activity (51, 52). Whereas $\alpha,\alpha$-dialkyl-$\beta$-aminoethanethiols are inactive (53, 54), some $\beta,\beta$-dialkyl-$\beta$-aminoethane thiosulfates and phosphorothioates (**5**) have

$$NH_2-\underset{\underset{R}{|}}{\overset{\overset{R}{|}}{C}}-CH_2SX \qquad X = SO_3^-, PO_3H^-$$

(**5**)

substantial activity (55). 2-Amino-1-pentanethiol and 2-amino-3-methyl-1-butanethiol also have good activity (55). C-Trialkyl derivatives of MEA (54), *sec*-mercaptoalkylamines (56), and 2-mercapto-2-phenethylamine (57) are inactive. Generally, the presence of a phenyl group in the MEA structure blocks activity (58). $\alpha,\alpha$-Dimethyl-2-aminoethanethiol, derived from penicillamine, is not protective but has radiosensitizing activity (59).

Alkylation or arylation of the mercapto group generally results in loss of activity. The *S*-benzyl derivative of MEA has some activity, probably resulting from *in vivo* debenzylation (60).

Attempts to determine whether the stereochemical structure of the aminoalkanethiols is important have revealed that a given stereoisomer may provide greater radioprotection than others. A small difference in activity was found for the *cis* and *trans* isomers of 2-aminocyclohexane-1-thiol (61). The *cis* forms of 2-mercaptocyclobutylamine and 2-mercapto-

cyclobutyl-*N*-methylamine have higher radioprotective activities in mice than those of the *trans* forms. No correlation could be found between protective activity and ability to protect against either induction of DNA single-strand breaks or inactivation of proliferative capacity of hamster cells *in vitro* (62), however. On the other hand, the *trans* forms were less toxic and somewhat more effective in competing for free radicals in DNA. The *D* and *L* isomers of 2-aminobutylisothiuronium bromide have been separated, and the *D* isomer was twice as active in mice as the *L* isomer (63). The optical isomers of dithiothreitol show a greater difference in protective ability, the *Dg* isomer protecting 50% of mice exposed to 650 rads, whereas the *Lg* isomer afforded no protection (64). The *Dg* isomer was also less toxic.

Other functional groups in the MEA structure have generally caused diminution or loss of protective ability. The presence of a carboxyl group frequently causes lower activity; cysteine, for instance, has the same dose reduction factor (1.7) in mice as that of MEA or MEG, but a much larger dose is required (65). This may be explained by the charge, or *Z* value, of the RSH molecule, which determines the concentration of thiol in the immediate vicinity of DNA, which has been shown to be in agreement with scavenging and chemical-repair reactions (66). The negative charge of a carboxyl group would thus be repelled by negative charges on DNA and prevent close accumulation of the thiol, necessary for DNA protection and repair.

N-Monosubstituted derivatives of MEA containing thioureide or sulfone substituents are inactive, although sulfonic acid zwitterions, $HS(CH_2)_2NH_2^+(CH_2)_3SO_3^-$, are strongly protective (67). The presence of hydroxyl often favors activity [e.g., *L*(+)-3-amino-4-mercapto-1-butanol gives good protection to mice (68)]. An additional thiol group diminished activity in a series of 2-alkyl-2-amino-1,3-propanedithiols, which showed little activity in mice (67). Dithiothreitol (Cleland's reagent, **6**)

$$HS-\underset{\underset{OH}{|}}{CH}-\underset{\underset{OH}{|}}{CH}-SH$$

(**6**)

has protective ability (64), and Carmack et al. (64) found no protection from the oxidized dithiane form. Its protective activity *in vivo* may be attributable to release of other nonprotein thiols from mixed disulfides. Also, this may reflect a requirement for a suitable redox potential. Dithiothreitol is so readily reduced that it quickly becomes oxidized in biological systems. It is a good protector *in vitro*, under conditions where it remains reduced. *N*-Carbamoyl ethyl derivatives of the phosphorothioate of MEA, however, had high protective activity (69). Sodium 2,3-dimercaptopropane sulfonate (Unithiol, **7**), which was studied in

$$HS-CH_2-CH-CH_2-SO_3Na$$
$$|$$
$$SH$$

**(7)**

Russia, was claimed to be more protective and less toxic than MEA (70).

S-Acylation of the MEA structure has provided some very active compounds, particularly where zwitterions have resulted. The thiosulfate (Bunte salt) (45), phosphorothioate (46), and trithiocarbonate (47) of MEA, all of which form zwitterions, have protective activities comparable to that of MEA. Corresponding zwitterions of mercaptoethylguanidine (MEG) also give protection to mice corresponding to that of MEG (47, 71). Of these S-acyl derivatives, the phosphorothioates have been particularly effective; S-(3-amino-2-hydroxypropyl)phos-phorothioate (**8**) and S-(2-aminopropyl)-phosphorothioate

$$NH_3^+-CH_2-CH-CH_2-SPO_3H^-$$
$$|$$
$$OH$$

**(8)**

have DRF values in mice of 2.16 and 1.86, respectively, compared to a DRF value of 1.84 for MEA (72). S-[2-(3-Aminopropylamino) ethyl]-phosphorothioate (**3**), better known as WR2721, from the screening program at the Walter Reed Army Institute of Research (73), has high antiradiation activity and has been studied in numerous investigations. 3-Aminopropylphosphorothioate (71), however, and

N-substituted derivatives of 2-aminoethylphosphorothioate are essentially inactive (71). Numerous publications concerning the synthesis and screening activities of the amino thiol derivatives submitted to the Walter Reed Army Institute of Research, leading to the selection of WR2721 as the most effective compound resulting from this screening program, have not been included here.

A comparison of the relative activities and toxicities of thiols with the corresponding thiosulfates showed the thiosulfates to be less toxic and comparable in activities (52, 74). In a series of 2-*N*-alkylaminoethanethiols, consisting of 66 compounds, the thiosulfates were generally superior to the corresponding thiols, disulfides, or thiazolidines (74), given intraperitoneally (i.p.) to mice. Another comparison of the relative effectiveness of thiols with the common mercapto-covering groups, the disulfide, thiosulfate, and phosphorothioate, was made with a series of 84 derivatives of 2-mercaptoacetamidine (75). Although generalities were not evident, by the i.p. route, the (3,5-dimethyl-1-adamantyl) methyl phosphorothioate (**9**) was the most effective com-

**(9)**

pound. Perorally, the disulfides appeared to be superior, the most effective compound of which was the 1-adamantyl-methyl disulfide. In a series of *N*-heterocyclic aminoethyl disulfides and aminoethiosulfuric acids, the thiosulfates were generally more active and less toxic than the disulfides, administered either i.p. or perorally (76). The most effective compound of this series was 2-(2-quinoxali-

**(10)**

nylamino) ethanethiosulfate (**10**) It is believed that the phosphorothioate group aids in cellular transport (77).

Two inorganic phosphorothioates, diammonium amidophosphorothioate (**11**) and diammonium thioamidodiphosphate (**12**) gave

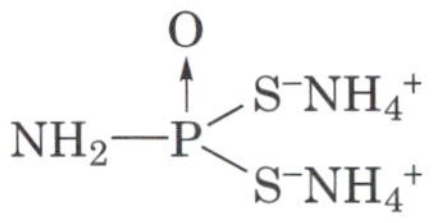

$$NH_2\!-\!\overset{\overset{\displaystyle O}{\uparrow}}{P}\!\!\overset{\displaystyle S^-NH_4^+}{\underset{\displaystyle S^-NH_4^+}{<}}$$

(**11**)

$$NH_2\!-\!\overset{\overset{\displaystyle O}{\uparrow}}{\underset{\displaystyle O^-NH_4^+}{P}}\!-\!S\!-\!\overset{\overset{\displaystyle O}{\uparrow}}{\underset{\displaystyle O^-NH_4^+}{P}}\!-\!NH_2$$

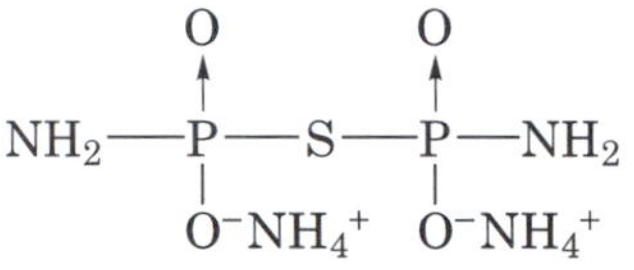

(**12**)

DRF values, respectively, of 2.30 and 2.16 at relatively low doses (72). Alkylation of the amidophosphorothioate lowered or eliminated activity, however (78).

In a series of straight-chain aliphatic thioesters of MEA, the best protection was found with the acetyl and octanoyl derivatives (79); the benzoyl ester was essentially inactive. *N*-Acetyl and *N,S*-diacetyl MEA showed minimal activity (80). In a series of hemimercaptals of MEA derived from glycolic acid, the most active protected mice at one-half the $LD_{50}$ dose, with activity comparable to that of MEA (81).

Other basic functional groups can replace the amino group in the MEA structure to provide protective thiols. The inclusion of the guanidino group has provided very active compounds, notably 2-mercaptoethylguanidine (MEG, **1**) and 2-mercaptopropylguanidine (MPG) (80). Solutions of these compounds were obtained by alkaline rearrangement of the aminoalkylisothiuronium (AET or APT; **13**) salts. When these compounds are em-

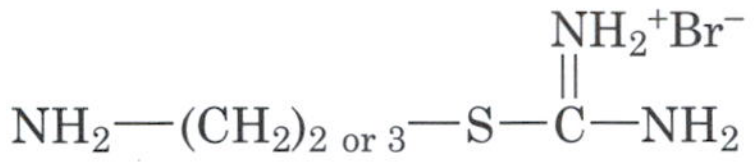

$$NH_2\!-\!(CH_2)_{2\ or\ 3}\!-\!S\!-\!\overset{\overset{\displaystyle NH_2^+Br^-}{||}}{C}\!-\!NH_2$$

(**13**)

ployed for radiation protection tests, the hydrobromides of the aminoalkylisothiuronium bromides are rearranged in neutral or alkaline media. This rearrangement has been termed "intratransguanylation." Thus, AET or APT yields solutions of MEG or MPG (Equation 4.1). These compounds are usually not isolated by this procedure, but they may be isolated as the sulfates (82) or the trithiocarbonate esters (47).

Although AET is not subject to air oxidation, as are most thiols, it is affected by moisture, resulting in conversion to 2-amino-2-thiazoline. The disulfide, bis(2-guanidinoethyl) disulfide (GED), is readily prepared, however, and is relatively stable. With more than three carbon atoms between the amino and isothiuronium functions, rearrangement

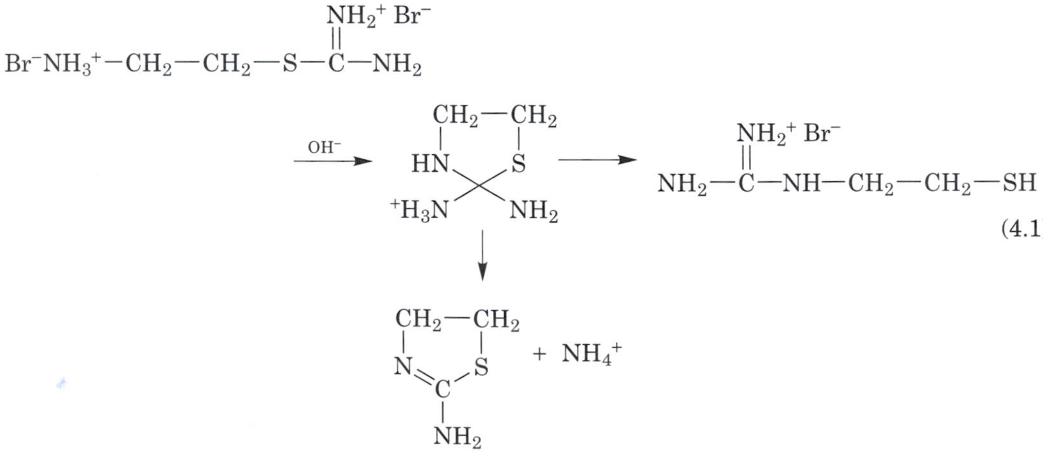

$$Br^-NH_3^+\!-\!CH_2\!-\!CH_2\!-\!S\!-\!\overset{\overset{\displaystyle NH_2^+\ Br^-}{||}}{C}\!-\!NH_2$$

$$\xrightarrow{OH^-}\quad \underset{^+H_3N}{\overset{HN}{\diagdown}}\!\!\overset{\overset{\displaystyle CH_2-CH_2}{|\qquad\ |}}{\underset{\diagdown}{\times}}\!\!\overset{S}{\underset{NH_2}{\diagup}}\quad\longrightarrow\quad NH_2\!-\!\overset{\overset{\displaystyle NH_2^+\ Br^-}{||}}{C}\!-\!NH\!-\!CH_2\!-\!CH_2\!-\!SH$$

(4.1)

$$\underset{\overset{\displaystyle |}{NH_2}}{N}\!\!\overset{\overset{\displaystyle CH_2-CH_2}{|\qquad\ |}}{\underset{\diagup}{\diagdown\!\!\!_{C}}}\!\!S\quad +\ NH_4^+$$

does not readily occur, and the isothiuronium salts give little protection. 2-Aminobutylthiopseudourea dihydrobromide, however, requires about one-fourth the molar quantity of AET for comparable protection in mice (63).

Replacement of the amino group by amidino has also resulted in compounds with good protective activity, particularly with Bunte salts of $a$-mercaptoacetamidines (14)

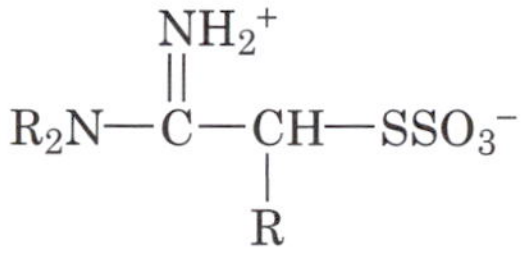

$$R_2N-\overset{\overset{\displaystyle NH_2^+}{\|}}{C}-\underset{\underset{\displaystyle R}{|}}{CH}-SSO_3^-$$

(14)

(83). Among the most effective of the amidinoalkylthiosulfuric acids are several terpene derivatives, including the bornyl (84). Other amidines related to MEA and MEG have been effective; 3,3′-dithiobis(propionamidine) (85) and propionamidines containing isothiuronium groups (86), for instance, have good activity.

Use of strongly basic nitrogen heterocycles having $pK_a$ values of 10–12.5 has also provided protective compounds having the dithiocarbamate group as the sulfur-containing function. Reaction of imino-$N$-alkyl pyridines, pyrimidines (87), quinaldines, and acridines (88) with carbon disulfide gave imino-$N$-carbodithioates (15) having moderate protective effects in mice.

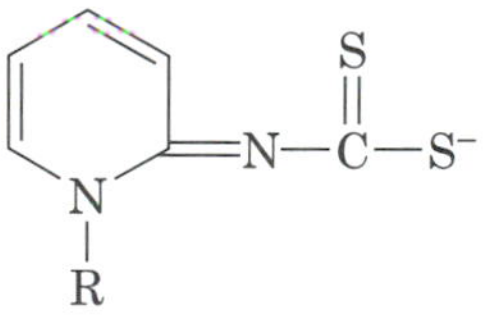

$$=N-\overset{\overset{\displaystyle S}{\|}}{C}-S^-$$

(15)

Substitution of the hydrazino group for amino has not provided many active compounds. Protection of mice has been reported for $N,N'$-bis(mercaptoacetyl) hydrazine (89), as well as for $N$-acetylthioglycolic hydrazide, $HSCH_2CONHNHCOCH_3$, and its disulfide (90).

Oxidation of the thiol group of the MEA structure has provided products with radio-

protective properties, particularly with the disulfides. The disulfides of MEA (cystamine) and MEG (GED) are as active as the parent thiols, although GED is more toxic than MEG (91). The argument has been advanced that the thiol is the active form of these compounds, given that some *in vitro* systems protected by MEA are not protected by cystamine (92), and the reduction of cystamine to MEA during irradiation of mice has been observed (93, 94). In the case of GED, appreciable amounts of this disulfide were found *in vivo* after administration of either MEG or GED (95). In theory, disulfides should be as effective as thiols because they are readily reduced to thiols intracellularly. The converse is also true, in that thiols are easily oxidized *in vivo* and therefore lose activity.

Cystine is nonprotective in mammals, probably because of its inability to penetrate some cellular membranes (93). Mixed disulfides of MEA have provided good protection, particularly those derived from $o$-substituted mercaptobenzenes, where zwitterions are formed with carboxyl, sulfonyl, or sulfinyl anions (16) (96). It is possible, however, that *in*

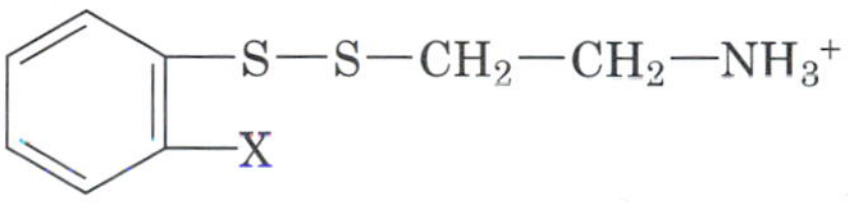

$$\text{S}-\text{S}-CH_2-CH_2-NH_3^+$$
$$\text{X}$$

(16) X = CO$_2^-$, SO$_3^-$, SO$_2^-$

*vivo* the unsymmetrical disulfides are disproportionating to the two symmetrical disulfides, giving rise to cystamine. Mixed disulfides containing $N$-decyl MEA are also effective (97), as is the mixed disulfide of thiolacetic acid and $N$-acetyl MEA (98). Disulfides lacking basic groups have generally been found inactive, although a bis(butanesulfinate) disulfide (17), derived from (18) by disproportionation, was highly active (99). The

$$NaO_2S-(CH_2)_4-S-S-(CH_2)_4-SO_2Na$$

(17)

$$CH_3CONH-(CH_2)_4-S-S-(CH_2)_4-SO_2Na$$

(18)

bis(butanesulfinate) trisulfide was also very active, protecting 100% of mice against a lethal dose of radiation . Two thiocarbamoyl disulfides (**19**; R = H, CH$_3$) also gave good protection (100).

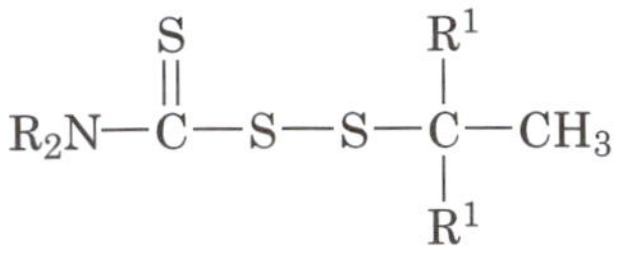

**(19)**

Higher oxidation states of the sulfur in MEA and MEG molecules have been obtained, and some protective activity has been found with these derivatives. The thiolsulfinates of both MEA (101) and MEG (102) have been prepared, as well as the corresponding thiolsulfonates (**20**) (103).

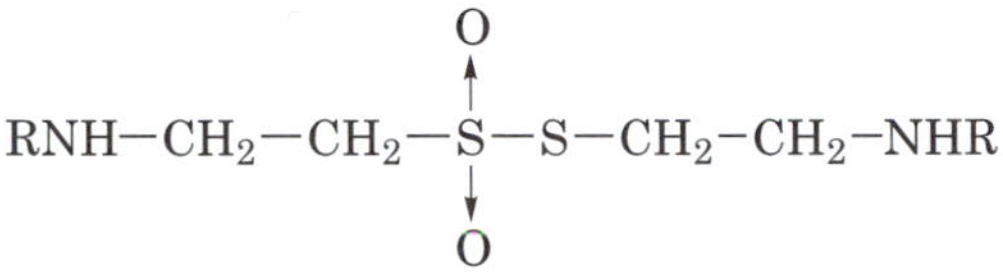

**(20)** R = H, C—NH$_2$

Protective activity has been reported for both the thiolsulfonate (103) and the thiolsulfinate of MEG (104), as well as for the thiolsulfonates of *N*-acetyl and *N*-decyl MEA. Taurine and hypotaurine (the SO$_3$H and SO$_2$H derivatives, respectively, of MEA), both metabolites of MEA in mice (105), provide essentially no protection (27).

Thiazolidines have been prepared from MEA or its N-substituted derivatives by reaction with aldehydes and ketones. A number of thiazolidines have shown good protective activity in mice, which has been attributed to ring opening *in vivo* to give the amino thiols (106). N-Substituted thiazolidines having oxy or thio cycloalkyl, aryl, or heterocyclic alkyl groups (**21**) have good activity (107). Thiazolidine-4-carboxylic acid, derived from cysteine, affords 40% protection to rats (108). Thiazolidines with particularly good activity are 2-propylthiazolidine (106), the 2-(3-phenyl propionate ester) derivative (**22**) (109), and the

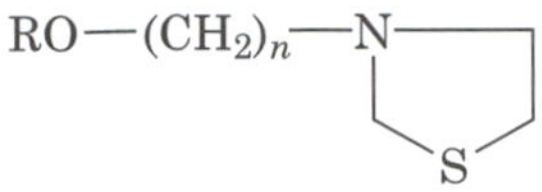

**(21)** R = cycloalkyl, aryl or heterocyclyl

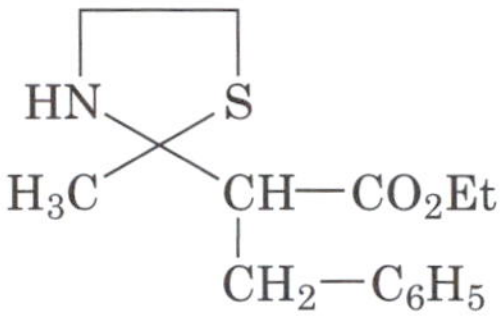

**(22)**

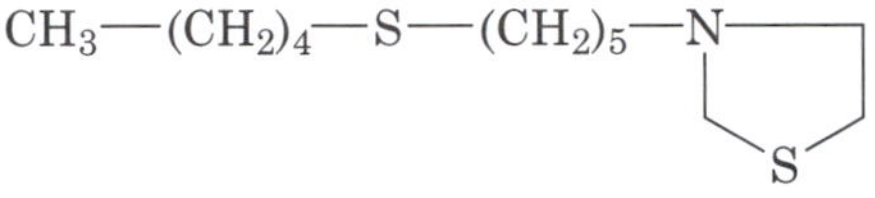

**(23)**

*N*-pentylthiopentyl derivative (**23**) (110). The latter compound was active orally. 2-Aminothiazoline, which is derived from AET at pH 2.5, has protective activity (111); it is probably converted to *N*-carbamyl cysteamine at pH 9.5 (112). 2-Mercaptothiazoline has been found active in two laboratories (16, 60); others have found it inactive (111, 113).

**1.4.2 Other Sulfur-Containing Compounds.** A number of dithiocarbamates have significant radioprotective effects, although the order of activity is less than that of MEA and its derivatives. The simplest compounds of this type, either with the nitrogen unsubstituted or bearing small alkyl groups, up to *n*-butyl, have shown the most activity (114, 115). 2-Methylpiperazinedithioformate (**24**), how-

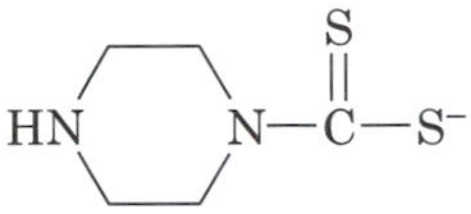

**(24)**

ever, provides protection more nearly comparable to that of MEA (116). The mechanism by which the dithiocarbamates protect is believed to differ from that of the aminothiols. Xanthates have not been found protective (117).

The related thiocarbamyl derivatives (25) and (26) have been reported to provide good protection (118).

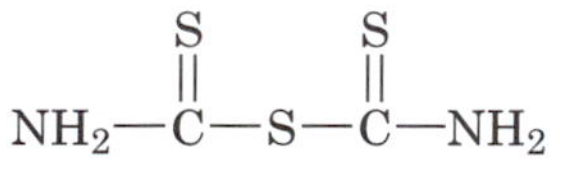

$$NH_2-\overset{\overset{\displaystyle S}{\|}}{C}-S-\overset{\overset{\displaystyle S}{\|}}{C}-NH_2$$

(25)

$$NH_2-\overset{\overset{\displaystyle S}{\|}}{C}-NH-\overset{\overset{\displaystyle S}{\|}}{C}-NH_2$$

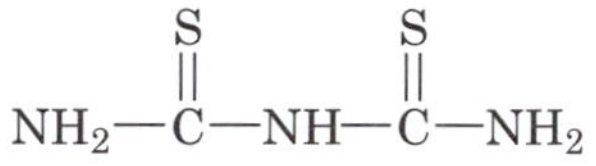

(26)

Reaction of cysteine with carbon disulfide gives the trithiocarbonate dithiocarbamate (27) (119), which is equivalent in protective

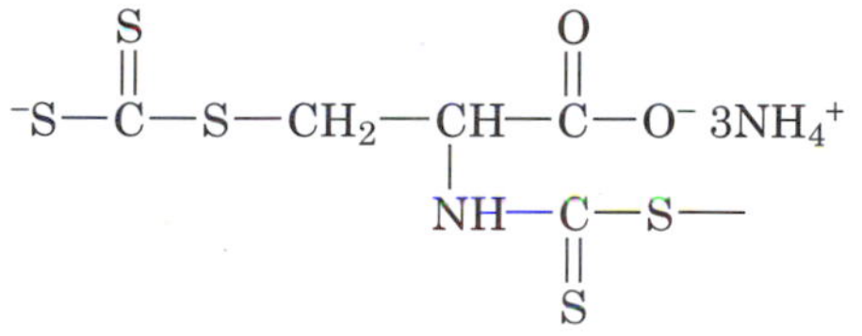

$$^-S-\overset{\overset{\displaystyle S}{\|}}{C}-S-CH_2-\underset{\underset{\displaystyle NH-\overset{\overset{\displaystyle }{\|}}{C}-S-}{|}}{CH}-\overset{\overset{\displaystyle O}{\|}}{C}-O^-\ 3NH_4^+$$

(27)

activity in mice to that of MEG but is only one-third as toxic. A metabolism study in mice showed the dithiocarbamate group to be appreciably stable *in vivo*, but the trithiocarbonate to be unstable (120). Trithiocarbonates of MEG and MPG and several derivatives of MEG (28) also provided good protection to mice against a lethal dose of X-radiation (47).

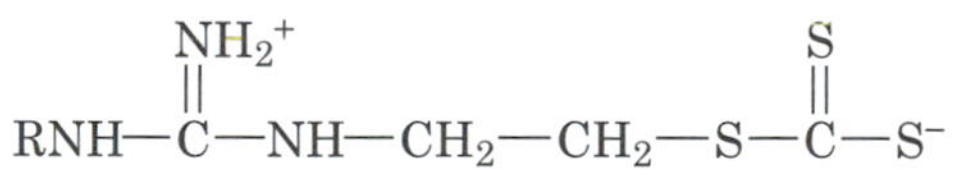

$$RNH-\overset{\overset{\displaystyle NH_2^+}{\|}}{C}-NH-CH_2-CH_2-S-\overset{\overset{\displaystyle S}{\|}}{C}-S^-$$

(28)

Thioureas and cyclic thioureas have shown only marginal or no protection. Thiourea itself protects mice only in massive doses (1800–2500 mg/kg) (5). S-Alkylisothioureas, with alkyls up to *n*-butyl, have shown moderate protective effects (121). Dithiooxamide is nonprotective, but symmetrical *N,N'*-dialkyldithiooxamides provide some protection (113). 1,5-Diphenyl-

thiocarbohydrazide and several derivatives have fair activity (122).

Simple, nonbasic thiols have no value as radiation protectors. Conflicting results have been reported for the dithiol BAL as well as for thioctic acid (27). 2,3-Dithiosuccinic acid is protective in mice vs. 700R (123), but most other dithiols are inactive, with the exception of dithiothreitol (64). 2-Mercaptoethanol protects bacteria (124) but not mice (125); it has also been found to be radiosensitizing (126).

Other sulfur-containing compounds with significant radioprotective ability include dimethyl sulfoxide (127) when given in large doses; other sulfoxides afford little or no protection. Large doses are required because the mechanism of protection may be hydroxyl-radical scavenging rather than hydrogen atom donation to radicals or DNA. Organic thiosulfates, other than those that liberate MEA or an active derivative of MEA, have generally failed to protect. Inorganic thiosulfate is a good protector of macromolecules *in vitro* or of the mucopolysaccharides of connective tissue *in vivo* (128); it does not protect animal cells, however, because of its inability to penetrate. Sodium cysteinethiosulfate (29), derived from

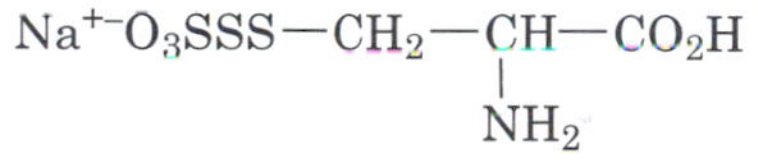

$$Na^+{}^-O_3SSS-CH_2-\underset{\underset{\displaystyle NH_2}{|}}{CH}-CO_2H$$

(29)

the cleavage of cystine with thiosulfate ion, has good activity, being protective of the intestines and kidneys of mice (129). The related S-sulfocysteine, having one less sulfur atom, is almost devoid of activity (130).

Mercapto acids have shown little protection, with the exception of thioglycolic acid, which is slightly protective, inactive, or sensitizing, depending on the system tested (27). The β-aminoethyl amide of thioglycolic acid, $HSCH_2CONHCH_2CH_2NH_2$, has good activity (81), however.

Monothio acids and their derivatives are generally inactive, although several dithio acid dianions, obtained by condensation of carbon disulfide with cyanomethylene compounds (30) show some protection of mice (131). The most active of this series is the dithio acid derived from 2-cyanoacryloylpyrrolidide (**31**), which provides 80% protection to mice against a lethal dose of γ-radiation (76). Dithio esters derived from pyridinium dithioacetic acid betaine (**32**) also show some protec-

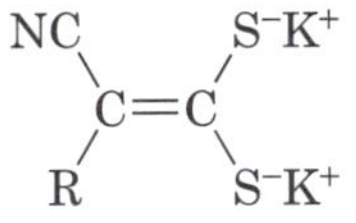

(**30**) R = CN, CO₂Et, C₆H₅CO

(**31**)

(**32**)

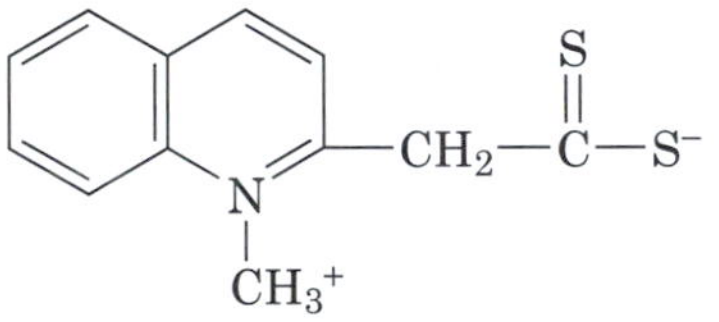

tive activity in both mice and bacteria (132). Quinolinium-2-dithioacetic acid zwitterions (**33**) provided good protection to mice, and the more soluble bis(methylthio) and methylthio amino derivatives (**34**) had slightly better effects (133). The corresponding pyridinium compounds had equally good activities, and both

(**33**)

(**34**) R = SCH₃, NR₂

the quinolinium and pyridinium compounds were protective at much lower dosage levels (2–18 mg/kg) than those for the aminothiols (150–600 mg/kg). Replacing the aromatic function with aliphatic in these compounds, yielding 3-amino-2-phenyldithiopropenoate esters, $R_2NCH = C(C_6H_5)CS_2CH_3$, resulted in only fair or poor protection (134).

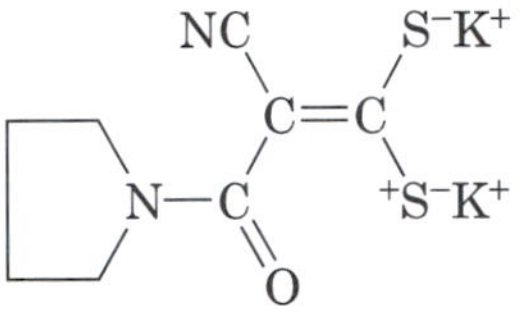

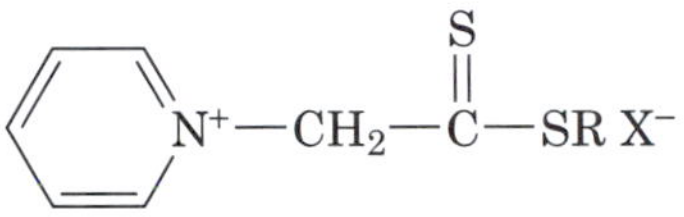

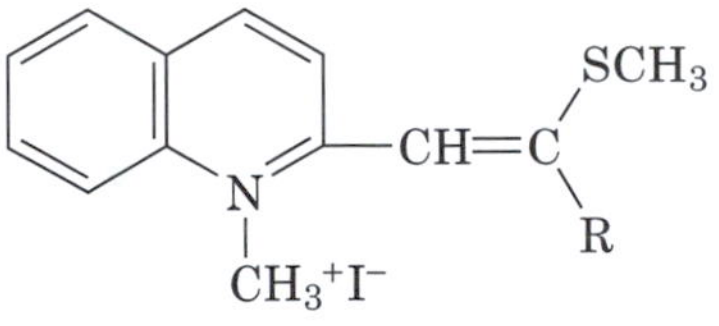

Thiols that occur naturally have not been found appreciably protective in animals, with the exception of glutathione (135), which has moderate activity. Pantoyl taurine apparently has some activity (136). Bacq (25) presented arguments that make it appear unlikely that coenzyme A is involved in radioprotection.

Selenium compounds have generally been ineffective in animal tests. Selenium analogs of the well-known radioprotectors, 2-aminoethaneselenol, 2-aminoethaneselenosulfuric acid (52), and 2-aminoethylselenopseudourea (137) are much more toxic than their sulfur analogs and are nonprotective. Sodium selenate (138) has been reported to be effective in rats administered postirradiation. Some selenium-containing heterocycles (e.g., selenoxanthene, selenoxanthone, and selenochromone) have been described as effective in rats (139). The investigation of organic selenium compounds as potential antiradiation agents has been reviewed by Klayman (140).

**1.4.3 Metabolic Inhibitors.** Cyanide ion has been found radioprotective in several laboratories (3, 141), but it must be administered immediately before irradiation because of its rapid detoxification (17). It has a number of biological properties in common with thiols, such as reduction of disulfide linkages and inhibition of copper-containing enzymes, but unlike thiols, it also inactivates cytochrome C oxidase, which controls oxygen consumption in mammals. Among other enzyme inhibitors, azide (142), hydroxylamine (143), and 3-amino-1,2,4-triazole (144) are weak protectors. The latter two compounds are inhibitors of catalase, but no relation between this effect and radioprotection was apparent.

Several organic nitriles show radioprotective effects; the most effective may be hydroxyacetonitrile (113). Fluoroacetate is protective (145) when sufficient time is allowed before irradiation for its conversion to fluorocitrate, an inhibitor of citrate metabolism. Other thiol group- or enzyme-inhibiting agents, such as iodoacetic acid, malonic acid, mercurials, and arsenicals have no protective ability, but some of these agents have radiosensitizing effects.

**1.4.4 Agents Involving Metal Ions.** A number of metal-binding agents are radioprotective and are also known to inhibit enzymes. Some metal complexes imitate the action of enzymes, such as copper complexes, which catalyze the decomposition of peroxides (146). In this aspect, the copper complex of 3,5-diisopropylsalicylate has shown activity against $\gamma$-radiation (147), presumably by mimicking the action of superoxide dismutase.

Copper complexes of the radioprotective pyridinium and quinolinium dithioacetic acid derivatives (133) also are able to mimic the action of superoxide dismutase (148). These effects may play a role in radiation protection.

Metal-binding agents already discussed include the dithiocarbamates as well as the aminothiols (149). EDTA protects mice only in very large doses (150), probably because very little EDTA enters the cells. 8-Hydroxyquinoline (oxine) is too toxic for animal studies, but was found highly protective in a polymer system (125). Other common metal-binding agents, such as $N$-nitroso-$N$-phenylhydroxylamine and nitrilotriacetate, show appreciable protection (151). Derivatives of 1,5-diphenylthiocarbohydrazide, avid metal binders, protect mice, rats, and dogs (122).

Some metal complexes have been tested and found to afford some protection. Iron complexes of polyamines (152) are active, as well as zinc complexes of MEA and MEG, the copper and iron complexes showing little or no activity (149). Copper complexes of diethyldithiocarbamate and dithiooxamide, however, give less protection than that of the uncomplexed ligands (125). Complexes of chlorophyllin, with Co, Mg, Mn, and V, are radioprotective in mice (153). Zinc aspartate has shown some protective properties (154).

**1.4.5 Hydroxyl-Containing Compounds.** Significant protection in animals by hydroxyl-containing compounds, including ethanol, was at first considered the result of antioxidant properties. Glycerol, however, is protective in mice as well as in other systems (5, 150). Phenols are protective in polymethacrylate tests (125), but many of them are too toxic for animal tests. The catecholamines provide some protection, possibly by lowering oxygen tension in the cells (27). A compound that increased catecholamine levels in irradiated rats, 1-acetylhydrazinylthiophenylformamidine, protected against a half-lethal dose of radiation (600 R) (154). The protective effects of gallic acid esters are attributed to inhibition of chain oxidation processes induced by radiation (155). Arachidoyl derivatives of pyrogallol and the naphthols have shown some protective activity (156). Ionol (2,6-di-$t$-butyl-4-methylphenol), injected after irradiation, prolongs the life of mice and alleviates intestinal damage (157).

Organic acids provide little or no protection, but the polycarboxylic acids, pyromellitic and benzenepentacarboxylic, are protective in mice (158). These polyionic substances are believed to protect by causing hypoxia from osmotic effects, rather than by chelating calcium ion, which also is believed to have an effect on radiation damage.

In a series of $S$-2-(3-aminopropyl-amino) alkylphosphorothioates, which are effective protectors in mice when given orally, the presence of hydroxyl groups in the alkyl chain generally lowers effectiveness for oral administration, but still allows good protection by i.p. injection (159).

**1.4.6 Heterocyclic Compounds.** Several relatively simple heterocyclic compounds provide significant protective activity in mice. In a series of imidazoles, imidazole itself, benzimidazole, and 1-naphthylmethylimidazole were the most effective compounds (160). Related imidazolidine-5-thiones were also protective. The cyclic analogs of AET, 2-aminoethyl- (**35**), and 2-aminopropylthioimidazoline, are moderately protective (161).

Of a large number of amine oxides tested for radiation protection, quinoxaline 1,4-di-$N$-oxide (**36**) (believed to act in part by radical trapping) was the most effective (162). It is protective in mice but radiosensitizing in the dog (163). A more recent $N$-oxide, 2,2,6,6-tetramethyl-4-hydroxypiperidine-$N$-oxide (Tem-

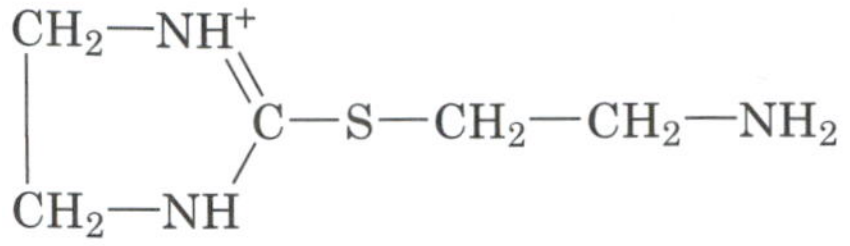

(35)

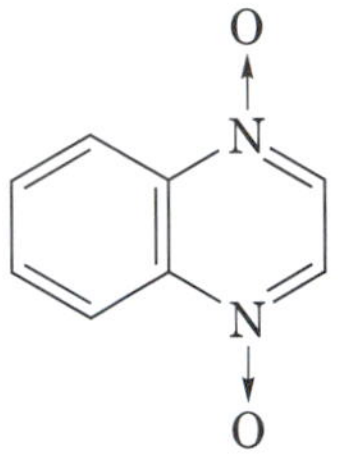

(36)

pol) and its reduced hydroxylamine are radio-protective in mice without lasting adverse effects (164). 3,5-Diamino-1,2,4-thiadiazole (165) and 3-(β-amino-ethyl)-1,3-thiazane-2,4-dione (166) (**37**) have some protective activity.

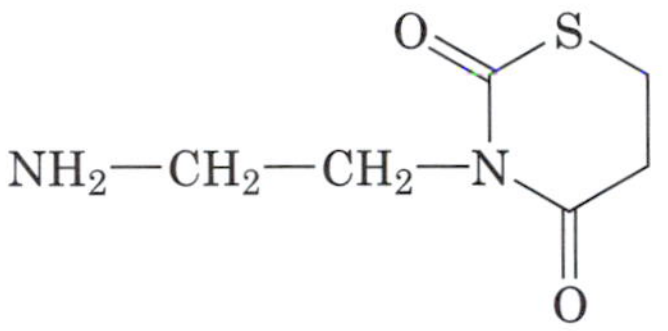

(37)

Aminoethyl and aminomethyl purines and pyrimidines gave one-third as much protection in mice as MEA (167). 8-Mercaptocaffeine and its S-β-aminoethyl and S-β-hydroxyethyl derivatives (**38**) had protective activity in mice

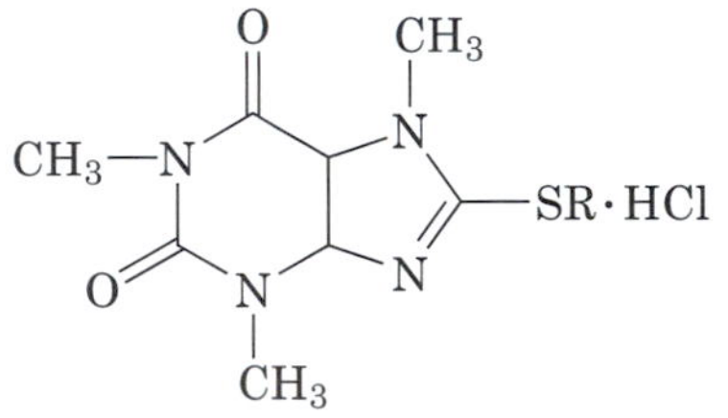

(38) R = H, $CH_2CH_2NH_2$, $CH_2CH_2OH$

similar to that of cystamine (168). These compounds also enhance hemopoiesis and decrease blood loss in irradiated animals.

Good protection was provided by 6-acyl-2,3-dimethyl-4,7-dimethoxybenzofurans (169), and fair protection was observed for several 2-dialkyl-1,3-oxathiolanes (170). In a large series of 1,3-dithiolanes tested, moderate protection was shown by 1,3-dithiolane itself and by its 2- and 4-methyl derivatives (171) (**39**). Although

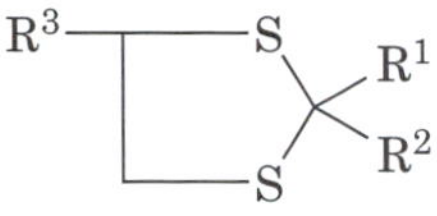

(39)

aryl 1,2-dithiole-3-thiones are known to raise glutathione levels in cells (172), 4-phenyl-1,2-dithiole-3-thione (**40**) and its S-methyl iodide

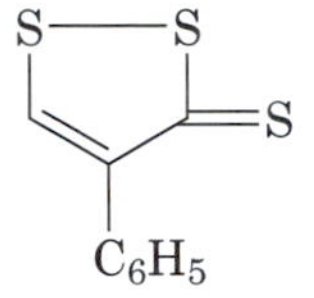

(40)

showed no activity in mice against a lethal dose of γ-radiation (134).

In a series of 2,1,3-benzothiadiazoles, the 4-hydroxy derivative (**41**) had the best protec-

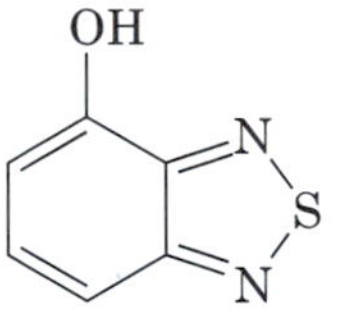

(41)

tive effect in mice (173), either i.p. or orally. Several aminothiazines, including 2-amino-4,6,6-trimethyl-1,3-thiazine (**42**), increased survival time in mice (174), without liberation of

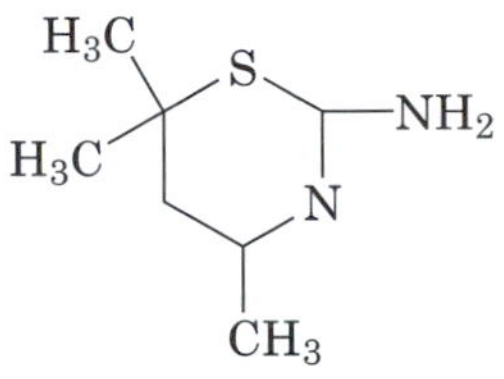

(42)

a thiol group. 2-Pyridinemethanethiol, structurally related to cysteamine, showed good protective potency in mice, but 2-pyrazinemethanethiol was inactive (175). The three isomeric mercaptopyridines also were active.

### 1.4.7 Physiologically Active Substances.

A number of familiar physiologic agents exert some radiation protection, which is generally of a lower order of activity than that provided by the amino thiols. Many of these agents are believed to be radioprotective by virtue of their ability to lower oxygen tension in the cells or to depress whole-body metabolism. Serotonin [5-hydroxytryptamine (5-ITT)] has been reported equal in activity to MEA (136). It is effective at a dose well below the toxic level (113). It has been most often used as the creatinine sulfate salt. Its activity has been attributed to its vasoconstrictor effect leading to hypoxia in radiosensitive tissues (176). Psilocybine (4-hydroxy-N,N-dimethyltryptamine), however, a more potent vasoconstrictor than serotonin, is inactive, suggesting that this effect may not be of major importance (177). There is evidence of central nervous system involvement in the activity of serotonin (178).

5-Hydroxytryptophan is comparable in activity to serotonin (179), and the 5-methyl ether of serotonin (mexamine) is also as effective, although higher alkyl ethers are ineffective (180). Numerous indole derivatives have been prepared as potential protectors, including 5-acetylindole, which has some activity (181), but none has exceeded serotonin or mexamine in potency. A series of acyl derivatives of 5-methoxytryptamine showed good protective effects in mice against a lethal dose of X-rays, the hexanoic and octanoic amides being the most potent (182). A synergistic radioprotective effect results from a combination of AET, ATP, and serotonin in mice and rats (183).

Central nervous system depressants have only small or moderate effects as radiation protectors. Chlorpromazine has been extensively studied, but exerts only a slight effect, which is most pronounced when administered 4.5 h before irradiation, when a state of hypothermia exists (184). Chlorprothixene is also most effective when body temperature and metabolism are depressed (185). Reserpine is effective when given 12–24 h before irradiation (186), possibly by release of serotonin and catecholamines (187).

Central nervous system stimulants generally are nonprotective. An exception is the magnesium complex of pemoline (2-imino-5-phenyl-4-oxazolidinone), which provides moderate protection to mice against 750 R (188). Of 21 analogs of imipramine, three showed significant activity (189).

The different classes of autonomic drugs provide some radiation protection; the causative factor is believed to be production of hypoxia by various mechanisms. Epinephrine provides some protection (190), but norepinephrine, which decreases oxygen tension in the spleen much less than does epinephrine, gives very little protection to mice (191). The cholinomimetic compounds arecoline, tremorine, and oxytremorine are also protective to a small extent in mice (192).

p-Aminopropiophenone (PAPP) apparently protects by induction of tissue hypoxia (193). It has been used in relatively small quantities in combination with other protective agents, such as MEA and AET (194, 195). The radioprotection afforded by PAPP is abolished by increased oxygen pressure during irradiation (196). PAPP and its ethylene ketal also gave good protection orally to mice against X-rays (197). p-Aminobenzophenone also provided good protection in this test. Monothio and dithio ketals of this compound gave little or no protection (198).

Physiological changes can probably account for the radioprotective action of some substances. Urethane (199), estrogens (200), and colchicine (201) can stimulate blood cell production by damaging bone marrow. If irradiation is carried out while there is an increased leucocyte/lymphocyte ratio in the blood, so that a greater percentage of more radioresistant cells are present, enhanced survival may result. The effect of colchicine may also be attributable to inhibition of mitosis, but there is evidence against this supposition (201). Colchicine is protective only when administered 2 or 3 days before irradiation, by which time mitotic inhibition has ceased. Urethane and the estrogens are similar in that they must be given a day or more before irradiation. The proestrogen tri-$p$-amisylchloro-ethylene is effective when given 5–30 days before irradiation (202). Other inhibitors of mitosis, however, can enhance survival; these include demecolcine (Colcemid), sodium arsenite, epinephrine, cortisone, and typhoid/paratyphoid vaccine (203). Tranquilizers and other psychotropic drugs possess only moderate radioprotective activities. These compounds probably are active by depression of whole-body metabolism through diminished oxygen uptake (35).

Procaine (204) and several derivatives of procaineamide, particularly the $p$-nitro derivative (205), have shown appreciable protective activity. 4-Hydroxybutyric acid and 6-phosphonogluconolactone, substances that stimulate turnover of NADP•H$_2$, a physiologic reducing agent, provide protection to mice (206). An antihistamine, thenaldine, affords moderate protection (204). Alloxan protects both mice (207) and the pancreatic ultrastructure of dogs (208).

### 1.4.8 Metabolites and Naturally Occurring Compounds.
A variety of compounds in these categories have been examined for radiation protection, but few effective protectants have been found. Some polysaccharides, such as dextran (209), those extracted from typhoid and proteus organisms (210), and a lipopolysaccharide from S. *abortus* (211), provide some protection for mice, possibly by inducing phagocytosis. Bacterial endotoxins, which are lipopolysaccharides of molecular weight around 1,000,000, show relatively good protective properties in both normal (200) and germ-free mice (212). Typhoid/paratyphoid vaccine shows similar protective properties (213).

Vitamins and coenzymes are not appreciably protective. Pyridoxal phosphate, however, has a moderate effect (214), which may be connected with a repair rather than a protective process (215). Several thiol-containing derivatives of vitamin B$_6$, including 5-mercaptopyridoxine (216), are also protective. Some of the naturally occurring pyrimidine bases and nucleotides (217), including ATP (218), have an effect in mitigating radiation damage, but their value may be ascribed to postirradiation repair. Protection from RNA, DNA, and derivatives has been claimed, but their effects are more likely attributable to repair processes (218–220). A protamine-ATP combination provides good protection to rats (221).

Among the commonly used antibiotics, the tetracyclines have shown the most favorable effects on survival rates of mice (222); this is believed to be the result of an increase in metabolic activity. A gallate-tannin complex (223) was active probably because of its antioxidant effect. 5,7-Dihydroxyisoflavones are effective when administered to mice percutaneously but not intraperitoneally (224), presumably because of protection of the capillaries. $O$-β-Hydroxyethylrutoside is also protective in mice, possibly by strengthening vascular walls and reducing bacterial invasion of the bloodstream (225). The radioprotective effect of rutin and other flavonoids has been controversial.

A series of extracts from the Chinese drug plants *Carthamus tinctorius*, *Sargentodoxae cuneata*, *Paeonia lactiflora*, *Salvia miltiorrizha*, and *Ligusticum chuanxiong* have shown significant protection in mice versus 7.5–8.0 Gy of γ-radiation (226). Their protective properties are believed to be related to their inhibitory effects on radiation-induced platelet hypercoagulation in the capillaries, which prevents excessive bleeding. Active constituents in the Ligusticum drug extract have been found to be harman alkaloids, including 1-(5-hydroxymethyl-2-furyl)-9 $H$-pyrido-[3,4-$b$]indole (227). Acetylsalicylic acid also provides moderate protection at this radiation dose.

### 1.4.9 Polymeric Substances.
A synthetic polymer prepared from $N$-vinylpyrrolidone and $S$-vinyl-(2,2-dimethylthiazolidyl)-$N$-monothiol carbamate (**43**) was found protective in mice, possibly by liberation of thiol groups (228). Other copolymers containing isothiouronium

(**43**)

salts, thiosulfates, and dithiocarbamate groups give appreciable protection when administered 24–48 h before irradiation (229). Polyinosinic-polycytidylic acid increases survival of mice, possibly by increasing the stem cell fraction in blood-forming tissues (230). Both poly(vinyl sulfate) (231) and heparin (232), a sulfated mucopolysaccharide, increase survival rates, possibly by affecting deoxyribonuclease activity.

**1.4.10 Miscellaneous Substances.** In a series of diethylsulfides and diethylsulfoxides, bis[2-$n$-butyrylamino)ethyl]-sulfoxide showed the greatest radioprotectant activity (233). Among 20 benzonitriles evaluated as radioprotectants in mice, 3,5-dinitrobenzonitrile showed significant activity (234). A series of N-substituted 1-$m$-hydroxyphenyl-2-aminoethanols, which are derivatives of phenylephrine, gave significant radioprotective effects (235). Cinnamonitriles and hydrocinnamonitriles produced survival rates of 43–58% in three derivatives of the former, the $p$-MeO, $p$-NO$_2$, and 2-chloro-$p$-MeO, after lethal doses of radiation (236). The $S$-phosphate of 5-mercaptomethylcytosine exhibited protective activity at a dose of 6 Gy (237). Radioprotective effects of two halogenated 1,3-perhydrothiazines have been described (238).

Radioprotective effects of isobombycol and its enanthate and cinnamate in mice have been reported (239). In a series of substituted anilines tested in mice versus a near-lethal dose of 6 mV photons, compounds with electronegative groups in the *meta* or *para* positions gave good protection to mice (240). No correlations between protective activity and a variety of molecular parameters could be found, however. Schiff bases of salicylaldehyde, 5-chlorosalicylaldehyde, and benzaldehyde with anilines reduced toxicity of the parent amines, but gave erratic protective results (241). The highest protection was observed for mixtures of $p$-aminopropiophenone with its Schiff bases or with the Schiff base of 1-($p$-aminophenyl)-1-propanol.

## 1.5 Mechanisms of Protective Action

The manner in which mammalian cells are protected from the damaging effects of ionizing radiation is not known in complete detail, although evidence is accumulating for several postulated pathways of radioprotection. Protection by means of radical trapping or antioxidant action, which can be demonstrated for simple systems, such as polymers, may be operative in animal cells as well. It is also probable that other mechanisms are more important in protection of cells, and possibly more than one mode of protection may be operative for a given type of agent. A number of the physiological agents that have been observed to be radioprotectors are believed to protect by producing various levels of anoxia; the evidence for this has been discussed (27).

**1.5.1 Protection by Anoxia or Hypoxia.** Protection by producing a state of cellular anoxia or hypoxia is based on the phenomenon of the "oxygen effect," the increase by two- to threefold of the damaging effects of radiation attributed to the presence of oxygen. A number of radioprotective drugs possess the physiological function of producing anoxia or severe hypoxia in various tissues; these include the catecholamines, histamine, choline esters, $p$-aminopropiophenone, morphine, ethyl alcohol, and nitrite. Other physiological effects, however, may contribute to their ability to protect, particularly with serotonin. Although the powerful protection afforded by this compound is not completely explained, a correlation between vasoconstrictive effects and radioprotection was found for a series of indolamines (242). Increasing the amount of oxygen available to radiosensitive tissues reversed the radioprotective effect of serotonin, histamine, and epinephrine, but caused much less reduction of protection by MEA or cysteine (243). The amino thiols, notably cysteine, MEA, and AET, can decrease oxygen consumption in the cells (244), but no appreciable hypoxia exists during the protective period (245). Enzymatic oxidation of radioprotective thiols has been reported (246), although the extent to which the resulting oxygen depletion contributes to radioprotection is uncertain.

A series of reports (reviewed in Ref. 247) provided evidence that WR2721 and its dephosphorylated thiol WR1065 have a major effect through their ability to cause local tissue hypoxia. The thiol also rapidly depletes the oxygen content of mammalian cell suspensions in culture (248). However, these results are put in question by the lack of adequate procedures to measure the oxygen concentration at critical sites for radiation protection (249).

### 1.5.2 Inhibition of Free-Radical Processes.

Mechanisms of protection involving "free-radical scavenging" are based on the assumption that the free radicals resulting from radiolysis of water are the main cause of radiation damage to the cells. Radioprotectors then would react with these radicals, such as $H^{\bullet}$, $HO^{\bullet}$, and $HO_2^{\bullet-}$, and prevent them from damaging biologically important molecules. This concept received support when a correlation was found between the protective action of about 100 substances in two systems: an aerated aqueous solution of polymethacrylate and the mouse (125). It is probable that radical scavenging is the primary event in the prevention of radiolytic fragmentation of the polymer (250), but it is probably not of equal importance in the cell. Radioprotection of mammalian cells by the $HO^{\bullet}$ scavenger dimethylsulfoxide requires concentrations of the order of 1 $M$, and the degree of protection is insensitive to oxygen concentration (251). Radioprotection of mammalian cells by thiols can be achieved at concentrations of the order of 1–10 m$M$ but this radioprotection can be reversed by exposure to molecular oxygen, indicating that it is not the result of $HO^{\bullet}$ scavenging (251). Diffusible radical scavenging could be radioprotective if mechanisms other than double-strand lesion formation are involved. In that case, it is more likely that less-reactive radicals than $HO^{\bullet}$ would be involved, given that lower concentrations of scavenger could be effective. Cytoplasmic irradiation (precluding direct damage to DNA) can be mutagenic, and there are indications that radical scavenging could be protective against this effect (252).

Reaction of sulfhydryl compounds with free radicals formed on macromolecules is considered more likely than reaction with $HO^{\bullet}$, to account for radioprotection in mammalian cells. Reaction rates with such radicals were measured for several radiation protectors: MEA, AET, APT, thiourea, cysteine, glutathione, propyl gallate, and diethyldithiocarbam-

ate (253). The fastest rates were observed for diethyldithiocarbamate, MEA, and cysteine. Cysteine and glutathione were found to accept electrons from irradiated proteins, whereas cystine and some nonsulfur compounds did not (254).

A number of antioxidant phenols, pyridines, and gallic acid esters are believed to be effective by virtue of their antioxidant action. A direct relation between radical inhibitory action and radiation protection has been observed (255). Protection by aliphatic alcohols, including glycerol, generally requires large concentrations for maximum protection (1–3 $M$) in cultured mammalian cells (256), but protective effects of radical scavengers have been found at much lower concentrations in bovine erythrocytes (257) and the erythrocyte membrane (258).

### 1.5.3 Mixed Disulfide Hypothesis.

This hypothesis of Eldjarn and Pihl (259) proposed that radioprotective thiols form mixed disulfides with thiol groups of proteins. The mixed disulfides provide protection to the protein thiols either by interfering with indirect radiation damage from radiolysis products of water or by facilitating energy transfer from the directly damaged protein to the administered thiol. Some arguments with this hypothesis have arisen; many thiols do not protect, and most thiols are capable of forming mixed disulfides (260). However, equilibrium constants for mixed disulfide formation are high for protective thiols but low for poor protectors (261).

In their original hypothesis, Eldjarn and Pihl proposed that the mixed disulfide bond would be cleaved by radical scavenging, but subsequent studies with protein solutions indicated that this may not be the case (262). Disulfide formation may also protect by moderating radiation-induced rearrangements (263). Radical scavenging may be an important function of the mixed disulfides (264), but mixed disulfide formation may also be a precursor for the liberation of cellular thiols, to be discussed under Section 1.5.4.

Another argument against this hypothesis is that many proteins are not damaged seriously by a dose of radiation that is lethal to mammals (265). Also, the nucleic acids, important target molecules of the cell nucleus, do not contain thiol or disulfide groups. The nuclear proteins involved in cell division have been proposed, however, as likely sites for mixed disulfide formation (266). RNA polymerase is particularly implicated for this process (267). Also, in favor of this hypothesis, $^{35}$S-MEA was found mainly bound to protein at the time of maximum protection (268). Certain enzymes, containing essential sulfhydryl groups, were found to be protected from X-ray damage by mixed disulfide formation (269).

**1.5.4 Biochemical Shock.** A number of biochemical and physiological disturbances take place in the cells after administration of thiols, and realization of the full extent of the cellular changes produced led to the postulation of the "biochemical shock" hypothesis of Bacq (270) and others. This states that protective thiols undergo mixed disulfide formations in the cells, leading to a series of disturbances including decreased oxygen consumption, decreased carbohydrate utilization, and mitotic delay by temporary inhibition of DNA and RNA synthesis, along with cardiovascular, endocrine, and permeability changes. The mitotic delay allows time for repair processes to restore normal nucleic acid synthesis.

Other metabolic effects observed after thiol administration include hypotension, hypothermia, and hypoxia (271). An increase in serotonin level has also been noted in rats after injection of amino thiols (272). Release of endogenous thiols is another metabolic effect of the radioprotective thiols. This has been caused not only by amino thiols but also by serotonin and hypoxia-causing compounds, as well as by the anoxic state (273). This increase in cellular thiol content is often 30- to 40-fold greater than the amount of thiol supplied by the protective agent. Protective effects of the amino thiols in Ehrlich ascites (274) and other tumor cells (275), as well as in mice (276), show direct correlations with the levels of nonprotein thiols. The natural radiosensitivity of mice was related to the concentration of thiol groups in the blood-forming tissues of the spleen (277), and development of radioresistance in cells was attributed to increased concentration of non-protein-bound thiols (278). Radioresistance in some tumor cells was believed to be attributed to protein thiol content

(279), although the level of hypoxia in tumor cells is also a factor affecting radiation sensitivity (280).

Further evidence of the importance of cellular thiol levels in radioprotection is found in the following situations. Protection of the chromosomal apparatus in Ehrlich ascites cells by MEA was associated with the increase in nonprotein thiol levels (281). Both MEA and cystamine increased plasma, liver, and spleen concentrations of free thiols and disulfides (282). Radioresistance of bacterial cells was believed to be the result of a repair system dependent on the thiol content of the cells (283). Revesz and coworkers considered that glutathione was a principal endogenous radioprotector released by administered thiols (284), but more recent research showed that only a small fraction of low molecular weight protein-bound thiols was identified as glutathione (285, 286).

**1.5.5 Control of DNA Breakdown.** The ability of the disulfides of the radioprotective amino thiols to bind reversibly to DNA, RNA, and nucleoproteins has been postulated as a result of *in vitro* studies (287). This, according to Brown (288), can result in two restorative effects: first, the loose ends of the helix resulting from single-strand rupture are held in place, so that shortening or alteration of the chain is prevented; and second, the replication rate of DNA is decreased or halted, so that repair can take place before radiation-induced alterations are replicated. This binding, together with either radical scavenging (289) or repair by proton donation, provides a possible route of protection of the nucleic acids by the amino thiols. It requires that the disulfide of the amino thiol be present for binding, and it also could explain why more than a three-carbon distance between amino and thiol functions leads to a sharp drop in protective ability. Portions of the DNA helix unprotected by histone have been found to accommodate an aliphatic chain of approximately 10 atoms; consequently, a disulfide with two or three carbons between the amino and disulfide functions would fit this exposed portion of the helix. Other strongly protective derivatives of MEA and MEG, such as the thiosulfate, phosphorothioate, trithiocarbonate, or acylthioesters, readily undergo disulfide formation. Arguments against this hypothesis include the fact that not all of the exposed areas of DNA are the same size. Also, in a series of protective disulfides and thiosulfates of MEA with N-heterocyclic substituents, the thiosulfates showed no binding ability for DNA, and whereas the disulfides have appreciable binding constants, there was no correlation between binding ability and protective activity (290).

Protective effects for DNA have been shown by thiourea and propyl gallate, as well as by cysteine and cystamine, apparently through antioxidant activity (291). Another proposed explanation for the protection of DNA by the amino thiols is that MEA renders cell membranes more resistant to radiation damage. Localization of repair enzymes and nucleases on the nuclear membrane makes it possible that radiation damage to the nuclear membrane could result in irreversible damage to DNA by nucleases, and interference with repair of DNA (292).

Other observations regarding the temporary inhibition of nucleoprotein synthesis by thiol protectors have been reported. Temporary inhibition of nuclear RNA synthesis in the radiosensitive tissue of rat thymus was found along with inhibition of thymidine phosphorylation for a short period (292). Radiosensitizers, such as penicillamine and β-mercaptoethanol, inhibited thymidine phosphorylation for a longer period. Some evidence for mixed disulfide formation with proteins (e.g., thymidine kinase) was also found. Inhibition of DNA synthesis in rat thymus, spleen, and regenerating liver by MEA and AET was believed to arise from a delay in the synthesis of relevant enzymes, nuclear RNA polymerase, and thymidine phosphorylating kinases (294). Although MEA decreases the frequency of radiation-induced single-strand breaks in DNA of mammalian cells (295), this was not considered to be the lesion responsible for the killing of *E. coli* cells by γ-radiation (296).

**1.5.6 Metabolic Effects.** Alteration of cellular metabolism can affect radiosensitivity either by changing oxygen concentration, and thus altering the extent of initial DNA damage, or by altering cell cycle progression, thus

allowing greater or lesser repair of potentially lethal damage. Even in cases where it can be demonstrated that a radioprotector can act by a purely radiochemical mechanism, it is possible that it could also act by altering metabolism (297) and the extent to which one mechanism dominates may be difficult to determine. Although a state of hypoxia considered sufficient to provide radiation protection is not brought about by most radioprotectors, some effect on oxygen availability and oxidation-reduction potential of the cells does result after their administration. A relation was observed between the duration of respiration inhibition and the radioprotective effect of cystaphos, the phosphorothioate of MEA (298). Several phosphorothioates were also found to induce vasodilation in the spleen, resulting in altered blood supply to the body, and decreasing tissue oxygen tensions (299). Aminoethyl and aminopropyl thiosulfates also decreased the oxidation-reduction potential in body tissues of rats and mice (300). They also increased serotonin and histamine levels, and decreased peroxide levels. Several heterocyclic compounds, including aryl derivatives of triazoline-2,5-dithione, decreased the oxygen tension in rat spleen, liver, and muscle (301); a correlation was observed between the decrease in oxygen tension and radioprotective effects of the compounds.

Radioprotective and radiosensitizing effects of various compounds have been related to an oxygen effect. A theory has been developed consisting of an "oxygen fixation hypothesis" (302), in which target free radicals react either with radical-reducing species, resulting in restoration, or with radical-oxidizing species, resulting in fixation of radical damage to a potentially lethal form. MEA and other thiols protect by adding to the pool of radical-reducing species, resulting in enhanced repair of free-radical damage. Electron-affinic compounds radiosensitize by adding to the pool of radical-oxidizing species, enhancing free-radical damage; *N*-ethylmaleimide has a similar effect. Metal ions, however, do not alter sensitivity to radiation inactivation of bovine carbonic anhydrase by oxidizing radicals, but do exert a protective effect against inactivation by reducing radicals (303).

An explanation of the protective effects of ethanol, and other hydroxy compounds, arose from the observation that ethanol adds to thymine under $\gamma$-irradiation (304). This prevents formation of thymine dimers, deleterious to the structure of DNA. It also explains the radiation resistance of bacterial spores, and protection of bacteria in glucose medium, where hydroxy compounds are in adequate supply to add to thymine.

Other cellular effects produced by the amino thiols may be involved in the complex process of radiation protection. Release of enzymes is one such effect, and various enzyme releases have been observed in rat plasma after introduction of either MEA or 5-mercaptopyridoxine, or by a state of hypoxia (305). Treatment with two nonprotective thiols, 2-mercaptoethanol and 4-mercaptopyridoxine, did not affect the plasma enzyme levels. The liberation of cellular thiols, discussed earlier, may be attributed to enzyme liberation, at least in part. Mixed disulfide formation may be a factor in this release, as suggested by the biochemical shock hypothesis.

The radioprotective thiols protect the erythropoietic system of animals, $^{59}$Fe uptake being used as the test for protection (306). MEA, AET, penicillamine, and 2-mercaptoethanol all inhibit phosphorylation of thymidine in rat thymus and spleen (307). This effect of the two protective agents MEA and AET is reversible, whereas that of the two sensitizing compounds is irreversible.

Addition of MEA to mitochondria first accelerates then slows respiration. A decrease in ATP synthesis was also noted, both in mitochondria and rat thymus nuclei, thus diminishing both respiration and phosphorylation coupling (308). Mixed disulfide formation is believed to be involved.

Bacq and Alexander proposed that a significant contribution to the radiobiological effects of ionizing radiation is attributed to cell membrane damage (15). The effect of X-rays on the permeability of Ehrlich ascites tumor cell membranes has since been studied by measuring loss of potassium from the cells (309). The radiosensitizing effects of Synkavit and excess oxygen were demonstrated by a marked loss of potassium from the irradiated cells, whereas the protective effects of MEA

and 2-amino-3-methylbutanthiol prevented this loss. Also, blocking of cell-surface amino and probably thiol groups with citraconic anhydride, dimethylmaleic anhydride, and diacetyl also modified radiation damage to the cell membrane. It was suggested that radioprotection may depend on a combination with cell-surface protein groups, which determine the surface charge and maintain the integrity of the cell membrane.

Radiation-protective effects were sought in a screening program involving more than a thousand kinds of Chinese herbs (310). Some of them raised the survival rate of dogs irradiated with a lethal dose of $\gamma$-rays by 30–40%. The ability to protect the hemopoietic and immune systems was believed to be the mechanism of protection.

Another class of radioprotectors, the nitroxides, are membrane permeable, stable free radicals, and protect both cell cultures and mice from radiation-induced cytotoxicity. A water-soluble nitroxide, Tempol, protected hamster cells from superoxide and other peroxides, and protected mice against an $LD_{50}$ dose of radiation (311). Potential mechanisms of protection include oxidation of reduced transition metals, superoxide dismutase-like activity, and scavenging of oxygen- and carbon-based free radicals.

Cytokines have been found to protect cells from the damaging effects of ionizing radiation. Interleukin 1 and tumor necrosis factor alpha (TNF-$\alpha$) protect mice from lethal doses of radiation, given before irradiation. At lower doses of radiation, hemopoietic growth factors, interleukins 1, -4, and -6, TNF-$\alpha$, interferon, and leukemia inhibitory factor promote recovery when administered after radiation, possibly by initiating autocrine/paracrine recovery and repair pathways (312).

Radioprotection by leukotrienes, especially with regard to hematopoietic stem cells, has been reviewed (313). *In vivo* radiation protection by prostaglandins and related compounds of the arachidonic acid cascade has also been noted (314). The role of mast cell mediators in radiation injury and protection has been discussed (315). Immunomodulators, either microbial agents (e.g., glucan) or recombinant cytokines can enhance hematopoietic and functional cell recovery after irradiation (316).

DNA-binding ligands, such as the bibenzimidazoles (317) and modulators of DNA repair and lesion fixation, such as 2-deoxy-D-glucose (318) have provided protection against radiation-induced cytogenetic damage. Also, calcium ion channel blockers, such as diltiazem (319) have increased survival rates of irradiated animals, alone or in combination with zinc aspartate (320). A postirradiation increase in calcium influx in mouse spleen lymphocytes (321) and an increase in cytosolic calcium in rat thymocytes, leading to apoptosis and necrosis, has been observed (322). Calcium ion influx activates a significant number of enzymes, including endonuclease (323), and $Ca^{2+}/Mg^{2+}$-dependent endonuclease mediates fragmentation of DNA. Lipid peroxidation is also reduced by calcium channel blockers, thus decreasing damage to cell membranes (324).

### 1.5.7 Use of Radioprotective Agents in Radiotherapy and Chemotherapy of Cancers.

The use of radioprotective agents to augment the effects of either radiotherapy or chemotherapy of various types of cancer has been investigated for 30 years or more, but until recently little positive benefit was observed. It is now clear that selective concentrations of protective agents can be realized in normal tissues, and that tumors are protected to a lesser extent. Favorable timing schedules for administering protective agents and anticancer therapies have played a significant role in the moderate success so far realized with radioprotector adjuvant treatments.

MEA gave favorable results when used in conjunction with X-rays and cyclophosphamide in rats with Geren's carcinoma (325). Cystamine decreased chromosomal aberrations in peripheral blood lymphocytes in uterine cancer patients (326). Although AET was found to penetrate normal and cancerous tissue of mice to the same extent, it prolonged the life spans of mice bearing ascites tumor cells (327). Favorable effects on irradiation of mice with Ehrlich carcinoma were reported for AET and DL-trans-2-amino-cyclohexanthiol (328). A combination of AET, serotonin, cysteine, and glutathione was definitely favorable to the survival of mice with Landschutz ascites tumors treated with 6000 R (329).

Differential distributions of MEA released from its phosphorothioate and thiosulfate in various tissues have been found (330); the phosphorothioate of MEA had a lower concentration in sarcoma M-1 than in the organs of mice (331). The phosphorothioate of MEA also diminished symptoms of radiation sickness in human patients undergoing radiation therapy for breast cancer (332). Also, in cancer patients, 2-mercapto-propionylglycine decreased the severity of lymphopenia and decreased the number of chromosome aberrations after irradiation (333). MEA, AET, 1-cysteine, and 1-cysteinyl-D-glucose restored the mitotic index in X-irradiated rats bearing Yoshida sarcoma (334); 5-fluorouracil was radiosensitizing in these experiments.

More recently, WR2721 (Amifostine, Ethyol) has exhibited activity as a chemoprotector. The compound requires activation by dephosphorylation to produce the free thiol. This process is catalyzed by capillary alkaline phosphatase, close to the desired site of protection. The neutral pH of normal tissues, compared with the slightly acidic pH of tumors, favors selective activation. Amifostine was able to reduce DNA platination when preincubated with cisplatin, but the effect was much weaker when given postincubation (335). Amifostine can also reduce the myelosuppression produced by cyclophosphamide, the combination of cyclophosphamide and cisplatin, and possibly carboplatin (336). The favorable effects noted with cisplatin therapy have been termed *cisplatin rescue* (337). Protection of hematopoietic stem cells from irradiation by amino thiols, synthetic polysaccharides, vitamins, and cytokines has been discussed (338). Possible mechanisms by which amifostine may protect normal hematopoietic stem cells from chemotherapeutic agents have been proposed, including direct binding to the alkylating agents, cisplatin and cyclophosphamide, or acceleration of the recovery of hematopoietic stem cells exposed to high doses of radiation (339).

Other radioprotective agents have shown promise as selective protective agents for normal tissue; dexrazoxane and mesna have been cited (340). Reviews on the effects of amifostine in protection of normal bone marrow stem cells (341) and as modulator of cisplatin- and carboplatin-induced side effects (342) have been published.

## 2  RADIOSENSITIZERS

### 2.1  Introduction

Radiosensitizers were developed for use in cancer therapy (37–39, 343). It is important for this purpose that they have a differential effect on the tumor vs. the normal tissue. Otherwise, nothing would be gained over increasing the dose of radiation. The relative increase in antitumor efficacy compared to normal tissue toxicity is referred to as *therapeutic gain*. Many of the compounds or strategies listed in this chapter have yet to be proven useful in cancer therapy, but are of mechanistic interest.

Two particular strategies have a clear rationale for selective radiosensitization of tumor cells and are currently in clinical use:

1. Sensitization of hypoxic cells by oxygen-mimetic drugs (344), or agents that alter oxygen delivery, because hypoxia occurs often in tumors but not in normal tissues, and about three times more radiation is required to kill hypoxic cells than oxygenated cells.

2. Incorporation of radiosensitizing thymidine analogs into DNA (345) because this will occur only in proliferating cells, and in some cases (e.g., brain tumors), the normal tissue surrounding the tumor is composed primarily of nonproliferating cells.

Strategies that take advantage of signal transduction differences between cancer cells and normal cells represent an exciting new class of radiosensitizers, although these strategies are still in the early stages of development.

Radiosensitizers are traditionally defined as agents that do not have a therapeutic effect of their own, but act to enhance the therapeutic effect of radiation. However, the term does apply to some antitumor agents, such as low dose cisplatin (346–349) and hyperthermia (350, 352), which appear to enhance radiation

damage in a truly synergistic manner. Otherwise, the combined use of radiation and antitumor agents is referred to as *combined modality therapy*, and relies on nonoverlapping normal tissue toxicities or on attacking different tumor cell populations to achieve a therapeutic gain. Combined modality therapy, although not discussed in this chapter, was reviewed recently by Phillips (353).

### 2.2  Radiosensitization by Alteration of Energy Absorption

The probability of ionization is essentially proportional to the number of electrons in the target molecule, regardless of chemical composition for the types of ionizing radiation that are normally used in radiation therapy (high energy photons or electrons). However, preferential energy absorption by particular elements can occur with certain energies or types of radiation. Two such cases are boron neutron capture and k-edge absorption of photons by atoms of high atomic weight.

**2.2.1 Boron Neutron Capture Therapy (BNCT).** Certain isotopes, such as $^{10}B$, capture low energy (thermal) neutrons very efficiently. This property is expressed as the thermal neutron cross section, in units of barns. The thermal neutron cross section for $^{10}B$ is 3837 barns. As early as 1936 (354), it was suggested that preferential incorporation of $^{10}B$ into tumors could be a useful strategy for selectively radiosensitizing tumor cells. $^{10}B$ itself is not radioactive, but neutron capture by $^{10}B$ is followed by radioactive decay of the resulting $^{11}B$ nucleus. $^{11}B$ splits into $^{4}He$ nuclei (alpha particles) and $^{7}Li$ ions, both of which are densely ionizing [i.e., high linear energy transfer (LET)], and consequently very cytotoxic. The ranges of these particles are such that their energy is deposited within one cell diameter of the neutron-capture event. These characteristics make $^{10}B$ a very desirable isotope for neutron-capture therapy, even though other isotopes have higher thermal neutron capture cross sections.

There are two challenges to implementation of this strategy. One is to deliver thermal neutrons to the tumor, given that low energy neutrons do not travel very far in tissue. The other is to design drugs that will selectively deliver $^{10}$B to tumors. The first challenge is being met with advances in instrumentation, such that BNCT is now regarded as a more realistic possibility than in the past (355). Neutrons of a sufficiently high energy to penetrate tissue (epithermal neutrons) are used in such a way that they become thermal neutrons at the depth corresponding to the tumor.

It is estimated that 10 parts per million (ppm) $^{10}$B would be enough to increase cytotoxicity twofold over that seen with neutrons alone (355). BSH ($Na_2B_{12}H_{11}SH$; **44**) was one

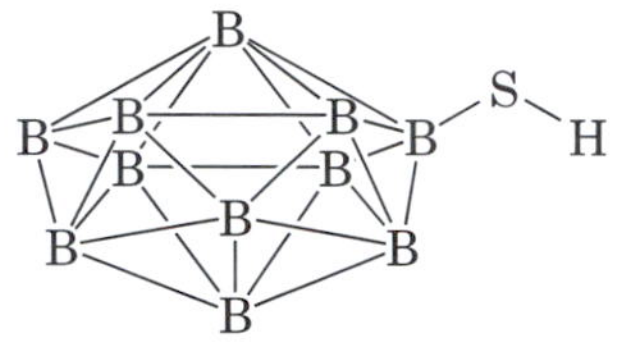

**(44)**

of the first compounds synthesized for this purpose (356). BSH and its disulfide, BSSB, are reported to accumulate in animal tumors, but the clinical biodistribution results from patients with brain tumors have been variable and unpredictable (357). In animal studies, very high tumor to normal brain tissue ratios were achieved, but there was no evidence of a therapeutic gain, suggesting that normal brain injury may be more related to the dose than to the vasculature (358). Clinical studies of brain tumor BNCT with sodium tetraborate and BSH did not result in improved survival and there was evidence of increased normal brain injury, consistent with preferential damage to the vasculature (359). Other types of BNCT agents are currently under development. Encouraging results were obtained with BPA (*p*-boronophenylalanine, **45**) in the treat-

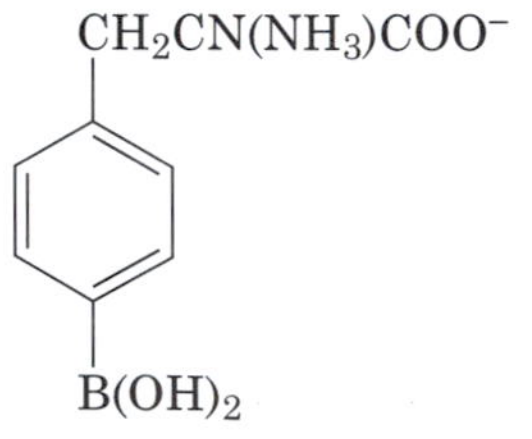

$$CH_2CN(NH_3)COO^-$$

$$B(OH)_2$$

**(45)**

ment of melanoma (360). $^{157}$Gd has been proposed as another isotope that could be useful in neutron-capture therapy (361). Although BNCT is a promising idea, the selective delivery of atoms of high neutron cross section (such as boron) to the target cells is still a limiting factor.

**2.2.2 k-Edge Absorption and Photoactivation of Elements of High Atomic Number.** The probability of absorption of a photon is highest when the energy of the photon is close to the binding energy of an electron in the target molecule (362). This effect is particularly noticeable for k-shell electrons of elements of high atomic number. A strategy for radiosensitization based on this effect is to incorporate atoms of high atomic number into DNA (363). The k-shell binding energy is characteristic for each element, and irradiation at energies just above this k-edge characteristic energy will result in selective absorption by a particular element. Iodine is particularly attractive for this purpose because: (*1*) the optimal energy of the activating photon is in a range that is reasonably achievable; (*2*) iodine is easily incorporated into compounds that can be delivered to tumors; and (*3*) photoactivation can occur, further enhancing the biological effect. Photoactivation is a process whereby inner shell electrons are ejected from an atom as a consequence of photon absorption, and cascading outer shell electrons fill the successively vacated orbitals, resulting in the emission of multiple low energy X-ray photons and electrons (364).

**2.2.3 Photodynamic Therapy.** Photodynamic therapy (PDT) consists of administration of a photosensitive compound and illumination of the tumor with visible light. Recent advances in light-delivery technology have provided methods for selective and thorough illumination of the tumor (365), although light delivery continues to be the principal limitation of this therapeutic approach.

PDT has been found to be effective in the treatment of several types of solid tumors in humans. Most of the clinical experience has been with hematoporphyrin derivative (HPD, **46**) or porfimer sodium (Photofrin), which is a derivative of HPD (366). A problem with these

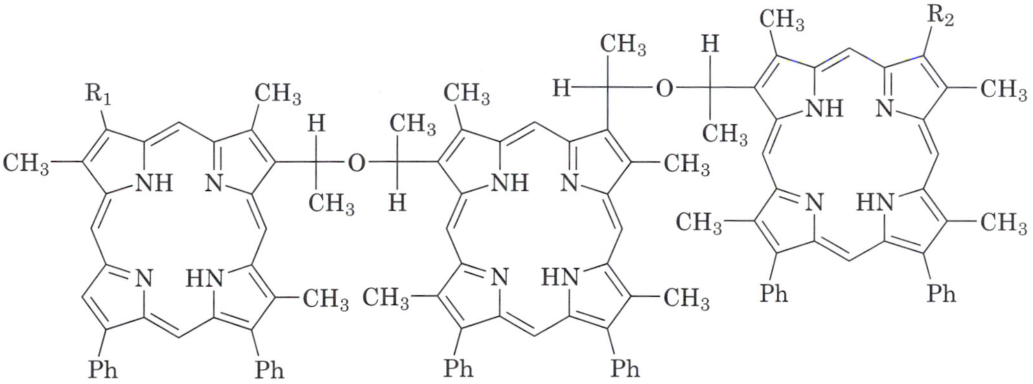

(46)

compounds is photosensitization of skin that persists for 6–8 weeks (367). *meta-tetra*(Hydroxyphenyl)chlorin (*m*-THPC) (368–370) has a shorter plasma half-life and a higher tumor/normal tissue ratio, and is currently in Phase I/II trials. 8-Aminolaevulinic acid (ALA) has been found to be effective clinically by topical application (371, 372) and in animal studies by systemic administration (373). ALA is converted *in vivo* to protoporphyrin IX, which is the active photosensitizer. Other tetrapyrroles that are being considered for PDT include purpurins (**47**) (374) and phthalocya-

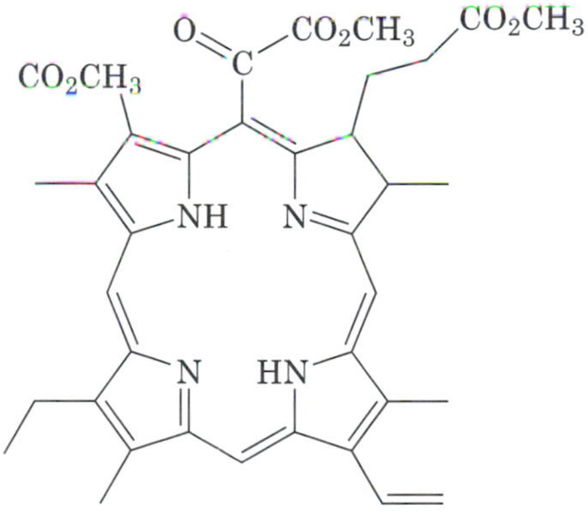

(47)

nines (375, 376). Cationic dyes have been investigated (377), given that there is evidence that a common feature of many types of tumor cells is the ability to concentrate moderately

lipophilic cationic dyes, attributed to differences in mitochondrial membrane potential (378).

The tumoricidal mechanism of PDT has two components: direct tumor cell killing and damage to the vasculature leading to tumor necrosis. The number of tumor cells from excised murine tumors that produce colonies *in vitro* decreases markedly with time between treatment and excision (379, 380), in support of the importance of damage to the vasculature. In contrast, tumor cell killing by ionizing radiation is evident when tumors are excised immediately after irradiation and plated for clonogenic assay (381). Hypoxia induced by PDT can be exploited by concurrent treatment with drugs that are metabolized under hypoxia to toxic species (380, 382).

## 2.3 Alteration of the Primary Radiolytic Products

Ionization of water is the most common consequence of irradiation of biological systems because they are composed mostly of water. Figure 4.1 shows the relative yields of products of water radiolysis (383). These reactive species differ considerably in their chemical properties. For example, H$^\bullet$ is a reducing radical and HO$^\bullet$ is highly oxidizing. A potential approach to radiosensitization is to convert the initial radiolytic products into more reactive or more selective species. This can be accomplished by including a substance that can react with the primary radiolytic products before they have a chance to react with other

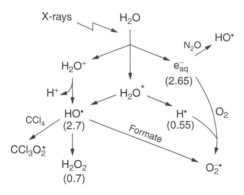

**Figure 4.1.** Products of radiolysis of water. Ionizing radiation causes an electron to be ejected from the water molecule (ionization), forming $H_2O^+$, which dissociates into $H^+$ and $HO^\bullet$. The electron becomes solvated ($e_{aq}^-$). Excitations also occur, and the excited water molecule ($H_2O^*$) can dissociate into $HO^\bullet$ and $H^\bullet$. Some of the reactions of these primary radicals are shown. The numbers in parentheses are the yields of the various species per 100 eV (G-values) (433).

molecules. For example, $O_2$ reacts readily with the aqueous electron ($e_{aq}^-$) and $H^\bullet$ to convert them to $O_2^{\bullet-}$ (Fig. 4.1). $N_2O$, on the other hand, can react with $e_{aq}^-$ to convert it to $HO^\bullet$ (Fig. 4.1). Hyperbaric $H_2$ can be used to convert $HO^\bullet$ radicals into $H^\bullet$. Isopropyl alcohol can be used to scavenge $HO^\bullet$ and $H^\bullet$, leaving the aqueous electron.

Such methods have been used to demonstrate differences in reactivity of these radiolytic products toward model biological targets, such as DNA (383). Studies with bacteria and mammalian cells have yielded variable results (384, 385), possibly because the modifiers can act by multiple mechanisms (385). In general, such studies have been consistent with lack of toxicity of the reducing radicals, suggesting that $HO^\bullet$ is the only diffusible primary radical that contributes to toxicity. One explanation is that multiple attacks on DNA must occur within a very small volume to produce a cytotoxic lesion (386). Only radicals with the reactivity of $HO^\bullet$ can contribute to these clustered lesions because other radicals diffuse away from the critical volume before reacting with DNA (387). For example, the conversion of $e_{aq}^-$ to $HO^\bullet$ by $N_2O$ more than $10^{-9}$ s after

irradiation should not affect cytotoxicity because the $HO^\bullet$ would be formed outside the critical volume (387).

A variation on this theme is to use modifiers that react with products of water radiolysis, to produce a species that is more selective in reacting with cellular targets (388) (Fig. 4.1). For example, Hiller et al. (389) found that the effectiveness of ionizing radiation in inactivating bacteriophage $T_2$ is five times greater if $HO^\bullet$ is converted to $CCl_3O_2^\bullet$ by inclusion of $CCl_4$ in the medium, whereas it is decreased by a factor of 20 by conversion of $HO^\bullet$ to $O_2^{\bullet-}$ by inclusion of formate. Similarly, uric acid can enhance radiation damage to alcohol dehydrogenase (in dilute solution), even though it protects lactate dehydrogenase (390). The reaction of $HO^\bullet$ scavengers with $HO^\bullet$ results in the formation of a more selective radical that may be more effective in damaging a specific cellular target. Dimethylsulfoxide (DMSO) reacts with $HO^\bullet$ to form $H_3C^\bullet$, or in the presence of oxygen, $CH_3OO^\bullet$. These are relatively nonreactive radicals, and DMSO is therefore considered to be an $HO^\bullet$ scavenger. DMSO protects against radiation-induced DNA damage. However, DMSO does not protect against $HO^\bullet$-initiated membrane damage, whereas other $HO^\bullet$ scavengers do (391).

## 2.4  Radiosensitization by Reaction with DNA Radicals

The principal mechanism of cell killing by ionizing radiation is the formation of clustered DNA lesions (386, 392, 393) by a combination of direct ionizations in the DNA molecule (the direct effect) and reaction of DNA with free radicals produced in the vicinity of DNA (the indirect effect). The reactions that produce the DNA radicals that are the precursors of these clustered lesions are complete within nanoseconds (387). However, the chemical reactions of these free radicals that result in damage fixation are not complete until 10 ms after irradiation (394), and there is an opportunity to alter the outcome of these reactions (Fig. 4.2). Damage fixation is a process that renders the damage nonrestorable by chemical protectors.

Reaction of DNA radicals with molecular oxygen results in damage fixation. This reaction occurs in competition with the restorative

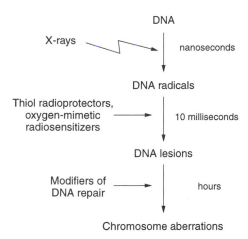

**Figure 4.2.** Time frame for radiochemical events. In living cells, DNA radicals are formed within nanoseconds by a combination of direct ionization of DNA and reaction of DNA with HO$^\bullet$ produced by the radiolysis of water. Protectors and sensitizers that modify initial lesion formation act in a millisecond time frame by reaction with these DNA radicals. Biological processes that can lead to repair or misrepair of the lesions take place over a period of hours.

reaction of DNA radicals with endogenous protectors (395). In experiments using chemical model systems (dilute solutions of macromolecules), sensitization by oxygen was observed only when radioprotectors were also present (395, 396). Thus, damage fixation will occur without a sensitizer (by internal bond rearrangement) if protective reactions are not fast enough. Either an increase in the concentration of oxygen-mimetic sensitizer or a decrease in the concentration of endogenous protector can result in radiosensitization.

The chemical property of oxygen that is the basis for oxygen mimetic radiosensitization is the one-electron redox potential, or electron affinity (397). Electron affinity correlates with hypoxic cell radiosensitization, for agents that sensitize by this mechanism, over a range of different chemical structures, with some exceptions (397, 398). Radiosensitization by an oxygen-mimetic mechanism occurs within milliseconds of irradiation, and techniques have been developed to distinguish between this mechanism and other mechanisms of sensitization by determining the time frame for the effect (399). For example, a rapid-mix ex-

periment was used to show that $N$-ethylmaleimide (**48**) can sensitize by an oxygen-mimetic mechanism, even though it can also sensitize

(**48**)

by reacting with cellular thiols (399). Whillans and Hunt (400) used a rapid-mix experiment to demonstrate that radiosensitization by misonidazole (**53**) does not occur if misonidazole is mixed with hypoxic cells more than 10 ms after irradiation. Some of the first compounds that were found to be radiosensitizers, such as $N$-ethylmaleimide (401) and menadione (402), are electron affinic, but were too toxic, too complicated mechanistically, or not sufficiently effective to be thoroughly developed for clinical use.

Attention in recent years has been concentrated on two general classes of electron-affinic radiosensitizers: quinones and nitroimidazoles (403). Highly electron affinic agents are effective *in vitro*, but not *in vivo*, presumably because they are so reactive they are depleted before they reach the target cells. The first nitro compounds that were found to be effective electron-affinic sensitizers *in vitro*, *p*-nitro-acetophenone (**49**) (404), *p*-nitro-3-dimethylamino-propiophenone (**50**) (405), and nitrofurazone (**51**) (406), were too toxic and

(**49**)

too metabolically unstable to be useful *in vivo*. Similarly, many quinones are effective *in vitro* but have been disappointing *in vivo* (407).

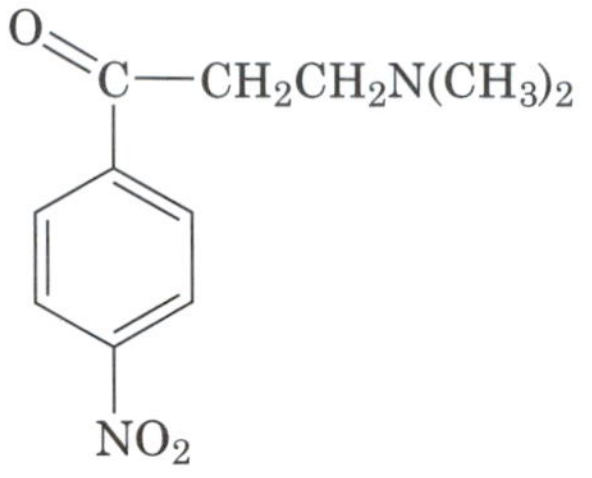

**(50)**

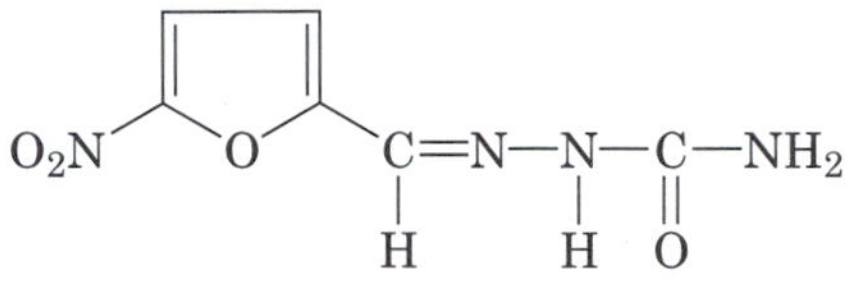

**(51)**

The first electron-affinic sensitizer based on the nitro functional group to be tested clinically (407) was metronidazole (1–13-hydroxyethyl-2-methyl-5-nitro-imidazole; Flagyl, **52**),

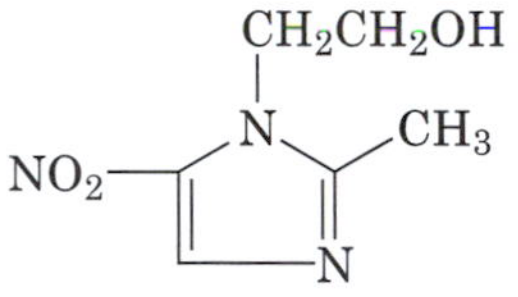

**(52)**

a 5-nitroimidazole that was already in clinical use as an antitrichimonal agent. The trial, conducted with patients with glioblastoma multiforme using nonstandard fractionation, was positive, in that the median survival for the sensitizer group (7 months) was superior to the median survival for the controls (3 months). However, the long-term survival of the sensitizer group was not superior to that of historical controls given standard fractionated radiotherapy. The 2-nitroimidazole, misonidazole (**53**), was tested more extensively

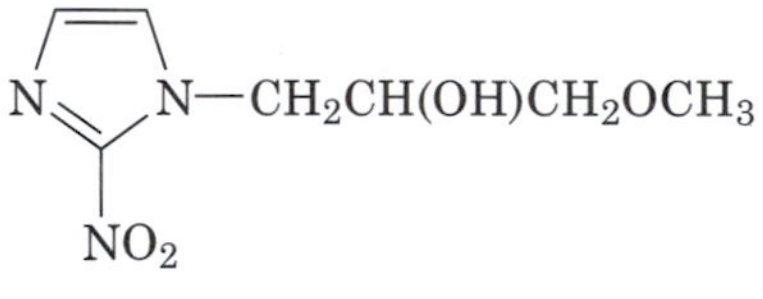

**(53)**

in clinical trials. Only 5 of 33 trials showed some possible benefit (408). The most promising result came from a large randomized trial of patients with pharyngeal cancer (409), with an overall disease-free survival of 46% for the misonidazole group vs. 26% for the controls.

The dose of misonidazole that can be administered is limited by peripheral neuropathy (410). In an effort to reduce this side effect, less lipophilic 2-nitroimidazoles were synthesized (411, 462). Desmethylmisonidazole (**54**) and etanidazole (**55**) are less neurotoxic than

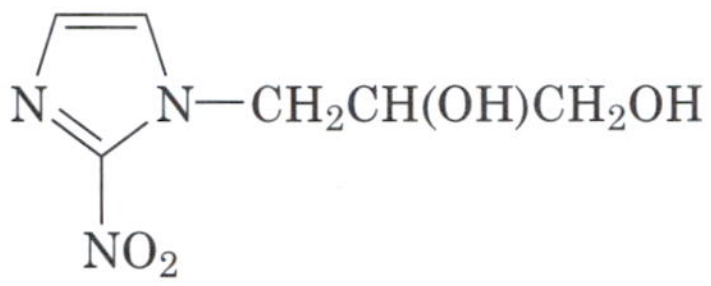

**(54)**

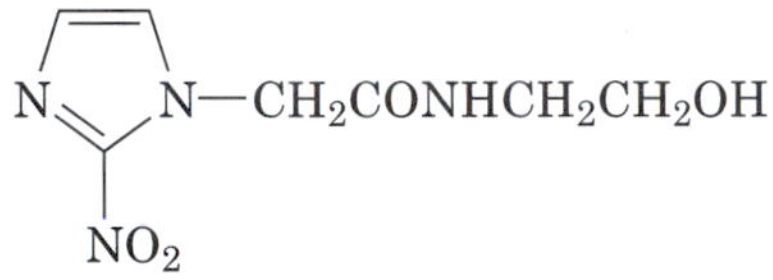

**(55)**

misonidazole, in keeping with their lower lipophilicity. In a phase I clinical trial (413), it was determined that 30% more desmethylmisonidazole than misonidazole could be administered, but this compound was not tested further. Etanidazole can be administered at about 4 times the dose of misonidazole (414), and peripheral neuropathy can be almost completely avoided by determination of individual patient pharmacokinetics and adjustment of the dosage accordingly (415). Efficacy data for etanidazole are not yet available, although several trials are nearing completion (415–418).

Nimorazole (**56**) (419), a 5-nitroimidazole, is less effective on a molar basis than the 2-nitroimidazoles, but its dose-limiting toxicity is different. The dose-limiting toxicity for nimorazole is nausea and vomiting, whereas it is peripheral neuropathy for the 2-nitroimidazoles. The toxicity of nimorazole is not cumulative, and it can therefore be given with each radiation fraction. A phase III trial of nimora-

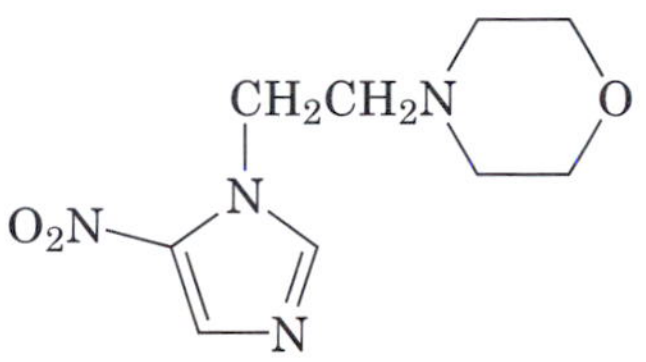

(56)

zole with 422 patients with squamous cell cancer of the larynx and pharynx (420, 421) showed an improvement in local control.

Pimonidazole (Ro 03–8799) (**57**) is a 2-ni-

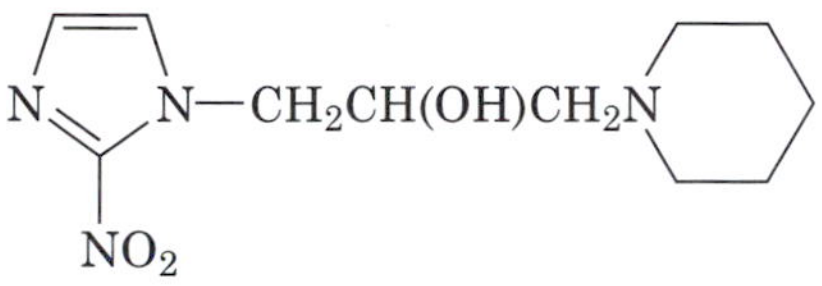

(57)

troimidazole with a basic side chain that has been found to accumulate in the acidic environment of tumors (422). Although pimonidazole is a better sensitizer than misonidazole *in vitro*, it is less effective than misonidazole in animal models, and clinical trials have indicated that it may be detrimental in combination with radiation therapy (423). A possible explanation for this adverse effect is suggested by a finding that pimonidazole can decrease blood perfusion in tumors (424). The idea of enhancing tumor uptake by attaching a basic side chain may still have merit, however (425–477).

Several new electron-affinic drugs have been developed, but have not yet been extensively tested. A series of 2-nitroimidazole derivatives were synthesized by attaching various amino acids at the 1-position through an amide bond (**58**) (428). Results with a series of nitroimidazoles with an acetohydroxamate

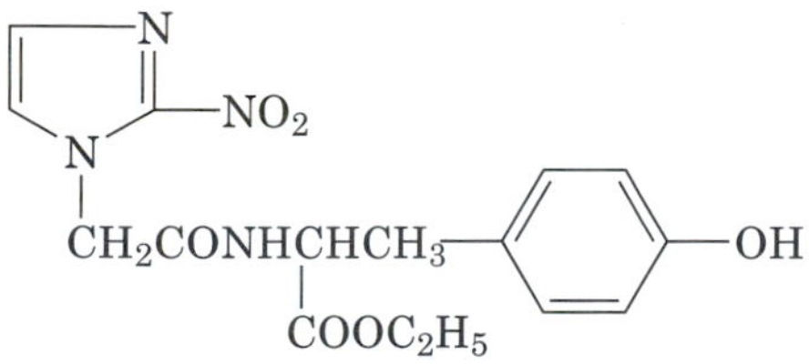

(58)

moiety on the side chain, of which KIN-804 (**59**) (429) and KIH-802 (**60**) (430) appear most promising, suggest that this functional

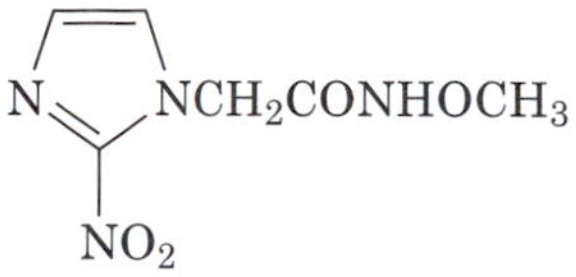

(59)

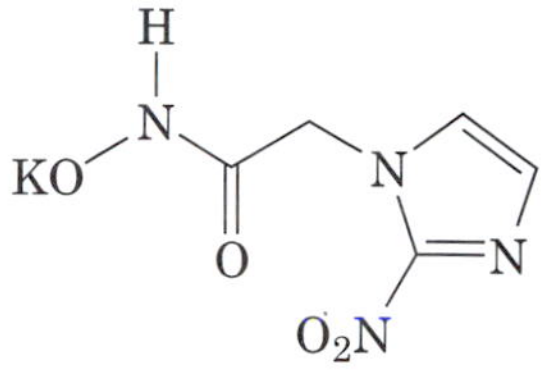

(60)

group may be useful for radiosensitizer design. AK-2123 (**61**) is a 3-nitrotriazole that is less toxic than misonidazole and is now in clinical trials (431, 432).

NLP-1 (5-[3-(2-nitro-1-imidazoyl)-propyl]-phenanthridinium bromide, **62**) was synthe-

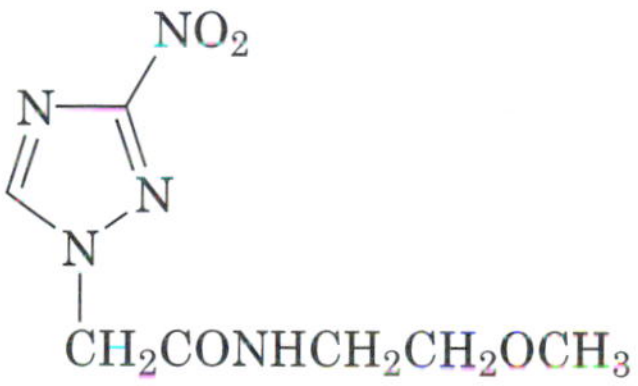

(61)

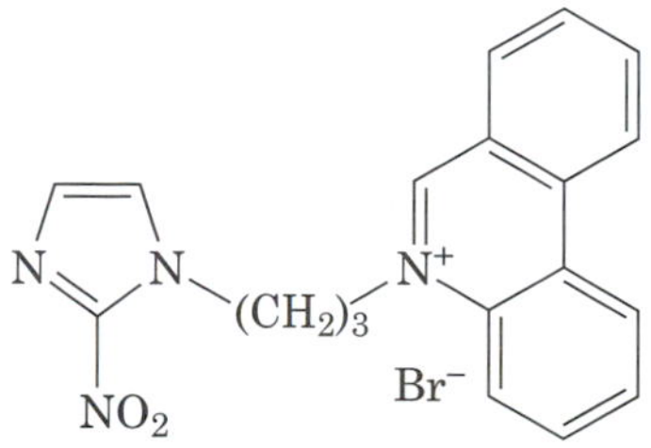

(62)

sized with the rationale of targeting the nitroimidazole to DNA through the intercalation of the phenanthridine ring (433). Nitracrine

[1-nitro-9-(dimethylaminopropyamino)acridine; (**63**)] is a DNA intercalating antitumor agent. Under conditions that minimize metab-

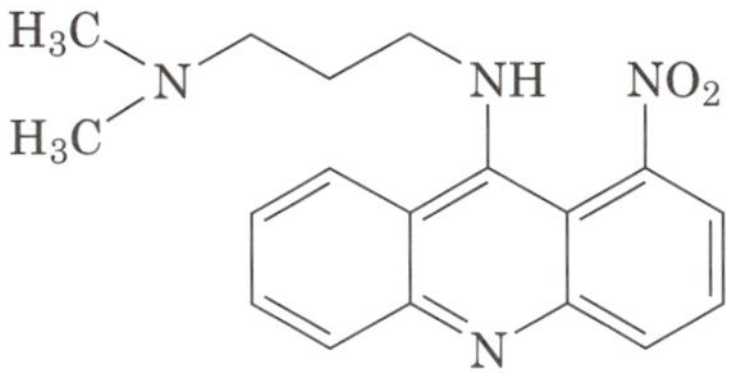

(**63**)

olism and cytotoxicity, nitracrine is a selective radiosensitizer of hypoxic cells, with an efficiency at least as great as expected from its electron affinity (434).

## 2.5 Additional Applications for Electron-Affinic Drugs in Cancer Therapy

The electron-affinity of hypoxic cell radiosensitizers confers additional biological properties to this class of drugs. They are readily reduced by cellular enzymes to reactive species that can cause biochemical alterations and cytotoxicity. The reversal of the first one-electron reduction, by reaction with oxygen, provides a mechanism for selectively affecting hypoxic cells. This biochemistry has been exploited for detection of hypoxia, additional radiosensitization, chemosensitization, and tumoricidal activity. Following this theme, new classes of drugs have been developed to optimize these mechanisms. Drugs that are selectively reduced under hypoxia include nitroaromatics, quinones, $N$-oxides, and transition-metal complexes.

### 2.5.1 Binding of Nitroimidazoles to Hypoxic Cells: Use in Detection of Hypoxia. Nitroimidazoles are reduced under hypoxic conditions to intermediates that bind to cellular macromolecules. This mechanism has been exploited in the detection of tumor hypoxia (435). Early studies made use of $^{14}$C-labeled misonidazole and autoradiography, and served to demonstrate that hypoxia does occur in patients' tumors (436). Another approach has been to use nitroimidazole derivatives that can be detected immunohistochemically (437–450). Fluorinated nitroimidazoles have

been synthesized for use in noninvasive detection of hypoxia (439–443). EF5 [2-(2-nitro-1$H$-imidazol-1-yl)-$N$-(2,2,3,3,3-pentafluorpropyl)acetamide, (**64**)] is a pentafluorinated

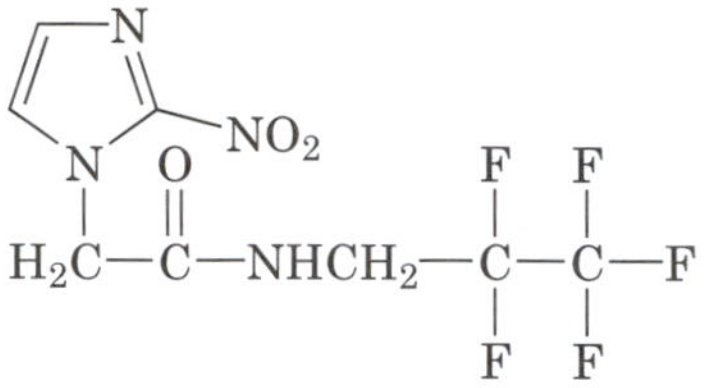

(**64**)

derivative of etanidazole that has been used for immunohistochemical analysis of hypoxia and is suitable for noninvasive imaging (440). KU-2285 (**65**) is a fluorinated etanidazole de-

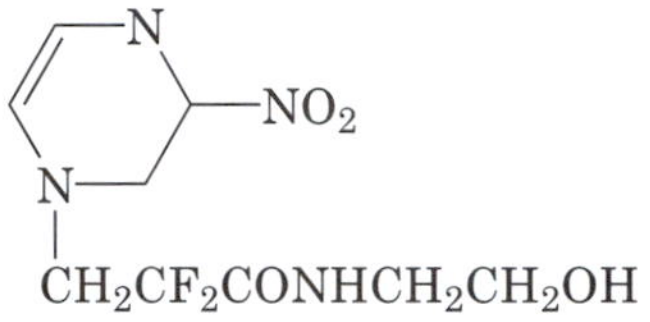

(**65**)

rivative that has a higher sensitizer efficiency than that of etanidazole and is in clinical trials as a hypoxic cell radiosensitizer (441). Pimonidazole (442) and IAZGP [iodinated-$\beta$-D-azomycin-galactopyranoside (443)] have also shown promise as hypoxia-imaging agents.

### 2.5.2 Additional Sensitization by Hypoxic Metabolism of Nitroimidazoles. Oxygen-mimetic radiosensitization by nitroimidazoles depends only on the intracellular concentration of the nitroimidazole at the time of irradiation. However, these same compounds are metabolized under hypoxic conditions to reactive intermediates that can cause radiosensitization by another mechanism. This additional sensitization is called the *preincubation effect* (444, 445). The effect is primarily a decrease in the shoulder of the radiation survival curve and is obtained even if the cells are reoxygenated before irradiation. Chemotherapeutic drug cytotoxicity is also enhanced by hypoxic preincubation with nitroimidazoles (446). The biochemical mechanism of the preincubation

effect has not been identified. One contribution to this effect is depletion of glutathione (447), but this cannot explain the full effect, and does not explain radiosensitization of cells that are reoxygenated before irradiation. When glutathione was depleted to a similar extent with another agent in a study by Taylor et al. that used Chinese hamster ovary cells (448), sensitization to melphalan was only a fraction of that observed with hypoxic preincubation with misonidazole.

**2.5.3 Bioreductive Drugs.** Hypoxic incubation of cells with nitroimidazoles can kill cells without the need for a second agent (444). The concentrations of 2-nitroimidazole radiosensitizers required to kill tumor cells are too high to allow their use as cytotoxic agents in the clinic. However, the chemistry involved is common to a class of agents, called bioreductive drugs, that have shown promise in the treatment of cancer, and represent a new drug design opportunity. Bioreductive drugs are drugs that are reduced under hypoxia to cytotoxic species. Because radiation spares hypoxic cells, the combined use of bioreductive drugs and radiation should provide a therapeutic benefit, even if the effects are strictly additive.

Tirapazamine (SR-4233; WIN 59075; 3-amino-1,2,4-benzotriazine 1,4-dioxide; **66**)

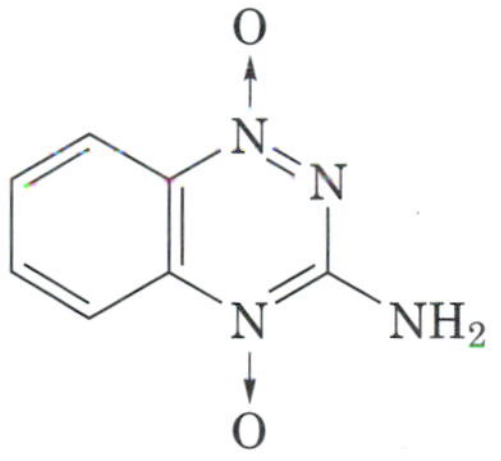

(66)

selectively kills and radiosensitizes hypoxic mammalian cells *in vitro* and murine tumor *in vivo* (449, 450). The selective killing of hypoxic cells by tirapazamine is attributed to its one-electron reduction to a reactive intermediate that produces DNA damage. Radiosensitization appears to be by a preincubation effect, given that hypoxic preincubation sensitizes reoxygenated cells to ionizing radiation, and

sensitization can be achieved when the drug is added after irradiation under hypoxic conditions, ruling out a radiochemical mechanism. Tirapazamine is currently in Phase I clinical trials, and muscle cramping has been noted as the most common side effect (451). A series of analogs of tirapazamine were synthesized to determine structure-activity relationships (450). One of these analogs, SR-4482 (**67**),

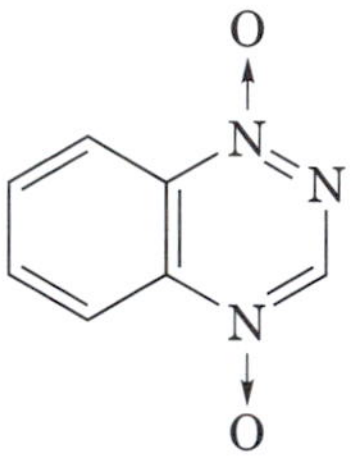

(67)

which contains no substituent on the 3-position of the triazine ring, is more toxic to hypoxic cells *in vitro*, but less toxic to mice, than tirapazamine. A relationship was found between the one-electron redox potential of these analogs and toxicity, although SR-4482 did not follow this pattern, and may represent a new class of benzotriazine di-*N*-oxides (450). RB90740 (**68**) is the lead compound selected

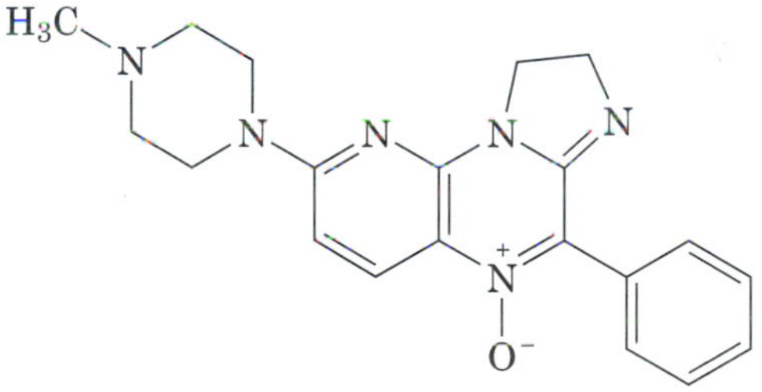

(68)

from a series of pyrazinemono-*N*-oxides that are selectively toxic to hypoxic mammalian cells (452). RB90740 has been found to accumulate in murine tumors, compared to normal tissue, providing additional selectivity.

RSU 1069 (**69**) is the lead compound in a series of 2-nitroimidazoles containing an alkylating moiety on the side chain (453). RSU 1069, which contains an aziridine ring as an alkylating moiety, is more potent than et-

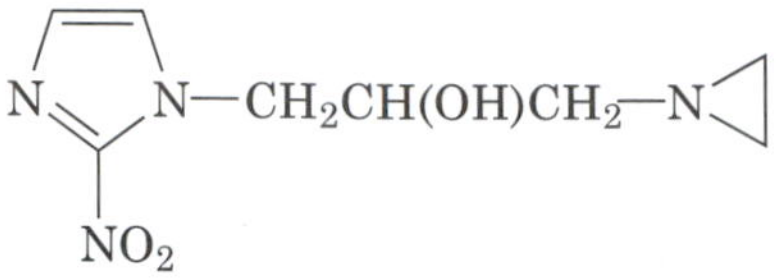

**(69)**

9,10-dione, **73**) is reduced under hypoxia to AQ4 (**74**), which binds to DNA (459). The terminal half-life for AQ4 is reported to be of the

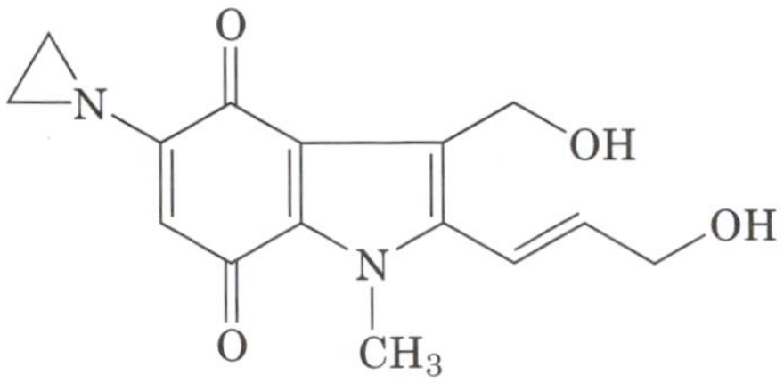

anidazole both as a hypoxic cell radiosensitizer and as a bioreductive drug. However, gastrointestinal toxicity limits the dose than can be administered clinically to levels that would not be expected to provide a benefit in cancer therapy. RB 6145 (**70**) is a prodrug of

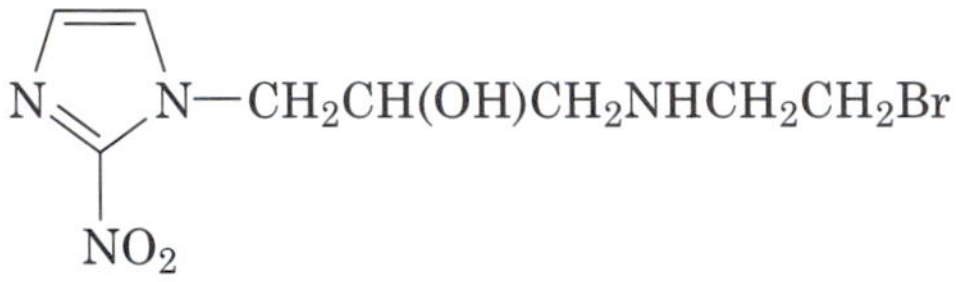

**(72)**

**(70)**

RSU-1069 that has lower toxicity with similar potency (454, 455).

Mitomycin C (**71**) is the prototype for qui-

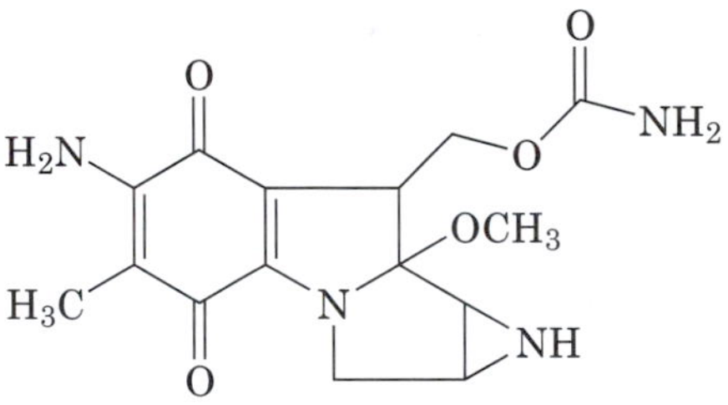

**(73)**

**(71)**

**(74)**

none-based bioreductive drugs (456). A clinical trial using mitomycin C and radiation for the treatment of squamous cell carcinoma of the head and neck showed an improvement in local control without enhancement of normal tissue toxicity (457). EO9 (**72**) is a mitomycin C analog containing an aziridine ring (458). A factor to consider is that the toxicities of bioreductive drugs depend considerably on the relative and absolute activities of cellular reductases; this is particularly a factor for substrates of DT-diaphorase, such as EO9, because DT-diaphorase activity can vary widely (458).

AQ4N (1,4-bis-{[2-(dimethylamino-$N$-oxide)ethylamino}5,8-dihydroxyanthracene-

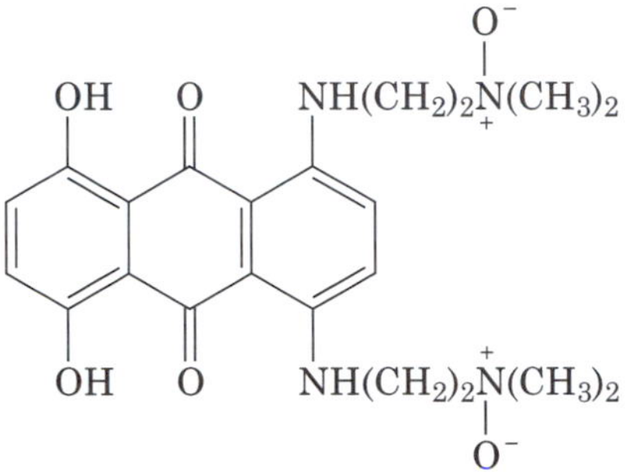

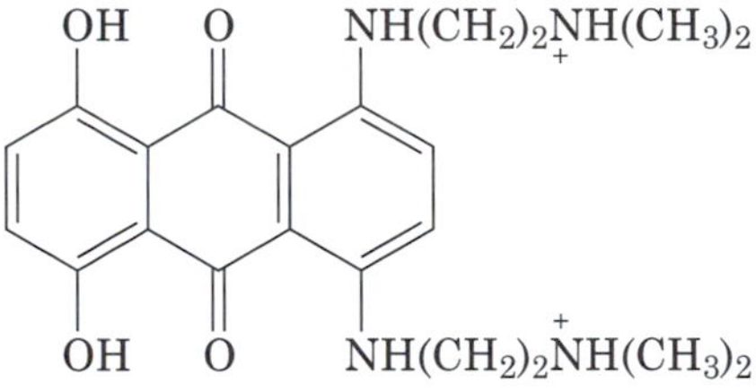

order of 24 h, whereas the terminal half-life for the parent drug AQ4N is 30 min (459). AQ4 is retained in hypoxic cells for days, and is cytostatic (460). Consequently, repopulation of the tumor may be suppressed by exposure of cells to AQ4 while they are hypoxic, even if they subsequently become reoxygenated. The benefit of combined use of AQ4N and radiation in animal tumor studies is obtained, even when the treatments are administered more than 24 h apart in either order, suggesting additivity rather than synergy.

## 2.6  Radiosensitization by Alteration of Oxygen Delivery

An alternative to using oxygen-mimetic radiosensitizers is to improve oxygen delivery to tumors. Tumor hypoxia can result from one of two mechanisms: growth of the tumor beyond the diffusion distance of oxygen (chronic hypoxia) (461) or intermittent impairment of tumor blood flow (acute or intermittent hypoxia) (462, 463).

The response of experimental tumors to radiation can be improved by increasing the oxygen content of the inspired air (464). Clinical trials with hyperbaric oxygen were carried out as early as the 1950s (465–467). Of nine prospective randomized trials, three showed a statistically significant benefit for hyperbaric oxygen (468). However, this strategy is technically difficult and impractical for widespread use in fractionated radiotherapy.

A prospective randomized trial using red blood cell transfusion to increase the hematocrit showed a benefit in the radiation therapy of cervical cancer (469). There is a limit beyond which this approach is counterproductive, in that increasing the hematocrit increases blood viscosity, resulting in poorer tumor perfusion. Furthermore, adaptation to chronically altered oxygenation results in reestablishment of tumor hypoxia (470). Adaptation to oxygenation status has been exploited in animal experiments to improve the tumor radiation response by adapting animals to low oxygen tensions (12% $O_2$) and then returning them to normal or higher than normal oxygen tensions just before irradiation (471).

Perfluorochemicals have been used to increase the oxygen-dissolving capacity of blood (472). Treatment of tumor-bearing rats with dodecafluoropentane and carbogen was recently reported to completely reverse the hypoxic cell radioresistance in this tumor model (473). Several clinical trials have been carried out with Fluosol (**75**) (Fluosol-DA is an emulsion containing perfluorodecalin and perfluorotripropylamine) (474). Mild liver dysfunction was noted as a side effect. A benefit from the use of Fluosol in combination with radiation was demonstrated in a trial of high grade brain tumors (475). Perflubron (perfluo-

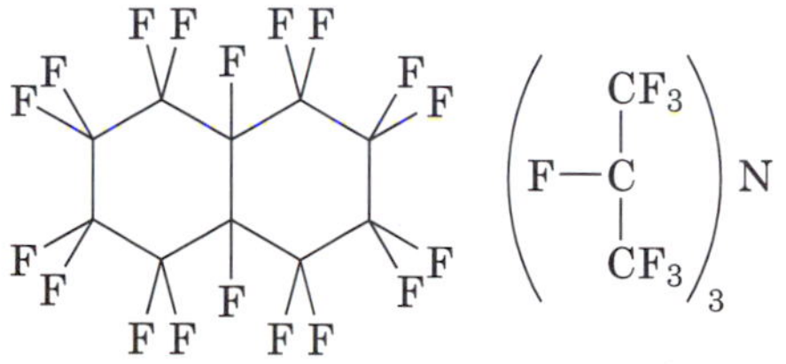

(75)

rooctyl bromide) is another perfluorochemical that has shown promise in animal model experiments (**76**) (476).

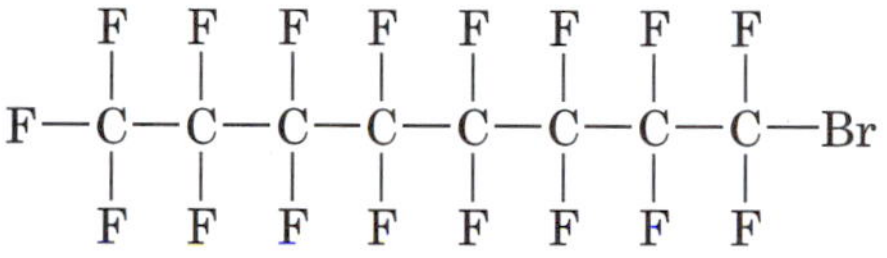

(76)

Various forms of hemoglobin have been used to increase the oxygen-carrying capacity of blood in animal experiments. Measurements using the Eppendorf oxygen needle electrode indicate that such treatments can effectively decrease tumor hypoxia (476). Radiobiological experiments in animal models have shown that tumor radiosensitivity is increased by perfluorochemicals or by ultrapurified polymerized bovine hemoglobin, by as much as an enhancement ratio of 3 (476), which is the theoretical maximum expected for total elimination of tumor hypoxia. Carbogen breathing (95% $O_2$, 5% $CO_2$) increases the effectiveness of these agents. Hemoglobin derivatized with polyethylene glycol (PEG-hemoglobin) has been demonstrated to improve tumor oxygenation in animal models (477). Derivatization with polyethylene glycol increases the circulating half-life of proteins, increases their solubility, and decreases their immunogenicity (478).

Allosteric modifiers of hemoglobin constitute another class of modifiers of tumor oxygenation (479). 2,3-Diphosphoglycerate (**77**) is a natural modifier of the oxygen affinity of hemoglobin and has been used to modify the tumor radiation response in experimental animals (480). Other allosteric modifiers of the oxygen affinity of hemoglobin that have been

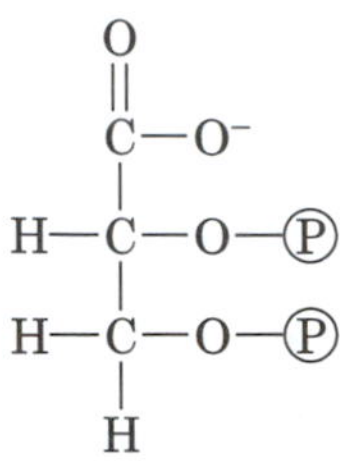

(77)

tested as radiation sensitizers include clofi-brate (**78**), bezafibrate, and gemfibnozil (481).

Pentoxifylline (**79**), a methylxanthine de-

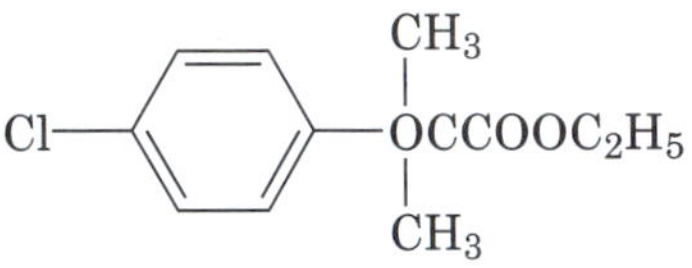

(78)

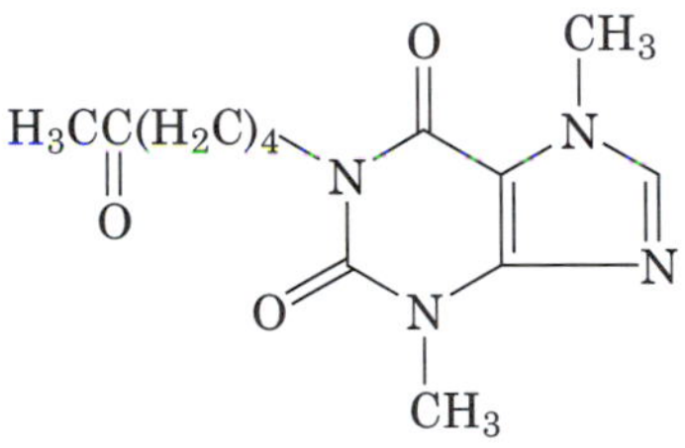

(79)

rivative, increases blood circulation by increasing blood cell deformability and decreasing blood viscosity (482). Pentoxifylline has been reported to increase murine tumor oxygenation and to enhance the radiation response of murine tumors (483). A consideration in the design of strategies for overcoming tumor hypoxia is the type of hypoxia (acute or chronic) that is targeted. Acute hypoxia is caused by intermittent cessation of blood flow. An agent that improves blood flow would be expected to be more effective against acute hypoxia than an agent that increases the oxygen content of blood. Nevertheless, the Eppendorf electrode data indicate that perfluorochemicals and hemoglobin solutions can overcome acute hypoxia, given that the oxygen electron should not have the spatial resolution to detect chronic hypoxia (narrow bands of hypoxia at a dis-

tance from a capillary corresponding to the diffusion limit for oxygen).

Nicotinamide (**80**) has been observed to en-

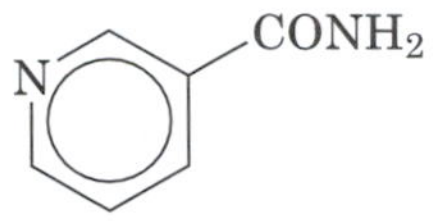

(80)

hance the radiation sensitivity of experimental murine tumors to a greater degree than normal tissues (484). This radiosensitization has been attributed to correction of acute hypoxia (485). In clinical studies, the combination of nicotinamide and carbogen (95% $O_2$, 5% $CO_2$) breathing has been shown to improve tumor oxygenation (486). A phase II study combining accelerated radiotherapy with carbogen breathing and nicotinamide ("ARCON") resulted in high local and regional control rates in head and neck cancer (487). Nitric oxide (endothelium-derived relaxing factor) is a potent vasodilator. Drugs that release NO, such as DEA/NO {$(C_2H_5)_2N[N(O)NO]^-$, (**81**)}

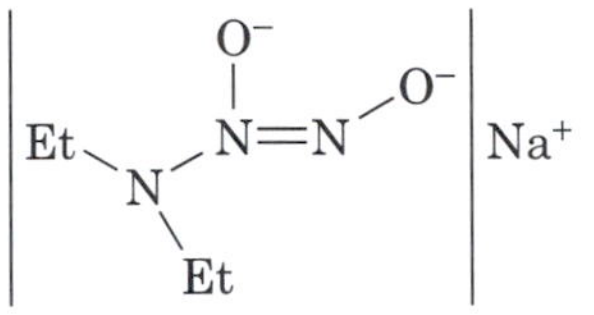

(81)

can sensitize tumors to radiation (488), although the mechanism appears to include a direct oxygen-mimetic effect (488). Nitric oxide synthase inhibitors can increase tumor hypoxia (488).

**2.6.1 RSR13.** RSR13 (*efaproxiral sodium*, **82**) is a synthetic allosteric modifier of hemoglobin, the first of a new class of pharmaceuti-

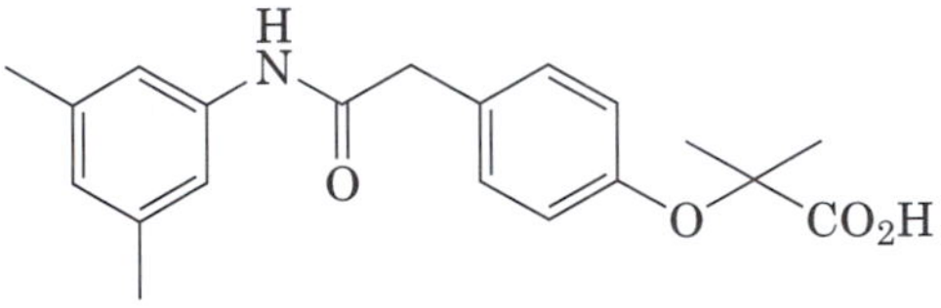

(**82**) RSR-13

cal agents. RSR13 is a small molecule that reduces hemoglobin oxygen-binding affinity and enhances the diffusion of oxygen from the blood to hypoxic tissues. RSR13 emulates the function of natural allosteric modifiers of hemoglobin such as hydrogen ions ($H^+$) and 2,3-diphosphoglycerate (2,3-DPG). This "turbocharging" of oxygen unloading from hemoglobin to tissue emulates and amplifies physiologic tissue oxygenation. This approach has broad clinical applicability in indications characterized by tissue hypoxia, including the use of RSR13 as an adjunct to radiation therapy (RT) and chemotherapy (CT), as well as cardiovascular, surgical, and critical care indications.

By increasing tissue oxygenation, RSR13 reduces tumor hypoxia and enhances the cytotoxic effects of RT and CT in animal models. These effects provide the rationale for RSR13 as an adjunct to RT and CT for the treatment of cancer patients with solid tumors. The goal of adjunctive RSR13 therapy is to achieve maximal concentrations of oxygen in the tumor tissue before administration of RT or during administration of CT, to decrease the hypoxic fraction of cells, and increase the radio- or chemoresponsiveness of malignant tumors. This hypothesis is supported by Phase Ib and II clinical efficacy data from studies using RSR13 combined with RT in the treatment of brain metastases, newly diagnosed glioblastoma multiforme (GBM), and locally advanced, inoperable non-small-cell lung cancer (NSCLC).

RSR13 has been studied for the prevention or treatment of conditions associated with tissue hypoxia in 17 phase I, II, and III clinical studies. RSR13 is being evaluated as a radioenhancement agent in patients receiving RT to treat brain metastases, GBM, or locally advanced, inoperable NSCLC. More than 400 cancer patients have received RSR13 in eight phase I–III radiation oncology studies. In addition, a study is ongoing to evaluate RSR13 as an adjunct to BCNU (carmustine) chemotherapy in patients with recurrent malignant glioma.

*2.6.1.1 Mechanism of Action.* RSR13 binds to a site on hemoglobin that is separate from the oxygen-binding site; therefore, it is referred to as an *allosteric modifier* (489, 490).

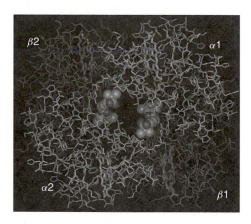

**Figure 4.3.** X-ray crystal structure of RSR13 bound to hemoglobin.

Natural and synthetic allosteric modifiers regulate oxygen-binding affinity by effecting a change of the hemoglobin tetramer from a high affinity to a low affinity structural conformation. Naturally occurring allosteric modifiers ($H^+$, $CO_2$, and 2,3-DPG) modulate hemoglobin oxygen-binding affinity and shift the oxygen equilibrium curve (OEC) to regulate oxygen unloading from hemoglobin to tissues under various physiologic conditions.

RSR13 binds to the central water cavity of the hemoglobin tetramer (Fig. 4.3). This site is different from the binding sites of natural allosteric modifiers, but the effects of RSR13 and the natural allosteric modifiers are similar, as shown in Fig. 4.4. Each molecule induces a rightward shift in the OEC, an effect that is described by an increased p50. The p50 is the oxygen pressure that results in 50% saturation of hemoglobin. Thus, an increase in p50 reflects a reciprocal decrease in oxygen-binding affinity of hemoglobin. Figure 4.4 depicts a p50 shift of 10 mmHg, a typical result in patients receiving a 100 mg/kg dose of RSR13.

The pharmacologic effect of RSR13 has broad clinical applicability in situations characterized by tissue hypoxia attributed to: (1) reduced blood flow (regional or global), (2) reduced oxygen carrying capacity, and/or (3) increased tissue oxygen demand. These therapeutic indications include the use of RSR13 as an adjunct to RT or CT, as well as cardiovascular, surgical, and critical care indications. Preclinical studies have shown that RSR13 can increase normal tis-

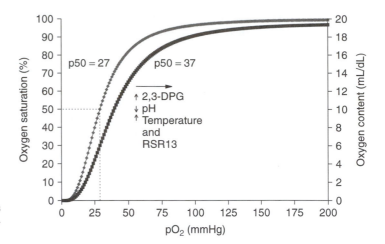

**Figure 4.4.** Allosteric modifiers of hemoglobin oxygen equilibrium curve for human blood.

sue oxygenation, reduce tumor hypoxia, improve the efficacy of RT and CT, and attenuate the functional and metabolic deficiencies attributed to myocardial and cerebral ischemia.

*2.6.1.2 Tumor Hypoxia.* Direct oxygen measurements in human tumors have confirmed tumor hypoxia (median partial pressure of oxygen [$pO_2$] < 20 mmHg) in many types of human tumors (491–495). Mechanisms for chronic or transient ischemic hypoxia may include obstruction of blood flow, inadequate or defective angiogenesis, and unregulated cellular growth that outstrips the capacity of the capillary blood supply (496–500). In general, tumor cells are oxygenated up to a distance of about 150 $\mu$m from a functional capillary; beyond that, tumor cells can become oxygen depleted and either die or survive in a hypoxic state. Calculations based on a three-dimensional simulation of oxygen diffusion from a network of vessels, with a geometry derived from observations of tumor microvasculature in the rat, have indicated that an increase in hemoglobin affinity for oxygen should have a beneficial effect on tumor radiotherapy (501). Although this should not be of much benefit in areas where the circulation is completely occluded (i.e., cells are completely anoxic), a recent report that a large proportion of the tumor cells in SiHa xenografts in mice are intermediately hypoxic (oxygen tension between 0.1 and 0.2%) (502) suggests that diffusion-limited hypoxia may be a significant contributor to tumor resistance.

The efficacy of RT can be affected by the extent of tumor oxygenation. Hypoxic tumors are more resistant to cell damage by radiation (503), and tumor hypoxia has been associated with a poor clinical prognosis of patients receiving RT (503–507). Oxygen measurements in human tumors have detected tumor hypoxia in GBM (491), brain metastases (491, 508), squamous cell carcinomas of the uterine cervix (493), and head and neck (492) and breast carcinomas (495). Because hypoxic tumors are substantially more resistant to radiation than oxygenated tumors, even small hypoxic fractions in a tumor may affect the overall response to RT and increase the probability that some tumor cells will survive RT.

Certain alkylating chemotherapeutic agents also require oxygen for maximal cytotoxicity. This may be related to effects of hypoxia on cellular metabolism, decreasing the cytotoxicity of anticancer drugs and enhancing genetic instability, which can lead to more rapid development of drug-resistant tumor cells (498–500, 509).

*2.6.1.3 Radiation Therapy Sensitization with RSR13.* Unlike most other radiosensitizing agents and strategies, the radiosensitizing effect of RSR13 is not dependent on its entry into the tumor. Instead, RSR13 enhances oxygen release from hemoglobin, thereby increasing the diffusion of oxygen from plasma and the vascular compartment to the hypoxic tumor cells. Enhanced tumor oxygenation is the basis for the radioenhancement and chemoenhancement effects of RSR13. The fact that RSR13 does not have to enter cells to increase the tumor sensitivity to RT or CT is an important differentiation between RSR13 and other attempts to improve the efficacy of cancer therapy. This is especially important in the setting of primary or metastatic brain tumors, where the blood-brain barrier acts to exclude or impede the entry of chemical agents into the brain parenchyma. Oxygen readily diffuses the blood-brain barrier and the cancer cell membrane to increase tumor oxygenation and, thereby, the effectiveness of therapy.

RSR13 is being developed as adjunctive therapy to standard RT for the treatment of solid tumors. Fractionated RT is administered once per day, 5 days per week. For most palliative indications, such as the treatment of brain metastases, RT is administered daily (Monday through Friday) for 2–3 weeks, for a total of 10–15 treatments. For other treatment regimens, such as the treatment of primary brain cancer and other solid tumors, RT is typically administered daily for 6–7 weeks, for a total of 30–35 treatments. In completed and ongoing radiation oncology studies, RSR13 has been administered as daily doses of 50–100 mg/kg, infused over 30–60 min through a central venous catheter, immediately before RT. This regimen results in the administration of a total of 10, 30, or 32 doses to patients with brain metastases, GBM, or NSCLC, respectively.

Phase II clinical studies suggest that RSR13 improves the efficacy of RT in patients with brain metastases, newly diagnosed GBM, and locally advanced, inoperable NSCLC.

Animal pharmacology studies have shown that RSR13 dose-dependently increases blood p50 (510–514), increases $pO_2$ in nontumor tissue (512, 515–517), and increases oxygen-diffusive transport in nontumor tissue (518). In rats bearing mammary carcinoma tumors, tumor $pO_2$ was measured using Eppendorf histograms with the tumor hypoxic fraction expressed as the percentage of readings ≤5 mmHg. In this model, RSR13 (150 mg/kg, i.v.) decreased the tumor hypoxic fraction from 36% (controls) to 0% (treated) and increased tumor oxygenation within 30 min after RSR13 dosing (519, 520).

By use of mice with subcutaneous human NSCLC xenographs and applying the noninvasive technique of blood oxygen level dependent magnetic resonance imaging (BOLD-MRI), RSR13 was shown to dose-dependently increase tumor $pO_2$. The maximum increase in tumor $pO_2$ was achieved after 200 mg/kg of RSR13 was administered by i.p. injection. In this human tumor xenograft model, RSR13 administered 30 min before 10-Gy radiation enhanced the radiation-induced tumor growth delay by a factor of 2.8. Without radiation, RSR13 had no effect on tumor growth delay (521).

In a mouse model with subcutaneous lung tumors, i.p. RSR13 dose-dependently enhanced the efficacy of fractionated RT (measured as an enhancement of tumor growth delay) by 22, 40, and 69% at 50, 100, and 200 mg/kg, respectively (520). In additional studies RSR13 decreased tumor cell survival when combined with fractionated radiation of mice bearing FSaII fibrosarcomas, squamous cell carcinomas (510), or mammary carcinoma tumors (522). The radioenhancement effect of RSR13 was shown to be oxygen dependent, with no direct cytotoxic effect on the tumor, bone marrow (520–522), or skin (522).

RSR13 was also shown to be an effective chemosensitizer of EMT-6 cells that were exposed to a variety of cytotoxic anticancer drugs in vitro (520). When tested as a single agent in clonogenic assays in vitro, RSR13 was not cytotoxic, and it did not alter the radiation response of bone marrow progenitor cells. These studies indicate that RSR13 has chemosensitizing activity when combined with alkylating/DNA-damaging agents.

The molecular basis of the chemoenhancement activity of RSR13 may involve both hemoglobin-dependent and -independent mechanisms. In vivo, RSR13 has demonstrated chemoenhancement in combination with various widely used agents. The effects of chemotherapeutic agents on tumor growth delay and development of lung metastases were potentiated by RSR13 in the Lewis lung carcinoma model (520–523) and the MB-49 bladder carcinoma model (520). In addition, RSR13 demonstrated a marked ability to decrease tumor volumes when given with BCNU in a 9-L gliosarcoma model in rats (524).

*2.6.1.4 Clinical Trials with RSR13.* A total of 17 Phase I-III clinical studies making use of RSR13 have been completed or are ongoing in patients with cancer (including studies specifically enrolling patients with brain metastases, newly diagnosed and recurrent GBM, and NSCLC), surgical patients, patients with cardiovascular disease, and healthy subjects. Phase II studies in patients with brain metastases, GBM, and NSCLC have been completed. One randomized phase III study in patients with brain metastases is ongoing and, in addition, a phase I/II study is ongoing to evaluate RSR13 as an adjunct to BCNU chemotherapy in patients with recurrent malignant glioma.

The RSR13 dosing in clinical trials, using the drug in combination with radiotherapy, is based on a phase I trial that studied escalation of both drug dosing and frequency of administration (525). According to this open-label study in patients undergoing palliative irradiation to 20–40 Gy in 1015 fractions, RSR13 could be administered daily in doses of up to 100 mg/kg for 10 consecutive treatments through central venous access and supplemental nasal oxygen at 4 L/min. The tolerance of the drug was supported by clinical monitoring of oxygen saturation and associated pharmacokinetic and pharmacodynamic studies. At 100 mg/kg the peak increase in p50 averaged 8.1 mmHg, consistent with the targeted physiological effect (525).

*2.6.1.4.1 Brain Metastases.* Nearly one-third of patients with systemic cancer develop brain metastases, a complication that profoundly affects the patients' quality of life and survival. In early studies, untreated patients with brain metastases had a median survival time of about 1 month. Without more aggressive treatment, nearly all patients died as a direct result of the brain metastases (526). Even with contemporary treatment, specifically earlier diagnosis, radiation therapy, and systemic chemotherapy, approximately 30–50% of brain metastases patients die as a direct result of the brain metastases (527). Expanding intracranial tumor masses lead to intractable headaches, nausea and vomiting, serious cognitive dysfunction, and one or more focal neurological deficits, including hemiparesis, seizures, visual, speech, and gait disturbances (528). Acute, catastrophic neurological complications, such as intracerebral hemorrhage and brainstem herniation, occur in 5–10% of brain metastases patients (529).

Study RT-008 was a phase II, open-label, multicenter study to assess the effect of RSR13 on enhancing radiation therapy in patients with brain metastases. Patients received a standard 10-day course of whole-brain radiation therapy (WBRT) (3 Gy in 10 fractions = 30 Gy) within 30 min of receiving RSR13 administered through a central venous access device. RSR13 administration began on the first day of WBRT and continued daily for a total of 10 doses.

Patient eligibility was based on histologically or cytologically confirmed breast, NSCLC, melanoma, genitourinary, or gastrointestinal primary carcinoma. Patients were stratified by recursive partitioning analysis (RPA), Class I or Class II, as previously described (529). A more recent analysis by the RTOG (Study 91–04) showed similar median survival times (MST) in each class. The analysis also indicated that no major change in the prognosis of brain metastases patients has been observed in the last 25 years, even with the advent of more aggressive multiagent CT regimens directed at both the primary tumor and extracranial metastases (530). At the time the study was closed there were 57 Class II patients enrolled.

The objective of the study was to compare MST in the study population to that from the RTOG Brain Metastases Database (BMD) through use of both the overall RTOG database and controls case-matched by prognostic factors. Exact case-matched controls were obtained for 38 patients [matching 5 of 5 criteria: age, Karnofsky Performance Status (KPS), extent of metastases, status of primary cancer, and location of primary tumor). RSR13-treated patients had significantly superior overall survival (6.4 months) compared to the historical BMD control group (4.1 months) by Kaplan-Meier estimates of MST ($P = 0.0269$ compared to the overall database) (Fig. 4.5) (531, 532). One-year survival was 23% for RSR13-treated patients compared to 15% for the overall BMD population, and 9% for all case-matched BMD controls.

Further improvements in survival were observed for RSR13-treated patients compared

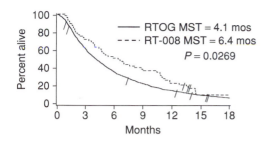

**Figure 4.5.** Survival results: study RSR13 RT-008 versus RTOG BMD (overall).

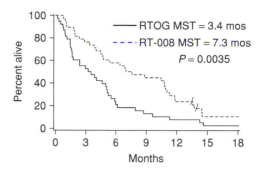

**Figure 4.6.** Survival results: study RSR13 RT-008 versus RTOG BMD (exact matches).

to exact case-matched controls (7.3 months versus 3.4 months, $P = 0.0035$, $n = 38$) (Fig. 4.6). One-year survival was 24% for RSR13-treated patients compared to 8% for exact case-matched BMD controls. All analyses used the log-rank test.

*2.6.1.4.2 Glioblastoma Multiforme.* Glioblastoma multiforme, or GBM, is a deadly form of primary brain cancer. This condition occurs in about 20% of all brain cancer patients in the United States, or approximately 3400 people per year. The median survival time of patients with GBM is approximately 9 to 10 months. Radiation therapy is the standard of care for GBM and is administered to most patients. The goal of radiation therapy is to prevent or reduce complications and improve survival time.

Patients with GBM received a standard 6-week course of cranial RT (2 Gy in 30 fractions = 60 Gy) after RSR13, at a dose of 100 mg/kg, was administered over 30 min through a central venous access device. RSR13 administration began on the first day of RT and continued daily for a total of 30 doses. RT was administered within 30 min after completion of each RSR13 infusion.

Patient eligibility was based on histologically confirmed supratentorial grade IV astrocytoma (GBM).

RT-007 was a study sponsored by the National Cancer Institute (NCI) and conducted by the New Approaches to Brain Tumor Therapy (NABTT) CNS Consortium. Survival results of the intent-to-treat analysis for this study have been published (533, 534). An intent-to-treat analysis was performed after all 50 patients had completed a minimum follow-up of 14 months. Survival data were compared to those from an NABTT historical database. The NABTT historical database consisted of 152 patients pooled from multiple studies. NABTT and RSR13 studies were performed in the same general time period (1994–2000).

Patients in the RSR13 and NABTT studies were predominantly male and Caucasian. Mean values for baseline parameters by study ranged from 53 to 62 years for age, from 83 to 88 for KPS, and from 62% to 83% for history of surgical resection. The distribution of patients by RPA class included 6 patients in Class III, 32 patients in Class IV, and 12 patients in Class V (535).

The MST was greater for RSR13-treated patients compared to the NABTT historical database: 12.3 months vs. 9.7 months and was statistically significant as determined by the Wilcoxon test ($P = 0.02$) and the log-rank test ($P = 0.04$) (Fig. 4.7).

*2.6.1.4.3 Non-Small-Cell Lung Cancer.* Non-small-cell lung cancer, or NSCLC, is a type of cancer that occurs in approximately 130,000 patients per year in the United States. RSR13 is currently being evaluating as a radiation enhancer for the treatment of patients with locally advanced, inoperable Stage IIIA and IIIB NSCLC. Radiation therapy for treatment of Stage III NSCLC is intended to prevent or reduce complications and control local tumor growth in the chest. The overall median survival time of patients with Stage III NSCLC is approximately 9 to 12 months.

Study RT-010 was a phase II, nonrandomized, open-label, multicenter efficacy and safety study. Patients with locally advanced

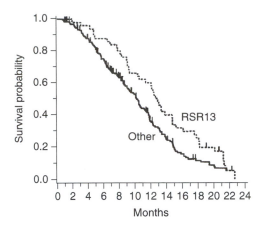

**Figure 4.7.** Kaplan-Meier survival distribution: study RSR13 RT-007 vs. NABTT historical database.

inoperable (Stage IIIA or IIIB) NSCLC received induction chemotherapy [paclitaxel (225 mg/m$^2$ i.v.) and carboplatin (area under the curve = 6, i.v.) every 3 weeks × two cycles] followed by standard thoracic RT with RSR13 over 6–7 weeks (up to 32 fractions/doses). RSR13 was administered at an initial dose of 75 mg/kg through a central venous access device just before daily RT. Dose reduction to 50 mg/kg or increase to 100 mg/kg was allowed per protocol depending on patient tolerance.

Objectives of this study were to evaluate complete and partial response rates in the chest (radiation portal), overall survival, progression-free interval in the chest, time to disease progression outside the radiation portal, and toxicities and adverse events associated with RSR13 and RT after induction chemotherapy.

The study was conducted in 47 evaluable patients with locally advanced, inoperable, Stage IIIA/IIIB non-small-cell lung cancer. The objectives of this study were to evaluate overall survival, progression-free interval in the chest, complete and partial response rates in the chest (radiation portal), and time to disease progression outside of the radiation portal. The patients received two courses of induction paclitaxel and carboplatin chemotherapy followed by daily RSR13 combined with chest radiation therapy for 32 doses.

The overall response rate was 89%, with complete and partial response rates of 9% and

80%, respectively (536). Median time to first progression was 9.9 months. Median tumor progression-free survival time in the radiation portal was 24.8 months, whereas median progression-free survival time outside the portal was 11.3 months. The median survival was 20.6 months, 1-year survival rate of 68%, and an estimated 2-year survival rate of 43% (537).

***2.6.1.5 Summary of RSR13 Results.*** The preclinical and clinical studies of RSR13 indicate that RSR13 increases oxygen release from red blood cells and increases the $pO_2$ in hypoxic tissues. Tumor hypoxia can limit the efficacy of radiation therapy; pretreatment of patients receiving radiation therapy for brain metastases, primary glioblastoma multiforme, and non-small-cell lung cancer with RSR13 appears to increase their chances for survival.

## 2.7 Radiosensitization by Depletion of Endogenous Protectors

Radiosensitization of hypoxic cells should be achievable by either increasing the concentration of oxygen-mimetic sensitizers or by decreasing the concentration of endogenous protectors, as discussed in Section 2.4. The concentration of endogenous protectors is too low to effectively compete with oxygen for reaction with DNA radicals, at oxygen concentrations found in well-oxygenated tissues. However, endogenous protectors become effective at low oxygen tensions and can undermine the effectiveness of electron-affinic sensitizers by competing with them for reaction with DNA radicals. Several human tumor cell lines have been reported to have high glutathione levels and to be correspondingly resistant to hypoxic radiosensitization with misonidazole (538). A strategy for selective radiosensitization of hypoxic cells is therefore to deplete endogenous radioprotectors.

Because glutathione (GSH) is the principal intracellular thiol, it has been the principal target for depletion. Selective depletion of GSH can be accomplished with weak electrophiles that are substrates for GSH *S*-transferases (539). Diethylmaleate (DEM, **83**) is one agent that meets these criteria (540). The enzyme-catalyzed reaction is substantially faster than chemical reactions with other endogenous nucleophiles at low DEM concentrations (540, 541). GSH depletion with low con-

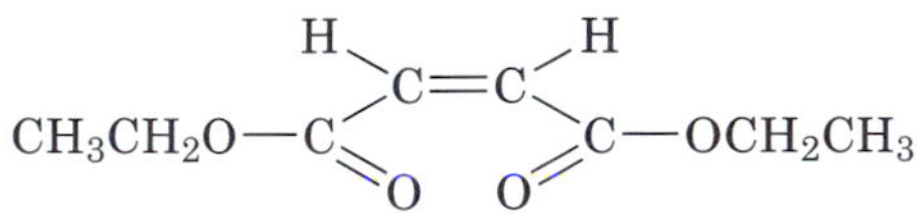

$$H-C=C-H$$

(structure) CH_3CH_2O—C(=O)—C=C—C(=O)—OCH_2CH_3

(83)

centrations of DEM sensitizes hypoxic cells to radiation and enhances the radiosensitizing effect of nitroimidazole radiosensitizers, without affecting aerobic cell radiosensitivity (541). Tumor sensitization can be achieved *in vivo* with DEM without enhancing normal tissue radiosensitivity, except for a slight enhancement of skin reactions (540). Enhancement of misonidazole tumor radiosensitization by DEM has also been demonstrated *in vivo* (36). The relative activities and substrate specificities of the various isozymes of GSH *S*-transferase (542, 543) in human tumors will be a consideration for optimization of this approach for human use. For example, it has been reported that GSH *S*-transferase π is overexpressed in many human tumors (544).

Another strategy for depletion of GSH is to inhibit its synthesis. *L,S*-Buthionine sulfoximine (*L,S*-BSO, **84**) is one of a series of sub-

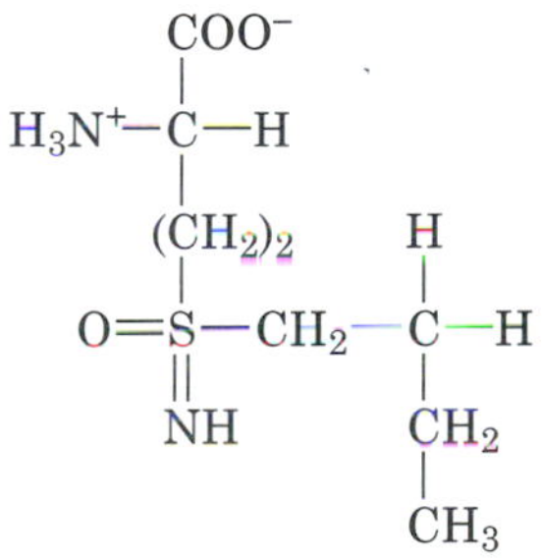

(structure 84)

COO⁻
|
H_3N⁺—C—H
|
(CH_2)_2      H
|             |
O=S—CH_2—C—H
‖             |
NH            CH_2
              |
              CH_3

(84)

strate analogs that inhibit γ-glutamyl cysteine synthetase, which is rate limiting for GSH synthesis (544, 545). GSH becomes depleted over a period of 4–24 h, depending on cell type, as GSH is lost to catabolism. *L(S,R)*-BSO is in clinical trials as a chemosensitizer (546). Early clinical results with bolus administration of *L(S,R)*-BSO indicated that tumor GSH depletion was not consistently achieved; however, continuous infusion trials appear to have been more successful in depleting tumor GSH (547–549). There have been no reports of clinical

toxicity with *L(S,R)*-BSO, although it does enhance the myelotoxicity of melphalan (547), and it has not yet been determined whether it produces a therapeutic gain as a chemosensitizer in the clinic. A possible explanation for the difficulty in achieving tumor GSH depletion in the clinic is that *L(S,R)*-BSO treatment can lead to increased expression of γ-glutamyl-cysteine synthetase messenger RNA (550). Tumor GSH depletion with *L,S*-BSO in a murine model system is less than expected from *in vitro* studies (551). Nevertheless, enhancement of etanidazole radiosensitization has been observed when GSH is depleted with *L(S,R)*-BSO in animal model systems (36). Enhancement of etanidazole-induced neurofilament degradation in a spinal cord organotypic model by *L,S*-BSO was not greater than the expected enhancement of hypoxic tumor cell radiosensitivity, suggesting a therapeutic gain may be achievable by the combination of *L,S*-BSO and etanidazole (552).

## 2.8 Radiosensitization by Inhibition of DNA Repair

The sensitivity of mammalian cells to ionizing radiation is very dependent on DNA double-strand break repair capacity (553), particularly at the radiation doses that are used in cancer therapy. Cell lines that are deficient in DNA double-strand break repair are radiosensitive (554, 555). Inhibition of DNA repair has great potential for tumor radiosensitization (556, 557), but strategies for selective sensitization of tumors remain elusive (558).

**2.8.1 PLD Repair Inhibitors.** One strategy is to inhibit a type of cell recovery from radiation damage that is called *potentially lethal damage repair* (PLDR). PLDR is not yet understood at the molecular level. PLDR is functionally defined as an increase in survival when cells are held under nutrient-deprived conditions, in comparison to the survival of cells that are immediately plated in fresh medium after irradiation (559).

Poly(ADP-ribose) polymerase (PARP) binds to single-strand breaks (SSB) in DNA (560), and is activated by DNA single-strand breaks. Inhibitors of this enzyme, such as 3-aminobenzamide (3AB, **85**) inhibit SSB repair and inhibit PLDR (561). Cleaver et al.

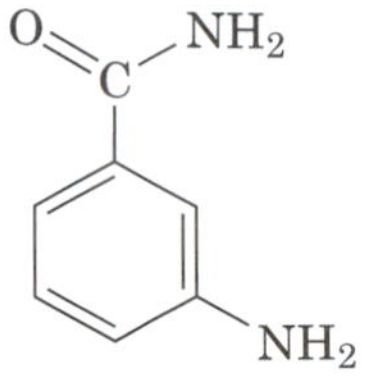

**(85)**

(562) cautioned that 3AB has other effects on the cell and may not be acting by inhibition of PARP, particularly considering that the concentrations required for inhibition of PLDR are much higher than the $K_i$ value for isolated PARP. Utsumi and Elkind (561) reported that two types of PLDR are inhibited by 3AB, at different concentrations of 3AB. A variety of compounds inhibit PARP (563) and there is currently intensive activity in the development of more effective PARP inhibitors. PD 128763 [3,4-dihydro-5-methyl-1(2$H$)-isoquinolinone; (**86**)] inhibits PARP at a concen-

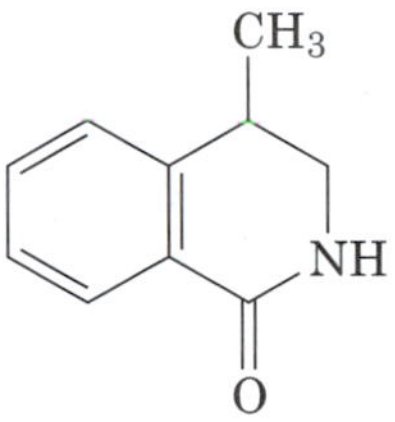

**(86)**

tration about 50 times lower than 3AB, and sensitizes both exponentially growing and stationary phase mammalian cells to ionizing radiation (564).

Other PLD repair inhibitors include actinomycin D (565), 9-$\beta$-D-arabinofuranosyladenine (ara-A) (566), 1-$\beta$-D-arabinofuranosylcytidine (ara-C) (567), cordycepin (3'-deoxyadenosine) (568), 3'-deoxyguanosine (569), and aniosotonic media (570). Irradiation of cells under aerobic conditions in the presence of high concentrations of misonidazole results in inhibition of PLD repair, even though misonidazole does not radiosensitize exponentially growing aerobic cells (571). Some DNA-repair inhibitors can also alter the survival curve of exponentially growing mammalian cells (572). Much remains to be understood about the mechanisms of radiosensitization by these

agents and about the differences in repair enzymology, if any, between tumors and normal tissue, before this approach can be used effectively in the clinic.

**2.8.2 Radiosensitization by Reaction with Protein Sulfhydryls.** Most exposed intracellular cysteine thiol groups normally occur in the reduced state (573), given that glutathione is kept highly reduced and enzymes catalyze redox reactions between intracellular glutathione and protein thiols/disulfides (574). In many cases oxidation or alkylation of these protein sulfhydryl groups results in loss of enzyme activity (575). Consequently, treatment of cells with thiol oxidants or thiol-binding reagents affects many metabolic processes (576).

It has long been recognized that thiol oxidants or thiol-binding reagents sensitize cells to ionizing radiation (577–580). Part of this effect is attributed to depletion of endogenous radioprotectors. However, this does not explain radiosensitization that is observed when thiol reagents are added after irradiation because endogenous chemical radioprotectors will have acted within 10 ms of irradiation (399). Recent studies have indicated that protein thiol loss results in inhibition of DNA double-strand break repair (581, 582), and that this can account for the postirradiation sensitization.

There are two approaches that can be taken to deplete protein thiols. One is to oxidize intracellular glutathione, which in turn will oxidize protein thiols. The other is to use thiol-binding reagents.

Diazenes have been used as reagents for selective oxidation of intracellular glutathione (582–584). Diamide [diazenedicarboxylic acid bis($N$-$N'$-dimethylamide), (**87**)] sensitizes mammalian cells to ionizing radiation (576, 585, 586). The predominant finding is that it decreases the shoulder of the radiation survival curve. Similar results were obtained with the diamide analogs DIP (**88**) and DIP + 1 (**89**) that penetrate cells more slowly, resulting in less drug toxicity (587) and with the diamide analog SR-4077 [diazenedicarboxylic acid bis($N$-$N'$-piperidide), (**90**)] that is similar in reactivity to diamide, but less cytotoxic (582). This is in contrast to the effect of glutathione

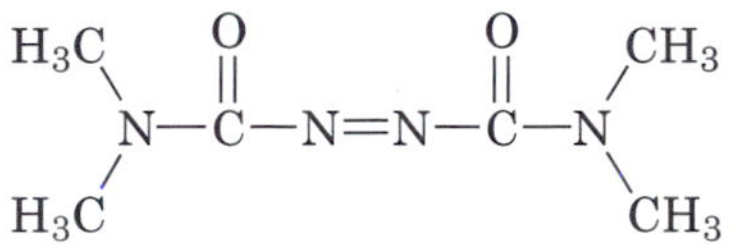

(87)

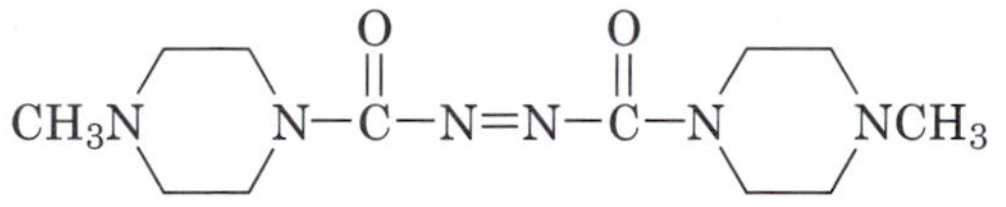

(88)

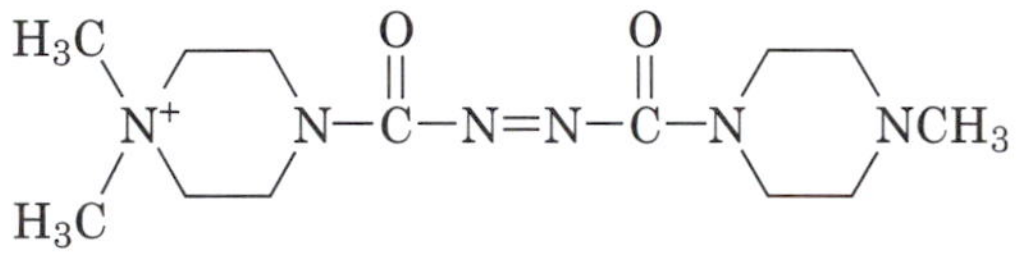

(89)

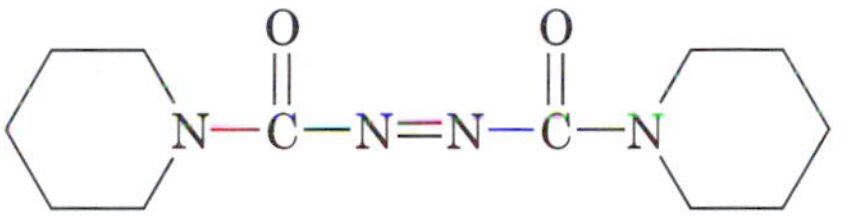

(90)

depletion, which is a change of the exponential slope of the survival curve (36). Radiosensitization by thiol oxidation can be achieved, even if the oxidant is added after irradiation, eliminating rapid radiation chemistry as a factor. Radiosensitization by glutathione oxidants is associated with inhibition of DNA double-strand break repair (581, 588).

Dimethyl fumarate (DMF, **91**) is a thiol-binding agent that depletes glutathione and protein thiols (589). DMF sensitizes hypoxic and oxygenated cells when added after irradiation and this effect is associated with inhibition of DNA repair. Additional sensitization of hypoxic cells is observed if DMF is added be-

(91)

fore irradiation, and this has been attributed to depletion of glutathione (589).

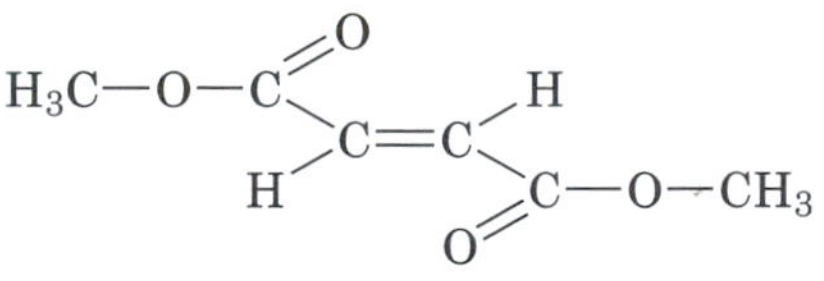

## 2.9 Radiosensitization by Perturbation of Cellular Metabolism

**2.9.1 Perturbation of Energy Metabolism.** Several studies have indicated that radiation sensitivity can be altered by treatment of mammalian cells with uncouplers or inhibitors of oxidative phosphorylation (590). A potential selective effect on hypoxic tumor cells is suggested by the observation that 5-thio-D-glucose can sensitize hypoxic cells to radiation (591), given that inhibition of glycolysis would be expected to have a greater impact on hypoxic cells than on oxygenated cells. Beyond selectivity toward hypoxic cells, 2-deoxy-D-glucose may selectively sensitize cancer cells that depend on aerobic glycolysis, while protecting bone marrow (592, 593).

The mechanism of radiosensitization by modulators of energy metabolism is not clear. No radiosensitization was observed when low concentrations of 2-deoxyglucose and rotenone were used in combination to achieve a steady-state decrease in the adenylate energy charge of Chinese hamster ovary cells, even though very low energy levels were maintained for up to 4 h and DNA repair was inhibited (594). This result suggests that a complete collapse of energy metabolism may be required to achieve radiosensitization, and that the mechanism may involve secondary events, such as failure to maintain ionic homeostasis, rather than simply a lack of ATP to carry out repair. It could be that the processes that result in damage fixation are equally inhibited by a low energy state, and that both damage fixation and damage repair recover when the ATP is restored. It is possible that under certain conditions, the balance could be shifted the other way, inhibiting damage fixation while permitting repair. This possibility is suggested by observations of radioprotection with uncouplers of oxidative phosphorylation (595).

An additional consideration arises *in vivo* because inhibition of oxidative phosphorylation will decrease oxygen consumption, providing a means of sensitization of chronically hypoxic cells by increasing the diffusion distance of oxygen.

**2.9.2 Abrogation of $G_2$ Delay.** Ionizing radiation induces cell-cycle arrest at both the $G_1$ and the $G_2$/M checkpoints (596). The purpose of cell-cycle arrest is presumed to be to allow time for DNA repair before progressing through the cell cycle (597), although the precise mechanism of "repair" involved is not clear. In the case of a $G_1$ arrest, abrogation of the arrest, for example, by transfection of a mutated form of the *p53* gene, does not result in increased cell killing (598), but has been associated with increased mutagenicity and carcinogenicity (599). Abrogation of the $G_2$/M arrest, on the other hand, can enhance radiation cell killing. The classic example of a radiosensitizer that acts by this mechanism is caffeine (**92**) (600–602).

(**92**)

$G_2$/M arrest appears to be an effect on cellular regulation, rather than simply a physical consequence of damage. Cyclin B1 mRNA and protein levels have been observed to decrease in correlation with radiation-induced $G_2$/M arrest (603, 604). Cyclin B1 is required for the $G_2$/M transition. Cyclin Bl forms a complex with p34$^{cdc2}$ kinase called *mitosis promoting factor* (MPF). MPF is kept inactive by phosphorylation. Caffeine can inhibit the protein kinase that phosphorylates p34$^{cdc2}$ (605). MPF is thought to trigger entry into mitosis upon its dephosphorylation by cdc25 (606, 607), by phosphorylating key proteins such as histones and lamins. Metting and Little (608) reported that failure to dephosphorylate the p34$^{cdc2}$-cydin B1 complex accompanies radiation-induced $G_2$ arrest in HeLa cells. Caffeine treatment results in increased p34$^{cdc2}$ kinase activity and increased histone mRNA in irradiated cells (609).

An interesting possibility is suggested by a study by Powell et al. (610), in which they found that p53$^-$ cells were sensitized more than p53$^+$ cells by caffeine. They suggested that this might be a basis for a therapeutic gain because many tumors have mutated forms of p53.

Other agents that abrogate the radiation-induced $G_2$/M arrest include staurosporine, 2-aminopunine, cordycepin, 6-dimethylaminopurine, theobromine, and theophilline (611–613). Staurosporine, a protein kinase inhibitor, can override both the $G_1$ and $G_2$/M blocks induced by irradiation (611). Hallahan et al. (614) reported radiosensitization of human squamous cell carcinoma cell lines with the protein kinase inhibitors sangivamycin and staurosporine.

**2.9.3 Radiosensitization by Growth Factors and Cytokines.** Radiation sensitivity can be altered with growth factors and cytokines, although it is likely that the effect of a particular agent of this type will be substantially dependent on the context of its use. Hallahan et al. (615) reported that TNF-$\alpha$ inhibits PLD repair in several human tumor cell lines when added 4–12 h before irradiation. Epidermal growth factor has been reported to radiosensitize human tumor cells that overexpress the EGF receptor (616, 617). Insulin and insulin-like growth factor 1 have been reported to inhibit PLD repair in human tumor cells (618).

**2.9.4 Halogenated Pyrimidines.** Substitution of a halogen atom for the hydrogen at the 5-carbon position of uracil or cytosine has produced a series of compounds that interfere with nucleotide metabolism. Radiosensitization by halogenated pyrimidines was noted in the late 1950s (619, 620). Radiosensitization by 5-iododeoxyuridine (IdUrd, **93**) requires incorporation into DNA and correlates with the percentage of thymidine replacement. The mechanism may involve enhancement of initial damage to DNA, inhibition of DNA repair, or alteration of cell-cycle kinetics, but has not been clearly established (621). A rationale for selective tumor radiosensitization is that the cells in some tumors proliferate more rapidly than the stem cells of the limiting normal tissue. Early clinical trials with Bromodeoxyuridine (BrdUrd) in the treatment of head and neck cancer resulted in considerable normal

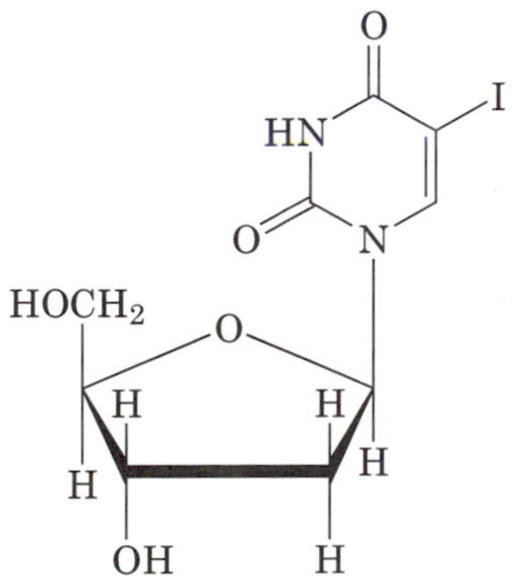

(93)

tissue toxicity, indicating that this rationale does not apply to all tumors (622). More recent trials have focused on brain tumors (623), where there appears to be a larger differential in proliferation rates between the tumor and the critical normal tissue cells (621). The limitation of this approach is that noncycling tumor cells will not be sensitized; the effectiveness of the treatment will depend on the percentage of cells that incorporate the analog. Combinations with biochemical modulators are being devised to enhance incorporation in tumor cells (621).

5-Fluorouracil (5-FU, **94**) is a cytotoxic

(94)

agent that is used in cancer chemotherapy. A number of clinical trials have shown a benefit for the combined use of 5-FU and ionizing radiation (621, 624), but it is not clear whether the effect is additive or synergistic, nor is it clear whether this represents a true therapeutic gain, in that increased normal tissue toxicity is also observed. 5-FU is thought to act principally by inhibition of thymidylate synthetase (TS), but it is also incorporated into RNA and can alter cell-cycle kinetics. Because inhibition of TS blocks *de novo* synthesis of

thymidine, 5-FU can also be used to enhance IdUrd incorporation into DNA (625). More specific TS inhibitors appear to retain radiosensitizing activity, suggesting that this is the dominant mechanism of radiosensitization (626).

Gemcitabine [2',2'-difluorodeoxycytidine (dFdC), (**95**)] radiosensitizes human tumor

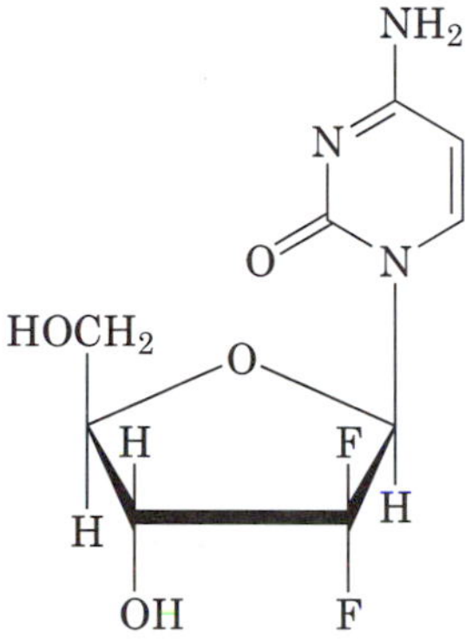

(95)

cells to radiation (627). This effect has been attributed to inhibition of ribonucleotide reductase, resulting in decreased deoxyribonucleotide pools (628). Selectivity could be achieved by differences in deoxycytidine kinase activity, given that this enzyme is required to activate this antimetabolite (629). Hydroxyurea (**96**), another inhibitor of ribo-

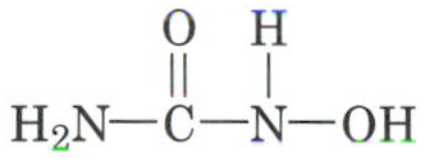

(96)

nucleotide reductase, has been tested as a radiosensitizer in clinical trials with some encouraging results (630).

Another strategy for inhibition of ribonucleotide reductase is to use 5-chloro-2'-de-oxycytidine or analogs thereof (631), which need to be metabolized by cytidine deaminase or deoxycytidylate deaminase to become active inhibitors, because it has been observed that a number of tumor cell lines express high levels of these enzymes (632). The most effective strategy appears to be to use this drug in combination with tetrahydrouridine (a cytidine deaminase inhibitor) to inhibit activation by cytidine deaminase because the greater differential between tumor and normal tissue appears to be in deoxycytidylate deaminase.

## 2.10 Radiosensitizers for which the Mechanism of Sensitization has not been Established

**2.10.1 Metal Ion Complexes.** A property of metal ions that could be useful in radiosensitizer design is the ability to readily exchange single electrons. Single-electron exchange could serve as a mechanism either for radioprotection or for radiosensitization. Some metal complexes that have been found to be hypoxic cell radiosensitizers are electron affinic. However, the mechanism of radiosensitization has generally not been established for this type of compound because metal ions also have the potential of interaction with thiol groups, disturbing DNA structure, and upsetting cellular metabolism (633).

Ferricenium salts (**97**) are very efficient ra-

(97)

diosensitizers of hypoxic mammalian cells, achieving an enhancement ratio of 2 at 10 $\mu M$, with little effect on oxygenated cells (634). Unlike other electron-affinic sensitizers, these compounds predominantly decrease the shoulder of the radiation survival curve. In that sense they resemble agents that inhibit DNA repair. Cu(I) salts have been reported to radiosensitize hypoxic bacterial or mammalian cells, but the mechanism has not been explored (635). Metalloporphyrins, and particularly Co(III) complexes, have been reported to preferentially radiosensitize hypoxic mammalian cells (636). Teicher et al. found various Co(III) and Fe(III) complexes to be hypoxic cell radiosensitizers (637). SN 24771 (**98**) is a cobalt (III)-nitrogen mustard complex that is selectively toxic to hypoxic cells (638).

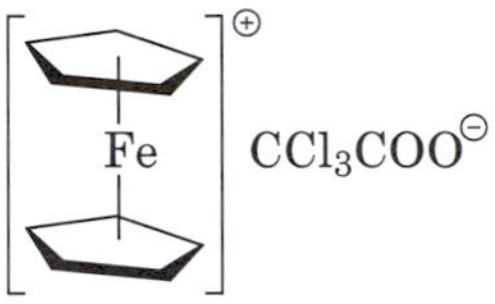

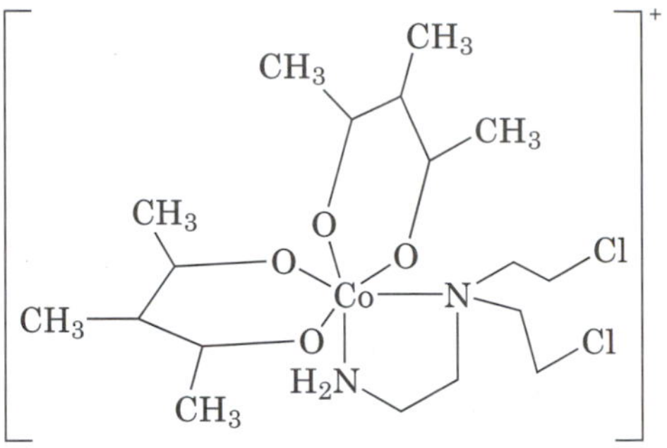

(98)

Cytotoxic platinum-containing drugs have been reported to have a more than additive effect on cell killing in combination with ionizing radiation (346–399, 639). Clinical trials with cisplatin and radiation in inoperable non-small-cell lung cancer have shown better tumor control (640). Carboplatin appears to be as effective as cisplatin in sensitizing human lung cell lines to radiation *in vitro*, and is less toxic clinically (639). Because Pt is not redox active, the mechanism of radiosensitization is most likely related to Pt binding to cellular macromolecules. Cisplatin cytotoxicity is related to its ability to produce DNA-DNA crosslinks. However, the mechanism of interaction with radiation has not yet been established.

Metal complexes of electron-affinic nitroheterocyclics have been synthesized, with the rationale that the ability of metals to bind to DNA will localize the electron-affinic agents to DNA and therefore improve their efficacy as hypoxic cell radiosensitizers (641, 642). Rh(II) complexes appear to be particularly effective.

**2.10.2 Thiols and Miscellaneous Compounds.** Although the best radioprotectors are thiols, several thiols, including isocysteine, 13-homocysteine, and D-penicillamine, have been reported to be radiosensitizers (643). Thioglycol and thioglycolic acid also cause sensitization (644). Thiamine diphosphate (645), riboflavin (646), and menadiol sodium phosphate (Synkovit) (647) act as sensitizers in animals. Demecolcine sensitizes mice when administered 12 h before irradiation, but is radioprotective when given 48 h before irradiation (648). Sensitization of mice or rats has also been reported for pentobarbital (649), nalorphine (650), butanone peroxide (651), methylhydrazine (652), and cupric salts (653).

**2.10.3 Bacterial Sensitizers.** Several compounds were found to sensitize bacterial cells to radiation in a series of studies in the 1960s in which the mechanism of sensitization is not clear. These include hadacidin (654), chloral hydrate and other halides (655), quaternary heterocycic salts, including phthalanilides, phenaziniums, and isoindoliniums (656), methylhydrazine (652), methylglyoxal (657), tetracyclines (658), and irradiated cupric salts (659).

# 3 SUMMARY AND PROSPECTS FOR FUTURE DEVELOPMENT OF CLINICAL RADIATION MODIFIERS

A variety of approaches have been developed for diminishing the effects of radiation on normal tissues or enhancing tumor cell killing by ionizing radiation. Nevertheless, it is not yet clear that these strategies will provide a therapeutic gain in radiation therapy. Results obtained with model systems do not always apply to more complicated biological systems. Many of these compounds have multiple pharmacological actions, which are sometimes antagonistic. Most often, agents have been tested clinically at suboptimal doses because of limiting toxicities, and the results have been inconclusive. The possibility that radioprotectors could protect the tumor makes it difficult to test radioprotectors in the clinic. Unless the rationale for selective radioprotection of normal tissues or radiosensitization of the tumor is foolproof, there is a risk in clinical testing because the clinical impact is often not known for several years. Nevertheless, several clinical trials are in progress, mostly with sensitizers, and substantial new information is expected from these trials within the next few years.

Several problems make it particularly difficult to develop clinical strategies for modification of radiation therapy. One problem is selective and effective drug delivery. Many of these compounds are metabolized or chemically altered before they reach the target cells.

Electron-affinic hypoxic cell radiosensitizers or bioreductive drugs that are excellent *in vitro* can be metabolically inactivated before they reach the hypoxic tumor cells. Reducing agents that might be excellent protectors *in vitro* can be inactivated by oxidation *in vivo*. Tumor vasculature is chaotic, creating problems in drug delivery (660). The intermittent vascular occlusion that creates problems in drug delivery is the same phenomenon that creates the hypoxic areas that need to be sensitized.

Selective drug delivery to the tumor is a difficult problem. Boron neutron-capture therapy, for example, depends entirely on selective uptake of the sensitizer in the tumor, and animal studies have suggested that this can be achieved, although the clinical results have not been as impressive. It is not enough to deliver more sensitizer to the tumor; it needs to be delivered to all the target cells in the tumor. The exponential relationship between dose and cell killing means that an equal increment in radiation dose is required for each log of cell kill. It is therefore of little use to sensitize 90% or 99% of the cells; this only decreases the dose necessary to kill the first log or two of cells, but has no impact on the remaining $10^7$ or $10^8$ cells. The situation in normal tissues is the reverse. It is not enough to protect some of the cells in the normal tissue because tissue damage can occur even if a small subset of these cells are killed, if their function is vital. In the case of early boron neutron-capture trials, it appears that accumulation of the sensitizer in the normal tissue vasculature negated the potential therapeutic gain that was expected from lower overall sensitizer concentrations in the normal tissue.

Despite the problems, the outlook for radiation modifiers is good. The radiation dose-response relationships for both tumor cure and normal tissue damage are steep, so that the challenge for modifiers is not too great; even a 20% shift in either response curve should have a large impact on clinical results. The classical view of the mechanism of action of ionizing radiation is that all of the biological effects can be accounted for by cell killing that results from clustered DNA lesions. That particular mechanism does not lend itself easily to differential modification of tumors and normal tissues, given the same radiation chemistry in both cases, with the exception that hypoxic tumor cells can be targeted with specific strategies. However, it is now recognized that ionizing radiation produces subtle changes in cell function, in addition to classical reproductive cell death, and that these other effects of radiation may be modifiable with agents that do not affect classical reproductive cell death (661).

Cytokine cascades persist for months after radiation, possibly contributing to tissue fibrosis, as a pathogenic mechanism (662, 663). Dittman et al. (664) suggested that radiation-induced differentiation of progenitor fibroblasts could be related to the development of tissue fibrosis, and they found that the Bowman-Birk proteinase inhibitor can inhibit radiation-induced premature differentiation of these cells. Delayed mutagenesis and cell death that occurs many cell divisions postirradiation could be attributed to an induced hypermutability (665). There are indications that mutagenesis may be inhibitable with strategies that would not affect the formation of clustered DNA lesions that are thought to be responsible for classical reproductive cell death (666). Apoptosis is a regulated mechanism of cell death that occurs in certain cell types more than others, and can be modified with agents that do not affect classical reproductive cell death (324, 667, 668). Vascular effects of radiation may be mediated in part by bioactive products that can be specifically antagonized (669–671).

The implication of these new findings is that new types of modifiers can be developed that target a specific aspect of the mechanism of action of ionizing radiation. If the mechanism is more important to normal tissue than to the tumor, or vice versa, the modifiers would not have to be delivered differentially and selectivity would be achieved by mechanistic differences.

## REFERENCES

1. W. M. Dale, L. H. Gray, and W. J. Meredith, *Philos. Trans. R. Soc. Lond.*, **242A**, 33 (1949).
2. R. Latarjet and E. Ephrati, *C. R. Soc. Biol.*, **142**, 497 (1948).
3. A. Herve and Z. M. Bacq, *C. R. Soc. Biol.*, **143**, 881 (1949).
4. H. M. Patt, E. B. Tyree, R. L. Straube, and D. E. Smith, *Science*, **110**, 213 (1949).
5. R. H. Mole, L. St. L. Philpot, and C. R. V. Hodges, *Nature*, **166**, 515 (1950).
6. Z. M. Bacq, A. Herve, L. Lecomte, P. Fischer, J. Blavier, G. Dechamps, H. LeBihan, and P. Rayet, *Arch. Int. Physiol.*, **59**, 442 (1951).
7. C. von Sonntag, *The Chemical Basis of Radiation Biology*, Taylor & Francis, London, 1987.
8. D. J. Pizzarello and R. L. Witcofski, *Basic Radiation Biology*, Lea and Febiger, Philadelphia, 1967.
9. A. P. Casarett, *Radiation Biology*, Prentice-Hall, Englewood Cliffs, NJ, 1968.
10. S. Okada, Ed., *Radiation Biochemistry*, Vol. **1**, Academic Press, New York, 1969.
11. H. Dertinger and H. Jung, *Molecular Radiation Biology*, Springer-Verlag, Berlin/New York, 1970.
12. P. Alexander and A. Charlesby in Z. M. Bacq and P. Alexander, Eds., *Radiation Symposium*, Butterworths, London, 1955, p. 49.
13. T. Henriksen, *Radiat. Res.*, **27**, 694 (1966).
14. S. Okada, Ed., *Radiation Biochemistry*, Vol. **1**, Academic Press, New York, 1969, p. 70.
15. Z. M. Bacq and P. Alexander, Eds., *Fundamentals of Radiobiology*, 2nd ed., Pergamon, Oxford, UK, 1961.
16. J. F. Thomson, *Radiation Protection in Mammals*, Reinhold, New York, 1962.
17. E. A. Bump and K. Malaker, *Radioprotectors: Chemical, Biological and Clinical Perspectives*, CRC Press, Boca Raton, FL, 1998.
18. H. L. Andrews, D. C. Peterson, and D. P. Jacobus, *Radiat. Res.*, **23**, 13 (1964).
19. Z. M. Bacq and P. Alexander, Eds., *Radiobiology Symposium*, Academic Press, New York, 1955.
20. W. D. Fischer, N. G. Anderson, and K. M. Wilbur, *Exp. Cell Res.*, **18**, 48 (1959).
21. G. Hotz, *Z. Naturforsch.*, **21b**, 148 (1966).
22. G. I. Gasanov, *Izv. Akad. Nauk Azerb, SSR, Ser. Biol. Med. Nauk*, **110** (1966); *Chem. Abstr.*, **66**, 5278 (1967).
23. K. S. Burdin, I. M. Parkomenko, Y. M. Petrusevich, and S. V. Shestakova, *Tr. Mosk. Obshch. Ispyt. Prir. Otd. Biol.*, **16**, 19 (1966); *Chem. Abstr.*, **66**, 62465 (1967).
24. J. F. Duplan and J. Fuhrer, *C. R. Soc. Biol.*, **160**, 1142 (1966).
25. Z. M. Bacq, *Sulfur Containing Radioprotective Agents*, Pergamon, Oxford, UK, 1975.

26. V. S. Balubukha, Ed., *Chemical Protection of the Body Against Ionizing Radiation*, Macmillan, New York, 1964.

27. O. F. Nygaard and M. G. Simic, Eds., *Radioprotectors and Anticarcinogens*, Academic Press, New York, 1983.

28. J. C. Livesey, D. J. Reed, and L. F. Adamson, *Radiation-Protective Drugs and Their Reaction Mechanisms*, Noyes, Park Ridge, NJ, 1985.

29. R. Huber and E. Spode, *Biologisch-Chemischer Strahlenschutz*, Akademie-Verlag, Berlin, 1963.

30. H. J. Melching and C. Streffer, *Prog. Drug Res.*, **9**, 11 (1966).

31. R. R. Overman and S. J. Jackson, *Annu. Rev. Med.*, **18**, 71 (1967).

32. E. F. Romantsev, *Radiation and Chemical Protection*, Atomizdat, Moscow, 1968.

33. W. O. Foye, *J. Pharm. Sci.*, **58**, 283 (1969).

34. D. L. Klayman and E. S. Copeland in E. J. Ariens, Ed., Drug Design, Vol. **VI**, Academic Press, New York, 1975.

35. V. G. Yashunskii and V. Yu. Kovtun, *Ospekhi Khim.*, **54**, 126 (1985).

36. E. A. Bump and J. M. Brown, *Pharmacol. Ther.*, **47**, 117–136 (1990).

37. R. Paoletti and R. Vertua, Eds., *Progress in Biochemical Pharmacology*, Vol. **1**, Butterworths, Washington, DC, 1965.

38. H. L. Moroson and M. Quintiliani, Eds., *Radiation Protection and Sensitization*, Barnes and Noble, New York, 1970.

39. Int. Confs. on Chemical Modifiers of Cancer Treatment, *Int. J. Radiat. Oncol. Biol. Phys.*, **8**, 323–815 (1982); *Int. J. Radiat. Oncol. Biol. Phys.*, **10**, 1161–1828 (1984); *Int. J. Radiat. Oncol. Biol. Phys.*, **12**, 1019–1551 (1986); *Int. J. Radiat. Oncol. Biol. Phys.*, **16**, 885–1353 (1989); *Int. J. Radiat. Oncol. Biol. Phys.*, **22**, 391–828 (1992); *Int. J. Radiat. Oncol. Biol. Phys.*, **29**, 229–642 (1994).

40. W. O. Foye, *Annu. Rep. Med. Chem.*, 324 (1966); *Annu. Rep. Med. Chem.*, 330 (1967).

41. E. R. Atkinson, *Annu. Rep. Med. Chem.*, 327 (1968); *Annu. Rep. Med. Chem.*, 346 (1970).

42. L. Giambarresi and A. J. Jacobs in L. J. Conklin and R. I. Walker, Eds., *Military Radiobiology*, Academic Press, Orlando, FL, 1987, pp. 265–301.

43. D. G. Doherty, W. T. Burnett Jr., and R. Shapira, *Radiat. Res.*, **7**, 13 (1957).

44. H. Langendorff and R. Koch, *Strahlentherapie*, **99**, 567 (1956).

45. B. Holmberg and B. Sorbo, *Nature*, **183**, 832 (1959).

46. S. Akerfeldt, *Acta Chem. Scand.*, **13**, 1479 (1959).

47. W. O. Foye, J. Mickles, R. N. Duvall, and J. R. Marshall, *J. Med. Chem.*, **6**, 509 (1963).

48. B. Sorbo, *Arch. Biochem. Biophys.*, **98**, 342 (1962).

49. A. F. Ferris, O. L. Salerni, and B. A. Schutz, *J. Med. Chem.*, **9**, 391 (1966).

50. J. R. Piper, C. R. Stringfellow Jr., and T. P. Johnston, *J. Med. Chem.*, **9**, 563 (1966).

51. D. W. van Bekkum and H. T. M. Nieuwerkerk, *Int. J. Radiat. Biol.*, **7**, 473 (1963).

52. D. L. Klayman, M. M. Grenan, and D. P. Jacobus, *J. Med. Chem.*, **12**, 510 (1969).

53. F. I. Carroll, J. D. White, and M. E. Wall, *J. Org. Chem.*, **28**, 1240 (1963).

54. G. W. Stacy, B. F. Barnett, and P. L. Strong, *J. Org. Chem.*, **30**, 592 (1965).

55. J. R. Piper, C. R. Stringfellow Jr., and T. P. Johnston, *J. Med. Chem.*, **9**, 911 (1966).

56. F. I. Carroll, J. D. White, and M. E. Wall, *J. Org. Chem.*, **28**, 1236 (1963).

57. K. Tonchev, Nauch, Tr. Vissh. Med. Inst., Sofia, **39**, 143 (1960); *Chem. Abstr.*, **55**, 19007 (1961).

58. L. I. Tank, *Med. Radiol.*, **5**, 34 (1960).

59. H. Langendorff, M. Langendorff, and R. Koch, *Strahlentherapie*, **107**, 121 (1958).

60. D. P. Jacobus and T. R. Sweeney, private communication.

61. H. Irie, *Strahlentherapie*, **110**, 456 (1959).

62. R. W. Hart, R. E. Gibson, J. D. Chapman, A. P. Reuvers, B. K. Sinha, R. K. Griffith, and D. T. Witiak, *J. Med. Chem.*, **18**, 323 (1975).

63. D. G. Doherty and R. Shapira, *J. Org. Chem.*, **28**, 1339 (1963).

64. M. Carinack, C. L. Kelley, S. D. Harrison Jr., and K. P. DuBois, *J. Med. Chem.*, **15**, 600 (1972).

65. R. L. Straube and H. M. Patt, *Proc. Soc. Exp. Biol. Med.*, **84**, 702 (1953).

66. J. A. Aguilera, G. L. Newton, R. C. Fahey, and J. F. Ward, *Radiat. Res.*, **130**, 194 (1992).

67. T. P. Johnston and C. R. Stringfellow Jr., *J. Med. Chem.*, **9**, 921 (1966).

68. G. R. Handrick and E. R. Atkinson, *J. Med. Chem.*, **9**, 558 (1966).

69. F. D. Carroll, M. B. Gopinathan, and A. Philip, *J. Med. Chem.*, **33**, 2501 (1990).

70. S. J. Arbusov, *Pharmazie*, **14**, 132 (1959).

71. S. Akerfeldt, *Acta Radiol. Ther. Phys. Biol.*, **1**, 465 (1963).

72. S. Akerfeldt, C. Ronnback, and A. Nelson, *Radiat. Res.*, **31**, 850 (1967).

73. J. R. Piper, C. R. Stringfellow Jr., R. D. Elliott, and T. P. Johnston, *J. Med. Chem.*, **12**, 236 (1969).

74. R. D. Westland, M. L. Mouk, J. L. Holmes, R. A. Cooley Jr., J. S. Hong, and M. M. Grenan, *J. Med. Chem.*, **15**, 968 (1972).

75. R. D. Westland, M. M. Merz, S. M. Alexander, L. S. Newton, L. Bauer, T. T. Conway, J. M. Barton, K. K. Khullar, and P. B. Devdhar, *J. Med. Chem.*, **15**, 1313 (1972).

76. W. O. Foye, Y. H. Lowe, and J. J. Lanzillo, *J. Pharm. Sci.*, **65**, 1247 (1976).

77. B. Shapiro, G. Kollmann, and D. Martin, *Radiat. Res.*, **44**, 421 (1970).

78. S. Akerfeldt, C. Ronnback, M. Hellström, and A. Nelson, *Radiat. Res.*, **35**, 61 (1968).

79. W. O. Foye, R. N. Duvall, and J. Mickles, *J. Pharm. Sci.*, **51**, 168 (1962).

80. D. O. Doherty and W. T. Burnett Jr., *Proc. Soc. Exp. Biol. Med.*, **89**, 312 (1955).

81. E. J. Jezequel, H. Frossard, M. Fatome, R. Perles, and P. Poutrain, *C. R. Acad. Sci., Ser. D*, **272**, 2826 (1971).

82. T. Taguchi, O. Komori, and M. Kojima, *Yakugaku Zasshi*, **81**, 1233 (1961).

83. L. Bauer and K. Sandberg, *J. Med. Chem.*, **7**, 766 (1964).

84. J. M. Barton and L. Bauer, *Can. J. Chem.*, **47**, 1233 (1969).

85. S. Robev, I. Baev, and N. Panov, *C. R. Acad. Bulg. Sci.*, **19**, 1035 (1966).

86. S. Robev, I. Baev, and N. Panov, *C. R. Acad. Bulg. Sci.*, **19**, 1143 (1966).

87. W. O. Foye and D. H. Kay, *J. Pharm. Sci.*, **57**, 345 (1968).

88. W. O. Foye, D. H. Kay, and P. R. Amin, *J. Pharm. Sci.*, **57**, 1793 (1968).

89. E. R. Atkinson, G. R. Handrick, R. J. Bruni, and F. E. Granchelli, *J. Med. Chem.*, **8**, 29 (1965).

90. F. L. Rose and A. L. Walpole, *Prog. Biochem. Pharmacol.*, **1**, 432 (1965).

91. E. E. Schwartz and B. Shapiro, *Radiat. Res.*, **13**, 780 (1960).

92. O. Vos, L. Budke, and A. J. Vergroesen, *Int. J. Radiat. Biol.*, **5**, 543 (1962).

93. P. Fischer and M. Goutier-Pirotte, *Arch. Int. Physiol.*, **62**, 76 (1954).

94. A. V. Titov, D. A. Golubentsov, and V. V. Mordukhovich, *Radiobiologiya*, **10**, 606 (1970); *Chem. Abstr.*, **73**, 127508 (1970).

95. B. Shapiro, E. E. Schwartz, and O. Kollman, *Radiat. Res.*, **18**, 17 (1963).

96. L. Field and H. K. Kim, *J. Med. Chem.*, **9**, 397 (1966); P. K. Srivastava, L. Field, and M. M. Grenan, *J. Med. Chem.*, **18**, 798 (1975).

97. L. Field, H. K. Kim, and M. Bellas, *J. Med. Chem.*, **10**, 1166 (1967).

98. L. Field and J. D. Buckman, *J. Org. Chem.*, **32**, 3467 (1967).

99. L. Field and Y. H. Khim, *J. Med. Chem.*, **15**, 312 (1972).

100. L. Field and J. D. Buckman, *J. Org. Chem.*, **33**, 3865 (1968).

101. D. L. Klayman and O. W. A. Milne, *J. Org. Chem.*, **31**, 2349 (1966).

102. W. O. Foye, A. M. Hebb, and J. Mickles, *J. Pharm. Sci.*, **56**, 292 (1967).

103. L. Field, A. Ferretti, R. R. Crenshaw, and T. C. Owen, *J. Med. Chem.*, **7**, 39 (1964).

104. R. I. H. Wang and A. T. Hasegawa, private communication.

105. R. A. Salvador, C. Davison, and P. K. Smith, *J. Pharmacol. Exp. Ther.*, **121**, 258 (1957).

106. A. Kaluszyner, P. Czerniak, and E. D. Bergmann, *Radiat. Res.*, **14**, 23 (1961).

107. R. D. Westland, R. A. Cooley Jr., J. L. Holmes, J. S. Hong, M. L. Lin, M. L. Zwiesler, and M. M. Grenan, *J. Med. Chem.*, **16**, 319 (1973).

108. R. Riemschneider, *Z. Naturforsch.*, **B16**, 75 (1961).

109. P. S. Farmer, C.-C. Leung, and E. M. K. Lui, *J. Med. Chem.*, **16**, 411 (1973).

110. R. D. Westland, M. H. Lin, R. A. Cooley Jr., M. L. Zwiesler, and M. M. Grenan, *J. Med. Chem.*, **16**, 328 (1973).

111. S. Shapira, D. O. Doherty, and W. T. Burnett Jr., *Radiat. Res.*, **7**, 22 (1957).

112. J. X. Khym, R. Shapira, and D. O. Doherty, *J. Am. Chem. Soc.*, **79**, 5663 (1957).

113. J. Doull, V. Plzak, and S. Brois, University of Chicago USAF Radiation Lab, Status Report, No. 2, August 1, 1961.

114. D. W. van Bekkum, *Acta Physiol. Pharmacol. Neerl.*, **4**, 508 (1956).

115. W. O. Foye and J. Mickles, *J. Med. Pharm. Chem.*, **5**, 846 (1962).

116. V. Palma, O. Galli, S. Garrattini, R. Paoletti, and R. Vertua, *Arzneim.-Forsch.*, **11**, 1034 (1961).

117. T. Stoichev, *Izv. Inst. Fiziol. Bulg. Akad. Nauk.*, **10**, 149 (1966); *Chem. Abstr.*, **67**, 29697 (1967).

118. V. O. Yakovlev and V. S. Mashtakov, *Khim. Zashch. Org. Ioniz. Izluch.*, 72 (1960); *Chem. Abstr.*, **55**, 27649 (1961).

119. R. I. H. Wang, W. Dooley Jr., W. O. Foye, and J. Mickles, *J. Med. Chem.*, **9**, 394 (1966).

120. W. O. Foye, R. S. F. Chu, K. A. Shah, and W. H. Parsons, *J. Pharm. Sci.*, **60**, 1839 (1971).

121. M. J. Ashwood-Smith and A. D. Smith, *Int. J. Radiat. Biol.*, **1**, 196 (1959).

122. A. A. Gorodetskii, R. O. Dubenko, P. S. Pel'kis, and E. Z. Ryabova, *Patog., Eksp. Profil. Ter. Luchevykh Porazhenii (Moscow), Sb.*, 179 (1964); *Chem. Abstr.*, **64**, 989 (1966).

123. F. Cugurra and E. Balestra, *Prog. Biochem. Pharmacol.*, **1**, 507 (1965).

124. A. Hollaender and C. O. Doudney, *Radiobiology Symposium, Liege, 1954*, Butterworths, London, 1955, p. 112.

125. P. Alexander, Z. M. Bacq, S. F. Cousens, M. Fox, A. Herve, and J. Lazar, *Radiat. Res.*, **2**, 392 (1955).

126. R. Koch in O. C. DeHevesy, Ed., *Advances in Radiobiology*, Oliver & Boyd, Edinburgh, 1957, p. 170.

127. M. J. Ashwood-Smith, *Int. J. Radiat. Biol.*, **3**, 41 (1961).

128. R. Brinkman, H. B. Lamberts, J. Wadel, and J. Zuideveld, *Int. J. Radiat. Biol.*, **3**, 205 (1961).

129. T. Zebro, *Acta Med. Pol.*, **7**, 83 (1966).

130. B. Hansen and B. Sorbo, *Acta Radiol.*, **56**, 141 (1961).

131. W. O. Foye and J. M. Kauffman, *J. Pharm. Sci.*, **57**, 1611 (1968).

132. W. O. Foye, Y. J. Cho, and K. H. Oh, *J. Pharm. Sci.*, **59**, 114 (1970).

133. W. O. Foye, R. W. Jones, P. K. Goshal, and B. Almassian, *J. Med. Chem.*, **30**, 57 (1987).

134. W. O. Foye, R. W. Jones, and P. K. Ghoshal, *J. Pharm. Sci.*, **76**, 809 (1987).

135. W. H. Chapman and E. P. Cronkite, *Proc. Soc. Exp. Biol. Med.*, **75**, 318 (1950).

136. Z. M. Bacq, *Acta Radiol.*, **41**, 47 (1954).

137. S.-H. Chu and H. O. Mautner, *J. Org. Chem.*, **27**, 2899 (1962).

138. Z. M. Hollo and S. Zlatarov, *Naturwissenschaften*, **47**, 328 (1960).

139. A. Breccia, R. Badiello, A. Trenta, and M. Mattii, *Radiat. Res.*, **38**, 483 (1969).

140. D. L. Klayman in D. L. Klayman and W. H. H. Gunther, Eds., *Organic Selenium Compounds; Their Chemistry and Biology*, John Wiley & Sons, New York, 1973, p. 727.

141. J. Schubert and J. F. Markley, *Nature*, **197**, 399 (1963).

142. Z. M. Bacq and A. Herve, *Br. J. Radiol.*, **24**, 618 (1951).

143. E. Boyland and E. Gallico, *Br. J. Cancer*, **6**, 160 (1952).

144. R. N. Feinstein and S. Berliner, *Science*, **125**, 936 (1957).

145. Z. M. Bacq, P. Fischer, and A. Herve, *Arch. Int. Physiol.*, **66**, 75 (1958).

146. H. Sigel and H. Erlenmeyer, *Helv. Chim. Acta*, **49**, 1266 (1966).

147. J. R. J. Sorenson, *J. Med. Chem.*, **27**, 1747 (1984).

148. W. O. Foye and K. Ghosh, *Trends Heterocyclic Chem.*, **1**, 9 (1991).

149. W. O. Foye and J. Mickles, *Prog. Biochem. Pharmacol.*, **1**, 152 (1965).

150. Z. M. Bacq, A. Herve, and P. Fischer, *Bull. Acad. Roy. Med. Belg.*, **18**, 226 (1953).

151. C. Corradi, R. Pagletti, and E. Pozza, *Atti Soc. Lomb. Sci. Med. Biol.*, **13**, 80 (1958); *Chem. Abstr.*, **52**, 17525 (1958).

152. O. J. Smirnova, *Radiobiologiya*, **2**, 378 (1962).

153. N. Kasugai, *J. Pharm. Soc. Jpn.*, **84**, 1152 (1964).

154. G. L. Floersheim, N. Chiodetti, and A. Biere, *Br. J. Radiol.*, **61**, 501 (1988).

155. S. S. Hasan, S. Mazumdar, P. M. Singh, and S. N. Pandey, *Nuklearmedizin*, **20**, 82 (1981).

156. A. A. Gorodetskii, V. A. Baraboi, and V. P. Chernetskii, *Vopr. Biofiz. Mekh. Deistoya Ioniz. Radiats.*, 159 (1964); *Chem. Abstr.*, **63**, 18617 (1965).

157. N. P. Buu-Hoi and N. D. Xuong, *J. Org. Chem.*, **26**, 2401 (1961).

158. V. S. Nesterenko, S. I. Suminov, Y. A. Gurvich, E. L. Styskin, and S. T. Kumok, *Farmakol. Toksikol. (Moscow)*, **37**, 443 (1974); *Chem. Abstr.*, **81**, 145897 (1974).

159. J. H. Barnes, *Nature*, **205**, 816 (1965).

160. J. R. Piper, L. M. Rose, T. P. Johnston, and M. M. Grenan, *J. Med. Chem.*, **18**, 803 (1975).

161. R. Rinaldi, Y. Bernard, and M. Guilhermet, *Compt. Rend.*, **261**, 570 (1965).

162. T. J. Haley, A. M. Flesher, and L. Mavis, *Arch. Int. Pharmacodyn.*, **138**, 133 (1962).

163. T. J. Haley, W. E. Trumbull, and J. A. Cannon, *Prog. Biochem. Pharmacol.*, **1**, 359 (1965).

164. S. M. Hahn, Z. Tochner, C. M. Krishna, et al., *Cancer Res.*, **52**, 1750 (1992).

165. K. Stratton and E. M. Davis, *Int. J. Radiat. Biol.*, **5**, 105 (1962).

166. E. Campaigne and P. K. Nargund, *J. Med. Chem.*, **7**, 132 (1964).

167. I. G. Krasnykh, V. S. Shaskkov, O. Y. Magidson, E. S. Golovchinskaya, and K. A. Chkhikvadze, *Farmakol. Toksikol. (Moscow)*, **24**, 572 (1961).

168. G. N. Krutovshikh, M. B. Kolesova, A. M. Rusanov, L. P. Vartanyan, and M. G. Shagoyon, *Khim.-Farm. Zh.*, **9**, 21 (1975); *Chem. Abstr.*, **83**, 108312 (1975).

169. H. Frossard, M. Fatome, R. Royer, J. P. Lechartier, J. Guillaumel, and P. Demerseman, *Chim. Ther.*, **8**, 32 (1973).

170. J. Bitoun, H. Blancou, M. Fatome, M. Flander, H. Frossard, R. Granger, P. Joyeux, R. Perles, and Y. Robbe, *Tray. Soc. Pharm. Mont. Fr.*, **33**, 147 (1973); *Chem. Abstr.*, **80**, 127 (1974).

171. G. Grassy, A. Terol, A. Belly, Y. Robbe, J. P. Chapat, R. Granger, M. Fatome, and L. Andrieu, *Eur. J. Med. Chem.-Chim. Ther.*, **10**, 14 (1975).

172. S. S. Ansher, P. Dolan, and E. Bueding, *Hepatology*, **3**, 932 (1983).

173. V. G. Vladimirov, Y. E. Strelnikov, I. A. Belenkaya, Y. L. Kostyukovskii, and N. S. Tsepova, *Radiobiologiya*, **14**, 766 (1974).

174. Y. Takagi, M. Shikata, and S. Akaboshi, *Radiat. Res.*, **15**, 116 (1974).

175. J. H. Barnes, M. Fatome, G. F. Esslemont, and C. E. L. Jones, *Eur. J. Med. Chem.-Chim. Ther.*, **18**, 515 (1983).

176. C. van der Meer and D. W. van Bekkum, *Int. J. Radiat. Biol.*, **4**, 105 (1961).

177. Z. Supek, M. Randic, and Z. Lovasen, *Int. J. Radiat. Biol.*, **4**, 111 (1961).

178. C. Streffer and G. Konermann in Ref. 38, p. 369.

179. S. Kobayashi, W. Nakamura, and H. Eto, *Int. J. Radiat. Biol.*, **11**, 505 (1966).

180. I. G. Krasnykh, P. G. Zherebchenko, V. S. Murashova, N. N. Suvorov, N. R. Sorokina, and V. S. Shashkov, *Radiobiologiya*, **2**, 156 (1962).

181. L. Andrieu, M. Fatome, R. Granger, Y. Robbe, and A. Terol, *Eur. J. Med. Chem.-Chim. Ther.*, **9**, 453 (1974).

182. R. T. Blickenstaff, S. M. Brandstadter, S. Reddy, and R. Witt, *J. Pharm. Sci.*, **83**, 216 (1994).

183. I. Baev and D. Benova, *Roentgenol. Radiol.*, **14**, 50 (1975).

184. M. Langendorff, H.-J. Melching, H. Langendorff, R. Koch, and R. Jacques, *Strahlentherapie*, **104**, 338 (1957).

185. A. Locker and H. Ellegast, *Strahlentherapie*, **129**, 273 (1966).

186. H. J. Melching and M. Langendorff, *Naturwissenschaften*, **44**, 377 (1957).

187. Z. M. Bacq and S. Liebecq-Hutter, *J. Physiol.*, **145**, S2p (1959).

188. H. Levan and D. L. Hebron, *J. Pharm. Sci.*, **57**, 1033 (1968).

189. G. Gansser, B. Marcot, C. Viel, M. Fatome, and J. D. Laval, *Ann. Pharm. Fr.*, **41**, 465 (1983).

190. J. L. Gray, E. J. Moulden, J. T. Tew, and H. Jensen, *Proc. Soc. Exp. Biol. Med.*, **79**, 384 (1952).

191. C. van der Meer, D. W. van Bekkum, and J. A. Cohen, Proceedings of the United Nations International Conference on the Peaceful Uses of Atomic Energy, 2nd ed., Vol. 23, 1958, p. 42.

192. A. H. Staib and K. Effier, *Naturwissenschaften*, **53**, 583 (1966).

193. J. B. Storer and J. M. Coon, *Proc. Soc. Exp. Biol. Med.*, **74**, 202 (1950); E. F. Romantsev and N. I. Bicheikina, *Radiobiologiya*, **4**, 743 (1964).

194. J. B. Storer, *Radiat. Res.*, **47**, 537 (1971).

195. L. T. Blouin and R. R. Overman, *Radiat. Res.*, **16**, 699 (1962).

196. P. R. Salerno and H. L. Friedell, *Radiat. Res.*, **1**, 559 (1954).

197. R. T. Blickenstaff, S. Reddy, and R. Witt, *Bioorg. Med. Chem.*, **2**, 1057 (1994).

198. R. T. Blickenstaff, S. M. Brandstadter, E. Foster, S. Reddy, and R. Witt, *Ind. Eng. Chem. Res.*, 32 (1993).

199. L. J. Cole and S. R. Gospe, *Radiat. Res.*, **11**, 438 (1959).

200. H. M. Patt, R. L. Straub, E. B. Tyree, M. N. Swift, and D. E. Smith, *Am. J. Physiol.*, **159**, 269 (1949).

201. W. W. Smith and I. M. Alderman, *Radiat. Res.*, **17**, 594 (1962).

202. K. Flemming and M. Langendorff, *Strahlentherapie*, **128**, 109 (1965).

203. W. E. Rothe and M. M. Grenan, *Science*, **133**, 888 (1961).

204. J. Cheymol, J. Louw, P. Chabrier, M. Adolphe, J. Seyden, and M. Selim, *Bull. Acad. Natl. Med.*, **144**, 681 (1960).

205. G. Arnaud, H. Frossard, J.-P. Gabriel, O. Bichon, J.-M. Saucier, J. Bourdais, and M. Guillerm, *Prog. Biochem. Pharmacol.*, **1**, 467 (1965).

206. H. Laborit, M. Dana, and P. Carlo, *Prog. Biochem. Pharmacol.*, **1**, 574 (1965).

207. D. P. Doolittle and J. Watson, *Int. J. Radiat. Biol.*, **11**, 389 (1966).

208. B. W. Volk, K. F. Wellmann, and S. S. Lazarus, *Am. J. Pathol.*, **51**, 207 (1967).

209. H. Blondal, *Br. J. Radiol.*, **30**, 219 (1957).

210. A. P. Duplischeva, K. K. Ivanov, and N. G. Silinova, *Radiobiologiya*, **6**, 318 (1966).

211. K. Flemming, *Biophys. Probl. Strahlenwirkung Jahrestag Tagungsber.*, **1965**, 138 (1966); *Chem. Abstr.*, **67**, 50865 (1967).

212. R. Wilson, G. D. Ledney, and T. Matsuzawa, *Prog. Biochem. Pharmacol.*, **1**, 622 (1965).

213. E. J. Ainsworth and F. A. Mitchell, *Nature*, **210**, 321 (1966).

214. H. Langendorff, H.-J. Melching, and H. Rosler, *Strahlentherapie*, **113**, 603 (1960).

215. H.-J. Melching, *Strahlentherapie*, **120**, 34 (1963).

216. R. Koch, *Acta Chem. Scand.*, **12**, 1873 (1958).

217. H. Langendorff, J.-J. Melching, and C. Streffer, *Strahlentherapie*, **118**, 341 (1962).

218. R. Wagner and E. C. Silverman, *Int. J. Radiat. Biol.*, **12**, 101 (1967).

219. M. A. Tumanyan, N. G. Sinilova, A. P. Duplischeva, and K. K. Ivanov, *Radiobiologiya*, **6**, 712 (1966).

220. M. E. Roberts, C. Jones, and M. A. Gerving, *Life Sci.*, **4**, 1913 (1965).

221. T. P. Pantev, N. V. Bokova, and I. T. Nikolov, *Eksp. Med. Morfol.*, **12**, 186 (1973).

222. T. Aleksandrov, I. Nikolov, D. Krustanov, and T. Tinev, *Roentgenol. Radiol.*, **5**, 45 (1966).

223. V. A. Barbaoi, *Biokhim. Progr. Tekhnol. Chai. Poizood. Akad. Nauk SSSR, Inst. Biokhim.*, 357 (1966); *Chem. Abstr.*, **66**, 2490 (1967).

224. J. M. Gazave, D. C. Modigliani, and G. Ligny, *Prog. Biochem. Pharmacol.*, **1**, 554 (1965).

225. V. Brueckner, *Strahlentherapie*, **145**, 732 (1973).

226. H.-F. Wang, X. Li, Y.-M. Chen, L. Yuan, and W. O. Foye, *J. Ethnopharmacol.*, **34**, 215 (1991).

227. X. Wang, Ph.D. Thesis, Massachusetts College of Pharmacy and Allied Health Sciences, 1993.

228. C. G. Overberger, H. Ringsdorf, and B. Avchen, *J. Med. Chem.*, **8**, 862 (1965).

229. J. Barnes, G. Esslemont, and P. Holt, *Makromol. Chem.*, **176**, 275 (1975).

230. O. V. Semina, A. G. Konnoplyannikov, and A. M. Pomerennyi, *Radiobiologiya*, **14**, 689 (1974).

231. N. P. Kharitanovich, G. A. Dokshina, V. A. Pegel, E. I. Yartsev, and V. G. Yashunskii, *Radiobiologiya*, **15**, 104 (1975).

232. V. Brueckner, *Umschau*, **73**, 607 (1973).

233. V. V. Zhamensky, A. D. Efremov, V. M. Bystrova, and T. P. Vasil'eva, *Khim. Farm. Zhur.*, **20**, 942 (1986).

234. J. P. Fernandez, Y. Robbe, J. P. Chapat, G. Cassanas, H. Sentenac-Roumanou, et al., *Farmaco Ed. Sci.*, **41**, 41 (1986).

235. T. A. Alpatova, V. I. Kulinskii, and V. G. Yashunskii, *Pharm. Chem. J. USSR*, **17**, 543 (1984).

236. S. A. Bol'shakova, Y. A. Goloshchapova, T. N. Tuzhilkova, A. D. Grebenyuk, and V. M. Myakishev, *Pharm. Chem. J. USSR*, **17**, 474 (1984).

237. G. Orzesyko and S. Bitny-Szlachto, *Acta Pol. Pharm.*, **43**, 237 (1988).

238. J. P. Fernandez, Y. Robbe, J. P. Chapat, R. Granger, M. Fatome, et al., *Farmaco Ed. Sci.*, **36**, 740 (1981).

239. Z. Li, Y. Wu, and Q. Tang, *Acta Pharm. Sinica*, **16**, 73 (1981).

240. R. T. Blickenstaff, S. M. Brandstadter, S. Reddy, R. Witt, and K. B. Lipkowitz, *J. Pharm. Sci.*, **83**, 219 (1994).

241. R. T. Blickenstaff, S. Reddy, R. Witt, and K. B. Lipkowitz, *Bioorg. Med. Chem.*, **2**, 1363 (1994).

242. P. G. Zherebchenko, G. M. Airapetyan, I. G. Krasnykh, N. N. Suvorov, and A. N. Shevchenko, *Patog. Eksp. Profil. Ter. Luchev. Porazenii (Moscow)*, 193 (1964); *Chem. Abstr.*, **64**, 2394 (1966).

243. H. A. Van der Breuk and D. Jamiesou, *Int. J. Radiat. Biol.*, **4**, 379 (1962).

244. L. Novak, *Bull. Cl. Sci. Acad. Roy. Belg.*, **52**, 633 (1966).

245. E. Y. Graevsky, I. M. Shapiro, M. M. Konstantinova, and N. F. Barakina in R. J. C. Harris, Ed., *The Initial Effects of Ionizing Radiations on Cells, A Symposium*, Academic Press, New York, 1961, p. 237.

246. C. C. Winterbourn, A. V. Peskin, and H. N. Parsons-Mair, *J. Biol. Chem.*, **18**, 1906 (2002).

247. J. Denekamp, A. Rojas, and F. A. Stewart in O. F. Nygaard and M. G. Simic, Eds., *Radio-

*protectors and Anticarcinogens*, Academic Press, New York, 1983, pp. 635–679.

248. J. W. Purdie, E. R. Inhabes, H. Schneider, and J. L. Labelle, *Int. J. Radiat. Biol.*, **43**, 517 (1983).

249. J. C. Livesey, D. J. Reed, and L. F. Adamson, *Radiation-Protective Drugs and Their Reaction Mechanisms*, Noyes, Park Ridge, NJ, 1985, p. 62.

250. P. Alexander and A. Charlesby, *Nature*, **173**, 578 (1954).

251. C. J. Koch in E. A. Bump and K. Malaker, Eds., *Radioprotectors: Chemical, Biological and Clinical Perspectives*, CRC Press, Boca Raton, FL, 1998, pp. 25–52.

252. L. J. Wu, G. Randers-Pehrson, A. Xu, C. A. Waldren, C. R. Geard, Z. Yu, and T. K. Hei, *Proc. Natl. Acad. Sci. USA*, **27**, 4959 (1999).

253. I. I. Sapezhinskii and E. G. Dontsova, *Biofizika*, **12**, 794 (1967).

254. W.-C. Hsin, C.-L. Chang, M.-J. Yao, C.-H. Hua, and H.-C. Chin, *Sci. Sinica (Peking)*, **15**, 211 (1966).

255. E. B. Burlakova, V. D. Gaintseva, L. V. Slepukhina, N. G. Khrapova, and N. M. Emanuel, *Dokl. Akad. Nauk SSSR*, **164**, 394 (1965); *Chem. Abstr.*, **54**, 2394 (1966).

256. R. Roots and S. Okada, *Int. J. Radiat. Biol.*, **21**, 329 (1972).

257. G. Bartosy and W. Leyko, *Int. J. Radiat. Biol.*, **39**, 9 (1981).

258. E. Gryelirslra, A. Bartkowiak, G. Bartosy, and W. Leyko, *Int. J. Radiat. Biol.*, **41**, 473 (1982).

259. L. Eldjarn and A. Pihl, *J. Biol. Chem.*, **225**, 499 (1965).

260. G. Gorin, *Prog. Biochem. Pharmacol.*, **1**, 142 (1965).

261. L. Eldjarn and A. Pihl, *Radiat. Res.*, **9**, 110 (1958).

262. E. A. Dickens and B. Shapiro, *US. Dep. Comm. Off Tech. Sew., P. B. Rep.*, **146**, 190 (1960); *Chem. Abstr.*, **56**, 13205 (1962).

263. G. Kollmann and B. Shapiro, *Radiat. Res.*, **27**, 474 (1966).

264. R. H. Bradford, R. Shapira, and D. G. Doherty, *Int. J. Radiat. Biol.*, **3**, 595 (1961).

265. D. Jamieson, *Nature*, **209**, 361 (1966).

266. T. Sanner and A. Pihl, *Scand. J. Clin. Lab. Invest.*, **22** (Suppl. 106), 53 (1968).

267. J. Sumegi, T. Sanner, and A. Pihl, *Int. J. Radiat. Biol.*, **20**, 397 (1971).

268. L. Eldjarn and A. Pihl, *J. Biol. Chem.*, **223**, 341 (1956).

269. A. Pihl and T. Sanner, *Biochem. Biophys. Acta*, **78**, 537 (1963).

270. Z. M. Bacq, *Bull. Acad. Roy. Med. Belg.*, **6**, 115 (1966).

271. C. Duyckaerts and C. Liebecq in Ref. 38, p. 429.

272. E. N. Goncharenko, Y. B. Kudryashov, and L. I. Alieva, *Dokl. Akad. Nauk SSSR*, **184**, 1437 (1969); *Chem. Abstr.*, **70**, 112161 (1969).

273. E. Y. Graevsky, M. M. Konstantinova, I. V. Nekrasoya, O. M. Sokolova, and A. G. Tarasenko, *Radiobiologiya*, **7**, 130 (1967).

274. E. V. Gusareva, A. G. Tarasenko, and E. Y. Graevsky, *Radiobiologiya*, **8**, 721 (1968).

275. L. Revesz, H. Modig, and P. Lindfors, *Tr. Kayakhsk Nauchn.-Issled. Inst. Onkol. Radiol.*, **6**, 36 (1969); *Chem. Abstr.*, **73**, 21940 (1970).

276. L. X. Thu, *Radiobiologiya*, **9**, 630 (1969).

277. L. X. Thu and E. Y. Graevsky, *Dokl. Akad. Nauk SSSR*, **182**, 965 (1968); *Chem. Abstr.*, **70**, 17403 (1969).

278. H. G. Modig and L. Revesz, *Atomkernenergie*, **14**, 214 (1969).

279. M. L. Efimov and N. I. Mironemko, *Radiobiologiya*, **9**, 359 (1969).

280. J. Denekamp, B. D. Michael, A. Rojas, and F. A. Stewart, *Br. J. Radiol.*, **54**, 1112 (1981).

281. A. G. Tarasenko, E. V. Gusareva, and E. Y. Graevsky, *Radiobiologiya*, **10**, 373 (1970).

282. P. Van Caneghem, *Strahlentherapie*, **137**, 231 (1969).

283. A. K. Bruce, P. A. Sansone, and T. J. MacVittie, *Radiat. Res.*, **38**, 95 (1969).

284. L. Revesz, M. Edgren, and T. Mishidai in T. Sugahara, Ed., *Modification of Radiosensitivity in Cancer Treatment*, Academic Press, New York, 1984, pp. 13–29.

285. R. Brigelius, C. Muchel, L. P. M. Akerboom, and H. Sies, *Biochem. Pharmacol.*, **32**, 2529 (1983).

286. J. C. Livesey and D. J. Reed, *Int. J. Radiat. Oncol. Biol. Phys.*, **10**, 1507 (1984).

287. E. Jellum, *Int. J. Radiat. Biol.*, **9**, 185 (1965).

288. P. E. Brown, *Nature*, **213**, 363 (1967).

289. G. Kollman, B. Shapiro, and D. Martin, *Radiat. Res.*, **31**, 721 (1967).

290. W. O. Foye, M. M. Karkaria, and W. H. Parsons, *J. Pharm. Sci.*, **69**, 84 (1980).

291. B. P. Ivannik and N. I. Ryabchenko, *Radiobiologiya*, **6**, 913 (1966).

292. P. Alexander, C. J. Dean, A. R. Lehmann, M. G. Ormerod, P. Felschreiber, and R. W. Serianni in Ref. 37, p. 15.

293. E. F. Romantsev, N. N. Koshcheenko, and I. V. Fillippovich in Ref. 38, p. 421.

294. R. Goutier and L. Baugnet-Mahieu in Ref. 38, p. 445.

295. S. Sawada and S. Okada, *Radiat. Res.*, **39**, 553 (1969).

296. D. M. Ginsberg and H. K. Webster, *Radiat. Res.*, **39**, 421 (1969).

297. J. Misustova, B. Hosek, and J. Kautska, *Strahlentherapie*, **156**, 790 (1980).

298. J. Misustova, L. Novak, and J. Kautsha, *Radiobiologiya*, **13**, 388 (1973).

299. J. M. Yuhas, J. O. Proctor, and L. H. Smith, *Radiat. Res.*, **54**, 222 (1973).

300. G. P. Bogatyrev, E. N. Goncharaenko, Y. Y. Chirkov, and Y. B. Kudryashov, *Biol. Nauki*, **16**, 50 (1973); *Chem. Abstr.*, **79**, 111626 (1973).

301. P. N. Kulyabko, N. G. Dubenko, and V. D. Konysheva, *Fiziol. Akt. Veshchestv.*, **4**, 87 (1972); *Chem. Abstr.*, **79**, 38542 (1973).

302. J. D. Chapman, A. P. Reuvers, J. Borsa, and C. L. Greenstock, *Radiat. Res.*, **56**, 291 (1973).

303. J. L. Redpath, R. Santus, J. Ovadia, and L. I. Grossweiner, *Int. J. Radiat. Biol.*, **28**, 243 (1975).

304. P. E. Brown, M. Calvin, and J. F. Newmark, *Science*, **151**, 68 (1966).

305. G. Plomteux, M. L. Beaumariage, Z. M. Bacq, and C. Heusghem in Ref. 38, p. 433.

306. P. V. Vittorio, E. A. Watkins, and S. Dziubalo-Blehm, *Can. J. Physiol. Pharmacol.*, **47**, 65 (1969).

307. I. V. Fillipovich and E. F. Romantsev, *Radiobiologiya*, **8**, 800 (1968).

308. Z. I. Zhulanova, E. F. Romantsev, E. E. Kalesnikow, and E. V. Kozyreva in Ref. 38, p. 453.

309. J. N. Mehrishi in Ref. 38, p. 265.

310. B. J. Wang, *Memorias do Instituto Oswaldo Cruz*, **86** (Suppl. 2), 165 (1991).

311. S. M. Hahn, C. M. Krishna, A. Samuni, W. DeGraff, D. O. Cuscela, P. Johnstone, and J. B. Mitchell, *Cancer Res.*, **54** (Suppl. 7), 2006s (1994).

312. R. Neta and J. J. Oppenheim, *Cancer Cells*, **3**, 391 (1991).

313. T. L. Walden Jr. and J. F. Kalinich, *Pharmacol. Ther.*, **39**, 379 (1988).

314. W. R. Hanson, K. A. Houseman, and P. W. Collins, *Pharmacol. Ther.*, **39**, 347 (1988).

315. M. A. Donlon, *Pharmacol. Ther.*, **39**, 373 (1988).

316. R. I. Walker, *Pharmacol. Ther.*, **39**, 13 (1988).

317. R. F. Martin and L. Denison, *Int. J. Radiat. Oncol. Biol. Phys.*, **23**, 579 (1992).

318. V. K. Jam, V. K. Kalia, R. Sharma, V. Maharajan, and M. Menon, *Int. J. Radiat. Oncol. Biol. Phys.*, **11**, 943 (1985).

319. P. J. Wood and D. G. Hirst, *Int. J. Radiat. Oncol. Biol. Phys.*, **16**, 1141 (1989).

320. G. L. Floersheim, *Radiat. Res.*, **133**, 80 (1993).

321. R. K. Kale and D. Samuel, *Indian J. Exp. Biol.*, **25**, 816 (1987).

322. H. Nakano and K. Shinohara, *Radiat. Res.*, **140**, 1 (1994).

323. B. Zhivotovsky, P. Nicotera, G. Bellono, K. Hanson, and S. Orreniius, *Exp. Cell Biol.*, **207**, 163 (1993).

324. R. E. Langley, S. T. Palayoor, C. N. Coleman, and E. A. Bump, *Radiat. Res.*, **136**, 320 (1993).

325. S. P. Sizenko and L. Y. Nesterovskaya, *Vopr. Eksp. Onkol. Sb.*, 263 (1965); *Chem. Abstr.*, **65**, 7868 (1966).

326. L. B. Berlin, *Dokl. Akad. Nauk SSSR*, **195**, 998 (1970); *Chem. Abstr.*, **74**, 72530 (1971).

327. J. R. Maisin, J. Hugon, and A. Leonard, *J. Belge Radiol.*, **47**, 871 (1964).

328. H. Yoshihara, *Nippon Igaku Hoshasen Gakkai Zasshi*, **23**, 1 (1963); *Chem. Abstr.*, **62**, 802 (1965).

329. J. R. Maisin and G. Mattelin, *Bull. Cancer*, **54**, 149 (1967).

330. J. J. Kelly, K. A. Herrington, S. P. Ward, A. Meister, and O. M. Friedman, *Cancer Res.*, **27A**, 137 (1967).

331. G. M. Airapetyan, G. A. Anorova, L. N. Kublik, G. K. Otarova, M. I. Pekaskii, V. N. Stanko, L. K. Eidus, and S. P. Yarmonenko, *Med. Radiol.*, **12**, 58 (1967).

332. E. Karosiene, *Sin. Izuch. Fiziol. Akt. Veshchestv., Mater. Konf*, **1971**, 47 (1971); *Chem. Abstr.*, **79**, 7385 (1973).

333. T. Sugahara, *Advan. Antimicrob. Antineoplastic Chemother., Proc. Int. Congr. Chemother., 7th*, **1971**, 825 (1972); *Chem. Abstr.*, **79**, 38697 (1973).

334. T. Ohshima and I. Tsykiyama, *Nippon Igaku Hoshasen Gakkai Zasshi*, **33**, 351 (1973); *Chem. Abstr.*, **81**, 45486 (1974).

335. W. J. van der Vijgh and G. J. Peters, *Semin. Oncol.*, **21** (5 Suppl. 11), 2 (1994).

336. G. T. Budd, V. Lorenzi, B. Ganapathi, D. Adelstein, R. Pelley, T. Olencki, D. McLain, and R. M. Bukowski, *Support. Care Cancer*, **2**, 380 (1994).

337. D. R. Gandara, E. A. Perez, V. Weibe, and M. W. DeGregorio, *Semin. Oncol.*, **18** (1 Suppl. 3), 49 (1991).

338. J. R. Yucali, *Leuk. Lymphoma*, **13**, 27 (1994).

339. M. Kalaycioglu and R. Bukowski, *Oncology*, **8**, 15, 19, 23 (1994).

340. C. Lewis, *Drug Saf.*, **11**, 153 (1994).

341. R. L. Capizzi, B. J. Scheffler, and P. S. Scheim, *Cancer*, **72**, 3A95 (1993).

342. M. Triskes and W. J. van der Vijgh, *Cancer Chemother. Pharmacol.*, **33**, 93 (1993).

343. C. N. Coleman, E. A. Bump, and R. A. Kramer, *J. Clin. Oncol.*, **6**, 709–733 (1988).

344. G. E. Adams, S. E. Dische, J. F. Fowler, and R. H. Thomlinson, *Lancet*, **1**, 186 (1976).

345. B. Djordjevic and W. Szybalski, *J. Exp. Med.*, **112**, 509–531 (1960).

346. M. Zak and J. Drobnik, *Strahlentherapie*, **142**, 112–115 (1971).

347. R. F. Kallman, D. Rapacchietta, and M. Zaghloul, *Int. J. Radiat. Oncol. Biol. Phys.*, **20**, 227–232 (1991).

348. M. Korbelik and K. A. Skov, *Radiat. Res.*, **119**, 145–156 (1989).

349. L. Yang, E. B. Douple, J. A. O'Hara, R. A. Crabtree, and A. Eastman, *Radiat. Res.*, **144**, 230–236 (1995).

350. G. C. Li, R. G. Evans, and G. M. Hahn, *Radiat. Res.*, **67**, 491–501 (1976).

351. K. J. Henle and D. B. Leeper, *Radiat. Res.*, **66**, 505–518 (1976).

352. B. A. Teicher and T. S. Herman in P. M. Mauch and J. S. Loeffler, Eds., *Radiation Oncology: Technology and Biology*, Saunders, Philadelphia, 1994, pp. 316–347.

353. T. Phillips in P. M. Mauch and J. S. Loeffler, Eds., *Radiation Oncology: Technology and Biology*, Saunders, Philadelphia, 1994, pp. 113–151.

354. G. L. Locher, *Am. J. Roentgenol.*, **36**, 1–13 (1936).

355. D. E. Wazer, R. G. Zamenhof, O. K. Harling, and H. Madoc-Jones in P. M. Mauch and J. S. Loeffler, Eds., *Radiation Oncology: Technology and Biology*, Saunders, Philadelphia, 1994, pp. 167–191.

356. A. H. Soloway, H. Hatanaka, and M. A. Davis, *J. Med. Chem.*, **10**, 714–717 (1967).

357. G. Stragliotto and H. Fankhauser in B. J. Allen, D. E. Moore, and B. V. Harrington, Eds., *Progress in Neutron Capture Therapy for Cancer*, Plenum, New York, 1992.

358. R. G. Fairchild, S. B. Kahl, B. H. Laster, J. Kalef-Ezra, and E. A. Popenoe, *Cancer Res.*, **50**, 4860–4865 (1990).

359. D. N. Slatkin, *Brain*, **114**, 1609–1629 (1991).

360. Y. Mishima, *Nucl. Sci. Applic.*, **4**, 349 (1991).

361. R. F. Martin, G. D. D'Cunha, M. Pardee, and B. J. Allen, *Int. J. Radiat. Biol.*, **54**, 205–208 (1988).

362. J. L. Humm and D. E. Charlton in K. F. Baverstock and D. E. Charlton, Eds., *DNA Damage by Auger Emitters*, Taylor & Francis, London, 1988, pp. 111–122.

363. R. G. Fairchild, A. B. Brill, and K. V. Ettinger, *Invest. Radiol.*, **17**, 407–416 (1982).

364. P. Auger, *J. Phys. (Paris)*, 204–206 (1925).

365. P. R. Almond in J. A. Purdy, Ed., *Advances in Radiation Oncology Physics: Dosimetry, Treatment Planning and Brachytherapy*, American Institute of Physics, Woodbury, NY, 1992, pp. 852–885.

366. S. L. Marcus in C. J. Gomer, Ed., *Future Directions and Applications in Photodynamic Therapy, 1S6*, SPIE Press, Bellingham, WA, 1990, pp. 5–56.

367. T. J. Dougherty, M. T. Cooper, and T. S. Mang, *Lasers Surg. Med.*, **10**, 485–488 (1990).

368. M. C. Berenbaum, S. L. Akande, R. Bonnet, H. Kaur, S. Ioannou, R. White, and U. Winfield, *Br. J. Cancer*, **54**, 717–725 (1986).

369. R. Bonnet, R. D. White, U. J. Winfield, and M. C. Berenbaum, *Biochem. J.*, **261**, 277–280 (1989).

370. H. B. Ris, H. J. Altermatt, R. Inderbitzi, R. Hess, B. Nachbur, J. C. M. Stewart, Q. Wang, C. K. Lim, R. Bonnet, M. C. Berenbaum, and U. Althaus, *Br. J. Cancer*, **64**, 1116–1120 (1991).

371. J. C. Kennedy and R. H. Pottier, *J. Photochem. Photobiol.*, **14**, 275–292 (1992).

372. P. Wolf, E. Rieger, and H. Kerl, *J. Am. Acad. Dermatol.*, **28**, 17–21 (1993).

373. L. A. Lofgren, A. M. Ronn, M. Noun, C. J. Lee, D. Yoo, and B. M. Steinberg, *Br. J. Cancer*, **72**, 857–864 (1995).

374. A. R. Morgan and S. H. Selman in D. Kessel, Ed., *Photodynamic Therapy of Neoplastic Disease*, Vol. **1**, CRC Press, Boca Raton, FL, 1990, pp. 247–262.

375. J. E. van Lier in D. Kessel, Ed., *Photodynamic Therapy of Neoplastic Disease*, Vol. **1**, CRC Press, Boca Raton, FL, 1990, pp. 279–290.

376. O. Peng and J. Moan, *Br. J. Cancer*, **72**, 565–574 (1995).

377. A. R. Oseroff, G. Ara, K. S. Wadwa, and T. Dahl in D. Kessel, Ed., *Photodynamic Therapy of Neoplastic Disease*, Vol. **1**, CRC Press, Boca Raton, FL, 1990, pp. 291–306.

378. K. K. Nadakavukaren, J. J. Nadakavukaren, and L. B. Chen, *Cancer Res.*, **45**, 6093–6099 (1985).

379. B. W. Henderson, N. W. Stephen, T. S. Mang, W. R. Potter, P. B. Malone, and T. J. Dougherty, *Cancer Res.*, **45**, 572–576 (1985).

380. I. P. J. Van Geel, H. Oppelaar, Y. G. Oussoren, J. J. Schuitmaker, and F. A. Stewart, *Br. J. Cancer*, **72**, 344–350 (1995).

381. G. W. Barendsen and J. J. Boerse, *Eur. J. Cancer*, **5**, 373–391 (1969).

382. S. Gonzales, M. R. Arnfield, B. E. Meeker, J. Tulip, W. H. Lakey, and J. D. Chapman, *Cancer Res.*, **46**, 2858–2862 (1986).

383. C. Von Sonntag, *The Chemical Basis of Radiation Biology*, Taylor & Francis, London, 1987, pp. 31–37.

384. R. Roots, A. Chatterjee, E. Blakely, P. Chang, K. Smith, and C. Tobias, *Radiat. Res.*, **92**, 245–254 (1982).

385. D. Ewing and D. S. Guilfoil, *Radiat. Res.*, **120**, 294–305 (1989).

386. J. F. Ward, W. F. Blakely, and E. I. Joner, *Radiat. Res.*, **103**, 383–392 (1985).

387. E. A. Bump, S. M. Pierce, L. Chin, C. Obcemea, and C. N. Coleman in M. G. Simic, K. A. Taylor, J. F. Ward, and C. Von Sonntag, Eds., *Oxygen Radicals in Biology and Medicine*, Plenum, New York, 1988, pp. 429–432.

388. R. L. Willson in H. Sies, Ed., *Oxidative Stress*, Academic Press, London, 1985, pp. 41–72.

389. K. O. Hiller, P. L. Hodd, and R. L. Willson, *Chem.-Biol. Interact.*, **47**, 293–305 (1983).

390. K. J. Kittridge and R. L. Willson, *FEBS Lett.*, **170**, 162–164 (1884).

391. J. A. Raleigh and W. Kremers, *Int. J. Radiat. Biol.*, **39**, 441–444 (1981).

392. D. J. Brenner and J. F. Ward, *Int. J. Radiat. Biol.*, **61**, 737–748 (1992).

393. J. F. Ward, *Int. J. Radiat. Biol.*, **57**, 1141–1150 (1990).

394. K. M. Prise, S. Davies, and B. D. Michael, *Int. J. Radiat. Biol.*, **61**, 721–728 (1992).

395. P. Howard-Flanders, *Nature*, **186**, 485–487 (1960).

396. F. Hutchinson, *Radiat. Res.*, **14**, 721–731 (1961).

397. G. E. Adams, I. R. Flockhart, C. E. Smithen, I. J. Stratford, P. Wardman, and M. E. Watts, *Radiat. Res.*, **67**, 9–20 (1976).

398. G. E. Adams and M. S. Cooke, *Int. J. Radiat. Biol. Relat. Stud. Phys. Chem. Med.*, **15**, 457–471 (1969).

399. G. E. Adams in H. Moroson and M. Quintiiani, Eds., *Radiation Protection and Sensitization*, Barnes & Noble, New York, 1970, pp. 3–14.

400. D. W. Whillans and J. W. Hunt, *Radiat. Res.*, **90**, 126–141 (1982).

401. B. A. Bridges, *Nature*, **188**, 415 (1960).

402. G. DiVita and D. H. Marrian in H. Moroson and M. Quintiiani, Eds., *Radiation Protection and Sensitization*, Barnes & Noble, New York, 1970, pp. 225–231.

403. G. E. Adams, *Int. J. Radiat. Oncol. Biol. Phys.*, **10**, 1181–1184 (1984).

404. G. E. Adams, *Br. Med. Bull.*, **29**, 48–53 (1973).

405. P. W. Sheldon and A. M. Smith, *Br. J. Cancer*, **31**, 81–88 (1975).

406. J. D. Chapman, A. P. Reuvers, J. Borsa, J. S. Henderson, and R. D. Migliore, *Cancer Chemother. Rep.*, **58** (Part 1), 559–570 (1974).

407. R. C. Urtason, P. Bond, and J. D. Chapman, *N. Engl. J. Med.*, **294**, 1364–1367 (1976).

408. S. E. Dische, *Radiother. Oncol.*, **3**, 97–115 (1985).

409. J. Overgaard, H. S. Hansen, and A. P. Anderson, *Int. J. Radiat. Oncol. Biol. Phys.*, **16**, 1065–1068 (1989).

410. T. H. Wasserman, J. S. Nelson, and D. VonGerichten, *Int. J. Radiat. Oncol. Biol. Phys.*, **10**, 1725–1730 (1984).

411. J. M. Brown and P. Workman, *Radiat. Res.*, **82**, 171–190 (1980).

412. D. M. Brown, E. T. Parker, and J. M. Brown, *Radiat. Res.*, **90**, 98–108 (1982).

413. C. N. Coleman, T. H. Wasserman, T. L. Phillips, J. M. Strong, R. C. Urtason, G. Schwade, R. J. Johnson, and G. Zagars, *Int. J. Radiat. Oncol. Biol. Phys.*, **8**, 371–375 (1982).

414. C. Beard, L. Buswell, M. A. Rose, L. Noll, D. Johnson, and C. N. Coleman, *Int. J. Radiat. Oncol. Biol. Phys.*, **29**, 611–616 (1994).

415. T. H. Wasserman, D. J. Lee, D. Cosmatos, N. Coleman, T. Phillips, L. Davis, V. Marcial, and J. Stetz, *Radiother. Oncol.*, **20** (Suppl.), 129–135 (1991).

416. D. J. Lee, T. L. Phillips, C. N. Coleman, D. Cosmatos, L. W. Davis, T. H. Wasserman, V. A. Marcial, and P. Rubin, *Int. J. Radiat. Oncol. Biol. Phys.*, **22**, 569–572 (1992).

417. D. Chassagne, H. Sancho-Garnier, I. Charreau, F. Eschwege, and E. P. Malaise, *Radiother. Oncol.*, **20** (Suppl.), 121–127 (1991).

418. D. Chassagne, I. Charreau, H. Sancho-Garnier, F. Eschwege, and E. P. Malaise, *Int. J. Radiat. Oncol. Biol. Phys.*, **22**, 581–584 (1992).

419. J. Overgaard, M. Overgaard, O. S. Nielsen, A. K. Pedersen, and A. R. Timothy, *Br. J. Cancer*, **46**, 904–911 (1982).

420. J. Overgaard, H. S. Hansen, B. Lindelov, M. Overgaard, K. Jorgensen, and B. Rasmusson, *Radiother. Oncol.*, **20** (Suppl.), 143–149 (1991).

421. J. Overgaard, H. S. Hansen, M. Overgaard, L. Bastholt, A. Berthelsen, L. Specht, B. Lindelov, and K. Jorgensen, *Radiother. Oncol.*, **46**, 135–146 (1998).

422. L. M. Cobb, J. Nolan, and S. A. Butler, *Br. J. Cancer*, **62**, 314–318 (1990).

423. S. Dische in W. C. Dewey, M. Edington, R. J. M. Fry, E. J. Hall, and G. F. Whitmore, Eds., *Radiation Research, a Twentieth-Century Perspective*, Academic Press, San Diego, 1992, pp. 584–594.

424. D. J. Chaplin and M. R. Horsman, *Int. J. Radiat. Oncol. Biol. Phys.*, **22**, 459–462 (1992).

425. M. F. Dennis, M. R. L. Stratford, P. Wardman, and M. E. Watts, *Int. J. Radiat. Biol.*, **47**, 629–643 (1985).

426. M. E. Watts and N. R. Jones, *Int. J. Radiat. Biol.*, **47**, 645–653 (1985).

427. N. M. Bleehen, H. F. V. Newman, T. S. Maughan, and P. A. Workman, *Int. J. Radiat. Oncol. Biol. Phys.*, **16**, 1093–1096 (1989).

428. K. C. Agrawal, C. A. Larroquette, and P. K. Garg, *Int. J. Radiat. Oncol. Biol. Phys.*, **10**, 1301–1305 (1984).

429. T. Tada, T. Nakajima, Y. Onoyama, C. Murayama, T. Mori, H. Nagasawa, H. Hori, and S. Inayama, *Int. J. Radiat. Oncol. Biol. Phys.*, **29**, 601–605 (1994).

430. H. Hon, C. Murayama, T. Mori, Y. Shibamoto, M. Abe, Y. Onoyama, and S. Inayama, *Int. J. Radiat. Oncol. Biol. Phys.*, **16**, 1029–1032 (1989).

431. A. H. Garcia-Angulo and V. T. Kagiya, *Int. J. Radiat. Oncol. Biol. Phys.*, **22**, 589–591 (1992).

432. L. C. Huan and B. Y. Hua, *Int. J. Radiat. Oncol. Biol. Phys.*, **29**, 607–610 (1994).

433. D. S. M. Cowan, R. Panicucci, R. A. McClelland, and A. M. Rauth, *Radiat. Res.*, **127**, 81–89 (1991).

434. W. R. Wilson and R. F. Anderson, *Int. J. Radiat. Oncol. Biol. Phys.*, **16**, 1001–1005 (1989).

435. A. J. Franko, *Int. J. Radiat. Oncol. Biol. Phys.*, **12**, 1195–1202 (1986).

436. R. C. Urtason, J. D. Chapman, J. A. Raleigh, A. J. Franko, and C. J. Koch, *Int. J. Radiat. Oncol. Biol. Phys.*, **12**, 1263–1267 (1986).

437. J. A. Raleigh, G. G. Miller, A. J. Franko, C. J. Koch, A. F. Fuciarelli, and D. A. Kelley, *Br. J. Cancer*, **56**, 395–400 (1987).

438. R. J. Hodgkiss, R. W. Middleton, J. Parrick, H. K. Rami, P. Wardman, and G. D. Wilson, *J. Med. Chem.*, **35**, 1920–1926 (1992).

439. J. A. Raleigh, A. J. Franko, E. O. Treiber, J. A. Lunt, and P. S. Allen, *Int. J. Radiat. Oncol. Biol. Phys.*, **12**, 1243–1245 (1986).

440. S. M. Evans, B. Joiner, W. T. Jenkins, K. M. Laughlin, E. M. Lord, and C. J. Koch, *Br. J. Cancer*, **72**, 875–882 (1995).

441. K. Sasai, H. Iwai, T. Yoshizawa, S. Nishimoto, Y. Shibamoto, N. Oya, T. Shibata, and M. Abe, *Int. J. Radiat. Oncol. Biol. Phys.*, **29**, 579–582 (1994).

442. P. F. Rijken, H. J. Bernsen, J. P. Peters, R. J. Hodgkiss, J. A. Raleigh, and A. J. van der Kogel, *Int. J. Radiat. Oncol. Biol. Phys.*, **48**, 571–582 (2000).

443. J. D. Chapman, R. F. Schneider, J. Urbain, and G. E. Hanks, *Semin. Radiat. Oncol.*, **11**, 47–57 (2001).

444. E. J. Hall and J. E. Biaglow, *Int. J. Radiat. Oncol. Biol. Phys.*, **2**, 521–530 (1977).

445. T. W. Wong, G. F. Whitmore, and S. Gulyas, *Radiat. Res.*, **75**, 541–555 (1978).

446. D. W. Siemann, *Int. J. Radiat. Oncol. Biol. Phys.*, **8**, 1029–1034 (1982).

447. C. J. Koch and K. A. Skov, *Int. J. Radiat. Oncol. Biol. Phys.*, **29**, 345–349 (1994).

448. Y. C. Taylor, E. A. Bump, and J. M. Brown, *Int. J. Radiat. Oncol. Biol. Phys.*, **8**, 705–708 (1982).

449. E. M. Zeman and J. M. Brown, *Int. J. Radiat. Oncol. Biol. Phys.*, **16**, 967–971 (1989).

450. E. M. Zeman, M. A. Baker, M. J. Lemmon, C. I. Pearson, J. A. Adams, J. M. Brown, W. W. Lee, and M. Tracy, *Int. J. Radiat. Oncol. Biol. Phys.*, **16**, 977–981 (1989).

451. N. Doherty, C. N. Coleman, L. Shulman, S. L. Hancock, C. Mariscal, R. Rampling, S. Senan, P. Workman, S. Kaye, and R. von Roemeling, *Int. J. Radiat. Oncol. Biol. Phys.*, **29**, 379–382 (1994).

452. B. M. Sutton, N. J. Reeves, M. A. Naylor, E. M. Fielden, S. Cole, G. E. Adams, and I. J. Stratford, *Int. J. Radiat. Oncol. Biol. Phys.*, **29**, 339–344 (1994).

453. G. E. Adams, I. Ahmed, P. W. Sheldon, and I. J. Stratford, *Br. J. Cancer*, **49**, 571–577 (1984).

454. J. S. Sebolt-Leopold, P. W. Vincent, K. A. Be-ningo, W. L. Elliott, W. R. Leopold, T. G. Hef-fner, J. N. Wiley, M. A. Stier, and M. J. Suto, *Int. J. Radiat. Oncol. Biol. Phys.*, **22**, 549–551 (1992).

455. D. W. Siemann, *Int. J. Radiat. Oncol. Biol. Phys.*, **29**, 301–306 (1994).

456. P. Workman, *Int. J. Radiat. Oncol. Biol. Phys.*, **22**, 631–637 (1992).

457. J. B. Weissberg, Y. H. Son, R. J. Papac, C. Saski, D. B. Fisher, R. Lawrence, S. Rockwell, A. C. Sartorelli, and J. J. Fischer, *Int. J. Radiat. Oncol. Biol. Phys.*, **17**, 3–9 (1989).

458. J. A. Plumb, M. Gerritsen, R. Milroy, P. Thom-son, and P. Workman, *Int. J. Radiat. Oncol. Biol. Phys.*, **29**, 295–299 (1994).

459. S. R. McKeown, M. V. Hejmadi, I. A. McIntyre, J. J. A. McAleer, and L. H. Patterson, *Br. J. Cancer*, **72**, 76–81 (1995).

460. S. R. McKeown, W. V. Hejmadi, O. P. Friery, J. J. A. McAleer, and L. H. Pattersonn, Pro-ceedings of the Ninth International Confer-ence on Chemical Modifiers of Cancer Treat-ment, Christ Church, Oxford, UK, August 22–26, 1995, Abstr. 1–20.

461. R. H. Thomlinson and L. H. Gray, *Br. J. Can-cer*, **9**, 539–549 (1955).

462. J. M. Brown, *Br. J. Radiol.*, **52**, 650–656 (1979).

463. M. J. Trotter, D. J. Chaplin, R. E. Durand, and P. L. Olive, *Int. J. Radiat. Oncol. Biol. Phys.*, **16**, 931–934 (1989).

464. C. Grau, M. R. Horsman, and J. Overgaard, *Int. J. Radiat. Oncol. Biol. Phys.*, **22**, 415–419 (1992).

465. I. Churchill-Davidson and C. Sanger, *Lancet*, **10**, 1091–1095 (1955).

466. H. D. Suit and O. C. Scott in K. H. Karcher, H. D. Kogelnik, and H. J. Meyer, Eds., *Progress in Radio-Oncology*, Springer-Verlag, Stuttgart, Germany, 1980, pp. 150–163.

467. L. W. Brady, H. P. Plenk, J. A. Hanley, J. R. Glassburn, S. Kramer, and R. G. Parker, *Int. J. Radiat. Oncol. Biol. Phys.*, **7**, 991–998 (1981).

468. S. Dische, *Radiother. Oncol.*, **20** (Suppl.), 9–11 (1991).

469. S. Dische, M. I. Saunders, and M. R. Wharbur-ton, *Int. J. Radiat. Oncol. Biol. Phys.*, **12**, 1335–1337 (1986).

470. D. G. Hirst, *Int. J. Radiat. Oncol. Biol. Phys.*, **12**, 2009–2017 (1986).

471. D. W. Siemmann, R. P. Hill, R. S. Bush, and P. Chhabra, *Int. J. Radiat. Oncol. Biol. Phys.*, **5**, 61–68 (1979).

472. J. J. Fischer, S. Rockwell, and D. F. Martin, *Int. J. Radiat. Oncol. Biol. Phys.*, **12**, 95–102 (1986).

473. C. J. Koch, P. R. Oprysko, A. L. Shuman, W. T. Jenkins, G. Brandt, and S. M. Evans, *Cancer Res.*, **62**, 3626–3629 (2002).

474. R. Lustig, L. N. McIntosh, C. Rose, J. Haas, S. Krasnow, M. Spaulding, and L. Prosnitz, *Int. J. Radiat. Oncol. Biol. Phys.*, **16**, 1587–1593 (1989).

475. R. G. Evans, B. F. Kimler, R. A. Morantz, and S. Batnitzky, *Int. J. Radiat. Oncol. Biol. Phys.*, **26**, 649–652 (1993).

476. B. A. Teicher in G. Fisher, Ed., *Hematology/Oncology Clinics of North America: Drug Re-sistance in Clinical Oncology and Hematology*, Vol. **9**, Saunders, Philadelphia, 1995, pp. 475–506.

477. B. A. Teicher, G. Ara, Y. Kakeji, M. Ikebe, Y. Maehara, and D. Buxton, Proceedings of the Forty-fourth Annual meeting of the Radiation Research Society, Chicago, April 14–17, 1996, Abstr. P04–70.

478. F. F. Davis, G. M. Kazo, and M. L. Nucci in V. H. L. Lee, Ed., *Peptide and Protein Drug Delivery*, Marcel Dekker, New York, 1991, Chapter 21.

479. D. J. Abraham, J. Kister, G. S. Joshi, M. C. Marden, and C. Poyart, *J. Mol. Biol.*, **248**, 845, 855 (1995).

480. D. G. Hirst and P. J. Wood, *Radiother. Oncol.*, **20** (Suppl.), 53–57 (1991).

481. D. G. Hirst, P. J. Wood, and H. C. Schwartz, *Radiat. Res.*, **112**, 164–172 (1987).

482. A. Ward and S. P. Clissold, *Drugs*, **34**, 50–97 (1987).

483. C. W. Song, C. M. Makepeace, R. J. Griffin, T. Hasegawa, J. L. Osborn, I. Choi, and B. S. Nah, *Int. J. Radiat. Oncol. Biol. Phys.*, **29**, 433–437 (1994).

484. M. R. Horsman, D. A. Chaplin, and J. M. Brown, *Radiat. Res.*, **109**, 479–489 (1987).

485. M. R. Horsman, P. E. G. Kristjansen, P. Mi-zuno, K. L. Christensen, D. J. Chaplin, B. Quis-torff, and J. Overgaard, *Int. J. Radiat. Oncol. Biol. Phys.*, **22**, 451–454 (1992).

486. V. M. Laurence, R. Ward, I. F. Dennis, and N. M. Bleehen, *Br. J. Cancer*, **72**, 198–205 (1995).

487. J. H. Kaanders, L. A. Pop, H. A. Marres, I. Bruaset, F. J. van den Hoogen, M. A. Merkx, and A. J. van der Kogel, *Int. J. Radiat. Oncol. Biol. Phys.*, **52**, 769–778 (2002).

488. J. Liebmann, A. M. DeLuca, D. Coffin, L. K. Keefer, D. Venzon, D. A. Wink, and J. B. Mitchell, *Cancer Res.*, **54**, 3365–3368 (1994).

489. D. J. Abraham, F. C. Wireko, R. S. Randad, et al., *Biochemistry*, **31**, 9141–9149 (1992).

490. R. S. Randad, M. A. Mahran, A. S. Mehanna, et al., *J. Med. Chem.*, **34**, 752–757 (1991).

491. R. Rampling, G. Cruickshank, A. D. Lewis, et al., *Int. J. Radiat. Oncol. Biol. Phys.*, **29**, 427–431 (1994).

492. R. A. Gatenby, H. B. Kessler, J. S. Rosenblum, et al., *Int. J. Radiat. Oncol. Biol. Phys.*, **14**, 831–838 (1988).

493. M. Hockel, K. Schlenger, C. Knoop, et al., *Cancer Res.*, **51**, 6098–6102 (1991).

494. P. Vaupel, F. Kallinowski, and P. Okunieff, *Cancer Res.*, **49**, 6449–6465 (1989).

495. P. Vaupel, K. Schlenger, C. Knoop, et al., *Cancer Res.*, **51**, 3316–3322 (1991).

496. R. K. Jain, *Cancer Res.*, **48**, 2641–2658 (1988).

497. S. Rockwell and C. S. Hughes, *Prog. Clin. Biol. Res.*, **354B**, 67–80 (1990).

498. J. E. Moulder and S. Rockwell, *Cancer Metastasis Rev.*, **5**, 313–341 (1987).

499. B. A. Teicher, *Hematol. Oncol. Clin. North Am.*, **9**, 475–506 (1995).

500. J. M. Brown and A. J. Giaccia, *Cancer Res.*, **58**, 1408–1416 (1998).

501. B. D. Kavanagh, T. W. Secomb, R. Hsu, P. S. Lin, J. Venitz, and M. W. Dewhirst, *Int. J. Radiat. Oncol. Biol. Phys.*, **53**, 172–179 (2002).

502. P. L. Olive, J. P. Banath, and R. E. Durand, *Radiat. Res.*, **158**, 159–166 (2002).

503. E. J. Hall, *The Oxygen Effect and Reoxygenation in Radiobiology for the Radiologist*, 3rd ed., Lippincott, Philadelphia, 1988.

504. A. W. Fyles, M. Milosevic, R. Wong, et al., *Radiother. Oncol.*, **48**, 149–156 (1998).

505. D. M. Brizel, G. S. Sibley, L. R. Prosnitz, et al., *Int. J. Radiat. Oncol. Biol. Phys.*, **38**, 285–289 (1997).

506. M. Nordsmark, M. Overgaard, and J. Overgaard, *Radiother. Oncol.*, **41**, 31–39 (1996).

507. P. Stadler, A. Becker, H. J. Feldmann, et al., *Int. J. Radiat. Oncol. Biol. Phys.*, **44**, 749–754 (1999).

508. M. De Santis, M. Balducci, L. Basilico, et al., *Rays*, **23**, 543–548 (1998).

509. B. A. Teicher, *Cancer Metastasis Rev.*, **13**, 139–168 (1994).

510. S. R. Khandelwal, P. S. Lin, C. E. Hall, et al., *Radiat. Oncol. Investig.*, **4**, 51–59 (1996).

511. M. P. Kunert, J. F. Liard, and D. J. Abraham, *Am. J. Physiol.*, **271**, 602–613 (1996).

512. M. P. Kunert, J. F. Liard, D. J. Abraham, et al., *Microvasc. Res.*, **52**, 58–68 (1996).

513. S. R. Khandelwal, B. D. Kavanagh, P. S. Lin, et al., *Br. J. Cancer*, **79**, 814–820 (1999).

514. P. S. Pagel, D. A. Hettrick, M. W. Montgomery, et al., *J. Pharmacol. Exp. Ther.*, **285**, 1–8 (1998).

515. M. Miyake, O. Y. Grinsberg, H. Hou, et al. in H. Swartz, Ed., *Oxygen Transport to Tissue*, 22nd ed., Plenum, New York, in press.

516. S. R. Khandelwal, R. S. Randad, P. S. Lin, et al., *Am. J. Physiol.*, **265**, 1450–1453 (1993).

517. O. Y. Grinsberg, M. Miyake, H. Hou, et al. in H. Swartz, Ed., *Oxygen Transport to Tissue*, 22nd ed., Plenum, New York, in press.

518. R. S. Richardson, K. Tagore, L. J. Haseler, et al., *J. Appl. Physiol.*, **84**, 995–1002 (1998).

519. N. P. Dupuis, T. Kusumoto, M. F. Robinson, et al., *Artif. Cells Blood Substit. Immobil. Biotechnol.*, **23**, 423–429 (1995).

520. B. A. Teicher, G. Ara, Y. Emi, et al., *Drug Dev. Res.*, **38**, 1–11 (1996).

521. G. P. Amorino, L. Haakil, G. E. Holburn, et al., *Radiat. Res.*, submitted (2001).

522. S. Rockwell and M. Kelley, *Radiat. Oncol. Investig.*, **6**, 199–208 (1998).

523. B. A. Teicher, J. S. Wong, H. Takeuchi, et al., *Cancer Chemother. Pharmacol.*, **42**, 24–30 (1998).

524. B. A. Teicher, data on file at Allos Therapeutics, 1997.

525. B. Kavanagh, S. Khandelwal, R. Schmidt-Ullrich, J. Roberts, E. Shaw, A. Pearlman, J. Venitz, K. Dusenberry, D. Abraham, and M. Gerber, *Int. J. Radiat. Oncol. Biol. Phys.*, **49**, 1133–1139 (2001).

526. P. Y. Wen and J. S. Loeffler, *Management of Brain Metastases*, Oncology, Huntington, 1999, pp. 941–961.

527. P. Davey, *Curr. Probl. Cancer*, **23**, 53–100 (1999).

528. R. A. Patchell, *Neurol. Clin. N. Am.*, **9**, 817–824 (1991).

529. L. Gaspar, C. Scott, M. Rotmar, et al., *Int. J. Radiat. Oncol. Biol. Phys.*, **37**, 745–751 (1997).

530. L. E. Gaspar, C. Scott, K. Murray, et al., *Int. J. Radiat. Oncol. Biol. Phys.*, **47**, 1001–1006 (2000).

531. E. G. Shaw, C. Scott, B. Stea, et al., 36th Proceedings of the American Society of Clinical Oncology, 19, 159a (Abstr. 615) (2000).

532. E. G. Shaw, C. Scott, B. Stea, et al., *Int. J. Radiat. Oncol. Biol. Phys.*, **48**, 183 (supplement) (2000).

533. L. R. Kleinberg, S. A. Grossman, K. Carson, et al., 36th Proceedings of the American Society of Clinical Oncology, 19, 158a (2000).

534. L. R. Kleinberg, S. A. Grossman, K. Carson, et al., *Int. J. Radiat. Oncol. Biol. Phys.*, **48**, 182 (supplement) (2000).

535. C. B. Scott, C. Scarantino, R. Urtasun, et al., *Int. J. Radiat. Oncol. Biol. Phys.*, **40**, 51–55 (1998).

536. H. Choy, A. Nabid, B. Stea, W. Roa, L. Souhami, F. Yunus, P. Roberts, and D. Johnson, 37th Proceedings of American Society of Clinical Oncology, 20, 313a (Abstr. 1248) (2001).

537. H. Choy, A. Nabid, B. Stea, W. Roa, L. Souhami, F. Yunus, P. Roberts, and D. Johnson, Proceedings of the 43rd Annual American Society for Therapeutic Radiology and Oncology, 51 (3 Suppl. 1), Abstr. 2254 (2001).

538. J. B. Mitchell, T. L. Phillips, W. DeGraff, J. Carmichael, R. K. Rajpal, and A. Russo, *Int. J. Radiat. Oncol. Biol. Phys.*, **12**, 1143–1146 (1986).

539. L. F. Chasseaud, *Adv. Cancer Res.*, **29**, 175–274 (1979).

540. E. A. Bump, N. Y. Yu, Y. C. Taylor, J. M. Brown, E. L. Travis, and M. R. Boyd in M. G. Simic and O. F. Nygaard, Eds., *Radioprotectors and Anticarcinogens*, Academic Press, New York, 1983, pp. 297–324.

541. E. A. Bump, N. Y. Yu, and J. M. Brown, *Science*, **217**, 544–545 (1982).

542. W. B. Jakoby, *Methods Enzymol.*, **113**, 495–499 (1985).

543. B. Ketterer and J. B. Taylor in K. Rackpaul and H. Rein, Eds., *Frontiers in Biotransformation*, Vol. **2**, Taylor & Francis, London, 1990, pp. 244–277.

544. D. J. Waxman, *Cancer Res.*, **50**, 6449–6454 (1990).

545. O. W. Griffith, *J. Biol. Chem.*, **257**, 13704–13712 (1982).

546. O. W. Griffith and A. Meister, *J. Biol. Chem.*, **254**, 7558–7560 (1979).

547. H. H. Bailey, R. T. Mulcahy, K. D. Tutsch, R. Z. Arzoomanian, D. Alberti, M. B. Tombes, G. Wilding, M. Pomplum, and D. R. Spriggs, *J. Clin. Oncol.*, **12**, 194–205 (1994).

548. P. J. O'Dwyer, T. C. Hamilton, R. C. Young, F. P. LaCreta, N. Carp, K. D. Tew, K. Padavic, L. Loomis, and R. F. Ozols, *J. Natl. Cancer Inst.*, **84**, 264–267 (1992).

549. G. H. Ripple, R. T. Mulcahy, K. D. Tutsch, R. Z. Arzoomanian, D. Alberti, C. Feirabend, D. Mahvi, J. Schink, M. Pomplun, G. Wilding, and H. H. Bailey, *Proc. 86th Annual Mtg. Am. Assoc. Cancer Res.*, Toronto, **36**, 238 (Abstr. 1416) (1995).

550. K. Yao, A. K. Godwin, R. F. Ozols, T. C. Hamilton, and P. J. O'Dwyer, *Cancer Res.*, **53**, 3662–3666 (1993).

551. K. Malaker, S. J. Hurwitz, E. A. Bump, O. W. Giffith, L. L. Lai, N. Riese, and C. N. Coleman, *Int. J. Radiat. Oncol. Biol. Phys.*, **29**, 407–412 (1994).

552. S. T. Palayoor, E. A. Bump, D. M. Saroff, J. R. Delfs, C. Geula, L. Menton-Brennan, S. J. Hurwitz, C. N. Coleman, and M. A. Stevenson, *Br. J. Cancer*, **27**, S117–S121 (1996).

553. M. Frankenberg-Schwager, *Radiother. Oncol.*, **14**, 307–320 (1989).

554. P. A. Jaggo, *Mutat. Res.*, **239**, 1–16 (1990).

555. R. Helbig, M. Z. Zdzienicka, and G. Speit, *Radiat. Res.*, **143**, 151–157 (1995).

556. E. Ben-Hur, *Radiat. Res.*, **88**, 155–164 (1981).

557. P. E. Bryant and D. Blocher, *Int. J. Radiat. Biol.*, **42**, 385–394 (1982).

558. J. F. Ward, *Int. J. Radiat. Biol.*, **12**, 1027–1032 (1986).

559. J. B. Little and H. Nagasawa, *Radiat. Res.*, **101**, 81–93 (1985).

560. J. Menissier-de Murcia, M. Molinete, G. Gradwhol, F. Simonin, and G. de Murcia, *J. Mol. Biol.*, **210**, 229–233 (1989).

561. H. Utsumi and M. M. Elkind, *Int. J. Radiat. Oncol. Biol. Phys.*, **29**, 577–578 (1994).

562. J. E. Cleaver, K. M. Milam, and W. F. Morgan, *Radiat. Res.*, **101**, 16–28 (1985).

563. M. Banasik, H. Komura, M. Shimoyama, and K. Ueda, *J. Biol. Chem.*, **267**, 1569–1575 (1992).

564. C. M. Arundel-Suto, S. V. Scavone, W. R. Turner, M. J. Suto, and J. S. Sebolt-Leopold, *Radiat. Res.*, **126**, 367–371 (1991).

565. A. Dritschio, A. Piro, and J. A. Belli, *Int. J. Radiat. Biol.*, **35**, 549–560 (1979).

566. G. Iliakis, *Radiat. Res.*, **83**, 537–552 (1980).

567. J. Ward, E. I. Joner, and W. F. Blakely, *Cancer Res.*, **44**, 59–63 (1984).

568. S. Nakatsugawa and T. Sugahara, *Radiat. Res.*, **84**, 265–275 (1980).

569. S. Nakatsugawa, T. Sugahara, and A. Kumar, *Int. J. Radiat. Biol.*, **41**, 343–346 (1982).

570. G. P. Raaphorst and W. C. Dewey, *Radiat. Res.*, **77**, 325–340 (1979).

571. D. M. Brown, C. Lionet, and J. M. Brown, *Radiat. Res.*, **97**, 162–170 (1984).

572. L. Kelland and G. G. Steel, *Int. J. Radiat. Biol.*, **54**, 229–244 (1988).

573. D. M. Ziegler, *Annu. Rev. Biochem.*, **54**, 305–329 (1985).

574. S. Eriksson, P. Askelof, K. Axelsson, J. Carlberg, C. Guntherburg, and B. Mannervik, *Acta Chem. Scand.*, **B28**, 922–930 (1974).

575. I. A. Cotgrave, M. Weis, L. Atzori, and P. Moldeus, *Pharmacol. Ther.*, **7**, 375–391 (1979).

576. J. W. Harris, *Pharmacol. Ther.*, **7**, 375–391 (1979).

577. M. R. Bianchi, M. Boccacci, M. Quintiliani, and E. Strom in R. Paoletti and R. Vertua, Eds., *Progress in Biochemical Pharmacology*, S. Karger, Farmington, Vol. I, 1965, pp. 384–391.

578. B. B. Singh, M. A. Shenoy, and A. R. Gopal-Ayengar in H. Moroson, D. S. Walker, and M. Quintiliani, Eds., *Radiation Protection and Sensitization*, Barnes & Noble, New York, 1969, pp. 159–162.

579. M. M. Klingerman and L. Schulhof, *Radiat. Res.*, **39**, 571–579 (1969).

580. G. E. Adams, *Br. Med. Bull.*, **29**, 48–53 (1973).

581. M. A. Baker and B. A. Hagner, *Biochim. Biophys. Acta*, **1037**, 39–47 (1990).

582. E. A. Bump, G. P. Jacobs, W. W. Lee, and J. M. Brown, *Int. J. Radiat. Oncol. Biol. Phys.*, **12**, 1533–1536 (1986).

583. N. S. Kosower, E. M. Kosower, and B. Wertheim, *Biochem. Biophys. Res. Commun.*, **37**, 593–596 (1969).

584. E. M. Kosower, N. S. Kosower, H. Kanety-Londner, and L. Levy, *Biochem. Biophys. Res. Commun.*, **59**, 347–351 (1974).

585. J. D. Chapman, A. P. Reuvers, J. Borsa, and C. Greenstock, *Radiat. Res.*, **56**, 291–306 (1973).

586. E. P. Clark, H. B. Michaels, E. C. Peterson, and E. R. Epp, *Radiat. Res.*, **93**, 479–491 (1983).

587. J. W. Harris, J. A. Power, N. S. Kosower, and E. M. Kosower, *Int. J. Radiat. Biol.*, **28**, 439–445 (1975).

588. J. F. Ward, E. I. Joiner, and W. F. Blakeley, *Cancer Res.*, **44**, 59–63 (1984).

589. K. D. Held, E. R. Epp, S. Awad, and J. E. Biaglow, *Radiat. Res.*, **127**, 75–80 (1991).

590. K. Nishizawa, C. Sato, and T. Morita, *Int. J. Radiat. Biol.*, **35**, 15–22 (1979).

591. W. A. Nagle, A. J. Moss Jr., H. G. Roberts Jr., and M. L. Baker, *Radiology*, **137**, 203–211 (1980).

592. V. K. Jam, V. K. Kalia, R. Sharma, V. Maharajan, and M. Menon, *Int. J. Radiat. Oncol. Biol. Phys.*, **11**, 943–950 (1985).

593. S. P. Singh, S. Singh, and V. Jam, *Int. J. Radiat. Biol.*, **58**, 791–797 (1990).

594. E. A. Bump, S. K. Calderwood, J. M. Sawyer, and J. M. Brown, *Int. J. Radiat. Oncol. Biol. Phys.*, **10**, 1411–1414 (1984).

595. F. Laval and J. B. Little, *Radiat. Res.*, **71**, 571–578 (1977).

596. D. B. Leeper, M. H. Schneiderman, and W. C. Dewey, *Radiat. Res.*, **50**, 401–417 (1972).

597. R. A. Tobey, *Nature*, **254**, 245–247 (1975).

598. W. J. Slichenmyer, W. G. Nelson, R. J. Slebos, and M. B. Kastan, *Cancer Res.*, **53**, 4164–4168 (1993).

599. K. Sankaranayanan and R. Chakraborty, *Radiat. Res.*, **143**, 121–143 (1995).

600. R. A. Walters, L. R. Gurley, and R. A. Tobey, *Biophys. J.*, **14**, 99–118 (1974).

601. P. M. Busse, S. K. Bose, R. W. Jones, and L. J. Tolmach, *Radiat. Res.*, **71**, 666–677 (1977).

602. K. E. Steinmann, G. S. Belinsky, D. Lee, and R. Schlegel, *Proc. Natl. Acad. Sci. USA*, **88**, 6843–6847 (1991).

603. R. J. Muschel, H. B. Zhang, G. Iliakis, and W. G. McKeena, *Cancer Res.*, **51**, 5113–5117 (1991).

604. A. Maity, W. G. McKeen, and R. J. Muschel, *Radiother. Oncol.*, **31**, 1–13 (1994).

605. C. Smythe and J. W. Newport, *Cell*, **68**, 787–797 (1992).

606. G. Draetta and D. Beach, *Cell*, **54**, 17–26 (1988).

607. J. B. A. Millar and P. Russell, *Cell*, **68**, 407–410 (1992).

608. N. F. Metting and J. B. Little, *Radiat. Res.*, **143**, 286–292 (1995).

609. J. Ham, N. E. A. Crompton, W. Burkart, and R. Jaussi, *Cancer Res.*, **53**, 1507–1510 (1993).

610. S. N. Powell, J. S. DeFrank, P. Connell, M. Eogan, F. Preffer, D. Dombkowski, W. Tang, and S. Friend, *Cancer Res.*, **55**, 1643–1648 (1995).

611. S. W. Tam and R. Chlegel, *Cell Growth Differ.*, **3**, 811–817 (1992).

612. S. P. Tomasovic and W. C. Dewey, *Radiat. Res.*, **74**, 112–128 (1978).

613. S. T. Palayoor, R. M. Macklis, E. A. Bump, and C. N. Coleman, *Radiat. Res.*, **141**, 235–243 (1995).

614. D. E. Hallahan, S. Virudachalam, J. L. Schwartz, N. Panje, R. Mustafi, and R. R. Weichselbaum, *Radiat. Res.*, **129**, 345–350 (1992).

615. D. E. Hallahan, M. A. Beckett, D. Kufe, and R. R. Weichselbaum, *Int. J. Radiat. Oncol. Biol. Phys.*, **19**, 69–74 (1990).

616. T. T. Kwok and R. M. Sutherland, *J. Natl. Cancer Inst.*, **81**, 1020–1024 (1989).

617. J. A. Bonner, N. J. Maihle, B. R. Folven, T. J. H. Christianson, and K. Spain, *Int. J. Radiat. Oncol. Biol. Phys.*, **29**, 243–247 (1994).

618. V. R. Jayanth, C. A. Belfi, A. R. Swick, and M. E. Varnes, *Radiat. Res.*, **143**, 165–174 (1995).

619. B. Djordjevic and W. Szybalski, *J. Exp. Med.*, **112**, 509–531 (1960).

620. C. Heidelberger, L. Griesbach, B. J. Montag, D. Mooren, O. Cruz, R. J. Schnitzer, and E. Grunberg, *Cancer Res.*, **18**, 305–317 (1958).

621. C. J. McGinn and T. J. Kinsella in P. M. Mauch and J. S. Loeffler, Eds., *Radiation Oncology: Technology and Biology*, Saunders, Philadelphia, 1994, pp. 90–112.

622. M. A. Bagshaw, R. L. S. Doggett, and K. C. Smith, *Am. J. Roentgenol.*, **99**, 886–894 (1967).

623. V. A. Levin, M. R. Prados, W. M. Wara, R. L. Davis, P. H. Gutin, T. L. Phillips, K. Lamborn, and C. B. Wilson, *Int. J. Radiat. Oncol. Biol. Phys.*, **32**, 75–83 (1995).

624. P. G. Gill, J. W. Denham, G. G. Jamieson, P. G. Devitt, E. Yeoh, and C. Olweny, *J. Clin. Oncol.*, **10**, 1037–1043 (1992).

625. T. S. Lawrence, M. A. Davis, J. P. Maybaum, P. L. Stetson, and W. D. Ensminger, *Int. J. Radiat. Oncol. Biol. Phys.*, **22**, 499–503 (1992).

626. S. H. Kim and J. H. Kim, *Proc. Am. Assoc. Cancer Res.*, **37**, Abstr. 4194 (1996).

627. T. S. Lawrence, E. Y. Chang, T. M. Hahn, L. W. Hertel, and D. S. Shewach, *Int. J. Radiat. Oncol. Biol. Phys.*, **34**, 867–872 (1996).

628. D. S. Shewach, D. Keena, L. Z. Rubsam, L. W. Hertel, and T. S. Lawrence, *Proc. Am. Assoc. Cancer Res.*, **37**, Abstr. 4196 (1996).

629. V. Gregoire, M. De Bast, J. F. Rosier, M. Beauduin, M. Bruniaux, B. De Coster, and P. Scalliet, *Proc. Am. Assoc. Cancer Res.*, **37**, Abstr. 4199 (1996).

630. W. Sinclair, *Int. J. Radiat. Oncol. Biol. Phys.*, **7**, 631–637 (1981).

631. L. M. Perez and S. Greer, *Int. J. Radiat. Oncol. Biol. Phys.*, **12**, 1523–1527 (1986).

632. G. Giusti, C. Mangoni, B. De Petrocellis, and E. Scarano, *Enzyme Biol. Clin.*, **11**, 375–383 (1970).

633. K. Skov, *Radiat. Res.*, **112**, 217–242 (1987).

634. A. M. Joy, D. M. L. Goodgame, and I. J. Stratford, *Int. J. Radiat. Oncol. Biol. Phys.*, **16**, 1053–1056 (1989).

635. I. Hesslewood, W. Cramp, D. McBnen, P. Williamson, and K. Lott, *Br. J. Cancer*, **37** (Suppl. III), 95–97 (1978).

636. J. A. O'Hara, E. B. Douple, M. J. Abrams, D. J. Picker, C. M. Giandomenico, and J. F. Vollano, *Int. J. Radiat. Oncol. Biol. Phys.*, **16**, 1049–1052 (1989).

637. B. A. Teicher, J. L. Jacobs, K. N. S. Cathcart, M. J. Abrams, J. F. Vollond, and D. H. Picker, *Radiat. Res.*, **109**, 36–46 (1987).

638. W. R. Wilson, J. W. Moselen, S. Cliffe, W. A. Denny, and D. C. Ware, *Int. J. Radiat. Oncol. Biol. Phys.*, **29**, 323–327 (1994).

639. H. J. M. Groen, S. Sleijfer, C. Meijer, H. H. Kampinga, A. W. T. Konings, E. G. E. De Vries, and N. H. Mulder, *Br. J. Cancer*, **72**, 1406–1411 (1995).

640. C. Schaake-Koning, W. van den Bogaert, O. Dalesio, J. Festen, J. Hoogenhout, P. van Houtte, A. Kirkpatrick, M. Koolen, B. Maat, A. Nijs, A. Renaud, P. Rodrigus, L. Schuster-Uitterhoeve, J. P. Sculier, N. van Zandwijk, and H. Bartelink, *N. Engl. J. Med.*, **326**, 524–530 (1992).

641. K. A. Skov and N. P. Farrell, *Int. J. Radiat. Biol.*, **57**, 947–958 (1990).

642. R. Chibber, I. J. Stratford, I. Ahmed, A. B. Robbins, D. Goodgame, and B. Lee, *Int. J. Radiat. Oncol. Biol. Phys.*, **10**, 1213–1215 (1984).

643. R. Koch in G. C. DeHevesy, Ed., *Advances in Radiobiology*, Oliver & Boyd, Edinburgh, 1957, p. 170.

644. R. Lange and A. Pihl, *Int. J. Radiat. Biol.*, **3**, 249–258 (1961).

645. H. Langendorff, R. Koch, and U. Hagen, *Strahlentherapie*, **99**, 375 (1956).

646. S. Grigorescu, C. Nedelcu, and M. Netase, *Radiobiologiya*, **6**, 819–821 (1965).

647. D. H. Maman, B. Marshall, and J. S. Mitchell, *Chemotherapia*, **3**, 225 (1961).

648. W. E. Rothe, M. M. Grenan, and S. M. Wilson, *Prog. Biochem. Pharmacol.*, **1**, 372 (1965).

649. C. Stuart and G. Cittadini, *Minerva Fisioter. Radiobiol.*, **10**, 337 (1965).

650. G. Caprino and L. Caprino, *Minerva Fisioter. Radiobiol.*, **10**, 311 (1965).

651. L. A. Tiunov, N. A. Kachurina, and O. L. Smirnova, *Radiobiologiya*, **6**, 343–348 (1966).

652. H. Moroson and D. Martin, *Nature*, **214**, 304–306 (1967).

653. V. G. Yakovlev, *Radiobiologiya*, **4**, 656–659 (1964).

654. R. F. Pittillo, E. R. Bannister, and E. P. Johnson, *Can. J. Microbiol.*, **12**, 17 (1966).

655. J. Burns, J. P. Garcia, M. B. Lucas, C. Moncrief, and R. F. Pittillo, *Radiat. Res.*, **25**, 460–469 (1965).

656. R. F. Pittillo, M. B. Lucas, and E. R. Bannister, *Radiat. Res.*, **29**, 549–567 (1966).

657. M. J. Ashwood-Smith, D. M. Robinson, J. H. Barnes, and B. A. Bridges, *Nature*, **216**, 137–139 (1967).

658. Z. M. Bacq, M. L. Beaumariage, and S. Liebecq-Hunter, *Int. J. Radiat. Biol.*, **9**, 175–178 (1965).

659. W. A. Cramp, *Radiat. Res.*, **30**, 221–236 (1967).

660. R. K. Jain, *Sci. Am.*, **271**, 58–65 (1994).

661. E. A. Bump, S. J. Braunhut, S. T. Palayoor, D. Medeiros, L. Lai, B. Cerce, R. E. Langley, and C. N. Coleman, *Int. J. Radiat. Oncol. Biol. Phys.*, **29**, 249–253 (1994).

662. W. H. McBride, *Int. J. Radiat. Oncol. Biol. Phys.*, **33**, 233–234 (1995).

663. P. Rubin, C. J. Johnston, J. P. Williams, S. McDonald, and J. N. Finkelstein, *Int. J. Radiat. Oncol. Biol. Phys.*, **33**, 99–109 (1995).

664. K. Dittmann, H. Loffier, M. Bamberg, and H. P. Rodemann, *Radiother. Oncol.*, **34**, 137–143 (1995).

665. J. B. Little, *Radiat. Res.*, **140**, 299–311 (1994).

666. J. S. Murley and D. J. Grdina in E. A. Bump and K. Malaker, Eds., *Radioprotectors: Chemical, Biological and Clinical Perspectives*, CRC Press, Boca Raton, FL, 1997.

667. D. E. McClain, D. E. Kalinich, and N. Ramakrishnan, *FASEB J.*, **9**, 1345–1354 (1995).

668. S. T. Palayoor, R. M. Macklis, E. A. Bump, and C. N. Coleman, *Radiat. Res.*, **141**, 231–239 (1995).

669. H. Kimura, N. Z. Wu, R. Dodge, D. P. Spencer, B. M. Klitzman, T. M. McIntyre, and M. W. Dewhirst, *Int. J. Radiat. Oncol. Biol. Phys.*, **33**, 627–633 (1995).

670. S. J. Braunhut in E. A. Bump and K. Malaker, Eds., *Radioprotectors: Chemical, Biological and Clinical Perspectives*, CRC Press, Boca Raton, FL, 1997.

671. S. N. Mooteri, J. L. Podolski, E. A. Drab, T. J. Saclarides, J. M. Onoda, S. S. Kantak, and D. B. Rubin, *Radiat. Res.*, **145**, 217–224 (1996).

672. E. B. Saloman, J. H. Hubbell, and J. H. Scofield, *Atomic Data and Nuclear Data Tables*, **38**, 1–197 (1988).

673. H. E. Johns and J. R. Cunningham, *The Physics of Radiology*, Charles C. Thomas, Springfield, IL, 1983, p. 88.

674. C. N. Coleman in V. T. deVita Jr., S. Hellman, and S. A. Rosenberg, Eds., *Cancer: Principles and Practice of Oncology*, 4th ed., Lippincott, Philadelphia, 1993, p. 2703.

# CHAPTER FIVE

# Synthetic Antiangiogenic Agents

 OREST W. BLASCHUK
Mc Gill University, Royal Victoria Hospital
Montreal, Quebec, Canada


J. MATTHEW SYMONDS
Adherex Technologies Inc.
Ottawa, Ontario, Canada

## Contents

1 Introduction, 216
2 Formation of Blood Vessels, 216
    2.1 Structure of Blood Vessels, 216
    2.2 General Description of Angiogenesis, 216
3 Compounds That Inhibit Angiogenesis, 216
    3.1 Inhibitors of Proteolysis, 216
    3.2 Inhibitors of Growth Factor Receptor
        Function, 217
    3.3 Inhibitors of Endothelial Cell Migration, 219
    3.4 Disruptors of the Endothelial Cell
        Cytoskeleton, 219
4 Summary and Future Directions, 220

*Burger's Medicinal Chemistry and Drug Discovery*
Sixth Edition, Volume 5: Chemotherapeutic Agents
Edited by Donald J. Abraham
ISBN 0-471-37031-2   © 2003 John Wiley & Sons, Inc.

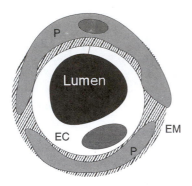

**Figure 5.1.** Structure of microvasculature. Endothelial cells (EC) form the wall of the microvessel and are surrounded by an extracellular matrix (EM) and pericytes (P).

## 1 INTRODUCTION

This chapter focuses on well-defined chemical compounds that antagonize specific targets believed to be critical regulators of the angiogenic process. These compounds have mainly been tested in the clinic for their ability to disrupt tumor-driven angiogenesis. It is noteworthy that very few antiangiogenic agents have thus far been successful in clinical trials.

## 2 FORMATION OF BLOOD VESSELS

### 2.1 Structure of Blood Vessels

Endothelial cells line the lumens of blood vessels (1, 2). In capillaries (i.e., small blood vessels, or microvessels), the endothelial cells are surrounded by a single layer of pericytes embedded in a basement membrane (also known as the basal lamina), whereas in large blood vessels (i.e., arteries and veins), endothelial cells rest on a lamina, which is encircled by multiple layers of smooth muscle cells (Fig. 5.1). Pericytes (which express $\alpha$-smooth muscle actin) and the smooth muscle cells are thought to regulate blood flow, as well as to stabilize the vasculature (1–3).

### 2.2 General Description of Angiogenesis

The process by which new vasculature arises from preexisting blood vessels is known as an-

giogenesis (4, 5) (Fig. 5.2). This process remains imperfectly understood. Endothelial cells are nonmigratory and quiescent (6). These cells become "activated" when the cumulative effects of proangiogenic stimuli are sufficient to overcome the influence of local antiangiogenic regulators (5). Endothelial cells of normal blood vessels rest on a basement membrane (1, 2) (Fig. 5.1). For angiogenesis to proceed in response to a stimulus, the endothelial cells must first detach from one another and degrade this basement membrane (4, 5, 7). Once released from these physical constraints, the endothelial cells begin migrating in a column toward the source of the proangiogenic stimulus. Cell division occurs at the front of the advancing column, whereas cell differentiation occurs in the rear. The differentiated endothelial cells adhere to one another and begin the process of forming a new capillary. Maturation of the capillaries is believed to be regulated by newly recruited pericytes, which are thought to suppress endothelial cell proliferation, as well as to collaborate with the endothelial cells in the synthesis of a basement membrane (2, 3, 8, 9).

## 3 COMPOUNDS THAT INHIBIT ANGIOGENESIS

### 3.1 Inhibitors of Proteolysis

The major structural components of blood vessel basement membranes are collagen IV, laminin, and heparan sulfate proteoglycans (HSPGs) (10–13). Endothelial cells use a group of zinc-binding endopeptidases, known as matrix metalloproteinases (MMPs), to degrade these components and breach the basement membrane (13–19). They also use these enzymes to migrate through the extracellular matrix of the perivascular stroma (which is predominantly composed of collagens I, III, and VII, fibronectin, and HSPGs) toward the proangiogenic stimulus (13, 14).

A number of synthetic substrate analog inhibitors of MMPs with antiangiogenic activity have been developed (20, 21) (Table 5.1). Two of the most potent broad-spectrum inhibitors

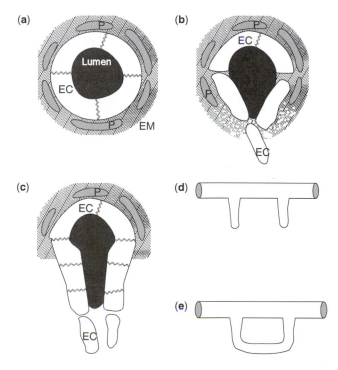

**Figure 5.2.** The angiogenic process. (a) The lumen of a mature microvessel is formed by endothelial cells (EC), which are supported by pericytes (P) and an extracellular matrix (EM). (b) After receiving an angiogenic signal, endothelial cells secrete enzymes (matrix metalloproteinases) that degrade the extracellular matrix, thus allowing them to migrate into the surrounding stroma. (c, d) The migrating endothelial cells proceed to form a microvascular column, or sprout. (e) Finally, the sprouts coalesce to from intact vessels. Further details concerning the angiogenic process are given in the text.

of MMP activity, marimastat and batimastat, are collagen peptidomimetics coupled to the zinc-binding group, hydroxamic acid (Fig. 5.3) (14, 15, 22, 23). These inhibitors bind to the active site of MMPs, which contains a zinc atom. They have been shown to be capable of inhibiting the formation of microvessels *in vitro* (24) and angiogenesis in animal models (25). Unfortunately, clinical trials using these and other MMP inhibitors to treat angiogenic-based diseases (such as cancer) have not been overly successful (20, 21). The reasons for the lack of success of MMP inhibitors in the clinic remain obscure, although it has become obvious that the roles played by MMPs during angiogenesis *in vivo* require further clarification (21).

## 3.2 Inhibitors of Growth Factor Receptor Function

The ability of receptor tyrosine kinases (RTKs) and their growth factor ligands to regulate angiogenesis has been extensively documented (5, 8, 9, 17, 26, 27). RTKs are integral membrane glycoproteins that exist as monomers on the cell surface (26, 27). They form dimers upon binding to their ligands. This triggers their intracellular kinase domain to catalyze the transfer of phosphate from ATP to protein substrates, thus initiating signaling cascades that ultimately cause changes in gene expression (26–30).

Vascular endothelial growth factor (VEGF), fibroblast growth factor (FGF), and platelet-

**Table 5.1  MMP Inhibitors with Demonstrated Antiangiogenic Activity**[a]

| Inhibitor | Class | Company |
|-----------|-------|---------|
| Marimastat | Peptidomimetic | British Biotech |
| Batimastat | Peptidomimetic | British Biotech |
| AG3340 | Nonpeptidomimetic | Aguoron Pharmaceuticals |
| Bay 12-9566 | Nonpeptidomimetic | Bayer Corporation |
| BMS-275291 | Nonpeptidomimetic | Bristol-Myers Squibb |
| CGS 27023A | Nonpeptidomimetic | Novartis Pharmaceuticals |

[a]Modified from Ref. 20.

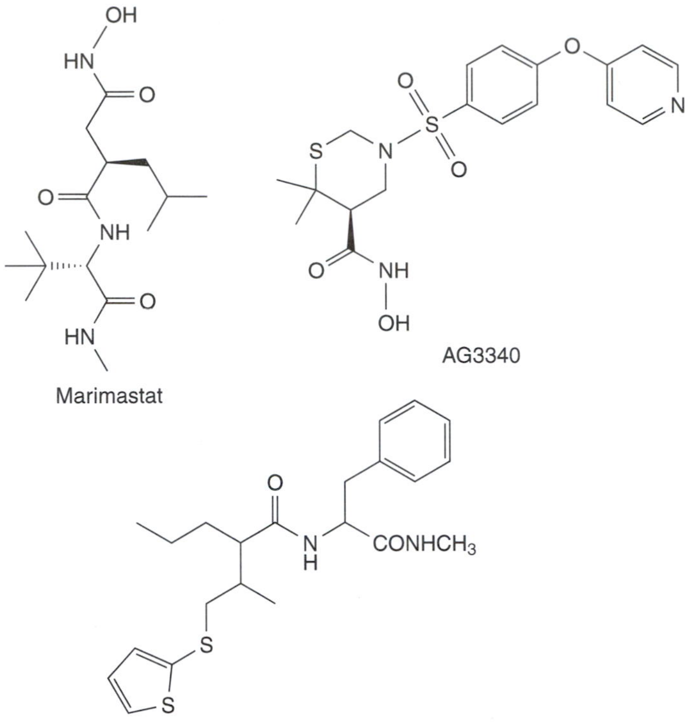

**Figure 5.3.** Structures of matrix metalloproteinase inhibitors.

derived growth factor (PDGF) are considered to be the most important positive regulators of angiogenesis (1, 5, 8, 9). VEGF and FGF stimulate endothelial cell migration, proliferation, and survival (1, 9, 17, 28, 31, 32), whereas PDGF regulates pericyte recruitment to immature microvessels (3, 8). In addition, VEGF is a potent inducer of vascular permeability, causing the detachment of endothelial cells from one another within microvessels (33, 34). This is likely to be the first step in the angiogenic process. Endothelial cells express the RTKs for VEGF, PDGF, and FGF (VEGF-R, PDGF-R, and FGF-R, respectively), whereas pericytes express FGF-R and PDGF-R (2, 35).

Two inhibitors of VEGF-R (also known as Flk-1/KDR) kinase activity are SU5416 and

ZD4190 (Table 5.2) (26, 36). SU5416 and ZD4190 have indolinone and quinazoline core structures, respectively (Fig. 5.4). A third broad-spectrum RTK inhibitor with an indolinone structure is SU6668 (37). Crystallographic studies have shown that SU6668 interacts with amino acids at the entrance to the ATP-binding site in the RTKs.

Because VEGF, FGF, and PDGF are all positive regulators of angiogenesis, it is not surprising that SU6668 is a more potent inhibitor of angiogenesis than agents (such as SU5416) that target only a single angiogenic factor (35, 37). SU6668 has been shown to inhibit VEGF- and FGF-induced endothelial cell proliferation *in vitro* and cause endothelial cell apoptosis within tumors *in vivo* (35, 37).

**Table 5.2   RTK Inhibitors with Demonstrated Antiangiogenic Activity**[a]

| Inhibitor | Target | Company | Clinical Trial |
|-----------|--------|---------|----------------|
| SU5416 | VEGF-R | Sugen | Phase III (colorectal cancer) |
| SU6668 | VEGF-R, FGF-R, PDGF-R | Sugen | Phase I |
| ZD4190 | VEGF-R | AstraZeneca | Discontinued |

[a]Modified from Ref. 26.

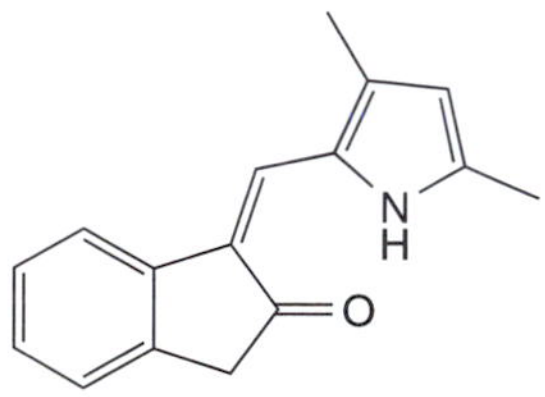

SU5416

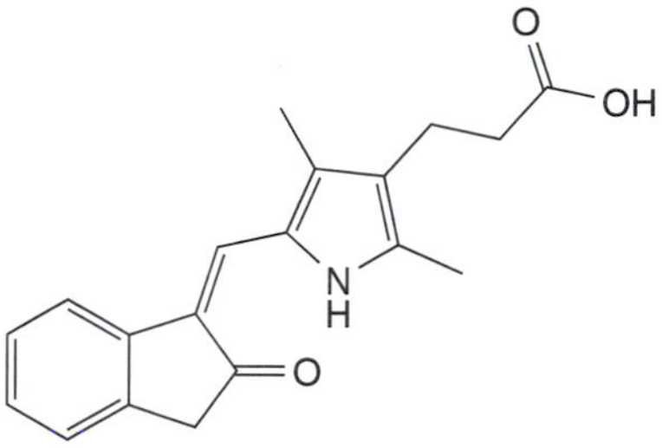

SU6668

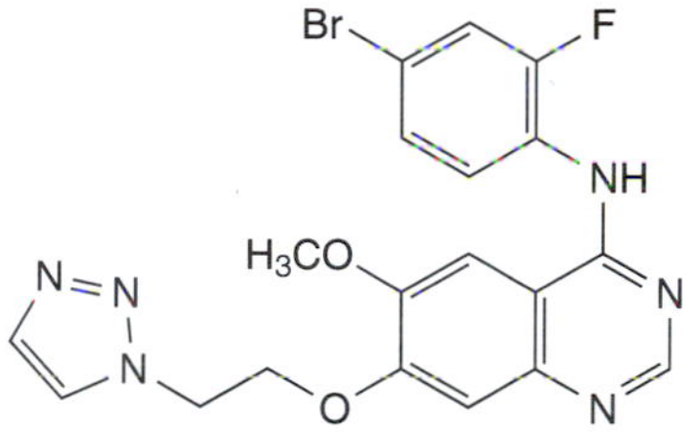

ZU4190

**Figure 5.4.** Structures of receptor tyrosine kinase inhibitors.

Furthermore, this agent also caused a decrease in pericyte vessel coverage within the newly formed vasculature of tumors *in vivo* (35). Collectively, these observations suggest that agents capable of simultaneously antagonizing the activity of multiple RTKs should be effective inhibitors of angiogenesis in the clinic.

### 3.3   Inhibitors of Endothelial Cell Migration

Integrins mediate the migration of endothelial cells through the extracellular matrix (38–41). These cell surface receptors are heterodimers composed of $\alpha$- and $\beta$-subunits (42, 43). The integrins $\alpha_v\beta_5$, $\alpha_v\beta_3$, and $\alpha_5\beta_1$ have been shown to play pivotal roles during angiogenesis (38–41). These integrins all recognize the amino acid sequence Arg-Gly-Asp (RGD), which is found within many extracellular ma-

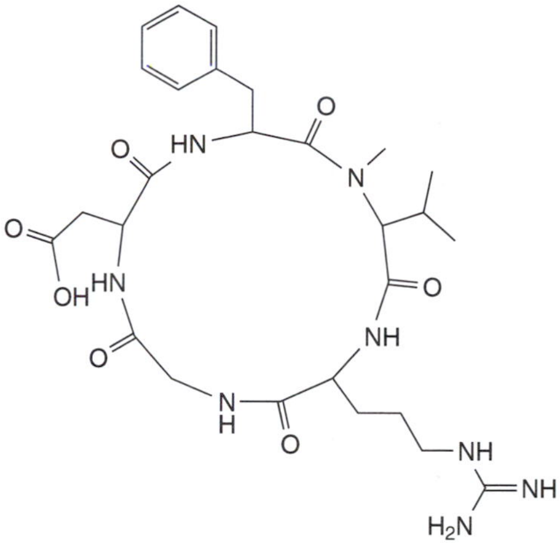

**Figure 5.5.** Structure of cyclo(RGDf-N(Me)V-).

trix proteins (such as fibronectin) (44). They bind to RGD-containing proteins with different affinities, thus making it possible to design antagonists that are specific for individual integrins (45–47).

Linear (44) and cyclic (41, 45, 46) peptides containing the RGD sequence have been developed that inhibit the interaction between integrins and various RGD-containing extracellular matrix proteins. Cyclic RGD-containing peptide antagonists of integrin $\alpha_v\beta_3$ have been shown to block angiogenesis (41, 45). Currently, the integrin $\alpha_v\beta_3$ antagonist cyclo (RGDf-N(Me)V-) (code EMD121974; Merck, Darmstadt, Germany) (Fig. 5.5) is in phase I/II clinical trials for Kaposi's sarcoma, brain tumors, and solid tumors (48).

### 3.4   Disruptors of the Endothelial Cell Cytoskeleton

Microtubules are an integral component of the endothelial cell cytoskeleton (49). They are in-

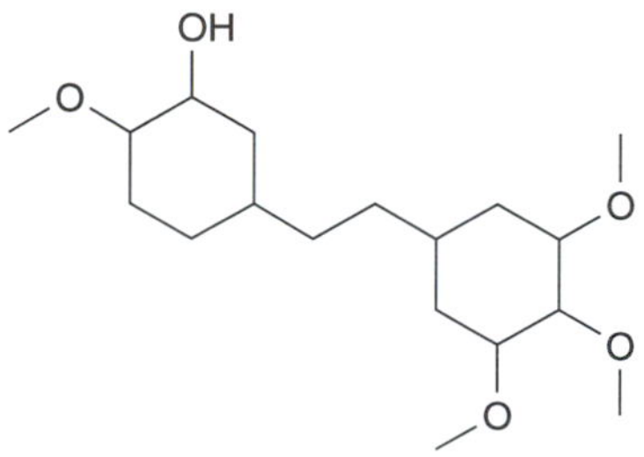

**Figure 5.6.** Structure of combretastatin A4.

**Table 5.3  Examples of Angiogenesis-Related Conditions**[a]

| | |
|---|---|
| Malignant tumors | Synovitis |
| Trachoma | Hemophilic joints |
| Meningioma | Psoriasis |
| Hemangioma | Pyogenic granuloma |
| Angiofibroma | Atherosclerotic plaques |
| Diabetic retinopathy | Hypertrophic scars |
| Neovascular glaucoma | Retrolental fibroplasias |
| Macular degeneration | Scleroderma |
| Vascular restenosis | Vascular adhesions |
| Corneal graft neovascularization | Dermatitis |
| Arteriovenous malformations | Endometriosis |
| Hemorrhagic telangiectasia | Rheumatoid arthritis |

[a]From Ref. 63.

volved in regulating endothelial cell shape and division (49–52). Microtubules are composed of tubulin polymers (53). Drugs capable of interfering with tubulin polymerization would therefore be expected to be effective antiangiogenic agents (54).

Combretastatin A4 (Fig. 5.6) is a tubulin-binding compound that was initially isolated from the South African tree *Combretan caffrum* (55, 56). This lipophilic *cis* stilbene destabilizes microtubules, thus affecting the cytoskeleton of dividing endothelial cells, causing apoptosis and inhibiting angiogenesis (53, 54, 56–58).

The solubility of combretastatin A4 was increased by attaching a phosphate moiety (59). The resulting compound, combretastatin A4 3-*o*-phosphate is a water-soluble prodrug whose phosphate moiety is cleaved *in vivo* by serum phosphatases to generate the active drug. This prodrug (developed by Oxigene Inc.) is currently in phase I clinical trials to evaluate its effects on tumor vasculature (54, 60).

## 4  SUMMARY AND FUTURE DIRECTIONS

Synthetic antiangiogenic compounds directed against four main groups of targets are currently being tested in the clinic: MMP, RTK, integrin, and microtubule antagonists. The preliminary clinical results obtained using most of these antagonists have been encouraging.

In view of the success of RTK antagonists in the clinic, a potentially promising area of investigation would be the development of compounds that antagonize the interaction between growth factors and their RTKs. Furthermore, in light of the clinical results obtained using antagonists of cell surface adhesion molecules (integrins), it would seem reasonable to target other adhesion molecules such as VE-cadherin and N-cadherin (1).

Only a few synthetic drugs have thus far been developed that are directed against specific proteins involved in angiogenesis. Although angiogenesis has been well described from a morphological perspective, there is a need to more precisely define the molecular mechanisms regulating this process (61, 62). Only then can new drugs be developed that are capable of modulating angiogenesis.

Most of the clinical trials to date have focused on determining the ability of antiangiogenic agents to prevent malignant tumor neovascularization. Angiogenesis has also been implicated in a wide variety of nonneoplastic diseases (Table 5.3) (63). Further clinical research is necessary to determine whether antiangiogenic agents will be useful in the treatment of these diseases.

## REFERENCES

1. O. W. Blaschuk and T. M. Rowlands, *Cancer Metastasis Rev.*, **19**, 1–5 (2000).

2. K. K. Hirschi and P. A. D'Amore, *Cardiovasc. Res.*, **32**, 687–698 (1996).

3. P. Lindahl, B. R. Johansson, P. Leveen, and C. Betsholtz, *Science*, **277**, 242–245 (1997).

4. G. W. Cockerill, J. R. Gamble, and M. A. Vadas, *Int. Rev. Cytol.*, **159**, 113–160 (1995).

5. D. Hanahan and J. Folkman, *Cell*, **86**, 353–364 (1996).

6. B. Hobson and J. Denekamp, *Br. J. Cancer*, **49**, 405–413 (1984).

7. D. H. Ausprunk and J. Folkman, *J. Microvasc. Res.*, **14**, 53–65 (1977).

8. M. Hellstrom, M. Kalen, P. Lindahl, A. Abramsson, and C. Betsholtz, *Development*, **126**, 3047–3055 (1999).

9. L. E. Benjamin, D. Golijanin, A. Itin, D. Pode, and E. Keshet, *J. Clin. Invest.*, **103**, 159–165 (1999).

10. R. Timpl, *Eur. J. Biochem.*, **180**, 487–502 (1989).

11. P. D. Yurchenco and J. C. Schittny, *FASEB J.*, **4**, 1577–1590 (1990).

12. M. Klagsbrun, *Curr. Opin. Cell Biol.*, **2**, 857–863 (1990).

13. J. T. Price, M. T. Bonovich, and E. C. Kohn, *Crit. Rev. Biochem. Mol. Biol.*, **32**, 175–253 (1997).

14. A. E. Yu, R. E. Hewitt, E. W. Connor, and W. G. Stetler-Stevenson, *Drugs Aging*, **11**, 229–244 (1997).

15. I. Massova, L. P. Kotra, R. Fridman, and S. Mobashery, *FASEB J.*, **12**, 1075–1095 (1998).

16. M. A. Moses, *Stem Cells*, **15**, 180–189 (1997).

17. S. Liekens, E. De Clerq, and J. Neyts, *Biochem. Pharmacol.*, **61**, 253–270 (2001).

18. J. Fang, Y. Shing, D. Wiederschain, L. Yan, C. Butterfield, G. Jackson, J. Harper, G. Tamvakopoulos, and M. A. Moses, *Proc. Natl. Acad. Sci. USA*, **97**, 3884–3889 (2000).

19. W. G. Stetler-Stevenson, *J. Clin. Invest.*, **103**, 1237–1241 (1999).

20. M. Hidalgo and S. G. Eckhardt, *J. Natl. Cancer Inst.*, **93**, 178–193 (2001).

21. S. Zucker, J. Cao, and W.-T. Chen, *Oncogene*, **19**, 6642–6650 (2000).

22. S. Odake, Y. Morita, T. Morikawa, N. Yoshida, H. Hori, and Y. Nagai, *Biochem. Biophys. Res. Commun.*, **199**, 1442–1446 (1994).

23. H. S. Rasmussen and P. P. McCann, *Pharmacol. Ther.*, **75**, 69–75 (1997).

24. W.-H. Zhu, X. Guo, S. Villaschi, and R. F. Nicosia, *Lab. Invest.*, **80**, 545–555 (2000).

25. G. Bergers, K. Javaherian, K. M. Lo, J. Folkman, and D. Hanahan, *Science*, **284**, 808–812 (1999).

26. J. M. Cherrington, L. M. Strawn, and L. K. Shawver, *Adv. Cancer Res.*, **79**, 1–38 (2000).

27. T. Pawson, *Nature*, **373**, 573–580 (1995).

28. W. Risau, *Nature*, **386**, 671–674 (1997).

29. H. D. Madhani, *Cell*, **106**, 9–11 (2001).

30. T. Pawson and T. M. Saxton, *Cell*, **97**, 675–678 (1999).

31. G. Seghezzi, S. Patel, C. J. Ren, A. Gualandris, G. Pintucci, E. S. Robbins, R. L. Shapiro, A. C. Galloway, D. B. Rifkin, and P. Mignatti, *J. Cell Biol.*, **141**, 1659–1673 (1998).

32. J. Folkman and P. A. D'Amore, *Cell*, **87**, 1153–1155 (1996).

33. W. G. Roberts and G. E. Palade, *Cancer Res.*, **57**, 765–772 (1997).

34. H. F. Dvorak, L. F. Brown, M. Detmar, and A. M. Dvorak, *Am. J. Pathol.*, **146**, 1029–1039 (1995).

35. R. M. Shaheen, W. W. Tseng, D. W. Davis, W. Liu, N. Reinmuth, R. Vellagas, A. A. Wieczorek, Y. Ogura, D. J. McConkey, K. E. Drazan, C. D. Bucana, G. McMahon, and L. M. Ellis, *Cancer Res.*, **61**, 1464–1468 (2001).

36. G. McMahon, L. Sun, C. Liang, and C. Tang, *Curr. Opin. Drug Disc. Dev.*, **1**, 131–146 (1998).

37. A. D. Laird, P. Vajkoczy, L. K. Shawver, A. Thurnher, C. Liang, M. Mohammadi, J. Schlessinger, A. Ullrich, S. R. Hubbard, R. A. Blake, T. A. T. Fong, L. M. Strawn, L. Sun, C. Tang, R. Hawtin, F. Tang, N. Shenoy, P. Hirth, G. McMahon, J. Cherrington, and M. Cherrington, *Cancer Res.*, **60**, 4152–4160 (2000).

38. B. P. Eliceiri and D. A. Cheresh, *J. Clin. Invest.*, **103**, 1227–1230 (1999).

39. R. O. Hynes, B. L. Bader, and K. Hodivala-Dilke, *Braz. J. Med. Biol. Res.*, **32**, 501–510 (1999).

40. S. Kim, K. Bell, S. A. Mousa, and J. A. Varner, *Am. J. Pathol.*, **156**, 1345–1362 (2000).

41. M. Friedlander, P. C. Brooks, R. W. Shaffer, C. M. Kincaid, J. A. Varner, and D. A. Cheresh, *Science*, **270**, 1500–1502 (1995).

42. K. Burridge and M. Chrzanowska-Wodnicka, *Annu. Rev. Cell Dev. Biol.*, **12**, 463–519 (1996).

43. R. O. Hynes, *Cell*, **69**, 11–25 (1992).

44. E. Ruoslahti, *Annu. Rev. Cell Dev. Biol.*, **12**, 697–715 (1996).

45. P. C. Brooks, A. M. P. Montgomery, M. Rosenfeld, R. A. Reisfeld, T. Hu, G. Klier, and D. A. Cheresh, *Cell*, **79**, 1157–1164 (1994).

46. M. Pfaff, K. Tangemann, B. Muller, M. Gurrath, G. Muller, H. Kessler, R. Timpl, and J. Engel, *J. Biol. Chem.*, **269**, 20233–20238 (1994).

47. K. M. Yamada, *J. Biol. Chem.*, **266**, 12809–12812 (1991).

48. M. A. Dechantsreiter, E. Planker, B. Matha, E. Lohof, G. Holzemann, A. Jonczyk, S. L. Goodman, and H. Kessler, *J. Med. Chem.*, **42**, 3033–3040 (1999).

49. A. I. Gottlieb, B. L. Langille, M. K. K. Wong, and D. W. Kim, *Lab. Invest.*, **65**, 123–137 (1991).

50. G. M. Tozer, V. E. Prise, J. Wilson, M. Cemazar, S. Shan, M. W. Dewhirst, P. R. Barber, B. Vojnovic, and D. J. Chaplin, *Cancer Res.*, **61**, 6413–6422 (2001).

51. A. M. Malek and S. Izumo, *J. Cell Sci.*, **109**, 713–726 (1996).

52. S. M. Galbraith, D. J. Chaplin, F. Lee, M. R. Stratford, R. J. Locke, B. Vojnovic, and G. M. Tozer, *Anticancer Res.*, **21**, 93–102 (2001).

53. A. Desai and T. J. Mitchison, *Ann. Rev. Cell Dev. Biol.*, **13**, 83–117 (1997).

54. J. Griggs, J. C. Metcalfe, and R. Hesketh, *Lancet Oncol.*, **2**, 82–87 (2001).

55. C. M. Lin, H. H. Ho, G. R. Pettit, and E. Hamel, *Biochemistry*, **28**, 6984–6991 (1989).

56. G. R. Petit, S. B. Singh, M. R. Boyd, E. Hamel, R. K. Petit, J. M. Schmidt, and F. Hogan, *J. Med. Chem.*, **38**, 1666–1672 (1995).

57. G. G. Dark, S. A. Hill, V. E. Prise, G. M. Tozer, G. R. Pettit, and D. J. Chaplin, *Cancer Res.*, **57**, 1829–1834 (1997).

58. S. Iyer, D. J. Chaplin, D. S. Rosenthal, A. H. Boulares, L.-Y. Li, and M. E. Smulson, *Cancer Res.*, **58**, 4510–4514 (1998).

59. G. R. Pettit and M. R. Rhodes, *Anticancer Drug Des.*, **13**, 183–191 (1998).

60. J. Adams, and P. J. Elliot, *Oncogene*, **19**, 6687–6692 (2000).

61. B. St. Croix, C. Rago, V. Velculescu, G. Traverso, K. E. Romans, E. Montgomery, A. Lal, G. J. Riggins, C. Lengauer, B. Vogelstein, and K. W. Kinzler, *Science*, **289**, 1197–1202 (2000).

62. E. B. Carson-Walter, D. N. Watkins, A. Nanda, B. Vogelstein, K. W. Kinzler, and B. St. Croix, *Cancer Res.*, **61**, 6649–6655 (2001).

63. P. E. Thorpe and R. A. B. Brekken, U.S. Pat. 6,342,219 (2002).

CHAPTER SIX

# Future Strategies in Immunotherapy

DOUGLAS F. LAKE
SARA O. DIONNE
University of Arizona Cancer Center
Tucson, Arizona

## Contents

1 Introduction, 224
   1.1 Innate Immunity, 224
   1.2 Adaptive Immunity, 224
2 Antibody-Directed Immunotherapy, 226
   2.1 History, 226
      2.1.1 Humanized Antibodies, 226
      2.1.2 Chimeric Antibodies, 226
   2.2 Current Antibodies on the Market, 227
      2.2.1 Therapeutic Antibodies, 227
      2.2.2 Trastuzumab, 230
      2.2.3 Rituximab, 230
      2.2.4 Gemtuzumab, 231
      2.2.5 Alemtuzumab, 233
   2.3 Anti-Inflammatory Therapeutic Antibodies, 233
      2.3.1 Infliximab, 233
      2.3.2 Etanercept, 234
3 Dendritic Cell Immunotherapy, 235
   3.1 Dendritic Cell Physiology, 235
   3.2 T-Cell Activation by Dendritic Cells, 235
   3.3 Immunotherapy Using Dendritic Cells, 236
   3.4 Dendritic Cell Immunotherapy Approaches for Undefined Tumor Antigens, 239
      3.4.1 Apoptotic Tumor Cells, 239
      3.4.2 Tumor Lysates, 239
      3.4.3 Tumor RNA, 240
      3.4.4 DC-Tumor Cell Fusion, 240
      3.4.5 Exosomes, 241
   3.5 Dendritic Cells Transduced with Viral Vectors, 241
      3.5.1 Adenoviral Transduction of DC, 242
      3.5.2 Pox Virus Transduction of DC, 242
4 Conclusions, 243
5 Abbreviations, 243

*Burger's Medicinal Chemistry and Drug Discovery*
Sixth Edition, Volume 5: Chemotherapeutic Agents
Edited by Donald J. Abraham
ISBN 0-471-37031-2　© 2003 John Wiley & Sons, Inc.

# 1   INTRODUCTION

The immune system is a complex, but remarkably precise cellular and molecular orchestra capable of responding to, eliminating, and providing protection from invading pathogens such as viruses, bacteria, fungi, and even tumors. The primary function of the immune system is to distinguish "self" from non-self. The term "self" encompasses host-derived tissue, whereas non-self includes infectious and non-infectious foreign agents. The immune system consists of many cell types, each specialized to fulfill an important role in the generation of a robust immune response on encounter of an invading agent and to provide life-long protection to the host.

The first section of this chapter will focus on the cellular and molecular components of the immune response and will be discussed in the context of future directions in immunotherapy. The remainder of the chapter will discuss current cutting edge and future immunotherapeutic strategies for cancer.

## 1.1   Innate Immunity

The two major components of immunity are the innate and adaptive (acquired) immune systems. As part of the innate immune system, physical barriers such as skin and mucosa provide the first line of defense against the outside environment (1). If skin or mucosal barriers (ocular conjunctiva, for example) are broken and an invading agent enters the eye or the blood stream, biochemical barriers such as lysozyme, present in tears, and serum complement provide protection from bacterial invasion and infection. Lysozyme breaks down bacterial cell walls and complement deposits on the surfaces of bacteria and viruses, ultimately resulting in their destruction. The breakdown of complement components also attracts inflammatory cells such as macrophages, neutrophils, and other granulocytes that are important cellular components of the innate immune response (2). These inflammatory cells bridge innate and adaptive immune responses. If these innate components are not successful, or if the antigenic insult is sufficiently large, adaptive immunity is called into play.

## 1.2   Adaptive Immunity

The adaptive immune system responds specifically to an enormous arsenal of antigens, discriminates between foreign and "self" antigens, and remembers previously encountered antigens so it can respond faster and more effectively to a second antigenic challenge. An adaptive immune response results in production of antigen-specific antibodies and T-cells. Antibodies provide protection against repeated invasion in an infectious disease setting and also coat, or opsonize, agents so that effector cells can destroy them (3). CD4-positive T-helper cells signal other cells in the immune system using a network of cytokines (4), whereas CD8-positive cytotoxic T-lymphocytes (CTL) lyse tumor cells or virally infected cells (5). Multiple gene segment rearrangements of B-cell receptors [immunoglobulins (Ig)] and T-cell receptors generate over a million different specificities in each B-cell and T-cell population to match the antigenic diversity found in nature (6).

Generation of specific immunity requires antigen-presenting cells (APC), including macrophages, dendritic cells (DC), skin-derived DC (Langerhans cells), and B-lymphocytes. These cells enzymatically digest protein antigens and present the derived peptides to the T-cell receptor (TCR) in association with class I and class II major histocompatibility complex (MHC) proteins (7). $CD4^+$ T-helper lymphocytes recognize peptides presented by MHC class II, whereas $CD8^+$ cytotoxic T-lymphocytes (CTL) recognize peptides presented by MHC class I. Class II peptides are derived from proteins outside the cell that have been engulfed by APCs into endosomes. These proteins are enzymatically digested into individual peptides, loaded onto MHC class II and carried to the surface for presentation to $CD4^+$ T-cells. In contrast to class II peptides, class I peptides are derived from within the cell. For example, cells that become infected with a virus present endogenous peptides in class I molecules to CD8$^+$ T-cells. Tumor cells also present endogenous tumor antigens in context of MHC class I. APC often become infected with viruses and present peptides derived from exogenous and endogenous sources to both CD4 and CD8$^+$ T-cells simultaneously. When an APC and T-cell interact, the TCR is responsible for the specificity of the interaction, and the interaction between a variety of co-stimulatory and adhesion molecules on both APC and the T-cell activate the T-cell, resulting in secretion of appropriate cytokines.

Two subsets of T-helper lymphocytes (Th1 and Th2) have been reported based on the cytokines they secrete. The Th1 subset produces interferon-$\gamma$ (IFN-$\gamma$) and interleukin-2 (IL-2) and enhances cell-mediated immunity by activation of macrophages, CTL, and natural killer (NK) cells. The Th2 subset produces IL-4, IL-5, and IL-6, which induce B-cell proliferation followed by differentiation into antibody-secreting plasma cells. These subsets down-regulate each other, such that Th2 cells produce IL-10, which inhibits cytokine production by Th1 lymphocytes. Conversely, IFN-$\gamma$ produced by Th1 cells inhibits the proliferation of Th2 cells. Although these subsets have been well characterized *in vitro*, it is still unclear whether they exist *in vivo* as distinct populations or as a single T-helper cell population that is capable of exhibiting either phenotype depending on the type of the antigen they encounter (8).

CD8$^+$ CTL are activated only after peptide antigen presentation and co-stimulation by APC. Then they become serial killers and no longer require co-stimulation by APC. Without initial co-stimulation, T-cells may become unresponsive to further antigenic stimulation (anergy). Activated CTL induce target cell death through lytic granule enzymes ("granzymes"), perforin, and the Fas-Fas ligand (Fas-FasL) apoptosis pathways. CD8$^+$ cells are thought to be a major component of anti-tumor responses and are the activation target of nearly all cellular immunotherapy protocols.

The immune system may eliminate tumors that arise spontaneously. Circumstantial evidence that the immune system is capable of tumor destruction is offered by the fact that individuals with natural or acquired immune deficiencies develop cancer at an increased incidence over the general population. For example, AIDS patients develop tumors at a much high rate than immunocompetent individuals (9). Spontaneous regression of tumors may result from hormonal fluctuations, tumor necrosis caused by abrogation of blood supply, elimination of carcinogens, differentiation, epigenetic mechanisms, apoptosis, and even psychological factors. However, the prevailing view is that immunological mechanisms play a leading role in spontaneous tumor regressions (10).

Although the innate immune system is clearly involved in prevention and early rejection of tumors, two general arms of the adaptive immune system participate in the elimination of established tumors. The humoral arm of the immune system consists of B-lymphocytes that primarily secrete antibodies. The cellular arm of the adaptive immune system is composed of T-cells responsible for tumor cell destruction. Both arms interact with each other intimately. For example, B-cells expressing surface immunoglobulin in peripheral secondary lymphoid organs may bind a soluble tumor protein antigen shed from a tumor (Her-2/neu, for example) (11). That protein is then internalized, processed into peptides and then presented as peptide-class II major histocompatibility complexes to CD4$^+$ T-helper cells that recognize each peptide-class II complex through their TCR. These T-helper cells then secrete cytokines, such as IL-4, that act directly on the B-cell that bound the antigen to help it differentiate into clonal antibody-secreting plasma cells producing high affinity antibody. Circulating soluble antibody may bind to antigen on the surfaces of tumor cells, marking the tumor cells for destruction by effector immune cells such as macrophages, NK cells, or neutrophils by antibody-dependent cellular cytotoxicity (ADCC) or by complement-mediated lysis.

One problem for the cellular arm of the immune system in fighting cancer is that most tumor cells express self-antigens. The immune system is tolerant to self (12), as T-cells strongly reactive to self-peptides are deleted in the thymus early in life (13–15). Therefore, when peptides from a tumor are presented to

T-cells, they do not recognize the peptides as avidly as they would if the cell were infected with a virus and presenting a foreign peptide. However, anti-tumor T-cells exist and can be isolated from peripheral blood and from tumors or tumor-draining lymph nodes (16–19). Tumor cells secrete cytokine that suppress immune cell function such as including transforming growth factor-$\beta$ and IL-10 (20–29). Several strategies have been employed to rescue and re-activate tumor-specific T-cells, including isolation of tumor-infiltrating lymphocytes (TIL), expansion of TIL, demonstration of autologous tumor lysis, and re-infusion of these TIL into the patient, indicating existence of tumor-specific CTL (30). Recent strategies for activating tumor-specific CTL have focused on the use of dendritic cells, which will be discussed in the latter section of this chapter. Next, we describe antibodies that have been humanized or made chimeric using genetic engineering techniques that are FDA-approved for cancer and anti-inflammatory treatment modalities.

## 2 ANTIBODY-DIRECTED IMMUNOTHERAPY

### 2.1 History

The development of "hybridoma" technology by Milstein and Köohler in the mid-1970s revolutionized the generation of specific antibodies for use in research and clinical applications (31). Hybridomas are made by fusing antibody-forming B-cells with an immortal, non–antibody-secreting plasma cell line resulting in a population of hybrid cells that are selected for secretion of an antibody specific for an antigen of interest. The secret to this technology is that the immortal plasma cell line does not secrete antibody and is deficient in a purine enzymatic salvage pathway, hypoxanthine phosphoribosyl transferase (HPRT). When these plasma cells are fused to B-cells and placed in medium containing hypoxanthine-aminopterin-thymidine (HAT), the aminopterin poisons the *de novo* purine synthesis pathway. Unfused cells die, and only the hybridomas survive in HAT, while the B-cell component of the hybridoma provide the purine salvage pathway and the plasma cells con-

tribute unlimited *in vitro* proliferation. Supernatants of bulk cultures of hybridomas are then screened for the presence of antibodies of interest, usually by ELISA. On a "hit" (a well containing hybridomas secreting antigen-reactive antibody), the hybridomas can be subcloned by limiting dilution such that they are monoclonal. Monoclonal hybridomas secreting an antibody of interest can then be mass-cultured for antibody production and re-tested for reactivity against its target and lack of reactivity against other tissues.

**2.1.1 Humanized Antibodies.** As one might imagine, mouse monoclonal antibodies far outnumber human monoclonal antibodies, because it is relatively easy to immunize mice and obtain splenic B-lymphocytes for hybridoma formation. The drawback of using mouse monoclonal antibodies for passive immunotherapy is that they induce "human anti-mouse antibody" (HAMA) responses when administered to humans (32). Therefore, many mouse antibodies have been "humanized" for therapeutic use in humans (33). Humanization of antibodies consists of exchanging mouse constant domains with their homologous human constant domains, thus decreasing the immunogenicity of the mouse antibody when administered to humans. A diagram of an antibody is shown in Fig. 6.1. To further humanize an antibody, mouse framework variable regions are replaced with human frameworks that provide the scaffolding for the complementarity determining regions (CDR). The problem with replacing mouse framework sequences with their homologous human sequences is that antibody specificity is often lost. There is a delicate balance between specificity and humanization. Computer modeling is most often used to overlay human amino acid sequences on mouse framework structures to determine what gene family of framework regions might least distort the CDRs. Then human framework region DNA is spliced frameworks 1–4, among mouse CDRs 1–3, cloned into an expression vector, and expressed as a recombinant protein from appropriate cells.

**2.1.2 Chimeric Antibodies.** Chimeric antibodies are less "human" than humanized an-

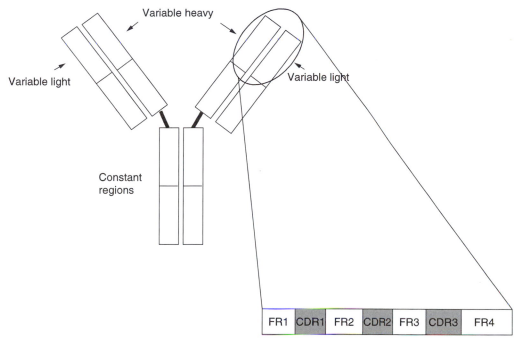

Variable domain structure

**Figure 6.1.** General structure of an antibody molecule. IgG1 contains three constant heavy chain domains, CH 1–3, and one variable domain. Each domain is approximately 110 amino acids in length. Variable light chains and constant light chains are noncovalently associated to the heavy chain variable and CH1 domains. Interchain disulfide bonding links the two heavy chains together in the hinge region, whereas intrachain disulfide bonding occurs within domains. The Fc portion of an antibody activates serum complement. The Fc portion binds Fc receptors on effector cells such as macrophages and neutrophils. The Fab contains variable heavy and light chains, and constant heavy and light chains. Fabs may be generated by papain digestion of IgG. Framework (FR) sequences interspersed between complementarity determining regions (CDR) may be exchanged among species, because CDR sequences impart specificity of the antibody.

tibodies, and as such, generally elicit more HAMA responses than a fully humanized antibody. Chimeric antibodies are generated by splicing DNA from murine heavy and light chain variable regions onto human IgG constant region DNA. This means that the entire variable region, including frameworks, is of murine origin, and only the effector, or Fc portion, of the antibody is human. The Fc portion of the antibody imparts effector function and allows antibody-dependent cellular cytotoxicity (ADCC) to occur after the antibody has bound to its target. Specificity is not usually a problem after generating a chimeric antibody, because it retains the natural framework scaffolds as the parent murine antibody.

## 2.2 Current Antibodies on the Market

**2.2.1 Therapeutic Antibodies.** Recently, the FDA has approved several humanized monoclonal antibodies for the treatment of various diseases. Because this chapter concentrates on future directions in immunotherapy for cancer, only chimeric or fully humanized antibodies will be described in detail. It should be recognized that antibodies for infectious diseases such as respiratory syncytial virus have been humanized (Palivizumab, Synagis; Medimmune) and are significant products for the treatment of those diseases. It is also important to note that many mouse monoclonal antibodies directed against human tumor an-

**Table 6.1 Therapeutic Antibodies**

| Trade Name (Generic Name) | Target | Manufacturer | Structure/ Chemical Class | Route of Administration | Efficacy/Potency | Dose | Potential Side Effects | Drug Interactions/ Contraindications | Absorption, Distribution, Metabolism, Eliminations |
|---|---|---|---|---|---|---|---|---|---|
| Herceptin (Trastuzumab) | Her-2/neu positive tumors | Genentech | Humanized mouse monoclonal antibody, $IgG_1$, $\kappa$ | IV | Mediates ADCC $K_d$ = 5 n$M$ | 4 mg/kg loading dose; 2 mg/kg weekly dose | Anaphylaxis; cardiotoxicity; infusion reactions; pulmonary events; anemia or leukopenia; diarrhea | None known | Half-life = 1.7–12 days, eliminated by RES |
| Rituxan (Rituximab) | CD20 positive B-cell neoplasms | IDEC and Genentech | Chimeric mouse/human monoclonal antibody, $IgG_1$, $\kappa$ | IV | Mediates direct apoptosis, ADCC and complement dependent lysis (CDL) $K_d$ = 8 n$M$ | 375 mg/m$^2$ weekly for 4 or 8 doses | Tumor lysis syndrome; severe infusion reactions; severe mucocutaneous reactions; pulmonary events; lymphopenia (B-cell) | Known IgE hypersensitivy to murine proteins, No drug interactions | Half-life = 76.3 h; eliminated by RES |
| Mylotarg (Gemtuzumab) | CD33 positive leukemias such as acute myelogenous leukemia | Wyeth-Ayerst | Humanized mouse monoclonal antibody, $IgG_4$, $\kappa$ 50% conjugated to calicheamicin | IV or central line | Internalization of antibody, Calicheamicin binds DNA and cause strand breaks | 9 mg/m$^2$ 2 doses, 14 days apart | Severe myelosuppression; hepatotoxicity; tumor lysis syndrome; infusion reactions; pulmonary events; Mucositis | Known hypersensitivity to calicheamicin, no drug interactions | Half-life = 45–100 h; eliminated by RES |

| Drug | Target | Company | Structure | Route | Mechanism | Dose | Adverse effects | Contraindications | Pharmacokinetics |
|---|---|---|---|---|---|---|---|---|---|
| Campath (Alemtuzumab) | CD52 positive leukemia such as B-chronic lymphocytic leukemia | Millenium and ILEX | Humanized rat monoclonal antibody, $IgG_1$, κ | IV | Mediates ADCC and CDL | Dose escalation: 3, 10, 30 mg as tolerated. Maintenance dose is 30 mg three times per week. | Hematologic toxicity; infusion reactions; opportunistic infections resulting in grade 3 or 4 sepsis | Known hypersensitivities to components of drug; pre-existing immunodeficiency; no drug interactions | Half-life = 12 days; eliminated by RES |
| Remicade (Infliximab) | TNF-α | Centocor | Chimeric mouse/human monoclonal antibody, $IgG_1$, κ | IV | Neutralizes the biological activity of TNF-α; $K_d$ = 0.1 n$M$ | 3 or 10 mg/kg every 4 or 8 weeks with concurrent methotrexate Tx | Risk of infection; infusion related reactions ANA formation | Known IgE hypersensitivity to murine proteins; no drug interactions | Half-life = 8–9.5 days; eliminated by RES |
| Enbrel (Etanercept) | TNF-α and TNF-β | Immunex (Amgen) | Dimeric fusion protein between TNF receptor and $IgG_1$ | Subcutaneous | Neutralizes the biological activity of TNF-α | Adults: 25 mg Peds: 0.4 mg/kg | Risk of infection; ANA formation; mild injection site reaction | Known hypersensitivity to components; sepsis | Half-life 102 h; eliminated by RES |

tigens have been approved by the FDA as radioconjugates for diagnostic imaging applications. These radiolabeled antibodies search out and bind primary and metastatic tumors, revealing their locations on scanning of cancer patients. Because these radiolabeled antibodies are of mouse origin and have not been humanized using genetic engineering techniques, they will not be discussed here.

Over the past 5–10 years, the promise of antibodies as "magic bullets" in treating disease has been partially realized. The FDA has approved 15 antibodies for use in humans as therapeutic treatments and/or imaging agents at the time of this writing. Therapeutic antibodies for cancer and rheumatoid arthritis are listed in Table 6.1.

**2.2.2 Trastuzumab.** Herceptin (Trastuzumab) is a prototype for future antibody-based immunotherapeutics. It is a recombinant DNA-produced, humanized monoclonal antibody ($IgG_1$-$\kappa$ chain) containing human framework regions and complementarity determining regions (CDRs) of mouse monoclonal antibody, 4D5 (34). It selectively binds to the extracellular domain of the human epidermal growth factor receptor-2 (EGF-2), also known as Her-2 with an affinity of 5 n$M$. The Her-2 protein is overexpressed on approximately 25–30% of primary human breast tumors as well as several other adenocarcinomas. Overexpression of Her-2 is often associated with increased tumor aggressiveness. Because Trastuzumab inhibits the proliferation of tumors overexpressing Her-2, it has been shown to be more effective against aggressive tumors in patients with an otherwise poor prognosis (35, 36). One of the effector mechanisms of Trastuzumab tumor cell killing is ADCC (37). ADCC occurs after multiple antibodies have bound to a tumor cell, exposing the Fc portion of the antibodies so that immune cells with IgG Fc receptors, such as macrophages, neutrophils, eosinophils, or NK cells, bind and either phagocytose the tumor cell or secrete lytic granules that result in tumor cell death. Another potential mechanism of tumor cell killing is induction of p27KIP1 and the Rb-related protein, p130, which results in a significant reduction in the number of cells in S-phase, thereby reducing tumor growth. Trastu-

zumab also induces phenotypic changes in tumors, which include down-modulation of the Her-2 receptor, increased cytokine susceptibility, restored E-cadherin expression, and reduced vascular endothelial growth factor production (38).

Because Her-2/neu is overexpressed on many adenocarcinomas, including breast tumors, Trastuzumab was tested for efficacy in humans with breast cancer. Phase III clinical trials designed to evaluate Trastuzumab in over 500 patients with metastatic breast cancer, either in combination with paclitaxel or as a single agent, demonstrated a significantly longer time to disease progression, a higher overall response rate, longer duration of response, and higher 1-year survival compared with chemotherapy alone. Trastuzumab is supplied lyophilized and reconstituted to 21 mg/mL with supplied diluent followed by further dilution in 0.95% sodium chloride. In studies using a loading dose of 4 mg/kg followed by weekly infusions of 2 mg/kg, a mean half-life of 5.8 days was observed (range, 1–32 days), with a mean serum concentration between 79 and 123 mg/ml between weeks 16 and 32, respectively. Trastuzumab is approved by the FDA for use in patients with metastatic breast cancer whose tumors overexpress Her-2 and who have previously received one or more chemotherapy regimens for their metastatic disease.

Adverse events from Trastuzumab administration are rare but can result in severe hypersensitivity, including systemic anaphylaxis, urticaria, bronchospasm, angioedema, or hypotension. A recent warning of cardiotoxicity has been issued for Trastuzumab, where its use in patients with cardiac dysfunction has resulted in congestive heart failure. This phenomenon is currently under further evaluation and investigation.

**2.2.3 Rituximab.** Rituxan (Rituximab) was the first monoclonal antibody to be approved by the FDA for cancer treatment. It is a genetically engineered monoclonal antibody that binds to CD20. CD20 is a B-lymphocyte lineage-restricted differentiation antigen found on normal and malignant B-lymphocytes, but not on other normal hematopoietic cells or antibody-producing plasma cells. CD20 has a molecular weight of 35 kDa and is a hydrophobic transmembrane protein that regulates cell cycle activation and differentiation during B-lymphocyte development (39). In contrast to Trastuzumab, Rituximab (MW 145 kDa) contains murine heavy and light chain variable regions genetically fused to human IgG1 heavy chain constant regions and human κ chain constant regions; it is a chimeric murine/human antibody (40, 41). The binding affinity for the CD20 antigen is approximately 8 n$M$. Rituximab-mediated killing of CD20-positive tumor cells *in vivo* is largely because of ADCC, and to a minor degree, induction of apoptosis by direct ligation of CD20 and complement-dependent lysis (40, 42).

Rituximab was initially approved for the treatment of low-grade or follicular, relapsed, or refractory CD20-positive B-cell non-Hodgkin's lymphoma. In clinical trials, patients were treated weekly with either four or eight doses of Rituximab at 375 mg/m$^2$ as an IV infusion at a concentration of 1–4 mg/mL in 0.9% sodium chloride or 5% dextrose in water. The overall response rate for four doses weekly was 48% with six complete responses; for eight doses weekly it was 57% with a 14% complete response rate. The mean serum half-life of Rituxan is dependent on dose, but also depends greatly on tumor burden and circulating CD20-positive tumor cells or B-lymphocytes. Peak and trough serum levels are inversely proportional to the number of circulating CD20-positive cells. After the third or fourth dose, normal B-cells remain depleted for 6–9 months after treatment, after which B-cell numbers returned to normal by 12 months. Retreatment can be attempted, because only 4 of 356 patients developed a human anti-chimeric antibody response. However, patients should be closely monitored on retreatment for serum sickness. Rituximab is currently under study for the treatment of other B-cell malignancies including chronic lymphocytic leukemia, Hodgkins disease, and other lymphoid malignancies which CD20 (43).

Adverse events caused by Rituximab infusion include severe infusion reactions, including hypotension, angioedema, hypoxia, bronchospasm, pulmonary infiltrates, myocardial infarction, ventricular fibrillation, and cardio-genic shock. Tumor lysis syndrome has been observed in patients with high levels of circulation CD20-positive tumor cells, in which rapid reduction of tumor volume is followed by acute renal failure, hyperkalemia, hypocalcemia, hyperuricemia, or hyperphosphatasemia. Patients with known IgE-mediated immediate hypersensitivity reactions to murine proteins should not receive Rituximab.

**2.2.4 Gemtuzumab.** Mylotarg (Gemtuzumab ozogamicin, CMA-676; Wyeth-Ayerst Laboratories) is an FDA-approved monoclonal antibody specific for CD33, a sialoadhesion protein found on leukemic blasts in 80–90% of patients with acute myelogenous leukemia (AML) (44). CD33 is also expressed on normal immature cells of the myelomonocytic lineage. The humanization of Gemtuzumab is similar to Trastuzumab in that it contains mouse CDRs and human framework and constant regions, such that over 98% of the antibody is human. Gemtuzumab is an IgG$_4$ with a κ light chain chemically linked to the cytotoxic agent, calicheamicin. In the formulation of the drug, 50% of the antibody is linked to $N$-acetyl-γ-calicheamicin through a bifunctional linker at between 4 and 6 moles of calicheamicin per mole of antibody, whereas the remaining 50% is not derivatized (Fig. 6.2) (45). Tumor cell internalization of Gemtuzumab linked to calicheamicin results in release of the cytotoxin from the antibody in the lysosome. Calicheamicin is then free to bind to the minor groove in DNA causing double strand breaks and cell death (46). The exact mechanisms of induction of leukemic cell death by Gemtuzumab without calicheamicin linkage are not yet known, but the binding of the antibody to CD33 on the leukemia cell surface is thought to induce apoptosis (47).

Gentuzumab was approved in May 2000 by the FDA for the treatment of patients 60 years and older in first relapse with CD33$^+$ AML who are not considered candidates for other types of cytotoxic chemotherapy. In combined phase II studies, 142 patients with CD33-positive AML in first relapse demonstrated a 30% overall response rate with Mylotarg therapy alone. Treated patients had relatively high incidences of myelosuppression, hyperbilirubinemia, and elevated hepatic transaminases

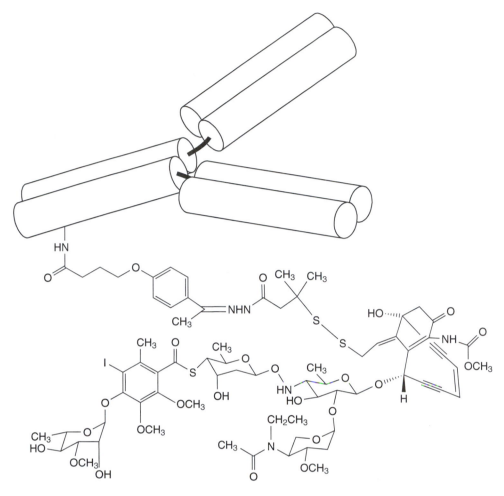

**Figure 6.2.** Diagram and chemical structure of Gemtuzumab ozogamicin. Calicheamicin (ozogamicin) is coupled to Gemtuzumab at an average loading of 4–6 moles per mole of antibody. The antibody is linked to N-acetyl γ calicheamicin. The molecular weight of the antibody-toxin conjugate is 151–153 kD. The injected formulation consists of 50% conjugated antibody and 50% unconjugated antibody.

(48). The incidences of severe mucositis and infections were low compared with mucositis resulting from conventional chemotherapeutic treatment. Sixteen percent of the patients had a complete response (CR), while 19% had a CR requiring platelet transfusions. The clinical data support the use of Gemtuzumab in AML patients with CD33-positive leukemia.

Adverse events caused by the administration of Gemtuzumab ozogamicin include severe myelosuppression, especially neutropenia, requiring careful hematological monitoring. Patients should be monitored for in-fection after Gemtuzumab administration and treated appropriately if necessary. Other adverse events associated with Gemtuzumab are hypersensitivity reactions, including anaphylaxis, infusion reactions, and pulmonary events including pulmonary edema, dyspnea, pleural effusions, hypoxia, and acute respiratory distress syndrome. Patients with greater than 30,000 leukemic cells/$\mu$L should be considered for leukoreduction to avoid tumor lysis syndrome, similar to that observed with Rituximab therapy.

**2.2.5 Alemtuzumab.** Campath (Alemtuzumab) is a recombinant DNA-derived humanized monoclonal antibody directed against CD52, a 21- to 28-kDa cell surface glycoprotein. CD52 is expressed on the surfaces of normal and malignant B- and T-lymphocytes, NK cells, monocytes, macrophages, a subset of granulocytes, and tissues of the male reproductive tract, but not on hematopoietic stem cells. Alemtuzumab is an IgG1 with a $\kappa$ light chain containing human variable framework and constant regions and (CDRs) from a rat monoclonal antibody (Campath-1G). CD52 is a glycosylphosphatidylinositol (GPI)-anchored protein with unknown function (49). Binding of Alemtuzumab to CD52 is thought to result in complement-dependent lysis and ADCC on tumor and normal cells (50).

Alemtuzumab is indicated for the treatment of B-cell chronic lymphocytic leukemia (B-CLL) in patients who have been treated with alkylating agents and who have failed fludarabine therapy. In a study where Alemtuzumab was infused intravenously once weekly for a maximum of 12 weeks over a range of doses, the overall average half-life during treatment was about 12 days. Alemtuzumab administered as a 30-mg intravenous infusion three times per week evaluated in CLL patients demonstrated variable peak and trough levels of the antibody during the first few weeks of treatment, but seemed to level off by week 6. Inter-patient variability was likely caused by tumor burden and circulating numbers of leukemia cells. However, increases in serum levels of Alemtuzumab corresponded to reduction in neoplastic cells. Clinical trials have shown that in previously treated B-CLL patients, partial responses occur at a rate of 40%, with 2–4% CRs. Responses are more likely in blood and bone marrow compared with lymph nodes. The median duration of response is 9–12 months. Because of the strong depleting activity on circulating lymphocytes, it has been used for purging residual disease in B-CLL, followed by autologous stem-cell transplantation (51, 52).

Alemtuzumab is provided as a sterile, clear, colorless, isotonic solution, pH 6.8–7.4, for injection. Each single use vial of antibody contains 30 mg Alemtuzumab, 24.0 mg sodium chloride, 3.5 mg dibasic sodium phosphate, 0.6 mg potassium chloride, 0.6 mg monobasic potassium phosphate, 0.3 mg polysorbate 80, and 0.056 mg disodium edetate. No preservatives are added. Infusions should be initiated at an initial dose of 3 mg with gradual escalation to 30 mg. Careful monitoring of blood pressure and hypotensive symptoms is recommended especially in patients with ischemic heart disease and in patients taking antihypertensive medications. If therapy is interrupted for 7 or more days, Alemtuzumab may be reinstituted with gradual dose escalation.

Adverse events are infusion related and are caused by tumor necrosis factor (TNF)-$\alpha$ and interleukin (IL)-6 release, usually during the first intravenous infusion, and include fever, rigor, nausea, vomiting, and hypotension that responds to steroids. Adverse events are usually less severe with subsequent infusions and can be prevented by with appropriate medication. Antihistamine and acetaminophen are recommended before infusion. Because CD52 is present on many types of leukocytes, immunosuppression resulting from depletion of normal B- and T-lymphocytes usually occurs, resulting in an increased risk for opportunistic infections.

## 2.3 Anti-Inflammatory Therapeutic Antibodies

Two recombinant antibodies, Infliximab and Etanercept, that bind to and neutralize TNF-$\alpha$ have been approved by the FDA for the treatment of rheumatoid arthritis (RA) and Crohn's disease.

**2.3.1 Infliximab.** Remicade (Infliximab) is a chimeric monoclonal antibody containing human constant and murine variable regions that inhibits TNF-$\alpha$ from binding to its receptor (53). It binds to TNF-$\alpha$ very strongly, with an association constant of 0.1 n$M$. TNF-$\alpha$ is secreted as a trimer by macrophages, T-cells, and NK cells. Biological activities of TNF-$\alpha$ include induction of pro-inflammatory cytokines such as IL-1 and IL-6 and increased leukocyte migration through up-regulation of endothelium permeability and adhesion molecules by both endothelial cells and leukocytes. TNF-$\alpha$ also activates eosinophils and neutrophils and induces acute phase proteins as well as enzymes such as matrix metalloproteinases involved in degradation of synoviocytes and chondrocytes in joint tissue. It is thought to play a major role in mediating tissue damage in RA and other autoimmune diseases. Infliximab has been shown to prevent RA in transgenic mice that develop polyarthritis caused by constitutive expression of TNF-$\alpha$ (54). When administered to mice after joint destruction had been established, damaged joints began to heal.

Infliximab in combination with methotrexate (MTX) is indicated for reducing the symptoms and inhibiting the progression of structural joint damage in patients with moderate-to-severe active rheumatoid arthritis who have had an inadequate response to methotrexate alone. It is also indicated for patients with Crohn's disease who have had inadequate responses to conventional therapy. Clinical studies with Infliximab in combination with MTX in 428 RA patients demonstrated serum half-lives of 8–9.5 days. In clinical trials, approximately 50% of patients receiving either 3 or 10 mg/kg of Infliximab every 4 weeks responded to treatment at a rate of approximately 50%, compared with placebo as measured by the American College of Rheumatology (ACR) response criteria (55). Treatment with Infliximab decreased inflammatory cell infiltration into inflamed areas in the joint, expression of adhesion molecules, E-selectin, intercellular adhesion molecule-1 (ICAM-1), and vascular cell adhesion molecule-1 (VCAM-1), chemoattractants such as IL-8 and monocyte chemotactic protein (MCP-1) and also inhibited expression of matrix metalloproteinases 1 and 3, which are involved in joint destruction (56). The treatment of Crohn's disease with Infliximab alone resulted in better than a 70% response rate compared with placebo within 4 weeks of receiving a single intravenous infusion according to the Crohn's Disease Activity Index (57).

Infliximab is provided as a sterile, lyophilized powder for reconstitution with 10 mL USP sterile water for injection such that the reconstituted material is 10 mg/mL followed by additional dilution into 250 mL of 0.9% sodium chloride. Recommended dose of Infliximab is 3 mg/kg given as an intravenous infusion between 0.4 and 4 mg/mL over a period of 2 h or more.

**2.3.2 Etanercept.** Enbrel (Etanercept), similar to Infliximab, binds to and neutralizes the biological activity of TNF-$\alpha$. Etanercept is novel in that it was constructed by fusing cDNA from the extracellular ligand-binding portion of the TNF receptor to cDNA from the Fc portion of IgG1 and has the same approximate molecular weight (150 kDa) as an IgG molecule (58). Because Etanercept contains the TNF receptor, it binds and neutralizes both TNF-$\alpha$ and TNF-$\beta$ (lymphotoxin) (59), in contrast to Infliximab, which binds only TNF-$\alpha$. However, Etanercept suppresses the same biologic and pathogenic mechanisms leading to RA as does Infliximab.

Like Infliximab, Etanercept is approved for use in adult patients with moderate-to-severe active RA. However, Etanercept was also approved to reduce the symptoms of moderate-to-severe polyarticular-course juvenile RA (JRA) in patients who have had inadequate responses to disease-modifying anti-rheumatic drugs (60) and for use in patients with psoriatic arthritis in combination with MTX who do not respond to MTX alone (61). Clinical evaluation of subcutaneous administration of Etanercept twice per week for 6 months demonstrated an overall 23% major clinical response, defined as maintenance of an ACR70 (70%) response over a 6-month period. Discontinuation of Etanercept generally resulted in return of symptoms within 1 month. If patients were retreated with Etanercept, they achieved the same response as the initial treatment.

Etanercept is supplied as 25 mg of lyophilized powder for reconstitution with 1 mL of USP bacteriostatic water for injection, resulting in a 25 mg/mL solution that may be self-injected by a patient or physician into the thigh, abdomen, or upper arm.

Risks associated with Infliximab and Etanercept treatment include increased risk of infections such as reactivation of latent tuberculosis, invasive fungal infections, and sepsis. Autoantibodies against DNA and other nuclear components were observed in 10% of patients treated with Infliximab or Etanercept, mimicking a lupus-like syndrome. Patients receiving placebo did not generate lupus-like antibodies. Infusion-related reactions to Infliximab included fever, chills, cardiopulmonary reactions, urticaria, and pruritus. Because Etanercept is injected subcutaneously, injection site reactions were limited to erythema, itching, pain, or swelling at the site of injection. Thirteen percent or fewer patients receiving Infliximab developed anti-Infliximab specific antibodies, while 11% of patients receiving Etanercept generated anti-TNF receptor-specific antibodies. These antibodies were all non-neutralizing and did not inhibit TNF-$\alpha$ from binding to its cell surface receptor. Administration of Infliximab and Etanercept is associated with increased infections (62), therefore immune responses to vaccines may be affected. Although not studied in-depth, patients may be immunized if necessary, with the exception of live or attenuated vaccines.

## 3  DENDRITIC CELL IMMUNOTHERAPY

The successful induction of cell-mediated anti-tumor immunity relies on the efficient capture of antigen by antigen-presenting cells, antigen processing, and presentation to T-lymphocytes. Antigen presenting cells (APC) such as B-cells, macrophages, and dendritic cells display antigenic peptides in the context of major histocompatibility complexes (MHC) to T-cells. Although B-cells and macrophages are capable of antigen uptake, processing, and antigen presentation, these APC are ineffective activators of naïve T-cells. In contrast, dendritic cells (DC) are considered the most effective, efficient, and potent APC of the immune system, because they are the only subset of APC that can present antigen to naïve T-cells resulting in T-cell activation (63). Cell-mediated tumor cytotoxicity is the desired outcome of tumor immunotherapy. Modalities aimed at the induction of specific anti-tumor T-cell responses by DC are extremely promising in the fight against cancer. The following sections emphasize pre-clinical and clinical studies employing DC to induce active, specific immunity against tumors.

### 3.1  Dendritic Cell Physiology

The first studies involving DC generation were performed by isolating CD34$^+$ progenitor cells from the bone marrow and culturing these cells in the presence of GM-CSF and TNF-$\alpha$ (64, 65). While this was an effective means of DC generation, a relatively small number of DC were obtained, because of the limited number of progenitor cells present in bone marrow. Subsequent studies demonstrated that DC could be generated from peripheral blood mononuclear cells (PBMC) directly isolated from blood. Monocyte-derived DC can be generated *ex vivo* by culturing monocytes with a cocktail of cytokines such as GM-CSF and IL-4 or IL-13, yielding an immature population of DC (66, 67). Immature DC can be driven to a mature phenotype with the addition of inflammatory cytokines such as IL-1$\beta$ and TNF-$\alpha$, bacterial derived products [lipopolysaccharide (LPS)], recombinant CD40L (costimulatory molecule found on T-cells), or double-stranded RNA (68–72). The ability to effectively generate DC *in vitro*, coupled with the fact that DC efficiently present antigens, makes this subset of immune cells attractive candidates for the treatment of cancer and other diseases.

### 3.2  T-Cell Activation by Dendritic Cells

Immature DC are efficient at antigen uptake through micropinocytosis, receptor-mediated endocytosis, or phagocytosis; however, they have low T-cell stimulatory capacity (73). Conversely, while mature DC are less efficient at antigen uptake and processing, they are very efficient presenters of antigenic peptide to T-cells, resulting in the initiation of immune responses (73, 74). Mature DC express high levels of MHC molecules, high levels of co-stimulatory molecules (such as CD80, CD86, and OX40L), and adhesion molecules (i.e., CD54 and CD11c) that bind to counter-receptors on T-cells (73, 75, 76). These multiple interactions between DC and T-cells play a major role in the activation of T-cells specific for antigenic peptide/MHC complexes displayed by DC (Fig. 6.3). DC also secrete a milieu of cytokines and chemokines involved in the recruitment and activation of T-cells. Chemokines are small molecules that serve as chemical attractants for lymphocytes. DC-CK and RANTES are two examples of such chemokines that attract naïve and memory T-cells toward DC displaying antigenic peptide (77). The production of cytokines such as interleu-

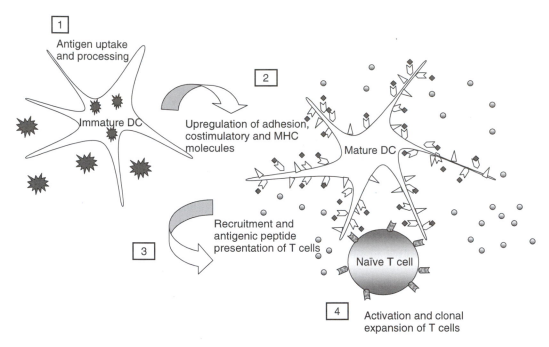

**Figure 6.3.** Antigen presentation and activation of naïve T-cells by mature DC.

kin-12 (IL-12) and IL-15 by DC are important cytokines for T-cell activation (78).

### 3.3 Immunotherapy Using Dendritic Cells

Recently, DC have been used in numerous clinical trials for the treatment of cancer. Immunization with DC is not toxic in either healthy subjects or cancer patients; no dose-limiting toxicity has been observed (79). The induction of tumor-specific T-cell responses has been detected in patients that have received DC immunotherapy. Several clinical trials are currently in progress investigating the safety and efficacy of immunotherapy of cancer with DC. *Ex vivo* incubation of DC with a source of tumor antigens is necessary to load tumor-derived antigenic epitopes on DC. Large numbers of DC generated *ex vivo* can be manipulated to enhance tumor antigen presentation and then re-administered to the patient to study the efficacy of DC immunotherapy. DC have been shown to induce strong anti-tumor immune responses both *in vitro* and *in vivo*. Early vaccination protocols involved DC pulsed with synthetic HLA-binding peptides. Since then, many other strategies involving DC have been investigated, such as

DC-tumor cell fusions, DC transfected with tumor RNA or viral vectors, and DC exposure to tumor apoptotic bodies or exosomes. We will discuss current DC immunotherapy strategies in further detail in the following sections. Table 6.2 provides a brief listing of the advantages and disadvantages of DC immunotherapy modalities.

Antigenic peptides are presented to T-cells by APC in the context of human leukocyte antigen (HLA) molecules. Numerous peptide epitopes have been identified and studied for both CD8+ and CD4+ tumor-specific T-cells (reviewed in ref. 80). Table 6.3 provides examples of HLA class I-binding peptide epitopes recognized by CD8+ CTL. Many tumor-associated antigens have been identified and may serve as targets for immunotherapy. Melanoma-specific cells can be propagated *in vitro*, therefore facilitating the identification of immunogenic melanoma antigens. However, there has been less success in the identification of tumor-associated antigens derived from other cancers because of the difficulty in generating tumor-specific CTL lines for *in vitro* studies. The use of peptides in immunotherapy is an attractive approach because

**Table 6.2   Advantages and Disadvantages of DC Immunotherapy Modalities**

|  | Advantages | Disadvantages |
|---|---|---|
| RNA transfection of DC | Simple technique; no requirement of tumor antigen identification | Difficult to measure transduction efficiency; potential oncogenic transformation of DC |
| DC-tumor cell fusion | Presentation full repertoire of tumor antigens possible with DC-derived costimulation; no requirement of tumor antigen identification | Low fusion efficiency; availability of autologous tumor cells; DC may not home to secondary lymphoid organs |
| Tumor cell lysate loading of DC | No requirement of tumor antigen identification; multiple epitopes presented by DC | Availability of autologous tumor cells |
| Apoptotic tumor cell loading of DC | May carry proteins that aid in loading peptides onto MHC; no requirement of tumor antigen identification; multiple epitopes presented by DC | Availability of autologous tumor cells |
| Peptide-pulsed DC | Ease of peptide synthesis; GMP manufacture straightforward known immunogen; multiple single peptides may be used to broaden T-cell response | Single antigenic epitope restricted by HLA haplotype |
| Viral vector transduction of DC | Long-lived, gene expression; may incorporate multiple defined antigens and/or costimulatory molecules | DC often resist transduction; safety: infectious agents |

peptides are easily synthesized to good manufacturing process (GMP) specifications. However, vaccination strategies of DC pulsed with peptide require more study to determine optimal dosing and route of administration.

Immunodominant peptides derived from tumor-associated antigens such as gp100 (melanoma), Her-2/neu (overexpressed on adenocarcinomas such as breast cancer), MAGE peptides, and carcinoembryonic antigens (CEA) have been used to vaccinate patients in several clinical trials (81–84).

A well-studied tumor antigen is gp100, a melanocyte lineage-restricted glycoprotein that is expressed by most melanoma cells (17). Bakker et al. (85) demonstrated that an HLA-A*0201-restricted gp100 immunodominant epitope (residues 209–217) was recognized by

**Table 6.3   Examples of HLA Class I Binding Tumor Peptides[*]**

| Classification | Gene | HLA-Allele | Peptide Epitope | Reference |
|---|---|---|---|---|
| Melanocyte differentiation antigens | $gp100_{(209-217)}$ | A2 | ITDQVPFSV | 85 |
|  | $tyrosinase_{(1-9)}$ | A2 | MLLAVLYLL | 86 |
|  | $MART-1_{(27-35)}$ | A2 | AAGIGILTV | 17 |
| Shared tumor/testis antigens | $MAGE-1_{(161-169)}$ | A1 | EADPTGHSY | 87 |
|  | $NY-ESO-1_{(53-62)}$ | A31 | ASGPGGGAPR | 88 |
| Non-mutated, overexpressed tumor antigens also expressed on normal tissue at low levels | $Her-2/neu_{(369-377)}$ | A2 | KIFGSLAFL | 89 |
|  | $CEA_{(61-70)}$ | A3 | HLFGYSWYK | 90 |
|  | $MUC-1_{(950-958)}$ | A2 | STAPPVHNV | 91 |
|  | $PSA_{(146-154)}$ | A2 | FLTPKKLQCV | 92 |
| Tumor-specific antigen caused by point mutation in normal gene | $P53_{(264-272)}$ | A2 | LLGRNSFEV | 93 |
|  | $\beta-catenin_{(29-37)}$ | A24 | SYLDSGIHF | 19 |
|  | $HSP70-2M_{(286-294)}$ | A2 | SLFEGIDIY | 94 |
| Viral gene products | $HPV\ E6_{(11-20)}$ | A2 | YMLDLQPETT | 95 |

[*]Table is not a complete listing of tumor antigens and peptides.

tumor infiltrating lymphocytes derived from melanoma patients. This gp100 epitope was modified to enhance peptide binding to HLA-A*0201 as an attempt to increase peptide immunogenicity (86). The modified peptide, g209 2M, enhanced CTL responses from patients with melanoma compared with the wild-type gp100 peptide *in vitro* (86). Early peptide immunotherapy trials involved g209 2M peptide administration in combination with IL-2 therapy (87). Vaccinations were well tolerated, and objective clinical responses were noted in 38% of patients (87). However, it was unclear whether these clinical responses were truly because of peptide therapy or a result of IL-2 administration. Future studies confirmed that peptide-specific CTL were induced in patients immunized with the g209 2M peptide postvaccination (88), demonstrating the feasibility of therapeutic peptide vaccines.

Recently, clinical trials involving DC-peptide therapy have begun. Patients with ovarian or breast cancer were vaccinated with DC pulsed with either Her-2/neu or MUC-1 derived peptides (89). Peptide-specific CTL responses were detected as measured by cytokine secretion and lysis of tumor cells. Another recent clinical trial involved pulsing autologous DC with the following melanoma peptides: g209 2M, MART-1 (AAGIGILTV), or tyrosinase (YMDGTMSQV) before administration in 16 patients with melanoma (90). Minimal toxicity was observed. Peptide-specific T-cell responses were generated *in vitro* against gp100 or tyrosinase from five patients; four of five patients demonstrated either tumor regression or stable disease. Immune responsiveness in this study correlated with clinical benefit. Transient increases in T-cell stimulation to MAGE peptides were also observed in six of eight patients enrolled in a separate DC clinical trial (91). These studies suggest that peptide-pulsed DC provide clinical benefit to melanoma patients.

Autologous DC were pulsed with MAGE-3A1 tumor peptide and administered to advanced stage IV melanoma patients who received five DC vaccinations (three intradermal and two intravenous injections). CTL specific for MAGE-3A1 were generated in 8 of 11 patients (92). Tumor regression of metastases was evident in 6 of 11 patients. Interestingly,

$CD8^+$ T-cells were absent from non-regressing lesions as well as the expression of MAGE-3 mRNA.

Several factors may play a role in the limited successes with DC-peptide immunotherapy. One reason may be that the optimal dose and route of administration are not yet known. Second, if tumor cells down-regulate the tumor antigen from which the peptide is derived, DC-peptide presentation to T-cells may be futile. Therefore, strategies using multiple peptides from multiple antigens are most likely to elicit effective responses with minimal tumor escape variants.

A current limitation of peptide immunotherapy is the applicability to a large patient population. Peptides are specific for HLA haplotypes, such that only cancer patients whose haplotype matches that of the tumor-derived peptide may be candidates for a given single peptide therapy. However, recent studies investigating peptide cross-reactivity for HLA molecules determined that a peptide with high affinity for one particular HLA class I allele can also bind to other alleles (93). Additionally, medical science is increasingly tailoring therapies to the individual.

Another problem facing immunotherapeutic approaches designed to stimulate T-cells involves self-tolerance. Thymic education of T-cells is a selection process in which T-cells that bind with high affinity to self-peptides are deleted through apoptosis (94). If these deleted T-cells were specific for a tumor antigen, then only low affinity T-cells would remain (13, 95, 96). These low affinity T-cells are incapable of responding strongly to tumor antigen without strong stimulation provided by DC. However self-specific T-cells can be found in the periphery (97–99). The existence of these self-specific T-cells, albeit with a low affinity, may initiate an anti-tumor immune response. These cells alone are insufficient to completely resolve tumor burden. Therefore, attempts have been made to increase both the HLA affinity and peptide immunogenicity of tumor-derived peptide similar to the modified gp100 peptide, g9 209 2M.

Several studies have shown that increased affinity of a peptide for MHC molecules is an important parameter in determining peptide immunogenicity (100–102). By modifying self-peptides, there is potential to activate T-cells that are either tolerant or have a weak affinity for self to initiate an immune response. Altered peptide ligands for p53, Her-2/neu, MAGE, carcinoembryonic antigen (CEA), and MART-1 tumor peptides have been successfully synthesized that have higher affinities for HLA molecules compared with wild-type, and elicit CTL responses *in vitro* (103, 104).

With the knowledge that tumor-derived peptides can be modified to increase HLA binding coupled with the potential that these peptides may cross-react with several HLA alleles, these findings may greatly impact the design of HLA-binding peptides for cancer immunotherapy applicable to a broad range of patients.

## 3.4 Dendritic Cell Immunotherapy Approaches for Undefined Tumor Antigens

An important issue in optimizing DC vaccines is choosing an ideal tumor antigen for DC loading. Tumor cell lysates, apoptotic tumor bodies, and tumor cells can be used as immunogens in DC cancer therapy for the development of anti-tumor strategies. For several tumor types, antigenic epitopes are unknown. In contrast to peptide immunotherapy, using tumor-derived products bypasses the need to consider HLA haplotypes and the identification of specific tumor-derived antigens. Several of these treatment modalities will now be reviewed, which demonstrate specific anti-tumor responses against tumors with undefined tumor antigens.

### 3.4.1 Apoptotic Tumor Cells. In an effort to overcome the limitations of HLA restriction and identification of specific tumor-associated antigens, whole tumor cells containing unidentified antigenic components as an antigen source have been investigated. Apoptotic tumor cells (ATC) have been studied to determine the effectiveness of these cells to serve as sources of antigen. Hoffmann et al. (105) generated a CD8$^+$ T-cell line specific for a squamous cell carcinoma of the head and neck cell line (SSCHN) to determine if immunogenic peptides could be presented by DC cultured with ATC. DC generated from healthy donors were tested for the ability to uptake ATC. Immature DC ingested ATC and were able to ef-

fectively present tumor antigens and induce a T-cell–mediated anti-tumor immune response more effectively than DC incubated with tumor-derived lysate preparations. An increase in CTL number and lytic function was observed when DC were loaded with pancreatic apoptotic tumor cells compared with cell lysate (106). The mechanism of enhanced DC priming using ATC is still unknown. Studies using ATC are ongoing in many laboratories and promise to yield data supporting their use in DC immunotherapy.

### 3.4.2 Tumor Lysates. Compared with vaccination approaches directed against a single tumor antigen, tumor cell lysates contain an array of tumor-derived antigens that have the potential of inducing broad T-cell responses against multiple antigens expressed by tumor cells, either previously identified antigens or unknown tumor-derived antigens. Although many vaccines employing tumor lysates have been prepared in the past, the addition of DC to the vaccine should lead to more robust immune responses than previously observed.

Tumor lysate-pulsed DC were tested in a phase I trial of pediatric patients with solid tumors (107). Patients were vaccinated with immature DC pulsed separately with either tumor cell lysates or keyhole limpet hemocyanin (KLH; an immunological tracer molecule). DC were administered intradermally every 2 weeks for a total of three vaccinations. Fifteen patients (10 of which completed all DC vaccinations) diagnosed with neuroblastoma, sarcoma, and renal malignances were immunized without observable toxicity. IFN-γ secreting T-cells were detected *in vitro* in 3 of 7 patients against tumor lysates and 6 of 10 patients for KLH. Five patients showed stable disease, including three who had minimal disease at time of vaccine administration and remained free of tumor 16–30 months post-vaccination. This trial demonstrated that tumor lysate or KLH-pulsed DC generated specific T-cell responses, thereby inducing regression of metastatic disease.

In another study, autologous tumor lysate-pulsed DC were administered intradermally to adult patients with stage IV solid malignancies every 2 weeks for three cycles with varying amounts of DC (108). The vaccine for this phase I clinical trial was composed of a mixture of lysate-pulsed DC and DC pulsed with KLH. No severe toxicity was reported in 14 patients that received three DC vaccinations. $CD4^+$ and $CD8^+$ T-cells were detected at the vaccination sites. Two patients with melanoma experienced a partial and a minor response, respectively. In other tumor types, phase I trials of patients with advanced gynecological malignancies and a malignant endocrine carcinoma demonstrated that vaccination with DC-autologous tumor lysates was safe, well-tolerated, and immunologically active with no significant adverse effects (109, 110). Tumor lysates were also derived from ovarian cancer cells and used to evaluate the potential of lysate-pulsed DC to induce tumor-specific T-cell responses against autologous tumors (111). DC-lysate stimulated proliferation of autologous T-cells and CTL-mediated lysis of autologous tumor cells. These effects were abrogated using anti-MHC class I and anti-CD8 antibodies. Furthermore, T-cells cultured with autologous tumor lysate-pulsed DC secreted cytokines that play an important role in immune cell recruitment, such as GM-CSF, TNF-$\alpha$, and IFN-$\gamma$. T-cells cultured with tumor lysates in the absence of DC resulted in no lytic activity, further confirming that DC play a major role in the initiation of anti-tumor immune responses.

In another study, DC generated from melanoma patients were loaded with either a melanoma peptide, gp100 (280–288), or melanoma tumor cell lysate to determine the ability to induce cytotoxic autologous T-cell activation (112). Weekly stimulations of PBMC were performed with the lysate and peptide DC preparations. $CD8^+$ CTL displayed strong lytic activity against melanoma cells irrelevant of the DC stimulations (peptide versus lysate). These findings indicate that a variety of DC immunotherapy strategies are likely to be effective at inducing CTL responses. Collectively, these studies demonstrate that the administration of tumor lysate-pulsed DC is non-toxic and effective at inducing immunological responses against autologous tumor.

**3.4.3 Tumor RNA.** Studies involving exposure of DC to purified tumor-derived RNA have shown that RNA is efficiently taken up by the DC, resulting in presentation of tumor antigens to T-cells. RNA extracted from metastatic colon cancer or lung cancer were loaded onto autologous DC (113). *In vitro* lysis of autologous tumor cells was observed when RNA-transfected DC were incubated with autologous T-cells. Loading of DC with RNA followed by potent stimulation of T-cells is an interesting contrast to the resistance of DC to transfection with plasmids.

A phase I trial involving patients with metastatic prostate cancer was conducted to determine the safety and efficacy of DC transfected with mRNA encoding for prostate-specific antigen (PSA) to induce specific anti-PSA T-cell responses (114). At all doses of mRNA-transfected DC administered, no evidence of dose-limiting toxicity or adverse effects were observed. Furthermore, PSA-specific T-cell responses were detected *in vitro* in all patients who received the vaccine.

This approach has also been applied to cervical cancer using autologous DC transfected with RNA encoding for E6 and E7 (115), oncoproteins constitutively expressed by many cervical carcinomas. Antigen-specific CTL responses were observed both *in vitro* and *in vivo*. Human cervical carcinoma cells expressing the E6 and E7 products were lysed by CTL.

These studies provide a rationale for the development of immunotherapy using DC transfected with RNA encoding for tumor antigens. DC transfected with tumor RNA may emerge as a method for inducing immune responses against tumor antigens. Clinical trials demonstrating vaccine safety are promising for the development of anti-tumor cellular vaccines.

**3.4.4 DC-Tumor Cell Fusion.** Hybrid cells composed of tumor cells fused to DC have been generated using either polyethylene glycol (PEG) or electrofusion techniques, in which these hybrid cells possess antigen presentation characteristics of DC as well as tumor antigens derived from tumor cells (Fig. 6.4). The distinct advantage of a fusion product between tumor cell and a DC is that the hybrid contains the co-stimulatory molecules that tumors rarely express. The hybrid also contains

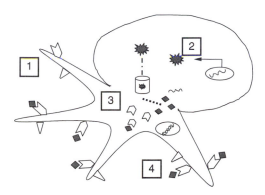

**Figure 6.4.** Tumor cell-dendritic cell hybrid. (1) On fusion, DC express costimulatory molecules, adhesion molecules, and MHC molecules necessary for T-cell activation. (2) Tumor cells transcribe genes that encode for tumor-derived antigens. (3) Tumor antigens can be processed by DC machinery and assembled onto MHC molecules. (4) Tumor-derived peptides can be presented in context of MHC molecules on the dendritic cell to tumor-specific T-cells.

multiple tumor antigens, including unidentified antigens that are endogenously processed and presented by MHC class I molecules.

Gong et al. (116) demonstrated that fusion between DC and autologous tumor cells is a feasible approach for immunotherapy. DC were fused with either autologous human breast or ovarian carcinoma cells (116, 117). These DC-tumor hybrids expressed both tumor-associated antigens and co-stimulatory molecules naturally expressed by DC. *In vitro*, hybrids induced both autologous T-cell proliferation and lysis of autologous tumor cells by responding CTL.

Fusion products of autologous renal cell carcinoma cells and allogeneic DC were tested in a phase I clinical trial (118). Post-vaccination with these fusion hybrids, four patients rejected all metastatic tumor lesions and two patients demonstrated tumor regression greater than 50% in strong partial responses.

Taken together, these studies indicate that the development of a "super" DC that can both stimulate and activate T-cells through DC-associated co-stimulatory molecules and presentation of tumor antigens derived from tumor cells may be a feasible approach for cancer immunotherapy. Currently, fusion efficiency is relatively low, ranging from 5–50% DC-tumor cell hybrids. This obstacle is currently being further addressed in several pre-clinical and clinical studies to maximize the clinical effectiveness of DC-tumor cell of hybrid cells.

**3.4.5 Exosomes.** Exosomes are membrane-bound vesicles released from either tumor cells or DC that have processed tumor cells. They are of endocytic origin ranging from 60 to 90 nm. Exosomes are thought to be secreted upon the fusion of multivesicular endosomes with the plasma membrane. Red blood cells, platelets, B- and T-cells, and DC have been found to secrete exosomes. Recent proteomic studies demonstrated that exosomes display MHC molecules (both class I and class II), co-stimulatory molecules, and several adhesion molecules (119). The mechanism of action of exosomes *in vivo* is not well defined. Exosomes may stimulate T-cells directly through surface MHC molecule expression or they may be ingested by other APC that could potentially further present peptide contained within the exosome. Wolfers et al. (120) demonstrated that upon DC uptake of tumor cell exosomes, presentation to CD8$^+$ T-cells resulted in potent anti-tumor effects in *in vivo* murine models, indicating that tumor-derived exosomes transfer antigenic tumor material to DC. Injection of DC-derived exosomes sensitized with tumor peptides induced anti-tumor immune responses in mice and eradicated previously established tumors (121). This anti-tumor effect of DC exosomes was dependent on CD8$^+$ T-cells as determined by T-cell depletion experiments *in vivo*. Recently, the first protein map of DC-derived exosomes was established (119), which may provide substantial insights into the exploitation of exosomes for immunotherapeutic use.

## 3.5 Dendritic Cells Transduced with Viral Vectors

One promising strategy for the future of immunotherapy is transduction of DC with recombinant, replication-incompetent viruses expressing multiple or single tumor antigens, cytokines, or co-stimulatory molecules. The goal of DC-based immunotherapy using viral vectors is to express one or more transgenes in DC so that endogenous and exogenous presentation of antigenic tumor peptides occurs in a favorable immunostimulatory environment using the complete repertoire of co-stimula-tory molecules present on DC. The advantages of viral transduction of DC are as follows. (*1*) Presentation of multiple epitopes derived from the same or multiple antigens may re-duce the probability that mutant tumor cells will escape immunological surveillance. (*2*) Antigen presentation by multiple MHC alleles provides even greater diversity of peptide epitopes. (*3*) Cytokines and co-stimulatory molecules can be co-expressed along with the tumor protein, eliciting stronger activation of T-cells. (*4*) Finally, the viral transduction pro-cedure itself may mature DC, enhancing anti-gen presentation capacity and activation of T-cells (122). Next, promising viral vector systems for active specific immunotherapy will be discussed.

### 3.5.1 Adenoviral Transduction of DC. DC

have been shown to be uniformly resistant to nearly all transfection techniques *in vitro* (123). Additionally, monocyte-derived DC do not undergo cell division and are therefore re-sistant to transduction by retroviral vectors (124). The fact that adenoviruses do not re-quire active cell division for infection is a dis-tinct advantage of the adenoviral system (125). Thus, it is possible to transduce DC with recombinant adenovirus. The problem is that DC express very low levels of the receptor for adenovirus, "coxsackie adenovirus receptor" (CAR) (126). One solution is to transduce DC with recombinant adenovirus encoding CAR as the transgene (127). In lieu of transducing DC with CAR so that they can be transduced again with the gene of interest, multiplicities of infection (MOI) as high as 1000 must be used to achieve high transduction efficiency (124, 128).

In a pre-clinical study, replication-deficient recombinant adenoviruses encoding human gp100 or MART-1 melanoma antigens were used to transduce human DC. The study dem-onstrated that human monocyte-derived DC could be infected at an MOI between 100 and 500 independent of the CAR. DC transduced with this replication-deficient adenovirus also elicited tumor-specific CTL *in vitro* from pa-tients with gp100$^+$ metastatic melanoma (129). These findings led to a clinical trial eval-uating the safety, dose-limiting toxicity, and maximum tolerated dose of autologous DC transduced with adenoviruses encoding the MART-1 and gp100 melanoma antigens with or without interleukin-2 therapy in patients with stage III or IV melanoma. Although the preclinical results are encouraging, no results have been published from this trial as of this writing.

### 3.5.2 Pox Virus Transduction of DC. Hu-

man DC have also been successfully trans-duced with various pox viruses including fowl pox and modified vaccinia ankara (MVA), a highly attenuated vaccinia virus. Pre-clinical work characterized the transduction effi-ciency of MVA encoding human MUC1 (tumor antigen that is aberrantly glycosylated on ma-lignant cells) and IL-2 with monocyte-derived DC. MUC1 expression in DC was from 27–54% of the cells, which also secreted biologically active IL-2. Over 72 h post-transduction, cyto-pathic effects of MVA resulted in decrease of the transgene expression. Transduction of both immature and mature DC did not alter expression of the MUC1 and IL-2 transgenes (130).

Other pre-clinical work demonstrated that human DC can be transduced with MVA ex-pressing tyrosinase (melanoma-associated an-tigen). These transduced DC stimulated anti-tyrosinase CTL responses from melanoma patients who had received smallpox vaccina-tion earlier in life (131). This data demon-strate the use of MVA transduced DC and ne-gate the argument that immunodominance by the pox vector itself will overwhelm any tumor antigen-specific immune response.

Another pre-clinical study that has led to a clinical trial in patients with carcinoembryonic antigen-positive (CEA) malignancies is evalua-tion of fowlpox transduced DC expressing three co-stimulatory molecules: B7-1, intercellular adhesion molecule-1, and leukocyte function-as-sociated antigen-3 (TRICOM). The goal of the study was to determine if DC infected with TRICOM would have an enhanced capacity to stimulate T-cell responses (132). Although the study was performed in a murine system, it showed that poxvirus vectors overexpressing of a triad of co-stimulatory molecules can sig-nificantly improve the efficacy of dendritic cells in priming specific immune responses.

These studies led to an NCI-sponsored clinical trial to determine the safety and feasibility of active immunotherapy comprising autologous DC infected with recombinant fowlpox-CEA-TRICOM vaccine in patients with locally recurrent or metastatic malignancies expressing CEA (133). In this ongoing study, autologous DC are harvested and infected with fowlpox-CEA-TRICOM vaccine. Patients receive their infected DC intradermally and subcutaneously followed by autologous DC mixed with CMV pp65 peptide and autologous DC mixed with tetanus toxoid SC and intradermally on day 1. Treatment was repeated every 3 weeks for a total of 4, 8, or 12 immunizations in the absence of unacceptable toxicity. CMV pp65 peptide and tetanus toxoid were administered so that immune responses to naïve and recall antigens can be monitored. We look forward to the results of this promising trial (134).

## 4 CONCLUSIONS

Many exciting developments have occurred in the field of immunotherapy in the past several years. This chapter has highlighted humanized antibody therapeutics, many of which have already been approved by the FDA, and the newest techniques in DC therapies for cancer. Clinical acceptance of humanized, chimeric, and diagnostic monoclonal antibodies has fostered growth in sales in 2000 of 1.8 billion dollars to nearly 3 billion dollars in 2001. We can expect to see many more humanized and chimeric antibodies in the immediate future.

Cellular immunotherapies using DC are now logistically and technologically feasible because closed system processing laboratories become more common. However, several parameters must be optimized to realize the full potential of DC in cancer immunotherapy. Vaccination schedules, route of DC administration, and the number of DC for effective immune responses must be resolved. The life span of DC *in vivo* post-vaccination and whether these cells effectively home to lymphoid tissues to encounter and activate T-cells also must be further investigated. Lastly, clinical trials involving DC immunotherapy need to combined with other treatment modalities

that are non-immunosuppressive to fully develop efficient anti-tumor therapies. Before poisoning the immune system with current cytotoxic chemotherapies, hopefully a paradigm shift will occur to activate and manipulate patients' own immune systems to therapeutic advantage, employing active specific immunotherapy. Therapeutic antibodies have now been accepted as viable treatments for cancer and other diseases. We are optimistic that some of the immunotherapeutic strategies outlined in this chapter will become front-line therapies in the future.

## 5 ABBREVIATIONS

ADCC  antibody dependent cell mediated cytotoxicity
APC  antigen presenting cell
CDR  complementarity determing region
CTL  cytotoxic T lymphocyte
DC  dendritic cell
HAMA  human anti-mouse antibody
HLA  human leukocyte antigen
IFN  interferon
Ig  immunoglobulin
IL  interleukin
MHC  major histocompatibility complex
NK  natural killer cell
PBMC  peripheral blood mononuclear cell
TCR  T cell receptor
TIL  tumor infiltrating lymophocyte
TNF  tumor necrosis factor

## REFERENCES

1. H. G. Boman, *Immunol. Rev.*, **173**, 5–16 (2000).

2. R. Barrington, M. Zhang, M. Fischer, and M. C. Carroll, *Immunol. Rev.*, **180**, 5–15 (2001).

3. L. L. Lanier and J. H. Phillips, *Immunol. Today*, **7**, 132–137 (1986).

4. Y. Kamogawa, L. A. Minasi, S. R. Carding, K. Bottomly, and R. A. Flavell, *Cell*, **75**, 985–995 (1993).

5. T. R. Mosmann, L. Li, and S. Sad, *Sem. Immunol.*, **9**, 87–92 (1997).

6. L. Rowen, B. F. Koop, and L. Hood, *Science*, **272**, 1755–1762 (1996).

7. C. A. Janeway, P. Travers, M. Walport, and J. D. Capra, Eds., *Immunobiology: The im-*

*mune system in health and disease*, 4th ed., Garland Publishing, London, UK, 1999, pp. 115–162.

8. R. Maldonado-Lopez and M. Moser, *Sem. Immunol.*, **13**, 275–282 (2001).

9. D. J. Straus, *Curr. Oncol. Rep.*, **3**, 260–265 (2001).

10. H. J. Deeg and G. Socie, *Blood*, **91**, 1833–1844 (1998).

11. S. M. Pupa, S. Menard, D. Morelli, B. Pozzi, G. De Palo, and M. I. Colnaghi, *Oncogene*, **8**, 2917–2923 (1993).

12. S. Nawrocki, P. J. Wysocki, and A. Mackiewicz, *Expert Opin. Biol. Ther.*, **1**, 193–204 (2001).

13. P. Kourilsky and N. Fazilleau, *Int. Rev. Immunol.*, **20**, 575–591 (2001).

14. S. J. Antonia, M. Extermann, and R. A. Flavell, *Crit. Rev. Oncol.*, **9**, 35–41 (1998).

15. J. W. Streilein, M. Takeuchi, and A. W. Taylor, *Hum. Immunol.*, **52**, 138–143 (1997).

16. T. Schwaab, A. R. Schned, J. A. Heaney, B. F. Cole, J. Atzpodien, F. Wittke, and M. S. Ernstoff, *J. Urol.*, **162**, 567–573 (1999).

17. Y. Kawakami, S. Eliyahu, K. Sakaguchi, P. F. Robbins, L. Rivoltini, J. R. Yannelli, E. Appella, and S. A. Rosenberg, *J. Exp. Med.*, **180**, 347–352 (1994).

18. P. Mulders, C. L. Tso, B. Gitlitz, R. Kaboo, A. Hinkel, S. Frand, S. Kiertscher, M. D. Roth, J. deKernion, R. Figlin, and A. Belldegrun, *Clin. Cancer Res.*, **5**, 445–454 (1999).

19. P. F. Robbins, M. El-Gamil, Y. F. Li, Y. Kawakami, D. Loftus, E. Appella, and S. A. Rosenberg, *J. Exp. Med.*, **183**, 1185–1192 (1996).

20. C. Asselin-Paturel, H. Echchakir, G. Carayol, F. Gay, P. Opolon, D. Grunenwald, S. Chouaib, and F. Mami-Chouaib, *Int. J. Cancer*, **77**, 7–12 (1998).

21. C. A. Bonham, L. Lu, R. A. Banas, P. Fontes, A. S. Rao, T. E. Starzl, A. Zeevi, and A. W. Thomson, *Transplant. Immunol.*, **4**, 186–191 (1996).

22. F. Geissmann, P. Revy, A. Regnault, Y. Lepelletier, M. Dy, N. Brousse, S. Amigorena, O. Hermine, and A. Durandy, *J. Immunol.*, **162**, 4567–4575 (1999).

23. J. J. Letterio and A. B. Roberts, *Annu. Rev. Immunol.*, **16**, 137–161 (1998).

24. J. A. McEarchern, J. J. Kobie, V. Mack, R. S. Wu, L. Meade-Tollin, C. L. Arteaga, N. Dumont, D. Besselsen, E. Seftor, M. J. Hendrix, E. Katsanis, and E. T. Akporiaye, *Int. J. Cancer*, **91**, 76–82 (2001).

25. R. S. Wu, J. J. Kobie, D. G. Besselsen, T. C. Fong, V. D. Mack, J. A. McEarchern, and E. T. Akporiaye, *Cancer Immunol. Immunother.*, **50**, 229–240 (2001).

26. P. Allavena, L. Piemonti, D. Longoni, S. Bernasconi, A. Stoppacciaro, L. Ruco, and A. Mantovani, *Eur. J. Immunol.*, **28**, 359–369 (1998).

27. C. Buelens, V. Verhasselt, D. De Groote, K. Thielemans, M. Goldman, and F. Willems, *Eur. J. Immunol.*, **27**, 756–762 (1997).

28. F. Wittke, R. Hoffmann, J. Buer, I. Dallmann, K. Oevermann, S. Sel, T. Wandert, A. Ganser, and J. Atzpodien, *Br. J. Cancer*, **79**, 1182–1184 (1999).

29. M. Yamamura, R. L. Modlin, J. D. Ohmen, and R. L. Moy, *J. Clin. Invest.*, **91**, 1005–1010 (1993).

30. S. A. Rosenberg, *Cancer J. Sci. Am.*, **6**, S2–S7 (2000).

31. G. Kohler and C. Milstein, *Nature*, **256**, 495–497 (1975).

32. R. Levy and R. A. Miller, *Ann. Rev. Med.*, **34**, 107–116 (1983).

33. J. S. Huston and A. J. George, *Hum. Antibodies*, **10**, 127–142 (2001).

34. R. M. Hudziak, G. D. Lewis, M. Winget, B. M. Fendly, H. M. Shepard, and A. Ullrich, *Mol. Cell. Biol.*, **9**, 1165–1172 (1989).

35. J. Baselga, L. Norton, J. Albanell, Y. M. Kim, and J. Mendelsohn, *Cancer Res.*, **58**, 2825–2831 (1998).

36. D. J. Slamon, W. Godolphin, L. A. Jones, J. A. Holt, S. G. Wong, D. E. Keith, W. J. Levin, S. G. Stuart, J. Udove, A. Ullrich, et al., *Science*, **244**, 707–712 (1989).

37. M. X. Sliwkowski, J. A. Lofgren, G. D. Lewis, T. E. Hotaling, B. M. Fendly, and J. A. Fox, *Sem. Oncol.*, **26**, 60–70 (1999).

38. A. J. Freebairn, A. J. Last, and T. M. Illidg, *Clin. Oncol.*, **13**, 427–433 (2001).

39. K. C. Anderson, M. P. Bates, B. L. Slaughenhoupt, G. S. Pinkus, S. F. Schlossman, and L. M. Nadler, *Blood*, **63**, 1424–1433 (1984).

40. M. E. Reff, K. Carner, K. S. Chambers, P. C. Chinn, J. E. Leonard, R. Raab, R. A. Newman, N. Hanna, and D. R. Anderson, *Blood*, **83**, 435–445 (1994).

41. O. W. Press, J. Howell-Clark, S. Anderson, and I. Bernstein, *Blood*, **83**, 1390–1397 (1994).

42. A. Demidem, T. Lam, S. Alas, K. Hariharan, N. Hanna, and B. Bonavida, *Cancer Biother. Radiopharm.*, **12**, 177–186 (1997).

43. A. J. Grillo-Lopez, E. Hedrick, M. Rashford, and M. Benyunes, *Sem. Oncol.*, **29**, 105–112 (2002).

44. P. Sorokin, *Clin. J. Oncol. Nurs.*, **4**, 279–280 (2000).

45. P. R. Hamann, L. M. Hinman, I. Hollander, C. F. Beyer, D. Lindh, R. Holcomb, W. Hallett, H. R. Tsou, J. Upeslacis, D. Shochat, A. Mountain, D. A. Flowers, and I. Bernstein, *Bioconjug. Chem.*, **13**, 47–58 (2002).

46. E. L. Sievers and M. Linenberger, *Curr. Opin. Oncol.*, **13**, 522–527 (2001).

47. V. H. van Der Velden, J. G. te Marvelde, P. G. Hoogeveen, I. D. Bernstein, A. B. Houtsmuller, M. S. Berger, and J. J. van Dongen, *Blood*, **97**, 3197–3204 (2001).

48. L. H. Leopold, M. S. Berger, and J. Feingold, *Clin. Lymphoma*, **2**, S29–S34 (2002).

49. C. Hale, M. Bartholomew, V. Taylor, J. Stables, P. Topley, and J. Tite, *Immunology*, **88**, 183–190 (1996).

50. J. Greenwood, S. D. Gorman, E. G. Routledge, I. S. Lloyd, and H. Waldmann, *Ther. Immunol.*, **1**, 247–255 (1994).

51. J. M. Flynn and J. C. Byrd, *Curr. Opin. Oncol.*, **12**, 574–581 (2000).

52. G. A. Pangalis, M. N. Dimopoulou, M. K. Angelopoulou, C. Tsekouras, T. P. Vassilakopoulos, G. Vaiopoulos, and M. P. Siakantaris, *Med. Oncol.*, **18**, 99–107 (2001).

53. D. M. Knight, H. Trinh, J. Le, S. Siegel, D. Shealy, M. McDonough, B. Scallon, M. A. Moore, J. Vilcek, P. Daddona, et al., *Mol. Immunol.*, **30**, 1443–1453 (1993).

54. S. A. Siegel, D. J. Shealy, M. T. Nakada, J. Le, D. S. Woulfe, L. Probert, G. Kollias, J. Ghrayeb, J. Vilcek, and P. E. Daddona, *Cytokine*, **7**, 15–25 (1995).

55. R. N. Maini, F. C. Breedveld, J. R. Kalden, J. S. Smolen, D. Davis, J. D. Macfarlane, C. Antoni, B. Leeb, M. J. Elliott, J. N. Woody, T. F. Schaible, and M. Feldmann, *Arthritis Rheum.*, **41**, 1552–1563 (1998).

56. M. Feldmann and R. N. Maini, *Annu. Rev. Immunol.*, **19**, 163–196 (2001).

57. S. R. Targan, S. B. Hanauer, S. J. van Deventer, L. Mayer, D. H. Present, T. Braakman, K. L. DeWoody, T. F. Schaible, and P. J. Rutgeerts, *N. Engl. J. Med.*, **337**, 1029–1035 (1997).

58. P. H. Wooley, J. Dutcher, M. B. Widmer, and S. Gillis, *J. Immunol.*, **151**, 6602–6607 (1993).

59. A. A. Grom, K. J. Murray, L. Luyrink, H. Emery, M. H. Passo, D. N. Glass, T. Bowlin, and C. Edwards, *Arthritis Rheum.*, **39**, 1703–1710 (1996).

60. D. J. Lovell, E. H. Giannini, A. Reiff, G. D. Cawkwell, E. D. Silverman, J. J. Nocton, L. D. Stein, A. Gedalia, N. T. Ilowite, C. A. Wallace, J. Whitmore, and B. K. Finck, *N. Engl. J. Med.*, **342**, 763–769 (2000).

61. P. J. Mease, B. S. Goffe, J. Metz, A. VanderStoep, B. Finck, and D. J. Burge, *Lancet*, **356**, 385–390 (2000).

62. K. Phillips, M. E. Husni, E. W. Karlson, and J. S. Coblyn, *Arthritis Rheum.*, **47**, 17–21 (2002).

63. P. Kalinski, C. M. Hilkens, E. A. Wierenga, and M. L. Kapsenberg, *Immunol. Today*, **20**, 561–567 (1999).

64. C. Caux, C. Dezutter-Dambuyant, D. Schmitt, and J. Banchereau, *Nature*, **360**, 258–261 (1992).

65. C. D. Reid, A. Stackpoole, A. Meager, and J. Tikerpae, *J. Immunol.*, **149**, 2681–2688 (1992).

66. N. Romani, D. Reider, M. Heuer, S. Ebner, E. Kampgen, B. Eibl, D. Niederwieser, and G. Schuler, *J. Immunol. Methods*, **196**, 137–151 (1996).

67. S. E. Alters, J. R. Gadea, B. Holm, J. Lebkowski, and R. Philip, *J. Immunother.*, **22**, 229–236 (1999).

68. F. Sallusto and A. Lanzavecchia, *J. Exp. Med.*, **179**, 1109–1118 (1994).

69. N. Romani, S. Gruner, D. Brang, E. Kampgen, A. Lenz, B. Trockenbacher, G. Konwalinka, P. O. Fritsch, R. M. Steinman, and G. Schuler, *J. Exp. Med.*, **180**, 83–93 (1994).

70. H. Jonuleit, U. Kuhn, G. Muller, K. Steinbrink, L. Paragnik, E. Schmitt, J. Knop, and A. H. Enk, *Eur. J. Immunol.*, **27**, 3135–3142 (1997).

71. M. Cella, M. Salio, Y. Sakakibara, H. Langen, I. Julkunen, and A. Lanzavecchia, *J. Exp. Med.*, **189**, 821–829 (1999).

72. M. A. Morse, H. K. Lyerly, E. Gilboa, E. Thomas, and S. K. Nair, *Cancer Res.*, **58**, 2965–2968 (1998).

73. J. Banchereau and R. M. Steinman, *Nature*, **392**, 245–252 (1998).

74. D. N. Hart, *Blood*, **90**, 3245–3287 (1997).

75. R. M. Steinman in W. E. Paul, Ed., *Dendritic Cells*, in *Fundamental Immunology*, Lippincott-Raven Publishers, Philadelphia, PA, 1999, pp. 547–573.

76. M. Rescigno, F. Granucci, and P. Ricciardi-Castagnoli, *Immunol. Cell Biol.*, **77**, 404–410 (1999).

77. J. G. Cyster, Chemokines and cell migration in secondary lymphoid organs. *Science*, **286**, 2098–2102 (1999).

78. J. Banchereau, F. Briere, C. Caux, J. Davoust, S. Lebecque, Y. J. Liu, B. Pulendran, and K. Palucka, *Annu. Rev. Immunol.*, **18**, 767–811 (2000).

79. R. M. Steinman and M. Dhodapkar, *Int. J. Cancer.*, **94**, 459–473 (2001).

80. N. Renkvist, C. Castelli, P. F. Robbins, and G. Parmiani, *Cancer Immunol. Immunother.*, **50**, 3–15 (2001).

81. M. L. Disis, K. H. Grabstein, P. R. Sleath, and M. A. Cheever, *Clin. Cancer Res.*, **5**, 1289–1297 (1999).

82. M. L. Disis, K. L. Knutson, K. Schiffman, K. Rinn, and D. G. McNeel, *Breast Cancer Res. Treat.*, **62**, 245–252 (2000).

83. K. L. Knutson, K. Schiffman, and M. L. Disis, *J. Clin. Invest.*, **107**, 477–484 (2001).

84. T. Itoh, Y. Ueda, I. Kawashima, I. Nukaya, H. Fujiwara, N. Fuji, T. Yamashita, T. Yoshimura, K. Okugawa, T. Iwasaki, M. Ideno, K. Takesako, M. Mitsuhashi, K. Orita, and H. Yamagishi, *Cancer Immunol. Immunother.*, **51**, 99–106 (2002).

85. A. B. Bakker, M. W. Schreurs, G. Tafazzul, A. J. de Boer, Y. Kawakami, G. J. Adema, and C. G. Figdor, *Int. J. Cancer.*, **62**, 97–102 (1995).

86. M. R. Parkhurst, M. L. Salgaller, S. Southwood, P. F. Robbins, A. Sette, S. A. Rosenberg, and Y. Kawakami, *J. Immunol.*, **157**, 2539–2548 (1996).

87. S. A. Rosenberg, J. C. Yang, D. J. Schwartzentruber, P. Hwu, F. M. Marincola, S. L. Topalian, N. P. Restifo, M. Sznol, S. L. Schwarz, P. J. Spiess, J. R. Wunderlich, C. A. Seipp, J. H. Einhorn, L. Rogers-Freezer, and D. E. White, *J. Immunol.*, **163**, 1690–1695 (1999).

88. K. H. Lee, E. Wang, M. B. Nielsen, J. Wunderlich, S. Migueles, M. Connors, S. M. Steinberg, S. A. Rosenberg, and F. M. Marincola, *J. Immunol.*, **163**, 6292–6300 (1999).

89. P. Brossart, S. Wirths, G. Stuhler, V. L. Reichardt, L. Kanz, and W. Brugger, *Blood*, **96**, 3102–3108 (2000).

90. R. Lau, F. Wang, G. Jeffery, V. Marty, J. Kuniyoshi, E. Bade, M. E. Ryback, and J. Weber, *J. Immunother.*, **24**, 66–78 (2001).

91. M. Toungouz, M. Libin, F. Bulte, L. Faid, F. Lehmann, D. Duriau, M. Laporte, D. Gangji, C. Bruyns, M. Lambermont, M. Goldman, and T. Velu, *J. Leukocyte Biol.*, **69**, 937–943 (2001).

92. B. Thurner, I. Haendle, C. Roder, D. Dieckmann, P. Keikavoussi, H. Jonuleit, A. Bender, C. Maczek, D. Schreiner, P. von den Driesch, E. B. Brocker, R. M. Steinman, A. Enk, E. Kampgen, and G. Schuler, *J. Exp. Med.*, **190**, 1669–1678 (1999).

93. J. Sidney, S. Southwood, D. L. Mann, M. A. Fernandez-Vina, M. J. Newman, and A. Sette, *Hum. Immunol.*, **62**, 1200–1216 (2001).

94. B. T. Ober, Q. Hu, J. T. Opferman, S. Hagevik, N. Chiu, C. R. Wang, and P. G. Ashton-Rickardt, *Int. Immunol.*, **12**, 1353–1363 (2000).

95. T. J. Pawlowski and U. D. Staerz, *Immunol. Today*, **15**, 205–209 (1994).

96. A. Martin-Fontecha, H. J. Schuurman, and A. Zapata, *Thymus*, **22**, 201–213 (1994).

97. A. Le Campion, C. Bourgeois, F. Lambolez, B. Martin, S. Leaument, N. Dautigny, C. Tanchot, C. Penit, and B. Lucas, *Proc. Natl. Acad. Sci. USA*, **99**, 4538–4543 (2002).

98. Q. Hu, C. R. Bazemore Walker, C. Girao, J. T. Opferman, J. Sun, J. Shabanowitz, D. F. Hunt, and P. G. Ashton-Rickardt, *Immunity*, **7**, 221–231 (1997).

99. A. Zippelius, M. J. Pittet, P. Batard, N. Rufer, M. de Smedt, P. Guillaume, K. Ellefsen, D. Valmori, D. Lienard, J. Plum, H. R. MacDonald, D. E. Speiser, J. C. Cerottini, and P. Romero, *J. Exp. Med.*, **195**, 485–494 (2002).

100. A. Sette, A. Vitiello, B. Reherman, P. Fowler, R. Nayersina, W. M. Kast, C. J. Melief, C. Oseroff, L. Yuan, J. Ruppert, et al., *J. Immunol.*, **153**, 5586–5592 (1994).

101. M. C. Feltkamp, M. P. Vierboom, W. M. Kast, and C. J. Melief, *Mol. Immunol.*, **31**, 1391–1401 (1994).

102. M. E. Ressing, A. Sette, R. M. Brandt, J. Ruppert, P. A. Wentworth, M. Hartman, C. Oseroff, H. M. Grey, C. J. Melief, and W. M. Kast, *J. Immunol.*, **154**, 5934–5943 (1995).

103. E. Keogh, J. Fikes, S. Southwood, E. Celis, R. Chesnut, and A. Sette, *J. Immunol.*, **167**, 787–796 (2001).

104. D. Valmori, J. F. Fonteneau, C. M. Lizana, N. Gervois, D. Lienard, D. Rimoldi, V. Jongeneel, F. Jotereau, J. C. Cerottini, and P. Romero, *J. Immunol.*, **160**, 1750–1758 (1998).

105. T. K. Hoffmann, N. Meidenbauer, G. Dworacki, H. Kanaya, and T. L. Whiteside, *Cancer Res.*, **60**, 3542–3549 (2000).

106. M. Schnurr, C. Scholz, S. Rothenfusser, P. Galambos, M. Dauer, J. Robe, S. Endres, and A. Eigler, *Cancer Res.*, **62**, 2347–2352 (2002).

107. J. D. Geiger, R. J. Hutchinson, L. F. Hohenkirk, E. A. McKenna, G. A. Yanik, J. E. Levine, A. E. Chang, T. M. Braun, and J. J. Mule, *Cancer Res.*, **61**, 8513–8519 (2001).

108. A. E. Chang, B. G. Redman, J. R. Whitfield, B. J. Nickoloff, T. M. Braun, P. P. Lee, J. D. Geiger, and J. J. Mule, *Clin. Cancer Res.*, **8**, 1021–1032 (2002).

109. J. Hernando, T. W. Park, K. Kubler, R. Offergeld, H. Schlebusch, and T. Bauknecht, *Cancer Immunol. Immunother.*, **51**, 45–52 (2002).

110. M. Schott, J. Feldkamp, D. Schattenberg, T. Krueger, C. Dotzenrath, J. Seissler, and W. A. Scherbaum, *Eur. J. Endocrinol.*, **142**, 300–306 (2000).

111. X. Zhao, Y. Q. Wei, and Z. L. Peng, *Immunol. Invest.*, **30**, 33–45 (2001).

112. Z. Abdel-Wahab, P. DeMatos, D. Hester, X. D. Dong, and H. F. Seigler, *Cell Immunol.*, **186**, 63–74 (1998).

113. S. K. Nair, M. Morse, D. Boczkowski, R. I. Cumming, L. Vasovic, E. Gilboa, and H. K. Lyerly, *Ann. Surg.*, **235**, 540–549 (2002).

114. A. Heiser, D. Coleman, J. Dannull, D. Yancey, M. A. Maurice, C. D. Lallas, P. Dahm, D. Niedzwiecki, E. Gilboa, and J. Vieweg, *J. Clin. Invest.*, **109**, 409–417 (2002).

115. C. Thornburg, D. Boczkowski, E. Gilboa, and S. K. Nair, *J. Immunother.*, **23**, 412–418 (2000).

116. J. Gong, D. Avigan, D. Chen, Z. Wu, S. Koido, M. Kashiwaba, and D. Kufe, *Proc. Natl. Acad. Sci. USA*, **97**, 2715–2718 (2000).

117. J. Gong, N. Nikrui, D. Chen, S. Koido, Z. Wu, Y. Tanaka, S. Cannistra, D. Avigan, and D. Kufe, *J. Immunol.*, **165**, 1705–1711 (2000).

118. A. Kugler, G. Stuhler, P. Walden, G. Zoller, A. Zobywalski, P. Brossart, U. Trefzer, S. Ullrich, C. A. Muller, V. Becker, A. J. Gross, B. Hemmerlein, L. Kanz, G. A. Muller, and R. H. Ringert, *Nature Med.*, **6**, 332–336 (2000).

119. C. Thery, M. Boussac, P. Veron, P. Ricciardi-Castagnoli, G. Raposo, J. Garin, and S. Amigorena, *J. Immunol.*, **166**, 7309–7318 (2001).

120. J. Wolfers, A. Lozier, G. Raposo, A. Regnault, C. Thery, C. Masurier, C. Flament, S. Pouzieux, F. Faure, T. Tursz, E. Angevin, S. Amigorena, and L. Zitvogel, *Nature Med.*, **7**, 297–303 (2001).

121. S. Amigorena, *Medicina (B Aires)*, **60**, 51–54 (2000).

122. A. Ribas, L. H. Butterfield, J. A. Glaspy, and J. S. Economou, *Curr. Gene Ther.*, **2**, 57–78 (2002).

123. M. Mincheff, I. Altankova, S. Zoubak, S. Tchakarov, C. Botev, S. Petrov, E. Krusteva, G. Kurteva, P. Kurtev, V. Dimitrov, M. Ilieva, G. Georgiev, T. Lissitchkov, I. Chernozemski, and H. T. Meryman, *Crit. Rev. Oncol. Hematol.*, **39**, 125–132 (2001).

124. J. F. Arthur, L. H. Butterfield, M. D. Roth, L. A. Bui, S. M. Kiertscher, R. Lau, S. Dubinett, J. Glaspy, W. H. McBride, and J. S. Economou, *Cancer Gene Ther.*, **4**, 17–25 (1997).

125. C. J. Kirk and J. J. Mule, *Hum. Gene Ther.*, **11**, 797–806 (2000).

126. D. Rea, F. H. Schagen, R. C. Hoeben, M. Mehtali, M. J. Havenga, R. E. Toes, C. J. Melief, and R. Offringa, *J. Virol.*, **73**, 10245–10253 (1999).

127. L. H. Stockwin, T. Matzow, N. T. Georgopoulos, L. J. Stanbridge, S. V. Jones, I. G. Martin, M. E. Blair-Zajdel, and G. E. Blair, *J. Immunol. Methods*, **259**, 205–215 (2002).

128. B. Gahn, F. Siller-Lopez, A. D. Pirooz, E. Yvon, S. Gottschalk, R. Longnecker, M. K. Brenner, H. E. Heslop, E. Aguilar-Cordova, and C. M. Rooney, *Int. J. Cancer*, **93**, 706–713 (2001).

129. G. P. Linette, S. Shankara, S. Longerich, S. Yang, R. Doll, C. Nicolette, F. I. Preffer, B. L. Roberts, and F. G. Haluska, *J. Immunol.*, **164**, 3402–3412 (2000).

130. K. T. Trevor, E. M. Hersh, J. Brailey, J. M. Balloul, and B. Acres, *Cancer Immunol. Immunother.*, **50**, 397–407 (2001).

131. I. Drexler, E. Antunes, M. Schmitz, T. Wolfel, C. Huber, V. Erfle, P. Rieber, M. Theobald, and G. Sutter, *Cancer Res.*, **59**, 4955–4963 (1999).

132. J. W. Hodge, A. N. Rad, D. W. Grosenbach, H. Sabzevari, A. G. Yafal, L. Gritz, and J. Schlom, *J. Natl. Cancer Inst.*, **92**, 1228–1239 (2000).

133. M. A. Morse, *Curr. Opin. Mol. Ther.*, **3**, 407–412 (2001).

134. National Cancer Institute, available online at http://www.nci.nih.gov/search/clinical_trials/, accessed on May 14, 2002.

# CHAPTER SEVEN

# Selective Toxicity

JOHN H. BLOCK
College of Pharmacy
Oregon State University
Corvallis, Oregon

*Burger's Medicinal Chemistry and Drug Discovery*
Sixth Edition, Volume 5: Chemotherapeutic Agents
Edited by Donald J. Abraham
ISBN 0-471-37031-2    © 2003 John Wiley & Sons, Inc.

## Contents

1 Introduction, 250
  1.1 Categorization of Therapeutic Agents, 250
  1.2 Principles of Selectivity, 252
    1.2.1 Comparative Distribution, 252
    1.2.2 Comparative Biochemistry, 253
    1.2.3 Comparative Cytology, 254
    1.2.4 Comparative Stereochemistry, 254
2 Examples of Selective Toxicity (9), 257
  2.1 Cancer Chemotherapy, 257
    2.1.1 Monoclonal Antibodies (*Comparative Cytology*), 257
    2.1.2 Imatinib (*Comparative Biochemistry*), 257
    2.1.3 Cisplatin/Carboplatin (*Comparative Distribution*), 257
    2.1.4 Newer (Not Commercially Available) Selective Approaches to Cancer Chemotherapy, 258
      2.1.4.1 Paclitaxel Antibody Conjugate (*Comparative Cytology*), 258
      2.1.4.2 Cationic Rhodacyanine Dye Analog (*Comparative Distribution*), 258
      2.1.4.3 Tricyclic Thiophene (*Comparative Distribution* and *Comparative Metabolism*), 258
      2.1.4.4 Diphtheria Toxin (*Comparative Biochemistry*), 259
    2.1.5 Summary, 259
  2.2 Dopaminergic Receptors, 259
    2.2.1 Receptor Destruction (*Comparative Cytology*), 259
  2.3 Anti-Infectives, 259
    2.3.1 Introduction, 259
    2.3.2 Examples Based on Chemical Class, 260
      2.3.2.1 β-Lactams (*Comparative Biochemistry*), 260
      2.3.2.2 Aminoglycosides (*Comparative Biochemistry* and *Comparative Distribution*), 260

2.3.2.3 Tetracyclines (*Comparative Biochemistry* and *Comparative Distribution*), 261
2.3.2.4 Macrolides (*Comparative Biochemistry*), 261
2.3.2.5 Quinolones/Fluroquinolones (*Comparative Biochemistry*), 263
2.3.2.6 Sulfonamides/Sulfanilamides and Diaminopyrimidines (*Comparative Biochemistry* and *Comparative Distribution*), 263
2.3.3 Examples Based on the Target Organism, 264
    2.3.3.1 Antiviral Drugs (*Comparative Biochemistry*), 266
    2.3.3.2 Antimycobacterial Drugs (*Comparative Biochemistry* and *Comparative Cytology*), 266
    2.3.3.3 Antifungal Drugs (*Comparative Biochemistry*), 266
    2.3.3.4 Drugs Used to Treat Parasitic Infections (*Comparative Biochemistry*), 268
3 Drug Chirality, 268
  3.1 Propoxyphene, 269
  3.2 Verapamil, 270
3.3 Warfarin, 271
3.4 Other Examples, 271
4 Additional Pharmacologically Active Agents, 273
  4.1 Bisphosphonates (*Comparative Distribution*), 273
  4.2 Inhibitors of Cyclooxygenase I (COX-1) and II (COX-2) (*Comparative Biochemistry* and *Comparative Distribution*), 274
  4.3 Antihistamines (*Comparative Biochemistry* and *Comparative Distribution*), 275
  4.4 Selective Estrogen Receptor Modulators (*Comparative Biochemistry*), 276
  4.5 Phosphodiesterase Inhibitors, 277
5 Insecticides, 277
  5.1 Neurotoxic Insecticides (*Comparative Biochemistry* and *Comparative Distribution*), 278
    5.1.1 Neonicotinoids, 278
    5.1.2 Spinosyns, 278
  5.2 Insect Growth Regulators (*Comparative Biochemistry*), 278
  5.3 Inhibitors of Oxidative Phosphorylation (*Comparative Biochemistry* and *Comparative Distribution*), 278
6 Conclusion, 279

# 1  INTRODUCTION

Selective toxicity was popularized by Professor Adrian Albert beginning with his lectures at University College London in 1948 and the first edition of his book in 1951. Subsequent editions appeared until the seventh edition in 1985 (1). The latter was reprinted in Japanese as two volumes in 1993 and 1994 (2, 3).

Today, the concept of selective toxicity largely is limited to chemotherapy and antibiotic therapy. How does one design a cancer chemotherapeutic agent that will kill a malignant cell and not interfere with mitosis in a benign cell? How does the biochemistry of a bacterial cell differ from that of a mammalian cell so that the antibiotic is toxic to the bacterium and "ignored" by the patient's cells?

It is important to realize that nearly all types of drug therapy can be thought of as selectively toxic. With the possible exception of some forms of hormonal replacement therapy, a drug's desired pharmacological response is actually an intervention in normal biochemical processes. Consider the nonste-

roidal anti-inflammatory drugs (NSAIDs) that inhibit cyclooxygenase (COX). Although reducing inflammation is the desired response for these agents, the drugs are toxic to the enzyme. $\beta$-Adrenergic agonists are exogenous ligands for the same receptors as epinephrine but are not subject to the same regulatory controls (biosynthesis, release, reuptake, metabolic disposal) as for the endogenous hormone. Herbicide and pesticide use in agriculture increasingly is based on the agent being selective for a specific plant or taking advantage of some specific property of the insect that differs from those of farm animals and humans.

## 1.1  Categorization of Therapeutic Agents

Focusing on humans, most drugs are agonists, antagonists, or replacement agents. Hormone replacement therapy is the basis for the use of insulin by a person with type 1 diabetes (formerly called insulin-dependent diabetes mellitus or IDDM) and levothyroxine for a patient with a thyroid deficiency. Insulin and thyroxine act only on specific receptors. As long as the dosing is correct, patients do not experience unpleasant adverse reactions. Excessive insulin causes severe, life-threatening hypoglycemia, leading to insulin shock. Because of the general distribution of thyroxine receptors throughout the body, excessive thyroxine accelerates intermediate metabolism in several organs.

This concept of replacement therapy can be extended to the administration of vitamins and minerals to a patient whose body stores have been depleted, or has a medical diagnosis for which increased administration of nutritional supplements is indicated (e.g., digestive disorders, malignancies, clinical depression, weight control with low calorie diets). The nutritional agents combine with specific receptors or, usually after a metabolic transformation, act as coenzymes (e.g., niacin-NAD/NADP, thiamine/thiamine pyrophosphate, folic acid/tetrahydrofolates, and cobalamin/adenosyl cobalamin). Again, when the dosing is correct, the patient does not experience adverse effects. Excessive doses of some vitamins can cause neuritis (pyridoxine/vitamin $B_6$), hypercalcemia (cholecalciferol/vitamin $D_3$ and ergocalciferol/vitamin $D_2$), and liver damage (retinol/vitamin A).

Agonists mimic the endogenous ligand when combining with the ligand's receptor. In theory, there should be excellent selectivity, but many times the receptors for the ligand are scattered widely throughout the body and located in many organs. Thus, an adrenergic agonist such as phenylephrine will constrict blood vessels while simultaneously increasing the heart rate. For some patients, the latter could lead to life-threatening tachycardia. Depending on the receptor, the active form of cholecalciferol (vitamin $D_3$), 1,25-dihydroxycholecalciferol (1) causes the synthesis of a calcium transport protein in the intestinal mucosa and regulates cell division. An analog of 1,25-dihydroxycholecalciferol, calcipotriene (Dovonex, 2) is used topically for psoriasis. If administered internally, the patient could experience severe hypercalcemia, leading to calcification of the soft tissues and blood vessel walls. In the context of selective toxicity, calcipotriene has poor selectivity.

Most drugs dispensed today are antagonists. (See other chapters for discussions of partial agonists and antagonists.) The desired

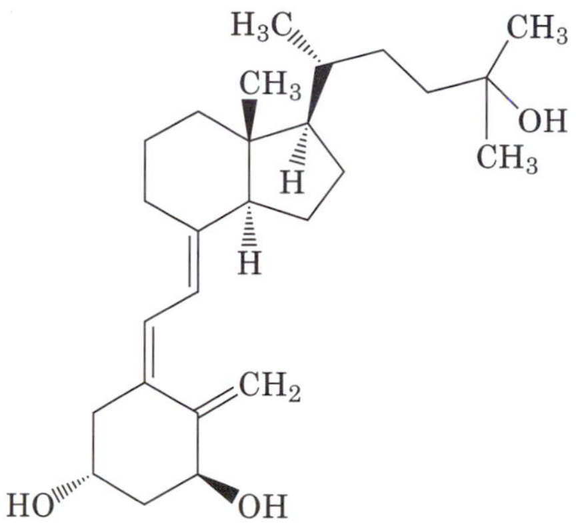

(1)

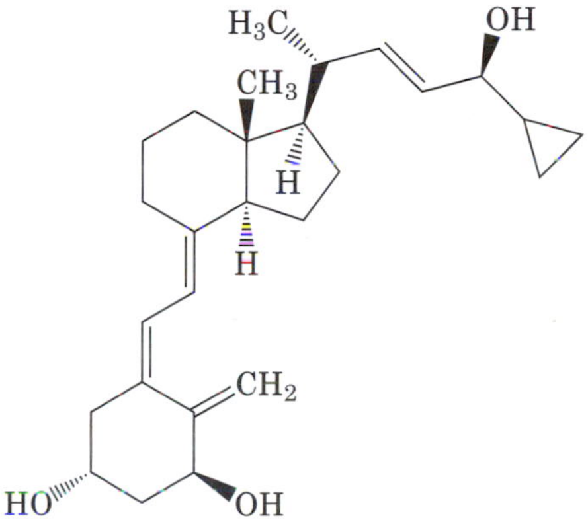

(2)

goal is to "block" responses in the cholinergic, adrenergic, rennin/angiotensin, and other integrated systems. Rarely are these biochemical systems localized in a specific organ. Inhibiting one enzyme can have unforeseen metabolic consequences downstream. A classic example is the angiotensin-converting enzyme (ACE) inhibitors that produce an irritating cough (to the patient and patient's family). Angiotensin-converting enzyme has more than one substrate: angiotensin I and bradykinin (Fig. 7.1). The desired target was inhibiting the formation of angiotensin II from angiotensin I and was the focus for the development of the ACE inhibitors (e.g., captopril, enalapril, and lisinopril). Angiotensin-converting enzyme also degrades bradykinin, one of the many peptides involved with pain production and inflammation. ACE inhibitors

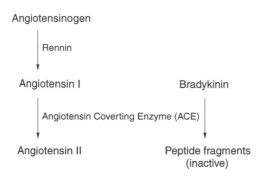

Angiotensinogen

|
Rennin

Angiotensin I                                          Bradykinin

|
Angiotensin Coverting Enzyme (ACE)

Angiotensin II                                          Peptide fragments
                                                        (inactive)

**Figure 7.1.** Rennin-angiotensin converting enzyme biochemistry.

cause production of excess bradykinin and may be the cause of the annoying dry cough characteristic of these drugs.

An important approach to increasing the selectivity of a pharmacological agent is to thoroughly understand the biochemistry of the system where drug therapy is indicated. By going downstream from the angiotensin-converting enzyme and employing an angiotensin II antagonist, there are fewer adverse reactions.

Asthma is a complex inflammatory disease of the respiratory system. One of the family of chemical mediators are the leukotrienes synthesized from arachindonic acid by the lipoxygenase enzyme complex (Fig. 7.2). The leukotrienes are one of the mediators of bronchoconstriction. In the treatments of asthma, the leukotriene $D_4$ ($LTD_4$) receptor antagonists, montelukast sodium (**3**) and zafirlukast (**4**), exhibit less adverse responses than the 5-lipoxygenase inhibitor, zileuton (**5**), which interferes with the formation of $LTA_4$, $LTC_4$, $LTD_4$, and the $LTE_4$ family of leukotrienes, all of which are chemical mediators of the immune response at several sites.

Distribution of enzyme types is another way to increase the selectivity. The cyclooxygenase (COX) isozymes are distributed unevenly among the various organs. What is now known as COX-1 is a constitutive enzyme found in many organs. COX-2 is inducible by cytokines and appears at sites when there is inflammation. Although not free of adverse reactions, celecoxib and rofecoxib do show fewer side effects in patients who must take these drugs on a chronic basis compared to the older COX-1 inhibitory nonsteroidal anti-inflammatory drugs (NSAIDs).

### 1.2 Principles of Selectivity

Professor Albert, after examining the various ways that selectivity of biological response is obtained from both natural products or synthetic agents, concluded that there are three possible ways that a pharmacologically active agent exerts selectivity: (*1*) comparative distribution, (*2*) comparative biochemistry, and (*3*) comparative cytology (*4*).

**1.2.1 Comparative Distribution.** Comparative distribution can be caused by differences in physical area where absorption occurs or differences in the drug's biodistribution. An insecticide accumulates in the insect because the insect has more exposed surface area relative to that of animals in the same environment. Radioactive iodine is used both to diagnose diseases of the thyroid and to destroy the thyroid gland because it accumulates in the thyroid as a result of specific iodine transport proteins. Parasites that remain in the intestinal tract are easier to treat because many of the newer drugs also remain in the intestinal tract. A rapidly dividing cell line will preferen-

(3)

(4)

(5)

tially incorporate a number of drugs, including those used in the treatment of cancer.

There are several examples where drugs cause adverse responses in patients because of poor comparative distribution. Most of the drugs used in the treatment of cancers do not differentiate between benign tissues, which are constantly dividing (bone marrow, intesti-

nal mucosa, hair follicles), and malignant cells (5). The result is immunosuppression and/or anemia, diarrhea or constipation, and loss of hair. A problem with the current antiviral drugs (see below) is that they cannot intercept free viruses but are effective only when the virus is inside a cell and dividing. These drugs cannot differentiate between virus-infected cells and cells free of virus.

**1.2.2 Comparative Biochemistry.** Comparative biochemistry is the basis for successful antibiotic therapy. The antibiotic class is one of the main reasons that human life expectancy has increased, producing a significant older population. The $\beta$-lactam antibiotics, penicillins and cephalosporins, inhibit

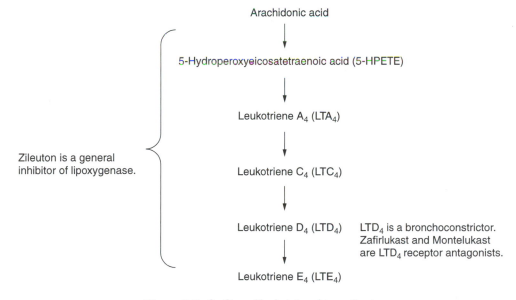

**Figure 7.2.** Outline of leukotriene biosynthesis.

bacterial transpeptidase, thus preventing the microorganism from completing the synthesis of its cell wall. Mammals do not have cell walls and, thus, are not affected by β-lactam antibiotics. Our lipid bilayers have completely different biochemistry from that of a bacterium. The various drugs that inhibit the biosynthesis of folic acid or its reduction to tetrahydrofolic acid (trimethoprim, sulfonamides) are bacteriocidal or bacteriostatic and do not affect mammals that cannot synthesize folic acid. (That is why folic acid is a vitamin.) Pyrimethamine shows a preference for plasmodia dihydrofolate reductase relative to the human enzyme and, therefore, has found use in the treatment of malaria (6).

In contrast, it has been difficult to develop antiviral and antifungal drugs that are selective for the infectious agent. Viruses are complete parasites in that they can reproduce only inside the host cell. Many viruses are dependent on the host cell's polymerases, ribosomes, and transfer RNAs. Most of today's antiviral drugs not only poorly differentiate between viral and human polymerases, but they also cannot distinguish between cells infected with the virus versus noninfected cells. Even inhibitors preferential for RNA-dependent DNA polymerase (reverse transcriptase), an enzyme not found in mammalian cells, can inhibit mammalian DNA-dependent DNA polymerase, leading to suppression of the patient's bone marrow.

Acyclovir and its analogs probably are the most successful antiviral drugs in use today. They are indicated for herpes simplex I and II and varicella zoster (chicken pox). The herpes virus has one of the more complicated viral genomes, coding for over 160 genes. One of these structural genes codes for thymidylate kinase, which is different from the mammalian kinase enzyme. The viral enzyme will phosphorylate inactive acyclovir (actually a prodrug), producing active acyclovir monophosphate (Fig. 7.3). The latter is phosphorylated next to the diphosphate and finally the triphosphate, which is the active antiviral drug. None of the mammalian kinases can significantly phosphorylate inactive acyclovir to the active form.

Fungal metabolism is more similar to mammalian metabolism than to bacterial metabolism (Fig. 7.4). Whereas fungi and plants produce ergosterol rather than cholesterol, fungi and mammals follow the same pathway to lanosterol from mevalonate and squalene (7). In contrast, plants produce cycloartenol from mevalonate and squalene.

Many of today's antifungal drugs take advantage of the fact that fungi produce ergosterol rather than cholesterol; however, ergosterol biosynthesis is very similar to that of cholesterol. Several of the antifungal drugs inhibit the C-14 demethylase that removes the 14-methyl group from lanosterol, thereby preventing the subsequent synthesis of ergosterol. For this reason, antifungal drugs that inhibit ergosterol's biosynthesis show poorer selectivity, with the result that many antifungal drugs can only be used topically, far away from the patient's organs that synthesize cholesterol.

**1.2.3 Comparative Cytology.** The third principle, comparative cytology, refers to the comparative taxonomic structure of cells. Examples include plant vs. animal cells; undifferentiated malignant cells vs. fully differentiated mature benign cells; the patient's immune system recognizing nonself cells, which produces the basis for monoclonal antibody therapy; cell wall vs. cell membrane; mitochondria vs. chloroplasts; and the presence of mitochondria in aerobes vs. their absence in anaerobes (8). There is overlap between comparative cytology and the first two principles. It is not uncommon to see selectivity caused by two and possibly all three principles, although usually one predominates. Penicillins and cephalosporins are selective for bacteria because of comparative cytology (cell wall vs. cell membrane) and comparative biochemistry (transpeptidase in bacteria).

**1.2.4 Comparative Stereochemistry.** The fourth principle of selectivity that is being used today in drug design is comparative stereochemistry. Most receptors are chiral and respond differently to drugs of different chirality. Increasingly, the Food and Drug Administration requires new drug applications to resolve racemic mixtures to determine which stereoisomer is biologically active. Drugs with chiral centers may exhibit more

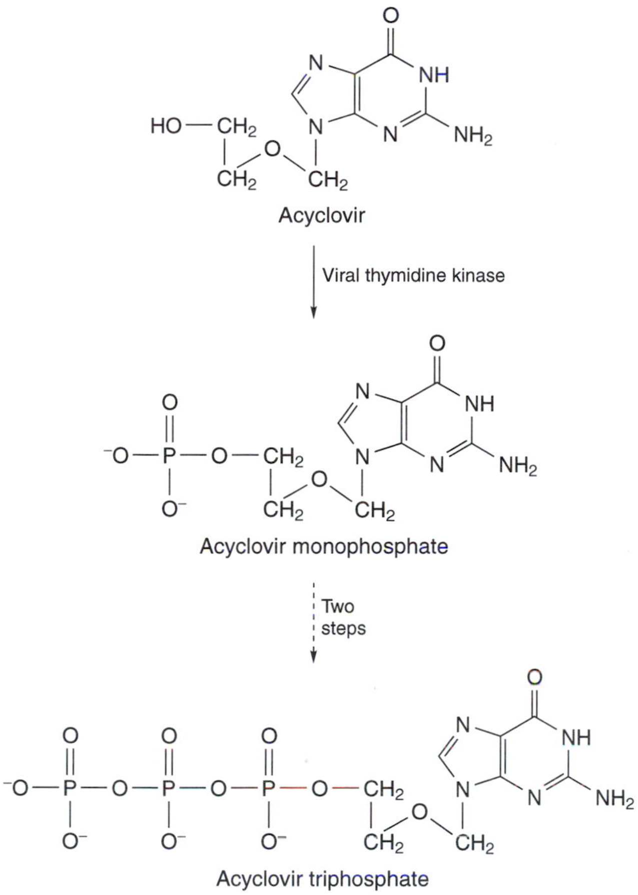

**Figure 7.3.** Acyclovir (acycloquanosine) activation.

complex selectivity. This can be caused by flexibility at the receptor, variations in transport across cell membranes to the site of action, and differences in metabolism for each stereoisomer. A classic example is the antitussive dextromethorphan (d-3-methoxy-N-methylmorphinan) (**6**), which is devoid of opiate ac-tivity compared to that of the levo isomer (l-3-methoxy-N-methylmorphinan) (**7**).

The remainder of this chapter consists of a brief review of strategies to enhance selectivity in biological response. Most examples consist of commercially available pharmacologically active agents, but there are descriptions

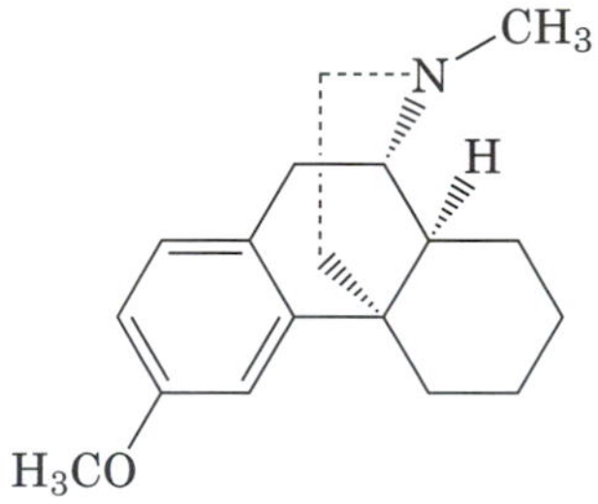

(6)

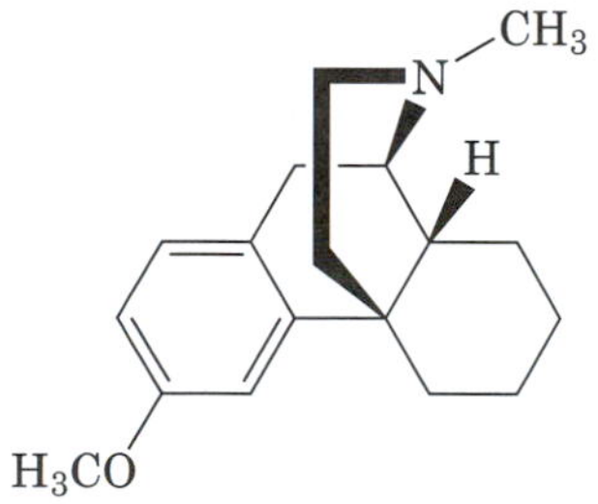

(7)

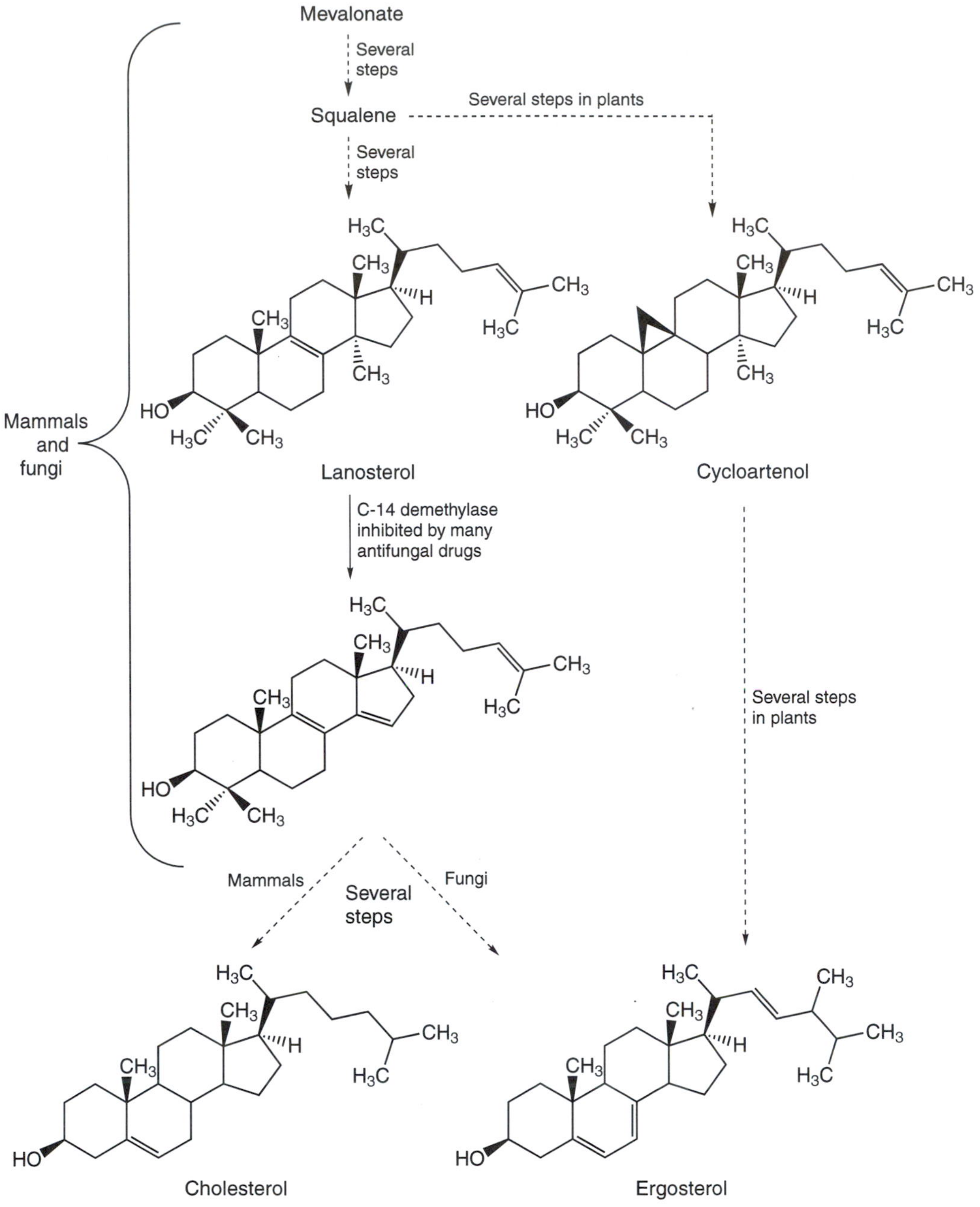

**Figure 7.4.** Cholesterol-ergosterol biosynthesis.

of recent research involving strategies that may or may not result in commercial products. Although Professor Albert's three principles are presented in each example, a standard pharmacological or toxicological classification is used.

## 2 EXAMPLES OF SELECTIVE TOXICITY (9)

### 2.1 Cancer Chemotherapy

**2.1.1 Monoclonal Antibodies (*Comparative Cytology*).** Monoclonal antibodies were going to be the fulfillment of Ehrlich's "magic bullets," particularly in the treatment of cancer where selective toxicity has been poor. They would carry the cytotoxic drug to the malignant cell where it would be released. The antibody would ignore benign cells. Alternatively, the antibody itself would attach to the malignant cell, providing the initial step for further response by the patient's immune system involving complement, macrophages, and T killer cells. This goal still has to be reached, but significant progress is being made.

The first impediment is the main source of monoclonal antibodies. They come from diverse nonhuman sources such as rodents and sheep. Repeated injections sensitize the patient to the animal protein. A partial solution as been "humanizing" the monoclonal antibody by using human genes to code for the constant region of the heavy and light chains and the animal genes for the variable regions. The result is that now approximately 70–75% of the immunoglobulin is human.

In addition, the malignant cell is not recognized by the patient's immune system as nonself because its surface contains essentially the same self-antigens as the patient's benign cells. Otherwise, the patient's immune system likely would have killed the malignant cells before they grew into a detectable mass. Therefore, the monoclonal antibodies developed in an animal where human cell surface antigens are foreign, also will respond to benign cells.

The most successful product to date that addresses this problem is trastuzumab (Herceptin) indicated for metastatic breast cancer containing cells that overexpress the HER2

proto-oncogene. This particular gene codes for a transmembrane receptor structurally related to the epidermal growth factor receptor. Trastuzumab binds to this receptor. This overexpression is seen in only 25–30% of primary breast cancers. Also, the gene is normally expressed in other cells, which means trastuzumab is not as selective as would be desired.

**2.1.2 Imatinib (*Comparative Biochemistry*).** The search for selective metabolism in malignant cells has focused on unique proteins synthesized by the cell that are involved in cell division. In the case of chronic myeloid leukemia, imatinib [Gleevec, (**8**)] inhibits a protein-tyrosine kinase formed by the Ber-Abl gene. The latter, and its protein product, are not found in normal cells. Although this drug does show selectivity, tyrosine kinases are common enzymes and imatinib also inhibits this tyrosine kinase receptor in platelet-derived growth factor and stem cell factor. Therefore, the patient can experience thrombocytopenia and neutropenia.

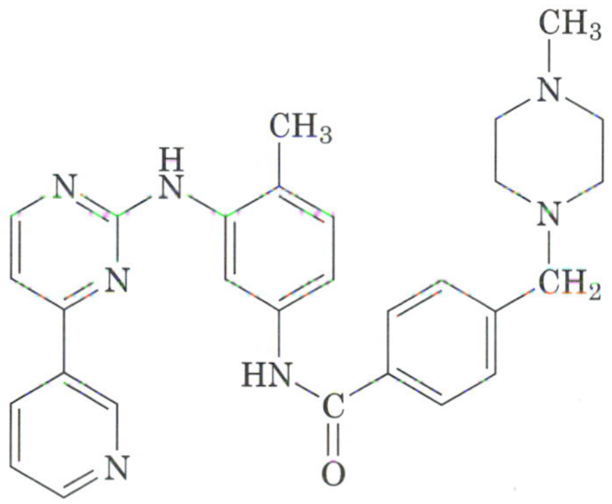

(8)

**2.1.3 Cisplatin/Carboplatin (*Comparative Distribution*).** It has been known almost from the time of its serendipitous discovery that the *cis* analogs of the platinum-containing cytotoxic agents, cisplatin (**9**) and carboplatin (**10**), are much more cytotoxic than the *trans* isomers. Both the *cis* and *trans* compounds enter the growing cell and form inter- and intrastrand crosslinks within the DNA double helix. The kinetics of this binding is very different between the two isomers. Further, DNA repair is more efficient with the *trans* isomers. The net result is that the *cis* isomer remains in the DNA long enough to be cytotoxic (10).

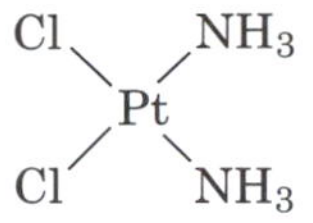

(9)

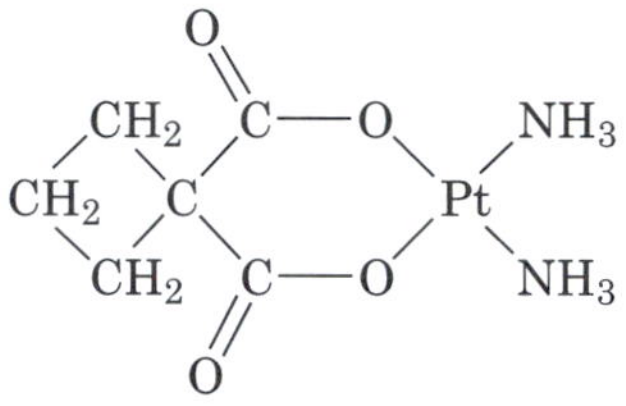

(10)

### 2.1.4 Newer (Not Commercially Available) Selective Approaches to Cancer Chemotherapy

*2.1.4.1 Paclitaxel Antibody Conjugate (Comparative Cytology).* The rationale for this project was to combine the best properties of two drug types and overcome each of their undesirable characteristics. Paclitaxel is a very effective anticancer drug but has poor solubility and lacks tumor specificity. Monoclonal immunoglobulins have good solubility properties and better tumor specificity, but lack good therapeutic efficacy. The goal was to develop a conjugate that would be inactive until attached to the malignant cell. The conjugate was produced by synthesizing 2′-glutaryl paclitaxel from glutaric anhydride. The latter's free carboxyl group was derivatized with *N,N*′-carbonyldiimidazole followed by addition of the monoclonal antibody (11). The binding character of the antibody was not altered by adding it to paclitaxel, and paclitaxel's cytotoxicity also was not affected when tested in rat neuroblastoma cell lines (11).

*2.1.4.2 Cationic Rhodacyanine Dye Analog (Comparative Distribution).* One cytotoxic target in a malignant cell is the cell's mitochondria, the membranes of which are negatively charged. Therefore, a positively charged molecule lipophilic enough to pass through the cell's hydrophobic membrane's lipid bilayer

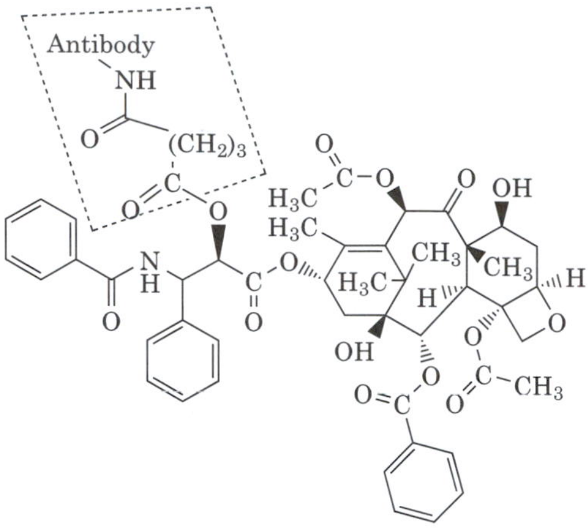

(11)

might be one way to obtain comparative distribution within the cell. It has been found that the positively charged rhodyacyanine dyes pass through the hydrophobic cell membrane lipid bilayer and concentrate in the negatively charged mitochondria membranes. Of course, both malignant and benign nucleated cells have mitochondria. The selectivity claimed for the rhodacyanine dyes (**12**) is based on the observation that carcinoma cells have a higher mitochondrial membrane potential relative to that of benign cells (12).

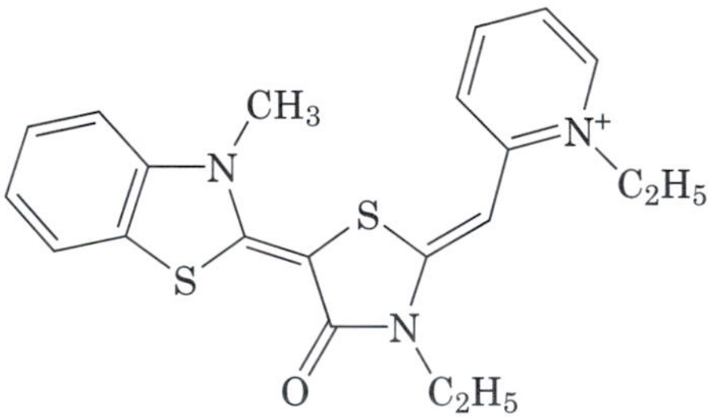

(12)

*2.1.4.3 Tricyclic Thiophene (Comparative Distribution and Comparative Metabolism).* The National Cancer Institute uses a panel of 60 human tumor cell lines to develop a "fingerprint" that, when analyzed using appropriate software, provides leads to mechanisms of cytotoxicity. An example is a tricyclic thiophene (**13**) that showed selective accumulation and metabolism in renal tumor cell lines.

Preliminary work in cell cultures and extracts indicate that a CYP 450 enzyme in the malignant cells may be responsible for forming an oxidized product that binds to proteins inside the cell. The latter leads to the cytotoxic response (13).

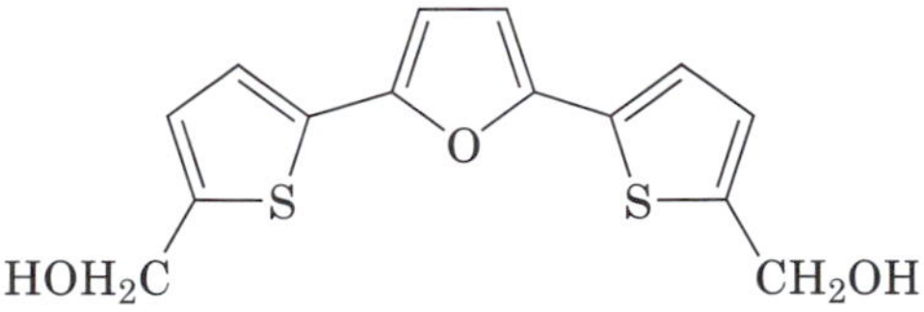

(13)

*2.1.4.4 Diphtheria Toxin (Comparative Biochemistry).* Rhabdomyosarcomas are characterized by a fusion protein formed from two abnormal genes that may contribute to the tumor cell's immortality, possibly by inhibiting apoptosis. By a process called protein swapping, the gene for diphtheria toxin A-chain is inserted into the malignant cell by a plasmid vector, thereby producing a fusion protein containing the toxin's A-chain and causing these tumor cells to undergo programmed cell death. Cells lacking the defective genes do not experience protein swapping and the gene for the toxin A-chain is not expressed (14).

**2.1.5 Summary.** The examples used in the treatment of cancers are attempts at trying to increase the selectivity in cancer chemotherapy using Professor Albert's three comparative principles. Nearly all of the drugs used today in cancer chemotherapy interfere with cell replication. Unfortunately, it has been difficult to specifically target malignant cells while sparing normal cells (i.e., gastric lining, skin, bone marrow). The end result is the use of very toxic drugs in the treatment of disease. If it were not for the lethality of most cancers, cancer chemotherapeutic drugs would not be approved by the regulatory authorities.

## 2.2  Dopaminergic Receptors

**2.2.1 Receptor Destruction (*Comparative Cytology*).** Although selective toxicity in drug design focuses on mechanisms that avoid adverse reactions, the concept can be applied to

models that help explain how receptors are destroyed during the disease process. Parkinson's disease involves destruction of dopaminergic receptors. Molecular probes that are very specific to dopaminergic receptors targeted in Parkinson's disease and do not affect other receptors in the brain are useful in determining how the disease progresses. One such probe is 1-methyl-4-phenylpyridinium [MPP$^+$, (**15**)] obtained by monoamine oxidase B oxidation of 1-methyl-1,2,3,6-tetrahydropyridine [MPTP, (**14**)]. MPP$^+$ inhibits mitochondrial complex I. It has been suggested that this inhibition results in less ATP for the neuron, depolarization of the mitochondria, and generation of reactive oxygen species, which leads to neuronal death. This model does not explain why dopaminergic neurons are more susceptible than other aerobic cells to MPP$^+$. In a series of experiments using another mitochondrial complex I inhibitor, rotenone, it appears that MPP$^+$ is specific to the structure of the dopaminergic receptor (15).

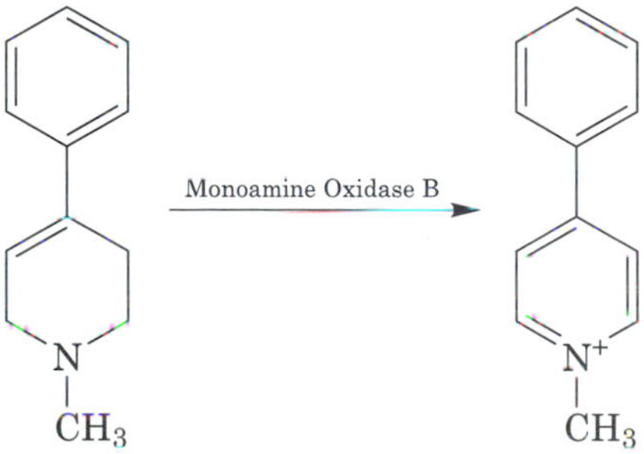

(14)                                              (15)

## 2.3  Anti-Infectives

**2.3.1 Introduction.** The approach to anti-infective therapy is based on the drug's being selectively toxic to the pathogen and, ideally, being "ignored" by the patient's cells. This group of widely prescribed drugs is covered in more detail in a separate chapter. Thus, this discussion is limited to a representative set of examples. Table 7.1 provides a summary of sites where antibiotics exert their antibacterial activity. The reader should refer to the anti-infectives chapters for discussions of structure-activity relationships (SARs), including how to reduce the

**Table 7.1   Summary of Sites Where Antibiotics Exert Their Antibacterial Effect**

| Site of Action | Antibiotic Class |
|---|---|
| Bacterial cell wall (peptidoglycan) synthesis | β-Lactams, Bacitracins, Vancomycin |
| Bacterial ribosomal units | Aminoglycosides, Tetracyclines, Macrolides |
| Organization of bacterial membrane | Polymixins, Colistins, Gramicidins |
| Bacterial DNA gyrase/topoisomerase | Quinolones |
| Folic acid biosynthesis | Sulfonamides |
| Dihydrofolate reductase | Trimethoprim |

risk of antibiotic resistance and improve the drug's pharmaceutical properties.

Most of the examples in this section are based on Professor Albert's principle of comparative biochemistry. Comparative cytology also plays a role, although most of the time this principle frustrates the drug designer and clinician. Many times antibiotics will be effective against Gram-positive but not Gram-negative bacteria. This difference often is attributed to the additional proteoglycan layer on the outside of the cell wall of Gram-negative bacteria. At the macro level, the additional barrier that keeps the gram stain from entering the bacterium also hinders the antibiotic from entering the cell.

### 2.3.2 Examples Based on Chemical Class

*2.3.2.1 β-Lactams (Comparative Biochemistry).* The β-lactam antibiotics consist of two major groups, penicillins (**16**) and cephaloporins (**17**), both of which are still the most widely prescribed drugs indicated for bacterial infections. They exhibit classic selective toxicity because they inhibit a key enzyme, transpeptidase, in the biosynthesis of the bacterial cell wall, a structure not found in the lipid bilayer of mammalian cell membranes. The inhibition is complex and involves the antibiotic co-

valently binding to "protein-binding proteins" (Fig. 7.5). The carbonyl carbon of the sterically strained β-lactam ring is attacked by the serine hydroxyl at the protein's active site, forming a stable covalent protein-antibiotic conjugate. In general, the β-lactam antibiotics are very safe. Adverse reactions are usually found in a small subset of patients who are allergic to this group of drugs.

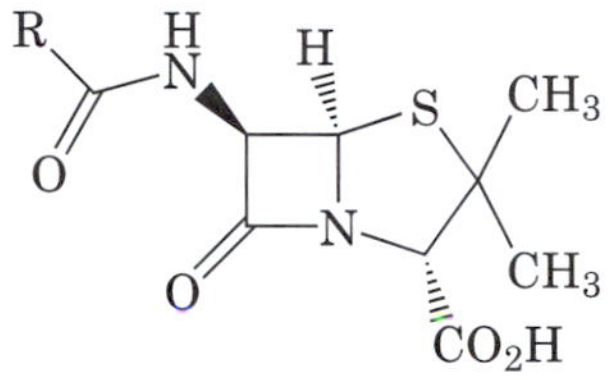

(16)

*2.3.2.2 Aminoglycosides (Comparative Biochemistry and Comparative Distribution).* In contrast with the β-lactam antibiotics, the aminoglycoside antibiotics have more structural diversity (Fig. 7.6). Examples include streptomycin, the gentamycin family, the kanamycin family, and the neomycin family. The one structural characteristic they have in common is one or two amino hexoses connected to a six-mem-

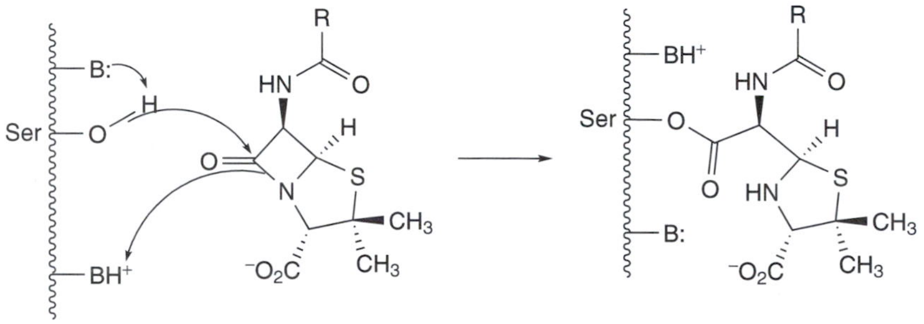

**Figure 7.5.** Penicillin mechanism.

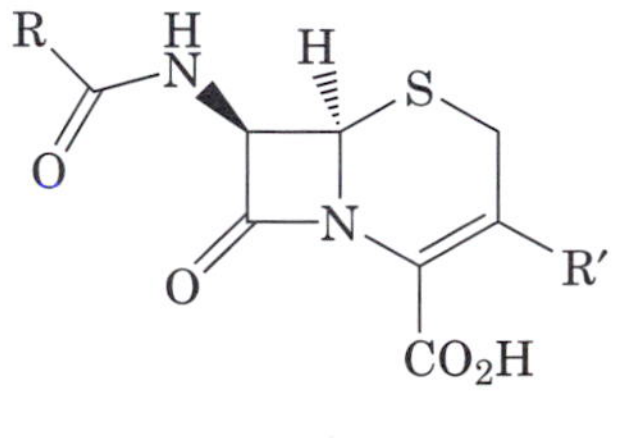

(17)

(18)

bered ring substituted with alcohols and amino moieties. Sometimes these are called aminocyclitols. Because the latter do not have an anomeric carbon, they are not sugars.

The aminoglycosides bind to the bacterial 30S ribosomal unit, causing inhibition of initiation of protein synthesis and sometimes misreading of the genetic code. Fortunately, the ribosomal units in mammalian cells are sufficiently different that the aminoglycosides do not readily bind to mammalian ribosomes.

Unfortunately, the aminoglycoside antibiotics do have severe toxicities. The neomycin family has such severe nephrotoxicity that these drugs usually are only administered topically. One of the most distressing and common adverse responses is ototoxicity, which can lead to permanent deafness. Although the biochemical mechanism of this toxicity is poorly understood, these antibiotics concentrate in the lymphatic tissue of the inner ear. The half-lives of the aminoglycosides are five to six times longer in the otic fluid compared to that of the plasma (16). This is an example in which comparative distribution increases the toxicity of the drug.

*2.3.2.3 Tetracyclines (Comparative Biochemistry and Comparative Distribution).* Like the aminoglycoside antibiotics, the tetracyclines (18) preferentially bind to the 30S ribosomal subunit, thus preventing elongation of peptide chains. In addition, the tetracyclines are actively transported into the bacterium, causing the drug to concentrate in the susceptible cell. Mammalian cells lack this active transport system for tetracyclines. Interestingly, some resistant bacteria have an active efflux system that can pump the tetracycline back out of the cell.

The tetracyclines are considered relatively nontoxic. Their ability to chelate di- and trivalent cations, particularly divalent calcium, can be a problem during certain periods of a patient's development. Tetracyclines are de-

posited in the bones of fetuses and growing children. They also are deposited in the developing teeth, producing discoloration. Therefore, tetracyclines normally are contraindicated for infants and pregnant women, although this problem is easily handled by prescribing an alternate antibiotic.

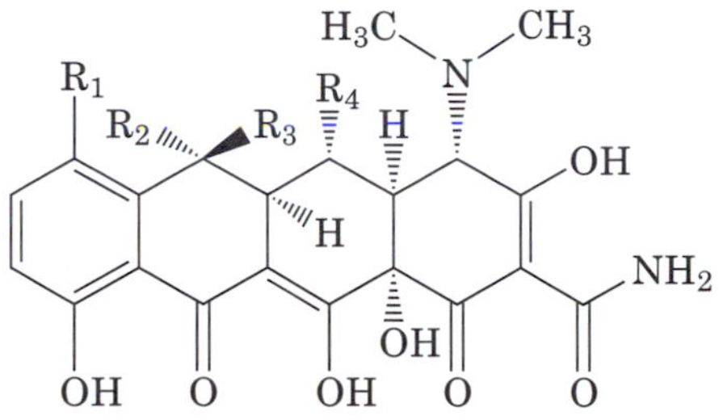

*2.3.2.4 Macrolides (Comparative Biochemistry).* Erythromycin (R = H) (19), clarithromycin (R = CH$_3$) (19), and azithromycin (20), produced by ring expansion of erythromycin, reversibly bind to the bacterial 50S ribosomal subunit. The result is inhibition of the growing protein chain. (Carbon 9 is labeled on erythromycin for reference.) Selective toxicity is achieved because these agents do not bind to mammalian ribosomes.

The macrolide antibiotics are not as selective as other anti-infective drugs. They can cause severe epigastric distress, possibly functioning as a motilin receptor agonist, which stimulates gas-

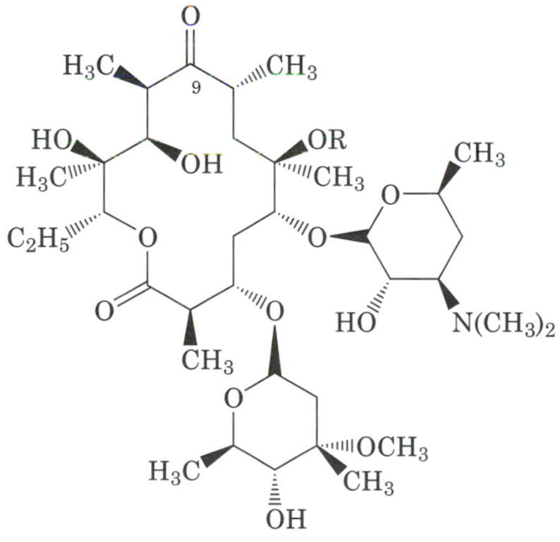

(19)

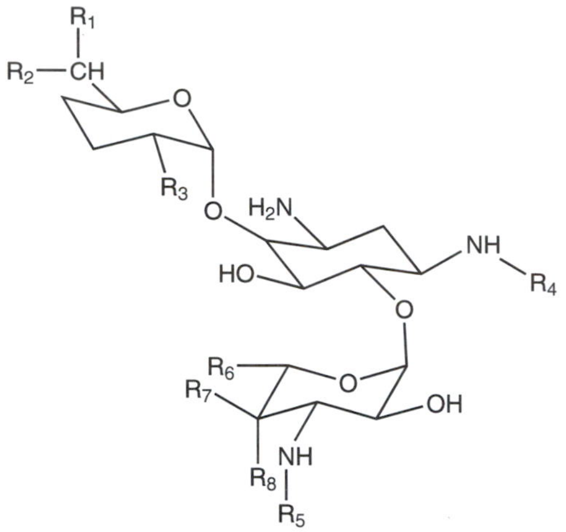

Gentamycin $C_1$: $R_1 = R_2 = R_5 = R_7 = CH_3$; $R_3 = NH_2$; $R_4 = H$; $R_8 = OH$
Kanamycin A: $R_1 = NH_2$; $R_2 = R_7 = H$; $R_3 = OH$; $R_4 = R_5 = R_8 = H$; $R_6 = CH_2OH$

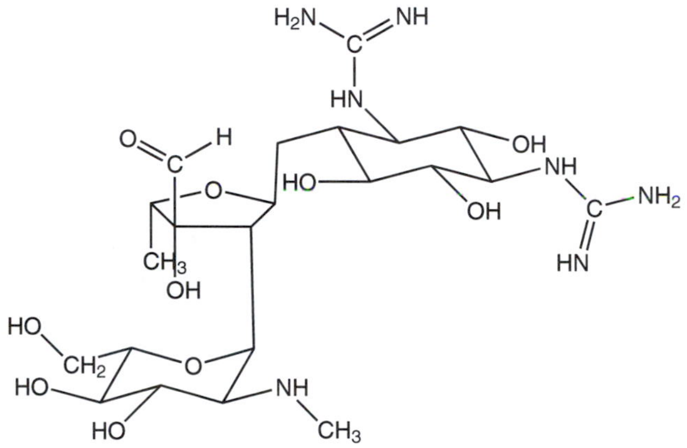

Streptomycin

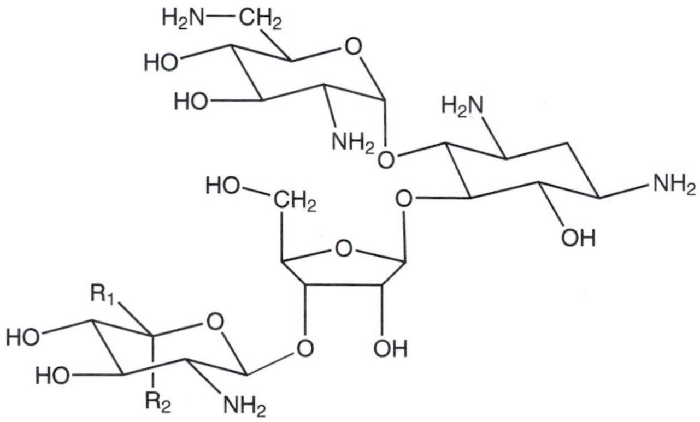

Neomycin B: $R_1 = H$; $R_2 = CH_2NH_2$
Neomycin C: $R_1 = CH_2NH_2$; $R_2 = H$

**Figure 7.6.** Representative aminoglycosides.

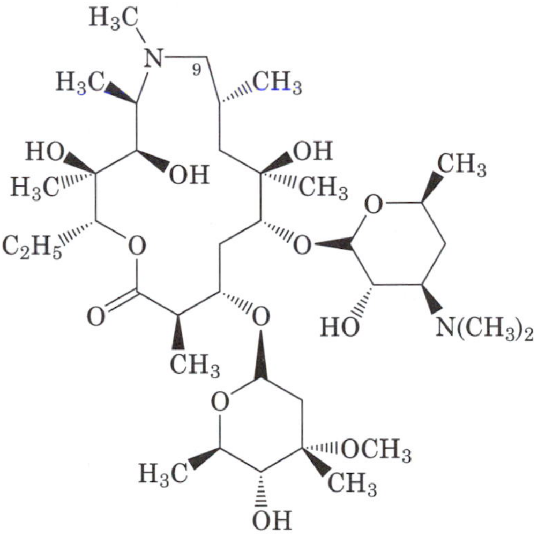

(20)

targets two bacterial DNA topoisomerase II enzymes, also known as DNA gyrases. DNA gyrase relaxes and reforms the DNA supercoil that is necessary for DNA to first be read and then reformed during replication. Selective toxicity arises because mammalian cells do not have DNA gyrases, although they do have a topoisomerase II. The latter requires a much higher dose of quinolones for inhibition to occur. Although no group of drugs administered internally is completely nontoxic, the quinolones show good selectivity.

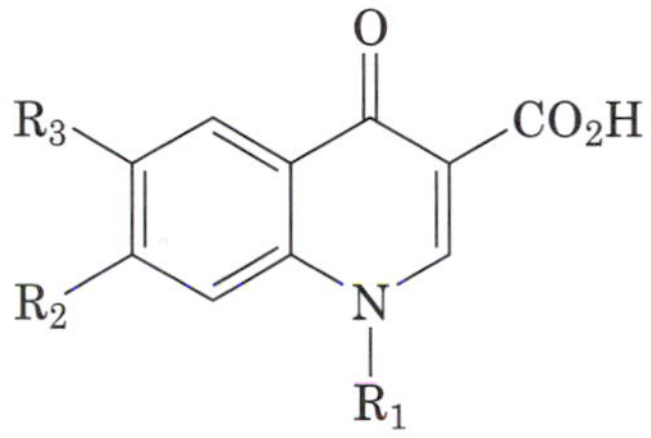

(21)

tric motility. This group also inhibits cytochrome P450 isozymes, CYP1A2 and CYP3A4. Table 7.2 lists the clinically relevant interactions. In each case, erythromycin causes an increase in serum concentrations of the drugs whose metabolism is inhibited by the antibiotic.

*2.3.2.5 Quinolones/Fluroquinolones (Comparative Biochemistry).* The quinolones/fluroquinolones (**21**) are one of the newer chemical classes of antibiotics and are completely synthetic. Their mechanism of action is unique among the antibiotics. The quinolone group

*2.3.2.6 Sulfonamides/Sulfanilamides and Diaminopyrimidines (Comparative Biochemistry and Comparative Distribution).* The discovery of sulfonamides (**22**) was the start of modern antibacterial chemotherapy and gave credence to the "magic bullet" ideal in drug design. The sulfonamide pharmacophore has several important applications in drug design. Although there are very few antibacterial sulfonamides still be-

**Table 7.2  Erythromycin–Drug Interactions**[a]

| ↑ Serum Levels | Remarks |
|---|---|
| CYP1A2 Inhibition | |
| Theophylline | Rifampin inhibits CYP1A2 metabolism of theophylline. |
| CYP3A4 Inhibition | |
| Warfarin | Recommend monitoring when beginning or stopping erythromycin. |
| Cisapride | Cisapride is contraindicated in patients taking erythromycin. |
| Alprazolam | Patients on these drugs should be monitored for increased sedation. |
| Diazepam | |
| Midazolam | |
| Triazolam | |
| Atorvastatin | Patients should be monitored for statin adverse reactions including liver and |
| Cerivastatin | muscle damage. |
| Lovastatin | |
| Simvastatin | |
| Sildenafil | Lower doses should be considered. |
| Astemizole | Astemizole is contraindicated in patients taking erythromycin. |
| Cyclosporine | Serum levels should be monitored. |

[a]*Pharmacists Letter,* Document 150401 (2001).

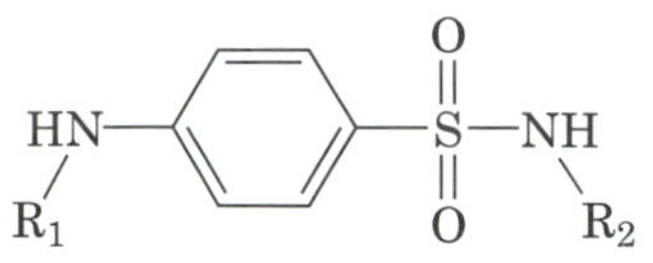

**(22)**

ing used in medicine, the sulfonamide moiety is found in diuretics (furosemide, thiazide, and thiazide-like, carbonic anhydrase inhibitors) and oral hypoglycemics (chlorpropamide, glyburide, glipizide, glimepiride, and other sulfonylureas).

The sulfonamides are selective for a key reaction found only in bacteria. Most bacteria synthesize their own folic acid. In contrast, humans obtain their folic acid from food and vitamin supplements. Therefore, in bacteria sulfonamides block folic acid biosynthesis by competitive inhibition of dihydropteroate synthase, which is the enzyme used by bacteria to incorporate p-aminobenzoic acid to form dihydropteroic acid (Fig. 7.7). Upon the addition of glutamic acid to the latter, the bacteria synthesize dihydrofolic acid ($FAH_2$, $FH_2$, DHF). A second antibiotic, trimethoprim (**23**), selectively inhibits bacterial dihydrofolate reduc-

tase. As long as the patient's folic acid status is adequate, there is minimal metabolic toxicity from the sulfonamides or trimethoprim.

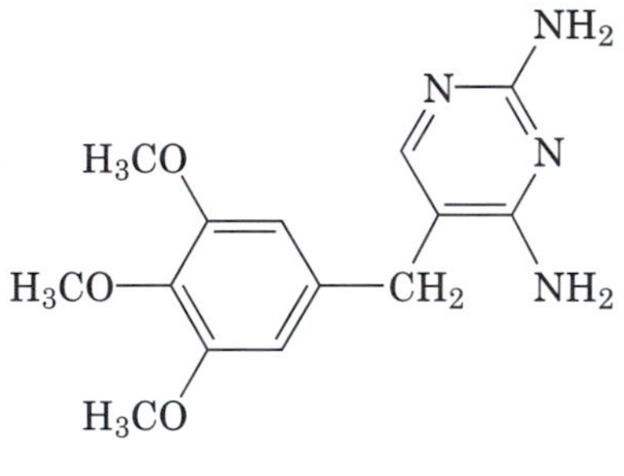

In contrast with these two antibacterial antibiotics, methotrexate (**24**) is one of the most used cytotoxic drugs for malignancies and, in lower doses, an immunosuppresive in autoimmune diseases (i.e., psoriasis and rheumatoid arthritis). Methotrexate inhibits mammalian dihydrofolate reductase (DHFR), which is found in every cell that uses one of the coenzyme forms of folic acid. Because of its poor comparative biochemical selectivity, it is common to administer one of the tetrahydrofolates as an antidote for methotrexate toxicity.

Dihydrofolate reductase also is a potential site for antifungal antibiotics. The problem was to find a drug that is selective for the fungal version of this enzyme. The result was pyrimethamine (**25**), which shows a preference for plasmodia dihydrofolate reductase relative to the mammalian enzyme.

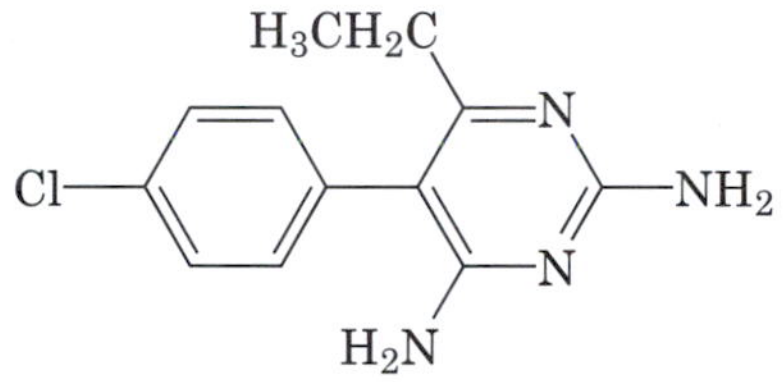

**(25)**

**2.3.3 Examples Based on the Target Organism.** When the pathogen's biochemistry and cell structure becomes more mammalian-like, it becomes more difficult to target a specific site that is significantly different from that found in the host. In some cases, the pathogen

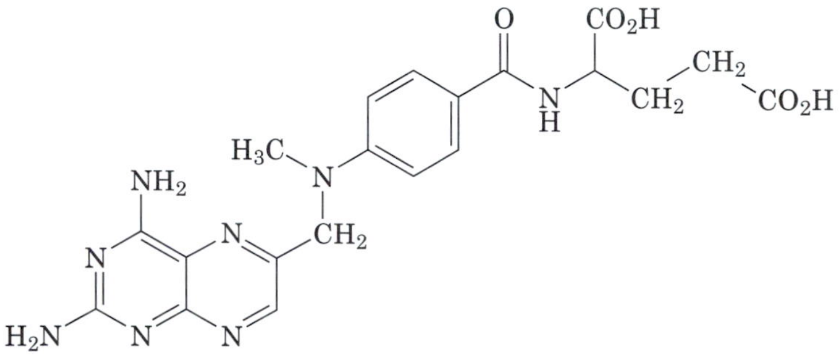

**(23)**

**(24)**

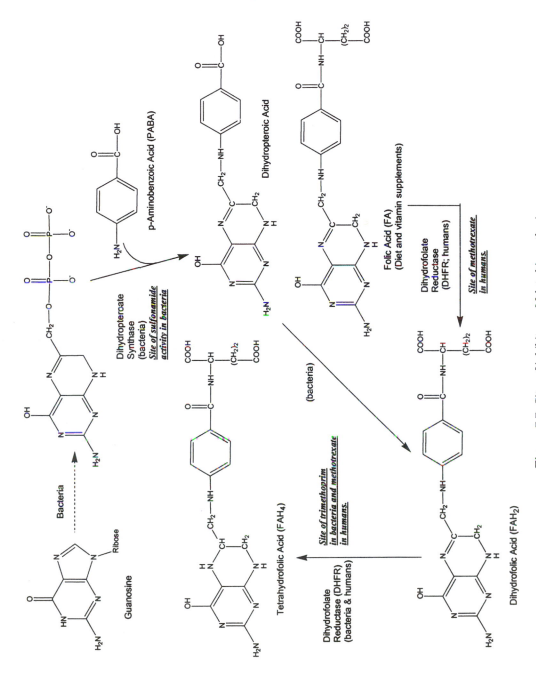

**Figure 7.7.** Sites of inhibitors of folate biosynthesis.

265

is able to hide from the host's immune system. This creates a problem getting the drug to the microorganism, which leads to reduced selectivity.

***2.3.3.1 Antiviral Drugs (Comparative Biochemistry).*** Viral biochemistry duplicates much of the host's biochemistry because it is a complete parasite. A virus cannot reproduce unless it is inside the host's cells where it uses many, if not all, of the cell's DNA and RNA polymerases, ribosomes, and t-RNAs. In contrast with bacteria where most antibiotics are effective against a wide variety of microorganisms, antiviral drugs are specific for narrow virus types. This is because the virus biochemistry tends to be specific for the virus type. Amantadine (**26**) and rimantadine (**27**) only block the uncoating of Influenza A, whereas oseltamivir (**28**) and zanamivir (**29**) inhibit viral neuraminidase found in influenzas A and B. Both sets of antiviral drugs show good selectivity for the virus.

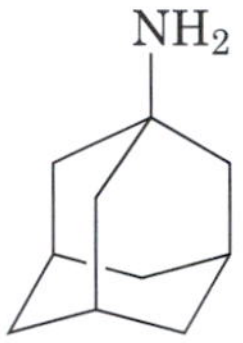

(26)

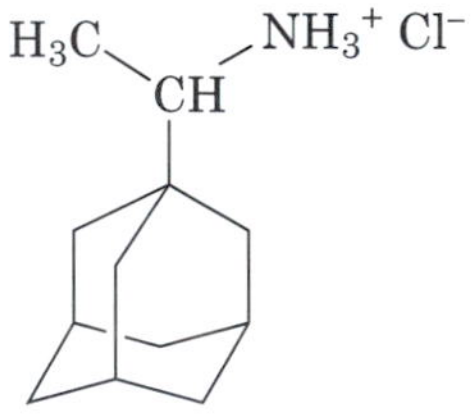

(27)

In contrast, the large number of drugs used to treat human immunodeficiency virus infections target enzymes (RNA-dependent DNA polymerase and HIV protease) specific to this virus and, in general, show poor selectivity for a variety of reasons. See Table 7.3 for examples and the antiviral chapter for more information.

***2.3.3.2 Antimycobacterial Drugs (Comparative Biochemistry and Comparative Cytology).***

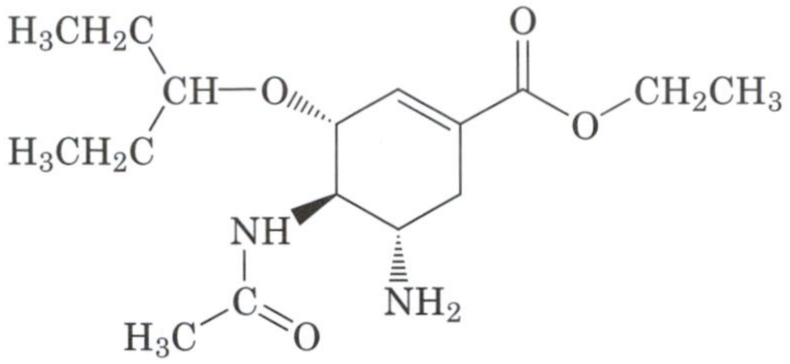

(28)

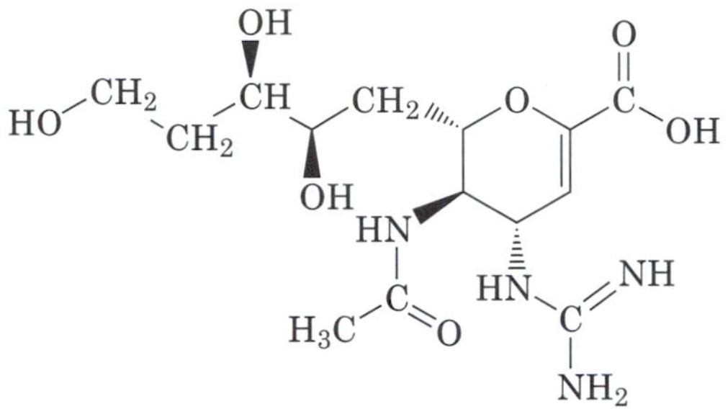

(29)

Mycobacteria-caused diseases are difficult to treat because the bacterium's complex outer membrane makes it difficult to get the drug inside the microorganism. It also is very slow growing, making it less susceptible to drugs that will stop the mycobacterial's reproduction. One of the standard drug treatments is isoniazid (**30**), which, after oxidative activation, inhibits mycolic acid biosynthesis. Mycolic acid is an essential part of the mycobacterium's cell wall. In contrast with the $\beta$-lactam antibiotics that show excellent selectivity, isoniazid has a boxed warning regarding adverse responses involving the liver, particularly in older patients whose lifestyle may have damaged this organ. Rifampin (**31**) shows a significant preference for DNA-dependent RNA polymerase. Even though the selectivity for the bacterial enzyme is very good, rifampin also induces the CYP1A2 and CYP3A4 isozymes and can be the cause of clinically significant drug-drug interactions (Table 7.4) that result in decreased serum concentrations of drugs metabolized by these isozymes.

***2.3.3.3 Antifungal Drugs (Comparative Biochemistry).*** Selective toxicity is more difficult with these targets. Most are structurally based on ergosterol, the central fungal sterol, and its

**Table 7.3   Representative Antiviral Drugs Classified by Virus**

| Virus | Drug | Site of Activity | Remarks |
|---|---|---|---|
| Influenza Type A | Amantadine Rimantadine | Blocks uncoating or entrance of the virus into the cell | Good selectivity. Restricted to Influenza Type A virus. |
| Influenza Types A and B | Oseltamivir Zanamivir | Viral neuraminidase | Good selectivity. Effective against both Influenza Types A and B virus. |
| Herpes Simplex I and II (HSV I and II), Varicella zoster (VZV) | Acyclovir Famciclovir | Substrate for herpes thymidine kinase | Good selectivity. This group of drugs are very poorly phosphorylated by the host cell's kinases. |
| Human Immunodeficiency Virus (HIV) | Zidovudine Didanosine Zalcitabine Lamivudine Stavudine Abacavir Delavirdine Nevirapine Efavirenz | HIV-RNA–dependent DNA polymerase (reverse transcriptase) | Although mammals do not have reverse transcriptase, the inhibitors also inhibit the host cell's DNA-dependent DNA polymerase. |
|  | Indinavir Ritonavir Saquinavir Nelfinavir Amprenavir Lopinavir | HIV protease | This group of drugs are very effective, but they have a large number of adverse reactions. The mechanisms of these complications are poorly understood. |
| Cytomegalovirus (CMV) | Foscarnet Cidofovir | Viral DNA–dependent DNA polymerase | There is a preference for the viral polymerase, but these drugs also inhibit the host cell's DNA polymerase. |

biosynthesis. Because ergosterol and its biosynthesis are so similar to that of cholesterol, antifungal drugs show poor selectivity when used internally. Thus, several antifungal drugs are limited to topical use. Several of the groupings are summarized in Table 7.5. More detailed information is to be found in the chapter on antifungals.

**Table 7.4   Rifampin–Drug Interactions[a]**

| ↓ Serum Levels | Remarks |
|---|---|
| CYP1A2 Induction | |
|   Theophylline | Rifampin induces CYP1A2 metabolism of theophylline. |
| CYP3A4 Induction | |
|   Warfarin | Recommend monitoring when beginning or stopping rifampin. |
|   Phenytoin | Recommend monitoring when beginning or stopping rifampin. |
|   Delavirdine | Depending on the clinician and the patient, rifampin can be |
|   Indinavir |    contraindicated for these anti-HIV drugs. |
|   Nelfinavir | |
|   Ritonavir | |
|   Saquinavir | |
|   Oral contraceptives | Nonhormonal methods of birth control or warn the patient to refrain from intercourse. |

[a]*Pharmacist's Letter*, Document 150401 (2001).

**Table 7.5  Antifungal Drugs Classified by Site of Action**

| Site of Drug's Action | Drug | Remarks |
|---|---|---|
| Ergosterol in fungal membrane | Amphotericins | Poor selectivity: Complexes with cholesterol (whose structure is very similar to that of cholesterol) in mammalian kidneys. Selectivity might be increased by forcing amphoteracin into an extended conformation.[a] |
| Fungal C-14$\alpha$-demethylase | Ketoconazole, Fluconazole, Itraconazole, Clotrimazole, Miconazole, Terconazole | Most are limited to topical use because of poor selectivity. The same enzyme is found in the biosynthesis of cholesterol in mammals. |
| Fungal squalene-2,3-epoxidase | Tolnaftate, Naftifine, Terbinafine | Most are limited to topical use because of poor selectivity. The same enzyme is found in the biosynthesis of cholesterol in mammals. |

[a]H. Resat, F. A. Sungur, M. Baginski, E. Borowski, and V. Aviyente, *J. Comput.-Aided Mol. Des.*, **14**, 689–703 (2000).

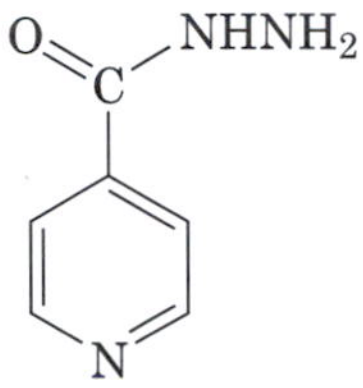

**(30)**

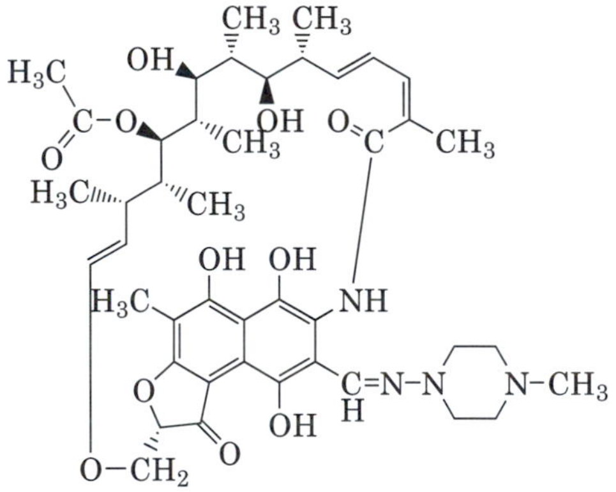

**(31)**

***2.3.3.4 Drugs Used to Treat Parasitic Infections (Comparative Biochemistry).*** The selectivity for receptor sites specific to parasites generally is not good. There are problems with patient compliance because of significant adverse reactions. Drugs must be taken over long periods of time and are costly for a patient population generally living in economically de-prived areas. The result is that diseases caused by parasites are difficult to treat. Malaria, the most common disease in this group, involves hundreds of millions of people. The quinine-based drugs, which are widely used to treat this disease, interfere with the last steps in the patient's heme biosynthesis that cause an increase in porphyrins, which may be toxic to the parasite. With the site of action being an essential component of the patient's metabolism, these drugs can be very toxic to the patient. See the chapter on drugs used to treat malaria and other parasitic infections for more information.

## 3  DRUG CHIRALITY

Many drugs are asymmetric. At least 25% are marketed as racemic mixtures (17). The United States Pharmacopeia recognizes this in defining the chemical standards for many drugs (18). Examples include:

Clomiphene Citrate, USP (Fig. 7.8)

Clomiphene citrate contains not less than 98.0% and not more than 102.0% of a mixture of the (*E*)- and (*Z*)-geometric isomers of $C_{20}H_{28}ClNOAC_6H_8O_7$, calculated on an anhydrous basis. It contains not less than 30.0% and not more than 50.0% of the *Z*-isomer.

Doxepin Hydrochloride, USP (Fig. 7.9)

Doxepin hydrochloride, an (*E*)- and (*Z*)-geometric isomer mixture, contains the equivalent of not less than 98.0% and not more than

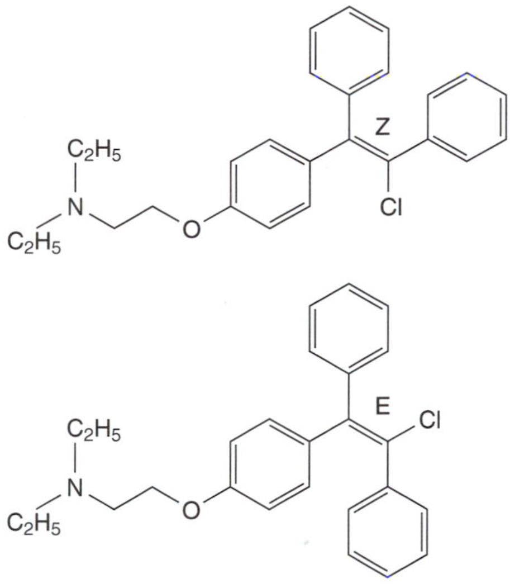

**Figure 7.8.** Clomiphene isomers.

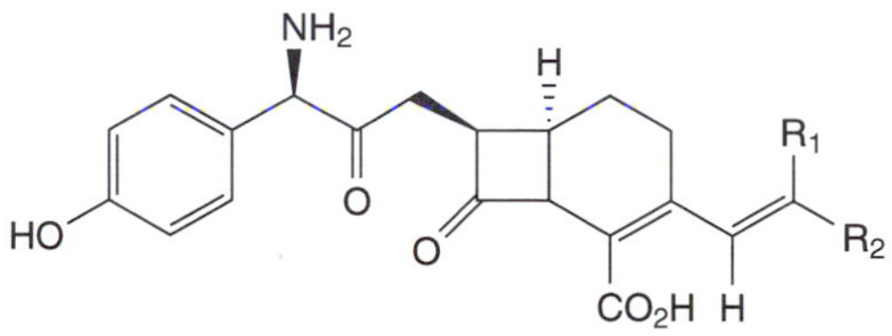

$R_1 = H; R_2 = CH_3$  E-Cefprozil
$R_1 = CH_3; R_2 = H$  Z-Cefprozil

**Figure 7.10.** Cefprozil isomers.

It now is realized that many times only one isomer contains the pharmacological activity. The routes of a drug's metabolic degradation also may vary with the isomer. In a few cases, the adverse responses may be stereoselective. Here are a few representative examples in which stereoselectivity can be significant (19).

### 3.1  Propoxyphene

Dextropropoxyphene [Darvon, (**32**)] is marketed as an analgesic, and levopropoxyphene (Novrad) as an antitussive (**33**). (Note that the

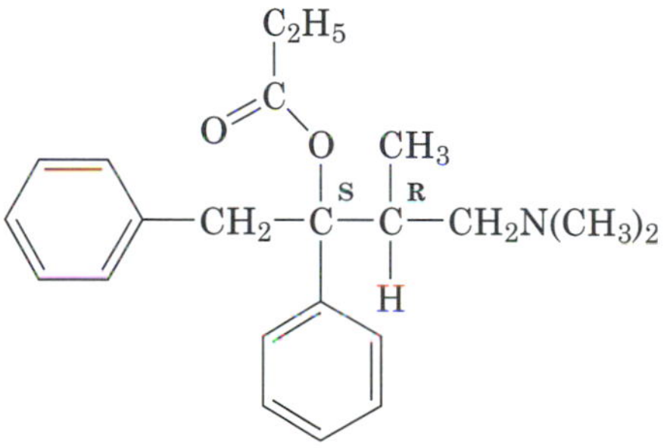

102.0% of doxepin ($C_{19}H_{21}NOAHCl$), calculated on a dried basis. It contains not less than 13.6% and not more than 18.1% of the (Z)-isomer and not less than 81.4% and not more than 88.2% of the (E)-isomer.

Cefprozil, USP (Fig. 7.10)

Although not specifically defined as a mixture, it is clear from the assay procedure that it is a mixture. The chromatographic procedure requires preparation of a resolution solution consisting of a mixture of equal volumes of the Cefprozil (Z)-isomer standard preparation and of the stock solution used to prepare the Cefprozil (E)-isomer standard preparation. The final line in the assay reads, "Calculate the quantity, in micrograms of cefprozil ($C_{18}H_{19}N_3O_3S$) in each mg of the Cefprozil taken by adding the values, in micrograms per mg, obtained for the cefprozil (Z)-isomer and for cefprozil (E)-isomer."

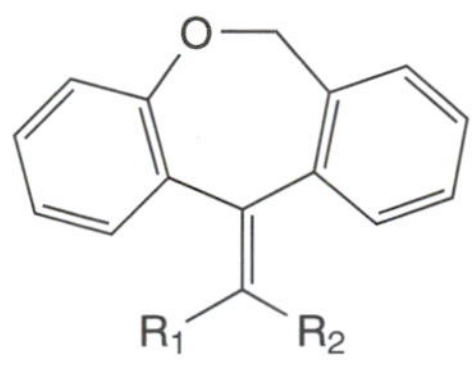

(32)

$R_1 = CH_2CH_2N(CH_3)_2; R_2 = H$  Z-Doxepin
$R_2 = H; R_2 = CH_2CH_2N(CH_3)_2$  E-Doxepin

**Figure 7.9.** Doxepin isomers.

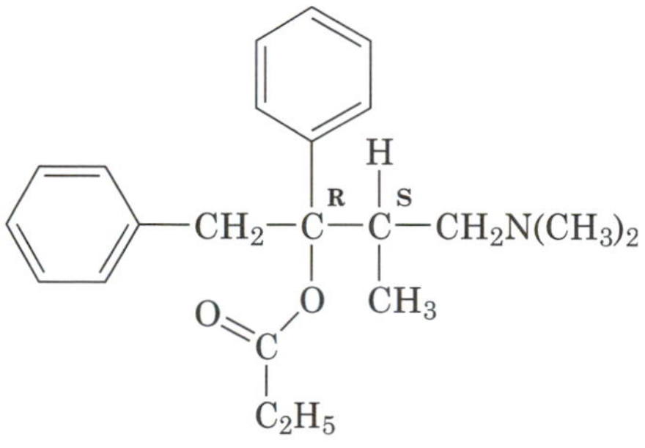

(33)

brand name *Novrad* is Darvon spelled backward and, therefore, the mirror image of Darvon.)

## 3.2 Verapamil

This calcium channel antagonist illustrates why it is difficult to conclude that one isomer is superior to the other. *S*-Verapamil (**34**) is the more active pharmacological stereoisomer than the less active *R*-verapamil (**35**), although the former is more rapidly metabolized by the first-pass effect. (First-pass refers to orally administered drugs that are extensively metabolized as they pass through the

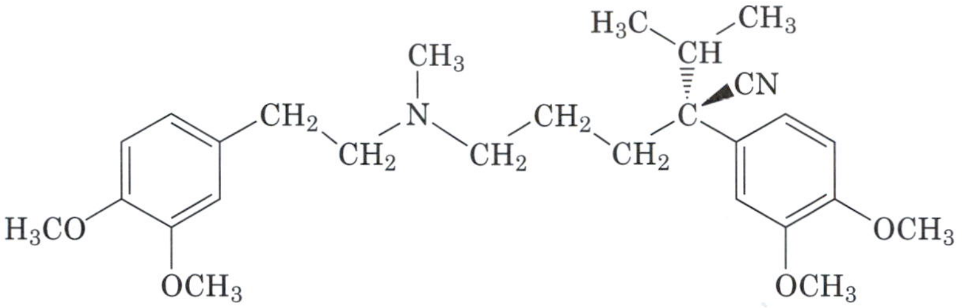

(34)

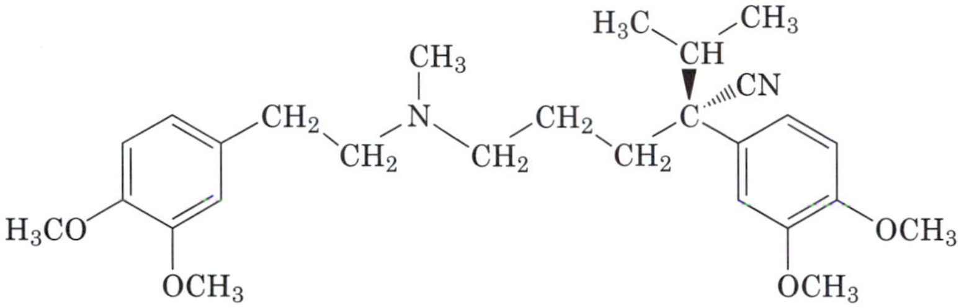

(35)

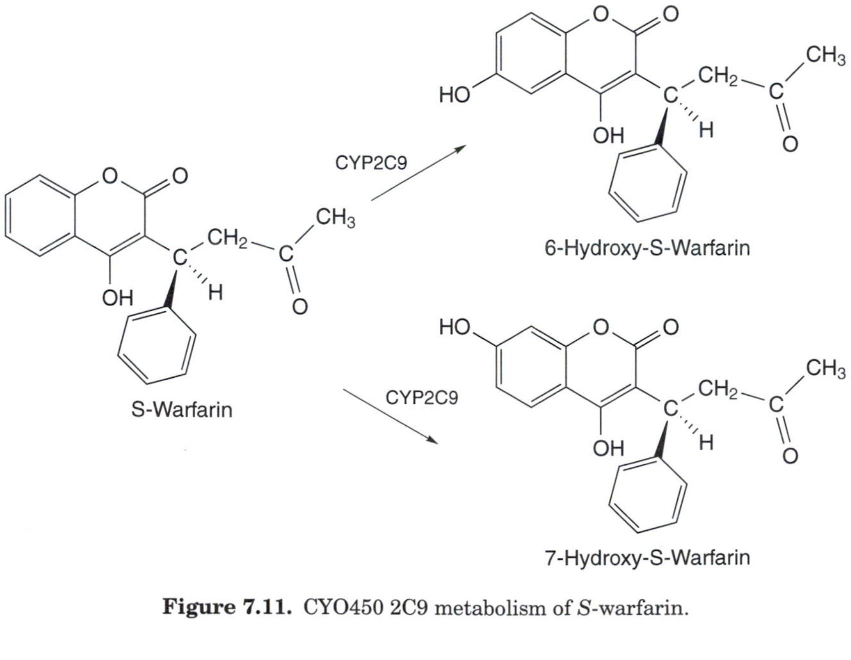

**Figure 7.11.** CYO450 2C9 metabolism of *S*-warfarin.

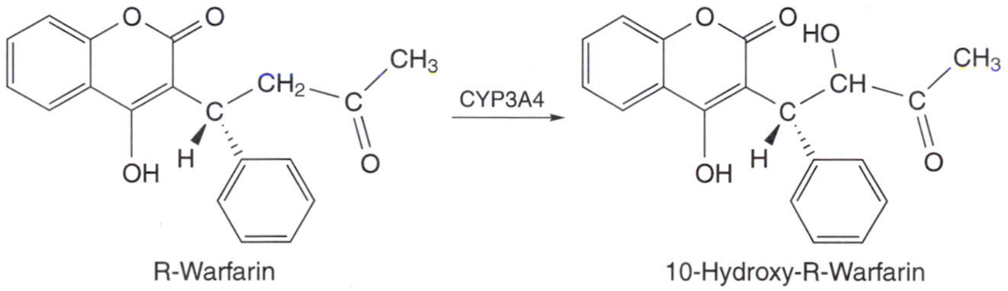

**Figure 7.12.** CYP450 3A4 metabolism of *R*-warfarin.

liver. It is not as significant when the drug is administered parenterally because the drug is dispersed before reaching the liver.) Therefore, intravenous administration of the racemic mixture of verapamil produces a longer duration of action than when administered orally because the more potent *S*-isomer will be metabolized more slowly.

### 3.3 Warfarin

The *S*-isomer of warfarin is more active and is metabolized by the CYP450 2C9 (CYP2C9) isozyme (Fig. 7.11), whereas the *R*-isomer is metabolized at a different position by CYP450 3A4 (CYP3A4) (Fig. 7.12) (20–22). The fact that two different CYP450 isozymes are required for warfarin metabolism and polymorphism is seen with CYP2C9 increases the chances of drug-drug interactions with this potent anticoagulant. Indeed, the package insert lists 22 different pharmacological classes of drugs that can alter warfarin's pharmacological response, as measured by prothrombin time/international normalized ratio (PT/INR). Specific examples are found in Table 7.6.

### 3.4 Other Examples

There are many synthetic products where the drug is marketed as a specific stereoisomer. Early SAR studies on the phenethyl amines showed that the substituent stereochemistry on both the α-carbon (amine carbon) and β-carbon (benzylic carbon) is crucial. Dextrorotatory substituents on the β-carbon increase central activity relative to that of the enantiomeric levorotatory analog. In contrast, levorotatory substituents on the β-carbon increase peripheral activity relative to that of the dextrorotatory analog. As already mentioned, the dextrorotatory opiates have antitussive activity and low abuse potential whereas the levorotatory opiates are analgesics with such high abuse potential that most are classified as Schedule II drugs by the Drug Enforcement Agency. Sometimes the generic name indicates the specific isomer: dextroamphetamine (Fig. 7.13), dexamethasone (**36**), dextromethorphan (**6**), levamisole (**37**), levobupivacaine (**38**), levothyroxine (Fig. 7.13), and levodopa (Fig. 7.13).

**Table 7.6  Potential Inhibitors and Inducers of Warfarin Metabolism**[a]

| Warfarin Isomer | CYP3A4 Inducers | CYP3A4 Inhibitors |
|---|---|---|
| *R*-warfarin | Carbamazepine, phenobarbital, dexamethasone, nevirapine, phenytoin, rifabutin, rifampin | Amiodarone, diltiazem, erythromycin, azole antifungals, norfloxacin, zafirlukast, zileuton |
| | **CYP2C9 Inducers** | **CYP2C9 Inhibitors** |
| *S*-warfarin | Carbamazepine, phenobarbital, phenytoin, primadone, rifampin | Amiodarone, cimetidine, azole antifungals, fluoxetine, isoniazid, sertraline, zafirlukast |

[a]*Pharmacists Letter,* Document 150401 (2001).

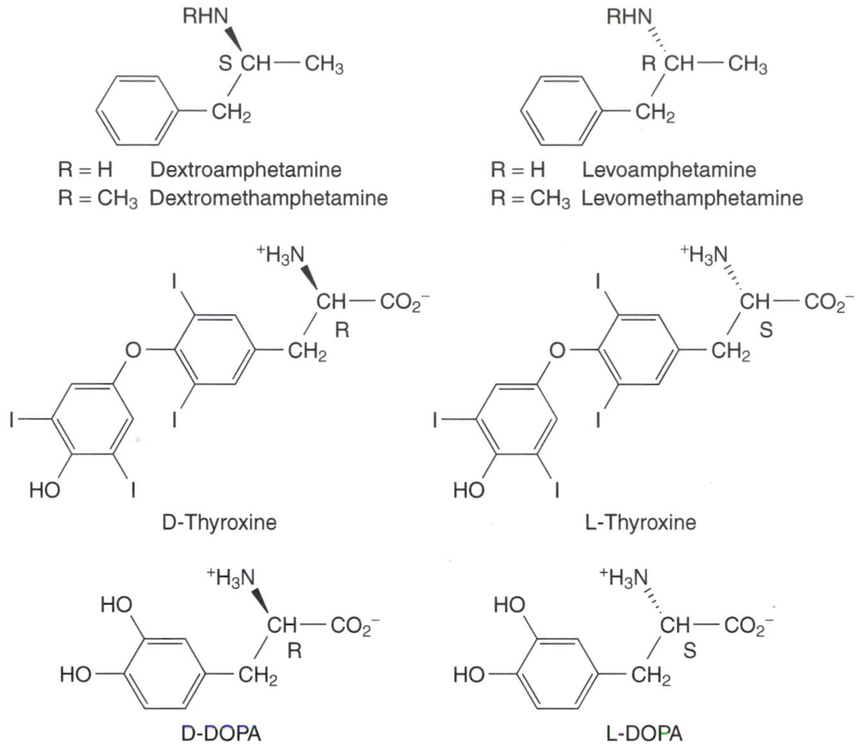

R = H      Dextroamphetamine
R = CH$_3$  Dextromethamphetamine

R = H      Levoamphetamine
R = CH$_3$  Levomethamphetamine

D-Thyroxine

L-Thyroxine

D-DOPA

L-DOPA

**Figure 7.13.** Stereoisomers based on L-phenylalanine.

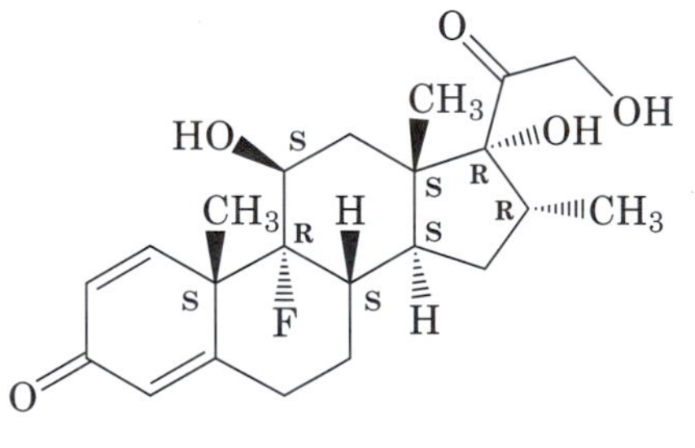

(36)

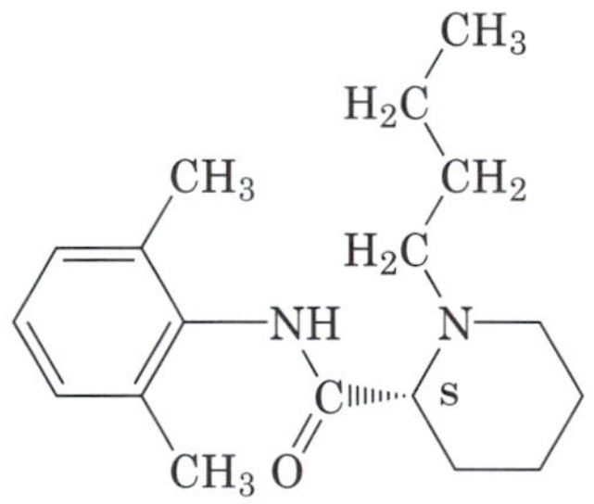

(38)

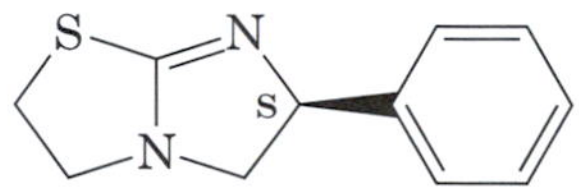

(37)

Other than its use as an analytical tool, the optical rotation does not provide information as to the molecule's shape, which, of course, determines fit at the receptor. Figure 7.13 contains drawings for a group of drugs based on L-phenyl-

alanine or L-tyrosine. *R*- and *S*-Amphetamine and *R*- and *S*-methamphetamine are centrally acting stimulants that exhibit complex pharmacology and high abuse potential. In general, the *R*-isomers (levo isomers) have less central nervous system (CNS) activity but stronger cardiovascular effects compared to that of their *S*-isomers (dextro isomers).

Steric differences in the two thyroxine isomers also cause differences in biochemical response. Both *S*- and *R*-thyroxine bind to thyroglobulin. The $S$-isomer (L-thyroxine) has about six times the antigoiter activity relative to that of $R$-thyroxine (D-thyroxine). The difference is more pronounced with the triiodothyronine isomers. $S$-Triiodothyronine has 13 times the antigoiter activity relative to that of $R$-triiodothyronine (23, 24). The fact that $R$-thyroxine is not devoid of hormonal activity is evidence for its being contraindicated in euthyroid patients who were prescribed the drug for hypercholesterolemia. ($R$-Thyroxine is no longer sold in the United States.)

Finally, there is L- or D-DOPA. This drug is the natural metabolite from hydroxylation of L-tyrosine. Keeping in mind that some peptide antibiotics contain D-amino acids, one would not expect D-DOPA to be a good substrate for DOPA decarboxylase. Further, repeated studies show that D-DOPA ($R$-isomer) does not have significant neurotrophic effects (25).

Levamisole [$S$-isomer; (37)] is an interesting drug. Its original indication was the treatment of worms in animals, possibly by blocking the parasite's cholinergic receptors. Later it was found that levamisole restores the immune response in humans when given with the anticancer drug 5-fluorouracil for the treatment of certain cancers of the colon. In contrast the $R$-isomer (dexamisole) shows little anthelmintic activity (26, 27).

A recent drug is the local anesthetic, levobupivacaine, the $S$-isomer (38) of bupivacaine. Many of the local anesthetics affect the neurons servicing the cardiac muscle. Sometimes this can be useful for the treatment of cardiac arrhythmias. On the other hand, particularly when administered in large amounts such that plasma levels become significant, some local anesthetics may cause depression of the myocardium, decreased cardiac output, heart block hypotension, bradycardia, and ventricular arrhythmias. Bupivacaine has been resolved with the $S$-isomer showing less cardiotoxic responses but still good local anesthetic activity (28).

# 4  ADDITIONAL PHARMACOLOGICALLY ACTIVE AGENTS

See the specific chapters for more complete discussions for each of the following pharmacological groups.

## 4.1  Bisphosphonates (*Comparative Distribution*)

Osteoporosis is an increasing problem in an aging population. Its mechanism is very complex because bone metabolism is intricate. Besides its structural or support role, bone can be considered as the body's calcium reservoir used to maintain calcium homeostasis. That means that there has to be a means for calcium to be stored as hydroxyapatite when not needed and removed from bone as blood levels begin to decrease. For purposes of this discussion, consider that osteoporosis results when the osteoblast (bone-forming) and osteoclast (bone-resorption) cells are not in balance.

Two widely used drugs (39), alendronate (R = 2-$n$-propylamine; Fosamax) and risedronate (R = 3-pyridyl; Actonel), used for re-

(39)

ducing the rate of calcium loss from bone tissue, belong to the bisphosphonates class. Both are administered orally and are widely distributed in the tissues and then concentrate onto the hydroxyapatite, where there are osteoclast cells. Although the mechanism of their toxicity is not understood, both drugs can cause local irritation and, for some patients, actual damage to the gastric linings. Even without damage, patients can experience severe gastric pain such that adherence is poor. Patients are advised to take either drug in an upright position with a full glass of water in the morning, 30 min before eating breakfast. It must be remembered that the bisphosphonates will be taken by the patients for the rest of their lives or until a better drug is discovered. Of the two drugs, risedronate appears to exhibit less adverse reactions, but it contains the same warning regarding taking it in an upright position with a glass of water. The selectivity of the current bisphosphonates is poor.

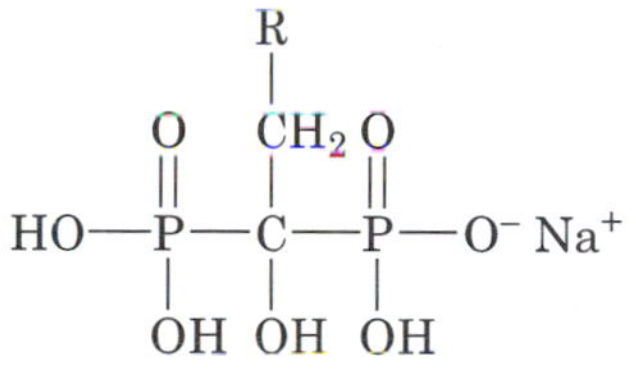

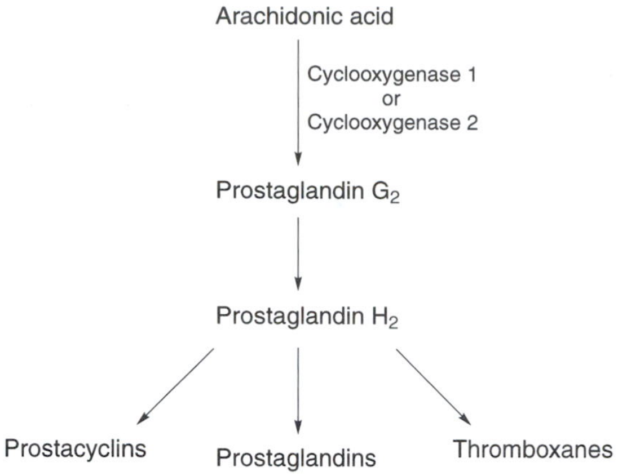

Arachidonic acid

Cyclooxygenase 1
or
Cyclooxygenase 2

Prostaglandin $G_2$

Prostaglandin $H_2$

Prostacyclins          Prostaglandins          Thromboxanes

**Figure 7.14.** End products from the cyclooxygenase cascade.

## 4.2  Inhibitors of Cyclooxygenase I (COX-1) and II (COX-2) (*Comparative Biochemistry* and *Comparative Distribution*)

Beginning with aspirin through today's nonsteroidal anti-inflammatory drugs (NSAIDs), the cyclooxygenase inhibitors have been very beneficial for the control of inflammation and fever. It now is realized that COX exists in at least two isozyme forms. COX-1 is constitutive and is found in many tissues. Its inhibition can reduce the integrity of the tissues where inhibition of COX-1 occurs. Note the outline shown in Fig. 7.14. Inhibition of the COX enzymes will reduce levels of prostacyclins, prostaglandins, and thromboxanes. Examples include the gastric lining and the kidney tubules. Although the NSAIDs are reasonably specific for cyclooxygenase, this enzyme is located at the beginning of an essential set of reactions, producing a diverse group of active compounds. There is an important difference between COX-1 and COX-2. Whereas COX-1 is constitutive, COX-2 is induced in tissues where inflammation is occurring. Therefore, selectivity is increased by designing NSAIDs that preferentially inhibit COX-2 in tissues experiencing inflammation without affecting COX-1 in healthy tissue (29, 30).

Fortunately, there are sufficient differences in the receptor sites of these two isozymes that compounds with reported good selectivity can be designed. The NSAIDs such as indomethacin (Indocin), (**40**) that inhibit COX-1 have a carboxylate group that binds to the guanidium residue of an arginine. COX-2 has a valine where

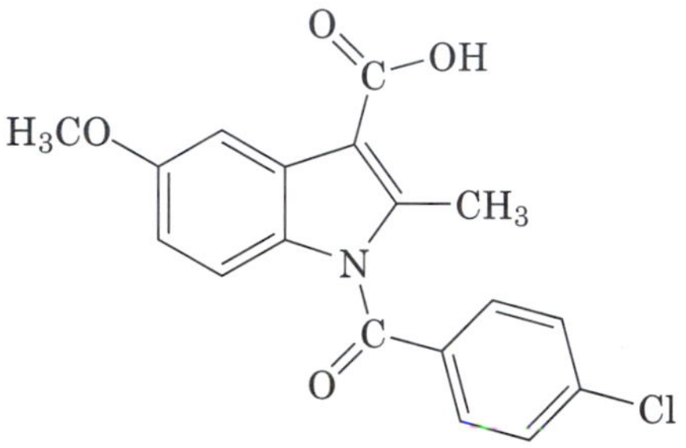

**(40)**

COX-1 has a bulkier isoleucine. The latter's larger hydrophobic side chain blocks entrance to a pocket that both enzymes contain. The net result is that the older acidic NSAIDs preferentially bind to COX-1 and the newer COX-2 inhibitors, rofecoxib [Vioxx, (**41**)] and celecoxib [Celebrex, (**42**)], do not bind well to the COX-1 site because the bulky aromatic ring on the COX-2 inhibitors cannot enter the cleft "guarded" by COX-1's isoleucine.

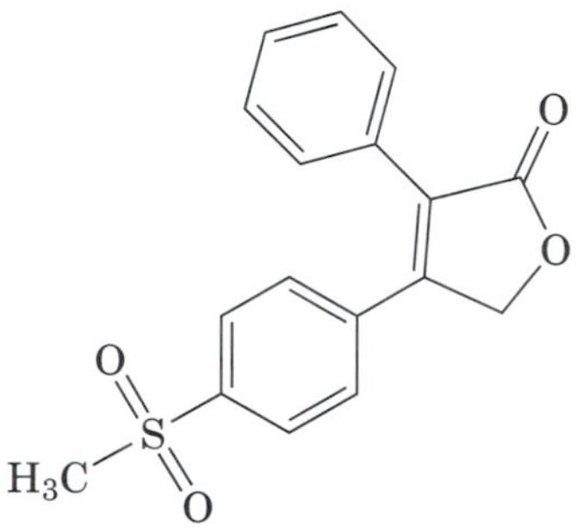

**(41)**

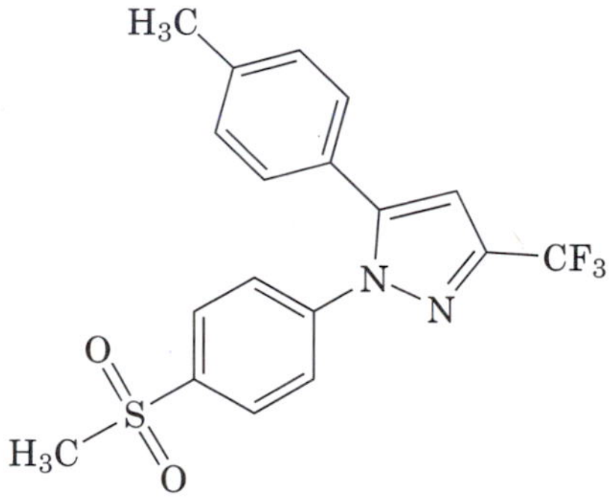

(42)

## 4.3 Antihistamines (*Comparative Biochemistry* and *Comparative Distribution*)

With two distinct classes of receptors and significant differences in their locations, antihistamines have good to excellent selectivity. First, there are the H1 and H2 receptor classes. Histamine is produced by a variety of cells. Patients taking the H1 antihistamines do so in response to the release of histamine from mast cells. These are located in the respiratory passages, skin, and gastrointestinal tract and are the cause of what patients refer to as an allergic response. The "first generation of H1 antihistamines" represented by diphenhydramine [Benadryl, (**43**)] and chlor-

(43)

pheniramine [Chlor-Trimeton, (**44**)] are effective, but they cross the blood-brain barrier, causing mild to significant sedation. Indeed, some of the early antihistamines are used as nonprescription sleep aids. They also could show anticholinergic effects. Sometimes the combination of crossing the blood-brain barrier and anticholinergic activity was put to good use in antinausea drugs, particularly when caused by motion sickness. The familiar

(44)

Dramamine is the chlorotheophylline salt of diphenhydramine (**43**). Thus, useful as these products are, the first generation of antihistamines show poor selectivity.

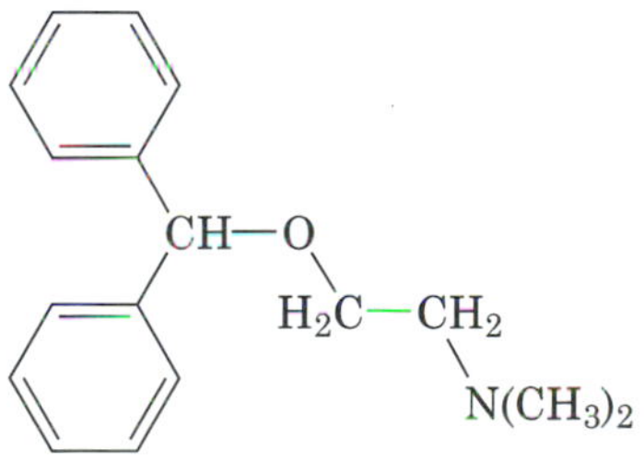

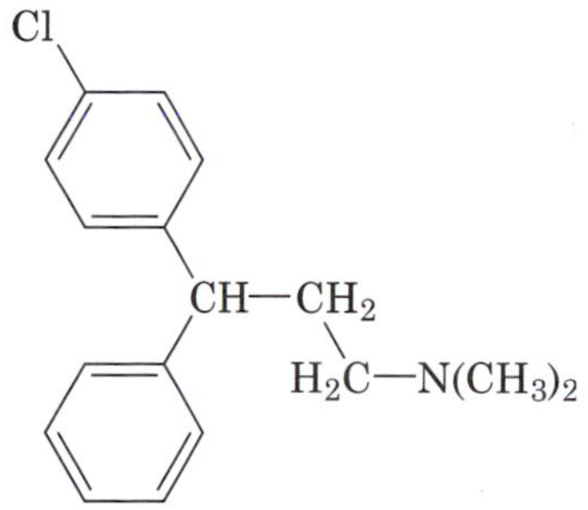

In contrast, the second generation of H1 antihistamines, loratidine [Claritin, (**45**)] and fexofenadine [Allegra, (**46**)] show better selec-

(45)

tivity because they are less likely to cross the blood-brain barrier, and, therefore, preferentially inhibit peripheral H1 receptors. As can be seen from their structures, they tend to be larger molecules. Nevertheless, their distribution is not an either-or situation. The second-generation H1 antihistamines still have central effects.

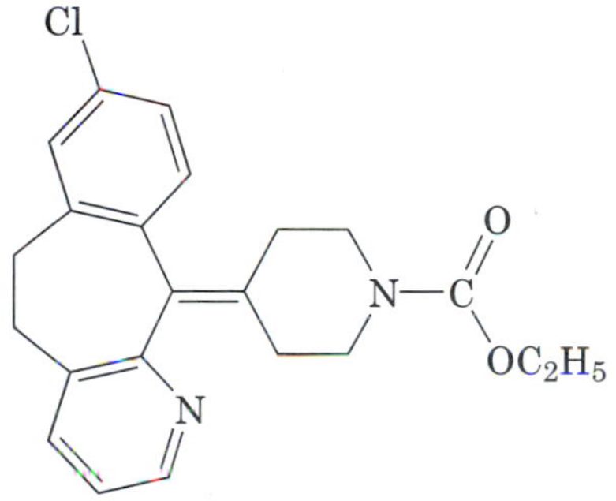

The H2 antihistamines, or blockers, are mainly used to reduce the secretion of gastric HCl. Their structure-activity relationships are significantly different from those of the H1 antagonists. Using cimetidine [Tagamet, (**47**)] as the prototypical molecule, H2 antagonist structure-activity relationships are based on the histamine structure. The selectivity of this group of drugs for the H2 receptor in the gas-

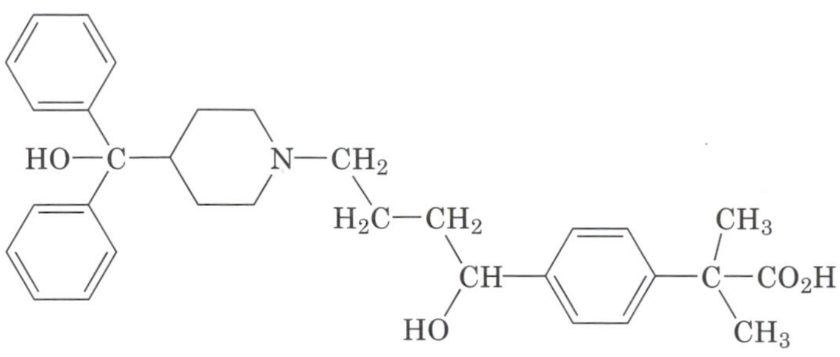

(46)

tric lining is remarkable when one considers how widely distributed are the H2 receptors. Although no drug is without adverse reactions, the H2 antagonists are approved for nonprescription use.

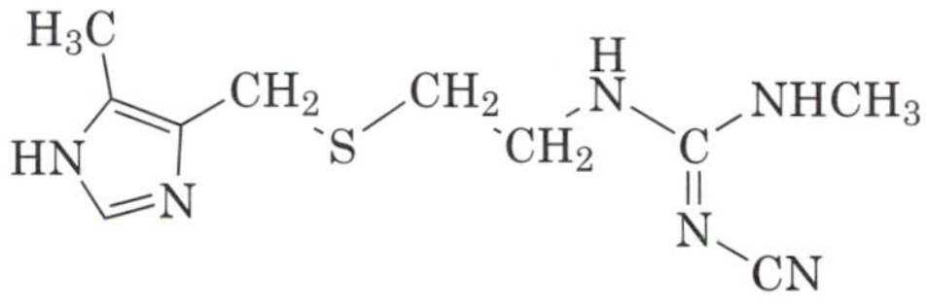

(47)

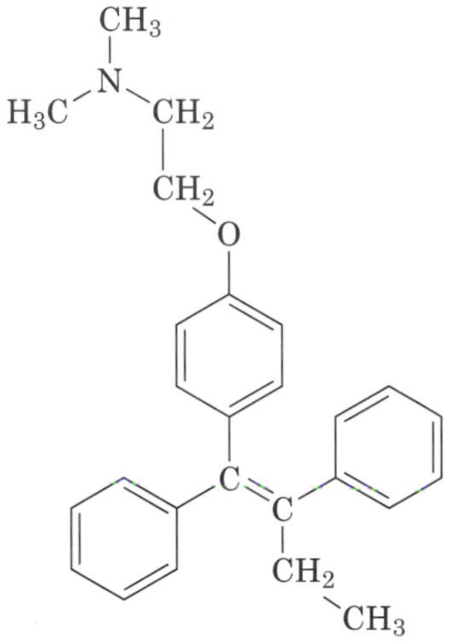

## 4.4 Selective Estrogen Receptor Modulators (*Comparative Biochemistry*)

The pharmacology of the selective estrogen receptor modulators (SERMs) is complex (see the chapter covering this group). They are not simple agonists or antagonists. Rather they can be considered variable agonists and antagonists. Their selectivity is very complex because it is dependent on the organ where the receptor is located.

This complexity can be illustrated with tamoxifen [Nolvadex, (48)], which is used for estrogen-sensitive breast cancer and reducing bone loss from osteoporosis (31). Prolonged treatment, however, increases the risk of endometrial cancer. Thus, tamoxifen is an estrogen antagonist in the mammary gland and an agonist in the uterus and bone. In contrast, raloxifene [Evista, (49)] does not appear to have many agonist properties in the uterus, but like tamoxifen is an antagonist in the breast and agonist in the bone.

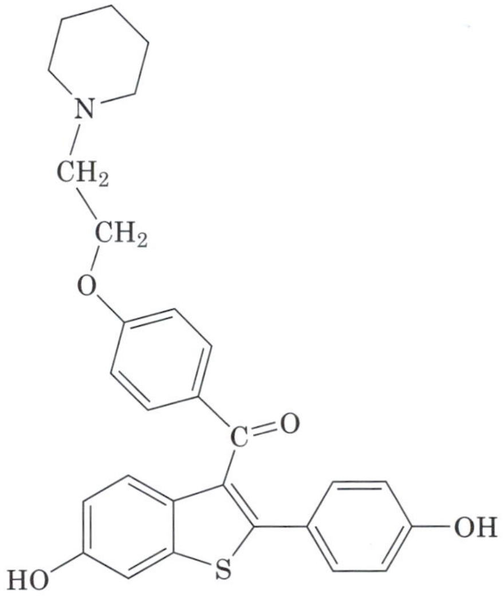

(48)

(49)

## 4.5 Phosphodiesterase Inhibitors

Inhibitors of phosphodiesterase type 5 have an important role in maintaining a desired lifestyle: treatment of erectile dysfunction caused by a variety of conditions. Originally developed for the treatment of angina (and not effective for this purpose), male test subjects reported the ease of having an erection, and the rest is history (32). A complex mechanism is involved. Nitric oxide (NO) activates guanylate cyclase, forming cyclic GMP (cGMP), which is hydrolyzed by a phosphodiesterase. Sildenafil [Viagra, (50)] and the newer compounds, cialis (51) and vardenafil (52), inhibit the phosphodiesterase. The selectivity is good, but it must be remembered that cyclic GMP,

like cyclic AMP, is ubiquitous and, therefore, the phosphodiesterases required to hydrolyze these chemical transmitters are also ubiquitous. There are a wide variety of phosphodiesterase isoforms and their distribution tends to be uneven. In other words, phosphodiesterase type 5 tends to be found in the corpus cavernosum and type 6 in the retina. Depending on the organ and enzyme isoform, sildenafil shows a 10 to 8500 times preference for the type 5 isoform (33, 34). More information on this group of drugs is found in another chapter of this series.

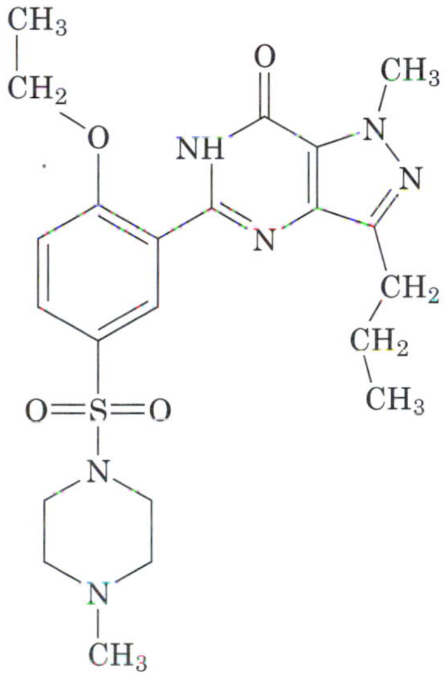

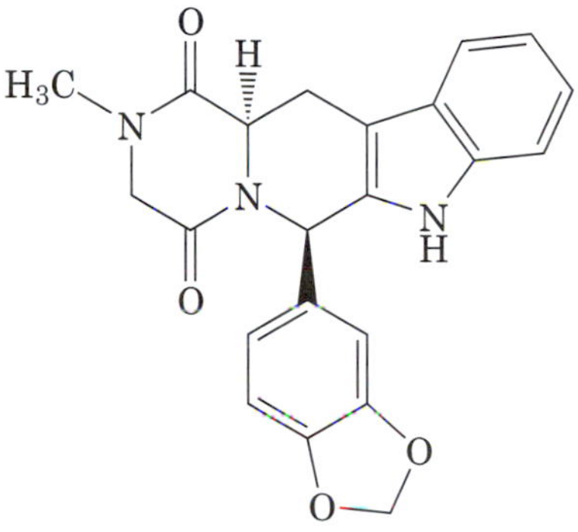

(52)

## 5 INSECTICIDES

This chapter closes with a brief description of the type of selectivity seen with insecticides. Today an insecticide must show good selectivity by targeting the insect and not affect the plant or mammal consuming the plant. Ideally, exploiting the principle of comparative biochemistry would be the goal, but many

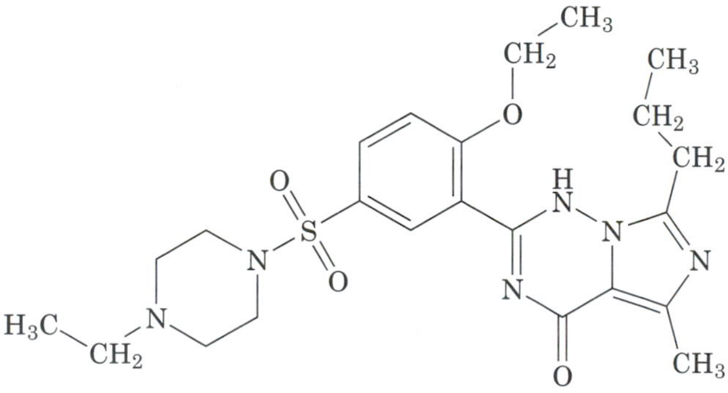

(50)

(51)

times the principle of comparative distribution is used by designing an insecticide to be nonsystemic and remain on the surface of the plant. Usually it can be washed off. Alternatively, a systemic insecticide is desired to kill sucking insects. Another approach is to design an insecticide that will degrade before the plant is harvested. Although the thrust is to protect humans and animals that consume the plants, safety of farm workers and food processors also is important (35, 36). The following classification is based on the target.

## 5.1 Neurotoxic Insecticides (*Comparative Biochemistry* and *Comparative Distribution*)

**5.1.1 Neonicotinoids.** The prototype compound that has wide use is imidacloprid (**53**). This group preferentially binds to the nicotinic acetylcholine receptors of insects relative to those found in animals (37). Their lipophilicity also tends to increase their safety for aquatic species.

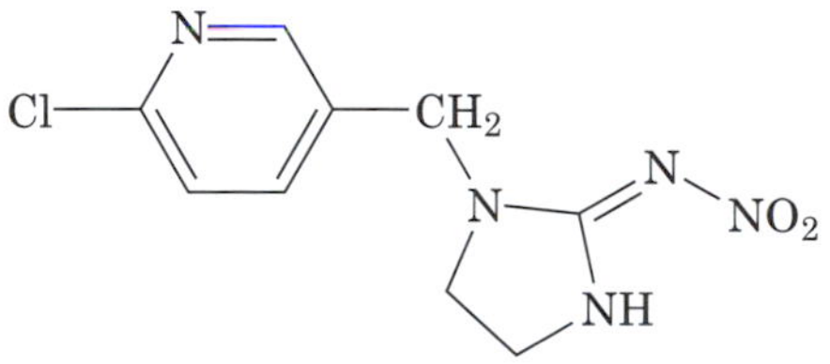

**5.1.2 Spinosyns.** A commercial mixture of the natural products spinosyns A and B (**54**) is reported to selectively activate insect acetylcholine receptors. They appear to have good selectivity for the insect receptor.

## 5.2 Insect Growth Regulators (*Comparative Biochemistry*)

Many of the insecticides in this class mimic insect juvenile hormone, including the insect-molting hormone, 20-hydroxyecdysone. This group of compounds are considered to have excellent selectivity because hormonal control of development in insects differs significantly from those of vertebrates. The growth regulators prevent the insect from undergoing normal maturation.

## 5.3 Inhibitors of Oxidative Phosphorylation (*Comparative Biochemistry* and *Comparative Distribution*)

This group tends to show specificity for either mitochondrial complex I or III. The basic selectivity is not as good as the neurotoxins and insect growth regulators. Mammals have similar mitochondrial complexes. Selectivity can be increased by administering the insecticide as a proinsecticide that is converted to the active chemical by an enzyme found only in the

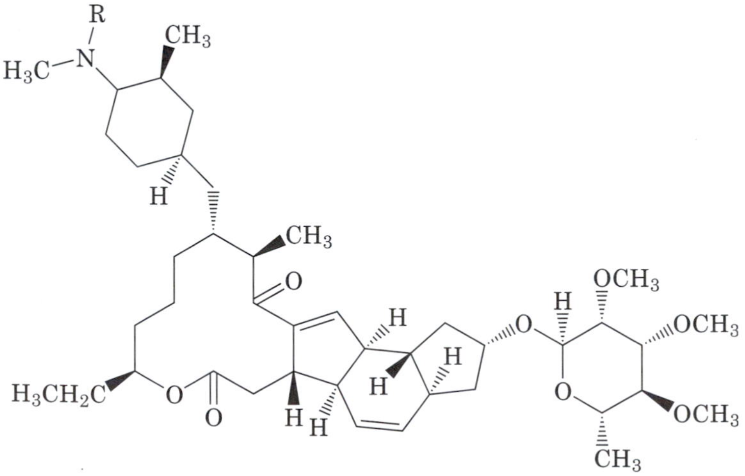

**(53)**

**(54)**

insect. Pharmacokinetic properties also may help increase selectivity.

## 6  CONCLUSION

Selective toxicity is an important goal and concept that must be used in designing a successful biologically active molecule. Today's drugs increasingly are being used to treat complex disease processes whose target receptors are found at several locations throughout the patient's body. There is selectivity within these receptors. The challenge continues to be to discover these differences and then use the principles of comparative biochemistry, distribution, cytology, and stereochemistry in designing new and better drugs.

## REFERENCES

1. A. Albert, *Selective Toxicity: The Physicochemical Basis of Therapy*, 7th ed., Chapman and Hall, New York, 1985.

2. A. Albert, *Selective Toxicity: The Physicochemical Basis of Therapy*, Vol. **1**, 7th ed. (*Sentaku Dokusel: Yakuzai Sayo no Seigyo to sono Bunshiteki Kiso*, Jo Kan. Dai 7 Han), Japan Scientific Societies Press, Tokyo, 1993.

3. A. Albert, *Selective Toxicity: The Physicochemical Basis of Therapy*, Vol. **2**, 7th ed. (*Sentaku Dokusel: Yakuzai Sayo no Seigyo to sono Bunshiteki Kiso*, Jo Kan. Dai 7 Han), Japan Scientific Societies Press, Tokyo, 1994.

4. See Ref. 1, pp. 17–20.

5. See Ref. 1, pp. 56–117.

6. See Ref. 1, pp. 118–174.

7. E. J. Corey, S. P. Matsuda, and B. Bartel, *Proc. Natl. Acad. Sci. USA*, **90**, 11628–11637 (1993).

8. See Ref. 1, pp. 175–205.

9. Information on approved drugs will be found in the package inserts. These are available in the Physicians Desk Reference and the Food and Drug Administration web site (www.fda.gov).

10. R. B. Ciccarelli, M. J. Solomon, A. Varshavsky, and S. J. Lippard, *Biochemistry*, **24**, 7533–7540 (1985).

11. V. Guillemard and H. U. Saragovi, *Cancer Res.*, **61**, 694–699 (2001).

12. R. Wadhwa, T. Sugihara, A. Yoshida, H. Nomura, R. R. Reddel, R. Simpson, H. Maruta,

and S. C. Kaul, *Cancer Res.*, **60**, 6818–6821 (2000).

13. M. J. Rivera, S. F. Stinson, D. T. Vistica, J. L. Jorden, S. Kenney, and E. A. Sausville, *Biochem. Pharmacol.*, **57**, 1283–1295 (1999).

14. E. S. Massuda, E. J. Dunphy, R. A. Redman, J. J. Schreiber, L. E. Nauta, F. G. Barr, I. H. Maxwell, and T. P. Cripe, *Proc. Natl. Acad. Sci. USA*, **94**, 14701–14706 (1997).

15. K. Nakamura, V. P. Bindokas, J. D. Marks, D. A. Wright, D. M. Frim, R. J. Miller, and U. J. Kang, *Mol. Pharmacol.*, **58**, 271–278 (2000).

16. P. T. B. Huy, A. Meulemans, M. Wassef, C. Manuel, O. Sterkers, and C. Amiel, *Antimicrob. Agents Chemother.*, **23**, 344–346 (1983).

17. A. J. Hutt and S. C. Tan, *Drugs*, **52** (Suppl. 5), 1–12 (1996).

18. *The United States Pharmacopeia 24-The National Formulary 19*, United States Pharmacopeial Convention, Washington, DC, 2000.

19. T. N. Riley, *U.S. Pharmacist*, **March**, 40–51 (1998).

20. B. K. Park, *Biochem. Pharmacol.*, **37**, 19–27 (1988).

21. L. S. Kaminsky and Zhi-Yi Zhang, *Pharmacol. Ther.*, **73**, 67–74 (1997).

22. A. R. Redman, *Pharmacotherapy*, **21**, 235–242 (2001).

23. E. C. Jorgensen in C. H. Li, Ed., *Hormonal Proteins and Peptides: Thyroid Hormones*, Vol. **6**, Academic Press, New York, 1978, p. 108.

24. E. C. Jorgensen in M. E. Wolff, Ed., *Burger's Medicinal Chemistry, Part 3*, 4th ed., Wiley, New York, 1981, p. 103.

25. M. A. Mena, V. Davila, and D. Sulzer, *J. Neurochem.*, **69**, 1398–1408 (1997).

26. J. A. Lewis, J. T. Fleming, S. McLafferty, H. Murphy, and C. Wu, *Mol. Pharmacol.*, **31**, 185–193 (1987).

27. A. H. M. Raeymaekers, L. F. C. Roevens, and P. A. J. Janssen, *Tetrahedron Lett.*, 1467–1470 (1967).

28. R. H. Foster and A. Markham, *Drugs*, **59**, 551–579 (2000).

29. R. G. Kurumbail, A. M. Stevens, J. K. Gierse, J. J. McDonald, R. A. Stegeman, J. Y. Pak, D. Gildehaus, J. M. Miyashiro, T. D. Penning, K. Seibert, P. C. Isakson, and W. C. Stallings, *Nature*, **384**, 644–648 (1996).

30. C. J. Hawkey, *Lancet*, **353**, 307–314 (1999).

31. M. Dutertre and C. L. Smith, *J. Pharmacol. Exp. Ther.*, **295**, 431–437 (2000).

32. D. Nachtshein, *West. J. Med.*, **169**, 112–113 (1998).

33. S. A. Ballard, C. J. Gingell, K. Tang, L. A. Turner, M. E. Price, and A. M. Naylor, *J. Urol.*, **159**, 2164–2167 (1998).

34. H. D. Langtry and A. Markham, *Drugs*, **57**, 967–989 (1999).

35. R. M. Hollingworth, New Insecticides: Mode of Action and Selective Toxicity, *ACS Symposium Series 774 (Agrochemical Discovery)*, American Chemical Society, Washington, DC, 2001, pp. 238–255.

36. T. Narahashi, *Pharmacol. Toxicol.*, **78**, 1–14 (1996).

37. K. Matsuda, S. D. Buckingham, J. C. Freeman, M. D. Squire, H. A. Baylis, and D. B. Sattelle, *Br. J. Pharmacol.*, **123**, 518–524 (1998).

# CHAPTER EIGHT

# Drug Resistance in Cancer Chemotherapy

AMELIA M. WALL
Division of Clinical Pharmacology and Therapeutics
Children's Hospital of Philadelphia
Cancer Pharmacology Institute
University of Pennsylvania
Philadelphia, Pennsylvania

## Contents

1 Introduction, 282
2 Mechanisms of Resistance, 282
   2.1 Alterations in Drug Targets, 282
   2.2 Alterations in Intracellular Retention
      of Drug, 285
   2.3 Alterations in Drug Detoxification, 286
   2.4 Increased DNA Repair, 287
   2.5 Defective Apoptosis, 287
   2.6 Epigenetic Changes, 289
3 Strategies to Overcome Resistance, 289
   3.1 Pharmacokinetic Monitoring, 289
   3.2 Pharmacogenetic Monitoring, 289
   3.3 Biologic Inhibition of Tumor Cell Properties
      (ABC Transporters), 290
4 Conclusions, 290
5 Acknowledgments, 290

*Burger's Medicinal Chemistry and Drug Discovery*
Sixth Edition, Volume 5: Chemotherapeutic Agents
Edited by Donald J. Abraham
ISBN 0-471-37031-2   © 2003 John Wiley & Sons, Inc.

# 1  INTRODUCTION

Chemotherapy is a mainstay of cancer therapies given to patients today. From the introduction of systemic chemotherapy in the late 1940s, an individual's response has been recognized to be dependent on many factors unique to the host, as well as the specific histologic and genetic subtype of malignancy. The design of rational, effective chemotherapeutic protocols involves using agents to which the tumor is susceptible that have different mechanisms of action, nonoverlapping dose-limiting toxicities, and that attack the malignant cells during different phases of the cell cycle. However, because of baseline genetic alterations in tumor cells and the fact that malignant cells can acquire pleiotropic changes in the presence of chemotherapy, tumors may still become refractory to both drugs they have been exposed to as well as to drugs with which they have never been treated. The latter case, termed multidrug resistance, can involve compounds with completely unrelated structures and mechanisms of action (1, 2). Clinical resistance occurs as the resistant clones are positively selected for during a course of chemotherapy (3). Despite the emergence of new classes of chemotherapeutic agents and the widespread use of rational protocol design, the study of the panoply of mechanisms of drug resistance continues to be essential for using this tool effectively for the most resistant subtypes of malignancies.

This chapter will attempt to describe some examples of acquired resistance in cancer cells, and discuss strategies employed therapeutically to overcome this resistance. Additionally, some somatic mutations also alter the susceptibility of a tumor to chemotherapy, and these will be discussed in brief. It is important to note, that although beyond the scope of this chapter, somatic polymorphisms that affect the amount of drug or dose intensity of a drug regimen given to a patient can ultimately impact the induction of resistance in tumor cells exposed to subclinical doses of drug and are also important clinical considerations concerning drug resistance.

# 2  MECHANISMS OF RESISTANCE

In 1984, Goldie and Coldman published a landmark paper attempting to use a mathematical model to follow the emergence of clinical drug resistance (4). This model attempted to describe clonal heterogeneity in tumors that provided the means for positive selection of resistant clones in the presence of a constant fraction of cell-kill during chemotherapy (Fig. 8.1). The Goldie-Coldman hypothesis of acquired mutations during therapy directly impacting resistance provides a basic framework for understanding drug resistance in the clinical setting.

The cancer cell exhibits unique molecular properties, which render the cell unable to halt replication in the presence of DNA damage. Thus, cancer cells can continue to survive despite DNA damage, can rapidly incorporate new molecular configurations that confer a survival benefit, and can replicate rapidly and efficiently, increasing the speed at which molecular mechanisms of resistance can be incorporated into the population of cells. These changes include increased gene copy, mutations, altered transcription, and epigenetic changes. The overall phenotype of a cancer cell that has incorporated these changes in its genome is a cell that either no longer accumulates drug, no longer makes the drug's target protein, or alters the target protein in such a way that the protein no longer binds or is affected by the drug (see Table 8.1).

## 2.1  Alterations in Drug Targets

One of the first mechanisms of acquired resistance to chemotherapy by the tumor cell was determined to be an alteration of the protein targeted by the drug, either by loss or gain of function. Additionally, some tumor cells can amplify copies of the genes encoding the drug's target and transcribe and translate more of the target molecules, overwhelming the drug's cytotoxic ability. Finally, some preexisting somatic mutations that are present in the tumor cell confer resistance to therapy and must be considered when choosing active agents for individuals.

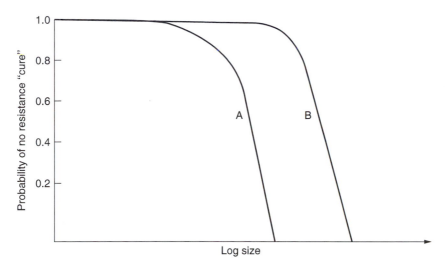

**Figure 8.1.** Goldie-Coldman model of tumor growth versus acquisition of drug-resistant clones measured by mutation rate. As acquired mutations increase over time, the probability of cure with chemotherapy decreases secondary to the emergence of drug resistance. Plot of two mutation rates, A and B, where A > B (function $p = e^{-\alpha N}$, where "the function defines the probability of there being zero resistant cells present for any given value of the tumor size and the mutation rate to resistance"). As tumors approach the steep portion of the curves, any increase in number of resistant cells significantly impacts the probability of success with single agent therapy, and combination chemotherapy should be considered. Adapted with permission from J. H. Goldie and A. J. Coldman AJ, The genetic origin of drug resistance in neoplasms: Implications for systemic therapy. *Cancer Res.*, **44**, 3643–3653 (1984).

A classic example of a tumor cell altering the target of drug therapy was the isolation of tumor cells that had increased levels of intracellular dihydrofolate reductase (DHFR) after exposure to the antifolate, methotrexate (5–7). DHFR is required to maintain intracellular pools of reduced folates, which serve as one-carbon donors for the pathways responsible for *de novo* purine synthesis, pyrimidine synthesis, and ultimately for the synthesis of amino acids. Because of its critical role in the folate pathway, this enzyme is the target of

**Table 8.1  Mechanisms for Tumor Cell Resistance to Chemotherapy**

| Mechanism for Resistance | Example |
|---|---|
| Alterations in drug targets | |
|   Increased expression of drug's target | DHFR, methotrexate |
|   Increased expression of biologic target | Bcr/abl, imatinib mesylate (Gleevac, STI571) |
|   Somatic genetic variation | Thymidylate synthase, 5-fluorouracil |
|   Tumor mutations in drug target | Topoisomerase II, epipodophyllotoxins |
| Alterations in intracellular retention of drug | |
|   Tumor decreases influx of drug | RFC, methotrexate |
|   Tumor increases efflux of drug | MDR, MRPs, BCRP and substrates |
|   Tumor affects pharmacologic path responsible for retention of drug | FPGS, methotrexate |
| Alterations in drug detoxification pathways | Cellular glutathione pathway, alkylators |
| Increased DNA repair | OGAT, nitrosureas |
| Defective apoptosis | *p53*, *bcl2* |
| Epigenetic changes | Methylation of genes involved in drug's cytotoxicity pathway |

the 4-amino folate analogs, or antifolate drugs (e.g., methotrexate). Some tumors that developed resistance to methotrexate over time were found to have amplified the *DHFR* gene, and subsequently increased levels of expressed DHFR enzyme (5, 6). These cells had increased DHFR activity, and an increased concentration of methotrexate was required to kill these cells. Mutation of DHFR has also been documented in some cell lines with acquired methotrexate resistance, resulting in either a DHFR molecule that is catalytically less active than wild-type DHFR or is less able to bind antifolates, but this has not yet been documented in patient samples (8, 9).

Another mechanism of resistance is to amplify genes encoding oncogenic fusion proteins. A new area of active investigation is the production of antitumor agents that target tumor tissue, avoiding damage to normal host tissue. One strategy is to target oncogenic fusion proteins, expressed in tumor cells with chromosomal translocations that result in an expressed protein. For example, BCR/ABL is a chimeric protein, whose expression is the result of the unbalanced translocation of chromosomes 9 and 22. BCR/ABL has been identified in both chronic and acute lymphocytic leukemias. The expressed protein is a tyrosine kinase. STI571 (Gleevac) was developed for its ability to inhibit the tyrosine kinase of BCR/ABL. *In vivo*, STI571 is able to successfully eradicate malignant lymphoblasts on initial exposure to the drug. However, patients uniformly relapse after therapy. Studies done with cells in culture show that these cells develop resistance due to either amplification of *bcr/abl* after prolonged exposure to STI571 or due to a single amino acid substitution in the threonine residue of the Abl portion of the fusion protein that is required for hydrogen bond formation with the drug molecule (10). Therefore, the early clinical results need to be viewed with caution, and additional investigation of combination therapies is currently underway.

Additional mechanisms of resistance occur when somatic variations of gene sequence affect a drug target. 5-Fluorouracil (5FU) is a chemical analog of uracil, and *in vivo*, is converted to its active metabolite 5-fluoro-2-deoxyuridine monophosphate (5-Fd-UMP). 5-Fd-UMP inhibits thymidylate synthase (TS), an enzyme required in the *de novo* pyrimidine synthesis pathway. Somatic polymorphisms in the enhancer/promoter region of (TS) have been documented, which involve a variable number of tandem repeats (VNTRs), and have been shown to influence cellular response to 5FU exposure (and potentially antifolate exposure) (11, 12). Cells whose promoters contain three repeats have higher TS activity than do cells with two repeats because of enhanced TS expression. Patients with a homozygous genotype for two tandem repeats have lower TS activity; therefore, tumor cells derived from this somatic background will respond well to 5FU therapy when compared with patients with a homozygous genotype for three repeats (13). Therefore, prospective pharmacogenetic screening of the host's somatic TS promoter region may help predict both tumor response to conventional doses of 5FU, as well as consider either alternative drug regimens or increasing 5FU dosing for patients with higher TS activity.

Finally, cancer cells can acquire structural changes in drug targets over time. Topoisomerase II is an enzyme targeted by several anticancer drugs because of its normal cellular role in affecting nuclear structure and function. Topoisomerases induce transient breaks in the phosphodiester backbone of DNA. Class II topoisomerases are necessary for the inducement of double-strand breaks during DNA synthesis, transcription, and chromosomal segregation, resolving torsional strain induced during the unwinding of DNA. Topoisomerase II inhibitors (epipodophyllotoxins and anthracyclines) stabilize the topoisomerase II-DNA covalent bond intermediate after scission has occurred, creating DNA damage that ultimately results in cell death. Multiple mutations in topoisomerase II have been identified in tumor cells. These mutations have been shown *in vitro* to induce resistance either through reduction in enzyme activity or alteration in protein structure that prevents binding of drug to the topoisomerase molecule (reviewed in Ref. 14).

## 2.2 Alterations in Intracellular Retention of Drug

By reducing the drug's ability to enter the cell or by rapid efflux of the drug from the cell, the cancer cell effectively removes the ability of the drug to exert its effect. This can be accomplished by either reducing copies of the genes for transporters required for the drug's influx into the cell or by increasing copy number of genes encoding transporters that rapidly efflux either the parent drug or the activated drug metabolite from the cell. Additionally, the cell can decrease metabolism of the parent drug, preventing the formation of toxic activated metabolites.

Altered influx of a drug as a mechanism of tumor resistance has been documented for therapy with methotrexate. The reduced folate carrier (RFC) is an ATP-dependent transporter that actively imports reduced folates into the cell against a concentration gradient for their use in DNA, RNA, and amino acid synthesis. Classic antifolates (e.g., methotrexate) are also transported via RFC into the cell. Mutations in RFC have been isolated in both human and rodent cells with acquired resistance to methotrexate (15). These mutations in RFC resulted in decreased methotrexate accumulation in the tumor cells, and therefore less cytotoxicity when these cells were exposed to increasing doses of methotrexate (15). Additionally, some human leukemia cells have been found to decrease expression of RFC, conferring marked resistance to antifolate therapy (16).

By increasing the efflux of drug molecules from the cell, resistance can be achieved. A subset of ATP binding cassette (ABC) transporters are localized to the plasma membrane of cells, and when expressed, confer resistance to multiple structurally unrelated drug compounds (this family of transporters is summarized with relevant references at http://www.nutrigene.4t.com/humanabc.htm) (17). These include the multidrug resistance protein (MDR), the multidrug resistance-related protein (MRP) and related family members, and the breast cancer resistance protein (BCRP). Expression of these gene products is often turned on in tumor cells after they acquire alterations or deletions in the regulator genes normally controlling their expression

(e.g., P53 or N-MYC) (18–21). Therefore, tumor cells can either increase expression of several ABC transporters after the transformation event, thus conferring *de novo* resistance to a broad spectrum of drugs or increase expression of these transporters after exposure to a drug *in vivo*. Because these transporters do transport a spectrum of compounds, up-regulation of ABC transporters in tumor cells not only yields cells that are resistant to drugs they have had previous exposure to, but also to drugs to which they have not yet been exposed. This limits the number of agents that can then be effectively used for eradication of these tumors *in vivo*.

MDR1, otherwise known as the P-glycoprotein (Pgp), was first identified in cells selected for multidrug resistance (22). It was subsequently discovered that cells that overexpress MDR1 acquire resistance to several compounds (primarily cationic and hydrophobic in structure), including etoposide, anthracyclines, taxanes, vinca alkaloids, steroids, and actinomycin D (23). A somatic mutation in *MDR1* has been identified (SNP in exon 26, C3435T) that confers a twofold decrease in duodenal expression of MDR1 and a fourfold reduction of activity in homozygous mutant individuals (24). The allele frequency for this mutation is ethnically distributed, with high frequency of the C allele found in African populations (25). The presence of somatic mutations in *MDR1* may have prognostic value for patients treated with chemotherapeutic agents that are substrates for this transporter.

MRP was first isolated in 1991 by Cole and colleagues, who were studying a small cell lung cancer cell line with acquired resistance unrelated to the expression of MDR1 (26–28). Since the discovery of MRP, many related family members have also been isolated (e.g., MRP2–6), each of which confer resistance to a signature spectrum of anionic compounds. (Table 8.2). Up-regulation of expression of MRPs in tumors, both at diagnosis and at relapse, have been documented. These tumors have been shown both *in vitro* and *in vivo* to be less responsive to chemotherapy (reviewed in Ref. 29).

Somatic mutations in *MRP2* that result in expression of an inactive transporter have

**Table 8.2  Multidrug Resistance Transporters and Chemotherapy Substrates (as Validated *In Vitro*)**

| | |
|---|---|
| MDR1 | Vinca alkaloids, anthracyclines, taxanes, actinomycin D; epipodophyllotoxins, steroids |
| MRP1 | Doxorubicin, daunomycin, vincristine, etoposide, methotrexate |
| MRP2 | Vinca alkaloids, cisplatin, CPT-11, methotrexate |
| MRP3 | Etoposide, teniposide, methotrexate, vincristine |
| MRP4 | Methotrexate, purine antimetabolites |
| MRP5 | Purine antimetabolites |
| BCRP | Mitoxantrone, camptothecins, anthracyclines |

been identified as the causative mutations in Dubin-Johnson Syndrome, an inherited disorder characterized by chronic mild hyperbilirubinemia (30–32). Patients who are heterozygotes for these mutations also excrete excess bilirubin byproducts in urine. It is suspected that mutations in this transporter may also contribute to hepatic toxicity with some anticancer drugs, and this is under current investigation.

BCRP was first identified in breast cancer cell lines with acquired resistance to mitoxantrone. Unlike MDR and the MRPs, BCRP is a "half molecule" transporter and is thought to function as either a homo- or heterodimer. This transporter also transports several anticancer agents (Table 8.2) (33, 34). Interestingly, a polymorphism has been identified in the third transmembrane domain of this protein that alters its substrate profile (35). At amino acid 482, three variants were isolated with nonredundant amino acid substitutions (arginine, glycine, and threonine). The frequency of the SNPs responsible for these substitutions is being investigated and may prove to be an additional somatic mutation that can alter individual response to chemotherapy. Additionally, the cell type in which BCRP is expressed may affect its substrate profile, especially if it is determined that this protein can function as a heterodimer. Conceivably, different cell types might express different "partners" for BCRP function, which would alter its substrate specificity profile.

Alterations in the pharmacologic retention of a drug molecule within the cell are also documented as a mechanism for resistance. Methotrexate is an anionic compound with a monoglutamate moiety, to which intracellular folylpolyglutamate synthase (FPGS) adds additional glutamate molecules. Polyglutamy-lated methotrexate is highly anionic and is no longer a substrate for either passive or active efflux from the cell. These polyglutamylated metabolites are active inhibitors of all proteins in the folate metabolism pathway normally inhibited by the parent molecule and provide prolonged toxicity when retained within the cell. In childhood acute lymphoblastic leukemia (ALL), patients with T lymphoblastic ALL have lower overall formation of polyglutamylated MTX than do patients with B lineage ALL. Because T lineage ALL is known to have a worse prognosis than B lineage ALL, the mechanism by which this occurs was investigated in primary ALL blasts. T lineage blasts were found to have lower FPGS activity than B lineage blasts, providing the mechanism by which less active metabolite was formed, and the tumor cells were more resistant to therapy (35–37).

## 2.3  Alterations in Drug Detoxification

Cells use secondary metabolism pathways (e.g., phase II metabolism) for the primary purpose of making drug molecules more water soluble in order to efflux them from the cell. By enhancing phase II metabolism, cancer cells can rapidly excrete drugs, thereby reducing their overall exposure time to the cytotoxic agents. For example, alterations in the glutathione conjugation pathway can alter a tumor cell's sensitivity to drugs that are significantly excreted via conjugation with glutathione.

γ-Glutamlycysteinylglycine, or reduced glutathione, is nucleophilic and conjugates with electrophilic atoms in three primary reactions: (1) nucleophilic addition of glutathione to electrophiles (e.g., detoxifies epoxides), (2) reduces lipid and DNA hyperoxides (e.g., thymine hydroperoxide), and (3) readily directly reduces some free radicals (e.g., hydroxy and carbon). Glutathione is present at high concentrations in tissues (0.1–10 m$M$) but is present in highest concentration in the liver (5–10 m$M$). At these high concentrations, glutathione can undergo nonenzymatic conjugation with nucleophiles. Cancer cells are noted for decreased glutathione concentrations when compared with the somatic tissues from which they are derived. Decreased intracellular glutathione pools prevent detoxification of nucleophiles. However, this decreased intracellular concentration of glutathione in some cell types is caused by overexpression of MRPs 1 and 2, which use glutathione to cotransport anticancer agents from the cell, even when these compounds are not conjugated to GSH (e.g., vincristine) (38). Conversely, some cancer cells have been shown to have elevated intracellular glutathione concentrations (two- to threefold), providing resistance to vinblastine, adriamycin, and VP-16 (39).

In tissues with low glutathione concentrations, glutathione-$S$-transferases are important to mediate glutathione conjugation to nucleophiles. There are multiple GST isozymes that have extensive substrate overlap. In humans, four cytosolic subfamilies of GSTs are important for drug detoxification: $\pi$, $\mu$, $\theta$, and $\alpha$. Although these enzymes are polymorphic, with $\mu$ and $\theta$ mutations comprising the majority of polymorphisms within the population, tumor cells usually up-regulate only the $\alpha$ and $\pi$ isoforms, thereby making these the most important predictors of tumor resistance. GSTs are active as either homo- or heterodimers. *In vitro* assays of GST activity in tumor cells has shown that *in vivo* tumor cells have increased GST activity, which renders them resistant to alkylating agents (38). Drugs with convincing data for the importance of glutathione conjugation *in vivo* include chlorambucil, melphalan, nitrogen mustard, acrolein, BCNU, hydroxyalkenals, ethacrynic acid, and steroids (40).

## 2.4 Increased DNA Repair

By increasing either the rapidity of DNA repair or decreasing the efficiency of repair, cancer cells can overcome some of the DNA damage exerted by certain chemotherapeutic agents. One example is $O6$-alkyl guanine alkyl transferase (OGAT), a constitutively expressed DNA repair protein, which removes alkyl groups from the $O6$-position of guanine in DNA. Tumor cells that exhibit high OGAT activity are resistant to agents that form $O6$-alkyl adducts, such as the nitrosureas (e.g., BCNU) and triazene compounds (e.g., procarbazine and temozolomide). Agents are currently in development that inactivate OGAT, rendering tumors sensitive to nitrosourea and triazene therapy (41).

## 2.5 Defective Apoptosis

Apoptosis, or programmed cell death, is an energy-dependent process by which cells undergo an orderly series of intracellular events leading to cell death. This process is initiated by a cell in response to specific stimuli, such as DNA damage. Apoptosis is required for maintaining appropriate function and structure of normal proliferating for renewable tissues. Disruption of the normal programmed cell death response prevents cells from self-destructing when irreversible damage takes place, and they survive with this damage. Because they do not apoptose, these cells can continue to replicate unchecked. Two examples of genes that are commonly mutated in cancer cells are *p53* and *bcl2*.

*p53* is a transcription factor, which in its wild-type form, both represses and initiates specific promoter transcription in addition to several other cellular functions. In normal cells, *p53* is activated in response to DNA damage, triggering either growth arrest or apoptosis depending on the current stage of the cell cycle (Fig. 8.2). Greater than 50% of tumors have been documented to contain functional inactivation of p53. The cancer cell genome is rendered unstable secondary to these alterations in *p53* or other changes that inactivate the apoptotic pathway (e.g., *bcl-2*). This allows for multiple secondary and tertiary molecular changes within the cancer cell genome (both in coding and noncoding regions), some of which are exploited in the presence of drug to allow for survival of the cell despite therapy. Cells that have mutant *p53* show a propensity to amplify DNA and show a distinct growth advantage to cells with wild-type *p53*. Therefore, these tumor cells can continue to divide, even under the selective pressure of chemotherapeutic agents whose mechanism of ac-

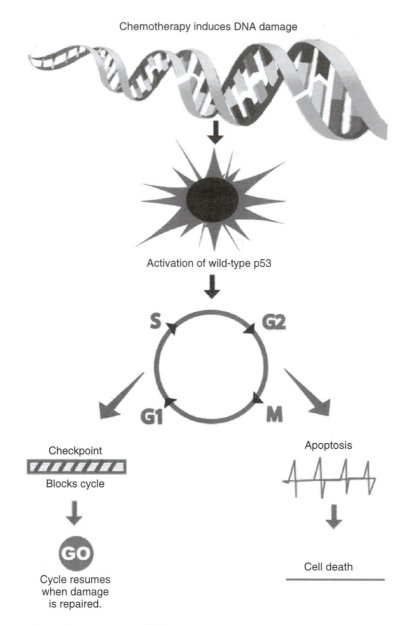

**Figure 8.2.** Chemotherapy induces DNA damage, stimulating activation of p53. Based on the stage of the cell cycle the cell is in when the damage takes place, different pathways are activated. When damage occurs early in the cell cycle, progression is arrested until DNA damage is repaired. Late in the cell cycle, apoptosis is stimulated.

tion is to damage DNA. Lowe et al. showed that normal mouse embryo fibroblasts lacking the *p53* gene were resistant to doxorubicin, 5-FU, and etoposide (42, 43). When the *p53* gene was transfected into these cells, the cells became sensitive to these drugs and died via the apoptotic pathway (44).

*bcl2* is an oncogene, the expression of which inhibits apoptosis. In tumor cells, *bcl2* is activated, allowing cell to survive in the face of cellular damage that would normally signal for programmed cell death, thus making these cells more resistant to chemotherapy-induced cytotoxicity. *In vitro* models of cell lines with *bcl-2* activation show that these cells continue to divide and grow despite high concentrations of drug in the media (45).

## 2.6 Epigenetic Changes

Changes in the cancer cell phenotype that do not alter the genotype (i.e., epigenetic changes) are an additional source of drug resistance. One example is the process of methylation/demethylation of DNA. Epigenetic silencing of tumor suppressor genes has been documented in human cancers. Methylation of DNA occurs in eukaryotic cells. DNA methyltransferase catalyzes the reaction, whereby a methyl group is transferred to the carbon 5 position of the cytosine ring within a 5′-CG-3′ sequence. Methylation of genes silences expression by preventing binding of transcription factors and recruiting binding of other proteins that limit the accessibility of the gene to the cell's transcription machinery. Demethylation of a gene activates its expression. Normal methylation patterns are disrupted in many tumor cells, with tumors exhibiting decreased methylation in areas of repetitive DNA and an increase in methylation of promoter CpG islands. Hypermethylation of tumor suppressor genes has been documented in many tumor cells, and tumor-specific patterns of gene methylation has identified a subset of patients whose tumors display widespread methylation abnormalities. These tumors are referred to as displaying the CpG island methylate phenotype (CIMP), and these patients tend to respond less well to chemotherapy. The exact mechanism by which this occurs is not known, but investigators are currently determining whether this phenomenon extends to drug targets or enzymes that catalyze drug activation/detoxification pathways.

## 3 STRATEGIES TO OVERCOME RESISTANCE

The future strategy to overcome resistance will be to individualize a patient's therapy prospectively, employing both knowledge of the patient's somatic mutations in drug metabolizing enzymes, drug receptors, and drug targets with knowledge of tumor-specific changes that affect cytotoxicity of agents normally given systemically to patients. Some strategies used clinically are pharmacokinetic-based strategies, pharmacogenetic-based strategies, and basic tumor biology-directed strategies for dosing. The future of individualization for patients lies in developing algorithms for dosing, based on a synthesis of all of these methods, to determine an overall phenotype for each patient prospectively.

## 3.1 Pharmacokinetic Monitoring

One strategy to overcome tumor cell resistance to chemotherapy is to employ pharmacokinetic strategies to further intensify doses of drugs by giving the maximally tolerated dose of drug as close together as possible. For example, by using 24-h methotrexate infusions clinically, one can presumably overcome the resistance phenotype seen in tumors overexpressing certain ABC transporters *in vitro*, whereby long exposure to methotrexate was able to overcome the high levels of resistance observed after short (4-h) exposure (e.g., MRP1–3) (46–48). Additionally, higher serum concentrations of methotrexate can overcome the resistance achieved by increased quantities of DHFR intracellularly or decreased expression of the reduced folate carrier. Unfortunately, not all interpatient variability can be determined prospectively for pharmacokinetic strategies to be completely effective prior to an individual's initial exposure to drug. Additionally, pharmacokinetic-guided dosing can be labor intensive for both the patient and medical staff.

## 3.2 Pharmacogenetic Monitoring

Pharmacogenetic-based strategies are currently used to identify monogenic traits in patients that would alter either a predisposition to toxicity or the efficacy of response. Currently, these strategies are not used prospectively in patients to alter dosing *a priori*, but rather once a patient experiences toxicity or doesn't respond to therapy, genotyping is performed. The future of pharmacogenetic monitoring, however, will be to develop dosing based on multiple genotypes in an individual (49–51), creating rational selection of drugs and doses for individual patients.

One example of a well-characterized polymorphism in drug metabolism that is currently used to screen patients treated with thiopurines is the SNPs present in thiopurine methyltransferase. 6-Mercaptopurine is a thiol-substituted analog of hypoxanthine, used to treat patients with acute lymphoblastic leukemia. TPMT methylates both the parent compound (6-MP) and the active phosphorylated metabolites of 6-MP, resulting in detoxification of the compound. Patients with homozygous mutations in (G238C) (TPMT*2) or (G460A and A719G) (TPMT*3A) experience profound neutropenia when exposed to thiopurines (52–54). Heterozygotes for these autosomal codominant mutations experience intermediate toxicity, with as many as one-third of these patients requiring a dose-reduction of thiopurines during therapy for ALL (55). Therefore, these somatic mutations alter the ability of patients to receive maximal doses potentially altering outcome in some studies.

### 3.3 Biologic Inhibition of Tumor Cell Properties (ABC Transporters)

Cyclosporine $A_1$ (CSA) and PSC833 are agents that block MDR1-mediated efflux of substrates *in vitro* and have been used systemically to inhibit MDR1 during chemotherapy, with mixed results (56). CSA has had some improvement in efficacy for systemic chemotherapy, probably because of the fact that it is a nonselective inhibitor and also inhibits BCRP-mediated efflux.

### 4 CONCLUSIONS

Drug resistance continues to provide an obstacle for administering chemotherapy to patients. Although some patients respond very well to chemotherapy, a significant number of treatment-refractory tumors are still under study. Ultimately, the combined knowledge of patient-specific factors that alter our ability to administer optimal doses of chemotherapy and tumor-specific factors that result in decreased efficacy of chemotherapy will be used to more efficiently dose patients. As new drugs

emerge for bioselective treatment of tumors, understanding induction of resistance continues to be essential for providing the best therapy to patients. With the powerful genomics tools that are becoming more available, it will soon be feasible to monitor both tumor and somatic variations in DNA sequence, expression of proteins, and *in vitro* response to therapy before initiating therapy in patients, thereby providing information required to reach the ultimate goals of reducing the incidence of side effects, improving response to therapy, and preventing the induction of resistant malignant clones *in vivo*.

### 5 ACKNOWLEDGMENTS

I would like to acknowledge Peter Adamson, M.D. (Children's Hospital of Philadelphia) for his editorial comments and Evan Katz (Children's Hospital of Philadelphia) for designing Fig. 8.2.

### REFERENCES

1. V. Ling, *Cancer Res.*, **57**, 2325 (1997).
2. I. B. Roninson and K. Choi, *Clin. Physiol. Biochem.*, **5**, 140 (1987).
3. J. H. Goldie, *Comp. Biochem. Physiol. C Pharmacol. Toxicol.*, **77**, 283 (1984).
4. J. H. Goldie and A. J. Coldman, *Cancer Res.*, **44**, 3643 (1984).
5. F. W. Alt, *Immunol. Today*, **13**, 306 (1992).
6. N. Denis, *Oncogene*, **6**, 1453 (1991).
7. H. B. Eastman, *Proc. Natl. Acad. Sci. USA*, **88**, 8572 (1991).
8. H. Miyachi, *Cancer Lett.*, **88**, 93 (1995).
9. R. L. Blakley, *Hum. Mutat.*, **11**, 259 (1998).
10. M. E. Gorre, *J. Appl. Microbiol.*, **90**, 788 (2001).
11. N. Horie, H. Aiba, K. Oguro, H. Hojo, and K. Takeishi, *Cell Struct. Funct.*, **20**, 191 (1995).
12. K. Kawakami, *Bioorg. Med. Chem.*, **7**, 2555 (1999).
13. E. Villafranca, *J. Clin. Oncol.*, **19**, 1779 (2001).
14. J. L. Nitiss and W. T. Beck, *Eur. J. Cancer*, **32A**, 958 (1996).
15. R. Zhao, *J. Biol. Chem.*, **273**, 19065 (1998).
16. R. Gorlick, *Blood*, **89**, 1013 (1997).
17. M. Mueller, *Human ABC Transporter Page [Online], Nutrition, Metabolism, and Genomics Group*, Wageningen University, The Netherlands, 2001.

18. M. E. Goldsmith, *J. Biol. Chem.*, **270**, 1894 (1995).

19. J. Sampath, *Nature Med.*, **7**, 1028 (2001).

20. J. V. Thottassery, *Proc. Natl. Acad. Sci. USA*, **94**, 11037 (1997).

21. Y. Gazitt, *Cancer Res.*, **52**, 2957 (1992).

22. R. L. Juliano, *Biochim. Biophys. Acta*, **455**, 152 (1976).

23. S. V. Ambudkar, *Ann. Rev. Pharmacol. Toxicol.*, **39**, 361 (1999).

24. S. Hoffmeyer, *Proc. Natl. Acad. Sci. USA*, **97**, 3473 (2000).

25. M. M. Ameyaw, *Pharmacogenetics*, **11**, 217 (2001).

26. S. P. Cole, G. Bhardwaj, J. H. Gerlach, J. E. Mackie, C. E. Grant, K. C. Almquist, A. J. Stewart, E. U. Kurz, A. M. Duncan, and R. G. Deeley, *Science*, **258**, 1650 (1992).

27. C. E. Grant, *Cancer Res.*, **54**, 357 (1994).

28. G. J. Zaman, M. J. Flens, M. R. van Leusden, M. de Haas, H. S. Mulder, J. Lankelma, H. M. Pinedo, R. J. Scheper, F. Baas, H. J. Broxterman, et al., *Proc. Natl. Acad. Sci. USA*, **91**, 8822 (1994).

29. S. P. Cole, *Bioessays*, **20**, 931 (1998).

30. J. Kartenbeck, *Hepatology*, **23**, 1061 (1996).

31. T. Kitamura, *Proc. Natl. Acad. Sci. USA*, **87**, 3557 (1990).

32. C. C. Paulusma, *Science*, **271**, 1126 (1996).

33. D. D. Ross, W. Yang, L. V. Abruzzo, W. S. Dalton, E. Schneider, H. Lage, M. Dietel, L. Greenberger, S. P. Cole, and L. A. Doyle, *J. Natl. Cancer Inst.*, **91**, 429 (1999).

34. L. A. Doyle, W. Yang, L. V. Abruzzo, T. Krogmann, Y. Gao, A. K. Rishi, and D. D. Ross, *Proc. Natl. Acad. Sci. USA*, **95**, 15665 (1998).

35. Y. Honjo, C. A. Hrycyna, Q. W. Yan, W. Y. Medina-Perez, R. W. Robey, A. van de Laar, T. Litman, M. Dean, and S. E. Bates, *Cancer Res.*, **61**, 6635 (2001).

36. Y. G. Assaraf, *J. Biol. Chem.*, **272**, 17460 (1997).

37. G. Jansen, *J. Biol. Chem.*, **273**, 30189 (1998).

38. C. M. Paumi, B. G. Ledford, P. K. Smitherman, A. J. Townsend, and C. S. Morrow, *J. Biol. Chem.*, **276**, 7952 (2001).

39. Y. Saito, Y. Nakada, T. Hotta, T. Mikami, K. Kurisu, K. Yamada, K. Kiya, K. Kawamoto, and T. Uozumi, *Int. J. Cancer*, **48**, 861 (1991).

40. K. D. Tew, *Cancer Res.*, **54**, 4313 (1994).

41. L. Tentori, L. Orlando, P. M. Lacal, E. Benincasa, I. Faraoni, E. Bonmassar, S. D'Atri, and G. Graziani, *Mol. Pharmacol.*, **52**, 249 (1997).

42. S. W. Lowe, *Curr. Opin. Oncol.*, **7**, 547 (1995).

43. S. W. Lowe, S. Bodis, A. McClatchey, L. Remington, H. E. Ruley, D. E. Fisher, D. E. Housman, and T. Jacks, *Science*, **266**, 807 (1994).

44. S. W. Lowe, H. E. Ruley, T. Jacks, and D. E. Housman, *Cell*, **74**, 957 (1993).

45. J. C. Reed, *Curr. Opin. Oncol.*, **7**, 541 (1995).

46. G. D. Kruh, H. Zeng, P. A. Rea, G. Liu, Z. S. Chen, K. Lee, and M. G. Belinsky, *J. Bioenerg. Biomembr.*, **33**, 493 (2001).

47. J. H. Hooijberg, H. J. Broxterman, M. Kool, Y. G. Assaraf, G. J. Peters, P. Noordhuis, R. J. Scheper, P. Borst, H. M. Pinedo, and G. Jansen, *Cancer Res.*, **59**, 2532 (1999).

48. M. Kool, M. van der Linden, M. de Haas, G. L. Scheffer, J. M. de Vree, A. J. Smith, G. Jansen, G. J. Peters, N. Ponne, R. J. Scheper, R. P. Elferink, F. Baas, and P. Borst, *Proc. Natl. Acad. Sci. USA*, **96**, 6914 (1999).

49. W. E. Evans, *Science*, **286**, 487 (1999).

50. W. E. Evans, *Annu. Rev. Genomics Hum. Genet.*, **2**, 9 (2001).

51. M. V. Relling and T. Dervieux, *Nature Rev. Cancer*, **1**, 99 (2001).

52. E. Y. Krynetski and W. E. Evans, *Pharm. Res.*, **16**, 342 (1999).

53. C. R. Yates, E. Y. Krynetski, T. Loennechen, M. Y. Fessing, H. L. Tai, C. H. Pui, M. V. Relling, and W. E. Evans, *Ann. Intern. Med.*, **126**, 608 (1997).

54. E. Y. Krynetski, H. L. Tai, C. R. Yates, M. Y. Fessing, T. Loennechen, J. D. Schuetz, M. V. Relling, and W. E. Evans, *Pharmacogenetics*, **6**, 279 (1996).

55. M. V. Relling, M. L. Hancock, G. K. Rivera, J. T. Sandlund, R. C. Ribeiro, E. Y. Krynetski, C. H. Pui, and W. E. Evans, *J. Natl. Cancer Inst.*, **91**, 2001 (1999).

56. A. F. List, K. J. Kopecky, C. L. Willman, D. R. Head, D. L. Persons, M. L. Slovak, R. Dorr, C. Karanes, H. E. Hynes, J. H. Doroshow, M. Shurafa, and F. R. Appelbaum, *Blood*, **98**, 3212 (2001).

# CHAPTER NINE

# Antiviral Agents, DNA

TIM MIDDLETON
TODD ROCKWAY
Abbott Laboratories
Abbott Park, Illinois

## Contents

1 Introduction, 294
2 The Viruses, 294
  2.1 Herpesviruses, 297
  2.2 HBV, 299
  2.3 Papillomaviruses, 301
  2.4 Polyomaviruses, 302
  2.5 Adenoviruses, 303
  2.6 Parvoviruses, 304
  2.7 Poxviruses, 305
3 Antiviral Compounds for DNA Viruses, 305
  3.1 Herpesviruses, 305
    3.1.1 Currently Approved Drugs, 305
      3.1.1.1 Range of Action-Modifications to ACV and Development of Unrelated Drugs, 307
      3.1.1.2 Prodrugs that Increase Oral Bioavailability, 309
      3.1.1.3 Resistance in Herpesviruses, 309
    3.1.2 Compounds Under Development, 310
      3.1.2.1 Nucleoside Analogs, 310
      3.1.2.2 Prodrugs Under Development, 318
      3.1.2.3 Nonnucleoside Inhibitors and Other Targets, 321
      3.1.2.4 Herpesvirus Protease Inhibitors, 324
      3.1.2.5 Immune Modulators, 325
  3.2 HBV, 326
    3.2.1 Pyrimidines, 326
    3.2.2 Purines, 330
    3.2.3 Resistance in HBV, 333
    3.2.4 Combination Treatments, 334
    3.2.5 Targeting Drugs to the Liver, 335
    3.2.6 Antisense Oligonucleotides, 337
  3.3 Papillomaviruses, 337
  3.4 Polyomaviruses, 339
  3.5 Adenoviruses, 340
  3.6 Poxviruses, 340
  3.7 Parvoviruses, 341
4 Conclusions, 342

*Burger's Medicinal Chemistry and Drug Discovery*
Sixth Edition, Volume 5: Chemotherapeutic Agents
Edited by Donald J. Abraham
ISBN 0-471-37031-2    © 2003 John Wiley & Sons, Inc.

# 1   INTRODUCTION

Antiviral chemotherapy is conceptually no different than chemotherapy of other infectious
agents, but the therapeutic targets are more
limited. Viruses are obligate intracellular parasites that rely heavily on their host cell to
produce progeny virions. The virus may supply as few as two nonstructural proteins to the
production line, but may borrow dozens of cellular proteins to gain access to the cell, produce the components of new virions, assemble
them, and export them from the cell. The genome sizes of viruses range from 25% to <1%
the size of typical pathogenic bacteria. Furthermore, because they borrow heavily from
cellular metabolic pathways, they tend to encode proteins that are similar to cellular proteins. Viruses have relatively few proteins
with unique activities not found in the host
cell. Thus, the difference between inhibiting a
viral protein and its cellular equivalent may be
small. With most viruses, one is forced to concentrate on a small number of targets and find
compounds that can exploit the sometimes
subtle differences between viral proteins and
cellular proteins with similar functions. Occasionally, a more effective approach has been to
not attack the virus directly, but rather to
stimulate the body's own defenses to fight it
more effectively. Some of the more effective
antiviral agents have been compounds that
stimulate the immune response against the virus.

# 2   THE VIRUSES

There are seven families of DNA viruses that
are pathogenic for humans. These pathogens
come from the Adenoviridae, Hepadnaviridae,
Herpesviridae, Polyomaviridae, Papilloma-
viridae, Parvoviridae, and Poxviridae families.
Herpesviruses, hepadnaviruses, and papillomaviruses are well established as human
health problems and as targets for antiviral
chemotherapy. The biology of adenoviruses,
parvoviruses, polyomaviruses, and poxviruses
has been intensively studied, but these viruses
have not been pursued as drug targets to the
extent of the other viral families.

DNA viruses have assumed more importance as pathogens as the prevalence of patients with suppressed immune function has
increased. This increased prevalence comes in
turn from two sources: greater use of organ
transplantation and the AIDS epidemic. Most
of the DNA viruses are ubiquitous in the human population, but several become a serious
health threat in the absence of a fully functional immune system. The lack of immune
surveillance allows them to propagate essentially unchecked. In particular, the overall
health threats posed by herpesviruses, hepatitis B virus (HBV), papillomaviruses, and polyomaviruses are worsened by immunodeficiency.

DNA viruses are defined as having their
genome in the form of DNA for the infectious
phase of their life cycle, although HBV goes
through an RNA intermediate during the replication cycle. Their genomes vary in size from
approximately 5000–300,000 base pairs (by
comparison, human cells contain about $3 \times 10^9$ base pairs). The biology and clinical details
of these viruses has been reviewed in detail
elsewhere (1); only the features most likely to
be important to drug development will be discussed here. We start with a general discussion of traits that apply to most viruses, followed by specific discussions of each family.

Most of the DNA viruses are chronic viruses—they infect their host and establish a
persistent infection in specific cell types that
may last for the lifetime of the host. Two requirements of such a long-lasting relationship
are a host cell in which it can remain indefinitely and a means of avoiding detection by the
immune system. For instance, $\alpha$-herpesvi-
ruses tend to infect nondividing cells in which
they can "lie low" by turning off the expression of most viral genes. On the other hand,
$\gamma$-herpesviruses infect dividing cells, but they
have evolved special mechanisms for persist-
ing in these cells and evading the immune system. Their genome is designed to be recognized similarly to host cell DNA, such that
cellular machinery replicates it when the cell
divides. They have also tailored the proteins
needed to maintain their presence in dividing
cells to evade the immune system. Some relationships are not as sophisticated; for in-

stance, most HBV and papillomavirus infections are destroyed by the immune system, but rates of production of virions are high enough to ensure long-term propagation of the viruses in a portion of the infected population. In this portion, a chronic infection will be established that will last for years to decades, thus providing a reservoir of virus in humans.

The following are the general steps needed by all viruses to produce progeny.

1. *Adsorption and entry.* Adsorption to the cell is mediated through interactions between ligands on the viral surface and specific cellular receptors, often involving multiple cellular proteins. Entry is usually by one of two routes, either fusion of a viral membrane with the cellular membrane or internalization by endocytosis and release of the virus from internal vesicles. These steps create the potential for good drug targets, and drugs to block retroviral binding have been developed. However, our understanding of the virus-receptor interactions for DNA viruses is not yet sufficient to be used for drug development.

2. *Uncoating and processing.* The extent of uncoating varies between species. In some cases, the DNA genome, with a minimal set of associated proteins, is transferred from the capsid to the nucleus. At the other extreme, some viruses, such as poxviruses, do a minimal amount of uncoating, and the entire replication process takes place in a specialized viral structure constructed by the virus in the cellular cytoplasm. Other steps may be required to prepare the genome for use, such as circularization of the linear herpesvirus genome after removal from the capsid or conversion of the partially double-stranded HBV genome to a completely double-stranded, covalently closed circular form. Again, uncoating and processing steps have not yet been exploited for drug development.

3. *Transcription of early proteins.* These include both regulatory proteins and proteins needed for replicating the viral genome. While critical to the production of progeny virions, achieving selectivity in inhibiting transcription of specific genes has posed a severe limitation, independent of whether the genes are cellular or viral. Herpesviruses provide an additional potential target, in that the productive cycle may be delayed indefinitely after infection (i.e., the virus becomes latent), in which case a reactivation step is needed to move the virus into the lytic cycle. Reactivation is a potential drug target, and in some cases, the reactivation pathways have been partially mapped. However, it is largely another form of transcriptional regulation and has not been profitably pursued as a drug target.

4. *Viral genome replication.* This is the primary focus of attention for all DNA viruses except papillomaviruses, polyomaviruses, and parvoviruses, which lack a virally encoded polymerase. Polymerase is the obvious target, but accessory proteins provide additional possibilities. Inhibitors of the helicase activity of papillomaviruses have been identified, but drugs have not yet been developed against this enzyme. Nucleoside kinases of herpesviruses represent another piece of the replication machinery that has played a central role in drug development, because they permit the use of drugs that are inactive in uninfected cells. These kinases perform the first addition of a phosphate to nucleoside analogs. Analogs that are poorly recognized by cellular kinases only get converted to the active form in infected cells. To the extent that the analog is selective for the viral kinase, toxicity is limited by the fact that the addition of the negative charge blocks the activated nucleoside analog from crossing the cellular membrane; therefore, uninfected cells are protected from exposure to the activated drug. This is a critical factor in providing specificity to several herpesvirus drugs and will be discussed in the context of those drugs.

Unlike RNA-dependent polymerases, DNA-dependent polymerases have relatively high fidelity, i.e., they are not as error-prone. This decreases the likelihood of resistance mutations accumulating in the viral population and has contributed to the emphasis on polymerase as a drug target. There are two types of exceptions to this generalization, however. The first is HBV, which has a polymerase with a high mutation rate. This is reflected in the relatively rapid appearance of strains of virus resistant to drugs. The second is loss of immune surveillance, which allows increased viremia with the accumulation of more mutations, even from viruses with a low overall error rate.

5. *Synthesis of late proteins.* Late in the viral life cycle, structural proteins are synthesized, including nucleocapsid proteins needed to bind to DNA, capsid/coat proteins that provide the shell of the virus, and in the case of lipid membrane–coated viruses, glycoproteins that serve as the ligands for cellular receptors. Whereas genes expressing these proteins are easy to identify in the viral genome and the proteins are usually synthesized in large quantities, there is no precedent for directly blocking their synthesis.

6. *Processing and assembly.* Two forms of processing have been exploited as drug targets. First, processing of the viral genome is needed in some cases. A set of herpesvirus proteins needed to cleave the genome to monomeric length and package it into capsids has been identified (2–4). Inhibitors of these proteins have recently been developed. Second, processing of the structural proteins is generally needed. Many of them are glycosylated by cellular glycosylation enzymes. Of more immediate interest, proteolytic cleavage is generally needed, either to cleave polyproteins to their individual proteins or to perform a maturation step, in which the morphology of the virus changes either after packaging is complete or after the virus has been released from the cell. Whereas the nature of this maturation step is not functionally well defined, protease inhibitors constitute one of the two major classes of currently licensed drugs for treatment of HIV-1, establishing a precedent for this approach. Recently, inhibitors of cytomegalovirus protease have also been developed.

7. *Egress of the virion from the cell.* Whereas the mechanisms by which viruses leave cells are understood to varying extents, this step has not been pursued for drug development.

The pathology of DNA viruses comes in three guises:

1. The destruction of the cell as an obligatory part of the production of progeny virus. All of the herpesviruses are capable of this, and it is the primary cause of damage by the α- and β-herpesviruses. This is also the primary cause of the pathology of adenoviruses, parvoviruses, and poxviruses.

2. Damage to the cell or organ resulting from the host immune response to the virus. This accounts for much of the pathology of the HBV; the loss of infected hepatocytes by T-cell–mediated destruction, coupled with high rates of viral infection, cause liver function to decrease to inadequate levels.

3. Stimulation of proliferation of the infected cell with the eventual generation of tumors. Tumorigenicity is an unintended consequence of the evolution of some viruses to regulate cellular proliferation, and it is a consequence from which the virus gains no benefit. This form of pathology is, in a sense, the most insidious of the three, because the tumors generally arise from the latent state of the virus or from a virus that has been unintentionally integrated into a cellular chromosome. In both cases, the viruses present very few protein targets for exploitation, either by the immune system or by drugs. The most serious clinical consequences of papillomaviruses and γ-herpesviruses are a result of this mechanism.

The classic examples of tumor viruses among DNA viruses are papillomaviruses, HBV, and γ-herpesviruses, particularly Epstein Barr virus (EBV) and human herpesvirus-8 (HHV-8, KHSV). EBV presents two types of proliferative disorders. The first is an acute, usually benign and self-limiting, proliferation of B-cells when infection occurs as an adult, which is associated with mononucleosis. The second generally occurs years after the initial infection and causes tumors to form in a variety of cell types, usually of lymphoid or epithelial origin. Papillomaviruses cause proliferation of epithelial cells, which becomes more aggressive over time, with some cells eventually becoming neoplastic. The role of hepatitis B in forming hepatocellular carcinoma is less well understood, but it may require the integration of the usually episomal genome into a host chromosome. This provides the potential for disruption of regulation of both viral and cellular gene synthesis. For the herpesviruses, and to a lesser extent the papillomaviruses, the quantitative significance of virally induced tumors has increased with the increase in depression of immune system function, either because of infection by HIV or because of the need prevent organ rejection after transplantation.

## 2.1 Herpesviruses

Herpesviruses are ubiquitous viruses, both in terms of the number of species that they infect and the proportion of human population infected. They are widely distributed throughout the animal kingdom, with about 100 species identified thus far, of which eight infect humans. An estimated 50–90% of adult humans are infected with at least one species of herpesvirus. This percentage points out the importance of latency in the life cycle of herpesviruses—they spend most of their time in what is functionally a state of dormancy, and they have effectively evolved to evade immune surveillance. The importance of evading the host immune system is shown by the fact that herpesviruses are among the more common pathogens when immune surveillance is removed or impaired.

Herpesviruses have large double-stranded linear genomes that range in size from approximately 120,000–230,000 kb. After infection, the genomes become circular, and they remain in the nucleus as an episome. The protein composition of herpesviruses is complex; typically

at least 30 proteins are found in the virion and at least twice that many are encoded in the genome. They are enveloped viruses, with several glycoproteins on the membrane surface. They encode an unusually large number of enzymes involved in replication of the viral DNA. There are seven proteins that are required to reconstruct herpes simplex virus-1 (HSV-1) DNA replication *in vitro* (5). In addition to these enzymes, additional enzymes involved in nucleotide metabolism and other steps are encoded. This wealth of viral machinery has made herpesviruses one of the most popular viral targets for drug development, with DNA replication receiving the majority of the attention. However, there are other targets that are now coming into their own, including a protease and enzymes required for processing the DNA and packaging it into capsids.

Understanding the entry of herpesviruses into cells has been complicated by the fact that there are multiple viral proteins involved in entry, and eliminating expression of them individually may not block infectivity (6–11). As pointed out by Roizman and Sears (12), this multiplicity of potential entry mechanisms may stem from the fact that herpesviruses infect multiple cell types, and different entry mechanisms may be needed for different cell types. If there is redundancy in entry pathways within the most commonly infected cell types, developing entry inhibitors of herpesviruses may be technically prohibitive.

DNA synthesis is carried out by viral enzymatic machinery in a manner that resembles that of host DNA replication. The exception is that the origin of replication is better defined for herpesviruses than for cellular DNA. The polymerases, helicases, primases, processivity factors, and single-stranded DNA binding proteins are nearly identical in function between virus and host (herpesviruses borrow their topoisomerase activity from the host cell). Fortunately, the structural differences in the proteins are sufficient to allow drugs to discriminate between the two. The need for herpesviruses to encode genes for enzymes involved in nucleotide metabolism may be a function of the amount of viral DNA that is synthesized per cell. Replication of viral DNA can result in an amount nearly equivalent to the amount of host cell DNA, thus placing a large burden on the cellular nucleotide pools.

Herpesviruses are categorized into three classes, $\alpha$, $\beta$, and $\gamma$. There is overall conservation of the machinery for nucleotide metabolism, DNA replication, and virion assembly and packaging between the classes (13–15). They differ in the genes that are the determinants of their tissue tropism.

The human $\alpha$ viruses are HSV-1, HSV-2, and varicella zoster virus (VZV). These viruses are characterized by a rapid reproductive cycle and rapid spread in cell culture. They are latent primarily in sensory ganglia. HSV-1 is classically associated with cold sores, although it can infect any organ. Cold sores are the most common form of recurrent HSV-1 infection, with 17–30% of humans having at least one recurrence per year (16, 17). The most common route of infection is through the mouth, with primary infection ranging from asymptomatic to mild mononucleosis (18, 19). HSV-2 is mostly associated with genital infections. Its common transmission route is sexual contact (20–22), and infection is usually delayed until the onset of sexual activity. Otherwise, the infections by the two viruses share many similarities. Approximately 20% of the adult population is predicted to be infected with HSV-2 (23), although this rate can vary widely depending on the level of sexual activity (24, 25).

VZV results in chicken pox (varicella) on initial infection and shingles (herpes zoster) if it is reactivated later in life. Approximately 95% of the population is predicted to be infected (26, 27). This proportion should decrease with the recent introduction of a vaccine against VZV (28, 29). The site of entry of VZV is likely to be respiratory mucosal epithelium. From there, it is thought to replicate in lymph nodes and infect reticuloendothelial cells, moving from these cells to cutaneous epithelium (30). It is at this site that the skin lesions occur. VZV infects T lymphocytes, epithelial cells, and dorsal root ganglia (31–33). The source of latent virus is cells of dorsal root ganglia (34–37). The cause of the reactivation from these ganglia, leading to herpes zoster, is not understood.

The mechanisms used for entry, transfer to the nucleus, and for DNA and protein synthesis and assembly seem to be similar to HSV.

Egress from the cell differs between the two, at least in cell culture. VZV is largely cell-associated, with little free virus found in cell culture (38–40). Whether this reflects a difference in viral spread *in vivo* is unclear.

The $\beta$-herpesviruses are more slowly replicating viruses. They include cytomegalovirus (HCMV), HHV-6, and HHV-7. $\beta$-Herpesviruses represent important opportunistic infections in cases of immunodeficiency. In a prospective study of 60 liver transplant patients, one-half were found to have active HCMV and HHV-7 infections, and a third had active HHV-6 infections (41).

HCMV is estimated to be responsible for about 8% of mononucleosis cases (42) and is a leading cause of birth defects caused by infection of the fetus (43). It is also the most frequent opportunistic viral infection found in HIV patients, with loss of vision because of retinitis the most common outcome (44). In a study conducted during the 1980s, 81% of AIDS patients were found to have signs of actively replicating HCMV at autopsy (45). Thirty to forty percent of AIDS patients with $CD4^+$ lymphocyte counts $<50/mm^3$ developed retinitis before the use of multidrug combinations for HIV infection (46).

The primary site of infection is mucosal epithelium. HCMV spreads throughout the body through infection of leukocytes, although the specific cell type is not well defined (47–52). HCMV infects cells of fibroblastic, endothelial, epithelial, macrophage, and muscle lineages (53). Myeloid cells are the most likely reservoir for latent infection (54–57).

The cellular receptors for $\beta$-herpesviruses have not been identified. Entry and transfer of the DNA to the nucleus seems to occur through pathways analogous to other herpesviruses (58). The synthesis of early proteins, DNA, and late proteins follows the pattern of $\alpha$-herpesviruses, but the timescale is longer. HCMV requires approximately three times longer in cell culture to complete the infectious cycle than HSV-1 (59).

HHV-6 and HHV-7 are genetically similar to HCMV and resemble other herpesviruses in their mode of entry, lytic production of progeny virions, and exit from the cell. HHV-6 infects several cell types, including T-cells, epithelial and endothelial cells, and neurons (60–

66). HHV-7 is found almost exclusively in T-cells (67–69). Infection rates for these two viruses have been estimated at 20–100% of the population (70–73).

The human γ-herpesviruses include EBV and HHV-8. The γ-herpesviruses have evolved two mechanisms for DNA replication, one used for production of progeny viruses (the lytic cycle) and the other simply for maintaining the presence of the viral DNA in the cell (latency). The latter is needed because these viruses primarily infect cells that either are programmed to divide or are reprogrammed by the virus to divide. The latent replication pathway maintains the copy number of viral genomes in the daughter cells. Note that this pathway is not entirely unique to γ-viruses. Other herpesviruses are replicated by cellular polymerases after infection to increase the viral genome copy number. However, latency is more complex in the γ-herpesviruses, and this is the only class that is associated with tumors.

The pathology of the γ-herpesviruses is counterintuitive, because whereas the other two classes do their damage by lysis of the cell when viruses are released, almost all of the diseases associated with γ-herpesviruses occur during the latent phase of their life cycle. With α- and β-herpesviruses, it is possible in principle to essentially eliminate clinical damage by preventing production of new viruses. With γ-viruses, clinical damage can be lessened by this approach in that blocking virogenesis decreases the number of infected cells carrying latent genomes. However, it is the latent cells themselves in which disease occurs, so treating the disease state directly requires eliminating the latent cells. This is a tall order, and thus far, drug development has focused on the productive phase of the life cycle rather than the latent phase.

EBV was first identified because of its capacity to induce tumors in B-cells. It is a ubiquitous virus, with nearly 100% of children being infected in developing countries (74). In more developed countries, infection is delayed, and about one-half of the children are infected. However, by adulthood, greater than 90% of the population is infected. EBV is one of the etiological agents associated with infectious mononucleosis, and this is the usual outcome of acute infection in adults. Latent infection is associated with a variety of malignancies, including B-cell lymphomas, Hodgkin's disease, leiomyosarcomas (muscle tumors), T-cell lymphomas, and nasopharyngeal carcinoma (75–84).

B-cells are the primary host cells for latent infection by EBV (85–87), although it is not clear whether the initial site of infection is B-cells or mucosal epithelium (88, 89). Binding of EBV to B-cells is accomplished through the viral glycoprotein gp350 (90, 91) and the cellular protein CD21 (92, 93). The virus is internalized through endocytosis, and the DNA is transferred to the nucleus (94, 95). The initial response is establishment of a latent infection, with only rare cells undergoing a lytic infection. During its latent phase, EBV synthesizes up to eight proteins and several small RNAs. These proteins activate resting B-cells to divide, and one of them, EBNA-1, participates in the latent replication of the viral DNA (96–101). The cascade of events that leads to the switch from latent infection to lytic infection is initiated by the activation of the viral protein BZLF1, and forcing the expression of this protein is sufficient to induce lytic infection (102, 103). While the details vary, the basic outline of the lytic production of progeny virus is the same as for the other classes of herpesviruses.

HHV-8 is the etiological agent of Kaposi's sarcoma (KS) (104, 105), the most frequently found neoplasm in AIDS patients (45). It is also associated with body cavity–based lymphoma and multicentric Castleman's disease. Retroviral therapy often results in resolution of KS and clearance of HHV-8. This does not seem to be a result of a direct effect of the HIV inhibitors on HHV-8, but rather a result of restoration of immune function after effective treatment of the HIV infection (106). The entry and exit of the latent phase are similar to EBV, and lytic production of daughter virions resembles other herpesviruses.

## 2.2 HBV

An estimated 350 million people worldwide are infected with HBV (107). The disease caused by HBV results not so much from the action of virus *per se*, but rather the damage that occurs to liver tissue as a result of the immune response to the virus. This liver damage can lead to hepatic fibrosis, cirrhosis, liver failure, and hepatocellular carcinoma (HCC), although HCC may have a more direct viral component (108–110). Chronic HBV infection is associated with an approximately 150-fold increase in the likelihood of developing HCC (111, 112). A vaccine is available for HBV, but difficulties of distribution and the large population of currently infected individuals have maintained a relatively high incidence of infection thus far (113). Thus, there is likely to remain a need for effective HBV treatment regimens in the foreseeable future.

Infection of adults is usually from sexual contact or exposure to contaminated blood (114, 115). The primary infection can be asymptomatic, or it may result in acute hepatitis. Most patients will resolve the primary infection (116, 117); about 95% of patients infected as adults clear the virus after an acute infection, although only about 10% of children infected before 4 years of age do so (118, 119). The remaining cases develop a persistent infection. This persistent infection can range from asymptomatic to high levels of viremia. Even for asymptomatic cases, there is a high incidence of HCC over the course of 30 years.

HBV is a DNA virus 3.2 kb in length with a primary target of the liver (120). In persistently infected individuals, about $10^{11}$ virions are released into the blood per day, and these virions have a half-life of about 1 day (121). The virus enters the cell in a partially double-stranded form and is converted to completely double-stranded, covalently closed, supercoiled DNA (cccDNA) in the nucleus (122–125). RNA is synthesized from this form, both for translation into viral proteins and for synthesis of genomic DNA. The combination of an RNA genomic intermediate and the cccDNA form presents two barriers to HBV antiviral development. First, the reverse transcriptase of HBV polymerase shares the feature of lack of fidelity with other reverse transcriptases. This lack of fidelity allows drug-resistant mutants to develop relatively easily. Second, the cccDNA form of the genome is quite stable. Drugs that block HBV polymerase activity have no effect on the cccDNA, and the infected cells remain and resume production of virus if treatment is stopped. At this time, the only way to remove the cccDNA is to destroy the infected cell by T-cell–mediated recognition of HBV epitopes. This is a slow process; HBV treatment generally requires time periods in excess of a year. The time required for removal of nuclear genomes is often longer than the time needed for resistance mutations to HBV inhibitors to develop, necessitating changes in treatment or additional treatments.

The proteins encoded by HBV are an important diagnostic tool in monitoring the state of the disease. HBV synthesizes four groups of proteins. The polymerase is both an RNA- and a DNA-dependent polymerase. The core protein (HBcAg) is the structural protein of the nucleocapsid. The core open reading frame also encodes a longer variant, referred to as HBeAg (126–131), whose function is unclear, but it is the best barometer of the outcome of an infection. A high level of HBeAg indicates a high level of viral replication. As an effective immune response is mounted against an HBV infection, HBeAg protein disappears from the serum, and antibodies to HBeAg appear; a process referred to as seroconversion. The strongest predictor of survival is seroconversion, with clearance of HBeAg and development of antibodies to HBeAg (132). The loss of HBeAg is not a perfect barometer, however, because mutations that result in the loss of expression of HBeAg can occur. Another indicator of the state of an infection is an increased presence of hepatocyte enzymes in serum. Serum alanine transferase (ALT) levels are used as an indication of the extent of liver cell necrosis.

HBV treatment presents a conundrum. On the one hand, the presence of elevated ALT and antibodies against HBeAg is an indication of liver damage resulting from an immune response to cells carrying surface HBV antigens. On the other hand, some amount of this response is necessary to clear the liver of cells carrying nuclear DNA copies of the genome. When effective treatments for HBV replication (e.g., lamivudine) are stopped, there is often a transient increase in ALT and viral DNA levels in serum. These symptoms usually clear without treatment and are probably a result of an immune response to cells where virion production resumed when the selection pressure of the drug was removed (133).

Stimulation of the immune response to HBV by interferon-$\alpha$ (IFN-$\alpha$) is an approved treatment method for HBV. However, only a small proportion of patients have responded well to modulating the immune response alone. Thus, small molecule inhibitors of HBV replication are critical to treating HBV. Furthermore, the development of strains resistant to monotherapy has dictated the need for additional anti-HBV drugs that can be used in combination therapy.

Several species play host to closely related viruses, and two of these have seen extensive use as models for human HBV infection. The first model is infection of duck hatchlings with duck hepatitis B virus (DHBV) (134). The second is chronic infection of woodchuck with woodchuck hepatitis virus (WHV). With the exception of interferon, comparison of compounds that have been studied both in clinical trials and in the woodchuck model shows good agreement between the two with regard to relative potencies of compounds (135).

### 2.3  Papillomaviruses

The papillomaviruses were initially grouped with the polyomaviruses in the Papovaviridae family. However, the genetic and phenotypic differences between them are sufficiently great that papillomaviruses and polyomaviruses are now placed in separate families. Papillomaviruses are small double-stranded DNA viruses which can exert dramatic effects on the proliferation of the infected cell. Over 70 distinct genotypes have been identified as infecting humans (136), and this variety is reflected in a variety of phenotypes observed. Phenotypic differences between strains are manifested in two ways: first, the cell type and anatomical location that is infected, and second, the severity of the infection. Papillomaviruses are highly restricted in the cell types that they infect. They infect epithelial cells (and in some cases, fibroblasts), but their tropism is more narrowly defined in that they infect only a specific type of epithelium, e.g. genitalia, bottoms of feet or hands, etc. About 40 of these strains infect mucosal, primarily anogenital, epithelium (137). The cellular proliferation that they support takes a number of morphological forms—warts, cysts, intraepithelial neoplasias, and papillomas, etc. The fi-

nal outcome of the proliferation is a function of the infecting strain. For the so-called low risk strains, these growths are usually benign, although they can be quite debilitating. For a subset of high risk strains (strains 16, 18, 31, 33, 45, others), these proliferating cells have a high propensity to develop into tumors (138, 139). About 95% of cervical cancers are caused by the high risk strains of HPV (140–142).

Papillomaviruses gain entrance to the body through breaks in the epithelium, which expose the basal epithelial cells (143). These cells are the sites of the initial infection, and they remain latently infected. They are constantly dividing, with one of the daughter cells migrating upwards toward the epithelial surface. As they move upward, the cells begin a complex differentiation program. Only when they are well along this differentiation pathway does virion production occurs. The progeny viruses are then shed at the skin surface. The primary route of transmission of anogenital viruses is sexual contact (144–146).

As is the case for other latent viruses, the balance between viral activity and immune surveillance is an important factor in the clinical outcome. Whereas the effect of immune suppression is not as dramatic for papillomaviruses as for herpesviruses, there is nevertheless an increase in pathological severity in immunosuppressed patients (147–149).

Papillomaviruses are non-enveloped viruses approximately 55 nm in diameter (150), with a genome of 7000–8000 base pairs. The capsid is composed predominantly of the L1 coat protein, with L2 as a minor substituent. Virus-like particles (VLPs) can be assembled from recombinant L1 and L2 in the absence of DNA. These VLPs are the primary focus of vaccine development. The genome consists of three parts: a regulatory region (long control region, LCR), an early region (E) encoding the nonstructural proteins, and a late region (L) encoding L1 and L2. The LCR contains binding sites for both E2 and a number of cellular transcription regulators. These factors are both positive and negative regulators, something that is very important with regard to progression to cancer.

Papillomaviruses encode three types of potential drug targets, two of which are currently of interest for medicinal chemistry.

1. E1 and E2. E2 is a regulatory protein for both transcription of viral genes and for viral DNA replication. It recognizes a 12-nucleotide binding site and activates transcription of viral genes (151, 152). A subset of the E2 binding sites is included in the origin of replication, and E2 binding is obligatory for efficient assembly of a replication complex at the origin (153, 154). E2 also has a role in ensuring that the viral genome gets distributed properly during mitosis and maintained in the nucleus during interphase (155). In addition to positively regulating transcription, E2 negatively regulates the synthesis of E6 and E7 RNAs. In tumor cells, this control is usually (though not always) abrogated by ligation of the viral DNA into a host chromosome at a point in the viral genome, such that the E2 binding sites are displaced from the E6 and E7 genes (156). Transcription of the E6 and E7 genes is consequently up-regulated, resulting in a cascade of events that lead to cellular proliferation. This mode of action of E2 complicates its use as a drug target, because inhibiting the negative regulation could actually stimulate E6- and E7-mediated proliferation. E1 is the only contribution of the virus to the DNA replication machinery. It is a site-specific DNA binding protein that has ATPase and DNA helicase activity (157, 158), and is responsible for recognition of the viral origin of DNA replication and helping recruit cellular replication proteins to this origin (153, 159). E1 then provides the essential helicase activity for the replication complex.

2. E6 and E7. The primary effect of these proteins is to inhibit cellular proteins controlling cell division. E7 binds the pRB protein to inactivate it (160, 161), and E6 binds p53 and activates a pathway that leads to its degradation (162, 163). E6 and E7 determine the carcinogenic potential of papillomaviruses. In high risk strains (e.g., 16, 18), they form strong associations with their cellular targets. In low risk strains (e.g. 6, 11), this association is weaker (164). The low risk strains efficiently cause cell proliferation, but cells become malignant less frequently than with high risk strains. The regulation of p53 activity alone by E6 is not sufficient to account for the activity of E6. An additional effect of E6 is to form a complex with the focal adhesion protein paxillin, disrupting the cytoskeletal structure (165).

3. The structural proteins L1 and L2. While these proteins are currently of interest for vaccine development, our understanding of papillomaviral entry has not lent itself to use of these proteins as targets for small molecule inhibitors.

These targets have proven remarkably elusive because papillomaviruses have developed such an intricate relationship with the cell. The pathogenesis of these viruses occurs during their latent phase, when viral DNA is being replicated but no infective virus is produced. Furthermore, the virologist's favorite target has been taken away—there is no viral polymerase. E1 helicase is an essential part of the replication machinery, but attempts to identify effective inhibitors of the helicase have not yet been successful. The end result has been that infection has been treated with blunt instruments, either by removing most of the infected cells surgically or with cytotoxins, or by stimulating an immune response against the virus. Success by these approaches has been limited, leaving room for small molecule inhibitors of viral activities.

## 2.4  Polyomaviruses

Of the 11 viruses currently in the Polyomaviridae family, two of them, BK and JC, infect humans. They are small double-stranded DNA viruses, with a diameter of about 40 nm and genome sizes of about 5000 base pairs. The most common route of infection is likely to be respiratory (166, 167). They seem to disseminate through the body through B-cells, and viral DNA can be detected in 5–40% of healthy subjects (168, 169). The tissue tropism is different for JC and BK viruses, with BK propagating in epithelial or fibroblastic cells and JC infecting neuronal cells.

JC virus causes progressive multifocal leukoencephalopathy (PML) in immunodeficient patients (170). This disease results from the destruction of the cells that produce the myelin sheaths that surround nerves. BK virus is associated with a wider variety of disease states, although it is quantitatively less important than the JC virus. These diseases include retinitis, nephritis, pneumonia, encephalitis, and kidney diseases associated with renal transplantation (171–177).

Polyomaviruses encode three capsid proteins and two nonstructural proteins. The nonstructural proteins provide both an enzymatic activity and regulatory functions. The smaller of the two (t antigen) is dispensable for lytic infection but interacts with cellular proteins involved in cell cycle regulation (178–181). The larger protein (T antigen) is a cornucopia of activities. It is a replication protein that recognizes the origin of replication, partially unwinds it on binding to it, and attracts cellular replication proteins to the site to initiate replication (182–188). It then becomes a replicative helicase, which remains associated with the cellular replication machinery to complete replication of the viral genome (189, 190). In addition, it combines two of the activities carried out by E7 and E6 of papillomaviruses: (*1*) it binds to Rb protein and the related p107 and p130 proteins, causing the release of the E2F transcription regulator, thus promoting progression of the cell into S phase (191, 192); and (*2*) it binds to p53, marking it for degradation (193, 194).

T antigen is the only protein that polyomaviruses contribute to the replication of their genome, thus, once again a viral polymerase is unavailable. However, experience with papillomaviruses has indicated that these viruses activate poorly understood pathways that increase the sensitivity of the cellular polymerases to certain polymerase inhibitors, the current example being cidofovir. This activity thus provides some selectivity of the drug for infected cells, even though it is targeting a cellular polymerase. The possibility exists that similar mechanisms will exist for polyomaviruses.

Binding and entry into the cell are poorly understood for polyomaviruses. JC virus seems to bind to a sialic acid–containing surface protein (195), and there is a report that BK virus may bind to phospholipids (196). If these observations are related to viral entry, it is unlikely that they will be the entire story;

there are likely to be multiple receptor factors. The bound virus enters the cell by endocytosis and is transferred to the nuclear membrane, where the DNA genome becomes uncoated and transferred into the nucleus. The nonstructural proteins are synthesized as early proteins, and DNA synthesis and late protein synthesis follow. Capsids are assembled in the nucleus (197, 198). Packaging of DNA and egress from the cell are not well understood for polyomaviruses.

## 2.5 Adenoviruses

Adenoviruses infect multiple tissue types and are the causative agent of an estimated 3% of respiratory disease in humans (199). Antibodies to adenoviruses can be found in about 50% of children (200), and adenoviruses have been detected in 5–50% of pediatric patients presenting with lower respiratory tract infections (201, 202). The virus enters its host through the mouth, nose, or eyes, and the primary sites of replication are respiratory epithelium, the eyes, and the gastrointestinal tract. Low levels of viral replication can persist for months to years, but the mechanisms supporting this persistency are not well understood (203–205).

Whereas adenoviruses are a less significant problem during immunosuppression than some other classes of viruses, immune dysfunction can exacerbate adenovirus infections. A study of 572 transplant patients indicated that 17% had an active adenovirus infection in the 150 days following transplantation; six of these infections were lethal (206). Adenovirus infection is more likely to lead to pneumonia in immunocompromised patients than in immunocompetent individuals (207–211).

Adenoviruses were discovered about 50 years ago and they grow well in cell culture, factors that led to their development as model viruses for studying both cellular and viral processes, including gene expression and regulation, RNA splicing, and cell cycle control. Adenovirus and papovavirus SV40 were key models for determining the enzymology by which semiconservative DNA replication occurs in eukaryotes (212). The discovery that human adenoviruses could cause malignant tumors in rodents led to intensive searches for adenovirus DNA in human tumors (213), but there is no evidence that adenovirus causes malignancy in the species from which it was derived. In recent years, much effort has gone into developing adenoviruses as gene therapy vectors (214–217). However, they have not received much attention with regard to development of chemotherapeutic agents.

Adenoviruses are non-enveloped viruses measuring 70–100 nm in diameter. They have an outer shell composed of seven peptides and an inner core with four peptides (218, 219). The linear double-stranded DNA genome consists of approximately 36,000 base pairs. They are rapidly replicating viruses, with a life cycle of about 24 h in cell culture. Binding and entry of adenoviruses is relatively well understood. Two cellular proteins, the CAR protein (a member of the immunoglobulin superfamily) and a member of the integrin family, form the cellular receptor (220, 221). The virus is internalized through endocytosis (222–224), released into the cytoplasm, and transported to nuclear pores, where the viral DNA is released into the nucleus.

Entry of the DNA into the nucleus is followed by early mRNA synthesis, the products of which both cause the cell to enter S phase and are needed for viral DNA synthesis. Adenoviruses encode their own DNA polymerase (225, 226). Replication requires minimally three viral proteins and two cellular proteins and differs significantly from the host cell DNA replication. Late protein synthesis is timed to begin at the time of DNA replication. The late gene products are assembled into capsids in the nucleus, after which a DNA molecule is packaged in the capsid (227). Enzymes for packaging the DNA into capsids and a virally encoded protease are both needed for assembly and maturation of the virions (228, 229). These are both steps that have been successfully employed as drug targets in herpesviruses.

## 2.6  Parvoviruses

The Parvoviridae family contains several significant animal pathogens, including one human pathogen, B19. Antibodies to B19 have been found in 30–50% of humans (230–233), indicating that exposure to the virus is much broader than the pathogenesis associated with

it. However, it is associated with several disease states, including red cell aplasia, fifth disease (often confused with measles in children), and fetal wastage (232, 234–238). Acute infection of adults results in nonspecific influenza-like symptoms, followed about a week later by symptoms of fifth disease (239–241). The primary cell types infected are of the erythroid lineage (242, 243). In immunocompromised patients, the most significant complication is red cell hypoplasia (244, 245).

Parvoviruses are the smallest of the human DNA viruses, with virions approximately 20–25 nm in diameter. The potential use of parvoviruses as gene therapy vectors (246, 247), as well as the existence of B19, has spurred interest in this family of viruses. Parvoviruses are non-enveloped viruses, with a capsid composed of either two or three proteins. This capsid contains one copy of the approximately 5000 base single-stranded genome (248). Both ends of the DNA contain palindromic sequences that are important for DNA replication (249–251). The DNA contains two large open reading frames, each of which codes for two to four proteins by a combination of differently spliced products and different translational start sites (248, 252–254).

As might be expected based on their size, parvoviruses are among the most dependent of viruses on host cell functions. They require that the host cell enter S phase to replicate, but unlike herpesviruses, papillomaviruses, or polyomaviruses, they do not encode a mechanism for activating the cell to do so (255–257). Two mechanisms are used by parvoviruses to bypass this block to productive viral infection, and the viruses fall into two genuses based on which approach they use. The first is to simply wait for the cell to enter S phase, the approach taken by the autonomous parvoviruses, including B19. The second is to depend on a helper virus, such as adenovirus or a herpesvirus, to activate the cell; this is the approach adopted by the helper virus–dependent parvoviruses. Adeno-associated virus (AAV) is the prototype for this genus of parvoviruses. AAV integrates into a site on chromosome 19 after infection, where it remains until it is rescued by an adenovirus or a herpesvirus (258, 259). Thus, it has a mechanism for maintaining latency. Whereas autonomous parvoviruses can develop a persistent infection, there is currently no evidence for a latent state.

Parvoviruses are relatively, though not entirely, species specific (260). Determinants on the capsid proteins needed for entry have been mapped, and for B19, the cellular receptor is globoside, the blood group P antigen (261). Entry is mediated through an endocytotic pathway, and the viral DNA is delivered to the nucleus. Replication of the viral DNA is through a mechanism that relies on the palindromic sequences for initiation of synthesis. The salient features of parvovirus replication from a medicinal chemistry perspective are as follows: (1) DNA synthesis is by the cellular DNA polymerases; (2) replication generates concatamers of the genome, which must be cleaved; and (3) the virus supplies the helicase activity. Both of the latter activities come from the NS1 protein in B19, or the Rep 78/68 protein in AAV (262–265). After translation of the coat proteins, they self-assemble in the nucleus, where the genomic DNA is packaged.

### 2.7 Poxviruses

Poxviruses are unique among human DNA viruses because they are complex DNA viruses that replicate entirely in the cytoplasm of host cells, in structures constructed by the virus (266). Their dimensions are approximately 350 by 270 nm, with genome sizes ranging from 130,000 to 300,000 base pairs (267). These genomes encode up to about 200 open reading frames (268–275). They contain virally encoded systems for carrying out both transcription and replication of viral genes and complex schemes for regulating these processes. In short, there are more potential drug targets encoded by poxviruses than for any other virus discussed in this chapter. However the extent of development of antipoxviral therapies has been limited, and their biology will therefore not be outlined in more detail here. It is reviewed by Moss (276).

Before the 19th century, poxviruses were a serious human health threat in the form of smallpox. With the introduction of vaccines and a campaign against smallpox on the part of the World Health Organization, smallpox was declared eradicated in 1980. Whereas laboratory stocks still exist, the last documented

infection was reported almost 25 years ago. Nevertheless, there are scenarios whereby poxviruses could become a health threat (277).

1. It would be a formidable biological weapon. A generation has grown up unvaccinated, and therefore susceptible, and current vaccine stocks are likely insufficient to stop its spread.
2. Related poxviruses may have the potential to become more virulent. As an example, human monkeypox outbreaks have occurred in central Africa within the last decade. Increased virulence or host range of this or a related poxvirus could have a significant health impact.
3. Recombinant vaccinia viruses have clinical potential as vehicles to deliver antigens from pathogens. However, in immunocompromised patients, vaccinia virus itself could become pathogenic.

While poxviruses may not generate much activity as drug targets in the near future, it is worth remembering that they are not extinct. They have generated significant human pathogens in relatively recent history. It is possible that others could again emerge from this family, a scenario played out from the retrovirus family with devastating consequences by HIV-1 in the 20th century.

## 3 ANTIVIRAL COMPOUNDS FOR DNA VIRUSES

The majority of approved drugs and compounds under development for DNA viruses are nucleoside analogs, although there are exciting developments in other target areas that will be discussed. This discussion is organized around the virus families. The currently available drugs are discussed first, followed by compounds under development. A factor of increasing concern with regard to treating DNA viruses is the development of resistance to drugs, and this topic will be discussed for currently available drugs.

### 3.1 Herpesviruses

**3.1.1 Currently Approved Drugs.** This section starts with acyclovir and its progeny, and follows with other nucleoside analogs. Foscarnet is not a nucleoside analog, but functionally it shares many of their features and will be discussed with them. Whereas a wide variety of antiherpesvirus nucleosides have been synthesized, much anti-herpesvirus drug development has focused on guanosine and thymidine analogs. The root of thymidine analog development was idoxuridine (5-iodo-2'-deoxyuridine, IDU) (**1**) (278), which was licensed for topical treatment of herpesviruses. Acyclovir (ACV) (**2**) is both the ancestor of and the par-

(**1**)

(**2**)

adigm for development of many of the purine nucleoside inhibitors. Discovered in the 1970s, it demonstrated a combination of metabolic stability and low toxicity that provided the springboard for development of a series of antiherpesviruses drugs (see Ref. 279 for a review). The relatively low toxicity derives from three sources. First, ACV is an analog of 2-deoxyguanosine, which must be activated by conversion to the triphosphate. Phosphorylation to ACV monophosphate is carried out efficiently by the thymidine kinases of $\alpha$- and $\gamma$-herpesviruses but very poorly by cellular en-

zymes. These thymidine kinases, unlike their cellular counterparts, bind a wide range of nucleosides in addition to thymidine. Second, the first phosphorylation results in localization of the monophosphorylated form to infected cells because the charged phosphate prevents translocation across the plasma membrane. Thus, ACV is concentrated in the infected cells. Third, ACV is recognized more efficiently by viral than cellular polymerases, providing another source of selectivity.

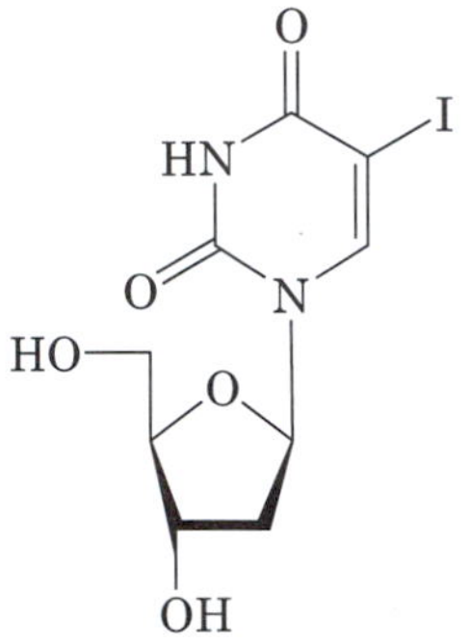

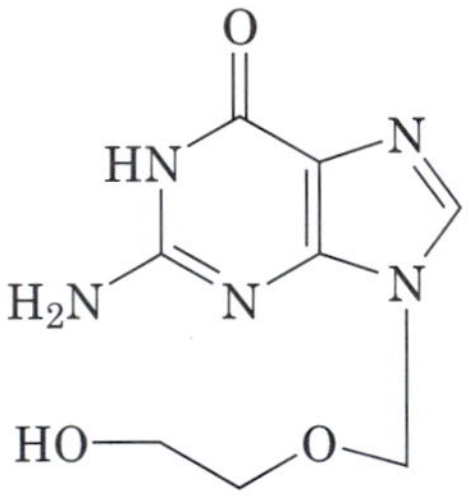

Further phosphorylation of ACV by cellular kinases leads to the triphosphate, which is incorporated into DNA by the viral DNA polymerase. Because it lacks an acceptor for the incoming phosphate of the next nucleotide, ACV acts as a chain terminator. Specifically, work of Reardon and Spector demonstrated that ACV is a competitive inhibitor with respect to deoxyguanosine triphosphate (dGTP), and that incorporation of ACV results in a large increase in $K_m$ of the next incoming base (280).

Acyclovir has been a very successful drug, with good potency against HSV and VZV and with few side effects. The attractive features of the ACV paradigm are delivery of the drug in an inactive form, which is activated primarily in infected cells and selectivity for the viral DNA polymerase over cellular enzymes. However, the specificity provided by the requirement for a viral kinase, along with the relatively low bioavailability of ACV, limit the diseases that it can be used to treat. ACV is a standard treatment for HSV-1, HSV-2, and VZV, but weakly effective against HCMV. HCMV does not encode a thymidine kinase, although a putative protein kinase, UL97, is capable of phosphorylating ACV at low levels. ACV is also ineffective against HSV and VZV strains that are deficient in thymidine kinase activity. The viral thymidine kinases are attractive drug targets because they are well studied, both functionally and structurally (reviewed in Ref. 281).

Additionally, ACV has a low oral bioavailability (15–30%). Increasing oral bioavailability of ACV would allow lower or less frequent doses. Prodrug forms of ACV and its analogs were developed that provide substantial increases in bioavailability. Valaciclovir (VACV) is a valyl ester of ACV that improves bioavailability to greater than 70%. The currently available progeny of ACV will be discussed with regard to improving on these two limitations of ACV.

*3.1.1.1 Range of Action-Modifications to ACV and Development of Unrelated Drugs.* Relatively small changes to the acyclic side chain of ACV have resulted in two additional drugs, ganciclovir (GCV) (**3**) and penciclovir (PCV) (**4**) that differ in their effectiveness

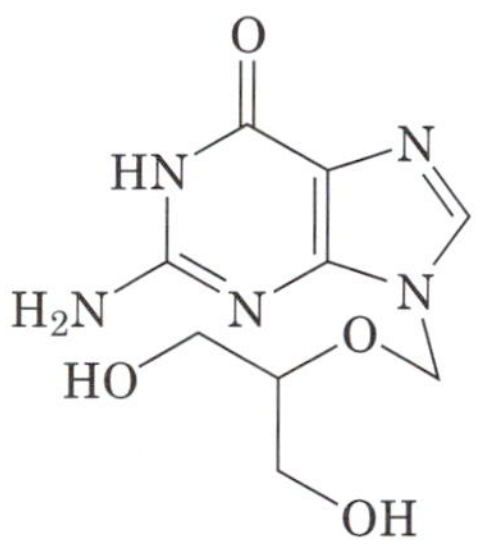

(3)

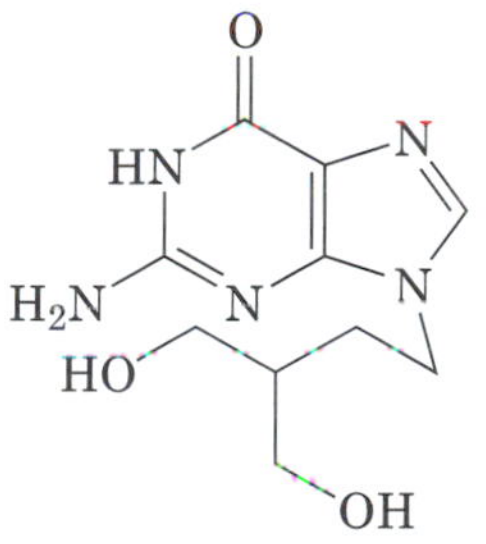

(4)

against herpeviruses. These differences result from changes in the efficiency of recognition by viral kinases and viral DNA polymerases.

Addition of a hydroxymethylene group to the acyclic side chain of ACV yielded GCV. GCV has antiviral activity against all herpesviruses, although its potency varies between viruses. GCV is more effective against HCMV than against HSV or VZV (282), the reverse of ACV. It is phosphorylated by both viral thymidine kinases (283) and by HCMV UL97 kinase (284, 285). HCMV phosphorylates GCV relatively efficiently, whereas it is very poor at phosphorylating ACV. HHV-6 and HHV-7 are β-herpesviruses with genetic similarity to HCMV. GCV is a potent inhibitor of HHV-6,

but it is a weak inhibitor of HHV-7, indicating that there are differences within the class of β-herpesviruses (286–290). GCV does have efficacy against thymidine kinase-deficient VZV strains, because of the fact that VZV encodes a UL97 homolog (ORF47) (291). GCV-triphosphate has a half-life sixfold longer than that of ACV-triphosphate, which results in higher intracellular concentrations of GCV-triphosphate than of ACV-triphosphate (reviewed in Ref. 292). This difference in triphosphate concentrations gives GCV efficacy against HSV, even though ACV-triphosphate is recognized by HSV DNA polymerase with higher affinity. GCV is also active in inhibiting HBV replication. The mechanism of action of GCV differs somewhat from ACV. Whereas ACV terminates the growing DNA chain, GCV seems to slow down the rate of elongation, while at the same time increasing the amount of repair.

The factors limiting GCV efficacy are its low bioavailability (<10%) and its side effect profile. Maintaining serum levels effective for treatment by oral administration of GCV is difficult; therefore, the initial treatment course involves intravenous injections twice daily for 2–3 weeks, followed by daily injections or oral dosing. The prodrug form valganciclovir improves the oral bioavailability of GCV (see Section 3.1.1.2), although it still has the toxicity profile of GCV (293). Despite the existence of at least four treatments for HCMV, there is still an unmet need for an effective, orally available drug with a limited side effect profile.

Intravenous delivery of GCV has been of value in treating HCMV retinitis, a complication of AIDS that often results in blindness. GCV is also efficacious for infections of the gastrointestinal tract and central nervous system. For HCMV retinitis, an additional option is to treat by implanting a slow release pellet in the eye. The pellet maintains high concentrations of GCV in the eye for a period of 7–8 months, with very low systemic concentrations. This results in effective treatment of the retinitis with low side effects. However, because HCMV causes a systemic infection, the localized treatment of the implant leaves the patient at risk for retinitis in the other eye or for gastrointestinal or neural complications.

Thus, treatment with implants is often combined with oral GCV to limit systemic viremia. Treatment options for HCMV retinitis are reviewed by Hoffman and Skiest (294).

The trade-off for the larger number of treatable viruses is toxicity—GCV is more cytotoxic than ACV and must be managed more carefully when used as a maintenance therapy. This aspect is critical because effective treatment requires long-term administration of the drug. The cytotoxicity is a function of the fact that GCV is phosphorylated more readily by cellular kinases than is ACV.

Penciclovir (**4**) shares structural similarity with GCV but is closer to ACV in its antiviral spectrum (295). It is phosphorylated by HSV thymidine kinase, and thus is inactive in kinase-deficient strains. However, the triphosphorylated form retains activity against ACV-resistant HSV polymerase. HSV thymidine kinase has a much higher affinity for PCV than ACV ($K_i$ = 1.5 versus 173 $\mu M$), but the triphosphate form is used about 40-fold less well by the HSV and VZV polymerases than ACV triphosphate (296–298). On the other hand, PCV triphosphate is approximately 10-fold more stable than ACV triphosphate, such that PCV-triphosphate levels are over 300-fold higher than ACV-triphosphate levels. In cell culture, the drugs were about equally inhibitory when present continuously [$EC_{50}$ = 3.8 and 4.2 $\mu g/mL$ against VZV (298)], but PCV retained antiviral activity for longer on withdrawal. The weak activity of PCV against HCMV suggests that it is not phosphorylated efficiently by the UL97 kinase.

PCV is activated at a low level by cellular kinases. This, combined with a greater than 4000-fold greater affinity for the HBV polymerase than the cellular polymerase, makes PCV an inhibitor of HBV replication. Oral bioavailability is about 5%; however, an effective prodrug form, famciclovir, has been developed (see below).

Cidofovir [9-(3-hydroxy-2-(phosphonomethoxy)propyl)cytosine, HPMPC] (**5**) is a pyrimidine phosphonate analog. Because it does not rely on a viral kinase to generate a monophosphate form, it is active against more viruses. In addition to HCMV and other herpesviruses, it is active against papillomaviruses, adenoviruses, and poxviruses. Because oral

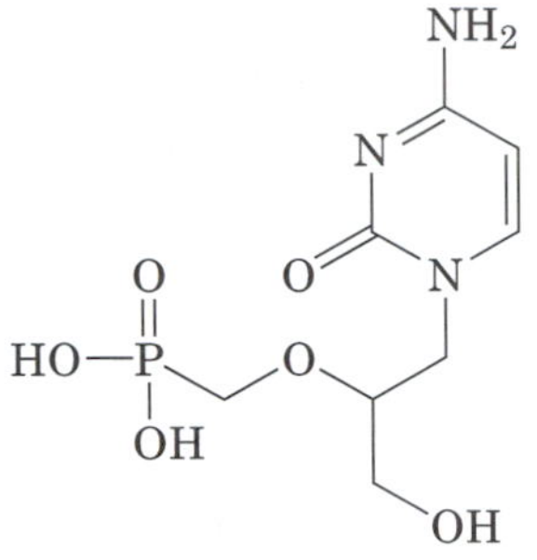

(5)

bioavailability is less than 5%, intravenous injection is required for systemic treatment (293). This is countered somewhat by its slow turnover, which allows longer periods between dosing. The primary factor limiting treatment with cidofovir is renal toxicity, which can be ameliorated somewhat by treatment with probenecid, hydration, and longer dosing intervals (299, 300). In a phase I/II study of its anti-HCMV effect, cidofovir was effective at inhibiting HCMV replication at a dose of 3 mg/kg weekly. The dose-limiting toxicity was renal damage (301).

Foscarnet (phosphonoformic acid) (**6**) is not a nucleoside analog, but rather an analog

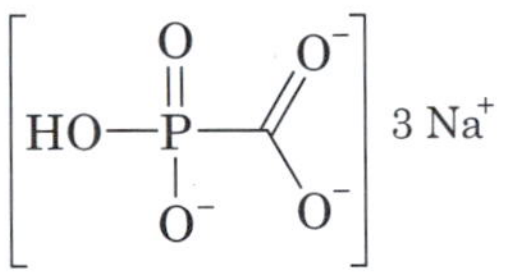

(6)

of the leaving group, pyrophosphate. It is active against all of the herpesvirus DNA polymerases but must be administered intravenously. Effectiveness against HCMV is equivalent to GCV, and the magnitude of side effects is similar. Side effects include renal toxicity and electrolyte imbalances (293). GCV has the advantage of a prodrug form, valganciclovir, with an oral bioavailability of 60%, whereas there is no prodrug from of foscarnet. However, foscarnet can be useful as a salvage therapy for HCMV infections that do not respond to GCV.

Brivudin (BVDU) (**7**) is a selective inhibitor of herpesviruses, with the selectivity deter-

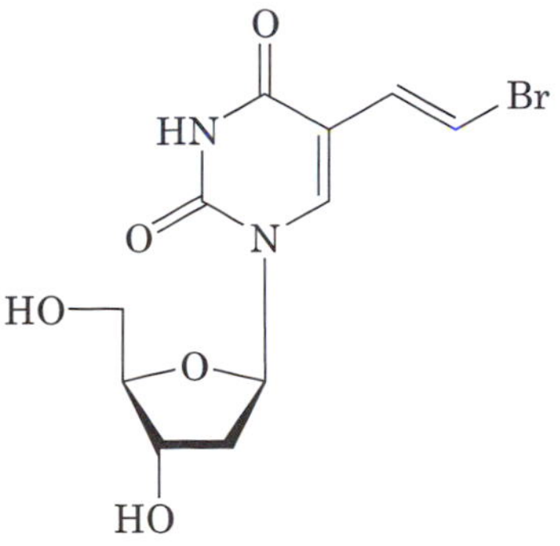

(7)

mined by the capacity of each viral kinase to convert it to the monophosphorylated form. BVDU is active against HSV-1, EBV, and is a very potent inhibitor of VZV, but is inactive against HSV-2, HCMV, HHV-6, and HHV-7 (302, 303). Orally administered brivudin was shown to be as effective as intravenous ACV in treating VZV in immunocompromised patients (304, 305). However, metabolic instability of BVDU has led to further development of analogs, discussed later.

Vidarabine (araA) is an adenosine analog, with the D-ribose replaced by D-arabinose (8).

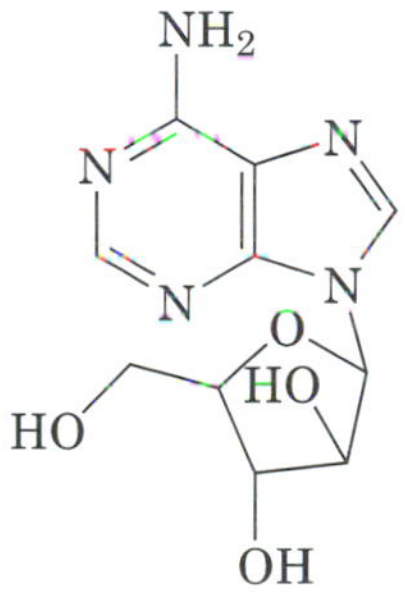

(8)

It is active against more viruses than ACV or GCV. However, it is also more toxic and less metabolically stable; it is prone to deamination by adenosine deaminase (303, 306–308). The deaminated product is at least 10-fold less potent than araA (309). Unlike ACV, araA is not preferentially phosphorylated in virus-infected cells. However, the half-life of the triphosphate form of araA (ara-ATP) is ap-

proximately threefold longer in HSV-infected cells compared with uninfected cells (310).

Resistance mutants of araA map to the DNA polymerase. AraA inhibits DNA polymerase activity through competitive inhibition with dATP. Ara-ATP is a substrate and can serve as an acceptor for the next base, but does so inefficiently, indicating that the mechanism of action of araA is both direct inhibition of viral DNA polymerase and a decrease in the rate of elongation by increasing the amount of excision repair needed (311). Because of its more rapid turnover and poorer toxicity profile, ACV and its derivatives have superseded ara A's use, but it remains a platform for further development.

Fomivirsen is a 21-nucleotide phosphorothioate oligonucleotide complementary to a region of the mRNA encoding immediate-early proteins of HCMV. Its mode of action is to bind to the mRNA and provide a substrate for a cellular RNase that destroys the mRNA. It has been approved for use in treating HCMV retinitis (312). In cell culture, it has an $EC_{50}$ between 0.03 and 0.2 $\mu M$ in fibroblasts. Using retinal pigment epithelial cells as the cell culture model rather than fibroblasts gave a sixfold increase in potency (313). Clearance studies in rabbits have indicated a half-life of 62 h; it is applied to patients intravitreally on a weekly basis. However, its use may be limited because of retinal toxicity (314).

*3.1.1.2 Prodrugs that Increase Oral Bioavailability.* The other limitation to ACV and the related drugs discussed earlier is their low oral bioavailability. A number of prodrug approaches have been used to increase oral bioavailability. Esterification of the acyclic side chain has been the most successful. For ACV, esterification with valine to give valciclovir (VACV) (9) resulted in a threefold improvement in oral bioavailability, with a spectrum of action and toxicity profile similar to ACV. Acetal derivatives have also led to efficacious compounds. The diacetal derivative of PCV, famciclovir (FCV) (10) is approved for treatment of HSV-1 and HSV-2 diseases. Oral bioavailability for FCV is 70%, compared with 5% for PCV. Other prodrug approaches are discussed in Section 3.1.2.

*3.1.1.3 Resistance in Herpesviruses.* Drug resistance of herpesviruses is not a significant

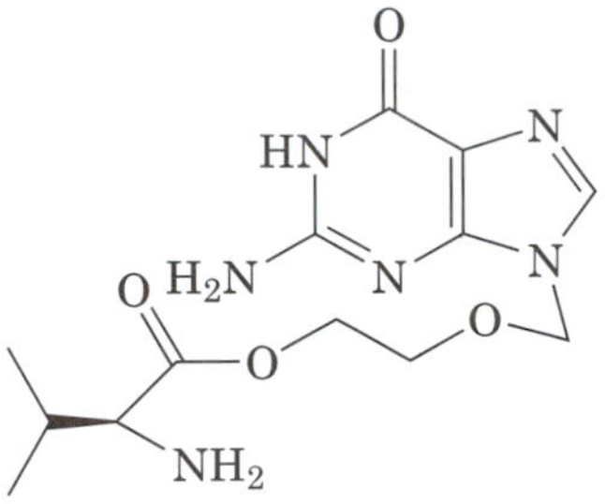

(9)

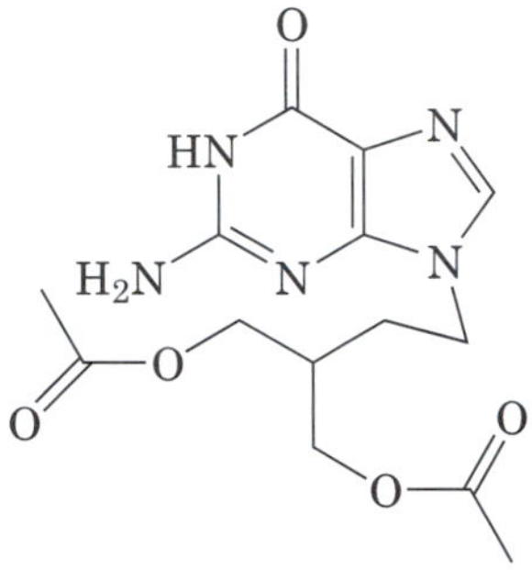

(10)

problem in immunocompetent hosts. Resistant strains do become a threat in immunocompromised patients, however, and this population is increasing in size because of suppression of the immune system in transplant patients and coinfection with HIV. This is particularly important for HCMV, where infection is largely asymptomatic except in immunocompromised patients.

Patterns of resistance mutations generated by currently available drugs have been well documented (315–317). Phosphorylation of cidofovir and foscarnet by UL97 is not required, so mutations occur in the DNA polymerase. The majority of isolates of GCV-resistant mutants map to the UL97 kinase. However, with long-term use of GCV, or coadministration with cidofovir or foscarnet, mutations in the DNA polymerase are noted. The mutations tend to confer resistance to both GCV and cidofovir, and occasional resistance to foscarnet is also observed (317).

As with GCV in HCMV, resistance to ACV (and V-ACV) in HSV or VZV is most often a result of mutations in the viral kinase. This resistance takes two forms: mutations that al-

ter the specificity of the kinase or loss of the expression of the enzyme. The same applies to PCV and FCV. ACV- and PCV-resistant strains are also generally cross-resistant, although the actual amino acid mutations selected by ACV and PCV differ (318).

Cross-resistance of drugs is an important issue in deciding salvage therapies when resistance occurs. Nucleoside analogs related to ACV (GCV, PCV, and their prodrugs) are cross-resistant, either because of a dependence on thymidine kinase for monophosphorylation or because they select for mutations in DNA polymerase that provide cross-resistance. Foscarnet is an alternative to these nucleoside analogs. Cidofovir also selects a different set of polymerase mutations, so that cidofovir-resistant strains remain sensitive to both foscarnet and ACV (319).

### 3.1.2 Compounds Under Development.

**3.1.2.1 Nucleoside Analogs.** ($R$)-9-(4-Hydroxy-2-(hydroxymethyl)butyl)guanine (H2G, omaciclovir) (**11**) is an acyclic guanosine analog

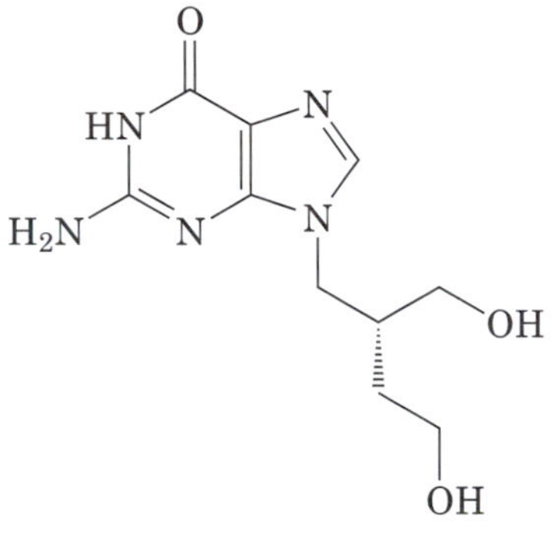

(11)

similar in structure to ACV. H2G is roughly comparable in activity to ACV against HSV-1, 20- to 400-fold more active against VZV, and somewhat less active against HSV-2 (320, 321), with a toxicity profile similar to ACV. H2G is efficiently phosphorylated by viral thymidine kinases (322), and the triphosphate form is maintained at levels more than 170-fold higher than for ACV in VZV-infected cells (321). The half-life of H2G ranged from 4 to 14 h, versus 1–2 h for ACV. H2G is a competitive inhibitor of dGTP incorporation by the viral DNA polymerases and is poorly recognized by human DNA polymerase $\alpha$. H2G seems to function by a mechanism analogous to GCV, in that it does not chain-terminate but rather inhibits progressive elongation by the polymerase. H2G is cross-resistant with ACV, and like ACV, resistance mutations accumulate first in the thymidine kinase (321). This compound is currently in phase II clinical development.

McGuigan et al. (323–325) constructed a potent and highly selective series of VZV inhibitors based on a bicyclic furopyrimidine base (**12**). When the R group was an 8–10 car-

(13)

(14)

(12)

bon alkyl group, these compounds were up to 100 times more potent in cell culture assays than ACV. A terminal halogen was tolerated with no loss of activity. Use of a phenyl group *para*-substituted with an alkyl chain increased potency a further 100-fold, with selectivity indexes of more than $10^6$. The optimal chain length decreased from 8–10 to 4–7 with the insertion of the phenyl group. These compounds were inactive against other herpesviruses and required the VZV thymidine kinase for activity.

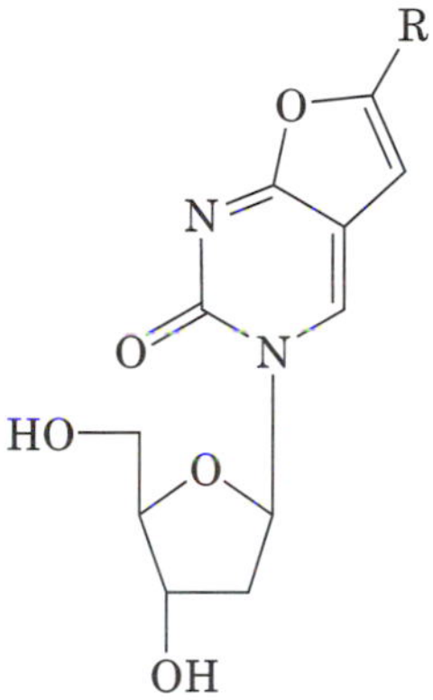

Modification of the side chain of ACV or GCV, forming the 9-alkoxy derivative (**13**), (**14**), produces isomers that retain activity against herpesviruses. The ACV isomer is three- to fivefold more active than ACV against HSV and VZV, whereas the GCV isomer is equivalent to ACV in activity (326). Bioavailability of these compounds is low, although it might be improved by using a

prodrug form. The length of the acyclic side chain is critical for recognition by the kinase, and 8-methyl purines are inactive, as are analogs containing substituents other than the hydroxymethylene group at the 2' position (327).

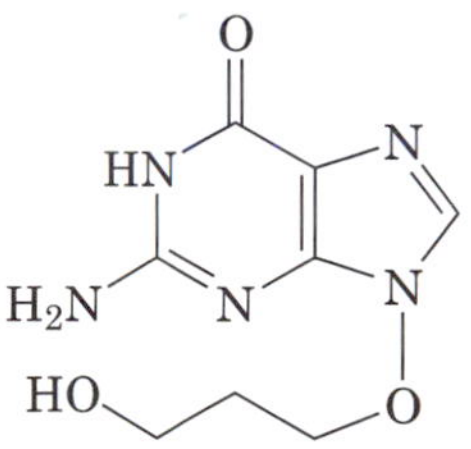

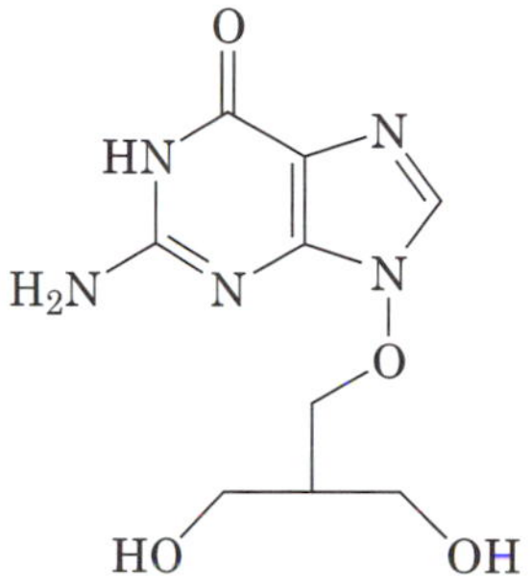

In contrast to L-nucleoside analogs, which have been shown to have activity against HIV reverse transcriptase and HBV polymerase with useful therapeutic indices, D-isomers have demonstrated potent antiviral activity, but with excessive toxicity. Variations on this theme, apio dideoxydidehydronucleosides (**15**) have been evaluated as herpesvirus inhibitors.

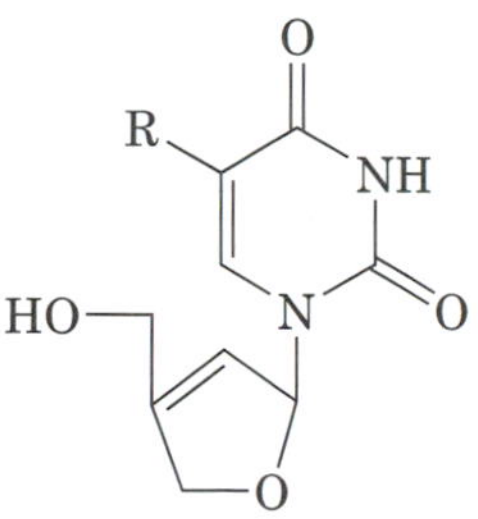

R = H, CH$_3$ or halogen

(15)

These compounds have the furanose oxygen and C2 transposed. They showed moderate activity against HCMV ($EC_{50}$ = 3–44 vs. 0.9 $\mu$g/mL for GCV) and were inactive against HSV-1 and HSV-2. Cytotoxicity was minimal.

A series of 2'-deoxy-2'-fluoro-4'-thioarabino-furanosyl nucleosides (**16**) were tested as HSV

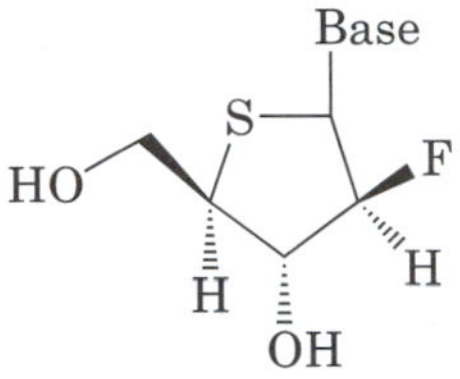

(16)

inhibitors (328). The $\beta$-anomers of pyrimidines were largely active against HSV-1 and HSV-2, with $EC_{50}$ against HSV-1 ranging from 0.012 to 27 $\mu$g/mL (ACV $EC_{50}$ = 0.32 $\mu$g/mL; Table 9.1). Of the purines, the 2,6-diaminopurine and guanine derivatives were highly active ($EC_{50}$ = 0.0041 $\mu$g/mL; HSV). The $\alpha$-anomers were largely inactive. Except for the thymine analog, selectivity indices were greater than 100.

The D- and L-enantiomers of cylohexenyl-guanine and -adenine (**17**) were synthesized as part of a study of how sugar analogs bind to nucleoside kinase and polymerase enzymes

(329). The adenosine analogs were inactive against HSV, HCMV, and VZV. Both guanosine enantiomers had activity comparable with ACV against HSV and VZV and activity comparable with GCV against HCMV. Selectivity indices for the D-enantiomer were >1000 for HSV and ranged from 7 to 40 for VZV and HCMV; the L-enantiomer showed two- to fivefold less selectivity than the D-enantiomer. Both compounds tended to be about fivefold less active against thymidine kinase–deficient strains, but they retained significant activity.

Borrowing from the observation of potency against HIV-reverse transcriptase of L-nucleoside analogs with lower toxicity than the D-isomers, synthesis of analogs of these compounds identified 2'-fluoro-5-methyl-$\beta$-L-arabinofuranosyluracil (L-FMAU) (**18**) as a compound with antiviral activity. It has been studied primarily as an anti-HBV inhibitor, but it is also an inhibitor of EBV (330, 331), with an $EC_{90}$ of 5 $\mu M$, equivalent to GCV (330). Its selectivity index was 180. L-FMAU showed no activity against HSV.

(1'S)(2'R)-9-{[1',2'-Bis(hydroxymethyl)-cycloprop-1'-yl]methyl}guanine (A-5021) (**19**) is a competitive inhibitor of HSV DNA polymerases with respect to dGTP (332). It is more potent than ACV against HSV-1, HSV-2, VZV,

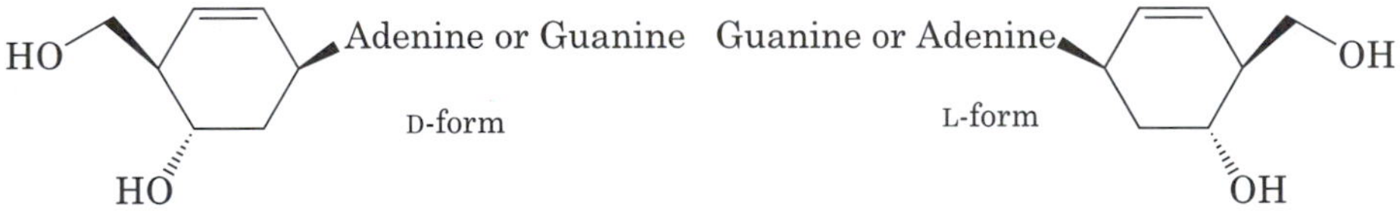

(17)

**Table 9.1   Inhibition of HSV-1 by 2'-deoxy-2'-fluoro-4'-thioarabinofuranosyl Nucleosides**

| Base | $EC_{50}$ ($\mu$g/mL) |
|---|---|
| Thymine | 0.32 |
| 5-Ethyluracil | 0.012 |
| 5-Iodouracil | 0.037 |
| 5-E-bromovinyluracil | 0.032 |
| 5-Chloroethyluracil | 0.11 |
| 5-Flurouracil | >27 |
| 5-Iodocytosine | 0.037 |
| 5-Ethyluracil | 1.0 |
| 5-Hydroxyethyluracil | 27 |
| ACV | 0.32 |

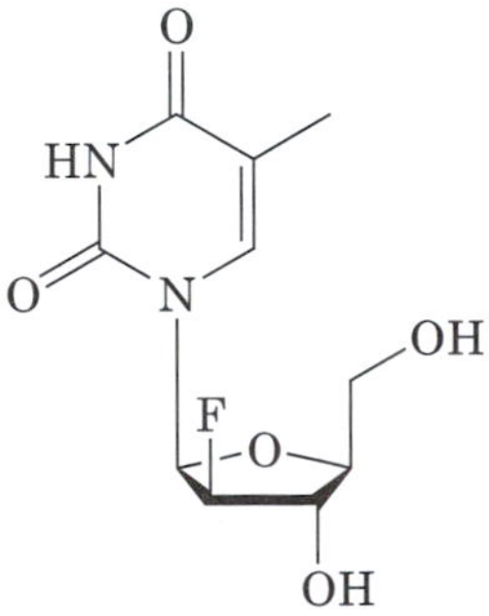

(18)

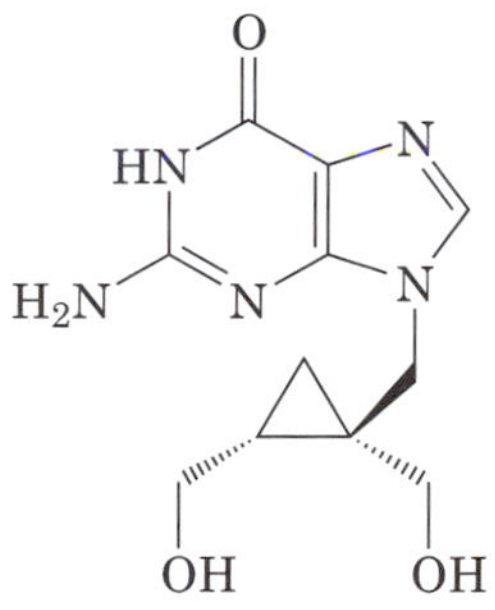

(19)

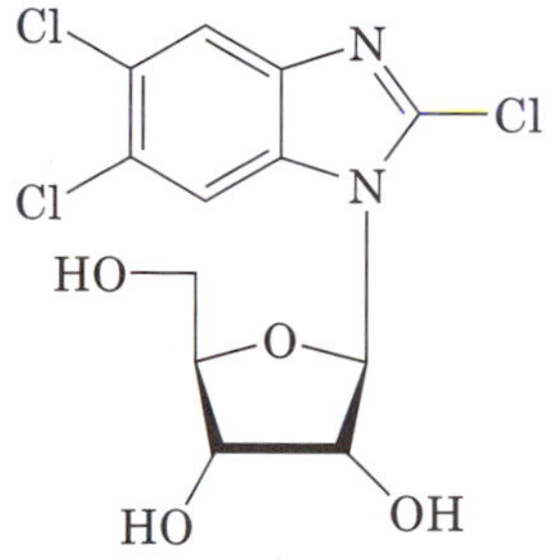

(20)

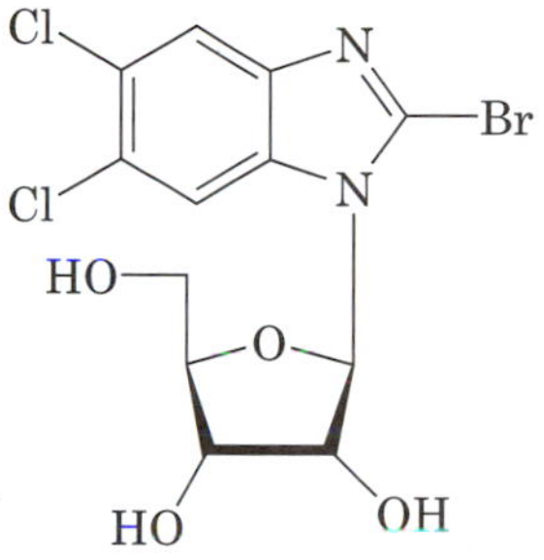

(21)

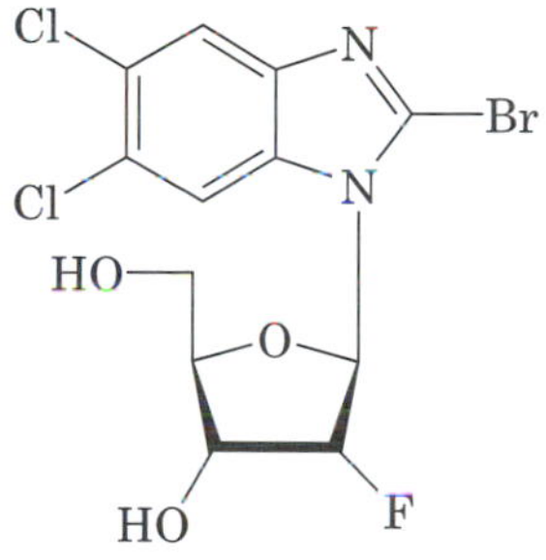

(22)

EBV, and HHV-6, but it is inactive against HHV-8. A-5021 was more effective than ACV at preventing intracutaneous and intracerebral HSV-1 infections in mice.

Modifications to the base of ACV have been made with varying degrees of success. Addition of a methyl group at N-1 of ACV reduces activity 5- to 20-fold. Etheno-ACV is inactive, but the 6-methyl- and 6-*tert*-butyl derivatives retain weak activity. Addition of a phenyl group at this position restores activity, particularly for the GCV analog, and results in a compound that shows activity against nucleoside kinase-deficient HSV and VZV. The biphenyl derivative show both reduced activity and narrower specificity, acting only on HSV. ACV analogs containing a 2-(3,5-dichloro anilino) substituent inhibit HSV-1 and HCMV replication but also inhibit cellular DNA polymerases at approximately the same concentration.

A series of benzimidazole and indole nucleosides have demonstrated activity against HCMV. 2,5,6-Trichloro-1–($\beta$-D-ribofuranosyl) benzimidazole (TCRB) (**20**) and 2-bromo-5,6-dichloro-1-$\beta$-D-ribofuranosyl benzimidazole (BDCRB) (**21**) are potent inhibitors of HCMV (333), but have half-lives of less than 1 h because of cleavage of the glycosidic bond. Introduction of a fluorine on the ribose to act as a stronger electron-withdrawing group (**22**) led to an approximately 10-fold loss of potency and an increase in cytotoxicity (334, 335). Removal of 3-N to give the corresponding indole nucleosides also caused a 10-fold loss in activity (336). A trichloro-tricyclic analog (**23**) was about as active as TCRB, but more cytotoxic,

suggesting that the tricyclic structure resulted in a relative increase in affinity for cellular enzymes (337). Replacement of the D-ribofuranose with a D-erythrofuranose gave a 10-fold increase in activity relative to TCRB and 40-fold greater potency than GCV. An iodine or amino group at the 2 position resulted in considerably less activity. BDCRB was inactive against EBV, but the related compound, 1263W94 was active (**24**), with $EC_{50}$ ranging from 0.15 to 1.1 $\mu M$ (338) 1263W94 also showed ninefold greater potency against clinical isolates of HCMV than GCV and was ef-

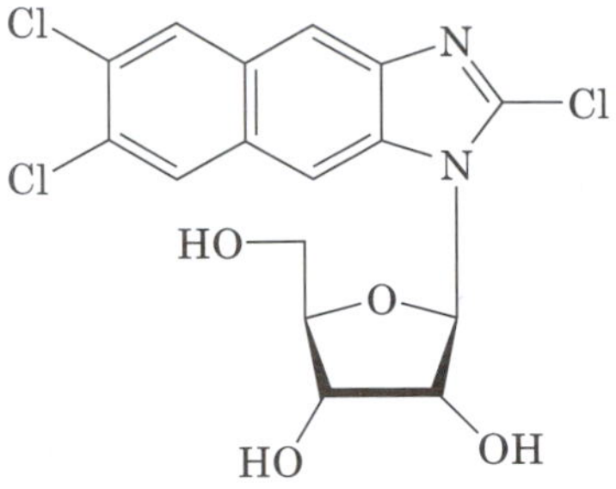

(23)

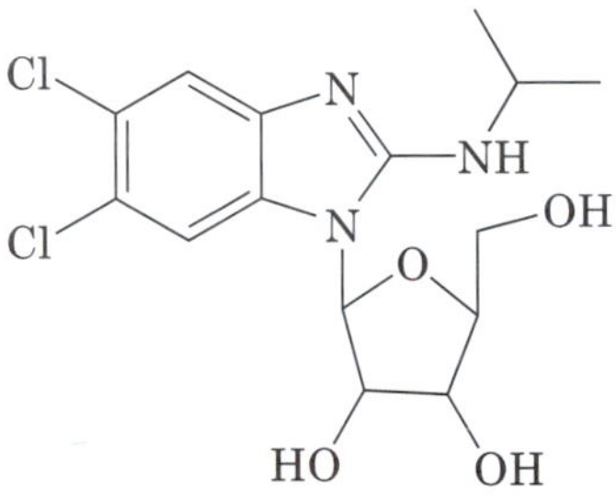

(24)

fective against GCV-resistant isolates (339). 1263W94 differs from BDCRB in having an L-ribose and a 2-isopropylamino group instead of a 2-bromine.

Several α-D- and α-L-lyxofuranosyl derivatives of 1263W94 and TCRB were synthesized and tested against HSV-1 and HCMV (340). None of the compounds were efficacious against HSV-1. The 2-halogen derivatives of both the α-lyxose and 5-deoxy-α-lyxose derivatives (e.g., 25) were active against HCMV, with the L-enantiomers being the most potent ($EC_{50}$ = 0.2–0.4 $\mu M$).

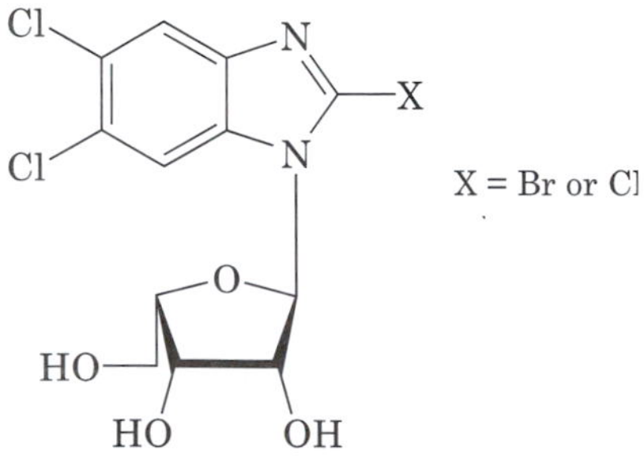

X = Br or Cl

(25)

The 2-halogenated compounds act through a different mechanism than most nucleoside analogs, and a target for them has been identified. Selection of strains resistant to TCRB or BDCRB identified mutations in the UL89 and UL56 genes, and these strains were also resistant to the 2-halo-erythrofuranose compounds (341–343). None of these four compounds generated resistance mutations in UL97. The 2-isopropylamino and 2-cyclopropylamino derivatives acted on the UL97 kinase, indicated by resistance mutations in UL97. UL89 and UL56 are required for postsynthetic DNA processing before packaging into virions. Studies with BDCRB indicated that assembly was blocked at the packaging step because the rolling circle replication product was not cleaved at the viral temini (343). Thus, these compounds identify another potential drug target in herpesviruses. Because the UL89 gene is highly conserved across herpesviruses, effective drugs that target this protein may have broad activity within the herpesvirus family.

UL6 is an HSV-1 gene required for DNA cleavage and encapsidation. Two thiourea compounds, N-(4-(3-(5-chloro-2,4-dimethoxyphenyl)-thioureido)-phenyl)-acetamide (CL-253824) (26) and its 2-fluoro-benzamide

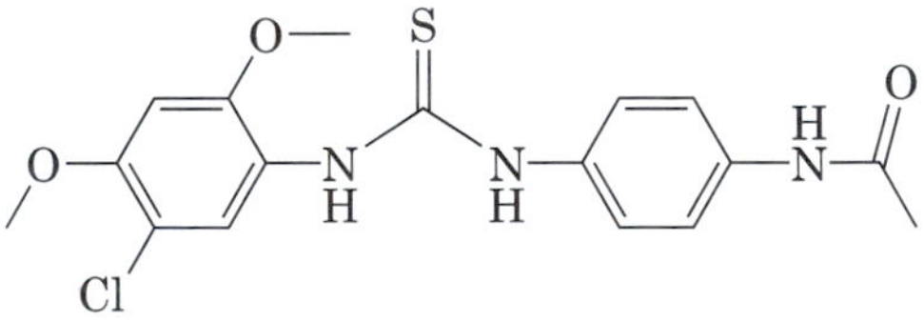

(26)

derivative (WAY-150138) (27), have been identified as inhibitors of cleavage and encapsidation (344). Resistance mutants generated with these compounds mapped to UL-6. CL-253824 is active against HSV-1 and weakly active against HSV-2 and HCMV ($EC_{50}$ values of 8–25 $\mu M$ against HSV-1). WAY-150138 is more active, with $EC_{50}$ = 0.2–1.5 $\mu M$ against HSV-1 and 6–19 $\mu M$ against HSV-2 and HCMV. Both inhibitors are inactive against VZV.

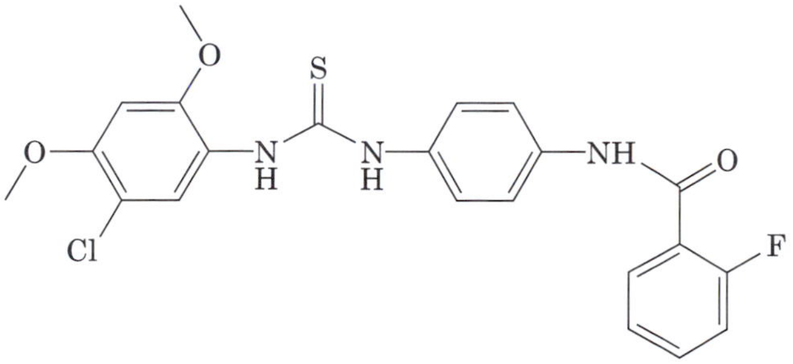

**(27)**

Vidaribine (araA) analogs are prone to metabolic degradation by deaminases. Modifications to the 6-N position (**28**) have proven par-

| $R_b$ |
|---|
| $NHCH_3$ |
| $N(CH_3)_2$ |
| $OCH_3$ |

**(28)**

**(29)**

$X = Cl, Br, I$

**(30)**

tially effective at blocking deamination (345). Methylation provided a 10- to 30-fold increase in potency against VZV relative to vidaribine and a decrease of about the same magnitude in cytotoxicity. Replacement of the amine with a methoxy group (araM) gives about 10-fold greater selectivity than araA against VZV. araM is inactive against other viruses because of an inability to be phosphorylated. Whereas these 6-N-substituted compounds afford a significant improvement, they still undergo relatively rapid catabolism. Their development as drugs will require prodrug forms that block this metabolism.

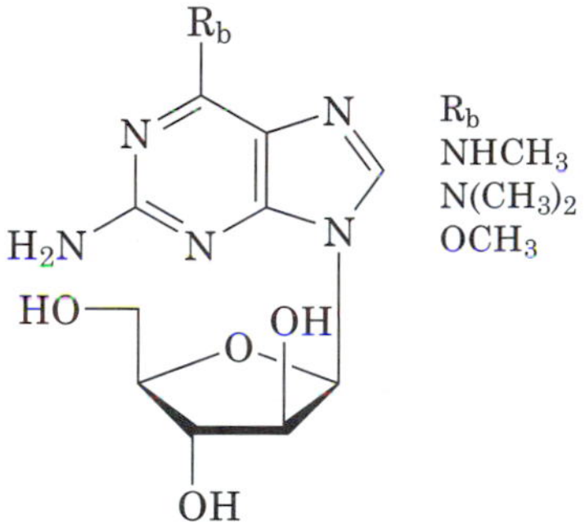

(E)-5-(2-bromovinyl)-2'-deoxyuridine (bivrudin, BVDU) (**7**) and 1-(B-D-arabinofuranosyl)-(E)-5-(2-bromovinyl)uracil (BVAU) (**29**) were shown to be potent inhibitors of VZV, although their metabolic instability made them impractical as drugs. Additional bromovinyluracil analogs (**30**) were made to circumvent

this catabolism (346). They contain a 2'-oxo moiety and one of three halogens on the vinyl group. These compounds were of equal or greater potency than ACV and PCV against VZV, with little toxicity. They were active in the order of Cl > Br > I. They also were active against EBV, but in the reverse order with regard to halogen. They showed activity against HSV-1 but not HSV-2.

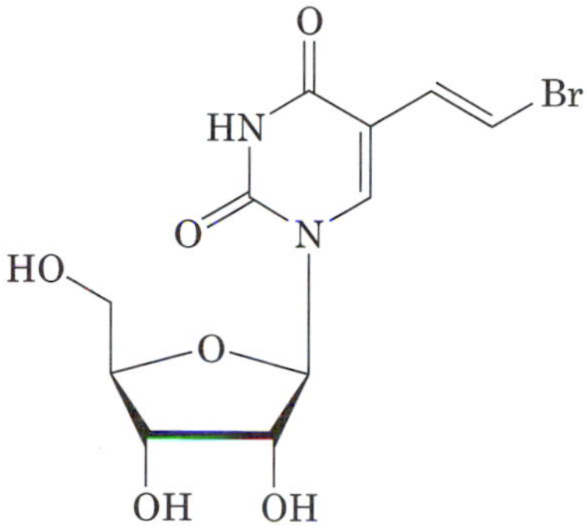

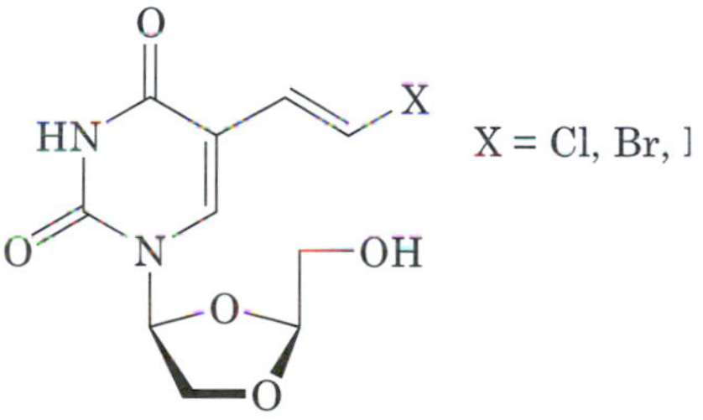

Out of a series of 5-substituted uracil nucleoside analogs containing a cyclopropyl side chain, bases with 5-halovinyl substitutions

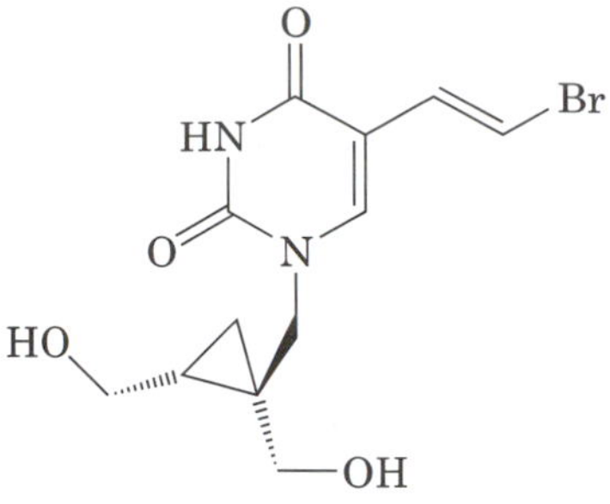

(31)

(31) gave activity up to 60-fold greater than ACV against VZV but were weaker than ACV against HSV-1 (347). The bromovinyl derivative had oral bioavailability of 69% in rats, and these compounds showed no detectable toxicity.

Similarly, L-dioxolane uracil nucleosides (32) were active against EBV, in the order X =

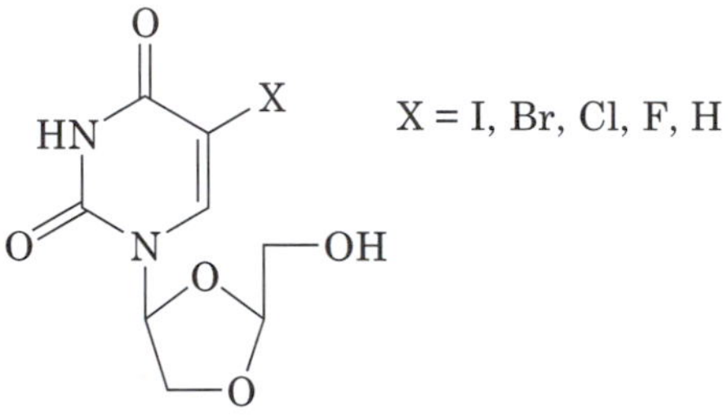

X = I, Br, Cl, F, H

(32)

I > Br > Cl (IC$_{50}$ = 0.033–0.6 $\mu M$), with little toxicity noted (348). Most L-nucleosides are phosphorylated by cellular kinases, but the activity of these dioxolane uracil analogs is dependent on the viral thymidine kinase to generate the monophosphorylated form (349, 350). Note that these compounds do not affect latent replication cycle of EBV, where the host cellular replication machinery replicates the viral DNA. As a result of this, virus DNA levels return to pretreatment levels by 50–60 days posttreatment because of reactivation of latent virions. The bromo- analog was 80-fold more active against VZV than was ACV, with no detectable toxicity (349).

Shigeta et al. (351) synthesized 20 analogues of 2-thio-pyrimidines and found that the 2-thiouracil and 2-thiothymidine arabinoside analogs were weakly to moderately active against HSV and VZV (>10-fold less active

than ACV). None of these compounds were active against HCMV or thymidine kinase–deficient HSV. 5-Chloro-2-thiocytosine and 5-bromo-2-thiocytosine (33) were relatively potent

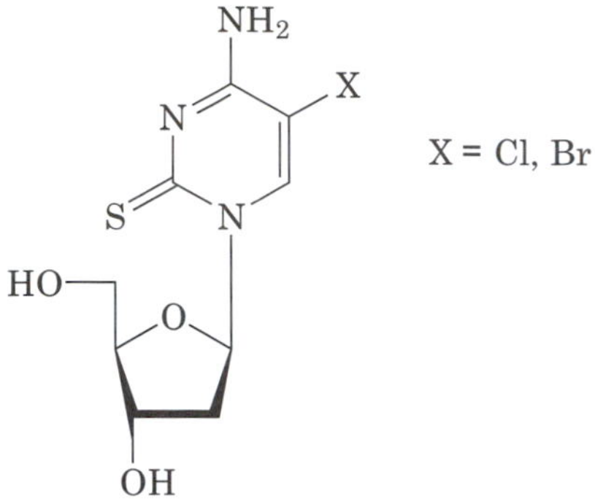

X = Cl, Br

(33)

inhibitors of VZV, with activities in the range of ACV.

9-[3-Hydroxy-2-(phosphonomethoxy)propyl] (HPMP)-purine and pyrimidine analogs have been synthesized for each of the four natural bases (352) (see (5) for HPMPC). HPMPG, the guanosine derivative, was about fivefold less active than ACV in cell culture but was more protective against HSV-1 infection in a mouse model. A related compound, 9-(3′-ethylphosphono-1′-hydroxymethyl-1′propyloxymethyl)-guanine (SR 3727A) (34) was inhibitory against

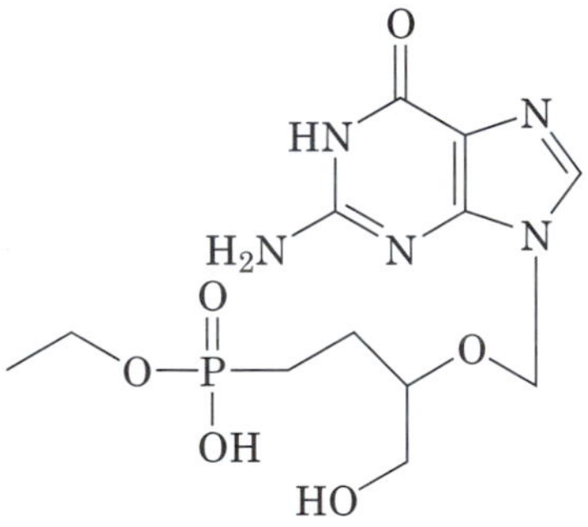

(34)

HCMV (EC$_{50}$ = 6–17 $\mu M$), with no toxicity noted in cell culture. It also provided protection against a lethal dose of murine CMV in mice (353). Studies of the efficacy of ACV phosphonate and GCV phosphonate in protecting against MCMV infection in mice showed the

GCV phosphonate derivative to be equivalent to GCV in potency in cell culture (354), and more effective at lengthening the time to death than the parent, GCV (355). However, it was less effective in cell culture and showed more renal toxicity than HPMPC (cidofovir). HPMPC has become the focus of development for this class of compounds.

HPMPC (cidofovir) (**5**) has shown broad-spectrum and potent activity against several DNA viruses, but nephrotoxicity limits its use (299, 300). Its cyclic congener (cHPMPC) (**35**) shows similar antiviral activity with reduced

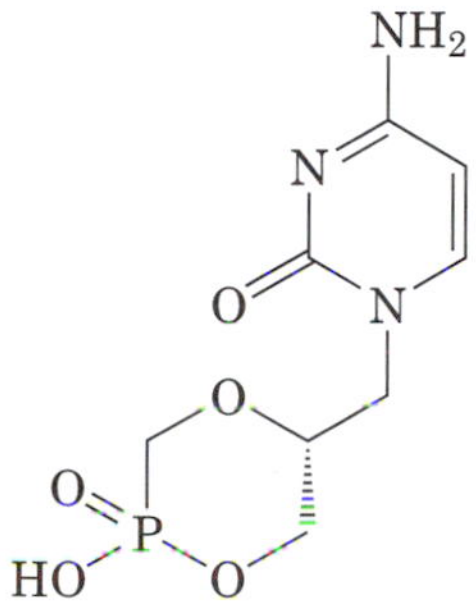

(35)

renal toxicity (299, 356–358). cHPMPC is converted after cellular uptake to cidofovir and shows similar potency and cytotoxicity in cell culture. In 14- and 30-day studies in rats, cHPMPC was >10-fold less toxic than cidofovir. Similar results were observed in guinea pigs and cynomolgus monkeys. These observations have also been extended to immunocompromised guinea pigs, which are especially sensitive to cidofovir toxicity.

S-2242 is a guanine analog substituted at the N9 position (**36**). It has potent activity

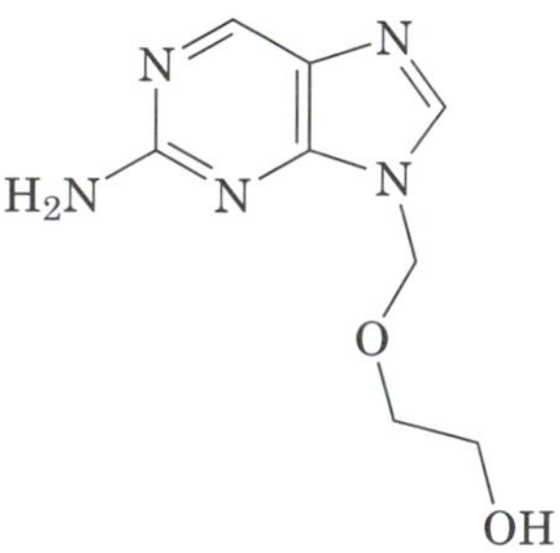

(36)

against HSV-1 and-2, VZV, HCMV, HHV-6, HHV-7, and HHV-8. $EC_{50}$ ranges from 0.01 to 0.4 $\mu$g/mL (359–361). Of 25 compounds tested, S-2242 was the most potent against HHV-7, with a selectivity index greater than 200. It also showed good potency and selectivity against HHV-8. S-2242 is about 10-fold less active against HSV than is acyclovir, but it is a good inhibitor of activity of thymidine-kinase-deficient strains of HSV and VZV. The toxicity profile in cell culture varied with cell type. In the majority of cell types it was comparable with or somewhat more toxic than GCV, though in HSB-2 and CEM cells it was more than 60-fold more toxic than GCV. Animal studies have shown toxicity profiles similar to GCV.

S-2242 has several characteristics that differ from currently available purine nucleoside drugs. Its activity against thymidine kinase–deficient HSV and VZV strains, its broad range of activity, and its toxicity profile indicate that phosphorylation does not involve a viral kinase. Instead, it is phosphorylated by a cellular deoxycytidine kinase (362). Blocking the phosphorylation of S-2242 by deoxycytidine kinase decreased its potency by 50- to 100-fold. S-2242 is not phosphorylated by HCMV UL97 (363), but infection with HCMV increased metabolites of S-2242 by 5- to 25-fold, indicating that HCMV either encodes an additional kinase capable of phosphorylating the compound or it activates a cellular kinase.

S-2242 resembles GCV and PCV in the stability of its triphosphate form, with a half-life of 3–6 h in CEM cells (362). When cells were treated for 24 h before infection with drug withdrawn at the time of infection, the apparent $EC_{50}$ of S-2242 increased from 0.45 $\mu$g/mL with continuous treatment to 0.7 $\mu$g/mL after the drug was withdrawn. By contrast, the apparent $EC_{50}$ of ACV increased from 0.03 to greater than 50 $\mu$g/mL when comparably treated (359).

Because a viral kinase is not required for activation of S-2242, the bulk of strains resistant to ACV, GCV, or PCV are sensitive to S-2242 (364). S-2242 shows cross-resistance with drugs generating certain DNA polymerase mutations (e.g., foscarnet), but it is not cross-resistant with drugs targeting other

polymerase amino acid mutations, e.g., cidofovir or bromovinyl-deoxyuridine.

### 3.1.2.2 Prodrugs Under Development.

There are three primary purposes that have
driven prodrug development:

1. To increase absorption of the drug. Low
   oral bioavailability is a characteristic of
   most nucleoside analogs, and lipophilic
   prodrug forms have provided dramatic increases in bioavailibility, as discussed earlier for ACV, GCV, and PCV.
2. To provide the drug in the monophosphate
   form. The need for an initial phosphorylation step to be carried out by a viral kinase
   provides a large degree of selectivity, but
   this approach is limited to viruses encoding
   such a kinase. One way to extend treatment to viruses lacking a suitable kinase is
   to develop nucleoside monophosphate analogs. In this case, the selectivity is limited
   to the difference in affinity for viral versus
   cellular polymerases, but if sufficient selectivity can be attained, the trade-off may be
   worthwhile for viruses where viral phosphorylation to the monophosphate is not
   an option (e.g., HBV), or for nucleoside kinase–deficient or –resistant herpesviruses.
   In this case, increasing the lipophilicity or
   masking the charge are essential considerations in developing the prodrug, because
   phosphorylated nucleosides are impermeable to cellular membranes.
3. To target the drug to a specific tissue. This
   approach may be particularly useful for diseases affecting the liver, such as HBV, or
   for certain cell types of the immune system.
   An example of this approach is compounds
   targeting the liver asialoglycoprotein receptor, discussed later.

A prodrug form of ACV, desciclovir (**37**), is
designed to increase uptake of the drug, which

is readily processed to ACV by xanthine oxidase (365). However, desciclovir is more toxic
than ACV. Modification of desciclovir by adding reducing sugars to the 2 position of adenine provided good water solubility, but poor
adsorption after oral dosing in rats. They were
also inefficiently processed to ACV (366).

A series of methylenecyclopropane analogs
of purine nucleosides and phosphoroalaninate
prodrug forms of these compounds (**38**) have
shown activity against several DNA viruses
(367–370). Note that the prodrug does not
need to undergo phosphorylation to the monophosphate. Most of these compounds were approximately as potent as GCV against HCMV,
the exceptions being 3, 5, 8, and 10, which
were 12- to 30-fold less active. Compounds 1–4
were also active against murine CMV in vivo
(371). The 2,6-diaminopurine prodrug was
equivalent to ACV in activity against HSV-1
and -2, and 100-fold more active than ACV
against VZV. Compounds 13 and 8 were very
potent against EBV. The diamino purine was
moderately active against HBV ($EC_{50} = 10$
$\mu M$) but the potency increased substantially
in prodrug (monophosphorylated) form ($EC_{50}$
$= 0.08 \mu M$). Overall, these compounds showed
good activity against the range of herpesviruses, but the potent compounds varied depending on the viral target, with no clear pattern emerging yet. Toxicity was minimal in a
variety of cell types.

Related analogs of ribose, spiropentane derivatives (**39**), were synthesized in four conformations, and conformation-dependent activity was observed (372). The *proximal* and
*medial-syn* adenosine analogs showed moderate activity against HCMV and EBV, although
the proximal analog was relatively toxic in one
cell line. The *distal* isomer of the guanosine
analog was also active against EBV, with no
toxicity noted. These compounds showed little
activity against HSV or VZV.

When the *proximal* isomer of the adenosine
analog was converted to the phenylphos-
phoalaninate prodrug form, it showed a 50-
fold increase in activity against HCMV ($EC_{50}$
$= 0.4 \mu M$), and gained 10-fold in potency
against EBV, although it was still cytotoxic to
Daudi cells ($EC_{50} = 2.8 \mu M$, $CC_{50} = 7.8 \mu M$). It
was also active against HBV and VZV ($EC_{50} =$
3.1 and 9.3 $\mu M$).

R = H (Desciclovir)
or sugar conjugates

(**37**)

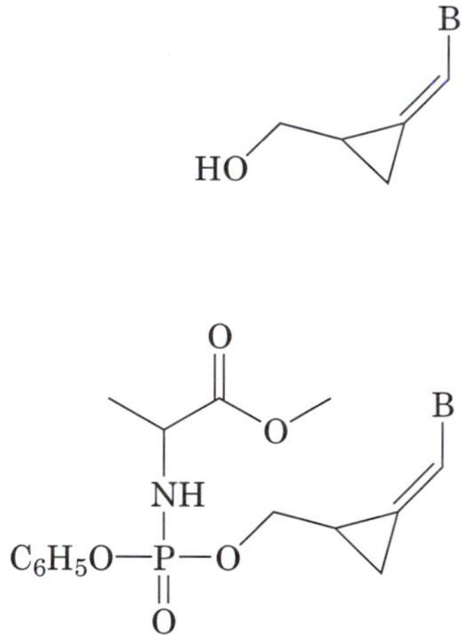

B

1 = Adenine
2 = Guanine
3 = 2-amino-6-methoxypurine
4 = 2,6-diaminopurine
5 = 2-amino-6-chloropurine
6 = 2-amino-6-cyclopropylaminopurine
13 = cytosine

7 = Adenine
8 = Guanine
9 = 2-amino-6-methoxypurine
10 = 2,6-diaminopurine
11 = 2-amino-6-chloropurine
12 = 2-amino-6-cyclopropylaminopurin

(38)

B = adenine or guanine

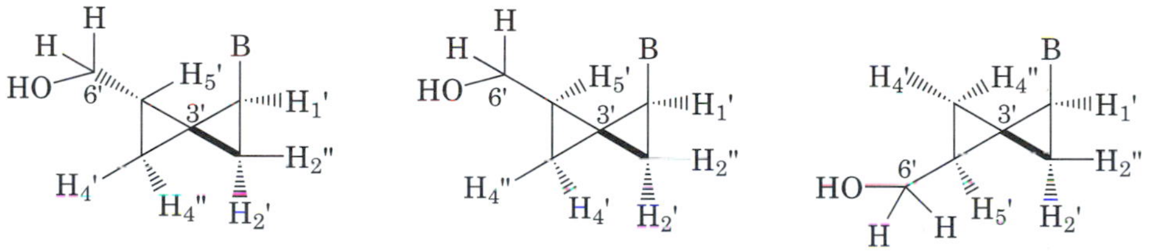

Proximal

Medial-syn

Distal

(39)

cycloSaligenyl nucleoside monophosphate derivatives were designed as an attempt to create a more efficacious tool for making monophosphate derivatives sufficiently lipophilic to enter cells (373). Derivatives of ACV and PCV (40) have been tested against HSV and EBV. The ACV derivatives were about equivalent to ACV in activity against a thymidine kinase-competent HSV strain and against EBV. However, in a thymidine kinase-deficient HSV strain, the cycloSaligenyl derivatives were 20- to 60-fold more active than ACV. This approach did not work with PCV derivatives, which were 32- to 95-fold less active than PCV.

Borrowing from experience with HIV, ACV-5'-(phenyl methoxy alaninyl) phosphate (41) was synthesized as a kinase-independent lipophilic ACV prodrug (374). This approach worked successfully for the HIV drug d4T, which was non-toxic and a potent HIV inhibitor. However, whereas the ACV derivative was not toxic, it had weak antiviral activity.

A phosphonomethyl ether analog of ACV (42) was fivefold less active against HSV-1 than ACV (352). However, as mechanistic studies would predict, the difference in efficacy for HSV and HCMV narrowed, with the analog being eight times more active against HCMV than ACV. Phosphonate ethers can also have an effect on cytotoxicity; a phosphono-derivative of GCV (43) is less toxic than GCV.

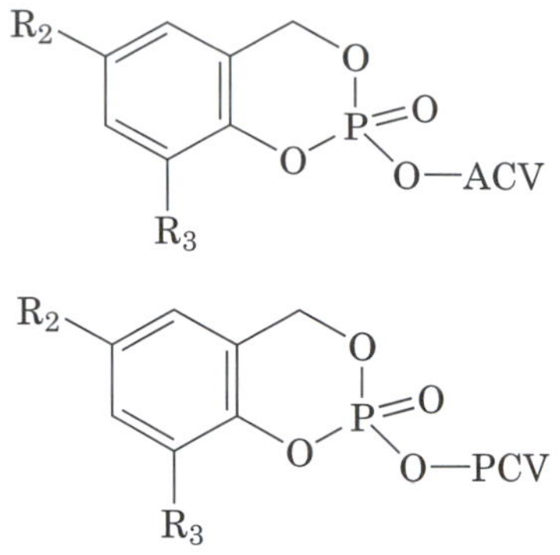

R_2 = H and R_3 = CH_3
R_2 = CH_3 and R_3 = H
R_2 = R_3 = H

R_2 = H and R_3 = CH_3
R_2 = CH_3 and R_3 = H
R_2 = R_3 = H

(40)

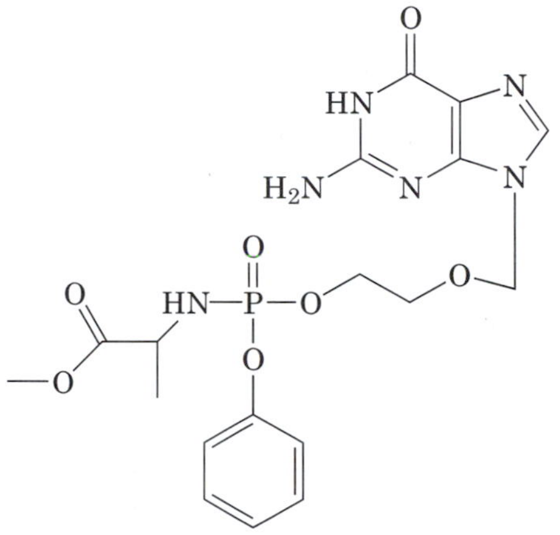

(41)

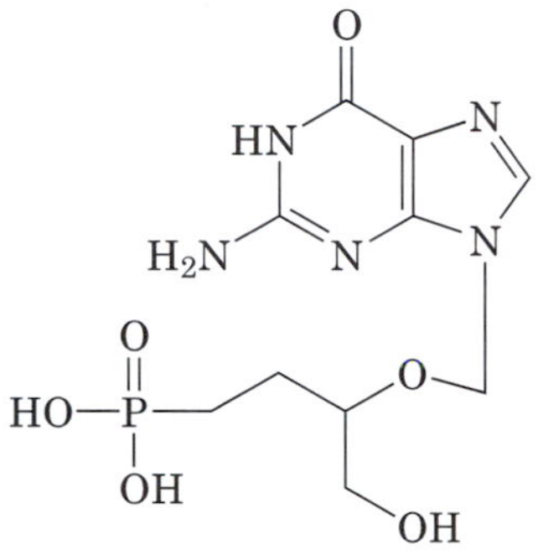

(43)

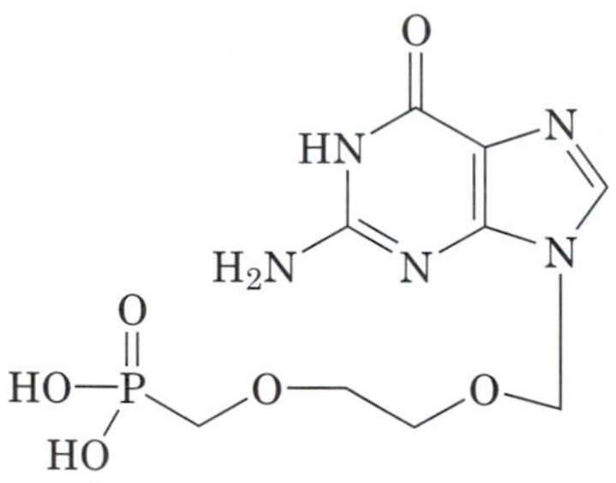

(42)

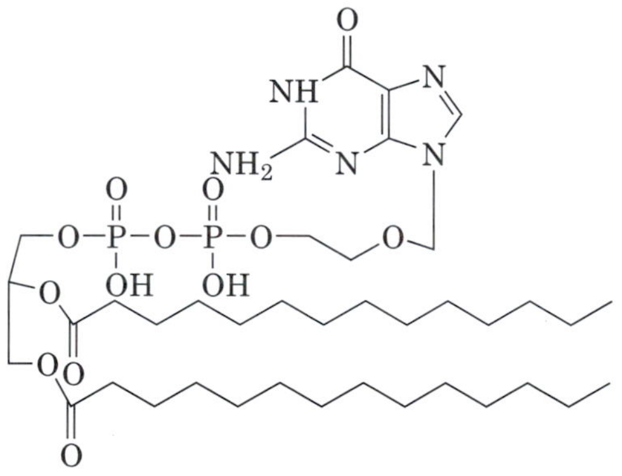

(44)

Another approach to treating thymidine kinase-deficient strains is to provide the phosphorylated form as a phospholipid, shown here for ACV (**44**) (375). ACV-diphosphate-dimyristoylglycerol was equivalent in activity to ACV against wild-type HSV-1 and HSV-2 strains. With thymidine kinase–deficient strains, ACV had no detectable activity, whereas the phospholipid prodrug was as active against them as against wild-type ($EC_{50}$ = 0.25–1 $\mu M$). In uninfected W138 lung fibroblasts, the levels of intracellular phosphorylated ACV were 56-fold higher using the phospholipid prodrug than those for ACV.

GCV has oral bioavailability of less than 10% in humans and is given intravenously for acute treatment of HCMV disease. It is active against HSV and VZV, but the poorer toxicity profile relative to ACV limits its use. An elaidic ester of GCV (E-GCV) (45) was 5- to 30-fold

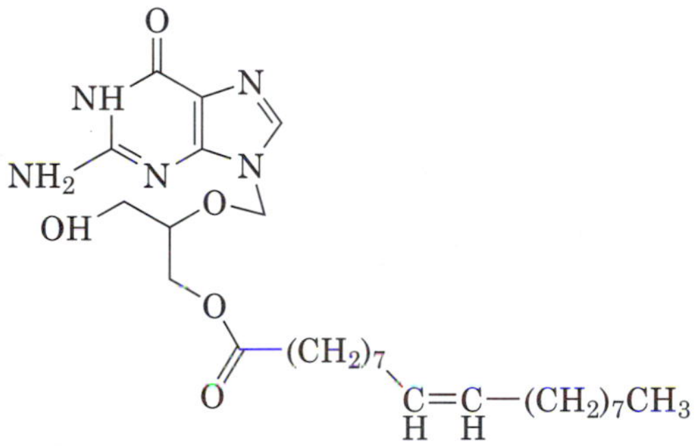

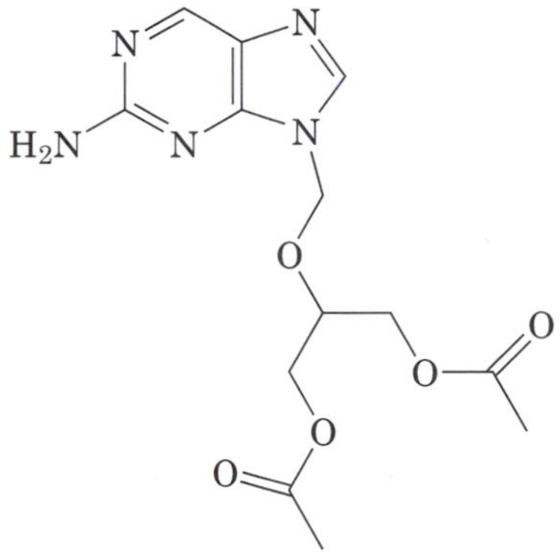

(46)

germinal epithelium, an effect that is also observed in mice treated with ganciclovir (277, 377).

3.1.2.3 Nonnucleoside Inhibitors and Other Targets. IMP dehydrogenase is the rate-limiting enzyme in the synthesis of guanine nucleotides, converting IMP to XMP. It is a target for two compounds, ribavirin (47) and myco-

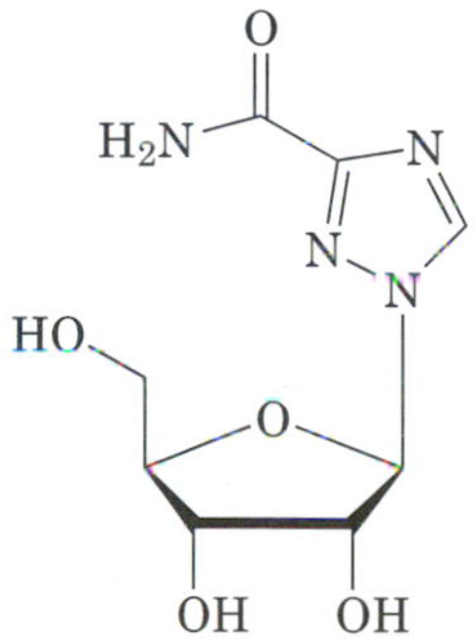

(45)

more potent in cell culture than GCV against HSV and 15- to 90-fold more potent than ACV (376). The ester was about equivalent to GCV in toxicity (less than twofold more toxic). E-GCV and GCV were of equivalent potency against VZV and HCMV ($EC_{50}$ = 0.7–4.5 $\mu M$ for both viruses). When E-GCV was used intraperitoneally to treat HSV-2 infection in mice, death rates were reduced to 10% under conditions where 100% of the mice died when treated with an equimolar amount of GCV. The activity profiles of GCV and E-GCV were similar with regard to potency in thymidine kinase–deficient strains or strains resistant to foscarnet or ACV.

H961 (46) (277) is the prodrug form of S-2242 (36). The poor oral bioavailability of S-2242 necessitates delivery by injection (377). However, the prodrug was been tested in mice infected with vaccinia virus and was effective when delivered orally. Oral administration of H961 at 100 mg/kg daily for 10 days completely blocked viral infection. The only toxic effect noted was atrophy of testicular

(47)

phenolic acid. Mycophenolate mofetil (MMF) is an immunosuppresant used for kidney transplant recipients. MMF is hydrolyzed to mycophenolic acid. Ribavirin both inhibits the activity of some RNA polymerases and increases their rate of nucleotide misincorporation, thus decreasing the fitness of the resulting virions (378–380). However, a second effect of both of these compounds is a depletion of GTP and dGTP pools through inhibition of IMP dehydrogenase activity. A synergy

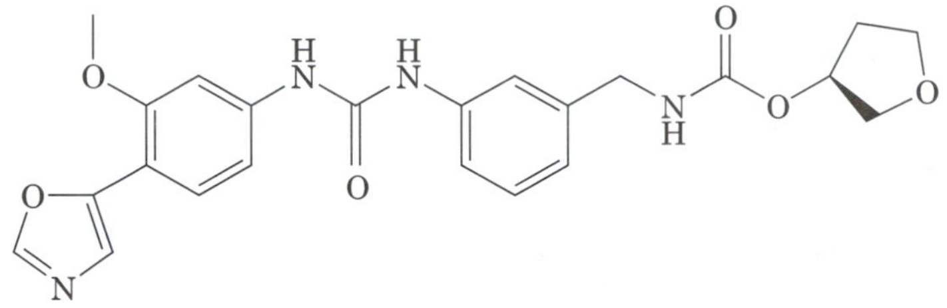

(48)

might be expected between these compounds and guanosine analogs, because a decrease in cellular dGTP would increase the rate of incorporation of the guanosine analog. This mode of action has been tested with MMF. A combination of MMF and acyclovir was more effective at preventing posttransplant lymphoproliferative disorders resulting from EBV reactivation or primary infection than was treatment with acyclovir alone (381). The combination of MMF and H2G was particularly effective, with a 10- to 150-fold increase in potency observed against HSV-1 and HSV-2 and VZV in cell culture (382). With a thymidine kinase–deficient HSV-1 strain, potency increased more than 2500-fold, such that H2G with MMF was more potent in the thymidine kinase–deficient strain than H2G alone in the wild-type strain. The effect in VZV strains lacking thymidine kinase was smaller than with HSV, but activity still approached that of H2G alone in a wild-type strain. Ribavirin caused similar effects, although smaller in magnitude. This approach may have more general applicability; a preliminary study in AIDS patients indicates that MMF may increase the effectiveness of currently used multidrug treatment regimens used against HIV (383).

VX-497 (48) is an uncompetitive inhibitor of human IMP dehydrogenase (384) and a broad-spectrum antiviral agent. Of the DNA viruses, it inhibits HBV and HCMV at $EC_{50}$ concentrations of 0.4–0.8 $\mu M$. HSV-1 is less potently inhibited ($EC_{50} = 6 \mu M$). VX-497 was 25- to 100-fold more potent than ribavirin against these viruses but was also more toxic in some cell lines, with a selectivity index of about 10 in those cell lines. This might be anticipated for a compound that targets an enzyme that is essential in rapidly dividing cell

types. The same constraint also applies to ribavirin, which had a smaller selectivity index in those cell types than did VX-497. Based on its mechanism of action, it is anticipated that VX-497 might share the synergies with guanosine analogs noted for MFF and ribavirin.

Hydroxyurea inhibits the activity of the cellular ribonucleotide reductase and potentiates the inhibitory effect of 2′,3′-dideoxynucleotide analogs against HIV-1, particularly didanosine (ddI) (385–387). The combination of hydroxyurea and ddI is currently being investigated as a much less expensive alternative for HIV treatment in those populations where cost makes triple drug regimens prohibitive. By analogy with the mechanism outlined above for interaction of MMF, ribavarin, and possibly VX-497 with guanosine analogs, hydroxyurea might act synergistically with nucleoside analogs targeting DNA viruses. When tested in combination with a variety of nucleoside analogs (ACV, GCV, PCV, H2G, cidofovir, and adefovir) for activity against HSV-1 and HSV-2, hydroxyurea had a moderate stimulatory effect (388). In thymidine kinase–deficient strains, where the triphosphate concentration of the nucleoside analogs would be much lower, the potentiation was larger. In these strains $EC_{50}$s were lowered from 20–100 to 0.2–5 $\mu g/mL$ by addition of hydroxyurea. The potentiating effect of hydroxyurea is not sufficient to warrant its general use for this purpose. However, given its potential for treating HIV infections, this combination of drugs could result when opportunistic herpesvirus infections are treated in AIDS patients.

Triciribine (TCN) (49) is a tricyclic nucleoside with antineoplastic and antiviral activity (389). It is phosphorylated by adenosine kinase to the monophosphate form but is not

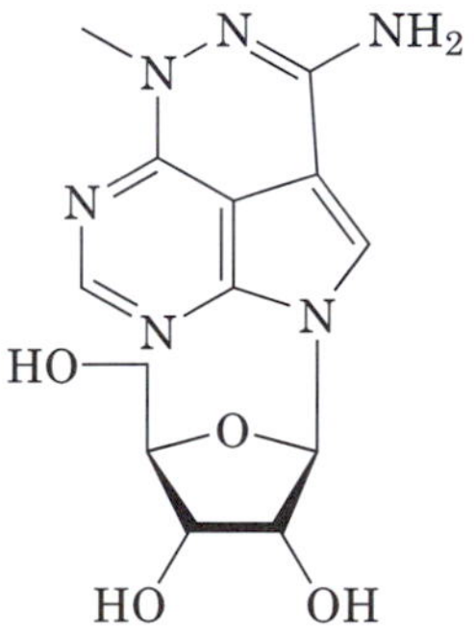

(49)

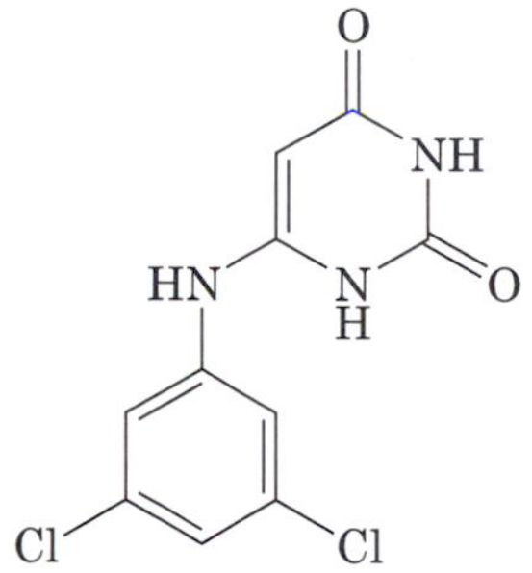

(51)

phosphorylated further, and it is not incorporated into DNA. In an effort to dissect its mechanism of action, analogs of TCN varying in the number of hydroxyl groups on the ribose were synthesized and tested for antiviral activity against HCMV. The parent TCN was active ($EC_{50}$ = 2.5 $\mu M$), but all of the analogs were inactive, indicating that hydroxyls at the 2′, 3′, and 5′ positions were essential for activity.

The helicase-primase complex of herpesviruses carries out two essential DNA replication activities: it provides a primer for DNA polymerase to initiate from, and it separates the two DNA strands. A series of dichloroanilino-purines and -pyrimidines (50), (51) were

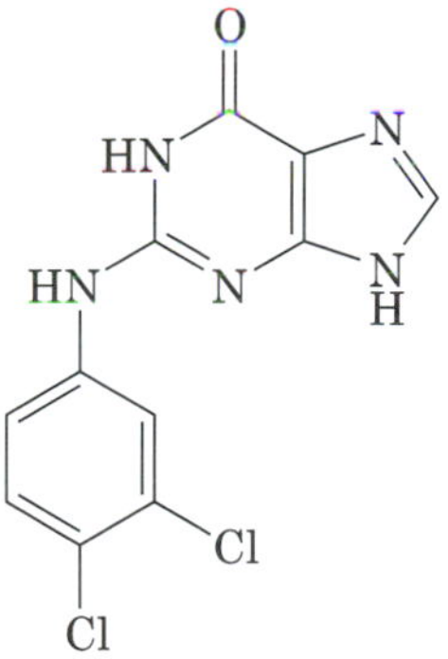

(50)

examined for their antiherpevirus activity, and several were found to inhibit not the polymerase but the helicase-primase complex (390, 391). They were weak inhibitors with

poor selectivity, but they demonstrate that this enzymatic activity is a potential drug target. Helicase-primase has proven a much more difficult target than polymerases, and thus far, no drugs targeting helicase have been approved for any disease. However, as mutants accumulate in the currently used targets, it may be necessary to broaden our approach to include targets such as helicase.

More recently, a 2-amino thiazole compound, T157602 (52) has been identified that

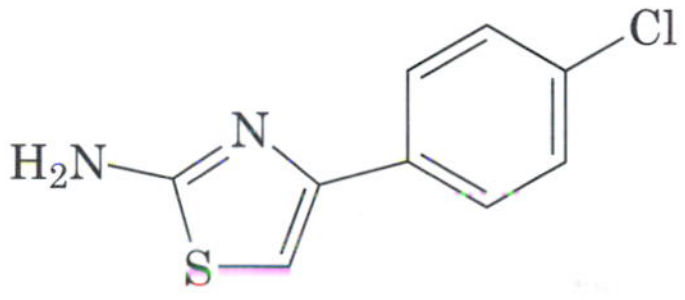

(52)

targets the UL5 protein of the HSV helicase-primase complex (392, 393). It has an $IC_{50}$ against the purified complex of 5 $\mu M$, an $EC_{50}$ in cell culture of 3 $\mu M$, and no toxicity at 100 $\mu M$. It is not active against helicase-primase complexes of VZV, HCMV, or EBV.

A 1,6 naphthyridine (53) and related 7,8-dihydroisoquinoline derivatives were identified as having potent activity against HCMV. Compound (54) was 39- to 220-fold more active than GCV. Strains resistant to GCV, foscarnet, cidofovir, and BDCRB were still sensitive to these compounds. In most cell lines, these compounds were substantially more cytotoxic than ACV or GCV, resulting in a poorer selectivity index than ACV and approximate equivalence to GCV.

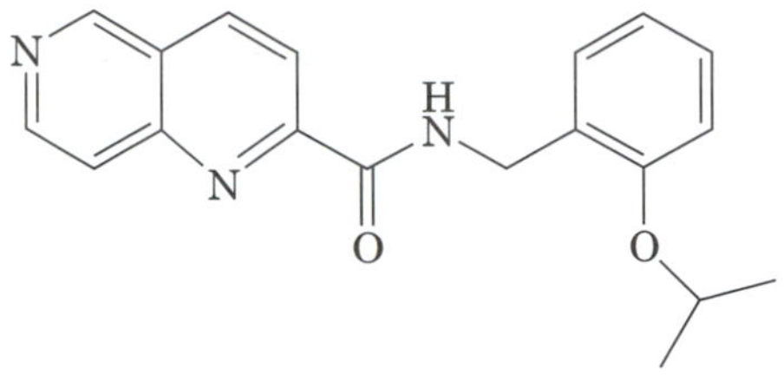

(53)

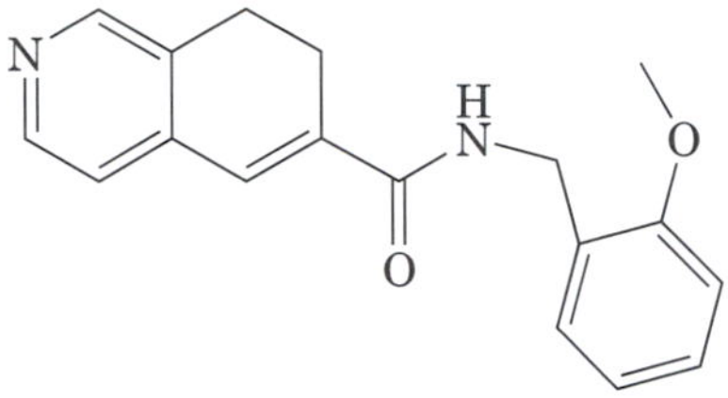

(54)

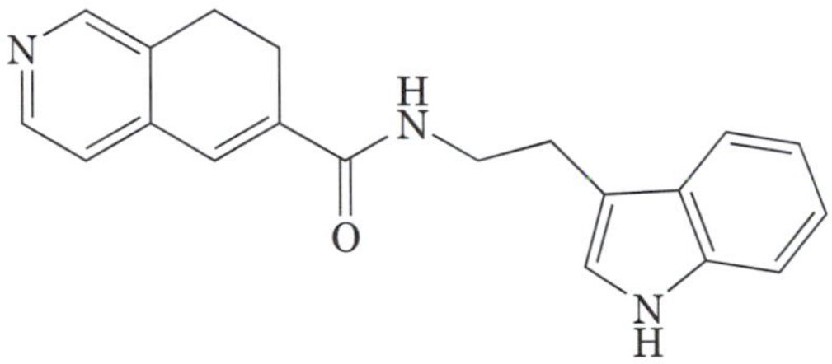

(55)

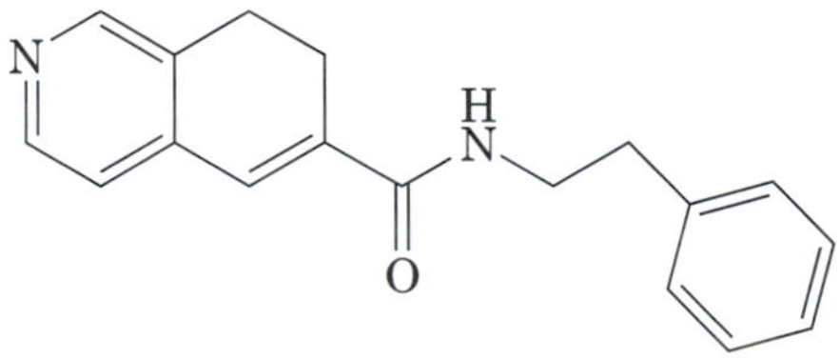

(56)

Dendrimers have been proposed as a topical microbicide, to be used to inhibit viral entry through mucosal cells (394). These molecules are highly branched polymers synthesized on a polyfunctional core, which are capped with a layer to give the desired surface properties. The choice of core, polymer, and capping group determine the shape, size, and charge characteristics of the molecule. Five dendrimers were tested for their capacity to

inhibit entry of HSV-1 and HSV-2. All five dendrimers showed antiviral activity in cell culture when they were added to cells before the addition of virus. Two of three dendrimers tested were effective in a mouse model in preventing infection when applied topically to the vagina. The compounds were applied as 15 $\mu$L of a 100 mg/mL solution and protected 15 of 16 mice from infection (16 of 16 control mice were infected). No toxic effects were noted from a single application, although multiple applications were not tested.

**3.1.2.4 Herpesvirus Protease Inhibitors.** Proteolytic activity is required to process herpesvirus structural protein precursors during assembly of the virion. The protease responsible is a serine protease encoded by the HSV-1 UL26 gene and the HCMV UL80 gene (reviewed in Refs. 395, 396). These herpesvirus proteases seem to be a distinct subfamily of serine proteases, with minimal structural similarity and minimal crossreactivity with inhibitors to other serine proteases. A benzimidazolylmethyl sulfoxide (**57**) inhibitor of HCMV

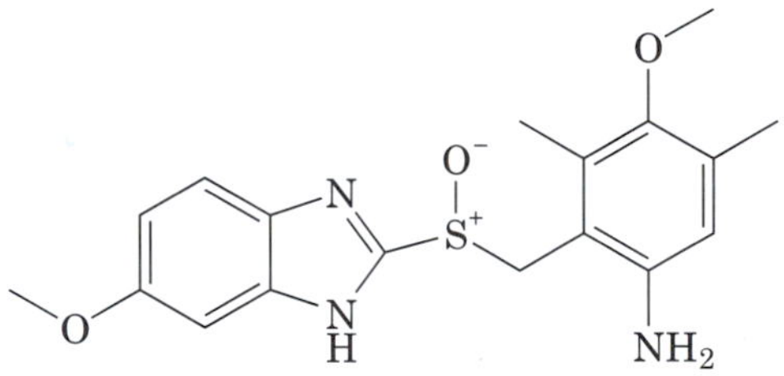

(57)

protease activity has been identified (397). This compound inactivates the protease through interactions with one or possibly two cysteines on its surface. It had an IC$_{50}$ in a biochemical assay with recombinant protease of 1.9 $\mu M$ and was functional in a cell culture assay, with an EC$_{50}$ of 18 $\mu M$. No toxicity noted at 100 $\mu M$. It also showed no activity at 100 $\mu M$ against chymotrypsin, trypsin, thrombin, factor Xa, plasmin, or kallikrein. Based on the lack of activity of this inhibitor against these other serine proteases and the amino acid sequence differences between the herpesvirus proteases and other serine proteases, the herpesvirus proteases seem to represent a subclass of the serine protease superfamily, an important determinant of the likelihood of identifying inhibitors with a useful selectivity index.

A series of thieno(2,3-d)oxazinone protease inhibitors (58) were demonstrated to have

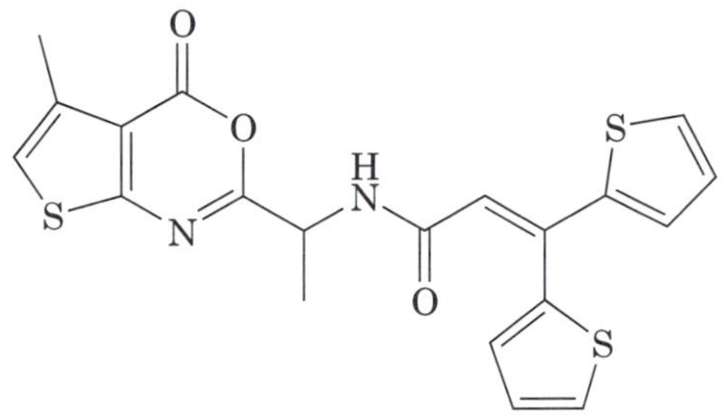

(58)

good potency against HSV-2, HCMV, and VZV ($EC_{50}$ from 0.02 to 0.9 $\mu M$) (398). Substitution of one or both of the thiophenes by phenyl groups had little effect on potency. These compounds were inactive at 100 $\mu M$ against elastase and trypsin but were moderately toxic ($CC_{50}$ = 20–100 $\mu M$).

A random screen for HSV-1 protease inhibitors yielded 1,4-dihydroxynaphthene (59) and three related naphthoquinones (60) (399).

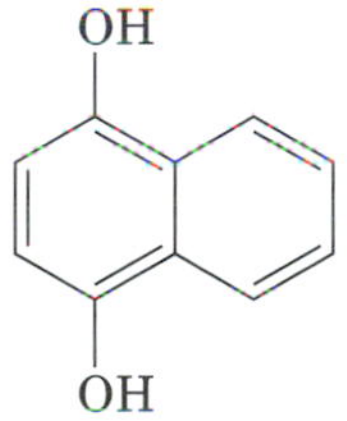

(59)

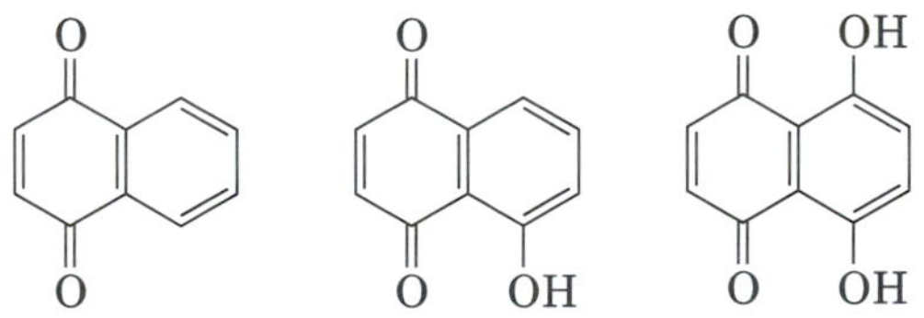

(60)

These compounds exhibited $IC_{50}$ values against the protease ranging from 6.4 to 16.9

$\mu M$. They were more active against HCMV protease, with $IC_{50}$ values of approximately 1 $\mu M$. While structurally similar, the naphthoquinones differed mechanistically from the naphthalene. Against HSV-1 protease, the naphthalene was a competitive inhibitor, whereas the naphthoquinones were noncompetitive. The reverse was the case against HCMV protease. These compounds were not active against trypsin, chymotrypsin, kallikrein, plasmin, thrombin, or factor Xa at 100 $\mu M$.

**3.1.2.5 Immune Modulators.** Imiquimod (61) has been tested as a topical treatment for herpesvirus and papillomavirus infection. A more potent analog, resiquimod (62), has been

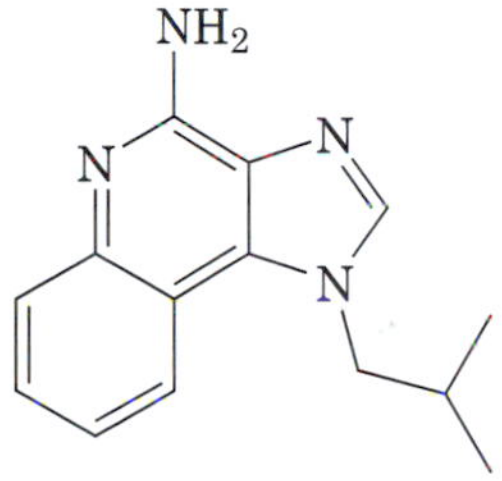

(61)

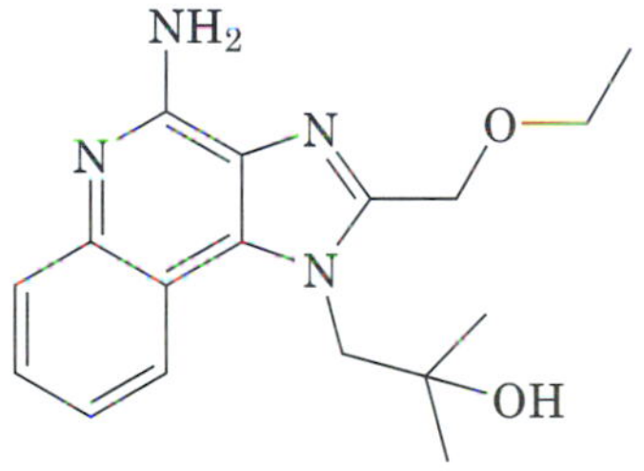

(62)

tested for anti-HSV-2 activity in a guinea pig model, where the endpoint was the number of recurrences over the test period (400). The number of recurrences was decreased by 65–75% by subcutaneous injection of resiquimod. Topical application of imiquimod also had a similar effect on recurrence rate (401). In a trial of 52 patients with frequently recurrent genital herpes, topical application of resiquimod one to three times per week (0.01% or 0.05%) for 3 weeks resulted in 32% of patients completing a 6-month observation period without a recurrence, compared with 6% in the control group (402). These compounds are immunomodulators whose effect is to increase interferon-$\alpha$ levels. Resiquimod was demonstrated to activate B-cells by a mode that resembled B-cell activation by the CD40 ligand (403). Given a mode of action that is quite distinct from other HSV-2 inhibitors, it may prove to be useful in combination therapy with inhibitors of viral targets such as polymerase.

## 3.2 HBV

Replication of HBV requires first transcribing the cccDNA genomes in the nucleus into RNA, followed by copying the RNA into DNA by the reverse transcriptase activity of HBV polymerase. Effective treatment of HBV has two important components. First the replication process of the virus must be interrupted, and second, the reservoir of cccDNA genomes in the nucleus must be eliminated. The replicative cycle can be effectively broken with HBV polymerase inhibitors, preventing production of new virions. However, it is common for plasma virus levels to be decreased to undetectable levels with polymerase inhibitors, only to have the viral replication rebound after removal of the drug because of the availability of the DNA genomes in the nucleus for transcription. These genome copies are stable and are only effectively removed by destroying infected cells. Thus, the second arm of the treatment scheme is to boost the capacity of the T-cell–mediated immune response to viral antigens expressed in infected cells.

The immunomodulatory approach was actually the first employed, with the use of interferon-$\alpha$ (IFN-$\alpha$). IFN-$\alpha$ is successful in stimulating T-cell–mediated removal of infected cells in a small proportion of patients (404). However, most patients do not respond and the side effects are often severe, so affecting the immune system alone has not proven a sufficient treatment (405, 406).

There is evidence that preventing reinfection with effective inhibitors of the replicative cycle is sufficient to allow eventual clearance of infected cells by the immune system, with or without immunomodulators. However, the course of treatment is long. A study of the clearance rate of virus showed a biphasic re-

sponse. In the initial phase, virus was removed with a half-life of 1.1 days, whereas in the second phase, the half-life was 18 days (407). Another study estimated a clearance rate for the second phase of 10–100 days (121). These two phases likely reflect the clearance of free virus and infected cells, respectively. If infected cells are cleared with a half-life of 18 days, removal of infected cells from the liver would be likely to take in excess of 2 years. The implications of this need for prolonged treatment are that drugs need to be well tolerated, with few side effects to keep patients on treatment, and that treatment regimens need to be able to deal with the inevitable accumulation of drug resistance mutations in the viral target.

Attempts to interrupt the replicative cycle have focused almost exclusively on the viral polymerase. Many of the compounds in use or under development as HBV polymerase inhibitors have their historical roots in either herpesviruses or HIV drug development. Most of the HBV inhibitors fall into two groups: (1) L-analogs of pyrimidine nucleosides or (2) purine analogs with modified sugar groups.

Before inhibitors of HBV polymerase are discussed, a note about nomenclature is needed. Historically, there has not been a universally accepted numbering system for HBV proteins. Mutations in HBV polymerase are reported by two numbering systems. For instance, the two most common mutations are M552I/V and M550I/V or I528M and I526M, depending on the source of the genome used. To standardize nomenclature, it has been recommended that a numbering system be used that starts with 1 at the consensus amino terminus of polymerase rather than at the beginning of the open reading frame, with each protein given a two-letter designation to identify it (408). With this system, the two mutations noted above become rtM204V/I and rtL180M. The numbering system used most commonly in the existing literature has been used in this chapter (M552I/V and I528M for the examples cited). Hopefully the proposed change in numbering will provide more clarity to the nomenclature in the future.

**3.2.1 Pyrimidines.** The most successful of the pyrimidine analogs in development thus far have been L-nucleoside analogs of active

D-nucleosides, including lamivudine (**63**), emtricitabine (FTC) (**64**), clevudine (L-FMAU) (**18**), and Fd4C (**65**) . These L-nucleo-

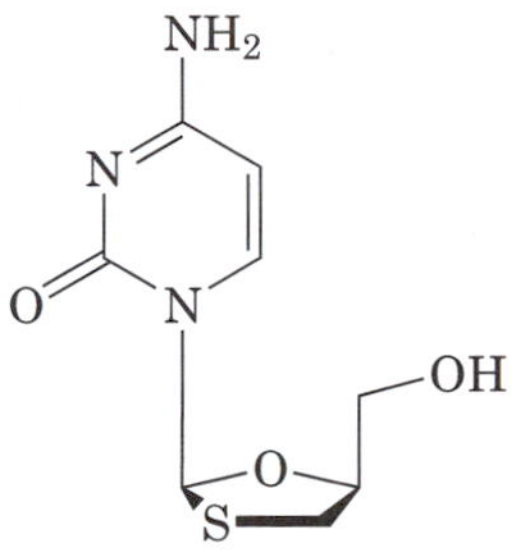

(**63**)

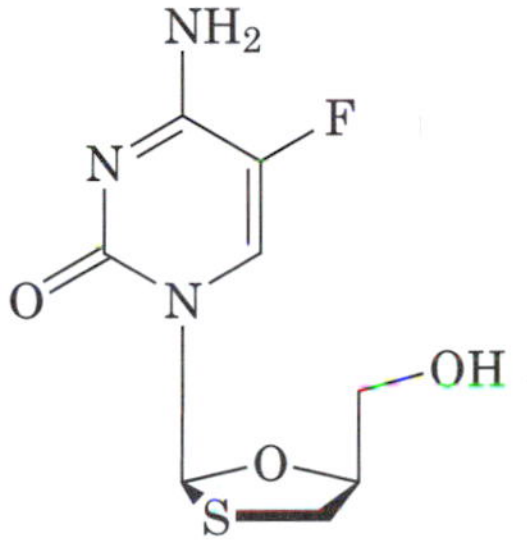

(**64**)

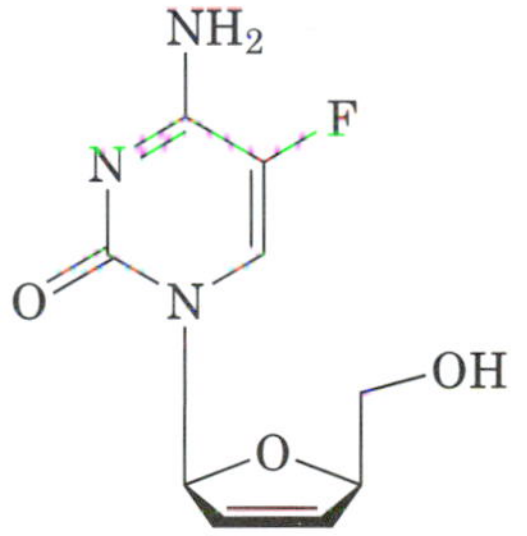

(**65**)

sides show potent activity while lacking the toxicity of the D-enantiomers. Some of them are quite specific for HBV, whereas other show activity against HIV and certain herpesviruses as well. The structure-activity relationship of these compounds as well as additional L-nucleosides have been elegantly examined by Bryant et al. (409), and the following conclusions can be drawn from their observations. (*1*) The specificity for HBV versus retroviruses derives primarily from the presence of a 3' hydroxyl on the ribose. (*2*) The lack of a 3' substituent resulted in retention of activity, but the specificity for HBV was lost. (*3*) Halogenating the 5 position on the pyrimidine ring reduced potency but did not affect specificity for HBV. (*4*) Retaining the 3' hydroxyl but changing its orientation (L-xylo-nucleoside) resulted in the loss of anti-HBV activity. These observations were made with cytidine and thymidine derivatives and were also extended to adenine derivatives.

Lamivudine is approved for treating hepatitis B patients. It is a chain terminator, and all three phosphorylation events to give the triphosphate are carried out by cellular enzymes. The first phosphorylation event is by a cellular deoxycytidine kinase rather than a viral kinase; despite this, cytotoxicity is low because all of the cellular polymerases have a very low affinity for lamivudine. Lamivudine triphosphate concentrations reach high levels, and the half-life of the triphosphate form is long, 17–19 h. Oral bioavailability is high (greater than 85%) (120).

Lamivudine is an effective initial treatment for HBV, but development of resistance is a major concern. HBV replicates through a genomic RNA intermediate, and the reverse transcriptase of HBV shares the characteristic of other reverse transcriptases of having poor fidelity. The incorporation of mismatches allows a rapid accumulation of mutations that result in resistance. For HBV polymerase, the most common mutation noted with lamivudine treatment is Met 552 to Val or Ile (M552I/V). This mutation changes a highly conserved motif (YMDD), and results in up to a 2-log increase in $EC_{50}$ of lamivudine (410). The price that the virus pays for this protective effect is an approximate 100-fold decrease in replication efficiency, although there are additional mutations (e.g., Leu[528] to Met) that partially compensate the loss of polymerase activity. Resistance is discussed further below.

Lamivudine has good potency against HBV in cell culture ($EC_{50} = 0.008 - 0.116 \mu M$) (411–415). The initial response to lamivudine at doses of 100–300 mg/day *in vivo* is good, with seroconversion and greater than 2-log loss in viral titers occurring over the first 6–12 months of treatment in greater than 80% of the patients (416). However, the response rate starts declining at some point between 1 and 2 years of treatment. In one study, the response rate was 96% at 12 months of treatment, but by 30–36 months it had dropped to 43% (416). ALT levels rebounded, indicative of liver damage, and virus titers increased. These changes were associated with a mutation in the YMDD motif (M552I/V).

For patients that respond successfully to lamivudine, both the length of treatment and how treatment is withdrawn seem to be important. In about 15% of patients, "hepatitis flares" were experienced after cessation of treatment, with significant increases in ALT and virus titers (133, 417). These flares may be a result of stimulation of clearance of infected cells carrying double-stranded DNA genomes (132, 406, 418, 419), and the majority of them resolve by themselves. However, in some cases, the flares may be severe enough to require readministration of lamivudine to limit the extent of the hepatitis. Possibly related to this phenomenon, an increase in the immune response to HBV was noted in 83% of patients treated with lamivudine during a time that corresponded with the decrease in viremia (420). Thus, whereas high levels of virus production have a depressive effect on T-cell–mediated HBV response, some level of virus synthesis may be needed to stimulate the immune system to remove infected cells.

The appearance of a mutation in the precore region that blocks expression of HBeAg is associated with a poor prognosis, presumably because lack of HBeAg results in less efficient T-cell–mediated destruction of infected cells. Lamivudine seems to select against precore mutants initially. In six patients that started lamivudine treatment expressing a precore mutant, all had reverted to wild-type by approximately 12 months of treatment. However, these mutants reappeared after prolonged therapy (421).

Organ transplantation poses an additional burden on treatment because of the immunosuppressive agents that are used to protect the new organ. Lamivudine has demonstrated potential as a prophylactic treatment before liver transplant. The extent of replication before a liver transplant is a predictor of recurrence of HBV (422). Lamivudine has been used to suppress HBV replication in liver transplant candidates both before and after transplantation. As an example, six patients with two to four times normal ALT levels were placed on daily 150-mg doses for 4 months, at which time all six were negative for HBV DNA and had ALT less than 1.2 times normal. After 6–15 months of treatment, no viral rebound had occurred (423). In another study, posttranplantation treatment with lamivudine resulted in loss of serum HbsAg in 64% of patients (424). Sixty percent of HBV-positive renal transplant patients treated with lamivudine for 1–4 months after transplantation responded to the treatment (425, 426), but when patients were followed after cessation of treatment, almost all of them rebounded (426). This is not a surprising result, because immunosuppression should result in poor clearance of infected cells.

Phosphorylation of lamivudine to the triphosphate form (3TCTP) is enhanced by hydroxyurea, methotrexate, or fludarabine, such that pool sizes increase by up to 3.5-fold (427). These compounds decreased the pool size of the endogenous competitor, dCTP. The result may be a moderate improvement in the efficacy of lamivudine. However, levels of 3TCTP needed to inhibit wild-type virus are readily achievable in the absence of this treatment, whereas the $EC_{50}$ against the M552V/I mutant is 50-fold higher than for wild-type. This is well out of the range of the effect of these compounds on 3TCTP levels.

L-dC (66), L-dT (67), and L-dA were tested for activity against a panel of 15 viruses, including herpesviruses, and were only active

(66)

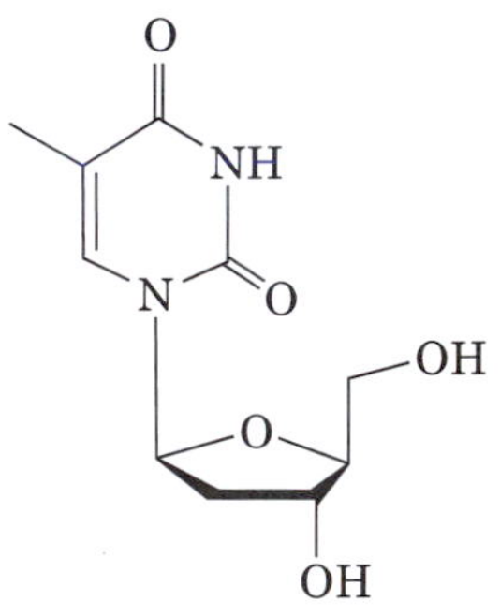

**(67)**

against HBV (409). They were also inactive against human DNA polymerases $\alpha$, $\beta$, and $\gamma$. Consistent with this observation, these nucleosides showed no cytotoxicity against a variety of cell lines. Finally, in infected woodchucks, L-dC and L-dT gave a 6- to 8-log reduction in serum virus titers after 3 weeks of treatment (10 mg/kg/day). No signs of toxicity were noted over 12 weeks of treatment. These pyrimidine analogs seem to be promising candidates for clinical development.

Another example of pursuit of an L-isomer of a D-nucleoside is 2′,3′-dideoxy-2′,3′-didehydro-$\beta$-L-5-fluorocyticine (Fd4C) (**65**) (413). Fd4C has potency in cell culture about 15- to 30-fold greater than lamivudine (EC$_{50}$ = 2–200 n$M$), with minimal cytoxicity (CC$_{50}$ = 7–20 vs. >50 $\mu M$ for lamivudine) (413, 428–430). The derivative lacking the flourine (d4C) was four- to fivefold less active than Fd4C (413). Similar results were noted in comparing lamivudine vs. Fd4C treatment of infected woodchucks (431). A 4-week treatment of HBV in a duckling model with Fd4C gave strong suppression of viremia and gave no evidence of toxicity (429). Higher levels of Fd4C triphosphate are maintained in cells, in part because conversion of Fd4C from the diphosphate to triphosphate form is more efficient than for lamivudine. Turnover of Fd4C metabolites is much slower than for lamivudine, and this is reflected in a longer time to reappearance of HBV DNA after removal of the drug for Fd4C (428). Fd4C has the potential for once daily dosing, which gives it an advantage over lamivudine. It suffers less loss of potency in viruses with mutations at amino acids 552 and 528 (432). The future of Fd4C as an

HBV antiviral may hinge on whether its increased potency, higher intracellular triphosphate levels than lamivudine, and better potency against resistance mutants allow it to retain effectiveness against these mutants.

An $S$-acyl-2 thioethyl-monophosphate (SATE) prodrug form of Fd4C (**68**) was tested

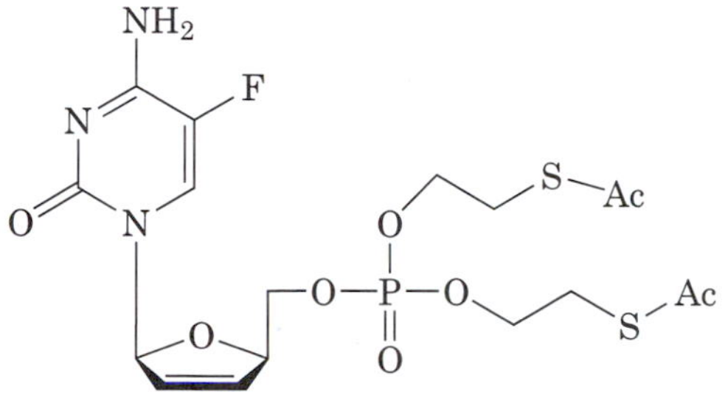

**(68)**

as a means of increasing the intracellular concentration of Fd4C-monophosphate. The prodrug was eightfold more potent than the parent and showed fourfold lower toxicity (430).

Additional members of this group include FTC (**64**) and L-FMAU (**18**). $\beta$-L-2′,3′-dideoxy-5-fluoro-3′-thiacytidine (emtricitabine, FTC) has good potency against HBV and low toxicity (cell culture EC$_{50}$ = 40 n$M$, TC$_{50}$ > 700 $\mu M$) (433). In woodchucks, it showed potency equivalent to lamivudine when administered at a dose of 30 mg/kg/day and no evidence of toxicity (434). 2′-Fluoro-5-methyl-$\beta$-L-arabinofuranosyluracil (L-FMAU) was a byproduct of the search for HIV antivirals. It was inactive against HIV, but inhibited HBV. Use of the D-enantiomer has been limited by both toxicity, primarily of the mitochondrial DNA polymerase (435) and by deamination (330). L-FMAU has 20-fold greater potency than the D-isomer (EC$_{50}$ = 0.1 versus 2.0 $\mu M$) and a selectivity index of >2000 versus 25 for the D-isomer (330, 436). In the duck model, oral administration of 40 mg/kg/day for 8 days gave a 72% decrease in peak viremia, with no short-term toxicity noted (437). Toxicity was also not noted in a 30-day study in mice or in 4-week or 3-month studies in woodchucks (438, 439). A short-term study in woodchucks demonstrated that the reduction in viral DNA synthesis was very rapid, with a 3-log reduction within 3 days of treatment. Viremia remained suppressed for a significant length of time after withdrawal of treatment, with one-half of the animals showing suppression of viral DNA at 10–12 weeks post-withdrawal (439). L-FMAU is a substrate for at least three cellular kinases, thymidine kinase, a mitochondrial deoxypyrimidine kinase, and deoxycytidine kinase (440). Related 2'-fluoro-5-methyl-β-L-arabinofuranosyluridine analogs were inactive, including 5 substitutions with halogens or alkyl groups. Cytosine and 5-iodocytosine showed weak to moderate activity, 15- and 50-fold less than FMAU, respectively (441).

2',3'-Dideoxy-3'-fluoro-D-ribonucleosides have also been shown to have potent antiviral activity, but excessive toxicity. Because other L-enantiomers have shown less toxicity while maintaining potency, L-enantiomers of these fluoro compounds were tested. They were inactive with the exception of the cytosine analog (69), which showed moderate activity

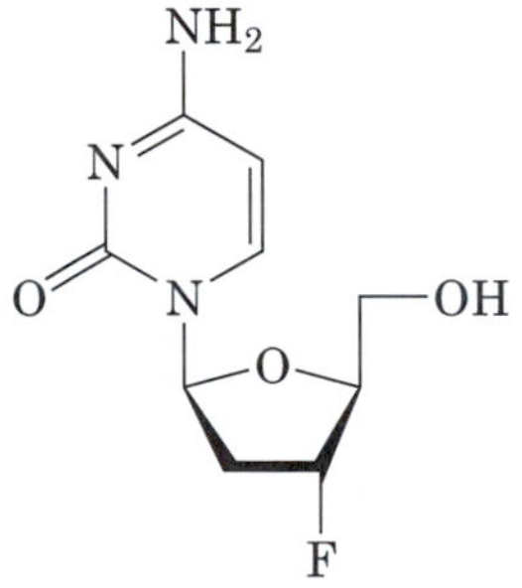

(69)

against HBV (442). Moving the fluorine to the 2' position gave a 5-fluorocytosine analog with potency against HBV about fourfold better than that of lamivudine (443). Several 2'-deoxy-2',2'-difluoro-L-erythro-pentofuranosyl nucleoside derivatives were synthesized to try to combine the antiviral effect and lack of toxicity of the L-nucleosides and the antiviral effect of 2'-deoxy-2',2'-difluoro nucleosides (444). While these compounds were not toxic, they did not show activity against HSV-1 or -2 or HBV.

Some β-D-2',3'-dideoxy-5-chloropyrimidine compounds have efficacy as HIV inhibitors, but the corresponding L-enantiomers do not. However, selected L-enantiomers of this class of compounds have shown activity against HBV. As an extension of this work, β-L-2',2'-dideoxy-5-chlorocytidine (70) and β-L-2',2'-dideoxy-3'-fluoro-chlorocytidine (71) were tested against HBV. They are active,

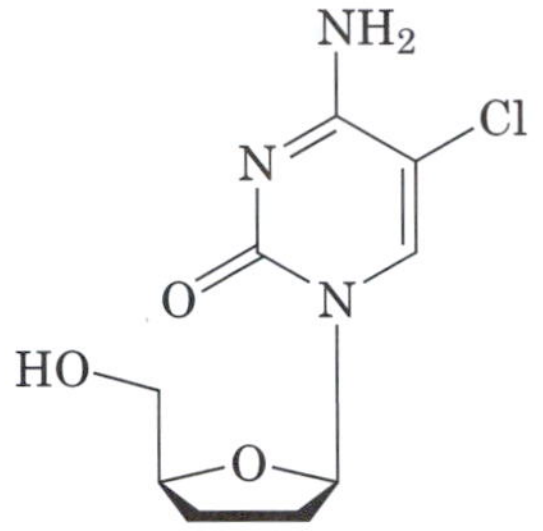

(70)

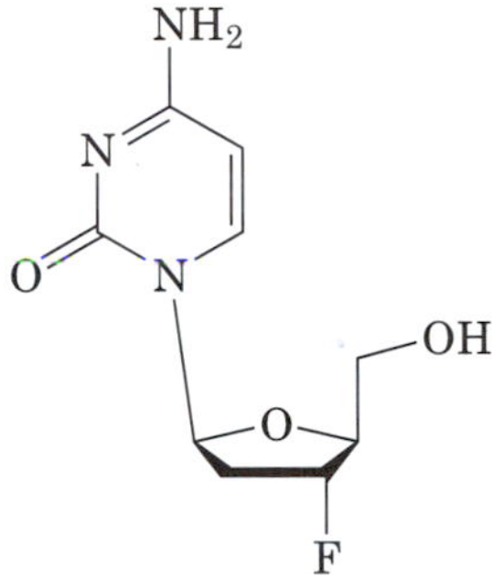

(71)

but were 100- to 800-fold weaker than lamivudine (445).

Cytallene (1-(4'-hydroxy-1',2'-butadienyl)-cytosine) (72) is a potent HBV inhibitor ($EC_{50}$ = 80 nM) (446). It is somewhat toxic ($TC_{50}$ = 12 μM), giving it a selectivity index of 150. It is efficiently phosphorylated by human deoxycytidine kinase (447). Cytallene is unique amongst these pyrimidines in being an acyclic analog.

**3.2.2 Purines.** GCV, PCV, and famciclovir (the prodrug form of PCV) have some use against HBV (448, 449). Whereas their potency approaches that of lamivudine, they have several disadvantages that limit their use. GCV and PCV require intravenous administration, and all three have poor side ef-

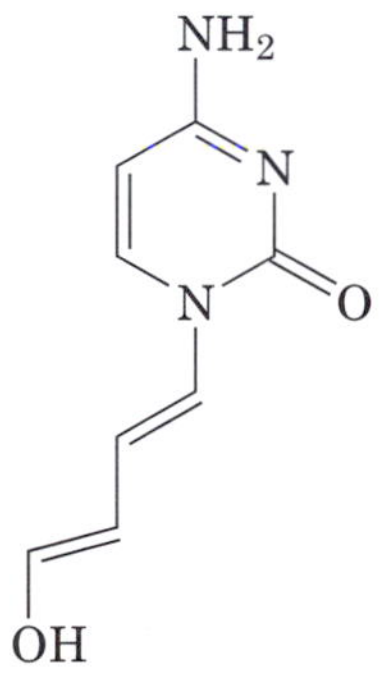

(72)

fect profiles relative to lamivudine. Further, as discussed below, they show cross-resistance to lamivudine. As monotherapies, they have no significant advantages over the use of lamivudine, but they may have use as part of a combination therapy regimen; this is discussed later.

ACV-triphosphate is an effective inhibitor of the HBV DNA polymerase ($IC_{50} < 1$ $\mu M$). However, ACV is ineffective against HBV because of the lack of phosphorylation of ACV to the monophosphate form. When ACV was supplied as a monophosphate, in the form of the orally bioavailable 1-O-hexadecyl propanediol-3-P-acyclovir (HDP-P-ACV) (73), it

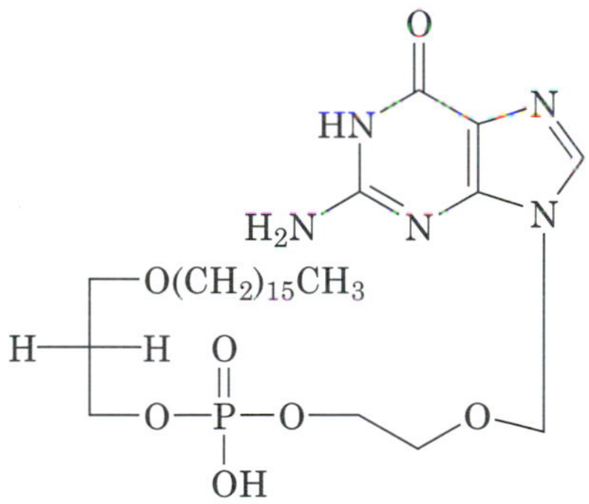

(73)

was effective in inhibiting virus production in woodchucks (450, 451). When HDP-P-ACV was given at 10 mg/kg twice daily, serum viral DNA was reduced by 95% after 4 weeks. ACV given at five times that molar dose had no effect on serum viral DNA levels. No evidence of

toxicity was noted for HDP-P-ACV at doses up to the maximum dose of 30 mg/kg twice daily.

An S-acyl-2 thioethyl-monophosphate (SATE) derivative of ACV (74) was also shown

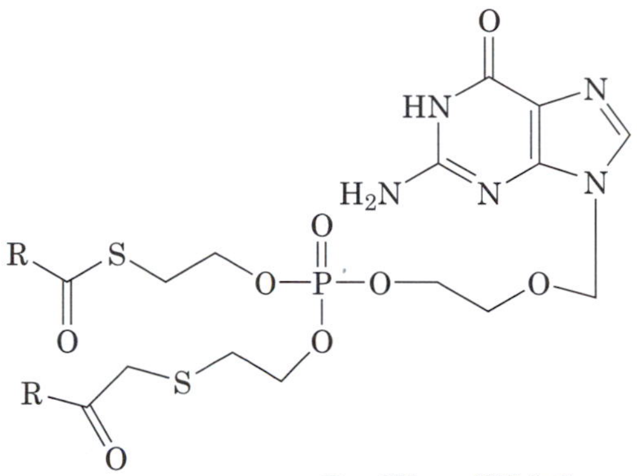

R = $CH_3$ or $(CH_3)_3C$

(74)

to have potent anti-HBV activity in the duck model in vitro and in vivo (452, 453). The methyl derivative exhibited an $EC_{50} = 30$ nM in infected primary duck hepatocytes, whereas the tert-butyl derivative exhibited an $EC_{50} = 0.6$ nM. The equivalent value for ACV was 100 nM. These compounds were more toxic than ACV, particularly the methyl derivative ($TC_{50} = 5$ $\mu M$). The tert-butyl derivative was less toxic ($TC_{50} = 120$ $\mu M$), and its high potency gave it a selectivity index of 200,000. The tert-butyl derivative was also effective in reducing viral titers in vivo.

Entecavir (BMS-200475) (75) is a guanosine analog that is approximately 10-fold more po-

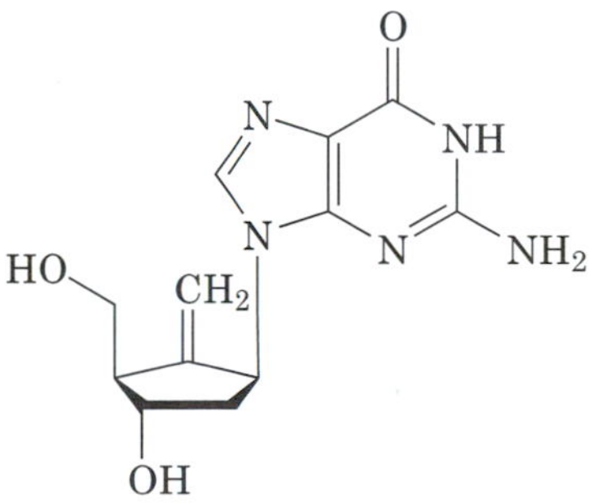

(75)

tent than lamivudine (454). It is a competitive inhibitor with respect to dGTP binding, and is capable of inhibiting the priming, reverse transcription, and DNA-dependent DNA synthesis steps of HBV polymerase. It is not a chain terminator; it causes termination two to three residues downstream of the site at which it is incorporated (454). In a 3-month study in woodchucks, doses of 0.1 or 0.5 mg entecavir/kg body weight reduced viral titers from $10^{10}$ to less than $10^3$ virions/mL (455). Similarly, in a 21-day study employing ducklings infected with duck hepatitis B virus, entecavir treatment resulted in a 2- to 3-log decrease in viral titer compared with less than a 1-log decrease with lamivudine (456).

A recent clinical study of entacavir in hepatitis B patients has been reported by de Man et al. (457). The 28-day double-blind placebo-controlled study tested four daily doses of entacavir (0.05, 0.1, 0.5, 1.0 mg). All doses of entacavir show a greater than 2-log mean reduction in viral load. Approximately 25% of the patients in the study (42 patients) exhibited HBV DNA levels below the levels of detection. The drug was well tolerated by all patients and there was no detectable change in ALT levels before or after the study. Patients receiving the 0.5 and 1.0 mg doses showed a slower return to baseline HBV DNA levels than the lower doses.

Adefovir (9,(-2-phosphonomethoxyethyl)-adenine, PMEA) (**76**), 1-$\beta$-2,6-diaminopurine

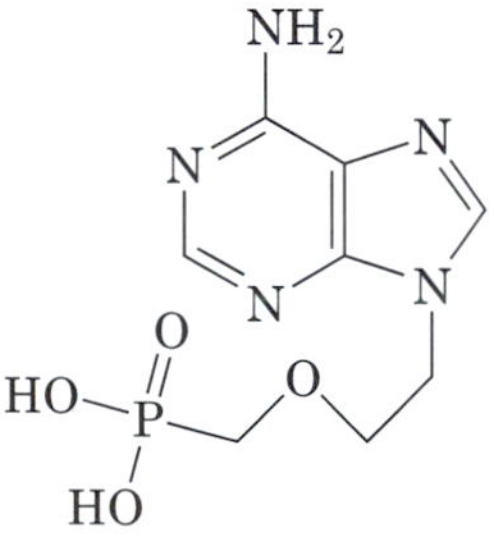

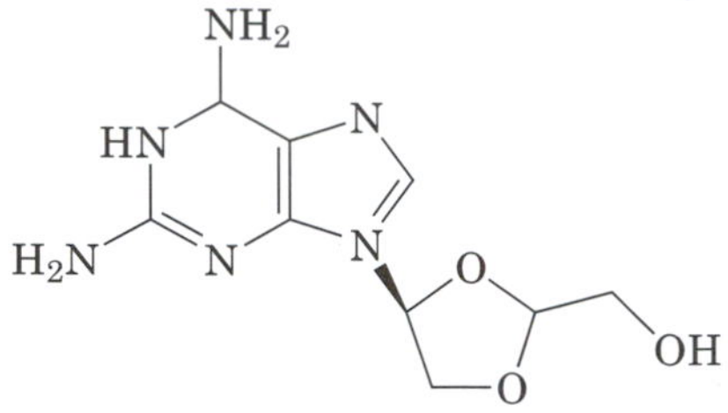

(77)

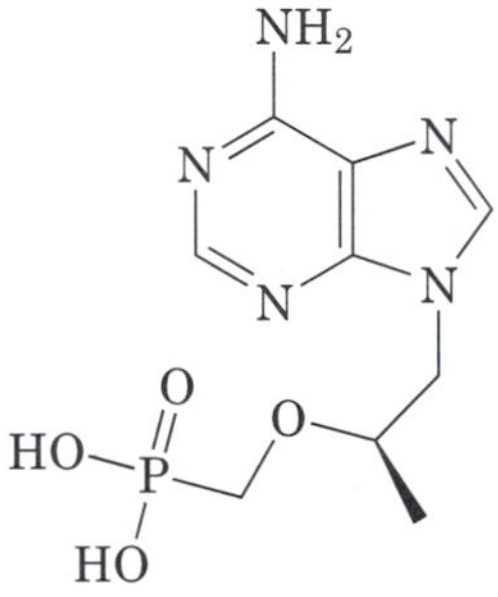

(78)

phonate analog of adenine monophosphate, which shows activity against herpesviruses and retroviruses, in addition to HBV (459). Adefovir is not orally bioavailable, but a prodrug form, adefovir dipivoxil (**79**) has 40% bio-

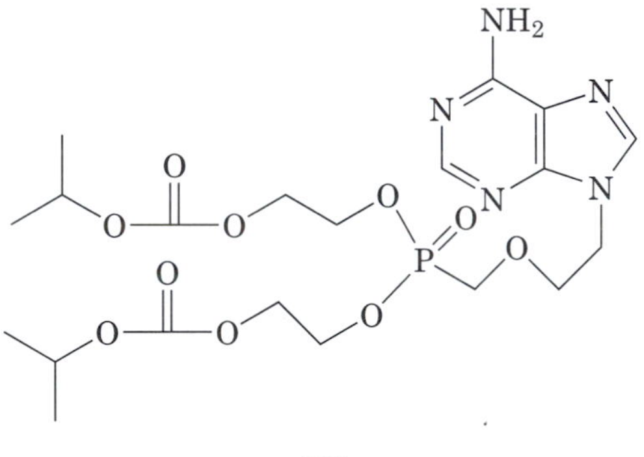

(76)

(79)

dioxalane (DAPD) (**77**), and 9,(3-hydroxy-2-phosphonomethoxypropyl)adenine (PMPA) (**78**) are purine analogs with potent anti-HBV activity (302, 458). Adefovir is an acyclic phos-

availability in humans (460). A 2-log drop in viral DNA secretion was noted from woodchuck hepatocytes infected with WHBV over 27 days of treatment (461). Treatment of woodchucks for 12 weeks with 15 mg adefovir dipixovil per kg body weight resulted in a 2.5-log decrease in serum viral DNA (462). A study of HBV production and clearance with adefovir dipivoxil treatment indicated that virus production was reduced to 0.7% of its pretreatment levels (407). In short-term studies in humans, adefovir dipivoxil caused an 1.8- to 4-log decrease in serum HBV DNA within 2 weeks (407, 460). A 48-week study in lamivudine-resistant patients resulted in a mean 4-log decrease in serum HBV DNA titers (463). HIV-positive patients experienced some liver toxicity, with elevations in ALT occurring in 50–60% of HIV-infected patients and in none of the HIV-negative patients (460, 464). Side effects were primarily associated with the gastrointestinal tract: abdominal pain, diarrhea, nausea, vomiting, etc. These side effects were measured in a 4-week study, which was not long enough to assess whether patients adapted to the drug with time.

Lobucivar (**80**) is a potent inhibitor of HBV replication, with some characteristics of adefovir. It has a shorter half-life than adefovir (459) but was 10-fold more potent than lamivudine against purified polymerase (454), and it was effective in short-term studies in woodchucks at 10–20 mg/kg/day (465). It does not show cross-resistance with lamivudine (466). However, it was withdrawn from clinical trials because of the appearance of tumors during carcinogenesis studies in rodents (467).

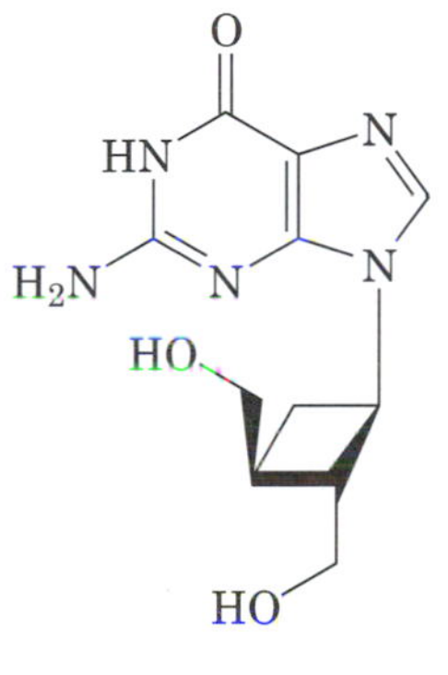

**3.2.3 Resistance in HBV.** Resistance mutants are the factor limiting successful treatment with lamuvidine (405, 468–471). A

(80)

study of 27 patients treated for 2–4 years showed the following pattern of response to lamivudine treatment (417): (1) a rapid decrease in viremia, with serum viral DNA levels decreasing 4–5 logs in the first year; (2) an improvement in liver histology over this time period; (3) appearance of resistant virus starting at about 8 months, which eventually appeared in one-half of the patients. The appearance of resistance was inversely correlated with the level of virus suppression-resistance mutants developed in 76% of the HbeAg$^+$ patients, but only 10% of the HbeAg$^-$ patients. The appearance of resistance mutations is also correlated with the phenotypic state of the liver, with high serum ALT levels being predictive of more rapid generation of drug resistance (472).

A mutation of Met$^{552}$ to Val or Ile (M552V/I), alone or in combination with Leu$^{528}$ to Met (L528M) increases the lamivudine EC$_{50}$ by 2–4 logs (473–475). The M552I/V mutation also causes a 1-log decrease in viral DNA replication (476). Because of this loss of fitness, discontinuation of lamuvidine treatment leads to reappearance of the wild-type virus, although if lamivudine treatment is restarted the resistant virus rapidly returns (417). This decrease in virus fitness is largely compensated for by L528M (476). The cause of the resistance because of the M552I/V mutation is decreased binding affinity of lamivudine by HBV polymerase, resulting from steric hindrance by the side chain of Val or Ile (477).

An additional mutation conferring resistance to lamivudine, A529T, has been reported in 3 of 23 patients included in a study of resistant patients (469). This mutation also introduces a stop codon in the gene coding HbsAg, which impairs its secretion. The effect on HbsAg may limit the spread of this mutation.

One of the lessons of drug resistance in other viruses, particularly HIV, is that treatment with combinations of drugs is likely to be important. Therefore, it is important to understand which combinations of drugs generate cross-resistance. In addition to lamivudine, the combination of M552I/V and L528M mutations also provide cross-resistance to FCV, PCV, ddC, FTC, AZT, FMAU, and to a lesser extent, Fd4C (477, 478). FCV generates an overlapping, but not identical set of resistance mutants; I528M, V521L, V555I, and T532S, but not mutations at M552, have been reported from both FCV- and lamivudine-resistant patients, (471, 479–481). It is likely that treatment of FCV-resistant patients with lamivudine would result in a more rapid appearance of M552V/I.

Adefovir is not cross-resistant with lamivudine (410, 432, 466, 482), thus providing an alternative therapy for resistant cases. A comparison of inhibition by lamivudine of wild-type HBV polymerase activity vs. mutant derivatives showed an increase in inhibition constants ($K_i$) of 8.0-, 19.6-, and 25.2-fold relative to wild-type for M552I, M552V, and M552V/L528M mutants, respectively. Increase in $K_i$ for adefovir was twofold or less for these mutants (410). Potency of lamivudine was decreased in cell culture by 100- to 10,000-fold by these mutants, whereas adefovir was only affected by 4- to 16-fold (476). In a study involving five patients who had developed resistance to lamivudine after 9–19 months of treatment, all responded to adefovir dipivoxil, with a 2- to >4-log decrease in viral titer (483). These decreases were maintained over the 11- to 15-month treatment period and were correlated with improvements in liver function.

The experience of HIV treatment indicates that additional therapies will be necessary to counteract developing resistance. HBV and HIV are similar with regard to the amount of virus that is produced per day and the fidelity of their polymerases. Combination therapy for HIV now typically involves three drugs, with alternative drugs available for patients who fail their initial therapy. The relatively limited selection of options currently available to treat HBV is unlikely to provide sufficient treatment.

**3.2.4 Combination Treatments.** Given the difficulty of maintaining effective long-term monotherapy, several combinations of drug treatments have been investigated. Treatment of patients that failed interferon-$\alpha$ monotherapy with interferon-$\alpha$ plus ribavirin showed improvements in a fraction of patients, but greater than one-half of the patients did not respond and the side effect profile was poor (484). When interferon-$\alpha$ was combined with lamivudine, there was a mod-

est but statistically insignificant improvement in the proportion of HBeAg seroconverters (485). The benefit of this combination seemed to depend on the extent of damage to the liver, because patients with lower baseline ALT levels showed greater improvement than those with higher ALT levels. Thus, it is possible that a subgroup of patients could be identified for which this combination of drugs could provide benefit.

In contrast to the previously mentioned beneficial effect of adding mycophenolate mofetil (MMF) in treatment of HSV and HIV, adding MMF to the treatment regimen of HBV-infected liver transplant patients who had failed lamivudine treatment did not provide any benefit (486). However, when mycophenolic acid or ribavirin were used in cell culture in combination with guanosine analogs, PCV and lobucavir, their potency was improved by as much as 10-fold (487). This effect could be reversed by adding exogenous guanosine, indicating that the increase in potency was a result of a decrease in dGTP pool size by mycophenolic acid or ribavirin. Because liver transplant patients are often treated with MMF, the prodrug form of mycophenolic acid, they represent a subset of HBV-infected patients that may derive more benefit from guanosine-based drugs than non-MMF–treated patients.

Relatively few studies have examined the effects of combination therapy in patients who were not already resistant to one of the treatments (412, 452, 488–490). A comparison of PCV, lamivudine, and adefovir in naive infection of duck hepatocytes showed that all combinations of the three treatments were at least additive (488). Cytotoxicity with all combinations was insignificant. Whereas PCV is not a useful follow-up to lamivudine in resistant patients caused by cross-resistance, the combination of PCV and lamivudine against drug-sensitive virus is more effective than either alone. Adefovir has the added advantage of suppressing the appearance of viruses resistant to either PCV or lamivudine, which should increase the length of time that those drugs will be useful. In a related study, PCV and lamivudine in combination were more effective at reducing the copy number of cccDNA than were either separately (490).

Similarly, the combination of famcyclovir (FCV) (the prodrug form of PCV) and lamivudine in woodchucks was at least additive relative to the individual drugs (489). As an example, treatment with 5 mg/kg of lamivudine reduced viremia by 75-fold, whereas 50 mg/kg of FCV showed a 10-fold decline. Combining the two treatments reduced viremia 10,000-fold.

In a short-term human trial, the combination of lamivudine and FCV was shown to be superior to lamivudine alone (491). After 12 weeks of treatment, the combination provided greater viral clearance than lamivudine alone. Viral DNA levels returned to pretreatment levels in 44% of the patients 16 weeks after termination of treatment but in none of the patients given the combination therapy. These combinations need examination in more long-term trials.

**3.2.5 Targeting Drugs to the Liver.** Two traits of the liver present an unusual opportunity for increasing potency and limiting the side effects of HBV treatments. First, hepatocytes have receptors capable of recognizing relatively simple molecules, which can be attached to make prodrug forms of drugs. Second, the liver is one of the first organs to which orally delivered compounds are exposed; thus, increasing uptake by the liver reduces the exposure of other organs to a drug. Several attempts have been made to take advantage of these features in designing anti-HBV drugs.

The most commonly used hepatocyte-specific receptor is the asialoglycoprotein receptor. Molecules with carbohydrates that terminate with galactose are recognized and internalized by this receptor. This subject is reviewed by Fiume et al. (492); several examples follow. Adenine arabinoside monophosphate (ara-AMP, vidariabine-MP) has been conjugated to human serum albumin (HSA) and used to treat patients for 1 and 4 weeks in separate studies. It was administered at a dose equivalent to 1.5 mg ara-AMP/kg/day, a dose that is 10- to 20-fold lower than is used for free ara-AMP. At this dose it was as effective as free ara-AMP at the higher concentration. However, with the free drug, by 4 weeks of treatment, neurotoxic and other side effects appeared; these side effects were absent with

the conjugated drug (493, 494). Similar results were obtained by conjugating ara-AMP to polylysine (495, 496). This conjugate was tested in the woodchuck model. Free ara-AMP administered at 2.5 mg/kg/day had no effect on viremia, whereas the conjugate delivered at the same molar concentration of drug decreased viremia to undetectable levels. Polylysine conjugates have the advantages over albumin conjugates of being synthetic rather than derived from a blood product and of being administered intramuscularly rather than intravenously. Both have the disadvantage of no oral bioavailability, however.

Adevofir is an effective drug with a slightly less favorable side effect profile than lamivudine. For treating HBV, it has the additional disadvantage that it is taken up poorly by hepatocytes. An approach conceptually similar to the one just discussed for targeting it to the liver was taken (497). Adefovir was derivatized with glycosides (**81**) with a high affinity for the asialoglycoprotein receptor. This modification resulted in a 10-fold increase in the amount of drug taken up by the liver (52–62% versus 5% for adefovir). This caused a decrease in the amount of drug taken up by extrahepatic tissues. The kidney removes about 10-fold more adefovir from serum than does the liver. However, the ratio of liver to kidney uptake shifted by 30- to 45-fold toward the liver with the prodrug forms. The end result was a fivefold increase in potency for the smaller glycoside and a 52-fold increase for the larger.

Another approach to targeting adefovir to the liver employed a lactosylated high-density lipoprotein (498). This protein, in combination with phospholipids, can bind lipophilic drugs. Adefovir is not sufficiently lipophilic, so it was derivatized by addition of a lithocholic acid-3α-oleate (referred to as PMEA-LO) (**82**). This derivative was mixed with the protein and phospholipids and injected intravenously into rats. It was taken up efficiently by the liver, with approximately 70% of the dose localized to the liver and less than 2% found in the kidneys. This ratio is 500 times greater than that noted with free adefovir. Of the liver dose, 88% was found in hepatocytes. The adefovir was efficiently released by acid hydrolysis in lysosomes.

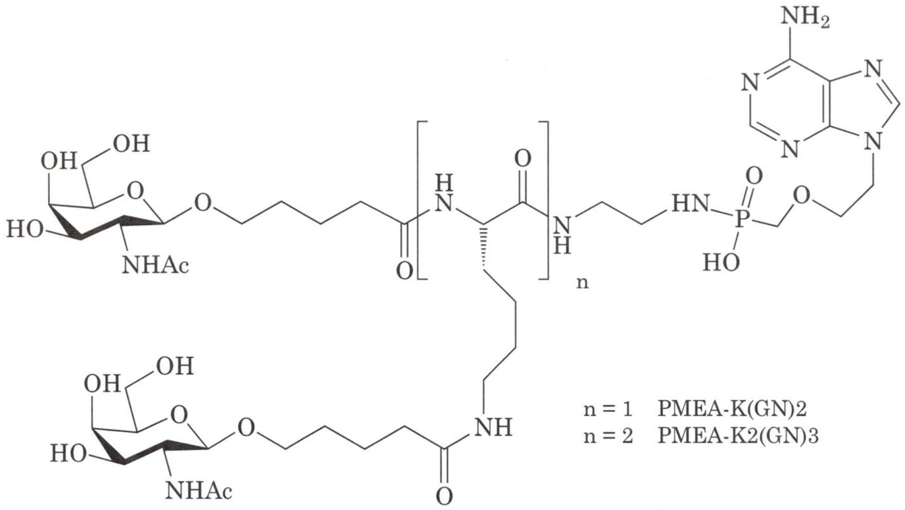

(81)

n = 1    PMEA-K(GN)2
n = 2    PMEA-K2(GN)3

A phosphatidyl derivative of lamivudine was synthesized that might be expected to be delivered preferentially to the liver (499). This lamivudine analog was comparable in activity in cell culture with the parent compound. Because lamivudine shows few toxic effects and an effective dose is attainable orally, there may be no advantage to such an approach with lamivudine. However, this compound may provide a model for targeting other compounds with less favorable characteristics.

If this liver-targeting approach can be combined with oral bioavailability, it may provide a powerful way to expand the selection of drugs that could be used without encountering harmful side effects. This drug delivery

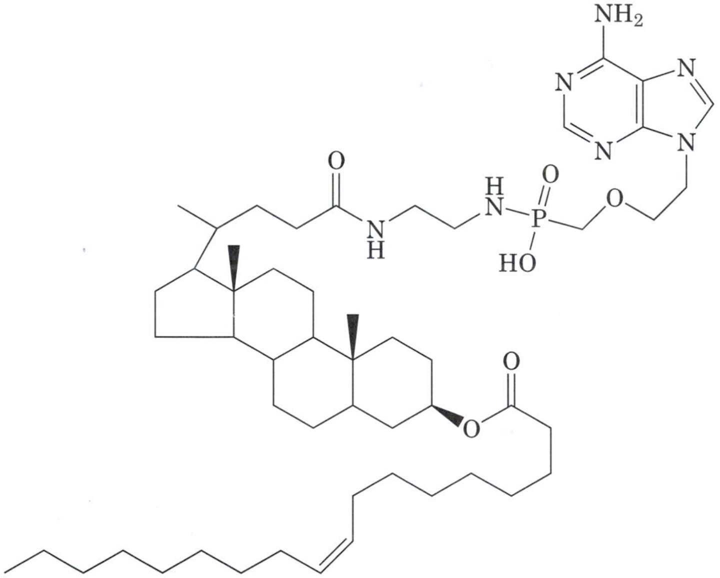

(82)

feature has a negative side, in that other cell types that can be infected by HBV may not recognize the prodrugs. These cells would then provide a reservoir of virus that is not eliminated. This aspect of these compounds merely provides another impetus to develop good combination therapies.

**3.2.6 Antisense Oligonucleotides.** Oligonucleotides that are complementary to a stretch of an RNA can, on binding to that RNA, either promote its degradation by a cellular RNase H or prevent movement of polymerases or ribosomes through that area. As an experimental tool, this approach has been used in dozens of model systems. This approach has also been applied by a few groups to the inhibition of HBV replication. The oligonucleotides face two hurdles. First, they are not orally bioavailable; second, they must avoid degradation by cellular nucleases. The first hurdle is currently cleared by intravenous injection and the second by modification of the oligonucleotide to make it more resistant to degradation.

A study of a series of antisense phosphorothioate oligonucleotides to the duck hepatitis virus genome identified two that showed effective inhibition of viral replication in isolated duck hepatocytes, with a 1- to 2-log decrease in viral DNA after 10 days of treatment (500). The most potent oligo was administered to infected ducks for 10 days to examine its potency in vivo. At a dose of 20 mg/kg body weight, viral DNA was reduced to nearly undetectable levels. No side effects were noted, although long-term treatment with phosphorothioate oligonucleotides can lead to cytoxicity. Treatment with the oligonucleotide before infection was able to block infection.

In an attempt to improve the membrane permeability and nuclease resistance of the oligonucleotides, oligoribonucleotides modified by addition of $2'$-$O$-(2,4-dinitrophenol) (DNP) at a 0.7 ratio of DNP to nucleotide were tested in the duck model (501). At ratios less than 0.5, oligonucleotides were easily degraded, whereas ratios >0.8 interfered with hybridization. The oligonucleotide inhibited purified polymerase with an $IC_{50}$ of 8–20 n$M$. The oligonucleotide was tested in ducklings with a 45-day treatment of 0.5–1 mg/kg/day.

Nine of nine treated ducks had undetectable viremia by day 25 of treatment. One of six control ducks spontaneously cleared the virus during this time. None of the treated ducks showed a reappearance of the virus during a 30-day follow-up period after treatment.

Antisense oligonucleotides offer some significant advantages. Their synthesis is routine and does not have to be reinvented for each new oligonucleotide. The number of potential targets is large, in that whereas the sequence and location in the genome affect the potency of the oligonucleotide, many sequences are likely to work. Thus, resistance can be overcome by moving to a new location; combination therapy could simply mean multiple oligonucleotides. Nevertheless, delivery of the oligonucleotide to the interior of the infected cell, especially in an orally bioavailable fashion, is still an imposing barrier to their use as drugs.

## 3.3 Papillomaviruses

Papillomaviruses have proven difficult to treat because of the paucity of targets in the latent phase. Thus, treatment regimens have focused on eliminating the infected cells rather than treating the virus directly. Methods for doing this have included surgery, cryotherapy or caustic agents applied to warts, and topical application of cytotoxins such as podophyllin (502, 503). These approaches have low success rates because of a high recurrence rate of the warts. Whereas the bulk of the wart can be removed by this method, the infected basal cells that provide the reservoir for continued viral genome replication are difficult to remove. One approach to overcoming this barrier comes from the observation that the warts may clear without treatment, presumably because of an immune response (504). This assertion is bolstered by observations that HPV-associated warts in HIV-positive patients responded to antiretroviral treatment (505, 506). Attempts to bolster the immune response led to treatment with interferon-$\alpha$ by injection into the infected tissue, but analogous to the response to interferon-$\alpha$ as an HBV treatment, interferon induction of an immune response has seen only limited success (507–509).

Induction of an interferon-mediated response by small molecule inducers of interferon has shown more success. Imiquimod (1-(2-methylpropyl)-1H-imidazo(4,5-c)quinolin-4-amine) (**62**) and the related compound S-28463 (**63**) induce an interferon-α-mediated response (510, 511). They have shown efficacy as immunomodulators against lesions caused by both herpesviruses and papillomaviruses (512, 513). In a placebo-controlled trial, 40% of patients experienced complete clearance of warts when treated with a 5% imiquimod cream for 8 weeks; no patients receiving placebo showed complete clearance (514). Three-quarters of the treated patients had a 50% reduction in wart area versus 8% with placebo. Of the patients who cleared the warts, 80% remained free of warts during a 10-week follow-up period. Greater than 50% of the patients reported mild-to-moderate localized side affects as a result of treatment, with no systemic effects reported. Similar results were noted in a 16-week study with 5% imiquimod, but 1% imiquimod was ineffective (14% versus 52% complete clearance) (515). Note that the improved results with imiquimod versus direct injection of interferon could be a result of an indirect effect of imiquimod on interferon induction. This effect is mediated by cytokines, which may also result in regulation of other factors that promote regression of the lesions. These trials demonstrated that a local interferon response is beneficial in a subset of papillomavirus-infected patients, but that it is not universally applicable. Factors that identify patients for whom this treatment is beneficial have yet to be identified.

A combination of vidaribine (**8**) and podophyllin (a DNA polymerase inhibitor and a cytotoxin) were tested in a 6-week topical application to 28 patients with cervical intraepithelial neoplasia (516). This treatment resulted in regression of the lesions and loss of detectable HPV DNA in 80% of the patients. However, 30% of patients for whom treatment resulted in regression of lesions relapsed during a 13-month follow-up period, indicating that longer treatments or additional combinations of drugs were needed for effective elimination of the proliferating cells.

5-fluorouracil (5-FU) (**83**) has been used as a topical application for treatment of warts

(**83**)

with mixed results. In two small studies, the majority of patients cleared the warts, with a low recurrence rate (517, 518). However, in a placebo-controlled trial of 40 subjects with weekly application of a cream containing the treatment for 4 weeks, fewer of the 5-FU treated patients showed regression of HPV at 4–6 months post-treatment than the placebo-treated patients, indicating that the 5-FU may have actually exacerbated the disease (519).

Photodynamic therapy mediated by 5-aminolevulinic acid (ALA) has been tested as a treatment for papillomavirus-induced low-grade cervical intraepithelial neoplasia (520). The basis of this approach is that the ALA accumulates in the proliferating tissue and sensitizes those cells to visible light. At 9 months post-treatment, 95% of the patients showed improvement in PAP smears, and 80% had no detectable virus. Side effects were minimal and transient.

Cidofovir (**5**) has shown efficacy in selectively blocking proliferation of HPV-infected cells. This is an attractive application of cidofovir because in treating papillomavirus infection it can be applied topically, which limits its negative side effect profile (521, 522). However, the mechanism for this inhibition is not immediately obvious because HPV does not encode a polymerase. Selectivity for HPV-infected cells implies a mechanism whereby the cellular polymerase of infected cells is more affected by cidofovir than that of uninfected cells. One mechanism has been identified by Johnson and Gangemi (523). In cell culture studies designed to determine the metabolic fate of cidofovir, the authors noted that in infected cells, cidofovir was readily converted to its fully phosphorylated form. However, in uninfected cells, most of the cidofovir was in the form of a choline adduct. Cidofovir inhibited proliferation of infected cells with an $EC_{50} =$

200 n$M$, whereas 1000 n$M$ inhibited normal cell growth by about 15%. This loss of proliferative capacity was maintained when drug was removed after a 1-week treatment. In uninfected cells, greater than 75% of the cidofovir was in the form of a choline adduct. In infected cells, greater than 75% of the cidofovir was phosphorylated. Thus, the concentration of cidofovir converted to polymerase substrate was considerably higher in infected cells. This factor, plus the ability to treat warts topically, may limit the toxic effects of cidofovir sufficiently to make it a practical treatment.

Using a rabbit model of papillomavirus infection, a 1% cidofovir cream applied for 18 days delayed the appearance of warts and decreased their size when applied within 1 week of infection (524). Its effectiveness decreased as the delay after infection increased. On the other hand, a related study where treatment was twice daily for 8 weeks found complete regression of the warts when treatment was delayed until 4 weeks after infection (525). Adefovir (76) also gave moderate activity. In about 50% of the warts treated with cidofovir, recurrences were noted with longer monitoring times. When this treatment was combined with vaccination with DNA coding for papillomavirus proteins E1, E6, and E7, the recurrence rate dropped from 53 to 15% (526). Vaccination alone was not an effective treatment in this study.

Treatment of 15 women with stage III cervical intraepithelial neoplasia with 1% cidofovir three times every other day resulted in removal of histological signs of the lesion in 7 of 15 patients. Four of those patients had undetectable levels of papillomavirus DNA. Two patients did not respond, and the remainder showed limited responses (527).

Respiratory papillomatosis is a rare but often debilitating and difficult to treat disease. A case has been reported where cidofovir alone was insufficient at reversing the growth of lesions in an advanced case (528). However a combination of treatment with cidofovir (5 mg/kg every 2 weeks) and interferon $\alpha$-2b led to a significant reduction in pulmonary lesions and complete regression of endobronchial lesions after 12 months of treatment.

The above studies concentrated on elimination of infected cells; few attempts to take on the virus directly have been reported. One such approach has been to identify compounds that cause release of zinc from the zinc binding sites if E6 (529, 530). Loss of zinc is associated with loss of capacity of E6 to bind to cellular proteins with which it interacts. In this case, the goal is not to block viral propagation but to interfere with the progression of the infected cell to malignancy. An initial screen identified compounds that were able to eject zinc from E6 and block binding of E6 to two of its cellular cofactors. One of these compounds, 4, 4'-dithiodimorpholine (84) was

(84)

weakly active in cell culture (EC$_{50}$ = 50–100 $\mu M$). No toxicity was noted, indicating that this compound showed specificity for E6 rather than having a general effect on zinc binding proteins. Further structure-activity relationship studies indicated that the monocyclic amines with the dithiobisamine moiety were important for activity. Decrease in ring size, moving the nitrogens out of the rings, or acyclic derivatives caused loss of specificity for E6, measured by increase in cytotoxicity. Modification of the N-S-S-N moiety resulted in loss of activity. Compounds that were incapable of electrophilic attack on the sulfur atoms coordinating the zinc or incapable of cleaving to generate a radical were also inactive.

## 3.4 Polyomaviruses

Whereas the disease states are different, the life cycles of polyomaviruses and papillomaviruses are similar. They both depend on the cellular DNA replication machinery, and they use similar mechanisms to stimulate cellular proliferation. These similarities are reflected in their currently available inhibitors. Murine polyomavirus and SV40, which are closely related to JC virus and BK virus, were inhibited in cell culture by cidofovir, and weakly inhibited by HPMPA ((S)-9(3-hydroxy-2-phospho-nylmethoxypropyl)adenine) (**85**) and PMEG (9-(2-phosphonylmethoxyethyl)-guanine) (**86**) (531). Other drugs tested in this study included

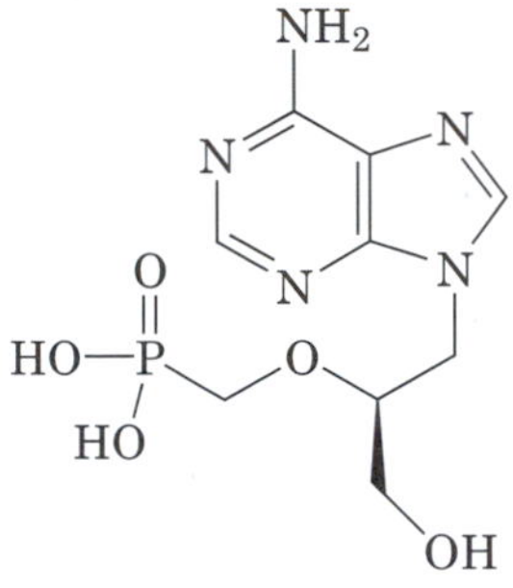

(85)

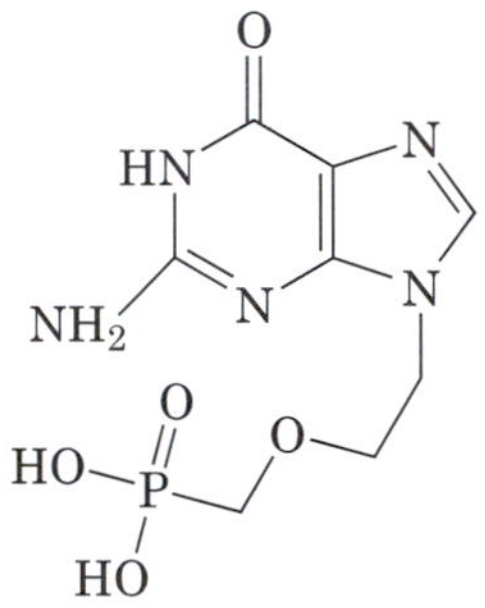

(86)

ACV (**2**), GCV (**3**), brivudine (**7**), ribavarin (**47**), and foscarnet (**6**), all of which were inactive. A neuroglial cell line persistently infected by JC virus was used to test ara-C (**87**) and cidofovir. ara-C had an antiviral effect at 10 $\mu$g/mL. Cido-

(87)

fovir showed only a minimal effect, but it was only tested up to 1 $\mu$g/mL (532). For comparison, the $EC_{50}$ for cidofovir against murine polyoma-virus was reported as 4–7 $\mu$g/mL (531).

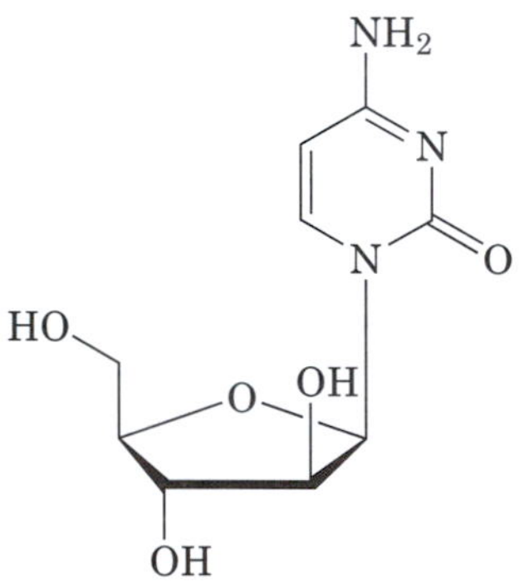

A hemangiosarcoma cell line derived by infection with mouse polyomavirus provided evidence that cidofovir was not exerting a direct antiviral effect (533). These tumor cells do not produce virus, yet they are susceptible to cido-fovir. When injected into SCID mice, tumor formation could be decreased to less than 5% of controls by injection of 100 mg/kg of cidofo-vir three times weekly. The metabolic fate of cidofovir in these cells was not examined, but it possible that its selectivity for the tumor cells is caused by an increased conversion to the active form as described for papillomavi-rus-infected cells (523). The case for the anti-polyomaviral activity of cidofovir is bolstered by observations of improvements in the outcome from AIDS-related multifocal leukoen-cephalopathy when cidofovir is included in the treatment regimen (534, 535).

## 3.5 Adenoviruses

Adenoviruses have received relatively limited study as drug targets. Again, cidofovir shows antiviral activity. Tested in a rabbit ocular model, topical application of 1% or 5% cidofo-vir cream significantly reduced adenoviral ti-ters (536). A small clinical study of seven pe-diatric bone marrow transplant patients with clinical signs of adenovirus infection showed efficacy for systemic cidofovir treatment (537). The patients were treated with 5 mg/kg weekly for 3 weeks and then every other week. Five of the seven responded to treatment with loss of detectable viral DNA and improved clinical symptoms. Comparison of cidofovir to S-2242 (**36**) and HPMPA (**88**) in a cell culture model showed that S-2242 was 30-fold more active than cidofovir, and HPMPA was sixfold more active (538).

## 3.6 Poxviruses

Poxviruses encode more of the machinery that they need to complete their life cycle than any of the other viruses discussed here. They therefore encode a wide variety of potential drug targets, but extensive efforts to develop anti-poxvirus drugs have not occurred. De

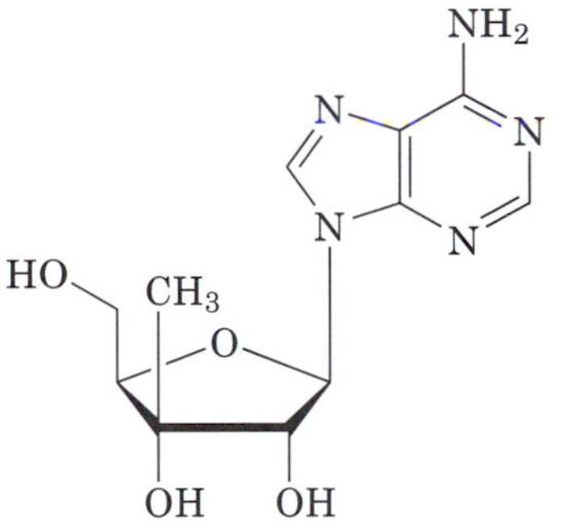

(88)

Clercq (539) has extensively reviewed the classes of known targets for which inhibitors exist. Whereas potent inhibitors of poxvirus replication have been identified, most of them validate their targets but are not practical drug candidates. The focus here will be on established drugs or likely candidates. In brief, some of the targets of effective inhibitors include IMP dehydrogenase, S-adenosyl homocysteine hydrolase, OMP decarboxylase and CTP synthetase, thymidylate synthase, and DNA polymerase. Detailed discussions of these inhibitors and their mechanism of action are given in Ref. 539.

Inhibition of IMP dehydrogenase by ribavirin effectively inhibits vaccinia virus replication ($EC_{50}$ = 4–20 $\mu$g/mL) (540). Ribavirin has shown efficacy in rabbit and mouse models applied topically to treat keratitis or injected in combination with cidofovir to treat systemic infection (541, 542). Nucleoside analogs showing activity include araA (543), 3'-C-methyl adenosine (88) (544), 8-methyl adenosine (89) (545), S-2242 (36), HPMPA (546), and cidofo-

vir (547). HPMPA inhibited vaccinia virus replication in cell culture with an $EC_{50}$ = 0.3–0.7 $\mu$g/mL (546). Cidofovir was 10-fold less potent; however, its improved side effect profile relative to HPMPA has resulted in its development and licensing as a drug. Cidofovir was also effective against parapoxviruses that infect humans and domestic animals (548).

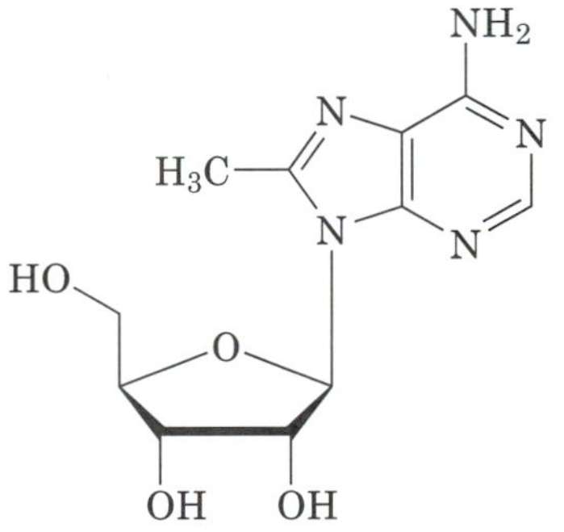

HPMPA and cidofovir were both active in preventing mortality caused by cowpox infection in mice when given subcutaneously at doses of 1–20 mg/kg/day (547, 549–551). In an infection regimen where untreated mice die between 6 and 8 days post-infection, cidofovir had some efficacy as a prophylactic treatment. When a single dose was given as much as 16 days before infection, survival rates increased from 0 to 50%. When treatment was given from day −4 to day 2, survival rates were greater than 90% (550). Another study designed to identify an easier delivery route for cidofovir showed that intranasal administration of 10–40 mg/kg of the drug was sufficient to provide greater than 90% protection from viral infection (551).

S-2242 showed potency equivalent to HPMPA in cell culture ($EC_{50}$ = 0.4 $\mu$g/mL) (359) and was more effective in infected mice. HPMPA and cidofovir were able to limit infection in normal mice, but were unable to prevent mortality in SCID mice. However H-961 (46), the prodrug of S-2242, provided complete protection against vaccinia virus infection in SCID mice (277).

## 3.7 Parvoviruses

The most common clinical manifestation of parvovirus infection is anemia caused by erythrocyte hypoplasia. Direct treatments of the virus have not been pursued, but the disease is usually self-limiting if the anemia is treated. The most common treatment is provision of intravenous immunoglobulin (244, 245). There is one report as well of treatment with erythropoietin to increase red blood cell counts (552). From the point of view of medicinal chemistry, the most effective treatments for parvoviruses have been treatments for HIV. Parvovirus-induced anemia in AIDS patients can be chronic rather than acute. Restoring an immune response by effective treat-

(89)

ment of the HIV infection is often sufficient remediation for the anemia (553–556).

## 4 CONCLUSIONS

A great deal has happened since the last edition of this review. Most of the developments have been in the expansion of useful nucleoside analogs, although other biological targets such as protease and accessory replication proteins are starting to come into their own. At the same time, the bar for treatment continues to rise higher with the appearance of resistance mutants to current effective treatments. Nucleoside analogs with novel binding modes (e.g., adefovir as an adjunct to lamivudine for HBV treatment) provide a partial answer to this problem. However, effective treatment is ultimately going to require additional targets other than the polymerases and nucleoside kinases, and such compounds are starting to appear. Even within the well-established class of nucleoside analogs, side effect profiles still curb the use of many of these compounds, so additional work remains to be done to remove this limitation.

For almost one-half of the virus families discussed here, few to no treatments have been devised that target the virus directly. These viruses (parvoviruses, papillomaviruses, polyomaviruses) pose major technical challenges because of the paucity of virus-encoded activities that they present as targets. Their treatment has benefited from unexpected modes of action (or unexpected metabolic fates) of nucleoside analogs, particularly cidofovir. However, attacking virally encoded targets is likely to be an important key to treating these virus families. Hopefully expansion of the investigation of drugs with novel modes of action for herpesviruses and hepatitis B virus will provide a boost to research on these viruses as well.

## REFERENCES

1. D. M. Knipe, P. M. Howley, D. E. Griffin, R. A. Lamb, M. A. Martin, B. Roizman, and S. E. Straus, *Fields Virology*, Lippincott Williams, and Wilkins, Philadelphia, PA, 2001.

2. B. F. Ladin, M. L. Blankenship, and T. Ben-Porat, *J. Virol.*, **33**, 1151–1164 (1980).

3. B. F. Ladin, S. Ihara, H. Hampl, and T. Ben-Porat, *Virology*, **116**, 544–561 (1982).

4. L. P. Deiss and N. Frenkel, *J. Virol.*, **57**, 933–941 (1986).

5. P. E. Boehmer and I. R. Lehman, *Annu. Rev. Biochem.*, **66**, 374–384 (1997).

6. J. D. Baines and B. Roizman, *J. Virol.*, **65**, 938–944 (1991).

7. A. Forrester, H. Farrell, G. Wilkinson, J. Kaye, N. Davis-Poynter, and T. Minson, *J. Virol.*, **66**, 341–348 (1992).

8. D. C. Johnson and M. W. Ligas, *J. Virol.*, **62**, 4605–4612 (1988).

9. D. C. Johnson, M. R. McDermott, C. Chrisp, and J. C. Glorioso, *J. Virol.*, **58**, 36–42 (1986).

10. R. Longnecker, S. Chatterjee, R. J. Whitley, and B. Roizman. *Proc. Natl. Acad. Sci. USA*, **84**, 4303–4307 (1987).

11. R. Longnecker and B. Roizman, *Science*, **236**, 573–576 (1987).

12. B. Roizman and A. E. Sears, in D. M. Knipe, B. N. Fields, P. M. Howley, Eds., *Fields Virology*, Lippincott-Raven, Philadelphia, PA, 1996, pp. 2231–2295.

13. R. Baer, A. T. Bankier, M. D. Biggin, P. L. Deininger, P. J. Farrell, T. J. Gibson, G. Hatfull, G. S. Hudson, S. C. Satchwell, and C. Seguin, *Nature*, **310**, 207–211 (1984).

14. A. J. Davison and J. E. Scott, *J. Gen. Virol.*, **67**, 1759–1816 (1986).

15. T. Kouzarides, A. T. Bankier, S. C. Satchwell, K. Weston, P. Tomlinson, and B. G. Barrell, *Virology*, **157**, 397–413 (1987).

16. S. K. Young, N. H. Rowe, and R. A. Buchanan, *Oral Surg. Oral Med. Oral Pathol.*, **41**, 498–507 (1976).

17. R. G. Douglas and R. B. Couch, *J. Immunol.*, **104**, 289–295 (1970).

18. W. P. Glezen, G. W. Fernald, and J. A. Lohr, *Am. J. Epidemiol.*, **101**, 111–121 (1975).

19. J. A. McMillan, L. B. Weiner, A. M. Higgins, and V. J. Lamparella, *Pediat. Infect. Dis. J.*, **12**, 280–284 (1993).

20. S. L. Deardourff, F. A. Deture, D. M. Drylie, Y. Centifano, and H. Kaufman, *J. Urol.*, **112**, 126–127 (1974).

21. W. E. Josey, A. J. Nahmias, Z. M. Naib, P. M. Utley, W. J. McKenzie, and M. T. Coleman, *Am. J. Obstet. Gynecol.*, **96**, 493–501 (1966).

22. W. E. Josey, A. J. Nahmias, and Z. M. Naib, *Obstet. Gynecolog. Surv.*, **27** (suppl), 295–302 (1972).

23. R. E. Johnson, A. J. Nahmias, L. S. Magder, F. K. Lee, C. A. Brooks, and C. B. Snowden, *N. Engl. J. Med.*, **321**, 7–12 (1989).

24. W. E. Rawls, H. L. Gardner, R. W. Flanders, S. P. Lowry, R. H. Kaufman, and J. L. Melnick, *Am. J. Obstet. Gynecol.*, **110**, 682–689 (1971).

25. W. E. Rawls and H. L. Gardner, *Clin. Obstet. Gynecol.*, **15**, 913–918 (1972).

26. S. Deguen, N. P. Chau, and A. Flahault, *J. Epidemiol. Commun. Health*, **52**, 46S–49S (1998).

27. C. K. Fairley and E. Miller, *J. Infect. Dis.*, **174**, S314–S319 (1996).

28. M. E. Halloran, S. L. Cochi, T. A. Lieu, M. Wharton, and L. Fehrs, *Am. J. Epidemiol.*, **140**, 81–104 (1994).

29. A. M. Arvin, J. F. Moffat, and R. Redman, *Adv. Vir. Res.*, **46**, 266–309 (1996).

30. C. Grose, *Pediatrics*, **68**, 735–737 (1981).

31. P. W. Annunziato, O. Lungu, C. Panagiotidis, J. H. Zhang, D. N. Silvers, A. A. Gershon, and S. J. Silverstein, *J. Virol.*, **74**, 2005–2010 (2000).

32. J. F. Moffat, M. D. Stein, H. Kaneshima, and A. M. Arvin, *J. Virol.*, **69**, 5236–5242 (1995).

33. A. F. Nikkels, S. Debrus, C. Sadzot-Delvaux, J. Piette, B. Rentier, and G. E. Pierard, *Neurology*, **45**, S47–S49 (1995).

34. Y. Furuta, T. Takasu, S. Fukuda, K. C. Sato-Matsumura, Y. Inuyama, R. Hondo, and K. Nagashima, *J. Infect. Dis.*, **166**, 1157–1159 (1992).

35. D. H. Gilden, A. Vafai, Y. Shtram, Y. Becker, M. Devlin, and M. Wellish, *Nature*, **306**, 478–480 (1983).

36. R. W. Hyman, J. R. Ecker, and R. B. Tenser, *Lancet*, **2**, 814–816 (1983).

37. S. E. Straus, W. Reinhold, H. A. Smith, W. T. Ruyechan, D. K. Henderson, R. M. Blaese, and J. Hay, *N. Engl. J. Med.*, **311**, 1362–1364 (1984).

38. B. Rentier, J. Piette, L. Baudoux, S. Debrus, P. Defechereux, M. P. Merville, C. Sadzot-Delvaux, and S. Schoonbroodt, *Vet Microbiol.*, **53**, 55–66 (1996).

39. J. I. Cohen, *Infect. Dis. Clinics North Am.*, **10**, 457–468 (1996).

40. C. Grose and T. I. Ng, *J. Infect. Dis.*, **166**, S7–S12 (1992).

41. P. D. Griffiths, M. Ait-Khaled, C. P. Bearcroft, D. A. Clark, A. Quaglia, S. E. Davies, A. K. Burroughs, K. Rolles, I. M. Kidd, S. N. Knight, S. M. Noibi, A. V. Cope, A. N. Phillips, and V. C. Emery, *J. Med. Virol.*, **59**, 496–501 (1999).

42. E. Klemola, R. Von Essen, G. Henle, and W. Henle, *J. Infect. Dis.*, **121**, 608–614 (1970).

43. W. J. Britt, R. F. Pass, S. Stagno, and C. A. Alford, *Transplantation Proc.*, **23**, 115–117 (1991).

44. J. E. Gallant, R. D. Moore, D. D. Richman, J. Keruly, and R. E. Chaisson, *J. Infect. Dis.*, **166**, 1223–1227 (1992).

45. R. McKenzie, W. D. Travis, S. A. Dolan, S. Pittaluga, I. M. Feuerstein, J. Shelhamer, R. Yarchoan, and H. Masur, *Medicine*, **70**, 326–343 (1991).

46. P. Pertel, R. Hirschtick, J. Phair, J. Chmiel, L. Poggensee, and R. Murphy, *J. Acquired Immune Deficiency Syndromes.*, **5**, 1069–1074 (1992).

47. M. G. Revello, M. Zavattoni, A. Sarasini, E. Percivalle, L. Simoncini, and G. Gerna, *J. Infect. Dis.*, **177**, 1170–1175 (1998).

48. J. A. Zaia, G. M. Schmidt, N. J. Chao, N. W. Rizk, A. P. Nademanee, J. C. Niland, D. A. Horak, J. Lee, G. Gallez-Hawkins, C. R. Kusnierz-Glaz, *Biology Blood Marrow Transplantation*, **1**, 88–93 (1995).

49. G. Gerna, M. Zavattoni, F. Baldanti, M. Furione, L. Chezzi, M. G. Revello, and E. Percivalle, *J. Med. Virol.*, **55**, 64–74 (1998).

50. G. Gerna, E. Percivalle, F. Baldanti, S. Sozzani, P. Lanzarini, E. Genini, D. Lilleri, and M. G. Revello, *J. Virol.*, **74**, 5629–5638 (2000).

51. T. M. Collins, M. R. Quirk, and M. C. Jordan, *J. Virol.*, **68**, 6305–6311 (1994).

52. C. A. Stoddart, R. D. Cardin, J. M. Boname, W. C. Manning, G. B. Abenes, and E. S. Mocarski, *J. Virol.*, **68**, 6243–6253 (1994).

53. C. Sinzger, A. Grefte, B. Plachter, A. S. Gouw, T. H. The, and G. Jahn. *J. Gen. Virol.*, **76**, 741–750 (1995).

54. K. Kondo, J. Xu, and E. S. Mocarski, *Proc. Natl. Acad. Sci. USA*, **93**, 11137–11142 (1996).

55. E. J. Minton, C. Tysoe, J. H. Sinclair, and J. G. Sissons, *J. Virol.*, **68**, 4017–4021 (1994).

56. G. Hahn, R. Jores, and E. S. Mocarski, *Proc. Natl. Acad. Sci. USA*, **95**, 3937–3942 (1998).

57. C. Soderberg-Naucler and J. Y. Nelson, *Int. Virol.*, **42**, 314–321 (1999).

58. T. Compton, R. R. Nepomuceno, and D. M. Nowlin, *Virology*, **191**, 387–395 (1992).

59. M. F. Stinski, in B. Roizman, Ed., *Herpesviruses*, Plenum Press, New York, 1983, pp. 67–113.

60. P. Lusso, P. D. Markham, E. Tschachler, F. di Marzo Veronese, S. Z. Salahuddin, D. V.

Ablashi, S. Pahwa, K. Krohn, and R. C. Gallo, *J. Exp. Med.*, **167**, 1659–1670 (1988).

61. K. Takahashi, S. Sonoda, K. Higashi, T. Kondo, H. Takahashi, M. Takahashi, and K. Yamanishi, *J. Virol.*, **63**, 3161–3163 (1989).

62. P. Lusso, M. S. Malnati, A. Garzino-Demo, R. W. Crowley, E. O. Long, and R. C. Gallo, *Nature*, **362**, 458–462 (1993).

63. P. Lusso, A. Garzino-Demo, R. W. Crowley, and M. S. Malnati, *J. Exp. Med.*, **181**, 1303–1310 (1995).

64. S. W. Aberle, C. W. Mandl, C. Kunz, and T. Popow-Kraupp, *J. Clin. Microbiol.*, **34**, 3223–3225 (1996).

65. R. F. Jarrett, D. A. Clark, S. F. Josephs, and D. E. Onions, *J. Med. Virol.*, **32**, 73–76 (1990).

66. M. Luppi, P. Barozzi, A. Maiorana, R. Marasca, R. Trovato, R. Fano, L. Ceccherini-Nelli, and G. Torelli, *J. Med. Virol.*, **47**, 105–111 (1995).

67. D. V. Ablashi, M. Handy, J. Bernbaum, L. G. Chatlynne, W. Lapps, B. Kramarsky, Z. N. Berneman, A. L. Komaroff, and J. E. Whitman, *J. Virol. Methods*, **73**, 123–140 (1998).

68. J. B. Black, D. A. Burns, C. S. Goldsmith, P. M. Feorino, K. Kite-Powell, R. F. Schinazi, P. W. Krug, and P. E. Pellett, *Vir. Res.*, **52**, 25–41 (1997).

69. P. Lusso, P. Secchiero, R. W. Crowley, A. Garzino-Demo, Z. N. Berneman, and R. C. Gallo, *Proc. Natl. Acad. Sci. USA*, **91**, 3872–3876 (1994).

70. P. H. Levine, J. Neequaye, M. Yadav, and R. Connelly, *Microbiol. Immunol.*, **36**, 169–172 (1992).

71. S. Ranger, S. Patillaud, F. Denis, A. Himmich, A. Sangare, S. M'Boup, A. Itoua-N'Gaporo, M. Prince-David, R. Chout, R. Cevallos, *J. Med. Virol.*, **34**, 194–198 (1991).

72. K. Tanaka-Taya, T. Kondo, T. Mukai, H. Miyoshi, Y. Yamamoto, S. Okada, and K. Yamanishi, *J. Med. Virol.*, **48**, 88–94 (1996).

73. T. Yoshikawa, Y. Asano, I. Kobayashi, T. Nakashima, T. Yazaki, S. Suga, T. Ozaki, L. S. Wyatt, and N. Frenkel, *J. Med. Virol.*, **41**, 319–323 (1993).

74. IARC Working Group on the Evaluation of Carcinogenic Risk to Humans, in *IARC Monograph*, Lyon, France, 1997.

75. J. L. Sullivan and B. A. Woda, *Immunodeficiency Rev.*, **1**, 325–347 (1989).

76. M. A. Nalesnik, *Springer Sem. Immunopathol.*, **20**, 325–342 (1998).

77. M. J. Boyle, W. A. Sewell, T. B. Sculley, A. Apolloni, J. J. Turner, C. E. Swanson, R. Penny, and D. A. Cooper, *Blood*, **78**, 3004–3011 (1991).

78. G. Gaidano, A. Carbone, and R. Dalla-Favera, *Am. J. Pathol.*, **152**, 623–630 (1998).

79. S. J. Hamilton-Dutoit, G. Pallesen, M. B. Franzmann, J. Karkov, F. Black, P. Skinhoj, and C. Pedersen, *Am. J. Pathol.*, **138**, 149–163 (1991).

80. R. Rubio, *Cancer*, **73**, 2400–2407 (1994).

81. E. S. Lee, J. Locker, M. Nalesnik, J. Reyes, R. Jaffe, M. Alashari, B. Nour, A. Tzakis, and P. S. Dickman, *N. Engl. J. Med.*, **332**, 19–25 (1995).

82. K. L. McClain, C. T. Leach, H. B. Jenson, V. V. Joshi, B. H. Pollock, R. T. Parmley, F. J. DiCarlo, E. G. Chadwick, and S. B. Murphy, *N. Engl. J. Med.*, **332**, 12–18 (1995).

83. H. Wolf, H. zur Hausen, and V. Becker, *Nature New Biol.*, **244**, 245–247 (1973).

84. J. F. Jones, S. Shurin, C. Abramowsky, R. R. Tubbs, C. G. Sciotto, R. Wahl, J. Sands, D. Gottman, B. Z. Katz, and J. Sklar, *N. Engl. J. Med.*, **318**, 733–741 (1988).

85. G. Wilson and G. Miller, *Virology*, **95**, 351–358 (1979).

86. N. Yamamoto, T. Katsuki, and Y. Hinuma, *JNCI*, **56**, 1105–1107 (1976).

87. M. Anvret and G. Miller, *Virology*, **111**, 47–55 (1981).

88. J. W. Sixbey, J. G. Nedrud, N. Raab-Traub, R. A. Hanes, and J. S. Pagano, *N. Engl. J. Med.*, **310**, 1225–1230 (1984).

89. M. A. Karajannis, M. Hummel, I. Anagnostopoulos, and H. Stein, *Blood*, **89**, 2856–2862 (1997).

90. G. R. Nemerow, C. Mold, V. K. Schwend, V. Tollefson, and N. R. Cooper, *J. Virol.*, **61**, 1416–1420 (1987).

91. J. Tanner, J. Weis, D. Fearon, Y. Whang, and E. Kieff, *Cell*, **50**, 203–213 (1987).

92. J. D. Fingeroth, J. J. Weis, T. F. Tedder, J. L. Strominger, P. A. Biro, and D. T. Fearon, *Proc. Natl. Acad. Sci. USA*, **81**, 4510–4514 (1984).

93. J. J. Weis, D. T. Fearon, L. B. Klickstein, W. W. Wong, S. A. Richards, A. de Bruyn Kops, J. A. Smith, and J. H. Weis, *Proc. Natl. Acad. Sci. USA*, **83**, 5639–5643 (1986).

94. J. C. Carel, B. L. Myones, B. Frazier, and V. M. Holers, *J. Biol. Chem.*, **265**, 12293–12299 (1990).

95. G. R. Nemerow and N. R. Cooper, *Virology*, **132**, 186–198 (1984).

96. C. Alfieri, M. Birkenbach, and E. Kieff, *Virology*, **181**, 595–608 (1991).

97. D. J. Moss, A. B. Rickinson, L. E. Wallace, and M. A. Epstein, *Nature.*, **291**, 664–666 (1981).

98. D. J. Moss, T. B. Sculley, and J. H. Pope, *J. Virol.*, **58**, 988–990 (1986).

99. S. Harrison, K. Fisenne, and J. Hearing, *J Virol.*, **68**, 1913–1925 (1994).

100. T. Middleton and B. Sugden, *J. Virol.*, **66**, 489–495 (1992).

101. T. Chittenden, S. Lupton, and A. J. Levine, *J. Virol.*, **63**, 3016–3025 (1989).

102. G. Miller, *Proc. Natl. Acad. Sci. USA*, **82**, 4085–4089 (1985).

103. L. Gradoville, E. Grogan, N. Taylor, and G. Miller, *Virology*, **178**, 345–354 (1990).

104. D. J. Blackbourn, D. Osmond, J. A. Levy, and E. T. Lennette, *J. Infect. Dis.*, **179**, 237–239 (1999).

105. R. Sarid, S. J. Olsen, and P. S. Moore, *Adv. Vir. Res.*, **52**, 139–232 (1999).

106. A. De Milito, M. Catucci, G. Venturi, L. Romano, L. Incandela, P. E. Valensin, and M. Zazzi, *J. Med. Virol.*, **57**, 140–144 (1999).

107. S. Davey, 76–82, World Health Organization, Geneva, Switzerland, 1996.

108. M. Omata, *N. Engl. J. Med.*, **339**, 114–115 (1998).

109. R. Idilman, N. De Maria, A. Colantoni, and D. H. Van Thiel, *J. Viral. Hepatitis*, **5**, 285–299 (1998).

110. R. P. Beasley, *Cancer*, **61**, 1942–1956 (1988).

111. R. P. Beasley, L. Y. Hwang, C. C. Lin, and C. S. Chien, *Lancet*, **2**, 1129–1133 (1981).

112. N. Gitlin, *Clin. Chem.*, **43**, 1500–1506 (1997).

113. V. Massari, P. Maison, J. C. Desenclos, and A. Flahault, *Eur J Epidemiol.*, **14**, 765–767 (1998).

114. D. Ganem, *Rev. Infect. Dis.*, **4**, 1026–1047 (1982).

115. J. H. Hoofnagle, *Annu. Rev. Med.*, **32**, 1–11 (1981).

116. A. R. Jilbert, T. T. Wu, J. M. England, P. M. Hall, N. Z. Carp, A. P. O'Connell, and W. S. Mason, *J. Virol.*, **66**, 1377–1388 (1992).

117. A. G. Redeker, *Am. J. Med. Sci.*, **270**, 9–16 (1975).

118. R. P. Perrillo, E. R. Schiff, G. L. Davis, H. C. Bodenheimer, K. Lindsay, J. Payne, J. L. Dienstag, C. O'Brien, C. Tamburro, I. M. Jacobson, *N. Engl. J. Med.*, **323**, 295–301 (1990).

119. L. Fu, and Y. C. Cheng, *Biochem. Pharmacol.*, **55**, 1567–1572 (1998).

120. B. Jarvis and D. Faulds, *Drugs*, **58**, 101–141 (1999).

121. M. A. Nowak, S. Bonhoeffer, A. M. Hill, R. Boehme, H. C. Thomas, and H. McDade, *Proc. Natl. Acad. Sci. USA*, **93**, 4398–4402 (1996).

122. F. Sattler and W. S. Robinson, *J. Virol.*, **32**, 226–233 (1979).

123. J. Summers, A. O'Connell, and I. Millman, *Proc. Natl. Acad. Sci. USA*, **72**, 4597–4601 (1975).

124. C. Seeger, D. Ganem, and H. E. Varmus, *Science*, **232**, 477–484 (1986).

125. H. Will, W. Reiser, T. Weimer, E. Pfaff, M. Buscher, R. Sprengel, R. Cattaneo, and H. Schaller, *J. Virol.*, **61**, 904–911 (1987).

126. P. D. Garcia, J. H. Ou, W. J. Rutter, and P. Walter, *J. Cell Biol.*, **106**, 1093–1104 (1988).

127. A. McLachlan, D. R. Milich, A. K. Raney, M. G. Riggs, J. L. Hughes, J. Sorge, and F. V. Chisari, *J. Virol.*, **61**, 683–692 (1987).

128. N. Ogata, R. H. Miller, K. G. Ishak, and R. H. Purcell, *Virology*, **194**, 263–276 (1993).

129. J. H. Ou, O. Laub, and W. J. Rutter, *Proc. Natl. Acad. Sci. USA*, **83**, 1578–1582 (1986).

130. M. J. Roossinck and A. Siddiqui, *J. Virol.*, **61**, 955–961 (1987).

131. D. N. Standring, J. H. Ou, F. R. Masiarz, and W. J. Rutter, *Proc. Natl. Acad. Sci. USA*, **85**, 8405–8409 (1988).

132. C. Niederau, T. Heintges, S. Lange, G. Goldmann, C. M. Niederau, L. Mohr, and D. Haussinger, *N. Engl. J. Med.*, **334**, 1422–1427 (1996).

133. P. Honkoop, R. A. de Man, H. G. Niesters, P. E. Zondervan, and S. W. Schalm, *Hepatology*, **32**, 635–639 (2000).

134. A. Nicoll, S. Locarnini, S. T. Chou, R. Smallwood, and P. Angus, *J. Gastroenterol Hepatol.*, **15**, 304–310 (2000).

135. B. E. Korba, P. Cote, W. Hornbuckle, B. C. Tennant, and J. L. Gerin, *Hepatology*, **31**, 1165–1175 (2000).

136. H. zur Hausen, *Biochim. Biophys. Acta.*, **1288**, F55–F78 (1996).

137. H. U. Bernard, S. Y. Chan, and H. Delius, *Curr. Topics Microbiol. Immunol.*, **186**, 33–54 (1994).

138. M. Durst, R. T. Dzarlieva-Petrusevska, P. Boukamp, N. E. Fusenig, and L. Gissmann, *Oncogene*, **1**, 251–256 (1987).

139. L. Pirisi, S. Yasumoto, M. Feller, J. Doniger, and J. A. DiPaolo, *J. Virol.*, **61**, 1061–1066 (1987).

140. A. J. van den Brule, C. J. Meijer, V. Bakels, P. Kenemans, and J. M. Walboomers, *J. Clin. Microbiol.*, **28**, 2739–2743 (1990).

141. B. C. Das, J. K. Sharma, V. Gopalkrishna, D. K. Das, V. Singh, L. Gissmann, H. zur Hausen, and U. K. Luthra, *J. Med. Virol.*, **36**, 239–245 (1992).

142. F. X. Bosch, M. M. Manos, N. Munoz, M. Sherman, A. M. Jansen, J. Peto, M. H. Schiffman, V. Moreno, R. Kurman, and K. V. Shah, *JNCI*, **87**, 796–802 (1995).

143. J. D. Oriel, *Br. J. Venereal Dis.*, **47**, 1–13 (1971).

144. L. T. Gutman, K. K. St Claire, V. D. Everett, D. L. Ingram, J. Soper, W. W. Johnston, G. G. Mulvaney, and W. C. Phelps, *J. Infect. Dis.*, **170**, 339–344 (1994).

145. E. Rylander, L. Ruusuvaara, M. W. Almstromer, M. Evander, and G. Wadell, *Obstet. Gynecol.*, **83**, 735–737 (1994).

146. A. B. Moscicki, J. Palefsky, J. Gonzales, and G. K. Schoolnik, *Pediat. Res.*, **28**, 507–513 (1990).

147. I. H. Frazer, G. Medley, R. M. Crapper, T. C. Brown, and I. R. Mackay, *Lancet*, **2**, 657–660 (1986).

148. J. M. Palefsky, J. Gonzales, R. M. Greenblatt, D. K. Ahn, and H. Hollander, *JAMA*, **263**, 2911–2916 (1990).

149. N. B. Kiviat, C. W. Critchlow, K. K. Holmes, J. Kuypers, J. Sayer, C. Dunphy, C. Surawicz, P. Kirby, R. Wood, and J. R. Daling, *AIDS*, **7**, 43–49 (1993).

150. J. T. Finch and A. Klug, *J. Mol. Biol.*, **13**, 1–12 (1965).

151. W. C. Phelps and P. M. Howley, *J. Virol.*, **61**, 1630–1638 (1987).

152. T. P. Cripe, T. H. Haugen, J. P. Turk, F. Tabatabai, P. G. Schmid, M. Durst, L. Gissmann, A. Roman, and L. P. Turek, *EMBO J.*, **6**, 3745–3753 (1987).

153. T. R. Sarafi and A. A. McBride, *Virology*, **211**, 385–396 (1995).

154. P. L. Winokur and A. A. McBride, *EMBO J.*, **11**, 4111–4118 (1992).

155. N. Bastien and A. A. McBride, *Virology*, **270**, 124–134 (2000).

156. M. M. Pater and A. Pater, *Virology*, **145**, 313–318 (1985).

157. L. Yang, I. Mohr, E. Fouts, D. A. Lim, M. Nohaile, and M. Botchan, *Proc. Natl. Acad. Sci. USA*, **90**, 5086–5090 (1993).

158. F. J. Hughes and M. A. Romanos, *Nucleic Acids Res.*, **21**, 5817–5823 (1993).

159. P. Park, W. Copeland, L. Yang, T. Wang, M. R. Botchan, and I. J. Mohr, *Proc. Natl. Acad. Sci. USA*, **91**, 8700–8704 (1994).

160. K. Munger, B. A. Werness, N. Dyson, W. C. Phelps, E. Harlow, and P. M. Howley, *EMBO J.*, **8**, 4099–4105 (1989).

161. J. D. Morris, T. Crook, L. R. Bandara, R. Davies, N. B. LaThangue, and K. H. Vousden, *Oncogene*, **8**, 893–898 (1993).

162. M. S. Lechner, D. H. Mack, A. B. Finicle, T. Crook, K. H. Vousden, and L. A. Laimins, *EMBO J.*, **11**, 3045–3052 (1992).

163. J. M. Huibregtse and S. L. Beaudenon, *Semin. Cancer Biol.*, **7**, 317–326 (1996).

164. J. M. Huibregtse and M. Scheffner, *Semin. Virol.*, **5**, 357–367 (1994).

165. X. Tong, and P. M. Howley, *Proc. Natl. Acad. Sci. USA*, **94**, 4412–4417 (1997).

166. J. Goudsmit, P. Wertheim-van Dillen, A. van Strien, and J. van der Noordaa, *J. Med. Virol.*, **10**, 91–99 (1982).

167. M. C. Monaco, P. N. Jensen, J. Hou, L. C. Durham, and E. O. Major, *J. Virol.*, **72**, 9918–9923 (1998).

168. T. Weber and E. O. Major, *Int. Virol.*, **40**, 98–111 (1997).

169. V. Dubois, H. Dutronc, M. E. Lafon, V. Poinsot, J. L. Pellegrin, J. M. Ragnaud, A. M. Ferrer, and H. J. Fleury, *J. Clin. Microbiol.*, **35**, 2288–2292 (1997).

170. S. Boubenider, C. Hiesse, S. Marchand, A. Hafi, F. Kriaa, and B. Charpentier, *J. Nephrol.*, **12**, 24–29 (1999).

171. B. G. Hedquist, G. Bratt, A. L. Hammarin, M. Grandien, I. Nennesmo, B. Sundelin, and S. Seregard, *Ophthalmology*, **106**, 129–132 (1999).

172. G. Bratt, A. L. Hammarin, M. Grandien, B. G. Hedquist, I. Nennesmo, B. Sundelin, and S. Seregard, *AIDS*, **13**, 1071–1075 (1999).

173. E. S. Sandler, V. M. Aquino, E. Goss-Shohet, S. Hinrichs, and K. Krisher, *Bone Marrow Transplant.*, **20**, 163–165 (1997).

174. J. F. Apperley, S. J. Rice, J. A. Bishop, Y. C. Chia, T. Krausz, S. D. Gardner, and J. M. Goldman, *Transplantation*, **43**, 108–112 (1987).

175. R. R. Arthur, K. V. Shah, S. J. Baust, G. W. Santos, and R. Saral, *N. Engl. J. Med.*, **315**, 230–234 (1986).

176. R. R. Arthur and K. V. Shah, *Prog. Med. Virol.*, **36**, 42–61 (1989).

177. R. R. Arthur, S. Dagostin, and K. V. Shah, *J. Clin. Microbiol.*, **27**, 1174–1179 (1989).

178. D. C. Pallas, L. K. Shahrik, B. L. Martin, S. Jaspers, T. B. Miller, D. L. Brautigan, and T. M. Roberts, *Cell*, **60**, 167–176 (1990).

179. B. Whalen, J. Laffin, T. D. Friedrich, and J. M. Lehman, *Exp. Cell Res.*, **251**, 121–127 (1999).

180. K. Rundell, S. Gaillard, and A. Porras, *Dev. Biol. Standard.*, **94**, 289–295 (1998).

181. A. Porras, R. Gaillard, and K. Rundell, *J. Virol.*, **73**, 3102–3107 (1999).

182. I. A. Mastrangelo, P. V. Hough, J. S. Wall, M. Dodson, F. B. Dean, and J. Hurwitz, *Nature*, **338**, 658–662 (1989).

183. W. S. Joo, H. Y. Kim, J. D. Purviance, K. R. Sreekumar, and P. A. Bullock, *Mol. Cell. Biol.*, **18**, 2677–2687 (1998).

184. K. A. Braun, Y. Lao, Z. He, C. J. Ingles, and M. S. Wold, *Biochemistry*, **36**, 8443–8454 (1997).

185. K. L. Collins, A. A. Russo, B. Y. Tseng, and T. J. Kelly, *EMBO J.*, **12**, 4555–4566 (1993).

186. S. G. Huang, K. Weisshart, I. Gilbert, and E. Fanning, *Biochemistry*, **37**, 15345–15352 (1998).

187. K. Wun-Kim, R. Upson, W. Young, T. Melendy, B. Stillman, and D. T. Simmons, *J. Virol.*, **67**, 7608–7611 (1993).

188. T. Melendy and B. Stillman, *J. Biol. Chem.*, **268**, 3389–3395 (1993).

189. F. B. Dean, P. Bullock, Y. Murakami, C. R. Wobbe, L. Weissbach, and J. Hurwitz, *Proc. Natl. Acad. Sci. USA*, **84**, 16–20 (1987).

190. H. Stahl, P. Droge, and R. Knippers. *EMBO J.*, **5**, 1939–1944 (1986).

191. J. A. DeCaprio, J. W. Ludlow, J. Figge, J. Y. Shew, C. M. Huang, W. H. Lee, E. Marsilio, E. Paucha, and D. M. Livingston, *Cell*, **54**, 275–283 (1988).

192. N. Dyson, K. Buchkovich, P. Whyte, and E. Harlow, *Cell*, **58**, 249–255 (1989).

193. D. P. Lane and L. V. Crawford, *Nature*, **278**, 261–263 (1979).

194. D. I. Linzer and A. J. Levine, *Cell*, **17**, 43–52 (1979).

195. C. K. Liu, G. Wei, and W. J. Atwood, *J. Virol.*, **72**, 4643–4649 (1998).

196. L. Sinibaldi, P. Goldoni, V. Pietropaolo, L. Cattani, C. Peluso, and C. Di Taranto, *Microbiologica*, **15**, 337–344 (1992).

197. J. Forstova, N. Krauzewicz, S. Wallace, A. J. Street, S. M. Dilworth, S. Beard, and B. E. Griffin, *J. Virol.*, **67**, 1405–1413 (1993).

198. D. H. Barouch and S. C. Harrison, *J. Virol.*, **68**, 3982–3989 (1994).

199. J. P. Fox, C. D. Brandt, F. E. Wassermann, C. E. Hall, I. Spigland, A. Kogon, and L. R. Elveback, *Am. J. Epidemiol.*, **89**, 25–50 (1969).

200. C. D. Brandt, H. W. Kim, A. J. Vargosko, B. C. Jeffries, J. O. Arrobio, B. Rindge, R. H. Parrott, and R. M. Chanock, *Am. J. Epidemiol.*, **90**, 484–500 (1969).

201. J. Y. Hong, H. J. Lee, P. A. Piedra, E. H. Choi, K. H. Park, Y. Y. Koh, and W. S. Kim, *Clin. Infect. Dis.*, **32**, 1423–1429 (2001).

202. R. J. Cooper, R. Hallett, A. B. Tullo, and P. E. Klapper, *Epidemiol. Infect.*, **125**, 333–345 (2000).

203. I. Nasz, G. Kulcsar, P. Dan, and K. Sallay, *J. Infect. Dis.*, **124**, 214–216 (1971).

204. J. van der Veen and M. Lambriex, *Infect. Immun.*, **7**, 604–609 (1973).

205. J. A. Mahr and L. R. Gooding, *Immunol. Rev.*, **168**, 121–130 (1999).

206. A. Baldwin, H. Kingman, M. Darville, A. B. Foot, D. Grier, J. M. Cornish, N. Goulden, A. Oakhill, D. H. Pamphilon, C. G. Steward, and D. I. Marks, *Bone Marrow Transplant.*, **26**, 1333–1338 (2000).

207. F. P. Siegal, S. H. Dikman, R. B. Arayata, and E. J. Bottone, *Am. J. Med.*, **71**, 1062–1067 (1981).

208. H. Stalder, J. C. Hierholzer, and M. N. Oxman, *J. Clin. Microbiol.*, **6**, 257–265 (1977).

209. H. J. Wigger and W. A. Blanc, *N. Engl. J. Med.*, **275**, 870–874 (1966).

210. J. M. Zahradnik, M. J. Spencer, and D. D. Porter, *Am. J. Med.*, **68**, 725–732 (1980).

211. L. R. Krilov, L. G. Rubin, M. Frogel, E. Gloster, K. Ni, M. Kaplan, and S. M. Lipson, *Rev. Infect. Dis.*, **12**, 303–307 (1990).

212. M. D. Challberg, and T. J. J. Kelly, *Proc. Natl. Acad. Sci. USA*, **76**, 655–659 (1979).

213. J. K. Mackey, P. M. Rigden, and M. Green, *Proc. Natl. Acad. Sci. USA*, **73**, 4657–4661 (1976).

214. T. Fujiwara, E. A. Grimm, T. Mukhopadhyay, W. W. Zhang, L. B. Owen-Schaub, and J. A. Roth, *Cancer Res.*, **54**, 2287–2291 (1994).

215. A. Mastrangeli, C. Danel, M. A. Rosenfeld, L. Stratford-Perricaudet, M. Perricaudet, A. Pavirani, J. P. Lecocq, and R. G. Crystal, *J. Clin. Invest.*, **91**, 225–234 (1993).

216. M. A. Rosenfeld, K. Yoshimura, B. C. Trapnell, K. Yoneyama, E. R. Rosenthal, W. Dalemans,

M. Fukayama, J. Bargon, L. E. Stier, L. Stratford-Perricaudet, *Cell*, **68**, 143–155 (1992).

217. W. C. Russell, *J. Gen. Virol.*, **81**, 2573–2604 (2000).

218. J. van Oostrum and R. M. Burnett, *J. Virol.*, **56**, 439–448 (1985).

219. M. A. Mirza and J. Weber, *Biochim. Biophys. Acta*, **696**, 76–86 (1982).

220. P. W. Roelvink, A. Lizonova, J. G. Lee, Y. Li, J. M. Bergelson, R. W. Finberg, D. E. Brough, I. Kovesdi, and T. J. Wickham, *J. Virol.*, **72**, 7909–7915 (1998).

221. T. J. Wickham, P. Mathias, D. A. Cheresh, and G. R. Nemerow, *Cell*, **73**, 309–319 (1993).

222. Y. Chardonnet and S. Dales, *Virology*, **40**, 462–477 (1970).

223. D. J. FitzGerald, R. Padmanabhan, I. Pastan, and M. C. Willingham, *Cell*, **32**, 607–617 (1983).

224. M. J. Varga, C. Weibull, and E. Everitt, *J. Virol.*, **65**, 6061–6070 (1991).

225. T. Enomoto, J. H. Lichy, J. E. Ikeda, and J. Hurwitz, *Proc. Natl. Acad. Sci. USA*, **78**, 6779–6783 (1981).

226. J. Field, R. M. Gronostajski, and J. Hurwitz, *J. Biol. Chem.*, **259**, 9487–9494 (1984).

227. L. Philipson, *Curr. Topics Microbiol. Immunol.*, **109**, 1–52 (1984).

228. J. Ding, W. J. McGrath, R. M. Sweet, and W. F. Mangel, *EMBO J.*, **15**, 1778–1783 (1996).

229. K. Tihanyi, M. Bourbonniere, A. Houde, C. Rancourt, and J. M. Weber, *J. Biol. Chem.*, **268**, 1780–1785 (1993).

230. B. J. Cohen, P. P. Mortimer, and M. S. Pereira, *J. Hygiene*, **91**, 113–130 (1983).

231. K. E. Brown and N. S. Young, *Adv. Pediatr. Iinfect. Dis.*, **13**, 101–126 (1997).

232. J. M. Shneerson, P. P. Mortimer, and E. M. Vandervelde, *Br Med. J.*, **280**, 1580 (1980).

233. L. Salvatori, C. Lavorino, A. Giglio, L. Alemanno, A. Di Carlo, F. Ameglio, and F. Caprilli, *New Microbiologica*, **22**, 181–186 (1999).

234. J. R. Pattison, S. E. Jones, J. Hodgson, L. R. Davis, J. M. White, C. E. Stroud, and L. Murtaza, *Lancet*, **1**, 664–665 (1981).

235. M. J. Anderson, S. E. Jones, S. P. Fisher-Hoch, E. Lewis, S. M. Hall, C. L. Bartlett, B. J. Cohen, P. P. Mortimer, and M. S. Pereira, *Lancet*, **1**, 1378 (1983).

236. M. J. Anderson, M. N. Khousam, D. J. Maxwell, S. J. Gould, L. C. Happerfield, and W. J. Smith, *Lancet*, **1**, 535 (1988).

237. G. J. Kurtzman, K. Ozawa, B. Cohen, G. Hanson, R. Oseas, and N. S. Young, *N. Engl. J. Med.*, **317**, 287–294 (1987).

238. M. Osaki, K. Matsubara, T. Iwasaki, T. Kurata, H. Nigami, H. Harigaya, and K. Baba, *Ann Hematol.*, **78**, 83–86 (1999).

239. M. J. Anderson, P. G. Higgins, L. R. Davis, J. S. Willman, S. E. Jones, I. M. Kidd, J. R. Pattison, and D. A. Tyrrell, *J. Infect. Dis.*, **152**, 257–265 (1985).

240. A. M. Garcia-Tapia, C. Fernandez-Gutierrez del Alamo, J. A. Giron, J. Mira, F. de la Rubia, A. Martinez-Rodriguez, M. V. Martin-Reina, R. Lopez-Caparros, R. Caliz, M. S. Caballero, *Clin. Infect. Dis.*, **21**, 1424–1430 (1995).

241. C. G. Potter, A. C. Potter, C. S. Hatton, H. M. Chapel, M. J. Anderson, J. R. Pattison, D. A. Tyrrell, P. G. Higgins, J. S. Willman, and H. F. Parry, *J. Clin. Invest.*, **79**, 1486–1492 (1987).

242. G. J. Kurtzman, B. Cohen, P. Meyers, A. Amunullah, and N. S. Young, *Lancet*, **2**, 1159–1162 (1988).

243. G. J. Kurtzman, P. Gascon, M. Caras, B. Cohen, and N. S. Young, *Blood*, **71**, 1448–1454 (1988).

244. P. R. Koduri, R. Kumapley, J. Valladares, and C. Teter, *Am. J. Hematol.*, **61**, 16–20 (1999).

245. T. W. Crook, B. B. Rogers, R. D. McFarland, S. H. Kroft, P. Muretto, J. A. Hernandez, M. J. Latimer, and R. W. McKenna, *Hum. Pathol.*, **31**, 161–168 (2000).

246. P. L. Hermonat and N. Muzyczka, *Proc. Natl. Acad. Sci. USA*, **81**, 6466–6470 (1984).

247. A. Srivastava, *Curr. Opin. Mol. Therapeut.*, **3**, 491–496 (2001).

248. A. Srivastava, E. W. Lusby, and K. I. Berns, *J. Virol.*, **45**, 555–564 (1983).

249. C. R. Astell, M. Smith, M. B. Chow, and D. C. Ward, *Cell*, **17**, 691–703 (1979).

250. V. Deiss, J. D. Tratschin, M. Weitz, and G. Siegl, *Virology*, **175**, 247–254 (1990).

251. E. Lusby, K. H. Fife, and K. I. Berns, *J. Virol.*, **34**, 402–409 (1980).

252. S. L. Rhode and P. R. Paradiso, *J. Virol.*, **45**, 173–184 (1983).

253. C. A. Laughlin, H. Westphal, and B. J. Carter, *Proc. Natl. Acad. Sci. USA*, **76**, 5567–5571 (1979).

254. S. Muralidhar, S. P. Becerra, and J. A. Rose, *J. Virol.*, **68**, 170–176 (1994).

255. P. Tattersall, *J. Virol.*, **10**, 586–590 (1972).

256. S. L. Rhode, *J. Virol.*, **11**, 856–861 (1973).

257. S. Wolter, R. Richards, and R. W. Armentrout, *Biochim. Biophys. Acta.*, **607**, 420–431 (1980).

258. R. M. Buller, J. E. Janik, E. D. Sebring, and J. A. Rose, *J. Virol.*, **40**, 241–247 (1981).

259. R. M. Kotin, J. C. Menninger, D. C. Ward, and K. I. Berns, *Genomics*, **10**, 831–834 (1991).

260. S. F. Cotmore and P. Tattersall, *Adv. Virol. Res.*, **33**, 91–174 (1987).

261. K. E. Brown, J. R. Hibbs, G. Gallinella, S. M. anderson, E. D. Lehman, P. McCarthy, and N. S. Young, *N. Engl. J. Med.*, **330**, 1192–1196 (1994).

262. D. S. Im and N. Muzyczka, *Cell*, **61**, 447–457 (1990).

263. R. O. Snyder, R. J. Samulski, and N. Muzyczka, *Cell*, **60**, 105–113 (1990).

264. J. R. Brister and N. Muzyczka, *J. Virol.*, **73**, 9325–9336 (1999).

265. R. H. Smith and R. M. Kotin, *J. Virol.*, **74**, 3122–3129 (2000).

266. C. Harford, A. Hamlin, and E. Riders, *Exp. Cell Res.*, **42**, 50–57 (1966).

267. J. Dubochet, M. Adrian, K. Richter, J. Garces, and R. Wittek, *J. Virol.*, **68**, 1935–1941 (1994).

268. G. Antoine, F. Scheiflinger, F. Dorner, and F. G. Falkner, *Virology*, **244**, 365–396 (1998).

269. S. J. Goebel, G. P. Johnson, M. E. Perkus, S. W. Davis, J. P. Winslow, and E. Paoletti, *Virology*, **179**, 247–266 (1990).

270. T. G. Senkevich, E. V. Koonin, J. J. Bugert, G. Darai, and B. Moss, *Virology*, **233**, 19–42 (1997).

271. C. Cameron, S. Hota-Mitchell, L. Chen, J. Barrett, J. X. Cao, C. Macaulay, D. Willer, D. Evans, and G. McFadden, *Virology*, **264**, 298–318 (1999).

272. D. O. Willer, G. McFadden, and D. H. Evans, *Virology*, **264**, 319–343 (1999).

273. C. L. Afonso, E. R. Tulman, Z. Lu, L. Zsak, G. F. Kutish, and D. L. Rock, *J. Virol.*, **74**, 3815–3831 (2000).

274. C. L. Afonso, E. R. Tulman, Z. Lu, E. Oma, G. F. Kutish, and D. L. Rock, *J. Virol.*, **73**, 533–552 (1999).

275. A. L. Bawden, K. J. Glassberg, J. Diggans, R. Shaw, W. Farmerie, and R. W. Moyer, *Virology*, **274**, 120–139 (2000).

276. B. Moss, in D. M. Knipe, P. M. Howley, et al, Eds., *Fields Virology*, Lippincott Williams, and Wilkins, Philadelphia, PA, 2001, pp. 2849–2883.

277. J. Neyts and E. De Clercq, *Antimicrob. Agents Chemother.*, **45**, 84–87 (2001).

278. W. H. Prusoff and D. C. Ward, *Biochem. Pharmacol.*, **25**, 1233–1239 (1976).

279. G. Darby, *J. Med. Virol.*, **Suppl 1**, 134–138 (1993).

280. J. E. Reardon and T. Spector, *J. Biol. Chem.*, **264**, 7405–7411 (1989).

281. J. Vogt, R. Perozzo, A. Pautsch, A. Prota, P. Schelling, B. Pilger, G. Folkers, L. Scapozza, and G. E. Schulz, *Proteins*, **41**, 545–553 (2000).

282. D. F. Smee, J. C. Martin, J. P. Verheyden, and T. R. Matthews, *Antimicrob. Agents Chemother.*, **23**, 676–682 (1983).

283. J. A. Fyfe, P. M. Keller, P. A. Furman, R. L. Miller, and G. B. Elion, *J. Biol. Chem.*, **253**, 8721–8727 (1978).

284. V. Sullivan, C. L. Talarico, S. C. Stanat, M. Davis, D. M. Coen, and K. K. Biron, *Nature*, **358**, 162–164 (1992).

285. E. Littler, A. D. Stuart, and M. S. Chee, *Nature*, **358**, 160–162 (1992).

286. D. C. Brennan, G. A. Storch, G. G. Singer, L. Lee, J. Rueda, and M. A. Schnitzler, *J. Infect. Dis.*, **181**, 1557–1561 (2000).

287. J. C. Mendez, D. H. Dockrell, M. J. Espy, T. F. Smith, J. A. Wilson, W. S. Harmsen, D. Ilstrup, and C. V. Paya, *J. Infect. Dis.*, **183**, 179–184 (2001).

288. N. Singh and D. L. Paterson, *Transplantation*, **69**, 2474–2479 (2000).

289. C. Manichanh, P. Grenot, A. Gautheret-Dejean, P. Debre, J. M. Huraux, and H. Agut, *Cytometry*, **40**, 135–140 (2000).

290. M. Yoshida, M. Yamada, T. Tsukazaki, S. Chatterjee, F. D. Lakeman, S. Nii, and R. J. Whitley, *Antiviral Res.*, **40**, 73–84 (1998).

291. S. Koyano, T. Suzutani, I. Yoshida, and M. Azuma, *Antimicrob. Agents Chemother.*, **40**, 920–923 (1996).

292. C. S. Crumpacker, *N. Engl. J. Med.*, **335**, 721–729 (1996).

293. S. Walmsley and A. Tseng, *Drug Safety*, **21**, 203–224 (1999).

294. V. F. Hoffman and D. J. Skiest, *Expert Opin. Invest. Drugs*, **9**, 207–220 (2000).

295. P. Ertl, W. Snowden, D. Lowe, W. H. Miller, P. Collins, and E. Littler. *Antivirol. Chem. Chemother.*, **6**, 89–97 (1995).

296. R. A. Vere Hodge and Y.-C. Cheng, *Antivirol. Chem. Chemother.*, **4**, 13–24 (1993).

297. D. L. Earnshaw, T. H. Bacon, S. J. Darlison, K. Edmonds, R. M. Perkins, and R. A. Vere Hodge, *Antimicrob. Agents Chemother.*, **36**, 2747–2757 (1992).

298. T. H. Bacon, J. Gilbart, B. A. Howard, and R-.C. Standring, *Antivirol. Chem. Chemother.*, **7**, 71–78 (1996).

299. N. Bischofberger, M. J. Hitchcock, M. S. Chen, D. B. Barkhimer, K. C. Cundy, K. M. Kent, S. A. Lacy, W. A. Lee, Z. H. Li, D. B. Mendel, *Antimicrob. Agents Chemother.*, **38**, 2387–2391 (1994).

300. M. A. Polis, K. M. Spooner, B. F. Baird, J. F. Manischewitz, H. S. Jaffe, P. E. Fisher, J. Falloon, R. T. Davey, J. A. Kovacs, R. E. Walker, *Antimicrob. Agents Chemother.*, **39**, 882–886 (1995).

301. J. P. Lalezari, W. L. Drew, E. Glutzer, C. James, D. Miner, J. Flaherty, P. E. Fisher, K. Cundy, J. Hannigan, J. C. Martin, *J. Infect. Dis.*, **171**, 788–796 (1995).

302. E. De Clercq, *Clin. Microbiol. Rev.*, **10**, 674–693 (1997).

303. H. Machida, M. Nishitani, Y. Watanabe, Y. Yoshimura, F. Kano, and S. Sakata, *Microbiol. Immunol.*, **39**, 201–206 (1995).

304. P. Wutzler, E. De Clercq, K. Wutke, and I. Farber, *J. Med. Virol.*, **46**, 252–257 (1995).

305. M. Heidl, H. Scholz, W. Dorffel, and J. Hermann, *Infection*, **19**, 401–405 (1991).

306. S. S. Cohen and W. Plunkett, *Ann. NY Acad. Sci.*, **255**, 269–286 (1975).

307. S. S. Cohen, *Med. Biol.*, **54**, 299–326 (1976).

308. S. S. Cohen, *Cancer*, **40**, 509–518 (1977).

309. C. Shipman, S. H. Smith, R. H. Carlson, and J. C. Drach, *Antimicrob. Agents Chemother.*, **9**, 120–127 (1976).

310. P. M. Schwartz, J. Novack, C. Shipman Jr., and J. C. Drach, *Biochem. Pharmacol.*, **33**, 2431–2438 (1984).

311. D. Derse and Y. C. Cheng, *J. Biol. Chem.*, **256**, 8525–8530 (1981).

312. M. D. de Smet, C. J. Meenken, and G. J. van den Horn, *Ocular Immunol. Inflam*, **7**, 189–198 (1999).

313. B. Detrick, C. N. Nagineni, L. R. Grillone, K. P. anderson, S. P. Henry, and J. J. Hooks, *Invest. Ophthal. Vis. Sci.*, **42**, 163–169 (2001).

314. M. Flores-Aguilar, G. Besen, C. Vuong, M. Tatebayashi, D. Munguia, P. Gangan, C. A. Wiley, and W. R. Freeman, *J. Infect. Dis.*, **175**, 1308–1316 (1997).

315. S. Chou, G. Marousek, S. Guentzel, S. E. Follansbee, M. E. Poscher, J. P. Lalezari, R. C. Miner, and W. L. Drew, *J. Infect. Dis.*, **176**, 786–789 (1997).

316. N. S. Lurain, K. D. Thompson, E. W. Holmes, and G. S. Read, *J. Virol.*, **66**, 7146–7152 (1992).

317. K. Harada, Y. Eizuru, Y. Isashiki, S. Ihara, and Y. Minamishima, *Arch. Virol.*, **142**, 215–225 (1997).

318. R. T. Sarisky, M. R. Quail, P. E. Clark, T. T. Nguyen, W. S. Halsey, R. J. Wittrock, J. O'Leary Bartus, M. M. Van Horn, G. M. Sathe, S. Van Horn, M. D. Kelly, T. H. Bacon, and J. J. Leary, *J. Virol.*, **75**, 1761–1769 (2001).

319. G. Andrei, R. Snoeck, E. De Clercq, R. Esnouf, P. Fiten, and G. Opdenakker, *J. Gen. Virol.*, **81**, 639–648 (2000).

320. D. M. Lowe, W. K. Alderton, M. R. Ellis, V. Parmar, W. H. Miller, G. B. Roberts, J. A. Fyfe, R. Gaillard, P. Ertl, and W. Snowden, *Antimicrob. Agents Chemother.*, **39**, 1802–1808 (1995).

321. T. I. Ng, Y. Shi, H. J. Huffaker, W. Kati, Y. Liu, C. M. Chen, Z. Lin, C. Maring, W. E. Kohlbrenner, and A. Molla, *Antimicrob. Agents Chemother.*, **45**, 1629–1636 (2001).

322. G. Abele, A. Karlstrom, J. Harmenberg, S. Shigeta, A. Larsson, B. Lindborg, and B. Wahren, *Antimicrob. Agents Chemother.*, **31**, 76–80 (1987).

323. C. McGuigan, C. J. Yarnold, G. Jones, S. Velazquez, H. Barucki, A. Brancale, G. Andrei, R. Snoeck, E. De Clercq, and J. Balzarini, *J. Med. Chem.*, **42**, 4479–4484 (1999).

324. C. McGuigan, H. Barucki, A. Carangio, S. Blewett, G. Andrei, R. Snoeck, E. De Clercq, J. Balzarini, and J. T. Erichsen, *J. Med. Chem.*, **43**, 4993–4997 (2000).

325. A. Brancale, C. McGuigan, G. Andrei, R. Snoeck, E. De Clercq, and J. Balzarini, *Bioorg. Med. Chem. Lett.*, **10**, 1215–1217 (2000).

326. M. R. Harnden, P. G. Wyatt, M. R. Boyd, and D. Sutton, *J. Med. Chem.*, **33**, 187–196 (1990).

327. S. Bailey, M. R. Harnden, and C. T. Shanks, *Antivirol. Chem. Chemother.*, **5**, 21–33 (1994).

328. Y. Yoshimura, K. Kitano, K. Yamada, S. Sakata, S. Miura, N. Ashida, and H. Machida, *Bioorg. Med. Chem.*, **8**, 1545–1558 (2000).

329. J. Wang, M. Froeyen, C. Hendrix, G. Andrei, R. Snoeck, E. De Clercq, and P. Herdewijn, *J. Med. Chem.*, **43**, 736–745 (2000).

330. C. K. Chu, T. Ma, K. Shanmuganathan, C. Wang, Y. Xiang, S. B. Pai, G. Q. Yao, J. P. Sommadossi, and Y. C. Cheng, *Antimicrob. Agents Chemother.*, **39**, 979–981 (1995).

331. G. Q. Yao, S. H. Liu, E. Chou, M. Kukhanova, C. K. Chu, and Y. C. Cheng, *Biochem. Pharmacol.*, **51**, 941–947 (1996).

332. N. Ono, S. Iwayama, K. Suzuki, T. Sekiyama, H. Nakazawa, T. Tsuji, M. Okunishi, T.

Daikoku, and Y. Nishiyama, *Antimicrob. Agents Chemother.*, **42**, 2095–2102 (1998).

333. L. B. Townsend, R. V. Devivar, S. R. Turk, M. R. Nassiri, and J. C. Drach, *J. Med. Chem.*, **38**, 4098–4105 (1995).

334. K. S. Gudmundsson, G. A. Freeman, J. C. Drach, and L. B. Townsend. *J. Med. Chem.*, **43**, 2473–2478 (2000).

335. L. B. Townsend, K. S. Gudmundsson, S. M. Daluge, J. J. Chen, Z. Zhu, G. W. Koszalka, L. Boyd, S. D. Chamberlain, G. A. Freeman, K. K. Biron, and J. C. Drach, *Nucleosides Nucleotides*, **18**, 509–519 (1999).

336. J. J. Chen, Y. Wei, J. C. Drach, and L. B. Townsend, *J. Med. Chem.*, **43**, 2449–2456 (2000).

337. Z. Zhu, B. Lippa, J. C. Drach, and L. B. Townsend, *J. Med. Chem.*, **43**, 2430–2437 (2000).

338. V. L. Zacny, E. Gershburg, M. G. Davis, K. K. Biron, and J. S. Pagano, *J. Virol.*, **73**, 7271–7277 (1999).

339. J. J. McSharry, A. McDonough, B. Olson, C. Talarico, M. Davis, and K. K. Biron, *Clin. Diag. Lab. Immunol.*, **8**, 1279–1281 (2001).

340. M. T. Migawa, J. L. Girardet, J. A. Walker, G. W. Koszalka, S. D. Chamberlain, J. C. Drach, and L. B. Townsend, *J. Med. Chem.*, **41**, 1242–1251 (1998).

341. P. M. Krosky, M. R. Underwood, S. R. Turk, K. W.-H. Feng, R. K. Jain, R. G. Ptak, A. C. Westerman, K. K. Biron, L. B. Townsend, and J. C. Drach, *J. Virol.*, **72**, 4721–4728 (1998).

342. K. S. Gudmundsson, J. Tidwell, N. Lippa, G. W. Koszalka, N. van Draanen, R. G. Ptak, J. C. Drach, and L. B. Townsend, *J. Med. Chem.*, **43**, 2464–2472 (2000).

343. M. R. Underwood, R. J. Harvey, S. C. Stanat, M. L. Hemphill, T. Miller, J. C. Drach, L. B. Townsend, and K. K. Biron, *J. Virol.*, **72**, 717–725 (1998).

344. M. van Zeijl, J. Fairhurst, T. R. Jones, S. K. Vernon, J. Morin, J. LaRocque, B. Feld, B. O'Hara, J. D. Bloom, and S. V. Johann, *J. Virol.*, **74**, 9054–9061 (2000).

345. G. W. Koszalka, D. R. Averett, J. A. Fyfe, G. B. Roberts, T. Spector, K. Biron, and T. A. Krenitsky, *Antimicrob. Agents Chemother.*, **35**, 1437–1443 (1991).

346. Y. Choi, L. Li, S. Grill, E. Gullen, C. S. Lee, G. Gumina, E. Tsujii, Y. C. Cheng, and C. K. Chu., *J. Med. Chem.*, **43**, 2538–2546 (2000).

347. T. Onishi, C. Mukai, R. Nakagawa, T. Sekiyama, M. Aoki, K. Suzuki, H. Nakazawa, N. Ono, Y. Ohmura, S. Iwayama, M. Okunishi, and T. Tsuji, *J. Med. Chem.*, **43**, 278–282 (2000).

348. J. S. Lin, T. Kira, E. Gullen, Y. Choi, F. Qu, C. K. Chu, and Y. C. Cheng, *J. Med. Chem.*, **42**, 2212–2217 (1999).

349. L. Li, G. E. Dutschman, E. A. Gullen, E. Tsujii, S. P. Grill, Y. Choi, C. K. Chu, and Y. C. Cheng, *Mol. Pharmacol.*, **58**, 1109–1114 (2000).

350. T. Kira, S. P. Grill, G. E. Dutschman, J. S. Lin, F. Qu, Y. Choi, C. K. Chu, and Y. C. Cheng, *Antimicrob. Agents Chemother.*, **44**, 3278–3284 (2000).

351. S. Shigeta, S. Mori, T. Kira, K. Takahashi, E. Kodama, K. Konno, T. Nagata, H. Kato, T. Wakayama, N. Koike, and M. Saneyoshi, *Antiviral. Chem. Chemother.*, **10**, 195–209 (1999).

352. C. U. Kim, P. F. Misco, B. Y. Luh, M. J. Hitchcock, I. Ghazzouli, and J. C. Martin, *J. Med. Chem.*, **34**, 2286–2294 (1991).

353. D. L. Barnard, J. H. Huffman, R. W. Sidwell, and E. J. Reist, *Antiviral Res.*, **22**, 77–89 (1993).

354. S. D. Chamberlain, K. K. Biron, R. E. Dornsife, D. R. Averett, L. Beauchamp, and G. W. Koszalka, *J. Med. Chem.*, **37**, 1371–1377 (1994).

355. D. F. Smee, S. T. Sugiyama, and E. J. Reist, *Antimicrob. Agents Chemother.*, **38**, 2165–2168 (1994).

356. E. R. Kern, J. Palmer, C. Hartline, and P. E. Vogt, *Antiviral Res.*, **26**, A329 (1995).

357. M. J. M. Hitchcock, S. A. Lacy, L. J. R., and E. R. Kern, *Antiviral Res.*, **26**, A358 (1995).

358. N. Bourne, F. J. Bravo, and D. I. Bernstein, *Antiviral Res.*, **47**, 103–109 (2000).

359. J. Neyts, G. Andrei, R. Snoeck, G. Jahne, I. Winkler, M. Helsberg, J. Balzarini, and E. De Clercq, *Antimicrob. Agents Chemother.*, **38**, 2710–2716 (1994).

360. D. Reymen, L. Naesens, J. Balzarini, A. Holy, H. Dvorakova, and E. De Clercq, *Antiviral Res.*, **28**, 343–357 (1995).

361. Y. Zhang, D. Schols, and E. De Clercq, *Antiviral Res.*, **43**, 23–35 (1999).

362. J. Neyts, J. Balzarini, G. Andrei, Z. Chaoyong, R. Snoeck, A. Zimmermann, T. Mertens, A. Karlsson, and E. De Clercq, *Mol. Pharmacol.*, **53**, 157–165 (1998).

363. A. Zimmermann, D. Michel, I. Pavic, W. Hampl, A. Luske, J. Neyts, E. De Clercq, and T. Mertens, *Antiviral Res.*, **36**, 35–42 (1997).

364. G. Andrei, R. Snoeck, and E. De Clercq, *Antimicrob. Agents Chemother.*, **39**, 1632–1635 (1995).

365. T. A. Krenitsky, W. W. Hall, P. de Miranda, L. M. Beauchamp, H. J. Schaeffer, and P. D. Whiteman, *Proc. Natl. Acad. Sci. USA*, **81**, 3209–3213 (1984).

366. S. D. Chamberlain, A. R. Moorman, T. C. Burnette, P. de Miranda, and T. A. Krenitsky, *Antiviral Chem. Chemother.*, **5**, 64–73 (1994).

367. Y. L. Qiu, M. B. Ksebati, R. G. Ptak, B. Y. Fan, J. M. Breitenbach, J. S. Lin, Y. C. Cheng, E. R. Kern, J. C. Drach, and J. Zemlicka, *J. Med. Chem.*, **41**, 10–23 (1998).

368. Y. L. Qiu, R. G. Ptak, J. M. Breitenbach, J. S. Lin, Y. C. Cheng, J. C. Drach, E. R. Kern, and J. Zemlicka, *Antiviral Res.*, **43**, 37–53 (1999).

369. Y. L. Qiu, A. Hempel, N. Camerman, A. Camerman, F. Geiser, R. G. Ptak, J. M. Breitenbach, T. Kira, L. Li, E. Gullen, Y. C. Cheng, J. C. Drach, and J. Zemlicka, *Nucleosides Nucleotides*, **18**, 597–598 (1999).

370. R. J. Rybak, C. B. Hartline, Y. L. Qiu, J. Zemlicka, E. Harden, G. Marshall, J. P. Sommadossi, and E. R. Kern, *Antimicrob. Agents Chemother.*, **44**, 1506–1511 (2000).

371. R. J. Rybak, J. Zemlicka, Y. L. Qiu, C. B. Hartline, and E. R. Kern, *Antiviral Res.*, **43**, 175–188 (1999).

372. H. P. Guan, M. B. Ksebati, Y. C. Cheng, J. C. Drach, E. R. Kern, and J. Zemlicka, *J. Organic Chem.*, **65**, 1280–1290 (2000).

373. A. Meerbach, R. Klocking, C. Meier, A. Lomp, B. Helbig, and P. Wutzler, *Antiviral Res.*, **45**, 69–77 (2000).

374. C. McGuigan, M. J. Slater, N. R. Parry, A. Perry, and S. Harris, *Bioorg. Med. Chem. Lett.*, **10**, 645–647 (2000).

375. K. Y. Hostetler, S. Parker, C. N. Sridhar, M. J. Martin, J. L. Li, L. M. Stuhmiller, G. M. van Wijk, H. van den Bosch, M. F. Gardner, K. A. Aldern, *Proc. Natl. Acad. Sci. USA*, **90**, 11835–11839 (1993).

376. G. Andrei, R. Snoeck, J. Neyts, M. L. Sandvold, F. Myhren, and E. De Clercq, *Antiviral Res.*, **45**, 157–167 (2000).

377. J. Neyts, G. Jahne, G. Andrei, R. Snoeck, I. Winkler, and E. De Clercq, *Antimicrob. Agents Chemother.*, **39**, 56–60 (1995).

378. S. Crotty, C. E. Cameron, and R. Andino, *Proc. Natl. Acad. Sci. USA*, **98**, 6895–6900 (2001).

379. S. Crotty, D. Maag, J. J. Arnold, W. Zhong, J. Y. Lau, Z. Hong, R. Andino, and C. E. Cameron, *Nature Med.*, **6**, 1375–1379 (2000).

380. D. Maag, C. Castro, Z. Hong, and C. E. Cameron, *J. Biol. Chem.*, **276**, 46094–46098 (2001).

381. S. A. Birkeland, H. K. Andersen, and S. J. Hamilton-Dutoit, *Transplantation*, **67**, 1209–1214 (1999).

382. J. Neyts, G. Andrei, and E. De Clercq, *Antimicrob. Agents Chemother.*, **42**, 3285–3289 (1998).

383. J. J. Coull, D. Turner, T. Melby, M. R. Betts, R. Lanier, and D. M. Margolis, *AIDS*, **26**, 423–434 (2001).

384. W. Markland, T. J. McQuaid, J. Jain, and A. D. Kwong, *Antimicrob. Agents Chemother.*, **44**, 859–866 (2000).

385. B. Clotet, L. Ruiz, C. Cabrera, A. Ibanez, M. P. Canadas, G. Sirera, J. Romeu, and J. Vila, *Antiviral Ther.*, **1**, 189–193 (1996).

386. A. Foli, F. Lori, R. Maserati, C. Tinelli, L. Minoli, and J. Lisziewicz, *Antiviral Ther.*, **2**, 31–38 (1997).

387. I. Sanne, R. A. Smego, and B. V. Mendelow, *Int. J. Infect. Dis.*, **5**, 43–48 (2001).

388. J. Neyts and E. De Clercq, *Antimicrob. Agents Chemother.*, **43**, 2885–2892 (1999).

389. A. R. Porcari, R. G. Ptak, K. Z. Borysko, J. M. Breitenbach, S. Vittori, L. L. Wotring, J. C. Drach, and L. B. Townsend, *J. Med. Chem.*, **43**, 2438–2448 (2000).

390. J. J. Crute, I. R. Lehman, J. Gambino, T. F. Yang, P. Medveczky, M. Medveczky, N. N. Khan, C. Mulder, J. Monroe, and G. E. Wright, *J. Med. Chem.*, **38**, 1820–1825 (1995).

391. M. Medveczky, T. F. Yang, J. Gambino, P. Medveczky, and G. E. Wright, *J. Med. Chem.*, **38**, 1811–1819 (1995).

392. F. C. Spector, L. Liang, H. Giordano, M. Sivaraja, and M. G. Peterson, *J. Virol.*, **72**, 6979–6987 (1998).

393. M. Sivaraja, H. Giordano, and M. G. Peterson, *Anal. Biochem.*, **265**, 22–27 (1998).

394. N. Bourne, L. R. Stanberry, E. R. Kern, G. Holan, B. Matthews, and D. I. Bernstein, *Antimicrob. Agents Chemother.*, **44**, 2471–2474 (2000).

395. W. Gibson and M. R. Hall, *Drug Design Discovery*, **15**, 39–47 (1997).

396. B. C. Holwerda, *Antiviral Res.*, **35**, 1–21 (1997).

397. D. L. Flynn, D. P. Becker, V. M. Dilworth, M. K. Highkin, P. J. Hippenmeyer, K. A. Houseman, L. M. Levine, M. Li, A. E. Moormann, A. Rankin, M. V. Toth, C. I. Villamil, A. J. Wittwer, and B. C. Holwerda, *Drug Design Discovery*, **15**, 3–15 (1997).

398. R. L. Jarvest, I. L. Pinto, S. M. Ashman, C. E. Dabrowski, A. V. Fernandez, L. J. Jennings, P.

Lavery, and D. G. Tew, *Bioorg. Med. Chem. Lett.*, **9**, 443–448 (1999).

399. M. Matsumoto, S. Misawa, N. Chiba, H. Takaku, and H. Hayashi, *Biol. Pharm. Bull.*, **24**, 236–241 (2001).

400. D. I. Bernstein, C. J. Harrison, M. A. Tomai, and R. L. Miller, *J. Infect. Dis.*, **183**, 844–849 (2001).

401. C. J. Harrison, R. L. Miller, and D. I. Bernstein, *Vaccine*, **19**, 1820–1826 (2001).

402. S. L. Spruance, S. K. Tyring, M. H. Smith, and T. C. Meng, *J. Infect. Dis.*, **184**, 196–200 (2001).

403. G. A. Bishop, Y. Hsing, B. S. Hostager, S. V. Jalukar, L. M. Ramirez, and M. A. Tomai, *J. Immunol.*, **165**, 5552–5557 (2000).

404. F. Bortolotti, P. Jara, C. Barbera, G. V. Gregorio, A. Vegnente, L. Zancan, L. Hierro, C. Crivellaro, G. M. Vergani, R. Iorio, M. Pace, P. Con, and A. Gatta, *Gut*, **46**, 715–718 (2000).

405. J. L. Dienstag, R. P. Perrillo, E. R. Schiff, M. Bartholomew, C. Vicary, and M. Rubin, *N. Engl. J. Med.*, **333**, 1657–1661 (1995).

406. D. K. Wong, A. M. Cheung, K. O'Rourke, C. D. Naylor, A. S. Detsky, and J. Heathcote, *Ann. Intern. Med.*, **119**, 312–323 (1993).

407. M. Tsiang, J. F. Rooney, J. J. Toole, and C. S. Gibbs, *Hepatology*, **29**, 1863–1869 (1999).

408. L. J. Stuyver, S. A. Locarnini, A. Lok, D. D. Richman, W. F. Carman, J. L. Dienstag, and R. F. Schinazi, *Hepatology*, **33**, 751–757 (2001).

409. M. L. Bryant, E. G. Bridges, L. Placidi, A. Faraj, A. G. Loi, C. Pierra, D. Dukhan, G. Gosselin, J. L. Imbach, B. Hernandez, A. Juodawlkis, B. Tennant, B. Korba, P. Cote, P. Marion, E. Cretton-Scott, R. F. Schinazi, and J. P. Sommadossi, *Antimicrob. Agents Chemother.*, **45**, 229–235 (2001).

410. X. Xiong, C. Flores, H. Yang, J. J. Toole, and C. S. Gibbs., *Hepatology*, **28**, 1669–1673 (1998).

411. S. F. Innaimo, M. Seifer, G. S. Bisacchi, D. N. Standring, R. Zahler, and R. J. Colonno, *Antimicrob. Agents Chemother.*, **41**, 1444–1448 (1997).

412. B. E. Korba and M. R. Boyd, *Antimicrob. Agents Chemother.*, **40**, 1282–1284 (1996).

413. T. S. Lin, M. Z. Luo, M. C. Liu, Y. L. Zhu, E. Gullen, G. E. Dutschman, and Y. C. Cheng, *J. Med. Chem.*, **39**, 1757–1759 (1996).

414. P. A. Furman, M. Davis, D. C. Liotta, M. Paff, L. W. Frick, D. J. Nelson, R. E. Dornsife, J. A.

Wurster, L. J. Wilson, J. A. Fyfe, *Antimicrob. Agents Chemother.*, **36**, 2686–2692 (1992).

415. J. Balzarini, O. Wedgwood, J. Kruining, H. Pelemans, R. Heijtink, E. De Clercq, and C. McGuigan, *Biochem. Biophys. Res. Commun.*, **225**, 363–369 (1996).

416. S. J. Hadziyannis, G. V. Papatheodoridis, E. Dimou, A. Laras, and C. Papaioannou, *Hepatology*, **32**, 847–851 (2000).

417. D. T. Lau, M. F. Khokhar, E. Doo, M. G. Ghany, D. Herion, Y. Park, D. E. Kleiner, P. Schmid, L. D. Condreay, J. Gauthier, M. C. Kuhns, T. J. Liang, and J. H. Hoofnagle, *Hepatology*, **32**, 828–834 (2000).

418. M. R. Brunetto, F. Oliveri, G. Rocca, D. Criscuolo, E. Chiaberge, M. Capalbo, E. David, G. Verme, and F. Bonino, *Hepatology*, **10**, 198–202 (1989).

419. P. Lampertico, E. Del Ninno, A. Manzin, M. F. Donato, M. G. Rumi, G. Lunghi, A. Morabito, M. Clementi, and M. Colombo, *Hepatology*, **26**, 1621–1625 (1997).

420. C. Boni, A. Bertoletti, A. Penna, A. Cavalli, M. Pilli, S. Urbani, P. Scognamiglio, R. Boehme, R. Panebianco, F. Fiaccadori, and C. Ferrari, *J. Clin. Invest.*, **102**, 968–975 (1998).

421. S. W. Cho, K. B. Hahm, and J. H. Kim, *Hepatology*, **32**, 1163–1169 (2000).

422. E. B. Keeffe, *J. Med. Virol.*, **61**, 403–408 (2000).

423. G. Macedo, J. C. Maia, A. Gomes, M. Beleza, A. Teixeira, and A. Ribeiro, *Transplant. Proc.*, **32**, 2642 (2000).

424. P. Andreone, P. Caraceni, G. L. Grazi, L. Belli, G. L. Milandri, G. Ercolani, E. Jovine, A. D'Errico, P. R. Dal Monte, G. Ideo, D. Forti, A. Mazziotti, A. Cavallari, and M. Bernardi, *J. Hepatol.*, **29**, 985–989 (1998).

425. C. Antoine, A. Landau, V. Menoyo, J. P. Duong, A. Duboust, and D. Glotz, *Transplant. Proc.*, **32**, 384–385 (2000).

426. D. Lewandowska, M. Durlik, K. KukuLa, T. Cieciura, R. Ciecierski, B. Walewska-Zielecka, J. Szmidt, W. Rowinski, and M. Lao, *Transplant. Proc.*, **32**, 1369–1370 (2000).

427. S. Kewn, P. G. Hoggard, S. D. Sales, M. A. Johnson, and D. J. Back, *Br. J. Clin. Pharmacol.*, **50**, 597–604 (2000).

428. Y. L. Zhu, D. E. Dutschman, S. H. Liu, E. G. Bridges, and Y. C. Cheng, *Antimicrob. Agents Chemother.*, **42**, 1805–1810 (1998).

429. F. Le Guerhier, C. Pichoud, S. Guerret, M. Chevallier, C. Jamard, O. Hantz, X. Y. Li,

S. H. Chen, I. King, C. Trepo, Y. C. Cheng, and F. Zoulim, *Antimicrob. Agents Chemother.*, **44**, 111–122 (2000).

430. X. Li, E. Carmichael, M. Feng, I. King, T. W. Doyle, and S. H. Chen, *Bioorg. Med. Chem. Lett.*, **8**, 57–62 (1998).

431. F. Le Guerhier, C. Pichoud, C. Jamard, S. Guerret, M. Chevallier, S. Peyrol, O. Hantz, I. King, C. Trepo, Y. C. Cheng, and F. Zoulim, *Antimicrob. Agents Chemother.*, **45**, 1065–1077 (2001).

432. L. Fu and Y. C. Cheng, *Antimicrob. Agents Chemother.*, **44**, 3402–3407 (2000).

433. R. F. Schinazi, G. Gosselin, A. Faraj, B. E. Korba, D. C. Liotta, C. K. Chu, C. Mathe, J. L. Imbach, and J. P. Sommadossi, *Antimicrob. Agents Chemother.*, **38**, 2172–2174 (1994).

434. B. E. Korba, R. F. Schinazi, P. Cote, B. C. Tennant, and J. L. Gerin, *Antimicrob. Agents Chemother.*, **44**, 1757–1760 (2000).

435. W. Lewis, E. S. Levine, B. Griniuviene, K. O. Tankersley, J. M. Colacino, J. P. Sommadossi, K. A. Watanabe, and F. W. Perrino, *Proc. Natl. Acad. Sci. USA*, **93**, 3592–3597 (1996).

436. S. Balakrishna Pai, S. H. Liu, Y. L. Zhu, C. K. Chu, and Y. C. Cheng, *Antimicrob. Agents Chemother.*, **40**, 380–386 (1996).

437. S. Aguesse-Germon, S. H. Liu, M. Chevallier, C. Pichoud, C. Jamard, C. Borel, C. K. Chu, C. Trepo, Y. C. Cheng, and F. Zoulim, *Antimicrob. Agents Chemother.*, **42**, 369–376 (1998).

438. C. K. Chu, F. D. Boudinot, S. F. Peek, J. H. Hong, Y. Choi, B. E. Korba, J. L. Gerin, P. J. Cote, B. C. Tennant, and Y. C. Cheng, *Antiviral Ther.*, **3**, 113–121 (1998).

439. S. F. Peek, P. J. Cote, J. R. Jacob, I. A. Toshkov, W. E. Hornbuckle, B. H. Baldwin, F. V. Wells, C. K. Chu, J. L. Gerin, B. C. Tennant, and B. E. Korba, *Hepatology*, **33**, 254–266 (2001).

440. S. H. Liu, K. L. Grove, and Y. C. Cheng, *Antimicrob. Agents Chemother.*, **42**, 833–839 (1998).

441. T. Ma, S. B. Pai, Y. L. Zhu, J. S. Lin, K. Shanmuganathan, J. Du, C. Wang, H. Kim, M. G. Newton, Y. C. Cheng, and C. K. Chu, *J. Med. Chem.*, **39**, 2835–2843 (1996).

442. B. K. Chun, R. F. Schinazi, Y. C. Cheng, and C. K. Chu, *Carbohydrate Res.*, **328**, 49–59 (2000).

443. S. H. Chen, Q. Wang, J. Mao, I. King, G. E. Dutschman, E. A. Gullen, Y. C. Cheng, and T. W. Doyle, *Bioorg. Med. Chem. Lett.*, **8**, 1589–1594 (1998).

444. L. P. Kotra, Y. Xiang, M. G. Newton, R. F. Schinazi, Y. C. Cheng, and C. K. Chu, *J. Med. Chem.*, **40**, 3635–3644 (1997).

445. C. Pierra, J. L. Imbach, E. De Clercq, J. Balzarini, A. Van Aerschot, P. Herdewijn, A. Faraj, A. G. Loi, J. P. Sommadossi, and G. Gosselin, *Antiviral Res.*, **45**, 169–183 (2000).

446. Y. L. Zhu, S. B. Pai, S. H. Liu, K. L. Grove, B. C. Jones, C. Simons, J. Zemlicka, and Y. C. Cheng, *Antimicrob. Agents Chemother.*, **41**, 1755–1760 (1997).

447. B. Kierdaszuk, C. Bohman, B. Ullman, and S. Eriksson, *Biochem. Pharmacol.*, **43**, 197–206 (1992).

448. F. E. Zahm, F. Bonino, R. Giuseppetti, and M. Rapicetta, *Ital. J. Gastroenterol Hepatol.*, **30**, 510–516 (1998).

449. R. A. de Man, P. Marcellin, F. Habal, P. Desmond, T. Wright, T. Rose, R. Jurewicz, and C. Young, *Hepatology*, **32**, 413–417 (2000).

450. K. Y. Hostetler, J. R. Beadle, G. D. Kini, M. F. Gardner, K. N. Wright, T. H. Wu, and B. A. Korba, *Biochem. Pharmacol.*, **53**, 1815–1822 (1997).

451. K. Y. Hostetler, J. R. Beadle, W. E. Hornbuckle, C. A. Bellezza, I. A. Tochkov, P. J. Cote, J. L. Gerin, B. E. Korba, and B. C. Tennant, *Antimicrob. Agents Chemother.*, **44**, 1964–1969 (2000).

452. C. Perigaud, G. Gosselin, J. L. Girardet, B. E. Korba, and J. L. Imbach, *Antiviral Res.*, **40**, 167–178 (1999).

453. O. Hantz, C. Perigaud, C. Borel, C. Jamard, F. Zoulim, C. Trepo, J. L. Imbach, and G. Gosselin, *Antiviral Res.*, **40**, 179–187 (1999).

454. M. Seifer, R. K. Hamatake, R. J. Colonno, and D. N. Standring, *Antimicrob. Agents Chemother.*, **42**, 3200–3208 (1998).

455. E. V. Genovesi, L. Lamb, I. Medina, D. Taylor, M. Seifer, S. Innaimo, R. J. Colonno, D. N. Standring, and J. M. Clark, *Antimicrob. Agents Chemother.*, **42**, 3209–3217 (1998).

456. P. L. Marion, F. H. Salazar, M. A. Winters, and R. J. Colonno, *Antimicrob. Agents Chemother.*, **46**, 82–88 (2002).

457. R. A. de Man, L. M. Wolters, F. Nevens, D. Chua, M. Sherman, C. L. Lai, A. Gadano, Y. Lee, F. Mazzotta, N. Thomas, and D. DeHertogh, *Hepatology*, **34**, 578–582 (2001).

458. R. A. Heijtink, J. Kruining, G. A. de Wilde, J. Balzarini, E. de Clercq, and S. W. Schalm, *Antimicrob. Agents Chemother.*, **38**, 2180–2182 (1994).

459. C. Ying, E. De Clercq, and J. Neyts, *J. Viral Hepatitis*, **7**, 79–83 (2000).

460. R. J. Gilson, K. B. Chopra, A. M. Newell, I. M. Murray-Lyon, M. R. Nelson, S. J. Rice, R. S. Tedder, J. Toole, H. S. Jaffe, and I. V. Weller, *J. Viral Hepatitis*, **6**, 387–395 (1999).

461. M. Dandri, M. R. Burda, H. Will, and J. Petersen, *Hepatology*, **32**, 139–146 (2000).

462. J. M. Cullen, D. H. Li, C. Brown, E. J. Eisenberg, K. C. Cundy, J. Wolfe, J. Toole, and C. Gibbs, *Antimicrob. Agents Chemother.*, **45**, 2740–2745 (2001).

463. Y. Benhamou, M. Bochet, V. Thibault, V. Calvez, M. H. Fievet, P. Vig, C. S. Gibbs, C. Brosgart, J. Fry, H. Namini, C. Katlama, and T. Poynard, *Lancet*, **358**, 718–723 (2001).

464. J. Kahn, S. Lagakos, M. Wulfsohn, D. Cherng, M. Miller, J. Cherrington, D. Hardy, G. Beall, R. Cooper, R. Murphy, N. Basgoz, E. Ng, S. Deeks, D. Winslow, J. J. Toole, and D. Coakley, *JAMA*, **282**, 2305–2312 (1999).

465. E. V. Genovesi, L. Lamb, I. Medina, D. Taylor, M. Seifer, S. Innaimo, R. J. Colonno, and J. M. Clark, *Antiviral Res.*, **48**, 197–203 (2000).

466. S. K. Ono-Nita, N. Kato, Y. Shiratori, K. H. Lan, H. Yoshida, F. J. Carrilho, and M. Omata, *J. Clin. Invest.*, **103**, 1635–1640 (1999).

467. G. C. Farrell, *Drugs*, **60**, 701–710 (2000).

468. D. Mutimer, D. Pillay, P. Shields, P. Cane, D. Ratcliffe, B. Martin, S. Buchan, L. Boxall, K. O'Donnell, J. Shaw, S. Hubscher, and E. Elias, *Gut*, **46**, 107–113 (2000).

469. C. T. Yeh, R. N. Chien, C. M. Chu, and Y. F. Liaw, *Hepatology*, **31**, 1318–1326 (2000).

470. T. Seta, O. Yokosuka, F. Imazeki, M. Tagawa, and H. Saisho, *J. Med. Virol.*, **60**, 8–16 (2000).

471. L. Fu, S. H. Liu, and Y. C. Cheng, *Biochem. Pharmacol.*, **57**, 1351–1359 (1999).

472. S. N. Si Ahmed, S. Ahmed, D. Tavan, C. Pichoud, F. Berby, L. Stuyver, M. Johnson, P. Merle, H. Abidi, C. Trepo, and F. Zoulim, *Hepatology*, **32**, 1078–1088 (2000).

473. M. I. Allen, M. Deslauriers, C. W. Andrews, G. A. Tipples, K. A. Walters, D. L. Tyrrell, N. Brown, and L. D. Condreay, *Hepatology*, **27**, 1670–1677 (1998).

474. M. M. Bartholomew, R. W. Jansen, L. J. Jeffers, K. R. Reddy, L. C. Johnson, H. Bunzendahl, L. D. Condreay, A. G. Tzakis, E. R. Schiff, and N. A. Brown, *Lancet*, **349**, 20–22 (1997).

475. K. P. Fischer and D. L. Tyrrell, *Antimicrob. Agents Chemother.*, **40**, 1957–1960 (1996).

476. S. K. Ono, N. Kato, Y. Shiratori, J. Kato, T. Goto, R. F. Schinazi, F. J. Carrilho, and M. Omata, *J. Clin. Invest.*, **107**, 449–455 (2001).

477. K. Das, X. Xiong, H. Yang, C. E. Westland, C. S. Gibbs, S. G. Sarafianos, and E. Arnold, *J. Virol.*, **75**, 4771–4779 (2001).

478. D. Mutimer, D. Pillay, P. Cook, D. Ratcliffe, K. O'Donnell, D. Dowling, J. Shaw, E. Elias, and P. A. Cane, *J. Infect. Dis.*, **181**, 713–716 (2000).

479. B. Seigneres, C. Pichoud, S. S. Ahmed, O. Hantz, C. Trepo, and F. Zoulim, *J. Infect. Dis.*, **181**, 1221–1233 (2000).

480. T. T. Aye, A. Bartholomeusz, T. Shaw, S. Bowden, A. Breschkin, J. McMillan, P. Angus, and S. Locarnini, *J. Hepatol.*, **26**, 1148–1153 (1997).

481. X. Xiong, H. Yang, C. E. Westland, R. Zou, and C. S. Gibbs, *Hepatology*, **31**, 219–224 (2000).

482. S. K. Ladner, T. J. Miller, M. J. Otto, and R. W. King, *Antiviral. Chem. Chemother.*, **9**, 65–72 (1998).

483. R. Perrillo, E. Schiff, E. Yoshida, A. Statler, K. Hirsch, T. Wright, K. Gutfreund, P. Lamy, and A. Murray, *Hepatology*, **32**, 129–134 (2000).

484. T. Cotonat, J. A. Quiroga, J. M. Lopez-Alcorocho, R. Clouet, M. Pardo, F. Manzarbeitia, and V. Carreno, *Hepatology*, **31**, 502–506 (2000).

485. S. W. Schalm, J. Heathcote, J. Cianciara, G. Farrell, M. Sherman, B. Willems, A. Dhillon, A. Moorat, J. Barber, and D. F. Gray, *Gut*, **46**, 562–568 (2000).

486. Z. Ben-Ari, R. Zemel, and R. Tur-Kaspa, *Transplantation*, **71**, 154–156 (2001).

487. C. Ying, E. De Clercq, and J. Neyts, *Antiviral Res.*, **48**, 117–124 (2000).

488. D. Colledge, G. Civitico, S. Locarnini, and T. Shaw, *Antimicrob. Agents Chemother.*, **44**, 551–560 (2000).

489. B. E. Korba, P. Cote, W. Hornbuckle, R. Schinazi, J. L. Gerin, and B. C. Tennant, *Antiviral Res.*, **45**, 19–32 (2000).

490. D. Colledge, S. Locarnini, and T. Shaw, *Hepatology*, **26**, 216–225 (1997).

491. G. K. Lau, M. Tsiang, J. Hou, S. Yuen, W. F. Carman, L. Zhang, C. S. Gibbs, and S. Lam, *Hepatology*, **32**, 394–399 (2000).

492. L. Fiume, G. Di Stefano, C. Busi, A. Mattioli, F. Bonino, M. Torrani-Cerenzia, G. Verme, M. Rapicetta, M. Bertini, and G. B. Gervasi, *J. Viral Hepatitis*, **4**, 363–370 (1997).

493. L. Fiume, G. Di Stefano, C. Busi, F. Incitti, M. Rapicetta, C. Farina, G. B. Gervasi, and F. Bonino, *J. Hepatol.*, **29**, 1032–1033 (1998).

494. M. R. Torrani Cerenzia, L. Fiume, C. Busi, A. Mattioli, G. Di Stefano, G. B. Gervasi, M. R. Brunetto, P. Piantino, G. Verme, and F. Bonino, *J. Hepatol.*, **20**, 307–309 (1994).

495. L. Fiume, G. Di Stefano, C. Busi, A. Mattioli, M. Rapicetta, R. Giuseppetti, A. R. Ciccaglione, and C. Argentini, *Hepatology*, **22**, 1072–1077 (1995).

496. L. Fiume, G. Di Stefano, C. Busi, A. Mattioli, G. Battista Gervasi, M. Bertini, C. Bartoli, R. Catalani, G. Caccia, C. Farina, A. Fissi, O. Pieroni, R. Giuseppetti, E. D'Ugo, R. Bruni, and M. Rapicetta, *J. Hepatol.*, **26**, 253–259 (1997).

497. E. A. Biessen, A. R. Valentijn, R. L. De Vrueh, E. Van De Bilt, L. A. Sliedregt, P. Prince, M. K. Bijsterbosch, J. H. Van Boom, G. A. Van Der Marel, P. J. Abrahams, and T. J. Van Berkel, *FASEB J.*, **14**, 1784–1792 (2000).

498. R. L. de Vrueh, E. T. Rump, E. van De Bilt, R. van Veghel, J. Balzarini, E. A. Biessen, T. J. van Berkel, and M. K. Bijsterbosch, *Antimicrob. Agents Chemother.*, **44**, 477–483 (2000).

499. H. Xie, M. Voronkov, D. C. Liotta, B. A. Korba, R. F. Schinazi, D. D. Richman, and K. Y. Hostetler, *Antiviral Res.*, **28**, 113–120 (1995).

500. W. B. Offensperger, C. Thoma, D. Moradpour, F. von Weizsacker, S. Offensperger, and H. E. Blum, *Methods Enzymol.*, **314**, 524–536 (2000).

501. W. Xin and J. H. Wang, *Antisense Nucleic Acid Drug Develop.*, **8**, 459–468 (1998).

502. A. Wikstrom and G. von Krogh, *Int. J. STD AIDS*, **9**, 537–542 (1998).

503. A. R. Kling, *Semin. Dermatol.*, **11**, 247–255 (1992).

504. S. Jablonska, and S. Majewski, in G. Gross, S. Jablonska, H. Pfister, and H. E. Stegner, Eds., *Genital Papillomavirus Infections*, Springer-Verlag, Berlin, 1990, pp. 263–281.

505. D. H. Spach and R. Colven, *J. Am. Acad. Dermatol.*, **40**, 818–821 (1999).

506. I. Heard, V. Schmitz, D. Costagliola, G. Orth, and M. D. Kazatchkine, *AIDS*, **12**, 1459–1464 (1998).

507. J. C. Vance, B. J. Bart, R. C. Hansen, R. C. Reichman, C. McEwen, K. D. Hatch, B. Berman, and D. J. Tanner, *Arch. Dermatol.*, **122**, 272–277 (1986).

508. L. J. Eron, F. Judson, S. Tucker, S. Prawer, J. Mills, K. Murphy, M. Hickey, M. Rogers, S. Flannigan, and N. Hien, *N. Engl. J. Med.*, **315**, 1059–1064 (1986).

509. A. E. Friedman-Kien, L. J. Eron, M. Conant, W. Growdon, H. Badiak, P. W. Bradstreet, D. Fedorczyk, J. R. Trout, and T. F. Plasse, *JAMA*, **259**, 533–538 (1988).

510. M. A. Tomai, S. J. Gibson, L. M. Imbertson, R. L. Miller, P. E. Myhre, M. J. Reiter, T. L. Wagner, C. B. Tamulinas, J. M. Beaurline, and J. F. Gerster, *Antiviral Res.*, **28**, 253–264 (1995).

511. M. A. Tomai, L. M. Imbertson, T. L. Stanczak, L. T. Tygrett, and T. J. Waldschmidt, *Cell. Immunol.*, **203**, 55–65 (2000).

512. C. J. Harrison, L. Jenski, T. Voychehovski, and D. I. Bernstein, *Antiviral Res.*, **10**, 209–223 (1988).

513. C. J. Harrison, L. R. Stanberry, and D. I. Bernstein, *Antiviral Res.*, **15**, 315–322 (1991).

514. K. R. Beutner and A. Ferenczy, *Am. J. Med.*, **102**, 28–37 (1997).

515. K. R. Beutner, S. K. Tyring, K. F. Trofatter Jr., J. M. Douglas Jr., S. Spruance, M. L. Owens, T. L. Fox, A. J. Hougham, and K. A. Schmitt, *Antimicrob Agents Chemother.*, **42**, 789–794 (1998).

516. A. Okamoto, C. D. Woodworth, K. Yen, J. Chung, S. Isonishi, T. Nikaido, T. Kiyokawa, H. Seo, Y. Kitahara, K. Ochiai, and T. Tanaka, *Oncol. Rep.*, **6**, 269–276 (1999).

517. G. L. Pride, *J. Reprod. Med.*, **35**, 384–387 (1990).

518. A. Ferenczy, *Obstet. Gynecol.*, **64**, 773–778 (1984).

519. M. M. Holmes, S. H. Weaver II, and S. T. Vermillion, *Infect. Dis. Obstet. Gynecol.*, **7**, 186–189 (1999).

520. F. Wierrani, A. Kubin, R. Jindra, M. Henry, K. Gharehbaghi, W. Grin, J. Soltz-Szotz, G. Alth, and W. Grunberger, *Cancer Detect. Prevent.*, **23**, 351–355 (1999).

521. K. C. Cundy, G. Lynch, and W. A. Lee, *Antiviral Res.*, **35**, 113–122 (1997).

522. G. Orlando, M. M. Fasolo, R. Beretta, R. Signori, B. Adriani, N. Zanchetta, and A. Cargnel, *AIDS*, **13**, 1978–1980 (1999).

523. J. A. Johnson and J. D. Gangemi, *Antimicrob. Agents Chemother.*, **43**, 1198–1205 (1999).

524. J. Duan, W. Paris, J. De Marte, D. Roopchand, T. L. Fleet, and M. G. Cordingley, *Antiviral Res.*, **46**, 135–144 (2000).

525. N. D. Christensen, M. D. Pickel, L. R. Budgeon, and J. W. Kreider, *Antiviral Res.*, **48**, 131–142 (2000).

526. N. D. Christensen, R. Han, N. M. Cladel, and M. D. Pickel, *Antimicrob. Agents Chemother.*, **45**, 1201–1209 (2001).

527. R. Snoeck, J. C. Noel, C. Muller, E. De Clercq, and M. Bossens, *J. Med. Virol.*, **60**, 205–209 (2000).

528. C. Armbruster, A. Kreuzer, H. Vorbach, and M. Huber, *Eur. Respir. J.*, **17**, 830–831 (2001).

529. W. Beerheide, H. U. Bernard, Y. J. Tan, A. Ganesan, W. G. Rice, and A. E. Ting, *JNCI*, **91**, 1211–1220 (1999).

530. W. Beerheide, M. M. Sim, Y. J. Tan, H. U. Bernard, and A. E. Ting, *Bioorg. Med. Chem.*, **8**, 2549–2560 (2000).

531. G. Andrei, R. Snoeck, M. Vandeputte, and E. De Clercq, *Antimicrob. Agents Chemother.*, **41**, 587–593 (1997).

532. J. Hou and E. O. Major, *J. Neurovirol.*, **4**, 451–456 (1998).

533. S. Liekens, E. Verbeken, E. De Clercq, and J. Neyts, *Int. J. Cancer*, **92**, 161–167 (2001).

534. A. De Luca, M. L. Giancola, A. Ammassari, S. Grisetti, A. Cingolani, M. G. Paglia, A. Govoni, R. Murri, L. Testa, A. D. Monforte, and A. Antinori, *AIDS*, **14**, F117–F121 (2000).

535. M. Herrero-Romero, E. Cordero, L. F. Lopez-Cortes, A. de Alarcon, and J. Pachon, *AIDS*, **15**, 809 (2001).

536. C. B. de Oliveira, D. Stevenson, L. LaBree, P. J. McDonnell, and M. D. Trousdale, *Antiviral Res.*, **31**, 165–172 (1996).

537. F. Legrand, D. Berrebi, N. Houhou, F. Freymuth, A. Faye, M. Duval, J. F. Mougenot, M. Peuchmaur, and E. Vilmer, *Bone Marrow Transplant.*, **27**, 621–626 (2001).

538. E. Kodama, S. Shigeta, T. Suzuki, and E. De Clercq, *Antiviral Res.*, **31**, 159–164 (1996).

539. E. De Clercq, *Clin. Microbiol. Rev.*, **14**, 382–397 (2001).

540. E. De Clercq, M. Cools, J. Balzarini, R. Snoeck, G. Andrei, M. Hosoya, S. Shigeta, T. Ueda, N. Minakawa, and A. Matsuda, *Antimicrob. Agents Chemother.*, **35**, 679–684 (1991).

541. R. W. Sidwell, L. B. Allen, G. P. Khare, J. H. Huffman, J. T. Witkowski, L. N. Simon, and R. K. Robins, *Antimicrob. Agents Chemother.*, **3**, 242–246 (1973).

542. D. F. Smee, K. W. Bailey, and R. W. Sidwell, *Antiviral. Chem. Chemother.*, **11**, 303–309 (2000).

543. M. Hasobe, J. G. McKee, D. R. Borcherding, and R. T. Borchardt, *Antimicrob. Agents Chemother.*, **31**, 1849–1851 (1987).

544. R. F. Nutt, M. J. Dickinson, F. W. Holly, and E. Walton, *J. Organic Chem.*, **33**, 1789–1795 (1968).

545. A. A. Van Aerschot, P. Mamos, N. J. Weyns, S. Ikeda, E. De Clercq, and P. A. Herdewijn, *J. Med. Chem.*, **36**, 2938–2942 (1993).

546. E. De Clercq, A. Holy, I. Rosenberg, T. Sakuma, J. Balzarini, and P. C. Maudgal, *Nature*, **323**, 464–467 (1986).

547. E. De Clercq, A. Holy, and I. Rosenberg, *Antimicrob. Agents Chemother.*, **33**, 185–191 (1989).

548. P. F. Nettleton, J. A. Gilray, H. W. Reid, and A. A. Mercer, *Antiviral Res.*, **48**, 205–208 (2000).

549. J. Neyts and E. De Clercq, *J. Med. Virol.*, **41**, 242–246 (1993).

550. M. Bray, M. Martinez, D. F. Smee, D. Kefauver, E. Thompson, and J. W. Huggins, *J. Infect. Dis.*, **181**, 10–19 (2000).

551. D. F. Smee, K. W. Bailey, M. Wong, R. W. Sidwell, M. Bray, M. Martinez, D. Kefauver, E. Thompson, and J. W. Huggins, *Antiviral Res.*, **47**, 171–177 (2000).

552. J. D. Bisognano, M. B. Morgan, B. D. Lowes, E. E. Wolfel, J. Lindenfeld, and L. S. Zisman, *Transplant. Proc.*, **31**, 2159–2160 (1999).

553. E. Mylonakis, B. P. Dickinson, M. D. Mileno, T. Flanigan, F. J. Schiffman, A. Mega, and J. D. Rich, *Am. J. Hematol.*, **60**, 164–166 (1999).

554. J. R. Arribas, J. M. Pena, and J. E. Echevarria, *Ann. Intern. Med.*, **132**, 1011 (2000).

555. A. J. Ware and T. Moore, *Clin. Infect. Dis.*, **32**, E122–E123 (2001).

556. M. Y. Chen, C. C. Hung, C. T. Fang, and S. M. Hsieh, *Clin. Infect. Dis.*, **32**, 1361–1365 (2001).

# CHAPTER TEN

# Antiviral Agents, RNA Viruses (Other than HIV), and Orthopoxviruses

Christopher Tseng
Catherine Laughlin
NIAID
National Institutes of Health
Bethesda, Maryland

## Contents

1 FDA Approved Antiviral Agents for Selected RNA Virus Infections, 360
  1.1 Introduction, 360
  1.2 Hepatitis C Virus, 361
    1.2.1 Interferon-$\alpha$, 362
    1.2.2 Interferon Combined With Ribavirin, 363
    1.2.3 Pegylated Interferon, 363
    1.2.4 Pegylated Interferon-$\alpha$ Combined With Ribavirin, 364
2 Discovery and Development of Inhibitors of RNA Viruses Other Than HIV, 364
  2.1 Influenza A and B Viruses, 364
    2.1.1 Inhibitors of Influenza Neuraminidase (Sialidase), 364
      2.1.1.1 Zanamivir (GG-167), 365
      2.1.1.2 Zanamivir Analogs, 368
      2.1.1.3 Oseltamivir (GS-4104), 368
      2.1.1.4 Oseltamivir Analogs, 370
      2.1.1.5 BANA Compounds, 372
      2.1.1.6 RWJ-270201 (BCX-1812, Peramivir), 374
      2.1.1.7 Pyrrolidine-Based Inhibitors, 376
    2.1.2 Inhibitors of Influenza Hemagglutinin, 377
      2.1.2.1 Monomeric Inhibitors of Influenza Hemagglutinin, 377
      2.1.2.2 Polymeric Carbohydrate-Based Inhibitors of Influenza Hemagglutinin, 380
    2.1.3 Inhibitors of RNA-Dependent RNA Polymerase (RNA Transcriptase), 384

*Burger's Medicinal Chemistry and Drug Discovery*
Sixth Edition, Volume 5: Chemotherapeutic Agents
Edited by Donald J. Abraham
ISBN 0-471-37031-2  © 2003 John Wiley & Sons, Inc.

2.1.3.1 Antisense Oligonucleotides, 384
2.1.3.2 DNAzymes, Ribozymes, and
        External Guide Sequences, 385
2.1.3.3 Short Capped Oligonucleotides,
        386
2.1.3.4 2,4-Dioxobutanoic Acid
        Derivatives, 386
2.1.3.5 2,6-Diketopiperazine
        Derivatives, 386
2.1.3.6 BMY-26270 and Analogs, 387
2.1.3.7 M1 Zinc Finger Peptides, 387
2.1.4 Inhibitors of Influenza M2 Protein, 388
2.1.5 Other Inhibitors of Influenza Viruses,
      388
2.1.5.1 Bisindolylmaleimides, 388
2.1.5.2 Pyrimidine Derivatives, 389
2.1.5.3 Natural Products, 389
2.1.5.4 Other Compounds, 390
2.2 Respiratory Syncytial Virus, Parainfluenza
Virus, and Measles Virus, 390
2.2.1 Inhibitors of RSV and
      Paramyxoviruses Fusion Proteins, 390
2.2.1.1 RFI-641 (WAY-154641), 390
2.2.1.2 VP-14637, 393
2.2.1.3 R170591, 393
2.2.1.4 NMSO3, 394
2.2.1.5 RD3-0028, 394
2.2.1.6 Benzanthrone Derivatives, 395
2.2.1.7 Immunoglobulins, 395
2.2.1.8 Peptides, 395
2.2.2 Oligonucleotides as Inhibitors of RSV,
      396
2.2.3 Other Inhibitors of RSV, 397
2.2.3.1 Natural Products, 397
2.2.3.2 Purines and Pyrimidines, 398
2.3 Picornaviruses, 398
2.3.1 Inhibitors of Picornaviral Attachment
      and Uncoating, 398
2.3.1.1 Long-Chain Compounds, 398
2.3.1.2 Isothiazoles, Dibenzofurans,
        and Dibenzosuberanes, 403
2.3.1.3 Soluble Intercellular Adhesion
        Molecules-1 (sICAM-1), 403
2.3.2 Inhibitors of Picornaviral Proteases,
      404

2.3.2.1 Peptidic Inhibitors, 404
2.3.2.2 Nonpeptidic Inhibitors, 409
2.3.3 Inhibitors of Picornaviral Replication
      (Enviroxime and Analogs), 411
2.3.4 Inhibitors of Picornaviral Protein 2C,
      413
2.3.5 Other Anti-Picornaviral Inhibitors, 413
2.3.5.1 Natural Products and Synthetic
        Analogs, 413
2.3.5.2 Other Compounds, 414
2.4 Hepatitis C Virus, 415
2.4.1 Inhibitors of HCV 5′ Untranslated
      Region and Core Gene, 415
2.4.1.1 Ribozymes and Antisense
        Oligonucleotides, 415
2.4.1.2 Inhibitors of HCV Internal
        Ribosome Entry Site, 417
2.4.2 HCV NS3 Protein, 417
2.4.2.1 Inhibitors of HCV NS3
        Protease, 418
2.4.2.2 Inhibitors of HCV NS3
        Helicase, 425
2.4.3 Inhibitors of NS5B RNA-Dependent
      RNA Polymerase, 425
2.4.4 VX-497, 426
2.4.5 Model Systems to Study HCV
      Replication, 427
2.5 Flaviviruses, 427
2.6 Arenaviruses, 430
2.7 Rotavirus, 430
2.8 Rubella Virus, 431
2.9 Broad-Spectrum Antiviral Compounds, 432
2.9.1 S-Adenosyl-L-Homocysteine Hydrolase
      Inhibitors, 432
2.9.2 Prostaglandins, 434
2.9.3 Polyoxometalates, 435
3 Orthopoxviruses, 435
3.1 Inhibitors of Orthopoxviruses, 437
3.1.1 Methisazone (Marboran), 437
3.1.2 Nucleoside Derivatives, 437
3.1.2.1 Cidofovir (HPMPC), 437
3.1.2.2 Ribavirin and Other Nucleoside
        Derivatives, 440

# 1 FDA APPROVED ANTIVIRAL AGENTS FOR SELECTED RNA VIRUS INFECTIONS

## 1.1 Introduction

The discovery and development of safe and effective antiviral therapies is inherently more difficult than the comparable effort to develop antibacterial agents. The primary difficulty is that viruses replicate inside the cells of their host and actually hijack host metabolic and replication processes and use them to replicate progeny viruses. Consequently, most compounds that inhibit viral replication are also toxic to the host. Most of the approved antiviral drugs target a specific viral function, such as amantadine's interaction with the influenza M2 ion channel. Accordingly, unlike antibiotics, most antiviral therapies are narrow spectrum.

Although there is an extensive and promising amount of new drug candidates in preclinical development, only nine compounds have been approved by the FDA for the treatment of infections caused by RNA viruses other than HIV. These are amantadine (**1**) and rimantadine (**2**) that interfere with the influ-

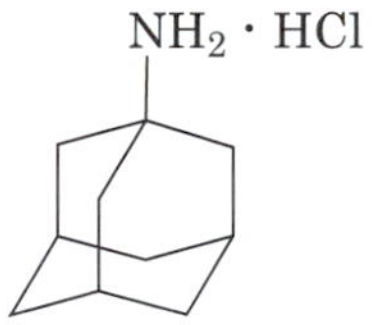

**(1)**

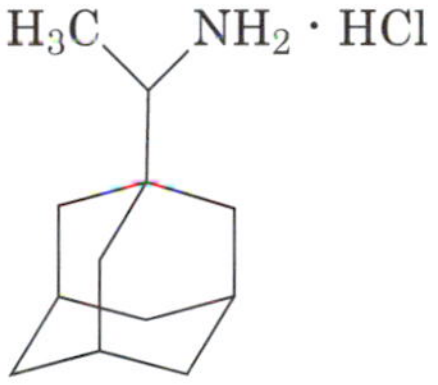

**(2)**

enza ion channel protein, and the newer inhibitors of the influenza neuraminidase, zanamivir (**3**) and oseltamivir (**4**). Respiratory

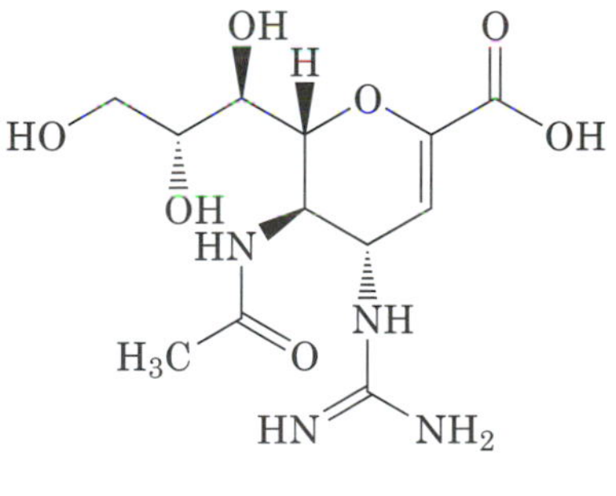

**(3)** Zanamivir

syncytial virus can be life threatening to infants, especially those with other underlying heart or lung problems. Aerosolized ribavirin (**5**) is approved as a therapy for this disease, and both a polyclonal antibody preparation, Respigam, and a monoclonal antibody, Syner-

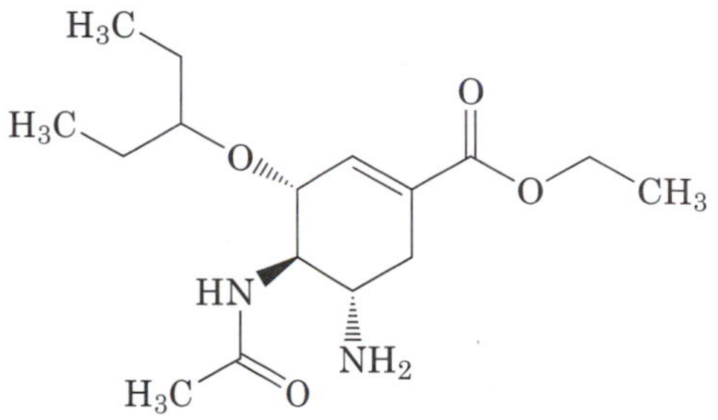

**(4)**

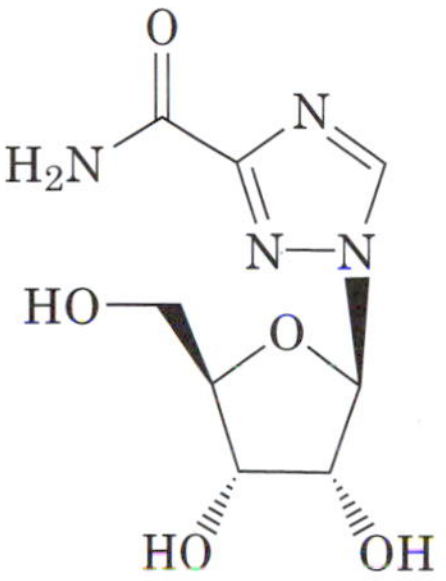

**(5)**

gis, are approved for prophylaxis in high-risk infants. Finally, several forms of interferon-$\alpha$, both natural and pegylated, and the combination of these interferons with ribavirin have been approved for the treatment of chronic hepatitis C infection.

This chapter will review the clinical data supporting the development of therapies for hepatitis C. In addition, the chapter includes a discussion of the historical efforts to discover a treatment for smallpox, a disease that has been eradicated from the world. However, interest in the development of treatments for this disease has been revived because of its potential for bioterrorist use.

## 1.2 Hepatitis C Virus

Hepatitis C virus (HCV) a member of the flavivirus genus, is a negative-stranded RNA virus that is estimated to have infected 170 million people globally and 4 million in the United States. The current HCV-associated mortality rate in the United States is almost 10,000 per year and is expected to triple in the next decade (1). With the implementation of screening blood products for HCV in the 1990s, the rate of new infections has decreased, but the virus is still spread as a consequence of intravenous drug use and blood to blood exposure (2, 3). The acute period of infection is rarely recognized, but in most people, progresses to an asymptomatic chronic infection that may last 20–30 years. Virus replicates at very high levels during this time, in the range of $10^{12}$ new particles per day (4), and eventually about 70% of chronic infections result in hepatitis and fibrosis and about one-fifth further progress to cirrhosis (1, 5). Those who develop cirrhosis are at high risk for further progression to hepatocellular carcinoma.

This high rate of viral turnover, combined with the high error rate of RNA-dependent RNA polymerases, results in the generation of genetically diverse quasispecies. This degree of diversity is correlated with outcome in that greater diversity in the acute phases is associated with progression to chronic disease (5) and response to interferon therapy is associated with a decrease in diverse quasispecies (2). Other prognostic factors that have been identified that generally are associated with a good outcome to therapy include the following: genotype 2 or 3, low baseline viral RNA, female gender, Caucasian race, age less than 40 years, and lack of cirrhosis (2, 6–10).

### 1.2.1 Interferon-α

*Discovery and Preclinical Findings.* Although the original report of the discovery of interferons and their antiviral potential was published in 1957 (11), and they have been actively studied between then and now, hepatitis B and C are the only viral infections for which they are widely used clinically. (Interferon is also approved, but not widely used, for the treatment of condyloma acuminata—a type of external genital wart). Interferons are a family of multiple low-molecular-weight cellular proteins that include the four types of α-interferon commercially available and in current clinical use. These are recombinant interferon alfa-2b (Intron A, Schering-Plough), recombinant interferon alfa-2a (Roferon, Hoffmann-La Roche), natural interferon alpha-n1 (Alferon N, Purdue Frederick), and recombinant consensus interferon (Infergen, InterMune).

*Mechanism of Action.* Interferons are thought to act at both immunomodulatory and antiviral levels. The classic antiviral activities of interferons are mediated through the interferon-induced cellular proteins 2′-5′-oligo adenylate synthetase and protein kinase R (PKR). The first activates an RNAse and thus causes the consequent degradation of viral and cellular RNAs. The PKR inhibits protein synthesis. A recent report demonstrated that interferon also inhibits HCV RNA translation through a PKR-independent pathway (12). Interferon's immunomodulatory actions are thought to involve enhancement of HLA class I antigen expression and signaling as well as stimulation of a Th-1 type immune response with production of γ-interferon and interleukin-2 (13).

Interferon seems to reduce the HCV viral load by decreasing the production of new virus from infected cells rather than by blocking the infection of new cells (4). The speed with which this effect is exerted, a 0.5–2.0 log reduction in 24 h, is astonishing (14, 15). This effect is most pronounced in patients infected with genotypes 2 or 3 as opposed to 1 and is likely to be correlated with their ability to respond to interferon treatment (6). Interestingly, the resistance of type 1 genotype HCV to interferon may be a property of the nonstructural NS5A viral protein that may somehow elude the interferon-induced inhibitory pathways. Another report similarly implicates the HCV E2 (16, 17).

*Monotherapy.* Two controlled clinical trials used interferon alfa-2b at 1–3 million units three times a week for 6 months. Complete response was defined as a return of liver enzyme levels to the normal range and was achieved by 50% of the enrolled patients. They also showed improvement in liver histology. However, virtually all patients relapsed after therapy was discontinued (18, 19). The strong association of response to therapy with the reduction of the viral load has led to the achievement of a sustained viral response, the reduction of viral load to undetectable for a period after cessation of therapy as the most easily studied reliable endpoint for clinical trials.

A standard course of therapy with interferon-α consists of 3 million units three times a week for 12–18 months. This usually results in a loss of virus load to the undetectable level in 40% while the patient is on therapy. However 50–90% of responders relapse and retreatment is usually not successful.

**1.2.2 Interferon Combined With Ribavirin.** The sustained viral response rate is improved if the nucleoside analogue ribavirin is combined with interferon-$\alpha$. One group reported a sustained rate of 35% after 4 weeks of therapy and 40% after 48 weeks (10).

Ribavirin is a purine nucleoside analog synthesized in 1970 with broad-spectrum antiviral activity *in vitro*. It was FDA approved in 1985 for aerosol treatment of RSV pneumonia in hospitalized infants and young children and for use as an oral medication in combination with injected interferon as a treatment for HCV in 1998. Other clinical targets of past or ongoing investigation include the arenavirus Lassa, the hanta bunyaviruses that cause hemorrhagic fever with renal syndrome and hantavirus pulmonary syndrome, and the encephalitis caused by the West Nile flavivirus.

*Mechanism of Action.* Ribavirin has been shown to have diverse activities in experimental systems and it is not clear which are relevant to its enhancement of interferon's activity against HCV (20). Ribavirin is a known inhibitor of inosine monophosphate dehydrogenase (IMPDH), an essential enzyme in the synthesis of guanosine triphosphate (GTP). IMPDH inhibition results in decreased intracellular GTP pools and a general inhibition of RNA synthesis. A second mechanistic possibility is that ribavirin may directly inhibit the HCV RNA polymerase. Ribavirin is phosphorylated by cellular kinases and ribavirin 5'-triphosphate, an analog of purine nucleotides, present in large amounts, would likely inhibit viral RNA synthesis. Consistent with this theory is the recent demonstration that ribavirin triphosphate is incorporated into RNA products synthesized by the HCV RNA-dependent RNA polymerase in cell-free systems (21). A variation on this proposed mechanism is derived from the demonstration that growth of poliovirus in high concentrations of ribavirin resulted in a high mutation rate with the generation of significant genomic diversity. Although viral replication was only moderately inhibited, viral fitness as measured by infec-

tivity was significantly decreased (22). The authors referred to this phenomenon "error catastrophe." Another hypothesized mechanism is interference with capping, the process in which a GTP is added to the 5' end of many viral and host mRNAs to provide protection from host nucleases. In another flavivirus, dengue, RTP competed with GTP for binding the viral enzyme responsible for capping (20). Finally, ribavirin has long been known to have immunomodulatory activity. Recent studies have suggested that the mechanism of this proposed mechanism is the stimulation of a Th-1 type cytokine response (2, 23).

Ribavirin is not effective as a monotherapy, although it did result in some improvement in liver enzymes and inflammation (2, 24). Side effects of ribavirin include anemia and exacerbation of cardiac disease. The anemia may be sufficiently severe as to require dose reduction in up to 10% of patients (2, 25). Furthermore, teratogenicity has been demonstrated in multiple animal species, and ribavirin is therefore contraindicated in both male and female partners to a pregnancy. The mechanism of ribavirin causing anemia is believed to be the result of the lack of ability of red blood cells to hydrolyze ribavirin-triphosphate. Consequently, ribavirin is concentrated in red blood cells leading to a depletion of ATP and damage to cellular membranes, culminating in removal by the reticuloendothelial system (26).

**1.2.3 Pegylated Interferon.** Recombinant interferon alfa-2b is linked to polyethylene glycol (PEG) a non-toxic water-soluble polymer to create PEG-Intron, the first pegylated interferon product approved by the U.S. FDA. A second pegylated product, Pegasys, is now available from another manufacturer and both are more effective clinically than their unmodified parents. Pegylation increases the half-life of interferon by increasing its molecular weight, which usually reduces elimination. In addition, pegylation stabilizes interferons to temperature and pH variation, protects them from the immune system and from degradation. For example the elimination half-life of the pegylated interferon alfa-2a is 77 h compared with 9 h for the unmodified interferon. Because of increased stability, pegylated interferons can be given once a week, instead of the thrice-weekly doses that are standard for unmodified interferon. This both improves compliance, and efficacy as serum levels are steady state rather than a fluctuating series of peaks and troughs (27). The PEG component of peginterferon alfa-2b is a 12-kDa linear molecule and the modified interferon is excreted renally. Peginterferon alfa-2a includes a 40-kDa branched-chain PEG component and is primarily cleared by the liver. Pegylation usually decreases the antiviral potency of interferon so determination of the optimum amount of modification requires balancing antiviral activity and the kinetics of elimination.

A large phase III study of pegylated interferon alfa-2b for 48 weeks showed a sustained viral response rate of 25% in patients receiving the pegylated form as opposed to 12% in patients receiving the non-pegylated parent interferons (28). This led to FDA approval of pegylated interferon alfa-2b for monotherapy in the United States. A similar study comparing peginterferon alfa-2a with its unmodified parent showed respective virologic response rate at 48 weeks of 69% versus 28% and final sustained viral response rates at week 72 of 39% versus 19% (27, 29). Commonly encountered side effects of pegylated interferons are the same as those of unmodified interferons, and include flu-like symptoms, injection site reactions, and psychiatric side effects, although neutropenia may be increased.

In general pegylated interferons are twice as effective as their unpegylated parents but do not decrease the relapse rate or alter genotype response sensitivities. Their future clinical role is likely to be as a component of combinations with ribavirin or as a monotherapy for ribavirin intolerant patients.

**1.2.4 Pegylated Interferon-$\alpha$ Combined With Ribavirin.** After the demonstration of the superiority of pegylated interferons to their unmodified parents as monotherapies, yet their inferiority to unmodified interferons combined with ribavirin, the next step was clearly to evaluate pegylated interferons in combination with ribavirin. This has been done with both interferon alfa-2b and alfa-2a. In the alfa-2b study, treatment for 48 weeks led to sustained viral response rates of 54%

with the pegylated combination compared with 47% for either the unpegylated combination or a lower dose pegylated combination (30). In this study, the response rate of patients infected with genotype 1 was increased from 33% with the non-pegylated combination to 42% with the pegylated combination. The rate of sustained viral response for non-group 1 genotypes was 82%. Compliance was demonstrated to be important, as the response rate of those who received more than 80% of both their pegylated interferon and ribavirin doses was 63% compared with 54% who did not. Pegylated alfa-2a combined with ribavirin was compared with pegylated alfa-2a alone or pegylated alfa-2b combined with ribavirin. The sustained viral response rates was 56% for the alfa-2a combination, and it seems clear that the combination of ribavirin and a pegylated interferon is currently the best therapeutic option for the treatment of chronic HCV infection (3, 31).

## 2 DISCOVERY AND DEVELOPMENT OF INHIBITORS OF RNA VIRUSES OTHER THAN HIV

This section provides updates on antiviral compounds that were published in the literature from 1996 to early 2002. Relevant reviews of earlier research are available and will be cited throughout the section.

### 2.1 Influenza A and B Viruses

**2.1.1 Inhibitors of Influenza Neuraminidase (Sialidase).** The approval of Relenza (zanamivir for inhalation) and Tamiflu (oseltamivir phosphate) by the U.S. FDA in 1999 (http://www.fda.gov/cder/approval/index.htm) marked the advent of clinically effective anti-influenza therapies achievable by structure-based drug design. Both drugs are inhibitors of influenza neuraminidase (NA). A third structure-based NA inhibitor, RWJ-270201 (32) (also known as BCX-1812), has also been in phase III trials in Europe (BioCryst News, August 10, 2001). There have been several excellent reviews of the work in this area published in recent years (33–39).

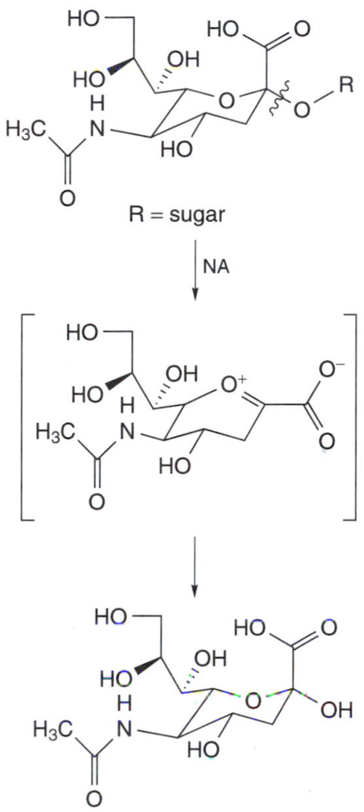

**Figure 10.1.** The catalytic mechanism of the neuraminidase-mediated cleavage of sialic acid.·

trum antiviral agents against both influenza A and B viruses, and (2) mutations of these conserved amino acids inactivates the enzyme, suggesting that the virus may not easily escape therapeutic intervention through mutation (41, 42).

Although there is no consensus on the mechanism of action of NA, it is accepted that the sialic acid cleavage by NA might proceed through an oxonium cation transition state intermediate adopting a half-chair conformation (Fig. 10.1) (38, 42, 43). It has been shown that a transition state mimic can be an inhibitor of a particular enzymic reaction. In this case, 2-deoxy-2,3-dehydro-$N$-acetylneuraminic acid (**6**) (Neu5Ac2en, also known as DANA) was the first mechanism-based analog synthesized, of which the pyranosidic ring adopts a similar planar structure to the putative sialosyl cation transition state intermediate. DANA shows potent NA inhibitory activity with $K_i$ in the micromolar range; however, this compound inhibits various viral, bacterial, and mammalian neuraminidases with similar affinity. It also failed to protect animals in experimental therapies against influenza virus (see the review articles cited above).

For convenience of discussions and comparisons, we use the numbering system of N2 subtype of influenza virus type A to denote the active site amino acid residues.

*2.1.1.1 Zanamivir (GG-167).* With the availability of the X-ray three-dimensional molecular structures of the NA active site with and without binding with DANA and sialic acid

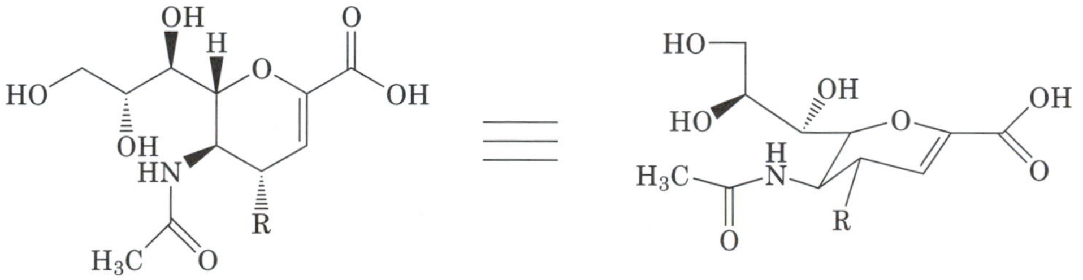

The premise for targeting NA (40) for antiinfluenza chemotherapy is based on the principles that (*1*) the active site amino acid residues are completely conserved across all known strains of influenza A and B NAs, suggesting that NA inhibitors can be broad-spec-

(**6**) DANA          R = OH

(**3**) Zanamivir  R = NH

NH

NH$_2$

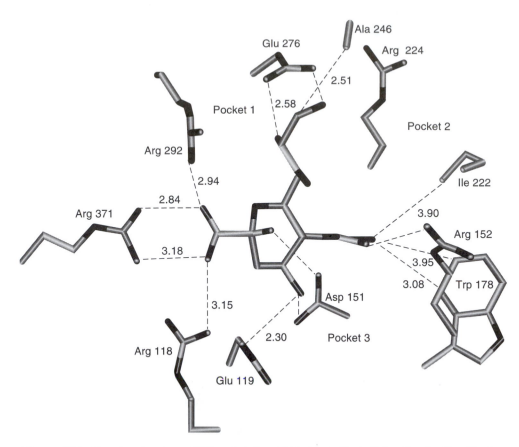

**Figure 10.2.** Complex structure of NA and sialic acid (dashes indicate H-bonding [red] and hydrophobic [black] interactions; some active site residues are omitted for clarity). Reproduced with permission of Dr. C. U. Kim (Gilead Sciences). See color insert.

(the natural substrate of NA), further drug design assisted by using the program GRID resulted in a rational-designed, potent transition state analog, 4-guanidino-2,4-dideoxy-N-acetylneuraminic acid (**3**) (zanamivir, also known as GG167) with binding affinity ($K_i \sim 10^{-11}\,M$) more than 100-fold tighter than that of DANA (43). It has been proposed that modifications at the $C_4$, $C_5$, and $C_6$ positions of sialic acid would lead to function-specific interactions with the NA active site; thereby, this might be the reason that the 4-guanidino group renders zanamivir influenza-specific (36). On the other hand, it might be because DANA has the same functional groups at these positions as that of sialic acid; DANA is a non-selective inhibitor.

The crystallographic structure of zanamivir complexed with NA is very similar to that found in the X-ray crystal structures of sialic acid and DANA complexed with NA (44). Figure 10.2 shows the key interactions between NA active site residues and sialic acid (37). The dihydropyran ring of zanamivir adopts a half-chair, near-flat conformation, and all $C_4$, $C_5$, and $C_6$ substituents on the ring are all equatorial on the same plane (45). The structures suggested strong charge-charge interactions of the carboxylate functional group with three arginine residues (Arg-292, -371, and -118). A quantitative structure-activity relationship (QSAR) model derived by COMparative BINding Energy (COMBINE) analysis, which concluded that the triarginyl cluster is the predominant factor for orienting and stabilizing inhibitor molecules (46), further supports this observation. A negatively charged group (e.g., carboxylate or phosphonate) that

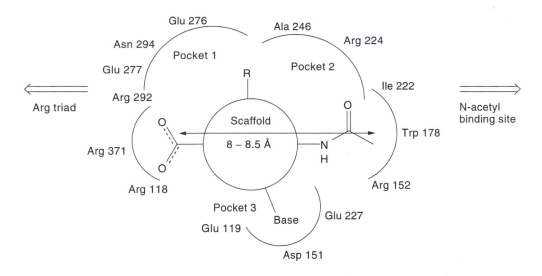

Pocket 1: Bifuntional
Pocket 2: Lipophilic, not utilized by sialic acid
Pocket 3: Cluster of negative charges

**Figure 10.3.** Basic principles for NA inhibitor design. Reproduced with permission from Dr. C. U. Kim (Gilead Sciences).

makes strong charge-charge interactions with the triarginyl pocket is highly favorable for binding (46, 47). These strong interactions and the interactions of the $N$-acetyl methyl group with the hydrophobic region formed by residues Ile-222 and Trp-178 and of the amide carbonyl oxygen with Arg-152 through a hydrogen bond (46, 48) are likely important for orientation of zanamivir and other inhibitors in the binding site (49, 50). Based on these observations, a model for anti-influenza NA inhibitor design has been suggested as shown in Fig. 10.3 (44, 51).

The X-ray crystal structures also show that the two terminal hydroxyls of the glycerol moiety form a bidentate hydrogen bond donor-acceptor interaction with the carboxylate of Glu-276 (Glu-276 re-orients on binding with inhibitors containing aliphatic substituents as discussed in the following sections) and the $C_8$ of the glycerol side-chain makes hydrophobic interactions with the hydrocarbon chain of Arg-224 (48). It is of interest to note that the $C_7$ hydroxyl of the glycerol moiety does not involve in the binding, suggesting that, in the sense of drug design, the $C_7$ hy-

droxyl could be eliminated and replaced with other functional group without compromising the affinity to the enzyme (48, 52).

Computational analysis predicted that replacement of the $C_4$ hydroxyl group in DANA by a positively charged amino group would be beneficial for binding affinity by the formation of a salt bridge with negatively charged Glu-119 (43). The analysis further predicted that even higher affinity could be achieved with a larger and more basic guanidino group, because the terminal nitrogens of the guanidino group seem to exhibit lateral binding to both Glu-119 and Glu-227 (43). In the crystal structure of zanamivir bound to NA, the predicted binding between one of the primary guanidinyl nitrogens and the carboxylate of Glu-227 does occur, whereas Glu-119, although slightly further removed than predicted, is found within a distance close enough for electrostatic interaction with the secondary guanidinyl nitrogen (34, 36), which also found to interact with the carboxylate of Asp-151. When the guanidino group occupies this binding pocket, it expulses the existing water and this replacement may contribute a favorable entropic factor to the binding (53). Because it needs to expulse a water molecule from the $C_4$ guanidino-binding pocket of the active site, zanamivir is a slow binding inhibitor (33, 40).

**2.1.1.2 Zanamivir Analogs.** In the search for further clinical candidates, a number of zanamivir analogs have been reported. Substitutions on the guanidino nitrogens generally resulted in much weaker inhibitors (34). 5-Trifluoroacetamido and 5-sulphonamide derivatives of zanamivir remained the activity approaching to that of zanamivir (54). The $C_6$ glycerol moiety has also been replaced by ether (7) (55), ketone (8) (55), carboxamide (9)

(53, 56–58), or a heterocycle such as triazole (10) (59). Interestingly, all of these $C_6$ modified compounds showed a strong selectivity against influenza A, with much worse activity against the type B virus. In the mouse model, despite the similar enzyme affinity and *in vitro* activity, the carboxamide analog (9) exhibited

a much lower intranasal efficacy compared to zanamivir in reducing viral titers in the infected animals (58).

It has been noticed that the carbohydrate and the dihydropyran rings have no direct interaction with the active site amino acids. In addition, unlike the active site of most other enzymes, the NA active site contains an unusually large number of polar or charged residues (37), implying that the ring structure might merely act as a structure frame (scaffold) to correctly orient the substituents to have proper electrostatic interactions with the active site amino acid residues. Therefore, novel inhibitors can be designed by constructing novel frame structures to place the interacting substituents in correct relative positions in the enzyme active site (32, 46, 50). A number of potent influenza NA inhibitors based on the ring structure of cyclohexene [e.g., oseltamivir (48)], benzene [e.g., BANA-206 (41)], cyclopantane [e.g., RWJ-270201 (32)], or pyrrolidine [e.g., ABT-675 (60)] have been reported. These compounds will be discussed in the following sections.

**2.1.1.3 Oseltamivir (GS-4104).** Taking into consideration that transition state forms the rather flat oxonium cation, which could be considered as an isostere of a double bond, the cyclohexene scaffold was selected as a replacement for the oxonium ring in the design of oseltamivir (4) (also known as GS-4104) and its related analogs. In addition, the carbocyclic system was expected to be more chemical and enzymic stable than the dihydropyran ring. A series of articles on the discovery and development of oseltamivir have been published by Kim at Gilead Sciences (37, 48, 61, 62). Oseltamivir is the prodrug of GS-4071 (11).

(11) GS 4071

| A/PR | $IC_{50} = 1$ nM |
| B/Lee | $IC_{50} = 3$ nM |

To treat influenza infection, oral administration is considered more convenient and economical for patient care. However, because of its extreme hydrophilicity, zanamivir is not orally bioavailable; it has to be given intranasally or by inhalation to treat patients. In the design of orally bioavailable drugs, balancing lipophilicity and water solubility could be critical for their absorption from the intestinal tract (37, 48). Gilead investigators decided to use a less polar amino group to replace the highly polar guanidino group, because this change might be beneficial for increasing oral absorption.

As does zanamivir, the cyclohexene-based inhibitors also contain the carboxylate and the acetamido moieties, located at $C_1$ and $C_4$ positions of the cyclohexene ring, respectively, which are anticipated to interact with the Arg triad (Arg-292, -371, and -118) on one end and with the amide recognition region (Trp-178, Ile-222, and Arg-152) on the opposite end, and thereby, help anchor the inhibitors into the NA binding side (Fig. 10.3).

It has been noticed that although the glycerol chain of zanamivir is polar in nature overall; the $C_8$ of the glycerol chain makes hydrophobic interactions with the hydrocarbon chain of Arg-224 (37, 48), suggesting that the optimization of this hydrophobic interaction would lead to new inhibitors with increased lipophilicity (48). The $C_7$ hydroxy group makes no direct interaction with the enzyme and is exposed to bulk solvent, suggesting that it may be replaced with other functional group (48, 52). These two notions provided the theoretical base for the design of GS-4071 (11) and its analogs by replacing the whole glycerol chain with a variety of alkoxy groups (at the $C_3$ position of the cyclohexene ring). Because the $C_3$ linker atom may not participate in the interaction with the enzyme, it can be replaced with nitrogen (63, 64), sulfur (65), or oxygen (37). The use of oxygen as the linkage in the design of GS-4071 was based on the need to reduce the electron density in the double bond in the cyclohexene ring, because the double bond in the sialosyl oxonium transition state intermediate is electron deficient, as well as the versatility in the synthesis of a variety of substituents. SAR studies showed that the

changes in length, size, and branching of the alkyl chains profoundly influence the NA inhibitory activity (48, 61, 62).

X-ray crystallographic analysis of GS-4071 (11) and its analogs bound to NA confirmed the existence of a hydrophobic region, which is corresponding to the glycerol-binding region for zanamivir, to accommodate bulky lipophilic groups (Fig. 10.4) (62). On binding of GS-4071 and related carbocyclic inhibitors, Glu-276 is forced to rotate to adopt an alternative conformation, which is stabilized by a strong charge–charge interaction with the nearby guanidino group of Arg-224, and expose its hydrophobic atoms to the aliphatic side chain of the inhibitor (Fig. 10.5). This reorientation enlarges the binding site and creates a much less polar environment, making this site possible to accommodate one branch of the hydrophobic pentyloxy group of GS-4071. The other branch of the pentyloxy group makes hydrophobic contacts with a larger, pre-existing binding region lined by hydrocarbon chains of Ile-222, Arg-224, and Ala-246 (Fig. 10.4) (48, 62). This latter pocket is significantly large enough to accommodate larger functional groups, such as a cyclohexyl ring (66).

Interestingly, the Glu-276 of type B NA undergoes a much smaller conformational change on binding to GS-4071 (Fig. 10.5) (62). It has been noticed that the region around Glu-276 is hydrophilic in type A NA, whereas it is hydrophobic and more crowded in type B NA (53). Therefore, the re-orientation creates distortions in the protein backbone near Glu-276 and in the second amino acid shell, which contains non-conserved amino acids compared with type A (67). As a consequence, the conformational rearrangement of type B NA residue Glu-276 occurs with energy penalties (53, 67), resulting in small changes in size and the polar nature of the pocket and decreased affinity for aliphatic side-chains (53, 62). Because of this, type B NA depends more likely on the second region, a larger hydrophobic pocket formed by Ile-222, Arg-224, and Ala-246, for inhibitor binding (32, 62); however, this region is very sensitive to the size of inhibitor bound and binding affinity is significantly affected by the increased bulk of the $C_3$ side-chain, suggested by the SAR studies with GS-

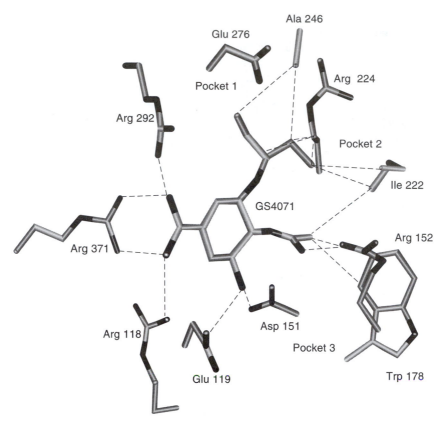

**Figure 10.4.** Complex structure of NA and GS 4071 (dashes indicate H-bonding [red] and hydrophobic [black] interactions; some active site residues are omitted for clarity). Reproduced with permission from Dr. C. U. Kim (Gilead Sciences). See color insert.

4071 analogs (61–63). In type A, the reorientation of Glu-276 resulted in minimal disruption of the second amino acid shell (41).

The NA of a clinical isolate (A/H1N1) recovered from a patient treated with oseltamivir was shown to be 400-fold more resistant to the drug than that of the wild-type virus (68). This variant carried a His-274-Tyr substitution. It was postulated that the large side-chain of Tyr at position 274 could interfere with the re-orientation of the side-chain of Glu-276 and, as a consequence, the successful binding of oseltamivir to the NA active site. Because no rearrangement of Glu-276 is required for zanamivir binding to the NA, this variant remained susceptible to zanamivir (68).

A similar rearrangement of Glu-276 has also been reported for the aforementioned carboxamide analogs of zanamivir when bound to

NA (53). This might explain the strong selective affinity for the NA of the type A virus.

GS-4071 (**11**) has a similar low bioavailability [~5% in rats (69)] compared with zanamivir. However, GS-4104 (**3**) improves the oral bioavailability after rapid conversion to the active form during gastrointestinal absorption. A high bioavailability was found in mice (~30%), dogs [~70% (69)], and humans [~80% (37, 70)]. Oral administration of GS-4104 results in high and sustained systemic absorption in animal tests with a half-life of 5 h in most tissues. In rats, a metabolite was isolated (71).

***2.1.1.4 Oseltamivir Analogs.*** The location of the double bond in the cyclohexene ring was found to be critical (37, 48, 62). Although molecular modeling analysis showed that (**11**) and (**12**) overlap quite well, their enzyme in-

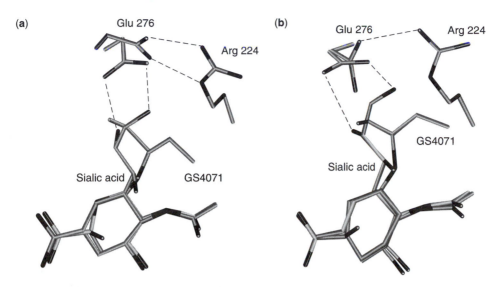

**Figure 10.5.** NA Glu-276 side-chain (a for type A neuraminidase complex; b for type B neuraminidase complex) can adopt alternative conformations on binding of sialic acid or GS4071 (dashes indicate H-bond; atoms are colored as following: blue for nitrogen, red for oxygen, brown for carbon in sialic acid complex, and green for carbon in GS4071 complex). Reproduced with permission from Dr. C. U. Kim (Gilead Sciences). See color insert.

**(12)**

| A/PR | $IC_{50} = 30$ nM |
|------|------------------|
| B/Lee | $IC_{50} = 500$ nM |

**(13)**

| A/Aichi | $IC_{50} = 17$ nM |
|---------|-------------------|
| B/Victoria | $IC_{50} = 23000$ nM |

hibitory activities were clearly different. For GS-4071 (**11**) the double bond is in a position analogous to the position of the oxonium ion double bond in the transition state. GS-4071 exhibited a significantly better activity than (**12**), especially against influenza B (62). Similarly, (**13**) was much more inhibitory than (**14**) (72).

The GS-4071 analog based on a tetrahydropyridazine ring, compound (**15**), was reported to have $IC_{50}$ values of 6 and 62 $\mu M$ against NA A/PR (H1N1) and B/Lee/40, respectively (73). Structural analyses of the GS-4071 analog revealed that the amino and acetamido groups

are in the pseudo-axial positions, rather than the preferred pseudo-equatorial positions as in the case of GS-4071, and that the 3-pentyloxy group points to the small hydrophobic pocket formed by the rearranged Glu-276, whereas in the case of GS-4071, the two ethyl moieties of the 3-pentyl side-chain bind in two different pockets as discussed above. The partial planar nature of the amide bond of the tetrahydropyridazine ring might cause the poor fit and energy penalty on binding to the enzyme (73).

The introduction of lipophilic substituents (chloro, methyl, and methylthio) at the $C_2$ position of GS-4071 resulted in a significant decrease of activity (~2000-fold reduction) (74). The $C_2$-fluoro analog of GS-4071 (**16**) re-

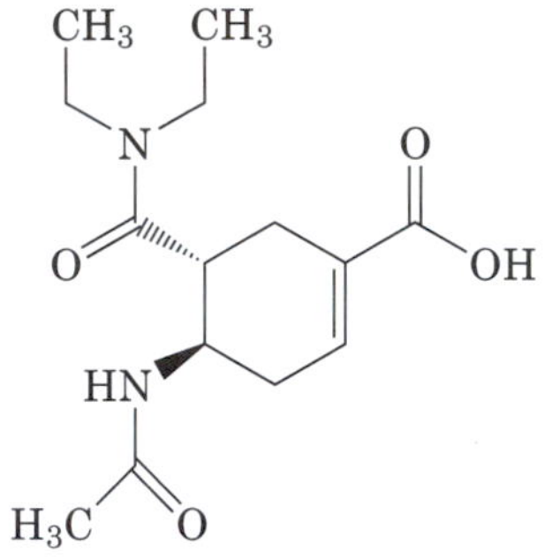

**(14)**

| A/Aichi | $IC_{50}$ = | 210 nM |
| B/Victoria | $IC_{50}$ = | 150000 nM |

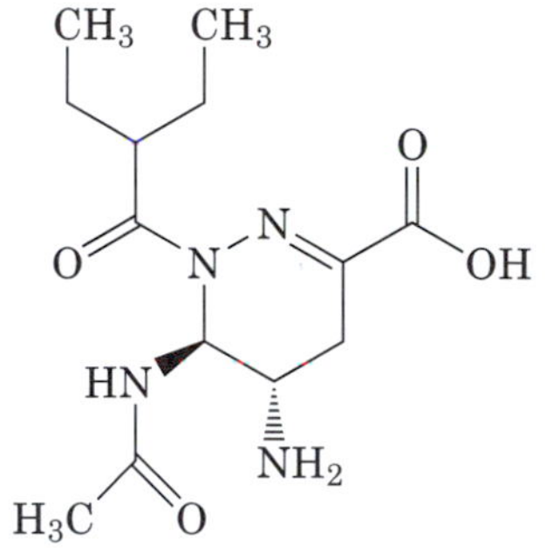

**(15)**

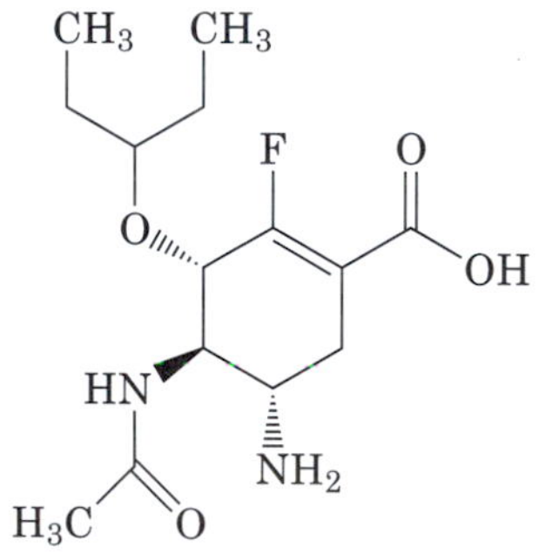

**(16)**

mained as potent as GS-4071 against NA A with an $IC_{50}$ of 3 n$M$; however, the activity against NA B was reduced by approximately 30-fold ($IC_{50}$, 90 n$M$) (61).

SAR studies by systematic modifications of substituents at the $C_3$, $C_4$, and $C_5$ positions of the cyclohexene ring indicated that oseltamivir seems to achieve the optimization of cyclohexene-based inhibitors in terms of biological and pharmacological activities, as well as synthetic practicality (66).

***2.1.1.5 BANA Compounds.*** The requirements of (*1*) a carboxyl group to interacts with the arginine triad (Arg-292, -371, and -118) in the NA active site, (*2*) a planar conformation near the carboxylate to resemble the transition-state-like structure, and (*3*) a proper spacing between the carboxylate and the acetamido groups for tight binding suggest that potent influenza NA inhibitors can be designed based on a benzoic acid template (scaffold) (50). During the early trials, the benzoic acid analog of zanamivir (**17**) was made, and unfortunately, was found to be devoid of NA

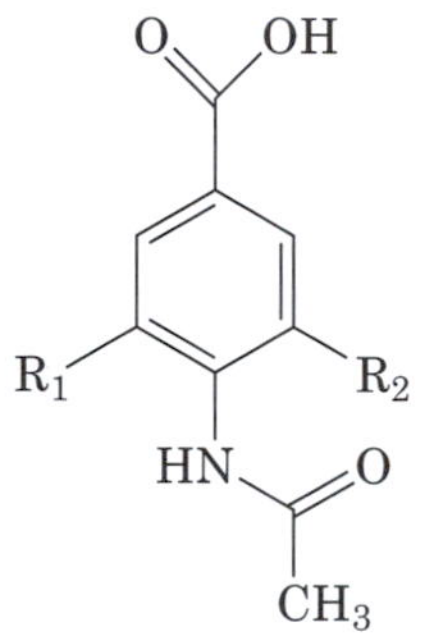

**(17)** $R_1$ = HC ...

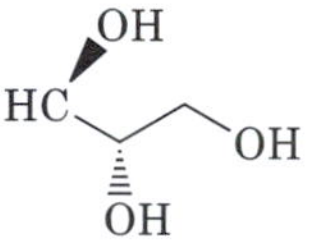

$R_2$ = ...

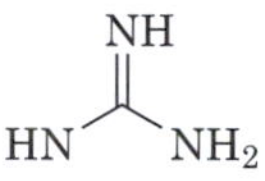

**(18)** BANA-113   $R_1$ = H   $R_2$ = ...

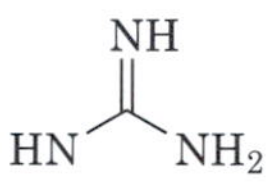

**(19)** $R_1$ = H   $R_2$ = ...

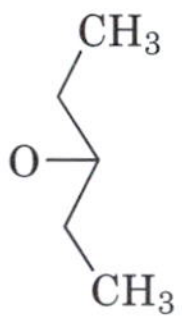

**(20)** $R_1$ 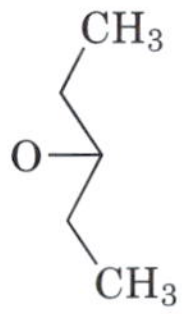 = ...   $R_2$ = $NH_2$

inhibitory activity (45, 75). Around the same time, BANA-113 (**18**) was discovered as a micromolar (IC$_{50}$ ~ 10 $\mu M$) inhibitor against both types A and B NAs (45, 49). BANA-113 contains a guanidino group, which is intended to mimic the C$_4$-guanidino of zanamivir to interact with the active site residues Glu-119 and Glu-227. In the crystal structure, it was surprising to find that the guanidino group of BANA-113 was 180° from the predicted position and formed a charge–charge interaction with residue Glu-276, the original glycerol-binding site of zanamivir (49, 50). Because BANA-113 is a symmetrical molecule, the guanidino group can select its more favorable binding pocket for interaction. Apparently, the active site pocket where the glycerol substituent of zanamivir bonds is the preferred binding pocket for the guanidino group. This might be the reason why the benzoic acid analog of zanamivir (**17**) was not active (75), because the benzene scaffold might not be able to provide right orientation for all substituents for the optimal binding to the NA active site (37, 76). When the guanidino group of BANA-113 was replaced with 3-pentyloxy (**19**), the inhibitory activity against type A NA remains, but not active against type B NA anymore (67). This is because the pentyloxy group binds to the hydrophobic pocket lined by Ile-222, Arg-224, Ala-246, and the re-oriented Glu-276, and the re-orientation of Glu-276 of type B NA is associated with energy penalties as discussed above.

The benzoic acid analog of GS-4071 (**20**) has also been reported (67). This compound showed similar inhibitory activity as that of BANA-113 against influenza A NA (~10 $\mu M$), but it was essentially inactive against influenza B NA (67). These observations suggest that the individual substituent contributions to overall binding to the active site cannot be considered to be additive; each substituent influences the overall interaction of the compound with the active site (66, 75). Therefore, a global consideration of binding energy is necessary to more reliably predict the binding energy of a designed compound (50).

Assisted with LeapFrog software, a *de novo* design program, Brouillette et al. postulated that for benzoic acid-based inhibitors, a small cyclic substituent containing side-chains, such

as a pyrrolidinone (76, 77), might be a suitable replacement for the N-acetyl grouping [this binding pocket, into which the N-acetyl grouping extends, is quite rigid and small (51, 76)]. When the N-acetylamino group of BANA-113 was replaced with a bis(hydroxymethyl)pyrrolidine-2-one ring, the resulting compound BANA-205 (**21**) exhibited IC$_{50}$ values similar

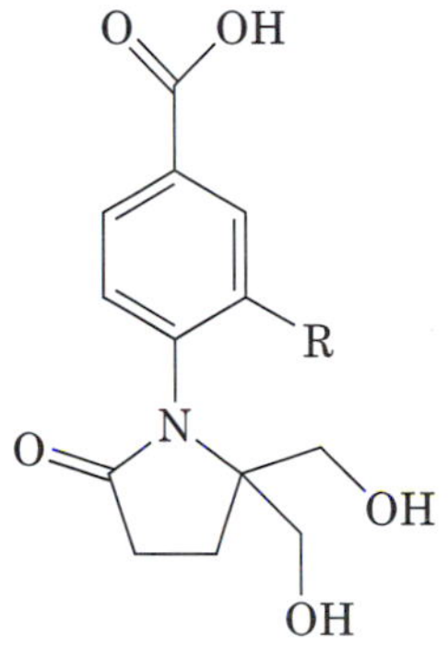

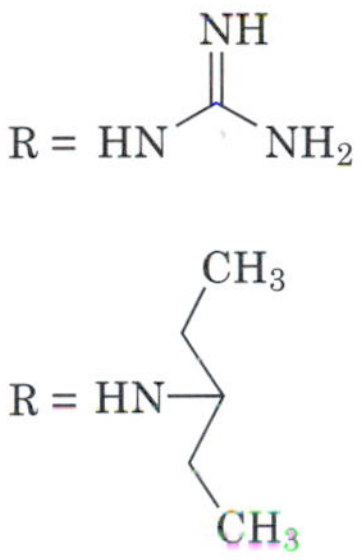

(**21**) BANA-205    R = HN (NH, NH$_2$ — guanidino)

(**22**) BANA-206    R = HN (3-pentylamino, CH$_3$ ... CH$_3$)

to that of BANA-113 (41, 77). Moreover, when both N-acetylamino and guanidino groups of BANA-113 were replaced with bis(hydroxymethyl)pyrrolidine-2-one and 3-pentylamino groups, respectively, the resulting compound BANA-206 (**22**) showed dramatic improvement in activity against influenza A NA reaching an IC$_{50}$ of 48 nM. However, its inhibitory activity against influenza B NA was significantly reduced (IC$_{50}$, 104 $\mu M$) (41, 77). For both BANA-205 and BANA-113, the guanidino group could interact with the active site residue Glu-276 in its native conformation; therefore, both compounds work equally well on both type A and B NA, whereas, for BANA-206, Glu-276 needs to adopt an alternative high-energy conformation to accommodate the hydrophobic pentylamino functional group, resulting in poorer inhibition constant (Fig. 10.6) (41, 77). The X-ray crystallography

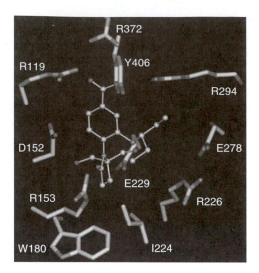

**Figure 10.6.** Complex structure of NA and BANA-206. Reproduced with permission from Dr. W. Brouillette (University of Alabama at Birmingham). See color insert.

also showed that the 2-pyrrolidinone and benzene rings are perpendicular to each other (76, 77). In conclusion, BANA-206 is a simple achiral benzoic acid derivative that has achieved nanomolar activity as an inhibitor of influenza A NA.

**2.1.1.6 RWJ-270201 (BCX-1812, Peramivir).** The investigators of BioCryst Pharmaceuticals compared the crystal structure complexes of $\alpha/\beta$-6-acetyl-amino-3,6-dideoxy-D-glycero-altro-2-nonulofuranosonic acid (**23**) (a

**(23)**

micromolar inhibitor) and DANA (**6**) and found that, regardless of their very different positions on the rings, the main functional groups (carboxylic acid, glycerol, acetamido

group, and $C_4$-hydroxy group) in both complexes have the same relative positions in the active site and have similar interactions with the enzyme (32). This finding suggested that a cyclopantane ring might be a suitable scaffold for novel NA inhibitors. Based on the literature data, simultaneous occupation of all binding regions by the four functionalities—carboxylate, guanidino, acetamido, and 3-pentyl—seems to be the basic requirements of a potent NA inhibitor. RWJ-270201 (**24**) was de-

**(24)** RWJ-270201

signed based on the premise that a cyclopantane ring could position these functionalities for optimal interaction with the four NA binding sites (32, 78).

Because this compound has five chiral centers, no efforts were made to fix the stereochemistry during the initial synthesis. Instead, it was synthesized as a mixture of isomers, followed by soaking a crystal of influenza NA in a solution of isomers to select the most active isomer from the mixture. RWJ-270201 was identified as the right isomer that bound to the active site (32).

X-ray crystallographic studies of RWJ-270201 bound to NA revealed that the carboxylic acid and 1'-acetylamino-2'-ethylbutyl group are trans to each other, whereas the guanidino and carboxylic acid groups are cis to each other (32). The guanidino group is bound to the same binding pocket as the guanidino group of zanamivir; however, they are oriented differently in the binding site. These differences in the orientation of the guanidino group might render RWJ-270201 active against the zanamivir-resistant strains containing Glu-119-Gly and Glu-119-Ala, which

**Table 10.1   Comparison of NA Inhibitory Activities of Zanamivir, Oseltamivir, and RWJ-270201 Against Various Influenza NAs (79)**

| | Mean $IC_{50}$ (n$M$) (range) | | | |
|---|---|---|---|---|
| | A/H1N1 | A/H2N2 | A/H3N2 | B |
| Zanamivir | 0.47 (0.3–0.8) | 1.23 (0.76–1.38) | 1.40 (0.68–2.32) | 4.75 (1.53–17.0) |
| Oseltamivir | 1.41 (0.69–2.24) | 0.48 (0.01–1.45) | 0.31 (0.21–0.56) | 9.68 (5.0–24.3) |
| RWJ-270201 | 0.35 (0.09–0.81) | 0.49 (0.17–1.39) | 0.36 (0.14–0.83) | 3.84 (0.6–10.8) |

$IC_{50}$, concentration of the compound required to inhibit enzyme activity by 50% in an acellular assay.

were selected by *in vitro* passages (32, 68, 79). In a matter similar to GS-4071, one of the two ethyl terminals of the 1′-acetylamino-2′-ethyl-butyl functionality point to the induced hydrophobic pocket created by the rearrangement of the Glu-276 side-chain, whereas the other ethyl moiety toward the pre-existing hydrophobic surface, which was formed by the hydrocarbon chains of Arg-224 and Ile-222 (32).

RWJ-270201 seemed to have a better *in vitro* activity against influenza A viruses than the other two drugs (Tables 10.1 and 10.2). In addition, all NA subtypes (N1-N9) of avian influenza viruses were also sensitive to RWJ-270201 (80–82). Exposure of cells to RWJ-270201 caused most of the virus to remain cell associated, with extracellular virus decreasing in a concentration-dependent manner (83). This seems in accordance with its effect as a neuraminidase inhibitor, which cause viral particles to be unable to release from cells but remain clumped at the cell surface.

Orally administered RWJ-270201 has been shown to be highly effective against experimentally induced influenza A (H1N1), A (H3N2), and B virus infection in mice (84, 85). When comparing the effects of the same doses, RWJ-270201 seemed to be more often efficacious than oseltamivir, although such differences were not seen in every experiment. Similar to oseltamivir, oral treatment could be delayed for up to 60 h post-infection, and the drug was still able to protect infected animals from death (84, 85). When administered intranasally, RWJ-270201 demonstrated better protection against lethality than oseltamivir and zanamivir at the same dose (79). Moreover, oral RWJ-270201 and oseltamivir protected mice against lethal challenge with A/Hong Kong/156/97 (H5N1) and A/quail/Hong Kong/G1/97 (H9N2); both viruses have been responsible for the 1997 Hong Kong outbreak (80, 81). Furthermore, pharmacodynamic evaluation of RWJ-270201 in mice predicted efficacy for once-daily dosing (82, 86). Indeed, in controlled clinical trials, oral, once-daily RWJ-270201 was well tolerated and effective in the treatment of experimental human influenza A and B infections (87).

Although RWJ-270201, oseltamivir, and zanamivir are structurally similar, they interact differently with residues of the NA active sites. In cross-resistance studies, RWJ-270201 was fully active against certain zanamivir-resistant enzymes and partially active against oseltamivir-resistant enzymes (68). A point mutation (Lys-189-Glu) in the hemagglutinin

**Table 10.2   *In Vitro* Antiviral Activities of Zanamivir, Oseltamivir, and RWJ-270201 on Influenza Virus Replication in MDCK cells (83)**

| | $EC_{50}$ ($\mu M$) | | | |
|---|---|---|---|---|
| | A/H1N1 (4 strains) | A/H3N2 (12 strains) | A/H5N1 (2 strains) | B (5 strains) |
| Zanamivir | 0.22–3.4 | <0.01–0.65 | 0.20–0.22 | 0.03–1.3 |
| Oseltamivir | 0.17–2.7 | <0.01–0.5 | 0.22–0.26 | 0.11–3.0 |
| RWJ-270201 | 0.09–1.5 | <0.01–0.19 | 0.01–0.02 | 0.06–3.2 |

$EC_{50}$, concentration of the compound required to inhibit viral-induced effect (cytopathic effect or plaque formation) or virus yield by 50% in cell culture.

The data on A/NWS/33 have been excluded from the table because both zanamivir and oseltamivir were inactive, and RWJ-270201 was only moderately active against this virus.

(HA) gene of A/Shangdong/H3N2 was selected in the presence of RWJ-270201 in cell culture (88). Based on virus challenge dose to infect mice, the resistant virus was approximately 10-fold less virulent than the wild-type virus. Mice infected with a lethal dose of the resistant virus could still be effectively treated with RWJ-270201 (88).

**2.1.1.7 Pyrrolidine-Based Inhibitors.** The investigators of Abbott Laboratories recently discovered that A-87380 (**25**) (IC$_{50}$, 50 $\mu M$ ver-

sus type A NA) could serve as a NA inhibitor lead after comparing it with zanamivir and by computer modeling (51, 89). A series of tri-substituted and tetra-substituted pyrrolidine analogs were then synthesized using high-throughput parallel synthetic combinatorial chemistry for SAR studies. Of all compounds synthesized, A-192558 (**26**) was the most potent inhibitor, with IC$_{50}$ values of 0.28 and 8 $\mu M$ against NA A and B, respectively (51). The preferential activity against NA A over NA B might be caused by the interaction between the urea functionality with the hydrophobic site by inducing a conformational change of Glu-276, and this change is known to be energetically less favorable for NA B. As anticipated, the carboxylate interacts with the pos-

itively charged arginine triad formed by Arg-118, Arg-292, and Arg-371 when the compound bound to the enzyme. The C$_2$ trifluoroacetamido group occupies the small hydrophobic pocket consisting of Ile-222 and Trp-178. Unexpectedly, the exocyclic amino group does not make close contact with all three acidic amino acid residues, Asp-151, Glu-119, and Glu-227.

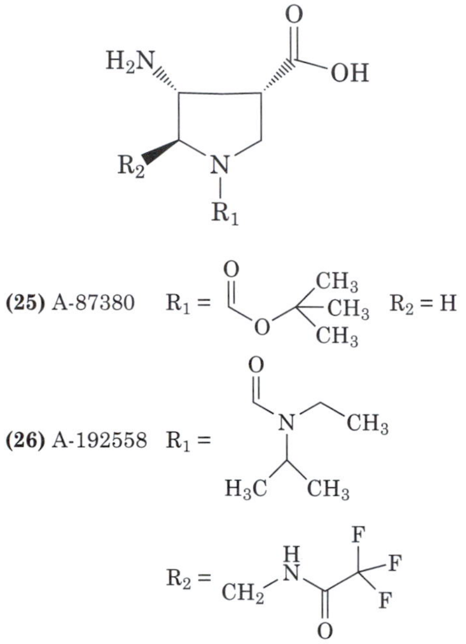

Subsequently, it was found that this traditional amine-binding pocket contains a previously unrecognized hydrophobic portion formed by Asp-151 and Leu-135 (90–92). This portion could be occupied by a cis-propenyl functional group through van der Waals force to achieve excellent affinity (90, 93). Moreover, studies with substituents pointing to the hydrophobic sub-site formed by Arg-224, Ile-222, and Ala-246 showed that a tertiary amine N-oxide (94) or ether (95) group could enhance a molecule's inhibitory activity. The enhanced activity comes from an intra-molecular H-bond involving the oxygen and the pyrrolidine nitrogen; this interaction serves to direct the aliphatic side-chains toward the hydrophobic surface in the active site of the enzyme. After an iterative structure-based method, the Abbott scientists identified ABT-675 (A-315675) (**27**) as a new potent broad-spectrum inhibitor

(**27**) ABT-675 (A-315675)

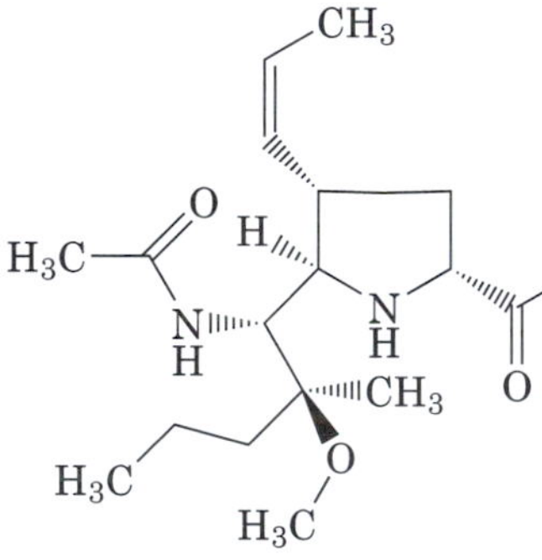

of influenza NA, with $K_i$ values ≤0.3 n$M$ (60). In MDCK cells, ABT-675 displayed comparable nanomolar activity as that of RWJ-270201 on the replication of both type A and B viruses. In the same study, GS-4071 was slightly less potent compared to ABT and RWJ compounds (60). The ethyl or isopropyl ester oral prodrug of the Abbott compound was equal or more active than GS-4104 against B/HK/5/72 in a BALB/c mouse model. For A/N2/Tokyo/3/67, the ethyl ester prodrug was more efficacious than GS-4104 (96). *In vitro* passage of A/N9/NWS/G70c, using high concentrations of A-315675, selected novel NA and HA mutations different from those selected by GS-4071. Variants selected with each drug were not highly cross-resistant (97).

In addition, ABT-675 dissociates from the enzyme about 18-fold more slowly than does GS-4071 ($t_{1/2}$ ~ 10 h for ABT-675, ~0.5 h for GS-4071) (98). The slower rate might mean a prolonged therapeutic effect, because the drug will stay bound to NA even if circulating drug has been removed from the site of influenza infection (98).

### 2.1.2 Inhibitors of Influenza Hemagglutinin

#### 2.1.2.1 Monomeric Inhibitors of Influenza Hemagglutinin

*2.1.2.1.1 BMY-27709 and Derivatives.* In a series of publications, Bristol-Myers Squibb investigators have identified BMY-27709 (**28**)

(**28**) BMY-27709

as a new lead for the development of influenza fusion inhibitors (99–101). In cell-based assays, this compound inhibited the growth of both H1 and H2 subtypes of influenza A virus (only when added at the early stage of infection) with $EC_{50}$ of 3–8 $\mu M$ (99). However, it is inactive against H3 subtype (99). To better understand the mechanism of action of the compound, 21 independent resistant viruses were selected (100). Two hot spots are identified. One is the methionine at position of 313 of the HA1. Another hot spot is the phenylalanine at position 110 in the HA2 subunit, which is mutated to either a serine (>100 resistance level) or a leucine (15–25 resistance level). Both H1 and H2 HAs contain a Phe-

110, while H3 HAs code for a Leu at this position. This may explain why H3 subtype viruses are not sensitive to BMY-27709 (99).

Most of the amino acid substitutions in the HAs of the resistant viruses are located in a region near the *N*-terminus of the HA2 subunit, suggesting that a binding pocket for BMY-27709 exists near this fusion peptide (100). It is known that HA2 subunit encodes the hydrophobic peptide believed to play a pivotal role in membrane fusion. A simulated H1 HA structure, constructed based on the known crystal structure of H3, also revealed a crevice in the region of Phe-110, in which BMY-27709 could be docked (100). Further photoaffinity-labeling experiments identified the covalent attachment site to be within HA2 amino acid residues 84–106, a region corresponding to part of the pocket proposed through molecular modeling (101). An interesting feature of the model is that HAs of H1 and H2 subtypes contain amino acid residues Glu-105 and Arg-106, which could form tight H-binding with the inhibitor. In contrast, these two respective positions in H3 are neutral Gln-105 and His-106 (100–102).

Through the use of reassortant viruses, drug-resistant variants, monoclonal antibody specificity, susceptibility to tryptic digestion, and transfectant viruses, it was concluded that BMY-27709 inhibits influenza virus infection by inhibiting the HA-mediated membrane fusion through blockage of the low-pH-induced conformational change of the native HA, which is a prerequisite for entry of the virus into host cells through membrane fusion (99–101).

A series of derivatives of BMY-27709 were synthesized in an attempt to illuminate the SARs associated with these quinolizidine salicylamides (103, 104). Variation of the substituents of the salicylic acid moiety suggested that the phenolic hydroxy group is essential for activity. This seems to be in agreement with the model, which shows the acid surrogate ketophenol moiety of BMY-27709 interacts with Arg-106 (100, 101). For substituents at the 5-position, small and non-polar groups are preferred, with optimal activity residing in the 5-halo and 5-methyl derivatives. The most active one is the 5-methyl phenol derivative (**29**), with an $EC_{50}$ of 0.25 $\mu g/mL$, which is

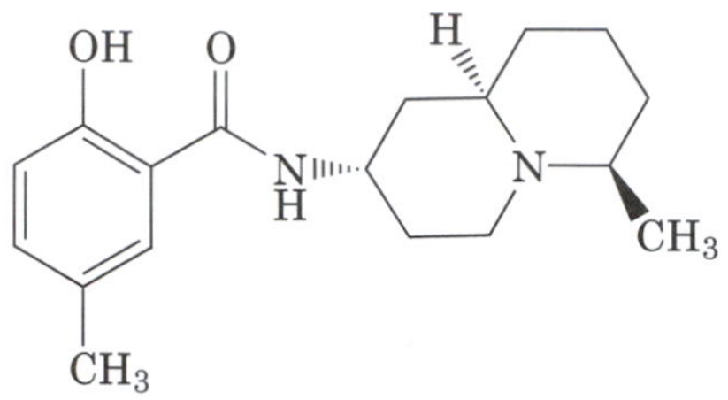

**(29)**

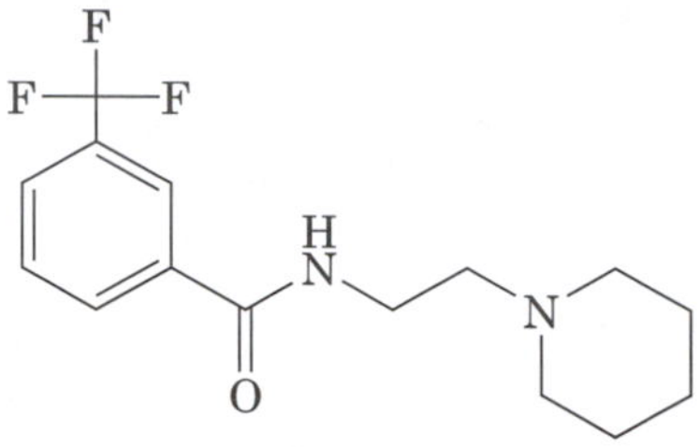

**(31)** CL-61917

fivefold more potent than BMY-27709 (103). The quinolizidine ring was then replaced with structurally simpler heterocycles—piperidines [closely resembles the structure of CL-61917 (102) discussed in the following section] or decahydroquinolines—to explore optimization of potency and spectrum of activity. The 2-methyl-*cis*-decahydroquinoline (**30**) showed

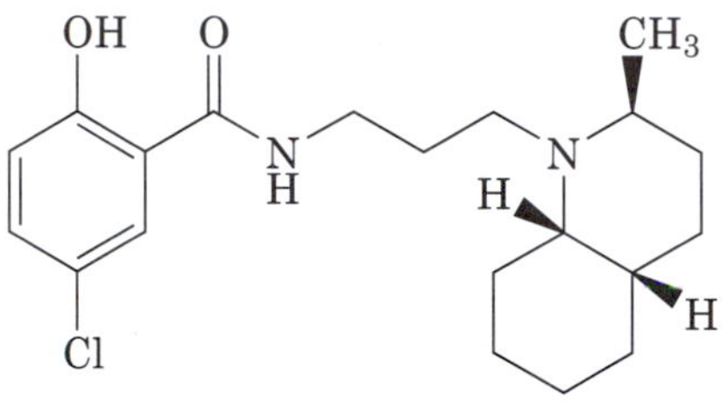

**(30)**

potent activity, with an $EC_{50}$ of 0.09 $\mu$g/mL in a plaque reduction assay against A/WSN/33 (H1N1) virus. However, this compound was moderately toxic to the MDCK cells (104). Disappointingly, none of the compounds in the series demonstrated significant activity against H3 subtype influenza virus type A.

H1 and H2 subtype HAs share nearly 70% overall sequence homology. However, H1 and H3 or H2 and H3 subtypes are with about 40% similarity overall (99).

*2.1.2.1.2 CL-61917 and Related Compounds.* By screening of a chemical library, Wyeth-Ayerst investigators also have identified several compounds that specifically inhibited replication of the H1 and H2 subtypes of influenza virus type A, particularly CL-61917 (**31**) (102). This compound showed $EC_{50} \sim 1$ $\mu$g/mL against replication of H1 and H2 subtypes; it was much less effective against H3 subtypes and virtually ineffective against influenza B virus. Interestingly, both CL-61917

and BMY-27709 are composed of a substituted benzamide linked to a nitrogen-containing heterocyclic ring structure. Both compounds demonstrate the ability to inhibit various manifestations of fusogenic activity of the representative strains of influenza A virus.

Computer-aided modeling and mutagenesis analysis suggested a putative docking site for CL-61917 in the middle of the stem region of the HA near the HA2 fusion protein. The docking surrounded by three (Phe-3, Asn-50, and Phe-110) of the four HA2 amino acid residues that are altered in the resistant mutants. The computer model illustrated two acid residues, Glu-105 and Asp-109, from one of the monomer chains of HA2, form charge–charge interaction with the piperidines nitrogen, whereas Arg-106, from a second HA2 chain, participates in a H-bonding with the amide carbonyl oxygen. The trifluorophenyl group points to a hydrophobic pocket lined partially with Phe-110 from the second monomer chain of HA2. In contrast, X-31 (a representative of H3 subtype) HA contains Gln-105, His-106, and Leu-110, yielding a significant poorer fit for CL-61917 (102).

Although there are close similarity between the CL and BMY compounds, each compound selects for some different mutations in different viral HAs, illustrating the need for caution in making generalization between these two families of compounds. Ultimately, clarification of the precise interactions between the inhibitors and HAs must await the outcomes of co-crystallization studies (102).

*2.1.2.1.3 Podocarpic Acid Derivative.* A group of Lilly investigators reported that a compound related to podocarpic acid, designed as 180299 (**32**) was identified as a specific inhibitor of influenza A viruses in tissue culture (105). Genetic analysis of reassortants be-

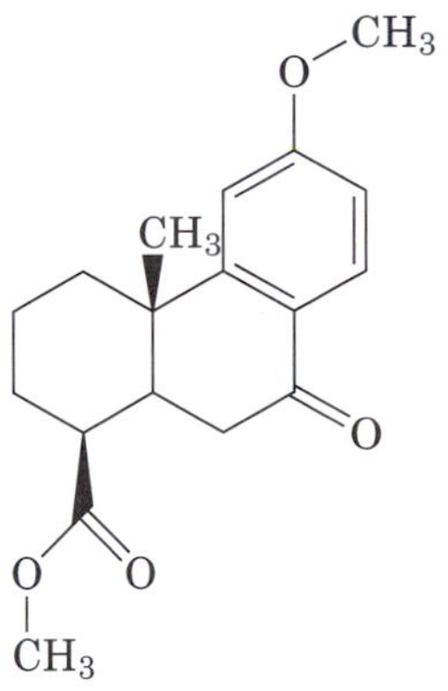

(32)

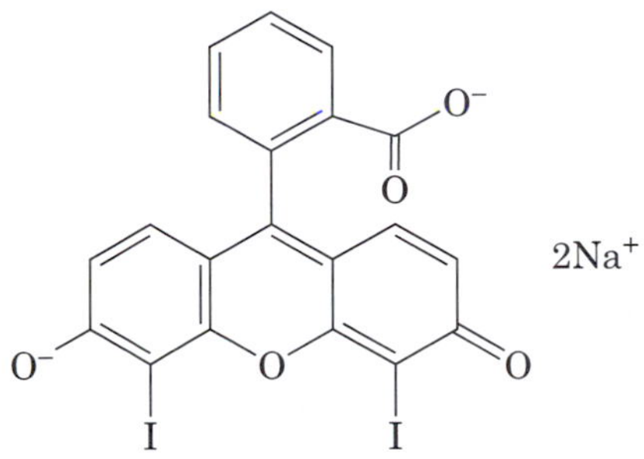

(33) C22

tween sensitive and resistant viruses, as well as independent isolates of mutant strains, showed that mutations are dispersed throughout the HA primary amino acid sequence and cluster in the interface between HA1 and HA2 and in a region near the fusion domain of HA2. Fusion of human erythrocytes and pH-of-inactivation studies suggested that, like the aforementioned CL and BMY compounds, 180299 interacts with the neutral pH conformation of influenza A HA and prevents the low-pH-induced change of HA to its fusogenic conformation (105). However, unlike the CL and BMY compounds, 180299 displayed no *in vitro* activity against S/WSN/33 (H1N1), a strain that displays an elevated pH-of-inactivation. It was found that naturally resistant influenza viruses [e.g., A/Aichi/68 (H3N2) and B/Lee/40] generally have an elevated pH-of-inactivation. In contrast, the most sensitive strain, A/Kawasaki/86 (an H1 subtype), has the lowest pH-of-inactivation (105).

*2.1.2.1.4 Diiodofluorescein.* White et al. used a computer-searching algorithm known as DOCK to conduct a series of structure-based inhibitor searching by targeting two sites surrounding HA2 54-81, the region of HA2 that undergoes conformational change at low pH (106). Of 12 new compounds selected by DOCK, diiodofluorescein (33) (also designated as C22) was identified as a new lead. This compound facilitates the conformational change with 50% effective concentration of approximately 8 $\mu M$, yet inhibits viral infectivity with an EC$_{50}$ of 8 $\mu M$. In the case of BMY-27709, the compound acts to stabilize HA dur-

ing the course of an infection and this stabilization impedes infectivity (100). In contrast, C22 inhibits viral fusion and infectivity by destabilizing HA; it acts as an irreversible facilitator of the conformational change (106).

*2.1.2.1.5 Stachyflin and Derivatives.* Stachyflin (34) a novel sesquiterpene deriva-

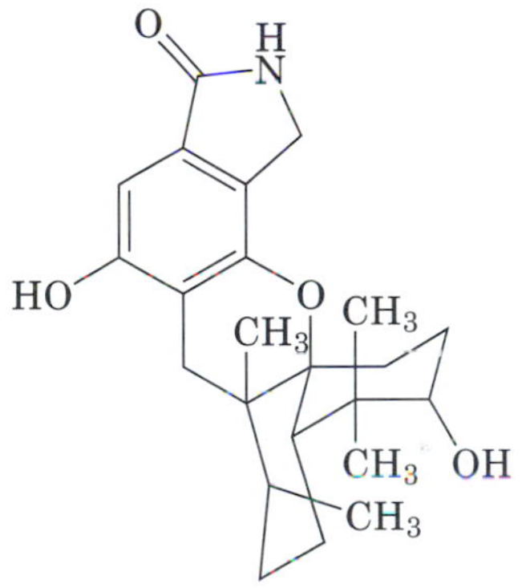

(34) Stachyflin

tive isolated from fungus, was shown as having H1 and H2 subtype-specific anti-influenza A virus activity by a group of investigators of Shionogi Labs (107). One-step virus growth experiment suggested that this compound interfered with the HA-mediated virus-cell membrane fusion process through the inhibition of the physiological HA conformational change induced by low pH.

Stachyflin is lipophilic and insoluble in water; therefore, it cannot be given orally to treat experimental infection in animals. However, when given intraperitoneally at doses 2–8 mg per mouse, twice a day, stachyflin showed approximately 70% reduction of virus [A/Kumamoto/5/67 (H2N2)] titers in the lungs of the infected animals compared with that of the control animals (108). Oral administration used either an aqueous solution of the phosphate ester prodrug or a solution of stachyflin in polyethylene glycol (PEG), PEG 400 or PEG 4000, could also achieve 60–70% reduction in the pulmonary virus titers (108). Unfortunately, concerns about its lack of activity against human clinical isolates and mutagenicity preclude stachyflin from further development (109).

The keto derivative of stachyflin III (**35**), as well as its phosphate prodrug III-Phos (**36**),

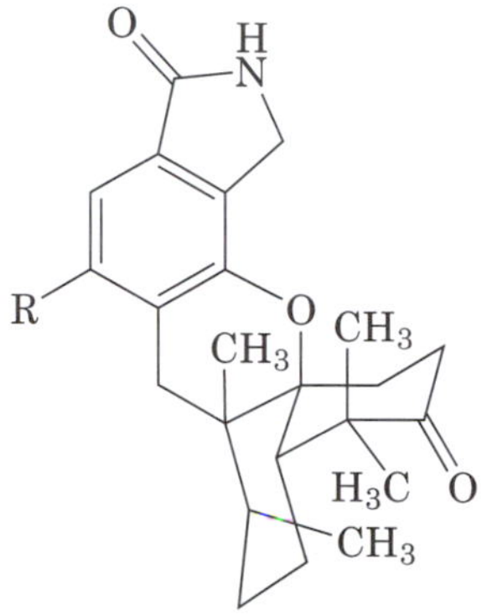

(**35**) R = OH

(**36**) R = O—P(=O)(OH)—OH

has been chosen for further development (109). Mice infected with mouse-adapted A/Kumamoto/5/67 (H2N2) were orally administered with III dissolved in PEG 4000 or with III-Phos dissolved in water (4 mg per mouse), twice a day for 2 days, resulting in approximately 85% virus inhibition in the lungs. Surprisingly, no effect was observed in ferrets infected with a fresh clinical isolate, A/Sendai/808/91 (H1N1), after oral administration of III, irrespective of the long-lasting high concentrations of the compound in the plasma. Nevertheless, intranasal administration of III-Phos could still inhibit viral replication in the nasal cavity and suppress fever. It was postulated that the accumulation of III and III-Phos on the surface of nasal membrane and good nasal absorption of III-Phos contribute to the *in vivo* efficacy after intranasal administration of III-Phos to the infected ferrets (109).

Both mouse and ferret are good animal models for the evaluation of experimental therapies of influenza infection. In mice, mouse-adapted influenza virus replicates in the lung and causes pneumonia without fever (110). In comparison, ferrets are susceptible to human influenza, which replicates in the upper respiratory tract and causes illness similar to humans (109, 111).

***2.1.2.2 Polymeric Carbohydrate-Based Inhibitors of Influenza Hemagglutinin.*** Multivalency (polyvalency) is the simultaneous binding of multiple ligands on one molecule to multiple receptors on another (112). An influenza virion presents approximately 200–300 copies of HA trimeric units on its surface; each subunit contains a binding site for sialic acid (SA; also termed Neu5Ac) at its outmost portion (113). Because the binding pocket is small and shallow, the interaction of a single HA binding site with a single SA is weak. Nevertheless, the binding of a viral particle to the surface of a cell is strong. This strong interaction reflects the interaction of multiple copies of HA on the viral surface simultaneously with multiple SA groups on the surface of the cell. Therefore, in principle, highly effective prevention of the attachment of influenza virus to the cell can be achieved with multivalent (polyvalent) inhibitors that present multiple copies of SA to the virus (114). Multivalent sialosides bearing multiple sialyl moieties tethered to various synthetic backbones of polymers, liposomes, or dendrimers have been reported (112, 113, 115).

*2.1.2.2.1 Sialic Acid-Containing Polymers.* In the X-ray crystal structures of HA complexed with SA derivatives, it has been shown that the 4-hydroxyl of SA does not interact with the enzyme (116, 117). Thus Watson et al. reported the synthesis of a multivalent SA-containing polymer (**37**) in which a sialoside has been coupled to polyacrylamide through a 4-*N*-linkage, as well as a 2-linked conjugate (**38**) (117). Both compounds showed potent activity (<0.5–4 $\mu M$ of SA units) against an influenza H3 subtype (X-31) and two H1 subtypes (A/Tokyo and G70C) by an HAI assay.

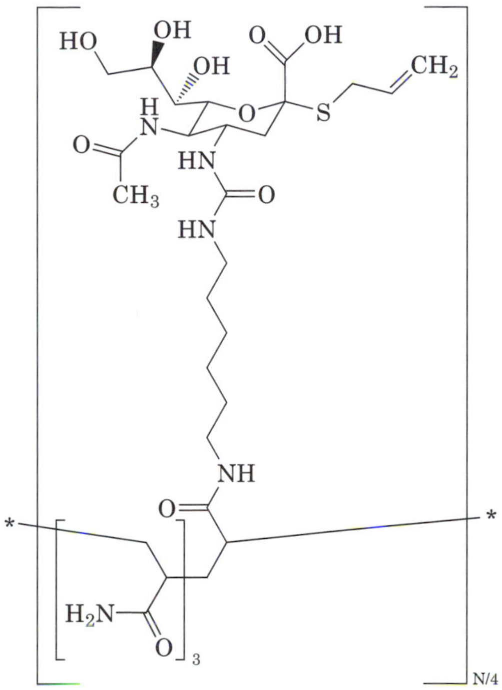

(37)

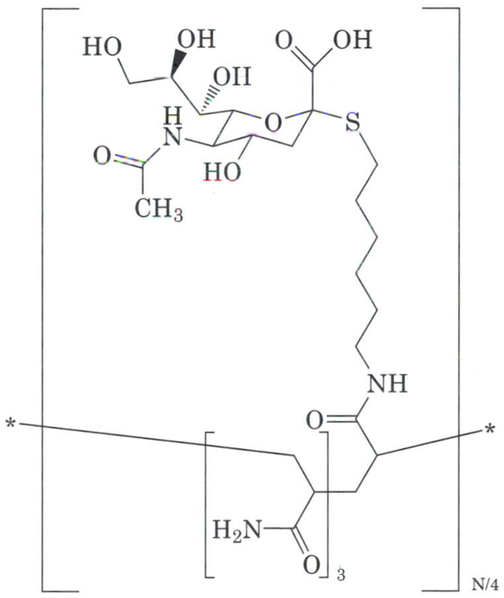

(38)

Generally, hemagglutination inhibition (HAI) assay is commonly used to measure the ability of a SA-containing polymer (glycopolymer)

that inhibits the attachment of influenza virus to erythrocytes (which serve as surrogate target cells). HAI assay is easy to perform, but limited to inhibition constants greater than 1 n$M$ (112). Therefore, inhibitors with the lowest concentration that inhibits hemagglutination smaller than 1 n$M$ seem to be equally effective in the HAI assay. Recently, Whitesides et al. reported a new method based on dual optical tweezers, termed OPTCOL, which is able to measure a single cell (e.g., erythrocyte) and a single microsphere coated with viral particles. The lowest limit of measurable inhibition constants is less than $10^{-18}$ M (112, 118). By using OPTCOL, a derivative of polyacrylamide pA(NeuAc, $\chi$) (**39**) with 35% of the side-chains ($\chi = 0.35$) tethered to SA by a short flexible linker was shown to prevent hemagglutination at concentration of 35 p$M$— $10^8$ times more effective than most monovalent derivatives of SA (e.g., $\alpha$-methylsialoside inhibits hemagglutination at concentration 2.5 m$M$) (118). Whitesides' group also reported, by using an enzyme-linked immunosorbent assay (ELISA), affinities of polyacrylamides bearing pendant $\alpha$-sialoside groups for the surface of influenza virus A X-31 (H3N2) ranged between $10^3$ and $>10^6$ greater than that of $\alpha$-methyl sialoside, on the basis of total sialic acid groups in solution (114).

The significantly enhanced ability of glycopolymers over monovalent SA derivatives at preventing the agglutination of red blood cells by influenza virus is thought as a result of high-affinity binding through polyvalency (because of the cooperative binding of multiple SA groups per inhibitor molecule) and steric stabilization (because of the steric prevention of virus from close approach to the cell by a water-swollen layer of the polymer) (119, 120). The latter mechanism has been further supported by studies that showed that the efficacy of SA-containing polymers in inhibiting hemagglutination got enhanced (by 2- to 20-fold) by adding potent monomeric NA inhibitors (121). The SA groups on the polymers bind the HA-binding pocket as well as the NA active site. The enhancement of inhibition by the polymers in the presence of NA inhibitors is probably caused by expansion of the adsorbed

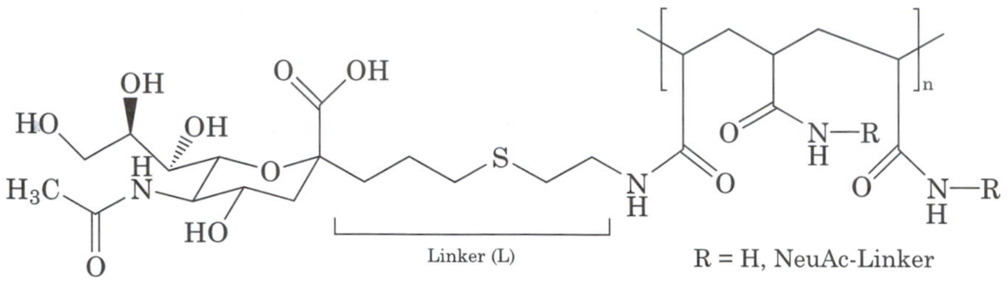

(39) pA(NeuAc); $\chi$ = mole fraction of NeuAc per side chain
= 0.05, 0.2, 0.35, 0.6, 1.0

polymer layer after the competitive release of SA groups originally bound to the NA active sites.

Furthermore, Whitesides et al. carried out both synthesis and the HAI assay, in the wells of microtiter plates, of libraries of polymers [poly(acrylic acid) (pAA)] bearing both sialoside and largely hydrophobic non-sialoside groups in random sequence (113). The *in situ* bioassay showed that new polymers containing both SA and certain non-sialoside groups [pAA(NeuAc-L; R)] were up to $10^4$-fold more potent than the parent polymer bearing only the SA residue [pAA(NeuAc-L)] (113). It is postulated that the non-sialoside groups R may be involved in non-specific binding to hydrophobic sites on the surface of virus, resulting in enhanced affinity of the new polymers for the viral surface. This strategy may serve generally for screening and obtaining leads in biological systems that involve multivalency.

Conceptually, if the non-sialoside groups can interfere with the conformational change of the membrane-fusion domain of HA, a polyvalent polymer can inhibit influenza virus not only during initial binding, but also later in the course of HA-mediated membrane fusion. To test the hypothesis, Wong et al. reported the synthesis and evaluation of a polymer of poly-L-glutamic acid conjugated with lysoganglioside GM$_3$ (40) (122). A fluorescent tag BODIPY was also attached. Using an ELISA assay, the polymer showed EC$_{50}$ against A/PR/8/34 (H1N1) at 7.5 p$M$ based on the sialic acid content. The influenza inhibitory activity of the polymer is enhanced by $10^3$-fold compared with that of lysoganglioside GM$_3$. Wong et al. thus proposed that the enhanced activity is a result from the formation of a stable poly-

mer/HA complex by a "chain lock" mechanism, which suggests that after the sugar groups bind to the active sites on the top of HA, the hydrophobic substituents (e.g., the sphingosine moiety) of lysoganglioside GM$_3$ wrap around the hydrophobic sites on the stem area providing further interaction with HA.

*2.1.2.2.2 Liposomes.* It has been reported that anti-influenza activity can be enhanced with a multivalent display of the sialoside on liposome. Wong et al. synthesized a series of conjugates of 3- OH or 3-F-substituted SA derivatives with distearoylphosphatidylethanolamine (DSPE) as the liposome (41) (123). As determined in an HAI assay against A/Aichi/2/68 (H3N2), the DSPE conjugates showed a $10^3$-fold increase in the inhibitory activity when compared with monomeric SA. However, the activity against A/PR/8/34 (H1N1) was not significantly enhanced.

Suzuki et al. demonstrated that selective activity for H3N2 subtype over H1N1 subtype was also observed with synthetic sialylphosphatidylethanolamine (sialyl PE) derivatives (42) (124). Because H3N2 subtype influenza A viruses (e.g., A/Aichi/2/68 and A/Memphis/1/71) preferentially bind to Neu5Ac$\alpha$2–6Gal and H1 subtype (e.g., A/PR/8/34) binds most effectively to Neu5Ac$\alpha$2–3Gal linkage of sialosugar chains on the cell membrane, Suzuki et al. speculated that the structural assembly of sialyl PE derivatives might exhibit some similarity to the Neu5Ac$\alpha$2–6Gal linkage.

*2.1.2.2.3 Dendritic Polymers.* Baker et al. have reported the synthesis of several SA-conjugated dendritic polymers with various architectures that included spheroidal polyamidoamine (PAMAM) dendrimers, comb-

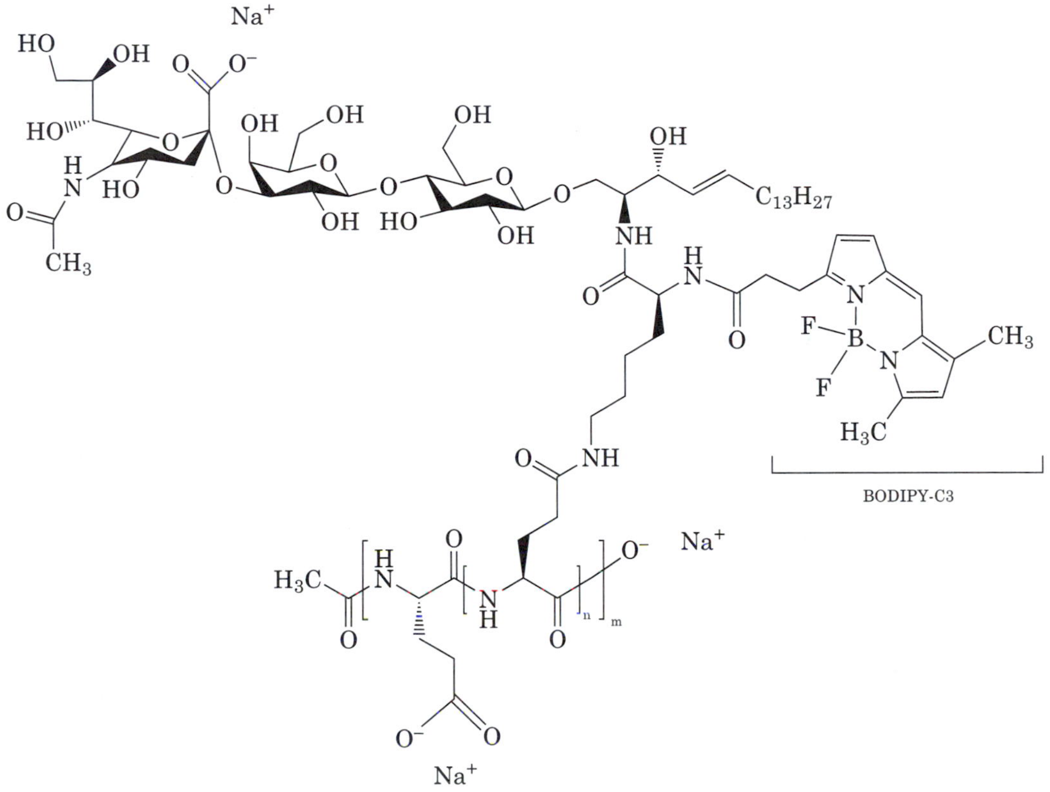

BODIPY-C3

(40) m = 270
n : m ≈ 1.5 : 100

branched and dendrigraft polymers, and linear-dendron architectural copolymers (125). Linear polyethyleneimine (PEI) was involved in the construction of the last three structures. Significant variation in susceptibility to these polymeric compounds was observed when tested for their ability to inhibit virus hemagglutination (HAI assay) and to block infection of MDCK cells (ELISA assay). Generally, the larger and more flexible linear-den-

dron architectural copolymers, as well as scaffolding-type comb-branched polymers and dendrigrafts are more efficient than the spheroidal dendrimers, which have a fixed rigid size and shape. Both the degree of SA conjugation and the polymer size seem to influence the bioactivity through polyvalent binding and steric hindrance. Influenza virus X-31 (H3N2) was shown to be more sensitive to these compounds than A/AA/6/60 (H2N2).

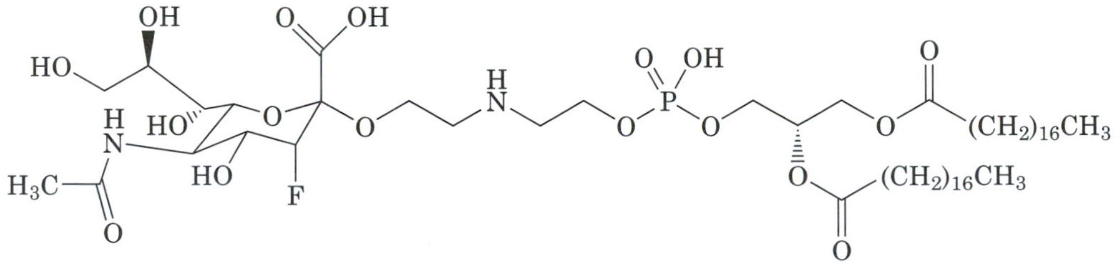

(41)

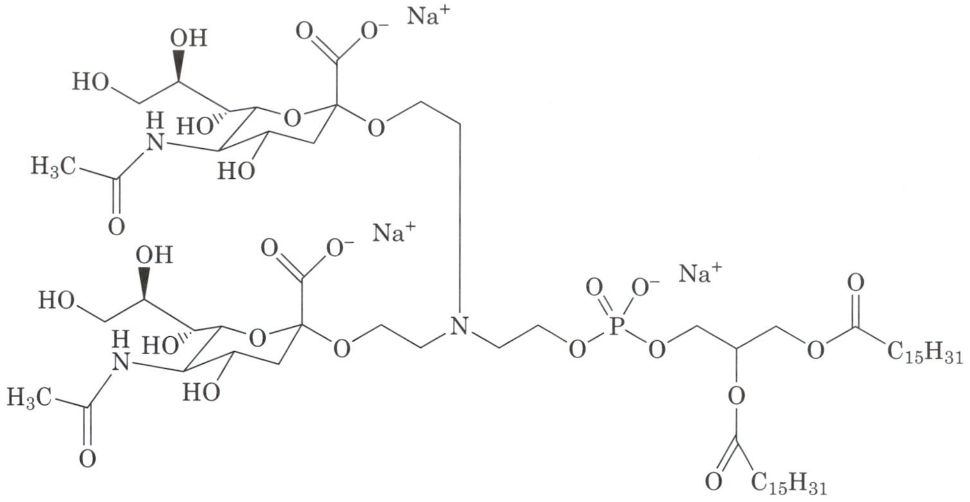

(42)

The most effective comb-branched and dendrigraft PEI polymers were up to $5 \times 10^4$-fold more effective than monomeric SA at inhibiting HA of the X-31 and sendai viruses.

A review on PAMAM dendrimers and their biomedical applications has been published recently (126).

### 2.1.3 Inhibitors of RNA-Dependent RNA Polymerase (RNA Transcriptase)

**2.1.3.1 Antisense Oligonucleotides.** It has been known that influenza viral mRNA synthesis is catalyzed by viral nucleocapsids, which consist of the individual viral RNAs (vRNAs) associated with four viral proteins: the nucleocapsid protein (NP) and the three P proteins (PB1, PB2, and PA). In a series of reports, Takaku et al. have demonstrated that antisense phosphodiester and phosphorothioate oligonucleotides that are complementary to the viral RNA polymerase (PB1, PB2, PA) and the nucleoprotein (NP) genes, specifically inhibited influenza A virus replication in MDCK cells (127, 128), enhanced survival of mice infected with influenza A virus (129, 130), and inhibited chloramphenicol acetyltransferase (CAT) protein expression in the clone 76 cell line (131–133). The clone 76 cell line is designed to express the influenza virus RNA polymerase (PB1, PB2, PA) and nucleoprotein (NP) genes in response to treatment with dexamethasone. The *in vitro* activities of

these oligonucleotides on the expression of the viral gene products were assessed on the basis of their inhibition of CAT protein expression with a CAT-ELISA method (131, 132).

The results revealed high inhibitory effects shown with the antisense oligonucleotides complementary to the sites of the PB2 AUG, PA AUG, and NP AUG initiation codons, with the best activity seen with the antisense oligonucleotides (ATATAAGTTATACCTTTCTT) targeting PB2 AUG. The antisense oligonucleotides targeted to the PB2 loop-forming site did not lead to efficient inhibition, and those targeted to PB1 AUG initiation codon and the loop forming sequence were considerably less effective. In addition, the inhibitory activities the free oligonucleotides could be increased significantly with liposomal encapsulation. It was shown that the endocapsulated antisense phosphorothioate oligonucleotides accumulated in the nuclear region of dexamethasone-treated clone 76 cells (132) or virus-infected cells (128); the endocapsulated antisense phosphodiester oligonucleotides were found within the cytoplasm. Intravenous administration of antisense phosphorothioate oligonucleotides containing the PB2 AUG initiation codon encapsulated with liposome to the mice infected with influenza A/PR/8/34 (H1N1) significantly inhibited viral growth in the lungs, prolonged the mean survival time in days and increased the survival rates of the infected mice (134). DMRIE-C liposome was more effective than Tfx-10. Because the PB2 mRNA sequences around the AUG initiation codons of influenza A and B viruses share very low homology, not surprisingly, these oligonucleotides failed to inhibit influenza B both *in vitro* and *in vivo* (130).

Treating dumbbell RNA/DNA chimeric oligonucleotides with RNase H can also generate antisense phosphodiester oligonucleotides, which in turn is bound to the target mRNA (133). These new class of oligonucleotides are consist of a sense RNA sequence and its complementary antisense DNA sequence, with two hairpin loop structures.

### 2.1.3.2 DNAzymes, Ribozymes, and External Guide Sequences.
In a recent report published by Toyoda et al., it was demonstrated *in vitro* that influenza virus replication was inhibited by RNA-cleaving DNA enzymes (135). Two oligonucleotides [DNzPB2 (14) and DNzPB2 (16)], which contain the DNA enzyme carrying the 10–23 catalytic sequence (136, 137), flanked with complementary sequences around the PB2 AUG initiation codon, were shown to be more effective than the same amount of antisense phosphorothioate oligonucleotides, which target AUG initiation codon sequences of PB2 mRNA. DNzPB2 (16) [TCTTTCCA**GGCTAGCTACAACGA**-ATTGAATA (sequence of 10–23 DNA enzyme is shown in bold italic)] was more efficient than DNzPB2 (14) (CTTTCCA**GGCTAGC-TACAACGA**ATTGAAT). One concern with these DNA enzymes is that their RNA-cleaving activity has not been able to optimize under physiological conditions.

Instead of targeting the *PB2* gene, Lazarev et al. selected a ribozyme gene directed at a specific cleavage of mRNA coding for PB1 protein (138). Because the *PB1* gene is one of the least variable *influenza A virus* genes, PB1-directed ribozyme may be expected to cleave mRNA of widely different virus strains. Oligodeoxyribonucleotides were synthesized corresponding to the *hammerhead ribozyme* gene (containing a 24-nucleotide catalytic domain), flanked with antisense sequences (12 nucleotides on both sides of the catalytic domain) complementary to a GUC site at position 1568 in PB1 mRNA of A/Kiev/59/79 (H1N1). Plasmids containing the oligodeoxyribonucleo-

tides were used to create ribozyme-expressing CV-1 cells, which along with the original CV-1 cells, were then infected with influenza A/Singapore/1/57 (H2N2) or A/WSN/33 (H1N1) for antiviral susceptibility. High levels (=90%) of inhibition of viral NP and NS1 proteins and influenza virus reproduction (by plaque assay) were noted in the cell lines expressing the functional ribozyme. Defective recombinant adenoviruses were also constructed carrying the genes of functional and non-functional ribozymes under the control of human cytomegalovirus promoter. Again, greater than 90% level of inhibition of the replication of influenza A/WSN/33 virus in CV-1 cells pre-infected with the recombinant ribozyme-expressing adenoviruses was observed compared with that in the non-infected cells. Analyzing the results with the cell line expressing the non-functional ribozyme suggested that the inhibition of influenza A virus reproduction with these ribozymes results mostly from the effect of RNA cleavage, and only to a small extent from the antisense effect of the flanking complementary sequences.

Ribonuclease P (RNase P) is an enzyme that cleaves tRNA precursors to generate the 5′ termini of mature tRNAs. Research conducted by Altman has shown that RNase P might target any RNA for specific cleavage provided that the RNA is associated with a custom-designed external guide sequence (EGS) RNA (139, 140). When the target RNA is complexed with EGS through hydrogen bonds, the resulting structure resembles a tRNA precursor and, therefore, is susceptible to cleavage by RNase P. Recently, Plehn-Dujowich and Altman reported that EGSs targeted to the influenza mRNA PB2 and NP genes effectively inhibited viral protein and particle production *in vitro* (141). Such inhibition was postulated as a consequence of a lowering of the amounts of the target mRNAs by the combined functions of EGSs and RNase P. In their design of EGSs, three sites in the PB2 transcript and two in the NP transcript that were assessable to digestion with RNase T1 were chosen to provide a G at the 3′ side of the putative cleavage site by RNase P in the target mRNA:EGS complex and a uracil 8 nts downstream, as is found in all tRNAs.

### 2.1.3.3 Short Capped Oligonucleotides.

To initiate influenza viral mRNA synthesis, the viral associated RNA-dependent RNA polymerase binds to the cap structure at the 5′ ends of host cell RNA polymerase II transcripts and then a virally encoded endonuclease cleaves the capped 5′-termini to provide caps for the 5′-termini of the viral mRNAs and to serve as primers for transcription by the viral RNA-dependent RNA polymerase (RNA transcriptase). The PB2 polymerase protein mediates both the binding and the endonucleolytic cleavage of capped mRNAs. Conceptually, the 5′-capped short RNA fragments by design are potential decoys of cap-dependent transcription. Takaku et al. reported the synthesis of short RNA molecules (8–13 ntds long) with a 5′-capped structure (m7GpppGm) using T7 RNA polymerase (141a). These short RNAs were tested, with or without liposomal encapsulation, for their inhibitory effect by a CAT-ELISA assay using the clone 76 cells, showing that the 9-ntd-long RNA molecule (m7GpppGmAAUACUCA) had the highest inhibitory activity. Furthermore, these RNA molecules exhibited higher inhibitory activity than that of the antisense phosphorothioate oligonucleotide complementary to the AUG initiation codon of PB2 mRNA.

Because the influenza virus employs the cap embezzled from the host cell, the virus is not sensitive to the inhibitory effect of S-adenosylhomocysteine (SAH) hydrolase inhibitors, which interfere S-adenosylmethionine (SAM)-dependent methylation reactions, leading to inhibition of a broad range of DNA and RNA viruses (142) (see Section 2.9.1).

### 2.1.3.4 2,4-Dioxobutanoic Acid Derivatives.

As discussed above, an intrinsic property of influenza virus RNA-dependent RNA polymerase (RNA transcriptase) is its cap-dependent endonucleolytic cleavage activity. Through a random screening, Tomassini and a group of Merck investigators have identified L-735882 (43) a 4-substituted 2,4-dioxobutanoic acid, as a specific inhibitor of cap-dependent endonuclease activity of the transcriptase and with antiviral activity against both influenza A and B viruses in cell culture (143). Compound (44), a synthetic analog of (43) with sub-micromolar antiviral activity, was found to be the most soluble in water and readily adsorbed into the

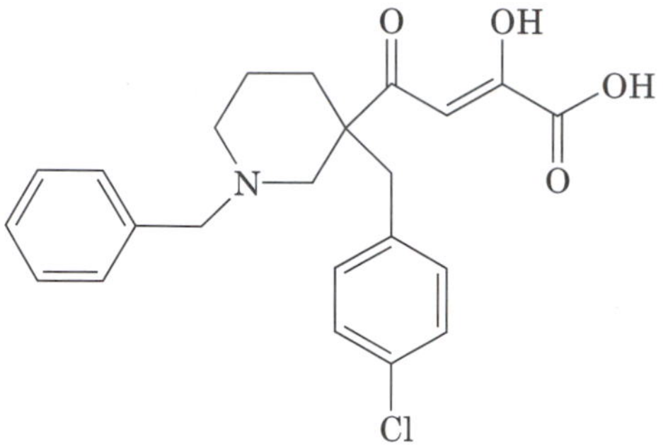

(43) L-735882

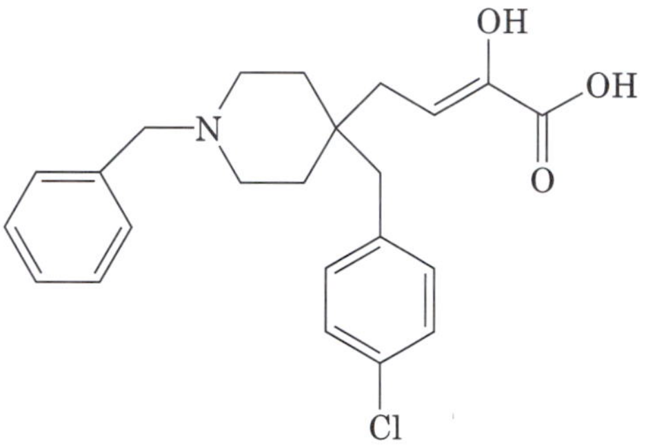

(44)

nasopharyngeal track in a mouse challenge model. When instilled intranasally into infected mice, (44), at its highest water-soluble dose, caused a 3.9-log reduction of the virus titers in nasal washes (143).

### 2.1.3.5 2,6-Diketopiperazine Derivatives.

Tomassini et al. also reported the finding of flutimide (45), a natural product isolated from

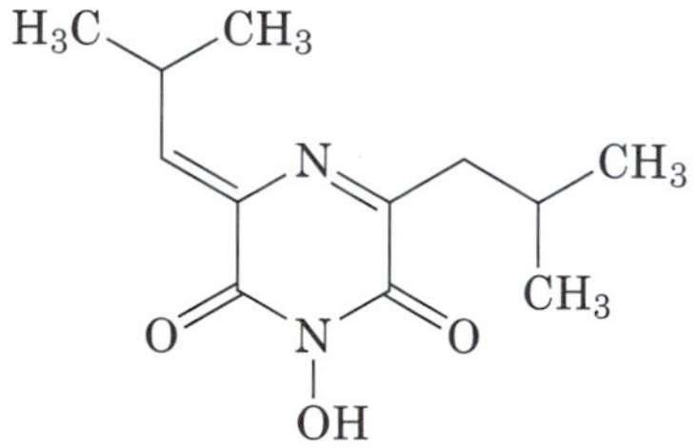

(45)

a fungus, which inhibited the cleavage of capped RNA by influenza virus endonuclease, with an IC$_{50}$ of 6.8 $\mu M$ (144). The SAR analysis with several synthetic analogs indicated that both the N-hydroxy and olefin groups were required for activity (145). The most potent analogs were compounds (46) and (47), both having IC$_{50}$ of ~0.9 $\mu M$. Flutimide inhibited

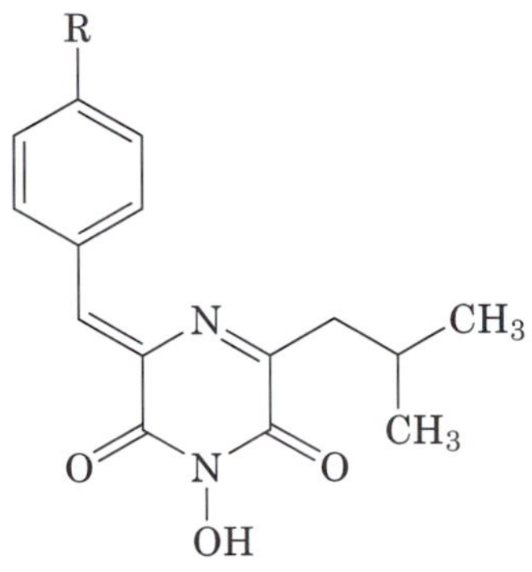

(46) R = OCH$_3$
(47) R = F

influenza virus infection of MDCK cells with an EC$_{50}$ of 5.9 $\mu M$ without any toxicity at 100 $\mu M$ concentrations. Compounds (46) and (47) seemed to be more potent in the antiviral assay; however, they showed cytotoxicity to the cells at >10 $\mu M$ concentrations (145).

**2.1.3.6 BMY-26270 and Analogs.** Krystal et al. of Bristol-Myers Squibb identified BMY-26270 (48) through a high-throughput *in vitro*

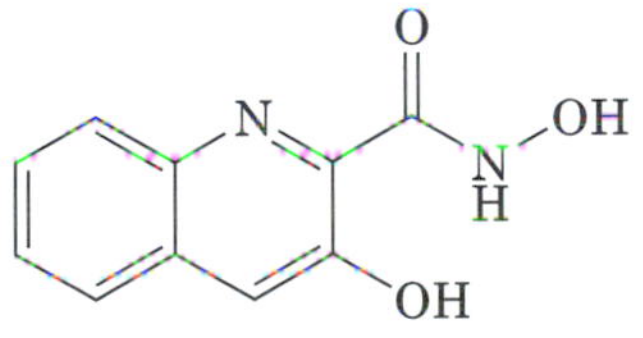

(48) BMY-26270

transcription assay from the company's chemical collection (146). This compound selectively inhibited influenza transcriptases of both A and B viruses with an IC$_{50}$ ~ 40 $\mu M$. When compared with related compounds selected from the chemical collection, it was suggested that the hydroxamic acid and phenol moieties, as well as their topological relationship, are essential for the activity. Two related *N*-hydroxy-imides, BMY-183355 (49) and BMY-183021 (50) where the relatively acidic amine NH presumably functions as an isostere of the phenolic hydrogen, also demonstrated an IC$_{50}$ ~ 50 $\mu M$. Notably, these

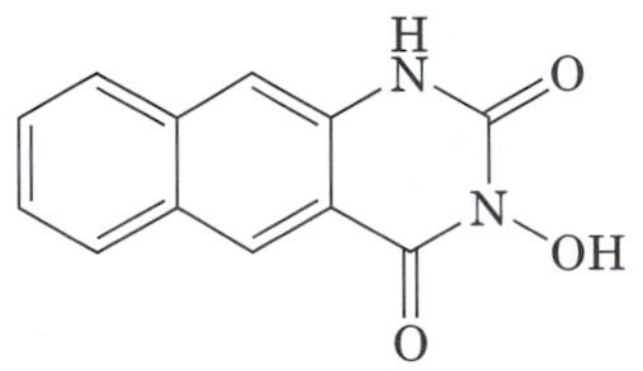

(49) BMY-183355

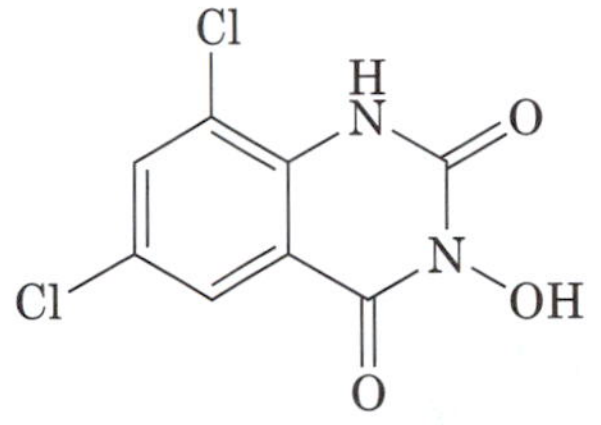

(50) BMY-183021

compounds inhibited endonuclease activity preferentially over capped RNA binding activity. However, they showed significant cytotoxicity in cell culture.

**2.1.3.7 M1 Zinc Finger Peptides.** The matrix protein (M1) is a major structural component of the influenza virion. M1 can bind to RNA directly and inhibit its own polymerase; it was proposed that this activity might be because of the presence of a zinc finger motif. Judd et al. reported that peptide 6, a synthetic peptide based on the zinc finger region of the M1 protein sequence of influenza A/PR/8/34 (H1N1) centered around residues 148–166, was 10$^3$-fold more effective in polymerase inhibition than was M1, and greater than 10$^3$-fold more effective, on a molar basis, than ribavirin and amantadine again virus A/PR/8/34 by measuring the inhibition of viral cytopathic effect in MDCK cells (147). Little or no *in vitro* antiviral activity could be seen if the peptide was added later than 1 h after virus challenge, suggesting that the peptide inhibits virus at an early stage in viral replication, presumably through inhibition of the polymerase. Pre-treatment with the peptide also significantly protected the cells from viral challenge. Because the M1 sequence representing peptide 6 is highly conserved among type A influenza viruses, peptide 6 exhibited *in vitro* antiviral activity against a wide range of type A influenza viruses representing

H1N1, H2N2, and H3N2 subtypes. Interestingly, it was also active against two type B influenza viruses (B/Lee/40 and B/Shanghai/4/94). Because peptide 6 shows therapeutic effect, it is possible that, with coordination of zinc, the peptide can assume a compact size and readily enters the cells. Derivatives with alternations in the finger loop, tail length, or residues involved in coordination of zinc showed reduced or abolished antiviral activity. When tested in a mouse model of influenza infections, peptide 6, administered intranasally beginning 4 h pre- or 8 h post-virus exposure to an H1N1 virus (A/PR/8/34) or an H3N2 virus (A/Victoria/3/75), was effective in preventing death, reducing the arterial oxygen decline, inhibiting lung consolidation, and reducing virus titers (titer reduction not seen with A/Victoria/3/75) in the lungs of infected animals (148).

**2.1.4 Inhibitors of Influenza M2 Protein.**
By employing a screen format that observes M2 expression in yeast cells, BL-1743 (**51**), a

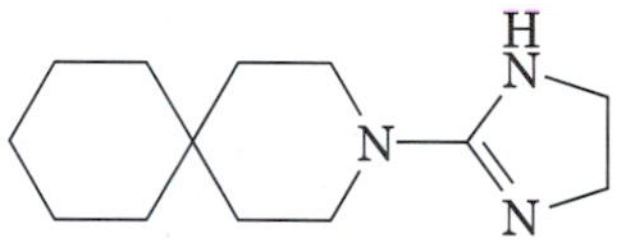

**(51) BL-1743**

spirene-containing lipophilic amine, was identified from the Bristol-Myers Squibb chemical collection as an inhibitor of influenza replication (149). In a plaque reduction assay, it showed a similar profile to that of amantadine in that it was active against an amantadine-sensitive strain (EC$_{50}$ ~ 2 $\mu M$ against A/Udorn/72) but not against amantadine-resistant A/WSN/33. The majority of BL-1743 resistant strains were also amantadine resistant. Inhibition of ion channel activity by BL-1743 differs from that with amantadine in that the inhibition with BL-1743 was reversible within the time frame of the experiment. The experimental results with amantadine on BL-1743-resistant strains indicated that two compounds interact differently with the M2 protein trans-membrane pore region. Because of the overlapping resistance profile of the two compounds and the higher apparent $K_i$ (4.7

$\mu M$ for BL-1743 and 0.3 $\mu M$ for amantadine), BL-1743 should not be regarded as a potential replacement of amantadine for the prophylaxis or treatment of influenza virus infections in humans.

Kolocouris et al. have also reported new derivatives of amantadine that exhibit antiviral activity presumably by inhibiting the M2 ion channel of influenza virus type A (150). Using a CPE assay in MDCK cells, compound (**52**),

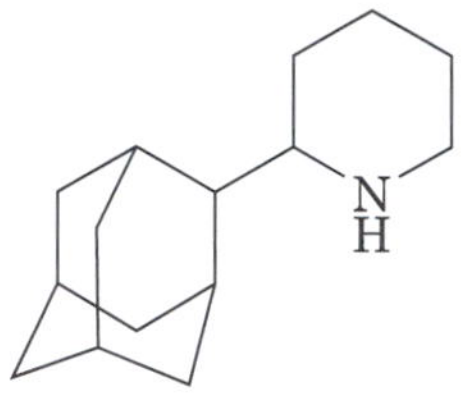

**(52)**

like amantadine and rimantadine, demonstrated antiviral effect on influenza virus A/Japan/305/57 (H2N2).

**2.1.5 Other Inhibitors of Influenza Viruses**
*2.1.5.1 Bisindolylmaleimides.* The effects of bisindolylmaleimide I (**53**), a potent inhibitor of protein kinase C (PKC), on the entry and

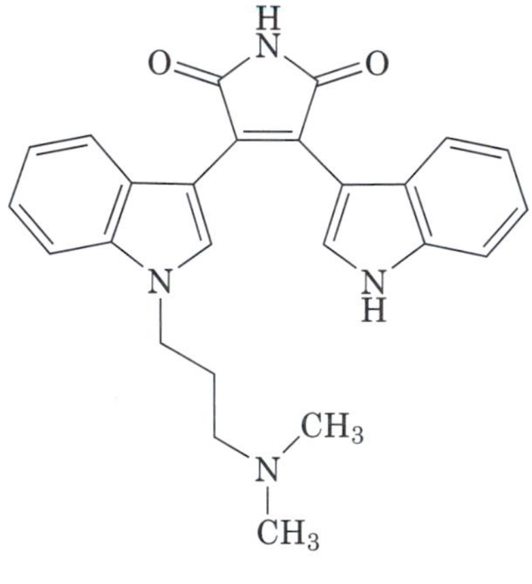

**(53)**

replication of influenza viruses were reported recently by Whittakar et al. (151). This compound inhibited *in vitro* replication of both influenza A and B viruses at micromolar concentrations in a dose-dependent and reversible manner. Further experiments showed that this compound blocked influenza virus entry within the first 60 min of infection, at some point (probably endocytosis) before entry of viral ribonucleoproteins (vRNPs) into the nucleus. Although this compound seemed to be acting during influenza virus entry, it was shown being not acting as a weak base. Therefore, it would be interesting to see if other pH-dependent viruses such as vesicular stomatitis virus and Semliki Forest virus are also sensitive to such PKC inhibitors.

**2.1.5.2 Pyrimidine Derivatives.** Efficient synthetic routes of 2-amino-4-($\omega$-hydroxy-alkylamino)pyrimidine derivatives (**54**) were

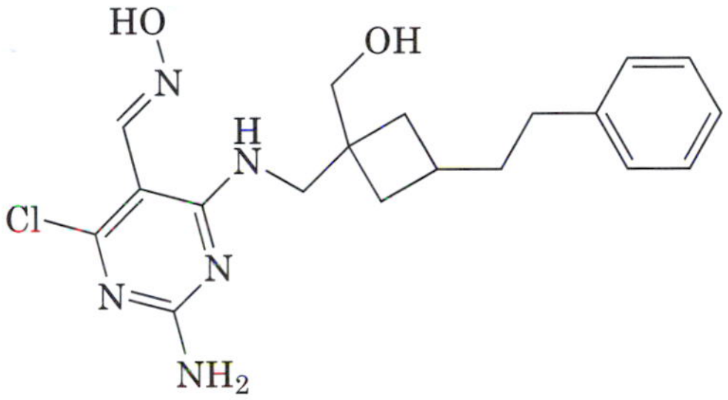

(54)

recently described (152). The compounds in which cyclobutyl group, which was further substituted by a phenylalkyl group at the 3'-position, were introduced to the $\beta$-position of the aminoalkyl side-chain were shown to be highly active in inhibiting both types of A and B influenza virus, with average $EC_{50}$ in the range of 0.1–0.01 $\mu M$. Although their toxicity for the stationary cells was not notable, they were extremely cytotoxic for the proliferating MDCK cells, suggesting that these compounds

might only be used for the topical treatment of influenza virus infection.

**2.1.5.3 Natural Products.** Certain biflavonoids isolated from *Rhus succedanea* and *Garcinia multiflora* exhibited strong inhibitory effects against influenza A and influenza B viruses through cell-based screens. These include robustaflavone (**55**) amentoflavone (**56**), and agathisflavone (**57**) (153). Robust-

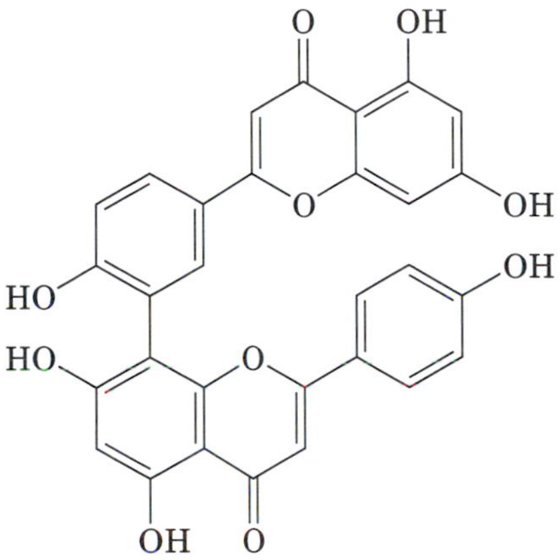

(56) Amentoflavone

aflavone was specifically active against influenza B/Panama/45/90, with an $EC_{50}$ of 0.23 $\mu g/mL$ and an SI of ~435. From the medicinal plant *Rhinacanthus nasutus*, both rhinacanthin-E (**58**) and rhinacanthin-F (**59**) were found to have *in vitro* activity against influenza type A virus, with $EC_{50}$ values of 7.4 and 3.1 $\mu g/mL$, respectively, in a CPE assay (154). Hirsutine (**60**), a Corynanthe-type monoterpenoid indole alkaloid found in the original plant of the Chinese "Kampo" medicine, was

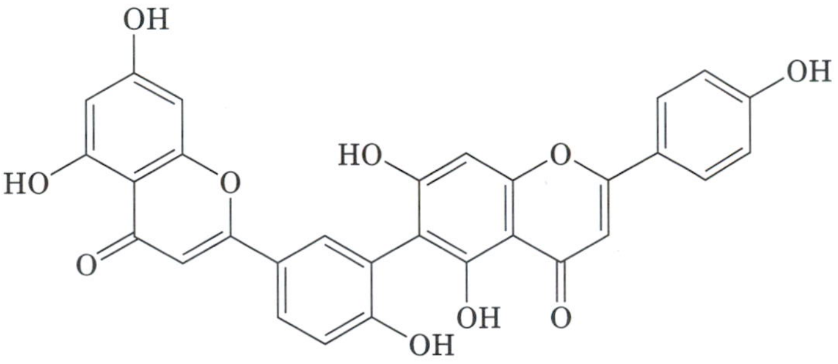

(55) Robustaflavone

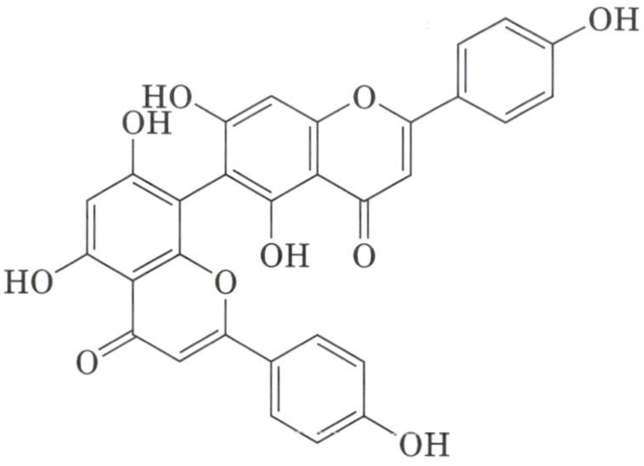

(57) Agathisflavone

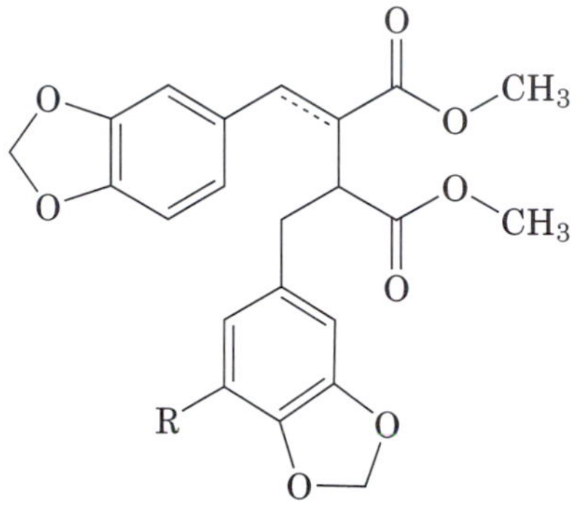

(58) Rinacanthin-E    R = $OCH_3$, Δ 7E
(59) Rinacanthin-F    R = $OCH_3$

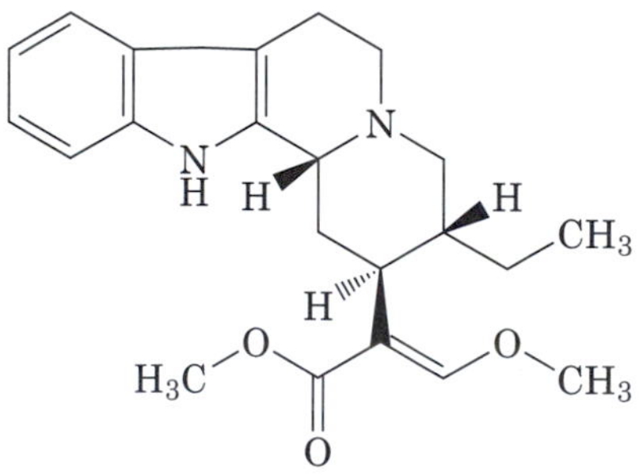

(60) Hirsutine

selectively active against strains of H3N2 viruses ($EC_{50} \sim 0.5$ μg/mL) (155). In a separate report, amentoflavone (56) from *Selaginella sinensis* displayed potent *in vitro* activity against respiratory syncytial virus (RSV) (156).

The following compounds are from microbial origins. Fattiviracin A1 (61) isolated from

the culture filtrate of *Streptomyces microflavus*, showed $EC_{50}$ of 2.05 μg/mL against an H1N1 subtype of influenza A virus measured by a plaque reduction assay (157). FR191512 (62), originated from a fungus, exhibited potent anti-influenza virus [A/PR/8/34 (H1N1)] activity *in vitro* using plaque inhibition assay (158, 159). Its activity was slightly less potent than that of zanamivir on a molar basis. When administered intranasally to mice infected with influenza A/PR/8/34, FR191512 prolonged the survival of infected mice. The $ED_{50}$ of FR191512, ribavirin, and zanamivir at day 7 after viral challenge was 3.2, 16.2, and 1.16 mg/kg, respectively.

**2.1.5.4 Other Compounds.** A potent inhibitor of all types of influenza A, B, and C viruses as well as the neuraminidase-resistant virus, termed T-705 (63), was reported at the 40th ICAAC (160). In an *in vitro* plaque reduction assay, the $EC_{50}$ ranged from 0.02 to 0.6 μg/mL (without showing cytotoxicity up to 500 μg/mL) in MDCK cells. In influenza A/PR/8/34 virus-infected mice, oral dosing of 100 mg/kg/day for 5 days significantly reduced both the mortality rate and the virus titers in the lungs. T-705 exhibited more potent therapeutic efficacy than oseltamivir in low and high dose infection.

Fullerene $C_{60}$, presented as complex with poly(*N*-vinylpyrrolidone) (PVP) in water, inhibited the reproduction of influenza A/Victory/35/72 (H3N2) *in vitro* in concentrations of 500 μg/mL or higher (161). The $C_{60}$-PVP complex caused possibly multiple effects in the viral replication cycle, because its efficacy was the same after its addition at different stages of the infection. Of interest, $C_{60}$ also was shown to inactivate Semliki Forest virus (an alphavirus) and vesicular stomatitis virus when illuminated with visible light (162).

## 2.2 Respiratory Syncytial Virus, Parainfluenza Virus, and Measles Virus

### 2.2.1 Inhibitors of RSV and Paramyxoviruses Fusion Proteins
**2.2.1.1 RFI-641 (WAY-154641).** In a series of reports, investigators at Wyeth-Ayerst Research described the discovery of a family of novel anti-RSV agents including CL-309623 (64), CL-387626 (65), and RFI-641 (66) (163–

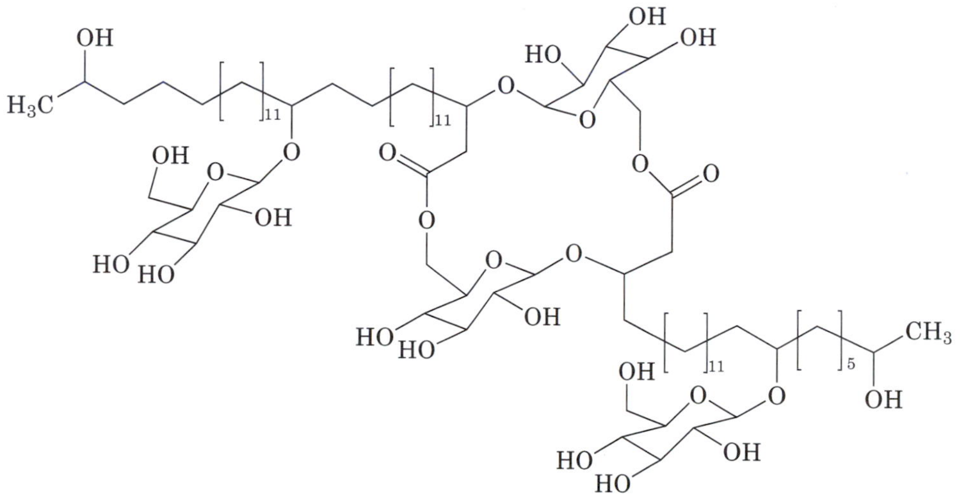

(61) Fattiviracin A1

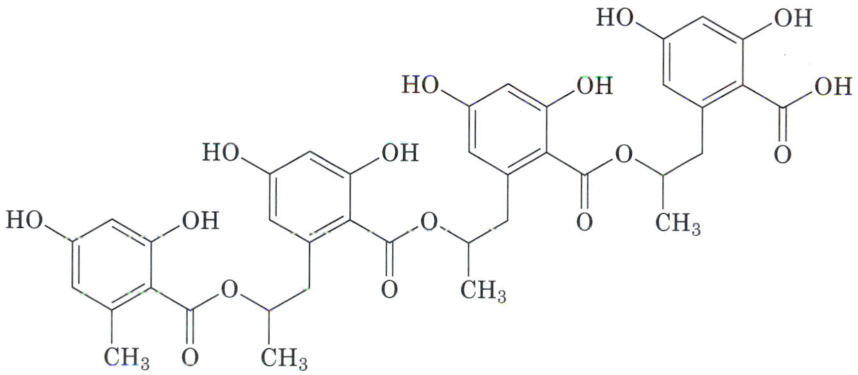

(62) FR191512

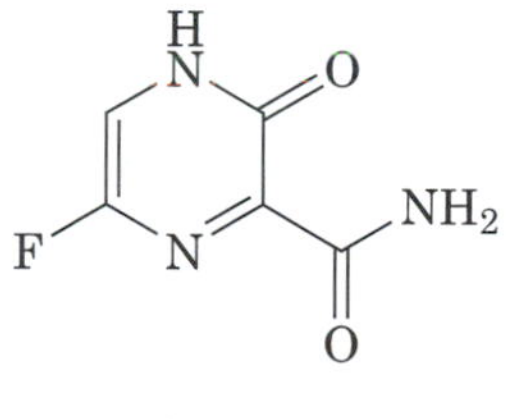

(63) T-705

166). The discovery began by screening of a compound library against a panel of viruses, and it turned out that CL-309623 was the only one to be remarkably effective and specific in inhibiting RSV infection. Interestingly, this compound was synthesized some 40 years ago as a brightener for industrial applications (165). Further refinement resulted in the biphenyl analogs CL-387626 and RFI-641, with improved anti-RSV potency (163, 166–169). Both of the biphenyl analogs inhibited laboratory and clinical isolates of RSV subtypes A and B *in vitro* in the range of sub-micromolar concentrations (e.g., 0.008–0.11 $\mu M$ for RFI-641) without notable cytotoxicity, and the antiviral activity was independent of the cell line used in the assay (167). By using a fluorescence-dequenching assay against a wild type (A2 strain) and a mutant (*cp*-52) virus that contains only the fusion (F) protein on its surface, it was shown that RFI-641 inhibited both

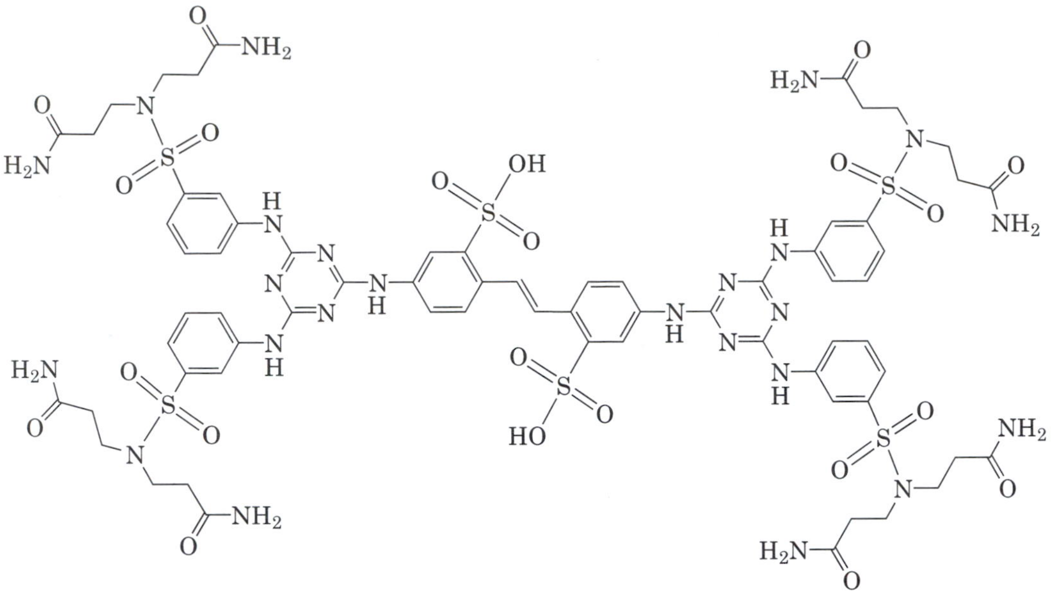

(64) CL-309623

virus-cell attachment and fusion events. Moreover, the fusion event is more sensitive than the attachment to the inhibitory effects of RFI-641 (170). Furthermore, because of heterogeneity of the aggregated F protein, multiple binding events happened when compound bound to the protein—an initial tight binding event followed by several weak binding events (163). Photoaffinity labeling experiments further suggested that the F1 subunit

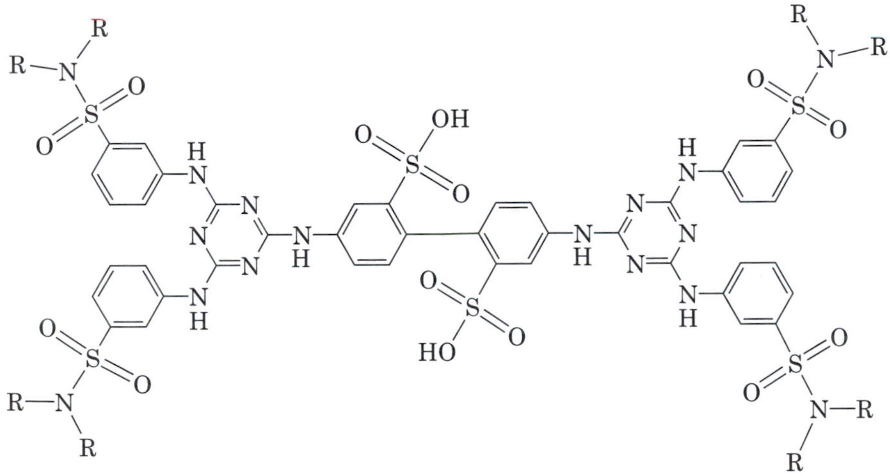

(65) CL-387626   R = CH$_2$ $\overset{O}{\underset{}{}}$ NH$_2$

(66) RFI-641   R = CH$_2$ NH$_2$

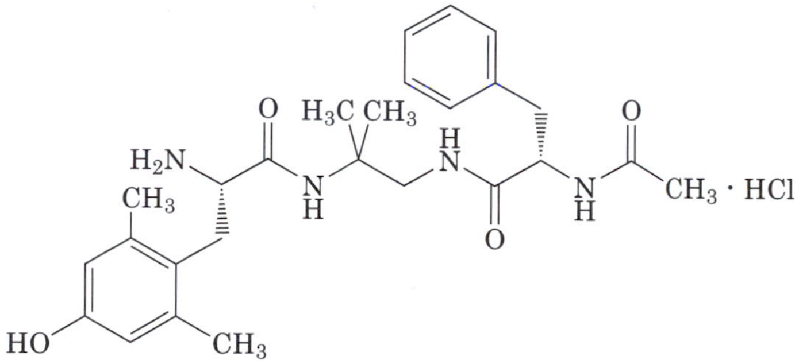

**(67)** VP14637          CAS#235106-62-4

of the fusion protein is the primary target for this series of compounds (170). The convergent chemical synthesis of these compounds involved the last step coupling of disubstituted monochlorotriazines with the biphenyl core by heating in the microwave at 105°C for 1 h (166).

The following structural features are required for strong antiviral potency (163, 165): (*1*) the core should be rigid and must bear two negatively charged groups, preferably in the 2,2′-positions; (*2*) sulfonic groups are better than carboxylic; (*3*) a molecule needs at least three aminobenzenesulfonamido fragments to show activity and the substituents on each of the fragment should be meta to each other for optimal activity; and (*4*) the outmost side-chains should have hydrogen bonding groups and two side-chains are better than one. Given the fact that multiple functional groupings are needed for activity by these rather bulky and symmetrical compounds, their interactions with F protein seem to involve multivalency.

When CL-387626 was given intranasally to RSV-infected cotton rats, it demonstrated significant prophylactic activity (164). Although the protection could last for 5–8 days after a single administration, the compound seemed to be more effective when it was administered closer to the time of virus challenge. The *in vivo* efficacy of RFI-641 was further determined in three animal models of RSV infection (mice, cotton rats, and African green monkeys) (166, 169). Prophylactic intranasal administration of RFI-641 significantly impacted on the establishment of infection and subsequent spread of RSV to the lungs. In addition, RFI-641 also showed therapeutic effect

because it could reduce viral loads in nasal and throat samples collected from infected monkeys after the compound was administered once daily, by the intranasal route, beginning 24 h after infection. RFI-641 has entered phase I clinical trials.

***2.2.1.2 VP-14637.*** VP-14637 (**67**) was a sub-nanomolar fusion inhibitor of RSV reported by ViroPharma Inc. recently (171–173). *In vitro*, this compound specifically inhibited RSV and displayed a remarkable anti-RSV potency against RSV A and B strains and a panel of clinical isolates as seen by using viral cytopathic effect (syncytial formation), antigen detection, and virus yield (virus production) assays. The plausible mechanism of action involves functions associated with the viral F protein, a highly conserved RSV protein that is essential for virus reproduction. In October 2000, ViroPharma initiated clinical trials with VP-14637. A special inhalation drug delivery device is used to administer the drug to the lungs (ViroPharma Press Release).

***2.2.1.3 R170591.*** Jansen Research Foundation recently identified R170591 (**68**), a

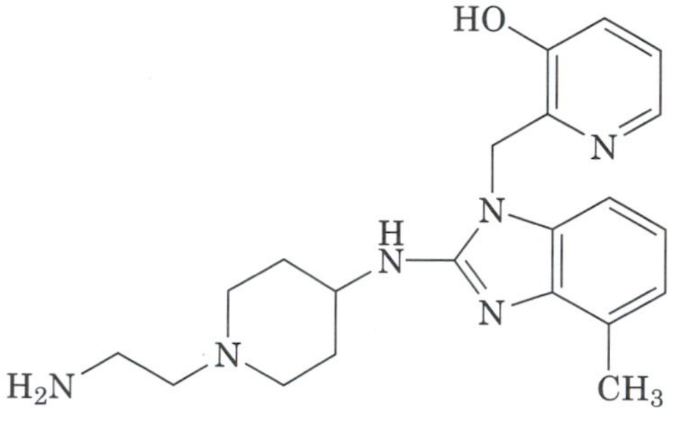

**(68)** R170591

benzimidazole derivative, through a cell-based assay as being capable of inhibiting fusion of RSV-infected HeLa cells (174, 175). Its *in vitro* efficacy (EC$_{50}$, 0.15 nM) was about $10^5$-fold more potent than that of ribavirin. It was active against human (subgroup A and B) and bovine RSV. Time-of-addition and mutagenesis studies suggested that it might interact with the F protein, leading to the inhibition of both virus–cell fusion early in the infection cycle and cell–cell fusion at the end of the replication cycle. Pretreatment of cotton rats by local (inhalation) or by systemic (intraperitoneal) application resulted in >90% reduction of pulmonary virus titers.

**2.2.1.4 NMSO3.** NMSO3 (**69**) is a nontoxic sulfated sialyl lipid that had been used to smooth the surface of instant noodles (176). This compound was found active in inhibiting RSV infection *in vitro* and *in vivo* (177). The compound did not show significant toxicity at the highest testing concentration in four different cell lines. Using ELISA, it was determined that the average EC$_{50}$ of NMSO3 against several selected laboratory strains and clinical isolates was 0.23 μM and that of ribavirin was 12.3 μM (in Hep-2 cells). However, the best efficacy was observed when it was added 0–1.5 h after the viral inoculation, during the time of virus adsorption and penetration. Therefore, NMSO3 might target RSV-F glycoprotein. This preliminary conclusion was further supported by the results of a temperature shift study with NMSO3 and anti-RSV (F) monoclonal antibodies. Nevertheless, it

was still active to some extent against virus replication, suggesting that NMSO3 might also inhibit some other later processes of RSV infection. Intraperitoneal administration of NMSO3 to RSV-infected cotton rats from 1 day before or 1 h after to 3 days, once a day every day, showed a significant reduction of RSV titer in the lungs. Its therapeutic effect against RSV growth in the lungs of infected animals was greater than that of ribavirin.

**2.2.1.5 RD3-0028.** Active anti-RSV benzodithiin derivatives were discovered through a random screening assay (178). Of these compounds, RD3–0028 (**70**) was the most active

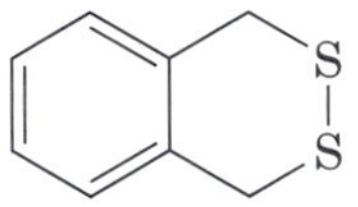

(**70**) RD3-0028

one, which showed activity against RSV subgroups A and B and clinical isolates, with EC$_{50}$ values in the range of 4.5–11 μM in HeLa cells. This compound inhibited RSV-induced syncytium formation even added to the culture up to 16 h post-infection, suggesting that it might target a late stage of viral replication (179). By further analysis of drug-resistant mutants, it was concluded that RD3–0028 treatment resulted in the production of defective viral particles by interfering with the intracellular synthesis or processing of the RSV F protein, or a step immediately thereafter (179, 180). When delivered by aerosol to virus-infected, cyclo-

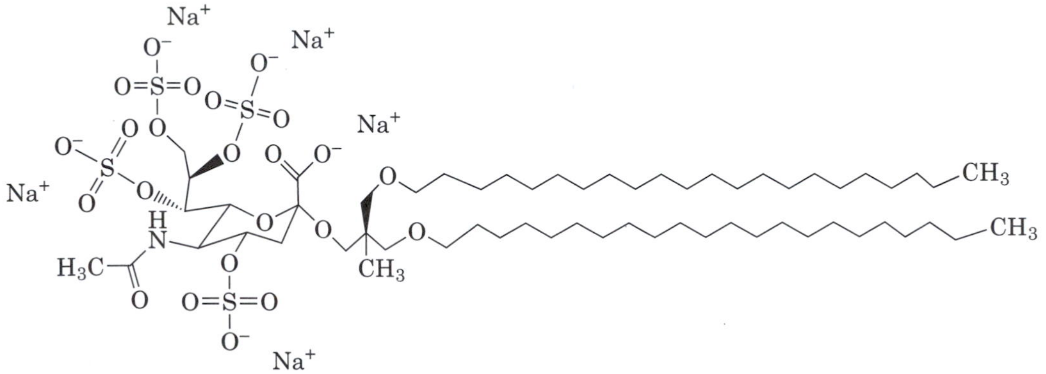

(**69**) NMSO3

phosphamide-treated immunosuppressed mice (181), RD3–0028 significantly reduced the pulmonary titers and protected the lungs of against tissue damage. RD3–0028 was not toxic for the mice at the therapeutic doses, and the minimal effective dose seemed to be much less than that of ribavirin (178).

***2.2.1.6 Benzanthrone Derivatives.*** Trimeris, Inc. has filed a patent (WO9839287) describing a series of RSV fusion-targeting inhibitors based on a benzanthrone skeleton (180, 182). Compound (**71**) displayed an $EC_{50}$ of 0.04 $\mu g/mL$ and a $CC_{50}$ of 4.15 $\mu g/mL$.

**(71)** Benzathrone

***2.2.1.7 Immunoglobulins.*** During the past few years, a wealth of data generated from prophylactic passive immunization on both experimental animals and humans have shown that parenteral administration of RSV-neutralizing antibodies could reduce the severity of RSV disease [see reviews by Mills (183) and Prince (184)]. Currently, there are two products manufactured by MedImmune—RespiGam (approved by the FDA in 1996) and Synagis (approved in 1998)—for RSV infections. RespiGam, a polyclonal respiratory syncytial virus immunoglobulin (RSVIG) derived from human plasma, is administered through intravenous infusions for the prevention of serious lower respiratory tract infection caused by RSV in high-risk human infants. Synagis (also known as palivizumab and MEDI-493), a humanized mouse IgG monoclonal antibody (mAb) directed against the RSV fusion protein, can be administered by intramuscular injections for use in a broader patient population than RespiGam, being suitable for prophylactic administration to pediatric patients at risk of RSV disease. In a recent study with cotton rats undergoing

prolonged immunosuppression with cyclophosphamide, both prophylaxis and therapy with RSVIG significantly reduced pulmonary RSV replication. In addition, the use of multiple therapeutic doses of RSVIG was able to prevent rebound viral replication, though virus was not completely eliminated (185).

A number of other mAb preparations directed against the F protein of RSV have also been actively pursued, including HNK20 (a mouse IgA mAb developed by OraVax) and RSHZ19 (a humanized mouse IgG mAb, also known as SB 209763, licensed from Scotgen to SmithKline Beecham for development). In preclinical studies in rodents and rhesus monkeys (186), intranasal delivery of HNK20 showed significant protective effect. However, in an international control trial conducted with 600 high-risk infants, HNK20 did not result in a significant decrease in the incidence of hospitalization associated with RSV lower respiratory tract infection (187). Similar to the case with HNK20, RSHZ19 showed prophylactic efficacy in mice and cotton rats (188) but failed to protect infants at risk for severe RSV disease (189).

Palivizumab (MEDI-493) and RSHZ19 recognize distinct neutralizing epitopes on the F protein of RSV. In series head-to-head experiments (190), it was clear that the F protein affinity of palivizumab was severalfold tighter than that of RSHZ19. Using ELISA to measure viral replication, palivizumab was approximately 5-fold more potent than RSHZ19 and 20-fold more potent than RSVIG to neutralize RSV [drugs were added either before (microneutralization assay) or after (fusion inhibition assay) attachment to Vero cells]. In a cotton rat prophylaxis model, palivuzumab was two- to fourfold more potent than RSHZ19 in inhibiting RSV replication in the lungs. Therefore, the difference in clinical efficacy seemed to relate to the greater potency of palivizumab compared with RSHZ19.

***2.2.1.8 Peptides.*** The F protein of paramyxoviruses and the transmembrane (TM) protein of retroviruses facilitate the fusion of the viral envelope or infected cell membranes with uninfected cell membranes. By recognizing that the fusion domains at the amino termini of RSV F1 subunit and HIV-1 TM show a high degree of sequence homology, Lambert et al. at Trimeris, Inc. have used a computer-searching strategy, based on the secondary structure characteristics to the DP-107 and DP-178 peptides of HIV-1 gp41, to identify conserved heptad repeat domains analogous to the DP-107 and DP-178 regions of HIV-1 gp-41 within the glycoproteins of paramyxoviruses, leading to the discovery of non-cytotoxic peptides T-118, T-205, and T-257 as nanomolar inhibitors against RSV, parainfluenza type 3 (PIV-3), and measles virus (MV), respectively, *in vitro* (191). Although these peptides were from domains that near the membrane anchor, the antiviral activity of these peptides was specific for the virus of origin. The sequence of T-118: Ac-$_{488}$FDASISQVNEKIN-QSLAFIRKSDELLHNVBAGKST$_{522}$-NH$_2$.

In separate studies, Yao and Compans showed that synthetic peptides containing the heptad repeat regions derived from the F proteins of human PIV-2 and PIV-3 could inhibit virus-induced cell fusion, virus entry, and spread of virus infection (192). Moreover, the inhibitory effects of these peptides were found to be virus-type specific. Similarly, Wild and Buckland reported that a peptide corresponding to the leucine zipper region (amino acids 455–490) of the MV F protein could block both MV entry and cell-to-cell fusion (193).

Graham et al. demonstrated that RhoA, a small cellular GTPase of the Ras superfamily, interacts with RSV F protein and facilitates virus-induced syncytium formation. This group further reported that RhoA$_{77-95}$; a peptide comprising amino acids 77–95 of RhoA, showed activity in inhibiting syncytium formation induced by RSV and PIV-3, both *in vitro* by inhibition of cell-to-cell fusion and *in vivo* by reduction of pulmonary virus titers in RSV-infected mice, when the peptide was administered intranasally to the animals immediately before or 2 h after RSV challenge (194). The authors suggested that when illness is mediated by the T-cell response and is not directly related to virus-induced cytopathology, antiviral therapy must be given early or combined with immunomodulators. The sequence of RhoA$_{77-95}$: TDVILMCFSIDSPDSLENI.

### 2.2.2 Oligonucleotides as Inhibitors of RSV.
The RSV genome encodes 10 viral proteins, which are, as shown in 3′ to 5′ order, NS1, NS2, N, P, M, NS3, G, F, M2, and L. Recently, Hybridon, Inc. reported the antiviral effects, by means of inhibiting RNA replication and transcription of NS2 mRNA, of an antisense phosphorothiolate oligonucleotide, v590 (5′-AAAAATGGGGCAAATAAATC-3′), which is complementary to two same 20-base sequences at the start of the *NS2* gene and the *P* gene, respectively (195). This oligonucleotide was 4- to 20-fold more potent than ribavirin in inhibiting RSV antigen and infectious virus yield. Treatment of cells with the compound specifically decreased the region of RSV RNA containing the v590 target sequence, suggesting that sequence-specific cleavage of RNA might have occurred (195). This cleavage might involve cellular ribonuclease H (RNase H), because RNase H has been shown to cleave RNA strand in oligonucleotide/RNA duplexes (195). Yet the all-phosphodiester-backbone and the G quartet–containing structural motifs might give rise nonspecific effects (196).

Targeting RNase L [(2–5A)-dependent endoribonuclease] to RSV RNA with 2′,5′-oligoadenylate (2–5A)-antisense chimeric oligonucleotides has been shown as an interesting antiviral approach (197). (For a comprehensive review of the 2–5A system, see Ref. 198.) The premise of this approach is that the 2–5A moiety of the chimeras attracts (or recruits) and activates ubiquitous intracellular latent RNase L, which causes the degradation of the RNA target bound to the antisense domain (197, 199).

Silverman and Torrence synthesized a variety of (2–5A)-antisense chimeras through covalent linkage of 3′,5′-antisense oligodeoxyribonucleotides (ODNs) and 2′,5′-oligoadenylate molecules through a linker (butanediols). Both termini were also chemically modified to protect the chimeras from enzymatic (e.g., 3′-exonuclease and phosphatase) degradation. Using a computer-assisted analysis of the secondary structure of the RSV RNA sequence, they identified several regions in the M2 and L mRNAs as the targets for the (2–5A)-antisense chimeras (200, 201). The chimeras, sp(5′A2′p)$_3$A-Bu$_2$-5′-ATGGTTATTT-GGGTTGTT-3′-3′T5′ [called spA$_4$-antiRSV3′-3′T/(8281–8299) or NIH8281, where sp = 5′-monothiophosphoryated], which targeted the sequence 8281–8299 in the ORF2 region of the

M2 mRNA, had the greatest antiviral potency in 9HTE cells. An approximately 75% reduction in viral yields (after 24 h post-infection) was observed when the compound was added twice to RSV-infected cells at 3.3 $\mu M$ per dose (201). This oligonucleotide was remarkably selective for the RSV M2 mRNA. Moreover, the selective and specific degradation of the *M2* gene was dependent on the presence of the tetrameric 2–5A moiety, thereby confirming the involvement of RNase L (201). Further studies with NIH8281 showed that this oligonucleotide inhibited several representative strains of both A and B RSV serotypes as well as bovine RSV with low cytotoxicity in a variety of antiviral assays (202). The compound was inhibitory only when added within 2 h of virus infection, and the activity was multiplicity of infection (MOI)-dependent. The compound did not inhibit measles or parainfluenza viruses (202).

In separate studies, Torrence et al. chose the conserved consensus sequences that occur in gene-start, intergenic, and gene–end signals within the RSV genome for inhibitor design. Of particular interest was the antisense 17-mer, 5′-AAAAATGGGGCAAATAA-3′, which could potentially target 10 sense targets, but with different hybridization efficiency, simultaneously (196). (This 17-mer was related in sequence to a 20-mer antisense phosphorothioate oligonucleotide v590 mentioned above.) To minimize the potential non-specific effects associated with the all-phosphorothioate, Torrence et al. used a "gapmer" approach, in which only three internucleotide linkages at the 5′ and 3′ termini of the antisense 17-mer were thiophosphorylated. The resultant (2–5A)-antisense gap-mer, NIH351 [(2–5A)-Bu$_2$-AsAsAsAATGGGGCAAAsTsAsA] inhibited RSV strain A2 virus yields (EC$_{50}$ and EC$_{90}$ were 0.3 and 1 $\mu M$, respectively, when added once, immediately before virus adsorption) 100-fold more potent than did ribavirin in Hep-2 and in MA-104 cells. This chimeric oligonucleotide was not toxic to cells and showed the most potent *in vitro* effects when given once a day for 3 days or twice a day for 2 days. Beside RSV A2 strain, NIH351 was a potent inhibitor of a number of representative members of both A and B strains of RSV whether assayed in human, monkey, or bovine (against bovine RSV) cells. When NIH351 was compared with NIH320, which has the same antisense design but without the 2–5A moiety, it was clear that NIH351 owed its 30-fold enhancement in antiviral activity to the involvement of the 2–5A system's RNase L (196).

### 2.2.3 Other Inhibitors of RSV

*2.2.3.1 Natural Products.* Phenylpropanoid glycosides (**72–75**) from the medicinal plant *Markhamia lutea Verbascoside* (203) and iri-

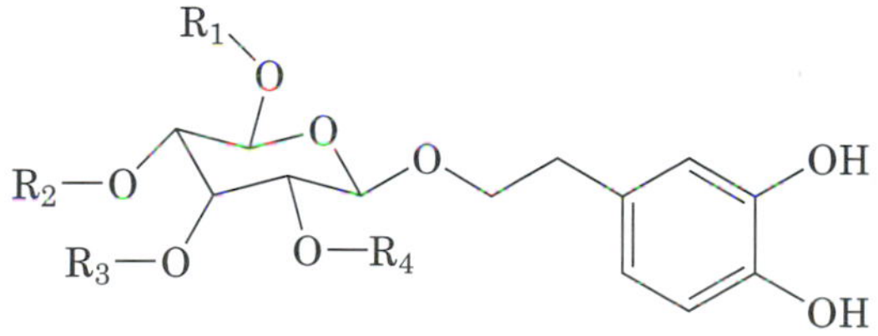

|      | R$_1$    | R$_2$ | R$_3$ | R$_4$ |
|------|----------|-------|-------|-------|
| (**72**) | H        | Caf   | Rha   | H     |
| (**73**) | Caf      | H     | Rha   | H     |
| (**74**) | COCH$_3$ | Caf   | Rha   | Api   |
| (**75**) | Caf      | H     | Rha   | Api   |

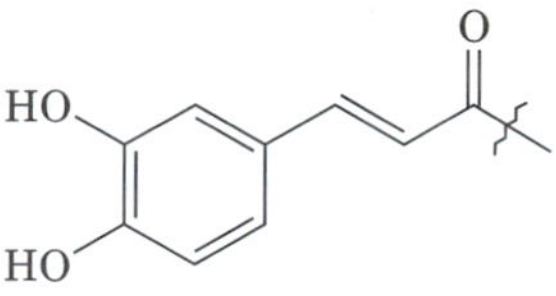

Caffeoyl 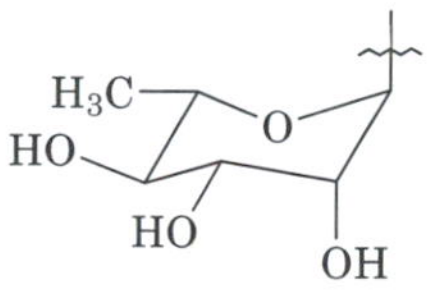                    Rhamnosyl 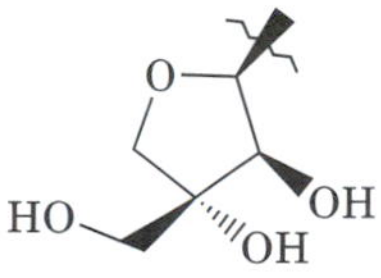                    Apiosyl

doid glycosides (**76** and **77**) from the medicinal plant *Barleria prionitis* (204) have been reported as antivirally active against RSV *in*

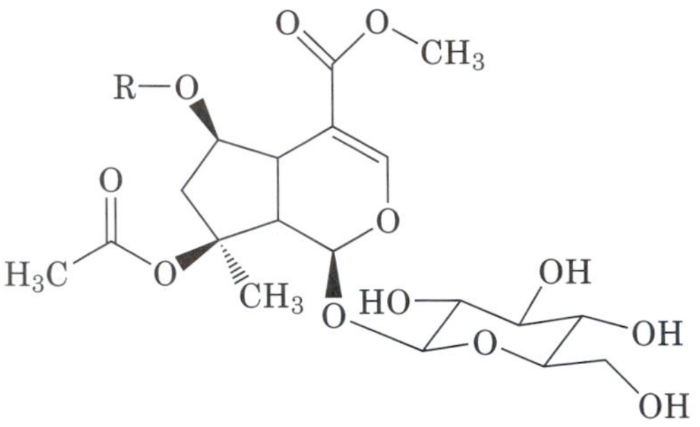

(76) R =

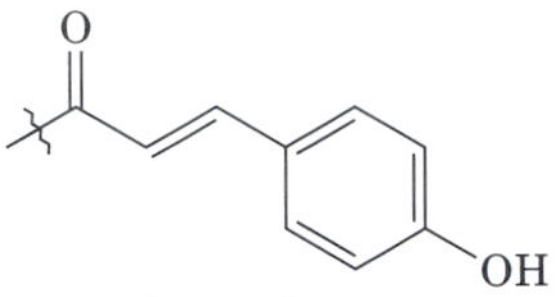

(77) R =

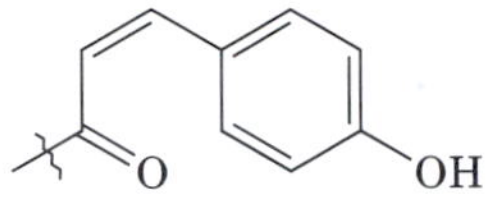

(78)

*vitro* in sub-micromolar concentrations, which were largely separated from their cytotoxic concentrations. The phenylpropanoid glycosides were active when added 3 h after virus infection of the cells.

**2.2.3.2 Purines and Pyrimidines.** Derivatives of purine and pyrimidine (e.g., EICAR, pyrazofurin, and cyclopentenyl cytosine) have been known for their potent anti-RSV activity both *in vitro* and *in vivo*. For a review of these and other anti-RSV compounds pre-dating 1996, see a recent publication by De Clercq (205).

There was a recent report showing that a synthetic guanine derivative (**78**) and its 2,6-diaminopurine analog both containing a common N9 cyclobutyl substituent, exhibited noteworthy *in vitro* activity against RSV with reasonable selectivity (206).

**2.3 Picornaviruses**

**2.3.1 Inhibitors of Picornaviral Attachment and Uncoating**

**2.3.1.1 Long-Chain Compounds.** Many picornaviruses share a common icosahedral capsid architecture constructed from 60 copies of

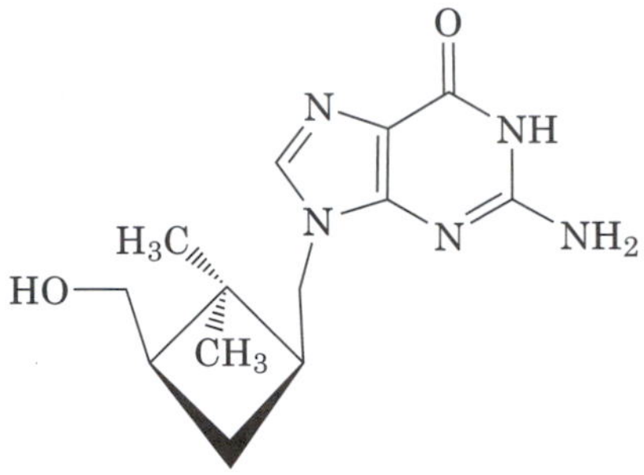

four proteins (VP1, VP2, VP3, and VP4) revealed by crystallographic studies of several human enteroviruses (namely, coxsackievirus B3, echovirus 1, poliovirus types 1, 2, and 3) and rhinovirus types 1A, 2 (207), 3, 14, and 16 (see references cited in Ref. 208). VP1, VP2, and VP3 compose the viral surface, whereas VP4 lays interior at the capsid/RNA interface (209).

In all of these structures, the virus surface reveal broad depressions, or canyons, formed by the junctions of VP1 and VP3. The canyon has been shown to be the site of receptor attachment for major group rhinoviruses (e.g., ICAM-1 for HRV-14) and for poliovirus. At the base, or floor, of the canyon there is a pore, which opens into the hydrophobic core, or pocket, within the VP1 protein (210). In most of the enteroviruses (comprise >60 serotypes) and rhinoviruses (>115 serotypes), the pocket is either empty (e.g., in HRV-14 and HRV-3) or occupied by a fatty acid-like pocket factor, of which the chemical identify remains unknown. A variety of diverse long-chain hydrophobic capsid-binding antiviral compounds (e.g., pleconaril) have been shown to displace the pocket factor and bind in the hydrophobic pocket. Drug binding has shown to lead to the inhibition of viral infectivity by stabilizing the viral particles and/or by preventing receptor attachment. The binding of antiviral drug not only causes local conformational changes in the drug-binding pocket but also stabilized the entire viral capsid against enzymatic degradation (209).

The drug binding-induced stabilization prevents the virion undergoes an irreversible conformational change (required for uncoating for RNA release) from the native 160S (or N) particle to the 135S (or A) particle, which facilitate cell entry. This stabilization effect has recently been shown to arise from higher entropy, and not through rigidification of the capsid as has been previously suggested (211–214). Binding of the antiviral compound increased the entropy (greater flexibility) of the HRV capsid and therefore reduced the free energy for uncoating. Further studies in the presence of soluble poliovirus receptor showed that capsid-binding compounds inhibited receptor-mediated N-to-A conversion through a combination of enthalpic and entropic effects (215).

Nevertheless, the variation in size and amino acid composition of the pockets, particularly, variations in the more hydrophobic end (the toe end) of the pocket among serotypes, might affect the pocket fit of these capsid-binding compounds. This factor might contribute, in part, to the different sensitivity of many of these serotypes to the compounds. In general, HRV-14, HRV-3, and poliovirus are more sensitive to longer compounds. Shorter compounds tend to be more effective against HRV-16, HRV-1A, and HRV-2 (216).

Several capsid-binding compounds have been investigated in clinical trials; only Viro-Pharma's pleconaril has advanced to phase III clinical studies (for a review of pleconaril and other drugs, see Refs. 217–223).

*2.3.1.1.1 Pleconaril.* Pleconaril (**79**) (registered as Picovir, also known as VP 63843) be-

(**79**) VP 63843    Pleconaril

longs to WIN series of compounds. It possess much improved potency, chemical and metabolic stability, pharmacokinetics (224, 225), and safety than its predecessors in the series. In a monkey liver microsomal assay, pleconaril was found to produce two minor met-

abolic products. In comparison, WIN 61893 (**80**) formed at least eight and WIN 54954 (**81**) formed 18 metabolic products. It was further

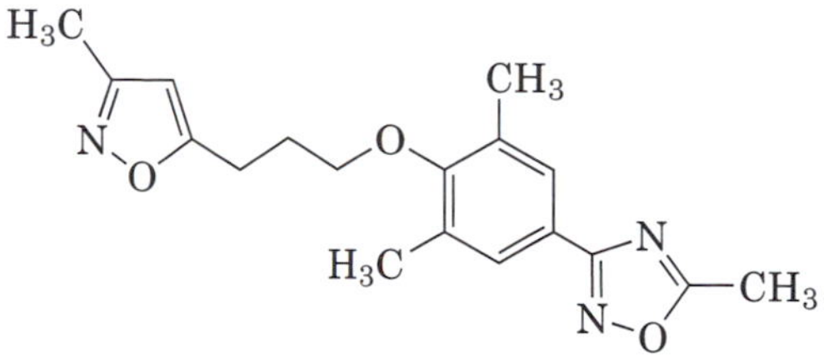

(**80**) WIN 61893

(**81**) WIN 54954

determined that the methyl groups on either end of WIN 54954 and WIN 61893 were the major sites of metabolism. Replacement of the methyl group on the oxadiazole ring of WIN 61893 with trifluoromethyl creates the metabolically stable and orally bioavailable (~70% in humans) pleconaril (221).

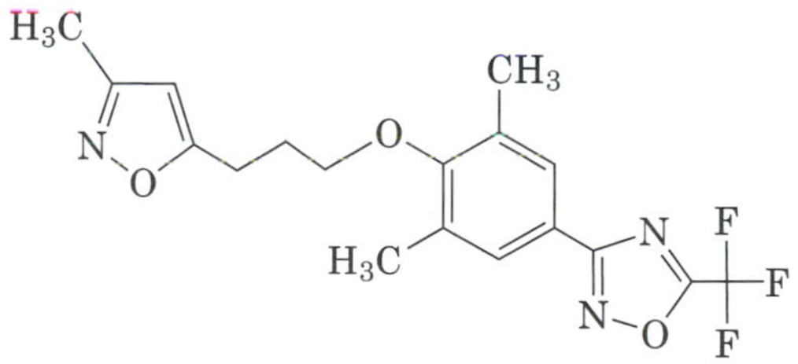

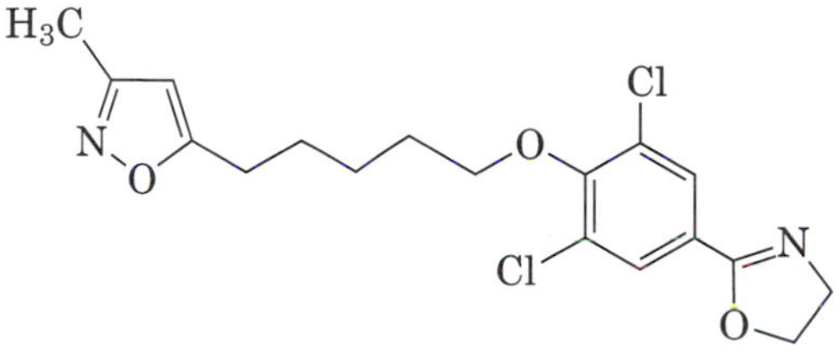

In preclinical studies, pleconaril has demonstrated a broad spectrum of activity against a wide range of rhinoviruses and non-polio enteroviruses both *in vitro* and *in vivo* (226). ViroPharma reported that pleconaril effectively inhibited the laboratory replication of 96% of the rhinovirus and enterovirus isolates from 322 human patients (http://www.viropharma. com/pipeline/pleconaril.htm). Significantly, echovirus 11, the most commonly isolated enteroviruses in the United States between 1970 and 1983, was the most sensitive serotype to pleconaril (226). This compound has also demonstrated excellent penetration into the central nerve system, liver, and nasal epithelium.

To date, pleconaril has shown clinical benefits for both adult and pediatric patients with enteroviral meningitis, viral respiratory infections, and potential life-threatening enteroviral infections in at-risk patient populations (e.g., patients with antibody-deficiency and

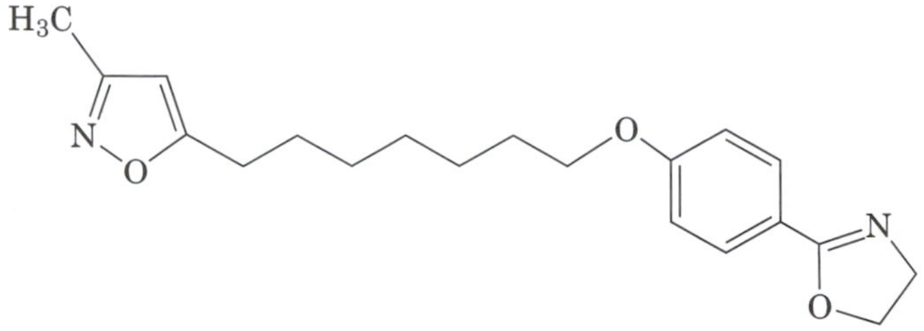

(82) WIN 51711     Disoxaril

bone marrow transplant recipients) (227, 228). ViroPharma issued two press releases in April 2000 and March 2001, respectively, reporting encouraging results from several phase III clinical studies of pleconaril in two disease indications: viral respiratory infection (common cold) in adults and viral meningitis in adults and children.

*2.3.1.1.2 Disoxaril Analogs.* In an attempt to search improved antipicornaviral compounds, based on the structure of disoxaril (WIN 51711) (82), Artico et al. synthesized a series of disoxaril analogs containing a terminal thiophene ring and a carbonyl group bound to the position 2 of thiophene (229). Although most of the analogs were equivalent to or more potent than disoxaril against HRV-14 and HRV-2, they were not broadly inhibitory to various HRVs as does by disoxaril. However, there were two broad-spectrum inhibitors. Compounds (83) and (84) were more potent than disoxaril when assayed against

HRV-2, and as potent as disoxaril against HRV-14 and the other 14 selected serotypes. When thiophene was replaced with benzene, the anti-rhinoviral activity deteriorated (230).

*2.3.1.1.3 Pirodavir, SCH 48973, and SDZ 880–061.* Pirodavir (R 77975) (85), developed by Janssen, is another capsid-binding inhibitor that has been in clinic evaluation. It was efficacious in experimentally induced HRV infection when the drug was administered intranasally before or after infection, but before onset of symptoms (219). However, no clinical benefit was seen in treating naturally occurring HRV colds by intranasal administration (231).

SCH 48973 (86) was identified at Schering-Plough through molecular modeling and in an assay designed to detect compounds that stabilize poliovirus to heat inactivation (232). When tested in cell culture, this compound demonstrated significant activity against a wide range of enteroviruses; it inhibited 80%

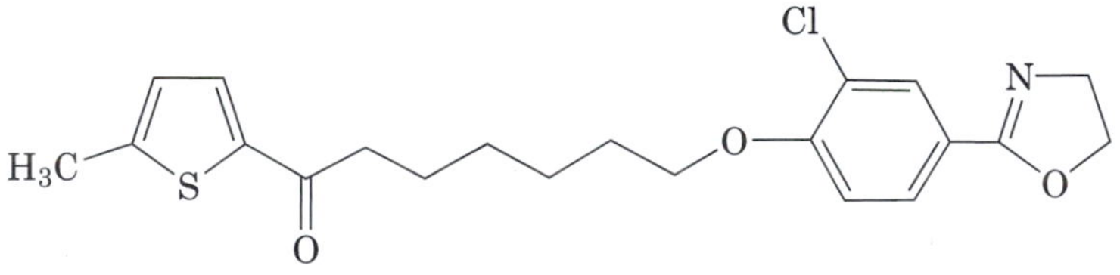

(83)

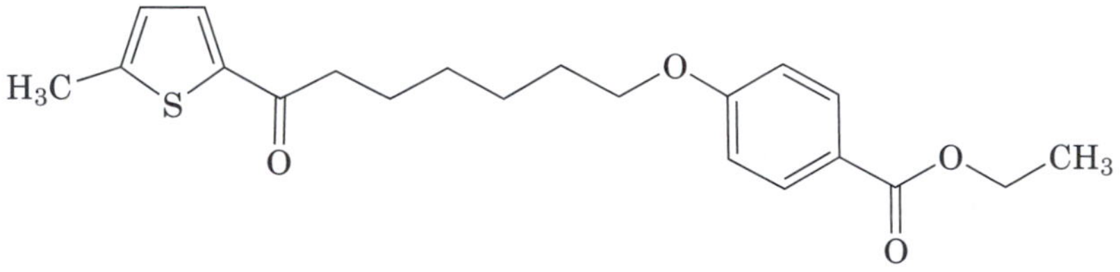

(84)

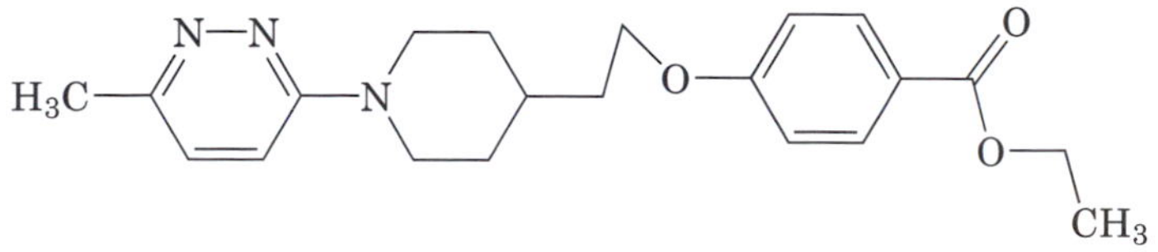

**(85)** Pirodavir (R 77975)

of 154 recent clinical isolates representing 15 common enterovirus serotypes with an average $EC_{50}$ of ~1 $\mu$g/mL. However, it was a poor inhibitor of HRV-14 (233).

Polioviruses exist in only three serotypes that are pathogenic for humans. Only poliovirus type 2 Lansing (PV2L) is neurovirulent in mice when injected intracerebrally (233). In poliovirus type 2–infected mice, therapeutic treatment with oral SCH 48973 (at dosages as low as 3 mg/kg/day) significantly reduced the viral titers in the brains and increased the survival of infected mice (232).

The crystal structure of PV2L complexed with SCH 48973 revealed that the compound was bound in a pocket within the $\beta$-barrel of VP1, in approximately the same position where the natural pocket factor binds to the virus. The structure also showed a surface depression located at the fivefold axis of PV2L capsid that is not present in the other two serotypes of poliovirus. In addition, unlike the other structures of enteroviruses, the entire PV2L VP4 is visible in the electron density,

and bases of the genomic RNA are observed stacking with conserved VP4 aromatic residues (233).

Based on the structural features of earlier SDZ series of compounds, investigators of Sandoz synthesized SDZ 880-061 (**87**) which possessed a relatively broad antiviral spectrum (234). In contrast to SCH 48973, SDZ 880-061 inhibited HRV-14 in the nanomolar concentrations. It inhibited 85% of 89 HRV serotypes tested at a concentration of $\leq$3 $\mu$g/mL. HRV-42 and HRV-68 were among those being most sensitive to SDZ 880-061, but were refractory to inhibition by pirodavir.

SDZ 880–061 was shown to bind to the same pocket, lying in the outer portion of the cavity (occupying 16 Å of the 21 Å length of the pocket) and cause similar, but less extensive, alternations of the HRV-14 VP1 backbone conformation compared with other capsid-binding antiviral agents. It might be that because it does not completely fill the hydrophobic pocket, SDZ 880-061 primarily interferes

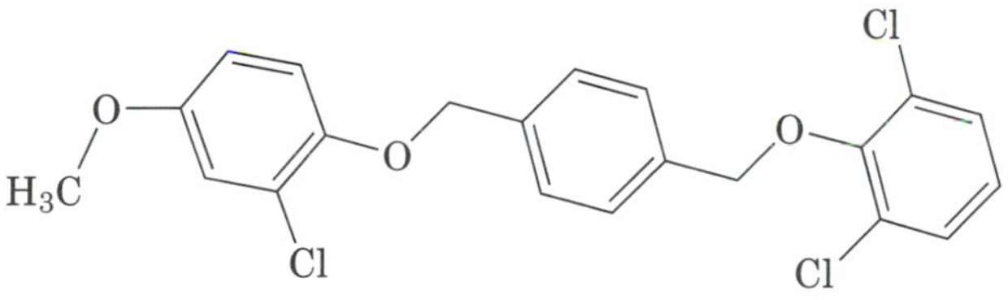

**(86)** SCH 48973

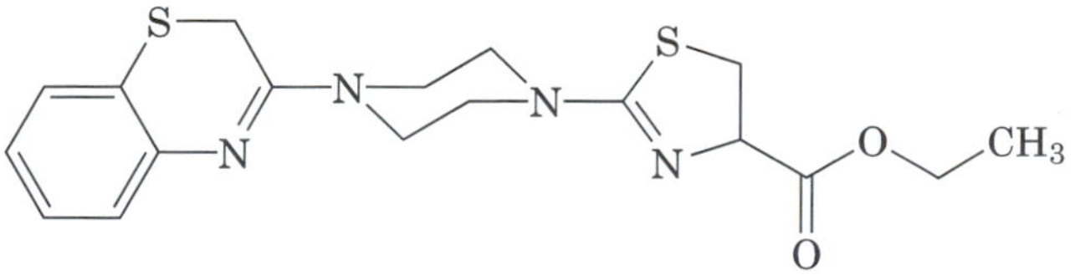

**(87)** SDZ 880-061

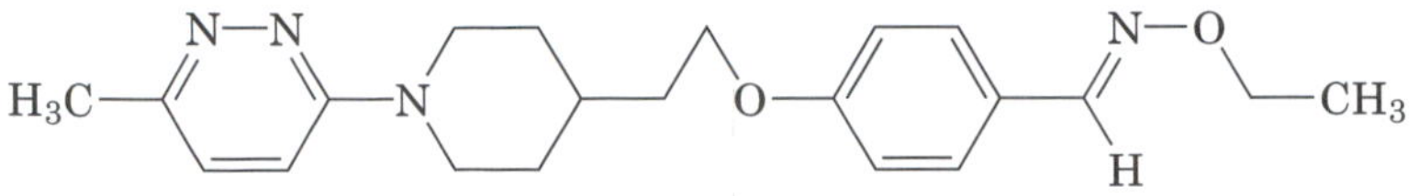

(88) BTA-188

with HRV-14 cellular attachment, and has only a marginal effect on uncoating (234).

*2.3.1.1.4 BTA-188.* Biota Holdings has recently developed BTA-188 (**88**) as a new potential candidate for the treatment of HRV disease. This compound, a capsid-binding inhibitor, was discovered through molecular modeling and SAR analyses (235). When tested in cell culture, BTA-188 inhibited 87 of 100 numbered HRV serotypes (median $EC_{50}$, 0.01 $\mu$g/mL, ranging from 0.0003 to >0.1 $\mu$g/mL) and all 40 clinical isolates (median $EC_{50}$, 0.004 $\mu$g/mL, ranging from <0.001 to 0.05 $\mu$g/mL) (236). In comparison, BTA-188 inhibited HRV-14 with an $EC_{50}$ of 1.0 ng/mL, whereas the $EC_{50}$ values were 30 and 3.2 ng/mL for pleconaril and pirodavir, respectively (235). BTA-188 and pirodavir are closely related to each other structurally. BTA-188 showed good oral bioavailability in rodents and dogs, and could be detect in the nasal epithelium of dogs with levels several times above the *in vitro* $EC_{90}$ for rhinoviruses. Its serum half-life was about 3 h (237).

*2.3.1.1.5 Other Long-Chain Compounds.* Hogle et al. have used a computational ligand design method called multiple copy simultaneous search (MCSS) to produce functionality maps of the drug binding sites of P3/Sabin poliovirus and rhinovirus-14 for the *de novo* design of new classes of picornavirus capsid-binding compounds (238). By simultaneously subjecting thousands of randomly placed copies of small molecular fragments, MCSS determined where specific functional (chemical) group have local potential energy minima in the binding site. Selected minima were clustered and connected with linkers [e.g., —(CH$_2$)$_n$—] to form candidate ligands. Their preliminary studies with fragment maps centered on the VP1 pocket suggested a template for a class of compounds that contain fused aromatic rings (e.g., benzimidazole) (238).

However, MCSS has its limitations because of the approximate methods being used. The same authors have recently reported a struc-

turally biased combinatorial approach (a combination of structure-based design and combinatorial chemistry) to overcome this difficulty (208). A small set of combinatorial libraries of ligands resembling the template suggested by MCSS (based on both P3/Sabin and P1/Mahoney poliovirus) were synthesized and screened by using a novel assay in which virus is incubated with crude libraries and the components that bind are identified by mass spectrometry. Potential binders were re-synthesized as members of smaller sub-libraries, which, in turn, were re-screened with the mass spectrometry assay and tested for reduction of rate constant for uncoating (N-to-A transition) with an immunoprecipitation assay. Promising leads were individually synthesized and re-tested. This iterative method identified three promising leads [compounds L367 (**89**), L383 (**90**) and L396 (**91**)] from a crude library containing 75 compounds (208, 239). All three compounds were micromolar

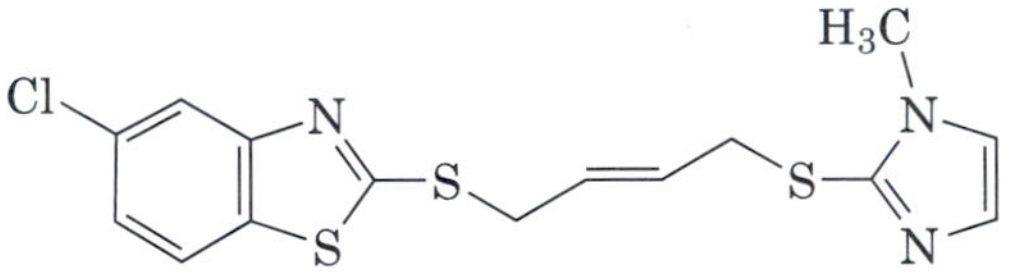

(89) L367

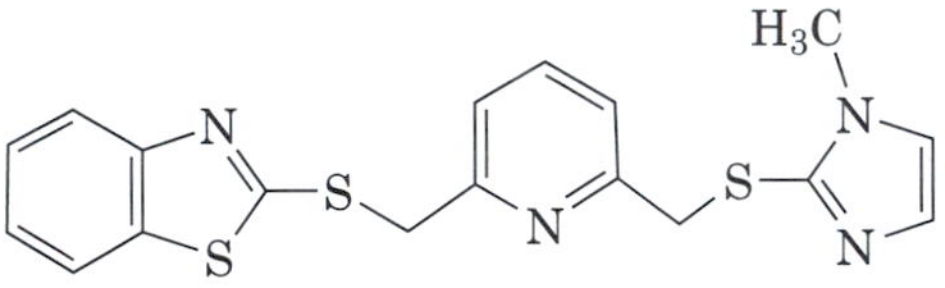

(90) L383

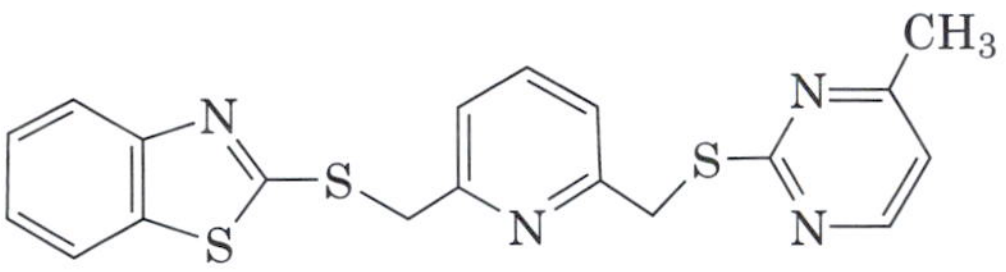

(91) L396

inhibitors against Mahoney strain of type 1 (P1/Mahoney) poliovirus uncoating and infectivity. However, they failed to show activity against the Sabin strain poliovirus in a Viro-Pharma assay system. L383 also inhibited HRV-14 and HRV-3 infectivity in cell culture with $EC_{50}$ of 0.8 and 0.55 $\mu M$, respectively. This work illustrates an advantage of designing ligand libraries, instead of individual compounds, as a means for drug design and discovery, because only enough information will be needed from the computational method, such as MCSS, to serve as a structural bias to guide (focus) the library design.

**2.3.1.2 Isothiazoles, Dibenzofurans, and Dibenzosuberanes.** The aforementioned compounds have a long-chain structure in common. The need to identify other chemical entities with activity against picornaviruses still exists. Among a series of 3,4,5-trisubstituted isothiazoles reported recently, IS-2 (**92**) showed the highest *in vitro* activity against poliovirus-1 ($EC_{50}$, 0.045 $\mu M$) and echovirus-9 ($EC_{50}$, 0.25 $\mu M$), if added to the cells within 1 h after poliovirus adsorption (240, 241). However, it was inactive against coxsackie B1 and rhinoviruses (242). In contrast, whereas IS-44 (**93**) and IS-50 (**94**) were inactive against both poliovirus-1 and echovirus-9, both compounds exhibited activity against rhinoviruses (242).

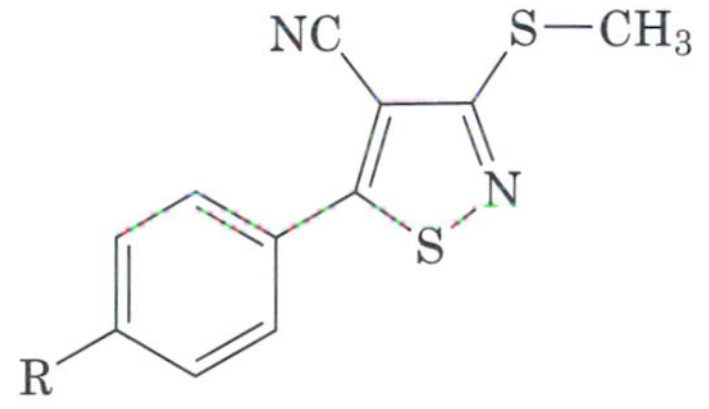

NC     S—CH₃

**(92)** IS-2     R = H

**(93)** IS-44     R = O—S(=O)(=O)—⟨ ⟩—CH₃

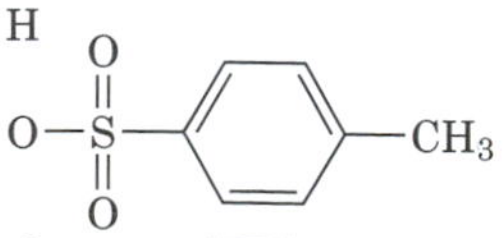

**(94)** IS-50     R = O—CH₂—⟨ ⟩

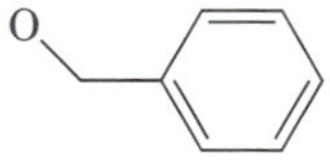

Of the 17 serotypes screened, 15 (88%) were sensitive to IS-50 ($EC_{50}$, 1–30 $\mu M$), including all of the group B serotypes screened. IS-44 was active against some group B rhinoviruses, with the lowest $EC_{50}$ values for HRV-2, -85, and -89 (0.3, 0.3, and 0.1 $\mu M$, respectively).

SAR showed that the active structural features seemed to have a short thioalkyl chain in the 3-position; a cyano or methylester group in the 4-position; and a not-substituted phenyl ring in the 5-position (241). Like the WIN compounds, IS-44 was shown to stabilize HRV-2 against thermal inactivation, suggesting that these isothiazoles might target the viral capsid.

Dibenzosuberenone (**95**) and 2-hydroxy-3-dibenzofuran carboxylic acid (**96**) are repre-

**(95)**

**(96)**

sentatives of other structurally distinct compounds that have shown to block rhinovirus (e.g., types 14 and 16) replication *in vitro* (243). Time-of-addition experiments showed that compounds work during an early stage of the viral infection cycle, probably on adsorption or uncoating. Although these compounds are not as extremely potent against rhinoviruses as many of the compounds discussed above, they can serve as leads for novel therapeutic agents because they also have an additional anti-inflammatory property. Combination antiviral and anti-inflammatory therapies may be of significant benefit for intervention of common cold; an illness may not result from direct virus-induced tissue damage, but rather from release of inflammatory mediators (243–245).

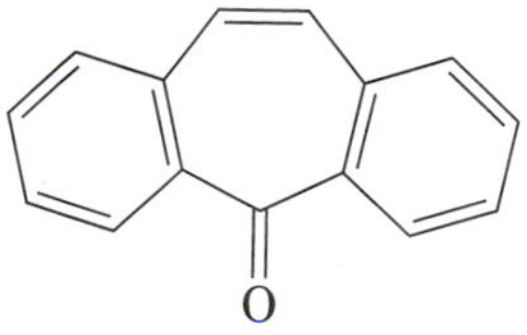

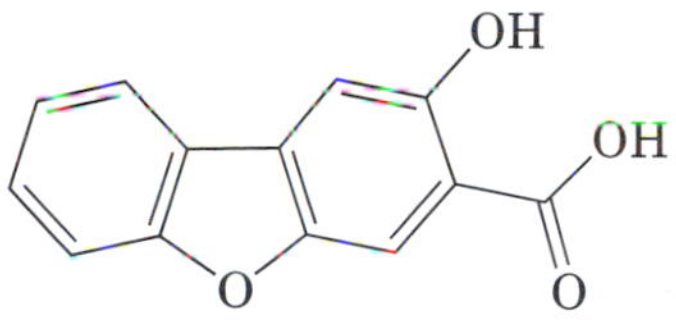

**2.3.1.3 Soluble Intercellular Adhesion Molecules-1 (sICAM-1).** Intercellular adhesion molecule 1 (ICAM-1, CD54) is a cytokine-inducible cell surface receptor that has also been shown to be the receptor for nearly 90% of the human rhinoviruses (the major group of human rhinoviruses). Truncated soluble ICAM-1 molecules could inhibit a broad spectrum of rhinovirus serotypes in a variety of different cell lines (222), both by acting as a competitive inhibitor and by irreversible disruption of the capsid with release of the viral RNA (246). A truncated from of sICAM-1, $tICAM_{453}$, was recently tested as an intranasal spray in preventing HRV-16 infection in chimpanzees (247). As chimpanzees do not show clinical manifestations of diseases, measuring anti-rhinovirus serum antibody responses and virus shedding were used to detect infection. By both of these measures, intranasal application of $tICAM_{453}$ was efficacious as prophylaxis against rhinovirus infection (247). Efficacy of tremacamra, a recombinant sICAM-1 developed by Boehringer Ingelheim, has been tested in controlled trials in humans (248). Tremacamra, as an inhaled solution or as a powder, was given intranasally either before or after inoculation with HRV-39 (but before onset of symptoms). The results indicated that tremacamra was effective in reducing the symptoms of experimental common colds, regardless of whether the drug was given before of after the challenge with virus. The mean virus titer and the concentration of interleukin-8 (IL-8) were significantly reduced in the lavage fluid from treated volunteers. This drug seemed to be well tolerated. It could neither penetrate through the nasal mucosa nor interfere with development of neutralizing antibody (248). It remains to be determined if tremacamra would be effective if given after onset of symptoms.

### 2.3.2 Inhibitors of Picornaviral Proteases

**2.3.2.1 Peptidic Inhibitors.** The genome of HRV contains a single open reading frame that can be translated into a large polyprotein, which undergoes further processing by two virally encoded proteases, designated 2A and 3C, to produce structural and functional proteins required for viral replication. In human rhinoviruses, the 2A protease separates the structural from the non-structural protein precursors, followed by the 3C protease, or its 3CD precursor, which carries out eight of the remaining nine proteolytic cleavage reactions, of which six cleavages occur at Gln-Gly bonds (249). The 2A and 3C proteases are a cysteine

protease, which contain a nucleophilic cysteine residue at the active site [Cys-172 in hepatitis A virus (HAV) 3C (250), Cys-147 in both human rhinovirus (HRV) 3C and poliovirus 3C (251, 252), or Cys-146 based on HRV-14 numbering (253)]; their tertiary structures are similar to the trypsin-like serine proteases.

Because the 3C protease (1) plays a critical role in replication and maturation of HRVs, (2) has high selectivity for substrates containing Gln-Gly bonds, (3) has conserved active site among the known HRV serotypes (254), and (4) has no known cellular homologs; this viral enzyme seems to be an attractive antiviral target.

The hexapeptide $H_2N$-Thr-Leu-Phe-Gln-Gly-Pro-$CO_2H$ (**97**) has been determined to be the minimal composition as an effective substrate for the HRV 3C protease, where Gln-Gly represents the scissile $P_1$-$P_{1'}$ bond (255). Peptide derivatives, where the scissile amide carbonyl was replaced with an electrophilic functionality (e.g., aldehyde, ketone, or Michael acceptor) that allows nucleophilic attack by the thiol group of the active site cysteine, have shown to be potent inhibitors of the HRV 3C protease (256, 257).

*2.3.2.1.1 AG7088 and Peptidyl Michael Acceptors.* A series of peptide-derived Michael acceptors, based on the X-ray structures of HRV-2 3C/inhibitors complexes, have been reported by investigators of Agouron Pharmaceuticals (257–263). Their investigations have led to the development of AG7088 (**98**), a potent and broad-spectrum anti-HRV compound currently in human clinical trials (administered as intranasal spray) for the treatment of common colds caused by the rhinovirus infection (Agouron press release, November 4, 1999).

The drug design by Dragovich et al. at Agouron began with Cbz-protected tripeptides containing a $P_1$-$P_{1'}$ equivalent *trans*-$\alpha,\beta$-unsaturated ester moiety (258). The representative methyl ester displayed relatively potent irreversible inhibition of HRV-14 3C protease; however, it exhibited moderate antiviral activity in H1-HeLa cells and was non-cytotoxic to the limits of its solubility. More importantly, it would not react readily with ubiquitous biological thiols (e.g., glutathione), suggesting

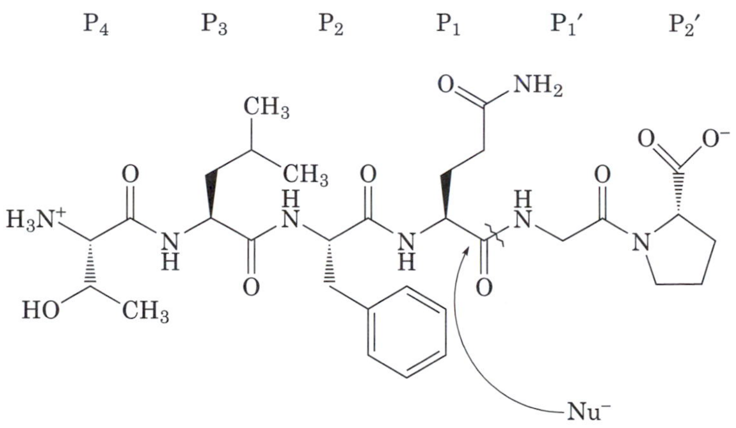

$$P_4 \qquad P_3 \qquad P_2 \qquad P_1 \qquad P_1' \qquad P_2'$$

**(97)**

that peptidyl Michael acceptors of this type could be developed into useful anti-HRV agents (258). Crystallographic analyses of enzyme-inhibitor complexes showed a covalent bond was formed between the 3C protease active site cysteine residue (Cys-147) and the β-carbon of the Michael acceptors, confirming the binding orientation of these compounds (258). Systematic SAR studies suggested the following criteria for bioactivity.

- At least three amino acids were required for effective binding of the inhibitor to the enzyme (261, 263).
- *trans*-α,β-Unsaturated esters were the optimal choice as the Michael acceptors (258, 261).
- The presence of a $P_1$ glutamine was essential (255, 263). Although a primary amide might not be required (264), *cis*-amide ge-

ometry was required for the γ-carboxamide side-chain, and incorporation of an (S)-γ-lactam moiety could impose the proper stereochemistry (260, 261).

- $P_2$ side-chains might be large and hydrophobic (255). Substitution of the $P_2$ phenyl ring with 4-fluorophenyl moderately enhanced the activity; other modifications generally resulted in reduced activity (261, 263).
- Replacement of the $P_2$-$P_3$ peptide bond with ketomethylene isostere slightly compromised the enzyme inhibitory activity but resulted in much improved antiviral properties (259, 261).
- Substitutions with a wide variety of functionally at $P_3$ were generally tolerated (255, 258, 261). Replacement of $P_3$ leucine with valine improved both enzyme inhibitory and antiviral properties (259).

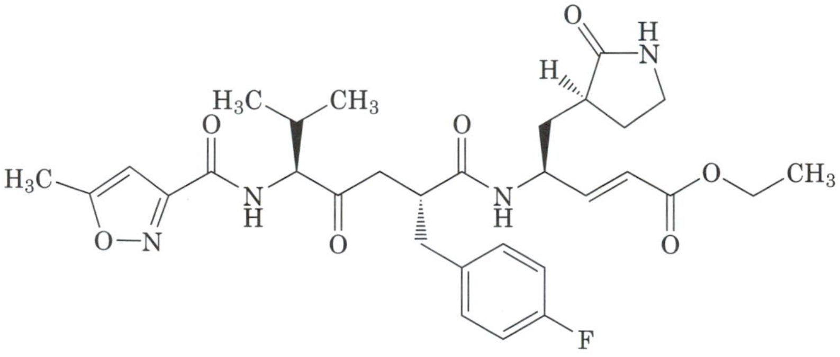

**(98)** AG 7088

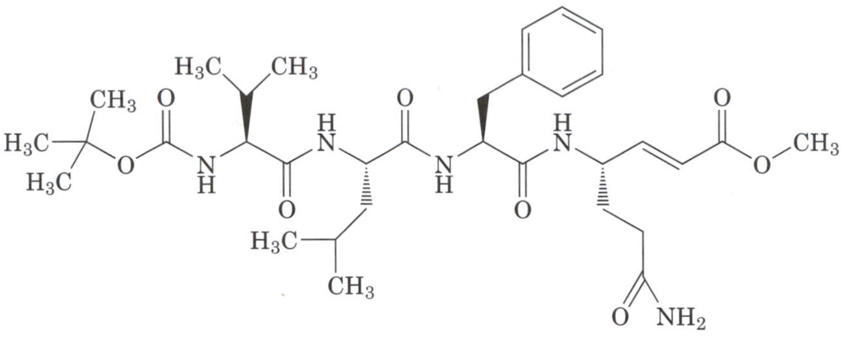

**(99)**

- $P_4$ should be small and hydrophobic (255). Replacement of $N$-terminal $P_4$ amide with a thiocarbamate or 5-methyl-isoxasole-3-carboxamide also improved both enzyme inhibitory and antiviral properties (259, 261, 263, 265).

  Interestingly, the structural features that made individual improvement in bioactivity could be combined in an additive manner (263), resulting in AG7088 (260, 261).

  AG7088 is a specific, potent, and irreversible inhibitor of HRV 3C protease (266). In cell cultures, it was non-toxic to the cells and inhibited a wide range of HRV serotypes, with a mean $EC_{50}$ and a mean $EC_{90}$ of 0.023 and 0.082 $\mu M$, respectively, and related echoviruses and enteroviruses tested (254, 266). Moreover, AG7088 seemed to be more potent and to have a broader anti-HRV spectrum than pleconaril when tested against HRV clinical isolates (267). AG7088 could be added up to 26 h after viral infection and still resulted in

significant reduction of infectious virus and the levels of inflammatory cytokines, IL-6, and IL-8 (267a).

Another series of potent anti-3C protease peptidyl Michael acceptors have also been reported by Kong et al. (255) These compounds, typified by (**99**), showed a very rapid, 1:1 stoichiometric, covalent inactivation of the enzyme as determined by electrospray mass spectrometry.

*2.3.2.1.2 Peptidyl Aldehydes and Ketones.* Peptide aldehydes in which an aldehyde moiety (e.g., a glutaminal) serves as the glutamine isostere at the $P_1$ position have been reported as reversible inhibitors 3C proteases of both HRV and HAV [(256, 264) and references cited therein]. For obtaining enzyme-inhibitor cocrystal structures for guiding structure-based drug design, Agouron investigators reported the synthesis of a series of tripeptide aldehyde inhibitors (represented by compound **100**) and their X-ray structures when they were covalently bound to HRV-2 3C protease (264).

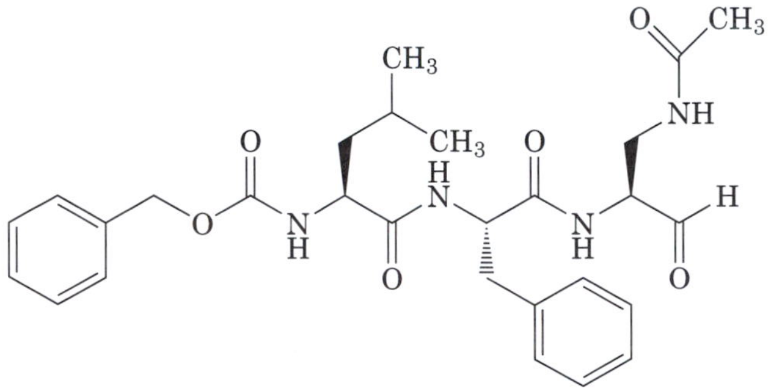

**(100)**

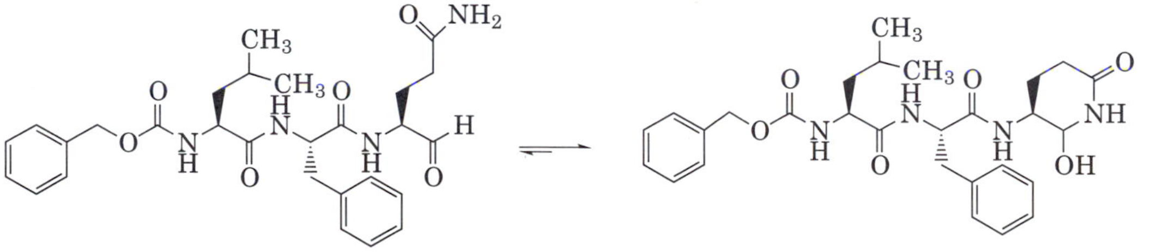

(101)

Only the *re* face of the aldehyde inhibitor was accessible to Cys-147. Isosteric replacement of P$_1$ glutaminal with *N*-acetyl-amino-alaninal not only prevented compound (**100**) from forming cyclic aminal, as in the case of compound (**101**), but also significantly improved the compound's enzyme inhibitory and anti-HRV properties over that of (**101**) (264). Independently, Lilly investigators reported that methionine sulfone residue could mimic the natural P$_1$ glutamine and demonstrated LY338387 (**102**) as the first dipeptide alde-

hyde with low micromolar enzyme inhibitory (reversible inhibition) and *in vitro* antiviral (HRV-14) activity (268).

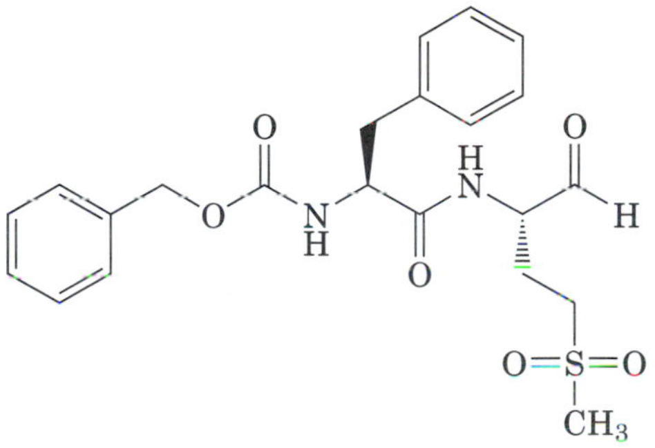

Because an aldehyde-containing compound is prone to have higher toxicity as well as less selectivity and stability, Dragovich et al. recently reported the preparation of a ketone-containing tripeptide (**103**) and showed that his compound displayed very potent levels of reversible 3C protease inhibition along with low *in vitro* cytotoxicity and sub-micromolar antiviral activity against HRV serotypes 14, 1A, and 10 (269). The benzothiazole nitrogen atom seemed to be important for hydrogen bonding interaction with the enzyme active site, because drastically reduced enzyme inhibitory activity was seen with an analog containing a 2-benzothiophene moiety (269). Independently, Vederas and Malcolm reported their work on a peptidyl monofluoromethyl ketone (**104**) as an irreversible inactivatior of HAV 3C (270). Enzyme inactivation might involve the formation of a (alkylthio)methyl ketone accompanied by liberation of fluoride ion

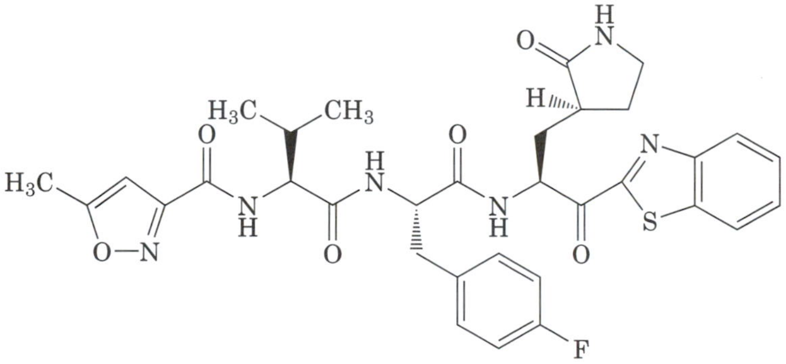

(102) LY 338387

(103)

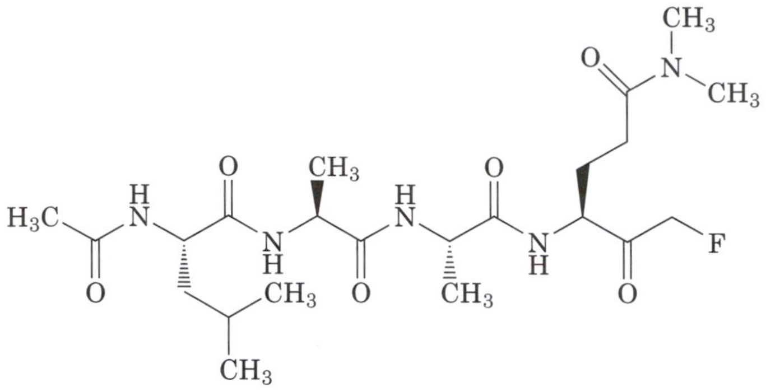

(104)

as monitored by the $C^{13}$ NMR spectrum of enzyme-inhibitor complex and $F^{19}$ NMR, respectively (270).

*2.3.2.1.3 Azapeptides.* Azapeptides, in which peptidic backbones contain hydrazine functionality, have also been known in the design of protease inhibitors (271, 272). Vederas and Malcolm, in extension of their previous work on peptidyl haloacetyl ketones, further reported that the haloacetyl azaglutamine peptides (105) and (106) as well as the sulfenamide azaglutamine derivative (107) irreversibly reacted with HAV 3C (250). Monitored by electrospray mass spectrometry, it was shown that there was displacement of halogen by the active site thiol of Cys-172. This mechanism of

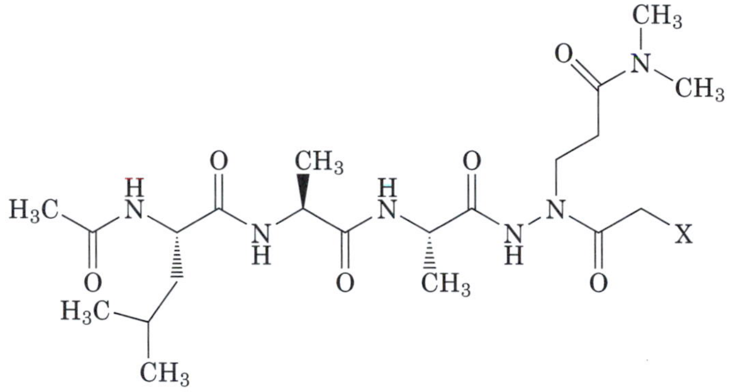

(105) X = Cl
(106) X = Br

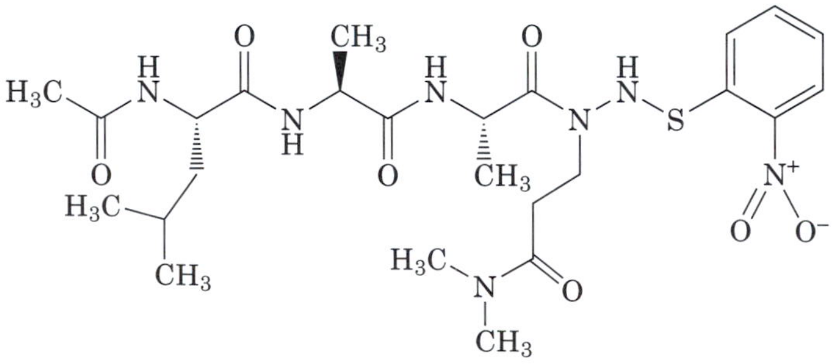

(107)

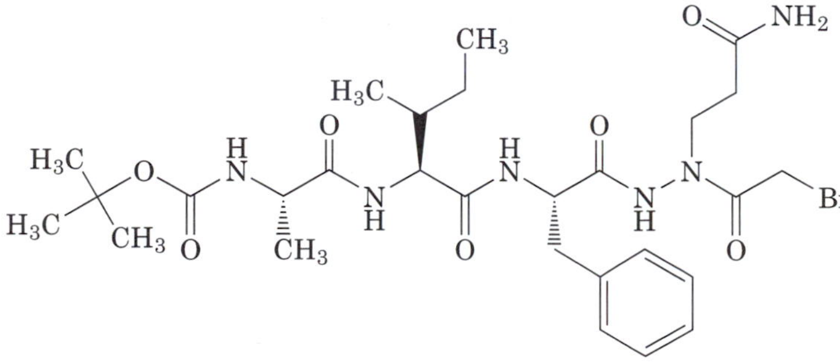

(108)

enzyme inactivation (i.e., formation of a covalent adduct between the enzyme and the inhibitor with the loss of a halogen) was further confirmed by the interaction of compound (108), prepared independently by Abbott Labs, with HRV-1B 3C protease (249).

Azodicarboxamide derivatives, represented by compound (109), were synthesized by Ved-

(110) Isatin

(109)

eras et al. as another series of irreversible 3C protease inhibitors, which formed covalent adducts with the enzyme active site thiol through Michael addition onto the azo moiety (273). Because these compounds react readily with extraneous thiols (e.g., dithiothreitol), they might not be appropriate drug candidates themselves. However, the authors argued that these compounds could be potential tools for probing the enzyme active site because recognition elements could readily be built onto either side of the azo moiety. Most of the current cysteine protease inhibitors recognize on the P region of the active site (273).

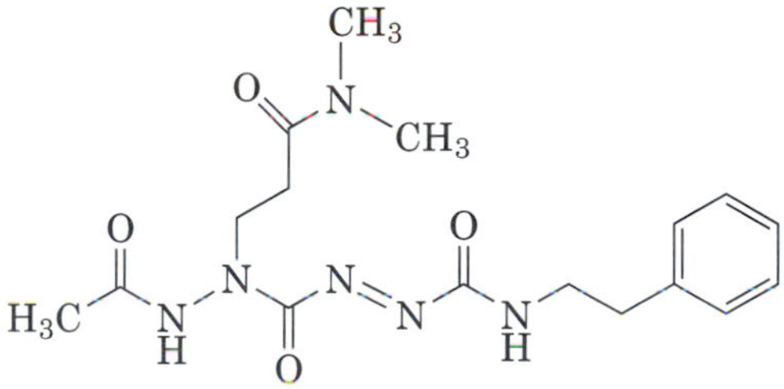

**2.3.2.2 Nonpeptidic Inhibitors.** Through a random screening effort, Lilly investigators identified the isatin (110) and the homophthalimides (111) as two interesting lead

compounds (274). Subsequent SAR studies resulted in compound LY353349 (112) and LY353352 (113) with much improved enzyme inhibitory activity. Molecular modeling and mass spectrometry studies with LY353349 as the model suggested that this compound is tightly bound to the 3C enzyme in a ratio of 1:1, presumably through the $C_3$ (non-benzylic) carbonyl under nucleophilic attack by the active site cysteine (274). LY353349 and LY353352 also displayed activity against HRV 2A protease (275). Because the proteolytic cleavage of viral polyprotein is carried out first by the 2A protease followed by the 3C, dual inhibition of both enzymes might result in cooperative inhibition of viral replication (275). This effect could be illustrated by LY353352, which had anti-HRV-14 2A and anti-HRV-14 3C $IC_{50}$ values of 63.3 and 55.4 $\mu M$, respectively; however, its anti-HRV-14 activity; exceeded the enzyme inhibitory activity, with an $EC_{50}$ of 15.8 $\mu M$. Of interest, a related compound, LY343814 (114), showed excellent *in vitro* anti-HRV-14 activity ($EC_{50}$, 4.2 $\mu M$), and was inhibitory to HRV-14 2A ($IC_{50}$, 20 $\mu M$), but not of HRV-14 3C ($IC_{50}$, >200 $\mu M$) (275).

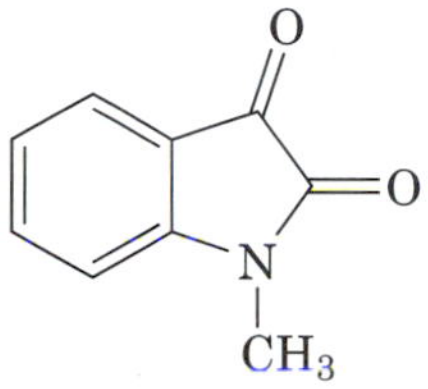

At the same time, Agouron investigators, who, based on the structural determinants

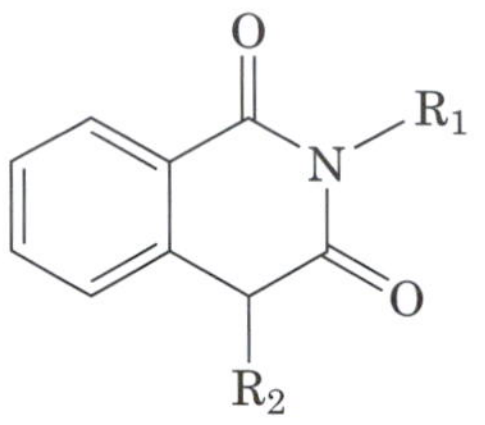

(111)            $R_1 = CH_3$                        $R_2 = CH_2$—[structure with 4-fluorophenyl ketone]

(112) LY 353349  $R_1 = CH_2$—[S(=O)(=O)—CH_3]     $R_2 = CH_2$—[phenyl ketone]

(113) LY 353352  $R_1 = CH_2$—[C(=O)—O—CH_2—CH_3]  $R_2 = CH_2$—[phenyl ketone]

(114) LY 343814  $R_1 = CH_2$—[propyl chain] CH_3   $R_2 =$ [C(=O)—CH_3]

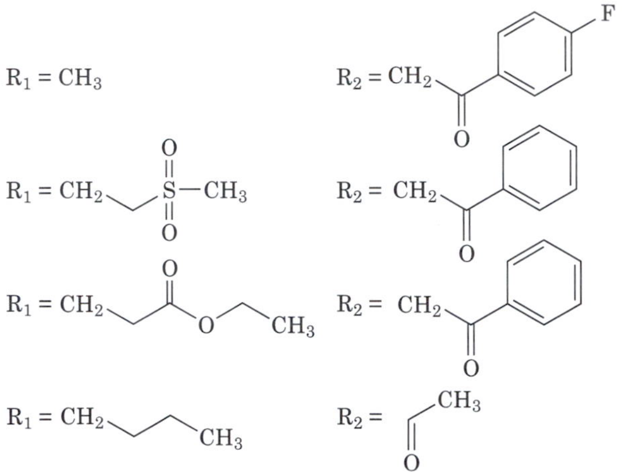

around the scissile cleavage, $P_1$ recognition, and $S_2$ sites, envisioned the cyclic $\alpha$-keto amide isatin structure as a good core for design of small molecule protease inhibitors, reported a series of synthetic isatins (2,3-dioxindoles) (253). Molecular modeling and SAR studies resulted in compound (115) as the

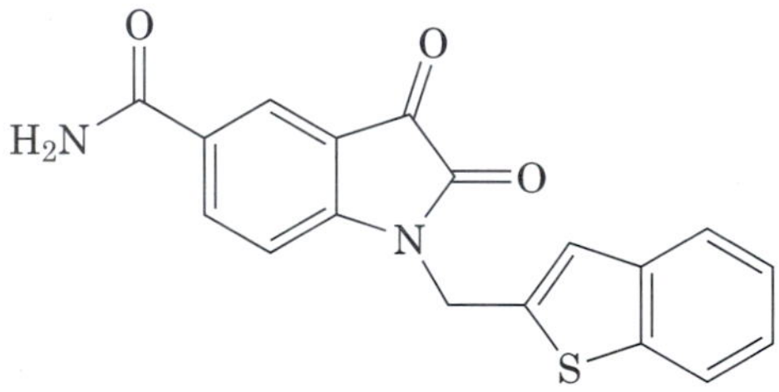

(115)

most potent inhibitor of HRV-14 3C protease ($K_i$, 2 n$M$) in the series. X-ray co-crystal structure further confirmed the existence of a covalent bond between Cys-147 and the electrophilic C3 of isatin and other important interactions between the enzyme and (115).

Disappointingly, because of its apparent cytotoxicity, no in vitro anti-HRV-14 activity was demonstrated (253).

In a continuous effort to find orally bioavailable small non-peptide 3C protease inhibitors, Agouron investigators used structure-based design and parallel synthesis on solid support to generate a structurally biased library of 5-substituted benzamides that contain a $C_3$ $\alpha,\beta$-unsaturated ester moiety (276). The benzamide-Michael acceptor core was designed to mimic the $P_1$ recognition element of the natural 3C protease substrate and the $C_5$ substituents were to optimize the binding of the inhibitors to the $S_3$-$S_4$ subsites of the enzyme. In such an arrangement, the $\alpha,\beta$-unsaturated ester group would be expected to undergo irreversible covalent 1,4-addition by the nucleophilic catalytic cysteine residue on the enzyme (confirmed later by a co-crystal structure). Surprisingly, these compounds generally showed very potent antiviral activity, despite their moderate enzyme inactivation rates. For the most potent compound (116), the antiviral EC$_{50}$ of 0.6 $\mu M$ was exceptional, given the modest $K_{obs}/I$ of 139 $M$/s (276).

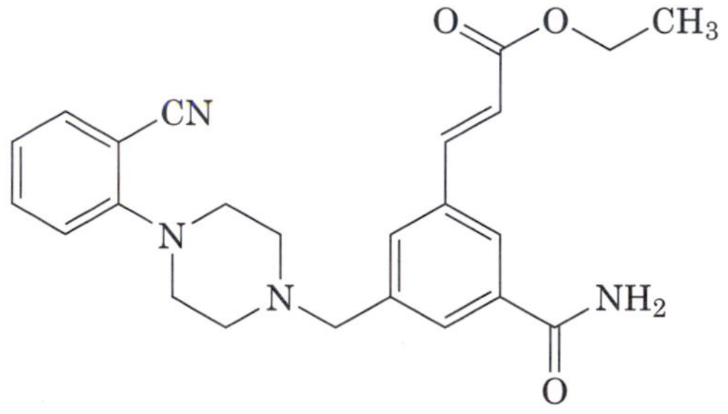

(116)

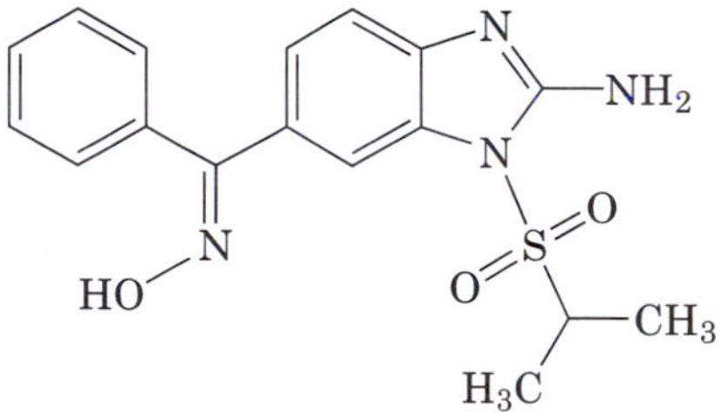

(118) Enviroxime

(119) Enviradene

Because nitric oxide (NO) or NO donors could inactivate many cysteine-containing enzymes, HRV 3C protease might also be susceptible to NO donors. To test this mechanism-based hypothesis, Wang et al. demonstrated that S-nitrosothiols [exemplified by GSNO (117)] could achieve a time- and concentra-

(117) GSNO

tion-dependent inactivation of HRV-14 3C protease with second-order rate constants (277). It was shown that the inhibition by these compounds was caused by the formation of an S-nitroso adduct and that the inactivation of the enzyme could be reversed by the addition of nucleophilic thiols, such as dithiothreitol, to the reaction mixture (277).

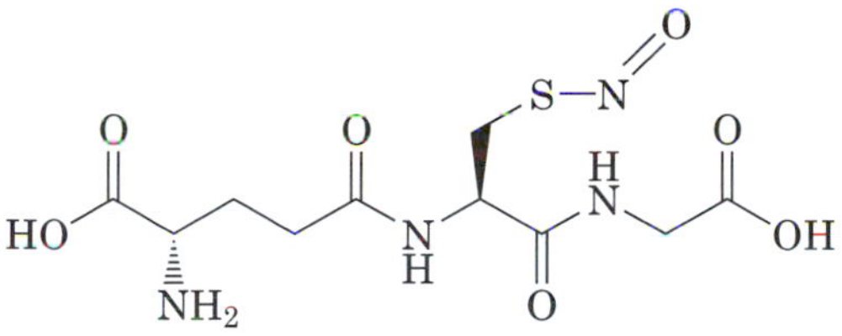

### 2.3.3 Inhibitors of Picornaviral Replication (Enviroxime and Analogs). In the early 1980s, two benzimidazole derivatives, enviroxime (118) and enviradene (119) were discovered and evaluated in the clinic by Lily Research Laboratories. However, both compounds failed in clinical studies because of poor oral bioavailability in humans. Enviroxime was also associated with emetic side effects (see references cited in Ref. 278).

Enviroxime possesses significant antiviral activity against both rhinoviruses and enteroviruses. Although a function involving

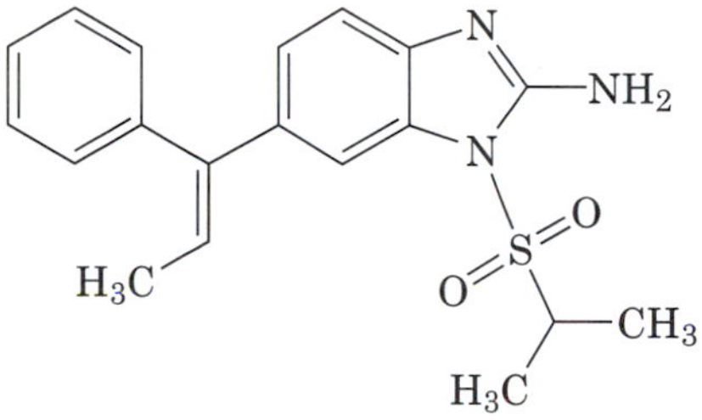

charged residues in the 3A region of the 3AB protein of HRV-14 has been suggested as a plausible target for enviroxime, the exact mechanism of action remains unclear. A mutant with an increased level of resistance to enviroxime showing mutations in multiple proteins or RNA sequences further suggested that enviroxime might target a complex of viral proteins and/or cellular factors (279). Interestingly, efforts to develop drug-resistant mutants mostly resulted in the selection of drug-sensitive mutants that exhibit no increase in their $EC_{50}$ values in standard plaque reduction assays (280). Despite extensive efforts, there was no clear evidence to show direct binding between enviroxime and a viral or a host protein. Such a complex mechanism of inhibition might explain the low levels of viral resistance to enviroxime and its related inhibitors (279).

Intrigued by the potential of enviroxime and its analogs, efforts to synthesize new analogs have recently been renewed with an emphasis on optimizing antiviral activity while maximizing oral bioavailability. It has been known that the vinyl oxime moiety of enviroxime is metabolically labile (281). In the case of enviradene, the vinyl methyl group also undergoes rapid allylic oxidation to a hydroxymethyl metabolite resulting in very low blood levels of enviradene in man and monkeys. The bioavailability improved when methyl was replaced with by acetylene, the resulting vinylacetylene derivative showed increased levels in blood in monkeys (282). Moreover, fluorine substitution on the left-hand aromatic ring further enhanced the oral blood level (280, 282). These compounds remained as potent and potentially broad-spectrum anti-picornaviral inhibitors as illustrated by compound (120), which inhibited poliovirus-1 (Mahoney)

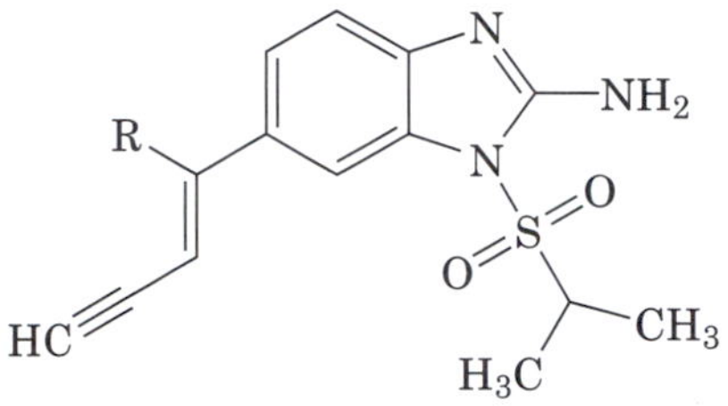

(120) R =

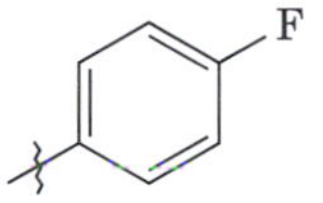

(121) R =

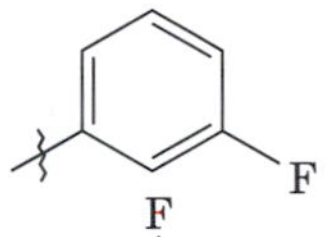

(122) R =

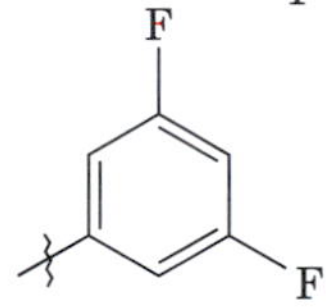

and HRV-14 with $EC_{50}$ of 0.04 and 0.13 $\mu$g/mL, respectively, in a plaque reduction assay. It was also efficacious by oral administration in treating a coxsackie A21 infection in CD-1 mice (278, 282).

In a study on cytochrome P450 function and hepatic porphyrin levels in mice, two fluoro-substituted vinylacetylene benzimidazoles, (121) and (122), which were shown to attain significant levels of plasma concentrations after oral dosing, caused a marked effect on liver enzymes and hepatic porphyrin levels. The multiple dose studies showed significant increases in liver weights as well as increases in serum levels of enzymes suggestive of hepatotoxicity (278). These hepatotoxic effects are most likely related to the acetylene moiety be-

cause structurally similar enviroxime and enviradene were free of these adverse effects during their respective preclinical and clinical evaluations. Nevertheless, there are marketed alkyne-containing drugs that do not have obvious detrimental effects on P450, its function, or hepatic porphyrin levels (278).

In studies with a series of C2 analogs of enviroxime, it was found that primary amino substitution, as in enviroxime, was the most antivirally active. The activity was reduced with those substituted with a larger group at $C_2$, which provided a repulsive steric interaction at $N_3$, resulting in less flexible conformation. In the case of enviroxime, a hydrogen atom of $C_2$ amino was shown to form an intramolecular hydrogen bond with the $N_1$ sulfonyl oxygen. This interaction might act to enhance the activity by holding the second hydrogen in a desirable orientation toward the enzyme active site (283). However, such internal hydrogen bonding might not be important in a series of 2-amino-3-substituted-6-[(E)-1-phenyl-2-(N-methylcarbamoyl)vinyl]imidazo[1,2-$\alpha$]pyridines, as evidenced by the reduction of anti-HRV-14 activity when the sulfite in compound (123) ($EC_{50}$, 0.17 $\mu$g/mL; by a plaque reduction assay) was replaced with corresponding sulfone as shown in (124) ($EC_{50}$, 0.64 $\mu$g/mL). The potency remained the same even the sulfite in (123) was replaced

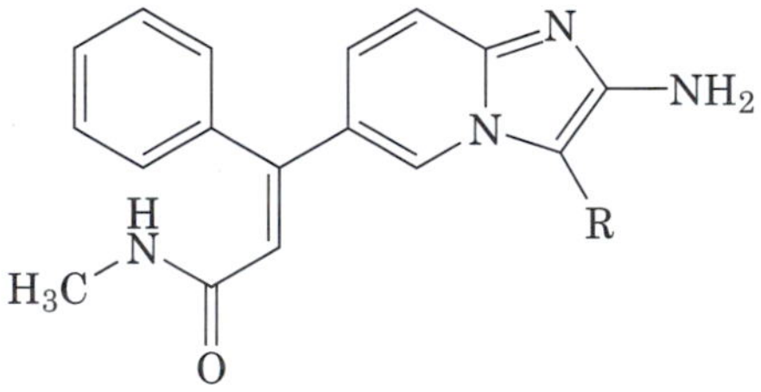

(123) R =

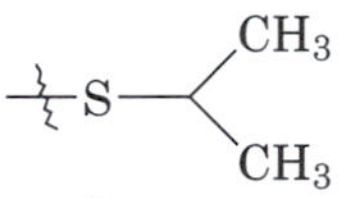

(124) R =

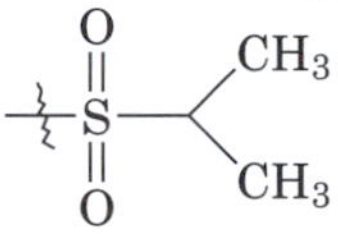

(125) R =

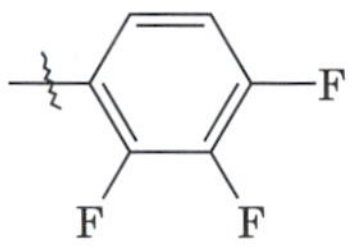

with a fluorine substituted aromatic ring, as demonstrated by compound (125) (284). When enviroxime (118) was combined with disoxaril (WIN 51711) (82), the combination exerted a significant synergistic inhibitory effect on poliovirus-1 (Mahoney) replication in FL cells, without concomitant synergic cytotoxic effect (285, 286). The combination also demonstrated synergistic *in vitro* antiviral effect in FL cells and *in vivo* protective effect in newborn mice infected with coxsackievirus B1 (287).

### 2.3.4 Inhibitors of Picornaviral Protein 2C.

The 2C protein of picornaviruses is a multifunctional and highly conserved non-structural protein involved in viral RNA replication, viral encapsidation, and other functions, such as membrane binding, RNA binding, protein–protein interactions, and NTPase activity (288, 289). Recently, mutational analyses with drug-resistant and drug-dependent variants revealed that two previously known benzimidazoles, HBB (126) (288, 290) and MRL-1237 (127) (291), exert their antiviral action

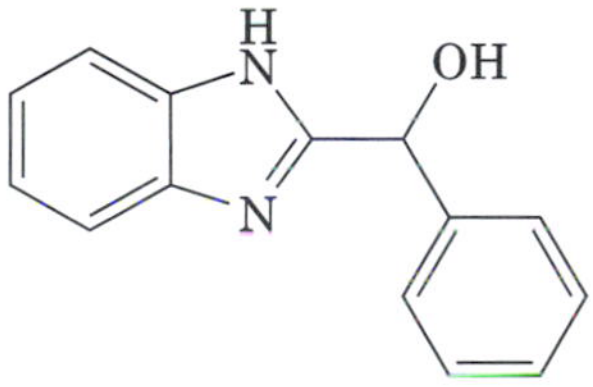

(126) HBB

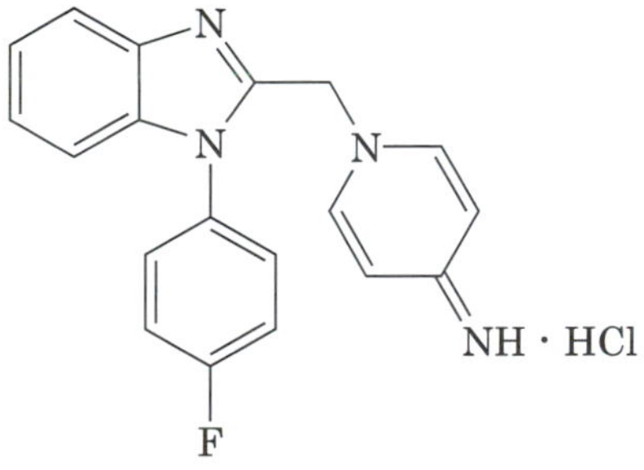

(127) MRL-1237

on the 2C protein. However, the exact mode of action of these compounds remains to be elucidated.

Similar studies with 5-(3,4-dichlorophenyl) methylhydantoin (128) suggested that the hydantoin derivative inhibits the encapsidation

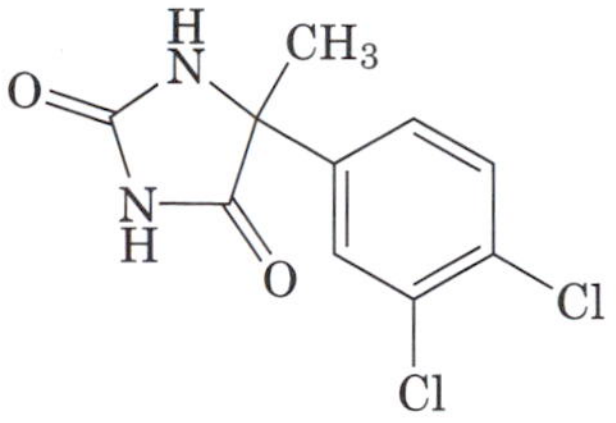

(128) 5-(3,4-Dichlorophenyl)methylhydantoin

function of the 2C protein (292). This finding was further confirmed with experiments in a cell-free system showing that this compound is an inhibitor of poliovirus assembly. It also inhibited the post-synthetic cleavage of poliovirus (293).

### 2.3.5 Other Anti-Picornaviral Inhibitors

#### 2.3.5.1 Natural Products and Synthetic Analogs.
By employing antiviral assay-guided fractionation of the ethanolic extract of *Pterocaulon sphacelatum*, a traditional medicinal plant used by the Australian aboriginal people, Semple et al. reported the isolation of the flavonoid chrysosplenol C (129), which

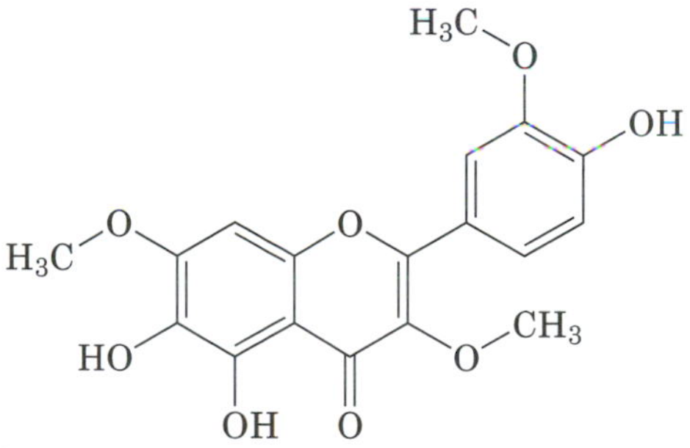

(129) Chrysosplenol C
(3,7,3'-Trimethoxy-5,6,4'-trihydroxyflavone)

showed *in vitro* activity against poliovirus with an $EC_{50}$ of 0.27 µg/mL (0.75 µM), and a maximum non-toxic concentration to proliferating Buffalo green monkey (BGM) kidney cells of 4 µg/mL (294). This was the first report of isolation of chrysosplenol C from the genus *Pterocaulon*.

The same group of investigators also discovered chrysophanic acid (**130**), isolated from another Australian aboriginal medicinal plant,

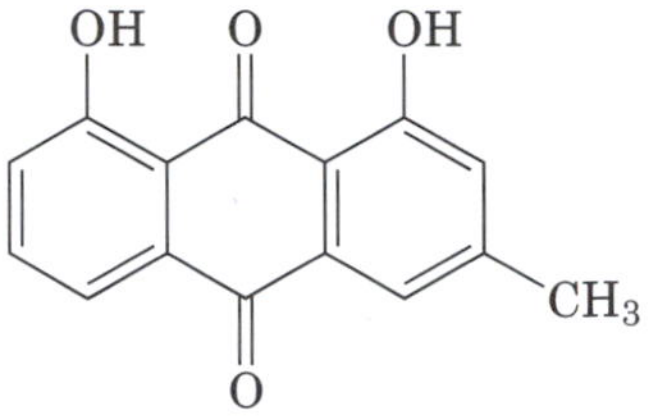

**(130)** Chrysophanic acid
(1,8-Dihydroxy-3-methylanthraquinone)

*Dianella longifolia*, to be inhibitory to the replication of poliovirus types 2 and 3 in BGM kidney cells, with $EC_{50}$ values of 0.21 and 0.02 $\mu$g/mL, respectively (295). The maximum nontoxic concentration for proliferating BGM cells was 12.5 $\mu$g/mL. This compound did not have an irreversible virucidal effect on poliovirus particle. Rather it might inhibit an early stage in the viral replication cycle as suggested by a time-of-addition study. Disappointingly, this compound did not show *in vitro* activity against coxsackievirus types A21 and B4 and human rhinovirus-2. Nevertheless, this was the first report of activity of an anthraquinone derivative against a non-enveloped virus (295). Structurally related anthraquinones have been shown to inhibit enveloped viruses by both virucidal and non-virucidal mechanisms (see references cited in Ref. 295).

A series of aporphinoid alkaloids were also shown active against poliovirus type 2 (296). Glaucine fumarate (**131**) was among the most active ones. It inhibited poliovirus replication

**(131)** Glaucine fumarate

in Vero cells with $EC_{50}$ and $CC_{50}$ of 9 and 142 $\mu$M (in a CPE assay), respectively. It was active even when added 1 h post-infection. This activity was further confirmed by reduction of virus yields after a single cycle of replication. The nature of the 1,2-substituents of the isoquinoline moiety was critical for activity and cytotoxicity; a methoxyl group at $C_2$ position seemed to be of most importance for anti-polioviral activity (296).

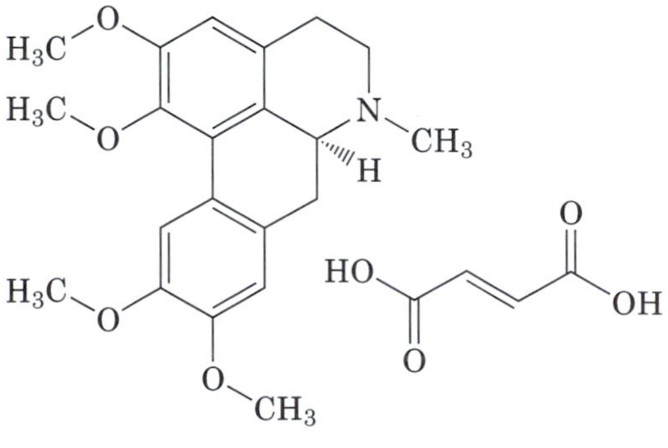

Desideri et al. have reported the anti-picornavirus activity of two series of synthetic flavonoids. First, 2-styrylchromones were evaluated against two selected human rhinoviruses, HRV-1B and HRV-14, by a plaque reduction assay in HeLa cells; it was shown that the majority of the compounds interfered with replication of both viruses. The most active compound was 4-nitro-2-styrylchromone (**132**),

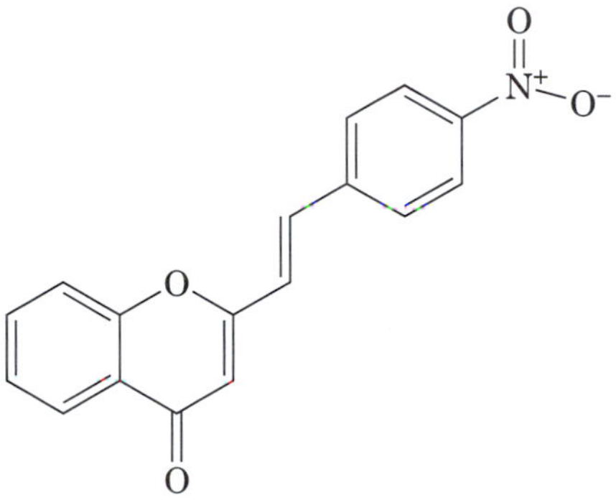

**(132)**

which had $EC_{50}$ values of 3.9 and 1.3 $\mu$M for HRV-1B and HRV-14, respectively. Its maximum non-toxic concentration for HeLa cells was 12.5 $\mu$M (297). In a second series of compounds, all synthetic homo-isoflavonoids were weakly effective against poliovirus-2, whereas they exhibited a variable degree of activity against HRVs 1B and 14 (298). Interestingly, the configuration of the chiral center in position 3, as illustrated by isoflavanone (**133**) did not seem to influence the activity against both rhinovirus serotypes, because the two enantiomers and the corresponding racemate were equipotent (299).

***2.3.5.2 Other Compounds.*** Although combinatorial chemistry has been recognized as a powerful tool for drug discovery, application of

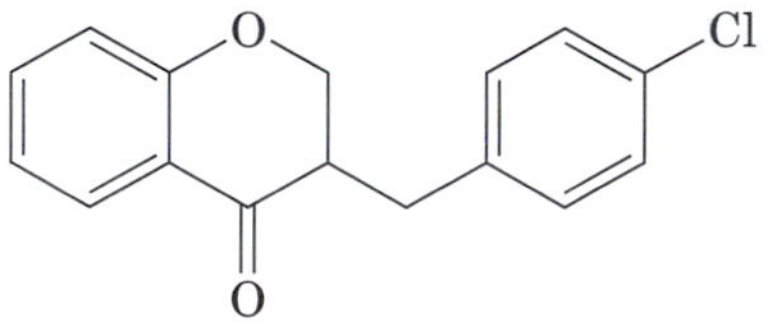

(133)

this technology has not been found widely used in antiviral research. One interesting work was recently reported in a communication by Lilly investigators who used a solution phase synthesis, coupled with a solid-supported aminomethylpolystyrene as "covalent scavenger" for removing isocyanate impurities, to yield equimolar mixtures of ureas for antirhinoviral testing in a whole cell assay (300). Subsequent deconvolution of hit mixtures led to two low cytotoxic leads, (134) and (135), with micromolar inhibitory activity against HRV-14.

## 2.4 Hepatitis C Virus

The HCV genome is a 9.5-kilobase, single-stranded, positive-sense RNA molecule, containing a single open reading frame that encodes for a polyprotein of 3010–3033 amino acids. This polyprotein undergoes maturational processing in the cytoplasm or in the endoplasmic reticulum (ER) of the infected cell to produce at least 10 mature proteins (C, E1, E2, P7, NS2, NS3, NS4A, NS4B, NS5A, and NS5B). In addition, an unusual feature of the HCV viral genome is the presence of two long and highly ordered untranslated regions at both 5′ and 3′ ends. All of these viral functions are potential target for therapeutic intervention; nevertheless, recent advances in anti-HCV drug development have largely focused on the 5′ untranslated region, the core, and the NS3 protein.

### 2.4.1 Inhibitors of HCV 5′ Untranslated Region and Core Gene

*2.4.1.1 Ribozymes and Antisense Oligonucleotides.* The 5′ UTR, which encodes the HCV internal ribosome entry site (IRES), and the core gene encoding the nucleocapsid protein of HCV are highly conserved among HCV isolates, making them attractive targets for ribozyme- and antisense oligonucleotide–based antiviral strategies (301).

Ribozyme Pharmaceuticals (RPI) reported their design and synthesis of hammerhead ribozymes targeting various conserved sites in the 5′ untranslated region (UTR). These ribozymes significantly reduced HCV 5′ UTR-mediated expression in a 5′ UTR-luceferase reporter system, as well as inhibited replication of an HCV-poliovirus chimera (302). Moreover, a nuclease resistant ribozyme, tar-

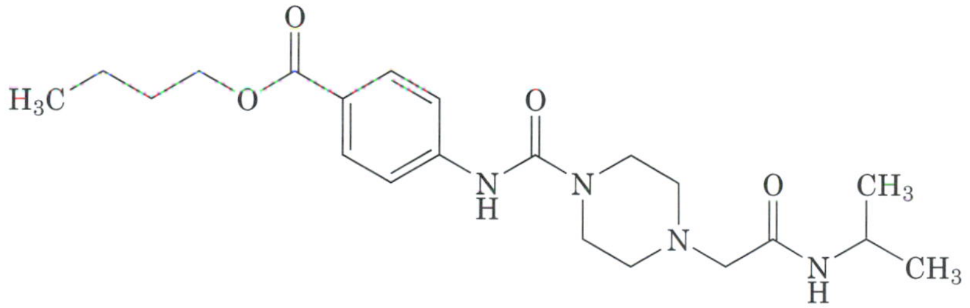

(134)

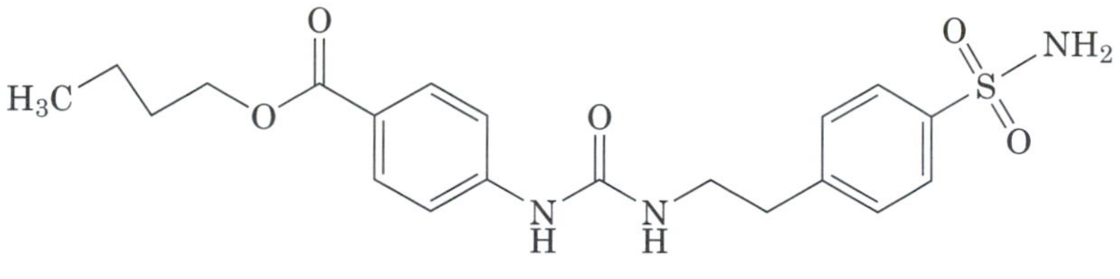

(135)

geting site 195, was selected for pharmacokinetics and tissue distribution studies after intravenous and subcutaneous administration in mice. The results showed that the ribozyme can be taken up and retained in the liver cells (303). RPI has recently completed their clinical trials of a 28-day safety and pharmacokinetic study of Heptazyme (LY 466700) following daily subcutaneous injections. Phase II clinical trials to study the drug's dose-ranging and efficacy makers in chronic HCV patients have been planned (RPI Press Release, September 28, 2000).

Wu et al. reported that two hammerhead ribozymes, which were designed to target just upstream of the start codon of the viral transcript and the core, respectively, were capable of suppressing HCV-luciferase reporter gene expression in a cell-free system and in transfected Huh7 cells (304). The same two regions were also targeted by DNA analogs of ribozymes (305). The third approach used by the group was to apply antisense oligonucleotides directed against a sequence in the 5'UTR IRES region and a region of the UTR overlapping the core protein translational start site of HCV. They reported that the antisense oligonucleotides, in the form of asialoglycoprotein-polylysine complexes, could be delivered by receptor-mediated endocytosis and caused specific inhibition of HCV-directed protein synthesis, as monitored by the expression of luciferase activity, in cells (306).

Patients infected with HCV genotype 1b have shown the poorest rate in response to interferon therapy. Kay et al. incubated total RNA from HCV 1b positive human livers, containing plus and minus strands, with a library of hammerhead ribozymes, and thereby isolated several effective ribozymes directed against a conserved region of the plus and minus strand of the HCV genome (307). The ribozymes were found to reduce or eliminate the respective plus or minus strand HCV RNAs in cultured cells and from primary human hepatocytes obtained from infected patients (307). Other studies by Hayashi et al. were targeting the core region of HCV 1b for the design of hammerhead ribozymes. They found that the ribozyme, whose cleavage site is located nearest to the initiation codon of the HCV ORF, showed the most efficient cleavage of the tar-

get RNA. On the other hand, the ribozyme with the cleavage site located farthest from the initiation codon blocked viral translation in a rabbit reticulocyte lysate most efficiently (308).

Hairpin ribozymes targeting HCV 5' UTR and capsid gene regions were reported by Barber et al. (309, 310). Because the 5' UTR contains considerable secondary structures, which could interfere with the cleavage activity of a ribozyme, the authors also prepared facilitator RNAs, trying to help to relax the secondary structure, and therefore, enhance the binding and activity of the ribozyme.

Different domains within the 5' UTR and core region have also been exploited as potential targets for inhibition of HCV translation by antisense oligonucleotides and oligodeoxynucleotides (ODNs). Isis Pharmaceuticals reported two phosphorothioate ODNs, ISIS 6095, which targeted a stem-loop structure within the 5' UTR known to be important for IRES function (nt 260–279), and ISIS 6547, which targeted sequences spanning the AUG used for initiation of HCV polyprotein translation (nt 330–349), effectively inhibited HCV gene expression as monitored in transformed hepatocytes (311). Reduction of RNA levels and the subsequent protein levels by these phosphorothioate ODNs was associated with RNase H cleavage of the RNA strand of the oligonucleotide–RNA duplex. On the other hand, 2'-modified (e.g., 2'-O-methoxyethyl) phosphodiesters oligonucleotides inhibited HCV core protein synthesis with comparable potency to phosphorothioate ODNs by an RNase H-independent mechanism (311, 312). In mice infected with an HCV-vaccinia virus recombinant, subcutaneous administration of ISIS 6547 and ISIS 14803 showed specific and dose-dependent inhibition of an HCV-luciferase reporter gene expression in the livers (313). (The two 20-base oligomers have the same sequence, but ISIS 14803 has 5-methylcytidine residues at all respective cytidine positions in ISIS 6547.) In March 2000, the company initiated clinical trials with ISIS 14803 in patients who failed interferon or interferon-plus-ribavirin therapy (314). In addition, investigators of Isis Pharmaceuticals reported a new probing strategy by using hybridization affinity screening and RNase H cleavage analysis on a fully randomized sequence DNA oligonucleotides (10-mer) library to identify energetically preferred hybridization sites on folded target RNA (315). The hypothesis is that binding-optimized shorter oligomers (10–15-ers) may have equivalent or greater affinity and specificity for hybridization site than longer ones (e.g., 20-mers).

Caselmann et al. reported that a phosphorothioate ODNs complementary to nucleotides 326–348 spanning the 3' end of the UTR and the start codon of the polyprotein precursor was very efficient in inhibiting HCV gene expression (316, 317). Moreover, strong inhibition of HCV gene expression was still remained with modified oligomers having methylphosphonate or benzylphosphonate modifications located at the termini (316, 318). Inhibition correlated with induction of RNase H activity. Furthermore, these oligomers could be coupled with cholesterol or bile acid to enhance their lipophilicity for improved liver specific delivery (319).

An independent investigation conducted by Vidalin et al. identified a domain within 5' UTR, which contains the conserved pyrimidine-rich tract (nt 103–138), as a new region susceptible to ODN inhibition (320). They also evaluated α-anomer phosphodiesters ODNs. It was found that α-ODNs inhibited HCV translation as efficiently as their β-ODNs counterparts. Wands et al. demonstrated that translation of HCV RNAs was efficiently inhibited by antisense RNA when the HCV core-luciferase cDNA was co-transfected with antisense RNA-producing contracts in Huh7 cells (321).

**2.4.1.2 Inhibitors of HCV Internal Ribosome Entry Site.** As a crucial RNA genomic structure required by HCV for initiation of translation, HCV internal ribosome entry site (IRES) has become an attractive target for therapeutic intervention. A small yeast RNA (a 60-nt-long RNA called inhibitor RNA or IRNA), which has previously shown to selectively block internal initiation of translation programmed by poliovirus RNA, has also shown to block HCV IRES-mediated translation in transient transfection of hepatoma cells (Huh-7) and a hepatoma cell line constitutively expressing IRNA. In these cells, it was further shown that replication of chimeric po-

liovirus containing the HCV IRES element was blocked (322). Site-directed mutagenesis studies suggested that the secondary structure of IRNA might be important in its competing with viral IRES structural elements for the binding of cellular proteins required for IRES-mediated translation (323).

Using an *in vitro* assay in which IRES-dependent translation of luciferase function could be selectively suppressed by the presence of inhibitor, BioChem Pharma and OSI Pharmaceuticals reported the findings of HCV IRES inhibition by phenazine and phenazine-like molecules by screening a compound library and fungal extracts (324). The hit compound (**136**) is also known as neutral red,

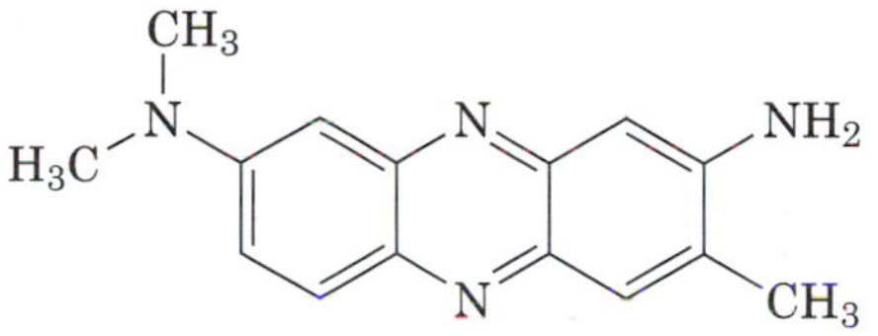

(136)

which is a dye used commonly in antiviral assays. The central ring seemed crucial for activity because the open ring analogs exhibited much lower or no activity. Polar substituents at positions 2 and 8 were also important for the inhibitory activity (324). Finally, Eisai Co. has patented several low molecular weight inhibitors, such as compounds (**137**) and (**138**) (see references in Refs. 324–326).

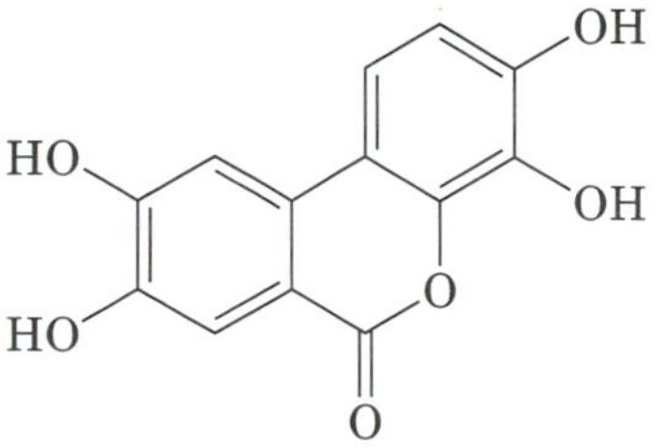

(137)

**2.4.2 HCV NS3 Protein.** NS3 is a multifunctional protein in which the *N*-terminal (ca. 180 amino acids) encodes a serine protease responsible for a distinct temporal hierarchy event of cleavage of NS3/4A, NS4A/4B, NS4B/5A, and NS5A/5B junctions, generating four

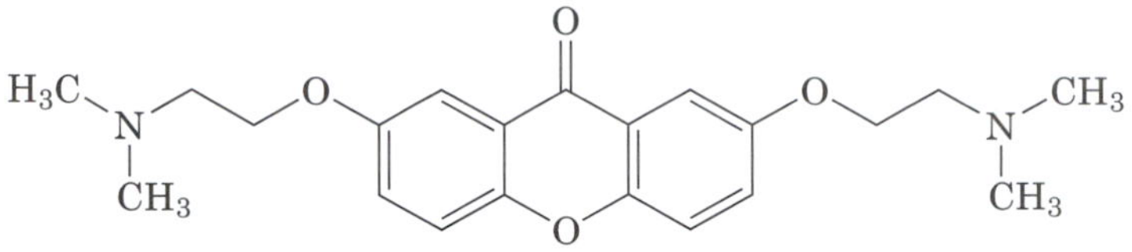

**(138)**

mature viral non-structural proteins, including NS4A, NS4B, NS5A, and NS5B. The remaining C-terminal two-thirds (ca. 450 amino acids) of the NS3 protein encode a nucleic acid–stimulated nucleoside triphosphatase (NTPase) and a helicase activities (327).

The interaction between the NS3 and its cofactor NS4A is required for the proteolytic activity of NS3. NS4A is a relatively small protein (54 amino acids). Binding of NS4A brings about several conformational changes, resulting in the stabilization of the conformation of the N-terminal domain of the protease and optimization of the alignment of the catalytic triad of His-57, Asp-81, and Ser-139 (327, 328). The structure of the isolated NS3 protease domain either in the presence or in the absence of the cofactor peptide was resolved by X-ray crystallography and by NMR spectroscopy.

A zinc ion is coordinated tetrahedrally by Cys-97, Cys-99, Cys-145, and His-149 at a site located remote from the active site. Interestingly, these residues are conserved in all known HCV genotypes. The zinc ion is believed to be required for structural integrity and activity of the enzyme (329, 330).

The NS3 serine protease has been regarded as one of the preferred targets for the development of anti-HCV agents because it presents three potential targets for antiviral design: (1) the enzyme active site, (2) the structural zinc-binding site, and (3) the NS4A binding site (327). However, the interaction between NS3

and NS4A is considered an unlikely target for the development of inhibitors because this region involves a very large surface area and the two components are tightly intercalated (328, 331).

#### 2.4.2.1 Inhibitors of HCV NS3 Protease

*2.4.2.1.1 Peptide-Based Inhibitors.* The NS3-dependent cleavage sites of the polyprotein have a consensus feature with cleavage occurring after cysteine (at the three intermolecular cleavage sites) or threonine (at NS3-4A intramolecular cleavage site) (Table 10.3). Other conserved features are an acid residue in the $P_6$ position (aspartic acid or glutamic acid), a small residue (serine or alanine) in $P_{1'}$, and a hydrophobic residue in the $P_{4'}$ position (332). Crystal structural studies have shown that the active site of the HCV enzyme is extended, shallow, with very little surface feature, and solvent exposed, requiring multiple weak interactions for binding of substrates and inhibitors. Moreover, NS3/4A is an induced-fit enzyme, requiring both the cofactor and the substrate to acquire its bioactive conformation (333). These characteristics pose significant challenge for the design of inhibitors. This challenge is demonstrated by the fact that common protease inhibitors, including serine protease inhibitors, are not effective against the HCV NS3 protease or are only active at high concentrations (334, 335).

The minimum length required for a peptide substrate is a decamer spanning from $P_6$ to $P_{4'}$ and incorporating preferentially all of these

**Table 10.3   Amino Acid Sequences of HCV Protease Cleavage Sites (331)**

| | $P_1 \sim P_{1'}$ |
|---|---|
| NS3–4A | H$_2$N $\cdots$ Asp Leu Glu Val Val Thr $\sim$ Ser Thr Trp Val $\cdots$ OH |
| NS4A-4B | H$_2$N $\cdots$ Asp Glu Met Glu Glu Cys $\sim$ Ala Ser His Leu $\cdots$ OH |
| NS4B-5A | H$_2$N $\cdots$ Asp Cys Ser Thr Pro Cys $\sim$ Ser Gly Ser Trp $\cdots$ OH |
| NS5A-5B | H$_2$N $\cdots$ Glu Asp Val Val Cys Cys $\sim$ Ser Met Ser Tyr $\cdots$ OH |

$\sim$, scissile bond.

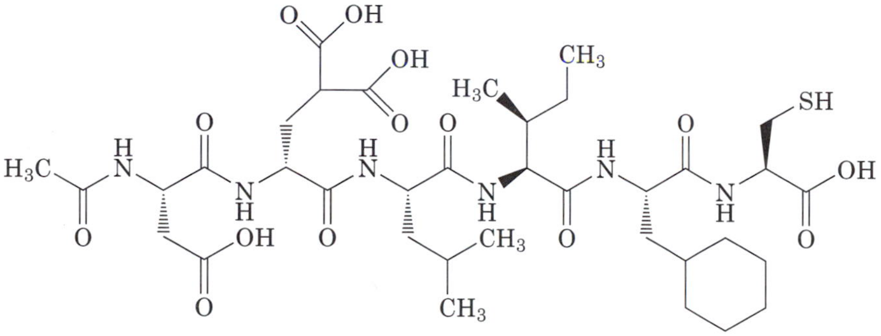

(139)

conserved features (336). Because it will be difficult to develop a large peptide substrate into therapeutic agent, recent research effects have been looking into smaller peptide-based inhibitors guided by SAR analyses of substrate specificities.

Investigations reported independently by Llinàs-Brunet et al. of Boehringer Ingelheim (Canada) (337) and Steinkühler et al. of IRBM in Italy (338) have shown that the NS3 protease undergoes inhibition by the $N$-terminal cleavage products of substrate peptides derived from the NS4A-4B [e.g., Asp-Glu-Met-Glu-Glu-Cys (338)], NS4B-5A, and NS5A-5B [e.g., Asp-Asp-Ile-Val-Pro-Cys (337)] cleavage sites.

Based on these hexapeptides, attempts have been tried to maximize the enzyme-inhibitor interaction through optimization of each amino acid residue by using single amino acid substitution and combinatory chemistry. The following are some substrate specificities observed in the SAR studies. (1) Efficient inhibitor binding requires two electrostatic interactions that involve both the $P_1$ carboxylic acid and a $P_6$ acidic residue. Moreover, the $P_1$ carboxylic acid functionality contributes most to the potency and specificity to these inhibitors (337, 339, 340). (2) $P_2$–$P_4$ sites prefer hydrophobic residues (332, 339, 341). (3) At the $P_5$ site, a negative charge is not an absolute requirement (339). Substitutions with D-amino acids generally result in enhanced potency (337, 339, 342).

After sequential optimization of the initial NS4A-4B inhibitor sequence (Ac-Asp-Glu-Met-Glu-Glu-Cys-OH; $IC_{50}$, 1 $\mu M$), Ingallinella et al. at IRBM reported a more than 600-fold increase in potency observed with Ac-Asp-D-Gla-Leu-Ile-Cha-Cys-OH (139) ($IC_{50}$, 1.5 n$M$) (339). It should be noted that the enzyme assay was based on the protease domain of NS3 protein. Shortening the length of the peptidic inhibitors has shown to have significant detrimental effect on the binding of inhibitors to the enzyme. However, it was later shown that the decrease in potency was much less when determined in the assay system that uses the full-length NS3 protein (protease-helicase/NTPase), suggesting that the helicase domain might have a significant influence on binding of protease inhibitor to the enzyme (343). It should also be noted that the $IC_{50}$ values vary according to different assay systems used by different groups of investigators.

A common strategy for the design of protease inhibitors is to incorporate an electrophilic carbonyl group like aldehyde (332, 340, 344), boronic acid (332, 345), trifluoromethyl ketone (340), ketoamide (340, 346, 347), ketoacid (344, 348, 349), and diketone (346) to the C-terminal (in place of the scissile amide bond) of a synthetic peptidic molecule as serine trap. The electrophilic carbonyl groups are anticipated to form a transition-state analog of the tetrahedral intermediate with the $\gamma$-OH nucleophile of the active site serine residue (340, 349).

However, because that there is a strong preference for cysteine at the $P_1$ position (337) and the cysteine's nucleophilic sulfhydryl side-chain is incompatible with the presence of an electrophile within the same molecule, this cysteine needs to be replaced with other small, hydrophobic amino acids to avoid this potential intramolecular interaction, which gener-

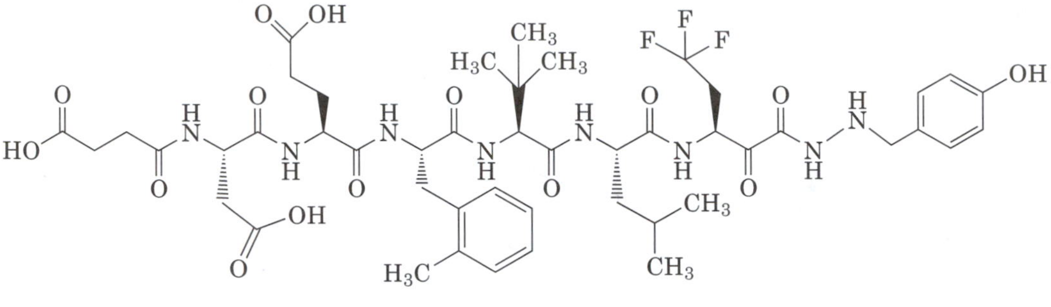

**(140)**

ally leads to inactivation of the inhibitor. α-Amino acids such as allylglycine, propargylglycine, aminobutyric acid, trifluroaminobutyric acid (332), difluoroaminobutyric acid (344, 348, 349), 1-amino-cyclopropylcarboxylic acid (342), and norvaline (340, 342) have been reported as the $P_1$ cysteine mimics.

Combining the optimal substitutions, Roche (Welwyn, UK) reported a series of nanomolar peptide-based inhibitors as represented by compound (**140**) (IC$_{50}$, 4 n$M$) (347) and compound (**141**) (IC$_{50}$, 80 n$M$) (345). Boehringer Ingelheim (Canada) reported compound (**142**) as a HCV serine protease-specific inhibitor (IC$_{50}$, 27 n$M$) (342). It is of interest to note that large aromatic group in $P_2$ resulted in stronger binding. $N$-terminal truncation yielded tetrapeptide (**143**), which is smaller but still remains potent (IC$_{50}$, 3.5 μ$M$) (342).

IRBM showed that hexapeptide α-ketoacid (**144**), incorporating difluoroaminobutyric acid in the $P_1$ position, acted as a potent, slow-binding inhibitor of the NS3 protease (IC$_{50}$, 1 n$M$) (349). The mechanism of action was thought to involve the rapid formation of a

reaction intermediate followed by a slow conversion into a tight 1:1 covalent complex (349). Similar mechanism of inactivation was also suggested for tripeptide (**145**) (IC$_{50}$, 1.4 μ$M$) (344, 349), and this was further supported by the crystal structure of the inhibitor bound to the HCV NS3–4A complex (344, 348). The same investigators have also incorporated the crucial α-ketoacid moiety into the preparation of peptidomimetics (e.g., compound (**146**); IC$_{50}$, 0.18 μ$M$) by *de novo* design (344).

Peptidic inhibitors described so far are mainly based on the determinants of ground-state substrate binding to the enzyme, which reside in the P region. The P′ region of the substrate is important for catalysis, whereas it contributes little to binding (351). Nevertheless, there are binding pockets in the S′ region that can be exploited for inhibitor binding. Ingallinella et al. of IRBM took non-cleavable decapeptides spanning $P_6$–$P_{4'}$, sequentially optimized the P′ residues, and generated sub-nanomolar inhibitors (351). For instance, the IC$_{50}$ for the decapeptide Ac-Asp-D-Glu-Leu-Ile-Cha-Cys-Pro-Cha-Asp-Leu-NH2 was <0.2

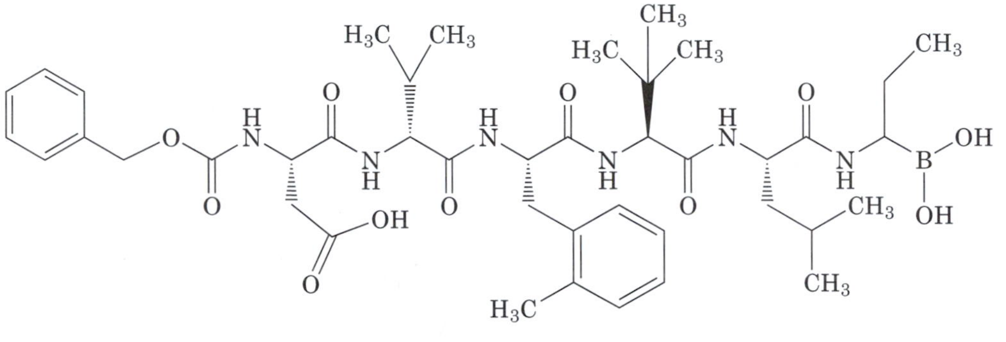

**(141)**

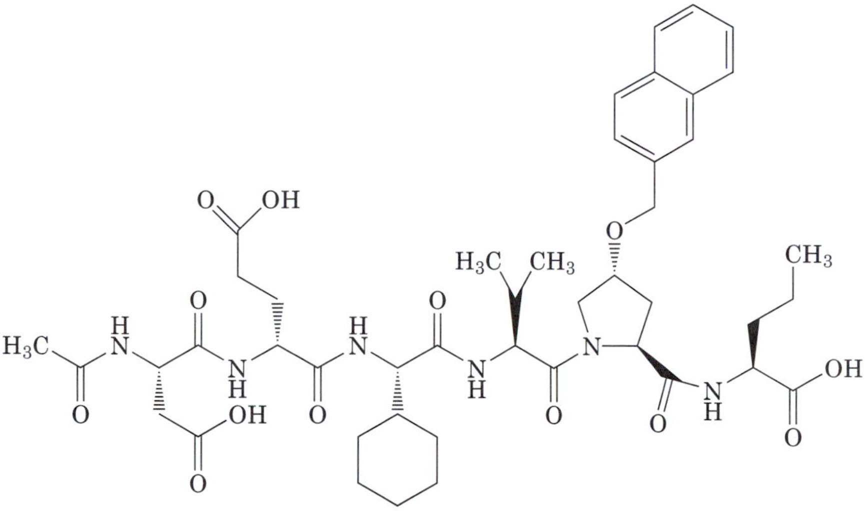

**(142)**

nM, whereas the IC$_{50}$ of hexapeptide Ac-Asp-D-Glu-Leu-Ile-Cha-Cys-OH, which is corresponding to P$_6$–P$_1$, was 15 nM (351).

*2.4.2.1.2 Low Molecular Weight Inhibitors.* Through a screening effort, Sudo et al. showed a number of benzamide derivatives as having activity against NS3-4A protease (352). The most active one in the series was RD3-4082 (**147**) (IC$_{50}$, 5.8 $\mu M$), which contains a long N-alkyl chain. However, this compound was also inhibitory to other seine pro-

teases. The most selective compound in the series against HCV serine protease was RD2-4039 (**148**), which has a N-phenyl moiety. Interestingly, RD2–4039 seemed to share a common structure with other HCV serine protease selective inhibitors, such as (**149**), reported independently by Kakiuchi (353). Compound (**149**) was found to be a non-competitive inhibitor, suggesting that these compounds do not bind to the substrate-binding pocket.

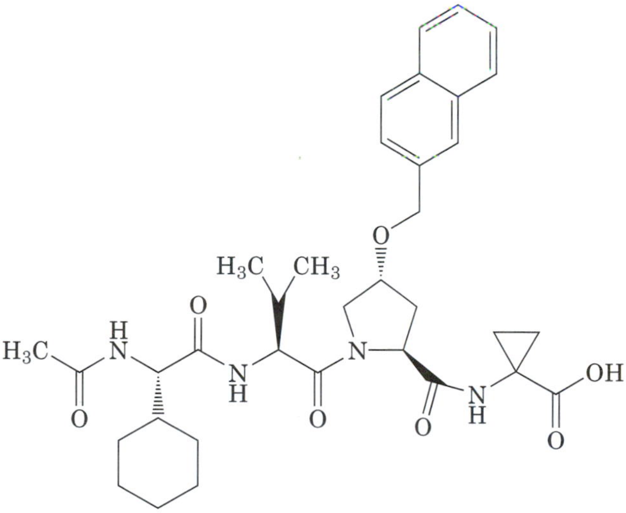

**(143)**

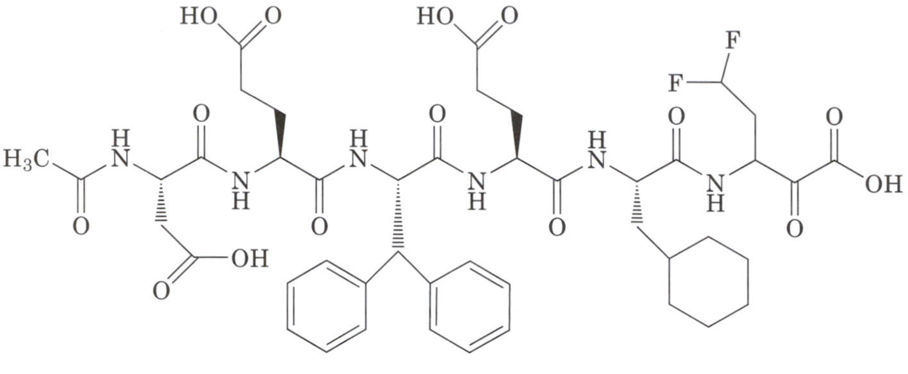

(144)

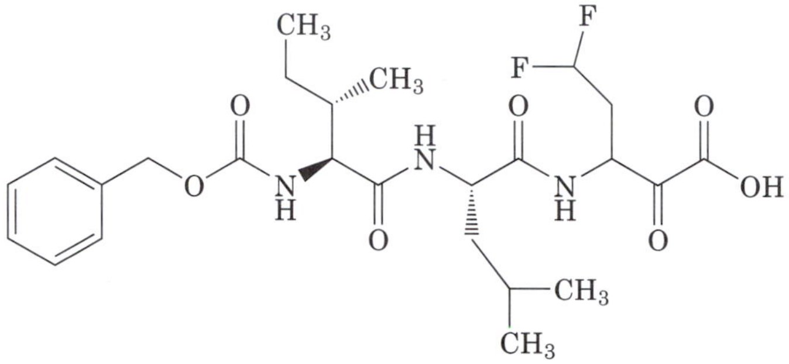

(145)

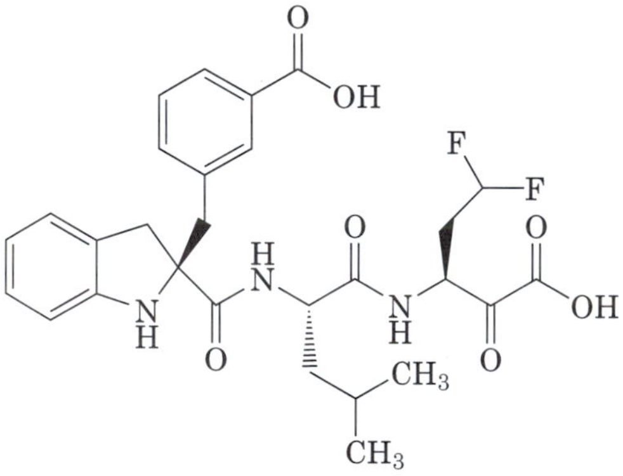

(146)

Derivatives of rhodanine have also been reported through screening, e.g., compound (150) (353) and RD4-6205 (151) (354), as well as through synthesis, e.g., compound (152) (355). Unfortunately, these compounds also showed to be non-selectively inhibitory to at least one other serine protease (chymotrypsin,

trypsin, plasmin, and elastase) in addition to their activity against HCV serine protease. Kinetic analysis revealed that RD4-6205 inhibits the HCV protease in a non-competitive manner.

HCV protease-guided screening of natural products yielded Sch 68631 (153) isolated from the fermentation culture broth of *Streptomyces* sp. (356), Sch 351633 (154), isolated from the fungus *Penicillium griseofulvum* (357), and mellein (155) from *Aspergillus ochraceus* (358).

Because HCV NS3 protease is a zinc-containing protein, Zn-ejecting compounds, such as 2,2'-dithiobis[(N-phenyl)benzamide] (156), which might interfere metal ligation could show enzyme inhibitory activity (328). However, zinc is essential for several cellular enzymes; ejecting zinc might cause non-selective interference with normal cell metabolism (331).

Other series of compounds, such as APC-6336 (157), might take advantage of forming a

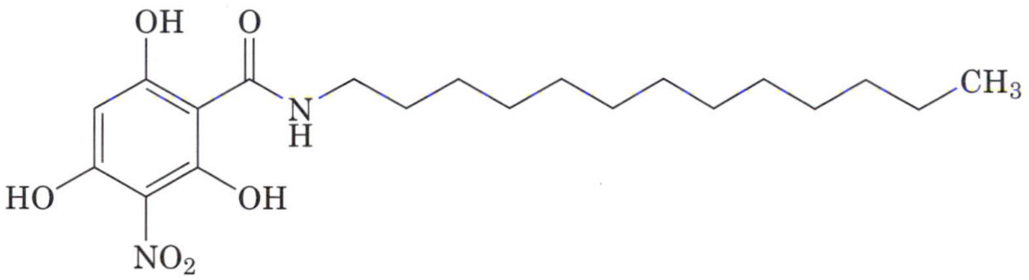

**(147) RD3-4082**

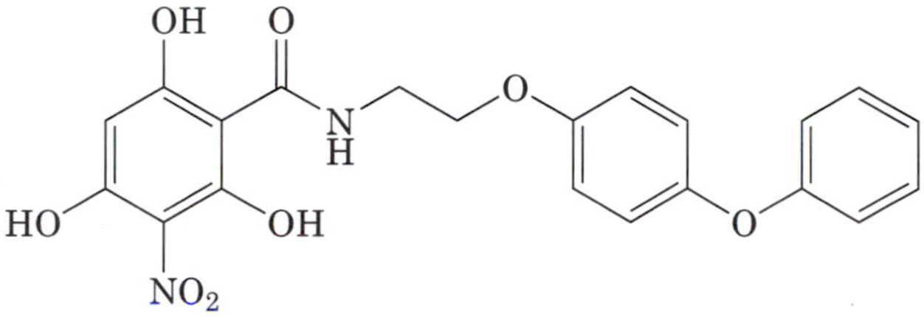

**(148) RD2-4039**

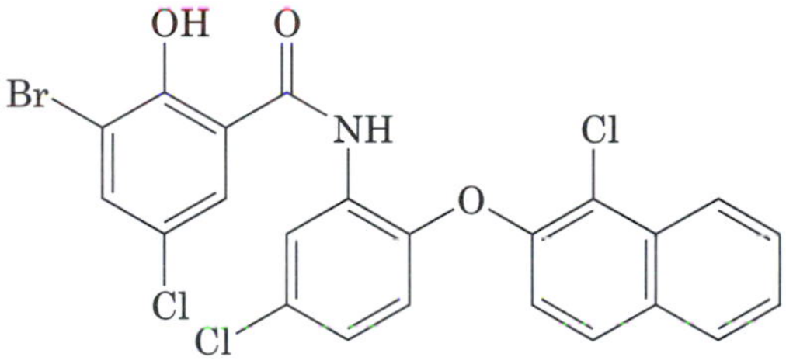

**(149)**

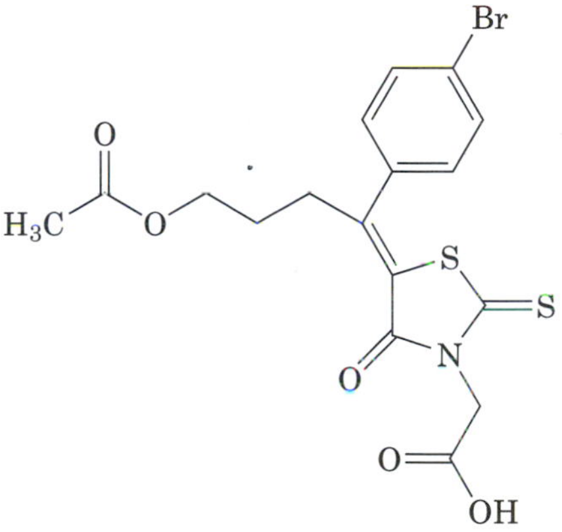

**(151) RD4-6205**

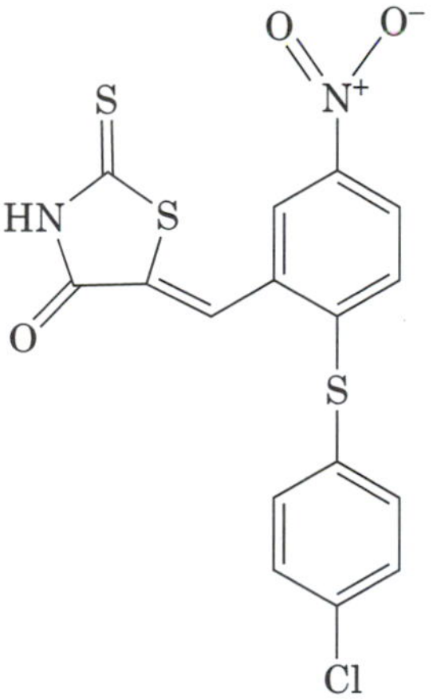

**(150)**

tetrahedral coordination with the zinc atom to help anchor their binding to the active site and thereby bring about enzyme inhibition. APC-6336 showed sub-micromolar inhibition against HCV NS3 in the presence of $Zn^{2+}$. The activity dropped more than 800-fold in the absence of $Zn^{2+}$ (359).

*2.4.2.1.3 Macromolecules.* Nishikawa et al. have used a genetic selection strategy, which involves repeated rounds of selection and amplification, to isolate NS3-binding RNA aptamers from pools of random RNA.

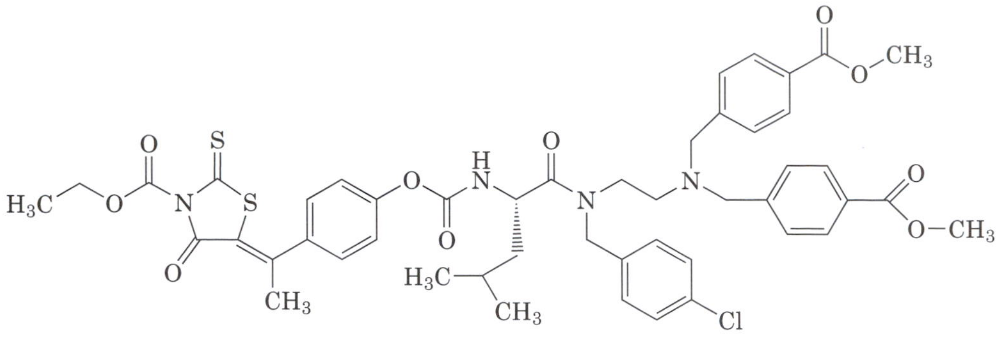

**(152)**

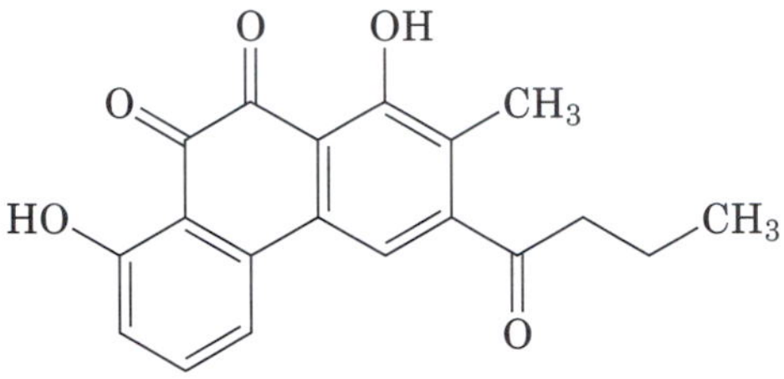

**(153)** SCH 68631

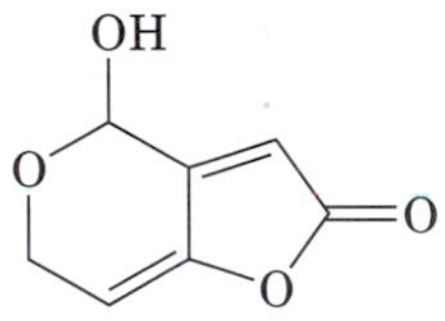

**(154)** SCH 351633

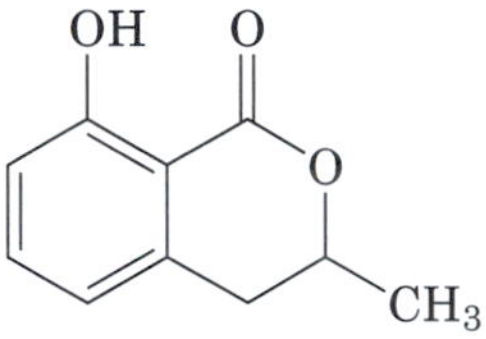

**(155)** Mellein

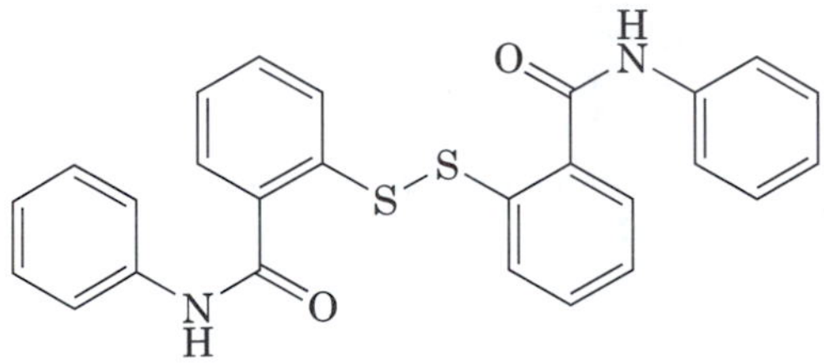

**(156)**

The aptamers so selected inhibited both the proteolytic and helicase activities of NS3 (360, 361). High affinity RNA aptamers selected by using truncated NS3 containing only the protease domain (ΔNS3) could bind to ΔNS3 with a binding constant of about 10 n$M$ and inhibit approximately 90% of the protease activity of ΔNS3 and a full-length NS3 protein fused with maltose-binding protein (362). Amino acid residues essential for aptamers binding have been revealed by surface plasmon resonance measurements (363).

Sollazzo et al. reported a similar affinity strategy for selection of macromolecular inhibitors, called minibodies ("minimized" antibody-like proteins), by phage display techniques (364). Based on known determinants for NS3 protease substrate recognition, affinity selection from a biased repertoire of minibody variants identified a competitive inhibitor of this enzyme (365). Moreover, characterization of the minibody inhibitor led to the synthesis of a cyclic hexapeptide mimicking the bioactive loop of the parent macromolecule. The same authors also reported the design of nanomolar NS3 protease inhibitors generated by reshaping the active site-binding loop of eglin c, which is a known potent inhibitor of several serine proteases isolated from *Hirudo medicinalis* (366).

More recently, Ueno et al. reported the development of a monoclonal antibody (Mab), 8D4, which recognized the active site of HCV NS3 protease (367). Interestingly, the variable fragment (Fv) of 8D4 had an inhibition profile almost identical to that of the parent IgG (368).

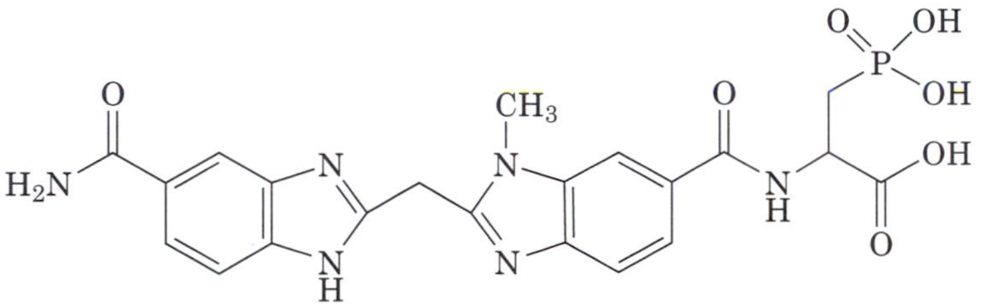

(157) APC-6336

### 2.4.2.2 Inhibitors of HCV NS3 Helicase.

The helicase activity associated with HCV NS3 protein has also been targeted for potential therapeutic intervention (369). The crystal structure of this enzyme has been determined to provide insights into the mechanism of unwinding (369–372). In addition to unwinding dsRNA, this enzyme also unwinds dsDNA (373). Using a DNA duplex substrate and recombinant HCV NS3 produced in *E. coli*, Biochem Pharma recently reported an assay system for HCV NS3 helicase activity that might be suitable for high-throughput screening of potential inhibitors (374). Other reported assay systems include an ELISA using a non-radioactive dsRNA substrate (375) and a scintillation proximity assay using radio labeled RNA/DNA hetero-duplex as the substrate (376). These assays can be amenable to high-throughput mode.

Notwithstanding, there have been only a few HCV helicase inhibitors reported. Viro-Pharma has patented two series of long-chain compounds as low micromolar inhibitors (**158** and **159**) (see references in Refs. 326 and 377). A preliminary SAR studied by Sim et al. showed that the essential elements for inhibitory activity were the NH group within the benzimidazole ring, the benzene group at the C$_2$ position of benzimidazole, and the nature of the linker (378).

By studying with ribavirin 5'-triphosphate, the ATP-binding domain of NTPase/helicase has been suggested as a potential antiviral target (379). Ribavirin 5'-triphosphate showed a competitive inhibitory mechanism with respect to ATP.

### 2.4.3 Inhibitors of NS5B RNA-Dependent RNA Polymerase.

The HCV NS5B protein, which encodes RNA-dependent RNA polymer-

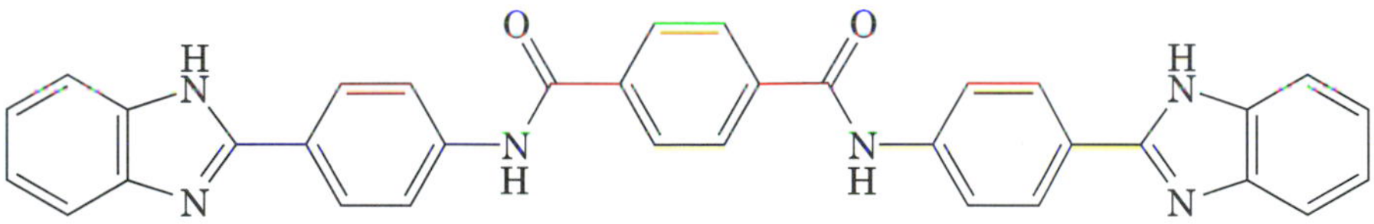

(158)

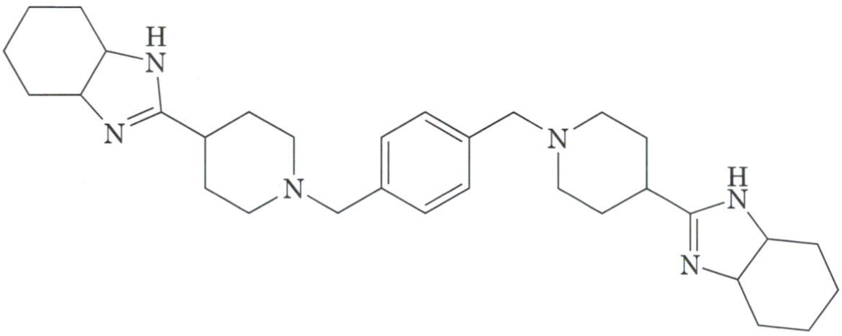

(159)

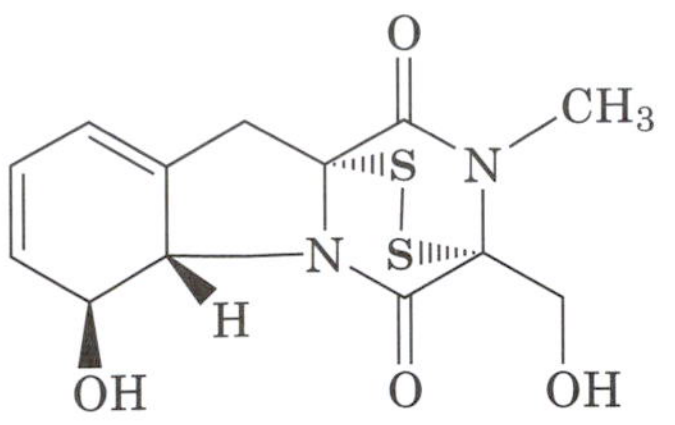

**(160)** Gliotoxin

**(161)**

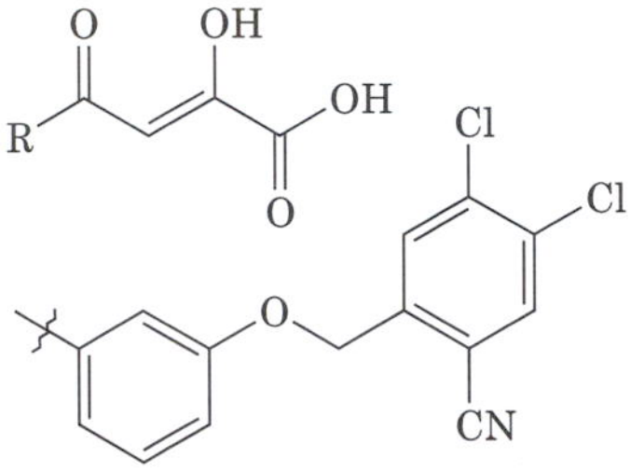

**(161a)** R =

**(161b)** R =

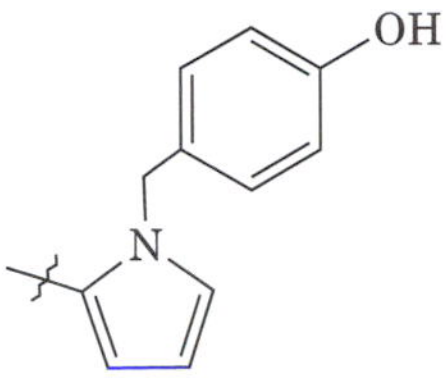

**(161c)** R =

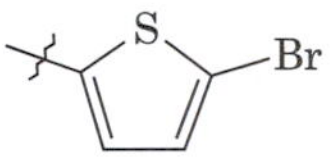

**(161d)** R =

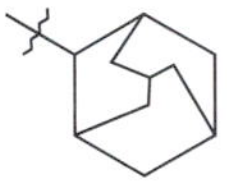

ase (RdRp) activity, involves in the synthesis of complementary (−)-RNA using the genome as the template and the subsequent synthesis of genomic RNA using this (−)-RNA. In addition to its important role in the viral RNA replication, the NS5B amino acid sequence is highly conserved among different HCV strains, this protein has been considered as an attractive target for antiviral therapy (377,

380). Biochemical and kinetic characterizations and the crystal structure of this enzyme have been reported (380–383). Gliotoxin (**160**), a known poliovirus 3D RdRp inhibitor, inhibited HCV NS5B RdRp in a dose-dependent manner (381); however, broad-spectrum antiviral agent ribavirin (in triphosphate form) did not show any effect (380). In a recent review article, there were several patented diketoacids (**161a-d**) cited as low nanomolar inhibitors (377).

**2.4.4 VX-497.** Although ribavirin may inhibit viral replication through multiple mechanisms of action, the major event is thought to be a depletion of the intracellular GTP and dGTP pools as a result of the inhibition of inosine 5′-monophosphate dehydrogenase (IMPDH) (384) (for a review of IMPDH and its inhibitors, see Refs. 385–388). Because (*1*) HCV infection involves both viral proliferation and liver inflammation and (*2*) blocking IMPDH could block the proliferation of certain cell types, such as lymphocytes (389), and the growth of viruses, including viruses closely related to HCV, such as BVDV (384), Vertex Pharmaceuticals investigators rationalized that inhibition of IMPDH might have potential to treat HCV infection (390). However, studies by Schering-Plough investigators pointed to the opposite that inhibition of IMPDH is unlikely to be effective approach, because the major mechanism of action of ribavirin is not related to its inhibition of host IMPDH (391). The hypothetical role of IMPDH in HCV life cycle is under scrutiny with the studies of VX-497 (**162**). VX-497 is a potent, reversible uncompetitive IMPDH inhibitor currently in clinical trials conducted by Vertex to treat HCV patients unresponsive to IFN (392) as well as in combination with IFN to treat patients who have not previously

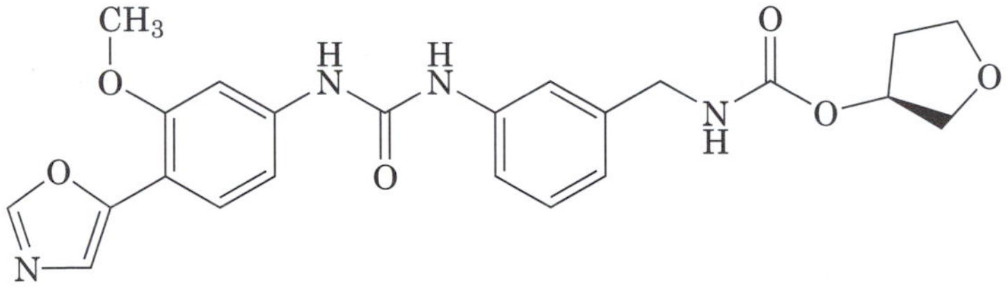

**(162)** VX-497

received HCV antiviral therapy (Vertex Pharmaceuticals Press Release, July 3, 2000). The company has also conducted clinical trials of VX-497 as a treatment for psoriasis, an autoimmune disease of skin.

The process leading to the discovery of VX-497 began with a molecular shape cluster-based screening of chemical library, followed by molecular modeling using DOCK program to model screening leads into the enzyme active site generated from the X-ray crystal structure of mycophenolic acid (MPA) bound to IMPDH (390). The crystal structure of VP-497 enzyme complex revealed several new interactions that are not observed in the binding of MPA (390).

VX-497 is a nanomolar inhibitor of both isoforms of human IMPDH ($K_i$ = 7 and 10 n$M$ against type I and II, respectively) (390). Its $K_i$ values are approximately 30-fold lower than that of ribavirin. In cultured cells, VX-497 showed broad-spectrum activity against several DNA and RNA viruses. Particularly, it was 17- to 186-fold more potent than ribavirin against HBV, HCMV, RSV, HSV-1, PIV-3, EMCV, and VEEV (384, 393). Moreover, the finding that both compounds' antiviral effects could be reversed by the addition of guanosine further strengthened the association of inhibition of IMPDH with the antiviral activities of both compounds (384).

**2.4.5 Model Systems to Study HCV Replication.** Robust and reliable cell-based *in vitro* screening systems are critical for the development of antiviral drugs. Unfortunately, no natural human liver cells are available for this purpose. Many cell lines have been described as methods for *in vitro* HCV replication; however, these systems are generally limited by the poor reproducibility and support low level of HCV replication (for a review of HCV cell culture systems, see Refs. 394–397). More recently, two independent research groups led by Bartenschlager (398) and by Rice (399), respectively, reported their efforts in the development of highly efficient cell culture systems based on the self-replication of engineered HCV sub-genomic RNAs (replicons) in transfected human hepatoma cell (Huh7) line. In these replicons, the neomycin phosphotransferase gene replaced the HCV structural genes. Using these systems high-level natural HCV RNA replication could be maintained stably for considerable periods of time and the replication was totally dependent on the activity of HCV NS5B RNA polymerase. Although these replicons do not represent a full natural viral replication and do not produce infectious virus particles, the systems seem to be suitable for the evaluation of HCV replication inhibitors because the replicons encode all the viral functions required for RNA replication (397, 400). Ultimately, an ideal system, which hopefully can happen in the future, is one that efficiently produces infectious HCV as well as a permissive cell line (395). Various animal models such as HCV SCID mouse model and HCV-trimera mouse model (401, 402) have been reported; however, the only reliable animal model for HCV infection is the chimpanzee (for a review, see Refs. 394, 396).

## 2.5 Flaviviruses

The family of flaviviridae contains three genera: hepacivirus [hepatitis C virus (HCV)], flavivirus [e.g., yellow fever virus (YFV), dengue virus (DENV), Japanese encephalitis virus (JEV), West Nile virus (WNV)], and pestivirus [e.g., bovine viral diarrhea virus (BVDV)]. This section will discuss the discovery and development of inhibitors of flaviviruses (for a review, see Ref. 403).

By screening compounds in cultured Vero cells, several known antiviral agents, such as interferon-$\alpha$, ribavirin, and 6-azauridine (**163**) (an OMP decarboxylase inhibitor), proved effective in reducing viral cytopathic effect (CPE) induced by flaviviruses (404). In a

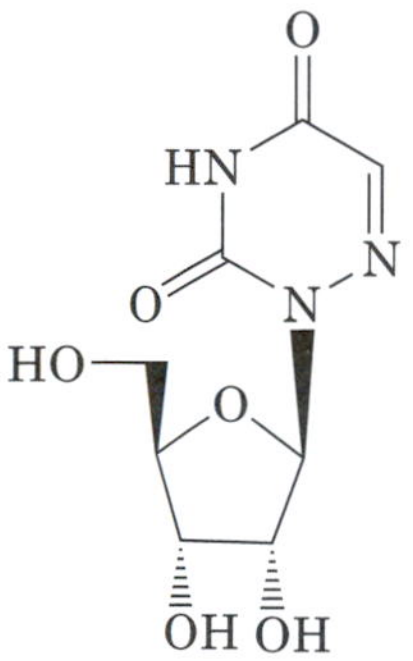

(**163**) 6-Azauridine

separate report, YFV (vaccine strain 17D) was shown to be highly sensitive to two IMPDH inhibitors, mycophenolic acid (**164**) and EICAR (**165**) as well as to a dihydrofolate reductase inhibitor, methotrexate (**166**) (405).

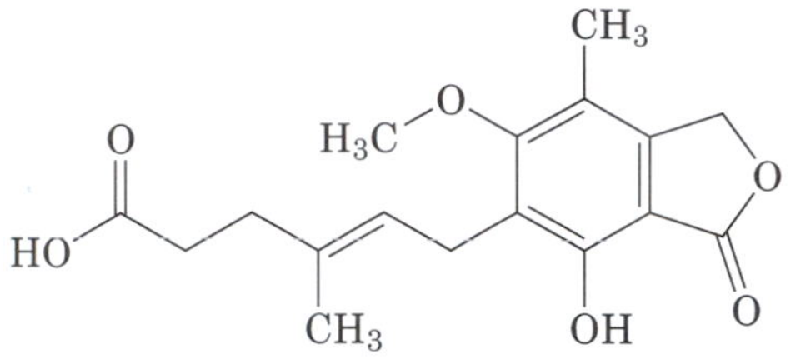

(**164**) Mycophenolic acid

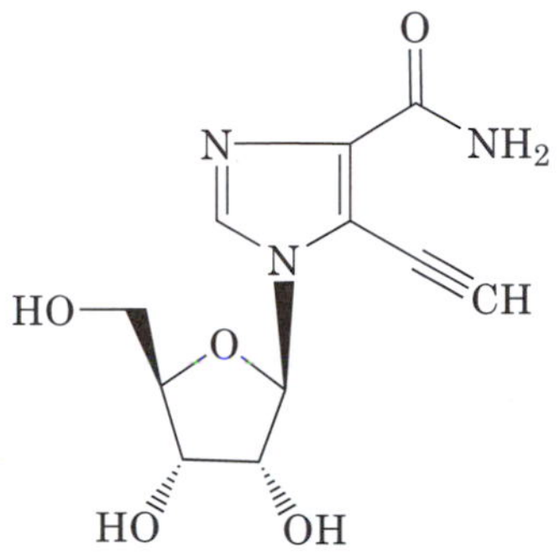

(**165**) EICAR

The recent outbreak of West Nile virus (WNV) in the United States has motivated a search of potential antiviral agents. Recent reports confirmed that WNV was susceptible to 6-azauridine (**163**) (406), cyclopentenyl cytosine (CPE-C) (**167**) (a CTP synthetase inhibi-

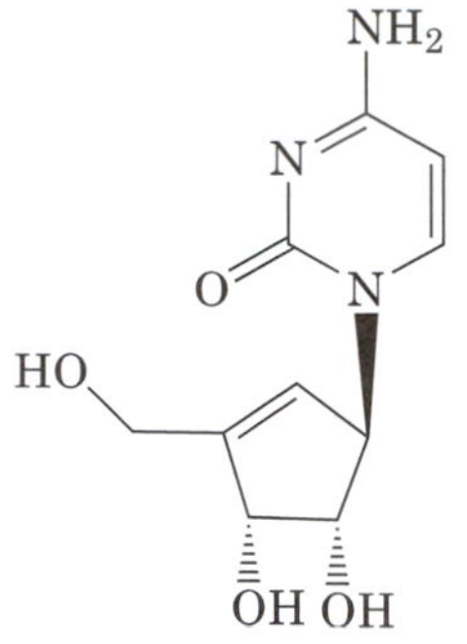

(**167**) CPE-C

tor) (406, 407), mycophenolic acid (**164**), and ribavirin (406, 408) in cell cultures. [Note: 6-azauridine was also inhibitory to sandfly fever Sicilian virus (a phlebovirus in the family of bunyaviridae) replication *in vitro* (409, 410).] Using cytopathic effect assay, ribavirin was active in MA-104 cells ($EC_{50}$ = 5 $\mu$g/mL), but it was not very active in Vero cells ($EC_{50}$ = 178 $\mu$g/mL) (411). This difference in activity might be because of differences in phosphorylation of ribavirin in different cell lines. Using virus yield reduction assay, the antiviral effect of 6-azauridine was seen at day 2, but not at day 6 post-infection (406). It seemed that the inhibition of virus replication by 6-azauridine was transient. On the other hand, the virus yield reduction result of ribavirin was essentially the same at day 2 and day 6 in MA-104 cells, suggesting the antiviral effect of ribavirin was sustained over a 6-day period (411). The modes of ribavirin action against WNV might involve both cellular metabolism and lethal mutagenesis of the viral genome. Riba-

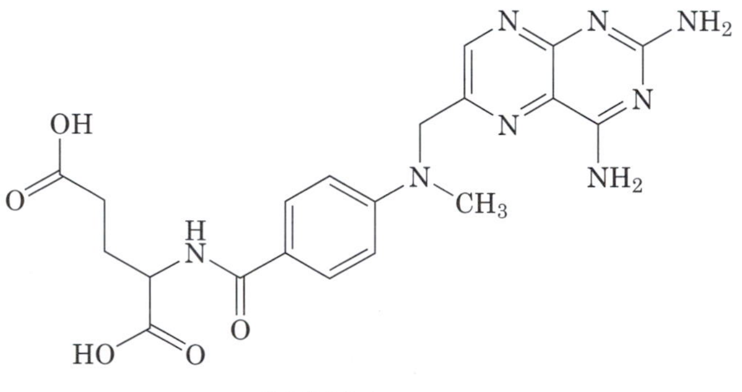

(**166**) Methotrexate

virin has been shown as an RNA virus muta-
gen that would cause RNA virus error catastrophe (22, 412).

The viral genome of mosquito-borne dengue virus encodes a single polyprotein, which
undergoes subsequent proteolytic processing
to form mature proteins by a combined action
of host proteases and a virus-encoded, two-
component protease, NS2B-NS3 serine protease (413). The crystal structure of the catalytic component of the NS3 at 2.1 Å resolution

was recently reported (414–416). It is anticipated that specific NS3 inhibitors could be designed based on this structure. Nevertheless,
several natural products with *in vitro* activity
against dengue virus have been reported recently; these included glabranine (**168**) isolated from *Tephrosia* sp (417), and gymnochrome D (**169**) isolated from the living fossil
crinoid *Gymnocrinus richeri* (418).

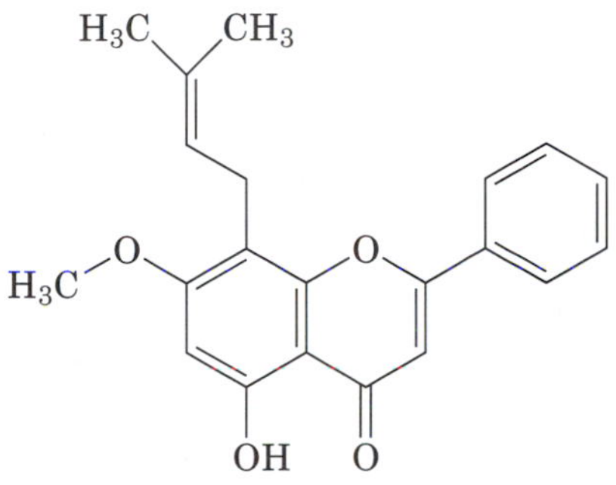

Bafilomycin A1 (**170**), a macrolide antibiotic isolated from the fermentation of *Streptomyces griseus*, was shown to inhibit the
growth of Japanese encephalitis virus (JEV)
in Vero cells (419). Viral inhibition correlated
with the disappearance of acidified cellular
compartments such as endosomes and lysosomes (referred to as ELS), suggesting that
bafilomycin A1-sensitive vacuolar-type proton
pumps are responsible for the acidification of
ELS and that the acidified compartments are
essential for the early phase of JEV infection
(419).

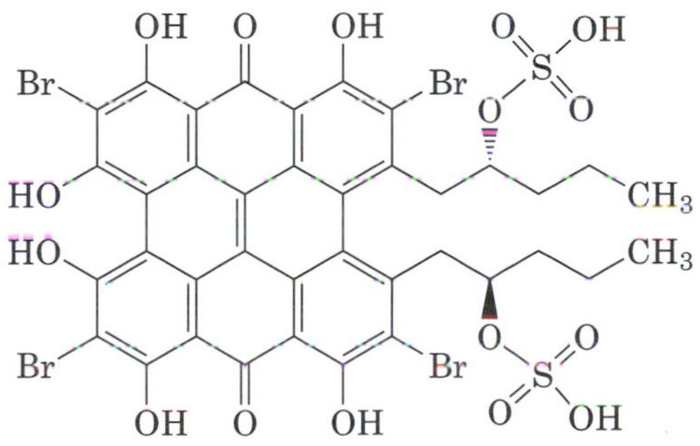

The effect of synthetic derivatives of natural furanonaphthoquinone on the replication
of JEV was also demonstrated in Vero cells
(420). The most active compound in the series,
FNQ3 (**171**), effectively inhibited the expression of viral proteins and also genomic RNA.

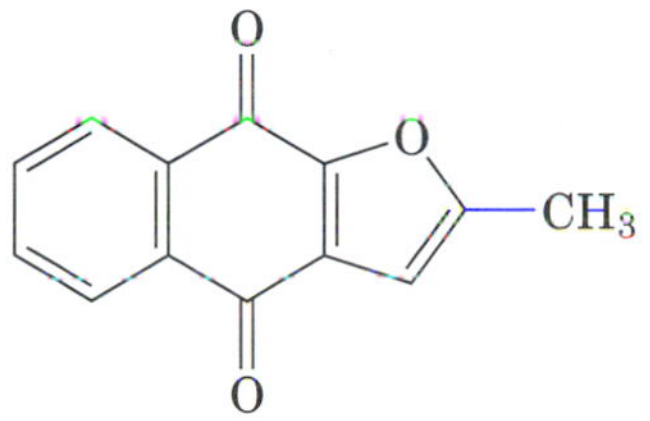

(168) Gabrauine

(169) Gymnochrome D

(171) FNQ3

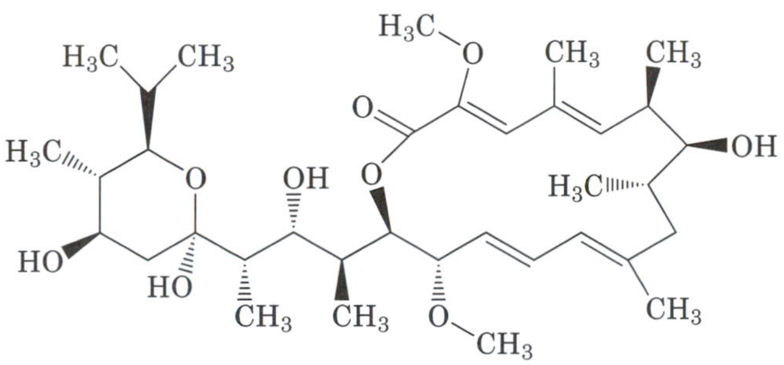

(170) Bafilomycin A 1

In addition, envelop protein E was much more inhibited than viral non-structural protein NS3 (420).

## 2.6   Arenaviruses

A number of anti-retroviral Zn-finger active compounds were tested *in vitro* against arenaviruses and the results demonstrated that (**172**), (**173**), and (**174**) have activities against

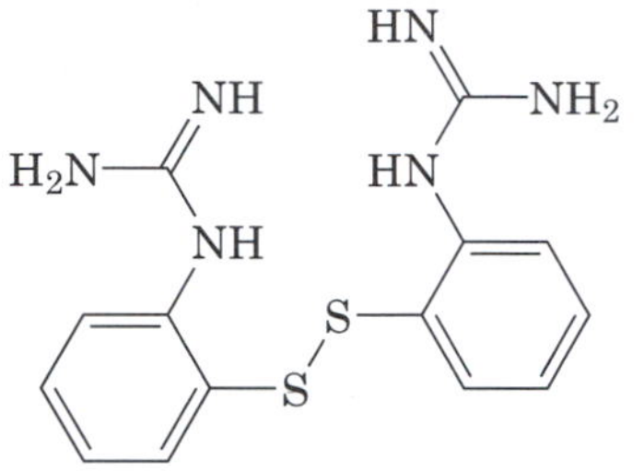

(**172**)

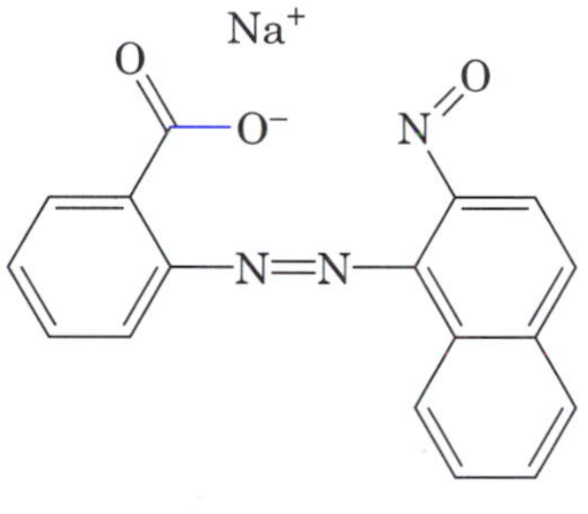

(**173**)

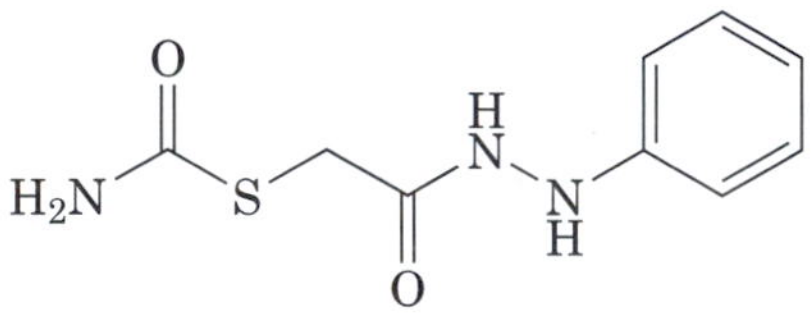

(**174**)

attenuated and pathogenic strains of Junin virus, the etiological agent of Argentine hemorrhagic fever, as well as antigenically related Tacaribe virus and Pichinde virus (421). Other pharmacologically active compounds reportedly showed *in vitro* activity against Junin, Tacaribe, and/or Pichinde virus included two phenotiazines [trifluoperazine (**175**) and chlorpromazine (**176**) (422)], two

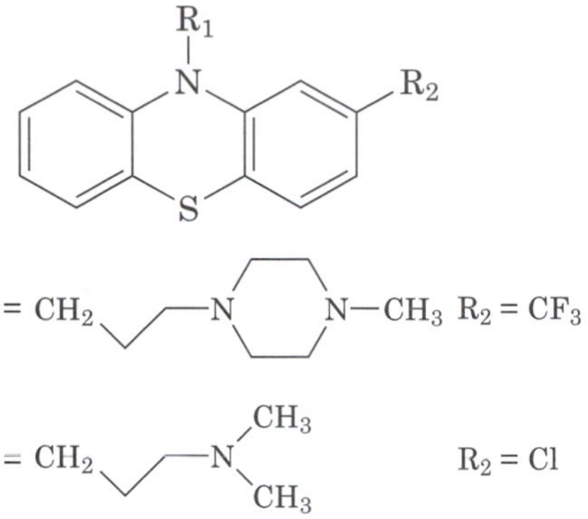

(**175**) $R_1 = CH_2$ ⟶N⟵N—CH$_3$  $R_2 = CF_3$

(**176**) $R_1 = CH_2$ ⟶N(CH$_3$)$_2$   $R_2 = Cl$

myristic acid analogs [2-hydroxymyristic acid (**177**) and 13-oxamyristic acid (**178**) (423)], and a brassinosteroid (**179**) (424). [Note: chlorpromazine also displayed strong inhibition of hepatitis A virus replication in BS-C-1 cells, possibly by preventing virus uncoating (425). Moreover, Prusiner et al. recently reported that chlorpromazine exhibited micromolar inhibition of formation of disease-causing isoform of the normal host prion protein in cultured cells chronically infected with prions (426).]

## 2.7   Rotavirus

Brefeldin A (BFA) (**180**) has been known as having specific activity in blocking protein transport from the endoplasmic reticulum (ER) to the Golgi complex. Because rotavirus uses ER for maturation, treatment with BFA was found to reduce progeny virus yield by 99% at 0.5 $\mu$g/mL (427). Electron microscopy analysis revealed that BFA interfered with the transition from the enveloped particle to the mature double-shelled rotavirus (427).

Sialic acid–containing surface glycoproteins of mature small intestine epithelial cells have been proposed as binding sites for animal rotaviruses. Therefore, as in the case of influenza virus, anti-rotaviral compounds might be derived from sialic acid-based compounds, particularly from multivalent sialic acid derivatives. Synthetic sialylphospholipid (**181**) was shown to exhibit dose-dependent inhibition against simian (SA-11 strain) and human (MO strain) rotaviruses in Rhesus monkey kidney cells (MA-104). The EC$_{50}$ values against SA-11 and MO were 4.4 and 16.1 $\mu M$, respectively

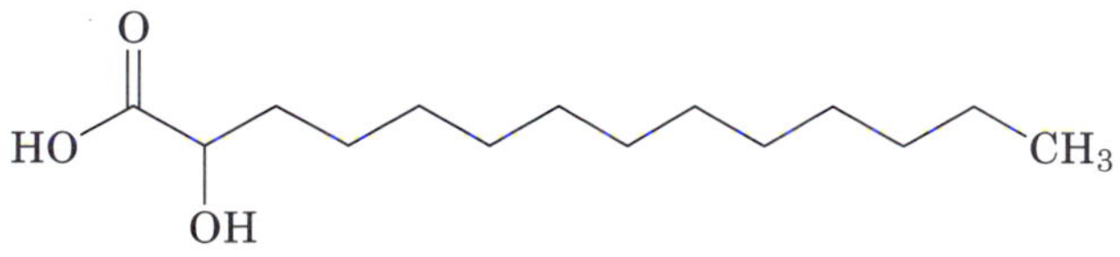

(177) 2-Hydroxytetradecanoic acid

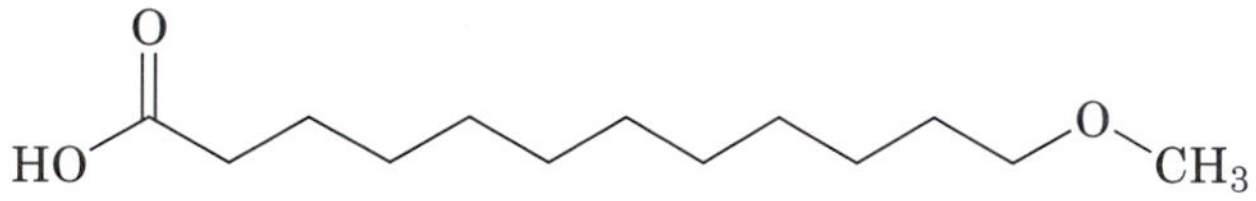

(178) 12-Methoxydodecanoic acid

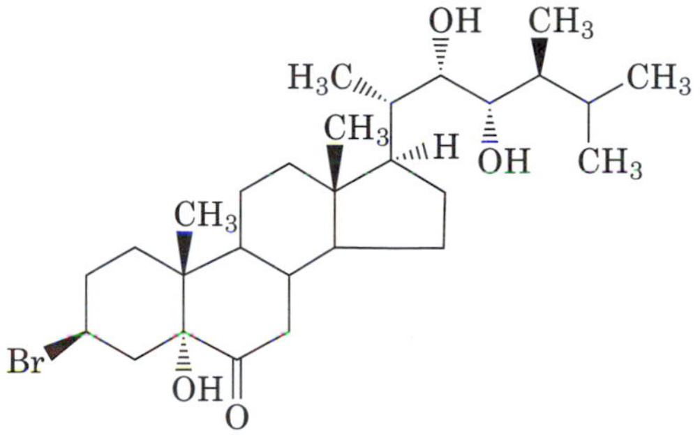

(179)

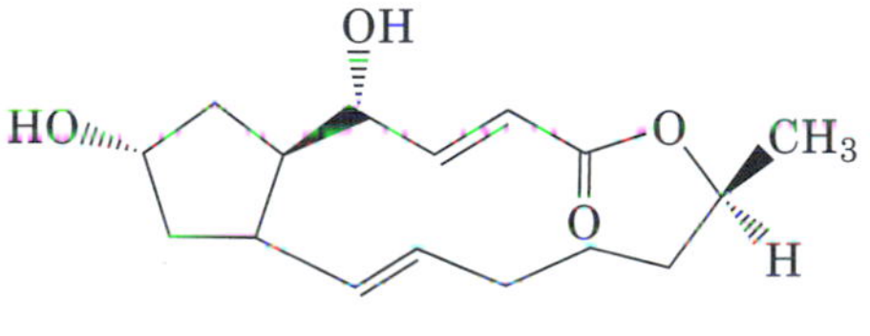

(180)

(428). A number of thioglycosides of sialic acid, exemplified by compound (182), have also been evaluated. Bovine (NCDV) rotavirus seemed to be the most sensitive serotype to these compounds, whereas human rotavirus (Wa strain) was not (429). More recently, sulfated colominic acid (183) was shown to exhibit suppressive effect on simian rotavirus SA-11 and human rotavirus MO infections, but not on Wa, which is a sialic acid–independent rotavirus (430). However, Wa rotavirus remained sensitive to other carbohydrate-containing compounds, such as neohesperidin (184) and hesperidin (185) isolated from the fruit of *Citrus aurantium* (431, 432). On the other hand, the aglycone, hesperetin, was not active.

## 2.8 Rubella Virus

To explore the chemical and biological relationships of 6-substituted uracil derivatives, a family of 2-methoxy- and 2-methylthio-6-[(2′-alkylamino)ethyl]-4(3H)-pyrimidinones, represented by compound (186), have shown various effects on virus yields in a plaque assay in Vero cells against vesicular stomatitis virus, sindbis virus, and rubella virus (433). The anti-rubella activity exhibited by (186) was

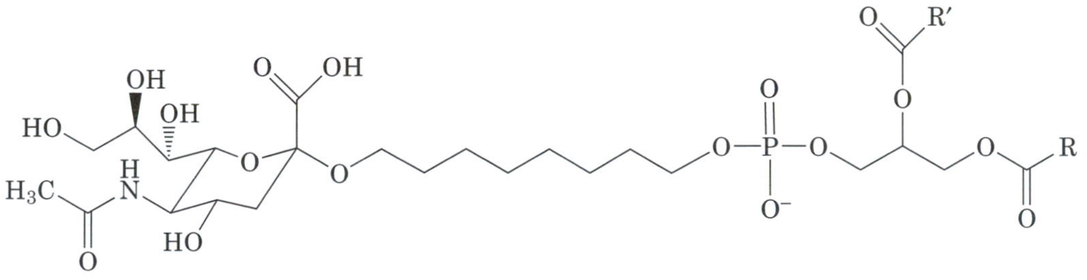

(181) Sialylphospholipid

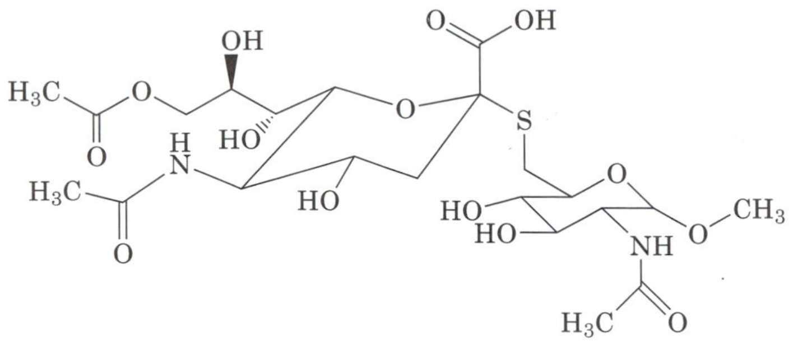

(182)

the most notable, with $EC_{50}$ and $CC_{50}$ of 4 and 250 μg/mL, respectively.

### 2.9 Broad-Spectrum Antiviral Compounds

**2.9.1 S-Adenosyl-L-Homocysteine Hydrolase Inhibitors.** S-adenosyl-L-homocysteine (SAH) hydrolase is an intracellular enzyme that regulates biological transmethylation in general. Because many animal viruses require SAH hydrolase in the methylation of the 5'-terminal residue of viral mRNA for forming the 5'-methylated cap structure necessary for viral protein translation and replication, this enzyme has been recognized as a suitable target for antiviral chemotherapy (434). Its inhibitors have generally shown broad-spectrum-activity against orthopox-, paramyxo-, rhabdo-, filo-, bunya-, arena-, and reoviruses (for a review, see Ref. 142). This enzyme has also been considered as an attractive target for parasite chemotherapy (435, 436) and other medical indications (437). More recent studies with

SAH hydrolase inhibitors on the replication of measles virus and Ebola virus are discussed in the respective sections.

Barnard et al. reported that both D-5'-noraristeromycin (**187**) and its L-isomer (**188**) (synthesized by Schneller et al.) were having potent *in vitro* anti-measles virus (MV) activity as determined by cytopathic effect reduction assay and by virus yield reduction assay (438). The D-like analog (**189**) was also potent MV inhibitors even when added to infected cells 24-h post-virus exposure, implying that an event occurring late in infection such as assembly or egress could be affected. When combined with ribavirin, these compounds demonstrated synergistic (additive) inhibition of MV replication at several concentrations.

Huggins et al. have recently established a lethal mouse model suitable for evaluation of prophylaxis and therapy of Ebola virus (a filovirus) (439). Intraperitoneal administration, thrice daily, of carbocyclic 3-deazaadenosine

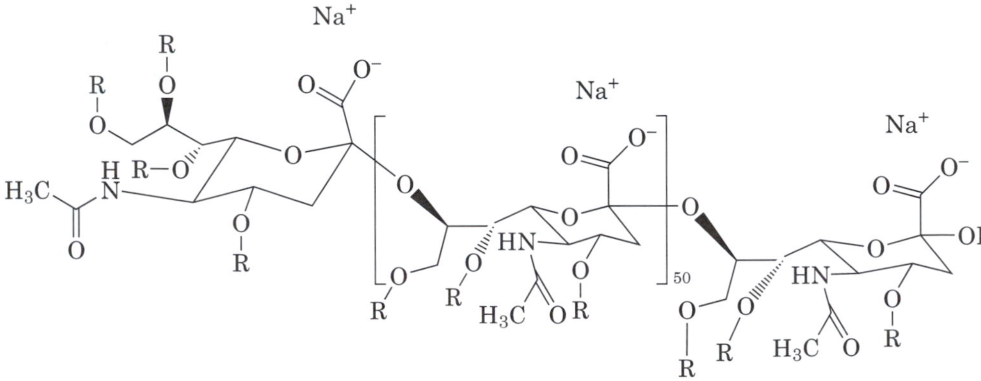

(183) R = H or SO$_3$Na

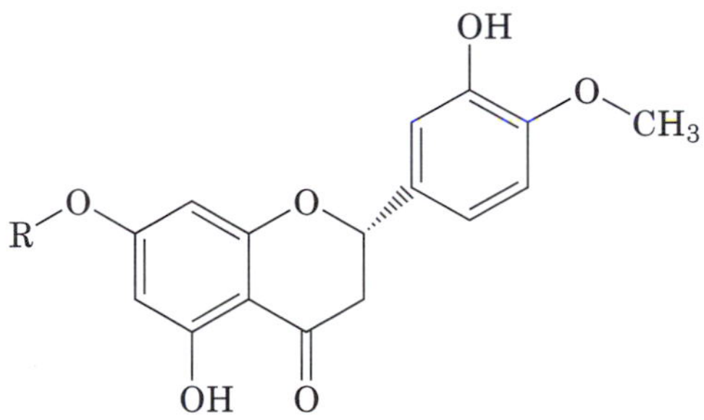

(184) Neohesperidin   R =

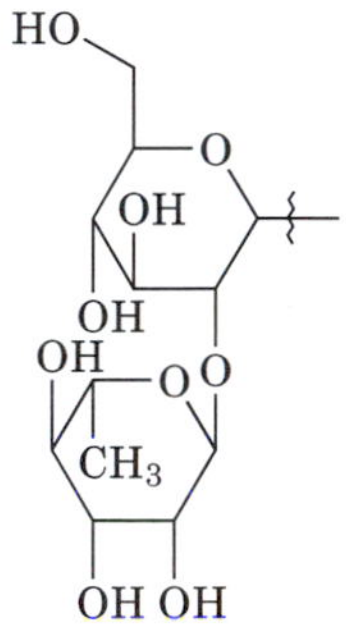

(185) Hesperidin   R =

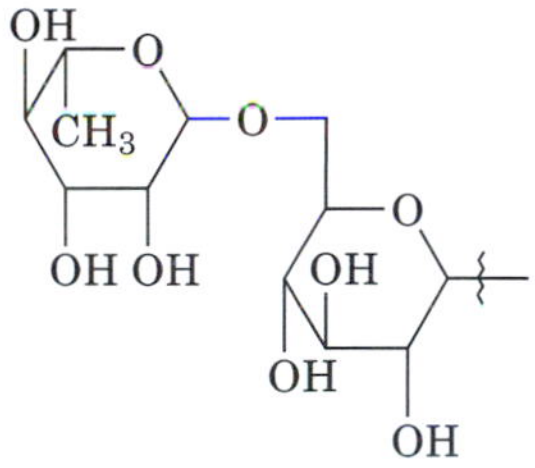

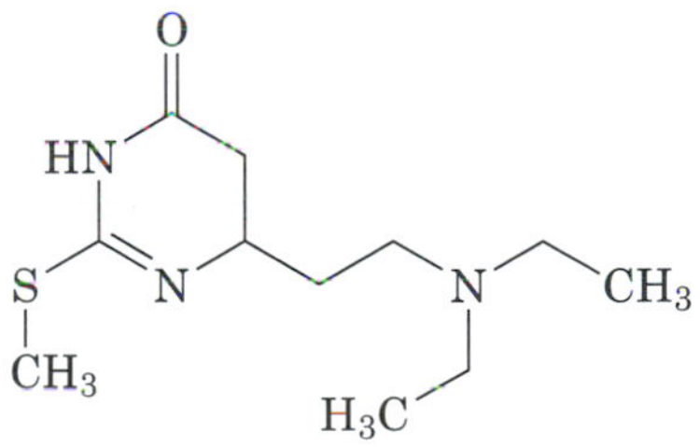

(186)

(187) R =

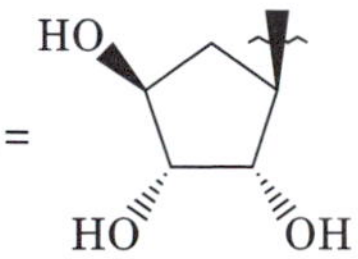

(188) R =

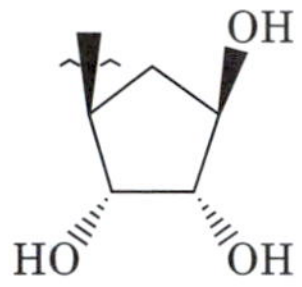

(189) R =

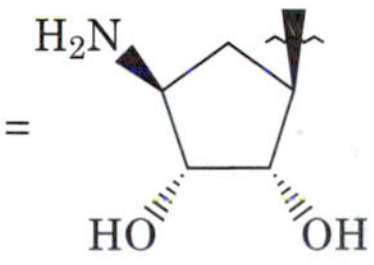

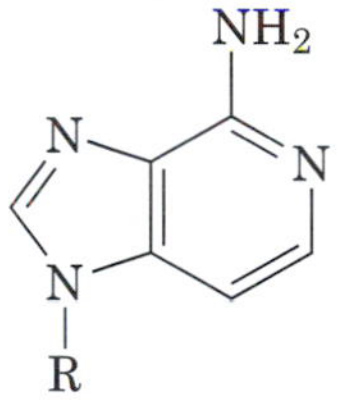

(190) R =

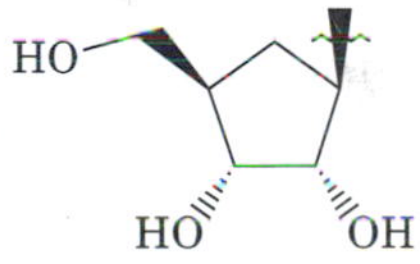

(191) R =

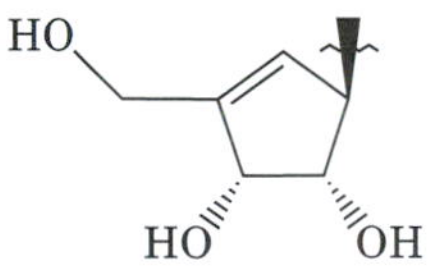

(190) significantly protected BALB/c mice from lethal infection with mouse-adapted Ebola Zaire virus, providing treatment was initiated on day −1, 0, or 1 day relative to time of virus challenge (440). Treatment with 2.2 mg/kg initiated on day 3 post-infection still resulted in 40% survival. In another study, a single subcutaneous dosing of 80 mg/kg or less of carbocyclic 3-deazaadenosine, or of 1 mg/kg or less of 3-deazaneplanocin A (191), provided equal or better protection, without causing toxicity (441). One dose of drug given on day 1 or 2 significantly reduced serum virus titers and resulted in survival of most or all animals. However, drug treatment given within 1 h after infection (day 0) was less effective. In SCID mice, single or multiple drug treatment suppressed Ebola replication, but did not prevent death (441). The prolonged efficacy of these two SAH hydrolase inhibitors demonstrated a

potential useful antiviral strategy in that drug treatment begins early in infection with high but non-toxic doses, to hold viral burden below the lethal threshold until the host immune system eliminates the infection (441).

### 2.9.2 Prostaglandins.

Cyclopentenone prostaglandins (PGs) have reportedly shown inhibitory activities against a variety of RNA and DNA viruses, including influenza A PR8, Sendai, poliovirus, VSV, Sindbis, rotavirus, HIV-1, vaccinia, HSV-1, and HSV-2 in cultured cells (reviewed by Santoro in Ref. 442). Only the cyclopentenone PGs were effective in inhibiting viral replication. It might be that the presence of an $\alpha,\beta$-unsaturated carbonyl group allows to form Michael adducts with proteinic nucleophiles and to bind covalently to the target proteins (443). The plausible mechanism of action of PGs includes the induction of cytoprotective heat shock protein (HSP) synthesis, notably of the 70K heat shock protein (HSP 70) in human cells, through activation of heat shock transcription factor 1 (HSF 1) (443). A recent study by Santoro et al. showed that $\Delta^{12}$-PGJ$_2$ (192) effec-

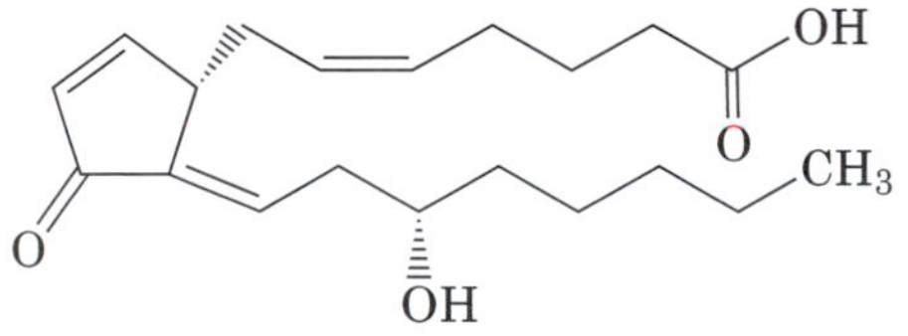

**(192) $\Delta^{12}$-PGJ$_2$**

tively inhibited viral protein synthesis of influenza virus A/PR8/34 (H1N1) as long as the host MDCK cells were synthesizing HSP 70, whose synthesis started 3 h after $\Delta^{12}$-PGJ$_2$ treatment and continued for at least 12 h in both uninfected and virus-infected cells (444). Nevertheless, the mechanism by which HSP can interfere with viral protein synthesis remains to be elucidated.

$\Delta^{12}$-PGJ$_2$ caused a dose-dependent reduction of influenza A/PR8/34 (H1N1) virus production in infected MDCK cells (greater than 95% at 6 $\mu$g/mL), and this antiviral effect could be sustained for at least 72 h post-viral infection. Drug treatment did not affect cell viability. In fact, the treatment actually pre-

vented the virus-induced inhibition of cellular RNA synthesis. Intraperitoneal administration of $\Delta^{12}$-PGJ$_2$ to PR8 virus-infected mice significantly reduced the virus titers in the lungs and increased the survival rates (50–60% of the animals that received a daily dosing of 5 $\mu$g/mouse for 7 days survived the infection). This compound was well tolerated by the animals (444).

At the same concentration of 6 $\mu$g/mL, PGA$_1$ (193) only modestly and transiently in-

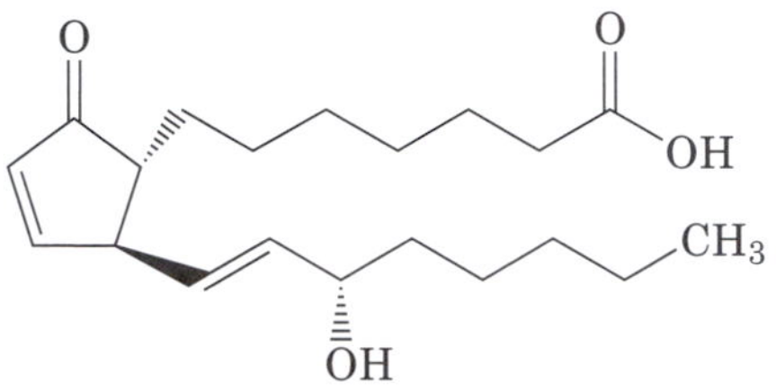

**(193) PGA$_1$**

hibited influenza PR8 replication in cultured cells. On the other hand, non-cyclopentenone PG of the E and D types, which are not able to induce HSP synthesis, did not affect PR8 virus replication (444). In an independent study, PGA$_1$ showed *in vitro* effect on the replication of avian influenza A, Ulster 73 (H7H1) (445).

Intriguingly, despite the fact that both PGA1 and $\Delta^{12}$-PGJ$_2$ inhibit poliovirus replication in a dose-dependent manner, infection with poliovirus seemed to inhibit, rather than induce, HSP 70 synthesis in PG-treated HeLa cells (446). Because cyclopentenone PGs lack of ability to induce heat shock response in polioviral-infected cells, poliovirus protein synthesis was not inhibited by PGs, suggesting that cyclopentenone PGs could interfere with a late event in the virus replication cycle, such as protein assembly and maturation of poliovirus virions (446).

Infection of monkey kidney MA104 cells with SA-11 simian rotavirus, a nonenveloped double-stranded RNA virus, was also shown to inhibit PGA$_1$-induced HSP 70 synthesis (447). Electron microscopic analysis revealed that, in the present of PGA$_1$, most of the virus particles remain in the membrane-enveloped intermediate form, and virus maturation is impaired. This effect might be caused by inhibition of glucosamine incorporation into the NSP4 glycoprotein as well as partial inhibition of VP7 and VP4 synthesis in PGA$_1$-treated cells.

Because PGs inhibit viral replication by acting on multiple cellular and viral targets, their potential use in the treatment of viral diseases remains as an open question.

**2.9.3 Polyoxometalates.** Polyoxometalates (POMs) are oligomeric aggregates of metal cations bridges by oxide anions that form by self-assembly processes (for a review on POMs in medicine, see Ref. 448). Previously, POMs have demonstrated broad-spectrum antiviral activity against many enveloped viruses (e.g., HIV, influenza, paramyxoviruses, herpesviruses, etc.) (448). More recent studies by Schinazi and Sidwell showed that a series of germanium- or silicon-centered POMs with the Barrel (e.g., JM2919), Keggin (e.g., JM2921), or double Keggin (e.g., JM2927) structure were highly inhibitory to influenza (both types of A and B) viruses (449) and respiratory syncytial virus (450). The *in vitro* anti-RSV results strongly suggested inhibition of virus attachment as the primary mode of action (450). In the case of influenza, greatest *in vitro* efficacy was also seen during the period of viral adsorption and penetration (449). A more precise mechanism of action was revealed by a study with a Keggin-type PM-523 by Shigeta and Schinazi, showing that inhibition of influenza replication in MDCK cells by POMs was not because of inhibition of virus binding to cells, but was associated with inhibition of fusion of the viral envelop to the cellular membrane (451).

Generally, recent clinical isolates of influenza A were more susceptible to these compounds than older, laboratory-adapted strains; H1N1 viruses were more sensitive to these effects than the H3N2 viruses (449). When PM-523 was combined with ribavirin, synergistic anti-influenza (H1N1) effects were demonstrated both *in vitro* and in mice (by intranasal administration) (452). In a separate investigation in mice, Liu et al. showed HPB-2 (given either orally or intraperitoneally) to be more effective than ribavirin in preventing deaths and lowering lung consolidation (453).

JM2919:
   $((CH_3)_3NH)_{10}(H)[Si_2Zr_3W_{18}O_{71}H_3]$
   $\cdot$ 10 H$_2$O
JM2921: $K_7[A\text{-}\alpha\text{-}GeNb_3W_9O_{40}] \cdot 18H_2O$
JM2927: $K8[A\text{-}\beta\text{-}Si_2Nb_6W_{18}O_{77}]$
PM-523: $(iPrNH3)_6(H)[PTi_2W_{10}O_{38}(O_2)_2]$
   $\cdot$ H$_2$O
HPB-2: $Ce_2H_3[BW_9{}^{VI}W_2{}^{V}Mn(H_2O)O_{39}]$
   $\cdot$ 12 H$_2$O

# 3   ORTHOPOXVIRUSES

Smallpox is presumably one of the most attractive pathogens to a potential bioterrorist because it meets the twin criteria of high transmissibility and high mortality. In addition, survivors are left with disfiguring sequelae. Historically drugs were tried both for treatment of smallpox and for prophylaxis of contacts but rarely in well-controlled clinical trials. Post-exposure prophylaxis with vaccinia immune globulin (VIG) demonstrated a modest anecdotal benefit when given to close contacts of smallpox patients along with revaccination, yet this scenario is not altogether relevant when an ever-increasing portion of the population has not received even a primary vaccination, and supplies of VIG are limited (454–456).

In 1963, post-exposure prophylaxis with marboran (N-methylisatin β-thiosemicarbazone) (**194**) was hailed as "the most significant

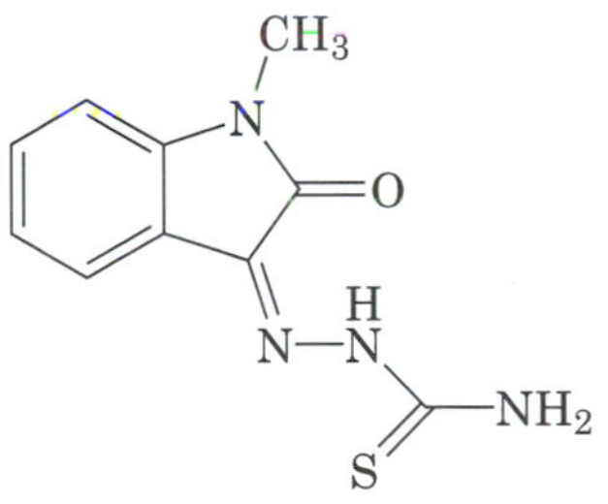

**(194)** Methisazone (Marboran)

advance in smallpox control since the days of JENNER." (457). However, this influential study was seriously flawed by current standards because most subjects were successfully vaccinated in infancy and revaccinated before receiving therapy. In addition, the study groups were not randomized and subject compliance with the dosing schedule was not adequately ascertained (458). This last point is especially relevant because marboran caused severe nausea and vomiting in approximately one-half of treated subjects. Other investigators conducted a well-designed post-exposure prophylaxis study of a similar thiosemicarbazone (4-bromo-3-methylisothiazole-5-carboxaldehyde thiosemicarbazone, M&B 7714) in India (459). Enrollees were limited to those without previous vaccination before their contact with an index case of smallpox. Although there was a small decrease seen in the incidence of smallpox in the treated group, there was no decrease in mortality of those who acquired smallpox. The authors concluded that post-exposure treatment was not appropriate for routine use as the benefit was small and the drug poorly tolerated—52% of those in the treatment group refused to complete their treatment course.

On the basis of encouraging data in animal models of poxvirus infections, a controlled clinical trial was undertaken in India to evaluate the treatment of smallpox with M&B 7714, the thiosemicarbazone described above that showed a small amount of benefit as a prophylactic agent. Unfortunately, this compound showed no therapeutic benefit in patients with or without prior vaccination (460).

Other antiviral agents were studied in small trials. An initial uncontrolled treatment study in Bangladesh of nine patients with cytosine arabinoside reported that eight survived compared with an expected 45% mortality in this area (461). This report was quickly followed by two randomized hospital-based controlled studies in Ethiopia and Bangladesh that showed no benefit for treatment (462, 463). The observation in one study that three treated patients who seemed to be improving died late in the course of infection is worrisome in view of cytosine arabinoside's immunosuppressive activity (463). Adenosine arabinoside was also studied in Bangladesh in a small double-blind placebo-controlled trial, and no differences were found between placebo and Ara-A treated groups in mortality, fever days, or duration of days of virus isolation.

Another historically important medical need was for prophylaxis and treatment of complications for vaccination for smallpox. Although vaccination is undeniably effective, it is a live vaccine and not sufficiently attenuated to prevent its unwanted replication in people with impaired immune systems. There are four complications of vaccination that are considered serious. Three of these involve viral replication and should potentially be responsive to antiviral therapy. These are as follows. (1) progressive vaccinia in which the original vaccination lesion gradually extends rather than resolves and new lesions appear at noncontiguous sites. This is almost always fatal. (2) Eczema vaccinatum in which lesions appear on previously normal as well as eczematous areas of skin. The prognosis is correlated with the extent of skin involved. (3) Generalized vaccinia, a generalized skin eruption, has a good prognosis. The fourth serious complication is post-vaccinial encephalitis and is thought to result from immunpathology rather than viral replication. In the absence of treatment progressive vaccinia is usually fatal (458), and the U.S. mortality rate for eczema vaccinatum in very young children was approximated as 33%. Fortunately treatment with VIG seems to significantly improve survival to about 93%, although complication rates are too low to conduct controlled clinical trials (455, 458).

Historical data on complication rates from the past will probably not be reliable predictors of future rates, should any government undertake the vaccination of large segments of the population to deter or ameliorate the consequences of a potential terrorist use of smallpox. The world's population has changed dramatically since the middle of the 20th century. Immunocompromised individuals comprise a much larger proportion of the overall population as a result of advances in transplantation and cancer treatment as well as the global devastation caused by HIV. In addition, the incidence of atopic dermatitis has dramatically increased in recent decades. As supplies of VIG are very limited, it may be as or even more important to identify an effective chemotherapeutic agent for the treatment of vaccinia complications as for the treatment of smallpox. Fortunately, because the viruses are closely related, most antiviral agents with activity against one of these viruses is likely to also inhibit the other.

New preclinical data support use of cidofovir (**195**) for both treatment of smallpox and of

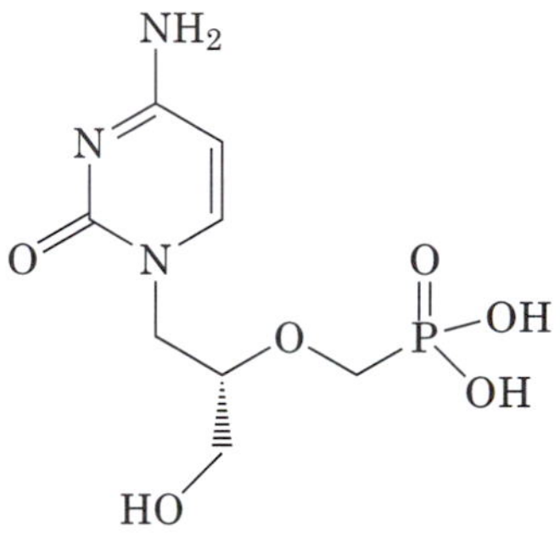

**(195)** Cidofovir

treatment of complications of vaccination. Accordingly, the Department of Health and Human Services has prepared and sponsored Investigational New Drugs (INDs) for both potential indications.

## 3.1 Inhibitors of Orthopoxviruses

Concerns about possible unnatural outbreak of smallpox (such as in a bioterrorist attack), prompted a renewed interest in the search for antiviral agents that might be useful to treat smallpox (variola). Because evaluation of anti-variola compounds cannot be done in a laboratory without a BSL-4 facility, and variola (as well as monkeypox) does not cause disease in adult mice (464), routine preclinical assessment of potential anti-variola compounds can only be studied in systems using surrogate viruses, such as vaccinia and cowpox viruses.

Very recently De Clercq published an influential review article to summarize the research on vaccinia virus inhibitors in his and others' laboratories (465). The inhibitors might serve as a paradigm for the chemotherapy of poxvirus infections. According to this review, the inhibitors could generally divide into two major categories—nucleoside derivatives and non-nucleoside organic molecules.

**3.1.1 Methisazone (Marboran).** Of the non-nucleoside compounds, methisazone (**194**), a thiosemicarbazone derivative of isatine (**110**),

is worthy of note. In the early 1960s, Bauer et al. first showed that methisazone protected infant mice from fatal encephalitis caused by intracerebral injection of variola virus (464). Around that time methisazone was used in a case of eczema vaccinatum and this appeared to be the first clinical use of antiviral drug in man (464). As discussed above, clinical experience with methisazone has also included the treatment of vaccinia gangrenosa, prophylaxis of vaccinia infection, prophylaxis of smallpox (the main indication for the use of the drug), and treatment of smallpox (466). In any event, methisazone could serve as the lead for the design of a next generation drug with much improved pharmacological properties and safety profile.

### 3.1.2 Nucleoside Derivatives

**3.1.2.1 Cidofovir (HPMPC).** To promptly identify an anti-poxvirus drug that could be immediately available in the event of a bioterrorism attack, initial attention has focused on currently approved antiviral agents. Recent preclinical studies against vaccinia and cowpox viruses have identified cidofovir (CDV) (**195**) as a promising candidate. Cidofovir was first described in the literature in 1987 by De Clercq and Holý (467) and was approved in 1996 by the U.S. FDA as an intravenous treatment for human cytomegalovirus (HCMV) retinitis in AIDS patients under the licensed name Vistide (468–470). Once inside the cells, cidofovir follows two-step phosphorylation by cellular enzymes first to cidofovir monophosphate, CDVp, (e.g., by pyrimidine nucleoside monophosphate kinase) then to cidofovir diphosphate, CDVpp (e.g., by pyruvate kinase) (471). The latter, structurally analogous to a nucleoside triphosphate, serves as a competitive inhibitor of dCTP and an alternative substrate for HCMV DNA polymerase (472, 473). Incorporation of a single cidofovir molecule causes a 31% decrease of in the rate of DNA elongation by HCMV DNA polymerase; incorporation of two consecutive molecules prohibits the DNA from further elongation (474). Furthermore, the intracellular cidofovir metabolites, namely CDVp, CDVpp, and CDVp choline, have very long half-life and these molecules confer a long-lasting antiviral response of cidofovir and infrequent dosing for antiviral therapy (469, 475, 476).

Cidofovir has broad-spectrum activity against various DNA virus, including polyomaviruses, papillomaviruses, adenoviruses, herpesviruses, and poxviruses (468, 469, 475, 476). Huggins et al. reported that, in Vero and BSC-40 cells, cidofovir inhibited vaccinia, cowpox, camelpox, and monkeypox viruses with $EC_{50}$ values in the range of 30–90 $\mu M$ and variola virus in the range of 10 $\mu M$ (Table 10.4) (477). It seemed that variola virus was most sensitive to cidofovir than the other orthopoxviruses in this particular study.

Several different animal models have been used to assess the therapeutic potential of cidofovir for the treatment of poxvirus infections. In earlier studies, De Clercq et al. had used intravenous injection of vaccinia virus to infect the mice and measured the suppression of tail lesion formation to assess a compound's antiviral effect (465). Similarly infected SCID mice also die from the disseminate vaccinia infection in addition to the development of tail lesions. In such infected SCID mice, cidofovir was shown to significantly delay the mean day of death using either treatment or prophylactic regimen (478). However, inoculation of virus by injection does not simulate the respiratory exposure that occurs in natural smallpox infection nor in a bioterrorist scenario, namely infection acquired by aerosol route. To mimic the natural infection, Bray et al. (479) and Smee et al. (480–482) demonstrated that aerosol or intranasal infection of BALB/c mice with vaccinia virus or cowpox virus caused the infected animal to develop pneumonia, lose weight, and eventually die from the disease. The efficacy of cidofovir observed in these new models can be summarized by the following.

- Cidofovir was active by intraperitoneal and intranasal routes against wild-type virus infections at non-toxic doses. It was not effective against infections caused by the cidofovir-resistant cowpox virus, and is not active orally.
- The efficacy of cidofovir against wild-type cowpox virus infections was similar to its activity against vaccinia virus infections.

- Mice could benefit from as little as a single treatment given a few days before or up to 4 days after virus exposure.
- Daily dosing with cidofovir was more beneficial than the single treatment regimen.

In a meeting presentation, Huggins reported that cynomolgus monkeys infected with monkeypox by small particle aerosol inoculation developed classical poxvirus lesions and pulmonary distress and that treatment of cidofovir, initiated on the day of infection, completely protected the animals from clinical and laboratory signs of disease (483). Topical cidofovir has been used to treat molluscum contagiosum (poxvirus) infections in AIDS patients (484).

In a bioterrorist attack, the number of exposed individuals is expected to be large; therefore, it may be technically difficult to administer cidofovir by injection. Because oral bioavailability of cidofovir is less than 5%, administration with an orally active prodrug form of cidofovir would be an ideal alternative under such circumstance.

Several reports by Hostetler et al. have shown that the oral bioavailability of nucleosides could be improved by conjugation with certain ether lipid groups, presumably by increasing oral absorption and cell membrane penetration (see references in Ref. 485). Cidofovir derivatives, HDP-CDV (196) and ODE-CDV (197), obtained by esterification of cido-

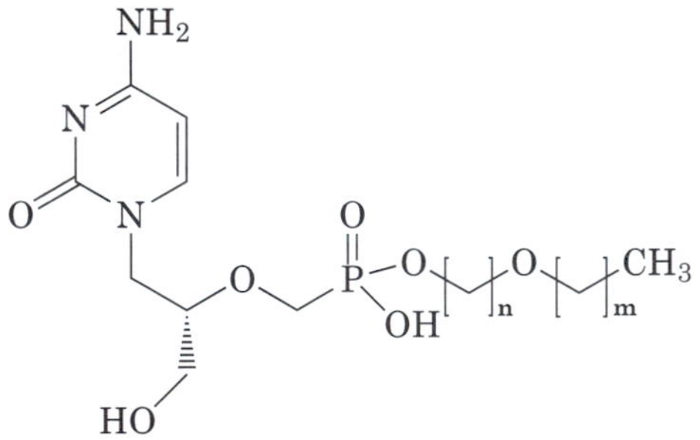

**(196)** HDP-CDV    n = 3, m = 15
**(197)** ODE-CDV    n = 2, m = 17

fovir with two long-chain alkoxyalkanols (3-hexadecyloxy-1-propanol and 3-octadecyloxy-1-ethanol, respectively) significantly enhanced both antiviral potency and selective indexes

Table 10.4  *In Vitro* Activity of Cidofovir, Ribavirin, Methisazone, Carbocyclic 3-Deazaadenosine, and 3-Deazaneplanocin A Against Vaccinia, Cowpox, Camelpox, Monkeypox, and Variola Viruses (477)

| | $EC_{50}$ ($\mu M$) in Indicated Cell Line | | | | | | | | | | | |
| | Vaccinia (Copenhagen) | | Cowpox (Brighton) | | Camelpox (Somalia) | | Monkeypox (Copenhagen) | | Variola (Bangladesh) | | Variola (Congo) | Variola (Garcia) |
| | Vero | BSC | Vero | BSC | Vero | BSC | Vero | BSC | Vero | BSC | BSC | BSC |
|---|---|---|---|---|---|---|---|---|---|---|---|---|
| Cidofovir (195) | 91 | 26 | 77 | 36 | 27 | 34 | 94 | 57 | 5 | 8 | 10 | 11 |
| Ribavirin (3) | 398 | 107 | 615 | 98 | 340 | 102 | 238 | 107 | 9 | 6 | 3 | 7 |
| Methisazone (194) | | | | | | | | | | 258 | | |
| Carbocyclic 3-deazaadenosine (190) | 2 | | >250 | | 30 | | 44 | | 1 | 1 | 2 | 4 |
| 3-Deazaneplanocin A (191) | 0.07 | | >250 | | 2 | | 2 | | 0.01 | 0.06 | 0.02 | 0.08 |

$EC_{50}$, concentration of the compound required to inhibit viral-induced effect (cytopathic effect or plaque formation) by 50% in cell culture.

**Table 10.5** *In Vitro* Activity of CDV, HDP-CDV, and ODE-CDV Against Vaccinia and Cowpox Viruses in HFF Cells (485)

| | Vaccinia (6 Strains) | | | Cowpox (Brighton) | | |
|---|---|---|---|---|---|---|
| | $EC_{50}$ $(\mu M)$ | $CC_{50}$ $(\mu M)$ | SI | $EC_{50}$ $(\mu M)$ | $CC_{50}$ $(\mu M)$ | SI |
| Cidofovir (CDV) (**195**) | 10.1–46.2 | 278 | 6.0–27.5 | 44.7 | 278 | 6.2 |
| HDP-CDV (**196**) | 0.2–1.2 | 31 | 25.8–155 | 0.6 | 31 | 50 |
| ODE-CDV (**197**) | 0.1–0.4 | 14.3 | 35.8–143 | 0.3 | 14.3 | 49 |

$EC_{50}$, concentration of the compound required to inhibit viral-induced effect (cytopathic effect or plaque formation) by 50% in cell culture; $CC_{50}$, concentration of the compound required to induce 50% normal cell morphological change or inhibit 50% normal cell growth in cell culture; SI, $CC_{50}/EC_{50}$.

over the parent compound in human foreskin fibroblast (HFF) cells using a plaque reduction assay (Table 10.5) (485). Importantly, the potency against variola was also increased by more than 100-fold with the alkoxyalkyl prodrugs [data from J. Huggins, (485)]. These long-chain alkoxyalkyl esters of cidofovir resemble readily absorbed dietary phospholipids as evident by the observation that cellular uptake of $^{14}$C-labeled HDP-CDV was severalfold greater than that observed with $^{14}$C-labeled CDV in human lung fibroblast cells (485).

**3.1.2.2 Ribavirin and Other Nucleoside Derivatives.** The other nucleoside derivatives discussed in De Clercq's review could attribute their mechanisms of action to the inhibition of viral DNA synthesis (e.g., adenine arabinoside) or one of the following enzymes: IMP dehydrogenase [e.g., ribavirin (**3**) and EICAR (**165**)] SAH hydrolase [e.g., 5'-noraristeromycin (**187**), carbocyclic 3-deazaadenosine (**190**), and 3-deazaneplanocin A (**191**)], OMP decarboxylase [e.g., 6-azauridine (**163**)], CTP synthetase [e.g., CPE-C (**167**)], and thymidylate synthase (e.g., 5-nitro-2'-deoxyuridine).

Ribavirin is another approved drug that may have the potential to treat poxvirus infections because its broad-spectrum antiviral activity also encompasses vaccinia virus (465). The anti-vaccinia activity could be attributable to the inhibition of IMP dehydrogenase by ribavirin 5'-monophosphate (486), as well as inhibition of the capping of vaccinia mRNA by ribavirin 5'-triphosphate (487). As shown in Table 10.4, although ribavirin was a weaker inhibitor than cidofovir against vaccinia, cowpox, camelpox, and monkeypox, it was comparable with cidofovir against variola virus. In a

separate study, Smee and Huggins reported that ribavirin inhibited plaque formation caused by camelpox, cowpox, monkeypox, or vaccinia viruses by 50% at approximately 2–12 $\mu M$ in mouse 3T3 cells and at 30–250 $\mu M$ in African green monkey kidney (Vero 76) cells (488, 489). The greater potency and increased toxicity of ribavirin in 3T3 cells was due, at least in part, to the higher amount of ribavirin taken into 3T3 and greater accumulation of mono-, di-, and triphosphate forms of the drug in 3T3 cells than in Vero cells (489). Ribavirin was marginally active ($EC_{50}$, 281 $\mu M$) against vaccinia and not active against cowpox virus when tested in HFF cells (485).

In animal models, ribavirin protected vaccinia tail lesion formation in mice (465). More recently Smee et al. showed that mice could not survive a high intranasal cowpox virus challenge, but ribavirin-treated animals lived several days longer than placebos (488, 490). However, under less severe cowpox virus infection, high dose of subcutaneous ribavirin (100 mg/kg/day) could completely protect the infected mice from death, and lower doses could also improve the survival rate. Ribavirin treatment followed sequentially by cidofovir significantly increased the mean time to death beyond that achieved with ribavirin alone (490).

Carbocyclic 3-deazaadenosine (**190**) and 3-deazaneplanocin A (**191**), two potent inhibitors of *S*-adenosylhomocysteine hydrolase, seemed to have *in vitro* activity against various poxviruses, except cowpox in Vero cells (Table 10.4). The sensitivity of poxviruses to this series of compounds in various cell lines is a subject of further investigations.

# REFERENCES

1. W. R. Kim, R. S. Brown Jr., N. A. Terrault, and H. El-Serag, *Hepatology*, **36**, 227–242 (2002).

2. A. M. Di Bisceglie, J. McHutchison, and C. M. Rice, *Hepatology*, **35**, 224–231 (2002).

3. A. Wasley and M. J. Alter, *Semin. Liver Dis.*, **20**, 1–13 (2000).

4. A. U. Neumann, N. P. Lam, H. Dahari, D. R. Gretch, T. E. Wiley, T. J. Layden, and A. S. Perelson, *Science*, **282**, 103–107 (1998).

5. P. Farci, A. Shimoda, A. Coiana, G. Diaz, G. Peddis, J. C. Melpolder, A. Strazzera, et al., *Science*, **288**, 339–344 (2000).

6. A. U. Neumann, N. P. Lam, H. Dahari, M. Davidian, T. E. Wiley, B. P. Mika, A. S. Perelson, and T. J. Layden, *J. Infect. Dis.*, **182**, 28–35 (2000).

7. N. N. Zein and J. Rakela, *Semin. Gastrointest. Dis.*, **6**, 46–53 (1995).

8. N. N. Zein, *Cytokines Cell Mol. Ther.*, **4**, 229–241 (1998).

9. K. R. Reddy, J. H. Hoofnagle, M. J. Tong, W. M. Lee, P. Pockros, E. J. Heathcote, D. Albert, et al., *Hepatology*, **30**, 787–793 (1999).

10. J. G. McHutchison, S. C. Gordon, E. R. Schiff, M. S. Shiffman, W. M. Lee, V. K. Rusgi, Z. D. Goodman, et al., *N. Engl. J. Med.*, **339**, 1485–1492 (1998).

11. A. Isaacs and J. Lindenmann, *Proc. R. Soc. Lond. [Biol.]*, **147**, 258–267 (1957).

12. J. Kato, N. Kato, M. Moriyama, T. Goto, H. Taniguchi, Y. Shiratori, and M. Omata, *J. Infect. Dis.*, **186**, 155–163 (2002).

13. N. N. Zein, *Expert Opin. Investig. Drugs*, **10**, 1457–1469 (2001).

14. J. E. Layden and T. J. Layden, *Hepatology*, **35**, 967–970 (2002).

15. N. P. Lam, A. U. Neumann, D. R. Gretch, T. E. Wiley, A. S. Perelson, and T. J. Layden, *Hepatology*, **26**, 226–231 (1997).

16. S. L. Tan and M. G. Katze, *Virology*, **284**, 1–12 (2001).

17. D. R. Taylor, S. T. Shi, P. R. Romano, G. N. Barber, and M. M. C. Lai, *Science*, **285**, 107–110 (1999).

18. A. M. Di Bisceglie, P. Martin, C. Kassiandes, et al., *N. Engl. J. Med*, **321**, 1506–1510 (1989).

19. G. L. Davis, L. A. Balart, E. R. Schiff, et al., *N. Engl. J. Med*, **321**, 1501–1506 (1989).

20. R. C. Tam, J. Y. N. Lau, and Z. Hong, *Antiv. Chem. Chemother.* **12**, 261–272 (2002).

21. D. Maag, C. Castro, Z. Hong, and C. E. Cameron, *J. Biol. Chem.*, **276**, 46094–46098 (2001).

22. S. Crotty, C. E. Cameron, and R. Andino, *Proc. Natl. Acad. Sci. USA*, **98**, 6895–6900 (2001).

23. R. C. Tam, B. Pai, J. Bard, C. Lim, D. R. Averett, U. T. Phan, and T. Milovanovic, *J. Hepatol.*, **30**, 376–382 (1999).

24. A. M. Di Bisceglie, H. S. Conjeevaram, M. W. Fried, R. Sallie, Y. Park, C. Yurdaydin, M. Swain, et al., *Ann. Intern. Med.*, **123**, 897–903 (1996).

25. W. C. Maddrey, *Semin. Liver Dis.*, **19**, 67–75 (1999).

26. L. De Franceschi, G. Fattovich, F. Turrini, K. Agi, C. Bragnara, F. Manzato, F. Noventa, and A. M. Stanzial. *Hepatology*, **31**, 997–1004 (2000).

27. A. Kozlowski, S. A. Charles, and J. M. Harris, *Biodrugs*, **15**, 419–429 (2001).

28. K. L. Lindsay, C. Trepo, T. Heintges, M. L. Shiffman, S. C. Gordon, J. C. Hoefs, E. R. Schiff, et al., *Hepatology*, **34**, 395–403 (2001).

29. S. Zeuzum, S. V. Feinman, J. Rasenack, E. J. Heathcote, M.-Y. Lai, E. Gane, J. O'Grady, et al., *N. Engl. J. Med.*, **343**, 1666–1672 (2000).

30. M. P. Manns, J. G. McHutchison, S. C. Gordon, V. K. Rustgi, M. Shiffman, R. Reindollar, Z. D. Goodman, et al., *Lancet*, **358**, 958–965 (2001).

31. M. W. Fried, M. L. Shiffman, K. Reddy, C. Smith, G. Marino, F. Goncales, D. Haeussinger, et al., *Gastroenterology*, **120**, A55 (2001).

32. Y. S. Babu, P. Chand, S. Bantia, P. Kotian, A. Dehghani, Y. El-Kattan, T. H. Lin, T. L. Hutchison, A. J. Elliott, C. D. Parker, S. L. Ananth, L. L. Horn, G. W. Laver, and J. A. Montgomery, *J. Med. Chem.*, **43**, 3482–3486 (2000).

33. R. C. Wade, *Structure*, **5**, 1139–1145 (1997).

34. M. J. Kiefel and M. von Itzstein, *Prog. Med. Chem.*, **36**, 1–28 (1999).

35. M. D. Dowle and P. D. Howes, *Expert Opin. Ther. Patents*, **8**, 1461–1478 (1998).

36. J. N. Varghese, *Drug Dev. Res.*, **46**, 176–196 (1999).

37. C. U. Kim, X. W. Chen, and D. B. Mendel, *Antivir. Chem. Chemother.*, **10**, 141–154 (1999).

38. G. M. Air, A. A. Ghate, and S. J. Stray, *Adv. Virus Res.*, **54**, 375–402 (1999).

39. D. P. Calfee and F. G. Hayden, *Drugs*, **56**, 537–553 (1998).

40. G. Taylor, *Curr. Opin. Struct. Biol.*, **6**, 830–837 (1996).

41. V. R. Atigadda, W. J. Brouillette, F. Duarte, S. M. Ali, Y. S. Babu, S. Bantia, P. Chand, N.

Chu, J. A. Montgomery, D. A. Walsh, E. A. Sudbeck, J. Finley, M. Luo, G. M. Air, and G. W. Laver, *J. Med. Chem.*, **42**, 2332–2343 (1999).

42. A. A. Ghate and G. M. Air, *Eur. J. Biochem.*, **258**, 320–331 (1998).

43. M. von Itzstein, J. C. Dyason, S. W. Oliver, H. F. White, W. Y. Wu, G. B. Kok, and M. S. Pegg, *J. Med. Chem.*, **39**, 388–391 (1996).

44. C. Kim, *Antiviral Res.*, **46**, Abstract 168, (2000).

45. M. Williams, N. Bischofberger, S. Swaminathan, and C. U. Kim, *Bioorg. Med. Chem. Lett.*, **5**, 2251–2254 (1995).

46. T. Wang and R. C. Wade, *J. Med. Chem.*, **44**, 961–971 (2001).

47. M. A. Williams, C. U. Kim, H. L. MacArthur, X. Wang, B. J. Chen, J. A. Graves, and W. G. Laver, *Antiviral Res.*, **50**, A84 (2001).

48. C. U. Kim, W. Lew, M. A. Williams, H. T. Liu, L. J. Zhang, S. Swaminathan, N. Bischofberger, M. S. Chen, D. B. Mendel, C. Y. Tai, W. G. Laver, and R. C. Stevens, *J. Am. Chem. Soc.*, **119**, 681–690 (1997).

49. S. Singh, M. J. Jedrzejas, G. M. Air, M. Luo, W. G. Laver, and W. J. Brouillette, *J. Med. Chem.*, **38**, 3217–3225 (1995).

50. M. Luo, G. Air, and W. J. Brouillette, *J. Infect. Dis.*, **176**, S62–S65 (1997).

51. G. Wang, Y. Chen, S. Wang, R. Gentles, T. Sowin, W. Kati, S. Muchmore, V. Giranda, K. Stewart, H. Sham, D. Kempf, and W. G. Laver, *J. Med. Chem.*, **44**, 1192–1201 (2001).

52. D. M. Andrews, P. C. Cherry, D. C. Humber, P. S. Jones, S. P. Keeling, P. F. Martin, C. D. Shaw, and S. Swanson, *Eur. J. Med. Chem.*, **34**, 563–574 (1999).

53. N. R. Taylor, A. Cleasby, O. Singh, T. Skarzynski, A. J. Wonacott, P. W. Smith, S. L. Sollis, P. D. Howes, P. C. Cherry, R. Bethell, P. Colman, and J. Varghese, *J. Med. Chem.*, **41**, 798–807 (1998).

54. P. W. Smith, I. D. Starkey, P. D. Howes, S. L. Sollis, S. P. Keeling, P. C. Cherry, M. von Itzstein, W. Y. Wu, and B. Jin, *Eur. J. Med. Chem.*, **31**, 143–150 (1996).

55. P. W. Smith, J. E. Robinson, D. N. Evans, S. L. Sollis, P. D. Howes, N. Trivedi, and R. C. Bethell, *Bioorg. Med. Chem. Lett.*, **9**, 601–604 (1999).

56. P. W. Smith, S. L. Sollis, P. D. Howes, P. C. Cherry, K. N. Cobley, H. Taylor, A. R. Whittington, J. Scicinski, R. C. Bethell, N. Taylor, T. Skarzynski, A. Cleasby, O. Singh, A.

Wonacott, J. Varghese, and P. Colman, *Bioorg. Med. Chem. Lett.*, **6**, 2931–2936 (1996).

57. S. L. Sollis, P. W. Smith, P. D. Howes, P. C. Cherry, and R. C. Bethell, *Bioorg. Med. Chem. Lett.*, **6**, 1805–1808 (1996).

58. P. W. Smith, S. L. Sollis, P. D. Howes, P. C. Cherry, I. D. Starkey, K. N. Cobley, H. Weston, J. Scicinski, A. Merritt, A. Whittington, P. Wyatt, N. Taylor, D. Green, R. Bethell, S. Madar, R. J. Fenton, P. J. Morley, T. Pateman, and A. Beresford, *J. Med. Chem.*, **41**, 787–797 (1998).

59. P. W. Smith, A. R. Whittington, S. L. Sollis, P. D. Howes, and N. R. Taylor, *Bioorg. Med. Chem. Lett.*, **7**, 2239–2242 (1997).

60. C. Maring, K. McDaniel, A. Krueger, C. Zhao, M. Sun, D. Madigan, D. DeGoey, H. J. Chen, C. Yeung, W. Flosi, D. Grampovnik, W. Kati, K. Stewart, V. Stoll, A. Saldivar, D. Montgomery, R. Carrick, K. Steffy, A. Molla, W. Kohlbrenner, A. Kennedy, T. Herrin, Y. Xu, and W. G. Laver, *Antiviral Res.*, **50**, A76 (2001).

61. C. U. Kim, W. Lew, M. A. Williams, H. W. Wu, L. J. Zhang, X. W. Chen, P. A. Escarpe, D. B. Mendel, W. G. Laver, and R. C. Stevens, *J. Med. Chem.*, **41**, 2451–2460 (1998).

62. W. Lew, X. W. Chen, and C. U. Kim, *Curr. Med. Chem.*, **7**, 663–672 (2000).

63. W. Lew, H. W. Wu, X. W. Chen, B. J. Graves, P. A. Escarpe, H. L. MacArthur, D. B. Mendel, and C. U. Kim, *Bioorg. Med. Chem. Lett.*, **10**, 1257–1260 (2000).

64. W. Lew, H. W. Wu, D. B. Mendel, P. A. Escarpe, X. W. Chen, W. G. Laver, B. J. Graves, and C. U. Kim, *Bioorg. Med. Chem. Lett.*, **8**, 3321–3324 (1998).

65. W. Lew, M. A. Williams, D. B. Mendel, P. A. Escarpe, and C. U. Kim, *Bioorg. Med. Chem. Lett.*, **7**, 1843–1846 (1997).

66. M. A. Williams, W. Lew, D. B. Mendel, C. Y. Tai, P. A. Escarpe, W. G. Laver, R. C. Stevens, and C. U. Kim, *Bioorg. Med. Chem. Lett.*, **7**, 1837–1842 (1997).

67. V. R. Atigadda, W. J. Brouillette, F. Duarte, Y. S. Babu, S. Bantia, P. Chand, N. M. Chu, J. A. Montgomery, D. A. Walsh, E. Sudbeck, J. Finley, G. M. Air, M. Luo, and G. W. Laver, *Bioorg. Med. Chem.*, **7**, 2487–2497 (1999).

68. L. V. Gubareva, R. G. Webster, and F. G. Hayden, *Antimicrob. Agents Chemother.*, **45**, 3403–3408 (2001).

69. W. X. Li, P. A. Escarpe, E. J. Eisenberg, K. C. Cundy, C. Sweet, K. J. Jakeman, J. Merson, W. Lew, M. Williams, L. J. Zhang, C. U. Kim, N.

Bischofberger, M. S. Chen, and D. B. Mendel, *Antimicrob. Agents Chemother.*, **42**, 647–653 (1998).

70. L. V. Gubareva, L. Kaiser, and F. G. Hayden, *Lancet*, **355**, 827–835 (2000).

71. W. Lew, P. A. Escarpe, D. B. Mendel, D. J. Sweeny, and C. U. Kim, *Bioorg. Med. Chem. Lett.*, **9**, 2811–2814 (1999).

72. S. A. Kerrigan, P. W. Smith, and P. J. Stoodley, *Tetrahedron Lett.*, **42**, 4709–4712 (2001).

73. L. J. Zhang, M. A. Williams, D. B. Mendel, P. A. Escarpe, X. W. Chen, K. Y. Wang, B. J. Graves, G. Lawton, and C. U. Kim, *Bioorg. Med. Chem. Lett.*, **9**, 1751–1756 (1999).

74. L. J. Zhang, M. A. Williams, D. B. Mendel, P. A. Escarpe, and C. U. Kim, *Bioorg. Med. Chem. Lett.*, **7**, 1847–1850 (1997).

75. P. Chand, Y. S. Babu, S. Bantia, N. M. Chu, L. B. Cole, P. L. Kotian, W. G. Laver, J. A. Montgomery, V. P. Pathak, S. L. Petty, D. P. Shrout, D. A. Walsh, and G. W. Walsh, *J. Med. Chem.*, **40**, 4030–4052 (1997).

76. W. J. Brouillette, V. R. Atigadda, F. Duarte, M. Luo, G. M. Air, Y. S. Babu, and S. Bantia, *Bioorg. Med. Chem. Lett.*, **9**, 1901–1906 (1999).

77. J. B. Finley, V. R. Atigadda, F. Duarte, J. J. Zhao, W. J. Brouillette, G. M. Air, and M. Luo, *J. Mol. Biol.*, **293**, 1107–1119 (1999).

78. P. Chand, P. L. Kotian, A. Dehghati, Y. El-Kattan, T. S. Lin, T. L. Hutchison, Y. S. Babu, S. Bantia, A. J. Elliott, and J. A. Montgomery, *J. Med. Chem.*, **44**, 4379–4392 (2001).

79. S. Bantia, C. D. Parker, S. L. Anath, L. L. Horn, K. Andries, P. Chand, P. L. Kotian, A. Dehghani, Y. El-Kattan, T. Lin, T. L. Hutchinson, J. A. Montgomery, D. L. Kellog, and S. Y. Babu, *Antimicrob. Agents Chemother.*, **45**, 1162–1167 (2001).

80. E. Govorkova, I. Leneva, K. Bush, and R. Webster, 40th Interscience Conference on Antimicrobial Agents and Chemotherapy, Am. Soc. Microbiol., Washington, DC, 2000, Abstract 1157.

81. E. A. Govorkova, I. A. Leneva, O. G. Goloubeva, K. Bush, and R. G. Webster, *Antimicrob. Agents Chemother.*, **45**, 2723–2732 (2001).

82. R. W. Sidwell and D. F. Smee, *Expert Opin. Investig. Drugs*, **11**, 859–869 (2002).

83. D. Smee, J. H. Huffman, A. C. Morrison, D. L. Barnard, and R. W. Sidwelll, *Antimicrob. Agents Chemother.*, **45**, 743–748 (2001).

84. R. W. Sidwell, D. F. Smee, J. H. Huffman, D. L. Barnard, K. W. Bailey, J. D. Morrey, and Y. S. Babu, *Antimicrob. Agents Chemother.*, **45**, 749–757 (2001).

85. R. W. Sidwell, D. F. Smee, J. H. Huffman, D. L. Barnard, J. M. Morrey, K. W. Bailey, W.-C. Feng, Y. S. Babu, and K. Bush, *Antiviral Res.*, **51**, 179–187 (2001).

86. G. L. Drusano, S. L. Preston, D. Smee, K. Bush, K. Bailey, and R. W. Sidwell, *Antimicrob. Agents Chemother.*, **45**, 2115–2118 (2001).

87. F. Hayden, J. Treanor, R. Qu, and C. Fowler, 40th Interscience Conference on Antimicrobial Agents and Chemotherapy, Am. Soc. Microbiol., Washington, DC, 2000, Abstract 1156.

88. D. F. Smee, R. W. Sidwell, A. C. Morrison, K. W. Bailey, E. Z. Baum, L. Ly, and P. C. Wagaman, *Antiviral Res.*, **52**, 251–259 (2001).

89. W. M. Kati, D. Montgomery, C. Maring, V. S. Stoll, V. Giranda, X. Chen, W. Graeme Laver, W. Kohlbrenner, and D. W. Norbeck, *Antimicrob. Agents Chemother.*, **45**, 2563–2570 (2001).

90. C. Zhao, C. Maring, M. Sun, K. Stewart, V. Stoll, Y. Xu, Y. Gu, A. Krueger, T. Herrin, H. Sham, W. G. Laver, D. Madigan, A. Kennedy, W. Kati, D. Montgomery, A. Saldivar, D. Kempf, and W. Kohlbrenner, *Antiviral Res.*, **46**, A53 (2000).

91. V. S. Stoll, K. McDaniel, C. Zhao, D. Madigan, M. H. Sun, A. Kennedy, A. Krueger, T. Herrin, W. Flosi, D. Grampovnik, D. DeGoey, H. Chen, C. Yeung, F. Mohammadi, C. Maring, L. L. Klein, D. Montgomery, A. C. Saldivar, W. Kati, K. Stewart, W. G. Laver, D. Kempf, and W. Kohlbrenner, 40th Interscience Conference on Antimicrobial Agents and Chemotherapy, Am. Soc. Microbiol., Washington, DC, 2000, Abstract 1847.

92. C. Maring, C. Zhao, V. Stewart, V. Stoll, W. Kati, D. Madigan, M. Sun, Y. Xu, Y. Gu, A. Krueger, T. Herrin, G. W. Laver, A. Saldivar, D. Montgomery, E. Muchmore, D. Kempf, and W. Kohlbrenner, *Antiviral Res.*, **46**, A59 (2000).

93. C. Zhao, M. Sun, C. Maring, A. Krueger, Y. Xu, D. Madigan, T. Herrin, Y. Hui, K. Stewart, V. Stoll, W. Kati, D. Montgomery, A. Saldivar, Y. Gu, G. W. Laver, K. Marsh, D. Kempf, and W. Kohlbrenner, 40th Interscience Conference on Antimicrobial Agents and Chemotherapy, Am. Soc. Microbiol., Washington, DC, 2000, Abstract 1849.

94. W. J. Flosi, D. Grampovnik, D. DeGoey, H. J. Chen, C. Yeung, L. L. Klein, V. Stoll, K. Stew-

art, D. Montgomery, A. Saldivar, W. Kati, W. G. Laver, D. Kempf, and W. Kohlbrenner, 40th Interscience Conference on Antimicrobial Agents and Chemotherapy, Am. Soc. Microbiol., Washington, DC, 2000, Abstract 1848.

95. A. Krueger, Y. Xu, C. Maring, D. Kempf, C. Zhao, M. Sun, T. Herrin, D. Madigan, K. McDaniel, A. Kennedy, W. Kati, D. Montgomery, A. Saldivar, W. Kohlbrenner, V. Stoll, K. Stewart, R. Carrick, and W. G. Laver, 40th Interscience Conference on Antimicrobial Agents and Chemotherapy, Am. Soc. Microbiol., Washington, DC, 2000, Abstract 1851.

96. N. A. Zielinski Mozny, S. T. Wilson, C. J. Maring, K. F. McDaniel, W. M. Kati, D. J. Kempf, R. J. Carrick, K. R. Steffy, A. M. Molla, W. E. Kohlbrenner, K. W. Mollison, and J. A. Meulbroek, 40th Interscience Conference on Antimicrobial Agents and Chemotherapy, Am. Soc. Microbiol., Washington, DC, 2001, Abstract H1583.

97. R. J. Carrick, N. Gusick, W. Kati, C. Maring, Y. Shi, D. Montgomery, T. Ng, K. Steffy, W. Kohlbrenner, and A. Molla, 40th Interscience Conference on Antimicrobial Agents and Chemotherapy, Am. Soc. Microbiol., Washington, DC, 2001, Abstract H1582.

98. W. Kati, D. Montgomery, K. McDaniel, C. Maring, A. Molla, and W. Kohlbrenner, *Antiviral Res.*, **50**, A83 (2001).

99. G. X. Luo, R. Colonno, and M. Krystal, *Virology*, **226**, 66–76 (1996).

100. G. X. Luo, A. Torri, W. E. Harte, S. Danetz, C. Cianci, L. Tiley, S. Day, D. Mullaney, K. L. Yu, C. Ouellet, P. Dextraze, N. Meanwell, R. Colonno, and M. Krystal, *J. Virol.*, **71**, 4062–4070 (1997).

101. C. Cianci, K. L. Yu, D. D. Dischino, W. E. Harte, M. Deshpande, L. Guangxiang, N. Meanwell, and M. Krystal, *J. Virol.*, **73**, 1785–1794 (1999).

102. S. J. Plotch, B. O'Hara, J. Morin, O. Palant, J. LaRocque, J. D. Bloom, S. A. Lang, M. J. DiGrandi, M. Bradley, R. Nilakantan, and Y. Gluzman, *J. Virol.*, **73**, 140–151 (1999).

103. K. L. Yu, E. Ruediger, G. X. Luo, C. Cianci, S. Danetz, L. Tiley, A. K. Trehan, I. Monkovic, B. Pearce, A. Martel, M. Krystal, and N. A. Meanwell, *Bioorg. Med. Chem. Lett.*, **9**, 2177–2180 (1999).

104. K. D. Combrink, H. B. Gulgeze, K. L. Yu, B. C. Pearce, A. K. Trehan, J. M. Wei, M. Deshpande, M. Krystal, A. Torri, G. X. Luo, C. Cianci, S. Danetz, L. Tiley, and N. A. Meanwell, *Bioorg. Med. Chem. Lett.*, **10**, 1649–1652 (2000).

105. K. A. Staschke, S. D. Hatch, J. C. Tang, W. J. Hornback, J. E. Munroe, J. M. Colacino, and M. A. Muesing, *Virology*, **248**, 264–274 (1998).

106. L. R. Hoffman, I. D. Kuntz, and J. M. White, *J. Virol.*, **71**, 8808–8820 (1997).

107. J. Yoshimoto, M. Kakui, H. Iwasaki, T. Fujiwara, H. Sugimoto, and N. Hattori, *Arch. Virol.*, **144**, 865–878 (1999).

108. S. Yagi, J. Ono, J. Yoshimoto, K. Sugita, N. Hattori, T. Fujioka, T. Fujiwara, H. Sugimoto, K. Hirano, and N. Hashimoto, *Pharm. Res.*, **16**, 1041–1046 (1999).

109. J. Yoshimoto, S. Yagi, J. Ono, K. Sugita, N. Hattori, T. Fujioka, T. Fujiwara, H. Sugimoto, and N. Hashimoto, *J. Pharm. Pharmacol.*, **52**, 1247–1255 (2000).

110. R. W. Sidwell and D. F. Smee, *Antiviral Res.*, **48**, 1–16 (2000).

111. P. D. Reuman, S. Keely, and G. M. Schiff, *J. Virol. Meth.*, **24**, 27–34 (1989).

112. M. Mammen, S. K. Choi, and G. M. Whitesides, *Angew Chem. Int. Edu.*, **37**, 2754–2794 (1998).

113. S. K. Choi, M. Mammen, and G. M. Whitesides, *J. Am. Chem. Soc.*, **119**, 4103–4111 (1997).

114. G. B. Sigal, M. Mammen, G. Dahmann, and G. M. Whitesides, *J. Am. Chem. Soc.*, **118**, 3789–3800 (1996).

115. K. J. Yarema and C. R. Bertozzi, *Curr. Opin. Chem. Biol.*, **2**, 49–61 (1998).

116. S. Watowich, J. Skehel, and D. Wiley, *Structure*, **2**, 719–731 (1994).

117. W. Y. Wu, B. Jin, G. Y. Krippner, and K. G. Watson, *Bioorg. Med. Chem. Lett.*, **10**, 341–343 (2000).

118. M. Mammen, K. Helmerson, R. Kishore, S. K. Choi, W. D. Phillips, and G. M. Whitesides, *Chem. Biol.*, **3**, 757–763 (1996).

119. L. WJ, A. Spaltenstein, J. E. W. Kinergy, and G. M. Whitesides, *J. Med. Chem.*, **37**, 3419–3433 (1994).

120. M. Mammen, G. Dahmann, and G. M. Whitesides, *J. Med. Chem.*, **38**, 4179–4190 (1995).

121. S. K. Choi, M. Mammen, and G. M. Whitesides, *Chem. Biol.*, **3**, 97–104 (1996).

122. H. Kamitakahara, T. Suzuki, N. Nishigori, Y. Suzuki, O. Kanie, and C. H. Wong, *Angew Chem. Int. Edu.*, **37**, 1524–1528 (1998).

123. X. L. Sun, Y. Kanie, C. T. Guo, O. Kanie, Y. Suzuki, and C. H. Wong, *Eur. J. Org. Chem.*, 2643–2653 (2000).

124. C. T. Guo, C. H. Wong, T. Kajimoto, T. Miura, Y. Ida, L. R. Juneja, M. J. Kim, H. Masuda, T. Suzuki, and Y. Suzuki, *Glycoconjugate J.*, **15**, 1099–1108 (1998).

125. J. D. Reuter, A. Myc, M. Hayes, Z. Gan, R. Roy, D. Quin, R. Yin, L. T. Piehler, R. Esfand, D. Tomalia, and J. Baker, *Bioconjugate Chem.*, **10**, 271–278 (1999).

126. E. Roseita and D. A. Tomalia, *Drug Discov. Today*, **6**, 427–436 (2001).

127. T. Abe, T. Hatta, K. Takai, H. Nakashima, T. Yokota, and H. Takaku, *Nucleosides Nucleotides*, **17**, 471–478 (1998).

128. T. Abe, S. Suzuki, T. Hatta, K. Takai, T. Yokota, and H. Takaku, *Antivir. Chem. Chemother.*, **9**, 253–262 (1998).

129. T. Mizuta, M. Fujiwara, T. Hatta, T. Abe, N. Miyano-Kurosaki, S. Shigeta, T. Yokota, and H. Takaku, *Nat. Biotechnol.*, **17**, 583–587 (1999).

130. T. Mizuta, M. Fujiwara, T. Abe, N. Kurosaki, T. Yokota, S. Shigeta, and H. Takaku, *Biochem. Biophys. Res. Commun.*, **279**, 158–161 (2000).

131. T. Hatta, Y. Nakagawa, K. Takai, S. Nakada, T. Yokota, and H. Takaku, *Biochem. Biophys. Res. Commun.*, **223**, 341–346 (1996).

132. T. Hatta, K. Takai, S. Nakada, T. Yokota, and H. Takaku, *Biochem. Biophys. Res. Commun.*, **232**, 545–549 (1997).

133. T. Abe, K. Takai, S. Nakada, T. Yokota, and H. Takaku, *FEBS Lett.*, **425**, 91–96 (1998).

134. T. Abe, T. Mizuta, S. Suzuki, T. Hatta, K. Takai, T. Yokota, and H. Takaku, *Nucleosides Nucleotides*, **18**, 1685–1688 (1999).

135. T. Toyoda, Y. Imamura, H. Takaku, T. Kashiwagi, K. Hara, J. Iwahashi, Y. Ohtsu, N. Tsumura, H. Kato, and N. Hamada, *FEBS Lett.*, **481**, 113–116 (2000).

136. S. W. Santoro and G. F. Joyce, *Proc. Natl. Acad. Sci. USA*, **94**, 4262–4266 (1997).

137. S. W. Santoro and G. F. Joyce, *Biochemistry*, **37**, 13330–13342 (1998).

138. V. N. Lazarev, M. M. Shmarov, A. N. Zakhartchouk, G. K. Yurov, O. U. Misurina, T. A. Akopian, N. F. Grinenko, N. G. Grodnitskaya, N. V. Kaverin, and B. S. Naroditsky, *Antiviral Res.*, **42**, 47–57 (1999).

139. A. C. Forster and S. Altman, *Science*, **249**, 783–786 (1990).

140. Y. Yaun and S. Altman, *Science*, **263**, 1269–1273 (1994).

141. D. Plehn-Dujowich and S. Altman, *Proc. Natl. Acad. Sci. USA*, **95**, 7327–7332 (1998).

141a. T. Hatta, M. Ishikawa, K. Takai, S. Nakada, T. Yokota, T. Hata, K. Miara, and H. Takaku, *Biochem. Biophys. Res. Commun.*, **249**, 103–106 (1998).

142. E. De Clercq, *Nucleosides Nucleotides*, **17**, 625–634 (1998).

143. J. C. Hastings, H. Selnick, B. Wolanski, and J. E. Tomassini, *Antimicrob. Agents Chemother.*, **40**, 1304–1307 (1996).

144. J. E. Tomassini, M. E. Davies, J. C. Hastings, R. Lingham, M. Mojena, S. L. Raghoobar, S. B. Singh, J. S. Tkacz, and M. A. Goetz, *Antimicrob. Agents Chemother.*, **40**, 1189–1193 (1996).

145. S. B. Singh and J. E. Tomassini, *J. Org. Chem.*, **66**, 5504–5516 (2001).

146. C. Cianci, T. D. Y. Chung, N. Meanwell, H. Putz, M. Hagen, R. J. Colonno, and M. Krystal, *Antivir. Chem. Chemother.*, **7**, 353–360 (1996).

147. E. H. Nasser, A. K. Judd, A. Sanchez, D. Anastasiou, and D. J. Bucher, *J. Virol.*, **70**, 8639–8644 (1996).

148. A. K. Judd, A. Sanchez, D. J. Bucher, J. H. Huffman, K. Bailey, and R. W. Sidwell, *Antimicrob. Agents Chemother.*, **41**, 687–692 (1997).

149. Q. Tu, L. H. Pinto, G. X. Luo, M. A. Shaughnessy, D. Mullaney, S. Kurtz, M. Krystal, and R. A. Lamb, *J. Virol.*, **70**, 4246–4252 (1996).

150. A. Kolocouris, D. Tataridis, G. Fytas, T. Mavromoustakos, G. B. Foscolos, N. Kolocouris, and E. De Clercq, *Bioorg. Med. Chem. Lett.*, **9**, 3465–3470 (1999).

151. C. N. Root, E. G. Wills, L. L. McNair, and G. R. Whittaker, *J. Gen. Virol.*, **81**, 2697–2705 (2000).

152. M. Hisaki, H. Imabori, M. Azuma, T. Suzutani, F. Iwakura, Y. Ohta, K. Kawanishi, Y. Ichigobara, M. Node, K. Nishide, I. Yoshida, and M. Ogasawara, *Antiviral Res.*, **42**, 121–137 (1999).

153. Y. M. Lin, M. T. Flavin, R. Schure, F. C. Chen, R. Sidwell, D. L. Barnard, J. H. Huffman, and E. R. Kern, *Planta Med.*, **65**, 120–125 (1999).

154. M. R. Kernan, A. Sendl, J. L. Chen, S. D. Jolad, P. Blanc, J. T. Murphy, C. A. Stoddart, W. Nanakorn, M. J. Balick, and E. J. Rozhon, *J. Nat. Prod.*, **60**, 635–637 (1997).

155. H. Takayama, Y. Iimura, M. Kitajima, N. Aimi, K. Konno, H. Inoue, M. Fujiwara, T. Mizuta, T. Yokota, S. Shigeta, K. Tokushisa, Y. Hanasaki, and K. Katsuura, *Bioorg. Med. Chem. Lett.*, **7**, 3145–3148 (1997).

156. S. C. Ma, P. P. H. But, V. E. C. Ooi, Y. H. He, S. H. S. Lee, S. F. Lee, and R. C. Lin, *Biol. Pharm. Bull.*, **24**, 311–312 (2001).

157. K. Yokomizo, Y. Miyamoto, K. Nagao, E. Kumagae, E. S. E. Habib, K. Suzuki, S. Harada, and M. Uyeda, *J. Antibiot.*, **51**, 1035–1039 (1998).

158. Y. Nishihara, E. Tsujii, S. Takase, Y. Tsurumi, T. Kino, M. Hino, M. Yamashita, and S. Hashimoto, *J. Antibiot.*, **53**, 1333–1340 (2000).

159. Y. Nishihara, E. Tsujii, Y. Yamagishi, T. Kino, M. Hino, M. Yamashita, and S. Hashimoto, *J. Antibiot.*, **53**, 1341–1345 (2000).

160. Y. Furuta, Y. Fukuda, M. Kuno, T. Kamiyama, K. Takahashi, N. Nomura, H. Egawa, S. Minami, Y. Watanabe, H. Narita, and K. Shiraki, 40th Interscience Conference on Antimicrobial Agents and Chemotherapy, Am. Soc. Microbiol., Washington, DC, 2000, Abstract 1852.

161. O. I. Kiselev, K. N. Kozeletskaya, E. Y. Melenevskaya, L. V. Vinogradova, E. E. Kever, S. I. Klenin, V. N. Zgonnik, M. A. Dumpis, and L. B. Piotrovsky, *Mol. Mat.*, **11**, 121–124 (1998).

162. F. Kasermann and D. Kempf, *Antiviral Res.*, **34**, 65–70 (1996).

163. W. D. Ding, B. Mitsner, G. Krishnamurthy, A. Aulabaugh, C. D. Hess, J. Zaccardi, M. Cutler, B. Feld, A. Gazumyan, Y. Raifeld, A. Nikitenko, S. A. Lang, Y. Gluzman, B. O'Hara, and G. A. Ellestad, *J. Med. Chem.*, **41**, 2671–2675 (1998).

164. P. R. Wyde, D. K. Moore-Poveda, B. O'Hara, W. D. Ding, B. Mitsner, and B. E. Gilbert, *Antiviral Res.*, **38**, 31–42 (1998).

165. A. Gazumyan, B. Mitsner, and G. A. Ellestad, *Curr. Pharm. Des.*, **6**, 525–546 (2000).

166. A. A. Nikitenki, Y. E. Raifeld, and T. Z. Wang, *Bioorg. Med. Chem. Lett.*, **11**, 1041–1044 (2001).

167. C. C. Huntley, B. Feld, W. Hu, S. Quartuccio, D. Grinberg, B. O'Hara, and A. Gazumyan, 40th Interscience Conference on Antimicrobial Agents and Chemotherapy, Am. Soc. Microbiol., Washington, DC, 2000, Abstract 187.

168. A. Gazumyan, B. Feld, W. Hu, S. Helwig, B. O'Hara, and C. C. Huntley, 40th Interscience Conference on Antimicrobial Agents and Chemotherapy, Am. Soc. Microbiol., Washington, DC, 2000, Abstract 1159.

169. W. J. Weiss, P. Wyde, G. Prince, and R.F.I. Project Team, 40th Interscience Conference on Antimicrobial Agents and Chemotherapy, Am. Soc. Microbiol, Washington, DC, 2000.

170. V. Razinkov, A. Gazumyan, A. Nikitenko, G. Ellestad, and G. Krishnamurthy, *Chem. Biol.*, **8**, 645–659 (2001).

171. D. C. Pevear, T. M. Tull, R. Direnzo, T. J. Nitz, and M. S. Collett, *Antiviral Res.*, **46**, A52 (2000).

172. T. J. Nitz, D. S. Pfarr, K. Bailey, C. W. Bender, C. W. Blackkedge, D. Cebzano, T. Draper, S. B. Ellis, J. Hehman, M. A. McWherter, D. H. Rys, J. Swestock, and N. Ye, 40th Interscience Conference on Antimicrobial Agents and Chemotherapy, Am. Soc. Microbiol., Washington, DC, 2000, Abstract 1853.

173. D. C. Pevear, T. M. Tull, R. Direnzo, N. Ma, T. J. Nitz, and M. S. Collett, 40th Interscience Conference on Antimicrobial Agents and Chemotherapy, Am. Soc. Microbiol., Washington, DC, 2000, Abstract 1854.

174. K. Andries, T. Gevers, R. Willebrords, J. Lacrampe, F. Janssens, and M. Moeremans, 40th Interscience Conference on Antimicrobial Agents and Chemotherapy, Am. Soc. Microbiol., Washington, DC, 2000, Abstract 1160.

175. K. Andries, M. Moeremans, T. Gevers, R. Willebrords, S. Sommen, J. Lacrampe, and F. Janssens, *Antiviral Res.*, **50**, A76 (2001).

176. S. Shigeta, *Expert Opin. Investig. Drugs*, **9**, 221–235 (2000).

177. K. Kimura, S. Mori, K. Tomita, K. Ohno, K. Takahashi, S. Shigeta, and M. Terada, *Antiviral Res.*, **47**, 41–51 (2000).

178. K. Sudo, W. Watanabe, K. Konno, R. Sato, T. Kajiyashiki, S. Shigeta, and T. Yokota, *Antimicrob. Agents Chemother.*, **43**, 752–757 (1999).

179. K. Sudo, K. Konno, W. Watanabe, S. Shigeta, and T. Yokota, *Microbiol. Immunol.*, **45**, 531–537 (2001).

180. N. A. Meanwell and M. Krystal, *Drug Discov. Today*, **5**, 241–252 (2000).

181. K. Sudo, W. Watanabe, S. Mori, K. Konno, S. Shigeta, and T. Yokota, *Antivir. Chem. Chemother.*, **10**, 135–139 (1999).

182. G. A. Prince, *Expert Opin. Ther. Patents*, **9**, 753–762 (1999).

183. J. Mills, *Antivir. Chemother.* **5**, 39–53 (1999).

184. G. A. Prince, *Expert Opin. Investig. Drugs*, **10**, 297–308 (2001).

185. M. Ottlini, D. Porter, V. Hemming, M. Zimmerman, N. Schwab, and G. Prince, *Bone Marrow Trans.*, **24**, 41–45 (1999).

186. R. Weltzin, V. Traina, K. Soike, J. Y. Zhang, P. Mack, G. Soman, G. Drabik, and T. P. Monath, *J. Infect. Dis.*, **174**, 256–261 (1996).

187. R. Weltzin and T. P. Monath, *Clin. Microbiol. Rev.*, **12**, 383–393 (1999).

188. P. R. Wyde, D. K. Moore, T. Hepburn, C. L. Silverman, T. G. Porter, M. Gross, G. Taylor, S. G. Demuth, and S. B. Dillon, *Pediatr. Res.*, **38**, 543–550 (1995).

189. H. C. Meissner, J. R. Groothuis, W. J. Rodriguez, R. C. Welliver, G. Hogg, P. H. Gray, R. Loh, E. A. F. Simoes, P. Sly, A. K. Miller, A. I. Nichols, D. K. Jorkasky, D. E. Everitt, and K. A. Thompson, *Antimicrob. Agents Chemother.*, **43**, 1183–1188 (1999).

190. S. Johnson, S. D. Griego, D. S. Pfarr, M. L. Doyle, R. Woods, D. Carlin, G. A. Prince, S. Koenig, J. F. Young, and S. B. Dillon, *J. Infect. Dis.*, **180**, 35–40 (1999).

191. D. M. Lambert, S. Barney, A. L. Lambert, K. Guthrie, R. Medinas, D. E. Davis, T. Bucy, J. Erickson, G. Merutka, and S. R. Petteway, *Proc. Natl. Acad. Sci. USA*, **93**, 2186–2191 (1996).

192. Q. Z. Yao and R. W. Compans, *Virology*, **223**, 103–112 (1996).

193. T. F. Wild and R. Buckland, *J. Gen. Virol.*, **78**, 107–111 (1997).

194. M. K. Pastey, T. L. Gower, P. W. Spearman, J. E. Crowe, and B. S. Graham, *Nat. Med.*, **6**, 35–40 (2000).

195. S. Jairath, P. B. Vargas, H. A. Hamlin, A. K. Field, and R. E. Kilkuskie, *Antiviral Res.*, **33**, 201–213 (1997).

196. M. R. Player, D. L. Barnard, and P. F. Torrence, *Proc. Natl. Acad. Sci. USA*, **95**, 8874–8879 (1998).

197. R. H. Silverman, J. K. Cowell, and P. F. Torrence, *Antisense Nucleic Acid Drug Dev.*, **9**, 409–414 (1999).

198. M. R. Player and P. F. Torrence, *Pharmacol. Ther.*, **78**, 55–113 (2001).

199. P. F. Torrence, W. Xiao, G. Li, H. Cramer, M. R. Player, and R. H. Silverman, *Antisense Nucleic Acid Drug Dev.*, **7**, 203–206 (1997).

200. W. Xiao, G. Y. Li, P. F. Torrence, N. M. Cirino, and R. H. Silverman, *Nucleosides Nucleotides*, **16**, 1735–1738 (1997).

201. N. M. Cirino, G. Li, W. Xiao, P. F. Torrence, and R. H. Silverman, *Proc. Natl. Acad. Sci. USA*, **94**, 1937–1942 (1997).

202. D. L. Barnard, R. W. Sidwell, W. Xiao, M. R. Player, S. A. Adah, and P. F. Torrence, *Antiviral Res.*, **41**, 119–134 (1999).

203. M. R. Kernan, A. Amarquaye, J. L. Chen, J. Chan, D. F. Sesin, N. Parkinson, Z. J. Ye, M. Barrett, C. Bales, C. A. Stoddart, P. Blanc, B. Sloan, C. Limbach, S. Mrisho, and E. J. Rozhon, *J. Nat. Prod.*, **61**, 564–570 (1998).

204. J. L. Chen, P. Blanc, C. Stoddart, M. Bogan, E. J. Rozhon, N. Parkinson, Z. Ye, R. Cooper, M. J. Balick, W. Nankorn, and M. R. Kernan, *J. Nat. Prod.*, **61**, 1295–1297 (1998).

205. E. De Clercq, *Int. J. Antimicrob. Agents*, **7**, 193–202 (1996).

206. J. M. Blanco, O. Caamano, F. Fernandez, X. Garcia-Mera, A. R. Hergueta, C. Lopez, J. E. Rodriguez-Borges, J. Balzarini, and E. De Clercq, *Chem. Pharm. Bull. (Tokyo)*, **47**, 1314–1317 (1999).

207. N. Verdaguer, D. Blaas, and I. Fita, *J. Mol. Biol.*, **300**, 1179–1194 (2000).

208. S. K. Tsang, J. Cheh, L. Isaacs, D. McCarthy, S. K. Choi, D. C. Pevear, G. M. Whitesides, and J. M. Hogle, *Chem. Biol.*, **8**, 33–45 (2001).

209. J. K. Lewis, B. Bothner, T. J. Smith, and G. Siuzdak, *Proc. Natl. Acad. Sci. USA*, **95**, 6774–6778 (1998).

210. T. J. Smith and T. Baker, *Adv. Virus Res.*, **52**, 1–23 (1999).

211. D. K. Phelps, B. Speelman, and C. B. Post, *Curr. Opin. Struct. Biol.*, **10**, 170–173 (2000).

212. S. K. Tsang, P. Danthi, M. Chow, and J. M. Hogle, *J. Mol. Biol.*, **296**, 335–340 (2000).

213. D. K. Phelps, P. J. Rossky, and C. B. Post, *J. Mol. Biol.*, **276**, 331–337 (1998).

214. D. K. Phelps and C. B. Post, *Protein Sci.*, **8**, 2281–2289 (1999).

215. S. K. Tsang, B. M. McDermott, V. R. Racaniello, and J. M. Hogle, *J. Virol.*, **75**, 4984–4989 (2001).

216. A. T. Hadfield, G. D. Diana, and M. G. Rossmann, *Proc. Natl. Acad. Sci. USA*, **96**, 14730–14735 (1999).

217. H. A. Rotbart, *Antivir. Chem. Chemother.*, **11**, 261–271 (2000).

218. J. M. Rogers, G. D. Diana, and M. A. McKinlay, *Antivir. Chemother.*, **458**, 69–76 (1999).

219. H. A. Rotbart, J. F. O'Connell, and M. A. McKinlay, *Antiviral Res.*, **38**, 1–14 (1998).

220. H. A. Rotbart, *Infect. Med.*, **17**, 488–494 (2000).

221. G. D. Diana and D. C. Pevear, *Antivir. Chem. Chemother.*, **8**, 401–408 (1997).

222. R. B. Turner, *Antiviral Res.*, **49**, 1–14 (2000).

223. H. A. Rotbart, *Pediatr. Infect. Dis. J.*, **18**, 632–633 (1999).

224. G. L. Kearns, S. M. Abdel-Rahman, L. P. James, D. L. Blowey, J. D. Marshall, T. G. Wells, and R. F. Jacobs, *Antimicrob. Agents Chemother.*, **43**, 634–638 (1999).

225. S. M. Abdel-Rahman and G. L. Kearns, *Antimicrob. Agents Chemother.*, **42**, 2706–2709 (1998).

226. D. C. Pevear, T. M. Tull, M. E. Seipel, and J. M. Groarke, *Antimicrob. Agents Chemother.*, **43**, 2109–2115 (1999).

227. G. M. Schiff and J. R. Sherwood, *J. Infect. Dis.*, **181**, 20–26 (2000).

228. M. H. Sawyer, *Curr. Opinion Ped.*, **13**, 65–69 (2001).

229. A. Mai, M. Artico, S. Massa, R. Ragno, A. De Montis, S. Corrias, M. G. Spiga, and P. La Colla, *Antivir. Chem. Chemother.*, **7**, 213–220 (1996).

230. A. Mai, M. Artico, G. Sbardella, S. Massa, A. DeMontis, I. Puddu, C. Musiu, and P. LaColla, *Antivir. Chem. Chemother.*, **8**, 235–242 (1997).

231. F. G. Hayden, G. J. Hipskind, D. H. Woerner, G. F. Eisen, M. Janssens, P. A. J. Janssen, and K. Andries, *Antimicrob. Agents Chemother.*, **39**, 290–294 (1995).

232. P. J. Buontempo, S. Cox, J. Wright-Minogue, J. L. DeMartino, A. M. Skelton, E. Ferrari, R. Albin, E. J. Rozhon, V. Girijavallabhan, J. F. Modlin, and J. F. O'Connell, *Antimicrob. Agents Chemother.*, **41**, 1220–1225 (1997).

233. K. N. Lentz, A. D. Smith, S. C. Geisler, S. Cox, P. Buontempo, A. Skelton, J. DeMartino, E. Rozhon, J. Schwartz, V. Girijavallabhan, J. OConnell, and E. Arnold, *Structure*, **5**, 961–978 (1997).

234. D. A. Oren, A. Zhang, H. Nesvadba, B. Rosenwirth, and E. Arnold, *J. Mol. Biol.*, **259**, 120–134 (1996).

235. K. G. Watson, W. Y. Wu, G. Y. Krippner, D. B. McConnell, B. Jin, P. C. Stanislawski, R. N. Brown, D. K. Chalmers, S. P. Tucker, R. Smith, S. Hamilton, A. Luttick, J. Ryan, and P. A. Reece, *Antiviral Res.*, **50**, A75 (2001).

236. F. G. Hayden, C. E. Crump, P. A. Reece, J. Watson, J. Ryan, R. Smith, and S. P. Tucker, *Antiviral Res.*, **50**, A75 (2001).

237. J. Ryan, R. W. Sidwell, D. Barnard, S. P. Tucker, and P. A. Reece, *Antiviral Res.*, **50**, A76 (2001).

238. D. Joseph-McCarthy, J. M. Hogle, and M. Karplus, *Proteins Struct. Funct. Genet.*, **29**, 32–58 (1997).

239. D. McCarthy, S. K. Tsang, D. J. Filman, J. M. Hogle, and M. Karplus, *J. Am. Chem. Soc.*, **123**, 12758–12769 (2001).

240. C. C. C. Cutri, A. Garozzo, M. A. Siracusa, M. C. Sarva, G. Tempera, E. Geremia, M. R. Pinizzotto, and F. Guerrera, *Bioorg. Med. Chem.*, **6**, 2271–2280 (1998).

241. C. C. C. Cutri, A. Garozzo, M. A. Siracusa, M. C. Sarva, A. Castro, E. Geremia, M. R. Pinizzotto, and F. Guerrera, *Bioorg. Med. Chem.*, **7**, 225–230 (1999).

242. A. Garozzo, C. C. C. Cutri, A. Castro, G. Tempera, F. Guerrera, M. C. Sarva, and E. Geremia, *Antiviral Res.*, **45**, 199–210 (2000).

243. M. A. Murray and L. M. Babe, *Antiviral Res.*, **44**, 123–131 (1999).

244. S. L. Johnston, *Trends Microbiol.*, **5**, 58–63 (1997).

245. K. McIntosh, *JAMA*, **281**, 1844 (1999).

246. J. M. Casasnovas, J. K. Bickford, and T. A. Springer, *J. Virol.*, **72**, 6244–6246 (1998).

247. E. D. Huguenel, D. Cohn, D. P. Dockum, J. M. Greve, M. A. Fournel, L. Hammond, R. Irwin, J. Mahoney, A. McClelland, E. Muchmore, A. C. Ohlin, and P. Scuderi, *Am. J. Respir. Crit. Care Med.*, **155**, 1206–1210 (1997).

248. R. B. Turner, M. T. Wecker, G. Pohl, T. J. Witek, E. McNally, R. George, B. Winther, and F. G. Hayden, *JAMA*, **281**, 1797 (1999).

249. W. M. Kati, H. L. Sham, J. O. McCall, D. A. Montgomery, G. T. Wang, W. Rosenbrook, L. Miesbauer, A. Buko, and D. W. Norbeck, *Arch. Biochem. Biophys.*, **362**, 363–375 (1999).

250. Y. T. Huang, B. A. Malcolm, and J. C. Vederas, *Bioorg. Med. Chem.*, **7**, 607–619 (1999).

251. S. C. Mosimann, M. M. Cherney, S. Sia, S. Plotch, and M. N. G. James, *J. Mol. Biol.*, **273**, 1032–1047 (1997).

252. S. Hata, T. Sato, H. Sorimachi, S. Ishiura, and K. Suzuki, *J. Virol. Methods*, **84**, 117–126 (2000).

253. S. E. Webber, J. Tikhe, S. T. Worland, S. A. Fuhrman, T. F. Hendrickson, D. A. Matthews, R. A. Love, A. K. Patick, J. W. Meador, R. A. Ferre, E. L. Brown, D. M. DeLisle, C. E. Ford, and S. L. Binford, *J. Med. Chem.*, **39**, 5072–5082 (1996).

254. S. L. Binford, J. Meador, F. Maldonado, M. A. Brothers, P. T. Weady, J. S. Isaacson, S. T. Maldonado, S. T. Worland, D. A. Matthews, and L. S. Zalman, 40th Interscience Conference on Antimicrobial Agents and Chemotherapy, Am. Soc. Microbiol., Washington, DC, 2000, Abstract 1162.

255. J. S. Kong, S. Venkatraman, K. Furness, S. Nimkar, T. A. Shepherd, Q. M. Wang, J. Aube, and R. P. Hanzlik, *J. Med. Chem.*, **41**, 2579–2587 (1998).

256. Q. M. Wang, *Expert Opin. Ther. Patents*, **8**, 1151–1156 (1998).

257. P. S. Dragovich, *Expert Opin. Ther. Patents*, **11**, 177–184 (2001).

258. P. S. Dragovich, S. E. Webber, R. E. Babine, S. A. Fuhrman, A. K. Patick, D. A. Matthews, C. A. Lee, S. H. Reich, T. J. Prins, J. T. Marakovits, E. S. Littlefield, R. Zhou, J. Tikhe, C. E. Ford, M. B. Wallace, J. W. Meador, R. A. Ferre, E. L. Brown, S. L. Binford, J. E. V. Harr, D. M. DeLisle, and S. T. Worland, *J. Med. Chem.*, **41**, 2806–2818 (1998).

259. P. S. Dragovich, T. J. Prins, R. Zhou, S. A. Fuhrman, A. K. Patick, D. A. Matthews, C. E. Ford, J. W. Meador, R. A. Ferre, and S. T. Worland, *J. Med. Chem.*, **42**, 1203–1212 (1999).

260. P. S. Dragovich, T. J. Prins, R. Zhou, S. E. Webber, J. T. Marakovits, S. A. Fuhrman, A. K. Patick, D. A. Matthews, C. A. Lee, C. E. Ford, B. J. Burke, P. A. Rejto, T. F. Hendrickson, T. Tuntland, E. L. Brown, J. W. Meador, R. A. Ferre, J. E. V. Harr, M. B. Kosa, and S. T. Worland, *J. Med. Chem.*, **42**, 1213–1224 (1999).

261. D. A. Matthews, P. S. Dragovich, S. E. Webber, S. A. Fuhrman, A. K. Patick, L. S. Zalman, T. F. Hendrickson, R. A. Love, T. J. Prins, J. T. Marakovits, R. Zhou, J. Tikhe, C. E. Ford, J. W. Meador, R. A. Ferre, E. L. Brown, S. L. Binford, M. A. Brothers, D. M. DeLisle, and S. T. Worland, *Proc. Natl. Acad. Sci. USA*, **96**, 11000–11007 (1999).

262. P. S. Dragovich, S. E. Webber, T. J. Prins, R. Zhou, J. T. Marakovits, J. G. Tikhe, S. A. Fuhrman, A. K. Patick, D. A. Matthews, C. E. Ford, E. L. Brown, S. L. Binford, J. W. Meador, R. A. Ferre, and S. T. Worland, *Bioorg. Med. Chem. Lett.*, **9**, 2189–2194 (1999).

263. P. S. Dragovich, S. E. Webber, R. E. Babine, S. A. Fuhrman, A. K. Patick, S. H. Reich, J. T. Marakovits, T. J. Prins, R. Zhou, J. Tikhe, E. S. Littlefield, T. M. Bleckman, M. B. Wallace, T. L. Little, C. Ford, J. Meador, E. L. Brown, S. L. Binford, D. M. DeLisle, and S. T. Worland, *J. Med. Chem.*, **41**, 2819–2834 (1998).

264. S. E. Webber, K. Okano, T. L. Little, S. H. Reich, Y. Xin, S. A. Fuhrman, D. A. Matthews, R. A. Love, T. F. Hendrickson, A. K. Patick, J. W. Meador, R. A. Ferre, E. L. Brown, C. E. Ford, S. L. Binford, and S. T. Worland, *J. Med. Chem.*, **41**, 2786–2805 (1998).

265. P. S. Dragovich, R. Zhou, D. J. Skalitzky, S. A. Fuhrman, A. K. Patick, C. E. Ford, J. W. Meador, and S. T. Worland, *Bioorg. Med. Chem.*, **7**, 589–598 (1999).

266. A. K. Patick, S. L. Binford, M. A. Brothers, R. L. Jackson, C. E. Ford, M. D. Diem, F. Maldonado, P. S. Dragovich, R. Zhou, T. J. Prins, S. A. Fuhrman, J. W. Meador, L. S. Zalman, D. A. Matthews, and S. T. Worland, *Antimicrob. Agents Chemother.*, **43**, 2444–2450 (1999).

267. L. Kaiser, C. E. Crump, and F. G. Hayden, *Antiviral Res.*, **47**, 215–220 (2000).

267a. L. S. Zalman, M. A. Brothers, P. S. Dragovich, R. Zhou, T. J. Prins, S. T. Worland, and A. K. Patick, *Antimicrob. Agents Chemother.*, **44**, 1236–1241 (2000).

268. T. A. Shepherd, G. A. Cox, E. McKinney, J. Tang, M. Wakulchik, R. E. Zimmerman, and E. C. Villarreal, *Bioorg. Med. Chem. Lett.*, **6**, 2893–2896 (1996).

269. P. S. Dragovich, R. Zhou, S. E. Webber, T. J. Prins, A. K. Kwok, K. Okano, S. A. Fuhrman, L. S. Zalman, F. C. Maldonado, E. L. Brown, J. W. Meador, A. K. Patick, C. E. Ford, M. A. Brothers, S. L. Binford, D. A. Matthews, R. A. Ferre, and S. T. Worland, *Bioorg. Med. Chem. Lett.*, **10**, 45–48 (2000).

270. T. S. Morris, S. Frormann, S. Shechosky, C. Lowe, M. S. Lall, V. Muller, R. Purcell, S. Ernerson, J. C. Vederas, and B. A. Malcolm, *Bioorg. Med. Chem.*, **5**, 797–807 (1997).

271. S. Venkatraman, J. S. Kong, S. Nimkar, Q. M. Wang, J. Aube, and R. P. Hanzlik, *Bioorg. Med. Chem. Lett.*, **9**, 577–580 (1999).

272. R. Xing and R. P. Hanzlik, *J. Med. Chem.*, **41**, 1344–1351 (1998).

273. R. D. Hill and J. C. Vederas, *J. Org. Chem.*, **64**, 9538–9546 (1999).

274. L. N. Jungheim, J. D. Cohen, R. B. Johnson, E. C. Villarreal, M. Wakulchik, R. J. Loncharich, and Q. M. Wang, *Bioorg. Med. Chem. Lett.*, **7**, 1589–1594 (1997).

275. Q. M. Wang, R. B. Johnson, L. N. Jungheim, J. D. Cohen, and E. C. Villarreal, *Antimicrob. Agents Chemother.*, **42**, 916–920 (1998).

276. S. H. Reich, T. Johnson, M. B. Wallace, S. E. Kephart, S. A. Fuhrman, S. T. Worland, D. A. Matthews, T. F. Hendrickson, F. Chan, J. Meador, R. A. Ferre, E. L. Brown, D. M. DeLisle, A. K. Patick, S. L. Binford, and C. E. Ford, *J. Med. Chem.*, **43**, 1670–1683 (2000).

277. M. Xian, Q. M. Wang, X. Chen, K. Wang, and P. G. Wang, *Bioorg. Med. Chem. Lett.*, **10**, 2097–2100 (2000).

278. M. J. Tebbe, C. B. Jensen, W. A. Spitzer, R. B. Franklin, M. C. George, and D. L. Phillips, *Antiviral Res.*, **42**, 25–33 (1999).

279. P. Brown-Augsburger, L. M. Vance, S. K. Malcolm, H. Hsiung, D. P. Smith, and B. A. Heinz, *Arch. Virol.*, **144**, 1569–1585 (1999).

280. M. J. Tebbe, W. A. Spitzer, F. Victor, S. C. Miller, C. C. Lee, T. R. Sattelberg, E. McKinney, and J. C. Tang, *J. Med. Chem.*, **40**, 3937–3946 (1997).

281. R. E. Stratford, M. P. Clay, B. A. Heinz, M. T. Kuhfeld, S. J. Osborne, D. L. Phillips, S. A. Sweetana, M. J. Tebbe, V. Vasudevan, L. L. Zornes, and T. D. Lindstrom, *J. Pharm. Sci.*, **88**, 747–753 (1999).

282. F. Victor, T. J. Brown, K. Campanale, B. A. Heinz, L. A. Shipley, K. S. Su, J. Tang, L. M. Vance, and W. A. Spitzer, *J. Med. Chem.*, **40**, 1511–1518 (1997).

283. F. Victor, R. Loncharich, J. Tang, and W. A. Spitzer, *J. Med. Chem.*, **40**, 3478–3483 (1997).

284. C. Hamdouchi, J. de Blas, M. del Prado, J. Gruber, B. A. Heinz, and L. Vance, *J. Med. Chem.*, **42**, 50–59 (1999).

285. L. Nikolaeva and A. S. Galabov, *Acta Virol.*, **43**, 303–311 (1999).

286. L. Nikolaeva and A. S. Galabov, *Acta Virol.*, **43**, 263–265 (1999).

287. L. Nikolaeva and A. S. Galabov, *Acta Virol.*, **44**, 73–78 (2000).

288. M. Klein, D. Hadaschik, H. Zimmermann, H. J. Eggers, and B. Nelsen-Salz, *J. Gen. Virol.*, **81**, 895–901 (2000).

289. T. Pfister and E. Wimmer, *J. Biol. Chem.*, **274**, 6992–7001 (1999).

290. D. Hadaschik, M. Klein, H. Zimmermann, H. J. Eggers, and B. Nelsen-Salz, *J. Virol.*, **73**, 10536–10539 (1999).

291. H. Shimizu, M. Agoh, Y. Agoh, H. Yoshida, K. Yoshi, T. Yoneyama, A. Hagiwara, and T. Miyamura, *J. Virol.*, **74**, 4146–4154 (2000).

292. L. M. Vance, N. Moscufo, M. Chow, and B. A. Heinz, *J. Virol.*, **71**, 8759–8765 (1997).

293. Y. Verlinden, A. Cuconati, E. Wimmer, and B. Rombaut, *Antiviral Res.*, **48**, 61–69 (2000).

294. S. J. Semple, S. F. Nobbs, S. M. Pyke, G. D. Reynolds, and R. L. P. Flower, *J. Ethnopharmacol.*, **68**, 283–288 (1999).

295. S. J. Semple, S. M. Pyke, G. D. Reynolds, and R. L. P. Flower, *Antiviral Res.*, **49**, 169–178 (2001).

296. J. Boustie, J. L. Stigliani, J. Montanha, M. Amoros, P. Payard, and L. Girre, *J. Nat. Prod.*, **61**, 480–484 (1998).

297. N. Desideri, C. Conti, P. Mastromarino, and F. Mastropaolo, *Antivir. Chem. Chemother.*, **11**, 373–381 (2000).

298. N. Desideri, S. Olivieri, M. L. Stein, R. Sgro, N. Orsi, and C. Conti, *Antivir. Chem. Chemother.*, **8**, 545–555 (1997).

299. M. G. Quaglia, N. Desideri, E. Bossu, R. Sgro, and C. Conti, *Chirality*, **11**, 495–500 (1999).

300. S. Kaldor, J. Fritz, J. Tang, and E. McKinney, *Bioorg. Med. Chem. Lett.*, **6**, 3041–3044 (1996).

301. K. J. Blight, A. A. Kolykhalov, K. E. Reed, E. V. Vagapov, and C. M. Rice, *Antivir. Ther.*, **3**, 71–81 (1998).

302. D. G. Macejak, K. L. Jensen, S. F. Jamison, K. Domenico, E. C. Roberts, N. Chaudhary, I. von Carlowitz, L. Bellon, M. J. Tong, A. Conrad, P. A. Pavco, and L. M. Blatt, *Hepatology*, **31**, 769–776 (2000).

303. P. A. Lee, L. M. Blatt, K. S. Blanchard, K. S. Bouhana, P. A. Pavco, L. Bellon, and J. A. Sandberg, *Hepatology*, **32**, 640–646 (2000).

304. N. Sakamoto, C. H. Wu, and G. Y. Wu, *J. Clin. Invest.*, **98**, 2720–2728 (1996).

305. M. Oketani, Y. Asahina, C. H. Wu, and G. Y. Wu, *J. Hepatol.*, **31**, 628–634 (1999).

306. C. H. Wu and G. Y. Wu, *Gastroenterology*, **114**, 1304–1312 (1998).

307. A. Lieber, C. Y. He, S. J. Polyak, D. R. Gretch, D. Barr, and M. A. Kay, *J. Virol.*, **70**, 8782–8791 (1996).

308. K. Ohkawa, N. Yuki, Y. Kanazawa, K. Ueda, E. Mita, Y. Sasaki, A. Kasahara, and N. Hayashi, *J. Hepatol.*, **27**, 78–84 (1997).

309. P. J. Welch, R. Tritz, S. Yei, M. Leavitt, M. Yu, and J. Barber, *Gene Ther.*, **3**, 994–1001 (1996).

310. P. J. Welch, S. P. Yei, and J. R. Barber, *Clin. Diagn. Virol.*, **10**, 163–171 (1998).

311. R. Hanecak, V. Brown-Driver, M. C. Fox, R. F. Azad, S. Furusako, C. Nozaki, C. Ford, H. Sasmor, and K. P. Anderson, *J. Virol.*, **70**, 5203–5212 (1996).

312. V. Brown-Driver, T. Eto, E. Lesnik, K. P. Anderson, and R. C. Hanecak, *Antisense Nucleic Acid Drug Dev.*, **9**, 145–154 (1999).

313. H. Zhang, R. Hanecak, V. Brown-Driver, R. Azad, B. Conklin, M. C. Fox, and K. P. Anderson, *Antimicrob. Agents Chemother.*, **43**, 347–353 (1999).

314. Anonymous, *Antiviral Agents Bull.*, **14**, 1 (2001).

315. W. F. Lima, V. Brown Driver, M. Fox, R. Hanecak, and T. W. Bruice, *J. Biol. Chem.*, **272**, 626–638 (1997).

316. W. H. Caselmann, S. Eisenhardt, and M. Alt, *Intervirology*, **40**, 394–399 (1997).

317. M. Alt, R. Renz, P. H. Hofschneider, and W. H. Caselmann, *Arch. Virol.*, **142**, 589–599 (1997).

318. M. Alt, S. Eisenhardt, M. Serwe, R. Renz, J. W. Engels, and W. H. Caselmann, *Eur. J. Clin. Invest.*, **29**, 868–876 (1999).

319. T. J. Lehmann and J. W. Engels, *Bioorg. Med. Chem.*, **9**, 1827–1835 (2001).

320. O. Vidalin, M. E. Major, B. Rayner, J. L. Imbach, C. Trepo, and G. Inchauspe, *Antimicrob. Agents Chemother.*, **40**, 2337–2344 (1996).

321. T. Wakita, D. Moradpour, K. Tokushihge, and J. R. Wands, *J. Med. Virol.*, **57**, 217–222 (1999).

322. S. Das, M. Ott, A. Yamane, W. M. Tsai, M. Gromeier, F. Lahser, S. Gupta, and A. Dasgupta, *J. Virol.*, **72**, 5638–5647 (1998).

323. A. Venkatesan, S. Das, and A. Dasgupta, *Nucleic Acids Res.*, **27**, 562–572 (1999).

324. W. Y. Wang, P. Preville, N. Morin, S. Mounir, W. Z. Cai, and M. A. Siddiqui, *Bioorg. Med. Chem. Lett.*, **10**, 1151–1154 (2000).

325. B. W. Dymock, P. S. Jones, and F. X. Wilson, *Antivir. Chem. Chemother.*, **11**, 79–96 (2000).

326. M. A. Walker, *Drug Discov. Today*, **4**, 518–529 (1999).

327. A. D. Kwong, J. L. Kim, G. Rao, D. Lipovsek, and S. A. Raybuck, *Antiviral Res.*, **41**, 67–84 (1999).

328. R. De Francesco, A. Pessi, and C. Steinkuhler, *J. Viral Hepatitis*, **6**, 23–30 (1999).

329. M. Stempniak, Z. Hostomska, B. R. Nodes, and Z. Hostomsky, *J. Virol.*, **71**, 2881–2886 (1997).

330. R. De Francesco, A. Urbani, M. C. Nardi, L. Tomei, C. Steinkuhler, and A. Tramontano, *Biochemistry*, **35**, 13282–13287 (1996).

331. R. Bartenschlager, *J. Viral Hepatitis*, **6**, 165–181 (1999).

332. M. R. Attwood, J. M. Bennett, A. D. Campbell, G. G. M. Canning, M. G. Carr, E. Conway, R. M. Dunsdon, J. R. Greening, P. S. Jones, P. B. Kay, B. K. Handa, D. N. Hurst, N. S. Jennings, S. Jordan, E. Keech, M. A. O'Brien, H. H. Overton, J. King-Underwood, T. M. Raynham, K. P. Stenson, C. S. Wilkinson, T. C. I. Wilkinson, and F. X. Wilson, *Antivir. Chem. Chemother.*, **10**, 259–273 (1999).

333. E. Bianchi, S. Orru, F. Dal Piaz, R. Ingenito, A. Casbarra, G. Biasiol, U. Koch, P. Pucci, and A. Pessi, *Biochemistry*, **38**, 13844–13852 (1999).

334. N. Kakiuchi, Y. Komoda, M. Hijikata, and K. Shimotohno, *J. Biochem.*, **122**, 749–755 (1997).

335. K. Sudo, H. Inoue, Y. Shimizu, K. Yamaji, K. Konno, S. Shigeta, T. Kaneko, T. Yokota, and K. Shimotohno, *Antiviral Res.*, **32**, 9–18 (1996).

336. C. Steinkuhler, A. Urbani, L. Tomei, G. Biasiol, M. Sardana, E. Bianchi, A. Pessi, and R. De Francesco, *J. Virol.*, **70**, 6694–6700 (1996).

337. M. Llinas-Brunet, M. Bailey, G. Fazal, S. Goulet, T. Halmos, S. Laplante, R. Maurice, M. Poirier, M. A. Poupart, D. Thibeault, D. Wernic, and D. Lamarre, *Bioorg. Med. Chem. Lett.*, **8**, 1713–1718 (1998).

338. C. Steinkuhler, G. Biasiol, M. Brunetti, A. Urbani, U. Koch, R. Cortese, A. Pessi, and R. De Francesco, *Biochemistry*, **37**, 8899–8905 (1998).

339. P. Ingallinella, S. Altamura, E. Bianchi, M. Taliani, R. Ingenito, R. Cortese, R. De Francesco, C. Steinkuhler, and A. Pessi, *Biochemistry*, **37**, 8906–8914 (1998).

340. M. Llinas-Brunet, M. Bailey, R. Deziel, G. Fazal, V. Gorys, S. Goulet, T. Halmos, R. Maurice, M. Poirier, M. A. Poupart, J. Rancourt, D. Thibeault, D. Wernic, and D. Lamarre, *Bioorg. Med. Chem. Lett.*, **8**, 2719–2724 (1998).

341. A. Urbani, E. Bianchi, F. Narjes, A. Tramontano, R. De Francesco, C. Steinkuhler, and A. Pessi, *J. Biol. Chem.*, **272**, 9204–9209 (1997).

342. M. Llinas-Brunet, M. Bailey, G. Fazal, E. Ghiro, V. Gorys, S. Goulet, T. Halmos, R. Maurice, M. Poirier, M. A. Poupart, J. Rancourt, D. Thibeault, D. Wernic, and D. Lamarre, *Bioorg. Med. Chem. Lett.*, **10**, 2267–2270 (2000).

343. C. J. Johansson, I. Hubatsch, E. Akerblom, G. Lindeberg, S. Winiwarter, U. H. Danielson, and A. Hallberg, *Bioorg. Med. Chem. Lett.*, **11**, 203–206 (2001).

344. C. Steinkuhler, U. Koch, F. Narjes, and V. Matassa, *Curr. Med. Chem.*, **8**, 919–932 (2001).

345. R. M. Dunsdon, J. R. Greening, P. S. Jones, S. Jordan, and F. X. Wilson, *Bioorg. Med. Chem. Lett.*, **10**, 1577–1579 (2000).

346. W. Han, Z. L. Hu, X. J. Jiang, and C. P. Decicco, *Bioorg. Med. Chem. Lett.*, **10**, 711–713 (2000).

347. J. M. Bennett, A. D. Campbell, A. J. Campbell, M. G. Carr, R. M. Dunsdon, J. R. Greening, D. N. Hurst, N. S. Jennings, P. S. Jones, S. Jordan, P. B. Kay, M. A. O'Brien, J. King-Underwood, T. M. Raynham, C. S. Wilkinson, T. C. I. Wilkinson, and F. X. Wilson, *Bioorg. Med. Chem. Lett.*, **11**, 355–357 (2001).

348. S. Di Marco, M. Rizzi, C. Volpari, M. A. Walsh, F. Narjes, S. Colarusso, R. De Francesco, V. G. Matassa, and M. Sollazzo, *J. Biol. Chem.*, **275**, 7152–7157 (2000).

349. F. Narjes, M. Brunetti, S. Colarusso, B. Gerlach, U. Koch, G. Biasiol, D. Fattori, R. De Francesco, V. G. Matassa, and C. Steinkuhler, *Biochemistry*, **39**, 1849–1861 (2000).

350. S. Ubol, C. Sukwattanapan, and Y. Maneerat, *J. Mol. Microbiol.*, **50**, 238–242 (2001).

351. P. Ingallinella, E. Bianchi, R. Ingenito, U. Koch, C. Steinkuhler, S. Altamura, and A. Pessi, *Biochemistry*, **39**, 12898–12906 (2000).

352. K. Sudo, Y. Matsumoto, M. Matsushima, K. Konno, K. Shimotohno, S. Shigeta, and T. Yokota, *Antivir. Chem. Chemother.*, **8**, 541–544 (1997).

353. N. Kakiuchi, Y. Komoda, K. Komoda, N. Takeshita, S. Okada, T. Tani, and K. Shimotohno, *FEBS Lett.*, **421**, 217–220 (1998).

354. K. Sudo, Y. Matsumoto, M. Matsushima, M. Fujiwara, K. Konno, K. Shimotohno, S. Shigeta, and T. Yokota, *Biochem. Biophys. Res. Commun.*, **238**, 643–647 (1997).

355. W. T. Sing, C. Lee, S. L. Yeo, S. P. Lim, and M. M. Sim, *Bioorg. Med. Chem. Lett.*, **11**, 91–94 (2001).

356. M. Chu, R. Mierzwa, I. Truumees, A. King, M. Patel, R. Berrie, A. Hart, N. Butkiewicz, B. DasMahapatra, T. M. Chan, and M. S. Puar, *Tetrahedron Lett.*, **37**, 7229–7232 (1996).

357. M. Chu, R. Mierzwa, L. He, A. King, M. Patel, J. Pichardo, A. Hart, N. Butkiewicz, and M. S. Puar, *Bioorg. Med. Chem. Lett.*, **9**, 1949–1952 (1999).

358. J. R. Dai, B. K. Carte', P. J. Sidebottom, A. L. S. Yew, S. B. Ng, Y. Huang, and M. S. Butler, *J. Nat. Prod.*, **64**, 125–126 (2001).

359. K. S. Yeung, N. A. Meanwell, Z. Qiu, D. Hernandez, S. Zhang, F. McPhee, S. Weinheimer, J. M. Clark, and J. W. Janc, *Bioorg. Med. Chem. Lett.*, **11**, 2355–2359 (2001).

360. P. T. Urvil, N. Kakiuchi, D. M. Zhou, K. Shimotohno, P. K. R. Kumar, and S. Nishikawa, *Eur. J. Biochem.*, **248**, 130–138 (1997).

361. P. K. R. Kumar, K. Machida, P. T. Urvil, N. Kakiuchi, D. Vishnuvardhan, K. Shimotohno, K. Taira, and S. Nishikawa, *Virology*, **237**, 270–282 (1997).

362. K. Fukuda, D. Vishnuvardhan, S. Sekiya, J. Hwang, N. Kakiuchi, T. Kazunari, K. Shimotohno, P. K. R. Kumar, and S. Nishikawa, *Eur. J. Biochem.*, **267**, 3685–3694 (2000).

363. J. Hwang, H. Fauzi, K. Fukuda, S. Sekiya, N. Kakiuchi, K. Shimotohno, K. Taira, I. Kusakabe, and S. Nishikawa, *Biochem. Biophys. Res. Commun.*, **279**, 557–562 (2000).

364. N. Dimasi, F. Martin, C. Volpari, M. Brunetti, G. Biasiol, S. Altamura, R. Cortese, R. De Francesco, C. Steinkuhler, and M. Sollazzo, *J. Virol.*, **71**, 7461–7469 (1997).

365. F. Martin, C. Steinkuhler, M. Brunetti, A. Pessi, R. Cortese, R. De Francesco, and M. Sollazzo, *Protein Eng.*, **12**, 1005–1011 (1999).

366. F. Martin, N. Dimasi, C. Volpari, C. Perrera, S. Di Marco, M. Brunetti, C. Steinkuhler, R. De Francesco, and M. Sollazzo, *Biochemistry*, **37**, 11459–11468 (1998).

367. T. Ueno, S. Misawa, Y. Ohba, M. Matsumoto, M. Mizunuma, N. Kasai, K. Tsumoto, I. Kumagai, and H. Hayashi, *J. Virol.*, **74**, 6300–6308 (2000).

368. N. Kasai, K. Tsumoto, S. Niwa, S. Misawa, T. Ueno, H. Hayashi, and I. Kumagai, *Biochem. Biophys. Res. Commun.*, **281**, 416–424 (2001).

369. N. Yao and P. C. Weber, *Antivir. Ther.*, **3**, 93–97 (1998).

370. N. Yao, T. Hesson, M. Cable, Z. Hong, A. D. Kwong, H. V. Le, and P. C. Weber, *Nat. Struct. Biol.*, **4**, 463 (2001).

371. H. S. Cho, N. C. Ha, L. W. Kang, K. M. Chung, S. H. Back, S. K. Jang, and B. H. Oh, *J. Biol. Chem.*, **273**, 15045–15052 (1998).

372. J. L. Kim, K. A. Morgenstern, J. P. Griffith, M. D. Dwyer, J. A. Thomson, M. A. Murcko, C. Lin, and P. R. Caron, *Structure*, **6**, 89–100 (1998).

373. C. L. Tai, W. K. Chi, D. S. Chen, and L. H. Hwang, *J. Virol.*, **70**, 8477–8484 (1996).

374. M. H. Alaoui-Ismaili, C. Gervais, S. Brunette, G. Gouin, M. Hamel, R. F. Rando, and J. Bedard, *Antiviral Res.*, **46**, 181–193 (2000).

375. C. C. Hsu, L. H. Hwang, Y. W. Huang, W. K. Chi, Y. D. Chu, and D. S. Chen, *Biochem. Biophys. Res. Commun.*, **253**, 594–599 (1998).

376. K. Kyono, M. Miyashiro, and I. Taguchi, *Anal. Biochem.*, **257**, 120–126 (1998).

377. Q. M. Wang, M. X. Du, M. A. Hockman, R. B. Johnson, and X. L. Sun, *Drugs Future*, **25**, 933–944 (2000).

378. C. W. Phoon, P. Y. Ng, A. E. Ting, S. L. Yeo, and M. M. Sim, *Bioorg. Med. Chem. Lett.*, **11**, 1647–1650 (2001).

379. P. Borowski, O. Mueller, A. Niebuhr, M. Kalitzky, L. H. Hwang, H. Schmitz, M. A. Siwecka, and T. Kulikowski, *Acta Biochim. Pol.*, **47**, 173–180 (2000).

380. V. Lohmann, A. Roos, F. Korner, J. O. Koch, and R. Bartenschlager, *J. Viral Hepatitis*, **7**, 167–174 (2000).

381. E. Ferrari, J. Wright-Minogue, J. W. S. Fang, B. M. Baroudy, J. Y. N. Lau, and Z. Hong, *J. Virol.*, **73**, 1649–1654 (1999).

382. V. Lohmann, A. Roos, F. Korner, J. O. Koch, and R. Bartenschlager, *Virology*, **249**, 108–118 (1998).

383. S. Bressanelli, L. Tomei, A. Roussel, I. Incitti, R. L. Vitale, M. Mathieu, R. De Francesco, and F. A. Rey, *Proc. Natl. Acad. Sci. USA*, **96**, 13034–13039 (1999).

384. W. Markland, T. J. McQuaid, J. Jain, and A. D. Kwong, *Antimicrob. Agents Chemother.*, **44**, 859–866 (2000).

385. K. W. Pankiewicz, *Expert Opin. Ther. Patents*, **11**, 1161–1170 (2001).

386. K. W. Pankiewicz, *Expert Opin. Ther. Patents*, **9**, 55–65 (1999).

387. B. M. Goldstein and T. D. Colby, *Curr. Med. Chem.*, **6**, 519–536 (1999).

388. J. O. Saunders and S. A. Raybuck, *Ann. Reports Med. Chem.*, **35**, 201–210 (2000).

389. J. Jain, S. J. Almquist, and D. Shlyakhter, *FASEB J.*, **14**, A1133 (2000).

390. M. D. Sintchak and E. Nimmesgern, *Immunopharmacology*, **47**, 163–184 (2000).

391. P. Ingravallo, E. Cretton-Scott, M. Xie, W. Zhong, B. Baroudy, J. Sommadossi, and J. Lau, American Society for Study of Liver Diseases, 1999, Abstract 991.

392. T. Wright, M. Shiffman, S. Knox, E. Ette, R. Kauffman, and J. Alam, American Society for Study of Liver Diseases, 1999, Abstract 990.

393. L. A. Sorbera, J. S. Silvestre, J. Castaner, and L. Martin, *Drugs Future*, **25**, 809–814 (2000).

394. R. F. Schinazi, E. Ilan, P. L. Black, X. J. Yao, and S. Dagan, *Antivir. Chem. Chemother.*, **10**, 99–114 (1999).

395. R. Bartenschlager and V. Lohmann, *Antiviral Res.*, **52**, 1–17 (2001).

396. R. Bartenschlager and V. Lohmann, *J. Gen. Virol.*, **81**, 1631–1648 (2000).

397. R. Bartenschlager and V. Lohmann, *Bailliere's Clin. Gastroentol.*, **14**, 241–254 (2000).

398. V. Lohmann, F. Körner, J. O. Koch, U. Herian, L. Theilmann, and R. Bartenschlager, *Science*, **285**, 110–113 (1999).

399. K. J. Blight, A. A. Kolykhalov, and C. M. Rice, *Science*, **290**, 1972–1974 (2000).

400. M. E. Major, K. Mihalik, J. Fernandez, J. Seidman, D. Kleiner, A. A. Kolykhalov, C. M. Rice, and S. M. Feinstone, *J. Virol.*, **73**, 3317–3325 (1999).

401. O. Nussbaum, E. Ilan, R. Eren, O. Ben-Moshe, Y. Arazi, S. Berr, I. Lubin, D. Shouval, E. Galun, Y. Reisner, and S. Dagan, *J. Hepatol.*, **32**, 172 (2000).

402. E. Ilan, R. Eren, O. Nussbaum, I. Lubin, D. Terkieltaub, O. Ben-Moshe, Y. Arazi, S. Berr, J. Gopher, A. Kitchinzky, D. Shouval, E. Galun, Y. Reisner, and S. Dagan, *Antiviral Res.*, **46**, 71 (2000).

403. P. Leyssen, E. De Clercq, and J. Neyts, *Clin. Microbiol. Rev.*, **13**, 67–82 (2000).

404. J. Crance, D. Gratier, H. Blancquaert, C. Rothlisberger, C. Bouvier, and A. Jouan, *S.S.A. Trav. Scient.*, **19**, 53–54 (1998).

405. J. Neyts, A. Meerbach, P. McKenna, and E. De Clercq, *Antiviral Res.*, **30**, 125–132 (1996).

406. J. D. Morrey, D. F. Smee, R. W. Sidwell, and C. Tseng, *Antiviral Res.*, **55**, 107–116 (2002).

407. G. Y. Song, V. Paul, H. Choo, J. Morrey, R. W. Sidwell, R. F. Schinazi, and C. K. Chu, *J. Med. Chem.*, **44**, 3985–3993 (2001).

408. I. Jordan, T. Briese, N. Fischer, J. Lau, and W. I. Lipkin, *J. Infect. Dis.*, **182**, 1214–1217 (2000).

409. J. M. Crance, D. Gratier, J. Guimet, and A. Jouan, *Res. Virol.*, **148**, 353–365 (1997).

410. J. M. Crance, D. Gratier, J. Guimet, and A. Jouan, *Trav. Scient.*, **55** (1998).

411. J. D. Morrey, personal communication, 2002.

412. S. Crotty, D. Maag, J. J. Arnold, W. Zhong, J. Y. N. Lau, Z. Hong, R. Anding, and C. E. Careron, *Nat. Med.*, **6**, 1375–1379 (2000).

413. R. I. Brinkworth, D. P. Fairlie, D. Leung, and P. R. Young, *J. Gen. Virol.*, **80**, 1167–1177 (1999).

414. H. M. K. Murthy, S. Clum, and R. Padmanabhan, *J. Biol. Chem.*, **274**, 5573–5580 (1999).

415. H. M. K. Murthy, K. Judge, L. DeLucas, S. Clum, and R. Padmanabhan, *Acta Cryst.*, **D55**, 1370–1372 (1999).

416. H. M. K. Murthy, K. Judge, L. DeLucas, and R. Padmanabhan, *J. Mol. Biol.*, **301**, 759–767 (2000).

417. I. Sanchez, F. Gomez-Garibay, J. Taboada, and B. H. Ruiz, *Phytother. Res.*, **14**, 89–92 (2000).

418. M. Laille, F. Gerald, and C. Debitus, *Cell. Mol. Life Sci.*, **54**, 167–170 (1998).

419. T. Andoh, H. Kawamata, M. Umatake, K. Terasawa, T. Takegami, and H. Ochiai, *J. Neurovirol.*, **4**, 627–631 (1998).

420. T. Takegami, E. Simamura, K. I. Hirai, and J. Koyama, *Antiviral Res.*, **37**, 37–45 (1998).

421. C. C. Garcia, N. A. Candurra, and E. B. Damonte, *Antivir. Chem. Chemother.*, **11**, 231–237 (2000).

422. N. A. Candurra, L. Maskin, and E. B. Damonte, *Antiviral Res.*, **31**, 149–158 (1996).

423. S. M. Cordo, N. A. Candurra, and E. B. Damonte, *Microbes Infect.*, **1**, 609–614 (1999).

424. M. B. Wachsman, E. M. F. Lopez, J. A. Ramirez, L. R. Galagovsky, and C. E. Coto, *Antivir. Chem. Chemother.*, **11**, 71–77 (2000).

425. N. E. Bishop, *Intervirology*, **41**, 261–271 (1998).

426. C. Korth, B. May, F. Cohen, and S. B. Prusiner, *Proc. Natl. Acad. Sci. USA*, **98**, 9836–9841 (2001).

427. A. Mirazimi, C. Henrik Von Bonsdorff, and L. Svensson, *Virology*, **217**, 554–563 (1996).

428. M. Koketsu, T. Nitoda, H. Sugino, L. R. Juneja, M. Kim, T. Yamamoto, N. Abe, T. Kajimoto, and C. H. Wong, *J. Med. Chem.*, **40**, 3332–3335 (1997).

429. M. J. Kiefel, B. Beisner, S. Bennett, I. D. Holmes, and M. von Itzstein, *J. Med. Chem.*, **39**, 1314–1320 (1996).

430. K. Konishi, Y. H. Gu, I. Hatano, and H. Ushijima, *Jpn. J. Infect. Dis.*, **53**, 62–66 (2000).

431. E. A. Bae, M. J. Han, M. Lee, and D. H. Kim, *Biol. Pharm. Bull.*, **23**, 1122–1124 (2000).

432. D. H. Kim, M. J. Song, E. A. Bae, and M. J. Han, *Biol. Pharm. Bull.*, **23**, 356–358 (2000).

433. M. Botta, F. Occhionero, R. Nicoletti, P. Mastromarino, C. Conti, M. Magrini, and R. Saladino, *Bioorg. Med. Chem.*, **7**, 1925–1931 (1999).

434. M. A. Turner, X. Yang, D. Yin, K. Kuczera, R. T. Borchardt, and P. L. Howell, *Cell Biochem. Biophys.*, **33**, 101–125 (2000).

435. Y. Kitade, A. Kozaki, T. Gotoh, T. Miwa, M. Nakanishi, and C. Yatome, *Nucleic Acids Symp. Ser.*, 25–26 (1999).

436. K. L. Seley and S. W. Schneller, *J. Med. Chem.*, **40**, 622–624 (1997).

437. C. S. Yuan, Y. Saso, E. Lazarides, R. T. Borchardt, and M. J. Robins, *Expert Opin. Ther. Patents*, **9**, 1197–1206 (1999).

438. D. L. Barnard, V. D. Stowell, K. L. Seley, V. R. Hegde, S. R. Das, V. P. Rajappan, S. W. Schneller, and D. F. Smee, *Antivir. Chem. Chemother.*, **12**, 241–250 (2001).

439. M. Bray, K. Davis, T. Geisbert, C. Schmaljohn, and J. Huggins, *J. Infect. Dis.*, **178**, 651–661 (1998).

440. J. Huggins, Z. X. Zhang, and M. Bray, *J. Infect. Dis.*, **179**, S240–S247 (1999).

441. M. Bray, J. Driscoll, and J. W. Huggins, *Antiviral Res.*, **45**, 135–147 (2000).

442. M. G. Santoro, *Trends Microbiol.*, **5**, 276–281 (1997).

443. A. Rossi, G. Elia, and S. Gabriella, *J. Biol. Chem.*, **271**, 32192–32196 (1996).

444. F. Pica, A. T. Palamara, A. Rossi, A. De Marco, C. Amici, and M. G. Santoro, *Antimicrob. Agents Chemother.*, **44**, 200–204 (2000).

445. G. Conti, P. Portincasa, S. Visalli, and C. Chezzi, *Virus Res.*, **75**, 43–57 (2001).

446. C. Conti, P. Mastromarino, P. Tomao, A. De-Marco, F. Pica, and M. G. Santoro, *Antimicrob. Agents Chemother.*, **40**, 367–372 (1996).

447. F. Superti, C. Amici, A. Tinari, G. Donelli, and M. G. Santoro, *J. Infect. Dis.*, **178**, 564–568 (1998).

448. J. T. Rhule, C. L. Hill, and D. A. Judd, *Chem. Rev.*, **98**, 327–357 (1998).

449. J. H. Huffman, R. W. Sidwell, D. L. Barnard, A. Morrison, M. J. Otto, C. L. Hill, and R. F. Schinazi, *Antivir. Chem. Chemother.*, **8**, 75–83 (1997).

450. D. L. Barnard, C. L. Hill, T. Gage, J. E. Matheson, J. H. Huffman, R. W. Sidwell, M. I. Otto, and R. F. Schinazi, *Antiviral Res.*, **34**, 27–37 (1997).

451. S. Shigeta, S. Mori, J. Watanabe, T. Yamase, and R. F. Schinazi, *Antivir. Chem. Chemother.*, **7**, 346–352 (1996).

452. S. Shigeta, S. Mori, J. Watanabe, S. Soeda, K. Takahashi, and T. Yamase, *Antimicrob. Agents Chemother.*, **41**, 1423–1427 (1997).

453. J. Liu, W. Mei, Y. Li, E. Wang, E. Ji, and P. Tao, *Antivir. Chem. Chemother.*, **11**, 367–372 (2000).

454. C. H. Kempe, G. Bowles, G. Meiklejohn, T. O. Berge, L. St. Vincent, B. V. Sundara Babu, S. Govindarajan, N. R. Ratnakanna, A. W. Downie, and V. R. Murthy, *Bull. Wld. Hlth. Org.*, **25**, 41–48 (1961).

455. C. H. Kempe, T. O. Berge, and B. England, *Pediatrics*, **18**, 177–188 (1956).

456. S. S. Marennikova, *Bull. Wld. Hlth. Org.*, **27**, 325–330 (1962).

457. D. J. Bauer, L. St. Vincent, C. H. Kempe, and A. W. Downie, *Lancet*, 494–496 (1963).

458. F. Fenner, D. A. Henderson, I. Arita, Z. Jezek, and I. D. Ladnyi, *Smallpox and Its Eradication*, World Health Organization, Bethesda, MD, 1988.

459. A. R. Rao and G. D. W. McKendrick, *Lancet*, **1**, 1072–1074 (1966).

460. A. R. Rao, J. A. McFadzean, and K. Kamalakshi, *Lancet*, **1**, 1068–1072 (1966).

461. M. S. Hossain, W. Hryniuk, J. Foerster, L. G. Isreals, A. S. Chowdhury, and M. K. Biswas, *Lancet*, **2**, 1230–1232 (1972).

462. D. T. Dennis, E. B. Doberstyn, S. Awoke, G. L. Royer Jr., and H. E. Renis, *Lancet*, **2**, 377–379 (1974).

463. K. A. Monsur, M. S. Hossain, F. Huq, M. M. Rahaman, and M. Q. Haque, *J. Infect. Dis.*, **131**, 40–43 (1975).

464. D. J. Bauer, *Br. Med. Bull.*, **41**, 309–314 (1985).

465. E. De Clercq, *Clin. Microbiol. Rev.*, **14**, 382–397 (2001).

466. D. J. Bauer, *Ann. NY Acad. Sci.*, **130**, 110–117 (1965).

467. E. De Clercq, T. Sakuma, M. Baba, R. Pauwels, J. Balzarini, I. Rosenberg, and A. Holý, *Antiviral Res.*, **8**, 261–272 (1987).

468. S. Safrin, J. Cherrington, and H. S. Jaffe, *Rev. Med. Virol.*, **7**, 145–156 (1997).

469. E. De Clercq, *Collect. Czech. Chem. Commun.*, **63**, 480–506 (1998).

470. G. L. Plosker and S. Noble, *Drugs*, **58**, 325–345 (1999).

471. T. Cihlar and M. S. Chen, *Mol. Pharmacol.*, **50**, 1502–1510 (1996).

472. X. Xiong, J. L. Smith, C. Kim, E. S. Huang, and M. S. Chen, *Biochem. Pharmacol.*, **51**, 1563–1567 (1996).

473. M. J. M. Hitchcock, H. S. Jaffe, J. C. Martin, and R. J. Stagg, *Antivir. Chem. Chemother.*, **7**, 115–127 (1996).

474. X. Xiong, J. L. Smith, and M. S. Chen, *Antimicrob. Agents Chemother.*, **41**, 594–599 (1997).

475. E. De Clercq, *Clin. Microbiol. Rev.*, **10**, 674–693 (1997).

476. L. Naesens, R. Snoeck, G. Andrei, J. Balzarini, J. Neyts, and E. De Clercq, *Antivir. Chem. Chemother.*, **8**, 1–23 (1997).

477. P. B. Jahrling, G. M. Zaucha, and J. W. Huggins, in W. M. Scheld, W. A. Craig, and J. M. Hughes, Eds, *Emerging Infections 4*, ASM Press, Washington, DC, 2000, pp. 187–200.

478. J. Neyts and E. De Clercq, *J. Med. Virol.*, **41**, 242–246 (1993).

479. M. Bray, M. Martinez, D. F. Smee, D. Kefauver, E. Thompson, and J. W. Huggins, *J. Infect. Dis.*, **181**, 10–19 (2000).

480. D. F. Smee, K. W. Bailey, M. H. Wong, and R. W. Sidwell, *Antiviral Res.*, **47**, 171–177 (2000).

481. D. F. Smee, K. W. Bailey, and R. W. Sidwell, *Antivir. Chem. Chemother.*, **12**, 71–76 (2001).

482. D. F. Smee, K. W. Bailey, M. H. Wong, and R. W. Sidwell, *Antiviral Res.*, **52**, 55–62 (2001).

483. J. W. Huggins, D. F. Smee, M. J. Martinez, and M. Bray, *Antiviral Res.*, **37**, A73 (1998).

484. E. G. Davies, A. Thrasher, K. Lacey, and J. Harper, *Lancet*, **353**, 2042 (1999).

485. E. R. Kern, C. Hartline, E. Harden, K. Keith, N. Rodriguez, J. R. Beadle, and K. Y. Hostetler, *Antimicrob. Agents Chemother.*, **46**, 991–995 (2002).

486. E. Katz, E. Margalith, and B. Winer, *J. Gen. Virol.*, **32**, 327–330 (1976).

487. B. B. Goswami, E. Borek, O. K. Sharma, J. Fujitaki, and R. A. Smith, *Biochem. Biophys. Res. Commun.*, **89**, 830–836 (1979).

488. D. F. Smee and J. W. Huggins, *Antiviral Res.*, **37**, A89 (1998).

489. D. F. Smee, M. Bray, and J. W. Huggins, *Antivir. Chem. Chemother.*, **12**, 327–335 (2001).

490. D. F. Smee, K. W. Bailey, and R. W. Sidwell, *Antivir. Chem. Chemother.*, **11**, 303–309 (2000).

# CHAPTER ELEVEN

# Rationale of Design of Anti-HIV Drugs

Ahmed S. Mehanna
Massachusetts College of Pharmacy and Health Sciences
Department of Pharmaceutical Sciences
School of Pharmacy
Boston, Massachusetts

## Contents

1 Introduction, 458
2 Human Retroviruses and AIDS Epidemiology, 459
3 HIV Infection and its Pathological Effects, 459
   3.1 Mode of Viral Transmission, 459
   3.2 Cellular Picture of the Infection, 459
   3.3 Clinical Picture of HIV Infection, 460
      3.3.1 Immunosuppressive Effects, 460
      3.3.2 Neurological Effects, 460
      3.3.3 Carcinogenic Effects, 460
4 HIV Structure and Molecular Biology, 460
   4.1 Virus Structure (7), 460
   4.2 HIV Genome, 461
5 HIV Life Cycle (10), 461
   5.1 Virus Binding to the Host Cell Membrane, 462
   5.2 Virus Fusion with Host Cell Cytoplasm and its Uncoating, 462
   5.3 Viral DNA Formation by the Reverse Transcriptase, 462
   5.4 Viral DNA Entry to the Host Cell Nucleus, 463
   5.5 Integration of Viral DNA into the Host Genome for Viral mRNA Production, 464
   5.6 Splicing of Viral mRNA from the Host Cell Genome, 464
   5.7 Migration of the Viral mRNA to the Cytoplasm, 464
   5.8 Assembly of Viral Proteins to Form the Virion, 464
   5.9 Viral Budding out of the Host Cell, 464
   5.10 Virus Maturation by HIV Protease (PR), 464
6 Targets for Drug Design of Anti-HIV Agents, 464
   6.1 An Overview, 464
   6.2 Inhibitors of Viral Entry, 465

*Burger's Medicinal Chemistry and Drug Discovery*
Sixth Edition, Volume 5: Chemotherapeutic Agents
Edited by Donald J. Abraham
ISBN 0-471-37031-2    © 2003 John Wiley & Sons, Inc.

6.2.1 HIV Vaccines, 465
6.2.2 Viral Adsorption Inhibitors, 465
    6.2.2.1 Polyanionic Compounds, 466
    6.2.2.2 Soluble CD4 Peptide
        Fragments, 466
6.2.3 Inhibitors of Gp120 Binding to the
    T-Cell Coreceptors, 467
    6.2.3.1 Gp120 Coreceptors, 467
    6.2.3.2 Gp120-CXCR4 Binding
        Inhibitors, 467
    6.2.3.3 Gp120-CCR5 Binding
        Inhibitors, 468
6.3 Inhibitors of Viral Fusion, 468
6.4 Inhibitors for Viral Uncoating, 469
6.5 Inhibitors of HIV Reverse Transcription, 469
    6.5.1 Molecular Aspects of HIV Reverse
    Transcriptase, 469
    6.5.2 HIV Reverse Transcriptase Inhibitors,
    470
        6.5.2.1 Nucleoside Reverse
        Transcriptase Inhibitors
        (NRTIs), 470
        6.5.2.2 Nonnucleoside Reverse
        Transcriptase Inhibitors
        (NNRTIs), 472

6.6 Inhibitors of HIV Ribonuclease (RnaseH),
    473
6.7 HIV Integrase Inhibitors, 473
    6.7.1 HIV Integrase (IN), 473
    6.7.2 Integrase Inhibitors, 473
6.8 Inhibitors of HIV Gene Expression
    (Transactivation Inhibitors), 473
6.9 Inhibitors of Virion Assembly, 474
6.10 Inhibition of Viral Maturation
    (HIV-1 Protease Inhibitors), 475
    6.10.1 HIV Protease, 475
    6.10.2 HIV Protease Inhibitors (PI), 475
7 HIV Drugs in Clinical Use, 477
8 The Need for New Anti-HIV Drugs, 478
    8.1 Development of Drug Resistance, 478
    8.2 Approaches to Overcome Drug Resistance
    (138–140), 478
        8.2.1 Switching the Drug Class, 478
        8.2.2 The Virus "Knock-Out" Strategy, 478
        8.2.3 Combination Therapy, 479
        8.2.4 Officially Approved Drug
        Combinations, 479
        8.2.5 Prodrug Conjugates: A New Trend in
        Combination Therapy, 479

# 1  INTRODUCTION

Acquired immunodeficiency syndrome or AIDS was first reported in the United States in 1981. The disease has killed more than 400,000 people in the United States and almost 21.8 million men, women, and children worldwide. Estimates by the Joint United Nations Program on HIV/AIDS (UNAIDS) and World Health Organization (WHO) indicate that, by the end of the last decade, over 36 million people were infected with the human immunodeficiency virus (HIV) that causes AIDS. Epidemiologists have shortened the name acquired immune deficiency syndrome to the chilling abbreviation AIDS. The disease is considered as the first catastrophic pandemic of the second half of the twentieth century and it was once described as the modern plague. Despite the fact that more is known now than ever before about how to prevent the spread of the epidemic, 5.3 million new HIV infections were reported during the year 2001 alone. Unless a cure is found or a life-prolonging therapy can be made more available throughout the world, the majority of those

now infected with the HIV will die within a decade. These deaths will not be the last because the virus continues to spread, causing 16,000 new infections a day.

Recently, Western countries have recorded a decrease in the death rate imputed to AIDS. This success has been largely attributed to the development and availability of chemotherapeutic agents that inhibit the infectivity of the HIV-1 virus, the causative agent of AIDS. The advancement in HIV research had led to unveiling of the details of the virus life cycle and the mechanism of its replication inside the immune system T-cells. The understanding of functions and molecular structures of the viral enzymes have contributed to the discovery of a large number of anti HIV drugs through rationale drug design. Several HIV inhibitors have been successfully introduced to the physician's arsenal to combat AIDS infections. All clinically approved anti-HIV drugs interfere with the virus life cycle by inhibiting one of two enzymes: the transcriptase or the protease. The two enzymes play a vital role in the process of viral replication inside the T-cell. Unfortunately, the efficacy of available drugs is always challenged by the development of drug resistance within short periods of clinical use (1). The development of drug resistance is attributed to the continuous evolution of mutant HIV strains. The problem of drug resistance has been solved, in part, by the combination therapy approach. Combination therapy requires administering more than one drug, each of which works by a different mechanism to suppress the virus ability to mutate. However, even with this approach, new viral resistant strains still emerge during therapy. The increasing development of resistance and cross-resistance to the currently available anti-HIV medications prompted researchers to seek alternative anti-HIV drug classes beyond the reverse transcriptase and protease inhibitors.

Efforts are directed now toward developing drugs capable of interfering with other steps of HIV life cycle including its adsorption, entry, fusion, uncoating, or integration with the T-cell lymphocytes. To understand the medicinal chemistry of anti-AIDS drugs and the rationale of their design, it is important for the reader to be acquainted with some information about both AIDS (the disease) and HIV (its causative organism). The present chapter covers various anti-HIV drug classes, the rationale of their design, and both the site and the molecular mechanism of action of each class. The clinical and cellular aspects of AIDS, together with its molecular biology, genetics nature, structure, and life cycle of the HIV, are also addressed.

## 2  HUMAN RETROVIRUSES AND AIDS EPIDEMIOLOGY

AIDS is a disease state that arises from infection with the human immune deficiency virus (abbreviated HIV). The virus belongs to a family of viruses called retroviruses. This group of viruses is characterized by having RNA as its genetic material. Retroviruses are capable of constructing DNA from RNA, a reversed process of the normal replication schemes, hence the classification of this group as retroviruses. Only three retroviruses are known to infect humans: HTLV I (Human T Leukemia Virus I), HTLV II (Human T Leukemia Virus II),

and HTLV III (Human T Lymphotropic Virus). Although HLTV I (first isolated in 1978) (2, 3) and HTLV II (first isolated in 1982) (4) are both known to cause leukemia, HTLV III was confirmed to be the etiological virus for the acquired immunodeficiency syndrome (AIDS). To distinguish HTLV III from the other two human retroviruses, HTLV I and HTLV II, it was given the name human immunodeficiency virus (HIV). HIV is further subcategorized into HIV-1 and HIV-2. The first causes AIDS, whereas the second causes lymphadenopathy (or swelling of the lymphatic glands). HIV-2 is sometimes called "lymphadenopathy associated virus," or LAV, to avoid name confusion with HIV-1, the most detrimental subtype to the immune system.

## 3  HIV INFECTION AND ITS PATHOLOGICAL EFFECTS

### 3.1  Mode of Viral Transmission

HIV is transmitted by several infection mechanisms, including unprotected sex, sharing of hypodermic needles for injection for drug use, from HIV-infected mother to her baby, human breast milk, or blood transfusion and coagulation products. The disease was found to be common among certain social groups who share a common lifestyle such as homosexuals and drug users.

### 3.2  Cellular Picture of the Infection

Once the virus is inside the body, it targets a certain type of cell called a T4 lymphocyte, a white blood cell that has a central role in regulating the immune system, specifically the CD4 helper T-cells (5). After the virus entry to the T4 cell, it may remain latent until the lymphocyte is immunologically stimulated by a secondary infection. Then the virus bursts into action, reproducing itself so furiously that the new virus particles escaping from the cell riddle the cellular membrane with holes and the lymphocyte ultimately dies. HIV starts its replication cycle in the host cell with the help of a viral enzyme called reverse transcriptase. The enzyme uses viral RNA as a template to assemble a double strand of viral DNA. The latter travels to the cell nucleus and inserts itself among the host's chromosomes, which provide the machinery for HIV-1 transcription and translation.

## 3.3 Clinical Picture of HIV Infection

### 3.3.1 Immunosuppressive Effects.
HIV infection results in a progressive decrease in the helper T-cell count, the hallmark of AIDS, from a normal value of 800–1300 cells/cm$^2$ of blood to below 200, which may give rise to a life-threatening illness (6). The suppressed immune system leaves the patient vulnerable to the so-called opportunistic infections by agents that would not harm a healthy person. The most common of such infections is pneumonia caused by *Pneumocystis carinii*, a widespread but generally harmless protozoan. Most of the clinical complications of AIDS patients result from such infections.

### 3.3.2 Neurological Effects.
Although most of the attention given to the HIV virus has gone to suppression of the immune system or AIDS, the virus is associated also with brain diseases and several types of cancer. The brain and spinal cord disease caused by HIV was first detected in brain and spinal cord tissues from AIDS patients in 1984. The chief pathologies observed in the brain, which appears to be independent of the immune deficiency, are an abnormal proliferation of the glial cells that surround the neurons and lesions resulting from loss of white matter (which is, along with gray matter, one of the two main types of brain tissue). This can ultimately give rise to a wide range of neurological symptoms such as dementia and multiple sclerosis.

### 3.3.3 Carcinogenic Effects.
Cancer is the third main type of HIV pathological manifestations. Whereas the neurological effects of HIV are distinct from the immune deficiency, cancer has a more ambiguous relation to the crippling process of the immune system. People infected with the virus have an increased risk of at least three types of human tumor. One is known as Kaposi's sarcoma, a rare tumor of blood vessel tissue in the skin or internal organs and had been known to exist mainly among older Italian and Jewish men and in Africa. The second type of cancer is carcinomas, including skin cancers, which are often seen in the mouth or rectum of infected homosexuals. The third major type of cancers observed with HIV infection is B-cell lymphomas (tumor originating in B-lymphocytes). The appearance of the same types of cancer among certain groups of young white middle-class males provided the first basis for how HIV infection is transmitted because this group turned out to have a history of homosexuality.

## 4  HIV STRUCTURE AND MOLECULAR BIOLOGY

### 4.1  Virus Structure (7)

The AIDS virus exists as a small particle called HIV virion. The particle is spherical in shape, with a diameter of roughly 1000-Å units (one ten thousandth of a millimeter). The particle is covered by a membrane, made up of lipid (fatty) bilayer material that is derived from the outer membrane of the host cell. Studding the membrane are glycoproteins (proteins with sugar chains attached). Each glycoprotein has two components: gp41 spans the bilayer membrane and gp120 extends beyond it. The combination of the bilayer membrane and its glycoproteins is called the envelope, which covers the virus core made up of proteins designated p24 and p18. The viral RNA is carried in the core, along with several copies of the enzyme reverse transcriptase, which catalyzes the assembly of the viral DNA. The envelope glycoproteins have an important role in HIV's entry to its host cell and also in the death of the host cell. The envelope proteins include some regions that are constant in all HIV strains, some of which are highly variable and others that are intermediate. The virus entry to the cell seems to depend on an interaction between one or more of the constant regions and certain molecules in the cell membrane. The envelope proteins are also involved in the process of budding of a new virus particle from the host cell, which may leave a hole in the cell's surface. On the other hand, T4 cells, which are the most significant target in HIV infection, selectively interact with the outer envelope of the virus, a process that sets up the mechanism for the virus entry into the

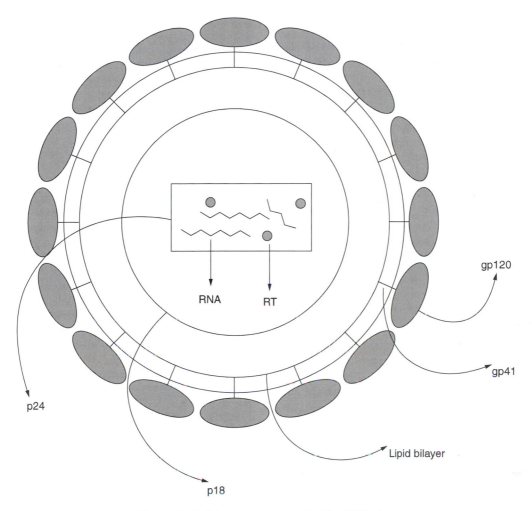

RNA          RT

gp120

gp41

p24

Lipid bilayer

p18

**Figure 11.1.** Schematic structure for the AIDS virus.

cell. Figure 11.1 depicts a schematic representation for HIV-1 virus structure.

### 4.2  HIV Genome

HIV is perhaps the most complicated retrovirus studied. In common with other retroviruses, it contains genes encoding the core structural protein (gag), the outer envelope glycoprotein (env), and viral DNA polymerase (pol) (the reverse transcriptase). It also contains sequences at either end of the genome called the long terminal repeats (LTRs), which contain regulatory elements associated with the virus replication (8). HIV is quite unusual, however, in containing five nonstructural genes, tat-III, art/trs, 3'-orf, sor, and r, some of

which are known to have important regulatory functions for viral replication (9). These unusual genes encode proteins that help to control the expression of viral genes to form its RNA. The latter is spliced to yield an array of messenger RNAs (mRNAs), from which all viral proteins are synthesized. The core proteins and reverse transcriptase are made from an mRNA corresponding to the entire genome. One splice yields the envelope protein mRNA and a second, the small mRNA from which the tat and trs proteins are made.

### 5  HIV LIFE CYCLE (10)

The overall process of HIV life cycle and replication starts with the virus binding to the host cell through specific surface receptor interactions. After the binding, the virus fuses itself with host cell cytoplasm through a very complex process, which involves a second set of surface protein interactions. After the virus fusion and its entry to the host cell cytoplasm, it makes use of the host cell machinery to express the genetic material necessary to produce its functional proteins. The last stage of the virus life cycle is the stage when the virus assembles itself inside the cell into new virus particles, followed by budding it out then its maturation, to become infective again. Figure 11.2 represents a diagram for the cycle of HIV replication, and the section below describes the details of each of these steps. Knowledge of HIV life cycle is essential to understand the rationale of design of various anti-HIV therapeutic agents. As shown in Fig. 11.2, the virus life cycle replication process can be described in 10 consecutive steps.

## 5.1   Virus Binding to the Host Cell Membrane

The plasma membrane of the host cell presents the first physical barrier to HIV infection. To overcome this barrier, the virus has in the structure of its envelope certain glycoproteins that catalyze binding of viral and cellular membranes. The viral binding and fusion machinery in HIV is contained in its outer envelope glycoproteins gp120 and gp41. Both glycoproteins are generated by proteolysis of the160-kDa precursor, gp160 (11–13). The glycoprotein gp120 binds with high affinity to another specific glycoprotein on the T-cell surface, called the CD4 receptor. This interaction represents the first step in the process of HIV infection.

## 5.2   Virus Fusion with Host Cell Cytoplasm and its Uncoating

Membrane fusion is another important step in the process of HIV infection. The fusion process allows HIV entry into the host cell cytoplasm. The process is mediated by the second envelope viral glycoprotein gp41. However, engagement of gp41 with the host cell membrane requires gp120, to bind, in addition to the CD4 receptors, to a second set of receptors on the T-cell surface, called coreceptors. These coreceptors are designated by certain abbrevi-

ations such as CXCR4 (the coreceptor for HIV-1 strains that infect T-cell, T-tropic, or X4 strains) and CCR5 (the coreceptor for HIV-1 strains that infect macrophages, M-tropic, or R5 strains). These coreceptors were recently discovered as domains on the T-cells' membrane, which normally act as receptors to certain chemicals released during inflammation, called chemokines (chemoattractant cytokines) (14, 15). The coreceptors were found to play a crucial role in the process of viral fusion to the host cell membrane and ultimately its entrance to the cytoplasm. Binding of gp120 to the coreceptors triggers structural changes in the viral transmembrane part of gp41, and this interaction results in uncovering of the outer end of gp41 (16). The exposed end of gp41 embeds itself in the host cell membrane through specific structures, leading to eventual fusion of the two membranes (17). The specific structural features that allow the gp41 to mediate fusion include two heptad-repeat (HR) regions (designated HR1 and HR2) in the gp41 domain and stretches of hydrophobic amino acids that constitute the fusion domain. In their most stable state, HR1 and HR2 form a "trimer-of-hairpins," in which the three HR2 regions pack in an antiparallel fashion into grooves present along a triple-stranded coiled-coil of the HR1 trimer. The HR regions are located between the fusion region and the transmembrane domain of gp41. Structural reorganization of HR1 and HR2 thus brings the fusion site and the transmembrane domains together and promotes mixing of the viral and cellular membranes. After HIV enters the cell, the virus loses its outer envelope in a process known as uncoating, then releases its RNA genome (encapsulated with the p24; see Fig. 11.1) into the cytoplasm.

## 5.3   Viral DNA Formation by the Reverse Transcriptase

After uncoating, a complementary strand of DNA is copied from viral RNA by HIV-DNA polymerase (reverse transcriptase). Subsequently, a second copy (positive strand) of DNA is made so that the genetic information is encoded in a double-strand form of DNA (18). The viral RNA (now bound to the newly formed DNA) is degraded by another ribonu-

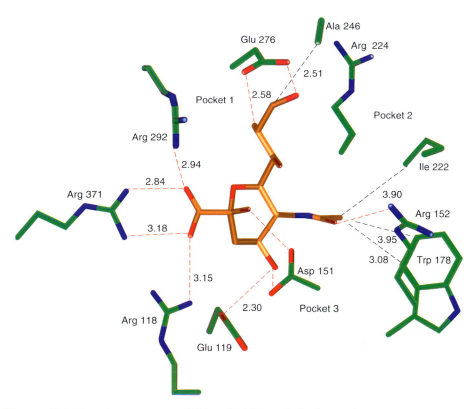

**Figure 10.2.** Complex structure of NA and sialic acid (dashes indicate H-bonding [red] and hydrophobic [black] interactions; some active site residues are omitted for clarity). Reproduced with permission of Dr. C. U. Kim (Gilead Sciences).

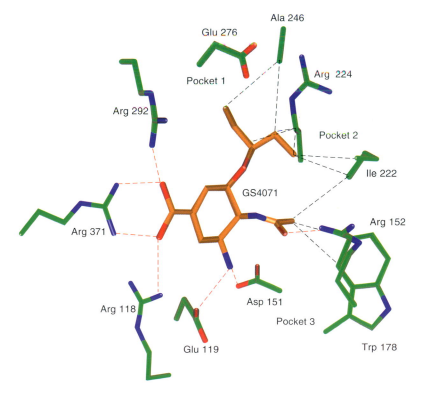

**Figure 10.4.** Complex structure of NA and GS 4071 (dashes indicate H-bonding [red] and hydrophobic [black] interactions; some active site residues are omitted for clarity). Reproduced with permission from Dr. C.U. Kim (Gilead Sciences).

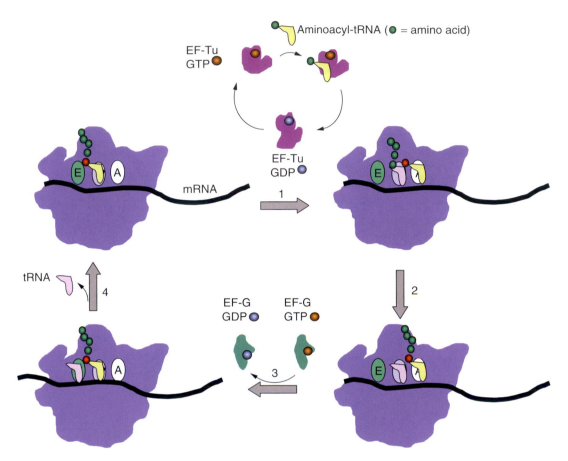

**Figure 15.1.** The bacterial protein synthesis elongation cycle. Step 1: Recognition and binding of the cognate aminoacyl-tRNA to the anticodon in the A-site. Step 2: Peptidyl transfer resulting in elongation of the peptide by one amino acid residue. Step 3: Translocation of the peptidyl-tRNA from the A-site to the P-site with concomitant movement of the unchanged tRNA to the E-site. Step 4: Exit of the unchanged tRNA completes the cycle.

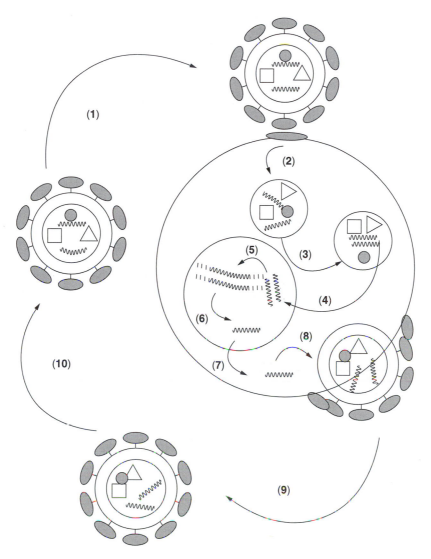

**Figure 11.2.** Schematic representation for HIV life cycle. (1) Binding of the virus to the T-cell through the gp120 and CD4 receptors. (2) Fusion through viral gp41 and loss of its envelope, the uncoating. (3) Viral DNA formation by reverse transcriptase followed by RNase. (4) Viral DNA entry to the host cell nucleus through its nuclear pores. (5) Viral DNA integration into host cell DNA by integrase. (6) Splicing of viral RNA by host RNA polymerase to produce viral mRNA. (7) Migration of viral RNA to the cytoplasm as mRNA to encode the synthesis of viral proteins. (8) Assembly of the virion containing the viral proteins as a single chain. (9) Viral budding through the host cell membrane with proteins as single chain. (10) Breakdown of the polyprotein precursor by the protease to give structural proteins and enzymes.

cleave enzyme (RnaseH), which cleaves RNA from the hybrid DNA, leading to the release of the viral DNA in the host cell cytoplasm. The newly formed viral DNA is called the provirus.

## 5.4   Viral DNA Entry to the Host Cell Nucleus

In the host cell HIV DNA, or the provirus, may stay in a free form or may enter the host cell nucleus, through pores in the nuclear membrane, and undergoes integration with the host cell genome.

## 5.5 Integration of Viral DNA into the Host Genome for Viral mRNA Production

Once the viral DNA is inside the nucleus, it integrates into the host chromosomes through the viral enzyme integrase, which is believed to be, like reverse transcriptase, a pol-encoded product (19). The linear viral DNA flanked by the two long terminal repeats (LTRs), found in the cytoplasm and nucleus of the infected cells, is the direct substrate for the viral integrase enzyme (20, 21). The enzyme inserts the linear double-stranded viral DNA at random into the host cell DNA (22, 23). Upon integration of viral DNA into the host genome, it starts to direct the transcription to produce its mRNA and other types of viral RNA.

## 5.6 Splicing of Viral mRNA from the Host Cell Genome

The transcription process of viral RNA occurs in the T-cell nucleus, as part of its own transcription. The splicing (cutting) of viral mRNA and other viral RNA molecules from the host DNA is also accomplished by the normal host cell RNA polymerase to produce an array of viral mRNA.

## 5.7 Migration of the Viral mRNA to the Cytoplasm

After viral mRNA is produced and spliced through use of the host cell genome, mRNA is released to the cytoplasm, where it directs the manufacture of various viral proteins, including core proteins, envelope proteins, regulatory proteins, and enzymes. Although the virus exploits the biochemical machinery of the host cell to produce its functional components, the synthesis of such products is regulated by the virus genome (24, 25).

## 5.8 Assembly of Viral Proteins to Form the Virion

The resulting viral proteins are produced as a single large polyprotein precursor, which is then transported, with the help of viral RNA, to the host cell membrane in preparation for its assembly into daughter particles. The assembly process occurs at the host cell membrane where the original uncoated viral envelope proteins exist. Each polyprotein precursor is enveloped with gp120 and gp41 glycoproteins after being encapsulated with the p18 and p24 viral core proteins. The particles are now ready to be expelled or budded out of the cell.

## 5.9 Viral Budding out of the Host Cell

The assembled particles are released out of the host cell membrane by a process called budding, which leaves behind a hole, or several holes, in the T4 cell membrane, which may contribute to the CD4 cell death. The host cells normally do not survive the invasion by HIV. They either disintegrate because of the large number of viruses budding off, or the body's immune system recognizes the viral envelope proteins in the cell membrane and destroys the damaged cells. This destruction of CD4 cells causes severe immunodeficiency because of the role of helper T-cells in mediating the system immunity. This paves the way for the so-called opportunistic infections such as candidiasis in the bronchi or lungs, cryptococcus, cytomegalovirus retinitis, herpes simplex infections, and pneumonia.

## 5.10 Virus Maturation by HIV Protease (PR)

During or shortly after budding, the polyprotein precursor, inside the daughter particles, undergoes cleavage by a specific viral enzyme known as HIV protease. The resulting degradation products are the viral functional enzymes and proteins necessary for its survival. The viral particles at this stage are called virions. The virus particles after the protease action have all the necessary constituents of mature virus and are capable of invading other T4 cells and repeating the life cycle (26, 27). The enzyme HIV protease constitutes, with the reverse transcriptase enzyme, the most extensively targeted steps in the virus life cycle to design anti-HIV agents.

## 6 TARGETS FOR DRUG DESIGN OF ANTI-HIV AGENTS

### 6.1 An Overview

The information available on HIV life cycle and its details provided medicinal chemists with several molecular and cellular sites for intervention to inhibit HIV replication (28). Virtually every step of the cycle has been targeted for drug design. Anti-HIV agents can be grouped into three groups based on the phase of their interference with the virus replication cycle, that is, the pretranscription, transcription, and posttranscription phases. Pretranscription inhibitors are those agents capable of blocking viral entry into the host cell. This class includes inhibitors of gp120 binding to CD4 (viral adsorption inhibitors), inhibitors of gp120 binding to coreceptors, and inhibitors of viral fusion and viral uncoating. The transcription phase is the central stage, through which viral DNA is transcribed from viral RNA under the control of the reverse transcriptase enzyme. Inhibitors for this enzyme represent a very important class of anti-HIV agents. The posttranscription inhibitors are agents that inhibit viral DNA integration into the host cell genome (integrase inhibitors) and agents that interfere with virus maturation by blocking the crucial step of producing viral functional proteins from the polyprotein precursor (protease inhibitors or viral maturation inhibitors).

All anti-HIV drugs currently approved for clinical use fall into the categories of reverse transcriptase and protease inhibitors. The reverse transcriptase inhibitors are further subclassified into two subclasses, the nucleoside reverse transcriptase inhibitors (NRTIs) and the nonnucleoside reverse transcriptase inhibitors (NNRTIs). The most challenging problem associated with HIV therapy is the development of drug resistance. The combination therapy approach has been broadly practiced in HIV treatment as a strategy to avoid such problem. However, despite the overall success of combination therapy in reducing viral load, morbidity, and mortality, it is mitigated by a variety of issues. The long-lived nature of the infection and the genetic plasticity inherent to this virus has led to the spread of multidrug-resistant variants of the virus. Therefore, the need for new treatment options beyond reverse transcriptase and protease inhibitors has emerged. Current efforts to develop new anti-HIV drugs are now directed more toward developing inhibitors for other steps in the virus life cycle, such as viral entry

into the T-cell, its integration with the host DNA, or its assembly into daughter particles. These proposed sites of intervention may provide additional options to maintain long-term suppression of the virus replication. The identification of specific inhibitors for each of these processes has recently validated these approaches as viable alternatives for the traditional classes of reverse transcriptase and protease inhibitors. Some of these new agents are in the preclinical trial phase. The section below addresses the latest in drug development in the area of anti-HIV therapeutics. The section is arranged in the same sequence of the virus life cycle shown in Fig. 11.2.

## 6.2 Inhibitors of Viral Entry

**6.2.1 HIV Vaccines.** The most ideal approach to inhibit HIV binding to the T-cells is to develop a vaccine that can neutralize the virus in circulation. Several attempts to develop antibodies to HIV as blocker for the viral binding to prevent infection have been reported (29, 30). However, because of the nature of the disease, people infected with HIV develop only low titers of neutralizing antibodies and that presents a problem for vaccine development. Knowledge of the detailed steps of the virus binding to the T-cell, together with discovery of the molecular structure of gp120, CD4, and the chemokine as coreceptors for gp120 binding, have directed vaccine development into more selective viral adsorption inhibitors (31).

**6.2.2 Viral Adsorption Inhibitors.** The recognition of the role of viral gp120 and gp41 glycoproteins in the processes of HIV binding and fusion with the T-cell surface flagged these sites as an attractive molecular target for intervention. In principle, agents affecting any of the viral entry events would be effective as inhibitors of HIV-1 replication. Targets of viral entry inhibition and fusion include the process of gp120 binding to CD4 receptors and other coreceptors (CCR5, CXCR4) on the T-cell membrane. The development of viral adsorption inhibitors and other anti-HIV agents together with molecular details of gp120 structure have been recently reviewed (32, 33) and are summarized below.

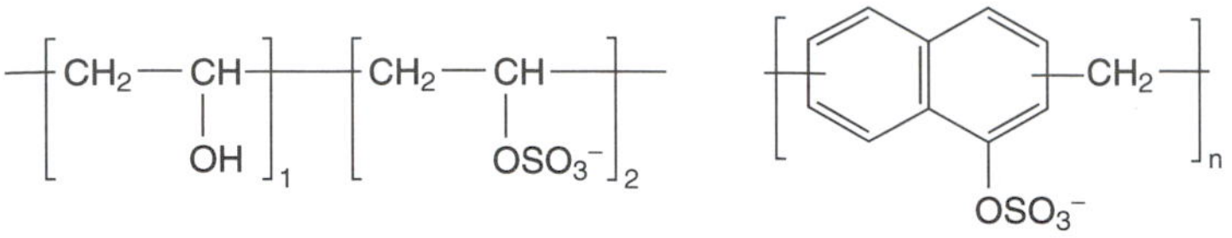

**Figure 11.3.** Promising adsorption inhibitors.

**6.2.2.1 Polyanionic Compounds.** Early reports showed that small peptide sequences for the env gp120 glycoprotein inhibit viral binding and replication (34). The polysulfated carbohydrate dextran sulfate was also found to inhibit HIV replication at a micromolar concentration range with a mechanism related to blocking the virus binding to the target cells (35, 36). Several other polyanionic compounds of natural and synthetic sources have been reported to inhibit gp120 binding to the cell membrane. Among these anionic compounds are polysulfates, polysulfonates, and polycarboxylates. All are believed to exert their antiviral activity through inhibiting gp120 binding to the CD4 receptor. The major molecular mechanism of action of these polyanionic substances in inhibiting gp120 binding is by shielding the positively charged sites on the V3 loop of the viral envelope glycoprotein gp120 (32). The V3 loop is one of five polypeptide loops that constitute the backbone of the glycoprotein gp120, found to be necessary for the initial viral attachment to the cell surface. Figure 11.3 depicts the structures of these promising viral adsorption inhibitors. The most promising of the polyanionic substances described as gp120 binding inhibitors are polyvinyl alcohol sulfate (PVAS, **1**), polynaphthalene sulfonate (**2**) and the polycarboxylate analog cosalane (**3**). The major role of polyanionic substances in general in the management of HIV infections may reside in the prevention of the sexual transmission of HIV. These compounds viewed as vaginal formulations may block HIV infection through both virus-to-cell and cell-to-cell contacts. Therefore, the merit of using these compounds as anti-HIV agents can be achieved if they are applied as vaginal microbicides (32).

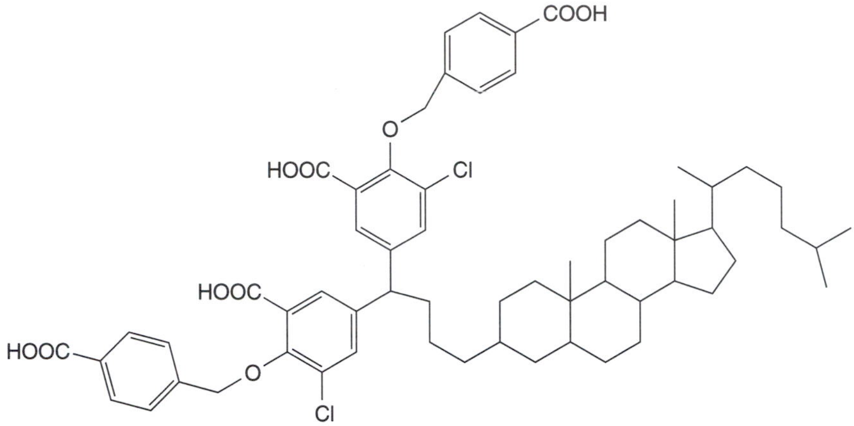

**6.2.2.2 Soluble CD4 Peptide Fragments.** Shortly after the discovery of CD4 as the primary receptor for HIV-1 binding on the T-cell membrane, soluble forms of CD4 (sCD4) were isolated and evaluated for clinical efficacy to suppress HIV infection (37–39). Although laboratory experiments had validated the overall approach, results from the initial clinical trials proved disappointing (38, 39). Recent efforts to use soluble CD4 forms have focused on CD4

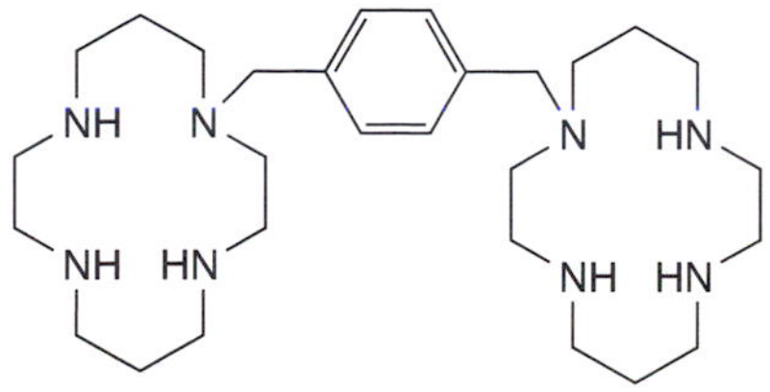

AMD3100

(4)

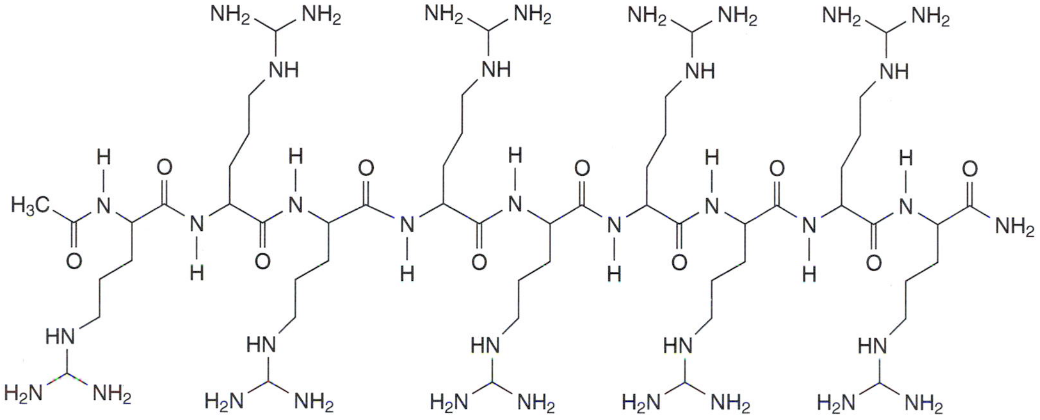

ALX40-4C

(5)

**Figure 11.4.** CXCR4 antagonists.

conjugates with toxins (40–42). Pro542 is a conjugation product of the CD4 soluble receptors (first two domains of CD4) with the constant regions (both heavy and light chains) of the human IgG2 (43, 44). Pro542 was found to be more effective than the soluble CD4 alone in blocking HIV transmission in the scid-hu-PBL mouse model (45). The results of clinical trials of Pro452 in HIV-infected adults and pediatric patients have demonstrated that the CD4 conjugate is both safe and effective (46).

### 6.2.3 Inhibitors of Gp120 Binding to the T-Cell Coreceptors

*6.2.3.1 Gp120 Coreceptors.* As explained under the virus entry part of the HIV life cycle, the coreceptors CXCR4 and CCR5 are newly discovered binding sites for gp120 on the T-cell membrane surface. These coreceptors naturally act as receptors for internal ligands released normally in cases of inflammation or immune responses. These ligands are generally designated as chemokines (*chemo*attrac-

tant cyto*kines*). Whereas CXCR4 has only one natural chemokine ligand called SDF-1 (stroma-cell-derived factor), CCR5 has two natural ligands named RANTES (regulated upon activation, normal T-cell expressed and secreted) and MIP-1 (macrophage inflammatory protein). MIP-1 is further classified into two subtypes, MIP-1$\alpha$ and MIP-1$\beta$. The importance of inhibiting gp120 binding to these coreceptors emerges from their role in the process of activating the second viral envelope protein gp41 binding to the T-cell surface, which ultimately leads to the virus entry.

*6.2.3.2 Gp120-CXCR4 Binding Inhibitors.* The most interesting member of this class is the compound bicyclam (AMD3100, **4**, Fig. 11.4). Bicyclam showed an *in vitro* anti-HIV-1 activity before the discovery of the HIV-1 coreceptors (47). Originally thought to inhibit a postentry step, however, AMD3100 was subsequently found to bind selectively to CXCR4 (48, 49). Biochemical evidence for this proposed mechanism of bicyclam action in-

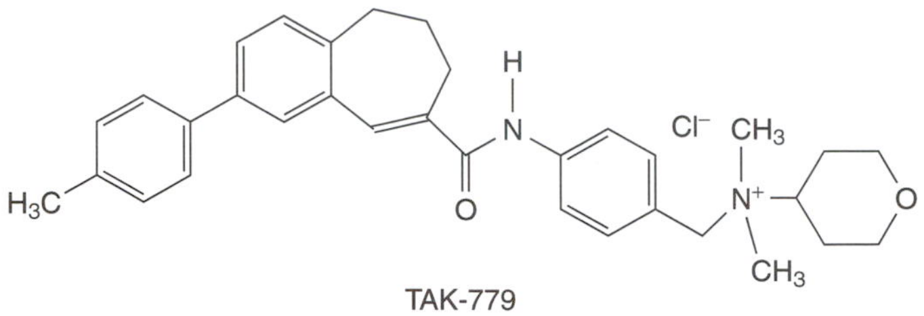

**Figure 11.5.** TAK-779, a potent
nonpeptide CCR5 antagonist.

TAK-779

(**6**)

cludes competition binding studies with both
the natural ligand of CXCR4 (SDF-1) and the
anti-CXCR4 monoclonal antibody 12G5. Bicyclam has shown modest antiviral activity in
the mouse model (50). The compound is reported to be the most specific and most potent
CXCR4 antagonist. Bicyclam is currently undergoing clinical evaluation. However, data
with respect to safety and efficacy are not yet
available. Additional CXCR4 inhibitors in preclinical studies include the polypeptide derivatives ALX40–4C (**5**, Fig. 11.4) (51), T22 (52),
T134, and T140 (53). There is a concern with
all of these compounds that long-term antagonism of CXCR4 function could be detrimental to the immune function, particularly in
the B-cell and granulocyte compartments (54–
56). Other agents that have been evaluated to
block gp120-CXCR4 binding included SDF-1,
the natural ligand for the CXCR4 receptor,
which was found to block a certain HIV strain
called T-tropic HIV (57). Figure 11.4 shows
the chemical structures for some of the
CXCR4 antagonists.

*6.2.3.3 Gp120-CCR5 Binding Inhibitors.* Several small molecules have been reported to inhibit binding of gp120 to CCR5. Some of these
agents are in preclinical development. These
CCR5 antagonists include TAK-779 (58, 59);
the 2-aryl-1-{*N*-(methyl)-*N*-(phenylsulfonyl)
amino}-4-(piperidin-1-yl)butanes (60); 1-
amino-2-phenyl-4-(piperidin-1-yl)butanes
(61); 1,3,4-trisubstituted pyrrolidines (62);
and PRO440, an anti-CCR5 monoclonal antibody (63). All of these agents were found to
block the binding of chemokines (e.g., MIP-1)
to CCR5. Consequently, they have the potential to block HIV binding to CCR5 and inhibit
replication of CCR5-dependent virus strains.
Although a promising class, an unresolved issue regarding the clinical utility of CCR5 in-

hibitors is the variability of gp120 structure
among different viral strains. The most attractive member of this class undergoing clinical evaluation is the quaternary ammonium
derivative TAK-779 (**6**, Fig. 11.5) (58, 59). It is
the first nonpeptide molecule that has been
described to block the replication of M-tropic
R5 HIV-1 strains at the CCR5 level. Other
agents undergoing evaluation as blockers for
gp120-CCR5 binding is the chemokine MIP-
1α, which has been identified as the most potent chemokine for inhibiting HIV infection
(64, 65). Figure 11.5 depicts the structure of
CCR5 antagonist TAK-779.

### 6.3 Inhibitors of Viral Fusion

The fusion of the virus envelope with the T-
cell membrane commences after the binding
of gp120 to the coreceptors CXCR4 and CCR5.
This binding triggers a spring-loaded action of
the glycoprotein gp41, which is normally covered by the larger gp120. The gp41 anchors
itself to the T-cell membrane through the
hairpin structures HR1 and HR2 (see virus
life cycle, fusion step). This initiates the fusion
of the two lipid bilayers of the virus and cell
membranes (66). Before the knowledge of
structural information of gp41, synthetic peptides overlapping the two HR regions HR1
[e.g., DP-107 peptide (67)] and HR2 [e.g., DP-
178 peptide (68)] were found to block HIV-1
replication in cell culture. The discovery that
these two regions constituted the fusion trigger subsequently provided a mechanistic basis
for their antiviral affects. The peptides act in a
dominant-negative fashion to prevent formation of the trimer-of-hairpins arrangement.
The first clinical trial of pentafuside (or DP-
178 or T-20 in the trial) was reported in 1998
(69). Pentafuside is a synthetic 36 amino acid
peptide that corresponds to the 127–162 resi-

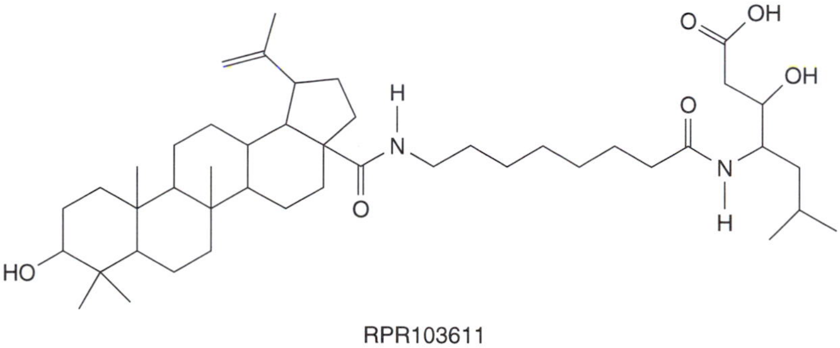

RPR103611

(7)

**Figure 11.6.** The nonpeptide fusion inhibitor betulinic acid derivative RPR103611.

dues of the of gp41 domain structure. Subsequent clinical trials have confirmed the overall safety and efficacy of T-20. Another polypeptide that is undergoing evaluation and believed to interact in a manner similar to that of T-20 is the antibiotic siamycin (a tricyclic 21 amino acid peptide isolated from Streptomyces) (70, 71). The betulinic acid derivative RPR103611 (**7**, Fig. 11.6) represents a nonpeptide inhibitor for gp41 fusion (72). Peptide and nonpeptide HIV fusion inhibitors have recently been reviewed (73). Figure 11.6 depicts the structure of the fusion inhibitor betulinic acid derivative RPR103611.

## 6.4 Inhibitors for Viral Uncoating

The viral uncoating process (loss of its nucleoprotein outer coat after fusion) results in the release of its RNA genome into the cytoplasm. Uncoating is controlled by the p7 nucleocapsid protein (NCp7), which is a peptide segment of the p18 protein. NCp7 is a zinc-containing protein, and zinc-displacing compounds were found to inhibit the virus uncoating process (74). The anti-influenza drug amantadine (**8**, Fig. 11.7) is an example that is reported to work by such a mechanism (75). Several other agents are reported to interfere with this step of viral nucleo-protein disassembly (i.e., uncoating). Among these agents are NOBA (3-nitrosobenzamide, **9**, Fig. 11.7) (76, 77), the dithiobenzamide-sulfonamide derivative DIBA (**10**, Fig. 11.7) (78, 79), SRR-SB3 (cyclic 2,2'-dithiobisbenzamide, **11**, Fig. 11.7) (80), dithiane (1,2-dithiane-4,5-diol,1,1-dioxide, **12**, Fig. 11.7) (81), and ADA (azadicarbonamide)

(82, 83). Figure 11.7 depicts the structures of viral uncoating inhibitors.

## 6.5 Inhibitors of HIV Reverse Transcription

The reverse transcription process has been one of the most thoroughly investigated steps in the virus life cycle. The transcription of HIV genetic material is initiated and mediated by the reverse transcriptase enzyme (RT). The enzyme is multifunctional, presenting both RNA and DNA polymerase as well as RnaseH activities. The reverse transcriptase is responsible for converting the single-strand RNA into the double-strand DNA. In addition, the enzyme has a catalytic unit for activating the RnaseH enzyme, which liberates the proviral DNA from RNA after transcription. Considering the crucial and complex role it plays in the synthesis of viral DNA, its inhibition has proved to be an effective approach in interrupting the HIV replication cycle. The enzyme has captivated the attention of many investigators striving to develop new anti-HIV drugs to overcome the continuously evolving HIV mutant strains.

**6.5.1 Molecular Aspects of HIV Reverse Transcriptase.** The mature HIV-1 RT is a heterodimeric enzyme composed of two subunits, p66 and p51 (84). X-ray crystal structures of the enzyme complexes with different inhibitors have been successively refined at 7 Å (85), 3.5 Å (86), 3.0 Å (87), and 2.2 Å (88). These crystal structure studies show that the polymerase regions of p66 and p51 are divided into four subdomains denoted *finger, palm, thumb,*

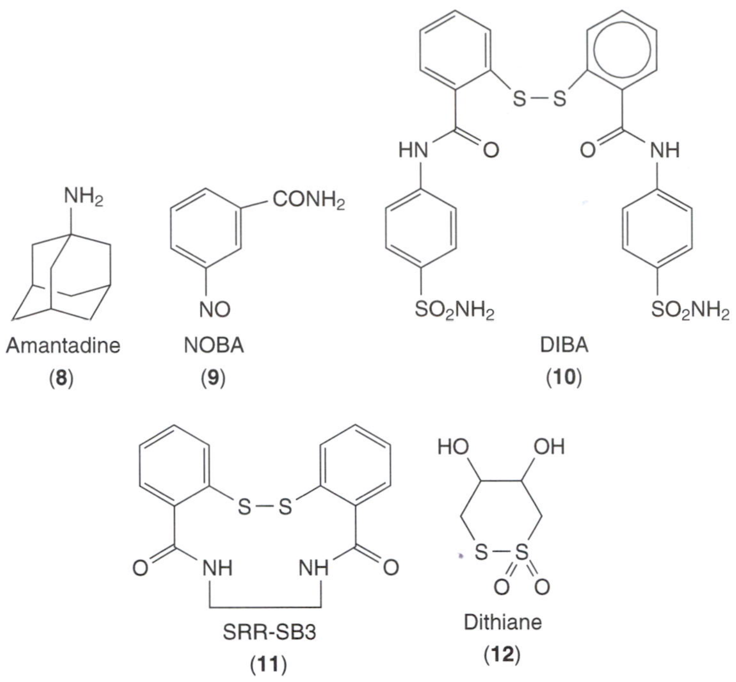

**Figure 11.7.** Viral uncoating inhibitors.

and *connection*. The p66 palm domain contains the polymerase active site. Three aspartic acid residues at positions 110, 185, and 186 constitute the catalytic triad (86, 87). A feature of HIV-1 RT that makes it such a difficult target for therapeutic intervention is the high degree of nucleotide sequence variation between different viral strains. The HIV-1 RT is highly error prone, with a predicted error rate of 5 to 10 per HIV-1 genome per round of replication *in vivo*. The mutations incorporated by retroviral RT include amino acid substitution and deletions of nucleotide sequence (89, 90).

**6.5.2 HIV Reverse Transcriptase Inhibitors.** Reverse transcriptase inhibitors are designed to inhibit viral DNA synthesis from viral RNA. The process of DNA synthesis involves incorporation, sequentially and one at a time, of activated nucleotides (as triphosphate) to the template RNA. The enzyme reverse transcriptase catalyzes this process by simultaneous binding to both the nucleotide and RNA at its catalytic active site. Inhibitors of HIV transcriptase act by impeding the nucleotide binding to the active site. The inhibi-

tion is achieved by two distinct mechanisms, either competitively or noncompetitively. The competitive inhibitors are nucleoside analogs or have nucleoside-like structures. The nucleoside inhibitors are further classified into purine and pyrimidine nucleoside inhibitors based on the nucleic acid base existing in the molecule. On the other hand, the noncompetitive inhibitors are not structurally related to the nucleotides and are referred to as allosteric inhibitors.

*6.5.2.1 Nucleoside Reverse Transcriptase Inhibitors (NRTIs).* Members of this class act as irreversible, competitive inhibitors for the HIV RT. Their description as competitive indicates the mechanism of action, given that they compete with normal substrates at the enzyme catalytic site. These normal substrates are different types of deoxyribonucleosides triphosphate (dNTP) or the deoxynucleotides. Therefore, members of this class, being nucleoside in nature (i.e., contain only the base and the sugar), require intracellular activation to the nucleoside triphosphate form (the nucleotide). This activation process requires three phosphorylation steps, whereby the compounds are converted successively to

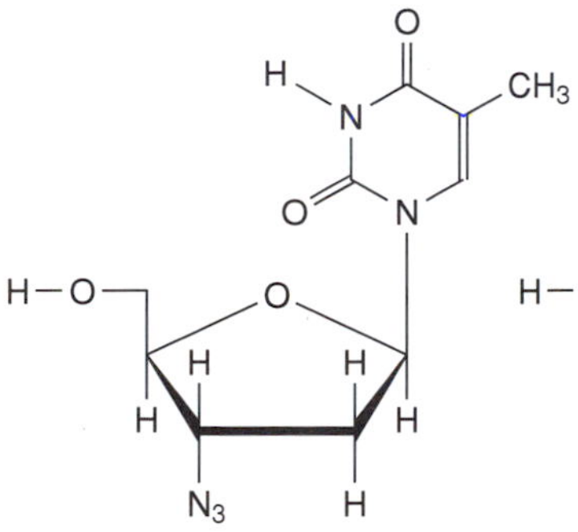

Zidovudine
(**13**)

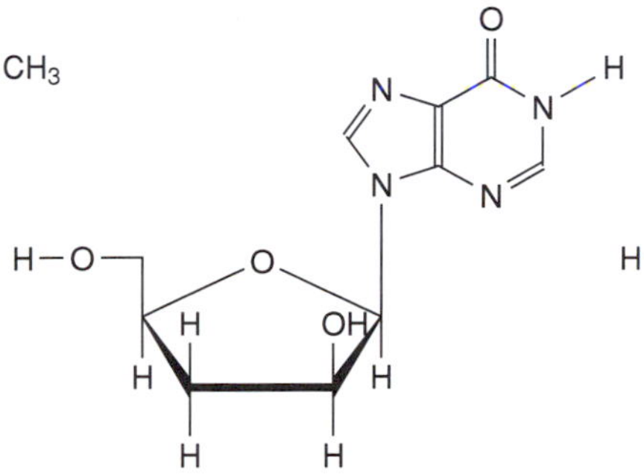

Didanosine
(**14**)

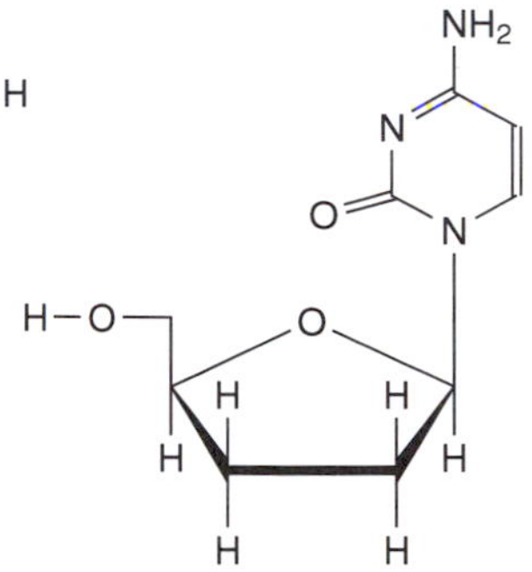

Zalcitabine
(**15**)

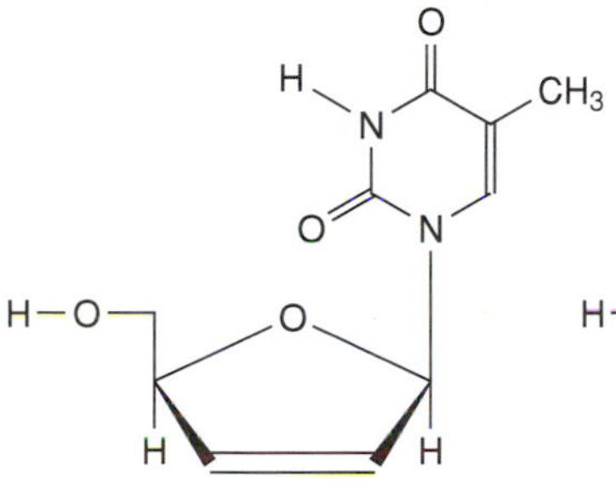

Stavudine
(**16**)

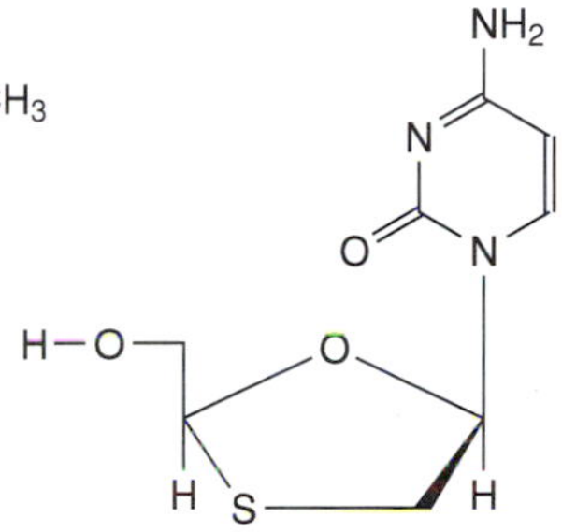

Lamivudine
(**17**)

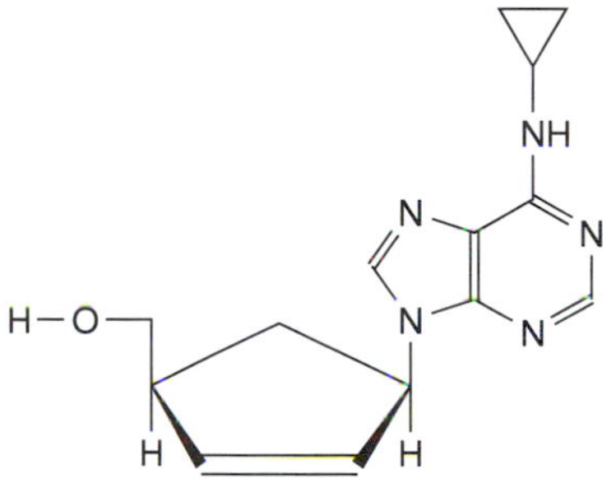

Abacavir
(**18**)

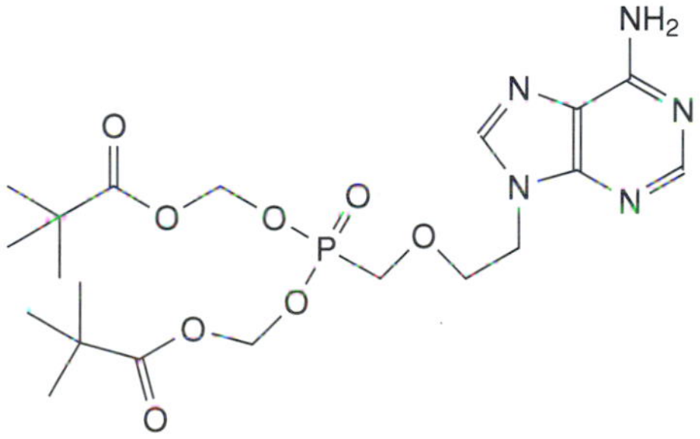

Adefovir dipivoxil
(**19**)

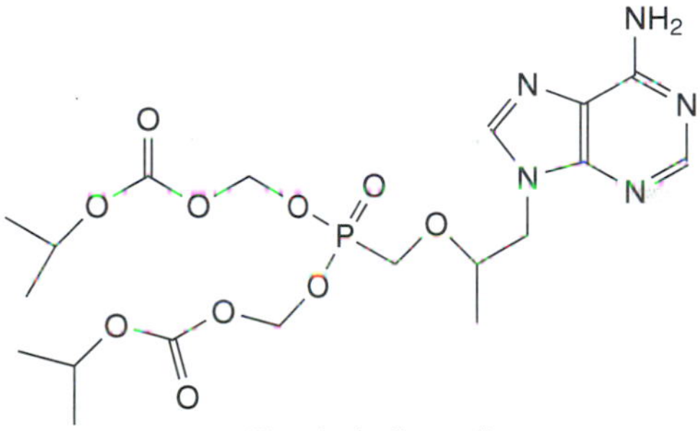

Tenofovir disproxil
(**20**)

**Figure 11.8.** Clinically approved and phase III clinical trial NRTI.

mono-, di-, and triphosphate by cellular kinases. The irreversible description of the mechanism of action of NRTIs is attributed to the fact that they are incorporated into the viral DNA through an irreversible covalent bond. The irreversible bond occurs through the activated 5-hydroxyl of the sugar. On the other hand, the structures of all NRTIs indicate their lack of the 3-hydroxyl group of the sugar. The absence of the 3-hydroxyl group results in a DNA intermediate product that does not elongate further through position 3. This process of blocking DNA elongation is commonly referred to as chain termination. Most drugs approved for treatment of HIV infections belong to the class of nucleoside inhibitors. These drugs include zidovudine (AZT, **13**, Fig. 11.8), which was the first member of

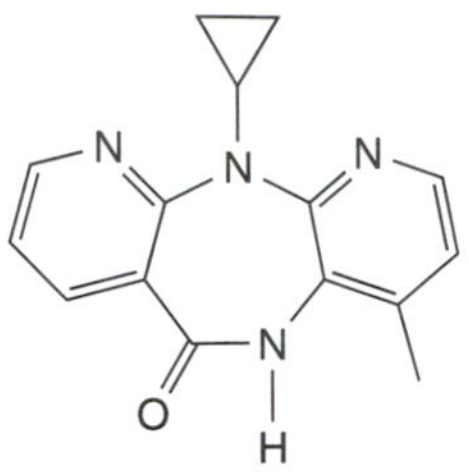

Nevirapine
(**21**)

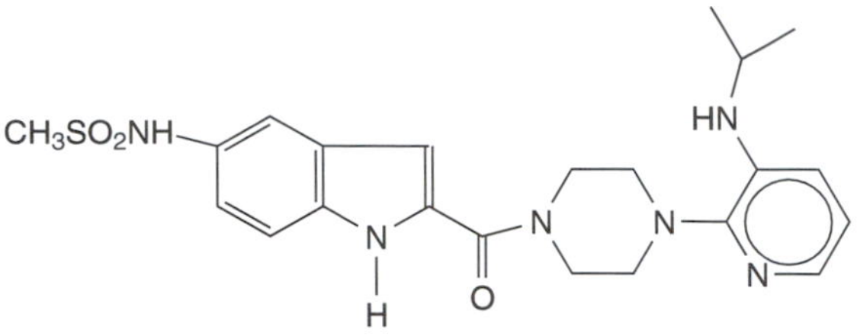


Delaviridine
(**22**)

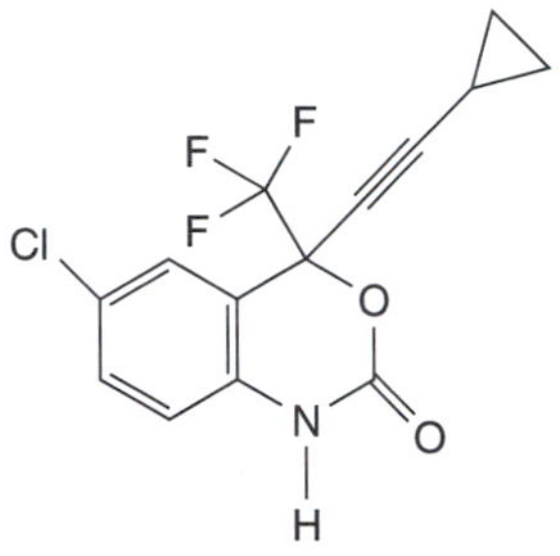

Efavirenz
(**23**)

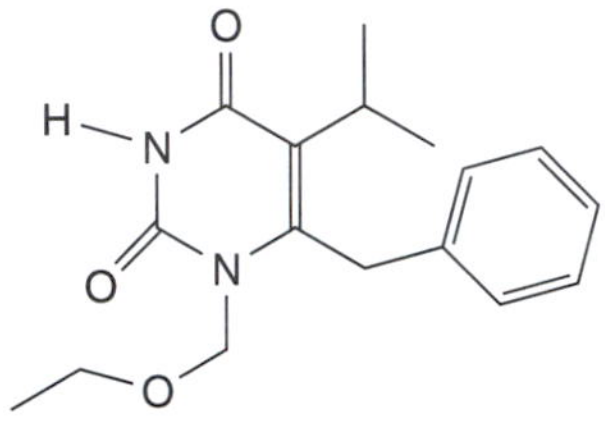

Emivirine
(**24**)

**Figure 11.9.** Clinically approved and phase III clinical trial NNRTI.

the class to be approved (91–93); didanosine (DDI, **14**, Fig. 11.8), zalcitabine (ddC, **15**, Fig. 11.8), stavudine (D4T, **16**, Fig. 11.8), lamivudine (3TC, **17**, Fig. 11.8), and abacavir (ABC, **18**, Fig. 11.8). In addition to acting as irreversible competitive inhibitors, NRTIs act as chain terminators by preventing DNA elongation. Some newer (second-generation) NRTIs have been developed and are undergoing preclinical trial phases, such as adefovir dipivoxil [bis(POC)-PMEA, **19**, Fig. 11.8] and tenofovir [bis(POC)-PMPA, **20**, Fig. 11.8], which is undergoing phase III clinical trials for the treatment of HIV infections (94). Trends in the design of nucleoside analogs such as anti-HIV drugs were recently reviewed (95). Figure 11.8 represents the structures of clinically approved nucleoside inhibitors and the phase III clinical trial inhibitors adefovir and tenofovir.

***6.5.2.2 Nonnucleoside Reverse Transcriptase Inhibitors (NNRTIs).*** In contrast to the NRTIs, the nonnucleoside reverse transcriptase inhibitors (NNRTIs) target the allosteric nonsubstrate binding sites. It has been suggested that the NNRTI binding sites may be functionally, and possibly also spatially, related to the substrate binding site (96, 97). Members of this class are described as noncompetitive, reversible inhibitors. The dissimilarity of their chemical structures to the natural substrate implies different binding sites, hence the classification as noncompetitive inhibitors. Furthermore, unlike NRTIs, members of this class are highly specific to HIV-1 RT without affecting the host DNA polymerases, which explains the low toxicity and side effects of these drugs. Compounds belonging to this class are classified as first-generation NNRTIs, which belong to more than 30 different classes of compounds (98) or second-generation NNRT inhibitors. Newer nonnucleoside reverse transcriptase inhibitors have recently been reviewed (99). Only three drugs belonging to the NNRTI class have been approved for clinical use: nevirapine (**21**, Fig. 11.9), delavirdine (**22**), and efavirenz (**23**, Fig. 11.9). The second-generation NNRTI emivirine (MK-442, **24**, Fig. 11.9) is in Phase III clinical trial. Figure 11.9 shows the structures of clinically approved NNRTIs and the Phase III clinical trial inhibitor emivirine.

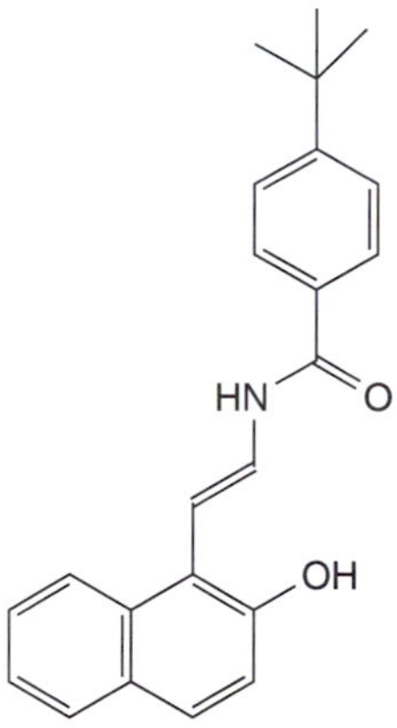

**BBNH**

**(25)**

**Figure 11.10.** The RNaseH inhibitor BBNH.

## 6.6 Inhibitors of HIV Ribonuclease (RnaseH)

The transcription of viral RNA copy to form the DNA molecule (under the control of RT) is followed by a second transcription of a positive strand of DNA to form a double helix. The tristranded product requires cleavage of the original RNA strand from the double-helix DNA molecule (the provirus). RnaseH (a viral ribonuclease) is the enzyme responsible for this step of removal of viral RNA from DNA. The activity of viral RNase is believed to be mediated by a special domain on the reverse transcriptase enzyme (100). Very little is known about inhibitors for this RT-associated ribonuclease H enzyme and research in this area is limited. This class of inhibitors is exemplified by the compound N-(4-t-butylbenzoyl)-2-hydroxy-1-naphthaldehyde hydrazone (BBNH, **25**, Fig. 11.10), which was reported to be a very potent inhibitor for RNase H (101). Figure 11.10 shows the structure of the Rnase H inhibitor BBNH.

## 6.7 HIV Integrase Inhibitors

**6.7.1 HIV Integrase (IN).** Integrase is one of the enzymes encoded by HIV-1. It catalyzes the insertion of the HIV-1 DNA into the host cell genome. Integration is required for stable maintenance of the viral genome and viral gene expression. The integration process requires that the IN protein recognizes linear viral DNA ends through its long terminal repeats (LTRs), followed by removal of two base pairs from the 3' ends of each provirus dinucleotide (102). These 3' processed viral DNA ends are then joined to the 5' ends of the cleaved host genome target site DNA by a one-step transesterification (103). Both the crystal structure of the IN catalytic domain and the solution structure of the DNA binding domain of HIV-1 IN have been determined (104, 105). The structural information has been helpful in the design of inhibitors at this crucial step in the HIV-1 replication cycle (106).

**6.7.2 Integrase Inhibitors.** Like other retroviruses, HIV cannot replicate without integration into a host chromosome. Accordingly, HIV integrase has been considered as an attractive target in the design of anti-HIV drugs. Numerous compounds have been described as inhibitors of HIV-1 integrase (107). The most promising examples of these inhibitors are L-chicoric acid (**26**, Fig. 11.11) and the diketo acid derivatives L-731,988 (**27**) and L-708,906 (**28**); all have been described as potent integrase inhibitors. The aspects of drug design for integrase inhibitors were the subject of a recent review as emerging therapeutic approaches to design inhibitors for HIV-1 infections (108). Figure 11.11 shows the structures of representative examples of the integrase inhibitors.

## 6.8 Inhibitors of HIV Gene Expression (Transactivation Inhibitors)

Agents belonging to this class inhibit the virus transcription phase at the host genome level after viral DNA integration. HIV uses the host genome to express its env, gag, and pol genes. These genes translate the production of the virus structural protein, for example, the envelope and functional proteins such as the reverse transcriptase, the integrase and, the protease. HIV gene expression occurs after the promotor LTR (long terminal repeat) ends of the gene are activated by binding to an activating viral protein called tat (transactivating). The activated LTR amplifies the process of gene expression. A number of compounds have been reported to inhibit HIV-1 replication through inhibiting this process of the gene transactivation through inhibiting tat

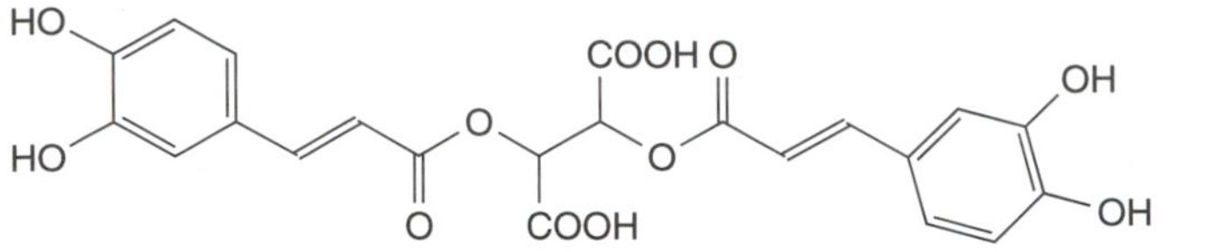

L. Chicoric acid

(**26**)

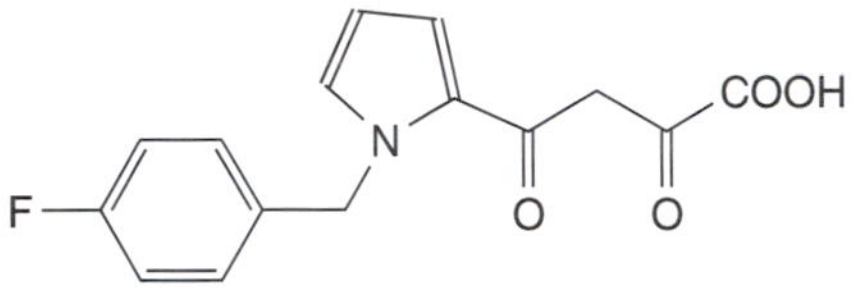

L.731,988

(**27**)

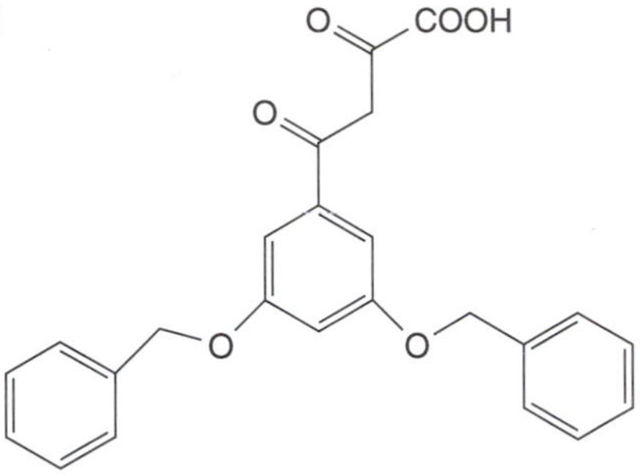

L708,906

(**28**)

**Figure 11.11.** Integrase inhibitors.

binding to the promoter LTR ends. Among these compounds are fluoroquinoline K-12 (**29**, Fig. 11.12) (109) and temacrazine (bistriazoloacridone, **30** Fig. 11.12). The latter was found to block HIV-1 RNA formation without interfering with the transcription of any cellular genes (110). Figure 11.12 shows

representative examples of HIV transcription transactivation inhibitors.

**6.9 Inhibitors of Virion Assembly**

The tripeptide Gly-Pro-Gly- $NH_2$, was found to have anti-HIV activity. This was originally attributed to an entry-inhibitory mechanism

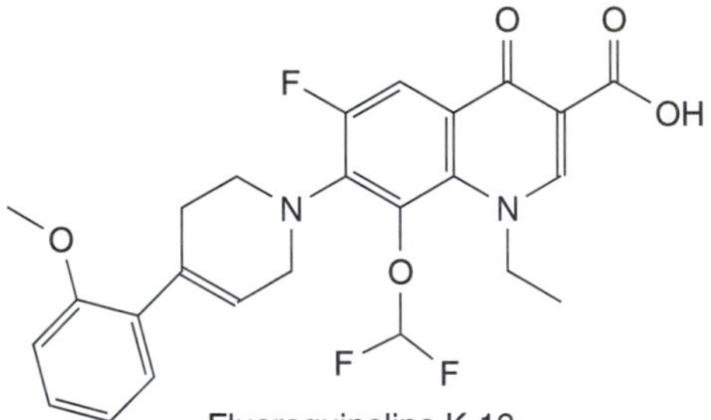

Fluoroquinoline K-12

(**29**)

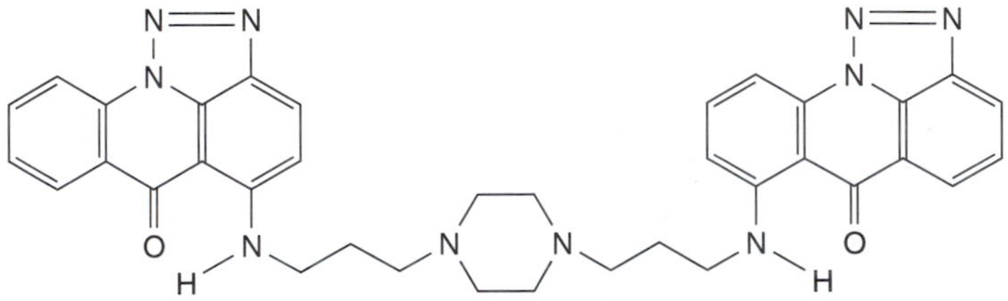

**Figure 11.12.** HIV transcription transactivation inhibitors.

Temacrazine

(**30**)

on the basis of the observation that this sequence is a highly conserved motif in the V3 loop of gp120 (111). However, recent studies have shown that GPG does not inhibit the early events in HIV-1 replication (because it affects the virus production from cells chronically infected with HIV-1; see Ref. 112). This suggested a novel mechanism of action that targets viral assembly and/or maturation. The mechanism of GPG is not yet fully understood, although this tripeptide may represent the first proof that small molecules can affect HIV replication by inhibiting the process of virion assembly.

## 6.10 Inhibition of Viral Maturation (HIV-1 Protease Inhibitors)

### 6.10.1 HIV Protease. HIV protease represents the second enzyme in the virus life cycle, after the reverse transcriptase, that has been extensively targeted for drug design. HIV protease is a proteolytic enzyme responsible for cleaving the large polyprotein precursor into biologically active protein products. HIV polyprotein precursor is encoded by the *gag* and *gag-pol* genes. These genes encode the precursor with HIV structural core proteins and various viral enzymes, including the reverse transcriptase, the RnaseH, the integrase, and the protease. HIV protease cleaves this polyprotein precursor during or shortly after viral budding to produce the mature virion. Therefore, inhibition of this posttranslational step leads to total arrest of viral maturation, thereby blocking infectivity of the virions (113). Structurally, the enzyme is classified as an aspartic protease. The X-ray crystal structure of HIV-1 protease (114–116) has revealed that the enzyme is a C2 symmetric homodimer of two identical subunits. Each of the two subunits is a 99 amino acid chain and each contributes a single aspartic acid residue to the active catalytic site, which exists in a cleft enclosed between two projecting loops from each subunit. The active site of the enzyme is a triad aspartyl-threonine-glycine residue, which is located within a loop whose structure is stabilized by hydrogen bonds. The loops of each monomer are interlinked by four hydrogen bonds that contribute to the stabilization the two subunits around the symmetry axis. These hydrogen bonds and the axiomatic structure of the enzyme have served as a foundation for designing inhibitors for the enzyme.

### 6.10.2 HIV Protease Inhibitors (PI). Three different approaches have been taken in the design of protease inhibitors. One is based on the transition-state mimetic approach. A large number of inhibitors were developed as competitive inhibitors for the natural substrate (the polyprotein precursor) binding to the enzyme. Eventually, all these competitive inhibitors are peptide in nature. The second approach of inhibitor design is based on disrupting the enzyme's twofold rotational C2-symmetry axis by forming specific hydrogen bonds and hydrophobic interactions with the amino acid residues involved in stabilizing the dimer. Agents belonging to this class are also peptide in nature. To improve the poor pharmacokinetic properties exhibited by peptide protease inhibitors, a third approach has emerged to develop inhibitors with fewer peptide characteristics. Accordingly, HIV protease inhibitors are classified as peptide inhibitors (first generation) and nonpeptide inhibitors (second generation). Members of both classes act by one of the above-mentioned mechanisms (i.e., competitive inhibition or dimer destabilizer). HIV protease inhibitors, peptides, and nonpeptides were previously (117) and recently (118) reviewed. At the molecular level of drug design, all first- or second-generation protease inhibitors are tailored after the target peptide linkages in the natural substrate (polyprotein precursor). The cleavage region has a phenylalanine-proline sequence at positions 167 and 168 of the gag-pol polyprotein. Inhibitors are designed with chemical functional groups that resist the cleavage. This is generally achieved by replacement of the scissile peptide bond of the substrate with a nonscissile structural moiety such as the hydroxyethylene group in the inhibitor.

All currently licensed protease inhibitors for the treatment of HIV infection (Fig. 11.13), that is, indinavir (**31**), ritonavir (**32**), saquinavir (**33**), nelfinavir (**34**), and amprenavir (**35**) (118, 119), share the same structural determinant, a hydroxyethylene group replacing the normal peptide bond. The hydroxyethlene

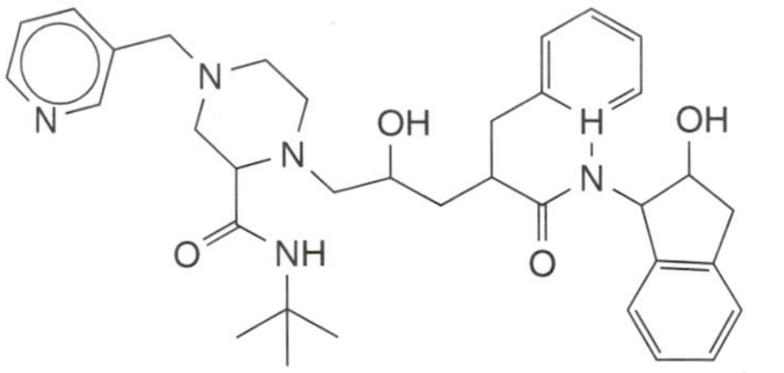

Indinavir
(**31**)

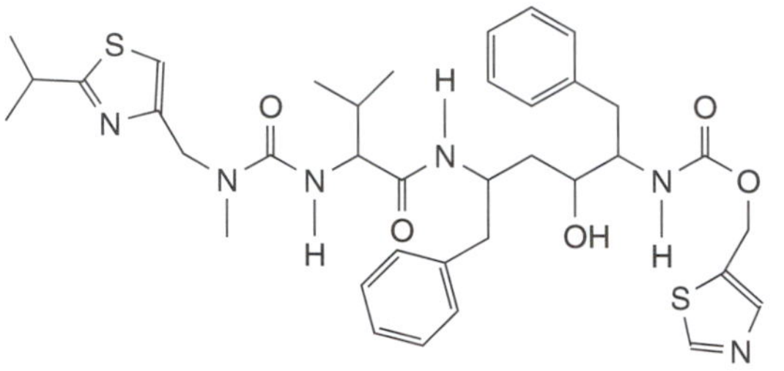

Ritonavir
(**32**)

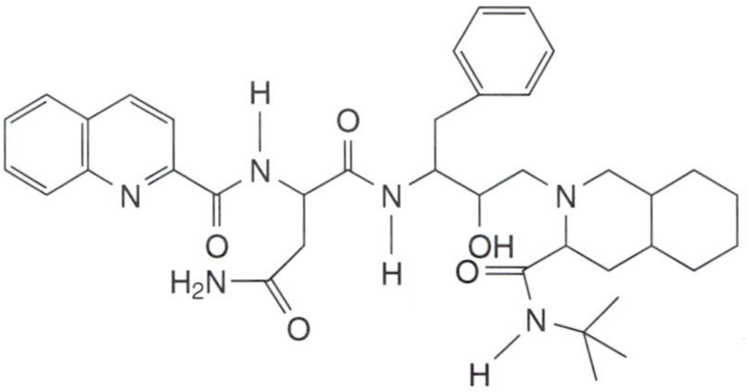

Saquinavir
(**33**)

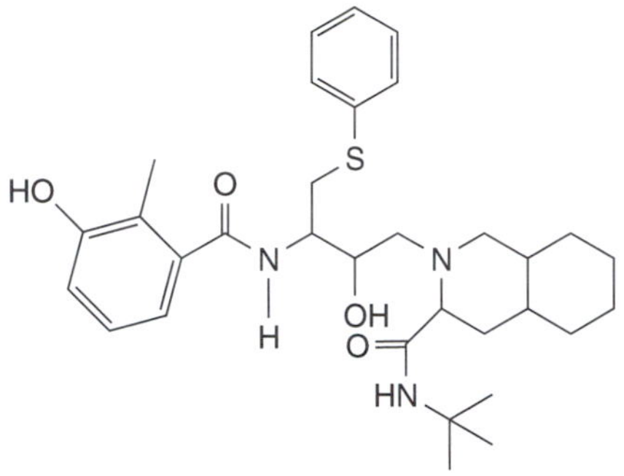

Neflinavir
(**34**)

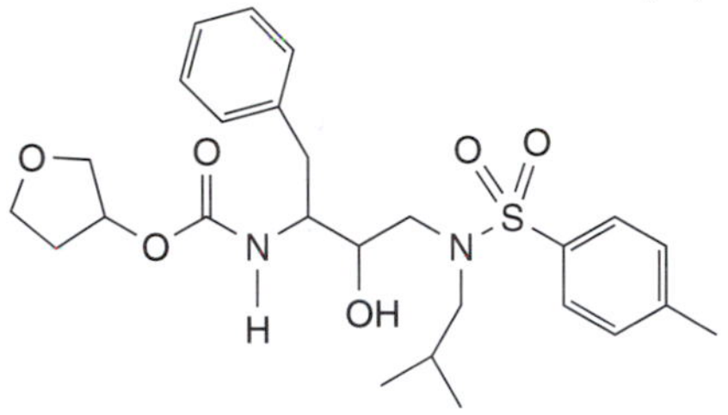

Amprenavir
(**35**)

**Figure 11.13.** Clinically approved protease inhibitors.

group has a dual purpose. In addition to making the molecule nonscissile, it allows the molecule to bind to the catalytic site by hydrogen bonding, which can destabilize the enzyme dimer. Other small size peptide inhibitors are reported to contain hydrophilic carboxyl groups as the hydrogen bonding destabilizer (120). Reports have appeared for peptidomimetic inhibitors with the sessile segment of the peptide substrate replaced with a 15-member macrocylic peptide with stable conformation (121). Certain monoclonal antibodies were also studied as noncompetitive inhibitors for the enzyme acting through suppressing the dimerization process (122). In general, protease inhibitors are designed with the aim of driving the molecule to the active site, where it acts either as a competitive inhibitor with stronger hydrophobic and or hydrogen bonding or as a destabilizer to the enzyme dimeric structure through similar forces of interaction. Figure 11.13 depicts the structure of the clinically approved protease inhibitors.

Some newer inhibitors with nonpeptide structure have been developed, such as lopinavir (ABT-378, **36**, Fig. 11.14) (119), the cyclic urea mozinavir (DMP 450, **37**, Fig. 11.14) (123), atazanavir (BMS-232632, **38**, Fig. 11.14) (124), tipranavir (PNU 140690, **39**, Fig. 11.14) (125), and the C2-symmetric protease

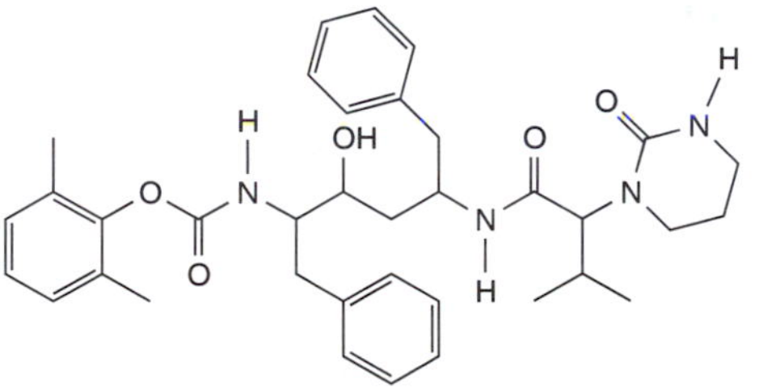

Lopinavir
(36)

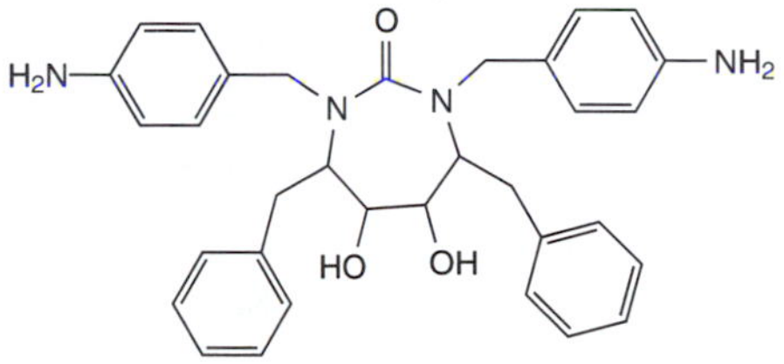

Mozenavir
(37)

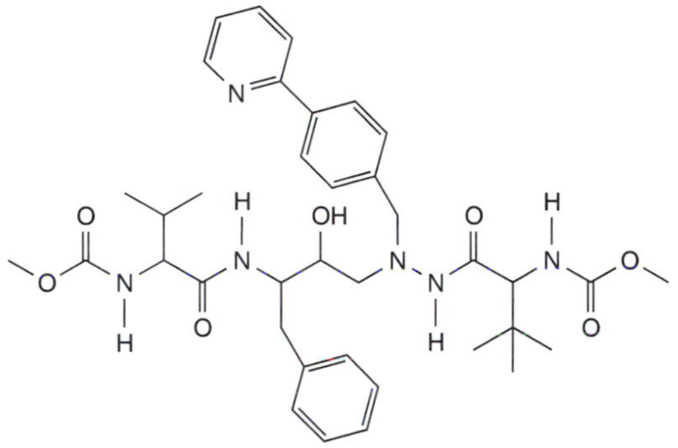

Atazanavir
(38)

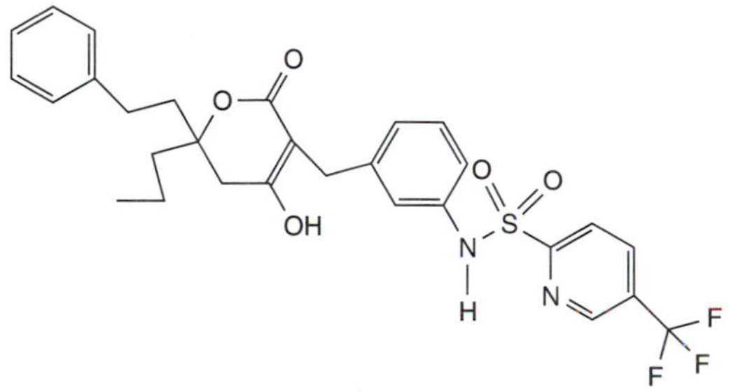

Tipranavir
(39)

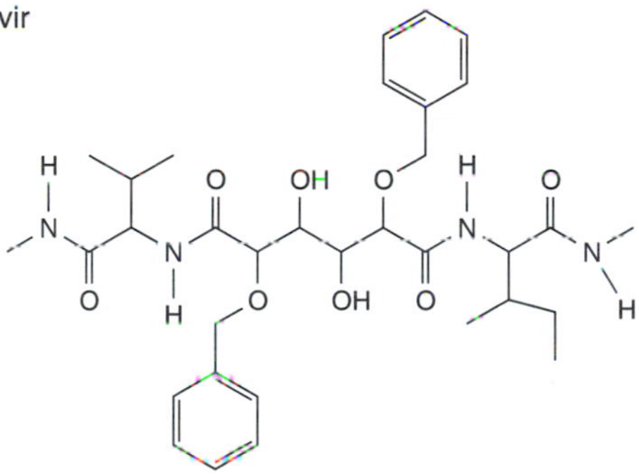

L-Mannaric acid
(40)

**Figure 11.14.** Newer protease inhibitors.

inhibitor L-mannaric acid (**40**, Fig. 11.14) (126). Figure 11.14 depicts the structures of these newer protease inhibitors. Recently, other potential sites were targeted for HIV protease inactivation by peptides and peptidomimetics based on the terminal sequence of the enzyme. These terminal sequences exist at the enzyme surface and are believed to be less prone to mutations. Cupric ion was described to bind to such surface sequences rich in histidine and cysteine amino acid residues, leading to enzyme inhibition (127). Computer-aided drug design and molecular modeling are used to analyze binding of the inhibitors to the enzyme and to develop new agents based on a rationale drug design approach (128–132).

## 7  HIV DRUGS IN CLINICAL USE

All currently available drugs for HIV therapy belong to one of three classes of inhibitors: the nucleoside reverse transcriptase inhibitors (NRTIs), the nonnucleoside reverse transcriptase inhibitors (NNRTIs), and the protease inhibitors (PIs). These drugs have gained a def-

**Table 11.1  HIV Approved Drugs for the Treatment of AIDS**

| Generic Name | Brand Name | Firm | Class |
|---|---|---|---|
| Zidovudine (AZT) | Retrovir | Glaxo Wellcome | NRTI |
| Didanosine (ddI) | Videx | Bristol-Myers Squibb | NRTI |
| Zalcitabine (ddC) | Hivid | Hoffman-La Roche | NRTI |
| Stavudine (d4T) | Zerit | Bristol-Myers Squibb | NRTI |
| Lamivudine (3TC) | Epivir | Glaxo Wellcome | NRTI |
| Abacavir (ABC) | Ziagen | Glaxo Wellcome | NRTI |
| Neveirapine | Viramune | Boehringer Ingelheim | NNRTI |
| Delavirdine | Rescriptor | Pharmacia | NNRTI |
| Efavirenz | Sustiva | Hoffman-La Roche | NNRTI |
| Idinavir | Crixivan | Mercke | PI |
| Ritonavir | Norvir | Abbott | PI |
| Saquinavir | Invirase | Hoffman-La Roche | PI |
| Nelfinavir | Viracept | Agouron Pharma | PI |
| Amprenavir | Agenerase | Glaxo Wellcome | PI |

inite place in the treatment of HIV-1 infections because they interfere with crucial events in the HIV replication cycle. NRTIs, which target the substrate binding site, include six drugs: zidovudine, didanosine, zalcitabine, stavudine, lamivudine, and abacavir. NNRTIs, which target nonsubstrate binding sites, include three drugs: nevirapine, delavirdine, and efavirenz. Protease inhibitors bind to the active site and act as either enzyme inhibitors or dimer-destabilizing factors; these include five drugs: indinavir, ritonavir, saquinavir, neflinavir, and amprenavir. Table 11.1 lists for each compound the generic name, brand name, the pharmaceutical firm that manufactures it, and its mechanistic classification.

## 8  THE NEED FOR NEW ANTI-HIV DRUGS

### 8.1  Development of Drug Resistance

In spite of having such an arsenal of drugs available for treatment of HIV infections, millions of dollars are being spent on AIDS research for developing new drugs. The development of drug-resistant HIV strains is a compelling reason for more efforts to develop newer inhibitors. Resistance arises from mutations in the viral genome, specifically in the regions that encode the molecular targets of therapy, HIV-1 RT and HIV-1 PR enzymes. These mutations alter the viral enzymes in such a way that the drug no longer inhibits the enzyme functions and the virus restores its

free replication power. In the case of reverse transcriptase inhibitors, drug resistance to nucleoside analogs develops after prolonged treatment and has been exhibited for all the nucleoside inhibitors (133–135). However, NNRTIs are notorious for rapidly eliciting resistance (136) because of the mutations that are clustered around the putative hydrophobic binding site of these drugs (137).

### 8.2  Approaches to Overcome Drug Resistance (138–140)

**8.2.1  Switching the Drug Class.**  Switching from one class of inhibitors to another (or, even within each class, from one compound to another) has proved to be effective in downplaying the resistance issue. Different NNRTIs do not necessarily lead to cross-resistance. For example, TSAO-resistant HIV-1 mutant strain containing [138]Glu to Lys mutation can be completely suppressed by another HIV-1-specific RT inhibitor like TIBO or nevirapine (141). Also α-APA derivative R89439 is very active against the [100]Leu to Ile mutant, which is highly resistant to the TIBO derivative (142).

**8.2.2  The Virus "Knock-Out" Strategy.**  This approach requires starting therapy with sufficiently high drug concentrations so as to "knock out" the virus and prevent breakthrough of any virus, whether resistant or nonresistant (143). Total knock-out can also be achieved by beginning with drug combinations that allow cells to clear from the virus at much lower drug concentrations than if the compounds were to be used individually.

**8.2.3 Combination Therapy.** HIV makes new copies of itself inside the infected cells at a very fast rate. Every day billions of new copies of HIV are made and millions of new cells die. One drug, by itself, can slow down the fast rate of infection. Two or more drugs can slow it down even more efficiently. Anti-HIV drugs from different drug groups attack the virus in different ways. For example, RT and protease are two different molecular targets. RT inhibitor will stop the HIV just after it enters the cell and a protease inhibitor stops it immediately before leaving it. Hitting at the different targets increases the chances of stopping HIV and protecting other cells from infection. In addition, different anti-HIV drugs can attack the virus in different cell types and different parts of the body. Combination therapy can reduce drug resistance to the virus, especially if the mutations counteract each other. A classic example is the $^{181}$Tyr to Cys mutation, which causes resistance to most of the NNRTIs. This mutation suppresses the $^{215}$Thr to Tyr mutation that causes resistance to AZT (144). In addition, the use of multidrug therapy allows low dosage of each drug. This minimizes the toxic effects exerted by individual drugs than if they were administered individually.

**8.2.4 Officially Approved Drug Combinations.** Among the formally recognized combination therapies are zidovudine and lamivudine (NRTI + NRTI); lopinavir and ritonavir (PI + PI); and abacavir, zidovudine, and lamivudine (NRTI + NRTI + NRTI). These combinations have been reported to achieve 1 to 3 log reductions in viral loads. Today, quadruple drug combinations are in vogue. These therapies have been reported to reduce the viral load to "undetectable levels." Such combinations are called highly active antiretroviral therapy (HAART). However, recent studies have shown that even after 30 months of HAART and undetectable viral load, patient-derived lymphocytes that were actively producing virus could be cultured *in vitro* (145). HIV remains a "sleeping giant."

**8.2.5 Prodrug Conjugates: A New Trend in Combination Therapy.** The recent trend in combination therapy has shifted lately into a new approach that combines the concepts of prodrug design as well as the combination therapy benefits. The approach involves the design of prodrug conjugates consisting of two of the desired inhibitors chemically conjoined though a linker molecule such as an amino acid or succinic acid. The prodrug is designed to spontaneously release the two drug components in the body. The prodrug conjugates not only provide single-dosing regimens, but have also proved to be of better anti-HIV activity than that of the individual drugs, probably because of the improved cell penetration properties (146–148).

## REFERENCES

1. B. A. Larder, G. Darby, and D. D. Richardman, *Science*, **234**, 1731–1734 (1989).

2. A. Saxon, R. H. Stevens, S. G. Quan, and D. W. Golde, *J. Immunol.*, **120**, 777 (1978).

3. B. J. Poiesz, F. W. Ruscetti, A. F. Gazdar, P. A. Bunn, J. D. Minna, and R. C. Gallo, *Proc. Natl. Acad. Sci. USA*, **77**, 7415 (1980).

4. F. Barre-Sinoussi, J. C. Chermann, F. Rey, M. T. Nugeyre, C. S. Chamaret, J. Gruest, C. Dauguest, C. Alex-Bin, F. Venzinet-Burn, C. Rouzioux, W. Rozenbaum, and L. Montagnier, *Science*, **220**, 868–871 (1983).

5. A. S. Fauci, *Science*, **239**, 617–622 (1988).

6. R. R. Redfield and D. S. Burke, *Sci. Am.*, **259**, 90–100 (1988).

7. R. C. Gallo, *Sci. Am.*, **256**, 47–56 (1987).

8. B. H. Hahn, G. M. Shaw, S. K. Arya, M. Popoviv, R. C. Gallo, and F. Wong-Staal, *Nature*, **321**, 412 (1984).

9. F. Wong-Staal, P. K. Chanda, and J. Ghrayeb, *AIDS Res. Hum. Retroviruses*, **3**, 33 (1987).

10. R. A. Kaltz and A. M. Shalka, *Annu. Rev. Biochem.*, **63**, 133–173 (1994).

11. R. Wyatt and J. Sodroski, *Science*, **280**, 1884–1888 (1998).

12. D. C. Chan and P. S. Kim, *Cell*, **93**, 681–684 (1998).

13. E. A. Berger, P. M. Murphy, and J. M. Farber, *Annu. Rev. Immunol.*, **17**, 657–700 (1999).

14. H. Choe, M. Farzan, Y. Sun, et al., *Cell*, **85**, 1135–1148 (1996).

15. B. J. Doranz, J. Rucker, Y. Yi, et al., *Cell*, **85**, 1149–1158 (1996).

16. O. J. Sattentau and J. P. Moore, *J. Exp. Med.*, **174**, 407–415 (1991).

17. D. J. Capon and R. H. Ward, *Annu. Rev. Immunol.*, **9**, 649–678 (1991).

18. F. Wong Staal and R. C. Gallo, *Nature*, **317**, 395 (1985).

19. C. M. Shaw, B. H. Hahn, S. K. Ayra, J. E. Groopman, R. C. Gallo, and F. Wong Staal, *Science*, **226**, 1165 (1984).

20. T. Fujiwara and K. Mizuuchi, *Cell*, **54**, 497–504 (1988).

21. P. O. Brown, B. Bowerman, H. E. Varmus, et al., *Proc. Natl. Acad. Sci. USA*, **86**, 2525–2529 (1989).

22. P. Lewis, M. Hensel, and M. Emerman, *EMBO J.*, **11**, 3053–3058 (1992).

23. H. Sakai, M. Kawamura, J. Sakuragi, et al., *J. Virol.*, **67**, 1169–1174 (1993).

24. C. A. Roses, I. G. Sodroski, W. C. Goh, A. L. Dayton, I. Lippkv, and W. A. Ilasettine, *Nature*, **319**, 555 (1986).

25. I. Sodroski, W. C. Got, C. Rosen, A. Dayton, F. Terwilliger, and W. Haseltine, *Nature*, **321**, 412 (1986).

26. I. Katoh and S. P. Goff, *J. Virol.*, **53**, 899–907 (1985).

27. N. E. Kohl, E. A. Emni, and W. A. Schleif, *Proc. Natl. Acad. Sci. USA*, **85**, 4686–4690 (1988).

28. J. J. Lipsky, *Lancet*, **348**, 800–803 (1996).

29. R. A. Wetss, P. R. Clapham, R. Cheingsong Popov, A. G. DalgLeish, C. A. Carne, I. V. D. Weller, and R. S. Tedder, *Nature*, **316**, 69 (1985).

30. M. Rohert Guruff, M. Brown, and R. C. Gallo, *Nature*, **316**, 72 (1985).

31. B. J. Doranz, *Emerg. Ther. Targets*, **4**, 423–437 (2000).

32. E. De Clercq, *Curr. Med. Chem.*, **8**, 1543–1572 (2001).

33. M. D. Miller and D. J. Hazuda, *Curr. Opin. Microbiol.*, **4**, 535–539 (2001).

34. C. B. Pert, J. M. Hill, M. R. Ruff, R. M. Berman, W. G. Robey, L. D. Arthur, F. W. Ruscetti, and W. L. Farrar, *Proc. Natl. Acad. Sci. USA*, **83**, 9254 (1986).

35. K. Ueno and S. Kuno, *Lancet*, **1**, 1379 (1987).

36. M. Ito, M. Baba, A. Sato, R. Pauwels, E. De Clercq, and S. Shigeta, *Antiviral Res.*, **7**, 361 (1987).

37. R. T. Schooley, T. C. Merigan, P. Gaut, M. S. Hirsch, M. Holodniy, T. Flynn, S. Liu, R. E. Byington, S. Henochowicz, E. Gubish, et al., *Ann. Intern. Med.*, **112**, 247–253 (1990).

38. J. O. Kahn, J. D. Allan, T. L. Hodges, L. D. Kaplan, C. J. Arri, H. F. Fitch, A. E. Izu, J. Mordenti, J. E. Sherwin, J. E. Groopman, et al., *Ann. Intern. Med.*, **112**, 254–261 (1990).

39. E. S. Daar, X. L. Li, T. Moudgil, and D. D. Ho, *Proc. Natl. Acad. Sci. USA*, **87**, 6574–6578 (1987).

40. P. E. Kennedy, B. Moss, and E. A. Berger, *Virology*, **192**, 375–379 (1993).

41. T. W. Pitts, M. J. Bohanon, M. F. Leach, T. J. McQuade, C. K. Marschke, J. A. Merritt, W. Wierenga, and J. A. Nicholas, *AIDS Res. Hum. Retroviruses*, **7**, 741–750 (1991).

42. R. T. Davey, C. M. Boenning, B. R. Herpin, D. H. Batts, J. A. Metcalf, L. Wathen, S. R. Cox, M. A. Polis, J. A. Kovacs, J. Falloon, et al., *J. Infect. Dis.*, **170**, 1180–1188 (1994).

43. A. Trkola, A. B. Pomales, H. Yuan, B. Korber, P. J. Maddon, G. P. Allaway, H. Katinger, C. F. Barbas, D. R. Burton, D. D. Ho, et al., *J. Virol.*, **69**, 6609–6617 (1995).

44. M. C. Gauduin, G. P. Allaway, P. J. Maddon, C. F. Barbas, D. R. Burton, and R. A. Koup, *J. Virol.*, **70**, 2586–2592 (1996).

45. M. C. Gauduin, G. P. Allaway, W. C. Olson, R. Weir, P. J. Maddon, and R. A. Koup, *J. Virol.*, **72**, 3475–3478 (1998).

46. J. M. Jacobson, I. Lowy, C. V. Fletcher, T. J. O'Neill, D. N. Tran, T. J. Ketas, A. Trkola, M. E. Klotman, P. J. Maddon, W. C. Olson, and R. J. Israel, *J. Infect. Dis.*, **182**, 326–329 (2000).

47. E. De Clercq, N. Yamamoto, R. Pauwels, J. Balzarini, M. Witvrouw, K. De Vreese, Z. Debyser, B. Rosenwirth, P. Peichl, R. Datema, et al., *Antimicrob. Agents Chemother.*, **38**, 668–674 (1994).

48. D. Schols, J. A. Este, G. Henson, and E. De Clercq, *Antiviral Res.*, **35**, 147–156 (1997).

49. G. A. Donzella, D. Schols, S. W. Lin, J. A. Este, K. A. Nagashima, P. J. Maddon, G. P. Allaway, T. P. Sakmar, G. Henson, E. De Clercq, and J. P. Moore, *Nat. Med.*, **4**, 72–77 (1998).

50. R. Datema, L. Rabin, M. Hincenbergs, M. B. Moreno, S. Warren, V. Linquist, B. Rosenwirth, J. Seifert, and J. M. McCune, *Antimicrob. Agents Chemother.*, **40**, 750–754 (1996).

51. B. J. Doranz, K. Grovit-Ferbas, M. P. Sharron, S. H. Mao, M. B. Goetz, E. S. Daar, R. W. Doms, and W. A. O'Brien, *J. Exp. Med.*, **186**, 1395–1400 (1997).

52. T. Murakami, T. Nakajima, Y. Koyanagi, K. Tachibana, N. Fujii, H. Tamamura, N. Yoshida, M. Waki, A. Matsumoto, O. Yoshie, et al., *J. Exp. Med.*, **186**, 1389–1393 (1997).

53. H. Tamamura, Y. Xu, T. Hattori, X. Zhang, R. Arakaki, K. Kanbara, A. Omagari, A. Otaka, T. Ibuka, N. Yamamoto, et al., *Biochem. Biophys. Res. Commun.*, **253**, 877–882 (1998).

54. Q. Ma, D. Jones, P. R. Borghesani, R. A. Segal, T. Nagasawa, T. Kishimoto, R. T. Bronson, and T. A. Springer, *Proc. Natl. Acad. Sci. USA*, **95**, 9448–9453 (1998).

55. Q. Ma, D. Jones, and T. A. Springer, *Immunity*, **10**, 463–471 (1999).

56. K. Kawabata, M. Ujikawa, T. Egawa, H. Kawamoto, K. Tachibana, H. Iizasa, Y. Katsura, T. Kishimoto, and T. Nagasawa, *Proc. Natl. Acad. Sci. USA*, **96**, 5663–5667 (1999).

57. A. D. Luster, *N. Engl. J. Med.*, **338**, 436–445 (1998).

58. M. Baba, O. Nishimura, N. Kanzaki, M. Okamoto, H. Sawada, Y. Iizawa, M. Shiraishi, Y. Aramaki, K. Okonogi, Y. Ogawa, et al., *Proc. Natl. Acad. Sci. USA*, **96**, 5698–5703 (1999).

59. M. Shiraishi, Y. Aramaki, M. Seto, H. Imoto, Y. Nishikawa, N. Kanzaki, M. Okamoto, H. Sawada, O. Nishimura, M. Baba, and M. Fujino, *J. Med. Chem.*, **43**, 2049–2063 (2000).

60. P. E. Finke, L. C. Meurer, B. Oates, S. G. Mills, M. MacCoss, L. Malkowitz, M. S. Springer, B. L. Daugherty, S. L. Gould, J. A. DeMartino, et al., *Bioorg. Med. Chem. Lett.*, **11**, 265–270 (2001).

61. C. P. Dorn, P. E. Finke, B. Oates, R. J. Budhu, S. G. Mills, M. MacCoss, L. Malkowitz, M. S. Springer, B. L. Daugherty, S. L. Gould, et al., *Bioorg. Med. Chem. Lett.*, **11**, 259–264 (2001).

62. J. J. Hale, R. J. Budhu, S. G. Mills, M. MacCoss, L. Malkowitz, S. Siciliano, S. L. Gould, J. A. DeMartino, and M. S. Springer, *Bioorg. Med. Chem. Lett.*, **11**, 1437–1440 (2001).

63. A. Trkola, T. J. Ketas, K. A. Nagashima, L. Zhao, T. Cilliers, L. Morris, J. P. Moore, P. J. Maddon, and W. C. Olson, *J. Virol.*, **75**, 579–588 (2001).

64. P. Menten, S. Struyf, E. Schutyser, A. Wuyts, E. DeClercq, D. Schols, P. Proost, and J. Damme, *J. Clin. Invest.*, **104**, R1–R5 (1999).

65. R. J. B. Nibbs, J. Yang, N. R. Landau, J. H. Mao, and G. J. Graham, *J. Biol. Chem.*, **274**, 17478–17483 (1999).

66. J. G. Sodroski, *Cell*, **99**, 243–246 (1999).

67. C. Wild, T. Oas, C. McDanal, D. Bolognesi, and T. Matthews, *Proc. Natl. Acad. Sci. USA*, **89**, 10537–10541 (1992).

68. C. T. Wild, D. C. Shugars, T. K. Greenwell, C. B. McDanal, and T. J. Matthews, *Proc. Natl. Acad. Sci. USA*, **91**, 9770–9774 (1994).

69. J. M. Kilby, S. Hopkins, T. M. Venetta, B. DiMassimo, G. A. Cloud, J. Y. Lee, L. Alldredge, E. Hunter, D. Lambert, D. Bolognesi, et al., *Nat. Med.*, **4**, 1302–1307 (1998).

70. S. Chokeijchai, E. Kojiama, S. Anderson, M. Nomizu, M. Tanaka, M. Machida, T. Date, K. Toyota, S. Ishida, K. Watanabe, H. Yoshioka, P. Roller, K. Murakami, and H. Mitsuya, *Antimicrob. Agents Chemother.*, **39**, 2345–2347 (1995).

71. H. Nakashima, K. Ichiyama, K. Inazawa, M. Ito, H. Hayashi, Y. Nishiara, E. Tsujii, and T. Kino, *Biol. Pharm. Bull.*, **19**, 405–412 (1996).

72. J. F. Mayaux, A. Bousseau, R. Pauwels, T. Huet, Y. Hénin, N. Dereu, M. Evers, F. Soler, C. Poujade, E. De Clercq, and J. B. Le Pecq, *Proc. Natl. Acad. Sci. USA*, **91**, 3564–3568 (1994).

73. S. Hang, Q. Zhao, and A. K. Debranth, *Curr. Pharm. Des.*, **8**, 563–580 (2002).

74. W. G. Rice, J. A. Turpin, L. O Arthur, and L. E. Henderson, *Int. Antiviral News*, **3**, 87–89 (1995).

75. L. Dolin, *Science*, **227**, 1296 (1985).

76. W. G. Rice, C. A. Schaeffer, B. Harten, F. Villinger, T. L. South, M. F. Summers, L. E. Henderson, J. W. Bess Jr., L. O. Arthur, J. S. McDougal, S. L. Orloff, J. Mendeleyev, and E. Kun, *Nature*, **361**, 473–475 (1993).

77. W. G. Rice, C. A. Schaeffer, L. Graham, M. Bu, J. S. McDougal, S. L. Orloff, F. Villinger, M. Young, S. Oroszlan, M. R. Fesen, Y. Pommier, J. Mendeleyev, and E. Kun, *Proc. Natl. Acad. Sci. USA*, **90**, 9721–9724 (1993).

78. W. G. Rice, J. G. Supko, L. Malspeis, R. W. Buckheit Jr., D. Clanton, M. Bu, L. Graham, C. A. Schaeffer, J. A. Turpin, J. Domagala, R. Gogliotti, J. P. Bader, S. M. Halliday, L. Coren, R. C. Sowder II, L. O. Arthur, and L. E. Henderson, *Science*, **270**, 1194–1197 (1995).

79. J. A. Turpin, S. J. Terpening, C. A. Schaeffer, G. Yu, G. J. Glover, R. L. Felsted, E. A. Sausville, and W. G. Rice, *J. Virol.*, **70**, 6180–6189 (1996).

80. M. Witvrouw, J. Balzarini, C. Pannecouque, S. Jhaumeer-Laulloo, J. A. Esté, D. Schols, P. Cherepanov, J. C. Schmit, Z. Debyser, A. M. Vandamme, J. Desmyter, S. R. Ramadas, and E. De Clercq, *Antimicrob. Agents Chemother.*, **41**, 262–268 (1997).

81. W. G. Rice, D. C. Baker, C. A. Schaeffer, L. Graham, M. Bu, S. Terpening, D. Clanton, R.

Schultz, J. P. Bader, R. W. Buckheit Jr., L. Field, P. K. Singh, and J. A. Turpin, *Antimicrob. Agents Chemother.*, **41**, 419–426 (1997).

82. M. Vandevelde, M. Witvrouw, J. C. Schmit, S. Sprecher, E. De Clercq, and J. P. Tassignon, *AIDS Res. Hum. Retroviruses*, **12**, 567–568 (1996).

83. W. G. Rice, J. A. Turpin, M. Huang, D. Clanton, R. W. Buckheit Jr., D. G. Covell, A. Wallqvist, N. B. McDonnell, R. N. DeGuzman, M. F. Summers, L. Zalkow, J. P. Bader, R. D. Haugwitz, and E. A. Sausville, *Nat. Med.*, **3**, 341–345 (1997).

84. F. D. Veronese, et al., *Science*, **231**, 1289 (1986).

85. E. Arnold, A. Jacobo-Molina, R. G. Nanni, R. L. Williams, X. Lu, J. Ding, A. D. Clark Jr., A. Zhang, A. L. Ferris, P. Clark, et al., *Nature*, **357**, 85–89 (1992).

86. L. A. Kohlstaed, J. Wang, J. M. Friedman, et al., *Science*, **256**, 1783–1790 (1992).

87. A. Jacobo-Molina, J. Ding, R. G. Nanni, A. D. Clark Jr., X. Lu, C. Tantillo, R. L. Williams, G. Kamer, A. L. Ferris, P. Clark, et al., *Proc. Natl. Acad. Sci. USA*, **90**, 6320–6324 (1993).

88. T. Unge, S. Knight, R. Bhikhabhai, S. Lovgren, Z. Dauter, K. Wilson, and B. Strandberg, *Structure*, **2**, 953–961 (1994).

89. V. K. Pathak and H. M. Temin, *Proc. Natl. Acad. Sci. USA*, **87**, 6019–6023 (1990).

90. V. K. Pathak and H. M. Temin, *Proc. Natl. Acad. Sci. USA*, **87**, 6024–6028 (1990).

91. H. Mitsuya, K. J. Weinhold, P. A. Furman, M. H. St Clair, S. Nusinofff-Lehrman, R. C. Gallo, D. Bolognesi, D. W. Barry, and S. Broder, *Proc. Natl. Acad. Sci. USA*, **82**, 7096–7100 (1985).

92. P. Herdewijn, J. Balzarini, E. De Clercq, R. Pauwels, M. Baba, S. Broder, and H. Vanderhaeghe, *J. Med. Chem.*, **30**, 1270–1278 (1987).

93. R. Yarchoan, C. Berg, P. Brouwers, M. A. Fischl, A. R. Spitzer, A. Wichman, J. Grafman, R. V. Thomas, B. Safai, A. Brunetti, C. F. Perno, P. J. Schmidt, S. M. Larson, C. E. Myers, and S. Broder, *Lancet*, **1**, 132 (1987).

94. A. Fridland, *Curr. Opin. Anti-Infect. Invest. Drugs*, **2**, 295 (2000).

95. H. M. ElKouni, *Curr. Pharm. Des.*, **8**, 581–593 (2002).

96. R. Esnouf, J. Ren, C. Ross, et al., *Nat. Struct. Biol.*, **2**, 303–308 (1995).

97. R. A. Spence, W. M. Kati, K. S. Anderson, and K. A. Johnson, *Science*, **267**, 988–993 (1995).

98. E. De Clercq, *Antiviral Res.*, **38**, 153–179 (1998).

99. G. Campiani, A. Ramumo, G. Maga, V. Nacci, C. Fattorusso, B. Galantotti, E. Morelli, and E. Novellino, *Curr. Pharm. Des.*, **8**, 615–657 (2002).

100. M. S. Johnson, M. A. McClure, D. F. Feng, J. Grey, and R. F. Doolittle, *Proc. Natl. Acad. Sci. USA*, **83**, 7648 (1986).

101. M. A. Parniak, D. Arion, and N. Sluis-Cremer, *Antiviral Ther.*, **5** (Suppl. 3), 3, Abstr. 2 (2000).

102. S. P. Goff, *Annu. Rev. Genet.*, **26**, 527–544 (1992).

103. A. Engelman, K. Mizuuchi, and R. Craigie, *Cell*, **62**, 829–837 (1990).

104. F. Dyda, A. B. Hickman, T. M. Jenkins, et al., *Science*, **266**, 1981–1986 (1994).

105. P. J. Lodi, J. A. Ernst, J. Kuszewski, et al., *Biochemistry*, **34**, 9826–9833 (1995).

106. K. Pfeifer, H. Merz, R. Steffen, W. E. G. Muller, S. Trumm, J. Schulz, E. Eich, and H. C. Schroder, *J. Pharm. Med.*, **2**, 75–97 (1992).

107. Y. Pommier, A. A. Pilon, K. Bajaj, A. Mazumder, and N. Neamati, *Antiviral Chem. Chemother.*, **8**, 463–483 (1997).

108. L. Tarrago-Litvak, M. L. Andreola, M. Fourneir, G. A. Nevinsky, V. Parissi, V. R. de Soultrait, and S. Litvak, *Curr. Pharm. Des.*, **8**, 595–614 (2002).

109. M. Baba, M. Okamoto, M. Makino, Y. Kimura, T. Ikeuchi, T. Sakaguchi, and T. Okamoto, *Antimicrob. Agents Chemother.*, **41**, 1250–1255 (1997).

110. J. A. Turpin, R. W. Buckheit Jr., D. Derse, M. Hollingshead, K. Williamson, C. Palamone, M. C. Osterling, S. A. Hill, L. Graham, C. A. Schaeffer, M. Bu, M. Huang, W. M. Cholody, C. Michejda, and W. G. Rice, *Antimicrob. Agents Chemother.*, **42**, 487–494 (1998).

111. J. Su, A. Palm, Y. Wu, S. Sandin, S. Hoglund, and A. Vahlne, *AIDS Res. Hum. Retroviruses*, **16**, 37–48 (2000).

112. J. Su, M. H. Naghavi, A. Jejcic, P. Horal, Y. Furuta, Y. P. Wu, S. L. Li, W. W. Hall, L. Goobar-Larsson, B. Svennerholm, and A. Vahlne, *J. Hum. Virol.*, **4**, 8–15 (2001).

113. C. Flexner, *N. Engl. J. Med.*, **338**, 1281–1292 (1998).

114. M. A. Navia, P. M. Fitzgerald, B. M. McKeever, et al., *Nature*, **337**, 615–620 (1989).

115. R. Lapatto, T. Blundell, A. Hemmings, et al., *Nature*, **342**, 299–302 (1989).

116. M. Miller, J. Schneider, B. K. Sathyanarayana, et al., *Science*, **246**, 1149–1152 (1989).

117. W. D. Nortbeck and D. J. Kempf, Annual Reports in Medicinal Chemistry: HIV Protease Inhibitors, Vol. **26**, Chap. 15, Academic Press, San Diego, CA, 1990, pp. 141–150.

118. S. Ren and E. J. Lien, *Prog. Drug Res.*, **151**, 1–31 (1998).

119. A. Molla, G. R. Granneman, E. Sun, and D. Kempf, *J. Antiviral Res.*, **39**, 1–23 (1998).

120. M. Hikaru, M. Takashi, N. Shingo, M. Takatoshi, K. Tooru, H. Yoshio, and K. Yoshilaki, *Bioorg. Med. Chem.*, **9**, 417–430 (2001).

121. D. R. March, G. Abbenante, D. Bergman, R. Brinkworth, W. Wickramsinghe, et al., *J. Am. Chem. Soc.*, **118**, 3375–3379 (1996).

122. P. Rezacova, J. Lescar, J. Brynda, M. Farby, M. Horejisi, J. Sedlacek, and G. A. Bentley, *Structure*, **9**, 887–895 (2001).

123. P. Y. S. Lam, Y. Ru, P. K. Jadhav, P. E. Aldrich, G. V. DeLucca, C. J. Eyermann, C. H. Chang, G. Emmett, E. R. Holler, W. F. Daneker, L. Li, P. N. Confalone, R. J. McHugh, Q. Han, R. Li, J. A. Markwalder, S. P. Seitz, T. R. Sharpe, L. T. Bacheler, M. M. Rayner, R. M. Klabe, L. Shum, D. L. Winslow, D. M. Kornhauser, D. A. Jackson, S. Erickson-Viitanen, and C. Hodge, *J. Med. Chem.*, **39**, 3514–3525 (1996).

124. G. Bold, A. Fässler, H. G. Capraro, R. Cozens, T. Klimkait, J. Lazdins, J. Mestan, B. Poncioni, J. Rösel, D. Stover, M. Tintelnot-Blomley, F. Acemoglu, W. Beck, E. Boss, M. Eschbach, T. Hürlimann, E. Masso, S. Roussel, K. Ucci-Stoll, D. Wyss, and M. Lang, *J. Med. Chem.*, **41**, 3387–3401 (1998).

125. R. M. W. Hoetelmans, *Curr. Opin. Anti-Infect. Invest. Drugs*, **1**, 241–245 (1999).

126. M. Alterman, M. Björsne, A. Mülman, B. Classon, I. Kvarnström, H. Danielson, P. O. Markgren, U. Nillroth, T. Unge, A. Hallberg, and B. Samuelsson, *J. Med. Chem.*, **41**, 3782–3792 (1998).

127. F. Lebon and M. Ledecq, *Curr. Med. Chem.*, **7**, 455–477 (2001).

128. S. Hanessian, N. Moitessier, and E. J. Therrien, *J. Comput.-Aided Mol. Des.*, **15**, 873–881 (2001).

129. V. Zoete, O. Michielin, and M. Karplus, *J. Mol. Biol.*, **315**, 21–52 (2002).

130. Y. Z. Chen, X. L. Gu, and W. Cao, *J. Mol. Graph. Modell.*, **19**, 560–570 (2001).

131. S. Wang, G. W. Milne, X. Yan, I. Posey, M. C. Nicklaus, et al., *J. Med. Chem.*, **39**, 2047–2054 (1996).

132. M. D. Varney, K. Appelt, V. Kalish, M. R. Reddy, J. Tatlock, et al., *J. Med. Chem.*, **37**, 2274–2284 (1994).

133. P. R. Harrigan, C. Stone, P. Griffin, et al., *J. Infect. Dis.*, **181**, 912–920 (2000).

134. D. J. Hooker, G. Tachedjian, A. E. Solomon, et al., *J. Virol.*, **70**, 8010–8018 (1996).

135. P. Kellam, C. A. Boucher, and B. A. Larder, *Proc. Natl. Acad. Sci. USA*, **89**, 1934–1938 (1992).

136. E. De Clercq, *Biochem. Pharmacol.*, **47**, 155–169 (1994).

137. C. Tantillo, J. Ding, A. Jacobo-Molina, R. G. Nanni, P. L. Boyer, S. H. Hughes, R. Pauwels, K. Andries, P. A. Janssen, and E. Arnold, *J. Mol. Biol.*, **243**, 369–387 (1994).

138. A. M. Vandamme, K. V. Vaerenbergh, and E. De Clercq, *Antiviral Chem. Chemother.*, **9**, 187–203 (1998).

139. E. De Clercq, *J. Med. Chem.*, **38**, 2491–2517 (1995).

140. E. De Clercq, *Med. Res. Rev.*, **16**, 125–157 (1996).

141. J. Balzarini, A. Karlsson, V. V. Sardana, E. A. Emini, M. J. Camarasa, and E. De Clercq, *Proc. Natl. Acad. Sci. USA*, **91**, 6599–6603 (1994).

142. R. Pauwels, M. P. de Bethune, K. Andries, P. Stoffels, et al., *Antiviral Res.*, **23**, 109 (1994).

143. J. Balzarini, M. Karlsson, M. J. Perez, M. J. Camarasa, and E. De Clercq, *Virology*, **196**, 157 (1993).

144. B. A. Larder, *Antimicrob. Agents Chemother.*, **36**, 4713–4717 (1992).

145. T. W. Chun, A. S. Fauci, L. Stuyer, S. B. Mizell, L. A. Ehler, J. M. Mican, M. Baseler, A. L. Lloyd, and M. A. Nowak, *Proc. Natl. Acad. Sci. USA*, **94**, 13193–13197 (1997).

146. H. Matsumoto, T. Kimura, T. Hamawaki, A. Kumaqai, T. Goto, H. Sano, and Y. Kiso, *Bioorg. Med. Chem.*, **9**, 1589–1600 (2001).

147. T. Kimura, H. Hikaru, T. Matsuda, T. Hamawaki, K. Akaji, and Y. Kiso, *Bioorg. Med. Chem. Lett.*, **9**, 803–806 (1999).

148. H. Matsumoto, T. Hamawaki, H. Ota, T. Kimura, T. Goto, K. Sano, Y. Hayashi, and Y. Kiso, *Bioorg. Med. Chem. Lett.*, **10**, 1227–1232 (2000).

CHAPTER TWELVE

# Organ Transplant Drugs

BIJOY KUNDU
Medicinal Chemistry Division
Central Drug Research Institute
Lucknow, India

**Contents**

1 Introduction, 486
    1.1 Role of T-Cells in Organ Rejection, 487
    1.2 Classification of Organ Transplant
        Rejections, 487
    1.3 Current Trend in the Management of Organ
        Transplantation, 487
2 Clinical Use of Agents, 489
    2.1 Current Drugs on the Market, 489
    2.2 Agents that Specifically Block T-Cell
        Function, 489
        2.2.1 Cyclosporine, 489
            2.2.1.1 History, 489
            2.2.1.2 Chemical Structure, 491
            2.2.1.3 Pharmacokinetics, 491
            2.2.1.4 Pharmacology, 492
            2.2.1.5 Structure-Activity Relationship,
                494
            2.2.1.6 Side Effects, 494
        2.2.2 FK 506, 494
            2.2.2.1 History, 494
            2.2.2.2 Chemical Structure, 495
            2.2.2.3 Pharmacokinetics, 495
            2.2.2.4 Pharmacology, 495
            2.2.2.5 Structure-Activity Relationship,
                499
            2.2.2.6 Side Effects, 500
        2.2.3 Sirolimus, 500
            2.2.3.1 History, 500
            2.2.3.2 Chemical Structure, 500
            2.2.3.3 Pharmacokinetics, 500
            2.2.3.4 Pharmacology, 500
            2.2.3.5 Structure-Activity Relationship,
                503
            2.2.3.6 Side Effects, 503
        2.2.4 15-Deoxyspergualin, 504
            2.2.4.1 History, 504
            2.2.4.2 Chemical Structure, 504
            2.2.4.3 Pharmacokinetics, 504
            2.2.4.4 Pharmacology, 504

*Burger's Medicinal Chemistry and Drug Discovery*
Sixth Edition, Volume 5: Chemotherapeutic Agents
Edited by Donald J. Abraham
ISBN 0-471-37031-2    © 2003 John Wiley & Sons, Inc.

2.2.4.5 Structure-Activity Relationship,
            505
     2.2.4.6 Side Effects, 506
 2.2.5 FTY 720, 506
     2.2.5.1 History, 506
     2.2.5.2 Chemical Structure, 507
     2.2.5.3 Pharmacokinetics, 507
     2.2.5.4 Pharmacology, 507
     2.2.5.5 Structure-Activity Relationship,
            508
     2.2.5.6 Side Effects, 509
 2.2.6 Monoclonal Antibodies, 509
     2.2.6.1 History, 509
     2.2.6.2 Pharmacokinetics, 510
     2.2.6.3 Pharmacology, 510
     2.2.6.4 Side Effects, 513
2.3 Agents that Block Nucleotide Synthesis, 513
 2.3.1 Azathioprine, 513
     2.3.1.1 History, 513
     2.3.1.2 Chemical Structure, 513
     2.3.1.3 Pharmacokinetics, 513
     2.3.1.4 Pharmacology, 513
     2.3.1.5 Structure-Activity Relationship,
            514
     2.3.1.6 Side Effects, 516
 2.3.2 Brequinar, 516
     2.3.2.1 History, 516
     2.3.2.2 Chemical Structure, 516
     2.3.2.3 Pharmacokinetics, 516
     2.3.2.4 Pharmacology, 516
     2.3.2.5 Structure-Activity Relationship,
            518
     2.3.2.6 Side Effects, 518

 2.3.3 Mycophenolate Mofetil, 518
     2.3.3.1 History, 518
     2.3.3.2 Chemical Structure, 519
     2.3.3.3 Pharmacokinetics, 519
     2.3.3.4 Pharmacology, 519
     2.3.3.5 Structure-Activity Relationship,
            520
     2.3.3.6 Side Effects, 521
 2.3.4 Mizoribine, 521
     2.3.4.1 History, 521
     2.3.4.2 Chemical Structure, 521
     2.3.4.3 Pharmacokinetics, 521
     2.3.4.4 Pharmacology, 521
     2.3.4.5 Structure-Activity Relationship,
            522
     2.3.4.6 Side Effects, 522
 2.3.5 Leflunomide, 522
     2.3.5.1 History, 522
     2.3.5.2 Chemical Structure, 522
     2.3.5.3 Pharmacokinetics, 522
     2.3.5.4 Pharmacology, 522
     2.3.5.5 Structure-Activity Relationship,
            523
     2.3.5.6 Side Effects, 523
3 Recent Developments, 523
 3.1 Stem Cells, 523
 3.2 Donor Bone Marrow as Antirejection
     Therapy, 524
 3.3 Antisense Technology, 524
 3.4 Molecules Capable of Inhibiting
     Costimulatory Signals, 525
4 Things to Come, 526
5 Web Site Addresses for Further Information, 526

# 1   INTRODUCTION

The immune response protects the body from potentially harmful substances (antigens) such as microorganisms, toxins, and cancer cells. The immune system distinguishes "self" from "foreign" by reacting to proteins on the surfaces of cells. The presence of foreign blood or tissue in the body triggers an immune response that can result in blood transfusion reactions and transplant rejection when antibodies are formed against foreign antigens on the transplanted or transfused material.

No two people (except identical twins) have identical tissue antigens. Therefore, in the absence of immunosuppressive drugs, organ and tissue transplantation would almost always cause an immune response against the foreign tissue (rejection), which would result in de-

struction of the transplant. Though tissue typing ensures that the organ or tissue is as similar as possible to the tissues of the recipient, unless the donor is an identical twin, no match is perfect and the possibility of organ/tissue rejection remains. Immunosuppressive therapy is used to prevent organ rejection.

Blocking the activated cells, which attack a transplanted organ without shutting down the rest of the immune system, allows the body to learn to tolerate the transplant and still defend itself against germs. Thus, the holy grail of transplantation biology is how to prevent organ-specific graft rejection in transplant recipients without compromising their ability to fight infections. Currently, transplant patients face a life-long battle to control their immune system; this requires that they use strong immunosuppressive drugs that may leave them unable to fight off infection by viruses or bacteria.

Historically, organ transplantation involving the kidney had been performed sporadically during the first half of the century (1, 2); however, planned programs for human organ transplantation started only in the late 1940s. The organs that are most commonly transplanted currently are kidney, liver, heart, lung, intestine, and pancreas.

## 1.1 Role of T-Cells in Organ Rejection

In experimental animals with congenital or induced T-cell deficiency, organ or tissue grafts survive indefinitely, indicating that T-cells play a central role in the specific immune response of allograft rejection. When a graft is introduced, it is accompanied by rejection because T-lymphocytes of the donor recognize the T-lymphocytes from the recipient as "foreign" bodies and react. Several studies have demonstrated that binding of alloantigen presented in the context of the major histocompatibility complex molecule (MHC) to the T cell receptor (TCR) on the surface of T-lymphocytes through direct and indirect recognition pathways (3) is the starting signal for T-cell activation (Fig. 12.1). Engagement of TCR with alloantigens recruits and activates a series of tyrosine kinases, including p56lck, p59fyn, and ZAP-70, followed by phosphorylation and activation of phospholipase C, and ultimately a rise in intracellular calcium (4). This leads to the activation of calcineurin enzyme, a serine-threonine phosphatase, that transduces signals to the nucleus to transcribe genes encoding cytokines relevant for the transition of the T-cells from the resting to the activated state.

Many studies have found that TCR stimulation alone is not enough to sustain full activation and that additional signals are required (5). Thus, antigen-presenting cells (APC) are able to deliver costimulatory signals to T-cells that, through independent intracellular pathways, synergize with TCR stimulation, leading to cytokine production and T-cell activation (6). The best-characterized costimulatory signal is that driven by CD28, expressed as a homodimer on the surface of T-cells (7). Engagement of TCR and activation of costimulatory signals allow transcription of genes for cytokines that induce T-cells to move from the resting G0 to the activated G1 state. G1 to S phase is regulated through a variety of receptors present on the T-cell surface, including IL-2 receptor, all acting on common intracellular elements in a pathway that controls key enzymes for the induction of cell division, such as cyclin/cyclin-dependent kinases. In this pathway *de novo* purine and pyrimidine synthesis is needed before lymphocytes can complete cell division. Thus, strategies to prevent T-cell activation or effector function can be potentially useful for antirejection therapy.

## 1.2 Classification of Organ Transplant Rejections

Clinically, organ transplant rejections can be classified into three categories and their management varies in accordance with the classification. Hyperacute rejection usually occurs within hours and is caused by antibodies already present in the recipient. Acute rejection is primarily T-cell based and can be controlled in most patients by the use of immunosuppressants. Chronic rejection is probably T-cell based and appears to be related to poor control of drug therapy, and usually requires another transplant. It is a slow, progressive process that usually begins inside the transplant organ's blood vessels, which are lined by donor cells that interact with host white blood cells in the bloodstream. Over time, as a result of inflammation and rejection reactions, scar tissue can accumulate inside these vessels, thus reducing or preventing blood flow into the filter and chemical plant portions of the kidney. If blockages become widespread, the organ becomes compromised because of the lack of oxygen and nutrients.

## 1.3 Current Trend in the Management of Organ Transplantation

In recent years, a greater understanding of the immunobiology of graft rejection, coupled with progressive improvements in the surgical and medical management, has revolutionized the field of clinical transplantation. Newer immunosuppressive schedules continue to be designed, not only to reduce the risk of rejection, but also to obtain a better therapeutic index that allows a further improvement of the graft

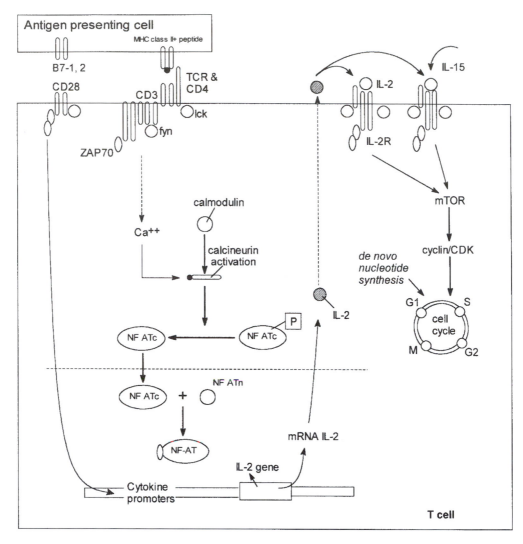

**Figure 12.1.** Principal pathways of T-cell activation that follow engagement of the alloantigen, presented by an APC in the context of an MHC molecule, with TCR on the surface of T-lymphocytes. APC, antigen-presenting cell; CD, cluster determinant; G1/G2, G1/G2 phases of cell cycle; Fyn, tyrosine kinase; lck, tyrosine kinase; IL, interleukin; IL-2R, interleukin receptor; M, M phase of cell cycle; MHC, major histocompatibility complex; mTOR, target for sirolimus protein; NF-AT, nuclear factor activating T-cell; NF-Atc, cytoplasmic subunit of NF-AT; NF-Atn, nuclear subunit of NF-AT; P, phosphate; S, S phase of cell cycle; TCR, T-cell receptor; ZAP 70, tyrosine kinase.

survival while minimizing the morbidity and drug-related toxicity. Presently, the optimal immunosuppressive strategy uses combinations of agents that act in synergistic fashion to provide the potency, freedom from toxic reactions, convenience of administration, and cost appropriate for the individual patient. This in turn has resulted in the significant in-

crease in the overall survival rates for transplant recipients, which is at an all-time high. Kidney recipients lead the 1-year patient survival rate with 94.3% followed by pancreas recipients at 91.1% (8). However, along with the success of organ transplantation, also comes the need for organs, which has grown at a much faster rate than the number of donated organs. In the United States alone the wait list is now more than 75,000, with approximately 5000 dying every year waiting for the organ transplantation (9). Nevertheless, an increase in the number of patients with end-stage kidney disease has also been observed at the rate of 7–8% each year. Currently, 12,000–14,000 kidney transplants alone are performed annually (10). Overall, organ transplants have risen by 5.4% in 2000 compared to 1999, according to preliminary data on U.S. organ donors released by the U.S. Department of Health and Human Services' Health Resources and Services Administration and the United Network for Organ Sharing. Organ transplants in 2000 totaled 22,827, an increase of 1172 over that of 21,655 transplants that occurred in 1999 (11). Thus, although multiple thousands of lives are being saved by the use of various immunosuppressive regimens, serious complications still occur as a result of treatment. This necessitates the ongoing search for novel, clinically efficacious, and nontoxic organ transplant drugs.

During the past 50 years, a large number of organ transplant drugs have been described. They have been successfully used under both allograft and xenograft conditions. These therapeutic agents interfere at different stages of T-cell activation and proliferation and can be classified into two main groups:

1. Agents that specifically block T-cell function, such as cyclosporine, tacrolimus, sirolimus, 15-deoxyspergualin, FTY 720, and monoclonal antibodies.
2. Agents that block nucleotide synthesis, such as azathioprine, brequinar, mycophenolate mofetil, mizoribine, and leflunomide.

The oldest class of immunosuppressants is the corticosteroids, which are extremely potent inhibitors of inflammation but when used by themselves do not adequately prevent organ rejection. The purpose of this review is to give an account of the organ transplant drugs, which are clinically used for inhibiting the rejection of allografts. Xenotransplantation using pig's organs, however, is gaining momentum, although its use under clinical conditions

has yet to be established (12). Many new experimental prototypes being studied for their antirejection properties in laboratory animals and in preclinical trials have been reviewed earlier (13) and will not be discussed here.

## 2 CLINICAL USE OF AGENTS

### 2.1 Current Drugs on the Market

Although cyclosporine, FK 506, and azathioprine, just to name a few of the most commonly used agents, are clinically available for treating organ transplants, their long-term use is associated with a high incidence of clinical complications, such as nephrotoxicity, hepatotoxicity, neurotoxicity, and gastrotoxicity. There is thus an ongoing requirement to develop immunosuppressive drugs with novel modes of action, with an improved ratio of desired activity to toxic effects that not only is exhibited by clinically used agents but also has an ability to prevent acute rejection and improve long-term graft function. Since the appearance of azathioprine in the 1960s, a large number of agents with immunosuppressive activity have been reported in the literature. Some of them are undergoing clinical trials and may soon enter into routine clinical use, whereas other pharmacological agents are currently being evaluated in laboratory animals. These compounds interfere at different stages of T-cell activation and proliferation and can be identified as inhibitors of nucleotide synthesis, growth factor signal transduction, and differentiation. In the last 5 years several new immunosuppressive drugs have been introduced in the market, out of which many new agents offer better therapeutic indexes and show promise in the treatment of allograft transplantation. A variety of structurally diverse immunosuppressive drugs used currently are summarized in Table 12.1.

### 2.2 Agents that Specifically Block T-Cell Function

#### 2.2.1 Cyclosporine

*2.2.1.1 History.* Cyclosporine (Novartis, Basel, Switzerland) is a cyclic undecapeptide isolated from the extracts of the fungus *Tolypocladium inflatum* by Jean-Francois Borel,

**Table 12.1  Current Drugs on Market**

| Drug Name | Code Numbers/ Other Names | Commercial Name/ Launching Year | Company Name | Injectable Formulation | Oral Capsule/Formulation |
|---|---|---|---|---|---|
| Cyclosporine | Ciclosporine | Sandimmun, 1983 | Novartis, Switzerland | 50 mg/mL | Soft gelatin capsules 25, 50, and 100 mg; oral solution 100 mg/mL |
| | | Neoral, 1995 | Novartis, Switzerland | | Soft gelatin capsules in 25 and 100 mg strengths; oral solution available in 100 mg/mL alcohol concentration |
| Tacrolimus | FK 506 | Prograf, 1994 | Fujisawa, USA | 5 mg/mL | Capsules of 0.5, 1, or 5 mg |
| Sirolimus | Rapamycin | Rapamune, 1999 | Wyeth-Ayerst Pharmaceuticals, USA | | Oral solution with 1 mg/mL, concentration |
| Deoxyspergualin | NSC 356894 and Gusperimus | Spanidin, 1994 | Nippon kayaku, Japan | | Tablet 100 mg |
| Monoclonal antibodies | Muromonab CD3 | Orthoclone (OKT3), 1986 | | Solution 5 mg/5 mL | |
| | Basiliximab | Simulect, 1998 | Novartis, Switzerland | 20 mg lyophilized powder | |
| | Daclizumab | Zenepax 1997 | Hoffman Laroche, USA | 25 mg/5 mL | |
| Azathioprine | B-W 322 | Imuran, 1968 | Burroughs Wellcome Lab | 10 mg/mL | 50 mg tablets |
| Mycophenolate | RS 61443 | Cell Cept, 1995 | Roche, USA | 6 mg/mL | 500 mg, 250 mg tablets or liquid 200 mg/mL |

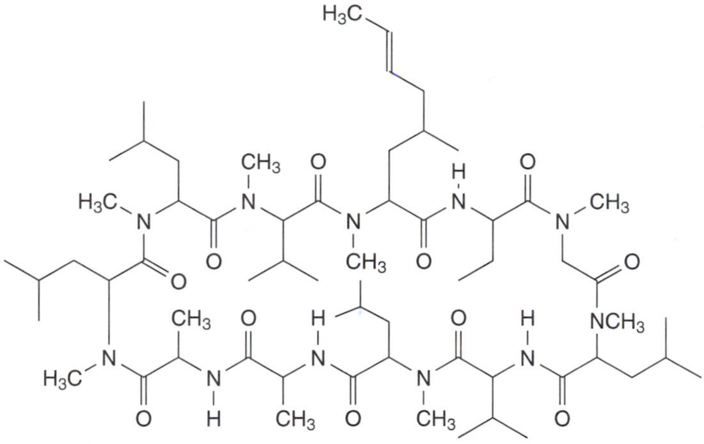

**Figure 12.2.** Structure of cyclosporine (**1**).

who worked in Switzerland for Novartis (formerly known as Sandoz) Pharmaceuticals. He discovered the immunosuppressant agent that ultimately moved transplantation from the realm of curiosity into routine therapy. Borel chose to examine a weak compound that was isolated from the soil fungus *Tolypocladium inflatum Gams* (subsequently renamed *Beauveria nivea*). The compound was thought to have little practical value, yet Novartis chemists continued to study and purify the compound because of its "interesting" chemical properties. Borel discovered that, unlike immunosuppressants that acted indiscriminately, this compound selectively suppressed the T-cells of the immune system (14). Excited by these characteristics, Borel continued his study and, in 1973, purified the compound called cyclosporine (also known as cyclosporin, ciclosporin). Cyclosporine was tested in humans for the first time in 1978 by Calne et al. (15) for kidney and by Powles et al. (16) for bone marrow. The results were startling: rejection was effectively inhibited in five of the seven patients receiving kidneys from mismatched deceased donors. Following extensive patient clinical trials, the U.S. Food and Drug Administration (U.S. FDA) granted clearance to Novartis to market cyclosporine under the brand name Sandimmune in November of 1983. This revolutionary therapy is indicated today for the prevention of organ rejection in kidney, liver, and heart transplant patients. Sandimmune (17) thus became the lifeline for thousands of transplant recipients

throughout the 1980s and into the next decade. However, the absorption of cyclosporine during chronic administration of Sandimmune soft gelatin capsules and oral solution was found to be erratic. Hence, continuous efforts were made to improve its bioavailability so as to minimize the toxicity and improve the risk-to-benefit ratio. Subsequently in 1995, a new oral formulation of cyclosporine in the form of cyclosporine microemulsion was introduced under the trade name Neoral by Novartis (18). With this formulation cyclosporine was found to disperse faster in the gut, thereby enhancing its absorption.

*2.2.1.2 Chemical Structure.* Chemically, cyclosporine (**1**) is a cyclic undecapeptide and designated as [R-[R*,R*-(E)]]-cyclic-(L-alanyl-D-alanyl-N-methyl-L-leucyl-N-methyl-L-leucyl-N-methyl-L-valyl-3-hydroxy-N,4-dimethyl-L-2-amino-6-octenoyl-L-aminobutyryl-N-methylglycyl-N-methyl-L-leucyl-L-valyl-N-methyl-L-leucyl). The structure is shown in Fig. 12.2.

*2.2.1.3 Pharmacokinetics.* Absorption of cyclosporine during chronic administration of Sandimmune soft gelatin capsules and oral solution was found to be erratic. However, Neoral was found to have increased bioavailability over that of sandimmune. The extent of absorption of cyclosporine was dependent on the individual patient, the patient population, and the formulation. The relationship between administered dose and exposure [area under the concentration vs. time curve (AUC)] was linear within the therapeutic dose range.

Intrasubject variability of AUC in renal transplant recipients was 9–21% for Neoral and 19–26% for Sandimmune. The disposition of cyclosporine from blood is generally biphasic, with a terminal half-life of approximately 8.4 h (range 5–18 h). After intravenous administration, the blood clearance of cyclosporine (assay: HPLC) was approximately 5–7 mL/min/kg in adult recipients of renal or liver allografts. After oral administration of Neoral, the time to peak blood cyclosporine concentrations ($T_{max}$) ranged from 1.5 to 2.0 h. Cyclosporine is distributed largely outside the blood volume. The steady-state volume of distribution during intravenous dosing has been reported as 3–5 L/kg in solid organ transplant recipients. In blood, the distribution is concentration dependent: approximately 33–47% is in plasma, 4–9% in lymphocytes, 5–12% in granulocytes, and 41–58% in erythrocytes. Cyclosporine is extensively metabolized by the cytochrome P450 III-A enzyme system in the liver, to a lesser degree in the gastrointestinal tract, and the kidney. At least 25 metabolites have been identified from human bile, feces, blood, and urine. The biological activity of the metabolites and their contributions to toxicity are considerably less than those of the parent compound. Blood cyclosporine clearance appears to be slightly slower in cardiac transplant patients [18]. Cyclosporine blood concentration needs to be monitored in transplant and rheumatoid arthritis patients taking Neoral, to avoid toxicity resulting from high concentrations. Dose adjustments also need to be made in transplant patients to minimize the possibility of organ rejection attributed to low concentration [18]. Only 0.1% of a cyclosporine dose is excreted unchanged in the urine. Elimination is primarily biliary, with only 6% of the dose (parent drug and metabolites) excreted in the urine [18].

### 2.2.1.4 Pharmacology.

It has been indicated in transplantation for the prophylaxis of organ rejection in kidney, liver, and heart transplants, and continues to be a cornerstone of immunosuppressive therapy. The immunosuppressive effect of cyclosporine depends on its inhibition of calcium-dependent T-cell activation [19]. After entering cells, cyclosporine binds to an intracellular receptor, cyclophilin [20]. The crystal structure of a complex between recombinant human cyclophilin A and CsA has been determined from a novel orthorhombic crystal form that contains only one monomer complex per asymmetric unit. The structure has been refined at 2.1-Å resolution to a crystallographic R-factor of 16.7%. The binding pocket of cyclophilin is a hydrophobic crevice defined by 13 residues that are within 4 Å of the bound CsA. In this high resolution structure, five direct hydrogen bonds and a network of water-mediated contacts stabilize the interactions between CsA and cyclophilin (Fig. 12.3). One of the water molecules plays a pivotal role and it is within a hydrogen bond distance of four atoms (MeBmt 1 O, W19, Gln63 NE2, and His54 NE2). The solvent molecule W19 is hydrogen bonded to this water molecule and to the carbonyl oxygen atom Asn71. The structure of the cyclophilin/CsA complex is consistent with a two-domain description of CsA. The cyclophilin binding domain composed of residues 1, 2, 3, 9, 10, and 11, and an exposed "effector domain" composed of residues 4, 5, 6, 7, and 8, which has higher flexibility and may be interacting with calcineurin. The accessible surface area of these five residues in this crystal complex is about 460 Å$^2$, compared with 230 Å$^2$ for the six residues making contact with cyclophilin. Many CsA analogs have shown a good agreement between cyclophilin binding and immunosuppressive activity based on the two-domain description of CsA. For example, substitution of MeVal[11] to MeAla[11] results in a 91.5% decrease in the immunosuppressive activity.

Once complexed with cyclophilin, the cyclosporine-cyclophilin complex inhibits the calcium- and calmodulin-dependent serine-threonine phosphatase activity of the enzyme calcineurin [21, 22], which in turn prevents the generation of the potent nuclear factor of activated T-cells. In fact, it acts in two different ways to inhibit the activation and proliferation of T-cytotoxic lymphocytes (white blood cells), the specific mediators of organ rejection. First, cyclosporine impedes the production and release of a protein IL-2 by T-helper white blood cells. Second, cyclosporine inhibits interleukin 2 receptor (IL-2R) expression on both T-helper and T-cytotoxic white blood

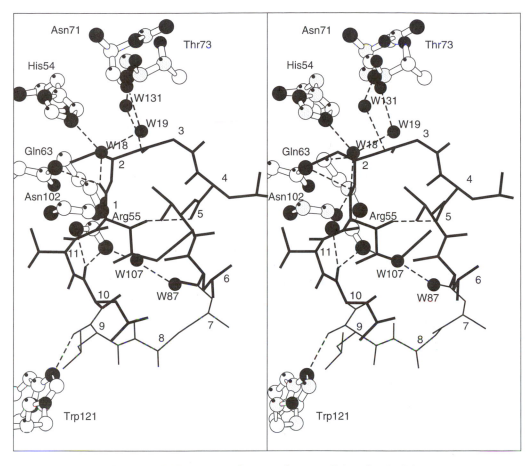

**Figure 12.3.** Stereoview of contacts between CsA and cyclophilin.

cells. These two actions effectively and selectively limit the differentiation and proliferation of T-cytotoxic white blood cells (22).

The superior selectivity and efficacy of cyclosporine compared with those of azathioprine were demonstrated by an increase in 1-year survival rates of renal allografts from cadaver donors to 80% for cyclosporine-based regimens and only 51% for azathioprine-based regimens (23). Furthermore, in contradistinction to the azathioprine-prednisone regimen, cyclosporine-prednisone therapy facilitated the successful clinical transplantation of hearts (24), livers (25), lungs (26), and combined heart and lung (24) allografts. Reports suggest that cyclosporine therapy may have an impact on chronic rejection and allograft coronary artery disease. The half-life for heart transplants improved from 5.4 years in the

1980–1985 period to 8.7 years in the 1986–1990 period (27). Similarly, an increase in renal allograft half-life from 16.6 to 23.04 years was observed when maintenance therapy was changed from azathioprine with high doses of steroids to cyclosporine and low dose steroids (28). In a prospective randomized study, the frequency of clinically defined chronic rejection after 4 years was 25% for the patients on azathioprine and steroids and only 9% for the patients on triple therapy including cyclosporine (29).

Cyclosporine dose and trough levels have been related to the development of chronic rejection in liver, kidney, and heart transplantation. Liver grafts in patients maintained on median cyclosporine levels (whole blood, trough level) of more than 175 $\mu$g/L in the first 28 days posttransplant had a significantly lower incidence of chronic rejection [2 of 49 versus 22 of 97 ($P = 0.002$)] (30). In a multivariate analysis of 587 kidney-alone transplants performed over a 5-year period, patients on less than 5 mg/kg of cyclosporine at 1 year were more prone to develop chronic rejection ($P = 0.007$) (31). Similarly, results from 225 heart transplant patients who had survived for 1 year after transplant demonstrated that actuarial 5-year survival in patients whose average cyclosporine dose was less than 3 mg/kg/day, was 60% compared to 77% in patients whose average dose was more than 3 mg/kg/day (32). Cyclosporine-based therapy significantly reduced patient morbidity and overcame at least some of the negative effects of the risk factors that were associated with azathioprine therapy. The risk factors included pretransplant blood transfusions, the possibility of retransplantation, and the requirement for matching of alleles of the human leukocyte antigen (HLA) system (33).

### 2.2.1.5 Structure-Activity Relationship.

Cyclosporine is the cornerstone of immunosuppressive therapy to prevent rejection of transplanted organs but it also causes kidney and liver damage. Thus there is a clear need for an alternative to cyclosporine with a reduced toxicity profile. This led researchers to carry out structure-activity relationship studies, to reduce or remove the toxic component. Studies with both naturally occurring modifications (34, 35) as well as with synthetically modified analogs of cyclosporine (36, 37) have been carried out. In the literature more than 3000 analogs of the drug have been synthesized, although none was found to be more potent than cyclosporine and was as toxic as cyclosporine. A strong correlation was observed between the immunosuppressive activity of cyclosporine analogs and the ability of the receptor-analog complex to bind to calcineurin *in vitro* and to inhibit its phosphatase activity against an artificial substrate *in vitro*. For example, even though the nonimmunosuppressive cyclosporine analog MeAla$^6$-cyclosporine binds to cyclophilin with an affinity similar to that of cyclosporine, the cyclophilin/MeAla$^6$-cyclosporine complex failed to inhibit calcineurin phosphatase activity (38). Recently, Isotechnika Biotech identified a novel analog ISAtx247 (the exact structure is currently

confidential), which provided better organ transplant survival and a superior toxicity profile to that of cyclosporine on a mg-for-mg basis in all animal models studied to date. Initial phase I single-dose human studies exhibited no significant adverse effects. Isotechnika has received approval from Canada's Health Protection Branch to begin phase I human clinical trials. An application for conducting human trials in the United States has also been filed (39).

### 2.2.1.6 Side Effects.

The major limitations of cyclosporine are its various toxicities; its nephrotoxicity, especially, is a serious one because several cell types, including those of the brain, heart, and kidney, require calcineurin to function. The nephrotoxicity of cyclosporine is associated with mucoid deposits in arterioles, focal fibrosis potentiated by vasoconstriction, a disturbed balance of ionic calcium, an altered balance of prostacyclin to thromboxane, and prorennin activation.

The other major limitation of cyclosporine is its inability to prevent chronic transplant nephropathy. The most common adverse events reported with Neoral in transplantation include renal dysfunction, tremor, hirsutism, hypertension, and gum hyperplasia (18). The risk increases with increasing doses of cyclosporine. High cyclosporine levels result in acute toxic symptoms such as nausea, headache, acute sensitivity of the skin, flushing, gum pain and bleeding, and a sensation of increased stomach size. High levels may also cause liver or kidney failure but this is temporary and no permanent failure is known (40). No clinically significant pharmacokinetic interactions occurred between cyclosporine and aspirin, ketoprofen, piroxicam, or indomethacin (18).

## 2.2.2 FK 506

### 2.2.2.1 History.

In 1984 scientists from Fujisawa isolated the macrolide FK 506 (tacrolimus) from a fermentation broth of *Streptomyces tsukubadensis* (soil sample, Tsukuba Japan) (41). The discovery was part of an extensive, ongoing screening for agents that can suppress or heighten the body's immune responses. Further tests conducted in laboratory animals by Ochiai at Chiba University in Chibashi, Japan, proved its ability to suppress

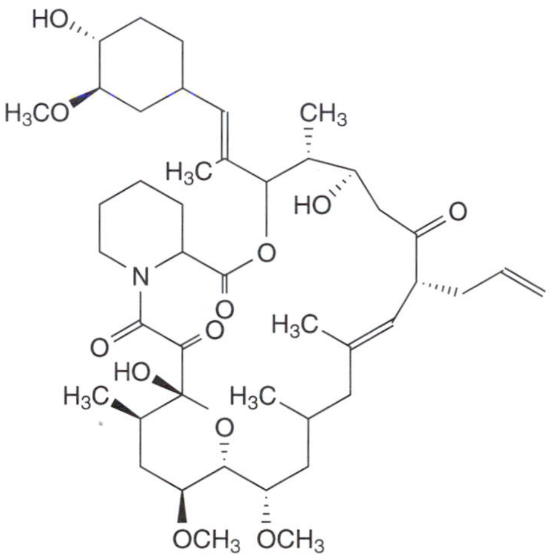

**Figure 12.4.** Structure of FK 506 (**2**).

the immune system. Unfortunately, further progress with this drug in Japan could not be pursued because of the cultural prohibition of cadaver transplantation. Subsequently, Starzl and his colleague carried out the drug's clinical development at the University of Pittsburgh Medical Center (Pittsburgh, PA). They found that the use of FK 506 decreased the incidence of rejection episodes while allowing a much lower dosage of steroids.

**2.2.2.2 Chemical Structure.** FK 506 is a 23-member macrolide designated as *[3S-[3R*[E(1S*, 3S*, 4S*)], 4S*, 5R*, 8S*, 9E, 12R*, 14R*, 15S*, 16R*, 18S*, 19S*, 26aR*]]-5,-6,-8,-11,-12,-13,-14,-15,-16,-17,-18,-19,-24,-25,-26,-26a-hexa-deca-hydro-5,19-di-hydroxy-3-[2-(4-hy-droxy-3-meth-oxy-cyclo-hexyl)-1-methyl-ethenyl]-14,16-di-meth-oxy-4,-10,-12,-18-tetra-methyl-8-(2-propenyl)-15,-19-epoxy-3H-pyrido[2,1-c][1,4]-oxa-azacyclo-tri-cosine-1,-7,-20,-21(4H,-23H)-tetrone monohydrate* (42). The structure (**2**; Fig. 12.4) was deduced on the basis of extensive chemical degradation and spectroscopic studies. The first report of total synthesis was published in 1989 by chemists at the Merck, Sharp & Dohme laboratories (43, 44).

**2.2.2.3 Pharmacokinetics.** Absorption of tacrolimus from the gastrointestinal tract after oral administration is incomplete and variable. The oral bioavailability in humans is variable and ranges from 5% to 67% in recipients of liver, small bowel, and kidney trans-

plantation. The absolute bioavailability of tacrolimus was $17 \pm 10\%$ in adult kidney transplant patients ($n = 26$), $22 \pm 6\%$ in adult liver transplant patients ($n = 17$), and $18 \pm 5\%$ in healthy volunteers ($n = 16$) (42). The tacrolimus maximum blood concentrations ($C_{max}$) were found to vary in different organ transplant settings: $29.7 \pm 7.2$ in normal healthy volunteers, $19.2 \pm 10.3$ in kidney transplant patients, and $68.5 \pm 30.0$ in liver transplant patients. In 11 liver transplant patients, Prograf administered 15 min after a high fat (400 kcal, 34% fat) breakfast, resulted in a decreased AUC ($27 \pm 18\%$) and $C_{max}$ ($50 \pm 19\%$), compared to that of a fasted state. The $t_{1/2}$ (h) of tacrolimus has been determined after intravenous (i.v.) and oral (p.o.) administration in healthy volunteers and both liver and kidney transplant patients. In healthy volunteers, $t_{1/2}$ is $34.8 \pm 11.4$ (p.o. 5 mg), in liver patients $t_{1/2}$ is $11.7 \pm 3.9$ (i.v. 0.05 mg/kg/12 h), and in kidney transplant patients $t_{1/2}$ is $18.8 \pm 16.7$ (i.v. 0.02 mg/kg/12 h) (42). The plasma protein binding of tacrolimus is approximately 99% and is independent of concentration over a range of 5–50 ng/mL. The distribution of tacrolimus between whole blood and plasma depends on several factors, such as hematocrit, temperature at the time of plasma separation, drug concentration, and plasma protein concentration. The mean clearance following i.v. administration of tacrolimus is 0.040, 0.083, and 0.053 L/h/kg in healthy volunteers, adult kidney transplant patients, and adult liver transplant patients, respectively. In humans, less than 1% of the dose administered is excreted unchanged in urine (42).

**2.2.2.4 Pharmacology.** Prograf has been used for the prophylaxis of rejection in liver transplants (since 1994) as well as for kidney transplants (since 1997). Tacrolimus was found to be at least 10–100 times more potent than cyclosporine in the *in vitro* models of immune suppression (45, 46). It binds competitively and with high affinity to a cytosolic receptor (immunophilin) termed as the FK binding protein (FKBP-12) (47). Studies have shown that FK 506 elicits its immunosuppressive activity by inhibiting the *cis-trans* peptidyl-prolyl isomerase activity of FKBP. The structure of the human FKBP complexed with

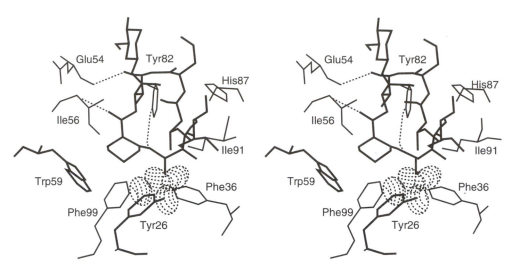

**Figure 12.5.** Stereoview of the binding region of FKBP-FK506 and selected residues of the hydrophobic pocket. Hydrogen bonds are shown between the C-1 ester carbonyl and the NH of Ile[56], the C-24 hydroxyl and the main-chain carbonyl of Glu[54], and the C-8 amide and the phenolic-OH of Tyr[82]. The C-9 carbonyl binding pocket is illustrated with van der Waals dot surfaces for the $\epsilon$-CHs of Tyr[26], Phe[36], and Phe[99]. The side chains from residues Tyr[26], Phe[36], Asp[37], Arg[42], Phe[46], Glu[54], Val[55], Ile[56], Trp[59], Tyr[82], His[87], Ile[91], and Phe[99] are within 4 Å of FK506.

immunosuppressant FK 506 has been determined to 1.7-Å resolution by X-ray crystallography (47). The conformation of the protein changes little upon complexation, but the conformation of FK 506 is markedly different in the bound and unbound forms. It binds in a shallow cavity between the $\alpha$-helix and the $\beta$-sheet with roughly 430 Å$^2$ (50%) of the ligand surface being buried at the protein-ligand interface and the remainder, encompassing the region around the allyl and the cyclohexyl groups, being exposed to the solvent (Fig. 12.5). Loops composed of residues 39–46, 50–56, and 82–95 flank the binding pocket, which is lined with conserved, aromatic residues. The side chain of Tyr[26], Phe[46], Phe[99], and Val[55]-Ile[56] make up the sides of the pocket, whereas the indole of Trp[59], in the $\alpha$-helix, is at the end of the pocket and serves as a platform for the pipecolinyl ring, the most deeply buried part of FK 506. There are five hydrogen bonds between FKBP and FK 506, out of which the fifth hydrogen bond, involving the C8 amide, is the most conspicuous because it is nearly orthogonal to the carbonyl plane and thus may be relevant to the mechanism of rotamase activity. Notably, complexed FK 506 more closely resembles free rapamycin

than it does free FK 506, which suggests that the higher affinity of rapamycin to FKBP ($K_d$ = 0.2 n$M$ for rapamycin versus 0.4 n$M$ for FK 506) reflects its greater preorganization. These investigations provide a structural framework to improve on the high affinity interactions of a clinically promising immunosuppressant with its predominant cytosolic receptor in the T-cell.

After binding with FKBP, the FK 506-FKBP complex binds with the catalytic A subunit of calcineurin and in turn inhibits protein phophatase activity of calcineurin, thereby blocking the activity of the cytoplasmic component of the nuclear factor of activated T-cells (48). This in turn prevents dephosphorylation of the cytoplasmic subunit of a transcription factor, nuclear factor of activated T-cells, which otherwise enters the nucleus and activates expression of T-cell activation lymphokine genes (49–51). Tacrolimus has been found to inhibit IL-2 production by T-cells in a fashion that is similar to that of cyclosporine (41). It prolonged the survival of the host and transplanted graft in animal transplant models of liver, kidney, heart, bone marrow, small bowel and pancreas, lung and trachea, skin, cornea, and limb, thereby suggesting its potential for clinical application to organ transplantation (52–54). In fact, in animals, tacrolimus has been demonstrated to suppress some humoral immunity and, to a greater extent, cell-mediated reactions such as allograft rejection, delayed-type hypersensitivity, collagen-induced arthritis, experimental allergic encephalomyelitis, and graft-vs.-host disease (GVHD).

In one of the first clinical trial studies carried out in liver transplant recipients (55), primary prevention with tacrolimus led to a significant improvement in both patient and graft survival. Rejection episodes and corticosteroid requirements were also reduced compared to those of the control receiving cyclosporine. In another trial performed by the same group (56), no significant difference was observed in patient survival between the tacrolimus and cyclosporine arms; however, freedom from rejection for patients treated with tacrolimus (61%) was much higher than that for patients treated with cyclosporine (18%). This was followed by extensive multicenter trials performed worldwide in liver and kidney transplant recipients, to compare the efficacy and tolerability of tacrolimus-based immunosuppressants with those of cyclosporine. In one of the trials (57) conducted in 529 recipients of liver transplants, the regimens based on tacrolimus and cyclosporine were comparable in terms of patient and graft survival. Similarly, in a European multicenter clinical trial involving 545 liver transplant recipients, the tacrolimus-treated group displayed an acute rejection episode rate of 43.4%, compared to 53.6% for the cyclosporine-treated group (58). In this study, both renal toxicity and neurological complications were more common among those patients who had been treated with tacrolimus. In the kidney transplant recipients, a European multicenter study failed to provide any evidence that the tacrolimus provides better immunosuppression than cyclosporine (59). Similarly, the 1-year analysis of the tacrolimus U.S. kidney transplant multicenter study demonstrated that immunosuppressive therapy with tacrolimus compared to cyclosporine led to statistically fewer rejections without affecting patient and graft survival and without major morbidity (60). Interestingly, in both liver and

kidney transplant settings, tacrolimus unlike cyclosporine is effective as a rescue therapy for acute rejection episodes resistant to conventional corticosteroid pulses (61, 62). Similar rescues after refractory rejection episodes have been reported for cardiac and combined pancreas-kidney transplants (63–65).

The efficacy of tacrolimus for the salvage treatment of chronic GVHD was evaluated in a single-arm, open-label phase II study. A total of 39 evaluable patients with chronic GVHD who failed previous immunosuppressive therapy with cyclosporine and prednisone were treated with tacrolimus, starting at a median of 20 months (range, 3–68 months) after transplantation (66). The Kaplan-Meier estimate of survival was 64% (95% confidence interval, 49–79%) at 3 years posttransplantation and the response rate was consistent with the earlier reports of salvage treatment for chronic GVHD. Further, because the elimination by *ex vivo* irradiation of mature lymphoid elements from the bowel allografts was known to eliminate the GVHD risk, the Pittsburgh group hypothesized that the infusion of donor bone marrow cells (BMC) in recipients of irradiated intestine may improve tolerogenesis without increasing the risk of GVHD. Their studies showed that, although recipients were protected from GVHD by irradiating intestinal allografts, the resulting leukocyte depletion led to chronic rejection of the transplanted bowel (67).

The safety and potential of tacrolimus are further evident by a growing number of pregnancies occurring in mothers receiving tacrolimus systemically. One hundred pregnancies in 84 mothers were recorded, out of which 71 progressed to delivery (68 live births, 2 neonatal deaths, and 1 stillbirth), 24 were terminated (12 spontaneous and 12 induced), and 3 were lost to follow-up. The most common complications in the neonate were hypoxia, hyperkalemia, and renal dysfunction, although they were transient in nature. Four neonates presented with malformations, without any consistent pattern of affected organ (68).

Recently, the efficacy of tacrolimus as monotherapy has also been evaluated in adult cardiac transplants. Forty-three patients received tacrolimus and prednisone as primary immunosuppression agents, without azathio-

**Table 12.2   Conversion to Tacrolimus from Cyclosporine in Various Organ Transplant Settings**

| Organ Transplant | Number of Patients | Results | Reference |
|---|---|---|---|
| Lung or heart–lung transplantation | 11 | Conversion to tacrolimus slows the decline of lung function in bronchiolitis obliterans syndrome. | 71 |
| Liver transplant | 21 | Conversion to tacrolimus improves hyperlipidemic states in stable liver transplant recipients. | 72 |
| Heart transplant | 10 Japanese | Postoperative follow-up after cardiac transplantation appears to be satisfactory with conversion to tacrolimus. | 73 |
| Renal transplant | 17 | Conversion from Cyclosporine to tacrolimus in stable renal transplant recipients may lead to attenuation of cardiovascular morbidity and chronic transplant nephropathy in the long term. | 74 |
| Heart transplant | 85 | Tacrolimus seems safe and effective for preventing rejection during the first year after heart transplant. It causes less high blood pressure and less high cholesterol, and has no more side effects in areas other than cyclosporine. | 75 |

prine or mycophenolate mofetil. Thirty-two of the 43 patients started on tacrolimus were weaned off steroids and maintained on monotherapy. Their results suggest that use of tacrolimus alone after steroid weaning provides effective immunosuppression with low incidence of rejection, cytomegalovirus infection, transplant arteriopathy, or posttransplant lymphoproliferative disease (69). In another study in heart transplant recipients, monotherapy with tacrolimus has been found to significantly reduce blood clot formation compared to that with cyclosporine, thereby suggesting that the drug may help prevent heart transplant vasculopathy (70).

In recent years a number of groups have recommended tacrolimus conversion as a means to antilymphocyte preparations for corticosteroid-resistant rejection in cyclosporine-based regimens. Some of the studies pertaining to tacrolimus conversions in a variety of organ transplant settings and their outcomes are summarized in Table 12.2. These findings suggest that conversion from a cyclosporine-based to a tacrolimus-based maintenance immunosuppression appears to be an effective and

safe approach to the management of patients with persistent or recurrent allograft rejection or those with cyclosporine intolerance.

In liver transplant patients, early chronic rejection episodes can be reversed by shifting from a cyclosporine-based protocol to a tacrolimus-based immunosuppressive regimen. This has been attributed to the fact that tacrolimus and transforming growth factor $\beta$-1 share the same binding site, the immunonophilin FKBP-12 (76). Thus, tacrolimus may interfere with the TGF $\beta$-1 signaling through competitive binding, and by antagonizing the fibrosis-promoting effect of this growth factor. Given the similar effect on T-lymphocytes, tacrolimus has also been proposed as an alternative to cyclosporine in the triple- or double-immunosuppressive regimen. Several groups have evaluated the efficacy of tacrolimus in combination with other clinically used immunosuppressants for the prevention of organ transplant rejections. In one of the studies, a randomized three-arm, parallel group, open label prospective study was performed at 15 North American centers to compare the three immunosuppressive regimens: (1) tacrolimus

+ azathioprine, (2) cyclosporine (Neoral) + mycophenolate mofetil, and (3) tacrolimus + mycophenolate mofetil. All patients were first cadaveric kidney transplants receiving the same maintenance corticosteroid regimen. All regimens yielded similar acute rejection rates and graft survival, but the tacrolimus + mycophenolate mofetil regimen was associated with the lowest rate of steroid-resistant rejection requiring antilymphocyte therapy (77). Tacrolimus plus mycophenolate mofetil has also been found to be effective in patients seeking second transplants. Low dose tacrolimus administered intravenously followed by oral tacrolimus, mycophenolate mofetil, and steroids provided effective immunosuppression in recipients ($n = 10$) of simultaneous pancreas-kidney transplants who had undergone previous transplants.

In this study, the investigators sought to decrease the morbidity associated with antibody therapy without increasing acute rejection rates. Two patients experienced episodes of acute rejection that responded to steroid treatment. After a mean follow-up of 20.4 months, the kidney and pancreas were found to be functioning in all patients. This is a new induction treatment for patients with previous transplants, sparing them of antibody induction therapy (78). Another group studied the efficacy of combined therapy involving tacrolimus plus mycophenolate mofetil in heart transplant recipients. The researchers enrolled 45 patients, 15 into phase I and 30 into phase II of the study. Intravenous tacrolimus was given for 2–3 days before switching to oral drugs. Target blood levels were 10–15 ng/mL. Treatment also included steroids and mycophenolate mofetil. During phase I, a 2 g/day dose of mycophenolate mofetil was given, whereas in phase II, doses were adjusted according to mycophenolic acid levels in the blood, with target levels of 2.5–4.5 $\mu$g/mL. Average follow-up was for 696 days in phase I and for 436 days in phase II. During phase I, patient survival was 100% and rejection was diagnosed in 67% of the patients. Data analysis suggests that an average mycophenolic acid plasma level of 3 $\mu$g/mL or more prevented rejections. Interestingly, during phase II, one patient died but rejection could be seen in only

10% of the patients. Steroids were successfully withdrawn from all patients who completed 6 months of treatment.

Thus, although combination therapy with tacrolimus and mycophenolate mofetil is effective for preventing rejection in heart transplant patients, routine drug level monitoring is critical (79). Similarly, in cadaveric kidney transplant recipients randomized to receive tacrolimus in combination with either azathioprine ($n = 59$) or mycophenolate mofetil 1 g/day ($n = 59$), or mycophenolate mofetil 2 g/day group ($n = 58$), tacrolimus in combination with an initial dose of mycophenolate mofetil 2 g/day was found to be a very effective and a safe regimen. The incidence of biopsy-confirmed acute rejection at 1 year was 32.2, 32.2, and 8.6% in the azathioprine, mycophenolate mofetil 1 g/day, and mycophenolate mofetil 2 g/day groups, respectively ($P < 0.01$). The overall incidence of posttransplant diabetes mellitus was 11.9%, with the lowest rate observed in the mycophenolate mofetil 2 g/day group (4.7%), and was reversible in 40% of the patients (80). In yet another study, the efficacy of double-drug protocol has also been evaluated in the hamster-to-rat xenotransplantation model involving heart and liver. Survival of heart and liver xenografts in the rats was 48 $\pm$ 4 and 63 $\pm$ 8 days, respectively, and after cessation of all immunosuppression, hearts were rejected after 18 $\pm$ 4 days and livers after 33 $\pm$ 8 days (81). Thus, the studies demonstrate improved graft survival and reduced rejection rates with the combined use of tacrolimus and mycophenolate mofetil in both allograft and xenograft settings.

***2.2.2.5 Structure-Activity Relationship.*** Several structurally related compounds of FK 506 have been isolated from *Streptomyces tsukubaensis:* methyl (FR900425), ethyl (FR900520), and proline (FR900525) analogs of FK 506. Because the FK 506 is the most active in this series of molecules, it has been studied most exclusively (82, 83). A novel analog of tacrolimus, L732531, a 32-*O*-1(1-hydroxyethylindol-5-yl) ascomycin derivative has been reported with potent immunosuppressive activity. It exhibited an improved therapeutic index compared to that of FK 506 in rodent models. Moreover, its biochemical properties were also found to be distinct from those of FK 506. These data suggest L732531 to be a potential candidate for treating organ transplant rejection (84).

*2.2.2.6 Side Effects.* Adverse effects associated with Prograf include tremor, hypertension, hypophosphatemia, infection, creatinine increase, headache, and diarrhea. Other toxicities include hyperglycemia, nephrotoxicity, chest pain, chest discomfort, palpitations, abnormal electrocardiogram, and abnormal distension (85). Prograf can cause neurotoxicity and nephrotoxicity, particularly when used in high doses. Nephrotoxicity has been reported in approximately 52% of kidney transplantation patients and in 40 and 36% of the liver transplant patients receiving Prograf in the U.S. and European randomized studies, respectively (42). New onset posttransplant diabetes mellitus (PTDM) was seen in 20% of the Prograf-treated kidney transplant patients and median time to the onset of PTDM was 68 days. Insulin dependency was reversible in 15% of these patients at 1 year and in 50% at 2 years posttransplantation (86). Prograf injection is contraindicated in patients with a hypersensitivity to HCO-60 (polyoxyl 60 hydrogenated castor oil).

### 2.2.3 Sirolimus

*2.2.3.1 History.* Another interesting immunosuppressant with a unique mechanism of action is sirolimus (rapamycin), which has been under development for more than 20 years before it gained FDA approval on September 15, 1999. It is also a microbial natural product and is produced by the actinomycete *Streptomyces hygroscopicus* isolated from Easter Island (Rapanui to its natives) soil samples in 1975. It emerged from an antifungal drug discovery program directed by Sehgal (87) at Ayerst Research in Montreal, Canada. The antifungal properties of the drug were not pursued when it became apparent that the drug caused involution of lymphoid tissue. Martel subsequently demonstrated that rapamycin suppresses experimental allergic encephalomyelitis and passive cutaneous anaphylaxis in the rat (88). It was not until the newly discovered structure of tacrolimus was found to be remarkably similar to that of sirolimus that led Cambridge group (89) to uncover the potential of the drug as an immunosuppressant.

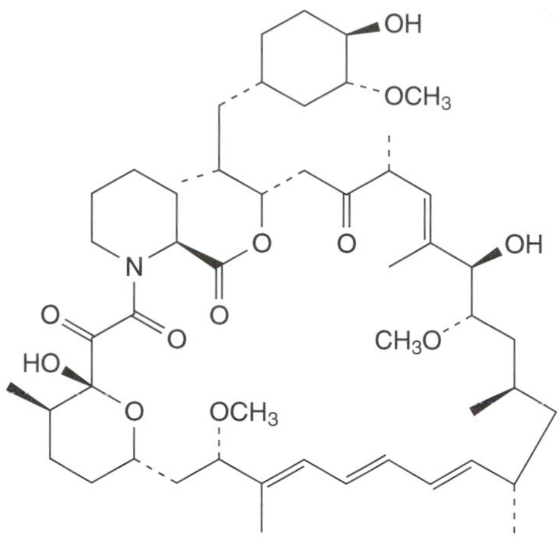

**Figure 12.6.** Structure of Rapamune (**3**).

*2.2.3.2 Chemical Structure.* Structural analysis of rapamycin (**3**; Fig. 12.6) revealed a macrocyclic triene lactone with strong structural similarity to FK 506.

*2.2.3.3 Pharmacokinetics.* For Rapamune the adult loading dose is 6 mg (equivalent to 3 × the maintenance dose). A loading dose is recommended because of the drug's long half-life; without a loading dose it may take ≥2 weeks to reach the steady state. Whereas the adult maintenance dose is 2 mg/day, the maintenance dose for children ≥13 years old is 1 mg/m$^2$/day, with a maximum dose of 2 mg/day. Though the oral absorption of sirolimus is rapid, with time-to-peak concentration of 2 h, the oral bioavailability is approximately 14%. Sirolimus is extensively metabolized in the liver by *O*-demethylation and/or by hydroxylation. Sirolimus is the major component in human whole blood and contributes to more than 90% of the immunosuppressive activity. The terminal elimination half-life ($t_{1/2}$) is 72 h in males and 61 h in females. The sirolimus trough concentrations (whole blood; steady-state; ng/mL) is 8.59 ± 4.01 for the 2 mg/day dose and 17.3 ± 7.4 for the 5 mg/day dose (90).

*2.2.3.4 Pharmacology.* Rapamycin, is a novel immunosuppressive agent developed for the prevention of organ rejection after renal transplantation. It allows doctors to eliminate cyclosporine, a standard component of typical multidrug regimens, at an early treatment stage. Like FK 506, rapamycin binds to and inhibits the isomerase activity of FKBPs and it

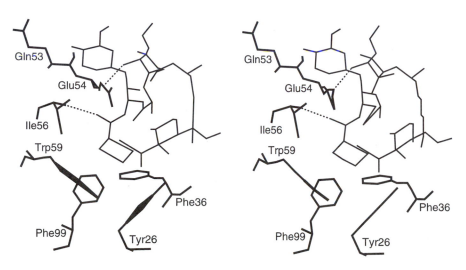

**Figure 12.7.** A stereodrawing of the binding pocket showing all of the bound rapamycin molecule and selected FKBP-12 residues.

is this rapamycin-FKBP complex that is responsible for eliciting immunosuppressive activity (91–93). Although it binds to the same intracellular binding protein or immunophilin known as FKBP-12 as FK 506, its unique mechanism of action and superior potency in blocking allograft rejection made it an entirely new class of immunosuppressants. The three-dimensional structure of the complex of human FKBP-12 and rapamycin, determined to 1.7-Å resolution by X-ray crystallographic techniques, provides a framework to interpret the effects of structural perturbation of either rapamycin or human FKBP-12 on signal transduction pathways (94). The protein component in the complex forms a five-stranded antiparallel β-sheet wrapping with a right-handed twist around a short α-helix, the same folding topology found in the complex of FKBP-12 with FK 506 and in uncomplexed FKBP-12. Rapamycin binds in a cavity between the β-sheet and α-helix with the pipecolinyl ring deeply buried in the protein (Fig. 12.7). The protein-ligand interface involves atoms from the pyranose ring through the C28 hydroxyl, with the remainder, including the C17-C22 triene exposed. The C1 ester, the pipecolinyl ring, the C8 and C9 carbonyls, and the pyranose ring adopt a conformation that is superimposable with the same groups in the FKBP-12/FK 506 complex. Three hydrogen

bonds between this region and FKBP-12 (Ile-56 NH to C1 carbonyl, Tyr-82 hydroxyl to C8 carbonyl, and Asp-37 carboxylate to C10 hydroxyl) and a C9 carbonyl binding pocket involving C—H—O interactions with ← hydrogens from Tyr-26, Phe-36, and Phe-99 are also identical with those found in the complex with FK 506, thus confirming the identical binding roles of the common structural elements in the two immunosuppressant ligands. Two additional hydrogen bonds are involved in rapamycin binding to FKBP-12. The first is from the Glu-54 main-chain carbonyl to C28 hydroxyl, which along the Ile-56 NH to C1 carbonyl-hydrogen bond may mimic the interaction of the dipeptide portion of a natural substrate with FKBP-12. The sirolimus-FKBP-12 complex, which has no effect on calcineurin activity, binds to and inhibits the activation of the mammalian target of rapamycin (mTOR), a key regulatory kinase in the signal transduction in T-cells.

The effect of rapamycin on immune cells *in vitro* markedly differs from the actions of other known immunosuppressants. Sirolimus was found to be an extremely potent inhibitor of both murine and human thymocyte proliferation induced by phytohemagglutinin and IL-2 and was at least 10–1000 times more potent than cyclosporine (95, 96). It strongly inhibits murine, porcine, and human T-lymphocyte proliferation when induced by antigens and cytokines (IL-2, IL-4, and IL-15), whereas cyclosporine and FK 506 block T-cell activation induced only by stimuli employing Ca-dependent pathways (97–100). Nevertheless, the effect of rapamycin could be seen even when the drug was added up to 12 h after the stimulation of T-cells, in contrast to cyclosporine and FK 506, which lose their effect after 2 h. These findings suggest that unlike cyclosporine and FK 506, which block $Ca^{2+}$-dependent cytokine transcription early in the G1 phase, rapamycin inhibits cytokine-mediated signal transduction pathways later in the G1 phase (97, 101).

Sirolimus inhibits the proliferation of B-cells induced by lipopolysaccharide through a Ca-independent pathway that is resistant to cyclosporine and FK 506 (101, 102). Further, its effect on the proliferation of purified peripheral blood mononuclear cells (PBMC) was also investigated in vitro and it was found to be at least 50–500 times more potent than cyclosporine in inhibiting PBMC proliferation induced by 0.1% PHA. Other biochemical events that are inhibited by rapamycin involve activation of p70S6kinase (103–105) and activation of the cdk2/cyclin E complex phosphorylation of retinoblastoma protein (106). In some animal studies, sirolimus-induced tolerization and the immunosuppressive effect lasted up to 6 months, even after the discontinuation of the therapy. Whether sirolimus can induce significant tolerization in humans is yet to be established.

Based on rapamycin's ability to inhibit both T- and B-cell proliferation in vitro, it was expected to exhibit potent inhibition in allograft rejection. This was evident in several animal models where rapamycin significantly prolonged survival of heart, kidney, and pancreas allografts in rats (107); nonvascularized fetal cardiac grafts in murine (108); and kidney allografts in pig (109) and primate (110). In addition it was also found to be effective in reversing advanced allograft rejection. Thus, rapamycin is a potent immunosuppressant with remarkable antirejection activity in animal models of organ transplantation. Further, because rapamycin and cyclosporine interfere at different stages and levels of immune response, combined use of these agents has been found to be very promising. In in vitro assays, the sirolimus-cyclosporine combination produced an inhibitory effect that was between 10- and 1000-fold greater than that produced by either agent alone (111). This was further evident in the phase I/II clinical trial studies when addition of sirolimus (0.5–7 mg/m²/day) was found to reduce overall incidence of rejection episodes from 32% to 7.5%, in patients who were treated with cyclosporine and prednisone alone (112). Similarly, a multicenter trial involving 149 renal transplant patients who had been treated with cyclosporine and prednisone confirmed that the addition of sirolimus (at 1, 3, or 5 mg/m²/day) reduced the incidence of acute rejection episodes (113); although within 6 months the cyclosporine-prednisone group showed a 40% incidence of rejection episodes, patients treated adjunctively with 1 or 3 mg/m²/day of sirolimus experienced only 10% incidence of rejection. The studies also showed that 1, 3, or 5 mg/m² of sirolimus did not reduce the number of rejection episodes in African-American patients who were treated with reduced cyclosporine doses. This finding was in contrast to what was found when this regimen was used to treat patients of non–African-American origin. However, the most recent clinical trial revealed that the adjustment of the cyclosporine-sirolimus regimen in African-American patients (using higher doses than those given to non–African-Americans) improved the 1-year survival rate of kidney allografts. The survival rate was increased to 97% from only 70% for those recipients who were treated with the cyclosporine-prednisolone regimen (114).

Finally, two phase III trials evaluating the efficacy of sirolimus have provided data indicating that sirolimus is safe and effective. The Rapamune U.S. Multicenter Study (115) included 719 renal transplant recipients who were randomized to receive either sirolimus (either 2 or 5 mg/day) or azathioprine (2–3 mg/kg/day). All patients also received cyclosporine and prednisone. Both sirolimus regimens were found to be significantly more effective than azathioprine at reducing the incidence of acute rejection. Graft and patient survival were similar in all treatment groups. In the Rapamune Global Study (116), conducted in

**Table 12.3   Results of phase III Clinical Trial Studies**

| | Sirolimus | | Azathioprine/ Placebo |
|---|---|---|---|
| | 2 mg | 5 mg | |
| Rapamune U.S. multicenter studies (719 renal transplant recipients) | | | |
| Incidence of biopsy-confirmed acute rejection at 6 months | 16.5% (47/284) | 11.3% (31/274) | 29.2% (47/161) |
| Graft survival at 12 months | 94.7% (269/284) | 92.7% (254/274) | 94.0% (151/161) |
| Patient survival at 12 months | 97.1% (276/284) | 96.0% (263/274) | 98.1% (158/161) |
| Rapamune worldwide studies (576 renal transplant recipients) | | | |
| Incidence of biopsy-confirmed acute rejection at 6 months | 24.7% (56/227) | 19.2% (42/219) | 41.5% (54/130) |
| Graft survival at 12 months | 89.9% (204/227) | 90.9% (199/219) | 87.7% (114/130) |
| Patient survival at 12 months | 96.5% (219/227) | 95.0% (208/219) | 94.6% (123/130) |

Australia, Canada, Europe, and the United States, 576 renal transplant recipients were stratified to receive either sirolimus (either 2 or 5 mg/day) or placebo. All patients also received cyclosporine and prednisone. As in the U.S. study, sirolimus was found to be superior to placebo at reducing the incidence of acute rejection within the first 6 months after transplantation. Graft and patient survival were similar in all treatment groups. The results of both phase III clinical trials are summarized in Table 12.3.

**2.2.3.5 Structure-Activity Relationship.** Structure-activity relationship studies with sirolimus led to the identification of SDZ-RAD (40-*O*-[2-hydroxyethyl]-rapamycin) as a new, orally active rapamycin derivative with potent immunosuppressive activity (117). Because of its synergistic interaction, SDZ-RAD is under clinical investigation as an immunosuppressant in combination with cyclosporine after organ transplantation (118). Coadministration of microemulsion cyclosporine (neoral) and the novel immunosuppressant SDZ-RAD potentiates the immunosuppressive efficacies of both drugs to suppress allograft rejection (119, 120). SDZ-RAD effectively ameliorates chronic renal allograft rejection in rats, probably mediated through the suppression of growth factors (121). SDZ-RAD exerts its

pharmacological effect by binding to a different effector protein, inhibits the S6p 70-kinase, and interrupts a different signal transduction pathway in contrast to tacrolimus. The pharmacokinetic data of the SDZ-RAD dose, normalized to 1 mg SDZ-RAD, were as follows: AUC (0, 24 h): $35.4 \pm 13.1$ $\mu$g/L/h; $C_{max}$: $7.9 \pm 2.7$ $\mu$g/L/h; and $t_{max}$: $1.5 \pm 0.9$ h (122). SDZ has also been evaluated in combination with FTY 720 and was found to be effective in maintaining heterotopic cardiac graft throughout the treatment period (123). Presently it is undergoing phase III clinical trials.

**2.2.3.6 Side Effects.** The most common side effects associated with Rapamune are hypercholesterolemia and hypertrigliyceridemia (hyperlipemia), thrombocytopenia, anemia, acne, abdominal pain, urinary tract infection, arthralgia, diarrhea, hypokalemia, lymphocele, and increased lactic dehydrogenase. Increased susceptibility to infection and the possible development of lymphoma may also result from immunosuppression (90). Specific adverse reactions associated with Rapamune administration occurring at a significantly higher frequency vs. controls were observed for both 2 and 5 mg/day schedules: hypercholesterolemia, hyperlipemia, hypertension, and rash in patients on the 5 mg/day schedule;

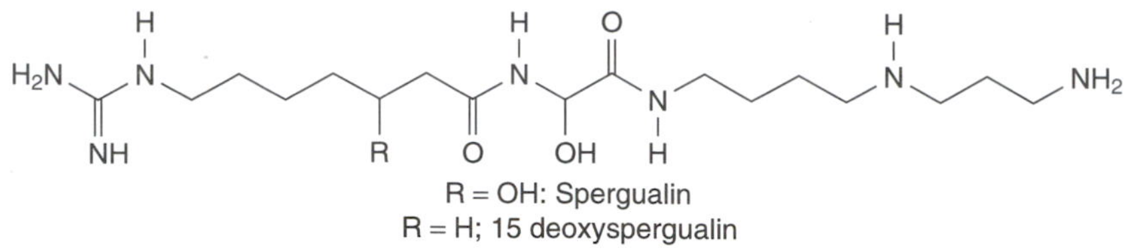

R = OH: Spergualin
R = H; 15 deoxyspergualin

**Figure 12.8.** Structure of spergualin (**4a**) and deoxyspergualin (**4b**).

anemia, arthralgia, diarrhea, hypokalemia, thrombocytopenia, and acne for patients on the 2 mg/day schedule. Elevations of triglycerides and cholesterol, and decreases in platelets and hemoglobin occurred in a dose-related manner (124). No significant interactions were observed with acyclovir, digoxin, glyburide, prednisolone, oral contraceptives, and co-trimoxazole (90).

### 2.2.4 15-Deoxyspergualin

**2.2.4.1 History.** In 1981 Umezawa isolated an antitumor and immunosuppressive antibiotic from the culture filtrates of *Bacillus laterosporus* (125, 126) called spergualin. Although it was originally identified because of its antitumor and antibiotic activity, the immunosuppressive activity was found to be more pronounced. Its structural instability, attributed to the presence of an $\alpha$-hydroxyglycine moiety, led to the synthesis of several hundred analogs in different laboratories, to identify a stable analog with potent immunosuppressive activity (127–129). These studies culminated in the identification of 15-deoxyspergualin (DSG), with strong immunosuppressive activity both *in vitro* and *in vivo* (130).

**2.2.4.2 Chemical Structure.** Spergualin, 1-amino-19-guanitido-11,15-dihydroxy-4,9, 12-triazathioprinenonadecane-10,13-dione (**4a**; Fig. 12.7) is a water-soluble peptide with a molecular weight of 496 Da. It is synthetically dehydroxylated to produce 15-deoxyspergualin (**4b**; Fig. 12.8).

**2.2.4.3 Pharmacokinetics.** DSG is a highly polar molecule because of which it has a poor (3–6%) oral bioavailability (131). When administered by intravenous bolus injection, it is rapidly cleared with a biphasic half-life in mice of about 1.9 and 11.6 min (132). The clearance rate was independent of dose and the predominant route of elimination of DSG was renal.

The maximum tolerated dose (MTD) in mice was 10.5 mg/kg when administered by intravenous bolus route. In mice, the peak plasma concentration after a single $LD_{10}$ dose was approximately 30 $\mu$g/mL.

Because the MTD of DSG is attributable to the peak plasma concentration, the generally accepted mode of administration to humans has been a 3-h intravenous infusion. Human pharmacokinetic studies of DSG given intravenously over 3 h for 5 consecutive days revealed rapid plasma clearance with $t_{1/2\alpha} = 37$ min and $t_{1/2\beta} = 9.2$ h. There was no accumulation of DSG in the plasma after each 3-h infusion for 5 days (132). Six metabolites of DSG have been identified using HPLC, but none of these metabolites exhibits the antitumor or immunosuppressive activity of the parent component. The area under the curve of 15-deoxyspergualin matched well with the administered dose (132, 133).

**2.2.4.4 Pharmacology.** *In vitro*, DSG moderately inhibits both the mitogen-stimulated proliferation of T-cells and the generation of cytotoxic T-cells (134, 135), without affecting IL-2 production. DSG inhibited growth of mouse EL-4 lymphoma cells with an $IC_{50}$ value of 0.02 $\mu$g/mL. Even though the cells were treated with DSG for only 4 h and then washed, the antiproliferative effect was long lasting, with an $IC_{50}$ value of 0.4 $\mu$g/mL. Studies have shown that DSG binds to the cells through its spermidine moiety instead of guanidine moiety and thus exerts its long-lasting antiproliferative effect (136). Because DSG blocks T-cell function *in vitro* when added 3–4 days after the stimulation of T-cells (134), it is believed to act at a late stage of the cell cycle of T-cells. However, this inhibitory effect can be reversed by the exogenous supplementation of IFN-$\gamma$, but not by IL-2 (137).

Although the molecular mode of action of DSG is not yet clearly understood, it binds specifically to two heat-shock proteins (Hsp), Hsp70 and Hsp90, which represent two members of a class of immunophilins. Although the exact binding site on Hsp70 for protein substrates is not yet known, a recent study has shown that the C-terminal four amino acids $^{647}$EEVD$^{650}$ might be playing a role in regulating ATPase activity and substrate binding. These four amino acids are also found at the C-terminus of Hsp90 and may be involved in a similar function (138). However, the role of the EEVD motif in biological processes has not yet been investigated.

Other effects of DSG include liposomal enzyme release, superoxide production, major histocompatibility antigen (MHC) class II up-regulation, and IL-1 production (139). DSG also reduced the production of antibodies in immunotoxin-treated mice (140), and in vitro reduced the expression of IgM on the surface of B-cells after the administration of lipopoly-saccharide or IFN-$\gamma$ to a mouse cell line of pre-B-cells (141). These studies may account for the immunosuppressive and antitumor properties of DSG and the pathway of immune suppression for DSG appears to be entirely different from that of other clinically used immunosuppressants.

The immunosuppressive activity of DSG has been demonstrated in many animal models of transplant rejection such as heart (142), liver (143), pancreas (144), pancreatic islet (144), and bone marrow (145). In one of the experiments, treatment of baboons with 4 mg/kg/day of 15-deoxyspergualin improved the survival of heart, but not renal, allografts (146). Similarly, treatment with DSG slightly delayed the rejection of heart allografts in rat recipients that had been presensitized with skin allografts (147). In preclinical models, DSG was found to be active both in prolonging allograft survival and in the reversal of graft rejection (148, 149). This is in contrast to clinically used cyclosporine, which is ineffective in reversing the ongoing rejection. DSG also demonstrated a rescue effect when intraperitoneal alginate-poly-L-lysine-alginate microencapsulated xenoislets induced cellular overgrowth that threatened the survival of the graft in streptozotocin-induced diabetic mice (150). Experiments in dogs showed that DSG treatment reversed acute rejection episode

(151), but caused significant gastrointestinal disturbances. In another study, DSG (40–220 mg/m$^2$) was highly effective in reversing rejection episodes in 27 of 34 (79.4%) recipients of renal transplants when used alone (152), and even more effective when administered in combination with methylprednisolone (87.5%; (153)). Similarly when DSG (3–5 mg/kg) was administered in daily 3-h intravenous infusions for 7 days, it reversed 76% of the rejection episodes in 260 patients within the first 6 months after transplantation (154). In addition, DSG also reversed 70% of steroid-resistant rejection episodes.

Its efficacy has been further demonstrated in xenograft models as well, where it prolonged the survival of xenogeneic kidney grafts in mongrel dogs receiving kidney of silver foxes (155). In fact, its activity was found to be markedly superior to that of cyclosporine and FK 506 in prolonging xenograft survival (156). Recently, several groups evaluated the efficacy of DSG alone or in combination with either T-cell monoclonal antibody or cyclosporine or FK 506 for the prolonged survival of xenografts in animal models (157–161). Coadministration of DSG and cyclosporine was effective and nontoxic and prolonged heart grafts in primate models (162). Further, it has also been reported that the use of DSG permits lowering of cyclosporine doses and reduction of nephrotoxicity when coadministered together (163).

In human transplantation, Amemiya et al. studied a variety of techniques for reversing rejection, using DSG alone, along with the use of DSG in rejections unresponsive to steroids (rescue therapy) (152, 164, 165). High rates of 70–90% of overall reversal were achieved in primary treatment, with an average rate of 81% for all patients studied. The rescue therapy similarly gave high rates of reversal with an average of 66% in these groups. Tanabe et al. showed that DSG was effective in allograft prolongation when used with a donor-specific transfusion protocol (166). These results suggest that DSG may be valuable as a prophylaxis for clinical human transplantation.

*2.2.4.5 Structure-Activity Relationship.* The structural instability of DSG resulting from the presence of an $\alpha$-hydroxyglycine moiety led to the synthesis of several hundred an-

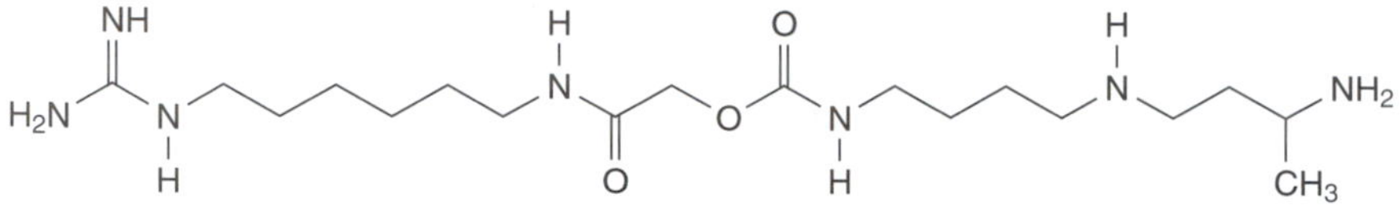

**Figure 12.9.** Structure of DSG analogue (**5**).

alogs in different laboratories, to identify a stable analog with potent immunosuppressive activity. Modifications were introduced in all three regions—spermidine moiety, α-hydroxyglycine moiety, and guanidinoheptanoic acid moiety—to clarify the roles of various functional moieties. Earlier structure-activity relationship studies carried out with DSG led to the conclusions that (*1*) the molecular length of the guanidino fattyacylaminoacyl moiety plays an important role in the expression of biological activity; (*2*) modification around the spermidine moiety plays an important role in biological activity; and (*3*) the 5-hydroxyl group is not essential for activity. None of the congeners was found to have a better activity profile than that of DSG; however, further optimization led to the identification of six novel analogs with strong immunosuppressive activity but all of them exhibited activity at a dose higher than 3 mg/kg, which is an effective dose for DSG (167). Later, Lebreton et al. (168, 169) carried out structure-activity relationship studies by synthesizing a series of novel analogs of DSG. They too introduced structural modifications in the hydroxyglycine region and in the spermidine region. The congeners were synthesized and tested in a graft-vs.-host disease (GVHD) model in mice to determine its optimum structure in terms of *in vivo* immunosuppressive activity. The studies led to the identification of a novel analog (**5**; Fig. 12.9), which demonstrated powerful activity at a dose of 0.3 mg/kg in the GVHD model and was much more potent than DSG in the demanding heart allotransplantation model in rats. The improvement of *in vivo* activity was related to an increase of the metabolic stability of the methylated analogs compared to that of the parent molecules. Because of its very low active dose, compatibility with subcutaneous administration in humans, and its favorable pharmacological and toxicological profile, it has been selected as a candidate for clinical evaluation.

*2.2.4.6 Side Effects.* Adverse effects associated with the treatment of 15-deoxyspergualin for allograft rejection includes numbness of the face, lips, and limbs (in 14% of patients), gastrointestinal toxicity (in 9%), and bone marrow suppression (in 54%). Because of its low bioavailability, 15-deoxyspergualin needs to be administered intravenously for only a short duration, thereby limiting its widespread application in organ transplantation. In rats treated with DSG at the active period, bone marrow suppression and damage of small intestine were significantly severe. The toxicity of DSG varied with the dosing time, whereas its efficacy did not (164).

### 2.2.5 FTY 720

*2.2.5.1 History.* In the 1990s, Yoshotomi Pharmaceuticals Ltd., Japan, in collaboration with Fujita of Kyoto University, began research on immunosuppressive substances from the products of vegetative wasp. From the culture broth of *Isaria sinclairii*, they isolated a potent immunosuppressant, ISP-I. It was found to be at least 10–100 times more potent than cyclosporine as an immunosuppressant of the immune response both *in vitro* and *in vivo* (170). Structurally, ISP-I was identical to myriocin (171, 172) and thermozymocidin (173), previously isolated from *M. albomyces* (ATCC 16425) and *M. sterillia* (ATCC 20349), respectively, as an antifungal agent. During structure-activity relationship studies of myriocin, it was observed that reduction of the ketone at C14 to methylene led to a congener with a 10-fold increase in the immunosuppressive activity (174), thereby suggesting that there is a strong possibility for the presence of compounds more potent than myriocin in the culture broth of myriocin-producing microorganisms. A search for new compounds having more potent activity than that of ISP-I led to the isolation of mycestericins D, F, E, and G (175–177) as minor analogs. These compounds were evaluated for their immunosup-

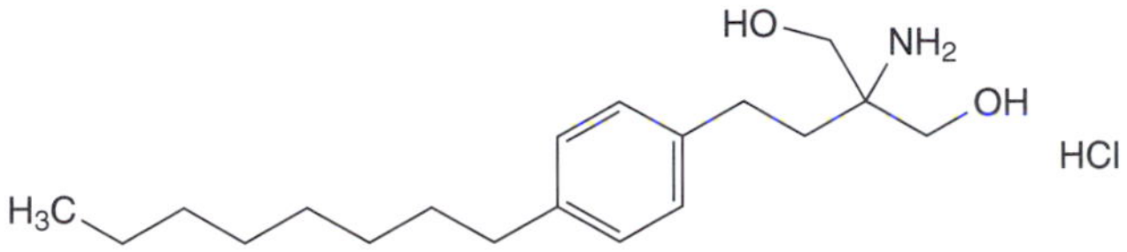

**Figure 12.10.** Structure of FTY 20 (**6**).

pressive activity in mouse allogenic MLR assay, where mycestericins D and E were found to be more or less equipotent to myriocin, whereas F and G were less active. A structure-activity relationship study on these natural variants by Fujita et al. led to the identification of 2-amino-2-tetradecylpropane-1,3-diol hydrochloride (ISP-I-55) as a promising lead for the development of potent organ transplant drugs (178). Finally, it was Adachi in 1995 who optimized these leads obtained from the synthetic as well as natural variants of ISP-I that led to the design and synthesis of the potent compound FTY 720 as a clinical candidate (179).

***2.2.5.2 Chemical Structure.*** The compound 2-amino-2[2-(4-octylphenyl)ethyl]-1,3-propane diol (**6**; Fig. 12.10) was first synthesized in 1995 in an effort to minimize the toxic *in vivo* properties of ISP-I.

***2.2.5.3 Pharmacokinetics.*** No data have yet been reported on the pharmacokinetic parameters of FTY 720.

***2.2.5.4 Pharmacology.*** FTY 720 exerts its immunosuppressive effect by inducing a drastic and selective decrease in blood lymphocytes, especially T-cells both *in vitro* and *in vivo*. *In vitro*, it causes cell death in lymphocytes and leukemia cells, whereas *in vivo* (rats and mice), a marked decrease in the number of blood lymphocytes is observed within 1 h after a single oral administration of FTY 720 at 5–10 mg/kg doses (180). It is presumed to be acting by altering lymphocyte trafficking/homing patterns through modulation of cell surface adhesion receptors and ligands in a manner that has yet to be elucidated. In the thymus, long-term daily administration of FTY 720 caused a three- to fourfold increase in the proportion of mature medullary thymocytes [CD4(+)CD8(−) and CD4(−)CD8(+)] as well as a slight decrease in the double-positive cell [CD4(+)CD8(+)] ratio, thereby suggesting that the immunosuppressive action of FTY 720, at least in part, could be attributable to its

inhibitory effect on T-cell emigration from the thymus to the periphery (181).

Thus, because of its distinct mechanism of action to reversibly alter lymphocyte homing patterns away from the graft, FTY 720 offers a unique reagent for both induction and maintenance of immunosuppression. FTY 720, when given i.v. or orally at 0.03 mg/kg or more, significantly prolonged skin allograft survival in a dose-dependent manner and showed more potent immunosuppressive activity than that of either cyclosporine or tacrolimus (182). To elucidate the mechanisms of the remarkable synergistic effect, mRNA expressions of Il-2 and γIFN and that of CD3, which reflects T-cell infiltration in allograft, were analyzed. The results suggest that the synergistic effect on the prolongation of allograft survival may be the result of the respective inhibitors of T-cell infiltration and cytokine production in grafts (183). The same group studied the significance of the timing of FTY 720 administration on the immunosuppressive effect to prolong rat skin allograft survival and found that it should be administered before the increase in T-cell infiltration into grafts so as to inhibit acute allograft rejection (184). Its clinical potential as an immunosuppressant has been demonstrated in a variety of allograft and xenograft transplantation models. It prolonged cardiac (185), pancreas (186), islet (187), liver (188), kidney (189), and renal (190, 191) allograft survival in laboratory animals without any drug-induced toxic side effects. Similarly, it was also able to prolong concordant xenograft survival in rats (192).

Although much research has yet to be done to unravel the nature of the mechanism of action of FTY 720, its efficacy has been sufficiently proved in numerous rodents and subhuman primate models, especially when administered in combination with one of the conventional immunosuppressants, such as cyclosporine, tacrolimus, or sacrolimus. FTY 720 potentiates the immunosuppressive effects of cyclosporine and/or sacrolimus both *in vitro* (by inhibiting of T-cell proliferative response) and *in vivo* (by inhibiting graft rejection).

Several groups have studied its efficacy in adjunct therapy with either cyclosporine or sacrolimus both in experimental allograft models involving renal (193), heart (194), liver (195, 196), and kidney (197) transplantation, and in a xenograft model involving islet transplantation in the pig-to-rat model (198). In one of the experiments carried out by Kahan et al. (199) a 14-day course of FTY 720 (0.05–8.0 mg/kg/day) by oral gavage prolonged heart allograft survival in a dose-dependent fashion. Although a 14-day oral course of cyclosporine (1.0 mg/kg/day) alone was ineffective (mean survival time = $7.0 \pm 0.7$ versus $6.4 \pm 0.6$ days in treated vs. untreated hosts), treatment with a combination of 1.0 mg/kg/day cyclosporine and 0.1 mg/kg/day FTY 720 extended allograft survival to $62.4 \pm 15.6$ days. Similarly, a 14-day oral course of 0.08 mg/kg/day sirolimus alone was ineffective ($6.8 \pm 0.6$ days; NS), but the combination of sirolimus with 0.5 mg/kg/day FTY 720 extended the mean survival time to $34.4 \pm 8.8$ days. The cyclosporine/sirolimus (0.5/0.08 mg/kg/day) combination acted synergistically with FTY 720 (0.1 mg/kg/day) to prolong heart survivals to >60 days (CI = 0.18). In experiments with other combinations, an attempt was made to define an effective range of FTY 720 doses that could be combined with a suboptimal dose (10 mg/kg) of cyclosporine for canine kidney allograft recipients. FTY 720, at a dose ranging from 0.1 to 3.0 mg/kg, significantly prolonged allograft survival in all groups receiving FTY 720 in combination with cyclosporine. None of the recipients died from the notable side effects of the drug, thereby suggesting that FTY 720 has a potent effect at an extremely low dose and a wide therapeutic window when combined with cyclosporine (197).

Recently, Fururkawa et al. (200) compared a single-dose study with FTY 720 at various doses and a combined-dose study with the conventional immunosuppressant cyclosporine in a canine liver allograft model. The median survival of untreated control animals was 9 days, whereas treatment with FTY 720 at a dose of 0.1 mg/kg/day prolonged graft survival

to 49.5 days. FTY 720 at 1 mg/kg/day showed a slight but insignificant prolongation to 16 days, but when the dose was increased to 5 mg/kg/day, the graft was rejected at 10 days. The combination of FTY 720 (0.1 mg/kg/day) with a subtherapeutic dose of cyclosporine (5 mg/kg/day) prolonged median animal survival from 40 days with cyclosporine alone to 74 days.

FTY 720 thus exhibits potent immunosuppressive activity and can be used synergistically with either cyclosporine, sacrolimus, or tacrolimus for antirejection therapy without enhancing their side effects. Although there is no information about the toxicity of the drug in humans, FTY 720 is metabolized by lysosomal and membrane enzymes as well as possibly by CYP 4A, but not by CYP 3A4, the enzyme that primarily biotransforms cyclosporine and sirolimus. These findings suggest that FTY 720 offers possibilities for the design of immunosuppressive drug combinations. Not only should the agent allow a further reduction in cyclosporine and/or sirolimus doses, but it should also facilitate the elimination of steroids from the immunosuppressive regimen. It is likely that in the near future FTY 720 will eventually prove to be an efficacious new weapon in the immunosuppressive armamentarium.

Recently, an antibody against FTY 720 has been raised by immunizing rabbits with an ovalbumin conjugate of 2-amino-2-(4-(4-mercaptobutyl)phenyl)ethylpropane-1,3-diol HCl (AMPD-4), which contains the essential structure of the novel immunosuppressant FTY 720. As the antibody reacted not only to AMPD-4 but also to FTY 720, it should be useful for the immunoassay of FTY 720 in body fluids, tissues, and cells (201).

***2.2.5.5 Structure-Activity Relationship.*** To investigate the structure-activity relationship, extensive modifications of ISP-I (**7**; Fig. 12.8) were conducted, and it was established that the fundamental structure possessing the immunosuppressive activity is a symmetrical 2-alkyl-2-aminopropane-1,3-diol. The tetradecyl, pentadecyl, and hexadecyl derivatives prolonged rat skin allograft survival in combination of LEW donor and F344 recipient and were more effective than cyclosporine. Among them, 2-amino-2-tetradecylpro-

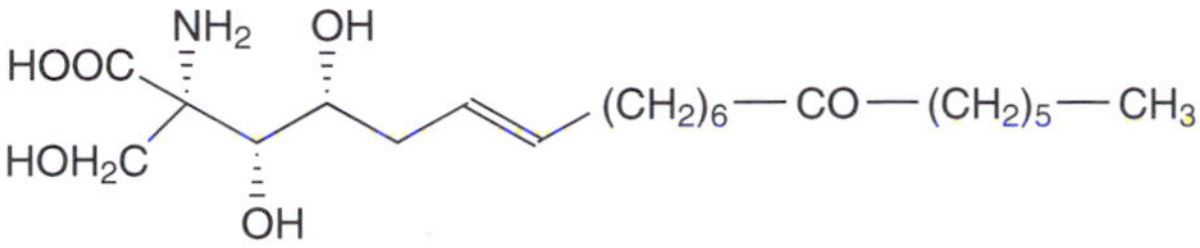

7: SP-I

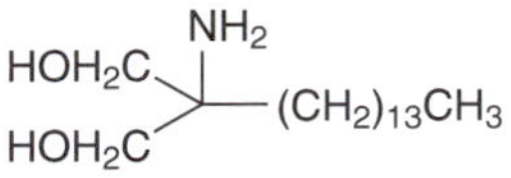

8: ISP-55

**Figure 12.11.** Structure of ISP-I (**7**) and ISP-55 (**8**).

pane-1,3-diol hydrochloride (ISP-I-55, **8**; Fig. 12.11) showed the lowest toxicity (178). Interestingly, each of the 2-amino-1,3-diol hydrochloride compounds is composed of a hydrophilic part (amino alcohol) and a lipophilic part (hydrocarbon chain), thereby suggesting that amphiphilicity should be one of the most important features of this class of compounds. Moreover, given that the lipophilic side chain contains a number of rotatable bonds, Adachi et al. hypothesized that by imposing restriction in the conformational flexibility, the activity could be improved. Accordingly, they introduced a phenyl ring in the lipophilic side chain because it is considered to be an effective template for introducing conformational constraints. Thus, eight compounds possessing a phenyl ring on a variety of positions within the side chain of ISP-I-55 were synthesized (179). The compounds displayed moderate to potent inhibitory activity; however, FTY 720 was the most potent compound and was at least three times more potent than ISP-I-55.

**2.2.5.6 Side Effects.** Administration of FTY 720 at an oral dose of 10 mg/kg for 14 days resulted in a prolongation of graft survival in rodents with a median mean time of 27.0 days without renal toxicity or other toxic sign. However, no information about the toxicity of the drug in humans is yet available.

### 2.2.6 Monoclonal Antibodies

**2.2.6.1 History.** Monoclonal antibodies offer enormous potential as saturation immunosuppressants because they can be used in prophylactic regimens to prevent early acute rejection. They are directed against various levels and stages of T-cell activation or effector function such as against CD4, adhesion molecules, cytokine receptor, transferrin receptor,

portion of T-cell receptor (CD3), and other targets as a means to alter the immune response to alloantigen in a more selective manner. Out of these, the most attractive approach is to target IL-2R with monoclonal antibodies, such as those raised in mice and reacting with the p55 light chain of the receptor, which show promise as immunosuppressive agents in experimental allograft models. Besides IL-2R, another interesting approach is to target CD3 antigen with anti-CD3 monoclonal antibody. Thus the drug works by targeting and blocking a specific region of immune system T-cells, thereby preventing the cells from multiplying in the transplanted organ. In other words, instead of attacking all of the immune cells and making the patient susceptible to infections, monoclonal antibodies can selectively target the cells that have recognized the foreign graft and have started to react to it.

Monoclonal IL-2R antibodies have been successful in pilot studies in preventing solid organ transplantation and in treating GVHD bone marrow transplantation. The monoclonal antibody directed against the p55 chain of the human IL-2R (BT 563) given for 10 days to 19 renal transplant recipients with corticosteroid-resistant graft rejection rescued 13 of 19 patients from rejection, and all had a functioning graft 43 months after treatment (202). Given the human anti-mouse antibody response, however, these antibodies were less effective in human transplantation than those in rodents (203, 204) because humans make antibodies of their own that neutralize these mouse proteins. As a consequence, mouse antibodies typically can be given only once, which is rarely enough to prevent or treat disease. Humanized monoclonals attempt to solve this problem. Using computer modeling and genetic engineering, the binding site of a mouse antibody, the small part of the antibody that attaches to its target, is combined with 90% of a human antibody to develop humanized antibodies. Although their availability was originally limited to murine antibodies produced by hybridoma technology, advances in the past few years have enabled production of "humanized" antibodies by the DNA method. They are reported to have decreased antigenicity and a prolonged plasma half-life, and can be used repeatedly for a longer period of time.

In contrast to monoclonal IL-2R antibodies, the anti-CD3 monoclonal antibody recognizes one of the seven protein chains that constitute the antigen receptor of T-cells and, by binding to this molecule, inhibits the function of T-cells (205). The target specificity of anti-CD3 provides considerable improvement over the use of polyclonal antibodies. Muromonab CD3 (trade name: orthoclone OKT3) has been introduced as an induction treatment in the attempt to block the initial insult to the target organ by T-lymphocytes, as well as to attenuate the frequency and severity of rejection episodes that threaten allograft survival (206). This product was developed at Ortho Biotech Inc., and was approved by the FDA in 1986. Commercially, OKT3 is available as an injectable solution (5 mg/5 mL).

Besides OKT3, two anti-IL-2R monoclonal antibodies, daclizumab and basiliximab (daclizumab, 10% murine and 90% human; basiliximab 40% murine and 60% human), have also been approved by the FDA for the prevention of acute renal transplant rejection (207, 208). Both are engineered human IgG monoclonal antibodies directed against the alpha-subunit (CD25) of the interleukin 2 receptor (IL-2R). Because they are specific for the alpha-subunit (Tac/CD25) of the interleukin 2 (IL-2) receptor on activated T-cells, they achieve immunosuppression by competitive antagonism of IL-2-induced T-cell proliferation. Basiliximab (trade name: Simulect) gained FDA approval in May 1998 and is marketed by Novartis Pharmaceutical Corporation (207). Daclizumab (trade name: Zenepax), developed and patented by Protein Design Laboratories, gained FDA approval in December 1997 and is marketed by Hoffman LaRoche (208). Com-

mercially, daclizumab is available in solution form (25 mg/5-mL vials) for injection, whereas basiliximab is available as a lyophilisate powder for injection (20 mg/vial).

*2.2.6.2 Pharmacokinetics.* The limited data available on the pharmacokinetics of intravenous daclizumab in renal transplant recipients indicate that a dosage of 1 mg/kg every 14 days for a total of five doses can maintain sufficient serum concentrations to provide immunosuppression for 3 months after transplantation. The first dose is given within 24 h before the transplant surgery. Daclizumab has a small volume of distribution (about 5.3 L), and systemic clearance is low (~15.1 mL/h, with interindividual variation of about 20%). The drug has a long terminal half-life (harmonic mean about 480 h), which is similar to that reported for human immunoglobulin G (207). With basiliximab, both single- and multiple-dose pharmacokinetic studies (ranging from 15 to 150 mg) have been carried out in kidney transplant recipients. The recommended regimen consists of two 20-mg doses given i.v. The first dose is given within 2 h before the transplant surgery. The second dose is given 4 days after the transplant. Pediatric (2–15 years of age) protocol is similar to that for adults except that the recommended dose is 12 mg/m$^2$ BSA, up to a maximum of 20 mg. Following i.v. infusion of 20 mg for 30 min, the peak mean ± SD serum concentration was 7.1 ± 5.1 mg/L. A dose-dependent increase in the $C_{max}$ and AUC value was observed up to the maximum tested single dose studied (i.e., 60 mg). The value of distribution is 8.6 ± 4.1 L, terminal half-life is 7.2 ± 3.2 days, and total body clearance is 41 ± 19 mL/h (208, 209).

For OKT3, each dose must be filtered through a low protein-binding 0.22-$\mu$ filter before administration, and is given by i.v. push over 20–40 s. The dose schedule for children less than 12 years old is 0.1 mg/kg/day for 10–14 days, whereas for children over 12 years and for adults it is 5 mg/day for 10–14 days. Immunosuppressive therapy should be either discontinued or substantially reduced during the course of OKT3 therapy. An adequate level of immunosuppression should be restored before the OKT3 course ends (206).

*2.2.6.3 Pharmacology.* In a randomized prospective trial conducted in 101 renal transplant recipients, induction with 2.5 mg OKT3 provided excellent rejection prophylaxis with fewer but persistent side effects. The use of mycophenolate diminishes the number of acute rejection episodes and the formation of anti-KT3 xenoantibody (210). Azathioprine, corticosteroids, and delayed addition of cyclosporine have also been used in the immunosuppressive regimen using prophylactic OKT3 to avoid overimmunosuppression and possible cyclosporine-related nephrotoxicity (211–213). As prophylaxis against renal allograft rejection, muromonab CD3 has been used in transplant recipients with delayed (214) and immediate graft function (215). Although OKT3 is used mainly in renal transplant patients, it is also effective in the treatment of heart, liver, and bone marrow transplants.

Daclizumab was studied in adult and pediatric renal allograft recipients, liver allograft recipients, and calcineurin-sparing protocols in renal transplant recipients (207). It directly and specifically interferes with IL-2 signaling at the receptor level by inhibiting the association and subsequent phosphorylation of the IL-2R beta- and gamma-chains induced by ligand binding (216). Daclizumab, when added to standard cyclosporine-based immunosuppressive therapy with or without azathioprine, significantly reduced the 6-month rate of acute rejection compared with that of placebo in two multicenter placebo-controlled phase III studies in renal transplant recipients. The mean number of rejection episodes was significantly reduced and the time to first acute rejection significantly increased in daclizumab vs. placebo recipients. Patient survival at 1 year after transplantation was significantly higher with daclizumab than with placebo in one study and showed a trend in favor of the drug in the other study (217). In a phase II study, acute rejection rates in patients treated with both daclizumab and mycophenolate mofetil (plus standard cyclosporine-based immunosuppression) were lower than those achieved with mycophenolate mofetil alone (218). A pooled analysis of two randomized, double-blind studies was performed on the efficacy and safety of daclizumab in renal transplant patients when combined with standard immunosuppression. Patients receiving their first cadaveric renal allograft were randomized to receive five doses of daclizumab ($n = 267$) or placebo ($n = 268$), starting preoperatively. Acute rejection at 1 year occurred less frequently with daclizumab ($n = 74$, 27.7%) than with placebo ($n = 116$, 43.3%) ($P = 0.0001$). Mean cumulative doses of corticosteroids were lower with daclizumab (4133 mg) than with placebo (4562 mg). One-year graft survival was 91.4% with daclizumab compared with 86.6% on placebo ($P = 0.065$), with patient survival of 98.5 and 95.1% for daclizumab and placebo, respectively ($P = 0.022$). Thus, therapy with daclizumab significantly reduces acute rejection in renal transplantation and improves patient survival without any increase in morbidity (219). Daclizumab was also found to exhibit substantial activity in patients ($n = 43$) with ongoing acute GVHD. There were no infusion-related reactions and no serious side effects related to daclizumab (220).

Basiliximab was studied in renal allograft recipients and subgroups of recipients of living-related and cadaveric transplants, and in patients with diabetes mellitus. Immunoprophylaxis with basiliximab has been demonstrated to significantly reduce the incidence of acute cellular rejection in adult renal allograft recipients (32% versus placebo, $P < 0.01$) (221). In a study of 732 patients treated with basiliximab, the percentage of biopsy-confirmed acute rejection episodes was decreased from 48% in those patients treated with placebo to 33% in the basiliximab group, which was statistically significant ($P < 0.0001$) (222). Although the risk of acute rejection was reduced, patients who received basiliximab had no adverse reactions and no increase in infectious complications or cancers compared to the placebo group.

Clinically, both daclizumab and basiliximab are used as part of an immunosuppression regimen that may include tacrolimus, cyclosporine, MMF, or corticosteroids. The results of some of the triple-drug regimens used under different organ transplant settings are summarized in Table 12.4. Although both new anti-IL-2R monoclonal antibodies, basiliximab and daclizumab, are additional agents for the prevention of acute renal transplant rejection, their general applicability and long-term therapeutic success remain to be proved.

**Table 12.4   Examples of Some Triple Drug Regimens Involving Monoclonal Antibodies**

| Monoclonal Antibody + Immunosuppressant Drug | Organ Transplantation Model ($n$ = number of patients) | Remarks | Reference |
|---|---|---|---|
| Daclizumab + MMF + steroids | Liver ($n$ = 25) | Daclizumab-based initial immunosuppression can be used safely to reduce the risk for infection with improved long-term graft and patient survival. | 223 |
| Tacrolimus + MMF + steroids + induction with monoclonal antibody | Solitary pancreas ($n$ = 23) | Improved graft survival and reduced rejection rates with the use of monoclonal antibody has been observed. | 224 |
| Daclizumab + sirolimus + MMF | Kidney transplantation ($n$ = 14) | A calcineurin inhibitor-sparing regimen appears to provide effective nonnephrotoxic immunosuppression for kidney transplantation without the need for a lymphocyte-depleting regimen. | 225 |
| Daclizumab + cyclosporine + MMF + prednisone | African–American and high risk Hispanic renal patients ($n$ = 49) | The addition of Daclizumab to an immunosuppressive regimen decreases acute rejection episodes in a high risk group of African–American and Hispanic renal transplant recipients. | 226 |
| Daclizumab + MMF + steroids | Renal transplantation ($n$ = 45) | Most of the rejections were moderate and easily reversible. Interestingly the actuarial 1-year graft survival was 95% with 100% patient survival without using cyclosporine. | 227 |
| Daclizumab + cyclosporine + MMF + prednisone | Cardiac transplantation ($n$ = 55) | Induction therapy with daclizumab safely reduces the frequency and severity of cardiac-allograft rejection during the induction period. | 228 |
| Tacrolimus + sirolimus + daclizumab | Islet transplantation in patients with type I diabetes mellitus ($n$ = 7) | Islet transplantation can result in insulin independence with excellent metabolic control. | 229 |
| Basiliximab + early initiation of cyclosporine therapy | Cadaveric or living-related donor renal transplants ($n$ = 138) | Basiliximab combined with early initiation of cyclosporine therapy resulted in low acute rejection rates similar to those achieved with ATG combined with delayed cyclosporine. Basiliximab therapy showed an excellent safety profile, with no increases in malignancies, infections, or deaths. | 230 |

2.2.6.4 *Side Effects.* Clinical experience from 2 phase III studies conducted to date indicates that daclizumab does not increase the incidence of adverse events when administered with standard cyclosporine-based dual or triple therapy to renal transplant recipients. The incidence of overall adverse events considered to be possibly or probably related to treatment is similar in the daclizumab and placebo treatment groups in both studies (222). For Simulect, severe acute (onset within 24 h) hypersensitivity reactions including anaphylaxis have been observed both on initial exposure and/or after reexposure after several months. In the case of severe hypersensitivity reaction, therapy with Simulect is permanently discontinued (208). The side effects of OKT3 may be life threatening, given that it is associated with markedly increased susceptibility to infection and leads to hypertension, hypotension, chest pain, dizziness, fainting, trembling, headache, and stiff neck (231).

## 2.3 Agents that Block Nucleotide Synthesis

### 2.3.1 Azathioprine

2.3.1.1 *History.* Several antimetabolites synthesized for cancer therapy were incidentally found to have immunosuppressive activity. In 1951 Elion for the first time synthesized 6-mercaptopurine (6-MP) as an inhibitor of nucleic acid base metabolism that significantly increased the life expectancy of leukemic children (232). To increase its efficacy, a large number of derivatives of 6-MP were synthesized and examined for their activities and metabolic fate. One of these was azathioprine, a prodrug of 6-MP, which temporarily protects 6-MP from catabolism and releases 6-MP inside leukemic cells. Though initially used in leukemia, 6-MP was found to suppress antibody formation and allograft rejection. The breakthrough came in 1959 when Schwartz and Dameshek (233) could prevent rabbits from producing antibodies to human serum albumin by treating them for 2 weeks with the antimetabolite 6-MP. Next, Calne et al. observed that bilaterally nephrotectomized dogs living on solitary renal allograft survived for years by treating them with derivatives related to 6-MP (234). Among several drugs pro-

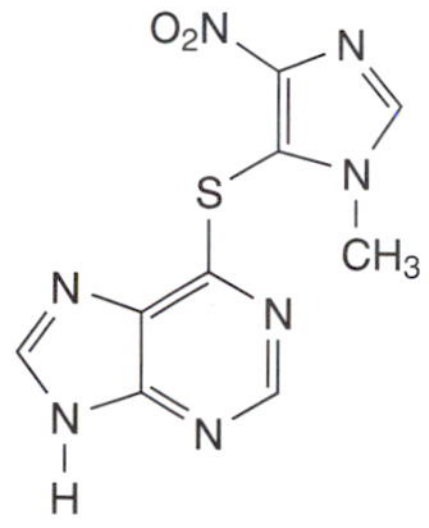

**Figure 12.12.** Structure of azathioprine (**9**).

vided by Hitchings and Elion, azathioprine was found to have the best therapeutic index. The first renal transplant recipient to receive azathioprine was an adult transplanted with an unrelated kidney in March 1961 (235). Finally, its clinical use was recommended in 1968 and since then it has remained the keystone of immunosuppressive treatment for renal transplantation. It is also used to treat severe cases of rheumatoid arthritis, systemic lupus, polymyositis, Crohn's disease, ulcerative colitis, and other autoimmune disorders.

2.3.1.2 *Chemical Structure.* Chemically, azathioprine is 6-(1-methyl-4-nitro-5-imidazyl) thiopurine (**9**; Fig. 12.12).

2.3.1.3 *Pharmacokinetics.* Oral azathioprine is well absorbed, and both azathioprine and its metabolite 6-mercaptopurine (6-MP) distribute throughout the body and are able to cross the placenta. Azathioprine is converted by hepatic xanthine oxidase to 6-MP, which is further metabolized to several compounds including 6-thiourate. These metabolites are excreted in the urine. The plasma half-life of azathioprine is <15 min, whereas the half-life of its active derivative 6-MP is 1–3 h.

2.3.1.4 *Pharmacology.* Azathioprine exerts its immunosuppressive and toxic properties through the release of 6-MP as the main metabolite *in vivo* (235, 236). The metabolic studies of azathioprine revealed that, after absorption, it is nonenzymatically cleaved by sulfhydryl-containing compounds (e.g., cysteine, red blood cells glutathione, etc.) to 6-MP. The latter is then enzymatically converted to ribonucleotide and thioinosinic acid. Interestingly, it is this thioinosinic acid that eventually interferes with the conversion of inosinic acid to guanylic and adenylic acids and gets itself converted to thioguanylic acid, which in turn affects the synthesis of DNA and polyadenylate-containing RNA (237, 238). The antiproliferative activity of azathioprine allows the drug to affect the dividing B- and T-lymphocytes during their proliferation cycle. Because T-cell stimulation by antigens causes cell proliferation, the predominant immunosuppressive activity of azathioprine is to block mitosis of activated cells by interfering with the nucleotide synthesis.

Earlier clinical transplant experiments demonstrated that the drug alone was relatively successful in preventing rejection, but soon it became evident that adjunctive maintenance corticosteroids were more effective. Studies suggested that low dose maintenance corticosteroids when used in combination with azathioprine could be as effective as the more generally accepted higher dose (239). Later randomized trials in renal transplant recipients of high dose vs. low dose corticosteroids in combination with azathioprine supported the concept, in that the low dose group exhibited fewer corticosteroid-related problems. Thus, low dose corticosteroids in combination with azathioprine became the common maintenance immunosuppressive regimen for kidney transplant patients (240). In fact, this was the breakthrough that allowed kidney transplantation to become a routine clinical approach.

However, after the introduction of cyclosporine in 1978, triple-drug therapy with cyclosporine, corticosteroids, and azathioprine became the most frequently used regimen for cadaver kidney recipients (241, 242). These three drugs are believed to complement each other in preventing graft rejection. One advantage of triple-drug therapy is that it allows more flexible immunosuppression, with the possibility of adjusting the dosage of individual components to minimize adverse effects, while maintaining adequate overall immunosuppression. The superior immunosuppressive efficacy of the azathioprine-based triple-drug regimen led to clinical trials with other triple-drug regimens consisting of recently introduced immunosuppressants, to identify the optimal combination of immunosuppressants for a variety of solid organ transplantation settings. Recently, several groups from the United States and Europe reported results of their single center/multicenter randomized trials by replacing azathioprine with other potent immunosuppressants: MMF, tacrolimus, and sirolimus in a variety of organ transplantation models. The results of comparative studies involving an azathioprine-based regimen with other regimens are summarized in Table 12.5. Substituting azathioprine with either MMF or sirolimus or tacrolimus resulted in an improved survival of graft and reduced occurrence and severity of acute rejection episodes in many organ transplantation models. Nevertheless, in another interesting experiment, conversion from azathioprine plus cyclosporine to MMF (20 mg/day) with consecutive reduction of cyclosporine in heart transplant recipients with cyclosporine-impaired renal function improved renal function to a significant extent (251). Thus, although recent studies point toward superior immunosuppression by use of a nonazathioprine-based triple-drug regimen, azathioprine still remains the keystone of immunosuppressive treatment.

*2.3.1.5 Structure-Activity Relationship.* Though it is evident that controlled release of 6-MP plays an important role in azathioprine's activity, studies from different laboratories suggest that the immunosuppressive effects of azathioprine may not be ascribed to the 6-MP alone. Crawford et al. (252) proposed that the secondary immunosuppressive effects of azathioprine might be attributable to the action of the methylnitroimidazolyl substituent. Based on this hypothesis, several analogs of azathioprine were designed and synthesized by replacing the 6-MP component with nontoxic thiols. In all, 24 such congeners were synthesized, out of which two compounds, **IXa** and **IXb** (Fig. 12.13), were found to be more effective than azathioprine in prolonging graft survival in mice. Toxicity studies with these two compounds showed that these analogs had no toxic effects at doses equivalent to that of azathioprine, which caused severe bone marrow depression. Biological effects that have been attributed to the methylnitroimidazolyl moiety of azathioprine involve interference with the processes such as antigen recognition, adherence, and cell-mediated cytotoxicity (253). These findings suggest that azathioprine may further inhibit cell proliferation by mechanisms

**Table 12.5  Comparison of Azathioprine (Aza)-Based Regimen with Other Potent Immunosuppressive Drugs in Different Organ Transplantation Settings**

| Organ Transplantation Model | Azathioprine-Based Treatment | Other Immunosuppressants | Results | Reference |
|---|---|---|---|---|
| Renal transplantation | Aza + cyclosporine + steroids (n = 161) | Sirolimus + cyclosporine + steroids (n = 558) | Use of sirolimus reduced occurrence and severity of acute rejection episodes with no increase in complications. | 114 |
| Pediatric renal transplant recipients | Cyclosporine + Aza + corticosteroids (n = 6) | Cyclosporine + MMF + corticosteroids (n = 16) | MMF leads to an improvement in the immunosuppression and renal function in children with ongoing rejection. | 243 |
| Renal recipients | Aza + cyclosporine + prednisone (n = 50) | MMF + cyclosporine + prednisone (n = 62) | MMF-based triple drug regimen results in fewer rejection episodes. | 244 |
| Renal allograft recipients | Aza + cyclosporine + prednisone (n = 26) | MMF + cyclosporine + prednisone (n = 22) | Graft function was excellent and similar in both groups during the first 6-month observation period. | 245 |
| Simultaneous kidney–pancreas recipients | Aza (n = 76) | MMF (n = 74) | Trends for most efficacy parameters favored MMF over Aza, and time to renal allograft rejection or treatment failure was statistically significantly longer for MMF. The use of MMF in the treatment of SPK recipients is a useful advance. | 246 |
| Liver allograft recipients | Aza + neoral + lymphocyte antibodies + steroid (n = 29) | MMF + neoral + lymphocyte antibodies + steroid (n = 28) | Primary immunosuppression with MMF is advantageous over Aza with regard to safety and efficacy. | 247 |
| Liver transplantation | Tacrolimus + Aza + antilymphocyte globulin + prednisolone (n = 56) | Tacrolimus + prednisone (n = 61) | Both tacrolimus-based dual and quadruple immunosuppressive induction regimens yield similar safety and effectiveness after liver transplantation. | 248 |
| Kidney transplantation | Aza + cyclosporine + steroid (n = 26) | MMF + cyclosporine + steroid (n = 25) | Graft survival demonstrated 12.5% graft losses in the Aza group vs. no kidney transplant losses in MMF group. | 249 |
| Pancreas transplantation | Aza + cyclosporine + steroid + antilymphocyte globulin (n = 13) | MMF + cyclosporine + steroid + antilymphocyte globulin (n = 12) | Patients treated with MMF required less frequent and less intensive treatment for acute rejection. However, its short- and long-term side effects should be further investigated. | 250 |

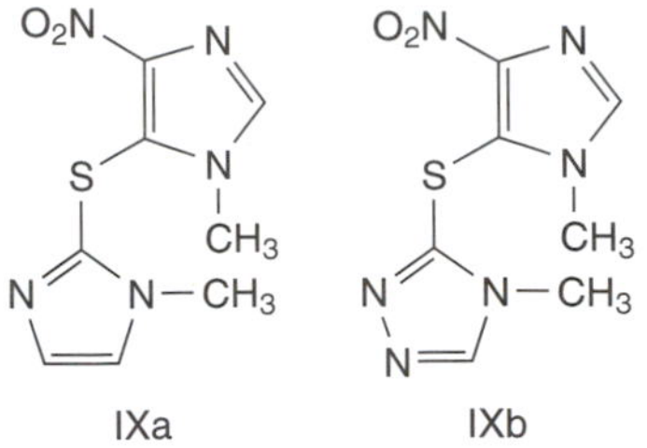

**Figure 12.13.** Structure of azathioprine analogues (9a & 9b).

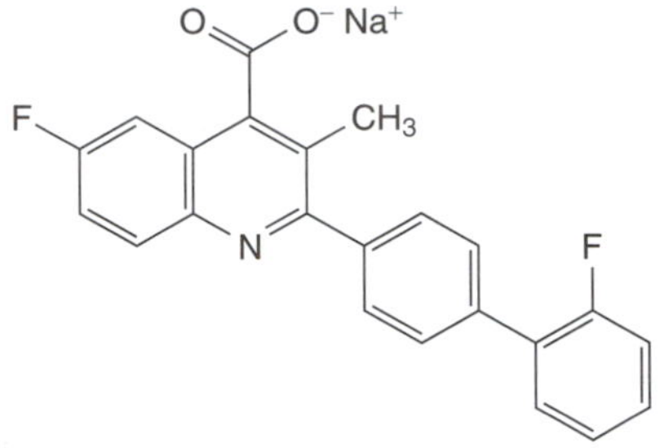

**Figure 12.14.** Structure of brequinar sodium (**10**).

independent of its effect on purine synthesis (254, 255). The drug inhibits the primary immune response with little effect on secondary responses, and is thus useful in preventing acute rejection but not in reversing the process, once started.

*2.3.1.6 Side Effects.* Major side effects associated with azathioprine are nausea, vomiting, mouth ulcers, or anorexia (60%). More serious, but less common, is bone marrow suppression, with reversible leucopenia occurring in up to 20% of the patients. This adverse reaction is usually dose dependent and can lead to further complications including infections and bleeding. Other unwanted side effects include headache, muscle aches, rash, pancreatitis, and, rarely, hepatotoxicity. The reported increased risk of lymphoproliferative disease and skin and urogenital cancers has not been confirmed but may be up to 4%. Azathioprine, however, does not appear to be excreted into breast milk but, in principle, its introduction during lactation should be avoided. Hepatotoxicity occurs in 2–10% of transplant patients receiving azathioprine. Metabolism of azathioprine is inhibited by Allopurinol, which potentiates the effect of azathioprine and increases the risk of myelosuppression. Combining it with other myelosuppressives may also increase the risk (256, 257).

### 2.3.2 Brequinar

*2.3.2.1 History.* Like other antiproliferative agents, brequinar was originally developed in 1985 as an antitumor agent (258). Later in 1993 it was found to exhibit potent and selective immunosuppressive activity (259).

*2.3.2.2 Chemical Structure.* The chemical structure of synthetic brequinar 6-fluoro-2-(2'-fluoro-1,1'-biphenyl-4-yl)-3-methyl-4-quinolinecarboxylic acid sodium salt (**10**) is shown in Fig. 12.14.

*2.3.2.3 Pharmacokinetics.* Brequinar is a water-soluble derivative with an oral bioavailability of 90%, and reaches peak concentrations in the plasma within 2–4 h of oral administration (260). The drug circulates in the peripheral blood tightly bound to serum proteins, with a half-life of approximately 15 h in humans and 17 h in rats. Once the drug is absorbed, it is distributed rapidly to peripheral organs, including liver and kidney (261). The areas under the curve (AUCs) for plasma levels of the drug increase linearly with the dose of the drug, and the plasma clearance is $19.2 \pm 7.7$ mL/min/mm$^2$ (260, 262). The extended half-life and the low plasma clearance allow its administration at an interval of 24–48 h. Brequinar is metabolized in the liver by the P450 cytochrome oxidase system, and is excreted primarily in feces (66%) and, to a lesser extent, in urine (23%). The oral administration of brequinar to rats for 30 days did not affect cyclosporine pharmacokinetics (263). The levels of the compound in the blood can be directly measured using high pressure liquid chromatography (264). The high level of bioavailability, the relative ease with which the compound can be administered, and the prolonged half-life are all features of the pharmacokinetics of brequinar that make the compound attractive for use in the clinical setting.

*2.3.2.4 Pharmacology.* Brequinar inhibits pyrimidine biosynthesis and noncompetitively blocks the activity of the enzyme dihydroorotate dehydrogenase (DHODH) (265, 266).

This enzyme is critical for the formation of uridine and cytidine, which are required for the synthesis of DNA and RNA (267). Studies have shown that brequinar sodium also has the ability to inhibit protein tyrosine phosphorylation and src-related protein tyrosine kinases, thereby suggesting that the activity of brequinar sodium may not be solely the result of the inhibition of pyrimidine nucleotide synthesis; inhibition of protein tyrosine phosphorylation may also be involved (268).

Although brequinar displays antitumor effects over a wide range of concentrations, it displays antilymphocyte effects over only a relatively narrow range (22–185 nmol/L) (269). In vitro, brequinar not only inhibits the proliferation of lymphocytes in a mixed lymphocyte culture but also a wide variety of cellular immune responses including alloantigen and mitogen-induced proliferation (270, 271).

In vivo, brequinar sodium has been found to be effective in suppressing graft-vs.-host responses and allograft rejections, which can be attributed to the potent inhibitory effect on T- and B-cell-mediated responses. Its efficacy as a primary immunosuppressive agent is evident by the effective inhibition of rejections in several models of vascular allografts in rodents. In rat transplant models, brequinar monotherapy has been shown to prolong the allograft survival of hearts, livers, and kidneys. A dose of 12 mg/kg brequinar administered three times a week for 30 days prolonged the survival of heart allografts to $45.5 \pm 12.26$ days, compared with a survival rate of $7.0 \pm 0.69$ days for untreated controls. The same brequinar protocol produced long-term survival ($>230$ days) in 12 out of 26 rat recipients of orthotopic liver (272). Furthermore, the oral administration of 4 mg/kg brequinar three times per week prolonged the survival of heterotopic cardiac allografts in nonhuman primates to $20.0 \pm 21.5$ days, compared to a survival rate of $8.0 \pm 0.5$ days for the controls (272).

Brequinar is also capable of suppressing xenograft rejection because it effectively prolonged survival of hamster-to-rat heart xenograft. In one experiment, brequinar (3 mg/kg/day for 90 days) prolonged the mean survival time of hamster heart xenotransplants in LEW recipients (an inbred strain of rats). The

survival time was increased to $24.5 \pm 42.2$ days (from $4.0 \pm 0.48$ days in the control animals); furthermore, four of the hearts continued to beat for more than 90 days (272). The hamster-to-rat heart xenograft is an example of an accelerated xenograft reaction and brequinar is the only agent capable of prolonging survival for this type of xenograft.

One of the most striking features of the immunosuppressive activity of brequinar sodium is its ability to synergistically interact with a number of other agents to prevent allograft and xenograft rejection. The combination of brequinar with cyclosporine and/or sirolimus was synergistic, as shown by the median effect analysis (273, 274). In primates, the brequinar-cyclosporine combination was able to prolong graft survival to a significant extent (275). Brequinar in combination with leflunomide or tacrolimus exhibited prolonged graft survival in a heterotopic rat cardiac allotransplantation model (276). Administration of BQR (3 mg/kg) with leflunomide (5 mg/kg) or FK 506 (0.5 mg/kg) exhibited prolonged graft survival in both drug combination groups, with a median survival time of 14 days compared to 5 days for controls. Similarly, brequinar in combination with cyclosporine also inhibited islet xenograft rejection in the pig-to-rat model (277). Thus the drug exhibits a number of characteristics that are considered desirable for inclusion in multidrug antirejection protocols.

In a phase I safety and pharmacokinetic study the efficacy of BQR in combination with cyclosporine was examined for the treatment and prophylaxis of rejection in organ transplant patients. The studies were performed in stable renal, hepatic, and cardiac transplant patients receiving cyclosporine and prednisone maintenance therapy for immunosuppression. In all three patient populations, the pharmacokinetics of BQR were characterized by a lower oral clearance (12–19 mL/min) than that seen in patients with cancer (approximately 30 mL/min at similar doses) and a long terminal half-life (13–18 h). This slower oral clearance for brequinar could be attributed either to a drug interaction between brequinar and cyclosporine or to altered clearance or metabolic processes in patients with transplants. Steady-state cyclosporine trough levels and the oral clearance of cyclosporine were not affected by brequinar coadministration. Among the three transplant populations, the cardiac transplant patients had lower oral clearance values for brequinar and of cyclosporine; however, the cause of this lower clearance is not yet clear. Safety results indicate that brequinar was well tolerated by this patient population (278).

**2.3.2.5 Structure-Activity Relationship.**
Structure-activity relationship studies of substituted cinchonic acid (279) and of some tetracyclic heterocycles (280) related to brequinar were studied to determine the structural features required for good activity and an optimal pharmacokinetic profile. In the first instance, compounds with various substituents at the 2-, 3-, 4-, and 6-position of the quinoline ring system were synthesized. The compounds were evaluated for their DHODase inhibitory activity as well as in the MLR assay, a standard model of cell-based immunity indicative of potential allograft rejection. The cinchonic acid core was found to be essential for activity, with only small lipophilic, electron-deficient substitution allowed on the benzo ring. The methyl group was optimal at position 3, although bridging with the 2-biphenyl retained potency. Among tetracyclic heterocycles series, a correlation between DHODase and MLR was observed. Compounds with an ethylene bridge or compounds with a thiomethylene moiety represent the best of the tetracyclic compounds synthesized. Molecular modeling showed the topology of all the tetracyclic ring systems to be similar. The differences in the activity observed with these topologically similar compounds have been in some cases attributed to changes in lipohilicity or basicity, or to the projection angle of the pendant aryl ring. Several of the compounds exhibited activity in DHODase and MLR assays comparable to that of brequinar, warranting further investigations for the development of potent and clinically useful immunosuppressants.

**2.3.2.6 Side Effects.** In a phase I safety trial, 45 cancer patients were administered brequinar by a single daily intravenous infusion for 5 days at a dose range of 36–300 mg/m$^2$/day; several side effects including transaminase elevations, thrombocytopenia, mucositis, phlebitis, and dermatitis were recorded (262). Subsequently, in a phase II efficacy study of brequinar, doses as high as 1800 mg/m$^2$ failed to reduce tumor growth in cancer patients (281). Once the adverse side effect appears, the noncompetitive nature of the enzyme inhibition provides the opportunity to reduce or withdraw treatment with the drug with rapid reversal of the drug-related effects.

### 2.3.3 Mycophenolate Mofetil

**2.3.3.1 History.** Mycophenolate mofetil (MMF; RS 61443) is a semisynthetic derivative of the antimetabolite mycophenolic acid. Mycophenolic acid (MPA) was initially derived from the cultures of the *Penicillium* species by Gosio in 1896 (282) and purified in 1913. Its antibacterial and antifungal activities were recognized in the 1940s. It was not until the 1980s that Nelson, Eugui, and Allison of Syntex, USA considered MPA for use as an immunosuppressant, as part of their search for selective immunosuppressants with a novel mode of action. They searched for a metabolic pathway more susceptible to inhibition in human lymphocytes than that in other cell types and their choice ultimately fell on two major pathways of purine synthesis. They postulated that depletion of GMP by inhibiting ionosine monophosphate dehydrogenase (IMPD) might result in an antiproliferative effect on lymphocytes to a greater extent than that on other cell types, given that lymphocytes use *de novo* purine synthesis, whereas other cells depend more on purine salvage. Thus, from several possible inhibitors, an antimetabolite mycophenolic acid was selected because it was found to be a potent noncompetitive reversible inhibitor of eukaryotic but not of prokaryotic IMP dehydrogenases (283, 284). In subsequent studies morpholinoethyl ester of MPA, mycophenolate mofetil, was selected from a number of derivatives on the basis of its structure, its ability to inhibit lymphocyte proliferation *in vitro*, its ability to inhibit antibody synthesis in mice, and its greater bioavailability when compared to MPA. Finally in 1987, Morris and colleagues at Stanford University (285) decided to evaluate MMF for use in transplantation.

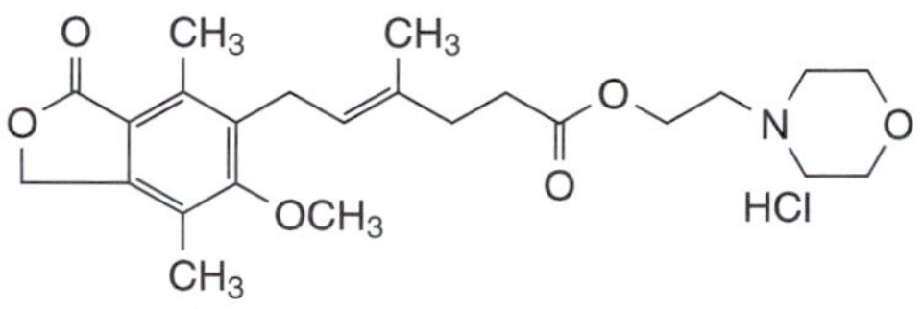

**Figure 12.15.** Structure of mycophenolate mofetil (**11**).

### 2.3.3.2 Chemical Structure.
Mycopheno-late mofetil (trade name: CellCept, **11**; Fig. 12.15) is a morpholinoethyl ester of mycophe-nolic acid.

### 2.3.3.3 Pharmacokinetics.
The bioavail-ability of MPA is only 43% to that of the ester, which is highly soluble at the lower pH in the upper GI tract and is absorbed more rapidly. The liver is the primary location for esterase-mediated hydrolysis of MMF into MPA. The liver is also the site of conversion of MPA to its primary metabolite, mycophenolate glucuro-nide (MPAG), a significant amount of which is secreted into the bile only to be recycled to the liver (enterohepatic recirculation), where it may be converted back to MPA. Thus, entero-hepatic recirculation is thought to contribute significantly to the MPA serum level. The high concentration of drug in the gut may account for the gastrointestinal side effects. Oral bio-availability (of MPA) varies from approxi-mately 95% in healthy volunteers to approxi-mately 50% in newly transplanted patients. The pharmacokinetic profiles of recipients of cadaveric-kidney transplants who had been treated with escalating doses of MMF (100–1750 mg, every second day) exhibited $C_{max}$ in the blood, 1.0–3.2 $\mu$g/mL, and the area under the concentration-time curve over a 24-h pe-riod ($AUC_{0-24}$), 6.4–37.6 $\mu$g/h/mL. The termi-nal half-life of the drug in the blood was 2.3–9.6 h (286).

### 2.3.3.4 Pharmacology.
Mycophenolate mofetil (MMF; RS 61443) is a semisynthetic derivative of the antimetabolite mycophenolic acid, with improved oral bioavailability and is rapidly deesterified *in vivo* to its active metab-olite, mycophenolic acid (MPA) (287). It was first introduced in 1995 for its clinical use in renal transplant patients. Later in 1998 it was approved for other solid organ transplant set-tings involving heart patients and recently as

a non-nephrotoxic immunosuppressant, for pediatric renal transplant recipients with chronic cyclosporine nephrotoxicity (288).

Both MMF and MPA inhibited prolifera-tion of T-cells and B-cells by blocking the pro-duction of guanosine nucleotides required for DNA synthesis. Mycophenolic acid noncom-petitively and reversibly inhibited the enzyme inosine monophosphate dehydegenase (IM-PDH), the rate-limiting enzyme for *de novo* purine synthesis during cell division. Transfer of fucose and mannose to glycoproteins was also found to be inhibited by mycophenolic ac-id-mediated depletion of GTP, which in turn may decrease the recruitment of lymphocytes into the sites of vascularized organ graft rejec-tion and inhibit the ongoing rejection (289).

Measurement of intracellular pools of GTP and dGTP in mitogen-activated PBMC and human T-lymphocytes in the presence or ab-sence of mycophenolic acid supports the hy-pothesis that antiproliferative effects of myco-phenolic acid mainly result from the depletion of GTP or dGTP. It was also found to com-pletely and reversibly suppress DNA synthesis in phytohemagglutinin-stimulated peripheral blood cells, thereby suggesting selectivity in the action of mycophenolic acid by not acting on other enzymes or metabolic functions or on thymidine transport (289).

Studies *in vivo* supported the *in vitro* find-ings with regard to the selectivity of MMF for lymphocytes than for other cell types. It inhib-ited generation of cytotoxic T-cells and rejec-tion of allogenic cells (290) and completely suppressed the formation of antibodies against xenogenic cells in rats (290, 291). The studies also provided theoretical justification for use of the drug in organ transplantation and autoimmune diseases.

In European phase III trials, 491 recipients of kidney transplants were randomized to re-ceive a placebo ($n = 166$) or 2 g/day MMF ($n = 165$) or 3 g/day MMF ($n = 160$), in combination with cyclosporine and prednisone therapy (292). The incidence of biopsy-proved rejec-tion episodes was reduced from 46.4% among placebo-treated patients to 17.0% for the 2 g/day MMF group and to 13.8% for the 3 g/day MMF group. However, 6 months after the transplant, the incidence of graft loss was 6.7% for the 2 g/day MMF group and 8.8% for the 3 g/day MMF group, compared to 10.2% for the placebo group. Although MMF reduced the thickening of the coronary vessels in a rat model, to date there is no clinical evidence that MMF affects the progression of graft-vessel disease or chronic rejection (293). These clinical studies show that MMF decreases the incidence and severity of acute rejection episodes, but has little effect on the overall long-term rate of graft survival.

In organ transplantation models MMF, when given in combination with corticosteroids or cyclosporine, reduces the frequency and severity of acute rejection episodes in kidney and heart transplants, improves patient and graft survival in heart allograft recipients, and increases renal allograft survival up to 3 years (294). The ability of MMF to facilitate sparing of other immunosuppressive agents, particularly in cyclosporine-related nephrotoxicity, is also promising. By permitting reduction in cyclosporine doses, MMF may stabilize or improve renal graft functions in patients with cyclosporine-related nephrotoxicity or chronic allograft nephropathy. Because MMF has a different mechanism of action to cyclosporine and corticosteroids, its immunosuppressive effect appears to be at least additive. The drug has therefore been proposed as an alternative to azathioprine for the prevention of graft rejection in kidney transplant patients. In patients shifted from cyclosporine to either azathioprine (2 mg/kg) or MMF (1 g twice daily), 1 year after transplantation, significantly fewer rejections occurred in patients converted to MMF than in patients converted to AZA (295). Conversion from azathioprine to MMF in pediatric renal transplant recipients with chronic rejection also led to significant improvement in the immunosuppression and renal function in children (243, 296). When used in combination with steroids and cyclosporine in heart transplant trials, CellCept reduced acute rejection and death better than did azathioprine. Thus, CellCept is considered to be a better replacement for azathioprine. Its use has been also favored in the prevention of renal rejections after primary simultaneous kidney-pancreas transplantation (246, 297).

Recently, daclizumab (monoclonal antibody) and MMF have been used as part of an immunosuppressive protocol, with the aim of inducing acceptance of ABO-incompatible mismatched liver allografts in humans (296). It has also been safely and effectively used for the treatment of GVDH in hematopoietic stem cell transplantation; however, the optimal dosage needs further investigation (297). Diabetic rats transplanted with adult porcine islets and immunosuppressed with MMF, cyclosporine, and leflunomide remained normoglycemic for up to 100 days (298).

Interestingly, CellCept has also been effective in reversing acute and resistant rejection episodes in liver and combined pancreas/kidney recipients. Early results of phase I and II clinical trials evaluating MMF therapy in liver, and in combined pancreas/kidney transplant recipients are encouraging (294). The main adverse effects associated with oral or intravenous MMF are gastrointestinal and hematologic in nature. Although the direct costs of using MMF versus azathioprine are higher, the decreased incidence and treatment of acute rejection in patients treated with MMF support its use as a cost-effective option during the first year after transplantation. Thus, MMF has become an important therapeutic tool in the transplant clinician's armamentarium.

*2.3.3.5 Structure-Activity Relationship.* Over a considerable period of time several research groups have attempted to obtain better therapeutic agents based on MPA by means of chemical modification (299), microbiological modification (300), or by latentiation (301). Nelson carried out synthesis and immunosuppressive activity of 12 side-chain variants of MPA (302). The compounds were synthesized either from MPA itself or from 5-(chloromethyl)-1,3-dihydroxy-6-methoxy-7-methyl-3-oxoisobenzofuran, a versatile intermediate for the synthesis of diverse side-chain variants. Replacement of the methylated E-bond of the natural product with a triple bond or a sulfur atom, with overall chain lengths equal to or greater than that of MPA led to compounds devoid of significant activity. Replacement of side-chain double bonds with difluoro, dobromo, or unsubstituted cyclopropane rings also removed most of the activity. Replacement of the double bond with an allenic link-

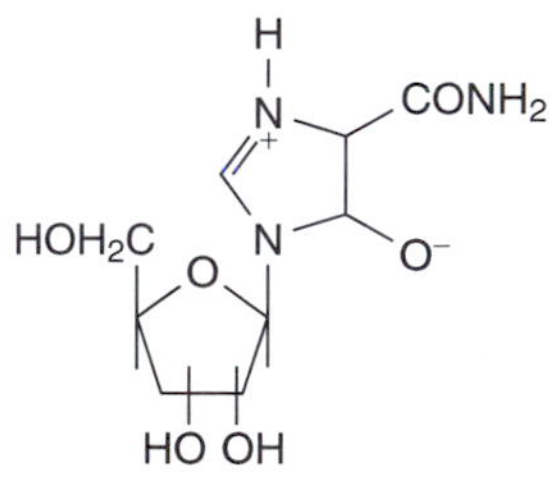

**Figure 12.16.** Structure of mizoribine (**12**).

age yielded a compound with about one-fifth of the immunosuppressive activity of MPA.

**2.3.3.6 Side Effects.** The side effects seen with CellCept use include diarrhea, leukopenia (reduction of white blood cells), sepsis, nausea, and vomiting (303). Allergic reactions to MMF have been observed; therefore, it is contraindicated in patients with a hypersensitivity to MMF, MPA, or any component of the drug product. Intravenous MMF is contraindicated in patients who are allergic to Polysorbate 80 (Tween). Antacids and oral Mg supplements markedly reduce MMF absorption. The administration of antibiotics such as metronidazole and fluoroquinolones and the resultant elimination of intestinal flora are associated with a 35–45% reduction in MPA bioavailability (303). In a phase I study of 48 recipients of cadaveric kidneys, MMF at doses of 100–3500 mg/day caused dose-dependent gastrointestinal toxicity (including gastritis and mild ileus) without any evidence of bone marrow suppression. In another trial, 21 recipients of renal transplants from cadaveric donors were administered 0.25–3.5 g/day of MMF in combination with a cyclosporine-prednisone regimen, and experienced limited side effects such as diarrhea, nausea, elevated liver enzymes, and an increased rate of infections with cytomegalovirus (304).

### 2.3.4 Mizoribine

**2.3.4.1 History.** Mizoribine is a naturally occurring molecule isolated from the filtrate of culture medium of *Eupenicillium brefeldianum* (M-2166) from the soil in 1974 by Mizuno et al. (305).

**2.3.4.2 Chemical Structure.** Mizoribine (MIZ) is a novel imidazole nucleoside (**12**; Fig. 12.16), with a strong immunosuppressive activity.

**2.3.4.3 Pharmacokinetics.** Mizoribine is mainly excreted by the kidney (306); therefore its plasma concentration depends greatly on renal function. Dosage adjustment is thus required when MIZ is given to patients with poor renal function (307). After absorption MIZ is metabolized in the liver into its active component mizoribine-5-P by adenosine kinase (308).

**2.3.4.4 Pharmacology.** Mizoribine has been approved in Japan and recommended as combination therapy along with glucocorticosteroids for renal transplant recipients. Like azathioprine, mizoribine is also a prodrug, which undergoes intracellular phosphorylation, affecting the synthesis of nucleic acid by antagonizing IMPDH and guanosine monophosphate (GMP) synthase, thus blocking one of the essential pathways of *de novo* purine synthesis in lymphocytes (309). It exerts its immunosuppressive activity by inhibiting DNA synthesis in the S phase of the cell cycle, thus preventing T-cell proliferation (308). Another possibility for the therapeutic effect of MIZ has been attributed to the regulation of glucocorticoid receptor function by binding to 14-3-3 proteins (310). It has also been found to suppress the expression of cyclin A mRNA in human B-cells by downregulating its stability, and thus downregulating their response (311). Studies to investigate specific binding proteins by use of the MIZ affinity column led to the identification of mammalian HSP 60 as a major target (312).

As an immunosuppressant, the drug is more potent than azathioprine with less bone marrow suppression (313). In experimental allograft models, mizoribine was found to exhibit a potent synergistic effect with cyclosporine. Therefore its use as an alternative to azathioprine has been recommended in the triple-drug combination therapy. The clinical efficacy of MIZ has been established by comparing standard triple therapy: cyclosporine + prednisone + azathioprine with a MIZ-based regimen in kidney transplant recipients (314–316). The MIZ group exhibited better renal function and graft survival than did the azathioprine group. MIZ also exhibited fewer adverse effects than did azathioprine. Therefore,

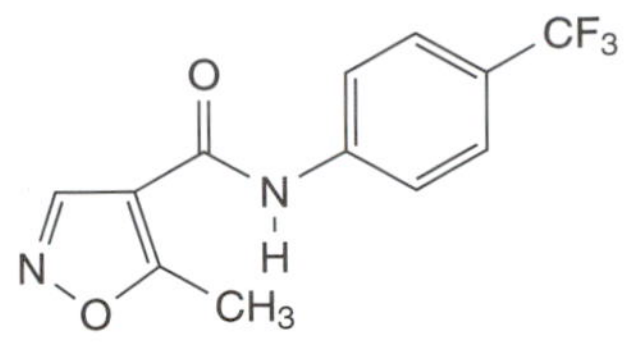

**Figure 12.17.** Structure of leflunomide (**13**).

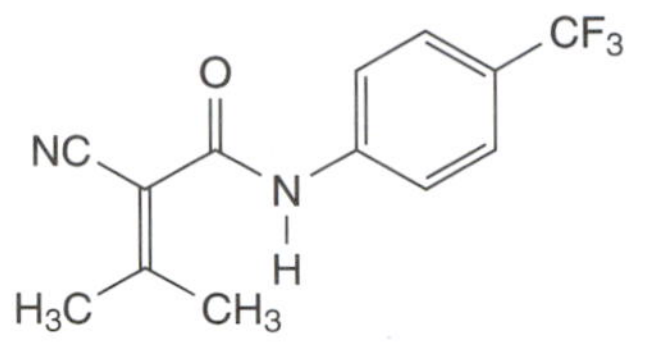

**Figure 12.18.** Structure of leflunomide analog A771726 (**13a**).

MIZ seems to be a much more useful immuno-suppressive agent than azathioprine for renal transplantation.

***2.3.4.5 Structure-Activity Relationship.*** No studies have been carried in this direction.

***2.3.4.6 Side Effects.*** MIZ has been shown in animal experiments to lack oncogenicity, and has been shown clinically to be associated with a low incidence of severe adverse reactions. An increased incidence of hepatic dysfunction and diabetes are some of the side effects observed in humans.

### 2.3.5 Leflunomide

***2.3.5.1 History.*** Leflunomide (LF), a derivative of isoxazole (HWA 486; Hoechst, Basel, Switzerland), inhibits various T-lymphocyte functions (317, 318) and was first described as an inhibitor of the T-cell-dependent antibody production by B-cells (319–321). In recent years several studies have demonstrated that the effect of LF is mediated through inhibition of *de novo* pyrimidine synthesis and tyrosine phosphorylation (322–324).

***2.3.5.2 Chemical Structure.*** Leflunomide (**13**; Fig. 12.17) is an isoxazole derivative [*N*-(4-trifluoro-methylphenyl)-5-methylisoxazol-4-carboxamide], with potent immuno-suppresive, anti-inflammatory, and anticancer activity (SU 101, Sugen Inc., USA).

***2.3.5.3 Pharmacokinetics.*** Although leflunomide has been extensively studied in animal models of transplantation, its clinical use has been initially approved as a disease-modifying antirheumatic drug (Arava) for the treatment of rheumatoid arthritis. This has been attributed to the long half-life (14 days) of leflunomide in humans, which can lead to a difficult situation with regard to its dose adjustment. In rodents, A771726 an active metabolite of LF (**13a**; Fig. 12.18), however, had a half-life of 10–30 h, which is 10 times shorter than its half-life in humans. Because the absorption of

leflunomide varies significantly between individuals and produces severe adverse effects, less-toxic analogs of leflunomide are needed.

***2.3.5.4 Pharmacology.*** After its administration, leflunomide is metabolized quickly to form an active metabolite, A771726, which has been identified as *N*-(4-trifluoromethyl-phenyl-2-cyano-3-hydroxy crotoamide. *In vitro*, A771726 is a potent inhibitor of protein tyrosine kinases (325) and human DHODH, an enzyme involved in pyrimidine biosynthesis (326). This in turn inhibits the proliferation of several immune and nonimmune cell lines and cell cycle progression (327, 328). Another important effect of A771726 is its ability to inhibit humoral-mediated responses by directly blocking T-cell-dependent and T-cell-independent B-cell proliferation and antibody production (298).

In spite of the chemical potential of leflunomide, its mechanism of action is not yet clear. It has been hypothesized that leflunomide inhibits T-cell activation by blocking the *lck* and *fyn* families of tyrosine kinases. These enzymes are known to be associated with the transduction of such growth factor receptor signals as IL-2, interleukin 3 (IL-3) and tumor necrosis factor alpha (TNF-$\alpha$) but not interleukin 1 (329). However, the most recent data show that leflunomide inhibits signal transduction after binding of interleukin 4 (IL-4) to the IL-4 receptor (330).

*In vivo*, leflunomide was found to prevent GVHD and prolonged allograft and xenograft survival in animal models. Moreover, leflunomide also suppressed antibody production in several animal models with allo and xenograft transplantation. In a rat model of GVHD, leflunomide not only was a powerful agent to prevent this otherwise terminal disorder, but was also effective when used as a therapy in an established GVHD. Its efficacy in preventing

GVHD has been attributed to the inhibition of uridine biosynthesis by leflunomide, which leads to a dual antiproliferative effect on both lymphocytes and smooth muscle cells (331, 332). It significantly prolonged survival of allografts: heart (333), pulmonary (334), and islet (335) in rat recipients. In the hamster-to-rat pulmonary model, although leflunomide displayed significant inhibition in xenograft rejection, it was accompanied by severe side effects (334, 335). However, subsequent studies with combination therapy using a variety of clinically used immunosuppressants such as leflunomide plus MMF plus cyclosporine (336), leflunomide plus DSG (337), leflunomide plus cyclosporine (338, 339), leflunomide plus tacrolimus (340), and leflunomide plus brequinar (276) have been reported to be very effective in preventing several xenograft and allograft rejections in animal models. In one of the experiments, the combination of cyclosporine (20 mg/kg/day by gavage) and leflunomide (5 mg/kg/day by gavage for 14–21 days) continuously from the day of transplant was able to completely inhibit the rejection of kidney, spleen, and pancreas xenografts in a hamster-to-rat xenotransplantation model. Only a transient treatment with leflunomide was necessary, and long-term graft survival could be maintained by cyclosporine alone. Histological examination of these grafts at >80 days posttransplantation indicated minimal signs of rejection (341). Leflunomide is currently undergoing clinical evaluation in a phase I/II clinical trial, to test its efficacy for preventing the rejection of kidney allografts.

*2.3.5.5 Structure-Activity Relationship.* Extensive structure-activity relationship studies with A771726 have been carried out to develop suitable analogs that will be clinically acceptable as immunosuppressants. A large number of compounds have been synthesized by different laboratories (342, 343), which suggests that substitution at position 4 of the aromatic ring results in potent congeners with enhanced activity compared to compounds with substitution at positions 2 and 3. Few malononitrilamide analogs (e.g., 279 and 715) have been identified that inhibit T- and B-cell proliferation, suppress immunoglobulin production, and interfere with cell adhesion.

The effects of these agents have been demonstrated in rat skin and cardiac allo- and xenotransplant models. The combined effect of MNA and tacrolimus in a high responder rat cardiac allotransplant model has also been investigated. Optimal doses of MNA or tacrolimus were found to have comparative effects on graft survival and histological changes, whereas a combination of the two drugs was beneficial with respect to both these parameters. Histological analysis of grafts have confirmed the benefit of the drug combination and no additional toxicity has been observed with combined therapy (344). Interestingly, a novel leflunomide analog, HMR 279, has been found to potentiate the immunosuppressive efficacy of microemulsion cyclosporine (Neoral) in rodent heart transplantation and also in a stringent allogeneic rodent lung transplant model. Combination therapies of Neoral (7.5 mg/kg/day) plus HMR 279 (10 mg/kg/day) or Neoral plus LFM were found to be most successful in preventing histologic allograft rejection compared to that of LFM monotherapy (345). Malononitrilamide analogs of A771726 are also being evaluated for immunosuppressive efficacy in preclinical models of transplantation.

*2.3.5.6 Side Effects.* In a phase II clinical trial, leflunomide showed high tolerability and efficacy in patients with advanced rheumatoid arthritis.

# 3  RECENT DEVELOPMENTS

## 3.1  Stem Cells

Stem cells are primitive cells that are capable of forming many different types of body cells. At the time of delivery, the blood in the baby's umbilical cord is quite rich in stem cells; however, as the child ages, these stem cells become less abundant and harder to find. Stem cells, induced to transform themselves into pancreatic cells, could overcome the shortage of donor pancreases and could possibly, when necessary, be genetically engineered to resist being rejected by the immune system. Liver tissue, damaged by infection or toxins such as alcohol, can similarly be replaced by stem cells differentiated into liver cells. Because the cells come from the patient's own body, there is no problem of rejection by the immune system, and no risk of introducing an infection that might have been present in a donor. However, the use of adult stem cells for therapy could certainly reduce, or even avoid, the practice of using stem cells obtained from human embryos or human fetal tissue. Adult stem cells, although present in only minute quantities, can be found throughout the body and are used to repair and regenerate certain types of tissue (346, 347).

Recently, exciting work has been performed in which cells exposed to appropriate inducing agents have been made to differentiate into a far larger number of cell types than was previously thought possible. Thus, bone marrow cells have been used to make nerve cells and nerve cells have been used to make liver cells.

Scientists from Stem Cells, Inc., demonstrated the production of mature liver cells from rigorously purified hematopoietic (blood) stem cells in mice. The study provides the first evidence that liver function can be restored from bone marrow cells in mice with a virulent form of liver failure and that highly purified blood stem cells can efficiently give rise to normal liver cells. Bone marrow is known to contain many cell types, including both mesenchymal (bone- and tissue-forming) and hematopoietic (blood-forming) stem cells. Different subsets of bone marrow cells were purified and each subset was tested by transplantation into mice. Only the subsets containing blood stem cells were able to produce hepatocytes. Furthermore, normal liver cells could be produced from as few as 50 of these highly purified hematopoietic stem cells. These remarkable results indicate that the hematopoietic stem cells are the only cells in the bone marrow responsible for the restoration of liver functions. This exciting development could greatly increase our ability to use adult stem cells therapeutically for the restoration of organs with end-stage disease instead of going for transplantation (348, 349).

### 3.2 Donor Bone Marrow as Antirejection Therapy

Organ rejection occurs less often and is less severe in patients who receive infusions of bone marrow from the same donor. A study of patients who received the extra boost of donor immune system cells indicates the procedure is safe and augments the cellular environment that is necessary for long-term acceptance of a transplanted organ. In other words, the presence of immune system cells from the donor help condition the recipient's system not to attack the new organ. The effect was found to be most dramatic in recipients of new hearts and lungs. The study included 268 patients who received donor bone marrow along with transplants of livers, pancreases, pancreatic islet cells, kidneys, intestines, hearts, and lungs. Of these, 229 received the bone marrow during their transplant operations. The remaining 39 received one infusion during surgery and two more infusions the first and second days after transplantation. The results were compared to 131 patients who received organ transplants without bone marrow. For heart transplant patients, acute cellular rejection occurred in 38% of those who received bone marrow, compared to 82% of those who did not. In lung transplant patients, the most significant result was a reduced incidence of chronic rejection. Obliterative bronchiolitis, the telltale marker for chronic rejection that prevents air exchange in the lung's bronchioles, occurred in 4.7% of the bone marrow patients and 31% in those who did not receive bone marrow. Similarly, the incidence of acute cellular rejection was much lower in the study kidney/pancreas recipients (350). Biopsies taken from the transplanted organs, lymph nodes, skin, and other tissues revealed that donor leukocytes, or white blood cells, had migrated from the transplanted organs to recipient tissues, where they persisted for years—in one case, for 29 years—after transplantation. Thus, simultaneous bone marrow infusion with solid organ transplantation significantly boosts the number of donor immune cells circulating with recipient cells, creating conditions considered essential for whole organ graft acceptance and the development of tolerance (351).

### 3.3 Antisense Technology

Antisense technology offers one of the most effective methods of drug design for selectively inhibiting the expression of a specific mRNA (352) involved during the process of organ rejection. It is hypothesized that the metabolic stages of a specific mRNA can be blocked by allowing an antisense oligonucleotide with a matching sequence to bind to a particular region of the specific mRNA. Despite several limitations, antisense technology offers a clear advantage over all of the previously described drugs. The advantages include: (1) a possibility to target any mRNA; (2) higher affinity and specificity compared with those of classical drugs; and (3) improved drug design, given that both the antisense oligo and the specific mRNA share the same chemistry. Thus, it is possible to target those molecules that are important in the rejection process or those that are used exclusively by T-cells or B-cells, thereby avoiding several nonspecific toxic effects. PS-oligos that are more stable than PD-oligos have been extensively used for antirejection therapy. Many studies have shown that antisense PS-oligos might inhibit protein expression by various mechanisms, that is, degradation of the targeted RNA, promotion of translational arrest, and/or inhibition of RNA processing. In one of the experiments, mouse antisense PS-oligos to ICAM-1 (IP-3082), which targets the 3' end of ICAM-1 mRNA, when delivered intravenously for 7 days, prolonged the survival of heart allografts. Similarly, a combination therapy of anti-ICAM-1 monoclonal antibody and anti-LFA-1 monoclonal antibody delivered intraperitoneally for 7 days induced tolerance to the transplantation of heart allograft (353). In another experiment, survival of pancreatic islets was prolonged to $78.3 \pm 16.5$ days by treating with combined IP-3082 and anti-LFA monoclonal antibody therapy.

Interestingly, the injection of donors intravenously with 10 mg/kg of an antisense oligo to rat ICAM-1 (IP-9125, which targets the 3'-untreated region of rat ICAM-1 mRNA) prolonged the survival of kidney allografts compared with that of controls (354). In yet another experiment, ex vivo perfusion of kidney allografts with 5, 10, or 20 mg of IP-0125 also prolonged the survival of kidney allografts. Thus, the PS-oligo IP-9125 blocks the rejection of organ allografts when used (1) for the postoperative treatment of a recipient, (2) for the pretransplant treatment of a donor, or (3) for the perfusion of grafts. Although these studies successfully demonstrate the efficacy

of antisense technology, further studies are needed to increase the resistance of oligos to degradation by nucleases and to improve the delivery of the oligos to the cells.

## 3.4 Molecules Capable of Inhibiting Costimulatory Signals

Induction of antigen-specific unresponsiveness to grafts is the ultimate goal for organ transplantation. In recent years blockade of the costimulatory signals has been reported to significantly prolong allograft survival. In the costimulation paradigm, the accessory signals generated by antigen-presenting cells are interrupted by distinct agents: the receptor conjugate CTLA4-immunoglobulin and anti-B7 or anti-CD40 ligand mAbs.

CTLA4-Ig is a genetically engineered fusion protein of human CTLA4 and the IgG 1 Fc region. It prevents T-cell activation by binding to human B7, which costimulates T-cells through CD28. Administration of CTLA4-Ig completely inhibited CD11a($-/-$) T-cell proliferation in response to alloantigens and significantly improved skin allograft survival in CD11a($-/-$) mice. Prolonged treatment of wild-type recipient mice with CTLA4-Ig and anti-LFA-1 increased median survival time to 45.5 days compared with 16 days after induction therapy, but it was not sufficient to induce indefinite allograft survival in this model (355). Repligen Corporation has entered into a clinical trial agreement with the National Cancer Institute (NCI) for a phase II trial of CTLA4-Ig.

Monoclonal antibodies to B7–1 and B7–2 have been used in vivo to examine the mechanisms underlying these processes and to evaluate costimulation antagonism as an approach to treatment of chronic autoimmune diseases. Administering monoclonal antibodies specific for these B7 ligands was able to delay the onset of acute renal allograft rejection in rhesus monkeys. The most durable effect resulted from simultaneous administration of both anti-B7 antibodies. The mechanism of action does not involve global depletion of T- or B-cells. Despite in vitro and in vivo evidence demonstrating the effectiveness of the anti-B7 antibodies in suppressing T-cell responsiveness to alloantigen, their use does not result in durable tolerance. Prolonged therapy with murine anti-B7 antibodies is limited by the development of neutralizing antibodies, but that problem was avoided when humanized anti-B7 reagents were used. Thus, anti-B7 antibody therapy may have use as an adjunctive agent for clinical allotransplantation (356).

Blockade of CD154-CD40 pathway with anti-CD154 antibodies has also been reported to prolong allograft survival in experimental transplantation models and to induce tolerance in some instances. However, anti-CD154 monotherapy is unable to induce tolerance in "stringent" models such as skin and islet transplantation and is not sufficient to prevent chronic graft vasculopathy in vascularized organ transplantation. Therefore, combined therapies of anti-CD154 antibodies plus donor-specific transfusion, bone marrow infusion, or B7 blockade by CTLA4-Ig have been tried, and synergistic effects for tolerance induction have been reported. Furthermore, the efficacy of CD154 blockade in primate models has been confirmed for islet and kidney transplantation. The mechanisms of CD154 blockade *in vivo* include CTLA4-dependent anergy or regulation, T-cell apoptosis, and induction of regulatory cells (357).

## 4   THINGS TO COME

In the near future, the continued development of organ transplantation will be shaped by at least four issues that have emerged with the recent maturation of this medical specialty.

1. The need to develop novel immunosuppressants that will offer more effective remedy for antirejection, especially xenograft rejection, without exacerbating the problems of toxic side effects or susceptibility to infection.
2. An emphasis on broadening the application of transplantation as a treatment modality to include patients, such as those presensitized to potential donors, that are not considered good candidates for transplantation.
3. The necessity to work out strategies to use organs from other species (xenograft) to provide a solution to an acute and growing shortage of human donor organs.
4. Development of artificial organs.

## 5   WEB SITE ADDRESSES FOR FURTHER INFORMATION

1. http://www.novartis-transplant.com/
2. http://www.upmc.edu
3. http://www.prohostonline.com
4. http://www.docguide.com
5. http://www.healthlink.mcw.edu
6. http://www.isotechnika.com
7. http://www.fujisawa.com
8. http://home.eznet.net
9. http://www.fp.com.pl/jkc/spanidin.html

## REFERENCES

1. F. D. Moore, *Transplant: The Give and Take of Tissue Transplantation*, Simon & Schuster, New York, 1972, pp. 66–79.
2. C. E. Groth, *Surg. Gynecol. Obstet.*, **134**, 33 (1972).
3. R. Colvin, *Annu. Rev. Med.*, **41**, 361 (1990).
4. M. Sayegh, B. Walschinger, and C. Carpenter, *Transplantation*, **57**, 1295 (1994).
5. R. Steinman and J. Young, *Curr. Opin. Immunol.*, **3**, 361 (1991).
6. R. Schwartz, *Cell*, **57**, 1073 (1989).
7. C. June, J. Lebettre, P. Linsley, and C. B. Thompson, *Immunol. Today*, **11**, 211 (1990).
8. HRSA Release (December 11 1997), HRSA press office, http://www.os.dhhs.gov/news/press/1997pres/971211b.html [Accessed 6/12/2000].
9. UNOS news release (March 9, 2001), United Network for Organ Sharing, http://www.unos.org/Newsroom/archive_newsrelease_20010309_75000.htm [Accessed 6/8/2001].
10. S. Hariharan, C. P. Johnson, B. A. Bresnahan, S. E. Taranto, M. J. Mcintish, and D. Stablein, *N. Engl. J. Med.*, **342**, 605 (2000).
11. HRSA/UNOS News Release (April 16, 2001), United Network for Organ Sharing, http://www.unos.org/Newsroom/archive_newsrelease_20010416_2000numbers.htm [Accessed 6/8/2001].
12. EID home page (February 9, 1996), http://www.cdc.gov/ncidod/EID/vol2no1/michler.htm [Accessed 6/12/2001].
13. B. Kundu and S. K. Khare, *Prog. Drug Res.*, **52**, 3 (1999).
14. Novartis Pharmaceutical Corporation (2000), http://www.novartis-transplant.com/history_of_transplant_treatment/history_sandimmune.html [Accessed 6/8/2001].

15. R. J. Calne, D. J. White, S. Thiru, D. B. Evans, P. McMaster, D. C. Dunn, G. N. Craddock, D. B. Pentlow, and K. Rolles, *Lancet*, **ii**, 1323 (1978).

16. R. L. Powles, A. J. Barrett, H. E. M. Clink, K. J. Sloane, and T. J. McElwain, *Lancet*, **ii**, 1327 (1978).

17. Novartis Pharmaceutical Corporation (November 1999), http://www.pharma.us.novartis.com/product/pi/pdf/sandimmune/pdf [Accessed 6/6/2001].

18. Novartis Pharmaceutical Corporation (November 1999), http://www.pharma.us.novartis.com/product/pi/pdf/neoral.pdf [Accessed 6/6/2001].

19. S. Schreiber, *Science*, **251**, 283 (1991).

20. V. Mikol, J. Kallen, G. Pflugl, and M. D. Walkinshaw, *J. Mol. Biol.*, **234**, 1119 (1993).

21. J. Liu, J. D. Farmer, W. S. Lane, J. Friedman, I. Weissman, and S. L. Schreiber, *Cell*, **66**, 807 (1991).

22. D. Fruman, C. B. Klec, B. E. Bierer, and S. J. Burakoff, *Proc. Natl. Acad. Sci. USA*, **89**, 3686 (1992).

23. B. A. Reitz, C. P. Bieber, A. A. Raney, J. L. Pennock, S. W. Jamieson, P. E. Oyer, and E. B. Stinson, *Transplant. Proc.*, **13**, 393 (1981).

24. T. E. Starzl, G. B. Klintmalm, K. A. Porter, S. Iwatsuki, and G. P. Schroter, *N. Engl. J. Med.*, **305**, 266 (1981).

25. N. R. Saunders, T. M. Egan, D. Chamberlain, and J. D. Cooper, *J. Thorac. Cardiovasc. Surg.*, **88**, 993 (1984).

26. J. D. Hosenpud, L. E. Bennett, B. M. Keck, B. Fiol, and R. J. Novick, *J. Heart Lung Transplant.*, **16**, 691 (1997).

27. Y. Vanrenterghem and J. Peeters, *Transplant. Proc.*, **26**, 2560 (1994).

28. R. D. Forbes, P. Cernacek, S. Zheng, M. Gomersall, and R. D. Guttmann, *Transplantation*, **61**, 791 (1996).

29. A. S. Soin, A. Rasmussen, N. V. Jamieson, C. J. Watson, P. J. Friend, D. G. Wight, and R. Y. Calne, *Transplantation*, **59**, 1119 (1995).

30. P. S. Almond, A. Matas, K. Gillingham, D. L. Dunn, W. D. Payne, P. Gores, R. Gruessner, and J. S. Najarian, *Transplantation*, **55**, 752 (1993).

31. H. Valantine, S. Hunt, P. Gamberg, J. Miller, and H. Luikart, *Transplant. Proc.*, **26**, 2710 (1994).

32. B. D. Kahan, C. T. van Buren, S. M. Flechner, M. Jarowenko, T. Yasumura, A. J. Rogers, N. Yoshimura, S. LeGrue, D. Drath, and R. H. Kerman, *Surgery*, **97**, 125 (1985).

33. R. Traber, H. R. Loosli, M. Hoffman, M. Kuhn, and A. Von Wartburg, *Helv. Chim. Acta*, **65**, 1655 (1982).

34. R. Traber, M. Kuhn, A. Rugger, H. Lichti, H. R. Loosli, and A. Von Wartburg, *Helv. Chim. Acta*, **60**, 1247 (1977).

35. R. M. Wenger, *Angew. Chem. Int. Ed. Engl.*, **24**, 77 (1985).

36. R. M. Wenger, *Helv. Chim. Acta*, **66**, 2672 (1983).

37. R. M. Wenger, *Helv. Chim. Acta*, **67**, 2672 (1984).

38. F. Dumont, M. Staruch, S. Koprak, J. Siekierka, C. Lin, R. Harrison, T. Sewell, V. Kindt, T. Beattie, M. Wyvratt, and N. Sigal, *J. Exp. Med.*, **176**, 751 (1992).

39. Canadian business homepage (April 3, 2000), http://www.isotechnika.com/corporate/main/isa_06282000.pdf [Accessed 6/6/2001].

40. Jon's Heart Transplant Update (July 11, 2000), http://www.jonsplace.org/CHF_tx/CHF transplant.htm [Accessed 12/24/2000].

41. T. Kino, H. Hatanaka, S. Miyata, N. Inamura, N. Nishiyama, T. Yajima, T. Goto, M. Okuhara, M. Kohsaka, H. Aoki, and T. Ochiai, *J. Antibiot.*, **40**, 1256 (1987).

42. N. Amaizi (March 3, 2001), The Drug Monitor, http://home.eznet.net/~webtent/tacro.html [Accessed 2/2/2001].

43. T. K. Jones, S. G. Mills, R. A. Reamer, D. Askin, R. Desmond, P. Volante, and I. Shinkai, *J. Am. Chem. Soc.*, **111**, 1157 (1989).

44. T. K. Jones, R. A. Reamer, P. Desmond, and S. G. Mills, *J. Am. Chem. Soc.*, **112**, 2998 (1990).

45. S. Sawada, G. Suzuki, Y. Kawase, and F. Takaku, *J. Immunol.*, **139**, 1797 (1987).

46. T. Kino, N. Inamura, F. Sakai, K. Nakahara, T. Goto, M. Okuhara, M. Kohsaka, H. Aoki, and H. Ochiai, *Transplant. Proc.*, **19**, 36 (1988).

47. G. D. Van Duyne, R. F. Standaert, P. A. Karplus, S. L. Schreiber, and J. Clardy, *Science*, **252**, 839 (1991).

48. M. W. Harding, A. Galat, D. E. Uehling, and S. L. Schreiber, *Nature*, **341**, 758 (1989).

49. W. M. Flanagan, B. Corthesy, R. J. Bram, and G. R. Crabtree, *Nature*, **352**, 803 (1991).

50. A. L. Defranco, *Nature*, **352**, 754 (1991).

51. S. L. Schreiber and G. R. Crabtree, *Immunol. Today*, **13**, 136 (1992).

52. A. L. Hoffman, L. Makowka, B. Banner, X. Cai, D. V. Cramer, A. Pacualone, S. Todo, and T. E. Starzl, *Transplantation*, **49**, 483 ( 1990).

53. Y. Yasunami, S. Ryu, and T. Kamei, *Transplantation*, **49**, 682 (1990).

54. K. Arai, T. Hotokebuchi, H. Miyahara, C. Arita, M. Mohtai, Y. Sugioka, and N. Kaibara, *Transplantation*, **48**, 782 (1989).

55. S. Todo, J. J. Fung, and A. J. Tzakis, *Transplant. Proc.*, **23**, 1397 (1991).

56. J. J. Fung, K. Abu-Elmagd, and A. Jain, *Transplant. Proc.*, **23**, 2977 (1991).

57. The U.S. Multicenter FK 506 Liver Study Group, *N. Engl. J. Med.*, **331**, 1110 (1994).

58. European FK 506 Multicentre Liver Study Group, *Lancet*, **344**, 423 (1994).

59. J. Van Hoff, Proceedings of the XVI International Congress of the Transplantation Society, August 25–30, 1996, Barcelona, Spain, p. 45.

60. FK 506 U.S. Kidney Transplant Multicenter Study Group. FK 506 in kidney transplantation results of the U.S. randomized, comparative, phase II study [Abstr.], Proceedings of the XVI International Congress of the Transplantation Society, August 25–30, 1996, Barcelona, Spain, p. 45.

61. A. Mathew, D. Talbot, E. Minford, D. Rix, G. Starkey, J. L. Forsythe, G. Proud, and R. M. Taylor, *Transplantation*, **60**, 1182 (1995).

62. M. L. Jordan, R. Naraghi, R. D. Smith, C. A. Vivas, V. P. Scantlebury, H. A. Gritsch, and J. McCauley, P. Randhawa, A. J. Demetris, J. McMichael, J. J. Fung, and T. E. Starzl, *Transplantation*, **63**, 223 (1997).

63. J. M. Armitage, R. L. Kormos, B. P. Griffith, R. L. Hardesty, F. J. Fricker, R. S. Stuart, G. C. Marrone, S. Todo, J. Fung, and T. E. Starzl, *Transplant. Proc.*, **23**, 1149 (1991).

64. P. McMaster and B. Dousset, *Transpl. Int.*, **5**, 125 (1992).

65. M. De Bonis, L. Reynolds, J. Barros, and B. P. Madden, *Eur. J. Cardiothorac. Surg.*, **19**, 690 (2001).

66. F. Carnevale-Schianca, P. Martin, K. Sullivan, M. Flowers, T. Gooley, C. Anasetti, J. Deeg, T. Furlong, P. McSweeney, R. Storb, and R. A. Nash, *Biol. Blood Marrow Transplant.*, **6**, 613 (2000).

67. N. Murase, Q. Ye, M. A. Nalesnik, A. J. Demetris, K. Abu-Elmagd, J. Reyes, N. Ichikawa, T. Okuda, J. J. Fung, and T. E. Starzl, *Transplantation*, **70**, 1632 (2000).

68. A. Kainz, I. Harabacz, I. S. Cowlrick, S. D. Gadgil, and D. Hagiwara, *Transplantation*, **70**, 1718 (2000).

69. D. A. Baran, L. Seguar, S. Kushwaha, M. Courtney, R. Correa, J. T. Fallon, J. Cheng, S. L. Lansman, and A. L. Gass, *J. Heart Lung Transplant.*, **20**, 59 (2001).

70. R. Freudenberger, J. Alexis, A. Gass, V. Fuster, and J. Badimon, *J. Heart Lung Transplant.*, **18**, 1228 (1999).

71. M. P. Revell, M. E. Lewis, C. G. Llewellyn-Jones, I. C. Wilson, and R. S. Bonser, *J. Heart Lung Transplant.*, **19**, 1219 (2000).

72. C. Manzarbeitia, D. J. Reich, K. D. Rothstein, L. E. Braitman, S. Levin, and S. J. Munoz, *Liver Transpl.*, **7**, 93 (2001).

73. M. Hachida, M. Nonoyama, N. Hanayama, M. Miyagishima, H. Hoshi, S. Saito, and H. Koyanagi, *Jpn. J. Thorac. Cardiovasc. Surg.*, **48**, 713 (2000).

74. G. Ligtenberg, R. J. Hene, P. J. Blankestijn, and H. A. Koomans, *J. Am. Soc. Nephrol.*, **12**, 368 (2001).

75. D. O. Taylor, M. L. Barr, B. Radovancevic, D. G. Renlund, R. M. Mentzer Jr., F. W. Smart, D. E. Tolman, O. H. Frazier, J. B. Young, and P. van Veldhuisen, *J. Heart Lung Transplant.*, **18**, 336 (1999).

76. T. Wang, P. Donahoe, and A. Zervas, *Science*, **265**, 674 (1994).

77. C. Johnson, N. Ahsan, T. Gonwa, P. Halloran, M. Stegall, M. Hardy, R. Metzger, C. Shield, L. Rocher, J. Scandling, J. Sorensen, L. Mulloy, J. Light, C. Corwin, G. Danovitch, M. Wachs, P. van Veldhuisen, K. Salm, D. Tolzman, and W. E. Fitzsimmons, *Transplantation*, **69**, 834 (2000).

78. G. Ciancio, J. Miller, and G. W. Burke, *Clin. Transplant.*, **15**, 142 (2001).

79. J. Miller, R. Mendez, J. D. Pirsch, and S. C. Jensik, *Transplantation*, **69**, 875 (2000).

80. B. M. Meiser, M. Pfeiffer, D. Schmidt, H. Reichenspurner, P. Ueberfuhr, D. Paulus, W. von Scheidt, E. Kreuzer, D. Seidel, and B. Reichart, *J. Heart Lung Transplant.*, **18**, 143 (1999).

81. D. G. Mollevi, Y. Ribas, M. M. Ginesta, T. Serrano, M. Mestre, A. Vidal, J. Figueras, and E. Jaurrieta, *Transplantation*, **71**, 217 (2001).

82. H. Hatanaka, T. Kino, S. Miyata, N. Inamura, A. Kuroda, T. Goto, H. Tanaka, and M. Okuhara, *J. Antibiot.*, **41**, 1592 (1988).

83. H. Hatanaka, T. Kino, M. Asano, T. Goto, H. Tanaka, and M. Okuhara, *J. Antibiot.*, **42**, 620 (1989).

84. L. B. Peterson, J. G. Cryan, R. Rosa, M. M. Martin, M. B. Wilsuz, P. J. Sinclair, F. Wong, J. N. Parsons, S. J. O'Keefe, and W. H. Parsons, *Transplantation*, **65**, 10 (1998).

85. Japanese FK 506 Study Group, *Transplant. Proc.*, **25**, 649 (1993).

86. K. A. Abraham, M. A. Little, A. M. Dorman, and J. J. Walshe, *Transpl. Int.*, **13**, 443 (2000).

87. S. N. Sehgal, H. Baker, and C. Vezina, *J. Antibiot.*, **28**, 727 (1975).

88. R. R. Martel, J. Klicius, and S. Galet, *Can. J. Physiol. Pharmacol.*, **55**, 48 (1977).

89. R. Calne, D. Collier, S. Lim, S. G. Pollard, A. Samaan, D. J. White, and S. Thiru, *Lancet*, **2**, 227 (1989).

90. N. Amaizi (January 1, 2001), The Drug Monitor, http://home.eznet.net/~webtent/sirolimus.html [Accessed 2/2/2001].

91. M. A. Wood and B. E. Bierer, *Perspect. Drug Discov. Res.*, **2**, 163 (1994).

92. E. J. Brown, M. W. Albers, T. B. Shin, K. Ichikawa, C. T. Keith, W. S. Lane, and S. L. Schreiber, *Nature*, **369**, 756 (1994).

93. R. E. Morris, *Transplant. Rev.*, **6**, 39 (1992).

94. G. D. Van Duyne, R. F. Standaert, P. A. Karplus, S. L. Schreiber, and J. Clardy, *Science*, **252**, 839 (1991).

95. L. M. Adams, L. M. Warner, W. L. Baeder, S. N. Sehgal, and J. Y. Chang, *J. Cell. Biol.*, **109**, 163A (1989).

96. G. A. Miller, C. C. Bansbach, and M. Rabin, *Transplantation*, **54**, 763 (1992).

97. F. J. Dumont, M. J. Staruch, S. L. Koprak, M. R. Melino, and N. H. Sigal, *J. Immunol.*, **144**, 251 (1990).

98. B. D. Kahan, S. Gibbons, N. Tejpal, S. M. Stepkowski, and T.-C. Chou, *Transplantation*, **51**, 232 (1991).

99. J. E. Kay, C. R. Benzie, M. R. Googier, C. J. Wick, and S. E. Doe, *Immunology*, **67**, 473 (1989).

100. C. H. June, J. A. Ledbetter, M. A. Gilespie, T. Lindsten, and A. J. Thompson, *Mol. Cell. Biol.*, **7**, 4472 (1987).

101. J. E. Kay, L. Kromwel, S. E. A. Doe, and M. Denyer, *Immunology*, **72**, 544 (1991).

102. L. S. Wicker, R. C. J. Boltz, V. Matt, E. A. Nichols, L. B. Peterson, and N. H. Sigal, *Eur. J. Immunol.*, **20**, 2277 (1990).

103. D. J. Price, J. R. Grove, V. Calvo, J. Arvuch, and B. E. Bierer, *Science*, **257**, 973 (1992).

104. C. J. Kuo, J. Chang, D. F. Fiorentino, W. M. Flanagan, J. Blenis, and G. R. Crabtree, *Nature*, **358**, 70 (1992).

105. J. Chung, C. J. Kuo, G. R. Crabtree, and J. Blenis, *Cell*, **69**, 1227 (1992).

106. T. Akiyama, T. Ohuchi, S. Sumida, K. Matsumoto, and K. Toyoshima, *Proc. Natl. Acad. Sci. USA*, **89**, 7900 (1992).

107. S. M. Stepkowski, H. Chen, P. Dloze, and B. D. Kahan, *Transplant. Proc.*, **23**, 507 (1991).

108. R. E. Morris, J. Wu, and R. Shorthouse, *Transplant. Proc.*, **22**, 1638 (1990).

109. D. S. Collier, R. Calne, S. Thiru, S. Lim, S. G. Pollard, P. Barron, M. Costa, and D. J. White, *Transplant. Proc.*, **22**, 1674 (1990).

110. D. S. Collier, R. Y. Calne, S. G. Pollard, P. J. Friend, and S. Thiru, *Transplant. Proc.*, **23**, 2246 (1991).

111. P. M. Kimball, R. H. Kerman, and B. D. Kahan, *Transplant. Proc.*, **23**, 1027 (1991).

112. B. D. Kahan, J. Podbielski, K. L. Napoli, S. M. Katz, H. U. Meier-Kriesche, and C. T. Van Buren, *Transplantation*, **66**, 1040 (1998).

113. B. D. Kahan, B. A. Julian, M. D. Pescovitz, Y. Vanrenterghem, and J. Neylan, *Transplantation*, **68**, 1526 (1999).

114. H. Podder, Proceedings of the 25th Annual American Society of Transplantation Surgeons, 1999.

115. B. D. Kahan, *Lancet*, **356**, 194 (2000).

116. A. S. MacDonald, *Transplantation*, **71**, 271 (2001).

117. W. Schuler, R. Sedrani, S. Cottens, B. Haberlin, M. Schulz, H. J. Schuurman, G. Zenke, H. G. Zerwes, and M. H. Schreier, *Transplantation*, **64**, 36 (1997).

118. G. A. Levy, D. Grant, K. Paradis, J. Campestrini, T. Smith, and J. M. Kovarik, *Transplantation*, **71**, 160 (2001).

119. H. J. Schuurman, J. Ringers, W. Schuler, W. Slingerland, and M. Jonker, *Transplantation*, **69**, 737 (2000).

120. B. Hausen, K. Boeke, G. J. Berry, U. Christians, W. Schuler, and R. E. Morris, *Ann. Thorac. Surg.*, **69**, 904 (2000).

121. O. Viklicky, H. Zou, V. Muller, J. Lacha, A. Szabo, and U. Heemann, *Transplantation*, **69**, 497 (2000).

122. G. I. Kirchner, M. Winkler, L. Mueller, C. Vidal, W. Jacobsen, A. Franzke, S. Wagner, S. Blick, M. P. Manns, and K. F. Sewing, *Br. J. Clin. Pharmacol.*, **50**, 449 (2000).

123. Z. Nikolova, A. Hof, Y. Baumlin, and R. P. Hof, *Transpl. Immunol.*, **8**, 115 (2000).

124. Doctor's Guide Global Edition (December 11, 2000), Docguide homepage, http://www.pslgroup.com/dg/1ED33A.htm [Accessed 6/6/2001].

125. T. Takeuchi, H. Iinuma, S. Kunimoto, M. Ishizuka, M. Takeuchi, H. Hamada, S. Naganawa, S. Kondo, and H. Umezawa, *J. Antibiot.*, **34**, 1619 (1981).

126. H. Umezawa, S. Kondo, H. Iinuma, S. Kunimoto, Y. Ikeda, H. Iwasawa, D. Ikeda, and T. Takeuchi, *J. Antibiot.*, **34**, 1622 (1981).

127. Y. Umeda, M. Moriguchi, H. Kuroda, T. Nakamura, H. Iinuma, T. Takeuchi, and H. Umezawa, *J. Antibiot.*, **38**, 886 (1985).

128. R. Nishizawa, Y. Takei, M. Yoshida, T. Tomiyoshi, T. Saino, K. Nishikawa, K. Nemoto, K. Takahashi, A. Fujii, and T. Nakamura, *J. Antibiot.*, **41**, 1629 (1988).

129. Y. Umeda, M. Moriguchi, H. Kuroda, T. Nakamura, A. Fujii, H. Inuma, T. Takeuchi, and H. Umezawa, *J. Antibiot.*, **40**, 1303 (1987).

130. Y. Umeda, M. Moriguchi, K. Ikai, H. Kuroda, T. Nakamura, A. Fujii, T. Takeuchi, and H. Umezawa, *J. Antibiot.*, **40**, 1316 (1987).

131. F. T. Thomas, M. A. Tepper, J. M. Thomas, and C. E. Haisch, *Ann. N. Y. Acad. Sci.*, **685**, 175 (1993).

132. J. F. Muindi, S. J. Lee, L. Baltzer, A. Jakubowski, H. I. Scher, L. A. Sprancmanis, C. M. Riley, D. Vander Velde, and C. W. Young, *Cancer Res.*, **51**, 3096 (1991).

133. P. Williams, H. Lopez, D. Britt, C. Chan, A. Ezrin, and R. Hottendorf, *J. Pharmacol. Toxicol. Methods*, **37**, 1 (1997).

134. S. Takahara, H. Jiang, Y. Takano, Y. Kokado, M. Ishibashi, A. Okuyama, and T. Sonoda, *Transplantation*, **53**, 914 (1992).

135. P. G. Kerr and R. C. Atkins, *Transplantation*, **48**, 1048 (1989).

136. M. Kawada, T. Someno, H. Inuma, T. Masuda, M. Ishizuka, and T. Takeuchi, *J. Antibiot.*, **53**, 705 (2000).

137. K. Nishimura and T. Tokunaga, *Int. J. Immunopharmacol.*, **10** (Suppl. 1), Abstr. WS21 (1988).

138. S. G. Nadler, D. D. Dischino, A. R. Malacko, J. S. Cleaveland, S. M. Fujihara, and H. Marquardt, *Biochem. Biophys. Res. Commun.*, **253**, 176 (1998).

139. A. M. Waaga, K. Ulrichs, M. Krzymanski, J. Treumer, M. L. Hansmann, T. Rommel, and W. Muller-Ruchholtz, *Transplant. Proc.*, **22**, 1613 (1990).

140. L. H. Pai, D. J. FitzGerald, M. Tepper, B. Schacter, G. Spitalny, and I. Pastan, *Cancer Res.*, **50**, 7750 (1990).

141. K. G. Sterbenz and M. A. Tepper, *Ann. N. Y. Acad. Sci.*, **685**, 205 (1993).

142. H. Reichenspurner, A. Hildebrant, P. A. Hunan, D. H. Boehm, A. G. Rose, J. A. Odell, B. Reichart, and H. U. Schorlemmer, *Transplantation*, **50**, 181 (1990).

143. K. Inagaki, Y. Fukuda, K. Sumimoto, K. Matsuna, H. Ito, M. Takahashi, and K. Dohl, *Transplant. Proc.*, **21**, 1069 (1989).

144. B. U. Von Specht, A. Dibelius, H. Konigsberger, G. Lodde, H. Roth, and W. Permanetter, *Transplant. Proc.*, **21**, 965 (1989).

145. K. Nemoto, M. Hagashi, J. Ito, Y. Sugawana, T. Mae, H. Fujii, F. Abe, A. Fujii, and T. Takeuchi, *Transplantation*, **51**, 712 (1991).

146. H. Reichenspurner, A. Hildebrandt, P. A. Human, D. H. Boehm, A. G. Rose, H. U. Schorlemmer, and B. Reichart, *Transplant. Proc.*, **22**, 1618 (1990).

147. H. U. Schorlemmer, G. Dickneite, and F. R. Seiler, *Transplant. Proc.*, **22**, 1626 (1990).

148. P. Walter, G. Dicknilite, G. Feilfel, and J. Thies, *Transplant. Proc.*, **19**, 3980 (1987).

149. R. Engemann, H. J. Gassel, E. Lafranz, C. Stoffregen, A. Thiede, and H. Hamelmann, *Transplant. Proc.*, **19**, 4241 (1987).

150. B. R. Hsu, F. H. Chang, J. H. Juang, Y. Y. Huang, and S. H. Fu, *Cell Transplant.*, **8**, 307 (1999).

151. H. Amemiya, S. Suzuki, S. Niiya, K. Fukao, N. Yamanaka, and J. Ito, *Transplant. Proc.*, **21**, 3468 (1989).

152. H. Amemiya, S. Suzuki, K. Ota, K. Takahashi, T. Sonoda, M. Ishibashi, R. Omoto, I. Koyama, K. Dohi, and Y. Fukuda, *Transplantation*, **49**, 337 (1990).

153. H. Amemiya, *Ann. N. Y. Acad. Sci.*, **685**, 196 (1993).

154. M. Kyo, Y. Ichikawa, Y. Fukunishi, T. Hanafusa, S. Nagano, S. Takahara, M. Ishibashi, A. Okuyama, H. Ihara, and N. Fujimoto, *Transplant. Proc.*, **27**, 1012 (1995).

155. D. Saumweber, T. Singer, C. Hammer, F. Krombach, D. Bohm, and J. Gokel, *Transplant. Proc.*, **21**, 542 (1989).

156. C. S. Wu, O. Korsgren, L. Wennberg, and A. Tibell, *Transplant. Int.*, **12**, 415 (1999).

157. M. Haga, H. Hirahara, M. Tsuchida, M. Watanabe, M. Takekubo, and J. Hayashi, *Transplant. Proc.*, **32**, 1006 (2000).

158. M. Haga, M. Tsuchida, H. Hirahara, T. Watanabe, J. I. Hayashi, H. Watanabe, Y. Matsumoto, T. Abo, and S. Eguchi, *Transplantation*, **69**, 2613 (2000).

159. Y. N. Tanabe, M. A. Randolph, A. Shimizu, and W. P. Lee, *Plast. Reconstr. Surg.*, **105**, 1695 (2000).

160. G. Wu, O. Korsgren, N. van Rooijen, L. Wennberg, and A. Tibell, *Xenotransplantation*, **6**, 262 (1999).

161. G. S. Wu, O. Korsgren, L. Wennberg, and A. Tibell, *Transpl. Int.*, **12**, 415 (1999).

162. D. M. Kapelanski, M. Perelman, L. Faber, D. E. Paez, E. F. Rose, and D. M. Behrendt, *J. Heart Transplant.*, **9**, 668 (1990).

163. H. Amemiya, K. Dohi, T. Otsubo, S. Endo, S. Nagano, M. Ishibashi, T. Hirano, K. Sato, T. Kurita, and K. Fukao, *Transplant. Proc.*, **23**, 1087 (1991).

164. H. Amemiya, S. Suzuki, K. Ota, K. Takahashi, T. Sonada, M. Ishibashi, R. Omoto, F. Koyama, K. Orita, and H. Takagi, *Int. J. Clin. Pharmacol. Res.*, **11**, 175 (1991).

165. I. H. Koyama, H. Amemiya, Y. Taguchi, T. Watanabe, and N. Nagashima, *Transplant. Proc.*, **23**, 1096 (1991).

166. K. Tanabe, K. Takahashi, M. Yasuo, K. Nemoto, M. Okada, Y. Kanetsuna, S. Oba, S. Fuchinoue, K. Ebihara, S. Onitsuka, T. Hayashi, S. Teraoka, H. Toma, T. Agishi, and K. Ota, *Transplant. Proc.*, **22**, 1620 (1990).

167. K. Maeda, Y. Umeda, and T. Saino, *Ann. N. Y. Acad. Sci.*, **685**, 123 (1993).

168. L. Lebreton, J. Annat, P. Derrepas, P. Dutartre, and P. Renaut, *J. Med. Chem.*, **42**, 277 (1999).

169. L. Lebreton, E. Jost, B. Carboni, J. Annat, M. Vaultier, P. Dutartre, and P. Renaut, *J. Med. Chem.*, **42**, 4749 (1999).

170. T. Fujita, K. Inoue, S. Yamamoto, T. Ikumoto, S. Sasaki, R. Toyama, K. Chiba, Y. Hoshino, and T. Okumoto, *J. Antibiot.*, **47**, 208 (1994).

171. D. J. Kluepfel, J. Bagli, H. Baker, M.-P. Charest, A. Kudelski, S. N. Sehgal, and C. Vezina, *J. Antibiot.*, **25**, 109 (1972).

172. J. F. Bagli, D. Kluepfel, and M. St-Jacques, *J. Org. Chem.*, **38**, 1253 (1973).

173. F. P. Aragozzini, P. L. Manachinin, R. Craveri, B. Rindone, and C. Scolastico, *Tetrahedron*, **28**, 5493 (1972).

174. T. Fujita, K. Inoue, S. Yamamoto, T. Ikumoto, S. Sasaki, R. Toyama, M. Yonata, K. Chiba, Y. Hoshino, and T. Okumoto, *J. Antibiot.*, **47**, 216 (1994).

175. S. Sasaki, R. Hashimoto, M. Kiuchi, K. Inoue, T. Ikumoto, R. Hirose, K. Chiba, Y. Hoshino, T. Okumoto, and T. Fujita, *J. Antibiot.*, **47**, 420 (1994).

176. T. Fujita, N. Hamamichi, T. Matsuzaki, Y. Kitao, M. Kiuchi, M. Node, and Y. Hirose, *Tetrahedron Lett.*, **36**, 8599 (1995).

177. T. Fujita, N. Hamamichi, M. Kiuchi, T. Matsuzaki, Y. Kitao, K. Inoue, R. Hirose, M. Yoneta, S. Saski, and K. Chiba, *J. Antibiot.*, **49**, 846 (1996).

178. T. Fujita, R. Horise, M. Yoneta, S. Sasaki, K. Inoue, M. Kiuchi, S. Hirase, K. Chiba, H. Sakamoto, and M. Arita, *J. Med. Chem.*, **39**, 4451 (1996).

179. K. Adachi, T. Kohara, N. Nakao, M. Arita, K. Chiba, T. Mishina, S. Sasaki, and T. Fujita, *Bioorg. Med. Chem. Lett.*, **5**, 853 (1995).

180. Y. Nagahara, S. Enosawa, M. Ikekita, S. Suzuki, and T. Shinomiya, *Immunopharmacology*, **48**, 75 (2000).

181. H. Yagi, R. Kamba, K. Chiba, H. Soga, K. Yaguchi, M. Nakamura, and T. Itoh, *Eur. J. Immunol.*, **30**, 1435 (2000).

182. K. Chiba, Y. Yanagawa, Y. Masubuchi, H. Kataoka, T. Kawaguchi, M. Ohtsuki, and Y. Hoshino, *J. Immunol.*, **160**, 5037 (1998).

183. Y. Yanagawa, K. Sugahara, H. Kataoka, T. Kawaguchi, Y. Masubuchi, and K. Chiba, *J. Immunol.*, **160**, 5493 (1998).

184. Y. Yanagawa, Y. Hoshino, and K. Chiba, *Int. J. Immunopharmacol.*, **22**, 597 (2000).

185. H. Nakajima, E. Sun, N. Mizuta, I. Fujiwara, and T. Oka, *Transplant. Proc.*, **30**, 2217 (1998).

186. K. Suzuki, T. Kazui, A. Kawabe, X. K. Li, N. Funeshima, H. Amemiya, and S. Suzuki, *Transplant. Proc.*, **30**, 3417 (1998).

187. T. Yamasaki, K. Inoue, H. Hayashi, Y. Gu, H. Setoyama, J. Ida, W. Cui, Y. Kawakami, M. Kogire, and M. Imamura, *Cell Transplant.*, **7**, 403 (1998).

188. K. Yamashita, M. Nomura, T. Omura, M. Takehara, T. Suzuki, T. Shimamura, A. Kishida, H. Furukawa, M. Murakami, T. Uede, and S. Todo, *Transplant. Proc.*, **31**, 1178 (1999).

189. T. Omura, T. Suzuki, T. Shimamura, M. B. Jin, R. Yokota, M. Fukai, J. Iida, M. Taniguchi, S. Magata, H. Horiuchi, K. Yamashita, M. Nomura, A. Kishida, M. Matsushita, H. Furukawa, and S. Todo, *Transplant. Proc.*, **31**, 2783 (1999).

190. S. Kunikata, T. Nagano, T. Nishioka, T. Akiyama, and T. Kurita, *Transplant. Proc.*, **31**, 1157 (1999).

191. K. Yuzawa, M. Otsuka, H. Taniguchi, Y. Takada, N. Sakurayama, Y. Jinzenji, S. Suzuki, and K. Fukao, *Transplant. Proc.*, **31**, 872 (1999).

192. I. Fujiwara, H. Nakajima, T. Matsuda, N. Mizuta, H. Yamagishi, and T. Oka, *Transplant. Proc.*, **31**, 2831 (1999).

193. P. Troncoso, S. M. Stepkowski, M. E. Wang, X. Qu, S. C. Chueh, J. Clark, and B. D. Kahan, *Transplantation*, **67**, 145 (1999).

194. S. M. Stepkowski, M. Wang, X. Qu, J. Yu, M. Okamoto, N. Tejpal, and B. D. Kahan, *Transplant. Proc.*, **30**, 2214 (1998).

195. A. Tamura, X. K. Li, N. Funeshima, S. Enosawa, H. Amemiya, and S. Suzuki, *Transplant. Proc.*, **31**, 2785 (1999).

196. A. Tamura, X. K. Li, N. Funeshima, S. Enosawa, H. Amemiya, M. Kitajima, and S. Suzuki, *Surgery*, **127**, 47 (2000).

197. S. Suzuki, T. Kakefuda, H. Amemiya, K. Chiba, Y. Hoshino, T. Kawaguchi, H. Kataoka, and F. Rahman, *Transpl. Int.*, **11**, 95 (1998).

198. J. Zhang, Z. Song, M. Wijkstrom, S. Bari, B. Sundberg, S. Nava, C. G. Groth, O. Korsgren, and L. Wennberg, *Transplant. Proc.*, **32**, 1017 (2000).

199. M. E. Wang, N. Tejpal, X. Qu, J. Yu, M. Okamoto, S. M. Stepkowski, and B. D. Kahan, *Transplantation*, **65**, 899 (1998).

200. H. Furukawa, T. Suzuki, M. B. Jin, K. Yamashita, M. Taniguchi, S. Magata, H. Ishikawa, K. Ogata, H. Masuko, T. Shimamura, M. Fukai, T. Hayashi, M. Fujita, K. Nagashima, T. Omura, A. Kishida, and S. Todo, *Transplantation*, **69**, 235 (2000).

201. T. Fujita, N. Matsumoto, S. Uchida, T. Kohno, T. Shimizu, R. Hirose, K. Yanada, W. Kurio, and K. Watabe, *Bioorg. Med. Chem. Lett.*, **10**, 337 (2000).

202. S. Carl, M. Wiesel, and V. Daniel, Proceedings of the XVI International Congress of the Transplantation Society, August 25–30, 1996, Barcelona, Spain, p. 1994.

203. J. Kupiec-Weglinski, T. Diamanstein, and N. Tilney, *Transplantation*, **46**, 785 (1988).

204. P. Neuhaus, W. Bechstein, G. Blumhardt, M. Weins, P. Lemmens, J. M. Langreber, R. Lohmann, R. Steffen, H. Schlag, and K. J. Slama, *Transplantation*, **55**, 1320 (1993).

205. T. W. Chang, P. C. Kung, and S. P. Gingras, and G. Goldstein, *Proc. Natl. Acad. Sci. USA*, **78**, 1805 (1981).

206. N. Amaizi (no date), The Drug Monitor homepage, http://home.eznet.net/~webtent/okt3.html [Accessed 2/2/2001].

207. N. Amaizi (January 1, 2001), The Drug Monitor homepage, http://home.eznet.net/~webtent/daclizumab.html [Accessed 2/2/2001].

208. N. Amaizi (January 1, 2001), The Drug Monitor homepage, http://home.eznet.net/~webtent/basiliximab.html [Accessed 2/2/2001].

209. Novartis Pharmaceutical Corporation (March 2001), Novartis homepage, http://www.pharma.us.novartis.com/product/pi/pdf/simulect.pdf [Accessed 6/8/2001].

210. S. M. Flechner, D. A. Goldfarb, R. Fairchild, C. S. Modlin, R. Fisher, B. Mastroianni, N. Boparai, K. J. O'Malley, D. J. Cook, and A. C. Novick, *Transplantation*, **69**, 2374 (2000).

211. D. Norman, L. Kahana, and F. J. Stuart, *Transplantation*, **55**, 44 (1993).

212. P. A. Todd and R. N. Brogden, *Drugs*, **37**, 871 (1989).

213. D. W. Hanto, M. D. Jendrisak, S. K. S. So, C. S. Mc Cullough, T. M. Rush, S. M. Michalski, D. Phelan, and T. Mohana Kumar, *Transplantation*, **57**, 377 (1994).

214. D. J. Cohen, A. L. Benevenisty, J. Cianci, and M. A. Hardy, *Am. J. Kidney Dis.*, **14** (Suppl. 2), 19 (1989).

215. J. M. Grino, A. M. Castelao, and D. Seron, *Am. J. Kidney Dis.*, **20**, 603 (1992).

216. J. Goebel, E. Stevens, K. Forrest, and T. L. Roszman, *Transpl. Immunol.*, **8**, 153 (2000).

217. L. R. Wiseman and D. Faulds, *Drugs*, **59**, 476 (2000).

218. L. R. Wiseman and D. Faulds, *Drugs*, **58**, 1029 (1999).

219. H. Ekberg, L. Backman, G. Tufveson, G. Tyden, B. Nashan, and F. Vincenti, *Transpl. Int.*, **13**, 151 (2000).

220. D. Przepiorka, N. A. Kernan, C. Ippoliti, E. B. Papadopoulos, S. Giralt, I. Khouri, J. G. Lu, J. Gajewski, A. Durett, K. Cleary, R. Champlin, B. S. Andersson, and S. Light, *Blood*, **95**, 83 (2000).

221. M. I. Lorber, J. Fastenau, D. Wilson, J. DiCesare, and M. L. Hall, *Clin. Transplant.*, **14**, 479 (2000).

222. J. Bell and J. Colaneri, *Nephrol. Nurs. J.*, **27**, 243 (2000).

223. S. Emre, G. Gondolesi, K. Polat, M. Ben-Haim, T. Artis, T. M. Fishbein, P. A. Sheiner, L. Kim-Schluger, M. E. Schwartz, and C. M. Miller, *Liver Transpl.*, **7**, 220 (2001).

224. J. S. Odorico, Y. T. Becker, M. Groshek, C. Werwinski, B. N. Becker, J. D. Pirsch, and H. W. Sollinger, *Cell Transplant.*, **9**, 919 (2000).

225. G. J. Chang, H. D. Mahanty, F. Vincenti, C. E. Freise, J. P. Roberts, N. L. Ascher, P. G. Stock, and R. Hirose, *Clin. Transplant.*, **14**, 550 (2000).

226. H. U. Meier-Kriesche, S. S. Palenkar, G. S. Friedman, S. P. Mulgaonkar, M. V. Goldblat, and B. Kaplan, *Transpl. Int.*, **13**, 142 (2000).

227. H. T. Tran, M. K. Acharya, D. B. McKay, M. H. Sayegh, C. B. Carpenter, H. J. R. Auchincloss, R. L. Kirkman, and E. L. Milford, *J. Am. Soc. Nephrol.*, **11**, 1903 (2000).

228. A. Beniaminovitz, S. Itescu, K. Lietz, M. Donovan, E. M. Burke, B. D. Groff, N. Edwards, and D. M. Mancini, *N. Engl. J. Med.*, **342**, 613 (2000).

229. A. M. Shapiro, J. R. Lakey, E. A. Ryan, G. S. Korbutt, E. Toth, G. L. Warnock, N. M. Kneteman, and R. V. Rajotte, *N. Engl. J. Med.*, **343**, 230 (2000).

230. H. Sollinger, B. Kaplan, M. D. Pescovitz, B. Philosophe, A. Roza, K. Brayman, and K. Somberg, *Transplantation*, **72**, 1915–1919 (2001).

231. F. Vincenti, R. Kirkman, S. Light, G. Bumgardner, M. Pescovitz, P. Halloran, J. Neylan, A. Wilkinson, H. Ekberg, R. Gaston, L. Backman, and J. Burdick, *N. Engl. J. Med.*, **338**, 161 (1998).

232. C. P. Rhodas, Ed., *Ann. N. Y. Acad. Sci.*, **60**, 183 (1954).

233. R. Schwartz and W. Dameshek, *Nature*, **183**, 1682 (1959).

234. R. Y. Calne, G. P. J. Alexandra, and J. E. Murray, *Ann. N. Y. Acad. Sci.*, **99**, 743 (1962).

235. J. E. Murray, J. P. Mcrrill, J. H. Harrison, R. E. Wilson, and G. J. Dammin, *N. Engl. J. Med.*, **268**, 1315 (1963).

236. G. B. Elion, *Science*, **244**, 41 (1989).

237. A. H. Chalmers, *Biochem. Pharmacol.*, **23**, 1891 (1974).

238. A. H. Chalmers, P. R. Knight, and M. R. Atkinson, *Aust. J. Exp. Biol. Med. Sci.*, **45**, 681 (1967).

239. M. G. McGeown, J. A. Kennedy, W. G. Loughridge, J. Douglas, J. A. Alexander, S. D. Clarke, J. McEvoy, and J. C. Hewitt, *Lancet*, **11**, 648 (1977).

240. J. R. Salaman, P. G. A. Griffin, and K. Price, *Transplant. Proc.*, **14**, 103 (1982).

241. R. Simmons, D. Canafax, and M. Strand, *Transplant. Proc.*, **17**, 266 (1985).

242. M. First, J. Alexander, and N. Wodhwa, *Transplant. Proc.*, **18**, 132 (1986).

243. J. R. Ferraris, M. L. Tambutti, M. A. Redal, D. Bustos, J. A. Ramirez, and N. Prigoshin, *Transplantation*, **70**, 297 (2000).

244. A. N. Arnold, D. G. Wombolt, T. V. Whelan, P. D. Chidester, I. Restaino, B. Gelpi, M. Stewart, R. L. Hurwitz, and T. R. McCune, *Clin. Transplant.*, **14**, 421 (2000).

245. R. P. Wuthrich, S. Cicvara, P. M. Ambuhl, and U. Binswanger, *Transplant*, **15**, 1228 (2000).

246. R. M. Merion, M. L. Henry, J. S. Melzer, H. W. Sollinger, D. E. Sutherland, and R. J. Taylor, *Transplantation*, **70**, 105 (2000).

247. M. Sterneck, L. Fischer, C. Gahlemann, M. Gundlach, X. Rogiers, and C. Broelsch, *Transplant*, **15**, 43 (2000).

248. P. Neuhaus, J. Klupp, J. M. Langrehr, U. Neumann, A. Gebhardt, J. Pratschke, S. G. Tullius, R. Lohmann, C. Radke, N. Rayes, R. Neuhaus, and W. O. Bechstein, *Transplantation*, **69**, 2343 (2000).

249. D. Di Landro, G. Sarzo, and F. Marchini, *Clin. Nephrol.*, **53**, 23 (2000).

250. P. Rigotti, R. Cadrobbi, N. Baldan, G. Sarzo, P. Parise, L. Furian, F. Marchini, and E. Ancona, *Nephrology*, **53**, 52 (2000).

251. I. Aleksic, M. Baryalei, T. Busch, B. Pieske, B. Schorn, J. Strauch, H. Sirbu, and H. Dalichau, *Transplantation*, **69**, 1586 (2000).

252. D. J. K. Crawford, J. L. Maddocks, D. N. Jones, and P. Szawlowski, *J. Med. Chem.*, **39**, 2690 (1966).

253. G. B. Elion, *Ann. N. Y. Acad. Sci.*, **685**, 401 (1993).

254. V. St Georges, *Ann. N. Y. Acad. Sci.*, **685**, 207 (1993).

255. B. S. Mitchell, J. S. Dayton, and L. A. Turka, *Ann. N. Y. Acad. Sci.*, **685**, 217 (1993).

256. N. Amaizi (January 1, 2001), The Drug Monitor homepage, http://home.eznet.net/~webtent/azathioprine.html [Accessed 2/2/2001].

257. Arthritis link (April 4, 2001), http://www.arthritislink.com/azadrweb.htm [Accessed 6/6/2001].

258. D. L. Dexter, D. P. Hesson, R. J. Ardecky, G. V. Gao, D. L. Tipett, B. A. Dusak, K. D. Paull, J. Plowman, B. M. Delarco, V. L. Narayanan, and M. Forbes, *Cancer Res.*, **45**, 5563 (1985).

259. B. D. Jaffe, E. A. Jones, S. E. Loveless, and S.-F. Chen, *Transplant. Proc.*, **25**, 19 (1993).

260. H. S. Shen, S.-F. Chen, D. Behrens, C. Whitney, D. Dexter, and M. Forbes, *Cancer Chemother. Pharmacol.*, **22**, 183 (1988).

261. B. D. Kahan, *N. Engl. J. Med.*, **321**, 1725 (1989).

262. C. L. Arteaga, T. Brown, and J. Kuhm, *Cancer Res.*, **49**, 4648 (1989).

263. T. B. Barnes, P. Campbell, I. Zajac, and D. Gayda, *Transplant. Proc.*, **25**, 71 (1993).

264. G. Hreha-Eiras, *Transplant. Proc.*, **25**, 32 (1993).

265. S.-F. Chen, L. M. Papp, R. J. Ardecky, G. V. Rao, D. P. Hesson, M. Forbes, and D. Dexter, *Biochem. Pharmacol.*, **40**, 709 (1990).

266. S.-F. Chen, R. Ruben, and D. Dexter, *Cancer Res.*, **46**, 5014 (1986).

267. P. Simon, R. M. Townsend, R. R. Harris, E. A. Jones, and B. D. Jaffee, *Transplant. Proc.*, **25**, 77 (1993).

268. X. Xu, H. Gong, L. Blinder, J. Shen, J. W. Williams, and A. S. Chong, *J. Pharmacol. Exp. Ther.*, **283**, 869 (1997).

269. J. J. Chen and M. E. Jones, *Arch. Biochem. Biophys.*, **176**, 82 (1976).

270. D. V. Cramer, F. A. Chapman, B. D. Jaffee, E. A. Jones, M. Knoop, G. Eiras-Hreha, and L. Makowka, *Transplantation*, **53**, 303 (1992).

271. G. Eiras-Hreha, D. V. Cramer, E. Cajulis, C. Cosenza, L. Mills, K. Hough, M. Frankland, F. A. Chapman, H. Wang, and I. Zajac, *Transplant. Proc.*, **25**, 708 (1993).

272. D. V. Cramer, F. A. Chapman, B. D. Jaffee, I. Zajac, G. Hreha-Eiras, C. Yasunaga, G. D. Wu, and L. Makowka, *Transplantation*, **54**, 403–408 (1992).

273. B. D. Kahan, N. Tejpal, S. Gibbons-Stubbers, Y. Tu, M. Wang, S. Stepkowski, and T. C. Chou, *Transplantation*, **55**, 894 (1993).

274. S. M. Stepkowski, Y. Tu, T. C. Chou, and B. D. Kahan, *Transplant. Proc.*, **26**, 3025 (1994).

275. T. Kawamura, D. A. Hullett, Y. Suzuki, W. O. Bechstein, A. M. Allison, and H. W. Sollinger, *Transplantation*, **55**, 691 (1993).

276. E. A. Antoniou, A. Deroover, A. J. Howie, K. Chondros, P. McMaster, and M. D'Silva, *Microsurgery*, **19**, 98 (1999).

277. L. Wennberg, Z. Song, M. Wijkstrom, J. Zhang, S. Bari, B. Sundberg, C. G. Groth, and O. Korsgren, *Transplant. Proc.*, **32**, 1026 (2000).

278. A. S. Joshi, S. Y. King, B. A. Zajac, L. Makowka, L. S. Sher, B. D. Kahan, A. H. Menkis, C. R. Stiller, B. Schaefle, and D. M. Kornhauser, *J. Clin. Pharmacol.*, **37**, 1121 (1997).

279. D. G. Batt, R. A. Copeland, R. L. Dowling, T. L. Gardner, E. A. Jones, M. J. Orwat, D. J. Pinto, W. J. Pitts, R. L. Magolda, and B. D. Jaffe, *Bioorg. Med. Chem. Lett.*, **5**, 1549 (1995).

280. W. J. Pitts, J. W. Jetter, D. J. Pinto, M. J. Orwat, D. G. Batt, S. R. Sherk, J. J. Petraitis, I. C. Jacobson, R. A. Copeland, R. L. Dowling, B. D. Jaffe, T. L. Gardner, E. A. Jones, and R. L. Magolda, *Bioorg. Med. Chem. Lett.*, **8**, 307 (1998).

281. J. Maroun, J. Ruckdeschel, R. Natalo, R. Morgan, B. Dallaire, R. Sisk, and J. Gyves, *Cancer Chemother. Pharmacol.*, **32**, 64 (1993).

282. B. Gosio, *Rev. Igiene Sanita Pubblica Ann.*, **7**, 825 (1896).

283. T. J. Franklin and J. M. Cook, *Biochem. J.*, **113**, 515 (1969).

284. R. Verham, T. D. Meek, L. Hedstrom, and C. C. Wang, *Mol. Biochem. Parasitol.*, **24**, 1 (1987).

285. R. E. Morris, E. G. Hoyt, and E. Eugui, *Surg. Forum*, **40**, 337 (1989).

286. H. W. Sollinger, M. H. Deierhoi, F. O. Belzer, A. G. Diethelm, and R. S. Kauffman, *Transplantation*, **53**, 428 (1992).

287. W. A. Lee, L. Gu, A. R. Miksztal, N. Chu, K. Leung, and P. H. Nelson, *Pharm. Res.*, **7**, 161 (1990).

288. G. Filler, J. Gellermann, M. Zimmering, and I. Mai, *Transpl. Int.*, **13**, 201 (2000).

289. A. C. Allison, S. J. Almquits, C. D. Muller, and E. M. Eugui, *Transplant. Proc.*, **23**, 10 (1991).

290. E. M. Eugui, A. Mirkovich, and A. C. Allison, *Scand. Immunol.*, **33**, 175 (1991).

291. N. K. Jerne and A. A. Nordin, *Science*, **140**, 405 (1963).

292. European Mycophenolate Mofetil Cooperative Study Group, *Lancet*, **345**, 1321 (1995).

293. C. Schmid, U. Heemann, H. Azuma, and N. L. Tilney, *Transplant. Proc.*, **27**, 438 (1995).

294. T. S. Mele and P. F. Halloran, *Immunopharmacology*, **47**, 215 (2000).

295. P. J. Smak Gregoor, T. van Gelder, N. M. van Besouw, B. J. van der Mast, J. N. IJzermans, and W. Weimar, *Transplantation*, **70**, 143 (2000).

296. W. C. Fang, J. Saltzman, S. Rososhansky, G. Szabo, S. O. Heard, B. Banner, R. Chari, and E. Katz, *Liver Transpl.*, **6**, 497 (2000).

297. M. G. Kiehl, M. Shipkova, N. Basara, and W. I. Blau, and A. A. Fauser, *Bone Marrow Transplant.*, **25** (Suppl. 2), S16 (2000).

298. L. Wennberg, Z. Song, J. Zhang, W. Bennet, S. Nava, S. Bari, B. Sundberg, O. Korsgren, and C. G. Groth, *Transplant. Proc.*, **32**, 1061 (2000).

299. D. F. Jones and S. D. Mills, *J. Med. Chem.*, **14**, 305 (1971).

300. D. F. Jones, R. H. Moore, and G. C. Crawley, *J. Chem. Soc. Perkin Trans.*, **1**, 1725 (1970).

301. K. Suzuki, S. Takaku, and T. Mori, *J. Antibiot.*, **29**, 275 (1976).

302. P. H. Nelson, E. Eugui, C. C. Wang, and A. C. Allison, *J. Med. Chem.*, **33**, 833 (1990).

303. N. Amaizi (December 12, 2000), The Drug Monitor homepage, http://home.eznet.net/~webtent/mmf.html [Accessed 2/2/2001].

304. M. H. Deierhoi, R. S. Kauffman, S. L. Hudson, W. H. Barber, J. J. Curtis, B. A. Julian, R. S. Gaston, D. A. Laskow, and A. G. Diethelm, *Ann. Surg.*, **217**, 476 (1993).

305. K. Mizuno, M. Tsujino, M. Takada, M. Hayashi, and K. Atsumi, *J. Antibiot. Tokyo*, **27**, 775 (1974).

306. S. A. Gruber, G. R. Erdmann, B. A. Burke, A. Moss, L. Bowers, W. J. Hrushesky, R. J. Cipolle, D. M. Canafax, and A. J. Matas, *Transplantation*, **53**, 12 (1992).

307. K. Sonda, K. Takahashi, K. Tanabe, S. Funchinoue, Y. Hayasaka, H. Kawaguchi, S. Teraoka, H. Toma, and K. Ota, *Transplant. Proc.*, **28**, 3643 (1996).

308. L. A. Turka, J. Dayton, G. Sinclair, C. B. Thompson, and B. S. Mitchell, *J. Clin. Invest.*, **87**, 940 (1991).

309. K. Sakaguchi, M. Tsujino, M. Hayashi, K. Kawai, K. Mizuno, and K. Hayano, *J. Antibiot. Tokyo*, **29**, 1320 (1976).

310. S. Takahashi, H. Wakui, J. A. Gustafsson, J. Zilliacus, and H. Itoh, *Biochem. Biophys. Res. Commun.*, **274**, 87 (2000).

311. S. Hirohata, K. Nakanishi, and T. Yanagida, *Clin. Exp. Immunol.*, **120**, 448 (2000).

312. H. Itoh, A. Komatsuda, H. Wakui, A. B. Miura, and Y. Tashima, *J. Biol. Chem.*, **274**, 35147 (1999).

313. H. Uchida, K. Yokota, N. Akiyama, Y. Masaki, K. Aso, M. Okubo, M. Okudaira, M. Kato, and N. Kashiwagi, *Transplant. Proc.*, **11**, 865 (1979).

314. Y. Kokado and M. Ishibashi, *Clin. Transplant.*, **22**, 1679 (1990).

315. K. Mita, N. Akiyama, T. Nagao, H. Sugimoto, S. Inoue, T. Osakabe, Y. Nakayama, K. Yokota, K. Sato, and H. Uchida, *Transplant. Proc.*, **22**, 1679 (1990).

316. K. Tanabe, T. Tokumoto, N. Ishikawa, A. Kanematsu, T. Oshima, M. Harano, M. Inui, T. Yagisawa, I. Nakajima, S. Fuchinoue, K. Takahashi, and H. Toma, *Transplant. Proc.*, **31**, (1999).

317. A. S. Chong, K. Rezaki, K. M. Gebol, A. Finnegan, P. Foster, X. Xu, and J. W. Williams, *Transplantation*, **61**, 140 (1996).

318. R. Lang, H. Wagner, and K. Heag, *Transplantation*, **59**, 382 (1995).

319. R. R. Bartlett, *Int. J. Immunopharmacol.*, **8**, 199 (1986).

320. S. Popovic and R. R. Bartlett, *Agents Actions*, **19**, 313 (1986).

321. R. R. Bartlett, S. Popovic, and R. X. Raiss, *Scand. J. Rheumatol. Suppl.*, **75**, 290 (1988).

322. T. Zielinski, D. Zeitter, S. Muller, and R. R. Bartlett, *Inflamm. Res.*, **44** (Suppl. 2), 5207 (1995).

323. X. Xu, J. W. Williams, E. G. Bremer, A. Finnegan, and A. S. Chang, *J. Biol. Chem.*, **270**, 12398 (1995).

324. H. M. Cherwinski, N. Byars, S. J. Ballaron, G. M. Nakano, J. M. Young, and J. T. Ransom, *Inflamm. Res.*, **44**, 317 (1995).

325. X. Xu, J. W. Williams, H. Gong, A. Finnegau, and A. S. Chong, *Biochem. Pharmacol.*, **52**, 527 (1996).

326. J. P. Davis, G. A. Cain, W. J. Pitts, R. L. Magolda, and R. A. Copeland, *Biochemistry*, **35**, 1270 (1996).

327. R. T. Elder, X. Xu, J. W. Williams, H. Gong, A. Finnegan, and A. S. Chong, *J. Immunol.*, **159**, 22 (1997).

328. H. M. Cherwinski, D. McCarley, R. Schtzman, B. Devens, and J. J. Ransom, *J. Pharmacol. Exp. Ther.*, **272**, 460 (1995).

329. R. R. Bartlett, M. Dimitrijevic, T. Mattar, T. Zielinski, T. Germann, E. Rude, G. H. Thoenes, C. C. Kuchle, H. U. Schorlemmer, and E. Bremer, *Agents Actions*, **32**, 10 (1991).

330. K. Siemasko, A. S. Chong, H. M. Jack, H. Gong, J. W. Williams, and A. Finnegan, *J. Immunol.*, **160**, 1581 (1998).

331. C. Mrowka, G. H. Thoenes, K. H. Langer, and R. R. Bartlett, *Ann. Hematol.*, **68**, 195 (1994).

332. R. V. Nair, W. Cao, and R. E. Morris, *Immunol. Lett.*, **48**, 77 (1995).

333. M. D'Silva, D. Candinas, O. Achilleos, S. Lee, E. Antoniou, A. DeRoover, A. Germenis, C. Stavropoulos, J. Buckels, and D. Mayer, *Transplantation*, **60**, 430 (1995).

334. Z. Guo, A. S. Chong, J. Shen, P. Foster, H. N. Sankary, L. McChesney, D. Mtal, S. C. Jensik, and J. W. Williams, *Transplantation*, **63**, 711 (1997).

335. D. D. Yuh, K. L. Gandy, R. E. Morris, G. Hoyt, J. Gutierrez, B. A. Reitz, and R. C. Robbins, *J. Heart Lung Transplant.*, **14**, 1136 (1995).

336. L. Wennberg, A. Karlsson-Parra, B. Sundberg, E. Rafael, J. Liu, S. Zhu, C. G. Groth, and O. Korsgren, *Transplantation*, **63**, 1234 (1997).

337. W. W. Hancock, T. Miyatake, N. Koyamada, J. P. Kut, M. Soares, M. E. Russell, F. H. Bach, and M. H. Sayegh, *Transplantation*, **64**, 696 (1997).

338. L. S. Yeh, C. R. Gregory, S. M. Griffey, R. A. Lecouter, S. M. Hou, and R. E. Morris, *Transplantation*, **64**, 919 (1997).

339. C. G. Groth, A. Tibell, L. Wennberg, and O. Korsgren, *J. Mol. Med.*, **77**, 153 (1999).

340. C. Rastellini, L. Cicalese, R. Leach, M. Braun, J. J. Fung, T. E. Starzl, and A. S. Rao, *Transplant. Proc.*, **31**, 646 (1999).

341. A. S. Chong, L. L. Ma, D. Yin, L. Blinder, J. Shen, and J. W. Williams, *Xenotransplantation*, **7**, 48 (2000).

342. C. A. Axton, M. E. J. Billingham, P. M. Bishop, P. T. Gallegher, T. A. Hicks, E. A. Kitchen, G. W. Mullier, W. M. Owton, and M. G. Parry, *J. Chem. Soc. Perkin Trans.*, **1**, 2203 (1992).

343. E. A. Kuo, P. Hambleton, D. P. Kay, P. L. Evans, S. S. Matharu, E. Little, N. McDowall, C. B. Jones, C. J. R. Hedgecock, C. M. Yea, A. W. E. Chan, P. W. Hairsine, I. R. Ager, W. G. Tully, R. A. Williamsoss, and R. Westwood, *J. Med. Chem.*, **39**, 4608 (1996).

344. Z. Qi, M. Simanaitis, and H. Ekberg, *Transpl. Immunol.*, **7**, 169 (1999).

345. B. Hausen, K. Boeke, G. J. Berry, J. F. Gummert, U. Christians, and R. E. Morris, *Transplantation*, **67**, 354 (1999).

346. E. Emerson (no date), USC Health Magazine homepage, http://www.usc.edu/hsc/info/pr/hmm/sum98/transplant.html [Accessed 5/23/2001].

347. C. H. Bebrovner (December 10, 2000), N.Y. Society for Ethical Culture, http://www.nysec.org/addresses/stem_cells.html [Accessed 1/12/2001].

348. E. Lagasse, H. Connors, M. Al-Dhalimy, M. Reitsma, M. Dohse, L. Osborne, X. Wang, M. Finegold, I. L. Weissman, and M. Grompe, *Nat. Med.*, **6**, 1212 (2000).

349. Backlash (November 1, 2000), Prohost online, http://www.prohostonline.com/_disc6/0000013a.htm [Accessed 1/12/2001].

350. UPMC home page (June 1, 2000), Lisa Rossi, http://www.upmc.edu/newsbureau/tx/lanbmt-bg.htm [Accessed 6/12/2001].

351. Doctor's Guide Global Edition (April 21, 1999), Docguide homepage, http://www.pslgroup.com/dg/F880A.htm [Accessed 6/6/2001].

352. R. M. Crooke, M. J. Graham, M. E. Cooke, and S. T. Crooke, *J. Pharmacol. Exp. Ther.*, **275**, 462 (1995).

353. M. Isobe, J. Suzuki, S. Yamazaki, Y. Yazaki, S. Horie, Y. Okubo, K. Maemura, Y. Yazaki, and M. Sekiguchi, *Circulation*, **96**, 2247 (1997).

354. S. M. Stepkowski, M. E. Wang, T. P. Condon, S. Cheng-Flournoy, K. Stecker, M. Graham, X. Qu, L. Tian, W. Chen, B. D. Kahan, and C. F. Bennett, *Transplantation*, **66**, 699 (1998).

355. H. Malm, M. Corbascio, C. Osterholm, S. Cowan, C. P. Larsen, T. C. Pearson, and H. Ekberg, *Transplantation*, **73**, 293 (2002).

356. A. D. Kirk, D. K. Tadaki, A. Celniker, D. S. Batty, J. D. Berning, J. O. Colonna, F. Cruzata, E. A. Elster, G. S. Gray, R. L. Kampen, N. B. Patterson, P. Szklut, J. Swanson, H. Xu, and D. M. Harlan, *Transplantation*, **72**, 377 (2001).

357. A. Yamada and M. H. Sayegh, *Transplantation*, **73**, S36 (2002).

# CHAPTER THIRTEEN

# Synthetic Antibacterial Agents

Nitya Anand
B-62, Nirala Nagar
Lucknow, India

William A. Remers
University of Arizona
College of Pharmacy
Tucson, Arizona

## Contents

1 Introduction, 538
2 Topical Synthetic Antibacterials, 539
  2.1 Terminology, 539
  2.2 Principles of Topical Antimicrobial Activity, 539
    2.2.1 Selective Toxicity, 539
    2.2.2 Cellular Targets, 543
    2.2.3 Mechanisms of Bacterial Resistance, 543
    2.2.4 Kinetics and Other Factors, 543
  2.3 Evaluation of Antimicrobial Activity, 544
    2.3.1 Methods of Testing Disinfectants, 544
    2.3.2 Methods for Testing Antiseptics, 544
  2.4 History, 545
  2.5 Halogens and Halophores, 545
    2.5.1 Chlorine and Chlorophores, 545
    2.5.2 Iodine, 547
  2.6 Alcohols, 547
  2.7 Phenols, 548
    2.7.1 Monophenols, 548
    2.7.2 Bisphenols, 549
  2.8 Epoxides and Aldehydes, 550
    2.8.1 Epoxides, 550
    2.8.2 Aldehydes, 551
  2.9 Acids, 551
  2.10 Oxidizing Agents, 551
  2.11 Heavy Metals, 552
    2.11.1 Mercury Compounds, 552
    2.11.2 Silver Salts, 553
  2.12 Dyes, 553
  2.13 Diarylureas, Amidines, and Biguanides, 554
  2.14 Surface-Active Agents, 555
    2.14.1 Cationic Surfactants, 555
    2.14.2 Anionic Surfactants, 556
    2.14.3 Amphoteric Surfactants, 557

*Burger's Medicinal Chemistry and Drug Discovery*
Sixth Edition, Volume 5: Chemotherapeutic Agents
Edited by Donald J. Abraham
ISBN 0-471-37031-2    © 2003 John Wiley & Sons, Inc.

3 Systemic Synthetic Antibacterials, 557
   3.1 Sulfonamides and Sulfones, 557
      3.1.1 Introduction, 557
      3.1.2 Clinical Use of Sulfonamides and
          Sulfones, 558
          3.1.2.1 Present Status in Therapeutics,
             558
          3.1.2.2 Adverse Reactions, 559
          3.1.2.3 Pharmacokinetics and
             Metabolism, 563
      3.1.3 Antimicrobial Spectrum, 565
      3.1.4 Mechanism of Action, 566
          3.1.4.1 Site of Folate Inhibition, 566
          3.1.4.2 Synergism with Dihydrofolate
             Reductase Inhibitors, 569
          3.1.4.3 Drug Resistance, 572
      3.1.5 Historical Background, 573
      3.1.6 Structure-Activity Relationship, 577
          3.1.6.1 Structure and Biological
             Activity, 577
          3.1.6.2 Quantitative Sar, 578
          3.1.6.3 Water/Lipid Solubility, 580
          3.1.6.4 Protein Binding, 581
   3.2 Quinolones, 582
      3.2.1 Introduction, 582

3.2.2 Antibacterial Activity, 583
      3.2.3 Structure-Activity Relationship, 584
      3.2.4 Mechanism of Action, 585
      3.2.5 Microbial Resistance, 586
      3.2.6 Pharmacokinetic Properties, 586
      3.2.7 Adverse Reactions, 586
   3.3 Oxazolidinones, 587
      3.3.1 Introduction, 587
      3.3.2 Linezolid, 588
          3.3.2.1 Antibacterial Activity, 588
          3.3.2.2 Mechanism of Action, 588
          3.3.2.3 Pharmacokinetics and
             Metabolism, 589
          3.3.2.4 Therapeutic Uses, 590
          3.3.2.5 Microbial Resistance, 590
          3.3.2.6 Toxicity and Adverse Effects,
             590
      3.3.3 New Oxazolidinones, 590
      3.3.4 Perspective, 593
   3.4 Other Systemic Synthetic Antibacterials, 594
      3.4.1 Nitrofurans, 594
      3.4.2 Methenamine, 595
      3.4.3 Cotrimoxazole, 596
   3.5 The Challenge of Antibacterial
      Chemotherapy, 596

# 1  INTRODUCTION

Synthetic antibacterial compounds are divided into two major classes: topical agents and systemic agents. The topical agents are termed disinfectants, antiseptics, and preservatives, depending on how they are used. Antiseptics and disinfectants differ from systemic agents in that they show little selective toxicity between the microbes and the host. Furthermore, most of them do not aid wound healing and may even impair it. Nevertheless, there are indispensable uses for disinfectants in hospital sanitation, including sterilization of surgical instruments, public health methods, and in the home. Antiseptics have important applications in the preoperative preparation of both surgeons and patients. They also are used in treating local infections caused by microorganisms refractive to systemic antimicrobial agents.

The development of systemic antibacterials had a strong dye-drug connection. Ehrlich in his classic studies on the selective uptake of chemicals by cells and tissues used dyes because they could be followed visually. His de-

velopment of selective staining methods for the identification of microorganisms led him to propose that dyes may also have selective toxicities for microbes. In fact many dyes were found to have antimicrobial activity and some are still used as germicides and disinfectants. This was the backdrop in which Ehrlich coined the term *chemotherapy* for selective killing of pathogenic microbes by chemicals. Discovery of the antibacterial activity of prontosil rubrum, a sulfonamide-azo dye, in a streptococcal infection in mice, led to the first effective chemotherapeutic agent employed for the treatment of systemic bacterial infections. The identification of sulfanilamide as the active component of prontosil led to the synthesis of a number of analogs called sulfonamides with improved activity.

The discovery of antibiotics and the development of bacterial resistance to sulfonamides reduced the use of sulfonamides. Nevertheless, they remain important for specific bacterial infections. The structurally related sulfones are the mainstay for treating leprosy. Other synthetic antibacterial agents have gained increasing attention in recent years.

This is especially true for the fluoroquinolones, which are highly effective systemically against Gram-positive and Gram-negative bacteria resistant to other agents. Very recently, the oxazolidinones have become another very promising class of synthetic antibacterial agents.

There are also some synthetic agents useful specifically for urinary tract infections. They include the nitrofurans and methenamine.

Systemic antibacterial agents are described in two different places in this volume. The antimycobacterial agents are discussed elsewhere, whereas the sulfonamides, sulfones, quinolones, oxazolidinones, and other compounds are discussed in the present chapter.

## 2 TOPICAL SYNTHETIC ANTIBACTERIALS

Table 13.1 summarizes the topical synthetic antibacterial agents in common use today. These agents possess a wide variety of chemical structures and properties, and they act by many different mechanisms to produce their antibacterial effects. Some of them are extremely toxic, which restricts their use to sterilizing surgical instruments and fumigating structures. Others are powerful agents that are irritating to skin. They are used to disinfect dairy barns, hospital areas, and the like. Less irritating agents are used as surgical scrubs. Still milder agents that possess potent antibacterial activity are used as antiseptics for treating wounds, or as mouthwashes. Very mild and nonirritating compounds may be used to kill bacteria on contact lenses. Thus there is a broad spectrum of needs for topical antibacterial agents, and a large number of chemical agents have been developed to meet these needs.

### 2.1 Terminology

The terms describing topical antibacterial agents are used rather loosely in everyday language, which sometimes creates confusion. Some of these terms have strict definitions by the U.S. Food and Drug Administration. Perhaps the two most important terms are disinfectant and antiseptic. A *disinfectant* is defined as a substance that destroys harmful microorganisms, although it may not kill bacterial spores. It is used when referring to agents applied to inanimate objects. Physical agents such as X-rays and ultraviolet light also are considered to disinfect. The term *antiseptic* is used for agents that kill or prevent the growth of microorganisms when used on living tissues. Antiseptics are used in soaps, mouthwashes, douches, and preparations for minor wounds and burns.

The ending *-cide* is used to denote killing action. Thus, a bactericide kills bacteria, a fungicide kills fungi, a virucide destroys viruses, a germicide kills various kinds of microorganisms, and a biocide kills all living organisms. Similarly, the ending *-stat* is used to describe an agent that prevents the growth of organisms, but does not necessarily kill them. The corresponding terms are bacteriostat, fungistat, and so forth.

Other commonly used terms include *antimicrobial*, which refers to an agent that kills or suppresses the growth of microorganisms, and *sanitizer*, which refers to an agent that reduces the number of harmful bacteria to an established safe limit. The criterion for sanitization is killing of 99.999% of specific test bacteria in 30 s, and it is commonly applied to eating and drinking implements and dairy equipment (1).

Terms used to describe processes include *asepsis*, which means the prevention of contamination by microorganisms; *decontamination*, which refers to the disinfection or sterilization of infected things; *sterilization*, which refers to killing all forms of life, especially microorganisms; and *fumigation*, which is the exposure of an area or object to disinfecting fumes.

### 2.2 Principles of Topical Antimicrobial Activity

**2.2.1 Selective Toxicity.** The ideal disinfectant would exert a rapidly lethal action against every pathogenic microorganism or spore and it would be inexpensive, stable, odorless, and nonstaining. Requirements for an ideal antiseptic are rapid and sustained lethality to microorganisms, activity in the presence of skin and bodily fluids, lack of irritation and allergenicity, lack of systemic toxicity when applied to skin and mucous membranes,

**Table 13.1 Topical Synthetic Antibacterials**

| Generic Name | Formula | Other Names | Use |
|---|---|---|---|
| **Chlorine and chlorophores** | | | |
| Chlorine | $Cl_2$ | | Drinking water; disinfection |
| Sodium chlorite | $NaClO_2$ | | Germicide |
| Sodium hypochlorite | $NaOCl$ | Dakin's solution | Wound disinfectant |
| Calcium hypochlorite dihydrate | $Ca(OCl)_2 \cdot 2H_2O$ | | Germicide |
| Lithium hypochlorite | $LiOCl$ | | Germicide |
| Chlorinated trisodium phosphate | $NaOCl \cdot 4Na_3PO_4$ | | Germicide |
| Chloramine T | **1** | Chlorazone, etc. | Drinking water; disinfection |
| Halazone | **2** | Pantocid | Sanitizer |
| Dichloromethylhydantoin | **3** | Halane, Dactin | Disinfect swimming pools |
| Trichloroisocyanuric acid | **4** | Symcolsine | Wound disinfectant |
| Dichlorocyanuric acid | **5** | | Wound disinfectant |
| Chlorazodin | **6** | Azochloramide | Wound disinfectant |
| **Iodine** | | | |
| Iodine | $I_2 + KI$ | Lugol's solution | Wound disinfectant |
| Iodine | $I_2 +$ ethanol | Iodine tincture | Wound disinfectant |
| Povidone–iodine | $I_2 +$ **7** | | Surgical antiseptic |
| **Alcohols** | | | |
| Ethanol | $C_2H_5OH$ | | Antiseptic |
| Isopropanol | $(CH_3)_2CHOH$ | | Antiseptic |
| Benzyl alcohol | $C_6H_5CH_2OH$ | | Pharmaceutical preservative |
| Phenethyl alcohol | **8** | | Pharmaceutical preservative |
| Chlorobutanol | **9** | Chloretone, etc. | Bacteriostatic for ophthalmology |
| Octoxynol | **10** | Triton X, etc. | Lens disinfectant |
| **Phenols** | | | |
| Phenol | $C_6H_5OH$ | Carbolic acid | Disinfectant |
| Cresols | $CH_3C_6H_5OH$ | | Disinfectant |
| 2-n-Amyl-5-methylphenol | **11** | | Oral antiseptic |
| Eugenol | **12** | | Mouthwash |
| 2-Benzyl-4-chlorophenol | **13** | Chlorophene | Disinfectant |
| 2-Phenylphenol | **14** | Dowicide 1 | Disinfectant |
| Triclosan | **15** | Irgasan | Disinfectant for cosmetics |
| Resorcinol | **16** | | Keratolytic |

540

| Name | Class | No. | Formula | Trade name / example | Use |
|---|---|---|---|---|---|
| Hexylresorcinol | | 17 | $HOC_6H_4CO_2alkyl$ | Crystoids, etc. | Mouthwash |
| Parabens | | | | | Pharmaceutical preservative |
| Hexachlorophene | Bisphenols | 18 | | PHisohex, etc | Topical antiseptic |
| Anthralin | | 19 | | Anthraderm, etc. | Keratolytic |
| Ethylene oxide | Epoxides and Aldehydes | 20 | | | Sterilization |
| Formaldehyde solution | | | $HCHO + water$ | Formalin | Sterilization |
| Glutaraldehyde | | 22 | | | Sterilization |
| Acetic acid | Acids | | $CH_3CO_2H$ | | Irrigation |
| Benzoic acid | | | $C_6H_5CO_2H$ | | Pharmaceutical preservative |
| Hydrogen peroxide | Oxidizing agents | | $H_2O_2$ | | Sterilization |
| Urea hydrogen peroxide | | | $H_2NCONH_2, H_2O_2$ | Hydrogen peroxide carbamide | Sterilization |
| Benzoyl peroxide | | 23 | | | Acne |
| Peracetic acid | | | $CF_3CO_3H$ | | Bactericide |
| Potassium permanganate | | | $KMnO_4$ | | Skin lesions |
| Ammoniated mercuric chloride | Heavy metals | | $Hg(NH_2)Cl$ | Ammoniated mercury | Skin infections |
| Mercuric oxide | | | $HgO$ | | Eye infections |
| Nitromersol | | 24 | | Metaphen | Bacterial antiseptic |
| Thiomersal | | 25 | | | Bacterial antiseptic |
| Phenylmercuric acetate | | 26 | | PMA, etc. | Bacterial antiseptic |
| Phenylmercuric nitrate | | 27 + 28 | | Phermernite | Bacterial antiseptic |
| Merbromin | | 29 | | Mercurichrome | Bacterial antiseptic |
| Silver nitrate | Silver salts | | $AgNO_3$ | Lunar caustic | Burns, ophthalmic |
| Toughened silver nitrate | | | $AgNO_3$ | | Wounds |

**Table 13.1** (Continued)

| Generic Name | Formula | Other Names | Use |
|---|---|---|---|
| Colloidal silver | Ag + protein | Mild silver protein | Wounds |
| Silver sulfadiazine | 30 | | Burns |
| | **Dyes** | | |
| Gentian violet | 31 | Methylrosaniline chloride, crystal violet, etc. | Topical antiseptic |
| Methylene blue | 32 | Urolene blue | Cystitis and urethritis, antiseptic |
| | **Diarylureas, amidines, and biguanides** | | |
| Triclocarban | 33 | Solubacter | Soaps and cosmetics |
| Propamidine | 34 | | Soaps and cosmetics |
| Dibromopropamidine | 35 | | Soaps and cosmetics |
| Chlorhexidine digluconate | 36 | Nolvasan, Sterilon | Wound cleansing, burns, surgical scrub |
| Polyhexamethylene biguanide | 37 | Polyaminopropyl biguanide | Contact lenses |
| Polyquaternium I | 38 | Polyquad | Contact lenses |
| | **Cationic surfactants** | | |
| | $[C_6H_5CH_2N(CH_3)_2\text{-alkyl}]^+/Cl^-$ | | |
| Benzalkonium chloride | 39 | Zephiran chloride, etc. Quatrachlor, etc. | Skin disinfectant |
| Benzethonium chloride | | | Skin disinfectant |
| Methylbenzethonium chloride | 40 | Diaperine chloride, Hyamine 10X | Diaper rash preventative |
| Cetylpyridinium chloride | 41 | Cepachol, etc. | Mouthwash and lozenges |
| Alkylbenzyldimethyl-ammonium chloride + alkyldimethyl(ethyl-benzyl)ammonium chloride | | BTC 2125M, Dual Quat | Biocide |
| Dimethyldioctyl-ammonium bromide | | Deciquam | Food preservative |
| Polyinones | 42 | Onamer M | Contact lenses |
| | **Anionic surfactants** | | |
| Sodium dodecylbenzene-sulfonate | 43 | Conoco C-50, etc. | Dairy disinfectant |
| Oxychlorosene | 43 + HClO | | Skin disinfectant |
| | **Amphoteric surfactants** | | |
| Dodecyl + tetradecyl di(aminoethyl)-glycines + tetradecyl-aminoglycines | | Tego 51 | Surgical disinfectant |

and no detrimental effect on wound healing. This ideal has not been realized completely (2). The need for selective toxicity increases from disinfectants, which are used on inanimate objects, through antiseptics, which are applied to skin and mucous membranes, to systemic antimicrobial agents. For antiseptics, a major concern is the *therapeutic index*, which is the ratio of the concentration that produces harmful effects to the concentration that is effective against microorganisms. The harmful effects include local tissue irritation and interference with the wound healing process. Hypersensitivity reactions and systemic toxicity resulting from absorption of the drug can be serious problems, as was the case when infants were commonly washed with hexachlorophene. A high degree of selective toxicity may be associated with a narrow antimicrobial spectrum and the emergence of resistance.

**2.2.2 Cellular Targets.** Chemical antimicrobial agents have a variety of cellular targets and, in some cases, multiple targets. Although the precise mechanisms of many agents remain unclear, there are a number of known interactions (3). For aldehydes, the target is the cell wall and the chemical mechanism is interaction with amino groups. The cell wall also is the target of anionic surfactants and the mechanism of action is lysis. Certain chelating agents such as EDTA form chelates with cations in the outer membrane of the cell wall and this process induces the release of lipopolysaccharides.

Phenols, quaternary ammonium compounds, biguanides, parabens, and hexachlorophene cause the leakage of low molecular weight compounds from cells and interfere with the normal proton flux. Phenol produces leakage, possible cell lysis, and proton flux changes; quaternary ammonium compounds cause leakage and protoplast lysis, and they interact with membrane phospholipids; chlorhexidine induces leakage and protoplast and spheroplast lysis; parabens cause leakage, transport inhibition, and selective inhibition of proton flux; and hexachlorophene causes leakage, protoplast lysis, and inhibition of respiration.

A variety of agents interfere with nucleic acid function. They include alkylation of DNA and RNA (and proteins) by ethylene oxide and formaldehyde, DNA intercalation by acridines and certain dyes, and inhibition of bacterial DNA gyrase (topoisomerase I) by quinolones.

Metal ions bind with sulfhydryl groups on enzymes or other proteins that may be associated with membranes.

**2.2.3 Mechanisms of Bacterial Resistance.** The outer membrane of Gram-negative bacteria is composed of lipopolysaccharide, proteins, and lipids (4). It presents a barrier to many chemical agents, including quaternary ammonium compounds and triphenylmethane dyes. In contrast, the cytoplasmic membrane of Gram-positive bacteria, excepting spores and mycobacteria, has greater permeability to most agents (5). Spores have a special coat and/or cortex that may not be permeable to hydrogen peroxide and chlorine disinfectants. Bacterial cells may also have efflux mechanisms that extrude agents such as quaternary ammonium compounds, dyes, and mercury compounds from their interiors (6).

**2.2.4 Kinetics and Other Factors.** Antibacterial action follows approximately first-order kinetics. The rate depends on concentration, pH, and the vehicle in which the agent is applied. The kinetics are most important when time is critical. An example of a rapid kinetics is the action of 70% ethanol on skin, which results in a 50% reduction in bacterial count in about 36 s. This ethanol concentration does not afford complete bactericidal action; only about 90% reduction in the bacterial count is obtained. Determination of kinetics is complicated by many factors, including diffusion, penetration, binding, and redistribution (7). Consequently, the rate of action often is not directly related to concentration and there is an optimal concentration. Thus, increasing the concentration of ethanol above 70% does not result in increased antibacterial activity. There is a general correlation of antibacterial activity and thermodynamic activity. The latter is related to the proportional saturation of the drug in the medium. This correlation breaks down when the capacity of the medium for the drug becomes a limiting factor because the amount of agent in the medium becomes depleted rapidly (8). One way to increase the capacity of water for poorly soluble agents such as iodine or hexachlorophene is to add a surfactant.

## 2.3 Evaluation of Antimicrobial Activity

### 2.3.1 Methods of Testing Disinfectants.
Disinfectants are required to have rapid and lethal antimicrobial effects, which means that bactericidal rather than bacteriostatic test methods are used. The test methods must be precise and reproducible, use standardized microbial strains, and have clearly defined inoculum and culture conditions. Standard bacterial strains are obtained from the American Type Culture Collection (ATCC, Manassas, VA), The activity of disinfectants is affected by pH, chelators and other metal ions, macromolecules and other organic matter, and detergent residues. Disinfectant activity depends on concentration and contact time according to the relationship $C^n t = k$, where $C$ is the disinfectant concentration, $t$ is the time required for lethal action, $n$ is the concentration exponent, and $k$ is a constant. Disinfectants with high exponents have a poor safety margin and require significantly increased contact time when they are diluted (9).

In the United States, the principal methods for testing disinfectants are published in the 16th edition of the *Official Methods of Analysis of the Association of Official Analytical Chemists International* (AOAC) (10). Detailed outlines may also be found in the 4th edition of *Disinfection, Sterilization, and Preservation* (11). Other countries have their own compendia of official methods. Three representative AOAC test methods are described briefly below. The literature indicated should be consulted for more details and further examples.

The phenol coefficient method compares bacterial activity of a disinfectant with that of phenol by a suspension test. A phenol coefficient number is the ratio of the greatest dilution killing test organisms in 10 min but not in 5 min to the greatest dilution of phenol giving the same result. The assay is run in broth culture at 20°C with contact times of 5, 10, and 15 min. It is used with specific strains of *Salmonella typhi*, *Staphylococcus aureus*, and *Pseudomonas aeruginosa*.

Available chlorine in disinfectants is measured by a capacity test in which disinfectant activity of a compound is compared with that of a standard sodium hypochlorite solution. The test organisms are *S. typhi* and *S. aureus*. Ten additions of inoculum are made at 1.5-min intervals and a subculture is made 1 min after each addition.

A carrier test is used for germicidal spray products. Broth cultures of test bacteria are spread on slides and dried 30–40 min at 37°C. The disinfectant is then sprayed onto the slides under standard conditions. After 10 min contact time at room temperature, material is transferred from the slides to subculture broth using suture loops or penicylinders. For a successful test, organisms must be killed in 10 out of 10 trials. The test organisms include *S. aureus*, *P. aeruginosa*, and *Salmonella choleraesuis*.

### 2.3.2 Methods for Testing Antiseptics.
The most important tests for disinfectants involve surgical hand scrubs. These tests incorporate the cup scrubbing procedure, in which a small area of skin is delineated by a glass cup. A wash solution (1 mL containing Triton X-100 in pH 7.9 phosphate buffer) is added by pipette and the skin area is scrubbed for 1 min. The procedure is repeated and an aliquot of the pooled sample is diluted and plated on agar. The plates are incubated 48 h at 37°C and colonies are counted (12).

The glove juice test is required by the FDA for surgical hand scrubs. In this test, loose gloves are placed on each hand and a sampling solution containing buffer and surfactant is added to one hand. This hand is massaged for 1 min and the sample is removed, plated on agar, and incubated to determine microbial growth. The opposite hand serves as a control and the test involves a group of people and statistical analysis (12).

In the modified Cade procedure, used for antimicrobial soaps, repeated hand washing under standard conditions is made in a series of basins and the bacteria in selected basins are plated on agar and incubated (12, 13). Hands are washed three times daily for 10 days for at least 10 consecutive days. Reductions in bacterial counts from baseline to the first basin and to the fourth and/or fifth basin are measured. At least 35 subjects are used in the test. Other surgical scrub tests are described in the literature (11).

## 2.4 History

As described in the interesting historical review by Block (14), the first disinfectant reported in the literature was in Homer's *The Odyssey*. Upon his return from the Trojan Wars, Odysseus burned sulfur to fumigate his house. The burning of sulfur (to sulfur dioxide) was used during the great plagues of the Middle Ages. Sulfur dioxide is still used as a disinfectant and preservative for fruit, fruit juices, and wine.

In the early 1500s, Paracelsus reformed the pharmacopeia and introduced compounds of mercury, lead, arsenic, copper, iron, and sulfur as disinfectants and antiseptics. Acidulated water, an antiseptic preparation containing wine or cider vinegar and cream of tartar, was used in the eighteenth century, as was mercuric chloride. Labarraque reported the use of calcium hypochlorite for wound dressings and general sanitation in 1825 (15) and Alcock recommended the use of chlorine to disinfect drinking water in 1827 (16). Tincture of iodine was admitted to the United States Pharmacopeia as an antiseptic in 1830. Richardson discovered the disinfectant activity of hydrogen peroxide in 1858. In 1887 an emulsion of coal tar creosote and soap was patented. A version of it was sold under the name Lysol.

Following the development of the principle of contagion by Holmes and Semmelweiss (17, 18), who independently reported the benefit of physicians washing their hands with calcium hypochlorite before conducting examinations and other procedures, Lister made a great advance in medicine by introducing antiseptic surgery (19). His techniques included the liberal use of phenol. Pasteur's pioneering studies in microbiology were followed by Koch's conclusive demonstration that bacteria invaded tissue and caused disease. In a comprehensive paper in 1881 he evaluated the ability of many chemicals to kill anthrax spores, reporting that halogens, mercuric chloride, and potassium permanganate were highly effective, but phenol was not (20).

Kronig and Paul established the basis for modern chemical disinfection in 1897 (21). They reported that chemical agents killed bacteria at a rate determined by concentration of the chemical and the temperature and they noted that disinfectants can be compared accurately only when they are tested under carefully controlled conditions. Rideal and Walker introduced the important phenol coefficient method for comparing disinfectants in 1903 (22). It is still used in standard assays.

The development of organic chemistry provided the basis for the synthesis of antibacterial agents of increasing potency and selectivity. Bechold and Ehrlich reported the high antibacterial activity of halogenated bisphenols in 1906 (23) and alkylresorcinols appeared in 1921. In 1935 Domagk showed that the addition of long alkyl chains to quaternary ammonium compounds greatly increased their disinfectant activity (24). Clorhexidine, a biguanide, was developed in 1965 by Rose and Swain through the structural modification of the antimalarial biguanides (25).

## 2.5 Halogens and Halophores

**2.5.1 Chlorine and Chlorophores.** Chlorine has been used since 1827 to disinfect drinking water and it remains the leading agent for this purpose because it is cheap and effective. Nevertheless, a variety of chlorine-generating compounds (chlorophores) and related compounds have been used in drinking water. Chlorine dioxide ($ClO_2$) has been used in recent years for drinking water disinfection and wastewater treatment. This highly reactive compound cannot be manufactured and shipped in bulk, but it is prepared at the site of consumption by treating a sodium chlorite solution with chlorine as indicated (Equation 13.1). Inorganic chloramines were used in the 1930s and early 1940s to improve the taste of drinking water, but their use was largely discontinued because of their inferior disinfectant properties. Combining chlorine and ammonia in water results in a mixture of species including $ClNH_2$, $Cl_2NH$, and $Cl_3N$. Approximately 25 times as much of these species is needed as is chlorine and the contact time required is 100 times as long (26). More recently, the addition of inorganic chloramines to water is being reconsidered because they prevent the formation of carcinogenic trihalomethanes from pollutants (27).

$$Cl_2 + 2NaClO_2 \rightarrow 2ClO_2 + 2NaCl \quad (13.1)$$

The germicidal species formed in water solutions of chlorine is hypochlorous acid (Equation 13.2). Based on this finding, a variety of inorganic hypochlorites were introduced as disinfectants. They include sodium hypochlorite, calcium hypochlorite dihydrate, lithium hypochlorite, and chlorinated trisodium phosphate [NaOCl·4(Na_3PO_4)]. The water solutions of sodium hypochlorite are available in different concentrations of available chlorine including the following: 0.4–0.5% (Dakin's solution used on wounds); 5.25% (Chlorox and related preparations); and 12–15% (liquid bleach).

$$Cl_2 + HOH \rightarrow HOCl + HCl \quad (13.2)$$

Organic compounds such as amides, sulfonamides, imides, and amidines form reasonably stable N-chloro derivatives, which may be produced in bulk quantities by treating appropriate nitrogen compounds with HOCl. In water they slowly release HOCl and regenerate the parent nitrogen compound (Equation 13.3). Among the N-chlorosulfonamides, chloramine-T (sodium p-toluenesulfonchloramide, 1) has been used for

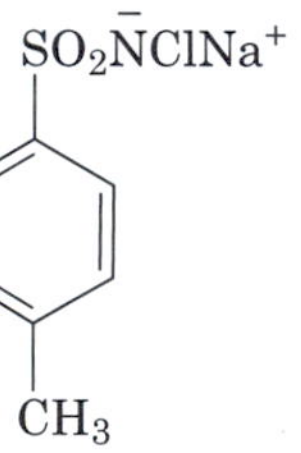

$$SO_2\overset{-}{N}ClNa^+$$

**(1)**

disinfecting wounds. It is permitted for use as a sanitizer in dairies and restaurants by the U.S. Public Health Service. Halazone (p-dichlorosulfamoylbenzoic acid, 2) is only slightly soluble in water, but its sodium salt is very soluble and has been used to disinfect drinking water (28). Heterocyclic chlora-

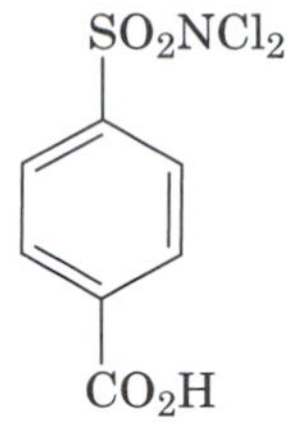

$$SO_2NCl_2$$

$$CO_2H$$

**(2)**

mines include dichlorodimethylhydantoin (halane, 3), which is used in some commercial sanitizing products and trichloroisocyanuric acid (trichloro-s-triazinetrione, 4),

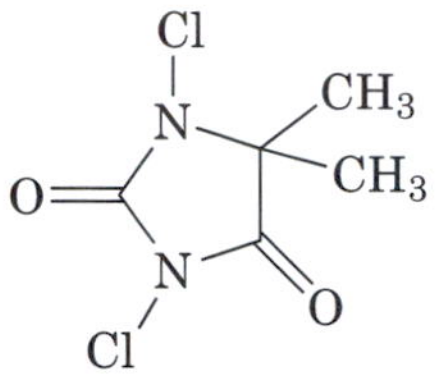

**(3)**

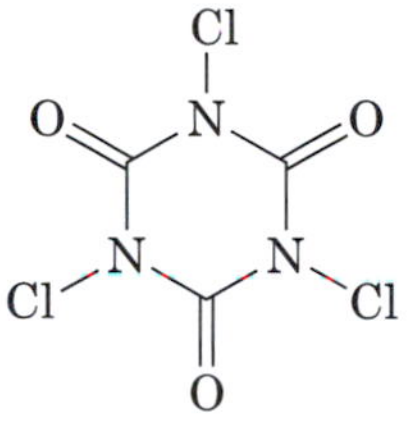

**(4)**

which is widely used to disinfect swimming pools. The sodium and potassium salts of dichloroisocyanuric acid (5) are used in laun-

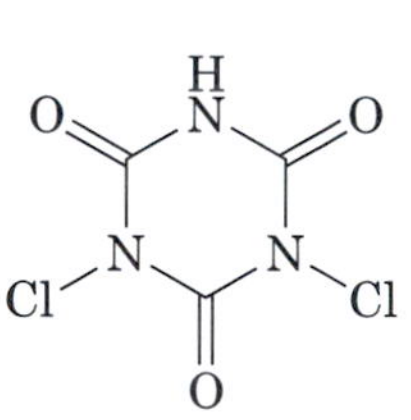

**(5)**

dry bleaches, dish-washing compounds, scouring powders, and industrial sanitizing products (29). Chloroazodin (N,N'-dichloroazodicarbonamidine, 6) is used for wound

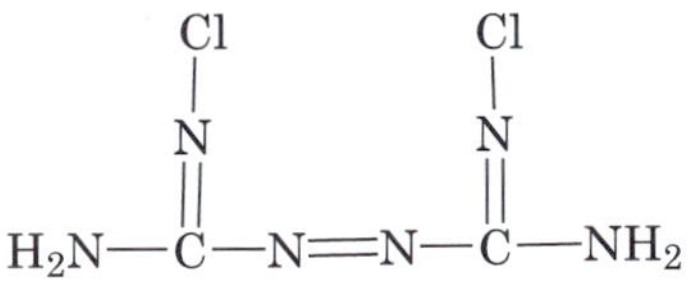

**(6)**

disinfection, packing for cavities, and lavage and irrigation. It has prolonged antiseptic action because of its relatively slow reaction with water.

$$RCONCl_2 + HOH \rightarrow RCONHCl + HOCl$$
$$(13.3)$$

Chlorine in minute amounts results in a destructive permeability of bacterial cell walls (30). Two different mechanisms have been proposed for the germicidal effects of hypochlorous acid. Both of them involve reactions with proteins: one is oxidation of sulfhydryl groups in the proteins and the other is chlorination of amide nitrogens (31). Although early workers considered that the germicidal action of N-chloramines was caused solely by hydrolysis to HOCl, it was suggested more recently that there is direct transfer of positive chlorine from N-chloramines to receptors in the bacterial cells (32).

**2.5.2 Iodine.** Molecular iodine $(I_2)$ is one of the oldest germicides and still one of the most useful. It is highly effective, economical, and minimally toxic at the usual concentrations. Water solutions of iodine are complex systems containing at least seven species, of which $I_2$ and HIO are considered to be the main disinfectants (33). Iodine preparations include the tincture (2% solution of $I_2$ in 50% ethanol containing iodide), Lugol's solution (5% $I_2$ in water with KI), and iodine solution (2% $I_2$ in water with NaI). The iodide salts solubilize $I_2$ and decrease its volatility. Preparations containing iodide ion have the triiodide ion $(I_3^-)$ as a major species. This ion is a weak disinfectant compared with $I_2$. Iodine as $I_2$ readily penetrates the cell walls of microorganisms. Suggested mechanisms of action include oxidation of the SH groups in proteins, iodination of amino groups, iodination of tyrosine residues,

and addition to the double bonds of unsaturated fatty acids (34). Povidone-iodine is a complex of $I_2$ with polyvinylpyrrolidone (7), a non-

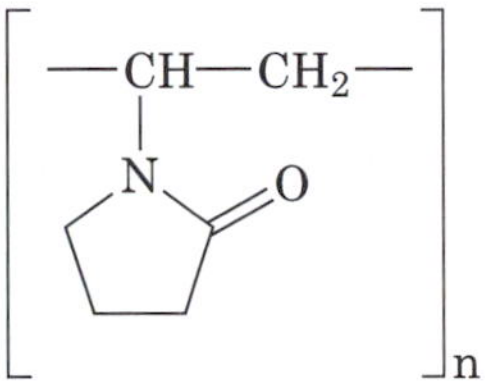

**(7)**

ionic surfactant polymer. This water-soluble complex contains approximately 10% of available $I_2$ (35), which it releases slowly. Povidone-iodine is nontoxic, nonirritating, and nonstaining. It is used as an antiseptic before surgery and injections and for treating wounds and lacerations. Available products include surgical scrubs, aerosols, ointments, antiseptic gauze pads, and mouthwashes.

## 2.6 Alcohols

Alcohols have desirable properties as disinfectants and antiseptics. They are bactericidal, inexpensive, and relatively nontoxic when used topically; however, they are more active against vegetative forms than against spores. Structure-activity relationships (SARs) among aliphatic alcohols have been established (36). Bactericidal activity increases as the homologous series of normal alcohols is ascended from methanol through ethanol, propanol, and so on to octanol. The phenol coefficients range from 0.026 for methanol to 21.0 for n-octanol when *Salmonella typhosa* is the organism. Against *Pseudomonas aeruginosa*, bactericidal activity of alcohols decreased in the order n-primary > i-primary > n-secondary > tertiary (36).

Ethanol and isopropanol are the only alcohols used routinely as antiseptics and disinfectants. They are most effective as 60–90% solutions in water and their mode of action is gross protein denaturation. The main use is skin disinfection, and an aerosol preparation is used to disinfect air.

Benzyl alcohol is commonly used as a preservative in vials of injectable drugs in concentrations of 1–4%. Its mild local anesthetic activity also is useful for injection. Further antiseptic uses for benzyl alcohol are in ointments and lotions. Phenylethyl alcohol (**8**) is

(**8**)

used primarily in perfumery. Among alcohols it has the unique property of greater potency against Gram-negative bacteria than against Gram-positive bacteria (37). It has been found to inhibit mRNA synthesis and DNA repair (38). Chlorobutanol (1,1,1-trichloro-2-methyl-2-propanol, **9**) is used as a bacteriostatic agent

(**9**)

in pharmaceutical preparations for injection and ophthalmic and intranasal administration.

Ethylene glycol, propylene glycol, and other glycols were used in the past as vapors to disinfect air. Good antimicrobial activity in this use depends on precise control of humidity.

Octoxynol (octylphenoxy polyethoxyethanol, **10**) consists of a mixture of compounds

(**10**)

containing 4–14 ethoxy units in addition to the terminal hydroxyethyl residue. Solutions containing this mixture are used to disinfect contact lenses. A single member of this set of compounds, octynol-9 ($n = 8$) is used as a spermatocide (39).

## 2.7 Phenols

**2.7.1 Monophenols.** Phenol was introduced by Lister as a surgical anesthetic in 1867. Originally, it was used undiluted and was toxic to tissue; however, it was considered to be less harmful than potential infections. Lister subsequently found that dilutions as low as 1:40 still gave effective antisepsis. The development of substituted phenols and bisphenols with superior antiseptic activity and lower toxicity has led to phenol being replaced as an antiseptic. It has limited use as a preservative for pharmaceuticals and a disinfectant for inanimate objects and excreta. Liquified phenol is phenol containing 10% water. This form is convenient for use in pharmaceutical preparations. Phenol and its derivatives act as gross cellular poisons at higher concentrations, penetrating cells and denaturating proteins. At lower concentrations, they inactivate essential enzymes. In 1% solution, phenol causes a pronounced leak of glutamic acid from cells. This finding suggests that the bactericidal activity of phenol results from physical damage to the permeability barrier in the bacterial cell wall (40).

Mixtures of the three isomeric methylphenols (cresols) are obtained from coal tar or petrolatum by alkaline extraction, acidification, and distillation. Cresol, NF is an inexpensive disinfectant with a phenol coefficient of 2.5. Compound cresol solution contains 50% cresol in saponified linseed or other suitable oil. It is soluble in water and is widely used for disinfecting inanimate objects and excreta.

The antibacterial potency of alkylphenols increases with increasing size of the alkyl group, which suggests that lipophilicity may be an important physical property for these compounds. *para*-Substituted alkylphenols increase in potency up to a chain length of six, then potency declines because of poor water solubility. Among the alkylphenols, 2-*n*-amyl-5-methylphenol (**11**) is used as an antiseptic in mouthwashes, gargles, and lozenges (41).

Eugenol (4-allyl-2-methoxyphenol, **12**) has a phenol coefficient of 14.4. It has local anesthetic activity in addition to antiseptic activity and these properties account for its use in mouthwashes.

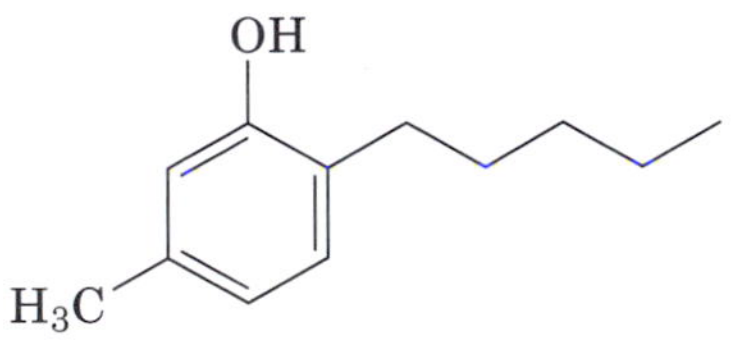

(11)

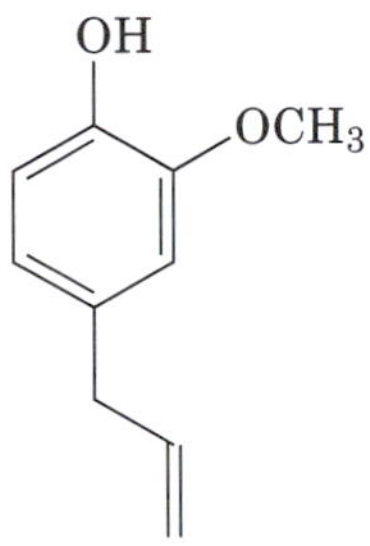

(12)

(13) R = H
(15) R = Cl

(14)

The antibacterial potency of phenols is increased by halogenation. *para*-Substituents are more effective than *ortho*-substituents. Addition of alkyl chains further increases potency and straight chains are more effective than branched ones. It is more effective to have the halogen *para* and the alkyl group *ortho* than in the reverse orientation. Increasing the molecular weight of the alkyl group usually increases the antibacterial potency (depending on the species) and decreases the toxicity (42, 43). A free hydroxyl group is required for antibacterial activity. Chloro and alkyl groups enhance potency by increasing lipophilicity and consequently reducing surface tension. Electron-withdrawing groups such as halogens increase the acidity of the phenol. Nitration increases antibacterial potency but also increases toxicity to higher species. Nitrophenols uncouple oxidative phosphorylation (44). The most widely used and effective substituted alkylhalophenols are chlorophene (2-benzyl-4-chlorophenol, **13**) and 2-phenylphenol (**14**). They are used against a broad spectrum of Gram-positive and Gram-negative bacteria as environmental disinfection in places such as hospitals, dairies, barns, poultry houses, and rest rooms. Triclosan (**15**) is a diphenylether substituted with one phenolic hydroxyl group and three chlorines. Its bacte-

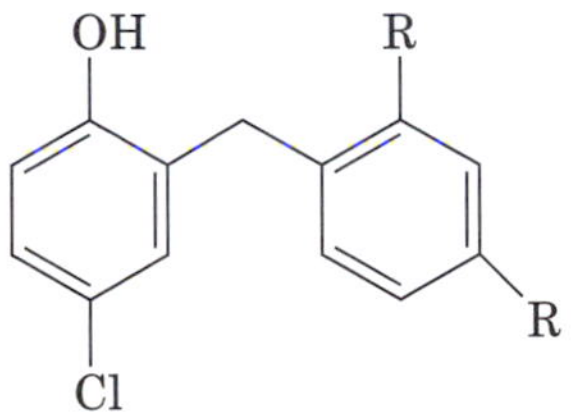

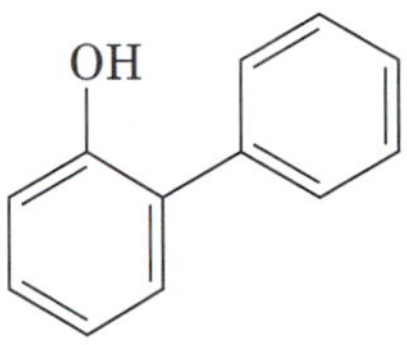

riostatic action on a broad spectrum of organisms makes it a useful disinfectant for cosmetic and detergent preparations.

The methyl, ethyl, propyl, and butyl esters of p-hydroxybenzoic acid (parabens) are used as preservatives for liquid dosage forms of pharmaceuticals and in cosmetics and industrial products. They are active against bacteria, yeasts and molds. Parabens are effective in low concentrations (0.1–0.3%) that are devoid of systemic effects, but as constituents of antibacterial ointments they can cause severe contact dermatitis.

**2.7.2 Bisphenols.** The parent bisphenol, resorcinol (**16**), has antibacterial and antifun-

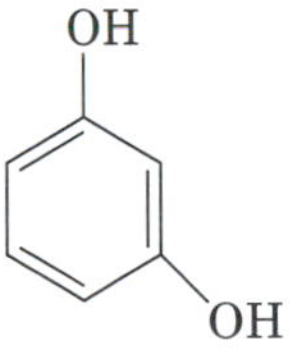

(16)

gal activity, but it is less potent than phenol. Nevertheless, its keratolytic properties make it useful in treating conditions such as acne, ringworm, eczema, and psoriasis. Resorcinol monoacetate is a prodrug that slowly liberates resorcinol. It has weaker but more prolonged action. As found with monophenols, the addition of alkyl substituents to bisphenols significantly increases antibacterial potency. Hexylresorcinol (4-*n*-hexylresorcinol, **17**) is an

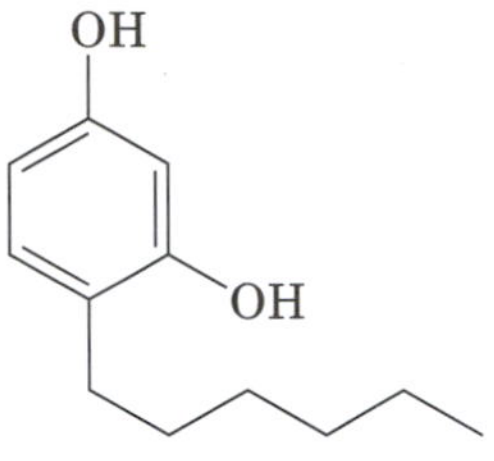

(17)

effective bactericidal agent that has a phenol coefficient of 313 against *Staphylococcus aureus* (45). It also has local anesthetic activity. Hexylresorcinol is used in 1:1000 water solution or glycerite in mouthwashes or pharyngeal antiseptic preparations.

Hexachlorophene is 2,2′-methylenebis-(3,4,6-trichlorophenol) (**18**). It was synthe-

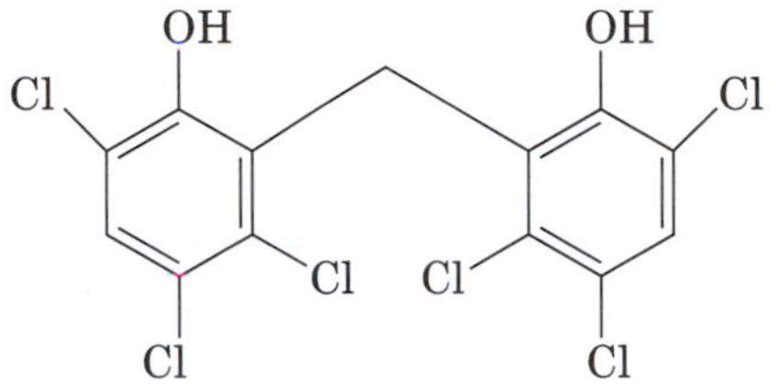

(18)

sized in 1941 by condensing 2 mol of trichlorophenol with 1 mol of formaldehyde in the presence of sulfuric acid (46). Hexachlorophene has high bacteriostatic activity, especially against Gram-positive bacteria: a 3% solution kills *S. aureus* in 15–30 s. It is less potent against Gram-negative bacteria and bactericidal activity requires longer contact time (24 h for some Gram-negative bacteria). Single applications of hexachlorophene are not more effective than ordinary soaps, but daily use results in a layer on the skin that confers prolonged bacteriostatic action. This property led to its widespread use as a topical antiseptic and in the late 1950s it successfully combatted virulent *Staphylococcus* infections in hospital nurseries throughout the world. Unfortunately, systemic toxicity can develop

from topical use, especially in infants, and more than 30 infants in France died from neurotoxicity resulting from exposure to baby powder containing 6% of hexachlorophene. In 1972 the FDA banned it from all over-the-counter (OTC) and cosmetic preparations, except at preservative levels of 0.1%. Prescription products bear warning labels concerning absorption and potential neurotoxicity (47). Hexachlorophene is still used for surgical scrubs, hand washing as part of patient care, and control of outbreaks of Gram-positive infections where other procedures have failed.

Anthralin (1,8-dihydroxyanthrone, **19**) is

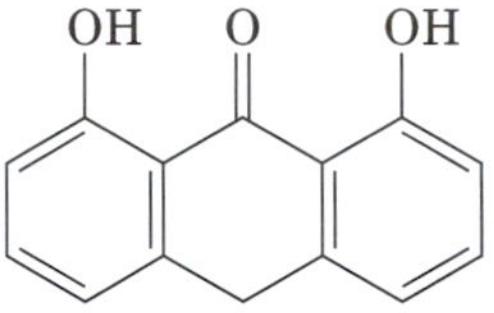

(19)

used against psoriasis and other chronic skin conditions because of its antiseptic, irritant, and keratolytic properties.

## 2.8 Epoxides and Aldehydes

Epoxides and simple aldehydes are highly reactive functionalities that readily alkylate nucleophilic groups such as amino, hydroxyl, and thiol on proteins and nitrogens on nucleic acids. Because of their high toxicity to higher organisms, they are used mainly to disinfect inanimate objects.

**2.8.1 Epoxides.** Ethylene oxide (oxirane, **20**) is a colorless flammable gas. It readily dif-

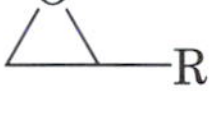

(20) X·R = H
(21) X·R = CH$_3$

fuses through porous materials and destroys all forms of microorganisms at room temperature (48). Ethylene oxide is very toxic and possibly carcinogenic. In concentrations of 3 to 80% v/v it forms explosive mixtures with air. This hazard may be eliminated by mixing it with carbon dioxide or fluorocarbons. For example, the product carboxide has 10% ethylene oxide and 90% carbon dioxide. Ethylene oxide has been used to sterilize temperature-sensitive medical equipment and certain pharmaceuticals that cannot be autoclaved. Propylene oxide is a liquid that also has been used as a sterilizing agent (49).

**2.8.2 Aldehydes.** Formaldehyde is a gas that readily undergoes oxidation to formic acid and polymerization to paraformaldehyde. It usually is used as formalin, an aqueous solution containing not less than 37% formaldehyde and 10–15% methanol to prevent polymerization. Formaldehyde exerts a slow but potent germicidal action thought to involve direct nonspecific alkylation of nucleophilic groups on proteins to form carbinol derivatives (Equation 13.4). It is used to disinfect surgical instruments. Formaldehyde is highly allergenic and a cancer suspect agent.

$$\text{Rnu} + \text{HCHO} \rightarrow \text{RnuCH}_2\text{OH} \quad (13.4)$$

Glutaraldehyde (1,5-pentanedial, **22**) is a reactive dialdehyde that readily undergoes self-condensation to form $\alpha,\beta$-unsaturated polymers (Equation 13.5). The commercial product is a stabilized solution containing 2% glutaraldehyde buffered to pH 7.5–8.0. Polymerization occurs above pH 8.5 (50). Glutaraldehyde Disinfectant Solution, NF is used to sterilize instruments and equipment that cannot be autoclaved. One special use is in disinfection of fiber-optic endoscopy equipment, especially to prevent the transmission of *Mycobacterium tuberculosis* between patients (51). Glutaraldehyde is effective against all microorganisms and its advantages over formaldehyde include less irritation and odor, although it can cause contact dermatitis. The mode of action of glutaraldehyde is based on its strong binding to outer cell layers, especially involving the $\varepsilon$-amino groups of proteins, rendering them impermeable (52). It may also inactivate cellular enzymes and alkylate nucleic acids (53).

## 2.9 Acids

Acetic acid in 1% water solution has been used in surgical dressings as a topical antimicrobial agent. In 0.25 to 2% solutions it is useful for infections of the external ear and for irrigation of the lower urinary tract. It is especially effective against *Pseudomonas* and other aerobic Gram-negative bacteria (54). Benzoic acid is employed externally as an antiseptic in lotions, ointments, and mouthwashes. As a preservative in food and pharmaceuticals, its effect depends on both pharmacokinetics and distribution between phases (55). It is more active when the pH is below its $pK_a$ value of 4.2.

## 2.10 Oxidizing Agents

Hydrogen peroxide ($H_2O_2$) is stable when it is pure, but small amounts of impurities promote rapid decomposition to water and oxygen. It is stabile in 3% water solution if deionized water and clean equipment are used. This solution is used for topical disinfection. Although it has potent activity against bacteria including anaerobes, it penetrates tissue poorly and the amount that does penetrate is rapidly decomposed by catalase. Thus, its antibacterial action in tissue is weak and brief. Hydrogen peroxide is more effective where living tissue is not present. It finds use in sterilizing milk, hospital water, food containers, and in the ultrasonic disinfection of dental and medical instruments. The lethal effect of hydrogen peroxide on microorganisms is thought to be attack on membrane lipids and

$$
\underset{\text{(22)}}{\overset{\displaystyle \text{O} \qquad\quad \text{O} \qquad\quad \text{O} \qquad\quad \text{CHO}}{\underset{\displaystyle \text{HC}(\text{CH}_2)_3\text{CH} \longrightarrow \text{HC}(\text{CH}_2)_3\text{CH}{=}\text{CH}(\text{CH}_2)_2\text{CHO}}{\parallel \qquad\quad \parallel \qquad\quad \parallel \qquad\quad |}}}
$$

$$
\longrightarrow \overset{\text{O}}{\overset{\parallel}{\text{HC}}}(\text{CH}_2)_3\text{CH}{=}\left(\overset{\text{CHO}}{\overset{|}{\text{C}}}(\text{CH}_2)_2\text{CH}{=}\right)_{\!n}\overset{\text{CHO}}{\overset{|}{\text{C}}}(\text{CH}_2)_2\text{CHO} \qquad (13.5)
$$

DNA. Decomposition to the highly reactive hydroxyl radical may be important in these processes (56). Concentrations of hydrogen peroxide used in sterilization (3% and higher) overcome the protective effect of catalase in bacterial cells.

Carbamide peroxide is a stable 1:1 complex of urea and hydrogen peroxide, which is provided as a 12.6% solution in anhydrous glycerine. It releases hydrogen peroxide when it is mixed with water.

Benzoyl peroxide (23) is chemically unstable and decomposes when heated. Its safety is improved by dilution with 30% water (Hydrous Benzoyl Peroxide, USP). Lotions contain 5–10% of hydrous benzoyl peroxide and they are stabilized by addition of dicalcium

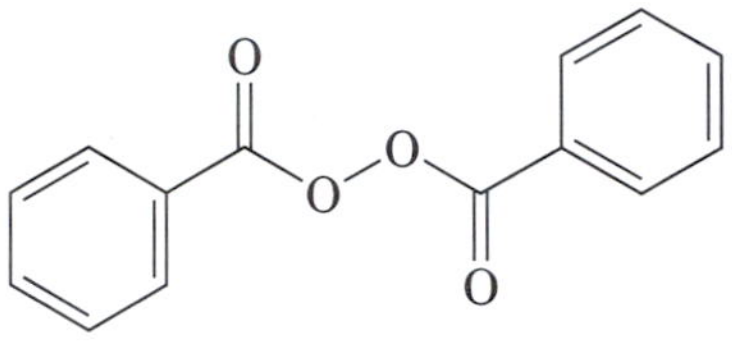

**(23)**

phosphate. Nonstabilized water solutions slowly decompose to hydrogen peroxide and benzoic acid. The main use of benzoyl peroxide is in treatment of acne, where it kills *Propionibacterium acnes*, decreases production of irritating fatty acids in sebum, and induces cell proliferation by its keratolytic action.

Peracetic acid is used widely in food processing and beverages. It is bactericidal at 0.001% and has the advantage of decomposing only to water, oxygen, and acetic acid (Equation 13.6).

$$2CH_3COOOH \rightarrow 2CH_3COOH + O_2 + H_2O$$

$$(13.6)$$

Potassium permanganate ($KMnO_4$) kills many microorganisms at a dilution of 1:10,000; however, this concentration is irritating to tissues and causes stains. It finds limited use in weeping skin lesions.

## 2.11 Heavy Metals

**2.11.1 Mercury Compounds.** In the past, mercuric chloride was used widely as an antisep-

tic, but its present use is limited to disinfecting instruments and occasional application to unabraided skin. The most significant inorganic mercury compound is ammoniated mercury ($Hg(NH_2)Cl$), which is used for skin infections such as impetigo. It is formulated as ammoniated mercury ointment, which contains 5 or 10% of the compound in liquid petrolatum and white ointment. Mercuric oxide (HgO) is used sometimes for inflammation of the eye.

There are two basic types of organomercurials: those in which the mercury is covalently bonded to carbon and those in which the mercury is bonded to a heteroatom such as oxygen, sulfur, or nitrogen. The latter type dissociates more readily than the former. Organomercurials are more bacteriostatic, less toxic, and less irritating than inorganic mercury compounds (57). Their mechanism of action is thought to result from binding with thiols in enzymes and other proteins. Thiols such as cysteine reverse their toxicity. Organomercurials are bacteriostatic and their potency is reduced substantially in serum because of the proteins present. They are not effective against spores. Among many organomercurials, nitromersol (**24**,

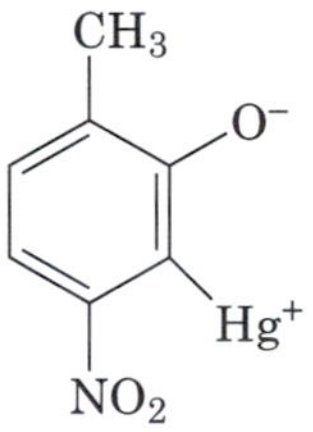

**(24)**

designated as the anhydride of 4-nitro-3-mercuri-2-methylphenol by the USP); thimerosal (merthiolate, **25**); phenylmercuric acetate (**26**); and

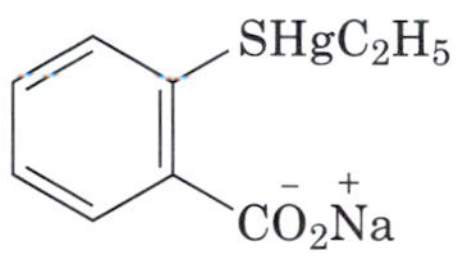

**(25)**

phenylmercuric nitrate [a mixture of phenylmercuric nitrate (**27**) and phenylmercuric hydroxide (**28**)] are marketed in many liquid and solid forms as bacteriostatic antiseptics. They also are used in biological products to prevent contamination (58).

Merbromin (mercurochrome, **29**) is used despite its very weak bacteriostatic action. Its brilliant red color may account for its popularity.

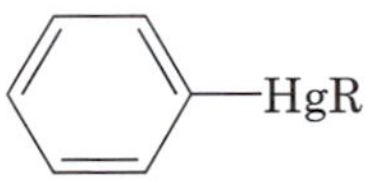

(**26**) R = OCOCH$_3$
(**27**) R = NO$_2$
(**28**) R = OH

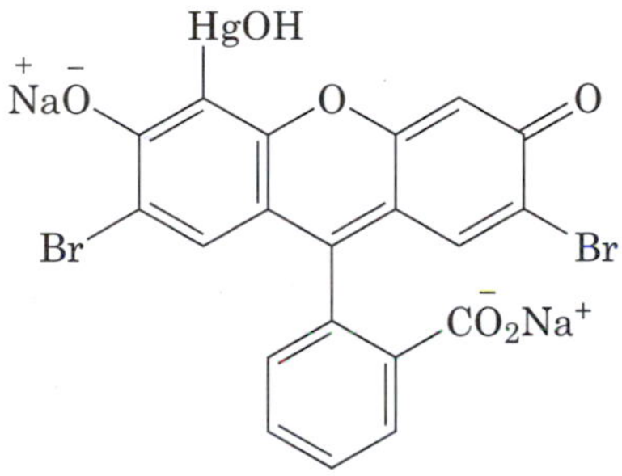

(**29**)

**2.11.2 Silver Salts.** Silver ions bind readily with biologically important functional groups including thiol, amine, phosphate, and carboxylate. Some of these interactions can alter the properties of bacterial proteins and cause them to precipitate. Other interactions may cause alterations in the bacterial cell wall and cytoplasmic membrane. These drastic changes result in an immediate bactericidal effect, and small amounts of silver ions subsequently liberated from silver-protein complexes provide sustained bacteriostatic action (56). Silver nitrate (AgNO$_3$) solutions are highly germicidal, de-

stroying most microorganisms at 0.1% concentration. Lower concentrations are bacteriostatic. Silver nitrate is particularly effective against gonococci and 1% solutions are used for the prophylaxis of *ophthalmia neonatorum*. In 5% solutions, silver nitrate is used, usually in conjunction with antibiotics, to treat extensive burns.

A solid form of silver nitrate known as lunar caustic (Toughened Silver Nitrate, USP) has been used to cauterize wounds.

Colloidal silver preparations retain substantial antibacterial activity and they are less injurious to tissues. One of these preparations, Mild Silver Protein, contains about 20% of elemental silver.

Silver sulfadiazine (**30**) is used in the topical treatment of extensive burns. It readily

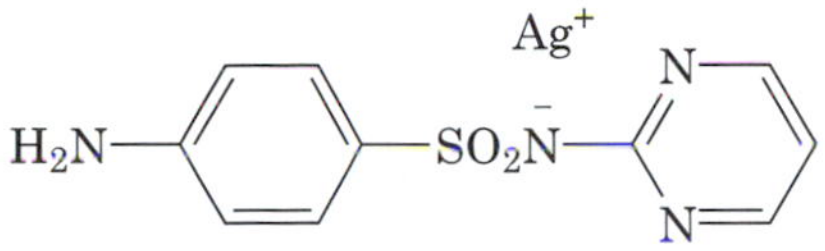

(**30**)

penetrates the eschar and the solubility is low enough that insufficient silver is released to precipitate proteins or chloride ions. It is effective against *Pseudomonas aeruginosa*.

## 2.12  Dyes

Organic dyes were used extensively as antimicrobial agents before the development of sulfonamides and antibiotics. Now only a few dyes such as gentian violet and methylene blue are used. Gentian violet (hexamethyl-*p*-rosaniline chloride, **31**) is a triphenylmethane

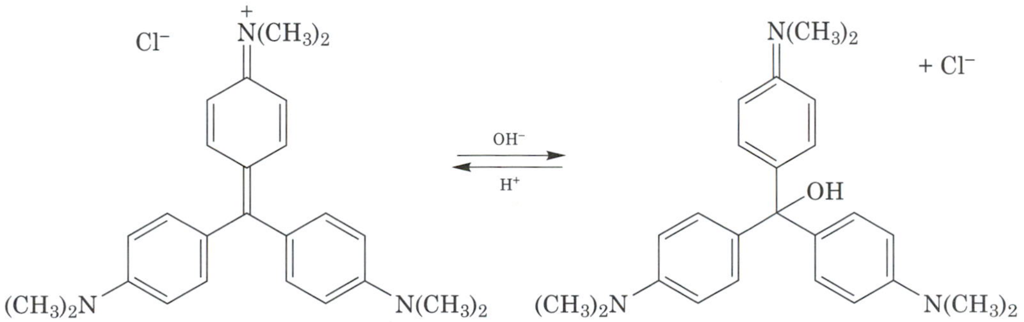

(**31**)                                                                     Leucobase

dye that is converted into a colorless form (leucobase) in alkaline solution (Equation 13.7). Cationic dyes such as gentian violet generally are active against Gram-positive bacteria and fungi; however, acid-fast and Gram-negative bacteria are resistant. Gentian violet has been used as a topical antibacterial agent, but its main use is as a topical agent for fungal infections. Gentian violet topical solution contains 1% of the agent and 10% of ethanol. Methylene blue (**32**) is a bacteriostatic agent that has

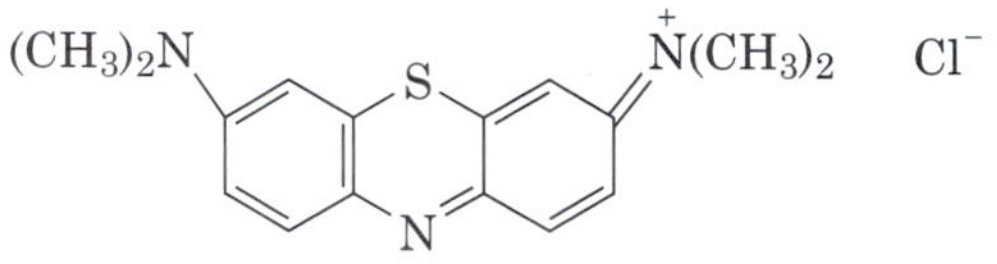

**(32)**

been used for cystitis and urethritis, infections associated with *E. coli* and *Neisseria gonorrhea*, respectively. Its redox properties make it useful for treatment of cyanide poisoning.

### 2.13 Diarylureas, Amidines, and Biguanides

Diarylureas (carbanilides) are potent antibacterial agents (60). Among a large number of these compounds, triclocarban (**33**) was cho-

**(33)**

sen for commercial development. It is used mainly in detergents, toilet soaps, and medicated cosmetics. Propamidine (**34**) and dibromopropamidine (**35**) are diamidines with activity against Gram-positive bacteria (61, 62). Their mode of action is not known.

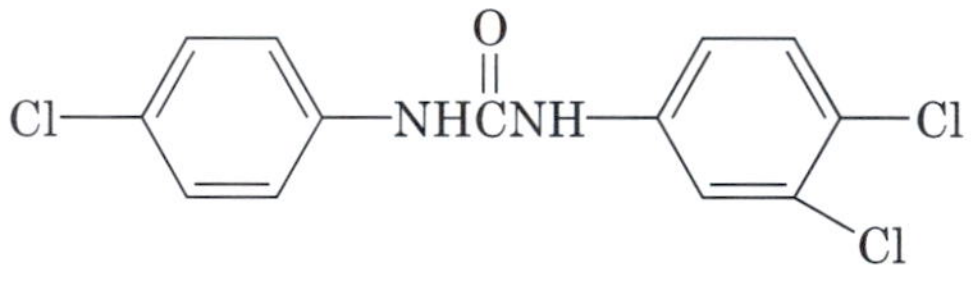

The antibacterial properties of biguanides were discovered by structural manipulation of earlier biguanides with antimalarial activity (25). They are strongly basic compounds that exist as dications at physiological pH. Their physical and antimicrobial properties resemble those of cationic surfactants, but they are not inactivated by anionic detergents unless the counter ions cause precipitation. The most important antibacterial biguanide is chlorhexidine (**36**), which is active against a broad spectrum of bacteria, except acid-fast bacteria and spores. It is not absorbed through skin or mucous membranes and it has no systemic toxicity or teratogenicity (63). The commercial product is chlorhexidine digluconate, which is highly water soluble. A 4% emulsion of it is used in wound cleansing (64), treatment of burns (65), and surgical scrub preparation of skin (66). In 0.2% solution it is used as a mouthwash to combat plaque-inducing bacteria (67).

Chlorhexidine acts by a sequence of events involving attraction to the bacterial cell, strong binding to certain phosphate-containing compounds on the bacterial surface, overcoming bacterial cell wall exclusion mechanisms, attraction to the cytoplasmic membrane, leakage of low molecular weight cytoplasmic components, and complexation with phosphated molecules such as ATP and nucleic acids (68).

Polymeric analogs of chlorhexidine, such as polyhexamethylene biguanide (also called polyaminopropyl biguanide, **37**), wherein the molecular weight is in the range 1000–3000 Da are used in disinfecting contact lenses because they have high antimicrobial potency, low binding to the lenses, and very low ocular toxicity (69). Another important polymeric quaternary compound is polyquaternium 1 (Polyquad, **38**), which has 2-butenyl chains

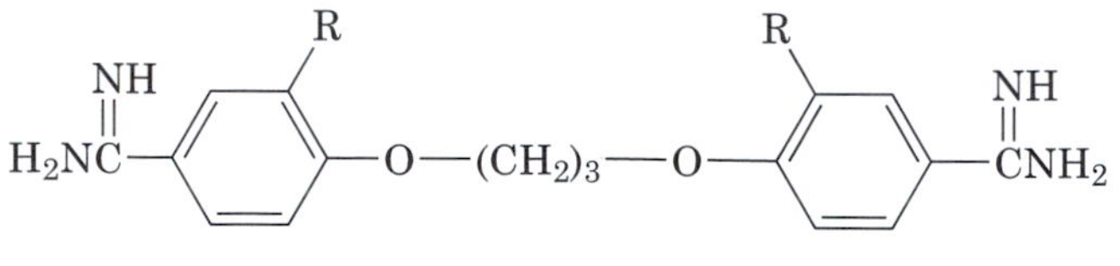

**(34)** R = H
**(35)** R = Br

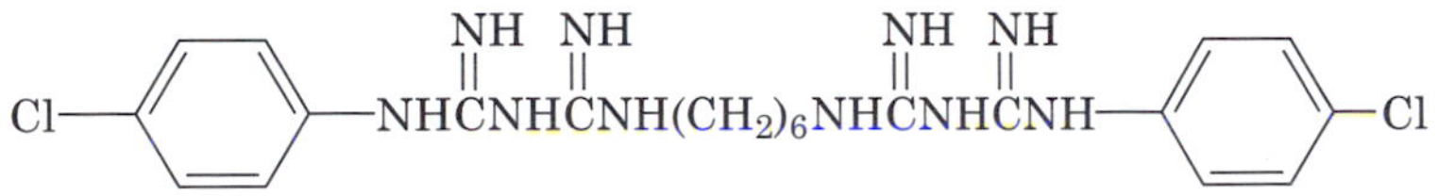

**(36)**

separating the quaternary nitrogens, and triethanolammonium groups at the chain ends (70).

### 2.14 Surface-Active Agents

**2.14.1 Cationic Surfactants.** Cationic surfactants are quaternary ammonium or pyridinium salts that are ionic in water and have surface-active properties. These properties are associated with the cationic head, which has high affinity for water, and a long hydrocarbon tail, which has high affinity for lipids. Cationic surfactants show potent activity against Gram-positive bacteria and lower activity against Gram-negative bacteria (71). *Pseudomonas* species and *Mycobacterium tuberculosis* are resistant.

The mechanism of action of cationic surfactants is association with cell wall protein followed by penetration and disruption of the cell membrane. The resistance of Gram-negative bacteria is attributed to difficulty in penetrating the outer membrane (72). There is no activity against bacterial spores. Desirable features of cationic surfactants include water solubility, low toxicity, relatively good tissue penetration, and freedom from stains and corrosion. Disadvantages include inactivation by anionic surfactants (all traces of soap must be removed from skin), reduced effectiveness in the presence of blood serum and pus, strong adsorption on fibrous material such as cotton, occasional allergic responses on prolonged use, and resistant organisms.

The important surfactants are benzalkonium chloride, benzethonium chloride, methybenzethonium chloride, and cetylpyridinium chloride. Benzalkonium chloride is a mixture of alkylbenzyldimethylammonium chlorides having the general formula $[C_6H_5CH_2N(CH_3)_2R]^+ Cl^-$, where R is a mixture of alkyl groups of which $C_{12}H_{25}$, $C_{14}H_{29}$, and $C_{16}H_{33}$ are the main components. It is used as an antiseptic for skin and mucosa in concentrations of 1:75 to 1:20,000. Other uses include irrigation and disinfection of surgical instruments. Benzethonium chloride (**39**) is also is used as a skin disinfectant and irrigant of mucous membranes. Methylbenzethonium chloride (Diaperene, **40**) is used specifically for control of diaper rash in infants caused by *Bacterium ammoniagenes*, a species that liberates ammonia.

Cetylpyridinium chloride (1-hexadecylpyridinium chloride, **41**) has its positively charged nitrogen as part of a pyridine ring. It finds use as a general anesthetic, irrigant for mucous membranes, and as a component of mouthwashes and lozenges.

Structure-activity relationships have been

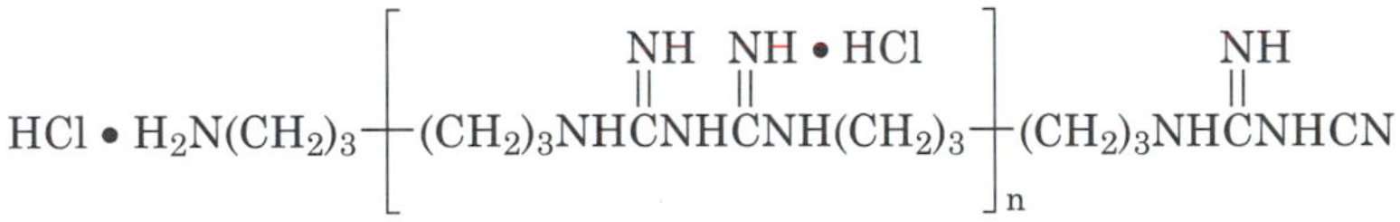

**(37)**

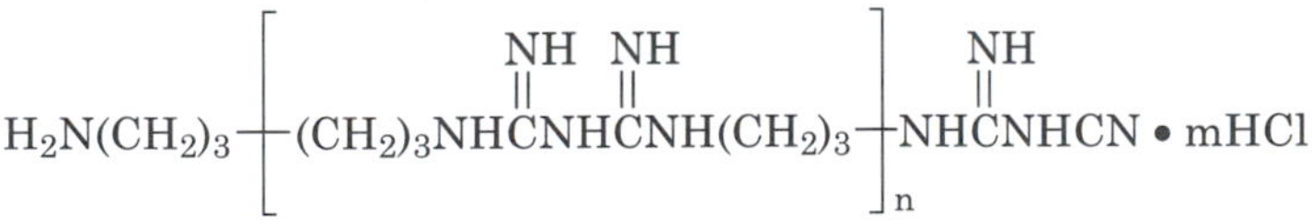

**(38)**

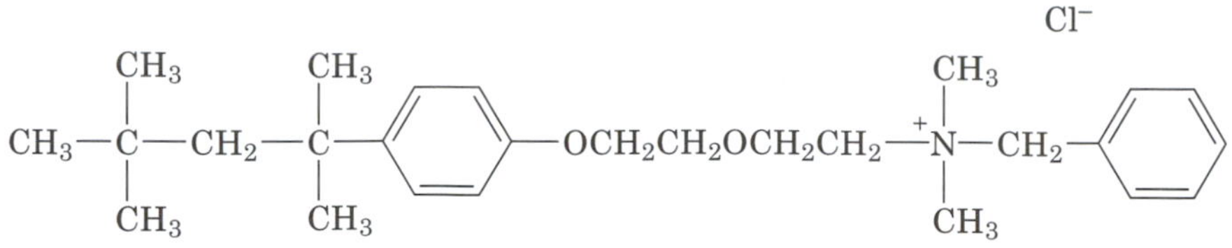

**(39)**

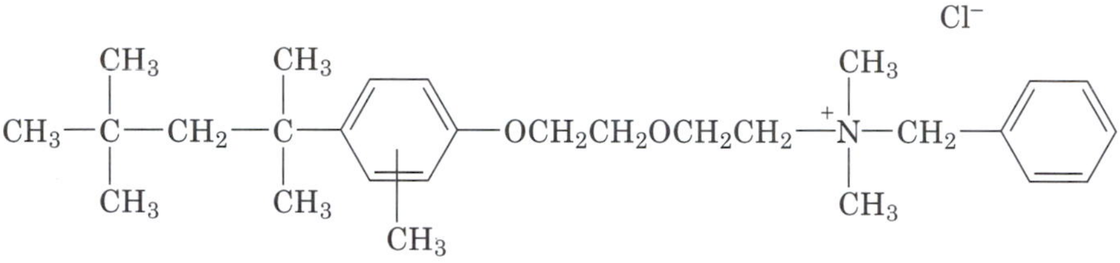

**(40)**

defined for cationic surfactants. For laurylpyridinium chlorides, antibacterial potency depends on the electronegativity of substituents on the pyridine ring with electron-releasing groups affording the highest potencies. There is a linear relationship between potency and partial charge on the pyridine nitrogen (73). Some highly potent substituted pyridines have not been commercialized because of their potential cost. Octanol-water partition coefficients were measured for a series of alkylbenzyldimethylammonium chlorides and used as the independent variable in a correlation with antibacterial potency. The parabolic relationship of the form $\log(1/C) = a + b \log P + c(\log P)^2$ was obtained, wherein $P$ is the partition coefficient and $C$ is the minimum inhibitory concentration (MIC). Maximum potency was found for the compound with the $C_{14}H_{29}$ side chain. Potency depended on the organism, with the MIC being 10 times as great for Gram-negative bacteria as for Gram-positive bacteria (74).

Mixtures of equal proportions of alkylbenzyldimethylammonium chloride and alky-

$$(CH_2)_{15}CH_3$$

**(41)**

ldimethyl(ethylbenzyl)ammonium chloride (alkyl = $C_{12}$ to $C_{18}$) are known as "dual quats." An example of these mixtures, BTC 2125M, has better biocidal activity than that of the individual species (75). Quaternary ammonium compounds having two long alkyl chains were made possible when catalytic amination of long-chain alcohols to give dialkylmethylamines became a commercial process. Quaternization of these compounds with methyl chloride provides products known as "twin-chain quats." An example of these compounds, dimethyldioctylammonium bromide (DECIQUAM) is used in the British food industry because of its low toxicity and good antibacterial potency (76). Polymeric quaternary ammonium compounds, named polyionenes, are milder and safer than monomeric quaternary ammonium salts (77). Among these compounds, Onamer M (**42**), is used as a preservative for contact lens solutions. It is less irritating than chlorhexidine.

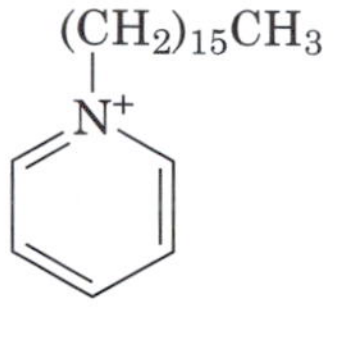

**2.14.2 Anionic Surfactants.** Mixtures of anionic surfactants with acids to lower the pH to 2–3 show rapid germicidal activity (78). Alkyarylsulfonates such as dodecylbenzenesulfonic acid (**43**) are the most effective surfactants and phosphoric acid is frequently used in the mixture. Possible modes of action are disorganization of the cellular membrane, denaturation of key enzymes and other proteins, and

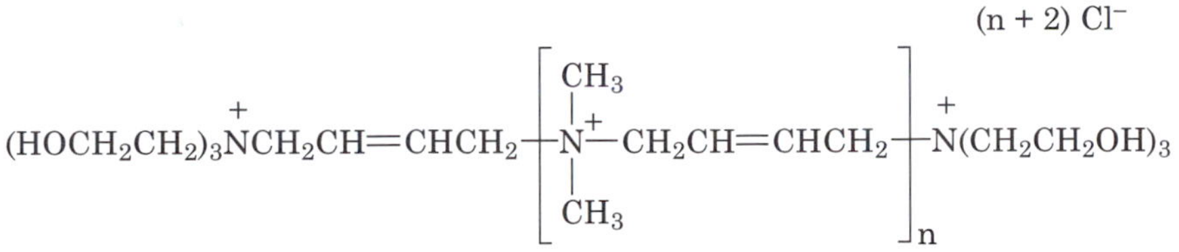

$$(\text{HOCH}_2\text{CH}_2)_3\overset{+}{\text{N}}\text{CH}_2\text{CH}=\text{CHCH}_2 - \left[ \begin{array}{c} \text{CH}_3 \\ | \\ \overset{+}{\text{N}} \\ | \\ \text{CH}_3 \end{array} - \text{CH}_2\text{CH}=\text{CHCH}_2 \right]_n \overset{+}{\text{N}}(\text{CH}_2\text{CH}_2\text{OH})_3 \qquad (n+2)\ \text{Cl}^-$$

(42)

$$\text{C}_{12}\text{H}_{25} - \underset{}{\bigcirc} - \overset{-}{\text{SO}}_3\text{Na}^+$$

(43)

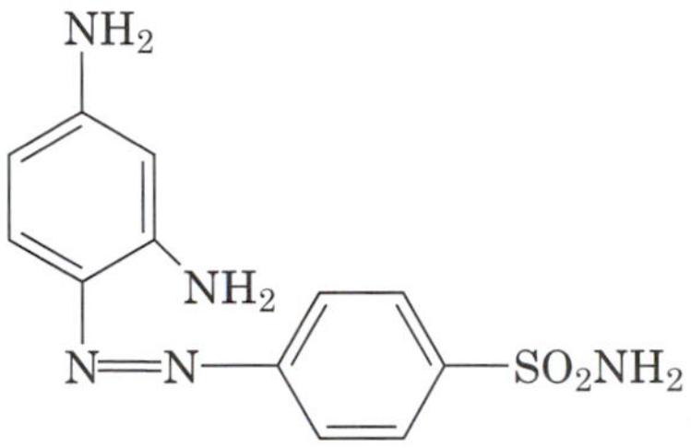

(44)

interruption of cellular transport (79). Products have been developed for disinfecting equipment in the food and dairy processing industry.

Oxychlorosene is a complex of sodium dodecylbenzenesulfonate and hypochlorous acid. It has a markedly rapid and complete cidal action against both Gram-positive and Gram-negative bacteria, fungi, yeasts, molds, viruses, and spores. It is applied as a disinfectant by irrigation, instillation, sprays, soaks, or wet compresses.

**2.14.3 Amphoteric Surfactants.** These preparations, known as ampholites, have been used as biocides in Europe for more than 40 years. They are based on mixtures of alkyldi(aminoethyl)glycines and other diaminoglycine derivatives. For example, Tego 51 contains dodecyl and tetradecyl di(aminoethyl)glycines plus dodecyl and tetradecyl aminoethylglycines (80). A 1% solution of this product kills many Gram-positive and Gram-negative bacteria within 1 min (81). Specific uses are in hand disinfection before surgery, disinfection of surgical instruments, and disinfection of rooms in hospitals and food-processing facilities.

# 3 SYSTEMIC SYNTHETIC ANTIBACTERIALS

## 3.1 Sulfonamides and Sulfones

**3.1.1 Introduction.** Prontosil rubrum (44), a sulfonamido-azo dye, was the first clinically useful systemic antibacterial agent to be dis-

covered. This discovery in the early 1930s and the development of sulfonamides and sulfones as a class of antibacterial agents, which followed, forms a fascinating chapter in the annals of medicinal chemistry. Their broad antimicrobial spectrum provided, for the first time, drugs for the cure and prevention of a variety of bacterial infections; their widespread clinical use brought about a sharp decline in morbidity and mortality of treatable infectious diseases, and thus proved of great medical and public health importance. Recognition of the inhibition of the action of sulfonamides by yeast extracts, which was shown to be attributed to the presence of p-aminobenzoic acid (PABA), required in folate biosynthesis, was the first clear demonstration of metabolite antagonism as a mechanism of drug action; this provided the long sought after mechanistic basis for drug action.

This led Fildes (1940) to propose his classic *theory of antimetabolites* as an approach to chemotherapy. The development of dihydrofolate reductase inhibitors as antimicrobial agents and their synergistic use in combination with sulfonamides, providing a unique rationale for the use of drug combinations, was a direct result of this interest generated in antimetabolites. That sulfanilamide (45) formed *in vivo* was responsible for the antibacterial action of prontosil focused attention on the

$$H_2N \overset{4}{-} \underset{}{\bigcirc} -\overset{1}{SO_2NH_2}$$

(45)

importance of drug metabolism and blood levels of the active species for drug action. Pharmacokinetic studies thus became an integral part of drug development. Carefully observed side effects in pharmacological and clinical studies of the early sulfonamides revealed new and unanticipated activities; successful exploitation of these leads opened up new areas in chemotherapy, such as oral antidiabetics, carbonic anhydrase inhibitors, and diuretics. This also highlighted the importance of side effects of drugs as a source of new leads in drug design. The rapidity with which new developments took place between 1933 and 1940, from the discovery of antibacterial activity of prontosil to the enunciation of the theory of antimetabolites by Fildes, indicates that the time was just ripe for major developments in drug research and needed only a catalyst, which was provided by the discovery of the antibacterial activity of sulfonamides. The discovery of sulfonamides thus was not only the beginning of the modern era of systemic synthetic antibacterials but also had a strong impact on developments in medicinal chemistry, which influenced later work in drug research in general and chemotherapy in particular.

The interest in sulfonamides and sulfones continues even seven decades after their discovery. Although no major new drug in this class has been added in the last three decades, and the addition of new classes of antibacterials has diminished the clinical use of the existing sulfonamides/sulfones, they still occupy a distinct place in the therapeutic armamentarium; for some conditions alone or in combination with trimethoprim they are still the drugs of choice. One major recent development in sulfonamides is the elucidation of their mode of action at the molecular level. With the identification of the pterin, PABA, and sulfonamide binding sites on the dihydropteroate synthase (DHPS) of the different classes of microbes, the stage is set for the design of a new generation of DHPS inhibitors with broad-based antimicrobial activity including pro-

karyotes and lower eukaryotes, as well as of agents with selective action against specific classes of microbes.

**3.1.2 Clinical Use of Sulfonamides and Sulfones.** About 20 sulfonamides and sulfones have been commonly used in clinical practice (Table 13.2). These vary widely in their absorption, distribution, and excretion patterns. Some remain largely unabsorbed after oral administration and are thus considered useful for gastrointestinal (GI) tract infections. Sulfonamides of another group characterized by high solubility, quick absorption, and rapid excretion, mainly in the unaltered form, are widely used in urinary tract infections. Those belonging to yet another group are absorbed rapidly but excreted slowly, or reabsorbed, resulting in maintenance of high blood levels for long periods; these sulfonamides require less frequent administration and are particularly useful for chronic conditions and for prophylaxis.

Sulfonamides and sulfones are widely used in clinical practice, even six decades after their discovery, because of the wide choice that provide of agents with greatly differing half-life and pharmacokinetic characteristics meeting the requirements of varied clinical situations, their wide antimicrobial spectrum, the benefits of synergistic action that their combination with dihydrofolate reductase inhibitors (DHFRI) provides, their highly selective action on the microbes with minimal effects on the host, their relative freedom from problems of superinfection, ease of administration, and favorable pharmacoeconomics. Furthermore, $p,p'$-diaminodiphenylsulfone (dapsone), remains the mainstay for the treatment of all forms of leprosy.

**3.1.2.1 Present Status in Therapeutics.** The number of conditions for which sulfonamides are drugs of first choice has declined because of a gradual increase in resistance to them and the addition of new classes of antimicrobials, although they still have a distinct and significant place in therapeutics (82–86).

Sulfonamides combined with trimethoprim are of value in the treatment of urinary tract infections, bacillary dysentery (particularly that caused by Shigella), salmonellosis, and chronic bronchitis. In meningococcal infections, sulfonamides are of value when the strains of *N. meningitides* or *H. influenzae* are sensitive to them. Sulfonamides are commonly used in preventing streptococcal infections and recurrence of rheumatic fever among susceptible subjects, especially in patients who are hypersensitive to penicillins. In methicillin-resistant staphylococcal and streptococcal infections and vancomycin-resistant enterococcal infections co-trimoxazole, a fixed-dose combination of sulfamethoxazole and trimethoprim is often considered as one of the treatment options.

Sulfacetamide sodium eye drops are employed extensively for the management of ophthalmic infections; a combination of topical and systemic application is of value in some conditions. Topical sulfonamides such as silver sulfadiazine and mafenide inhibit enterobacteriacease, *P. aeruginosa*, staphylococci and streptococci, and they are extensively used to reduce the bacterial load in burn eschars.

Sulfonamides have been found useful for the treatment of infections resulting from *Listeria monocytogenes*, especially in penicillin-allergic patients. They are commonly used for the prophylaxis and treatment of otitis media in children.

Sulfonamides, alone or combined with trimethoprim, are the drugs of choice in the treatment of infections resulting from *Nocardia* species, including cerebral nocardiosis; sulfisoxazole, sulfamethoxazole, and sulfadiazine are the commonly used drugs (87). Some clinicians prefer to use sulfonamides alone to avoid the greater risk of hemolytic toxicity observed more commonly with combination therapy. A pyrimethamine-sulfadiazine combination is commonly used for the treatment of all forms of toxoplasmosis, including maternofetal toxoplasmosis. Co-trimoxazole and fansidar (fixed-dose combinations of sulfadeoxine and pyrimethamine) are commonly used for the prophylaxis and treatment of *Pneumocystis carinii* infection, a common sequelae in patients with AIDS.

Sulfamethoxine, sulfamethoxypyridazine, and dapsone combined with pyrimethamine find use for the prophylaxis and treatment of chloroquine-resistant falciparum malaria.

Dapsone remains the of drug choice for all forms of leprosy and is an essential component of all multidrug therapy regimens. Dapsone has also been reported to cure some cases of Crohn's disease, which may have a mycobacterial origin. Though sulfonamides in general are not very effective against tuberculosis, some are active against nontuberculous mycobacterial infections, which have acquired importance in immunocompromised subjects such as in cases of AIDS; co-trimoxazole has been found effective in patients with *Mycobacterium marinum* infection, whereas sulfisoxazole has been used successfully for *M. fortuitum* infections (88).

*3.1.2.1.1 Other Conditions.* Shortly after the introduction of sulfa drugs, sulfapyridine was found to have a unique beneficial effects on some inflammatory conditions, especially dermatologic, unrelated to their antibacterial activity (89). Later, dapsone was found to share the same properties at a much lower dose and with improved therapeutic index (90). The disorders that respond are dermatitis herpetiformis (DH), pyoderma gangrenosum, subcorneal pustular dermatosis, acrodermatitis continua, impetigo herpetiformis, ulcerative colitis, and cutaneous lesions of patients with lupus erythematosus. Dapsone is the drug of choice for the treatment of DH and other similar inflammatory conditions that are characterized by neutrophil infiltration. These disorders are characterized by edema followed by granulocytic inflammation or by vesicle or bullae formation (91). Coleman et al. described useful *in vitro* test systems for the study of inhibition of neutrophil function, which could help in picking up more leads and active compounds in this area (92). The mechanism of action is not fully understood, but it has been proposed that these drugs enter or influence the protein moiety of glycosaminoglycans and decrease tissue viscosity, resulting in prevention of edema, dilution of tissue fluid, and decrease in inflammation and vesicle and bullae formation. It is likely that this additional action of DDS may in part account for the extraordinary sensitivity of lepra bacilli to it. Salicylazosulfapyridine is the treatment of choice for ulcerative colitis.

*3.1.2.2 Adverse Reactions.* The sulfonamides are generally safe drugs, even though

# Table 13.2 Characteristics of Commonly Used Sulfonamides and Sulfones [a]

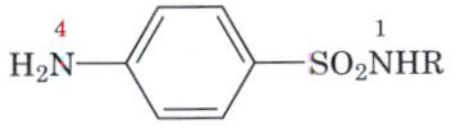

Structure: $H_2N$—(benzene ring, position 4 and 1)—$SO_2NHR$

| Generic Name | R | Common Proprietary Names | In Vivo Activity [b] Against E. coli (μmol/L) | Water Solubility [c] (mg/100 mL at 25°C) | p$K_a$ | Liposolubility [d] (%) | Protein Binding at (1.0 μmol/ml, % bound) | Plasma Half-Life/h (human) | % $N^4$-Metabolite in Urine [e] (human) |
|---|---|---|---|---|---|---|---|---|---|
| Phthalylsulfathiazole [f] | | Thalazole, Sulfthaladine | Poorly absorbed locally acting | Insoluble | Acid | | | | |
| Sulfaguanidine | C(=NH)NH₂ | Guanicil, Resulfon | 4 | 100 | Base | | | 5 | |
| | | | Well absorbed, rapidly excreted | | | | | | |
| Sulfamethizole Sulfamethylthiadiazole | 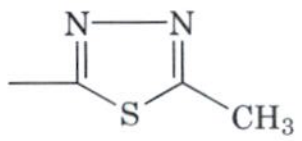 | Methisul, Lucosil | | 25 (pH 6.5) | 5.5 | | 22 | 2.5 | 6 |
| | | Ultrasul | | | | | | | |
| Sulfathiazole | 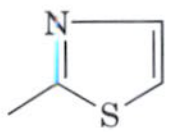 | Cibazole, Thiazamide | 1.6 | 60 (pH 6) | 7.25 | 15.3 | 68 | 4 | 30 (40) |
| Sulfisoxazole | 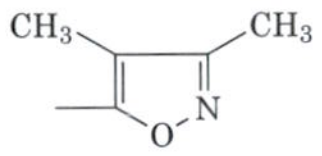 | Gantrisin, Urosulfin | 2.15 | 350 (pH 6) | 5.0 | 4.8 | 76.5 | 6.0 | 16 (30) |
| Sulfisomidine | 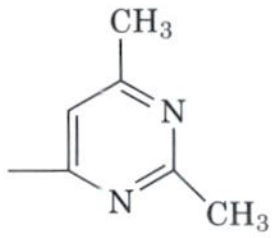 | Elkosin, Aristamid | 1.5 | 300 (30°) | 7.4 | 19.0 | 67 | 7.5 | 4 |

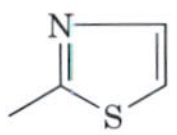

| Compound | Substituent | Trade names | | | | | | | |
|---|---|---|---|---|---|---|---|---|---|
| Sulfacetamide | -COCH₃ | Albucid | 2.3 | 670 | 5.4 | 2.0 | 9.5 | 7 | 5 |
| Sulfapyridine | 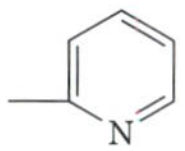 | Bubasin Dagenan MB693 | 4.8 | 30 | 8.4 | 14 | 70 | 9 | 30 |
| Sulfanilamide | H | Prontalbin | 128 | 750 | 10.5 | 71 | 9 | 9 | |
| *Readily absorbed, medium rate of excretion* | | | | | | | | | |
| Sulfaphenazole | 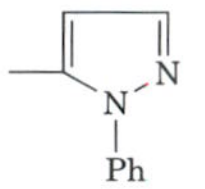 | Orisul | 1.0 | 150 | 6.09 | 69 | 87.5 | 10 | 20 (80) |
| Sulfamethoxazole | 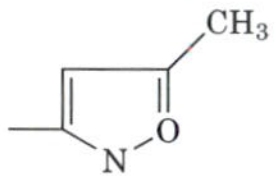 | Ganthanol | 0.8 | Sparingly soluble | 6.0 | 20.5 | 60 | 11 | 60 (14) |
| Sulfadiazine | 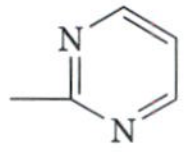 | Deberal Pyrimal | 0.9 | 8 | 6.52 | 26.4 | 37.8 | 17 | 25 |
| *Readily absorbed, slowly excreted* | | | | | | | | | |
| Sulfamethyldiazine | 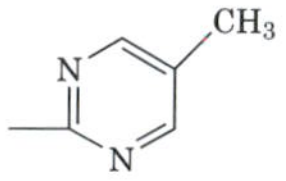 | Pallidin | 1.0 | 40 (pH 5.5) | 6.7 | 69.6 | 74.0 | 35 | |
| Sulfamethoxydiazine Sulfamonomethazine | 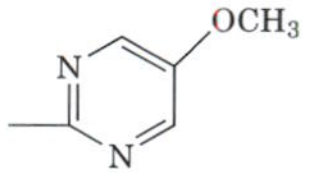 | Sulfameter Durenat | 2.0 | Very sparingly soluble | 7.0 | 64.0 | 74.2 | 37 | 20$^c$ (30) |
| Sulfamethoxypyridazine | 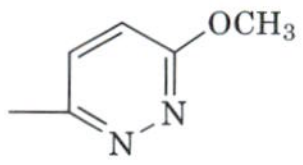 | Lederkyn Kynex | 1.0 | 147 (pH 6.5) | 7.2 | 70.4 | 77 | 37 | 50 (15) |

**Table 13.2** *(Continued)*

| Generic Name | R | Common Proprietary Names | In Vivo Activity[b] Against *E. coli* (μmol/L) | Water Solubility[c] (mg/100 mL at 25°C) | $pK_a$ | Liposolubility[d] (%) | Protein Binding at (1.0 μmol/ml, % bound) | Plasma Half-Life/h (human) | % $N^4$-Metabolite in Urine[e] (human) |
|---|---|---|---|---|---|---|---|---|---|
| Sulfadimethoxine | 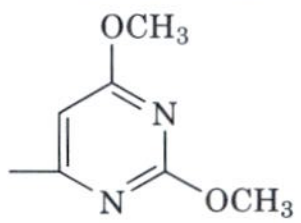 OCH₃ structure | Madribon | 0.7 | 29.5 (pH 6.7) | 6.1 | 78.7 | 92.3 | 40 | 15 (70) |
| Sulfamethoxpyrazine Sulfametopyrazine | 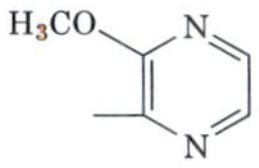 structure | Sulfalene Kelfizina | 1.85 | Very sparingly soluble | 6.1 | | 65 | 65 | 65 |
| Sulformethoxine | 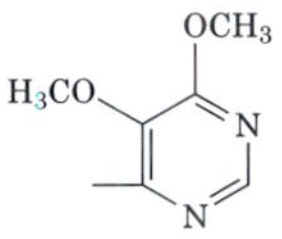 structure | Sulfadoxine Fanasil | 0.8 | — | 6.1 | 5 | 95 | 150 | 60[a] (10) |
| Diaminodiphenylsulfone | H | Dapsone Avolosulphon | 44 | 14 | Base | 13 | 50,[c] ii | 20 | |
| Diacetamidodiphenyl-sulfone | COCH₃ | Acedapsone | | 0.3 | | | | 43 days, i.m. | |

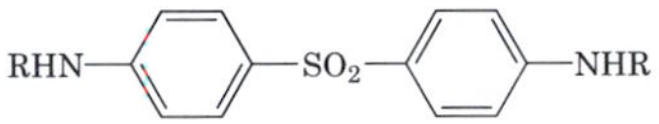

$$RHN-\!\!\!\bigcirc\!\!\!-SO_2-\!\!\!\bigcirc\!\!\!-NHR$$

[a] Arranged in order of increasing plasma half-life. Unless otherwise stated the data are from Rieder (115).
[b] From Struller (296).
[c] From Ref. 297.
[d] Determined by partition between ethylene dichloride and sodium phosphate buffer.
[e] From William and Park (100).
[f] $N^4$-Phthalyl.

the list of possible adverse reactions is long (85, 86). The most common side effects are related to allergic skin reactions, which vary from relatively minor skin rashes, maculopapular rashes, and urticarial reactions to severe, even life-threatening reactions such as erythema multiform, Stevens-Johnson syndrome, and toxic epidermal necrolysis (TEN). The severe hypersensitivity reactions occur most commonly after treatment with long-acting sulfonamides, whether used alone or in combination with pyrimethamine as for malaria prophylaxis or treatment. The skin eruption is also frequent in patients with AIDS being treated with pyrimethamine-sulfadoxine for *P. carinii* pneumonia, and is associated with pancytopenia in some patients, and may be severe enough to require discontinuation of drugs (90). Individuals seropositive for HIV are more susceptible to developing adverse reactions to sulfonamides/sulfones: 40–80% in patients with AIDS compared to 5% in patients with other immune deficiencies (93, 94). Photosensitivity reactions are also relatively common with sulfonamides.

Hemolytic adverse reactions may occur occasionally, and when they occur, drug administration may need to be discontinued. These include hematologic reactions such as methemoglobinemia, agranulocytosis, thrombocytopenia, kernicterus in the newborn, and hemolytic anemia in patients with G6PD deficiency. Kernicterus can result from administration of sulfonamides to the mother or to the newborn because sulfonamides displace bilirubin from albumin in the newborn. Therefore pregnant women near term or newborns should not be given sulfonamides. Hemolytic anemia is relatively more common with sulfone therapy in leprosy patients, and most often is related to the undernourished status of these patients; discontinuation of treatment is often not necessary and only supplemental therapy is required.

The minor adverse reactions reported include GI reactions such as nausea, vomiting, and diarrhea; and neurologic effects, such as peripheral neuritis, insomnia, and headache. Crystalluria, one of the earliest serious toxic reactions reported with sulfonamides, has been more or less overcome with the discovery of agents that are highly soluble at the pH of urine, or are excreted mainly as water-soluble metabolites.

By binding to albumin sites, sulfonamides may displace drugs such as warfarin, methotrexate, and hypoglycemic sulfonylurea drugs, and may thus potentiate the action of these drugs. Sulfonamide concentrations are increased by indomethacin, salicylates, and probenecid.

It has been suggested that adverse reactions of sulfonamides may be attributable to the formation of reactive hydroxylamine metabolites, together with a deficient glutathione system needed for scavenging these reactive molecules (95). It has been suggested that the covalent adducts formed by the $N^4$-hydroxylamine metabolites with human epidermal keratinocytes are very likely responsible for the initiation and propagation of the cutaneous hypersensitivity reaction observed with these drugs (96). This is supported by *in vitro* experiments in which sulfamethoxazole hydroxylamine has been found to be cytotoxic for lymphocytes, whereas the parent compound was not (97, 98). With dapsone it has been shown that its hydroxylamine metabolite seems to be responsible for methemoglobinemia; when dapsone is combined with cimetidine, an inhibitor of $N$-oxidation, the increase of methemoglobinemia is reduced (99).

### 3.1.2.3 Pharmacokinetics and Metabolism

*3.1.2.3.1 Sulfonamides.* The sulfonamide drugs vary widely in their pharmacokinetic properties (Table 13.2). Those that are highly ionized are not absorbed from the GI tract after oral administration, leading to a high local concentration of the drug, and were thus considered useful for enteric infections. A majority of the sulfonamides, however, are well absorbed, mainly from the small intestine, and insignificantly from the stomach. Absorption occurs of the un-ionized part, related to their lipid solubility. In rate and extent of absorption most sulfonamides behave similarly within the $pK_a$ range 4.5–10.5. After absorption they are fairly evenly distributed in all the body tissues. High levels are achieved in pleural, peritoneal, synovial, and ocular fluids that approximate 80% of serum levels; CSF levels are effective in meningeal infections. Those that are highly soluble do not, in general, attain a high tissue concentration, show no tendency to crystallize in the kidney, are more readily excreted, and are useful in treating genitourinary infections. The relatively less soluble ones build up high levels in blood, tissues, and extravascular fluids and are useful for treating systemic infections. This wide range of solubilities and pharmacokinetic characteristics of different sulfonamides permits the access of one or the other member of the group to almost any site in the body, thus adding greatly to their usefulness as chemotherapeutic agents. The free, non-protein-bound drugs and their metabolic products are ultrafiltered in the glomeruli, then partly reabsorbed. Tubular secretion also plays an important role in the excretion of sulfonamides and their metabolites. The structural features of the compounds have a marked effect on these processes and determine the rate of excretion. The renal clearance rates of the metabolites are generally higher than those of the parent drugs.

Metabolism of sulfonamides takes place primarily in the liver and involves mainly $N^4$-acetylation, to a lesser extent glucuronidation, and to a very small degree, C-hydroxylation of phenyl and heterocyclic rings and of alkyl substituents and O- and ring N-dealkylation. Variation of the substituents markedly influences the metabolic fate of the sulfonamides (Table 13.2); the metabolism also differs markedly in different animal species (100–105). Some of the sulfonamides, such as sulfisomidine, are excreted almost unchanged; in most of them $N^4$-acetylation occurs to a substantial degree; but some of the newer sulfonamides, such as sulfadimethoxine and sulfphenazole, are excreted mainly as the glucuronide. The metabolites in human urine of the commonly used sulfonamides reveal the wide variation in their metabolic disposition (106).

Fujita (107) has carried out regression analyses on the rates of metabolism and renal excretion of sulfonamides in terms of their substituent constants. Equations showing the best correlation indicate that the most important factor governing the rate-determining step of the hepatic acetylation is the hydrophobicity of the drug and that $pK_a$ does not play a significant role. The excretion pattern seems to be more complex and would have to take into consideration additional parameters to give an acceptable correlation.

*3.1.2.3.2 Sulfones.* Dapsone is well-absorbed after oral administration and is evenly distributed in almost all the body tissues. It is excreted mainly through the kidneys; less than 5% is excreted unchanged, very little N-acetylation takes place, and most of it is present as the mono-N-glucuronide (108–110). It has been shown that there are marked animal species differences in the metabolism of dapsone; humans are relatively slow acetylators compared to rhesus monkey (111, 112). Dapsone has a half-life of about 20 h. Acedapsone following intramuscular injection is very slowly absorbed and deacetylated. It has a half-life of about 42.6 days. There are marked animal species differences in the metabolism of acedapsone also; mice deacetylate acedapsone efficiently, but rats do not (113).

*3.1.2.3.3 Half-Life.* The half-lives of sulfonamides is of importance because the dosage regimen is related to it; dose schedule is a function of the pharmacokinetic parameters. Kruger-Thiemer and his associates have reported a mathematical model for correlating these parameters with dose schedules (114).

The half-life of different sulfonamides in clinical use vary widely, from 2.5 to 150 h (Table 13.2) and also show marked differences in different animal species. Reider (115) correlated the $pK_a$, liposolubility, surface activity, and protein binding of a group of 21 sulfonamides with their half-life in humans. It was reported that long-acting sulfonamides were, in general, more lipid soluble than were the short-acting compounds, but no clear-cut relationship could be established; factors such as tubular secretion and tubular reabsorption are also involved. In 2-sulfapyrimidines; a 4-$CH_3$ group increases the half-life, 4,6-$(CH_3)_2$ reduces it to less than one-half, the corresponding methoxy derivatives have a much longer half-life, and both 5-$CH_3$ and 5-$OCH_3$ prolong half-life to the same extent. Similarly, in 4-sulfapyrimidines, the 2,6-$(OCH_3)_2$ analog is the most persistent sulfonamide known; sulfamethoxypyridazine has a half-life about twice as long as that of sulfapyrazine. Thus although no clear-cut pattern of relationship between structure and half-life is discernible,

**Table 13.3  Antimicrobial Spectrum of Sulfonamides and Sulfones**

| Gram-Positive/Acid Fast | Gram-Negative | Others |
|---|---|---|
| | Highly sensitive | |
| *Bacillus anthracis* (some strains) | *Calymmatobacterium granulomatis* | *Actinomyces bovis* |
| *Corynebacterium diphtheria* (some strains) | *Hemophilus ducreyi* | *Chlamyia trachomatis* |
| | *H. influenzae* | *Coccidia* |
| *Mycobacterium leprae* (to sulfones) | *Listeria monocytogenes* | *Lymphogranuloma venereusm* virus |
| *Staphylococcus aureus* | *Neisseria gonorrheae* | |
| *Streptococcus pneumoniae* | *N. Meningitidis* | *Plasmodium falciparum* |
| *S. pyogenes* (Group A) | *Pasteurella pestis* | *P. malariae* |
| | *Proteus mirabilis* | *Pneumocystis carinii* |
| | *Shigella flexneri* | *Nocardia* species |
| | *S. sonnei* | *Toxoplasma* |
| | *Vibrio cholerae* | *Trachoma* viruses |
| | Weakly susceptible | |
| *Clostridium welchii* | *Aerobacter aerogenes* | *Plasmodium vivax* |
| *Mycobacterium tuberculosis* | *Brucella abortus* | |
| *Mycobacterium avium* | *Escherichia coli* | |
| *Mycobacterium intracellulare* | *Klebsiella pneumoniae* | |
| *Streptococcus viridans* | *Pseudomonas aeruginosa* | |
| | *Salmonella* | |

the methoxy and methyl groups in general seem to prolong half-life.

**3.1.3 Antimicrobial Spectrum.** After the initial dramatic results obtained with sulfonamides in the treatment of streptococcal infections, studies with these drugs were extended to other microorganisms including bacteria, viruses, protozoa, and fungi. It was found that many Gram-positive and Gram-negative bacteria, mycobacteria, some large viruses, protozoa, and fungi are susceptible to the action of sulfonamides and sulfones (Table 13.3). In almost all cases their action is related to PABA antagonism.

The sulfonamides and sulfones have a relatively broad antibacterial spectrum. Individual sulfonamides do differ in their antibacterial spectrum, but these differences are more quantitative than qualitative. The bacteria most susceptible to sulfonamides include pneumococci, streptococci, meningococci, staphylococci, some coliform bacteria, and shigellae. Lepra bacilli are susceptible to sulfones. One limitation of sulfonamides is their weak activity against bacteria responsible for typhoid fever, diphtheria, and subacute bacterial endocarditis. They have prac-

tically no activity against *P. aeruginosa*. Another limitation with sulfonamides is the rising incidence of resistant isolates in the community. Synergism of their action by dihydrofolate reductase inhibitors, and their introduction to combination therapy has to some extent helped to remedy this situation. Sulfamethoxazole has been shown to have impressive *in vitro* activity against *Mycobacterium avium* and *M. intracellulare* (116).

Sulfonamides and sulfones are also active against malarial parasites, although parasites vary greatly in their sensitivity to them; *P. falciparum* is very sensitive and *P. vivax* is less so. Their effect is potentiated by pyrimethamine (117). The action of sulfones and sulfonamides is mainly against the blood forms, with marginal activity against primary (preerythrocytic) tissue forms and no activity against sexual forms and latent tissue forms.

Sulfonamides have been shown to be highly active against *Eimeria* (118–120), Toxoplasma (121, 122), and *Nocardia* spp. (123, 124), and in combination with pyrimethamine are widely used for coccidiosis (125), toxoplasmosis (126), and nocardiosis (87, 127, 128).

McCallum and Findlay (129) showed that

experimental *Lymphogranuloma venereum* virus infection in mice was cured by sulfonamides. Later, other Chlamydiae were also found to be inhibited by sulfonamides, which led to the successful clinical use of these drugs in the treatment of trachoma (130).

Sulfonamides have also high activity against *Pneumoystis carinii* (131) and combined with trimethoprim are largely used for the treatment of *P. carinii* pneumonia in AIDS patients.

### 3.1.4 Mechanism of Action

***3.1.4.1 Site of Folate Inhibition.*** The antimicrobial action of sulfonamides is characterized by a competitive antagonism with *p*-aminobenzoic acid (PABA), an essential growth factor vital to the metabolism of the microorganisms. Evidence for this antagonism started coming soon after the discovery of sulfonamides. It was found that substances antagonizing the action of sulfonamides were present in peptones (132), various body tissues, and fluids, especially after autolysis or acid hydrolysis (133), pus (134), bacteria (135, 136), and yeast extract (137, 138). Woods (137) obtained evidence that PABA is the probable antagonistic agent in yeast extract, and showed that synthetic PABA could completely reverse the bacteriostatic activity of sulfanilamide against various bacteria *in vitro*. Selbie (139) and Findlay (140) soon after found that PABA could antagonize the action of sulfonamides *in vivo* as well. Blanchard (141), McIllwain (142), and Rubbo et al. (143) finally isolated PABA from these sources. This led Woods (137) to suggest that, because of its similarity of structure with that of PABA, sulfanilamide interfered with the utilization of PABA by the enzyme system necessary for the growth of bacteria. Based on these observations, a more general and clear enunciation of the theory of metabolite antagonism to explain the action of chemotherapeutic agents was given by Fildes in 1940 (144) in his classic paper entitled "A rational approach in chemotherapy."

Further studies showed that the inhibition of growth by sulfonamides in simple media can be reversed not only competitively by PABA, but also noncompetitively by a number of compounds not structurally related to PABA, such as L-methionine, L-serine, glycine, adenine,

guanine, and thymine (145, 146). The relationship of sulfonamides to purine was uncovered by the finding that sulfonamide-inhibited cultures accumulated 4-amino-5-imidazolecarboxamide ribotide (147), a compound later shown by Shive et al. (148) and Gots (149) to be a precursor of purine biosynthesis.

With the concurrent knowledge gained in the field of bacterial physiology and metabolism, these isolated facts could be gradually fit into a pattern. The determination of the structure of folic acid by Angier et al. (150) and Mowat et al. (151) revealed that PABA was an integral part of its structure. After this, Tschesche (152) made the suggestion that folic acid is formed by the condensation of PABA or *p*-aminobenzoylglutamic acid (PABG) with a pteridine and that sulfonamides compete in this condensation. Soon the structure of the active coenzyme form of folic acid, leucovorin (folinic acid, citrovorum factor), was established and its involvement in biosynthetic steps where one-carbon units are added was elucidated (153, 154); the amino acids, purines, and pyrimidines that are able to replace or spare PABA are precisely those whose formation requires one-carbon addition catalyzed by folic acid.

Direct evidence of the inhibition of folic acid synthesis by sulfonamides was soon obtained by studies on bacterial cultures. It was already known that a number of organisms could use PABA and folic acid as alternative essential growth factors (155). Lampen and Jones (156, 157) found that the growth of some strains of *Lactobacillus arabinosus* and *L. plantarum* in media containing PABA was inhibited competitively by sulfonamides, whereas folic acid caused a noncompetitive type of reversal of this inhibition, suggestive of its being the product of the inhibited reaction. Nimmo-Smith et al. (158) reported a similar inhibition of folic acid synthesis by sulfonamides and its competitive reversal by PABA in nongrowing suspensions of *L. plantarum*. Inhibition of folic acid synthesis by sulfonamides was also demonstrated in a PABA-requiring mutant, in the parent wild strain of *E. coli* (159, 160), and in cultures of *Staphylococcus aureus*.

The demonstration of the enzymic synthesis of dihydropteroate (DHP) and dihydrofolate (DHF) (Fig. 13.1) in cell-free extracts of a number of organisms (161–165) set the stage for examining the action of sulfonamides at the enzyme level. It was soon demonstrated that the synthesis of DHP from PABA is sensitive to inhibition by sulfonamides, and that the relation between a sulfonamide and PABA remained strictly competitive as long as the two compounds were added simultaneously. If the enzyme and sulfonamide are preincubated with a low concentration of pteridine, subsequent addition of PABA failed to reverse the inhibition; if, however, a high pteridine concentration is used, preincubation results in a much lesser degree of inhibition. Brown (166) showed that the enzyme was not irreversibly inactivated. These results were suggestive of sulfonamide incorporation. It was soon realized that sulfonamides could act as alternate substrates for the enzymes (165–168), resulting in the formation of sulfa-pteroates. Roland et al. (169), however, showed that dihydropterinsulfonamide thus formed did not inhibit DHPS or other folate enzymes. Consequently, this incorporation was not of physiological significance.

Brown (166) observed that the enzymic synthesis was much more sensitive to inhibition by sulfonamides than to bacterial growth, suggestive of impeded permeability of the intact organisms to sulfonamides compared to that of PABA. The more potent inhibitors of folate biosynthesis were, in general, better growth inhibitors also. Hotchkiss and Evans (170) suggested that differences in the response of various organisms to sulfonamides may be attributed to quantitative differences in the ability of individual isoenzymes to produce folic acid from PABA in the presence of sulfonamides.

In a more recent study of the enzymic mechanism and sulfonamide inhibition of DHPS from *Streptococcus pneumoniae* (219), it has been shown that the sulfonamides were capable of displacing PABA in a competetive manner, with equilibrium binding constants that were significantly higher than the equivalent $K_i$ values deduced from steady-state kinetic measurements, indicating that the target for sulfonamide inhibition of *S. pneumoniae* DHPS is the enzyme-DHPP binary

complex, rather than the apoprotein form of the enzyme.

Richey and Brown (171) purified dyhydropteroate synthetase/synthase ($H_2$-pteroate synthase; DHPS) from *E. coli*, and showed that it could use *p*-aminobenzoylglutamate (PABG) also as the substrate to form dihydrofolate directly (Fig. 13.1). PABA is, however, not the natural substrate for this enzyme except in a few bacteria such as *M. tuberculosis*, which forms dihydrofolate directly from dihydropterin pyrophosphate. Shiota et al. (172) and Ortiz and Hotchkiss (173) have shown that the utilization of both substrates, PABA and PABG, is competitively inhibited by sulfonamides.

The cell-free $H_2$-pteroate synthesizing system isolated from *E. coli* has become a very useful tool for studying structure-activity correlations among agonists and antagonists of PABA and the inhibitory effect of sulfonamides (173–176).

The mechanism of action of dapsone (and other sulfones) is similar to that of sulfonamides, in that the action is antagonized by PABA in mycobacteria (175–177), other bacteria (178), and protozoa (179). The exceptionally high antibacterial activity of DDS against *M. leprae* has attracted special attention (180). There is evidence that, as with sulfonamides (167), DDS is also incorporated to form an analog of dihydropteroate, although this also may be of no physiological importance. In *M. kansasii*, Panitch and Levy found a 14- to 15-fold accumulation of DDS within the bacterial cells after 8 days of treatment (181); there may also be similar accumulation within the *M. leprae* bacilli. Additional sites of action outside the folate-synthesizing enzyme system have also been proposed. DDS has unique beneficial effects on some dermal inflammatory conditions (loc. cit.), and it is likely that this action may contribute to its activity against *M. leprae*.

A similar mode of action of sulfonamides and sulfones has been demonstrated in most of the other classes of microbes tested that are susceptible to their action. In the case of chlamydia it has been shown that the sulfonamide-sensitive members of this group, such as trachoma inclusion conjunctivitis viruses, have a folic acid metabolism similar to that of bacte-

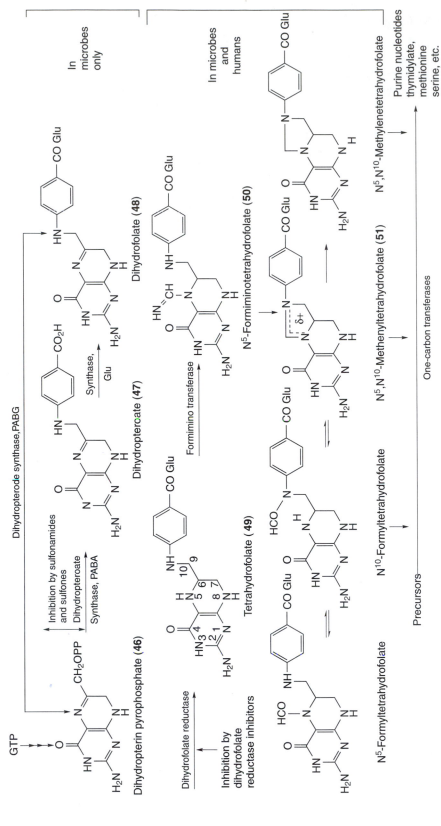

**Figure 13.1.** Folate metabolism. Sites of action of sulfonamides/sulfones and dihydrofolate reductase inhibitors.

ria, and that the action of sulfonamides against them is competitively antagonized by PABA (182–184).

*3.1.4.1.1 Selectivity of Action.* The presence of the folate-synthesizing system has been demonstrated in a variety of bacteria (164, 167, 173, 175, 185), protozoa (186–188), yeasts (189), and plants (190–192), and this serves to explain the broad spectrum of action of sulfonamides. Higher organisms (e.g., mammals) do not possess this biosynthesis system and require preformed folic acid and are thus unaffected by sulfonamides. This selective action on the parasite based on the difference in the metabolic pathway between the microbes and humans makes sulfonamides "ideal" chemotherapeutic agents.

**3.1.4.2 Synergism with Dihydrofolate Reductase Inhibitors.** The discovery by Hitchings et al. in 1948 (193) of certain diaminopyrimidines showing good antimicrobial activity through antifolate mechanism, and the reports that antifolates acted synergistically with sulfonamides (194–196), added a new dimension to the therapeutic use of sulfonamides (197). The elucidation of the folic acid pathway and the demonstration of its inhibition by both sulfonamides and dihydrofolate reductase inhibitors (Fig. 13.1) elucidated the mechanism of this synergism. It is a consequence of the sequential occurrence of the twin loci of inhibition in the *de novo* folic acid biosynthesis. Factors resulting from this combination that contribute to its usefulness include a severalfold increase in chemotherapeutic indices, better tolerance of the drugs, ability to delay development of resistance, and ability to produce cures where the curative effects of the individual drugs are minimal (198).

Recent crystallographic studies of DHPSs and DHFRs from different organisms have greatly helped to understand the structure of the ligand-binding sites on these enzymes and the molecular basis of their action and synergism (loc. cit.). The choice of the individual drugs used in the combinations is based on the best pharmacokinetic fit (199). For example, trimethoprim with sulfamethoxazole, both having a half-life of about 11 h, is a commonly used antibacterial combination. Dihydrofolate reductases from various sources differ mark-

edly in their binding ability to various inhibitors; pyrimethamine is bound much more strongly to the enzyme from plasmodia than from bacteria, and the converse is true for trimethoprim (200, 201). This explains the choice of trimethoprim for bacterial infections and of pyrimethamine for antimalarial chemotherapy. Thus, although pronounced differences exist in the affinity of DHF inhibitors for dihydrofolate reductases of different origins, the structural requirements for inhibitors of the dihydropteroate synthases for the various species studied are somewhat similar (198). Co-trimoxazole, a fixed-dose combination of trimethoprim and sulfamethoxazole (TMP/SMX), is a very commonly used drug for a variety of bacterial infections.

**(52)**

*3.1.4.2.1 Dihydropteroate Synthase (DHPS).* The gene encoding the DHPS from a number of organisms has been cloned, sequenced, and expressed (203–211). Although DHPS is a monofunctional enzyme in prokaryotes, including *Mycobacterial* spp. in plants (210) and protozoa (208), it is part of a bifunctional enzyme and in yeasts (209), it is part of a trifunctional enzyme combining the preceding one and two steps, respectively, of the folate biosynthetic pathway (203). DHPS is reported to be a homodimer in most prokaryotes, including *E. coli* (205), *S. aureus* (207), and *M. tuberculosis/leprae* (203), whereas eukaryotic bifunctional DHPS is reported to be either a dimer or a trimer. The DHPSs from *E. coli* (205), *S. aureus* (207), and *M. tuberculosis* (203) have now been crystallized and their high resolution crystal structures determined. Based on the information available from the crystal-structure studies and that of the distribution of known sulfonamide/sulfone resistance mutations, the binding sites for the substrates could be located, and the possible

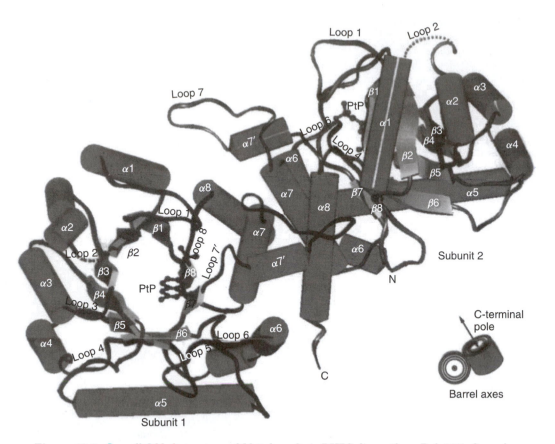

**Figure 13.2.** Overall folded structure of *M. tuberculosis* DHPS dimer (from Ref. 203). See color insert.

mechanism of action of PABA, sulfonamides, and sulfones proposed (203). The implications of this model for the catalytic mechanisms and the likely geometry of the transition state have also been proposed (203).

There are individual variations in the structure of the DHPSs from different organisms, but overall there are many common structural features and a unified picture of the site and mode of binding of the substrates and the inhibitors has emerged. The DHPS consists of 282 amino acids in the case of *E. coli* (205, 206), of 267 amino acids in *S. aureus*, and of 280 amino acids in *M. tuberculosis* (203). The DHPSs belong to the TIM-barrel class of protein structures.

The polypeptide chain is folded into an eight-loop α/β-barrel with a distorted cylindrical shape. It has eight α-helices stacked around the outside of an inner cylinder of par-

allel β-strands. The residues constituting the outer eight helices, the inner parallel β-sheets, and the α,β-connections have been identified. The intermolecular contacts within the crystal structure suggest a dimeric structure for the enzyme, the interface deriving from the proximity of extensive shallow concave areas of each monomer.

The overall TIM-barrel fold and dimerization interface of DHPSs of *M. tuberculosis* (*Mtb*), *E. coli*, and *S. aureus* are similar, with 38% sequence identity (203). The folded *Mtb* DHPS dimer structure as obtained from crystal structure studies is shown in Fig. 13.2.

It has been shown that the pterin-binding pocket of DHPS of *Mtb* is formed by the side chains of 12 amino acid residues (Fig. 13.3), and that this binding pocket occurs in a deep cleft in the barrel (203). Each hydrogen-bond donor/acceptor group of the pterin moiety is

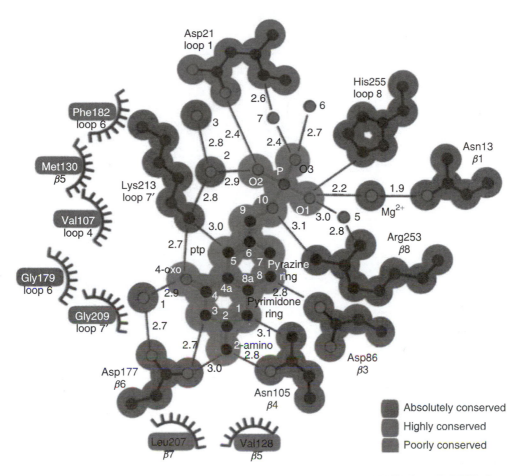

**Figure 13.3.** The pterin monophosphate binding site in *M. tuberculosis* DHPS (from Ref. 203). See color insert.

engaged in interactions with hydrogen-bond donors/acceptors provided by the DHPS amino acid side chains. The residues involved in forming polar bonds and hydrophobic interactions have also been identified. The residues that form the DHP binding site are highly conserved in a number of bacterial and even fungal species, and the residues involved in forming the binding site of the phosphate moiety are somewhat different between different organisms. This has been attributed to different conformations the enzyme assumes during catalysis. The loop 1 containing the C-terminal pole of the β-barrel has been suggested to play a crucial and dynamic function during the catalytic action of the enzyme, and it undergoes extensive conformational changes to place the functionally relevant residues, such as Asp21,

in the proper position for catalysis. It thus appears that a high level of sequence conservation with relative conformational flexibility of loop 1 seem to be important for the catalytic reaction mechanism of DHPS. It has been suggested that the loops of DHPS, which serve an important functional role, are flexible and can assume different conformations (203).

Baca et al. (203) proposed a mechanism in which both loops 1 and 2 play an important role in catalysis by shielding the active site from bulk solvent and allowing PP transfer to occur. Based on these data the transition state geometry, as shown in Fig. 13.4, has been proposed for the catalytic site in which the 4-amino group of the attacking nucleophile displaces the pyrophosphate from the opposite side of the 6-methyl carbon atom.

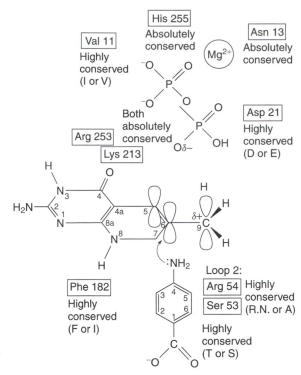

**Figure 13.4.** Possible geometry of catalytic transition state.

The main picture that emerges from these crystallographic studies is of a highly conserved pterin-binding pocket spanning both prokaryotic and lower enkaryotic DHPSs from a wide range of species. Inhibitors designed to fit this site should have a broad spectrum of activity against a variety of microbial pathogens. The fact that this site is highly conserved would imply that drug-resistance mutations are less likely to occur, which would minimize the resistance development problems in inhibitors based on this target. Further, having no counterpart in the mammalian host specificity would not be a concern in the design of inhibitors of this class and selectivity is ensured. This target is not very well exploited and offers good scope for design of antimicrobial agents, including against mycobacteria. The only concern may be overlap with dihydrofolate reductase inhibitors, which have a mammalian counterpart. Sulfones/sulfonamides (and PABA) bind in a less conserved site, which offers greater scope for selectivity of action for different classes of microbes. Most of the sulfonamides so far designed belong to this class. The knowledge gained about the PABA/

sulfone/sulfonamide binding sites of DHPS of different organisms would offer a better scope for design of inhibitors selective to the microbes. These current developments in the understanding of molecular mechanism of action of sulfonamides/sulfones offers a suitable scope for design of a second generation of dihydropteroate synthase inhibitors.

**3.1.4.3 Drug Resistance.** Emergence of drug-resistant strains is a serious problem with sulfonamides, as with many other antimicrobials. Because of the long-time use of this group of drugs, the incidence of drug-resistant isolates in the community has become quite alarming. The majority of isolates of *N. meningitides* of serogroups B and C in the United States and of group A isolated from other countries are now resistant. A similar situation prevails with respect to Shigella and strains of *E. coli* isolated from patients. Given that the mode of action of all sulfonamide/sulfone antimicrobials involves the same basic mechanisms, different sulfonamides usually show cross-resistance, but not to antimicrobials of other classes acting by different mechanisms.

Resistance can arise by one or more of the following mechanisms: (1) increased production of PABA by the pathogen (212, 213); (2) mutation in the DHPS gene, resulting in altered sensitivity of the enzyme, making it selectively more sensitive to the natural substrate (214–219); (3) gene amplification of the enzyme so that more enzyme is produced, thus rendering its saturation by antagonist difficult; (4) a bypass mechanism by which the microorganism develops an ability to utilize more effectively the folic acid present in the host (220); and (5) reduction in permeability of the cell to sulfonamides so that less drug is transported into the cells (214).

In bacteria the first two mechanisms (i.e., overproduction of PABA and causing point mutations that reduced sensitivity of the DHPS to sulfonamides) seem to be the most common. Resistant mutants develop by random mutation and selection or by transfer of resistance factors (R-factors) by plasmids (217). With the advent of molecular cloning techniques and sequence analysis of the DHPS gene/enzyme from resistant isolates the development of resistance and resistance transfer have now been studied at the molecular level. It has been shown that a single point mutation can confer resistance to sulfonamides and sulfones. Most of the high level sulfonamide resistance in Gram-negative bacteria appears to be accounted for by only two plasmid-borne genes, *Sul-I* and *Sul-II* (221). It has been known for many years that multiple drug resistance involving streptomycin, chloramphenicol, tetracycline, and sulfonamides could be transferred between *Shigella* and *E. coli* in mixed cultivation in the host (222). Drug resistance acquired in this manner is usually persistent and irreversible and can be transferred to other sensitive strains indefinitely. However, in meningococci the sulfonamide resistance has been reported to be chromosomally located. Similarly with *S. aureus*, it has been reported that of the nine resistant clinical isolates from different geographical locations analyzed, all had mutations in their chromosomal gene leading to an altered DHPS (207). These isolates differed in 15 amino acid residues from the wild-type sequence, and in this way DHPS appears to be different from DHFR (223). In *M. leprae* resistance to dapsone has been shown to be attributed to mutation at highly conserved amino acid residues 53 or 55 (224).

In the case of plasmodia a bypass mechanism (i.e., ability to use preformed folic acid) seems to be more commonly operative. Bishop has described strains resistant to both sulfonamides and pyrimethamine (220, 225) that can presumably utilize the reduced forms of folic acid available in the host erythrocytes.

**3.1.5 Historical Background.** The story of sulfonamides goes back to the early years of this century when Orlein, Dressel, and Kethe of I.G. Farbenindustrie in Germany (226) found that introduction of a sulfamyl group imparted fastness to acid wool dyes, thus indicating affinity for protein molecules. However, none of these sulfonamides was investigated for antibacterial activity. The interest in dyes as possible antimicrobials was prompted by Ehrlich's studies on the relationship between selective staining of cells by dyes and their antiprotozoal activity, which also led to the testing of azo dyes for antibacterial activity; some of them indeed showed such activity. In an attempt to improve the antimicrobial activity of quinine derivatives, Heidelberger and Jacobs (227) prepared dyes based on dihydrocupreine, which included *p*-aminobenzenesulfonamido-hydrocupreine. Although the latter was reported to have bactericidal activity, it did not arouse much interest because the activity, having been tested *in vitro*, was of a low order and no further work was published on these compounds.

Mietzch and Klarer at I.G. Farbenindustrie synthesized a variety of azo dyes, a continuation of Ehrlich's interest in imparting to azo dyes the property of specific binding to bacterial proteins, comparable to the binding to wool proteins. Mietzsch and Klarer (228) synthesized a group of such dyes containing a sulfonamide radical, which included prontosil rubrum (1). Domagk, also at I.G. Farbenindustrie, carried out the antibacterial testing of these dyes. Realizing the lack of correlation between *in vitro* and *in vivo* screening, Domagk decided to do the testing in mice, a very fortunate decision, because otherwise the fate of sulfonamides might have been different. Domagk (229) observed in 1932 that prontosil protected mice against streptococcal infections and rabbits against staphylococcal infections, though it was without action *in vitro* on bacteria. Anecdotal reports say that the first patient to be successfully treated with prontosil was Hildegarde Domagk, the daughter of its discoverer, who had septicemia attributed to a stitching needle prick. Foerster (230) published the first clinical success with prontosil in a case of staphylococcal septicemia in 1933.

These studies aroused worldwide interest and further developments took place at a very fast rate. One of the earliest systematic investigations on sulfonamides was by Trefouel, Nitti, and Bovet (231) working at the Pasteur Institute in Paris. Under a program of structural modification of this class of compounds, they prepared a series of azo dyes by coupling diazotized sulfanilamides with phenols with or without amino or alkyl groups. They observed that variation in the structure of the phenolic moiety had very little effect on antibacterial activity, whereas even small changes in the sulfanilamide component abolished the activity. These observations pointed to the benzenesulfonamide residue as the active structural unit, and led to considering *p*-aminobenzenesulfonamide (sulfanilamide, **45**) as the putative metabolite responsible for the antibacterial activity. They suggested that prontosil was converted to sulfanilamide in animals and showed that sulfanilamide was as effective as the parent dyestuff in protecting mice infected with streptococci. They also showed that sulfanilamide exerted a bacteriostatic effect *in vitro* on susceptible organisms. Soon after, Colebrook and Kenny (232) observed that although prontosil was inactive *in vitro*, the blood of patients treated with it had bacteriostatic activity. They also reported the dramatic cure of 64 cases of puerperal sepsis by prontosil, whereas Buttle, Gray, and Stephenson (233) showed that sulfanilamide could cure streptococcal and meningococcal infections in mice. Fuller's (234) demonstration of the presence of sulfanilamide in the blood and its isolation from urine of patients (and mice) under treatment with prontosil firmly established that prontosil is reduced in the body to form sulfanilamide (**64**), a compound synthesized as early as 1909 by Gelmo

(235). Fuller concluded that the therapeutic action of prontosil was very likely the result of its reducton *in vivo* to sulfanilamide. Among the early patients to be treated with sulfanilamide was Franklin D. Roosevelt Jr., the son of the President of the United States. His recovery from a streptococcal throat infection helped to overcome early reservations on the medicinal value of antibacterial chemotherapy with sulfonamides. Ehrlich's concept of a relationship between the affinity of dyes for a parasite and their antimicrobial activity, which focused attention on sulfonamide azo dye, was found to be irrelevant to the activity of the latter. Nevertheless, sulfanilamide proved to be the "magic bullet" of Ehrlich and heralded the era of *chemotherapy*, a term coined by Ehrlich to emphasize the concept of selective action of chemicals on microbes as opposed to the action on host cells. The era of modern chemotherapy had now begun. Domagk was awarded the Nobel Prize for Medicine in 1939, primarily for the discovery of the antibacterial activity of sulfonamides.

*Earlier Sulfonamides.* These discoveries had a tremendous impact not only on the development of sulfonamides as antimicrobials, but also on developments in chemotherapy in general. Sulfanilamide, being easy to prepare, cheap, and not covered by patents, became available for widespread use and brought a new hope for the treatment of microbial infections. Recognizing the potential of sulfonamides, almost all major research organizations the world over initiated research programs for the synthesis and study of analogs and derivatives of sulfanilamide, particularly to improve its antimicrobial spectrum, therapeutic ratio, and pharmacokinetic properties. New sulfonamides were introduced in quick succession. Sulfapyridine (M&B 693, **66**) (236), reported in 1938, was one of the earliest of the new sulfonamides to be used in clinical practice for the treatment of pneumonia and remained the drug of choice until it was replaced by sulfathiazole. Sulfapyridine was used on Winston Churchill to cure pneumonia in 1943 during his trip to Africa.

Sulfathiazole (**67**) (237) was the second sulfonamide to be introduced in clinical practice. It replaced sulfapyridine because of its wider antibacterial spectrum and higher therapeutic

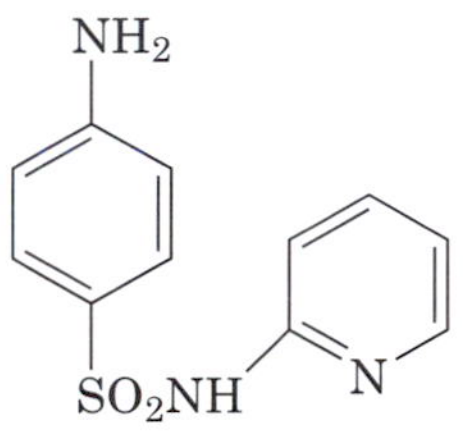

(53)

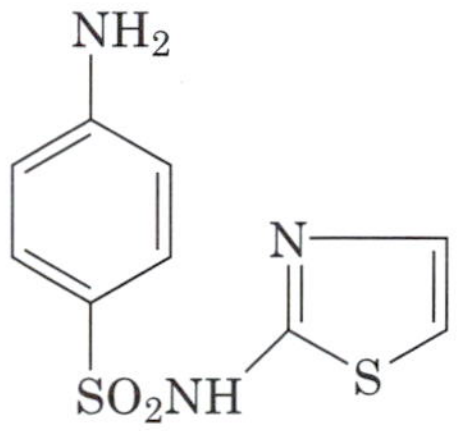

(54)

index. Substitution of the thiazole ring by alkyl groups did not improve the activity, whereas a 4-phenyl residue enhanced both the activity and the toxicity.

Some of the other important sulfonamides introduced in clinical practice during this period were sulfacetamide, the corresponding $N^4$-pthallyl and succinyl derivatives, sulfadiazine, sulfamerazine, sulfamethazine, sulfisomidine, sulfamethizole, and sulfisoxazole (Table 13.2). These compounds differed widely in their pharmacokinetic profile and helped in enlarging the scope of therapeutic use of sulfonamides. This work was described in 1948 in a very exhaustive monograph by Northey, which may be consulted for research work of this period (238).

*Later Sulfonamides.* This widespread interest in new sulfonamides continued until about 1945, when interest gradually shifted to antibiotics after the introduction of penicillin. However, after about a decade of the use of penicillins, problems encountered with antibiotics, such as emergence of resistant strains, superinfection, and allergic reactions, brought about a revival of interest in sulfonamides. The knowledge gained during this period about the selectivity of action of sulfonamides on the pathogens, the relationship between their solubility and toxicity, and their pharmacokinetics gave a new direction to further

developments and a second generation of sulfonamides began to appear with improved characteristics.

A major advance in sulfonamide therapy came with the proper appreciation of the role of pharmacokinetic studies in determining the dosage schedule of these drugs. It was realized that some of the "earlier" sulfonamides such as sulfadiazine (**55**) and sulfamerazine (**56**)

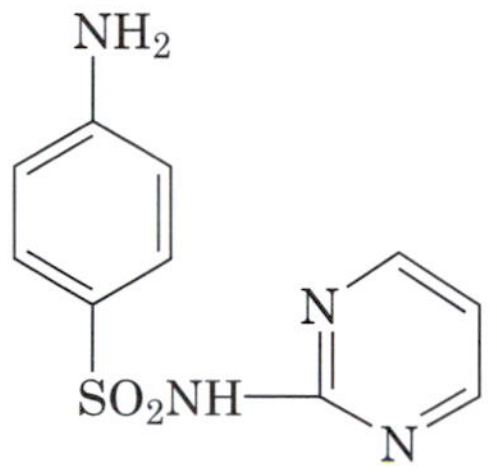

(55)

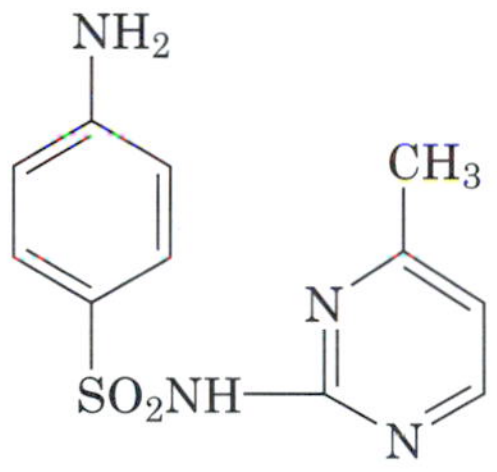

(56)

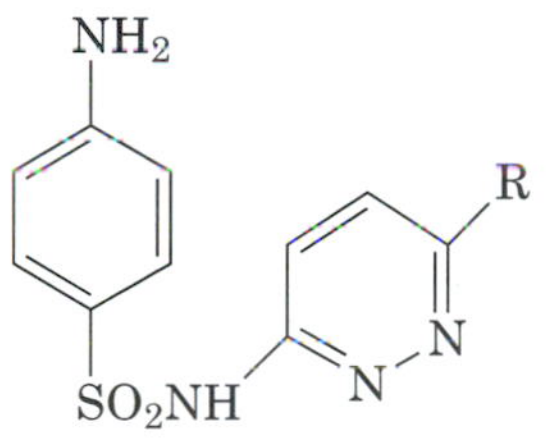

(57)

had a long half-life (17 and 24 h, respectively) and required less frequent administration than was normally prescribed. The era of newer, long-acting sulfonamides started in 1956 with the introduction of sulfamethoxypyridazine (**57**) (239) having a half-life of around 37 h, the longest known at that time, which needed to be administered only once a day (240). In 1959 sulfadimethoxine was introduced with a half-life of approximately 40 h

(241–243). A related 4-sulfonamidopyrimi-
dine, sulfomethoxine (**58**) (244, 245), having
the two methoxyl radicals in the 5,6-positions,

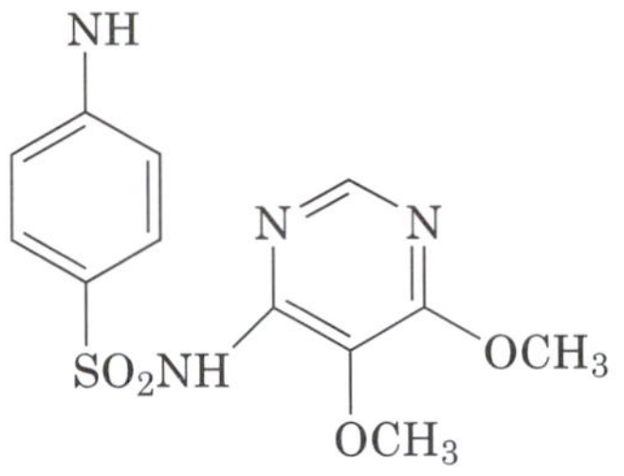

(**58**)

was soon introduced. It has by far the longest
half-life, about 150 h, and needs administration only once a week.

Some of the other sulfonamides introduced
in clinical practice in this period were sulfa-
methyldiazine, sulfaphenazole, sulfamoxol,
and sulfamethoxazole (**59**)(246) (Table 13.2).

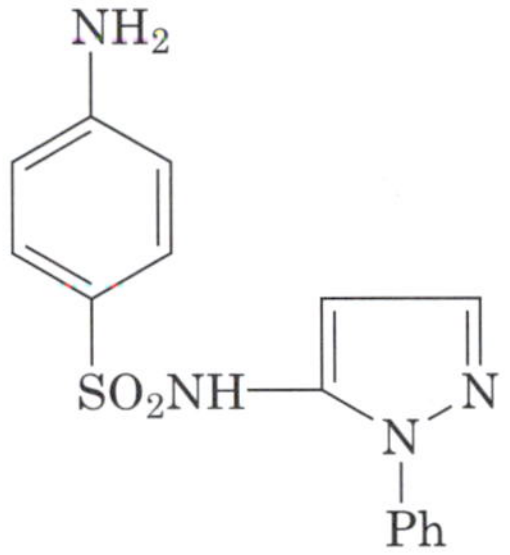

(**59**)

Sulfamethoxazole is a particularly important
sulfonamide of this period in view of its well-
matched half-life (~11 h) with that of trimethoprim, and a fixed-dose combination of
the two, co-trimoxazole is widely used in clinical practice.

*Sulfones.* The demonstration that experimental tuberculosis could be controlled by 4,4'-
diaminodiphenylsulfone (dapsone, **60**)(247) and
disodium 4,4'-diaminodiphenylsulfone-*N,N*'-
didextrose sulfonate (promin, **61**)(248) was a
major advance in the chemotherapy of mycobacterial infections. Although dapsone and
promin proved disappointing in the therapy of
human tuberculosis, the interest aroused in

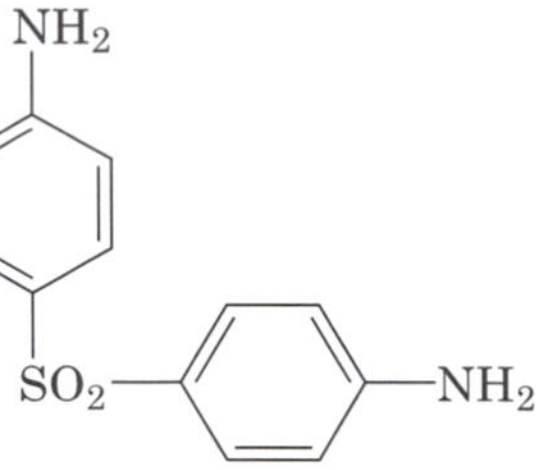

(**60**)

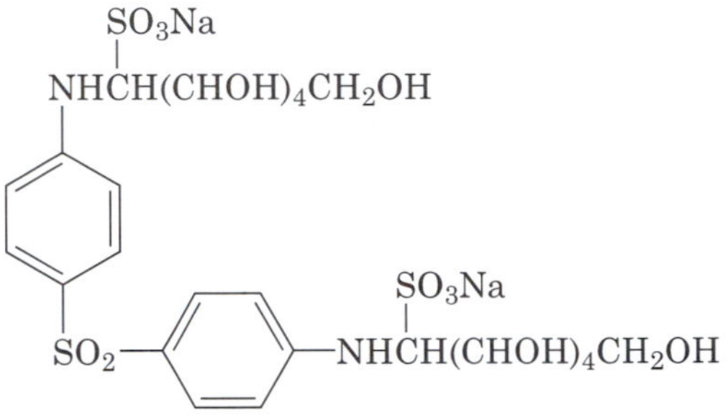

(**61**)

the possibility of treatment of mycobacterial
infections with sulfones led to the demonstration of the favorable effect of promin in rat
leprosy (249). This was soon followed by the
successful treatment of leprosy patients, first
with promin and later with dapsone itself.
Since then dapsone has remained the mainstay for the treatment of all forms of human
leprosy (250). It has now been shown that *Mycobacterium leprae* is unusually sensitive to
dapsone (251) and that its growth can be inhibited by very low concentration of the latter.

An important advance in the use of dapsone took place with the demonstration that
*N,N*'-diacyl derivatives and certain Schiff
bases of dapsone are prodrugs and have a repository effect and release dapsone slowly;
*N,N*'-diacetamidodiphenylsulfone (acedapsone, **62**) and the Schiff base DSBA (**63**) are
particularly useful as repository forms (252).
After a single intramuscular injection of 225
mg of acedapsone, a therapeutic level of DDS
(20–25 μg/mL) is maintained in the blood for
as long as 60–68 days, and it is useful in the
prophylaxis and treatment of leprosy.

To improve on the antimycobacterial and
antiprotozoal (especially antimalarial) activi-

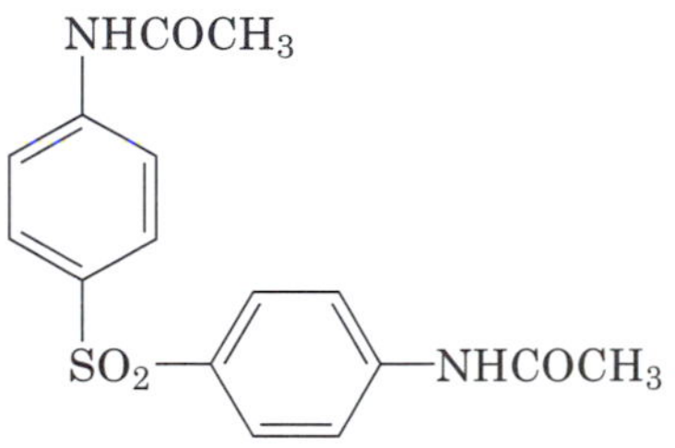

NHCOCH$_3$

SO$_2$—⟨ ⟩—NHCOCH$_3$

**(62)**

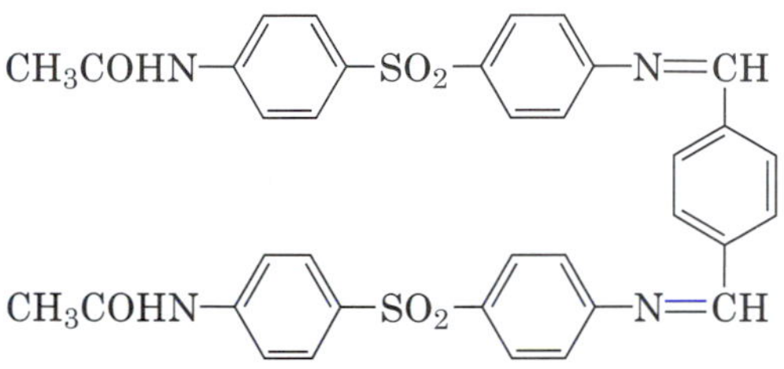

CH$_3$COHN—⟨ ⟩—SO$_2$—⟨ ⟩—N=CH

CH$_3$COHN—⟨ ⟩—SO$_2$—⟨ ⟩—N=CH

**(63)**

ties of dapsone a variety of substituted sulfones have been prepared.

Overall, none of the substituted diaminodiphenylsulfones was significantly more active *in vivo* than DDS to offer much advantage (253). Even after almost 50 years of use in clinical practice, DDS alone or in combination with other drugs (as in multidrug therapy) continues to be the mainstay of chemotherapy of leprosy.

### 3.1.6 Structure-Activity Relationship
**3.1.6.1 Structure and Biological Activity.** Although the story of sulfonamides started with the discovery of their antimicrobial action, susequent studies established their usefulness as carbonic anhydrase inhibitors, diuretics (saluretics), and antidiabetics (insulin-releasers), and more recently also as endothelin antagonists. Compounds with each type of action possess certain specific common structural features. The present discussion, however, is restricted only to the antimicrobial sulfonamides and sulfones, characterized by their ability to inhibit the *de novo* biosynthesis of folic acid by competing with PABA for 7,8-dihydro-6-hydroxymethylpterin pyrophosphate at the active site of dihydropteroate synthase (254–256).

*3.1.6.1.1 Sulfonamides.* Because sulfanilamide (**45**) is a rather small molecule and there are not too many variations that can be carried out without changing the basic nucleus, the following generalizations regarding structure-activity relationships were arrived at quite early in the development of sulfonamides, which guided the subsequent work on molecular modification, and these generalizations still hold:

1. The amino and sulfonyl radicals on the benzene ring should be in 1,4-disposition for activity; the amino group should be unsubstituted or have a substituent that is removed readily *in vivo*.

2. Replacement of the benzene ring by other ring systems, or the introduction of additional substituents on it, decreases or abolishes the activity.

3. Exchange of the SO$_2$NH$_2$ by SO$_2$C$_6$H$_4$-*p*-NH$_2$ retains the activity, whereas exchange by CONH$_2$ by COC$_6$H$_4$-*p*-NH$_2$ markedly reduces the activity.

4. $N^1$-Monosubstitution results in more active compounds with greatly modified pharmacokinetic properties. $N^1$-Disubstitution in general leads to inactive compounds.

5. The $N^1$-substitution should be such that its p$K_a$ value would approximate the physiological pH.

The presence of a *p*-aminobenzensulfonyl radical thus seems inviolate for maintaining good activity and practically all the attention was focused on $N^1$-substituents. These substituents seem to affect mainly the physicochemical and the pharmacokinetic characteristics of the drugs.

*3.1.6.1.2 Sulfones.* The following broad generalizations hold for the SAR of sulfones:

1. One *p*-aminophenylsulfonyl residue is essential for activity; the amino group in this moiety should be unsubstituted or have a substituent that is removed readily *in vivo*.

2. The second benzene ring should preferably have small substituents that will make this

ring electron rich (such as $CH_3$, $OCH_3$, OH, $NH_2$, $NHC_2H_5$); *p*-substitution is most favorable for activity.

3. Replacement of the second phenyl ring by heterocycles does not improve the activity.

Diaminodiphenyl sulfone (DDS) has retained its preeminent position, even after 50 years of use.

*3.1.6.2 Quantitative Sar.* Studies to find a correlation between physicochemical properties and bacteriostatic activity of sulfonamides have been pursued almost since their discovery. The substituents that attracted the attention of investigators quite early were the amino and the sulfonamido groups in the molecule, and several groups of investigators almost simultaneously noted a correlation between the bacteriostalic activity and degree of ionization of sulfonamides. The primary amino group in sulfonamides apparently plays a vital part in producing bacteriostasis, given that any substituent on it causes complete loss of activity. Seydel et al. (257, 258), from a study of infrared (IR) spectra and activity of a number of sulfonamides, concluded that the amount of negative charge on the aromatic amino group is important for the activity. However, variation in activity within a series of compounds could not be related to a change in base strength, given that all the active sulfonamides (and sulfones) have a basic dissociation constant of about 2, which is close to that of PABA. Foernzler and Martin (259) computed the electronic characteristics of a series of 50 sulfonamides by the combination of linear atomic and molecular orbital methods (LCAO-MO methods) and found that the electronic charge on the *p*-amino group did not vary significantly with a change in the $N^1$-substituent.

Thus attention was focused mainly on the $N^1$-acidic dissociation values, which vary widely from about 3 to 11. Fox and Rose (260) noted that sulfathiazole and sulfadiazine were about 600 times as active as sulfanilamide against a variety of microorganisms, and that approximately 600 times as much PABA was required to antagonize their action as to antagonize the action of sulfanilamide; however, the same amount of PABA was required to

antagonize the MIC of each drug. This suggested that the active species in both cases were similar, and that the increase in bacteriostatic activity was ascribed to the presence of a larger proportion of the drug in an active (ionized) form. They found that the concentration of the ionized form of each drug at the minimum effective concentration was of the same order. Thus if only the ionized fraction at pH 7 was considered instead of the total concentration, the PABA/drug ratio was reduced to 1:1.6–6.4. They also observed that with a 10-fold increase in ionization of sulfanilamide on altering the pH from 6.8 to 7.8, there was an eightfold increase in bacteriostatic activity. On the basis of these observations, Fox and Rose suggested that only the ionized fraction of the MIC is responsible for the antibacterial action. Schmelkes et al. (261) also noted the effect of pH of the culture medium on the MIC of sulfonamides and suggested that the active species in a sulfonamide solution is anionic.

Bell and Roblin (262), in an extensive study of the relationship between the $pK_a$ of a series of sulfonamides and their *in vitro* antibacterial activity against *E. coli*, found that the plot of log 1/MIC against $pK_a$ was a parabolic curve, and that the highest point of the curve lay between $pK_a$ values of 6 and 7.4; the maximal activity was thus observed in compounds whose $pK_a$ approximated the physiological pH. Because the $pK_a$ values are related to the nature of the $N^1$-substituent, the investigators emphasized the value of this relationship for predicting the MIC of new sulfonamides. The $pK_a$ of most of the active sulfonamides discovered since then, and particularly of the long-acting ones, falls in this range (Table 13.2). Bell and Roblin correlated Woods and Fildes's hypothesis regarding the structural similarity of a metabolite and its antagonist with the observed facts of ionization. They emphasized the need of polarization of the sulfonyl group of active sulfonamides, so as to resemble as closely as possible the geometric and electronic characteristics of the *p*-aminobenzoate ion, and postulated the "the more negative the $SO_2$ group of $N^1$-substituted sulfonamides, the more bacteriostatic the compound will be." The acid dissociation constants were considered to be an indirect measure of the negative character of the $SO_2$ group. The hypothesis of Bell and Roblin stated that the un-ionized molecules had a bacteriostatic activity, too, although weaker than that of the ionized form. Furthermore, it was supposed that increasing the acidity of a compound decreased the negativity of the $SO_2$ group, thus reducing the bacteriostatic activity of the charged and uncharged molecules.

Cowles (263) and Brueckner (264), in a study of the effect of pH of the medium on the antibacterial activity of sulfonamides, found that the activity increased with increase in pH of the medium only up to the point at which the ionization of the drug was about 50%, and then it decreased. Brueckner assumed different intra- and extracellular pH values to explain the observed effects. Cowles suggested that the sulfonamides penetrate the bacterial cell in the un-ionized form, but once inside the cell, the bacteriostatic action is attributed to the ionized form. Hence for maximum activity, the compounds should have a $pK_a$ value that gives the proper balance between the intrinsic activity and penetration; the half-dissociated state appeared to present the best compromise between transport and activity. This provided an alternative explanation for the parabolic relationship observed by Bell and Roblin between $pK_a$ and MIC.

Seydel et al. (257, 265, 266) and Cammarata and Allen (267) have cited examples of active sulfonamides whose $pK_a$ values lie outside the optimal limits given by Bell and Roblin, and showed that if a small homologous series is used, a linear relationship of the $pK_a$ to the MIC is obtained.

Seydel and associates in a study of sulfanilides and $N^1$-(3-pyridyl)-sulfanilamides extrapolated the electron density on the 1-NH group from the study of IR and NMR data and Hammett sigma values of the parent anilines and have correlated the data with the MIC against *E. coli*. Anilines were used for studying the IR spectra because they could be dissolved in nonpolar solvents, thus giving more valid data; this was not possible with sulfanilamides because of low solubility in such solvents. Seydel (268) and Garrett et al. (269) found an approximately linear relationship between bacteriostatic activity, Hammett sigma value, and electron density of the $N^1$-nitrogen of a group of *m*- and *p*-substituted

sulfanilides and emphasized the possibility of predicting the *in vitro* antibacterial activity of sulfanilamides by use of this relationship. Later, Seydel (270) included in this study 3-sulfapyridines, carried out regression analyses of the data, and obtained a very acceptable correlation coefficient.

The functional relationship between acid dissociation constant and the activity of sulfonamides has not been questioned since the investigations just cited were published. This, however, does not imply that the ions of different sulfonamides are equally active; other factors also have an influence and account for the observed differences in activity of different sulfonamides, such as affinity for the concerned enzyme. The $pK_a$ value is related to solubility, distribution and partition coefficients, permeability across membranes, protein binding, tubular secretion, and reabsorption in the kidneys.

Fujita and Hansch (271), in a multiparameter linear free-energy approach, correlated the $pK_a$, hydrophobicity constant, and Hammett sigma values of a series of sulfanilides and $N^1$-benzoyl and $N^1$-heterocyclic sulfanilamides with their MIC data against Gram-positive and Gram-negative organisms and their protein-binding capacity. They devised suitable equations by regression analyses for the *meta*- and *para*-substituted compounds; the correlation for the *para*-substituted compounds was rather poor. The hydrophobicity of the compounds was found to play a definite role in the activity. It was shown that keeping the lipophilicity of the substituents unchanged, the logarithmic plot of activity against the dissociation constant gives two straight lines with opposite slopes, the point of intersection of which corresponds to the maximal activity for a series of sulfanilamides. They suggested the optimal values of the dissociation constant and hydrophobicity for maximum activity against the organisms studied.

Yamazaki et al. (272), in their study of the relationship between antibacterial activity and $pK_a$ of 14 $N^1$-heterocyclic sulfanilamides, considered separately the activities of the compounds in terms of the concentrations of their ionized and un-ionized forms and their total concentrations in the culture medium.

They found that, when the relationship between $pK_a$ and ions is considered, it is linear for ionized and un-ionized states, giving two lines having opposite slopes and intersecting each other. The point of intersection corresponds to the pH of the culture medium. They found the $pK_a$ for optimal activity to be between 6.6 and 7.4.

In these studies it was noticed that some of the sulfonamides had lower antibacterial activity than expected, possibly because of their poor permeation. To define the role of permeability in the antibacterial activity of sulfonamides, Seydel et al. (273) extended these investigations to cell-free folate synthesizing systems and correlated the inhibitory activity of these compounds on this enzyme system and on the intact organisms to their $pK_a$, Hammett sigma, chemical shift, and $\pi$ values. The rate-determining steps for sulfonamide action in the cell-free system and a whole-cell system were found to have similar substituent dependencies. From a comparison of the linear free-energy relationships obtained in the two systems, they suggested that the observed parabolic dependency of the antibacterial activity indicates that it is not the extracellular ionic concentration, which, in turn, is limited by the permeation of un-ionized compounds, thus supporting Cowles and Brueckner's postulates (loc. cit.). They concluded that the lipophilic factors are not important in the cell-free system or for *in vitro* antibacterial activity when permeability is not limited by ionization.

Intensive subsequent work in this field over the last four decades has fully supported the views expressed quite early in the development of sulfonamides of the predominant role of ionization for their antibacterial activity, and that the degree of ionization determines the antibacterial activity, given that the ionized form is more potent than the un-ionized form.

*3.1.6.2.1 Sulfones.* Ever since the discovery of the antimycobacterial activity of sulfones in the 1940s and that they share a common biological mode of action with sulfonamides (competitive antagonism of PABA), the question of their structural similarity with sulfonamides, which enables them to inhibit dihydropteroate synthase, has attracted much

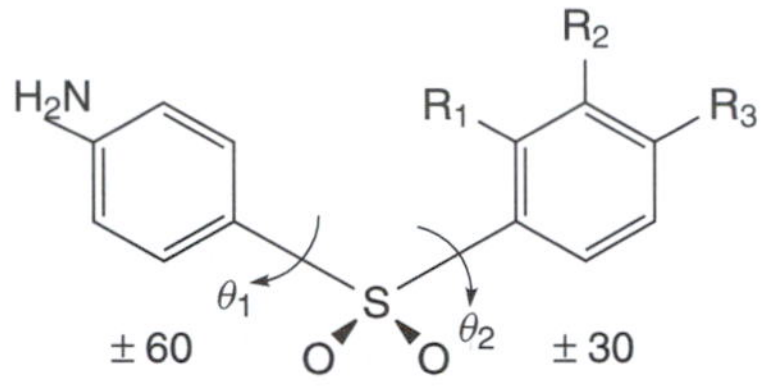

**Figure 13.5.** Proposed active conformation of 4-aminodiphenylsulfones.

attention. It was realized quite early in these studies that as with sulfonamides, $4\text{-}NH_2\text{-}C_6H_4\text{-}SO_2$ was inviolate for optimal activity and the substituents in the second phenyl could modulate the activity. A number of QSAR studies have been reported (274–282) on 4-aminodiphenylsulfones to analyze the contribution of these substituents to the biological activity using linear free-energy, molecular modeling, and conformational analysis methods. It has been shown that electronic and steric effects have the decisive role both on binding to the enzyme and the overall biological activity. The electronic effects were rationalized in terms of electronic charge perturbations that are transmitted from the multisubstituted aryl ring to the essential structural moiety, $4\text{-}NH_2\text{-}C_6H_4\text{-}SO_2$, through the $SO_2$ group mainly through hyperconjugation (279). In conformational analysis using the MINDO semiemperical molecular orbital method, it was found that 4-aminodiphenylsulfones show multiple conformational energy minima, mainly attributed to the torsional freedom of the sulfur-carbon bond of the substituted aryl ring with the $4\text{-}NH_2\text{-}C_6H_4\text{-}SO_2$. The highly active derivatives were in general shown to be less flexible and inhibition potency increased as entropy decreased (280, 281); a butterfly-type structure (Fig. 13.5) was considered to best represent the active conformations.

*3.1.6.3 Water/Lipid Solubility.* The clinically used sulfonamides being weak acids are, in general, soluble in basic aqueous solutions. As the pH is lowered, the solubility of their $N^1$-substituted sulfonamides decreases, usually reaching a minimum in the pH range of 3 to 5. This minimum corresponds to the solubility of the molecular species in water (Table

13.2). With a further decrease in pH corresponding to that of a moderately strong acid, the sulfa drugs dissolve as cations.

The solubility of sulfonamides is of clinical and toxicological significance because damage to kidneys is caused by crystallization of sulfonamides or their $N^4$-acetyl derivatives. Their solubility in the pH range of urine (i.e., pH 5.5–6.5) is thus of practical interest. One of the significant advances in the first phase of sulfonamide research was the development of compounds with greater water solubility, such as sulfisoxazole, which helped to overcome the problem of crystallization in the kidney of earlier sulfonamides. However, apart from the solubility of the parent compounds, the solubility of their $N^4$-acetyl derivatives, which are the main metabolic products, is of great importance because these are generally less soluble than the parent compounds. For example, sulfathiazole, which itself is unlikely to be precipitated, is metabolized to its $N^1$-acetyl derivative, which has a poor solubility that is likely to lead to its crystallization in the kidney. The solubility of sulfonamides and their principal metabolites in aqueous media, particularly in buffered solutions and body fluids, therefore, has been the subject of many studies aimed at enhancing our understanding of their behavior in clinical situations (284, 285).

An important factor affecting the chemotherapeutic activity of sulfonamides and their *in vivo* transport is lipophilicity of the undissociated molecule. The partition coefficients measured in solvents of different dielectric constants have been used to determine the lipid solubility and hydrophobicity constant (269–284). Chromatographic $R_1$ values in a number of thin-layer chromatography systems have also been used as an expression of the lipophilic character of sulfonamides and found to correspond well with the Hansch values in an isobutyl alcohol-water system (286).

Table 13.2 gives the percentage of various sulfonamides passing from aqueous phase into ethylene chloride as determined by Rieder (284). Lipid solubility of different sulfonamides varies over a considerable range. These differences unquestionably influence their pharmacokinetics and antibacterial activity. It has been noted by Rieder (284) that long-acting sulfonamides with a high tubular reabsorption are generally distinguished by a high degree of lipid solubility. The antibacterial activity and the half-life are also related to lipid solubility. Although a precise relationship between these factors has not been established, it has been shown in general, that as the lipid solubility increases, so does the half-life and *in vitro* activity against *E. coli*.

**3.1.6.4 Protein Binding.** A particularly important role in the action of sulfonamides is played by their binding to proteins. Protein binding, in general, blocks the availability of sulfonamides as of many other drugs (the bound drug is chemotherapeutically inactive), and reduces their metabolism by the liver. The binding is reversible; thus the active free form is liberated as its level in the blood is gradually lowered. The sulfonamide concentration in other body fluids is also dependent on its protein binding. Thus the unbound fraction of the drug in the plasma seems to be significant for activity, toxicity, and metabolism, whereas protein binding appears to modulate the availability of the drug and its half-life. The manner and extent of binding of sulfonamides has been the subject of many studies (284, 287–289), and the important characteristics of the binding are now reasonably clear. The binding affinity of different sulfonamides varies widely with their structure (Table 13.2) as also with the animal species and the physiological status of the animal (284, 290). In plasma the drug binds predominantly to the albumin fraction. The binding is weak (4–5 kcal) and is easily reversed by dilution. It appears to be predominantly hydrophobic, with ionic binding being relatively less significant (271, 289). Thus the structural features that favor binding are the same as those that increase lipophilicity, such as the presence of alkyl, alkoxy, or arly groups (261, 287, 291). $N^4$-Acetyl derivatives are more strongly bound than the parent drugs. Introduction of hydroxyl or amino groups decreases protein binding, and glucuronidation almost abolishes it. Seydel (292), in a study of the effect of the nature and position of substituents on protein binding and lipid solubility, has shown that among isomers, *ortho*-substituted compounds have the lowest protein binding. This would indicate that steric factors have a role in protein binding and that $N^1$-nitrogen atom of the sulfonamide is involved. The binding seems to take place with the basic centers of arginine, lysine, and histidine in the proteins (284). The locus of binding of several sulfonamides to serum albumin has been shown by high resolution NMR spectral studies to involve the benzene ring more than the heterocycle (293).

There have been attempts to establish correlations between physicochemical properties of sulfonamides, their protein binding, and their biological activity. Martin (294) established a functional relationship between excretion and distribution and binding to albumin, and Kruger-Thiemer et al. (288) derived a mathematical relationship. Moriguchi et al. (295) observed a parabolic relationship between protein binding and *in vitro* bacteriostatic activity in a series of sulfonamides, and suggested that too strong an affinity between sulfonamides and proteins would prevent them from reaching their site of action in bacteria; with too low an affinity, they would not be able to bind effectively with enzyme proteins to cause bacteriostasis, assuming that affinity for enzyme proteins is paralleled by affinity to bacterial proteins. In a multiparameter study of a series of $N^1$-heterocyclic sulfonamides, Fujita and Hansch (271) considered that in the free state, sulfonamides exist as two different species, neutral and ionized, whereas in the bound state they exist in only one form. They developed suitable equations by regression analysis and showed that for the series of sulfonamides of closely related structure, whose $pK_a$ value does not vary appreciably, the binding is governed mainly by the $N^1$-substituent, which supported the earlier results (287).

The implications of protein binding for chemotherapeutic activity are not fully understood. The factors favoring protein binding are also those that would favor transport across membranes, tubular reabsoption, and increased binding to enzyme protein. $N^1$-Acetyl derivatives are more strongly bound to proteins and yet are better excreted. No universally applicable relationship has been found between half-life of sulfonamides and protein binding, although it has been established in general that protein binding modulates bioavailability and prolongs the half-life of sulfonamides, as of other drugs.

## 3.2 Quinolones

The 4-quinolones have a number of advantages over other classes of antibacterial agents. They are effective against many organisms, well-absorbed orally, well-distributed in tissues, and they have relatively long serum half-lives and minimal toxicity. Because of deep-tissue and cell penetration, they are useful for urinary tract infections, prostatitis, infections of the skin and bones, and penicillan-resistant sexually transmitted diseases.

**3.2.1 Introduction.** The first quinolone of commercial importance, nalidixic acid (**64**), was prepared in 1962 by Lesher (298). Norfloxacin, a fluoroquinolone with a broad spectrum of antibacterial activity, was patented in 1978 by Irikura (299). Chemical modifications based on their structures have since led to thousands of new analogs, some of which have significantly improved effectiveness. The quinolone antibacterial group contains 4-quinoline-3-carboxylic acids such as ciprofloxacin (**65**), lomefloxacin (**66**), norfloxacin

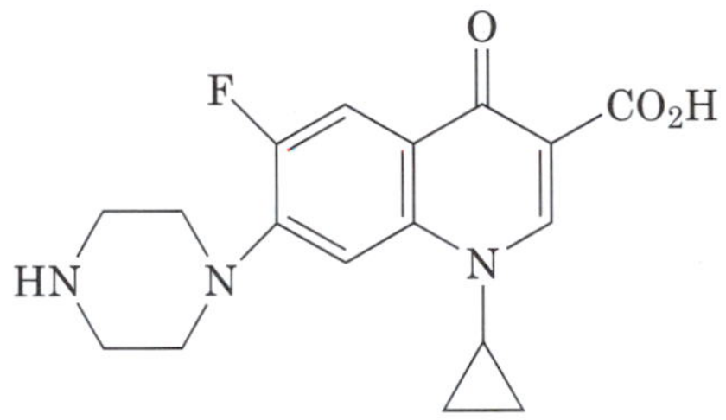

(65)

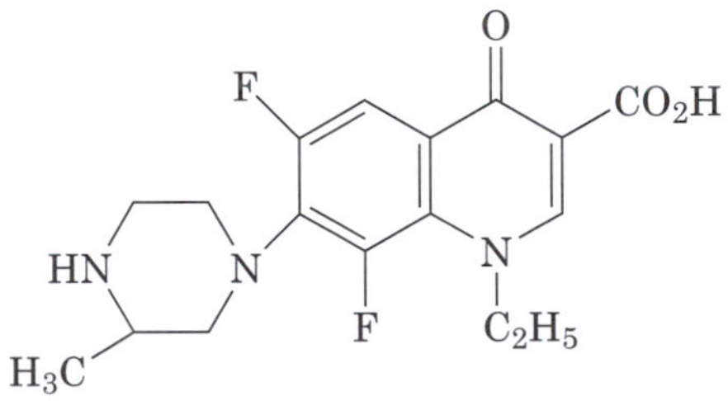

(66)

(**67**), pefloxacin (**68**), sparfloxacin (**69**), gatifloxacin (**70**), and moxifloxacin (**71**); and the tricyclic analog ofloxacin (**72**), for which the

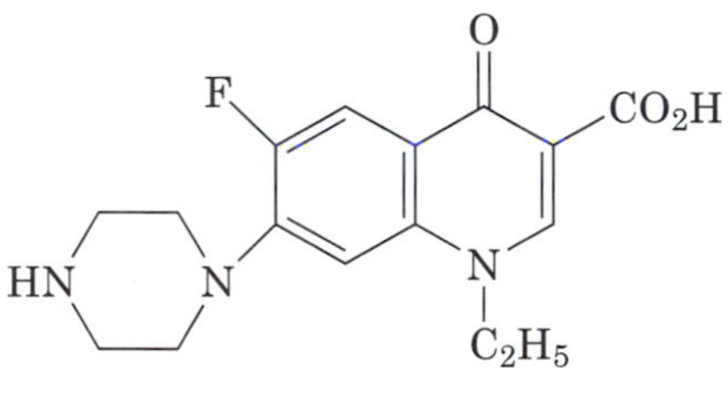

(67)

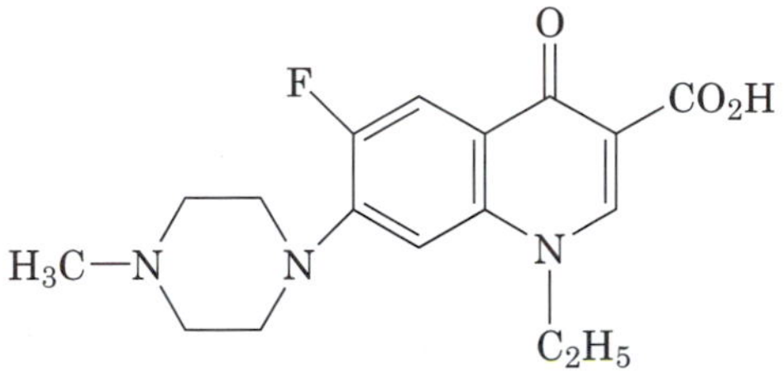

(68)

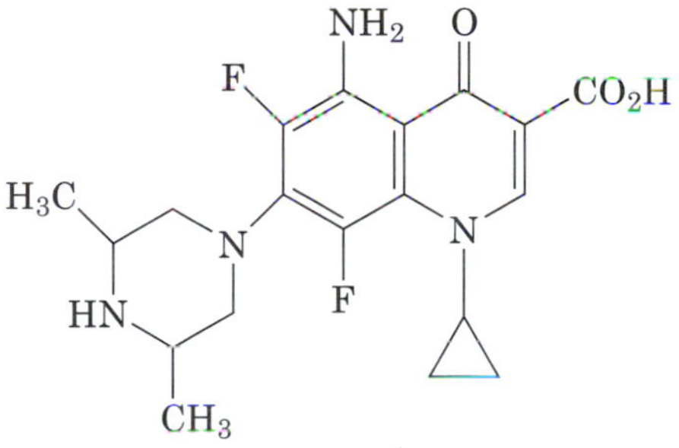

(69)

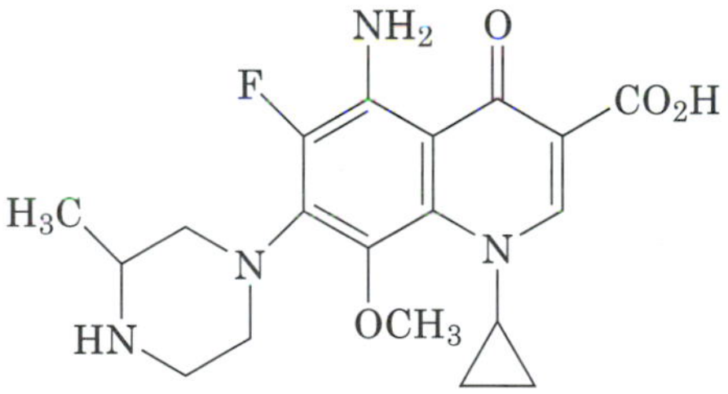

(70)

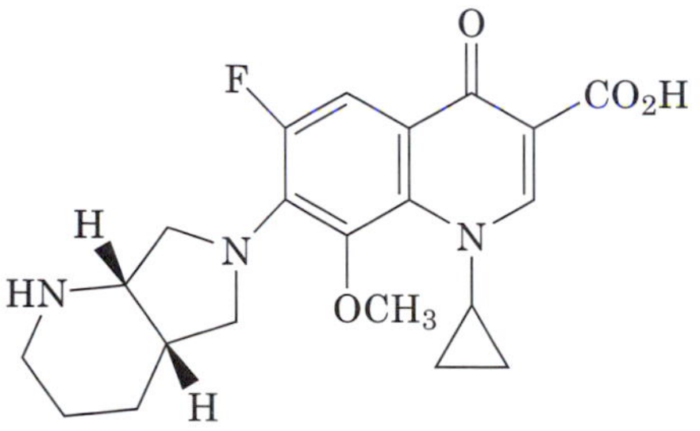

(71)

**3.2.2 Antibacterial Activity.** Fluoroquinolones are important broad-spectrum antibacterial agents. All of them are active against species such as *Enterobacter cloacea, Proteus*

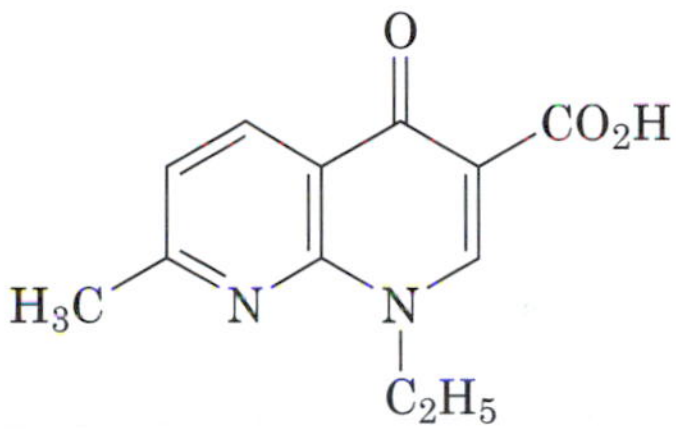

(64)

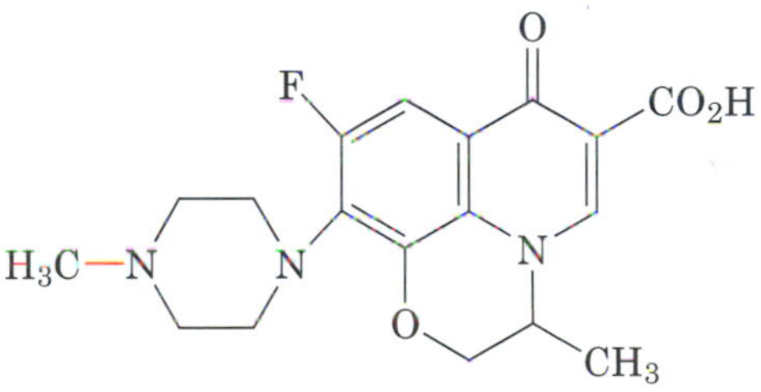

(72)

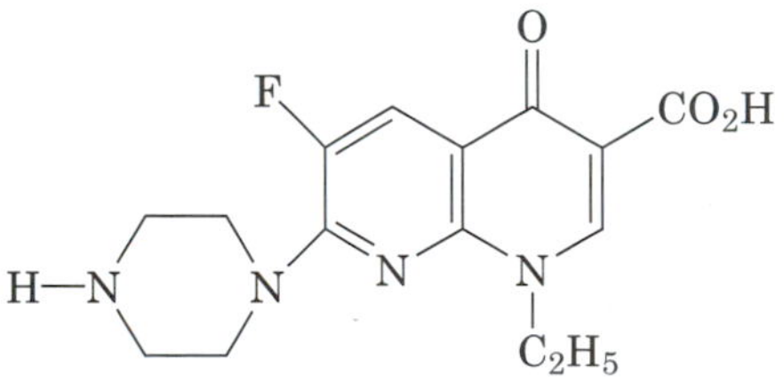

(73)

(*S*)-enantiomer (levofloxacin) is marketed separately. It also contains 1,8-naphthyridine-4-one-3-carboxylic acids such as nalidixic acid (enoxacin, **73**), and trovafloxacin/alatrofloxacin (**74**), as well as cinnolin-4-one-3-carboxylic acid (cinoxacin, **75**).

*mirabilis, Morganella morganii,* and *Staphylococcus epidermis.* Furthermore, most of

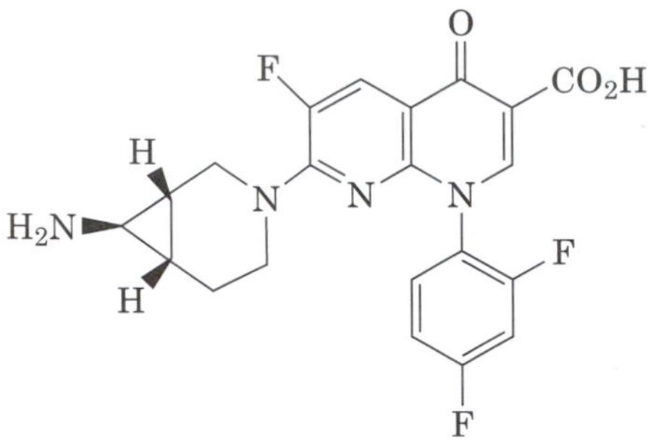

(74)

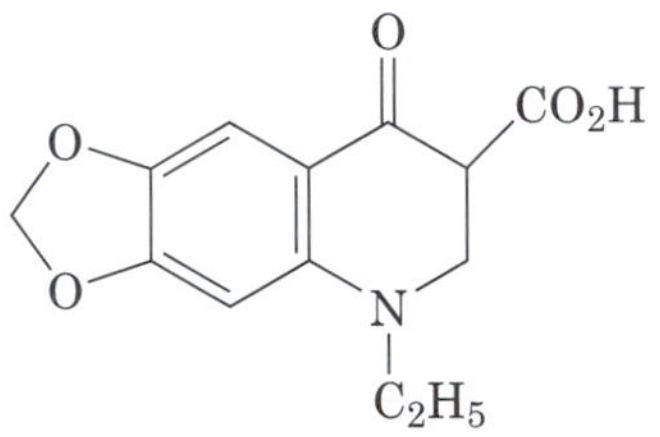

(75)

them are active against *Hemophilus influenzae*, *Providencia rettgeri*, *Pseudomonas aeruginosa*, *Serratia marcesans*, *Staphylococcus aureus*, *Enterococcus fecalis*, *Mycoplasma pneumonia*, *Chlamidia pneumonia*, and *Nisseria gonorrhoea*. When anthrax infections resulting from terrorist activities in the United States emerged in October of 2001, ciprofloxacin became the principal drug for treating this bacterium. Although other antibacterial agents such as penicillin and doxycycline are active

against anthrax infections, ciprofloxacin is effective against more strains of anthrax. Clinical indications for the quinolones and fluoroquinolones are given in Table 13.4.

**3.2.3 Structure-Activity Relationship.** The minimum pharmacophore required for significant antibacterial activity consists of the 4-pyridone ring with a 3-carboxylic acid group (Fig. 13.6). Reduction of the 2,3-double bond eliminates activity. Most of the highly active quinolones have a fluorine atom at C6 (fluoroquinolones) because it increases lipophilicity, which facilitates penetration into cells. Many analogs have piperazino groups on C7 because they broaden the spectrum, especially to include Gram-negative organisms such as *Pseudomonas aeruginosa*; however, they also increase affinity for the GABA receptor, which contributes to CNS side effects. This receptor binding affinity can be reduced by adding a methyl or ethyl group to the piperazine or by placing a bulky substituent on N1 (300). Substitutents on the piperazine ring can shift excretion of the compound from kidney to liver and they extend its half-life. Quinolones with greater amounts of liver metabolism and biliary excretion are useful in patients with impaired renal function (301). Replacement of C8 by nitrogen, to give a naphthyridine (e.g., enoxacin and trovafloxacin) increases bioavailability (302), whereas a methoxy group in place of hydrogen or fluorine on C8 provides greater stability to ultraviolet light and less phototoxicity in mice (303). Compounds such

**Table 13.4  Therapeutic Indications for Quinolones**

| Compound | Indications |
|---|---|
| Nalidixic acid | Urinary tract infections |
| Cinoxacin | Urinary tract infections |
| Ciprofloxacin | Acute sinusitis, lower respiratory tract infections, nosocomial pneumonia, skin infections, bone/joint infections, urinary tract infections, gonorrhea, anthrax |
| Enoxacin | Gonorrhea, urinary tract infections |
| Gatifloxacin | Chronic bronchitis, acute sinusitis, urinary tract infections, pyelonephritis |
| Levofloxacin | Chronic bronchitis, acute sinusitis, urinary tract infections, pneumonia, skin infections |
| Norfloxacin | Urinary tract infections, gonorrhoea, chronic bacterial prostatitis |
| Ofloxacin | Liver cirrhosis, epididmytis, gonorrhea, chlamydia |
| Sparfloxacin | Pneumonia, chronic bronchitis |
| Trovofloxacin/alatrofloxacin | Gynecological and pelvic infections |

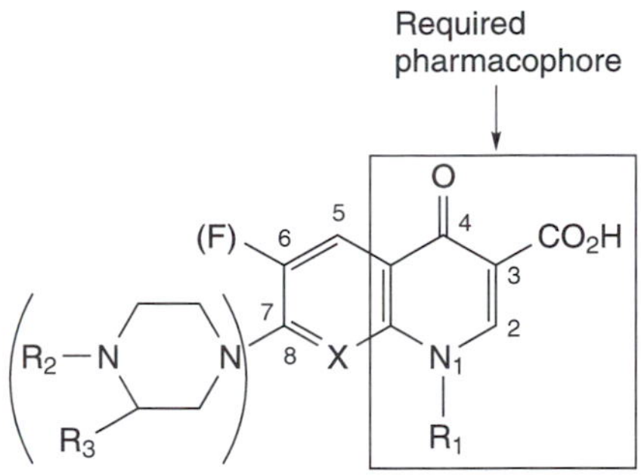

**Figure 13.6.** Quinolone pharmacophore and substituents.

as sparfloxacin (**69**), with an amino group at C5, also have reduced phototoxicity.

Isosteric replacement of nitrogen for C2 provides cinnolines, such as cinoxacin (**75**), which have good antibacterial activity and pharmacokinetic properties. Isomeric naphthyridines including the 1,5 and 1,6 isomers retain antibacterial activity. Compounds with ring fusions at 1,8 (ofloxacin, **72**); 5,6; and 7,8 also are effective antibacterials.

The stereochemistry of the methyl group in the third ring of ofloxacin is important to antibacterial activity. The (S)-enantiomer (marketed separately as levofloxacin) is 10-fold more potent than the (R)-enantiomer and it is less selective for topoisomerase II (304, 305). Interestingly, the corresponding methylene analog (which flattens the ring to allow stronger intercalation between base pairs) has 20-fold greater topoisomerase activity than that of either of the methyl enantiomers (306).

Fluorine at C6 enhances inhibition of DNA gyrase and provides activity against *Staphylococci*, whereas a piperazine substituent at C7 affords the best activity against Gram-negative bacteria. Addition of a second fluorine at C8 increases absorption and half-life. Ring alkylation improves gram-posotive potency and half-life. A number of newer quinolones such as ciprofloxacin, gatifloxacin (307), and moxifloxacin (308) have cyclopropyl sunstituents at N1. This substituent, or the combination of an amine at C5 and fluorine at C8 (sparfloxacin), increases potency against mycoplasma and chlamydia species.

Earlier antibacterial quinolones such as nalidixic acid and cinoxacin have only the 3-carbox-

ylic acid as an ionizable group. Its relatively high $pK_a$ in the range of 5.4 to 6.4 is thought to result from an acid-weakening hydrogen bond with the 4-carbonyl group (309). More recent quinolones with a piperazine or other substituent with a basic nitrogen at C7 have a second $pK_a$ in the range of 8.1 to 9.3. Consequently, significant fractions of these compounds exist as zwitterions at physiological pH values. Decreased solubility in urine of higher pH presents a potential problem for these compounds.

The 4-carbonyl group and 3-carboxylic acid functionalities of quinolones provide an excellent site for chelation with divalent of trivalent metals. Quinolones can form 1:1, 2:1, or 3:1 chelates, depending on the particular metal ion, relative concentration of the quinalone, and the pH. The relative insolubility of these chelates causes incompatibilities with antacids ($Ca^{2+}$, $Mg^{2+}$, and $Bi^{3+}$), hematinics ($Fe^{2+}$), and mineral supplements ($Zn^{2+}$).

It has been noted that, if just 20 different substituents are taken two at a time for the seven available positions on quinolones, there would be 84,000 possible compounds to synthesize and test (310). The use of computer-assisted quantitative structure-activity relationships has helped narrow the search for new compounds of this type. Koga and coworkers developed an equation that related the potency of compounds against *E. coli* (MIC) to the length of the substituent on N1, the size of substituents on C8, an enhancement factor for C7 substituents, and a factor for the detrimental effect of attachment of C7 substituents by an NCO function (311).

**3.2.4 Mechanism of Action.** The mechanism of action of quinolones is prevention of the detachment of gyrase from DNA. This enzyme is the bacterial form of topoisomerase, which allows the relaxation of supercoiled DNA that is required for normal transcription. Bacterial gyrase is different enough from mammalian topoisomerase so that 78 of 90 quinolones tested were selective only to bacteria. Those compounds with poorer selectivity had a propyl group at N1 and two fluorines at C6 and C8 (312). The antibacterial activity of quinolones is antagonized by chloramphenicol and rifampin, which suggests that protein synthesis is required for killing (313). This

**Table 13.5  Pharmacokinetic Properties of Quinolones**[a]

| Compound | Single Oral Dose (mg) | Peak Serum Concentration, ($\mu$g/mL) | Half-Life (h) | Protein Binding (%) | Urinary Recovery (% unchanged) |
|---|---|---|---|---|---|
| Nalidixic acid | 1000 | 20–40 | 6 | 93–97 | 3 |
| Ciprofloxacin | 200 | 0.8 | 4–6 | 40–50 | 20–40 |
| Enoxacin | 200 | 1.0 | 5 | 40 | 20–40 |
| Lomefloxacin | 200 | 0.7 | 3–4 | 10 | 65 |
| Norfloxacin | 100 | 1.0 | 5 | 10–15 | 26–32 |
| Gatifloxacin | 200 | 2.0 | 7.8 | 20 | 74 |
| Ofloxacin | 200 | 1.5 | 9 | 32 | 65–80 |
| Sparfloxacin | 400 | 1.3 | 20 | 46 | 10 |
| Trovafloxacin/alatrofloxacin | 100 | 1.0 | 9.1 | 76 | 6 |

[a]Abstracted from *Drug Facts and Comparisons.*

mechanism is consistent with apoptosis rather than necrosis. Fluoroquinolines also inhibit topoisomerase IV, an enzyme that functions in partitioning of the chromosomal DNA during bacterial cell division.

**3.2.5 Microbial Resistance.** Drug resistance was observed in the early clinical trials with nalidixic acid (314). One mechanism of resistance is by mutation at various locations on the gyrase gene. These mutations confer cross-resistance to all other quinlones. Another mechanism is by mutation in the genes that code for porins, which are membrane proteins by which quinolones enter Gram-negative cells. These mutations raise tolerance fourfold. Other mutations of serious concern are those that reduce membrane lipopolysaccharides to afford cross-resistance with antibacterial agents of other chemical classes (315). Considering the problem of resistance to quinolones, Moellering has concluded that, "The future viability of the quinolones will in large part depend on the ability of the medical community to use them wisely" (314). In particular, the increasing use of quinolones for oral periodental treatment could bring about an expansion of resistance to quinolones that would limit their value in treating deep-tissue infections.

**3.2.6 Pharmacokinetic Properties.** Some phamacokinetic properties of quinolones are listed in Table 13.5. Bioavalibility after oral administration is good, but it is substantially reduced by magnesium or aluminum antacids. Distribution to tissues is superior to that of most other drugs because there is little binding to plasma proteins. Quinolones achieve tissue-to-serum ratios of over 2 to 1, in contrast to less than one-half for $\beta$-lactams and aminoglycosides (316). Clearance is by kidneys (ofloxacin), or by liver (pefloxacin and difloxacin), or by both (norfloxacin, ciprofloxacin, enoxacin, and fleroxacin). Renal clearance correlates with creatinine clearance; consequently, reduced drug dosage is appropriate for patients having creatinine clearance under 30 mL/min. The renal clearance rates of ciprofloxacin and norfloxacin exceed the glomerular filtration rates, indicating net renal tubular secretion. Lomafloxacin has the advantage of a relatively long half-life and so can be administered once daily.

Metabolism of quinolones is primarily by glucuronide conjugation at the 3-carboxyl group and is inactivating. The piperazine ring is readily metabolized in the compounds that have this functionality, and this metabolism reduces antimicrobial activity (316). Approximately one-eighth of enoxacin is cleared as the oxo metabolite. Six metabolites with modifications in the piperazine ring were found for norfloxacin (301).

**3.2.7 Adverse Reactions.** Adverse reactions, which are mostly mild and reversible, include headache, dizziness, joint swelling, and leukopenia. Lomefloxacin, sparfloxacin, ofloxacin, and trovafloxacin/alatrofloxacin cause photosensitization. Quinolones can affect the central nervous system by two mechanisms: (1) accumulation of ingested xanthines, including caffeine and theophylline

(317); and (2) blockade of GABA receptors, which can cause convulsions (318). A study on proconvulsant effects of quinolones in a strain of DBA/2 mice susceptible to sound-induced seizures showed an incidence in the order pefloxacin > enoxacin > rufloxacin > norfloxacin > cinoxacin > ciprofloxacin > nalidixic acid (319). Another study based on the induction of fatal convulsions by a combination of nonsteroidal anti-inflammatory drugs and quinolones gave the following order of potency when fenbufen was the anti-inflammatory agent: enoxacin > lomefloxacin > norfloxacin, with ofloxacin and ciprofloxacin causing no deaths (320).

### 3.3 Oxazolidinones

**3.3.1 Introduction.** The oxazolidinones are a new class of synthetic antibacterial agents with activity against a broad spectrum of Gram-positive pathogens, including those resistant to currently used antibacterials (321). Following the lead of activity discovered in a series of 5-(halomethyl)-3-aryl-2-oxazolidinones against plant pathogens described in a patent issued to them in 1978, scientists at E. I. DuPout de Nemours (Wilmington, DE) observed antibacterial activity in ($R$)-5-hydroxymethyl-3-aryl-2-oxazolidinone (S-6123) against human pathogens (322). Further optimization led to the emergence of two highly active antibacterial drug candidates (322): ($S$)-[(3-(4-methylsulfinylphenyl)-2-oxo-5-oxazolidinyl)methyl]acetamide (DUP-105, **76**)

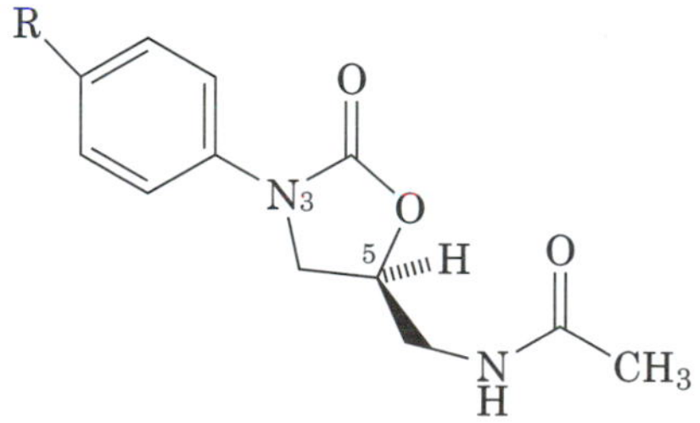

**(76)** DUP 105, R = CH$_3$SO
**(77)** DUP 721, R = CH$_3$CO

and ($S$)-[(3-(4-acetylphenyl)-2-oxo-5-oxazolidinyl)methyl]acetamide (DUP-721, **77**) (323, 324), with a number of special features, including:

- activity against a number of therapeutically important multidrug resistant Gram-positive organisms (322–326)
- equally active when administered by oral or parenteral routes, showing good oral absorption (323, 327)
- a novel mechanism of action, and consequently less likely to have cross-resistance with existing antimicrobials (322, 324, 328, 329).

Furthermore, the prototype structure consisting of distinct structural units offered much scope for molecular modification.

Although the development of these two agents was subsequently discontinued consequent upon DUP-721 exhibiting toxicity in rodents (330), the special features of their antibacterial activity attracted much attention and prompted studies on oxazolidinones in a number of laboratories. Pharmacia-Upjohn (Piscataway, NJ) scientists were able to identify two drug candidates for human studies, eperezolid (PNU-100592, **78**) and linezolid

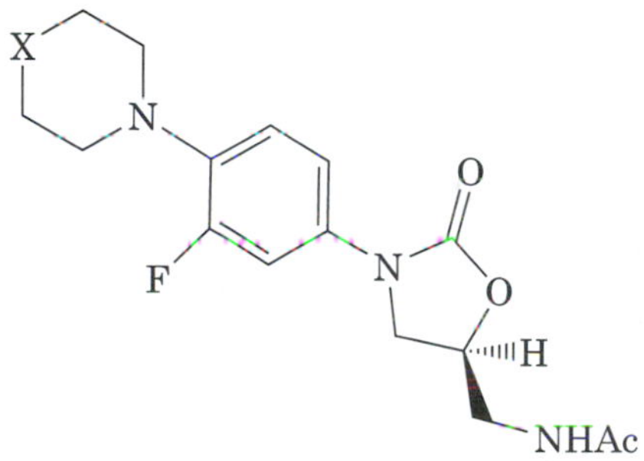

**(78)** PNU 100592, X = NCOCH$_2$OH
**(79)** PNU 10076, X = O

(PNU-100766, **79**) (330–332). Both compounds were active *in vitro* and *in vivo* (experimental mouse models) against methicillin-resistant *S. aureus* (MRSA) and *S. epidermidis* (MRSE), against penicillin and cephalosporin-resistant *S. pneumoniae*, comparing favorably with vancomycin activity, and against vancomycin-sensitive and -resistant *Enterococcus* spp. (VSE and VRE). Both eperezolid and linezolid went successfully through phase I human trials without any significant safety concerns. Linezolid has undergone more extensive phase II and III clinical evaluation (333).

### 3.3.2 Linezolid.

Linezolid (Zyvox, Zyvoxam, **79**) is the first of this new class of oxazolidinone antibacterials to be approved in the United States, the United Kingdom, and Canada for the treatment of Gram-positive infections. Specific indications include complicated and uncomplicated skin and soft-tissue infections (SSTIs), community and clinically acquired pneumonia, and vancomycin-resistant enterococcal infections (334–337).

*Synthesis.* Linezolid and related oxazolidinones have been prepared through a novel asymmetric synthesis involving the reactions of *N*-lithiocarbamates of the appropriate aniline with (*R*)-glycidylbutyrate as the key step, and the resultant (*R*)-5-(hydroxymethyl)oxazolidinone converted to linezolid in a few steps in excellent yield and high enantiomeric purity (330, 334).

#### 3.3.2.1 Antibacterial Activity.

Linezolid has inhibitory activity against a broad range of Gram-positive bacteria, including methicillin-resistant *Staphylococcus aureus* (MRSA), glycopeptide-intermediate *S. aureus* (GISA), vancomycin-resistant enterococci (VRE), and penicillin/cephalosporin-sensitive, -intermediate, and -resistant streptococci and pneumococci (332). Linezolid is in general bacteriostatic and displays bactericidal activity only against some strains, which include some pneumococci *Bacteroides fragilis* and *Clostridium perfringens* (332, 333). Initial breakpoint criteria established for MICs of linezolid are ≤4 $\mu$g/mL for susceptibility and ≥16 $\mu$g/mL for resistance (334, 335, 338).

*3.3.2.1.1 Gram-Negative Bacteria.* Linezolid is significantly less active against most Gram-negative organisms. It has only moderate activity against *Moraxella catarrhalis*, *Haemophilus influenzae*, *Legionella* spp., and *Bordatella pertussis* and practically no activity against enterobacteriaceae, *Klebsiella*, *Proteus*, and *Pseudomonas aeruginosa*. The few Gram-negative organisms against which linezolid has good activity are *Flavobacterium meningosepticum* and *Pasteurella multicida* with MIC values of 2 and 4 $\mu$g/mL, respectively (332, 336, 337).

*3.3.2.1.2 Activity Against Anaerobes.* Linezolid demonstrated activity comparable to that of vancomycin (MIC: 1–2 $\mu$g/mL) against *Clostridium difficile* and *C. perfringens*. It also showed good activity against Gram-negative anaerobes including *Bacteroids* spp. (MIC: 4 $\mu$g/mL), *Fusobacterium nucleatum* (MIC: 0.5 $\mu$g/mL), *F. meningosepticum* (MIC: 2–4 $\mu$g/mL), and *Prevotella* spp. (MIC: 1–2 $\mu$g/mL) (336, 338, 339).

*3.3.2.1.3 In Vivo Antibacterial Activity.* In murine bacteremia models linezolid was more active than vancomycin against methicillin-sensitive *Staphylococcus aureus* (MSSA), and displayed comparable activity against MRSA, though less active for MRSE. In addition linezolid displayed consistent *in vivo* activity against pneumococci, including against multidrug-resistant strains and vancomycin-resistant *E. faecium* but was less active than vancomycin against aminoglycoside-resistant *E. faecalis* (340).

Of particular interest is the report by Cynamon et al. (341) that oral linezolid (25, 50, and 100 mg/kg) demonstrated efficacy against *M tuberculosis* in a murine model, though it was somewhat less active than isoniazid. Subsequently, PNU-100480, the thiomorpholine analog of linezolid, has been reported to be as active as INH against *M. tuberculosis*. This provides a new lead for the design of antimycobacterial agents. Linezolid also showed promising activity in a rat experimental endocarditis model (342) and in an experimental model of acute otitis media (343) produced by a multidrug-resistant pneumococcal isolate.

#### 3.3.2.2 Mechanism of Action.

It has been established that linezolid (and related oxazolidininones) act through inhibition of the initiation phase of bacterial protein synthesis. Although the exact mode of action at the molecular level is not fully elucidated, linezolid has been reported to bind directly to a site on 23S ribosomal RNA of the bacterial 50S ribosomal subunit, thereby preventing the formation of the functional 70S-initiation complex (344, 345), formed with 30S ribosomal subunit, mRNA, initiation factors, fMet-tRNA, and 50S ribosome, which is an essential step of the bacterial translation process (Fig. 13.7). In a subsequent study it has been reported that mutations in the central loop of domain V of

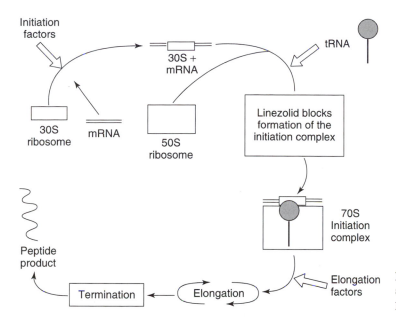

**Figure 13.7.** Oxazolidinones: schematic representation of the proposed mechanism of action.

23S rRNA, a component of the ribosomal peptidyl transferase center, conferred resistance to linezolid (346). It would thus appear that linezolid and other oxazolidinones disrupt the interaction of fMet-tRNA with the 50S subunit during formation of the preinitiation complex, and affect the translocation step.

The mechanism for the poor activity against Gram-negative bacteria is not clear. One possible reason is the operation of the selective efflux mechanism of Gram-negative bacteria.

This inhibition of the protein synthesis at an early stage is a new site/mechanism of action, and therefore there is little chance of development of cross-resistance between linezolid and other existing antibacterial agents; in fact, none has been reported so far (347).

*3.3.2.2.1 Effect On Virulence Factors.* Linezolid inhibited the expression of virulence factors from *S. aureus* and *Streptococcus pyogenes in vitro* at concentrations ranging from 12.5 to 50% of the MIC for the organism; production of α-hemolysin and coagulase by *S. aureus* and of streptolysin O and DNAase by *S. pyogenes* was markedly inhibited at these concentrations (348).

**3.3.2.3 Pharmacokinetics and Metabolism.** Linezolid follows a similar pharmacokinetic (PK) profile by both oral and intravenous routes of administration. It is rapidly and completely absorbed in humans after oral administration, with a mean absolute availability of almost 100%; the peak plasma concentration was reached in 1–2 h, with elimination half-life of 5.5 h and volume of distribution of 45 L (349, 350). The circulating drug is moderately (31%) bound to plasma proteins (351). At steady state after 15 days of twice-daily administration of 375 and 625 mg of linezolid in 24 volunteers, $C_{max}$ values were 12 and 18 $\mu$g/mL, respectively, whereas the $MIC_{90}$ for susceptible pathogens were $\geq 4$ $\mu$g/mL (349). With intravenous linezolid (500 or 625 mg b.i.d. for 16 doses) the minimum plasma concentrations were 3.5 and 3.8 $\mu$g/mL, with plasma concentration exceeding 4 $\mu$g/mL for $\geq 75\%$ of the dosage interval, above the MIC of most of the pathogens targeted (350).

The pharmacokinetic properties of linezolid do not seem to be influenced by age. The clearance was somewhat lower in females, both young and old, but this was not of much significance and dose adjustments are not warranted on the grounds of age or gender (351).

*3.3.2.3.1 Metabolism.* Linezolid appears to be metabolized by oxidation of the morpholine ring to form two inactive carboxylic acid metabolites (352). In volunteers, unchanged linezolid accounted for 90% of the circulating dose, with the major metabolite accounting for <6%. After a single 500 mg oral dose, 80–85% of linezolid was recovered in urine and 7–12% in feces over a 7-day period. In elimination, 35% of the drug appeared unchanged in the urine and 50% appeared as one of the two major inactive metabolites (352, 353).

**3.3.2.4 Therapeutic Uses.** The impressive antimicrobial activity against organisms that are resistant to other antimicrobial agents, high bioavailability after oral administration, favorable metabolic stability, low/no incidence of cross-resistance with other antimicrobials, low propensity to developing microbial resistance, and favorable safety profile make linezolid an attractive antimicrobial agent for the treatment of problem Gram-positive infections. Oral and intravenous linezolid have been reported to be equally effective in the treatment of plain or complicated SSTIs (354, 355), urinary tract infections (UTIs) caused by methicillin-sensitive or -resistant *Stephylococcus* spp., and pneumonia, including nosocomial and community-acquired pneumonia, which require hospitalization (356). The clinical response had a parallel favorable microbiological response. In these studies, linezolid was as effective as established treatments, including third-generation cephalosphorins, oxacillin, flucloxacillin, and clarithromycin.

*3.3.2.4.1 Other Serious Gram-Positive Bacterial Infections.* Linezolid has also been evaluated in a variety of other infections (e.g., bacteremia, intra-abdominal abscesses, osteomyelitis, lower and upper respiratory tract infections) caused by vancomycin-resistant *E. faecium/faecalis*, MRSA or MRSE, with clinical and microbiological cures ranging from 70% to 90% (357, 358). Linezolid thus appears an effective treatment option for a variety of serious, multidrug-resistant Gram-positive bacterial infections.

**3.3.2.5 Microbial Resistance.** There are as yet very few definitive reports of the development of resistance to linezolid, but these are bound to appear as the use increases. Development of resistance to linezolid was reported first in two patients who had *E. faecium* bacteremia with long-standing indwelling devices that could not be removed and had more than 4 weeks of i.v. linezolid; the MIC, which was

initially 2 μg/mL dose, rose to 16 and 32 μg/mL (347), respectively, after treatment. Five more cases of linezolid resistance have been reported more recently out of 45 patients under treatment, one of which developed resistance during treatment out of 45 linezolid-treated patients (348).

**3.3.2.6 Toxicity and Adverse Effects**
*3.3.2.6.1 Preclinical Toxicology.* In dogs, the "no observed adverse effect level" (NOAEL) of linezolid was 20 mg/kg for both sexes when administered orally for 28 days. Doses of 40 and 50 mg/kg/day were well tolerated with only mild effects. At toxic levels hypocellularity of the bone marrow (reversible decrease in white blood cells and platelets) and atrophy of the lymphoid tissue were observed.

*3.3.2.6.2 Human Safety.* Linezolid was well tolerated in human volunteers after oral or i.v. administration of daily doses up to 625 mg b.i.d. (349, 350, 360, 361). The most common adverse effects were nausea (5.4%), diarrhea (5.2%), or oral cavity symptoms (tongue discoloration, 2.5%; oral monilia, 2.2%). Serious drug-related adverse events (e.g., elevated liver enzymes, atrial fibrillation, or worsening renal failure) occurred in <1% of cases. In the linezolid compassionate-use trial of patients with significant, resistant Gram-positive infections, the overall adverse event rate was about 33%, of which approximately 6% were considered serious events (360). The most frequent adverse events reported were thrombocytopenia (2.6%) and dermatological reaction (2.5%) (361). Overall, at the end of therapy, linezolid was well tolerated by about 78% of patients (361).

Myelosuppression (including anemia, leukopenia, pancytopenia, and thrombocytopenia) has been reported in three patients receiving linezolid for 2 and 6 weeks, and 4 months, respectively. With discontinuation of linezolid treatment, the affected hematological parameters rose toward pretreatment levels (362). The U.S. FDA has recommended that complete blood counts should be monitored weekly, particularly in patients who receive linezolid for longer than 2 weeks.

**3.3.3 New Oxazolidinones.** The special features of the antibacterial activity of oxazolidinones highlighted earlier attracted much

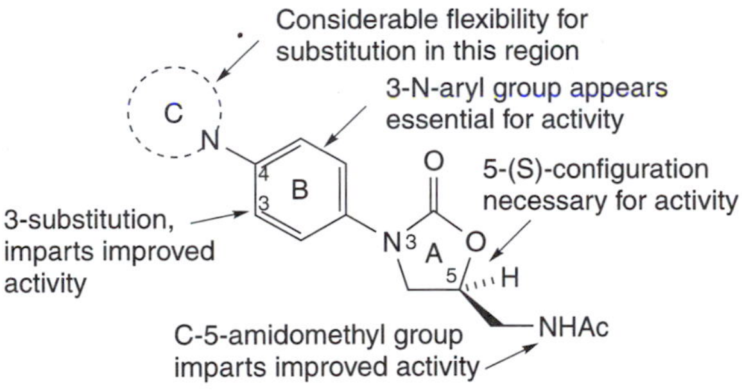

Figure 13.8. Oxazolidinones as antibacterials: important structural features.

attention. A number of reports have appeared describing new oxazolodinones with improved antibacterial activities. On the basis of the early structure-activity studies (321, 330), some structural features that appeared important for good antibacterial activity in oxazolidinones could be identified and are presented in Fig. 13.8 (363). The diagrammatic representation in Fig. 13.8 acted as a useful guide for subsequent studies and broadly still holds. The structure-activity relationship of the reported oxazolidinones (364) is presented in the context of this picture of the pharmacophore.

*A-Ring Modification.* The oxazolidinone ring structure seems essential for good antibacterial activity and relatively few successful modification of the A-ring have been reported (364). The tricyclic analogs having a methylene bridge between rings A and B have been reported, with the transfused *trans* homolog (80) having good activity (365, 366), although the exact linezolid analog (81) was only weakly active. The 3-pyridyl analog (82) is the most active compound in this series (365).

*B-Ring Modification.* The replacement of the B-aromatic residue by heteroaromatic rings as in (83–87) gave compounds that had

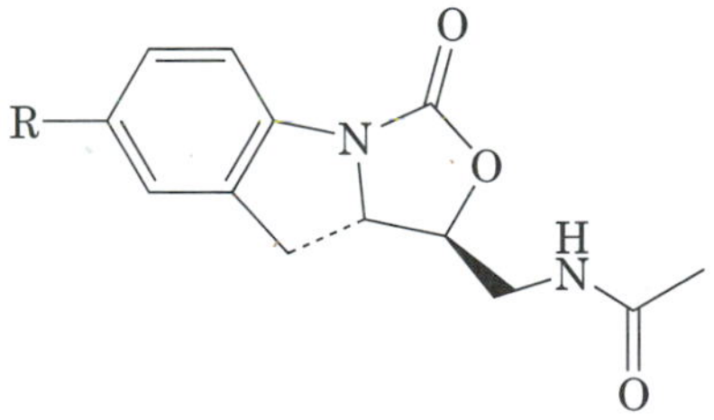

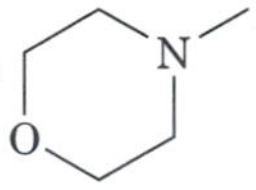

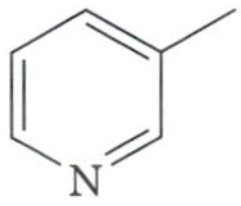

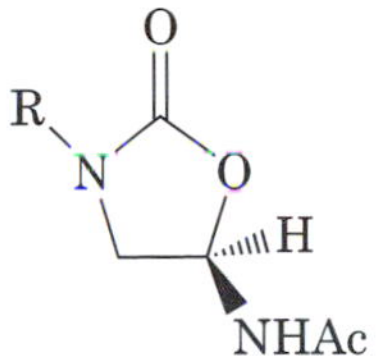

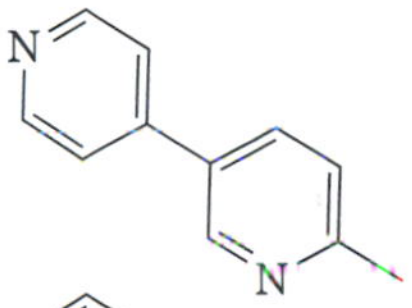

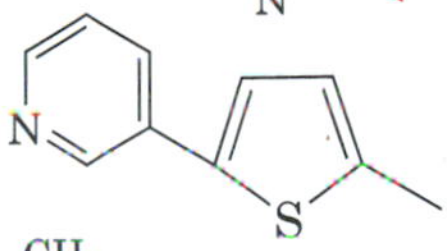

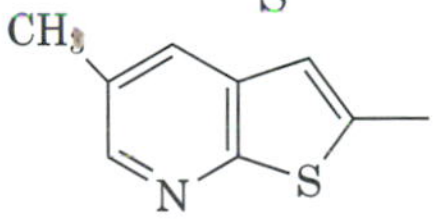

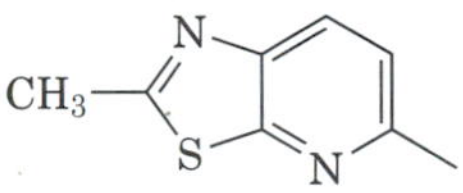

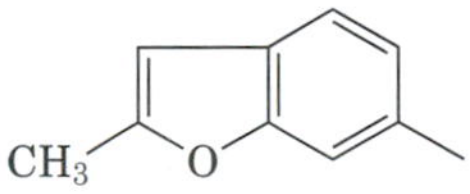

modest to good activity by subcutaneous administration (367); only the benzofuran analog (87) had good Gram-positive antibacterial activity (368). Thus the replacement of ring B did not seem to result in any significant improvement in the antibacterial profile.

Analogs that incorporate a fused hetero ring on aromatic ring B with a lactam residue, as in (**88–92**), have been reported to exhibit good an-

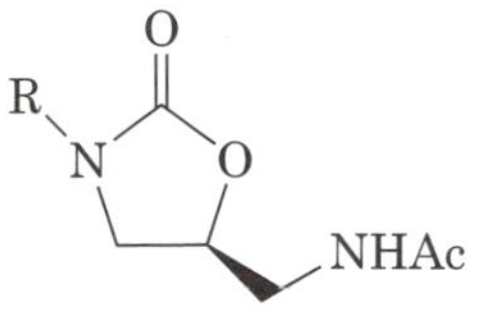

(88) R =

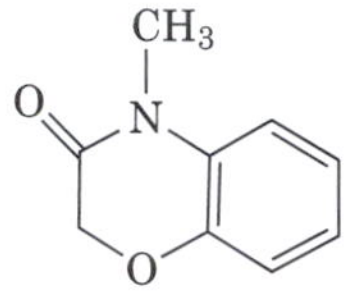

(89) R =

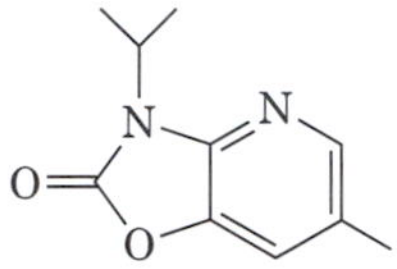

(90) R =

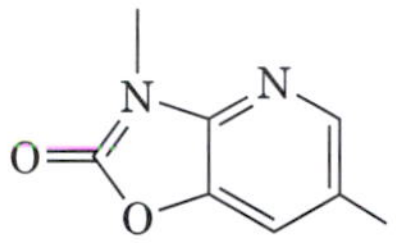

(91) R =

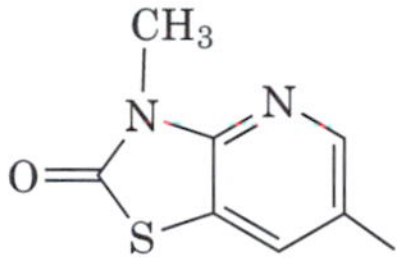

(92) R =

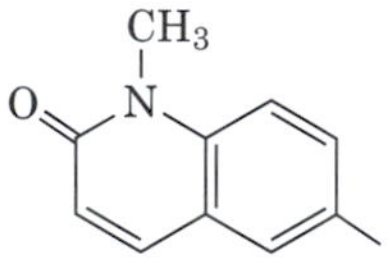

tibacterial activity. The benzoxazinone (**88**) has *in vitro* activity comparable to that of linezolid, with better i.v. PK parameters (369, 370). The benzoxazolone (**89**) and benzothiazolone (**91**) exhibit greater potency than that of linezolid (MIC range 0.25–2 μg/mL) (371). Other active compounds reported in this group include the oxazolopyridine analog (**15**) and the quinolone analog (**92**) (372, 373).

*C-Ring Modification.* A great deal of flexibility exists around the site C. A series of oxazolidinones in which the distal nitrogen on the aromatic ring is replaced by a five-membered heteroaromatic residue were reported

to have good activity. This included the pyrrole analog PNU-107922 (**93**), the cyanopyrrole analog (**94**), and the cyanopyrazole analog PNU-172576 (**95**) (374). These analogs, apart from being more active than linezolid against Gram-positive pathogens, are also active against fastidious Gram-negative organisms; (**94**) and (**95**) showed significant activity against *H. influenzae* and *M. catarrahalsis* and represent the first oxazolidinones with potentially useful activity against Gram-negative pathogens; MIC values of <0.125–0.5 μg/mL against Gram-positive organisms and of 1–4 μg/mL against Gram-negative bacteria have been reported. These compounds are also orally active in mice. The cyanothiazole (**96**) and 5-cyanothiophene (**97**) analogs were even more potent against both Gram-positive and fastidious Gram-negative organisms (MIC > 0.125 μg/mL) and were orally active (375). Similar activity is reported for cyanoethylthiadiazole (**98**) (376). The pyrazoles (**99**) and (**100**) were also active but were less potent than other members of this group (375).

*B- and C-Ring Fused Analogs.* A number of analogs have been reported wherein the B and C rings are bridged by one or two atoms to form a rigid tricyclic system (377–379). The bridged pyrido-benzofuran (**101**), the imidazobenzoxazinyl (**102**), the pyrazinoindolyl (**103**), and pyrazinobenzoxazinyl analogs (**104**) have been reported and shown to have potent *in vitro* activity against both Gram-positive and Gram-negative bacteria, but have poor oral activity attributed to unfavorable pharmacokinetics.

*D-Ring Analogs.* These oxazolidinones are quite flexible for substitution around ring C and can tolerate a ring on the distal nitrogen of the piperazinyl ring. The analogs (**105–111**) have been reported to have good *in vitro* activity against Gram-positive organisms and are also orally active *in vivo* mouse model (380). The isooxazolylpiperazines (**110**) have even better activity (381). A QSAR analysis of these compounds has been carried out and suggests that steric factors are the most important determinants of activity in this class (382).

*Acetamide Side-Chain Modification.* The tolerance limits for change around the acetamide side chain are rather limited. A study with the benzthiazolones (**112**) and the

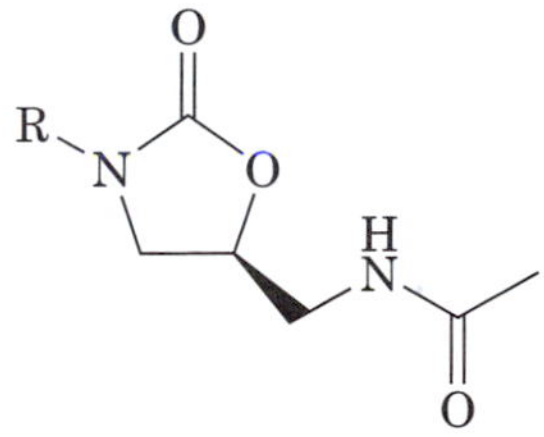

(101) R =

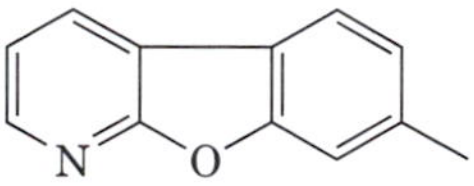

(102) R =

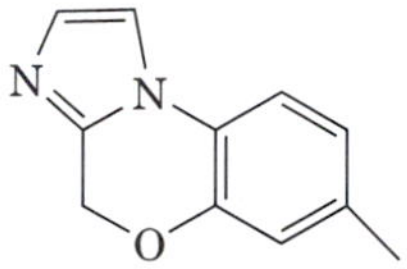

(103) R = H₃C

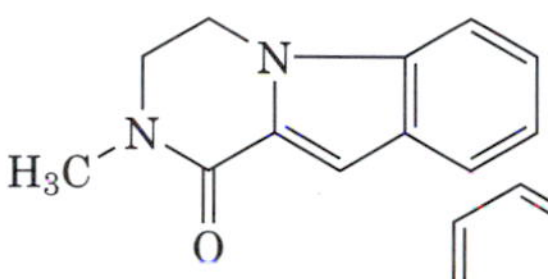

(104) R = HO

features; their unique mode of action provides low likelihood of developing cross-resistance with existing antibacterials; excellent oral bio-availability offers good flexibility in patient

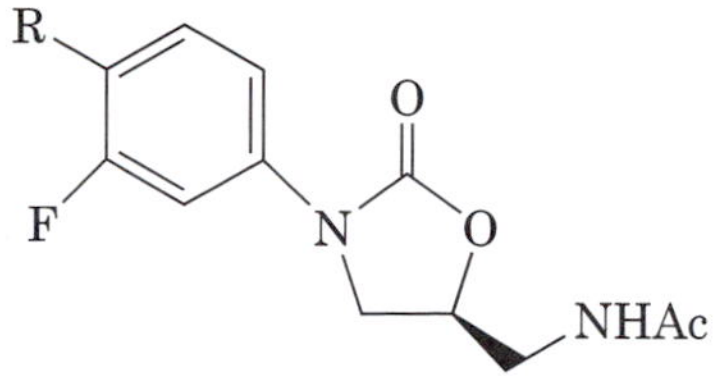

(93) R =

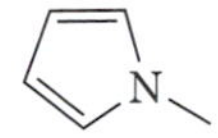

(94) R =

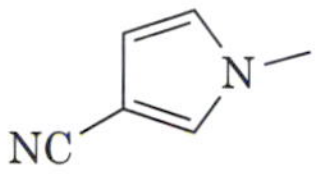

(95) R =

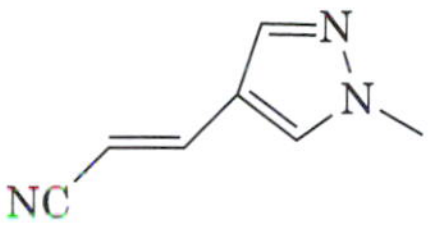

(96) R =

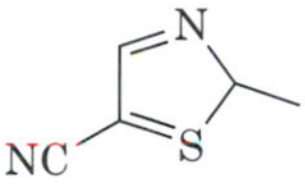

(97) R =

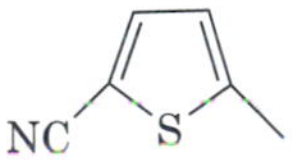

(98) R =

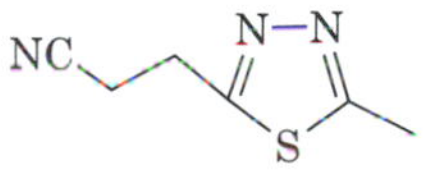

(99) R =

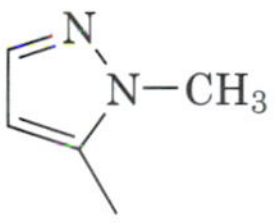

(100) R =

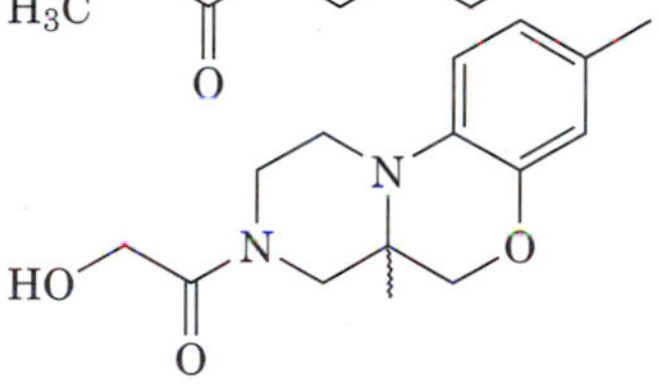

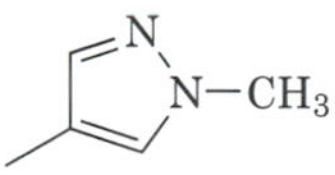

piperazinoindoles (**113**) showed that good antimicrobial activity was retained when R was small alkyl or alkoxy, or corresponding thioamide or thiourea derivatives; the thioamides in some cases resulted in improvement of activity. The thioamide (**113**) is fourfold more active than the corresponding carboxamide, with MIC values against staphylococci, streptococci, and enterecocci ranging from 0.12 to 0.5 μg/mL and with good activity against *H. influenzae* with MIC of 0.5–1 μg/mL (370, 383). Tokuyama et al., in a recent SAR study on 5-substituted oxazolidinones, have also shown that elongation of the methylene chain and conversion of the acetamido moiety into a guanidino group decreased the antibacterial activity (384). However, replacement of carbonyl by thiocarbonyl or thiocarbamate groups greatly enhanced the *in vitro* antibacterial activity, and some of the compounds have stronger activity than that of linezolid (384).

**3.3.4 Perspective.** Oxazolidinones as a new class of antibacterial agents offer some special

management from hospital to ambulatory setting; their chemical structure offers good scope for molecular modification to improve their therapeutic profile, reduce toxicity, and improve safety. The structure-activity studies carried out so far indicate the promise that this class holds.

Linezolid is the first of this new class of antibacterial drugs. Its broad spectrum of activity

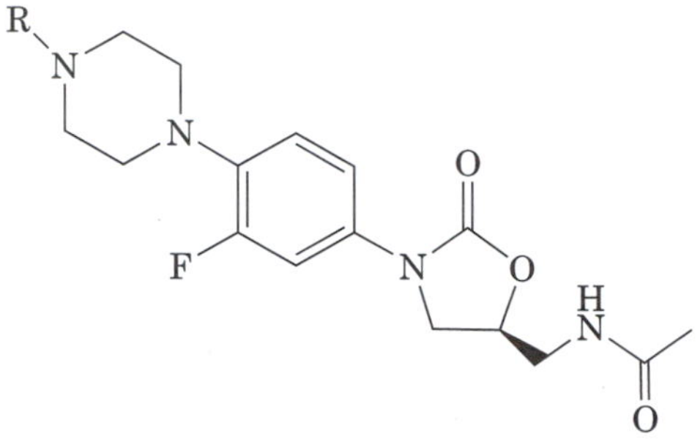

(105) R =

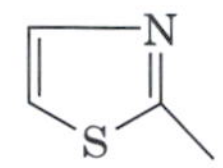

(106) R =

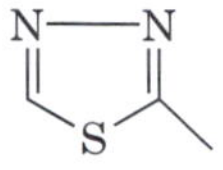

(107) R =

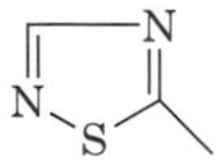

(108) R =

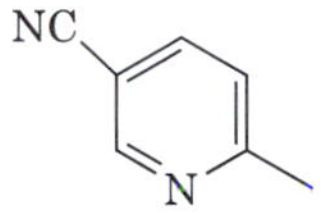

(109) R =

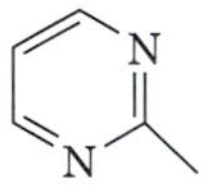

(110) R =

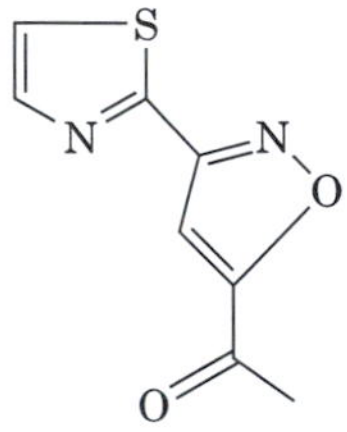

(111) R =

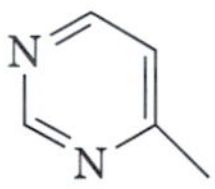

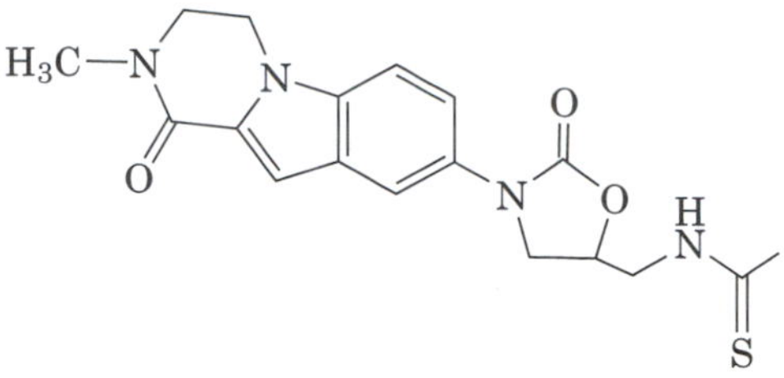

(113)

## 3.4 Other Systemic Synthetic Antibacterials

### 3.4.1 Nitrofurans.

Thousands of nitrofurans have been synthesized and tested for antibacterial activity, but only nitrofurazone, nitrofurazolidone, and nitrofurantoin have been approved for human use in the United States. Nitrofurazone was described by Dodd and Stillman in 1944 (385) and nitrofurantoin was patented in 1952. Nihydrazone and furaltadone are used in animal feeds.

Nitrofurazone (**114**) is the semicarbazone of 5-nitro-2-furaldehyde (386). It has a broad spectrum of antibacterial activity in the range of 1:100,000 to 1:200,000, although it is not effective against *Pseudomonas aeruginosa* or fungi. It is used topically in the treatment of burns and in prevention of bacterial infection in skin graft procedures.

Furazolidone (**115**) is the hydrazone formed from 5-nitro-2-furaldehyde and 3-amino-2-oxazolidinone (387). It has a broad spectrum of bactericidal activity against intestinal pathogens, including various species of *Salmonella*, *Shigella*, *Proteus*, and *Enterobacter*, as well as *E. coli* and *Vibrio cholerae*. It is useful for the treatment of bacterial or protozoal diarrhea because it is effective and only a small fraction of an oral dose is absorbed.

The hydrazone prepared from 5-nitro-2-furaldehyde and 1-aminohydantoin is known as nitrofurantoin (**116**). It is active against

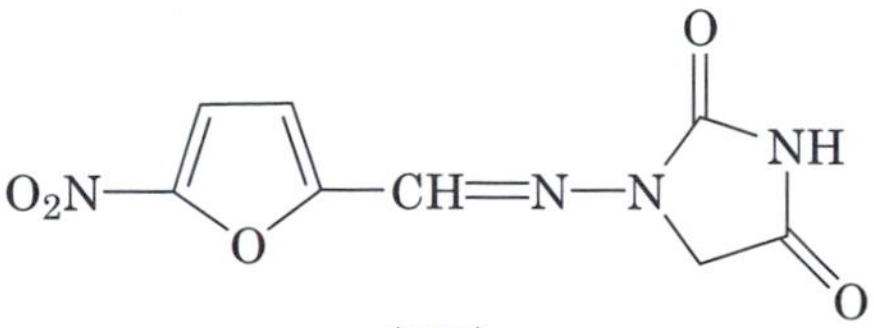

against many problem Gram-positive infections makes it a useful addition to therapeutics. It may also find use in combination therapy with other antibacterials, including in mycobacterial infections. However, there have been some reports of myelosuppression in persons treated for longer than 2 weeks, which calls for continuous monitoring during treatment.

(116)

many Gram-positive and Gram-negative bacteria at concentrations of 5–10 $\mu$g/mL. Upon oral administration, it is rapidly absorbed, but cleared so rapidly that concentrations adequate for an antibacterial effect in plasma cannot be obtained; however, it accumulates in urine in sufficient concentration for treatment of urinary tract infections.

Structure-activity relationships for nitrofuran derivatives indicate that a 5-nitro group

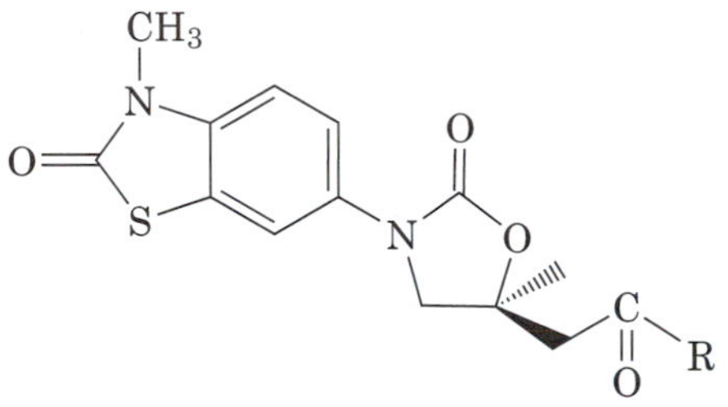

(112)

is required for antibacterial activity. Substitutents at position 2 can be varied widely, with the most potent compounds having azomethine (C=N), vinyl, or heterocyclic groups. Few compounds with 3 or 4 substituents have been prepared (388).

Products found in the urine of rats fed nitrofurazone included 4-cyano-2-oxobutyraldehyde semicarbazone (117) and other compo-

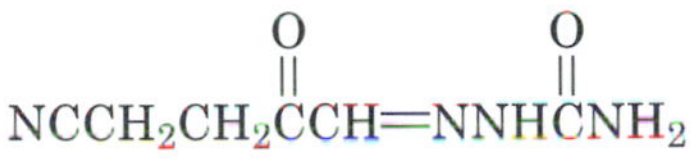

(117)

nents thought to include amino or hydroxylamino-substituted furaldehyde semicarbazone (389). 4-Cyano-2-oxobutyraldehyde semicarbazone also was found when nitrofurazone was incubated with small intestinal mucosa or liver homogenates from rats (390). Chickens and rats fed nitrofurantoin produced metabolites in urine from which a small amount of the corresponding 4-hydroxy derivative (118) was isolated (391).

It has been suggested that nucleic acids, especially tRNA, are the primary target in nitrofuran mutagenesis and carcinogenesis (392). Thus, incubation of labeled 2-amino-4-(5-nitro-2-furyl)-thiazole with rodent liver

preparations led to covalent attachment of metabolites to added yeast tRNA. Enzymatic hydrolysis of the product gave two covalent adducts (393). Another factor in mutagenesis is thought to be protein binding. When nitrofurazone-sensitive strains if *E. coli* were exposed to labeled nitrofurazone, radioactive species were tightly bound to proteins in the trichloroacetic acid-insoluble fraction, whereas very little radioactivity was found in proteins from resistant mutants (394). A similar result was obtained when labeled nitrofurylthiazole derivatives were incubated with mammalian tissues. Addition of thiols to the incubation mixtures substantially decreased protein binding (395). The structure of the carcinogen-protein adducts is not known; however, the reactive form of nitrofurazone is thought to be a *N*-hydroxylamine based on an experiment in which albumen binding was observed when nitrofurazone was reduced electrochemically at $-0.8$ V. This potential reduces nitro groups, but not hydroxylamines (394).

**3.4.2 Methenamine.** Methenamine (hexamethylenetetramine, **119**) is prepared by evaporating a mixture of formaldehyde and strong ammonia water. It was patented by Missner and Schwiedessen in 1956 (396). The free base has almost no antibacterial activity, but acidification results in the liberation of formaldehyde, which is strongly bactericidal (see Section 2.7.2). Acidification is provided by formulating methenamine as a mandelate or hippurate salt, or by administering ammonium chloride or sodium biphosphate to acidify the urine. Methenamine is sometimes used in long-term suppression of bacterial urinary tract infections, although it is not a treatment of choice for corresponding acute infections (397). Certain bacteria are resistant to methenamine because they liberate urease, an enzyme that hydrolyzes urea to ammonia and thereby raises urinary pH. This problem can

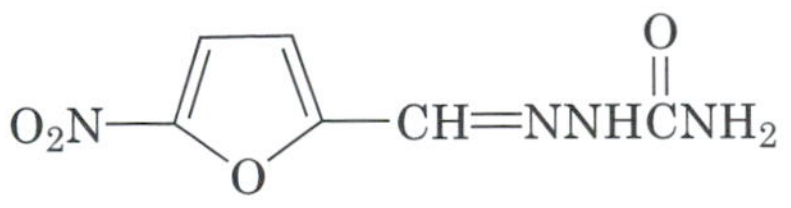

(114)

be overcome by giving acetohydroxamic acid (120), which inhibits urease.

### 3.4.3 Cotrimoxazole. Reduced folates are carriers of one-carbon fragments in the bio-

$$O_2N-\text{(furan)}-C=N-N\text{(oxazolidinone)}$$

**(115)**

synthesis of purines and pyrimidines. Agents that inhibit dihydrofolate reductase prevent the formation of important tetrahydrofolates, which results in decreased nucleotide synthesis and cell death. The use of dihydrofolate reductase inhibitors in chemotherapy of cancer and malaria is discussed in the appropriate chapters. There are species differences in dihydrofolate reductases (398), and these differences have been used to discover compounds that are lethal to bacteria but relatively innocuous to mammals (399). The selective inhibitor of bacterial dihydrofolate reductase used clinically is trimethoprim [5-(3,4,5-trimethoxybenzyl)-2,4-diaminopyrimidine, (52)](400). It is potent against a broad spectrum of Gram-positive and Gram-negative bacteria, although *P. aeruginosa* is resistant and resistant strains of *E. coli* have been isolated from patients (401). Trimethoprim in not usually used alone, but in combination with sulfamethoxazole, a sulfonamide with a similar half-life. This combination, called cotrimazole, has a synergistic effect that prevents the development of bacterial resistance (402). Cotrimazole is used mainly in the treatment of urinary tract infections, but it has been effective against gonorrhea, chest infections caused by pneumococci and *Hemophilis influenzae*, and enteric *Salmonella* infections. Side effects occur mainly in patients with impaired folate metabolism and include megaloblastosis, leukopenia, and thrombocytopenia (403).

$$CH_3\overset{O}{\overset{\|}{C}}ONH_2$$

**(120)**

### 3.5 The Challenge of Antibacterial Chemotherapy

There is a dramatic increase in the incidence of bacterial infections resistant to most commonly used antibacterials. In a few reports, infections in some patients were not susceptible to any known antibacterial, and the patients eventually died. These tragedies sound an alarm bell for antibacterial chemotherapy.

$$HO-\text{, } O_2N-\text{(furan)}-CH=N-N\text{(imidazolidinedione)}$$

**(118)**

A certain propensity to resistance development could be expected in bacteria exposed to antibiotics because of their rapid rate of multiplication, which would promote the selection of less susceptible (resistant) mutants continuously from a population with mixed sensitivities. This effect, coupled with overprescription and indiscriminate use of antibacterial agents, is to a large measure responsible for the emergence of this alarming situation.

This situation was aggravated further by a decline in antibacterial research, which resulted in very few new classes of antibacterial agents being discovered over the last few decades. The big successes achieved in antibacterial chemotherapy by sulfonamides and antibiotics created the feeling that bacterial infections were conquered. Complacency set in, and for many years little attention was paid to new agents for antibacterial chemotherapy, despite the inherent nature of infectious agents to develop resistance, and the fact that the developing world still had significant morbidity and mortality from bacterial infections. The serious challenge to public health caused by the emergence of bacterial resistance has highlighted the need for a continuous effort to develop new antibacterial agents. Recent advances in molecular biology, genomics, and drug design approaches offer many new opportunities to discover antibacterial agents. The recent introduction of linezolid, a syn-

**(119)**

thetic antibacterial with a new mechanism and activity against multidrug-resistant Gram-positive organisms, demonstrates that this can be achieved. Folate antagonists, fluoroquinolines and related prototypes, and oxazolidinones offer good prospects for development of further new analogs, as discussed above, and they need to be fully explored more fully. Even more important is the discovery of new classes of antibacterial agents with novel modes of action, which would be less likely to have cross-resistance to known antibacterials. There is a need to maintain constant pressure in the effort to develop new antibacterial agents.

The activity of linezolid has aroused worldwide interest, and many research laboratories around the world have been working in this class of compounds. The compounds that deserve special mention are AZD 2563 and RBx 7644. Clinical trials of AZD 2563, developed by AstraZeneca, have started; whereas RBx 7644, developed by Ranbaxy Laboratories, India, is awaiting clinical trial permission. AZD 2563 has a novel heterocyclic replacement for the C5 N-acetate group. Worldwide data presented (404) show that it is active against Gram-positive bacteria, including multidrug-resistant strains, and may have the added convenience of potential once-daily administration. RBx 7644 (405) is a novel compound that retains linezolid's excellent activity against both sensitive and resistant Gram-positive pathogens; moreover, it is distinctly more active against all anaerobes (Gram-positive or Gram-negative) and has significant, improved inhibitory activity on slime-producing and glass-adherent bacteria, thus extending the indications for its use and decreasing the chances of resistance development.

## REFERENCES

1. C. W. Chambers, *J. Milk Food Technol.*, **19**, 183 (1956).

2. A. R. Martin in J. N. Delgado and W. A. Remers, Eds., *Wilson and Gisvold's Textbook of Organic Medicinal and Pharmaceutical Chemistry*, 9th ed., Lippincott, Philadelphia, 1991, p. 129.

3. S. S. Block in S. S. Block, Ed., *Disinfection, Sterilization, and Preservation*, 4th ed., Lea & Febiger, Philadelphia, 1991, p. 3.

4. J. W. Costerton, J. M. Ingraham, and K.-J. Cheng, *Bacteriol. Rev.*, **38**, 87 (1974).

5. S. M. Hammond, P. A. Lambert, and A. N. Rycroft, *The Bacterial Cell Surface*, Croom Helm, London, 1984.

6. I. G. Jones and M. Midgley, *FEMS Microbiol. Lett.*, **28**, 355 (1985).

7. S. C. Harvey in A. G. Gilman, L. S. Goodman, and A. Gilman, Eds., *The Pharmacological Basis of Therapeutics*, 6th ed., Macmillan, New York, 1980, p. 966.

8. N. A. Allawala and S. Riegelman, *J. Am. Pharm. Assoc. Sci. Ed.*, **42**, 267 (1953).

9. S. W. V. Sutton in J. M. Ascenzi, Ed., *Handbook of Disinfectants and Antiseptics*, Marcel Dekker, New York, 1966, pp. 43–62.

10. P. Cunniff, Ed., *Official Methods of Analysis of the Association of Official Analytical Chemists International*, Vol. **1**, 16th ed., AOAC International, Arlington, VA, 1995.

11. M. K. Bruch in S. S. Block, Ed., *Disinfection, Sterilization, and Preservation*, 4th ed., Lea & Febiger, Philadelphia, 1991, pp. 1028–1046.

12. A. R. Cade, *J. Soc. Cosmet. Chem.*, **2**, 281 (1951).

13. S. S. Block in S. S. Block, Ed., *Disinfection, Sterilization, and Preservation*, 4th ed., Lea & Febiger, Philadelphia, 1991, p. 3.

14. A.-G. Labarraque, *The Use of Sodium and Calcium Hypochlorites*, Huzard, Paris, 1825.

15. T. Alcock, *Lancet*, **11**, 643 (1827).

16. O. W. Holmes, *N. Engl. Q. J. Med. Surg.*, **1**, 503 (1843).

17. I. P. Semmelweiss, *The Etiology, Conception, and Prophylaxis of Childbed-fever*, Hartlehen, Pest/Vienna/Leipzig, 1861.

18. J. Lister, *Lancet*, **2**, 353 (1867).

19. R. Koch, *Mittheilungen aus dem Kaiserlichen Gesundheitsamte*, **1**, 234 (1881).

20. B. Kronig and T. L. Paul, *A. Hyg. Infekt.*, **25**, 1 (1897).

21. S. Rideal and J. T. A. Walker, *J. R. Sanit. Inst.*, **24**, 424 (1903).

22. H. Bechold and P. Ehrlich, *Z. Physiol. Chem.*, **47**, 173 (1906).

23. G. Domagk, *Dtsch. Med. Wochenschr.*, **61**, 829 (1935).

24. F. L. Rose and G. Swain, *J. Chem. Soc.*, 4422 (1956).

25. J. Cremieux and J. Fleurette in S. S. Block, Ed., *Disinfection, Sterilization, and Preservation*, 4th ed., Lea & Febiger, Philadelphia, 1991, pp. 1009–1027.

26. J. Race, *Chlorination of Water*, John Wiley & Sons, New York, 1988, pp. 1–132.

27. T. S. Norman, L. L. Harms, and R. W. Looyegna, *J. Am. Water Works Assoc.*, **72**, 176 (1980).

28. G. F. Reddish and A. W. Pauley, *Bull. Nat. Formul. Comm.*, **13**, 11 (1945).

29. J. S. Thompson, *Soap Chem. Spec.*, **40**, 45 (1964).

30. L. Friberg, *Acta Pathol. Microbiol. Scand.*, **40**, 67 (1957).

31. W. E. Knox, P. K. Stumpf, D. E. Green, and V. H. Auerbach, *J. Bacteriol.*, **55**, 451 (1948).

32. M. Kosugi, J. J. Kaminski, S. H. Selk, I. H. Pitman, N. Bodor, and T. Higuchi, *J. Pharm. Sci.*, **65**, 1743 (1976).

33. S. L. Chang, *J. Sanit. Eng. Div. Proc. ASCE*, **97**, 689 (1971).

34. K. Apostolov, *J. Hygiene*, **84**, 381 (1980).

35. H. A. Shelanski and M. V. Shelanski, *J. Int. Coll. Surg.*, **25**, 727 (1956).

36. F. W. Tilley and J. M. Schaeffer, *J. Bacteriol.*, **12**, 303 (1926).

37. B. D. Lilly and J. H. Brewer, *J. Am. Pharm. Assoc.*, **52**, 6 (1953).

38. C. K. K. Nair, D. S. Pradham, and A. Sreenivasan, *J. Bacteriol.*, **121**, 392 (1965).

39. FDA Advisory Panel on Nonprescription Contraceptive Products, Fed. Reg. 45, 82014, Dec. 12, 1980.

40. J. Judis, *J. Pharm. Sci.*, **52**, 126 (1963).

41. E. M. Richardson and E. E. Reid, *J. Am. Chem. Soc.*, **62**, 413 (1940).

42. E. G. Klarman, V. A. Shternov, and L. W. Gates, *J. Am. Chem. Soc.*, **55**, 2576 (1933).

43. E. G. Klarman, L. W. Gates, V. A. Shternov, and P. H. Cox, *J. Am. Chem. Soc.*, **55**, 4657 (1933).

44. C. M. Suter, *Chem. Rev.*, **28**, 209 (1941).

45. D. O. O'Connor and J. R. Rubino in S. S. Block, Ed., *Disinfection, Sterilization, and Preservation*, 4th ed., Lea & Febiger, Philadelphia, 1991, p. 205.

46. W. S. Gump, U.S. Pat. 2,250,480 (1941).

47. R. E. Bambury in M. E. Wolff, Ed., *Burger's Medicinal Chemistry, Part II*, 4th ed., John Wiley & Sons, New York, 1979, p. 49.

48. G. L. Gilbert, V. M. Gambill, D. R. Spiner, R. K. Hoffmann, and C. R. Phillips, *Appl. Microbiol.*, **12**, 496 (1964).

49. A. Hart and M. W. Brown, *Appl. Microbiol.*, **28**, 1069 (1974).

50. S. Margel and A. Rembaum, *Macromolecules*, **13**, 19 (1980).

51. G. A. Ayliffe, J. R. Babb, and C. R. Bradley, *J. Hosp. Infect.*, **7**, 295 (1986).

52. R. C. Hughes and P. F. Thurman, *Biochem. J.*, **119**, 925 (1970).

53. D. Hopwood, *Histochem. J.*, **7**, 267 (1975).

54. E. R. Jawetz in B. G. Katzung, Ed., *Basic and Clinical Pharmacology*, 6th ed., Appleton and Lange, Norwalk, CT, 1995, p. 748.

55. E. R. Garrett and O. R. Woods, *J. Am. Pharm. Sci.*, **42**, 736 (1953).

56. I. Fridovich, *Am. Sci.*, **63**, 54 (1975).

57. A. R. Martin in J. N. Delgado and W. A. Remers, Eds., *Wilson and Gisvold's Textbook of Organic Medicinal and Pharmaceutical Chemistry*, 9th ed., Lippincott, Philadelphia, 1991, p. 140.

58. E. R. Jawetz in B. G. Katzung, Ed., *Basic and Clinical Pharmacology*, 6th ed., Appleton and Lange, Norwalk, CT, 1995, p. 749.

59. S. C. Harvey in A. G. Gilman, L. S. Goodman, and A. Gilman, Eds., *The Pharmacological Basis of Therapeutics*, 6th ed., Macmillan, New York, 1980, p. 976.

60. D. J. Beaver, D. P. Roman, and P. J. Stoffel, *J. Am. Chem. Soc.*, **79**, 1236 (1957).

61. W. R. Thrower and F. C. Valentine, *Lancet*, **1**, 133 (1943).

62. S. S. Berg and G. Newberry, *J. Chem. Soc.*, 642 (1949).

63. M. J. Winrow, *J. Periodontal Res.*, **12** (Suppl.), 45 (1973).

64. S. I. Hnatko, *Can. Med. Assoc. J.*, **117**, 223 (1977).

65. H. G. Smylie, J. R. C. Logie, and G. Smith, *Br. Med. J.*, **4**, 586 (1973).

66. A. F. Peterson, A. Rosenberg, and S. D. Alatary, *Surg. Gynecol. Obstet.*, **146**, 63 (1978).

67. M. Addy, *J. Clin. Periodontol.*, **13**, 957 (1986).

68. P. M. Woodcock in K. P. Payne, Ed., *Industrial Biocides*, John Wiley & Sons, Chichester, UK, 1988, pp. 19–36.

69. C. R. Enyaert in M. J. Schick, Ed., *Nonionic Surfactants*, Marcel Dekker, New York, 1967.

70. M. J. Miller in J. M. Ascenzi, Ed., *Handbook of Disinfectants and Antiseptics*, Marcel Dekker, New York, 1996, pp. 83–107.

71. A. R. Martin in J. N. Delgado and W. A. Remers, Eds., *Wilson and Gisvold's Textbook of Organic Medicinal and Pharmaceutical Chemistry*, 9th ed., Lippincott, Philadelphia, 1987, p. 137.

72. D. W. Blois and J. Swarbrick, *J. Pharm. Sci.*, **61**, 393 (1972).

73. H. Kourai, *J. Antibact. Antifung. Agents*, **13**, 245 (1985).

74. C. Hansch and A. Leo, *Exploring QSAR: Fundamentals and Applications in Chemistry and Biology*, American Chemical Society, Washington, DC, 1995, pp. 418–419.

75. D. F. Green and A. N. Petrocci, *Soap Cosmet. Chem. Spec.*, **8**, 33 (1980).

76. A. N. Petrocci, H. A. Green, J. J. Merianos, and B. Like, C.S.M.A. Proceedings of the 60th Mid-Year Meeting, May, 1974, pp. 78–79.

77. R. L. Stark, U.S. Pat. 4,525,346 (1985).

78. H. N. Prince and R. N. Prince, *Chemical Times and Trends*, July, 1987, p. 8.

79. R. D. Swisher, *Surfactant Biodegradation*, Marcel Dekker, New York, 1987, pp. 181–209.

80. A. Schmitz, U.S. Pat. 2,684,946 (1954).

81. G. Sykes, *Disinfection and Sterilization*, 2nd Ed., Lippincott, Philadelphia, 1965, pp. 377–378.

82. L. Weinstein, M. A. Madoff, and C. M. Samet, *N. Engl. J. Med.*, **263**, 793, 842, 900, 952 (1960).

83. M. I. Fischl, G. M. Dickinson, and L. La Voie, *J. Am. Med. Assoc.*, **259**, 1185 (1988).

84. FDA Drug Efficacy Reports, through *J. Am. Pharm. Assoc.*, **NS9**535 (1969).

85. R. Berkow, Ed., *The Merck Manual*, Merck, Rahway, NJ, 1987, pp. 43–44.

86. G. L. Archer and R. E. Polk in *Harison's Principles of Internal Medicine*, 15th ed., McGraw-Hill, New York, 2001, p. 877.

87. M. C. Bach, L. D. Sabath, and M. Finland, *Antimicrob. Agents Chemother.*, **3**, 1 (1973).

88. M. D. Iseman in S. L. Gorbach, J. G. Bartlest, and N. R. Blacklow, Eds., *Infectious Diseases*, 2nd ed., Saunders, Philadelphia, 1998, pp. 1513–1528.

89. M. J. Costells, *Arch. Dermatol. Syph.*, **42**, 161 (1940).

90. (a) A. L. Lorinez and R. W. Pearson, *Arch. Dermatol.*, **85**2 (1962); (b) J. J. Zone, *Curr. Probl. Dermatol.*, **3**, 4 (1991).

91. O. J. Sione, *Med. Hypotheses*, **31**, 99 (1990).

92. M. D. Coleman, J. K. Smith, A. D. Perris, S. Buck, and J. K. Seydel, *J. Pharm. Pharmacol.*, **49**, 53 (1997).

93. A. J. M. Van Der Ven, P. P. Koopmans, T. B. Vree, and J. W. M. Van Der Mer, *J. Antimicrob. Chemother.*, **34**, 1 (1994).

94. J. A. Kovacs, J. W. Hiemens, and A. M. Macher, *Ann. Intern. Med.*, **100**, 663 (1984).

95. N. H. Sheer, S. P. Spelberg, D. M. Grant, B. K. Tang, and W. Kalow, *Ann. Intern. Med.*, **105**, 179 (1986).

96. T. P. Reilly, L. H. Lash, M. A. Doll, et al., *J. Invest. Dermatol.*, **114**, 1164 (2000).

97. M. J. Rieder, J. Uetrecht, N. H. Shear, and S. P. Spielberg, *J. Pharmacol. Exp. Ther.*, **244**, 724 (1988).

98. M. J. Rieder, E. Sisson, I. Bird, and W. Y. Almawi, *Int. J. Immunopharmacol.*, **14**, 1175 (1992).

99. M. D. Coleman, A. K. Scott, A. M. Breckenridge, and B. K. Park, *Br. J. Clin. Pharmacol.*, **30**, 761 (1990).

100. R. T. Williams and D. V. Parke, *Annu. Rev. Pharmacol.*, **4**, 85 (1964).

101. H. Nogami, A. Hasegawa, M. Hanano, and K. Imaoka, *Yakugaku Zasshi*, **88**, 893 (1968).

102. M. Yamazaki, M. Aoki, and A. Kamada, *Chem. Pharm. Bull. (Tokyo)*, **16**, 707 (1968).

103. K. Kakemi, T. Arita, and K. Kakemi, *Arch. Pract. Pharm.*, **25**, 22 (1965).

104. T. Koizumi, T. Arita, and K. Kakemi, *Chem. Pharm. Bull. (Tokyo)*, **12**, 428 (1964).

105. R. H. Adamson, J. W. Bridges, M. R. Kibby, S. R. Walker, and R. T. Williams, *Biochem. J.*, **118**, 41 (1970).

106. G. Zbinden in R. F. Gould, Ed., *Molecular Modificaton in Drug Design*, American Chemical Society, Washington, DC, 1964, p. 25.

107. T. Fujita in R. F. Gould, Ed., *Biological Correlations The Hansch Approach*, American Chemical Society, Washington, DC, 1972, p. 80.

108. G. A. Ellard, *Br. J. Pharmacol.*, **26**, 212 (1966).

109. S. R. M. Bushby and A. J. Woiwood, *Am. Rev. Tuberc. Pulm. Dis.*, **72**, 123 (1955).

110. S. R. M. Bushby and A. J. Woiwood, *Biochem. J.*, **63**, 406 (1956).

111. H. B. Hucker, *Annu. Rev. Pharmacol.*, **10**, 99 (1970).

112. G. R. Gordon, J. H. Peters, R. Gelber, and L. Levy, *Proc. West Pharmacol. Soc.*, **13**, 17 (1970).

113. P. E. Thompson, *Int. J. Leprosy*, **35**, 605 (1967).

114. (a) E. Kruger-Thiemer and P. Bunger, *Arzneim.-Forsch.*, **16**1431 (1961); (b) E. Kruger-Thiemer and P. Bunger, *Arzneim.-Forsch.*, **11**, 867 (1961).

115. J. Reider, *Arzneim.-Forsch.*, **13**, 81, 89, 95 (1963).

116. W. V. J. Raszka, L. P. Skillman, and P. L. McEvoy, *Diagn. Microbiol. Infect. Dis.*, **18**, 201 (1994).

117. W. Peters, *Adv. Parasitol.*, **12**, 69 (1974).

118. P. P. Levine, *Cornell Vet.*, **29**, 309 (1939).

119. P. P. Levine, *J. Parasitol.*, **26**, 233 (1940).

120. L. P. Joyner, S. F. M. Davies, and S. B. Kendall in R. J. Schnitzer and F. Hawking, Eds., *Experimental Chemotherapy*, Academic Press, New York, 1963, p. 445.

121. A. B. Sabin and J. Waren, *J. Bacteriol.*, **41**, M50, 80 (1941).

122. E. Biocca, *Arg. Biol. (Sao Paulo)*, **7**, 27 (1943).

123. R. E. Strauss, A. M. Kilgman, and D. M. Pilsbury, *Am. Rev. Tuberc.*, **63**, 441 (1951).

124. R. G. Connar, T. B. Ferguson, W. C. Sealy, and N. F. Conant, *J. Thorac. Surg.*, **22**, 424 (1951).

125. S. B. Kendall and L. P. Joyner, *Vet. Record*, **70**, 632 (1958).

126. H. Werner, *Biol. Chil. Parasitol.*, **25**, 65 (1970).

127. R. A. Smego, M. B. Moeller, and H. A. Gallis, *Arch. Intern. Med.*, **143**, 711 (1983).

128. G. K. Herkes, J. Fryer, R. Rushworth, et al., *Aust. N. Z. J. Med.*, **19**, 475 (1989).

129. F. O. McCallum and G. M. Findlay, *Lancet*, **2**, 136 (1938).

130. W. G. Forster and J. R. McGibony, *Am. J. Ophthalmol.*, **27C**, 1107 (1944).

131. M. A. Fischl., G. M. Dickinson, and L. Law Voie, *JAMA*, **259**, 1185 (1988).

132. J. S. Lockwood, *JAMA*, **111**, 2259 (1938).

133. C. M. MacLod, *J. Exp. Med.*, **72**, 217 (1940).

134. D. A. Boroff, A. Cooper, and J. G. M. Bullowa, *J. Immunol.*, **43**, 341 (1942).

135. T. C. Stamp, *Lancet*, **2**, 10 (1939).

136. H. N. Green, *Br. J. Exp. Pathol.*, **21**, 38 (1940).

137. D. D. Woods, *Br. J. Exp. Pathol.*, **21**, 74 (1940).

138. S. Ratner, M. Blanchard, A. F. Coburn, and D. E. Green, *J. Biol. Chem.*, **155**, 689 (1944).

139. F. R. Selbie, *Br. J. Exp. Pathol.*, **21**, 90 (1940).

140. G. M. Findlay, *Br. J. Exp. Pathol.*, **21**, 356 (1940).

141. K. C. Blanchard, *J. Biol. Chem.*, **140**, 919 (1941).

142. H. McIllwain, *Br. J. Exp. Pathol.*, **23**, 265 (1942).

143. S. D. Rubbo, M. Maxwell, R. A. Fairbridge, and J. M. Gillespie, *Aust. J. Exp. Biol. Med. Sci.*, **19**, 185 (1941).

144. P. Fildes, *Lancet*, **1**, 955 (1940).

145. E. A. Bliss and P. H. Long, *Bull. Johns Hopkins Hosp.*, **69**, 14 (1941).

146. E. E. Snell and H. K. Mitchell, *Arch. Biochem.*, **1**, 93 (1943).

147. M. R. Stetten and C. L. Fox Jr., *J. Biol. Chem.*, **161**, 333 (1945).

148. W. Shive, W. W. Ackermann, M. Gordon, M. E. Getzendaner, and R. E. Eakin, *J. Am. Chem. Soc.*, **69**, 725 (1947).

149. J. S. Gots, *Nature*, **172**, 256 (1953).

150. R. B. Angier, J. H. Boothe, B. L. Hutchings, J. H. Mowat, J. Semb, E. L. R. Stokstad, Y. Subba Row, C. W. Waller, D. B. Cosulich, M. J. Fahrenbach, M. E. Hultquist, E. Kuh, E. H. Northey, D. R. Seeger, J. P. Sickless, and J. M. Smith Jr., *Science*, **103**, 667 (1946).

151. J. H. Mowat, J. H. Boothe, B. L. Hutchings, E. L. R. Stokstad, C. W. Waller, R. Angier, J. Semb, D. B. Consulich, and Y. Subba Row, *Ann. N. Y. Acad. Sci.*, **48**, 279 (1946–1947).

152. R. Tschesche, *Z. Naturforsch.*, **26b**, 10 (1947).

153. A. D. Welch and C. A. Nichol, *Annu. Rev. Biochem.*, **21**, 633 (1952).

154. M. Friedkin, *Annu. Rev. Biochem.*, **32**, 185 (1963).

155. D. D. Woods in *Chemistry and Biology of Pteridines (Ciba Foundation Symposium)*, Little Brown, Boston, 1954, p. 220.

156. J. O. Lampen and M. J. Jones, *J. Biol. Chem.*, **166**, 435 (1946).

157. J. O. Lampen and M. J. Jones, *J. Biol. Chem.*, **170**, 133 (1947).

158. A. K. Miller, *Proc. Soc. Exp. Biol. Med.*, **57**, 151 (1944).

159. A. K. Miller, P. Bruno, and R. M. Berglund, *J. Bacteriol.*, **54**, G20, 9 (1947).

160. J. Lascelles and D. D. Woods, *Br. J. Exp. Pathol.*, **33**, 288 (1952).

161. R. H. Nimmo-Smith, J. Lasceles, and D. D. Woods, *Br. J. Exp. Pathol.*, **29**, 264 (1948).

162. T. Shiota and M. M. Disraely, *Biochim. Biophys. Acta*, **52**, 467 (1961).

163. T. Shiota, M. N. Disraely, and M. P. McCan, *J. Biol. Chem.*, **236**, 2534 (1961).

164. G. M. Brown, R. A. Weisman, and D. A. Molnar, *J. Biol. Chem.*, **236**, 2534 (1961).

165. R. Weisman and G. M. Brown, *J. Biol. Chem.*, **239**, 326 (1964).

166. G. M. Brown, *J. Biol. Chem.*, **237**, 536 (1962).

167. L. Bock, G. H. Miller, K. J. Schaper, and J. K. Seydel, *J. Med. Chem.*, **17**, 23 (1974).

168. G. Swedberg, S. Castensson, and O. Skold, *J. Bacteriol.*, **1**, 129, 137 (1979).

169. S. Roland, R. Ferone, R. J. Harvey, Y. L. Styles, and R. W. Morrison, *J. Biol. Chem.*, **254**, 10337 (1979).

170. R. D. Hotchkiss and A. H. Evans, *Fed. Proc.*, **19**, 912 (1960).

171. D. P. Richey and G. M. Brown, *J. Biol. Chem.*, **244**, 1582 (1969).

172. T. Shiota, C. M. Baugh, R. Jackson, and R. Dillarde, *Biochemistry*, **8**, 5022 (1969).

173. P. J. Ortiz and R. D. Hotchkiss, *Biochemistry*, **31**, 174 (1969).

174. P. J. Ortiz, *Biochemistry*, **9**, 355 (1970).

175. B. L. Toth-Martinez, S. Papp, Z. Dinya, and F. J. Hernadi, *Biosystems*, **7**, 172 (1975).

176. H. H. W. Thjssen, *J. Med. Chem.*, **20**, 233 (1977).

177. G. Brownice, A. F. Green, and M. Woodbine, *Br. J. Pharmacol.*, **3**, 15 (1948).

178. C. Levaditti, *Compte. Rend. Soc. Biol.*, **135**, 1109 (1941).

179. S. P. Ramakrishnan, P. C. Basu, H. Singh, and N. Singh, *Bull. World Health Organ.*, **27**, 213 (1962).

180. J. K. Seydel, M. Richter, and E. Wemple, *Int. J. Leprosy*, **48**, 18 (1977).

181. M. L. Panitch and L. Levy, *Lepr. Rev.*, **49**, 131 (1978).

182. R. J. Cenedella and J. J. Jarrell, *Am. J. Trop. Med. Hyg.*, **19**, 592 (1970).

183. H. R. Morgan, *J. Exp. Med.*, **88**, 285 (1948).

184. J. W. Moulder, *The Biochemistry of Intracellular Parasitism*, University of Chicago Press, Chicago, 1962, p. 105.

185. L. P. Jones and F. D. Williams, *Can. J. Microbiol.*, **14**, 933 (1968).

186. R. Ferone, *J. Protozool.*, **20**, 459 (1973).

187. R. D. Walter and E. Konigk, *Hoppe-Seyler's Z. Physiol. Chem.*, **355**, 431 (1974).

188. J. L. McCullough and T. H. Maren, *Mol. Pharmacol.*, **10**, 140 (1974).

189. L. Jaenicke and P. H. C. Chan, *Angew. Chem.*, **72**, 752 (1960).

190. O. Okinaka and K. Iwai, *Anal. Biochem.*, **31**, 174 (1969).

191. H. Mitsuda and Y. Suzuki, *J. Vitaminol. (Kyoto)*, **14**, 106 (1968).

192. K. Iwai and O. Okinaka, *J. Vitaminol. (Kyoto)*, **14**, 170 (1968).

193. B. Roth in *Inhibition of Folate Metabolism in Chemotherapy*, Springer-Verlag, Berlin, Germany, 1983, p. 107.

194. J. Greenberg, *J. Pharmacol. Exp. Ther.*, **97**, 484 (1949).

195. J. Greenberg and E. M. Richeson, *J. Pharmacol. Exp. Biol. Med.*, **99**, 444 (1950).

196. J. Greenberg and E. M. Richeson, *Proc. Soc. Exp. Biol. Med.*, **77**, 174 (1951).

197. L. P. Garrod, D. G. James, and A. A. G. Lewis, *Postgrad. Med. J.*, **45** (Suppl.), 1 (1969).

198. G. H. Hitchings, *Med. J. Aust.*, **1** (Suppl.), 5 (1973).

199. J. K. Seydel and E. Wempe, *Chemotherapy*, **21**, 131 (1975).

200. S. R. M. Bushby and G. H. Hitchings, *Br. J. Pharmacol. Chemother.*, **33**, 72 (1968).

201. J. J. Burchall and G. H. Hitchings, *Mol. Pharmacol.*, **1**, 126 (1965).

202. G. P. Wormser and G. T. Kausch in *Inhibition of Folate Metabolism in Chemotherapy: The Origins and Uses of Co-trimoxazole*, Springer-Verlag, Berlin, Germany, 1983, p. 1.

203. A. M. Baca, R. Sivaraporn, S. Turlly, W. Sirawaraporne, and W. G. J. Hol, *J. Mol. Biol.*, **302**, 1193 (2000).

204. V. Nopponpunth, W. Sirawaraporn, P. J. Greene, and D. V. Santi, *J. Bacteriol.*, **181**, 6814 (1999).

205. A. Achari, D. O. Somers, J. N. Champness, P. K. Bryant, J. Rosemond, and D. K. Stammers, *Nat. Struct. Biol.*, **4**, 490 (1997).

206. W. S. Dallas, J. E. Gowen, P. D. Ray, M. J. Cox, and I. K. Dev, *J. Bacteriol.*, **174**, 5961 (1992).

207. I. C. Hampele, A. D'Arcy, G. E. Dale, D. Kostrewa, J. Nielsen, C. Oefner, M. G. F. Page, H. J. Schonfeld, D. Stuber, and R. L. Then, *J. Mol. Biol.*, **268**, 21 (1997).

208. T. Triglia and A. F. Cowman, *Proc. Natl. Acad. Sci. USA*, **91**, 7149 (1997).

209. F. Volpe, M. Dyer, J. G. Scaife, G. Darby, D. K. Stammers, and C. J. Delves, *Gene*, **112**, 213 (1992).

210. F. Rebeille, D. Macherel, J. M. Mouillon, J. Garin, and R. Douce, *EMBO J.*, **16**, 947 (1997).

211. D. R. Brooks, P. Wang, M. Read, W. M. Watkins, P. F. Sims, and J. E. Hyde, *Eur. J. Biochem.*, **224**, 397 (1994).

212. M. Landy, N. W. Larkun, E. J. Oswald, and F. Strighoff, *Science*, **97**, 265 (1943).

213. P. J. White and D. D. Woods, *J. Gen. Microbiol.*, **40**, 243 (1965).

214. M. L. Pato and G. M. Brown, *Arch. Biochem. Biophys.*, **103**, 443 (1963).

215. R. Ho and L. Cormen, *Antimicrob. Agents Chemother.*, **5**, 388 (1974).

216. E. M. Wise Jr., and M. M. Abou-Donia, *Proc. Natl. Acad. Sci. USA*, **72**, 2621 (1975).

217. A. Bishop, *Biol. Rev.*, **34**, 445 (1959).

218. P. Huovinen, L. Sundstrom, G. Swedberg, and O. Skold, *Antimicrob. Agents Chemother.*, **39**, 279 (1995).

219. H. G. Vinnicombe and J. P. Derrick, *Biochem. Biophys. Res. Commun.*, **258**, 752 (1996).

220. A. Bishop, *Parasitology*, **53**, 10 (1963).

221. P. Radstrom, G. Swedberg, and O. Skold, *Antimicrob. Agents Chemother.*, **35**, 1840 (1991).

222. T. Watanabe, *Bacteriol. Rev.*, **27**, 87 (1963).

223. G. E. Dale, C. Broger, A. D'Arcy, P. G. Hartman, et al., *J. Mol. Biol.*, **266**, 23 (1997).

224. M. Kai, M. Matsuoka, N. Nakata, S. Maeda, M. Gidoh, Y. Maeda, K. Hashimoto, K. Kobayashi, and Y. Kashiwabara, *FEMS Microbiol. Lett.*, **177**, 231 (1999).

225. A. Bishop in L. G. Goodwin and R. N. Nimmo-Smith, Eds., *Drug Parasites and Hosts*, Little Brown, Boston, 1962, p. 98.

226. Horlein, Dressel, and Kethe quoted by F. Mietzsch, *Chem. Ber.*, **71A**, 15 (1938).

227. M. Hedelberger and W. A. Jacobs, *J. Am. Chem. Soc.*, **41**, 2131 (1939).

228. F. Mietzsch and J. Kiarer, Ger. Pat. 607,537 (1935); *Chem. Abstr.*, **29**, 4135 (1935).

229. G. Domagk, *Dtsch. Med. Wochenschr.*, **61**, 250 (1935).

230. R. Foerster, *Zbt. Haut-u Geschlechiskr.*, **45**, 549 (1933).

231. J. Trefouel, Mme. J. Trfouel, F. Nitti, and D. Bovel, *C. R. Soc. Biol.*, **120**, 756 (1935).

232. L. Colebrook and M. Kenny, *Lancet*, **1**, 1279 (1936).

233. G. A. H. Buttle, W. H. Grey, and D. Stephenson, *Lancet*, **1**, 1286 (1936).

234. A. T. Fuller, *Lancet*, **1**, 194 (1937).

235. P. Gelmo, *J. Prakt. Chem.*, **77**, 369 (1908); *Chem. Abstr.*, **2**, 2551 (1908).

236. L. E. H. Whitby, *Lancet*, **2**, 1210 (1938).

237. R. J. Fosbinder and L. A. Walter, *J. Am. Chem. Soc.*, **61**, 2032 (1939).

238. E. H. Northey, *The Sulfonamides and Allied Compounds*, American Chemical Society Monograph Services, Reinhold, New York, 1948.

239. R. L. Nichols, W. F. Jones Jr., and M. Finland, *Proc. Soc. Exp. Biol. Med.*, **92**, 637 (1956).

240. L. Weinstein, M. A. Madoff, and C. M. Samet, *N. Engl. J. Med.*, **263**, 793, 842, 900, 952 (1960).

241. L. Neipp and R. L. Meyer, *Ann. N. Y. Acad. Sci.*, **69**, 447 (1957).

242. W. Klotzer and H. Bretschneider, *Monatsh. Chem.*, **87**, 136 (1956); *Chem. Abstr.*, **51**, 15610 (1957).

243. B. Fust and E. Bohni, *Antibiot. Med. Clin. Ther.*, **6** (Suppl. 1), 3 (1959).

244. G. Hitzenberger and K. H. Spitzky, *Med. Klin.*, **57**, 310 (1962); *Chem. Abstr.*, **57**, 5274 (1962).

245. M. Reber, G. Rutshauser, and H. Tholen in *3rd International Congress on Chemotherapy, Stuttgart, Germany, 1963*, Vol. **1**, Thieme, Stuttgart, 1964, p. 648.

246. B. Fust and E. Bohni, *Schweiz. Med. Wochenschr.*, **92**, 1599 (1962).

247. N. Rist, *Nature*, **146**, 838 (1940).

248. W. H. Feldman, H. C. Hinshaw, and H. E. Moses, *Am. Rev. Tuberc.*, **45**, 303 (1942).

249. E. V. Cowdry and C. Ruangsiri, *Arch. Pathol.*, **32**, 632 (1941).

250. S. G. Browne, *Adv. Pharmacol. Chemother.*, **7**, 211 (1969).

251. C. C. Shepard, D. H. McRae, and J. A. Habas, *Proc. Soc. Exp. Biol. Med.*, **122**, 893 (1966).

252. E. F. Elslager, *Prog. Drug Res.*, **18**, 99 (1974).

253. N. Anand in M. E. Wolff, Ed., Burger's Medicinal Chemistry and Drug Discovery, Vol. **2**, 5th ed., John Wiley & Sons, New York, 1996, p. 538.

254. J. K. Seydel in E. J. Ariens, Ed., *Physico-Chemical Aspects of Drug Action*, Pergamon, New York, 1988, p. 169.

255. P. G. De Beneditti, *Adv. Drug Res.*, **16**, 227 (1987); *Prog. Drug Res.*, **36**, 361 (1991).

256. P. G. Sammes, Comprehensive Medicinal Chemistry, Vol. **II**, 1st ed., Pergamon, Oxford, UK, 1990, p. 255.

257. J. K. Seydel and E. Wempe, *Arzneim.-Forsch.*, **14**, 705 (1964).

258. J. K. Seydel, E. Kruger-Thiemer, and E. Wempe, *Z. Naturforsch.*, **15b**, 620 (1960).

259. E. C. Foernzler and A. N. Martin, *J. Pharm. Sci.*, **56**, 608 (1967).

260. C. L. Fox Jr., and H. M. Rose, *Proc. Soc. Exp. Biol. Med.*, **50**, 142 (1942).

261. F. C. Schmelkes, O. Wyss, H. C. Marks, B. J. Ludwig, and F. B. Stranskov, *Proc. Soc. Exp. Biol. Med.*, **50**, 145 (1942).

262. P. H. Bell and R. O. Roblin Jr., *J. Am. Chem. Soc.*, **64**, 2905 (1942).

263. P. B. Cowles, *Yale J. Biol. Med.*, **14**, 599 (1942).

264. A. H. Brueckner, *Yale J. Biol. Med.*, **15**, 813 (1943).

265. J. K. Seydel, *Arzneim.-Forsch.*, **16**, 1447 (1966).

266. J. K. Seydel, *J. Pharm. Sci.*, **57**, 1455 (1967).

267. A. Cammarata and R. C. Allen, *J. Pharm. Sci.*, **56**, 640 (1967).

268. J. K. Seydel, *Mol. Pharmacol.*, **2**, 259 (1966).

269. E. R. Garrett, J. B. Mielck, J. K. Seydel, and H. J. Kessler, *J. Med. Chem.*, **14**, 724 (1971).

270. J. K. Seydel, *J. Med. Chem.*, **14**, 724 (1971).

271. T. Fujita and C. Hansch, *J. Med. Chem.*, **10**, 991 (1967).

272. M. Yamazaki, N. Kakeya, T. Morishita, A. Kamada, and A. Aoki, *Chem. Pharm. Bull. (Tokyo)*, **18**, 702 (1970).

273. G. H. Miller, P. H. Doukas, and J. K. Seydel, *J. Med. Chem.*, **15**, 700 (1972).

274. A. Koch, J. K. Seydel, A. Gasco, C. Tirani, and R. Fruttero, *Quant. Struct.-Act. Relat.*, **12**, 373 (1993).

275. P. G. De Beneditti, D. Iarossi, U. Folli, C. Frassineti, M. C. Menziani, and C. Cennamo, *J. Med. Chem.*, **32**, 396 (1989).

276. P. G. Deneditti, D. Iarossi, U. Folli, C. Frassineti, M. C. Menziani, and C. Cennamo, *J. Med. Chem.*, **30**, 459 (1987).

277. M. Wiese, J. K. Seydel, H. Pieper, G. Kruger, K. R. Noll, and J. Keck, *Quant. Struct.-Act. Relat.*, **6**, 164 (1987).

278. M. G. Koehler, A. J. Hopfinger, and J. K. Seydel, *J. Mol. Struct. (Theochem)*, **179**, 319 (1988).

279. P. G. De Beneditti, *Prog. Drug Res.*, **36**, 361 (1991).

280. (a) R. L. Lopez de Compadre, R. A. Pearlstein, A. J. Hopfinder, and J. K. Seydel, *J. Med. Chem.*, **30**, 900 (1987); (b) A. J. Hopfinder, R. L. Lopez de Compadre, M. G. Koshler, S. Emery, and J. K. Seydel, *Quant. Struct.-Act. Relat.*, **6**, 111 (1987).

281. P. G. De Beneditti, *Prog. Drug Res.*, **36**, 361 (1991).

282. M. Cocchi, D. Iarossi, M. C. Menziani, P. G. De Beneditti, and C. Frassineti, *Struct. Chem.*, **3**, 129 (1992).

283. Y. U. Sokolov, M. C. Menziani, M. Cocchi, and P. G. Beneditti, *Theo. Chem.*, **79**, 293 (1991).

284. J. Rieder, *Arzneim.-Forsch.*, **13**, 81, 89, 95 (1963).

285. D. Lehr, *Ann. N. Y. Acad. Sci.*, **69**, 417 (1957).

286. G. L. Biagi, A. M. Barbaro, M. C. Guerra, G. C. Forti, and M. E. Fracasso, *J. Med. Chem.*, **17**, 28 (1974).

287. W. Scholtan, *Arzneim.-Forsch.*, **14**, 348 (1964); *Arzneim.-Forsch.*, **18**, 505 (1968).

288. E. Kruger-Thiemer, W. Diller, and P. Bunger, *Antimicrob. Agents Chemother.*, 183 (1965).

289. K. Irmscher, D. Gabe, K. Jahnke, and W. Scholtan, *Arzneim.-Forsch.*, **16**, 1019 (1966).

290. W. Scholtan, *Chemotherapia*, **6**, 180 (1963).

291. J. A. Shannon, *Ann. N. Y. Acad. Sci.*, **44**, 455 (1943).

292. J. K. Seydel, Ed., *Drug Design*, Vol. **1**, Academic Press, New York, 1971, p. 343.

293. O. Jardetzky and N. G. Wade-Jardetzky, *Mol. Pharmacol.*, **1**, 214 (1965).

294. B. K. Martin, *Nature*, **207**, 274 (1965).

295. I. Moriguchi, S. Wada, and T. Nishizawa, *Chem. Pharm. Bull. (Tokyo)*, **16**, 601 (1968).

296. T. Struller, *Prog. Drug Res.*, **12**, 389–457 (1968).

297. S. Budavari, Ed., *The Merck Index*, 12th ed., Merck, Rahway, NJ, 1996.

298. G. Y. Lesher, E. D. Froelich, M. D. Grant, J. H. Bailey, and R. P. Brundage, *J. Med. Pharm. Chem.*, **5**, 1063 (1962).

299. T. Irikura, U.S. Pat. 4,146,719 (1978).

300. A. Bryskier and J. F. Chantot, *Drugs*, **49** (Suppl. 2), 16–18 (1995).

301. G. L. Drusano in J. S. Wolfson and D. C. Hooper, Eds., *Quinolone Antimicrobial Agents*, American Society for Microbiology, Washington, DC, 1989, pp. 71–105.

302. B. B. Gooding and R. N. Jones, *Antimicrob. Agents Chemother.*, **37**, 349 (1993).

303. K. Marutani, M. Matsumoto, Y. Otabe, M. Nagamuta, K. Tanaka, A. Miyoshi, T. Hasegawa, H. Nagana, S. Matsubara, R. Kamide, T. Yokata, F. Matsumoto, and Y. Ueda, *Antimicrob. Agents Chemother.*, **37**, 2217 (1993); M. Matsumoto, K. Kojima, N. Nagano, S. Matsubara, and T. Yokota, *Antimicrob. Agents Chemother.*, **36**, 1715 (1992).

304. I. Hayakawa, et al., *Antimicrob. Agents Chemother.*, **29**, 163 (1986).

305. L. A. Mitscher, et al., *J. Med. Chem.*, **30**, 2283 (1987).

306. K. Koshino, K. Sato, T. Ure, and Y. Osada, *Antimicrob. Agents Chemother.*, **33**, 1816 (1989).

307. J. P. Sanchez, et al., *J. Med. Chem.*, **38**, 4478 (1995).

308. U. Petersen, et al., *Curr. Opin. Invest. Drugs*, **1**, 45 (2000).

309. H. A. Shelanski and M. V. Shelanski, *J. Int. Coll. Surg.*, **25**, 727 (1956).

310. C. Hansch and A. Leo, *Exploring QSAR: Fundamentals and Applications in Chemistry and Biology*, American Chemical Society, Washington, DC, 1995, p. 440.

311. Y. Yamashita, T. Ashizawa, M. Morimoto, J. Hosomi, and H. Nakano, *Cancer Res.*, **52**, 2818 (1992).

312. A. R. Ronald, M. Turck, and R. G. Petersdorf, *N. Engl. J. Med.*, **275**, 1081 (1966).

313. D. C. Hooper and J. S. Wolfson in D. C. Hooper and J. S. Wolfson, Eds., *Quinolone Antimicrobial Agents*, American Society for Microbiology, Washington, DC, 1989, pp. 249–271.

314. R. C. Moellering in J. S. Wolfson and D. C. Hooper, Eds., *Quinolone Antimicrobial Agents*, American Society for Microbiology, Washington, DC, 1989, pp. 273–283.

315. C. C. Sanders in W. E. Sanders and C. C. Sanders, Eds., *Fluoroquinolines in the Treatment of Infectious Diseases*, Physicians and Scientists Publ., Glenview, IL, 1990, pp. 1–27.

316. J. J. Schentag and D. E. Nix in W. E. Sanders and C. C. Sanders, Eds., *Fluoroquinolines in the Treatment of Infectious Diseases*, Physicians and Scientists Publ., Glenview, IL, 1990, pp. 5–34.

317. J. S. Wolfson, D. C. Hooper, and M. N. Swartz in J. S. Wolfson and D. C. Hooper, Eds., *Quinolone Antimicrobial Agents*, American Society for Microbiology, Washington, DC, 1989, pp. 5–34.

318. P. S. Leitman, *Drugs*, **49** (Suppl. 2), 159 (1995).

319. M. DeSarro, A. Zappala, S. Chimirri, S. Grasso, and G. B. DeSarro, *Antimicrob. Agents Chemother.*, **37**, 1497 (1993).

320. S. Murayama, Y. Hara, A. Ally, T. Suzuki, and M. Tamagawa, *Nippon Yakurigaku Zasshi*, **99**, 13 (1992).

321. C. W. Ford, J. C. Hamel, D. Stapert, et al., *Trends Microbiol.*, **5**, 196 (1997).

322. J. S. Daly, G. M. Eliopoulos, S. Willey, and R. C. Moellering Jr., *Antimicrob. Agents Chemother.*, **32**, 1341 (1988).

323. A. M. Slee, M. A. W. Uonola, R. J. Mcripley, et al., *Antimicrob. Agents Chemother.*, **31**, 1791 (1987).

324. J. S. Daly, G. M. Eliopoulos, E. Reiszner, and R. C. Moellering Jr., *J. Antimicrob. Chemother.*, **21**, 721 (1988).

325. H. C. Neu, A. Novelli, G. Saha, et al., *Antimicrob. Agents Chemother.*, **32**, 580 (1988).

326. W. Brumfitt and J. M. T. Hamilton-Miller, *J. Antimicrob. Chemother.*, **21**, 711 (1988).

327. G. M. Zajac, H. E. Lam, and A. M. Hoffman, 27th Interscience Conference on Antimicrobial Agents and Chemotherapy, New York, Abstr. 247 (1987).

328. D. C. Eustice, P. A. Feldman, and A. M. Slee, *Biochem. Biophys. Res. Commun.*, **150**, 965 (1988).

329. D. C. Eustice, P. A. Feldman, I. Zajac, and A. M. Slee, *Antimicrob. Agents Chemother.*, **32**, 1218 (1988).

330. S. J. Brickner, D. K. Hutchinson, and M. R. Barbachyn, *J. Med. Chem.*, **39**, 673 (1996).

331. S. J. Brickner, *Curr. Pharmacol. Design*, **2**, 175 (1996).

332. G. E. Zurenko, B. H. Yagi, R. D. Schaadt, et al., *Antimicrob. Agents Chemother.*, **40**, 839 (1996).

333. D. J. Biedenbach and R. N. Jones, *J. Clin. Microbiol.*, **35**, 3198 (1997).

334. D. J. Diekema and R. N. Jones, *Drugs*, **59**, 7 (2000).

335. Y.-Q. Xiong, M. R. Yeaman, and A. S. Bayer, *Drugs Today*, **36**, 631 (2000).

336. C. M. Perry and B. Jarvis, *Drugs*, **61**, 525 (2001).

337. M. C. Di Pentima, E. O. Mason Jr., and S. L. Kaplan, *Clin. Infect. Dis.*, **26**, 1169 (1998).

338. E. J. C. Goldstein, D. M. Citron, and C. V. Merriam, *Antimicrob. Agents Chemother.*, **43**, 1469 (1999).

339. C. Edlund, H. Oh, and C. E. Nord, *Clin. Microb. Infect.*, **5**, 51 (1999).

340. C. W. Ford, J. C. Hamel, D. M. Wilson, et al., *Antimicrob. Agents Chemother.*, **40**, 1508 (1996).

341. M. H. Cynamon, S. P. Klemens, C. A. Sharpe, et al., *Antimicrob. Agents Chemother.*, **43**, 1189 (1999).

342. M. J. Gehman, P. G. Pitsakis, S. V. Mallela, et al., Proceedings of the 37th Annual Meeting of the American Society of Infectious Diseases, Nov. 18–21, 1999, Philadelphia, Abstr. 192 (1999).

343. S. I. Pelton, M. Figueira, R. Albut, et al., *Antimicrob. Agents Chemother.*, **44**, 654 (2000).

344. S. M. Swaney, H. Aoki, and M. C. Ganoza, *Antimicrob. Agents Chemother.*, **42**, 3251 (1998).

345. A. H. Lin, R. W. Murray, T. J. Vidmar, and K. R. Marotti, *Antimicrob. Agents Chemother.*, **41**, 2127 (1997).

346. P. Kloss, L. Xiong, D. L. Shinabarger, et al., *J. Mol. Biol.*, **294,** 93 (1999); L. Xiong, P. Kloss, S. Douthwaite, et al., *J. Bacteriol.*, **182**, 5325 (2000).

347. G. E. Zurenko, W. M. Todd, B. Hafkin, et al., *Antimicrob. Agents Chemother.*, **43**, Abstr. C848 (1999).

348. C. G. Gemell and C. W. Ford, Proceedings of the 39th Interscience Conference on Antimicrobial Agents and Chemotherapy, San Francisco, B-118 (1999).

349. D. J. Stalker, C. P. Wajszczuk, D. H. Batts, et al., *Antimicrob. Agents Chemother.*, **41**, Abstr. A115 (1997).

350. D. J. Stalker, C. P. Wajszczuk, D. H. Batts, et al., *Antimicrob. Agents Chemother.*, **41**, Abstr. A116 (1997).

351. Pharmacia and Upjohn Company, Zyvox (linezolid): General review (Data on file Kalamazoo, MI, 1999).

352. I. C. Wienkers, M. A. Wynakla, K. L. Feenstra, et al., Proceedings of the 39th Interscience Conference on Antimicrobial Agents and Chemotherapy, San Francisco, A-68 (1999).

353. K. I. Feenstra, J. G. Slatter, D. J. Stalker, et al., Proceedings of the 38th Interscience Conference on Antimicrobial Agents and Chemotherapy, San Diego, 17 (1998).

354. S. K. Cammarata, B. Hafkin, D. M. Demke, et al., *Clin. Microb. Infect.*, **5**, Abstr.133 (1999).

355. R. C. Moellering Jr., *Ann. Intern. Med.*, **130**, 155 (1999).

356. S. K. Cammarata, B. Hafkin, W. M. Todd, et al., *Am. J. Respir. Crit. Care Med.*, **159**, A844 (1999).

357. M. C. Birmingham, G. S. Zimmer, B. Hafkin, et al., Proceedings of the 38th Interscience Conference on Antimicrobial Agents and Chemotherapy, San Diego, Abstr. MN26 (1998).

358. J. W. Chien, M. L. Kucia, and R. A. Salata, *Clin. Infect. Dis.*, **30**, 146 (2000).

359. R. D. Gonzales, P. C. Schreckenberger, M. B. Graham, et al., *Lancet*, **357**, 1179 (2001).

360. N. E. Wilks, M. A. McConnell-Martin, T. H. Oliphant, et al., *Antimicrob. Agents Chemother.*, **43**, Abstr. 1763 (1999).

361. M. C. Birmingham, G. S. Zimmer, B. Hafkin, et al., *Antimicrob. Agents Chemother.*, **43**, Abstr. 1098 (1999).

362. S. L. Green, J. C. Maddox, and E. D. Huttenbach, *JAMA*, **285**, 1291 (2001).

363. G. E. Zurenko, C. W. Ford, D. K. Hutchinson, S. J. Brickner, and M. R. Barbachyn, *Exp. Opin. Invest. Drugs*, **6**, 151 (1997).

364. R. Gadwood and D. A. Shinabarger, *Annu. Rep. Med. Chem.*, **35**, 136 (2001).

365. D. M. Gleave, S. J. Brickner, P. R. Manninen, et al., *Bioorg. Med. Chem. Lett.*, **8**, 1231 (1998).

366. M. D. Gleave and S. J. Brickner, *J. Org. Chem.*, **61**, 6470 (1996).

367. S. Bartel, W. R. Endermann, W. Guamieri, et al., Proceedings of the 37th Interscience Conference on Antimicrobial Agents and Chemotherapy, Toronto, Ontario, Canada, Abstr. F18 (1997).

368. B. Riedl, D. Habich, A. Stolle, et al., U.S. Pat. 5,684,023 (1997).

369. D. Habich, S. Bartel, R. Endermann, et al., Proceedings of the 39th Interscience Conference on Antimicrobial Agents and Chemotherapy, San Francisco, Abstr. F-566 (1999).

370. S. Bartel, W. Guamieri, D. Habich, et al., WO 9937641 (1999).

371. D. Habich, S. Bartel, R. Endermann, et al., Proceedings of the 38th Interscience Conference on Antimicrobial Agents and Chemotherapy, San Diego, Abstr. F-129 (1998).

372. S. Bartel, R. Endermann, W. Guamieri, et al., Proceedings of the 38th Interscience Conference on Antimicrobial Agents and Chemotherapy, San Diego, Abstr. F-130 (1998).

373. A. Stolle, D. Habich, B. Riedl, et al., U.S. Pat. 5,869,659 (1999).

374. M. J. Gein, D. K. Hutchinson, D. A. Alwine, et al., *J. Med. Chem.*, **41**, 5144 (1996).

375. R. C. Gadwood, L. M. Thomasco, E. A. Weaver, et al., Proceedings of the 39th Interscience

Conference on Antimicrobial Agents and Chemotherapy, San Francisco, Abstr. 571 (1999).

376. R. C. Gadwood, L. M. Thomasco, and D. J. Anderson, U.S. Pat. 5,977,373 (1999).

377. S. Bartel, R. Endermann, W. Guamieri, et al., Proceedings of the 37th IUPAC Congress, Berlin, Germany, Abstr. SYN-2-140 (1999).

378. S. Raddetz, S. Bartel, W. Guamieri, et al., WO 9940094 (1998).

379. S. Bartel, W. Guamieri, D. Habich, et al., WO 9937652 (1999).

380. J. A. Tucker, D. A. Allwine, K. C. Grega, et al., J. Med. Chem., 41, 3727 (1998).

381. A. N. Pae, H. Y. Kim, H. J. Joo, et al., Bioorg. Med. Chem. Lett., 9, 2679 (1999).

382. A. N. Pae, S. Y. Kim, H. Y. Kim, et al., Bioorg. Med. Chem. Lett., 9, 2685 (1999).

383. S. Bartel, R. Endermann, W. Guamieri, et al., Proceedings of the 39th Interscience Conference on Antimicrobial Agents and Chemotherapy, San Francisco, Abstr. F-565 (1999).

384. R. Tokuyama, Y. Takahashi, et al., Chem. Pharm. Bull. (Tokyo), 49, 347, 353, 361 (2001).

385. M. C. Dodd, W. B. Stillman, N. Roys, and C. Crosby, J. Pharmacol. Exp. Ther., 82, 11 (1944).

386. W. B. Stillman and A. B. Scott, U.S. Pat. 2,416,234 (1947).

387. G. D. Drake and K. J. Hayes, U. S. Pat. 2,759,931 (1956).

388. R. E. Bambury in M. E. Wolff, Ed., Burger's Medicinal Chemistry, 4th ed., John Wiley & Sons, New York, 1979, p. 65.

389. H. E. Paul, V. R. Ells, F. Kopko, and R. C. Bender, J. Med. Chem., 2, 563 (1960).

390. K. Tatsumi, T. Ou, H. Yoshimura, and H. Tsukamoto, Chem. Pharm. Bull., 19, 330 (1971).

391. J. Olivard, G. M. Rose, G. M. Klein, and J. P. Heotis, J. Med Chem., 19, 729 (1976).

392. P. L. Olive and D. R. McCalla, Cancer Res., 35, 781 (1975).

393. S. Swaminathan, R. C. Wong, G. M. Lower, and G. T. Bryan, Proc. Am. Assoc. Cancer Res., 18, 187 (1977).

394. D. R. McCalla, A. Ruvers, and C. Kaiser, J. Bacteriol., 104, 1126 (1970).

395. S. Swaminathan, G. M. Lower, and G. T. Bryan, Proc. Am. Assoc. Cancer Res., 19, 150 (1978).

396. F. Meissner and E. Schwiedersen, U.S. Pat. 2,762,799 (1956).

397. M. J. Katul and I. N. Frank, J. Urol., 101, 320 (1970).

398. E. A. Falco, G. H. Hitchings, P. B. Russell, and H. VanderWerf, Nature, 164, 107 (1949).

399. J. J. Burchall and G. H. Hitchings, Mol. Pharmacol., 1, 126 (1965).

400. B. Roth, E. A. Falco, G. Hitchings, and S. Bushby, J. Med. Pharm. Chem., 5, 1103 (1962).

401. A. P. Ball and E. T. Wallace, J. Int. Med. Res., 18 (1974).

402. G. H. Hitchings and S. R. M. Bushby, 5th Int. Congr. Biochem., 1961, p. 165.

403. V. Herbert, J. Infect. Dis., 128 (Nov. Suppl.), 433 (1973).

404. Proceedings of the 41st Interscience Congress on Antimicrobial Agents and Chemotherapy (ICAAC), September 22–25, 2001, Chicago, IL, Posters F-1023 to F-1039

405. A. Mehta, S. Arora, B. Das, A. Ray, S. Rudra, and A. Rattan, PCT WO 02/06278 (2001).

# CHAPTER FOURTEEN

# β-Lactam Antibiotics

Daniele Andreotti
Stefano Biondi
Enza Di Modugno
GlaxoSmithKline Research Center
Verona, Italy

## Contents

1 Introduction, 608
2 Mechanism of Action, 610
  2.1 Bacterial Cell Wall of Gram-Positive and Gram-Negative Strains, 610
  2.2 Action of β-Lactams, 611
3 Mechanism of Resistance to β-Lactams, 613
  3.1 Acquisition and Spread of Resistance to β-Lactams, 613
  3.2 Modification of Native PBPs: Mosaic Gene Formation, 616
    3.2.1 Methicillin-Resistant Staphylococci (MRS): Mechanism and Genetics, 617
  3.3 β-Lactamase Enzymes: Biological Evolution of PBP and β-Lactamases, 618
  3.4 Classification and Relevance of β-Lactamases in Clinics, 618
    3.4.1 Hydrolysis of Different Classes of β-Lactams, 619
  3.5 β-Lactam Resistance Attributed to Decreased Outer Membrane Permeability and Active Efflux, 621
    3.5.1 Porin Loss in Enterobacteria, 621
    3.5.2 Intrinsic Resistance to β-Lactams in Pseudomonas aeruginosa, 621
4 Clinical Application (Hospital and Community), 623
  4.1 Pharmacokinetic Properties, 623
  4.2 Side Effects and Drug Interactions, 624
  4.3 Indications for Use, 625
5 History and Discovery of β-Lactams, 628
  5.1 Penicillin, 628
    5.1.1 Synthesis of Penicillin, 629
    5.1.2 Derivatives of 6-Amino-Penicillanic Acid (6-APA): Biological Activity and Structure-Activity Relationship (SAR), 630
      5.1.2.1 C6-Substituted Penicillins, 634
    5.1.3 Other Modified Penicillins, 637

*Burger's Medicinal Chemistry and Drug Discovery*
Sixth Edition, Volume 5: Chemotherapeutic Agents
Edited by Donald J. Abraham
ISBN 0-471-37031-2 © 2003 John Wiley & Sons, Inc.

5.2 Cephalosporins and Synthetic Analogs:
Oxacephems, Carbacephems, 637
    5.2.1 Derivatives of C7-
        Aminocephalosporanic Acid, 638
    5.2.2 Classification, Biological Activity, and
        Structure-Activity Relationship of Ceph-
        alosporins, 639
    5.2.3 The Total Synthesis of Cephalosporin
        C, 645
        5.2.3.1 The Penicillin Sulfoxide-
            Cephalosporin Conversion, 646
    5.2.4 C7-α-Substituted Cephalosporins, 649
        5.2.4.1 C7-α-Methoxylated Derivatives,
            650
        5.2.4.2 C7-α-Formamido Derivatives,
            651
    5.2.5 Oxacephalosporins (1-Oxadethia-
        cephems), 652
    5.2.6 Carbacephalosporins (Carbacephems), 654
5.3 Penems, 654
    5.3.1 Synthesis, 654
        5.3.1.1 Woodward's Phosphorane
            Route, 654
        5.3.1.2 Extension of the Phosphorane
            Route, 654
        5.3.1.3 Alternative Methods for the
            Synthesis of Penems, 656
    5.3.2 Biological Properties, 658
5.4 Monobactams and Nocardicins, 659
    5.4.1 Nocardicins, 660
        5.4.1.1 Derivatives of 3-Amino
            Nocardicinic Acid (3-ANA) and
            the Synthesis of Nocardicins, 660
        5.4.1.2 Biological Activity and SAR, 663
        5.4.1.3 Formadicins, 664
    5.4.2 Monobactams, 664
        5.4.2.1 Structure Determination and
            Synthesis, 665

5.4.2.2 Biological Activity of Natural
    Products and Synthetic
    Derivatives, 665
5.4.2.3 Alternative N1-Activating
    Groups, 666
5.5 Carbapenems and Trinems, 668
    5.5.1 Discovery of Carbapenems, 668
        5.5.1.1 Natural Products: Occurrence,
            Structural Variations, and
            Chemistry, 670
        5.5.1.2 Synthesis of Natural
            Carbapenems, 671
        5.5.1.3 Synthesis of
            1β-Methylcarbapen-
            ems, 676
    5.5.2 Trinems and Policyclic Carbapenems,
        684
    5.5.3 Biological Properties of Carbapenems
        and Trinems, 690
5.6 β-Lactamase Inhibitors, 694
    5.6.1 Discovery of Clavulanic Acid, 695
    5.6.2 Sulbactam, 697
    5.6.3 Tazobactam, 697
    5.6.4 6-Heterocyclylmethylene Penems, 698
    5.6.5 Carbapenems, 699
    5.6.6 Penams and Penam Sulfones, 700
    5.6.7 Miscellaneous β-Lactams, 700
6 Recent Developments in β-Lactams, 702
    6.1 Anti–Gram-Positive β-Lactams, 703
        6.1.1 Cephalosporins, 703
        6.1.2 Carbapenems and Trinems, 703
    6.2 Antibacterial Broad-Spectrum β-Lactams, 708
    6.3 Orally Active Carbapenems/Trinems, 709
    6.4 β-Lactamase Inhibitors, 710
7 Current Trends Driving Industry, 712
    7.1 Inhibitors of Protein Export in Bacteria, 712
8 Web Site Addresses and Recommended Reading
    for Further Information, 714

# 1  INTRODUCTION

The history of β-lactam antibiotics is generally considered to begin with Alexander Fleming and his observation in 1928 that a strain of the mould *Penicillium* produced an antibacterial agent that was named penicillin (1).

β-Lactam antibiotics have been in clinical use for over 50 years. These pharmaceuticals remain to the present the most commonly used antibiotics, and their introduction for the clinical use is one of the most important, if not the most important, medical developments of the century.

The success of penicillin G had an unexpected effect: an increasing number of *Staphylococcus aureus* isolates resistant to penicillin G in London hospitals spread progressively all around the world. It was discovered that certain bacterial enzymes, the β-lactamases, had the ability to hydrolyze the β-lactam ring of these antibiotics, thus diminishing their effectiveness as antibacterial agents (2). Based on the rapidity of the widespread resistance to penicillins, it became obvious that the hydrolytic enzymes could potentially destroy the utility of this potent class of antibiotics. The pharmaceutical industry has invested to iden-

Evolution of the β-lactam class

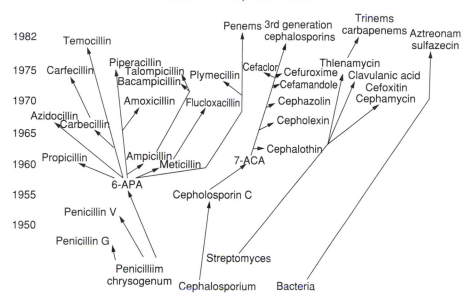

tify novel β-lactams over the past 40 years in an attempt to keep ahead of the continuous evolution of new β-lactamases with altered hydrolytic properties.

Two approaches were undertaken: development of agents stable to hydrolysis by the major β-lactamases, and identification of potent inhibitors for these enzymes. The number of compounds, even when limited to therapeutically important compounds, based on the β-lactam ring is now huge. As new compounds have been discovered and developed they have been described by their trivial names, penicillins, cephalosporins, clavulanic acid, and so on, although the situation has become so complicated that it has become necessary to group them according to their chemical structures.

The agents having current or potential therapeutic use can be represented by seven principal chemical skeletons:

- *Penams*. Penicillins are bicyclic structures where the β-lactam ring is fused with a five-membered thiazolidine ring.
- *Penems*. These differ from penams by the presence of a double bond between the C2 and C3 positions. No natural penems have so far been described, but many have been

synthesized in the expectation that they would combine desirable properties of penicillins and carbapenems.

- *Carbapenams and carbapenems*. These compounds differ from penams and penems by the presence of a carbon atom at position 1. Many natural and synthetic members of the group have been described, and some of them are currently used in the clinics.
- *Cephems, oxacephems, and carbacephems*. These compounds are characterized by the presence of a β-lactam ring fused with a six-membered unsaturated ring, having at position 1 a sulfur, an oxygen, or a carbon atom, respectively. In particular, the Cephem class has been very prolific in generating good antibiotics, which have found extensive application in the treatment of bacterial infections.
- *Oxapenams*. This class is generally characterized by a very weak antibacterial activity but has found therapeutic applications as inhibitors of bacterial β-lactamases. Its skeleton differs from that of penams by the presence of an oxygen atom at position 1.
- *Monobactams*. The monocyclic β-lactam is the simplest structure still retaining antibacterial activity.

β-Lactam skeletons

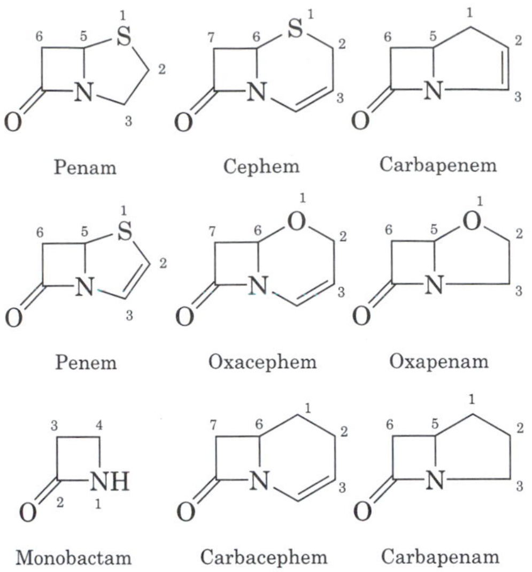

Penam            Cephem            Carbapenem

Penem            Oxacephem            Oxapenam

Monobactam            Carbacephem            Carbapenam

**Scheme 14.1.**

## 2 MECHANISM OF ACTION

### 2.1 Bacterial Cell Wall of Gram-Positive and Gram-Negative Strains

β-Lactams inhibit the synthesis of peptidoglycan (3, 4), which is the major polymer of the bacterial cell wall. Peptidoglycan maintains the cell shape and protects against osmotic forces. That capability depends on its netlike structure, composed of long sugar chains crosslinked by peptides.

In Gram-positive organisms, the peptidoglycan typically forms a thick layer external to the cytoplasmic membrane and accounts for 50% of the dry weight of the bacterium, whereas in Gram-negative bacteria and mycobacteria, peptidoglycan is a thinner layer (5), or hydrated gel (6), sandwiched between the outer and cytoplasmic membranes (Fig. 14.1). To manufacture peptidoglycan, bacteria first synthesize precursor molecules of uridine diphosphate (UDP) linked to N-acetylmuramic acid pentapeptide (7). The sequence of the pentapeptide varies among species, but the two terminal residues are D-alanine, and the third amino acid, usually L-lysine or m-diaminopimelic acid, bears a free amino group, which may be substituted with a "bridge" of additional amino acids (Table 14.1).

These precursors that are produced in the cytoplasm, are then transferred from UDP to an isopropenoid carrier located in the cytoplasmic membrane. An N-acetylglucosamine residue is added to give "disaccharide pentapeptide," which is transported across the membrane and inserted into the existing sacculus by transglycosylation and transpeptidation. Transglycosylation extends sugar chains by attaching the muramyl residue of a new precursor to a free N-acetylglucosamine residue on the existing peptidoglycan. Transpeptidation crosslinks adjacent sugar chains through their pentapeptides (Fig. 14.1). Peptidoglycan transglycosylase and D-alanyl-D-alanine transpeptidase activities often reside at separate sites on the same proteins, and most bacteria possess multiple peptidoglycan transglycosylases/transpeptidases, each with a different role (9, 10). These enzymes contain lipophilic sequences, which anchor them to the cytoplasmic membrane and allow particular activities to be localized. Their interplay ensures the maintenance of cell shape (10, 11) and their relative amounts may vary with the growth rate (12). Additional enzymes hydrolyze crosslinked peptidoglycan and are collectively called autolysins or peptidoglycan hydrolases (13). D-Alanyl-D-alanine carboxypeptidases hydrolyze D-alanine groups from pentapeptides that have served as amino donors (Fig. 14.2) and probably have a regulatory role in peptidoglycan synthesis.

More drastic hydrolysis of peptidoglycan are undertaken by (1) D,D-endopeptidases, which cleave the crosslink synthesized by D-alanyl-D-alanine transpeptidases; (2) N-acetylmuramyl-L-alanine amidase, which cleaves peptides from muramyl residue; and (3) lytic glycosylase, β-N-acetylglucosaminidases, and β-N-acetylmuraminidases, all of which hydrolyze the sugar backbone of the peptidoglycan (Fig. 14.2).

Many species have multiple autolysins (14, 15); for example, Escherichia coli has at least 11 different enzymes (16), catalyzing five modes of cleavage. The transglycosylases, transpeptidases, and autolysins are often (but not always) membrane bound, allowing their activities to be localized. Their main role is to

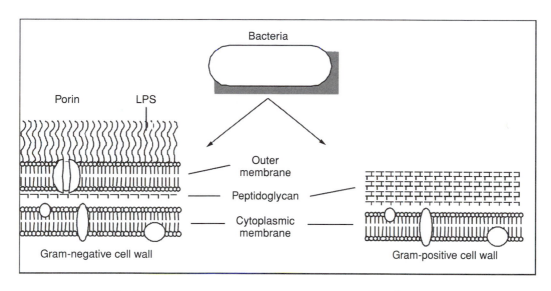

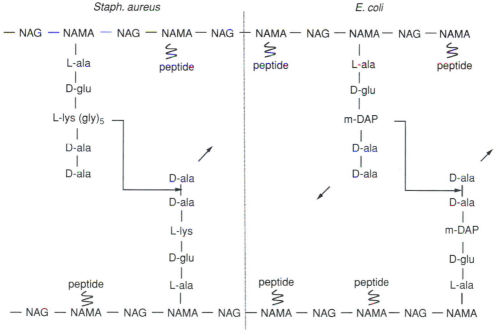

NAG = N-acetylglucosamine
NAMA = N-acetylmuramic acid

**Figure 14.1.** Bacterial cell wall of Gram-positive and Gram-negative strains.

provide sites for insertion of new peptidogly-cans (17), but they may also assist in daughter cell cleavage, autolysis, and flagellar extrusion. Degraded peptidoglycan fragments are absorbed and recycled.

## 2.2 Action of β-Lactams

β-Lactams inhibit D-alanyl-D-alanine transpeptidase activity by acylation, forming stable esters with the opened lactam ring attached to

**Table 14.1  Bridge in the Peptidoglycans of Bacteria**[a]

| Organism | Amino Acid 3, Chain 1[b] | Bridge[b] | Amino Acid 4, Chain 2[b] |
|---|---|---|---|
| *Staphylococcus aureus* | L-Lys | (Gly)$_5$ | D-Ala |
| *Staphylococcus epidermidis* | L-Lys | (Gly)$_{4-5}$–L-Ala | D-Ala |
| *Streptococcus agalacticae* | L-Lys | L-Ala (or L-Ser)–L-Ala | D-Ala |
| *Streptococcus* group G | L-Lys | L-Ala (or L-Ser)–L-Ala | D-Ala |
| *Streptococcus salivarius* | L-Lys | Gly–L-Thr | D-Ala |
| *Streptococcus bovis* | L-Lys | L-Ser–L-Ala–L-Thr | D-Ala |
| *Enterococcus faecium* | L-Lys | D-Asn | D-Ala |
| *Enterococcus faecalis* | L-Lys | (L-Ala)$_{2-3}$ | D-Ala |
| *Streptococcus* spp. | L-Lys | Gly–L-Ala | D-Ala |
| Gram-negative bacilli | *m*-Dap[c] | Direct; no bridge | D-Ala |

[a]See Ref. 8.

[b]Positions of amino acids 3 and 4 and of the bridge are indicated in Fig. 14.1.

[c]*m*-DAP, *m*-diaminopimelic acid.

the hydroxyl group of the enzyme's active site (Fig. 14.3). These esters mostly have hydrolysis half-lives of several hours, so inactivation is effectively permanent (18). β-Lactams have no effect on transglycosylation. They do inhibit the D-alanyl-D-alanine carboxypeptidases and some peptidoglycan endopeptidases, but these are regulatory enzymes whose inactivation is of little consequence.

The ability of β-lactams to inhibit the D-alanyl-D-alanine trans- and carboxypeptidase depends on conformational similarity between the amide (O=C—N) bond of the β-lactam ring and the peptide link of D-alanyl-D-alanine (3, 19). Considerable effort was put into comparing the bond angles of these structures and showing how they resembled one another. However, D-alanyl-D-alanine transpeptidases are more versatile than was once thought, and attempts to find conformational identity now seem misplaced. In particular, many D-alanyl esters, as well D-alanyl-D-alanine, are substrates, despite the fact that these have very different molecular shapes (20). Inhibition of transpeptidases alone theoretically may cause bacteriolysis, by yielding a wall that cannot withstand osmotic forces (21, 22). In practice, however, autolysins accelerate lysis by destroying the existing wall (13, 17, 23). Evidence for this view comes from the occurrence

**Figure 14.2.** Action of autolysins. 1, β-*N*-acetylglucosaminidase; 2, lytic transglycosylase and β-*N*-acetylmuraminidase (lysozyme); 3, D,D-endopeptidase; 4, *N*-acetylmuramyl-L-alanine amidase; 5, L,D-carboxypeptidase; 6, D,D-carboxypeptidase. Although lytic transglycosylase and β-*N*-acetylmuraminidase (lysozyme) cleave the same target bond, the former enzyme additionally catalyzes transfer of the glycosyl bond to the 6'-hydroxyl group of the same muramic acid, yielding 1,6-anhydromuramic acid. NacGlc, *N*-acetylglucosamine; NacMur, *N*-acetylmuramic acid; *m*-DAP, *m*-diaminopimelic acid.

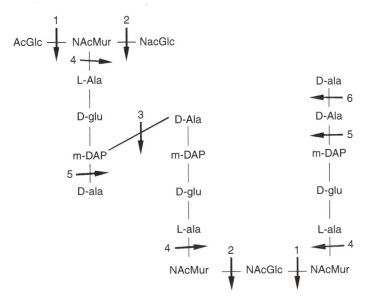

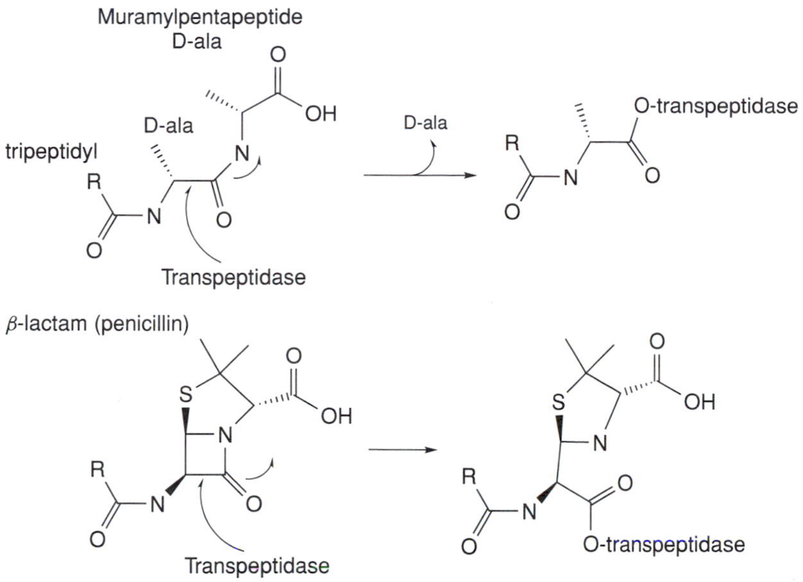

**Figure 14.3.** Mechanism of action of β-lactams.

of "tolerant" mutants, particularly of Gram-positive cocci that are inhibited but not killed by β-lactams (24–28). These organisms lack various autolysins (29), although no single loss is consistently associated with the phenotype (13); a few tolerant isolates have minor penicillin-binding protein (PBP) changes (30) (See Tables 14.2 and 14.3). β-Lactams, or compounds accumulating through their action, may "trigger" deregulation of autolysins, thus initiating lysis. Tomasz (13) suggested that lipoteichoic acid was the trigger in pneumococci, but the general relevance of this observation is uncertain. Insight into the complexity of the relationships that arise is provided by the realization that mutational loss of cytoplasmic N-acetylmuramyl-L-alanine amidase autolysins is the cause of β-lactamase de-repression in C. freundii (31).

## 3  MECHANISM OF RESISTANCE TO β-LACTAMS

Among the various mechanisms of acquired resistance to β-lactam antibiotics, resistance attributed to production of β-lactamases by the cell is the most prevalent. Alteration in the preexisting PBPs; acquisition of a novel PBP

insensitive to β-lactams; changes in the outer membrane proteins of Gram-negative organisms; and active efflux, which prevent these compounds from reaching their targets, can also confer resistance.

### 3.1  Acquisition and Spread of Resistance to β-Lactams

Bacteria can become resistant to some antibiotics mutating existing genes, but in many cases a new gene must be acquired.

Although bacteria can acquire new genes by bacteriophage transduction or by transformation (uptake of DNA from the external environment), these types of transfer tend to occur mainly between members of the same species. Such narrow host range resistance transfer can be important clinically. For example, transformation is probably spreading PBP-type resistance genes between clinical isolates of *Streptococcus pneumoniae*, a common cause of bacterial pneumonia. A major clinical problem is the transfer of resistance genes across genus and species lines. Such broad host range transfer is most likely to be mediated by conjugation (transfer of DNA through a pore formed in the fused membranes of two bacteria). There are two type of

**Table 14.2  PBPs of Gram-Negative Bacteria**

| PBP | Molecular Mass (kDa)[a] | Activity[b] | Essentiality | Reference(s) |
|---|---|---|---|---|
| Gram-negative rods | | | | 18, 34 |
| 1a | 88–118 | Tg/Tp | Yes | 35 |
| 1b | 81–103 | Tg/Tp | Yes | 35–37 |
| 2 | 63–78 | Tg/Tp | Yes | 38, 39 |
| 3 | 59–66 | Tg/Tp | Yes | 35 |
| 4 | 44–51 | D,D-Cp | No | 40, 41 |
| 5 | 40–44 | D,D-Cp | No | 40, 42 |
| 6 | 38–41 | D,D-Cp | No | 42, 43 |
| *Neisseria* spp. | | | | 44–47 |
| 1 | 90–108 | —[c] | Yes | |
| 2 | 59–63 | — | Yes | |
| 3 | 44–48 | — | No | |
| *Haemophilus influenzae* | | | | 32–34 |
| $1^d = 1a^e$ | 90 | — | — | |
| $2^d = 1b^e$ | 84 | — | — | |
| $3^d = 2^e$ | 75 | — | — | |
| $4^d = 3a^e$ | 68 | — | Yes | |
| $5^d = 3b^e$ | 64 | — | Yes | |
| $6^d = 4^e$ | 48 | — | — | |
| $7^d = 5^e$ | 41 | — | — | |
| $8^d = 6^e$ | 27 | — | — | |
| *Acinetobacter baumannii* | | | | 34, 48 |
| 1 | 94 | — | — | |
| 2 | 64 | — | — | |
| 3 | 57 | — | — | |
| 4 | 51 | — | — | |
| 5 | 38 | — | — | |

[a]Ranges indicate values found by different workers and for different species within the group.

[b]Activity: Tg/Tp, transglycosylase/D-alanyl-D-alanine transpeptidase, with these activities residing at different sites on the protein; D,D-Cp, D-alanyl-D-alanine carboxypeptidase. Catalytic activities of these enzymes are shown in Figs. 14.2 and 14.3.

[c]—, Not determined.

[d]Nomenclature of Ref. 32.

[e]Nomenclature of Ref. 33.

conjugative elements, plasmids and chromosomal elements known as conjugative transposons.

*Plasmids.* The best-studied type of conjugative element is the plasmid. Plasmids that transfer themselves by conjugation must carry a number of genes encoding proteins needed for the conjugation process itself (*tra* genes). Thus, self-transmissible plasmids are usually at least 25 kb (kilobases). Some plasmids that cannot transfer themselves can still be transferred by conjugation because they are mobilized by self-transmissible plasmids. Such plasmids are called mobilizable plasmids and can be much smaller than self-transmissible plasmids because they need only one or

two genes (*mob* genes) that allow to them to take advantage of the transfer machinery provided by the other plasmids. Self-transmissible or mobilizable plasmids can acquire and transmit multiple antibiotic resistance genes. There are two ways plasmids can acquire multiple resistance genes. One way is to acquire sequential transposon insertions. However, most multiresistance plasmids apparently did not arise in this way. A newly discovered type of integrating element, called integron, is probably responsible for evolution of many of the plasmids that carry multiple resistance genes (Fig. 14.4).

Integrons, like transposons, are linear DNA segments that insert into DNA. Unlike

**Table 14.3   PBPs of Gram-Positive Bacteria**

| PBP | Molecular Mass (kDa) | Activity[a] | Essentiality | Notes | Reference(s) |
|---|---|---|---|---|---|
| *Enterococcus faecalis* | | | | | 49 |
| 1 | 115 | —[b] | — | Overproduced in resistant mutants | |
| 2 | — | — | — | | |
| 3 | — | — | — | | |
| 4 | — | — | — | | |
| 5 | 74 | — | — | | |
| 6 | 43 | D,D-Cp | — | | |
| *Enterococcus faecium/Enterococcus hirae* | | | | | 49 |
| 1 | 126 | — | | Overproduction causes resistance to penicillins in clinical isolates | 50–53 |
| 2 | 92.5 | — | | | |
| 3 | 90 | Tg/Tp | Yes | | |
| 4 | 86 | — | — | | |
| 5 | 79–81[c] | Tg/Tp | Yes[d] | | |
| 6 | 40 | | | | |
| *Rhodococcus equi* | | | | | 54 |
| 1 | 59 | — | — | Replaced by 3′ in imipenem mutants | 54 |
| 2 | 56 | — | — | | |
| 3 | 43 | — | — | | |
| 4 | 26 | — | — | | |
| *Staphylococci* | | | | | 55–57 |
| 1 | 85 | | Debated | | 57 |
| 2 | 81 | Tg/Tp | Yes | Resolves into two components | 56, 57 |
| 3 | 75 | Tg/Tp | Yes | | 56 |
| 4 | 45 | D,D-Cp | No | | 58–60 |
| 2′ = 2a (MRS only) | 76.4 | Tg/Tp | —[d] | Not related to PBP-2 | 61–63 |
| *Streptococcus pneumoniae* | | | | | 47, 64–66 |
| 1a | 98[e] | — | Yes | PBP-1C is low affinity form of PBP-1A | |
| 1b | 95 | — | Yes | | |
| 2a | 81[e] | — | Yes | | |
| 2b | 79[e] | — | Yes | | |
| 2x | 85[e] | — | Yes | | 67, 68 |
| 3 | 43 | D,D-Cp | No | | 69 |
| *Streptococcus pyogenes* | | | | | 30, 70 |
| 1 | 95 | — | — | | |
| 2 | 83 | — | — | | |
| 3 | 76 | — | — | | |
| 4 | 48 | — | — | | |

[a]Activity: Tg/Tp, transglycosylase/D-alanyl-D-alanine transpeptidase, with these activities residing at different sites on the protein; D,D-Cp, D-alanyl-D-alanine carboxypeptidase. Catalytic activities of these enzymes are shown in Figs. 14.2 and 14.3.

[b]—, Not determined.

[c]Range of values reported by different authors.

[d]PBP-2′ of MRS and PBP-5 of *E. faecium* can functionally replace all other PBPs in these species.

[e]Molecular weights vary in resistant isolates where the PBPs are encoded by mosaic genes.

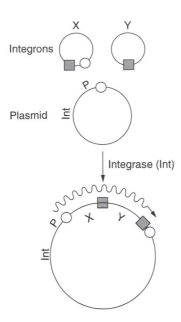

**Figure 14.4.** Integration of two integrons, carrying promoterless resistance genes X and Y, into a plasmid. The plasmid supplies the promoter (P) and integrase (Int). ○, 8-bp consensus site; ■, 59- to 60-bp shared sequence.

transposons, however, integrons integrate at a single site and do not encode a transposase. The first integron inserted into a plasmid is inserted at an 8-bp consensus site on the plasmid. Integration is mediated by an integrase encoded on plasmid that receives the integron. The plasmid must also provide a promoter because most integrons contain promoterless antibiotic resistance genes.

*Conjugative Transposons.* A second type of conjugative element has been discovered and characterized only within the past decade: the conjugative transposon. Conjugative transposons are usually located in the bacterial chromosome and can transfer themselves from the chromosome of the donor to the chromosome of the recipient. They can also integrate into plasmids. Their mechanism of transfer is different from that of other known gene transfer elements. They excise themselves from the donor genome to form a covalently closed circle that does not replicate. This circular intermediate transfers with a mechanism similar to that of plasmids. In the recipient, the circular intermediate integrates

in the chromosome by a mechanism that does not duplicate the target site. Conjugative transposons are probably responsible for at least as much resistance gene transfer as plasmids, especially among Gram-positive bacteria, and they have a very broad host range. Conjugative transposons can transfer not only among species within the Gram-positive group or within the Gram-negative group but also between Gram-positive and Gram-negative bacteria. Conjugative transposons were overlooked for a long time because they are located in the chromosome and thus cannot be detected as easily as plasmids. Moreover, one of the few groups of bacteria in which conjugative transposons have not been found is the group of species closely related to *E. coli*. Like self-transmissible plasmids, conjugative transposons can mediate the transfer of the other DNA. They can mobilize coresident plasmids, and some can also mediate the transfer of unlinked segments of chromosomal DNA.

## 3.2 Modification of Native PBPs: Mosaic Gene Formation

Modification of the normal PBPs is the sole cause of resistance to β-lactams in pneumococci (71–73) and other α-hemolytic streptococci (74) and, together with impermeability, is a major component of non-β-lactamase-mediated "intrinsic" resistance in *Neisseria* spp. (44, 45, 47) and *Haemophilus influenzae* (32, 75). Target modification may contribute to β-lactam resistance in some *Acinetobacter* isolates (76, 77), but its general importance is uncertain in this genus, where β-lactamases and impermeability are also involved (78). Although modifications of normal PBPs are very rare, in some instances they have been thought to cause resistance in Staphylococci, enterococci, enterobacteria, and pseudomonas (79, 80). Critically, the organisms in which PBP modification has proved important (i.e., α-hemolitic streptococci, hemophili, and *Neisseria* spp.) are transformable with naked DNA (47). They can acquire fragments of PBP genes from other organisms with inherent or acquired β-lactam resistance and insert this into their own PBP genes. The resulting "mosaic" gene encodes β-lactam-resistant PBPs. The transformation can occur when sequence divergence between the incoming DNA and native material is less than 25–305 bp, permitting interspecies recombination (47). Restriction and mismatch repair systems do not limit exchange as much as might be anticipated and may even serve to correct errors during replication. Mosaic gene formation is understood for *Neisseria meningitidis*, where penicillin-insensitive strains have alteration only in PBP-2. In particular, the PBP-2 gene has regions of normal meningococcal sequence interspersed with inserts with 14 to 23% divergences (81, 82). These inserts closely resemble the corresponding section of the PBP-2 genes of *Neisseria flavescens* and *Neisseria cinerea*, which are throat commensals that are inherently unsusceptible to penicillin.

Penicillin-resistant gonococci have mosaic PBP-2 genes with inserts identical to those in resistant meningococci (46, 82). In addition, however, gonococci with high level penicillin resistance commonly have modifications of PBP-1 and reduced permeability (44, 45, 83, 84). Mosaic gene formation has also been extensively studied in pneumococci (47). High level benzylpenicillin resistance [minimum inhibitory concentration (MIC) values of more than 1 μg/mL, compared to MIC values of ≤0.06 μg/mL for "normal" isolates] requires modification, through mosaic gene formation, of each PBP-1a, -2x, -2a, and -2b, although low level resistance (MIC values of 0.12 to 1 μg/mL) arise when only three of these are altered (47). Mosaic PBP-1a, -2x, and -2b genes of different resistant isolates have diverse inserts, whereas corresponding PBP gene sequences from susceptible isolates are highly conserved (65, 67, 68, 85). It is concluded that resistance has evolved on many separate occasions. The source of the foreign DNA is uncertain. Recombination events between penicillin-resistant *Streptococcus pneumaniae* and *Streptococcus mitis* are implicated, but their direction is uncertain (71). Mosaic gene formation has not yet been proved in *H. influenzae* isolates with non-β-lactamase-mediated ampicillin resistance but seems likely, inasmuch as the resistance is associated with reduced PBP affinity (32, 75, 86, 87) and is readily transformable in the laboratory (88). The PBPs affected are 68- and 63-kDa proteins, numbered 3a and 3b or 4 and 5, respectively.

Alternative routes of peptidoglycan synthesis allow a different form of target-mediated resistance, compared with that described above, and are important in Staphylococci and *E. faecium*.

### 3.2.1 Methicillin-Resistant Staphylococci (MRS): Mechanism and Genetics.

Staphylococci normally have two essential PBPs, 2 and 3 (55). These components are inhibited by many β-lactams, including methicillin. MRS retain β-lactam-sensitive PBP-1 to PBP-3 but manufacture an additional transglycosylase/transpeptidase, PBP-2′ (PBP-2a), encoded by the *mecA* gene (61, 62, 89). PBP-2′ continues to function when PBP-1, -2, and -3 have been inactivated and, by itself, can yield a stable peptidoglycan, albeit with many fewer crosslinks than that of the normal wall (90). *mecA* by a chromosomal insert called the *mec* determinant (91, 92) is probably originated outside Staphylococci, perhaps through fusion between a β-lactam gene and a PBP gene (93). The *mec* determinant is not readily self-transmissible but has spread within *S. aureus* and several coagulase-negative species, indicating that some horizontal transfer is possible (94–96).

Expression of *mecA* and, consequently, of resistance varies among and within staphylococcal strains (63).

Many strains show "hetero-resistance," with only a small proportion of the cell being obviously resistant, whereas the huge majority remain apparently susceptible (97–99). This behavior reflects the genic background of the host staphylococcus, not genes on the *mec* determinant itself (100). Several chromosomal genes have been implicated in the control of *mecA*, including those in the *fem* cluster, particularly *femA* (101) as well as a recently described determinant, *chr** (102). Curiously, loss of β-lactamase from MRS causes *mecA* to be homogeneously expressed; the relation is unclear but implies a linkage in the control of the two-resistance determinant (103). Expression of *mecA* and of resistance is induced by β-lactams and is also influenced by the environmental conditions, being promoted by high osmolality, low temperature, and neutral to alkaline pH (104–106). The latter points bear on laboratory detection, which

Evolution of PBP and β-lactamases

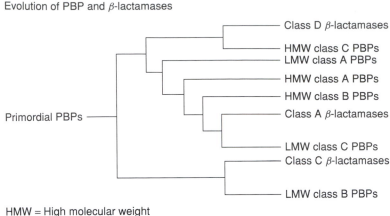

HMW = High molecular weight
LMW = Low molecular weight

is best achieved by testing with methicillin or oxacillin at 30°C on Mueller-Hinton or IsoSensitest agar supplemented with 5–7% NaCl. Tests with other β-lactams, at 37°C, or media with normal osmolality often fail to detect resistance, especially in coagulase-negative species.

*PBP-5 in Enterococcus Faecium.* Bypass resistance is also important in *E. faecium*, although rare in *E. faecalis*. *E. faecium* has six PBPs, with PBP-3 serving as the main D-alanyl-D-alanine transpeptidase. This protein is inhibited by penicillins and carbapenems, not by cephalosporins and monobactams. At low temperature (32°C), or following mutation, *E. faecium* can switch to using PBP5, a transpeptidase that has minimal affinity for any β-lactam. Acting alone, PBP-5 can manufacture a stable peptidoglycan (51, 107). Unlike PBP-2′ of MRS, PBP-5 is universal in *E. faecium* and its permanent expression requires only a regulatory mutation, not aquisition of foreign DNA. Many isolates of *E. faecium* have undergone this mutation and are consequently (and obviously) resistant to all β-lactams (52, 108).

### 3.3  β-Lactamase Enzymes: Biological Evolution of PBP and β-Lactamases

Study (109) of the evolution of β-lactamases and PBPs based on the multiple amino acid sequence alignment resulted in a scheme that revealed the relationship among these enzymes. The diagram "Evolution of PBP and β-lactamases" shows that, among various

PBPs, the low molecular weight enzymes are the most related to β-lactamases.

The diverse classes of β-lactamases are more closely akin to the different PBP classes from which they arose than they are to each other. This is a further clear indication that β-lactamases originated from different groups of PBPs rather than from an immediate shared common ancestor. Class A β-lactamases clustered together with the class C low molecular weight PBPs and the class C β-lactamases group with the class B low molecular weight PBPs. It would appear that the PBPs with the simpler domain structure (low molecular weight) were used as templates for β-lactamase evolution. On the other hand, class D for β-lactamases represents an exception to this observation. They cluster together with the high molecular weight class C PBPs. The unique role of these PBPs in bacteria is in signal transduction, to initiate the synthesis of PBPs with low affinity for β-lactams and class A β-lactamases. Therefore, the functional and structural relationships of these PBPs to the other classes of PBPs are quite distant, as was also demonstrated by multiple-sequence analysis (109).

### 3.4  Classification and Relevance of β-Lactamases in Clinics

From a clinical perspective, the most important β-lactamases are those that threaten the use of the most beneficial β-lactam–containing antimicrobial agents. These agents argu-

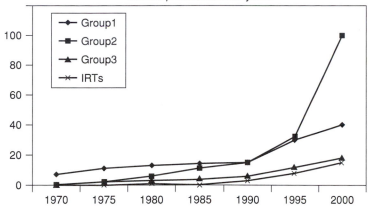

**Figure 14.5.** Frequency of occurrence of β-lactamases in clinical isolates.

ably include the oral cephalosporins and the β-lactamase inhibitor combination, amoxicillin-clavulanic acid, that are used frequently for community infections such as otitis media, and the more potent agents such as the expanded-spectrum cephalosporins and carbapenems that are used extensively for nosocomial infections. Most of the β-lactam agents introduced after 1980 were selected for pharmaceutical development, in part, because they were stable to hydrolysis by the major β-lactamases. However, enzymes that hydrolyzed these agents emerged shortly after introduction of new members of these classes of agents. Among the more relevant groups of enzymes recently identified are the ESBLs (extended-spectrum β-lactamases) group 1, plasmid-mediated cephalosporines group 2, the inhibitor-resistant TEM (IRT) and SHV variants, and the group 3 (class B) metallo-β-lactamases. The frequency with which they have appeared is shown in Fig. 14.5, again emphasizing the almost exponential growth of ESBLs.

**3.4.1 Hydrolysis of Different Classes of β-Lactams.** Many attempts have been made to classify β-lactamases. Most have been based on phenotypic properties, with great weight being placed on whether cephaloridine is hydrolyzed more rapidly than benzylpenicillin or vice versa; whether enzyme inhibition is achieved by cloxacillin, clavulanate, aztreonam, or p-chloro-mercuribenzoate; and whether the enzyme is chromosomally encoded or plasmid-mediated. Phenotypic features are useful and convenient, in that they can be related to an-

tibiogram data from clinical laboratories, because they permit all β-lactamases to be classified, and because they allow "weighting" of minor structural changes that critically alter the hydrolytic spectrum. The first phenotypic classification that achieved wide acceptance was proposed by Richmond and Sykes in 1973 (110) (Table 14.4).

Superimposed on this there grew to be a convention whereby plasmid-mediated β-lactamases (which fell into classes III and V of the Richmond-Sykes scheme) were given a three-letters and one-number code (111). By 1984, these plasmid-mediated types could be clustered into three groups, according to whether they were most active against benzylpenicillin, oxacillin, or carbenicillin (112). Broad-spectrum enzymes corresponded to Richmond-Sykes class III, but the oxacillinases and carbenicillinases, although very different from each other, both fell into class V. Other problems also emerged in the Richmond-Sykes scheme (113). Class I, it was realized, included two different classes of cephalosporinases: (1) the clavulanate-insensitive, molecular weight 40-kDa, chromosomally encoded enzymes from most enterobacteria; and (2) the 30-kDa cefuroxime, hydrolyzing β-lactamases from *Proteus vulgaris* (114, 115). Finally, many enzymes placed in class II were proved to be transposon-mediated and accordingly were transferred to classes III and V. These deficiencies prompted Bush to propose a major organization in 1989 (113), with a revision in 1995 (116).

**Table 14.4  Correspondence Between Molecular and Phenotypic Classification of β-Lactamases**

| Structural Class (Ambler) | Functional Group (Bush) | Richmond–Sykes class | Activity[a] | | | | | | | Inhibition[b] | | |
|---|---|---|---|---|---|---|---|---|---|---|---|---|
| | | | Penicillin | Carbenicillin | Oxacillin | Cephaloridine | Cefotaxime | Aztreonam | Imipenem | Clavulanate | Aztreonam | EDTA |
| Serine β-lactamases | | | | | | | | | | | | |
| A | 2a | NL[c] | +++ | + | – | ± | – | – | – | ++ | – | – |
| | 2b | II and III | +++ | + | + | ++ | – | – | – | ++ | – | – |
| | 2be | IV | +++ | + | + | ++ | ++ | ++ | – | ++ | – | – |
| | 2br | NL | +++ | + | + | + | – | – | – | – | – | – |
| | 2c | II and V | ++ | +++ | + | + | – | – | – | + | – | – |
| | 2e | Ic | ++ | ++ | – | ++ | ++ | ++ | – | ++ | – | – |
| | 2f | NL | ++ | + | ? | + | + | ++ | ++ | + | – | – |
| C | 1 | I, except Ic | ++ | ± | – | +++ | + | – | – | – | – | – |
| D | 2d | V | ++ | + | +++ | + | – | – | – | V | ++ | – |
| Undetermined[d] | 4[d] | NL | ++ | ++ | ++ | V | V | – | – | – | – | – |
| Zinc β-lactamases | | | | | | | | | | | | |
| B | 3 | NL | ++ | ++ | ++ | ++ | ++ | – | ++ | – | – | ++ |

[a]Activity, +++, preferred substrate (highest $V_{max}$); ++, good substrate; +, hydrolyzed; ±, barely hydrolyzed; –, stable; V, varies within group; ?, uncertain.
[b]Inhibition, ++, strong inhibitor of all members of class; +, moderate inhibition; V, inhibition varies within the class; –, negligible inhibition.
[c]NL, not listed.
[d]None of Bush's group 4 enzymes has yet been sequenced. They are assumed to be serine types because they lack carbapenemase activity and are not inhibited by EDTA.

The Bush numbering classifies β-lactamases based on their relative hydrolysis of penicillin, oxacillin, carbenicillin, cephaloridine, expanded-spectrum cephalosporins, and imipenem and on their susceptibility to inhibition by clavulanate, aztreonam, and ethylenediaminetetraacetic acid (EDTA) (Table 14.4). Four "functional groups" are recognized, with group 2 enzymes (clavulanate-sensitive, aztreonam-resistant penicillinases) being divided into six subgroups and with the enzymes in group 4 only partially characterized. This scheme is now to be preferred over the Richmond-Sykes classification. Increasingly, however, β-lactamases are classified by primary structure and sequence homology, as first proposed by Ambler (117), rather than by their phenotypic properties (Table 14.4).

Sequence-based classifications reflect fundamental relationships, unlike phenotypic classification, and cannot be distorted by mutations that alter substrate specificity and inhibitor susceptibility. Only four molecular classes, A to D, are recognized, each with distinct sequence motifs (118). Classes A, C, and D constitute serine active-site enzymes, and class B has the zinc types. The distinction in chromosomally encoded and plasmid-mediated β-lactamases is dubious to biochemists because each molecular class and functional group contains plasmidic enzymes as well as those normally encoded by chromosomes (116). Nevertheless, the distinction between chromosomal β-lactamases and those coded by plasmids (or other inserts) remains fundamental to the clinical microbiologist; chromosomal β-lactamases are ubiquitous in species, whereas plasmid-mediated types are less universal but cross interspecies lines. In Fig. 14.6 is reported a schematic representation of the mechanism of catalysis of serine β-lactamases.

## 3.5  β-Lactam Resistance Attributed to Decreased Outer Membrane Permeability and Active Efflux

In contrast to Gram-positive organisms, in which β-lactams have unhindered access to their PBP targets, in Gram-negative bacteria the outer membrane acts as a barrier to these compounds. The balance between antibiotic influx and clearance, whether ascribed to hydrolysis or trapping, determines the susceptibility or resistance of the cell, given that permeability barriers alone rarely produce significant levels of resistance. In *E. coli* and other Gram-negative bacteria, β-lactams diffuse across the outer membrane primarily through water channels consisting of a specific class of proteins, termed porins. *E. coli* produces at least two porin types, OmpF and OmpC. Mutations that cause reduced expression or alternation of OmpF and/or OmpC result in decreased susceptibility to many β-lactams (119).

**3.5.1 Porin Loss in Enterobacteria.** Permeability mutants of enterobacteria can easily be isolated in the laboratory and are found to lack one or more porins (120–124). For *E. coli*, MICs of penicillins, cefoxitin as well as narrow- and extended-spectrum cephalosporins, increase in a stepwise manner as first OmpC and OmpF are lost (125). Loss of OmpF has the greater effect because it forms larger pores than OmpC (126). MICs of carbapenems aminothiazolyl cephalosporins are only scarcely affected by these losses, probably because disruption in the periplasm is minimal. Despite this ease of selection, porin-deficient enterobacteria are rarely found in clinical settings, probably because they are nutritionally disadvantaged (122). Nevertheless, such mutants have been selected during cefoxitin therapy of infections caused by *Klebsiella pneumoniae*, producing ESBLs (127, 128) and during imipenem therapy of an infection caused by a β-lactamase-depressed *Enterobacter cloacae* strain (129).

In some cases, the combination of impermeability with weak β-lactamase production confers clinical resistance, whereas the β-lactamase alone fails to do so (128, 130, 131). Porin-deficient mutants of *Serratia marcescens*, lacking a 40-kDa protein, seem to be selected more often than analogous mutants of other enterobacteria (122).

**3.5.2 Intrinsic Resistance to β-Lactams in *Pseudomonas aeruginosa*.** Impermeability-mediated resistance is widely perceived as a greater problem in *P. aeruginosa* than in enterobacteria but is not completely understood. Most *P. aeruginosa* isolates are 32- to 48-fold less susceptible to penicillins and cephalosporins than hy-

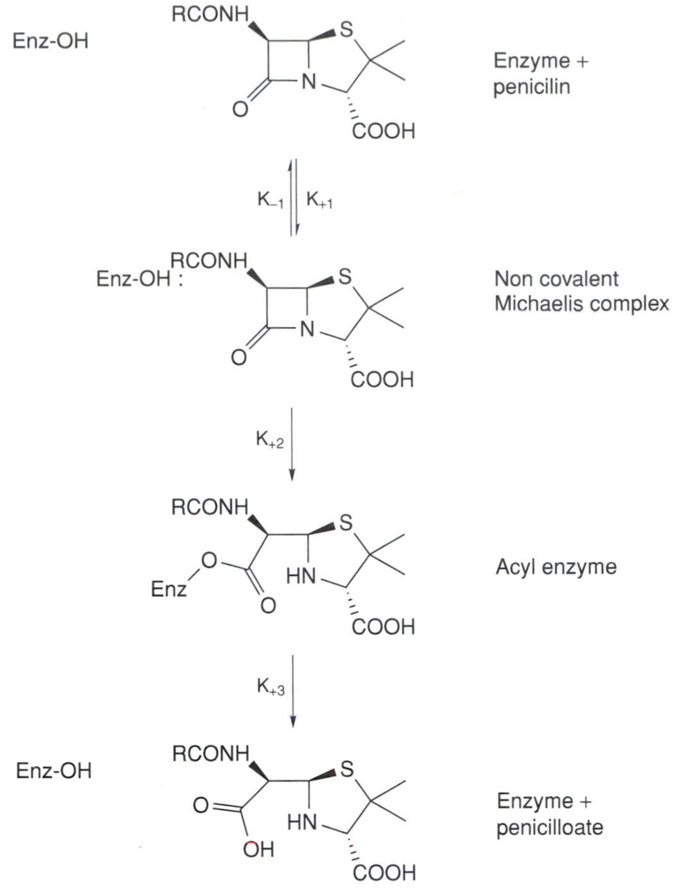

Enz-OH

Enzyme + penicilin

$K_{-1} \| K_{+1}$

Enz-OH :

Non covalent Michaelis complex

$K_{+2}$

Enz

Acyl enzyme

$K_{+3}$

Enz-OH

Enzyme + penicilloate

$K_{+1}$, $K_{-1}$, $K_{+2}$, $K_{+3}$ are the rate costants; $K_{+3}$ is extremely slow.

**Figure 14.6.** Mechanism of action of serine β-lactamases.

Release of the β-lactam, is often in a fragmented form, as shown. Enz-OH = serine-enzyme.

persusceptible laboratory mutants, which are inferred, from their lack of other mechanisms, to be hyperpermeable (132, 133).

P. aeruginosa is characterized by an intrinsic resistance to a wide variety of antimicrobial agents, including β-lactams, β-lactam inhibitors (134), tetracyclines, quinolones, and chloramphenicol. Although this resistance has most often been caused by a highly impermeable outer membrane, it is now recognized to result from the synergy between a unique tripartite energy-driven efflux system (MexAB-OprM) with wide substrate specificity (135–137) and low outer membrane permeability. Recently, it has been shown that MexAB-OprM directly contributes

to β-lactam resistance through efflux of this class of antibiotics (138). The emergence in P. aeruginosa of resistance specific to imipenem is associated with the loss of the specific channel OprD (139). It appears that imipenem preferentially uses this channel to cross the outer membrane, which under normal circumstances probably transports basic amino acids (140).

The carbapenem uptake is reduced irrespective of whether OprD porin of P. aeruginosa is lost. This causes an increase of the MIC values of imipenem from 1–2 to 8–32 μg/mL and those of meropenem from 0.25–0.5 to 2–4 μg/mL (141). This loss has no effect on resistance to β-lactam classes (142). The greater effect on imipenem than on meropenem reflects the slight liability of imipenem to the class C β-lactamase of *P. aeruginosa*. This enzyme gives greater protection when permeability is reduced by porin loss (143). Meropenem has the capacity to escape the β-lactamase (141, 144) and so is less affected by the porin loss. Selection of OprD-deficient mutants is a problem in imipenem therapy, occurring in 15 to 20% of patients treated for *Pseudomonas* infections (145, 146).

## 4  CLINICAL APPLICATION (HOSPITAL AND COMMUNITY)

### 4.1  Pharmacokinetic Properties

*Penicillins.* For most penicillins, renal excretion is the major route of elimination. With few exceptions, renal clearance tends to exceed the glomerular filtration rate. In all cases studied, there is evidence of active renal tubular secretion, in that the coadministration of probenecid invariably causes a decrease in renal clearance, an increase in serum concentrations, and/or a prolongation in serum-life. By the same token, renal dysfunction has a profound effect on elimination rate, serum levels, and drug accumulation. Although homodialysis or peritoneal dialysis can be effective in some patients with end-stage renal failure, the rate of drug removal is not simply related to serum protein binding. Even though antimicrobial activity in bile is usually high compared to that in serum, biliary excretion contributes only insignificantly to the elimination of most penicillins. Notable exceptions are metampicillin, nafcillin, and the ureidopenicillins. β-Lactam cleavage appears to be a common metabolic pathway among penicillins. Penicilloic acid formations, although variable, may represent as much as 50% of total clearance of a given agent. Part of this may occur in the gastrointestinal lumen or during initial transit through the gut wall and liver after oral administration. Although there is evidence of N-deacylation for some penicillins, only traces of 6-aminopenicillanic acid or penicilloic acid are found in human urine.

*Cephalosporins.* In general, the cephalosporin and cephamycin antibiotics are a family of parenterally administered antimicrobial agents. Notable orally absorbed cephalosporins include cephaloglycine, cephalexin, cephradine, cefadroxil, cefaclor, and cefratrizine. Urinary excretion is the primary route of elimination for all cephalosporins, although biliary excretion and metabolism contribute to the elimination of several. For most, urinary excretion involves both glomerular filtration and renal tubular secretion. As might be expected, kidney dysfunction or the coadministration of probenecid causes dramatic changes in drug elimination. Cephalosporin antibiotics are remarkably stable *in vivo*, with few exceptions. The metabolism of cephalosporins has generally not been studied with discriminating assay technique. Those that have been studied in this fashion have shown insignificant metabolism.

The pharmacological properties of the parenteral penicillins and cephalosporins are summarized in Table 14.5 (147–149). In general, β-lactam antibiotics penetrate most areas of the body except the eye, prostate (except aztreonam), and uninflamed meninges. Although entry into the cerebrospinal fluid (CSF) is satisfactory with most of the penicillins and carbapenems, only certain cephalosporins reach therapeutic levels in the CSF. Protein binding varies from 2% to 98%. The serum half-life of penicillins is short, ranging from 0.5 to 1.5 h. The half-life of cephalosporins, and especially the third-generation cephalosporins, tends to be longer. The half-life of ceftriaxone, the longest of all these agents, is 8 h. Most penicillins, cephalosporins, and aztreonam are removed intact by renal excretion, and those require dose modification in patients with renal impairment. However, the isoxazolyl penicillins (nafcillin and oxacillin), the ureidopenicillins, and a few cephalosporins (ceftriaxone and cefoperazone) have partial biliary excretion. Dose modification is needed less often for the ureidopenicillins and these cephalosporins, and is not needed at all for the isoxazolyl penicillins.

*Carbapenems.* Imipenem is hydrolyzed in the renal tubules, producing nephrotoxic metabolites, and requires coadministration of cilastain, an inhibitor of the dehydropeptidase enzyme (DHP-1). This prolongs its half-life and preserves renal function. Meropenem is

**Table 14.5  Pharmacokinetic Properties of the β-Lactams**

| Antimicrobial | Protein Binding | $t_{1/2}$ (h) | Cerebrospinal Fluid Penetration | Route of Excretion |
|---|---|---|---|---|
| Penicillin G | 55 | 0.5 | Yes | Renal |
| Nafcillin | 87 | 0.5 | Yes | Biliary and renal |
| Oxacillin | 93 | 0.5 | Yes | Biliary and renal |
| Ampicillin | 17 | 1 | Yes | Renal |
| Carbenicillin | 50 | 1.1 | Yes | Renal |
| Ticarcillin | 50 | 1.2 | Yes | Renal |
| Piperacillin | 50 | 1.3 | Yes | Renal and biliary |
| Mezlocillin | 50 | 1.1 | Yes | Renal and biliary |
| Azlocillin | 20 | 0.8 | Yes | Renal and biliary |
| Aztreonam | 56 | 1.8 | Yes | Renal |
| Imipenem/cilastatin | 20/40 | 1 | Yes | Renal |
| Meropenem | 2 | ~1 | Yes | Renal |
| Cephalothin | 71 | 0.6 | | Renal |
| Cefazolin | 80 | 1.8 | | Renal |
| Cephradine | 10 | 0.7 | | Renal |
| Cefamandole | 75 | 0.8 | | Renal |
| Cefonicid | 98 | 4.5 | | Renal |
| Cefuroxime | 35 | 1.3 | Yes | Renal |
| Cefoxitin | 70 | 0.8 | | Renal |
| Cefotetan | 90 | 3.5 | | Renal |
| Cefotaxime | 35 | 1 | Yes | Renal |
| Ceftizoxime | 30 | 1.7 | Yes | Renal |
| Ceftriaxone | 90 | 8 | Yes | Renal and biliary |
| Cefoperazone | 90 | 2 | | Renal and biliary |
| Ceftazidime | 17 | 1.8 | Yes | Renal |
| Cefepime | 20 | 2.1 | Yes[a] | Renal |

[a]Animal studies only.

not susceptible to renal hydrolysis and therefore does not require coadministration of an inhibitor of hydrolysis (150). Both carbapenems require dose reductions in patients with renal impairment.

## 4.2  Side Effects and Drug Interactions

A remarkably favorable chemotherapeutic index has been one of the most important qualities of the animal and human pharmacology of the first penicillins. Aside from problems of local irritation and of neurotoxicity with extremely large doses, these antibiotics have minimal direct toxicity. For nearly 20 years, allergic or immune-mediated reactions represented the only common complications of the use of penicillins. However, the development of newer penicillins and cephalosporins, although providing some of the safest and most important antibiotics currently used, has introduced new direct toxicity problems.

Pain and sterile inflammatory reactions at the site of intramuscular injection are among the most common local effects of therapy with the penicillins and cephalosporins. With benzylpenicillin these reactions appear to be related to the concentration of the antibiotic, and this also may be the case for the other penicillins (151).

Phlebitis develops in many patients receiving intravenous penicillins or cephalosporins (152). The relative risk of phlebitis among the different cephalosporins is unsettled (153–155).

Gastrointestinal side effects are among the most common adverse reactions to oral treatment with the β-lactam antibiotics. Some degree of gastric irritation is reported in 2–7% of patients (156). The discomfort occasionally results in discontinuation of the drug, although serious consequences are rare. Diarrhea is also a relatively common problem, which may be of mild or life-threatening severity. At least three types are distinguishable: (1) nonspecific diarrhea, (2) pseudomembranous colitis, and (3) presumed ischemic colitis.

Some of the most important reactions to β-lactam antibiotics result from the involvement of humoral or cellular immunity. Almost all patients receiving benzylpenicillin develop antibodies to it (151, 157). The frequency of allergic reactions to penicillins has been reported as 0.7–10%, and that to cephalosporins has been estimated to be 0.8–7.5% (151, 158, 159). Humoral hypersensivity produces some of the most serious adverse effects, including urticaria, angioedema, bronchospasm, and anaphylaxis. The frequency of anaphylactic reactions to penicillin has been estimated to be 0.045–0.15% (151, 160). A history of previous exposure to penicillin is common but not invariable (151). Skin testing may be helpful in distinguishing those patients at risk for anaphylaxis, especially if testing is conducted with both major and minor antigenic determinants of the penicillin (158, 161).

Skin rashes during the administration of a penicillin or cephalosporin are relatively common, occurring in approximately 2% of cases (156, 162, 163).

Aztreonam, however, is felt to be safe for administration to penicillin-allergic patients. Imipenem produces seizures in 0.4–1.5% of patients, predominantly affected by renal insufficiency with underlying central nervous system (CNS) disease. Meropenem, however, appears to have a much lower potential to cause seizures (164–167).

### 4.3 Indications for Use

Table 14.6 summarizes some of the more important reasons for using β-lactam agents. In addition to being the drugs of choice for treatment of a variety of infections caused by specific bacteria, the β-lactam drugs are the stalwart agents for empirical treatment of febrile neutropenic patients (168). Unfortunately, the choice of empirical therapy today is less straightforward than in the past, and no obviously best regime exists. Important factors influencing the choice of empirical therapy for febrile neutropenia include the type of cancer

chemotherapy being used, expected severity and duration of neutropenia, the presence of an indwelling long-term catheter, previous use of prophylactic antibiotics or gut decontamination, the patients' symptoms, and the bacterial resistance pattern of the hospital.

Although there has been a striking increase in Gram-positive infections, the morbidity and mortality caused by Gram-negative infections (especially Pseudomonas) suggest that empirical therapy continues to cover these organisms. Supportive evidence exists for the empiric use of β-lactam drugs in combination with an aminoglycoside. Piperacillin, mezlocillin, azlocillin, ticarcillin, and ceftazidine have shown similar efficacy, with response rates of 55–88% (169–177). Also, a convenient regimen of once-daily dosing of ceftriaxone and amikacin proved as effective as multiple daily-dosing regimens (178). Toxicity for aminoglycoside use led to trials with combinations of two β-lactam agents. Good results were achieved with multiple regimens, including carbenicillin plus cephalotin, cefoperazone plus aztreonam, cefoperazone plus mezlocillin, ceftazidime plus piperacillin, and ceftazidime plus ticarcillin/clavulanate (169–171). However, double β-lactam therapy remains controversial because of the possibilities of increased selection of resistant organisms, drug antagonism, prolongation of neutropenia, and potentiation of bleeding disorders.

β-Lactam monotherapy has also been investigated. Supportive data exist for the use of ceftazidime and imipenem (170–172, 179–183). However, the increase in frequency of resistant Gram-negative infections, and the finding by some workers that ceftazidime treatment is inadequate, suggest cautions in using ceftazidime monotherapy (170–174). Aztreonam in combination with an agent covering Gram-positive organisms (e.g., vancomycin) is of great utility in patients with penicillin allergy (184, 185). β-Lactam antibiotics have played a key role in improving the care of both immunocompromised and nonimmunocompromised patients. These agents are bactericidal, well tolerated, widely distributed throughout the body, and, most important, clinically effective.

**Table 14.6  Recommendations for the Use of β-Lactams**

| β-Lactam | Recommendation | Organism or Condition | Notes |
|---|---|---|---|
| Penicillin G | Primary indication | *Streptococcus pyogenes, Streptococcus pneumoniae,* and enterococcal infections | |
| | Drug of choice | Treponemal infection, prevention of rheumatic fever. | |
| | Can be used | Puerperal infection: anaerobic streptococci, *Streptomyces agalactiae,* clostridial infection, and infection attributed to mouth flora: Gram-positive cocci, Gram-negative cocci, and *Actiomyces* | |
| Penicillinase-resistant penicillins | Can be used | Susceptible *Staphylococcus aureus* infection | |
| Aminopenicillins | Recommended | Prevention of endocarditis | |
| | Can be used | Infection of respiratory tract in areas with low prevalence of β-lactamase and *Haemophilus influenzae.* Urinary tract infection | |
| Extended-spectrum | Primary indication | *Pseudomonas* spp. | |
| | Can be used | Infection of urinary tract, respiratory tract, and bone with Gram-negative bacilli and mixed aerobic/anaerobic infections | |
| Combination drugs | Can be used | Mixed bacterial infection: community and hospital acquired pneumonia, especially if aspiration, intra-abdominal and gynecological infections, osteomyelitis, and skin-structure infection | |
| Timentin (ticarcillin and clavulanic acid) | | | |
| Unasyn (ampicillin and sulbactam) | Can be used | Mixed bacterial infection: intra-abdominal: obstetric, gynecologic, soft-tissue, bone infection | |
| Zosyn (Piperacillin and Tazobactam) | Can be used | Mixed bacterial infection: lower respiratory tract, intra-abdominal, skin, and soft tissue infection | |
| Monobactams Aztreonam | Can be used | Urinary tract, lower respiratory tract, skin-structure, and intra-abdominal infections, patients with penicillin allergy | |
| Carbapenems | Can be used | Resistant Gram-negative bacilli infection with ESBL. | |
| Imipenem | Drug of choice | Nosocomial infection, when multiresistant Gram-negative bacilli or mixed infections are suspected | |
| Meropenem | Can be used | Nosocomial infection when multiresistant Gram-negative bacilli or mixed infections are suspected. Meningitis in pediatric population older than 3 months | |

| Agent | Recommendation | Indications | Comments |
|---|---|---|---|
| First-generation cephalosporins | Drug of choice | Prophylaxis of surgical procedures | |
| | Can be used | Infection attributed to *S. aureus* or nonenterococcal streptococci (e.g, skin and soft tissue infections, pharyngitis) | |
| Second-generation cefuroxime | Can be used | Respiratory tract infections: pneumonia epiglottis, complicated sinusitis, soft-tissue infections, bacteremia | Good *H influenzae* coverage |
| Cefoxitin | Drug of choice | Pelvic inflammatory disease (+ doxycicline) | |
| | Recommended | Prophylaxis of colorectal surgery | |
| | Can be used | Mixed aerobic/anaerobic infections: intra-abdominal infections, skin and soft-tissue infections, including diabetic foot infections and decubitus ulcers | |
| Cefotetan | | Mixed aerobic/anaerobic infections: intra-abdominal infections, skin and soft-tissue infections, including diabetic foot infections and decubitus ulcers | Slightly less anaerobic coverage but better Gram-negative coverage than cefoxitin |
| Third-generation Ceftriaxone | Drug of choice | *Neisseria gonhorroea* chancroid, Lyme disease if neurological involvement, carditis, arthritis, or refractory late constitutional symptoms | Once-daily dosing, good CSF penetration |
| | Recommended | Meningitis attributed to *H. influenzae*, *Neisseria* meningitis, and penicillin-resistant *S. pneumoniae* | |
| | Can be used | Nosocomial infection caused by a sensitive Gram-negative baccilli: pneumoniae, wound, and complicated urinary tract infections Home treatment of chronic infections | |
| Ceftazidime | Should be used | Infections that are likely attributable to *Pseudomonas aeruginosa* | |
| | Can be used | Empiric treatment of febrile neutropenia | |
| | Drug of choice | Meningitis attributed to *P. aeruginosa* | |
| Cefoperazone | Can be used | Empiric treatment of febrile neutropenia | Moderate antipseudomonal activity |
| Cefepime | Can be used | Infections of lower respiratory tract, urinary tract, skin and soft tissue, and in female reproductive tract | |

627

# 5  HISTORY AND DISCOVERY OF β-LACTAMS

One of the fundamental milestones in medicinal chemistry is represented by the knowledge acquired during the studies carried out on a small and only apparently simple four-membered ring, the β-lactam. The history of β-lactam can be compared to a sky full of stars and the following paragraphs describe the discovery of the most important β-lactam classes that have had a remarkable impact in the antibacterial field, either as antibacterials or as β-lactamase inhibitors.

## 5.1  Penicillin

The antibiotic era began in 1940 with the demonstration made by Florey and his colleagues (186) that penicillin, the substance discovered by Fleming in 1929, was a chemotherapeutic agent of unprecedented potency (187). This story has been recounted many times (188–191) and is here presented only as a brief overview. The observation by Alexander Fleming that partial lysis of Staphylococci colonies occurred on plates contaminated with *Penicillium notatum* was made in 1928 (192).

Although Fleming cultured the mould and named the active "mould broth filtrate" penicillin, its therapeutic potential was not recognized at that time.

The life-saving capacity of penicillin was first demonstrated in animals (186), and in humans the following year, when the small quantities of penicillin available necessitated the recovery of the antibiotic from the urine of the patient for reuse (193). Once penicillin had been shown effective in humans, the search for improved production was initiated. War-torn Britain was thought unsuitable for such research, so key scientists moved to the United States to continue the effort alongside their American colleagues. Here, replacement of the surface cultures of the low yielding *P. notatum* with an aerated deep fermentation of a high yielding strain of *Penicillium chrysogenus*, and the use of corn-steep liquor as growth promoter, led to greatly improved penicillin production. This provided sufficient material for the treatment of infections associated with serious battle casualties during the end of

World War II and so confirmed the important role of penicillin in the saving of life.

Soon after the demonstration of the extraordinary efficacy in treating infections, penicillin was the subject of in-depth studies by groups on both sides of the Atlantic to solve the structure of penicillin. The correct structure of benzylpenicillin (Pen-G) was eventually identified as the fused β-lactam-thiazolidine bicyclic structure (1), as opposed to an

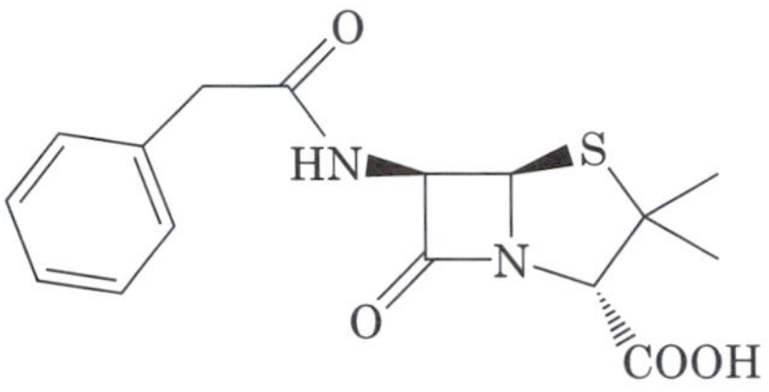

(1) Penicillin G (1929–1940)

oxazolone alternative, by an X-ray crystallographic analysis in 1945 (194).

Isolation of reasonably pure samples of penicillin revealed differences in chemical behavior and biological properties between the materials studied by different groups. This arose as a consequence of the different acyl side-chains (RCO) attached to the C6-amino group.

The source of the acyl side chains is the carboxylic acids present in the culture media producing that penicillin; thus, the phenylacetic acid in corn-steep liquor resulted in the phenylacetamido group (RCO = $PhCH_2CO$) of penicillin G (1). From among the many penicillins that have been isolated from *Penicillium* spp. fermentations, either unaided or resulting from the incorporation of added precursor acids ($RCO_2H$), several of the better-known examples are shown (Table 14.7) with their British and U.S. designations.

The next major advance was the isolation and characterization of the penicillin nucleus, 6-amino-penicillanic acid (6-APA) (2), in 1957 by the Beecham group at Brockham Park (203). This major breakthrough provided a source of chiral material for the preparation of semisynthetic penicillins.

Concurrently with identification of 6-APA (2) from biochemical investigations, Sheehan and Henerey-Logan concluded their total syn-

## Table 14.7  Biosynthetic Penicillins

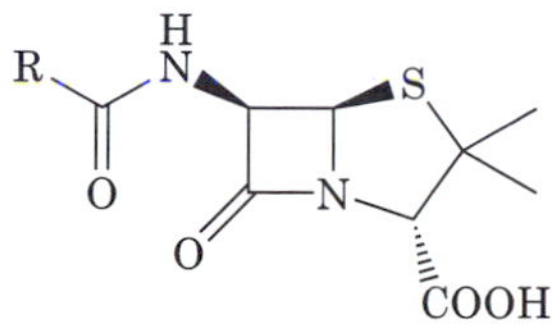

| (1) | Name | British Name | U.S. Name | R | Ref. |
|---|---|---|---|---|---|
| a | 2-Pentenyl penicillin | I | F | $CH_3CH_2CH{=}CHCH_2{-}$ | 195, 196 |
| b | Pentyl penicillin | Dihydro I | Dihydro F | $CH_3(CH_2)_4{-}$ | 196, 197 |
| c | Heptyl penicillin | IV | K | $CH_3(CH_2)_6{-}$ | 196 |
| d | Benzyl penicillin | II | G | $PhCH_2{-}$ | 198 |
| e | p-Hydroxybenzyl penicillin | III | X | $4\text{-HO-}PhCH_2{-}$ | 198 |
| f | (D)-4-Amino-4-carboxybutyl penicillin |  | N | $(\text{D})\text{-}HO_2C\text{-}CH(NH_2)(CH_2)_3{-}$ | 199 |
| g | (L)-4-Amino-4-carboxybutyl penicillin |  | iso-N | $(\text{L})\text{-}HO_2C\text{-}CH(NH_2)(CH_2)_3{-}$ | 200 |
| h | Phenoxymethyl penicillin |  | V | $PhOCH_2{-}$ | 197 |
| i | Butylthiomethyl penicillin | BT |  | $C_4H_9SCH_2{-}$ | 201 |
| j | Allylthiomethyl penicillin | AT | O | $CH_2{=}CHCH_2SCH_2{-}$ | 201 |
| k | p-Aminobenzyl penicillin |  | T | $4\text{-}H_2N\text{-}PhCH_2{-}$ | 202 |
| l | p-Nitrobenzyl penicillin |  |  | $4\text{-}O_2N\text{-}PhCH_2{-}$ | 202 |

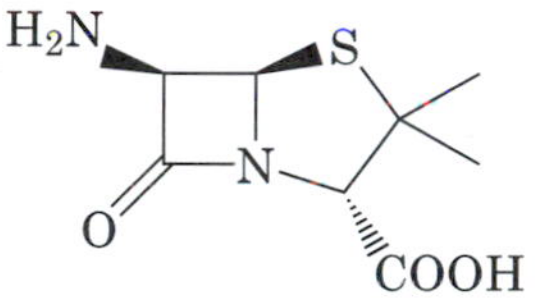

(2) 6-amino-penicillanic acid (6-APA)

thesis of penicillin (1) (204, 205) (see Section 5.1.1) and 6-APA (206, 207). However, this synthesis gave a low overall yield and lacked stereospecificity, so it could not compete with the biochemical route as a practical source of material for chemical modifications. Today, 6-APA (2), obtained from penicillin G by enzymatic cleavage of the phenylacetyl side chain, is a readily available bulk chemical that is used for the preparation of semisynthetic penicillins and related antibiotics.

### 5.1.1 Synthesis of Penicillin.
From the considerable synthetic effort of many groups that began in the 1940s, three major synthetic approaches to penicillin have been reported. The first one resulted from the already mentioned pioneering work of Sheehan, in which the β-lactam ring was formed by cyclization in

very low yield of the penicilloic methyl ester (3) with dicyclohexylcarbodiimide (DCCI) (204, 205).

In the Bose approach, formation of the β-lactam ring through cycloaddition of the dihydro-thiazoline (4) and the ketene derived

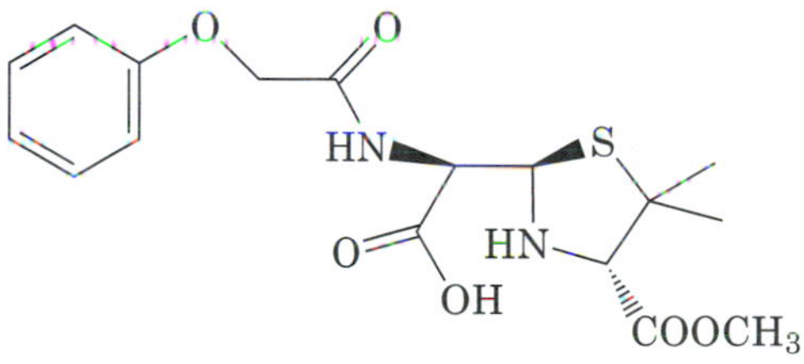

(3)

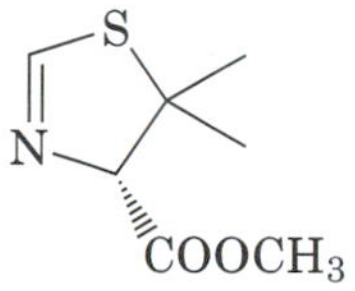

(4)

from azidoacetyl chloride gave the thermodynamically favored, unnatural trans β-lactam penicillin structure (5) (208).

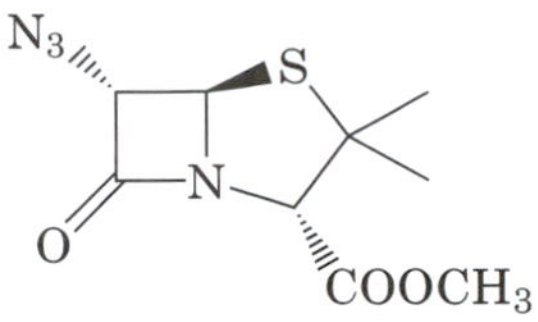

(5)

Subsequently, however, a method was reported for equilibration of the C6-stereochemistry, to give some of the *cis* β-lactam possessing the natural stereochemistry, by kinetic quenching of the anion generated from the derived Schiff base (6) (209).

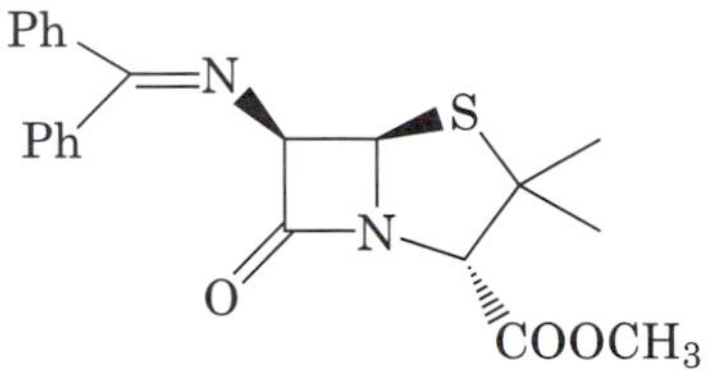

(6)

It was not until 1976 that the only stereoselective synthesis of penicillin was described by Baldwin (210). Thus, the peptide precursor (7), derived from cysteine and D-isodehydrova-

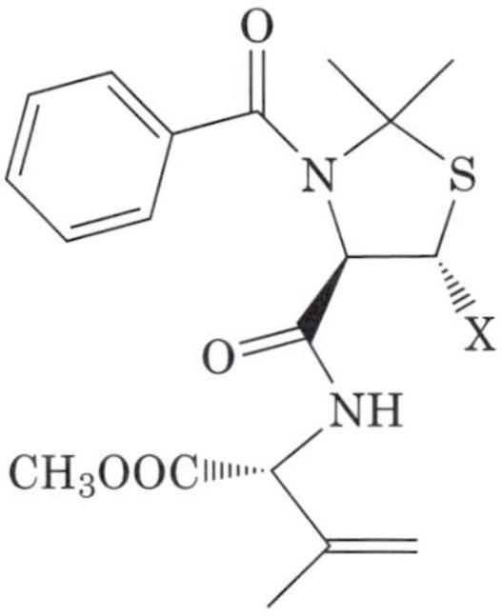

(7) X = H
(8) X = Cl

line methyl ester, was cyclized through the chloride (8) to the β-lactam (9).

Conversion of the latter by a multistep process then generated the sulfenic acid (10), which cyclized to provide the sulfoxide (11).

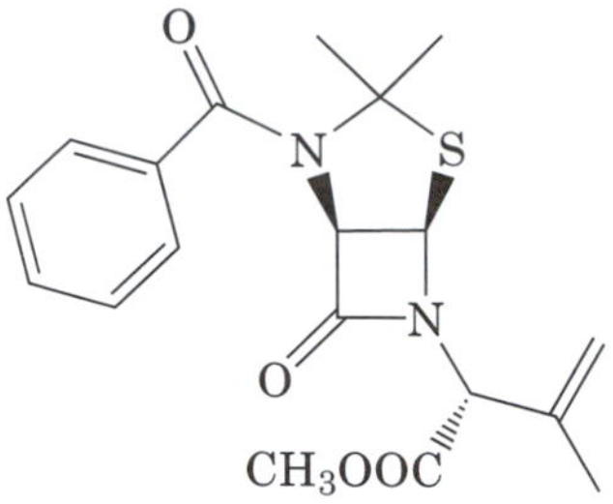

(9)

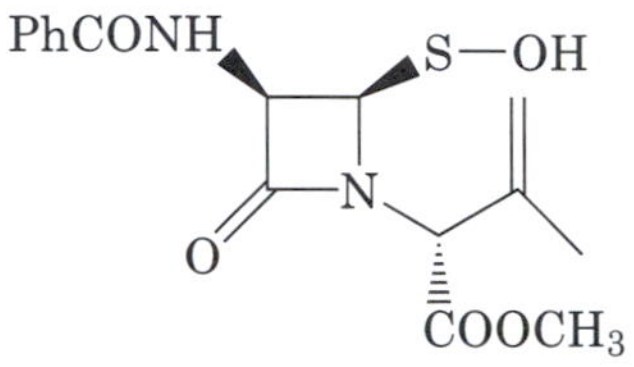

(10)

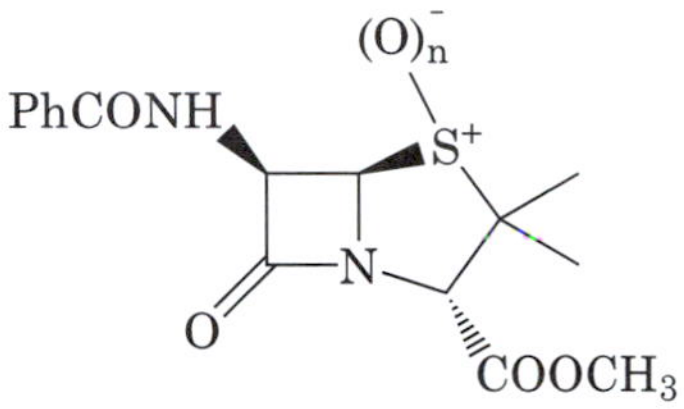

(11) n = 1
(12) n = 0

Finally, deoxygentation gave the penicillin ester (12).

**5.1.2 Derivatives of 6-Amino-Penicillanic Acid (6-APA): Biological Activity and Structure-Activity Relationship (SAR).** The availability of 6-APA (2) led to the preparation of many thousands of semisynthetic penicillins, through N-acylation of the C6-amino group. Most, if not all, of the methods used in peptide chemistry have been employed for this process (211). This increase in the scope of side-chain variation allowed significant improvements in the biological properties of the derived analogs, including greater stability to penicillinase (β-lactamase) and an expanded spectrum of antibacterial activity. Some early examples of semisynthetic penicillins to be studied in humans were substituted derivatives of the acid-stable penicillin V (**13**), such as phenethicillin (**14**) and propicillin (**15**), which, despite being

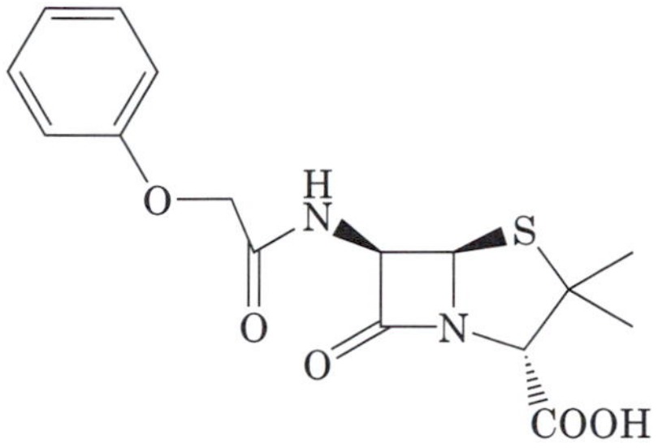

(**13**) Penicillin V

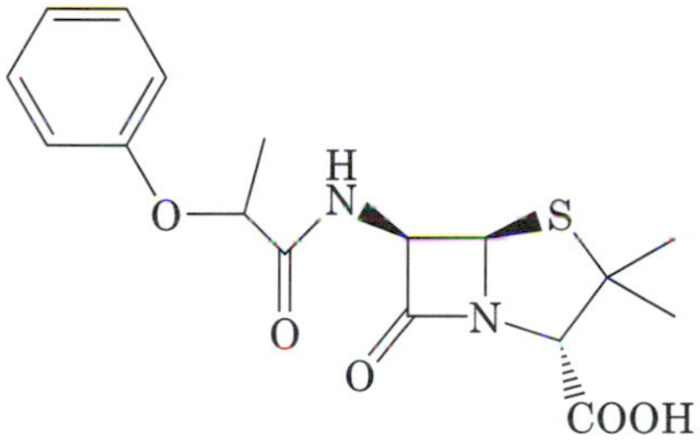

(**14**) Phenethicillin

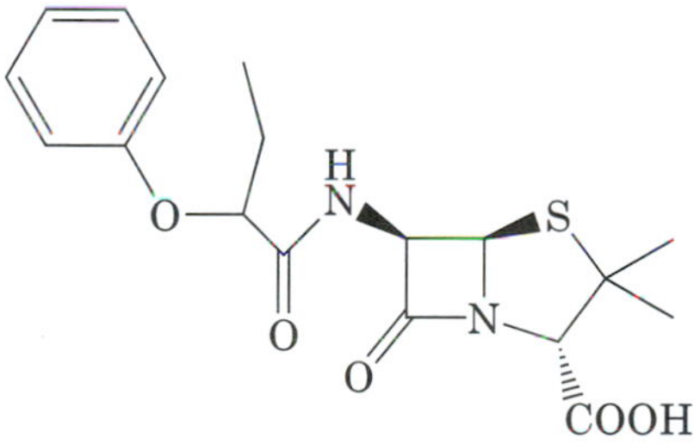

(**15**) Propicillin

mixtures of diastereoisomers, offered increased oral absorption, thus giving higher and more prolonged blood levels than those of penicillin V (212).

The fast widespread use of penicillin G and probably the misuse or abuse of this drug had already resulted in an increase in the occurrence of penicillinase (β-lactamases)-produc-

ing strains of *S. aureus*. By 1960 this was becoming a worldwide clinical problem (213). Often these strains of *S. aureus* were virulent and resistant to most other antibiotics, thus making them difficult to eradicate. Much early research was therefore directed at identifying new penicillins resistant to inactivation by these penicillinases. Methicillin (**16**) was in-

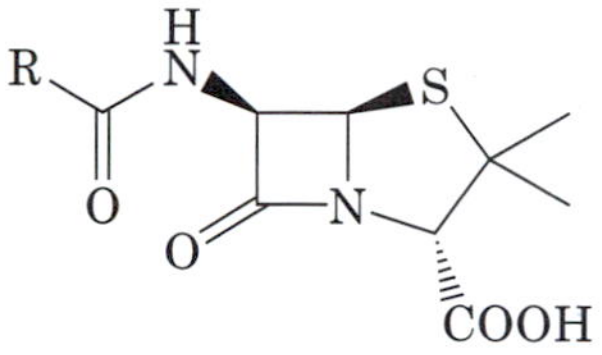

(**16**) R =

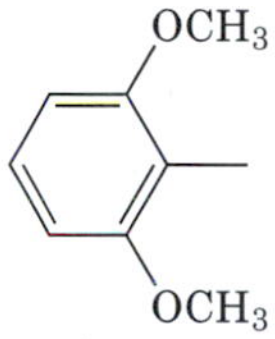

(**17**) R =

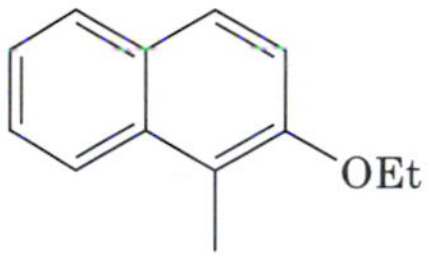

troduced into clinical usage in 1960 (214), followed by nafcillin (**17**) (215) and the isoxazole penicillins, oxacillin (**18a**) (216), cloxacillin

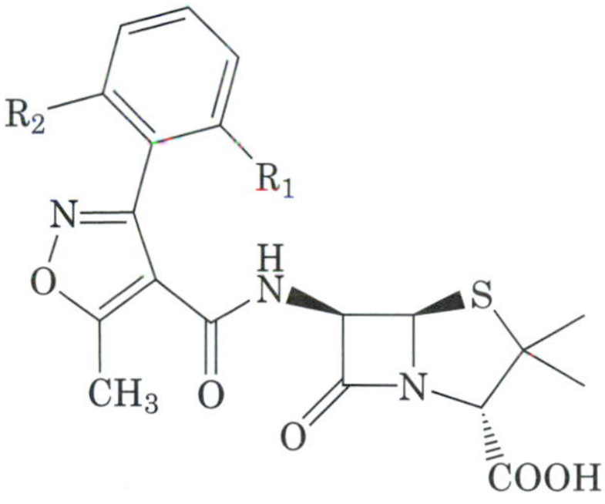

(**18a**) $R_1 = R_2 = H$
(**18b**) $R_1 = H, R_2 = Cl$
(**18c**) $R_1 = R_2 = Cl$
(**18d**) $R_1 = F, R_2 = Cl$

(**18b**) (217), dicloxacillin (**18c**) (218), and flucloxacillin (**18d**) (219). In all these cases, the steric bulk of the side-chain group adjacent to the amide carbonyl group is believed to protect the β-lactam ring from unwanted hydrolysis by the bacterial enzyme penicillinase.

The aminopenicillins such as ampicillin (**19**) (220), epicillin (**20**) (221), amoxycillin

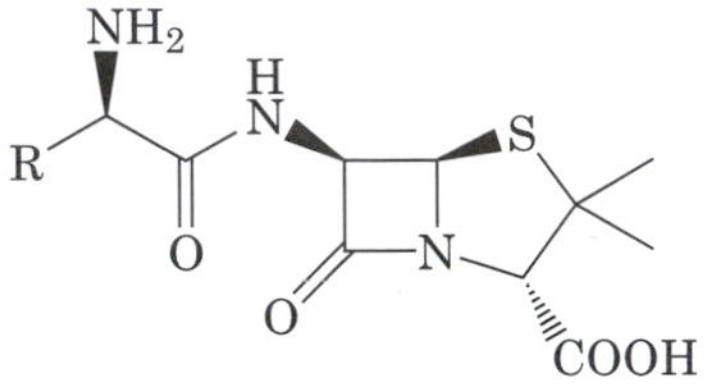

(**19**) R =

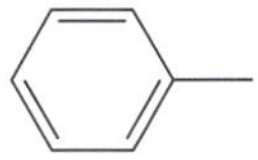

(**20**) R =

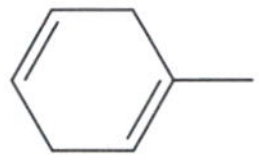

(**21**) R = HO—

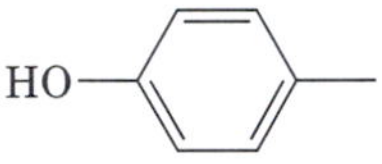

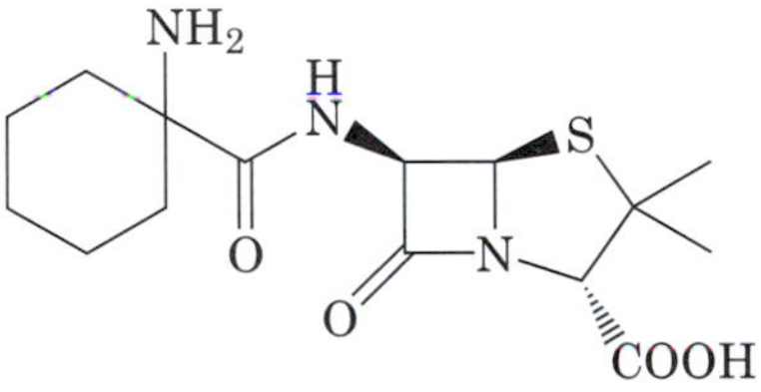

(**22**)

(**21**) (222), and cyclacillin (**22**) (223) introduced into the clinic between 1961 and 1972, are characterized by their broad spectrum of antibacterial activity against both Gram-positive and Gram-negative bacteria and good oral absorption.

Subsequently the prodrug esters talampicillin (**23**) (224), bacampicillin (**24**) (225), pivampicillin (**25**) (226), and lenampicillin (**26**) (227), which all release ampicillin (**19**) once absorbed, have been developed as agents that improve the oral absorption of ampicillin (**19**). With these esters, the oral absorption of ampicillin approaches the excellent level achieved by amoxycillin (**21**).

One of the next challenges to be addressed by researchers was to extend the spectrum of antibacterial acitivity to cover the opportunistic pathogen *Pseudomonas aeruginosa*. Car-

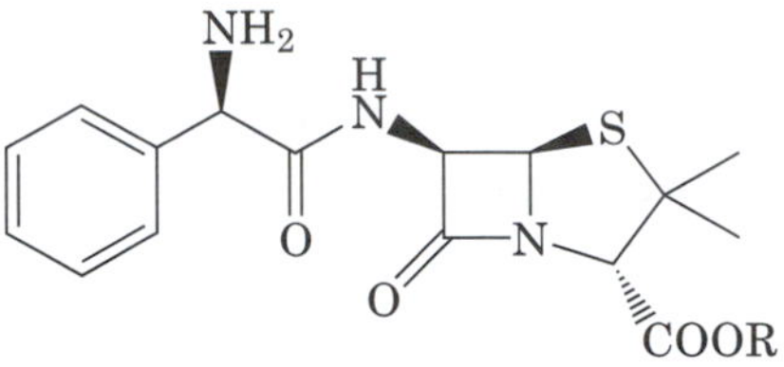

(**23**) R =

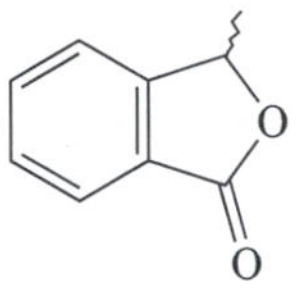

(**24**) R = CH(CH₃)OCO₂Et

(**25**) R = CH₂OCOC(CH₃)₃

(**26**) R =

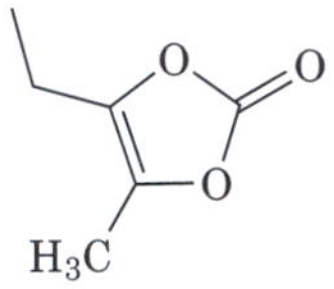

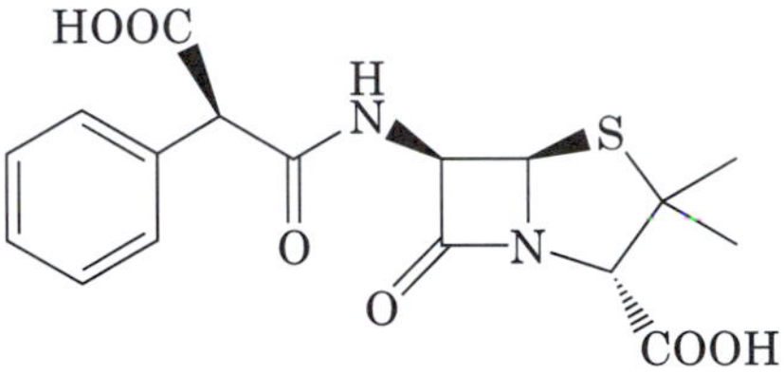

(**27**)

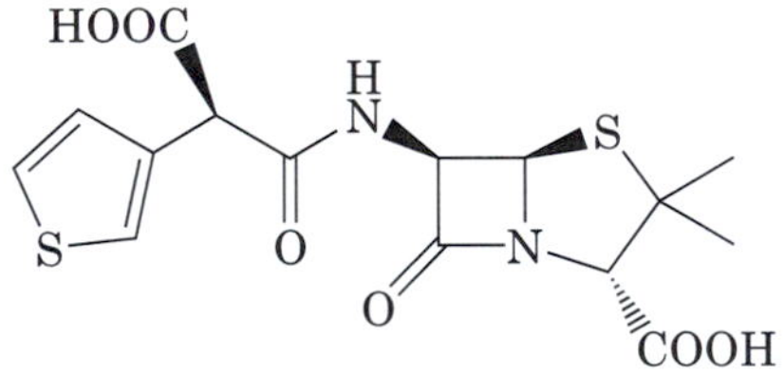

(**28**)

benicillin (**27**) (228), ticarcillin (**28**) (230), and sulbenicillin (**29**) (229), all derivatives with an acidic group in the acylamino side chain, are parenteral agents with adequate antipseudomonas potency combined with good safety tolerance that have found clinical use.

A second generation of penicillins active against *Pseudomonas* species was obtained by acylation of the side-chain amino group of ampicillin (**19**) or amoxycillin (**21**). A selection from the many *N*-acylated aminopenicillin an-

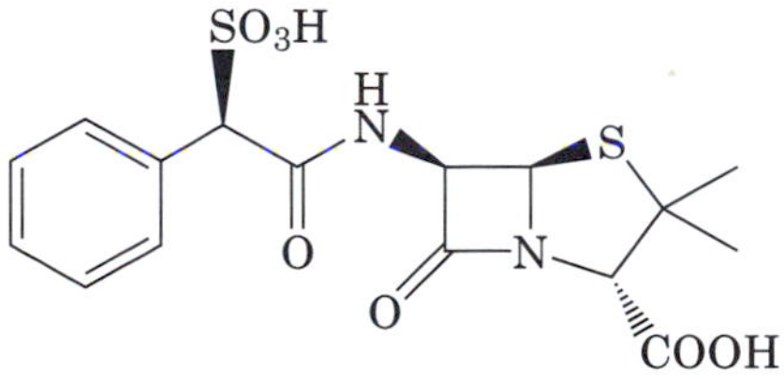

SO$_3$H

(29)

alogs (**30–35**) that have been considered for development as antipseudomonas agents is shown in Table 14.8 (**30–35**). Of these the most successful has been piperacillin (**34**), which together with ticarcillin (**28**), remains the penicillin of choice for treating infections caused by this organism.

In the 1970s the amidino penicillin, mecillinam (**36**), which is unusual in that it has a nonacyl side chain and antibacterial activity

**Table 14.8   *N*-Acylated Aminopenicillins with Activity Against *Pseudomonas aeruginosa***

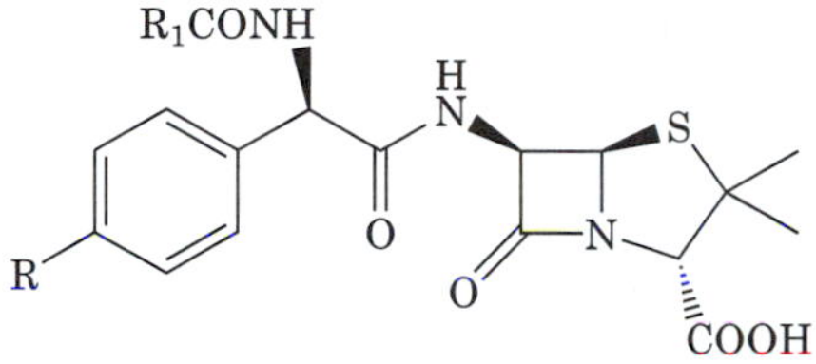

R$_1$CONH

COOH

| Compound | Name | R | R$_1$ | Reference |
|---|---|---|---|---|
| (**30**) | Apalcillin | H | 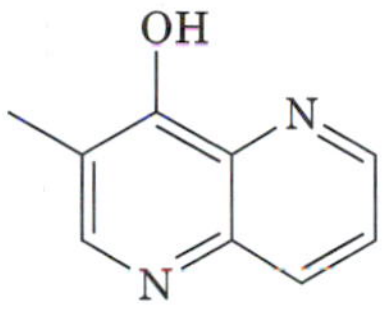 *(structure: 3-methyl-4-hydroxy-1,5-naphthyridine)* | 231 |
| (**31**) | Timoxicillin | OH | 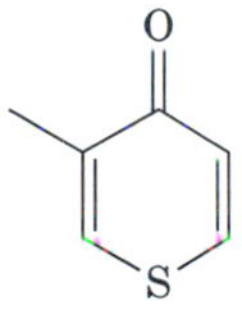 *(structure: 3-methyl-4-oxo-thiopyran)* | 232 |
| (**32**) | Azlocillin | H | 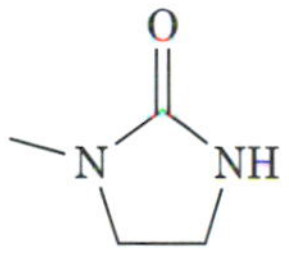 *(structure: 2-oxoimidazolidine)* | 233 |
| (**33**) | Mezlocillin | H | 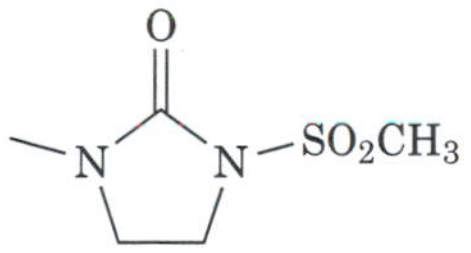 *(structure: N-SO$_2$CH$_3$ substituted 2-oxoimidazolidine)* | 234 |
| (**34**) | Piperacillin | H | 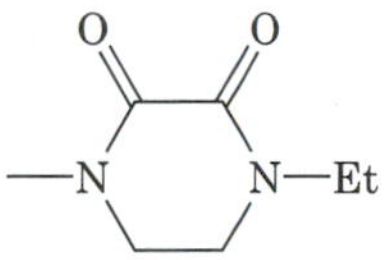 *(structure: 4-ethyl-2,3-dioxopiperazine)* | 235 |
| (**35**) | (D)-Asparagine derivative | OH | 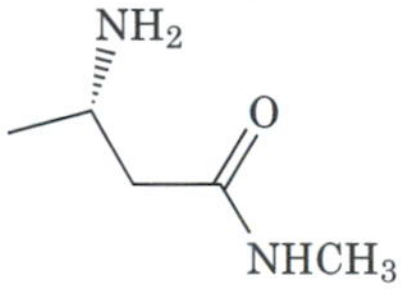 *(structure: NH$_2$ and C(O)NHCH$_3$)* | 236 |

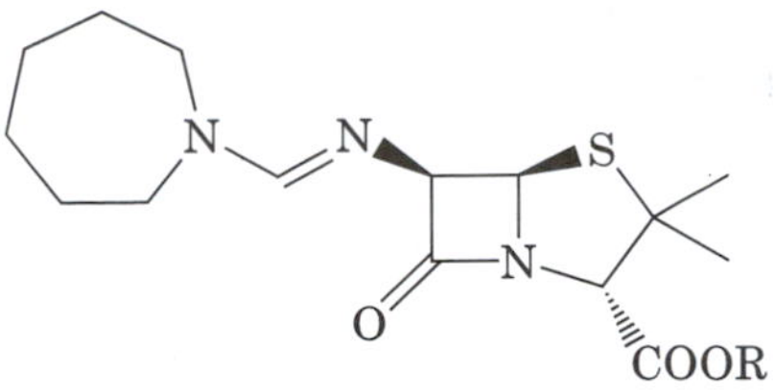

(36) R = H
(37) R = CH$_2$OCOC(CH$_3$)$_3$
(38) R = CH(CH$_3$)OCO$_2$Et

limited to Gram-negative organisms, was reported (237). Also, the oral pivaloyloxymethyl (238) and carbonate prodrug esters have been described (239).

A recent publication describes a further range of C6-acylamino β-lactamase stable penicillins from which BRL-44154 (39) was

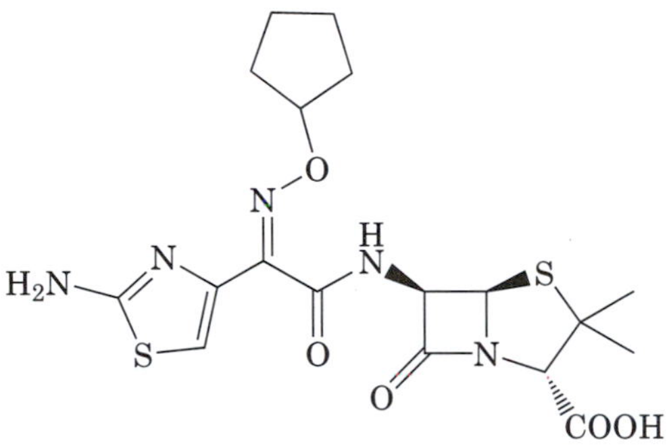

(39) BRL-44154

identified as exhibiting high activity against Gram-positive organisms including methicillin-resistant Staphylococci (MRSA) (240).

Table 14.9 shows a comparison of the *in vitro* antibacterial activities of representative penicillins, including C6-α-substituted derivatives described below (see Section 5.1.2.1). Further information on the penicillin described above (and many more that have not been included) is available from the references in the data review of β-lactam antibiotics by Rolinson and more recently by Wright (241, 242).

**5.1.2.1 C6-Substituted Penicillins.** Early interest in C6-substituted penicillins arose from the idea that the presence of a C6-α-methyl group in a penicillin molecule would provide a closer mimic of the D-alanyl-D-ala-

nine component of the cell wall (244, 245). Although this was not shown to be the case (246, 247), the discovery of the naturally occurring ring C7-α-methoxy cephalosporin derivatives (cephamycin, Section 6.2), turned attention to the synthesis of penicillins having this and other substituents in the C6-position.

In many cases the methodology for the introduction of the methoxy substituent parallels that developed for the cephalosporins (248). A method for direct methoxylation of a C6-acylaminopenicillin relies on the addition of methoxide to an acylimine (40) (249, 250),

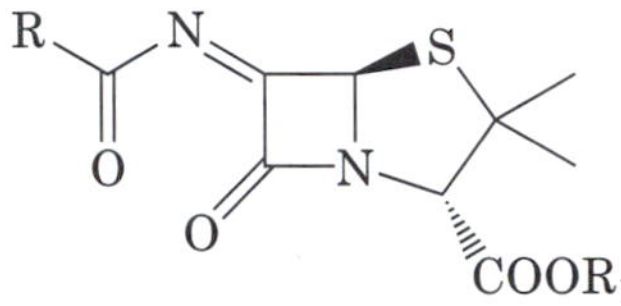

(40)

whereas an indirect procedure involves the addition of methanol to the putative acylamine formed from the C6-α-methylthiopenicillin (41) in the presence of a mercuric salt (251).

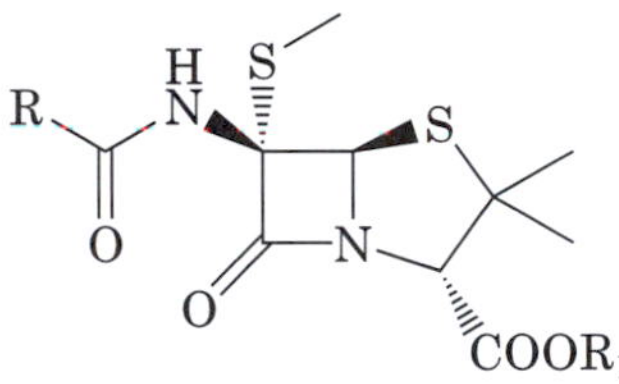

(41)

The C6-α-methylthio substituent, as well as alkyl and substituted alkyl, is available from the anion of the C6-Schiff base (42) with a suitable reagent (247, 251, 252). The C6-α-stereochemistry of the new substituents is en-

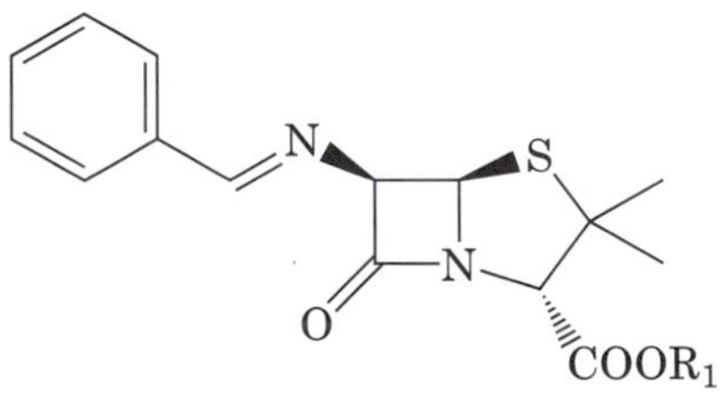

(42)

**Table 14.9  *In Vitro* Antibacterial Activity of Penicillins[a,b]**

| Organism | Pen G | Meth | Clox | Amox | Ticar | Pip | Mecill | BRL-44154 | 6a-Substituted Temo | 6a-Substituted Form |
|---|---|---|---|---|---|---|---|---|---|---|
| *E. coli* | 50->100 | >100 | >100 | 2->100 | 0.8->100 | 0.06->64 | <0.4-50 | 1->64 | 2-16 | <0.03-0.5 |
| *K. pneumoniae* | 0.5-100 | >100 | >100 | 16-100 | 25->100 | 1->64 | <0.4->100 | 8->64 | 1->100 | <0.03-2 |
| *Enterobacter* sp. | 50->100 | >100 | >100 | 12.5-100 | 0.8->100 | 1->64 | <0.4->100 | 16->64 | 1-32 | <0.03-0.12 |
| *Citrobacter* sp. | >100 | >100 | >100 | 5->100 | <1.6->100 | 1->64 | 0.5->100 | 8->32 | 2-16 | <0.03-16 |
| *S. marcescens* | >100 | >100 | >100 | >100 | 4->100 | 2->64 | 0.8->100 | 16->64 | 8->100 | <0.03-8 |
| *P. mirabilis* | 16-32 | >100 | >100 | 0.8->100 | <1.6->100 | 0.06-64 | 3.2->100 | 2->64 | 0.5-16 | 0.06-0.25 |
| *Proteus* sp (In. +) | 0.4->100 | 6.3->100 | 50->100 | 0.4->100 | <1.6-100 | 0.06->64 | 1.6->100 | 2->64 | 1-4 | <0.03-2 |
| *P. aeruginosa* | >100 | >100 | >100 | >64 | 3.2->100 | 1->64 | 12.5->100 | 16->64 | >100 | 0.12-16 |
| *H. influenzae* | 0.08->100 | — | 2->100 | 0.05->64 | 0.25->100 | <0.03-2 | 16->100 | 0.06-2 | 0.5-4 | 0.06-0.5 |
| *B. fragilis* | 0.08->25 | >100 | >100 | 0.12->32 | <1.6->100 | 3-100 | >100 | 8->64 | 16-128 | 32 |
| *S. aureus* | <0.005->100 | 0.4-2.5 | 0.12-2 | 0.05->100 | 0.4-25 | 0.8-100 | 1.6->100 | 0.25-1 | >100 | >100 |
| *S. pyogenes* | 0.005-50 | 0.1-0.8 | <0.03-0.06 | <0.03-0.12 | 0.2-0.8 | <0.015-0.12 | 0.8-6.3 | 0.03-0.06 | >100 | 4 |
| *S. pneumoniae* | 0.006-2 | 0.1-0.2 | 0.12 | <0.03-4 | 1.25-2 | 0.015-0.06 | 0.8-12.5 | <0.03-8 | >100 | 4 |
| *N. gonorrhoeae* | 0.003->100 | 0.05-2 | 0.12->64 | 0.01->16 | 0.02-16 | 0.015-1 | 0.03-8 | — | 0.03-8 | <0.008-0.06 |

[a]Minimum inhibitory concentrations (range mg/L). Refs. 241, 265, 243, and GSK in house data.

[b]Abbreviations: Pen G, benzylpenicillin; Meth, methicillin; Clox, cloxacillin; Amox, amoxicillin; Ticar, ticarcillin; Pip, piperacillin; Mecill, mecillinam; Temo, temocillin; Form, Formidacillin.

sured, given that the incoming group always approaches from the less-hindered α-face of the β-lactam ring (252).

Other methods for introduction of methoxy and a variety of alkyl or substituted alkyl variations make use of diazo-intermediates (253) or isonitrile chemistry (254).

Alternative reactive intermediates for the insertion of a C6-α-substituent have been generated by way of keteneimines (255), sulfenimines (256), and quinolone methide intermediates (257).

After the methoxy series a wide variety of the other C6-α-substituents were investigated on the penicillin ring system, and it was found that a C6-α-formamidino group with an appropiate side chain provided a series of highly active antibiotics (258). At the time of this discovery naturally occurring β-lactams possessing this substituent were not known. Introduction of the C6-α-formamido substituent followed a similar pattern to that of the methoxy group. Initially, this was obtained by displacement of the methylthio group from (41) by ammonia in the presence of a mercury salt, followed by formylation of the C6-α-amino group. Subsequently, a more direct addition used N,N-bis trimethylsilylformamide (BSF) (258, 259). In both the methoxy and formamido series it is also possible to generate the C6-amino nucleus (43), which can be ac-

(44)

(45)

antibacterial potency is compromised in comparison to the C6-α-hydrogen analogs. Only one derivative, temocillin (46) (262), has pro-

(46)

(43) X = OCH₃, NHCHO

ylated with the appropriate side-chain acid as required (259).

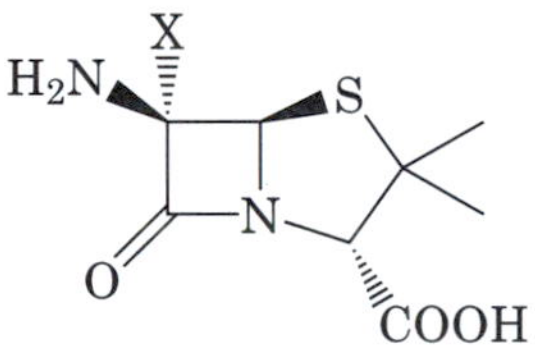

Other methods for the introduction of the formamido group make use of the N-trifluoromethylsulfonylamino penicillin (44) (260) or the C6-α-methylsulfinyl penicillin (45) (261) to generate the acylimine, which is then trapped with BSF.

A feature of the 6,6-disubstituted penicillins is the stability of the β-lactam ring to hydrolysis by β-lactamases, but usually the

gressed to the clinic but, in spite of the prolonged blood levels in humans ($t_{1/2}$ = 4.5–5 h), has found limited utility because its spectrum of antibacterial activity is restricted to Gram-negative organisms only. When the thienyl carboxylic acid moiety was replaced by a catecholic sulfonic acid, to give (47), the spectrum of antibacterial activity was expanded to include P. aeruginosa but not the Gram-positive organisms (263).

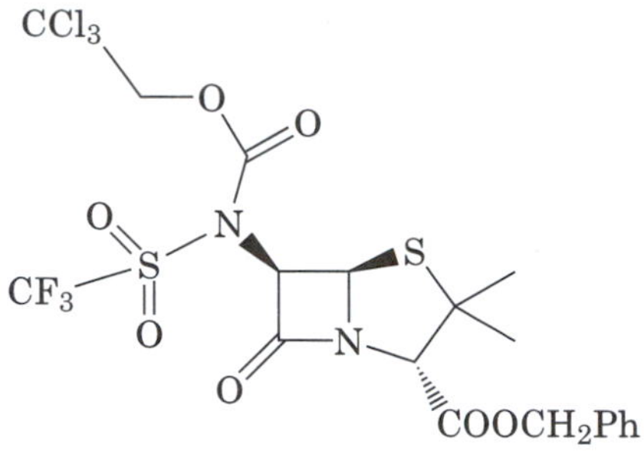

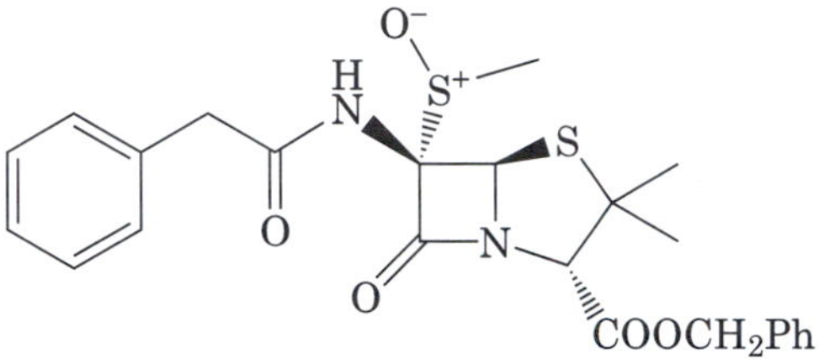

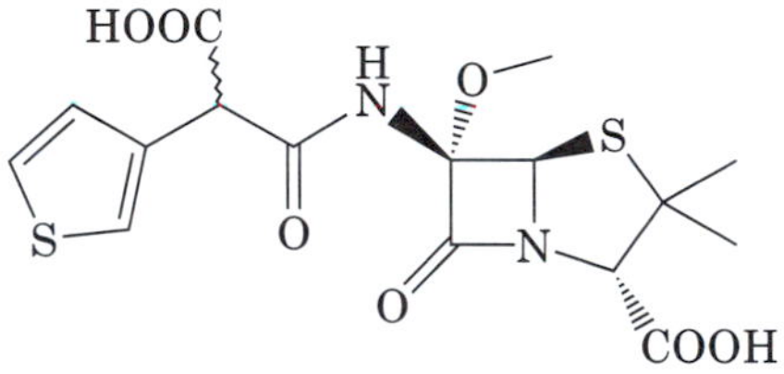

After an extensive research program (264) the combination of the N-acylated dihydroxyphenylglycyl C6-α-formamidino group produced formidacillin (BRL-36650, 48) (265). This derivative was highly potent against

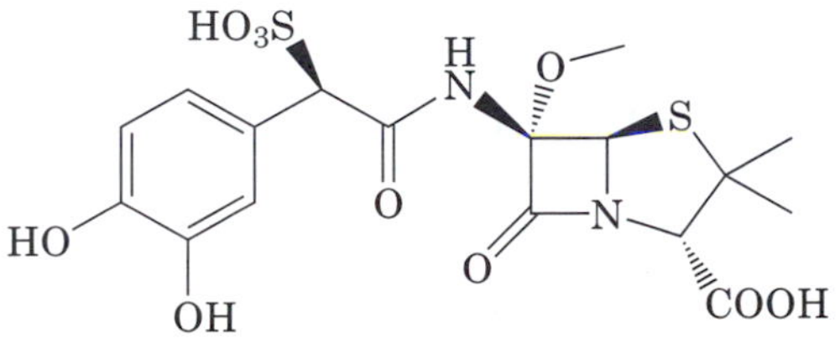

(47)

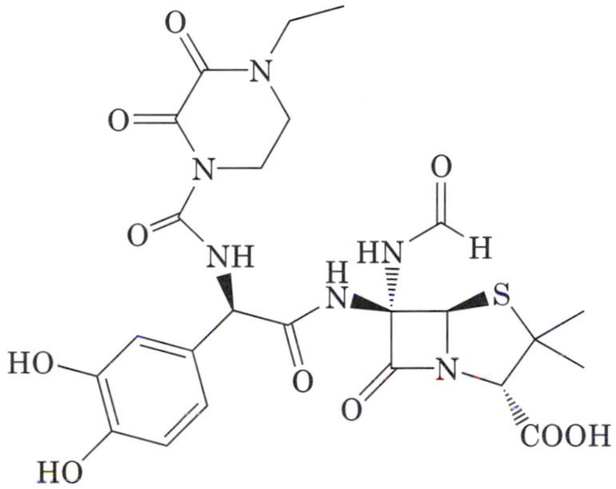

(48) BRL-36650

(49) T5575   R = —N

(50) T5578   R = —N

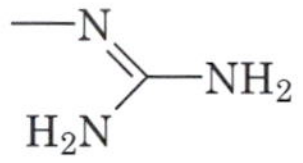

this class followed a similar story. In fact, the antibacterial activity of a metabolite produced by *Cephalosporium acremonium* was originally detected in 1945 by Brotzu, an Italian scientist in Sardinia who isolated a strain from sewage outfall (270). Subsequent work by Abraham at Oxford resulted in the isolation and characterization of the active drug, in particular cephalosporin C (51) (271, 272). Con-

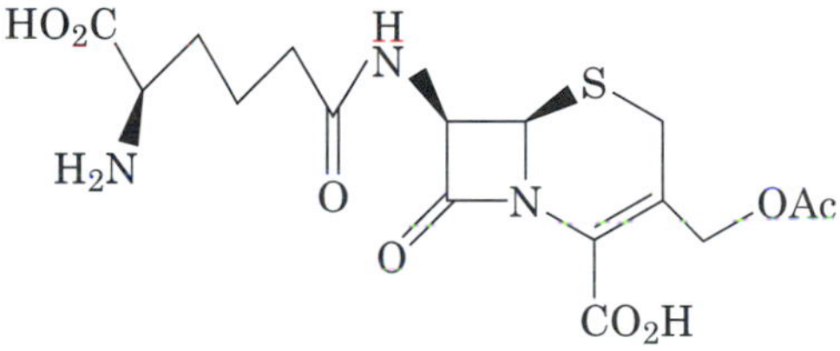

(51) Cephalosporin C (1948)

Gram-negative organisms, including *Pseudomonas* species, and possessed some activity against *Streptococci* but not against the *Staphylococci*.

### 5.1.3 Other Modified Penicillins.

Chemical modification of most of the other positions (S1, C2, C3, and C6) of the penicillin molecule have been reported and summarized elsewhere (211). More recent reports have described novel 2-carboxypenam analogs T5575 (49) and T5578 (50), in which both the C2- and C3-substituents have been changed from those of the natural penicillins (266–269). These compounds displayed potent antibacterial activity against Gram-negative organisms, including *Pseudomonas aeruginosa*, with stability to β-lactamases generally greater than that observed for the natural penicillin structure.

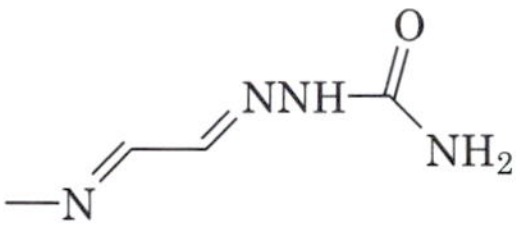

### 5.2   Cephalosporins and Synthetic Analogs: Oxacephems, Carbacephems

After penicillins, cephalosporins represent the best-known β-lactam class. The discovery of

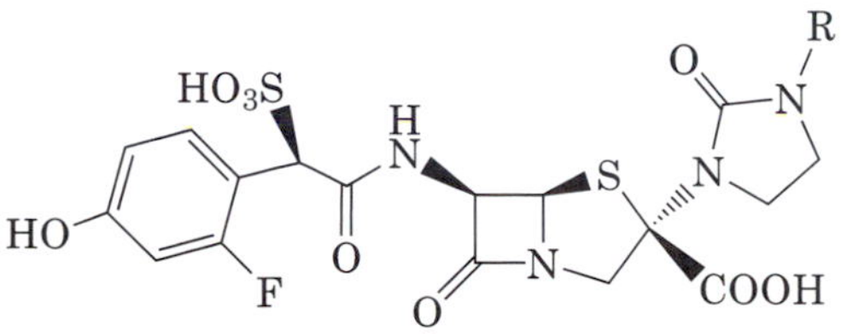

firmation of its structure was later obtained by X-ray crystallography (273).

Cephalosporin C (51) attracted the attention of scientists mainly because of its stability to penicillinases. Cephalosporins (52) are characterized by a bicyclic ring system, in which the β-lactam is fused to a six-membered

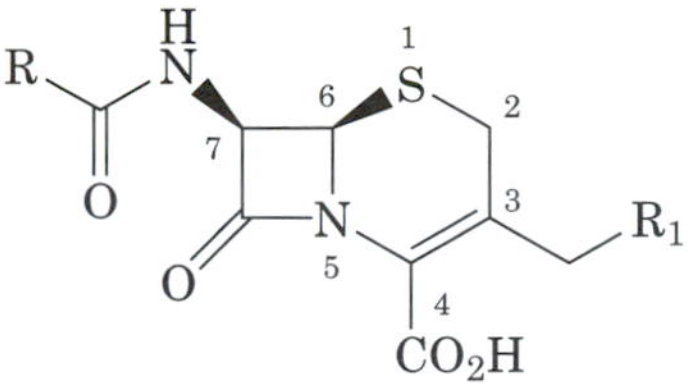

(52) cephalosporins

ring bearing at position 1 a sulfur atom; oxacephalosporins (**53**), an oxygen atom; and carbacephalosporins (**54**), a carbon atom.

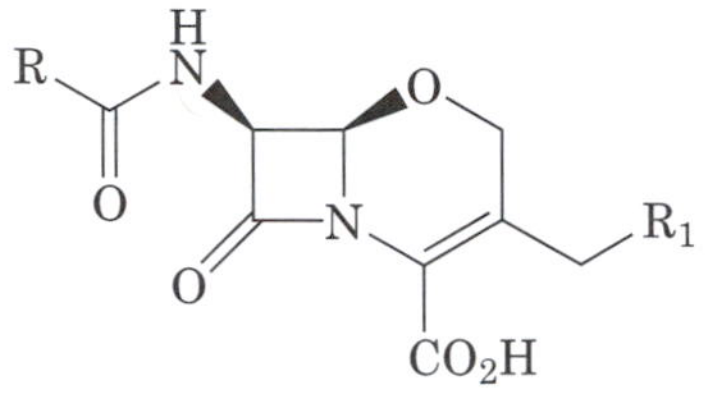

(**53**) oxacephalosporins

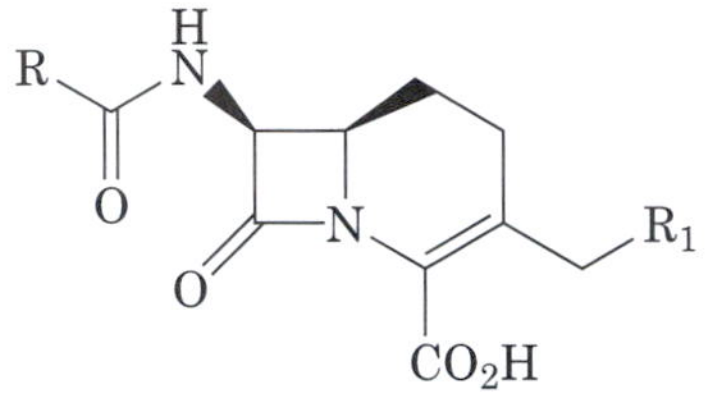

(**54**) carbacephalosporins

The isolation of the first cephalosporin was followed by the discovery of other natural metabolites from fungi and actinomycetes, including those with a C7-$\alpha$-methoxy substituent (**55**) (274) and a C7-$\alpha$-formamido group (**56**) (275).

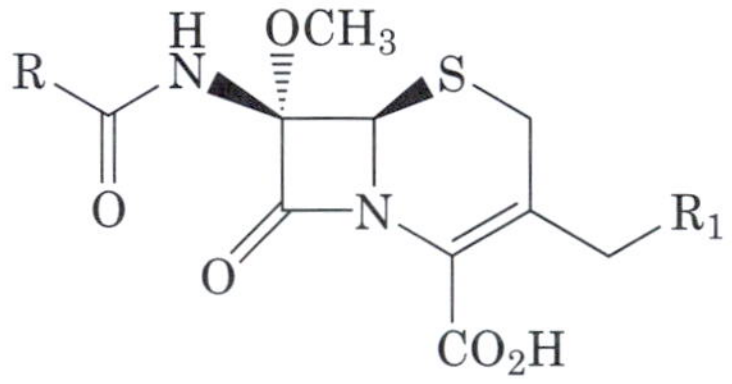

### 5.2.1 Derivatives of C7-Aminocephalosporanic Acid.

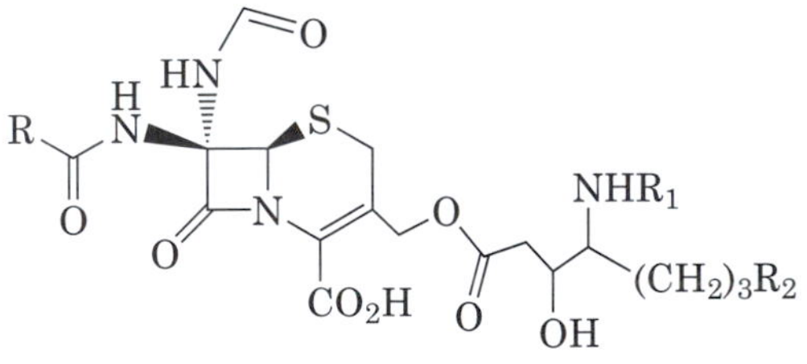

The structure of naturally occurring members of this series is characterized by

(**55**) cephamycins

(**56**) cephabacins

the $\alpha$-aminoadipic side chain at C7 (271, 272, 274, 276–280). Some variations have been isolated that have the side chain derived from glutaric acid (281) or 5-hydroxy-5-carboxy-valeric acid (282, 283).

The natural products exhibit a low level of antibacterial activity. Cleavage of the amide bond of the aminoadipoyl side chain of cephalosporin C (**51**) is a high yielding process (284) that affords 7-amino-cephalosporanic acid (7-ACA, **57**), ideally suited for the synthesis of a

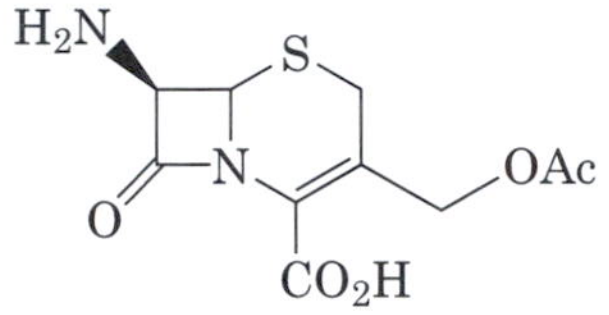

(**57**) 7-amino-cephalosporanic acid (7-ACA)

wide range of semisynthetic cephalosporins by acylation of the C7 amino group.

In cephalosporins there are two positions available for chemical manipulation, C3 and C7. A wide variety of amine acylation methods have been used for the production of C7-acylamino derivatives by the use of acyl chlorides, mixed anhydrides, active esters, and carbodiimides (285). To improve the chemical reactivity of 7-ACA (**57**), the solubility in organic solvents is increased by conversion of the carboxylic acid at C4 of 7-ACA (**57**) into an ester such as *tert*-butyl dimethylsilyl, benzhydryl, *p*-nitrobenzyl, or *p*-methoxybenzyl. However, ester derivatives of 7-ACA (**57**) become very sensitive to basic conditions and the $\Delta^3$ double bond isomerizes to the unwanted $\Delta^2$ isomer (286).

Reactions at C3 have mainly involved displacement of 3′-acetate, to give substitution with heteroatoms, especially sulfur and nitrogen (287), whereas elimination followed by catalytic hydrogenation (288) or acid-catalyzed reduction by use of trialkylsilanes leads to 3-methyl cephalosporins (285). An alternative to the displacement of the 3′-acetoxy for the synthesis of 3′-heteroatom cephalosporins is through functionalization of a 3-methyl derivative of a cephalosporin sulfoxide by allylic bromination to the 3′-bromomethyl compound suitable for displacement reactions (286). Hydrolytic deacylation by base or ester-

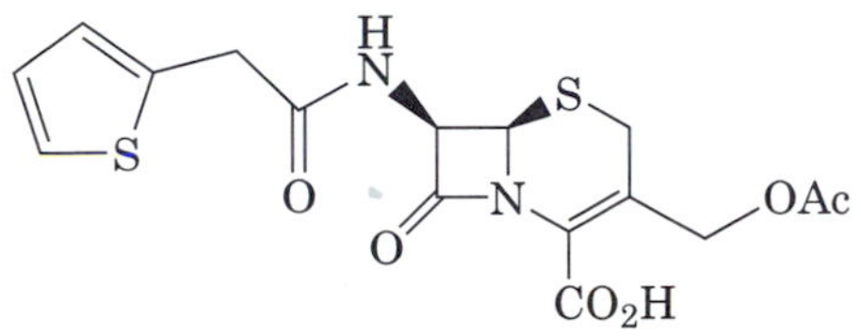

**(58)** Cephalothin (first generation)

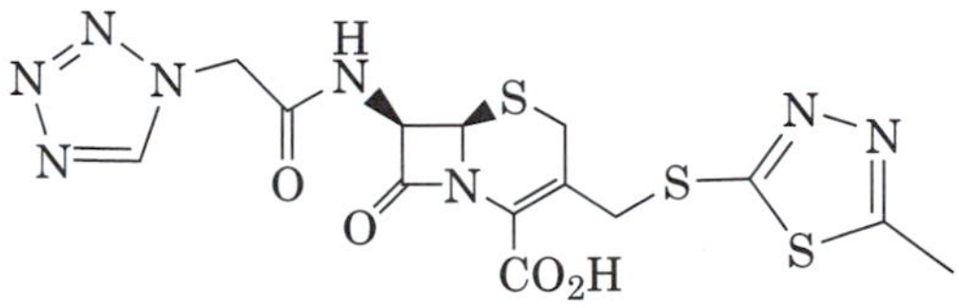

**(59)** Cefazolin (first generation)

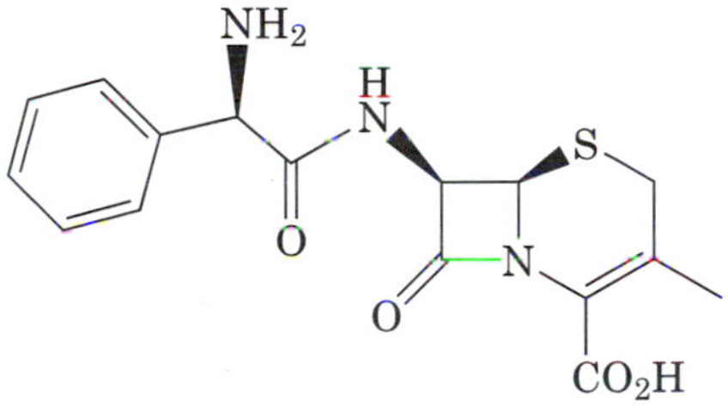

**(60)** Cephalexin (first generation)

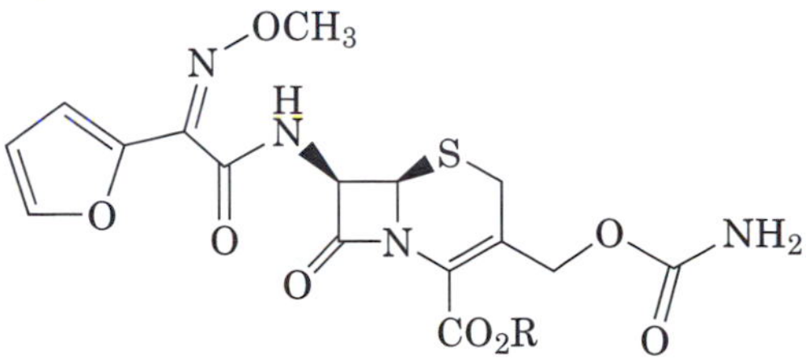

**(62a)** Cefuroxime (second generation)  R = H
**(62b)** Cefuroxime axetil      R = CH(CH₃)OAc

(292). A summary of many of the common reactions and derivatives of 7-ACA (**57**) is given in Fig. 14.7.

**5.2.2 Classification, Biological Activity, and Structure-Activity Relationship of Cephalosporins.** Some 50 different cephalosporins are in clinical use or at an advanced stage of development (241, 285, 293, 294) and many attempts have been made to classify these based on stability to β-lactamases, potency, antibacterial spectrum, and pharmacological properties. The most common approach has been to divide the group into various generations primarily based on their antibacterial spectrum, with both parenteral and oral agents being covered (285, 295, 296), and also on their level of chemical sophistication. First-generation derivatives such as cephalothin (**58**) (297), cefazolin (**59**) (298), and the orally absorbed cephalexin (**60**) (299) possess activity against Gram-positive bacteria, but a relatively narrow spectrum against Gram-negative strains attributed in part to their susceptibility to β-lactamases.

Second-generation compounds such as cefamandole (**63**) (300) and cefuroxime (**62a**) (301) have a broader spectrum of activity, with enhanced activity against *H. influenzae* and

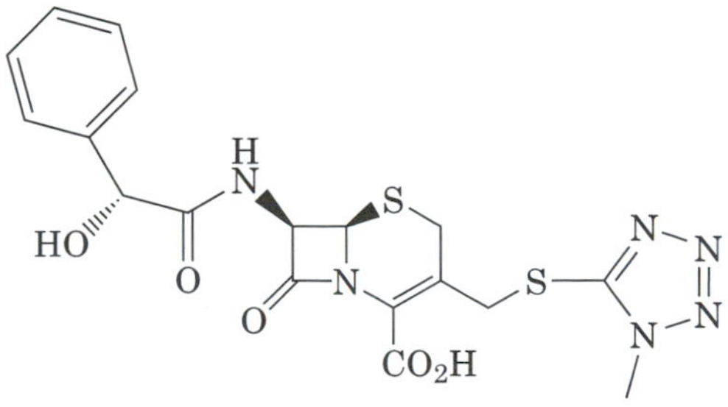

ase gives 3'-hydroxymethyl cephalosporins (289), which can be converted into carbamoyl compounds such as cefuroxime (**62a**) (290).

Other important intermediates are 3-exomethylene derivatives that, by ozonolysis of the exocyclic double bond, provide an entry to cephalosporins having a heteroatom attached directly at C3 as in the commercially important cefaclor (**61**) (291). The exomethylene compounds are also easily converted to 3'-bromomethyl derivatives with bromine and base

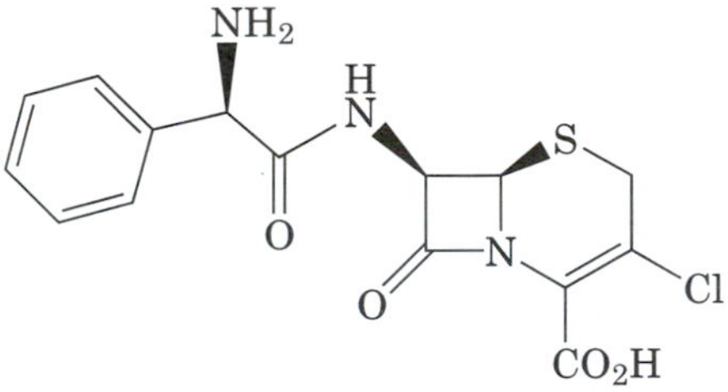

**(61)** Cefaclor (second generation)

**(63)** Cefamandole (second generation)

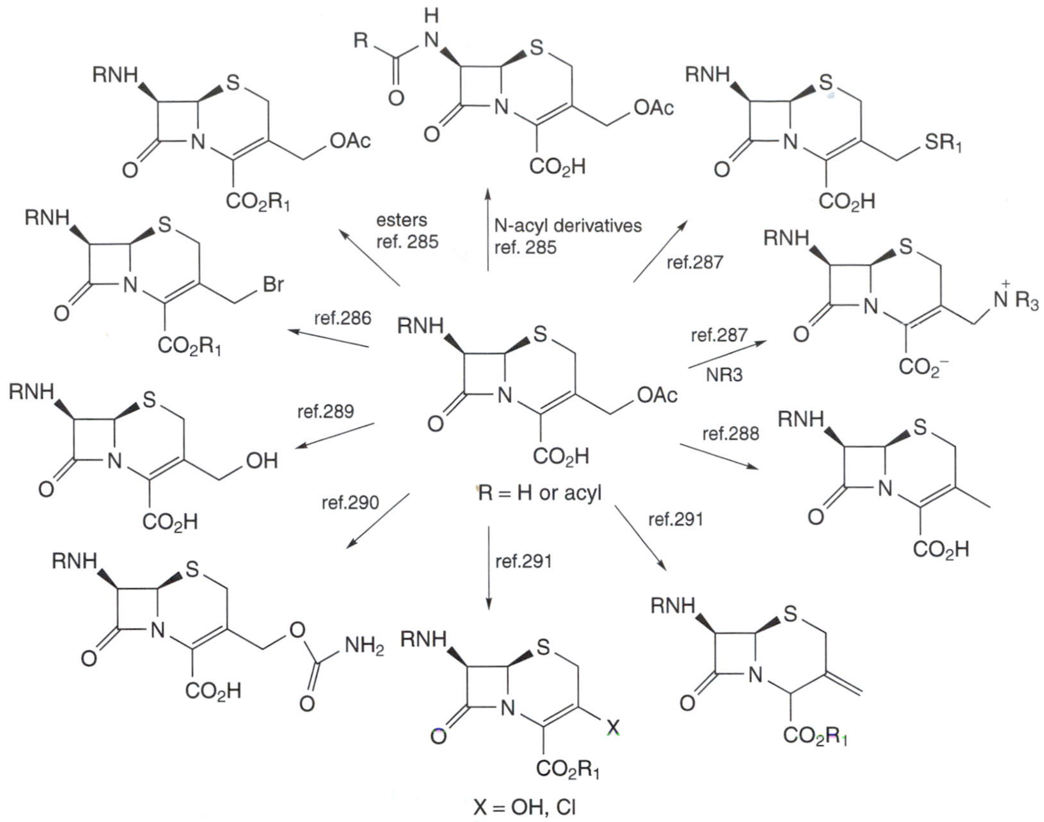

**Figure 14.7.** Summary of some common reactions and derivatives of 7-ACA.

the Enterobacteriaceae resulting from an increased stability to β-lactamases. Oral agents in this group are cefaclor (**61**) (302) and the axetil ester of cefuroxime (**62b**) (303). The highly β-lactamase stable cephamycins such as cefoxitin (**64**) and cefotetan (**65**) are also

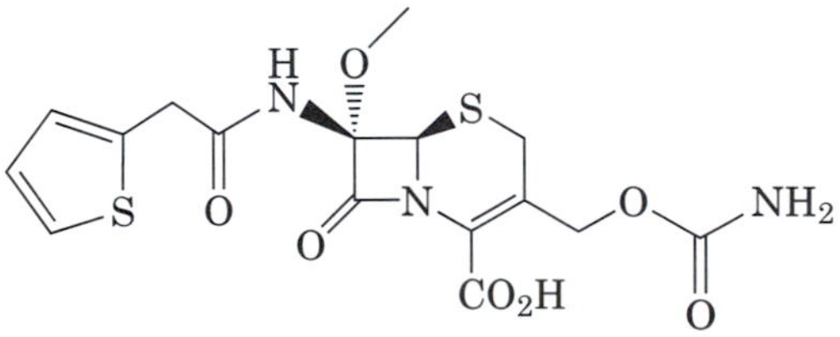

(**64**) Cefoxitin (second generation)

included in this group and show excellent activity against *Bacteroides fragilis* (300, 304).

The third-generation cephalosporin compounds, which originated with cefotaxime (**66**)

(305, 306), have a broader spectrum of activity, especially against the Enterobacteriaceae. They show several important advantages over the first- and second-generation compounds, including increased resistance to many plasmid and chromosomally mediated β-lactamases (296). Ceftriaxone (**67**) (305, 307) is an example that has a very prolonged half-life ($t_{1/2}$ = 7–8 h in humans), making it suitable for once-daily administration. Compounds such as cefoperazone (**68**) (308) and ceftazidime (**69**) (309) show improved activity against *Pseudomonas aeruginosa*, which is also a predominant feature of cefsulodin (**70**) (305, 310). The only oral agent of this group is cefixime (**71**) (311). The third-generation cephalosporins show potent anti-Gram-negative but modest anti-Gram-positive activity, being inferior to the first-generation agents in this respect, and like the majority of cephalo-

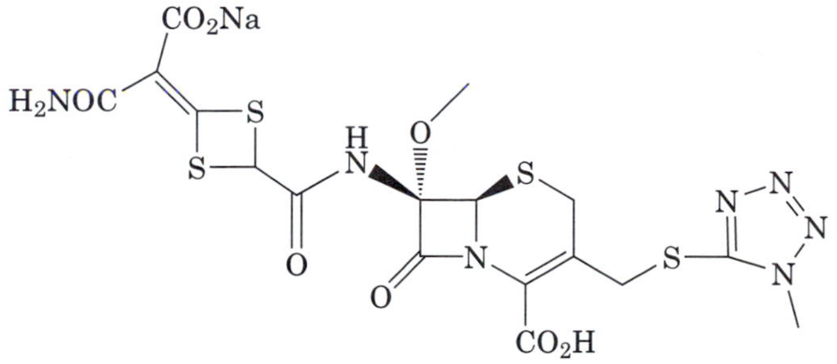

(65) Cefotetan (second generation)

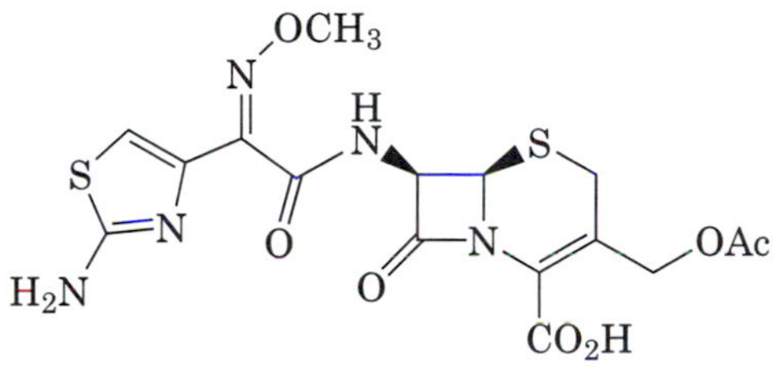

(66) Cefotaxime (third generation)

sporins described so far, show no activity against methicillin-resistant Staphylococci (MRS) (296).

Compounds such as cefpirome (72) (312) and cefepime (73) (313), and others undergoing clinical development (314, 315), constitute examples of the fourth-generation cephalosporins (285), which show some slight further advantages. However, although recent research effort in the cephalosporin field has identified new derivatives able to treat methicillin-resistant *Staphylococcus aureus* strains (see Sec-

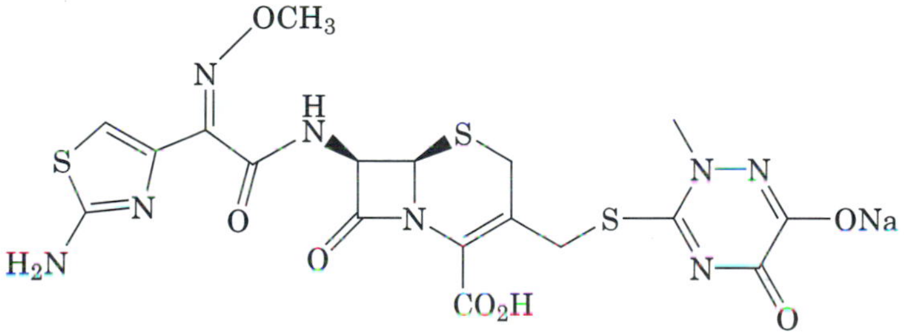

(67) Ceftriaxone (third generation)

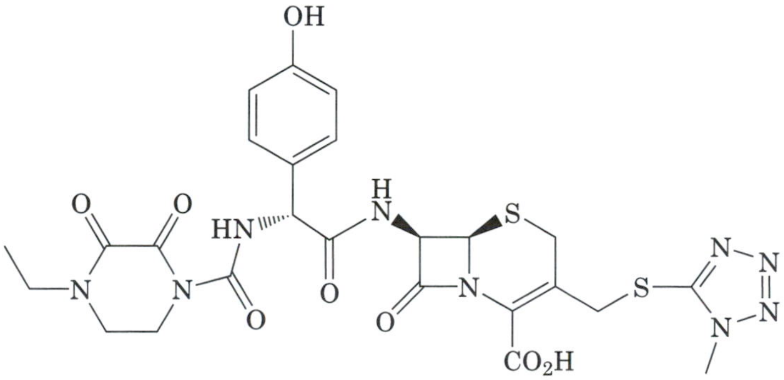

(68) Cefoperazone (third generation)

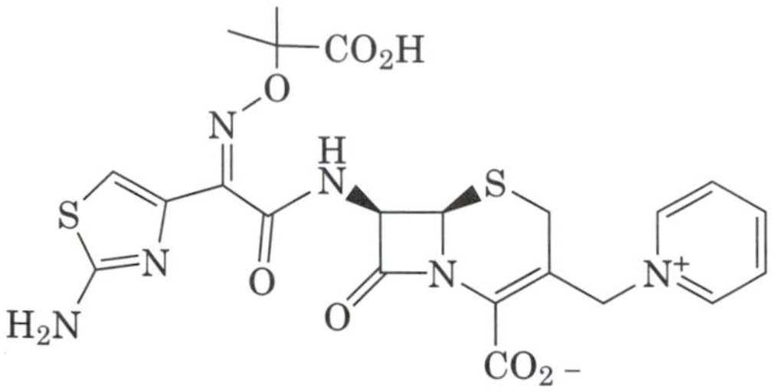

(69) Ceftazidime (third generation)

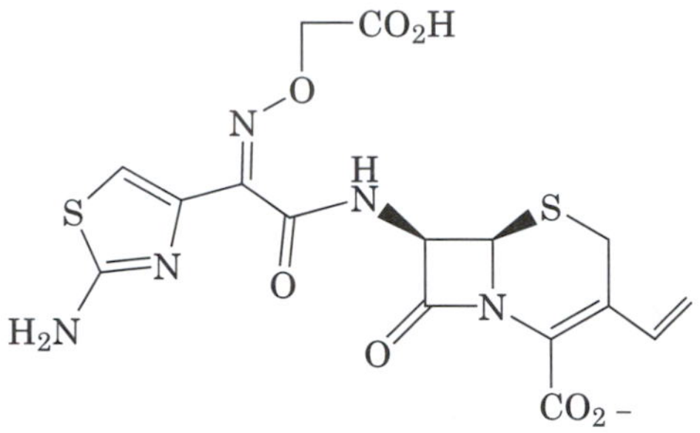

(71) Cefixime (third generation)

tion 6.1.1), problems with extended-spectrum β-lactamases (ESBLs) still undermine their potential use as broad-spectrum agents.

Table 14.10 shows a comparison of the *in vitro* antibacterial activities of representative cephalosporins of these groups.

Ultimately, the biological activity of the cephalosporins depends on their affinity and interaction with the target enzymes (PBPs). However, many other factors such as penetration to the target site, β-lactamase stability, pharmacokinetic parameters, and metabolic stability will all influence the final antibacterial effectiveness of these agents. Most structure-activity relationships (SARs) have been derived from *in vitro* activities against whole cells by use of MIC values to compare structural variations.

Major factors influencing these properties result from changes in the C7-acylamino substituent or C3 variations, although virtually all positions of the cephalosporin nucleus have been modified (296), as illustrated in Fig. 14.8.

Several highly active derivatives have resulted from replacement of the sulfur atom with oxygen or carbon or by transposition of the heteroatom from position 1 to position 2 ("isocephem" derivatives), whereas substitution at C7 strongly affects β-lactamase stability.

A large number of acyl groups have been introduced at C7 and showed significant changes in both potency and spectrum of activity (285). Introduction of the aminothiazole group improves the activity against Gram-negative strains (316), and in combination with the *syn* oximino grouping confers resistance to β-lactamases (317). The nature of the C3-substituent predominantly influences the pharmacokinetic and pharmacological properties but also the antibacterial activity. Thus C3 methyl compounds with a C7 phenylglycine side chain are orally absorbed, whereas C3 vinyl derivatives, even with the C7 oximino side chain, also show promise as oral agents (285). Some interesting synthetic methodologies have been developed for these vinyl compounds by use of allenyl azetidinone intermediates and a variety of organometallic reagents (318, 319).

The synthesis of potent antipseudomonal β-lactam derivatives can be achieved by introducing siderophore-like moieties that allow the antibiotic to use the *ton*B transport system, overcoming in that way the mechanism of resistance caused by a reduced permeation barrier. In general, siderophores are high affinity iron-binding compounds that the bacte-

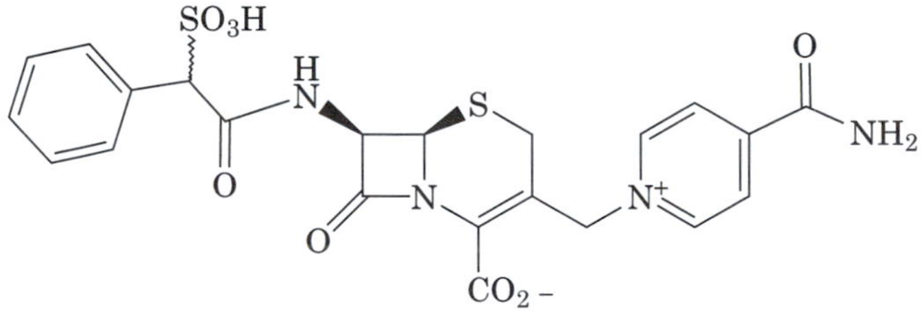

(70) Cefsulodin (third generation)

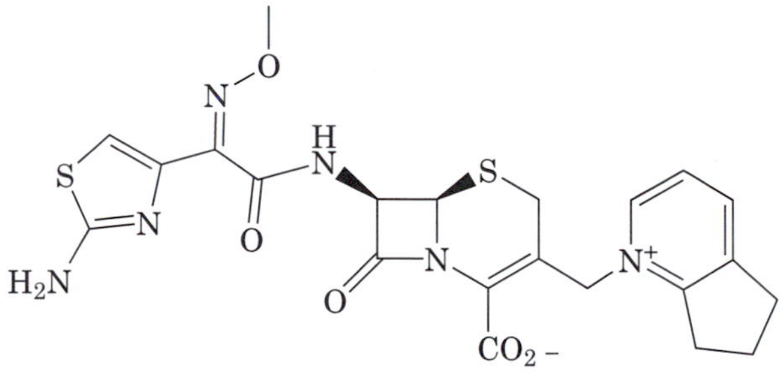

(72) Cefpirome (fourth generation)

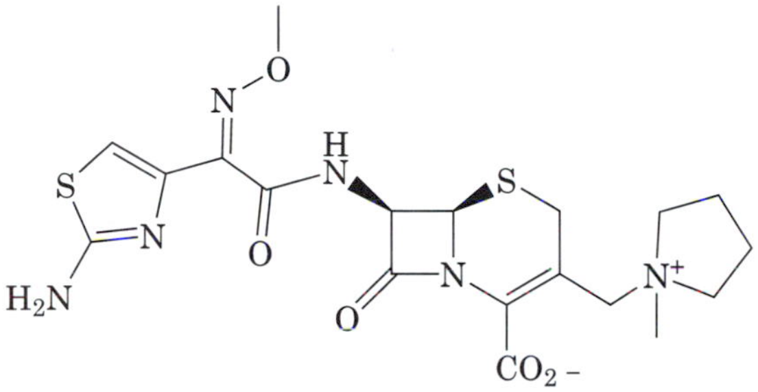

(73) Cefepime (fourth generation)

ria release in their surroundings (320, 321). These bacterial products, which are capable of chelating ferric iron, are actively transported into the bacteria.

Catechol groups have been used to enhance the activity, particularly against *Pseudomonas aeruginosa*, and many examples have been described (294). The effect of the catechol group seems to be similar irrespective of whether it is incorporated in the C7 acylamino side chain, as in M-14659 (74) (322), or the C3 position, as illustrated by (76) (323). Noteworthily, BRL-57342 (75) has been described (324), which not only exhibits antipseudomonal activity but also retains some anti-Grampositive activity that is an unusual feature for cephalosporins bearing a catechol group. Evidence has been presented to show that these compounds penetrate into the cell through use of the bacterial *ton*B-dependent iron-transport system (325, 326).

Other examples of cephalosporin derivatives bearing a catechol moiety are GR69153 (77) (327) and KP-736 (78) (328). However, it is very common that bacteria can very rapidly develop resistance to siderophore-like drugs.

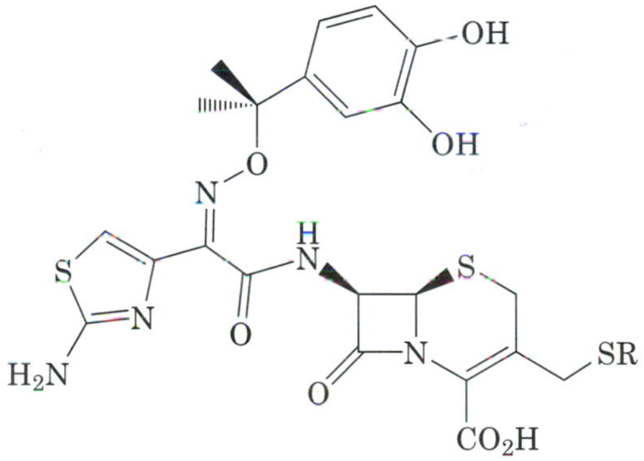

(74) M-14659    R =

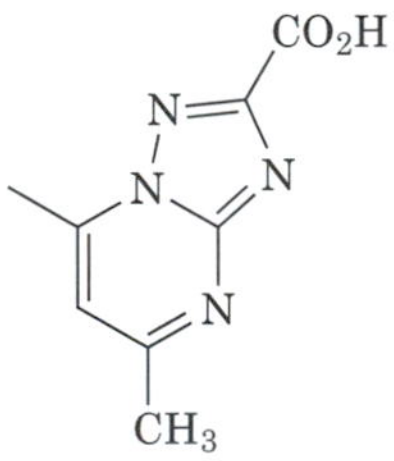

(75) BRL 57342   R =

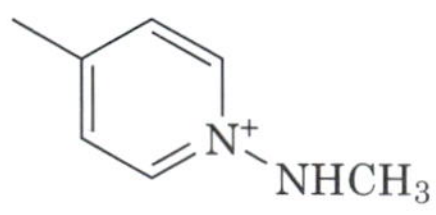

**Table 14.10  Antibacterial Activity (mg/L) of Some Representative Cephalosporins[a,b]**

| Organism | 1st | | | 2nd | | | 3rd | | | | 4th | |
|---|---|---|---|---|---|---|---|---|---|---|---|---|
| | CET | CEZ | CEX (os) | CMD | CXM | CFL (os) | CTX | CTR | CAZ | CFM (os) | CPR | CPM |
| E. coli | 4->64 | 2->128 | 4->128 | 1-32 | 1-8 | 1->128 | 0.03-1 | 0.12-1 | 0.12-1 | 0.4-32 | 0.03-0.25 | 0.02-1 |
| K. pneumoniae | 4->128 | 2->128 | 4->128 | 0.5->128 | 2-4 | 1->128 | 0.03-0.12 | 0.06-0.12 | 0.06-0.25 | 0.05-0.4 | 0.06-NA | 0.01-2 |
| Enterobacter sp. | >128 | >128 | >128 | 32->128 | 16->128 | >128 | 0.12->64 | 0.25-NA | 0.25->64 | 64->100 | 0.12-4 | 0.01-32 |
| Citrobacter sp. | 64 | >128 | >128 | 8 | 8 | >128 | 0.25 | 0.5 | 0.5 | 64->100 | 0.12 | 0.01-0.5 |
| S. marcescens | >128 | >128 | >128 | >64 | 64->128 | >128 | 0.12->64 | 0.25-NA | 0.12-2 | 2-100 | 0.12-4 | 0.03-3.2 |
| P. mirabilis | 2-4 | 4-8 | 8 | 1-2 | 1-2 | 1 | 0.03 | 0.03-NA | 0.06 | 0.05 | 0.03-NA | 0.02-0.25 |
| Proteus sp. (In +) | >128 | >128 | >128 | 8 | 8 | >128 | 0.12 | 0.12 | 0.12 | NA | 0.25 | NA |
| P. aeruginosa | >128 | >128 | >128 | >128 | >128 | >128 | 32->128 | 32-64 | 2-16 | NA | 4-16 | 0.5-16 |
| H. influenzae | 4 | 8 | 8 | 1->81->8 | 0.5 | 1-2 | 0.03-0.06 | 0.03-0.06 | 0.12 | 0.12 | 0.03 | 0.02-0.06 |
| B. fragilis | 64 | 64 | 64 | 64 | 32 | >128 | 32 | 64 | 64 | NA | 32 | 8->125 |
| S. aureus | 0.25-0.5 | 0.25-1 | 2-4 | 0.5-1 | 1 | 1-4 | 2 | 4 | 4-8 | 8->100 | 0.5 | 1-4 |
| S. pyogenes | 0.12 | 0.12 | 0.5 | 0.06 | 0.03 | 0.25 | 0.03 | 0.03 | 0.12 | 0.2 | 0.03 | <0.01-0.06 |
| S. pneumoniae | 0.12 | 0.12 | 2 | 0.25 | 0.12 | 1 | 0.12 | 0.25 | 0.25 | 0.4 | 0.12 | 0.01-0.25 |
| N. gonorrhoeae | 0.5 | 0.5-2 | 2 | 0.5 | 0.06 | 0.12-0.5 | <0.01-0.03 | <0.01-0.01 | 0.06-0.12 | NA | <0.01 | <0.01-0.06 |

[a] Refs. 241, 311, 313.

[b] Abbreviations: CET, cephalotin; CEZ, cefazolin; CEX, cephalexin; CMD, cefamandole; CXM, cefuroxime; CFL, cefaclor; CTX, cefotaxime; CTR, ceftriaxone; CAZ, ceftazidime; CFM, cefixime; CPR, cefpirome; CPM, cefepime.

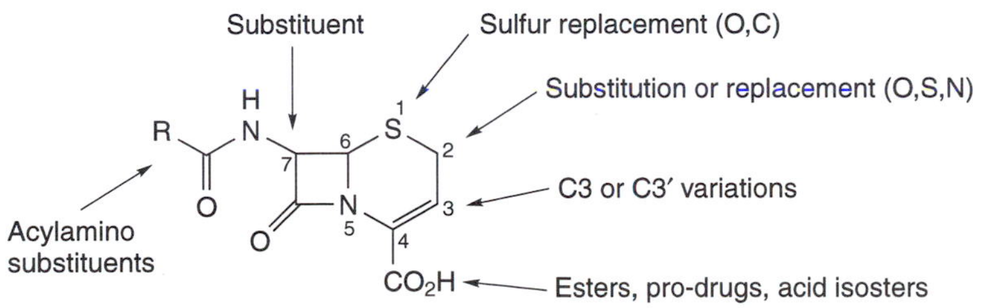

**Figure 14.8.** Structural variations of the cephalosporin ring system.

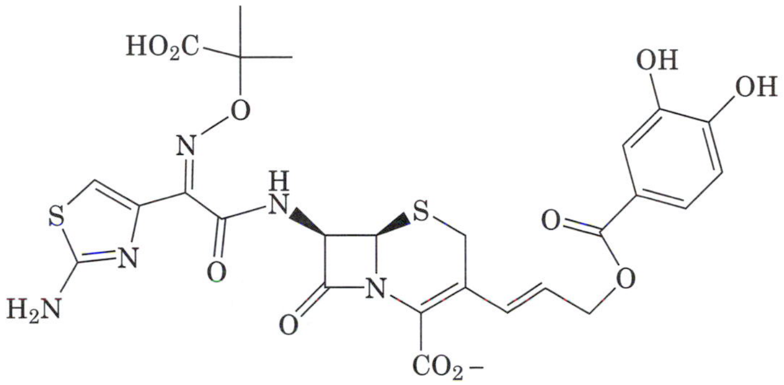

(76)

### 5.2.3 The Total Synthesis of Cephalosporin C.

Intermediates such as (**79**) (329, 330) have been used to provide the cephalosporin lactone ring system (**80**) and deacetylcephalothin (331). Later, by use of a [2 + 2] cycloaddition reaction between the thiazine (**81**) and the ketene derived from azidoacetyl chloride, the total racemic synthesis of cephalothin was achieved (332–334).

To date, however, the only complete synthesis of cephalosporin C is that described by

(77) GR 69153

Woodward in his Nobel lecture of 1965 and published in 1966 (335, 336) (Figs. 14.9 and 14.10). Protection of the nitrogen, sulfur, and carboxylic acid of L(+) cysteine (**82**) provided the cyclic intermediate (**83**). Introduction of the hydrazino group gave (**84**) that, by oxidation with lead tetracetate and treatment with sodium acetate, was converted into the *trans*-hydroxy ester (**85**). Formation of the mesylate, inversion of the stereochemistry by displacement with azide, and reduction provided the *cis*-amino ester (**86**) that gave the key β-lactam intermediate (**87**), of which, on cyclization, the absolute stereochemistry was confirmed by X-ray crystallography.

This β-lactam derivative was reacted with dialdehyde (**88**) in a Michael addition manner to yield (**89**). Treatment with trifluoroacetic acid removed both nitrogen- and sulfur-protecting groups and resulted in cyclization to the cephalosporin precursor (**90**), in which the amino group was then acylated with the suitably protected D-α-aminoadipic acid side chain, forming (**91**). Reduction, acetylation, and equilibration provided the cephalosporin C ester (**92**), from which the protecting groups

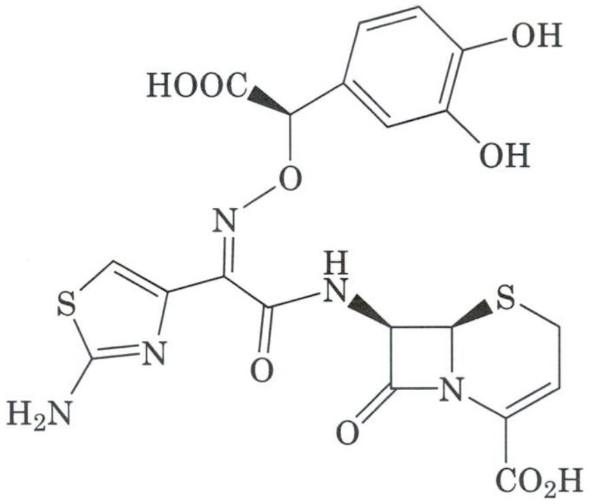

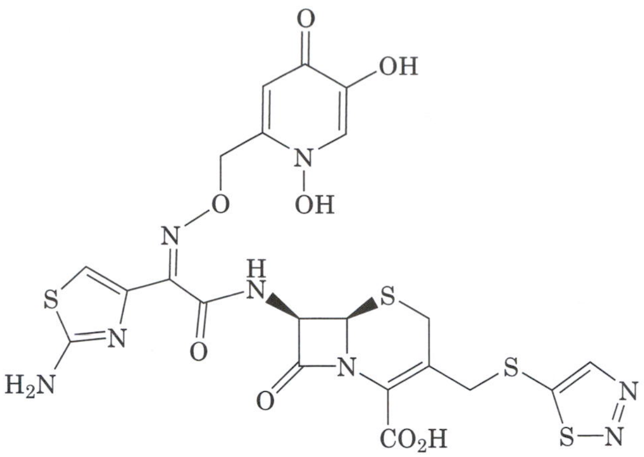

(78) KP-736

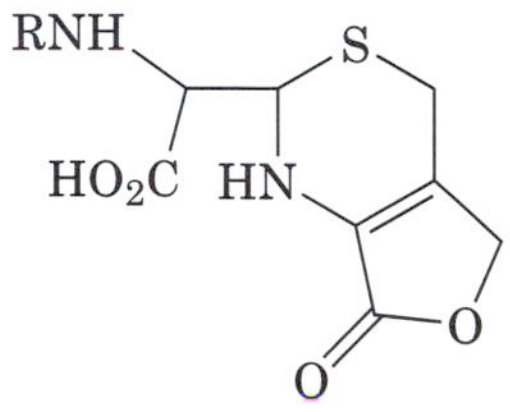

(79)

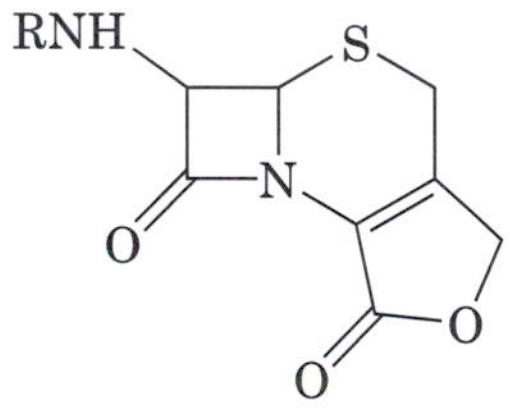

(80)

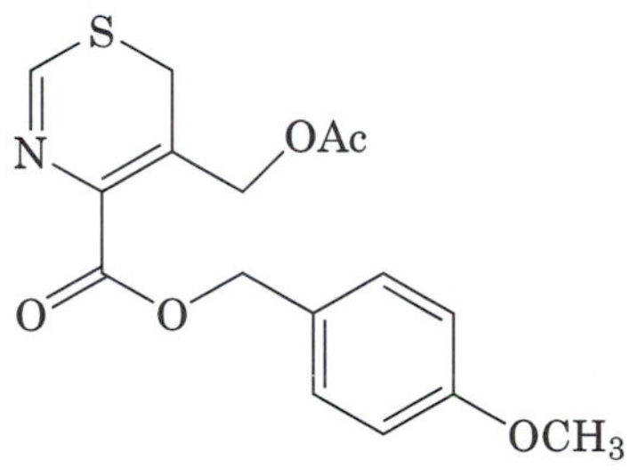

(81)

were removed with zinc and acetic acid to give the free acid (93), which was identical to a sample of authentic material.

*5.2.3.1 The Penicillin Sulfoxide-Cephalosporin Conversion.* Early work on the chemistry of penicillin sulfoxide by Morin (337) demonstrated the thermal rearrangement of (94) in the presence of acid to the deacetoxycephalosporin ring system (96). Speculation that this transformation occurred through the sulfenic acid intermediate (95) was later confirmed by its isolation in a crystalline form (338). The rearrangement of (94) to (96) afforded an attractive route to cephalosporins lacking the C3-acetoxymethyl group of the natural derivative, starting from a relatively cheap chiral starting material while retaining the stereochemical integrity of the β-lactam ring. The chemistry of the penicillin sulfoxides and their use for the interconversion of β-lactam antibiotics have been extensively reviewed by Cooper and coworkers (339–341). Here is reported only the process leading to the commercially available cephalexin (60) and cefaclor (61) (Fig. 14.11).

In the case of cephalexin (62) the trichloroethyl ester of penicillin V sulfoxide (94) was successfully rearranged to the deacetoxycephalosporin, followed by cleavage of the C7 side chain and acylation with a suitably protected D-α-phenylglycine. Removal of the amine- and acid-protecting groups gave a synthesis that could be adapted for the production of cephalexin (62) on a multikilogram scale (342). The

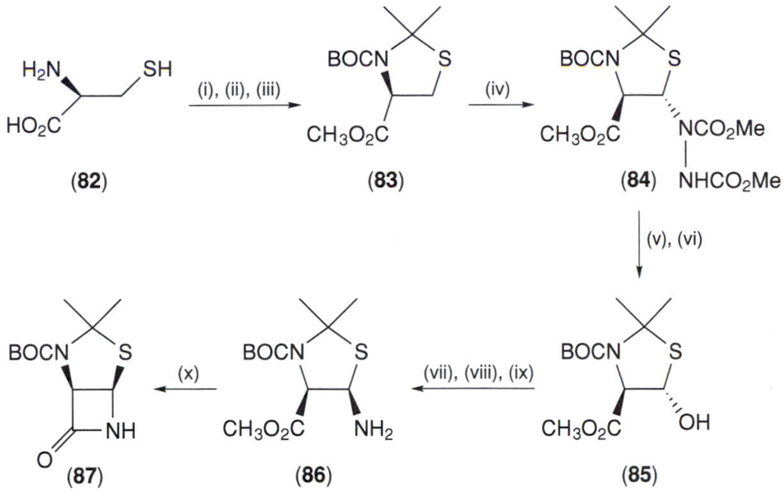

Reagents: (i) acetone; (ii) BOCCl, pyridine; (iii) $CH_2N_2$; (iv) $MeO_2CN = NCO_2Me$; (v) $Pb(OAc)_4$; (vi) NaOAc, MeOH; (vii) $MeSO_2Cl$, DIPEA; (viii) $NaN_3$; (ix) Al/Hg; (x) triisobutylaluminium.

**Figure 14.9.** Woodward's synthesis of cephalosporin C.

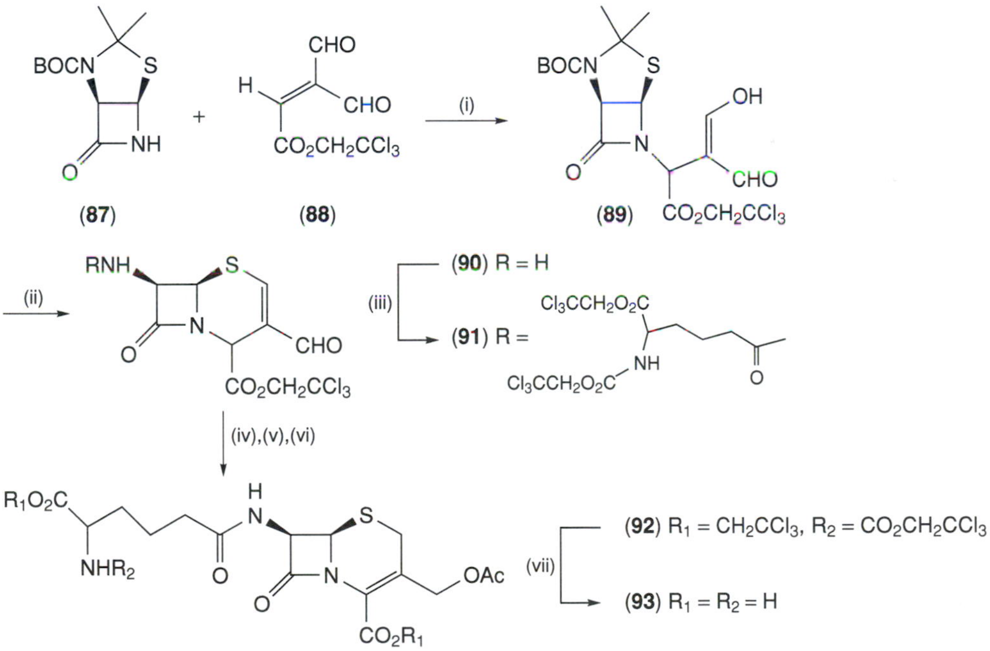

Reagents: (i) 80°C; (ii) $CF_3CO_2H$; (iii) acylation; (iv) diborane; (v) $Ac_2O$, pyridine; (vi) pyridine; (vii) Zn, $CH_3CO_2H$.

**Figure 14.10.** Woodward's synthesis of cephalosporin C.

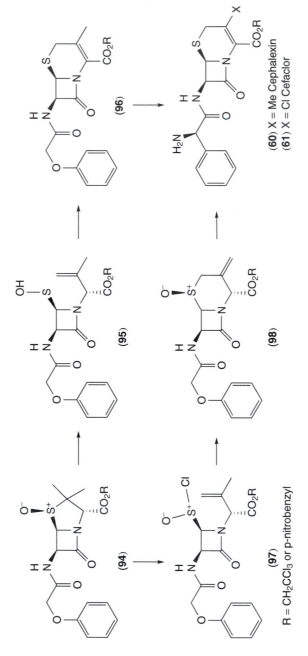

**Figure 14.11.** Rearrangement of penicillin sulfoxide to cephalosporins.

(60) X = Me Cephalexin
(61) X = Cl Cefaclor

R = CH₂CCl₃ or p-nitrobenzyl

648

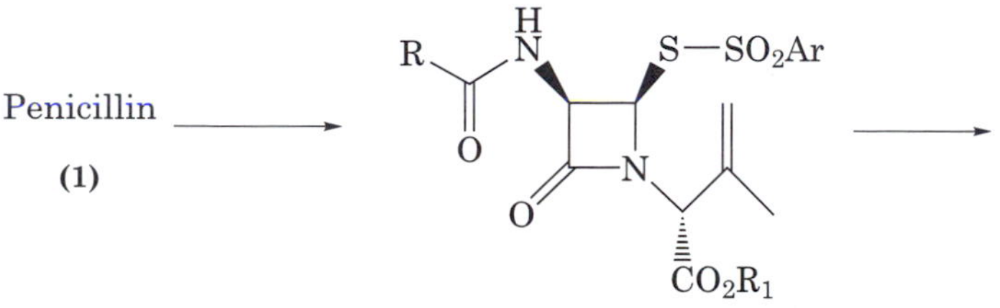

Penicillin ⟶

(1)

(99)

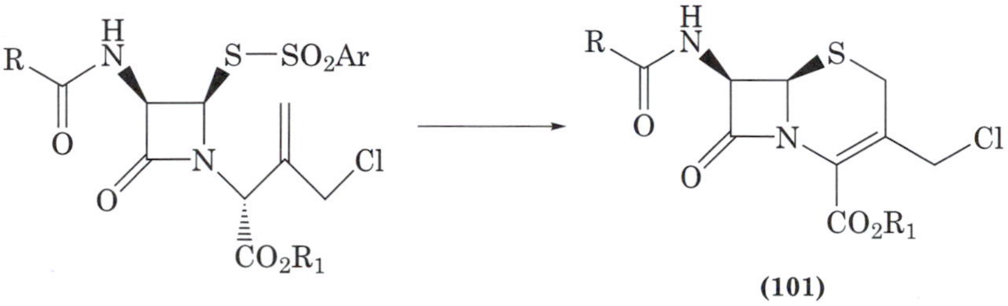

(100) ⟶ (101)

synthesis of cefaclor (59) stems from work that showed that penicillin sulfinyl chlorides (97) generated from (94) afforded 3-exomethylene cepham sulfoxides (98) on treatment with tin chloride (IV) or other Lewis acids (343). Reduction of the sulfoxide and ozonolysis of the double bond gave the C3-hydroxycephem, which was converted to the corresponding chloride. Side-chain exchange and deprotection provided a viable synthesis of cefaclor (59) (291). 4-Arylsulfonylthioazetidin-4-ones (99) have been efficiently converted (yields of 70–90%) to 3-chloromethyl cephalosporins (101) through an electrolytic ene-type chlorination to (100) and subsequent base-catalyzed ring closure (344). This procedure has provided the basis for the successful development of a commercially viable process for the synthesis of the key cephalosporin intermediate 7-amino-3-chloromethyl-3-cephem-4-carboxylic acid, p-methoxybenzyl ester (ACLE), as well as other derivatives, by the Otsuka Chemical Company of Japan. Variations of the process have also afforded 3-hydroxy cephalosporins (345).

**5.2.4 C7-α-Substituted Cephalosporins.** In 1971 two naturally occurring cephalosporins possessing a C7-α-methoxy substituent were isolated from a Streptomyces strain, and

shown to be C7-α-methoxycephalosporin C (102) and the C3-carbamate (103) (274).

Subsequently, several further examples of this type of natural product were discovered and characterized (285, 346). All possess the

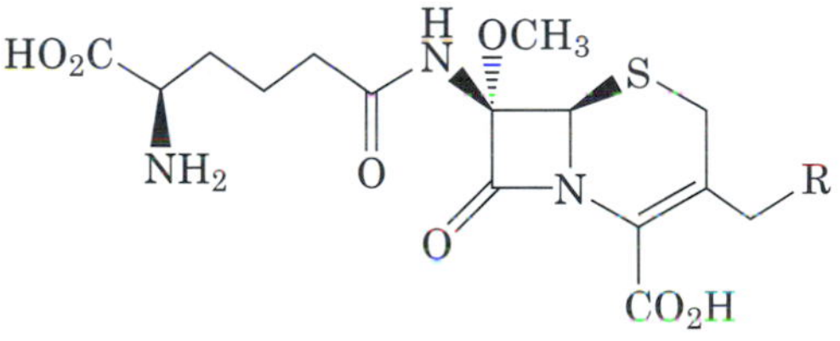

(102) R = OCOCH₃
(103) R = OCONH₂

(D)-α-aminoadipic acid side chain but differ in the nature of the C3-substituent. Interest in this class of compounds arises primarily from their intrinsically higher resistance to hydrolysis by β-lactamases. Collectively, they have been called the cephamycins and their chemistry and biology have been comprehensively reviewed (248). Later, a range of C7-α-methoxylated cephalosporins (104), of bacterial origin possessing oligopeptide side chains at C3 and known as the cephabacins-M, have been reported (347). In addition, a new structural class incorporating a C7-α-formylamino sub-

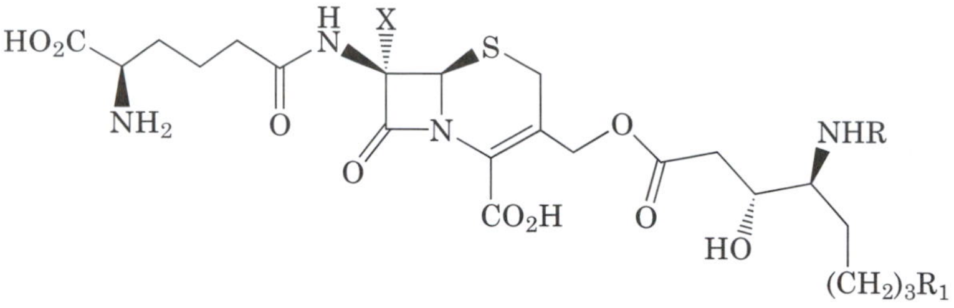

(104) Cephabacins-M    X = OCH$_3$
(105) Cephabacins-F    X = NHCHO

stituent (105) has been isolated (275, 348, 349). These have been named as the cephabacins-F or alternatively as the chitinovorins A-D (275, 349, 350).

### 5.2.4.1 C7-α-Methoxylated Derivatives.

As with the cephalosporins, the first area of SAR to be investigated was the replacement of the α-aminoadipic acid side chain with other acylamino variations. Methods involving acyl-exchange reactions (351) and removal of the side chain using imines and imidoyl chlorides (352–354) or through an oxamic acid derivative (355) have all been described, together with the effect on antibacterial activity (248). Numerous approaches have been directed toward the chemical introduction of the 7α-methoxy, and other substituents into the cephalosporin ring system (248). Many have also found application in the penicillin field (Section 5.1). Of the range of groups introduced, however, it is the methoxy and recently the formamido group that have been found to provide the best combination of stability to β-lactamases and antimicrobial potency (356, 357).

Although initial approaches to methoxylation used C7-diazo cephalosporins (253), as in the penicillins, the most widely used methods rely on addition of methoxide to acylimine intermediates such as (106) (358), sulfenimines

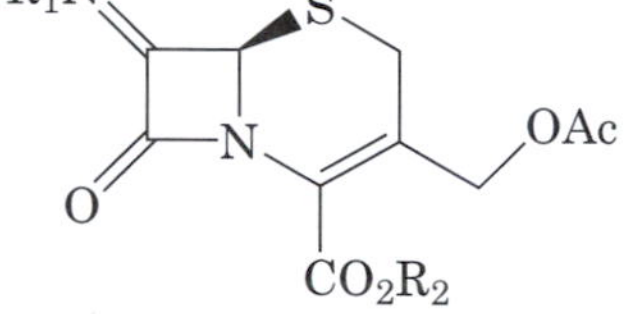

(107) (256), or ketenimines (359). Addition to the quinone methide derivative (108) is a method that has found application in the synthesis of cefmetazole (109) (360, 361), a semisynthetic cephamycin.

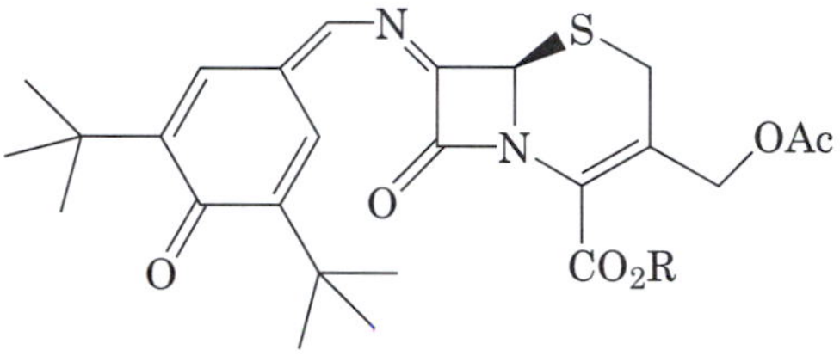

(108)

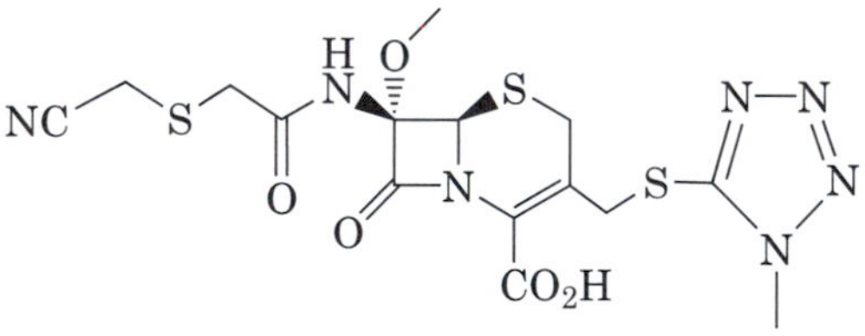

(109) Cefmetazole

Again, in a manner similar to that of the penicillins, an alternative approach to acylimine formation uses carbanion formation α to the β-lactam carbonyl followed by reaction with an electrophile, to give a substrate suitable for the introduction of the methoxy group. Thus, generation of the anion from the Schiff base (110) and reaction with methyl methanethiosulfate gave the corresponding C7-α-thiomethyl cephalosporin (111), which can be converted to the acylamino derivative (112). Solvolysis in methanol in the presence of mercury salts then leads to C7-α-methoxy cephalosporin (251).

(106) R$_1$ = RCO
(107) R$_1$ = RS

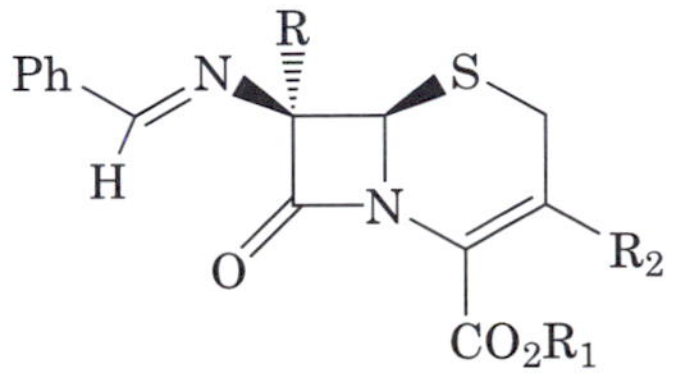

(110) R = H
(111) R = SCH$_3$

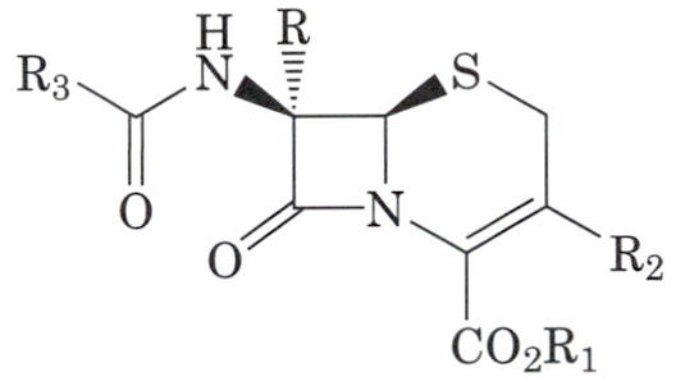

(112) R = SCH$_3$
(113) R = OCH$_3$

Factors related to the methylthiolation of cephalosporins have been extensively studied, particularly by workers at Squibb (362). Approaches to the total synthesis of methoxylated cephalosporins have been described by Merck researchers (332–334) and Kishi (363). The first member of this group to be used clinically was cefoxitin (64), which is very stable to β-lactamases, whereas the carbamoyl group provides stability to esterases (300). Cefoxitin (64) has a fairly broad spectrum of activity, including anaerobic bacteria such as *Bacteroides fragilis*, a common pathogen resulting from abdominal surgery (364). Examples of newer cephamycins having a C3-*N*-methyltetrazolylthiomethyl substituent are cefmetazole (109) (361, 365) and cefotetan (65) (304), reported as showing improved Gram-positive activity or pharmacokinetic properties over those of cefoxitin (64).

*5.2.4.2 C7-α-Formamido Derivatives.* Before the discovery of the natural products, the Beecham group had already shown the utility of the formamido substituent with the synthesis of several highly active, β-lactamase stable C6/C7-α-substituted penicillins and cephalosporins (258, 259). One of the initial approaches to the conversion of the unsubstituted cephalosporin ring system into the formamido nucleus was by way of (114), readily available from work on the methoxy

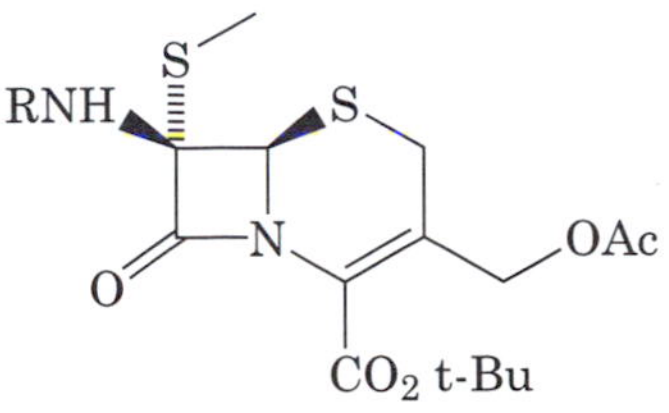

(114) R = H
(115) R = R$_1$CO

series. Acylation provides (115), from which the methylthio group can be displaced with ammonia in the presence of a mercury salt followed by formylation and acid deprotection, to give the appropriate C7-α-formamido cephalosporin acid (116). Alternatively, the form-

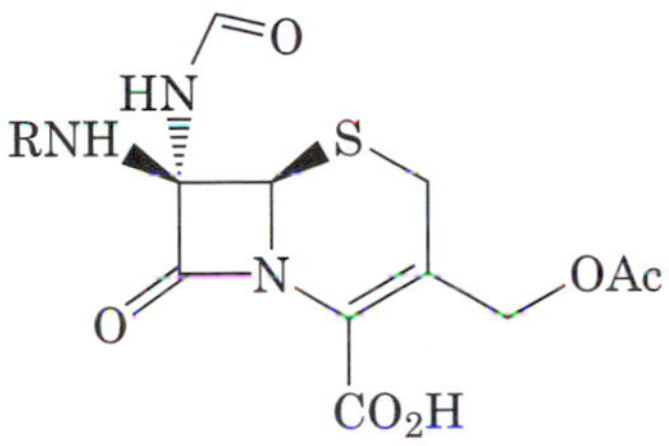

(116) R = R$_1$CO
(117) R = H

amido group can be introduced directly by treatment of (115) with *N,N*-bis(trimethylsilyl)formamide in the presence of mercuric acetate (259, 366).

Other methods for incorporating the formamido substituent have been reported (260, 261), as well as high yielding "one-pot" large-scale preparation of the formamido nucleus by use of silyl protection and quinone methide methodology (367).

In a series of structure-activity studies varying both the C7-acylamino side chain and C3-substituent it was demonstrated that the formamido cephalosporins in many cases showed advantages over the cephamycins in conferring high β-lactamase stability without compromising antimicrobial activity (368, 369). In particular the catecholic derivative (118, BRL 41897) was identified as a broad spectrum agent active against both Gram-negative and Gram-positive bacteria with high potency against *Pseudomonas aeruginosa*, and which reached the stage of single dose volun-

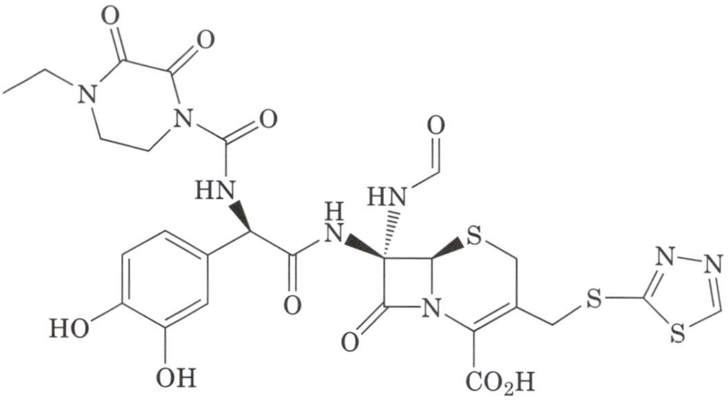

(118) BRL-41897

teer studies, before adverse toxicological effects caused its withdrawal from development (370).

### 5.2.5 Oxacephalosporins (1-Oxadethiacephems).

The synthesis of 1-oxadethiaceph-3-em-4-carboxylic acids (1-oxacephems) was initially carried out by the Merck group in 1974 with the synthesis of racemic 1-oxacephalothin (119) (371). This showed activity comparable to cephalothin (58) itself.

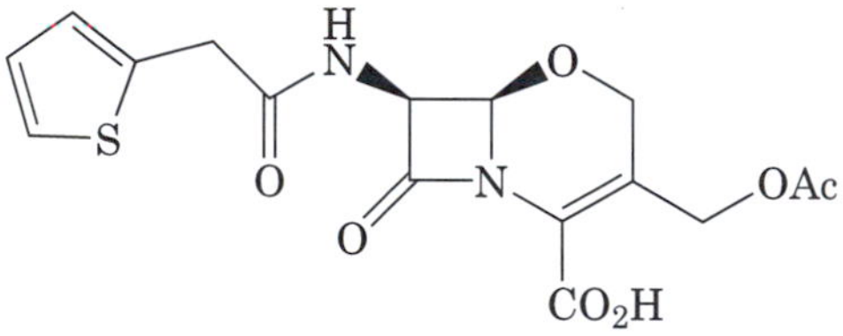

(119)

Subsequent examples also demonstrated the potential utility of these nonnatural 1-oxa analogs over the natural cephalosporins (372–375). Extensive SAR studies by the Shionogi group led to the identification and development of moxalactam (latamoxef, 120) as a broad-spectrum clinical agent equal to or better than many third generation cephalosporins (376). However, moxalactam possesses poor activity against S. aureus, while the manifestation of side effects, including problems associated with blood clotting due to decreased vitamin K synthesis, have considerably restricted its clinical use (377, 378). Fur-

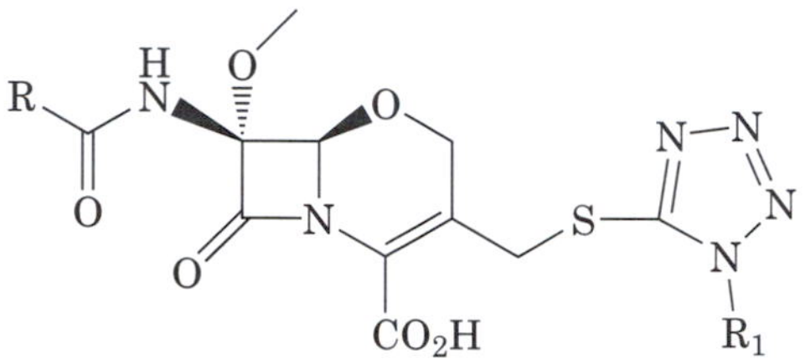

(120) Moxalactam

R = 

R₁ = CH₃

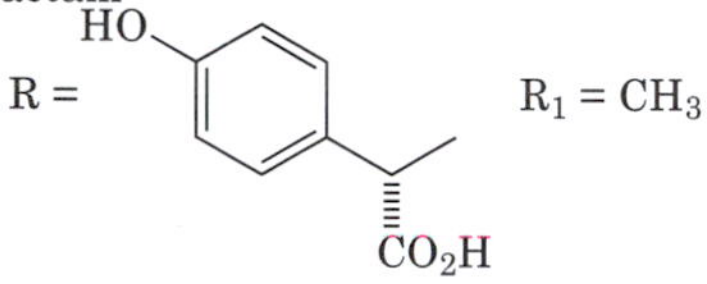

(121) Flomoxef  R = F₂CHSCH₂
R₁ = CH₂CH₂OH

ther studies identified flomoxef (121) with improved Gram-positive activity as an agent to overcome these problems (379).

The identification of moxalactam (120) and flomoxef (121) as clinical candidates led to an intensive effort by Shionogi to establish a stereocontrolled, commercially viable process for the production of the compound, and other 1-oxacephems. The culmination of this effort can be illustrated (Fig. 14.12) by the conversion of the readily available penicillin nucleus (6-APA) to the 1-oxacephem nucleus (129) used for the production of both moxalactam (120) and flomoxef (121) (378, 380).

Protection, oxidation, and epimerization of the benzoyl derivative of 6-APA (122) readily gave the epipenicillin β-sulfoxide (123), which was converted to the key epi-oxazolinone azetidinone intermediate (124). Transformation

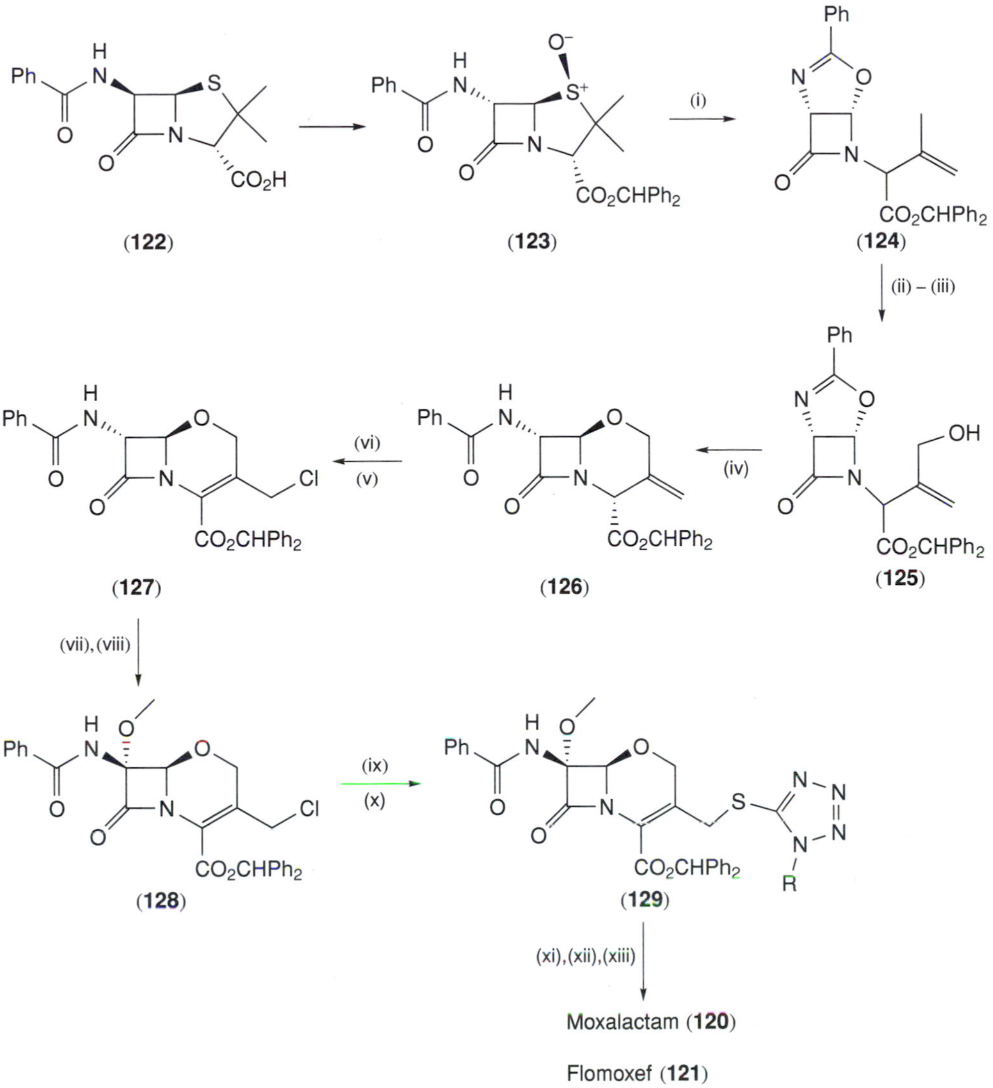

Reagents: (i) Δ/Ph₃P; (ii) Cl₂/Base, (iii) I₂/CuO/DMSO/H₂O; (iv) BF₃-Et₂O; (v) Cl₂/hν; (vi) DBN; (vii) t-BuOCl/LiOMe; (viii) H⁺; (ix) Na₂S₂O₃; (x) NaS-Tetrazole; (xi) PCl₅/Pyridine; (xii) MeOH; (xiii) Acylation.

**Figure 14.12.** Conversion of penicillin to 1-oxacephems.

to the allylic alcohol (125) was followed by Lewis acid catalyzed cyclization to the oxacepham ring system (126), and then photochemical halogenation to the chloromethyl 1-oxo-cephem ester (127). Introduction of the methoxy group and appropriate tetrazole substituent afforded (128). Deacylation then gave in high overall yield the ester (129), which is a suitable substrate for conversion to moxalac-

tam (120) or flomoxef (121) on a commercially viable scale. A range of 2-methyl analogs (e.g., 130) have been reported but no clinical candidates were selected (381).

Subsequent to the discovery of the C7-α-formamido-substituted cephalosporins the synthesis of a series of analogous 1-oxacephem antibiotics was described (370). Many of these show a level of antibacterial activity and β-lac-

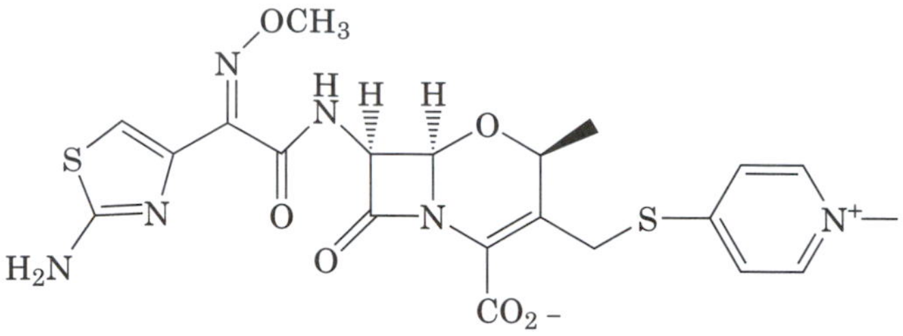

(130)

tamase stability comparable to that of moxalactam or cephalosporins such as ceftazidime (23) (382).

### 5.2.6 Carbacephalosporins (Carbacephems).
Another series of nonnatural cephalosporins of recent interest are the 1-carbacephems, particularly with the discovery of the oral compound loracarbef (131), which has a spec-

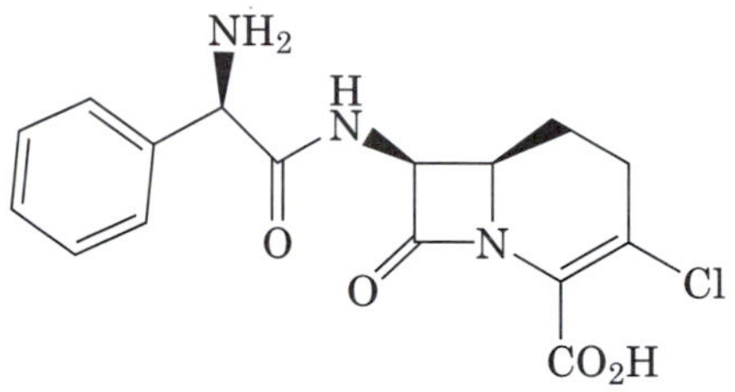

(131) Loracarbef

trum of activity similar to that of cefaclor, but with greater chemical stability, a longer half-life, and better oral absorption (383, 384). The additional chemical stability conferred by the carbacephem nucleus has been further demonstrated by the synthesis of directly linked quaternary derivatives such as (132), which have not been reported in the cephalosporin

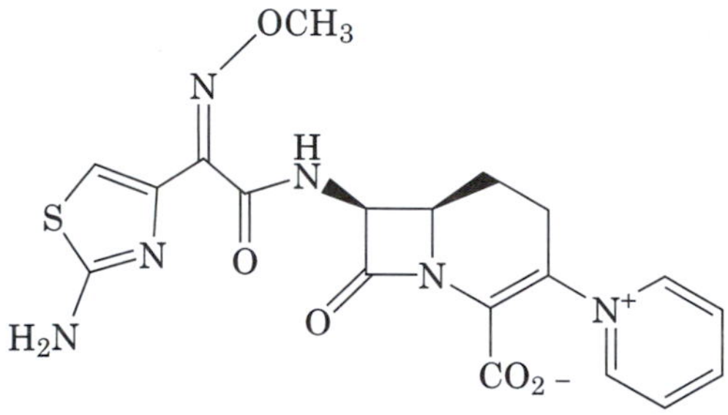

(132)

series (385). The Lilly group has also described detailed SAR leading to the identification of agents, exemplified by (133), with activity against methicillin-resistant Staphylococci (386).

### 5.3 Penems

Penems, unlike penams, carbapenems, and cephalosporins (Section 5.5), have not been found in nature. However, penem development has clearly been influenced by the structures of closely related β-lactam natural products. In fact, when first described by Woodward, penems were viewed as hybrid molecules that combined the chemical structural features responsible not only for β-lactam reactivity of penams and cephalosporins (387) but also, so they hoped, for their biological properties.

### 5.3.1 Synthesis
#### 5.3.1.1 Woodward's Phosphorane Route.
The first penem synthesis to be described was that of the 6-phenoxyacetamido penem (140) (Fig. 14.13) (387, 388). The strategy adopted involved a semisynthetic approach based on the penicillin derived Kamiya disulfide (134) and employed an intramolecular Wittig reaction as the key step (135 to 139) in the formation of the fused thiazoline ring system. The demonstration of antibacterial activity for the penem (140), despite its limited stability, led to a proliferation of studies on this new β-lactam system.

#### 5.3.1.2 Extension of the Phosphorane Route.
Although alternative routes have been developed (Section 5.3.1.3), the most versatile routes to penems remain those involving the phosphorane strategy or closely related methods involving 2,3-double bond formation. To-

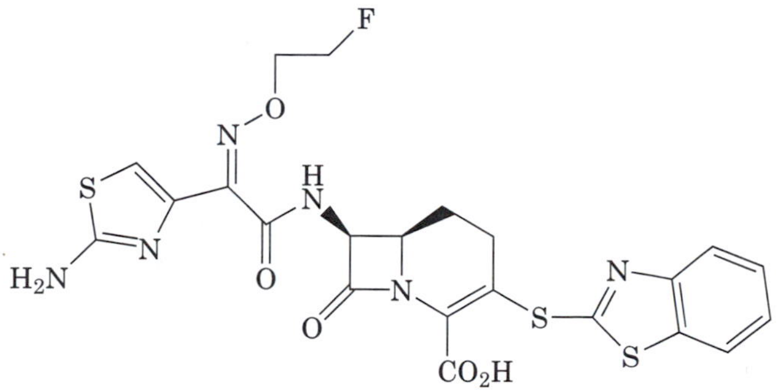

(133)

tal synthesis embodying this strategy has been used to prepare a range of 2- and 6-substituted racemic penems. Nucleophilic displacement of the acetate function of 4-acetoxyazetidinone (141) (389) provided azetidinones (142–144) that were elaborated to give the penems (145) (390), (146) (391), and (147) (392).

Similar treatment of 3-substituted (or disubstituted) acetoxyazetidinones provided examples of 2-substituted 6-alkyl and 6,6-dialkyl-penems (393). Acetoxyazetidinone (141) also

provided the starting point for the preparation of the racemic mixture (148) of a penem containing the 1-hydroxyethyl substituent found in the naturally occurring carbapenems (394). The demonstration that, in common with the carbapenems, maximal antibacterial activity resided in the racemate with the *trans*-6-[1(*R*)-hydroxyethyl] group together with the earlier indication that it was also associated with 5(*R*)-stereochemistry (395), provided the impetus for the development of routes to chiral penems.

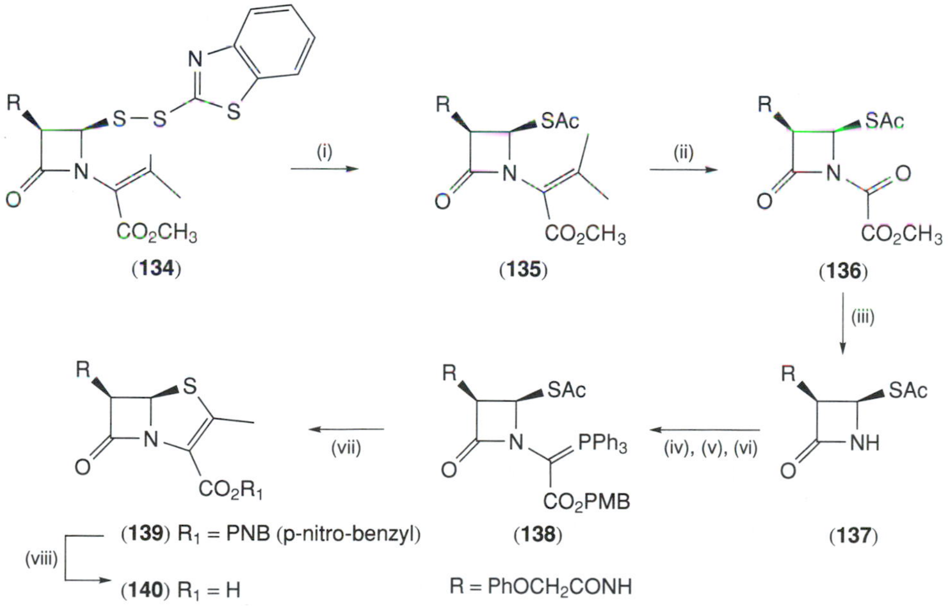

Reagents: (i) Ph$_3$P, Ac$_2$O, AcOH, Pyridine; (ii) O$_3$; (iii) MeOH; (iv) p-nitro-benzyl glyoxylate ethyl hemiacetal; (v) SOCl$_2$, base; (vi) Ph$_3$P, base; (vii) Δ, Toluene; (viii) H$_2$, Pd/C.

**Figure 14.13.** Woodward's synthesis of penems.

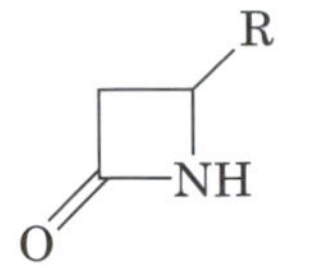

(141) R = OAc
(142) R = SCOCH₃
(143) R = SC(=S)SEt
(144) R = SC(=S)OEt

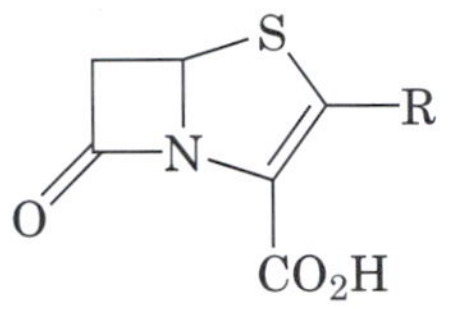

(145) R = CH₃
(146) R = SEt
(147) R = OEt

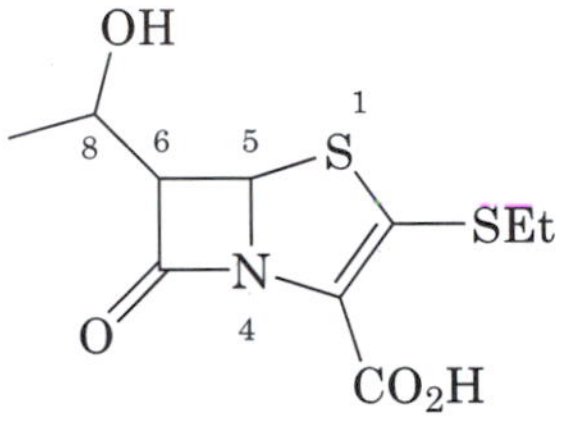

(148)

A publication describing the stereochemical outcome of alcohol condensations of the readily available 6,6-dibromopenicillanates (149) with acetaldehyde (396), made them attractive starting materials for the synthesis of chiral 6-(1-hydroxyethyl) penems (Fig. 14.14).

Reductive removal of the halogen atom of the major product (150), having the desired 8(R) stereochemistry, provided the (5R,6S,8R) penam (151). A number of strategies have been developed for opening the thiazolidine ring of the suitably protected penams (151) to give intermediates suitable for elaboration to penems. For example, chlorinolysis of (151) provided access to the chloro azetidinone (152) (397), whereas cleavage using mercuric acetate gave the acetoxy azetidinone (153) after oxidative removal of the nitrogen substituent (398). Reaction of either (152) (397) or (153) (399) with sodium trithiocarbonate proceeded with retention of configuration to give azetidi-

nones (154), which were converted to the desired trans-8(R)-penems (155). In these studies it was noted that the high temperatures required to cyclize the intermediate phosphoranes was also responsible for a novel C5-epimerization of the penem, possibly involving the intermediacy of a betaine leading to the unwanted cis-penems (156) (399).

Other methods for the cleavage of the thiazolidine ring of penicillanates retaining the stereochemistry of the carbon bearing the sulfur atom, to produce suitable intermediates for penem synthesis, have been described. Silver-assisted cleavage afforded mercaptides (400), whereas the sulfenic acids generated by the thermolysis of penicillanate sulfoxides have been trapped by acetylenes (401) and thiols (402). Most of the more efficient synthesis reported use the commercially available tert-butyldimethylsilyl protected acetoxy azetidinone (153), which is prepared by total synthesis.

### 5.3.1.3 Alternative Methods for the Synthesis of Penems.

Figure 14.15 summarizes the more widely applicable alternative strategies, which have been used to synthesize penems (163). Although demonstrated, syntheses involving S-C2 and N-C3 ring closures have been little used. The diazoketone/carbene insertion route widely used for the synthesis of carbapenems has not been demonstrated for penems (see Section 5.5) (403). In fact, lack of reactivity appears related to the interaction of the sulfur atom with the carbenoid derived from the diazo species.

The reactivity of the oxalimide carbonyl group provides two useful alternatives to the original phosphorane route for the formation of the 2,3-double bond. Treatment of thioesters (160a) (404, 205) and trithiocarbonates (161a) (405, 406) with trialkylphosphites at high temperatures gave the corresponding carbon- and sulfur-substituted penems (163). In a detailed study (407) it was demonstrated that for the thioesters, carbene generation and interception gave the intermediate trialkoxyphosphoranes (160b), which underwent Wittig-type cyclization. In contrast, it has been proposed that cyclization of the trithiocarbonates proceeds through insertion of the carbene into the more reactive thiocarbonyl group, to form a tricyclic episulfide

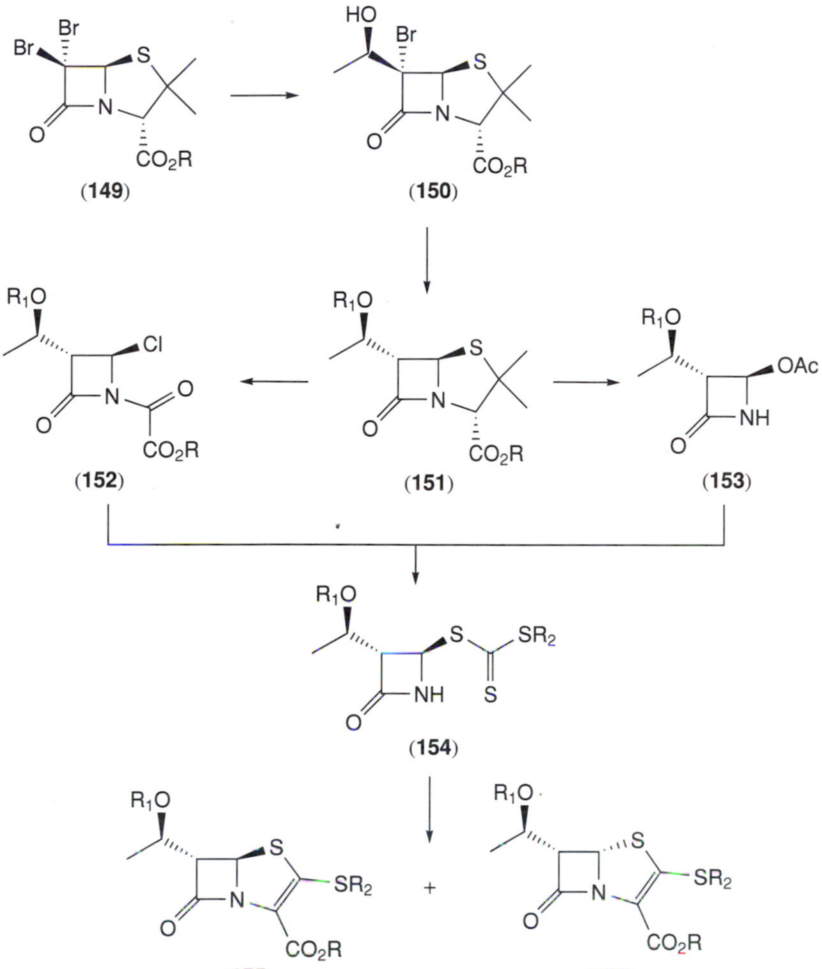

**Figure 14.14.** Synthesis of chiral penems.

(**162**), which is readily desulfurized to give the penem (**163**) (405, 406). The lower temperatures and shorter reaction times required for cyclization of trithiocarbonates, by use of this procedure, avoided the C5 epimerization experienced with the conventional phosphorane route. The high reactivity of the oxalimide carbonyl has also been exploited in the cyclization of "inverse phosphoranes" of the type (**161b**) (408). Sulfur-C5 bond formation was employed for the synthesis of the first examples of 2-aryloxy penems (**163**, Y = OAryl) (409).

Stereospecific cyclization of an intermediate of the type (**158**), in which the leaving group (Z) was chlorine, proceeded with inversion of configuration to give the 5 (R)-penem (**163**). Ring contraction of 2-thiacephems of the type (**157**) has been the subject of extensive studies (410). The unpredictable stereoselectivity of desulfurizations that use phosphorus (III) reagents was overcome by oxidation to the 1,1-dioxides, which underwent stereospecific thermal desulfonylation to give the corresponding 5-(R)-penems. Despite this, it was concluded that the route offered no practical advantage over routes that used 2,3-double bond-forming strategies. Penems possessing leaving groups at the two positions have

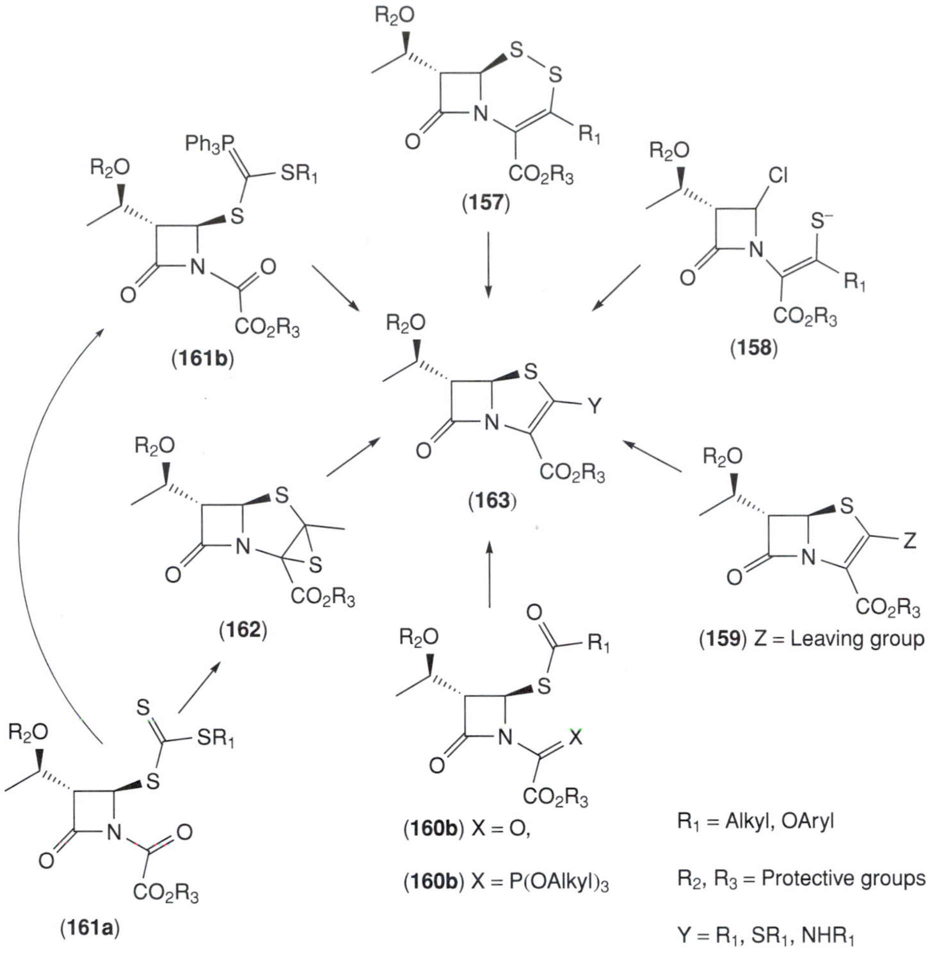

**Figure 14.15.** Alternative methods for the synthesis of penems.

proved to be useful intermediates for the synthesis of a large number of analogs. Regioselective oxidation of the sulfide provided 2-ethylsulfinylpenems [(**159**), Z = S(O)Et], which underwent displacement reactions with thiolates, to give a range of thiosubstituted penems (411). In a conceptually similar approach, amines have been shown to displace phenolic-leaving groups from penems of the type (**159**) (Z = OAr) to provide a general route to 2-substituted aminopenems (**163**, Y = NHR$_1$) (412).

Later, the generality of this approach has been realized with the demonstration of displacement of the triflate group of penems

(**159**, Z = OSO$_2$CF$_3$) by thiolates and cuprates, to give both sulfur- and carbon-substituted penems (413).

**5.3.2 Biological Properties.** Unlike the antibacterially inactive penicillanic and cephalosporanic acids, the 6-unsubstituted penems exhibited good activity against Gram-positive bacteria but displayed more modest potency against Gram-negative strains (390). Penems such as (**140**), incorporating a 6β-acylamino side chain, were of limited stability and showed only weak antibacterial properties (388). In contrast to methoxylation, 6α-methylation of (**140**) improved stability, although the compound still exhibited only weak antibacterial activity (414). Similarly, 6α-methylpenems were also reported to be more stable than the methoxy analogs, whereas 6,6-dialkylpenems have been shown to be extremely stable but devoid of antibacterial activity (415).

Incorporation of the 1(*R*)-hydroxyethyl moiety found in the naturally occurring carbapenem thienamycin proved crucial for providing penems with broad-spectrum of activity and stability to β-lactamases. As with the carbapenems, activity was markedly affected by the relative stereochemistry at the three chiral centers. Further evaluation of the components within the racemic mixture of penems (**148**) revealed that the *trans*-8(*R*)-isomer was 20- to 30-fold more potent than the *trans*-8(*S*)-isomer, with the two *cis*-isomers being of intermediate potency (416). The same study showed that removal or alkylation of the 8-hydroxy group resulted in loss of β-lactamase stability and reduced potency against non-β-lactamase-producing bacteria. Increased bulk resulting from an additional methyl group or substitution of the C8-methyl group was not tolerated; 6-hydroxy-isopropyl-, 6-(1-hydroxy-1-propyl)-, and 6-(1-hydroxy-2-phenylethyl)-penems were all devoid of useful antibacterial activity (417). With the exception of the 6-(substituted) methylenepenems, which showed potent β-lactamase inhibitory properties (Section 5.7), all the penems that have progressed have 5(*R*),6(*S*),8(*R*)-stereochemistry. The relatively less demanding nature of the 2-substituent is amply demonstrated by the wide range of 6-(1-hydroxyethyl) penems, which have been described as having potent antibacterial activity. Typically, these compounds show excellent activity against Gram-positive bacteria including β-lactamase-producing strains, whereas activity against Gram-negative organisms is more modest and generally not as good as that seen with carbapenems. Except for compounds possessing a basic amino function in the 2-substituent, penems are devoid of useful activity against *Pseudomonas* species (418). Changes in 2-substituents are often accompanied by changes in susceptibility to β-lactamase and human renal dehydropeptidase-1 enzymes as well as pharmacokinetics (e.g., oral absorption); these properties have proved important in the selection of candidates for progression.

Table 14.11 lists some of the compounds that have been selected for development as potential antibacterial agents. Some of these penems [e.g., SCH 29482 (**164**) and faropenems (**169**, SUN 5555)] have oral absorption properties as acids, whereas others [e.g., ritipenem (**166**, FCE 22101)] have required the use of the prodrug approach to give good blood levels by this route.

However, development of the orally active β-lactam antibiotic SCH 29482 (**164**) has been discontinued by Schering-Plough because of an odor problem (427). This compound was in Phase II studies for treatment of gonorrhea, pneumonia, and UTI in the United States, United Kingdom, Ireland, and other European markets. In humans, it was rapidly absorbed when administered orally, with half-life of 1.5–2 h. No significant clinical or biochemical changes were seen. It was highly active (MIC < 8 μg/mL) against Gram-positive and Gram-negative bacteria, including some strains resistant to third-generation cephalosporins, but had no antipseudomonal activity.

Faropenem (**169**) is another example of an oral and injectable penem antibacterial agent with a unique spectrum of activity against aerobic and anaerobic Gram-positive and Gram-negative bacteria, excluding *Pseudomonas aeruginosa*. It is also highly stable against various β-lactamases. Faropenem was launched in Japan in 1997.

Ritipenem acoxil (FCE 22891) is the oral acetoxymethyl ester of ritipenem (**166**, FCE 22101), an injectable penem derivative for the treatment of urinary tract and lower respiratory tract infections.

## 5.4  Monobactams and Nocardicins

The isolation of the nocardicins in 1976 (428) and the monobactams in 1981 (429, 430) revealed for the first time the potential for antimicrobial activity in simple monocyclic β-lactam structures as opposed to the fused-ring systems of the penicillins and cephalosporins. This provided great impetus for new ideas on the structural features necessary for β-lactam

**Table 14.11   Penems Selected for Detailed Investigation**

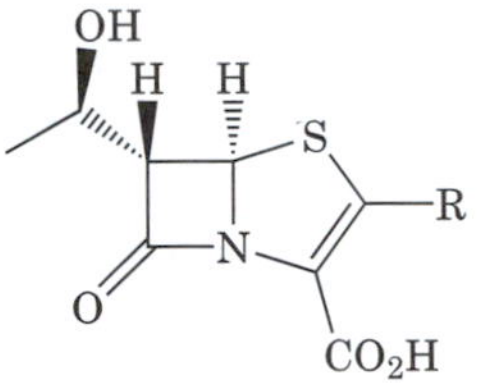

| Compound | Name/Code | R | Reference |
|----------|-----------|---|-----------|
| (164) | SCH 29482 | $SCH_2CH_3$ | 419 |
| (165) | SCH 34343 | $S(CH_2)_2OCONH_2$ | 420 |
| (166) | FCE 22101 | $CH_2OCONH_2$ | 421 |
| (167) | FCE 29464 | $CH_2OCH_3$ | 422 |
| (168) | HRE 664 | 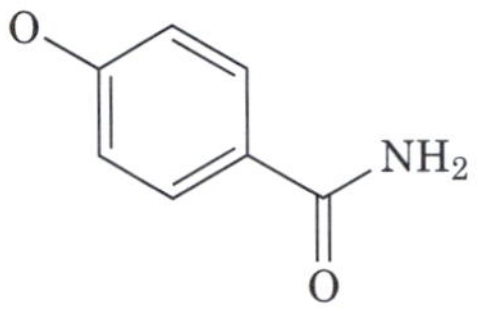 | 423 |
| (169) | SUN 5555 | 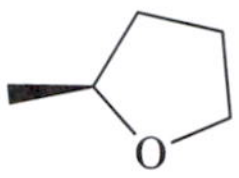 | 424 |
| (170) | CGP 37697 | 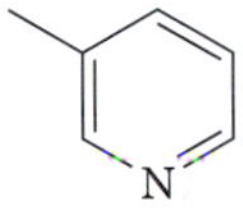 | 425 |
| (171) | CP 70429 | 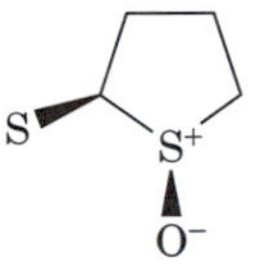 | 426 |

recognition and activation, together with the development of a wide range of synthetic methods for the construction of both natural products and analogs (431).

**5.4.1 Nocardicins.** The nocardicins were detected in a fermentation broth of a strain of *Nocardia uniformis* by a screening procedure through the use of a mutant strain of *E. coli* supersensitive to β-lactam antibiotics. The first compound to be characterized was no-cardicin A, which shows a modest level of antibacterial activity *in vitro* against Gram-negative bacteria (428). Structural determination was from spectroscopic evidence and by degradation experiments (432, 433). Several related compounds (nocardicins B-G) have also been characterized (432, 434), whereas the most re-

cent variation is the chloro-substituted derivative from a Streptomyces species (435). The structures of the natural nocardicins are shown in Table 14.12.

**5.4.1.1 Derivatives of 3-Amino Nocardicinic Acid (3-ANA) and the Synthesis of Nocardicins.** As with the penicillins and cephalosporins initial approaches to improve the potency of the natural products revolved around changing the acylamino side chain by way of the 3-amino norcardicinic acid (3-ANA) nucleus (180). Initially, this was prepared by deacylation of nocardicin C by use of microbial amidases or chemical methods (436, 437). A more practical approach made use of the reaction of the oxime group of nocardicin A with a large excess of di-*t*-butyl dicarbonate, to give

**Table 14.12   Structures of Natural Nocardicins**

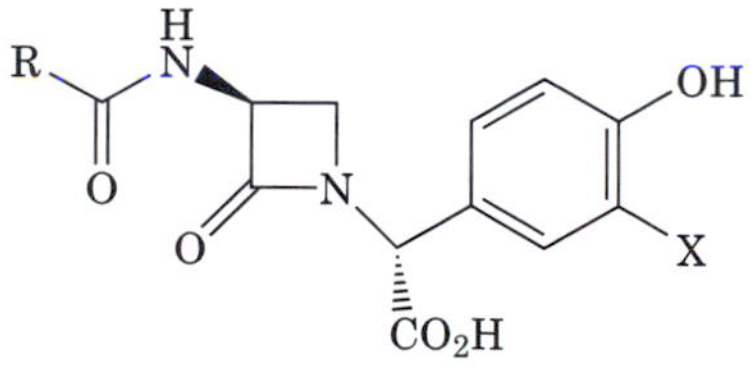

R—C(=O)—NH— (azetidinone β-lactam ring) —N— CH(CO$_2$H) — phenyl(OH)(X)

| Compound | Nocardicin | X | R | Reference(s) |
|----------|-----------|---|---|--------------|
| (172) | A | H | 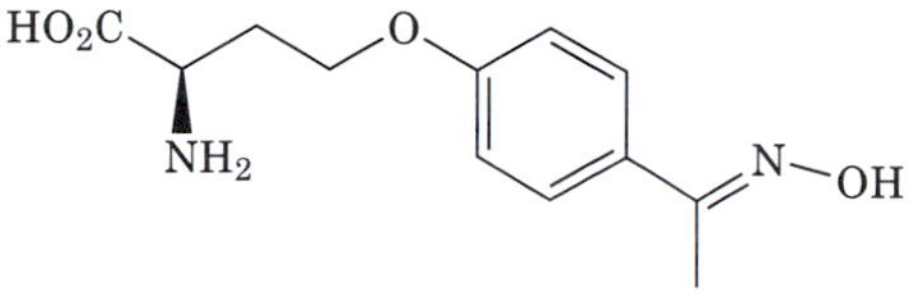 HO$_2$C—CH(NH$_2$)—CH$_2$CH$_2$—O—C$_6$H$_4$—C(=N—OH)CH$_3$ | 432, 433, 435 |
| (173) | B | H | 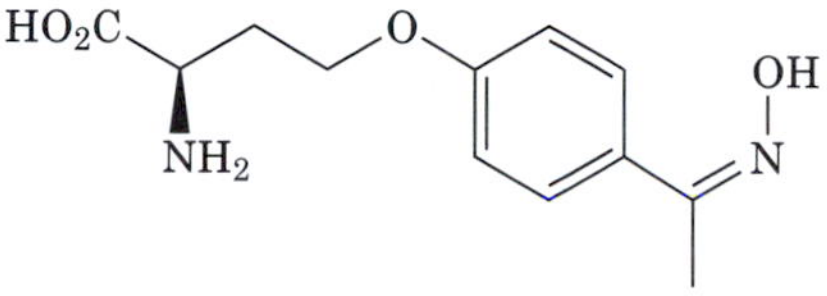 HO$_2$C—CH(NH$_2$)—CH$_2$CH$_2$—O—C$_6$H$_4$—C(=N—OH)CH$_3$ | 432, 435 |
| (174) | C | H | 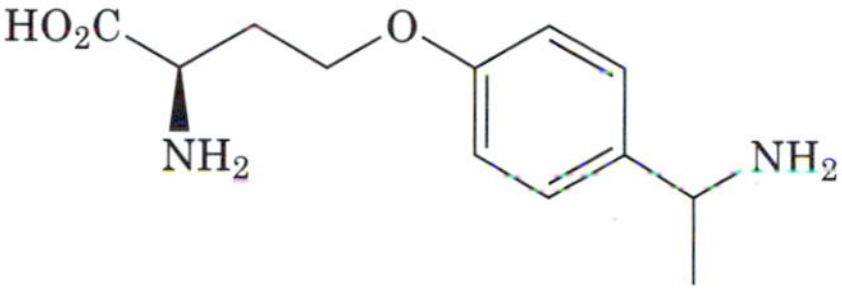 HO$_2$C—CH(NH$_2$)—CH$_2$CH$_2$—O—C$_6$H$_4$—CH(NH$_2$)CH$_3$ | 432, 435 |
| (175) | D | H | 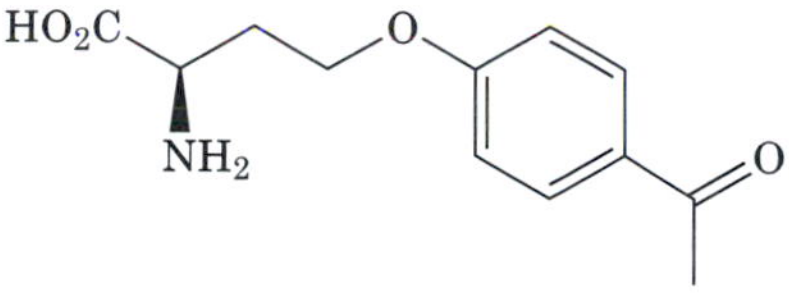 HO$_2$C—CH(NH$_2$)—CH$_2$CH$_2$—O—C$_6$H$_4$—C(=O)CH$_3$ | 432, 435 |
| (176) | E | H | 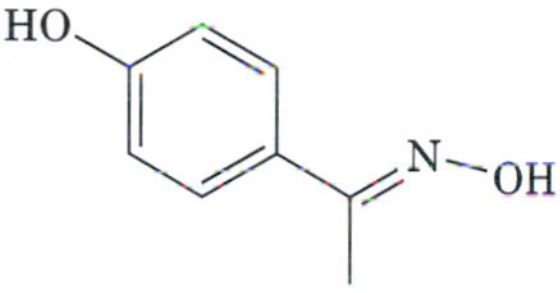 HO—C$_6$H$_4$—C(=N—OH)CH$_3$ | 432, 435 |
| (177) | F | H | 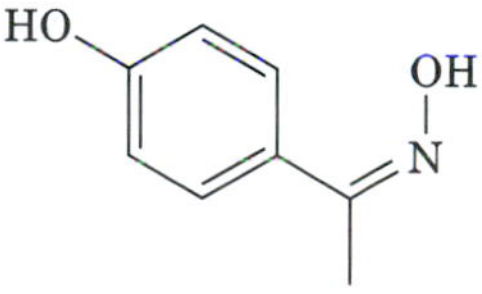 HO—C$_6$H$_4$—C(=N—OH)CH$_3$ | 432, 435 |
| (178) | G | H | 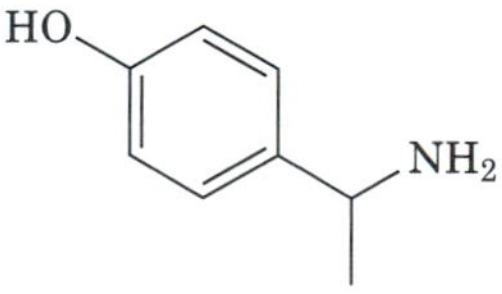 HO—C$_6$H$_4$—CH(NH$_2$)CH$_3$ | 432, 435 |
| (179) | Chlorocardicin | Cl | 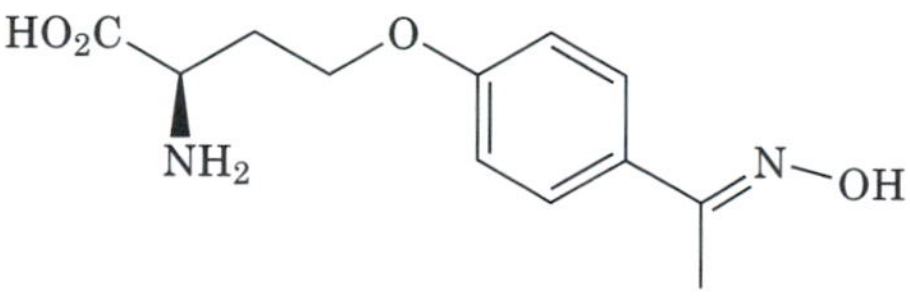 HO$_2$C—CH(NH$_2$)—CH$_2$CH$_2$—O—C$_6$H$_4$—C(=N—OH)CH$_3$ | 436 |

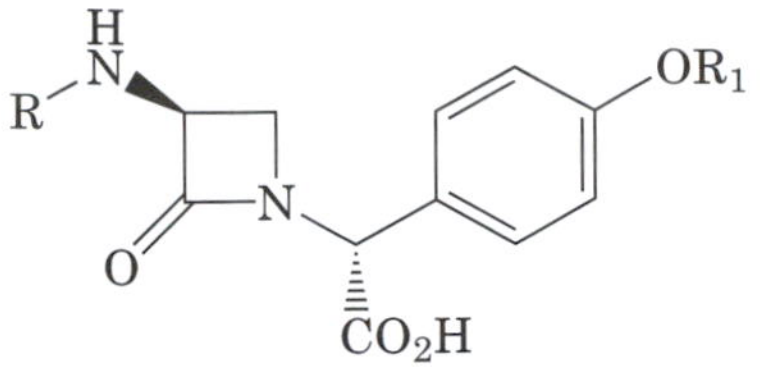

(180) R = R₁ = H (3-ANA)
(181) R = R₁ = CO₂ t-But

(184)

(181), which on treatment with trifluoroacetic acid afforded 3-ANA in excellent yield (438).

Although semisynthetic approaches to the nocardicins from penicillin-derived β-lactams have been reported (437, 439), by far the greatest effort has been directed toward totally synthetic methods that afford a versatile range of structural types. One early approach made use of the cycloaddition reaction of the thioimidate (182) with the ketene generated

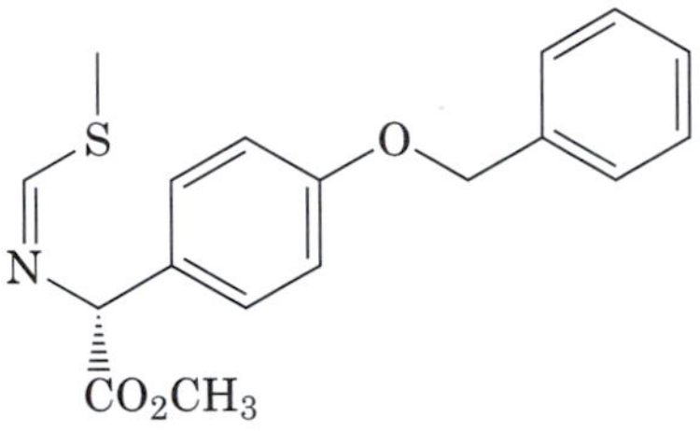

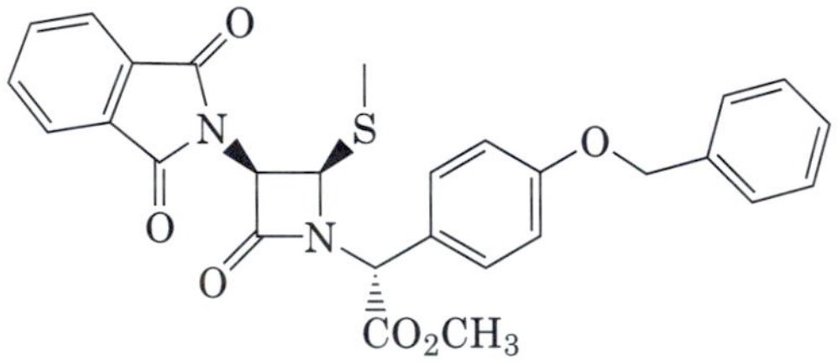

first examples of N1-C4 ring closure to form the β-lactam ring (441).

Undoubtedly, however, the most widely used approach involving cyclizations of this type is the hydroxamate method developed by Miller, which provides the opportunity to use readily available amino acids in a high yielding process to afford chiral β-lactams of virtually any description without racemization or elimination taking place (442). Thus, use of the Mitsunobu procedure or base-catalyzed cyclization of β-halo hydroxamates (185) provides good yields of the cyclic product (186),

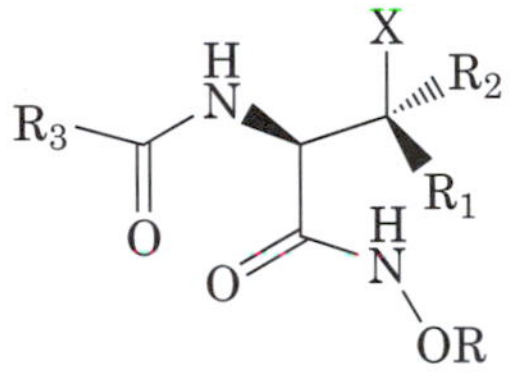

(182)

(185)

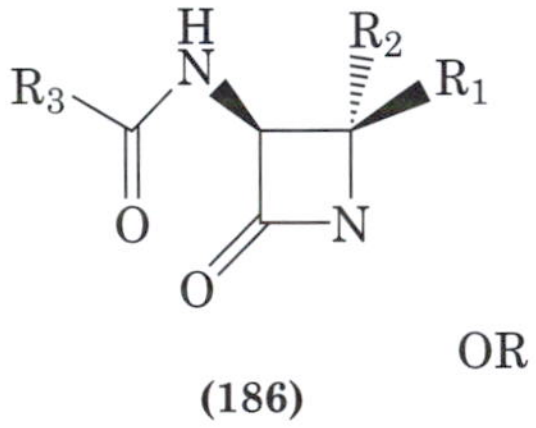

(183)

(186)

from phthalimidoacetyl chloride. This gave the cis-substituted β-lactam (183), which was readily converted to 3-ANA, and then nocardicin A and other natural side-chain derivatives (440).

Another synthesis of nocardicin A reported by the Lilly group used the L-cysteine-derived thiazolidine (184), where intramolecular displacement of the chlorine provided one of the

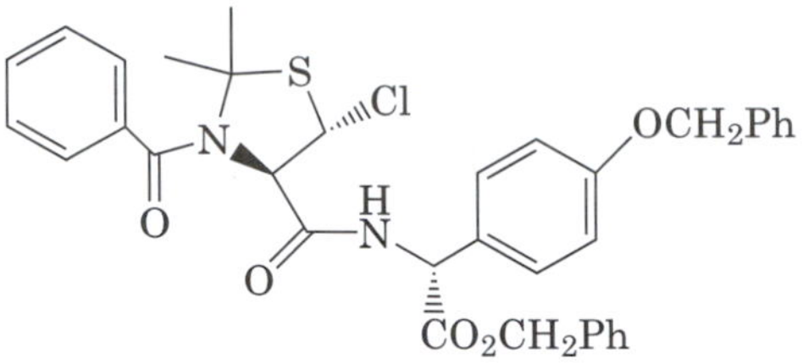

from which the β-lactam (187) is obtained by reduction of the free N-hydroxy β-lactam (186; R = H) with titanium trichloride (443).

Miller used this method for the synthesis of the nocardicin ring system starting from tBoc-L-serine, to give the β-lactam (188). This was followed by introduction of the phenylglycine residue by alkylation or diazo insertion using

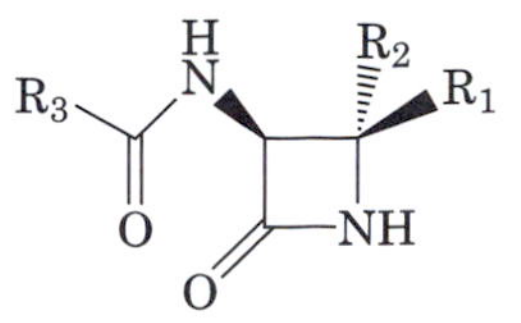

(187)

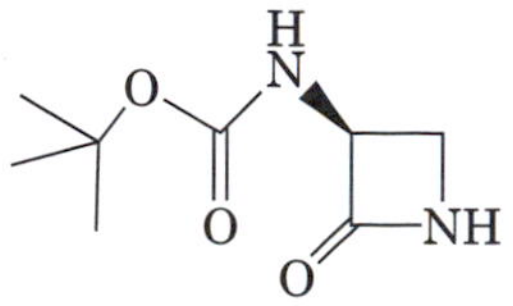

(188)

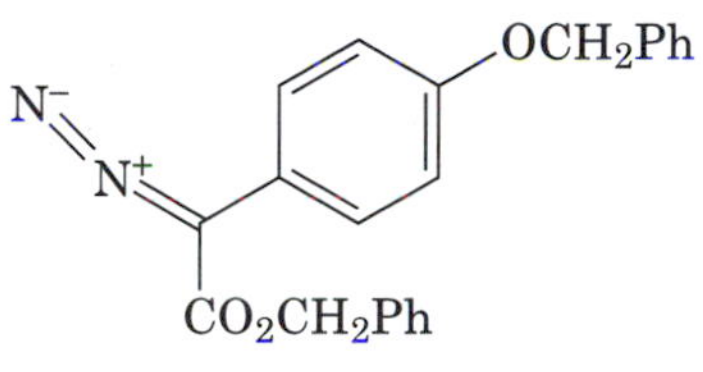

(189)

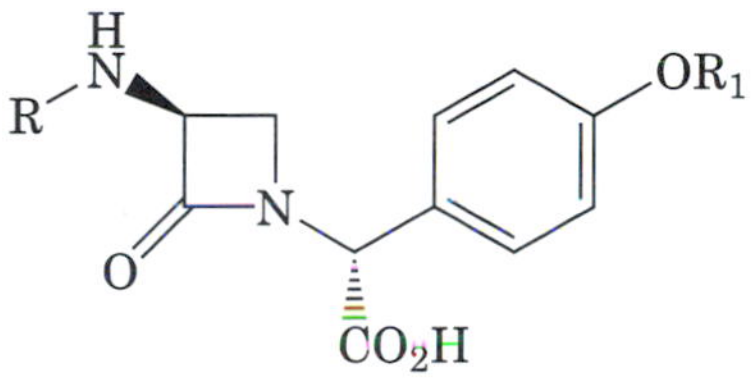

(190)

(191)

(189), giving an overall yield of 45% of (190), a fully protected version of 3-ANA (444).

This biomimetic-like ring closure has also been used by Townsend (446). Intermediate (191) was cyclized in a modified cyclodehydration procedure by use of triethylphosphite rather than triphenylphosphine, which avoided any problems of epimerization at C5. After suitable deprotection and acylation with the appropriate side chains, nocardicins A-G were synthesized in good yields; for example, nocardicin A was produced in an overall yield of 22% from L-serine and D-(p-hydroxyphenyl)-glycine. Other approaches to the nocardicin skeleton have made use of α-methylene β-lactams (446) as well as novel β-lactam-forming

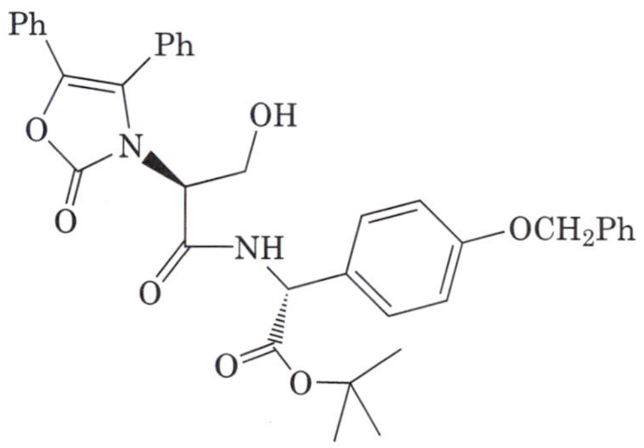

reactions (447), whereas general aspects of the synthesis of monocyclic β-lactams and 3-amino-2-azetidinones have been reviewed (431, 448, 449).

*5.4.1.2 Biological Activity and SAR.* Nocardicin A shows *in vitro* activity (3.13–50 μg/mL) against several Gram-negative organisms, including *Pseudomonas aeruginosa*, *Proteus* species, *Serratia marcescens*, and strains of *Neisseria*. No activity of significance is seen against *Staphylococci* or *E. coli* (450). Nocardicins C-E show weak activity, whereas nocardicins F and G are inactive, illustrating the importance of the *syn*-oxime function and the homoserine residue for activity in these natural products (451). Functional group modifications of the side chain (oxime, ketone, amine) of nocardicins A, C, and D have been carried out, but almost all the compounds show reduced activity. Acetamido derivatives of 3-ANA show weak activity, but replacement of the p-hydroxyphenyl residue by other aromatic or heteroaromatic groups generally does not reduce potency. After extensive SAR studies involving the preparation of several hundred compounds, it was concluded that the only useful antibiotic in the series was nocardicin A and, to maintain an effective level of activity, only limited modification of the natural product is possible (451). The most interesting property of nocardicin A relates to its mode of action *in vivo*, where it was found to act synergistically with serum bactericidal factors against *P. aeruginosa* and with polymorphonuclear leukocytes (PMNs) against *P. aeruginosa*, *E. coli*, and *P. vulgaris*. Unlike most antibiotics the bactericidal activity of nocardicin A increased markedly against these

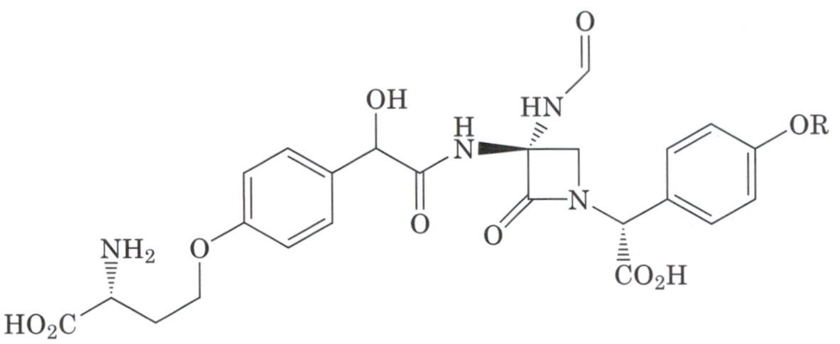

**(192)** R = D-glucuronic acid; Formadicin A
**(193)** R = H; Formadicin C

organisms in the presence of fresh serum and PMNs, an effect that was reflected in a more potent *in vivo* activity than would have been anticipated from *in vitro* MIC values (450, 451).

**5.4.1.3 Formadicins.** The most recently described natural products having the nocardicin skeleton are the formacidins A-D, isolated from a species of *Flexibacter* (452).

Structural determination showed that formadicins A (**192**) and C (**193**) possessed the formamido substituent in the C3-position of the β-lactam ring (453). These natural products show a fairly narrow spectrum of activity that is similar to that of nocardicin A, of which those with the formamido substituent were significantly more stable to hydrolysis by β-lactamases. No synthetic analogs have yet been reported.

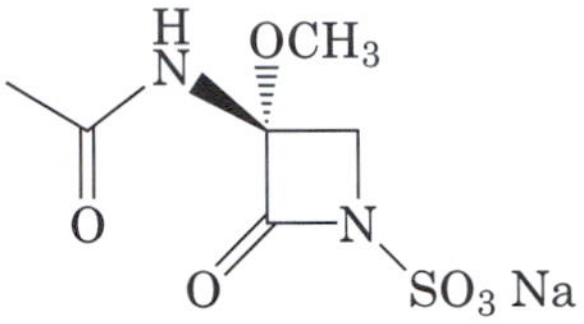

**5.4.2 Monobactams.** In 1980 groups at Takeda and Squibb reported on the isolation of a new class of monocyclic β-lactams from bacteria (429, 430). Up to that time, the only producers of β-lactam antibiotics were the fungi and actinomycetes. Characterized by the presence of an N1-sulfonic acid grouping in the β-lactam ring, they have collectively become known as the monobactams (454). The simplest member of the family is the acetamido derivative SQ 26180 (**194**) (455), whereas other members of the series are sulfazecin (**195**) (456), isosulfazecin (**196**) (457), and several related compounds based on structure (**197**) (458), together with other examples having oligopeptide side chains (459–461). One natural compound (**198**) has been reported to have a C4-methyl group (462).

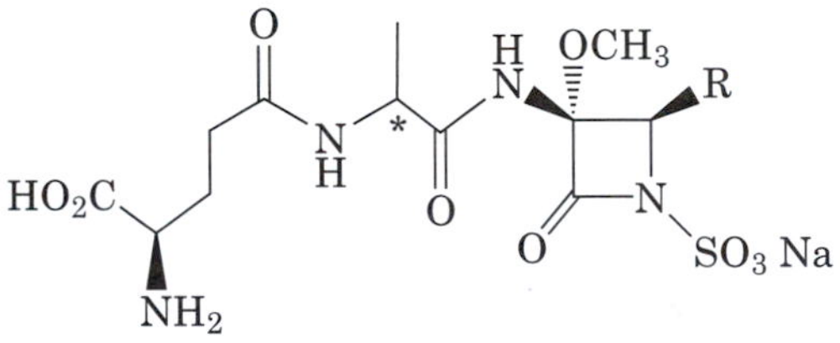

**(194)** SQ 26180

**(195)** R = H; Sulfazecin; * = D
**(196)** R = H; Isosulfazecin; * = L
**(197)** R = $CH_3$; * = D

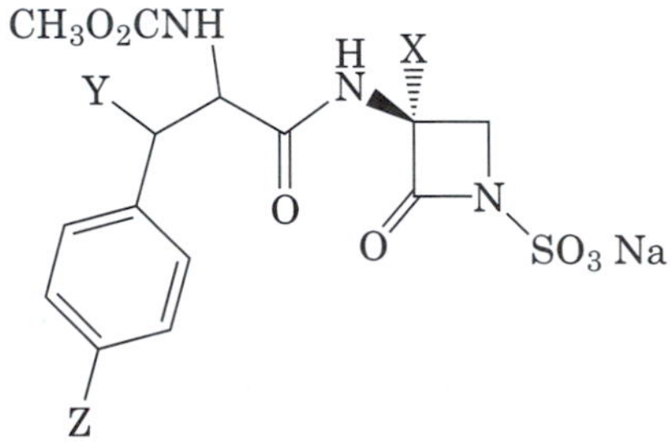

**(198)** X = $OCH_3$
         Y = $OSO_3M$
         Z = H, OH, $OSO_3M$

**5.4.2.1 Structure Determination and Synthesis.** The structure of SQ 26180 (**194**) was readily determined from spectroscopic data. Confirmation was made by degradation of the thiazine ring of an appropriately substituted methoxylated cephalosporin to a simple acetamido β-lactam. N-Sulfonation then gave material identical with the natural product, thus establishing that the stereochemistry at the C3 position of SQ 26180 (**194**) was the same as found in the cephalosporins and cephamycins (455). The structure and stereochemistry of sulfazecin (**195**) was unambiguously established by X-ray crystallography (463), whereas that of the other natural monobactams was obtained by a combination of spectroscopic properties and hydrolytic experiments to determine the nature of the peptide side chains. Isolation of the monobactams has led to the synthesis of many acylamino derivatives of 3-amino monobactamic acid (3-AMA) (**199**) or

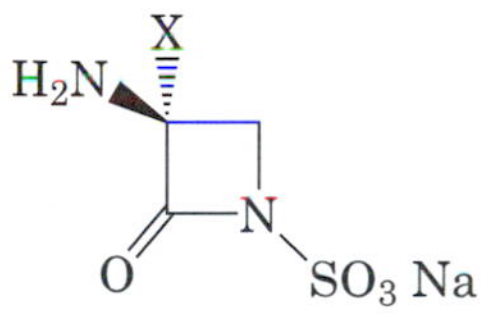

(**199**) X = H
(**200**) X = OCH$_3$

the methoxy nucleus (**200**). Because deacylation of the natural products was not satisfactory, β-lactams (**199**) and (**200**) were initially prepared by degradation of 6-APA (**2**) followed by sulfonation of the derived β-lactam; in the case of the methoxy β-lactam (**200**), only racemic material was obtainable (454, 464). Total synthesis has also been extensively used and is now almost always the method of choice. This was first achieved by direct base-catalyzed cyclization of the acyl sulfamate (**201**) (465).

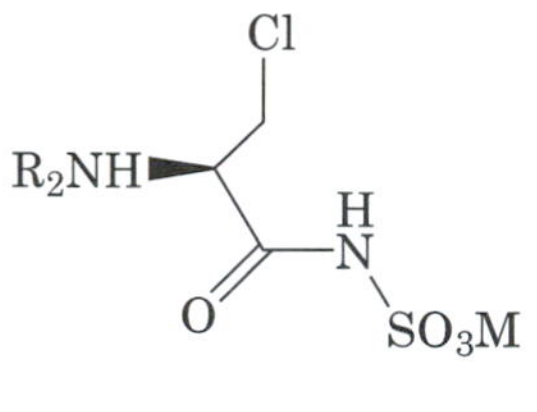

(**201**) R$_1$ = H, CH$_3$
R$_2$ = Acyl derivative

Subsequently, a wide range of methods have been developed, starting from β-hydroxy amino acids such as serine to intermediate (**202**) (R$_1$ = H) or threonine, giving (**203**) (R$_1$

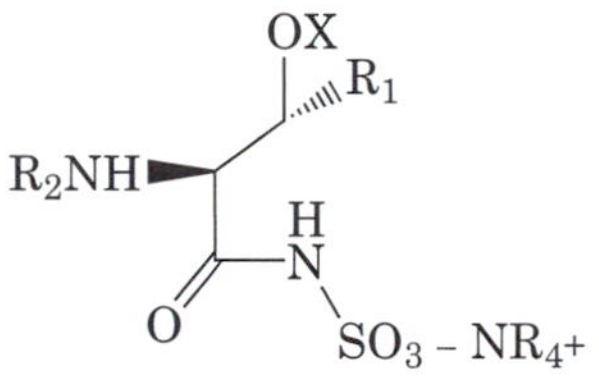

(**202**)

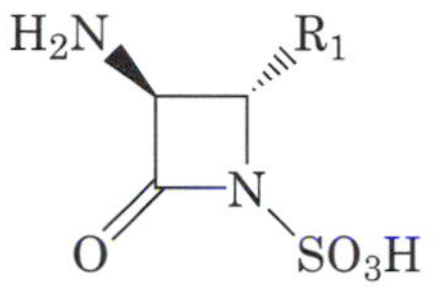

(**203**)

= Me), to provide a ready route to the nucleus (**203**) (R$_1$ = H or Me), ideally suitable for derivatization by acylation of the C3 amino group (466).

**5.4.2.2 Biological Activity of Natural Products and Synthetic Derivatives.** Like the penicillins, cephalosporins, and other β-lactam antibiotics, the monobactams interfere with the synthesis of bacterial cell walls by binding to PBPs. Aztreonam (see below), one of the clinically efficacious agents, binds specifically to PBP-3 (467). All of the natural products exhibit poor activity, but extensive modification of the C3-amido side chain by acylation of 3-AMA has provided several highly potent compounds. In addition, structural modification at C4 and N1 of the β-lactam ring can profoundly affect antibacterial activity and β-lactamase stability (468). The importance of the C3-stereochemistry was demonstrated by the synthesis of the enantiomers of (**204**). Only the S-isomer (MIC < 0.05 µg/mL against *P. rettgeri*; R-isomer > 100 µg/mL) corresponding to that found in the penicillin and cephalosporin family of natural products was active (466). Compared to benzyl penicillin the analogous monobactam (**205**) shows a similar pattern of activity, but with reduced potency. With the carbenicillin, side-chain activity was reduced, but with ureido derivatives activity

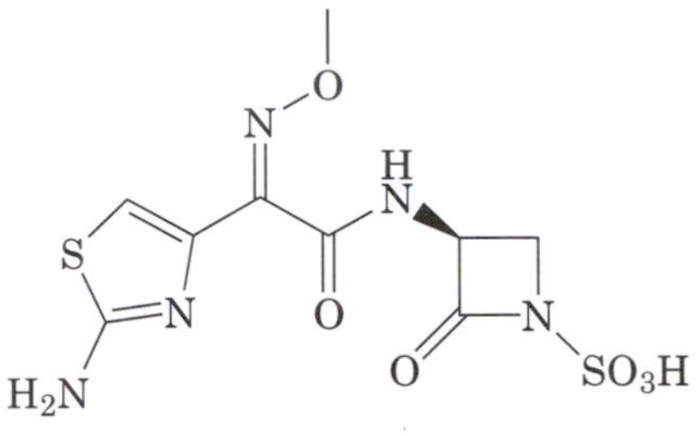

(204)

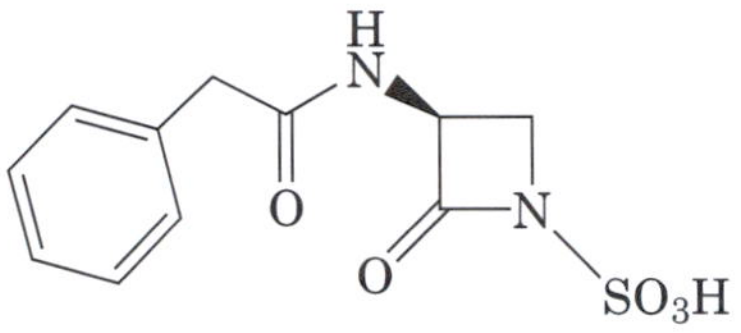

(205)

was similar to that of the penicillins. A similar pattern was apparent with the corresponding analogs of the cephalosporins (468). One early observation revealed that C4-alkyl substituents could considerably influence both antimicrobial potency and β-lactamase stability. Extensive SAR studies resulted in the identification of the totally synthetic derivative aztreonam (206) as a product with a potent and

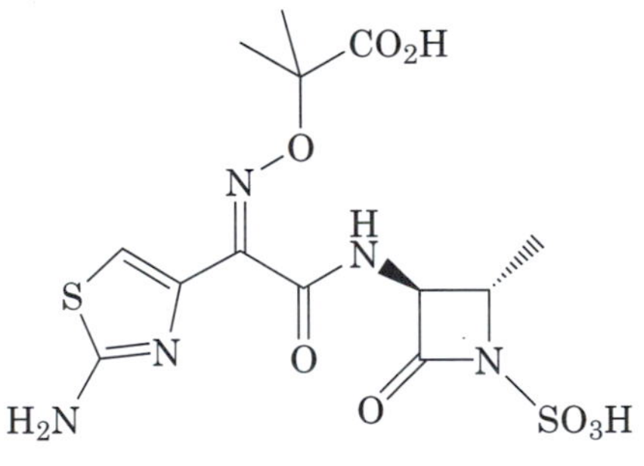

(206) Aztreonam

useful spectrum of clinical activity against Gram-negative bacteria, but with virtually no activity against Gram-positive pathogens (466, 469, 470). A second synthetic monobactam to reach the clinic is carumonam (207), which shows some improvement in activity over that of aztreonam against the Enterobacteriaceae family of bacteria (471). Table 14.13

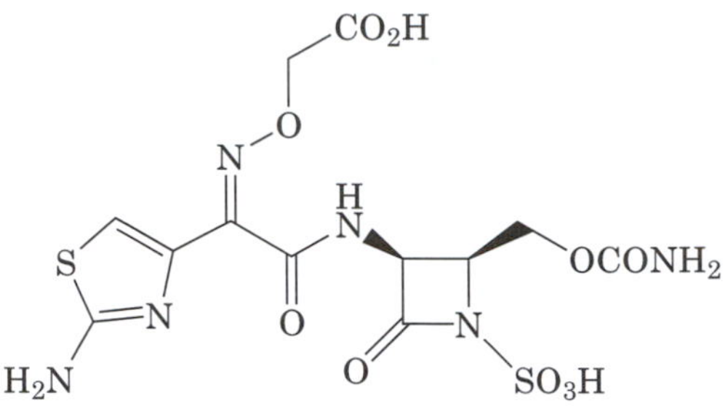

(207) Carumonam

shows some representative activities of a number of monobactams and related analogs.

**5.4.2.3 Alternative N1-Activating Groups.** The sulfamate residue found in the monobactams is considered both to activate the β-lactam ring for interaction with the active site serine of a PBP and to provide the anionic charge required for binding, as postulated by Knox (466, 473). This has stimulated the effort to synthesize monocyclic derivatives that have new N1-activating substituents, which could act in the same way. Several of them have been discovered that exhibit the properties necessary to afford antibacterial activity, and are exemplified by structures (208–211),

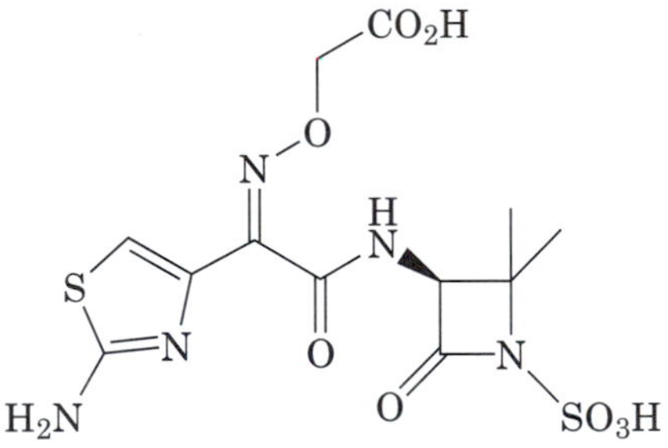

(208) Tigemonam

all of which exhibit various levels of antibacterial activity (Table 14.12) and β-lactamase stability (466). Tigemonam (208) is a potent anti-Gram-negative agent with good stability to β-lactamases. It is also orally absorbed, which is unusual among the monobactams, and encouraging results have been obtained in phase I and phase II clinical studies (466). Another example of heteroatom activation is seen in the oxamazin series exemplified by oximonam (211). The t-butyl glycolate ester prodrug (gloximonam) of (211) is also absorbed by the

**Table 14.13  *In Vitro* Antibacterial Activity of Monobactams and Analogs[a,b]**

| Organism | Pen G | CAZ | (204) | AZTR (205) | CAR (206) | TIG (207) | (208) | (209) | OXIM (210) |
|---|---|---|---|---|---|---|---|---|---|
| *E. coli* | >100 | 0.4 | 50->100 | 0.1–0.2 | 0.1–0.2 | 0.4–0.8 | 0.4–0.8 | 0.4–0.8 | 0.25–0.5 |
| *K. aerogenes* | 12.5->100 | 0.1–1.6 | 50 | 0.2->100 | | 0.4–0.8 | 0.4–0.8 | 0.4->100 | |
| *E. cloacae* | 25->100 | 0.2->100 | >100 | 0.01–50 | 0.25–16 | 0.8–50 | 0.8->100 | 0.8–50 | 1->64 |
| *P. rettgeri* | 3.1 | <0.05 | 25 | 0.05 | 0.06–0.12 | <0.05 | 0.2 | <0.05–0.4 | 0.12 |
| *P. aeruginosa* | >100 | 1.6 | >100 | 3.1 | 1–4 | 1.6->100 | 0.3->100 | 0.8–12.5 | >64 |
| *S. aureus* | 0.05–3.1 | 12.5 | 3.1–6.3 | >100 | >128 | | | | >64 |

[a]Minimum inhibitory concentrations (range mg/L). Refs. 466, 468, 471, 472.
[b]Abbreviations: Pen G, benzyl penicillin; CAZ, ceftazidime; AZTR, aztreonam; CAR, carumonam; TIG, tigemonam; OXIM, oximonam.

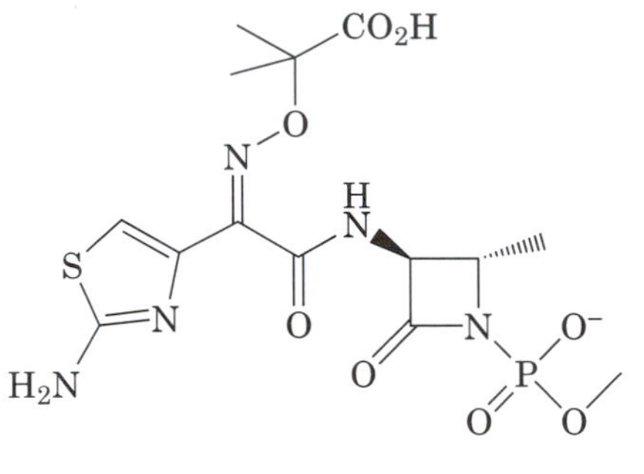

**(209)**

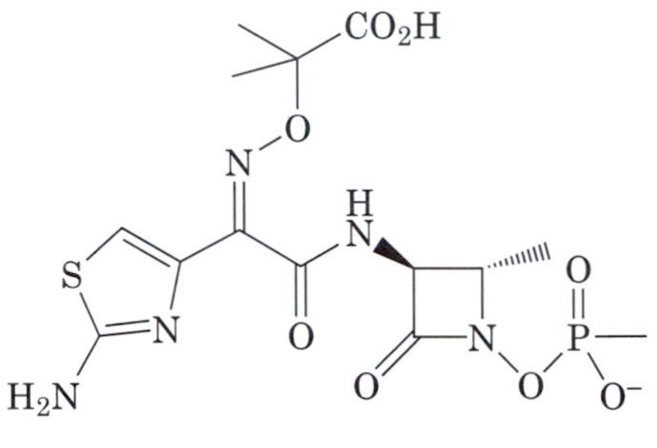

**(210)**

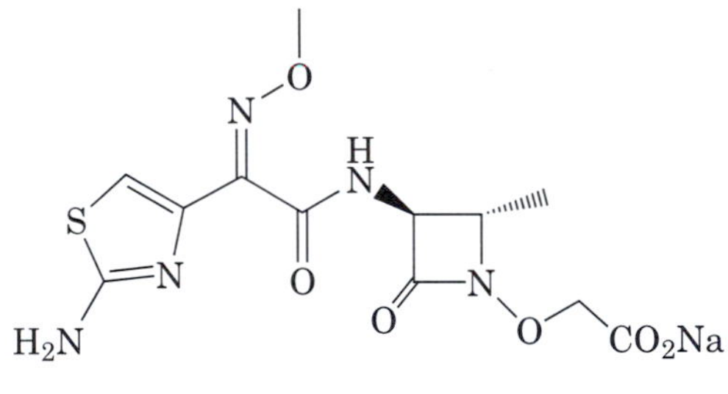

**(211) Oximonam**

oral route, giving good therapeutic levels of the parent drug (466). Efforts to improve potency against *Pseudomonas aeruginosa* led to the discovery of SQ 83360 (**212**), having the 3-hydroxy-4-pyridone residue, which acts as a catechol bioisoter, leading to uptake by the bacterial *ton*B-dependent iron transport pathway in a manner similar to that of the catecholic *β*-lactams (474). A somewhat different series is that in which a tetrazole has replaced the sulfonate group to give, for example, RU 44790 (**213**), which has been shown to compare favorably in activity with aztreonam (475).

In contrast, N1-sulfonated monobactams having the (*R*)-1-hydroxyethyl of the carbapenems (Section 5.4) at the C3-position showed only weak activity (476). The most recent class of compounds reported as being active are C3-alkylidene derivatives (**214**) with, surprisingly, a neutral N-acyl activating group on the N1-position of the *β*-lactam ring (477).

## 5.5 Carbapenems and Trinems

Although structurally related, carbapenems and trinems have different origins. The carbapenems were discovered as natural products during the screening of microbial metabolites, whereas the trinems were conceived as synthetic analogs designed to improve the biological properties of existing classes of *β*-lactam antibiotics.

**5.5.1 Discovery of Carbapenems.** Imipenem (MK 0787, **215**) and meropenem (SM 7338, **216**) can be considered as the most significant examples of *β*-lactams. The rationale by which it has been possible to discover these agents

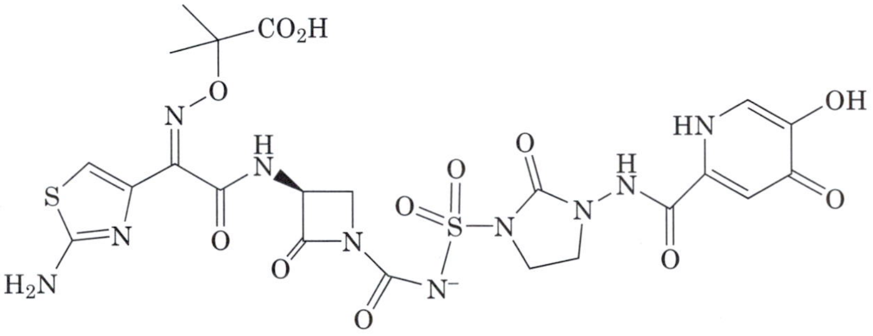

**(212) SQ 83360**

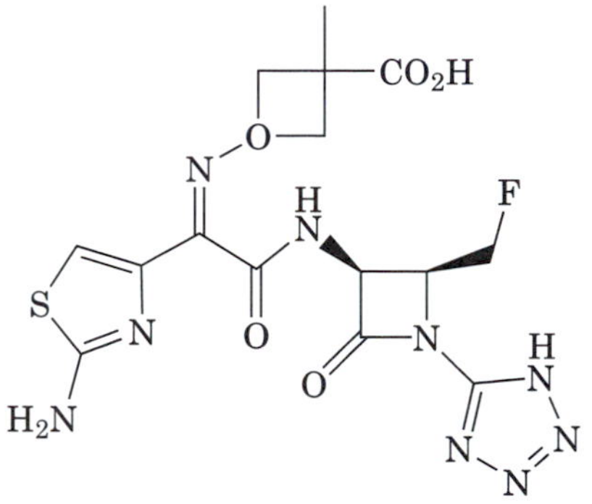

(213) Ru 44790

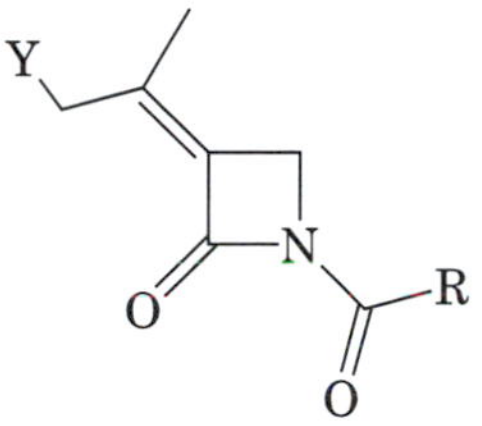

(214)

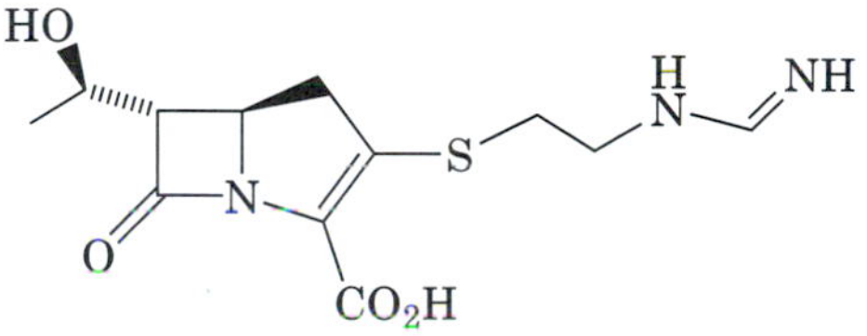

(215) Imipenem MK 0787 (1982)

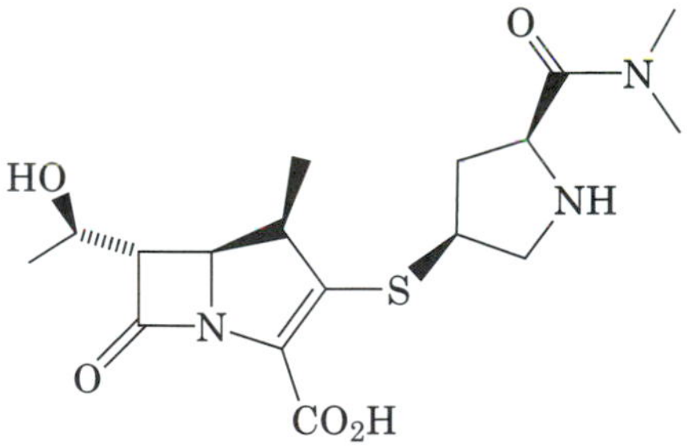

(216) Meropenem SM7338 (1987)

elements, to adjust the spectrum of activity and the enzymatic and chemical stability. Direct comparison of thienamycin (217) with

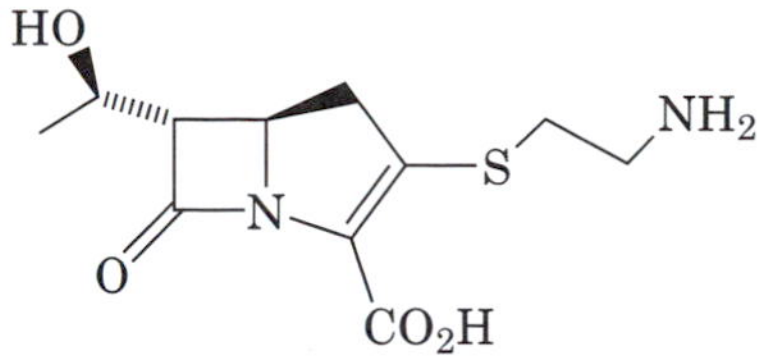

(217) Thienamycin (1977)

meropenem (216) clearly shows this process. Introduction of a 1-β-methyl group and formation of a conformationally constrained $SCH_2CH_2NH_2$ system resulted in one of most potent and effective β-lactam antibiotics currently used.

In the mid-1970s during the course of screening for inhibitors of cell wall biosynthesis the Merck group isolated thienamycin (217) from *Streptomyces cattleya*, followed by other members of the family of carbapenems (478).

At the same time a Beecham group, while screening for inhibitors of β-lactamases (Section 5.7), isolated from *Streptomyces olivaceus* a group of interrelated metabolites that they called the olivanic acids (479, 480). All these compounds were characterized by the presence of the 5-(*R*)-7-oxo-1-azabicyclo[3.2.0]hept-2-ene-2-carboxylic acid ring system, generally referred to as 1-carbapen-2-em-3-carboxylic acid (218), and which has been isolated as a natural

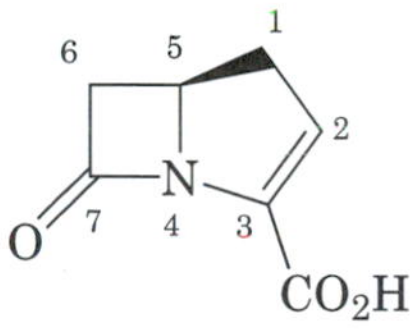

(218)

product from certain species of *Serratia* and *Erwinia* (481).

All the other natural products possess substituents at C2 and C6 of this nucleus. Over 50 variations are known and many of them have been listed (346, 482). The discovery of this novel series of highly active bicyclic compounds, so markedly different from the peni-

can be described as an elegant and successful example of applied medicinal chemistry. The research has faced and resolved several issues related to chemical and biological properties of these molecules optimizing the key stuctural

cillins and cephalosporins, revitalized ideas concerning the structural and stereochemical features required for biological activity and stability among the β-lactams.

**5.5.1.1 Natural Products: Occurrence, Structural Variations, and Chemistry.** While a variety of *Streptomyces* spp. produce the carbapenems, only *S. cattleya* and *S. penemfaciens* have been reported to produce thienamycin (**217**) (470, 482). Several research groups have been involved in the search for new carbapenems (483–487). Often the natural products are found as a mixture of stereoisomers, usually in low yield (1–20 μg/L), and much effort has been expended on strain improvement and the optimization of fermentation conditions (488). Thienamycin (**217**) has the (R)-configuration at the C8-hydroxy group with a *trans* arrangement of protons about the β-lactam ring (489). The history of the discovery, structural elucidation, and chemistry of thienamycin (**217**) and related carbapenem antibiotics has been well documented (490). The first metabolites isolated from *S. olivaceus* were a series of sulfated derivatives such as MM13902 (**219a**; R = SO$_3$H), the corresponding sulfoxide MM4550 (**220a**), and the side-chain-saturated compound MM17880 (**221**) (480, 486, 491).

Later, several nonsulfated members (e.g., **219b**, **220b**, **221**, **222** and **223**) of this group

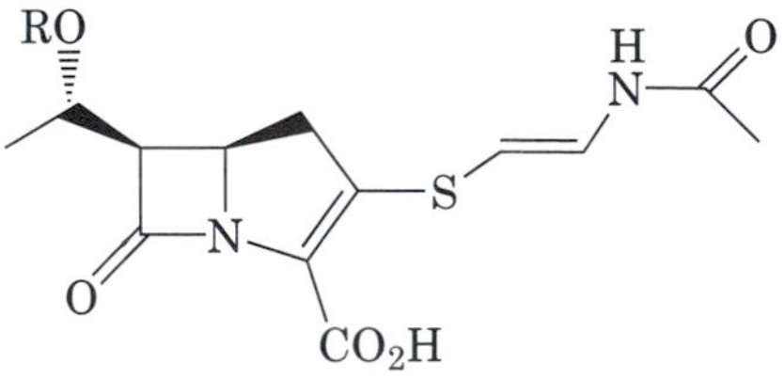

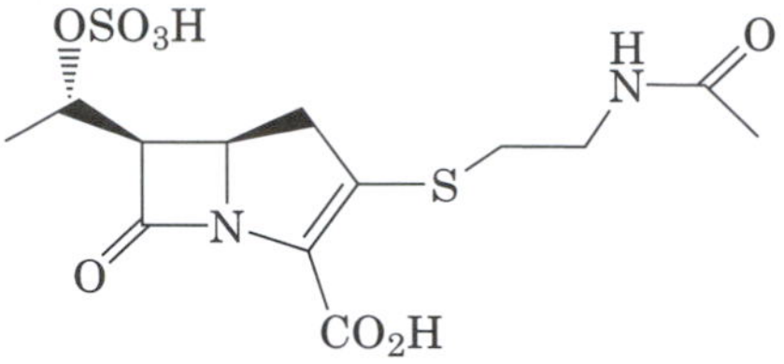

**(221)** MM17880

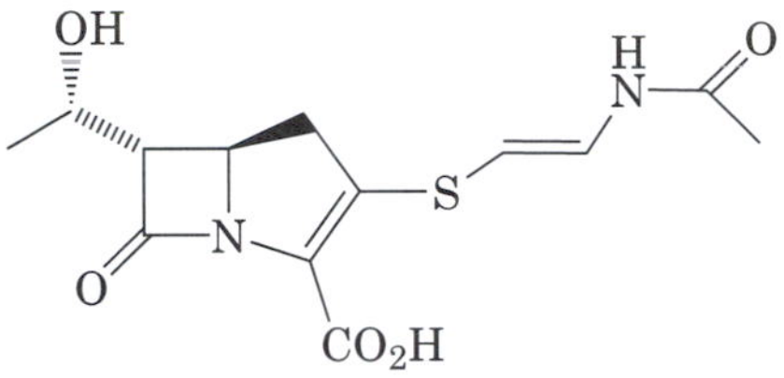

**(222)** MM22383

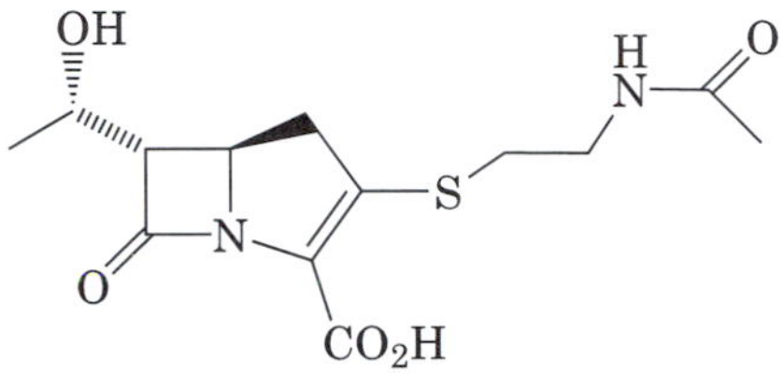

**(223)** MM22381

| **(219a)** MM13902 | R = SO$_3$H |
| **(219b)** MM22380 | R = H |

| **(220a)** MM4550 | R = SO$_3$H |
| **(220b)** MM22382 | R = H |

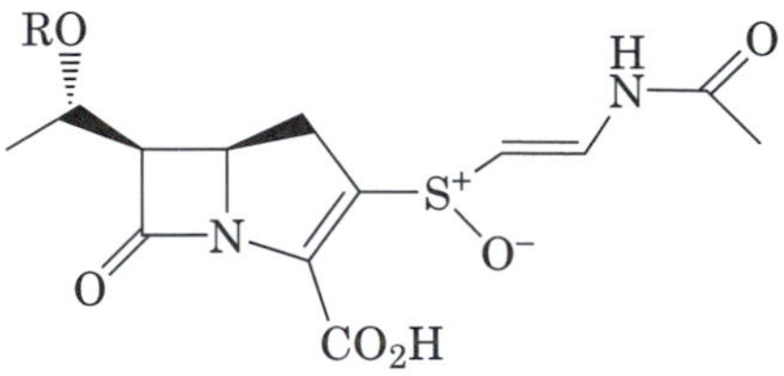

were also obtained from *S. olivaceus* (492, 493) and have shown to correspond to a series of thienamycins and epi-thienamycins analogs isolated from *S. parvogriseus* (490). In the olivanic acids the stereochemistry is 8(S) with a *cis*-substituted β-lactam in the sulfated series, whereas both *cis*- and *trans*-β-lactams are found in the case of the C8-hydroxy compounds (494, 495). Following reports of thienamycin (**217**) and the olivanic acids (**219–223**) a number of structurally related carbapenems having differing C2- and C6-substituents were reported. Representative examples of these compounds are PS-5 (**224**), PS-6 (**225**) (496), carpetimycin A (**226**) (497), asparenomycin A (**227**) (290), and pluracidomycin A (**228**) (487), isolated from a wide variety of sources.

The carbapenems often show chemical instability and can be extremely sensitive to reaction conditions. Nevertheless, a wide range of functional group modifications have been carried out on the natural products. Many de-

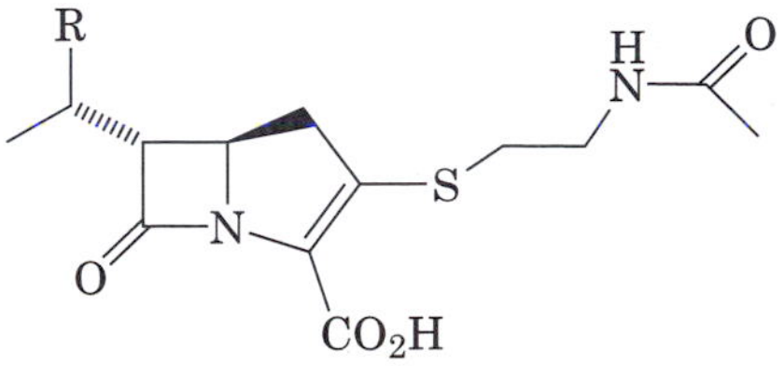

**(224)** PS-5    R = H
**(225)** PS-6    R = CH₃

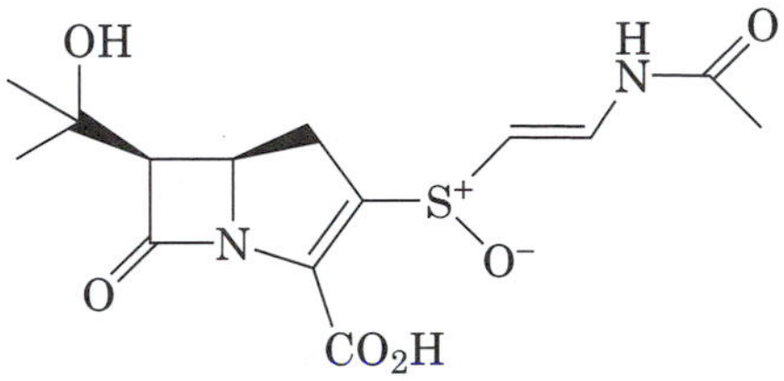

**(226)** Carpetimycin A

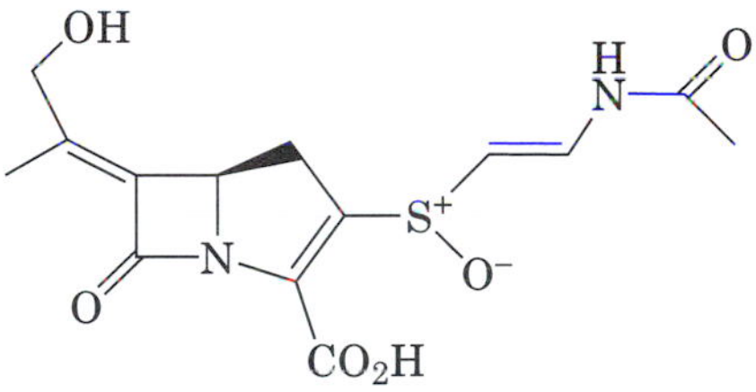

**(227)** Asparenomycin A

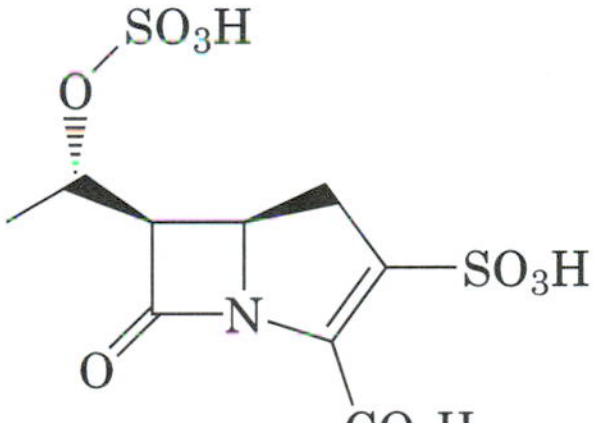

**(228)** Pluracidomycin A

rivatives of the amino, hydroxy, and carboxy groups of thienamycin have been obtained for SAR studies; other reactions cover oxidation or removal of the cysteaminyl side chain and isomerization of the double bond (490, 499). The largest class of derivatives are those obtained by modification of the amino group, particularly amidines, which greatly improve chemical stability without compromising po-

tency. It was this series that gave *N*-formimidoyl thienamycin (imipenem, **215**) (500), which eventually became the first clinically used member of the carbapenem family (501).

In the olivanic acids the side-chain double bond can be isomerized to the (*Z*)-isomer (502). It also reacts readily with hypobromous acid to produce a bromohydrin, which readily breaks down to the key thiol intermediate (**229**), ideally suited for the synthesis of new C2-alkenyl or alkyl analogs (503).

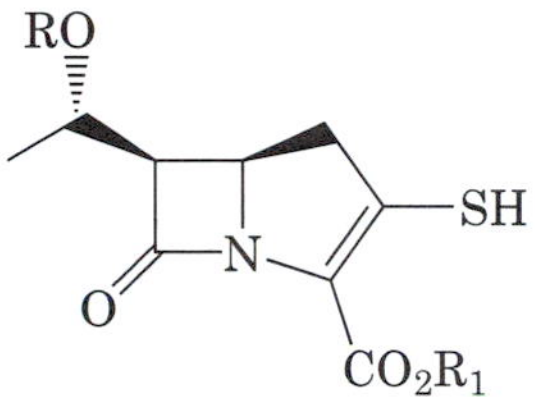

**(229)**

With the C8-hydroxyolivanic acids, inversion of the (*S*)-stereochemistry affords an entry into the 8(*R*) thienamycin series (504). Another useful reaction for introducing C2-sulfur substituents was developed with PS-5 (**224**) by displacement of the *S*-oxide with other thiols (505).

Many wide-ranging and detailed reviews of the early and most recent chemistry of the carbapenems have been published (370, 482, 490, 499, 502, 506–513).

*5.5.1.2 Synthesis of Natural Carbapenems.*
The novelty and potent biological properties of the carbapenems, together with low fermentation yields, resulted in an intensive synthetic effort at a relatively early stage, to produce not only the natural products but also, particularly, synthetic analogs having improved chemical and metabolic stability (see Section 5.5.1.3) over that of the natural materials (489, 514–516). The most common approach has been to construct an appropriately functionalized and stereochemically correct β-lactam, followed by cyclization, to form the more highly strained bicyclic system in the first steps of the synthesis. Variations of this approach have been applied to produce virtually all of the natural products (517). The unsubstituted ring system (**218**) was synthesized,

before the discovery of the natural material, in racemic or chiral form by several groups, starting from azetidinone (230) or (231), ob-

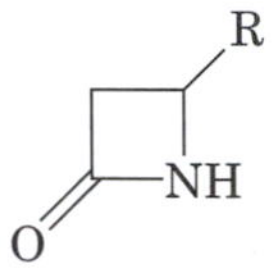

(230) R = (structure with OAc)

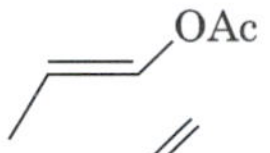

(231) R = (structure)

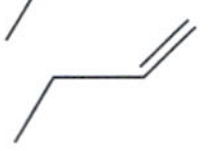

tained from chlorosulfonyl isocyanate (CSI) and an appropriate olefin (518–521).

Progression to the phosphorane (232 or 233) was followed by oxidation to the aldehyde

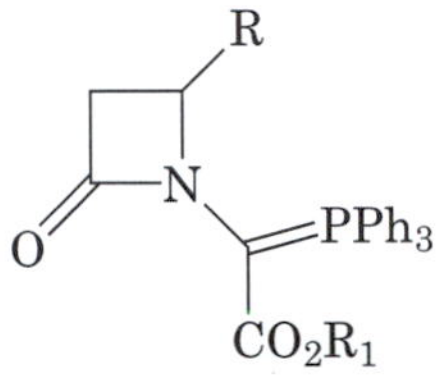

(232) R = (structure with OH)

(233) R = (structure)

(234) R = (structure with O)

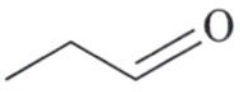

(234), which readily cyclized by an intramolecular Wittig reaction. Removal of the acid-protecting group provided the unstable sodium salt of (218). The formation of the [2,3] double bond by this procedure has been extensively used both for analogs and for natural product synthesis (522). Since its discovery, thienamycin (217) has been the focal point for innumerable synthetic studies reported in the carbapenem field (490, 512, 517). Many of the methods originated from the Merck group are still widely used. Their first synthesis of racemic thienamycin (217) also made use of (230), introduction of the hydroxyethyl side chain being by way of an aldol reaction giving a *trans* β-lactam, which was elaborated by a lengthy process to the dibromide (235).

Cyclization and decarboxylation, followed by elimination and deprotection, then gave

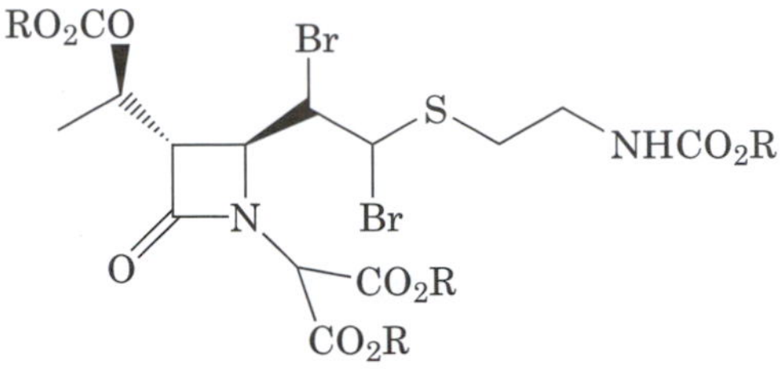

(235) R = p-nitro-benzyl

(±)-thienamycin (523). A chiral synthesis involving [3,4] bond formation and, starting from the L-aspartate-derived β-lactam, (236) soon followed (524). Stereocontrolled introduction of the side chain and elaboration to the dithiane derivative (237) was followed by introduction of the hydroxyethyl side chain (238) and then the diazo intermediate (239) (Fig. 14.16).

Through the use of a catalytic amount of dirhodium tetra acetate, cyclization proceeded extremely efficiently to the bicyclic keto-ester (240), which has become the essential intermediate for the synthesis of a wide range of analogs of thienamicyn. Introduction of the C2-cysteaminyl side chain through the enol phosphate (241) and final deprotection afforded thienamycin (217), whereas using a protected amidine side chain imipenem (215) is directly accessible (525). By following a similar methodology, several 2-thio-substituted derivatives were obtained, including cyclic amidines (242), aryl and heteroarylthio (243)

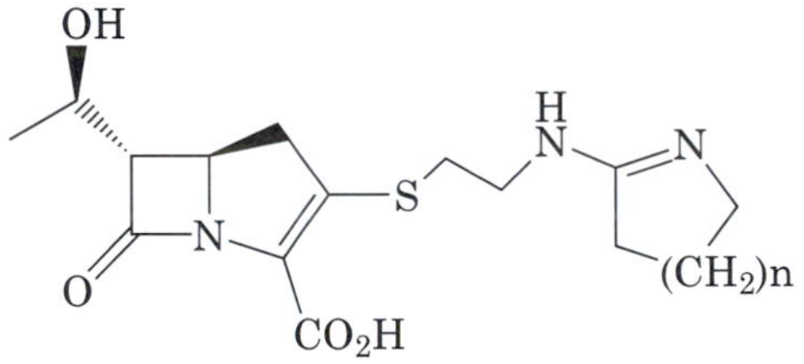

(242)

carbapenems (526), and compounds bearing a quaternary ammonium group (244) (527).

A similar approach that used the stable enol triflate (245) allowed the synthesis of aryl and vinyl carbapenems such as (246–248) through the use of an organometallic reagent-mediated coupling methodology (528).

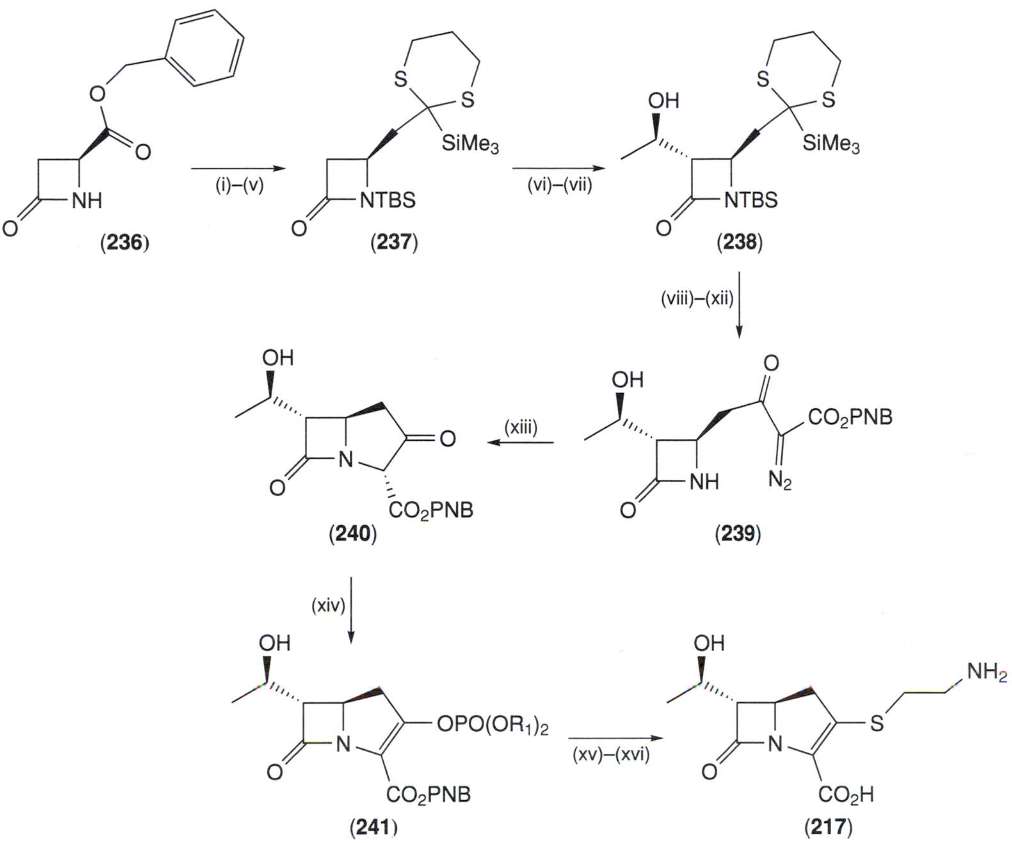

**Figure 14.16.** A stereocontrolled synthesis of thienamycin.

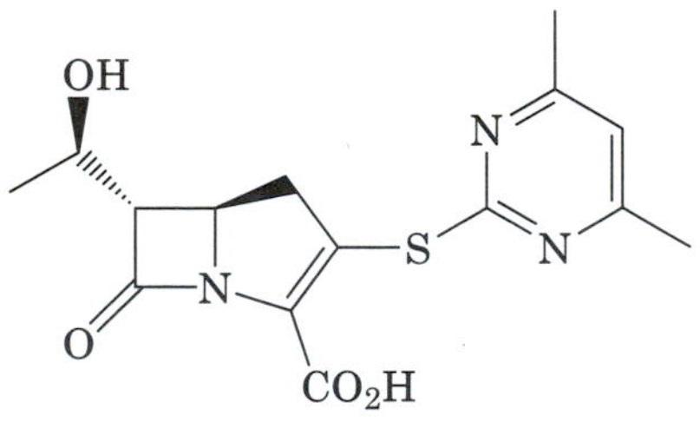

Reagents: (i) NaBH$_4$, MeOH; (ii) MeSO$_2$Cl, TEA, DCM 0°C; (iii) NaI, Acetone, Δ; (iv) TBSCl, TEA, DMF

(v) 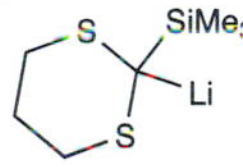 [structure] Li  , THF, –78°C; (vi) LDA, N Acethyl imidazole, THF, –78°C; (vii) K selectride, KI, Ether;

(viii) HgCl$_2$, HgO MeOH, H$_2$O, Δ; (ix) H$_2$O$_2$, MeOH, H$_2$O; (x) CDI, THF, then Mg—[structure]  ;

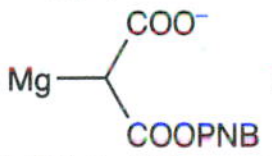

(xi) HCl, MeOH; (xii) p-carboxy benzene sufonyl azide, TEA, CH$_3$CN; (xiii) Rh(OAc)$_4$, Toluene, 80°C;

(xiv) ClPO(OPh)$_2$, DIPEA, CH$_3$CN, 0°C; (xv) HSCH$_2$CH$_2$NHCOOPNB, DIPEA, CH$_3$CN, –5°C;

(xvi) H$_2$, Pd/C, 3.5 atm.

(243) 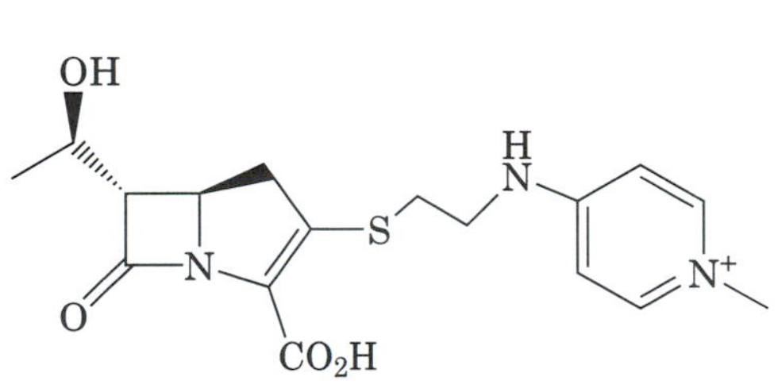                                                  (244)

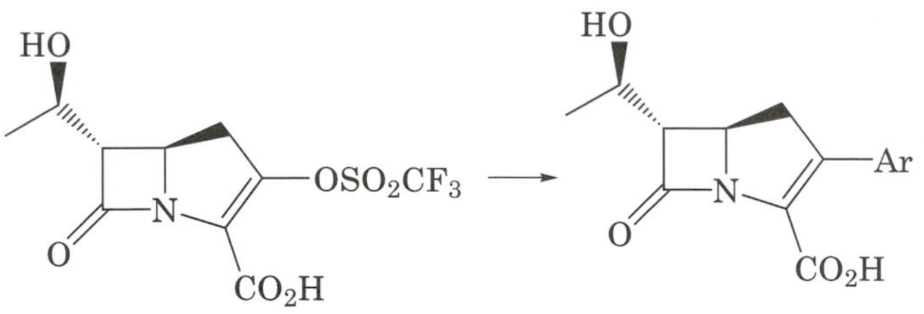

**(245)**

**(246)** Ar =

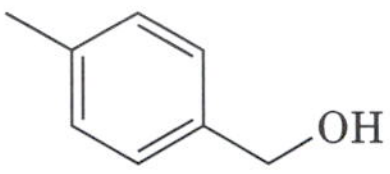

**(247)** Ar =

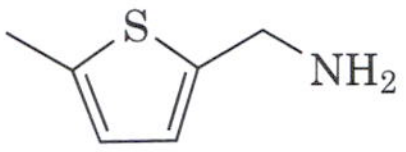

**(248)** Ar =

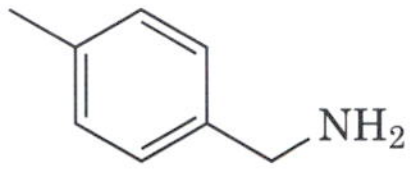

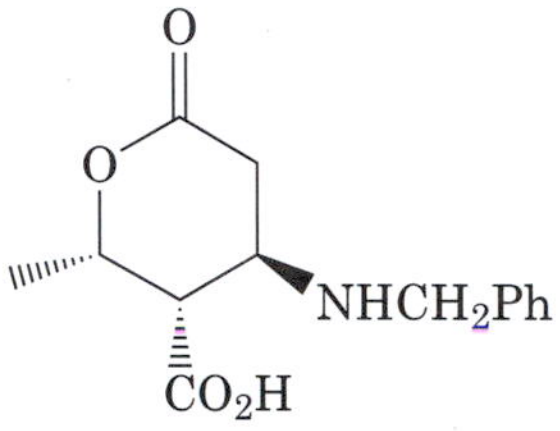

**(249)**

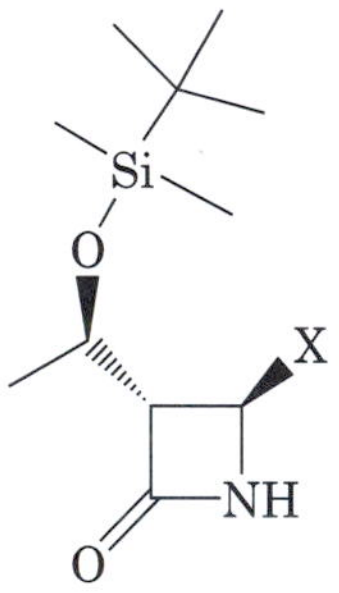

**(250)** X = Cl
**(251)** X = OAc

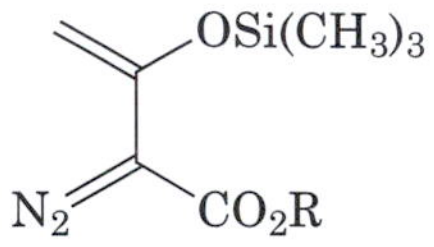

**(252)**

Another approach starts from the (±)-lactone (**249**) obtainable from acetone dicarboxylic acid (516). Initially carried out in the racemic series, this route afforded a practical synthesis of thienamycin, although conversion of the side-chain stereochemistry from (*S*) to (*R*) was required. Subsequently, an enantioselective route was developed from enantiomerically pure lactone (**249**) (529). Displacement of the 4-acetoxy or chloro group from azetidinones such as (**250**) and (**251**), respectively, with the trimethylsilyl enol ether (**252**) provided a further improvement in refining the synthesis of thienamycin (**217**) and its analogs (530, 531).

The success of this approach has been recognized by the commercial development of 4-acetoxy azetidinone (**251**) for penem, carbapenem, and trinem synthesis. Of several reported syntheses, the two most successful use simple chiral starting materials derived from 3-hydroxybutyric or lactic acid and a cycloaddition procedure to construct the β-lactam ring (532, 533). Other syntheses in the thienamycin series using β-lactams derived from penicillin (530, 534, 535), carbohydrates (536, 537), amino acids (538, 539), isoxazolidines (540), and organo-iron or cobalt complexes (541, 542) have been described together with alternative methods for bicyclic ring construction by use of the Dieckman cyclization (543) or an intramolecular Michael reaction (544).

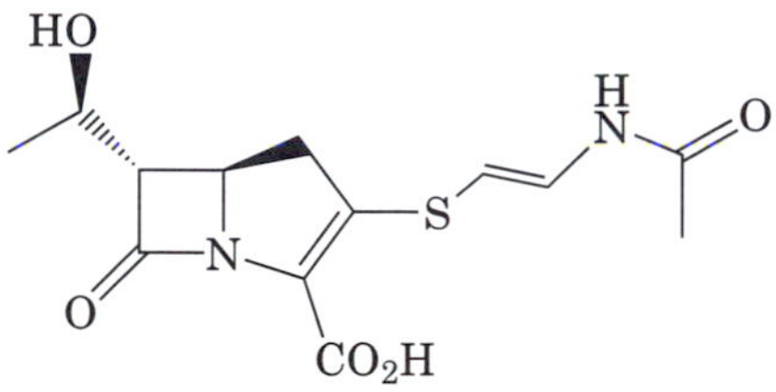

**(253)**

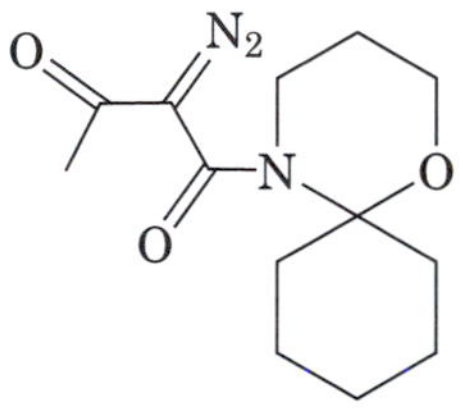

**(254)**

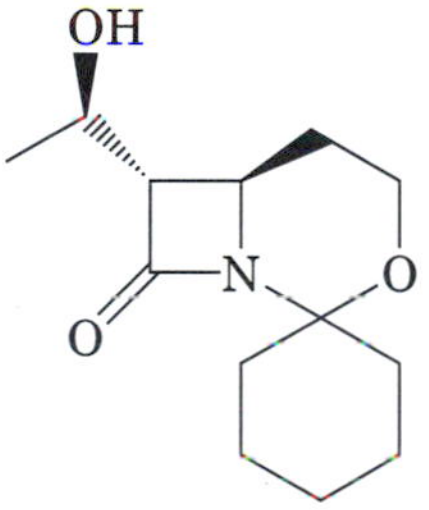

**(255)**

**(256)**

**(257)**

The discovery that thiol esters could participate in the Wittig cyclization procedure to give C2-substituted derivatives provided the basis for the total synthesis of the olivanic acid MM22383 (**222**) (493) and *N*-acetyldehydrothienamycin (**253**) (490).

In this case, formation of the β-lactam ring was obtained by means of dirhodium tetra acetate-catalyzed cyclization of the diazo-intermediate (**254**), to give the *trans*-β-lactam (**255**), as a mixture of hydroxy epimers after reduction of the carbonyl group.

Progression to the thio-ester (**256**) was followed by cyclization in boiling toluene to give the two epimers of the cyclic product (**257**). These were separated and deprotected to afford the racemic natural products (522, 545).

Another route to the olivanic acids was by way of (**258**) (546), which readily reacted with acetamidoethanethiol to form the carbapenam (**259**). Reintroduction of the double bond gave the ester of (+) MM22381 (**260**) (547).

This method has also been used for the synthesis of PS-5 (**224**), although most approaches toward this natural product have used the chiral (3*R*,4*R*) 3-ethyl, 4-acetoxy azetidinone (**261**) and the diazo-ketone (**262**). In the case of the carpetimycins such as (**226**), the major problem is to obtain the thermodynamically less favored *cis*-arrangement of protons around the β-lactam ring in precursors such as (**263**). One attractive method involved a directed aldol condensation using (**264**) and

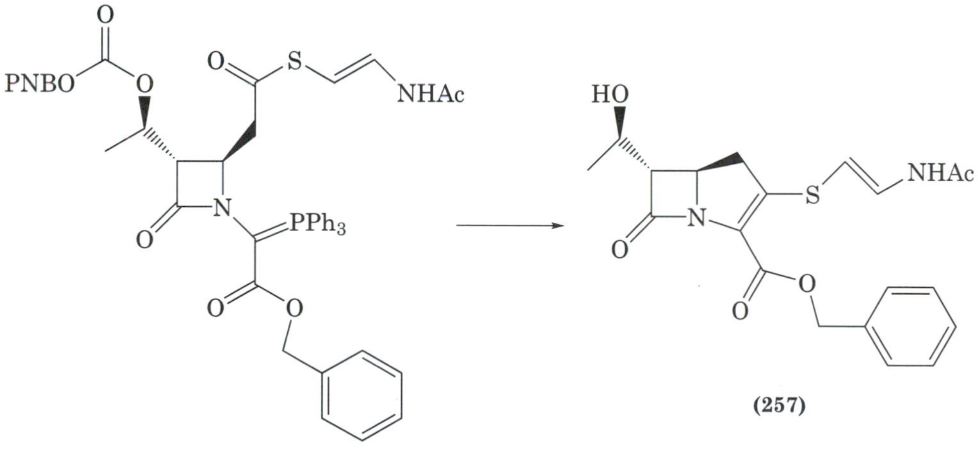

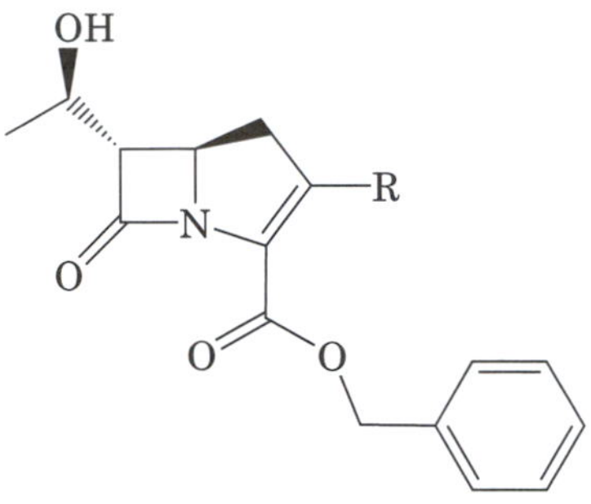

(258) R = H

(260) R = S⌒⌒NHAc

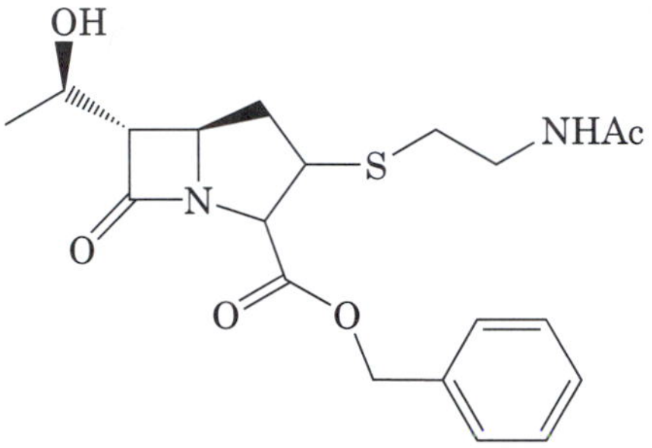

(259)

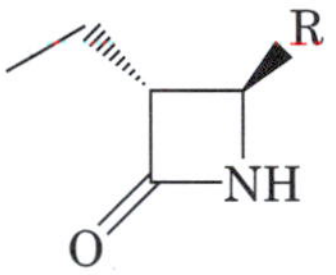

(261) R = OAc

(262) R =

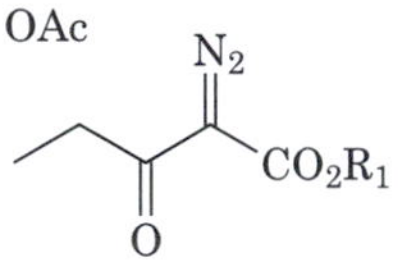

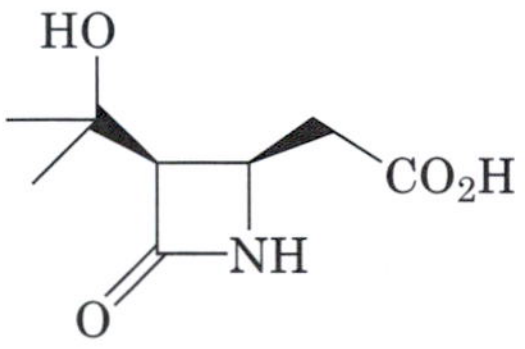

(263)

acetone, where metal ion chelation of the β-lactam enolate with the neighboring methoxyethoxymethoxy (MEM) group in the pres-

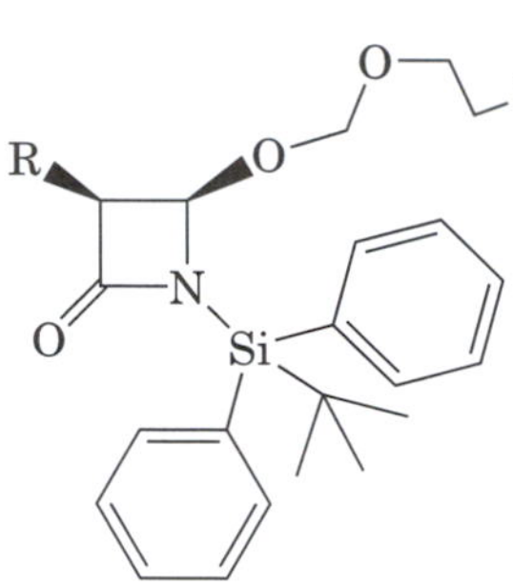

(264) R = H

(265) R =

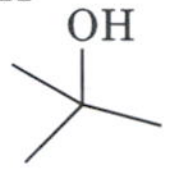

ence of the bulky silyl residue resulted predominantly in addition on the β-face, giving the *cis*-product (265) (548).

The asparenomycin natural products such as (227) have an alkylidene substituent at the C6-position. Synthesis in this series has followed a similar pattern by way of a bicyclic keto-ester derived from an appropriately functionalized monocyclic precursor. Many elegant routes to all these natural products and synthetic intermediates have been reported and reviewed (513, 517).

*5.5.1.3 Synthesis of 1β-Methylcarbapenems.* In 1984 the Merck group reported the synthesis of the 1β-methyl-substituted carbapenem (266), which showed improved chemi-

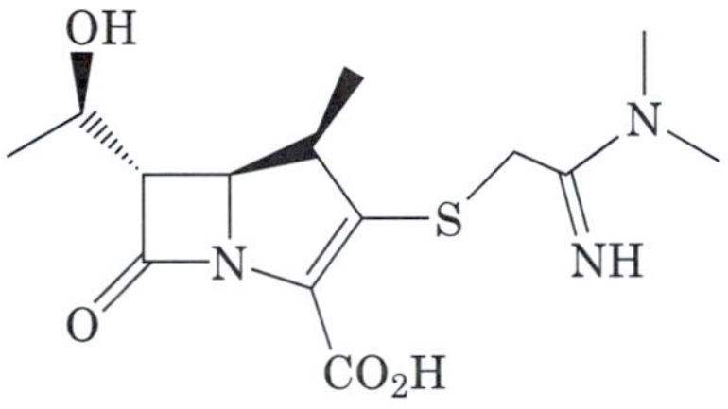

(266)

cal and metabolic stability while retaining the antimicrobial activity of imipenem (215) (549). The synthesis was achieved by alkylation of the methyl ester of (235) (R = TBS), followed by elaboration of the acid (267) to the keto-ester (268), although little stereocontrol was achievable by this route.

The advantages offered by the 1β-methyl series has led to a proliferation of methods for

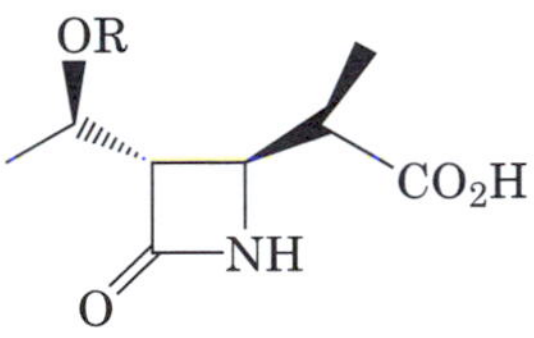

(267)

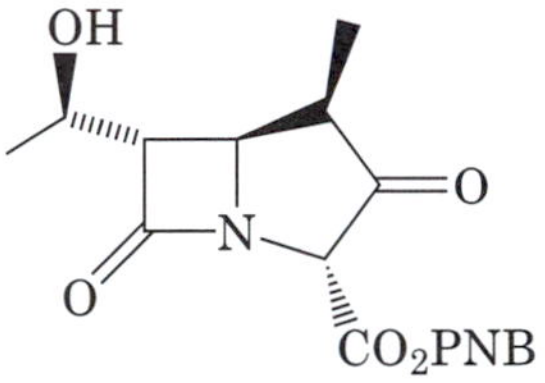

(268)

the stereoselective synthesis of (262) and other intermediates leading to (263). Most often, this method has made use of displacement of the acetoxy group from the chiral 4-acetoxy azetidinone (251) with tin or boron enolate of general structure (269), the product being

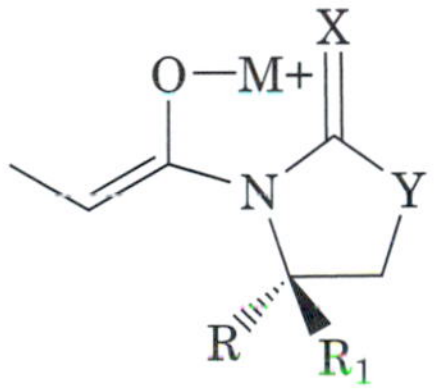

(269) X, Y, = O, S
      R, R₁ = alkyl

readily convertible to the acid. Yields are generally good, with a ratio of β:α isomers ranging from 24:1 with (270) (550) to greater than

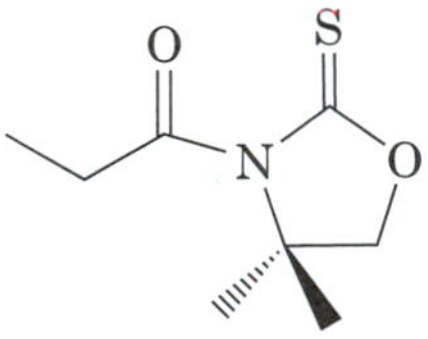

(270)

90:1 with (271) (551); many other variations have been reported (506). Direct incorporation of the diazo side chain is also possible (552).

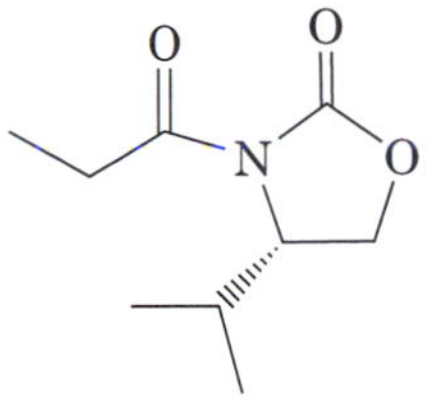

(271)

Alternative variations for constructing the 1β-methyl carbapenem ring system have made use of (R)- or (S)-methyl 2-methylhydroxypropionate (553, 554), the alcohol (272)

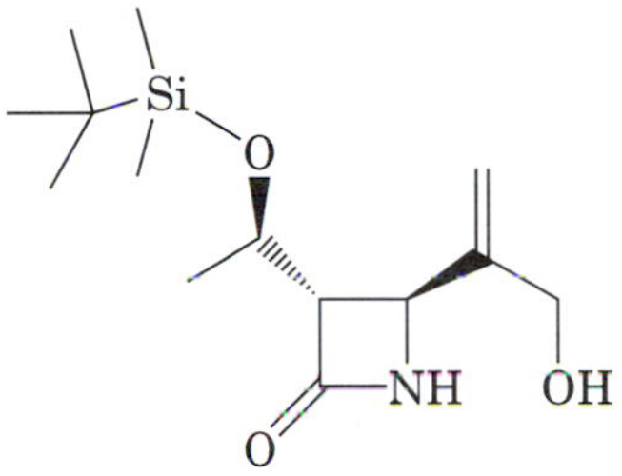

(272)

(555), malonic acid derivative (273) (556), and an enzymatic approach starting from (274) (557). Numerous synthetic efforts have fo-

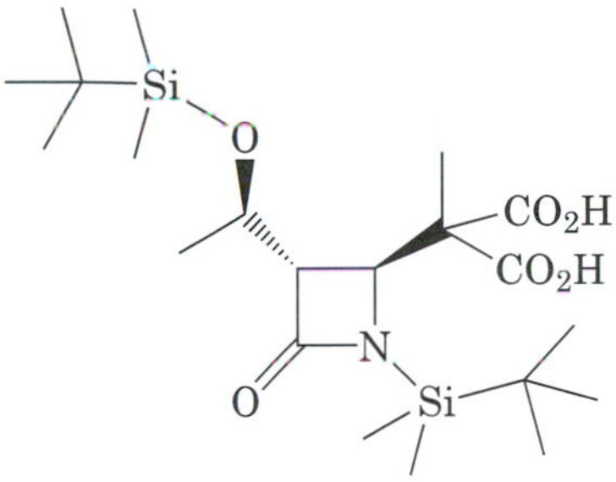

(273)

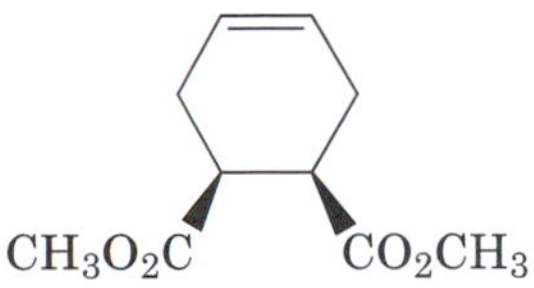

(274)

cused on the synthesis of 1β-methyl carbapenems in the search for a broad-spectrum agent suitable to treat severe infections. Compounds (**275–286**) are representative examples taken from the literature (558–563).

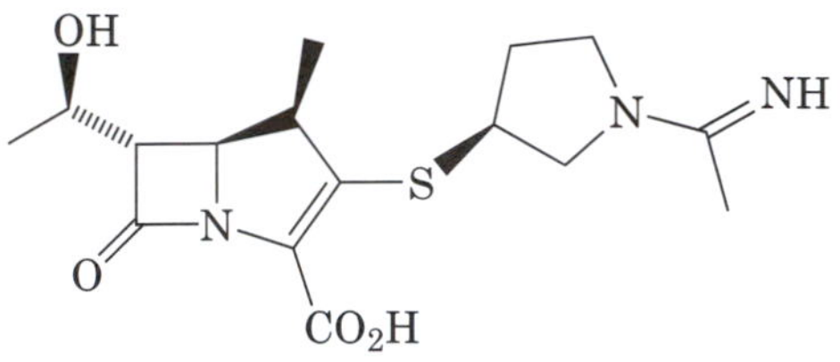

(**275**) Panipenem

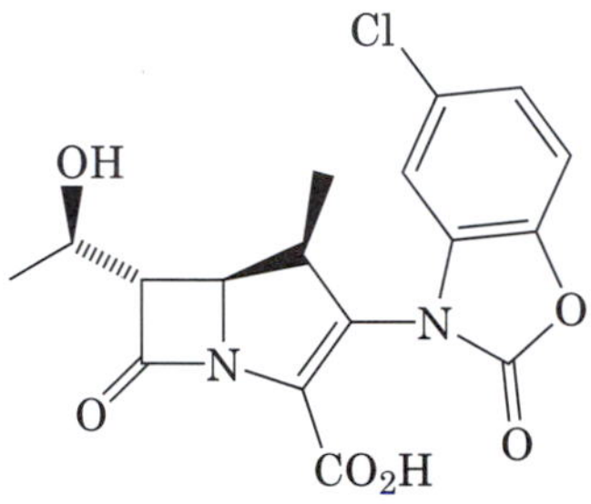

(**276**)

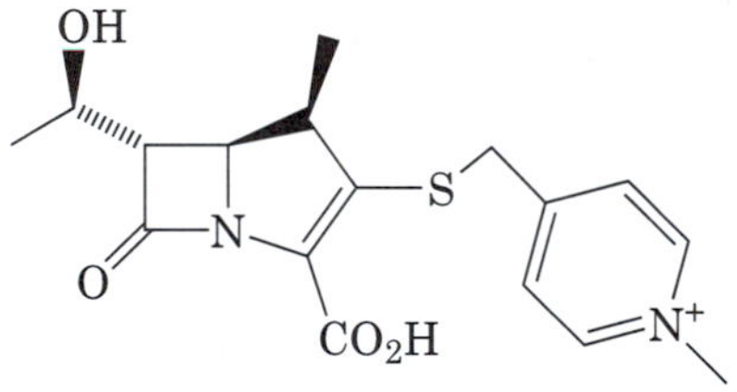

(**277**)

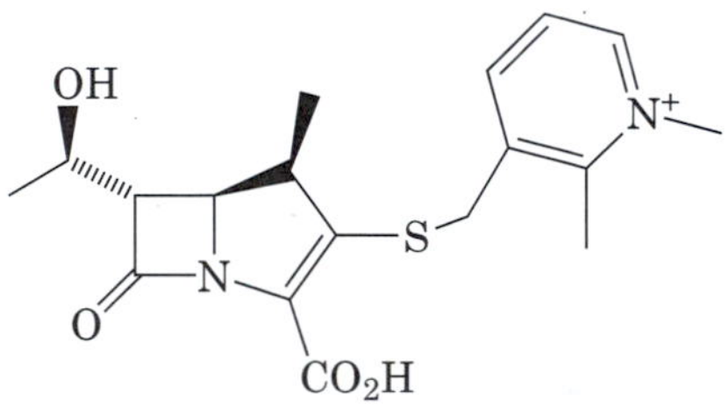

(**278**)

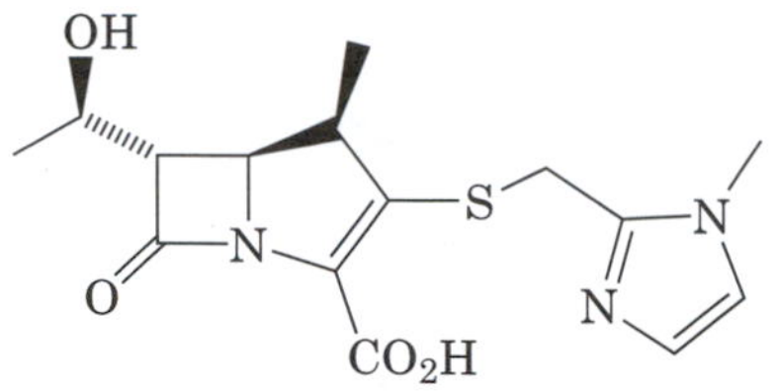

Effective substituents replacing the 1β methyl were extensively explored by several research groups, although compounds having

(**279**)

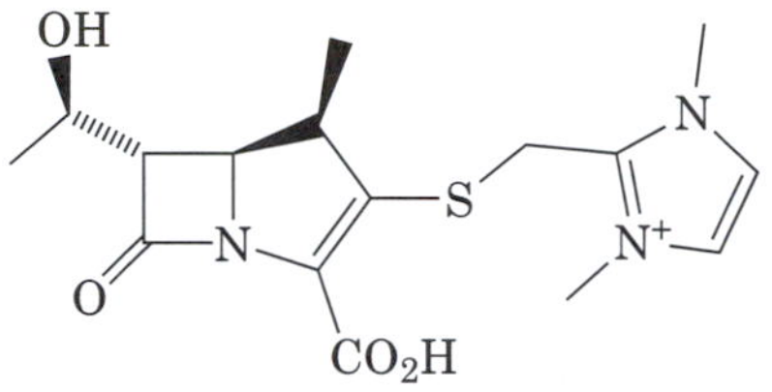

(**280**)

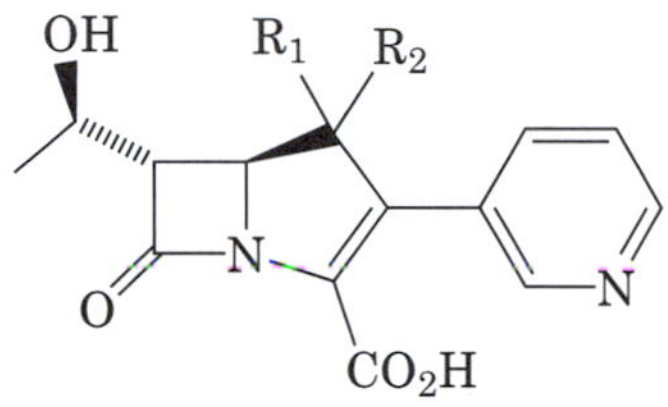

(**281**) $R_1$, $R_2$ = H or $CH_3$

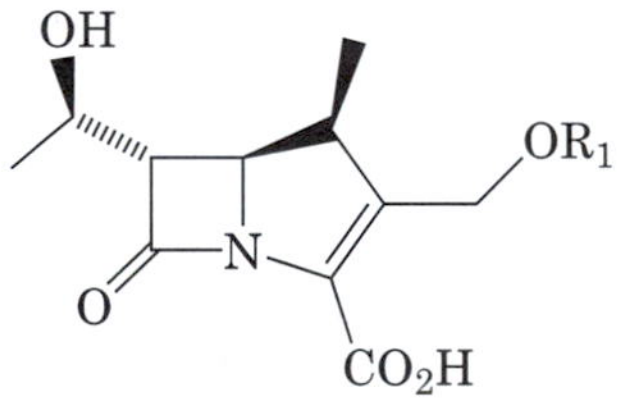

(**282**) $R_1$ = H, Ac, $CONH_2$, $CON(CH_3)_2$, CONHAr, $CONH(CH_2)_2Py+$

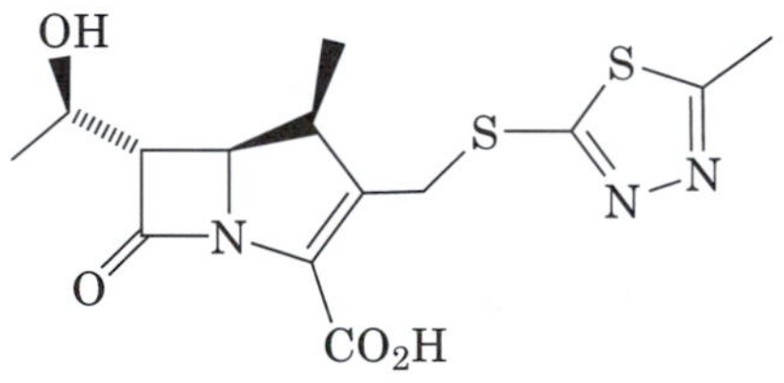

(**283**)

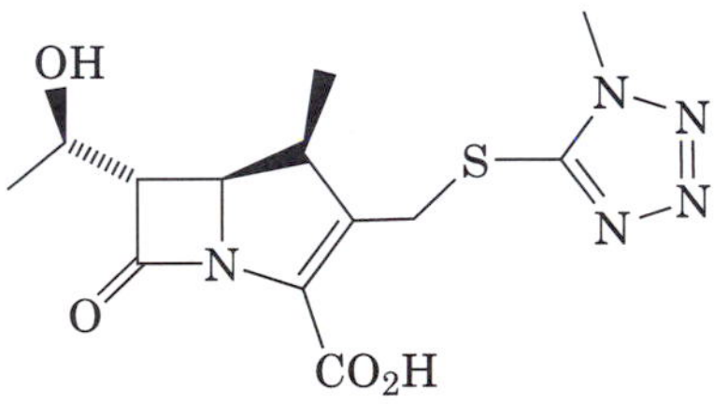

(284)

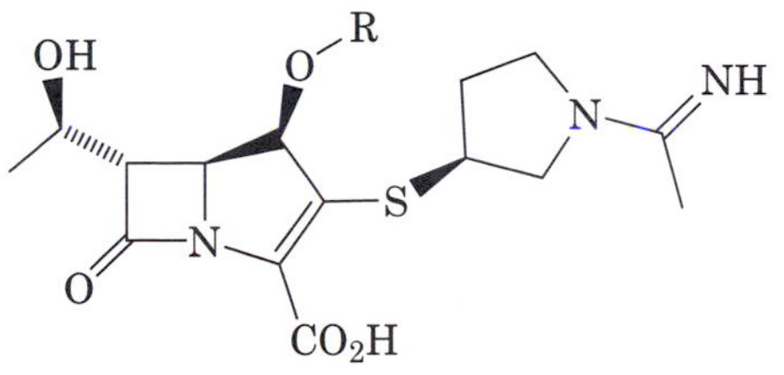

(288) R = H
(289) R = CH₃

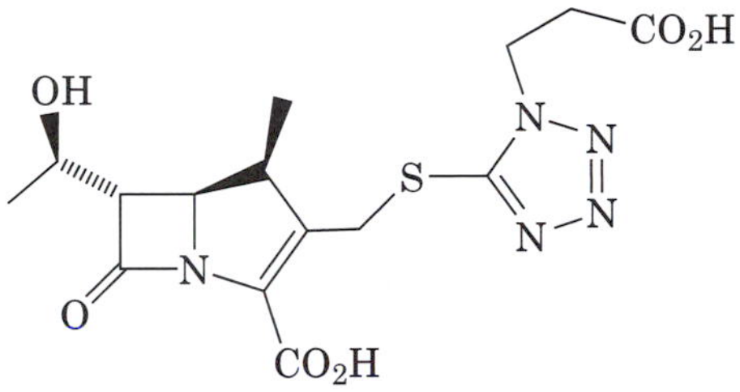

(285)

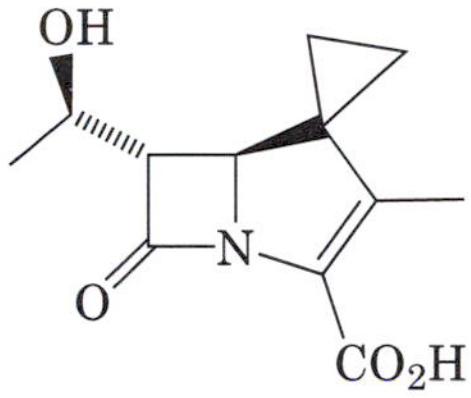

(290)

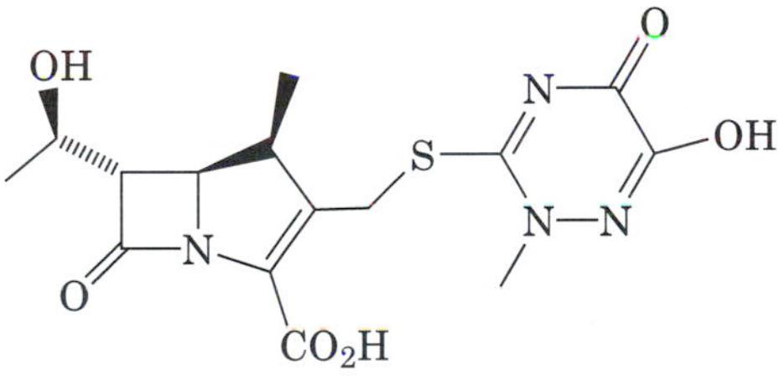

(286)

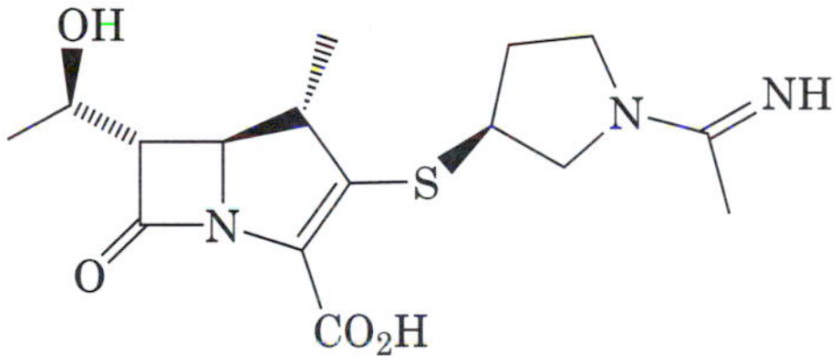

(291)

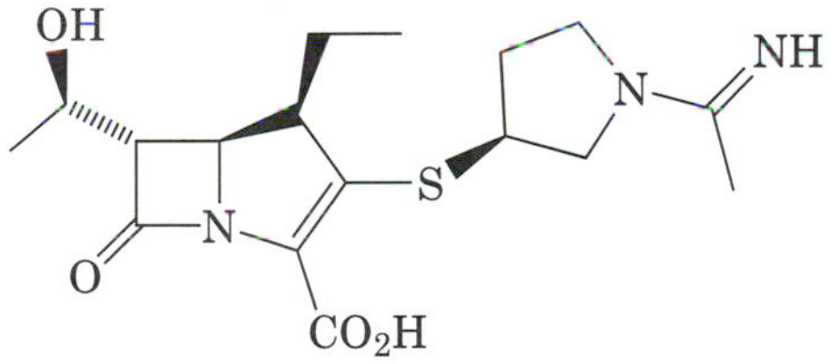

(287)

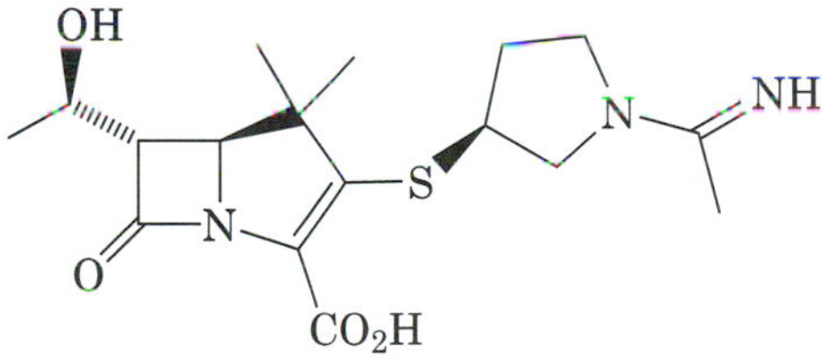

(292)

ethyl (**287**), hydroxy (**288**), methoxy (**289**), or 1–1′-*spiro*-cyclopropane (**290**) in place of methyl are less active (564, 565). Other compounds prepared by Sankyo include 1α-methyl and 1,1-dimethyl panipenem derivatives (**291**) and (**292**) (566–569). Further modifications were studied by Roche, which eventually obtained Ro 19–8928 (**293**) (567), a compound

with good activity against *Pseudomonas aeruginosa*. Moreover, fluorination at position 1, reported by Merck, resulted in very unstable compounds such as (**294**) (570).

A series having 1β-aminoalkyl substituents has aroused some interest (571). The Bristol-Myers Squibb group extensively explored this class of derivatives and compounds (**295**) and (**296**) are examples of such modifications (572–577).

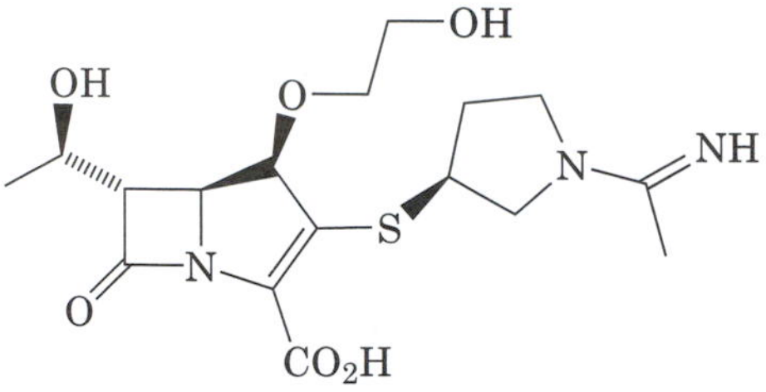

(293) Ro 19-8928

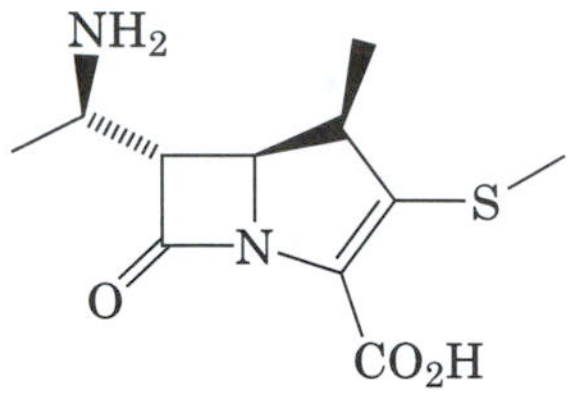

(297) BMS-40383

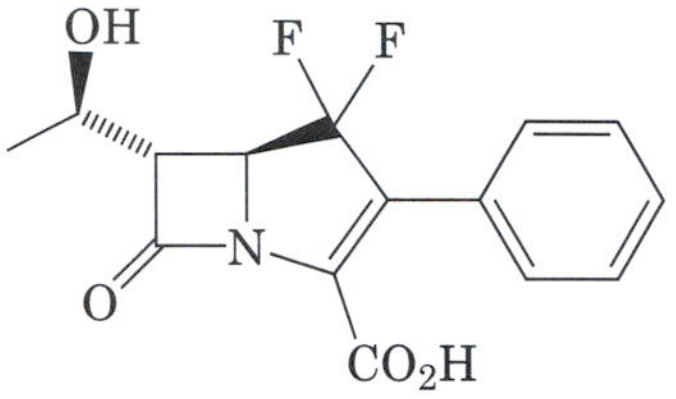

(294)

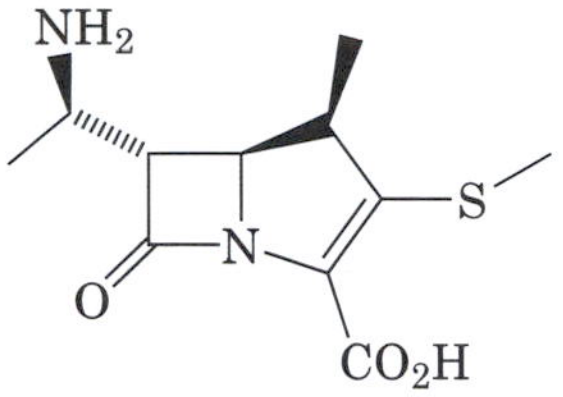

(298) BMS-40591
Carbapenems bearing a basic group at C6 position

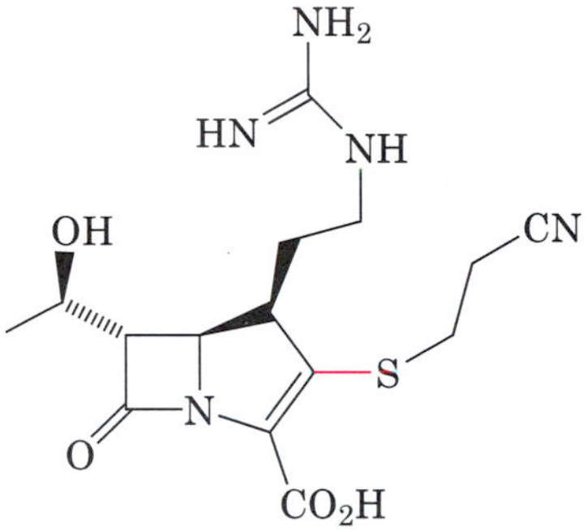

(295) BMS-181139

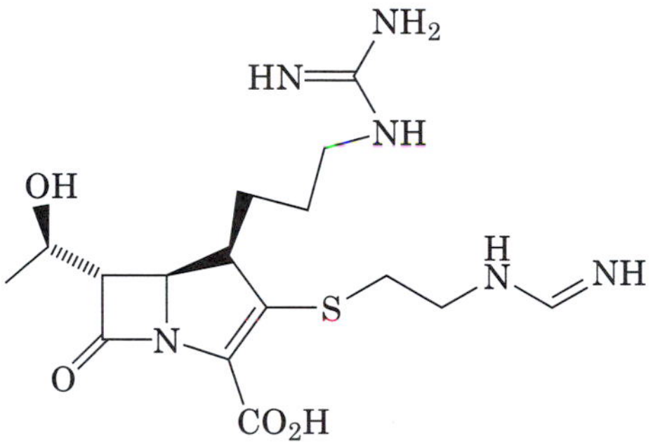

(299) BMS-45742

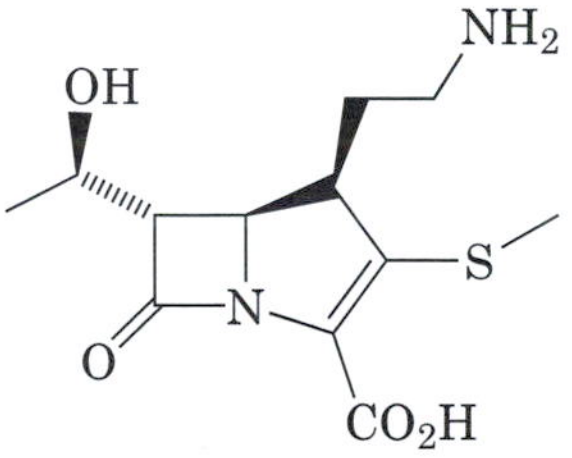

(296) BMS-40383

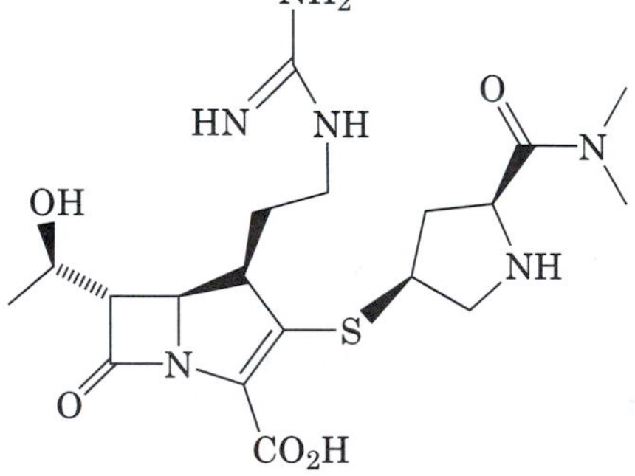

(300) BMS-45047
Carapenems bearing a basic group
at C1 and C2 positions

They also synthesized carbapenems (297–302) that bore the amino moiety either at C6 or at the C1 position and made a comparison with other derivatives that bore the hydroxy-

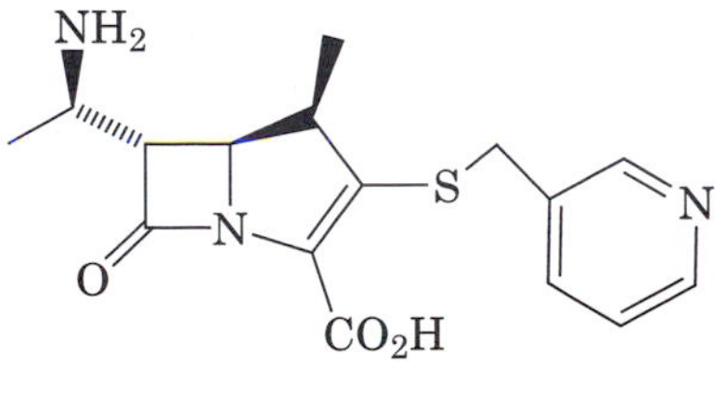

(301) BMY-40732

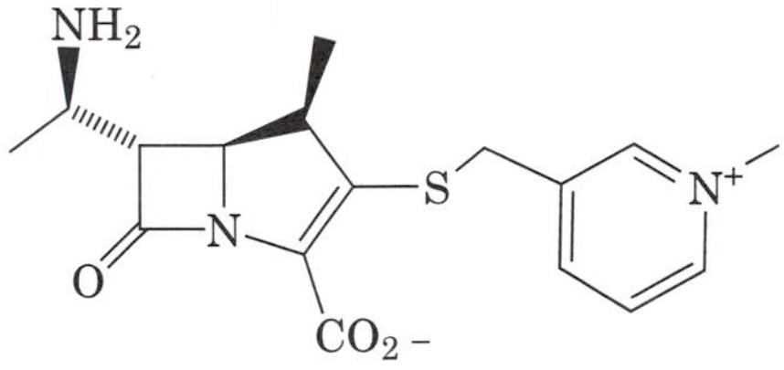

(302) BMY-40886
Carbapenems bearing a cationic or basic
group at C2 and C6 position

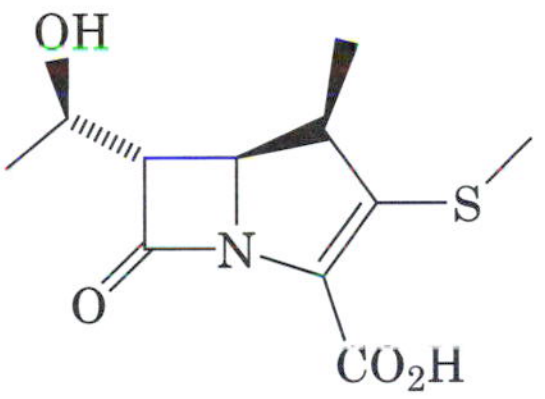

(303) BMY-27946

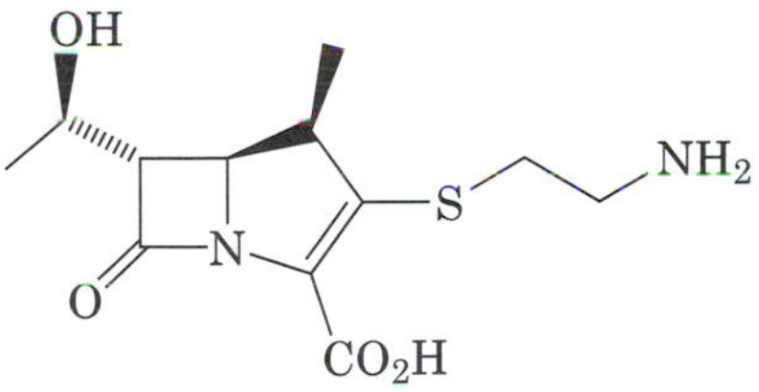

(304) BMS-182880

ethyl side chain (**303** and **304**), reporting
three major findings (571, 578, 579):

1. The presence of a cationic group was reconfirmed as essential to retain antipseudomonal activity.
2. The antipseudomonal activity was observed, regardless of the position of the cationic group at C1, C2, or C6.

3. The presence of a second basic group at the
   C1 or C6 position of a carbapenem already
   bearing a cationic center at C2 position allows the molecule to exert its antipseudomonal activity without being taken up
   through porin protein D2, thereby overcoming one of the mechanisms of resistance
   acquired by *Pseudomonas aeruginosa*.

Besides the introduction of a basic functional group, Bristol-Myers Squibb also investigated other modifications of the hydroxyethyl side chain (580). From their studies it
was reconfirmed that the presence of an electron-withdrawing group at C6 was difficult because of the low chemical stability of the resulting molecules (**305–307**) and that the

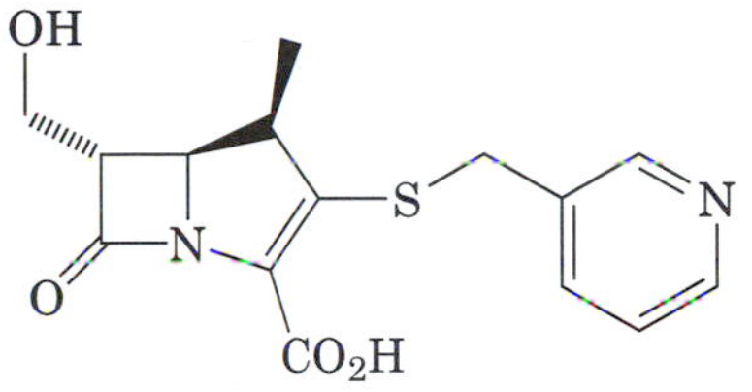

(305)

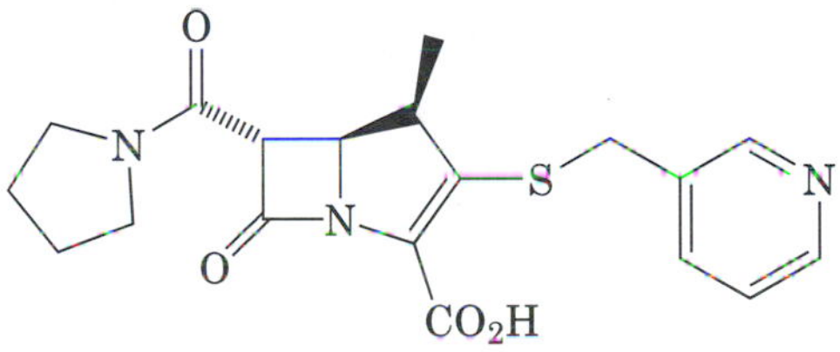

(306)

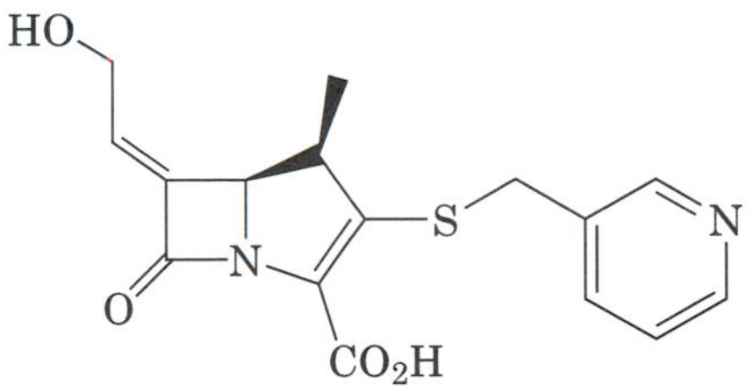

(307)

hydroxyethyl side chain is the best compromise between chemical stability and microbiological activity (581–584).

Another interesting example regarding modification at C6 has been reported by Nagao (585), where the synthesis of 6-methylthiocarbapenems (**308**) and (**309**), by reaction of the

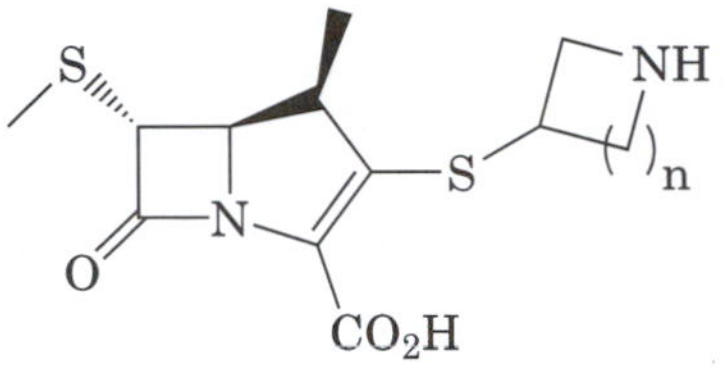

(**308**) n = 1
(**309**) n = 2

anion at position C6 of the carbapenem scaffold, is described.

In 1987 meropenem (**216**) was presented by Sumitomo as a drug candidate (586), featuring improved efficacy and safety over those of other types of carbapenem antibiotics. As soon as it became clear that meropenem was endowed with very potent antibacterial activity and a wide spectrum of action against Gram-positive and Gram-negative organisms including *Pseudomonas* spp., synthesis of a variety of analogs having the 1β-methylcarbapenem skeleton and 2,4-disubstituted pyrrolidine moiety at C2 was reported by many research groups (587–590).

Methylation of the nitrogen atom in the pyrrolidine ring and modification of the carboxamide group enhanced the DHP-I stability (591) and, in particular, the introduction of hydrophilic groups (**310**, **311**) on the amido moiety increased the antipseudomonal activity (592).

Addition of a second cationic (basic) center was explored to improve the biological profile

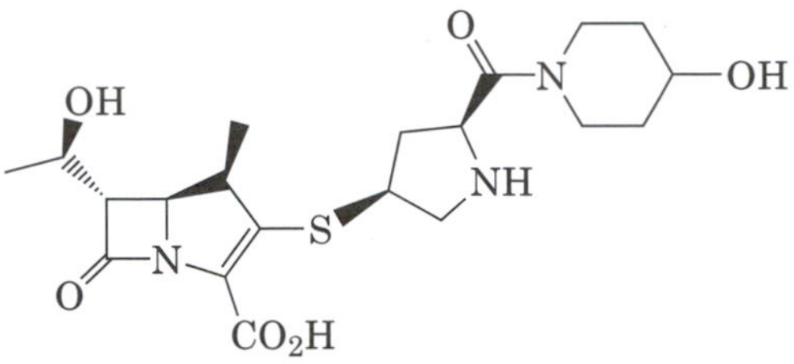

(**310**)

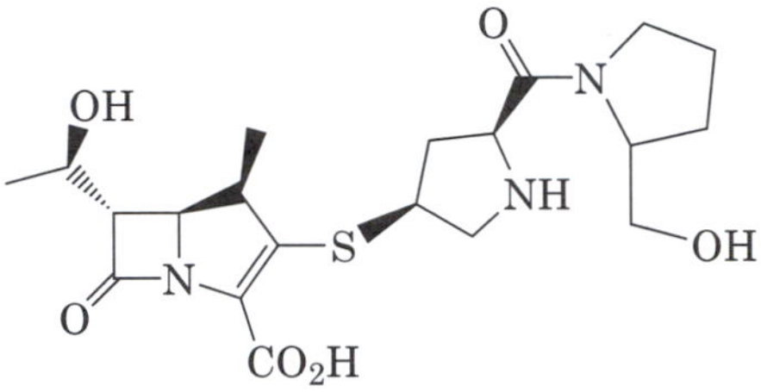

(**311**)

of the drug. Following this line of research, Nishi reported (593) the piperazine derivative (**312**), which retained the antipseudomonal activity.

Subsequently, other piperazine derviatives (**313**, **314**) were reported by Lee (594); interestingly, only these examples showed good activity against *Pseudomonas aueruginosa*. Introduction of a quaternary ammonium moiety in the C2 proline side chain was particularly efficacious at improving the half-life of the drug by reduction of the renal clearance (**315**) (595). Alternatively, it has been shown by Oh (596) that the use of a sulfonium moiety can also provide the extra cationic center (**316**). Zeneca and Merck groups achieved a remarkable improvement in terms of half-life with

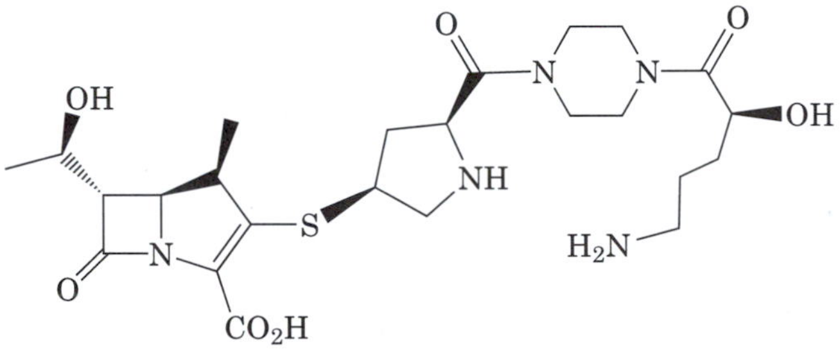

(**312**) DX-8739

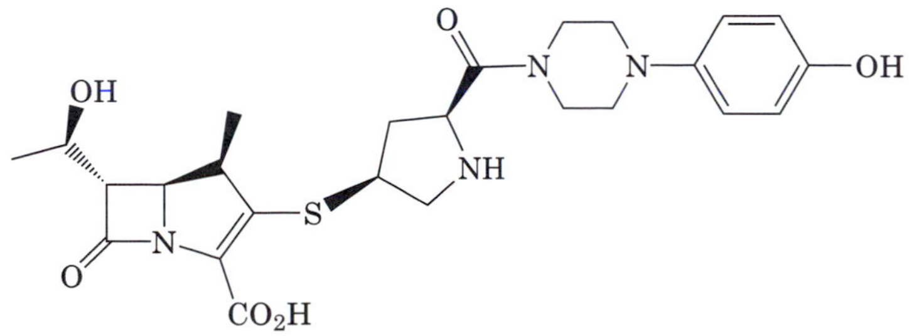

(313)

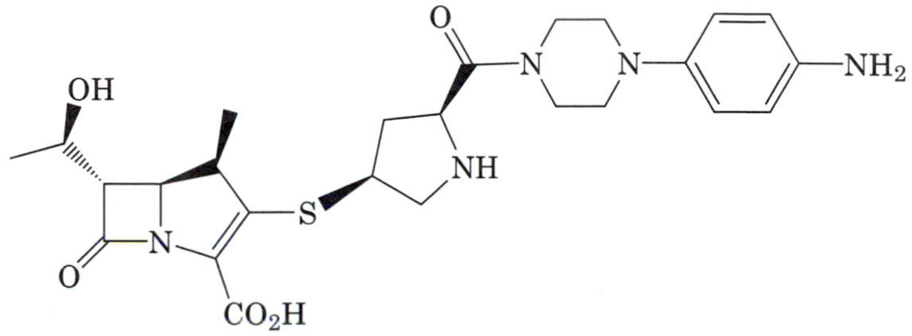

(314)

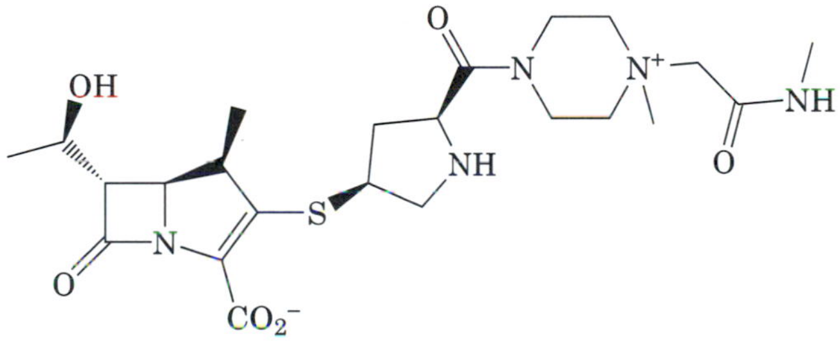

(315)

the synthesis of MK-826 (**317**), a result that has been confirmed in humans. The compound is highly serum bound, approximately 95%, and has a mean terminal $t_{1/2}$ of about 4.5 h after intravenous administration (597,

(316)

598). However, the presence of an overall negative charge on the molecule has limited the broadness of spectrum of action of the antibiotic. In fact, it lacks activity against *P. aueruginosa* and also against penicillin-resistant *S. pneumoniae* (599).

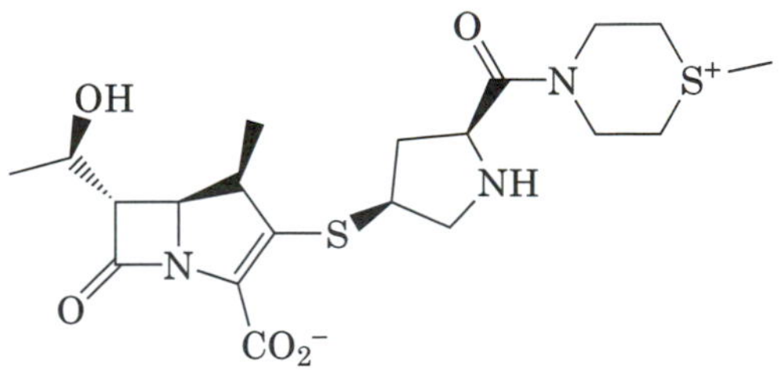

Another approach to prepare meropenem mimics was undertaken by introduction of a catechol moiety in a manner similar to what has been done in cephalosporin chemistry (see Section 5.2.2). This modification confirmed that it is possible to improve the *in vivo* antipseudomonal activity (**318**) (600).

Although most compounds exhibited an antibacterial profile similar to that of meropenem, Banyu, first with J-111,347 (**319**) and

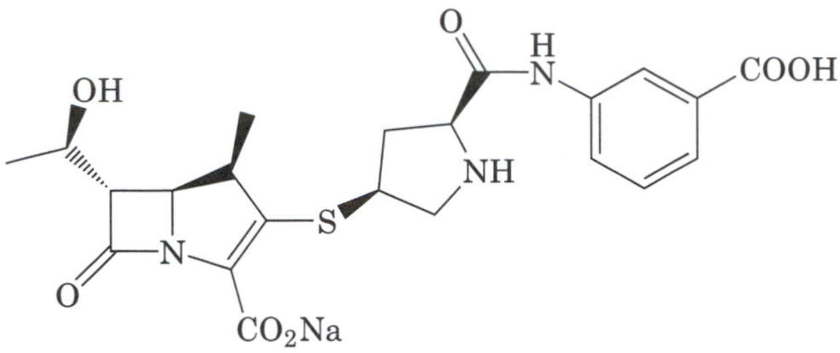

**(317)** MK-826

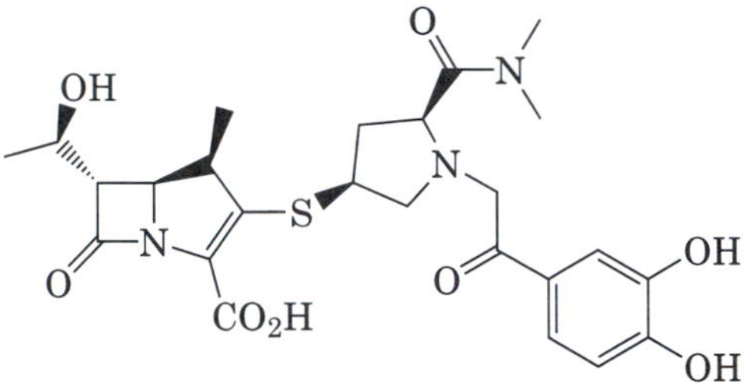

**(318)**

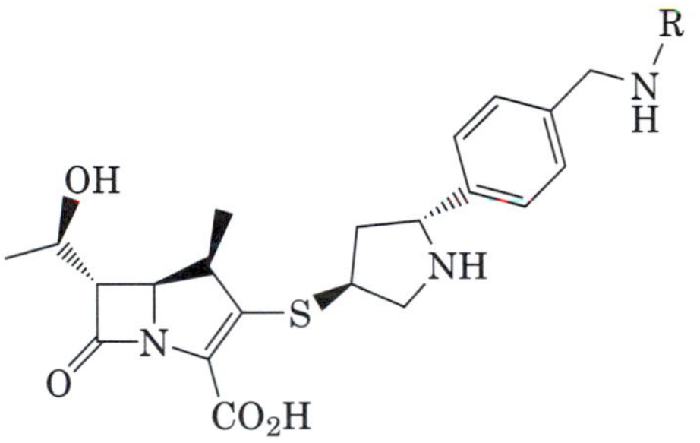

**(322)** J-114,871

**(319)** J-111,347   R = H
**(320)** J-111,225   R = CH$_3$

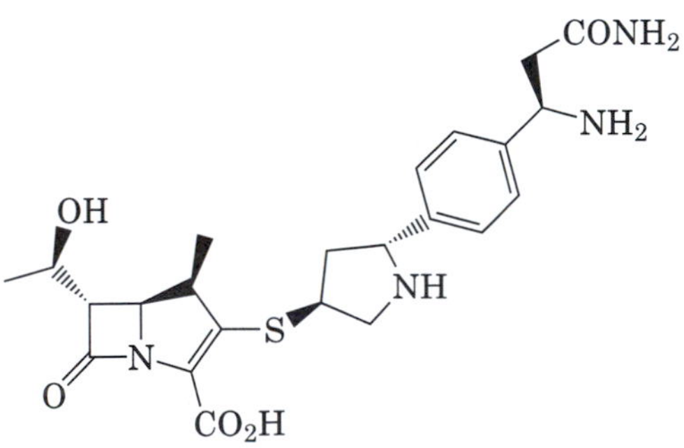

**(321)** J-114,870

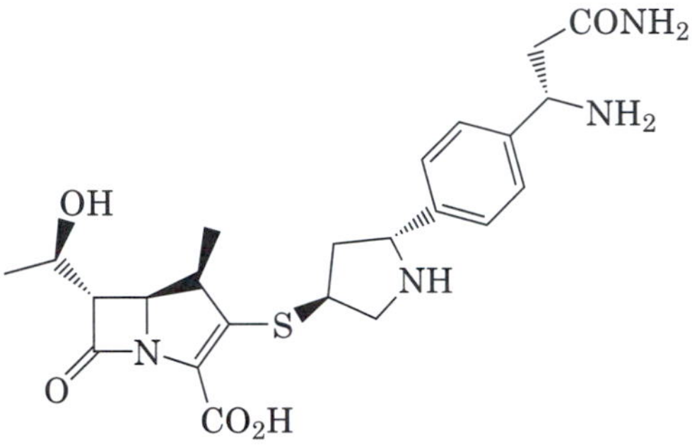

then with J-111,225 (**320**), J-114,870 (**321**), and J-114,871 (**322**), demonstrated that it is possible to further expand the antibacterial spectrum of meropenem mimics, including anti-methicillin-resistant *Staphylococcus aureus* (anti-MRSA) activity (601–604).

Further evaluation of the prototype compound J-111,347 (**319**) has been suspended because of its epileptogenicity. This has been eliminated by N-methylation (**320**) or carbamoylmethyl substitutions (**321, 322**) in the other three compounds. *In vitro* and *in vivo* data are reported in Tables 14.14 and 14.15, respectively (604).

**5.5.2 Trinems and Policyclic Carbapenems.** Further developments in seeking to advance the properties of the carbapenems have seen the synthesis of both tricyclic and polycyclic analogs. Starting from azetidinone (**323**) and using the phosphorane route, the Glaxo group prepared a series of so-called *tribactams* (and lately renamed *trinems*). Belonging to this class the 4-methoxy trinem, GV104326 (sanfetrinem, **324**) (605, 606), showed a very potent microbiological activity. Its conversion to

**Table 14.14   MIC 90 (μg/mL) of Imipenem Derivatives**

| Organism | J-111,225 | J-114,870 | J-114,871 | Imipenem |
|---|---|---|---|---|
| *S. aureus* (MSSA) | 0.016 | 0.016 | 0.016 | 0.016 |
| *S. aureus* (MRSA) | 4 | 4 | 4 | 128 |
| Coagulase-negative staphylococci MR | 4 | 4 | 4 | 128 |
| *S. pyogenes* | ≦0.008 | ≦0.008 | ≦0.008 | ≦0.008 |
| *S. pneumoniae* (incl. PenR strains) | 0.25 | 0.25 | 0.25 | 0.25 |
| *E. faecalis* | 8 | 8 | 8 | 4 |
| Enterobacteriaceae | 0.032–1 | 0.016–2 | 0.032–1 | 0.125–4 |
| *H. influenzae* | 0.125 | 0.125 | 0.125 | 1 |
| *P. aeruginosa* Imi-S | 4 | 4 | | 4 |
| *P. aeruginosa* Imi-R | 16 | 16 | 16 | 32 |
| *Acinetobacter* spp. | 0.125 | 0.25 | 0.125 | 0.125 |
| *C. difficile* | 2 | 2 | 2 | 8 |

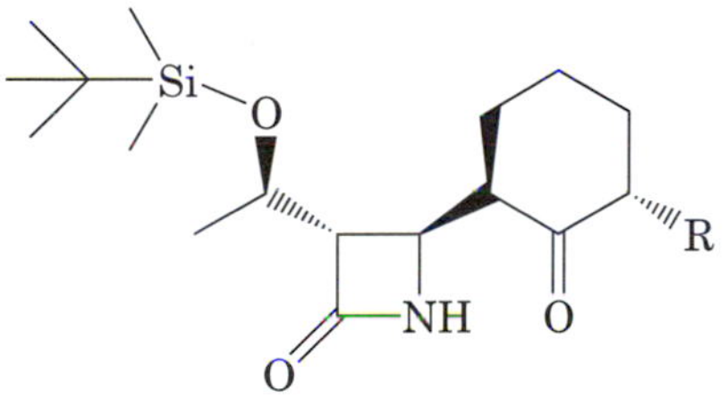

**(323)**

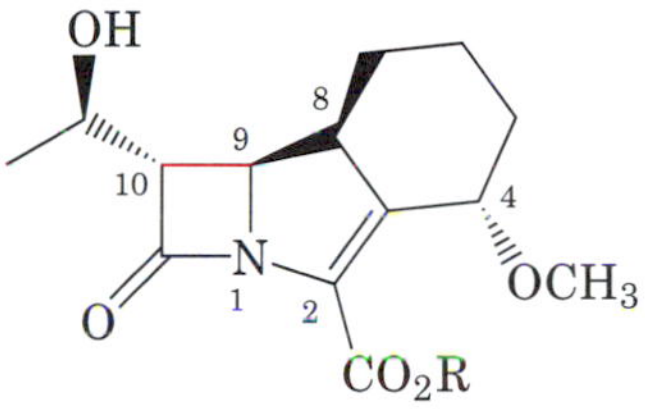

**(324)** Sanfetrinem   R = Na

**(325)** Sanfetrinem cilexetil

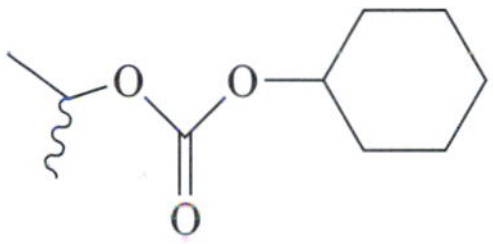

the corresponding metabolically labile pro-drug ester (**325**) demonstrated a safety profile and pharmaceutical stability to warrant further progression into the clinics (607–610). The biological properties of trinems include a broad spectrum of activity versus Gram-positives and Gram-negatives, either aerobic and anaerobic, and stability to clinically relevant β-lactamases and human dehydropeptidases (DHP-I) (611).

Structure-activity relationship studies showed that the best biological profile was ob-

served in those compounds bearing a hetero-atom attached at C4 and having absolute configuration (4S,8S). The construction of the trinem backbone is very challenging because it contains five stereogenic centers that need to be built in a stereospecific manner. A great deal of work has involved the stereoselective

**Table 14.15   *In Vivo* Efficacy in Systemic Murine Infections Expressed as ED$_{50}$ (mg/kg)$^{a,b}$**

| Systemic infection | J-111,225 | Imipenem | Vancomycin |
|---|---|---|---|
| MS *S. aureus* | 0.040 | 0.060 | 0.523 |
| MR *S. aureus* BB6221 normal mice | 5.42 | 94.73 | 2.56 |
| MR *S. aureus* BB6221 immunosuppressed mice | 8.53 | 93.76 | 8.83 |
| *Pseudomonas aeruginosa* BB5746 | 0.483 | 0.242 | — |

[a]MS, methicillin-susceptible; MR, methicillin-resistant; PR, penicillin-resistant.

[b]In a model of thigh infection caused by MR *S. aureus* BB6294 in immunosuppressed mice, bacterial counts were significantly decreased with J-111,225 versus controls and imipenem- or vancomycin-treated mice ($P < 0.01$) at 4 h post-therapy.

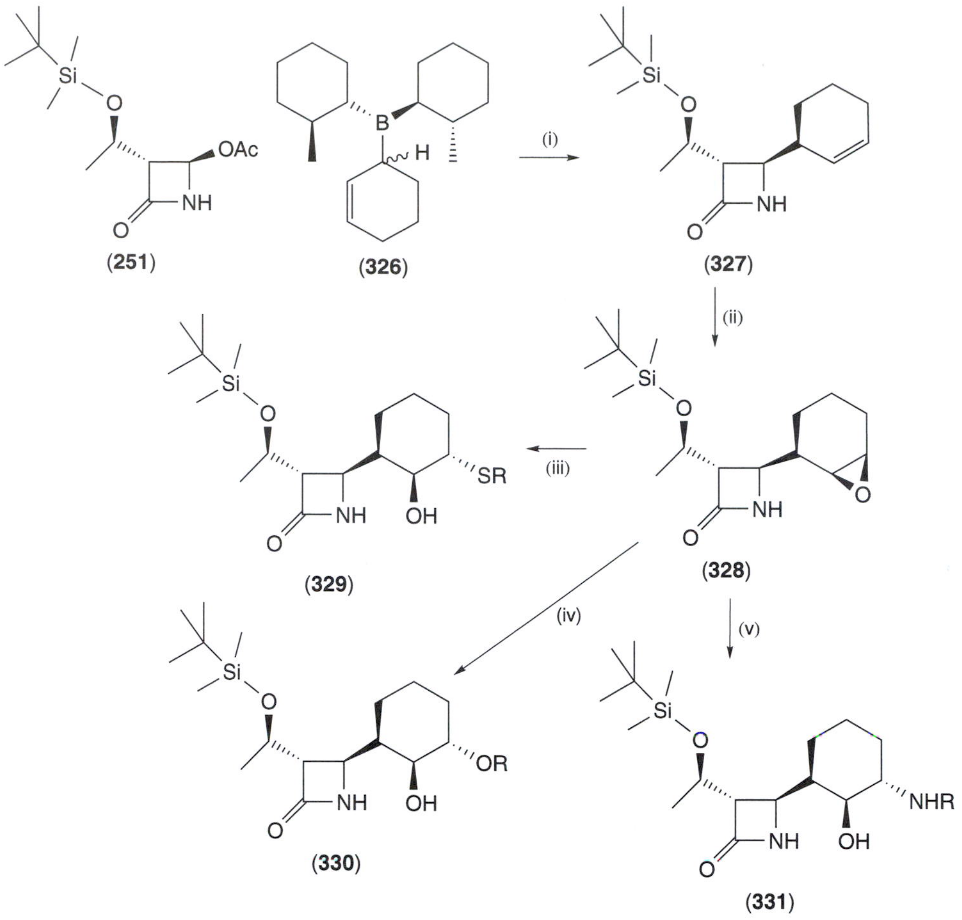

Reagents: (i) ZnEt₂, THF, 25C; (ii) Magnesium monoperoxyphthalate, CH₂Cl₂; (iii) RSH, H⁺; (iv) ROH, H⁺; (v) RNH₂, H⁺.

**Figure 14.17.** Highly diastereoselective synthesis of intermediates to trinems.

synthesis of advanced intermediates such as epoxide (**328**) and epoxyphosphate (**334**) that have allowed the introduction at C4 of sulfur-, oxygen-, and nitrogen-containing functional groups, as shown in Figs. 14.17 and 14.18. Reaction of 4-acetoxy azetidinone (**251**) with 2-cyclohexenylborane (**326**) gave cyclohexenyl derivative (**327**) in high yield and selectivity (612). This intermediate was converted into epoxide (**328**) that underwent nucleophilic addition, to afford sulfides (**329**), alkoxides (**330**), and amines (**331**) that were further progressed to derivatives of general formula (**323**).

Another diastereoselective synthesis in-

volved the ketoazetidinone (**332**) that was protected at the nitrogen of the β-lactam ring and reacted with diethylchlorophosphate to give (**333**). Epoxidation gave the advanced intermediate (**334**) that, upon reaction with nucleophiles, was converted into azetidinone (**323**) (613, 614).

A synthetic improvement has been introduced with the direct condensation methodology (615–617), in which the enantiomerically enriched silyl enol ether of 2-methoxycyclohexanone (**335**) (618, 619) is reacted with the 4-acetoxy azetidinone (**251**), to yield the 6'-methoxy ketoazetidinone (**336**) with high stereoselectively. After introduction of the cyclo-

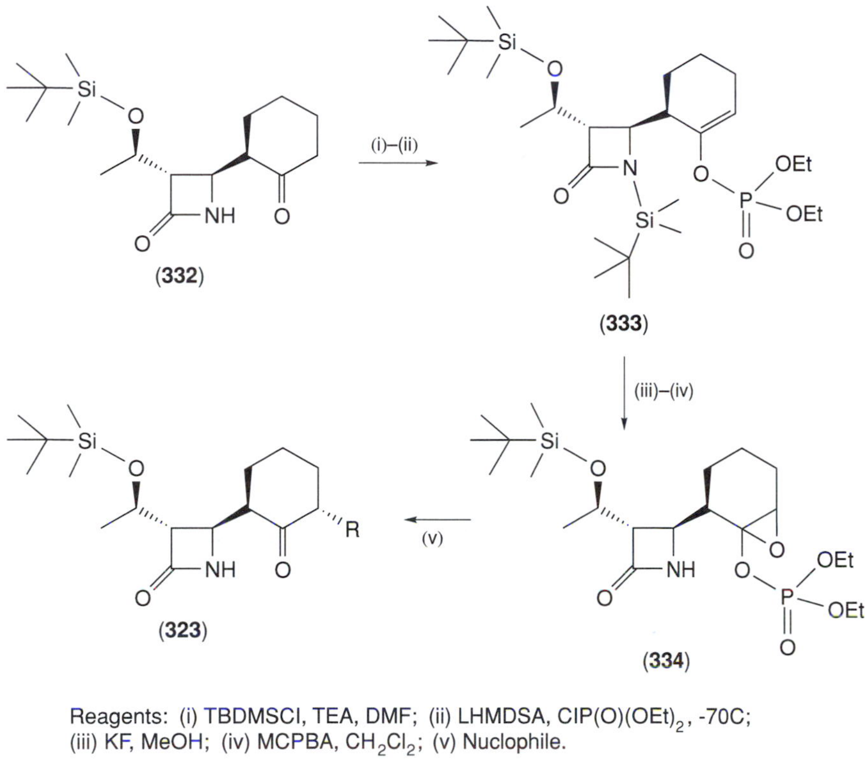

Reagents: (i) TBDMSCl, TEA, DMF; (ii) LHMDSA, ClP(O)(OEt)$_2$, -70C;
(iii) KF, MeOH; (iv) MCPBA, CH$_2$Cl$_2$; (v) Nuclophile.

**Figure 14.18.** Synthesis and use of epoxyphosphate 334.

hexyl ring, intramolecular cyclization of an oxalimido derivative (**337**) produced the tricyclic system (**338**), and removal of the protecting groups gave sanfetrinem (**324**). A highly diastereoselective and practical synthesis of sanfetrinem (**324**), in which a diasterometic mixture of 6'-methoxy ketoazetidinone is enolized and the zinc enolate protonated with diethylmalonate, to give diastereomerically pure (**336**), has been reported by Hanessian (620). Figure 14.19 shows a route that has been applied on large scale to produce multikilogram quantities of sanfetrinem (**324**).

Furthermore, the synthesis of 4-N-substituted trinems such as (**339**) has allowed the preparation of ureas (**340**) (621), amides (**341**) (622), and, most important, 4-N-methylformamidino compound GV129606X (**342**) (623–625). GV129606X (**342**) is an injectable trinem with a broad spectrum of activity, including *Pseudomonas* spp., whose development has been terminated after preclinical studies showed toxicity problems.

Through use of similar chemistry, both Hoechst and Bayer described tetracyclic analogs (626, 627). Thus, compound (**343**) was elaborated to afford the tetracyclic carbapenem (**344**) (627). Ring size, stereochemistry, and heteroatom position were modified (**345–348**), although these studies were not fruitful from the viewpoint of antibacterial activity.

Introduction of an aryl or heteroaryl substituted 4-exomethylenyl substituent (628) provided a series of compounds that showed high potency against resistant Gram-positive strains such as penicillin-resistant Pneumococci and methicillin-resistant Staphylococci (MRS). In particular, GV143253X (**349**) showed a broad spectrum, including vancomycin-resistant Enterococci and *H. influenzae*, good *in vivo* efficacy against MRSA in septicemia, and thigh infections in the mouse.

Sankyo Laboratories synthesized a series of 4-substituted trinems (**350–352**) bearing a

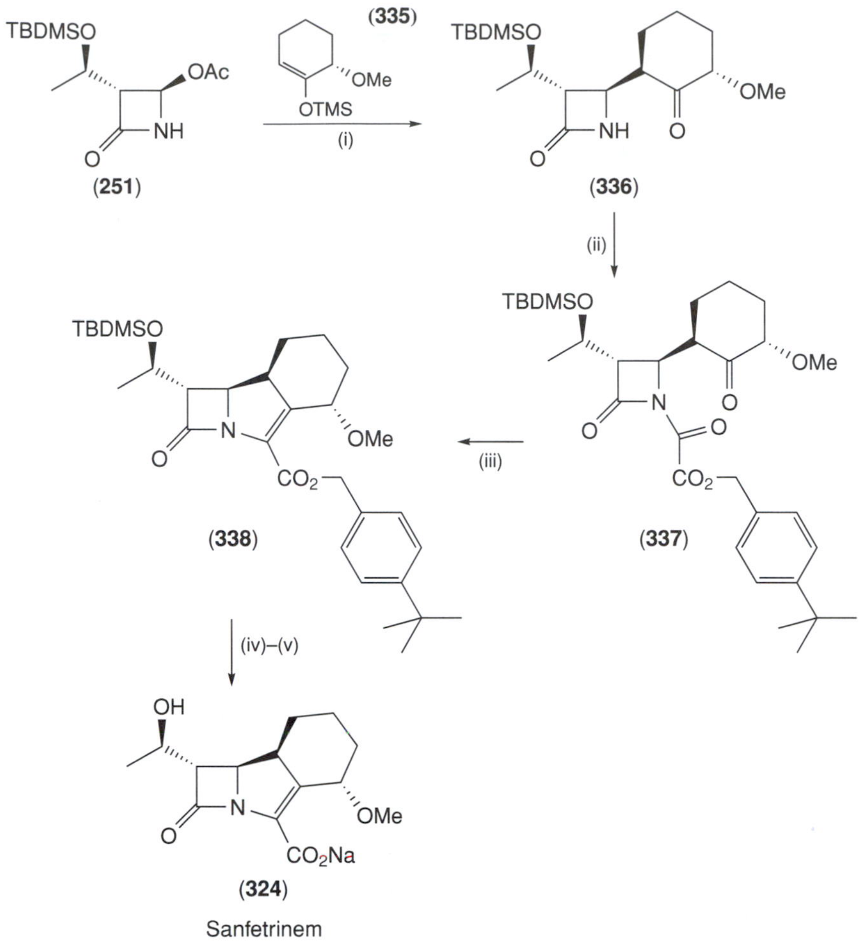

Reagents (i) SnCl$_4$; (ii) ClCOCO$_2$R, Pyridine, CH$_2$Cl$_2$; (iii) P(OEt)$_3$, xylene, reflux; (iv) TBAF, AcOH, THF; (v) H$_2$, Pd/C.

**Figure 14.19.** Highly diastereoselective and practical synthesis of sanfetrinem.

heteroaryl substituent with an antibacterial spectrum similar to that of 4-exomethylenyl trinems (629).

GlaxoWellcome also reported the synthesis of pentacyclic β-lactams (353) that were conceived as conformationally constrained analogs of 4-exomethylenyl trinems (630). This class is characterized by antibacterial activity and affinity to PBP2a similar to those of 4-exomethylenyl trinems but with an inferior pharmacokinetic profile (631).

Another interesting series of anti-MRSA trinems bearing a pyrrolidinyl moiety was recently studied by Sankyo (632, 633), who pre-

pared a series of derivatives starting from epoxide (328) (Fig. 14.20), or hydroxymethyl intermediate (358) (Fig. 14.21). Specifically, epoxide (328) was converted in two steps into ketoazetidinones (354) and (355), which upon standard cyclization and deprotection procedures gave trinems (356) and (357). Alcohol (358) was mesylated to give (359), which was reacted with thiol (360), giving rise to both 6′ isomers (362), possibly through elimination of mesylate to form intermediate (361) and subsequent Michael addition. Standard procedures resulted in the formation of the corresponding trinems (363) and (364), which are

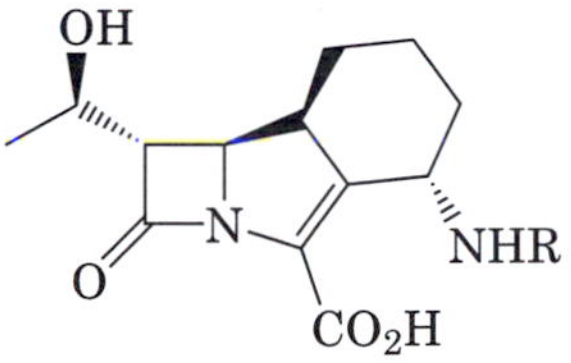

(339)

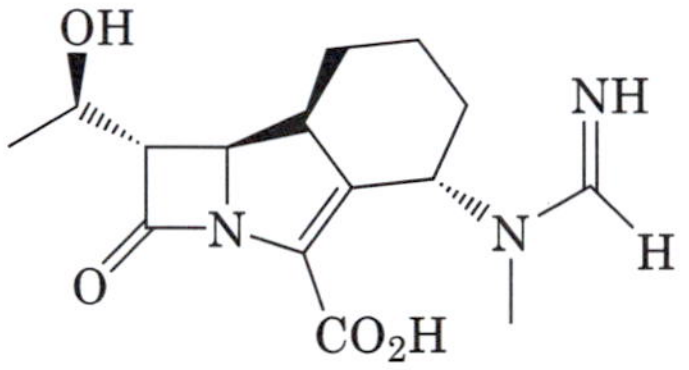

(342) GV129606X

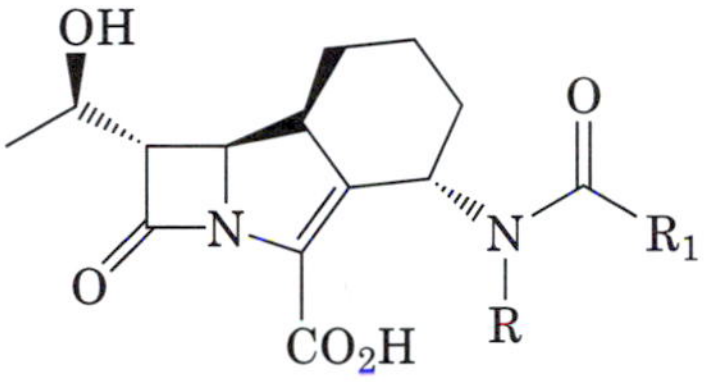

(341)

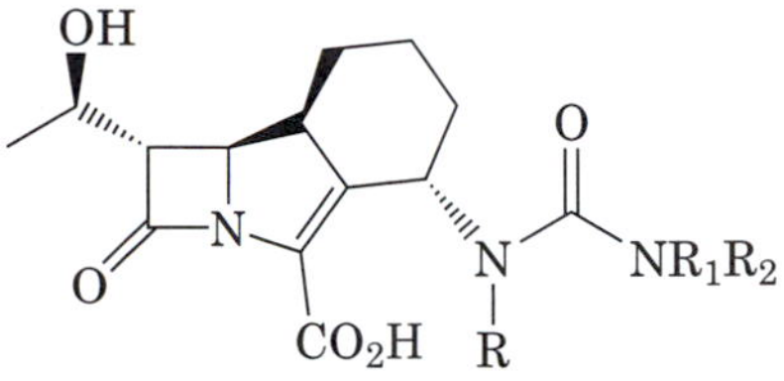

(340) R = H, CH₃

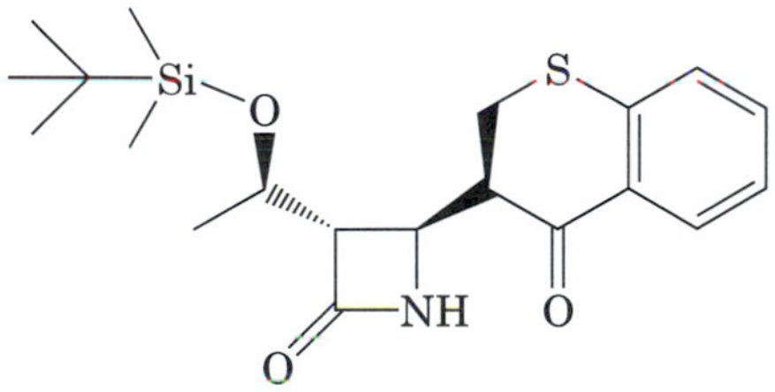

(343)

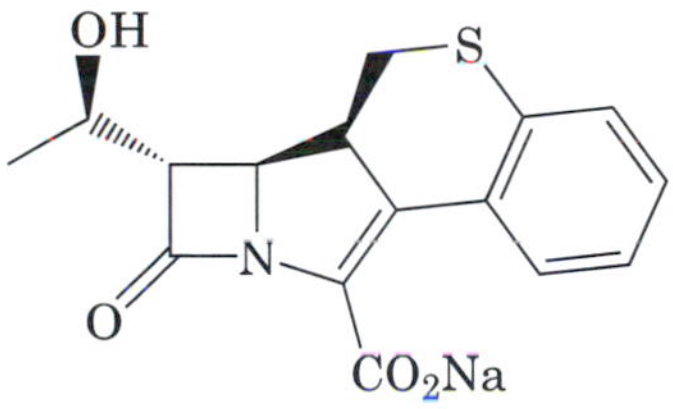

(344)

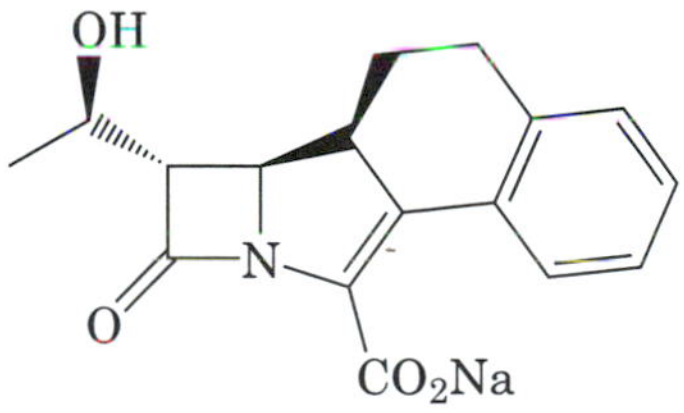

(345)

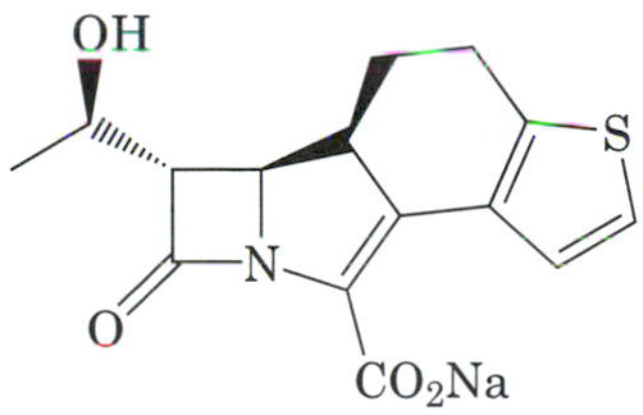

(347)

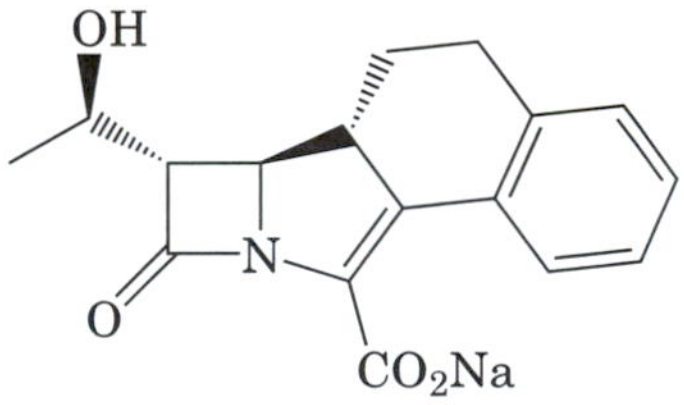

(346)

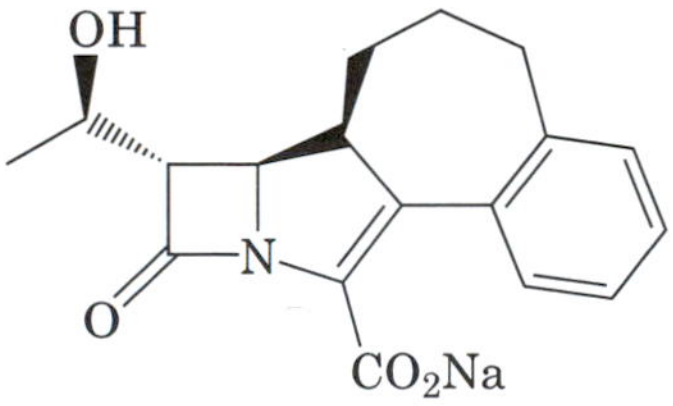

(348)

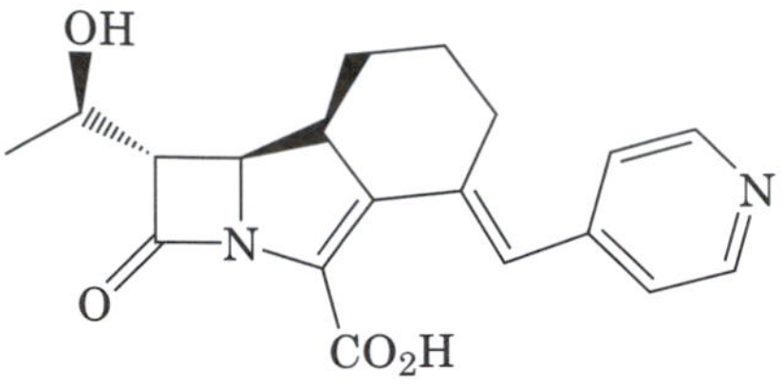

(349) GV143253X

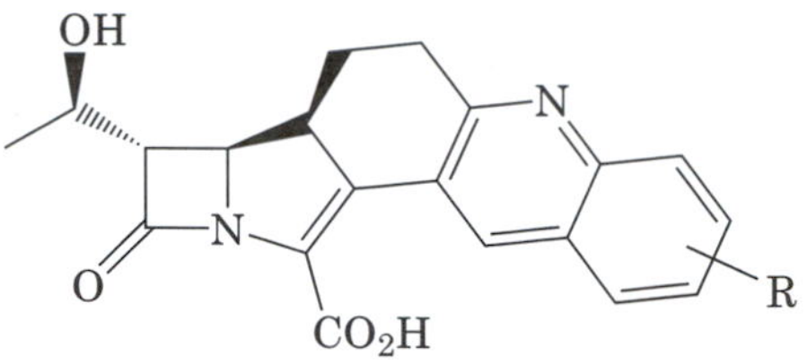

(353)

sulfur regioisomeric variants of their counter-parts, (356) and (357). Interestingly, they found out that the pyrrolidinyl group properly oriented could result in a good antibacterial activity against Gram-positive strains includ-

(350) X = NH₂

(351) X = HN

(352) X = S

ing MRSA. Best results were obtained with compound (364), which was comparable to vancomycin.

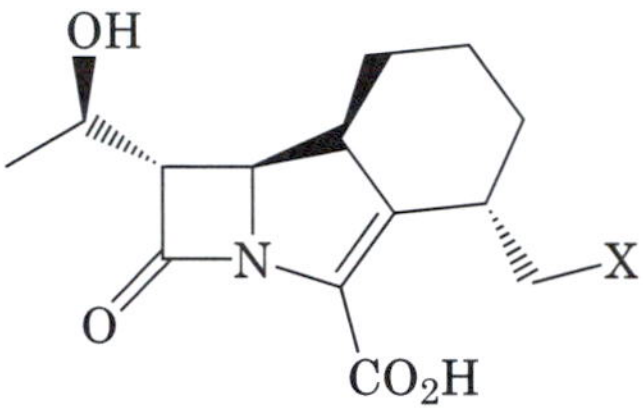

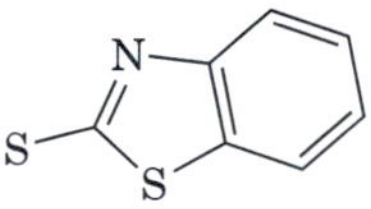

**5.5.3 Biological Properties of Carbapenems and Trinems.** Thienamycin (217), the olivanic acids, and the majority of carbapenems are broad-spectrum antibacterial agents showing good stability to β-lactamases. Thienamycin (217) is the most potent of the natural products having activity against a wide range of Gram-positive and Gram-negative bacteria including *Pseudomonas aeruginosa* (634). Structure-activity studies indicate that the latter results derive from the presence of the basic C2-cysteaminyl residue, given that *N*-acylated derivatives show much reduced activity (499). The olivanic acid MM13902 (219a; R = SO₃H) is also a broad-spectrum agent, but lacks significant activity against *Pseudomonas* spp. (635). In many cases, the carbapenems, particularly the olivanic acid

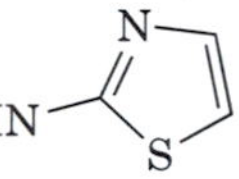

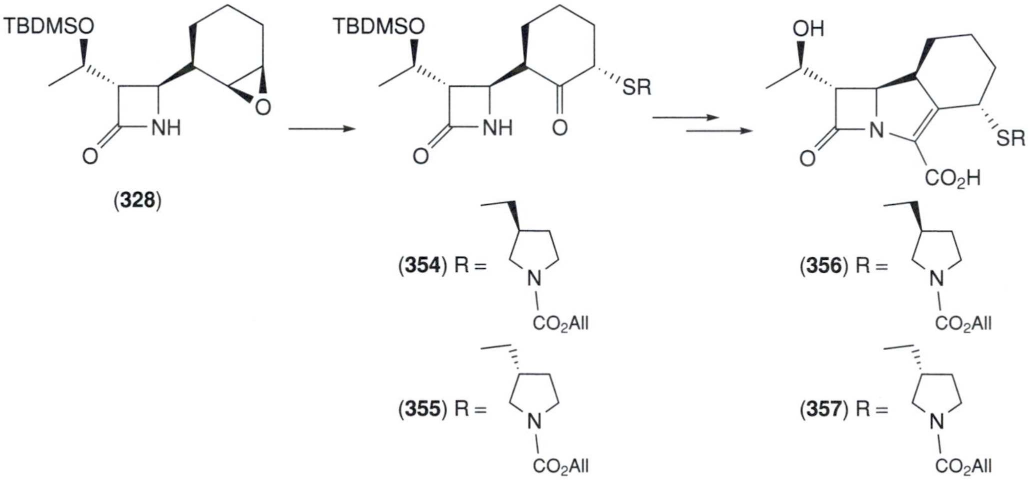

(328)

(354) R =

(355) R =

(356) R =

(357) R =

**Figure 14.20.** Preparation of derivatives starting from epoxide (328).

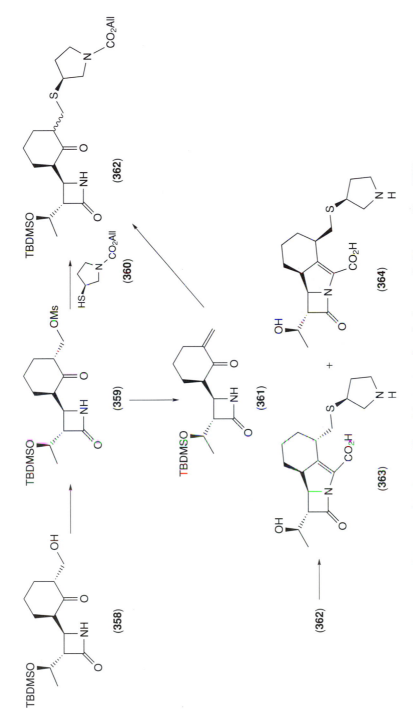

**Figure 14.21.** Preparation of derivatives from hydroxymethyl intermediate (**358**).

**Table 14.16  In Vitro Antibacterial Activity of Carbapenems [MIC (μg/mL)]$^a$**

| Organism | Thienamycin (213) | Imipenem (224) | MM13902 (216a) | PS-5 (219) | Asparenomycin A (222) | Pluracidomycin A (223) | Meropenem (214) | Biapenem (361) | Sanfetrinem (319a) |
|---|---|---|---|---|---|---|---|---|---|
| E. coli | 0.2–0.4 | 0.25 | 0.2–1.6 | 1.56 | 0.39–3.13 | 6.3 | 0.03 | 0.25 | 0.5 |
| K. pneumoniae | 0.4 | 0.25 | 0.4–3.1 | 3.13 | 0.78 | 6.3 | 0.06 | 0.25 | 2 |
| Enterobacter spp. | 1.6 | 0.5–1 | | | | 12.5 | 0.12 | 0.25 | 2 |
| S. marcescens | 1.6 | 0.5–2 | | | | 12.5 | 0.12 | 2 | 16 |
| P. mirabilis | 3.1 | 2 | 0.2 | 6.25 | 3.13 | 12.5 | 0.12 | 2 | 2 |
| P. aeruginosa | 3.1 | 2–4 | 25–50 | 50–100 | 25 | >100 | 8 | 4 | >32 |
| H. influenzae | — | 0.25–0.5 | 0.1 | — | — | — | 0.12 | 1 | 0.25 |
| B. fragilis | 0.4 | 0.25 | 0.4 | — | — | — | 1 | 0.5 | — |
| S. aureus | 0.04 | 0.03–0.06 | 1.6 | 0.025–0.39 | 1.56 | 25–50 | 0.12 | 0.06 | 0.12 |
| S. pyogenes | 0.01 | 0.01 | 0.2 | 0.08 | 1.56 | 25–50 | 0.12 | 0.008 | 0.015 |
| S. pneumoniae | 1.6 | 0.03 | 6.2 | 0.02 | — | — | 0.06 | 0.03 | 0.007 |

$^a$Refs. 241, 487, 506, 605, 635, 636, 637.

sulfoxide MM4550 (**220a**; R = SO$_3$H) (480), are good inhibitors of isolated β-lactamases, (Section 5.7). The *in vitro* activity of some representative natural products is shown in Table 14.16.

Unlike the penicillins and cephalosporins a rigid adherence to the *cis*-stereochemistry of substituents around the β-lactam ring is not necessary for activity. In the olivanic acids the des-sulfate derivatives are not as active as the corresponding sulfates, and in this series the *cis*-isomers are better than the *trans*-isomers (635). Thienamycin (**217**) is the most stable to β-lactamases, which is attributed to the presence of the 8(*R*)-hydroxyethyl side chain, given that synthetic examples lacking the C6-substituent are much more susceptible to hydrolysis by these enzymes (638). Overall, in the nonsulfated hydroxyethyl, series SAR would indicate the order of potency as *trans*-β-lactam with 8(*R*) stereochemistry > *cis* 8(*S*) > *trans* 8(*S*) (499). Binding studies indicate that in *E. coli* thienamycin has the greatest affinity for PBP-2, whereas most of the newer cephalosporins bind to PBP-3 (639). Although thienamycin (**217**) was not suitable for further development because of its chemical instability both in concentrated solution and in the solid state, the corresponding *N*-formimidoyl derivative, imipenem (**215**) was a crystalline product with much improved stability and suitable for progression (500).

Imipenem (**215**) exhibits an outstanding spectrum of activity against aerobic and anaerobic Gram-positive and Gram-negative bacteria, showing a potent bactericidal effect against *Pseudomonas aeruginosa*, *Serratia*, *Bacteroides fragilis*, Enterococci, and many other species (640). Against some 800 clinical isolates, imipenem inhibited the majority of organisms at concentrations below 1 μg/mL and was not hydrolyzed by plasmid or chromosomal β-lactamases, although both *Pseudomonas maltophilia* and *P. cepacia* were resistant (641). The plasma half-life of imipenem (1 h) in humans was considered satisfactory, although urinary recoveries were quite variable (6–40%) (642). This was attributed to extensive metabolism by the renal tubular brush border dipeptidase (DHP-1), to which all of the natural carbapenems are susceptible (643). To overcome this obstacle, Merck researchers developed imipenem in combination with an inhibitor (cilastatin, **365**) of DHP-1, to give an acceptable urinary recovery of the antibiotic

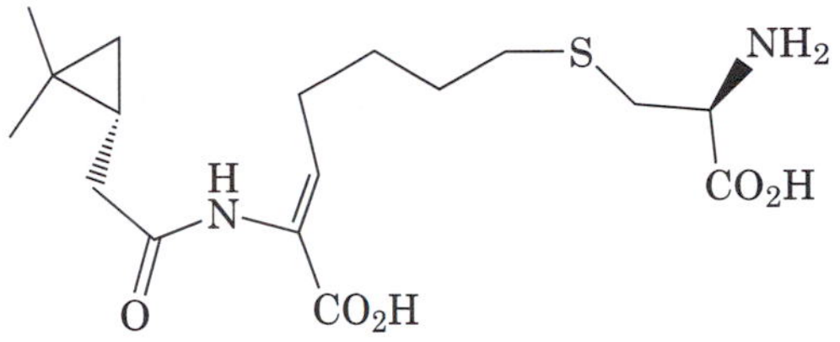

**(365)** Cilastatin

(644); the formation of any potential nephrotoxic degradation products was also reduced. After successful clinical trials it was this combination that was developed as the broad-spectrum parenteral agent primaxin (640, 645).

Although large numbers of analogs of imipenem (**215**) have been synthesized, particularly with modified C2-sulfur-linked substituents, the only other compound that has been developed in this series is panipenem (**275**), which, although a broad-spectrum agent, still requires coadministration with betamipron (**366**), a DHP-I inhibitor (646). One series of

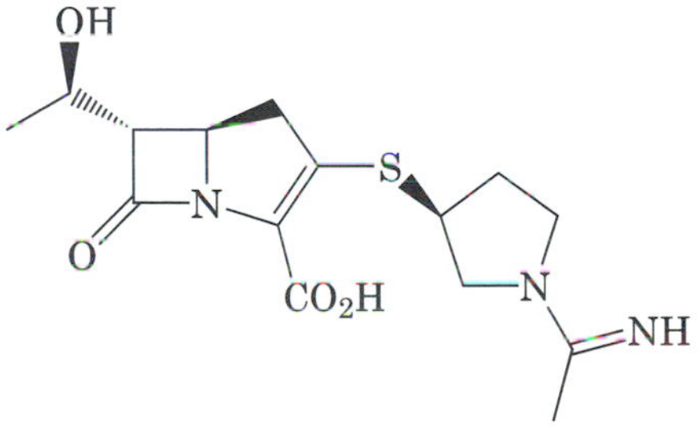

**(275)** Panipenem

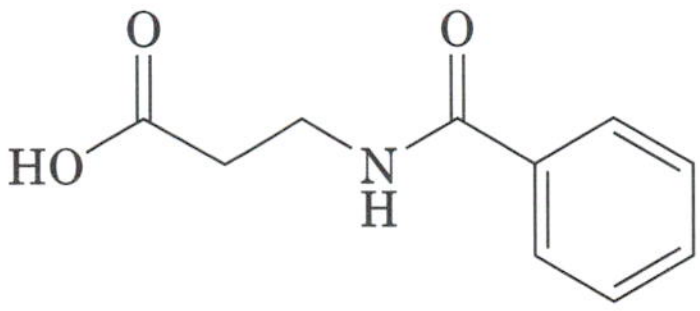

**(366)** Betamipron

thienamycin analogs of some interest are those with an aryl group directly attached at C2. These have been reported as having good activity and stability (647) and have evolved into a very interesting class of anti-MRSA agents (see Section 6.1).

The 1β-methyl series has seen the development of meropenem (**216**), a broad-spectrum agent comparable to imipenem (**215**) (Table 14.16), and which is sufficiently stable to DHP-I to allow administration without the necessity of an inhibitor (587, 648). Another compound that has been considered as a clinical candidate is the triazolium derivative biapenem (**367**) (506). Sanfetrinem (**324**) is not

**(367)** Biapenem

active against *Pseudomonas* spp. but it has a broad spectrum of antimicrobial activity (Table 14.16) and is stable to DHP-I. Of particular interest is the cilexetil prodrug ester (**325**), which is orally absorbed in humans (506). Oral absorbption was also demonstrated with the pivaloyloxymethyl ester of the tetracyclic β-lactam derivative (**344**) (627).

The antibacterial activity of trinems can be highly influenced by the kind of substituent at C4 (649), and could be targeted toward broad-spectrum agents including *Pseudomonas* spp. (623), as shown by GV129606X (**342**) or highly resistant Gram-positive bacteria like MRSA and penicillin-resistant Streptococci like GV143253X (**349**).

## 5.6 β-Lactamase Inhibitors

Widespread use of β-lactams, the largest family of antibiotics in current clinical use, has inevitably led to the emergence of resistant bacteria. The most commonly encountered mechanism of resistance is that attributable to the production of β-lactamases, a group of enzymes capable of catalyzing the hydrolysis of the β-lactam ring. The existence of these enzymes was recognized as early as 1940 soon after the isolation of penicillin, and fears relat-

ing to their plasmid-mediated spread throughout the bacterial population have been fully realized (2, 650).

The β-lactamases constitute a large and diverse family of enzymes, classification of which has been the subject of a number of publications. As stated earlier (Section 3) the most recent and comprehensive of these and its relationship to previous classification schemes (651–655) have been described by Bush and coworkers (656). However, for simplicity, the earlier classification originally proposed by Ambler, which is based on amino acid sequence homology, will be used in this review (657). In this classification the enzymes are divided into three groups (classes A, C, and D), which are active-site serine β-lactamases and one group (class B) of metallo β-lactamases containing at least one $Zn^{2+}$ ion per active subunit.

Class A contains many of the clinically important enzymes, including those from Gram-positive bacteria and the widely found TEM-type enzymes from Gram-negative organisms. The majority of them are transmissible, plasmid-mediated enzymes, often referred to as penicillinases because their preferred substrates are the penicillins.

Class B contains the broad-spectrum metallo-enzymes, which are capable of hydrolyzing most classes of β-lactams including the carbapenems. The earliest reported metallo-enzymes were confined to a small group of organisms with little clinical relevance. However, recent isolation of plasmid-mediated metallo β-lactamases has led to concern that there is potential for rapid dissemination of these enzymes, which are not susceptible to any of the clinically used inhibitors (658).

Class C contains predominantly chromosomally mediated enzymes from Gram-negative bacteria whose preferred substrates are the cephalosporins and are thus often referred to as cephalosporinases. Many producers of this class of enzymes are regulated by an inducible mechanism of control in response to the presence of antibiotics (659). Class D includes enzymes capable of hydrolyzing the more β-lactamase stable isoxazolyl penicillins.

Two strategies have evolved to combat β-lactamase-mediated resistance. The first of these, involving the development of classes of β-lactam antibiotic with improved stability is

**Table 14.17  β-Lactamase Inhibitory Activity (IC$_{50}$ μg/mL)$^a$**

| Organism | Enzyme class | Clavulanic acid (362) | Sulbactam (369) | Tazobactam (372) | BRL 42715 (380) |
|---|---|---|---|---|---|
| S. aureus | A | 0.063 | 1.4 | 0.27 | 0.016 |
| E. coli (TEM-1) | A | 0.055 | 1.7 | 0.028 | 0.002 |
| E. coli (SHV-1) | A | 0.035 | 13.0 | 0.14 | 0.001 |
| Enterobacter cloacae (P99) | C | >50 | 5.0 | 0.93 | 0.002 |
| E. coli (OXA-1) | D | 0.71 | 2.2 | 1.1 | 0.001 |

$^a$IC$_{50}$ determined after 5-min preincubation of enzyme and inhibitor.

dealt with in the appropriate sections of this chapter. The second approach has been the identification of β-lactamase inhibitors for co-administration with the antibiotic, thereby protecting it from hydrolysis by β-lactamase-producing organisms.

The search for inhibitors has involved both natural product screening and the creative ingenuity of the medicinal chemist. The early developments in the field of β-lactamase inhibitor research have been reviewed by Cole (660) and Cartwright and Waley (661). In this section we cover the clinically used β-lactamase inhibitors and the important developments since these early reviews were written. None of the reported non-β-lactam inhibitors is likely to have any clinical impact and will not be discussed because they fall outside the scope of this chapter.

**5.6.1  Discovery of Clavulanic Acid.** Clavulanic acid (368) can be considered as the most

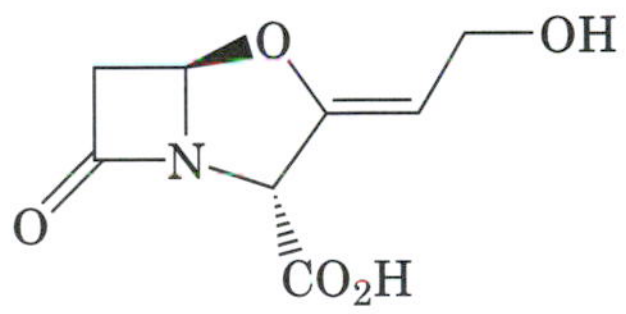

(368) Clavulanic acid (1976)

important and representative among the inhibitors of β-lactamases (662).

Clavulanic acid (368), the first clinically useful β-lactamase inhibitor, was identified as a natural product from a strain of *Streptomyces clavuligerus* through use of an assay designed to detect inhibitors by their ability to protect penicillin G from hydrolysis by a β-lactamase-producing strain of *Klebsiella aerogenes* (663).

X-ray analysis of the 4-nitrobenzyl ester revealed a novel fused β-lactam containing an oxygen atom instead of sulfur and lacking the acylamino side chain present in penicillins and cephalosporins (662). Clavulanic acid is a potent inhibitor of a wide range of clinically important class A β-lactamases, with more modest activity against class D and none against the class B and C enzymes (664). Comparative inhibition data for clavulanic acid and other inhibitors, which will be discussed in detail, are shown in Table 14.17 (665).

Clavulanic acid (368) possesses only a poor level of antibacterial activity in its own right, but is capable of synergizing the activity of a number of β-lactam antibiotics against a range of β-lactamase-producing strains of both Gram-positive and Gram-negative bacteria (666). Comparative synergy data from a study in which low inhibitor concentrations

**Table 14.18  Amoxycillin MIC (μg/mL) in the Presence of μg/mL of Inhibitor**

| Organism | Enzyme class | No. inhibitor | Clavulanic acid (362) | Sulbactam (369) | Tazobactam (372) | BRL 42715 (380) |
|---|---|---|---|---|---|---|
| Proteus mirabilis | A | >512 | 16 | 64 | 16 | 2 |
| E. coli (TEM-1) | A | >512 | 8 | 128 | 8 | 2 |
| K. pneumoniae | A | 256 | 4 | 64 | 16 | 2 |
| Enterobacter cloacae | C | 512 | >512 | 256 | 256 | 1 |
| E. coli (OXA-1) | D | >512 | >512 | >512 | >512 | 2 |

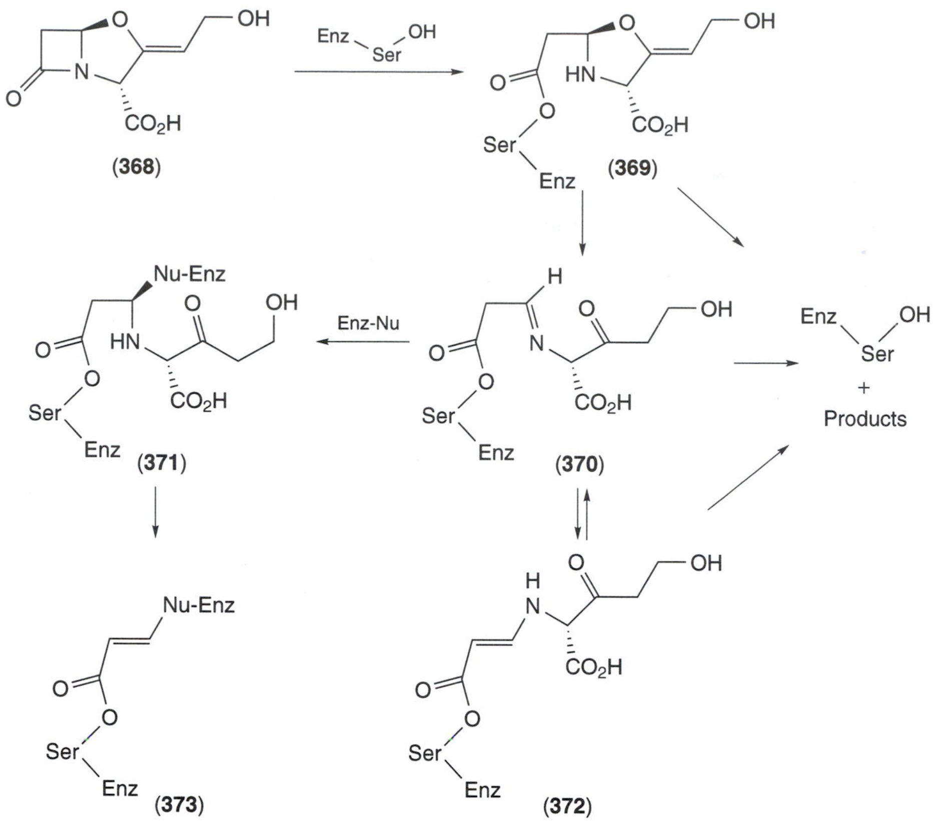

**Figure 14.22.** Mechanism of action of clavulanic acid.

were used are shown in Table 14.18 (667). Although greater synergy can be demonstrated at higher levels these data serve to illustrate the differences in spectrum and potency of the four inhibitors featured in this review. Clavulanic acid (**368**) markedly reduces the MIC of amoxycillin (**21**) against the organisms producing class A β-lactamase. However, at this inhibitor level, its more modest potency against the class D enzyme is not translated into whole-cell activity.

Marketed as Augmentin (potassium clavulanate plus amoxycillin) and Timentin (potassium clavulanate plus ticarcillin), clavulanic acid has found widespread use, particularly as the former formulation, with sales in excess of $1500 million in 2000. The pharmacokinetic and early clinical trial data for amoxycillin/clavulanic acid combinations were summarized by Cole (660) and further clinical experience was reviewed in 1989 (668).

Despite a general increase in the frequency of β-lactamase-producing strains observed in the period since its introduction, two recent and extensive reviews of published clinical data concluded that there had been no significant increase in resistance to Augmentin (669, 670). Clavulanic acid (**368**) belongs to a class of enzyme inhibitors often referred to as "suicide" or "mechanism-based" inhibitors and, because of its clinical importance, has been the subject of many kinetic and mechanistic studies (671–674). These studies have led to a complex and still incomplete model for the interaction of clavulanic acid with β-lactamases, the salient features of which are shown in Fig. 14.22.

Acylation of the active site serine followed by ring opening of the intermediate oxazolidine (**369**) provides the imine (**370**), which may undergo tautomerization to the enamine derivative (**372**). Although all three of these intermediates may undergo hydrolysis, leading to turnover and release of active enzyme, the anticipated hydrolytic stability of (**372**) (a stable β-aminoacrylate) led to its assignment as the structure for the so-called transiently inhibited species (673). It has been proposed that the irreversible inhibition results from the trapping of a second nucleophile of the enzyme by the imine species (**370**) to afford (**371**), followed by β-elimination to give (**373**). Although it has been suggested that the second nucleophilic group might be provided by a lysine, some modeling studies based on enzyme X-ray crystal structure led to the hypothesis that it is a second and conserved serine that is captured by the imine species (**370**) (674). The biosynthesis (675) and chemistry (370, 676) of clavulanic acid have also received much attention. Despite extensive modifications, especially those involving derivatization of the allylic alcohol, no inhibitors derived from clavulanic acid have progressed to the clinic.

**5.6.2 Sulbactam.** Diazotization/bromination of 6-APA (**2**), followed by oxidation, gave the 6,6-dibromopenicillanic acid sulfone (**374**), which on catalytic hydrogenation provided sulbactam (**375**) (677).

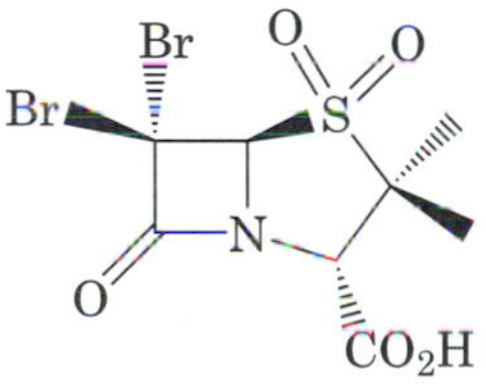

(**374**)

Sulbactam (**375**) was shown to be an irreversible inhibitor of several β-lactamases

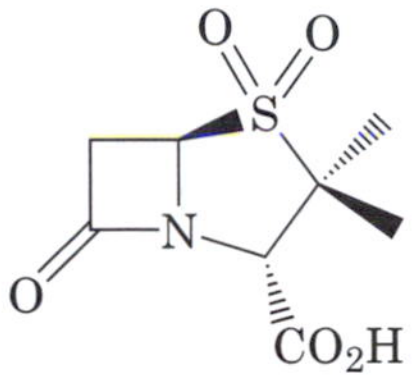

(**375**) Sulbactam

(678, 679). Compared with clavulanic acid (Table 14.17) sulbactam is a modest inhibitor of the class A enzymes but shows improved potency against those of class C, although at a level considered to be of little clinical use. The levels of synergy achieved in whole cells with sulbactam (**375**) (Table 14.18) reflect its poorer potency against the class A enzymes. Its mode of action is believed to be essentially similar to that of clavulanic acid (**368**), with a cascade following β-lactam ring opening, eventually leading to the same type of inactivated species (**373**) (673). Demonstration of synergy in combination with ampicillin (**19**) led to the development of a 1:1 formulation, which is marketed as Unasyn for parenteral use. Poor absorption of sulbactam (**375**) precluded the use of this combination by the oral route and led to the development of the mutual prodrug sultamicillin (**376**). After absorption from the gut this double ester undergoes rapid cleavage by nonspecific esterases, to give good serum levels of both ampicillin (**19**) and sulbactam (**375**) (680).

**5.6.3 Tazobactam.** Renewed interest in penicillanic acid sulfones stimulated by the discovery of the properties of sulbactam (**375**) led to the investigation of the effects of substitution in the 2-methyl group. Exploitation of the reactivity of the azido function of the 2-β-

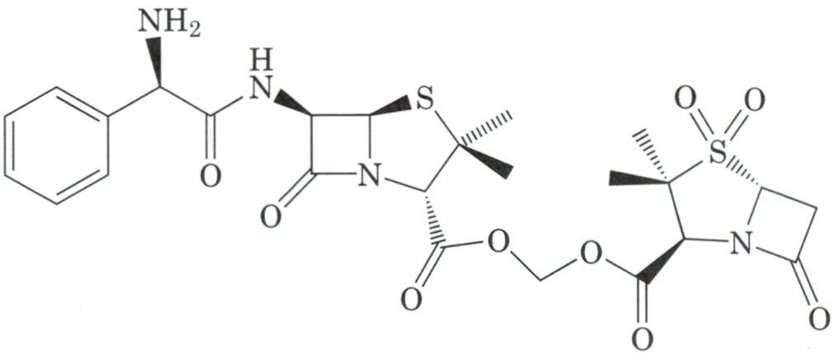

(**376**) Sultamicillin

azidomethylpenam (**377**) to cycloaddition reactions with acetylenes provided, after deprotection, a series of 2β-(triazolyl)methylpenams (**378**) with potent β-lactamase inhibitory activity (681).

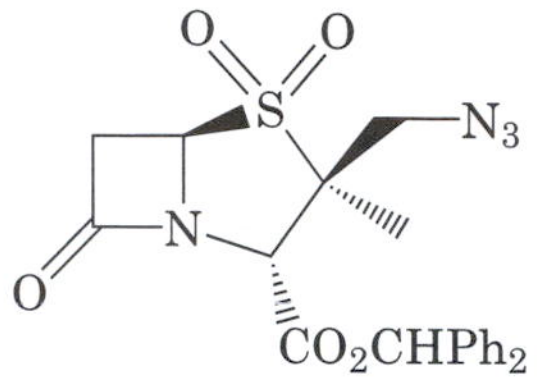

(**377**)

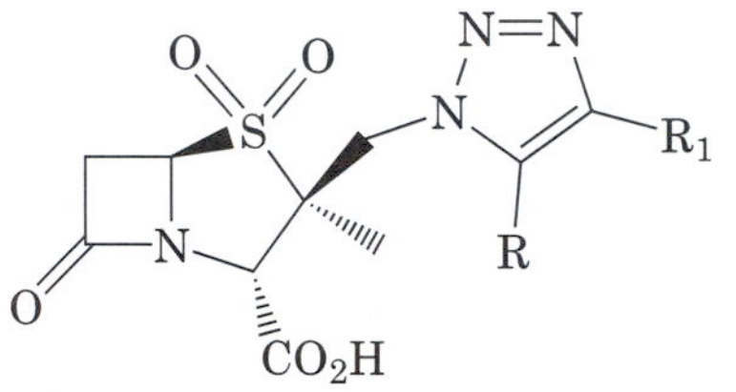

(**378**) Tazobactam

The unsubstituted triazole (**378**; R = R$_1$ = H), which was also available through the reaction of (**374**) with vinyl acetate, originally selected for its potency and ease of preparation, has found clinical use as tazobactam. Tazobactam (**378**) is more potent than sulbactam (Table 14.17), with good activity against the clinically important TEM-1 enzymes. It also shows a modest level of activity against the class C enzymes, which are not inhibited by clavulanic acid (**368**). In combination with amoxycillin levels of synergism similar to those of clavulanic acid are observed for organisms producing class A β-lactamases (Table 14.18). At higher inhibitor levels synergy against class C-producing organisms has been demonstrated for tazobactam (**378**) as well as its ability to synergize the activity of a range of β-lactam antibiotics (682, 683). Among these, piperacillin (**34**) was selected as the partner antibiotic for the marketed parenteral formulation; experience of its clinical use was reviewed in 1993 (684). The degradation of tazobactam (**378**) has been studied extensively and its mechanism of action was considered similar to that of sulbactam (**375**) (685, 686).

**5.6.4  6-Heterocyclylmethylene Penems.** Sch-29482 (**379**) has been reported to be an inhib-

(**379**)

itor of class C and class D β-lactamases (687). However, fewer additional data are available for penems bearing the 6-(1-hydroxyethyl) substituent, which have been studied widely for their antibacterial activity. Dehydration of (**379**) gave the E- and Z-isomers of the 6-ethylidenepenem (**380**), both of which were

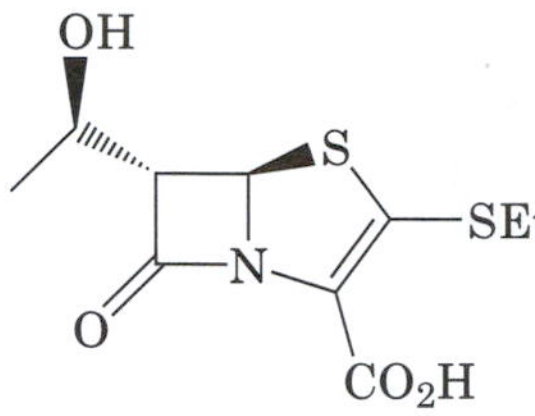

(**380**)

shown to be potent broad-spectrum β-lactamase inhibitors with much weaker antibacterial activity than that of the parent penem (688).

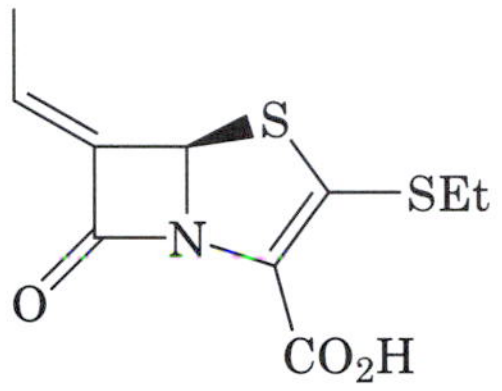

Extensive structure-activity relationship studies involving modification at both the C2 and C8 positions have been described (689).

From these studies the Z-triazolylmethylenepenem (**386**, BRL-42715) was selected for further evaluation based on its overall activity and stability to human renal dehydropeptidase (DHP-I), which has the ability to degrade penems in a manner similar to that of the carbapenems (690). An atom-efficient, stereocontrolled synthesis from 6-amino penicillanic acid (**2**) has been described, the key steps of which are shown in Fig. 14.23 (691).

Reaction of the anion derived from the bromopenem (**381**) with 1-methyl-1,2,3-triazole-4-carboxaldehyde provided a mixture of the lithium salts (**382**), *in situ* acetylation of which gave a mixture of the bromoacetates (**383**). Reductive elimination of this mixture

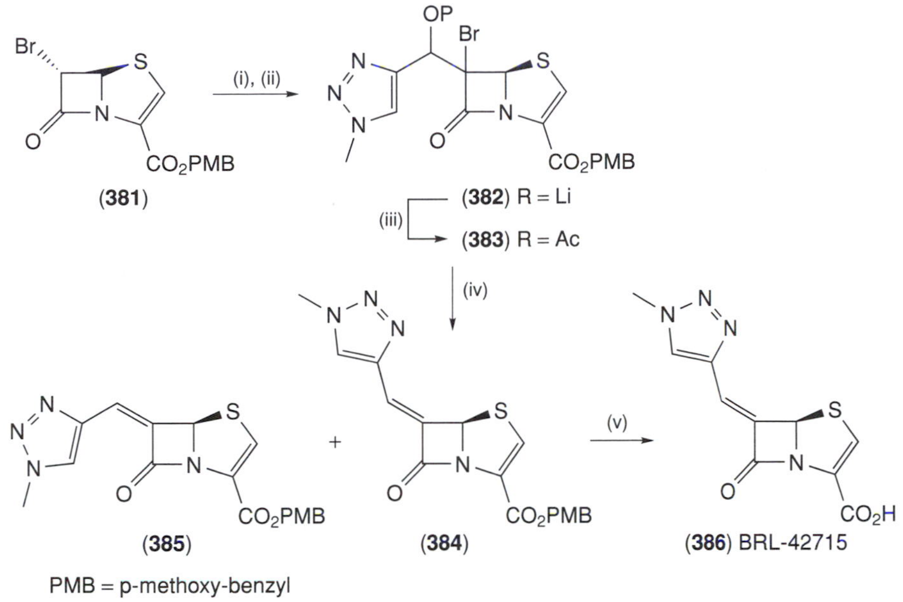

PMB = p-methoxy-benzyl

Reagents: (i) Ph$_2$NLi, THF, -78°C; (ii) 1-methyl-1,2,3-triazole-4-carboxaldehyde; (iii) Ac$_2$O, -78°C to 20°C; (iv) Zn, N,N,N′,N′-tetramethylethylethylenediammine dihydrochloride, NH$_4$Cl, DMF; (v) AlCl$_3$, anisole.

**Figure 14.23.** Synthesis of BRL-42715.

gave the desired *Z*-triazolylmethylenepenem ester (**384**) as the major product, which was deprotected to provide the acid (**386**). BRL-42715 (**386**) is a potent broad-spectrum inhibitor capable of synergizing the activity of β-lactam antibiotics against a wide range of β-lactamase-producing organisms (665). Tables 14.17 and 14.18 reveal that BRL-42715 (**386**) is a more potent inhibitor than clavulanic acid (**368**), sulbactam (**375**), and tazobactam (**378**) against class A, C, and D enzymes and that this potency is translated into

(**387**)

more impressive synergism in whole-cell assays. Detailed kinetics confirming the efficiency of inactivation have been described for a number of β-lactamases (692, 693), and a mechanism of action involving a novel rearrangement to a dihydrothiazepine (**387**) proposed (694).

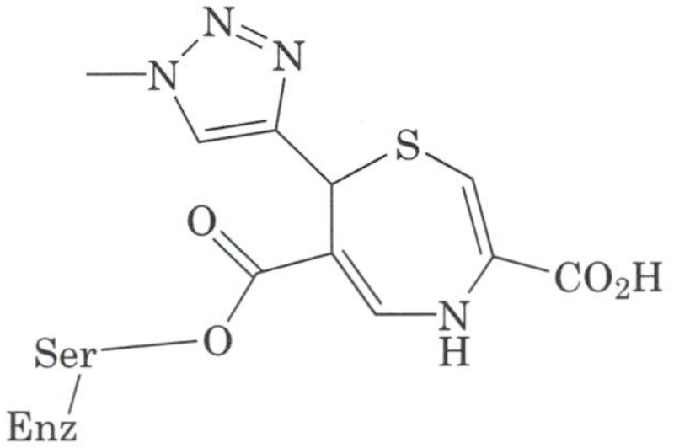

Despite its impressive inhibitor profile, development of BRL-42715 (**386**) was terminated because of its failure to fulfill other desired technological features. A second inhibitor, SB-206999 (**388**), possessing an activity profile similar to that of BRL-42715, is currently under development (695).

**5.6.5 Carbapenems.** Many of the carbapenems described in Section 5.5 are potent β-lactamase inhibitors with a spectrum that includes the class C enzymes not inhibited by clavulanic acid (665).

The mechanism of action of carbapenems has been studied (673) and the rearrangement to a Δ$^1$-pyrroline (**389**) after β-lactam ring opening was proposed to explain stabilization

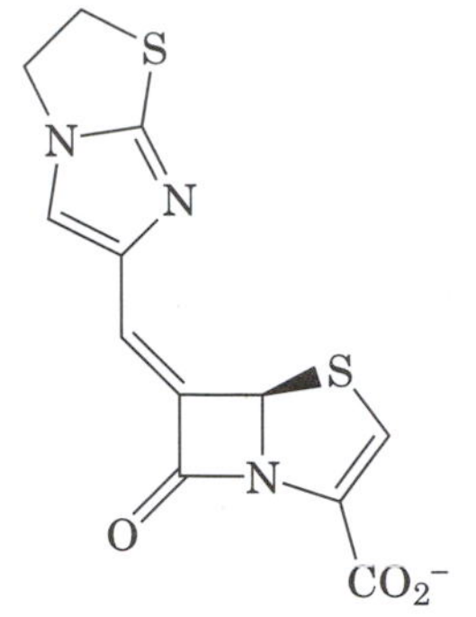

(388) SB-206999

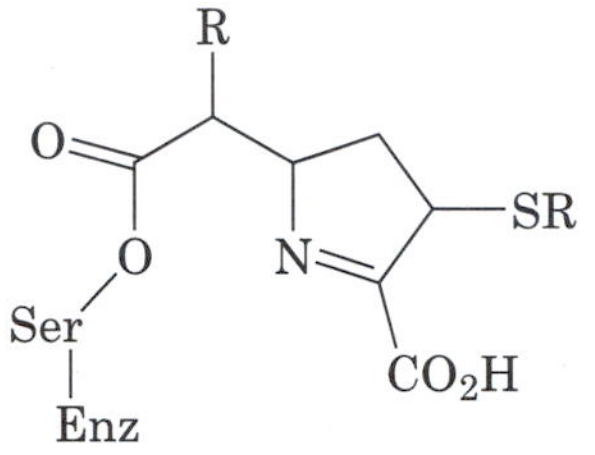

(389)

of the enzyme-bound species. Limitations on the level of carbapenem used, imposed by their potent antibacterial activities, have often precluded the demonstration of significant synergy and no compound of this class has progressed as an inhibitor.

**5.6.6 Penams and Penam Sulfones.** 6β-Bromopenicillanic acid (390), often referred to as

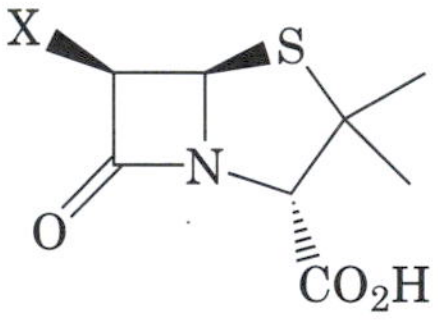

(390) X = Br
(391) X = I
(392) X = Cl

brobactam, was described as a β-lactamase inhibitor as early as 1978 (696). Extensive studies of its mechanism of action have provided powerful evidence for an active-site serine-bound dihydrothiazine (393) in the inactivated enzyme (697).

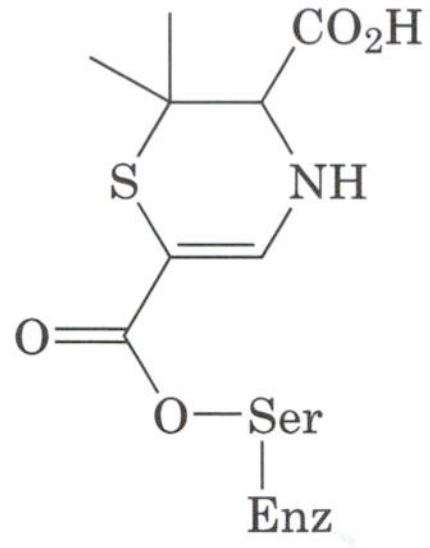

(393)

It has been noted that this dihydrothiazine (393) and the dihydrothiazepine (387) share the same enamine moiety seen in the proposed structure of the transiently inhibited species (372) for clavulanic acid (694, 698). Despite the demonstration of good synergistic properties and a favorable pharmacokinetic profile in humans, brobactam (390) has not found clinical application (699, 700). Comparable activity has been reported for 6β-iodopenicillanic acid (391) (701), whereas the chloro analog is less potent (392) (702). Stimulated by the discoveries of sulbactam (375), tazobactam (378), and brobactam (390), modification of the penam nucleus has provided a large number of β-lactamase inhibitors. Tables 14.19 and 14.20, which are by no means comprehensive, illustrate the structural types of penam and penam sulfones that have been reported to possess these properties.

**5.6.7 Miscellaneous β-Lactams.** β-Lactamase inhibitory properties have been reported for a large number of other β-lactam-containing structures including cephems and monocyclic β-lactams (660, 661). Among the more recent reports are those of activity for 7α-(1-hydroxyethyl)-cephems (including their sulfoxides and sulfones) and oxacephems (394) bearing electron-withdrawing groups at the 3-position (721). Activity has also been reported for other cephems bearing nonclassical 7-substituents [e.g., the 7-allene functionality and alkylidene cephems (395) and (396) (722)].

Stabilization of the oxapenem ring system by the introduction of bulky 2-substituents has allowed the demonstration of β-lac-

**Table 14.19  Penam Inhibitors of β-Lactamases**

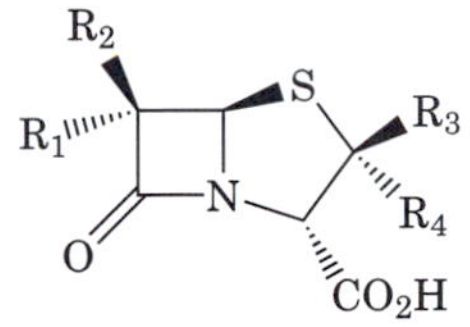

| $R_1$ | $R_2$ | $R_3$ | $R_4$ | Reference |
|---|---|---|---|---|
| H | Br | $CH_2X$<br>X = F, $N_3$, OAc,<br>OMe, SCN | Me | 703 |
| H | $(CF_3SO_2)_2N$ | Me | Me | 704 |
| H | $HOSO_2NH$ | Me | Me | 705 |
| H | 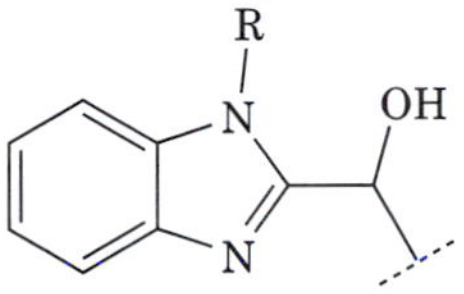 R | Me | Me | 706 |
| | 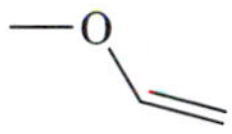 —O | Me | Me | 707 |
| | 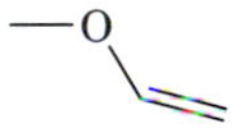 —O | Me | Me | 708 |

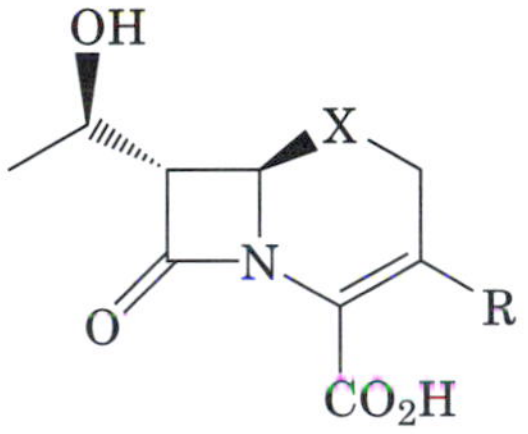

**(394)** X = S, SO, $SO_2$, O
R = electron withdrawing group

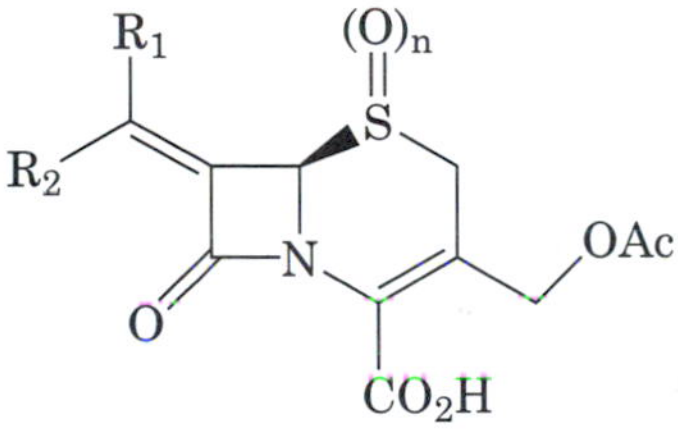

**(396)** n = 0, 2

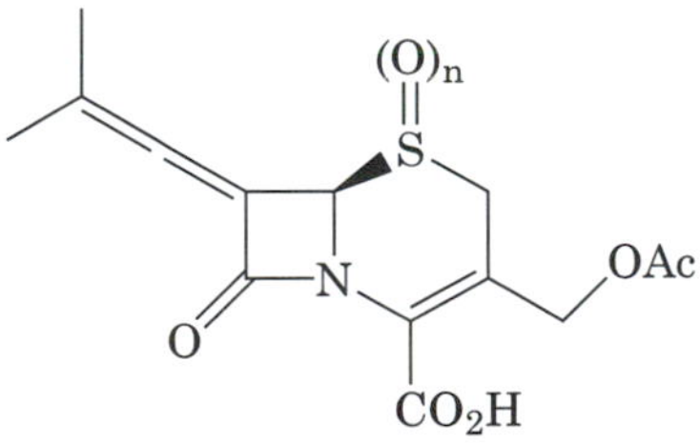

**(395)** n = 0, 2

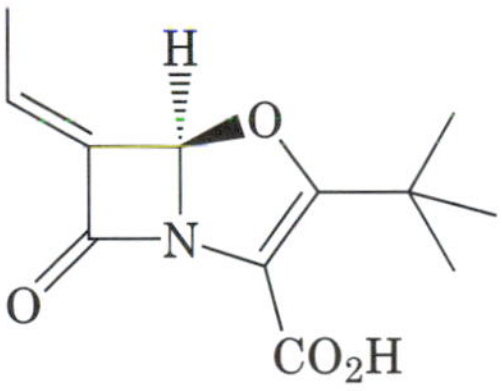

**(397)**

tamase inhibition for members of this highly ring strained class of β-lactam derivatives. Activity has been demonstrated for the 2-*tert*-butyloxapenems [(**397**) (723) and (**398**) (724)] and the 2-isopropyloxapenem (**399**) (725), the latter in particular showing potent inhibition of class C enzymes. Recently, selective and potent inhibitors of class C enzymes have been reported for a series of bridged monobactams (**400**) (726).

**Table 14.20  Penam Sulfone Inhibitors of β-Lactamases**

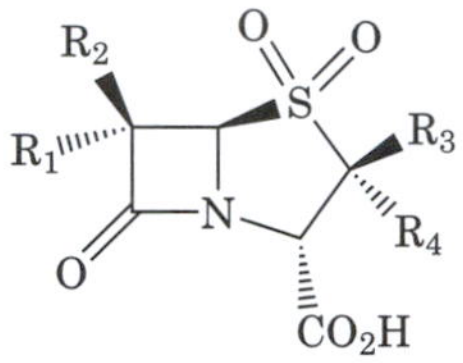

| $R_1$ | $R_2$ | $R_3$ | $R_4$ | Ref. |
|---|---|---|---|---|
| H | H | $CH_2Cl$ | Me | 709 |
| H | $CF_3SO_2NH$ | Me | Me | 710 |
| H | PhCH(OH) | Me | Me | 711 |
| H | X = OAc, F, OH 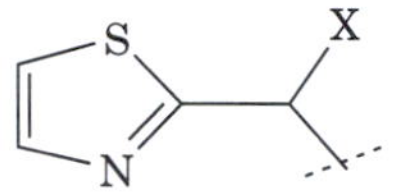 | Me | Me | 712 |
| H | H | 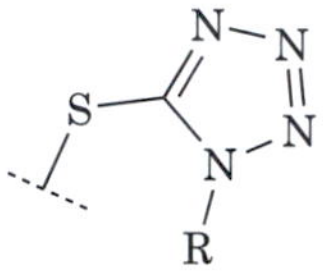 R = Me, Ph | Me | 713 |
| H | H | $CH_2SCN$ | Me | 713 |
| H | H | 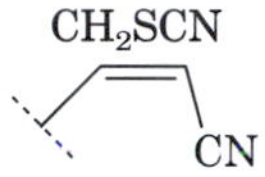 | Me | 714 |
| H | $RCOCH_2$ R = Me, $CH_2OPh$, CN, $CO_2Et$ | Me | Me | 715 |
| H | 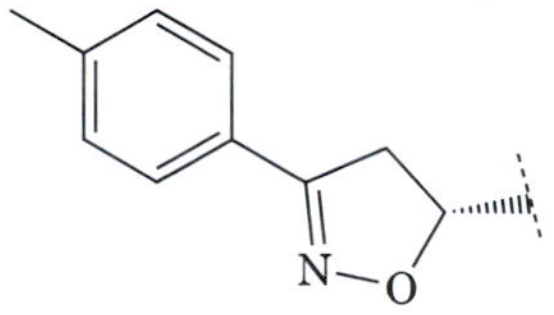 | Me | Me | 716 |
| | 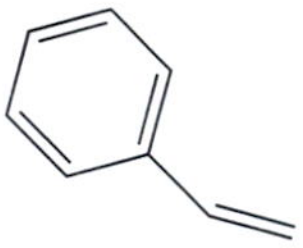 | Me | Me | 717 |
| | 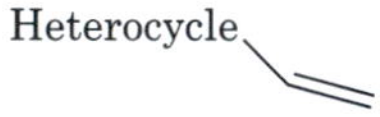 Heterocycle | Me | Me | 718, 719 |
| | 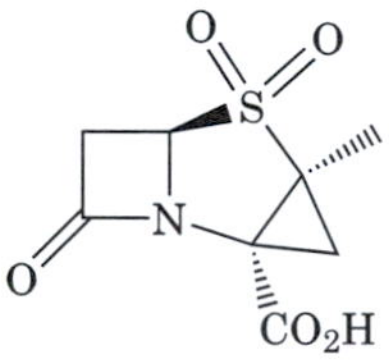 | | | 720 |

# 6  RECENT DEVELOPMENTS IN β-LACTAMS

The pharmaceutical companies that are still actively involved in the research of new β-lactam antibiotics after almost 60 years are encouraged by the safety profile that has been proved by extensive use throughout the years and by the continuous launch of new agents. In

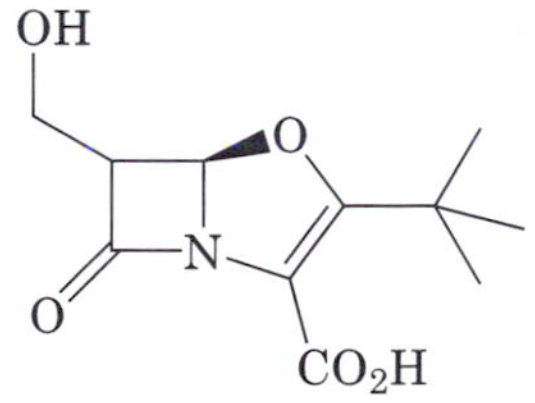

(398)

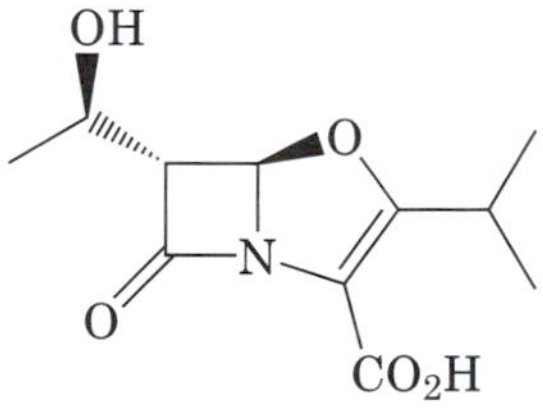

(399)

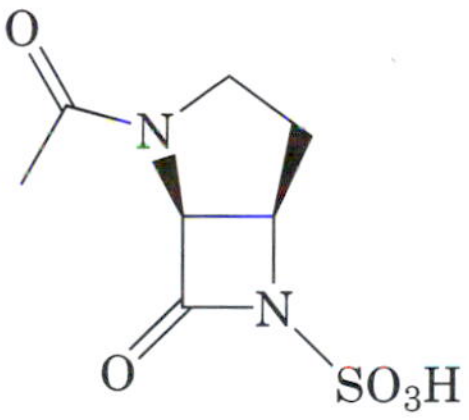

(400)

particular, the main efforts are directed toward the discovery of agents active against dangerous emerging resistant strains, and a more favorable pharmacodynamic and tolerability, reducing the dosage, number of daily administrations, length of treatment, and side effects.

## 6.1 Anti–Gram-Positive β-Lactams

MRSA (methicillin-resistant *Staphylococcus aureus*) was first described in the early 1960s and has since become one of the major nosocomial infections in the hospital setting (727). MRSA still remains as an important target for the development of anti-infective agents. It has been shown that these strains are resistant to penicillins, cephalosporins, and also carbapenems. Although effective, therapy with glycopeptides such as vancomycin or teicoplanin is relatively limited because of side effects shown by this class of antibacterial agents. Therefore, potent anti-MRSA agents

with low levels of side effects would be highly desirable. Nearly all strains of MRSA share a common feature of carrying the *mecA* gene encoding for an additional, but with low affinity for, β-lactam penicillin binding protein, PBP2a (728, 729).

### 6.1.1 Cephalosporins.
A new generation of cephalosporins with enhanced affinity for PBP2a was identified during the late 1990s. The most interesting derivatives are characterized by the presence of a highly basic or ionized moiety at the C3-position bearing at C7 the amino thiazole ring or a closely related structure.

Microcide was particularly active in this field with the identification of several products: MC-02331 (**401**), MC-02363 (**402**), and MC-02479 (**403**); RWJ-54428 (**404**); RWJ-333441 and its prodrugs (730) (Microcide Pharmaceuticals/RW Johnson Pharmaceutical Research Institute). Other cephalosporin derivatives of particular interest were also identified: Ro-63–9141 (**405**, F. Hoffmann-La Roche), TOC-39 (**406**, Taiho) and TOC 50 (**407**, Taiho). All these compounds have shown a promising anti-MRSA activity with the potential to overcome the recent issues associated with the development of bacterial resistance.

### 6.1.2 Carbapenems and Trinems.
Several pharmaceutical companies have reported a series of carbapenems and trinems showing excellent activity against MRSA and vancomycin-resistant Enterococci (731–735). These compounds (**408–411**) showed very good affinity for PBP2a, typical of MRSA and PBP5, which is overproduced by resistant Enterococci.

The presence of an aromatic moiety, as in the 2-arylcarbapenems or aryl-4-exomethylenyl trinems, was recognized as a necessary key feature in the most active β-lactam anti-MRSA agents. Most of these carbapenems were derivatives in which the C2 side chain was linked by a C-C bond instead of a C-S bond. Moreover, the presence of a quaternary nitrogen on the side chain conferred satisfactory water solubility, an essential property for an injectable drug.

The carbapenem class was extensively investigated and several potent compounds

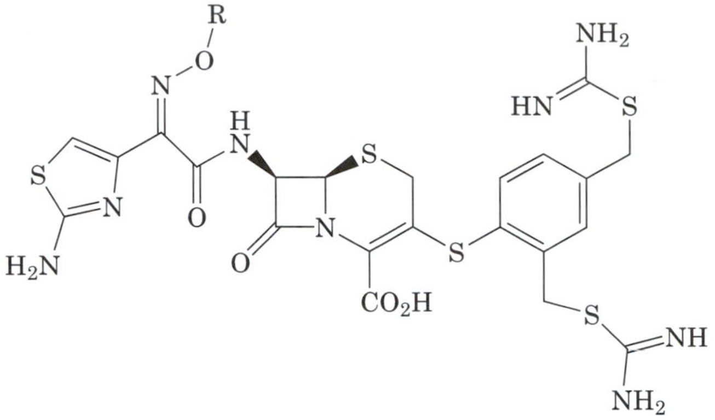

(401) MC-02331    R = H,

(402) MC-02363    R =

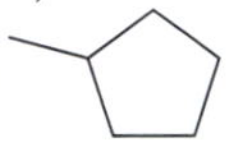

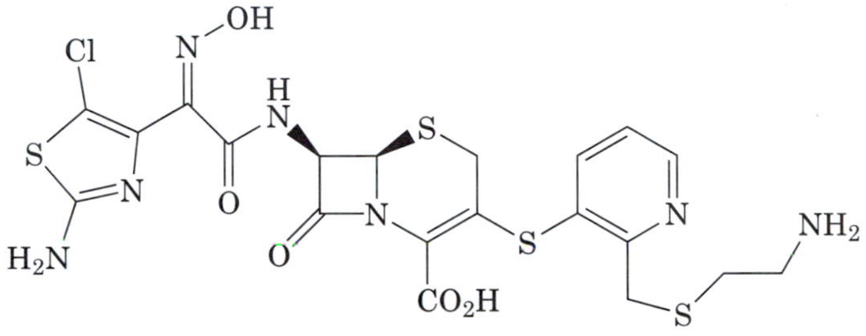

(403) MC-02479 or RWJ-54428

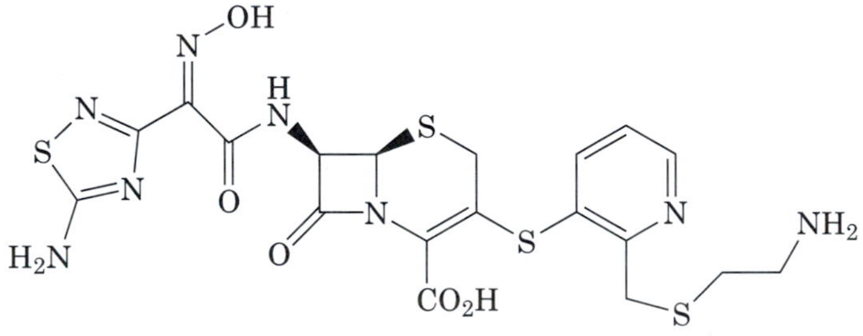

(404) RWJ-333441

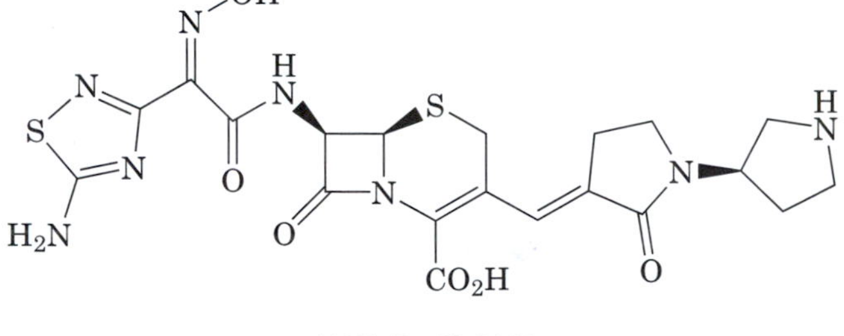

(405) Ro-63-9141

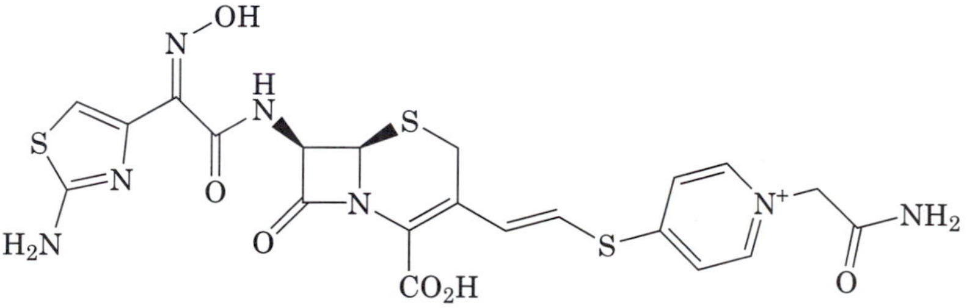

(406) TOC-39

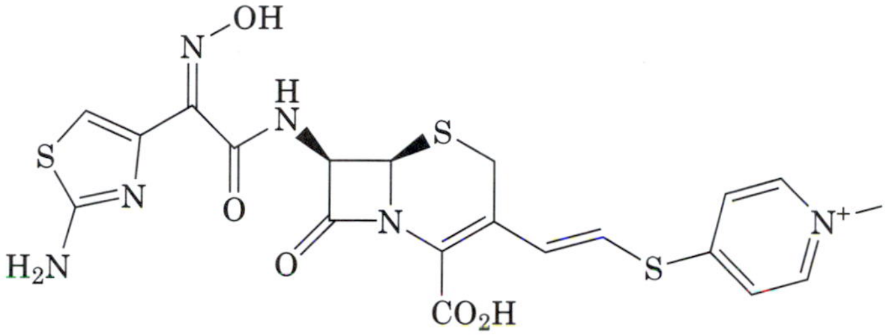

(407) TOC-50

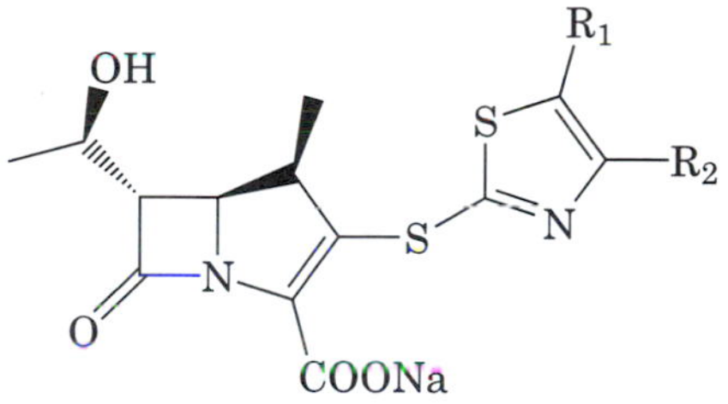

Sumitomo

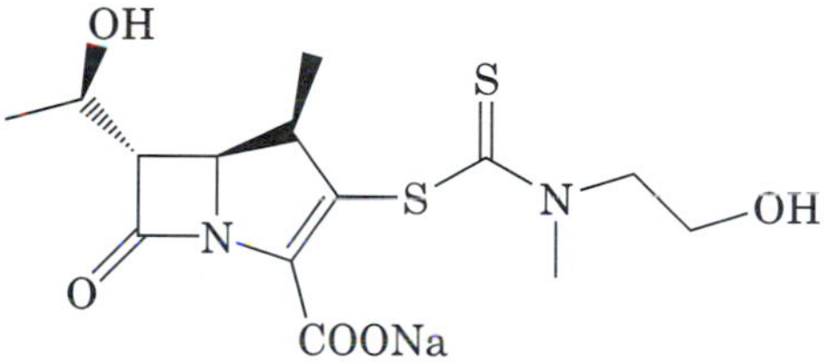

(411) BO-3482 Banyu

(408) R₁ = H,         R₂ = —[pyridinium]—N⁺—

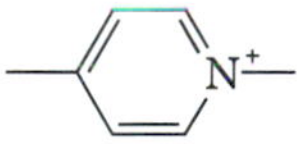

(409) R₁ = CH₃,       R₂ = CH₃

(410) R₁ = —[phenyl]   R₂ = H

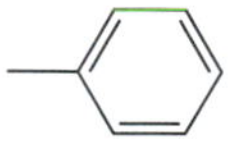

were identified: L-741–462 (**412**), L-696–256 (**413**), L-742–728 (**414**), and L-786–392 (**415**). All these compounds underwent clinical trials but for different reasons their clinical development was discontinued.

Clinical development of L-741–462 (**412**) was discontinued because of an unexpected immunotoxic reaction observed in Rhesus monkeys (736), which was thought to be the result of a specific acylation of lysine residues on the cell surface followed by recognition of the appended hapten (Fig. 14.24). The carbapenem L-786–392 (**415**) was designed to overcome this problem, given that acylation of lysine by β-lactam would release the sulphonamido moiety, thereby avoiding its recognition as hapten. However, despite its excellent *in vitro* and *in vivo* activity, this compound failed to meet the required safety profile.

GV143253X (**349**, GlaxoWellcome) was the first reported example of trinem with a very interesting *in vitro* and *in vivo* antimicrobial profile that is comparable to that of vancomycin. This compound belongs to a relatively novel class of antibiotics and in this class the presence of the double bond has been shown to be important for the anti-MRSA activity.

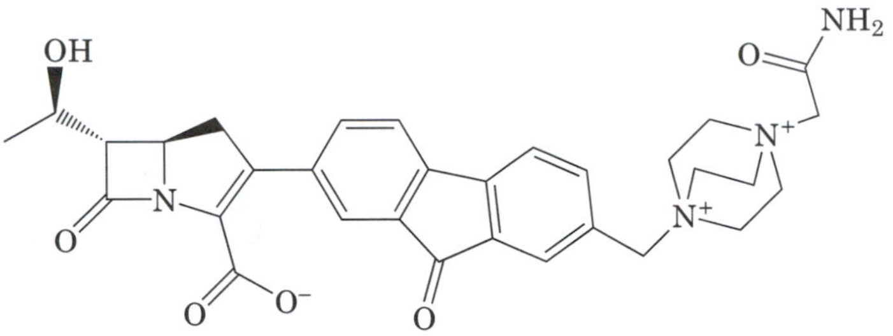

(412) L-741-462

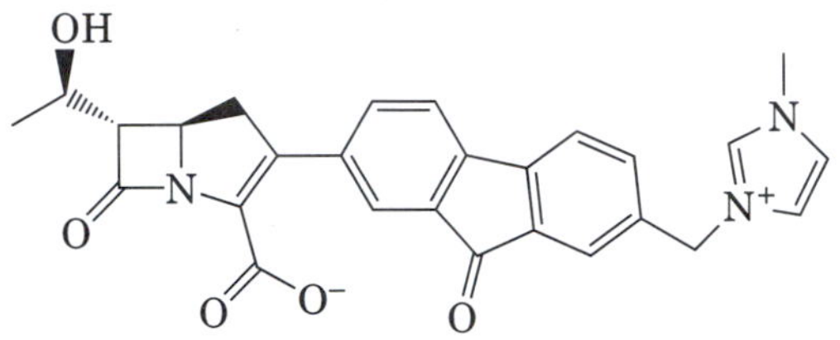

(413) L-696-256

To substantiate the hypothesis, the cyclopropanation of the exo-double bond resulted in a compound (420) still microbiologically active, but lacking anti-MRSA activity. Additionally, inclusion of the double bond into an aromatic ring (419) retained the anti-MRSA activity (630, 631).

However, the unsatisfactory pharmacokinetic profile shown by GV143253X (349) in healthy volunteers suggested discontinuing further studies on this compound.

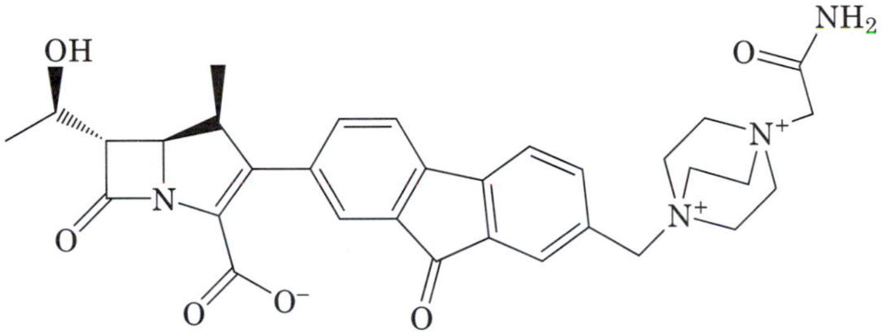

(414) L-742-728

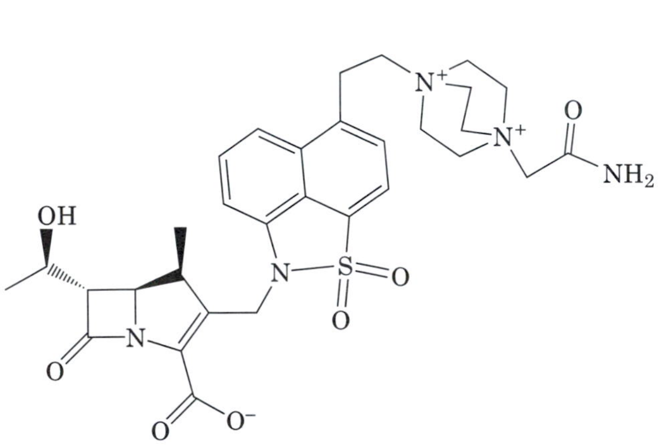

(415) L-786-392

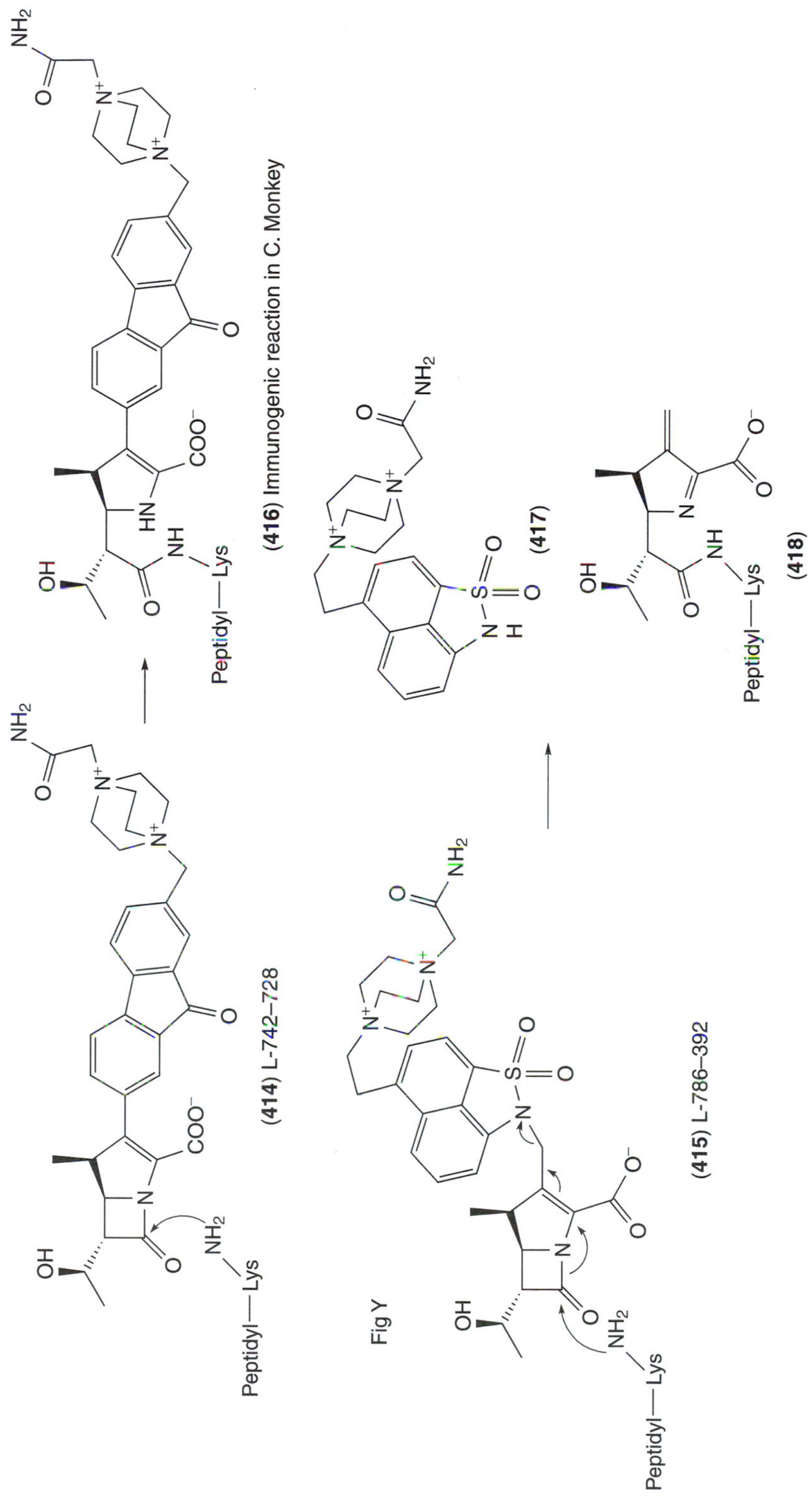
(414) L-742-728
(416) Immunogenic reaction in C. Monkey
Fig Y
(415) L-786-392
(417)
(418)
Peptidyl—Lys
COO$^-$
NH$_2$

**Figure 14.24.** Acylation of cell surface protein by L-741-762 and L-786-392.

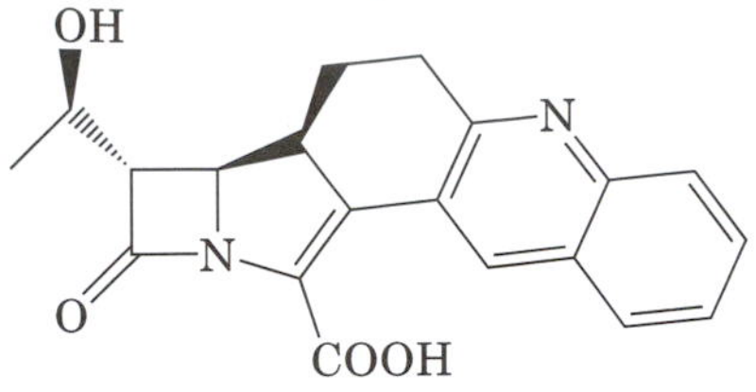

(419)

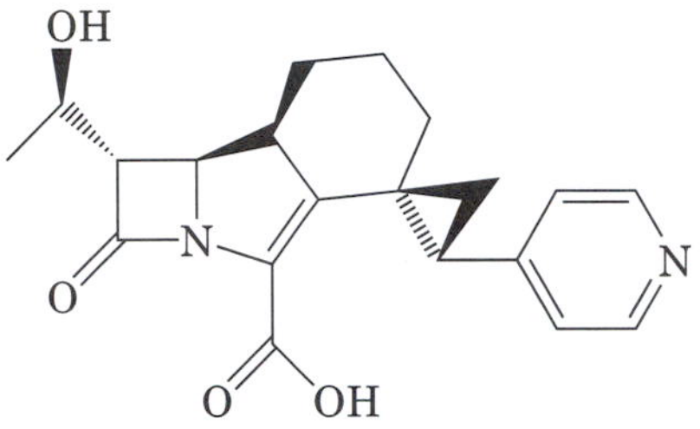

(420)

## 6.2 Antibacterial Broad-Spectrum β-Lactams

The intrinsic potency of carbapenems and trinems has encouraged pharmaceutical companies to investigate these classes of β-lactams to find broad-spectrum agents for the treatment of severe hospital infections.

A series of compounds that are characterized by the presence of the 3-thiopyrrolidine group, also present in meropenem (216), have been reported. The Korea Institute of Science and Technology (KIST) has synthesized DK-35C (421) (737, 738), a compound that exhib-

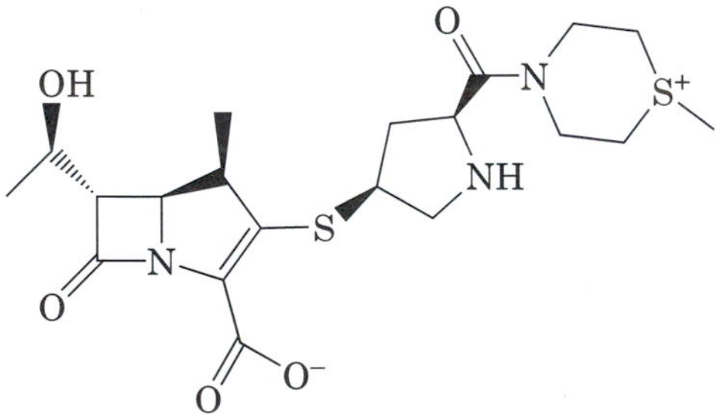

(421) DK-35C

ited greater potency than that of imipenem (215) or meropenem (216). Intracerebroventricular injection in male rats (739) showed a proconvulsive activity greater than that of

meropenem but weaker than that of imipenem or cefazolin (59); however, it is not known whether DK-35C (421) possesses a greater capability than that of other carbapenems to penetrate the blood-brain barrier.

Another interesting broad-spectrum compound is DA-1131 (422) (740–744) that, simi-

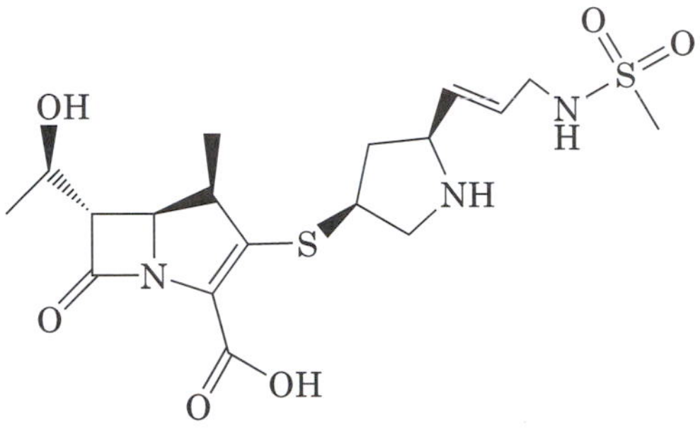

(422) DA-1131

larly to panipenem [(275), marketed by Sankyo in combination with betamipron (366) under the trade name Carbenin], needs to be coadministered with a nephroprotector to avoid renal impairment caused by drug accumulation in the renal cortex (742).

Shionogi has reported S-4661 (423) (745, 746), a carbapenem with an antibacterial activity comparable to that of meropenem (216)

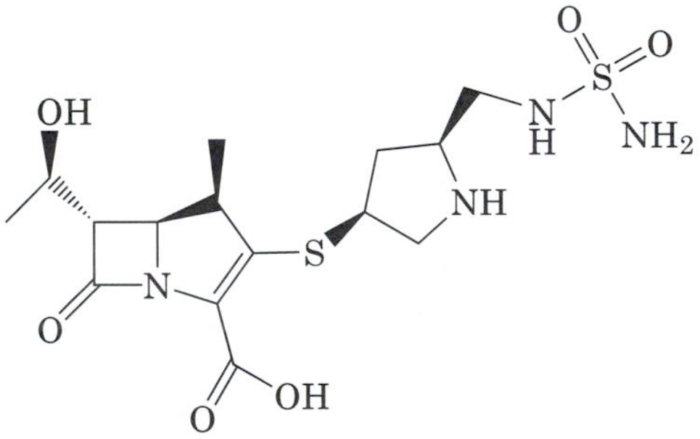

(423) S-4661

but with a lower frequency of antibiotic-induced resistance. Clinical trials have shown its good efficacy for urinary tract infections (UTI) and respiratory tract infections (RTI), but there was evidence also of an increase in the levels of hepatic enzymes (GOT/GPT) (747, 748).

Other recent compounds include MK-826 (**317**) (Merck and Zeneca, Daloxate), a compound with a prolonged half-life and excellent activity against β-lactamase producer strains (749–752), and J-111,225 (**424**) (Banyu Phar-

(**424**) J-111,225

maceutical Co.) (753), which is an ultra-broad-spectrum agent including MRSA and *Pseudomonas* species.

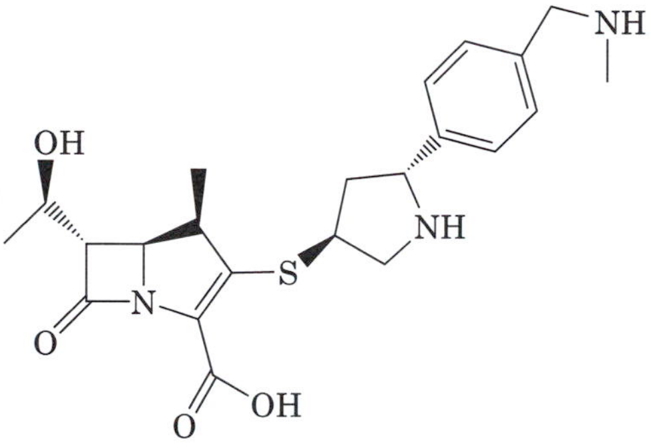

Trinems were also studied to obtain compounds with a broad spectrum of action, including *Pseudomonas* spp. GV129606X (**342**, GlaxoWellcome), which features a 4-*N*-methylformimidoyl group, showed activity against β-lactamase-producing resistant strains, good efficacy and pharmacokinetic profile in animal models.

## 6.3  Orally Active Carbapenems/Trinems

Pharmaceutical companies have put an intensive effort to produce orally active 1β-methyl carbapenems and trinems for the treatment of community-aquired infections. Wyeth-Ayerst has published a study in which CL-191121 (**425**), an aminomethyl tetrahydrofuranyl carbapenem, was reacted with amino acids to give

(**425**) CL-191121

the corresponding peptidoyl derivatives (754, 755). This work led to the identification of OCA-983 (**426**) (756), which is a valinyl pro-

(**426**) OCA-983

drug of CL-191121 (**425**) with improved oral bioavailability and *in vivo* efficacy against Gram-positive and Gram-negative bacteria (757, 758). OCA-983 (**426**) represents the first example of a peptidic prodrug of carbapenems orally absorbed by active transport.

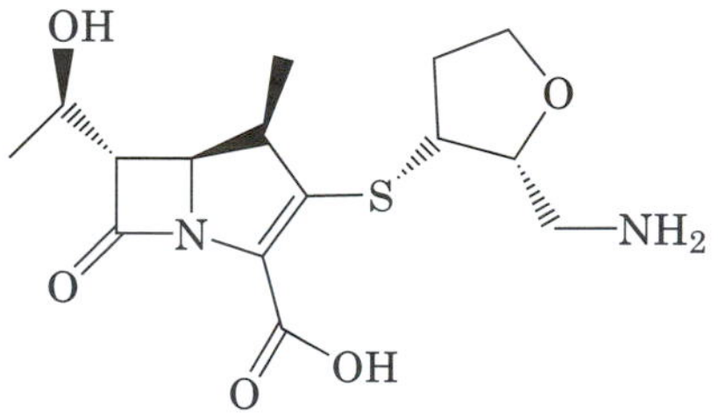

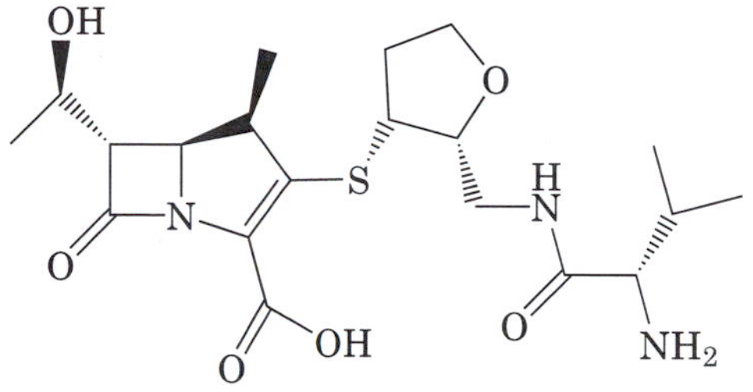

Other examples of broad-spectrum oral carbapenems for the community are reported by Lederle, with L-084 (**428**) (759, 760), a prodrug of LJC-11036 (**427**) and by Sankyo, with

(**427**) LJC-11036

CS-834 (**430**) (761–763), a prodrug of R-95867 (**429**).

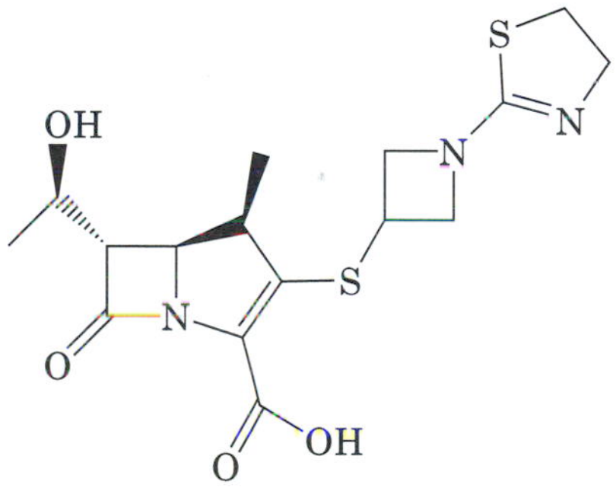

OCA-983 (**426**), L-084 (**428**), and CS-834 (**430**) showed very similar antimicrobial profiles and are suitable for the treatment of respiratory tract infections, with good stability to ESBL and broad-spectrum β-lactamases. Two major drawbacks could be envisaged for such compounds:

• The dosing regimen (t.i.d. for L-084 and CS-834) required maintenance of the concen-

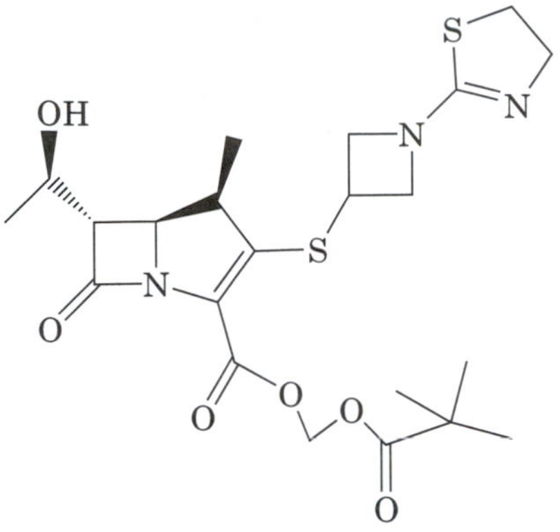

**(428)** L-084

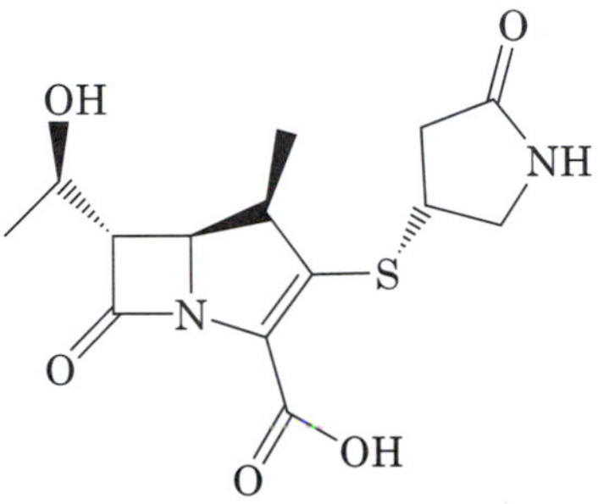

**(429)** R-95867

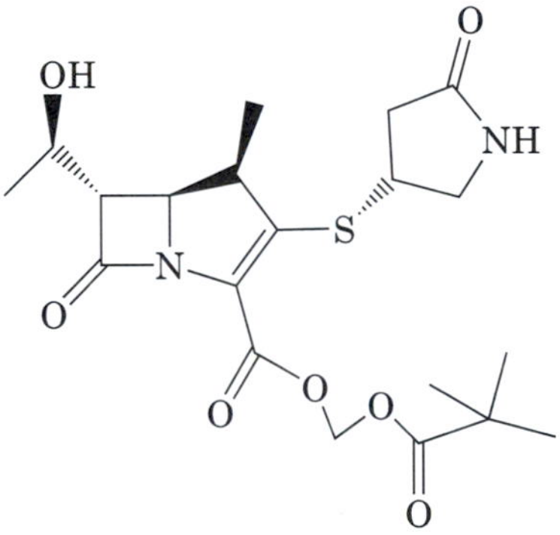

**(430)** CS-834

## 6.4  β-Lactamase Inhibitors

Most recently, the problem of inactivation of β-lactams by β-lactamases has been carefully investigated, as it is evident from the literature (764–767). Several sulfone derivatives such as compound (**375**), obtained from chemical modification of 6-APA (**2**), have been obtained (768). In addition, mechanistic and crystallographic studies with mono and bis penicillanic acids (**432**, **433**) and TEM-1-β-lac-

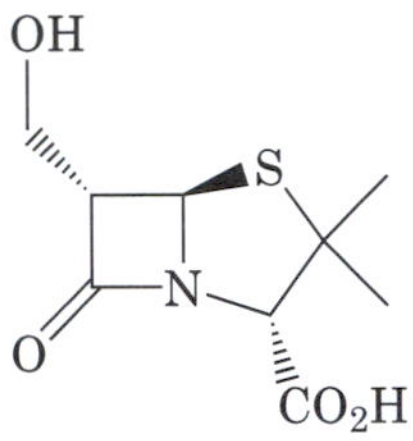

**(432)**

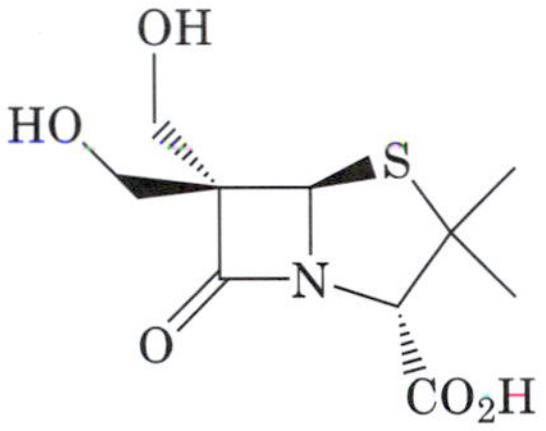

**(433)**

tamase from *E. coli*, have been reported (769–771). In particular, bis hydroxymethyl penicillanic acid (**433**) showed inhibitory activity on both class A and class C β-lactamases.

Scientists from Hoffmann-La Roche used the β-hydroxymethyl derivative (**434**) to prepare a series of alkenyl penam sulfones (772) that inhibited TEM and SHV, among which Ro 48–1220 (**435**) showed the best inhibitory properties (773), comparable to those of clavulanic acid (**368**) and tazobactam (**378**). Through use of a similar procedure, Taiho prepared a series of amides, starting from the carboxylic acid (**436**), among which compound (**437**) showed better synergy than that of tazobactam (**378**) *in vitro* against class C β-lactamases (774).

Starting from the chloro methyl (**438**), Taiho also reported the synthesis of an inter-

tration of the circulating compound above the MIC value for a sufficient time in the 24 h.

- The relatively high cost of treatment compared to that of other drugs already available in this overcrowded market.

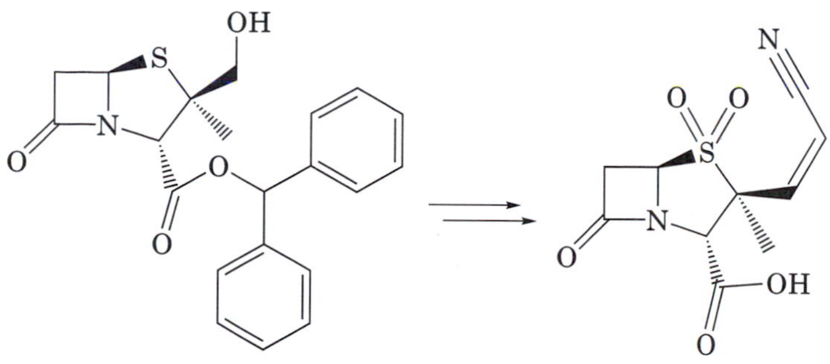

(434)                                         (435) Ro 48-1220

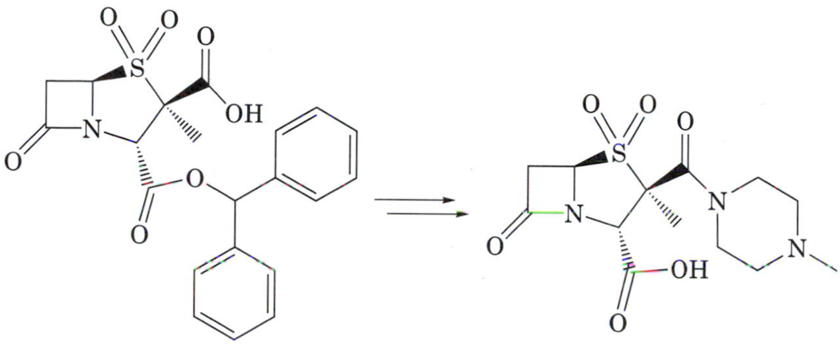

(436)                                         (437)

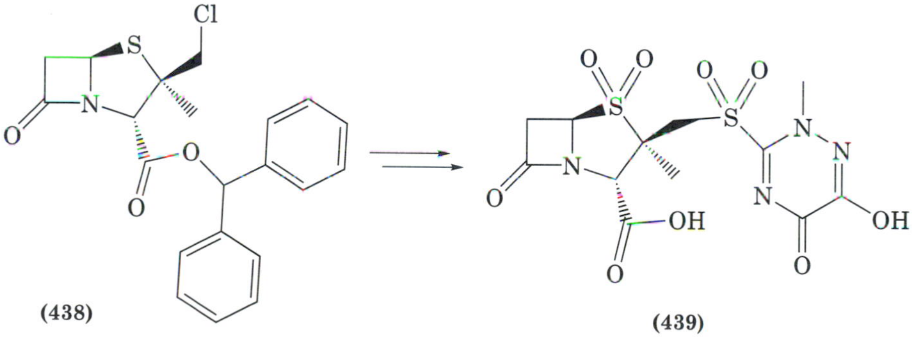

(438)                                         (439)

esting compound (**439**) bearing the same heteroaromatic moiety of ceftriaxone. This compound possesses an excellent *in vitro* activity against cephalosporinases (775).

Introduction of a double bond conjugated with the carbonyl groups of the β-lactam ring (alkylidene derivatives) has been extensively studied in several classes of β-lactams other than penems (see Section 5.6.4), producing potent inhibitors. (*Z*)-Triazol-4-yl-methylene penicillanic acid sulfones (**440–442**) showed

very good inhibitory activity against TEM β-lactamases. Compounds (**440–442**) were 2–3 times more active than tazobactam (**378**) against cephalosporinases from *Enterobacter cloacae* (776).

Inhibition of class C β-lactamases has been observed for 7-alkylidene cephalosporanic acid sulfones such as compound (**443**). This gave excellent inhibitory activity against *Enterobacter cloacae* P99 and *Enterobacter cloacae* SC-12368 (722).

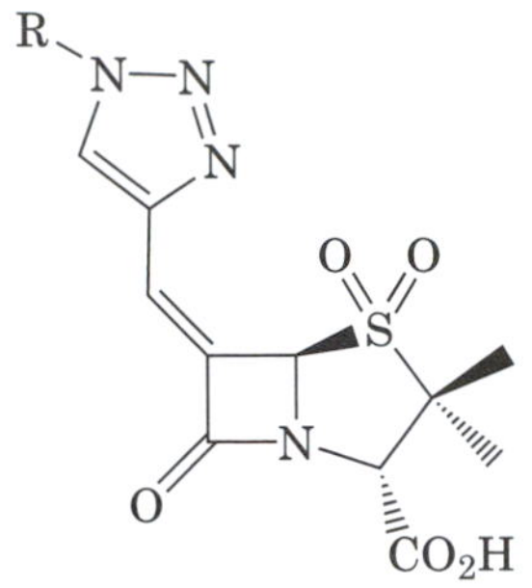

(440) R = CH$_3$
(441) R = CH$_2$CH$_2$OH

(442) R = CH$_2$CH$_2$—N$^+$ ⟨pyridinium⟩

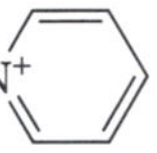

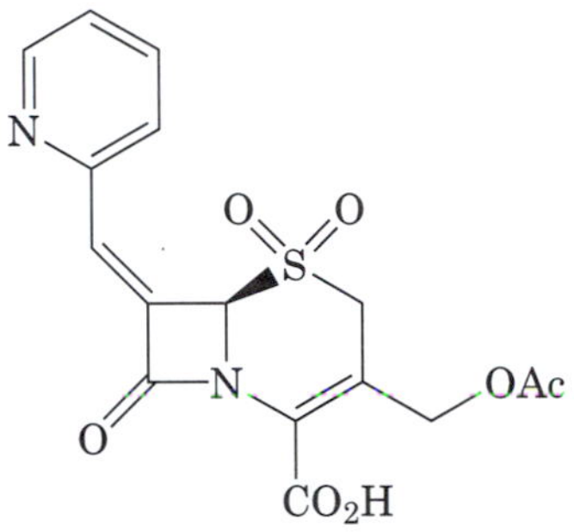

(443)

## 7  CURRENT TRENDS DRIVING INDUSTRY

Almost 300 antibacterial products are available on the world market, but successful treatment of bacterial disease is becoming increasingly problematic as the number of elderly and immunocompromised patients increases, the pathogens undergo changes, and resistance to current agents become more widespread. These factors are driving the pharmaceutical industry to develop new and more powerful agents from existing classes of antibacterial drugs and to reevaluate their approaches to antibacterial drug discovery. Bacterial genomics, supported by fundamental research on bacterial physiology, environmental adaptation, and host-pathogen interactions are expected to supply the drug targets of the future. Although products of this research are some way off, new and improved agents continue to emerge from established screening and chemical modification programs.

The antibacterial market is increasingly well supplied with effective oral and parenteral agents that can provide treatment for the majority of infectious diseases. The criteria of novelty and competitive advantage for new agents therefore become increasingly difficult to attain. The relentless spread of antibacterial resistance to agents that were initially thought to have overcome all resistance problems is now offering necessary scope for novelty and innovation. The message coming from many opinion leaders in the medical profession is also one of strong encouragement for the pharmaceutical industry to pursue the quest for new and better antibacterials. Drug resistance is seen as a serious potential threat to the continuing effectiveness of current antibacterial agents, and a new compound that can address an important resistance problem is likely to find a place in the crowded market.

The most problematic forms of resistance being encountered are penicillin resistance in *S. pneumoniae*, glycopeptide resistance in enterococci, methicillin and multidrug resistance in Staphylococci, extended-spectrum β-lactamase resistance in MTB, and metronidazole resistance in *Helicobacter pylori*.

Other potential areas for innovation relate to the changing pattern of infectious disease brought about by the AIDS epidemic and the increasing frequency of organ transplantation, implanted prosthetic devices, and other developments in invasive medical procedures, together with a demographic shift to a more elderly and institutionalized population in the developed world. Antibiotics are now increasingly required to be effective in patients who are immunocompromised or debilitated through age-underlying disease.

### 7.1  Inhibitors of Protein Export in Bacteria

Signal peptidase enzymes represent a novel class of antibacterial targets (777) involved in the export of proteins across the cytoplasmic membrane in bacteria (778, 779). Proteins that are processed through this pathway are synthesized as preproteins with an extra domain, the signal sequence, attached to the amino terminus (780). This signal sequence allows the preprotein to cross the cytoplasmic

Current medical needs

| Organism | Infection | Resistance |
|---|---|---|
| **Nosocomial** | | |
| MRSA/E | Bacteremia, Pneumonia Sugical wound infections, Endocarditis | Quinolones, β-Lactams, Rifampin, Tetracyclines |
| Enterococci | Bacteremia, UTI, Surgical wound infections | Vancomycin, Quinolones Aminoglycosides; β-Lactams |
| *Pseudomonas spp* | Bacteremia, Pneumonia UTI, Burns | Quinolones, Aminoglycosides Tetracyclines, β-Lactams |
| **Community** | | |
| *S pneumoniae* | Meningitis, Pneumonia, Arthritis, Bacteremia | Macrolides, β-Lactams |
| **Emerging pathogens** | | |
| *Stenotrophomonas malt.* *Burkholderia cepacia* | Cystic fibrosis, burns | Carbapenems |
| *Mycobacterium tuberc.* | Tuberculosis | Isonyazide |

membrane by means of a multiprotein complex, the Sec proteins. When it is outside, the signal peptidase enzyme (781) cleaves the signal sequence, thus liberating the protein, which is now able to fold into its active conformation. It is believed that by inhibiting this pathway, the proteins remain functionally inactive and eventually should accumulate in the cytoplasmic membrane, compromising its integrity and leading to cell death. A weak inhibition of signal peptidases was obtained with polysubstituted azetidinones (**444–446**) (782), in which was observed a stereochemical preference for (4S) isomers.

A screening of serine protease inhibitors against the signal peptidase from *E. coli* LP1 allowed identification of several classes of active bicyclic β-lactams. Esters of clavulanic acid (**447**) confirmed this preference for (5S) stereochemistry and were more active than the corresponding thioclavam analogs (**448**). Oxidation to the corresponding sulphone regained the activity in the decarboxylated series ($R_2$ = H).

The best compounds were obtained in the 5(S) penem series, allowing identification of benzyl ester (**449**), which has the opposite stereochemistry to that required for inhibition of bacterial PBPs and showed an $IC_{50}$ value of 10 $\mu M$ (783, 784). The hit compound was studied in depth and led to the synthesis of the penem

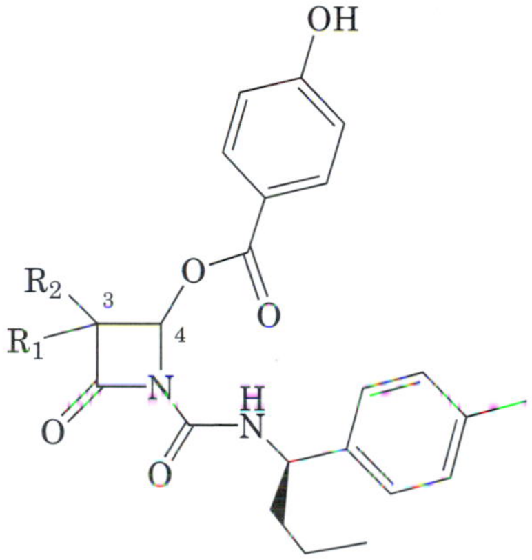

(**444**) (4S) $R_1$ = Et, $R_2$ = Et
(**445**) (4S) $R_1$ = $CH_3$, $R_2$ = $CH_3$
(**446**) (4R) $R_1$ = $CH_3$, $R_2$ = $CH_3$

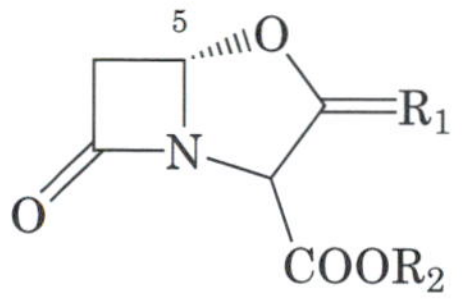

allyl ester derivative (**450**), which showed complete inhibition of processing in a permeable strain of *E. coli*, with an $IC_{50}$ value of 0.07 $\mu M$. Although the activity could not be man-

(**447**)

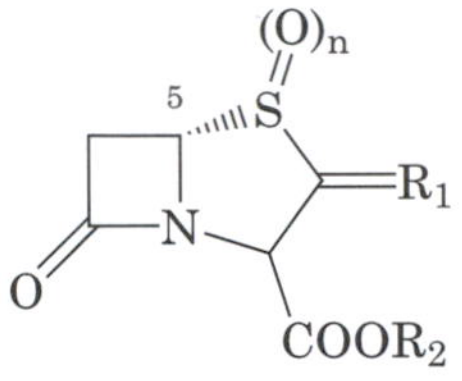

**(448)**

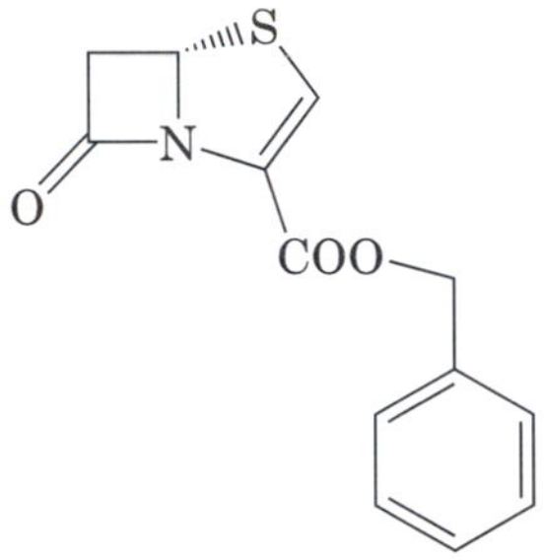

**(449)**

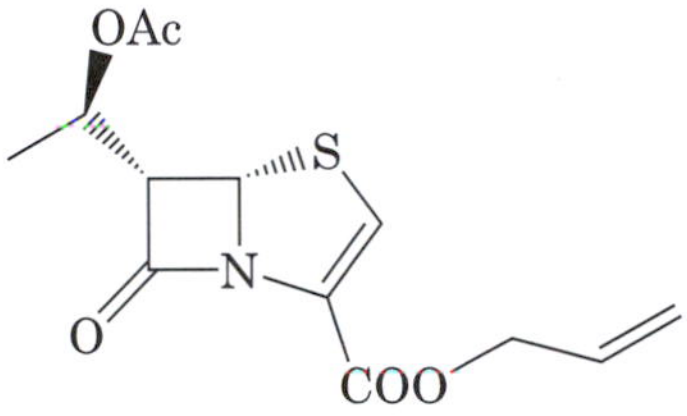

**(450)**

tained in standard and clinical strains, this work has highlighted a new valuable target for the discovery of novel β-lactam antibiotics (785).

## 8  WEB SITE ADDRESSES AND RECOMMENDED READING FOR FURTHER INFORMATION

If someone just needs quick information related to the β-lactam antibiotics and their use, nowadays it is sufficient to use a popular research engine searching for "lactam antibiotics" and retrieve thousands of Web pages. Here are listed only a few of them that contain interesting and precise specific information:

• The links http://www.microbe.org, http://www.idlinx.com, http://www.bact.wisc.edu/

bact330/330Lecturetopics.htm, http://www.ches.siu.edu/fix/medmicro/index.htm can be consulted to get an overview of the infection disease treatments, bacteria morphology, mechanism of action of antibiotics, and mechanism of resistance to the current antibacterial agents.

• The link http://wizard.pharm.wayne.edu/medchem/betalactam.htm leads to the Web page entitled: Chemistry of Beta Lactam Antibiotics PHA 421—Infectious Disease Module, a Medicinal Chemistry Tutorial on Beta Lactam Antibiotics.

• An interesting course of Chemotherapy can be found at the following address: http://www.vet.purdue.edu/depts/bms/courses/chmrx/bms45_96.htm, which leads to Drug Groups, in which the major antibacterial classes are extensively discussed.

• At http://www.grunenthal.com/knowledge-Base/kb_start.htm there is the Grünenthal Knowledge Base, a library of Knowledge-Modules in which are reported the antibiotics, including Beta-Lactam Antibiotics. A fast entry-level reference guide.

• A very good microbiology textbook can be found at the following link: http://www.bact.wisc.edu/microtextbook/index.html, in which, by clicking on http://www.bact.wisc.edu/microtextbook/ControlGrowth/antibiotic.html, a chapter on antibiotics is reported.

## REFERENCES

1. A. Fleming, *Br. J. Exp. Pathol.*, **10**, 226 (1929).
2. E. P. Abraham and E. Chain, *Nature*, **146**, 837 (1940).
3. D. J. Tipper and J. L. Strominger, *Proc. Natl. Acad. Sci. USA*, **54**; 1133 (1965).
4. D. J. Waxman and J. L. Strominger, *Annu. Rev. Biochem.*, **52**, 825 (1983).
5. H. Nikaido in J. M. Ghuysen and R. Hakenbeck, Eds., *Bacterial Cell Wall*, Elsevier Science BV, New York, 1994, pp. 547–558.
6. J. A. Hobot, E. Carlemalm, W. Villiger, and E. Kellenberger, *J. Bacteriol.*, **160**, 143 (1984).
7. M. Matsuhashi in J. M. Ghuysen and R. Hakenbeck, Eds., *Bacterial Cell Wall*, Elsevier Science BV, New York, 1994, pp. 55–71.
8. K. H. Schleifer and O. Kandler, *Bacteriol. Rev.*, **36**, 407 (1972).

9. J. M. Ghuysen and G. Dive in J. M. Ghuysen and R. Hakenbeck, Eds., *Bacterial Cell Wall*, Elsevier Science BV, New York, 1994, pp. 103–129.

10. F. B. Wientjes and N. Nanninga, *Res. Microbiol.*, **142**, 333 (1991).

11. B. Glauner and J. V. Holtje, *J. Biol. Chem.*, **265**, 18988 (1990).

12. E. Tuomanem and R. Cozens, *J. Bacteriol.*, **169**, 5308 (1987).

13. A. Tomasz, *Ann. Rev. Microbiol.*, **33**, 113 (1979).

14. G. Bernadsky, T. J. Beveridge, and A. J. Clarke, *J. Bacteriol.*, **176**, 5225 (1994).

15. G. D. Shockman and J. V. Holtje in J. M. Ghuysen and R. Hakenbeck, Eds., *Bacterial Cell Wall*, Elsevier Science BV, New York, 1994, pp. 131–166.

16. J. V. Holtje and E. I. Tuomanem, *J. Gen. Microbiol.*, **137**, 441 (1991).

17. R. J. Doyle, J. Chaloupka, and V. Vinter, *Microbiol. Rev.*, **52**, 554 (1988).

18. B. G. Spratt, *Eur. J. Biochem.*, **72**, 341 (1977).

19. D. J. Waxman, R. R. Yocum, and J. L. Strominger, *Philos. Trans. R. Soc. London Ser. B*, **289**, 257 (1980).

20. D. Cabaret, J. Liu, M. Wakselman, R. F. Pratt, and Y. Xu, *Bioorg. Med. Chem.*, **2**, 757 (1994).

21. T. D. McDowell and K. E. Reed, *Antimicrob. Agents Chemother.*, **33**, 1680 (1989).

22. B. G. Spratt and A. B. Pardee, *Nature*, **254**, 516 (1975).

23. D. V. Edwards and W. D. Donachie in M. A. De Pedro, Ed., *Bacterial Growth and Lysis*, Plenum, New York, 1993, pp. 369–374.

24. E. Grahn, S. E. Holm, and K. Roos, *Scand. J. Infect. Dis.*, **19**, 421 (1987).

25. L. Gutman and A. Tomasz, *Antimicrob. Agents Chemother.*, **22**, 128 (1982).

26. S. Handwergwer and A. Tomasz, *Rev. Infect. Dis.*, **7**, 368 (1985).

27. T. D. McDowell and C. L. Lemanski, *J. Bacteriol.*, **170**, 1783 (1988).

28. A. Tomasz and J. V. Holtje in D. Schlesinger, Ed., *Microbiology*, American Society for Microbiology, Washington, DC, 1977, pp. 202–215.

29. E. Tuomanen, D. T. Durack, and A. Tomasz, *Antimicrob. Agents Chemother.*, **30**, 521 (1986).

30. G. J. Van Asselt, G. De Kort, and J. A. M. Van De Klundert, *J. Antimicrob. Chemother.*, **35**, 67 (1995).

31. C. Jacobs, L. J. Huang, E. Bartowsky, S. Normark, and J. T. Park, *EMBO J.*, **13**, 4684 (1994).

32. P. M. Mendelman, D. O. Chaffin, and G. Kalaitzoglou, *J. Antimicrob. Chemother.*, **25**, 525 (1990).

33. T. R. Parr Jr., R. A. Moore, L. V. Moore, and R. E. Hancock, *Antimicrob. Agents Chemother.*, **31**, 121 (1987).

34. N. H. Georgopapadakou and F. Y. Liu, *Antimicrob. Agents Chemother.*, **18**, 148 (1980).

35. T. Tamura, H. Suzuki, Y. Nischimura, J. Mizoguchi, and Y. Hirota, *Proc. Natl. Acad. Sci. USA*, **77**, 4499 (1980).

36. J. Nacagawa, S. Tamake, and M. Matsuhashi, *Agric. Biol. Chem.*, **43**, 1379 (1979).

37. J. Nacagawa, S. Tamaki, S. Tomioka, and M. Matsuhashi, *J. Biol. Chem.*, **259**, 13937 (1984).

38. S. J. Curtis and J. L. Strominger, *J. Bacteriol.*, **145**, 398 (1981).

39. F. Ishino, W. Park, S. Tomioka, et al., *J. Biol. Chem.*, **261**, 7024 (1986).

40. M. A. De Pedro, U. Schwarz, Y. Nischimura, and Y. Hirota, *FEMS Microbiol. Lett.*, **9**, 219 (1980).

41. M. Matsuhashi, Y. Takagaki, I. N. Maruyama, et al., *Proc. Natl. Acad. Sci. USA*, **74**, 2976 (1977).

42. H. Amanuma and J. L. Strominger, *J. Biol. Chem.*, **255**, 11173 (1980).

43. J. K. Broome-Smith and B. G. Spratt, *J. Bacteriol.*, **152**, 904 (1982).

44. T. J. Dougherty, *Antimicrob. Agents Chemother.*, **30**, 649 (1986).

45. T. J. Dougherty, A. E. Koller, and A. Tomasz, *Antimicrob. Agents Chemother.*, **18**, 730 (1980).

46. B. G. Spratt, *Nature*, **332**, 173 (1988).

47. B. G. Spratt, *Science*, **264**, 388 (1994).

48. C. Urban, E. Go, N. Mariano, et al., *J. Infect. Dis.*, **167**, 448 (1993).

49. R. Williamson, L. Gutmann, T. Horaud, F. Delbos, and J. F. Acar, *J. Gen. Microbiol.*, **132**, 1929 (1986).

50. R. Fontana, M. Alderighi, M. Ligozzi, H. Lopez, A. Sucari, and G. Satta, *Antimicrob. Agents Chemother.*, **38**, 1980 (1994).

51. R. Fontana, R. Cerini, P. Longoni, A. Grossato, and P. Canepari, *J. Bacteriol.*, **155**, 1343 (1983).

52. M. L. Grayson, G. M. Eliopoulos, C. B. Wennersten, K. L. Ruoff, P. C. De Girolami, N. J. Fer-

raro, and R. C. Noellering, Jr, *Antimicrob. Agents Chemother.*, **35**, 2180 (1991).

53. R. Williamson, C. Le Bouguenec, L. Gutmann, and T. Horaud, *J. Gen. Microbiol.*, **131**, 1933 (1985).

54. P. Nordmann, M. H. Nicolas, and L. Gutmann, *Antimicrob. Agents Chemother.*, **37**, 1406 (1993).

55. H. F. Chambers, M. J. Sachdeva, and C. J. Hackbarth, *Biochem. J.*, **301**, 139 (1994).

56. N. H. Georgopapadakou, B. A. Dix, and Y. R. Mauriz, *Antimicrob. Agents Chemother.*, **29**, 333 (1986).

57. P. E. Reynolds in P. Actor, L. DaneoMoore, M. L. Higgins, M. R. J. Salton, and G. D. Schockman, Eds., *Antibiotic Inhibition of Bacterial Cell Surface Assembly and Function*, American Society for Microbiology, Washington, DC, 1988, pp. 343–351.

58. N. H. Georgopapadakou, L. M. Cummings, E. R. LaSala, J. Unowsky, and D. L. Pruess in P. Actor, L. DaneoMoore, M. L. Higgins, M. R. J. Salton, and G. D. Schockman, Eds., *Antibiotic Inhibition of Bacterial Cell Surface Assembly and Function*, American Society for Microbiology, Washington, DC, 1988, pp. 597–602.

59. J. W. Kozarich and J. L. Strominger, *J. Biol. Chem.*, **253**, 1272 (1978).

60. A. W. Wyke, J. B. Ward, M. V. Hayes, and N. A. Curtis, *Eur. J. Biochem.*, **119**, 389 (1981).

61. B. J. Hartman and A. Tomasz, *J. Bacteriol.*, **158**, 513 (1984).

62. P. E. Reynolds and C. Fuller, *FEMS Microbiol. Lett.*, **33**, 251 (1986).

63. K. Ubukata, R. Nonoguchi, M. Matsuhashi, and M. Konno, *J. Bacteriol.*, **171**, 2882 (1987).

64. L. J. Chalkley and H. K. Koornhof, *J. Antimicrob. Chemother.*, **22**, 791 (1988).

65. Z. Markiewicz and A. Tomasz, *J. Clin. Microbiol.*, **27**, 405 (1989).

66. S. Zighelboim and A. Tomasz, *Antimicrob. Agents Chemother.*, **17**, 434 (1980).

67. M. Jamin, C. Damblon, S. Millier, R. Hakenbeck, and J. M. Frere, *Biochem. J.*, **292**, 735 (1993).

68. M. Jamin, R. Hakenbeck, and J. M. Frere, *FEBS Lett.*, **331**, 101 (1993).

69. R. Hakenbeck and M. Kohiyama, *Eur. J. Biochem.*, **127**, 231 (1982).

70. S. Yan, P. M. Mendelman, and D. L. Stevens, *FEMS Microbiol. Lett.*, **110**, 313 (1993).

71. C. G. Dowson, T. J. Coffey, C. Kell, and R. A. Whiley, *Mol. Microbiol.*, **9**, 635 (1993).

72. R. Kakenbeck, M. Tarpay, and A. Tomasz, *Antimicrob. Agents Chemother.*, **17**, 364 (1980).

73. S. Handwerger and A. Tomasz, *Antimicrob. Agents Chemother.*, **30**, 57 (1986).

74. C. G. Dowson, A. Hutchinson, N. Woodford, A. P. Johnson, R. C. George, and B. G. Spratt, *Proc. Natl. Acad. Sci. USA*, **87**, 5858 (1990).

75. T. R. Parr Jr. and L. E. Bryan, *Antimicrob. Agents Chemother.*, **25**, 747 (1984).

76. M. Gehrlein, H. Leyng, W. Culmann, S. Wendt, and W. Opferkuch, *Chemotherapy*, **37**, 405 (1991).

77. C. Urban, E. Go, M. Mariano, and J. J. Rahal, *FEMS Microbiol. Lett.*, **125**, 193 (1995).

78. M. Obara and T. Nakae, *J. Antimicrob. Chemother.*, **28**, 791 (1991).

79. N. H. Georgopapadakou, *Antimicrob. Agents Chemother.*, **37**, 2045 (1993).

80. A. Tomasz, H. B. Drugeon, H. M. De Lencastre, D. Jabes, L. McDougall, and J. Billie, *Antimicrob. Agents Chemother.*, **33**, 1869 (1989).

81. L. D. Bowler, Q. Y. Zhang, and B. G. Spratt, *J. Bacteriol.*, **176**, 333 (1994).

82. B. G. Spratt, L. D. Bowler, Q. Y. Zhang, J. Zhou, and J. M. Smith, *J. Mol. Evol.*, **34**, 115 (1992).

83. H. Faruki and P. F. Sparling, *Antimicrob. Agents Chemother.*, **30**, 856 (1986).

84. W. Pan and B. G. Spratt, *Mol. Microbiol.*, **11**, 769 (1994).

85. G. Laible, B. G. Spratt, and R. Hakenbeck, *Mol. Microbiol.*, **5**, 1993 (1991).

86. P. M. Mendelman, D. O. Chaffin, T. L. Stull, C. E. Rubens, K. D. Mack, and A. L. Smith, *Antimicrob. Agents Chemother.*, **26**, 235 (1984).

87. D. A. Serfass, P. M. Mendelman, D. O. Chaffin, and C. A. Needham, *J. Gen. Microbiol.*, **132**, 2855 (1986).

88. M. Powell and D. M. Livermore, *J. Antimicrob. Chemother.*, **26**, 741 (1990).

89. K. Ubukata, N. Yamashita, and M. Konno, *Antimicrob. Agents Chemother.*, **27**, 851 (1985).

90. B. L. De Jonge, Y. S. Chang, D. Gage, and A. Tomasz, *J. Biol. Chem.*, **267**, 11255 (1992).

91. W. D. Beck, B. Berger-Bachi, and F. H. Kayser, *J. Bacteriol.*, **165**, 373 (1986).

92. P. R. Matthews, K. C. Reed, and P. R. Stewart, *J. Gen. Microbiol.*, **133**, 1919 (1987).

93. M. D. Song, M. Wachi, M. Doi, F. Ishino, and M. Matsuhashi, *FEBS Lett.*, **221**, 167 (1987).

94. W. C. Gaisford and P. E. Reynolds, *Eur. J. Biochem.*, **185**, 211 (1989).

95. J. M. Musser and V. Kapur, *J. Clin. Microbiol.*, **30**, 2058 (1992).

96. E. Suzuki, K. Hiramatsu, and T. Yokota, *Antimicrob. Agents Chemother.*, **36**, 429 (1992).

97. B. J. Hartman and A. Tomasz, *Antimicrob. Agents Chemother.*, **29**, 85 (1986).

98. P. R. Matthews and P. R. Stewart, *FEMS Microbiol. Lett.*, **22**, 161 (1984).

99. S. J. Seligman, *J. Gen. Microbiol.*, **42**, 315 (1966).

100. K. Murakami and A. Tomasz, *J. Bacteriol.*, **171**, 874 (1989).

101. H. Midhof, B. Reinicke, P. Blimel, B. Berger-Bachi, and H. Labischinski, *J. Bacteriol.*, **173**, 3507 (1991).

102. C. Ryffel, A. Strassle, F. H. Kayser, and B. Berger-Bachi, *Antimicrob. Agents Chemother.*, **38**, 724 (1994).

103. J. M. Boyce and A. A. Medeiros, *Antimicrob. Agents Chemother.*, **31**, 1426 (1987).

104. D. I. Anner, *Med. J. Aust.*, **1**, 444 (1969).

105. H. F. Chambers and C. J. Hackbarth, *Antimicrob. Agents Chemother.*, **31**, 1982 (1987).

106. B. Inglis, P. R. Matthews, and P. R. Stewart, *J. Gen. Microbiol.*, **134**, 1465 (1988).

107. P. Canepari, L. M. Del Mar, G. Cornaglia, R. Fontana, and G Satta, *J. Gen. Microbiol.*, **132**, 625 (1986).

108. B. E. Murray, *Clin. Microbiol. Rev.*, **3**, 46 (1990).

109. I. Nassova and S. Mobashery, *Antimicrob. Agents Chemother.*, **42**, 1 (1998).

110. M. H. Richmond and R. B. Sykes, *Adv. Microbiol. Physiol.*, **9**, 31 (1973).

111. M. Matthew, *J. Antimicrob. Chemother.*, **5**, 349 (1979).

112. P. M. Mendelman, D. O. Chaffin, and G. Kalaitzoglou, *J. Antimicrob. Chemother.*, **25**, 525 (1990).

113. K. Bush, *Antimicrob. Agents Chemother.*, **33**, 259 (1989).

114. A. Aspiotis, W. Cullmann, W. Dick, and M. Stieglitz, *Chemotherapy*, **32**, 236 (1986).

115. Y. Yang, and D. M. Livermore, *Antimicrob. Agents Chemother.*, **32**, 1385 (1988).

116. K. Bush, G. A. Jacoby, and A. A. Medeiros, *Antimicrob. Agents Chemother.*, **39**, 1211 (1995).

117. R. P. Ambler, *Philos. Trans. R. Soc. London Ser. B*, **289**, 321 (1980).

118. J. M. Ghuysen, *Annu. Rev. Microbiol.*, **45**, 37 (1991).

119. D. M. Livermore, *Rev. Infect. Dis.*, **10**, 691 (1988).

120. H. Y. Chen and D. M. Livermore, *J. Antimicrob. Chemother.*, **32** (Suppl. B), 63 (1993).

121. N. A. C. Curtis, R. L. Eisenstad, K. A. Turner, and A. J. White, *J. Antimicrob. Chemother.*, **15**, 642 (1983).

122. H. Nikaido, *Antimicrob. Agents Chemother.*, **33**, 1831 (1989).

123. H. Nikaido and S. Normark, *Mol. Microbiol.*, **1**, 29 (1987).

124. A. Raimondi, A. Traverso, and H. Nikaido, *Antimicrob. Agents Chemother.*, **35**, 1174 (1991).

125. A. Jaffe, Y. A. Chabbert, and O. Semonin, *Antimicrob. Agents Chemother.*, **22**, 942 (1982).

126. H. Nikaido and M. Vaara, *Microbiol. Rev.*, **49**, 1 (1985).

127. L. Martinez-Martinez, S. Hernandez-Alléo, S. Albertí, J. M. Tomás, V. J. Benedi, and G. Λ. Jacoby, *Antimicrob. Agents Chemother.*, **40**, 342 (1996).

128. B. Pangon, C. Bizet, A. Bure, et al., *J. Infect. Dis.*, **159**, 1005 (1989).

129. E. H. Lee, M. H. Nicolas, M. D. Kitzis, G. Pialoux, E. Collatz, and L. Gutmann, *Antimicrob. Agents Chemother.*, **35**, 1093 (1991).

130. P. Ledent, X. Raquet, B. Joris, J. Van Beeumen, and J. M. Frere, *Biochem. J.*, **292**, 255 (1993).

131. D. A. Weber, C. C. Sanders, J. S. Bakken, and J. P. Quinn, *J. Infect. Dis.*, **162**, 460 (1990).

132. X. Z. Li, D. Ma, D. M. Livermore, and H. Nikaido, *Antimicrob. Agents Chemother.*, **38**, 1742 (1994).

133. D. M. Livermore, *J. Med. Microbiol.*, **18**, 261 (1984).

134. X. Z. Li, L. Zang, R. Srikumar, and K. Poole. *Antimicrob. Agents Chemother.*, **42**, 399 (1998).

135. T. Kohler, M. Kok, M. Hamzehpour, P. Plesiat, N. Gotoh, T. Nishino, L. K. Curty, and J. C. Pechere, *Antimicrob. Agents Chemother.*, **40**, 2288 (1996).

136. X. Z. Li, H. Nikaido, and K. Poole, *Antimicrob. Agents Chemother.*, **39**, 1948 (1995).

137. H. Nikaido, *Science*, **264**, 382 (1994).

138. R. Srikumar, T. Kon, N. Gotoh, and K. Poole, *Antimicrob. Agents Chemother.*, **42**, 65 (1998).

139. J. Trias, J. Dufresne, R. C. Levesque, and H. Nikaido, *Antimicrob. Agents Chemother.*, **33**, 1201 (1989).

140. T. Fukuoka, S. Ohya, T. Narita, M. Katsuta, M. Iijima, N. Masuda, H. Yasuda, J. Trias, and H. Nikaido, *Antimicrob. Agents Chemother.*, **37**, 322 (1993).

141. D. M. Livermore and Y. J. Yang, *J. Antimicrob. Chemother.*, **24** (Suppl. A), 149 (1989).

142. J. Trias, J. Dufresne, R. C. Levesque, and H. Nikaido, *Antimicrob. Agents Chemother.*, **33**, 1202 (1989).

143. D. M Livermore, *Antimicrob. Agents Chemother.*, **36**, 2046 (1992).

144. Y. J. Yang and D. M. Livermore, *J. Antimicrob. Chemother.*, **24** (Suppl. A), 207 (1989).

145. G. Calandra, F. Ricci, C. Wang, and K. Brown, *Lancet*, **2**, 340 (1986).

146. J. P. Quinn, E. J. Dudek, C. A. Di Vincenzo, D. A. Lucks, and S. A. Lerner, *J. Infect. Dis.*, **154**, 289 (1986).

147. H. F. Chambers and H. C. Neu in G. L. Mandell, J. B. Bennett, and R. Dolin, Eds., *Principles and Practice of Infectious Diseases*, 4th ed., Churchill Livingstone, New York, 1995, pp. 233–246.

148. H. F. Chambers and H. C. Neu in G. L. Mandell, J. B. Bennett, and R. Dolin, Eds., *Principles and Practice of Infectious Diseases*, 4th ed., Churchill Livingstone, New York, 1995, pp. 264–272.

149. A. W. Karchmer in G. L. Mandell, J. B. Bennett, and R. Dolin, Eds., *Principles and Practice of Infectious Diseases*, 4th ed., Churchill Livingstone, New York, 1995, pp. 247–264.

150. R. D. Pryka and G. M. Haig, *Ann. Pharmacother.*, **28**, 1045 (1994).

151. G. L. Mandell and M. A. Sande in A. G. Gilman, L. S. Goodman, and A. Gilman, Eds., *The Pharmacologic Basis of Therapeutics*, Macmillan, New York, 1980, p. 1126.

152. A. Svedhem, K. Alestig, and M. Jertborn, *Antimicrob. Agents Chemother.*, **18**, 349 (1980).

153. J. Carrizosa, M. E. Levison, and D. Kaye, *Antimicrob. Agents Chemother.*, **3**, 306 (1973).

154. S. Berger, E. C. Ernst, and M. Barza, *Antimicrob. Agents Chemother.*, **9**, 575 (1976).

155. B. Trollfors, K. Alestig, and R. Norrby, *Scand. J. Infect. Dis.*, **11**, 315 (1979).

156. T. Bergan, *Infection*, **7** (Suppl.), 507 (1979).

157. L. D. Petz in N. R. Rose and H. H. Friedman, Eds., *Manual of Clinical Immunology: Auto-immune and Drug-Immune Hemolitic Ane-*

158. L. D. Petz, *J. Infect. Dis.*, **134** (Suppl.), S74 (1978).

159. G. T. Stewart, *Ann. Rev. Pharmacol.*, **13**, 309 (1973).

160. J. Porter and H. Jick, *Lancet*, **1**, 587 (1977).

161. B. B. Levine and D. M. Zolov, *J. Allergy*, **43**, 231 (1969).

162. C. Van Winzum, *J. Antimicrob. Chemother.*, **4** (Suppl.), 91 (1978).

163. S. Ahalsted, B. Ekstrom, P. O. Svard, B. Sjoberg, A. Kristofferson, and B. Ortengren, *CRC Crit. Rev. Toxicol.*, **7**, 219 (1980).

164. Anonymous, *Med. Lett.*, **38**, 88 (1996).

165. S. R. Norrby, *Med. Clin. North. Am.*, **79**, 745 (1995).

166. A. de Sarro, D. Ammendola, M. Zappala, S. Grasso, and G. B. de Sarro, *Antimicrob. Agents Chemother.*, **39**, 232 (1995).

167. S. R. Norrby, P. A. Newell, K. L. Faulkner, and W. Lesky, *J. Antimicrob. Chemother.*, **36**, 207 (1995).

168. W. T. Hughes, D. Armstrong, G. P. Bodey, A. E. Brown, J. E. Edwards, R. Feld, P. Pizzo, K. V. Rolston, J. L. Shenep, and L. S. Young, *Clin. Infect. Dis.*, **25**, 551 (1997).

169. G. P. Bodey, *Clin. Infect. Dis.*, **17**, S378 (1993).

170. H. Giamarellou, *Med. Clin. North Am.*, **79**, 559 (1996).

171. P. A. Pizzo, *N. Engl. J. Med.*, **328**, 1323 (1993).

172. J. W. Sanders, N. R. Powe, and R. D. Moore, *J. Infect. Dis.*, **164**, 907 (1991).

173. D. J. Winston, W. G. Ho, L. S. Young, W. L. Hewitt, and R. P. Gale, *Arch. Intern. Med.*, **142**, 1663 (1982).

174. J. C. Wade, S. C. Schimpff, K. A. Newman, C. L. Fortner, H. C. Standiford, and P. H. Wiernik, *Am. J. Med.*, **71**, 983 (1981).

175. R. D. Lawson, L. O. Gentry, G. P. Bodey, M. J. Keating, and T. L. Smith, *Am. J. Med. Sci.*, **287**, 16 (1984).

176. J. Klastersky, M. P. Glauer, S. C. Schimpff, S. H. Zinner, and H. Gaya, *Antimicrob. Agents Chemother.*, **29**, 263 (1986).

177. Anonymous, *N. Engl. J. Med.*, **317**, 1692 (1987).

178. Anonymous, *Ann. Intern. Med.*, **119**, 584 (1993).

179. P. A. Pizzo, J. W. Hathron, J. Himenez, M. Browne, J. Commers, D. Cotton, J. Gress, D.

Longo, D. Marshall, and J. McKnight, *N. Engl. J. Med.*, **315**, 552 (1986).

180. G. P. Bodey, M. E. Alvarez, P. G. Jones, K. V. Rolston, L. Steelhammer, and V. Fainstein, *Antimicrob. Agents Chemother.*, **30**, 211 (1986).

181. J. A. Miller, T. Butler, R. A. Beveridge, A. N. Kales, R. A. Binder, L. J. Smith, W. M. Ueno, G. Milkovich, S. Goldwater, and A. Marion, *Clin. Ther.*, **15**, 486 (1993).

182. M. J. Leyland, K. F. Bayston, J. Choen, R. Warren, A. C. Newland, A. J. Bint, C. Cefai, D. G. White, S. A. Murray, and D. Bareford, *J. Antimicrob. Chemother.*, **30**, 843 (1992).

183. A. G. Freifeld, T. Walsh, D. Marshall, D. J. Gress, S. M. Steinberg, J. Hathorn, M. Rubin, P. Jarosinski, V. Gill, and R. C. Young, *J. Clin. Oncol.*, **13**, 165 (1995).

184. K. V. I. Rolston, P. Berkey, G. P. Bodey, E. J. Anaissie, N. M. Khardori, J. H. Joshi, M. J. Keating, F. A. Holmes, F. F. Cabanillas, and L. A. Elting, *Arch. Intern. Med.*, **152**, 283 (1992).

185. P. G. Jones, K. V. I. Rolston, V. Fainstein, L. Elting, R. S. Walters, and G. P. Bodey, *Am. J. Med.*, **81**, 243 (1986).

186. E. Chain, H. W. Florey, A. D. Gardner, N. G. Heatley, M. A. Jennings, J. Orr-Ewing, and A. G. Sanders, *Lancet*, 226 (1940).

187. E. H. Flynn, M. H. Mc Cormick, M. C. Stamper, H. De Valeria, and C. W. Godzeski, *J. Am. Chem. Soc.*, **84**, 4594 (1962).

188. H. T. Clarke, J. R. Johnson, and R. Robinson, *The Chemistry of Penicillin*, Princeton University Press, Princeton, NJ, 1949.

189. D. Wilson, *In Search of Penicillin*, Knopf, New York, 1976.

190. R. Hare, *The Birth of Penicillin*, Allen & Unwin, London, 1970.

191. H. W. Florey, E. Chain, N. G. Heatley, M. A. Jennings, A. G. Sanders, E. P. Abraham, and M. E. Florey, Antibiotics, Vol. **2**, Oxford University Press, London, 1946.

192. E. P. Abraham in A. L. Demain and N. A. Solomon, Eds., Antibiotics Containing the $\beta$-Lactam Structure I, Vol. **67**, Springer-Verlag, Berlin/Heidelberg/New York/Tokyo, 1983, Chapter 1.

193. E. P. Abraham, E. Chain, A. D. Gardner, N. G. Heatley, M. A. Jennings, and H. W. Florey, *Lancet*, 177 (1941).

194. D. Hodgkin Crowfoot, C. W. Bunn, B. W. Rogers-Low, and A. Tuner-Jones, *The Chemis-*

*try of Penicillin*, Princeton University Press, Princeton, NJ, 1949, pp. 310–367.

195. H. T. Clarke, J. R. Johnson, and R. Robinson, *The Chemistry of Penicillin*, Princeton University Press, Princeton, NJ, 1949.

196. J. A. Thorn and M. J. Johnson, *J. Am. Chem. Soc.*, **72**, 2052 (1950).

197. D. C. Mortimer and M. J. Johnson, *J. Am. Chem. Soc.*, **74**, 4098 (1952).

198. O. K. Behens, J. J. Corse, R. G. Jones, M. J. Mann, Q. F. Soper, F. R. Van Abeele, and Ming-Chien Chiang, *J. Biol. Chem.*, **175**, 771 (1948).

199. E. P. Abraham and G. G. F. Newton, *Biochem. J.*, **58**, 94 (1954).

200. E. H. Flynn, M. H. McCormick, M. C. Stamper, H. De Valeria, and C. W. Godzeski, *J. Am. Chem. Soc.*, **84**, 4594 (1962).

201. Q. F. Soper, C. W. Whitehead, O. K. Behrens, J. J. Corse, and R. G. Jones, *J. Am. Chem. Soc.*, **70**, 2849 (1948).

202. A. L. Tosoni, D. G. Glass, and L. Goldsmith, *Biochem. J.*, **69**, 476 (1958).

203. F. R. Batchelor, F. P. Doyle, J. H. C. Nayler, and G. N. Rolinson, *Nature*, **183**, 257 (1959).

204. J. C. Sheehan and K. R. Henery-Logan, *J. Am. Chem. Soc.*, **79**, 1262 (1957).

205. J. C. Sheehan and K. R. Henery-Logan, *J. Am. Chem. Soc.*, **81**, 3089 (1959).

206. J. C. Sheehan and K. R. Henery-Logan, *J. Am. Chem. Soc.*, **81**, 2912 (1959).

207. J. C. Sheehan and K. R. Henery-Logan, *J. Am. Chem. Soc.*, **84**, 2983 (1962).

208. A. K. Bose, G. Spiegelman, and M. S. Manhas, *J. Am. Chem. Soc.*, **90**, 4506 (1968).

209. R. A. Firestone, N. S. Maciejewicz, R. R. W. Ratcliffe, and B. G. Christensen, *J. Org. Chem.*, **39**, 437 (1974).

210. J. E. Baldwin, M. A. Christie, S. B. Haber, and L. I. Kruse, *J. Am. Chem. Soc.*, **98**, 3045 (1976).

211. R. J. Ponsford, Kirk-Othmer Encyclopedia of Chemical Technology: The Penicillins, 4th ed., Vol. **3**, John Wiley & Sons, New York, 1992, pp. 129–158.

212. G. M. Williamson, J. K. Morrison, and K. J. Stevens, *Lancet*, **1**, 847 (1961).

213. M. Ridley, D. Barrie, K. Lynn, and K. C. Stead, *Lancet*, **1**, 230 (1970).

214. F. P. Doyle, K. D. Hardy, J. H. C Nayler, M. J. Soulal, E. R. Stove, and H. R. J. Waddington, *J. Chem. Soc.*, 1453 (1962).

215. J. A. Yurchenco, M. W. Hooper, and G. H. Warren, *J. Antibiot. Chemother.*, **12**, 534 (1962).

216. F. P. Doyle, A. A. W. Long, J. H. C. Nayler, and E. R. Stove, *Nature*, **192**, 1183 (1961).

217. J. H. C. Nayler, A. A. W. Long, D. M. Brown, P. Acred, G. N. Rolinson, F. R. Batchelor, S. Stevens, and R. Sutherland, *Nature*, **195**, 1264 (1962).

218. J. V. Bennett, C. F. Gravenkamper, J. L. Brodie, and W. M. M. Kirby, *Antimicrob. Agents Chemother.*, 257 (1965).

219. R. Sutherland, E. A. P. Croydon, and G. N. Rolinson, *Br. Med. J.*, **4**, 455 (1970).

220. G. N. Rolinson and S. Stevens, *Br. Med. J.*, **2**, 191 (1961).

221. J. E. Dolfini, H. E. Applegate, G. Bach, H. Basch, J. Bernstein, J. Schwartz, and F. L. Weisenborn, *J. Med. Chem.*, **14**, 117 (1971).

222. R. N. Brogden, T. M. Speight, and G. S. Avery, *Drugs*, **9**, 88 (1975).

223. S. B. Roseman, L. S. Weber, G. Owen, G. H. Warren, H. E. Alburn, D. E. Clark, H. Fletcher, and N. H. Grant, *Antimicrob. Agents Chemother.*, 590 (1967).

224. J. P. Clayton, M. Cole, S. W. Elson, and H. Ferres, *Antimicrob. Agents Chemother.*, **5**, 670 (1974).

225. N. O. Bodin, B. Ekström, U. Forsgren, L. P. Jalar, L. Magni, C. H. Ramsay, and B. Sjöberg, *Antimicrob. Agents Chemother.*, **8**, 518 (1975).

226. E. L. Foltz, J. W. West, I. H. Breslow, and H. Wallick, *Antimicrob. Agents Chemother.*, 422 (1970).

227. F. Sakamoto, S. Ikeda, and G. Tsukamoto, *Chem. Pharm. Bull.*, **32**, 2241 (1984).

228. P. Acred, D. M. Brown, E. T. Knudsen, G. N. Rolinson, and R. Sutherland, *Nature*, **215**, 25 (1967).

229. S. Morimoto, H. Nomura, T. Fugono, T. Azuma, I. Minami, M. Hori, and T. Masuad, *J. Med. Chem.*, **15**, 1108 (1972).

230. R. Sutherland, J. Brunett, and G. N. Rolinson, *Antimicrob. Agents Chemother.*, 390 (1971).

231. H. Noguchi, Y. Eda, H. Tobiki, T. Nakagome, and T. Komatsu, *Antimicrob. Agents Chemother.*, **9**, 262 (1976).

232. T. Kashiwagi, I. Isaka, N. Kawahara, K. Nakano, A. Koda, T. Osaza, I. Souza, Y. Murakami, and M. Murakami, Proceedings of the 15th Interscience Conference on Antimicrobial Agents and Chemotherapy, Washington, DC, 1975, Abstr. 255.

233. D. Stewart and G. P. Bodey, *Antimicrob. Agents Chemother.*, **11**, 865 (1977).

234. G. P. Bodey and T. Pan, *Antimicrob. Agents Chemother.*, **11**, 74 (1977).

235. K. P. Fu and H. C. Neu, *Antimicrob. Agents Chemother.*, **13**, 358 (1978).

236. M. Wagatsuma, M. Seto, T. Miyagishima, M. Kawazu, T. Yamaguchi, and S. Ohshima, *J. Antibiot.*, **36**, 147 (1983).

237. F. J. Lund in J. Elks, Ed., *Recent Advances in the Chemistry of β-Lactam Antibiotics*, Special Publication No. 28, Royal Chemistry Society, London, 1977, pp. 25–45.

238. K. Roholt, *J. Antimicrob. Chemother.*, **3** (Suppl. B), 71 (1977).

239. K. Josefsson, T. Bergan, L. Magni, B. G. Pring, and D. Westrlund, *Eur. J. Clin. Pharmacol.*, **23**, 249 (1982).

240. P. Brown, S. H. Calvert, P. C. A. Chapman, S. C. Cosham, A. J. Eglington, R. L. Elliot, M. A. Harris, J. D. Hinks, J. Lowther, D. J. Merrikin, M. J. Pearson, R. J. Ponsford, and J. V. Sims, *J. Chem. Soc. Perkin Trans.* **1**, 881 (1991).

241. G. N. Rolinson, *J. Antimicrob. Chemother.*, **17**, 5 (1986).

242. A. J. Wright, *Mayo Clin. Proc.*, **74**, 290 (1999).

243. V. Lorian, Ed., *Antibiotics in Laboratory Medicine*, 2nd ed., Williams & Wilkins, Baltimore, 1986.

244. D. J. Tipper in D. J. Tipper, Ed., *Mode of Action of β-Lactam Antibiotics: Antibiotic Inhibitors of Bacterial Cell Wall Biosynthesis*, Pergamon, Elmsford, NY, 1987, pp. 133–170.

245. J. L. Strominger and D. J. Tipper, *Am. J. Med.*, **39**, 708 (1965).

246. E. H. W. Bohme, H. E. Applegate, B. Teoplitz, and J. Z. Gougoutas, *J. Am. Chem. Soc.*, **93**, 4324 (1971).

247. R. A. Firestone, N. Schelechow, D. B. R. Johnson, and B. G. Christensen, *Tetrahedron Lett.*, **13**, 375 (1972).

248. E. M. Gordon and R. B. Sykes in R. B. Morin and M. Gorman, Eds., *The Chemistry and Biology of β-Lactam Antibiotics*, Vol. **1**, Academic Press, New York, 1982, pp. 199–370.

249. J. E. Baldwin, F. J. Urban, R. D. G. Cooper, and F. L. Jose, *J. Am. Chem. Soc.*, **95**, 2401 (1973).

250. G. A. Koppel and R. E. Koehler, *J. Am. Chem. Soc.*, **95**, 2403 (1973).

251. W. A. Slusarchyk, H. E. Applegate, P. Funke, W. Koster, M. S. Puar, M. Young, and J. E. Dolfini, *J. Org. Chem.*, **38**, 943 (1973).

252. T. Jen, J. Frazee, and R. E. Hoover, *J. Org. Chem.*, **38**, 2857 (1973).

253. L. D. Cama, W. J. Leanza, T. R. Beattie, and B. G. Christensen, *J. Am. Chem. Soc.*, **94**, 1408 (1972).

254. P. H. Bentley and J. P. Clayton, *J. Chem. Soc. Chem. Commun.*, 278 (1974).

255. A. W. Taylor and G. Burton, *Tetrahedron Lett.*, **18**, 3831 (1977).

256. E. M. Gordon, H. W. Chang, and C. M. Cimarusti, *J. Am. Chem. Soc.*, **99**, 5504 (1977).

257. H. Yanagisawa, M. Fukushima, A. Ando, and H. Nakao, *J. Antibiot.*, **24**, 969 (1976).

258. R. J. Ponsford, M. J. Basker, G. Burton, A. W. Guest, F. P. Harrington, P. H. Milner, M. J. Pearson, T. C. Smale, and A. V. Stachulski in A. G. Brown and S. M. Roberts, Eds., *Recent Advances in the Chemistry of β-Lactam Antibiotics*, Special Publication No. 52, Royal Chemistry Society, London, 1985, pp. 32–57.

259. P. H. Milner, A. W. Guest, F. P. Harrington, R. J. Ponsford, T. C. Smale, and A. V. Stachulski, *J. Chem. Soc. Chem. Commun.*, 1335 (1984).

260. C. L. Branch, M. J. Pearson, and T. C. Smale, *J. Chem. Soc. Perkin Trans.*1, 2865 (1988).

261. A. C. Kaura and M. J. Pearson, *Tetrahedron Lett.*, **26**, 2597 (1985).

262. B. Slocombe, M. J. Basker, P. H. Bentley, J. P. Clayton, M. Cole, K. R. Comber, R. A. Dixon, R. A. Edmondson, D. Jackson, D. J. Merrikin, and R. Sutherland, *Antimicrob. Agents Chemother.*, **20**, 38 (1981).

263. G. Burton, D. J. Best, R. A. Dixon, R. F. Kenyon, and A. G. Lashford, *J. Antibiot.*, **39**, 1419 (1986).

264. A. W. Guest, F. P. Harrington, P. H. Milner, R. J. Ponsford, T. C. Smale, A. V. Stachulski, M. J. Basker, and B. Slocombe, *J. Antibiot.*, **39**, 1498 (1986).

265. M. J. Basker, R. A. Edmondson, S. J. Knott, R. J. Ponsford, B. Slocombe, and S. J. White, *Antimicrob. Agents Chemother.*, **26**, 734 (1984).

266. H. Ochiai, Y. Murotani, H. Yamamoto, O. Yoshino, H. Morisawa, K. Momomoi, and Y. Watanabe, Proceedings of the 32nd Interscience Conference on Antimicrobial Agents and Chemotherapy, Anaheim, CA, 1992, Abstr. 383.

267. Y. Watanabe, S. Minami, T. Hayashi, H. Araki, and H. Ochiai, Proceedings of the 32nd Interscience Conference on Antimicrobial Agents and Chemotherapy, Anaheim, CA, 1992, Abstr. 384.

268. N. Matsumura, S. Minami, and S. Mitsuhashi, *J. Antimicrob. Chemother.*, **39**, 31 (1997).

269. Y. Watanabe, S. Minami, T. Hayashi, H. Araki, K. Harumi, R. Kitayama, and H. Ochiai, *Antimicrob. Agents Chemother.*, **39**, 2787 (1995).

270. E. H. Flynn, *Cephalosporins and Penicillins*, Academic Press, London/New York, 1970.

271. G. G. F. Newton and E. P. Abraham, *Nature*, **175**, 548 (1955).

272. E. P. Abraham and G. G. F. Newton, *Biochem. J.*, **79**, 377 (1961).

273. D. C. Hodgkin and E. N. Masten, *Biochem. J.*, **79**, 393 (1961).

274. R. Nagarajan, L. D. Boeck, M. Gorman, R. L. Amil, C. E. Higgens, M. M. Hoehn, W. M. Stark, and J. G. Whitney, *J. Am. Chem. Soc.*, **93**, 2308 (1971).

275. S. Tsubotani, T. Hida, F. Kasahara, Y. Wada, and S. Harada, *J. Antibiot.*, **37**, 1546 (1984).

276. F. M. Huber, R. H. Baltz, and P. G. Caltrider, *Appl. Microbiol.*, **16**, 1011 (1968).

277. C. E. Higgins, R. L. Hamill, T. H. Sands, M. M. Hoehn, N. E. Davis, N. Nagarajan, and L. D. Boeck, *J. Antibiot.*, **27**, 298 (1974).

278. P. Traxler, H. J. Treichler, and J. Nuesch, *J. Antibiot.*, **38**, 605 (1975).

279. T. Kaznaki, T. Fukita, H. Shirafuji, and Y. Fujisana, *J. Antibiot.*, **27**, 361 (1974).

280. T. Kaznaki, T. Fukita, K. Kitano, K. Katamoto, K. Nara, and Y. Nakao, *J. Ferment. Technol.*, **54**, 720 (1976).

281. K. Kitano, Y. Fujusawa, K. Katamoto, K. Nara, and Y. Nakao, *J. Ferment. Technol.*, **54**, 712 (1976).

282. J. Shoji, R. Sakazaki, K. Matsumoto, T. Tanimoto, Y. Terui, S. Kozuki, and E. Kondo, *J. Antibiot.*, **36**, 167 (1983).

283. K. A. Alvi, C. D. Reeves, J. Peterson, and J. Lein, *J. Antibiot.*, **48**, 338 (1995).

284. L. D. Hatfield, W. H. W. Lunn, B. G. Jackson, L. R. Peters, L. C. Blaszczac, J. W. Fisher, J. P. Gardiner, and J. M. Dunigan in G. I. Gregory, Ed., *Recent Advances in the Chemistry of β-Lactam Antibiotics*, Special Publication No. 38, Royal Chemistry Society, London, 1981, pp. 109–124.

285. J. Roberts, Kirk-Othmer Encyclopedia of Chemical Technology: The Cephalosporins, Vol. **3**, John Wiley & Sons, New York, 1992, pp. 28–82.

286. C. F. Murray and J. A. Webber in E. H. Flynn, Ed., *Cephalosporins and Penicillins, Chemistry and Biology*, Academic Press, London/New York, 1972, pp. 134–182.

287. L. D. Hatfield, J. W. Fisher, J. M. Durrigan, R. W. Burchfield, J. M. Greene, J. A. Webber, R. T. Vasileff, and M. D. Kinnick, *Philos. Trans. R. Soc. Lond. B.*, **289**, 173 (1980).

288. R. J. Stedman, K. Swered, and J. R. Hoover, *J. Med. Chem.*, **7**, 117 (1964).

289. T. Takaya, H. Takasugi, T. Murakawa, and H. Nakano, *J. Antibiot.*, **34**, 1300 (1981).

290. E. M. Wilson, *Chem. Ind.*, 217 (1984).

291. S. Kukolja and R. R. Chauvette in R. B. Morin, Ed., *Lactam Antibiotics*, Vol. **1**, Academic Press, New York, 1982, pp. 93–198.

292. G. A. Koppel, M. D. Kinnick, and L. J. Nummy, *J. Am. Chem. Soc.*, **99**, 2821 (1977).

293. J. Elks, *Drugs*, **34** (Suppl. 2), 240 (1987).

294. C. H. Newall and P. D. Hallam in P. G. Sammes and J. B. Taylor, Eds., Comprehensive Medicinal Chemistry, Vol. **2**, Pergamon, Elmsford, NY, 1990, pp. 609–653.

295. N. C. Klein and B. A. Cunha, *Adv. Ther.*, **21**, 83 (1995).

296. W. Dürkheimer, F. Adam, G. Fisher, and R. Kirrstetter in B. Testa, Ed., Advances in Drug Research, Vol. **17**, Academic Press, New York, 1988, pp. 61–234.

297. R. R. Chauvette, B. G. Jackson, E. R. Lavagnino, R. B. Morin, R. A. Mueller, R. P. Pioch, R. W. Roeske, C. W. Ryan, J. L. Spencer, and E. Van Heyningen, *J. Am. Chem. Soc.*, **84**, 3401 (1962).

298. Symposium, *J. Infect. Dis.*, **128** (Suppl.), 312–424 (1973).

299. C. W. Ryan, R. L. Simon, and E. M. Van Heyningen, *J. Med. Chem.*, **12**, 310 (1969).

300. H. G. Adams, G. A. Stilwell, and M. Turk, *Antimicrob. Agents Chemother.*, **9**, 1019 (1976).

301. R. N. Brogden, R. C. Heel, T. M. Speight, and G. S. Avery, *Drugs*, **17**, 233 (1979).

302. H. C. Neu and K. P. Fu, *Antimicrob. Agents Chemother.*, **13**, 584 (1978).

303. P. E. O. Williams and S. M. Harding, *J. Antimicrob. Chemother.*, **13**, 191 (1984).

304. Symposium, *J. Antimicrob. Chemother.*, **11** (Suppl. A), 1–239 (1983).

305. H. L. Muytens and J. van der Ros-van de Repe, *Antimicrob. Agents Chemother.*, **21**, 925 (1982).

306. Symposium, *Rev. Infect. Dis.*, **4** (Suppl.), 281–488 (1982).

307. D. M. Richards, R. C. Heel, R. N. Brogden, T. M. Speight, and G. S. Avery, *Drugs*, **27**, 469 (1984).

308. R. N. Brogden, A. Carmine, R. C. Heel, P. A. Morley, T. M. Speight, and G. S. Avery, *Drugs*, **22**, 432 (1981).

309. Symposium, *J. Antimicrob. Chemother.*, **8** (Suppl. B), 1–358 (1981).

310. K. Tsuchiya, M. Kondo, and H. Nagamote, *Antimicrob. Agents Chemother.*, **13**, 137 (1978).

311. R. N. Brogden and D. M. Campoli-Richards, *Drugs*, **38**, 524 (1989).

312. R. N. Jones, C. Thornsberry, and A. L. Berry, *Antimicrob. Agents Chemother.*, **25**, 710 (1984).

313. R. E. Kessler, M. Bies, R. E. Buck, D. R. Chisholm, T. A. Pursiano, M. Misiek, K. E. Price, and F. Leitner, *Antimicrob. Agents Chemother.*, **27**, 207 (1985).

314. M. L. Hammond, *Ann. Rep. Med. Chem.*, **28**, 119 (1993).

315. M. R. Jefson, S. J. Hecker, and J. P. Dirlam, *Ann. Rep. Med. Chem.*, **29**, 113 (1994).

316. M. Numata, I. Minamida, M. Yamaoka, M. Shiraishi, T. Miyawaki, H. Akimoto, K. Naito, and M. Kida, *J. Antibiot.*, **31**, 1262 (1978).

317. P. C. Cherry, M. C. Cook, M. W. Foxton, M. Gregson, G. I. Gregory, and G. B. Webb in J. Elks, Ed., *Recent Advances in the Chemistry of β-Lactam Antibiotics*, Special Publication No. 28, Royal Chemical Society, London, 1977, pp. 145–152.

318. J. Kant and V. Farina, *Tetrahedron Lett.*, **33**, 3563 (1992).

319. J. Kant, J. A. Roth, C. E. Fuller, D. G. Walker, D. A. Benigni, and V. Farina, *J. Org. Chem.*, **59**, 4956 (1994).

320. J. L. Martinez, A. Delgado-Iribarren, and F. Baquero, *FEMS Microbiol. Rev.*, **75**, 45 (1990).

321. M. J. Miller and F. Maouin, *Acc. Chem. Res.*, **26**, 241 (1993).

322. H. Mochizuki, Y. Oikawa, H. Yamada, S. Kasukabe, T. Shiihara, K. Murakami, K. Kato, J. Ishiguro, and H. Kosuzume, *J. Antibiot.*, **41**, 377 (1988).

323. T. Okita, K. Imae, T. Hasegawa, S. Imura, S. Masuyoshi, H. Kamachi, and H. Kamei, *J. Antibiot.*, **46**, 833 (1993).

324. R. G. Adams, E. G. Brian, C. L. Branch, A. W. Guest, F. P. Harrington, L. Mizen, J. E. Neale, M. J. Pearson, I. N. Simpson, H. Smulders, and I. I. Zomaya, *J. Antibiot.*, **48**, 417 (1995).

325. N. Watanabe, T. Nagasu, K. Katsu, and K. Kitoh, *Antimicrob. Agents Chemother.*, **31**, 497 (1987).

326. I. A. Critchley, M. J. Basker, R. A. Edmonson, and S. J. Knott, *J. Antimicrob. Chemother.*, **28**, 377 (1991).

327. P. Silley, J. W. Griffiths, D. Monsey, and A. M. Harris, *Antimicrob. Agents Chemother.*, **34**, 1806 (1990).

328. Y. Tatsumi, T. Maejima, and S. Mitsuhashi, *Antimicrob. Agents Chemother.*, **39**, 613 (1995).

329. J. E. Dolfini, J. Schwartz, and F. Weisenborn, *J. Org. Chem.*, **34**, 1582 (1969).

330. R. Heymes, G. Amiard, and G. Nominee, *Bull. Soc. Chim. Fr.*, 563 (1974).

331. S. L. Neidleman, S. C. Pan, L. A. Last, and J. E. Dolfini, *J. Med. Chem.*, **13**, 386 (1970).

332. R. W. Ratcliffe and B. G. Christensen, *Tetrahedron Lett.*, **14**, 4645 (1973).

333. R. W. Ratcliffe and B. G. Christensen, *Tetrahedron Lett.*, **14**, 4649 (1973).

334. R. W. Ratcliffe and B. G. Christensen, *Tetrahedron Lett.*, **14**, 4653 (1973).

335. R. B. Woodward, *Science*, **153**, 487 (1966).

336. R. B. Woodward, K. Heusler, J. Gostelli, P. Naegeli, W. Oppolzer, R. Ramage, S. Ranganthan, and H. Vorbruggen, *J. Am. Chem. Soc.*, **88**, 852 (1966).

337. R. B. Morin, B. G. Jackson, R. A. Mueller, E. R. Lavagnino, W. B. Scanlon, and S. L. Andrews, *J. Am. Chem. Soc.*, **85**, 1896 (1963).

338. T. S. Chou, J. R. Burgtorf, A. I. Ellis, S. R. Lammert, and S. Kukolja, *J. Am. Chem. Soc.*, **96**, 1609 (1974).

339. R. D. G. Cooper and D. O. Spry in E. H. Flynn, Ed., *Cephalosporins and Penicillins: Chemistry and Biology*, Academic Press, London/New York, 1972, pp. 183–254.

340. R. D. G. Cooper and G. A. Koppel in R. B. Morin and M. Gorman, Eds., *Chemistry and Biology of β-Lactam Antibiotics*, Vol. **1**, Academic Press, New York, 1982, pp. 2–92.

341. R. D. G. Cooper, L. D. Hatfield, and D. O. Spry, *Acc. Chem. Res.*, **6**, 32 (1973).

342. R. R. Chauvette, P. A. Pennington, C. W. Ryan, R. D. G. Cooper, F. L. Jose, I. G. Wright, E. M. Van Heyningen, and G. W. Huffman, *J. Org. Chem.*, **36**, 1259 (1971).

343. S. Kukolja, S. R. Lammert, M. R. Gleissner, and A. I. Ellis, *J. Am. Chem. Soc.*, **98**, 5040 (1976).

344. S. Torii, H. Tanaka, N. Saitoh, T. Siroi, H. Sasaoka, and J. Nokami, *Tetrahedron Lett.*, **23**, 2187 (1982).

345. H. Tanaka, M. Taniguchi, Y. Kameyama, M. Monnin, M. Sasaoka, T. Shiroi, S. Nagao, and S. Torii, *Chem. Lett.*, 1867 (1990).

346. R. Southgate and S. Elson, *Progress in the Chemistry of Organic Natural Products*, Springer-Verlag, Berlin/New York, 1985, pp. 1–106.

347. Y. Nozaki, N. Katayama, S. Tsubotani, H. Ono, and H. Okazaki, *J. Antibiot.*, **38**, 1141 (1985).

348. P. D. Singh, M. G. Young, J. H. Johnson, C. M. Cimarusti, and R. B. Sykes, *J. Antibiot.*, **37**, 773 (1984).

349. J. Shoji, T. Kato, R. Sakazaki, W. Nagata, Y. Terui, Y. Nakagawa, M. Shiro, K. Matsumoto, T. Hattori, T. Yoshida, and E. Kondo, *J. Antibiot.*, **37**, 1486 (1984).

350. J. Shoji, R. Sakazaki, T. Kato, Y. Terui, K. Matsumoto, T. Tanimoto, T. Hattori, K. Hirooka, and E. Kondo, *J. Antibiot.*, **38**, 538 (1985).

351. S. Karady, S. H. Pines, L. M. Weinstock, F. E. Roberts, G. S. Brenner, A. M. Hoinowski, T. Y. Cheng, and M. Sletzinger, *J. Am. Chem. Soc.*, **94**, 1410 (1972).

352. W. H. W. Lunn, R. W. Burchfield, T. K. Elzey, and E. V. Mason, *Tetrahedron Lett.*, **15**, 1307 (1974).

353. S. Karady, J. S. Amato, L. M. Weinstock, and M. Sletzinger, *Tetrahedron Lett.*, **19**, 407 (1978).

354. H. E. Applegate, C. M. Cimarusti, and W. A. Slusarcyk, *J. Chem. Soc. Chem. Commun.*, 293 (1980).

355. M. Shiokazi, N. Ishida, K. Iino, and T. Hiraoka, *Tetrahedron*, **36**, 2735 (1980).

356. E. O. Stapley, J. Birnbaum, K. A. Miller, W. Hyman, D. Endlin, and B. Woodruff, *Rev. Infect. Dis.*, **1**, 73 (1979).

357. K. Okonogi, A. Sugiura, M. Kuno, H. Ono, S. Harada, and E. Higashide, *J. Antibiot.*, **38**, 1555 (1985).

358. G. A. Koppel and R. E. Koehler, *J. Am. Chem. Soc.*, **102**, 1690 (1980).

359. T. Saito, Y. Sugimura, Y. Iwano, K. Iino, and T. Hiraoka, *Tetrahedron Lett.*, **17**, 1310 (1976).

360. H. Yanagisawa, M. Fukushima, A. Ando, and H. Nakao, *Tetrahedron Lett.*, **16**, 2705 (1975).

361. H. Nakao, H. Yanagisawa, B. Shimizu, M. Kaneko, M. Nagano, and S. Sugawara, *J. Antibiot.*, **29**, 554 (1976).

362. H. E. Applegate, C. M. Cimarusti, J. E. Dolfini, P. Funke, W. H. Koster, M. S. Puar, W. A. Slusarchyk, and M. G. Young, *J. Org. Chem.*, **44**, 811 (1979).

363. S. Nakatsuka, H. Tanino, and Y. Kishi, *J. Am. Chem. Soc.*, **97**, 5008 (1975).

364. R. N. Brogden, R. C. Heel, T. M. Speight, and G. S. Avery, *Drugs*, **17**, 1 (1979).

365. M. Benlloch, A. Torres, and F. Soriano, *J. Antimicrob. Chemother.*, **10**, 347 (1982).

366. A. W. Guest, C. L. Branch, S. C. Finch, A. C. Kaura, P. H. Milner, M. J. Pearson, R. J. Ponsford, and T. C. Smale, *J. Chem. Soc. Perkin Trans I*, 45 (1987).

367. P. D. Berry, A. G. Brown, J. C. Hanson, A. C. Kaura, P. H. Milner, C. J. Moores, K. Quick, R. N. Saunders, R. Southgate, and N. Whittall, *Tetrahedron Lett.*, **32**, 2683 (1991).

368. M. J. Basker, C. L. Branch, S. C. Finch, A. W. Guest, P. H. Milner, M. J. Pearson, R. J. Ponsford, and T. C. Smale, *J. Antibiot.*, **39**, 1788 (1986).

369. C. L. Branch, M. J. Basker, S. C. Finch, A. W. Guest, F. P. Harrington, A. C. Kaura, S. J. Knott, P. H. Milner, and M. J. Pearson, *J. Antibiot.*, **40**, 646 (1987).

370. R. Southgate, C. Branch, S. Coulton, and E. Hunt in G. Lukacs and M. Ohno, Eds., *Recent Progress in the Chemical Synthesis of Antibiotics*, Springer-Verlag, Berlin/New York, 1993, pp. 622–675.

371. L. D. Cama and B. G. Christensen, *J. Am. Chem. Soc.*, **96**, 7584 (1974).

372. R. A. Firestone, J. L. Fahey, N. S. Maciejewicz, G. S. Patel, and B. G. Christensen, *J. Med. Chem.*, **20**, 551 (1977).

373. M. Narisada, M. Onoue, and H. Nagata, *Heterocycles*, **7**, 839 (1977).

374. C. L. Branch and M. J. Pearson, *J. Chem. Soc. Perkin Trans. I*, 2268 (1979).

375. T. Ogasa, H. Saito, Y. Hashimoto, K. Sato, and T. Hirata, *Chem. Pharm. Bull.*, 315 (1989).

376. M. Narisada, T. Yoshida, M. Onoue, M. Ohtani, T. Okada, T. Tsuji, I. Kikkada, N. Haga, H. Satoh, H. Itani, and W. Nagata, *J. Med. Chem.*, **22**, 757 (1979).

377. J. J. Lipsky, J. C. Lewis, and W. J. Novick, *Antimicrob. Agents Chemother.*, **25**, 380 (1985).

378. M. Narisada and T. Tsuji in G. Lukacs and M. Ohno, Eds., *Recent Progress in the Chemical Synthesis of Antibiotics*, Springer-Verlag, Berlin/New York, 1990, pp. 705–725.

379. T. Tsuji, H. Satoh, M. Narisada, Y. Hamashima, and T. Yoshida, *J. Antibiot.*, **36**, 466 (1985).

380. W. Nagata, *Pure Appl. Chem.*, **61**, 325 (1989).

381. T. Okonogi, S. Shibahara, Y. Murai, T. Yoshida, S. Inouye, S. Kondo, and B. G. Christensen, *J. Antibiot.*, **43**, 357 (1990).

382. C. L. Branch, M. J. Basker, and M. J. Pearson, *J. Antibiot.*, **39**, 1792 (1986).

383. R. D. G. Cooper in M. I. Page, Ed., *The Chemistry of β-Lactams*, Chapman & Hall, London, 1992, pp. 272–303.

384. H. C. Neu in M. I. Page, Ed., *The Chemistry of β-Lactams*, Chapman & Hall, London, 1992, p. 118.

385. G. K. Cook, J. H. McDonald III, W. Alborn Jr., D. B. Boyd, J. A. Eudaly, J. M. Indelicato, R. Johnson, J. S. Kasher, C. E. Pasini, D. A. Preston, and E. C. Y. Wu, *J. Med. Chem.*, **32**, 2442 (1989).

386. R. J. Ternansky, S. E. Draheim, A. J. Pike, W. F. Bells, S. J. West, C. L. Jordan, E. C. Y. Wu, D. A. Preston, W. Alborn Jr., J. S. Kasher, and B. L. Hawkins, *J. Med. Chem.*, **36**, 1971 (1993).

387. R. B. Woodward in J. Elks, Ed., *Recent Advances in the Chemistry of β-Lactam Antibiotics*, Special Publication No. 28, Royal Chemical Society, London, 1977, pp. 167–180.

388. I. Ernst, J. Gosteli, C. W. Greengrass, W. Holick, D. E. Jackman, H. R. Pfaendler, and R. B. Woodward, *J. Am. Chem. Soc.*, **100**, 8214 (1978).

389. K. Claues, D. Grimm, and G. Prossel, *Justus Liebigs Ann. Chem.*, 539 (1974).

390. M. Lang, K. Prasad, W. Holick, J. Gosteli, I. Ernest, and R. B. Woodward, *J. Am. Chem. Soc.*, **101**, 6296 (1979).

391. M. Lang, K. Prasad, J. Gosteli, and R. B. Woodward, *Helv. Chim. Acta*, **63**, 1093 (1980).

392. E. G. Brain (to Beecham Group Ltd.), Br. Pat. 2,042,508A (1980).

393. J. Gosteli, I. Ernest, M. Lang, and R. B. Woodward, Eur. Pat. 0,000,258 (1978).

394. S. W. McCombie, A. K. Ganguly, V. M. Girijavallabhan, P. D. Jeffrey, S. Lin, P. Pinto, and A. T. McPhail, *Tetrahedron Lett.*, **22**, 3489 (1981).

395. H. R. Pfaendler, J. Gosteli, and R. B. Woodward, *J. Am. Chem. Soc.*, **101**, 6306 (1979).

396. F. DiNinno, T. R. Beattie, and B. G. Christensen, *J. Org. Chem.*, **42**, 2960 (1977).

397. V. M. Girijavallabhan, A. K. Ganguly, S. W. McCombie, P. Pinto, and R. Rizvi, *Tetrahedron Lett.*, **22**, 3485 (1981).

398. A. Yoshida, T. Hayashi, N. Takeda, S. Oida, and E. Ohki, *Chem. Pharm. Bull.*, **29**, 2899 (1981).

399. T. Hayashi, A. Yoshida, N. Takeda, S. Oida, S. Suguwara, and E. Ohki, *Chem. Pharm. Bull.*, **29**, 3158 (1981).

400. M. Alpegiani, A. Bedeschi, F. Giudici, E. Perrone, and G. Franceschi, *J. Am. Chem. Soc.*, **107**, 6398 (1985).

401. M. Foglio, C. Battistini, F. Zarini, C. Scarafile, and G. Franceschi, *Heterocycles*, **20**, 1491 (1983).

402. M. Alpegiani, A. Bedeschi, M. Foglio, F. Giudici, and E. Perrone, *Tetrahedron Lett.*, **24**, 1627 (1983).

403. J. Marchand-Brynaert, J. Vekemans, S. Bogdan, M. Cossement, and L. Ghosez in G. I. Gregory, Ed., *Recent Advances in the Chemistry of β-Lactam Antibiotics*, Special Publication No. 38, Royal Chemistry Society, London, 1981, pp. 269–280.

404. C. Battistini, C. Scarafile, M. Foglio, and G. Franceschi, *Tetrahedron Lett.*, **25**, 2395 (1984).

405. A. Yoshida, T. Hayashi, N. Takeda, S. Oida, and E. Ohki, *Chem. Pharm. Bull.*, **31**, 768 (1983).

406. A. Alfonso, F. Hon, J. Weinstein, and A. K. Ganguly, *J. Am. Chem. Soc.*, **104**, 6138 (1982).

407. E. Perrone, M. Alpegiani, A. Bedeschi, F. Giudici, and G. Franceschi, *Tetrahedron Lett.*, **25**, 2399 (1984).

408. A. J. Baker, M. M. Campbell, and M. J. Jenkins, *Tetrahedron Lett.*, **30**, 4359 (1990).

409. M. D. Cooke, K. W. Moore, B. C. Ross, and S. E. Turner, *J. Chem. Soc. Chem. Commun.*, 1005 (1983).

410. G. Franceschi, E. Perrone, M. Alpegiani, A. Bedeschi, C. Della Bruna, and F. Zarini in P. H. Bentley and R. Southgate, Eds., *Recent Advances in the Chemistry of β-Lactam Antibiotics*, Special Publication No. 70, Royal Chemistry Society, London, 1988, pp. 222–246.

411. F. DiNinno, D. A. Muthard, R. W. Ratcliffe, and B. G. Christensen, *Tetrahedron Lett.*, **23**, 3535 (1982).

412. A. J. Baker, M. R. Teall, and G. Johnson, *Tetrahedron Lett.*, **28**, 2283 (1987).

413. D. Phillips and B. T. O'Neill, *Tetrahedron Lett.*, **31**, 3291 (1990).

414. J. Banville, P. Lapointe, B. Belleau, and M. Menard, *Can. J. Chem.*, **66**, 1390 (1980).

415. J. Gosteli, W. Holick, M. Lang, and R. B. Woodward in G. I. Gregory, Ed., *Recent Advances in the Chemistry of β-Lactam Antibiotics*, Special Publication No. 38, Royal Chemistry Society, London, 1981, pp. 359–367.

416. V. M. Girijavallabhan, A. K. Ganguly, S. W. McCombie, P. Pinto, and R. Rizvi, Proceedings of the 21st Interscience Conference on Antimicrobial Agents and Chemotherapy, Chicago, 1981, Abstr. 830.

417. M. Menard and A. Martel (to Bristol Myers), Br. Pat. Appl. 2042515A (Sept. 1980).

418. O. Zak, M. Lang, R. Cozens, E. A. Konopka, H. Mett, P. Schneider, W. Tosch, and R. Scartazzini, *J. Clin. Pharmacol.*, **28**, 128 (1988).

419. I. Phillips, R. Wise, and H. C. Neu, Eds., *J. Antimicrob. Chemother.*, **9** (Suppl. C) (1982).

420. R. Wise and I. Phillips, Eds., *J. Antimicrob. Chemother.*, **15** (Suppl. C) (1985).

421. D. Reeves, D. Speller, R. Spencer, and P. J. Daly, Eds., *J. Antimicrob. Chemother.*, **23** (Suppl. C) (1989).

422. M. Alpegiani, A. Bedeschi, F. Zarini, C. Della Bruna, D. Jabes, E. Perrone, and G. Franceschi, *J. Antibiot.*, **45**, 797 (1992).

423. N. Klesel, D. Isert, M. Lambert, G. Seibert, and E. Schrinner, *J. Antibiot.*, **40**, 1184 (1987).

424. T. Nishino, Y. Maeda, E. Ohtsu, S. Koizuka, T. Nishihara, H. Adachi, K. Okamoto, and M. Ishiguro, *J. Antibiot.*, **42**, 977 (1989).

425. A. Badeschi, G. Visentin, E. Perrone, F. Giuducu, F. Zarini, G. Franceschi, G. Meinardi, P. Castellani, D. Jabes, R. Rossi, and C. Della Bruna, *J. Antibiot.*, **43**, 306 (1990).

426. M. Minamimura, Y. Taniyama, E. Inoue, and S. Mitsuhashi, *Antimicrob. Agents Chemother.*, **37**, 1547 (1993).

427. Source, Pharmaproject, 13 October 2000, Analysts Meeting, March 1983.

428. H. Aoki, H. Sakai, M. Kohsaka, T. Konami, J. Hosoda, Y. Kubochi, E. Iguchi, and H. Imanaka, *J. Antibiot.*, **29**, 492 (1976).

429. A. Imada, K. Kitano, K. Kintaka, M. Muroi, and M. Asai, *Nature*, **289**, 590 (1981).

430. R. B. Sykes, C. M. Cimarusti, D. P. Bonner, K. Bush, D. M. Floyd, N. H. Georgopapadakou, W. H. Koster, W. C. Liu, W. L. Parker, P. A. Principe, M. L. Rathnum, W. A. Slusarchyk, W. H. Trejo, and J. S. Wells, *Nature*, **291**, 489 (1981).

431. R. C. Thomas in G. Lukacs and M. H. Ohno, Eds., *Recent Progress in the Chemical Synthesis of Antibiotics*, Springer-Verlag, Berlin, 1990, pp. 533–564.

432. M. Hashimoto, T. Komori, and T. Kamiya, *J. Antibiot.*, **29**, 890 (1976).

433. M. Hashimoto, T. Komori, and T. Kamiya, *J. Am. Chem. Soc.*, **98**, 3023 (1976).

434. M. Kurita, K. Jomon, T. Komori, N. Miyairi, H. Aoki, S. Kuge, T. Kamiya, and H. Imanaka, *J. Antibiot.*, **29**, 1243 (1976).

435. T. Kamiya, H. Aoki, and Y. Mino in R. B. Morin and M. Gorman, Eds., Chemistry and Biology of β-Lactam Antibiotics, Vol. **2**, Academic Press, New York, 1982, pp. 166–226.

436. L. J. Nisbet, R. J. Mehta, Y. Oh, C. H. Pan, R. G. Phelan, M. J. Polansky, M. C. Shearer, A. J. Giovenella, and S. F. Grappel, *J. Antibiot.*, **38**, 133 (1985).

437. T. Komori, K. Kunigita, K. Nakahara, H. Aoki, and H. Imanaka, *Agric. Biol. Chem.*, **42**, 1439 (1978).

438. T. Kamiya in J. Elks, Ed., *Recent Advances in the Chemistry of β-Lactam Antibiotics*, Special Publication No. 28, Royal Chemical Society, London, 1977, pp. 281–294.

439. K. Shaffner-Sabba, B. W. Muller, R. Scartazzini, and H. Wehrli, *Helv. Chim. Acta*, **63**, 321 (1980).

440. M. Foglio, G. Franceschi, P. Lombardi, C. Scarafile, and F. Arcamone, *J. Chem. Soc. Chem. Commun.*, 1101 (1978).

441. T. Kamiya, M. Hashimoto, O. Nakaguchi, and T. Oku, *Tetrahedron*, **35**, 323 (1979).

442. G. A. Koppel, L. McShane, F. Jose, and R. D. G. Cooper, *J. Am. Chem. Soc.*, **100**, 3933 (1978).

443. M. J. Miller, *Acc. Chem. Res.*, **19**, 49 (1986).

444. P. G. Mattingley and M. J. Miller, *J. Org. Chem.*, **45**, 410 (1980).

445. P. G. Mattingley and M. J. Miller, *J. Org. Chem.*, **46**, 1557 (1981).

446. G. M. Salituro and C. A. Townsend, *J. Am. Chem. Soc.*, **112**, 760 (1990).

447. K. Chiba, M. Mori, and Y. Ban, *Tetrahedron*, **41**, 387 (1985).

448. H. H. Wasserman, D. J. Hlasta, A. W. Tremper, and J. S. Wu, *J. Org. Chem.*, **46**, 2999 (1981).

449. F. H. van der Steen and G. van Koten, *Tetrahedron*, **47**, 7503 (1991).

450. G. I. George, Ed., *BioMed. Chem. Lett.*, **3**, 2337 (1993).

451. M. Nishida, Y. Mine, S. Nonoyama, and H. Koji, *J. Antibiot.*, **30**, 917 (1977).

452. N. Katayama, Y. Nozaki, K. Okonogi, H. Ono, S. Harada, and H. Okazaki, *J. Antibiot.*, **38**, 1117 (1985).

453. T. Hida, S. Tsubotani, N. Katayama, H. Okazaki, and S. Harada, *J. Antibiot.*, **38**, 1128 (1985).

454. W. H. Koster, C. M. Cimarusti, and R. B. Sykes in R. B. Morin and M. Gorman, Eds., Chemistry and Biology of β-Lactam Antibiotics, Vol. **3**, Academic Press, New York, 1982, pp. 339–375.

455. W. L. Parker, W. H. Koster, C. M. Cimarusti, D. M. Floyd, W. Liu, and M. L. Rathnum, *J. Antibiot.*, **35**, 189 (1982).

456. I. Asai, K. Haibara, M. Muroi, K. Kintaka, and T. Kishi, *J. Antibiot.*, **34**, 621 (1981).

457. K. Kintaka, K. Haibara, M. Asai, and A. Imada, *J. Antibiot.*, **34**, 1081 (1981).

458. W. L. Parker and M. L. Rathnum, *J. Antibiot.*, **35**, 300 (1982).

459. P. D. Singh, J. H. Johnson, P. C. Ward, J. S. Wells, W. H. Trejo, and R. B. Sykes, *J. Antibiot.*, **36**, 1246 (1983).

460. R. Cooper, K. Bush, P. A. Principe, W. H. Trejo, J. S. Wells, and R. B. Sykes, *J. Antibiot.*, **36**, 1252 (1983).

461. T. Kato, H. Hinoo, Y. Terui, J. Nishikawa, Y. Nakagawa, Y. Ikenishi, and J. Shoji, *J. Antibiot.*, **40**, 139 (1987).

462. S. J. Box, A. G. Brown, M. L. Gilpin, M. N. Gwynn, and S. R. Spear, *J. Antibiot.*, **41**, 7 (1988).

463. K. Kamiya, M. Takamoto, Y. Wada, and M. Asai, *Acta Crystallogr.*, **B37**, 1626 (1981).

464. C. M. Cimarusti, H. E. Applegate, H. W. Chang, D. M. Floyd, W. M. Koster, W. A. Slusarchyk, and M. G. Young, *J. Org. Chem.*, **47**, 179 (1982).

465. D. M. Floyd, A. W. Fritz, and C. M. Cimarusti, *J. Org. Chem.*, **47**, 176 (1982).

466. K. R. Linder, D. P. Bonner, and W. H. Koster, Kirk-Othmer Encyclopedia of Chemical Technology: The Monobactams, 4th ed., Vol. **3**, John Wiley & Sons, New York, 1992, pp. 107–129.

467. N. H. Georgopapadakou, S. A. Smith, and R. B. Sykes, *Antimicrob. Agents Chemother.*, **21**, 950 (1982).

468. D. P. Bonner and R. B. Sykes, *J. Antimicrob. Chemother.*, **14**, 313 (1985).

469. Symposium, *Rev. Infect. Dis.*, **7** (Suppl. 4) (1985).

470. R. B. Sykes and D. P. Bonner, *Am. J. Med.*, **78** (Suppl. A), 2 (1985).

471. P. Angehrn, *Chemotherapy*, **31**, 440 (1985).

472. S. R. Woulfe and M. J. Miller, *J. Med. Chem.*, **28**, 1447 (1985).

473. J. B. Bartolone, G. J. Hite, J. A. Kelly, and J. R. Knox in A. G. Brown and S. M. Roberts, Eds., *Recent Advances in the Chemistry of β-Lactam Antibiotics*, Special Publication No. 52, Royal Chemistry Society, London, 1985, pp. 327–337.

474. H. Nikaido and E. Y. Rosenberg, *J. Bacteriol.*, **172**, 1361 (1990).

475. J. F. Chantot, M. Klitch, G. Teutsch, A. Bryskier, P. Collette, A. Markus, and G. Seibert, *Antimicrob. Agents Chemother.*, **36**, 1756 (1992).

476. K. Obi, Y. Ito, and S. Terashima, *Chem. Pharm. Bull.*, **38**, 917 (1990).

477. S. J. Brickner, J. J. Gaikema, L. J. Greenfield, G. E. Zurenko, and P. R. Manninen, *BioMed. Chem. Lett.*, **3**, 2241 (1993).

478. J. S. Kahan, F. M. Kahan, R. T. Goegelman, S. A. Currie, M. Jackson, E. O. Stapley, T. W. Miller, A. K. Miller, D. Hendlin, S. Mochales, S. Hernandez, H. B. Woodruff, and J. Birnbaum, *J. Antibiot.*, **32**, 1 (1979).

479. D. Butterworth, M. Cole, G. Hanscomb, and G. N. Rolinson, *J. Antibiot.*, **32**, 287 (1979).

480. J. D. Hood, S. J. Box, and M. S. Verrall, *J. Antibiot.*, **32**, 295 (1979).

481. W. L. Parker, M. L. Rathnum, J. S. Wells, W. H. Trejo, P. A. Principe, and R. B. Sykes, *J. Antibiot.*, **35**, 653 (1982).

482. R. Southgate and N. F. Osborne, Kirk-Othmer Encyclopedia of Chemical Technology: Carbapenems and Penems, 4th ed., Vol. **3**, John Wiley & Sons, New York, 1992, pp. 1–27.

483. K. Okamura, S. Hirata, Y. Okumura, Y. Fukagawa, Y. Shimauchi, K. Kouno, and T. Ishikura, *J. Antibiot.*, **31**, 480 (1978).

484. M. Nakayama, A. Iwasaki, S. Kimura, T. Mizoguchi, S. Tanabe, A. Murakami, I. Watanabe, M. Okuchi, H. Itoh, Y. Saino, F. Kobayashi, and T. Mori, *J. Antibiot.*, **33**, 1388 (1980).

485. K. Tanaka, J. Shoji, Y. Terui, N. Tsuji, E. Kondo, M. Miyama, Y. Kawamura, T. Hattori, K. Matsumoto, and T. Yoshida, *J. Antibiot.*, **34**, 909 (1981).

486. A. G. Brown, D. F. Corbett, A. J. Eglington, and T. T. Howarth, *J. Chem. Soc. Chem. Commun.*, 523 (1977).

487. N. Tsuji, K. Nagashima, M. Kobayashi, Y. Terui, K. Matsumoto, and E. Kondo, *J. Antibiot.*, **35**, 536 (1982).

488. D. Butterworth, J. D. Hood, and M. S. Verrall in A. Mizrahi, Ed., *Advances in Biotechnological Processes*, Vol. **1**, A. R. Liss, New York, 1982, pp. 252–282.

489. G. Albers-Schonberg, B. H. Arison, T. O. Hensens, K. Hirshfield, R. W. Ratcliffe, E. Walton, L. J. Rushwinkle, R. B. Morin, and B. G. Christensen, *J. Am. Chem. Soc.*, **100**, 6491 (1978).

490. R. W. Ratcliffe and G. Albers-Schonberg in R. B. Morin and M. Gorman, Eds., Chemistry and Biology of β-Lactam Antibiotics, Vol. **2**, Academic Press, New York, 1982, pp. 227–313.

491. D. F. Corbett, A. J. Eglington, and T. T. Howarth, *J. Chem. Soc. Chem. Commun.*, 953 (1977).

492. S. J. Box, J. D. Hood, and S. R. Spear, *J. Antibiot.*, **32**, 1239 (1979).

493. A. G. Brown, *Kirk-Othmer Encyclopedia of Chemical Technology*, 3rd ed. (Suppl.), John Wiley & Sons, New York, 1984, pp. 83–131.

494. A. G. Brown, D. F. Corbett, A. J. Eglington, and T. T. Howarth, *J. Antibiot.*, **32**, 961 (1979).

495. A. G. Brown, D. F. Corbett, A. J. Eglington, and T. T. Howarth, *Tetrahedron*, **39**, 2551 (1983).

496. K. Yamamoto, T. Yoshioda, Y. Kato, N. Shibamoto, K. Okamura, Y. Shimanuchi, and T. Ishikura, *J. Antibiot.*, **33**, 796 (1980).

497. Y. Nakayama, S. Kimura, S. Tanabe, T. Mizoguchi, I. Watanabe, T. Mori, K. Mori, K. Miyahara, and T. Kawasaki, *J. Antibiot.*, **34**, 818 (1981).

498. N. Tsuji, K. Nagashima, M. Kobayashi, J. Shoji, T. Kato, Y. Terui, H. Nakai, and M. Shiro, *J. Antibiot.*, **35**, 24 (1982).

499. W. J. Leanza, K. J. Wildonger, J. Hannah, D. H. Shih, R. W. Ratcliffe, L. Barash, E. Walton, R. A. Firestone, G. F. Patel, F. M. Kahan, J. S. Kahan, and B. G. Christensen in G. I. Gregory, Ed., *Recent Advances in the Chemistry of β-Lactam Antibiotics*, Special Publication No. 38, Royal Chemistry Society, London, 1981, pp. 240–245.

500. W. J. Leanza, K. J. Wildonger, T. W. Miller, and B. G. Christensen, *J. Med. Chem.*, **22**, 1435 (1979).

501. D. W. Graham, W. T. Ashton, L. Barash, J. E. Brown, R. D. Brown, L. F. Canning, A. Chen, J. P. Springer, and E. F. Rogers, *J. Med. Chem.*, **30**, 1074 (1987).

502. A. G. Brown, D. F. Corbett, A. J. Eglington, and T. T. Howarth in G. I. Gregory, Ed., *Recent Advances in the Chemistry of β-Lactam Antibiotics*, Special Publication No. 38, Royal Chemistry Society, London, 1981, pp. 255–268.

503. D. F. Corbett, *J. Chem. Soc. Chem. Commun.*, 803 (1981).

504. D. F. Corbett, S. Coulton, and R. Southgate, *J. Chem. Soc. Perkin Trans I*, 3011 (1982).

505. K. Yamamoto, T. Yoshioka, Y. Kato, K. Ishiki, M. Nishino, F. Nakamura, Y. Shimauchi, and T. Ishikura, *Tetrahedron Lett.*, **23**, 897 (1982).

506. S. Coulton and E. Hunt in G. P. Ellis and D. K. Luscombe, Eds., Progress in Medicinal Chemistry, Vol. **33**, Elsevier Science, Amsterdam/New York, 1996.

507. T. Kametani, K. Fukumoto, and M. Ihara, *Heterocycles*, **17**, 463 (1982).

508. W. Durkeheimer, J. Blumbach, R. Lattrell, and K. H. Scheunemann, *Angew. Chem. Int. Ed. Engl.*, **24**, 180 (1985).

509. A. H. Berks, *Tetrahedron*, **52**, 331 (1996).

510. A. Sasaki and M. Sunagawa, *Chem. Heterocycl. Compd.*, **34**, 1249 (1999).

511. M. Sunagawa and A. Sasaki, *Heterocycles*, **54**, 497 (2001).

512. T. Nagahara and T. Kametani, *Heterocycles*, **25**, 729 (1987).

513. C. Palomo in G. Lukacs and M. Ohno, Eds., *Recent Progress in the Chemical Synthesis of Antibiotics*, Springer-Verlag, Berlin, 1990, pp. 562–612.

514. D. B. R. Johnston, S. M. Schmitt, F. A. Bouffard, and B. G. Christensen, *J. Am. Chem. Soc.*, **100**, 313 (1978).

515. R. W. Ratcliffe, T. N. Salzmann, and B. G. Christensen, *Tetrahedron Lett.*, **21**, 31 (1980).

516. D. G. Melillo, I. Shinkai, T. Liu, K. M. Ryan, and M. Sletzinger, *Tetrahedron Lett.*, **21**, 2783 (1980).

517. R. Southgate, *Contemp. Org. Synth.*, **1**, 417 (1994).

518. L. D. Cama and B. G. Christensen, *J. Am. Chem. Soc.*, **100**, 8006 (1978).

519. A. J. G. Baxter, K. H. Dickinson, P. M. Roberts, T. C. Smale, and R. Southgate, *J. Chem. Soc. Chem. Commun.*, 236 (1979).

520. J. H. Bateson, A. J. G. Baxter, P. M. Roberts, T. C. Smale, and R. Southgate, *J. Chem. Soc. Perkin Trans I*, 3242 (1981).

521. H. R. Pfaendler, J. Gosteli, R. B. Woodward, and G. Rihs, *J. Am. Chem. Soc.*, **103**, 4526 (1981).

522. J. H. Bateson, A. J. G. Baxter, K. H. Dickinson, R. I. Hickling, R. J. Ponsford, P. M. Roberts, T. C. Smale, and R. Southgate in G. D. Gregory, Ed., *Recent Advances in the Chemistry of β-Lactam Antibiotics*, Special Publication No. 38, Royal Chemistry Society, London, 1981, pp. 291–313.

523. S. M. Schmitt, D. B. R. Johnston, and B. G. Christensen, *J. Org. Chem.*, **45**, 1142 (1980).

524. T. N. Salzmann, R. W. Ratcliffe, B. G. Christensen, and F. A. Bouffard, *J. Am. Chem. Soc.*, **102**, 6161 (1980).

525. I. Shinkai, R. A. Reamer, F. W. Hartner, T. Liu, and M. Sletzinger, *Tetrahedron Lett.*, **23**, 4903 (1982).

526. A. J. G. Baxter, P. Davis, R. J. Ponsford, and R. Southgate, *Tetrahedron Lett.*, **21**, 5071 (1980).

527. J. Hannah, C. R. Johnson, A. F. Wagner, and E. Walton, *J. Med. Chem.*, **25**, 457 (1982).

528. L. D. Cama, K. J. Wildonger, R. N. Guthikonda, R. W. Ratcliffe, and B. G. Christensen, *Tetrahedron*, **39**, 2531 (1983).

529. D. G. Melillo, R. J. Cuetovich, K. M. Ryan, and M. Sletzinger, *J. Org. Chem.*, **51**, 1498 (1986).

530. S. Karady, J. S. Amato, R. A. Reamer, and L. M. Weinstock, *J. Am. Chem. Soc.*, **103**, 6745 (1981).

531. P. J. Reader and E. J. J. Grabowski, *Tetrahedron Lett.*, **23**, 2293 (1982).

532. G. I. George, J. Kant, and H. S. Gill, *J. Am. Chem. Soc.*, **109**, 1129 (1987).

533. Y. Ito, Y. Kobayashi, T. Kawabata, M. Takase, and S. Terashima, *Tetrahedron*, **45**, 5767 (1989).

534. H. Maruyama and T. Hiraoka, *J. Org. Chem.*, **51**, 399 (1986).

535. G. J. Quallich, J. Bordner, M. L. Elliot, P. Morrisey, R. A. Volkmann, and M. M. Wroblesoska-Adam, *J. Org. Chem.*, **55**, 367 (1990).

536. M. Miyashita, M. N. Chida, and A. Yoshikoshi, *J. Chem. Soc. Chem. Commun.*, 1454 (1982).

537. A. Knierzinger and A. Vasella, *J. Chem. Soc. Chem. Commun.*, 9 (1985).

538. H. Maruyama, M. Shiozaki, and T. Hiraoka, *Bull. Chim. Soc.*, **58**, 3264 (1985).

539. M. Shiozaki, N. Ishida, H. Maruyama, and T. Hiraoka, *Tetrahedron*, **39**, 2399 (1983).

540. T. Kamata, T. Nagahara, and T. Honda, *J. Org. Chem.*, **50**, 2327 (1985).

541. S. T. Hodgson, D. M. Hollinshead, and S. V. Ley, *Tetrahedron*, **41**, 5871 (1985).

542. G. Pattenden and S. J. Reynolds, *J. Chem. Soc. Perkin Trans. I*, 379 (1994).

543. A. I. Meyers, T. J. Sowin, S. Sholz, and Y. Ueda, *Tetrahedron Lett.*, **28**, 5103 (1987).

544. S. Hanessian, D. Desilets, and Y. L. Bennani, *J. Org. Chem.*, **55**, 3098 (1990).

545. R. J. Ponsford and R. Southgate, *J. Chem. Soc. Chem. Commun.*, 1085 (1990).

546. J. H. Bateson, P. M. Roberts, T. C. Smale, and R. Southgate, *J. Chem. Soc. Perkin Trans I*, 1541 (1990).

547. J. H. Bateson, R. I. Hickling, T. C. Smale, and R. Southgate, *J. Chem. Soc. Perkin Trans. I*, 1793 (1990).

548. M. Shibasaki, Y. Ishida, and N. Okabe, *Tetrahedron Lett.*, **26**, 2217 (1985).

549. D. H. Shih, F. Baker, L. Cama, and B. G. Christensen, *Heterocycles*, **21**, 29, 1984.

550. R. Deziel and D. Faureau, *Tetrahedron Lett.*, **27**, 5687 (1986).

551. L. M. Fuentes, I. Shinkai, and T. Salzmann, *J. Am. Chem. Soc.*, **108**, 4675 (1986).

552. R. Deziel and M. Endo, *Tetrahedron Lett.*, **29**, 61 (1988).

553. F. Shirai and H. Nakai, *Chem. Lett.*, 445 (1989).

554. M. Ihara, M. Takahashi, M. Fukumoto, and T. Kametani, *J. Chem. Soc. Perkin Trans I*, 2215 (1989).

555. M. Kitumuro, K. Nagai, Y. Hsiao, and R. Noyori, *Tetrahedron Lett.*, **31**, 5686 (1990).

556. W. B. Choi, H. R. O. Churchill, J. E. Lynch, A. S. Thompson, C. R. Humphrey, R. P. Volante, R. P. Raider, and I. Shinkai, *Tetrahedron Lett.*, **35**, 2275 (1994).

557. H. Kaga, S. Kobayashi, and M. Ohno, *Tetrahedron Lett.*, **30**, 113 (1989).

558. T. Shibata and Y. Sugimura, *J. Antibiot.*, **42**, 374 (1989).

559. C. U. Kim, B. Y. Luh, P. F. Misco, and M. J. M. Hitchcock, *J. Med. Chem.*, **32**, 601 (1989).

560. R. N. Guthikonda, L. D. Cama, M. L. Quesada, M. F. Woods, T. N. Salzmann, and B. G. Christensen, *J. Med. Chem.*, **30**, 871 (1987).

561. S. M. Schmitt, T. N. Salzmann, D. H. Shih, and B. G. Christensen, *J. Antibiot.*, **41**, 780 (1988).

562. T. N. Salzmann, F. P. DiNinno, M. L. Greenlee, R. N. Guthikonda, M. L. Quesada, S. M. Schmitt, J. J. Hermann, and M. F. Woods in P. H. Bentley and R. Southgate, Eds., *Fourth International Symposium on Recent Advances in the Chemistry of β-Lactam Antibiotics*, Royal Chemistry Society, London, 1989, p. 171.

563. J. Haruta, K. Nishi, K. Kikuchi, S. Matsuda, Y. Tamura, and Y. Kita, *Chem. Pharm. Bull.*, **37**, 2338 (1989).

564. A. Andrus, F. Baker, F. A. Bouffard, L. D. Cama, B. G. Christensen, R. N. Guthikonda, J. V. Heck, D. B. R. Johnston, W. J. Leanza, R. W. Ratcliffe, T. N. Salzmann, S. M. Schmitt, D. H. Shih, N. V. Shah, K. J. Vildonger, and R. R. Wilkening in A. G. Brown and S. M. Roberts, Eds., *Recent Advances in the Chemistry of β-Lactam Antibiotics*, Special Publication No. 52, Royal Chemistry Society, London, 1985, pp. 86–99.

565. C. U. Kim, P. F. Misco, and B. Y. Luh, *Heterocycles*, **26**, 1193 (1987).

566. T. Shibata, K. Iino, T. Tanaka, T. Hashimoto, Y. Kameyama, and Y. Sugimura, *Tetrahedron Lett.*, **26**, 4739 (1985).

567. E. Goetschi, P. Anghern, and R. Then, Proceedings of the 27th Interscience Conference on Antimicrobial Agents and Chemotherapy, New York, 1987, Abstr. 759.

568. Y. Nagao, T. Abe, H. Shimizu, T. Kumagai, and Y. Inoue, *J. Chem. Soc. Chem. Commun.*, 821 (1989).

569. R. L. Rosati, L. V. Kapili, P. Morrissey, and J. A. Retsema, *J. Med. Chem.*, **33**, 291 (1990).

570. N. V. Shah and L. D. Cama, *Heterocycles*, **25**, 221 (1987).

571. M. Menard, J. Banville, A. Martel, J. Desiderio, J. Fung-Tomc, and R. A. Partyka in P. H. Bentley and R. J. Ponsford, Eds., *Recent Advances in the Chemistry of Anti-Infective Agents*, Special Publication No. 119, Royal Chemistry Society, London, 1993, pp. 3–20.

572. A. Martel, J. Banville, C. Bachand, G. Bouthillier, J. Corbeil, J. Fung-Tomc, P. Lapointe, H. Mastalerz, V. S. Rao, R. Remillard, B. Turmel, M. Menard, R. E. Kessler, and R. A. Partyka, Proceedings of the 31st Interscience Conference on Antimicrobial Agents and Chemotherapy, Chicago, 1991, Abstr. 827.

573. J. Banville, C. Bachand, J. Corbeil, J. Desiderio, J. Fung-Tomc, P. Lapointe, A. Martel, V. S. Rao, R. Remillard, M. Menard, R. E. Kessler, and R. A. Partyka, Proceedings of the 31st Interscience Conference on Antimicrobial Agents and Chemotherapy, Chicago, 1991, Abstr. 828.

574. C. Bachand, J. Banville, G. Bouthillier, J. Corbeil, J. P. Daris, J. Desiderio, J. Fung-Tomc, P. Lapointe, A. Martel, H. Mastalerz, R. Remillard, Y. Tsai, M. Menard, R. E. Kessler, and R. A. Partyka, Proceedings of the 31st Interscience Conference on Antimicrobial Agents and Chemotherapy, Chicago, 1991, Abstr. 829.

575. V. S. Rao, C. Bachand, J. Banville, G. Bouthillier, J. Desiderio, J. Fung-Tomc, P. Lapointe, A. Martel, H. Mastalerz, R. Remillard, A. Michel, M. Menard, R. E. Kessler, and R. A. Partyka, Proceedings of the 31st Interscience Conference on Antimicrobial Agents and Chemotherapy, Chicago, 1991, Abstr. 830.

576. R. E. Kessler, J. Fung-Tomc, B. Minassian, B. Kolek, E. Huczko, and D. P. Bonner, Proceedings of the 34th Interscience Conference on Antimicrobial Agents and Chemotherapy, Orlando, FL, 1994, Abstr. F42.

577. J. Fung-Tomc, E. Huczko, J. Banville, M. Menard, B. Kolek, E. Gradelski, R. E. Kessler, and D. P. Bonner, *Antimicrob. Agents Chemother.*, **39**, 394 (1995).

578. V. S. Rao, R. Remillard, and M. Menard, *Heteroat. Chem.*, **3**, 25 (1992).

579. J. Banville, Eur. Pat. Appl. EP372582 A2 (1990).

580. E. H. Ruediger and C. Solomon, *J. Org. Chem.*, **56**, 3183 (1991).

581. H. Mastalerz and M. Menard, *Heterocycles*, **32**, 93 (1991).

582. G. Bouthillier, H. Mastalerz, M. Menard, and J. Fung-Tomc, *J. Antibiot.*, **45**, 240 (1992).

583. H. Mastalerz, M. Menard, E. H. Ruediger, and J. Fung-Tomc, *J. Med. Chem.*, **35**, 25 (1992).

584. V. S. Rao, R. Remillard, and M. Menard, *Heteroat. Chem.*, **3**, 25 (1992).

585. Y. Nagao, T. Abe, H. Shimizu, T. Kumagai, and Y. Inoue, *Heterocycles*, **33**, 523 (1992).

586. M. Sunagawa, H. Matsumura, T. Inoue, M. Fukasawa, and M. Kato, Proceedings of the 27th Interscience Conference on Antimicrobial Agents and Chemotherapy, New York, 1987, Abstr. 752.

587. M. Sunagawa, H. Matsumura, T. Inoue, M. Fukasawa, and M. Kato, *J. Antibiot.*, **43**, 519 (1990).

588. M. Sunagawa, H. Matsumura, T. Inoue, M. Fukasawa, and M. Kato, *J. Antibiot.*, **44**, 459 (1991).

589. M. Sunagawa, H. Matsumura, T. Inoue, H. Yamaga, and M. Fukasawa, *J. Antibiot.*, **45**, 971 (1992).

590. M. Sunagawa, H. Matsumura, and M. Fukasawa, *J. Antibiot.*, **45**, 1983 (1992).

591. M. Sunagawa and A. Sasaki, *Heterocycles*, **54**, 497 (2001).

592. C. H. Oh, S. Y. Hong, and J. H. Cho, *Korean J. Med. Chem.*, **2**, 7 (1992).

593. T. Nishi, M. Takemura, T. Saito, Y. Ishida, K. Sugita, M. Houmura, H. Ishi, K. Sato, I. Hayakawa, and T. Hayano, Proceedings of the 32nd Interscience Conference on Antimicrobial Agents and Chemotherapy, Orlando, FL, 1992, Abstr. 131.

594. H. W. Lee, E. N. Kim, K. K. Kim, H. J. Son, J. K. Kim, C. R. Lee, and J. W. Kim, *Korean J. Med. Chem.*, **4**, 101 (1994).

595. M. Sunagawa, A. Sasaki, H. Yamaga, H. Shinagawa, Y. Sumita, and H. Nouda, *J. Antibiot.*, **47**, 1337 (1994).

596. C. H. Oh and J. H. Cho, *J. Antibiot.*, **47**, 126 (1994).

597. C. J. Gill, J. J. Jackson, L. S. Gerckens, B. A. Pelak, R. K. Thompson, J. G. Sundelof, H. Kropp, and H. Rosen, *Antimicrob. Agents Chemother.*, **42**, 1996 (1998).

598. J. G. Sundelof, R. Hajdu, C. J. Gill, R. Thompson, H. Rosen, and H. Kropp, *Antimicrob. Agents Chemother.*, **41**, 1743 (1997).

599. I. Odenholt, E. Lowdin, and O. Cars, *Antimicrob. Agents Chemother.*, **42**, 2365 (1998).

600. M. Sunagawa, A. Sasaki, H. Yamaga, H. Shinagawa, M. Fukasawa, and Y. Sumita, *J. Antibiot.*, **47**, 1354 (1994).

601. A. Shimizu, H. Sugimoto, S. Sakuraba, H. Imamura, H. Sato, N. Ohtake, R. Ushijima, S. Nakagawa, C. Suzuki, T. Hashizume, and H. Morishima, Proceedings of the 38th Interscience Conference on Antimicrobial Agents and Chemotherapy, San Diego, 1998, Abstr. F-52.

602. Y. Adachi, R. Nagano, K. Shibata, Y. Kato, T. Hashizume, and H. Morishima, Proceedings of the 38th Interscience Conference on Antimicrobial Agents and Chemotherapy, San Diego, 1998, Abstr. F-54.

603. R. Nagano, Y. Adachi, T. Hashizume, and H. Morishima, Proceedings of the 38th Interscience Conference on Antimicrobial Agents and Chemotherapy, San Diego, 1998, Abstr. F-55.

604. R. Nagano, K. Shibata, Y. Adachi, H. Imamura, and T. Hashizume, *Antimicrob. Agents Chemother.*, **44**, 489 (2000).

605. A. Perboni, B. Tamburini, T. Rossi, D. Donati, G. Tarzia, and G. Gaviraghi in P. H. Bentley and R. J. Ponsford, Eds., *Recent Advances in the Chemistry of Anti-Infective Agents*, Special Publication No. 119, Royal Chemistry Society, London, 1993, pp. 21–35.

606. B. Tamburini, A. Perboni, T. Rossi, D. Donati, D. Andreotti, G. Gaviraghi, S. Biondi, and C. Bismara, Eur. Pat. Appl. EP0416953 (1991).

607. E. Di Modugno, I. Erbetti, L. Ferrari, G. L. Galassi, S. M. Hammond, and L. Xerri, *Antimicrob. Agents Chemother.*, **38**, 2362 (1994).

608. R. Wise, J. M. Andrews, and N. Brenwald, *Antimicrob. Agents Chemother.*, **40**, 1248 (1996).

609. J. Ngo and J. Castaner, *Drugs Future*, **21**, 1238 (1996).

610. S. Tamura, S. Miyazaki, K. Takeda, A. I. Y. Ohno, T. Matsumoto, N. Furuya, and K. Yamaguchi, *Antimicrob. Agents Chemother.*, **42**, 1858 (1998).

611. G. Gaviraghi, *Eur. J. Med. Chem.*, **30** (Suppl.), 467S (1995).

612. T. Rossi, S. Biondi, S. Contini, R. J. Thomas, and C. Marchioro, *J. Am. Chem. Soc.*, **117**, 9604 (1995).

613. A. Perboni, Eur. Pat. Appl. EP0502488 A2 (1992).

614. S. Biondi, G. Gaviraghi, and T. Rossi, *Bioorg. Med. Chem. Lett.*, **6**, 525 (1996).

615. C. Ghiron, E. Piga, T. Rossi, and R. J. Thomas, *Tetrahedron Lett.*, **37**, 3891 (1996).

616. G. Kennedy, T. Rossi, and B. Tamburini, *Tetrahedron Lett.*, **37**, 7441 (1996).

617. T. Matsumoto, T. Murayama, S. Mitsuhashi, and T. Miura, *Tetrahedron Lett.*, **40**, 5043 (1999).

618. P. Stead, H. Marley, M. Mahmoudian, G. Webb, D. Noble, I. P. Yam, E. Piga, T. Rossi, S. Roberts, and M. Dawson, *Tetrahedron* Asymmetry, **7**, 2247 (1996).

619. C. Fuganti, P. Grasselli, M. Mendozza, S. Servi, and G. Zucchi, *Tetrahedron*, **53**, 2617 (1997).

620. S. Hanessian and M. J. Rozema, *J. Am. Chem. Soc.*, **118**, 9884 (1996).

621. S. Gehanne, E. Piga, D. Andreotti, S. Biondi, and D. A. Pizzi, *Bioorg. Med. Chem. Lett.*, **6**, 2791 (1996).

622. D. Andreotti, S. Biondi, and D. Donati, *Chem. Heterocycl. Compd. (NY)*, **34**, 1324 (1999).

623. S. Biondi, A. Pecunioso, F. Busi, S. A. Contini, D. Donati, M. Maffeis, D. A. Pizzi, L. Rossi, T. Rossi, and F. M. Sabbatini, *Tetrahedron*, **56**, 5649 (2000).

624. E. Di Modugno, R. Broggio, I. Erbetti, and J. Lowther, *Antimicrob. Agents Chemother.*, **41**, 2742 (1997).

625. A. Perboni, D. Donati, and G. Tarzia, Eur. Pat. 502468 A1 (1992).

626. T. Wollmann, U. Gerlach, R. Horlein, N. Krass, R. Lattrell, M. Limbert, and A. Markus in P. H. Bentley and R. J. Ponsford, Eds., *Recent Advances in the Chemistry of Anti-Infective Agents*, Special Publication No. 119, Royal Chemistry Society, London, 1993, pp. 50–66.

627. G. Shmidt, W. Schrock, and R. Endermann, *Bioorg. Med. Chem. Lett.*, **3**, 2193 (1993).

628. T. Rossi, D. Andreotti, G. Tedesco, L. Tarsi, E. Ratti, A. Feriani, D. A. Pizzi, G. Gaviraghi, S. Biondi, and G. Finizia, Int. Pat. WO-09821210 (1998).

629. O. Kanno and I. Kawamoto, *Tetrahedron*, **56**, 5639 (2000).

630. D. Andreotti, S. Biondi, D. Donati, S. Lociuro, and G. Pain, *Can J. Chem.*, **78**, 772 (2000).

631. D. Andreotti and S. Biondi, *Curr. Opin. Anti-Infect. Invest. Drugs*, **2**, 133 (2000).

632. O. Kanno, Y. Shimoji, S. Ohya, and I. Kawamoto, *J Antibiot.*, **53**, 404 (2000).

633. M. Mori and O. Kanno, *Annu. Rep. Sankyo Res. Lab.*, **52**, 15 (2000).

634. H. C. Neu, *Am. J. Med.*, **78** (Suppl. 6A), 33 (1985).

635. M. J. Basker, R. J. Boon, and P. A. Hunter, *J. Antibiot.*, **33**, 378 (1980).

636. M. Sakamoto, H. Iguchi, K. Okumura, S. Hori, Y. Fukagawa, and T. Ishikura, *J. Antibiot.*, **32**, 272 (1979).

637. Y. Kimura, K. Motokawa, H. Nagata, Y. Kameda, S. Matsuura, M. Madama, and T. Yoshida, *J. Antibiot.*, **35**, 32 (1982).

638. M. J. Basker, J. H. Bateson, A. J. G. Baxter, R. J. Ponsford, P. M. Roberts, R. Southgate, T. C. Smale, and J. Smith, *J. Antibiot.*, **34**, 1224 (1981).

639. B. G. Spratt, V. Jobanputra, and W. Zimmerman, *Antimicrob. Agents Chemother.*, **12**, 406 (1977).

640. F. M. Kahan, H. Kropp, J. G. Sundelof, and J. Birnbaum, *J. Antimicrob. Chemother.*, **12** (Suppl. D), 1 (1983).

641. H. C. Neu and P. Labthavibul, *Antimicrob. Agents Chemother.*, **21**, 180 (1982).

642. S. R. Norrby, B. Bjornegard, F. Ferber, and K. H. Jones, *J. Antimicrob. Chemother.*, **12** (Suppl. D), 109 (1983).

643. H. Kropp, J. G. Sundelof, R. Hajdu, and F. M. Kahan, *Antimicrob. Agents Chemother.*, **22**, 62 (1982).

644. S. R. Norrby, K. Alestig, B. Bjornegard, L. A. Burman, F. Ferber, J. L. Huber, K. H. Jones, F. M. Kahan, J. S. Kahan, H. Kropp, M. A. P. Meisinger, and J. G. Sundelof, *Antimicrob. Agents Chemother.*, **23**, 23 (1983).

645. J. Birnbaum, F. M. Kahan, H. Kropp, and J. S. MacDonald, *Am. J. Med.*, **78** (Suppl. 6A), 3 (1985).

646. A. Kurihara, H. Naganuma, M. Hisaoka, H. Tokiwa, and Y. Kawaraha, *Antimicrob. Agents Chemother.*, **36**, 1810 (1992).

647. F. DiNinno, D. A. Muthard, and T. N. Salzmann, *BioMed. Chem. Lett.*, **3**, 2187 (1993).

648. P. Davey, A. Davies, D. Livermore, D. Speller, and P. J. Daly, Eds., *J. Antimicrob. Chemother.*, **24** (Suppl. A), 1–311 (1989).

649. S. Biondi in P. H. Bentley and R. J. Ponsford, Eds., *Recent Advances in the Chemistry of Anti-Infective Agents*, Special Publication No. 119, Royal Chemistry Society, London, 1993, pp. 86–100.

650. N. Datta and P. Kontamichalou, *Nature*, **208**, 239 (1965).

651. M. H. Richmond and R. B. Sykes, *Adv. Microb. Physiol.*, **9**, 31–88 (1973).

652. R. B. Sykes and M. J. Matthew, *J. Antimicrob. Chemother.*, **2**, 115–157 (1976).

653. K. Bush, *Antimicrob. Agents Chemother.*, **33**, 259–263 (1989).

654. K. Bush, *Antimicrob. Agents Chemother.*, **33**, 264–270 (1989).

655. K. Bush, *Antimicrob. Agents Chemother.*, **33**, 271–276 (1989).

656. K. Bush, G. A. Jacoby, and A. A. Medeiros, *Antimicrob. Agents Chemother.*, **39**, 1211–1233 (1995).

657. R. P. Ambler, *Philos. Trans. R. Soc. London*, **289**, 321 (1980).

658. D. Payne, *J. Med. Microbiol.*, **39**, 93 (1993).

659. W. Cullman and W. Dick, *Chemotherapy*, **35**, 43 (1989).

660. M. Cole, *Drugs Future*, **6**, 697 (1981).

661. S. J. Cartwright and S. G. Waley, *Med. Res. Rev.*, **3**, 341 (1983).

662. T. T. Howarth, A. G. Brown, and J. King, *J. Chem. Soc. Chem. Commun.*, 266 (1976).

663. A. G. Brown, D. Butterworth, M. Cole, G. Hanscomb, J. D. Hood, and C. Reading, *J. Antibiot.*, **29**, 668 (1976).

664. C. Reading and M. Cole, *Antimicrob. Agents Chemother.*, **11**, 852 (1977).

665. K. Coleman, D. R. J. Griffin, J. W. J. Page, and P. A. Upshon, *Antimicrob. Agents Chemother.*, **33**, 1580 (1989).

666. P. A. Hunter, K. Coleman, J. Fisher, and D. Taylor, *J. Antimicrob. Chemother.*, **6**, 455 (1980).

667. J. S. Bennett, G. Brooks, J. P. Broom, S. H. Calvert, K. Coleman, and I. Francois, *J. Antibiot.*, **44**, 969 (1991).

668. D. C. E. Speller, L. O. White, P. J. Wilkinson, and P. J. Daly, Eds., *J. Antimicrob. Chemother.*, **24** (Suppl. B) (1989).

669. H. C. Neu, A. P. R. Wilson, and R. N. Gruneberg, *J. Chemother.*, **5**, 67 (1993).

670. G. N. Rollinson, *J. Chemother.*, **6**, 283 (1994).

671. C. Reading and P. Hepburn, *Biochem. J.*, **179**, 67 (1979).

672. C. Reading and T. Farmer, *Biochem. J.*, **199**, 779 (1981).

673. J. R. Knowles, *Acc. Chem. Res.*, **18**, 97 (1985).

674. U. Imtiaz, E. Billings, J. R. Knox, E. K. Manavathu, S. A. Lerner, and S. Mobashery, *J. Am. Chem. Soc.*, **115**, 4435 (1993).

675. N. H. Nicholson, K. H. Baggaley, R. Cassels, M. Davison, S. W. Elson, M. Fulston, J. W. Tyler, and S. R. Woroniecki, *J. Chem. Soc. Chem. Commun.*, 1281 (1994).

676. G. Brooks, G. Bruton, M. J. Finn, J. B. Harbridge, M. A. Harris, T. J. Howarth, E. Hunt, I. Stirling, and I. I. Zomaya in A. G. Brown and S. M. Roberts, Eds., *Recent Advances in the Chemistry of β-Lactam Antibiotics*, Special Publication No. 52, Royal Chemistry Society, London, 1985, pp. 222–241.

677. R. A. Volkman, R. D. Carroll, R. B. Drolet, M. L. Elliott, and B. S. Moore, *J. Org. Chem.*, **47**, 3344 (1982).

678. A. R. English, J. A. Retsema, A. E. Girard, J. E. Lynch, and W. E. Barth, *Antimicrob. Agents Chemother.*, **14**, 414 (1978).

679. J. A. Retsema, A. R. English, and A. E. Girard, *Antimicrob. Agents Chemother.*, **17**, 615 (1980).

680. H. A. Friedel, D. M. Campoli-Richards, and K. L. Goa, *Drugs*, **37**, 491 (1989).

681. R. G. Micetich, S. N. Maiti, P. Spevak, T. W. Hall, S. Yamabe, N. Ishida, M. Tanaka, T. Yamazaki, A. Nakai, and K. Ogawa, *J. Med. Chem.*, **30**, 1469 (1987).

682. L. Gutmann, M. D. Kitzis, S. Yamabe, and J. F. Acar, *Antimicrob. Agents Chemother.*, **29**, 955 (1986).

683. F. Higashitani, A. Hyodo, N. Ishida, M. Inoue, and S. Mitsuhashi, *J. Antimicrob. Chemother.*, **25**, 567 (1990).

684. D. Greenwood and R. G. Finch, Eds., *J. Antimicrob. Chemother.*, **31** (Suppl. A) (1993).

685. T. Marunaka, E. Matsushima, Y. Minami, K. Yoshida, and R. Azuma, *Chem. Pharm. Bull.*, **36**, 4478 (1988).

686. E. Matsushima, K. Yoshida, R. Azuma, Y. Minami, and T. Marunaka, *Chem. Pharm. Bull.*, **36**, 4593 (1988).

687. J. C. Pechere, R. Letarte, R. Guay, C. Asselin, and C. Morin, *J. Antimicrob. Chemother.*, **9** (Suppl. C), 123 (1982).

688. M. J. Basker and N. F. Osborne, *J. Antibiot.*, **43**, 70 (1990).

689. N. J. P. Broom, S. Coulton, I. Francois, J. B. Harbridge, J. H. C. Nayler, and N. F. Osborne in P. H. Bentley and R. Southgate, Eds., *Recent Advances in the Chemistry of β-Lactam Antibiotics*, Special Publication No. 70, Royal Chemistry Society, London, 1989, pp. 247–258.

690. K. Coleman, D. R. J. Griffin, and P. A. Upshon, *Antimicrob. Agents Chemother.*, **35**, 1748 (1991).

691. N. F. Osborne, R. J. Atkins, N. J. P. Broom, S. Coulton, J. B. Harbridge, M. A. Harris, I. Stirling-Francois, and G. Walker, *J. Chem. Soc. Perkin Trans I*, 179 (1994).

692. T. H. Farmer, J. W. J. Page, D. J. Payne, and D. J. C. Knowles, *Biochem. J.*, **303**, 825 (1994).

693. A. Bulychev, I. Massova, S. A. Lerner, and S. Mobashery, *J. Am. Chem. Soc.*, **117**, 4797 (1995).

694. N. J. P. Broom, T. H. Farmer, N. F. Osborne, and J. W. Tyler, *J. Chem. Soc. Chem. Commun.*, 1663 (1992).

695. K. Coleman, *Exp. Opin. Invest. Drugs*, **4**, 693 (1995).

696. R. F. Pratt and M. J. Loosemore, *Proc. Natl. Acad. Sci. USA*, **75**, 4145 (1978).

697. R. T. Aplin, C. V. Robinson, C. J. Shofield, and S. G. Waley, *J. Chem. Soc. Chem. Commun.*, 121 (1993).

698. B. S. Orlek, P. G. Sammes, V. Knott-Hunziger, and S. G. Waley, *J. Chem. Soc. Chem. Commun.*, 962 (1979).

699. H. C. Neu, *Antimicrob. Agents Chemother.*, **23**, 63 (1983).

700. R. Wise, N. O'Sullivan, J. Johnson, and J. M. Andews, *Antimicrob. Agents Chemother.*, **36**, 1002 (1992).

701. R. Wise, J. M. Andews, and N. Patel, *J. Antimicrob. Chemother.*, **7**, 531 (1981).

702. W. von Daehne, *J. Antibiot.*, **33**, 451 (1980).

703. W. von Daehne, E. T. Hansen, and N. Rastrup-Anderson in A. G. Brown and S. M. Roberts, Eds., *Recent Advances in the Chemistry of β-Lactam Antibiotics*, Special Publication No. 52, Royal Chemistry Society, London, 1985, pp. 375–380.

704. P. F. S. Mezes, R. W. Frieson, T. Viswanatha, and G. I. Dmitrienko, *Heterocycles*, **19**, 1207 (1982).

705. M. Yamashita, S. Hashimoto, M. Ezaki, M. Iwami, T. Komori, M. Koshaka, and H. Imanaka, *J. Antibiot.*, **36**, 1774 (1983).

706. Y. L. Chen, K. Hedberg, K. Guarino, J. A. Retsema, M. Anderson, M. Manousos, and J. F. Barrett, *J. Antibiot.*, **44**, 870 (1991).

707. N.-X. Chin, M. J. McElrath, and H. C. Neu, *Chemotherapy*, **34**, 318 (1988).

708. D. G. Brenner and J. R. Knowles, *Biochemistry*, **23**, 5839 (1984).

709. W. J. Gottstein, L. B. Crast Jr., R. G. Graham, U. J. Haynes, and D. N. McGregor, *J. Med. Chem.*, **24**, 1531 (1981).

710. G. I. Dmitrienko, C. R. Copeland, L. Arnold, M. E. Savard, A. J. Clarke, and T. Viswanatha, *Bioorganic Chem.*, **13**, 34 (1985).

711. G. C. Knight and S. G. Waley, *Biochem. J.*, **225**, 435 (1985).

712. Y. L. Chen, K. Hedberg, J. F. Barrett, and J. A. Retsema, *J. Antibiot.*, **41**, 134 (1988).

713. H. Tanaka, M. Tanaka, A. Nakai, S. Yamada, N. Ishida, T. Otani, and S. Torii, *J. Antibiot.*, **41**, 579 (1988).

714. P. Angehrn, W. Cullmann, P. Hohl, C. Hubschwerlen, H. Richter, and R. L. Then, Proceedings of the 34th Interscience Conference on Antimicrobial Agents and Chemotherapy, Orlando, FL, 1994, Abstr. F-153.

715. S. Adam, R. Then, and P. Angehern, *J. Antibiot.*, **46**, 64 (1993).

716. C. B. Ziegler, Y. Yang, P. Fabio, D. A. Steinberg, W. Weiss, M. J. Wildey, and K. Bush, Proceedings of the 34th Interscience Conference on Antimicrobial Agents and Chemotherapy, Orlando, FL, 1994, Abstr. C-54.

717. C. D. Foulds, M. Kosmirak, and P. G. Sammes, *J. Chem. Soc. Perkin Trans I*, 963 (1985).

718. Y. L. Chen, C.-W. Chang, K. Hedberg, K. Guarino, W. M. Welch, L. Kiessling, J. A. Retsema, S. L. Haskell, M. Anderson, M. Manousos, and J. F. Barrett, *J. Antibiot.*, **40**, 803 (1987).

719. C. Im, S. N. Maiti, R. G. Micetich, M. Daneshtalab, K. Atchison, O. A. Phillips, and C. Kunugita, *J. Antibiot.*, **47**, 1030 (1994).

720. D. D. Keith, J. Tengi, P. Rossman, L. Todaro, and M. Weigele, *Tetrahedron*, **39**, 2445 (1983).

721. S. Nishimura, N. Yasuda, H. Sasaki, Y. Matsumoto, T. Kamimura, K. Sakane, and T. Takaya, *J. Antibiot.*, **43**, 114 (1990).

722. J. D. Buynak, K. Wu, B. Bachmann, D. Khasnis, L. Hua, H. K. Nguyen, and C. L. Carver, *J. Med. Chem.*, **38**, 1022 (1995).

723. H. Wild and W. Hartwig, *Synthesis*, 1099 (1992).

724. H. S. Pfaendler, T. Neumann, and R. Bartsch, *Synthesis*, 1179 (1992).

725. M. Murakami, T. Aoki, M. Matsuura, and W. Nagata, *J. Antibiot.*, **43**, 1441 (1990).

726. A. D'Arcy, K. Gubernator, E.-M. Gutnecht, I. Heinze-Krauss, C. Hubschwerlen, M. Kania, C. Oefner, M. G. P. Page, J.-L. Specklin, and F. Winkler, Proceedings of the 35th Interscience Conference on Antimicrobial Agents and Chemotherapy, San Francisco, 1995, Abstr. F-150.

727. O. G. Brakstad and J. A. Maeland, *AMPIS*, **105**, 264 (1997).

728. M. Michel and L Gutmann, *Lancet*, **349**, 1901 (1997).

729. H. F. Chamber, *Clin. Microbiol. Rev.*, **10**, 781 (1997).

730. N. Jiraskova, *Curr. Opin. Invest. Drugs*, **2**, 209 (2001).

731. R. R. Wilkening, R. W. Ratcliffe, K. J. Wildonger, L. D. Cama, K. D. Dykstra, F. P. DiNinno, T. A. Blizzard, M. L. Hammond, J. V. Heck, K. L. Dorso, E. St. Rose, J. Kohler, and G. G. Hammond, *Bioorg. Med. Chem. Lett.*, **9**, 673 (1999).

732. N. Yasuda, C. Yang, K. M. Wells, M. S. Jensen, and D. L. Hughes, *Tetrahedron Lett.*, **40**, 427 (1998).

733. D. P. Rotella, *Chemtracts: Org. Chem.*, **12**, 675 (1999).

734. L. D. Cama, R. W. Ratcliffe, R. R. Wilkening, K. J. Wildonger, and W. Sun, Int. Pat. WO-09920627 (1999).

735. L. D. Cama, R. W. Ratcliffe, R. R. Wilkening, K. J. Wildonger, and W. Sun, Int. Pat. WO-09920628 (1999).

736. B. Chan, Proceedings of the 38th Interscience Conference on Antimicrobial Agents and Chemotherapy (Part 1), IDdb Meeting Report, September 24–27, 1998.

737. C.-H. Oh, S. C. Lee, S.-J. Park, I.-K. Lee, K. H. Nam, K.-S. Lee, B.-Y. Chung, and J.-H. Cho, *Arch. Pharm. Pharm. Med. Chem.*, **332**, 111 (1999).

738. J. H. Woo, G. Bae, and J. Ryu, *J. Antimicrob. Chemother.*, **44** (Suppl. A), 74 (1999).

739. C. Jin, I. Jung, H. J. Ku, J. Yook, D. H. Kim, M. Kim, J. H. Cho, and C. H. Oh, *Toxicology*, **138**, 59 (1999).

740. S. H. Kim, J. W. Kwon, W. B. Kim, and M. G. Lee, *Antimicrob. Agents Chemother.*, **43**, 2524 (1999).

741. S. H. Kim, W. B. Kim, and M. G. Lee, *Drug Metab. Dispos.*, **27**, 710 (1999).

742. S. H. Kim, W. B. Kim, J. W. Kwon, and M. G. Lee, *Biopharm. Drug Dispos.*, **20**, 125 (1999).

743. C. Lee, C. Cho, and J. Rhee, *Korean J. Med. Chem.*, **8**, 2 (1999).

744. S. H. Kim, W. B. Kim, and M. G. Lee, *Biopharm. Drug Dispos.*, **19**, 231 (1998).

745. J. H. Epstein-Toney, *Curr. Opin. Anti-Infective Invest. Drugs*, **1**, 64 (1999).

746. M. Tsuji, Y. Ishii, A. Ohno, S. Miyazaki, and K. Yamaguchi, *Antimicrob. Agents Chemother.*, **42**, 94 (1998).

747. S. Arakawa, S. Kamidono, T. Inamatsu, and J. Shimada, *Proceedings of the 37th Interscience Conference on Antimicrobial Agents and Chemotherapy*, San Diego, CA, F-218 (1997).

748. A. Saito, T. Inamatsu, and J. Shimada, *Proceedings of the 37th Interscience Conference on Antimicrobial Agents and Chemotherapy*, San Diego, CA, F-219 (1997).

749. M. Kume, H. Ooka, and H. Ishitobi, *Tetrahedron*, **53**, 1635 (1997).

750. P. C. Fuchs, A. L. Barry, and S. D. Brown, *J. Antimicrob. Chemother.*, **43**, 703 (1999).

751. J. Kohler, K. L. Dorso, K. Young, G. G. Hammond, H. Rosen, H. Kropp, and L. L. Silver, *Antimicrob. Agents Chemother.*, **43**, 1170 (1999).

752. M. L. Van Ogtrop, *Curr. Opin. Anti-Infective Invest. Drugs*, **1**, 74 (1999).

753. D. Andes, *Curr. Opin. Anti-Infective Invest. Drugs*, **1**, 111 (1999).

754. Y. I. Lin, P. Bitha, S. M. Sakya, S. A. Lang, Y. Yang, W. J. Weiss, P. J. Petersen, K. Bush, and R. T. Testa, *Bioorg. Med. Chem. Lett.*, **7**, 3063 (1997).

755. P. J. Petersen, S. A. Lang, Y. Youjun, N. Bhachech, W. J. Weiss, Y. Lin, N. V. Jacobus, P. Bitha, K. Bush, L. Zhong, R. T. Testa, S. M. Sakya, and T. W. Strohmeyer, *Bioorg. Med. Chem. Lett.*, **7**, 1811 (1997).

756. B. Panayota, L. Zhong, and I. L. Yang, *J. Antibiot.*, **52**, 643 (1999).

757. W. J. Weiss, P. J. Petersen, N. V. Jacobus, P. Bitha, Y. I. Lin, and R. T. Testa, *Antimicrob. Agents Chemother.*, **43**, 454 (1999).

758. W. J. Weiss, S. M. Mikels, P. J. Petersen, N. V. Jacobus, P. Bitha, Y. I. Lin, and R. T. Testa, *Antimicrob. Agents Chemother.*, **43**, 460 (1999).

759. M. Hikida, K. Itahashi, A. Igarashi, T. Shiba, and M. Kitamura, *Antimicrob. Agents Chemother.*, **43**, 2010 (1999).

760. D. Andes, *Curr. Opin. Anti-Infective Invest. Drugs*, **1**, 101 (1999).

761. M. L. Van Ogtrop, *Curr. Opin. Anti-Infective Invest. Drugs*, **1**, 70 (1999).

762. (a) E. Sakagawa, M. Otsuki, T. Ou, and T. Nishino, *J. Antimicrob. Chemother.*, **42**, 427–437 (1998); (b) Erratum to document cited in CA130:107445, *J. Antimicrob. Chemother.*, **43**, 169 (1999).

763. (a) K. Yamaguchi, H. Domon, S. Miyazaki, K. Takeda, A. Ohno, Y. Iishii, and N. Furuya, *Antimicrob. Agents Chemother.*, **42**, 555–563 (1998); (b) Erratum to document cited in CA128:303664, *Antimicrob. Agents Chemother.*, **42**, 1527 (1998).

764. O. A. Mascaretti, G. O. Danelon, E. L. Setti, M. Laborde, and E. G. Mata, *Curr. Pharmaceut. Des.*, **5**, 939 (1999).

765. S. M. Maiti, O. A. Phillips, R. G. Micetich, and D. M. Livermore, *Curr. Med. Chem.*, **5**, 441 (1998).

766. E. Di Modugno, and A. Felici, *Curr. Opin. Anti-Infective Invest. Drugs*, **1**, 25 (1999).

767. S. N. Maiti and O. Phillips, *Curr. Opin. Anti-Infective Invest. Drugs*, **1**, 40 (1999).

768. T. P. Smyth, M. E. O'Donnell, and M. J. O'Connor, *J. Org. Chem.*, **63**, 7600 (1998).

769. K. Miyashita, I. Massova, P. Taibi, and S. Mobashery, *J. Am. Chem. Soc.*, **117**, 11055 (1995).

770. L. Maveyraud, I. Massova, C. Birck, K. Miyashita, J. P. Samana, and S. Mobashery, *J. Am. Chem. Soc.*, **118**, 7435 (1996).

771. T. Nagase, D. Golemi, A. Ishiwata, and S. Mobashery, *Bioorg. Chem.*, **29**, 140 (2001).

772. H. G. F. Richter, P. Anghern, C. Hubschwerlen, M. Kania, M. G. P. Page, J. L. Specklin, and F. K. Winkler, *J. Med. Chem.*, **39**, 3712 (1996).

773. L. S. Tzouvalekis, M. Gazouli, E. E. Prinarakis, E. Tzelepi, and N. I. Legakis, *Antimicrob. Agents Chemother.*, **41**, 475 (1997).

774. N. A. V. Reddy, E. L. Setti, O. A. Phillips, D. P. Czajkowski, H. Atwal, K. Atchison, R. G. Micetich, and S. N. Maiti, *J. Antibiot.*, **50**, 276 (1997).

775. E. L. Setti, O. A. Phillips, N. A. V. Reddy, C. Kunugita, A. Hyodo, R. G. Micetich, and S. N. Maiti, *Heterocycles*, **42**, 645 (1996).

776. P. Eby, M. D. Cummings, O. A. Phillips, D. P. Czajkowski, M. P. Singh, P. Spevak, R. G. Micetich, and S. N. Maiti, *Heterocycles*, **42**, 653 (1996).

777. M. T. Black and G. Bruton, *Curr. Pharmaceut. Des.*, **4**, 133 (1998).

778. R. E. Dalbey, *Mol. Microbiol.*, **5**, 2855 (1991).

779. M. Inouye and S. Halegoua, *CRC Crit. Rev. Biochem.*, **7**, 339 (1980).

780. W. Wickner, A. J. M. Driessen, and F.-U. Hartl, *Ann. Rev. Biochem.*, **60**, 101 (1991).

781. J. M. Van Dijl, A. De Jong, J. Vehmaanpera, G. Venema, and S. Bron, *EMBO J.*, **11**, 2819 (1992).

782. D. Kuo, J. Weidner, P. Griffin, S. K. Shan, and W. B. Knight, *Biochemistry*, **33**, 8347 (1994).

783. A. E. Allsop, G. Brooks, G. Bruton, S. Coulton, P. D. Edwards, I. K. Hatton, A. C. Kaura, S. D. McLean, N. D. Pearson, T. Smale, and R. Southgate, *Bioorg. Med. Chem. Lett.*, **5**, 443 (1995).

784. A. E. Allsop, G. Brooks, P. D. Edwards, A. C. Kaura, and R. Southgate, *J. Antibiot.*, **49**, 921 (1996).

785. A. E. Allsop, M. J. Ashby, G. Brooks, G. Bruton, S. Coulton, P. D. Edwards, S. A. Elsmere, I. K. Hatton, A. C. Kaura, S. D. McLean, M. J. Pearson, N. D. Pearson, C. R. Perry, T. C. Smale, and R. Southgate in P. H. Bentley and P. J. O'Hanlon, Eds., *Anti-Infectives: Recent Advances in Chemistry and Structure-Activity Relationships*, Special Publication No. 198, Royal Chemistry Society, London, 1997, pp. 61–72.

# CHAPTER FIFTEEN

# Tetracycline, Aminoglycoside, Macrolide, and Miscellaneous Antibiotics

GERARD D. WRIGHT
Antimicrobial Research Centre
Department of Biochemistry
McMaster University
Hamilton, Ontario, Canada


DANIEL T. W. CHU
Chiron Corporation
Emeryville, California

## Contents

1 Introduction, 738
2 Tetracyclines, 739
   2.1 Clinical Use of Tetracycline Antibiotics and Currently Used Drugs, 739
   2.2 Side Effects, Toxicity, and Contraindications of Tetracycline Antibiotics, 741
   2.3 Pharmacology and Mode of Action of the Tetracycline Antibiotics, 741
   2.4 History and Biosynthesis of the Tetracycline Antibiotics, 742
   2.5 Tetracycline Resistance, 743
   2.6 Recent Developments in the Tetracycline Antibiotic Field, 746
3 Aminoglycoside Antibiotics, 747
   3.1 Clinical Use of Aminoglycoside Antibiotics and Currently Used Drugs, 747
   3.2 Side Effects, Toxicity, and Contraindications of the Aminoglycoside Antibiotics, 748
   3.3 Pharmacology and Mode of Action of the Aminoglycoside Antibiotics, 749
   3.4 History and Biosynthesis of the Aminoglycoside Antibiotics, 751
   3.5 Aminoglycoside Antibiotic Resistance, 753
   3.6 Recent Developments in the Aminoglycoside Antibiotic Field, 757
4 Macrolide Antibiotics, 758
   4.1 Clinical Use of Macrolide Antibiotics and Currently Used Drugs, 758

*Burger's Medicinal Chemistry and Drug Discovery*
Sixth Edition, Volume 5: Chemotherapeutic Agents
Edited by Donald J. Abraham
ISBN 0-471-37031-2   © 2003 John Wiley & Sons, Inc.

4.2 Side Effects, Toxicity, and Contraindications of the Macrolide Antibiotics, 761
4.3 Pharmacology and Mode of Action of the Macrolide Antibiotics, 762
4.4 History, Biosynthesis, and Structure-Activity Relationships of the Macrolide Antibiotics, 765
    4.4.1 C-9 Ketone Modification, 769
    4.4.2 C-8 Modification, 770
    4.4.3 C-6 Modification, 770
    4.4.4 C-11 and C-12 Modification, 771
    4.4.5 Cladinose Modification, 772
    4.4.6 Core–Skeleton Modification, 772
4.5 Resistance to Macrolide Antibiotics, 774
4.6 Recent Developments in the Macrolide Antibiotic Field, 776
    4.6.1 Novel Macrolides to Overcome Bacterial Resistance, 776
        4.6.1.1 Ketolides, 776

        4.6.1.2 Anhydrolides and Acylides, 781
        4.6.1.3 4″-Carbamate Macrolides, 782
    4.6.2 Non-Antibacterial Activity of Macrolides, 782
        4.6.2.1 Gastrokinetic Effects, 783
        4.6.2.2 Effects on Cytokines, 783
        4.6.2.3 Immunomodulatory Effects, 784
        4.6.2.4 Antitumor Effects, 784
        4.6.2.5 Antiparasitic Activity, 784
4.7 Future Prospects in the Macrolide Field, 784
5 Other Antibiotics That Target the Bacterial Ribosome, 785
    5.1 Streptogramins, 785
    5.2 Lincosamides, 787
    5.3 Chloramphenicol, 789
    5.4 Fusidic Acid, 792
    5.5 Oxazolidinones, 792
6 Future Prospects, 793

# 1 INTRODUCTION

Infections caused by bacteria are among the leading causes of death worldwide and antibiotics have made a dramatic impact in our capacity to intervene in these diseases. The tetracycline, aminoglycoside, and macrolide antibiotics were among the first classes of antibiotics discovered and have been mainstays of antibacterial therapy for half a century. These antibiotics all target the bacterial ribosome and interfere in the process of translation of the messenger RNA into protein and thus block a fundamental process in bacterial metabolism. For the most part, these compounds have little effect on the translational machinery of eukaryotic organisms, and thus this differentiation is the basis for the selectivity and consequent lower toxicity of these natural products. These antibiotics continue to be essential for the treatment of bacterial infections and in many cases are drugs of choice.

The bacterial ribosome consists of both a large (50S) and small (30S) subunits that come together to form the intact, 70S ribosome. Each subunit consists of both protein and RNA (rRNA) components: 31 proteins, 23S and 5S rRNAs for the 50S subunit, and 21 proteins and 16S rRNA for the small subunit. Although a complete description of the translational process is beyond the scope of this chapter, a brief outline of the key steps is pre-

sented in Fig. 15.1. The active site of the ribosome is composed of three subsites: the aminoacyl site (A), the peptidyltransfer site (P), and the exit site (E). Briefly, translation is initiated with the recognition and binding of messenger RNA (mRNA), which encodes a specific protein, to the 30S subunit in the presence of an initiator transfer RNA (tRNA) and of initiation factor 2 (IF2). Binding of the initiator tRNA in the P-site is followed by binding of the 50S subunit, generating a translation competent ribosome particle. The elongation cycle then consists of rounds of presentation of specific tRNAs charged with their cognate amino acids by elongation factor Tu and binding of these aminoacyl-tRNAs to the A-site with synchronized hydrolysis of GTP (Step 1, Fig. 15.1). Once in place on the ribosome, the aminoacyl-tRNA in the A-site participates in peptide bond formation by nucleophilic attack of the amino acid $\alpha$-amino group at the activated carboxyl group of the growing peptide chain linked to the tRNA in the P-site (Step 2, Fig. 15.1). The result is the elongation of the peptide by one amino acid residue in the A-site followed subsequently by a translocation step consisting of a shift in the mRNA register in an elongation factor G–dependent manner (Step 3, Fig. 15.1). This includes binding of the now uncharged tRNA in the E-site followed by its ejection, and binding of the peptidyl-tRNA to the P-site, leaving the A-site

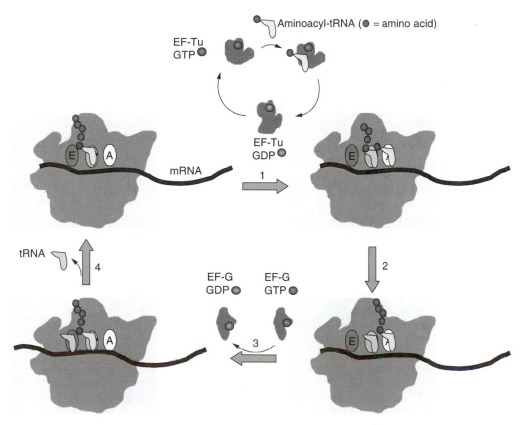

**Figure 15.1.** The bacterial protein synthesis elongation cycle. Step 1: Recognition and binding of the cognate aminoacyl-tRNA to the anticodon in the A-site. Step 2: Peptidyl transfer resulting in elongation of the peptide by one amino acid residue. Step 3: Translocation of the peptidyl-tRNA from the A-site to the P-site with concomitant movement of the uncharged tRNA to the E-site. Step 4: Exit of the uncharged tRNA completes the cycle. See color insert.

free again to bind an appropriate aminoacyl-tRNA (Step 4, Fig. 15.1). The complexities of the binding and enzymatic events as well as the multitude of proteins and RNAs involved in the translation process have been fertile ground for the evolution of antimicrobial agents, and all of the antibiotics discussed in this chapter exert an impact directly on critical aspects of translation.

## 2  TETRACYCLINES

### 2.1  Clinical Use of Tetracycline Antibiotics and Currently Used Drugs

The tetracycline antibiotics (see Table 15.1) were first introduced over 50 years ago as the first of the broad-spectrum antibiotics, effica-

cious against aerobic and anaerobic Gram-positive and Gram-negative bacteria. They are generally orally administered and have found extensive use in the treatment of infectious diseases and continue to be widely used, but are now being supplanted by other agents such as the quinolones. Nonetheless, the tetracyclines remain first-line drugs in the treatment of infections caused by pathogens of the family Rickettsiae (causative agents of Rocky Mountain spotted fever, typhus, Q fever, ehrlichiosis), *Chlamydia pneumoniae*, *Mycoplasma pneumoniae*, *Chlamydia trachomatis*, *Borrelia burgdorferi* (Lyme disease), members of the genus *Brucella* that cause brucellosis, *Calymmatobacterium granulomatis* (granuloma inguinale), *Vibrio cholera* (cholera), and

**Table 15.1   Tetracycline Antibiotics in Current Use**

| Generic Name | Trade Name | Structure | Route | Dose |
|---|---|---|---|---|
| Tetracycline | Achromycin | 1 | Oral | 250 mg, 4 times daily |
| Chlortetracycline | Aureomycin | 2 | Ophthalmic or topical ointment | |
| Oxytetracycline | Vibramycin | 3 | Oral i.v. | 200 mg/day |
| Demethylchlortetracyline | Declomycin | 4 | Oral | 3.25–3.75 mg/kg |
| Minoclycine | Minocin | 5 | Oral | 100 mg every 12 h |
| Doxycycline | Doxycin | 6 | Oral | 50–100 mg every 12 h |

others. The tetracyclines are also often used in combination with other agents for the treatment of peptic ulcer disease caused by *Helicobacter pylori* infection. Furthermore, tetracyclines can be used in the treatment of a number of conditions including acne vulgaris, legionellosis, syphilis, tularemia, and plague caused by *Yersinia pestis*. Doxycycline (**6**) is also used for chemoprophylaxis of malaria under certain conditions. The tetracycline antibiotics are therefore of great clinical importance and utility, especially aginst infections caused by Gram-negative bacteria.

As the name suggests, tetracyclines possess four rings that form the minimal naphthacene tetracycle (Fig. 15.2). Positions at the "bottom" of the molecule (10, 11, 1) and most of ring A (positions 2, 3, and 4) represent the invariant pharmacophore region of the mole-

cule, where modifications are not tolerated without loss of antibiotic activity. The remaining positions where substitution is permitted have been exploited by nature and synthetic chemists in the development of new tetracyclines. Table 15.1 lists a number of tetracyclines and their structures. The natural products tetracycline (**1**), chlortetracycline (**2**), oxytetracycline (**3**), demethylchlortetracyline (**4**), and the semisynthetic minocycline (**5**) and doxycycline (**6**) are frequently used in the clinic. The glycylcyclines such as GAR-936 (**7**) are new semisynthetic derivatives that link modified glycines to 9-amino derivatives of various tetracyclines (1, 2).

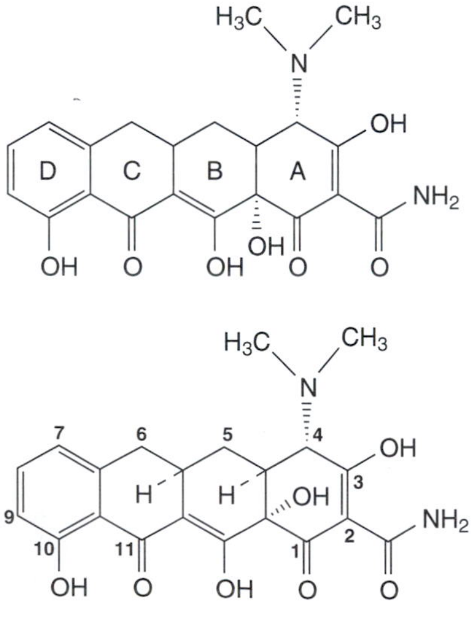

(1)

(2)

**Figure 15.2.** Tetracycline pharmacophore and numbering.

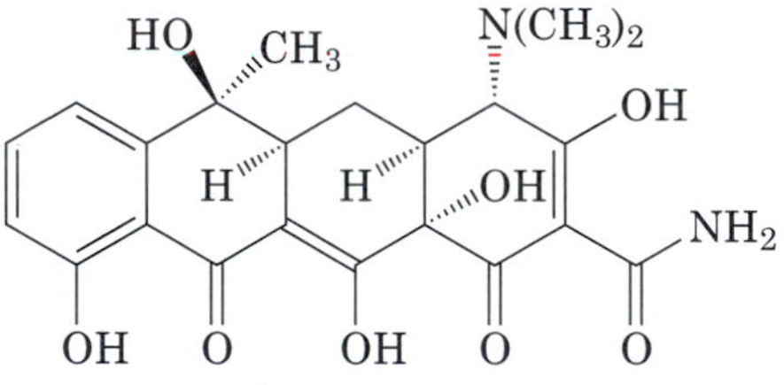

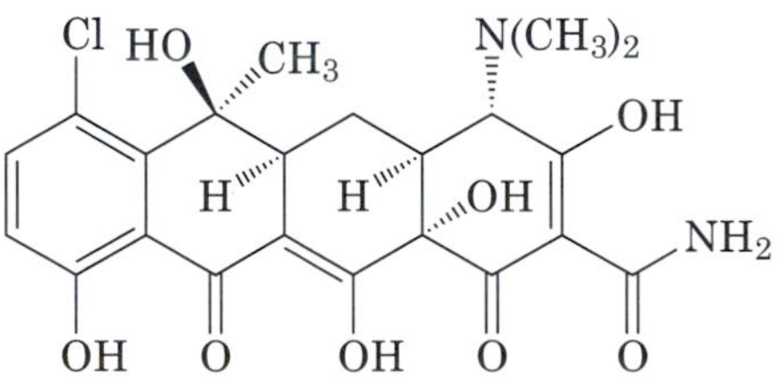

The tetracyclines avidly bind metals through the hydroxyl and carbonyl groups of

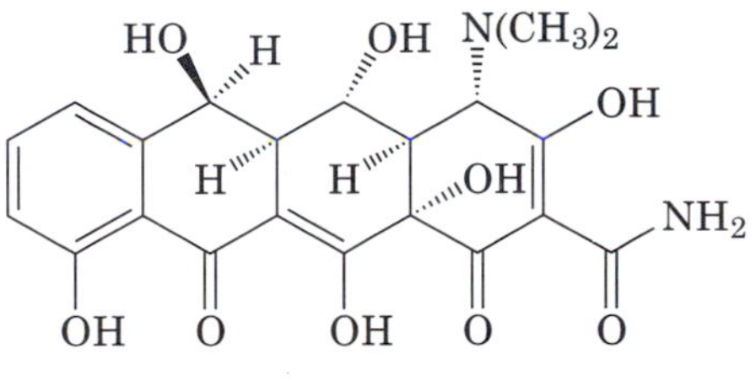

(3)

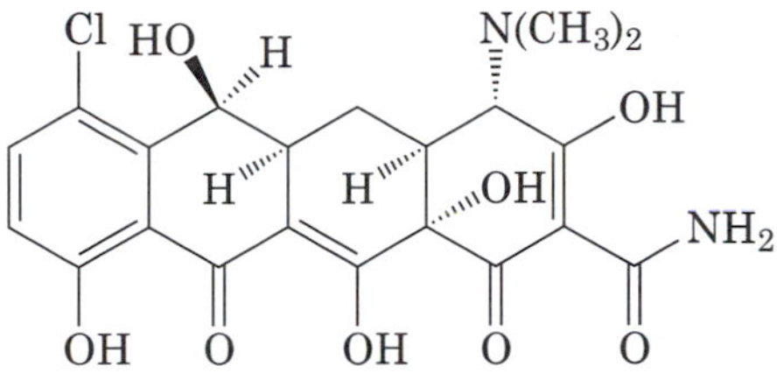

(4)

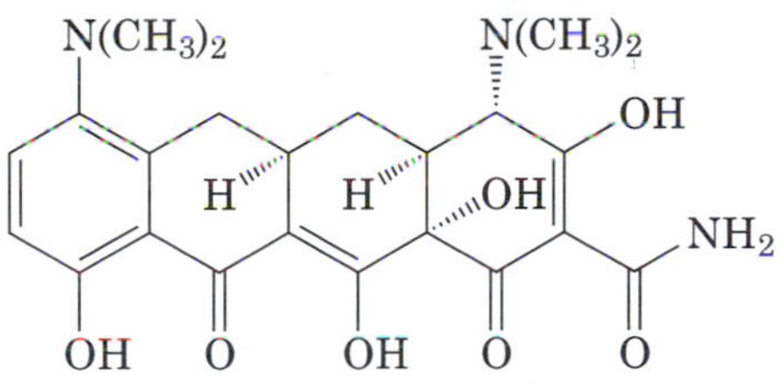

(5)

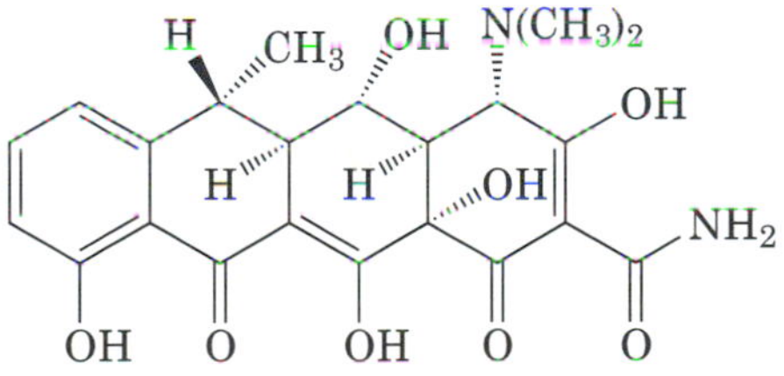

(6)

Metal

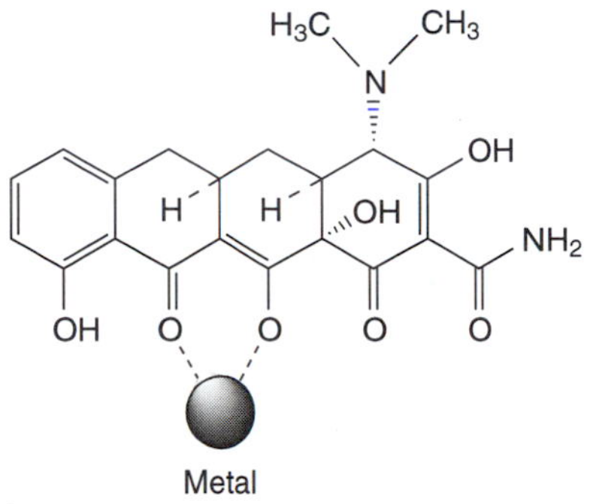

**Figure 15.3.** Binding of metals by tetracycline antibiotics.

affinity for divalent metals including $Ca^{2+}$, tetracyclines can stain the teeth and bones and are therefore not given to younger children or during pregnancy. The affinity for metals also means that coadministration with certain foods such as milk or other agents such as $Ca^{2+}$- or $Mg^{2+}$-containing antacids is not appropriate because it reduces bioavailability. $Ca^{2+}$ binding and subsequent precipitation is also thought to be the reason that injection of tetracyclines is painful and therefore not a preferred route of administration. Tetracyclines can also affect the usefulness of oral contraceptives.

Side effects tend to be minor, including dizziness or lightheadedness for minocycline, some gastrointestinal discomfort, and candidal overgrowth, which is common to virtually all antibiotics, and under rare conditions esophageal ulceration. High levels of tetracycline have been associated with hepatotoxicity, which is exacerbated by preexisting renal impairment or coadministration with other hepatotoxic agents.

### 2.3 Pharmacology and Mode of Action of the Tetracycline Antibiotics

Tetracyclines are readily absorbed in the gastrointestinal tract and are subsequently widely distributed in most tissues. These antibiotics also cross the placenta and are present in breast milk and therefore are not recommended for pregnant or lactating women. Most tetracycline antibiotics are eliminated through the kidney and are therefore not recommended for patients with renal problems

the B and C rings (Fig. 15.3) and this appears to be the biologically active form of the molecules (see Section 2.3).

### 2.2 Side Effects, Toxicity, and Contraindications of Tetracycline Antibiotics

The tetracyclines are in general quite safe and well tolerated. They are all associated with photosensitivity, and patients should avoid direct exposure to the sun. As a result of their

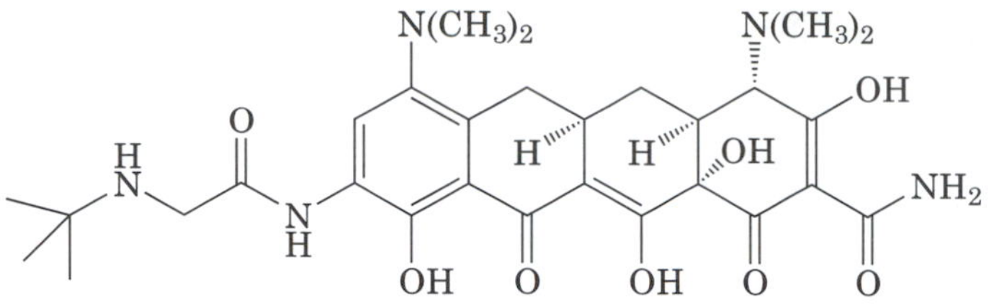

(7)

because it may lead to toxicity. The exception is doxycycline (**6**), which is excreted through the bile. Serum half-lives are long (6 to >20 h, depending on the agents) and thus once or twice daily dosing is recommended on the order of 100 mg per dose (Table 15.1). Metabolism of these antibiotics is minimal.

The tetracycline antibiotics in current clinical use are generally bacteriostatic, inhibiting cell growth but not actually killing the cell. Thus, if the antibiotic is removed from the medium, the cells recover and resume growth. Tetracycline freely passes through lipid bilayers and enters the cell in a diffusion-controlled manner and no transport proteins appear to be necessary (3, 4). The mechanism of antibiotic action involves binding to the bacterial ribosome and arresting translation. Various methods have determined (*1*) that the 30S subunit is the preferred site of binding and (*2*) that there is one high affinity binding site, located in the A-site of the ribosome, and several lower affinity sites. The crystal structure of tetracycline bound to the 30S subunit of the *Thermus thermophilus* 30S subunit has been determined to 3.4 Å (5). In this structure, two binding sites for the antibiotic were determined, one in the A-site as predicted and another in the body of the ribosome. The first site is likely the more clinically relevant site, in that it agrees well with several decades of binding and biochemical data describing the tetracycline–ribosome interaction. The key features of the binding of tetracycline to the A-site are shown in Fig. 15.4. All of the interactions between the antibiotic and the ribosome occur with the 16S rRNA, and no contacts with proteins are observed. Tetracycline binds in a 20 × 7-Å pocket above the A-site binding pocket for the aminoacyl tRNA. As predicted by structure–activity studies, all

four rings of tetracycline participate in binding to the ribosome and, in particular, Ring D stacks with the pyrimidine ring of C1054. The entire pharmocophore region of the molecule (see Fig. 15.2) is involved in specific interactions with the 16S rRNA. Not surprisingly, a $Mg^{2+}$ ion is chelated by the antibiotic and the rRNA phosphate backbone as predicted by the affinity of tetracycline for divalent ions and the requirement for $Mg^{2+}$ ion in tetracycline binding. This $Mg^{2+}$ is also present in the structure of the ribosome in the absence of the antibiotic and may represent a key conserved binding element. The tetracycline-binding region is poorly conserved between eukaryotes and bacteria and helps to explain the low toxicity and specificity of these antibiotics.

Modeling of a tRNA molecule in the structure of the tetracycline–30S complex reveals a steric clash between the tRNA and the antibiotic. Furthermore, because tetracycline binds on the opposite side of the codon–anticodon binding pocket, it is possible that tetracycline does permit presentation of the aminoacyltRNA by EF-Tu, which would trigger GTP hydrolysis (see Fig. 15.1). Therefore, tetracyclines may act in two ways, first by preventing occupancy of the A-site by the aminoacyltRNA and thus arresting translation, and second in a catalytic fashion, depleting GTP stores in the cell.

## 2.4 History and Biosynthesis of the Tetracycline Antibiotics

The first tetracycline antibiotic to be isolated was chlortetracycline (**2**) from cultures *of Streptomyces aureofaciens* in 1948, which was followed in the same year by oxytetracycline (**3**) from *Streptomyces rimosus*. Chemical deoxygenation of oxytetracycline at C6 yielded doxycycline (**6**); minocycline (**5**) was prepared

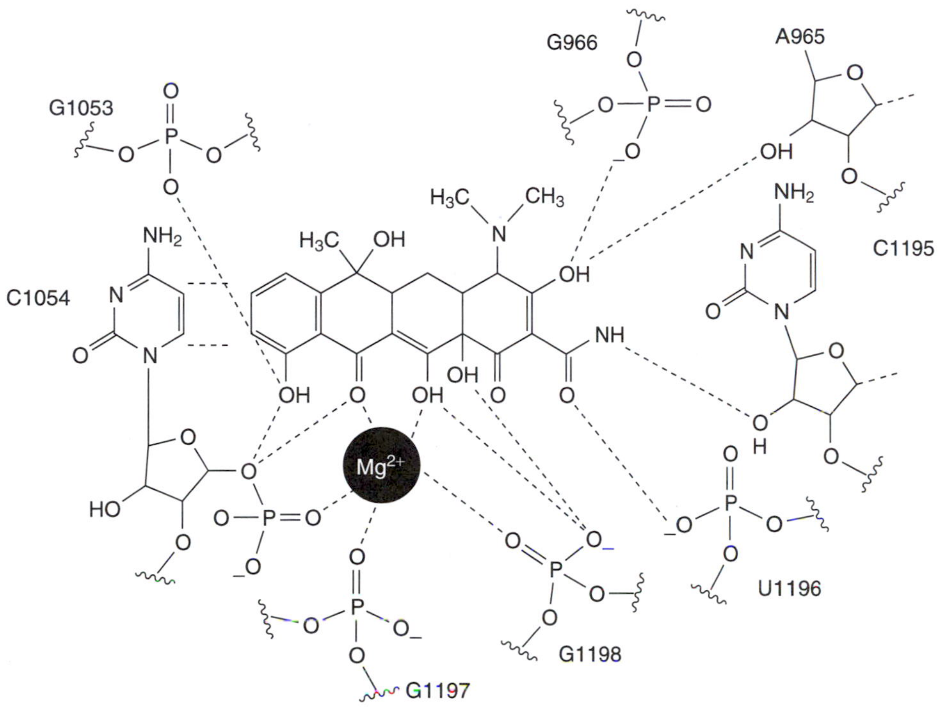

**Figure 15.4.** Interaction of tetracycline with the 30S ribosomal subunit (redrawn from Ref. 5). Base numbers are from *E. coli*.

from demethylchlortetracyclinc in the mid 1960s. The synthesis of the newer glycylcyclines was first described in the early 1990s.

The tetracycline antibiotics are polyketide natural products and therefore derived from the sequential linkage of acetyl units that form the core polyketide. The biosynthetic clusters encoding the genes necessary for the biosynthesis of chlortetracycline and oxytetracycline have been cloned and sequenced (6–8). From these gene sequences and chemical intuition, a plausible biosynthetic pathway can be constructed, beginning with the starter unit malonamyl-CoA, which provides the requisite amide at C2 (Fig. 15.5). Eight successive condensations with malonyl CoA, each providing a C2 unit, generate the linear polyketide. This is followed by cyclization to yield 6-methylpretetramid [possibly by the *otcD1* gene product of *S. rimosus* (9)], hydroxylation, oxidation, and reductive transamination at C4a as well as hydroxylation at C12a to generate 4-amino-anhydrotetracyline. *S*-Adenosylmethionine–dependent methylation followed by

two successive hydroxylation events and a reduction generates oxytetracycline. The *otcC* gene product of *S. aureofaciens* has been shown to be the anhydrotetracyline oxidase, hydroxylating anhydrotetracycline at position 6 (10). The gene encoding the enzyme required for chlorination of tetracycline at position 7 to give chlortetracycline has also been cloned (11). The precise gene products for the remainder of the biosynthetic steps have not yet been fully characterized and therefore the details of this pathway remain probable, but speculative.

## 2.5  Tetracycline Resistance

Tetracycline antibiotics are no longer used for empirical treatment of many Gram-negative and Gram-positive infections as a result of the dissemination and prevalence of resistance, which has played an important role in decreased use of these otherwise potent antibiotics over the past years. Clinically relevant resistance occurs primarily by two mechanisms: active efflux and ribosomal protection,

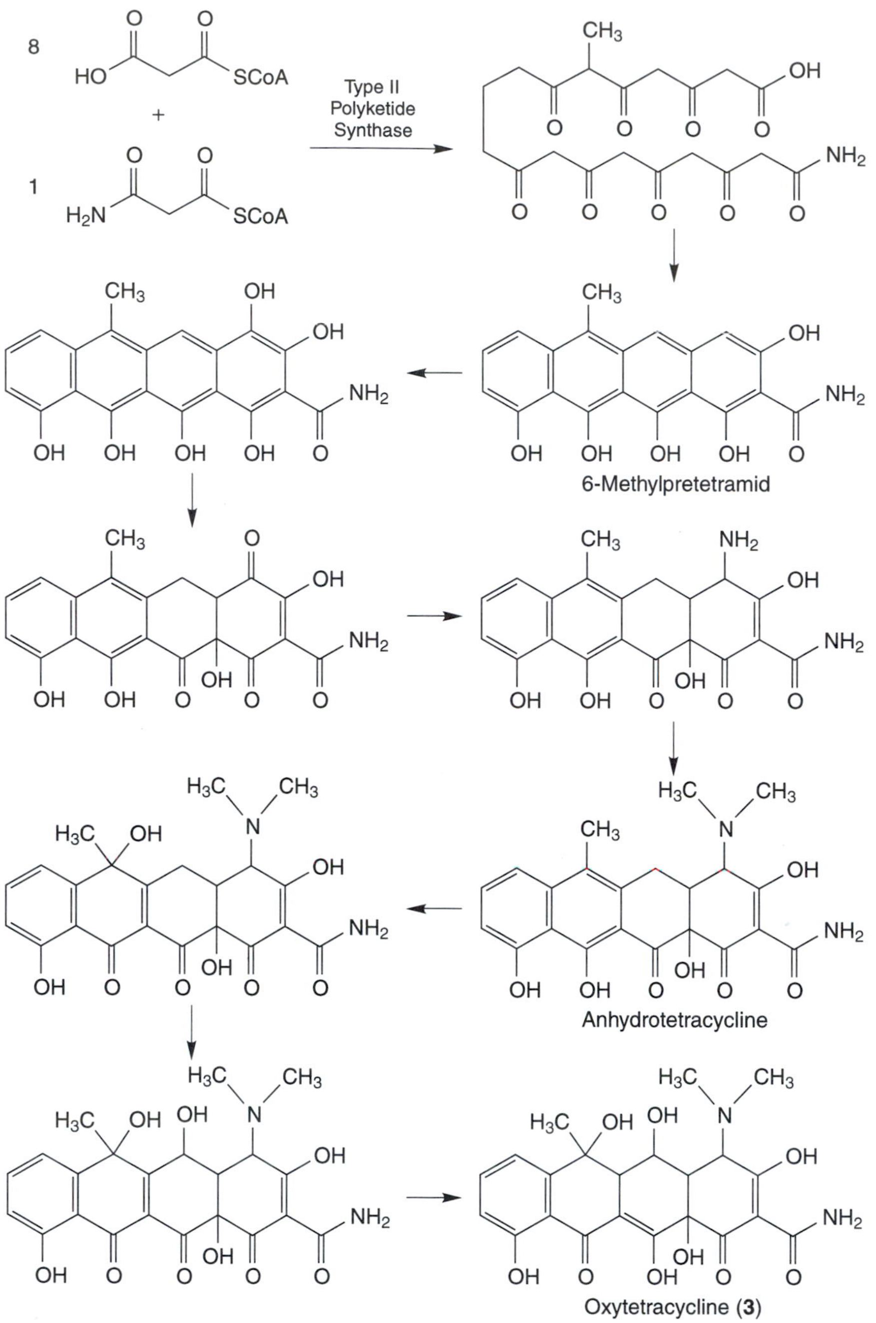

**Figure 15.5.** Predicted oxytetracycline biosynthetic pathway.

although enzymatic modification and other unknown mechanisms have been documented (see reviews in Refs. 12–14). Ribosomal mutations, a major source of resistance for other inhibitors of translation, are rarely implicated in tetracycline resistance, though mutation at position 1058 of the 16S rRNA (*Escherichia coli* numbering) has been associated with tet-

**Table 15.2    Tetracycline Efflux Determinants**

| Group | Proteins | Bacterial Genera Associated with Resistance Type | Number of Transmembrane Helices |
|---|---|---|---|
| 1 | Tet(A) | Numerous including *Pseudomonas, Serratia, Escherichia, Klebsiella, Vibrio* | 12 |
| | Tet(B) | Numerous including *Escherichia, Providencia, Moraxella, Shigella, Vibrio* | |
| | Tet(C) | Numerous including *Salmonella, Proteus, Pseudomonas, Escherichia, Vibrio* | |
| | Tet(D) | *Aeromonas, Yersinia, Enterobacter, Pasteurella* | |
| | Tet(E) | *Aeromonas, Serratia, Escherichia, Vibrio* | |
| | Tet(G) | *Vibrio* | |
| | Tet(H) | *Pasteurella* | |
| | Tet(Z) | *Corynebacterium* | |
| 2 | Tet(K) and Tet(L) | *Staphylococcus, Clostridium, Enterococcus, Streptococcus* | 14 |
| 3 | OtrB and Tcr3 | *Streptomyces, Mycobacterium* | 14 |
| 4 | TetA(P) | *Clostridium* | 12 |
| 5 | Tet(V) | *Mycobacterium* | 10–11 |
| 6 | TetAB | *Corynebacterium* | ATP binding cassette transporter |

racycline resistance in *Propionibacterium acnes* (15, 16). The efflux proteins all appear to be members of the major facilitator superfamily of efflux proteins that export small molecules in a proton-dependent fashion (17). They are integral membrane proteins of approximate molecular mass of 46 kDa and can be subdivided into six groups based on amino acid sequence similarity (Table 15.2) (18).

At present there is no available three-dimensional (3D) structure available for a tetracycline efflux protein, but there has been a significant amount of biochemical research describing the topology and antibiotic specificity of some of these proteins. The preponderance of work has come from the Yamaguchi laboratory describing the group 1 protein Tet(A) (19–22). The work from this lab has suggested that the 12 helices form a water-filled channel flanked by transmembrane helices TM1, TM2, TM4, TM5, TM8, TM10, and TM11 with a flexible intracellular loop between TM6 and TM7. This channel presumably facilitates the export of tetracycline from inside the cell, resulting in resistance.

The second prevalent mechanism of tetracycline resistance is ribosomal protection (23). There are a number of genes encoding these proteins of approximately 70 kDa (Table 15.3), and the Tet(M) and Tet(O) proteins have been the best studied (24, 25). Amino acid sequence analysis shows similarity of these proteins with elongation factors G and Tu, especially in the N-terminal GTP binding domain. GTP binding and perhaps hydrolysis are necessary for binding of these tetracycline resistance determinants to the ribosome, which triggers release of tetracycline from the A-site. In the presence of nonhydrolyzable GTP analogs, stable complexes between this class of resistance protein and the ribosome can be formed. Cryoelectron microscopic studies have been used to study the Tet(O)–ribosome complex, and the 3D structure was determined to be 16 Å (26). This structure demonstrates that Tet(O) has a shape similar to that of EF-G and binds analogously to the A-site region of the ribosome. However, unlike EF-G, which causes a significant structural change in the ribosome (27), binding of Tet(O) does not sig-

**Table 15.3   Ribosomal Protection Tetracycline Resistance Determinants**

| Protein | Bacterial Genera Associated with Resistance Type |
|---|---|
| Tet(M) | Numerous including *Enterococcus, Streptococcus, Staphylococcus, Neisseria* |
| Tet(O) | *Campylobacter, Enterococcus, Lactobacillus* |
| Tet(Q) | *Bacteroides, Streptococcus, Lactobacillus* |
| Tet(S) | *Enterococcus, Lactococcus, Listeria* |
| Tet(T) | *Streptococcus* |
| Tet(W) | *Butyrivibrio* |
| TetP(B) | *Clostridium* |
| Otr(A) | *Streptomyces, Mycobacterium* |

nificantly affect the ribosome conformation. These studies provide a possible mechanism of action, in which Tet(O)–GTP binding decreases the affinity of tetracycline for the ribosome and subsequent GTP hydrolysis releases Tet(O), generating a translation competent ribosome.

A final mechanism of tetracycline resistance is enzymatic degradation. The only enzyme described with this activity is TetX, an $O_2$-dependent oxidase that paradoxically was identified in the anaerobe *Bacteroides fragilis* (28, 29). Under aerobic conditions in *E. coli*, TetX modifies tetracycline in an unknown fashion to generate inactive antibiotics. This resistance element has not yet been identified as a clinical problem.

## 2.6   Recent Developments in the Tetracycline Antibiotic Field

Although the use of tetracycline antibiotics has diminished over the past years, they continue to be of great importance in the treatment of infectious disease. The broad dissemination of resistance throughout both Gram-negative and Gram-positive populations has had a significant impact on their clinical utility. The glycylcyclines, tetracyclines substituted on position 9 [e.g., GAR-936 (**7**) (1)], do not appear to be susceptible to the common ribosome protection or efflux resistance (30). Some molecules are in advanced clinical trials for the treatment of drug-resistant infections (31, 32). Modified tetracyclines [e.g., 13-cyclopentylthio-5-hydroxy-6-deoxytetracycline (**8**)] have also been shown to block some tetracycline efflux pumps such as TetA and TetB, resulting in a reversal of antibiotic resistance (33–35).

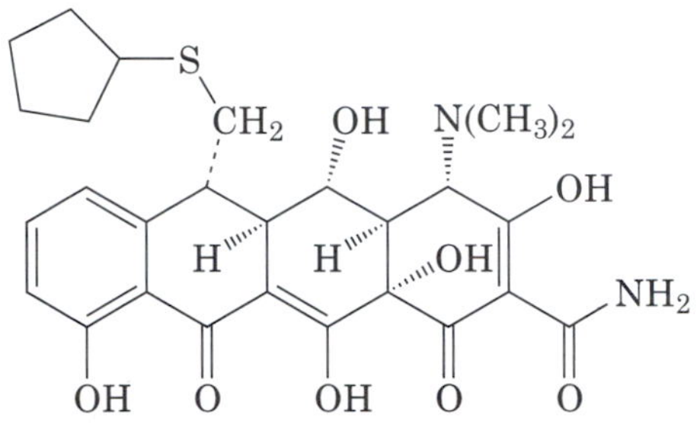

(8)

Tetracyclines and chemically modified tetracyclines have been investigated for properties other than their antimicrobial activity (36). For example, tetracyclines are known to have anti-inflammatory properties. These molecules have therefore been pursued as chemotherapeutic agents with promise in the treatment of inflammatory disease such as osteo- and rheumatoid arthritis. Tetracyclines exert a number of effects related to inflammation. These include the reduction of levels of mRNA encoding NO synthase (37), increasing the levels of cyclooxygenase-2 and subsequent increase in prostaglandin E2 levels (38), and the inhibition of enzymes important to connective tissue biology such as matrix metalloproteinases (MMPs) and gelatinases. There is a rich literature on the inhibition of MMPs by antimicrobial and chemically modified tetracyclines [e.g., CMT-3 (**9**)] and this property could have application in osteoarthritis (39), periodontitis (40, 41), corneal diseases (42, 43), and cancer (44).

Thus, in addition to the well-established antimicrobial activity of tetracyclines, there are several new therapeutic areas on which

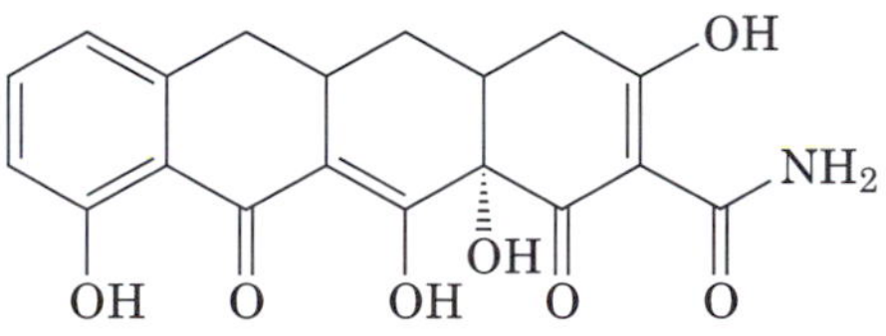

(9)

these compounds, or their derivatives, may prove to have a profound impact.

## 3 AMINOGLYCOSIDE ANTIBIOTICS

### 3.1 Clinical Use of Aminoglycoside Antibiotics and Currently Used Drugs

The aminoglycoside antibiotics find use as broad-spectrum agents for the treatment of infections caused by aerobic Gram-negative and Gram-positive bacteria including *Klebsiella pneumoniae*, *Pseudomonas aeruginosa*, *E. coli*, *Proteus* sp., *Serratia marcescens*, and *Staphylococci* (45). Aminoglycosides are also used in combination therapy with penicillins for the treatment of enterococcal infections. Some compounds such as paromomycin have been employed in treatment of protozoal infections (46). Streptomycin (10) was the first aminoglycoside isolated and the first antibiotic with potent activity against *Mycobacterium tuberculosis* and this antibiotic continues to be used to treat tuberculosis, but as a result of the development of resistance, now in combination therapy with other antibiotics (45). Streptomycin can also be used for the treatment of tularemia, plague and leprosy. The aminoglycosides are highly water soluble and poorly absorbed orally. These antibiotics are therefore primarily delivered by intramuscular injection or intravenously. Topical use of aminoglycosides also is frequently used including gentamicin (11), a complex of gentamicin C1, gentamicin C1a, and gentamicin C2, in drops to treat ophthalmic infections, neomycin (12) in topical formulation with corticosteroids, and tobramycin (13), aerosolized to treat respiratory infections.

The aminoglycoside antibiotics are more properly termed aminoglycoside-aminocyclitols because they not only incorporate aminosugars, but also six-membered aminocyclitol

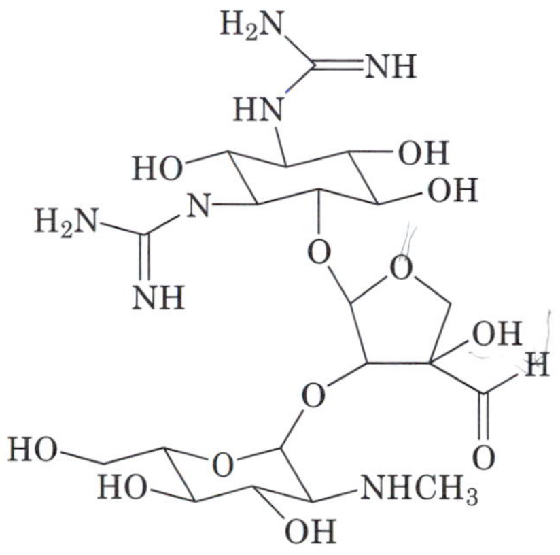

(10)

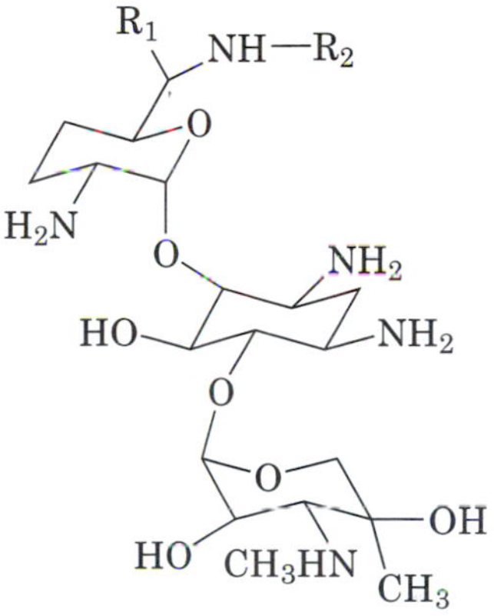

Gentamicin (11)
C1   R1 = $CH_3$   R2 = $CH_3$
C1a  R1 = H      R2 = H
C2   R1 = $CH_3$   R2 = H

rings, cyclic carbon rings functionalized with amino and hydroxyl groups (47). However, the name aminoglycoside is more frequently used, despite the fact that some members of the group such as spectinomycin do not have any sugar component. The clinically important aminoglycosides have been generally subdivided into two broad groups based on structure: compounds that contain a 2-deoxystreptamine aminocyclitol ring and those that do not. The 2-deoxystreptamine–containing antibiotics are further divided into two groups based on whether the aminocyclitol is substituted with amino sugars at positions 4 and 5 or 4 and 6. The former are also characterized by the presence of a pentose at position 5 [e.g.,

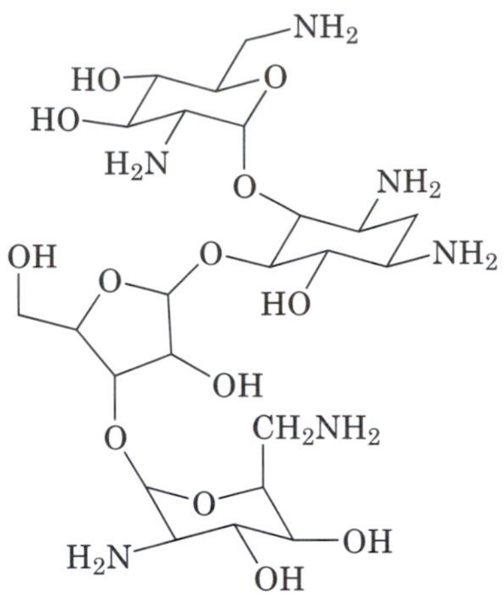

(12)

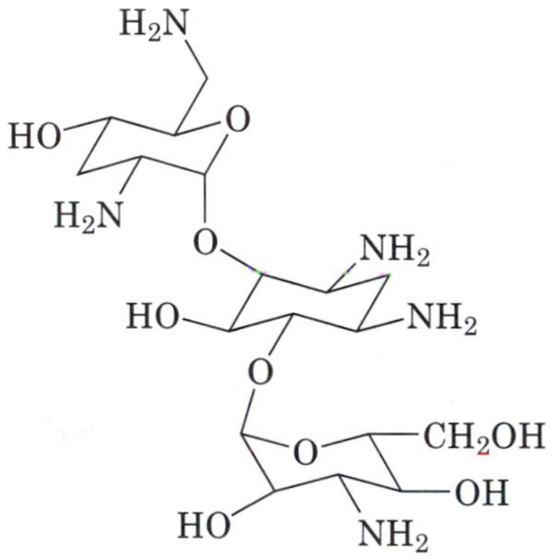

(13)

neomycin (**12**)]. The preponderance of the best-tolerated antibiotics are the 4,6-disubstituted 2-deoxystreptamine compounds such as gentamicin (**11**). The second broad class of antibiotics includes the clinically used antibiot-

ics streptomycin (**10**) and spectinomycin, which differ in their modes of action (see Section 3.3).

The characteristics of the most commonly used aminoglycosides can be found in Table 15.4. The presence of numerous amine groups in all compounds, positively charged at neutral pH, ensures an overall cationic nature and substantial water solubility.

## 3.2 Side Effects, Toxicity, and Contraindications of the Aminoglycoside Antibiotics

Most aminoglycosides show oto- and nephrotoxic effects and the risk of damage is higher in patients with impaired renal function or after high doses of these antibiotics. They are therefore contraindicated with other drugs that can impair renal function such as some diuretics and cephalosporins. Ototoxicity is the result of damage to the sensory hair cells and the eighth cranial nerve and is irreversible (48). Nephrotoxicity is the result of accumulation of the antibiotics in the proximal renal tubule and this has a number of effects including hydroxyl radical formation, inhibition of phospholipases, increases in thomboxane synthesis, and others (reviewed in Ref. 49). As a result of this ubiquitous problem, once-daily regimens have been established that account for patient body weight; close monitoring of serum concentration is recommended (45). Aminoglycosides cross the placenta and are thus used in treatment of pregnant women only in life threatening circumstances. Small amounts of the antibiotics are present in breast milk and therefore have the potential to harm the infant; however, this must be balanced with the knowledge that oral absorption

**Table 15.4  Aminoglycoside Antibiotics in Clinical Use**

| Generic Name | Trade Name | Structure | Route | Dose[a] |
|---|---|---|---|---|
| Streptomycin | Streptomycin | **10** | i.m. | 15 mg/kg/day |
| Amikacin | Amikin | **14** | i.v. or i.m. | 15 mg/kg |
| Tobramycin | TOBI | **13** | Inhalation | 300 mg/12 h |
| Gentamicin C | Garamycin Parenteral | **11** | i.v. or i.m. | 3 mg/kg/day in three 1 mg/kg doses |
| Gentamicin C | Garamycin Ophthalmic | **11** | Drops | |
| Netilmicin | Netromycin | **15** | i.v. or i.m. | 4 mg/kg/day in two 2 mg/kg doses |
| Neomycin | Neosporin (and others) | **12** | Topical | |

[a]Adult patients with normal renal function.

is poor. Rapid infusion by i.v. and coadministration with muscle relaxants can lead to neuromuscular blockade.

## 3.3 Pharmacology and Mode of Action of the Aminoglycoside Antibiotics

As noted above, as a consequence of the poor oral availability of aminoglycosides, parenteral administration is required. This poor absorption through the gastrointestinal tract has been taken advantage of for the sterilization of the gut before abdominal surgery. Upon injection, the serum half-life of aminoglycosides in patients is on the order of 2 h, but this is highly variable, especially in cases of renal failure (45). There is little metabolism of the aminoglycosides, which are excreted by glomerular filtration. However, there have been reports of the generation of an uncharacterized metabolite of gentamicin and other aminoglycosides after incubation with hepatic microsomes that is toxic to outer hair cells (50, 51). The details of the structure and mechanism of these metabolites and their relationship with clinical toxicity remain to be explored.

The aminoglycosides are generally bactericidal antibiotics and the main site of action of these antibiotics is the ribosome. Identification of the ribosome as the primary target is supported by the observation that high level resistance to aminoglycosides can occur through ribosomal modification by specific methyltransferases or by point mutation in rRNA and certain ribosomal proteins. Chemical footprinting studies have identified the 16S rRNA as the primary site of binding for the aminoglycoside antibiotics, particularly within the mRNA decoding A-site (52, 53). However, simple interference with translation is rarely sufficient to achieve cell death, and the mechanism of bactericidal activity of aminoglycosides has not been completely resolved. Aminoglycosides enter the cell in a multiphasic process (54), consisting of first an energy-independent accumulation of antibiotic on the cell surface, likely as a result of interaction of the cationic drugs with the anionic components of the cell wall (teichoic acids in Gram-positive bacteria), outer membrane (lipopolysaccharides in Gram-negative bacteria), and cellular membrane (phospholipids). This is followed

by an energy-dependent transport of the antibiotic across the cell membrane. This transport requires the $\Delta\psi$ portion of the membrane potential and explains why anaerobic bacteria are less sensitive to these antibiotics and how mutations in electron-transport components or the presence of inhibitors of this process interfere with antibiotic access to the cell (55, 56). The initial entry phase is followed by a second energy-dependent accumulation of antibiotic within the cell. This has numerous effects other than on translation, including membrane damage resulting in membrane "leakiness" and inhibition of DNA synthesis (57–59).

A key element in the process of cell death appears to reside in the fact that bactericidal aminoglycosides cause mistranslation of the genetic code (60–62). This results in the biosynthesis of miscoded proteins. Davis has proposed that these aberrant proteins, with associated unpredictable folding, play a role in membrane damage, which precipitates cell death (63, 64). Miscoding associated with aminoglycosides has long been recognized and in fact has been used to suppress point mutations (65).

The specific details of the interactions of aminoglycosides with the target 16S rRNA are now being elucidated. A series of elegant model studies using NMR techniques by the group of Puglisi have suggested that binding of antibiotics such as paromomycin and gentamicin results in a conformational change in the rRNA in this region (66–68). Recently, the structures of the 30S ribosomal subunit of *T. thermophilus* in complex with streptomycin, spectinomycin, and paromomycin have been solved at a resolution of 3 Å, providing unprecedented molecular insight into their mode of action (69). Although all these aminoglycoside antibiotics interfere with translation, they have different effects. Both streptomycin and 2-deoxystreptamine aminoglycosides such as paromomycin, which is prototypical of this group, cause mistranslation and cell death but footprint to different areas of the 16S rRNA. Spectinomycin, on the other hand, does not cause mistranslation but blocks the translocation of the tRNA from the A-site to the P-site after peptidyltransfer, by freezing the ribosome in an inactive conformation, and is

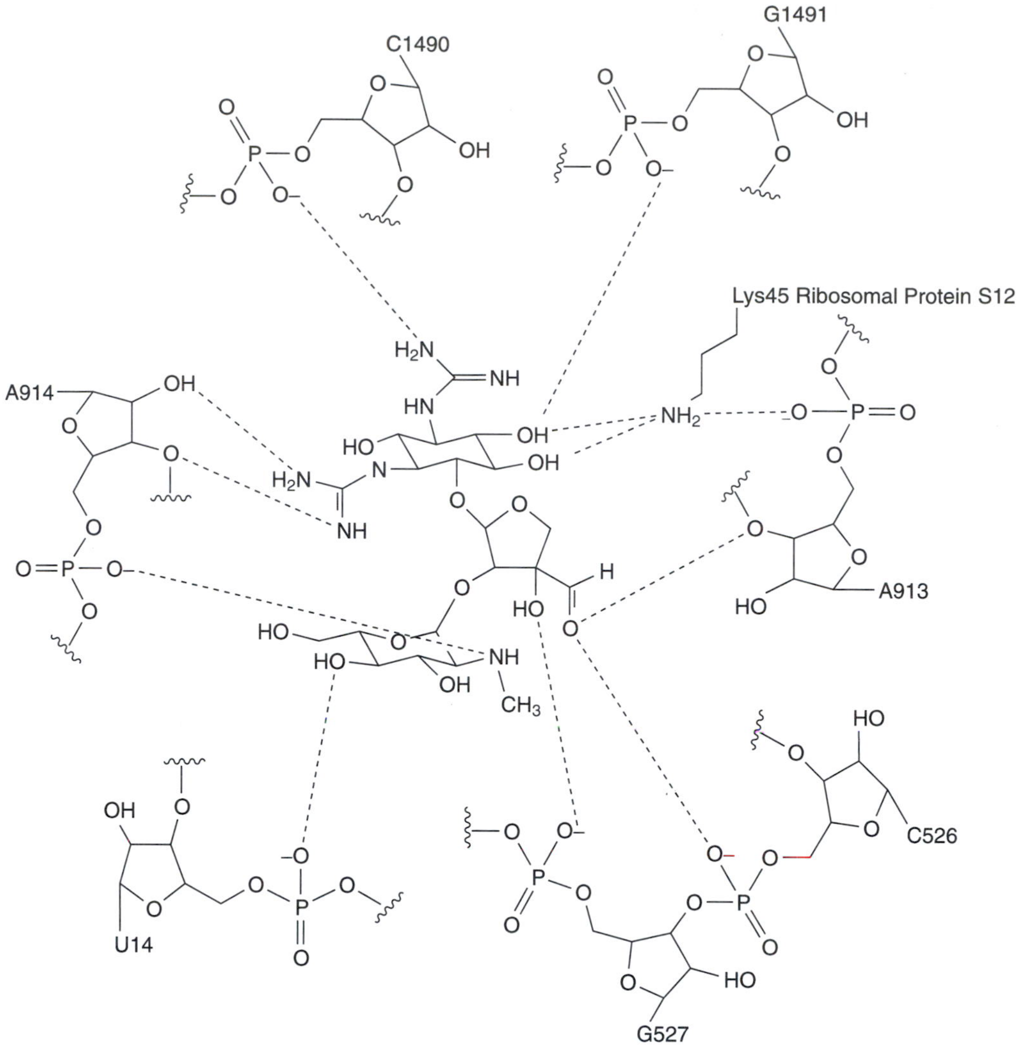

**Figure 15.6.** Interaction of streptomycin with the 30S ribosomal subunit (redrawn from Ref. 69). *E. coli* numbering of 16S rRNA is shown.

bacteriostatic (70–72). The structure of the spectinomycin–30S complex confirms the predicted (70) direct interactions with G1064 and C1192 of the 16S rRNA and places the rigid spectinomycin molecule at the juncture of 16S rRNA helices H34 and H35. This interaction prevents either movement or a conformation change in H34 during translocation.

Streptomycin, on the other hand, binds not only to the 16S rRNA through a number of contacts between H27, H44, and H18 but also directly with ribosomal protein S12 (Fig.

15.6). This interaction stabilizes a conformation of the ribosome, termed *ram* (ribosomal ambiguity), that increases tRNA affinity in the A-site. Stabilization of this conformation likely results in the loss of tRNA selectivity in the A-site and consequently increases translation errors, which contribute to cell death. The paromomycin–30S complex (Fig. 15.7) reveals that the antibiotic binds to a pocket formed by the major grove of H44 and forms direct interactions between a number of A-site 16S bases including A1408 and G1494 (69). Importantly,

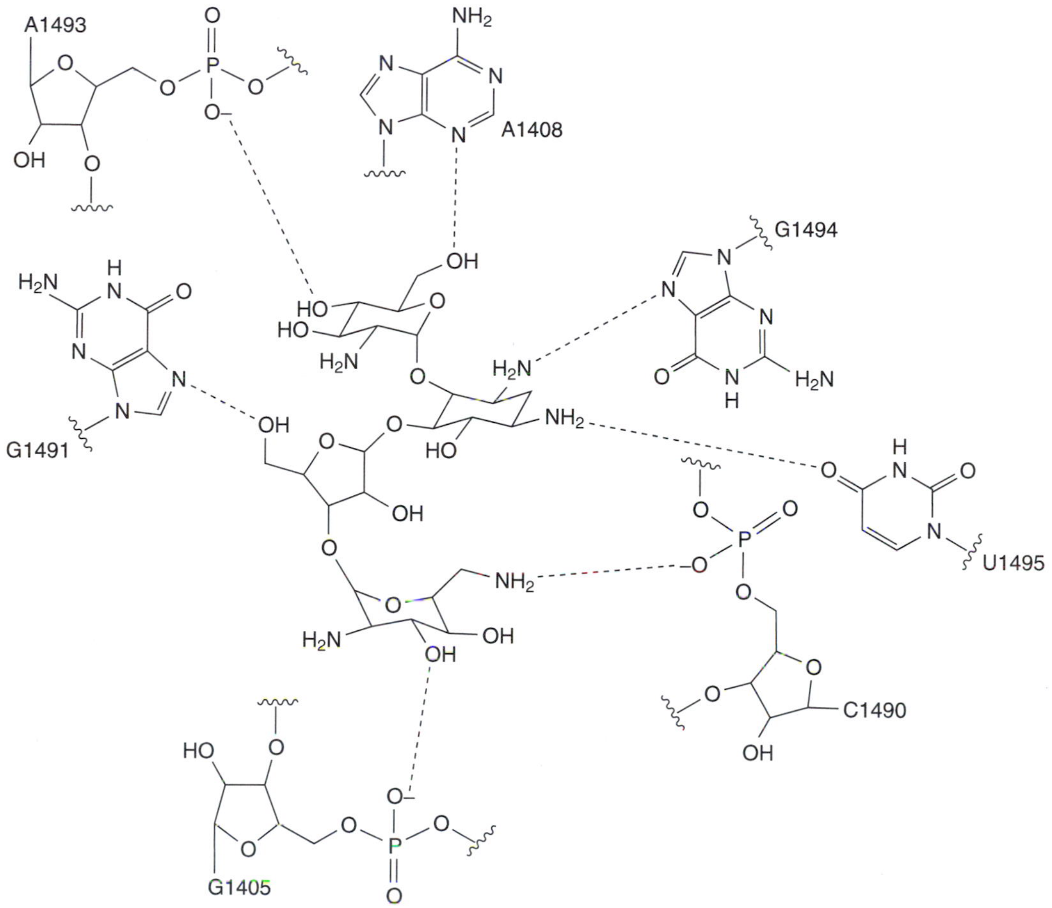

**Figure 15.7.** Interaction of paromomycin with the 30S ribosomal subunit (redrawn from Ref. 69). *E. coli* numbering of 16S rRNA is shown.

bases A1492 and A1493 of H44 are flipped out toward the mRNA binding region upon binding of paromomycin. The structure of the paromomycin–30S complex is now complemented by the structure of the complex of the 30S subunit with the anticodon loop of tRNA$^{Phe}$ and a U$_6$ hexanucleotide (73). In this structure, 16S rRNA bases A1492 and A1493 are also flipped out toward the mRNA, interacting with the minor grove of the codon–anticodon helix. Paromomycin therefore induces a similar conformational change in the absence of a cognate codon–anticodon interaction, and this may be the molecular basis for mRNA misreading caused by binding of the 2-deoxystreptamine aminoglycosides to the ribosomal A-site.

## 3.4 History and Biosynthesis of the Aminoglycoside Antibiotics

Walksman and colleagues discovered the first aminoglycoside antibiotics, streptomycin in 1944 and neomycin in 1949, using systematic searches for antimicrobial compounds derived from soil organisms (74, 75). These antibiotics were rapidly incorporated into clinical use, streptomycin being the first antibiotic with potent antimycobacterial activity used for the treatment of tuberculosis. The discovery of these antibiotics was followed by the identification of numerous additional members of the group including kanamycin in 1957 by Umezawa's group in Japan (76), and gentamicin (1963) and sisomicin (1970) both from Schering-Plough in the United States (77, 78).

Although new aminoglycosides continued to be identified and characterized from natural sources after this initial phase of discovery, clinically directed drug discovery in the 1970s saw a focus on the semisynthetic derivatization of existing antibiotics. This effort was guided by known natural derivatives that showed less sensitivity to aminoglycoside resistance by chemical modification, which was rapidly emerging as a grave clinical problem (see Section 3.5). For example, modification of N1 through amination by an $\alpha$-hydroxy-4-aminobutyryl (AHB) group naturally found in the aminoglycoside butirosin was mimicked in the synthesis of amikacin (N1-AHB-kanamycin A) (14), netilimicin (N1-ethylsisomicin) (15), isepamicin (N1-$\alpha$-hydroxy-$\gamma$-aminopropionyl-gentamicin B), and arbekacin [N1-AHB-3′-4′-dideoxykanamycin B (dibekacin, 16)], all of which found, and continue to find, clinical use. Despite the identification of a number of aminoglycosides derived from both natural sources and through semisynthesis, there has not been a new antibiotic of this class introduced into clinical practice for over 2 decades, with the exception of arbekacin (16) in Japan. The reasons for this may reflect the increased availability of newer broad-spectrum agents such as the fluoroquinolones, which lack the general toxicity of aminoglycosides.

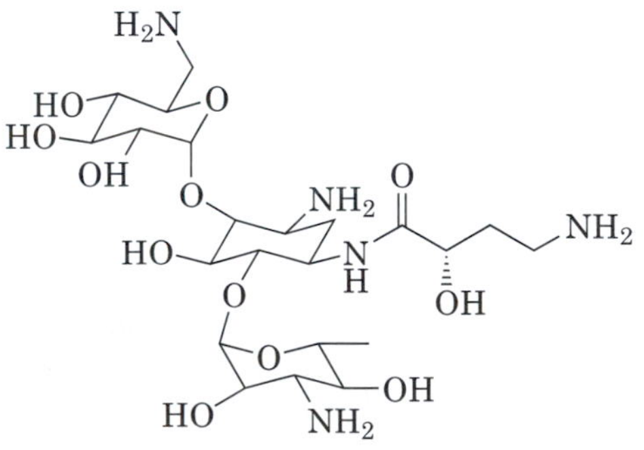

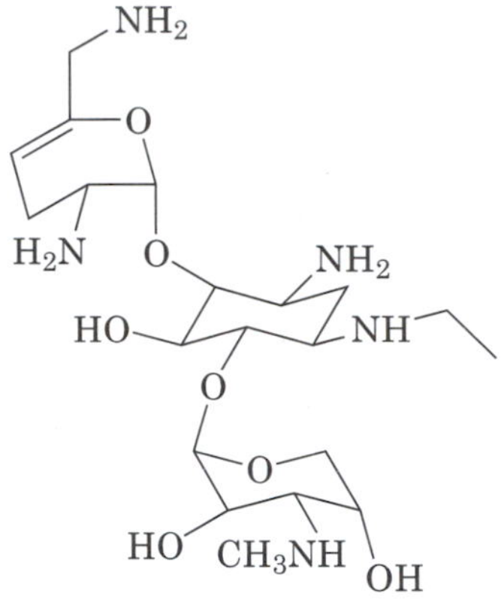

(15)

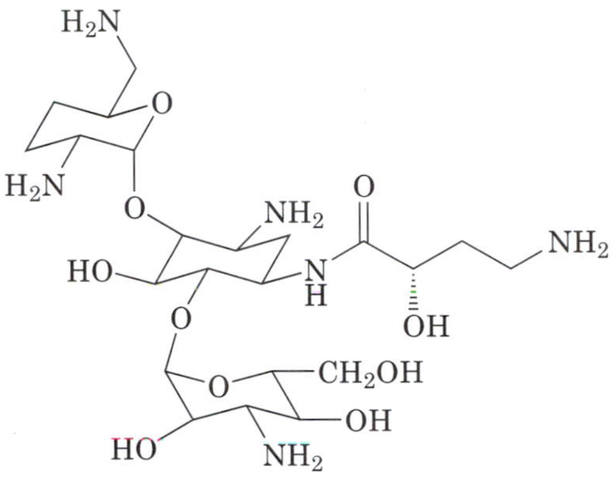

(16)

(14)

The aminoglycoside antibiotics are synthesized primarily by actinomycete bacteria from the genera *Streptomyces*, *Micromonospora*, and *Saccharopolyspora*, which produce a

number of antibiotics, but also from bacteria of the genera *Bacillus* and *Pseudomonas* (79). The distinguishing aminocyclitol ring is derived from inositide biosynthesis from glucose-6-phosphate, which, although essential to eukaryotic membrane structure and signal transduction, is generally thought to be rare within the prokaryotes. Nonetheless, it has been shown that actinomycetes and a few other bacteria have inositol-containing lipids in their membranes and cell walls (see references in ref. 80). Piepersburg has noted that the available biosynthetic data point to two major routes for the synthesis of the aminocyclitol rings: (1) through a D-*myo*-inositol-3-phosphate synthase generating *scyllo*-inosamine, which is necessary in the formation of streptomycin and spectinomycin; and (2) through a dehydroquinate synthaselike path-

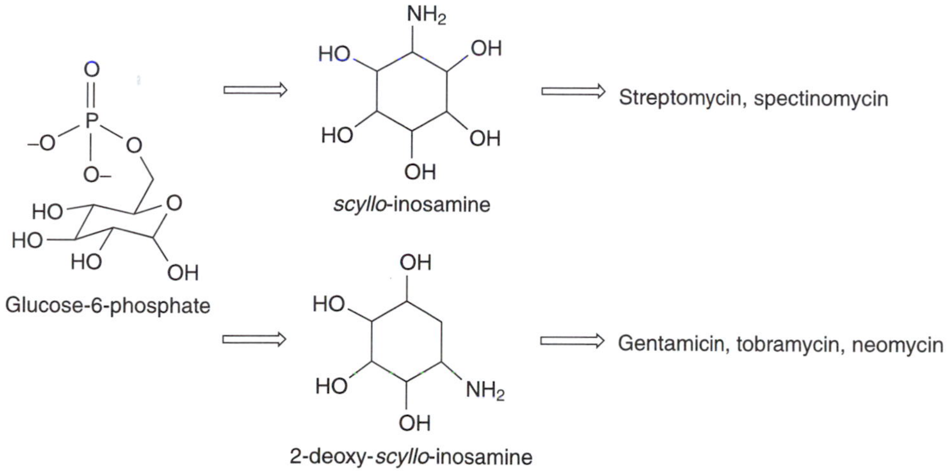

**Figure 15.8.** Biosynthetic strategies for the generation of aminocyclitol rings.

way generating 2-deoxy-*scyllo*-inosamine, which is essential for the formation of all the important 2-deoxystreptamine antibiotics such as kanamycin and gentamicin (Fig. 15.8) (79). These then are the fundamental steps in aminoglycoside biosynthesis. The specific sugar components that are characteristic of each individual aminoglycoside arise both by modification before and after transfer to the aminocyclitol rings, and the precise pathways appear to be unique to individual antibiotics with few common intermediate steps. An exception is the biosynthesis of paromamine, which is a central element in the formation of the majority of 2-deoxystreptamine antibiotics (Fig. 15.9).

The precise details of aminoglycoside biosynthesis await complete sequencing of biosynthetic genes and the accompanying detailed biochemical analysis of gene product function, which are lacking for most important aminoglycosides, with the exception of streptomycin and fortimicin (reviewed in Ref. 79) and more recently butirosin (81). Research on the biosynthesis of streptomycin has formed the basis of our understanding of aminoglycoside biosynthesis and has been essential in deciphering the molecular strategy used by the microorganisms to generate the antibiotic. Of particular interest is the fact that streptomycin is prepared as the 6-phospho-derivative, which does not have antimicrobial activity. This therefore protects the cell

against suicide during antibiotic production (82). A 6-phospho-streptomycin–specific exporter, StrV/StrW, facilitates passage from the cytosol (83). Once on the outside of the cell, an extracellular 6-phospho-streptomycin phosphatase, StrK, removes the phosphate, revealing the active streptomycin antibiotic in the extracellular medium (84). This clever approach of biosynthesizing antibiotic precursors that are activated only once outside the cell may prove to be a general approach in avoiding toxicity in this class of antibiotics.

### 3.5  Aminoglycoside Antibiotic Resistance

Resistance to the aminoglycosides occurs primarily through the production of enzymes that modify the antibiotics in either an ATP- or acetylCoA-dependent fashion, although other resistance mechanisms are known including drug efflux, alteration of ribosome components, and altered uptake of the drugs. Efflux has thus far been of minor clinical importance and restricted to nonspecific pumps in only a few organisms [e.g., *P. aeruginosa* (85, 86), *Burkholderia pseudomallei* (87), *Mycobacterium fortuitum* (88), and *E. coli* (89, 90)]. Importantly, the presence of the active MexXY–OprM efflux system in *P. aeruginosa* has recently been linked to suppression of aminoglycoside activity in this organism by cations such as $Mg^{2+}$, suggesting that inhibitors of efflux pumps may potentiate aminoglycoside action in this organism (91). Altered up-

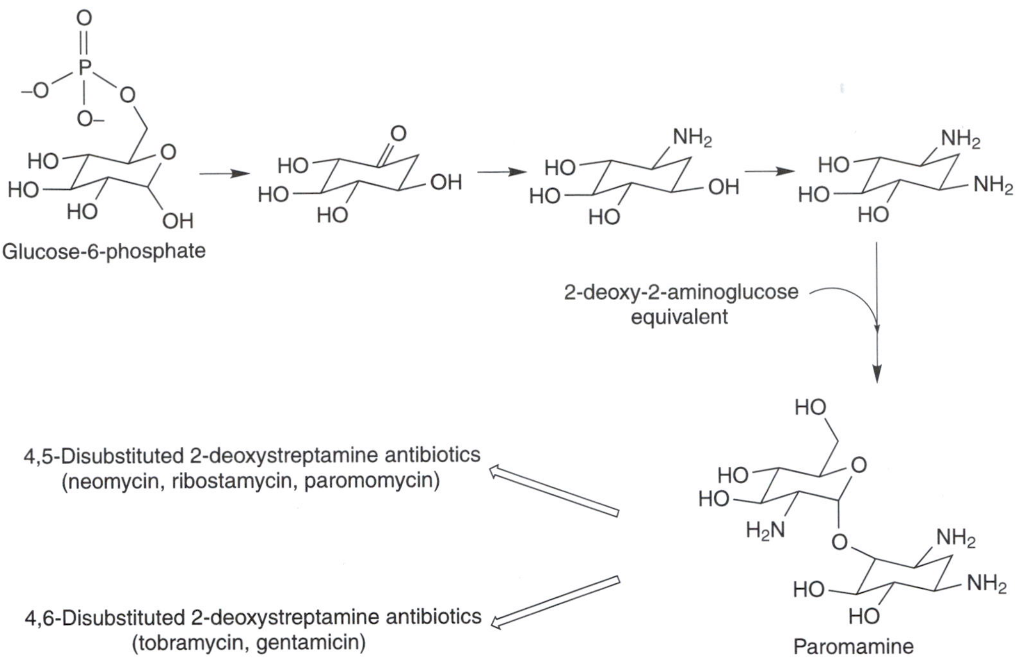

**Figure 15.9.** Paromamine is a common biosynthetic intermediate for several 2-deoxystreptamine aminoglycoside antibiotics.

take of aminoglycosides is also infrequently associated with clinical resistance, though as noted above, anaerobic conditions and mutations that affect the membrane potential necessary for aminoglycoside entry into the cell are known.

Target modification by mutation or chemical modification of the ribosome also results in aminoglycoside resistance. Point mutations in the target 16S rRNA are not uncommon in some clinically important bacteria such as *M. tuberculosis*, where streptomycin resistance is associated with a variety of mutations in the 16S rRNA gene *rrs* (92). The structure of the streptomycin–30S subunit complex (69) clarifies the molecular basis of resistance, demonstrating ionic and H-bond interactions between the antibiotic and mutation-susceptible bases. Resistance to streptomycin can also arise frequently from point mutation of Lys43 of the ribosomal protein S12 (92), which in the crystal structure of the streptomycin–30S subunit complex binds directly to streptomycin (69). Resistance to the 2-deoxystreptamine aminoglycosides, which do not bind in the

same area of the A-site as streptomycin, can also arise from 16S rRNA mutations, though these are generally less frequently associated with clinical resistance. Modification of the 16S rRNA target by methylation is a prevalent mechanism of high level resistance in aminoglycoside-producing bacteria, where S-adenosylmethionine–dependent methylation of N7 of G1405 and/or A1408 results in resistance (93). These bases interact directly with paromomycin in the crystal structure of the paromomycin–30S subunit complex (see Fig. 15.7) (69).

The most frequent means of aminoglycoside resistance is through enzyme-catalyzed detoxification of the antibiotics. Unlike other antibiotics such as the β-lactams, where enzymatic destruction is also the most relevant mechanism of resistance, aminoglycoside resistance does not occur through cleavage of the molecule. Rather, the aminoglycosides are covalently modified by phosphorylation, adenylation, or acetylation on key hydroxyl or amine groups that result in steric blocking of the antibiotic binding site on the ribosome.

These group transfer reactions require ATP or acetylCoA co-substrates and thus antibiotic inactivation occurs only inside the cell and consequently aminoglycoside modifying enzymes are intracellular. There are numerous individual aminoglycoside resistance enzymes and these are classified by the type of modification: APH for aminoglycoside phosphotransferases, ANT for aminoglycoside adenyltransferases, and AAC for acetyltransferases. The standard nomenclature used to describe aminoglycoside resistance enzymes consists first of designation of the type of transfer (APH, ANT, or AAC) followed by the site of modification in parentheses (e.g., position 3, 3', 2''), followed by a designation of the resistance profile by a Roman numeral (I, II, III, . . .), and finally a lowercase letter designating the specific gene, which is assigned in incremental fashion as it is reported (94). For example, spectinomycin is modified by ANT(9)-Ia, an adenyltransferase (ANT) that modifies spectinomycin only (I) at position 9, and the gene was cloned from *Staphylococcus aureus* (a); AAC(6')-Ic is an acetyltransferase(AAC) that acetylates kanamycin, tobramycin, amikacin, and neomycin (I) at position 6', and the gene was cloned from *Serratia marcescens*. Figure 15.10 shows the structures of gentamicin C2 and streptomycin and the predominant sites of modification by resistance enzymes typically found in clinical isolates. (For a more comprehensive description of aminoglycoside resistance genes and the associated regiospecificity of group transfer see Refs. 47, 94, and 95.) Three-dimensional structures of members of each class of resistance enzyme (APH, ANT, and AAC) are now known and have served to illuminate the molecular basis of resistance.

Aminoglycoside phosphotransferases constitute a large family of aminoglycoside resistance enzymes that require ATP for activity (96). APHs modify aminoglycosides by phosphorylation of hydroxyl groups with release of ADP:

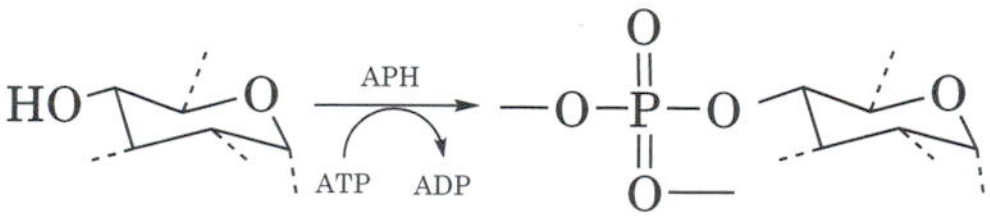

(15.1)

Enzymes of this family share significant primary sequence homology in the C-terminus, which is now known to contain many of the residues important to catalysis. The 3D structure of one aminoglycoside kinase, APH(3')-IIIa, has been determined in the absence of bound nucleotides and in the presence of ADP and the nonhydrolyzable ATP analog AMPPNP (97, 98). These structures revealed close similarity to the fold of the Ser/Thr/Tyr protein kinase superfamily, despite the fact that only 5 of 264 amino acids are conserved between APH(3')-IIIa and protein kinases. Not surprisingly, these residues play important roles in ATP binding and phosphoryl transfer, which is common to both groups of enzymes (97, 99). The structural relationship between both aminoglycoside and protein kinases has been expanded by studies indicating that these enzyme families share sensitivity to inhibitors (100, 101) and by the fact that aminoglycoside kinases do have weak, but demonstrable protein kinase activity (102).

There are fewer known aminoglycoside adenyltransferases than either APHs or AACs; nonetheless, two of these, ANT(4')-Ia from *S. aureus* and ANT(2'')-Ia from *Enterobacteriaceae* are significant clinical problems conferring resistance to most clinically important 2-deoxystreptamine aminoglycosides including tobramycin, gentamicin C, and amikacin. These enzymes catalyze the transfer of an AMP group to aminoglycoside hydroxyls with release of pyrophosphate:

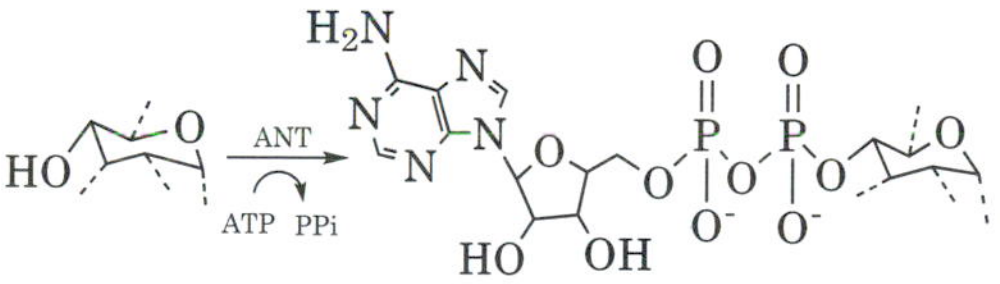

(15.2)

The 3D structure of dimeric ANT(4')-Ia in the absence of substrates and in the presence of kanamycin and the nonhydrolyzable ATP analog AMPCPP have been determined (103, 104). The enzyme active site is at the interface of the dimer with residues from both subunits contributing to substrate binding and catalysis. Although there is minimal primary sequence homology among the ANTs, several key residues are conserved, indicating that the

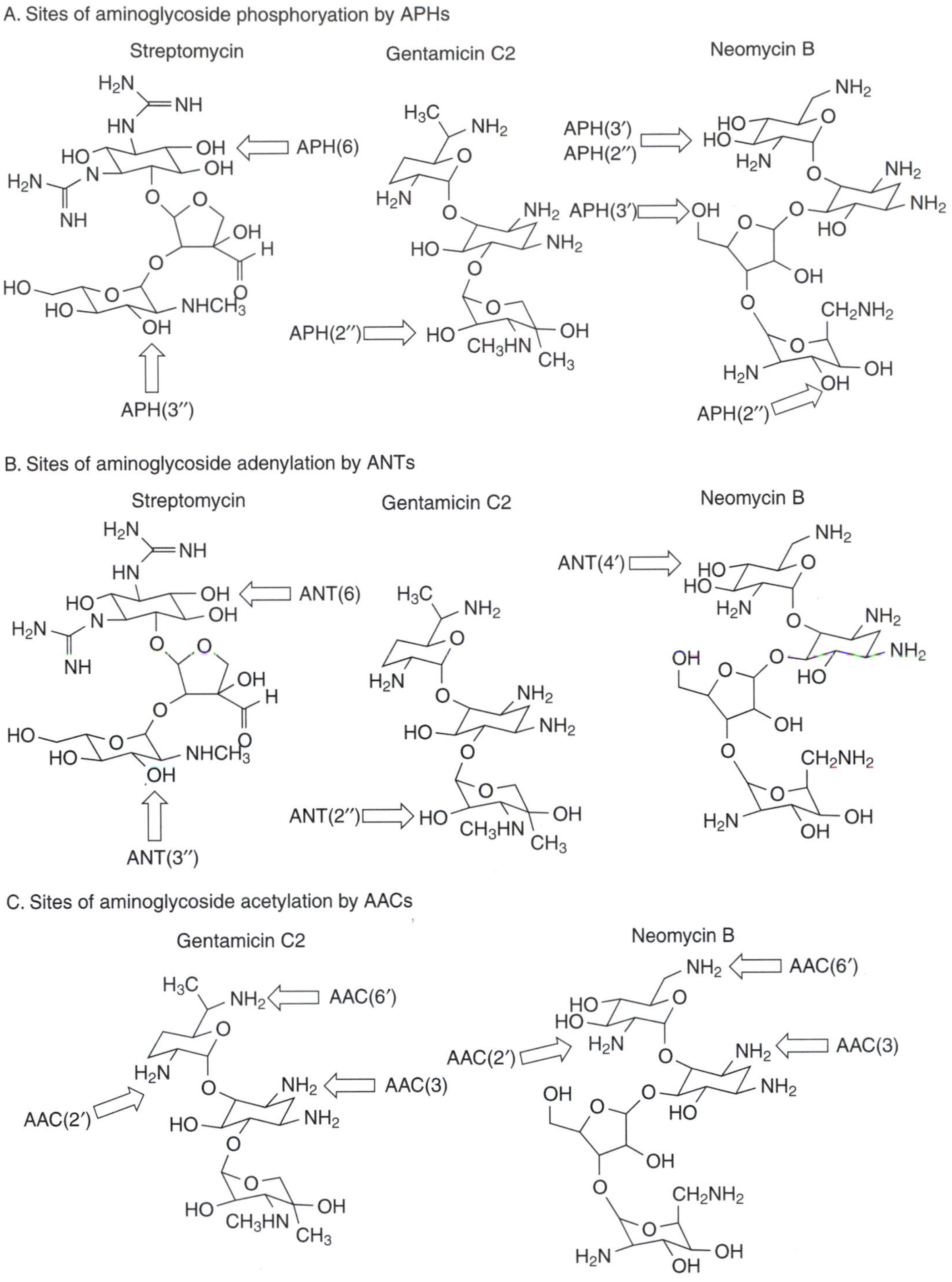

A. Sites of aminoglycoside phosphoryation by APHs

B. Sites of aminoglycoside adenylation by ANTs

C. Sites of aminoglycoside acetylation by AACs

**Figure 15.10.** Major sites of aminoglycoside modification catalyzed by resistance enzymes.

general structure and active site are largely conserved among the enzymes, though the specific details of aminoglycoside recognition are divergent.

The aminoglycoside acetyltransferases represent the largest group of resistance aminoglycoside enzymes. These enzymes N-acetylate positions 2', 6', 3, or, less frequently, position 1 of 2-deoxystreptamine aminoglycosides:

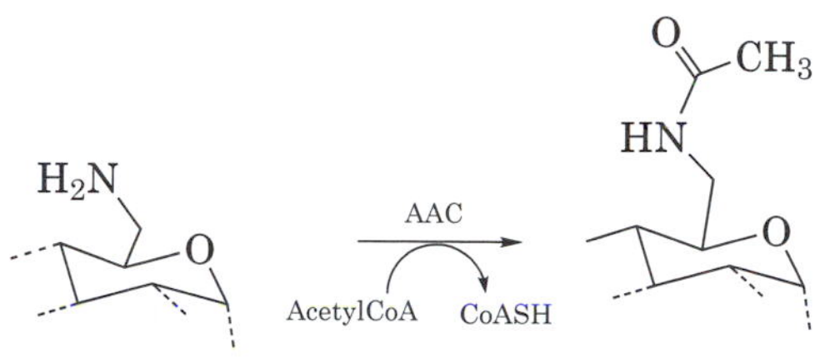

$$(15.3)$$

Whereas N-acetylation is clearly predominant, low level O-acetylation can occur with at least one enzyme (105). AACs are widely distributed in both Gram-positive and Gram-negative bacteria and are important causes of aminoglycoside resistance in the clinic, where AAC(3) isozymes are frequently associated with gentamicin resistance and AAC(6') with resistance to amikacin and tobramycin. The 3D structures of members of both of these families have been characterized: AAC(6')-Ii, a chromosomally encoded enzyme from *Enterococcus faecium* (106), and the plasmid-encoded AAC(3)-Ia from *S. marcescens* (107). Although these enzymes show little amino acid homology, the fold of the proteins is conserved and shared with other enzymes that belong to the GCN5 superfamily of acyltransferases (108).

## 3.6  Recent Developments in the Aminoglycoside Antibiotic Field

Research in the development of new antimicrobial aminoglycosides has not been extensive over the past decade but recently several new reports have appeared that examine these antibiotics and their derivatives in a creative fashion. The group of Chi-Huey Wong at the Scripps Research Institute has prepared a se-

ries of *O*-acyl and dimeric aminoglycosides based on the minimal neamine (**17**) core structure (109, 110). Some of these compounds such as (**18**) possess not only good antimicrobial activity but also inhibit some aminoglycoside resistance enzymes (111). A series of modified aminoglycosides, in which the amino groups at positions 1, 3, 2', 6' of neamine and kanamycin were replaced by hydrogen atoms, have been synthesized by Mobashery and colleagues at Wayne State University (112). Some of these compounds retain antibacterial activity and are not substrates for a subset of aminoglycoside resistance enzymes (112, 113). The same group also reported the synthesis and characterization of mechanism-based inactivators of aminoglycoside 3'-phosphotransferases by incorporating a nitro group at position 2', such as compound (**19**), that generates an alkylating species upon enzymatic phosphorylation at position 3' (114).

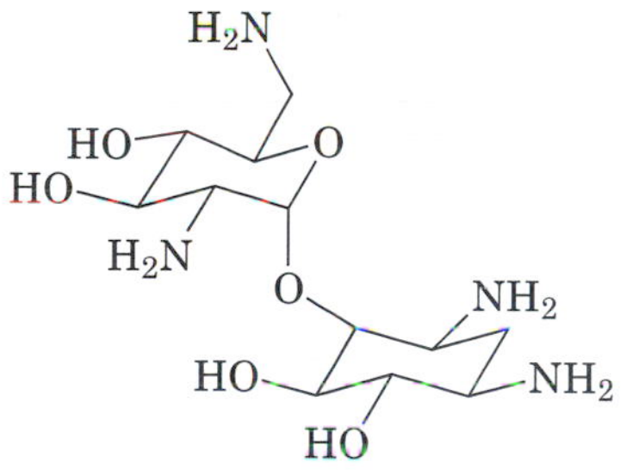

(**17**)

The affinity of aminoglycosides for RNA has proven useful in numerous studies (reviewed in Ref. 115). This includes selection of RNA aptamers with high and specific affinity for aminoglycoside using methods such as SELEX (systematic evolution of ligands by exponential enrichment). Aminoglycosides have also been shown to be inhibitors of ribozyme action such as Group I intron splicing (123–125), hammerhead (123–125), human hepatitis delta virus (126), ribonuclease P (127, 128), and others. These experiments raise the possibility of using aminoglycoside antibiotics and their derivatives to target specific RNA molecules directly in a therapeutically useful fashion.

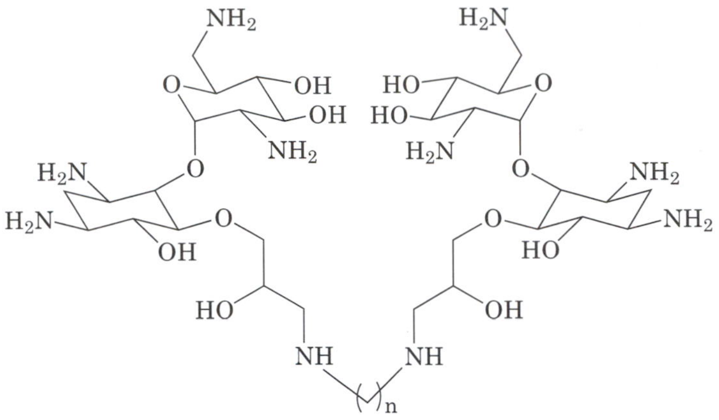

(18)

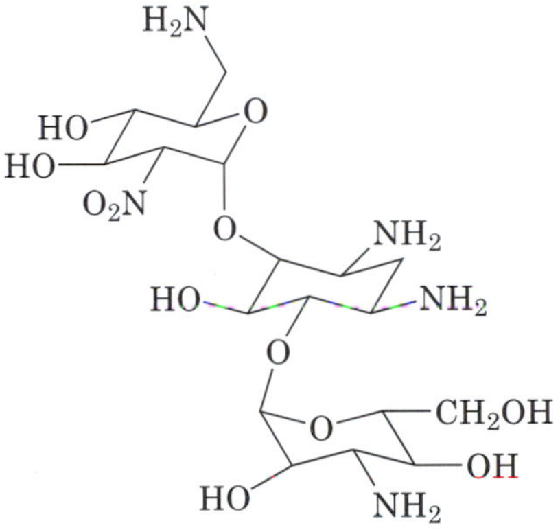

(19)

## 4 MACROLIDE ANTIBIOTICS

### 4.1 Clinical Use of Macrolide Antibiotics and Currently Used Drugs

Macrolide antibiotics belong to the family of macrocyclic antibiotics and are a well-established class of antibacterial agents for both human and veterinary applications. They have been used for the treatment of various bacterial infections in both out-patient and in-patient settings for more than 40 years, and as a result of their very good safety profile, they are extensively prescribed to children. The term macrolide means large macrocyclic lactone and was first used by R. B. Woodward for the class of natural products produced by *Streptomyces* species (129). Macrolide antibiotics are generally lipophilic and consist of a central highly substituted lactone ring (termed an aglycone) functionalized with various carbohydrate residues. This chapter covers antibacterial macrolides of clinical significance having aglycone of 12–16 atoms, with one or more sugars (or aminosugars) attached to the lactone core. The currently marketed macrolide antibiotics play a very important role in treating bacterial infections and have gained wide acceptance for the treatment of both upper and lower respiratory tract infections, as well as cutaneous infections. Because of the extensive medical use of this class of antibiotics, the term macrolides is now generally synonymous with macrolide antibiotics in the medical-related scientific communities. A book on the chemistry, pharmacology, and clinical uses of approved macrolides was published in 1993 (130); this chapter therefore presents a brief summary of information on these three areas.

Naturally occurring macrolide antibiotics are grouped into three major groups of 12-, 14-, and 16-membered macrolides with the aglycone consisting of 12-, 14-, and 16-atom cyclic lactone rings, respectively. For example, erythromycin A (**20**) is a 14-membered macrolide (a 14-atom cyclic lactone ring) and possesses desosamine and cladinose glycosidically linked to C-5 and C-3, respectively.

In addition to natural macrolides, many semisynthetic macrolides have been made and developed for clinical use. A 15-membered macrolide, the azalide azithromycin (**21**), that incorporates an additional nitrogen atom in

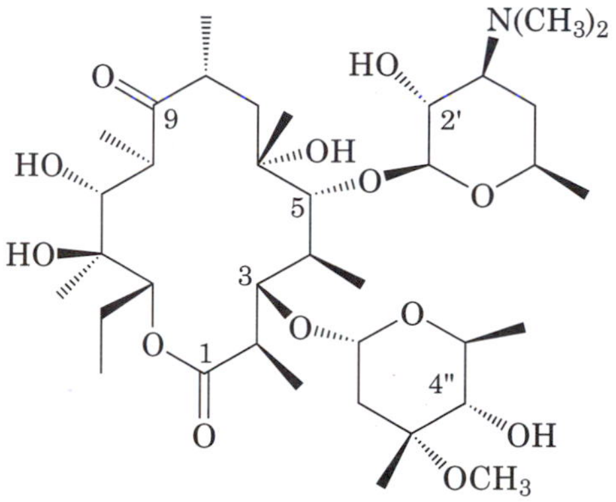

(20)

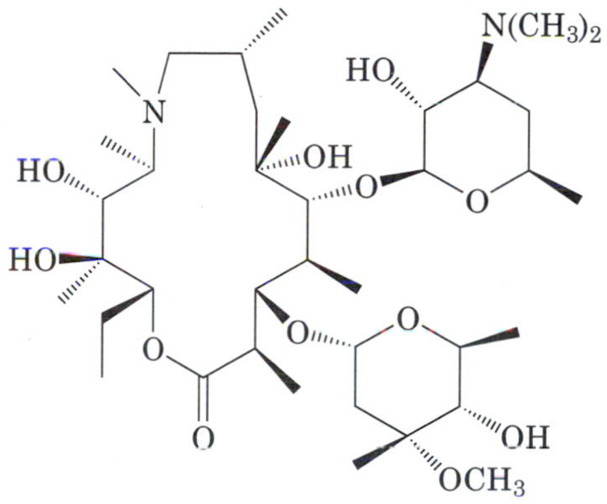

(21)

the aglycone has been successfully synthesized and developed for human use.

Among the 14-membered naturally occurring macrolides, erythromycin A is by far the most useful and possesses excellent antibacterial activity against Gram-positive bacteria and mycoplasmas. Erythromycin was developed initially for the treatment of staphylococcal infections for patients allergic to penicillin; however, after more than 40 years of use, most of the staphylococci isolated in hospitals are erythromycin resistant. Erythromycin and its semisynthetic derivatives are now used to treat lower and upper respiratory infections as well as skin and soft tissue infections, and are administered either orally or parenterally. Some of the disadvantages of erythromycin A are low bioavailability, a narrow spectrum of activity, and high gastrointestinal side effects.

These liabilities have prompted research laboratories throughout the world to search for better semisynthetic erythromycin derivatives. These derivatives possess improved therapeutic properties such as enhanced antibacterial activity and/or broadened spectrum (including Gram-negative bacteria such as *Haemophilus influenzae*), or improved pharmacokinetic properties as well as reducing gastrointestinal side effects.

Several semisynthetic derivatives of erythromycin A, including clarithromycin (**22**), dirithromycin (**23**), flurithromycin (**24**), and roxithromycin (**25**), have been successfully developed and are currently in clinical use. These semisynthetic newer macrolides are used for the following indications: community acquired pneumonia, acute bacterial exacerbation of chronic bronchitis, acute bacterial sinusitis, tonsillitis/pharyngitis, otitis media, skin and soft-tissue infections, and ophthalmologic infections. They possess good antibacterial activity against the common respiratory pathogens such as *Streptococcus pyogenes*, *Streptococcus pneumoniae*, *H. influenzae*, and *M. catarrhalis* as well as atypical bacteria such as *L. pneumophila*, *M. pneumoniae*, and *C. pneumoniae*. In addition to these derivatives of erythromycin A, an oleandomycin (**26**) analog called triacetyloleandomycin (**27**) has also been developed for limited clinical use.

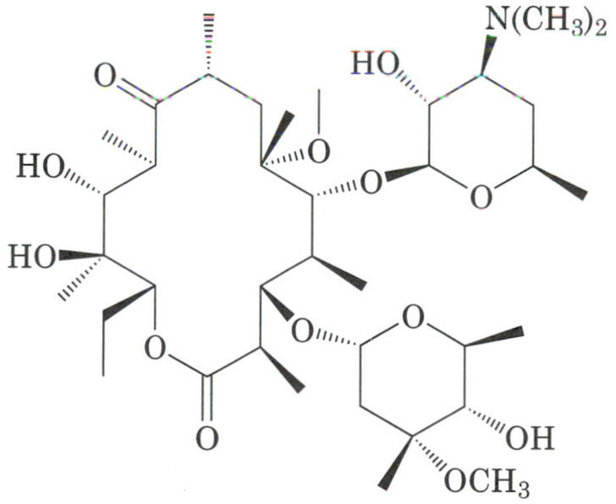

(22)

The 16-membered macrolides possess a wide diversity of structures and are usually subdivided into the leucomycin- and tylosin-related groups, with derivatives from the first

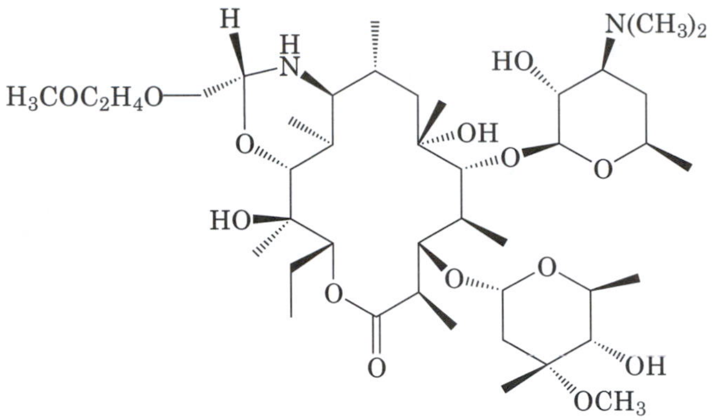
H
H
N
N(CH3)2
H3COC2H4O
HO
OH
O
O
HO
O
O
O
O
OH
OCH3

(23)

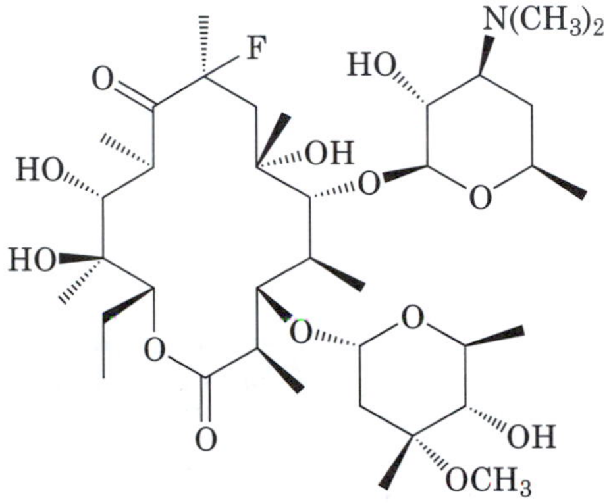
F
N(CH3)2
O
HO
HO
OH
HO
O
O
O
O
O
O
OH
OCH3

(24)

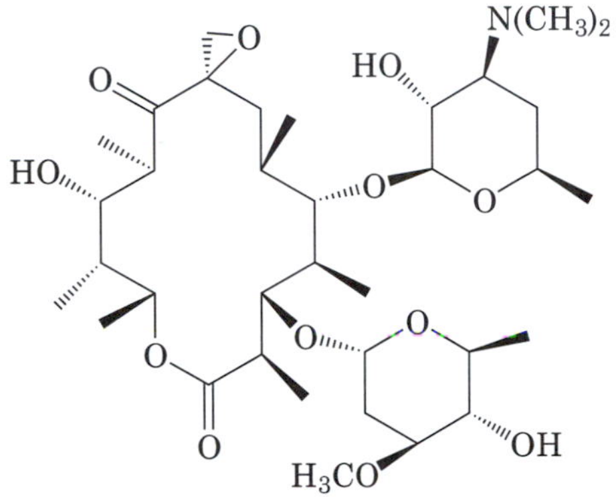
O
N(CH3)2
O
HO
HO
O
O
O
O
O
O
H3CO
OH

(26)

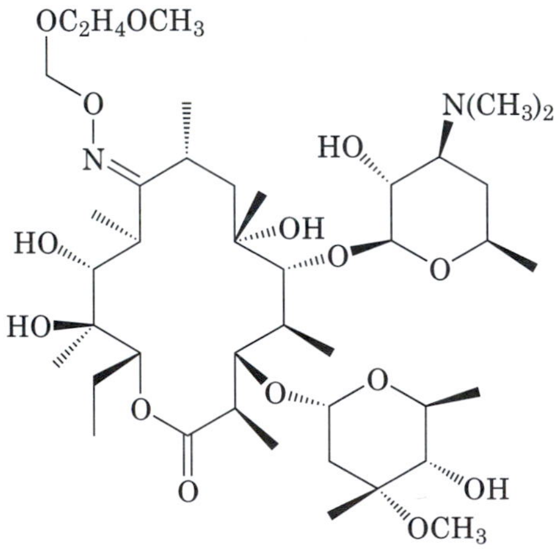
OC2H4OCH3
O
N
N(CH3)2
HO
OH
HO
HO
O
O
O
O
O
O
OH
OCH3

(25)

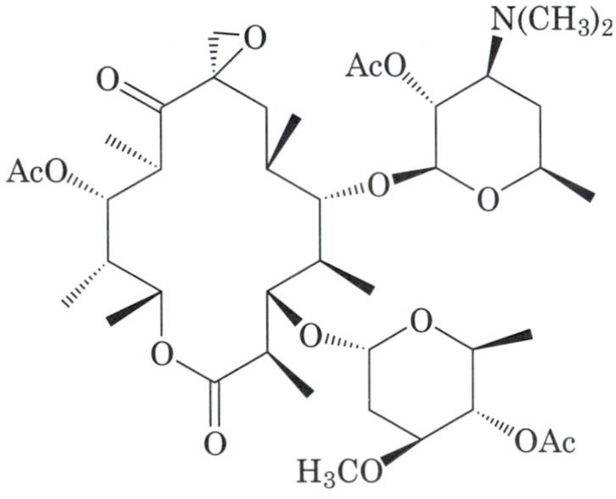
O
N(CH3)2
O
AcO
AcO
O
O
O
O
O
O
H3CO
OAc

(27)

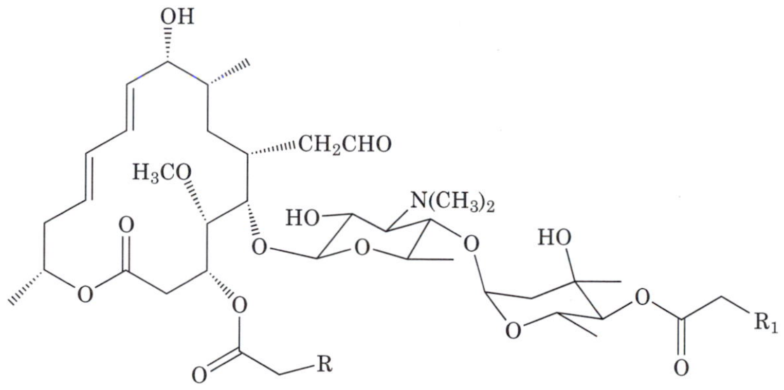

**(28)** R = H      R₁ = isopropyl
**(29)** R = CH₃    R₁ = CH₃

group developed for human use and the latter group for veterinary purposes. These
compounds, natural or semisynthetic, have
achieved limited success in human medicine
and include josamycin (**28**), midecamycin
(**29**), miokamycin (**30**), rokitamycin (**31**), and
spiramycin (**32**).

A list of macrolides currently on the market
for various clinical uses is summarized in Table 15.5.

## 4.2 Side Effects, Toxicity, and Contraindications of the Macrolide Antibiotics

The majority of side effects associated with
macrolide antibiotics are mild and transient

in nature. The most frequently reported
events in adults are diarrhea, nausea, abnormal taste, dyspepsia, abdominal pain (substantially less with newer macrolides), and
headache. Like other kinds of antibiotics,
pseudomembranous colitis has been reported
with macrolide use, ranging in severity from
mild to life-threatening. There have been isolated reports of transient central nervous system side effects such as confusion, hallucinations, seizures, and vertigo associated with
erythromycin use.

Macrolides are known to interact with cytochrome P450-dependent monooxygenases
that affect the metabolism and elimination of

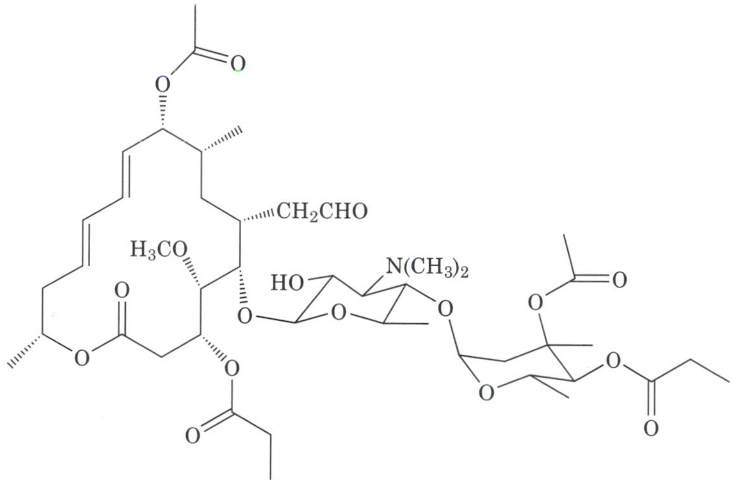

**(30)**

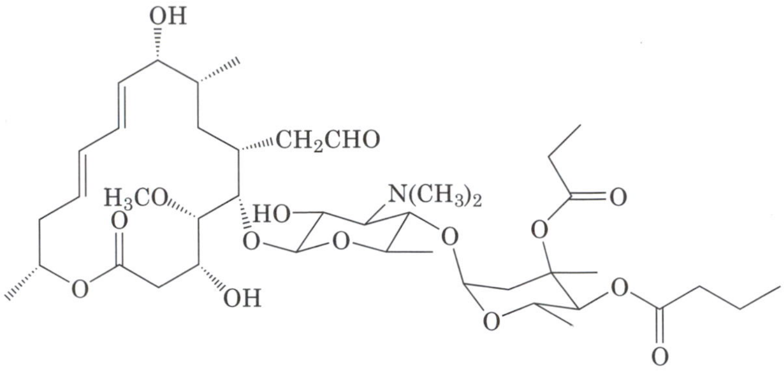

(31)

other drugs (131). Thus, macrolides may have an impact on the drug level of other drugs using these enzymes for drug metabolism. For example, combined therapy with a macrolide and theophylline (a bronchodilator) or carbamazepine (a psychotropic drug) causes an increase in the serum levels of the latter drugs. Cardiac arrhythmias such as ventricular tachycardia have been reported in patients receiving erythromycin A therapy. Thus, macrolides are contraindicated in patients receiving terfenadine therapy who have preexisting cardiac abnormalities (arrhythmia, bradycardia, QT interval prolongation, ischemic heart disease and congestive heart failure). Hepatotoxicity (132), ototoxicity (133), dermatologic effects (134), pancreatitis (135), cardiovascular toxicity such as QT prolongation (136) and induced hypotension (137), and hemolytic

anemia (138) have been reported less commonly for some earlier macrolides.

## 4.3 Pharmacology and Mode of Action of the Macrolide Antibiotics

Macrolide antibiotics are primarily administered orally. They are readily absorbed from the gastrointestinal tract (139, 140). Because macrolides are weakly basic, they are predominantly absorbed in the alkaline intestinal environment. Erythromycin A is acid unstable. Thus, its absorption among different patients is highly variable. The drug degrades differently during its passage through the acid environment of the stomach of different patients. The lower bioavailability of earlier macrolides is the result of acidic instability, incomplete absorption, and a first-pass effect. Many different water-insoluble salts or esters

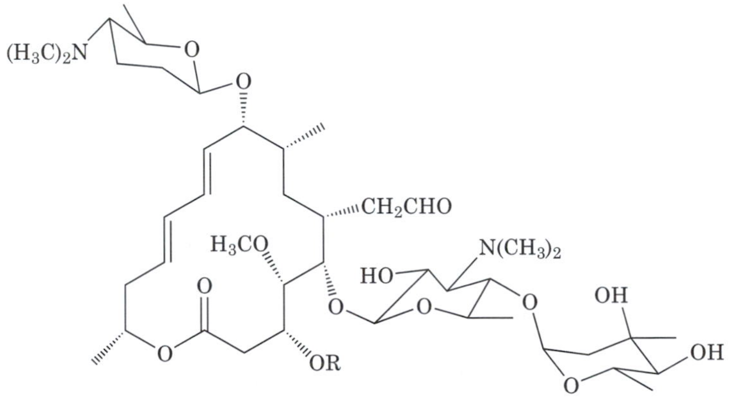

(32)

I R = H; II R = acetyl; III R = propionyl

**Table 15.5  Selected Commercial Clinical Macrolides**

| Generic Name | Trade Name | Originator | Class | Route |
|---|---|---|---|---|
| Erythromycin (enteric-coated tablet) | Ery-Tab | Abbott | 14 | Oral |
| Erythromycin (polymer-coated particles) | PEC Dispertab | Abbott | 14 | Oral |
| Erythromycin topical gel | Emgel | Glaxo Wellcome | 14 | Topical |
| Erythromycin ophthalmic ointment | Ilotycin ointment | Dista | 14 | Topical |
| Erythromycin + benzoyl peroxide gel | Benzamycin | Dermik | 14 | Topical |
| Erythromycin acistrate | Erasis (Finland) | Orion | 14 | Oral |
| Erythromycin ethylsuccinate | E.E.S. | Abbott | 14 | Oral |
| Erythromycin ethylsuccinate + sulfisoxazole acetyl | Erythromycin ethylsuccinate + sulfisoxazole acetyl | Lederle | 14 | Oral |
| Erythromycin estolate | Ilosone | Dista | 14 | Oral |
| Erythromycin latobionate | Erythrocin lactobionate | Abbott | 14 | i.v. |
| Erythromycin gluceptate | Ilotycin glucoheptonate | Dista | 14 | i.v. |
| Erythromycin stearate | Erythrocin stearate | Abbott | 14 | Oral |
| Erythromycin stinoprate | Eritrocist (Italy) | Edmond Pharma | 14 | Oral |
| Erythromycin-11,12-carbonate | Davercin (Poland) | Tarchomin | 14 | Oral |
| Azithromycin | Zithromax | Pfizer | 15 | Oral |
|  | Sunamed (Yugoslavia) | Pliva |  |  |
| Clarithromycin | Biaxin | Abbott | 14 | Oral |
| Dirithromycin | Dynabac | Sanofi | 14 | Oral |
| Flurithromycin ethylsuccinate | Flurizic (Italy) | Pierrel | 14 | Oral |
| Roxithromycin | Rulid (France) | HMR | 14 | Oral |
| Triacetyloleandomycin | TAO | Roerig | 14 | Oral |
| Josamycin | Josamycin (Japan) | Yamanouchi | 16 | Oral |
| Midecamycin | Medemycin (Japan) | Meiji Seika | 16 | Oral |
| Miokamycin | Miocamycin (Japan) | Meiji Seika | 16 | Oral |
| Rokitamycin | Ricamycin (Japan) | Toyo Jozo | 16 | Oral |
| Spiramycin | Rovamycine (France) | RPR | 16 | Oral |
| Telithromycin (pending) | Ketek | Aventis | Ketolide | Oral |

of erythromycin A have been prepared to protect it from acid degradation. A polymeric enteric-coated erythromycin A has also been made. The enteric coating can protect it from the degradation by the acidic environment of the stomach. Newer semisynthetic macrolides are more acid stable and retain good oral bioavailability.

The pharmacokinetics of a single-dose administration of roxithromycin was found to be nonlinear in the dose range of 150 to 450 mg given orally (141). The $C_{max}$ values were 7.9 and 12.4 mg/L, respectively. The $C_{max}$ and AUC values were decreased with food intake. The oral absorption of dirithromycin was rapid, having an absolute bioavailability of 10% (142). Upon oral administration of 500 mg of dirithromycin, the $C_{max}$ values ranged

from 0.1 to 0.5 $\mu$g/mL, with an elimination serum half-life of 44 h and the oral bioavailability ranged from 6 to 14% (143). The absorption is not significantly affected by food (144). The oral bioavailability of azithromycin is 37%. A regimen of 2 × 500 mg on the first day followed by the maintenance dose of 500 mg daily gave a mean $C_{max}$ of 0.62 $\mu$g/mL (145). The apparent half-life between 8 and 24 h was 11–14 h. In humans, clarithromycin was nearly completely absorbed. Over a dose range of 100–1000 mg administered orally, the pharmacokinetics of clarithromycin was dose dependent, with a terminal half-life ranging from 2.3 to 6 h (146).

Macrolides bind to plasma and interstitial proteins and binding to plasma proteins varies widely from 10 to 93% at therapeutic concentrations. Erythromycin A and roxithromycin bind specifically to $\alpha$-1-acid glycoprotein (AGP) and, to a minor extent, nonspecifically to albumin (147). Roxithromycin (300 mg daily), clarithromycin (500 mg daily), and azithromycin (500 mg daily) for 12 days provided steady-state maximum and trough plasma concentration values of approximately 10 and 1.9 $\mu$g/mL; 2.1 and 0.4 $\mu$g/mL; and 0.19 and 0.05 $\mu$g/mL, respectively (148). The terminal elimination half-life for roxithromycin was significantly prolonged in patients with severely impaired renal function (15.5 h) compared with that of the group with normal renal function (7.9 h) (149). Clarithromycin and its 14-OH active metabolite were found to be extensively distributed into human lung tissue of patients undergoing lung resection. A mean calculated ratio of concentrations of 11.3 in lung tissue and 2.4 in plasma were found (150). The pharmacokinetics of clarithromycin suspension in infants and children has been reported (151). The drug absorption was rapid, reaching a mean peak plasma concentration within 3 h. The mean $C_{max}$ of 4.58 and 1.26 $\mu$g/mL for its metabolite were obtained upon a single dose of 7.5 mg/kg clarithromycin suspension to children. After a single oral dose of 500 mg, concentrations of azithromycin in all tissues were higher than that in serum (152). The pharmacokinetics of a 3-day (500 mg/day) and 5-day (500 mg on day 1, followed by 250 mg/day on days 2–5) regimens of azithromycin were found to be similar, having identical plasma profiles ($C_{max}$ 0.37 versus 0.31 $\mu$g/mL) on day 1. The accumulation of azithromycin in plasma was higher with the 3-day regimen than that with the 5-day regimen (0.31 versus 0.18 $\mu$g/mL) on the last day (153). Once-a-week azithromycin in AIDS patients on zidovudine did not alter the disposition of zidovudine (154). The concentration of azithromycin was elevated in middle ear effusion in children, with effusion concentration of 1.02 $\mu$g/mL at 12 h after the oral administration of 10 mg/kg (155). Concentrations of azithromycin in gallbladder and liver were usually within twofold of each other and ≥20-fold greater than those in serum (156).

As a result of their appropriate antibacterial spectrum as well as good tissue distribu-tion profiles, macrolides are excellent therapeutic agents for upper and lower respiratory infections. A high lung-to-plasma ratio is found with newer macrolides such as azithromycin, clarithromycin, and roxithromycin. The ability of macrolides to concentrate in phagocytes and alveolar macrophages (9 to >20 times extracellular levels) accounts for the high tissue concentrations. Clarithromycin attained high and balanced concentrations both intracellularly (to levels severalfold higher than that of serum) and extracellularly (to levels approximately equal to or higher than that of serum). The peak concentration in $\mu$g/mL or $\mu$g/g of clarithromycin after oral dose of 250 mg bid for bronchial secretion, lung (500 mg bid), tonsil, nasal mucosa, and saliva were 3.98, 13.5, 5.34, 5.92, and 2.22, respectively (157).

The primary site of metabolism of macrolides is the liver and, to a lesser degree, the kidneys and lungs (139). The common metabolic pathways for 14-membered macrolides include the $N$-demethylation of the desosamine by the P450 3A and, to a lesser extent, the hydrolysis of the neutral sugar cladinose. Roxithromycin does not undergo extensive metabolism, with the parent compound found in both urine and feces. Minor metabolites, the mono- and di-$N$-demethyl, and the descladinosyl derivatives were found in the urine (158). Transformations specific for individual macrolides are given below. Erythromycin A undergoes an intramolecular cyclization reaction under acidic conditions to form the 8,9-anhydroerythromycin-6,9-hemiketal (**33**) and subsequently the erythromycin-6,9;9,12-spiroketal (**34**) (159).

Erythromycylamine (**35**) is the principal metabolite of dirithromycin found in both urine and feces (160). Azithromycin does not metabolize extensively, with the parent compound accounting for 75% of the excreted drug-related substances (161). Minor metabolites are derivatives resulting from various reactions or combinations thereof, such as 3'-$N$-demethylation, 9a-$N$-demethylation, hydrolysis of the cladinose, 3"-$O$-demethylation, hydroxylation, and hydrolysis of the lactone core. 14-($R$)-Hydroxyclarithromycin (**36**) was the major metabolite found in the plasma in humans together with as many as eight minor

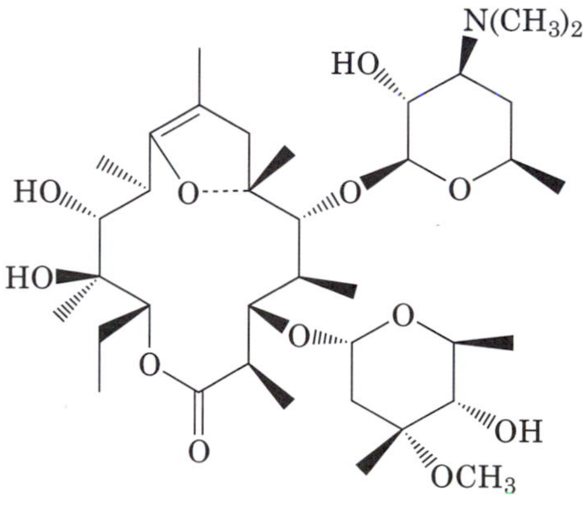

(33)

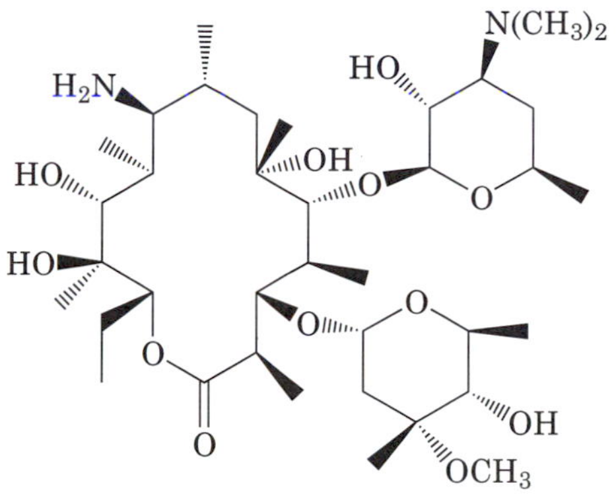

(35)

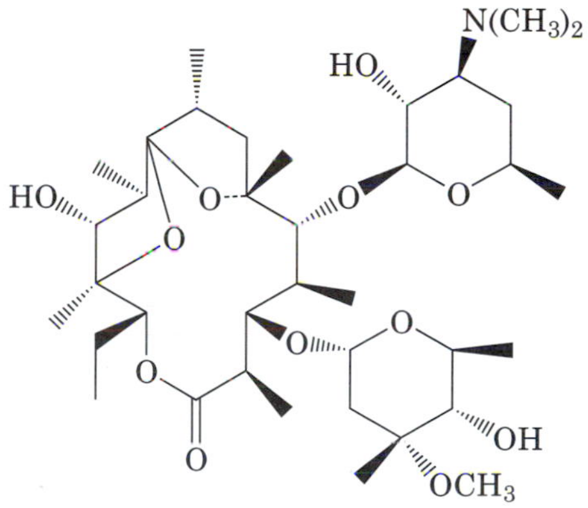

(34)

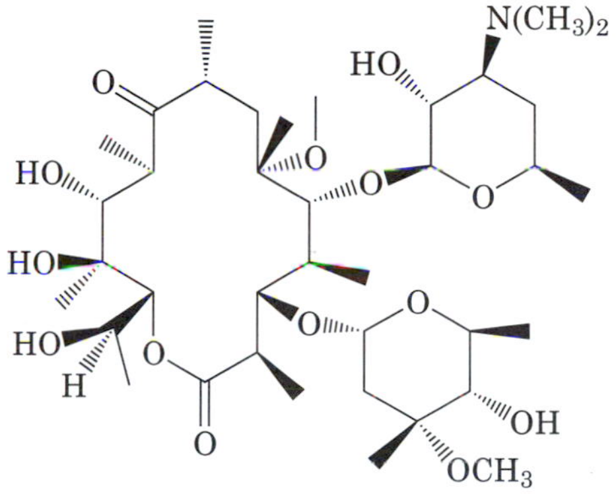

(36)

metabolites (162). Combination of the 14-(R)-hydroxyclarithromycin and the parent clarithromycin produced a synergetic effect against *H. influenzae* (163). As for the metabolism of 16-membered macrolides, deacylation of the esters and hydroxylation by enzymatic oxidative reaction are more important than *N*-demethylation (164).

Macrolide antibiotics inhibit protein biosynthesis through interaction with the 50S subunit of the bacterial ribosome. Erythromycin A possesses high specificity and affinity for the bacterial 50S subunit, having a dissociation constant of approximately $10^{-8}$ *M*, as measured by equilibrium dialysis (165) and footprinting (166). Specifically, this class of antibiotics binds to the ribosomal P-site. The exit of this site includes a tunnel through

which the newly formed peptide emerges after peptide bond formation (167, 168). The atomic structures of the macrolides erythromycin, clarithromycin, and roxithromycin bound to the 50S ribosomal subunit of *Deinococcus radiodurans* have been determined and have revealed that these antibiotics block the entrance to the peptide elongation tunnel, thereby providing the molecular rationale for inhibition of translation by these antibiotics (Fig. 15.11) (169).

## 4.4 History, Biosynthesis, and Structure-Activity Relationships of the Macrolide Antibiotics

The most widely used macrolide antibiotic, erythromycin A was first identified in the fermentation products of a strain of *Saccharopolyspora erythraea* (formerly known as

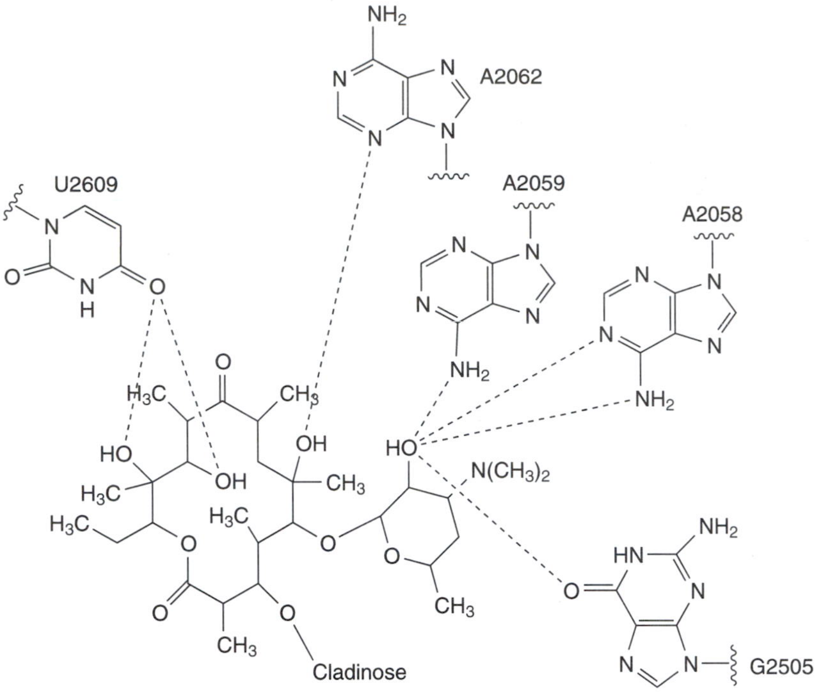

**Figure 15.11.** Interaction of erythromycin with the 50S ribosomal subunit of *D. radiodurans* (redrawn from Ref. 169). The equivalent *E. coli* numbering is shown and the interaction with G2057 is predicted based on the interaction with A2040 for the *D. radiodurans* 23S rRNA.

*Streptomyces erythreus*) isolated from a soil sample from the Philippines (170). The various aglycones (12-, 14-, and 16-monocyclic lactones) are produced through polyketide biosynthetic pathways. Thus, many macrolides with their core lactones share similar substituents and stereochemistry predicted by the polyketide biosynthetic mechanisms (170, 171). The biosynthesis of the macrolides can best be illustrated by the synthesis of erythromycin A (Fig. 15.12). Erythromycin A is composed of a 14-membered aglycone, to which are attached 6-deoxysugars, D-desosamine at C-5, and L-cladinose at C-3. The aglycone deoxyerythronolide (**37**) is assembled by a polyketide synthase complex, deoxyerythronolide B synthase (DEBS) encoded by the *eryAI*, *eryAII*, and *eryAIII* genes corresponding to DEBS1, DEBS2, and DEBS3, respectively (172). After the completion of the synthesis of (**37**) by DEBS, it is hydroxylated at C-6 by the P450 hydroxylase EryF to produce erythronolide B (**38**) (173, 174). Addition of L-mycarose by the gene products of the EryB locus yields 3-α-mycarosylerythronolide B (**39**). Subsequent addition of D-desosamine by EryC-associated enzymes gives the first bioactive macrolide in the biosynthetic pathway erythromycin D (**40**). Conversion of (**40**) to erythromycin A is accomplished by two enzymatic actions: (1) a P450 hydroxylase encoded by *eryK* gene hydroxylates the C-12 of erythromycin D [to give erythromycin C (**41**)] (175) and (2) an *O*-methyltransferase encoded by *eryG* gene methylates the hydroxyl at C''-3 on the mycarose moiety of erythromycin C [to yield erythromycin A (**20**)] (176).

The 14-membered macrolides are in general more potent, having lower MIC values than those of the 12- or 16-membered macrolides (177). Currently, the 14- and 15-membered macrolides are by far the most useful in terms of use as human therapeutics. Their comparative *in vitro* antibacterial activities are given in Table 15.6.

Many chemical modifications of erythromycin have been performed over the years. The orientation of the sugars relative to the

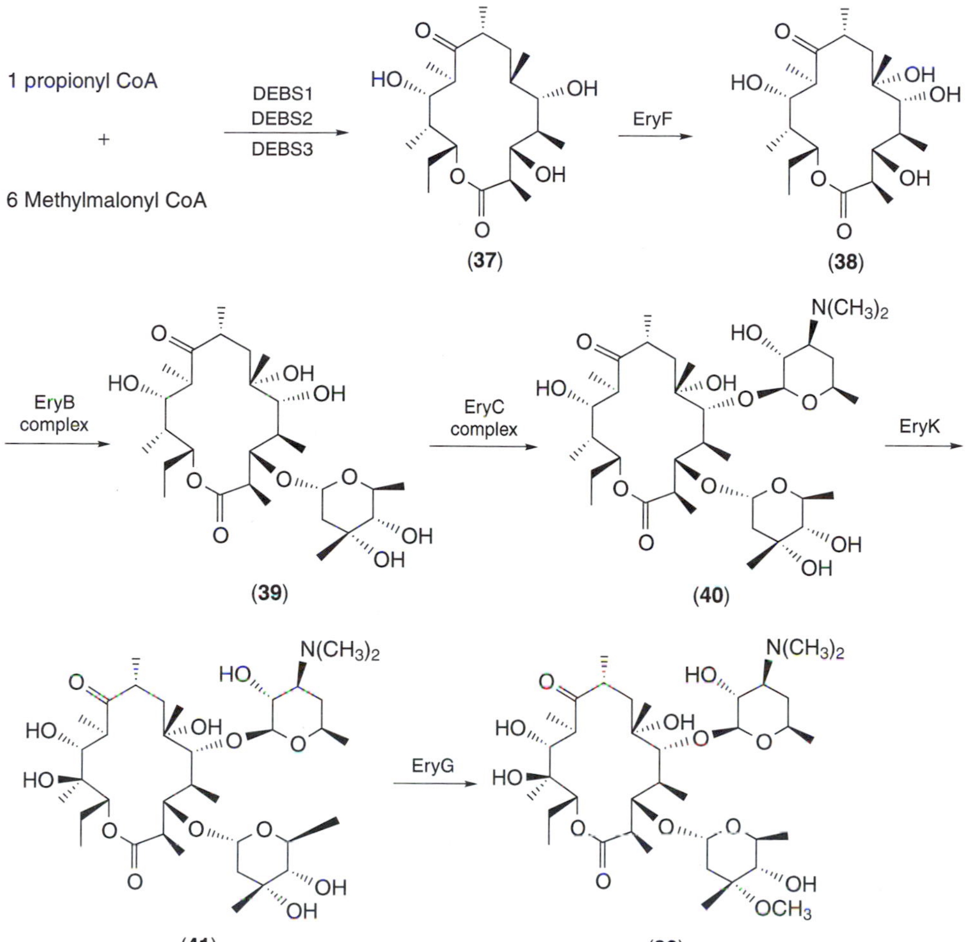

**Figure 15.12.** Biosynthesis of erythromycin by *S. erythraea.*

aglycone macrolactone in erythromycin A is considered to be important to ribosomal binding and hence antibacterial activity. The conformation of the erythronolide is therefore important because of its influence on the interaction of erythromycin A with the bacterial ribosome (178). Erythromycin A is acid labile, making it readily susceptible to degradation in the stomach. The acid-conversion products (33) and (34) lack antibacterial activity. Research efforts in the past several years were focused on strategies to prevent this undesirable acid degradation.

In the early years, relatively water-insoluble, acid stable salts, esters, and formulations were developed to protect erythromycin A's

passage through the stomach. An additional benefit from this approach was that these compounds mask the bitter taste associated with macrolides. Ester derivatives of erythromycin A were prepared by the acylation of the 2'-hydroxyl group. These esters (such as erythromycin propionate, acetate, and ethyl succinate) are prodrugs and are converted back to the active parent erythromycin A by hydrolysis in the body. Triacetyloleandomycin (27) possesses improved oral bioavailability and taste over that of oleandomycin.

The initial erythromycin A acid-degradation product (33) is formed by the intramolecular cyclization of the 9-ketone and C-6-hydroxyl groups. The subsequent production of a

**Table 15.6  Comparative *In Vitro* Activities of Selected Macrolides[a]**

| Organism | MIC$_{90}$ (μg/mL) of | | | | | | | |
|---|---|---|---|---|---|---|---|---|
| | E | C | A | R | EA | D | F | AZ |
| *S. aureus* | >128 | >128 | >128 | >128 | >128 | >128 | >128 | >128 |
| MRSA | >128 | >128 | >128 | >128 | >128 | >128 | >128 | >128 |
| *S. epidermidis* | >128 | >128 | >128 | >128 | >128 | >128 | >128 | >128 |
| *S. pyogenes* | 0.03 | 0.015 | 0.03 | 0.06 | 0.12 | 0.12 | 0.06 | 0.12 |
| *S. pneumoniae* | 0.03 | 0.015 | 0.015 | 0.03 | 0.06 | 0.12 | 0.06 | 0.12 |
| *S. agalactiae* | 0.06 | 0.06 | 0.06 | 0.25 | 0.12 | 0.25 | 0.12 | 0.12 |
| *Corynebacterium* spp. | 16 | 4 | 8 | 16 | 8 | 8 | 16 | 128 |
| *L. monocytogenes* | 0.5 | 0.25 | 0.5 | 1 | 1 | 2 | 0.5 | 2 |
| *B. catarrhalis* | 0.25 | 0.25 | 0.12 | 1 | 0.25 | 0.25 | 0.25 | 0.06 |
| *N. gonorrhoeae* | 0.5 | 0.5 | 0.5 | 1 | 2 | 4 | 1 | 0.06 |
| *C. jejuni* | 1 | 2 | 2 | 4 | 1 | 0.25 | 1 | 0.12 |
| *L. pneumophila* | 2 | 0.25 | 0.5 | 0.5 | 8 | 16 | 2 | 2 |
| *H. influenzae* | 4 | 8 | 4 | 8 | 8 | 8 | 8 | 0.5 |
| *B. pertussis* | 0.03 | 0.03 | 0.06 | 0.25 | 0.03 | 0.03 | 0.06 | 0.06 |
| *B. fragilis* | 4 | 2 | 1 | 32 | 32 | >128 | 8 | 2 |
| *C. perfringens* | 1 | 0.5 | 0.5 | 2 | 1 | 4 | 2 | 0.25 |
| *P. acnes* | 0.03 | 0.03 | 0.06 | 0.06 | 0.5 | 0.5 | 1 | 0.03 |
| *P. streptococcus* spp. | 4 | 4 | 4 | 32 | >128 | >128 | 16 | 2 |
| *Enterococcus* spp. | >128 | >128 | >128 | >128 | >128 | >128 | >128 | >128 |

[a]Data taken from Ref. 177. E, erythromycin; C, clarithromycin; A, A-62671; R, roxithromycin; EA, erythromycylamine; D, dirithromycin; F, flurithromycin; AZ, azithromycin; MRSA, methicillin-resistant *Staphylococcus aureus*.

6,9;9,12-spiroketal (**34**) involves the action of C-12 hydroxyl group on the C-8–C-9 olefin. Chemical modification aimed at these functional groups (i.e., at positions C-9, C-8, C-6, and C-12) was attempted in search of stable derivatives with antibacterial activity.

**4.4.1 C-9 Ketone Modification.** Reduction of erythromycin gave 9-dihydroerythromycin, which is more stable than erythromycin A but has less potent antibacterial activity (179). However, reduction of the oxime or hydrazone of erythromycin yielded 9-(*S*)-erythromycylamine (**35**) with antibacterial activity similar to that of erythromycin A but now with decreased oral bioavailability (180). Five series of 9-oximino-ether derivatives of erythromycin A have been prepared and evaluated (181). These are aliphatic-, aromatic-, oximino-ethers containing nitrogen, oxygen, or sulfur atoms. The oxime with *E* stereochemistry is more active than the isomer with *Z* configuration. The more important of these derivatives is roxithromycin (**25**), a derivative having a methoxyethoxyether side-chain (182). It possesses slightly less antibacterial activity than that of erythromycin A, with the exception of *Ureaplasma urealyticum*, against which it is more potent. Its antibacterial spectrum is similar to that of erythromycin A (183). The 11-*O*-alkyl-9-oximino ether ER 42859 (**42**) was also prepared and is more acid stable than erythromycin (184).

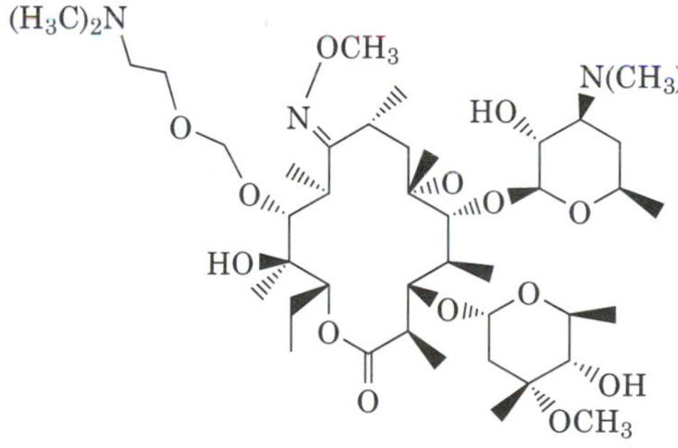

(**42**)

The 9-(*S*)-erythromycylamine was found to be poorly absorbed after oral administration in phase I clinical study (185). To improve its oral bioavailability, a series of 9-*N*-alkyl derivatives of erythromycylamine were prepared and were found to have potent *in vitro* and *in vivo* antibacterial activity. The *N*-(1-propyl)-9-(*S*)-erythromycylamine (LY 281389) (**43**) has *in vitro* activity similar to that of erythromycin A with onefold better activity against *H. influenzae*. Its *in vivo* activity is substantially better and it is more acid stable with superior oral bioavailability, achieving higher plasma and tissue levels (186). Although the *S*-isomer is more active in the monosubstituted 9-amino erythromycylamine, for the dialkylamino series, the 9-(*R*) isomer is more active. A-69991 (**44**) and A-70310 (**45**) are the two 9-(*R*) azacyclic erythromycin derivatives with superior *in vivo* efficacy and pharmacokinetic parameters than those of erythromycin A (187). They are as active as erythromycin A *in vitro* (188).

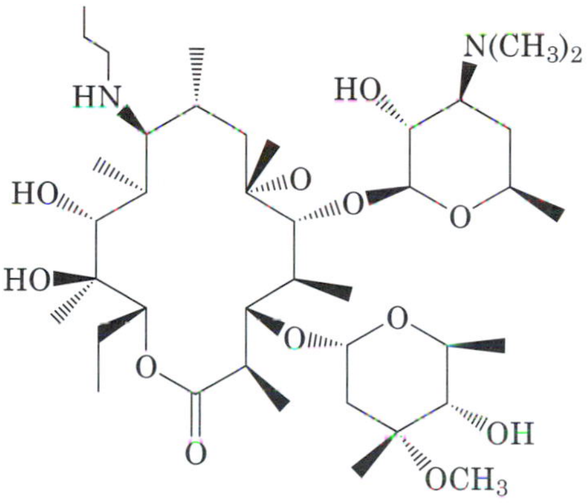

(**43**)

Another approach to improve the oral bioavailability of (**35**) is to search for a prodrug of erythromycylamine. Dirithromycin rapidly nonenzymatically hydrolyzes under acid condition back to erythromycylamine. When given subcutaneously in mice and rats, the serum levels of dirithromycin and its major metabolite erythromycylamine were found to be higher than the corresponding values obtained for erythromycin A (189).

A series of 9-deoxo-12-deoxy-9,12-epoxy erythromycin derivatives were prepared. A-69334 [(9*S*,11*S*)-11-amino-9-deoxo-11,12-dideoxy-9,12-epoxyerythromycin A (**46**)] possesses antibacterial activity similar to that of

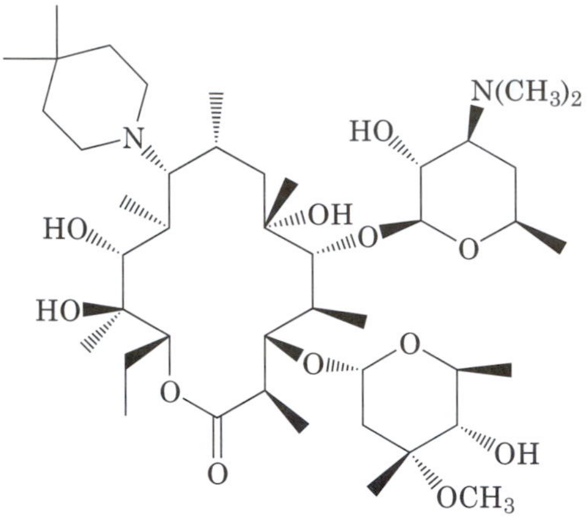

(44)

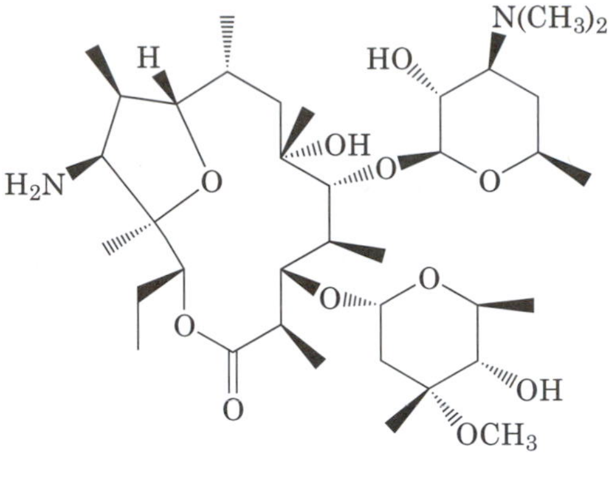

(46)

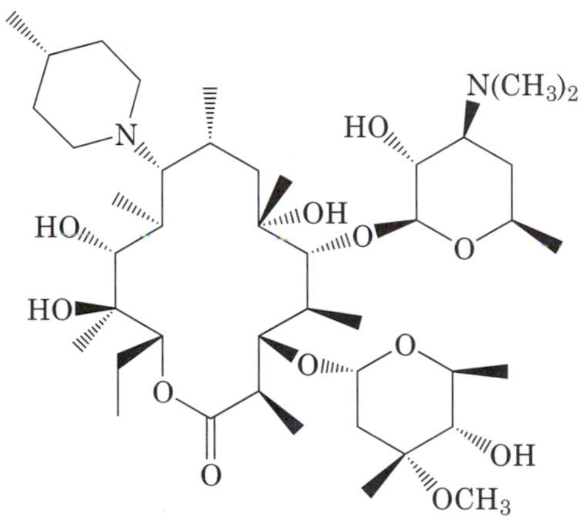

(45)

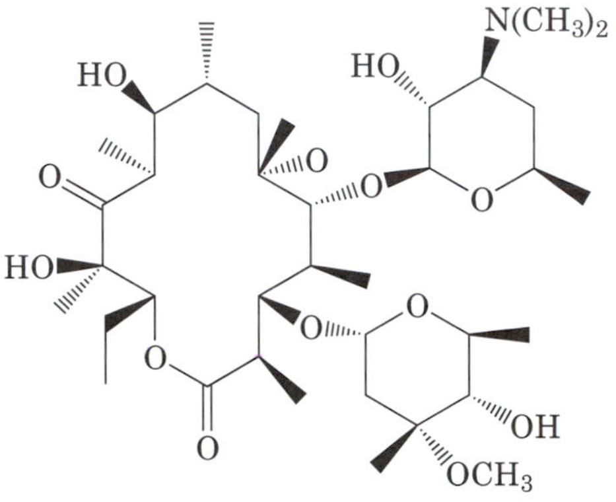

(47)

erythromycin A, with twofold more potent activity *in vitro* against *S. pyogenes* C203 and *S. pneumoniae* 6303 (190). It showed superior oral efficacy over that of erythromycin in *H. influenzae*–induced otitis media in a gerbil model (191). An erythromycin derivative with 9,11 functional groups interchanged, (9S)-11-dehydroxy-9-deoxo-9-hydroxy-11-oxo-erythromycin A (**47**), was made and found to be slightly less active than erythromycin A (192).

**4.4.2 C-8 Modification.** To prevent the formation of anhydrohemiketal erythromycin (**33**), the 8-fluoro-erythromycin A, flurithromycin, was prepared by the use of both chemical and biotransformation procedures (193,

194). Although its activity against staphylococci, streptococci, *H. influenzae*, and *M. catarrhalis* was similar to that of erythromycin A, it was more potent against anaerobes (195).

**4.4.3 C-6 Modification.** A family of 6-deoxyerythromycins was prepared by using a genetically engineered strain of *S. erythraea*, in which the *eryF* gene (which encodes the P450, C-6–specific hydroxylase) had been inactivated (196). These 6-deoxy derivatives are less potent than their corresponding erythromycins against bacteria *in vitro*. However, 6-deoxyerythromycin A (**48**) is as potent as erythromycin A in mouse protection tests against *S. aureus*, *S. pyogenes*, and *S. pneumoniae* (197). 6-Deoxyerythromycin A loses antibacterial activity slowly, although there is no C-6 hydroxyl group participation in the degradation process. A loss of the cladinose moiety was observed when it was treated with 10% acetic acid (198). These results suggest that the C-6 hydroxyl group (or another substituent at this position) may be important for biological activity. This hypothesis has been confimed with the availability of the 3D structure of erythromycin bound to the 50S ribosomal subunit, where this hydroxyl is within hydrogen-bonding distance with N of A2062 (*E. coli* numbering) (Fig. 15.11) (169).

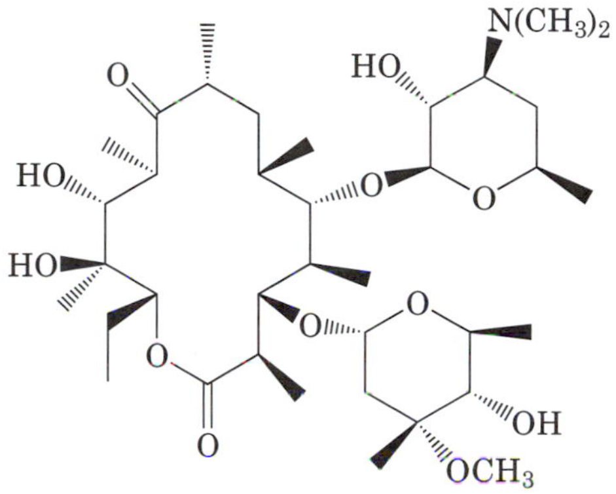

(48)

To define the importance of the 6-hydroxyl group on the biological activity of the 14-membered macrolide, a series of 6-*O*-alkyl derivatives of erythromycin have been synthesized (199). Alkylation by small alkyl groups at the 6-hydroxyl was found to have a minimal impact on the biological activity. However, this modification makes the molecule more acid stable, improving its pharmacokinetic profiles. 6-*O*-Methylerythromycin A, known as clarithromycin, has been developed and is now marketed worldwide (200). 6-*O*-Methyl derivatives of erythromycin A are much more acid stable than 6-deoxyerythromycin A. Although clarithromycin's antibacterial spectrum is similar to that of erythromycin A (201), it is about half as active as erythromycin A against *H. influenzae*. Its 14-hydroxy human metabolite, 14-(*R*)-hydroxyclarithromycin, possesses the same MIC against *H. influenzae* as erythromycin (163). Combining it with clarithromycin produced a synergistic effect against *H. influenzae* both *in vitro* (202) and *in vivo* (203). Clarithromycin was found to be safe and effective upon combination with other antimyco-

bacterial agents for the treatment of disseminated *M. avium* complex infection in AIDS patients (204). Clarithromycin was also found to be effective when combined with a proton pump inhibitor for the treatment of metronidazole-resistant *H. pylori*–positive gastric ulcer (205).

**4.4.4 C-11 and C-12 Modification.** A series of erythromycin-11,12-methylene cyclic acetals was prepared to test for antibacterial activity. Although the erythromycin-11,12-methyleneacetal (**49**) is more active than erythromycin A *in vitro*, it possesses similar efficacy to that of erythromycin A in treating experimental infections in mice (206).

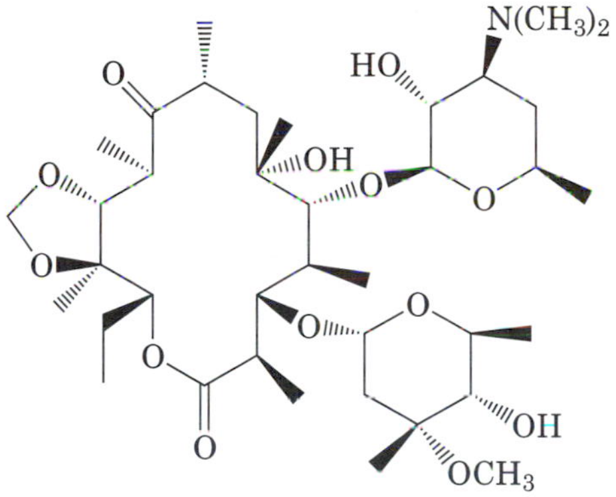

(49)

Erythromycin-11,12-carbonate (**50**) was prepared to avoid the spiroketal formation during the degradation process of erythromycin A in acid medium (207). It is twice as active *in vitro* and more stable, with better pharmacokinetics than that of erythromycin A (208). However, it has higher hepatotoxicity potential and its development was discontinued (209, 210). The 3″-fluoro-11,12-carbonate derivatives of erythromycins (**51**) and (**52**) possess about half the *in vitro* activity as that of erythromycin A (211).

A series of 11-deoxy-11-(carboxyamino)-6-*O*-methylerythromycin A 11,12-cyclic carbamate derivatives (**53**) have been synthesized (212). They possess *in vitro* antibacterial activity comparable to that of clarithromycin.

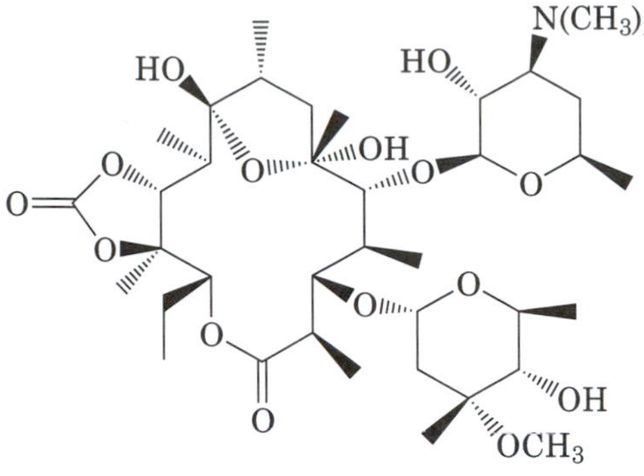

(50)

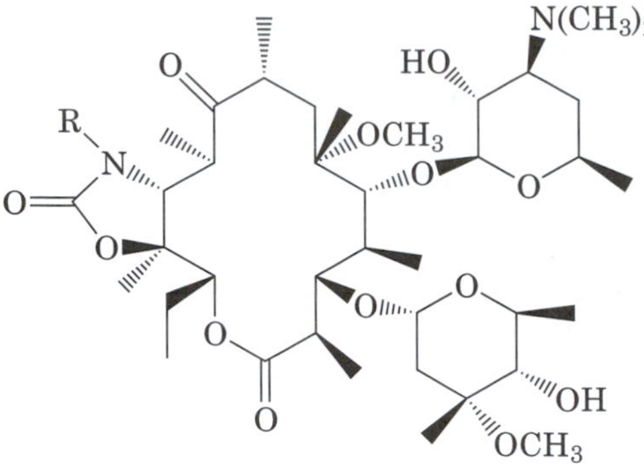

(53)

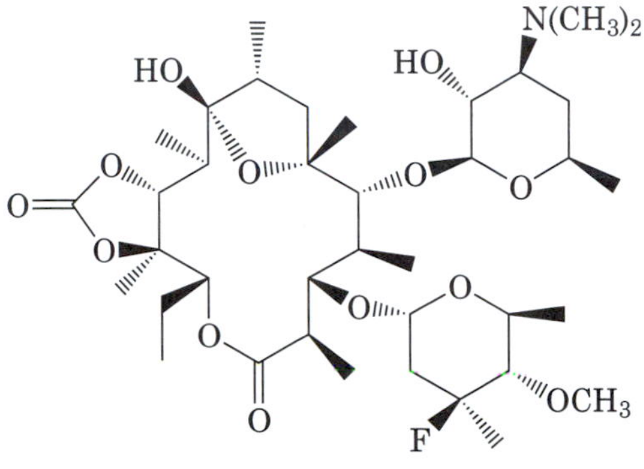

(51)

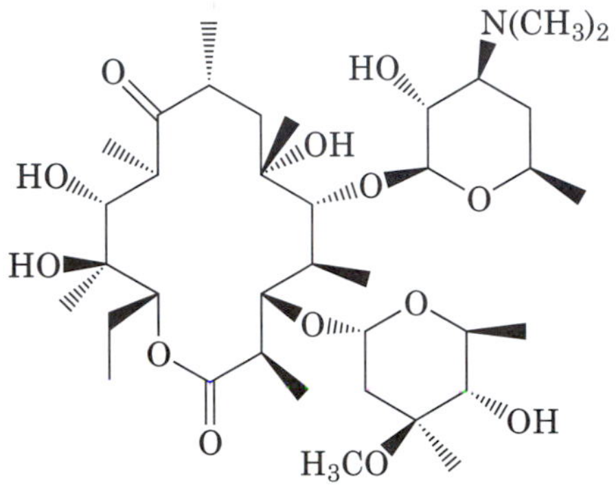

(54)

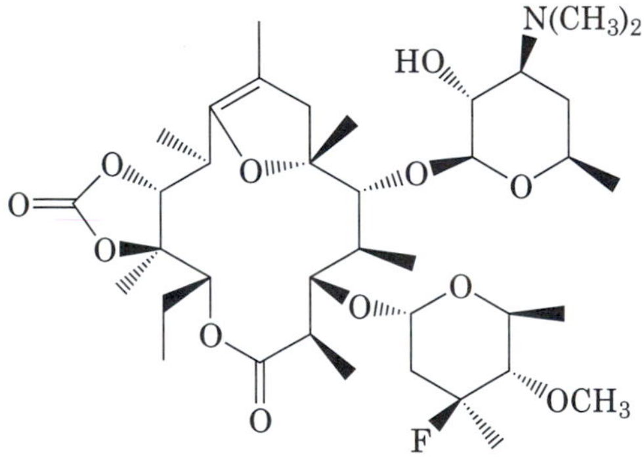

(52)

**4.4.5 Cladinose Modification.** Modification on the cladinose sugar has been attempted. The 3″-epi-erythromycin A (**54**) has *in vitro* antibacterial activity comparable to that of erythromycin A, indicating that the stereo-

chemistry at the 3″ position does not have great influence on the antibacterial activity (192).

**4.4.6 Core–Skeleton Modification.** Natural antibacterial macrolides have 12-, 14-, and 16-membered ring core structures by virtue of their biosynthetic process. The 15-membered macrolides, however, are made synthetically by a Beckmann rearrangement of erythromycin-9-oxime, to produce the ring-expanded derivatives in which an extra amino group is embedded in the 14-membered ring skeleton. These derivatives are named azalides. Azithromycin (**21**) is made by a Beckmann rearrangement of erythromycin-9-($E$)-oxime, followed by reduction of the imino ether and subsequent $N$-methylation (213), and has antibacterial activity comparable to that of erythromycin A. Even though it has a 15-membered ring structure, erythromycin-resistant staphylococci show complete cross-resistance to it (214). Azithromycin is more acid stable and the bioavailability in animals was found to be two- or threefold higher than that of erythromycin A (213).

A series of 8a-azalides (9-deoxo-8a-aza-8a-homoerythromycins) was prepared to compare their activities with the 9a-azalide, azithromycin, by Beckmann rearrangement of erythromycin-9-(Z)-oxime followed by reduction of the imino ether and subsequent N-methylation (215). Simple 8a-alkyl derivatives exhibited good antibacterial activity similar to that of azithromycin. Polar 8a–side chain 8a–azalide derivatives, however, have less antibacterial activity. Replacement of the 4″-hydroxyl group with an amino group resulted in a two- to eightfold increase in activity against Gram-negative bacteria but two- to eightfold decrease in activity against Gram-positive bacteria. 9-Deoxo-8a-methyl-8-aza-8a-homoerythromycin A (55) and 4″-deoxy-4″-amino-9-deoxo-8a-methyl-8a-aza-8a-homoerythromycin A (56) were found to be more efficacious than azithromycin and clarithromycin in mouse protection tests because they were more acid stable and had better pharmacokinetic profiles (216, 217).

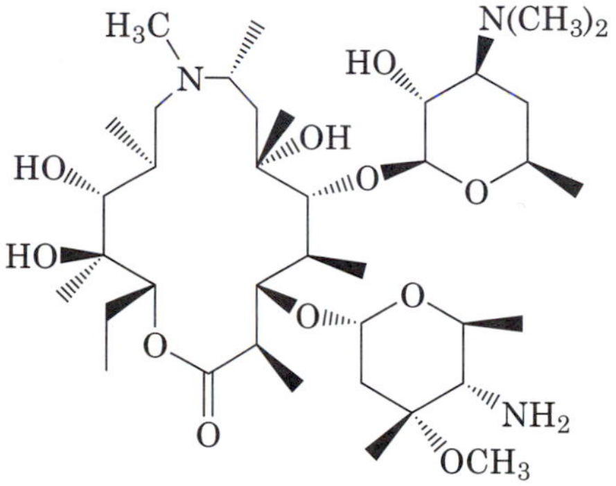

(56)

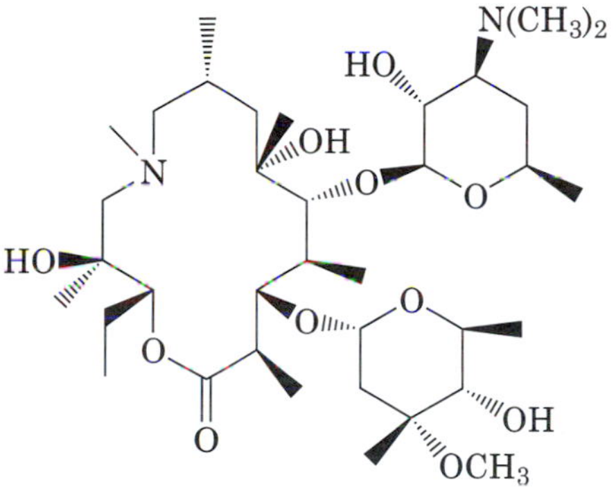

(57)

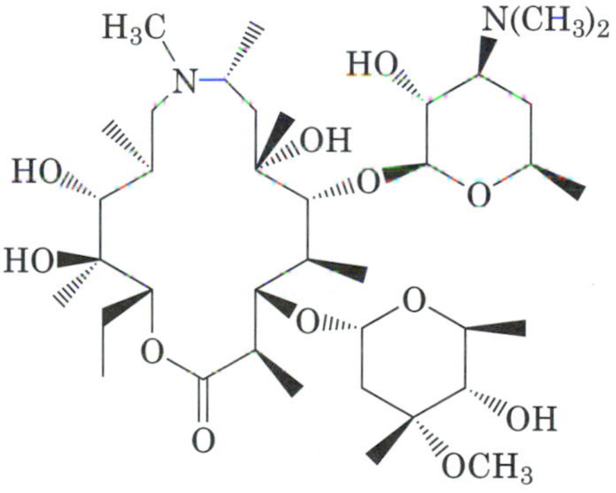

(55)

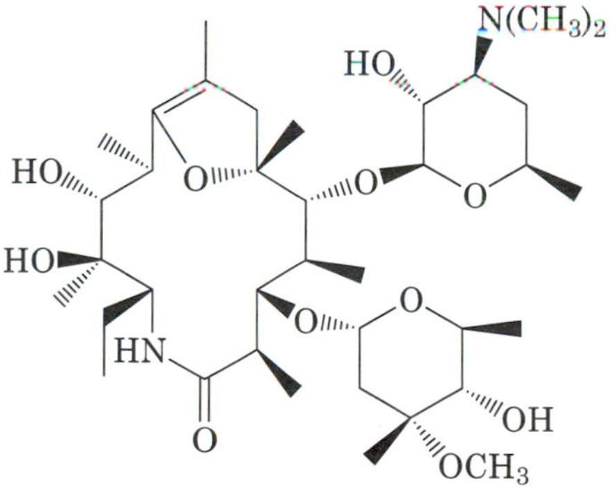

(58)

A series of novel 14-membered azalides was prepared. The 10-aza-10-methyl-9-deoxo-11-deoxyerythromycin (57) has an antibacterial spectrum similar to that of erythromycin A but was less active than either erythromycin A or azithromycin (218, 219). The 14-membered

lactam (58) was prepared but found to have minimal antibacterial activity but good gastrointestinal motor-stimulating activity (220).

## 4.5  Resistance to Macrolide Antibiotics

Bacterial resistance to macrolide antibiotics is emerging as a significant problem. Thirty-seven percent of 1113 *S. pneumoniae* clinical isolates collected between 1996 and 1997 in Spain were found to be resistant to macrolides (221). Based on the susceptibility study on 302 *S. pneumoniae* isolates from central Italian patients, the erythromycin A resistance rate increased from 7.1% in 1993 to 32.8% in 1997 (222). In a 1996–1997 winter U.S. surveillance study, among 1276 *S. pneumoniae* clinical isolates, 23% were resistant to erythromycin A, azithromycin, and clarithromycin (223). Although the *S. pneumoniae* resistance to macrolides is high in the United States as well as in most of Europe, the resistance rates in Norway and Canada are rather low (about 3–5%) (224, 225).

With respect to *S. pyogenes*, 14.5% clinical isolates identified in a Finnish surveillance study were erythromycin resistant (226). In Spain, 27% of the β-hemolytic group A and 12% of group C streptococci isolates collected between May 1996 and April 1997 were erythromycin resistant (221). In France, the erythromycin resistance rate in viridans group streptococci was 40% (227), whereas in Taiwan, 55% of *S. oralis* isolates were erythromycin resistant (228). Antimicrobial susceptibility data from the SENTRY antimicrobial surveillance program conducted in the United States and Canada indicated that *M. catarrhalis* and *H. influenzae* isolates remain highly susceptible to macrolides with less than 1% and 5% resistance, respectively (223, 229).

The molecular mechanisms of resistance to macrolides in pathogenic bacteria has been summarized and reviewed (230). Resistance to macrolides is found to occur by one of the three mechanisms: target-site modification, active efflux, and enzyme-catalyzed antibiotic inactivation.

Target-site modification is by far the most clinically significant resistance mechanism for macrolides. Two types of target-site modification resistance for macrolides have been described: (*1*) posttranscriptional modification of the 23S rRNA by adenine-$N^{6,6}$-dimethyltransferase ($MLS_B$ resistance), and (*2*) site-specific mutations in the 23 rRNA gene. The $MLS_B$ resistance confers resistance to macrolide, lincosamide, and streptogramin B classes of antibiotics. This resistance is mediated by plasmid or transposon-encoded genes called *erm* (erythromycin-resistant methylases), encoding enzymes that catalyze the $N^{6,6}$-dimethylation of the adenosine-2058 residue of bacterial 23S rRNA (*E. coli* numbering) (231, 232). The determination of the 3D structures of macrolide antibiotics bound to the ribosomal 50S subunit provides insight into the molecular basis of Erm-mediated resistance (169). In these structures, the hydroxyl group of the macrolide desosamine sugar forms hydrogen bounds with N1 and N6 of A2058; methylation of N6 therefore generates a steric block of this interaction, resulting in resistance (Fig. 15.13).

The $N^{6,6}$-dimethyltransferases have been found in numerous organisms including staphylococci, streptococci, enterococci, *Clostridium difficile*, *E. coli*, and *Bacillus* strains. The *erm*-mediated resistance phenotype can be constitutive or induced (233, 234). Enzyme induction is affected by exposure of the organism to both 14- and 15-membered, but not 16-membered, macrolides. Over two dozen *erm* genes have been identified encoding proteins of about 29 kDa in size (235). These genes are widely distributed in Gram-positive bacteria and, paradoxically, in many Gram-negative bacteria, which are generally not susceptible to macrolides as a result of the impermeable outer membrane. The presence of *erm* genes also confers reduced susceptibility to azithromycin (236).

The 3D structures of ErmC′ and ErmAM have been determined by X-ray crystallography and NMR methods, respectively (237, 238). These enzymes are approximately 50% identical and, not surprisingly, then show very similar folds consisting of an amino-terminal *S*-adenosyl-methionine binding domain and a helical C-terminal RNA binding domain. The structures of ErmC′, in complex with the substrate *S*-adenosyl-methionine, the product *S*-adenosyl-homocysteine, and the methyl transferase inhibitor sinefungin, have also been reported (239). These structures therefore provide the opportunity to develop inhibitors of Erm activity that could be used to block resistance. Several Erm inhibitors have in fact

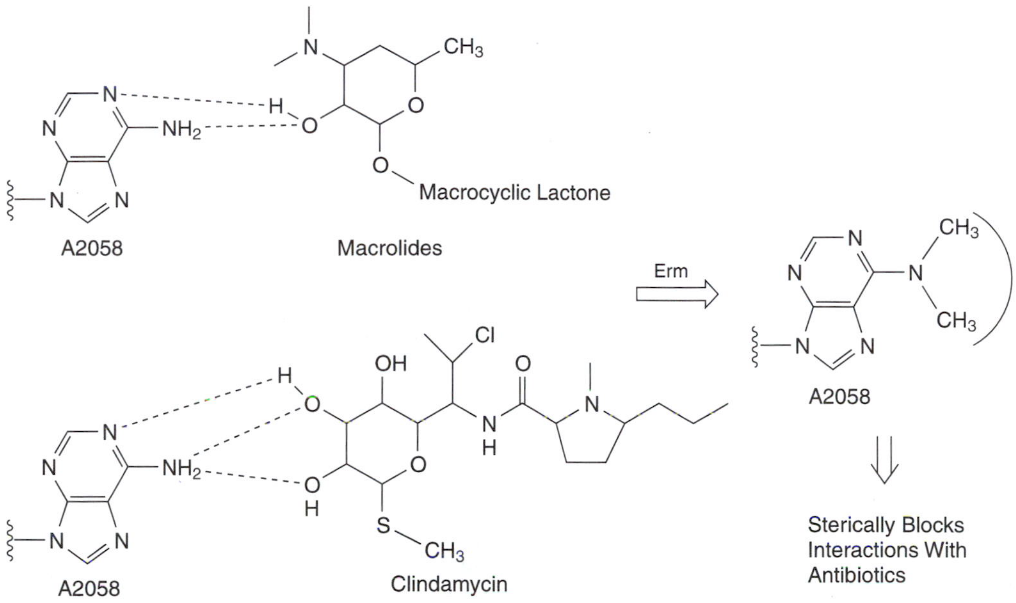

**Figure 15.13.** Outcome of methylation of A2058 by Erm resistance enzymes.

been identified by high throughput screening (240) and the SAR by NMR (241) methods, and phage display has also been used to discover inhibitory peptides (242). These studies indicate that it may be possible to identify molecules that block Erm-catalyzed methylation of the ribosomes.

The second target-site modification mechanism conferring macrolide resistance is through site-specific mutation in the 23S rRNA gene (reviewed in ref. 243). In *H. pylori*, A2142G mutation was linked with high level cross-resistance to all MLS$_B$ antibiotics. The A2143G mutation gave rise to an intermediate level of resistance to clarithromycin and clindamycin, but not streptogramin B (244, 245). Other examples of site-specific mutations include A2063G or A2064G in *M. pneumoniae*; A2059G or C2611A and G in *S. pneumoniae*; A2058C, G, and U in *M. avium*; and A2048G, G2057A, or A2059G in *Propionibacteria* (243). The number used here corresponds to the position in *E. coli* 23S rRNA. Novel mutations in either 23S rRNA alleles or ribosomal protein L4 were recently found to be responsible for macrolide resistance in *S. pneumoniae* isolates passaged with azithromycin (246).

Active efflux of macrolides is a fairly common mechanism for erythromycin A resistance in *S. pyogenes* and *S. pneumoniae* (247). In some of these bacteria, this type of resistance is attributed to the presence of the *mefA* gene, which encodes a membrane-associated efflux protein that confers resistance to macrolides but not to lincosamides or streptogramin B (248). The *mefE* gene (90% sequence identity to *mefA* in *S. pyogenes*) was found to be present in erythromycin-resistant M phenotype (macrolide resistant but clindamycin and streptogramin B susceptible) in *S. pneumoniae* (249). The macrolide efflux gene *mreA*, distinct from the *mefA* in *S. pneumoniae* and *S. pyogenes* and the multicomponent *mrsA* in *S. aureus*, was found in a strain of *S. agalactiae*, which displayed resistance to 14-, 15-, and 16-membered macrolides (250). Several *S. agalactiae* clinical isolates with the M phenotype harbor *mefA* and *mefE* genes (251). The efflux system is the major contributor to the macrolide resistance in *Burkholderia pseudomallei* (252). The *mef* gene has been identified in clinical isolates of *Acinetobacter junii* and *Neisseria gonorrhoeae*. These strains could transfer the *mef* gene into one or more of the following recipients: Gram-negative *M. ca-*

*tarrhalis*, *N. perflava*, and *N. mucosa* and Gram-positive *E. faecalis* (253). The *mtrCDE*-encoded efflux pump has been suggested to be one of the resistant mechanisms in *N. gonorrhoeae*. The *mtrCDE* genes constitute a single transcriptional unit that is negatively regulated by the adjacent *mtrR* gene product. *N. gonorrhoeae* clinical isolates with mutation in *mtrR* transcriptional repressor gene possessed reduced azithromycin susceptibility (253). The *mrsA* and *mrsB* genes are also efflux systems that are mostly isolated from *S. aureus* and *S. epidermidis* (254); a homolog *mrsC* was isolated from *Enterococcus faecium* (255).

Macrolide resistance by antibiotic inactivation can also occur through phosphorylation (256, 257), deesterification of the lactone ring (256, 257), and glycosylation (258–260). Two glycosyltransferases and a glycosidase are involved in oleandomycin modification (261). A glycosyl transferase inactivating macrolides encoded by *gimA* from *S. ambofaciens* was found to locate downstream of *srmA*, a gene that confers resistance to spiramycin and has a high degree of similarity to *S. lividans* glycosyl transferase, which inactivates macrolides (262). The *E. coli* clinical isolate BM2506 is highly resistant to macrolides by having macrolide 2'phosphotransferase II [MPH(2')II]. This strain is found to harbor two plasmids, pTZ3721 (84kb) and pTZ3723 (24kb). It appears that the *mphB* gene is located on these two plasmids in BM2506 and can be transferred to other strains of *E. coli* by conjugation or mobilization (263).

A minireview on new nomenclature for macrolide and macrolide–lincosamide–streptogramin B (MLS$_B$) resistance determinants was recently published (264). The common macrolide phenotypes are: MLS$_B$ for strains that carry Erm methylase; M for strains resistant to macrolide but are clindamycin and streptogramin B susceptible; ML for strains resistant to 14-, 15-, and 16-membered macrolide and lincosamides; and MS for strains resistant to macrolide and streptogramins.

## 4.6 Recent Developments in the Macrolide Antibiotic Field

### 4.6.1 Novel Macrolides to Overcome Bacterial Resistance. Several macrolide derivatives having potent antibacterial activity against erythromycin-sensitive strains and at the same time with moderate activity against erythromycin-resistant *S. pyogenes* were first reported in 1989 (265). The 11,12-carbamate clarithromycin analogs and the 11,12-carbonate erythromycin analogs with additional modifications at the 4' position of the cladinose were found to be active against both inducible and constitutive resistant *S. pyogenes*. The two naturally occurring descladinosyl 14-membered macrolide derivatives, narbomycin (**59**) and picromycin (**60**), with the presence of a 3-keto group, were isolated in the early 1950s (266, 267). Although these two compounds are poorly active, they are not inducers of macrolide resistance. With these new insights provided by the above information, recent chemical modifications of erythromycin have generated several new classes of macrolide derivatives having potent activity against erythromycin-sensitive and -resistant bacteria.

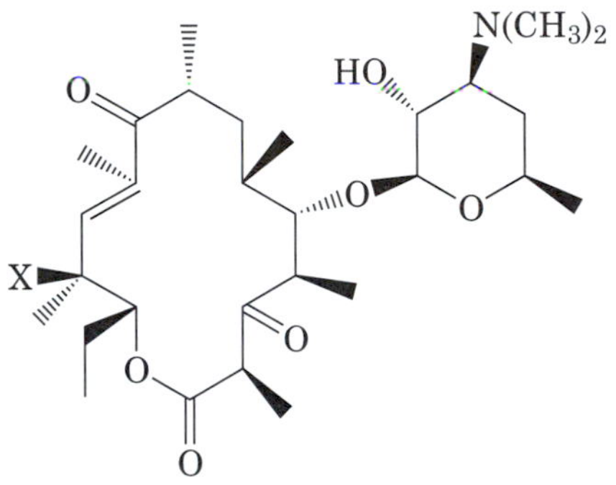

(59)    X = H
(60)    X = OH

### 4.6.1.1 Ketolides. Apart from the natural products narbomycin and picromycin, certain recently synthesized 3-keto macrolide derivatives were found to be active against both penicillin-resistant and erythromycin-resistant *S. pneumoniae*. Like narbomycin and picromycin, these derivatives do not induce MLS$_B$ resistance in staphylococci and streptococci (268). These 3-descladinoxyl-3-oxo-11,12-cyclic carbamate derivatives of erythromycin or clarithromycin are called ketolides. So far, three ketolides have undergone clinical development: HMR 3004 (**61**), HMR 3647 (telithro-

mycin, (**62**), and ABT-773 (**63**). The ketolides are very active against respiratory pathogens, including erythromycin-resistant strains (269, 270).

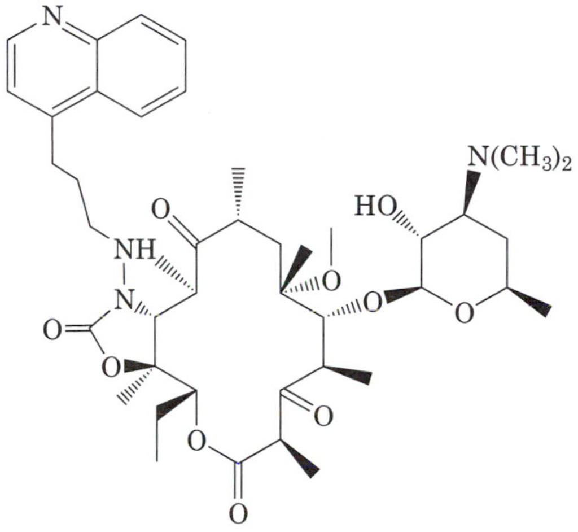

(**61**)

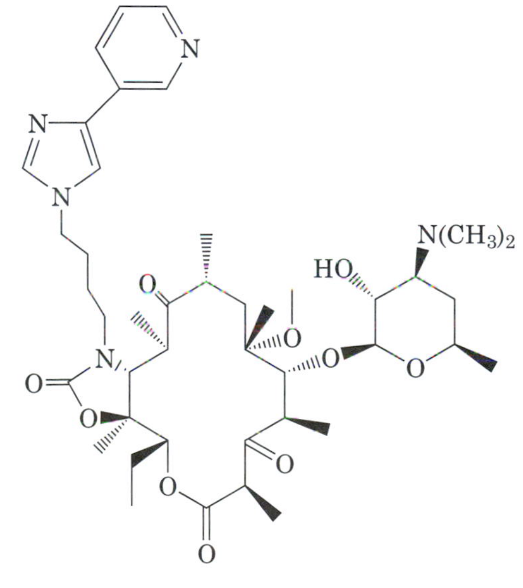

(**62**)

HMR 3004 is very active against penicillin- and erythromycin-resistant pneumococci with a MIC$_{90}$ value of 0.25 $\mu$g/mL (271, 272) and is very efficacious against respiratory infections in animal studies (273). It was found to be

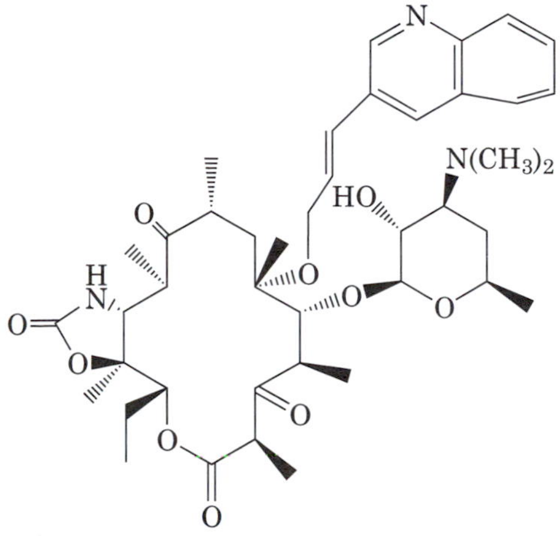

(**63**)

active against $\beta$-lactamase–producing *H. influenzae* in a murine model of experimental pneumonia and more active than azithromycin, ciprofloxacin, clarithromycin, erythromycin, and pristinamycin (274).

Another ketolide being developed is telithromycin (HMR 3647). The U.S. FDA issued an approval letter for telithromycin in 2001 for the following indications: community-acquired pneumonia, acute bacterial exacerbation of chronic bronchitis, and acute bacterial sinusitis. Against *H. influenzae*, telithromycin is as active as azithromycin, with a MIC$_{90}$ value of 4 $\mu$g/mL (275). When tested against *M. catarrhalis*, it has MIC$_{50}$/MIC$_{90}$ values of 0.06/0.125 $\mu$g/mL (276). It has MIC$_{90}$ and MBC$_{90}$ values of 0.25 $\mu$g/mL (277). The pharmacodynamic properties of telithromycin demonstrated by time-kill kinetics and post-antibiotic effect on enterococci and *Bacteroides fragilis* were found to be similar to those obtained with macrolides (278). Although it is found to be active against MLS$_{B}$-resistant pneumococci, telithromycin did not bind to the methylated ribosomes isolated from the MLS$_{B}$-resistant strain (279). In a murine model of experimental pneumonia, telithromycin was effective against $\beta$-lactamase–producing *H. influenzae* (274). It is also effective for the treatment of *L. pneumophila* in a guinea pig pneumonia experimental model (280). In a mouse peritonitis model of enterococci infection, telithromycin was found to be highly active against erythromycin-susceptible and -intermediate strains (281). HMR 3004 and telithromycin were found to be highly active *in vitro* against *M. pneumoniae* (282), *E. faecalis* (283), and *Bordetella pertussis* (284). The oral antibacterial activities of telithromycin in different infections induced in mice by *S. aureus*, *S. pneumoniae*, streptococci, enterococci, and *H. influenzae* were found to be excellent (285).

A series of 6-*O*-substituted ketolides was reported to possess excellent activity against inducible resistant *S. aureus* and *S. pneumoniae* as well as constitutively resistant *S. pneumoniae* (286). Derivatives with an aryl group at the 6-*O* position with a propenyl spacer together with an 11,12-cyclic carbamate and a 3-keto group exhibited the best activity. In this study, the unsubstituted 11,12-cyclic carbamate analogs were more active than either the carbazate or tricyclic analogs [ABT-773 (**63**) versus A-201316 (**64**) versus 197579 (**65**)].

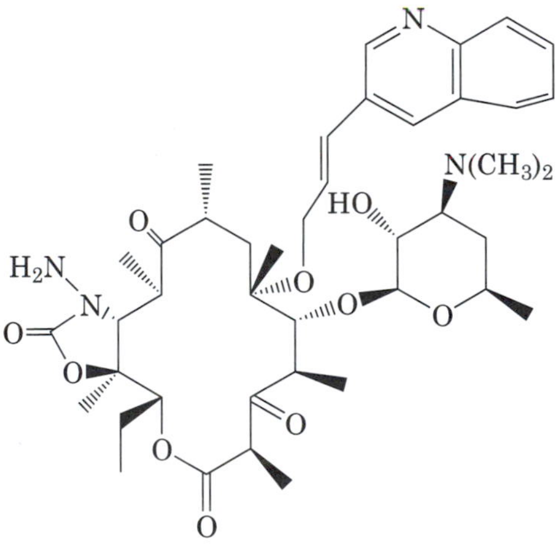

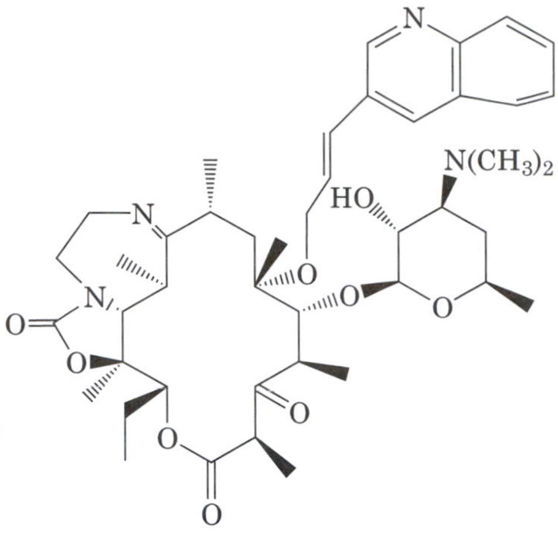

(65)

(64)

ABT-773 is currently in late-stage clinical trials. It has a potent antibacterial spectrum including activity against penicillin- and macrolide-resistant Gram-positive bacteria (270, 287). ABT-773 was found to be the most active compound tested among other macrolides against *S. pneumoniae* with $MIC_{90}$ value of 0.03 $\mu$g/mL. It is as potent as azithromycin but

more potent than clarithromycin and erythromycin A against *H. influenzae* ($MIC_{90}$ value = 4 $\mu$g/mL) and *M. catarrhalis* ($MIC_{90}$ value = 0.06 $\mu$g/mL). It also has good activity against Gram-negative and atypical respiratory tract pathogens and *H. pylori* ($MIC_{90}$ value = 0.06 $\mu$g/mL against macrolide-susceptible strain) (288). ABT-773 is active against anaerobic bacteria. For *B. fragilis*, telithromycin is one to two dilution levels less active than ABT-773. For all anaerobic tested strains, ABT-773 was found to be more active than erythromycin A by four or more dilution levels (289). A comparative study on the *in vitro* antibacterial activity of ABT-773 against 207 aerobic and 162 anaerobic antral sinus puncture isolates showed that erythromycin-resistant pneumococci strains were susceptible to ABT-773 with $MIC_{90}$ value of 0.125 $\mu$g/mL (290). ABT-773 had superior activity against macrolide-resistant *S. pneumoniae* than that of telithromycin with $MIC_{90}$ value in $\mu$g/mL: erm strain 0.015 versus >0.12 and mef strain 0.12 versus 1. Against different *S. pyogenes* strains ABT versus telithromycin has the following $MIC_{90}$ values in $\mu$g/mL: erm strain 0.5 versus >8 and mef strain 0.12 versus 1 (291–294). The high activity of ABT-773 against macrolide-resistant *S. pneumoniae* is probably the result of several factors. For example, ABT-773 was demonstrated to bind tighter than erythromycin to the ribosome target (295). ABT-773 was also shown (*1*) to accumulate in macrolide-sensitive *S. pneumoniae* at a higher rate than that of erythromycin A; (2) to bind to methylated ribosomes, although at lower affinities than those of wild-type ribosomes; and (3) to accumulate in *S. pneumoniae* strains with the efflux-resistant phenotype.

ABT-773 possesses a favorable pharmacokinetic profile, with plasma elimination half-lives averaging 1.6, 4.5, 3.0, and 5.9 h after i.v. dosing in mouse, rat, monkey, and dog, respectively. Peak plasma concentrations averaged 1.47, 0.52, 0.56, and 0.84 µg/mL, having bioavailabilities of 49.5, 60.0, 35.8, and 44.1% after oral dosing in the same animal species, respectively. The lung concentration of ABT-773 was >25-fold higher than plasma concentration after oral dosing in rat (296). In animal studies, ABT-773 was found to be fivefold more efficacious than azithromycin against *mefE*-bearing strains and had excellent efficacy against *ermAM*-bearing strains, whereas azithromycin was inactive (297). ABT-773 was shown to have 3- to 16-fold improved efficacy over that of telithromycin against macrolide-resistant *S. pneumoniae* (both *ermAM* and *mefE*) in rat pulmonary infection (298).

HMR 3562 (**66**) and HMR 3787 (**67**) are two 2-fluoro ketolides having potent *in vitro* antibacterial activity against inducible resistant *S. aureus* and *S. pneumoniae* as well as constitutively resistant *S. pneumoniae*. They are more potent than their corresponding nonfluorinated parent ketolides against *H. influenzae* (299, 300). Both compounds display high therapeutic efficacy in mice infected by different common respiratory pathogens, such as multidrug-resistant pneumococci and *H. influenzae* (301). A series of C-2 halo-substituted tricyclic ketolides were prepared, showing that the 2-fluoro-ketolide analog is more potent than the 2-bromo derivative. A-241550 (2-F analog of TE-802) (**68**) was found to be more potent and efficacious than TE-802 (the C2-H parent), particularly against *H. influenzae* (302).

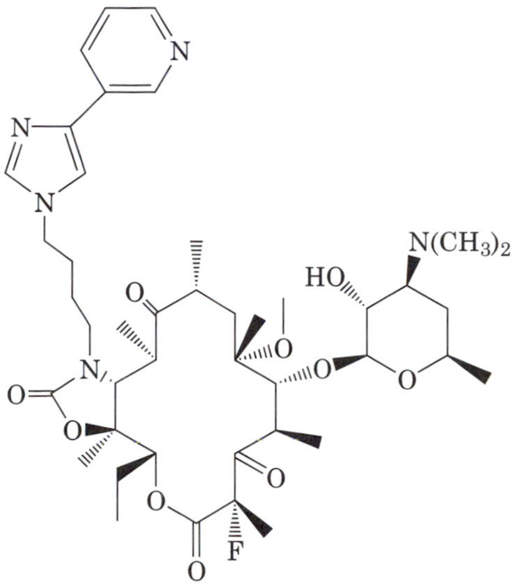

A C-2 fluorinated ketolide with the 9-keto group replaced by an oxime, designated as CP-654,743 (**69**), was shown to be potent *in vitro* against macrolide-resistant respiratory pathogens, including *S. pneumoniae*, *S. pyogenes*, and *H. influenzae* (303). The MIC$_{90}$ values in

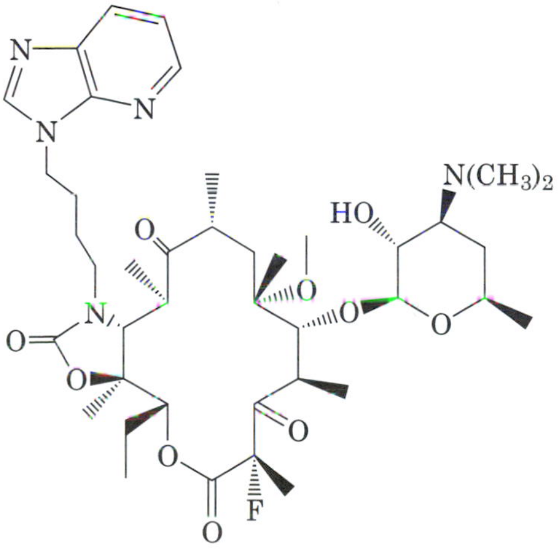

(66)

(67)

µg/mL against *S. pneumoniae*, *S. pneumoniae* (*ermB*$^+$), *S. pneumoniae* (*mefA*$^+$), *H. influenzae*, *S. pyogenes* (*ermB/ermA*$^+$), and *S. pyogenes* (*mefA*$^+$) for CP-654,743 are 0.004, 0.125, 0.25, 2.0, 6.3, and 1.0, respectively. This improved *in vitro* potency appeared to translate into improved *in vivo* activity, especially with improved activity against MLS$_B$ pneumococci (304). CP-654,743 has a large volume of

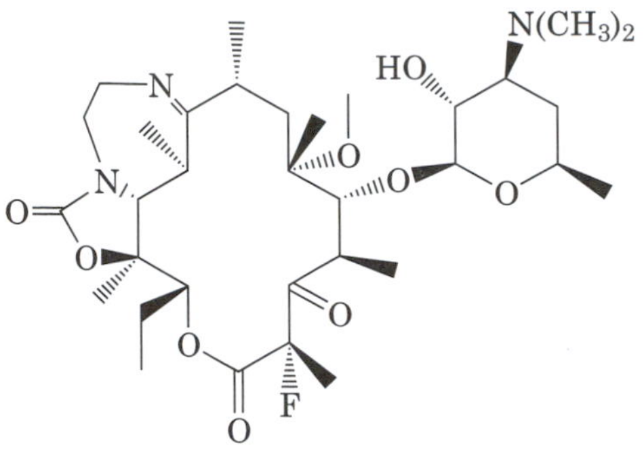

**(68)**

distribution and moderately high clearance in rats, dogs, and monkeys and the liver microsomal data predict moderate plasma hepatic clearance in humans. It is moderately bound to serum proteins (305).

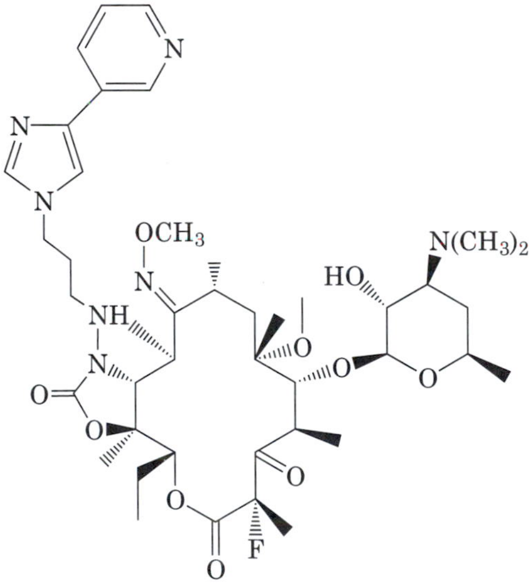

**(69)**

The *in vitro* activity of the ketolide HMR3832 (**70**), with a butyl imidazoyl anilino moiety substituted on the *N*-carbamate position, was reported (306). This compound is about one to two times less active than clarithromycin against macrolide susceptible *S. pneumoniae*, but at the same time retains activity against strains harboring macrolide-re-

sistant determinants. The $MIC_{90}$ values in μg/mL against *H. influenzae*, *M. catarrhalis*, *N. meningitis*, *B. pertussis*, *C. pneumoniae*, and *M. pneumoniae* are 2–4, 0.12, 0.12, 0.12, 0.25, and 0.0005, respectively.

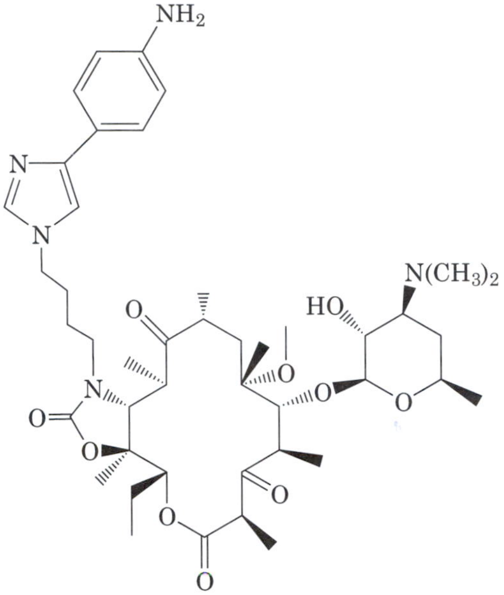

**(70)**

Apart from the synthesis of ketolides from erythromycin A or its analog clarithromycin, a series of C-13–modified ketolides (**71**), in which the C-13 ethyl group is replaced by other alkyl or vinyl groups (R = alkyl or vinyl), were prepared. These were synthesized from a modified erythromycin A template produced in a genetically engineered *Streptomyces coelicolor* strain (307). The *S. erythraea* polyketide synthase genes were engineered to contain an inactive ketosynthase 1 domain ($KS1^0$) and then expressed in *S. coelicolor*. The strain was fermented with the appropriate diketide thioester precursor, to yield C-13–modified 6-deoxyerythronolide analogs with C-13 being a methyl or vinyl group. These derivatives were then purified and fermented with a $KS1^0$ *S. erythraea* to glycosylate at the O-3 and O-5 positions and hydroxylate at the C-6 and C-12. Further chemical modification of these novel core templates produced the C-13–modified ketolide (**71**). The *in vitro* activity of the C-13–modified ketolides compared to that of telithromycin (R = ethyl) is given in Table 15.7. They have activity slightly less than that of telithromycin, showing that certain C-13–modified ketolides retain good activity against macrolide-resistant bacteria. Thus, this biosynthetic strategy may offer an opportunity to generate potent and useful novel macrolide derivatives.

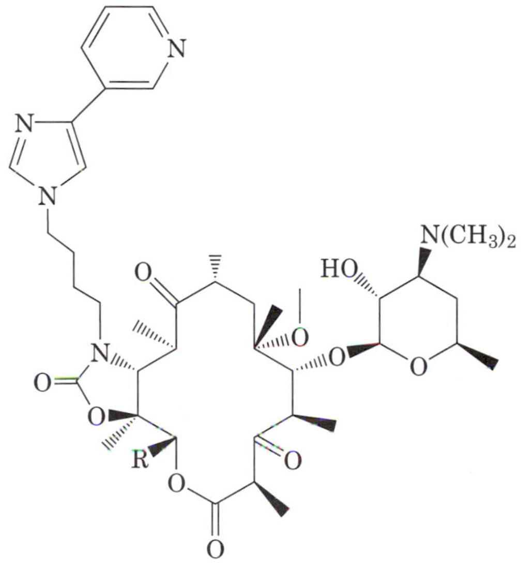

ketolides) (308, 309). A-179461 (**72**), an anhydrolide, having a structure similar to that of HMR 3004, was found to have similar antibacterial activity *in vitro* to that of HMR 3004 against Gram-positive organisms. Thus, functionality other than a keto group at the C-3 position can be used as cladinosyl group replacement to produce potent novel macrolide derivatives (310).

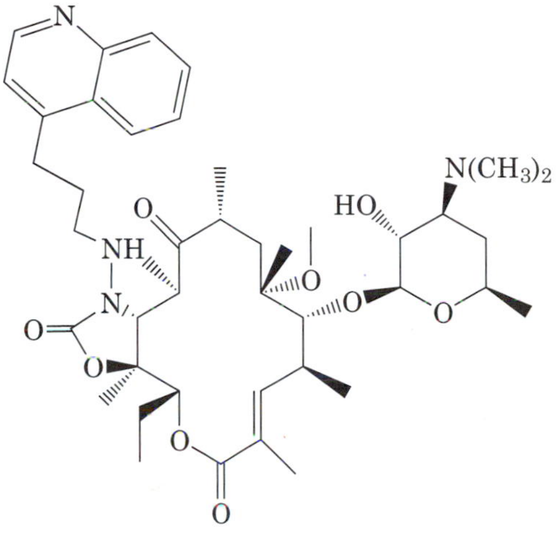

(**71**)

(**72**)

**4.6.1.2 Anhydrolides and Acylides.** A series of 2,3-anhydroerythromycin derivatives, termed anhydrolides, were synthesized. They possess a carbon–carbon double bond at the C-2–C-3 position, having an $sp^2$ carbon at position C-3 (the same position as the 3-keto group in the

A series of 3-descladinosyl-3-acylated macrolide derivatives with potent antibacterial activity were prepared and called acylides. The 3-O-acyl-5-desosaminyl-erythronolide 11,12-carbonate TEA 0769 (**73**) and 3-O-acyl-5-desosaminyl-erythronolide 6,9;11,12-biscarbonate FMA 122 (**74**) have potent *in vitro*

**Table 15.7  Comparative *In Vitro* Activity of C-13–Modified Ketolides (71)**

| Organism | Resistance Mechanism | N | Median MIC ($\mu$g/mL) Telithromycin Derivatives | | |
| --- | --- | --- | --- | --- | --- |
| | | | R = Ethyl | R = Methyl | R = Vinyl |
| *M. catarrhalis* | | 3 | 0.25 | 8 | 0.25 |
| *H. influenzae* | | 12 | 4 | 8 | 8 |
| *S. aureus* (MSSA, MRSA) | | 3 | 0.25 | 0.5 | 0.5 |
| Staphylococci | $erm_j$ | 5 | 0.5 | 16 | 4 |
| | $erm_c$ | 5 | ≥16 | ≥16 | ≥16 |
| | msr | 2 | 1 | 2 | 2 |
| Enterococci (VRE) | | 5 | 0.06 | 0.12 | 0.12 |
| Enterococci | | 11 | ≤0.015 | 0.03 | 0.03 |
| *S. pneumoniae* | erm | 5 | 0.03 | 0.12 | 0.12 |
| | mef | 8 | 0.25 | 0.5 | 1 |

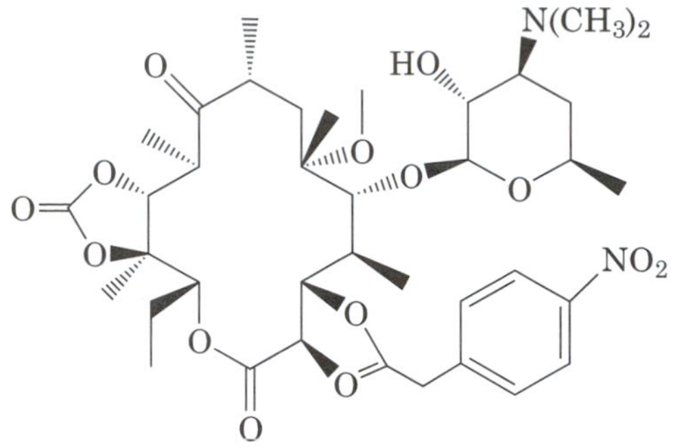

(73)

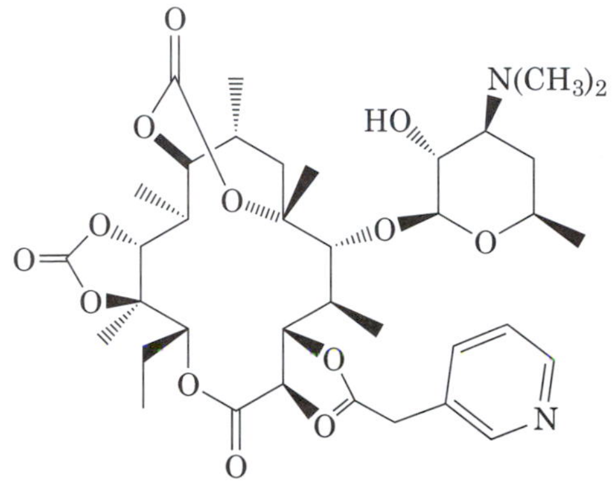

(74)

antibacterial activity against Gram-positive organisms similar to that of telithromycin (311–313). TEA 0769 was found to be twofold more active against staphylococci (MIC, *S. aureus* 209P: 0.05 $\mu$g/mL) and 16-fold more active against enterococci (MIC, *E. faecalis*: 0.05 $\mu$g/mL) than was clarithromycin. It is also active against erythromycin-resistant streptococci. The antibacterial activity of FMA 122 compared with that of azithromycin given in $\mu$g/mL against *H. influenzae*, *S. aureus* 209P, *S. pneumoniae* IID533 (erythromycin sensitive), *S. pneumoniae* 210 (erythromycin-resistant efflux), and *S. pneumoniae* 205 (erythromycin-resistant-*erm*) is as follows: 0.78 versus 1.56, 0.10 versus 0.39, 0.025 versus 0.10, 0.10 versus 0.78, and 6.25 versus 100, respectively. Thus, the acylides represent a potent macrolide series with good anti–*H. influenzae* activity while having activity against macrolide-resistant bacteria.

**4.6.1.3 4″-Carbamate Macrolides.** A series of 4″-carbamates of 14- and 15-membered macrolides were prepared, in which the 4″ hydroxyl group was replaced by a carbamate group. They have potent *in vitro* activity against Gram-positive and Gram-negative respiratory pathogens including the macrolide-resistant *S. pneumoniae* (314). CP-544,372 (**75**) possesses good *in vitro* antibacterial activity comparable to that of telithromycin. Its MICs in $\mu$g/mL against *S. pyogenes mefA*, *S. pneumoniae ermB*, and *S. pneumoniae mefE* are 0.5, 0.16, and 0.08, respectively (315). Its maximum drug concentration ($C_{max}$) in the lung is higher than that in serum. In mice, CP-544,372, given at 1.6–100 mg/kg orally, produced the serum $C_{max}$ of 0.001–0.66 $\mu$g/mL, AUC of 0.13–5.5 $\mu$g h$^{-1}$ mL$^{-1}$, and a mean terminal half-life of 6.5 h (316). In murine acute pneumonia models of infection induced by macrolide-sensitive and -resistant strains of pneumococci or *H. influenzae*, CP-544,372 was shown to be orally active (317). It has either more potent or similar *in vivo* efficacy against macrolide-resistant *S. pneumoniae* compared to telithromycin or HMR 3004. Thus, modification at the 4″-position of erythromycincylamine led to 4″-carbamate macrolide derivatives with increased *in vitro* and oral *in vivo* antibacterial activity against macrolide-resistant *S. pneumoniae*, while maintaining potent *in vitro* and *in vivo* activity against macrolide-sensitive *S. pneumoniae* and *H. influenzae*.

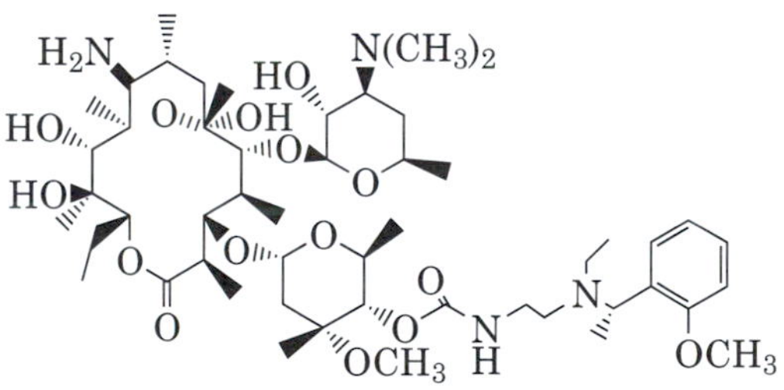

(75)

**4.6.2 Non-Antibacterial Activity of Macrolides.** Other than having antibacterial activity, macrolide antibiotics also have diverse non-antibiotic properties. They have a broad range of biological response-modifying effects on inflammation, tumor cells, airway secretions, and host defense. Other effects such as their gastrokinetic effect are also described below.

**4.6.2.1 Gastrokinetic Effects.** The gastrokinetic effect of macrolides such as erythromycin A and oleandomycin has been known for some time. They induced strong macular contraction in the gastrointestinal tract of dogs. Erythromycin A, when given at 0.03 mg/kg intravenously to dogs, induces a pattern of migrating contractions in the gastrointestinal tract, mimicking the motilin effect on gastrointestinal contractile activity (318). At an intravenous dose of 7 mg/kg, an immediate increase in contractile activity in the whole length of the intestinal tract was observed (319). Thus, erythromycin A acts as a nonpeptide agonist of motilin (320, 321). Erythromycin A administered intravenously to normal volunteers at a dose of 200 mg after a meal induced powerful peristaltic contractions, improved antroduodenal coordination, and phase three–like activity (322).

Motilides are macrolide derivatives with reduced antibacterial activity and have the activity to induce gastrointestinal contractions acting as a motilin mimic (323). A detailed study of macrolide derivatives on their potency in displacing motilin and contractility identified EM-523 (**76**) for clinical development as a gastrointestinal tract prokinetic agent (324, 325). 8,9-Anhydro-4″-deoxy-3′-*N*-desmethyl-3′-ethylerythromycin B-6,9-hemiacetal, ABT-229 (**77**) is another potent prokinetic agent (326), with >300,000 times more potency than that of erythromycin A. It is also 400-fold more active than its 4″,12-dihydroxy congener EM-523. It has good oral bioavailability with 39% compared to 1.4% for EM-523. Thus the 4″-hydroxyl group is not a major contributor to the motilin agonistic activity, whereas the 2′-hydroxyl group seems necessary (327).

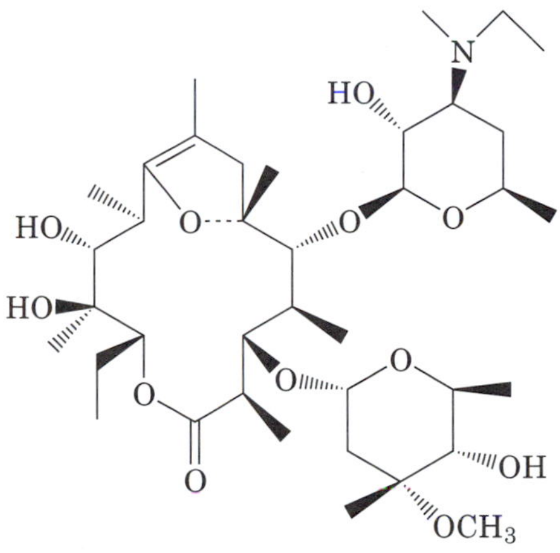

**4.6.2.2 Effects on Cytokines.** Macrolides have the ability to inhibit the production of proinflammatory cytokines and reactive oxygen species by airway neutrophils. In whole blood, erythromycin A was found to cause a dose-dependent decrease in heat-killed *S. pneumoniae*–induced production of tumor necrosis factor alpha (TNF-$\alpha$) and interleukin 6 (IL-6) *in vitro* (328). In patients with chronic

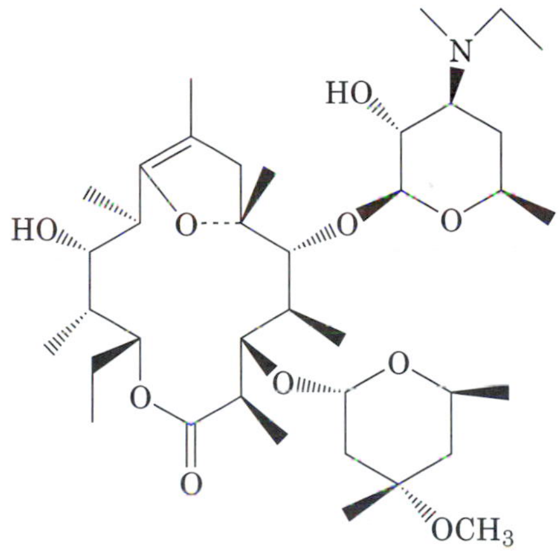

(76)

(77)

airway diseases, administration of erythromycin had an inhibitory effect on IL-8 release (329, 330). In mice, HMR 3004 was found to downregulate the pneumococcus-induced IL-6 and IL-1$\beta$ and nitric oxide in bronchoalveolar lavage fluid. It limited the neutrophil recruitment to the lung tissue and alveoli without interference on phagocytosis. It also abrogated lung edema (331, 332). Roxithromycin was also found to suppress the production of IL-8, IL-6, and granulocyte-macrophage colony-stimulating factor *in vitro* in human bronchial epithelial cells. It also inhibited neutrophil adhesion to epithelial cells. Roxithromycin's efficacy in airway disease may be attributable to its effects in modulating local recruitment of inflammatory cells (333). Clarithromycin was also reported to suppress the IL-8 release in human bronchial epithelial cell line BRAS-2B (334) and inhibit NF-κB activation in human peripheral blood mononuclear cells and pulmonary cells (335). Azithromycin, clarithromycin, and roxithromycin were reported to suppress the edema production in carrageenin-injected rats (336). Macrolides are also shown to inhibit mRNA expression of IL-1 (337), IL-8 (338), endothelin-1 (339), and inducible NO synthase (340).

**4.6.2.3 Immunomodulatory Effects.** Macrolide antibiotics such as clarithromycin decrease neutrophil oxidative burst and proinflammatory cytokine generation and release. Thus they are shown to possess immunomodulatory effects and can be used as biological response modifiers. Macrolides may protect airway epithelium against oxidative damage by phospholipid-sensitized phagocytes (341). They can decrease mucus hypersecretion both *in vitro* (342) and *in vivo* (343). Clinical efficacy of macrolides is high upon administration of macrolides to patients with diffuse panbronchiolitis (DPB), a disease with airflow limitation, sinusitis, sputum expectoration, dyspnea, as well as infection with *P. aeruginosa* (344, 345). The 5-year survival rate for patients with DPB was 26% in 1984. However, after the introduction of macrolide therapy, the 10-year survival rate has increased to 94%, an impressive improvement (346). Erythromycin A, clarithromycin, and azithromycin were found to suppress the expression of *P. aeruginosa* and *P. mirabilis* flagelin dose dependently (347). Flagella are among the virulence factors of Gram-negative rods and have a role in the initiation of biofilm formation. Macrolide therapy is found to be beneficial for patients with macrolide-resistant *P. aeruginosa* (DPB) and cystic fibrosis, although the improvement was not associated with the disappearance of *Pseudomonas* (348–350). A beneficial effect was also seen for patients on macrolide therapy for chronic sinusitis (351).

**4.6.2.4 Antitumor Effects.** Macrolides have been reported to have antitumor effects in mice (352). Clarithromycin prolonged the survival of patients with nonresectable, nonsmall cell lung cancer (353). It inhibited tumor angiogenesis, growth, and metastasis of mouse B16 melanoma cells. Although clarithromycin showed no direct cytotoxicity to the tumor cell *in vitro*, spleen cells obtained from the tumor-bearing rats receiving this compound showed a stronger tumor-neutralizing activity (354). Using a mouse dorsal air sac model, roxithromycin was found to inhibit tumor angiogenesis in a dose-dependent manner (355).

**4.6.2.5 Antiparasitic Activity.** Megalomicin (**78**) is a 14-membered macrolide with three sugar substituents (one neutral sugar and two aminosugars) isolated from *Micromonospora megalomicea* (356). It was found to have antiparasitic activity. It inhibits vesicular transport between the medial- and *trans*-Golgi, resulting in the undersialylation of the cellular protein (357). It was found to be active in an *in vivo* animal model with complete protection of the BALA/c mice from death caused by acute *Trypanosoma brucei* infection and significantly reducing the parasitema.

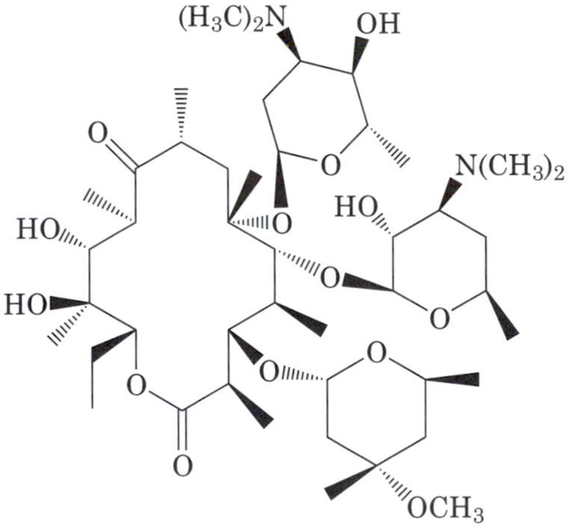

(78)

### 4.7 Future Prospects in the Macrolide Field

Novel macrolides can now be prepared by either chemical modification of the natural macrolides or gene manipulation of the producing organisms. The current novel macrolides have activity against macrolide-resistant bacteria. Telithromycin, a ketolide, has been successfully developed for the treatment of bacterial infections. Further clinical study shall no doubt show its efficacy for the treatment of respiratory infections caused by macrolide-resistant pneumococci. Additional ketolides are expected to come to market in the near future. This will provide physicians with additional tools to combat the threat of bacterial resistance. The recently expanding knowledge of mechanisms of bacterial resistance to macrolides will accelerate the discovery of new and useful macrolide derivatives other than the ketolide class. The new findings on many non-antibacterial activities of macrolide will open doors for the design and development of novel agents for the treatment of inflammatory diseases, parasitic diseases, cancers, and gastrointestinal motility disorders. Investigation directed toward study of the mechanism of action of these nonbacterial effects will no doubt be intensified.

## 5 OTHER ANTIBIOTICS THAT TARGET THE BACTERIAL RIBOSOME

### 5.1 Streptogramins

The streptogramin antibiotics are natural products, derived from soil bacteria, that inhibit bacteria protein synthesis, thus resulting in cell death. These antibiotics were discovered in the 1950s but did not find extensive clinical use as a result of poor water solubility. Recently, the preparation of semisynthetic water-soluble derivatives of streptogramins has facilitated their use in the treatment of infections caused primarily by Gram-positive bacteria such as staphylococci, streptococci, and enterococci as well as some Gram-negative organisms including *Neisseria* and *Legionella* (358, 359). Streptogramins are composed of two distinct chemical classes, group A and group B. Group A streptogramins are polyunsaturated macrolactones derived from acetate (polyketide synthesis) and amino acids [e.g., pristinamycin-IIA (**79**)]. Group B streptogramins are cyclic hexadepsipeptides [e.g., pristinamycin-IA (**80**)], cyclized through an ester linkage between the C-terminal carboxylate and the hydroxyl of an *N*-terminal Thr. The streptogramin-producing bacteria, such as the pristinamycin producer *Streptomyces pristinaespiralis*, simultaneously generate

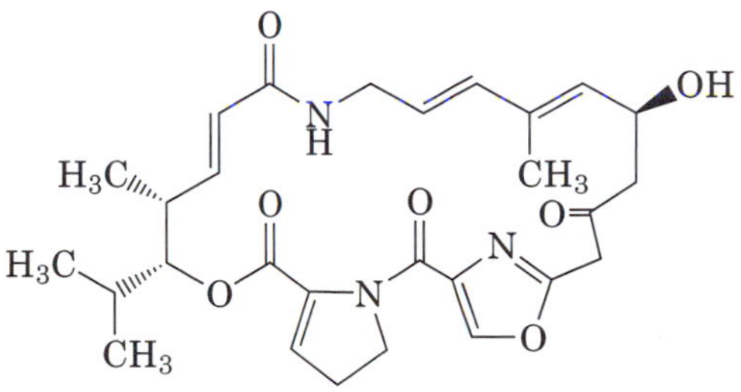

(79)

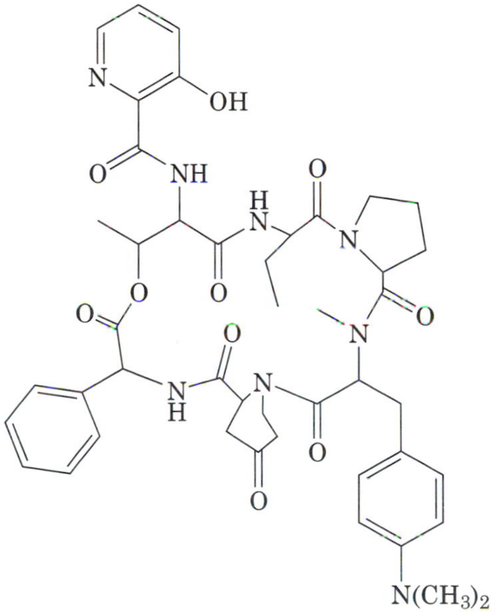

(80)

both a group A and a group B antibiotic in a molar ratio of approximately 60:40.

Individually, the group A and B streptogramins are generally bacteriostatic, but become bactericidal when coadministered. This synergistic effect has been exploited in the use of the semisynthetic compounds dalfopristin (**81**) and quinupristin (**82**), water-soluble derivatives of pristinamycin IIA and pristinamycin IA, respectively, that together in a 7:3 ratio form the drug Synercid. This formulation was recently approved in North America for the treatment of serious infections caused by antibiotic-resistant Gram-positive pathogens. The route of administration is by i.v. in a dose of 7.5 mg/kg every 8–12 h. Synercid inhibits cytochrome P450 3A4 activity, which is implicated in the metabolism of numerous other drugs including cyclosporine, midazolam, and others (for reviews, see refs. 360 and 361), and thus caution should be used when administering Synercid with other drugs metabolized by this route.

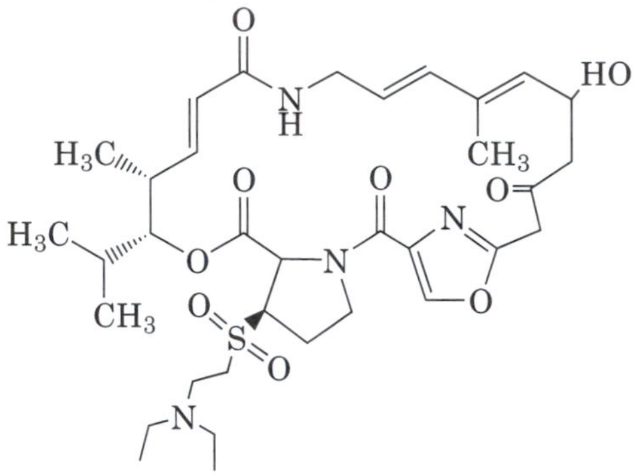

(81)

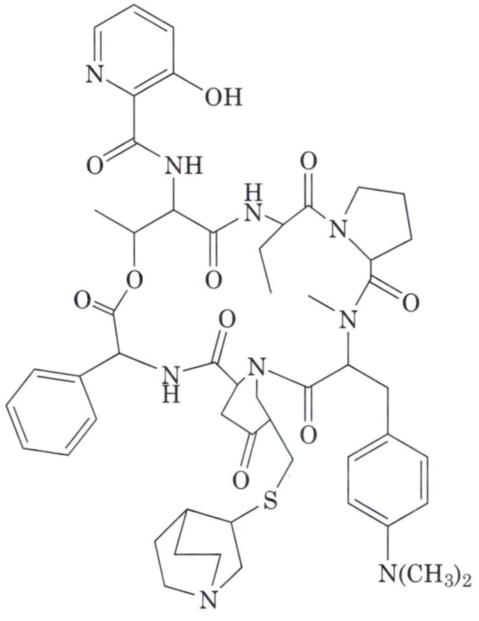

(82)

Both group A and group B streptogramins bind to the 50S ribosomal subunit, but in distinct sites (362). Group A antibiotics inhibit both aminoacyl-tRNA binding to the A-site and peptide bond formation by binding tightly

to 50S and free 70S ribosomal particles (363, 364). This tight interaction is also accompanied by a conformational change in the 50S subunit (365, 366). Group B streptogramins bind to the 50S subunit, which blocks peptide elongation and induces premature chain termination (367, 368). The B streptogramins overlap the macrolide–lincosamide binding sites (369), which explains the cross-resistance observed by Erm ribosomal methyltransferases (232). Binding of group A streptogramins to the ribosome facilitates binding of a group B streptogramin, which is the likely basis for synergy between the molecules (369–371).

Resistance to the streptogramins can occur by way of distinct efflux, chemical modification, or target protection/alteration mechanisms for both A and B group streptogramins. On the other hand, resistance to the synergistic mixture of A and B group streptogramins requires an A group resistance determinant, not necessarily coupled with one for group B streptogramins (372).

Ribosomal protection by Erm methyltransferases at position A2058 of the 23S rRNA, which also confers resistance to the macrolide and lincosamide antibiotics (see section above), provides resistance to the group B streptogramins. This results in the $MLS_B$-resistance phenotype. In the case of inducible *erm* gene expression, group B streptogramins are not inducers, unlike the 14-membered macrolides. The presence of an Erm methyltransferase does not confer resistance to the group A streptogramins and thus Synercid is still effective against these strains (359, 373, 374). Point mutations in the 23rRNA can also result in resistance to the group B streptogramins [e.g., C2611U (375), C2611A, C2611G, and A2058G (246)].

Active efflux mechanisms specific to either the A or B group streptogramins have been identified. The Vga proteins from *S. aureus*, members of the ABC (ATP-binding cassette) family of efflux proteins, are associated with group A streptogramin resistance (376–378). Efflux-mediated resistance to the group B streptogramins has not been associated thus far with clinically relevant resistance. Recently, a virginiamycin S (group B streptogramin) resistance gene, *varS*, was cloned

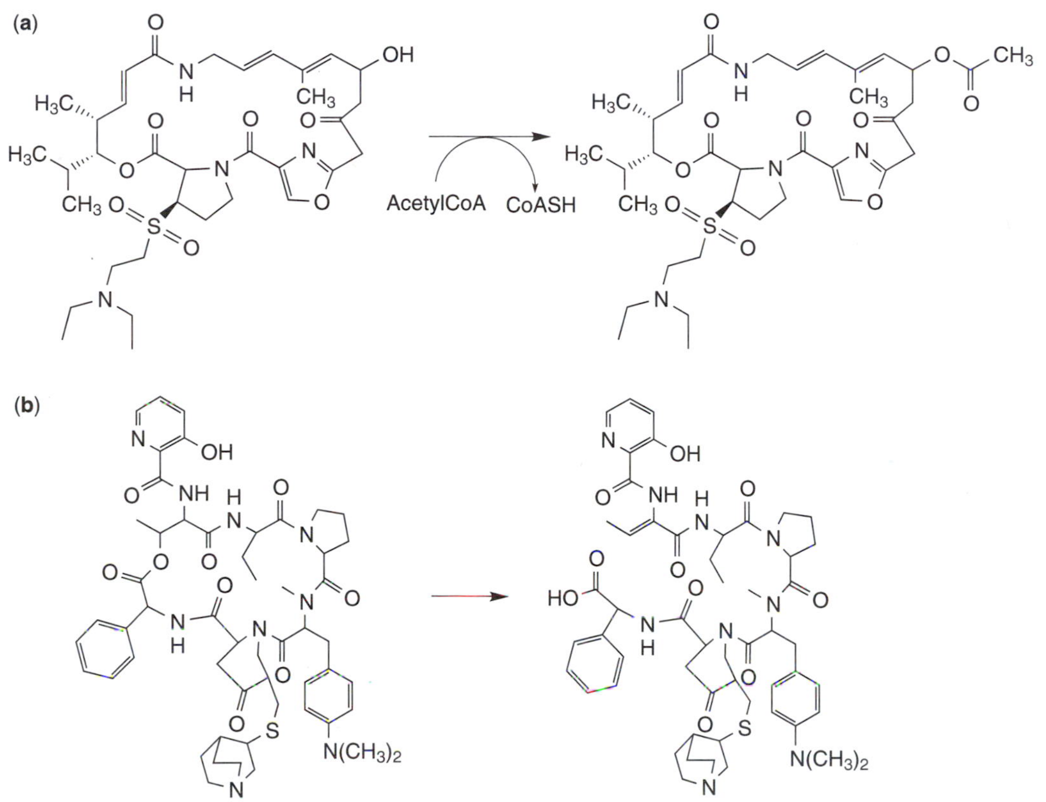

**Figure 15.14.** Modification of streptogramin by resistance enzymes. (a) Group A streptogramin acetyltransferase modification of dalfopristin. (b) Group B streptogramin lysase inactivation of quinupristin.

from the virginiamycin producer *Streptomyces virginiae* (379). The VarS protein is a homolog of other drug-resistant ABC transporters and heterologous expression in *Streptomyces lividans* resulted in virginiamycin S resistance.

Enzymatic inactivation of the streptogramins is a predominant mechanism of resistance. Group A streptogramins are inactivated by O-acetylation catalyzed by acetyl-CoA-dependent acetyltransferases (Fig. 15.14a). Several acetyltransferases termed Sat or Vat have been identified both in group A streptogramin-resistant Gram-positive clinical isolates (e.g., 380–382) and in the chromosomes of a number of bacteria (383). The crystal structure of VatD from *E. faecium* has been determined in the presence and absence of virginiamycin M1 and CoA (384). Inactivation of group B streptogramins occurs by linearization of the cyclic depsipeptide by the enzyme

Vgb. The mechanism of ring opening had been thought to occur by hydrolysis; thus, Vgb has been termed a hydrolase or lactonase (385–387). However, recent results indicate that this enzyme does not use water to open the ring and in fact operates by an elimination reaction (Fig. 15.14b) (388). Homologs of this enzyme are also found in the chromosomes of numerous bacteria (388), and thus potential streptogramin inactivating enzymes are widespread.

A new orally active streptogramin, RPR-106972 (**83**), has been reported with an activity profile similar to that of Synercid but with increased activity toward the respiratory pathogens *H. influenzae* and *M. catarrhalis* (389).

## 5.2 Lincosamides

Lincomycin (**84**), a lincosamide, is produced by *Streptomyces lincolnensis* and was first dis-

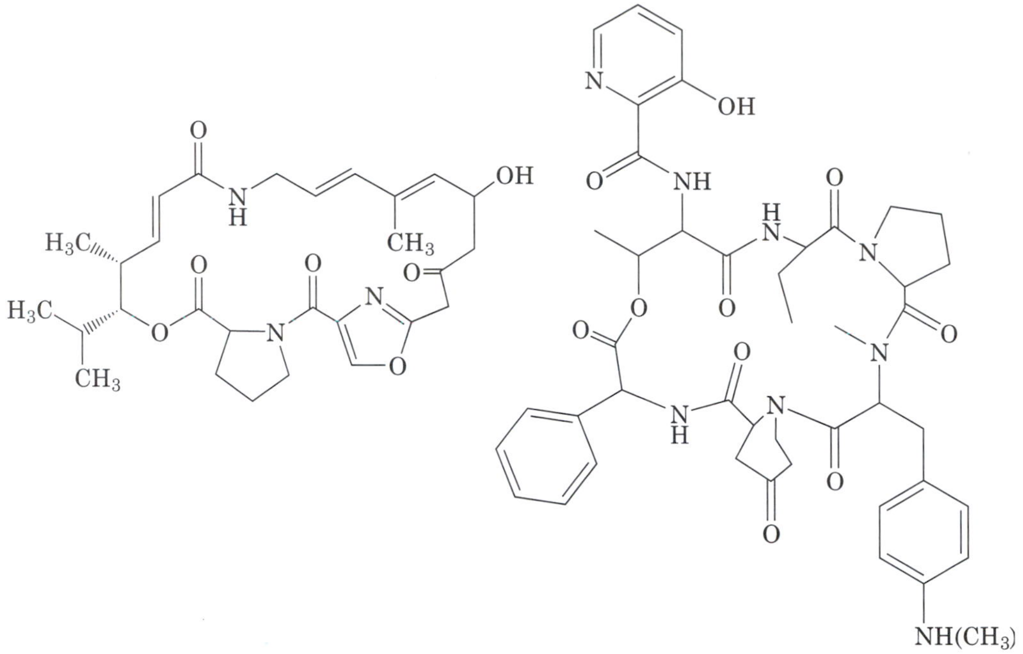

**(83)**

covered in 1962 (390). This antibiotic is composed of an amino acid portion, *trans-N*-methyl-4-*N*-propyl-L-proline, linked through an amide bond to an unusual sugar, 6-amino-6,8-dideoxy-1-thio-D-erythro-α-D-galactoocto-pyranoside. Clindamycin (**85**), a chlorinated semisynthetic derivative of lincomycin, shows better absorption and antimicrobial activity than that of the parent compound. These antibiotics are useful for the treatment of non-CNS infections caused by Gram-positive bacteria such as streptococci, staphylococci, and enterococci. Clindamycin is especially important in the treatment of infections caused by susceptible anaerobes, including species of *Bacteroides* and *Clostridium* (391) [but not *C. difficile*, where clindamycin treatment can cause *C. difficile*–associated colitis (392)]. Clindamycin is also useful in combination with primaquine for the treatment of *Pneumocystis carinii* infection in AIDS patients (393, 394).

Clindamycin is orally active and efficiently absorbed with an adult dose of 140–150 mg every 4 h. The mode of action of the lincosamide antibiotics includes binding to the 50S

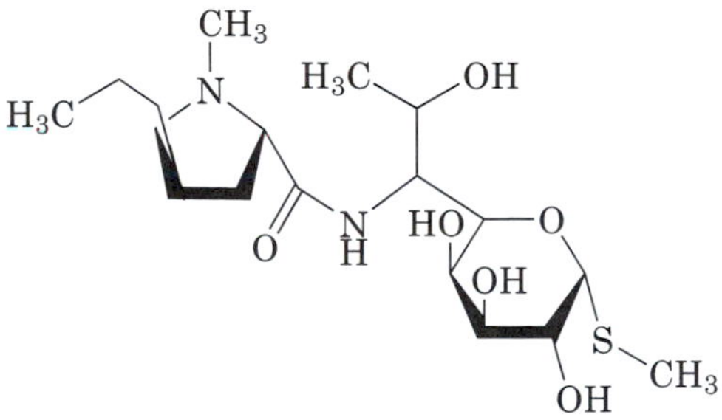

**(84)**

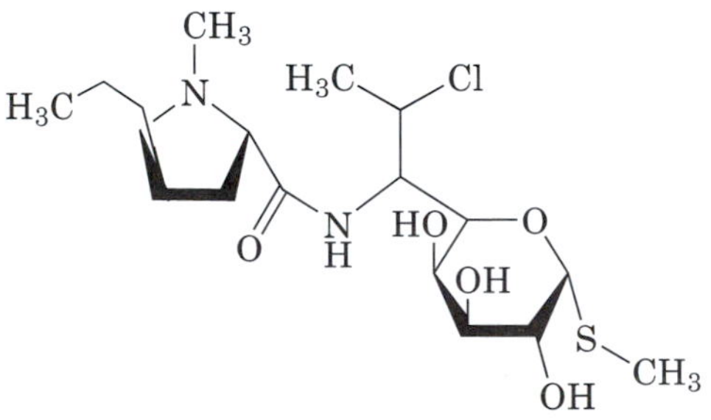

**(85)**

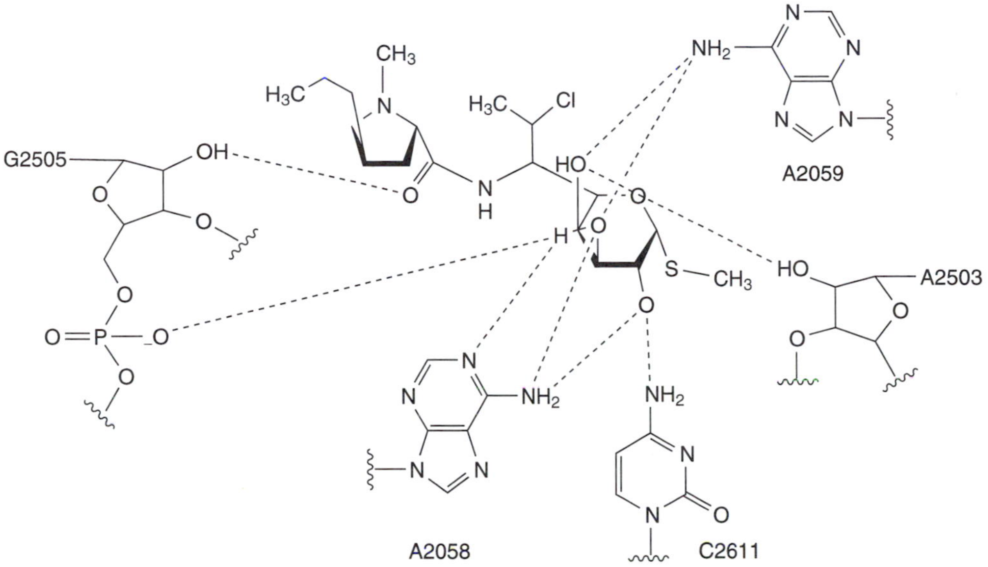

**Figure 15.15.** Interaction of clindamycin with the 50S ribosomal subunit of *D. radiodurans* (redrawn from Ref. 169). The equivalent *E. coli* numbering is shown and the interaction with C2611 is predicted based on the interaction with U2590 for the *D. radiodurans* 23S rRNA.

ribosomal subunit, which blocks aminoacyl-tRNA binding, thereby inhibiting protein synthesis (395). The 3D structure of clindamycin bound to the 50S ribosomal subunit has been reported (169). The molecular structure demonstrates that the antibiotic binds in a region overlapping the A- and P-sites of the ribosome exclusively through contacts with the 23S rRNA (Fig. 15.15), resulting in blocking of the peptide elongation tunnel and interfering with peptidyltransfer.

The complete lincosomycin biosynthetic gene cluster from *S. lincolnensis* has been cloned by the group of Piepersburg (396). Biosynthetic studies have indicated that the propyl proline moiety is derived from tyrosine (397) and that the sugar is derived either from glucose 1-phosphate or octulose-8-phosphate (395).

Clindamycin resistance can occur through ribosomal methylation by the *erm* gene products (MLS$_B$ phenotype), thereby confirming the site of action of the antibiotic and suggesting overlapping binding sites on the ribosome with the macrolide and group B streptogramins (232). Lincosamide resistance can also occur through mutations in the 23S rRNA (e.g., at positions A2058 and A2059) (246,

398). Chemical modification of lincosamides by phosphorylation and adenylation has also been detected, with the latter mechanism being associated with resistance in *S. aureus*, *S. haemolyticus*, and *E. faecium* (399–401). Surprisingly, the *linA* and *linA'* gene products from *S. haemolyticus* and *S. aureus*, respectively, modify lincomycin at position 3 of the sugar moiety and modify clindamycin at position 4 (Fig. 15.16) (400). The LinB enzyme from *E. faecium*, on the other hand, modifies lincosamides exclusively at position 3 (399).

## 5.3 Chloramphenicol

Chloramphenicol (**86**) is an antibiotic produced by *Streptomyces venezuelae* and other soil bacteria that was first discovered in 1947 (402) and is now exclusively produced synthetically. Chloramphenicol has a broad spectrum of activity against both aerobic and anaerobic Gram-positive and Gram-negative bacteria but is now infrequently used because of toxicity (see below). Nonetheless, chloramphenicol (or its prodrugs, the palmitate or succinate esters) is still indicated for the treatment of acute typhoid fever (*Salmonella typhi*) and as an alternative choice in the treatment of bacterial meningitis (*H. influenzae*, *S. pneu-*

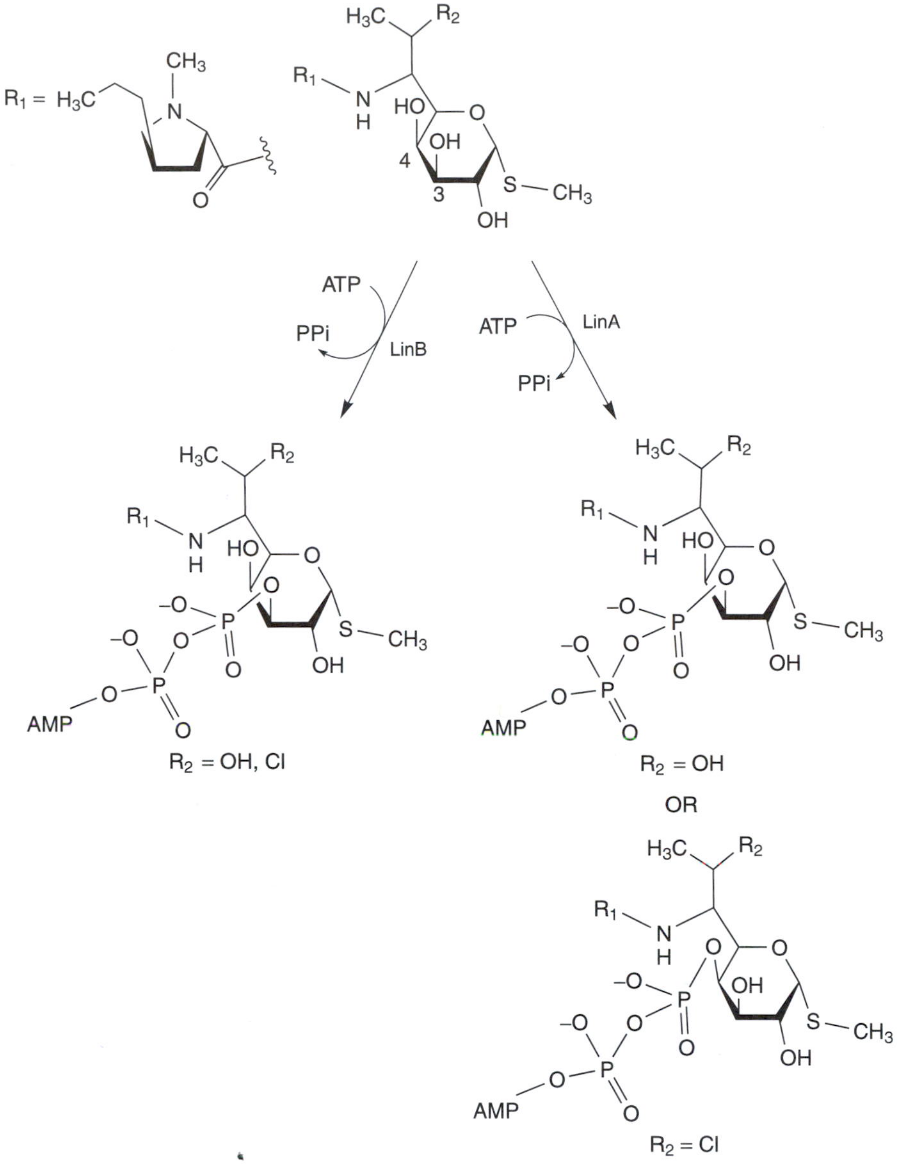

**Figure 15.16.** Modification of lincosamide antibiotics by adenyltransferases.

*moniae, N. menigitidis*) and other serious infections where β-lactam allergy or resistance is a problem. The antibiotic is well distributed in all tissues including the brain, which enables its use in the treatment of meningitis (391). Dosage is typically 50 mg/kg/day orally administered at 6-h intervals and peak serum levels are achieved 1–3 h after an oral dose.

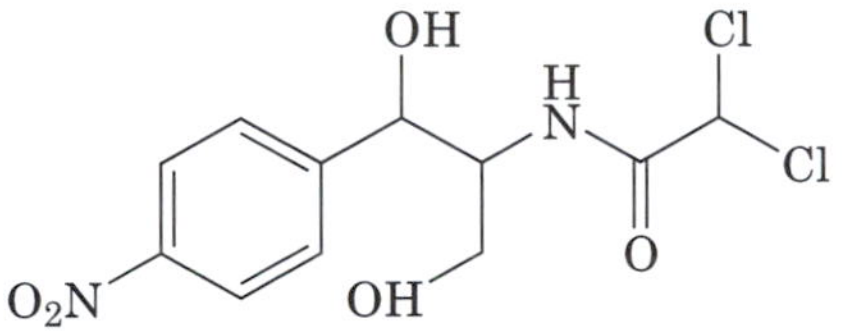

(86)

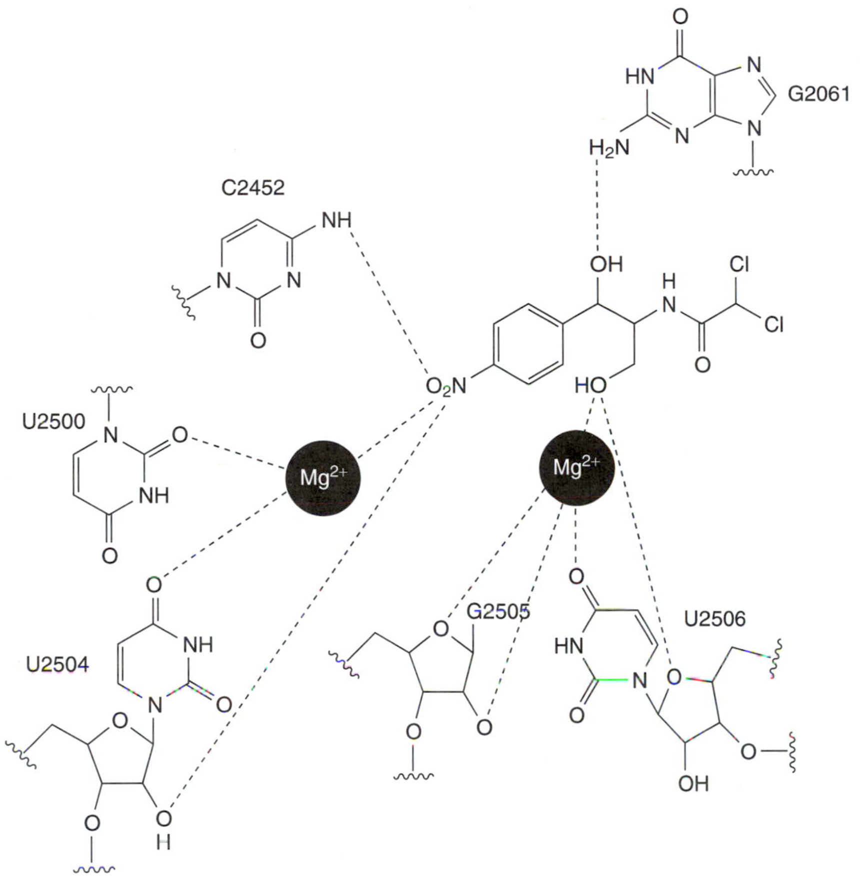

**Figure 15.17.** Interaction of chloramphenicol with the 50S ribosomal subunit of *D. radiodurans* (redrawn from Ref. 169). The equivalent *E. coli* numbering of the 23S rRNA is shown.

Chloramphenicol is generally bacteriostatic and binds to the P-site of the 50S ribosomal subunit, thus inhibiting translation (403). Affinity labeling has indicated that chloramphenicol binds to ribosomal protein L16 (404) and chemical footprinting studies have shown the 23S rRNA to also be a binding site (405). The 3D structure of chorampheni col bound to the 50S ribosomal subunit has been reported, confirming the interaction with the 23S rRNA, although not the predicted contact with L16 (169). Interestingly, key contacts between the primary alcohol at C3 and the nitro group with the rRNA are mediated through contacts with putative $Mg^{2+}$ ions (Fig. 15.17).

The major adverse effect of chloramphenicol is a risk of fatal irreversible aplastic ane-

mia that occurs after therapy and does not appear to be related to dose or administration route (406–408). Large doses in premature and neonatal infants can result in "gray baby syndrome," which is associated with vomiting, failure to feed, distended abdomen, and blue-gray skin color and can be rapidly fatal (391, 409). Reversible bone marrow suppression and several other adverse effects including gastrointestinal problems, headache, and mild depression have also been noted (391, 406).

Resistance to chloramphenicol can be the result of mutation in the 23S rRNA [e.g., G2032 (410) or C2452 (411) (*E. coli* numbering), altered permeability (412), and efflux (413–417)], but is generally the result of modifying enzymes that acetylate the antibiotic at the hydroxyl at position 3 in an acetylCoA-

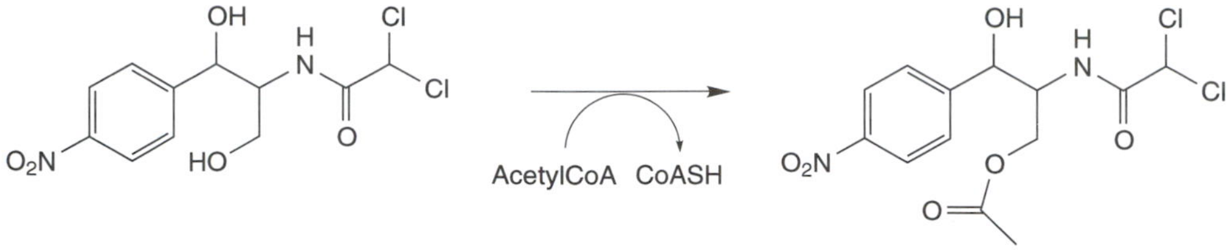

**Figure 15.18.** Action of chloramphenicol acetyltransferases.

dependent fashion (Fig. 15.18) (418, 419). The 3D structure of chloramphenicol bound to the 50S ribosomal subunit reveals contacts between the the cytosine base at position 2452 and the antibiotic nitro group, and the primary hydroxyl at position 3 with nucleotides, both direct and through a $Mg^{2+}$ ion, demonstrating the molecular basis of resistance (169).

There are more than 20 chloramphenicol acetyltransferase (CAT) isozymes known, all of which are trimers composed of monomers of approximately 25 kDa. These enzymes form part of a larger acetyltransferase family that also includes the streptogramin A acetyltransferases (418). The $CAT_{III}$ isozyme from *E. coli* has been studied most extensively and the 3D structure has been observed at high resolution (1.75 Å) (420). The structure and subsequent enzyme studies have revealed that the enzyme active site lies at the interface of the subunits of the trimer, with amino acid side chains from each partner contributing to substrate binding and catalysis (reviewed in Ref. 419). The $CAT_I$ isozyme is the most prevalent in clinical settings and also confers resistance to the antibiotic fusidic acid (421).

### 5.4 Fusidic Acid

Fusidic acid (**87**) is a steroidal antibiotic produced by the fungus *Fusidium coccineum* that finds use largely as an ophthalmic and topical agent for the treatment of wound infections caused by staphylococci and streptococci. The antibiotic can also be introduced by the i.v. route in cases of staphylococcal infection that has not responded to other therapies. Fusidic acid has no effect on *P. aeruginosa* and other Gram-negative bacteria because of the lack of outer membrane permeability (422). This antibiotic acts by binding to EF-G on the ribo-

some after GTP hydrolysis (see Fig. 15.1). EF-G undergoes a dramatic conformational change upon hydrolysis of GTP (423, 424), and fusidic acid stabilizes the complex, thus preventing dissociation of the elongation factor (425, 426). Resistance to fusidic acid is predominantly the result of point mutations in EF-G (427, 428), although binding to chloramphenicol acetyltransferase I can also contribute to resistance (421, 429).

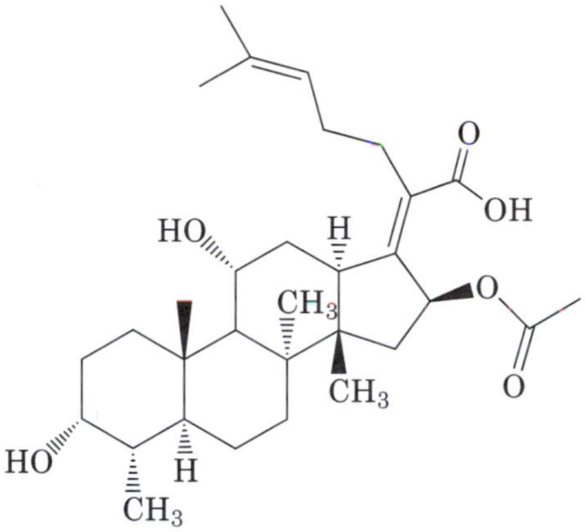

(**87**)

### 5.5 Oxazolidinones

The oxazolidinones are a relatively new class of antibiotics that inhibit bacterial protein synthesis. These compounds are not derived from natural products and represent one of the few completely synthetic antibiotics available, along with the sulfonamides, trimethroprim, and the fluoroquinolones. The oxazolidinones were first reported in the late 1980s (430) and one member of this class, linezolid (Zyvox) (**88**), is now approved for use in the treatment of drug-resistant and life-threatening infections caused by the Gram-positive staphylococci, streptococci, and enterococci (431–433). These antibiotics are bacteriostatic against enterococci and staphylococci but bactericidal against streptococci. Linezolid can be administered either orally or by i.v. injection in a dose of 600 mg every 12 h. Oral absorption is rapid with a $T_{max}$ of 1–1.5 h, generally. The oxazolidinones bind to the 50S ribosomal subunit with micromolar affinity and impede protein synthesis, possibly by interfering with the initiation step (434), although a consensus mechanism of action has not yet been established; recent evidence indicates that the 23S rRNA is the main site of interaction (435). Resistance to the oxazolidinones has been observed in the laboratory associated with mutations in the 23S rRNA [e.g., G2447U and G2576U (436, 437)].

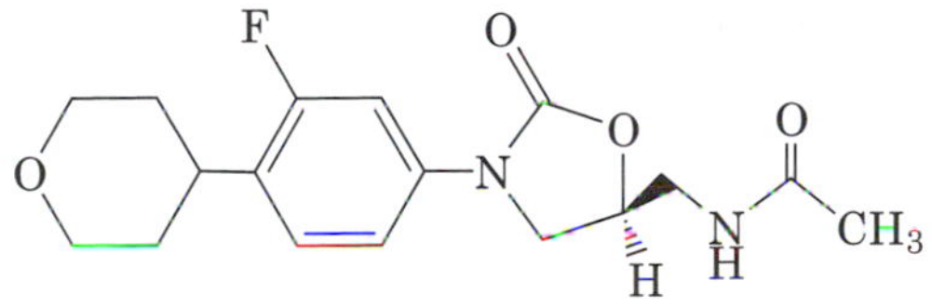

(88)

## 6 FUTURE PROSPECTS

Inhibition of the bacterial protein synthesis remains, along with DNA synthesis and cell wall synthesis, a favorite and proven target for antibacterial agents. The past 50 years have seen the discovery and clinical use of numerous antibacterial agents that inhibit translation and the highly conserved 16S and 23S rRNAs serve as the molecular target for most of these agents. The need for new antibacterial agents in the face of a growing antibiotic-resistance problem means that derivatives of known antibiotics that evade resistance or the synthesis of inhibitors of resistance mechanisms that can be coadministered with an "old" antibiotic are routes that must be explored (438). New antibiotics that inhibit translation at sites that are not susceptible to existing resistance are also being discovered. For example, evernimycin (89) is an oligosaccharide antibiotic active against Gram-positive bacteria (439) that acts at the initiation factor 2 binding site (440, 441).

The example of Synercid and other streptogramin antibiotics provides precedent for a reexamination of known compounds with antimicrobial activity as starting places for new drug development programs. Other antibiotics that interfere with translation have not been exploited, such as the peptide edeine [for which there is now a structure of the antibiotic bound to the 30S ribosome (442)], and the compound sparsomycin, largely attributed to indiscriminant inhibition of eukaryotic and bacterial translation, may therefore serve as templates for further compound screening and development. Other strategies such as the catalytic ribotoxins produced by aspergilli that specifically cleave rRNAs may also prove useful under some conditions (443).

The increasing number of high resolution crystal structures of ribosomes and their complexes with antibiotics now provides an unprecedented opportunity to rationally undertake structure and function analyses on inhibitors of translation and to leverage this information in the development of new antibiotics. The proven utility of bacterial protein

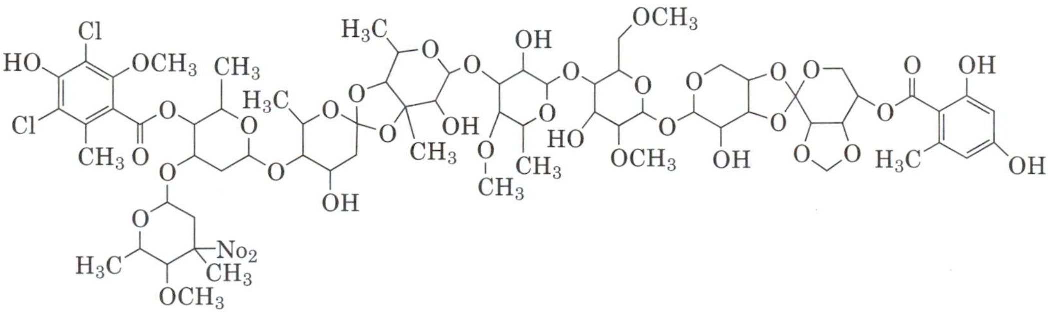

(89)

synthesis as an outstanding target for antibiotics will no doubt continue to be exploited in the future.

# REFERENCES

1. P. E. Sum, V. J. Lee, R. T. Testa, J. J. Hlavka, G. A. Ellestad, J. D. Bloom, Y. Gluzman, and F. P. Tally, *J. Med. Chem.*, **37**, 184–188 (1994).

2. P. E. Sum and P. Petersen, *Bioorg. Med. Chem. Lett.*, **9**, 1459–1462 (1999).

3. M. Argast and C. F. Beck, *Arch. Microbiol.*, **141**, 260–265 (1985).

4. M. Argast and C. F. Beck, *Antimicrob. Agents Chemother.*, **26**, 263–265 (1984).

5. D. E. Brodersen, W. M. Clemons Jr., A. P. Carter, R. J. Morgan-Warren, B. T. Wimberly, and V. Ramakrishnan, *Cell*, **103**, 1143–1154 (2000).

6. T. Dairi, T. Nakano, T. Mizukami, K. Aisaka, M. Hasegawa, and R. Katsumata, *Biosci. Biotechnol. Biochem.*, **59**, 1360–1361 (1995).

7. M. J. Butler, E. J. Friend, I. S. Hunter, F. S. Kaczmarek, D. A. Sugden, and M. Warren, *Mol. Gen. Genet.*, **215**, 231–238 (1989).

8. C. Binnie, M. Warren, and M. J. Butler, *J. Bacteriol.*, **171**, 887–895 (1989).

9. H. Petkovic, A. Thamchaipenet, L. H. Zhou, D. Hranueli, P. Raspor, P. G. Waterman, and I. S. Hunter, *J. Biol. Chem.*, **274**, 32829–32834 (1999).

10. I. Vancurova, J. Volc, M. Flieger, J. Neuzil, J. Novotna, J. Vlach, and V. Behal, *Biochem. J.*, **253**, 263–267 (1988).

11. T. Dairi, T. Nakano, K. Aisaka, R. Katsumata, and M. Hasegawa, *Biosci. Biotechnol. Biochem.*, **59**, 1099–1106 (1995).

12. M. C. Roberts, *FEMS Microbiol. Rev.*, **19**, 1–24 (1996).

13. D. Schnappinger and W. Hillen, *Arch. Microbiol.*, **165**, 359–369 (1996).

14. I. Chopra and M. Roberts, *Microbiol. Mol. Biol. Rev.*, **65**, 232–260; second page, table of contents (2001).

15. J. I. Ross, E. A. Eady, J. H. Cove, and W. J. Cunliffe, *Antimicrob. Agents Chemother.*, **42**, 1702–1705 (1998).

16. J. I. Ross, A. M. Snelling, E. A. Eady, J. H. Cove, W. J. Cunliffe, J. J. Leyden, P. Collignon, B. Dreno, A. Reynaud, J. Fluhr, and S. Oshima, *Br. J. Dermatol.*, **144**, 339–346 (2001).

17. I. T. Paulsen, M. H. Brown, and R. A. Skurray, *Microbiol. Rev.*, **60**, 575–608 (1996).

18. L. M. McMurray and S. B. Levy in V. A. Fischetti, R. P. Novick, J. J. Ferretti, D. A. Portnoy, and J. I. Rood, Eds., *Gram-positive Pathogens*, ASM Press, Washington, DC, 2000, pp. 660–677.

19. N. Tamura, S. Konishi, S. Iwaki, T. Kimura-Someya, S. Nada, and A. Yamaguchi, *J. Biol. Chem.*, **276**, 20330–20339 (2001).

20. Y. Someya, T. Kimura-Someya, and A. Yamaguchi, *J. Biol. Chem.*, **275**, 210–214 (2000).

21. Y. Kubo, S. Konishi, T. Kawabe, S. Nada, and A. Yamaguchi, *J. Biol. Chem.*, **275**, 5270–5274 (2000).

22. S. Iwaki, N. Tamura, T. Kimura-Someya, S. Nada, and A. Yamaguchi, *J. Biol. Chem.*, **275**, 22704–22712 (2000).

23. D. E. Taylor and A. Chau, *Antimicrob. Agents Chemother.*, **40**, 1–5 (1996).

24. D. E. Taylor, L. J. Jerome, J. Grewal, and N. Chang, *Can. J. Microbiol.*, **41**, 965–970 (1995).

25. V. Burdett, *J. Bacteriol.*, **178**, 3246–3251 (1996).

26. C. M. Spahn, G. Blaha, R. K. Agrawal, P. Penczek, R. A. Grassucci, C. A. Trieber, S. R. Connell, D. E. Taylor, K. H. Nierhaus, and J. Frank, *Mol. Cell.*, **7**, 1037–1045 (2001).

27. J. Frank and R. K. Agrawal, *Nature*, **406**, 318–322 (2000).

28. B. S. Speer and A. A. Salyers, *J. Bacteriol.*, **170**, 1423–1429 (1988).

29. B. S. Speer and A. A. Salyers, *J. Bacteriol.*, **171**, 148–153 (1989).

30. S. J. Projan, *Pharmacotherapy*, **20**, 219S–223S; discussion 224S–228S (2000).

31. T. M. Murphy, J. M. Deitz, P. J. Petersen, S. M. Mikels, and W. J. Weiss, *Antimicrob. Agents Chemother.*, **44**, 3022–3027 (2000).

32. M. L. van Ogtrop, D. Andes, T. J. Stamstad, B. Conklin, W. J. Weiss, W. A. Craig, and O. Vesga, *Antimicrob. Agents Chemother.*, **44**, 943–949 (2000).

33. M. L. Nelson and S. B. Levy, *Antimicrob. Agents Chemother.*, **43**, 1719–1724 (1999).

34. M. L. Nelson, B. H. Park, and S. B. Levy, *J. Med. Chem.*, **37**, 1355–1361 (1994).

35. M. L. Nelson, B. H. Park, J. S. Andrews, V. A. Georgian, R. C. Thomas, and S. B. Levy, *J. Med. Chem.*, **36**, 370–377 (1993).

36. R. Greenwald and L. Golub, *Curr. Med. Chem.*, **8**, 237–242 (2001).

37. A. R. Amin, M. G. Attur, G. D. Thakker, P. D. Patel, P. R. Vyas, R. N. Patel, I. R. Patel, and

S. B. Abramson, *Proc. Natl. Acad. Sci. USA*, **93**, 14014–14019 (1996).

38. R. N. Patel, M. G. Attur, M. N. Dave, I. V. Patel, S. A. Stuchin, S. B. Abramson, and A. R. Amin, *J. Immunol.*, **163**, 3459–3467 (1999).

39. M. E. Ryan, R. A. Greenwald, and L. M. Golub, *Curr. Opin. Rheumatol.*, **8**, 238–247 (1996).

40. M. E. Ryan, S. Ramamurthy, and L. M. Golub, *Curr. Opin. Periodontol.*, **3**, 85–96 (1996).

41. B. R. Rifkin, A. T. Vernillo, and L. M. Golub, *J. Periodontol.*, **64**, 819–827 (1993).

42. R. A. Ralph, *Cornea*, **19**, 274–277 (2000).

43. D. Dursun, M. C. Kim, A. Solomon, and S. C. Pflugfelder, *Am. J. Ophthalmol.*, **132**, 8–13 (2001).

44. M. Hidalgo and S. G. Eckhardt, *J. Natl. Cancer Inst.*, **93**, 178–193 (2001).

45. R. S. Edson and C. L. Terrell, *Mayo Clin. Proc.*, **74**, 519–528 (1999).

46. J. E. Rosenblatt, *Mayo Clin. Proc.*, **74**, 1161–1175 (1999).

47. G. D. Wright, A. M. Berghuis, and S. Mobashery, *Adv. Exp. Med. Biol.*, **456**, 27–69 (1998).

48. R. E. Brummett and K. E. Fox, *Antimicrob. Agents Chemother.*, **33**, 797–800 (1989).

49. B. H. Ali, *Gen. Pharmacol.*, **26**, 1477–1487 (1995).

50. S. A. Crann, M. Y. Huang, J. D. McLaren, and J. Schacht, *Biochem. Pharmacol.*, **43**, 1835–1839 (1992).

51. S. A. Crann and J. Schacht, *Audiol. Neurootol.*, **1**, 80–85 (1996).

52. D. Moazed and H. F. Noller, *Nature*, **27**, 389–394 (1987).

53. J. Woodcock, D. Moazed, M. Cannon, J. Davies, and H. F. Noller, *EMBO J.*, **10**, 3099–3103 (1991).

54. H. W. Taber, J. P. Mueller, P. F. Miller, and A. S. Arrow, *Microbiol. Rev.*, **51**, 439–457 (1987).

55. L. E. Bryan and S. Kwan, *Antimicrob. Agents Chemother.*, **23**, 835–845 (1983).

56. S. Gilman and V. A. Saunders, *J. Antimicrob. Chemother.*, 37–44 (1986).

57. D. T. Dubin and B. D. Davis, *Biochim. Biophys. Acta*, **52**, 400–402 (1961).

58. R. Hancock, *J. Bacteriol.*, **88**, 633–639 (1961).

59. N. Tanaka, K. Matsunaga, H. Yamaki, and T. Nishimura, *Biochem. Biophys. Res. Commun.*, **122**, 460–465 (1984).

60. J. Davies, L. Gorini, and B. D. Davis, *Mol. Pharmacol.*, **1**, 93–106 (1965).

61. J. Davies, D. S. Jones, and H. G. Khorana, *J. Mol. Biol.*, **18**, 48–57 (1966).

62. J. Davies and B. D. Davis, *J. Biol. Chem.*, **243**, 3312–3316 (1968).

63. B. D. Davis, L. L. Chen, and P. C. Tai, *Proc. Natl. Acad. Sci. USA*, **83**, 6164–6168 (1986).

64. B. D. Davis, *Microbiol. Rev.*, **51**, 341–350 (1987).

65. A. Singh, D. Ursic, and J. Davies, *Nature*, **277**, 146–148 (1979).

66. S. Yoshizawa, D. Fourmy, and J. D. Puglisi, *EMBO J.*, **17**, 6437–6448 (1998).

67. D. Fourmy, S. Yoshizawa, and J. D. Puglisi, *J. Mol. Biol.*, **277**, 333–345 (1998).

68. D. Fourmy, M. I. Recht, S. C. Blanchard, and J. D. Puglisi, *Science*, **274**, 1367–1371 (1996).

69. A. P. Carter, W. M. Clemons, D. E. Brodersen, R. J. Morgan-Warren, B. T. Wimberly, and V. Ramakrishnan, *Nature*, **407**, 340–348 (2000).

70. M. F. Brink, G. Brink, M. P. Verbeet, and H. A. de Boer, *Nucleic Acids Res.*, **22**, 325–331 (1994).

71. E. P. Bakker, *J. Gen. Microbiol.*, **138**, 563–569 (1992).

72. O. Jerinic and S. Joseph, *J. Mol. Biol.*, **304**, 707–713 (2000).

73. J. M. Ogle, D. E. Brodersen, W. M. Clemons Jr., M. J. Tarry, A. P. Carter, and V. Ramakrishnan, *Science*, **292**, 897–902 (2001).

74. A. Schatz, E. Bugie, and S. A. Waksman, *Proc. Soc. Exp. Biol. Med.*, **55**, 66–69 (1944).

75. S. A. Waksman and H. A. Lechevalier, *Science*, **109**, 305–307 (1949).

76. H. Umezawa, M. Ueda, K. Maeda, K. Yagishita, S. Kando, Y. Okami, R. Utahara, Y. Osato, K. Nitta, and T. Kakeuchi, *J. Antibiot.*, **10**, 181–189 (1957).

77. M. J. Weinstein, G. M. Luedemann, E. M. Oden, G. H. Wagman, J. P. Rosselet, J. A. Marquez, C. T. Coniglio, W. Charney, H. L. Herzog, and J. Black, *J. Med. Chem.*, **6**, 463–464 (1963).

78. M. J. Weinstein, J. A. Marquez, R. T. Testa, G. H. Wagman, E. M. Oden, and J. A. Waitz, *J. Antibiot.*, **23**, 551–554 (1970).

79. W. Piepersberg in W. R. Strohl, Ed., *Biotechnology of Antibiotics*, Marcel Dekker, New York, 1997, pp. 81–163.

80. M. Jackson, D. C. Crick, and P. J. Brennan, *J. Biol. Chem.*, **275**, 30092–30099 (2000).

81. Y. Ota, H. Tamegai, F. Kudo, H. Kuriki, A. Koike-Takeshita, T. Eguchi, and K. Kakinuma, *J. Antibiot. (Tokyo)*, **53**, 1158–1167 (2000).

82. K. Pissowotzki, K. Mansouri, and W. Piepersberg, *Mol. Gen. Genet.*, **231**, 113–123 (1991).

83. S. Beyer, J. Distler, and W. Piepersberg, *Mol. Gen. Genet.*, **250**, 775–784 (1996).

84. K. Mansouri and W. Piepersberg, *Mol. Gen. Genet.*, **228**, 459–469 (1991).

85. J. R. Aires, T. Kohler, H. Nikaido, and P. Plesiat, *Antimicrob. Agents Chemother.*, **43**, 2624–2628 (1999).

86. S. Westbrock-Wadman, D. R. Sherman, M. J. Hickey, S. N. Coulter, Y. Q. Zhu, P. Warrener, L. Y. Nguyen, R. M. Shawar, K. R. Folger, and C. K. Stover, *Antimicrob. Agents Chemother.*, **43**, 2975–2983 (1999).

87. R. A. Moore, D. DeShazer, S. Reckseidler, A. Weissman, and D. E. Woods, *Antimicrob. Agents Chemother.*, **43**, 465–470 (1999).

88. J. A. Aínsa, M. C. Blokpoel, I. Otal, D. B. Young, K. A. De Smet, and C. Martin, *J. Bacteriol.*, **180**, 5836–5843 (1998).

89. R. Edgar and E. Bibi, *EMBO J.*, **18**, 822–832 (1999).

90. E. Y. Rosenberg, D. Ma, and H. Nikaido, *J. Bacteriol.*, **182**, 1754–1756 (2000).

91. W. Mao, M. S. Warren, A. Lee, A. Mistry, and O. Lomovskaya, *Antimicrob. Agents Chemother.*, **45**, 2001–2007 (2001).

92. S. Sreevatsan, X. Pan, K. E. Stockbauer, D. L. Williams, B. N. Kreiswirth, and J. M. Musser, *Antimicrob. Agents Chemother.*, **40**, 1024–1026 (1996).

93. A. A. Beauclerk and E. Cundliffe, *J. Mol. Biol.*, **93**, 661–671 (1987).

94. K. J. Shaw, P. N. Rather, R. S. Hare, and G. H. Miller, *Microbiol. Rev.*, **57**, 138–163 (1993).

95. G. D. Wright, *Curr. Opin. Microbiol.*, **2**, 499–503 (1999).

96. G. D. Wright and P. R. Thompson, *Front. Biosci.*, **4**, D9–D21 (1999).

97. W. C. Hon, G. A. McKay, P. R. Thompson, R. M. Sweet, D. S. Yang, G. D. Wright, and A. M. Berghuis, *Cell*, **89**, 887–895 (1997).

98. D. L. Burk, W. C. Hon, A. K. Leung, and A. M. Berghuis, *Biochemistry*, **40**, 8756–8764 (2001).

99. D. D. Boehr, P. R. Thompson, and G. D. Wright, *J. Biol. Chem.*, **276**, 23929–23936 (2001).

100. D. M. Daigle, G. A. McKay, and G. D. Wright, *J. Biol. Chem.*, **272**, 24755–24758 (1997).

101. D. D. Boehr, W. S. Lane, and G. D. Wright, *Chem. Biol.*, **8**, 791–800 (2001).

102. D. M. Daigle, G. A. McKay, P. R. Thompson, and G. D. Wright, *Chem. Biol.*, **6**, 11–18 (1998).

103. J. Sakon, H. H. Liao, A. M. Kanikula, M. M. Benning, I. Rayment, and H. M. Holden, *Biochemistry*, **32**, 11977–11984 (1993).

104. L. C. Perdersen, M. M. Benning, and H. M. Holden, *Biochemistry*, **34**, 13305–13311 (1995).

105. D. M. Daigle, D. W. Hughes, and G. D. Wright, *Chem. Biol.*, **6**, 99–110 (1999).

106. L. E. Wybenga-Groot, K. Draker, G. D. Wright, and A. M. Berghuis, *Struct. Fold Des.*, **7**, 497–507 (1999).

107. E. Wolf, A. Vassilev, Y. Makino, A. Sali, Y. Nakatani, and S. K. Burley, *Cell*, **94**, 439–449 (1998).

108. F. Dyda, D. C. Klein, and A. B. Hickman, *Annu. Rev. Biophys. Biomol. Struct.*, **29**, 81–103 (2000).

109. W. A. Greenberg, E. S. Priestley, P. Sears, P. B. Alper, C. Rosenbohm, M. Hendrix, S.-C. Hung, and C.-H. Wong, *J. Am. Chem. Soc.*, **121**, 6527–6541 (1999).

110. C. H. Wong, M. Hendrix, E. S. Priestley, and W. A. Greenberg, *Chem. Biol.*, **5**, 397–406 (1998).

111. S. J. Sucheck, A. L. Wong, K. M. Koeller, D. D. Boehr, K.-A. Draker, P. Sears, G. D. Wright, and C.-H. Wong, *J. Am. Chem. Soc.*, **122**, 5230–5231 (2000).

112. J. Roestamadji, I. Grapsas, and S. Mobashery, *J. Am. Chem. Soc.*, **117**, 11060–11069 (1995).

113. G. A. McKay, J. Roestamadji, S. Mobashery, and G. D. Wright, *Antimicrob. Agents Chemother.*, **40**, 2648–2650 (1996).

114. J. Roestamadji, I. Grapsas, and S. Mobashery, *J. Am. Chem. Soc.*, **117**, 80–84 (1995).

115. R. Schroeder, C. Waldsich, and H. Wank, *EMBO J.*, **19**, 1–9 (2000).

116. A. D. Ellington, *Curr. Biol.*, **4**, 427–429 (1994).

117. S. M. Lato, A. R. Boles, and A. D. Ellington, *Chem. Biol.*, **2**, 291–303 (1995).

118. Y. Wang and R. R. Rando, *Chem. Biol.*, **2**, 281–290 (1995).

119. Y. Wang, J. Killian, K. Hamasaki, and R. R. Rando, *Biochemistry*, **35**, 12338–12346 (1996).

120. S. M. Lato and A. D. Ellington, *Mol. Divers.*, **2**, 103–110 (1996).

121. L. Jiang, A. K. Suri, R. Fiala, and D. J. Patel, *Chem. Biol.*, **4**, 35–50 (1997).

122. L. Jiang, A. Majumdar, W. Hu, T. J. Jaishree, W. Xu, and D. J. Patel, *Struct. Fold Des.*, **7**, 817–827 (1999).

123. U. von Ahsen, J. Davies, and R. Schroeder, *Nature*, **353**, 368–370 (1991).

124. U. von Ahsen and R. Schroeder, *Nucleic Acids Res.*, **19**, 2261–2265 (1991).

125. U. von Ahsen, J. Davies, and R. Schroeder, *J. Mol. Biol.*, **226**, 935–941 (1992).

126. J. Rogers, A. H. Chang, U. von Ahsen, R. Schroeder, and J. Davies, *J. Mol. Biol.*, **259**, 916–925 (1996).

127. N. E. Mikkelsen, M. Brannvall, A. Virtanen, and L. A. Kirsebom, *Proc. Natl. Acad. Sci. USA*, **96**, 6155–6160 (1999).

128. A. Tekos, A. Tsagla, C. Stathopoulos, and D. Drainas, *FEBS Lett.*, **485**, 71–75 (2000).

129. R. W. Woodward, *Angew. Chem.*, **69**, 50 (1957).

130. A. J. Bryskier, J. P. Butzler, H. C. Neu, and P. M. Tulkens, Eds., *Macrolides: Chemistry, Pharmacology and Clinical Uses*, Arnette Blackwell, Paris, 1993.

131. T. Miura, M. Iwasaki, M. Komori, H. Ohi, M. Kitada, H. Mitsui, and T. Kamataki, *J. Antimicrob. Chemother.*, **24**, 551–559 (1989).

132. D. F. Johnson and W. H. Hall, *Engl. J. Med.*, **265**, 1200 (1961).

133. P. D. Kroboth, M. A. McNeil, A. Kreeger, J. Dominguez, and R. Rault, *Arch. Intern. Med.*, **143**, 1263–1265 (1983).

134. J. E. Slater, *Ann. Allergy*, **66**, 193–195 (1991).

135. C. R. E. Hawworth, *Br. Med. J.*, **298**, 190 (1989).

136. S. Nattel, S. Ranger, M. Talajic, R. Lemery, and D. Roy, *Am. J. Med.*, **89**, 235–238 (1990).

137. C. J. Fichtenbaum, J. D. Babb, and D. A. Baker, *Conn. Med.*, **52**, 135–136 (1988).

138. S. J. Nance, S. Ladisch, T. L. Williamson, and G. Garratty, *Vox Sang.*, **55**, 233–236 (1988).

139. P. Periti, T. Mazzei, E. Mini, and A. Novelli, *Clin. Pharmacokinet.*, **16**, 193–214 (1989).

140. P. Periti, T. Mazzei, E. Mini, and A. Novelli, *Clin. Pharmacokinet.*, **16**, 261–282 (1989).

141. D. Tremblay, H. Jaeger, J. B. Fourtillan, and C. Manuel, *J. Clin. Pract.*, **22**(Suppl. 55), 49 (1988).

142. G. Bozler, G. Henizel, U. Lechner, K. Schumacher, and U. Busch, Proceedings of the 28th Interscience Conference on Antimicrobial Agents and Chemotherapy, Abstr. 924 (1988).

143. G. D. Sides, B. J. Cerimele, H. R. Black, U. Busch, and K. A. DeSante, *J. Antimicrob. Chemother.*, **31**(Suppl. C), 65–75 (1993).

144. A. M. Lepore, G. Bonardi, and G. C. Maggi, *Int. J. Clin. Pharmacol. Res.*, **8**, 253–257 (1988).

145. G. Foulds, R. M. Shepard, and R. B. Johnson, *J. Antimicrob. Chemother.*, **25**(Suppl. A), 73–82 (1990).

146. S. Y. Chu, L. T. Sennello, S. T. Bunnell, L. L. Varga, D. S. Wilson, and R. C. Sonders, *Antimicrob. Agents Chemother.*, **36**, 2447–2453 (1992).

147. G. A. Dette, H. Knothe, and G. Koulen, *Drugs Exp. Clin. Res.*, **13**, 567–576 (1987).

148. O. Nilsen, Proceedings of the 2nd International Conference on the Macrolides, Azalides and Streptogramins, Abstr. 211 (1994).

149. C. E. Halstenson, J. A. Opsahl, M. H. Schwenk, J. M. Kovarik, S. K. Puri, I. Ho, and G. R. Matzke, *Antimicrob. Agents Chemother.*, **34**, 385–389 (1990).

150. D. N. Fish, M. H. Gotfried, L. H. Danziger, and K. A. Rodvold, *Antimicrob. Agents Chemother.*, **38**, 876–878 (1994).

151. V. N. Gan, S. Y. Chu, H. T. Kusmiesz, and J. C. Craft, *Antimicrob. Agents Chemother.*, **36**, 2478–2480 (1992).

152. J. A. Retsema, A. E. Girard, D. Girard, and W. B. Milisen, *J. Antimicrob. Chemother.*, **25**(Suppl. A), 83–89 (1990).

153. A. Wildfeuer, H. Laufen, M. Leitold, and T. Zimmermann, *J. Antimicrob. Chemother.*, **31**(Suppl. E), 51–56 (1993).

154. J. P. Chave, A. Munafo, J. Y. Chatton, P. Dayer, M. P. Glauser, and J. Biollaz, *Antimicrob. Agents Chemother.*, **36**, 1013–1018 (1992).

155. J. Pukander, Proceedings of the 2nd International Conference on the Macrolides, Azalides and Streptogramins, Abstr. 199 (1994).

156. G. Foulds, R. Ferraina, H. Fouda, R. B. Johnson, A. Kamel, R. M. Shepard, R. Falcone, and C. Hanna, Proceedings of the 33rd Interscience Conference on Antimicrobial Agents and Chemotherapy, Abstr. 738 (1993).

157. F. Scaglione, G. Demartini, and F. Fraschini, Proceedings of the 2nd International Conference on the Macrolides, Azalides and Streptogramins, Abstr. 207 (1994).

158. A. McLean, J. A. Sutton, J. Salmon, and D. Chatelet, *Br. J. Clin. Pract.*, **42**(Suppl. 55), 52 (1988).

159. G. S. Duthu, *J. Liq. Chromatogr.*, **7**, 1023 (1984).

160. F. T. Counter, A. M. Felty-Duckworth, H. A. Kirst, and J. W. Paschal, Proceedings of the 28th Interscience Conference on Antimicrobial Agents and Chemotherapy, Abstr. 920 (1988).

161. R. M. Shepard, H. Fouda, D. F. Johnson, R. Ferraina, and M. A. Mullins, Proceedings of

the International Congress for Infectious Diseases, Abstr. 184 (1990).

162. T. Adachi, S. Morimoto, H. Kondoh, T. Nagate, Y. Watanabe, and K. Sota, *J. Antibiot. (Tokyo)*, **41**, 966–975 (1988).

163. D. J. Hardy, R. N. Swanson, R. A. Rode, K. Marsh, N. L. Shipkowitz, and J. J. Clement, *Antimicrob. Agents Chemother.*, **34**, 1407–1413 (1990).

164. T. Osono and H. Umezawa, *J. Antimicrob. Chemother.*, **16**(Suppl. A), 151–166 (1985).

165. S. Pestka and R. A. Lemahieu, *Antimicrob. Agents Chemother.*, **6**, 479–488 (1974).

166. S. Douthwaite and C. Aagaard, *J. Mol. Biol.*, **232**, 725–731 (1993).

167. A. Yonath, K. R. Leonard, and H. G. Wittmann, *Science*, **236**, 813–816 (1987).

168. P. Nissen, J. Hansen, N. Ban, P. B. Moore, and T. A. Steitz, *Science*, **289**, 920–930 (2000).

169. F. Schlunzen, R. Zarivach, J. Harms, A. Bashan, A. Tocilj, R. Albrecht, A. Yonath, and F. Franceschi, *Nature*, **413**, 814–821 (2001).

170. R. Bentley and J. W. Bennett, *Annu. Rev. Microbiol.*, **53**, 411–446 (1999).

171. D. O'Hagan, *Nat. Prod. Rep.*, **12**, 1–32 (1995).

172. S. Donadio, M. J. Staver, J. B. McAlpine, S. J. Swanson, and L. Katz, *Science*, **252**, 675–679 (1991).

173. J. F. Andersen and C. R. Hutchinson, *J. Bacteriol.*, **174**, 725–735 (1992).

174. J. M. Weber, J. O. Leung, S. J. Swanson, K. B. Idler, and J. B. McAlpine, *Science*, **252**, 114–117 (1991).

175. D. Stassi, S. Donadio, M. J. Staver, and L. Katz, *J. Bacteriol.*, **175**, 182–189 (1993).

176. J. M. Weber, B. Schoner, and R. Losick, *Gene*, **75**, 235–241 (1989).

177. D. J. Hardy, D. M. Hensey, J. M. Beyer, C. Vojtko, E. J. McDonald, and P. B. Fernandes, *Antimicrob. Agents Chemother.*, **32**, 1710–1719 (1988).

178. P. A. Lartey and J. J. Perun, Eds., *Studies in Natural Products Chemistry*, Vol. **13**, Elsevier Science, Amsterdam/New York, 1993.

179. P. F. Wiley, K. Gerzon, E. H. PFlynn, M. V. Sigal, O. Weaver, U. C. Quarck, R. R. Chauvette, and R. Monahan, *J. Am. Chem. Soc.*, **79**, 6062 (1957).

180. E. H. Massey, B. S. Kitchell, L. D. Martin, and K. Gerzon, *J. Med. Chem.*, **17**, 105–107 (1974).

181. J. C. Gasc, S. G. d'Ambrieres, A. Lutz, and J. F. Chantot, *J. Antibiot. (Tokyo)*, **44**, 313–330 (1991).

182. I. Phillips, J. Pechers, A. Davies, and D. Speller, *J. Antimicrob. Chemother.*, **20**(Suppl. B), 1 (1987).

183. T. Barlam and H. C. Neu, *Antimicrob. Agents Chemother.*, **25**, 529–531 (1984).

184. J. M. Wilson, P. C. Hannan, C. Shillingford, and D. J. Knowles, *J. Antibiot. (Tokyo)*, **42**, 454–462 (1989).

185. H. Sakakabara and S. Omura, Eds., *Macrolide Antibiotics: Chemistry, Biology and Practice*, Academic Press, Orlando, FL, 1984.

186. H. A. Kirst, J. A. Wind, J. P. Leeds, K. E. Willard, M. Debono, R. Bonjouklian, J. M. Greene, K. A. Sullivan, J. W. Paschal, J. B. Deeter, et al., *J. Med. Chem.*, **33**, 3086–3094 (1990).

187. C. Maring, L. Klein, R. Pariza, P. A. Lartey, D. Grampovnik, C. Yeung, M. Buytendrop, and D. J. Hardy, Proceedings of the 29th Interscience Conference on Antimicrobial Agents and Chemotherapy, Abstr. 1023 (1989).

188. D. J. Hardy, C. Vojtko, J. M. Beyer, D. M. Hensey, C. Maring, D. C. Klein, R. Pariza, P. A. Lartey, and J. J. Clement, Proceedings of the 29th Interscience Conference on Antimicrobial Agents and Chemotherapy, Abstr. 1024 (1989).

189. F. T. Counter, P. W. Ensminger, D. A. Preston, C. Y. Wu, J. M. Greene, A. M. Felty-Duckworth, J. W. Paschal, and H. A. Kirst, *Antimicrob. Agents Chemother.*, **35**, 1116–1126 (1991).

190. L. Freiberg, C. Edwards, D. Bacino, L. Seif, P. A. Lartey, and D. Whittern, Proceedings of the 29th Interscience Conference on Antimicrobial Agents and Chemotherapy, Abstr. 1029 (1989).

191. J. J. Clement, N. L. Shipkowitz, R. N. Swanson, and P. A. Lartey, Proceedings of the 30th Interscience Conference on Antimicrobial Agents and Chemotherapy, Abstr. 814 (1990).

192. M. Nakata, T. Tamai, Y. Miura, M. Kinoshita, and K. Tatsuta, *J. Antibiot. (Tokyo)*, **46**, 813–826 (1993).

193. L. Toscano, G. Fioriello, R. Spagnoli, L. Cappelletti, and G. Zanuso, *J. Antibiot. (Tokyo)*, **36**, 1439–1450 (1983).

194. L. Toscano, G. Fioriello, S. Silingardi, and M. Inglesi, *Tetrahedron*, **40**, 2177 (1984).

195. G. Gialdroni Grassi, R. Alesina, C. Bersani, A. Ferrara, A. Fietta, and V. Peona, *Chemioterapia*, **5**, 177–184 (1986).

196. J. B. McAlpine, R. N. Swanson, D. Whitten, A. Burko, and J. M. Weber, Proceedings of the

30th Interscience Conference on Antimicrobial Agents and Chemotherapy, Abstr. 810 (1990).

197. J. J. Clement, C. Hanson, N. L. Shipkowitz, D. J. Hardy, and R. N. Swanson, Proceedings of the 30th Interscience Conference on Antimicrobial Agents and Chemotherapy, Abstr. 811 (1990).

198. R. Faghin, T. Pagano, J. McAlpine, K. Tanaka, J. Plattner, and P. Lartey, *J. Antibiot. (Tokyo)*, **46**, 698–700 (1993).

199. S. Morimoto, Y. Misawa, T. Adachi, T. Nagate, Y. Watanabe, and S. Omura, *J. Antibiot. (Tokyo)*, **43**, 286–294 (1990).

200. S. Morimoto, Y. Takahashi, Y. Watanabe, and S. Omura, *J. Antibiot. (Tokyo)*, **37**, 187–189 (1984).

201. P. B. Fernandes, R. Bailer, R. Swanson, C. W. Hanson, E. McDonald, N. Ramer, D. Hardy, N. Shipkowitz, R. R. Bower, and E. Gade, *Antimicrob. Agents Chemother.*, **30**, 865–873 (1986).

202. J. H. Jorgensen, L. A. Maher, and A. W. Howell, *Antimicrob. Agents Chemother.*, **35**, 1524–1526 (1991).

203. E. Vallee, E. Azoulay-Dupuis, R. Swanson, E. Bergogne-Berezin, and J. J. Pocidalo, *J. Antimicrob. Chemother.*, **27**(Suppl. A), 31–41 (1991).

204. G. Notario and D. Henry, Proceedings of the 2nd International Conference on the Macrolides, Azalides and Streptogramins, Abstr. 297 (1994).

205. R. Logan, H. Schaufelbrger, P. Gummett, J. Barton, and J. Misewicz, Proceedings of the 2nd International Conference on the Macrolides, Azalides and Streptogramins, Abstr. 317 (1994).

206. E. Hunt, D. J. Knowles, C. Shillingford, and I. I. Zomaya, *J. Antibiot. (Tokyo)*, **41**, 1644–1648 (1988).

207. W. Slawinski, H. Bojarska-Dahlig, T. Glabski, I. Dziegielewska, M. Biedrzycki, and S. Neperty, *Recl. Trav. Chim. Pays-Bas*, **94**, 236 (1975).

208. J. Kolwas, *Drugs Today*, **17**, 26 (1981).

209. P. Mannisto and M. Viluksela, *J. Antimicrob. Chemother.*, **25**, 476 (1990).

210. H. Bojarska-Dahlig, *J. Antimicrob. Chemother.*, **25**, 475–477 (1990).

211. C. Edwards, P. Kartey, and T. Pagano, Proceedings of the 206th Meeting of the American Chemical Society, Abstr. MED-164 (1993).

212. W. Baker, J. Clark, R. Stephens, and K. Kim, *J. Org. Chem.*, **53**, 2340 (1988).

213. G. M. Bright, A. A. Nagel, J. Bordner, K. A. Desai, J. N. Dibrino, J. Nowakowska, L. Vincent, R. M. Watrous, F. C. Sciavolino, A. R. English, et al., *J. Antibiot. (Tokyo)*, **41**, 1029–1047 (1988).

214. K. T. Dunkin, S. Jones, and A. J. Howard, *J. Antimicrob. Chemother.*, **21**, 405–411 (1988).

215. R. Wilkening, R. Ratcliffe, M. Szymonifka, K. Shankaran, A. May, T. Blizzard, J. Heck, C. Herbert, A. Graham, and K. Bartizal, Proceedings of the 33rd Interscience Conference on Antimicrobial Agents and Chemotherapy, Abstr. 426 (1993).

216. C. Gill, G. Abruzzo, A. Flattery, J. Smith, H. Kropp, and K. Bartizal, Proceedings of the 33rd Interscience Conference on Antimicrobial Agents and Chemotherapy, Abstr. 427 (1993).

217. B. Pelak, L. Gerckens, and H. Kropp, Proceedings of the 33rd Interscience Conference on Antimicrobial Agents and Chemotherapy, Abstr. 428 (1993).

218. A. Jones, *J. Org. Chem.*, **57**, 4361 (1992).

219. A. B. Jones and C. M. Herbert, *J. Antibiot. (Tokyo)*, **45**, 1785–1791 (1992).

220. R. Faghih, H. N. Nellans, and J. J. Plattner, *Drug Future*, **23**, 861 (1998).

221. F. Baquero, J. A. Garcia-Rodriguez, J. G. de Lomas, and L. Aguilar, *Antimicrob. Agents Chemother.*, **43**, 178–180 (1999).

222. P. Oster, A. Zanchi, S. Cresti, M. Lattanzi, F. Montagnani, C. Cellesi, and G. M. Rossolini, *Antimicrob. Agents Chemother.*, **43**, 2510–2512 (1999).

223. C. Thornsberry, P. T. Ogilvie, H. P. Holley Jr., and D. F. Sahm, *Antimicrob. Agents Chemother.*, **43**, 2612–2623 (1999).

224. T. Bergan, P. Gaustad, E. A. Hoiby, B. P. Berdal, G. Furuberg, J. Baann, and T. Tonjum, *Int. J. Antimicrob. Agents*, **10**, 77–81 (1998).

225. N. J. Johnston, J. C. De Azavedo, J. D. Kellner, and D. E. Low, *Antimicrob. Agents Chemother.*, **42**, 2425–2426 (1998).

226. J. Kataja, P. Huovinen, M. Skurnik, and H. Seppala, *Antimicrob. Agents Chemother.*, **43**, 48–52 (1999).

227. C. Arpin, M. H. Canron, J. Maugein, and C. Quentin, *Antimicrob. Agents Chemother.*, **43**, 2335–2336 (1999).

228. L. J. Teng, P. R. Hsueh, Y. C. Chen, S. W. Ho, and K. T. Luh, *J. Antimicrob. Chemother.*, **41**, 621–627 (1998).

229. G. V. Doern, R. N. Jones, M. A. Pfaller, and K. Kugler, *Antimicrob. Agents Chemother.*, **43**, 385–389 (1999).

230. Y. Nakajima, *J. Infect. Chemother.*, **5**, 61–74 (1999).

231. R. Leclercq and P. Courvalin, *Antimicrob. Agents Chemother.*, **35**, 1273–1276 (1991).

232. R. Leclercq and P. Courvalin, *Antimicrob. Agents Chemother.*, **35**, 1267–1272 (1991).

233. D. Dubnau, *CRC Crit. Rev. Biochem.*, **16**, 103–132 (1984).

234. B. Weisblum, *Br. Med. Bull.*, **40**, 47–53 (1984).

235. B. Weisblum, *Antimicrob. Agents Chemother.*, **39**, 577–585 (1995).

236. M. C. Roberts, W. O. Chung, D. Roe, M. Xia, C. Marquez, G. Borthagaray, W. L. Whittington, and K. K. Holmes, *Antimicrob. Agents Chemother.*, **43**, 1367–1372 (1999).

237. D. E. Bussiere, S. W. Muchmore, C. G. Dealwis, G. Schluckebier, V. L. Nienaber, R. P. Edalji, K. A. Walter, U. S. Ladror, T. F. Holzman, and C. Abad-Zapatero, *Biochemistry*, **37**, 7103–7112 (1998).

238. L. Yu, A. M. Petros, A. Schnuchel, P. Zhong, J. M. Severin, K. Walter, T. F. Holzman, and S. W. Fesik, *Nat. Struct. Biol.*, **4**, 483–489 (1997).

239. G. Schluckebier, P. Zhong, K. D. Stewart, T. J. Kavanaugh, and C. Abad-Zapatero, *J. Mol. Biol.*, **289**, 277–291 (1999).

240. J. Clancy, B. J. Schmieder, J. W. Petitpas, M. Manousos, J. A. Williams, J. A. Faiella, A. E. Girard, and P. R. McGuirk, *J. Antibiot.*, **48**, 1273–1279 (1995).

241. P. J. Hajduk, J. Dinges, J. M. Schkeryantz, D. Janowick, M. Kaminski, M. Tufano, D. J. Augeri, A. Petros, V. Nienaber, P. Zhong, R. Hammond, M. Coen, B. Beutel, L. Katz, and S. W. Fesik, *J. Med. Chem.*, **42**, 3852–3859 (1999).

242. R. B. Giannattasio and B. Weisblum, *Antimicrob. Agents Chemother.*, **44**, 1961–1963 (2000).

243. B. Vester and S. Douthwaite, *Antimicrob. Agents Chemother.*, **45**, 1–12 (2001).

244. G. Wang and D. E. Taylor, *Antimicrob. Agents Chemother.*, **42**, 1952–1958 (1998).

245. Y. J. Debets-Ossenkopp, A. B. Brinkman, E. J. Kuipers, C. M. Vandenbroucke-Grauls, and J. G. Kusters, *Antimicrob. Agents Chemother.*, **42**, 2749–2751 (1998).

246. A. Tait-Kamradt, T. Davies, M. Cronan, M. R. Jacobs, P. C. Appelbaum, and J. Sutcliffe, *Antimicrob. Agents Chemother.*, **44**, 2118–2125 (2000).

247. J. Sutcliffe, A. Tait-Kamradt, and L. Wondrack, *Antimicrob. Agents Chemother.*, **40**, 1817–1824 (1996).

248. J. Clancy, J. Petitpas, F. Dib-Hajj, W. Yuan, M. Cronan, A. V. Kamath, J. Bergeron, and J. A. Retsema, *Mol. Microbiol.*, **22**, 867–879 (1996).

249. A. Tait-Kamradt, J. Clancy, M. Cronan, F. Dib-Hajj, L. Wondrack, W. Yuan, and J. Sutcliffe, *Antimicrob. Agents Chemother.*, **41**, 2251–2255 (1997).

250. J. Clancy, F. Dib-Hajj, J. W. Petitpas, and W. Yuan, *Antimicrob. Agents Chemother.*, **41**, 2719–2723 (1997).

251. C. Arpin, H. Daube, F. Tessier, and C. Quentin, *Antimicrob. Agents Chemother.*, **43**, 944–946 (1999).

252. V. A. Luna, S. Cousin Jr., W. L. Whittington, and M. C. Roberts, *Antimicrob. Agents Chemother.*, **44**, 2503–2506 (2000).

253. L. Zarantonelli, G. Borthagaray, E. H. Lee, and W. M. Shafer, *Antimicrob. Agents Chemother.*, **43**, 2468–2472 (1999).

254. J. Sutcliffe, T. Grebe, A. Tait-Kamradt, and L. Wondrack, *Antimicrob. Agents Chemother.*, **40**, 2562–2566 (1996).

255. K. V. Singh, K. Malathum, and B. E. Murray, *Antimicrob. Agents Chemother.*, **45**, 263–266 (2001).

256. N. Noguchi, A. Emura, H. Matsuyama, K. O'Hara, M. Sasatsu, and M. Kono, *Antimicrob. Agents Chemother.*, **39**, 2359–2363 (1995).

257. N. Noguchi, J. Katayama, and K. O'Hara, *FEMS Microbiol. Lett.*, **144**, 197–202 (1996).

258. E. Cundliffe, *Antimicrob. Agents Chemother.*, **36**, 348–352 (1992).

259. M. S. Kuo, D. G. Chirby, A. D. Argoudelis, J. I. Cialdella, J. H. Coats, and V. P. Marshall, *Antimicrob. Agents Chemother.*, **33**, 2089–2091 (1989).

260. J. Sasaki, K. Mizoue, S. Morimoto, and S. Omura, *J. Antibiot. (Tokyo)*, **49**, 1110–1118 (1996).

261. L. M. Quiros, I. Aguirrezabalaga, C. Olano, C. Mendez, and J. A. Salas, *Mol. Microbiol.*, **28**, 1177–1185 (1998).

262. A. Gourmelen, M. H. Blondelet-Rouault, and J. L. Pernodet, *Antimicrob. Agents Chemother.*, **42**, 2612–2619 (1998).

263. J. Katayama, H. Okada, K. O'Hara, and N. Noguchi, *Biol. Pharm. Bull.*, **21**, 326–329 (1998).

264. M. C. Roberts, J. Sutcliffe, P. Courvalin, L. B. Jensen, J. Rood, and H. Seppala, *Antimicrob. Agents Chemother.*, **43**, 2823–2830 (1999).

265. P. B. Fernandes, W. R. Baker, L. A. Freiberg, D. J. Hardy, and E. J. McDonald, *Antimicrob. Agents Chemother.*, **33**, 78–81 (1989).

266. H. Brackmann and W. Henekl, *Chem. Ber.*, **84**, 284 (1951).

267. L. Corbaz, L. Ettlinger, E. Gaumann, W. Keller, F. Kradolfer, E. Kyburz, L. Neip, V. Prelog, R. Reusser, and H. Zahner, *Helv. Chim. Acta*, **35**, 935 (1955).

268. A. Bonnefoy, A. M. Girard, C. Agouridas, and J. F. Chantot, *J. Antimicrob. Chemother.*, **40**, 85–90 (1997).

269. K. Malathum, T. M. Coque, K. V. Singh, and B. E. Murray, *Antimicrob. Agents Chemother.*, **43**, 930–936 (1999).

270. A. B. Brueggemann, G. V. Doern, H. K. Huynh, E. M. Wingert, and P. R. Rhomberg, *Antimicrob. Agents Chemother.*, **44**, 447–449 (2000).

271. L. M. Ednie, S. K. Spangler, M. R. Jacobs, and P. C. Appelbaum, *Antimicrob. Agents Chemother.*, **41**, 1037–1041 (1997).

272. C. Jamjian, D. J. Biedenbach, and R. N. Jones, *Antimicrob. Agents Chemother.*, **41**, 454–459 (1997).

273. C. Agouridas, A. Bonnefoy, and J. F. Chantot, *Antimicrob. Agents Chemother.*, **41**, 2149–2158 (1997).

274. K. E. Piper, M. S. Rouse, J. M. Steckelberg, W. R. Wilson, and R. Patel, *Antimicrob. Agents Chemother.*, **43**, 708–710 (1999).

275. A. L. Barry, P. C. Fuchs, and S. D. Brown, *Antimicrob. Agents Chemother.*, **42**, 2138–2140 (1998).

276. G. A. Pankuch, D. B. Hoellman, G. Lin, S. Bajaksouzian, M. R. Jacobs, and P. C. Appelbaum, *Antimicrob. Agents Chemother.*, **42**, 3032–3034 (1998).

277. P. M. Roblin and M. R. Hammerschlag, *Antimicrob. Agents Chemother.*, **42**, 1515–1516 (1998).

278. F. J. Boswell, J. M. Andrews, and R. Wise, *J. Antimicrob. Chemother.*, **41**, 149–153 (1998).

279. C. Agouridas, P. Collette, P. Mauvais, and J. F. Chantot, Proceedings of the 35th Interscience Conference on Antimicrobial Agents and Chemotherapy, Abstr. F170 (1995).

280. P. H. Edelstein and M. A. Edelstein, *Antimicrob. Agents Chemother.*, **43**, 90–95 (1999).

281. K. V. Singh, K. K. Zscheck, and B. E. Murray, *Antimicrob. Agents Chemother.*, **44**, 3434–3437 (2000).

282. C. M. Bebear, H. Renaudin, M. D. Aydin, J. F. Chantot, and C. Bebear, *J. Antimicrob. Chemother.*, **39**, 669–670 (1997).

283. D. B. Hoellman, G. Lin, M. R. Jacobs, and P. C. Appelbaum, *Antimicrob. Agents Chemother.*, **43**, 166–168 (1999).

284. J. E. Hoppe and A. Bryskier, *Antimicrob. Agents Chemother.*, **42**, 965–966 (1998).

285. A. Bonnefoy, M. Guitton, C. Delachaume, P. Le Priol, and A. M. Girard, *Antimicrob. Agents Chemother.*, **45**, 1688–1692 (2001).

286. Y. S. Or, R. F. Clark, S. Wang, D. T. Chu, A. M. Nilius, R. K. Flamm, M. Mitten, P. Ewing, J. Alder, and Z. Ma, *J. Med. Chem.*, **43**, 1045–1049 (2000).

287. K. L. Credito, G. Lin, G. A. Pankuch, S. Bajaksouzian, M. R. Jacobs, and P. C. Appelbaum, *Antimicrob. Agents Chemother.*, **45**, 67–72 (2001).

288. A. M. Nilius, M. H. Bui, L. Almer, D. Hensey-Rudloff, J. Beyer, Z. Ma, Y. S. Or, and R. K. Flamm, *Antimicrob. Agents Chemother.*, **45**, 2163–2168 (2001).

289. D. M. Citron and M. D. Appleman, *Antimicrob. Agents Chemother.*, **45**, 345–348 (2001).

290. E. J. Goldstein, G. Conrads, D. M. Citron, C. V. Merriam, Y. Warren, and K. Tyrrell, *Antimicrob. Agents Chemother.*, **45**, 2363–2367 (2001).

291. D. Shortridge, N. Ramer, J. Beyer, Z. Ma, Y. S. Or, and R. K. Flamm, Proceedings of the 39th Interscience Conference on Antimicrobial Agents and Chemotherapy, Abstr. 2136 (1999).

292. M. H. Bui, L. Almer, D. M. Hensey, Z. Ma, Y. S. Or, A. M. Nilius, and R. K. Flamm, Proceedings of the 39th Interscience Conference on Antimicrobial Agents and Chemotherapy, Abstr. 2138 (1999).

293. M. M. Neuhauser, J. L. Prause, D. H. Li, R. Jung, N. Boyea, J. M. Hackleman, L. H. Danziger, and S. L. Pendland, Proceedings of the 39th Interscience Conference on Antimicrobial Agents and Chemotherapy, Abstr. 2139 (1999).

294. D. B. Hoellman, G. Lin, S. Bajaksouzian, M. R. Jacobs, and P. C. Appelbaum, Proceedings of the 39th Interscience Conference on Antimicrobial Agents and Chemotherapy, Abstr. 2140 (1999).

295. J. O. Capobianco, Z. Cao, V. D. Shortridge, Z. Ma, R. K. Flamm, and P. Zhong, *Antimicrob. Agents Chemother.*, **44**, 1562–1567 (2000).

296. L. Hernandez, N. Sadrsadeh, S. Krill, Z. Ma, and K. Marsh, Proceedings of the 39th Interscience Conference on Antimicrobial Agents and Chemotherapy, Abstr. 2148 (1999).

297. J. Meubroek, M. Mitten, K. W. Mollison, P. Ewing, J. Alder, A. M. Nilius, R. K. Flamm, Z. Ma, and Y. S. Or, Proceedings of the 39th

Interscience Conference on Antimicrobial Agents and Chemotherapy, Abstr. 2151 (1999).

298. M. Mitten, J. Meubroek, L. Paige, P. Ewing, J. Alder, K. W. Mollison, A. M. Nilius, R. K. Flamm, Z. Ma, and Y. S. Or, Proceedings of the 39th Interscience Conference on Antimicrobial Agents and Chemotherapy, Abstr. 2150 (1999).

299. A. Denis, C. Agouridas, A. Bonnefoy, F. Bretin, C. Fromentin, and A. Bonnet, Proceedings of the 39th Interscience Conference on Antimicrobial Agents and Chemotherapy, Abstr. 2152 (1999).

300. A. Bonnefoy, A. Denis, F. Bretin, C. Fromentin, and C. Agouridas, Proceedings of the 39th Interscience Conference on Antimicrobial Agents and Chemotherapy, Abstr. 2153 (1999).

301. A. Bonnefoy, A. Denis, F. Bretin, C. Fromentin, and C. Agouridas, Proceedings of the 39th Interscience Conference on Antimicrobial Agents and Chemotherapy, Abstr. 2156 (1999).

302. L. T. Phan, Y. S. Or, Y. C. Chen, D. T. Chu, P. Ewing, A. M. Nilius, M. H. Bui, P. M. Raney, D. Hensey-Rudloff, M. Mitten, D. Henry, and J. Plattner, Proceedings of the 38th Interscience Conference on Antimicrobial Agents and Chemotherapy, Abstr. F127 (1998).

303. T. Kaneko, W. McMillen, J. Sutcliffe, J. Duigan, and J. Petitpas, Proceedings of the 40th Interscience Conference on Antimicrobial Agents and Chemotherapy, Abstr. 1815 (2000).

304. D. Girard, H. W. Mathieu, S. M. Finegan, C. R. Cimochowski, T. Kaneko, and W. McMillen, Proceedings of the 40th Interscience Conference on Antimicrobial Agents and Chemotherapy, Abstr. 1816 (2000).

305. D. Girard, H. W. Mathieu, R. M. Shepard, S. Yee, T. Kaneko, and W. McMillen, Proceedings of the 40th Interscience Conference on Antimicrobial Agents and Chemotherapy, Abstr. 1817 (2000).

306. D. Flemingham, M. J. Robbins, I. L. Mathias, C. Dencer, H. Salman, G. L. Ridgway, and A. Bryskier, Proceedings of the 40th Interscience Conference on Antimicrobial Agents and Chemotherapy, Abstr. 2172 (2000).

307. K. Bush, D. Abbanat, G. Ashley, E. Baum, J. Carney, M. Fardis, B. Foleno, E. Grant, T. Henninger, J. J. Hilliard, Y. Li, P. Licardi, M. Loeloff, M. Macielag, J. Melton, E. Wira, and D. J. Hlasta, Proceedings of the 40th Inter-

science Conference on Antimicrobial Agents and Chemotherapy, Abstr. 1821 (2000).

308. R. L. Elliott, D. Pireh, G. Griesgraber, A. M. Nilius, P. J. Ewing, M. H. Bui, P. M. Raney, R. K. Flamm, K. Kim, R. F. Henry, D. T. Chu, J. J. Plattner, and Y. S. Or, J. Med. Chem., **41**, 1651–1659 (1998).

309. G. Griesgraber, M. J. Kramer, R. L. Elliott, A. M. Nilius, P. J. Ewing, P. M. Raney, M. H. Bui, R. K. Flamm, D. T. Chu, J. J. Plattner, and Y. S. Or, J. Med.Chem., **41**, 1660–1670 (1998).

310. R. L. Elliott, D. Pireh, A. M. Nilius, P. M. Johnson, R. K. Flamm, D. T. Chu, J. J. Plattner, and Y. S. Or, Biorg. Med. Chem. Lett., **7**, 641 (1997).

311. T. Asaka, M. Kashimura, T. Tanikawa, T. Ishii, A. Matsuura, K. Matsumoto, K. Suzuki, N. K., T. Akashi, and S. Morimoto, Proceedings of the 37th Interscience Conference on Antimicrobial Agents and Chemotherapy, Abstr. F262 (1997).

312. T. Asaka, M. Kashimura, A. Manka, T. Tanikawa, T. Ishii, K. Suzuki, H. Sugiyama, T. Akashi, H. Saito, T. Adachi, and S. Morimoto, Proceedings of the 39th Interscience Conference on Antimicrobial Agents and Chemotherapy, Abstr. 2159 (1999).

313. T. Asaka, M. Kashimura, T. Tanikawa, K. Suzuki, H. Sugiyama, T. Akashi, T. Adachi, and S. Morimoto, Proceedings of the 39th Interscience Conference on Antimicrobial Agents and Chemotherapy, Abstr. 2160 (1999).

314. Y. J. Wu, R. Wons Jr., D. Durkin, M. Goldsmith, W. G. Su, J. Rainville, K. Smyth, B. V. Yang, M. Massa, J. Kane, K. Brighty, K. Blair, R. Monahan, J. Sutcliffe, L. Brennan, J. Duigan, J. Petitpas, A. Tait-Kamradt, R. Linde II, S. Miller, T. Kaneko, M. Keaney, N. Birsner, S. Brickner, and J. Roache, Proceedings of the 38th Interscience Conference on Antimicrobial Agents and Chemotherapy, Abstr. F123 (1998).

315. Y. J. Wu, K. Smyth, J. Rainville, T. Kaneko, J. Sutcliffe, L. Brennan, J. Duigan, A. E. Girard, S. M. Finegan, and C. R. Cimochowski, Proceedings of the 38th Interscience Conference on Antimicrobial Agents and Chemotherapy, Abstr. F122 (1998).

316. D. Girard, H. W. Mathieu, A. E. Girard, C. A. Menard, W. G. Su, K. Smyth, and J. Rainville, Proceedings of the 38th Interscience Conference on Antimicrobial Agents and Chemotherapy, Abstr. F120 (1998).

317. D. Girard, H. W. Mathieu, A. E. Girard, S. M. Finegan, C. A. Menard, W. G. Su, K. Smyth, and J. Rainville, Proceedings of the 38th Inter-

science Conference on Antimicrobial Agents and Chemotherapy, Abstr. F121 (1998).

318. Z. Itoh, M. Nakaya, T. Suzuki, H. Hiral, and K. Wakabayashi, *Am. J. Physiol. Gastrointest. Liver Physiol.*, **10**, G688 (1984).

319. G. P. Zara, H. H. Thompson, M. A. Pilot, and H. D. Ritchie, *J. Antimicrob. Chemother.*, **16**(Suppl. A), 175–179 (1985).

320. T. Peeters, G. Matthijs, I. Depoortere, T. Cachet, J. Hoogmartens, and G. Vantrappen, *Am. J. Physiol. Gastrointest. Liver Physiol.*, **257**, G470–G474 (1989).

321. T. L. Peeters, *Gastroenterology*, **105**, 1886–1899 (1993).

322. V. Annese, J. Janssens, G. Vantrappen, J. Tack, T. L. Peeters, P. Willemse, and E. Van Cutsem, *Gastroenterology*, **102**, 823–828 (1992).

323. Z. Itoh and S. Omura, *Dig. Dis. Sci.*, **32**, 915 (1987).

324. I. Depoortere, T. Peeters, G. Matthijs, T. Cachet, J. Hoogmartens, and G. Vantrappen, *J. Gastrointest. Motil.*, **1**, 150 (1989).

325. I. Depoortere, T. L. Peeters, and G. Vantrappen, *Peptides*, **11**, 515–519 (1990).

326. P. A. Lartey, H. N. Nellans, R. Faghih, A. Petersen, C. M. Edwards, L. Freiberg, S. Quigley, K. Marsh, L. L. Klein, and J. J. Plattner, *J. Med. Chem.*, **38**, 1793–1798 (1995).

327. H. Koga, T. Sato, K. Tsuzuki, and H. Takanashi, *Biorg. Med. Chem. Lett.*, **4**, 1649 (1994).

328. M. J. Schultz, P. Speelman, S. Zaat, S. J. van Deventer, and T. van der Poll, *Antimicrob. Agents Chemother.*, **42**, 1605–1609 (1998).

329. K. Oishi, F. Sonoda, S. Kobayashi, A. Iwagaki, T. Nagatake, K. Matsushima, and K. Matsumoto, *Infect. Immun.*, **62**, 4145–4152 (1994).

330. H. Takizawa, M. Desaki, T. Ohtoshi, S. Kawasaki, T. Kohyama, M. Sato, M. Tanaka, T. Kasama, K. Kobayashi, J. Nakajima, and K. Ito, *Am. J. Respir. Crit. Care Med.*, **156**, 266–271 (1997).

331. M. Duong, M. Simard, Y. Bergeron, N. Ouellet, M. Cote-Richer, and M. G. Bergeron, *Antimicrob. Agents Chemother.*, **42**, 3309–3312 (1998).

332. M. Duong, M. Simard, Y. Bergeron, and M. G. Bergeron, *Antimicrob. Agents Chemother.*, **45**, 252–262 (2001).

333. S. Kawasaki, H. Takizawa, T. Ohtoshi, N. Takeuchi, T. Kohyama, H. Nakamura, T. Kasama, K. Kobayashi, K. Nakahara, Y. Morita, and K. Yamamoto, *Antimicrob. Agents Chemother.*, **42**, 1499–1502 (1998).

334. T. Sunazuka, H. Takizawa, M. Desaki, K. Suzuki, R. Obata, K. Otoguro, and S. Omura, *J. Antibiot. (Tokyo)*, **52**, 71–74 (1999).

335. T. Ichiyama, M. Nishikawa, T. Yoshitomi, S. Hasegawa, T. Matsubara, T. Hayashi, and S. Furukawa, *Antimicrob. Agents Chemother.*, **45**, 44–47 (2001).

336. F. Scaglione and G. Rossoni, *J. Antimicrob. Chemother.*, **41**(Suppl. B), 47–50 (1998).

337. T. Miyanohara, M. Ushikai, S. Matsune, K. Ueno, S. Katahira, and Y. Kurono, *Laryngoscope*, **110**, 126–131 (2000).

338. M. Desaki, H. Takizawa, T. Ohtoshi, T. Kasama, K. Kobayashi, T. Sunazuka, S. Omura, K. Yamamoto, and K. Ito, *Biochem. Biophys. Res. Commun.*, **267**, 124–128 (2000).

339. H. Takizawa, M. Desaki, T. Ohtoshi, S. Kawasaki, T. Kohyama, M. Sato, J. Nakajima, M. Yanagisawa, and K. Ito, *Eur. Respir. J.*, **12**, 57–63 (1998).

340. J. Tamaoki, M. Kondo, K. Kohri, K. Aoshiba, E. Tagaya, and A. Nagai, *J. Immunol.*, **163**, 2909–2915 (1999).

341. C. Feldman, R. Anderson, A. J. Theron, G. Ramafi, P. J. Cole, and R. Wilson, *Inflammation*, **21**, 655–665 (1997).

342. K. Okamoto, C. Kishioka, J. S. Kim, C. D. Wegner, J. A. Muebroek, and B. K. Rubin, *J. Resp. Crit. Care Med.*, **159**, A35 (1999).

343. J. Tamaoki, K. Takeyama, E. Tagaya, and K. Konno, *Antimicrob. Agents Chemother.*, **39**, 1688–1690 (1995).

344. H. Takeda, H. Miura, M. Kawahira, H. Kobayashi, S. Otomo, and S. Nakaike, *Kansenshogaku Zasshi*, **63**, 71–78 (1989).

345. S. Kudoh, T. Uetake, K. Hagiwara, M. Hirayama, L. H. Hus, H. Kimura, and Y. Sugiyama, *Jpn. J. Thorac. Dis.*, **25**, 632–642 (1987).

346. H. Nagai, H. Shishido, R. Yoneda, E. Yamaguchi, A. Tamura, and A. Kurashima, *Respiration*, **58**, 145–149 (1991).

347. K. Kawamura-Sato, Y. Iinuma, T. Hasegawa, T. Horii, T. Yamashino, and M. Ohta, *Antimicrob. Agents Chemother.*, **44**, 2869–2872 (2000).

348. M. Sawaki, R. Mikami, K. Mikasa, M. Kunimatsu, S. Ito, and N. Narita, *Kansenshogaku Zasshi*, **60**, 45–50 (1986).

349. T. Fujii, J. Kadota, K. Kawakami, K. Iida, R. Shirai, M. Kaseda, S. Kawamoto, and S. Kohno, *Thorax*, **50**, 1246–1252 (1995).

350. N. Nakanishi, N. Ueda, M. Kitade, and T. Moritaka, *Jpn. J. Thorac. Dis.*, **33**, 771–774 (1995).

351. C. S. Rhee, Y. Majima, S. Arima, H. W. Jung, T. H. Jinn, Y. G. Min, and Y. Sakakura, *Ann. Otol. Rhinol. Laryngol.*, **109**, 484–487 (2000).

352. K. Hamada, E. Kita, M. Sawaki, K. Mikasa, and N. Narita, *Chemotherapy*, **41**, 59–69 (1995).

353. K. Mikasa, M. Sawaki, E. Kita, K. Hamada, S. Teramoto, M. Sakamoto, K. Maeda, M. Konishi, and N. Narita, *Chemotherapy*, **43**, 288–296 (1997).

354. K. Sassa, Y. Mizushima, T. Fujishita, R. Oosaki, and M. Kobayashi, *Antimicrob. Agents Chemother.*, **43**, 67–72 (1999).

355. J. Yatsunami, N. Tsuruta, N. Hara, and S. Hayashi, *Cancer Lett.*, **131**, 137–143 (1998).

356. M. J. Weinstein, G. H. Wagman, J. A. Marquez, R. T. Testa, E. Oden, and J. A. Waitz, *J. Antibiot. (Tokyo)*, **22**, 253–258 (1969).

357. P. Bonay, I. Duran-Chica, M. Fresno, B. Alarcon, and A. Alcina, *Antimicrob. Agents Chemother.*, **42**, 2668–2673 (1998).

358. G. Bonfiglio and P. M. Furneri, *Expert Opin. Invest. Drugs*, **10**, 185–198 (2001).

359. D. E. Low, *Microb. Drug Resist.*, **1**, 223–234 (1995).

360. G. Delgado Jr., M. M. Neuhauser, D. T. Bearden, and L. H. Danziger, *Pharmacotherapy*, **20**, 1469–1485 (2000).

361. D. R. Allington and M. P. Rivey, *Clin. Ther.*, **23**, 24–44 (2001).

362. B. T. Porse and R. A. Garrett, *J. Mol. Biol.*, **286**, 375–387 (1999).

363. G. Chinali, P. Moureau, and C. G. Cocito, *J. Biol. Chem.*, **259**, 9563–9568 (1984).

364. C. Cocito, M. Di Giambattista, E. Nyssen, and P. Vannuffel, *J. Antimicrob. Chemother.*, **39**(Suppl. A), 7–13 (1997).

365. C. Cocito, F. Vanlinden, and C. Branlant, *Biochim. Biophys. Acta*, **739**, 158–163 (1983).

366. P. Moureau, M. Di Giambattista, and C. Cocito, *Biochim. Biophys. Acta*, **739**, 164–172 (1983).

367. G. Chinali, E. Nyssen, M. Di Giambattista, and C. Cocito, *Biochim. Biophys. Acta*, **949**, 71–78 (1988).

368. G. Chinali, E. Nyssen, M. Di Giambattista, and C. Cocito, *Biochim. Biophys. Acta*, **951**, 42–52 (1988).

369. M. Di Giambattista, A. P. Thielen, J. A. Maassen, W. Moller, and C. Cocito, *Biochemistry*, **25**, 3540–3547 (1986).

370. A. Contreras and D. Vazquez, *Eur. J. Biochem.*, **74**, 549–551 (1977).

371. M. Di Giambattista, E. Nyssen, A. Pecher, and C. Cocito, *Biochemistry*, **29**, 9203–9211 (1990).

372. N. el Solh and J. Allignet, *Drug Resist. Updates*, **1**, 169–175 (1998).

373. D. H. Bouanchaud, *J. Antimicrob. Chemother.*, **39**(Suppl. A), 15–21 (1997).

374. D. H. Bouanchaud, *J. Antimicrob. Chemother.*, **30**(Suppl. A), 95–99 (1992).

375. P. Vannuffel, M. Di Giambattista, E. A. Morgan, and C. Cocito, *J. Biol. Chem.*, **267**, 8377–8382 (1992).

376. J. Allignet, V. Loncle, and N. El Solh, *Gene*, **117**, 45–51 (1992).

377. J. Allignet and N. El Solh, *Gene*, **202**, 133–138 (1997).

378. J. Haroche, J. Allignet, C. Buchrieser, and N. El Solh, *Antimicrob. Agents Chemother.*, **44**, 2271–2275 (2000).

379. C. K. Lee, Y. Kamitani, T. Nihira, and Y. Yamada, *J. Bacteriol.*, **181**, 3293–3297 (1999).

380. R. Rende-Fournier, R. Leclercq, M. Galimand, J. Duval, and P. Courvalin, *Antimicrob. Agents Chemother.*, **37**, 2119–2125 (1993).

381. J. Allignet and N. El Solh, *Antimicrob. Agents Chemother.*, **39**, 2027–2036 (1995).

382. G. Werner and W. Witte, *Antimicrob. Agents Chemother.*, **43**, 1813–1814 (1999).

383. A. Seoane and J. M. Garcia Lobo, *Antimicrob. Agents Chemother.*, **44**, 905–909 (2000).

384. M. Sugantino and S. L. Roderick, *Biochemistry*, **41**, 2209–2216 (2002).

385. C. H. Kim, N. Otake, and H. Yonehara, *J. Antibiot.*, **27**, 903–908 (1974).

386. F. Le Goffic, M. L. Capmau, J. Abbe, C. Cerceau, A. Dublanchet, and J. Duval, *Ann. Microbiol. (Paris)*, **128B**, 471–474 (1977).

387. J. Allignet, V. Loncle, P. Mazodier, and N. El Solh, *Plasmid*, **20**, 271–275 (1988).

388. T. A. Mukhtar, K. P. Koteva, D. W. Hughes, and G. D. Wright, *Biochemistry*, **40**, 8877–8886 (2001).

389. A. King, J. May, and I. Phillips, *J. Antimicrob. Chemother.*, **42**, 711–719 (1998).

390. D. J. Mason, A. Deietz, and C. DeBoer in J. C. Sylvester, Ed., *Antimicrobial Agents and Chemotherapy, 1962*, American Society for Microbiology, Ann Arbor, MI, 1963, pp. 554–559.

391. M. J. Kasten, *Mayo Clin. Proc.*, **74**, 825–833 (1999).

392. J. Freeman and M. H. Wilcox, *Microbes Infect.*, **1**, 377–384 (1999).

393. E. Toma, S. Fournier, M. Poisson, R. Morisset, D. Phaneuf, and C. Vega, *Lancet*, **1**, 1046–1048 (1989).

394. J. A. Fishman, *Antimicrob. Agents Chemother.*, **42**, 1309–1314 (1998).

395. S.-T. Chung, J. J. Manis, S. J. McWethy, T. E. Patt, D. F. Witz, H. J. Wof, and M. G. Wovcha in W. R. Stohl, Ed., *Biotechnology of Antibiotics*, Marcel Dekker, New York, 1997, pp. 165–186.

396. U. Peschke, H. Schmidt, H. Z. Zhang, and W. Piepersberg, *Mol. Microbiol.*, **16**, 1137–1156 (1995).

397. D. Neusser, H. Schmidt, J. Spizek, J. Novotna, U. Peschke, S. Kaschabeck, P. Tichy, and W. Piepersberg, *Arch. Microbiol.*, **169**, 322–332 (1998).

398. G. Wang and D. E. Taylor, *Antimicrob. Agents Chemother.*, **42**, 1952–1958 (1998).

399. B. Bozdogan, L. Berrezouga, M. S. Kuo, D. A. Yurek, K. A. Farley, B. J. Stockman, and R. Leclercq, *Antimicrob. Agents Chemother.*, **43**, 925–929 (1999).

400. A. Brisson-Noel, P. Delrieu, D. Samain, and P. Courvalin, *J. Biol. Chem.*, **263**, 15880–15887 (1988).

401. A. Brisson-Noel and P. Courvalin, *Gene*, **43**, 247–253 (1986).

402. J. Ehrlich, Q. R. Bartz, R. M. Smith, D. A. Joslyn, and P. R. Burkholder, *Science*, **106**, 417 (1947).

403. D. Drainas, D. L. Kalpaxis, and C. Coutsogeorgopoulos, *Eur. J. Biochem.*, **164**, 53–58 (1987).

404. O. Pongs, R. Bald, and V. A. Erdmann, *Proc. Natl. Acad. Sci. USA*, **70**, 2229–2233 (1973).

405. D. Moazed and H. F. Noller, *Biochimie*, **69**, 879–884 (1987).

406. A. A. Yunis, *Am. J. Med.*, **87**, 44N–48N (1989).

407. B. C. West, G. A. DeVault Jr., J. C. Clement, and D. M. Williams, *Rev. Infect. Dis.*, **10**, 1048–1051 (1988).

408. S. A. Rayner and R. J. Buckley, *Drug Saf.*, **14**, 273–276 (1996).

409. M. Knight, *J. Clin. Pharmacol.*, **34**, 128–135 (1994).

410. S. Douthwaite, *J. Bacteriol.*, **174**, 1333–1338 (1992).

411. B. Vester and R. A. Garett, *EMBO J.*, **7**, 3577–3587 (1988).

412. J. L. Burns, C. E. Rubens, P. M. Mendelman, and A. L. Smith, *Antimicrob. Agents Chemother.*, **29**, 445–450 (1986).

413. K. Poole, K. Krebes, C. McNally, and S. Neshat, *J. Bacteriol.*, **175**, 7363–7372 (1993).

414. H. Okusu, D. Ma, and H. Nikaido, *J. Bacteriol.*, **178**, 306–308 (1996).

415. K. Poole, N. Gotoh, H. Tsujimoto, Q. Zhao, A. Wada, T. Yamasaki, S. Neshat, J. Yamagishi, X. Z. Li, and T. Nishino, *Mol. Microbiol.*, **21**, 713–724 (1996).

416. T. Kohler, M. Michea-Hamzehpour, U. Henze, N. Gotoh, L. K. Curty, and J. C. Pechere, *Mol. Microbiol.*, **23**, 345–354 (1997).

417. J. Kieboom and J. de Bont, *Microbiology*, **147**, 43–51 (2001).

418. I. A. Murray and W. V. Shaw, *Antimicrob. Agents Chemother.*, **41**, 1–6 (1997).

419. W. V. Shaw and A. G. Leslie, *Annu. Rev. Biophys. Biophys. Chem.*, **20**, 363–386 (1991).

420. A. G. Leslie, P. C. Moody, and W. V. Shaw, *Proc. Natl. Acad. Sci. USA*, **85**, 4133–4137 (1988).

421. A. D. Bennett and W. V. Shaw, *Biochem. J.*, **215**, 29–38 (1983).

422. M. Vaara, *Antimicrob. Agents Chemother.*, **37**, 354–356 (1993).

423. M. Laurberg, O. Kristensen, K. Martemyanov, A. T. Gudkov, I. Nagaev, D. Hughes, and A. Liljas, *J. Mol. Biol.*, **303**, 593–603 (2000).

424. R. K. Agrawal, P. Penczek, R. A. Grassucci, and J. Frank, *Proc. Natl. Acad. Sci. USA*, **95**, 6134–6138 (1998).

425. K. Burns, M. Cannon, and E. Cundliffe, *FEBS Lett.*, **40**, 219–223 (1974).

426. G. R. Willie, N. Richman, W. P. Godtfredsen, and J. W. Bodley, *Biochemistry*, **14**, 1713–1718 (1975).

427. I. Nagaev, J. Bjorkman, D. I. Andersson, and D. Hughes, *Mol. Microbiol.*, **40**, 433–439 (2001).

428. J. Turnidge and P. Collignon, *Int. J. Antimicrob. Agents*, **12**(Suppl. 2), S35–S44 (1999).

429. G. N. Proctor, J. McKell, and R. H. Rownd, *J. Bacteriol.*, **155**, 937–939 (1983).

430. A. L. Barry, *Antimicrob. Agents Chemother.*, **32**, 150–152 (1988).

431. D. I. Diekema and R. N. Jones, *Drugs*, **59**, 7–16 (2000).

432. J. F. Plouffe, *Clin. Infect. Dis.*, **31**(Suppl. 4), S144–S149 (2000).

433. J. F. Barrett, *Curr. Opin. Invest. Drugs*, **1**, 181–187 (2000).

434. A. H. Lin, R. W. Murray, T. J. Vidmar, and K. R. Marotti, *Antimicrob. Agents Chemother.*, **41**, 2127–2131 (1997).

435. L. Xiong, P. Kloss, S. Douthwaite, N. M. Andersen, S. Swaney, D. L. Shinabarger, and A. S. Mankin, *J. Bacteriol.*, **182**, 5325–5331 (2000).

436. P. Kloss, L. Xiong, D. L. Shinabarger, and A. S. Mankin, *J. Mol. Biol.*, **294**, 93–101 (1999).

437. L. Xiong, P. Kloss, S. Douthwaite, N. M. Andersen, S. Swaney, D. L. Shinabarger, and A. S. Mankin, *J. Bacteriol.*, **182**, 5325–5331 (2000).

438. G. D. Wright, *Chem. Biol.*, **7**, R127–R132 (2000).

439. A. K. Ganguly, *J. Antibiot. (Tokyo)*, **53**, 1038–1044 (2000).

440. W. S. Champney and C. L. Tober, *Antimicrob. Agents Chemother.*, **44**, 1413–1417 (2000).

441. L. Belova, T. Tenson, L. Xiong, P. M. McNicholas, and A. S. Mankin, *Proc. Natl. Acad. Sci. USA*, **98**, 3726–3731 (2001).

442. M. Pioletti, F. Schlunzen, J. Harms, R. Zarivach, M. Gluhmann, H. Avila, A. Bashan, H. Bartels, T. Auerbach, C. Jacobi, T. Hartsch, A. Yonath, and F. Franceschi, *EMBO J.*, **20**, 1829–1839 (2001).

443. R. Kao and J. Davies in R. A. Garrett, S. R. Douthwaite, A. Liljas, A. T. Matheson, P. B. Moore, and H. F. Noller, Eds., *The Ribosome: Structure, Function, Antibiotics and Cellular Interactions*, ASM Press, Washington, DC, 2000, pp. 451–473.

# CHAPTER SIXTEEN

# Antimycobacterial Agents

Piero Sensi
University of Milan
Milan, Italy

Giuliana Gialdroni Grassi
University of Pavia
Pavia, Italy

## Contents

1 Introduction, 809
2 The Mycobacteria, 813
3 Pathogenesis and Epidemiology, 821
  3.1 Tuberculosis, 821
  3.2 Leprosy, 825
4 Screening and Evaluation of Antimycobacterial Agents, 826
5 Current Drugs on the Market, 829
  5.1 Synthetic Products, 829
    5.1.1 Sulfones, 829
      5.1.1.1 History, 829
      5.1.1.2 SARs, 829
      5.1.1.3 Activity and Mechanism of Action, 830
      5.1.1.4 Pharmacokinetics, 830
      5.1.1.5 Adverse Effects, 830
    5.1.2 *Para*-amino-salicylic Acid, 830
      5.1.2.1 History, 830
      5.1.2.2 SARs, 830
      5.1.2.3 Activity and Mechanism of Action, 831
      5.1.2.4 Pharmacokinetics, 831
      5.1.2.5 Adverse Effects, 831
    5.1.3 Thioacetazone, 832
      5.1.3.1 History, 832
      5.1.3.2 SARs, 832
      5.1.3.3 Activity and Mechanism of Action, 832
      5.1.3.4 Pharmacokinetics, 832
      5.1.3.5 Adverse Effects, 832
    5.1.4 Isoniazid, 833
      5.1.4.1 History, 833
      5.1.4.2 SARs, 833
      5.1.4.3 Activity and Mechanism of Action, 833

*Burger's Medicinal Chemistry and Drug Discovery*
Sixth Edition, Volume 5: Chemotherapeutic Agents
Edited by Donald J. Abraham
ISBN 0-471-37031-2   © 2003 John Wiley & Sons, Inc.

5.1.4.4 Pharmacokinetics, 835
5.1.4.5 Adverse Effects, 835
5.1.5 Ethionamide, 835
    5.1.5.1 History and SARs, 835
    5.1.5.2 Activity and Mechanism of
        Action, 836
    5.1.5.3 Pharmacokinetics, 836
    5.1.5.4 Adverse Effects, 836
5.1.6 Pyrazinamide, 836
    5.1.6.1 History, 836
    5.1.6.2 SARs, 836
    5.1.6.3 Activity and Mechanism of
        Action, 837
    5.1.6.4 Pharmacokinetics, 837
    5.1.6.5 Adverse Effects, 837
5.1.7 Clofazimine, 837
    5.1.7.1 History, 837
    5.1.7.2 Activity and Mechanism of
        Action, 838
    5.1.7.3 Pharmacokinetics, 838
    5.1.7.4 SARs, 839
    5.1.7.5 Adverse Reactions, 839
5.1.8 Ethambutol, 839
    5.1.8.1 History, 839
    5.1.8.2 SARs, 839
    5.1.8.3 Activity and Mechanism of
        Action, 840
    5.1.8.4 Pharmacokinetics, 840
    5.1.8.5 Adverse Effects, 840
5.2 Antibiotics, 840
  5.2.1 Streptomycin, 840
    5.2.1.1 History, 840
    5.2.1.2 SARs, 840
    5.2.1.3 Activity and Mechanism of
        Action, 841
    5.2.1.4 Pharmacokinetics, 842
    5.2.1.5 Adverse Effects, 842
  5.2.2 Kanamycin, 842
    5.2.2.1 History, 842
    5.2.2.2 Activity and Mechanism of
        Action, 843
    5.2.2.3 Pharmacokinetics, 843
    5.2.2.4 Adverse Effects, 844
  5.2.3 Amikacin, 844
    5.2.3.1 History, 844
    5.2.3.2 Activity, 844
  5.2.4 Viomycin and Capreomycin, 844
    5.2.4.1 History, 844

5.2.4.2 Activity and Mechanism of
    Action, 844
5.2.4.3 Pharmacokinetics, 845
5.2.4.4 Adverse Reactions, 845
5.2.5 Cycloserine, 846
    5.2.5.1 History, 846
    5.2.5.2 SARs, 846
    5.2.5.3 Activity and Mechanism of
        Action, 846
    5.2.5.4 Pharmacokinetics, 846
    5.2.5.5 Adverse Reactions, 846
5.2.6 Rifamycins, 846
    5.2.6.1 History, 846
    5.2.6.2 SARs, 847
    5.2.6.3 Mechanism of Action and
        Resistance, 850
5.2.7 Antimycobacterial Rifamycins, 850
    5.2.7.1 Rifampicin, 850
    5.2.7.2 Rifabutin, 852
    5.2.7.3 Rifapentine, 852
    5.2.7.4 Rifalazil, 852
    5.2.7.5 Rifamycin T9, 853
5.3 Drugs Under Investigation, 853
  5.3.1 Fluoroquinolones, 853
    5.3.1.1 Historical Development, 853
    5.3.1.2 Antimycobacterial Activity and
        Mechanism of Action, 853
    5.3.1.3 Clinical Uses in Mycobacterial
        Infection, 855
    5.3.1.4 Clinical Doses and Adverse
        Effects, 855
  5.3.2 Macrolides, 855
    5.3.2.1 Historical Development, 855
    5.3.2.2 Clarithromycin, 856
    5.3.2.3 Azithromycin, 857
  5.3.3 $\beta$-Lactams and $\beta$-Lactamase Inhibitors,
    858
  5.3.4 Nitroimidazopyrans, 858
  5.3.5 Tryptanthrin, 859
  5.3.6 Oxazolidinones, 859
  5.3.7 Thiolactomycins, 859
  5.3.8 Rifazalil and Rifamycin T9, 860
6 Recent Developments and Present Status of
Chemotherapy, 860
  6.1 Tuberculosis, 860
  6.2 Mycobacteriosis from NTM, 864
  6.3 Leprosy, 865
7 Things to Come, 866

# 1 INTRODUCTION

Some species of mycobacteria are pathogenic for several animal species and are responsible for two important human chronic diseases, tuberculosis and leprosy, as well as for other less widespread but severe infections, traditionally called atypical mycobacterioses. *Mycobacterium leprae* (identified by Hansen in 1871) and *M. tuberculosis* (identified by Koch in 1882) were among the first bacteria recognized as causative agents, respectively, of leprosy and tuberculosis. The dramatic importance of these two still-present illnesses is well known; over the past 200 years, tuberculosis was responsible for the death of 1000 million people (1), and leprosy has been one of the most terrifying diseases since antiquity, and its stigma persists in virtually all cultures in some form (2).

More recently, diseases caused by other mycobacteria, such as *M. avium* complex in immunodepressed patients, have become of increasing importance (3). Despite the early discovery of the etiological agents of the infections, only in the past 5 decades have drugs that are highly effective in the treatment of mycobacterial diseases been discovered (Table 16.1).

Tuberculosis probably appeared in humans about 8000 years ago. Evidence of human tuberculosis, such as bone deformities suggesting Pott's disease, has been found in mummies from pre-dynastic Egypt and pre-Columbian Peru. Paintings, drawings, and, in successive eras, written descriptions witness that a disease similar to what we now define as tuberculosis was well known among ancient peoples. For a long time the popular opinion considered tuberculosis a hereditary disease, although Aristotle (384–322 B.C.), Galen (A.D. 131–201), and Avicenna (980–1037) supported the theory of a contagious etiology.

The first clinical description of phthisis is attributed to Hyppocrates (460–370 B.C.), who, besides the clinical signs, considered the inspection of sputum of great value for diagnostic and prognostic purposes. This view was shared by Aretaeus (4, 5).

For several centuries, knowledge of the disease did not progress, and treatment was based on superstitious practices (bloodletting,

administration of wolf's liver boiled in wine, or boiled crocodile, "the king's touch"), as well as old Hindu remedies, also quoted in the Old Testament (milk from a lactating woman, many meats and vegetables, avoidance of fatigue).

When it became possible to perform autopsies, in the late 1400s, the morbid anatomy allowed for better insight in the pathology of disease as well as allowing correlation with clinical symptoms. Vesalius (1514–1564) was the initiator of modern anatomy, whereas about the same period, Fracastorus (1478–1553) laid the foundation of the epidemiology of infectious diseases with his book "De contagione," where he described three ways of infection: direct contact, fomites, and air. He postulated the existence of "seminaria," imperceptible particles that could exist outside the body for several years and still infect. Among other clever intuitions, we can quote that for phthisis, he postulated that "seminaria" could come not only from outside, but could arise within the body from "putrefaction of the humors." Moreover, he observed that infection is selective in affecting some organs and sparing some others, and he suggested that sterilization at the initial stage of the disease could be efficacious, eradicating the infecting organisms ("germs").

The concept of the contagious nature of phthisis was widely accepted in the following centuries, particularly in Southern Europe, while supporters of the hereditary or constitutional nature persisted in Northern Europe. The evidence of the contagious spreading of the disease induced some governments to take measures of disinfections, decontamination of houses, goods, clothes, personal objects, and some attempts of compulsory report-keeping of affected patients. The Republic of Lucca (Italy) was the first government to take these measures. Its example was later followed by Florence, Naples, and Spain. However, financial considerations and the fear that these laws could give rise to prejudices against tuberculous patients led to their demise by the end of 18th century. However, in the 19th century and the early 20th century, the number of cases and deaths for phthisis showed a continuous increase, particularly in overcrowded cities, where poverty, malnutrition, and poor

**Table 16.1  Antimycobacterial Drugs**

| Generic Name | Trade Name | Name of Chemical Class | Who Marketed the Non-Generic Substance | Efficacy/Potency[a] | Route of Administration | Dose[b] |
|---|---|---|---|---|---|---|
| **Synthetic products** | | | | | | |
| p-aminosalicylic acid | Paser | Aminosalicylates | Jacobus (USA) | 1 μg/mL | Oral | 10–12 g/day |
| Clofazimine | Lamprene | Riminophenazines | Geigy (USA) | 0.1–1 μg/g[e] | Oral | 300 mg/once a month + 50 mg/day |
| Dapsone | Alvosulfon | Sulfones | Wyeth Ayerst (Canada) | 10 ng/mL for M. leprae[f] | Oral | 100 mg/day |
| Ethambutol | Myambutol | Alkylenediamines | Lederle (USA) | 0.5–8 μg/mL | Oral | 15–25 mg/kg/day (max. 2.5 g) |
| Ethionamide | Trecator | Nicotinamide | Wyeth-Ayerst (USA) | 0.6–10 μg/mL | Oral | 15 mg/kg/day (max. 1 g) |
| Isoniazid | Lamiazid and Nydrazid | Isonicotinic acids | Lannet (USA) | 0.02–0.2 μg/mL | Oral | 5 mg/kg/day (max. 300 mg/day) |
| Prothionamide[c] | Trevintix[c] | Nicotinamides | CSL (Australia) | 1–10 μg/mL | Oral | 15 mg/kg/day (max. 1 g) |
| Pyrazinamide | Tebrazid and Zinamide | Nicotinamides | ICN (Canada) Merck Sharp and Dohme (UK) | 5–20 μg/mL (at pH 5.5) | Oral | 20–35 mg/kg/day (max. 3 g) |
| Thiacetazone[c] | | Thiosemicarbazones | | 1 μg/mL | Oral | 150 mg/day (or 2.5 mg/kg/day) |
| Thiocarlide | Isoxyl | Thioureas | Inibsa (Spain) | 0.1–5 μg/mL | Oral | 3–6 g/day |
| **Antibiotics** | | | | | | |
| Amikacin[d] | Amikin | Aminoglycosides | Apothecon (USA) Bristol-Myers Squibb (UK) | 4 μg/mL | IM or IV infusion | 15 mg/kg/day (max. 1.5 g) |

| Drug | Trade name | Composition/source | Manufacturer | MIC[a] | Route | Dosage[b] |
|---|---|---|---|---|---|---|
| Capreomycin | Capstat | Polypeptides complex from Streptomyces capreolus | Lilly (USA) | 10 µg/mL | IM | 1 g/day |
| Cycloserine | Seromycin | Produced by Streptomyces orchidaceus or S. garophalus | Lilly (USA) | 5–20 µg/mL | Oral | 0.5–1 g/day |
| Kanamycin[d] | Kantrex | Aminoglycosides | Aphothecon (USA) | 0.5–10 µg/mL | IM | 15 mg/kg/day (max. 1.5 g) |
| Rifabutin | Mycobutin | Rifamycins or ansamycins | Adria (USA) Pharmacia and Upjohn (Italy) | 0.1–1 µg/mL | Oral | 150–450 mg/day |
| Rifamycin SV[d] | Rifocin | Rifamycins or ansamycins | Lepetit (Italy) Chepharin (Aust.) | 0.1 µg/mL | IM or IV infusion | 0.5–0.75 g/day |
| Rifampicin or Rifampin[d] | Rifadin and Rimactane | Rifamycins or ansamycins | Marion-Merrel Dow (USA) Ciba (USA) | 0.1–2 µg/mL | Oral (IV infusion) | 600 mg/day |
| Rifapentine[g] | | Rifamycins or ansamycins | | 0.1–2 µg/mL | Oral | |
| Streptomycin[d] | Streptomycin injection | Aminoglycosides | Official preparation (USP23) | 1–8 µg/mL | IM | 15–20 mg/kg/day (max. 1 g/day) |
| Viomycin[c] | | Polypeptides from Streptomyces pumiceus or S. floridae | | 2–16 µg/mL | IM | 1 g/day |

[a]MIC, minimum inhibitory concentration in vitro (µg/mL) for M. tuberculosis.
[b]Dosage is merely indicative. It depends on the type of therapeutic regimen in which it is included.
[c]Not yet released.
[d]Drug active also against other bacterial species.
[e](Foot-pad technique) for M. leprae.
[f](In vitro mouse tissue culture) for M. leprae.
[g]At present not commercially available.

work conditions in the era of initial industrialization favored contagion and spreading of the disease.

At the same time, some progresses in the clinical diagnosis could be registered: the introduction of percussion of the thorax by Auenbrugger and of auscultation by means of the stethoscope by Laennec (1819) allowed for a better diagnosis as well as a correlation of the clinical with postmortem findings. We owe to Laennec also the concept of the unity of tuberculosis, namely that tubercles present in various organs in different stages of evolution were an expression of the same disease. He reached this conclusion on the basis of the description of the pathologic lesions found at the autopsies of tuberculous patients performed by Bayle, who described miliary nodules, cavitary, and extrapulmonary lesions, but was unable to differentiate them from bronchiectasis or lung abscess. Laennec's concept of the etiological unity of the different tuberculous lesions was not uniformly accepted by the scientific community and was credited more than one century after, when the microbiologic demonstration of tubercle bacilli in lesions was possible. According to Scholnlein (1839), tubercle was to be considered the fundamental lesion of "tuberculosis," so that this word should substitute the old definitions of phthisis or consumption. The empiricism and the disputes on the etiology of tuberculosis ended in 1882 with the discovery and isolation of tubercle bacillus by Robert Koch, who could also reproduce the disease in animals.

A considerable progress in the diagnosis of the disease and its clinical stages was represented by the discovery of X-rays by Roentgen in 1885. It took some years to realize how valuable X-rays were in the clinical knowledge of the course of tuberculosis.

From the middle of 19th century, while scientific research progressed, new attitudes toward treatment emerged. The superstitious practices of the past were discarded, and the observation that rest, healthy climate, such as a long stay on the mountains, at the seashore, and spas with mineral waters and proper nutrition exerted a favorable effect on the course of the disease led to the opening of numerous sanatoria. They were based on completely different concepts on which the previous institutions were founded, that is with charitable purposes for poor people and with the aim of stopping the spread of infection in crowded towns. Sanatoria greatly contributed to ameliorate the outcome of tuberculosis and also to examine closely its clinical course, its stages, and its different manifestations. The first sanatoria were set up in Europe, particularly in Germany, Switzerland, and Italy, but quite soon they proliferated in the United States and in the rest of the world. Some controversies arose about the benefits of the sanatorium system: according to somebody's opinion, it could represent a kind of imprisonment, possibly lasting some months, some years, or the whole lifetime, with a negative influence on the patient's personality. For others, it represented the possibility to escape a situation of isolation or even rejection from community life. On the whole, considering the lack of specific therapies of the tuberculosis at that time, the experience of the sanatorium system could be considered positive.

At the same time, the awareness of the contagious character of the disease as well as of the efficacy of preventive measures led numerous national organizations to promote a campaign to educate people about the communicability of tuberculosis and about the use of hygienic measures to prevent its dissemination. A further objective was to improve the epidemiological situation of tuberculosis through legislation, research, better patient care, and establishment of hospitals specializing in tuberculosis patients. The initiative gained great success and was adopted by many countries throughout the world. In 1902, an International Central Bureau for the campaign against tuberculosis was established with an office in Berlin and then in Geneva. Since then, the campaign has reached significant achievements under the auspices of the International Union against Tuberculosis and Lung Disease and World Health Organization as well as other national institutions.

After the discovery of tubercle bacillus and the observation that guinea pigs inoculated with living or dead tubercle bacilli had a different reaction from the latter inoculation that allowed animals to survive (Koch phenomenon), Koch inferred that a treatment with a sterile filtrate of cultured organisms

(Old Tuberculin or OT) could act as a remedy for tuberculosis in humans. Unfortunately, the assumption was not correct, and most of the patients treated with this preparation died. Later, tuberculin was revealed to be one of the major methods to diagnose tuberculosis.

Before the introduction of the specific chemotherapy of tuberculosis, the main therapies were collapsotherapy and surgical resection. Collapsotherapy was based on the idea that allowing the lung to rest would hasten the healing of tuberculous lesions. The lung rest could be obtained by introducing air in the pleural space. Carlo Forlanini in Pavia (1894) performed the first successful pneumothorax, a practice that was widely adopted everywhere.

The presence of pleural adhesion prevented the effective collapse of the lung. The invention of the thoracoscope by Jacobs allowed the adhesions to be cut by introducing the apparatus through the thoracic wall into the pleural space. Also, surgery was used to obtain the collapse of the lung. It consisted in removing ribs (thoracoplasty) to bring the chest wall down to the lung; it was first applied by de Carenville in 1885. Results were quite favorable, significantly prolonging the life of patients. Resection of the diseased lung was also used. Lobectomy and pneumonectomy achieved some successes, but the incidence of complications and mortality was high. At present, surgery is uncommon; it is only applied in selected patients where the disease was caused by multiresistant mycobacteria.

Another approach against the disease was the search for a vaccine. In 1921, Calmette and Guerin of the Pasteur Institute of Paris succeeded in preparing a vaccine from an attenuated strain of *Mycobacterium bovis* (BCG) that proved to be effective in preventing tuberculosis and was widely adopted some years later.

A first step in specific chemotherapy of tuberculosis was represented by Domagk's discovery of sulfones, which showed a certain activity on tuberculosis; however, they were toxic.

The real beginning of the modern era of antituberculous chemotherapy was the discovery of streptomycin (Waksman, 1944), which was extremely active against tubercle bacillus *in vitro* and *in vivo*, both in animal

infection and in human disease. Unfortunately, after a brief period of treatment, mycobacteria became resistant to the adopted drug, and therapy was no longer effective. The discovery of other antituberculous agents showed that the resistance problems could be bypassed by the use of multidrug associations. The discovery of isoniazid (1952), and later of rifampicin (1966), the two most powerful antimycobacterial agents with high bactericidal activity, allowed the adoption of multidrug regimens, together with other agents (ethambutol, pyrazinamide, ethionamide, cycloserine, *p*-aminosalycilic acid, and thioacetazone) capable of eradicating *M. tuberculosis* from sputum in 1–2 months and curing the disease in 6–12 months (6, 7).

Since then the course of tuberculosis has been profoundly changed, and the old measures and remedies have been discarded. In industrialized countries, only a few sanatoria are now open for people whose social conditions do not insure a proper compliance to therapy at home. Collapse procedures are no longer applied, and surgical resections are limited to few selected indications.

Unexpectedly, in 1985, the number of cases of tuberculosis in western countries, which were always in constant decline, particularly after the introduction of chemotherapy, showed a sharp increase. This alarming phenomenon prompted a series of studies and surveys to find the cause. Since 1990, this trend now seems to be stable or in decline. In the United States, even if the overall number of cases has declined, the relative proportion of cases occurring among foreign-born individuals increased from 27% to 42% in 1998 (8, 9).

The resurgence of tuberculosis has been attributed mostly to the spread of HIV infection and increasing waves of immigration. Moreover, multidrug resistant (MDR) strains of *M. tuberculosis* are emerging, rendering older therapies largely ineffective. New strategies have been developed or are under study to withstand this situation.

## 2 THE MYCOBACTERIA

The genus *Mycobacterium* belongs to the order Actinomycetales and the family Mycobacteriaceae; it is characterized by nonmotile, nonsporulating rods that resist decolorization with acidified organic solvents and alcohol (10). For this reason, they are also called acid-fast bacteria. Some mycobacterial species are pathogenic for humans: the so-called *Mycobacterium* complex includes *M. tuberculosis, M. bovis*, and *M. africanum*. *M. microti* is pathogenic for voles, field mice, and other rodents, but it was considered nonpathogenic in humans. However, in recent years, it has been demonstrated that it can cause disease in immunosuppressed, and particularly, in HIV$^+$ patients (11). *M. leprae*, pathogenic for man, was considered of uncertain taxonomy for some time, but now is definitely classified among mycobacteria. The study of its genome has contributed to clarify its position in this genus (12).

Other mycobacterial species that resemble *M. tuberculosis* in some morphologic aspects and cultural requirements, but show a characteristic reaction pattern in a battery of biochemical tests and little or no pathogenic effect in humans (at least in absence of some underlying conditions), were improperly defined in the past as "atypical mycobacteria." They were classified by Runyon (13) into four groups according to their growth rates and pigment production. This classification, however, is inadequate to define the different species in a clear-cut way. The development of recombinant DNA technology has allowed us to define better the species belonging to these groups, but the general definition still does not seem satisfactory. A common definition in the recent years has been "mycobacteria other than tuberculosis" (MOTT), but now "Non-Tuberculous Mycobacteria" (NTM) is preferred. The most recent and appropriate grouping of these organisms, proposed by the American Thoracic Society, is based on the type of clinical disease they produce: pulmonary disease, lymphadenitis, cutaneous disease, and disseminated disease (Table 16.2). The term mycobacteriosis is proposed for the diseases caused by these organisms (14, 15).

*M. tuberculosis* is a nonmotile bacillus 1–2 $\mu$m long and 0.3–0.6 $\mu$m wide. It can be demonstrated in pathologic specimens by means of specific staining procedures, the most widely used being the carbol fuchsin or Zichl-Neelsen stain or fluorescent acid-fast staining methods, using auramine as primary fluorochrome. *In vitro* culture of tubercle bacilli is slow on solid media and sometimes difficult. The nutritional requirements are not particularly complex, but the content of the medium greatly influences the composition of the mycobacterial cell. The most common media used for the isolation of *M. tuberculosis* from pathological specimens and for its maintenance are solid media, with egg yolk as a base (Petragnani, Lowenstein-Jensen, and IUTM media) or agar (Middlebrook 7H10, 7H11, and 7H12). In these media, the culture begins to appear 12–15 days after inoculation, but full growth is obtained after 30–40 days. When inoculation is made with pathological material from patients, observation must be prolonged. The most widely used maintenance liquid media are synthetic media containing albumin (Dubos, Youmans, and 7HT media). They usually allow rapid growth (8–10 days), and addition of Tween 80 makes possible to obtain uniformly dispersed growth (16, 17).

In addition to culture in solid and liquid media, recent laboratory diagnostic methods for mycobacterial infections and for susceptibility patterns to antimicrobial drugs include radiometric methods that measure the release of $^{14}CO_2$ from a C-labeled substrate (BACTEC), antigen demonstration by enzyme-linked immunosorbent assays (ELISA), DNA probes, nucleic acid amplification methods, and restriction fragment length polymorphism (RFLP) analysis of genomic DNA (18).

*In vivo* pathogenic activity of mycobacteria was demonstrated in guinea pigs. The rabbit, which is susceptible to *M. bovis* infection but scarcely or not at all to *M. tuberculosis*, was used to differentiate between the two infections. *M. leprae*, or Hansen bacillus, is resistant to acid and alcohol and requires the Ziehl-Neelsen method of staining to be recognized. It is found in lepromatous lesions, where it is arranged mostly in clumps. It is 1–8 $\mu$m long, nonmotile, and nonsporeforming, with a doubling time of 14 days. The most difficult problem in accumulating information about the biology, the susceptibility to antibiotics, and the epidemiology of the disease was the impossibility of cultivating this mycobacterium *in vitro*. Tests in animals have been improved in

**Table 16.2    Classification of the Non-Tuberculous Mycobacteria (NTM) Recovered From Humans**

| Clinical Disease | Common Etiologic Species | Geographical Distribution | Species Rarely Causing Disease in Humans |
|---|---|---|---|
| Pulmonary disease | • *M. avium* complex<br>• *M. kansasii*<br>*M. abscessus*<br>• *M. xenopi*<br>• *M. malmoense* | Worldwide<br>USA, coal mining regions Europe<br>Worldwide, mostly in USA<br>Europe Canada<br>UK, Northern Europe | • *M. simiae*<br>• *M. szulgai*<br>*M. fortuitum*<br>*M. celatum*<br>*M. shimiodie*<br>*M. haemophilum*<br>*M. smegmatis*<br>*M. microti* |
| Lymphadenitis | • *M. avium* complex<br>• *M. scrofulaceum*<br>• *M. malmoense* | Worldwide<br>Worldwide<br>UK, Northern Europe (especially Scandinavia) | *M. fortuitum*<br>*M. chelonae*<br>*M. abscessus*<br>• *M. kansasii*<br>*M. nonchromogenicum*<br>*M. smegmatis*<br>• *M. haemophlum* |
| Cutaneous disease | • *M. marinum*<br>*M. fortuitum*<br>*M. chelonae*<br>*M. abscessus*<br>• *M. ulcerans* | Worldwide<br>Worldwide, mostly USA<br>Australia, tropics, Africa, SE Asia | • *M. avium* complex<br>• *M. kansasii*<br>*M. nonchromogenicum*<br>*M. haemophilum* |
| Disseminated disease | • *M. avium* complex<br>• *M. kansasii*<br>*M. chelonae*<br>*M. haemophilum* | Worldwide<br>USA<br>USA<br>USA, Australia | *M. abscessus*<br>*M. xenopi*<br>• *M. malmoense*<br>*M. genavense*<br>• *M. simiae*<br>*M. conspicum*<br>• *M. marinum*<br>*M. fortuitum* |

•Slowly growing species.

Adapted from American Thoracic Society (ATS), Diagnosis and treatment of disease caused by nontuberculous Myco-bacteria, *Am. J. Resp. Crit. Care Med.* **156**, 51 (1997).

recent years with introduction of the infection of foot-pads in mice and an infection model in the armadillo, but they are complex and can be performed only in the specialized laboratories (19).

The biochemical constitution of mycobacteria is complex and an enormous amount of work has been done in this field. Novel chemical structures have been discovered, but the relationship between these and the pathogenic and biological activities of mycobacteria have not yet been satisfactorily elucidated (19–24). Information about metabolism of mycobacteria is extremely voluminous, but the overall picture of the mycobacterial metabolism is far from complete.

Some interesting differences exist in the metabolic properties of tubercle bacilli grown *in vivo* and *in vitro*, which must be considered when the practical value of antimycobacterial agents designed and tested in laboratory is assessed in practice. Populations of *M. tuberculosis* H37 Rv (a virulent strain) grown in the lungs of mice and populations grown *in vitro* show two different phenotypes, Phe I and Phe II, with marked differences in the metabolism of certain energy sources, in the production of detectable sulfolipids, and in immunogenicity (Phe II is a better immunogen than Phe I) (25–27). Because the shift from Phe I to Phe II is readily reversible, it can be deduced that the genome of H37 Rv remains the same (28); probably, a modification in its surface occurs.

A great deal of effort has been focused on research of the constituents of the mycobacterial cell wall, because they are responsible for many of the pathogenic effects of tubercle bacilli. The core of the mycobacterial cell wall is the mycolylarabinogalactanpeptidoglycan constituted by three covalently attached macromolecules (peptidoglycan, arabinogalactan, and mycolic acid). A schematic structure, simplified from (29), is outlined in (1). The pepti-

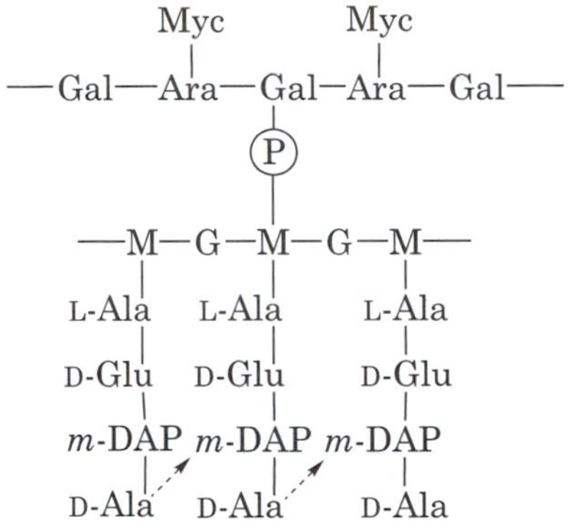

$$
\begin{array}{ccc}
 & \text{Myc} & \text{Myc} \\
 & | & | \\
-\text{Gal}-\text{Ara}- & \text{Gal}-\text{Ara} & -\text{Gal}-- \\
 & | & \\
 & \textcircled{P} & \\
 & | & \\
-\text{M}-\text{G}- & \text{M}-\text{G} & -\text{M}- \\
| & | & | \\
\text{L-Ala} & \text{L-Ala} & \text{L-Ala} \\
| & | & | \\
\text{D-Glu} & \text{D-Glu} & \text{D-Glu} \\
| & | & | \\
m\text{-DAP} & m\text{-DAP} & m\text{-DAP} \\
| & | & | \\
\text{D-Ala} & \text{D-Ala} & \text{D-Ala}
\end{array}
$$

(1)

doglycan (or murein) consists of a repeating disaccharide unit, in which N-acetyl-D-glucosamine (G) is linked in a 1–4 linkage to N-glycolyl-D-muramic acid (M) attached to L-alanine-D-glutamic acid-NH$_2$-meso-diaminopimelic acid-NH$_2$-D-alanine. This unit is linked to a polysaccharide unit, the arabinogalactan, through a disaccharide phosphate (rhamnose-galactose phosphate). The arabinogalactan is connected to a glycolipidic region constituted of esters of mycolic acids.

Mycolic acids are $\alpha$-branched, $\beta$-hydroxylated long-chain fatty acids, of which three principal groups are known: the corynomycolic acids, the nocardic acids, and the mycobacterial mycolic acids (30, 31). Mycolic acids can also be detected in the skin lesions of patients suffering from lepromatous leprosy, indicating that the agent of leprosy is a mycobacterium containing it (32). The chemical structures of methoxylated mycolic acid and $\beta$-mycolic acid extracted from M. tuberculosis sp. hominis are shown in structures (2) and (3) (33). The mycolic acids are linked through their carboxy groups to the end terminal OH groups of the D-arabinofuranose (Araf) molecules, branches of the arabinogalactan of the cell wall (34–36). Mycolic acids are known to be acid-fast, because they bind fuchsin and the binding is acid-fast. Thus, it seems that the acid fastness of mycobacteria depends on two mechanisms: the capacity of the mycobacterial cell to take fuchsin into its interior and the capacity of mycolic acids to form a complex with the dye (37–39).

In addition to the lipid murein part of the rigid structure, there is a series of soluble lipid compounds that seem to be located in or on the outer part of the cell wall: lipoarabinomannan, waxes D, cord factor, mycosides, sulfolipids, and phospholipids (20–22). Lipoarabinomannan (LAM) is an essential part of the cell envelope, being a component of the plasma membrane, which lacks covalent association with the cell core. In this macromolecule, arabinan chains are attached to a mannan back-

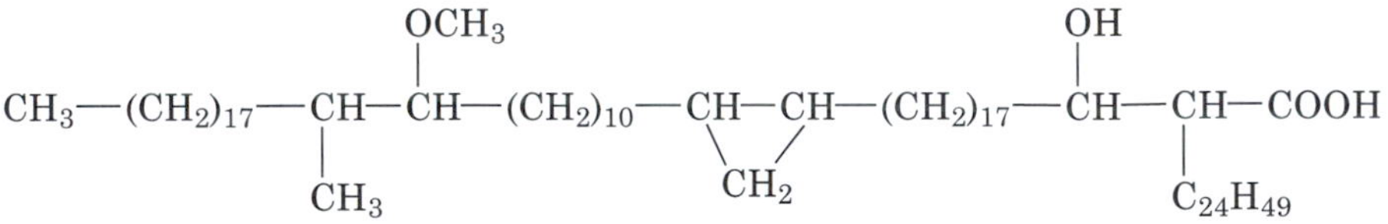

$$
\begin{array}{ccccc}
 & \overset{\displaystyle \text{OCH}_3}{\underset{\displaystyle |}{}} & & \overset{\displaystyle \text{OH}}{\underset{\displaystyle |}{}} & \\
\text{CH}_3-(\text{CH}_2)_{17}-\text{CH}-\text{CH}-(\text{CH}_2)_{10}-\text{CH}-\text{CH}-(\text{CH}_2)_{17}-\text{CH}-\text{CH}-\text{COOH} & & & & \\
 & \underset{\displaystyle \text{CH}_3}{|} & \underset{\displaystyle \text{CH}_2}{\diagdown\diagup} & & \underset{\displaystyle \text{C}_{24}\text{H}_{49}}{|}
\end{array}
$$

(2)

$$
\begin{array}{ccccc}
 & \overset{\displaystyle \text{O}}{\underset{\displaystyle \|}{}} & & \overset{\displaystyle \text{OH}}{\underset{\displaystyle |}{}} & \\
\text{CH}_3-(\text{CH}_2)_{17}-\text{CH}-\text{C}-[\text{C}_{17}\text{H}_{34}]-\text{CH}-\text{CH}-(\text{CH}_2)_{19}-\text{CH}-\text{CH}-\text{COOH} & & & & \\
 & \underset{\displaystyle \text{CH}_3}{|} & \underset{\displaystyle \text{CH}_2}{\diagdown\diagup} & & \underset{\displaystyle \text{C}_{24}\text{H}_{49}}{|}
\end{array}
$$

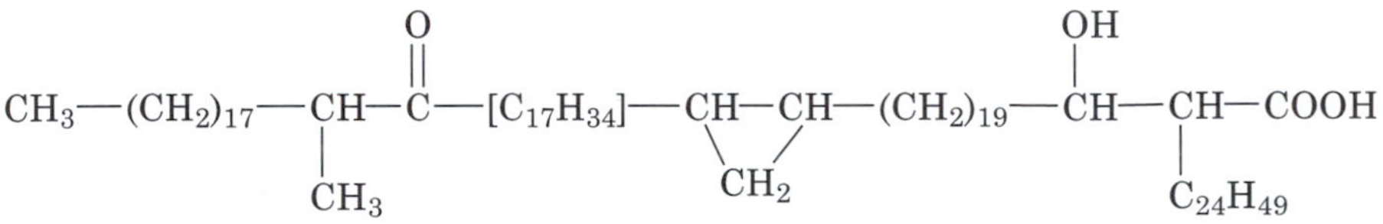

(3)

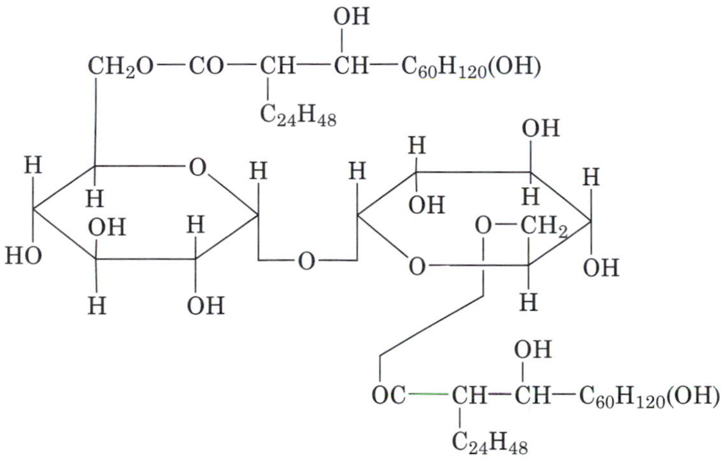

(4)

bone, which is, in turn, attached to a phosphatidylinositol esterified with fatty acids like palmitate and 10-methyloctadecanoate (tuberculostearate).

The so-called waxes D are ether-soluble, acetone-insoluble, and chloroform-extractable peptidoglycolipid components of the mycobacterial cell, probably autolysis products of the cell wall (40). Because they are esters of mycolic acids with arabinogalactan linked to a mucopeptide containing $N$-acetylglucosamine, $N$-glycolylmuramic-acid, L- and D-alanine, meso-diaminopimelic acid, and D-glutamic acid, it has been suggested that they are materials synthesized in excess of those needed for insertion into the cell wall (22, 39–41). The constitution of wax D differs in different varieties and strains of mycobacteria.

Cord-factor is a toxic glycolipid (6, 6′-dimycolate of trehalose) structure (4), which has been deemed to be responsible for the phenomenon of cording (42, 43) (i. e., the capacity of M. tuberculosis to grow in serpentine cords), a property that is correlated with its capacity to kill guinea pigs (44, 45). The detergent properties of cord factor and its location on the outer cell wall have led to the suggestion that it may play a role in facilitating the penetration of certain molecules necessary for growth of mycobacteria (39).

Mycosides are glycolipids and peptidoglycolipids type-specific of mycobacteria (46) that often have in common terminal saccharide moieties containing rhamnoses $O$-methylated in various positions (47). They can be divided into two main categories (31): phenolicglycolipids with branched-chain fatty acids and peptidoglycolipids consisting of a sugar moiety, a short peptide, and a fatty acid. The biological activity of mycosides is still obscure. They probably have a role in cellular permeability (48). Glycolipids or peptidoglycolipids are responsible for the ropelike appearance that is evident in one of the outer layers of mycobacteria when they are visualized by the technique of negative staining (39).

The sulfolipids (2, 3, 6, 6′-tetraesters of trehalose; **5**) (22–28, 49), to which are attributed the cytochemical neutral-red fixing activity of viable, cordforming tubercle bacilli, seem to

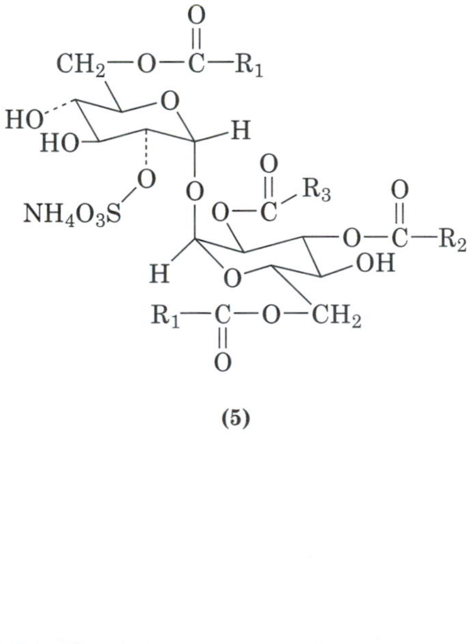

(5)

play a role in conferring virulence to tubercle bacilli and influencing their pathogenicity (49), acting synergically with the cord factor (50).

The phospholipids (cardiolipin, phosphatidylethanolamine-glycosyl diglyceride, and phosphatidylinositol-monomannosides and -oligomannosides) were considered to be antigenic substances elaborated by *M. tuberculosis*, but the most purified preparations have been shown to behave only as haptens (22, 30, 51).

Even though some suggestions have been made about the biological activities of the lipids of the tubercle bacillus, a clear structure-function relationship has not yet been delineated. Nor has it been determined which structural features can produce favorable or detrimental effects. Also, the biosynthetic pathways leading to the formation and assembly of the cell wall have not been clarified, and the enzymes involved have not been purified.

Other interesting substances isolated from mycobacteria are the mycobactins, which are a group of bacterial growth factors that occur only in the genus *Mycobacterium* (52). The isolation of the mycobactins was followed by the identification of growth factors in other microorganisms, the sideramines, which differ from mycobactins but share with them some common properties, the most relevant being strong chelating capacity for ferric ions. At least nine mycobactin groups have been isolated from different mycobacteria. They have the same basic constitution, with some variations in details of structure. They consist of an octahedral iron-binding site (containing two secondary hydroxamate groups, an oxazoline ring, and a phenolic hydroxyl group) and a hydrophobic chain (containing up to 20 carbon atoms). Mycobactin P (**6**), isolated in 1946, was the first example of a natural product with an exceptional iron-chelating activity, and its structure was the first to be determined. Mycobactin S (**7**) is the most active of these factors, showing growth stimulation at concentrations as low as 0.3 ng/mL. Mycobactin M (**8**) is a representative of the structure of M-type factors. The most biochemically unusual product in the structure of mycobactins is $N^6$-hydroxylysine, which is present in the molecule in both acyclic and cyclic forms. All the

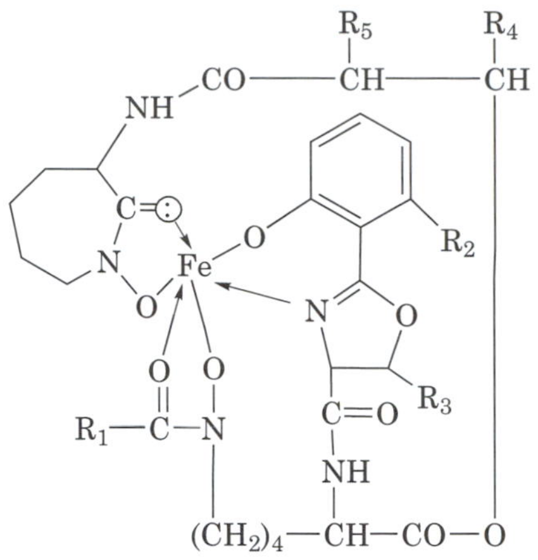

(**6**) $R_1 = C_{17}H_{34}$, $R_2 = CH_3$, $R_3 = H$, $R_4 = C_2H_5$, $R_5 = CH_3$
(**7**) $R_1 = C_{17}H_{34}$, $R_2 = H$, $R_3 = H$, $R_4 = CH_3$, $R_5 = H$
(**8**) $R_1 = CH_3$, $R_2 = H$, $R_3 = CH_3$, $R_4 = C_{17}H_{34}$, $R_5 = CH_3$

known mycobactins contain either a salicylic acid or a 6-methylsalicylic acid moiety. The oxazoline rings derive from 3-hydroxy amino acids, either serine or threonine. The mycobactins are powerful iron chelators with an association constant of $>10^{30}$ and are essential growth factors for the mycobacteria. Therefore, it seems likely that they are involved in mycobacterial iron metabolism (53). Mycobacteria respond to iron deficiency by producing salicylic or 6-methylsalicylic acid, together with mycobactins that have a great affinity for ferric iron. It has been suggested that in the mycobacteria, salicylic or methylsalicylic acid mobilizes the iron in the environment and that mycobactins are involved in the active transport of iron into the cell.

Mycobacterial proteins have been submitted to extensive research since the days of the discovery of Robert Koch and the preparation of the old tuberculin (OT), a heat inactivated concentrate of proteins released to the growth medium from stationary culture of *M. tuberculosis*. This product termed "tuberculin" elicits a reaction in the skin, that is an expression of delayed-type hypersensitivity (DHT), and has been used for the diagnosis of tuberculosis. At present, an ammonium sulfate precipitable protein fraction of the culture filtrate, termed "tuberculin purified protein derivative" (PPD) is currently used to determine the "tuberculin reaction" (54).

Numerous proteins with different locations have been identified in *M. tuberculosis*. They can be cytoplasmic, cell-membrane, and cell-wall associated or secreted to the surrounding medium. They have been sequenced, but in most cases, their function has not been identified.

Cytoplasmic proteins are involved in amino acid and protein biosynthesis, and sometimes they can show a different behavior in virulent and avirulent strains. For instance, the virulent strains (H37 Rv) possess one type of aspariginase, and the avirulent one (H37 Ra) possesses two asparaginases. The avirulent strain has an aspartotransferase that transfers the aspartyl moiety of asparagine to hydroxylamine, whereas the avirulent strain does not (55–56).

Other proteins show a significant homology with the so called "heat shock proteins" or "stress proteins." They have many functions, particularly in assisting the microrganism in adaption to environmental changes (57). Proteins synthesized in the cytoplasm may be destined to function outside the cytoplasm either in association with the cytoplasmic membrane or cell wall or can be exported to the surrounding medium. Proteins secreted by *M. tuberculosis* are very potent in generating a cellular immune response. The vaccine of Calmette-Guerin (BCG), an attenuated strains of *M. bovis*, is in position to confer protection against disease only if it is alive. It has been suggested that only in this situation, where the secretion of proteins is continuous, will the reaction of T-cells responsible for recognition of the infected macrofage be elicited and lead to the control of infection at an early stage. Proteins associated with the cell membrane or the cell wall are lipoproteins, which besides fulfilling functional activities such as binding some nutrients, could be immunogenic. Proteins exported to the exterior have some distinct functions. As example we can quote superoxide dismutase (SOD), an enzyme that can paralise the host defense mechanism by inactivating the toxic oxide radicals generated by the activated macrophage (58).

The functions of mycobacterial proteins that have been found in the past years could now be confirmed, elucidated, and completed with those of numerous other proteins and heterogeneous compounds constituting the mycobacterial cells through the informations obtained by the complete sequencing of *M. tuberculosis* genome (virulent strain H37Rv).

This achievement, reported in 1998, represents a milestone in knowledge of biology of *M. tuberculosis*. The genome results of a composite sequence of 4.411.529 base pairs, with a guanine + cytosine (G + C) content of 65.6%, which is relatively constant throughout the genome. It contains 3924 proteins encoding genes. It is the second largest bacterial genome after *E. coli*. By the analysis of the proteins encoded in genes, a precise function could be attributed to a 40% of them and some information was obtained for another 44%. The remaining 16% do not resemble any known proteins and may account for specific mycobacterial functions (59, 60).

From the composition of the genome sequence, it can be inferred that *M. tuberculosis* has the capacity to synthetize all essential amino acids, vitamins, and enzymes, necessary co-factors for major metabolic pathways present in the cell (lipid metabolism glycolysis, pentosephosphate pathways, tricarboxylic acid, glyoxylate cycles) as well as for a number of others.

Two major important findings emerged by the genome sequence: the identification and characterization of repeat elements as well as of regulatory genes. The presence of repeat elements, that vary from strain to strain, allows the strain subtyping by molecular techniques. The expression of regulatory genes is governed by 13 putative sigma factors, which promote the recognition of the subunit of mycobacterial RNA polymerase. More than 100 regulatory proteins are predicted, and other transcription regulators and regulatory protein pairs are present that may transmit signals from outside of the bacterial cell into the cytoplasm.

Two new families of acidic glycine-rich proteins have been recognized, the PE and PPE families, which represent about 10% of the coding capacity of the genome. Their genes are clustered. Their function is unknown except for a lipase. These proteins include one of the major antigens, the serine-rich antigen, present in leprosy patients. Even if at present there is no proof, some immunological function could be possibly attributed to them and, according to some authors, also some role as virulence factors (61).

The complex lipidic constitution of *M. tuberculosis* cells led to the presumption of the presence of numerous genes encoding for lipolitic and lipogenic activity. In fact, over 250 genes for biosynthesis of various lipid classes were found in *M. tuberculosis* in contrast with the approximately 50 genes of the same class in *E. coli*. For fatty acid degradation, a process that leads to release lipids within mammalian cells and tubercles, there is a series of enzymes that are involved such as 36 acylCoA-synthases and a family of 36 related enzymes that could catalyze the first step in fatty acid degradation. The successive degradation steps are catalyzed by the enoyl CoA hydratase/isomerase superfamily of enzymes, enzymes converting 3-hydroxy fatty acid into 3-keto fatty acid, the acetyl-CoA, C-acetyltransferases, and the Fad A/Fad B β-oxidation complex. For fatty acid synthesis, two distinct families of enzymes are involved: fatty acid synthase I and II (FAS I on FAS II), which contain seven enzyme activities. Two subunits of FAS II system, the enoylreductase synthase corresponding to the enzyme *Inh*A and β-ketoacylsynthase proteins *Kat*A are target for isoniazid. Numerous other genes are contributed to the fatty acid biosynthesis (carboxylase system, Fab A–like β-hydroxyacyl-ACP-dehydrase, S-adenosyl-L-methionine-dependent enzymes, etc.).

It was found that mycobacteria also synthetize polyketides by several different mechanisms, similar to those involved in erythromycin biosynthesis. There are four classes of polyketide synthases. Their function in mycobacteria is unknown, but it is of potential interest because some polyketides such as rapamycin show antibacterial activities or immunological properties.

From genome sequencing, useful information is expected to clarify some aspects of virulence, pathogenicity, and immunology of *M. tuberculosis*. At present little information has been obtained about virulence. Before the completion of the genome sequence three virulence factors were known: catalase-peroxidase, with a protective activity against reactive oxygen species produced by the phagocyte; *mce* that encodes macrophage colonizing factor; and a sigma factor gene, *sig*A, mutations, which can lead to virulence attenuations. After genome sequence a series of phospholipases C, lipases, and esterases, which attack cellular or vacuolar membranes, could be individuated as further virulence factors.

To identify factors of virulence, the genomic analysis of *M. tuberculosis* H37Rv and of the closely related attenuated strains H37Ra has allowed to individuate some genetic differences that is two polimorphisms: a 480-kb fragment in *M. tuberculosis* H37Rv was replaced by two fragments of 220 and 260 kb in *M. tuberculosis* H37Ra, while there was a fragment of restriction endonuclease *Dra*I in *M. tuberculosis* H37Ra that had no counterpart in *M. tuberculosis* H37Rv (62). It has been suggested that these polymorphisms were the results of transposition of an insertion sequence (IS6110), which may have inactivated virulence genes. However complementation of *M. tuberculosis* H37Ra with the restriction fragment from *M. tuberculosis* H37Rv that encompasses the IS6110 did not restore the virulence as it could be demonstrated by infection in mice. Therefore, the role played by the genomic variation demonstrated in the two strains has not been elucidated.

A further approach to identify genes for putative virulence has been to perform comparative genomics of the members of *M. tuberculosis* complex, an approach facilitated by the paucity of genetic differences among them at the nucleotide level (63). The genome sequence of *M. bovis* is almost completed (64). A comparative study of *M. bovis* and *M. bovis* BCG (a strain of attenuated virulence used for vaccination) has shown that three regions, designated as RD1, RD2, and RD3, are deleted in *M. bovis* BCG relative to *M. bovis*, but it is not clear if these deletion regions play a role in the attenuation of *M. bovis* BCG. Other deletion regions were found from BCG relative to *M. tuberculosis*. Loss of RD5 locus removed the genes for three phospholipase C enzymes, other deletions concerned Esat6 protein, (a potent T-cell antigen), the genes involved in lipopolysaccharide biosynthesis. Although definitive comparison of PE and PPE is not possible as the genome sequence of BCG is not completed, analysis so far performed revealed remarkable differences in the sequence of PE and PPE proteins, which could have a role in virulence. Therefore, it seems that comparative genomics may offer some indications to understand the attenuation of *M. bovis* BCG.

The genome of *M. leprae*, which has recently been fully sequenced, is smaller that that of *M. tuberculosis*, containing 3.268.182 base pairs. Its GC content circa 58% and the number of genes is also inferior (about 1800 vs. 3966 of *M. tuberculosis* H37Rv) (12). Assuming that the genome of *M. leprae* was once equivalent and similar in size to other mycobacteria, it can be inferred that extensive downsizing and rearrangement may have occurred during evolution. The leprosy bacillus may have lost more than 2000 genes. Comparison with the genome of *M. tuberculosis* revealed a mosaic arrangement, where areas (about 65 segments) showing synteny are flanked by unrelated genomic segments, Many sequences are noncoding and demonstrated to be pseudogenes, namely inactivated version of genes that are still functional in *M. tuberculosis*. The present arrangement denotes that extensive recombination events between dispersed repetitive sequences occurred. Gene deletion and decay led to the elimination or decrease of many important metabolic activities such as siderophore production, oxidative microaerophilic, and anaerobic respiratory chains as well as of many catabolic systems. On the other hand, among the outer lipids of the leprosy bacillus there is a phenolic glycolipid 1, a component not found in the envelope of *M. tuberculosis*. Several enzymes don't have any counterpart in *M. tuberculosis*, such as uridine-phosphorilase and adenylate cyclase, and two transport systems, one for sugars the other for divalent metal ion uptake. The composition of genome of *M. leprae* that shows an extended loss of functions in comparison to *M. tuberculosis* may account for some peculiar biological properties of this organism: extremely long doubling time, non-cultivability *in vitro*, obligate intracellular location, and pronounced tropism for the Schwann cells of the peripheral nervous system in humans.

Further insight in gene functions and comparison with genomes of other mycobacteria hopefully will reveal the mechanisms of pathogenicity and virulence and provide the basis for the design of new chemotherapeutic agents and vaccines.

# 3 PATHOGENESIS AND EPIDEMIOLOGY

## 3.1 Tuberculosis

Tuberculosis is sometimes an acute but more frequently a chronic communicable disease that derives its character from several properties of the tubercle bacillus, which in contrast with many common bacterial pathogens, multiplies very slowly, does not produce exotoxins, and does not stimulate an early reaction from the host. The tubercle bacillus is also an intracellular parasite, living and multiplying inside macrophages.

The structure and evolution of lesions caused by tubercle bacilli are determined by host aspecific defense system, immunologic response, and genetic factors. Most commonly mycobacteria reach the lung alveoli by inhalation. If the number of bacilli is low, the strain is of moderate virulence and the intrisic microbicidal activity (genetically determined) of resident alveolar macrophages is of good level, the tubercle bacilli ingested are rapidly killed and no further evolution of infection follows. However mycobacteria may evoke a variable immunological response leading the host organism either to a state of resistance or to a state of disease, which depends on the number of infecting organisms and the already quoted host defense capacity and genetic factors.

To schematize the events that follow the penetration of tubercle bacilli in the organism, four stages have been described in the pathogenesis of tuberculosis, mainly based on the researches of Lurie (65) and Dannenberg (66) in rabbits. The results of animal research can be only partially applied to human disease. However they have contributed to a better understanding of the main aspects of its course (67).

The first stage has been described above, and because of the rapid destruction of the infecting organism, it can be self-limiting. However if the number of bacilli is high or if the macrophage microbicidal activity is inadequate to kill them, bacilli multiply inside the cell, causing its burst. This event, through the production of chemotactic factors from the released bacilli attracts monocytes from the circulation and gives origin to the second stage.

Blood-born monocytes are not yet activated, so they cannot destroy the bacilli, which grow within the cell at a logarithmic rate. At this time, the immunological response is triggered; mycobacterial antigens processed by macrophages elicit the subset lymphocytes (CD+ T-cells) primarily involved in cell-mediated immunity, the main mechanism of protective immunity in tuberculosis. These cells secrete a number of lymphokines (particularly interferon-$\gamma$) capable of inducing macrophages to kill intracellular mycobacteria (68, 69). At this stage the key event in the immunological response to mycobacterial antigens, i.e., the formation of granuloma (tubercle), is initiated.

Stage 3 starts when the logarithmic phase of bacillary growth stops. It is characterized by the emergence of an other subset of lymphocytes, CD8+ T-cells, that probably play some role also in the late phases of stage 2 and through their action caseous necrosis is produced. CD8+ T-cells are endowed with cytotoxic activity and mediate the delayed-type hypersensitivity, which leads to caseous necrosis and to interruption of logarithmic bacillary growth. CD8+ T-cells destroy the nonactivated macrophages by eliminating the intracellular environment favorable to bacillary growth. The released bacilli will be then phagocytized by activated macrophages (whose activation is accelerated by previously sensitized lymphocytes). It is still debated how CD8+ T-cells contribute to the production of caseous necrosis; among different ideas, it has been suggested that CD8+ T-cells induce local sensitivity to tumor necrosis factor (TNF) (69–71).

Further research has been dedicated to investigate the role of apoptosis (programmed cell death) of T-lymphocytes in the pathogenesis of TB. Conflicting results have been obtained. A recent hystopathologic finding in biopsy material of TB patients seems to throw some light on this subject (72, 73). Large numbers of apoptotic T-cells and macrophages

were seen in areas surrounding caseous foci. Macrophages are negative for anti-apoptotic bcl-2 protein, but positive the pro-apoptotic bax protein, whereas the associated T-cells express interferon (IFN-$\gamma$) and FasL, data that indicate that caseation is associated with apoptosis of macrophages and T lymphocytes. If these findings are confirmed, apoptosis should represent a mechanism for tissue destruction and consequently for TB transmission (74, 75).

In its typical elementary structure, granuloma shows a solid caseous center surrounded by immature macrophages (allowing intracellular bacillary growth) and activated macrophages (that can differentiate in epithelioid cells and Langhans cells, giant polynuclear cells), which kill the bacilli. The development or the arrestment of the disease depends on the balance between the two types of macrophages. Caseous material is not favorable to the bacillary growth because of acidic pH, low oxygen tension, and presence of inhibitory fatty acids; mycobacteria do not multiply but can survive for years in this environment (76–77).

The disease proceeds when stage 4 is reached, in which liquefaction of the caseous center occurs, often followed by formation of cavities where bacilli multiply extracellularly reaching very high density. They can be discharged into the airways, diffusing in other parts of the lung and in the environment. Liquefaction and cavitation as markers of disease progression occur in human disease, but not in infection model in rabbits (67). However, the schematized patterns of infection in animals find some counterpart in the course of human infection and disease. In fact, the first contact of the human organism with tubercle bacilli, which usually takes place in infancy or adolescence, normally does not produce any clinical manifestation. The anatomic lesion induced by proliferation of mycobacteria and the reactive regional adenitis are called "primary complex." At that moment the subject shows a positive tuberculin test [a cutaneous reaction obtained by injection or percutaneous application of culture filtrates of mycobacteria or of their purified protein content (PPD)], which indicates a state of hypersensitivity to the tubercle bacillus, not necessarily a state of immunity. Usually the primary complex remains clinically silent, but it can also progress and evolve to a state of disease.

According to the definition of the American Thoracic Society (ATS, 1981), tuberculosis infection does not mean disease (78). The state of disease is defined by the appearance of clinically, radiologically, and bacteriologically documentable signs and symptoms of infection.

Chronic pulmonary tuberculosis in adults may be caused by reactivation of the primary infection or to exogenous reinfection. Some aspects of the state of latency of *M. tuberculosis* in lung tissue, cause of possible reactivation of the disease, have been recently elucidated that could have some impact in clinical settings. It has been demonstrated that some mutants are not capable to maintain the state of latency; this property is associated to inactivation of cyclopropane synthetase (79) or isocytrate lyase (80), which is required for the metabolism of fatty acids. Moreover a mutant has been discovered that cannot proliferate in the lungs, although it proliferates normally in other tissues. This strain is unable to synthetize or transport the cell wall associated phtiocerol, dimycocerosate, a lipid found only in pathogenic mycobacteria (81).

A typical characteristic of tuberculosis is the formation in the infected tissue of tubercles, nodular formations, which can have different sizes and different modes of diffusion, giving rise to various clinical forms called miliary, infiltrate, lobar tuberculosis, and so on. The disease progresses by means of ulceration, caseation, and cavitation, with bronchogenic spread of infectious material. Healing may occur at any stage of the disease by processes of resolution, fibrosis, and calcification.

Control of the disease has been achieved in part through mass vaccination with BCG (the bacillus of Calmette and Guèrin, an attenuated strain of *M. tuberculosis bovis*), but above all through correct application of active chemotherapeutic agents. Chemotherapeutic treatment now available enables to stop the propagation of the disease in a high percentage of cases, by killing the pathogenic bacilli, thus permitting the organism to repair or to confine the pathological alterations. Another important consequence of chemotherapeutic treatment is the prevention of dissemination of virulent bacilli to other persons.

Despite the efficacious drugs now available for the treatment of tuberculosis, this illness is present all over the world. According to recent estimates by the World Health Organization (WHO), 8.4 million new cases occurred in 1999, up from 8.0 million in 1997. The rise affected mostly the African countries, which had a 20% increase in TB incidence attributed to the continuous spread of HIV/AIDS. The total number of smear positive cases was 1.4 million both in 1998 and 1999 (82). In developing countries, the majority of deaths occurs in young adults, between 15 and 40 years and represents 98% of all deaths caused by tuberculosis (83). Twenty-two are in the highest burden countries and more than one-half the cases occurred in India, China, Indonesia, Bangladesh, and Pakistan. The highest incidence rates pro capita were in sub-Saharan Africa (82, 84).

In developed countries after the First World War, the trend to a decrease in tuberculosis morbidity has been constant, probably related to better general hygienic conditions, preventive measures, and in the last decades, to the introduction of effective chemotherapy. However, in 1985, the reports of WHO indicated a resurgence of tuberculosis, even in advanced countries of Europe and North America. In the United States, where the incidence of tuberculosis has been declining at an average rate of 6% each year until 1985, the case rate has risen from 9.3 for 100,000 in 1985 to 10.3 in 1990 (85–88).

In April 1993, the WHO took the extraordinary step of declaring tuberculosis a "global emergency" (88). However, after this increase, case rates started to become stable or declining. The most recent data, issued in 1998, indicated that the incidence of tuberculosis in the United States was 6.8 per 100,000 per year, a decrease of 31% since 1992. However the relative proportion of cases occurring among foreign-born individuals increased from 27% in 1992 to 42% in 1998 (84). It is well known that this phenomenon occurs in poor areas of major cities in industrialized countries. A dramatic increase in the number of tuberculosis cases was notified since the early 1990s in some Eastern European countries, particularly Georgia, Kazakhstan, Kyrgystan, Latvia, Lithuania, Romania, and the Russian Federation. In Russia, the number of cases more than doubled in the 5 years between 1992 and 1997 (89).

Bolivia, Haiti, and Peru have notification rates greater that 100 per 100,000. In Peru, where the notification rates reached 225 per 100,000 in 1994, the intensified control and the prevention and therapy measures led to a decrease in notification rates of 3–4% per year, while the treatment success rates now exceed 90%. China and Vietnam reported a remarkable increase in number of TB cases during the 1990s. Now the two countries have implemented a control program that seems to give good results. In the United States, the highest percentage of TB cases are registered among blacks and Hispanics between the ages of 25 and 44 years (89–90). In Western Europe, the Gulf States, and Australia, the majority of cases are among foreign-born immigrants. In 1997, immigrants accounted for more than one-half of tuberculosis cases in these areas.

Why the highest rate of tuberculosis is in immigrants is not fully understood, even if many factors are known to contribute to this behavior. Poor living conditions, poor socio-economic status, and malnutrition surely play an important role.

Moreover it must be noted that the spread of HIV infection/AIDS has coincided with the recrudescence of TB (91). AIDS is the highest risk factor that favors the progression of latent infection to active tuberculosis. In HIV-infected subjects, cell-mediated immunity is depressed, because the major targets of this virus are CD4+ T lymphocytes, whose number reduction and functional impairment prevents them to mount the essential immunological response to *M. tuberculosis* challenge. At present it is estimated that the risk of developing tuberculosis in HIV infected adults is 3–8% annually throughout the world, with rates varying widely among geographical areas within the same country. The majority of cases are concentrated in sub-Saharan African countries where the AIDS epidemic has caused about an 100% increase in reported tuberculosis cases (92).

HIV-infected subjects may develop tuberculosis by direct progression of exogenous infection to disease or by recrudescence of previously acquired latent infection. They are at least 30-fold more likely to develop reactivation tuberculosis as HIV infection progresses. Severe outbreaks of tuberculosis have been reported when groups of HIV-infected persons have been exposed to a person with infectious tuberculosis (93–95).

Tuberculosis tends to occur in the early stages of HIV infection. Sometimes it is the first manifestation of the disease caused by HIV. The clinical course of tuberculosis is determined by the degree of immunosuppression. Very often it is a severe disease, frequently with extrapulmonary localization (92).

A further consequence of spreading HIV infection has been an increase of infections caused by mycobacteria other than tuberculous (MOTT) or non-tuberculous mycobacteria (NTM), previously defined as atypical mycobacteria. They only partially resemble the tubercle bacillus and cause diseases less frequently than it. In some geographical areas, their incidence is rather high, such as *M. kansasii* in central United States and *M. ulcerans* in Australia and South Africa. Unfortunately the drugs developed for the treatment of tuberculosis have a poor efficacy for these infections. Among NTM, the *M. avium intracellulare* complex (MAC), which in the past was very seldomly recognized as cause of infection, showed to be a frequent cause of severe disseminated infection in AIDS patients occurring most frequently in the late stages of the disease. However, at present, a growing number of patients have MAC infection as the initial manifestation of AIDS. In the last years, MAC has been isolated from 20% to 80% of patients fully followed from initial diagnosis of AIDS to death (96–98).

The resurgence of tuberculosis has been accompanied by another phenomenon of importance, that is the emergence of multiple drug-resistant (MDR) strains of *M. tuberculosis* in the etiology of tuberculosis. MDR organisms demonstrate *in vitro* resistance to at least two major antituberculosis drugs, usually isoniazid and rifampicin.

Cases caused by these strains are difficult to treat. Multidrug-resistant TB has emerged particularly in the states of Eastern Europe and Russia, probably as a consequence of the collapse of public health services, associated with growing poverty, malnutrition, and war. Percentages of 9.4% for primary MDR-TB were reported from Russia in 1998, whereas percentages as high as 55% and 23% were reported from patients in Russian prisons (90, 99).

## 3.2 Leprosy

Leprosy is a chronic disease caused by *M. leprae*, a mycobacterium that multiplies even more slowly than the tubercle bacilli, having a doubling time of about 15 days. It is an obligate intracellular organism that can not be cultured *in vitro*. Mice foot-pad infection and a model of infection in armadillo are the means to grow this mycobacterium (100).

For a long time, humans were considered the only natural reservoir of *M. leprae*, but actually it has been demonstrated that about 15% of wild armadillos in Louisiana and Texas and some primates can be infected by *M. leprae* (19, 101) The mode of transmission of leprosy is not fully elucidated. Household contacts with patients with borderline leprosy and lepromatous leprosy whose nasal mucosa is heavily infected are at risk of acquiring the infection because of dissemination of infected secretions in the air, whereas dissemination through skin lesions and contaminated soil seems less important (102–104). *M. leprae* causes granulomatous lesions that differ according to the immune status of the patient. Granulomatous lesions with epithelioid cells, many lymphocytes, and few *Mycobacterium leprae* are characteristic of tuberculoid leprosy, the form of disease occurring in subjects with a good immunologic reactivity. In lepromatous leprosy, the form of disease occurring in subjects with impaired cell-mediated immunity, granuloma shows a massive infiltration of tissues with large macrophages filled with mycobacteria. Lymph nodes have hyperplastic germinal centers and in the paracortical areas there are very few lymphocytes.

The clinical manifestations of tuberculoid leprosy consist of skin macules with clear center that are insensitive to pain stimuli because of the involvement of peripheral nerves. The skin lesions of lepromatous leprosy consist of numerous, symmetric, small, hypopigmented,

or erythematous papules in the initial phase, whereas nodules develop later in the course of the disease. Then the lesions diffuse to the eyes, upper respiratory tract, and many other organs and tissues. The damage to peripheral nerves lead to loss of sensation, deformity, and mutilation. The borderline forms of leprosy show lesions that have characteristics intermediate between the two main forms of the disease.

The majority of people exposed to leprosy will not develop it. Those who develop it, after an incubation period that can range from 1 to $\geq$40 years (usually 5–7 years), may have only single lesion (indeterminate leprosy) that is often self-healing or may progress to the paucibacillary or multibacillary stages. The progression and evolution of the infection is influenced by many factors: socioeconomic status, immune status of the host, genetic factors, and concomitant infections. For instance, a previous infection with *M. tuberculosis* may boost the immune system, diminishing the chances of developing leprosy (105). BCG vaccination seems to provide protection against leprosy, but the figures proving it are quite different, ranging from 20% to 80% (106). Conflicting results are reported also about the influence of concomitant HIV infection on the development of leprosy. So far there is no indication that the HIV epidemic is causing an increase in the number of leprosy patients. It is possible that the incubation period for leprosy is too long to discover the relationship between the two diseases (102).

The influence of genetics is documented by familial clustering (107) by the link between NRAMP1 gene (108) and susceptibility to leprosy, as well between HLA genes and the type of leprosy (109, 110).

The Ridley and Jopling classification (111) correlates the clinical manifestations of leprosy with cell-mediated immune response. The types of disease defined as indeterminate (I) and tuberculoid (TT) occur in patients who exhibit a relatively good cell-mediated immunity against *M. leprae*, borderline (BT), mid-borderline (BB), and borderline lepromatous (BL) in patients with less effective cell-mediated immunity, and lepromatous (LL) in patients anergic to *M. leprae*. In the last classes of patients, the disease is widespread and involves the skin and upper respiratory tract, the anterior chamber of the eye, the testes, the lymph nodes, the periosteum, and the superficial sensory and motor nerves. The WHO classification is simpler. Leprosy is subdivided in two types: paucibacillary corresponding to I, TT, and BT and multibacillary corresponding to BB, BL, and LL (112–114). Paucibacillary leprosy is defined by five or fewer skin lesions with no bacilli on skin smears. The bacterial index can range from 0 (none found in 100 oil-immersion fields) to 6+ (over 1000 bacilli per field). Multibacillary cases have six or more skin lesions and may be skin-smear positive (100).

The use of polymerase chain reaction (PCR) for *M. leprae* DNA in skin biopsies to increase sensitivity and specificity of diagnosis has proved to be of limited use. Sensitivity is high in multibacillary cases but low in paucibacillary cases. The number of false positives is also high (115–117). At present, it represents an interesting research tool for some particular investigations.

Leprosy was an epidemic disease in Europe in medieval times but now is confined to some tropical areas, especially India, the Philippines, South America, and tropical Africa. Social and economic poverty is the main reason for the prevalence of this disease, which has the greatest distribution in underdeveloped countries. Chemotherapy offers a great possibility for eradication of the disease and multidrug therapy gave very favorable results.

The success of the multidrug therapeutic regimens recommended by WHO in 1981 led the WHO Assembly in 1991 to set a goal of elimination of leprosy as a public health problem by the year 2000 (118). Elimination was defined as a prevalence of ≤1 case for 10,000 population, with a case being defined as a patient receiving or requiring chemotherapy.

This goal has not been reached and it has been pushed back to 2005, but a significant decline in incidence has occurred in some countries such as China and Mexico.

The estimates of total cases of leprosy have fallen from 10–12 million to about 5.2 million in 1991. In 1999 about 800,000 cases were registered for treatment worldwide. About the same number of cases was detected in 1998. About 75% of registered patients live in South-

East Asia, particularly in India, where the rate of cases per 100,000 population is 5.3 and the absolute number of patients is >500,000. However there are still 32 countries that have a leprosy prevalence of >1 case per 10,000 population (104, 112–114).

## 4  SCREENING AND EVALUATION OF ANTIMYCOBACTERIAL AGENTS

In the search for new antimycobacterial agents, demonstration of *in vitro* activity against the virulent strain of *M. tuberculosis* H37 Rv is one of the simplest preliminary tests. Although much more predictive than other *in vitro* models using avirulent or fast-growing mycobacteria (*M. smegmatis, M. phlei*), the *in vitro* test with *M. tuberculosis* gives a large number of false positives, and unfortunately, also some false negative results. Despite these limitations, the *in vitro* test, with various modifications of the inoculum size, the culture media, and the observation time, is still used in many laboratories for the blind primary screening of a large number of compounds. It is also used in antibiotic screening, where thousands of fermentation broths must be tested, and a primary screen using *in vivo* models is a practical impossibility.

Rapid susceptibility testing of *M. tuberculosis* can be performed using the radiometric assay of $^{14}CO_2$ produced by the action of the microorganism on $^{14}C$-palmitic acid substrate. Results from this method, which are useful for screening new chemical compounds, are available after 4–10 days of incubation instead of the 20–30 days required by conventional culture media (119, 120).

To avoid the use of radioactive culture medium, the firefly bioluminescence is used to detect ATP during mycobacterial metabolism, comparing the ATP production in drug-containing broths with that of the control broths (121).

Another test, devised to shorten the time required for sensitivity testing, uses mycobacteria infected with a specific reporter phage expressing the firefly luciferase gene. The photoreaction produced by the reaction of luciferin with ATP allows the measurement of bacterial growth or inhibition by drugs. The assays requires small amounts of test compounds and is designed to screen a large number of antimycobacterial products (122).

The *in vitro* tests give only an indication of activity; quantitative evaluation of the potential usefulness of the new products must be obtained through *in vivo* tests. These are performed, generally speaking, by inoculating virulent mycobacteria strains into laboratory animals, administering the product to a group of them, and comparing the course of infection in treated and untreated animals. There are a variety of procedures for performing these tests, differing with respect to animal species (mouse, guinea pig, rabbit, etc.), mycobacterial strain and size of inoculum, route of product administration, and evaluation of the results. The most current procedure for the evaluation of antituberculous drugs uses mice infected with the human virulent mycobacterial strain, evaluating the results in terms of $ED_{50}$, survival time, the pathology of the lung, and bacterial count. The products active in the mice are then evaluated in other *in vivo* tests using more sophisticated techniques.

The best species for extrapolation of the results to humans is the rhesus monkey, *Macaca mulata* (123, 124). Experimental tuberculosis in this species closey parallels the human disease, and despite the difficulties in terms of time, space, and cost of the test, it is advisable to perform it, especially when doubtful results have been obtained from other species. In any case, extrapolation to the human disease of the results obtained in animals requires comparison of the kinetics and metabolism of the product in the different animal species and in humans. Differences in activity are sometimes clearly related to differences in the metabolic behaviors.

In the case of leprosy, for a long time no screening or evaluation models were available using the pathogenic agent *M. leprae*, which could not be cultivated *in vitro* or transmitted to animals. Therefore, the experimental infection of rodents with *M. lepraemurium* was used for evaluating the effect of potential drugs, although this model shows a low predictivity for activity in humans. For example, dapsone is inactive and isoniazid very active in this test, whereas the opposite is true for human leprosy. Only in 1960 was local infection in the mouse footpad with *M. leprae* set up (125). This model has been successfully used, with various procedural modifications, for screening and evaluation of drugs (126, 127). Thymectomy and body irradiation of mice inoculated with *M. leprae* provokes dissemination of bacilli, and this may be a model for a generalized infection (128). Another model of experimental leprosy was set up using the armadillo, which develops a severe disseminated lepromatoid disease several months after inoculation with a suspension of human leprosy bacilli (129). Limitations in the use of this model for chemotherapeutic evaluation are derived not only from its cost, but also from the importance to reserve the infected animals as a source of bacillary material for the development of a vaccine.

All the available animal models for the evaluation of antileprotic agents are too time-consuming for the screening of a large number of compounds. The search for short-term models is necessary. More recently it was reported that an *in vitro* culture system for *M. leprae* in cells harvested from livers of nine banded armadillos, inoculated earlier with human-derived microorganism had been created. The system seems to be adequate for initial screening of the drugs (129).

Activity of products against NTM is generally tested *ad hoc*, *in vitro*, or *in vivo*, and the compounds selected for this purpose are initially those showing antitubercular action, although their activity is quite variable against the various mycobacterial species. In recent years there has been increasing interest in identifying reliable *in vitro* susceptibility test methods against *M. avium* complex because of the severity of the infections caused by this microorganism and the increase of their frequency (130).

In addition to the standard *in vitro* assays, some *in vitro* studies have been carried out in presence of macrophages to determine the activity of drugs against *M. avium* complex and other NTM when they are located intracellularly (131–133). A model of chronic disseminated *M. avium* complex infection, developed in beige mice, revealed to be a useful tool in studying the *in vivo* activity of antimycobacterial drugs (134).

The inclusion of a number of NTM in the large antibacterial screening programs of new compounds could provide leads for finding new, more active chemotherapeutic agents for the diseases caused by these microorganisms. On the other hand, there is certainly need for more knowledge on the biochemistry and physiology of NTM and on their pathogenic behavior in laboratory animals.

Using the laboratory models available for testing and evaluating products, there are two possible approaches in the search for antimycobacterial drugs. The first approach is the blind screening of a large number of compounds, which permits the detection of a certain number of structures endowed with antimycobacterial activity. Chemical modification of these "lead" structures, accompanied by careful studies of structure-activity relationships (SARs), can yield the optimal derivative for therapeutic use. The second approach, more challenging from the scientific point of view, is based on designing drugs to act selectively on biochemical targets specific for the particular microorganism.

The development of new antimycobacterial drugs is made difficult by a number of factors (135). First, large scale screening systems for the detection of new antimycobacterial agents are particularly time-consuming and entail some problems related to the handling of the pathogens. Second, the development of an antimycobacterial drug takes more time and human resources than the development of other antimicrobial agents. Third, and probably most important, tuberculosis and leprosy are predominant in developing countries with low economic resources, and industrial laboratories are reluctant to invest in research for new products to be used in those geographic areas, where an additional drawback is the lack of patent protection. However, the actual situation makes the search for new and more effective antimycobacterial drugs necessary.

Although screening for antimycobacterial agents will continue to be a possible way to discover useful new drugs, the increasing knowledge of the biochemistry of the mycobacteria makes possible a more rational approach to the problem. In particular, the studies of the biosynthesis of the unique constituents of the mycobacterial cell will indicate the targets for inhibitors of the biosynthetic pathways present specifically in the microorganisms of the mycobacterium species.

The complete sequencing of *M. tuberculosis* genoma opens new opportunities to find the basic targets of the activity of antimycobacterial agents and possibly the way to design drugs specifically directed to inhibit essential enzymes. However precise functions have been attributed to about 40% of the encoded proteins, whereas there is only some information for 44%. Therefore, the horizon for screening unexploited antimicrobial targets is broad. The discovery of selective inhibitors will allow to have products ineffective on the eukariotic cells of the host, therefore, potentially with little toxicity and no activity on the normal microbial flora. At present it has been demonstrated that the fatty acid synthase (FAS II) may be the target of isoniazid (136).

A novel approach to identify new antimicrobial agents has been the application of combinatorial chemistry, which creates molecules starting from a rational building block. It represents a powerful tool for a rapid and automated screening of active components and for the creation of libraries including enormous series of derivatives.

Although some important clues have appeared in the field of specific biochemical pathways inside the mycobacteria and of the mechanism of action of antimycobacterial drugs, little progress has been made till now in overcoming the problem of drug resistance in mycobacteria. *M. tuberculosis* mutates spontaneously and randomly to resistance to isoniazid, streptomycin, ethambutol, and rifampicin (137), and that is in a genotype form.

The knowledge of the mechanism of mycobacterial resistance to a drug could indicate the direction in which to search for new products specifically overcoming this mechanism. In the case of aminoglycoside antibiotics, the mechanism of resistance in *Enterobacteriaceae* and some *Pseudomonas* strains has been extensively studied, and the inactivating enzymes have been identified. This information has allowed a rational design of chemical modifications of aminoglycoside antibiotics at the site of attack of inactivating enzymes giving new products that were active against resistant strains (138).

Unfortunately, strides toward understanding the molecular basis of mycobacterial resistance to the various drugs have been made only rather recently. The status of knowledge of the resistance mechanisms is not sufficient to conceive a rational design of molecules acting on the mutants resistant to the specific drugs. Also the knowledge of the genetic mechanisms and genotypical systems governing resistance in mycobacteria is limited because of the difficulty of using mycobacteria in classical genetic techniques (slow growth, lipid-rich cell walls impermeable to DNA uptake). Only during the last decade, application of recombinant DNA technologies to the mycobacteria has opened new doors to obtain information on some basic mechanisms of resistance to drugs. However, at present, the delay or prevention of the evolution toward resistance of mycobacterial flora is accomplished only through combination therapy, based on the complementary action of the constituents. It is well known that in this case the probability of a concomitant mutation toward resistant strains is much lower than the probability of mutation to resistance to a single drug.

## 5  CURRENT DRUGS ON THE MARKET

Among the several thousand compounds screened for antimycobacterial activity, only a few have had therapeutic indices sufficient to warrant introducing them into clinical use (Table 16.1). These drugs are described in some detail, together with general information about the chemical analogs, mechanism of action, pharmacokinetics and metabolism, clinical use, and effects. The information, in summary form, is intended to cover the aspects that are useful for understanding the role of each product in current therapy, the limitations of use, and the need for improvement or for further studies.

The drugs have been subdivided arbitrarily into synthetic products and antibiotics, and within these two categories they are listed in a quasi-chronological order, with products having structural similarities grouped with the representative first introduced into therapy.

## 5.1  Synthetic Products

### 5.1.1  Sulfones

*5.1.1.1 History.*  The sulfones were synthesized by analogy to sulfonamides, which had no antimycobacterial activity. The first sulfones prepared, 4, 4′-diaminodiphenylsulfone (dapsone) (**9**) and its glucose bisulfite deriva-

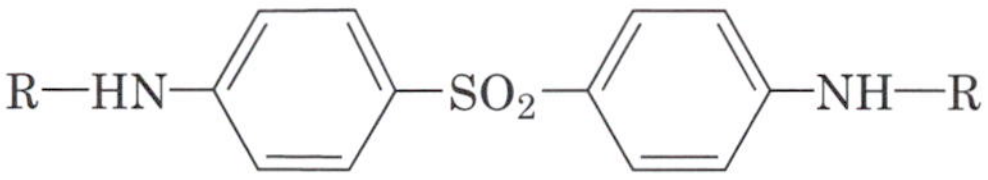

$$R-HN-\underset{}{\bigcirc}-SO_2-\underset{}{\bigcirc}-NH-R$$

(**9**)   R = H

(**10**)  R = CH(CHOH)$_4$CH$_2$OH
         |
         SO$_3$Na

(**11**)  R = CH$_2$SO$_2$Na

(**12**)  R = CH—CH$_2$—CH—C$_6$H$_5$
             |              |
             SO$_3$Na        SO$_3$Na

(**13**)  R = COCH$_3$

tive, glucosulfone sodium (**10**), were found to be active in suppressing experimental tubercle infections (139–140). The usefulness in the chemotherapy of human tuberculosis was limited, but the discovery of some effect of compound (**10**) in leprosy experimentally induced in rats (141) opened the way to their successful introduction into the treatment of human leprosy.

*5.1.1.2 SARs.*  Because it was thought that the activity of compound (**10**) was caused by its metabolic conversion into dapsone, intensive studies have been carried out that vary the structure of the latter to find optimal activity and to improve its low solubility. Various substitutions on the phenyl ring yielded products less active than dapsone. The only product of this type that has some clinical use is acetosulfone sodium, which contains one SO$_2$N(Na)—COCH$_3$ group in the ortho position to the sulfone group. Its antibacterial effect seems to be caused by the unchanged drug. Substitutions in both the amino groups gave rise to products that are in general active only if they are converted metabolically into the parent dapsone.

Substitution on the amino groups to improve solubility yielded products such as the above-mentioned glucosulfone, the methanesulfonic acid derivative, sulfoxone sodium, aldenosulfone (**11**), and the cinnamaldehyde-sodium bisulfite addition product, sulfetrone sodium, solasulfone (**12**), which have limited use in leprosy treatment. They act through their metabolic conversion to dapsone in the body. Several methanesulfonic acid derivatives of 4, 4'-diamino-diphenyl sulfone have been tested for their ability to be metabolized to dapsone (142).

The 4,4'-diacetyldiaminodiphenylsulfone, acedapsone (**13**), has low activity *in vitro* but is used as an injectable depot sulfone, which releases dapsone at a steady rate over several weeks.

Although a large number of sulfones have been synthesized and tested as potential antileprotic agents, dapsone continues to be the basic therapeutic agent for *M. leprae* infections, often as a component of multidrug programs.

### 5.1.1.3 Activity and Mechanism of Action.
Dapsone is weakly bactericidal against *M. leprae* at concentration estimated to be of the order of 0.003 $\mu$g/mL (143), when the microorganism has been isolated from untreated patients. *M. leprae* becomes resistant to dapsone and its congeners after chronic administration.

It is assumed that dapsone interferes with incorporation of *p*-aminobenzoic acid into dihydrofolate, in analogy with the action of sulfonamides in other bacterial systems This mechanism of action of dapsone has been proved for *E. coli* (144), and the finding of cross-resistance to sulfones and sulfonamides in *Mycobacterium* species 607 indirectly confirms their similarity of action (145). Unfortunately, the inability to cultivate *M. leprae in vitro* makes it difficult ultimately to verify that dapsone acts through the proposed mechanism on this organism.

### 5.1.1.4 Pharmacokinetics.
Dapsone is usually administered orally at a daily dose of 100 mg (118). It is nearly completely absorbed from the gastrointestinal (GI) tract, well distributed into all tissues, and excreted in high percentage in the urine as mono-*N*-sulfamate and other unidentified metabolites (143). It is

monoacetylated in humans. The characteristics of dapsone acetylation parallel those of isoniazid and sulfametazine, thereby establishing genetic polymorphism for the acetylation of dapsone in humans (146). Two metabolic factors, greater acetylation and greater clearance of dapsone from the circulation, may contribute to emergence of dapsone-resistant *M. leprae* (147).

Acedapsone is administered intramuscularly in oily suspension at a dose of 225 mg every 7 weeks, The product is slowly released from the site of injection, and the plasma contains mainly dapsone and its monoacetyl derivative; the ratio of these two products depends on whether the patient is a slow or rapid acetylator, as in the case of dapsone. The advantage of this type of depot usage was also considered for the prophylaxis of people exposed to risk (148).

### 5.1.1.5 Adverse Effects.
Dapsone and its derivatives may be administered for years with certain precautions: gradual increase of doses, rest periods to prevent cumulative effects, and laboratory and clinical supervision. Among the most common side effects are gastrointestinal and central nervous system disturbances. Also, the most common effect is hemolysis in varying degrees. There are indications that individuals with a glucose-6-phosphate dehydrogenase deficiency are more susceptible to hemolysis during sulfone therapy, although this is controversial (149).

## 5.1.2 Para-amino-salicylic Acid
### 5.1.2.1 History.
The observation that benzoates and salicylates have a stimulatory effect on the respiration of mycobacteria (150) suggested that analogs of benzoic acid might interfere with the oxidative metabolism of the bacilli. When various compounds structurally related to benzoic acid were tested, it was found that some of them had limited antituberculous activity; *p*-aminosalicylic acid (PAS) (**14**) was the most active (151). The discovery of PAS cannot be quoted as an example of biochemically oriented chemotherapeutic research, because in fact, its mechanism of action is not by way on the respiration of mycobacteria.

### 5.1.2.2 SARs.
The *in vitro* and *in vivo* antimycobacterial activity of a simple molecule

COOR
OH

NHR'

| | R | R' |
|------|----------|------------------------|
| (14) | H | H |
| (15) | $C_6H_5$ | H |
| (16) | H | $COC_6H_5$ |
| (17) | H | $COCH_3$ |
| (18) | H | $COCH_2NH_2$ |

such as PAS stimulated the synthesis and testing of many derivatives. This extensive research failed to give rise to better drugs but did provide knowledge of the structural requirements for activity in the series.

Modification of the position of the hydroxy and amino groups with respect to the carboxy group resulted in a sharp decrease of activity. The amino group confers a distinct pharmacodynamic property to the molecule, eliminating the antipyretic and analgesic activities of salicylic acid and giving the specific tuberculostatic activity. Further nuclear substitution and replacement of the amino, hydroxy, or carboxy groups with other groups yielded inactive or poorly active products. Also, functional derivatives on the amino, hydroxy, and carboxy groups of PAS are generally inactive, unless they are converted *in vivo* into the active molecule. Among the latter, the phenylester (15) and benzamidosalicylic acid (16) must be mentioned.

### 5.1.2.3 Activity and Mechanism of Action.
PAS has bacteriostatic activity *in vitro* on *M. tuberculosis* at a concentration of the order of 1 $\mu$g/mL. It is active only against growing tubercle bacilli, being inactive against intracellular organisms. Other microorganisms are not affected by the compound. Most of the other mycobacteria are insensitive to the drug. Although active against *M. leprae* in the mouse footpad test (152), is not used in the current treatment of leprosy.

The mechanism of action of PAS is not clear. It was demonstrated earlier that *p*-aminobenzoic acid antagonizes the antibacterial activity of PAS *in vitro* and *in vivo*, and therefore, it could act by competitively blocking the synthesis of dihydrofolic acid. On the other hand the formation of mycobactin, an ionophore for iron transport, is strongly inhibited by PAS, and the bacteriostatic activity of PAS might be caused by the inhibition of the metabolic pathway for iron uptake (153–155).

### 5.1.2.4 Pharmacokinetics.
PAS is readily absorbed by the gastrointestinal tract and well distributed throughout the body, but it does not penetrate into cerebrospinal fluid (CSF) of patients with uninflamed meninges. Peak serum levels of 7–8 $\mu$g/mL are reached 1–2 h after oral administration of 60 mg/kg of PAS. The half-life time of serum levels is very short (0.75 h) It is quickly eliminated through the urine in form of inactive metabolites, the *N*-acetyl (17) and the *N*-glycyl (18) derivatives. The common daily dose for adults is around 10–20 g, generally divided in two or four doses administered after meals, according to patient tolerance. In children the usual daily dose is 0.2–0.5 g/kg. Probenecid interferes with renal tubular secretion of PAS and increases the serum levels of the drug.

Although PAS easily undergoes decarboxylation *in vitro*, it seems quite stable in the body. The *in vivo* acetylation of PAS competes with acetylation of isoniazid, resulting in higher plasma concentration of free isoniazid than when isoniazid is given alone. This fact has no proven clinical relevance.

### 5.1.2.5 Adverse Effects.
PAS has been used for many years in combination with streptomycin and isoniazid for the treatment of all forms of tuberculosis.

Gastrointestinal irritation is a very common side effect, with manifestations of various degrees of intensity and severity, which may lead to discontinuation of treatment. Hypersensitivity reactions occur in about 5–10% of patients, usually with manifestations like rash, fever, and pruritus, rarely followed by exfoliative dermatitis or hepatitis of allergic nature. PAS should be used with caution in patients with renal diseases, because it is largely excreted in the urine.

The therapeutic use of PAS is declining because of the high incidence of side effects and poor patient acceptance and the introduction of more potent and safer antituberculous drugs. However it is still used in developing countries because it is inexpensive.

### 5.1.3 Thioacetazone

*5.1.3.1 History.* The synthesis of a series of thiosemicarbazones as intermediates in the preparation of analogs of sulfathiadiazole (156–157), which has weak antituberculous activity (158), led to the discovery of the thiosemicarbazone of p-acetamidobenzaldehyde (thioacetazone, amithiozone, **19**), the most ac-

$$H_2NCSNHN=HC-\phantom{xx}-NHCOCH_3$$

**(19)**

tive substance *in vitro* and *in vivo* of the series (159–161).

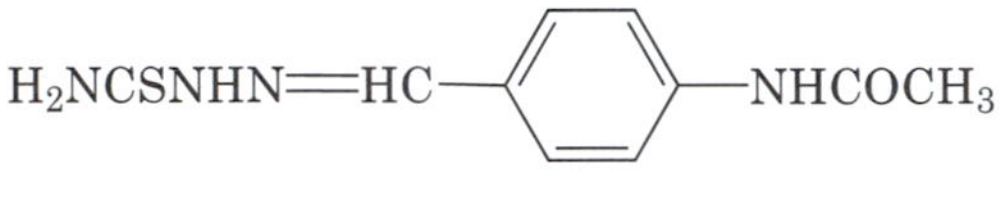

*5.1.3.2 SARs.* In the search for better products, many modifications of the thioacetazone molecule have been made. The studies clearly indicate that the activity resides in the thiosemicarbazone structure of the aromatic aldehydes. In fact, several products with modified aromatic moieties, including some heterocyclic nuclei, have been found to be active. Some of them have been tested clinically with positive results, but thioacetazone is the only thiosemicarbazone still in clinical use.

Several thioureas that are structurally related to thiosemicarbazones have been found to be active *in vitro* and *in vivo* against *M. tuberculosis*, the most active being the diphenylthioureas with p-alkoxy groups in one or both the aromatic nuclei. Of these, thiocarlide (**20**) and thiambutosine (**21**) have been studied clinically with disappointing results. Thiambutosine was recommended for leprosy pa-

tients who do not tolerate dapsone (162), but the last WHO report on chemotherapy of leprosy does not mention thiambutosine among the currently available antileprosy drugs (118).

*5.1.3.3 Activity and Mechanism of Action.* Thioacetazone is active against *M. tuberculosis* at concentration of 1 $\mu$g/mL. Its activity in tuberculous animal infections is comparable with that of streptomycin. Its action is bacteriostatic, and the development and frequency of resistant strains is rather high. Some of the thioacetazone-resistant strains are also resistant to ethionamide (163). The mechanism of action of thioacetazone is not known. Its tuberculostatic activity is not counteracted by p-aminobenzoic acid (164), and there is a partial cross-resistance to antituberculous thioureas.

Its good activity *in vitro* and *in vivo*, and the lack of cross-resistance to isoniazid and streptomycin, indicated the use of thioacetazone as a drug to combine with these drugs to delay bacterial resistance.

*5.1.3.4 Pharmacokinetics.* Thioacetazone is well absorbed from the gastrointestinal tract (165). Oral administration of 150 mg of thioacetazone gives serum levels of 1.6 $\mu$g/mL at peak and of 0.2 $\mu$g/mL after 2 days. Large amounts of thioacetazone are excreted in the urine, but there is not enough information on its metabolism.

*5.1.3.5 Adverse Effects.* Despite relatively low toxicity in laboratory animals, thioacetazone has limitations in clinical use because of serious side effects (such as gastrointestinal disorders, liver damage, and anemia) when administered to humans at the initial proposed daily dose of 300 mg. A review of the side ef-

$$\text{iso-}C_5H_{11}O-\phantom{xx}-NHCSNH-\phantom{xx}-OC_5H_{11}\text{-iso}$$

**(20)**

$$n\text{-}C_4H_9\text{-}O-\phantom{xx}-NHCSNH-\phantom{xx}-N(CH_3)_2$$

**(21)**

fects and efficacy of thioacetazone in relation to the dosage indicates that the drug has an activity comparable with that of PAS and an acceptable toxicity when administered at lower dosage (166).

A dose of 300 mg of isoniazid plus 150 mg of thioacetazone is an inexpensive and acceptable combination for long-term therapeutic treatment after the initial treatment with three drugs. This schedule is used in developing countries (167), although there are considerable differences in side effects among patients from different geographic areas (168). The reported effectiveness in leprosy (169) notwithstanding, thioacetazone is of limited usefulness in the treatment of this disease because of its side effects and the emergence of resistance of *M. leprae* to the drug.

### 5.1.4 Isoniazid

*5.1.4.1 History.* After the early report that nicotinamide possesses tuberculostatic activity (170–171), several compounds related to it were examined. Attention was aimed at the isonicotinic acid derivatives, and in view of the already established antituberculous activity of thiosemicarbazones, the thiosemicarbazone of isonicotinyl aldehyde was prepared. When isonicotinyl hydrazide (isoniazid, **22**), described in the chemical literature in 1912 (172), was prepared as an intermediate in the synthesis of the aldehyde and tested, it proved to be a potent antitubercular agent *in vitro* and *in vivo* (173–175).

*5.1.4.2 SARs.* The outstanding antituberculous activity of isoniazid in experimental infections, confirmed by the clinical trials, stimulated the study of chemical modifications of this simple molecule. At least 100 analogs were prepared, but structural changes caused a reduction in or loss of activity. Among the modified forms that retained appreciable activity, the *N2*-alkyl derivatives should be mentioned. In particular the *N2*-isopropyl derivative (iproniazid, **23**), was found to be active *in vivo*. Extensive clinical trials proved the therapeutic effectiveness of iproniazid and revealed its psychomotor stimulant effect, caused by the inhibition of monoamine oxidase (176). Use of iproniazid in the treatment of tuberculosis or of psychotic and neurotic depression was discontinued because of the

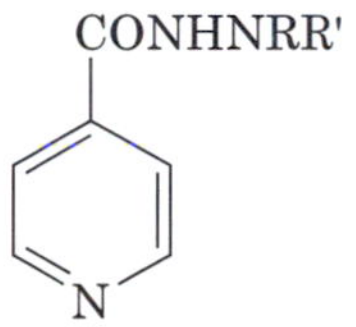

|        | R | R'            |
|--------|---|---------------|
| (22)   | H | H             |
| (23)   | H | $CH(CH_3)_2$  |

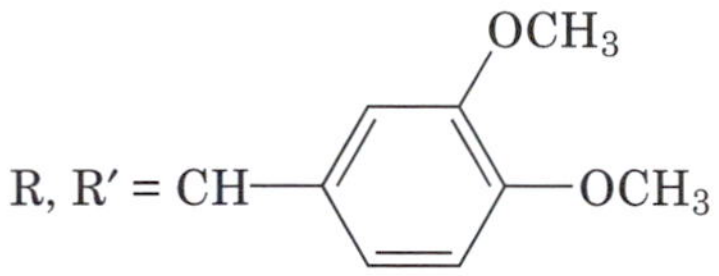

(24)

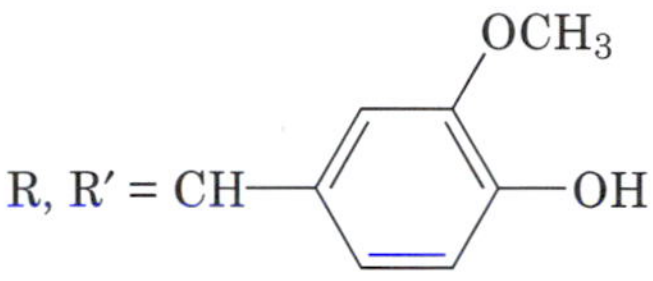

(25)

hepatic toxicity of the drug. Although acetyl isoniazid is inactive, its hydrazones constitute a group of isoniazid congeners that have activity of the same order of the parent compound. The activity of these compounds generally is related to the rate of their hydrolysis to the parent compound. Some hydrazones have been introduced into therapeutic use, such as the 3,4-dimethoxybenzilidene (verazide, **24**) and the 3-methoxy-4-hydroxy-benzilidene (phthivazid, **25**) derivatives; however, their use is questionable, and products of this kind now have limited or no application.

*5.1.4.3 Activity and Mechanism of Action.* Isoniazid has bacteriostatic and bactericidal activity *in vitro* against *M. tuberculosis* and also against strains resistant to other antimycobacterial drugs. The minimal inhibitory concentration for the human strain is about 0.05 μg/mL. It acts on growing cells and not on resting cells and is also effective against intracellular bacilli. Its effect on the non-tuberculous mycobacteria is marginal or non-existent. Isoniazid is active in various models of experimental tuberculosis in animals. It shows limited activity against *M. leprae* in the mouse footpad test (177), and it is essentially inactive in human leprosy (178).

Resistance of *M. tuberculosis* to isoniazid develops rapidly if it is used alone in the treatment of clinical infection and can be prevented or delayed by combination with other antimycobacterial agents. Resistance does not seem to be a problem when isoniazid is used alone in prophylaxis, probably because the bacillary load is low. It is recommended as a single drug in tuberculosis chemoprophylaxis, especially in household contacts and close associates of patients with isoniazid-sensitive tuberculosis.

Several hypotheses have been put forward concerning the mechanism of action of isoniazid, taking into account that the activity of this drug is very specific against mycobacteria at low concentrations. Investigations in this direction have indicated that the action of isoniazid is on the biosynthetic pathway to the mycolic acids (179–180). Apparently, isoniazid blocks the synthesis of fatty acids longer than C26 in chain length (181).

Another hypothesis suggests that it is the isonicotinic acid the responsible for the inhibitory effect of isoniazid on mycobacteria (182–

184). Isoniazid is said to penetrate the cell, where it is hydrolyzed enzymatically to isonicotinic acid, which at the intracellular pH is nearly completely ionized and cannot return across the membrane and accumulates inside the cell. Isonicotinic acid is then quaternized and competes with nicotinic acid through the formation of an analog of the nicotinamide-adenine dinucleotide, which does not have the activity of the natural coenzyme. Alteration of the metabolic functions of the cell follows, particularly with respect to lipid metabolism.

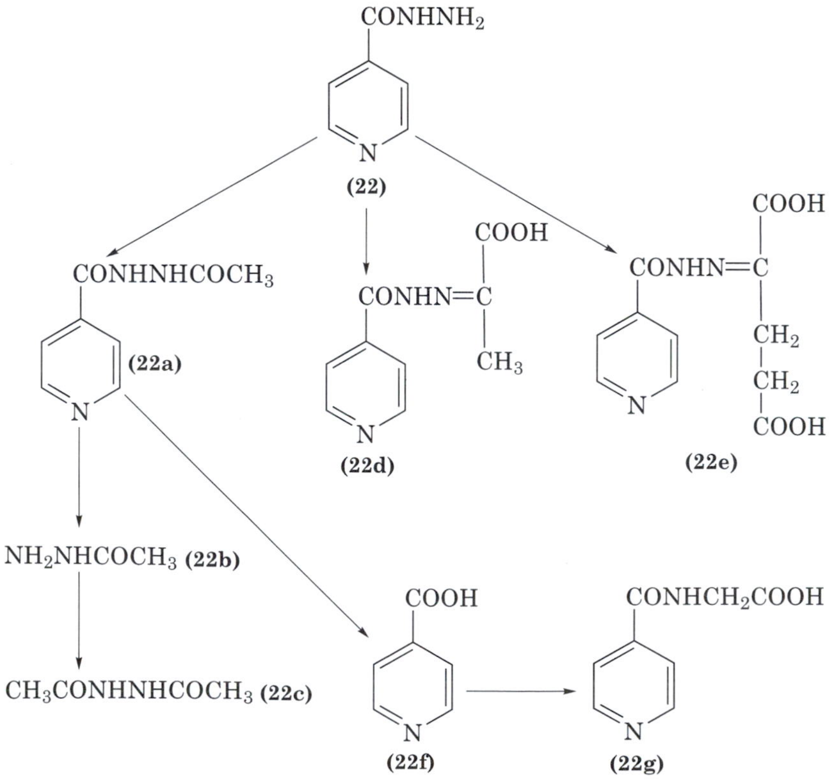

More recently it has been discovered that the activity of isoniazid results from a peroxidative reaction catalysed by a catalase-peroxidase, which is encoded by the *kat G* gene (185). In fact, clinical isolates of *M. tuberculosis* highly resistant to isoniazid lack *kat G* (186). The most common and potent mechanism of resistance to INH involves loss of *Kat G* activity to prevent activation of the drug. The *kat G*–mediated oxidation of INH to a potent electrophile proceeds at the expense of hydrogen peroxide, which acts as an electron sink for the reaction (187). The activated INH electrophile is then free to interact with any number of cellular nucleophiles poisoning the nicotinamide biosynthetic pool. The discovery of a novel gene, *inh A*, may complete the understanding of the mechanism of action of isoniazid. The product of *inh A* has a certain analogy with the enterobacterial *env M* enzyme, which is associated with the biosynthesis of fatty acids, phospholipids, and lipopolysaccharides (188). The inh A protein could be the primary target of action of isoniazid, but there is also the possibility that *inh A* requires NAD(H) as a coenzyme and that its activity can be affected by the incorporation of iso-NAD produced by action of catalase-peroxidase on isoniazid, leading to a block of the synthesis of mycolic acid and loss of acid fastness. It is still to be clarified the fact that *inh A* and *kat G* are also present in the non-tubercular mycobacteria naturally resistant to isoniazid. In *M. tuberculosis*, the two genes are altered in isoniazid resistant isolates (189).

*5.1.4.4 Pharmacokinetics.* Isoniazid is readily absorbed from the gastrointestinal tract in humans. Peak serum levels of the order of 5 $\mu$g/mL are obtained 1–2 h after administration of a 5-mg/kg dose. After absorption, isoniazid is well distributed in body fluids and tissues, including the cerebrospinal fluid, and penetrates into the macrophages. Isoniazid is excreted mainly in the urine in unchanged form, together with various inactive metabolites: *N*-acetyl isoniazid (**22a**), monoacetyl hydrazine (**22b**), diacetyl hydrazine (**22c**), isoniazid hydrazones of pyruvic acid (**22d**), and $\alpha$-ketoglutaric acid (**22e**), isonicotinic acid (**22f**), and isonicotinylglycine (**22g**).

The primary metabolic route that determines the rate at which isoniazid is eliminated from the body is acetylation in the liver to acetyl isoniazid. There are large differences among individuals in the rate at which isoniazid is acetylated. The acetylation rate of isoniazid seems to be under genetic control (190–191), and individuals can be slow or rapid acetylators of the drug. The serum half-lives of isoniazid in a large number of subjects show a bimodal distribution; the isoniazid half-lives of rapid metabolizers range from 45 to 110 min and those of slow metabolizers from 2 to 4.5 h (192). The rate of acetylation seems to be conditioned by ethnic background. The proportion of slow acetylators varies from 10% among the Japanese and Eskimos to 60% among blacks and Caucasians. The isoniazid acetylator status of tuberculosis patients treated with isoniazid-containing regimens seems to be relevant only for once-weekly treatments with the drug (193).

*5.1.4.5 Adverse Effects.* Isoniazid is usually well tolerated for a long period of treatment. Hepatic side effects consist of frequent subclinic asymptomatic enzyme abnormalities (increase of SGOT and bilirubin), which occasionally cause severe clinical hepatitis, especially in patients with previous hepatobiliary diseases and in alcoholics. The risk of hepatitis is age-related, being very rare in children or young people. Neurological side effects, such as peripheral neuritis, are rare at the daily dose of 5 mg/kg, but more frequent at doses of 10 mg/kg. Administration of pyridoxine to patients receiving high doses of isoniazid generally prevents neurological disturbances. Other rare effects are at the GI level and moderate hypersensitivity reactions.

### 5.1.5 Ethionamide

*5.1.5.1 History and SARs.* As mentioned before, the discovery of isoniazid was the consequence of research based on the weak antitubercular activity of nicotinamide. This lead was pursued in various directions, and among the earliest modifications, thioisonicotinamide (**26**) (194–196) seemed to have the in-

CSNH₂

(26)  R = H
(27)  R = $C_2H_5$
(28)  R = *n*-$C_3H_7$

triguing property of an *in vivo* efficacy superior to that expected from the *in vitro* activity. The hypothesis that some metabolic product of the drug was responsible for the activity *in vivo* stimulated the synthesis as well the testing of a series of potential thioisonicotinamide metabolites and various other derivatives. Among the latter, increased activity was observed for the 2-alkyl derivatives (197). 2-ethyl thioisonicotinamide (ethionamide, **27**) and the 2-*n*-propyl analog (prothionamide, **28**) were selected for clinical use. Of these two drugs, ethionamide has been more extensively studied, although prothionamide seems to possess biological properties similar to it.

**5.1.5.2 Activity and Mechanism of Action.** At concentrations around 0.6–2.5 $\mu$g/mL, ethionamide is active *in vitro* against *M. tuberculosis* strains, either sensitive or resistant to isoniazid, streptomycin, and *p*-aminosalicylic acid. It also shows activity against other mycobacteria, especially *M. kansasii*. Administered orally, ethionamide is effective in the treatment of experimental tuberculosis in animals. Although activity against *M. leprae* in animal infections has been reported, ethionamide is rarely used in the therapeutic treatment of leprosy. Bacterial resistance develops quickly when ethionamide is given alone; therefore, it is used in combination with other antimycobacterial drugs.

The antibacterial action of ethionamide seems to be caused by an inhibitory effect on mycolic acid synthesis, with a concomitant effect on nonmycolic acid-bound lipids (198). This pattern is like that shown by isoniazid. However, other studies indicate that ethionamide disturbs the synthesis of mycolic acid in both resistant and susceptible mycobacteria, whereas isoniazid inhibits the synthesis of all kinds of mycolic acid in the same way in all susceptible strains and has no effect on mycolic acid synthesis in resistant strains (199). On the other hand, a missense mutation within the mycobacterial *inh A* gene was shown to confer resistance to both isoniazid and ethionamide in *M. smegmatis* and *M. bovis* (188).

**5.1.5.3 Pharmacokinetics.** In case of tuberculosis, ethionamide is given orally at doses varying from 125 mg to a maximum of 1 g daily. It is rapidly absorbed and widely distributed in the body tissues and fluids, including the CSF. It has a short half-life and is rapidly excreted in the urine, but only a minor percentage is in the form of unaltered product. A series of metabolites has been found in the urine: the active sulfoxide, the 2-ethyl isonicotinic acid and amide, and the corresponding dihydropyridine derivatives (200).

**5.1.5.4 Adverse Effects.** GI side effects, sometimes very severe, are common and constitute the major cause for discontinuation of the treatment with ethionamide. Neurotoxicity with manifestations of mental disturbances is also relatively frequent during the treatment with ethionamide.

**5.1.6 Pyrazinamide**

**5.1.6.1 History.** In studies of chemical modifications of the nicotinamide structure, other heterocyclic nuclei have been investigated, and the 2-carboxamidopyrazine (pyrazinamide, **29**) was synthesized and tested for

**(29)** R = H
**(30)** R = CH$_2$- N ⟨ ⟩ O

antituberculous activity (201–203). The *in vitro* activity of pyrazinamide against *M. tuberculosis* was found to be negligible at neutral pH and of the order of 5–20 $\mu$g/mL at pH 5.5. The best activity of pyrazinamide is against intracellular mycobacteria in monocytes, probably because of the low intracellular pH, which favors its activity.

**5.1.6.2 SARs.** Compounds with other substitutions on the pyrazine nucleus or other carboxamido heterocycles were found to be inactive or less active than pyrazinamide. The only active analog developed because of its potential superiority over pyrazinamide was morphazinamide (*N*-morpholinomethylamide of pyrazinoic acid, **30**) (204). Interest in this drug ceased when it was ascertained that the activity and toxicity parallel those of pyrazinamide, to which morphazinamide is converted *in vivo* (205). Some pyrazinoic acid esters (206) and some *N*-pyrazinylthyoureas (207) were found to be more active *in vitro* against

(29)

*M. tuberculosis* than pyrazinamide, but no data *in vivo* have been reported.

**5.1.6.3 Activity and Mechanism of Action.** The activity of pyrazinamide against intracellular mycobacteria parallels the fact that pyrazinamide is inactive in guinea pig tuberculosis, predominantly extracellular, and very active in murine tuberculosis, which has an important intracellular component. Apart from these observations, the mechanism of action of pyrazinamide remains unknown. According to some authors, the activity of pyrazinamide is caused by its intracellular conversion into pyrazinoic acid (208), but the mechanism of action of the latter is unknown. Recently, the gene *pncA*, encoding the pyrazinamidase (PZase) of *M. tuberculosis* has been identified (209); mutations in *pncA* have been shown to be associated with pyrazinamide resistance. Recent studies suggest that the mutations found in *pncA* occur preferentially in conserved regions of *pncA* that might be very important for the binding and processing of PZA (210). Recent work has revealed that the target of pyrazinamide is the eukaryotic-like fatty acid sinthetase 1 (FAS-1) of *M. tuberculosis* (211).

Pyrazinamide was introduced in therapy in 1952 for the treatment of tuberculosis. After a period of declining use, a renewed interest has been shown on this drug for its potential role in the short-course chemotherapy regimens. Because of its bactericidal effect on the intracellular mycobacteria, pyrazinamide is recommended especially in the first 2 months of treatment of tuberculosis in combination with rifampicin and isoniazid. The usual daily dos-

ages are 20–35 mg/kg given orally in three or four equally spaced doses.

**5.1.6.4 Pharmacokinetics.** Pyrazinamide (**29a**) is well absorbed from the gastrointestinal tract. Peak serum concentrations occur about 2 h after a dose by mouth and have been reported to be about 35 $\mu$g/mL after 1.5 g, and 66 $\mu$g/mL after 3 g. Pyrazinamide is widely distributed in body fluids and tissues and diffuses into the CSF. The half-life time has been reported to be about 9–10 h. It is metabolized primarily in the liver by hydrolysis to the major active metabolite pyrazinoic acid (**29a**) which is subsequently converted into the major excretory product 5-hydroxypyrazinoic acid (**29b**) and pyrazinuric acid (**29c**). It is excreted through the kidney mainly by glomerular filtration. About 70% of a dose appears in the urine within 24 h, mainly as metabolites and 4–14% as unchanged drug.

**5.1.6.5 Adverse Effects.** Hepatotoxicity is the most common and serious side effect of pyrazinamide and is related with the dose and length of treatment. Liver functions should be checked before the administration of the drug and at frequent intervals during the therapy. Another side effect observed with pyrazinamide is arthralgia, caused by elevation of plasma uric acid levels.

### 5.1.7 Clofazimine

**5.1.7.1 History.** Clofazimine belongs to a peculiar class of phenazines called riminophenazines. Studies on these compounds derived from the original observation that treating a solution of 2-aminodiphenylamine with ferric chloride produced a red crystalline precipitate that completely inhibited the growth of tubercle bacilli H37Rv strain *in vitro* and was not inactivated by human serum (212–214). The *in vivo* activity was moderate and the toxicity low. Such a compound, named B-283 (**31**), was the leading structure for a se-

|        |  R           |  R'          |  R"          | R''' |
|--------|--------------|--------------|--------------|------|
| (**31**) | H          | H            | H            | H    |
| (**32**) | *p*-ClC$_6$H$_4$ | CH(CH$_3$)$_2$ | *p*-ClC$_6$H$_4$ | H |
| (**33**) | C$_6$H$_5$ | CH(CH$_2$–CH$_2$)(CH$_2$–CH$_2$)CH$_2$ | C$_6$H$_5$ | Cl |

ries of riminophenazines, which are alkyl or arylimino derivatives. Among the first compounds synthesized, B-663, later named clofazimine (**32**), was the most active (215, 216).

### 5.1.7.2 Activity and Mechanism of Action.

Clofazimine has an *in vitro* inhibitory activity against *M. tuberculosis* at concentrations of 0.1–0.5 $\mu$g/mL. Strains resistant to isoniazid, and/or streptomycin, PAS, and thioacetazone, are susceptible to the drug. Some MOTT are also susceptible to this compound. Minimal inhibitory concentrations for *M. avium* complex (MAC) are 0.125–1 $\mu$g/mL (217). This activity has rendered the drug eligible for the treatment of infections caused by this organism.

Clofazimine is not only bacteriostatic, but also bactericidal (but the latter action is rather slow) and only on multiplying mycobacteria. The mechanism of action of riminophenazines has not been elucidated, mainly because of the low solubility of these compounds in aqueous media. The activity seems to be correlated with the *p*-quinoid system; in fact, when this is removed by reductive acylation, activity disappears. The mycobacteria under anaerobic conditions reduce the quinoid system. It has been shown that 20% of the respiratory hydrogen

can be transferred from a respiratory enzyme to clofazimine (214). Its action has been also related to iron chelation, with resulting production of nascent oxygen radicals intracellularly (218).

In the treatment of murine tuberculosis, clofazimine was found to be very active, but much higher doses were necessary to achieve a therapeutic effect in guinea pigs and monkeys. Trials in chronic human pulmonary tuberculosis indicated that clofazimine had no significant effect on the disease at doses up to 10 mg/kg. In experimental infections with mycobacteria, clofazimine was found to be much more active than isoniazid against *M. kansasii* (219). Other studies have shown that clofazimine is also active in experimental infections with *M. johnei* (220), *M. ulcerans* (221), *M. lepraemurium* (222, 223), and *M. leprae* (224–231). In particular, *M. leprae* seems to be about 10 times more susceptible to clofazimine than *M. tuberculosis* (177). The marked activity against leprosy was confirmed in clinical trials (224–230). Generally speaking, the activity of clofazimine is similar to that of dapsone. Dapsone-resistant mutants are susceptible to clofazimine. During treatment of lepromatous leprosy with clofazimine, the characteristic inflammatory reaction erythema nodosum leprosum (ENL) seldom develops; if clofazimine is combined with dapsone, the latter agent no longer causes ENL. This makes the concurrent use of corticosteroids unnecessary. It was suggested that clofazimine would have a corticosteroids like anti-inflammatory action (231). In a dye-hyaluronidase spreading test, clofazimine had a hyaluronidase inhibitory effect, after a single oral administration of 100–200 mg in humans (232). In agreement with these results, it was found that clofazimine (50–100 mg/kg per day) inhibited rat adjuvant arthritis and the inflammatory paw swelling following an adjuvant injection (233). It did not inhibit the primary antibody response to sheep erythrocytes or the tuberculin skin response. Thus clofazimine seems to have anti-inflammatory but not immunosuppressive activity.

### 5.1.7.3 Pharmacokinetics.

Clofazimine is an effective alternative drug in the therapy of leprosy. It is most active when administered twice weekly or daily, but can be administered at monthly intervals as well, permitting monitoring of the treatment. The daily dose should not exceed 100 mg. Clofazimine has a peculiar pharmacokinetic pattern, characterized by slow absorption, low blood concentration, and extremely slow excretion. All riminophenazines are soluble in fats, and in micronized form, are well absorbed by the intestine. Riminophenazines, once absorbed by the intestine, are carried by lipoproteins in the blood and ingested by macrophages. After continued oral administration, the macrophages appear as red-orange phagosomes. Therefore, the compounds have a diffusion that is mainly intracellular. They are stored in the body for a long time, and this confers some prophylactic action on them (219).

*5.1.7.4 SARs.* To obtain compounds with better pharmacodynamic and pharmacokinetic properties, which would be more useful than clofazimine in the treatment of mycobacterial infections, a number of riminophenazines were prepared and tested. All the derivatives with hydrophilic groups are either less active or inactive. Compound B-1912 (**33**) was selected for further investigation because it showed higher serum levels and lower tissues levels (except than in lipids) than clofazimine. In *M. leprae* infections experimentally induced in mice, it was shown that clofazimine and B-1912 have the same activity (234). Newer riminophenazines, such as B746 and B4157, showed increased antimycobacterial activity and produced less skin pigmentation, which is the main drawback of this group of compounds, and both need further investigation (235).

*5.1.7.5 Adverse Reactions.* In laboratory animals, clofazimine has low acute and subacute toxicity (236). In clinical use, treatment with clofazimine is not accompanied by relevant toxicity. The main and sometimes unacceptable side effect is a red-purple coloration of skin, particularly within skin lesions (231). Gastrointestinal intolerance may also occur.

### 5.1.8 Ethambutol

*5.1.8.1 History.* Extensive studies of the chemical and biological properties of compounds related to alkylenediamine were carried out after the discovery, during the screening of randomly selected compounds, of the antimycobacterial activity of $N,N'$-diisopropylethylenediamine (**34**) (237, 238). This compound was found to be active both *in vitro* and *in vivo*, with a therapeutic index of the same order of streptomycin.

$$RNHCH_2CH_2NHR$$

(**34**)   $R = CH(CH_3)_2$

(**35**)   $R = \underset{\displaystyle CH_2OH}{CHC_2H_5}$

(**36**)   $R = \underset{\displaystyle CHO}{CHC_2H_5}$

(**37**)   $R = \underset{\displaystyle COOH}{CHC_2H_5}$

*5.1.8.2 SARs.* Chemical modifications of the compound (**34**), attempted with the aim of obtaining the product with the highest therapeutic index, gave indications of the structural requirements for antimycobacterial activity. More relevant are the following: the presence of two basic amine centers, the distance between the two carbons, and the presence of a simple, small branched alkyl group on each nitrogen. A correlation between metal chelation of compounds of this structure and antimycobacterial activity suggested the use of synthesizing products with hydroxy substitution of the $N$-alkyl groups as more effective metal chelators and possibly more active antimycobacterial agents The most active of these derivatives was the *dextro* isomer of $N,N'$-bis-(1-hydroxy-2-butyl)ethylenediamine (ethambutol, **35**). The *meso* isomer is less active and the *levo* almost inactive. Furthermore, hydroxy substitution on other alkyl groups (isopropyl, *t*-butyl) or in other positions of the butyl group gave inactive products. These data seem in contrast with the working hypothesis of a correlation between metal chelation and antimycobacterial activity. Various other modifications of the structure of ethambutol gave inactive products, with a few exceptions. The $OCH_3$, $OC_2H_5$, and $HNCH_3$ derivatives have the same activity *in vivo* as the parent compound because dealkylation occur in the body. In addition, the monohydroxy unsymmetrical analog has activity equal to ethambutol but is more toxic.

### 5.1.8.3 Activity and Mechanism of Action.

Ethambutol inhibits *in vitro* the growth of most of the human strains of *M. tuberculosis* at concentrations around 1 μg/mL. Strains resistant to other antimycobacterial agents are just as sensitive to ethambutol. Among the other mycobacteria, *M. bovis*, *M. kansasii*, and *M. marinum* are usually susceptible. The MICs for *M. avium intracellulare* range between 0.95 and 15 μg/mL; 63% of strains is inhibited by 1.9 μg/mL or less (239, 240). Ethambutol is not active in the experimental mouse infection with *M. leprae* (177) and is not used in the treatment of human leprosy. The efficacy of ethambutol against *M. tuberculosis in vivo* was proved in various experimental models in animals and confirmed in the clinical trials in human tuberculosis.

The effect of ethambutol on mycobacteria is primarily bacteriostatic, with a maximum inhibitory effect at neutral pH. The primary mechanism of action is not understood. The growth inhibition by ethambutol is largely independent of concentration, being more related to the time of exposure. It seems that most of ethambutol taken up by mycobacteria has not direct role in growth inhibition, and there is no information on the subcellular components responsible for critical ethambutol binding (241, 242). Treatment of mycobacteria with ethambutol results in inhibition of protein and DNA synthesis, and it was proposed that ethambutol interferes with the role of polyamines and divalent cations in ribonucleic acid metabolism (243–245). Other authors have found an inhibitory effect of ethambutol on phosphorylation of specific compounds of intermediary metabolism, under conditions of endogenous respiration (246). Other proposed that primary sites of inhibition include arabinogalactan biosynthesis (247) and glucose metabolism (248). However, further studies are necessary to clarify the primary action of the drug.

### 5.1.8.4 Pharmacokinetics.

In current therapeutic use, the drug is administered orally at daily doses of 15–25 mg/kg, in combination with other antitubercular agents, to prevent emergence of resistant strains (249, 250). The drug is well absorbed from the gastrointestinal tract. Except for the cerebrospinal fluid, it is widely distributed to most tissues and fluids. About one-half the ingested dose is excreted as active drug in the urine, where there are also minor quantities of two inactive metabolites: the dialdehyde (**36**) and the dicarboxylic acid (**37**) derivatives (251, 252).

### 5.1.8.5 Adverse Effects.

Ethambutol is rather well tolerated. The main side effect of concern is ocular toxicity, consisting of retrobulbar neuritis with various symptoms, including reduced visual acuity, constriction of visual fields, and color blindness. Ocular toxicity seems to be dose-related (253). At a daily dose of 25 mg/kg, visual impairment occurs in about 3% of patients, rising to 20% at doses higher than 30 mg/kg per day.

## 5.2 Antibiotics

### 5.2.1 Streptomycin

#### 5.2.1.1 History.

Streptomycin was discovered in 1944 as a fermentation product of *Streptomyces griseus* (254). It belongs to the family of aminoglycoside antibiotics, which includes kanamycin, gentamicin, neomycin, amikacin, nebramycin, paromomycin, ribostamycin, tobramycin, sisomicin, dibekacin, netilmicin, kasugamycin, and spectinomycin. In terms of chemical structure, they are aminocyclitols (cyclohexane with hydroxyl and amino or guanidino substituents) with glycosyl substituents at one or more hydroxyl groups. Streptomycin is an *N*-methyl-L-glucosaminidostreptosidostreptidine made up of three components: streptidine, streptose, and *N*-methyl-L-glucosamine (**38**). The intact molecule is necessary for antibacterial action.

Mannosidostreptomycin (**39**) is another antibiotic, produced together with streptomycin by *S. griseus*, which has not found clinical application because it is less active than streptomycin itself. Hydroxystreptomycin (**40**), produced by *S. griseocarneus*, has biological properties similar to those of streptomycin, with no advantages over it.

#### 5.2.1.2 SARs.

In the attempt to improve activity and/or decrease toxicity of streptomycin, some chemical modifications have been performed (e.g., on aldehyde or guanidino functions), which yielded generally less active

| | $R^1$ | $R^2$ | $R^3$ |
|---|---|---|---|
| (38) | —CHO | H | H |
| (39) | —CHO | H | (see structure) |
| (40) | —CHO | —OH | H |
| (41) | $CH_2OH$ | H | H |

products. One chemical derivative of strepto-mycin, dihydrostreptomycin (**41**), obtained by catalytic hydrogenation of the carbonyl group of streptose, has almost the same antibacterial activity of the parent compound and investi-gators hoped that it would differ from the par-ent in having lower toxicity. Later clinical ex-perience did not confirm this hope.

*5.2.1.3 Activity and Mechanism of Action.*
Streptomycin is both bacteriostatic and bacte-ricidal for the tubercle bacillus *in vitro*, according to the concentration of the antibi-otic. Concentrations of streptomycin around 1 $\mu$g/mL inhibit the growth of *M. tuberculosis H37Rv*. NTM are not susceptible to strepto-mycin. The antibacterial activity of strepto-mycin is not restricted to *M. tuberculosis* but includes a variety of Gram-positive and Gram-

negative bacteria. The most important clinical use of streptomycin is in the therapy of tuber-culosis and it was the first really effective drug for this disease. Its importance declined after the introduction of other powerful oral antitu-berculous agents. Since the introduction of other broad-spectrum antibiotics, the use of streptomycin in the treatment of infections is limited to diseases in which other alternatives are lacking and the sensitivity of the infecting organism indicates the choice and eventually the use of this drug in combination with other antibiotics. Thus, it is still a drug of first choice for enterococcal endocarditis (in combination with penicillin or ampicillin), in brucellosis (in combination with tetracycline), in plague, and in tularemia.

The investigation on the mechanism of ac-tion of streptomycin has involved a number of elegant studies in microbiological chemistry and molecular biology that have led to a suc-cession of hypotheses and to a continuous in-crease in knowledge not only of the mode of action of the antibiotic but also of the biology of the bacteria. After a series of preliminary hypotheses, it was finally ascertained that the drug is a specific inhibitor of protein biosyn-thesis in intact bacteria (255, 256). The ribo-some, and particularly its 30S subunit, is the site of action of the antibiotic (257, 258), and after careful disassemble of 30S ribosomes, a protein designated P10 was determined to be the genetic locus responsible for the pheno-typic expression of sensitivity and resistance and dependence on streptomycin (259). The antibiotic induces a misreading of the genetic code, demonstrated through studies of the er-roneous incorporation of amino acids in cell-free ribosome systems (260). It was deduced that the misreading *in vivo* was the cause of the bactericidal effect of streptomycin, be-cause it resulted in "flooding the cell" with erroneous, non-functional proteins. However, it was subsequently demonstrated that this could not be the case because in the intact bacteria the antibiotic inhibits the synthesis of proteins (261). The ultimate mode by which streptomycin exerts its bactericidal activity is not yet clear. Two hypotheses have been put forward: one suggesting that streptomycin specifically inhibits initiation of protein syn-thesis (262) (this is supported by the involve-

ment of protein P10, the site of action of streptomycin, in the initiation reaction) and the
other suggesting that it inhibits peptide chain
elongation, that is, the synthesis of peptide
bonds at any time during the growth of the
peptide chain (263, 264). As noted before, sensitivity and resistance to and dependence on
streptomycin all seem to be expressed in the
ribosome and apparently are multiple alleles
of a single genetic locus. Streptomycin-resistant mutant cells arise spontaneously in a bacterial culture, with a frequency of the order
$1 \cdot 10^{-6}$ (265).

In the phenomenon of streptomycin dependence, bacteria require streptomycin to grow;
these bacteria also arise by spontaneous mutation (266), and the mechanism of their behavior is also related to the reading of codons.
This can be done correctly only in the presence
of streptomycin, which overcomes an undis-
criminating restriction (caused by mutation),
leading to a mutant in which the ribosomal
screen does not allow normal translation for
growth (261–267). In addition, resistance to
streptomycin can be transferred by means of
R-factors or plasmids, namely, by extra-chro-
mosomal DNA carrying multiple antibiotic resistance (268).

The mode of transmission of resistance is
particularly frequent among enterobacteria.
Enzymatic inactivation is a frequent cause of
resistance to streptomycin in eubacteria. The
aminoglycoside-inactivating enzymes are phosphotransferases, acetyl-transferases, and ade-
nyl-transferases. Because they act by inactivating a chemical group that is common to
different aminoglycosides, bacterial strains
that produce only one of them can be resistant
to all aminoglycosides possessing the same
chemical group (cross-resistance). Streptomycin can be inactivated by some streptomycin-
adenyltransferase and streptomycin-phospho-
transferase, which usually do not affect other
aminoglycosides except spectinomycin (172).

In mycobacteria, mutations in the *rpsL*
gene, which encodes the ribosomal protein
S12 have been shown to confer resistance to
streptomycin. Analysis of the primary structure of the ribosomal protein S12 in *M. tuberculosis* has revealed that mutations in the
gene replacing Lys[43] or Lys[88] by arginine are
frequently associated with resistance to strep-

tomycin (269–272). A second type of mutation
conferring resistance has been identified in
streptomycin-resistant strains of *M. tuberculosis* that have a wild-type *rpsL* gene. These
strains have point mutations in the 16S rRNA
clustered in two regions around nucleotides
530 and 915 (273, 274). In those isolates with a
wild-type 16S rRNA and *rpsL* gene, other
mechanism of drug resistance can be hypothesized, such as modifications of other components of the ribosome or alteration in cellular
permeability.

**5.2.1.4 Pharmacokinetics.** Streptomycin,
like all other aminoglycoside antibiotics, is not
absorbed from the gastrointestinal tract, and
therefore, it must be administered parenterally. Serum peak levels are reached in 1–2 h,
and the values are 9–15 $\mu$g/mL after administration of 1 g. Its half-life is 2–3 h. The serum
protein binding of streptomycin is 25–35%
(275). Streptomycin diffuses slowly into the
pleura and better into the peritoneal, pericar-
dial, and synovial fluids. It does not penetrate
into spinal fluid, unless the meninges are inflamed. Urinary elimination is rapid, and 70%
of the drug is excreted in unmodified form in
the first 24 h.

**5.2.1.5 Adverse Effects.** The most important toxic effects of streptomycin involve the
peripheral and central nervous system. The
eighth cranial nerve is the most frequently in-
jured by prolonged administration of streptomycin, especially in its vestibular portion,
causing equilibrium disturbances to appear.
Treatment with 2–3 g/day for 2–4 months
produces this type of side effect in about 75%
of patients, but the incidence is much less at
doses of 1 g/day. Other side effects are hypersensitivity reactions and renal damage.

## 5.2.2 Kanamycin

**5.2.2.1 History.** Another aminoglycoside
antibiotic, used in the therapy of tuberculosis
and named kanamycin, was isolated in 1957 as
a fermentation product of *Streptomyces kana-
myceticus*. It consists of three components:
kanamycin A, B, and C (**42–44**). Kanamycin A
(**42**) is the largest part of the mixture (98%).
Structural studies indicated that the molecule
contains deoxystreptamine (instead of the
streptidine present in the streptomycin molecule) and two amino sugars: kanosamine and

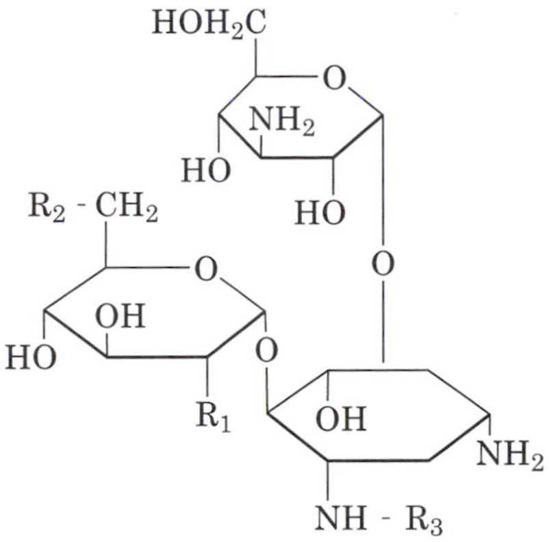

| R₁ | R₂ | R₃ |
|---|---|---|
| (42) OH | NH₂ | H |
| (43) NH₂ | NH₂ | H |
| (44) NH₂ | OH | H |
| (45) NH₂ | NH₂ | CO−CHOH—CH₂−CH₂—NH₂ |

6-glucosamine. It is water soluble and stable at both acid and basic pH as well as at high temperature.

**5.2.2.2 Activity and Mechanism of Action.**
Kanamycin has a quite broad spectrum of activity, including Gram-positive cocci and Gram-negative bacteria, as well as *M. tuberculosis* and some other mycobacteria. Its activity against *M. tuberculosis* is weaker than that of streptomycin (276). The bactericidal concentrations are close to the bacteriostatic ones, but they are hard to achieve *in vivo* (277). Kanamycin is used in therapy of infections caused by penicillin-resistant *Staphylococcus aureus* (now less frequently used because the availability of penicillinase-resistant penicillins and other antistaphylococcal antibiotics) or by Gram-negative bacilli, such as *E. coli, Klebsiella, Enterobacter,* and *Proteus*. It has no activity against *Pseudomonas*. It has been used in the therapy of tuberculosis, in combination with other antituberculous drugs. The common dose is 15 mg/kg per day, but a total dose of 1.5 g must not be exceeded.

The mechanism of action of kanamycin is similar to that of streptomycin, because it produces a misreading of the genetic code, interacting with 30S ribosomal subunit in more than one site (whereas streptomycin is bound only to one site), and inhibits protein synthesis (278–280). All aminoglycoside antibiotics that contain a 2-deoxystreptamine moiety cause miscoding. Kanamycin, like streptomycin and other aminoglycosides, blocks both the initiation and elongation of peptide chains. This mechanism was confirmed in mycobacteria. *In vitro* studies on cell-free preparations of *M. bovis* have shown that kanamycin inhibits polypeptide synthesis, followed by breakdown of polysomes and detachment of mRNA (281). A kanamycin-induced increase in $^{14}$C-isoleucine incorporation by poly-U-directed ribosomes indicated a misreading of the genetic code, but this did not seem to be directly related to the bactericidal action of the antibiotic.

Resistance to kanamycin can be acquired *in vitro* in a stepwise fashion by subculturing bacteria in increasing concentrations of antibiotic. In addition to chromosomal resistance, resistance to kanamycin can be acquired by conjugation, through the transfer of extrachromosomal DNA, the so-called R-factors or plasmids, and coding aminoglycoside-inactivating enzymes (phosphotransferases, acetyltransferases, and nucleotidyltransferases). Kanamycin A can be inactivated by neomycin-kanamycin phosphotransferases I and II, kanamycin acetyltransferase, and gentamicin adenyltransferase. Cross-resistance will appear to any other aminoglycoside antibiotic that may be inactivated by the same enzyme (269). Cross-resistance is total with neomycin and paromomycin. With streptomycin, a "one-way resistance" is observed, namely strains resistant to kanamycin and neomycin are usually resistant to streptomycin, whereas streptomycin-resistant strains are usually susceptible to kanamycin and gentamicin. It has been suggested that this is caused by different sites of action of the antibiotics on the ribosomes (282). Thus, in therapy, it is advisable to administer streptomycin before kanamycin.

**5.2.2.3 Pharmacokinetics.** From a pharmacokinetic point of view, kanamycin behaves similarly to the other aminoglycosides: it is not absorbed when given by the oral route, but it is rapidly absorbed after intramuscolar administration, reaching high peak serum levels 2–3 h after administration (283). There is almost no serum protein binding (275). Diffusion into cerebrospinal fluid is poor when meninges are normal but increases when they are inflamed. Kanamycin diffuses quite well into pleural, peritoneal, synovial, and ascitic fluids (284–285), but poorly into bile, feces, amniotic fluid, and so on. It is excreted by the kidney, mainly by glomerular filtration (50–80%) and in unmodified form.

**5.2.2.4 Adverse Effects.** Drawbacks to the use of kanamycin are its ototoxicity and nephrotoxicity (286). Cochlear and vestibular functions are damaged in about 5% of patients, but the percentage increases proportionally to the total dose administered. Thus, in prolonged treatments, as in the case of tuberculosis, patients must be closely followed. Nephrotoxicity can be prevented if dosage to patients with impaired renal function is reduced in accord with the decrease in creatinine clearance or the increase in creatininemia (284). Kanamycin can produce neurotoxicity, with curare-like effects caused by neuromuscular blockade.

### 5.2.3 Amikacin

**5.2.3.1 History.** Amikacin (**45**) is a semisynthetic analog of kanamycin, in which the C-1 aminogroup is amidated with a 2-hydroxy-4-amino-butirric acid moiety. The synthesis of amikacin was suggested by the observation that butirosin B (an antibiotic produced by B. circulans differing from ribostamycin only in having an α-hydroxy-γ-aminobutyric acid substituent on the amine in position 1 of 2-deoxystreptamine) was active against Pseudomonas and other inactivating strains, unlike ribostamycin. The different aminoglycosides obtained by acylation with α-hydroxy-γ-aminobutyric acid were found to be insensitive to the enzymes phosphorylating the oxygen in position 3', to those acylating the nitrogen in position 3, and to those determining the nucleotide attachment to the oxygen in position 2.

**5.2.3.2 Activity.** Amikacin, the derivative obtained by acylating the nitrogen on C-1 of kanamycin A with α-hydroxy-γ- aminobutyric acid, is the aminoglycoside with the broadest spectrum of activity and can be used especially in cases of infections caused by bacteria resistant to other aminoglycosides. In fact, amikacin is not inactivated by many of the R-factor mediated enzymes that attack kanamycin. In

particular, useful activity against Pseudomonas aeruginosa results. In in vitro and animal trials, amikacin is among the most active, if not the most active, aminoglycoside against M. tuberculosis (287).

Amikacin is not used for the initial treatment of susceptible tuberculosis, mainly because of its cost, but it seems to have merit as an alternative drug for the retreatment of resistant M. tuberculosis infections. Amikacin seems to have a role in the treatment of NTM infections, especially if the infections are caused by rapidly growing mycobacteria (M. fortuitum and M. chelonae) (288). It is one of the most bactericidal agent against M. avium complex both in vitro and in beige mouse models (289–290). In some clinical studies with AIDS patients, M. avium intracellulare complex bacteremia was cleared by combination of amikacin with other drugs (clarithromycin and ciprofloxacin) (291).

### 5.2.4 Viomycin and Capreomycin

**5.2.4.1 History.** Viomycin and capreomycin are two structurally similar polypeptide antibiotics. Viomycin (**46**) was discovered independently by two groups of investigators in 1951 (292, 293) from an actinomyces named Streptomyceus puniceus by one group and S. floridae by the other. Capreomycin is a polypeptide complex isolated in 1960 from Streptomyces capreolus (294) and the structure of the same components of the complex (**47**) indicated its similarity with viomycin.

**5.2.4.2 Activity and Mechanism of Action.** Viomycin is relatively more active against the mycobacteria than against other bacteria. It inhibits protein synthesis (295) but has little or no miscoding activity (279). Viomycin-resistant mutants isolated from M. smegmatis have altered ribosomes (296): one of these mutants had altered 50S subunits and others have 30S subunits. The genetic locus for viomycin-capreomycin resistance (vic locus) in M. smegmatis consisted of two groups: vic A and vic B. It is likely that alterations in the 30S subunit conferred by vic B and alterations in the 50S subunit conferred by vic A interact in response to viomycin (297). There is a one-way cross-resistance with kanamycin: viomycin-resistant strains may retain their susceptibility to kanamycin, but kanamycin-

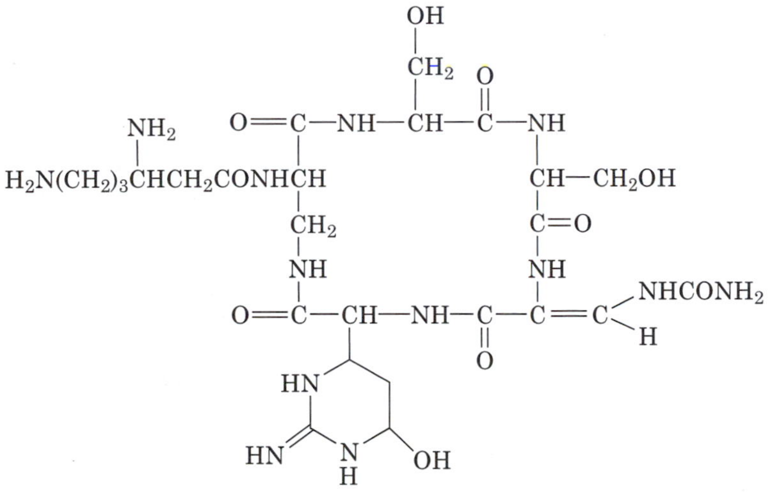

$$
\begin{array}{c}
\text{OH} \\
|\\
\text{CH}_2 \quad \text{O} \\
\\
\text{NH}_2 \quad\quad \text{O}=\text{C}-\text{NH}-\text{CH}-\overset{\|}{\text{C}}-\text{NH} \\
| \\
\text{H}_2\text{N(CH}_2)_3\text{CHCH}_2\text{CONHCH} \quad\quad \text{CH}-\text{CH}_2\text{OH} \\
| \quad\quad\quad\quad\quad\quad\quad\quad\quad\quad\quad | \\
\text{CH}_2 \quad\quad\quad\quad\quad\quad\quad\quad \text{C}=\text{O} \\
| \quad\quad\quad\quad\quad\quad\quad\quad\quad\quad | \\
\text{NH} \quad\quad\quad\quad\quad\quad\quad\quad \text{NH} \quad\quad \text{NHCONH}_2 \\
| \quad\quad\quad\quad\quad\quad\quad\quad\quad\quad | \\
\text{O}=\text{C}-\text{CH}-\text{NH}-\overset{\|}{\underset{\text{O}}{\text{C}}}-\text{C}=\text{C}\diagdown_{\text{H}} \\
| \\
\text{HN} \\
\\
\text{HN}\quad\quad\text{N}\quad\text{OH} \\
\overset{}{\underset{\text{H}}{}}
\end{array}
$$

(46)

resistant strains are also resistant to viomycin. It is interesting to note that in *M. smegmatis*, the genetic locus for neomycin-kanamycin resistance (*nek* locus) is not linked to *str* locus (for streptomycin resistance) as in *E. coli* but is linked to *vic* locus (297).

Capreomycin is active only against *M. tuberculosis* and some other mycobacteria, particularly *M. kansasii*. Cross-resistance to kanamycin, neomycin, and viomycin has been described, and a phenomenon of partial "one-way resistance" to kanamycin has been demonstrated. Capreomycin-resistant strains are not always fully resistant to kanamycin, but kanamycin-resistant strains are always resistant to capreomycin.

**5.2.4.3 Pharmacokinetics.** This type of antibiotics behaves like the aminoglycosides. They are not absorbed by the gastrointestinal tract. Peak serum concentrations of capreomycin are achieved 1–2 h after intramuscular administration of 1 g of the drug. The half-life is 3–6 h. The drug is eliminated unchanged in the urine.

**5.2.4.4 Adverse Reactions.** Side effects produced by viomycin are severe and frequent.

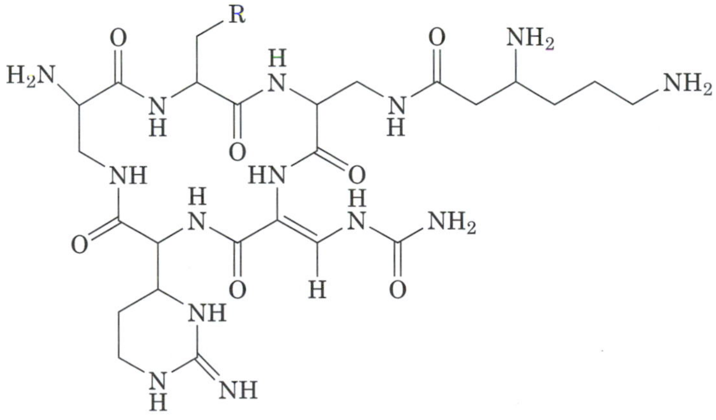

(47) R = H, OH

Vestibular and auditory impairment, renal damage, and disturbance in the electrolyte balance have a high incidence during viomycin therapy. For these reasons the drug is now seldom used.

Renal damage is the most consistent and significant toxic effect of capreomycin, which is also potentially toxic to the eighth cranial nerve. Capreomycin is usually reserved for re-treatment regimens, when the primary anti-tuberculous drugs cannot be used because of toxicity or the presence of resistant bacilli.

### 5.2.5 Cycloserine

***5.2.5.1 History.*** D-Cycloserine was isolated in 1955 independently by several groups of workers from cultures of *Streptomyces gary-phalus, S. orchidaceus,* and *S. lavendulae* (298–300). On the basis of degradation studies and physicochemical properties, the structure of D-4-amino-3-isoxazolidone (**48**) was as-

(**48**)

signed to this antibiotic (301, 302). The structure was confirmed by synthesis (303), and various methods of preparation have been reported subsequently, which have also been used to prepare L-cycloserine from L-serine.

***5.2.5.2 SARs.*** No improvement on the antibacterial activity of D-cycloserine was obtained through chemical variation of its structure. Only the synthetic L-cycloserine possesses some antibacterial activity, but probably with a different mechanism of action.

***5.2.5.3 Activity and Mechanism of Action.*** Cycloserine inhibits *M. tuberculosis* at concentrations of 5–20 $\mu$g/mL. Strains resistant to other antimycobacterial drugs have the same sensitivity to cycloserine. The antibiotic is also active *in vitro* against a variety of Gram-positive and Gram-negative microorganisms, but only in cultures media-free of D-alanine, which results to be an antagonist of the antibacterial activity of cycloserine.

Concerning the mechanism of action of cycloserine, it has been proven in some bacterial species that this antibiotic interferes with the synthesis of the cell wall. In fact, cycloserine induces the formation of protoplasts in *E. coli.* Microorganisms treated with cycloserine accumulate a muramic-uridine-nucleotide peptide, which differs from that produced by penicillin in the absence of the terminal D-alanine dipeptide. The inhibition of alanine racemase, which converts L-alanine into D-alanine, is probably the primary action of cycloserine (304–306). The mechanism of action in mycobacteria is likely the same.

***5.2.5.4 Pharmacokinetics.*** Cycloserine possesses different pharmacokinetic properties in the various animal species and this explains the different responses in animal infections. *In vivo*, cycloserine was ineffective against experimental tuberculosis in mice and marginally effective in guinea pigs, but some activity was found against the disease induced in the monkey. The drug is more effective in humans than in animals. When given orally to humans, cycloserine is well and quickly absorbed from the gastrointestinal tract and is well distributed in the body fluids and tissues. The usual dose for adults is 250 mg twice a day orally, always in combination with other effective tuberculostatic agents. About one-half of the ingested dose is excreted unchanged in the urine in 24 h. A part of the antibiotic is metabolized into products not yet identified.

***5.2.5.5 Adverse Reactions.*** Cycloserine produces severe side effects in the central nervous system that can also generate psychotic states with suicidal tendencies and epileptic convulsions. Therefore, its use is limited only to cases in which other drugs cannot be used.

### 5.2.6 Rifamycins

***5.2.6.1 History.*** The rifamycin antibiotics were discovered in 1959 as metabolites of a microorganism originally considered to belong to the genus *Streptomyces* and subsequently reclassified as a *Nocardia* (*Nocardia mediterranea*) (307–309) and more recently as *Amycolatopsis mediterranea.* The crude material extracted from the fermentation broths contained several rifamycins (rifamycin complex) (310). Only rifamycin B was isolated as a pure crystalline substance, and it is essentially the only component found when sodium diethyl-

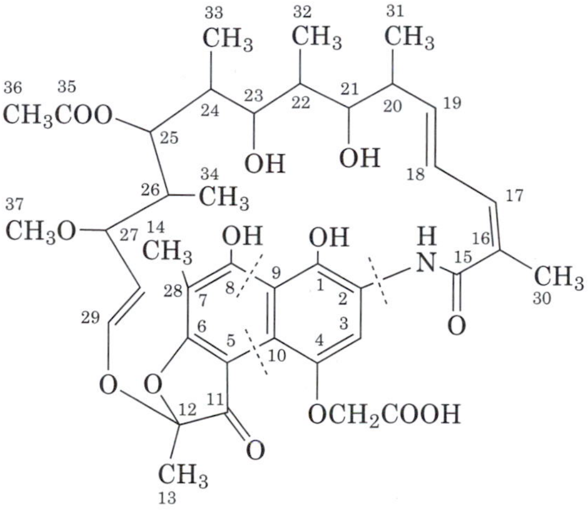

$$CH_3COO$$ (36, 35) ... $$CH_3O$$ (37) ...

(49)

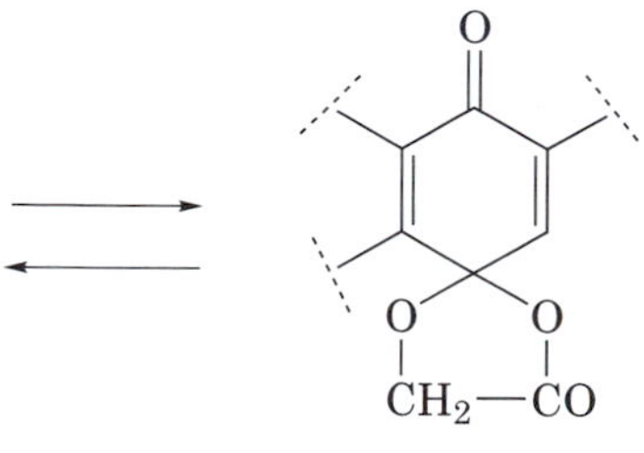

(50)

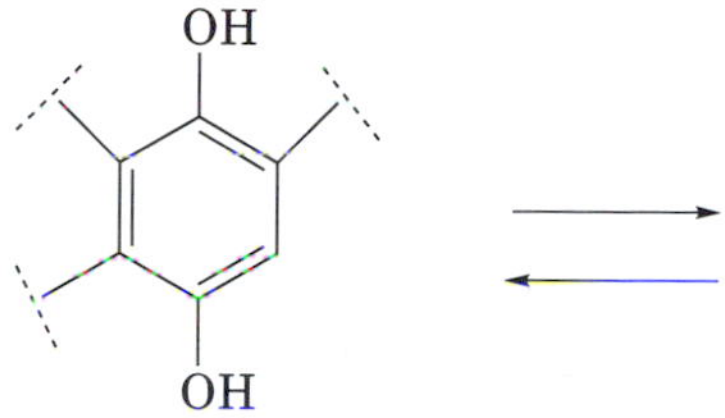

(52)

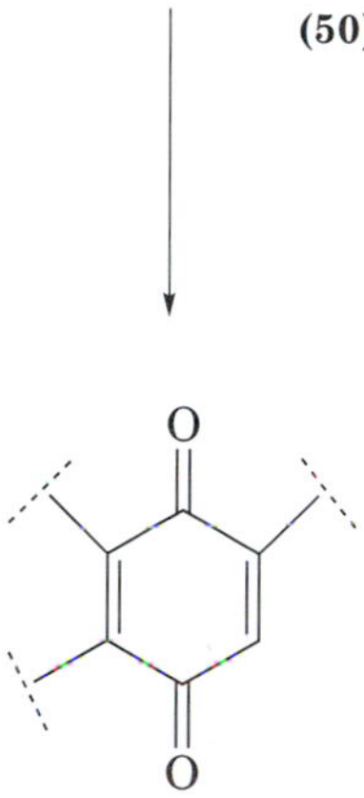

(51)

barbiturate is added to the fermentation media (311).

Rifamycin B (49) has the unusual property that, in oxygenated aqueous solutions, it tends to change spontaneously into other products with greater antibacterial activity (rifamycin O, 50; rifamycin S, 51). Rifamycin SV (52) was obtained from rifamycin S (310–313) by mild reduction. The structures of rifamycin B and of the related compounds involved in the "activation" process have been elucidated by chemical and X-ray crystallographic methods (314–316).

The rifamycins are the first natural substances to have been assigned an *ansa* structure consisting of an aromatic moiety spanned by an aliphatic bridge. At present, several natural substances with *ansa* structures are known, and for those that are metabolites of actinomycetales, (e.g., streptovaricins, tolypomycins, and halomycins), the general name of "ansamycins" has been proposed (317).

Among the first rifamycins, the sodium salt of rifamycin SV was introduced into therapeutic use in 1963 and is currently used in various countries for the parenteral and topical treatment of infections caused by Gram-positive bacteria and infections of the biliary tract also caused by Gram-negative bacteria. Systematic studies of the chemical modifications of the natural rifamycins were planned with the aim of obtaining a derivative presenting the following advantages over rifamycin SV: oral absorption, more prolonged therapeutic blood levels, and higher activity in the treatment of mycobacterial and Gram-negative bacterial infections (318). Several hundred rifamycin derivatives have been prepared and carefully evaluated for their potential therapeutic efficacy. These efforts brought forth rifampicin (synthesized in 1964 and introduced in therapeutic use in 1968), rifapentine in 1978, rifabutin in 1983, rifalazil (KRM-1648) in 1993, and rifamycin T9 in 1995.

*5.2.6.2 SARs.* Extensive chemical modifications have been performed on the various parts of the rifamycin molecule: the glycolic chain of rifamycin B, the aliphatic *ansa*, and the chromophoric nucleus. A great deal of information about the structure-activity relationships has been obtained (318–321). All rifamycins possessing a free carboxy group are partially or totally inactive because they do not enter into the bacterial cell. Requirements for activity seem to be the presence of two free hydroxyls in positions C21 and C23 on the *ansa* chain and of two polar groups (either free hydroxyl or carbonyl) at positions C1 to C8 of the naphtoquinone nucleus, together with a conformation of the *ansa* chain that results in certain specific geometrical relationship among these four functional groups.

This conclusion is based on the following data. First, substitution or elimination of the two free hydroxyls in position C21 and C23 gives inactive products. Inversion of the configuration at C23 leads to an inactive compound, and the inversion at C21 strongly reduces the activity (322, 323). Second, modifications of the *ansa* chain that alter its conformation (e.g., C16–C17 and C18–C19 monoepoxy and diepoxy derivatives) give inactive or less active products. Also, the stepwise hydrogenation of the ansa chain double bonds results in a gradual decrease in activity as a consequence of the increase in flexibility of the *ansa* diverging from the most active conformation. Third, the oxygenated functions at C1 and C8 must be either free hydroxyl or carbonyl to maintain biological activity. Fourth, the four oxygenated functions at C1, C8, C21, and C23 not only must be unhindered and underivatized, but must display well-defined relationships with one another. The absolute requirements for these four functions to be in a correct geometrical relationship suggest that they are involved in the noncovalent attachment of the antibiotic to the bacterial target enzyme. This contention is supported by the observation that in the active derivatives these four functional groups lie on the same side of the molecule and almost in the some plane with identical interatomic distances between the four oxygens (324), as seen in the spatial model derived from X-ray studies (325). However, conformational differences have been observed at the junction of the *ansa* chain to the naphtoquinone chromophore, according to the nature of the substituents in position 3 (326). Protonic nuclear magnetic

resonance ($^1$H NMR) studies of rifamycin S have confirmed that the conformation of the molecule in solution corresponds well to that obtained in the solid state (327).

Finally, all modifications at C3 and/or C4 position that do not interfere with the previous requirements do not affect the general activity of the products (318–321).

A great number of active derivatives have been obtained by modifications on the glycolic moiety of rifamycin B and on position C3 and/or C4 of the aromatic nucleus of rifamycin S or SV, leaving unaltered the structure and the conformation of the *ansa* chain. Among the earlier derivatives of the glycolic side chain, the diethylamide of rifamycin B, rifamide (**53**), had a better therapeutic index

(**53**)

than rifamycin SV (328), but it still suffers from most of the limitations of use of rifamycin SV (329).

Chemical modifications of the chromophoric nucleus of rifamycin on C3 and/or C4 positions gave a large number of derivatives. The nature of the substituents at positions C3 and/or C4 influences the physicochemical properties of the derivatives, especially lipophilicity. The various derivatives show a minor degree of variation in antibacterial activity against intact cells, because the transport through bacterial wall and membrane is the major factor affected by these substituents (330). Other biological characteristics influenced by the various modifications of the positions C3 and/or C4 are the absorption from the gastrointestinal tract and the kinetics of elimination. Rifamycin derivatives with substitutions at position C4 include a series of 4-aminoderivatives. Modifications at positions C3 and C4 include rifamycins with heterocyclic nuclei fused on these positions (e.g., 3,4-phenazinerifamycins, phenoxazinerifamycins, pyrrolerifamycins, thiazolerifamycins, and imidazorifamycins) with various groups on the heterocyclic nuclei. Rifamycins with substituents only in position C3 are represented by 3-thioethers, 3-aminomethylrifamycins, 3-carboxyrifamycins, 3-aminorifamycins, and 3-hydrazinorifamycins. When the 3-formyl-rifamycin SV (54) was prepared by oxidation of N-dialkyl-aminomethyl rifamycin SV (331), it was found that

many of its N,N-disubstituted hydrazones had high activity, both *in vitro* and *in vivo*, against *M. tuberculosis* and Gram-positive bacteria, and moderate activity against Gram-negative bacteria. For some of these derivatives, the *in vivo* activity in animal infections was of the same order whether the products were administered orally or parenterally, indicating good absorption from the gastrointestinal tract (318). The hydrazone of 3-formylrifamycin SV

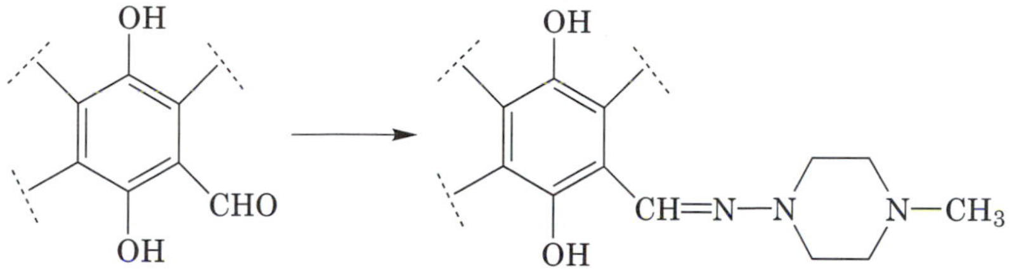

(54)

(55)

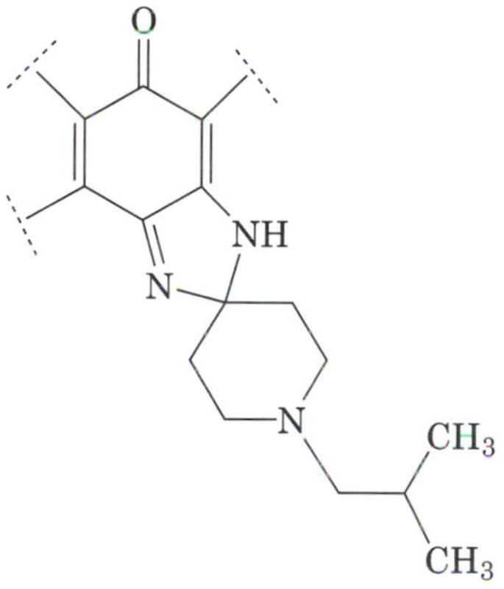

(56)

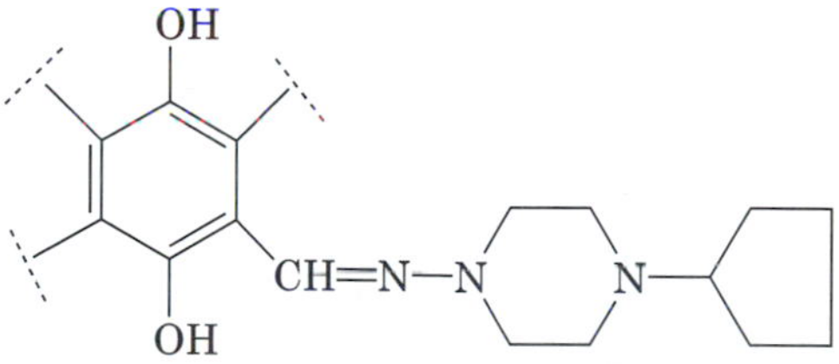

(57)

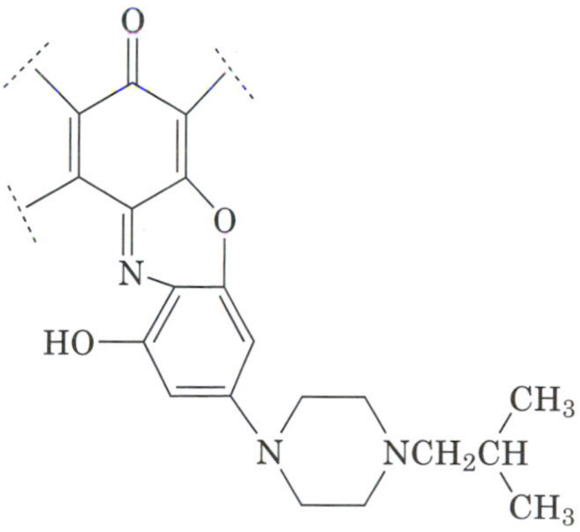

(58)

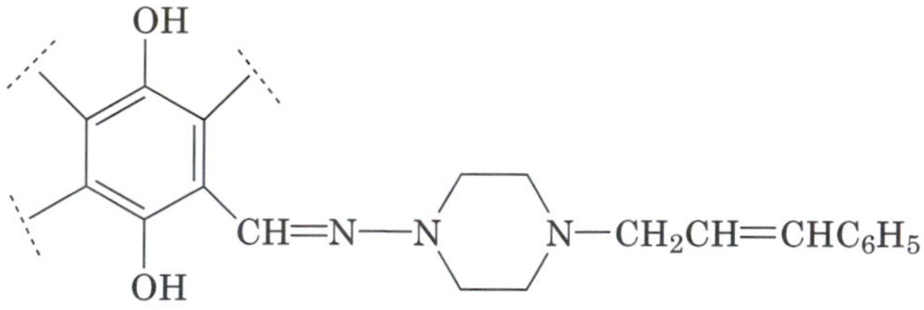

(59)

with $N$-amino-$N'$-methylpiperazine (rifampicin, **55**) (318, 332) was the most active *in vivo* and was selected for clinical use.

The knowledge of the SARs in the rifamycins was the basis for further development of this family of antibiotics, leading to other products in clinical use [rifabutin (**56**), rifapentine (**57**), rifalazil (**58**), and rifamycin T9 (**59**)].

**5.2.6.3 Mechanism of Action and Resistance.**
Studied mainly for rifampicin, the mechanism of antibacterial action is the same for all the active rifamycins and resides in the specific inhibition of the activity of the enzyme DNA-directed RNA polymerase (DDRP). The mammalian DDRP is resistant even to high concentration of rifamycins (333, 334). The inhibition of the action of the bacterial RNA polymerase by rifampicin is caused by the formation of a rather stable, noncovalent complex between the antibiotic and the enzyme with a binding constant of $10^{-9}$ $M$ at 37°C (355). One molecule of rifampicin (mol. wt. 823) is bound with one molecule of the enzyme (mol. wt. 455,000) (336). The drug is not bound covalently to the protein because the complex dissociates in presence of guadinium chloride (337). The binding site of rifampicin to the RNA polymerase has been particularly well studied in *E. coli* (338–340). The inhibitory effect of rifampicin on DDRP as the cause of its bactericidal activity has been verified on several mycobacterial species, e.g., *M. smegmatis*, *M. bovis BCG*, and *M. tuberculosis* (341–342).

The DDRP is comprised of five subunits ($\alpha$, $\alpha'$, $\beta$, $\beta'$, and $\sigma$) and rifampicin binds to the $\beta$ subunit. When rifampicin is bound to the enzyme, the complex can still attach to the DNA template and catalyze the initiation of RNA synthesis with the formation of the first phosphodiester bond, e.g., of the dinucleotide pppApU. However, the formation of a second phosphodiester bond, and therefore, the synthesis of long chain RNAs, is inhibited. The action of rifampicin is to lead to an abortive initiation of RNA synthesis (344, 345).

Resistance to rifampicin arises spontaneously in strains not exposed previously to the antibiotic at a rate of one mutation per $10^7$ to $10^8$ organisms. In *E. coli*, but also in *M. leprae* and in *M. tuberculosis*, resistance to rifampicin results from missense mutations in the *rpoB* gene, which encodes the $\beta$ subunit of RNA polymerase. These mutations are all located in a short region of 27 codons near the centre of *rpoB* and consist predominantly of point mutations, although in-frame deletions and insertions also occur. In most of *M. tuberculosis* rifampicin resistant clinical isolates, changes have occurred in the codons for Ser 531, or His 526, or Asp 516 (346–351). The analysis of genetic alterations in the *rpoB* gene has been suggested for predicting rifampicin and other rifamycins susceptibility (352). In addition to the mechanisms caused by the *rpoB* mutations, the lower susceptibility of fast growing *Mycobacterium* spp. to rifampicin may be partly caused by antibiotic inactivation mechanisms, consisting in the 23-*O*-ribosylation of rifampicin (353).

Although the main mechanism of resistance is the modification of the target enzyme, some mutagenic treatments yield resistant mutants in which the RNA polymerase is still highly sensitive to the drug, but the rate of rifampicin uptake is reduced. The mechanism of this permeability mutation is not yet clear (354).

**5.2.7 Antimycobacterial Rifamycins**

**5.2.7.1 Rifampicin.** Rifampicin (Rifampin, USAN, **55**) is the 3-[(4-methyl-1-piperazinyl)iminomethyl]-rifamycin SV. It has an amphoteric nature (p$Ka$ 1.7, 7.9) and is soluble in most organic solvents, slightly soluble in water at neutral pH, but more soluble in acidic and alkaline solutions.

Rifampicin is active *in vitro* against *M. tuberculosis* at concentrations below 1 $\mu$g/mL in semisynthetic media. It is active against other Gram-positive bacteria at lower concentrations and against Gram-negative bacteria at concentrations of 1–20 $\mu$g/mL. Rifampicin is active at the same concentrations against strains resistant to other antibiotics and antimycobacterials.

The bactericidal activity of rifampicin is demonstrated at concentrations close to the static ones. It is possible to isolate strains resistant to rifampicin from mycobacterial cultures exposed to the antibiotic, but the frequency of resistant mutants to rifampicin in sensitive populations of *M. tuberculosis* is lower than that to other antimycobacterial drugs (355).

Rifampicin is active against *M. leprae*, suppressing the multiplication and the viability of the bacilli in laboratory animal infections (128, 356–359). It is also active against many other mycobacteria, although at concentrations generally higher than those effective against *M. tuberculosis*. Among the various mycobacteria tested, *M. kansasii* and *M. marinum* are the most sensitive (360). There is no evidence of cross-resistance among rifampicin and other antibiotics or antituberculous drugs (361). Transfer of resistance to rifampicin could not be obtained.

Preliminary *in vivo* studies (362) showed that the antituberculous efficacy of rifampicin in experimental infections in mice was comparable with that of isoniazid and markedly superior to that of streptomycin, kanamycin, and ethionamide. In guinea pigs, it was comparable with streptomycin. Rifampicin was also remarkably active in experimental infections caused by Gram-positive and some Gram-negative bacteria.

The excellent antituberculous activity of rifampicin *in vivo* was confirmed by other experiments in many laboratories using various animal models (mice, guinea pigs, rabbits) and various schedules of treatment and criteria of evaluation (363–373). The overall results indicate that rifampicin has a high bactericidal effect and a therapeutic efficacy of the same order of isoniazid and superior to all the other antituberculous drugs. The combination of rifampicin plus isoniazid has been shown to produce a more rapid, complete, and durable sterilization of infected animals than other combination (366).

Many clinical trials have confirmed the efficacy of rifampicin in a variety of bacterial infections (374). In particular, rifampicin is largely used for the treatment of human tuberculosis both in newly diagnosed patients and in patients whose primary chemotherapy has failed. Normally, rifampicin is administered orally in a dose of 600 mg daily in combination with other antituberculous drugs. The short-course chemotherapy (SCT) of tuberculosis is a very effective treatment based on the combined administration of rifampicin

with isoniazid and other antituberculous drugs for 6–9 months. In human leprosy (128, 356–359, 375, 376), treatment with rifampicin alone or in combination with other antileprosy drugs gave favorable results in almost all the patients. In particular, the variations of the morphological index, and when carried out, the mouse footpad inoculation, showed the rapid and constant bactericidal action of rifampicin. The same effect was observed in many patients who had become resistant to other antileprosy drugs.

*5.2.7.1.1 Pharmacokinetics.* After oral administration, rifampicin is well absorbed in animals and humans (377). After administration of 150 and 300 mg to humans, serum levels reach maximum values around h 2 and persist as appreciable values beyond h 8 and h 12, respectively. When the dose is increased, serum levels are high and long lasting. Generally, the serum levels found at the beginning of treatment are higher than the levels that gradually set in as treatment continues. This phenomenon occurs during the first few weeks of treatment (378, 379). The half-life of rifampicin is around 3 h and increases in patients with biliary obstruction or liver disease (380–382). Rifampicin shows an extensive distribution in the tissues and crosses the blood-brain barrier. It reaches good antibacterial levels in cavern exudate and in pleural fluid. It penetrates inside the macrophages, where the tubercle bacillus, as an intracellular parasite, can live and multiply.

Rifampicin is eliminated through both the bile and the urine. It appears rapidly in the bile, where it reaches high concentrations that last even when the antibiotic is not measurable in the serum. After reaching the threshold of hepatic eliminatory capacity, biliary levels do not increase with an increased dosage, but serum and urinary levels do. In humans, rifampicin is mainly metabolized to 25-*O*-desacetylrifampicin (383), which is only slightly less active than the parent drug against *M. tuberculosis*, but considerably less active against some other bacteria. Both rifampicin and desacetylrifampicin are excreted in high concentrations in the bile. Rifampicin is reabsorbed from the gut, forming an enterohepatic cycle, but the desacetyl derivative is poorly absorbed and thus is excreted with the feces. Rifampicin has a stimulating effect on microsomal drug-metabolizing enzymes (384), which leads to a decreased half-life for a number of compounds, including prednisone, norethisterone, digitoxin, quinidine, ketoconazole, and the sulfonylureas.

*5.2.7.1.2 Adverse Reactions.* Rifampicin has a low toxicity according to acute, subacute, and chronic toxicity studies in several animal species (385, 386). Results of animal and human studies showed no toxic effect on the ear and eye.

In humans, adverse reactions to daily rifampicin are uncommon and usually trivial. The most frequent ill effects are cutaneous reactions and gastrointestinal disturbances. Rifampicin can also disturb liver function, but the risk of its causing serious or permanent liver damage is limited, particularly among patients with no previous history of liver disease.

In the case of intermittent therapy, when the drug is given three times, or once a week, a high incidence of severe side effects (such as the "flu-like" syndrome and cytotoxic reactions) may result. These adverse reactions seem to be associated with rifampicin-dependent antibodies, suggesting an immunological basis. However, with proper adjustment of the size of each single dose, the interval between doses, and the length of treatment, it has been possible to develop intermittent regimens with rifampicin that are highly effective and acceptably safe (386).

*5.2.7.2 Rifabutin.* Rifabutin (**56**) belongs to the group of spiroimidazorifamycins obtained by condensation of 3-amino-4-iminorifamycin SV with *N*-butyl-4-piperidone (387–389). Rifabutin has an antibacterial spectrum similar to rifampicin, but seems to possess incomplete cross-resistance with rifampicin *in vitro*. In fact some strains resistant to rifampicin are still moderately sensitive to rifabutin. Another characteristic of rifabutin is that its activity against *Mycobacterium avium* complex is higher than that of rifampicin (390–394).

Rifabutin is rapidly absorbed by mouth, but its oral bioavailability is only 20%, and there are considerable interpatient variations. The elimination half-life is long (35–40 h), but as a result of a large volume of distribution,

average plasma concentrations remain relatively low after repeated administration of standard doses. Binding to plasma proteins is about 70%. Rifabutin is slowly but extensively metabolized, possibly to more than 20 compounds in humans, the 25-desacetyl derivative being the main metabolite (395).

Rifabutin has proven value in preventing or delaying mycobacterial infections in immunocompromised patients (396). It has been approved in the United States and in other countries for the prevention of MAC infections in AIDS patients.

*5.2.7.3 Rifapentine.* Rifapentine (**57**) is the 3-(4-cyclopentyl-1-piperazinyl-iminomethyl)-rifamycin SV and therefore is an analog of rifampicin in which a cyclopentyl group substitutes for a methyl group on the piperazine ring (397). Its activity is similar to that of rifampicin, but is slightly superior against mycobacteria, including MAC (398, 408). It is more lipophilic and has a serum half-life about five times longer than rifampicin. This is because of its stronger binding to serum proteins than rifampicin (401, 402). Rifapentine is excreted primarily as intact drug in the feces; less than 5% of it and its metabolites are excreted in the urine. The primary metabolite in bile and feces is 25-desacetyl-rifapentine, with smaller amounts of the degradation byproducts, 3-formyl-rifapentine and 3-formyl-desacetyl-rifapentine, formed in the gut.

In experimental tuberculous infections, rifapentine administered once a week has practically the same therapeutic efficacy as rifampicin administered daily (397). It is in clinical trial as a drug for the therapy of tuberculosis and leprosy at a lower dosage and frequency of administration than rifampicin. However, relapse with rifamycin monoresistant tuberculosis occurred among HIV-seropositive tuberculosis patients treated with a once-weekly isoniazid/rifapentine continuation phase regimen (403).

*5.2.7.4 Rifalazil.* Rifalazil (previously known as KRM-1648) (**58**) is the 3'-hydroxy-5'-(4-isobutyl-1-piperazinyl) benzoxazinorifamycin, selected among various benzoxazine rifamycins (404–406). Rifalazil is more potent *in vitro* and *in vivo* against *M. tuberculosis* and *M. avium* complex than rifampicin. Rifalazil demonstrated excellent *in vivo* efficacy against *M. tuberculosis* infections produced by rifampicin-sensitive organisms, having activity in the murine model superior to rifampicin (407). The potent antimycobacterial activity of rifalazil is largely because of its increased ability to penetrate the mycobacterial cell walls (408). The high tissue drug levels and long plasma half-life of rifalazil significantly contributes to its excellent *in vivo* efficacy. Various experiments with rifalazil in combination with isoniazid demonstrated its potential for short-course treatment of *M. tuberculosis* infections. Rifalazil plus isoniazid for a minimum of 10 weeks was necessary to maintain a non-culturable state through the observation period (409). Rifalazil in combination with pyrazinamide and ethambutol has sterilizing activity comparable with that of the combination isoniazid plus rifampicin, but significantly better with respect to relapse of infection (410). If rifalazil will be developed for human therapy, because of its remarkable activity, it can, in combination with isoniazid, significantly shorten the duration of therapy. This suggests that ultra–short-course therapy is an attainable goal.

*5.2.7.5 Rifamycin T9.* Rifamycin T9 (**59**) is the 3-(4-cinnamyl-1-piperazinyl-iminomethyl) rifamycin SV, synthesized and evaluated for its antimycobacterial activity in the Chemical Pharmaceutical Research Institute of Sofia, Bulgaria. Investigations conducted by the developers of the drug indicated good therapeutic activity, lack of toxicity, and favorable pharmacokinetic and bioavailability in experimental animals. The *in vitro* activity of T9 against *M. tuberculosis* and MAC and its chemotherapeutic activity in experimental tuberculosis in mice gave encouraging results. The excellent *in vitro*, intracellular, and *in vivo* activities of T9, as well as its promising bioavailability, warrant its potential usefulness in the treatment of mycobacterial infections (411). Results of other studies show that T9 is twice as active as rifabutin and four times more active than rifampicin in inhibiting the growth of rifampicin-sensitive strains of *Mycobacterium leprae*. Furthermore, there is a demonstrated synergy between T9 and ofloxacin, suggesting that combination of T9 and ofloxacin would be an ideal formulation in multidrug regimen for leprosy (412).

## 5.3 Drugs Under Investigation

In recent years the need of new drugs to meet the problems connected with the emergence of multidrug resistant (MDR) tuberculosis and mycobacteriosis particularly in AIDS patients, failing any really new antituberculous compound, has led to carefully evaluate the antituberculous activity of antimicrobial agents developed for their activity against common Gram-positive and Gram-negative bacteria.

Some promising drugs are present in the groups of fluoroquinolones, macrolides, and β-lactams (in combination with β-lactamase inhibitors), but also in a new series of products, such as azaindoloquinazolindione alkaloids and nitroimidazopyrans.

### 5.3.1 Fluoroquinolones

*5.3.1.1 Historical Development.* These products represent a development of the earlier analogs (nalidixic acid, oxolinic acid, pipemidic acid, and cinoxacin) in being more potent *in vitro* and in having broader antibacterial spectrum, which includes Gram-positive and Gram-negative organisms.

They have also improved pharmacokinetic properties; whereas the old agents distributed very poorly in body tissues and fluids, so that they can be employed only as urinary antiseptics, the new derivatives distribute much better in all body districts, penetrate cells, and can be efficaciously used in the treatment of systemic infections. Their oral bioavailability is excellent (413).

Key points in determining these characteristics are the attachment of a fluorine atom at C-6 and of piperazinyl or *N*-methylpiperazinyl groups at C-7 and of alkyl or cycloalkyl groups at *N*-1 of the 1,4-dihydro-4-oxo-3-quinoline-carboxylic acid (414).

*5.3.1.2 Antimycobacterial Activity and Mechanism of Action.* Derivatives recognized with activity against mycobacteria are ofloxacin (**60**) and ciprofloxacin (**61**), which have been studied at a larger extent (415–421), whereas sparfloxacin (419, 420), levofloxacin, which is the levoisomer of ofloxacin (421), lomefloxacin (422, 423), WIN 5723 (424), and AM-1155 (425) have been more recently submitted to investigation and are under development. Another fluoroquinolone, moxifloxacin, recently introduced in clinical practice, shows a good *in*

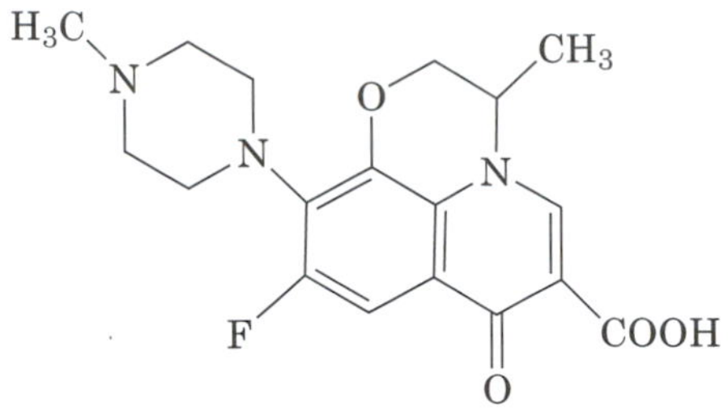

(60)

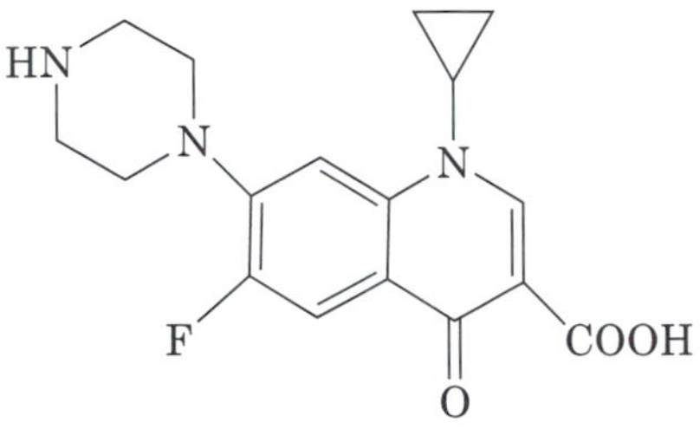

(61)

vitro activity against *M. tuberculosis* and *M. avium* complex and does not cause phototoxicity (426).

MICs of ofloxacin and ciprofloxacin for *M. tuberculosis* range from 0.12 to 2 μg/mL. Both drugs are bactericidal, the MBC/MIC ratio being from 2 to 4. Levofloxacin, the levo-isomer of ofloxacin, is twice as active as the parent drug; the bactericidal concentration corresponds to MIC. Ciprofloxacin and ofloxacin are also active against *M. avium* complex. MICs of ciprofloxacin for 50% of strains range from 1 to 16 μg/mL and MIC for 90% from 2 to 16 μg/mL (with data from some series exceeding 100 μg/mL). MICs of ofloxacin are a little higher, ranging from 2 to 16 μg/mL for 50% of strains and from 8 to 16 μg/mL for 90% of strains (again with values from a few studies exceeding 100 μg/mL). The MBC/MIC ratio most commonly ranges from 1 to 8. In addition, ciprofloxacin and ofloxacin show a quite good *in vitro* activity against *M. bovis*, *M. kansasii* and *M. fortuitum*, but not against *M. chelonae*.

The targets of quinolone action in bacterial cell are topoisomerase II, a DNA girase that contains two A subunits (Gyr A) and two B subunits, encoded by the *Gyr A* gene. DNA girase functions within viable bacterial cell include introduction of negative supercoils and Iopoisomerase IV, involved with decatenation of linked DNA molecules. They are essential for DNA recombination and repair. The main effects of fluoroquinolones are the inhibition of DNA supercoiling and the damage to DNA, whose synthesis is rapidly interrupted. At high quinolone concentrations, RNA and protein synthesis is also inhibited and cell filamentation occurs. It has been suggested, however, that quinolones may have other effects in the bacterial cells, for instance, the formation of an irreversible complex of drug, DNA, and enzyme, functioning as a "poison." In fact, nalidixic acid reduces the burst size of bacteriophages T7 at permissive temperature in *E. coli* strain containing thermosensitive Gyr A subunits, but not at non-permissive elevated temperature, suggesting that the inhibitory action of the drug on phage T7 depends on DNA inhibition even if the girase function is not required for phage growth (427, 428).

Moreover, quinolone concentrations that inhibit the DNA supercoiling and decatenating activity of purified DNA gyrase are several-fold higher than those required to inhibit bacterial growth. Again this discrepancy has suggested that other targets for quinolones exist in bacterial cell and several interpretations of this behavior have been put forward.

Activity of fluoroquinolone on susceptible bacterial species is bactericidal, but the inhibition of DNA synthesis does not seem sufficient to explain bacterial killing. All steps of this effect have not yet been elucidated; cells in logarithmic phase of growth are rapidly killed and the rate of killing increases with the increase of drug concentration, up to a maximum above which, paradoxically, the killing is reduced (429–430). This effect was already observed with penicillin and is known as the "Eagle" effect (431) from the name of the author who first described it. It is possible that this effect is a result of the inhibition of protein synthesis occurring in presence of high concentrations of quinolones. This hypothesis is supported also by the observation that bacteria exposed to antibiotics that inhibits protein synthesis (chloramphenicol, rifampicin) or to aminoacid starvation are less efficiently killed by fluoroquinolones.

*In vitro* association with other antituberculous drugs has given variable results. Combination of ciprofloxacin with major antituberculous drugs did not show any synergistic effect, but only indifference against *M. tuberculosis*. Synergism between ciprofloxacin or ofloxacin and ethambutol could be put in evidence for some strains of *M. avium*, whereas the effect was less evident when rifampicin replaced ethambutol. The combination of sparfloxacin with ethambutol, but not with rifampicin, had a synergistic effect against *M. avium*. Triple combination of sparfloxacin, rifampicin, and ethambutol resulted in synergism (415, 432, 433).

Triple combinations including isoniazid, rifabutine plus either ciprofloxacin, ofloxacin or norfloxacin, and isoniazid, rifapentine plus either pefloxacin or ciprofloxacin were the most active ones against resistant strains of *M. tuberculosis*. Rifampicin in combination with ciprofloxacin and amikacin in combination with isoniazid and ciprofloxacin gave the best results against *M. fortuitum* (433).

The synergistic interaction of antituberculous drugs *in vitro*, when obtained with *in vivo* achievable concentrations, may give some indications on the possible clinical efficacy.

Further information for the clinical application of fluoroquinolones can be drawn also from the activity of ofloxacin (434), ciprofloxacin (435), sparfloxacin (436), and levofloxacin (437), both singly or in combination with other antituberculous drugs in some experimental models of *M. tuberculosis* infection in mice.

Ofloxacin was shown to be highly active against *M. leprae* (MIC = 1.5 $\mu g/mL$), and the combination of ofloxacin and rifabutin or rifampicin has a synergistic effect (438). The combination of ofloxacin and rifabutin deserves further attention in the multidrug therapy of leprosy.

### 5.3.1.3 Clinical Uses in Mycobacterial Infection.
Fluoroquinolones have been introduced in several multidrug regimens, particularly in retreatment regimens, of MDR tuberculosis both in immunocompetent and AIDS patients and in MAC disease. Ofloxacin, ciprofloxacin, and more recently, sparfloxacin have been administered either with ethambutol, pyrazinamide, or with isoniazid and rifabutin and in other combinations. Even if

some results seem promising, no definite conclusion on the clinical value of these new approaches to treatment can be drawn.

### 5.3.1.4 Clinical Doses and Adverse Effects.
The daily dosage varies according to the different derivatives as follows: ofloxacin, 400 mg b.i.d.; ciprofloxacin, 750 mg o.d. or b.i.d. or 500 mg t.i.d.; sparfloxacin, 200–400 mg o.d. and moxifloxacin, 400 mg o.d. (439, 440).

Fluoroquinolones are usually well tolerated. The most common side effects are gastrointestinal reactions, central nervous disturbances, and skin and hypersensitivity reactions, particularly photosensitivity reactions (441). So far the quoted side effects have been observed during therapies of limited duration. Concerning the photosensitizing potential of the fluoroquinolones, it varies considerably between different agents. The likelihood of phototoxicity has probably been a limiting factor for the use of sparfloxacin in tuberculosis patients. There is an indication that a methoxy group at position 8 confers reduced phototoxicity, and this is the case of moxifloxacin. So far, however, no data support an increase of adverse events connected with the use of fluoroquinolones in the treatment of tuberculosis and mycobacteriosis.

Sporadic cases of hepatotoxicity have been reported, probably because of the association with other hepatotoxic drugs. It is suggested to monitor closely patients with concomitant liver impairment to stop medication if necessary (442, 443).

### 5.3.2 Macrolides
### 5.3.2.1 Historical Development.
Macrolides are a group of antibiotics characterized by a large lactonic structure of 12, 14, or 16 atoms. The lactone ring has hydroxyl, alkyl, and ketone groups in various positions. In the 16-membered ring macrolides, an aldehydic group can also be present. Sometimes two of three hydroxyl groups are substituted by sugars, which can be neutral (6-deoxy-hexoses), bound through a $\alpha$-glycosidic linkage, or basic (3-amino-sugars) bound through a $\beta$-glycosidic linkage.

The first derivative introduced in therapy was erythromycin, which was isolated in 1952 from the fermentation broths of *Streptomyces erythreus* and followed by other natural derivatives, such as spiramycin and oleandomycin, produced by other *Streptomyces* species. For some decades no other derivative was added to this antibiotic family until the 1980s, when josamycin and myocamycin, a semi-synthetic product, entered clinical use. Since then, new semi-synthetic derivatives have become available, namely roxithromycin, dirithromycin, flurithromycin, clarithromycin, and azithromycin. They belong to the 14-membered ring derivatives except the last one, which has a 15-membered ring with an *N*-methyl group between C9 and C10; for this reason, the name "azalides" has been proposed for azithromycin and other 15-membered ring macrolides (444).

Macrolides are usually administered by oral route, but because of their physico-chemical properties, they are poorly and erratically absorbed. Passage through the stomach can result in degradation and loss of activity. To obviate to this inconvenience, different salts and esters as well as adequate pharmaceutical formulations have been prepared (445). Once absorbed, they diffuse well in tissues and penetrate cells, so that they can be active in intracellular infections (446). They are metabolized in liver to a different extent according to derivatives, and because of their inducing activity on cytochrome P-450, they can interfere in the activity of other drugs metabolized by the same enzyme such as teophylline, antipyrine, and methylprednisolone (447). Macrolides are eliminated mainly by gastrointestinal route.

Among side effects, gastrointestinal disturbances prevail, and some derivatives show a certain degree of hepatotoxicity. However they are considered to be among the best-tolerated antibiotics (448).

*5.3.2.1.1 Antimycobacterial Activity.* The general spectrum of activity of macrolides includes Gram-positive cocci, some Gram-negative species, and intracellular pathogens (449).

The most recent macrolide derivatives show better pharmacokinetic characteristics, particularly longer half-life, and higher tissue and intracellular concentration than old derivatives. In addition, they show better activity against some Gram-negative species, such as *H. influenzae*, and in addition, clarithromycin and azithromycin have some action against mycobacteria (449).

It was known that erythromycin could be active in infections caused by rapidly growing mycobacteria, *M. fortuitum* and *M. chelonae*, and probably also in *M. smegmatis* infection. New macrolides (clarithromycin, azithromycin, and roxithromycin) have a similar or higher activity (416, 450).

At present, the interest is focused on determining which derivatives can have some clinically exploitable activity against MAC and *M. tuberculosis*.

*5.3.2.2 Clarithromycin.* Clarithromycin (**62**), as well as its hydroxy-metabolite (14-hydroxy-clarithromycin), are active *in vitro* against

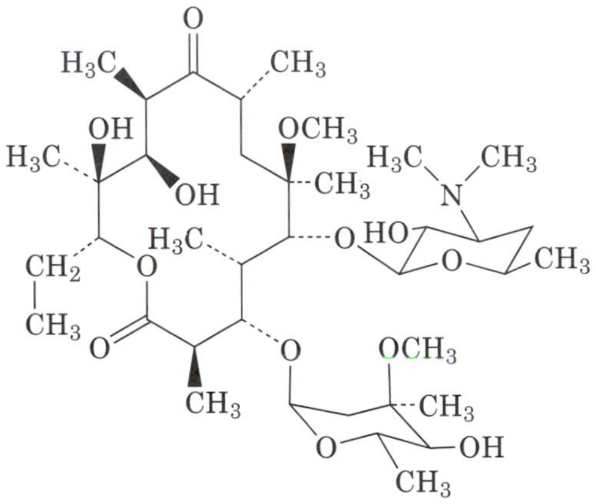

(62)

MAC. However, MICs are quite variable from study to study (451–454). Several *in vitro* experiments were carried out in presence of human macrophages, where clarithromycin accumulates (455–466). MIC is 1 $\mu$g/mL and MBC range from 16 to 64 $\mu$g/mL (456): these values are similar to those obtained in broth cultures at pH 7.4 and lower than those found in broth culture at pH 6.8 and 5. Intracellularly, mycobacteria grow at pH 6.8 and pH 5 in phagosomes and phagolysosomes, respectively. The fact that MIC and MBC against them were lower than those found in broth cultures is an indirect indication of concentrations of clarithromycin exceeding those in extracellular medium. In fact it was previously demonstrated that clarithromycin accumulates in cells at concentration 10- to 16-fold higher than those in the extracellular fluid (457–459).

Moreover, it was showed that single 2-h pulsed exposure of macrophages infected with *M. avium* to clarithromycin at 3 µg/mL completely inhibited the intracellular bacterial growth during the first 4 days of observation. This finding suggests that *in vivo*, the intracellular bacteria can be inhibited after a short period of exposure to the high concentrations of drug that can be reached at the time of blood peak level, rather then to prolonged exposure to low concentrations (460).

*In vivo* experiences confirmed the *in vitro* results; clarithromycin showed a good activity in the beige mouse model of disseminated *M. avium* infection, inducing a dose-related reduction in spleen and liver microbial cell counts for treatment with doses of 50, 100, and 200 mg/kg. Also, in patients with infection caused by *M. avium*, susceptible to 2 µg/mL or less, a dramatic decline in intensity of bacteremia up to a negative blood culture was achieved after 4–10 weeks of therapy (416, 453).

Combination of clarithromycin with amikacin, ethambutol, or rifampicin did not result in activity superior to that of clarithromycin alone, whereas the combinations with clofazimine or rifabutin were more active than clarithromycin alone. The triple combinations clarithromycin-rifampicin-clofazimine, clarithromycin-rifampicin-clofazimine, or clarithromycin-clofazimine-ethambutol were significantly more active than the macrolide alone (461).

Clarithromycin has been recommended by the U.S. Public Health Service Task Force on Prophylaxis and Therapy for MAC as a primary agent for the treatment of disseminated infection caused by *M. avium*. It should be used in combination with other antimycobacterial drugs that have shown activity against MAC, including ethambutol, clofazimine, and rifampicin (467).

The activity of some macrolides has been recently tested against *M. leprae in vitro* using ATP assay and *in vivo* using a food-pad model. Erythromycin was active *in vitro* but not *in vivo* (466, 468). Clarithromycin was active both *in vitro* (MIC = 0.1–2 µg/mL) and *in vivo*, demonstrating also a bactericidal activity (469). Similar data were obtained with roxithromycin. The acid stability of these new macrolides together with their long half-life, better penetration, and longer persistance in tissues are probably the reason for their activity, which deserves further investigation (466–469).

**5.3.2.3 Azithromycin.** Among the new macrolide derivatives, azithromycin (**63**) has a pe-

(63)

culiar place because of its chemical structure, consisting of a 15-membered lactone ring, and its pharmacokinetic characteristics. In fact, blood levels are very low, whereas very high concentrations are reached in tissues and cells; in polymorphonuclear neutrophyles, they can be up to 200- to 300-fold higher than extracellular concentrations. From cells, the antibiotic is slowly released, a phenomenon that assumes a particular relevance at the infectious focus where PMN accumulate. Half-life is about 60 h (470, 471). Because of these features, azithromycin can be administered for 3 days, assuring therapeutic tissue and cell concentrations for about 7–10 days (472–474). Azithromycin is endowed with good activity against some rapidly growing mycobacterial species and MAC. Its MICs on the latter species vary quite widely; the values range between 17 and 94 µg/mL (475, 476), but the $MIC_{90}$ (62 µg/mL) exceeds concentrations in tissues after oral dosing (477–478). It is bactericidal against *M. avium* and *M. xenopi* in macrophages. The addition of another active antimycobacterial drug such as amikacin or rifabutin enhanced the intracellular killing (479). Uptake of azithromycin by macrophages was increased by the addition of TNF or inter-feron-$\gamma$, but it is not known if the stimulation of azithromycin uptake accounts by itself for the increased killing of the macrolide (480).

The activity of azithromycin against MAC was confirmed in infection of rats treated with cyclosporine (481) and in the beige mouse model of *M. avium* infection (476, 482). The drug was efficacious when administered intermittently. Azithromycin given in combination with rifapentine on a once weekly basis for 8 weeks showed promising activity (483).

Combination with amikacin and clofazimine (484) or with clofazimine and ethambutol enhanced the efficacy of treatment; combination with rifabutin did not seem to be significantly superior to rifabutin alone (482).

In the same experimental model in beige mice, azithromycin had activity comparable with clarithromycin and rifampicin against *M. kansasii, M. xenopi, M. simiae*, and *M. malmoense*. The combinations with amikacin and clofazimine were more effective than the single drug (485).

Azithromycin was less active than clarithromycin and roxithromycin against *M. leprae* (118).

A few clinical trials have been carried out with azithromycin in *M. avium* infection in AIDS patients; after a few weeks of treatment (0.5–0.6 g once daily), a consistent reduction of bacteremia and improvement of clinical signs were observed (486). As already noted, macrolides cannot be employed alone in the therapy of mycobacteriosis; they must be included in polychemotherapeutic regimens.

**5.3.3 β-Lactams and β-Lactamase Inhibitors.** The resistance of *M. tuberculosis* to β-lactams has been attributed to the production of β-lactamases, with characteristics of both penicillinases and cephalosporinases (487).

One of the ways that was useful in overcoming the problem of β-lactamase resistance in other microorganisms has been to associate an inhibitor of the enzyme to penicillin, restoring the activity of the antibiotic susceptible to the enzymatic action. Inhibitors of β-lactamase such as clavulanic acid, sulbactam, and tazobactam have been associated to amoxycillin, ampicillin, ticarcillin, and piper-

acillin, and have been proven to be active not only *in vivo* but also in clinical setting, in the treatment of infections caused by β-lactamase–producing strains (488, 489). A number of β-lactamases have been characterized; they have different properties and different susceptibilities to β-lactamase inhibitors (490). β-lactamase produced by *M. tuberculosis* is susceptible to clavulanic acid; therefore, interest has been focused to ascertain whether its combination with a penicillin derivative could have some activity against *M. tuberculosis* (491). *In vitro*, the combination amoxycillin-clavulanic acid was remarkably more active than amoxycillin alone and had bactericidal activity (492, 493).

Two patients with multidrug resistant tuberculosis and poorly responding to multidrug regimens including ethionamide, capreomycin, and cycloserine, in addition to streptomycin, ethambutol, or pyrazinamide, were successfully treated following the addition of amoxycillin-clavulanic acid to these regimens (494).

Imipenem and meropenem, two carbapenems highly resistant to β-lactamases, were shown to have antimycobacterial activity only when a peculiar technique of *in vitro* dosing (daily addition of the drug) was adopted, compensating for the loss of activity of the antibiotics during the test incubation period. In fact the instability of the two compounds, particularly of imipenem, in the culture medium can hide their antimycobacterial activity (495).

Other β-lactams resistant to the β-lactamase action should be investigated for activity against mycobacteria and for their possible inclusion in multidrug regimens.

**5.3.4 Nitroimidazopyrans.** It has been reported recently the synthesis and the biological properties of a series of compounds containing a nitroimidazopyran nucleus that possess antitubercular activity (496). After activation by a mechanism dependent on *M. tuberculosis* F 420 cofactor, nitroimidazopyrans inhibited the synthesis of proteins and cell-wall lipids. In contrast to current antitubercular drugs, nitroimidazopyrans exhibited bactericidal activity against both replicating and static *M. tuberculosis*. Structure-activity relationship studies focusing on antitubercular activity revealed substantial variety in the tolerated substituent at C3, but optimal activity was achieved with lipophilic groups. The stereochemistry at C3 was important for activity, as the *S*-enantiomers were generally at least 10-fold more active than the *R*-enantiomers.

Lead compund PA-824 (**64**) showed potent bactericidal activity against multidrug resistant *M. tuberculosis* and promising oral activity in animal infection models. The nitroimidazopyrans represent a class of antitubercular compounds that act by a new mechanism but share some interesting parallels and contrast with INH. Like INH, the PA-824 prodrug requires bacterial activation, albeit by a different F420 dependent mechanism. PA-824 also inhibit a step in the synthesis of cell-wall mycolates, but at a more terminal step than INH.

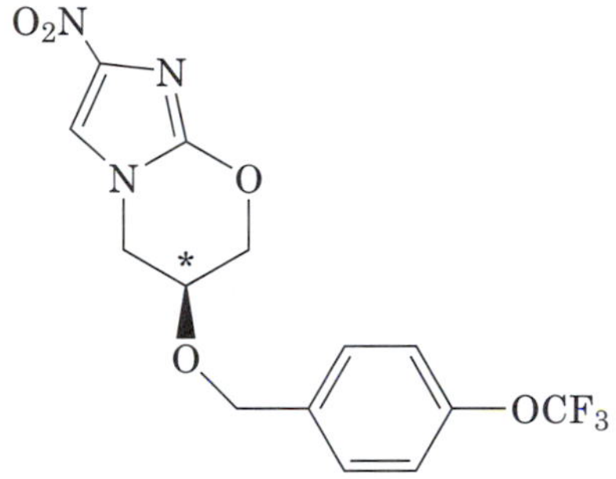

centration of 1 mg/L and against MAC at 2 mg/L, and it retained its activity against a panel of multiple drug-resistant strains.

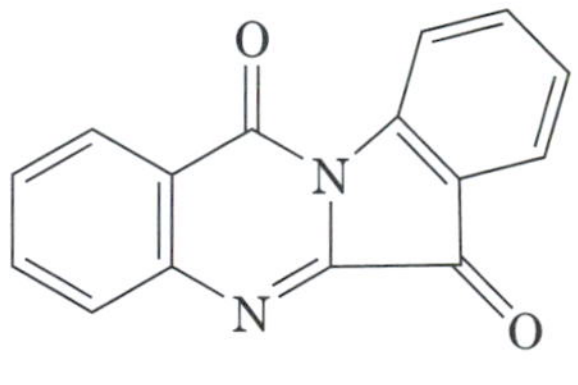

Extensive structure-activity studies have been performed on trytanthrin, with synthesis of many analogs using two parts of the molecule suitable for combinatorial chemistry. After examination of the various features of many analogs, PA-505 (the 2-aza-8-isooctyl analog of tryptanthrin (**66**) was considered the best and tested in detail. Its MIC against drug-sensitive and drug-resistant *M. tuberculosis* was 0.015 μg/mL and against MAC 19075 was 0.06 μg/mL. However, the *in vivo* activity in infected mice was unsatisfactory and the research is continuing on another series of derivatives (497).

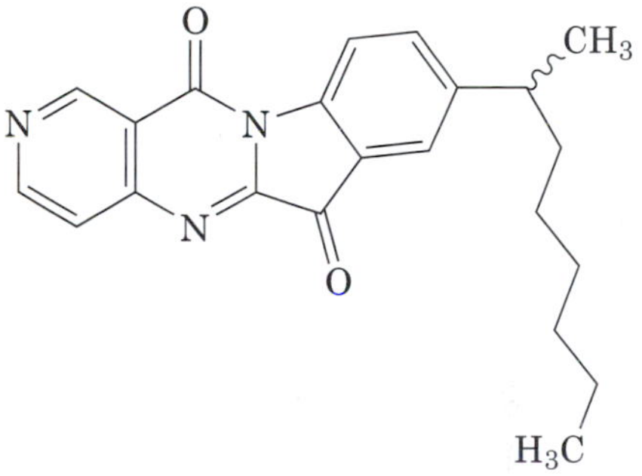

(64)

(66)

### 5.3.5 Tryptanthrin.

The indolo-quinazolinone alkaloid tryptanthrin (**65**) has been

(65)

known about since 1915. Isolated from a Chinese-Taiwanese medicinal plant, *Strobilantes cusia*, it was pursued intensively because of its very attractive *in vitro* properties, simple structure, and ease of synthesis. It retained its activity against *M. tuberculosis* H37Rv at con-

### 5.3.6 Oxazolidinones.

The oxazolidinones represent a class of products, with a general skeleton indicated in (**67**), extensively studied for their activity against Gram-positive and Gram-negative pathogens resistant to several antibacterial drugs. Some of them have been selected for their effect against mycobacteria. In particular, U-100480 (**68**) and DuP-721 (**69**) showed very low MIC against drug-sensitive and drug-resistant *M. tuberculosis* isolates and deserve further development (498, 499).

### 5.3.7 Thiolactomycins.

Thiolactomycins (TLM) (**70**) have been isolated as a metabolite of a soil *Nocardia* species. It is a unique thiolactone that exhibits antimycobacterial activity by specifically inhibiting fatty acids and mycolic acid biosynthesis. TLM targets two

| | R' | R'' |
|---|---|---|
| **(67)** | H | H |
| **(68)** | (thiomorpholine–N–phenyl, F) | NHCOCH$_3$ |
| **(69)** | CH$_3$CO– | NHCOCH$_3$ |

**(70)**

β-ketoacyl-carrier protein synthases, KasA and KasB, consistent with the fact that both enzymes belong to the fatty acid synthases type II system involved in fatty acid and mycolic acid biosynthesis. The design and synthesis of several TLM derivatives have led to compounds more potent both *in vitro* against fatty acids and mycolic acid biosynthesis and *in vivo* against *M. tuberculosis*. A three-dimensional structural model of Kas has been generated to improve understanding of the catalytic site of mycobacterial Kas proteins and to provide a more rational approach to the design of new drugs (500).

**5.3.8 Rifazalil and Rifamycin T9.** See Section 5.2.6.

# 6   RECENT DEVELOPMENTS AND PRESENT STATUS OF CHEMOTHERAPY

## 6.1   Tuberculosis

Chemotherapy has been the most potent and useful tool for modifying the evolution and prognosis of tuberculosis and has almost reached the goal of eradicating tuberculosis from some western countries. This positive tendency has been interrupted around 1985 when the incidence of tuberculosis started to increase for different reasons, outlined earlier. Moreover the rate of disease caused by resistant strains of mycobacteria, steadily falling in previous decades (being about 5% in industrialized countries), has risen to a varying extent according to countries and to socioeconomic status of some population classes. Data have not been systematically monitored in each country or region: therefore they have only an indicative value.

Recent data (1996–1999) reported by WHO and International Union Against Tuberculosis and Lung Disease, which undertook a global project on anti-tuberculous drug resistance surveillance in 35 countries, indicate that among patients not submitted to previous treatment, a median of 10.7% (range, 1.7–36.9%) of *M. tuberculosis* strains were resistant to at least one drug among the four first-line drugs taken into account (streptomycin, isoniazid, rifampicin, and ethambutol). The prevalence increased in Estonia from 28.2% in 1994 to 36.9% in 1998, and in Denmark from 9.9% in 1995 to 13.1% in 1998. In Western Europe, the median prevalence of primary multidrug resistance was less than 1%, whereas in Eastern Europe it was increasing, after a period of decline. In Estonia, it was 14.1% and in Latvia it was 9%. In the United States, the prevalence of primary resistance to any drug was 12.3%, whereas it was 1.6% for multidrug resistance. In patients who had prior treatment for more that 1 month, the prevalence of resistance for any drug ranged from 5.3% to 100% and that for MDR mycobacteria from 0% to 54% (501, 502). When there was a concomitant resistance to INH and RMP, the failure rate was over 50% in immunocompetent patients and about 100% with mortality of 50–90% in AIDS patients.

Data from many countries, China, India, and most African countries, are still lacking and are being collected together with those of other 40–50 countries. High rates of resistance were found in the countries of the former Soviet Union, The Domenican Republic, and Argentina. Of great concern is the very high incidence of 89% of non-responding patients and 24% of patients with newly diagnosed tuberculosis among the prison population of some countries of the former Soviet Union, whose disease was caused by MDR tubercle bacilli. It is evident that there is an enormous threat for the transmission of the drug-resistant strains to the community (503–505).

The success of chemotherapy is strictly related to the use of drug combination (to avoid the emergence of resistance) and to the rigorous adherence to the prescribed therapy regimens. Unfortunately, despite the introduction of simpler and shorter course of antimycobacterial chemotherapy, the compliance of patients is poor, particularly in developing countries. In the attempt to avoid the disastrous consequences of this behavior, after limited experiences in same countries, by the early 1980s, WHO implemented the directly observed therapy short-course (DOTS), i.e., patients had to be administered drugs (standard drugs for 6 months) under the direct observation of a supervisor (health workers) along with other measures to favor the accomplishment of the therapy (506, 507). Undeniably, DOTS achieved remarkable successes in some countries such as China (508). The first experiences, however, have shown that, even if the rate of implementation increases, tuberculosis mortality will not be halved (509) before 2030, and that the DOTS program is unable to control tuberculosis in countries with high levels of HIV infection and MDR tuberculosis. In these settings, where the DOTs regimen is expected to be ineffective, a new control strategy has been proposed, the "DOTs-plus" program, introducing either standardized or individualized regimens with second-line drugs. Second-line drugs are expensive, and the organization of this program is complex. Therefore, it seems that "DOTS-plus" strategy will be hard to implement on a large scale in low socioeconomic countries where MDR tuberculosis is prevalent (510, 511).

A turning point for the polychemotherapy of tuberculosis was represented by the introduction in clinical use of rifampicin in 1968. The addition of this potent drug to the association of the other active drugs already in use allowed the application of short-course treatment by reducing the period of treatment from 12–24 months to 6 months (or a maximum of 9 months). At present the choice of the therapeutic regimen is based on the knowledge of the epidemiology of resistant strains in the area where a patient is treated to apply the most adequate therapeutic regimen. This commonly must be prescribed on empirical basis, because the results of susceptibility tests require some weeks to be available. In the case of a presumably susceptible population of tubercle bacilli, an initial treatment with an association of three drugs is advised, whereas if a multiple resistance is suspected, a four- or five-drug regimen is recommended.

Drugs of first choice for the initial treatment are rifampicin, isoniazid, pyrazinamide, and ethambutol. Drugs should be administered daily for a 2-month period. In the meantime, the results of susceptibility tests should be available, which together with the clinical observation of the patient, will give indications for further treatment. A 4- to 6-month period follows in which drugs can be administered intermittently, two or three times a week. If the strain is susceptible to isoniazid and rifampicin, the maintenance treatment can continue with these two drugs (512, 513). However, many other options are to be considered, according to patient condition and requirements. An array of regimens have been suggested and applied with success for maintenance period: one of the most commonly employed regimen includes isoniazid + rifampicin + pyrazinamide + streptomycin or ethambutol administered twice or thrice a week. The same therapeutic regimen can be followed in HIV patients, but it should be extended to at least 9 months, and a fourth drug should be added initially for severe disease. When isoniazid-resistant tuberculosis is suspected, a three-drug regimen can be adopted for the first 2 months (rifampicin + ethambutol + pyrazinamide) followed by 7 months of daily rifampicin and ethambutol. An alternative treatment is a four-drug regimen (rifampicin + isoniazid + pyrazinamide and ethambutol or streptomycin), applied for the first 2 months, followed by rifampicin + ethambutol twice a week for 7 months or more. For patients responding to treatment in which tubercle bacilli turn out to be susceptible to all administered drugs, ethambutol (or strepto-mycin) can be withdrawn and isoniazid and rifampicin are continued for 4 or 6 months.

For patients living in areas where tuberculosis sustained by multidrug-resistant mycobacteria is widespread, a five-drug regimen is recommended initially, including both ethambutol and rifampicin. Successively, the treatment will be modified on the basis of results of susceptibility tests. In such patients, other antituberculous drugs, the second-line drugs, ethionamide, cycloserine, capreomycin, kanamycin, para-aminosalycilic acid, and sometimes clofazimine, are variably combined with each other and with first-line drugs with the aim to overcome resistance. The activity of other drugs showing *in vitro* activity against *M. tuberculosis* are under clinical investigation: amikacin, ciprofloxacin, and ofloxacin. Rifabutin has been included in some drug combinations, and it seems to have some efficacy also in the treatment of cases associated with low-level rifampicin resistance. Multidrug-resistant tuberculosis often requires prolonged treatment up to 24 months to reach sputum conversion (514–518).

The common dosages of antituberculous drugs and the schemas of some possible regimens are given in Tables 16.3 and 16.4.

The side effect of antituberculous therapy reflects the toxicity of the drug that is associated in the different combination regimens. Hepatotoxicity is quite frequent because drugs, and particularly the first-line drugs, are endowed with liver toxicity. Hepatitis with liver enzymes elevation can occur and it should be determined which is the responsible drug to discontinue it. Usually, after 1 week, liver enzymes return to baseline levels, and therapy can be resumed, but in some cases, the drug responsible of hepatotoxicity must be substituted with another better-tolerated agent. Other severe side effects are represented by hematologic derangements and neuropathy.

Antituberculus drugs, particularly rifampicin and isoniazid, caused by the induction of citochrome P-450 system, may interfere with the metabolism of other drugs, metabolized through the same enzymatic pathway and the reverse can be true. In Table 16.5, the principal drug interactions of antituberculous drugs are indicated.

So far the prevention of TB has been put into practice by the vaccination with BCG since the late 1940s. The evaluation of its activity is still controversial (519). The explanation of the disparate results obtained in different countries has been attributed to variation between strains of BCG and/or to the different policy adopted for vaccination (vaccination has been prescribed to neonates, to school age children, to select high-risk groups, to tuberculin negative subjects). Repeated doses of BCG have been administered in countries of the ex-Soviet Union. Unfortunately there have been almost no comparative evaluations of the effectiveness of the different policies. It has been observed that poor results are obtained in populations that are exposed to many different "environmental" mycobacteria, which could itself provide a kind of "vaccination" so that BCG could not improve greatly the response. This assumption seems confirmed by the results of a large trial performed in 1968 in South India, where the exposure to environmental mycobacteria was high. Neither of the two vaccines tested that were obtained from different BCG strains offered protection against tuberculosis. On the other hand, in front of continue decline in TB incidence in the community, countries of northern Europe (Sweden and England) showed a trend to discontinue the routine use of BCG and to restrict its use to high-risk groups (TB contacts and recent immigrants).

At present, about 100 million children receive BCG annually throughout the world. Most countries now follow the policies of the Expanded Program on Immunization (EPI)/ Global Program on Vaccines of the WHO, which recommends only a single dose of BCG be given at birth or at the earliest contact with a health service.

Another form of TB prevention is chemoprophylaxis (520), i.e., the administration of isoniazid to individuals not yet infected (tuberculin-negative) but at risk to acquire tuberculosis because of close contacts with TB patients. Chemoprophylaxis should be also prescribed to infected subjects, i.e., tuberculin-positive (under 35 years of age), to prevent the progression of the infection to the active disease when they live or have close contacts with high-incidence groups (nursing home,

**Table 16.3  Common Dosages of Available Antimycobacterial Drugs**

| | Dose and Route of Administration in Adults | | |
| --- | --- | --- | --- |
| | Daily Dose | Twice Weekly Dose | Thrice Weekly Dose |
| First-line drugs | | | |
| Isoniazid (INH) | 5 mg/kg os (max. 300 mg) | 15 mg/kg os (max. 900 mg) | 15 mg/kg os (max. 900 mg) |
| Rifampicin (RMP) | 10 mg/kg os (max. 600 mg) | 10 mg/kg os (max. 600 mg) | 10 mg/kg os (max. 600 mg) |
| Rifabutin (RTB) | 300 mg os | 300 mg os | 150 mg os |
| Ethambutol (EMB) | 15–25 mg/kg os (max. 2.5 g) | 50 mg/kg os | 25–30 mg/kg os |
| Streptomycin (SM) | 15 mg/kg IM (max. 1 g) | 25.30 mg/kg IM (max. 1.5 mg) | 25–30 mg/kg IM (max. 1.5 g) |
| Pyrazinamide | 15–30 mg/kg os (max. 2 g) | 30 mg/kg os (max. 2 g) | 30 mg/kg os[a] |
| Second-line drugs[b] | | | |
| p-Aminosalicylic acid (PAS) | 150 mg/kg os (max. 10–12 g) | | |
| Ethionamide | 15–20 mg/kg os (max. 1 g) | | |
| Prothionamide | 15–20 mg/kg os (max. 1 g) | | |
| Thioacetazone | 150 mg os | | |
| Capreomycin | 15–30 mg/kg IM (max. 1 g) | | |
| Viomycin | 1 g IM | | |
| Cycloserine | 15–20 mg/kg os (max. 1 g) | | |
| Kanamycin | 15–30 mg/kg IM (max. 1 g) | | |

[a]Doses higher than 2 g are not well tolerated.
[b]Second-line drugs are seldom used at present. Some have been withdrawn from the market in many countries. The dosage in intermittent regimens is usually the same as for initial treatment.

**Table 16.4   Tuberculosis Chemotherapy Regimens**

| Treatment (months) | Regimen[a] | |
| --- | --- | --- |
| | Initial | Maintenance |
| 8 | 2SHRZ | 6HR |
| 8 | 2HRE | $6H_3R_3Z_3E_3$ (or S) |
| 6 | 2HRE | $4H_3R_3$ |
| 6 | 2HRZ ($\pm$S or E) | $4H_3R_3$ |
| 6 | 6HRS | |
| 6 | 2HRZ | $4H_2R_2$ |
| 6 | $2S_3H_3R_3Z_3$ | $2S_3H_3R_32H_3R_3$ |
| 6 | $6H_3R_3Z_3$(S or E) | |
| 6 | 0.5SHRZ $1.5S_2H_2R_2Z_2$ | $4H_2R_2$ |

[a]A numeral preceeding letters indicates the number of months that drug combination is given, e.g., 4RH indicates 4 months of daily isoniazid and rifampicin. A numeral in subscript following a letter indicates the number of doses per week in intermittent regimens, e.g., $5H_3R_3E_3$ indicates rifampicin, isoniazid, and ethambutol each given three times weekly for 5 months. If there is no subscript, the drug is given daily.

E, ethambutol; H, isoniazid; R, rifampicin (rifampin); S, streptomycin; Z, pirazinamide.

mental institution, correctional or long-term facilities, etc.). The recommended doses of INH are 10 mg/kg for children and adolescents and 5 mg/kg daily for adults (maximum daily dose = 300 mg). For subjects requiring DOT, a twice-weekly dose of 15 mg/kg to a maximum of 900 mg is suggested when daily DOT is not possible. If the contact is likely to be infected with INH-resistant strain, rifampicin should substitute INH in chemoprophy-laxis at the dosage of 10 mg/kg up to 600 mg daily. The duration of the preventive therapy ranges from a minimum of 6 months up to 12 months.

## 6.2   Mycobacteriosis from NTM

These mycobacteriosis are particularly frequent in AIDS patients; about 90% of them are affected. The most frequent opportunistic infection in these patients is now caused by the

**Table 16.5   Interaction of Some Antimycobacterial Drugs with Other Agents**

| Antimycobacterial Drug | Interaction with |
| --- | --- |
| Ethambutol | $\uparrow$ Chloroquine, chloramphenicol, didanosine, disulfiram, isoprinosine, thiazide diuretics, zalcitabine |
| Isoniazid | $\uparrow$ Anticoagulants, acetaminophen, barbiturates, carbamazepin, cyclosporine, corticosteroids, diazepam, didanosine, disulfiram, flucitosine, ketoconazole, methyldopa, phenitoin, rifampicin,[a] pyrazinamide,[a] theophylline, vidarabine, zalcitabine |
| Pyrazinamide | $\uparrow$ Didanosine, isoniazid,[a] ketokonazole, viridazole, thiazine diuretics |
| Rifampicin | $\downarrow$ Analgesic, anticoagulants, anticonvulsant, azathioprine, barbiturates, beta-adrenergic blockers, chloramphenicol, oral contraceptives, corticosteroids, cyclosporine, dapsone, diazepam, digoxin, disopyramide, estrogens, haloperidol, methadone, protease inhibitors (saquinavir, ritonavir, indinavir, nelfinavir), quinidine, zidovudine |
| | $\uparrow$ Didanosine, isoniazid,[a] zalcitabine |
| Rifabutin | Less interaction than rifampicin. |
| | $\downarrow$ clarithromycin, cyclosporine, protease inhibitors |
| Streptomycin | $\uparrow$ Acyclovir, amphothericin, nephrotoxic beta-lactams, carboplatinum, cisplatinum, cyclosporine, loop diuretics, 5-fluorocytosine neuromuscolar blocking agents, NSAID, vancomycin |

$\uparrow$ Possibile increase of activity and/or toxicity.
$\downarrow$ Possibile decrease of activity.
[a]Antimycobacterial drug that contributes to the increased hepatotoxicity of therapeutic regimens.

*M. avium* complex. The disease may be localized in lungs, lymph nodes, and skin while in patients with less >50 CD4/mm$^3$ the disease is disseminated. *M. kansasii* and *M. xenopi* cause chronic pulmonary infections, *M. genavense* is responsible for disseminated infections, and *M. haemophilum* causes septic lesions involving skin and joints. Except *M. kansasii*, NTM are scarcely susceptible to classic antituberculous drugs. Therapeutic regimens (521) applied in these infections are summarized in Table 16.6.

## 6.3 Leprosy

The therapy of leprosy has reached significant achievements since 1982, when WHO promoted new protocols of treatment, providing a short-term multidrug therapy (MDT) (522, 523). In this period, the estimates of the total number of leprosy cases in the world has been falling downward. This tendency has not to be attributed only to implementation of MDT but to a series of other factors, such the revision in the definition of patients. Patients who have completed their course of chemotherapy are no longer counted as registered cases even if they have residual disabilities. In addition, better living conditions and better nutrition greatly contributed to the decline of incidence of the disease in some countries. There has been a reduction of 42% in cases of leprosy in the world since 1985; the global estimates in 1991 were 5.5 million cases (of which 2 million were not receiving treatment) compared with 10–12 million in the past (522–524).

The regimens that have been followed according to WHO recommendations in developing countries are mainly two, one directed to treat paucibacillary disease and the other for multibacillary disease. The former consist in the administration of dapsone 100 mg daily plus rifampicin 600 mg monthly for 6 months, the latter in the administration of dapsone 100 mg plus clofazimine 50 mg daily plus rifampicin 600 mg and clofazimine 300 mg monthly for a minimum of 2 years (522, 523). Relapse rates have been considered low enough for WHO to recommend the reduction of the therapy to 1 year (525). In some cases a daily treatment with rifampicin and dapsone is privileged for 1 year in paucibacillary leprosy and for 2 years with the addition of clofazimine in multibacillary disease (522, 523).

On the basis of a multicenter trial, the WHO Committee on Leprosy concluded that single-lesion paucibacillary disease could be treated with a single dose of therapy consisting of 600 mg rifampicin, 400 mg ofloxacin, and 100 mg minocycline. This regimen seems to be as effective as the standard paucibacillary regimen previously approved by WHO (526), that is, dapsone 100 mg daily plus rifampicin monthly for 6 months. However the results are still under debate.

During therapy, about 25% of all borderline and lepromatous patients develop acute inflammatory reactions probably caused by the formation of immune complexes caused by the high antigen load released by dying mycobacteria, with a consequent secretion of TNF-$\alpha$ from macrophages (527). Reactions are of two main types, namely reversal reaction and erythema nodosum leprosum. The former consists in new and increased inflammation in preexisting skin lesions, with or without involvement of nerve trunks; the latter consists in the formation of small tender erythematous subcutaneous nodules, accompanied by fever, arthralgia, vasculitis, adenopathy, and further inflammatory reaction in other organs. These reactions are usually controlled with corticosteroids. In the most severe cases thalidomide, azathioprine, and cyclosporine have been employed, but because of serious side effects, their use has to be restricted and closely monitored (528).

A field which attracts a particular interest is the possibility to develop a vaccine to prevent the disease. Three vaccines have been prepared combining BCG with *M. leprae* or other mycobacterial species and are currently under investigation, but results of these trials will only be available in 8–10 years (529). A preliminary report on a trial performed in Venezuela with a vaccine constituted by BCG and killed *M. leprae* does not evidentiate any advantage over the use of BCG alone (530). In fact, BCG alone, particularly if repeated, has shown some protective activity against both tuberculosis and leprosy (531).

**Table 16.6  Regimens for the Treatment and Prophylaxis of Some NTM Infections**

| Pathology | Therapy | Duration of Therapy |
| --- | --- | --- |
| *M. kansasii:* pulmonary disease | Isoniazid 300 mg, + rifampicin 600 mg, + ethambutol 25 mg/kg for 2 months, (then 15 mg/kg), daily. Possible alternative drugs: clarithromycin, sulfamethoxazole, amikacin, newer quinolones, rifabutin | 18 Months, with a minimum of 12 months sputum negativity |
| *M. avium* complex: pulmonary disease | Clarithromycin 500 mg, twice daily (or azithromycin 250 o 500 mg three time a week) + rifabutin 300 mg (or rifampicin 600 mg) + ethambutol 25 mg/kg daily for 2 months, then (15 mg/kg) + streptomycin 0.750-1 g three times a week for 2–3 months | Until culture negative for at least 12 months |
| *M. avium* complex: disseminated disease | Clarithromycin 500 mg twice a day (or azithromycin 250 mg three times a week) + rifampicin 600 mg (or rifabutin 300 mg + ethambutol 25 mg/kg for 2 months, (then 15 mg/kg) Streptomycin 0.75-1 g 2–3 times a week for the first 8 weeks should be considered if tolerated. | Until culture negative or lifelong |
| | Prophylaxis in adults with AIDS with CD4 < 50 cells: Rifabutin 300 mg daily or clarithromycin 500 mg twice daily; or azithromycin 1200 mg once weekly; or azithromycin 1200 mg once weekly + rifabutin 300 mg daily | Lifelong |
| *M. avium, M. scrofulaceum, M. malmoense:* cervical lymphadenitis | Possibly susceptible to clarithomycin, amikacin, ciprofloxacin, doxicycline, clofazimine. Surgical resection | 4–6 Months |
| *M. marinum:* cutaneous diseases | Clarythromycin 500 mg twice daily; or minocycline or doxycycline 100 mg twice daily; or TMP/SMX 160/800 mg twice daily: or rifampicin 600 mg + ethambutol 15 mg/kg daily | 3 Months |
| *M. malmoense:* pulmonary disease | Ethambutol, rifampicin, streptomycin (in combination according to susceptibility tests) | Variable |
| *M. simiae:* pulmonary and disseminated disease | Clarithromycin, ethambutol, rifabutin, streptomycin (4 drug in combination according to susceptibility test) | Variable |

## 7  THINGS TO COME

What is the future of tuberculosis? What is the way to overcome the many problems that this disease still poses?

The future projections of the global tuberculosis epidemic, as estimated by WHO, are not optimistic. If effective national tuberculosis programs are not implemented, it is likely that the epidemic will worsen. Several factors,

**Table 16.6**   (*Continued*)

| Pathology | Therapy | Duration of Therapy |
|---|---|---|
| Rapidly growing mycobacteria: *M. fortuitum* | Amikacin, tobramycin, cefoxitin, imipenem, newer macrolides, quinolones, doxycycline, minocycline, sulfonamides (2 drug combination therapy, according to susceptibility test) | 6–12 Months |
| *M. abscessus* | Amikacin, cefoxitin, imipenem, newer macrolides (2 drug combination therapy, according to susceptibility test) | 2–4 Weeks (curative course probably of 4–6 months) |

Adapted from American Thoracic Society (ATS), Diagnosis and treatment of disease caused by nontuberculous Mycobacteria, *Am. J. Resp. Crit. Care Med.*, **156**, 51 (1997).

mostly particular to developing countries, contribute to confirm this hypothesis: war, natural disaster, poverty, malnutrition, demographic changes, increasing antituberculous drug resistance, and expansion of HIV pandemic, especially in Southeast Asia. It is estimated that the global HIV-related tuberculosis cases will increase from 700,000 in 1995 to 1.4 million in 2000 (507, 532).

However, some lines of research have to be pursued to find possibly innovative solutions to the problem of tuberculosis.

It is a common feeling and it seems to be a realistic approach that the global problem of tuberculosis will be solved when an efficacious vaccine is (will be!) available. A first goal is achieving a better evaluation of the effectiveness of BCG vaccine, determining the factors that influence it, and identifying the immunological correlates of its behavior in different populations, when the effectiveness of the vaccine is known. A better knowledge of its mechanism of action will allow some genetic modifications leading to an enhanced activity. This approach has been attempted, and encouraging results have been obtained. Strains of BCG secreting cytokines or overexpressing specific antigens showed enhanced immunogenecity (533, 534).

DNA vaccines have raised great interest; a recombinant antigen when administered as DNA vaccine is immunoprotective as the results obtained in animal models demonstrated (535, 536). That means that any of the 4000 genes identified in the *M. tuberculosis* genoma could be rapidly tested for their potential use as subunit vaccines. This possibility greatly enhances the projects for vaccine development (537). The steps for developing an improved tuberculosis vaccine have been indicated (538). More than 190 candidates vaccines have been screened in animal models (539).

There are already at least 40 potential vaccines that could be considered for clinical trials, but the enormous complexity and the cost of the organization of such trials delays or prevents their realization (537).

From the point of view of treatment, immunotherapy could offer a solution to the problem of MDR tuberculosis as well of other of mycobacterioses scarcely susceptible to the classical antituberculous therapy. Cytokines could be directly administered in patients systemically or by aerosol. Experiments that have been performed in animals and some limited trials in humans with IL-2 and IFN-$\gamma$ seem to give some promising results (540, 541).

The goal to be reached in the search of new antituberculous chemotherapeutic agents should be the development of drugs with novel mechanism of action and new targets, high antimycobacterial activity, improved pharmacokinetic characteristics, and better tolerability. Possibly, the ideal antimycobacterial agent should be capable to eradicate the infection by a very short course of therapy, better if by single-drug administration. This result has been practically reached by the use the single dose of the association rifampicin, ofloxacin, and minocycline in the treatment of certain forms of leprosy (542).

Improvements in the implementation of the DOT program, which in past years has been deceiving in many countries, will be of value. It was estimated that 25% of the world's population may have access to DOTS in mid-2000, in comparison with about 10% in 1995. To further improve the results, it is necessary to increase substantially case detection and notifications. The WHO targets for TB control are 85% treatment success rate among smear positive cases and detection of 70% of all such cases. These targets had not been met by the start of the year 2000, but could be achieved by 2005. To achieve this program, it is necessary that the largest countries, particularly India, which have a high incidence rate of TB, extend the DOTS procedure to the whole population by 2004 (507, 532).

In the near future new and old tools (drugs, vaccines, diagnostic methods) can be combined and used in different ways to improve the outcome of tuberculosis, but it is necessary that the recent discoveries and hopefully some new therapeutic agents will provide us with the means to apply better strategies to the therapy and prevention of tuberculosis.

## REFERENCES

1. L. S. Young and G. P. Wormser, *Scand. J. Infect. Dis.*, **93**(suppl), 9 (1994).

2. W. H. Iopling, *Lepr. Rev.*, **62**, 1 (1991).

3. C. B. Inderlied, C. A. Kemper, and L. E. M. Bermudez, *Clin. Microbiol. Rev.*, **6**, 266 (1993).

4. F. Ayvazian, in L. B. Reichman, E. S. Hershfield, Eds., *Tuberculosis. A Comprehensive International Approach*, Marcel Dekker, Inc., New York, 1993, p. 1.

5. A. L. Davis, in L. B. Reichman, E. S. Hershfield, Eds., *Tuberculosis. A Comprehensive International Approach*, 2nd ed., Marcel Dekker, Inc., New York, 2000, p. 3.

6. P. Sensi, *Rev. Infect. Dis.*, **5**(suppl 3), S402, (1983).

7. A. Sakula, *Br. J. Dis. Chest*, **82**, 23 (1988).

8. Anonymous, *MMRW Morbid. Mortal. Wkly. Rep.*, **48**, 732 (1999).

9. C. Dye, et al., *J. Am. Med. Ass.*, **282**, 677–686 (1999).

10. G. S. Wilson and A. A. Miles, *Topley and Wilson's Principles of Bacteriology and Immunity*, Arnold, London, 1975.

11. D. van Soolingen, A. G. M. van der Zanden, P. E. W. de Haas, et al., *J. Clin. Microbiol.*, **36**, 1840 (1998).

12. S. T. Cole, K. Eiglmeier, J. Parkhill, K. D. James, N. R. Thomson, P. R. Wheeler, N. Honoré, T. Garnier, et al., *Nature*, **409**, 1007 (2001).

13. E. H. Runyon, *Adv. Tuberc. Res.*, **14**, 235 (1965).

14. American Thoracic Society, *Am. J. Am. Resp. Crit. Care*, **156**, 51 (1997).

15. G. D. Roberts, E. Koneman, and Y. K. Kim, in A. B. Balows, W. J. Hausler, K. L. Hermann, H. D. Isenberg, H. J. Shadomy, Eds., *Manual of Clinical Microbiology*, 5th ed., ASM, Washington DC, 1991, p. 304.

16. R. Buttiaux, H. Beerens, and A. Tacquet, *Manuel de Techniques Bactériologiques*, 4th ed., Flammarion Medécine Sciences, Paris, 1974.

17. G. Canetti and J. Grosset, *Techniques et Indications des Exames Bactériologiques en Tuberculose*, Editions de la Tourelle, St. Mandè, 1968.

18. B. Watt, A. Kayner, and G. Harris, *Rev. Med. Microbiol.*, **4**, 97 (1993).

19. G. A. Snow, *Bacteriol. Rev.*, **34**, 99 (1970).

20. E. Lederer, *Pure Appl. Chem.*, **25**, 135 (1971).

21. E. Lederer, A. Adam, R. Ciorbarn, J. F. Petit, and J. Wietzerbin, *Mol. Cell. Biochem.*, **7**, 87 (1975).

22. M. B. Goren, *Bacteriol. Rev.*, **36**, 33 (1972).

23. T. Ramakrishnan, M. Suryanarayana Murthy, and K. P. Gopinathan, *Bacteriol. Rev.*, **36**, 65 (1972).

24. L. P. Macham and C. Ratledge, *J. Gen. Microbiol.*, **89**, 379 (1975).

25. W. Segal and U. Block, *J. Bacteriol.*, **72**, 132 (1956).

26. W. Segal and U. Block, *Am. Rev. Tuberc. Pulm. Dis.*, **75**, 495 (1957).

27. W. Segal and W. T. Miller, *Proc. Soc. Exp. Biol. Med.*, **118**, 613 (1965).

28. W. Segal, *Proc. Soc. Exp. Biol. Med.*, **118**, 214 (1965).

29. M. R. McNeil and P. J. Brennan, *Res. Microbiol.*, **142**, 451 (1991).

30. C. Asselineau, *The Bacterial Lipids*, Hermann, Paris and Holden Day, San Francisco, 1966, p. 176.

31. E. Lederer, *Chem. Phys. Lipids*, **1**, 294 (1967).

32. A. H. Etémadi and J. Convit, *Infect. Immun.*, **10**, 236 (1974).

33. A. H. Etémadi, *Exp. Ann. Biochem. Med.*, **28**, 77 (1967).

34. J. Azuma and Y. Yamamura, *J. Biochem.*, **52**, 200 (1962).

35. J. Azuma and Y. Yamamura, *J. Biochem.*, **53**, 275 (1963).

36. N. P. V. Acharya, M. Senn, and E. Lederer, *C. R. Acad. Sci. Paris Ser. C.*, **264**, 2173 (1967).

37. J. W. Berg, *Proc. Soc. Exp. Biol. Med.*, **84**, 196 (1953).

38. J. W. Berg, *Yale J. Biol. Med.*, **26**, 215 (1953).

39. L. Barksdale and K. S. Kim, *Bacteriol. Rev.*, **41**, 217 (1977).

40. J. Markovits, E. Vilkas, and E. Lederer, *Eur. J. Biochem.*, **18**, 287 (1971).

41. F. Kanetsuma, *Biochim. Biophys. Acta*, **98**, 476 (1965).

42. H. Bloch, *J. Exp. Med.*, **91**, 197 (1950).

43. H. Noll, H. Bloch, J. Asselineau, and E. Lederer, *Biochim. Biophys. Acta*, **20**, 299 (1956).

44. E. Dàrzins and G. Fahar, *Dis. Chest.*, **30**, 642 (1956).

45. G. Middlebrook, R. J. Dubos, and C. Pierce, *J. Exp. Med.*, **88**, 521 (1948).

46. D. W. Smith, H. M. Randall, A. P. McLennan, and E. Lederer, *Nature*, **186**, 887 (1960).

47. A. P. McLennan, D. W. Smith, and H. M. Randall, *Biochem. J.*, **80**, 309 (1961).

48. G. Lanéelle and J. Asselineau, *Eur. J. Biochem.*, **5**, 487 (1968).

49. J. E. Clark-Curtiss, in J. J. McFadden, Ed., *Molecular Biology of the Mycobacteria*, Surrey University Press, London, 1990, p. 77.

50. P. R. S. Gangadharam, M. L. Cohn, and G. Middlebrook, *Tubercle*, **44**, 452 (1963).

51. M. Kato and M. B. Goren, *Japan J. Med. Sci. Biol.*, **27**, 120 (1974).

52. P. Pigretti, E. Vilkas, E. Lederer, and H. Bloch, *Bull. Soc. Chim. Biol.*, **47**, 2039 (1965).

53. V. Braun, and K. Hantke, in L. Ninet, P. E. Bost, P. M. Bouanchaud, J. Florent, Eds., *The Future of Antibiotherapy and Antibiotic Research*, Academic Press, London, 1981, 285.

54. K. Kanai, E. Wiegeshaus, and D. W. Smith, *Japan J. Med. Sci. Biol.*, **23**, 327 (1970).

55. H. N. Jayaram, T. Ramakrishnan, and C. S. Vaidyanathan, *Arch. Biochem. Biophys.*, **126**, 165 (1968).

56. H. N. Jayaram, T. Ramakrishnan, and C. S. Vaidyanathan, *Indian J. Biochem.*, **6**, 106 (1969).

57. D. D. Young and T. R. Garbe, *Infect. Immun.*, **59**, 3086 (1991).

58. Y. Zang, R. Lathigne, D. Garbe, et al., *Microbiology*, **5**, 381 (1991).

59. S. T. Cole, R. Brosch, J. Parkill, C. Garnier, C. Churcher, D. Harris, S. V. Gordon, K. Eiglmeier, et al., *Nature*, **393**, 537 (1998).

60. Y. C. Manabe, A. M. Dannenberg, and W. R. Bishai, *Int. J. Tuberc.*, **4**(suppl 21), S18 (2000).

61. I. Ramakrishnan, N. A. Federspice, and S. Falkow, *Science*, **288**, 1436 (2000).

62. R. Brosch, W. J. Philipp, E. Stavropoulos, M. J. Colston, S. T. Cole, and S. V. Gordon, *Infect. Immunol.*, **67**, 5768 (1999).

63. R. Brosch, S. V. Gordon, K. Eiglmeier, T. Garnier, and S. T. Cole, *Res. Microbiol.*, **151**, 135 (2000).

64. S. V. Gordon, K. Eiglmeier, T. Garnier, R. Brosch, J. Parkhill, B. Barrell, S. T. Cole, and R. G. Hewinson, *Tubercle*, **81**, 157 (2001).

65. M. B. Lurie, *Resistance to Tuberculosis: Experimental Studies in Native and Acquired Defensive Mechanisms*, Harvard University Press, Cambridge, MA, 1964.

66. A. M. Dannenberg, *Immunol. Today*, **12**, 228 (1991).

67. E. A. Nardell, in L. B. Reichman and E. S. Hershfield, Eds., *Tuberculosis. A Comprehensive International Approach*, M. Dekker, Inc., New York, 1993, p. 103.

68. J. L. Flinn, J. Chan, K. J. Triebold, D. K. Dalton, T. A. Stewart, and B. R. Bloom, *J. Exp. Med.*, **178**, 2249 (1993).

69. J. M. Orme, *Curr. Opin. Immunol.*, **5**, 497 (1993).

70. L. E. M. Bermudez, and L. S. Young, *J. Immunol.*, **140**, 3006 (1988).

71. M. Dems, *Clin. Exp. Immunol.*, **4**, 66 (1991).

72. A. Fayyazi, B. Eichimeyer, A. Soruni, et al., *J. Pathol.*, **191**, 417 (2000).

73. A. Gaw Rook and A. Zumnia, *Curr. Opin. Pulm. Med.*, **7**, 116 (2001).

74. A. C. Fratazzi, R. D. Arbeit, C. Carini, et al., *J. Leukocyte Biol.*, **66**, 763 (1999).

75. R. Placido, G. Mancino, A. Amendola, et al., *J. Pathol.*, **181**, 31 (1997).

76. D. Maiti, Bhattacharyya, and J. Basu, *J. Biol. Chem.*, **276**, 329 (2000).

77. M. Rojas, L. F. Garcia, J. Nigou, et al., *J. Infect. Dis.*, **182**, 240 (2000).

78. American Thoracic Society, *Am. Rev. Respir. Dis.*, **123**, 343 (1981).

79. M. S. Glicknoan, J. S. Cox, and W. R. Jacobs, *Mol. Cell*, **5**, 717 (2000).

80. J. D. McKinney, K. H. ZuBentrup, E. J. Munoz-Elias, et al., *Nature*, **406**, 735 (2000).

81. J. S. Cox, B. Chen, M. McNeil, et al., *Nature*, **402**, 79 (1999).

82. World Health Organization, *Global Tuberculosis Report. WHO Report 2001*, Geneva, Switzerland, 2001.

83. C. J. Murray, K. Stiblo, and A. Rouillon, *Bull. Int. Union Tuberc. Lung Dis.*, **65**, 24 (1990).

84. J. L. Johnson and J. J. Ellner, *Curr. Opin. Pulm. Med.*, **6**, 189 (2000).

85. G. W. Comstock and G. M. Cauthen, in Ref. 61, p. 23.

86. Center for Disease Control, *MMWR Morbid. Mortal. Wkly. Rep.*, **39**, 7 (1990).

87. Center for Disease Control, *Tuberculosis Statistics in United States 1989—HH Publication No. (CDC) 91–8322*, Public Health Service, Atlanta, GA, 1989.

88. WHO, Assembly Emphasizes Tuberculosis Crisis, WHA/14, WHO, Geneva, 1993.

89. D. Bleed, C. Dye, and M. C. Raviglione, *Curr. Opin. Pulm. Med.*, **6**, 174 (2000).

90. A. Pablos-Mendez, M. C. Raviglione, and A. Laszlo, et al., *N. Engl. J. Med.*, **338**, 641 (1998).

91. P. F. Barnes and S. A. Barrows, *Ann. Intern. Med.*, **119**, 400 (1993).

92. P. C. Hopwell, in Ref. 61, p. 369.

93. G. Di Perri, M. C. Danzi, G. De Cecchi, et al., *Lancet*, **2**, 1502 (1989).

94. C. L. Daley, P. M. Small, G. F. Schecter, et al., *N. Engl. J. Med.*, **326**, 231 (1992).

95. Center for Disease Control, *MMWR Morbid. Mortal. Wkly. Rep.*, **39**, 718 (1990)

96. C. R. Horsburgh Jr., *N. Engl. J. Med.*, **324**, 1332 (1991).

97. L. E. Bermudez, C. B. Interlied, and L. S. Young, *Curr. Clin. Topics Infect. Dis.*, **12**, 257 (1992).

98. L. S. Young, *J. Antimicrob. Chemother.*, **32**, 179 (1993).

99. R. Coninx, C. Mathieu, M. Debacker, et al., *Lancet*, **353**, 969 (1999).

100. R. R. Jacobson and J. L Krakenbuhl, *Lancet*, **353**, 655 (1999).

101. C. R. Valverde, D. Canfield, R. Tarara, et al., *Int. J. Lepr.*, **66**, 140 (1998).

102. S. M. Van Beers, M. Y. L. De Wit, and P. R. Katser, *FEMS Microbiol. Lett.*, **136**, 121 (1996).

103. S. M. Van Beers, M. Hatta, and P. R. Klatser, *Int. J. Lepr.*, **67**, 119 (1999).

104. J. Visschedijk, J. Van De Broek, P. Eggens, et al., *Tropic Med. Int. Health*, **5**, 388 (2000).

105. T. Lietman, T. Porco, and S. Blower, *J. Publ. Health*, **87**, 1923 (1997).

106. P. E. M. Fine, *Lancet*, **346**, 1339 (1995).

107. M. R. Chakravarti and F. Vogel, *A Twin Study on Leprosy*, George Thieme, Stuttgart, Germany, 1973.

108. I. Abel, F. O. Sanchez, J. Oberti, et al., *J. Infect. Dis.*, **177**, 133 (1998).

109. R. R. P. de Vries and T. H. M. Ottenholf, in R. C. Hastings, Ed., *Leprosy*, 2nd ed, Churchill Livingston, Edinburgh, 1994, p. 113.

110. W. W. Ooi and S. L. Moschetta, *Clin. Infect. Dis.*, **32**, 930 (2001).

111. D. S. Ridley and W. H. Jopling, *Int. J. Lepr.*, **31**, 255 (1966).

112. WHO Expert Committee on Leprosy, *WHO Tech Report Series No. 874*, World Health Organization, Geneva, Switzerland, 1998.

113. WHO, *Wkly. Epidemiol. Rec.*, **73**, 153 (1998).

114. WHO, *Wkly. Epidemiol. Rec.*, **73**, 313 (1999).

115. J. Wichitwechkarn, S. Karnjan, S. Shuntawuttisettee, et al., *J. Clin. Microbiol.*, **33**, 45 (1995).

116. M. Nisimura, K. S. Kwon, K. Shibuta, et al., *Mod. Pathol.*, **7**, 253 (1994).

117. J. Jamil, J. T. Keer, S. B. Lucas, et al., *Lancet*, **342**, 264 (1993).

118. WHO, *Technical Reports Series No 847*, World Health Organization, Geneva, Switzerland, 1994.

119. G. Vinckè, O. Yegers, H. Vanachter, P. A. Jenkins, and J. P. Butzler, *J. Antimicrob. Chemother.*, **10**, 351 (1982).

120. A. Laszlo, P. Gill, V. Handzel, M. M. Hodgkin, and D. M. Helbecque, *J. Clin. Microbiol.*, **8**, 1335 (1983).

121. L. E. Nillson, S. E. Hoffner, and S. Anséhn, *Antimicrob. Agents Chemother.*, **32**, 208 (1988).

122. R. C. Cooksey, J. T. Crawford, W. R. Jacobs, and T. M. Shinnick, *Antimicrob. Agents Chemother.*, **37**, 1348 (1993).

123. L. H. Schmidt, *Am. Rev. Tuberc.*, **74**, 138 (1956).

124. L. H. Schmidt, *Ann. NY Acad. Sci.*, **135**, 747 (1966).

125. C. C. Shepard, *J. Exp. Med.*, **112**, 445 (1960).

126. C. C. Shepard and Y. T. Chang, *Proc. Soc. Exp. Biol. Med.*, **109**, 636 (1962).

127. R. J. W. Rees, *J. Exp. Pathol.*, **45**, 207 (1964).

128. R. J. W. Rees, J. M. H. Pearson, and M. F. R. Waters, *Br. Med. J.*, **1**, 89 (1970).

129. E. E. Storrs, G. P. Walsh, H. P. Burchfield, and C. H. Binford, *Science*, **183**, 851 (1974).

130. C. B. Inderlied, L. S. Young, and J. K. Yamada, *Antimicrob. Agents Chemother.*, **31**, 1697 (1987).

131. A. J. Crowle, A. Y. Tsang, A. E. Vatter, and M. H. May, *J. Clin. Microbiol.*, **24**, 812 (1986).

132. L. E. Bermudez and L. S. Young, *Braz. J. Med. Biol. Res.*, **20**, 191 (1987).

133. C. B. Interlied, L. Barbara-Burnham, M. Wu, L. S. Young, and L. E. M. Bermudez, *Antimicrob. Agents Chemother.*, **38**, 1838 (1994).

134. L. E. Bermudez, P. Stevens, P. Kolonowski, M. Wu, and L. S. Young, *J. Immunol.*, **143**, 2996 (1989).

135. P. Sensi, *Rev. Infect. Dis.*, **11**(Suppl 2), S467 (1989).

136. K. Mdluli, *Science*, **280**, 1607 (1998).

137. H. L. David, *Appl. Microbiol.*, **20**, 810 (1970).

138. K. E. Price, D. R. Chrisholm, M. Misiek, et al., *J. Antibiot. (Tokyo)*, **25**, 709 (1972).

139. N. Rist, *C. R. Soc. Biol.*, **130**, 972 (1939).

140. W. H. Feldman, H. C. Hinshaw, and H. E. Moses, *Proc. Staff Meet. Mayo Clin.*, **15**, 695 (1940).

141. E. V. Cowdry and C. Ruangsiri, *Arch. Pathol.*, **32**, 632 (1941).

142. Y. Kurono, K. Ikeda, and K. Uekama, *Chem. Pharm. Bull. (Tokyo)*, **22**, 1261 (1974).

143. C. C. Shepard, *Ann. Rev. Pharmacol.*, **9**, 37 (1969).

144. J. L. McCullough and T. H. Moren, *Antimicrob. Agents Chemother.*, **3**, 665 (1973).

145. W. T. Colwell, G. Chan, V. H. Brown, et al., *J. Med. Chem.*, **17**, 142 (1974).

146. R. Gelber, J. H. Peters, G. R. Gordon, A. J. Glazko, and L. Levy, *Clin. Pharmacol. Ther.*, **12**, 225 (1971).

147. J. H. Peters, *Am. J. Trop. Med. Hyg.*, **23**, 222 (1974).

148. Editorial, *Lancet*, **2**, 534 (1971).

149. J. H. S. Pettit and J. Chin, *Lancet*, **2**, 1014 (1969).

150. F. Bernheim, *J. Bacteriol.*, **41**, 385 (1941).

151. J. Lehmann, *Lancet*, **1**, 15 (1946).

152. R. J. Rees, *Trans. Roy. Soc. Trop. Med. Hyg.*, **61**, 581 (1957).

153. C. Ratledge and B. J. Marshall, *Biochim. Biophys. Acta*, **279**, 58 (1972).

154. C. Ratledge and K. A. Brown, *Am. Rev. Respir. Dis.*, **106**, 74 (1972).

155. K. A. Brown and C. Ratledge, *Biochim. Biophys. Acta*, **385**, 207 (1975).

156. R. Behnisch, F. Mietzsch, and H. Schmidt, *Am. Rev. Tuberc.*, **61**, 1 (1950).

157. R. Behnisch, F. Mietzsch, and H. Schmidt, *Angew. Chem.*, **60**, 113 (1948).

158. G. Domagk, R. Behnisch, F. Mietzsch, and H. Schmidt, *Naturwissenschaften.*, **33**, 315 (1946).

159. G. Domagk, *Am. Rev. Tuberc.*, **61**, 8 (1950).

160. R. Donovik and J. Bernstein, *Am. Rev. Tuberc.*, **60**, 539 (1949).

161. D. M. Spain, W. G. Childress and J. S. Fischer, *Am. Rev. Tuberc.*, **62**, 144 (1950).

162. WHO, *Technical Report—Comité OMS d'Experts de la Lèpre*, World Health Organization, Geneva, Switzerland, 1970, p. 459.

163. Anonymous, *Br. Med. J.*, **1**, 33 (1975).

164. G. Domagk, *Schweiz. Z. Pathol. Bakteriol.*, **12**, 575 (1949).

165. G. A. Gallard, J. M. Dickinson, and D. A. Mithchison, *Tubercle*, **55**, 41 (1974).

166. H. Blaha, *Med. Welt.*, **25**, 915 (1974).

167. WHO, *Comitè OMS d'Experts de la Tuberculouse Technical Report*, World Health Organization, Geneva, Switzerland, 1974, p. 552.

168. Anonymous, *Bull. WIIO*, **47**, 211 (1972).

169. J. Lowe, *Lancet*, **2**, 1065 (1954).

170. H. H. Fox, *Science*, **116**, 129 (1952).

171. D. McKenzie, L. Malone, S. Kushner, J. J. Oleson, and Y. Subbarow, *J. Lab. Clin. Med.*, **33**, 1249 (1948).

172. H. Meyer and J. Mally, *Chem. Abstr.*, **6**, 2073 (1912).

173. V. Chorine, *Compt. Rend.*, **220**, 150 (1945).

174. J. Bernstein, W. A. Lott, B. A. Steinberg, and H. L. Yale, *Am. Rev. Tuberc.*, **65**, 357 (1952).

175. H. A. Offe, W. Siekfen, and G. Domagk, *Z. Naturforsch*, **446**, 462 (1952).

176. E. A. Zeller, J. Barsky, J. R. Fouts, W. F. Kirchheimer, and L. S. van Orden, *Experientia*, **8**, 349 (1952).

177. C. C. Shepard and Y. T. Chang, *Int. J. Lepr.*, **2**, 260 (1964).

178. J. A. Doull, J. N. Rodriguez, A. R. Davidson, et al., *Int. J. Lepr.*, **25**, 173 (1957).

179. F. G. Winder and P. B. Collins, *Am. Rev. Respir. Dis.*, **100**, 101 (1969).

180. K. Takayama, *Ann. NY Acad. Sci.*, **4**, 26 (1974).

181. K. Takayama and H. K. Schones, *Fed. Proc.*, **33**, 1425 (1974).

182. J. K. Seydel, E. Wempe, and H. J. Nestler, *Arzneim-Forsch.*, **18**, 362 (1968).

183. H. J. Nestler, *Arzneim-Forsch.*, **16**, 1442 (1966).

184. J. K. Seydel, K. J. Schaper, E. Wempe, and H. P. Cordes, *J. Med. Chem.*, **19**, 483 (1976).

185. D. B. Young, *Curr. Biol.*, **4**, 351 (1994).

186. Y. Zhang, B. Heym, B. Allen, D. Young, and S. Cole, *Nature*, **358**, 591 (1992).

187. K. Johnson and P. G. Schultz, *J. Am. Chem. Soc.*, **116**, 7425 (1995).

188. A. Banerjee, E. Dubman, A. Quemard, et al., *Science*, **263**, 227 (1994).

189. Y. Zhang and D. Young, *J. Antimicrob. Chemother.*, **34**, 313 (1994).

190. W. Mandel, D. A. Heaton, W. F. Russel, and G. Middlebrook, *J. Clin. Invest.*, **38**, 1356 (1959).

191. S. Sunahara, M. Urano, and M. Ogawa, *Science*, **134**, 1530 (1961).

192. H. Tiitinen, *Scand. J. Respir. Dis.*, **50**, 110 (1969).

193. G. A. Ellard, *Clin. Pharmacol. Ther.*, **19**, 610 (1976).

194. T. S. Gardner, E. Wenis, and J. Lee, *J. Org. Chem.*, **19**, 753 (1954).

195. R. I. Meltzer, A. D. Lewis, and J. A. King, *J Am. Chem. Soc.*, **77**, 4062 (1955).

196. N. Rist, F. Grumbach, and D. Libermann, *Am. Rev. Tuberc.*, **79**, 1 (1959).

197. F. Grumbach, N. Rist, D. Libermann, et al., *Compt. Rend.*, **242**, 2187 (1956).

198. G. Winder, P. B. Collins, and D. Whelan, *J. Gen. Microbiol.*, **66**, 379 (1971).

199. A. Quemard, G. Lanéelle, and C. Lacave, *Antimicrob. Agents Chemother.*, **36**, 1316 (1992).

200. A. Bieder, P. Brunel, and L. Mazeau, *Ann. Pharm. Fr.*, **24**, 493 (1966).

201. S. Kushner, H. Dalalian, J. L. Sanjurio, et al., *J Am. Chem. Soc.*, **74**, 3617 (1952).

202. L. Malone, A. Schurr, H. Lindh, et al., *Am. Rev. Tuberc.*, **65**, 511 (1952).

203. E. F. Rogers, W. J. Leanza, H. J. Becker, et al., *Science*, **116**, 253 (1952).

204. E. Felder, D. Pitrè, and U. Tiepolo, *Minerva Med.*, **53**, 1699 (1962).

205. L. Trnka, J. Kuska, and A. Havel, *Chemotherapia*, **9**, 158 (1965).

206. L. Heifets and P. Lindholm-Levy, *Am. Rev. Respir. Dis.*, **145**, 1223 (1992).

207. K. Wisterowicz, M. Foks, M. Janowiec, and Z. Zwolska-Krewk, *Acta Pol. Pharmacol.*, **46**, 101 (1989).

208. K. Konno, F. M. Feldmann, and W. McDermott, *Am. Rev. Respir. Dis.*, **95**, 461 (1967).

209. A. Scorpio and Y. Zhang, *Nat. Med.*, **2**, 662 (1996).

210. N. Lemaitre, W. Sougaloff, C. Truffot-Pernot, and V. Yarlier, *Antimicrobial. Agents Chemother.*, **43**, 1761 (1999).

211. O. Zimong, J. S. Cox, J. T. Welch, C. Vilchèze, and W. R. Jacobs, *Nat. Med.*, **6**, 1043 (2000).

212. V. C. Barry, J. G. Belton, J. F. O'Sullivan, and D. Twomey, *J. Chem. Soc.*, **896**, (1956).

213. V. C. Barry, M. L. Conalty, and E. E. Graffney, *J. Pharm. Pharmacol.*, **8**, 1089 (1956).

214. V. C. Barry, J. G. Belton, M. L. Conalty, and D. Twomey, *Nature*, **162**, 622 (1948).

215. V. C. Barry, *Sci. Proc. Roy. Dubl. Soc., Ser. A.*, **3**, 153 (1969).

216. W. A. Vischer, *Arzneim-Forsch.*, **18**, 1529 (1968).

217. P. J. Lindholm-Levy and L. B. Heifets, *Tubercle*, **69**, 186 (1988).

218. Y. Niwa, T. Sakance, Y. Miyachi, and M. Ozaki, *J. Clin. Microbiol.*, **20**, 837 (1984).

219. W. A. Vischer, *Arzneim-Forsch.*, **20**, 714 (1970).

220. N. J. L. Gilmour, *Br. Vet. J.*, **122**, 517 (1966).

221. H. F. Lunn and R. J. W. Rees, *Lancet*, **1**, 247 (1964).

222. Y. T. Chang, *Antimicrob. Agents. Chemother.*, **2**, 294 (1962).

223. Y. T. Chang, *Int. J. Lepr.*, **34**, 1 (1966).

224. Y. T. Chang, *Int. J. Lepr.*, **35**, 78 (1967).

225. R. Y. W. Rees, *Int. J. Lepr.*, **33**, 646 (1965).

226. J. M. Gangas, *Lepr. Rev.*, **38**, 225 (1967).

227. J. H. S. Pettit and R. J. W. Rees, *Int. J. Lepr.*, **34**, 391 (1966).

228. J. H. S. Pettit, R. J. W. Rees, and D. S. Ridley, *Int. J. Lepr.*, **35**, 25 (1967).

229. A. G. Warren, *Lepr. Rev.*, **39**, 61 (1968).

230. F. M. Imkamp, *Lepr. Rev.*, **39**, 119 (1968).

231. S. G. Browne, *Adv. Pharmacol. Chemother.*, **7**, 211 (1969).

232. H. Mathies and U. Ress, *Arzneim-Forsch.*, **20**, 1838 (1970).

233. H. L. F. Currey and P. Fowler, *Br. J. Pharmacol.*, **45**, 676 (1972).

234. C. C. Shepard, L. L. Walker, R. M. van Landingham, and M. A. Redus, *Proc. Soc. Exp. Biol. Med.*, **137**, 728 (1971).

235. V. M. Reddy, J. F. O'Sullivan, and P. R. I. Gangadharam, *J. Antimicrob. Chemother.*, **43**, 615 (1999).

236. E. G. Steuger, L. Aeppli, E. Peheim, and P. E. Thomann, *Arzneim-Forsch.*, **20**, 794 (1970).

237. J. P. Thomas, C. O. Baughn, R. G. Wilkinson, and R. G. Shepherd, *Am. Rev. Respir. Dis.*, **83**, 891 (1961).

238. E. G. Wilkinson, M. B. Cantrall, and R. G. Sheperd, *J. Med. Chem.*, **5**, 835 (1962).

239. A. G. Karlson, *Am. Rev. Respir. Dis.*, **84**, 905 (1961).

240. L. B. Heifets, M. D. Iseman, and P. J. Lindholm-Levy, *Antimicrob. Agents Chemother.*, **30**, 927 (1986).

241. W. H. Beggs and F. A. Andrews, *Am. Rev. Respir. Dis.*, **108**, 691 (1973).

242. W. H. Beggs and F. A. Andrews, *Antimicrob. Agents Chemother.*, **5**, 234 (1974).

243. M. Forbes, N. A. Kuck, and E. A. Peets, *J. Bacteriol.*, **84**, 1099 (1962).

244. M. Forbes, N. A. Kuck, and E. A. Peets, *J. Bacteriol.*, **89**, 1299 (1965).

245. M. Forbes, N. A. Kuck, and E. A. Peets, *Ann. NY Acad. Sci.*, **135**, 726 (1966).

246. H. Reutgen and H. Iwainsky, *Z. Naturforsch.*, **27**, 1405 (1972).

247. K. Takayama and J. O. Kilburn, *Antimicrob. Agents Chemother.*, **33**, 1493 (1989).

248. G. Silve, P. Valero-Guillen, A. Quemard, et al., *Antimicrob. Agents Chemother.*, **37**, 1536 (1993).

249. R. F. Corpe and F. A. Blalcock, *Dis. Chest.*, **48**, 305 (1965).

250. I. D. Bobrowitz and K. S. Go Kulanathan, *Dis. Chest.*, **48**, 239 (1965).

251. V. A. Place and J. P. Thomas, *Am. Rev. Respir. Dis.*, **87**, 901 (1963).

252. E. A. Peets, W. M. Sweeney, V. A. Place, and D. A. Buyske, *Am. Rev. Respir. Dis.*, **91**, 51 (1965).

253. J. E. Leihold, *Ann. NY Acad. Sci.*, **135**, 904 (1966).

254. A. Schatz., E. Bugie, and S. A. Waksman, *Proc. Soc. Exp. Biol. Med.*, **55**, 66 (1944).

255. B. D. Davis, *Microbiol. Rev.*, **51**, 341 (1987).

256. C. R. Krishna Murthi, *Biochem. J.*, **76**, 362 (1960).

257. C. R. Spotts and R. Y. Stamier, *Nature*, **192**, 633 (1961).

258. E. C. Cox, J. R. White, and J. G. Flakes, *Proc. Nat. Acad. Sci. USA*, **51**, 703 (1964).

259. M. Ozaki, S. Mizushima, and M. Nomura, *Nature*, **222**, 333 (1969).

260. J. Davies, D. S. Jones, and H. G. Khorana, *J. Mol. Biol.*, **18**, 48 (1966).

261. D. Elseviers and L. Gorini in S. Mitsuhashi, Ed., *Drug Action and Drug Resistance in Bacteria, Vol. 2, Aminoglycoside Antibiotics*, University Park Press, Baltimore, MD, 1975, p. 147.

262. L. Luzzato, D. Apirion, and D. Schlessinger, *Proc. Nat. Acad. Sci. USA*, **60**, 873 (1968).

263. J. Modollel and B. D. Davis, *Proc. Nat. Acad. Sci.USA*, **61**, 1270 (1968).

264. J. Modollel and B. D. Davis, *Nature*, **224**, 345 (1969).

265. H. B. Newcombe and M. H. Nyholm, *Genetics*, **35**, 603 (1950).

266. H. B. Newcombe and R. Haxizko, *J. Bacteriol.*, **57**, 565 (1949).

267. J. G. Flakes, E. C. Cox, M. L. Witting, and J. R. White, *Biochem. Biophys. Res. Commun.*, **7**, 390 (1962).

268. H. Umezawa. M. Okamishi, S. Kondo, et al., *Science*, **157**, 1559 (1967).

269. S. Mitsuhashi, H. Kawabe in A. Whelton and H. R. Neu, Eds., *The Aminoglycosides. Microbiology, Clinical Use and Toxicology*, Marcel Dekker Inc., New York, 1982, p. 97.

270. M. Finken, P. Kirschner, A. Meier, and E. C. Böttger, *Mol. Microbiol.*, **9**, 1239 (1993).

271. J. Nair, D. A. Rouse, G. H. Bai, and S. L. Morris, *Mol. Microbiol.*, **10**, 521 (1993).

272. N. Honorè and S. T. Cole, *Antimicrob. Agents Chemother.*, **38**, 238 (1994).

273. J. Douglas and L. M. Stein, *J. Infect. Dis.*, **167**, 1506 (1993).

274. A. Meier, P. Kirschner, F. C. Bange, et al., *Böttger, Antimicrob. Agents Chemother.*, **38**, 228 (1994).

275. R. C. Gordon, C. Regamey, and W. M. M. Kirby, *Antimicrob. Agents Chemother.*, **2**, 214 (1972).

276. E. M. Weyer, *Ann. NY Acad. Sci.*, **132**, 771 (1996).

277. E. M. Yow and H. Abu-Nassar, *Antibiot. Chemother.*, **11**, 148 (1963).

278. D. Apirion and D. Schlessinger, *J. Bacteriol.*, **96**, 768 (1968).

279. J. Davis, L. Gorini, and D. B. Davies, *Mol. Pharmacol.*, **1**, 93 (1965).

280. H. Masukawa, N. Tanaka, and H. Umezawa, *J. Antibiot. (Tokyo)*, **21**, 517 (1968).

281. N. Tanaka, Y. Yoshida, K. Sashikata, H. Yamaguchi, and H. Umezawa, *J. Antibiot. (Tokyo)*, **19**, 65 (1966).

282. K. Konno, K. Oizumi, N. Kumano, and S. Oka, *Am. Rev. Respir. Dis.*, **108**, 101 (1973).

283. D. H. Starkey and E. Gregory, *Can. Med. Assoc. J.*, **105**, 587 (1971).

284. J. T. Doluisio, L. W. Dittert, and J. C. La Piana, *J. Pharmacokinet. Biopharmaceut.*, **1**, 253 (1973).

285. L. L. McDonald and J. W. St. Geme, *Antimicrob. Agents Chemother.*, **2**, 41 (1972).

286. E. Kuntz, *Klin. Wochenschr.*, **40**, 1107 (1962).

287. W. E. Sanders, R. Cacciatore, H. Valdez, et al., *Am. Rev. Respir. Dis.*, **113**(suppl 4), 59 (1976).

288. W. E. Sanders Jr., C. Hartwig, N. Schneider, et al., *Tubercle*, **63**, 201 (1982).

289. L. B. Heifets and P. Lindholm-Levy, *Antimicrob. Agents Chemother.*, **33**, 1298 (1989).

290. C. B. Inderlied, P. T. Kolonoski, M. Wu, and L. S. Young, *Antimicrob. Agents Chemother.*, **33**, 176 (1989).

291. F. De Lalla, R. Maserati, P. Scarpellini, et al., *Antimicrob. Agents Chemother.*, **36**, 1567 (1992).

292. A. C. Finlay, G. I. Hobby, F. Hochstein, et al., *Am. Rev. Respir. Dis.*, **63**, 1 (1951).

293. Q. R. Bartz, J. Ehrlich, J. D. Mold, et al., *Am. Rev. Tuberc. Pulm. Dis.*, **63**, 4–6 (1951).

294. E. B. Herr Jr. and M. O. Redstone, *Ann. NY Acad. Sci.*, **135**, 940 (1966).

295. N. Tanaka and S. Igusa, *J. Antibiot. (Tokyo)*, **1**, 239 (1968).

296. T. Yamada, K. Masuda, K. Shoj, and M. Hari, *J. Bacteriol.*, **112**, 1 (1972).

297. T. Yamada, K. Masuda, Y. Mizuguchi, and K. Suga, *Antimicrob. Agents Chemother.*, **9**, 817 (1976).

298. R. L. Hamed, P. H. Hidy, and E. K. La Baw, *Antibiot. Chemother.*, **5**, 204 (1955).

299. D. A. Harris, R. Ruger, M. A. Reagan, et al., *Antibiot. Chemother.*, **5**, 183 (1955).

300. G. Shull and J. Sardinas, *Antibiot. Chemother.*, **5**, 398 (1955).

301. F. A. Kuehl, F. J. Wolf, N. R. Trenner, et al., *J. Am. Chem. Soc.*, **77**, 2344 (1955).

302. P. H. Hidy, E. B. Hodge, V. V. Young, et al., *J. Am. Chem. Soc.*, **77**, 2345 (1955).

303. C. H. Stammer, A. N. Wilson, and C. F. Spencer, *J. Am. Chem. Soc.*, **79**, 3236 (1957).

304. J. L. Strominger, E. Ito, and R. H. Threnn, *J. Am. Chem. Soc.*, **82**, 998 (1960).

305. J. L. Strominger, R. H. Threnn, and S. S. Scott, *J. Am. Chem. Soc.*, **181**, 3083 (1959).

306. F. C. Neuhaus and J. L. Lynch, *Biochemistry*, **3**, 471 (1964).

307. P. Sensi, P. Margalith, and M. T. Timbal, *Farmaco Ed. Sci.*, **14**, 146 (1959).

308. P. Margalith and G. Beretta, *Mycopathol. Mycol. Appl.*, **8**, 321 (1960).

309. J. E. Thiemann, G. Zucco, and G. Pelizza, *Arch. Microbiol.*, **67**, 147 (1969).

310. P. Sensi, A. M. Greco, and R. Ballotta. *Antibiot. Annu.* (1959–1960) ASM, Washington DC, 1960, p. 252.

311. P. Margalith and H. Pagani, *Appl. Microbiol.*, **9**, 325 (1961).

312. P. Sensi, M. T. Timbal, and G. Maffii, *Experientia*, **16**, 412 (1960).

313. P. Sensi, R. Ballotta, A. M. Greco, and G. G. Gallo, *Farmaco Ed. Sci.*, **16**, 165 (1961).

314. W. Oppolzer, V. Prelog, and P. Sensi, *Experientia*, **20**, 336 (1964).

315. M. Brufani, W. Fedeli, G. Giacomello, and A. Vaciago, *Experientia*, **20**, 339 (1964).

316. J. Leitich, W. Oppolzer, and V. Prelog, *Experientia*, **20**, 343 (1964).

317. W. Oppolzer and V. Prelog, *Helv. Chim. Acta*, **56**, 2287 (1973).

318. P. Sensi, N. Maggi, S. Furesz, and G. Maffi, *Antimicrob. Agents Chemother.*, **6**, 669 (1966).

319. P. Sensi, *Pure Appl. Chem.*, **35**, 383 (1973).

320. G. Lancini, and W. Zanichelli in D. Perlman, Ed., *Structure-Activity Relationship Among the Semisynthetic Antibiotics*, Academic Press, New York, 1977, p. 531.

321. P. Sensi and G. Lancini in C. Hansch Ed., *Comprehensive Medicinal Chemistry*, Vol. **2**, Pergamon Press, New York, 1990, p. 793.

322. M. Brufani, G. Cecchini, L. Cellai, et al., *J. Antibiot.*, **38**, 259 (1985).

323. M. Brufani, L. Cellai, L. Cozzella, et al., *J. Antibiot.*, **38**, 1359 (1985).

324. S. K. Arora, *Mol. Pharmacol.*, **23**, 133 (1983).

325. M. Brufani, S. Cerrini, W. Fedeli, and A. Vaciago, *J. Mol. Biol.*, **87**, 409 (1974).

326. M. Brufani, L. Cellai, S. Cerrini, et al., *Mol. Pharmacol.*, **21**, 394 (1981).

327. G. G. Gallo, E. Martinelli, V. Pagani, and P. Sensi, *Tetrahedron*, **30**, 3093 (1974).

328. P. Sensi, N. Maggi, R. Ballotta, et al., *J. Med. Chem.*, **7**, 596 (1964).

329. G. Maffii and P. Schiatti, *Toxicol. Appl. Pharmacol.*, **8**, 138 (1966).

330. G. Pelizza, G. C. Lancini, G. C. Allievi, and G. G. Gallo, *Farmaco Ed. Sci.*, **28**, 298 (1973).

331. N. Maggi, R. Pallanza, and P. Sensi, *Antimicrob. Agents Chemother.*, **5**, 765 (1965).

332. N. Maggi, C. R. Pasqualucci, R. Ballotta, and P. Sensi, *Chemotherapia*, **11**, 285 (1966).

333. G. Hartmann, K. O. Honikel, F. Knussel, and J. Nuesch, *Biochim. Biophys. Acta*, **145**, 843 (1967).

334. H. Umezawa, S. Mizuno, H. Yamazaki, and K. Nitta, *J. Antibiot. (Tokyo)*, **21**, 234 (1968).

335. W. Wehrli, *Eur. J. Biochem.*, **80**, 325 (1977).

336. R. J. White and G. C. Lancini, *Biochim. Biophys. Acta*, **240**, 429 (1971).

337. U. J. Lill and G. R. Hartmann, *Eur. J. Biochem.*, **38**, 336 (1973).

338. S. Riva and L. G. Silvestri, *Ann. Rev. Microbiol.*, **26**, 199 (1972).

339. W. Wehrli and M. Staehelin in J. W. Corcoran and H. Hahn, Eds., *Antibiotics, Vol. 3, Mechanism of Action of Antimicrobial and Antitumor Agent*, Springer-Verlag, Berlin, 1975, p. 252.

340. W. Stender, A. A. Stutz, and K. H. Scheit, *Eur. J. Biochem.*, **56**, 129 (1975).

341. R. J. White, G. C. Lancini, and L. Silvestri, *J. Bacteriol.*, **108**, 737 (1971).

342. K. Konno, K. Oizumi, and S. Oka, *Am. Rev. Respir. Dis.*, **107**, 1006 (1973).

343. K. Konno, K. Oizumi, F. Arji, et al., *Am. Rev. Respir. Dis.*, **107**, 1002 (1973).

344. W. Schulz and W. Zillig, *Nucleic Acids Res.*, **9**, 6889 (1981).

345. C. Kessler, M. Huaifeng, and G. R. Hartmann, *Eur. J. Biochem.*, **122**, 515 (1982).

346. N. Honorè and S. T. Cole, *Antimicrob. Agents Chemother.*, **37**, 414 (1993).

347. A. Telenti, P. Imboden, F. Marchesi, et al., *Lancet*, **341**, 647 (1993).

348. L. Miller, J. T. Crawford, and T. M. Shinnick, *Antimicrob. Agents Chemother.*, **38**, 805 (1994).

349. P. Matsiota-Bernard, G. Vrion, and P. Marinis, *J. Clin. Microbiol.*, **36**, 201 (1998).

350. V. Sintchenko, P. J. Jelfs, W. K. Cheu, and G. L. Gilbert, *J. Antimicrob. Chemother.*, **44**, 294 (1999).

351. D. L. Williams, L. Spring, L. Collins, et al., *Antimicrob. Agents Chemother.*, **42**, 1833 (1998).

352. B. Yang, H. Koga, H. Ohno, et al., *J. Antimicrob. Chemother.*, **42**, 621 (1998).

353. E. R. Dabbs, K. Yazawa, Y. Mikami, et al., *Antimicrob. Agents Chemother.*, **39**, 1007 (1995).

354. G. R. Hartmann, P. Heinrich, and M. C. Kollenda, *Angew. Chem. Int. Ed. Engl.*, **24**, 1009 (1985).

355. L. Verbist and A. Gyselen, *Am. Rev. Respir. Dis.*, **98**, 923 (1968).

356. G. R. F. Hilson, D. K. Banerjee, and J. B. Holmes, *Int. J. Lepr.*, **39**, 349 (1971).

357. C. C. Shepard, L. L. Walker, R. M. Van Landingham, and M. A. Redus, *Am. J. Trop. Med. Hyg.*, **20**, 616 (1971).

358. S. R. Pattyn, *Int. J. Lepr.*, **41**, 489 (1973).

359. S. R. Pattyn and E. J. Saerens, *Ann. Soc. Belg, Med. Trop.*, **54**, 35 (1974).

360. J. K. McClatchy, R. F. Wagonner, and W. Lester, *Am. Rev. Respir. Dis.*, **100**, 234 (1969).

361. P. W. Steinbruck, *Acta Tuberc. Pneumol. Belg.*, **60**, 413 (1969).

362. R. Pallanza, V. Arioli, S. Furesz, and G. Bolzoni, *Arzneim-Forsch.*, **17**, 529 (1967).

363. F. Grumbach and N. Rist, *Rev. Tuberc. (Paris)*, **31**, 749 (1967).

364. L. Verbist, *Acta Tuberc. Pneumol. Belg.*, **60**, 390 (1969).

365. F. Grumbach, G. Canetti, and M. Le Lirzin, *Tubercle*, **50**, 280 (1969).

366. F. Grumbach, G. Canetti, and M. Le Lirzin, *Rev. Tubercle Pneumol.*, **34**, 312 (1970).

367. F. Kradolfer and R. Schnell, *Chemotherapy*, **15**, 242 (1970).

368. F. Kradolfer, *Am. Rev. Respir. Dis.*, **98**, 104 (1968).

369. F. Kradolfer, *Antibiot. Chemother.*, **16**, 352 (1970).

370. V. Nitti, E. Catena, A. Ninni, and A. Di Filippo, *Arch. Tisiol.*, **21**, 867 (1966).

371. V. Nitti, E. Catena, A. Ninni, and A. Di Filippo, *Chemotherapia*, **12**, 369 (1967).

372. V. Nitti, *Antibiot. Chemother.*, **16**, 444 (1970).

373. M. Lucchesi and P. Mancini, *Antibiot. Chemother.*, **16**, 431 (1970).

374. G. Binda, E. Domenichini, A. Gottardi, et al., *Arzneim-Forsch.*, **21**, 796 (1971).

375. D. L. Leiker, *Int. J. Lepr.*, **39**, 462 (1971).

376. R. J. W. Rees, M. F. R. Waters, H. S. Helmy, and J. M. H. Pearson, *Int. J. Lepr.*, **41**, 682 (1973).

377. S. Furesz, R. Scotti, R. Pallanza, and E. Mapelli, *Arzneim-Forsch.*, **17**, 726 (1967).

378. G. Curci, A. Ninni, and F. Iodice, *Acta Tuberc. Pneumol. Belg.*, **60**, 276 (1969).

379. G. Acocella, V. Pagani, M. Marchetti, et al., *Chemotherapy*, **16**, 356 (1971).

380. G. Acocella, *Clin. Pharmacokinet.*, **3**, 108 (1978).

381. G. Acocella, L. Bonollo, M. Garimoldi, et al., *Gut*, **13**, 47 (1972).

382. L. Dettli and F. Spina, *Farmaco Ed. Sci.*, **23**, 795 (1968).

383. N. Maggi, S. Furesz, R. Pallanza, and G. Pelizza, *Arzneim-Forsch.*, **19**, 651 (1969).

384. A. M. Baciewicz, T. H. Self, and W. B. Bekemeyer, *Arch. Intern. Med.*, **147**, 565 (1987).

385. S. Furesz, *Antibiot. Chemother.*, **16**, 316 (1970).

386. P. Kluyskens, *Acta Tuberc. Pneumol. Belg.*, **60**, 323 (1969).

387. L. Marsili, C. R. Pasqualucci, A. Vigevani, et al., *J. Antibiot.*, **34**, 1033 (1981).

388. C. Della Bruna, G. Schioppacassi, D. Ungheri, et al., *J. Antibiot.*, **36**, 1502 (1983).

389. D. Ungheri, C. Della Bruna, and A. Sanfilippo, *G. Ital. Chemother.*, **31**, 211 (1984).

390. C. L. Woodley and J. O. Kilbum, *Am. Rev. Respir. Dis.*, **126**, 586 (1982).

391. H. Saito, K. Sato, and H. Tomioka, *Tubercle*, **69**, 187 (1988).

392. L. B. Heifets and M. D. Iseman, *Am. Rev. Respir. Dis.*, **132**, 710 (1985).

393. M. H. Cynamon, *Antimicrob. Agents Chemother.*, **28**, 440 (1985).

394. S. P. Klemens, M. A. Grossi, and M. H. Cynamon, *Antimicrob. Agents Chemother.*, **38**, 234 (1994).

395. M. H. Skinner and T. F. Blaschke, *Clin. Pharmacokinet.*, **28**, 115 (1995).

396. S. D. Nightingale, D. W. Cameron, and F. M. Gordin, *N. Engl. J. Med.*, **329**, 828 (1993).

397. V. Arioli, M. Berti, G. Carniti, et al., *J. Antibiot.*, **34**, 1026 (1981).

398. J. M. Dickinson and D. A. Mitchison, *Tubercle*, **68**, 113 (1987).

399. C. S. F. Easmon and J. P. Crane, *J. Antimicrob. Chemother.*, **13**, 585 (1984).

400. S. P. Klemens and M. H. Cynamon, *J. Antimicrob. Chemother.*, **29**, 555 (1992).

401. A. Assandri, T. Cristina, and L. Moro, *J. Antibiot.*, **31**, 894 (1978).

402. A. Assandri, A. Perazzi, and M. Berti, *J. Antibiot.*, **30**, 409 (1977).

403. A. Vernon, W. Burman, D. Benator, et al., *Lancet*, **353**, 1843 (1999).

404. T. Yamane, T. Hashizume, K. Yamashita, et al., *Chem. Pharm. Bull.*, **40**, 2707 (1992).

405. T. Yamane, T. Hashizume, K. Yamashita, et al., *Chem. Pharm. Bull.*, **41**, 148 (1993).

406. S. P. Klemens, M. A. Grossi, and M. H. Cynamon, *Antimicrob. Agents Chemother.*, **38**, 2245 (1994).

407. K. Fuji, H. Saito, H. Tomioka, T. Mae, and K. Hosoe, *Antimicrob Agents Chemother.*, **39**, 1489 (1995).

408. A. M. Dhople and M. A. Ibanez, *J. Antimicrob. Chemother.*, **35**, 463 (1995).

409. C. M. Shoen, S. E. Chase, M. S. De Stefano, et al., *Antimicrob. Agents Chemother.*, **44**, 1458 (2000).

410. A. M. Lenaerts, S. E. Chase, and M. H. Cynamon, *Antimicrob. Agents Chemother.*, **44**, 3167 (2000).

411. V. M. Reddy, G. Nadadhu, D. Daneluzzi, et al., *Antimicrob. Agents Chemother.*, **39**, 2320 (1995).

412. A. M. Dhople and V. Dimova, *Arzneim-Forsch/Drug Res.*, **46**, 210 (1996).

413. N. Karabalut, and G. Drusano in D. C. Hooper, J. S. Wolfson, Eds., *Quinolone Antimicrobial Agents*, 2nd ed., American Society for Microbiology, Washington DC, 1993, p. 195.

414. L. A. Mitscher, P. Devasthole, and R. Zadov in Ref. 420, p. 3.

415. G. M. Eliopoulos and C. T. Eliopoulos in Ref. 420, p. 161.

416. L. B. Heifets, *Semin. Respir. Infect.*, **9**, 84 (1994).

417. N. Khardory, K. Rolston, B. Rosenbaum, et al., *J. Antimicrob. Chemother.*, **24**, 667 (1989).

418. J. A. Garcia-Rodriguez and A. C. Gomez-Garcia, *J. Antimicrob. Chemother.*, **32**, 797 (1993).

419. N. Rastogi and K. S. Goh, *Antimicrob. Agents Chemother.*, **35**, 1933 (1991).

420. B. Ji, C. Truffot-Pernot, and J. Grosset, *Tubercle*, **72**, 181 (1991).

421. N. Mor, J. Vanderkolk, and L. Heifets, *Antimicrob. Agents Chemother.*, **38**, 1161 (1993).

422. C. Piersimoni, V. Morbiducci, S. Bornigia, et al., *Am. Rev. Respir. Dis.*, **146**, 1445 (1992).

423. D. K. Benerjee, J. Ford, and S. Markanday, *J. Antimicrob. Chemother.*, **30**, 236 (1992).

424. L. B. Heifets and P. J. Lindholm-Levy, *Antimicrob. Agents Chemother.*, **34**, 770 (1990).

425. H. Tomioka, H. Saito, and K. Sato, *Antimicrob. Agents Chemother.*, **37**, 1259 (1993).

426. I. Man, J. Murphy, and J. Ferguson, *J. Antimicrob. Chemother.*, **43**(suppl B), 77 (1999).

427. T. D. Gootz, K. E. Brighty in V. T. Andriole, Ed., *The Quinolones*, 2nd ed., Academic Press, San Diego, CA, 1998, p. 29.

428. K. N. Kreuzer and N. R. Cozzarelli, *J. Bacteriol.*, **140**, 424 (1979).

429. G. C. Crumplin and J. T. Smith, *Nature*, **260**, 643 (1976).

430. J. T. Smith, *Pharm. J.*, **233**, 299 (1984).

431. H. Eagle and A. D. Musselmann, *J. Exp. Med.*, **88**, 99 (1948).

432. M. Casal, J. Gutierrez, J. Gonzales, and P. Ruiz, *Chemioterapia*, **6**, 437 (1987).

433. M. Casal, F. Rodriguez, J. Gutierrez, and P. Ruiz, *Rev. Infect. Dis.*, **11**(suppl 5), S1042 (1989).

434. M. Tsukamura, *Am. Rev. Respir. Dis.*, **132**, 915 (1985).

435. M. Chadwich, G. Nicholson, and H. Gaya, *Am. J. Med.*, **87**(suppl 5A), 35S (1989).

436. V. Lalande, C. Truffot-Pernot, A. Paccaly-Moulin, et al., *Antimicrob. Agents Chemother.*, **37**, 407 (1993).

437. S. P. Klemens, C. A. Sharpe, M. C. Roggie, and M. H. Cynamon, *Antimicrob. Agents Chemother.*, **38**, 1476 (1994).

438. A. M. Dhople, M. H. Ibanez, and G. D. Gardner, *Arzneim-Forsch/Drug Res.*, **43**, 384 (1993).

439. Hong Kong Chest Service and British Medical Research Council, *Tubercle Lung Dis.*, **73**, 59 (1992).

440. R. J. O'Brien, in Ref. 410, p. 207.

441. D. C. Hooper and J. S. Wolfson, *Quinolone Antimicrobial Agents*, 2nd ed., AMS, Washington DC, 1993, p. 489.

442. P. Ball, *Rev. Infect. Dis.*, **11**(suppl 5), S1365 (1989).

443. N. Kennedy, R. Fox, L. Viso, et al., *J. Antimicrob. Chemother.*, **32**, 897 (1993).

444. A. Bryskier, J. Gasc, and E. Agoridas in A. Bryskier, J. P. Butzler, H. C. Neu, and P. M. Tulkens, Eds., *Macrolides: Chemistry, Structure, Activity*, Arnette Blackwell, Oxford, 1993, p. 5.

445. H. Lode, M. Boeckh, and T. Schaberg, in Ref. 457, p. 409.

446. M. T. Labro, in Ref. 457, p. 379.

447. D. Mansuy and M. Delaforge, in Ref. 457, p. 635.

448. R. E. Polk and D. Israel, in Ref. 457, p. 647.

449. H. C. Neu, in Ref. 457, p. 167.

450. C. B. Inderlied, L. M. Bermudez, and L. S. Young, in Ref. 457, p. 285.

451. L. B. Heifets, in L. B. Heifets, Ed., *Drug Susceptibility in the Chemotherapy of Mycobacterial Infections*, CRC Press, Inc., Boca Raton, FL, 1991.

452. C. Truffot-Pernot and J. B. Grosset, *Antimicrob. Agents Chemother.*, **35**, 1677 (1991).

453. S. Naik and R. Ruck, *Antimicrob. Agents Chemother.*, **33**, 1614 (1989).

454. P. B. Fernandes, D. J. Hardy, D. McDaniel, et al., *Antimicrob. Agents Chemother.*, **33**, 1531 (1989).

455. Y. Cohen, C. Perronne, C. Truffot-Pernot, et al., *Antimicrob. Agents Chemother.*, **36**, 2104 (1992).

456. N. Mor and L. Heifets, *Antimicrob. Agents Chemother.*, **37**, 111 (1993).

457. N. Mor, J. Vanderkolk, and L. Heifets, *Pharmacotherapy*, **14**, 100 (1994).

458. R. Anderson, G. Joone, and C. E. J. Van Rensburg, *J. Antimicrob. Chemother.*, **22**, 923 (1988).

459. M. Ishiguro, H. Koga, S. Kohno, et al., *J. Antimicrob. Chemother.*, **24**, 719 (1989).

460. N. Mor and L. Heifets, *Antimicrob. Agents Chemother.*, **37**, 1380 (1993).

461. S. P. Klemens, M. S. De Stefano, and M. H. Cynamon, *Antimicrob. Agents Chemother.*, **36**, 2413 (1992).

462. R. E. Chaisson, C. Benson, M. Dube, et al., Progr. Abst. 32nd Intersci. Conf. Antimicrob. Agents Chemother. ASM, Washington, DC, 1992, abst. 891.

463. L. B. Heifets, N. Mor, and J. Vanderkolk, *Antimicrob. Agents Chemother.*, **37**, 2364 (1993).

464. P. Prokocima, M. Dellerson, C. Crafs, et al., Progr. Abst. 30th Intersci. Conf. Antimicrob. Agents Chemother. ASM, Washington, DC, 1990, abst. 634.

465. B. Dautzenberg, C. Truffot-Pernot, S. Legris, et al., *Am. Rev. Respir. Dis.*, **144**, 564 (1991).

466. S. G. Franzblau and R. C. Hasting, *Antimicrob. Agents Chemother.*, **32**, 1758 (1988).

467. H. Masur, *N. Engl. J. Med.*, **329**, 898 (1993).

468. R. Gelber, *Prog. Drug Res.*, **34**, 421 (1990).

469. N. Ramasel, J. Krahenbuhl, and R. C. Hasting, *Antimicrob. Agents Chemother.*, **33**, 657 (1989).

470. H. Lode, *Eur. J. Clin. Microbiol. Infect. Dis.*, **10**, 807 (1991).

471. P. J. McDonald and H. Pruul, *Eur. J. Clin. Microbiol. Infect. Dis.*, **10**, 828 (1991).

472. A. I. M. Hoepelman, A. P. Sips, J. L. M. Van Helmond, et al., *J. Antimicrob. Chemother.*, **31**(suppl E), 147 (1993).

473. F. Bradbury, *J. Antimicrob. Chemother.*, **31**(suppl E), 153 (1993).

474. J. Myburgh, G. J. Nagel, and E. Petschel, *J. Antimicrob. Chemother.*, **31**(suppl E), 163 (1993).

475. O. G. Berlin, L. S. Young, S. A. Floyd-Reising, and D. A. Bruckner, *Eur. J. Clin. Microbiol. Infect. Dis.*, **6**, 486 (1987).

476. C. B. Inderlied, P. T. Kolonoski, M. Wu, and L. S. Young, *J. Infect. Dis.*, **159**, 994 (1989).

477. A. E. Girard, D. Girard, and J. A. Retsema, *J. Antimicrob. Chemother.*, **25**(suppl A), 61 (1990).

478. J. A. Retsema, A. E. Girard, D. Girard, and W. B. Milisen, *J. Antimicrob. Chemother.*, **25**(suppl A), 83 (1990).

479. M. J. Gevaudan, C. Bollet, M. N. Mallet, et al., *Phatol. Biol.*, **35**, 413 (1990).

480. L. E. Bermudez, C. Interlied, and L. S. Young, *Antimicrob. Agents Chemother.*, **35**, 2625 (1991).

481. S. T. Brown, F. F. Edwards, E. M. Bernard, et al., *Antimicrob. Agents Chemother.*, **37**, 398 (1993).

482. M. H. Cynamon and S. P. Klemens, *Antimicrob. Agents Chemother.*, **36**, 1611 (1992).

483. S. P. Klemens and M. H. Cynamon, *Antimicrob. Agents Chemother.*, **38**, 1721 (1994).

484. P. T. Kolonoski, J. Martinelli, M. L. Petrosky, et al., 89th Annual Meeting of the American Society for Microbiology, New Orleans, LA, **51**, 163 (1989).

485. R. H. Gelber, P. Siu, M. Tsang, and L. P. Murray, *Antimicrob. Agents Chemother.*, **35**, 760 (1991).

486. L. S. Young, L. Wiviott, M. Wu, et al., *Lancet*, **338**, 1107 (1991).

487. J. E. Kasic, *Am. Rev. Respir. Dis.*, **91**, 117 (1965).

488. H. C. Neu, A. P. R. Wilson, and R. N. Grunberg, *J. Chemother.*, **5**, 67 (1993).

489. K. Bush, *Antimicrob. Agents. Chemother.*, **33**, 259 (1989).

490. Y. Zhang, V. A. Steingube, and R. J. Wallace Jr., *Am. Rev. Respir. Dis.*, **145**, 657 (1992).

491. G. N. Rolinson, *J. Chemother.*, **6**, 283 (1994).

492. M. H. Cynamon and G. S. Palmer, *Antimicrob. Agents Chemother.*, **24**, 429 (1983).

493. M. Casal, F. Rodriguez, M. Benavente, and M. Luna, *Eur. J. Clin. Microbiol.*, **5**, 453 (1986).

494. J. P. Nadler, J. Berger, J. A. Nord, R. Cofsky, and M. Saxena, *Chest*, **99**, 1025 (1991).

495. B. Watt, J. R. Edwards, A. Rayner, et al., *Tubercle Lung Dis.*, **73**, 134 (1992).

496. C. K. Stover, P. Warrener, D. R. VanDevanter, et al., *Nature*, **405/22**, 962 (2000).

497. L. A. Mitsher and W. Baker, *Med. Chem. Rev.*, **18**, 363 (1998).

498. M. R. Barbachyn, S. J. Brickner, D. K. Hutchinson, et al., Progr. Abst. 35th Intersci. Conf. Antimicrob. Agents Chemother. ASM, Washington, DC, 1995, Abst. F227.

499. D. R. Ashtekan, R. Costa-Pereira, R. Ajjer, N. Vishanatan, and W. Rittel, *Diagn. Microbiol. Infect. Dis.*, **14**, 465 (1991).

500. L. Kremer, J. D. Douglas, A. R. Baulard et al., *J. Biol. Chem.*, **275**, 6857 (2000).

501. M. A. Espinal, A. Laszlo, L. Simonsen, et al., *N. Engl. J. Med.*, **344**, 1294 (2001).

502. P. A. Willcox, *Curr. Opin. Pulmon. Med.*, **6**, 198 (2000).

503. R. Cominx, C. Mathien, M. Debacker, et al., *Lancet*, **363**, 969 (1999).

504. M. Kimerling, H. Kluge, N. Vezhnina, et al., *Int. J. Tuberc. Lung Dis.*, **3**, 451 (1999).

505. P. Willcox, *Curr. Opin. Pulmon. Med.*, **6**, 198 (2000).

506. R. Bayer and D. Wilkinson, *Lancet*, **345**, 1545 (1995).

507. P. Nunn, *Scand. J. Infect. Dis.*, **33**, 329 (2001).

508. China Tuberculosis Control Collaboration, *Lancet*, **347**, 358 (1996).

509. World Global Tuberculosis Programme, *WHO Report 1998*, World Health Organization, Geneva, Switzerland, 1998.

510. P. Farmer and J. Y. Kim,. *Br. Med. J.*, **517**, 671 (1998).

511. A. Gbayisomore, A. A. Lardizabal, and L. B. Reichman, *Curr. Opin. Infect. Dis.*, **13**, 155 (2000).

512. R. J. O'Brien in L. B. Reichman and E. S. Hershelfield, Eds., *Tuberculosis*, Marcel Dekker Inc., New York, 1993, p. 207.

513. C. Grassi, *Medit. J. Infect. Parasit. Dis.*, **7**, 17 (1992).

514. M. D. Iseman and J. A. Sbarbaro, *Curr. Clin. Topics Infect. Dis.*, **12**, 188 (1992).

515. American Thoracic Society, *Am. J. Respir. Crit. Care Med.*, **149**, 1359 (1994).

516. R. F. Jacobs, *Clin. Infect. Dis.*, **19**, 1 (1994).

517. M. D. Iseman, *N. Engl. J. Med.*, **329**, 784 (1993).

518. S. Agrawal, N. S. Thomas, A. B. Dhanicula, et al., *Curr. Opin. Pulmon. Med.*, **7**, 142 (2001).

519. P. E. M. Fine in L. B. Reichman and E. S. Herschfield, Eds., *Tuberculosis*, 2nd ed., Marcel Dekker, New York, 2000, p 503.

520. L. J. Geiter, in Ref. 519, p. 241.

521. C. J. Wallace, J. Glassroth, D. E. Griffith, et al., *Am. J. Crit. Care Med.*, **156**, 51 (1997).

522. L. J. Yoder, *Curr. Opin. Infect. Dis.*, **4**, 302 (1991).

523. H. J. Yoder, *Curr. Opin. Infect. Dis.*, **6**, 349 (1993).

524. S. K. Nordeen, *Lepr. Rev.*, **62**, 72 (1991).

525. B. Ji, *Lepr Rev.*, **69**, 106 (1998).

526. Single-Lesion Multicentre Trial Group, *Indian J. Lepr.*, **69**, 121 (1997).

527. P. F. Barnes, D. Chatterjee, P. J. Brennan, et al., *Infect. Immunol.*, **60**, 1441 (1992).

528. W. J. Britton, *Lepr. Rev.*, **69**, 225 (1998).

529. M. D. Gupta, *Indian J. Lepr.*, **63**, 342 (1991).

530. J. Convit, C. Sampson, M. Zuniga, et al., *Lancet*, **339**, 446 (1992).

531. J. M. Ponninhans, P. E. M. Fine, J. A. C. Sterne, et al., *Lancet*, **339**, 6363 (1992).

532. D. Maher, M. C. Raviglione in D. Schlossberg, Ed., *Tuberculosis and Nontuberculosis Mycobacterial Infection*, 4th ed, WB Saunders Co, Philadelphia, PA, 1999, p. 104.

533. P. J. Murray, A. Aldovini, R. A. Young, *Proc. Natl. Acad. Sci. USA*, **93**, 934 (1996).

534. C. K. Stover, G. P. Bansal, M. S. Hanson, et al., *J. Exp. Med.*, **178**, 197 (1993).

535. R. E. Tascon, M. J. Colston, S. Ragno, et al., *Nat. Med.*, **2**, 893 (1996).

536. K. Huygen, J. Content, O. Denis, et al., *Nat. Med.*, **2**, 893 (1996).

537. A. S. Pym and S. T. Cole, *Lancet*, **353**, 1004 (1999).

538. A. M. Ginsberg, *Clin. Infect. Dis.*, **30**(suppl 3), S233 (2000).

539. P. M. Small and P. I. Fujiwara, *N. Engl. J. Med.*, **345**, 189 (2001).

540. G. A. W. Rook, G. Seah, and A. Ustianowski, *Eur. Resp. J.*, **17**, 537 (2001).

541. S. M. Holland, *Semin. Resp. Infect.*, **16**, 47 (2001).

542. Single Lesion Multicenter Trial Group, *Lepr. Rev.*, **68**, 341 (1997).

# CHAPTER SEVENTEEN

# Antifungal Agents

WILLIAM J. WATKINS
THOMAS E. RENAU
Essential Therapeutics
Mountain View, California

## Contents

1 Introduction, 882
  1.1 Fungal Diseases and Pathogens, 882
    1.1.1 *Candida* spp, 882
    1.1.2 *Cryptococcus neoformans*, 884
    1.1.3 Opportunistic Filamentous Fungi, 884
  1.2 Trends in Incidence of Fungal Infection, 884
  1.3 Epidemiology of Resistance, 884
  1.4 Diagnostic and Microbiological Issues, 885
  1.5 Selectivity, 886
    1.5.1 The Biosynthesis and Cellular
        Functions of Ergosterol, 886
    1.5.2 The Fungal Cell Wall, 888
2 Current Therapeutic Options, 889
  2.1 Current Drugs, 889
    2.1.1 Drugs for Treatment of Systemic
        Mycoses, 889
    2.1.2 Drugs for Treatment of Superficial
        Cutaneous Mycoses, 889
3 Antifungal Chemical Classes, 889
  3.1 Polyenes: Amphotericin B and Nystatin, 889
    3.1.1 Overview and Mode of Action, 889
    3.1.2 Structure-Activity Relationship, 889
    3.1.3 Resistance, 889
    3.1.4 Side Effects, 889
    3.1.5 Absorption, Distribution, Metabolism,
        Excretion (ADME), 892
    3.1.6 Drug Interactions, 892
    3.1.7 Polyene Liposomal Formulations, 892
    3.1.8 Things to Come, 893
  3.2 Azoles, 893
    3.2.1 Overview and Mode of Action, 893
    3.2.2 History of Azole Discovery, 895
    3.2.3 Structure-Activity Relationship, 895
    3.2.4 Resistance, 896
    3.2.5 Side Effects, 896
    3.2.6 ADME, 897
    3.2.7 Drug Interactions, 897
    3.2.8 Things to Come, 897
      3.2.8.1 Voriconazole, 897

*Burger's Medicinal Chemistry and Drug Discovery*
Sixth Edition, Volume 5: Chemotherapeutic Agents
Edited by Donald J. Abraham
ISBN 0-471-37031-2    © 2003 John Wiley & Sons, Inc.

3.2.8.2 Posaconazole and
Ravuconazole, 898
3.3 Allylamines, 899
3.3.1 Overview and Mode of Action, 899
3.3.2 Structure-Activity Relationship, 899
3.3.3 Resistance, 900
3.3.4 Side Effects, 900
3.3.5 ADME, 900
3.3.6 Drug Interactions, 900
3.3.7 Things to Come, 900
3.4 Candins, 900
3.4.1 Overview and Mode of Action, 900
3.4.2 Structure-Activity Relationship, 901
3.4.3 Side Effects, 902
3.4.4 ADME, 902
3.4.5 Things to Come, 902
3.4.5.1 Anidulafungin (LY303366, V-
Echinocandin, V-002), 902
3.4.5.2 Micafungin (FK-463), 903
3.5 Miscellaneous, 903
3.5.1 Other Inhibitors of Ergosterol
Biosynthesis, 903
3.5.1.1 Thiocarbamates, 903
3.5.1.2 Morpholines, 903
3.5.1.3 Other Azoles, 903
3.5.2 Flucytosine, 903

3.5.2.1 Overview and Mode of Action,
903
3.5.2.2 Resistance, 904
3.5.2.3 Side Effects, 904
3.5.2.4 ADME, 904
3.5.2.5 Drug Interactions, 905
3.5.3 Griseofulvin, 905
3.5.4 Other Topical Agents in Clinical Use,
906
3.5.5 Other Classes of Medicinal Interest,
906
3.5.5.1 Polyoxins and Nikkomycins,
906
3.5.5.2 Aureobasidins, 906
3.5.5.3 Sordarins, 907
3.5.5.4 Pradimicins and Benanomycins,
908
3.5.5.5 N-Myristoyl Transferase
Inhibitors, 908
3.5.5.6 Fungal Efflux Pump Inhibitors,
909
4 New Trends in Antifungal Research, 910
5 Web Sites and Recommended Reading, 911
5.1 Web Sites, 911
5.2 Other Texts, 911

# 1 INTRODUCTION

Of the five fundamental Kingdoms of Life, the Kingdom Fungi is arguably the most diverse and prevalent. Unlike the Kingdom Monera (containing bacteria), fungi are eukaryotic organisms whose cellular functions consequently resemble those of plants and animals more closely. Thus the issue of selectivity predominates in the quest for safe and effective chemotherapeutic remedies for diseases caused by fungi. As with all chemotherapy, there is a risk-reward ratio to be taken into account; in the context of fungal infections, this ratio may vary greatly, from minor irritations such as athlete's foot to life-threatening systemic infections such as those caused by *Aspergillus fumigatus*. This chapter addresses medicinal aspects of the treatment of fungal diseases of all types, but because most recent research has been directed toward the treatment of systemic infections, emphasis is placed on this aspect.

## 1.1 Fungal Diseases and Pathogens

**1.1.1 *Candida* spp.** Invasive candidiasis is the most common nosocomial mycosis, per-

haps because the causative organism is a component of the endogenous flora of the human alimentary tract. There has been debate over the significance of positive blood cultures (candidemia) in the progression of fungal disease. Given the high mortality rates (up to 75%) in cases of invasive candidiasis, the current consensus is that all high risk patients with candidemia should receive therapy (1).

*C. glabrata, C. krusei, C. tropicalis*, and *C. parapsilosis* have emerged in recent years as troublesome organisms, challenging the supremacy of *C. albicans* in candida infections. The most common manifestation of these infections, particularly in AIDS patients, is oral or esophageal candidiasis ("thrush": white plaques that cause pain or difficulty upon swallowing); studies suggest that up to 90% of those suffering from AIDS have had at least one such episode (2). Catheter-related candida infections are also very common in hospitalized patients and, if not treated adequately, may lead to disseminated disease in which virtually any organ may be affected. This is particularly a risk in patients undergoing treat-

**Table 17.1 Common Fungal Infections**[a]

| Disease | Etiologic Agents | Main Tissues Affected |
|---|---|---|
| | **Contagious, Superficial Disease** | |
| Dermatophytoses (ringworm/tinea) | *Epidermophyton, Microsporum, Trichophyton* spp. | Skin, hair, nails |
| | **Noncontagious, Systemic Diseases** | |
| Aspergillosis | *Aspergillus* spp. | External ear, lungs, eye, brain |
| Blastomycosis | *Blastomyces dermatitidis* | Lungs, skin, bone, testes |
| Candidiasis | *Candida* spp. | Respiratory, gastrointestinal and urogenital tracts; skin |
| Chromomycosis | *Cladosporium, Fonsecaea,* and *Phialophora* spp. | Skin |
| Coccidioidomycosis | *Coccidioides inimitis* | Lungs, skin, joints, meninges |
| Cryptococcosis | *Cryptococcus neoformans* | Lungs, meninges |
| Histoplasmosis | *Histoplasma capsulatum* | Lungs, spleen, liver, adrenals, lymph nodes |
| Mucormycosis | *Absidia, Mucor, Rhizopus* spp. | Nasal mucosa, lungs, blood vessels, brain |
| Paracoccidioidomycosis | *Paracoccidioides brasiliensis* | Skin, nasal mucosa, lungs, liver, adrenals, lymph nodes |
| Pneumocystosis | *Pneumocystis carinii* | Lungs |
| Pseudoallescheriasis | *Pseudoallescheria boydii* | External ear, lungs, eye |
| Sporotrichosis | *Sporothrix schenkii* | Skin, joints, lungs |

[a]Reproduced with permission from *Burger's Medicinal Chemistry and Drug Discovery: Therapeutic Agents*, Vol. 2, 5th ed., Chapter 35, p. 639.

ment for leukemia, with prolonged periods of bone marrow dysfunction and neutropenia.

### 1.1.2 Cryptococcus neoformans.

This organism, which survives in the feces of pigeons and is probably acquired in humans by aerosolization and inhalation, emerged at the height of the AIDS pandemic as a life-threatening pathogen. In recent years incidence of these infections has abated somewhat, but an episode of cryptococcal meningitis remains a serious cause for concern, and after therapy a life-long antifungal regimen is required in AIDS patients to prevent relapse.

### 1.1.3 Opportunistic Filamentous Fungi.

These are a group of organisms that cause very serious infections, particularly in neutropenic patients. Infection attributed to *Aspergillus* spp. is the most common. After the inhalation of spores, pulmonary disease develops; in severe cases, the disease spreads to any of several other organs. Mortality in cases of invasive pulmonary aspergillosis remains high, despite some improvement in early diagnosis, and the need for effective, less toxic therapies remains acute (3). The situation is complicated by the incidence of pulmonary infection resulting from other moulds that may be refractory to agents that have some effect on *Aspergillus* spp.

### 1.2 Trends in Incidence of Fungal Infection

The greatest use of antifungal agents, by far, is in treatment of dermatophytic infections—dandruff, athlete's foot, and toenail infections, for example—that occur in otherwise healthy individuals. The incidence of such disease is relatively constant. Because these are infections of the skin, many of the available agents are applied topically, although a few notable oral alternatives are now available. Despite the relative triviality of the infections, however, their eradication is often problematic and requires many weeks or months of therapy.

Far more serious, but much less prevalent, is the morbidity caused by systemic mycoses. In populations with a fully functioning immune system, the incidence of invasive fungal infections is low, but in the immunocompromised host the situation is very different. Be-

cause the proportion of the population that is immunocompromised has increased sharply in recent years, so the incidence of life-threatening fungal infections has risen dramatically. Data from the National Nosocomial Infections Surveillance System conducted in the United States showed a 487% increase in candida bloodstream infections between 1980 and 1989 (4). The increase in numbers of immunocompromised patients is attributed principally to three factors: (1) advances in transplant technology, necessitating a period of deliberate immunosuppression to avoid complications resulting from rejection of the foreign organ; (2) chemotherapeutic regimens in cancer that kill rapidly dividing cells such as key parts of the immune system; and (3) the spread of the AIDS pandemic. By the early 1990s, fungemia was recognized as the cause of 10% of nosocomial infections (5). The advent of highly active antiretroviral therapy (HAART) in the treatment of AIDS has ameliorated the number of fungal infections in this subpopulation, but fungal infections remain a significant cause of morbidity and mortality, particularly in the nosocomial setting (6).

### 1.3 Epidemiology of Resistance

As with all other areas of anti-infective drug therapy, the phenomenon of resistance to currently available antifungal therapies is a major concern, and the biochemical means by which such resistance to various drug classes is manifest in the clinic is discussed in section 3. However, it is appropriate at this juncture to compare and contrast the epidemiological circumstances by which such resistance develops in the antifungal (as opposed to antibacterial or antiviral) context.

A major cause of concern in antibacterial chemotherapy is the rapid spread of resistant genes within a bacterial strain (or even across species) by the exchange of genetic material on plasmids, a phenomenon that has been exacerbated in the West by the growth in air travel, ensuring that any mechanism by which a population of bacteria becomes resistant to a particular agent will be shared across much of the world in a relatively short time. Fortunately, in the fungal context (where life-threatening systemic diseases are not contagious), the spread of resistance by such means occurs at a negligible rate. Here, there are two major ways in which resistance becomes manifest. The first arises when initial antifungal therapy (combined with a defective immune system) fails to eradicate the organism from the patient. Under these circumstances, repeated sublethal exposure to the drug allows a variety of resistance mechanisms to be induced, in much the same way as resistant strains are deliberately generated in the laboratory. The second mechanism for the emergence of infections that are refractory to therapy arises (like all resistance phenomena) from the use of antifungal agents: inevitably, any agent is less efficacious for some fungal species than for others. Thus, the use of any drug is bound to present an environmental stress on the most susceptible strains, allowing the less susceptible to become more prevalent. In the clinical context, the most obvious manifestation of this phenomenon has been the recent identification of *Candida* spp. that are *intrinsically* resistant to fluconazole in cases of invasive candidiasis (7). These are strains that ordinarily make up only a small proportion of the fungal burden, but emerge as predominant pathogens when the population of *C. albicans* is reduced through antifungal therapy.

Thus, although antifungal therapy is not bedeviled by the facile mechanisms for the spread of resistance that are such a problem in the antibacterial or antiviral contexts, nonetheless there are ways in which drugs become less efficacious over time, and the consequent clinical limitations should not be underestimated (5).

## 1.4 Diagnostic and Microbiological Issues

Although the pharmaceutical industry commonly classifies antifungal with antibacterial (and antiviral) research and development as a single therapeutic area (anti-infectives), there are some important clinical differences that affect the way in which antifungal agents are used, and that have a significant impact on the discovery process. First, considerably fewer clinical microbiological data are available to allow meaningful analysis of the pathogenic organisms involved. This is partly because less routine identification of fungal species (much

less individual strains) is performed in the general hospital setting. The generation time of most fungi is longer than that of bacteria, and consequently the common practice is to rule out bacteria as origins of an infection and proceed to antifungal therapy on an empiric basis. In some extreme cases, such as for invasive aspergillus infections, the incubation time is comparable with the time it takes for the disease to progress from presentation with fever to death. This, combined with the reduction in postmortem clinical microbiological investigations, may contribute to significant underreporting of the incidence of serious systemic fungal infections in general.

Second, only recently have standard methods begun to emerge for routine susceptibility testing, further confounding attempts to interpret historical data in a comparative manner (8). The variable morphology of important pathogens (the dimorphism of *Candida* spp. between yeast and hyphal forms, or the vastly different morphology of *Cryptococcus neoformans in vitro* and *in vivo*, for example), the differing correlations of growth time *in vitro* with clinical outcomes across species, "trailing" effects with important classes of agents confounding the definition of a minimum inhibitory concentration (MIC), and difficulties in defining interpretative breakpoints for use in a predictive manner clinically, all combine to render the definition of broadly useful methods extremely problematic; the need for different methods for different organisms further increases the burden on clinical microbiology laboratories.

Given the limited success in defining microbiological susceptibility testing conditions that correlate with clinical outcomes, it is only to be expected that trends in the emergence of resistance may be hard to define. However, particularly for azoles, approximately 50% of azole-resistant isolates correlate with clinical failure (9, 10), and the increase in MIC upon sequential isolation from individual patients provides clear evidence of such trends (11).

In this context, it is not surprising that the field of antifungal research is littered with examples where tests *in vitro* were not adequate to predict effects *in vivo*, and many of these were false negatives, examples where the compounds turned out to be active *in vivo* despite apparently poor *in vitro* antifungal activity. Two very prominent and significant classes of antifungals (the azoles and the echinocandins) were significantly underestimated by conventional *in vitro* methods; the use of pharmacokinetic and *in vivo* efficacy tests at an early stage in the discovery process is therefore particularly appropriate in this field of medicinal discovery (12).

### 1.5 Selectivity

The search for antifungal agents that are clinically useful may fairly be characterized as the search for selective ways to kill one eukaryotic organism within another, and for systemic mycoses, it is also fair to say that until recently clinicians have been forced to use agents whose mammalian toxicity has been the limiting factor in achieving successful outcomes. Indeed, amphotericin B, the gold standard of therapy for serious invasive mycoses for over 40 years, causes nephrotoxicity in over 80% of patients who receive it, and a primary concern in the design of therapeutic regimens (both in amount and frequency of dosing, as well as in the search for new formulations) is to avoid permanent defects in renal function (see Sections 3.1.4, 3.1.7, and 3.1.8).

The two primary differences between fungi and higher eukaryotes that have been exploited in the identification of antifungal agents with clinical utility are the constitution of the cell membrane and the architecture of the cell wall.

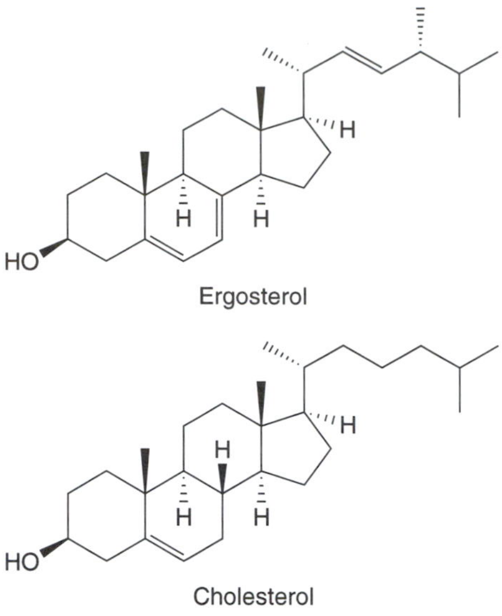

#### 1.5.1 The Biosynthesis and Cellular Functions of Ergosterol.

In virtually all pathogenic fungi, the principal sterol is ergosterol (Fig. 17.1), which differs from its mammalian counterpart cholesterol in several structural respects. Although none of the classes of currently available antifungal agents was discovered by a rational search based on this fact, a remarkable number of them turn out to depend on it for their utility.

The principal route for the biosynthesis of ergosterol from squalene in *C. albicans* is shown in Schemes 17.1 and 17.2. Not surprisingly, there are many parallels with the biosynthesis of cholesterol in mammals; even when the substrates differ, many of the chem-

**Figure 17.1.** Ergosterol and cholesterol.

ical transformations are analogous. Furthermore, it is important to appreciate that the route to ergosterol in the latter stages is not identical in all fungal species, nor is there necessarily a unique path in each (13). For example, the order of methylenation at $C_{24}$ and demethylations at $C_4$ and $C_{14}$ may vary, as may the isomerizations and adjustments in oxidation state that lead thereafter to the appropriate installation of olefins. Thus inhibition of one of these later steps may cause not only the depletion of cellular reserves of ergosterol, but also the accumulation of aberrant sterols as intermediates in the pathway.

Ergosterol may constitute 10% of the dry weight of a fungal cell. Using *S. cerevisiae* as a convenient model organism and studying growth patterns using surrogate sterols, it has been shown to be important for the growth and survival of fungi in a variety of ways (14). At least four sterol functions have been implicated: sparking, critical domain, domain, and bulk (15). Of these, the sparking function (associated with cellular proliferation) is the most structurally demanding and requires a sterol with a planar $\alpha$-face (i.e., no methyl group at $C_{14}$) and bearing a $C_{5-6}$ olefin, and the bulk function is the least demanding, requiring only a sterol bearing a $3\beta$-hydroxy group.

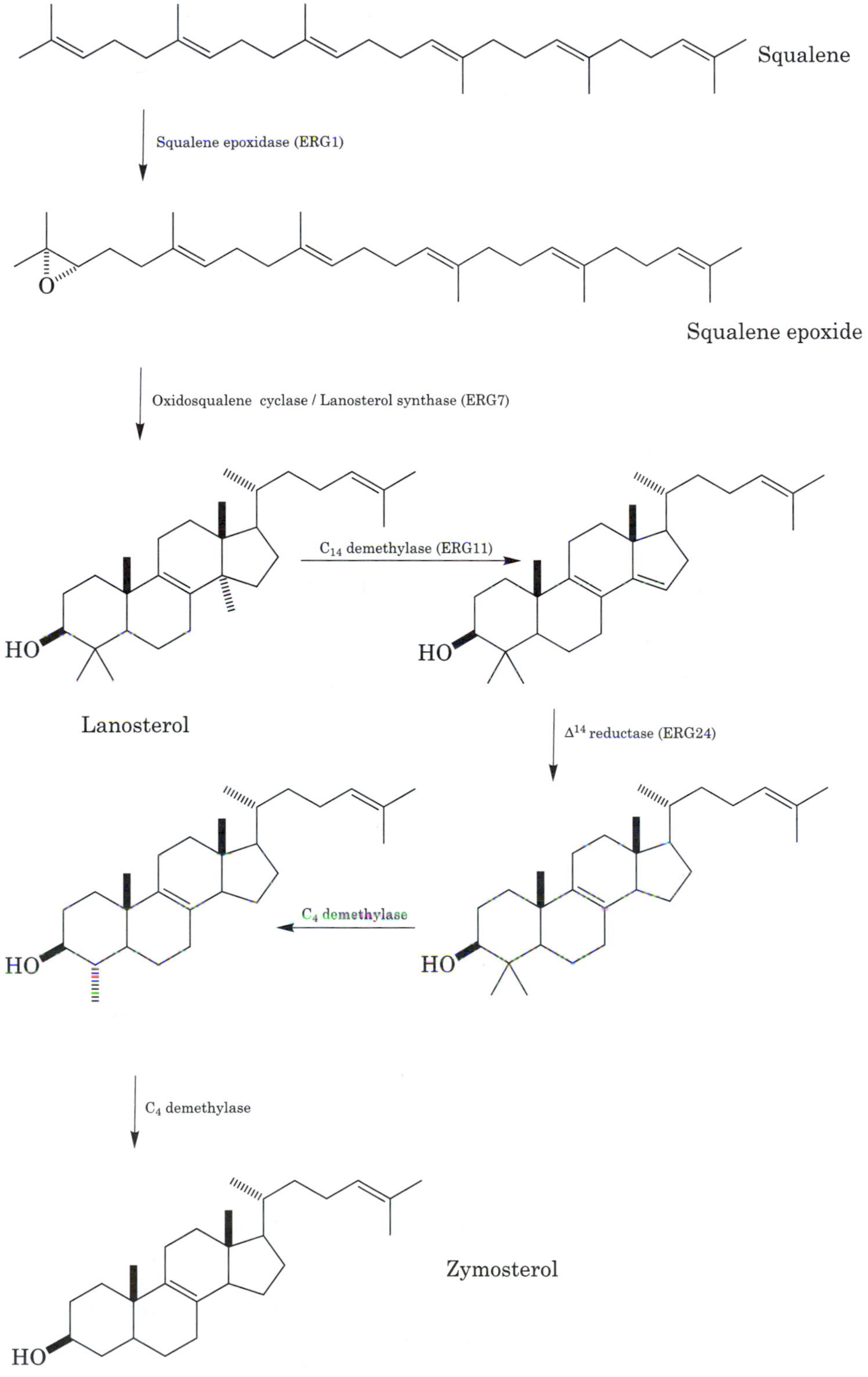

**Scheme 17.1.** Biosynthetic pathway from squalene to zymosterol. For further stereochemical detail, see Fig. 17.1.

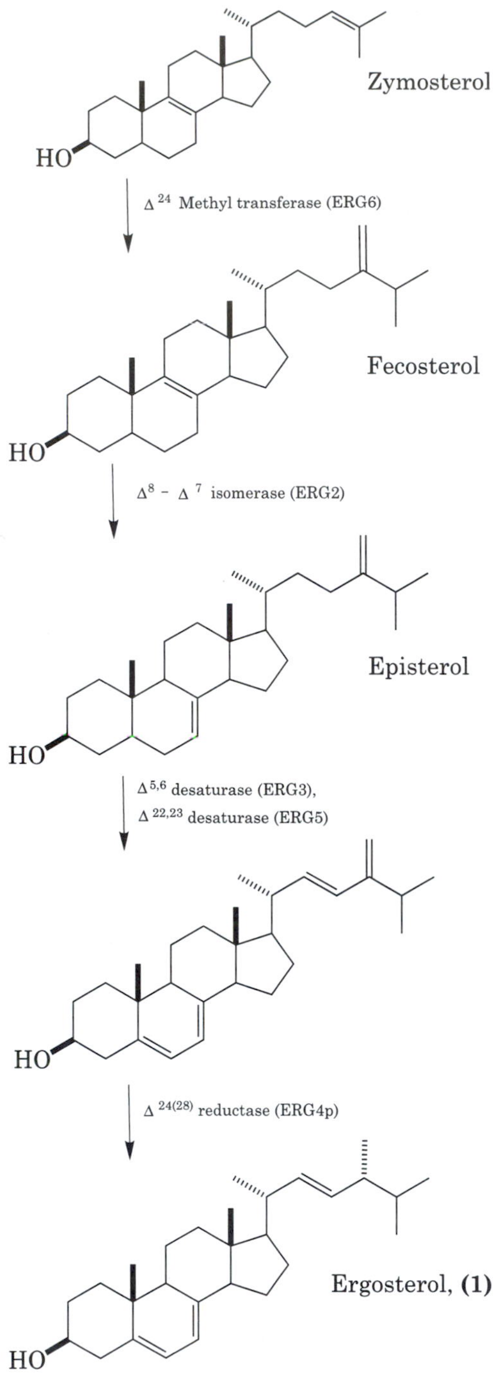

**Scheme 17.2.** Biosynthetic pathway from zymosterol to ergosterol (in C. albicans). For further stereochemical detail, see Fig. 17.1.

### 1.5.2 The Fungal Cell Wall.

The fungal cell wall is a complex structure external to the cell membrane containing a high degree of polysaccharide content. It is essential for fungal viability, maintaining organism osmotic integrity and shape during its growth. In pathogenic fungi, adhesion to the mammalian host appears to be regulated by the cell wall. Given that there are no mammalian counterparts to the fungal cell wall, compounds that target it should be selective and inherently fungicidal, making them quite attractive for clinical development. The clinical success of antibacterial agents such as the β-lactams that target the bacterial cell wall suggest that a functional equivalent in fungi would have good clinical potential (16).

The composition of the fungal cell wall varies among species but is generally composed of glucans, chitin, and mannoproteins. Chitin and glucan fibrils form the scaffolding, whereas embedded mannoproteins form the matrix that is responsible for cell wall porosity and hydrophobicity.

Glucans, which constitute nearly 60% of the fungal cell wall, are variously linked β-1,3-, α-1,3-, and β-1,6-glucose polymers composed of approximately 1500 monomer units with side chains averaging 150 monomer units in length. Glucan is evenly distributed over the entire cell surface but is primarily deposited at sites of new cell wall growth. Caspofungin (see Section 3.5), an inhibitor of β-1,3-glucan synthase, was recently approved to treat patients with refractory invasive aspergillosis. Other glucan synthase inhibitors of the candin class that are currently in clinical development include Micafungin (FK-463) and Anidulafungin (LY303366, V-002).

Chitin, a homopolymer of monomer units of N-acetyl-D-glucosamine, is a minor but essential cell wall component. Its proportion in the fungal cell wall varies among species (i.e., 2% in yeasts, higher in moulds). Chitin synthesis occurs on the cytoplasmic surface of the plasma membrane with the translocation of the linear polymer through the membrane to the cell surface. Chitin synthesis is competitively inhibited by the naturally occurring nucleoside-peptides polyoxins and nikkomycins (see Section 3.5.1.2).

Mannoproteins, which are *O*- and *N*-glyco-sylated proteins that constitute approximately 40% of the cell wall, are linked at their manno-oligosaccharide to β-1,3-glucan by a phosphodiester linkage. Because glycosylation of fungal mannoproteins is generally similar to that of mammalian mannoproteins, disruption of this target is considered less attractive.

## 2 CURRENT THERAPEUTIC OPTIONS

### 2.1 Current Drugs

The entries in the following tables are listed according to mode of action, with those affecting ergosterol biosynthesis or function (the vast majority) grouped together.

The pharmaceutical requirements (and markets) of agents for use in treating systemic and superficial cutaneous mycoses differ greatly. Consequently, compounds for these different uses are listed in separate tables below. Where significant use for a single agent is found in both contexts, it is listed in both tables.

**2.1.1 Drugs for Treatment of Systemic Mycoses.** See page 890.

**2.1.2 Drugs for Treatment of Superficial Cutaneous Mycoses.** See page 891.

## 3 ANTIFUNGAL CHEMICAL CLASSES

### 3.1 Polyenes: Amphotericin B and Nystatin

**3.1.1 Overview and Mode of Action.** Amphotericin B (Fig. 17.2), produced by *Streptomyces nodosus*, was discovered in 1955 by Gold and coworkers. It is the archetypal polyene antifungal and was the first systemic agent available to treat invasive fungal infections. The primary indications of its current therapy include invasive candidiasis, cryptococcosis, aspergillosis, and histoplasmosis. It is the treatment of choice for empirical therapy in febrile neutropenic patients.

The polyenes act by binding to ergosterol, a sterol present in the membrane of sensitive fungi. The interaction with ergosterol forms pores within the fungal cell membrane, result-ing in leakage of cellular contents such as sodium, potassium, and hydrogen ions. Polyenes display little or no activity against bacteria, as these organisms lack membrane-bound sterols. Although amphotericin B binds approximately 10 times more strongly to fungal cell membrane components than to mammalian cell membrane cholesterol, the interaction with and disruption of mammalian cells by the compound can result in adverse side effects. Consequently, polyenes such as nystatin that have a higher affinity for cholesterol have correspondingly greater toxicities for mammalian cells.

**3.1.2 Structure-Activity Relationship.** Approximately 60 polyene antifungal compounds have been described to date. They are amphoteric, poorly soluble in water, and somewhat unstable because of extensive unsaturation. They consist of a large, macrocyclic core (38-membered in the case of amphotericin B), where a hydrophilic, polyhydroxylated chain ($C_1$–$C_{13}$ in amphotericin B) is bound to a lipophilic polyene fragment ($C_{20}$–$C_{33}$ in amphotericin B) by two poly-substituted alkyl links ($C_{34}$–$C_{37}$ and $C_{14}$–$C_{19}$ in amphotericin B, where the latter exists in a pyranose hemiacetal form). Diversity among the polyenes arises principally from the number and configurational arrangement of hydroxyl groups and double bonds in the macrocycle. X-ray crystallography has shown amphotericin B to be rod-shaped, with the hydrophilic hydroxyl groups of the macrolide forming an opposing face to the lipophilic polyene portion.

**3.1.3 Resistance.** Clinical resistance to polyene treatment remains rare; however, resistant strains have been isolated under laboratory conditions. These strains typically demonstrate alterations in the nature of the sterols or in the amount of sterols present in the membrane.

**3.1.4 Side Effects.** The clinical use of polyenes is limited, principally because of dose-limiting nephrotoxicity. The primary manifestations of nephrotoxicity include decreased glomerular filtration, loss of urinary concentrating ability, renal loss of sodium and potassium ions, and renal tubular acidosis. Ampho-

**Table 17.2  Drugs for Treatment of Systemic Mycoses[a]**

| Chemical Class[b] | Generic Name | Trade Name | Originator | Formulations | Indications, Dose |
|---|---|---|---|---|---|
| Polyenes | Amphotericin B (AmB) | Fungizone | Bristol-Myers Squibb | Deoxycholate | Systemic fungal infections; no more than 70 mg over 2 days |
| | AmB Lipid Complex | ABELCET | The Liposome Company | Dimyristoyl phosphatidylcholine, dimyristoyl phosphatidylglycerol | Systemic fungal infections; 5 mg/kg/day |
| | AmB Colloidal Dispersion | Amphocil Amphotec | InterMune | | Invasive aspergillosis; 3–6 mg/kg/day |
| | Liposomal AmB | Ambisome | Fujisawa | Liposomal membranes | Empiric therapy, 3 mg/kg/day; systemic fungal infections, 3–6 mg/kg/day |
| | AmB Oral suspension | Fungizone Oral Suspension | Bristol-Myers Squibb | Oral, 100 mg/mL suspension | Systemic fungal infections; 2 g/day |
| Azoles | Fluconazole | Diflucan | Pfizer | Oral, i.v. | Candidiasis, cryptococcal meningitis; up to 400 mg/day |
| | Itraconazole | Sporanox | Janssen | Oral, i.v. | Blastomycosis, histoplasmosis, aspergillosis; up to 400 mg/day |
| | Ketoconazole | Nizoral | Janssen | Oral | Candidiasis; 400 mg/day |
| | Voriconazole | Vfend | Pfizer | Oral, i.v. | Acute invasive aspergillosis; 4 mg/kg/day i.v.[c] |
| Allylamines | Terbinafine | Lamisil | Sandoz (Novartis) | Oral | Candidiasis; 250 mg/day |
| Candins | Caspofungin | Cancidas | Merck | i.v. infusion | Refractory invasive aspergillosis; 50 mg/day |
| Others | Flucytosine | Ancobon | Roche | Oral | Candidiasis, cryptococcal meningitis; 50–150 mg/kg/day |

[a]Reproduced with permission from www.doctorfungus.org.
[b]For structures, see relevant sections.
[c]FDA advisory committee recommendations; FDA approval still pending.

**Table 17.3  Drugs for Treatment of Superficial Cutaneous Mycoses**[a]

| Chemical Class[b] | Generic Name | Trade Name | Originator | Formulations[c] | Indications[c] |
|---|---|---|---|---|---|
| Polyenes | Amphotericin B | Fungizone | Bristol-Myers Squibb | C, L, O | CC |
|  | Nystatin | Various | Various | C, O, OS, P, VT, T | CC, OC, VC |
| Azoles | Butoconazole | Femstat | Syntex (Roche) | C | VC |
|  | Clotrimazole | Various | Bayer | C, L, S, T, VT | D, CC, OC, VC |
|  | Econazole | Spectazole | Janssen | C | D, CC |
|  | Itraconazole | Sporanox | Janssen | S | D |
|  | Ketoconazole | Nizoral | Janssen | C, S | D, CC |
|  | Miconazole | Micatin, Monistat | Janssen | C, L, S, P, VS | D, CC, VC |
|  | Oxiconazole | Oxistat | Roche | C, L | D, CC |
|  | Sulconazole | Exelderm | Syntex (Roche) | C, S | D, CC |
|  | Terconazole | Terazole | Janssen | C, VS | VC |
|  | Tioconazole | Vagistat | Pfizer | C, VO | VC |
| Allylamines | Naftifine | Naftin | Sandoz (Novartis) | C, O, P | D |
|  | Terbinafine | Lamisil | Sandoz (Novartis) | C, S | D |
|  | Butenafine | Mentax | Kaken | C | D |
| Miscellaneous | Amorolfine | Loceryl | Roche | C?? | D |
|  | Ciclopirox | Loprox | Aventis | C, L | D, CC |
|  | Griseofulvin | Various | Various | Oral | D |
|  | Haloprogin | Halotex | Bristol-Myers Squibb | C | D, CC |
|  | Tolnaftate | Aftate, NP-27, Tinactin, Ting | Schering-Plough | C, S, P | D |
|  | Undecylenate | Cruex, Desenex | Fisons | C, P, O, S | D |

[a]Reproduced with permission from www.doctorfungus.org.
[b]For structures, see relevant sections.
[c]C, cream; L, lotion; O, ointment; OS, oral suspension; P, powder; S, solution/spray; VO, vaginal ointment; VS, suppository; T, troche; VT, vaginal tablet; D, dermatophytosis; CC, cutaneous candidiasis; OC, oropharyngeal candidiasis; VC, vulvovaginal candidiasis.

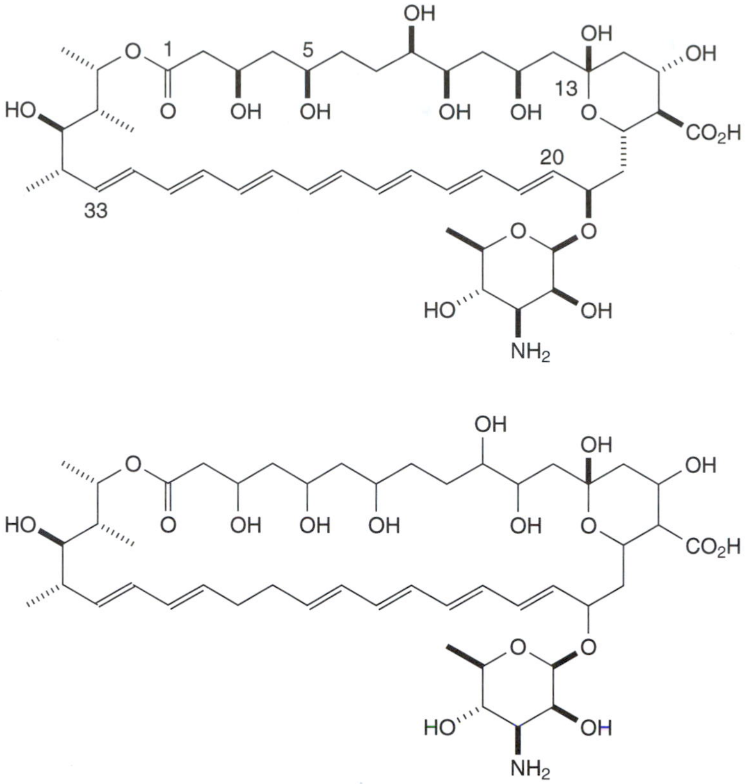

**Figure 17.2.** Structures of polyenes.

tericin B-induced renal impairment may result in the premature or temporary cessation of treatment or a reduction in dosage, leading to a worsening of the infection or prolonged hospitalization. Complications in addition to nephrotoxicity include hypokalemia and anemia, caused by reduced erythropoetin synthesis.

**3.1.5 Absorption, Distribution, Metabolism, Excretion (ADME).** Amphotericin B is poorly absorbed when given orally, and therefore is administered only by intravenous injection. Serum concentrations of the compound plateau at 2.5 mg/L, with little increase for dosages >1 mg/kg/day. Side effects appear more related to the total cumulative dose than serum concentrations of the drug. The distribution of amphotericin B is believed to follow a three-compartment model, with a reported total volume of distribution of 4 L/kg; the compound is distributed to various organs of the body such as lungs, kidneys, and the

spleen. Little of the agent penetrates into cerebrospinal fluid (CSF). The metabolism of the compound is not well understood, and the drug is eliminated primarily by way of renal and biliary routes.

**3.1.6 Drug Interactions.** Because multiple drug therapy is typically required for patients receiving amphotericin B, the risk of drug interactions is quite high. Caution is advised with concomitant use of other nephrotoxic agents such as aminoglycosides.

**3.1.7 Polyene Liposomal Formulations.** In an effort to overcome the side effects associated with the use of amphotericin B, a number of lipid based formulations have been developed, each with its own composition and pharmacokinetic behavior (17, 18). There are three currently available lipid formulations of amphotericin B that include: amphotericin B lipid complex (ABLC, Abelcet), amphotericin

**Table 17.4  Lipid-Based Formulations of Amphotericin B$^a$**

| Commercial Name | Lipid Formulation | Lipid Configuration | Vehicle Lipid |
|---|---|---|---|
| Abelcet | ABLC | Ribbonlike | DMPC, DMPG |
| Amphocil, Amphotec | ABCD | Disklike | Cholesteryl sulfate |
| AmBisome | Liposomal amphotericin B | Vesicle | HSPC, DSPG, cholesterol |

$^a$ABLC, amphotericin B lipid complex; ABCD, amphotericin B colloidal dispersion; DMPC, dimyristoyl phosphatidylcholine; DMPG, dimyristoyl phosphatidylglycerol; DSPG, distearoyl phosphatidylglycerol; HSPC, hydrogenated soy phosphatidylcholine.

B colloidal dispersion (ABCD, Amphotec), and liposomal amphotericin B (L-AMB, AmBisome) (Table 17.4). Comparative physicochemical properties of amphotericin B and its lipid formulations are shown in Table 17.4. Clinical use of lipid-based formulations of amphotericin B has shown that antifungal activity is maintained while the rate of nephrotoxicity is appreciably reduced. One of the limitations to the use of polyene liposomal formulations is their cost. Ongoing evaluation of these compounds should address the pharmacoeconomic issues by analyzing the cost-benefit ratios for their wider use.

**3.1.8 Things to Come.** Like amphotericin B, the use of nystatin has been limited because of its nephrotoxic side effects. Lipid formulations of nystatin have displayed decreased toxicity with the maintenance of antifungal activity; liposomal nystatin is a newer formulation whose pharmacokinetics, efficacy, toxicity, and comparative advantages to lipid formulations of amphotericin B are currently being evaluated (19–25).

**3.2  Azoles**

**3.2.1 Overview and Mode of Action.** Azoles have emerged as the preeminent class of antifungal agents for all except the most life-threatening infections, and even in these cases highly active new variants are being developed. The most significant agents in clinical use are shown in Fig. 17.3. Chlormidazole, miconazole, clotrimazole, and econazole are administered topically; ketoconazole, itraconazole, and fluconazole are useful in the treatment of systemic infections. The discussion of the pharmacology of various members of the class below is limited to systemic agents.

Azoles interfere with ergosterol biosynthesis by inhibiting demethylation at $C_{14}$ of lanosterol (see Scheme 17.1), thereby causing accumulation of 14$\alpha$-methylated sterols that disrupt the various sterol functions in the cell. Except at very high concentrations, the agents are fungistatic, a feature that is of some clinical concern in highly immunocompromised patients, where eradication of the pathogen is a very desirable outcome of therapy. The target protein ($C_{14}$ demethylase, also known as CYP51) is a cytochrome P450 (CyP450) enzyme that achieves its effect by three sequential radical-mediated hydroxylations, and spectroscopic studies were used to show that the triazole or imidazole moiety in the inhibitor coordinates at the sixth ligand binding site of the ferrous ion (26). Because the protein is membrane bound, X-ray crystallographic studies have not been readily forthcoming, despite the many years of intensive research in this area. Very recently, however, the structure of a complex of a soluble CYP51 ortholog from *Mycobacterium tuberculosis* with fluconazole has been solved (27).

Despite the presence of an analogous enzyme for the $C_{14}$-demethylation of lanosterol in mammalian cells, the fungal enzymes are typically 100- to 1000-fold more susceptible to inhibition by azoles. For at least one agent, inhibition of the fungal enzyme was shown to be noncompetitive with respect to the substrate, whereas in the corresponding mammalian enzyme the inhibition was competitive (and therefore potentially reduced by increases in substrate concentration); this may provide a further explanation for the relative lack of effect in mammalian systems (28).

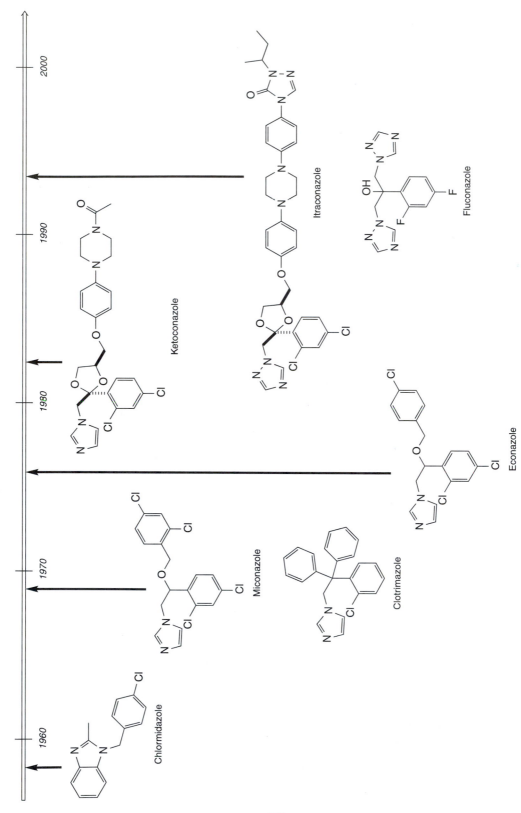

**Figure 17.3.** Chronology of discovery of clinically significant azoles.

894

**3.2.2 History of Azole Discovery.** The azoles were the first purely synthetic class of antifungal agents to be exploited clinically, and their discovery was facilitated by the search for agents for use in the control of fungi in an agrochemical context, where they have had an equally dramatic impact. Thus early antifungal screens soon established that simple imidazole derivatives had activity, building on the first (serendipitous) observation by Woolley in 1944 that benzimidazole affected fungal growth *in vitro* (29). Of the very many agents that have been reported, the chronology of discovery of the most significant clinically is highlighted below.

The first azole to become available for clinical use (as a topical agent) was chlormidazole (Fig. 17.3), introduced by Chemie Gruenenthal in 1958. It was followed in 1969 by Janssen's miconazole (30) and Bayer's clotrimazole (31), and econazole (30) was launched by Janssen in 1974. Even today, the latter three agents remain the mainstay of topical therapy for many dermatophytoses.

In the late 1970s and 1980s, the emphasis in research in anti-infectives was toward the identification of antibacterial agents, in that the clinical need was particularly evident in that area. The only significant event in the fungal context was the introduction by Janssen in 1981 of ketoconazole, the first azole with utility in the treatment of systemic mycoses (32). Like its forebears, this agent is an imidazole derivative, albeit of significantly greater structural complexity; the dioxolane ring was first incorporated in much simpler acetophenone derivatives that were reported as part of the miconazole discovery effort (33). It proved a significant advance in the treatment of serious fungal disease in that it provided an orally available, less toxic alternative to amphotericin B, and for nearly 10 years was the only azole available for this purpose. As such, it saw widespread use, despite limitations with respect to bioavailability and pharmacokinetic interactions (see below).

The finding that imidazole could be replaced by triazole as the $C_{14}$-demethylase heme-coordinating ligand was a major breakthrough because it provided both greater selectivity for the fungal enzyme and greater metabolic stability *in vivo*. The issue of selectivity was a particular concern, given the ubiquity of CyP450-mediated enzymes in mammalian systems and the attendant association of azoles with hepatic side effects and effects on hormone synthesis. Janssen's itraconazole made use of this key change (33), although the extended ketoconazole-like side chain rendered the compound extremely insoluble, presenting formidable challenges in identifying formulations that allowed parenteral administration or gave reliable oral bioavailability. Thus, despite the appearance on the market of an oral capsule form in 1992, an oral solution in cyclodextrin was not launched until 1997, and it took until 1999 for an intravenous formulation (in hydroxypropyl-$\beta$-cyclodextrin) to become available.

The discovery of fluconazole at Pfizer was an interesting study in the optimization of molecular properties important for good activity *in vivo*. As with most azole discovery programs at this time, a mouse survival model involving challenge with a lethal, systemic dose of *C. albicans* was being used at an early stage in the evaluation process to differentiate analogs. The use of the triazole moiety was based on the reduced propensity for first-pass metabolism, and the emphasis on analogs containing a tertiary alcohol (as opposed to dioxolanes or tetrahydrofurans) was also based on the superior activity observed *in vivo*. A conscious attempt was made to focus on polar analogs, particularly with regard to the side chain, thereby minimizing serum protein binding and maximizing the amount of unbound drug available at the site of infection; it was this consideration that led to the incorporation of a second azole moiety into the molecule. Finally, systematic variation of the halo-substituted aromatic group, combined with pharmacokinetic studies in mice, led to the selection of the 2,4-difluorophenyl analog, which was notable for its high urinary recovery (indicative of metabolic stability) (34). This discovery strategy has been fully vindicated: the solubility, safety, and predictable pharmacokinetics of fluconazole, launched in 1991, have driven its success as the world's predominant prescription antifungal agent.

**3.2.3 Structure-Activity Relationship.** From the many series of azoles that have been re-

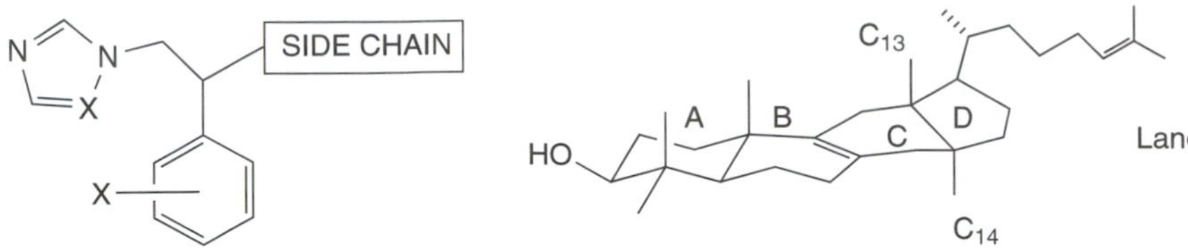

**Figure 17.4.** Azole pharmacophore.

ported by diverse groups, several common structural features emerge: an imidazole or triazole heme-coordinating group, a halo-substituted aromatic separated from the azole moiety by two atoms, and a side chain (see Fig. 17.4). The latter represents the feature of greatest diversity across the family.

Early postulates that the halo-substituted aromatic residue binds in the pocket normally occupied by the A/B-ring region of lanosterol (see conformational depiction in Fig. 17.4) now seem less likely in the light of recent modeling and crystallographic studies (27, 35); indeed, it may be that this residue binds in the same hydrophobic cleft as the $C_{17}$ side chain of the substrate. The structural diversity and variable length of the side chains explored by various groups suggests that this part of the pharmacophore may extend beyond the substrate binding site, perhaps into the substrate access channel. This general binding model is further substantiated by the more recent discovery that incorporation of a methyl group $\beta$ to the aromatic ring gives added potency (particularly against moulds), but only for one of the four possible stereoisomers (36). This may be because the methyl group occupies the binding pocket filled by the $C_{13}$ methyl group on the $\beta$ face of lanosterol; it is also likely that its presence favors conformations in which the side chain is antiperiplanar to the aromatic ring (37).

**3.2.4 Resistance.** Resistance to azoles has become a significant problem since their use in the treatment of systemic fungal infections has increased (38). Target modification is clearly a common contributor to clinical resistance to azole therapy and has been implicated directly for *C. neoformans* (39), *C. albicans* (40–42), *C. glabrata* (43), and by inference for other *Candida* spp. (44). Different mutations have been documented, making it difficult to

rationally design new agents that are less prone to this resistance mechanism. Azole resistance also occurs when compensatory changes exist in other enzymes in the ergosterol biosynthesis pathway, notably inactivation of $\Delta^{5,6}$-desaturase. This leads to the build-up of nontoxic 14$\alpha$-methylated steroids in fungal membranes. The change in membrane composition can lead to cross-resistance to polyenes (45, 46).

Probably the most prevalent cause of resistance to azole therapy is caused by reduction in intracellular drug concentrations ascribed to active efflux. In *C. albicans*, two types of efflux pump have been shown to be clinically relevant: a member of the Major Facilitator superfamily known as MDR1 (or BEN), and ATP-binding cassette (ABC)-type transporters CDR1 (47) and CDR2 (48). Fluconazole (Fig. 17.3) is a substrate for all three of these pumps, whereas other azoles are affected only by CDR1/2. Homologs of the ABC-type transporters have been shown to confer resistance in clinical isolates of *C. glabrata* (CgCDR1/2) (49, 50) and *A. fumigatus* (ADR1) (51). Tools for the convenient characterization of such resistance mechanisms in clinical isolates are becoming available (52).

**3.2.5 Side Effects.** The most common adverse effects of ketoconazole (Fig. 17.3) treatment are dose-dependent nausea and vomiting, which occur in approximately 20% of patients receiving a 400 mg daily dose. Better tolerance is achieved by administration with food. Hepatic abnormalities are common, with approximately 5–10% of patients experiencing a mild, asymptomatic elevation of liver transaminases, which return to normal after cessation of treatment. Liver function tests are indicated in patients with hepatic abnormalities. Hepatic insult by the azoles is mechanism based, and the selectivity improves from ketoconazole to fluconazole.

Itraconazole is typically well tolerated, with few limiting adverse effects in patients receiving up to 400 mg of the compound. Most adverse effects are transient and can be dealt with by reduction in dose.

Of the three systemic azoles, fluconazole demonstrates the least adverse effects. Daily doses up to 1200 mg are well tolerated. Marked increases in hepatic transaminases are seen with less than 1% of patients treated with fluconazole, and more severe hepatic toxicity or hepatitis is rare.

**3.2.6 ADME.** Ketoconazole was the first successful orally active azole useful for fungal infections. An acidic environment is required for its dissolution; therefore, its absorption is decreased in patients with decreased gastric acidity, such as those receiving antacids or $H_2$-histamine blockers. After oral doses of 200, 400, and 800 mg the peak plasma concentrations of the agent are approximately 4, 8, and 20 $\mu$g/mL, respectively. The half-life of ketoconazole increases with dose and can be as long as 8 h (after an 800 mg dose). The compound is extensively metabolized, with inactive metabolites excreted mainly through the bile. Concentrations of the active drug or its metabolites in the urine are low.

Like ketoconazole, itraconazole is soluble at low pH. Hence, its absorption is affected by patients in the fasting state or those individuals receiving concurrent treatment with $H_2$ receptor antagonists or antacids. Plasma levels vary widely among patients; however, clinically useful levels can be achieved when the compound is taken with food or an acidic drink. Peak plasma concentrations are reached 1.5–4 h after drug administration. The compound is extensively metabolized in the liver and is excreted as inactive metabolites into the bile and urine. One metabolite, hydroxyitraconazole, is as active as the parent. Although it is eliminated more rapidly, hydroxyitraconazole's maximum plasma concentration is approximately 2 times higher than that of the parent.

Fluconazole is well absorbed; its bioavailability is greater than 90%. Unlike ketoconazole and itraconazole, absorption is not af-

fected by intragastric pH. The serum protein binding of fluconazole is 12%, whereas both ketoconazole and itraconazole are >95% bound. The volume of distribution of fluconazole approximates that of total body water and the drug penetrates well into nearly all body tissues, including the CSF. In further contrast to ketoconazole and itraconazole, fluconazole is relatively stable to metabolism; its terminal half-life ranges from 27 to 37 h. More than 90% of a therapeutic dose is excreted through the kidneys, 80% of which is recovered unchanged in the urine.

**3.2.7 Drug Interactions.** As a class, the azoles are notable for their propensity for drug-drug interactions. Most have measurable affinity for human CyP450 enzymes. Interactions of ketoconazole with other agents, however, are much more frequent than with newer compounds, especially fluconazole. Compounds such as rifampicin increase ketoconazole clearance through the induction of hepatic enzymes.

Itraconazole-induced inhibition of CyP450 enzymes may lead to potentially toxic concentrations of coadministered drugs. As with ketoconazole, compounds that affect the CyP450 enzyme systems can possibly increase or decrease the metabolism of the compound.

The types of drug interactions observed with fluconazole are generally similar to those of ketoconazole or itraconazole, but the number and severity of relevant interactions reported with fluconazole are much lower than those with the other two agents.

**3.2.8 Things to Come.** With the success of fluconazole and the increased incidence of fungal infections in general, there have been many attempts to develop novel agents with increased activity, particularly against organisms that cause serious disease in neutropenic patients, notably *Aspergillus* spp. The three azoles (Fig. 17.5) that are currently under FDA review or in late-stage development are reviewed briefly below.

*3.2.8.1 Voriconazole.* Voriconazole (Vfend), discovered by Pfizer and previously known as UK-109,496 (53), is under FDA review. Unlike fluconazole, this derivative shows potent activity against a wide variety of fungi, in-

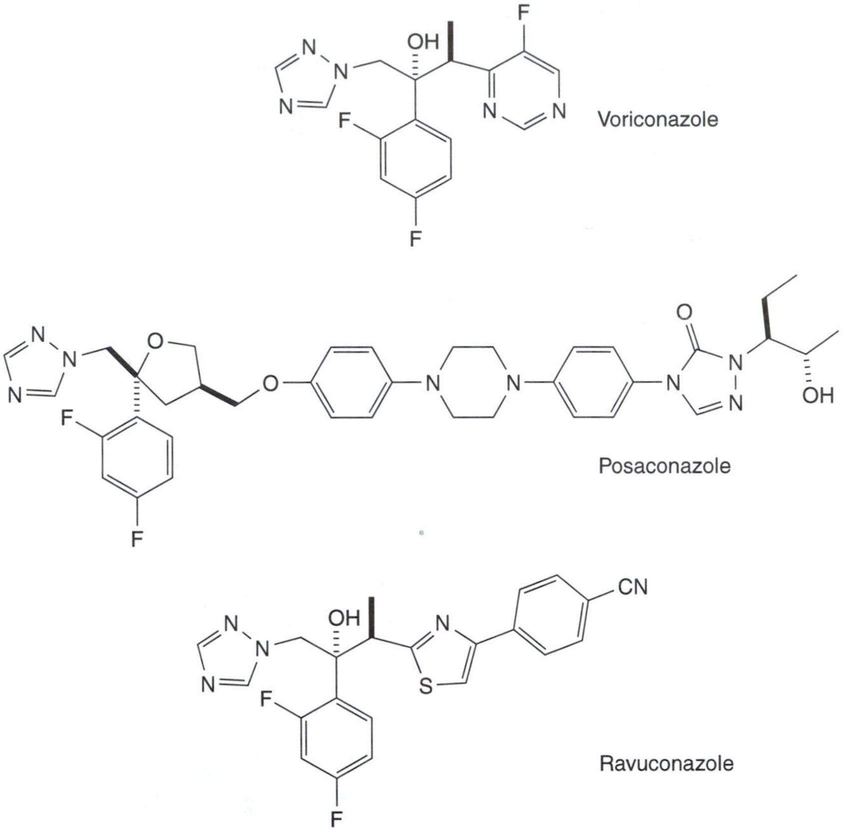

**Figure 17.5.** Azoles in advanced clinical development.

cluding all the clinically important pathogens (54–60). Voriconazole is clearly more potent than itraconazole against *Aspergillus* spp. (59) and is comparable to posaconazole and ravuconazole in its activity against *C. albicans*; however, it appears slightly more potent against *C. glabrata* (61). In general, *Candida* spp. that are less susceptible to fluconazole also exhibit higher MICs to voriconazole and other azoles (62). Despite this, the voriconazole $\text{MIC}_{90}$ for 1300 bloodstream isolates was 0.5 μg/mL (61).

Of all the azoles in current use or under development, voriconazole most closely resembles fluconazole in its pharmacokinetic profile: in humans, bioavailability is high (up to 90%), protein binding (58%) and volume of distribution (2 L/kg) are low. The mean half-life of about 6 h is, however, considerably lower than that of fluconazole, which is attrib-

utable to extensive metabolism; pharmacokinetics are nonlinear, possibly suggesting saturable clearance mechanisms (63). In phase II clinical trials the compound was shown to be efficacious in acute or chronic invasive aspergillosis in both neutropenic and nonneutropenic patients (64, 65), and in oropharyngeal candidiasis (66). Voriconazole has also been shown to be a suitable alternative to amphotericin B preparations for empirical antifungal therapy in patients with neutropenia and persistent fever (67).

**3.2.8.2 Posaconazole and Ravuconazole.** The *in vitro* activity of posaconazole (Schering-Plough) (Fig. 17.5) appears similar to that of voriconazole and ravuconazole (55, 61, 68). Currently in phase III clinical trials, the compound has been shown to be efficacious in a wide variety of infections in animal models, including those caused by rarer pathogens (69–74).

Posaconazole is subject to enterohepatic re-circulation and is eliminated primarily by way of the bile and feces (75, 76). The compound exhibits high oral bioavailability in animal models (77), but its low solubility precludes convenient formulation for intravenous use. A program aimed at identifying a more soluble prodrug has identified SCH 59884, in which the hydroxyl group of posaconazole is derivatized as a butyryl ester bearing a γ-phosphate, as a potential candidate (78, 79). This agent is inactive *in vitro*, but is dephosphorylated *in vivo* to produce the active 4-hydroxybutyrate ester of posaconazole. This, in turn, is hydrolyzed to the parent compound in human serum.

Ravuconazole (Esai/Bristol-Myers Squibb) has *in vitro* activity that is broadly similar in spectrum and potency to that of voriconazole and posaconazole (55, 80–82). Like posaconazole, the compound exhibits good oral bioavailability and a long serum half-life in both rats and dogs (83, 84) (in contrast, the half-life of voriconazole is significantly shorter). The activity of the agent in animal models of systemic and pulmonary infections attributed to *Candida*, *Cryptococcus*, and *Aspergillus* spp. has been demonstrated (83, 85, 86).

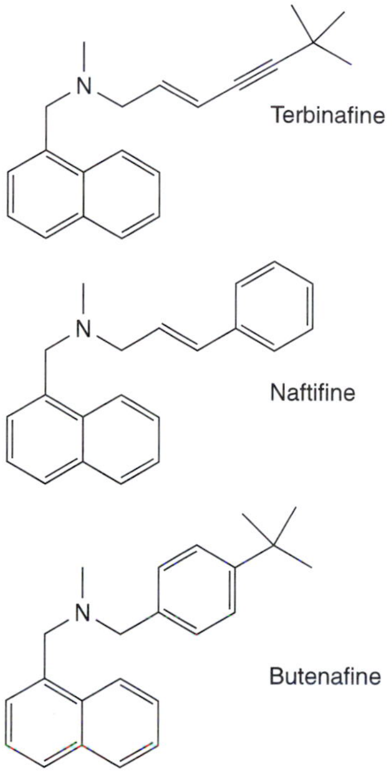

## 3.3 Allylamines

### 3.3.1 Overview and Mode of Action.
The allylamines are the most prominent of a number of antifungal classes (see also Section 3.5.1.1) that exert their activity by inhibition of squalene epoxidase; the intracellular accumulation of squalene that results is thought to be the primary cause of the fungicidal consequences of exposure to the drug (87). The predominant example of this class of antifungal agent is terbinafine (Fig. 17.6), which is one of the mainstays for the treatment of dermatophytosis. In the treatment of onychomycosis, in particular, it is safer, more efficacious, and requires shorter duration of therapy than griseofulvin (the previous standard, but inadequate, agent); its efficacy in this indication is enhanced by the propensity of the compound to accumulate in nails and hair. It has seen some use, particularly in combination with fluconazole, in the treatment of oropharyngeal infections arising from *Candida* spp. (88,

**Figure 17.6.** Allylamines.

89). The other significant member of the class is naftifine, which is an older agent whose use has been in the topical treatment of ringworm (tinea cruris, tinea corporis) (90). Butenafine (91) is used for the same purpose.

The inhibition of squalene epoxidase by the allylamines is reversible and noncompetitive with respect to squalene, NADPH, and FAD (92), and the agents have no effects on other enzymes in the ergosterol biosynthetic pathway (93). In cell-free rat liver extracts, terbinafine has been shown to inhibit cholesterol synthesis, also by specific inhibition of squalene epoxidase, although the concentration required to do so is at least 1000-fold greater than that required in the analogous assay derived from fungal cells (93).

### 3.3.2 Structure-Activity Relationship.
The moniker "allylamine" was originally applied to the class because the tertiary (*E*)-allylamine structural element was perceived to be essential for antifungal activity (94). In more recent studies, however, it has been shown that this feature can be replaced by a suitably substi-

tuted benzylamine, or by a homopropargylamine. Thus, the structural requirements for potent activity are represented in the broadest sense by two lipophilic domains linked to a central polar moiety by spacers of appropriate length; for good activity, the polar moiety is a tertiary amine, and one of the lipophilic domains consists of a bicyclic aromatic ring system such as naphthalene or benzo[b]thiophene (95).

### 3.3.3 Resistance.
Clinical resistance to terbinafine has not yet been documented. However, it has been shown that inhibitors of CDR-type pumps in *C. albicans* can reduce the MIC to terbinafine (96); thus, the potential for the emergence of efflux-mediated resistance is established. Allylamines are widely used in the agrochemical context, and resistance has been documented in the corn pathogen *Ustilago maydis* (97), in which both decreased intracellular concentrations and decreased affinity for the target enzyme were implicated.

### 3.3.4 Side Effects.
In general, the adverse events after oral terbinafine treatment are mild and transient. Rare cases of liver failure have occurred with the use of the compound; however, in the majority of the cases reported, the patients had an underlying hepatic condition.

### 3.3.5 ADME.
Terbinafine is well absorbed (>70%) after oral administration, and the bioavailability of about 40% is attributable largely to first-pass metabolism. Nonetheless, after a single oral dose of 250 mg, peak plasma concentrations of only 1 $\mu$g/mL are achieved, despite >99% plasma protein binding. Such relatively low plasma concentrations are ascribed to the extensive distribution to the sebum and skin with a terminal half-life of 200–400 h, reflecting the slow elimination of the compound from the body. Terbinafine is extensively metabolized, although none of the metabolites observed to date possesses antifungal activity. Approximately 70% of the administered dose is eliminated in the urine.

### 3.3.6 Drug Interactions.
Drug-drug interactions conducted in healthy volunteers showed that terbinafine does not affect the clearance of antipyrine or digoxin but does decrease the clearance of caffeine by 19%. Terbinafine clearance is increased 100% by rifampin, a CyP450 enzyme inducer, and decreased by a factor of 3 by cimetidine, a CyP450 inhibitor.

### 3.3.7 Things to Come.
The growing awareness of toenail onychomycosis should be a catalyst for the clinical use of terbinafine. Several large open-labeled, multicenter trials of terbinafine treatment have been conducted and continue to demonstrate the efficacy of the compound in the treatment of this particular infection (98). The potential utility of the agent in the treatment of systemic infections has also recently been explored (89).

## 3.4 Candins

### 3.4.1 Overview and Mode of Action.
Echinocandins, natural products discovered in the 1970s, and the related pneumocandins, discovered in the 1980s by researchers at Merck, are presumed to act as noncompetitive inhibitors of (1,3)-$\beta$-D-glucan synthase, an enzyme complex that forms glucan polymers in the fungal cell wall [16, 99, 100; see Section 1.5.2]. There are at least two subunits of this enzyme: one a catalytic subunit in the plasma membrane, the other a GTP-binding protein that activates the catalytic subunit. Analysis of various resistant mutants of *S. cerevisiae* led to the cloning of the echinocandin target gene *ETG1*. The gene encodes for a 215-kDa protein that contains 16 putative transmembrane domains. Whereas disruption of *ETG1* alone does not lead to lethality in *S. cerevisiae*, disruption of *ETG1* together with a homologous gene, *FKS2*, leads to fungal cell death (100).

Derivatives of the echinocandins and pneumocandins have been explored, and three prominent analogs have emerged. One of these, Cancidas (caspofungin acetate, MK-0991), was approved in early 2001 for the treatment of refractory invasive aspergillosis in patients who do not respond or cannot tolerate other therapies such as amphotericin B, lipid formulations of the polyenes, and/or itraconazole. A summary of the pharmacology,

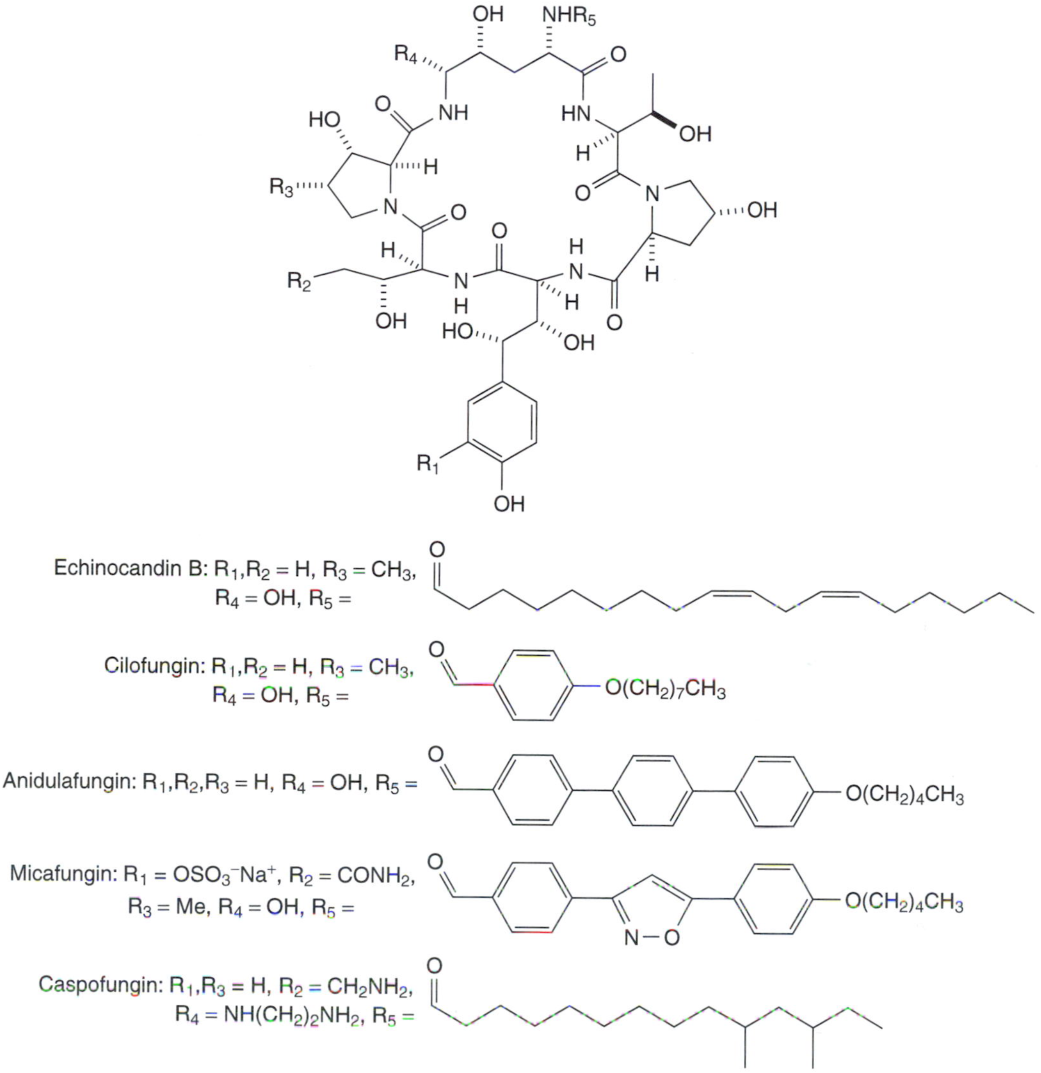

**Figure 17.7.** Notable candins.

pharmacokinetics, clinical efficacy, and adverse effects of caspofungin has been reported (101).

Caspofungin and the two related compounds in clinical development (anidulafungin and micafungin, see below) have potent activity against a variety of *Candida* and *Aspergillus* spp., including azole-resistant strains.

**3.4.2 Structure-Activity Relationship.** Echinocandin B (Fig. 17.7) is the archetype of the family of glucan synthase inhibitors. The

structural features common to these lipopeptides include a cyclic peptide composed of six amino acids, one of which (an ornithine) is acylated at the $\alpha$-N with a fatty acid containing 14–18 carbons. The peptide is cyclized by an aminal linkage connecting 3-hydroxy-4-methylproline to the $\delta$-amino group of dihydroxyornithine. This group is readily cleaved at basic pH, resulting in ring opening and complete loss of antifungal activity.

The clinical limitations of echinocandin B include poor solubility, lack of oral bioavailability, and a propensity to lyse red blood cells *in vitro*. Because chemical modifications implicated the lipophilic acyl tail in the lytic potential of echinocandin B, various analogs of the natural product were prepared in which the cyclic peptide nucleus was acylated with modified fatty acid tails. Antifungal activity was maintained in these compounds when a lipophilic tail of at least 12 carbon atoms in length was incorporated, with $C_{18}$ unbranched fatty acids being optimal. Cilofungin (Fig. 17.7), echinocandin B with a *p*-octyloxybenzoyl side chain, was prepared in this fashion and chosen for clinical evaluation, but its development was halted because of its poor solubility, narrow spectrum of activity, and nephrotoxicity (thought to be attributed to the dosing vehicle).

A related analog (anidulafungin, LY303366, V-002), containing the echinocandin B nucleus with a terphenyl side chain with a tail of five carbon atoms, is in late-stage clinical development (102). Recent structure-activity relationship (SAR) studies with anidulafungin have focused principally on (*1*) improving the water solubility of the compounds by introducing phosphonate and phosphate ester prodrugs on the phenolic hydroxy group (103), and (*2*) addressing the instability of the compound under strongly basic conditions by incorporating nitrogen-containing ethers at the hemiaminal hydroxy group (104).

In general the echinocandins display poor intrinsic water solubility; however, the isolation of the natural product FR901379 was notable in that it was the first example of an intrinsically soluble echinocandin-like cyclic hexapeptide. Side-chain modification of this natural product led to the discovery of micafungin (FK-463), which is in late-stage clinical trials (105). The target compound was prepared by condensation of an active ester of 4-(5-(4-pentoxyphenyl)isoxazol-3-yl)benzoic acid with the cyclic peptide nucleus, obtained by enzymatic deacylation of the natural product.

The related pneumocandins differ in structure from the echinocandin B class of lipopeptides by the replacement of threonine with hydroxyglutamine, along with a modified proline in place of the 3-hydroxy-4-methylproline in the cyclic peptide nucleus. The glutamine residue provides a handle for chemical modification to improve both activity and solubility. The lipid tail of pneumocandin $B_0$, a component of the pneumocandins produced by the fungus *Glarea lozoyensis,* has 10,12-dimethylmyristate in place of echinocandin B's linoleoyl side chain. In contrast to echinocandin B, pneumocandin $B_0$ did not cause lysis of red blood cells. The aim of early medicinal chemistry focused on the improvement of solubility. One strategy employed a prodrug approach, the best candidate from which was the phosphate derivative of pneumocandin $B_0$, L-693,989 (106). In an effort to identify an intrinsically water-soluble and less labile analog, further chemical modification focused on the hemiaminal and the hydroxyglutamine residue (107). This effort ultimately afforded caspofungin (Fig. 17.7).

Various programs have focused on identifying new glucan synthase inhibitors that may overcome several of the problems associated with the echinocandins and pneumocandins. For example, cyclic amino hexapeptides of echinocandin B have been reported (108), as well as the discovery of novel classes of (1,3)-β-D-glucan synthase inhibitors (109).

**3.4.3 Side Effects.** Adverse effects after caspofungin treatment have been minimal and are typically related to histamine-mediated symptoms such as rash, facial swelling, and/or the sensation of warmth. A more complete side-effect profile of caspofungin will emerge as its use becomes more widespread.

**3.4.4 ADME.** Caspofungin is not orally bioavailable and is therefore administered parenterally. After a 1-h intravenous infusion, the compound is extensively distributed into tissues, and is slowly metabolized by phase I transformations such as hydrolysis and *N*-acetylation. Elimination of caspofungin and its metabolites occurs equally through the feces and urine; only a small percentage of the parent compound is detected in the latter.

**3.4.5 Things to Come.** As noted above there are two compounds in addition to caspofungin that are in late-stage clinical trials:

*3.4.5.1 Anidulafungin (LY303366, V-Echinocandin, V-002).* This agent is a pentyloxyterphenyl side-chain derivative of echinocandin B (Fig. 17.7). The agent is in late-stage phase II and early phase III clinical trials (110). Like caspofungin, the compound is active *in vitro* against *Candida* and *Aspergillus* spp. (111) and its potent activity in animal models of disseminated candidiasis and pulmonary aspergillosis has been described (112–114). Anidulafungin is also active in animal models of esophageal candidiasis and aspergillosis using organisms that are resistant to fluconazole and itraconazole, respectively (115, 116). The pharmacokinetics of the compound in healthy and HIV-infected volunteers after single-dose administration have been reported (117).

***3.4.5.2 Micafungin (FK-463).*** Micafungin (Fig. 17.7), like anidulafungin and caspofungin, has potent *in vitro* activity against a variety of *Candida* (MIC range $\leq 0.004-2 \ \mu g/mL$) and *Aspergillus* ($\leq 0.004-0.03 \ \mu g/mL$) spp. (118–121), and is also active in a number of animal efficacy models (122–124). The compound has favorable pharmacokinetics and was well tolerated in a single-dose phase I study in healthy volunteers (125). In a phase II study in an AIDS population, micafungin was effective in improving or clearing the clinical signs and symptoms of esophageal candidiasis at 12.5, 25, and 50 mg once daily for up to 21 days (126). In addition, once-daily dosing for 14-21 days revealed no safety-related concerns.

### 3.5 Miscellaneous

#### 3.5.1 Other Inhibitors of Ergosterol Biosynthesis

***3.5.1.1 Thiocarbamates.*** The clinically most prominent class of antifungal ergosterol biosynthesis inhibitors not mentioned above are the thiocarbamates. Like the allylamines, these are reversible, noncompetitive inhibitors of squalene epoxidase (127), with inherent selectivity for the fungal enzymes over mammalian (128). The most significant mem-

ber of the class is tolnaftate (Fig. 17.8). Unlike terbinafine, its spectrum is restricted primarily to dermatophytes; the lack of effect on *Candida* spp. is attributed to poor penetration of the cell envelope (128). Consequently, it is not used in the treatment of systemic disease and its pharmacokinetic properties dictate topical administration for the treatment of skin infections, such as athlete's foot, where cure rates are around 80% (cf. 95% for miconazole) (90).

***3.5.1.2 Morpholines.*** Discovered in the 1970s, amorolfine (Fig. 17.9) has seen some use topically in the treatment of nail infections. Like fenpropimorph (Fig. 17.9), which is a prominent antifungal agrochemical, the compound inhibits both $\Delta^{14}$ reductase and $\Delta^7-\Delta^8$ isomerase. Antifungal activity is attributable primarily to inhibition of the former, which is an essential enzyme in *S. cerevisiae* (129). The toxicity of amorolfine precludes its systemic use (127).

***3.5.1.3 Other Azoles.*** The structures of azoles listed in Table 17.3, but not discussed in the text, are shown in Fig. 17.10.

#### 3.5.2 Flucytosine.

***3.5.2.1 Overview and Mode of Action.*** Flucytosine (Fig. 17.11) is a fluorinated pyrimidine related to the anticancer agent fluorouracil. It was originally developed in 1957 as an antineoplastic agent, but reports in 1968 of its being used to treat candida and cryptococcal infections in humans led to its use as an antifungal agent. As with many other antifungal compounds, *in vitro* susceptibility testing of flucytosine has correlated poorly with

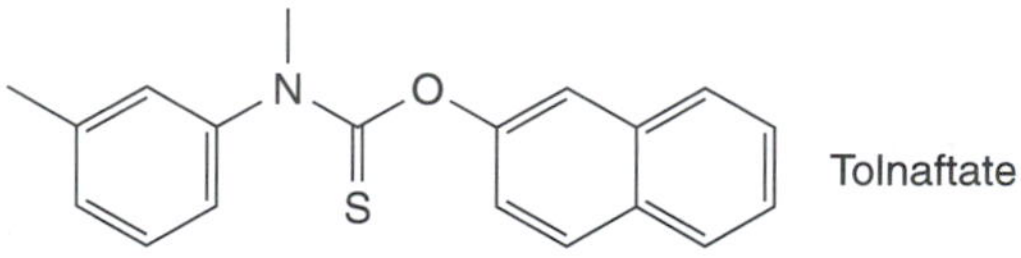

Amorolfine

Fenpropimorph

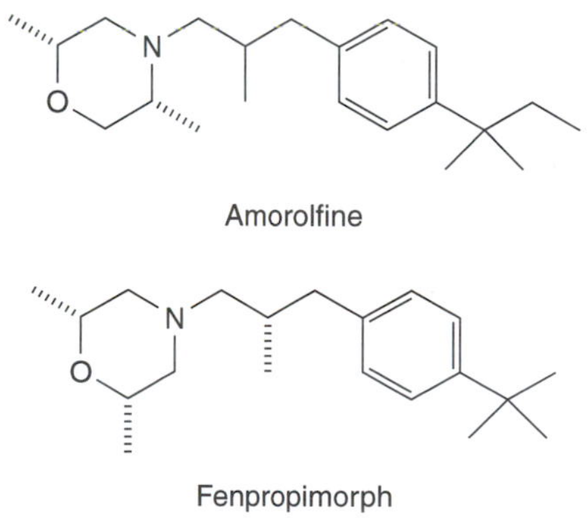

**Figure 17.9.** Amorolfine and fenpropimorph.

Tolnaftate

**Figure 17.8.** Tolnaftate.

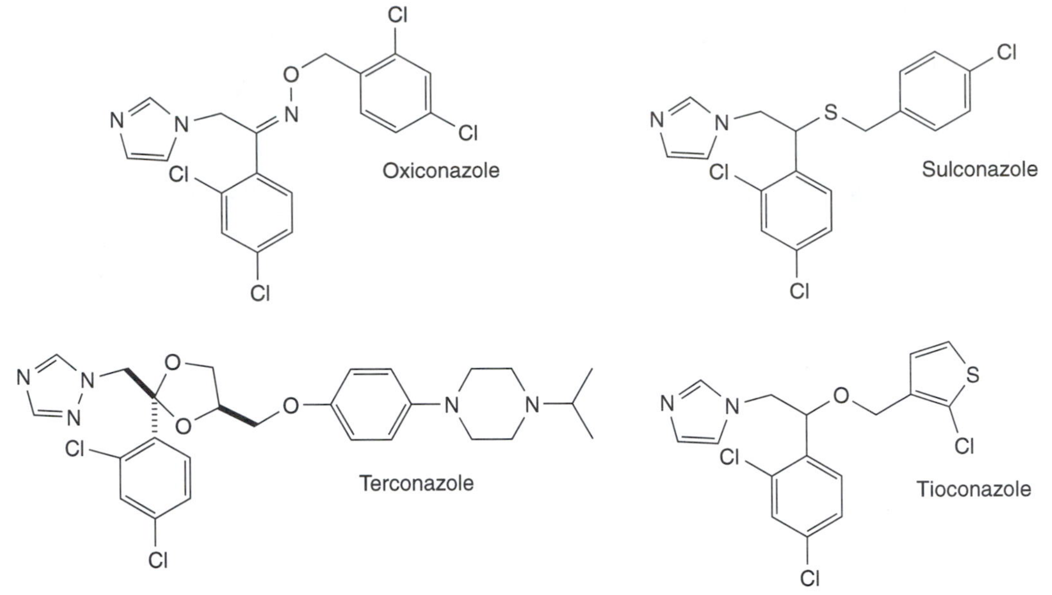

**Figure 17.10.** Other azoles for topical use.

clinical outcome. Therapeutically, flucytosine is used predominantly in combination with amphotericin B to treat cryptococcal meningitis and infections caused by *Candida* spp. In these cases, lower doses of amphotericin B can be used.

Susceptible fungi, such as *Crytococcus neoformans* and *Candida* spp., deaminate flucytosine by way of cytosine deaminase to 5-fluorouracil, which is ultimately converted into metabolites that inhibit DNA synthesis (Scheme 17.3). 5-Fluorouracil is also incorporated into fungal RNA, thereby disrupting transcription and translation. Selectivity is achieved because mammalian cells are unable to convert flucytosine to fluorouracil.

***3.5.2.2 Resistance.*** Resistance to flucytosine is common and can be caused by the loss of the permease that is necessary for com-

pound transport or decreased activity of either cytosine deaminase or UMP pyrophosphorylase (130). Strategies to prevent the emergence of resistant isolates have focused on maintaining sufficient drug concentration at the infection site as well as combining the compound with other antifungal agents such as amphotericin B.

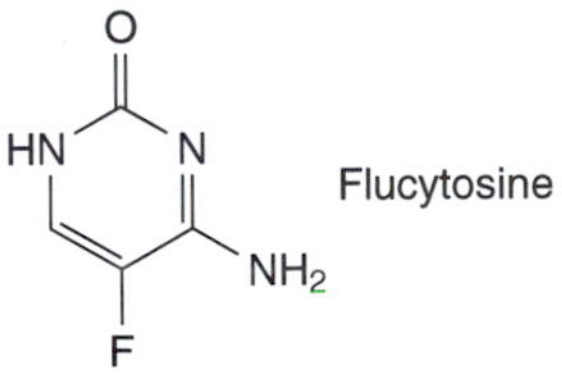

***3.5.2.3 Side Effects.*** The most serious adverse effects associated with flucytosine therapy are hematological, manifested as leukopenia and thrombocytopenia. Patients that are more prone to this complication, such as those having an underlying hematologic disorder, require careful monitoring after treatment. The hematological toxicities are thought to result from the conversion of flucytosine to the antimetabolite 5-fluorouracil by bacteria in the host.

***3.5.2.4 ADME.*** Approximately 80–90% of a dose of flucytosine is absorbed after oral administration. It is widely distributed in the body with a volume of distribution that approximates total body water. Flucytosine levels in the CSF are approximately 80% of the simultaneous serum levels. Approximately 80% of a given dose is excreted unchanged in the urine, with the half-life of the drug in healthy individuals ranging from 3 to 6 h.

**Figure 17.11.** Flucytosine.

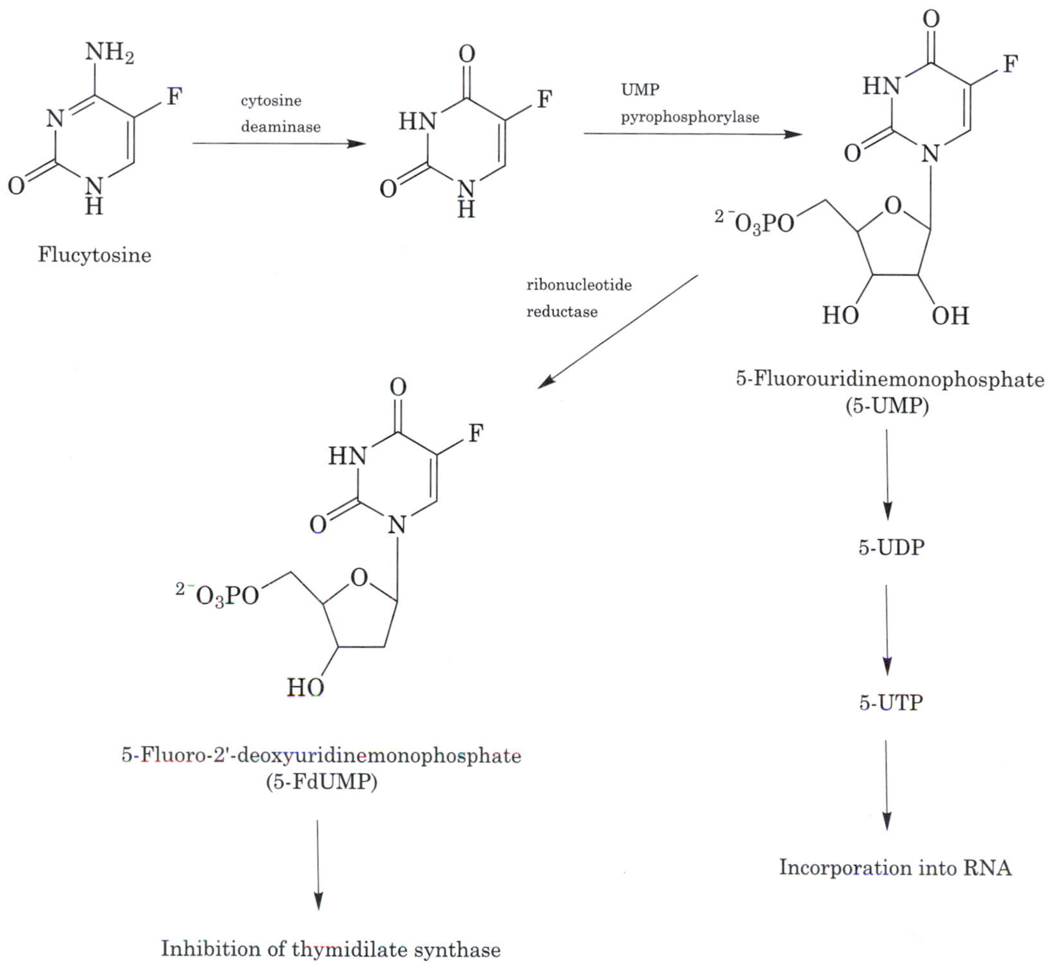

**Scheme 17.3.** Metabolism of flucytosine.

***3.5.2.5 Drug Interactions.*** Concomitant administration of agents that result in toxicities similar to those observed with flucytosine, such as the treatment of patients with cryptococcal meningitis with AZT and/or ganciclovir, requires caution. Drugs known to cause renal dysfunction, such as amphotericin B, may change the elimination profile of flucytosine, resulting in elevated flucytosine levels.

**3.5.3 Griseofulvin.** Griseofulvin (Fig. 17.12) is an antifungal agent of great historical significance. A natural product first isolated in 1939 (131), it was introduced as an oral agent for the treatment of skin and nail infections in 1958. Despite limited efficacy, untoward side effects (particularly headache), and therapeutic regimens lasting up to 12 months (132), griseofulvin served as the first-line drug for treatment of dermatophytosis for many years, only recently being displaced by itraconazole and terbinafine. The compound acts in a fungi-

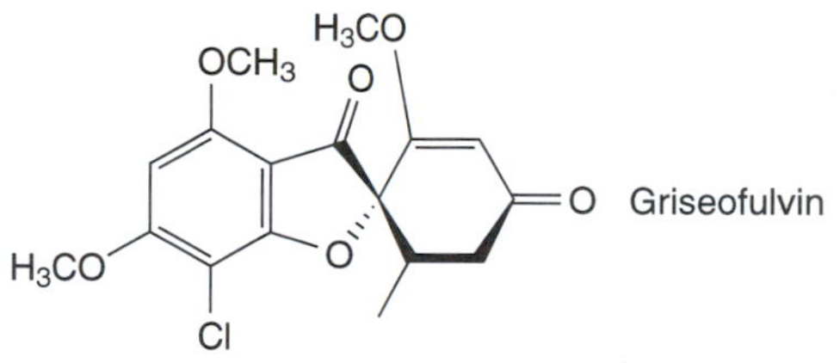

**Figure 17.12.** Griseofulvin.

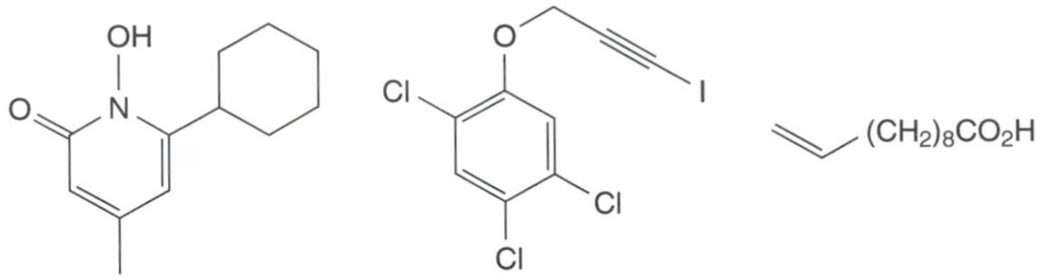

**Figure 17.13.** Other topical agents.

Ciclopirox          Haloprogin          Undecylenate

static manner by inhibiting the formation of microtubules in the process of cell division (133).

### 3.5.4 Other Topical Agents in Clinical Use.

The structures of other agents referred to in Table 17.3, but not discussed in the text, are shown in Fig. 17.13. Ciclopirox is used as the ethanolamine salt and is a broad-spectrum agent with some antibacterial activity (134). Haloprogin (135, 136) also has antibacterial activity. Undecylenate is commonly marketed as the zinc salt.

### 3.5.5 Other Classes of Medicinal Interest.

There are very many natural products and totally synthetic materials that have antifungal activity *in vitro*, but only a few in which the mode of action has been determined in sufficient detail to justify a focused medicinal development program. Those that have not been described above are discussed briefly in this section.

*3.5.5.1 Polyoxins and Nikkomycins.* These are naturally occurring nucleoside peptide antibiotics that inhibit chitin synthase, an enzyme that catalyzes the polymerization of *N*-acetylglucosamine, a major component of the fungal cell wall (see Section 1.5.2). A comprehensive review of synthetic efforts and subsequent biological studies on these agents has been reported (137).

Nikkomycin Z (Fig. 17.14) (138), the most advanced of these agents, has demonstrated additive and synergistic interactions with ei-

ther fluconazole or itraconazole against *C. albicans* and *C. neoformans in vitro* (139). Marked synergism was also observed with nikkomycin Z and itraconazole against *A. fumigatus*. Nikkomycin Z is active against the less common endemic mycoses such as histoplasmosis, where pronounced synergistic interactions with fluconazole have been observed both *in vitro* and *in vivo* (140). An apparent drawback of nikkomycin Z is that it appears to inhibit only weakly the most abundant chitin synthase found in fungi.

A series of nikkomycin analogs has recently been prepared; among them, an analog containing a phenanthrene moiety at the terminal amino acid possessed potent antichitin synthase activity (141). A structurally distinct chitin synthase inhibitor, Ro-09-3143 (Fig. 17.15) arrests cell growth in *C. albicans* (142).

*3.5.5.2 Aureobasidins.* Aureobasidin A (Fig. 17.16), a cyclic depsipeptide produced by *Aureobasidium pullulan,* inhibits inositol phosphorylceramide synthase (IPC synthase), an enzyme essential and unique in fungal sphingolipid biosynthesis (143). Target modification and efflux have both been implicated as a mode of resistance to the compounds (144, 145).

The syntheses of Aureobasidin A and several related cyclopeptide derivatives have been reported (146). Aureobasidin derivatives with modifications at amino acid position 6, 7, or 8 were prepared as part of a study to elaborate the SAR of the natural product. Whereas

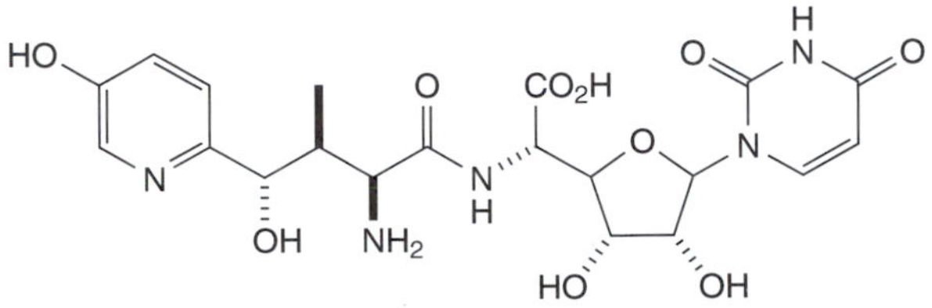

Nikkomycin Z

**Figure 17.14.** Nikkomycin Z.

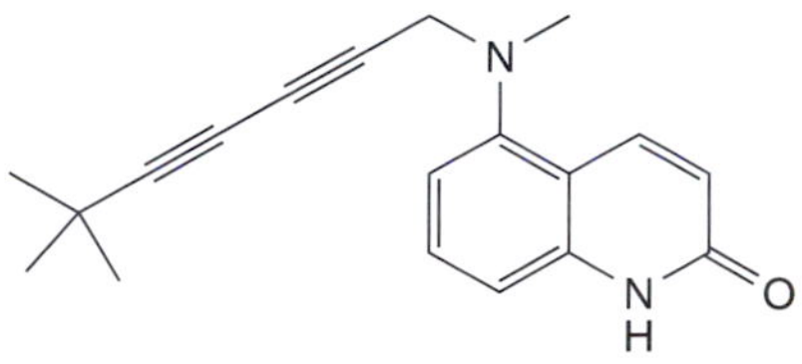

Ro-09-3143

**Figure 17.15.** Ro-09-3143.

analogs having L-glutamic acid at position 6 or 8 showed weak activity, esterification of the γ-carboxyl group with benzyl or shorter alkyl ($C_4$–$C_6$) alcohols significantly enhanced the potency (147). Introduction of a longer $C_{14}$ alkyl chain resulted in total loss of antifungal activity. SAR for the inhibition of the ABC-transporter P-glycoprotein (P-gp) by various aureobasidins has recently been disclosed. Although several compounds demonstrated potent inhibition of this efflux pump, the P-gp activity and the SAR for antifungal activity were not correlated (148).

**3.5.5.3 Sordarins.** Both fungal and mammalian cells require two proteins, elongation factors 1 (EF1) and 2 (EF2), for ribosomal translocation during protein synthesis. Fungi also have a third elongation factor, EF3, which is not present in mammalian cells, and as such is a potential target for antifungal drug development. A family of selective EF2 inhibitors, derived from the tetracyclic diterpene glycoside natural product sordarin (Fig. 17.17), has

been identified and shown to possess activity *in vitro* against a wide range of pathogenic fungi, including *Candida* spp., *Cryptococcus neoformans,* and *P. carinii* (149–153). Target modification has been implicated as a mode of resistance to the sordarins (151).

Using a sordarin derivative, the localization of EF2 in the 80S ribosome fraction of *Saccharomyces cerevisiae* has been determined (154). In addition, the presumed binding site of sordarin in *C. albicans* has been identified by use of a photaffinity label of a sordarin derivative (155).

*In vivo*, sordarin derivatives have shown efficacy against systemic infections in mice caused by fluconazole-sensitive and -resistant *Candida albicans,* with $ED_{50}$ values ranging from 10 to 25 mg/kg (156). The activities of the sordarins in experimental models of aspergillosis and pneumocystosis have also been reported (157). The toxicological properties of the new sordarin derivatives have been evaluated in several *in vitro* and *in vivo* preclinical studies (158, 159). Overall, the compounds have demonstrated no evidence of genotoxicity in the Ames test, are not clastogenic in cultured human lymphocytes, and are well tolerated in rats and dogs.

Modification of the sugar unit affords the tetrahydrofuran derivative GM 237354 and structurally related analogs (150). Researchers at Merck have also reported the prepara-

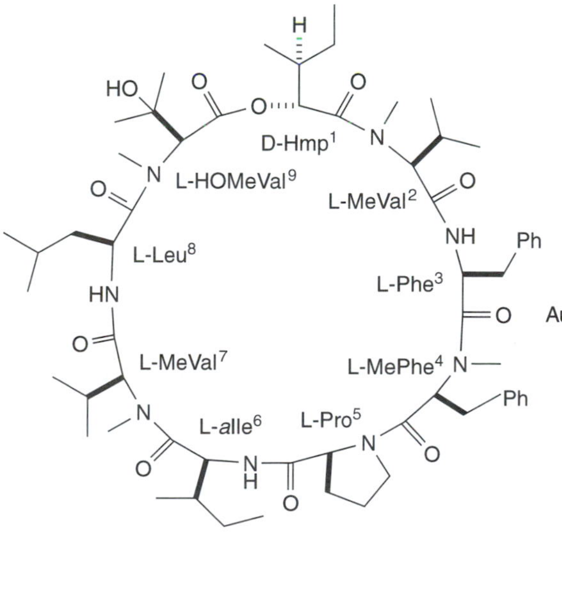

Aureobasidin A

**Figure 17.16.** Aureobasidin A.

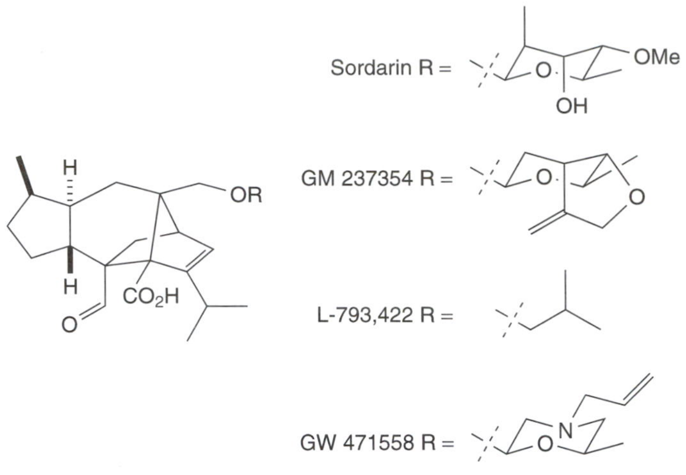

**Figure 17.17.** Sordarin and notable derivatives.

tion and evaluation of a variety of alkyl-substituted derivatives such as L-793,422 (160). Compounds of this type clearly demonstrate that a certain degree of lipophilicity on the side chain is important for optimal antifungal activity. Biological studies have confirmed that L-793,422 and GM 237354 share the same mode of action (161).

An enantio-specific synthesis of the monocyclic core of sordarin has been achieved through the conversion of (+)-3,9-dibromocamphor into a 1,1,2,2,5-penta-substituted cyclopentane bearing all of the key functionalities present in the natural product (162). In addition, preliminary reports describing the SAR of the sordarin class of antifungal agents have recently appeared (163, 164).

Recently, efforts have been directed toward the identification of new sordarin agents that display improved activity against *Candida* spp. other than *C. albicans,* and that possess improved pharmacological properties such as increased efficacy and decreased toxicity. To that end, a new class in which the traditional sugar moiety is replaced by a 6-methylmorpholin-2-yl group with *N*-4' substituents, known as the azasordarins, has been identified. These compounds, highlighted by GW 471558 (Fig. 17.17), have the advantage of being easier to synthesize from fermentation-derived starting materials than from the parent class (165, 166).

*3.5.5.4 Pradimicins and Benanomycins.* These compounds are dihydrobenzonaphthacene quinones conjugated with a D-amino acid

and a disaccharide side chain (167). They bind to cell wall mannoproteins in a calcium-dependent manner that causes disruption of the plasma membrane and leakage of intracellular potassium. Spectroscopic studies on the interaction of BMS 181184 (Fig. 17.18), a water-soluble pradimicin derivative, suggest that two molecules of the compound bind one $Ca^{2+}$ ion, and each compound binds two mannosyl residues (168, 169). BMS 181184 possesses activity toward *Aspergillus* spp. *in vitro*, but is less potent than itraconazole or amphotericin B (170). In a model of invasive pulmonary aspergillosis in persistently neutropenic rabbits, daily doses of 50 and 150 mg/kg of BMS 181184 were as effective as amphotericin B at 1 mg/kg/day (171).

*3.5.5.5 N-Myristoyl Transferase Inhibitors.* *N*-myristoyl transferase (NMT) is a cytosolic enzyme that catalyzes the transfer of myristate from myristoylCoA to the N-terminal glycine amine of a variety of eukaryotic proteins, thereby facilitating protein-protein or protein-lipid interactions involved in intracellular signal transduction cascades. The enzyme has been shown to be essential for the viability of both *C. albicans* and *Cryptococcus neoformans* (172, 173).

An approach to inhibit NMT by exploiting the peptide binding site has been reported. Remarkably, it proved possible to mimic four terminal aminoacids (ALYASKLS-NH₂) of a weak octapeptide inhibitor of the substrate by use of an 11-aminoundecanoyl motif. Initial optimization of this lead gave a highly potent

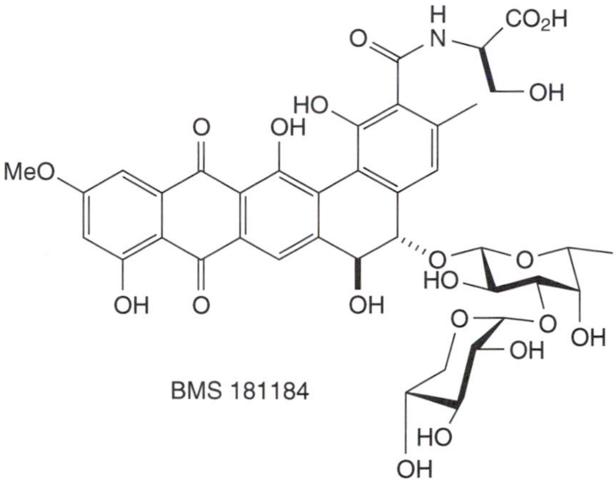

**Figure 17.18.** BMS 181184.

BMS 181184

and selective agent (**1**, Fig. 17.19), which was shown to be a competitive inhibitor of *C. albicans* NMT with respect to the octapeptide substrate GNAASARR-NH$_2$, with a $K_{i(app)}$ value of 70 n$M$, and to exhibit 400-fold selectivity over the human enzyme (174). However, no whole-cell activity was observable. Addition of a second carboxylic acid (**2**) ameliorated the potency against the enzyme, but gave a compound with weak fungistatic activity, as did replacement of the remaining peptidic residues (**3**) (175). The crystal structure of an inhibitor in this series bound to NMT from *S. cerevisiae* has been published (176).

Two alternative structure-based drug design approaches, based on a lead identified through high throughput screening, have also been reported. In the first, a compound with weak but selective activity (**4**, Fig. 17.19; IC$_{50}$ = 0.98 $\mu M$; 200-fold selectivity over the human enzyme) (177) was refined to afford highly potent agents with activity against *C. albicans* both *in vitro* and in a rat model of systemic candidiasis (**5**, **6**) (178). Like compounds 1–3, these inhibitors interact with the key carboxylate residue (Leu451) implicated in the acyl transfer reaction.

In the second structure-based drug design approach (179, 180), early leads based on screening hits that were competitive with the peptide substrate (e.g., **7**, Fig. 17.19; IC$_{50}$ = 0.5 $\mu M$) were potent enzyme inhibitors but lacked whole-cell antifungal activity. Based on the premise that this was attributed to excessive hydrophilicity, lipophilic substituents were added to the

primary amino group. This effort culminated in the identification of a highly potent compound (**8**; IC$_{50}$ = 86 n$M$; *C. albicans* MIC 0.09 $\mu$g/mL) that was also shown to be fungicidal. A surprising lack of activity against *Aspergillus fumigatus* was attributed to a single change (Phe to Ser) in the binding pocket of the benzothiazole-carboxamide, and this was corroborated by site-directed mutagenesis studies in *Candida*.

**3.5.5.6 Fungal Efflux Pump Inhibitors.** It will be apparent from the above discussions on resistance that the inhibition of efflux pumps in pathogenic fungi would be expected to have a significant effect on the susceptibility of several clinically problematic *Candida* spp. toward several classes of antifungal drugs. The first reports of inhibitors of ABC-type pumps in *C. albicans* and *C. glabrata* have recently appeared (181). The agents lack antifungal activity and were characterized by their ability to increase intrinsic susceptibility to known pump substrates (azoles, terbinafine, rhodamine 6G), but not to agents not subject to efflux (amphotericin B). In a fluorescence assay, the compounds were shown to increase intracellular accumulation of rhodamine 6G. Such compounds can reverse CDR-mediated azole resistance in *C. albicans* (64- to 128-fold reduction in MIC of fluconazole or posaconazole) and reduce intrinsic resistance in *C. glabrata* (8- to 16-fold reduction in MIC). A representative of the class, milbemycin $\alpha$-9 (MC-510,027, Fig. 17.20), was shown to dramatically reduce the MIC$_{90}$ of a broad panel of clinical isolates of *Candida* (182).

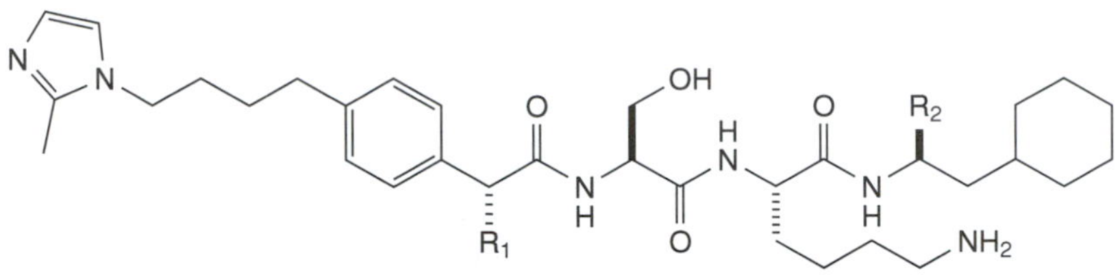

**(1)**, R$_1$ = Me, R$_2$ = H; **(2)**, R$_1$ = H, R$_2$ = CO$_2$H

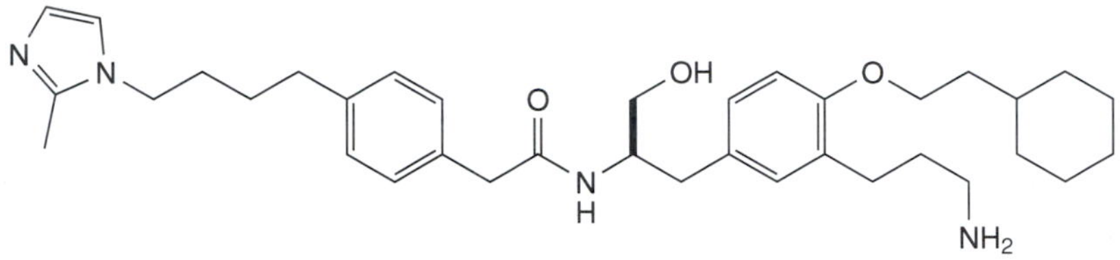

**(3)**

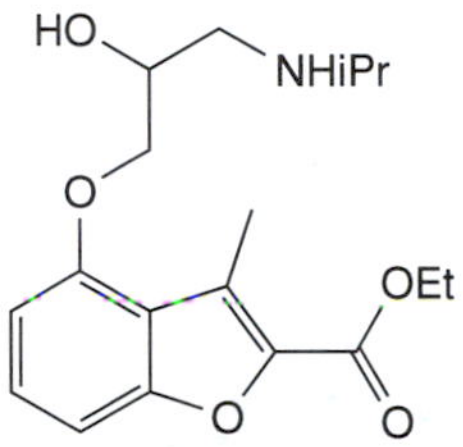

**(4)**

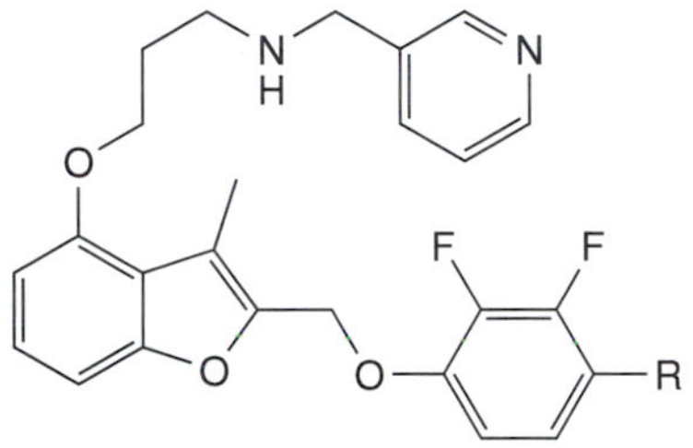

**(5)**, R = H; **(6)**, R = F

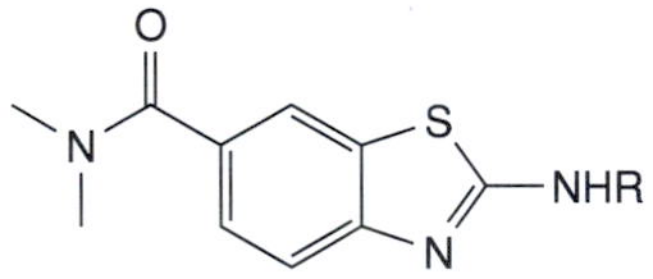

**(7)**, R =

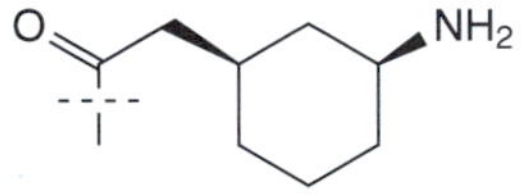

**(8)**, R =

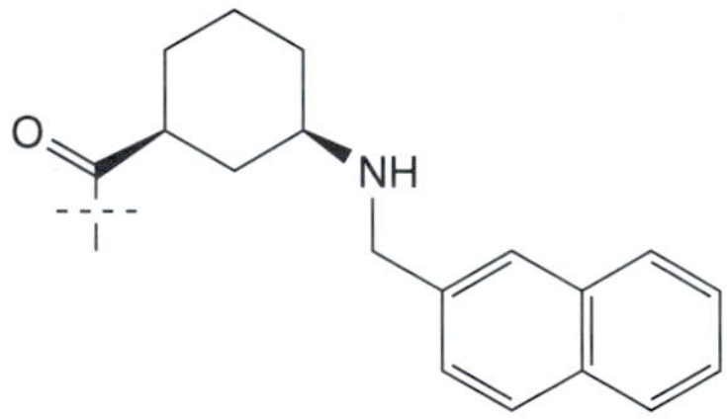

**Figure 17.19.** NMT inhibitors.

## 4  NEW TRENDS IN ANTIFUNGAL RESEARCH

The sequencing of microbial genomes is revolutionizing the discovery of novel antifungal drugs, providing the tools for the rational identification of novel targets and compounds. High throughput genomic sequencing, combined with fragment assembly tools, has delivered a cornucopia of sequence information to assist in the search of new targets.

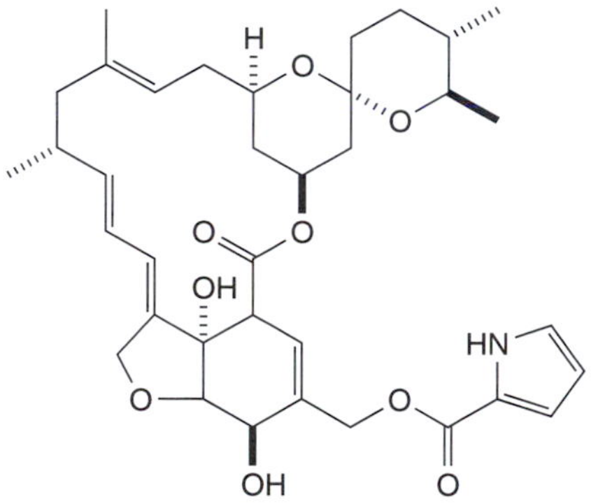

MC-510,027

**Figure 17.20.** MC-510,027.

Genomic information, combined with the ability to selectively delete or modify genes of interest, is proving useful in evaluating the selectivity of a target and its essentiality for growth. In addition, comparative genomics allows the identification of potential targets shared across fungal species. For instance, entire biochemical pathways can be reconstructed and compared in different pathogens. Sequence comparisons may also provide some indication of potential mammalian toxicity if proteins of similar sequence exist in mammalian sequence databases.

Numerous databases, available over the Internet and easily downloaded onto local servers, are now available that contain both sequence and functionality information. In addition, certain commercial databases are available for nonexclusive use by commercial subscribers.

Overall, the genomic revolution is expected to have a profound impact on antifungal drug discovery, with the potential for the identification of new agents with novel mechanisms of action (183).

## 5 WEB SITES AND RECOMMENDED READING

### 5.1 Web Sites

A number of very informative Web sites serve as excellent starting points for further information on various aspects of the material covered in the chapter. The following are good sources for general information:

- http://www.doctorfungus.org/ This site is maintained principally by clinicians and others at the University of Texas Medical School in Houston, and is an excellent and authoritative source, particularly on clinical and microbiological aspects of fungal disease. The site also contains links to the manufacturer's Web sites for clinically significant antifungal agents.

- http://www.aspergillus.man.ac.uk/ This site is associated with the center of clinical expertise in systemic aspergillosis at the University of Manchester, UK, and is designed to provide information on pathogenic *Aspergilli* for clinicians and scientific researchers.

- http://www.mic.ki.se/Diseases/c1.html This site, based in Sweden's Karolinska Institute, is an excellent source of background information on both bacterial and fungal diseases, and contains many links to other informative sites.

- http://genome-www.stanford.edu/Saccharomyces/ This is a scientific database of the molecular biology and genetics of the yeast *Saccharomyces cerevisiae* (baker's yeast), which serves as a genetically manipulable surrogate model for pathogenic organisms. It contains many links to other sources of proteomic and biochemical information concerning this organism.

### 5.2 Other Texts

Several recent reviews on clinical aspects of fungal disease and chemotherapeutic options are available, and are cited at particularly relevant points above. In particular, the reader's attention is drawn to the following texts:

- V. L. Yu, T. C. Merigan, and S. L Barriere, Eds., *Antimicrobial Therapy and Vaccines*, Williams & Wilkins, Baltimore, 1999. This book provides comprehensive information on clinical and microbiological aspects of antifungal chemotherapy.

- J. Sutcliffe and N. H. Georgopapadakou, Eds., *Emerging Targets in Antibacterial and Antifungal Chemotherapy*, Chapman & Hall, London, 1992. Although nearly a decade old, this book contains a series of chapters that concisely cover the clinical context and, particularly, biochemical aspects of current antifungal agents.

## REFERENCES

1. J. E. Edwards Jr., G. P. Bodey, R. A. Bowden, T. Büchner, B. E. de Pauw, S. G. Filler, M. A. Ghannoum, M. Glauser, R. Herbrecht, C. A. Kauffman, S. Kohno, P. Martino, F. Meunier, T. Mori, M. A. Pfaller, J. H. Rex, T. R. Rogers, R. H. Rubin, J. Solomkin, C. Viscoli, T. J. Walsh, and M. White, *Clin. Infect. Dis.*, **25**, 43–59 (1997).

2. W. G. Powderly, *AIDS Res. Hum. Retroviruses*, **10**, 925–929 (1994).

3. D. W. Denning and D. A. Stevens, *Clin. Inf. Dis.*, **12**, 1147–1201 (1990).

4. J. H. Rex, M. G. Rinaldi, and M. A. Pfaller, *Antimicrob. Agents Chemother.*, **39**, 1–8 (1995).

5. T. J. Walsh in J. Sutcliffe and N. Georgopadakou, Eds., *Emerging Targets in Antibacterial and Antifungal Chemotherapy*, Chapman & Hall, New York/London, 1992, pp. 349–373.

6. M. B. Edmond, S. E. Wallace, D. K. McClish, M. A. Pfaller, R. N. Jones, and R. P. Wenzel, *Clin. Inf. Dis.*, **29**, 239–244 (1999).

7. F. C. Odds, *J. Antimicrob. Chemother.*, **31**, 463–471 (1993).

8. National Committee for Clinical Laboratory Standards: (a) Reference Method for Broth Dilution Antifungal Susceptibility Testing of Yeasts (Approved standard), NCCLS document M27-A, 1997; (b) Reference Method for Broth Dilution Antifungal Susceptibility Testing of Conidium-Forming Filamentous Fungi (Proposed standard), NCCLS document M38-P, 1998, National Committee for Clinical Laboratory Standards, Wayne, PA.

9. P. Marichal, *Curr. Opin. Anti-Infect. Inv. Drugs*, **1**, 318–333 (1999).

10. D. A. Stevens and K. Holmberg, *Curr. Opin. Anti-Infect. Inv. Drugs*, **1**, 306–317 (1999).

11. R. Franz, S. L. Kelly, D. C. Lamb, D. E. Kelly, M. Ruhnke, and J. Morschhauser, *Antimicrob. Agents Chemother.*, **42**, 3065–3072 (1998).

12. J. F. Ryley and K. Barrett-Bee in J. Sutcliffe and N. Georgopapadakou, Eds., *Emerging Targets in Antibacterial and Antifungal Chemotherapy*, Chapman & Hall, New York/London, 1992, pp. 546–567.

13. K. Barrett-Bee and N. Ryder in J. Sutcliffe and N. Georgopapadakou, Eds., *Emerging Targets in Antibacterial and Antifungal Chemotherapy*, Chapman & Hall, New York/London, 1992, pp. 410–436.

14. L. W. Parks, R. T. Lorenz, and W. M. Casey in J. Sutcliffe and N. Georgopapadakou, Eds., *Emerging Targets in Antibacterial and Antifungal Chemotherapy*, Chapman & Hall, New York/London, 1992, pp. 393–409.

15. R. J. Rodriguez, C. Low, C. D. K. Bottema, and L. W. Parks, *Biochim. Biophys. Acta*, **837**, 336–343 (1985).

16. N. H. Georgopapadakou, *Exp. Opin. Invest. Drugs*, **10**, 269–280 (2001).

17. S. Arikan and J. H. Rex, *Curr. Pharm. Des.*, **7**, 393–415 (2001).

18. A. Wong-Beringer, R. A. Jacobs, and B. J. Guglielmo, *Clin. Infect. Dis.*, **27**, 603–618 (1998).

19. R. Powles, S. Mawhorter, and T. Williams, Proceedings of the 39th Interscience Conference on Antimicrobial Agents and Chemotherapy, Abstr. LB-4 (1999).

20. A. H. Williams and J. E. Moore, Proceedings of the 39th Interscience Conference on Antimicrobial Agents and Chemotherapy, Abstr. 1420 (1999).

21. F. C. J. Offner, R. Herbrecht, D. Engelhard, H. F. L. Guiot, G. Samonis, A. Marinus, R. J. Roberts, and B. E. De Pauw, Proceedings of the 40th Interscience Conference on Antimicrobial Agents and Chemotherapy, Abstr. 1102 (2000).

22. A. J. Carillo-Munoz, G. Quindos, C. Tur, M. T. Ruesga, Y. Miranda, O. Del Valle, P. A. Cossum, and T. L. Wallace, *J. Antimicrob. Chemother.*, **44**, 397–401 (1999).

23. K. L. Oakley, C. B. Moore, and D. W. Denning, *Antimicrob. Agents Chemother.*, **43**, 1264–1266 (1999).

24. R. T. Mehta, R. L. Hopfer, L. A. Gunner, R. L. Juliano, and G. Lopez-Berestein, *Antimicrob. Agents Chemother.*, **31**, 1897–1900 (1987).

25. M. A. Hossain and M. A. Ghannoum, *Exp. Opin. Invest. Drugs*, **9**, 1797–1813 (2000).

26. J. B. Schenkman, H. Remmer, and R. W. Estabrook, *Mol. Pharmacol.*, **3**, 113–123 (1967).

27. L. M. Podust, T. L. Poulos, and M. R. Waterman, *Proc. Natl. Acad. Sci. USA*, **98**, 3068–3073 (2001).

28. K. Barrett-Bee, J. Lees, P. Pinder, J. Campbell, and L. Newboult, *Ann. N. Y. Acad. Sci.*, **544**, 231–244 (1988).

29. D. W. Wooley, *J. Biol. Chem.*, **152**, 225–232 (1944).

30. E. F. Godefroi, J. Heeres, J. Van Cutsem, and P. A. J. Janssen, *J. Med. Chem.*, **12**, 784–791 (1969).

31. M. Plempel, K. Bartmann, K. H. Buchel, and E. Regel, *Antimicrob. Agents Chemother.*, **1969**, 271–274 (1969).

32. J. Heeres, L. J. J. Backx, J. H. Mostmans, and J. Van Cutsem, *J. Med. Chem.*, **22**, 1003–1005 (1979).

33. J. Heeres, J. J. Backx, and J. Van Cutsem, *J. Med. Chem.*, **27**, 894–900 (1984).

34. K. Richardson, K. Cooper, M. S. Marriott, M. H. Tarbit, P. F. Troke, and P. J. Whittle, *Ann. N. Y. Acad. Sci.*, **544**, 4–11 (1988).

35. H. Ji, W. Zhang, Y. Zhou, M. Zhang, J. Zhu, Y. Song, J. Lu, and J. Zhu, *J. Med. Chem.*, **43**, 2493–2505 (2000).

36. R. P. Dickinson, A. S. Bell, C. A. Hitchcock, S. Narayanaswami, S. J. Ray, K. Richardson, and P. F. Troke, *Bioorg. Med. Chem. Lett.*, **6**, 2031–2036 (1996).

37. J. Bartroli, E. Turmo, M. Alguero, E. Boncompte, M. L. Vericat, J. Garcia-Rafanell, and J. Forn, *J. Med. Chem.*, **38**, 3918–3932 (1995).

38. F. C. Odds, *J. Antimicrob. Chemother.*, **31**, 463–471 (1993).

39. K. Venkateswarlu, M. Taylor, N. J. Manning, M. G. Rinaldi, and S. L. Kelly, *Antimicrob. Agents Chemother.*, **41**, 748–751 (1997).

40. D. Sanglard, F. Ischer, L. Koymans, and J. Bille, *Antimicrob. Agents Chemother.*, **42**, 241–253 (1998).

41. K. Asai, N. Tsuchimori, K. Okonogi, J. R. Perfect, O. Gotoh, and Y. Yoshida, *Antimicrob. Agents Chemother.*, **43**, 1163–1169 (1999).

42. D. C. Lamb, D. E. Kelly, T. C. White, and S. L. Kelly, *Antimicrob. Agents Chemother.*, **44**, 63–67 (2000).

43. D. Sanglard, F. Ischer, D. C. Calabrese, P. A. Majcherczyk, and J. Bille, *Antimicrob. Agents Chemother.*, **43**, 2753–2765 (1999).

44. T. Fukuoka, Y. Fu, and S. G. Filler, Proceedings of the 38th Interscience Conference on Antimicrobial Agents and Chemotherapy, Abstr. J-90 (1998).

45. S. L. Kelly, D. C. Lamb, D. E. Kelly, J. Loeffler, and H. Einsele, *Lancet*, **348**, 1523 (1996).

46. F. S. Nolte, T. Parkinson, D. J. Falconer, S. Dix, J. Williams, C. Gilmore, R. Geller, and J. R. Wingard, *Antimicrob. Agents Chemother.*, **41**, 196–199 (1997).

47. G. D. Albertson, M. Nimi, R. D. Cannon, and H. F. Jenkinson, *Antimicrob. Agents Chemother.*, **40**, 2835–2841 (1996).

48. D. Sanglard, F. Ischer, M. Monod, and J. Bille, *Microbiology*, **143** (Pt. 2), 405 (1997).

49. D. Sanglard, F. Ischer, L. Koymans, and J. Bille, *Antimicrob. Agents Chemother.*, **42**, 241–253 (1998).

50. D. Sanglard, F. Ischer, and J. Bille, Proceedings of the 38th Interscience Conference on Antimicrobial Agents and Chemotherapy, Abstr. C-148 (1998).

51. J. W. Slaven, M. J. Anderson, D. Sanglard, G. K. Dixon, J. Bille, I. S. Roberts, and D. W. Denning, Proceedings of the 39th Interscience Conference on Antimicrobial Agents and Chemotherapy, Abstr. 447 (1999).

52. S. Maesaki, P. Marichal, H. Vanden Bossche, D. Sanglard, and S. Kohno, *J. Antimicrob. Chemother.*, **44**, 27–31 (1999).

53. R. P. Dickinson, A. S. Bell, C. A. Hitchcock, S. Narayanaswami, S. J. Ray, K. Richardson, and P. F. Troke, *Bioorg. Med. Chem Lett.*, **6**, 2031–2036 (1996).

54. G. M. Gonzalez, R. Tijerina, D. A. Sutton, and M. G. Rinaldi, Proceedings of the 39th Interscience Conference on Antimicrobial Agents and Chemotherapy, Abstr. 1511 (1999).

55. M. A. Pfaller, S. A. Messer, S. Gee, S. Joly, C. Pujol, D. J. Sullivan, D. C. Coleman, and D. R. Soll, *J. Clin. Microbiol.*, **37**, 870–872 (1999).

56. M. Cuenca-Estrella, B. Ruiz-Diez, J. V. Martinez-Suarez, A. Monzon, and J. L. Rodriguez-Tudela, *J. Antimicrob. Chemother.*, **43**, 149–151 (1999).

57. M. A. Ghannoum, I. Okogbule-Wonodi, N. Bhat, and H. Sanati, *J. Chemother.*, **11**, 34 (1999).

58. G. Quindos, S. Bernal, M. Chavez, M. J. Gutierrez, M. M. E. Strella, and A. Valverde, Pro-

ceedings of the 38th Interscience Conference on Antimicrobial Agents and Chemotherapy, Abstr. J-6 (1998).

59. F. Marco, M. A. Pfaller, S. A. Messer, and R. N. Jones, *Med. Mycol.*, **36**, 433 (1998).

60. M. H. Nguyen and C. Y. Yu, *Antimicrob. Agents Chemother.*, **42**, 471–472 (1998).

61. M. A. Pfaller, S. A. Messer, R. J. Hollis, R. N. Jones, G. V. Doern, M. E. Brandt, and R. A. Hajjeh, *Antimicrob. Agents Chemother.*, **42**, 3242–3244 (1998).

62. C. J. Clancy, A. Fothergill, M. G. Rinaldi, and M. H. Nguyen, Proceedings of the 39th Interscience Conference on Antimicrobial Agents and Chemotherapy, Abstr. 1518 (1999).

63. D. J. Sheehan, C. A. Hitchcock, and C. M. Sibley, *Clin. Microbiol. Rev.*, **12**, 40–79 (1999).

64. D. Denning, A. del Favero, E. Gluckman, D. Norfolk, M. Ruhnke, S. Yonren, P. Troke, and N. Sarantis, Proceedings of the 35th Interscience Conference on Antimicrobial Agents and Chemotherapy, Abstr. F-80 (1995).

65. B. Dupont, D. Denning, H. Lode, S. Yonren, P. Troke, and N. Sarantis, Proceedings of the 35th Interscience Conference on Antimicrobial Agents and Chemotherapy, Abstr. F-81 (1995).

66. P. F. Troke, K. W. Brammer, C. A. Hitchcock, S. Yonren, and N. Sarantis, Proceedings of the 35th Interscience Conference on Antimicrobial Agents and Chemotherapy, Abstr. F-73 (1995).

67. T. J. Walsh, P. Pappas, D. J. Winston, H. M Lazarus, F. Petersen, J. Raffalli, S. Yanovich, P. Stiff, R. Greenberg, G. Donowitz, and J. Lee, *N. Engl. J. Med.*, **346**, 225–234 (2002).

68. F. Barchiesi, D. Arzeni, A. W. Fothergill, L. F. Di Francesco, F. Caselli, M. G. Rinaldi, and G. Scalise, *Antimicrob. Agents Chemother.*, **44**, 226–229 (2000).

69. M. Lozano-Chiu, S. Arikan, V. L. Paetznick, E. J. Anaissie, D. Loebenberg, and J. H. Rex, *Antimicrob. Agents Chemother.*, **43**, 589–591 (1999).

70. L. Najvar, J. R. Graybill, R. Bocanegra, and H. Al-Abdely, Proceedings of the 38th Interscience Conference on Antimicrobial Agents and Chemotherapy, Abstr. J-68 (1998).

71. P. Melby, J. R. Graybill, and H. Al-Abdely, Proceedings of the 38th Interscience Conference on Antimicrobial Agents and Chemotherapy, Abstr. B-58 (1998).

72. L. Najvar, J. R. Graybill, R. Bocanegra, and H. Al-Abdely, Proceedings of the 38th Inter-

science Conference on Antimicrobial Agents and Chemotherapy, Abstr. J-69 (1998).

73. J. Molina, O. Martins-Filho, Z. Brener, A. J. Romanha, D. Loebenberg, and J. A. Urbina, *Antimicrob. Agents Chemother.*, **44**, 150–155 (2000).

74. W. R. Kirkpatrick, R. K. McAtee, A. W. Fothergill, D. Loebenberg, M. G. Rinaldi, and T. F. Patterson, *Antimicrob. Agents Chemother.*, **44**, 780–782 (2000).

75. C. V. Magatti, D. Hesk, M. J. Lauzon, S. S. Saluja, and X. Wang, *J. Label. Comp. Radiopharm.*, **41**, 731 (1998).

76. P. Krieter, J. Achanfuo-Yeboah, M. Shea, M. Thonoor, M. Cayen, and J. Patrick, Proceedings of the 39th Interscience Conference on Antimicrobial Agents and Chemotherapy, Abstr. 1199 (1999).

77. A. Noemir, P. Kumari, M. J. Hilbert, D. Loebenberg, A. Cacciapuoti, F. J. Menzel, E. J. Moss, R. Hare, G. H. Miller, and M. N. Cayen, Proceedings of the 35th Interscience Conference on Antimicrobial Agents and Chemotherapy, Abstr. F-68 (1995).

78. A. A. Nomeir, P. Kumari, S. Gupta, D. Loebenberg, A. Cacciapuoti, R. Hare, C. C. Lin, and M. N. Cayen, Proceedings of the 39th Interscience Conference on Antimicrobial Agents and Chemotherapy, Abstr. 1934 (1999).

79. D. Loebenberg, F. Menzel Jr., E. Corcoran, C. Mendrick, K. Raynor, A. F. Cacciapuoti, and R. S. Hare, Proceedings of the 39th Interscience Conference on Antimicrobial Agents and Chemotherapy, Abstr. 1933 (1999).

80. J. C. Fung-Tomc, E. Huczko, B. Minassian, and D. P. Bonner, *Antimicrob. Agents Chemother.*, **42**, 313–318 (1998).

81. D. J. Diekema, M. A. Pfaller, S. A. Messer, A. Houston, R. J. Hollis, G. V. Doern, R. N. Jones, and the Sentry Participants Group, *Antimicrob. Agents Chemother.*, **43**, 2236–2239 (1999).

82. C. B. Moore, C. M. Walls, and D. W. Denning, *Antimicrob. Agents Chemother.*, **44**, 441–443 (2000).

83. K. Hata, J. Kimura, H. Miki, T. Toyosawa, T. Nakamura, and K. Katsu, *Antimicrob. Agents Chemother.*, **40**, 2237–2242 (1996).

84. T. Nakamura, R. Sakai, J. Somoda, T. Katoh, T. Kaneko, and T. Horie, Proceedings of the 35th Interscience Conference on Antimicrobial Agents and Chemotherapy, Abstr. F94 (1995).

85. K. Hata, J. Kimura, H. Miki, T. Toyosawa, M. Moriyama, and K. Katsu, *Antimicrob. Agents Chemother.*, **40**, 2243–2247 (1996).

86. K. Shock, S. Marino, V. Andriole, and T. Baumgartner, Proceedings of the 38th Interscience Conference on Antimicrobial Agents and Chemotherapy, Abstr. J-54 (1998).

87. N. S. Ryder, *Ann. N. Y. Acad. Sci.*, **544**, 208–220 (1988).

88. M. A. Ghannoum and B. Elewski, *Clin. Diagn. Lab. Immunol.*, **6**, 921–923 (1999).

89. J. Vazquez, A. Lamaraca, R. Schwartz, R. Ramirez, L. Smith, R. Pollard, J. Gill, A. Fothergill, L. Ince, J. Wirzman, A. Perez, and J. Felser, Proceedings of the 40th Interscience Conference on Antimicrobial Agents and Chemotherapy, Abstr. 1418 (2000).

90. J. E. Bennett in J. G. Hardman and L. E. Limbird, Eds., *Goodman & Gilman's The Pharmacological Basis of Therapeutics*, 9th ed., McGraw-Hill, 1996, p. 1187.

91. T. Arika, M. Yokoo, T. Hase, T. Maeda, K. Amemiya, and H. Yamaguchi, *Antimicrob. Agents Chemother.*, **34**, 2250–2253 (1990).

92. N. S. Ryder and M. C. Dupont, *Biochem. J.*, **230**, 765–770 (1985).

93. N. S. Ryder, *Antimicrob. Agents Chemother.*, **27**, 252–256 (1985).

94. A. Stutz and G. Petranyi, *J. Med. Chem.*, **27**, 1539–1543 (1984).

95. P. Nussbaumer, I. Leitner, and A. Stutz, *J. Med. Chem.*, **37**, 610–615 (1994).

96. O. Lomovskaya, M. Warren, A. Mistry, A. Staley, J. Galazzo, H. Fuernkranz, M. Lee, G. Miller, and D. Sanglard, Proceedings of the 39th Interscience Conference on Antimicrobial Agents and Chemotherapy, Abstr. 1269 (1999).

97. A. B. Orth, M. J. Henry, and H. D. Sisler, *Pestic. Biochem. Physiol.*, **37**, 182–191 (1990).

98. R. Pollak and S. A. Billstein, *J. Am. Podiatr. Med. Assoc.*, **91**, 127–131 (2001).

99. M. N. Kurtz and J. H. Rex, *Adv. Protein Chem.*, **56**, 423–475 (2001).

100. D. W. Denning, *J. Antimicrob. Chemother.*, **40**, 611–614 (1997).

101. A. Hoang, *Am. J. Health Syst. Pharm.*, **58**, 1206–1214 (2001).

102. M. Debono, W. W. Turner, L. LaGrandeur, F. J. Burkhardt, J. S. Nissen, K. K. Nichols, M. J. Rodriguez, M. J. Zweifel, D. J. Zeckner, R. S. Gordee, J. Tang, and T. R. Parr Jr., *J. Med. Chem.*, **38**, 3271–3281 (1995).

103. M. J. Rodriquez, V. Vasudevan, J. A. Jamison, P. S. Borromeo, and W. W. Turner, *Bioorg. Med. Chem. Lett.*, **9**, 1863–1868 (1999).

104. J. A. Jamison, L. M. LaGrandeur, M. J. Rodriquez, W. W. Turner, and D. J. Zeckner, *J. Antibiot.*, **51**, 239–242 (1998).

105. M. Tomishima, H. Ohki, A. Yanada, H. Takasugi, K. Maki, S. Tawara, and H. Tanaka, *J. Antibiot.*, **52**, 674–676 (1999).

106. J. M. Balkovec, R. M. Black, M. L. Hammond, J. V. Heck, R. A. Zambias, G. Abruzzo, K. Bartizal, H. Kropp, C. Trainor, R. E. Schwartz, D. C. McFadden, K. H. Nollstadt, L. A. Pittarelli, M. A. Powles, and D. M. Schmatz, *J. Med. Chem.*, **35**, 194–198 (1992).

107. F. A. Bouffard, R. A. Zambias, J. F. Dropinski, J. M. Balkovec, M. L. Hammond, G. K. Abruzzo, K. F. Bartizal, J. A. Marrinan, M. B. Kurtz, D. C. McFadden, K. H. Nollstadt, M. A. Powles, and D. M. Schmatz, *J. Med. Chem.*, **37**, 222–225 (1994).

108. L. L. Klein, L. Li, H-J. Chen, C. B. Curty, D. A. DeGoey, D. J. Grampovnik, C. L. Leone, S. A. Thomas, C. M. Yeung, K. W. Funk, V. Kishore, E. O. Lundell, D. Wodka, J. A. Meulbroek, J. D. Adler, A. M. Nilius, P. A. Lartey, and J. J. Plattner, *Bioorg. Med. Chem.*, **8**, 1677–1696 (2000).

109. J. Onishi, M. Meinz, J. Thompson, J. Curotto, S. Dreikorn, M. Rosenbach, C. Douglas, G. Abruzzo, A. Flattery, L. Kong, A. Cabello, F. Vicente, F. Pelaez, M. T. Diez, I. Martin, G. Bills, R. Giacobbe, A. Dombrowski, R. Schwartz, S. Morris, G. Harris, A. Tsipouras, K. Wilson, and M. B. Kurtz, *Antimicrob. Agents Chemother.*, **44**, 368–377 (2000).

110. S. Hawser, *Curr. Opin. Anti-Infect. Invest. Drugs*, **1**, 353–360 (1999).

111. K. L. Oakley, C. B. Moore, and D. W. Denning, *Antimicrob. Agents Chemother.*, **42**, 2726–2730 (1998).

112. R. Petraitiene, V. Petraitis, A. H. Groll, M. Candelario, T. Sein, A. Bell, C. A. Lyman, C. L. McMillian, J. Bacher, and T. J. Walsh, *Antimicrob. Agents Chemother.*, **43**, 2148–2155 (1999).

113. V. Petraitis, R. Petraitiene, A. H. Groll, A. Bell, D. P. Callender, T. Sein, R. L. Schaufele, C. L. McMillian, J. Bacher, and T. J. Walsh, *Antimicrob. Agents Chemother.*, **42**, 289–2905 (1998).

114. J. Roberts, K. Schock, S. Marino, and V. T. Andriole, *Antimicrob. Agents Chemother.*, **44**, 3381–3388 (2000).

115. L. K. Najvar, R. Bocanegra, S. E. Sanche, and J. R. Graybill, Proceedings of the 39th Inter-

science Conference on Antimicrobial Agents and Chemotherapy, Abstr. 2002 (1999).

116. V. Petraitis, R. Petraitiene, A. H. Groll, T. Sein, R. L. Schaufele, C. A. Lyman, A. Francesconi, J. Bacher, S. C. Piscitelli, and T. J. Walsh, *Antimicrob. Agents Chemother.*, **45**, 471–479 (2001).

117. L. Ni, B. Smith, B. Hathcer, M. Goldman, C. McMillian, M. Turik, I. Rajman, L. J. Wheat, and V. S. Watkins, Proceedings of the 38th Interscience Conference on Antimicrobial Agents and Chemotherapy, Abstr. J-134 (1998).

118. R. A. Fromtling and J. Castaner, *Drugs Future*, **23**, 1273–1278 (1998).

119. S. Tawara, F. Ikeda, K. Maki, Y. Morishita, K. Otomo, N. Teratani, T. Goto, M. Tomishima, H. Ohki, A. Yamada, K. Kawabata, H. Takasugi, K. Sakane, H. Tanaka, F. Matsumoto, and S. Kuwahara, *Antimicrob. Agents Chemother.*, **44**, 57–62 (2000).

120. H. Mikamo, Y. Sato, and T. Tamaya, *J. Antimicrob. Chemother.*, **46**, 485–487 (2000).

121. K. Uchida, Y. Nishiyama, N. Yokota, and H. Yamaguchi, *J. Antibiot. (Tokyo)*, **53**, 1175–1181 (2000).

122. S. Maesaki, M. A. Hossain, Y. Miyazaki, K. Tomono, T. Tashiro, and S. Kohno, *Antimicrob. Agents Chemother.*, **44**, 1728–1730 (2000).

123. F. Ikeda, Y. Wakai, S. Matsumoto, K. Maki, E. Watabe, S. Tawara, T. Goto, Y. Watanabe, F. Matsumoto, and S. Kuwahara, *Antimicrob. Agents Chemother.*, **44**, 614–618 (2000).

124. S. Matsumoto, Y. Wakai, T. Nakai, K. Hatano, T. Ushitani, F. Ikeda, S. Tawara, T. Goto, F. Matsumoto, and S. Kuwahara, *Antimicrob. Agents Chemother.*, **44**, 619–621 (2000).

125. J. Azuma, I. Yamamoto, M. Ogura, T. Mukai, H. Suematsu, H. Kageyama, K. Nakahara, K. Yoshida, and T. Takaya, Proceedings of the 38th Interscience Conference on Antimicrobial Agents and Chemotherapy, Abstr. F-146 (1998).

126. K. Pettengell, J. Mynhardt, T. Kluyts, and P. Soni, Proceedings of the 39th Interscience Conference on Antimicrobial Agents and Chemotherapy, Abstr. 1421 (1999).

127. N. Georgopapadakou and T. J. Walsh, *Antimicrob. Agents Chemother.*, **40**, 279–291 (1996).

128. N. S. Ryder, I. Frank, and M. C. Dupont, *Antimicrob. Agents Chemother.*, **29**, 858–860 (1986).

129. C. M. Marcireau, D. Guyonnet, and F. Karst, *Curr. Genet.*, **22**, 267–272 (1992).

130. H. Vanden Bosche, P. Marichal, and F. C. Odds, *Trends Microbiol.*, **2**, 393 (1994).

131. S. Budavari, Ed., *The Merck Index*, 12th ed., Merck & Co., Inc., Rahway, NJ, 1996, p. 775.

132. J. E. F. Reynolds, Ed., *Martindale, The Extra Pharmacopoeia*, 31st ed., Royal Pharmaceutical Society of Great Britain, London, 1996, p. 407.

133. K. Gull and A. P. J. Trinci, *Nature*, **244**, 292–294 (1973).

134. See Ref. 123, p. 380.

135. See Ref. 123, p. 785.

136. E. F. Harrison, P. Zwadyk, R. J. Bequette, E. E. Hamlow, P. A. Tavormina, and W. A. Zygmunt, *Appl. Microbiol.*, **19**, 746–750 (1970).

137. D. Zhang and M. J. Miller, *Curr. Pharm. Des.*, **5**, 73–99 (1999).

138. N. H. Georgopapadakou, *Curr. Opin. Anti-Infect. Invest. Drugs*, **1**, 346–352 (1999).

139. R. K. Li and M. G. Rinaldi, *Antimicrob. Agents Chemother.*, **43**, 1401–1405 (1999).

140. J. R. Graybill, L. K. Najvar, R. Bocanegra, R. F. Hector, and M. F. Luther, *Antimicrob. Agents Chemother.*, **42**, 2371–2374 (1998).

141. K. Obi, J. Uda, K. Iwase, O. Sugimoto, H. Ebisu, and A. Matsuda, *Bioorg. Med. Chem. Lett.*, **10**, 1451–1454 (2000).

142. M. Sudoh, T. Yamazaki, K. Masubuchi, M. Taniguchi, N. Shimma, M. Arisawa, and H. Yamada-Okabe, *J. Biol. Chem.*, **275**, 32901–32905 (2000).

143. W. Zhong, M. W. Jeffries, and N. H. Georgopadakou, *Antimicrob. Agents Chemother.*, **44**, 651–653 (2000).

144. M. Kuroda, T. Hashida-Okado, R. Yasumoto, K. Gomi, I. Kato, and K. Takesako, *Mol. Gen. Genet.*, **261**, 290–296 (1999).

145. A. Ogawa, T. Hashika-Okado, M. Endo, H. Yoshioka, T. Tsuruo, K. Takesako, and I. Kato, *Antimicrob. Agents Chemother.*, **42**, 755–761 (1998).

146. U. Schmidt, A. Schumacher, J. Mittendorf, and B. Riedl, *J. Pept. Res.*, **52**, 143–154 (1998).

147. T. Kurome, T. Inoue, K. Takesako, and I. Kato, *J. Antibiot.*, **51**, 359–367 (1998).

148. F. Tiberghien, T. Kurome, K. Takesako, A. Didier, T. Wenandy, and F. Loor, *J. Med. Chem.*, **43**, 2547–2556 (2000).

149. D. Gargallo-Viola, *Curr. Opin. Anti-Infect. Invest. Drugs*, **1**, 297–305 (1999).

150. E. Herreros, C. M. Martinez, M. J. Almela, M. S. Marriott, F. G. de las Heras, and D. Gargallo-Viola, *Antimicrob. Agents Chemother.*, **42**, 2863–2869 (1998).

151. L. Capa, A. Mendoza, J. L. Lavandera, F. G. de las Heras, and J. F. Garcia-Bustos, *Antimicrob. Agents Chemother.*, **42**, 2694–2699 (1998).

152. J. M. Dominquez, M. G. Gomez-Lorenzo, and J. J. Martin, *J. Biol. Chem.*, **274**, 22423–22427 (1999).

153. M. Shastry, J. Nielsen, T. Ku, M. Hsu, P. Liberator, J. Anderson, D. Schmatz, and M. Justice, *Microbiology*, **147** (Pt. 2), 383–390 (2001).

154. M. G. Gomez-Lorenzo, C. M. Spahn, R. K. Agrawal, R. A. Grassucci, P. Penczek, K. Chakraburtty, J. P. Ballesta, J. L. Lavandera, J. F. Garcia-Bustos, and J. Frank, *EMBO J.*, **19**, 2710–2718 (2000).

155. J. M. Dominquez and J. Martin, *J. Biol. Chem.*, **276**, 31402–31407 (2001).

156. A. Martinez, E. Jiminez, P. Aviles, J. Caballero, F. G. de las Heras, and D. Gargallo-Viola, Proceedings of the 39th Interscience Conference on Antimicrobial Agents and Chemotherapy, Abstr. 294 (1999).

157. A. Martinez, P. Aviles, E. Jimenez, J. Caballero, and D. Gargallo-Viola, *Antimicrob. Agents Chemother.*, **44**, 3389–3394 (2000).

158. E. Herreros, M. J. Almela, S. Lozano, C. M. Martinez, and D. Gargallo-Viola, Proceedings of the 39th Interscience Conference on Antimicrobial Agents and Chemotherapy, Abstr. 158 (1999).

159. D. G. Gatehouse, T. C. Williams, A. T. Sullivan, G. H. Apperley, S. P. Close, S. R. Nesfield, A. Martinez, and D. Gargallo-Viola, Proceedings of the 38th Interscience Conference on Antimicrobial Agents and Chemotherapy, Abstr. J-75 (1998).

160. B. Tse, J. M. Balkovec, C. M. Blazey, M-J. Hsu, J. Nielsen, and D. Schmatz, *Bioorg. Med. Chem. Lett.*, **8**, 2269–2272 (1998).

161. M. C. Justice, M-J. Hsu, B. Tse, T. Ku, J. Balkovec, D. Schmatz, and J. Nielsen, *J. Biol. Chem.*, **273**, 3148–3151 (1998).

162. J. C. Cuevas and J. L. Martos, *Tetrahedron Lett.*, **39**, 8553–8556 (1998).

163. E. M. Arribas, J. Castro, I. R. Clemens, J. C. Cuevas, J. Chicharro, M. T. Fraile, S. Garcia-Ochoa, F. G. de las Heras, and J. R. Ruiz, *Bioorg. Med. Chem Lett.*, **12**, 117–120 (2002).

164. J. M. Bueno, J. C. Cuevas, J. M. Fiandor, S. Garcia-Ochoa, and F. G. de las Heras, *Bioorg. Med. Chem Lett.*, **12**, 121–124 (2002).

165. M. Cuenca-Estrella, E. Mellado, T. M. Diaz-Guerra, A. Monzon, and J. L. Rodriquez-Tudela, *Antimicrob. Agents Chemother.*, **45**, 1905–1907 (2001).

166. E. Herreros, M. J. Almela, S. Lozano, F. G. de las Heras, and D. Bargallo-Viola, *Antimicrob. Agents Chemother.*, **45**, 3132–3139 (2001).

167. J. C. Fung-Tomc, B. Minassian, E. Huczko, B. Kolek, D. P. Bonner, and R. E. Kessler, *Antimicrob. Agents Chemother.*, **39**, 295–300 (1995).

168. K. Fujikawa, Y. Tsukamoto, T. Oki, and Y. C. Lee, *Glycobiology*, **8**, 407–414 (1998).

169. M. Hu, Y. Ishizuka, Y. Igarashi, T. Oki, and H. Nakanishi, *Spectrochim. Acta A Mol. Biol. Spectrosc.*, **56A**, 181–191 (2000).

170. K. L. Oakley, C. B. Moore, and D. W. Denning, *Int. J. Antimicrob. Agents*, **12**, 267–269 (1999).

171. C. E. Gonzalez, A. H. Groll, N. Giri, D. Shetty, I. Al-Mohsen, T. Sein, E. Feuerstein, J. Bacher, S. Piscitelli, and T. J. Walsh, *Antimicrob. Agents Chemother.*, **42**, 2399–2404 (1998).

172. R. A. Weinberg, C. A. McWherter, S. K. Freeman, D. C. Wood, J. I. Gordon, and S. C. Lee, *Mol. Microbiol.*, **16**, 241–250 (1995).

173. J. K. Lodge, E. Jackson-Machelski, D. L. Toffaletti, J. R. Perfect, and J. I Gordon, *Proc. Natl. Acad. Sci. USA*, **91**, 12008–12012 (1994).

174. B. Devadas, S. K. Freeman, M. E. Zupec, H-F. Lu, S. R. Nagarajan, N. S. Kishore, J. K. Lodge, D. W. Kuneman, C. A. McWherter, D. V. Vinjamoori, D. P. Getman, J. I. Gordon, and J. A. Sikorski, *J. Med. Chem.*, **40**, 2609–2625 (1997).

175. B. Devadas, S. K. Freeman, C. A. McWherter, N. S. Kishore, J. K. Lodge, E. Jackson-Machelski, J. I. Gordon, and J. A. Sikorski, *J. Med. Chem.*, **40**, 996–1000 (1998).

176. R. S. Bhatnagar, K. Futterer, T. A. Farazi, S. Korolev, C. L. Murray, E. Jackson-Machelski, G. W. Gokel, J. I. Gordon, and G. Waksman, *Nat. Struct. Biol.*, **5**, 1091 (1998).

177. M. Masubuchi, K. Kawasaki, H. Ebiike, Y. Ikeda, S. Tsujii, S. Sogabe, T. Fujii, K. Sakata, Y. Shiratori, Y. Aoki, T. Ohtsuka, and N. Shimma, *Bioorg. Med. Chem. Lett.*, **11**, 1833–1837 (2001).

178. H. Ebiike, M. Masubuchi, P. Liu, K. Kawasaki, K. Morikami, S. Sogabe, M. Hayase, T. Fujii, K. Sakata, H. Shindoh, Y. Shiratori, Y. Aoki, T. Ohtsuka, and N. Shimma, *Bioorg. Med. Chem. Lett.*, **12**, 607–610 (2002).

179. D. R. Armour, A. S. Bell, M. I. Kemp, M. P. Edwards, and A. Wood, Proceedings of the 221st American Chemical Society National Meeting, Paper 349 (2001).

180. A. S. Bell, D. R. Armour, M. P. Edwards, M. I. Kemp, and A. Wood, Proceedings of the 221st American Chemical Society National Meeting, Paper 350 (2001).

181. O. Lomovskaya, M. Warren, A. Mistry, A. Staley, J. Galazzo, H. Fuernkranz, M. Lee, G. Miller, and D. Sanglard, Proceedings of the 39th Interscience Conference on Antimicrobial Agents and Chemotherapy, Abstr. 1269 (1999).

182. S. Chamberland, J. Blais, D. P. Cotter, M. K. Hoang, J. Galazzo, A. Staley, M. Lee, and G. H. Miller, Proceedings of the 39th Interscience Conference on Antimicrobial Agents and Chemotherapy, Abstr. 1270 (1999).

183. D. T. Moir, K. J. Shaw, R. S. Hare, and G. F. Vovis, *Antimicrob. Agents Chemother.*, **43**, 439–446 (1999).

# CHAPTER EIGHTEEN

# Antimalarial Agents

DEE ANN CASTEEL
Department of Chemistry
Bucknell University
Lewisburg, Pennsylvania

## Contents

1 Introduction, 920
  1.1 The Disease, 920
  1.2 The Parasite, 921
  1.3 Parasite Biochemistry and Genetics, 924
  1.4 Global Incidence, 926
  1.5 Resistance, 928
  1.6 Immunity and Prophylaxis, 930
  1.7 Vector Control, 931
  1.8 Economic and Political Issues, 932
2 Antimalarial Agents for Chemotherapy and
  Prophylaxis: Current Drugs in Use, 933
  2.1 Quinine and Quinidine, 933
  2.2 Chloroquine, 938
  2.3 Mefloquine, 943
  2.4 Halofantrine, 946
  2.5 Primaquine, 949
  2.6 Antifolates, 954
    2.6.1 Proguanil and Chlorproguanil, 954
    2.6.2 Pyrimethamine, 956
    2.6.3 Sulfonamides and Sulfones, 957
    2.6.4 Combinations, 958
    2.6.5 Related Compounds, 960
  2.7 Artemisinin and Its Derivatives, 960
    2.7.1 Background, Isolation, and Chemistry, 960
    2.7.2 Structure-Activity Relationships, 963
    2.7.3 Mechanism of Action, 966
    2.7.4 First Generation Artemisinins, 970
      2.7.4.1 Artemisinin, 971
      2.7.4.2 Dihydroartemisinin, 972
      2.7.4.3 Artesunate, 972
      2.7.4.4 Artemether, 973
      2.7.4.5 Arteether, 974
      2.7.4.6 Artelinic Acid, 975
  2.8 Other Agents, 976
    2.8.1 Atovaquone and Malarone, 976
    2.8.2 Lumefantrine, 978
    2.8.3 Pyronaridine, 979
    2.8.4 Amodiaquine, 981
    2.8.5 Antibiotics, 982

*Burger's Medicinal Chemistry and Drug Discovery*
Sixth Edition, Volume 5: Chemotherapeutic Agents
Edited by Donald J. Abraham
ISBN 0-471-37031-2   © 2003 John Wiley & Sons, Inc.

2.8.6 Yingzhaosu A, Yingzhaosu C, and
    Arteflene (Ro 42-1611), 984
3 Antimalarial Agents for Chemotherapy and
    Prophylaxis: Experimental Agents, 985
    3.1 Synthetic Experimental Antimalarial
        Agents, 985

3.2 New Targets for Antimalarial
    Chemotherapy, 993
3.3 Antimalarial Natural Products, 993
4 Resources, 999

"If we take as our standard of importance the greatest harm to the greatest number, then there is no question that malaria is the most important of all infectious diseases (1)."

"Ah, poor heart! he is so shaked of a burning quotidian tertian, that it is most lamentable to behold (2)."

# 1 INTRODUCTION

Malaria is one of the most serious, complex, and refractory health problems facing humanity this century. Some 300–500 million of the world's people are infected by the disease, presenting over 120 million clinical cases annually. It is estimated that between 1.5 and 2.7 million people die from malaria every year, either directly or in association with acute respiratory infections and anemia, and up to 1 million of those deaths are among children younger than 5 years old. Malaria is a leading cause of morbidity and mortality in the developing world, particularly in tropical Africa, and it remains an outstanding tropical disease control priority.

## 1.1 The Disease

Human malaria is caused by four species of protozoan parasites of the Plasmodium genus. These are *Plasmodium falciparum*, *P. vivax*, *P. ovale*, and *P. malariae*, each of which presents slightly different clinical symptoms. *P. falciparum* is the most widespread of the four geographically and the most pernicious, causing the majority of malaria-related morbidity and mortality. Other Plasmodia species specifically infect a variety of birds, reptiles, amphibians, and mammals.

Parasites are transmitted from one person to another by an insect vector, the female anopheline mosquito. In most malarious areas, several species are able to transmit the parasite and the exact species responsible vary from region to region. Male mosquitoes do not transmit the disease. These mosquitoes are present in almost all countries in the tropics and subtropics, and they bite during nighttime hours, from dusk to dawn. It has been demonstrated that transmission can occur from transfusion of infected blood or from mother to child *in utero*, although these instances are very rare when compared with mosquito inoculation. The parasites develop in the gut of the mosquito and are passed on in the saliva when an infected mosquito bites a person. Uninfected mosquitoes become infected by taking a blood meal from an infected human. The parasites are carried by the blood to the victim's liver. After 9–16 days, the parasites have multiplied greatly and then return to the blood and penetrate the red cells. Inside erythrocytes, the parasite begins its replication cycle, a cycle of differing duration depending on the infecting species. Rupture of the infected erythrocyte and release of the merozoites begins a new cycle of red cell infection and parasite replication.

The signs and symptoms of malaria illness are variable, but most patients experience fever. Other symptoms often include headache, back pain, chills, muscle ache, increased sweating, malaise, nausea, and sometimes vomiting, diarrhea, and cough. Early stages of malaria may resemble the onset of the flu. Between paroxysms, the patient may remain febrile or may become asymptomatic. Early in an infection, the cyclic patterns of fever may not be noticeable, but later, a clear cyclic trend with symptoms recurring at regular intervals occurs. Of the four species of parasite, only falciparum malaria can progress rapidly to the cerebral stage, where infected red cells obstruct the blood vessels in the brain. Cerebral malaria is a medical emergency best managed in an intensive care unit. Untreated cases can progress to coma, renal failure, liver failure, pulmonary edema, convulsions, and death. Although infections with *P. vivax* and *P. ovale* often cause less serious illness, parasites may remain dormant in the liver for many months, causing a reappearance of symptoms months or even years later.

Malaria is diagnosed by the clinical symptoms and by microscopic examination of the blood. Stained thick and thin blood smears are used to diagnose malaria and to quantify the level of parasitemia. Giemsa-stained thin smears are used to differentiate between the species of parasite. Clinical symptoms are an inaccurate means of diagnosis by themselves, although in the absence of adequate laboratory facilities, as is the case in many malarious regions, it is the only means available. Malaria can normally be cured by antimalarial drugs. The symptoms quickly disappear once the parasites are killed. The standard measures of clinical antimalarial drug efficiency are fever clearance time and parasite clearance time. In certain geographic regions, however, the parasites have developed resistance to antimalarial agents, particularly chloroquine. Patients in these areas require treatment with newer, often more expensive drugs.

In the last several years, it has been noted that microscopic examination of blood is an inadequate method for detecting low levels of parasitemia. The lack of sensitivity seldom affects treatment and diagnosis in acute cases but it does limit understanding of the degree to which malaria is chronic. In an endemic area, polymerase chain reaction studies revealed that more than 90% of the exposed population at any one time was chronically infected with *P. falciparum* (3).

In treating malaria, curing patients is often difficult to define. The relief of symptoms of a malaria attack is a "clinical cure." Should parasites remain, even after symptoms have resolved, either in blood cells or in liver tissue, then recrudescences and/or relapses may result in the re-establishment of the infection. A "radical cure" is when the parasites are completely eliminated from the body so that relapses cannot occur. Obviously, a radical cure is the ideal therapeutic endpoint.

The choice of antimalarial agent(s) for treatment in each particular case is determined by a multiplicity of factors including the parasite species causing the infection, the acquired immune status of the patient, the susceptibility of the parasite strain to antimalarial agents, the facilities and resources available for health care, and the genetic make-up of the patient. Rapid onset and a relatively long duration of antimalarial action to cover three to four parasite life cycles are deemed essential for radical therapy. During pregnancy, women are at high risk of death from falciparum malaria. Also at risk are children who are prone to severe attacks until they develop partial immunity. Non-immune travelers to malarious areas are similarly vulnerable.

## 1.2 The Parasite

The four species of human malaria parasites are evolutionarily, morphologically and clinically distinct. *P. vivax*, *P. malariae*, and *P. ovale* are closely related on an evolutionary basis to a number of simian malarias. When comparing small subunit ribosomal RNA gene sequences, *P. vivax* is closer to *P. fragile*, a parasite of toque monkeys, than to either *P. ovale* or *P. malariae*. It has been suggested that *P. malariae* was derived from a West African chimpanzee malaria. And a plasmodium of New World monkeys, *P. brasilianum*, may in fact be *P. malariae* that has adapted to a new host over the last few hundred years. These parasites most likely arose alongside the primate hosts an estimated 30 million years ago. *P. falciparum* seems more closely related to avian malarias and is of more recent origin (4, 5), perhaps corresponding to the rise of agriculture (6).

Recently, a new species, dubbed *P. vivax*-like parasite, has been described that infects humans (7). Its morphological characteristics show it to be similar to *P. vivax*, but analysis of its DNA indicates that the sequence for the circumsporozoite protein (CS) gene is quite different from that of *P. vivax*. Rather, its CS gene seems to be identical to that of a parasite isolated from toque monkeys, *P. simiovale*. The CS protein is the major surface protein of the sporozoite stage of the parasite and has been studied as a source of malaria vaccine antigens.

The life cycle of the parasite in both mosquitoes and humans is complex (Fig. 18.1) as is

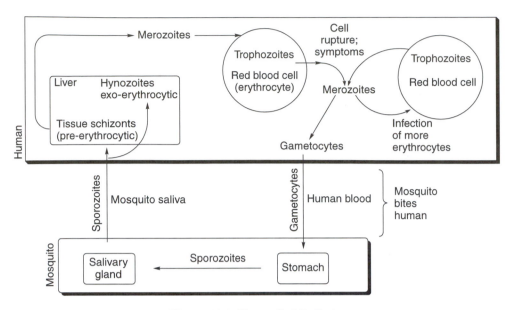

**Figure 18.1.** Plasmodia Life-Cycle.

the terminology of the various parasite development stages (8). When an infected mosquito bites, sporozoites are injected into the blood stream of the human victim and then travel to liver tissue where they invade parenchymal cells, a process involving receptor ligand mediated adhesion. During development and multiplication in the liver, known as the pre-erythrocytic stage, the host is asymptomatic. After a variable period of time, 5–7 days for falciparum, 6–8 days for vivax, 9 days for ovale, and 12–16 days for malariae, merozoites (5000–40,000 per sporozoite) are released from the liver and the parasites take up residence in the red blood cells (erythrocytic stage) again by way of a receptor mediated process. In non-relapsing malarias (falciparum and malariae), no parasites are left in the liver; the infection moves entirely into the blood stage. In relapsing malarias, some of the merozoites (or sporozoites) differentiate into a dormant non-dividing stage (hypnozoites), providing a reservoir of parasites in the liver that can be activated for up to 5 years after the initial infection. Invasion of the red cell by a merozoite results in the development of the trophozoite stage. The parasite feeds on the protein portion of hemoglobin and a waste product, hemozoin, accumulates in the host cell cytoplasm. After the parasite undergoes

nuclear divisions, the erythrocyte bursts and merozoites, parasite waste, and cell debris are released. The presence of the debris is the cause of the episodes of fever and chills associated with malaria. The merozoites released by the red cell rupture go on to infect more erythrocytes. Time intervals between cell rupture (fever), infection of other erythrocytes, and then their rupture (new bouts of fever) are characteristic of the parasite species. A few merozoites become differentiated into male and female gametocytes, forms that are dormant in humans. When a mosquito takes a blood meal from the infected human, the gametocytes begin sexual reproduction in the digestive track of the mosquito. Ultimately, sporozoites form and reside in the mosquito saliva, ready for a new round of infection.

In discussing malaria, the terms recrudescence and relapse are used to described the return of disease symptoms from different reservoirs of residual parasites. When a patient has been symptom-free for a period of time greater than the usual periodicity of the paroxysms and then clinical symptoms of malaria return, the situation is termed "recrudescence" if the re-established infection is a result of surviving erythrocytic forms of the parasite. If symptoms reappear because of the continuing presence of parasites in liver tissue, the term "relapse" is used. Because only *P. vivax* and *P. ovale* have hypnozoites that reside in the liver, these are the true relapsing malarias.

*Plasmodium vivax*, benign tertian malaria, or simply tertian malaria, presents acute symptoms that recur every second day; each erythrocytic cycle is completed within 41–45 h. Infections with *P. vivax* were termed benign, or uncomplicated, in distinction from the severe morbidity associated with *P. falciparum* infections. The pathology of vivax infections is almost invariably transient and with no detectable tissue-damaging consequences, and vivax parasites do not sequester to any detectable degree at any stage in their development. As a true relapsing malaria, vivax malaria is characterized by a prolonged or secondary tissue development stage that can remain dormant for very long periods in the liver. Relapse can occur months or even years after clearance of the initial blood-stage infection and different strains have characteristic periods between primary infection and late relapses. This allows vivax to remain endemic in areas that experience cold winters with periods of no chance for transmission by mosquito. Treatment with drugs that kill only the erythrocytic forms of the parasite will not effect a radical cure of a vivax infection. Vivax is widely spread geographically with the notable exception of tropical Africa. This has been explained by the absence of a red cell surface antigen in most black Africans that vivax requires for cell penetration (9, 10).

*Plasmodium ovale* is much less widespread than the other three species, with patchy distributions where it does occur in tropical Africa and in islands of the western Pacific. A relapsing malaria that is rarely the cause of morbidity even in children, its features are similar to that of vivax with an erythrocytic cycle of 49–50 h. Ovale and vivax overlap very little geographically. *P. vivax* and *P. ovale* infect only young erythrocytes. The fever of *P. malariae* or quartan malaria recurs every third day or 72 h and is noted for persistent reappearance of symptoms. Even so, it is not a true relapsing malaria; *P. malariae* has extremely long-lasting erythrocytic forms that can persist in an infected host for decades at very low parasite densities. Mature red blood cells are the target of *P. malariae* infection. Its geographic distribution is broad but irregular.

The three species of malaria above cause a comparatively mild form of the disease. Given the selectivities for age of the erythrocyte infected, the degree of total parasitemia is limited. Red cells are destroyed in the peripheral capillaries and anemia results. Although in non-immune children and travelers these symptoms can be quite severe, the global incidence of mortality from vivax, ovale, and malariae is extremely small. The possibility of relapse or deep recrudescence, however, makes the treatment and monitoring of patients critical.

In contrast, *P. falciparum*, subtertian malaria, or tropical malaria can lead to serious and life-threatening conditions when untreated. Erythrocytes of all ages can become infected by *P. falciparum*, and thus, a high percentage of red cells can become parasitized. The severe pathology associated with falciparum malaria involves the adhesion of parasitized erythrocytes to the capillary endothelium. The cytoadherence seems to involve parasite-derived proteins that are presented in the membranes of infected red cells (10, 11). The resulting sequestration in the post-capillary microvasculature leads to nearly continuous aggravation of the endothelial tissue in every organ affected and often irreversible tissue damage. Microcirculatory arrest occurs and when this happens in the brain, the condition is termed cerebral malaria; delirium, coma, convulsion, and death may ensue. Falciparum malaria is characterized by erythrocytic cycles of 48 h, but it does not relapse because it forms no hypnozoites in the liver. It is the most serious form of the infection, most widespread geographically, and accounts for the vast majority of malaria deaths. In areas of intense transmission, persons may be infected by more that one of these species at a time, causing complications in treatment.

Humans are not the only vertebrates that are parasitized by Plasmodia species. Several other parasites and their hosts have been used extensively for research in malaria. *P. berghei*, *P. vinckei*, and *P. yoelii* are useful models for malaria in mice and rats, while *P. cynomolgi* and *P. knowlesi* are studied in various monkey species. The human parasites can be studied in owl monkey (aotus) and in the splenecto-mized chimpanzee. Many avian and reptilian malarias are known, but their value as exper-imental models of human malaria is not as great as the rodent and primate malarias.

The most common method for evaluating the antimalarial activity of drugs and experi-mental compounds is the microdilution tech-nique introduced by Desjardins and co-work-ers in 1979 (12) and modified by Milhous et al. (13). Cultured intra-erythrocytic asexual forms of *P. falciparum* are treated with serial dilutions of compounds to be tested. Inhibition of uptake of [G-$^3$H]hypoxanthine by the para-sites serves as the indicator of antimalarial activity. [2,8-$^3$H]Adenosine may also be used as the radiolabel in these assays (14). For some types of studies, standard cultures are often unsuitable because of the high proportion of uninfected erythrocytes. A new procedure for producing highly concentrated cultures of *P. falciparum* from the ring stage provides a new tool in malaria research (15).

## 1.3 Parasite Biochemistry and Genetics

The biology, biochemistry, and genetics of Plasmodia are topics of wide interest. The in-formation from those fields that more directly impacts on the design of new antimalarial agents is presented briefly.

For the majority of their life cycle in hu-mans, malaria parasites live in red blood cells. Within the erythrocytes, the parasites feed on hemoglobin, digesting the protein as a source of amino acids and releasing heme [Fe(II)pro-toporphyrin IX]. Disruption of hemoglobin ca-tabolism results in parasite death. Hemoglo-bin digestion by *P. falciparum* proceeds by an ordered metabolic pathway (16). The initial events are the endocytosis of hemoglobin from the host cytoplasm and transport to the para-site's food vacuole. The tetramer structure of hemoglobin does not spontaneously dissociate in the food vacuole. Ample evidence exists that malaria parasites degrade hemoglobin in a stepwise fashion, a process mediated by a se-ries of proteases (17, 18). Initial cleavage of native hemoglobin occurs specifically at 33Phe-34Leu of the $\alpha$ chain mediated by an aspartic protease, plasmepsin I, which has been cloned and characterized from *P. falcipa-rum* (19, 20). Cleavage at this site can be ex-

pected to unravel the tetramer, making it susceptible to further proteolysis. The plas-mepsin I gene has striking homology to hu-man renin and cathepsin D. A second aspartic protease, plasmepsin II, prefers to act on acid-denatured globin but also has some overlap of specificity with plasmepsin I. The biosynthesis of the plasmepsins has been explored (21). The crystal structures of recombinant plasmepsin II (22) and plasmepsin II complexed with a known aspartic protease inhibitor, pepstatin A, are available (23). The active sites of plas-mepsins cloned from *P falciparum*, *P. vivax*, and *P. malariae* have been compared (24). The literature on plasmepsins I and II from *P. fal-ciparum* has been reviewed (25). The third en-zyme in the degradation pathways is a cys-teine protease, falcipain. Falcipain cleaves denatured protein but does not act on intact hemoglobin. A fourth enzyme, falcilysin, a novel metallopeptidase, acts downstream of the others (26). In addition to a role in globin degradation, these enzymes may function in dissociation of the hemoglobin tetramer and the release of heme from globin (27).

The other by-product of hemoglobin me-tabolism is heme. Between 25% and 75% of the erythrocyte hemoglobin is digested during growth of the parasite in the red cell (17). Be-cause erythrocytes contain 310–350 mg/ml of hemoglobin, the concentration of released heme is ~20 mM. However, if one considers that the heme would be released in the small volume of the parasite's food vacuole, the con-centrations could reach 200–500 mM (28). The heme then undergoes an autoxidation to produce toxic hematin [aquaFe(III)protopor-phyrin IX]. The parasite detoxifies hematin by conversion to an insoluble material known as hemozoin or malaria pigment. An X-ray dif-fraction study has shown that hemozoin is chemically and structurally identical with $\beta$-hematin, a synthetic product (29). The structure of $\beta$-hematin (and hence of hemo-zoin) was shown to be a crystalline dimer of hematin in which two protoporphyrin systems coordinate with one another by association of the propionate side-chains with the Fe(III) center of the opposite hematin (30). This is in contrast with an earlier theory that hemozoin was a polymer, rather than a dimer, of hema-tin.

The mechanism of hemozoin/β-hematin formation *in vivo* or *in vitro* remains unclear, even though a number of researchers have explored this question. Early workers suggested than the process was enzyme catalyzed (31) and it was reported in 1992 that cell-free preparations from infected erythrocytes caused hematin to precipitate in its insoluble form (32, 33). Extensive efforts to identify an enzyme responsible for this process, a "heme polymerase," have not been successful. Others have suggested that the process of hemozoin formation is not dependent on protein and that it was an essentially spontaneous, physicochemical process (34), although subsequent workers were not able to detect autocatalysis (35). A lipid catalyst has also been proposed (36, 37). Finally, histidine-rich proteins such as Pf-hrp-2 (*P. falciparum* histidine-rich protein 2) may play a role in hemozoin formation (38, 39); specifically, they may serve as initiators or scaffolds for hematin binding before sequestration as hemozoin (40). New studies have indicated that the mechanism of the chemical formation of β-hematin can best be understood as a biomineralization process, a view that may shed light on the biological formation of hemozoin (41).

Malaria parasites evade the human immune system through a process of continuous variation in a specific protein, erythrocyte membrane protein 1 (EMP-1) (42–45). During infection of the red blood cell, *P. falciparum* synthesizes EMP-1 that presents on the surface of the infected cell. EMP-1 serves to bind infected cells to blood vessels in the brain and in other organs. The presence of the EMP-1 protein would also be expected to notify the immune system of an infectious agent present in the cell. But, the parasite carries as many as 150 genes for EMP-1, each encoding a slightly different protein. New variants of EMP-1 allow the parasite to avoid destruction by immune processes. Studies have estimated that about 1 in 50 of each new generation of parasites secretes a different EMP-1 protein.

The conventional view is that Plasmodia use *de novo* folate synthesis because they lack folate salvage pathways. Inhibitors of key enzymes in the folate pathway, especially dihydrofolate reductase (DHFR) and dihydropteroate synthase (DHPS), have been shown to be clinically valuable antimalarials. A more complex picture is emerging, however, because it has been shown that some strains of *P. falciparum* are able to use exogenous folate, thus circumventing any blockage to *de novo* synthesis provided by antimalarial drugs.

Plasmodia synthesize dihydrofolate by a pathway unique to microorganisms. Para-aminobenzoic acid (PABA) is linked with a pteridine to form dihydropteroate by the enzyme dihydropteroate synthetase (DHPS), an enzyme not present in mammals. Then, conjugation of dihydropteroate with glutamate forms dihydrofolate (dihydropteroylglutamate). In contrast, mammalian cells obtain dihydrofolate through reduction of dietary folic acid. Sulfonamides and sulfones, inhibitors of DHPS, are selectively toxic to the parasite and relatively safe in the human host.

Malaria parasites are unable to use preformed pyrimidines using "salvage pathways" as mammalian cells do. Rather, plasmodia synthesize pyrimidines *de novo*. An important enzymatic target is dihydroorotate dehydrogenase (DHOD), which catalyzes the conversion of dihydroorotate to orotate, an intermediate in the pyrimidine biosynthetic pathway. Some compounds with antimalarial action such as atovaquone have been found to be inhibitors of DHOD. In a step further along in pyrimidine biosynthesis, tetrahydrofolate is a required cofactor. Compounds that inhibit dihydrofolate reductase (DHFR) effectively cut off the supply of tetrahydrofolate. Thus, compounds such as pyrimethamine and proguanil that inhibit DHFR are effective antimalarial agents.

There exist other metabolic pathways in Plasmodia that are potential sites for drug action. The shikimate pathway, a series of enzymatic conversions that produces aromatic cofactors and aromatic amino acids, was detected in apicomplexan parasites including plasmodia (46). Because the shikimate pathway is absent in mammals, inhibition of enzymes in this pathway provides an excellent prospect for drug design efforts. Plasmodia, as with other members of the phylum Apicomplexa, has been shown to possess an unusual organelle, a plastid that seems to have been acquired from algae at some point in its evolutionary history (47–49). Replication of this apicomplexan plastid, termed the apicoplast, is essential for parasite survival (50). An apicoplast metabolic pathway has been identified that is not found in animals, the mevalonate-independent pathway of isoprenoid biosynthesis (51), which is also known as the 1-deoxy-D-xylulose 5-phosphate (DOXP) pathway. Sequence data provided by the malaria genome project suggested the presence of two genes encoding important enzymes from the DOXP path, DOXP synthase and DOXP reductoisomerase.

Erythrocytes infected by plasmodia suffer oxidant damage from the parasite; the parasite causes measurable oxidation to the host red blood cell (52). The cell may be placed under oxidant stress from parasite-generated oxidants and from a weakening of the defense mechanisms of the cell itself. Increases in methemoglobin formation and lipid peroxidation have been documented in infected cells.

Two of the 14 chromosomes of the *P. falciparum* genome, roughly 7% of the total genome, have been sequenced. The sequence of chromosome 2 was published in 1998 (53) and that of chromosome 3 in 1999 (54). Approximately 215 protein-coding genes were identified on chromosome 2 and 209 on chromosome 3, giving an estimate of 6500 genes. Sequencing the *P. falciparum* genome required overcoming the technical difficulties of high A + T content. A genome-wide high-resolution linkage map of *P. falciparum* has been published (55) as has a shotgun optical map (56). The WHO maintains a database of gene protein information on malaria parasites (57). It is anticipated that new targets for drug development will arise from examination of the gene sequence.

## 1.4  Global Incidence

Malaria is endemic in 101 countries and territories; 45 of these are in Africa, 21 in the Americas, 4 in Europe, 14 in the Eastern Mediterranean, 8 in Southeast Asia, and 9 in Western Asia. There are small pockets of transmission occurring in a further 12 countries (Table 18.1). Although *P. falciparum* is the predominant parasite, falciparum malaria occurs only sporadically or does not exist in 19 of those countries. For comparison, in 1955, there were 140 countries or areas where malaria

was endemic. Global statistical information on malaria is presented in Table 18.2 and is available from the WHO web site (58).

An indication of the magnitude of the health problem that malaria poses can be obtained from WHO's Malaria Fact Sheet (60):

• Malaria is a public health problem for more than 2400 million people, 40% of the world population (Table 18.2).
• There are an estimated 300–500 million clinical cases/year.
• 1,086,000 deaths were reported as attributable to malaria in 1999 (61).
• Approximately 90% of both malaria incidence and mortality occur in sub-Saharan Africa.
• The mortality is estimated to be greater than 1 million; the vast majority of these are young children in Africa, especially in rural areas with limited health care available.
• One child dies every 30 s or 3000 children per day under the age of 5 years.
• In affected countries, 3 of 10 hospital beds are occupied by malaria victims.

Historically, malaria was endemic throughout much of the continental United States; an estimated 600,000 cases were reported in 1914. During the late 1940s, a number of factors combined to successfully interrupt malaria transmission including improved socioeconomic conditions, water management programs, insect-control efforts, and case management approaches. Since then, surveillance has been maintained to detect reintroduction of transmission. In recent decades, almost all cases of malaria in the United States were imported by travelers from regions of the world where malaria transmission is known to occur. In 1997, the CDC reported 1544 cases of malaria in the United States, the highest overall incidence since 1980 and the highest number of civilian cases since 1968. This figure represents a 10.9% increase over 1996. Of these cases, only five were acquired in the United States. Six persons died of malaria (62).

Africa has the highest levels of endemicity in the world; in very large areas transmission is intense and perennial. In areas with alti-

**Table 18.1   Countries Where Malaria Transmission Occurs (59)**

| | | |
|---|---|---|
| Afghanistan[a,f] | Algeria[a,d] | Angola[c,f,g] |
| Argentina[b] | Armenia[b] | Azerbaijan[a,d] |
| Bangladesh[f,g] | Belize[a,e] | Benin[c,f] |
| Bhutan[f,g] | Bolivia[a,f,g] | Botswana[c,f] |
| Brazil | Burkina Faso[c,f] | Burundi[c,f] |
| Cambodia[c,f,g,h] | Cameroon[c,f,g] | Cape Verde[d] |
| Central Afr. Rep.[c,f,g] | Chad[c,f] | China |
| Colombia[f,g] | Comoros[c,f] | Congo[c,f] |
| Costa Rica[b] | Côte d'Ivoire[c,f,g] | Dem. People's Rep. Korea[b,d] |
| Dem. Rep. Congo[c,f,g] | Djibouti[c,f] | Dominican Repepublic[c,d,e] |
| Ecuador[f] | Egypt[d] | El Salvador[b,d] |
| Equatorial Guinea[c,f,g] | Eritrea[c,f] | Ethiopia[c,f] |
| French Guiana[c,i] | Gabon[c,f,g] | Gambia[c,f] |
| Georgia[b] | Ghana[c,f] | Guatemala[a] |
| Guinea[c,f] | Guinea-Bissau[c,f] | Guyana[f] |
| Haiti[c,e] | Honduras[a] | India[f] |
| Indonesia[f,g,j] | Iran, Islamic Rep.[f] | Iraq[b] |
| Kenya[c,f,g] | Lao People's Dem. Rep.[c,f] | Liberia[c,f,g] |
| Madagascar[c,f] | Malawi[c,f,g] | Malaysia[c,d,f] |
| Mali[c,f] | Mauritania[c] | Mauritius[b] |
| Mayotte[c] | Mexico[b] | Morocco[b,d] |
| Mozambique[c,f,g] | Myanmar[c,f,g] | Namibia[c,f] |
| Nepal[a,f] | Nicaragua[a,e] | Niger[c,f] |
| Nigeria[c,f] | Oman[d,f] | Pakistan[f] |
| Panama[a,f] | Papua New Guinea[c,f,g,j] | Paraguay[b] |
| Peru[f,g] | Philippines[f] | Republic of Korea[b,d] |
| Rwanda[c,f,g] | Sao Tome & Principe[c,f] | Saudi Arabia[c,f] |
| Senegal[c,f] | Sierra Leone[c,f] | Solomon Islands[c,f] |
| Somalia[c,f] | South Africa[c,f] | Sri Lanka[a,f] |
| Sudan[c,f] | Suriname[c,f,k] | Swaziland[c,f] |
| Syrian Arab Republic[b] | Tajikistan[b,f] | Thailand[f,g,h,i,k] |
| Togo[c,f] | Turkey[b] | Turkmenistan[b] |
| Uganda[c,f] | United Arab Emirates[d] | United Rep. Tanzania[c,f,g] |
| Vanuatu[c,f,g,j] | Venezuela[f] | Vietnam[c,f,g] |
| Yemen[c,f] | Zambia[c,f] | Zimbabwe[c,f] |

[a]Predominately *P. vivax;* [b]exclusively *P. vivax;* [c]predominately *P. falciparum;* [d]limited; [e]no resistance reported; [f]chloroquine-resistant falciparum reported; [g]sulfadoxine/pyrimethamine resistance reported; [h]mefloquine resistance reported; [i]multi-drug resistance reported; [j]chloroquine-resistant vivax reported; [k]lessened sensitivity to quinine reported.

tudes over 1500 m and rainfall below 1000 mm/year, endemicity decreases, and the potential for epidemic outbreaks increases. Ecological, demographical, and meteorological factors including quasi-cyclic occurrence of heavy rains have led to epidemics or serious exacerbations of endemicity, especially in Botswana, Burundi, Ethiopia, Kenya, Madagascar, Rwanda, Sudan, Swaziland, Zaire, and Zambia.

Excluding Africa, of the total number of cases reported annually to the WHO, more than two-thirds are concentrated in only six countries: India, Brazil, Sri Lanka, Afghanistan, Vietnam, and Colombia. In other parts of the world, the distribution of malaria varies greatly from country to country and from region to region within countries. For example, in India, the majority of reported cases occur in only a handful of states. Three states of the Amazonian Basin in Brazil account for close to 80% of all cases while representing only 6.1% of the country's overall population.

Inadequate and irregular reporting, particularly in areas known to be highly endemic and often out of the reach of established health services, make it difficult to obtain accurate information on the incidence of malar-

**Table 18.2   Reported Cases of Malaria, Endemic Countries Only, 1982–1997 (58)**

| Year | World Total (Countries Reporting) | Africa | Americas | Asia | Oceania |
|------|-----------------------------------|--------|----------|------|---------|
| 1982 | 14,530,337 (92) | 8,045,746 (39) | 713,878 (21) | 5,580,381 (29)[a] | 190,332 (3) |
| 1983 | 9,024,790 (80) | 4,351,945 (26) | 829,546 (21) | 3,606,587 (30) | 236,712 (3) |
| 1984 | 10,514,053 (83) | 5,483,965 (29) | 929,891 (21) | 3,849,728 (30) | 250,469 (3) |
| 1985 | 19,271,828 (91) | 14,557,717 (37) | 909,162 (21) | 3,556,744 (30) | 248,205 (3) |
| 1986 | 23,829,689 (90) | 18,567,724 (36) | 948,906 (21) | 4,088,518 (30)[a] | 224,541 (3) |
| 1987 | 26,633,220 (90) | 21,294,523 (36) | 1,016,327 (21) | 4,058,717 (30) | 263,653 (3) |
| 1988 | 30,887,818 (91) | 25,602,604 (37) | 1,118,132 (21) | 3,992,351 (30) | 174,731 (3) |
| 1989 | 38,634,970 (90) | 33,284,881 (36) | 1,111,732 (21) | 4,013,520 (3) | 224,837 (3) |
| 1990 | 27,238,125 (85) | 21,903,265 (32) | 1,055,897 (21) | 4,029,017 (29) | 249,946 (3) |
| 1991 | 27,208,754 (86) | 21,570,180 (33) | 1,230,155 (21) | 4,163,317 (29) | 245,102 (3) |
| 1992 | 26,689,817 (84) | 21,371,102 (33) | 1,186,053 (20) | 3,939,559 (28) | 193,103 (3) |
| 1993 | 32,657,473 (86) | 27,603,152 (35) | 982,040 (20) | 3,868,984 (28) | 203,297 (3) |
| 1994 | 38,433,333 (88) | 32,619,436 (36) | 1,113,407 (21) | 3,935,507 (28) | 764,983 (3)[b] |
| 1995 | 27,378,851 (85) | 21,512,357 (35) | 1,279,651 (20) | 4,409,796 (27) | 177,047 (3) |
| 1996 | 24,273,550 (76) | 18,096,602 (24) | 1,138,453 (20) | 4,877,033 (29) | 161,462 (3) |
| 1997 | 18,715,601 (73) | 12,260,126 (22) | 1,054,230 (20) | 5,288,916 (28) | 112,329 (3) |

[a]Not all cases from China confirmed by laboratory tests.
[b]Not all cases from Papua New Guinea confirmed by laboratory tests.

ial disease in many areas. Under ideal circumstances, each case would be confirmed clinically by microscopic examination of a blood smear. In practice, malaria is most often defined in association with disease symptoms rather than with microscopic confirmation. And priority is given to reporting of severe and complicated cases and malaria deaths rather than total number of cases. Especially in Africa, strict reporting is fragmentary and based on clinical signs and symptoms.

## 1.5   Resistance

Drug resistance in malaria has been defined as the "ability of a parasite strain to survive and/or multiply despite the administration and absorption of a drug given in doses equal to, or higher than, those usually recommended but within the limits of tolerance of the subject" (63). Later modification of this definition stated that "the form of the drug active against the parasite must gain access to the parasite, or the infected blood cell, for the duration of time necessary for its normal action." The modified definition makes clear that for parasites to be termed resistant, the drug must be bioavailable. The phenomenon of drug resistance has become common in many infectious diseases: tuberculosis, staphylococci, streptococci, HIV, and syphilis. Drug re-

sistance was perhaps first recognized in the treatment of malaria. Early in this century, it was noted that some cases of malaria responded much more poorly to quinine, the only drug available at the time, and that larger doses of drug had to be administered to effect a cure. At present, the increasing occurrence of chloroquine-resistant strains of *P. falciparum* has inspired global drug-design and development efforts in an attempt to identify new agents for the treatment of resistant strains of Plasmodia. The sensitivity of a parasite strain to a drug can be classified into four general groups: sensitive (S), in which clearance of asexual parasitemia occurs within 7 days of the initiation of treatment without subsequent recrudescence; slightly resistant (RI), in which asexual parasitemia is cleared as in sensitive cases but recrudescence follows; moderately resistant (RII), in which the level of asexual parasitemia is reduced markedly but not cleared fully; and highly resistant (RIII), in which little or no reduction in asexual parasitemia occurs. A higher resistance to a particular agent is not necessarily correlated with higher virulence. A single isolate of *P. falciparum* from an infected individual consists of parasites with differing drug responses.

For descriptions of the history of the emergence of malarial drug resistance, see Ref. 64.

The overall picture is that malaria parasites have developed resistance to each and every therapeutic agent in areas where drug selection pressure has been applied. Drug pressure is considered to be the main causative factor of selecting and propagating resistant *P. falciparum* in a particular locality or area. Chloroquine has been the agent of choice against *P. falciparum* for decades. When first brought into wholesale use for mass chemotherapy in the 1950s, its effectiveness led to the hope of global malaria eradication. Such broad usage for chloroquine, however, induced resistance beginning in South America and Southeast Asia. Even so, for compounds such as chloroquine, prolonged drug exposure seems to be required before resistance began to develop (65). Chloroquine resistance emerged much later in Africa than in South America and Southeast Asia. The first case from East Africa was documented in non-immune travelers returning from that area in 1979 (66). Since that time, chloroquine resistance has spread rapidly in Africa, and it is estimated that more than 60% of the *P. falciparum* strains that infect non-immune travelers are resistant to chloroquine. Significantly, chloroquine still largely retains is therapeutic efficacy in semi-immune populations. Among the countries where falciparum malaria exists, only those of Central America have not recorded the resistance of falciparum malaria to chloroquine. An excellent set of maps tracking the development of chloroquine resistance globally is shown in Ref. 63.

The resistance of *P. vivax* to chloroquine is not uncommon in Indonesia and Papua New Guinea, although the extent of the spread is not known. There have been sporadic reports of chloroquine-resistant vivax from Myanmar, Thailand, Borneo, India, and Brazil (67). Although not widespread as yet, primaquine-resistant strains of *P. vivax* have been reported (68, 69), although others have raised concerns about the definition of primaquine resistance (70).

Chloroquine is not the only agent that has lost effectiveness against *P. falciparum* because of the development of drug resistance. Resistance to sulfadoxine/pyrimethamine has developed in Southeast Asia, South America, and Africa. For the dihydrofolate reductase inhibitors proguanil and pyrimethamine, resistance can be induced by a single large dose. In Thailand, there are indications that more than 50% of cases in certain border areas no longer respond to mefloquine therapy, while the sensitivity to quinine is also diminishing in areas of Thailand and Vietnam. In cases of resistance, quinine, sulfadoxine/pyrimethamine, mefloquine, and now the artemisinin-type agents, are being used therapeutically.

Thailand has served as a laboratory for the development of drug-resistant *P. falciparum* (71). Beginning in the early 1950s, resistance to pyrimethamine arose as a result of its use in prophylaxis and in presumptive treatment (treatment pending laboratory diagnosis). Initial experiences with emerging chloroquine resistance involved recrudescence on administration of standard doses in the late 1950s, although its took until the mid-1970s until chloroquine resistance became so frequent and problematic that alternate therapies were introduced. Sulfadoxine/pyrimethamine as a routine treatment was useful only for a short time before resistance emerged. By 1982, the recommended treatment regimen was changed again to a 7-day course of quinine/tetracycline. The inconvenience and poor compliance associated with a 7-day course necessitated the move to the triple combination mefloquine/sulfadoxine/pyrimethamine in 1985. Treatment failures with mefloquine were observed beginning in 1991.

Multi-drug resistance may be defined as resistance of *P. falciparum* to more than one operational class of antimalarial agents. Clinically, the problem of multi-drug resistant strains is limited to the Indochina Subcontinent, with some foci in Sabah, Malaysia, Papua New Guinea, and West Irian, Indonesia. Some sporadic occurrences of multi-drug resistant organisms have been seen in tropical Africa and South America. The border areas of Thailand with Cambodia and Myanmar are the areas of hardcore multi-drug resistance. *P. falciparum* in these areas are resistant to chloroquine and other 4-aminoquinolines, sulfadoxine/pyrimethamine, and quinine (if used as a single agent).

Given these observations, resistance to any new therapeutic agent can be expected, and strategies are being developed to ensure an appreciable effective lifetime for each new drug. One important tactic has been the employment of combination therapy (72); the useful lifetime of pyrimethamine has been extended by an estimated 15 years because of its use in combination with sulfadoxine (73). For a discussion of the factors that influence the emergence of resistance and a consideration of the outcomes when antimalarial drugs are used in combinations, see Refs. 72, 74. Many therapies are being studied with the new artemisinin agents in combination with older agents. Another tactic is the avoidance of initiating widespread therapy with new agents in areas where other, older agents remain effective. The hope is that the newer agents might be held in reserve until absolutely necessary. Researchers have been studying the dynamics involved in the emergence of drug-resistant malaria strains and are attempting to develop models that will assist in the creation of strategies for deployment of newer agents in the field (75).

Current efforts at analyzing the molecular mechanisms of resistance are offering new approaches for drug resistance (76, 77). Mutations in key molecular targets had been identified for antifolate resistant strains. Mechanisms that confer resistance to chloroquine seem to be more complex, but new information is helping to clarify the variables. Rapid efflux of chloroquine from resistant parasites can be reversed by co-administration of certain agents. Additional research on the molecular basis of resistance may provide improved therapeutic strategies for use in resistant organisms.

## 1.6  Immunity and Prophylaxis

The average age of first infection with malaria is less than 1 year old for persons in most endemic areas. Estimates are that between 75,000 and 200,000 infant deaths each year are associated with malaria infection in pregnancy (78). Exposure does not induce lifelong immunity to further malaria infections. As children age, they acquire a functional immunity that provides fewer clinical attacks and lessened clinical symptoms. Adults can obtain sufficient immunity so that a substantial reduction in infection rates is observed. Asymp-tomatic parasitemia may occur among persons who have been long-term residents of malaria-endemic areas. As a result, children have the highest levels of mortality and morbidity resulting from malaria. It had long been assumed that these data were the result of a high transmission rate of the parasite. New studies have indicated, however, that most endemic areas contain parasites with a high degree of antigenic diversity (79). Each infection does in fact provide immunity to that particular strain; new infections are caused by parasites with different antigenic profiles. A calculation of transmission rates incorporating these considerations yields numbers an order of magnitude lower than previously estimated.

Pregnant women have a unique susceptibility to malaria, such that "maternal malaria," a distinct clinical entity, causes serious pregnancy-related complications in endemic areas. Increases in miscarriage, premature delivery, retardation of fetal growth, anemia, low birth weight, and mother and infant mortality rates are observed when the mother contracts malaria during pregnancy. Adult women acquire the same degree of immunity to the common strains of parasites as do adult men; however, after becoming pregnant, this acquired immunity diminishes markedly. With successive pregnancies, the loss of immunity is less pronounced. To explain the apparent loss of acquired immunity in pregnancy, researchers studied the binding of infected red blood cells to placental tissue (80). A subpopulation of *P. falciparum* parasites preferentially binds to placental tissue and multiplies there. The binding site in placenta normally binds chondroitin sulfate A (CSA). The severity of the infection must be caused by the initial appearance of placental tissue that harbors the infection. As the woman develops immunity to this subpopulation of parasites, the frequency and severity of malaria in a second or third pregnancy is lessened. With the rising incidence of HIV/AIDS in Africa and concern about maternal transmission of the virus, the problems of simultaneous multiple drug therapies emerge. For a review of the therapeutic and safety issues in the concurrent use of anti-HIV agents with antimalarial agents in pregnant patients, see Ref. 81.

Glucose-6-phosphate dehydrogenase (G6PD) deficiency is a genetic condition most prevalent among persons living in malaria-endemic areas. The most common African form of G6PD deficiency was associated with a 46–58% reduction in the risk of severe malaria (82). Parasites inflict oxidant damage to the erythrocytes that are infected and erythrocytes that are deficient in G6PD are especially susceptible to oxidant damage. In addition, it has been shown that phagocytosis of parasiticized G6PD-deficient erythrocytes occurs earlier than does phagocytosis of infected normal erythrocytes, perhaps accounting for the observed protective effect (83).

Other inherited red cell disorders contribute an immunity to malaria infections, at least partially as a result of sensitization to oxidant stress. These include the well-known example of heterozygous hemoglobin S (sickle cell trait) (84) as well as thalassemia (85), persistence of fetal hemoglobin, and hemoglobin E. $\alpha$-Thalassemia seems to increase susceptibility to malaria in early childhood and then offers immunity later in life (86). An explanation of the effect involving hemoglobin/membrane interactions had been proposed (87). Southeast Asian ovalocytosis is another genetic disorder that confers marked protection against cerebral malaria (88). It is caused by a deletion in the gene for the erythrocyte membrane band 3. The band 3 protein is responsible to the cytoadherence of parasitized cells. The mutation occurs in high frequency in the western Pacific.

Non-immune travelers, with no prior exposure to malaria, are at greater risk of acquiring serious infections. If the infection presents clinical symptoms only after return from a malarious area, there may be a delay in diagnosis by physicians unfamiliar with the disease. Travelers to endemic areas should seek advice about the use of chemoprophylactic regimens and stand-by treatment. Further protection is gained by preventing exposure to mosquitoes using insect repellents on clothes and skin and staying indoors at night. For reviews on malaria chemoprophylaxis, see Refs. 89, 90.

Chemoprophylaxis demands a nearly impossible level of safety, especially when used over long periods. Drugs are given to healthy individuals, most often to prevent a disease of low probability. The most successful agent to date has been chloroquine, with a 50-year safety record that is matchless among anti-infective agents. Malaria prophylactic regimens are generally highly efficacious and low in cost, especially when compared with the costs of treatment including hospitalization. Even so, compliance has remained an issue among U.S. travelers to malarious areas; only 21% of travelers used recommended malaria prophylaxis appropriate for their area of travel in 1997 (62). International travel recommendations are available from both the WHO (59, 91) and the CDC (92, 93).

## 1.7 Vector Control

Systematic malaria control began after the discovery of the malaria parasite by Laveran in 1880 (for which he was awarded the Nobel Prize for medicine in 1907), and the demonstration by Ross in 1897 that the mosquito was the vector of malaria (94). Attempts to control mosquitoes have paralleled therapeutic and chemoprophylactic approaches in battling the disease (95). In 1955, the World Health Assembly initiated a malaria eradication program. The stated goal was the interruption of malaria transmission by reducing mosquito populations that fed on humans; complete elimination of the vector was not attempted. Household spraying with DDT was the main tactic. The approach was most successful in temperate regions where transmission was unstable. However, in some areas when spraying was halted, mosquito populations returned but were now resistant to DDT.

An alternative to reducing or eliminating vector populations is the prevention of contact between humans and mosquitoes. The use of insecticide-impregnated bed nets seems to be effective in limiting malaria in a number of different areas: the Solomon Islands, where malaria prevalence rates are among the highest in the world (96); Guatemala (97); Kenya (98); Tanzania (99); and The Gambia (100). Continual governmental support for such programs will be needed if any long-term benefits are to be ultimately accrued from such an approach. The CDC gives detailed instructions for travelers about avoiding contact with potentially infectious mosquitoes, including the use of bed nets and insect repellents (92).

A recent response to vector control has been studies of the replacement of normal mosquito populations with strains that cannot support normal parasite development (101). Success will require identification of the mosquito genes that inhibit parasite growth, the introduction of such genes in the mosquito genome, and dispersing the trait through natural populations. Efforts to sequence the *Anopheles gambiae* genome are underway (102). Fortunately, there are a number of points in the parasite life cycle within the mosquito that provide opportunities for manipulation.

It has been suggested both in the scientific literature and in the popular press that certain tropical diseases, especially those that involve a mosquito vector, are likely to increase in incidence with the increase in global temperatures. However, a statistical study incorporating a number of critical factors including temperature did not forecast global malaria expansion (103).

## 1.8 Economic and Political Issues

Today, malaria is becoming a greater health problem than before in many parts of the world. Epidemics are occurring in areas where transmission had been eliminated. These outbreaks are generally associated with deteriorating social and economic conditions, and many victims are in underprivileged rural populations. Demographic, economic, and political pressures compel entire populations (seasonal workers, nomadic tribes, and farmers migrating to newly developed urban areas or new agricultural and economic developments) to leave malaria-free areas and move into endemic zones. These people are often non-immune and at high risk of severe disease. Unfortunately, these population movements and the intensive urbanization are not always accompanied by adequate development of sanitation and health care.

Malaria is now mainly confined to the poorer tropical areas of Africa, Asia, and Latin America, but the problems of controlling malaria in these countries are aggravated by inadequate health structures and poor socioeconomic conditions. Moreover, in many areas, conflict, economic crises, and administrative disorganization result in the disruption of health services. As a result, control efforts are interrupted and more people are put at risk. The absence of adequate health services frequently results in a recourse to self-administration of drugs often with incomplete treatment. This is a major factor in the increase in resistance of the parasites to drugs.

Malaria thus has disastrous social consequences and is a heavy burden on personal and national economic development. On the level of the individual, an episode of the disease was estimated to cost US$8.67 in Ghana (reported in 1997), although costs would vary depending on the costs of the drugs (104). Short-term costs—loss of work time, losses associated with child morbidity and mortality, costs of treatment and prevention—are only part of the story. A high proportion of the cost of malaria care is the opportunity cost to the caretakers. Children in particular may suffer from chronic anemia and malnutrition as a result of repeated bouts of malaria, leading to altered physical and cognitive development.

In the aggregate, costs are staggering. According to 1997 estimates, direct and indirect costs of malaria in sub-Saharan Africa exceed $2 billion (60). In contrast, the average cost for each nation in Africa to implement programs to control malaria is estimated at $300,000/year or approximately US$0.06 per person for a country of 5 million persons.

Areas with malaria are almost exclusively poor, with low rates of growth (105). But is this a cause or an effect? Successful elimination of malaria has historically required well-organized and well-financed programs. Even so, some of the most effective control efforts have used few resources beyond labor. Sustaining such efforts over long periods has not been successful in many locations, however. Endemic countries may find trade impeded, lowered interest in international investment and commerce, and fewer opportunities for tourism. Evidence suggests that overall economic development is hindered by malaria. In places where malaria has been eradicated, economic growth accelerated after eradication. Growth rates in endemic countries are over 1% per year less than that in comparable states without malaria. Over a period of decades, the economic burden can become immense (106). And so while countries do not become prosperous by controlling malaria alone, it is certainly made easier.

A final issue is that of the economics of drug development of antimalarial agents and drugs to treat parasite diseases more generally. Because the populations affected are found mostly in developing countries with limited health-care resources, pharmaceutical companies have little financial incentive to invest in the development of new agents to treat these diseases (107).

On May 13, 1998, the Director General of the WHO announced the Roll Back Malaria Campaign with the goal of cutting global incidence by 50% by 2010. Another 50% reduction was targeted for 2015. Details of the program may be found at WHO web sites (108) as well as reports on the progress made to date (109, 110). As part of the Roll Back Malaria effort, a program called Medicines for Malaria Venture has been established (111).

## 2 ANTIMALARIAL AGENTS FOR CHEMOTHERAPY AND PROPHYLAXIS: CURRENT DRUGS IN USE

The therapeutic agents in use against malaria are summarized in Table 18.3. The available drugs represent a wide variety of structural types and modes of action. One of the drugs, quinine, has been in use for hundreds of years. Several were discovered and approved very recently. Some find uses in other infectious diseases while many are specifically used in malaria. The agents may be used in treatment or for chemoprophylaxis; some are valued in both applications. Each of the drugs is described in detail below.

### 2.1 Quinine and Quinidine

Quinine (1) may be claimed without exaggeration as the drug to have relieved more human suffering than any other in history (112). For 300 years, it was the only known effective treatment for a life-threatening infectious disease. Only a handful of other treatments, emetine for amoebic dysentery, mercury for syphilis, chaulmoogra oil for leprosy, and herbal anthelmintics, were effective as specific anti-infective agents until this century. The fanciful story of the miraculous cure of the

Countess of Chinchón, wife of the Viceroy of Peru, by administration of a native remedy produced from tree bark is charming but very far from fact. For a scholarly discussion of the early history of cinchona, see Ref. 113. Tree bark from the cinchona tree, *Cinchona officinalis* and other Cinchona species, a native plant from South America, was the source for an effective treatment of recurrent fevers. In the 19th century, the active principles, the cinchona alkaloids, quinine (1), quinidine (2), cinchonine (3), and cinchonidine (4), were isolated and purified. The formal synthesis of quinine by Woodward and Doering in 1944–1945 was a landmark in modern synthetic chemistry (114, 115). The first stereoselective total synthesis of quinine was recently reported by Stork and co-workers (116). Included in their paper is a brief history of synthetic efforts toward quinine.

The principal areas producing cinchona are central Africa, India, and Indonesia. Commercial formulations of quinine have approximately 10% dihydroquinine as an impurity. The two compounds have nearly equivalent antimalarial activity, however, so that efficacy is not affected. The preparation Quinimax is a mixture of cinchona alkaloids, predominately quinine, and is reported to be more effective than quinine alone.

The stereochemical differences among the cinchona alkaloids result in differences in potency, and the stereoelectronic features have been examined (117). Conformational differences between the diastereomers apparently lead to differing ability to form critical hydrogen bonds. Quinidine is two- to threefold more active than quinine in both chloroquine-sensitive and chloroquine-resistant strains of *P. falciparum* (118). Likewise, cinchonine is more active than cinchonidine *in vitro* (119). The differences in activity based on stereochemistry are greater for those compounds like quinine and quinidine, which have a rigid quinuclidine moiety, than for synthetic compounds such as mefloquine, which bear a piperidine ring (120).

The main metabolite of quinine is 3-hydroxyquinine, which is produced by the action of CYP4503A4 (121). Certain drugs were found to inhibit the metabolism of quinine, including tetracycline, doxycycline, omeprazole,

**Table 18.3 Drugs Currently in Use**

| Generic Name | Structure | Trade Names | Year | Originator | Chemical Class | Dosage Forms |
|---|---|---|---|---|---|---|
| Quinine | 1 | Quinamm | | | Quinoline alcohol | Many sizes capsules and tablets, IV form no longer available in U.S. |
| Quinidine | 2 | | | | Quinoline alcohol | IV solution |
| Chloroquine | 5 | Aralen, Avloclor, Nivaquin, Resochin | | | 4-Aminoquinoline | 500-mg Tablet (300-mg base), oral |
| Hydroxychloroquine | 6 | Plaquenil | | | 4-Aminoquinoline | 200-mg (155-mg base) Tablet oral |
| Mefloquine | 10 | Mephaquine Lariam | 1985 1989 | Mepha Hoffmann-LaRoche | Quinoline alcohol | 250-mg Tablet, oral |
| Halofantrine | 13 | Halfan | 1988 | Smith Kline & French | Phenanthrenemethanol | 250-mg Tablet, oral |
| Primaquine | 14 | | | | 8-Aminoquinoline | 26.3-mg (15-mg base) Tablet, oral |
| Bulaquine | 21 | Aablaquine | 2000 | Central Drug Research Institute | 8-Aminoquinoline | Aablaquine = bulaquine + chloroquine combination pack |
| Proguanil | 23 | Paludrine | | | Biguanide | 100-mg Tablet oral |
| Chlorproguanil | 24 | | | | Biguanide | |
| Pyrimethamine | 27 | Daraprim | | | Pyrimidine | |
| Sulfadoxine | 31 | | | | Sulfanilamide | |
| Pyrimethamine/ sulfadoxine | | Fansidar | | | Combination | 500 mg Sulfadoxine, 25 mg pyrimethamine tablet oral |
| Mefloquine/ pyrimethamine/ sulfadoxine | | Fansimef | | | Combination | |
| Dapsone | 33 | | | | Sulfone | |
| Pyrimethamine/dapsone | | Maloprim Deltaprim | | | Combination | 25 mg Pyrimethamine, 100 mg dapsone tablets |
| Trimethoprim | 28 | | | | Pyrimidine | |
| Sulfamethoxazole | 34 | Gantanol | | | Sulfanilamide | 500-mg Tablet |
| Sulfamethoxazole/ trimethoprim | | Bactrim Cotrim | | | Combination | |
| Sulfisoxazole | 35 | Gantrisin | | | Sulfanilamide | |

934

| Drug name | No. | Brand names | Year | Company | Class | Formulation/Notes |
|---|---|---|---|---|---|---|
| Sulfalene | 32 | | | | Sulfanilamide | |
| Artemisinin | 42 | | 1987 | Ping Hau Sau Research Group | Artemisinin | |
| Artemether | 50 | Paluther | 1994 | | Artemisinin | |
| Arteether | 51 | Artemotil | 2000 | Central Drug Research Institute | Artemisinin | |
| Dihydroartemisinin | 47 | | | | Artemisinin | |
| Artesunate | 48 | Arsumax, Plasmotrim | 1996 | | Artemisinin | |
| Atovaquone | 68 | Mepron, Wellvone | 1992 | Wellcome | Hydroxynaphthoquinone | 250-mg Tablet, also IV solution |
| Pyronaridine | 70 | | | China | 4-Aminobenzo[a]1,5-naphthyridine | |
| Amodiaquine | 72 | Achromycin and many others | | | 4-Aminoquinoline | |
| Tetracycline | 76 | | | | Tetracycline | |
| Doxycycline | 77 | Vibramycin, Doxycyclin | | Pfizer | Tetracycline | 100-mg Tablet, oral |
| Azithromycin | 78 | Sunamed, Zithromax | 1988 | Pliva, Pfizer | Macrolide | |
| Clindamycin | 83 | Cleocin | | | Linosamide | |
| Ciprofloxacin | 79 | | | | Fluoroquinolone | |
| Atovaquone/proguanil | | Malarone | 1997 | GlaxoSmith Kline | Combination | 250 mg Atovaquone, 100 mg proguanil oral |
| Lumefantrine (benflumatol) | 69 | | | | Fluorene alcohol | |
| Lumafantrine/artemether (coartemether) | | Coartem, Riamet | 1992 | China | Combination | |
| | | | 1999 | Novartis | | |

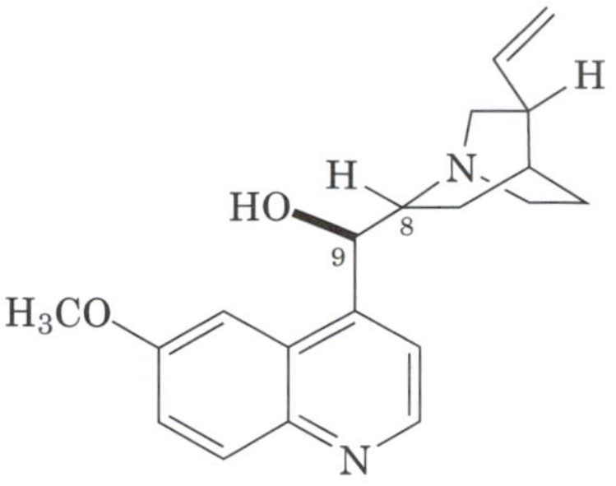

**(1)**
quinine
(–)-8S, 9R

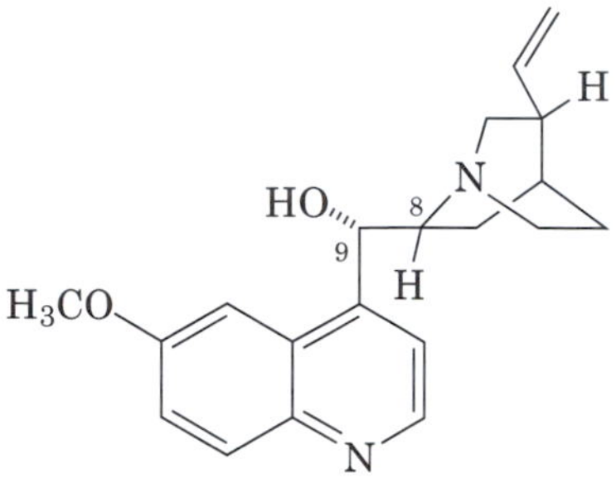

**(2)**
quinidine
(+)-8R, 9S

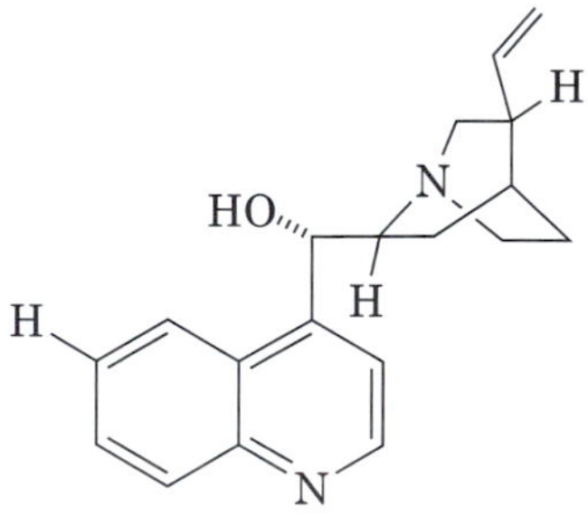

**(3)**
cinchonine

**(4)**
cinchonidine

ketoconazole, troleandomycin, and erythromycin (122, 123). In contrast, chloroquine, pyrimethamine, proguanil and its metabolites, norfloxacin, and dapsone did not appreciably inhibit quinine metabolism.

Since the development of potent synthetic agents, quinine had fallen into disuse as a first-line antimalarial. The small difference between therapeutic and toxic doses is a de-

cided drawback, and longer treatments over several days are required for good cure rates. But with the advent of multi-drug resistant strains, quinine has returned as an important agent in severe cases involving drug-resistant parasites. It is almost always used in combination with tetracycline. Effective treatment usually involves a 7-day course of the combination (124). Therefore, it is not as convenient as single-day treatments, and compliance can become a problem. In addition, tetracycline cannot be given to children.

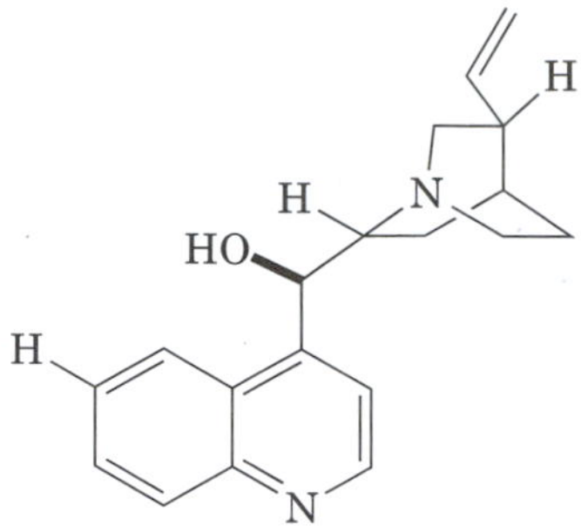

Quinine has been examined in combinations with other agents. A study suggested that co-administration of allopurinol, a potent inhibitor of purine biosynthesis, brings about faster eradication of *P. falciparum* and clinical remission than with quinine alone (125). Combinations of omeprazole, a proton pump inhibitor, with quinine were synergistic *in vitro* (126).

Several studies have examined the viability of rectal routes of quinine administration. Intra-rectal formulations may be given in the presence of vomiting and nausea and may be especially important in areas where the safety of parenteral administration cannot by assured. Quinine or Quinimax given by rectal administration was a viable alternative to parenteral routes (127, 128). Despite the benefits of rectal administration, concerns remain about the low melting point of the vehicle, especially for use in warm climates.

There is an increased interest in the use of Quinimax as an alternative to quinine. Like quinine, 7-day dosing regimens are required for acceptable cure rates (129). The pharmacokinetics of Quinimax suppositories were studied in children (130). A comparison of intra-rectal, intramuscular, and intravenous routes of administration showed that peak plasma concentrations and areas under the plasma concentration-time curve were similar; the time to peak plasma concentration was shorter for the intra-rectal route, however.

When used for monotherapy, sensitivity to quinine was declining over time; in Thailand, for example, the cure rates were 94% in 1978–1979, 86% in 1979–1980, and 76% in 1980–1981 (131). By combining quinine with tetracycline, the cure rates were improved to >95%. Tetracycline serves to increase the plasma quinine levels above those seen with equivalent dosing of quinine alone. In a study involving highly drug-resistant parasites, clinical response to quinine/tetracycline did not decline between 1981 and 1990. African strains of P. falciparum are generally susceptible to quinine, although a few reports indicate a lessening of its effectiveness (132, 133); clinical cases of strains resistant to quinine have been documented in East Africa (134) and Brazil (135). In the Brazil study, data from the sequence analysis for the pfmdr1 gene indicated that the mechanism for quinine resistance may differ from that of either chloroquine or mefloquine. Some researchers have gathered evidence that mutations in Pgh1, the protein product of pfmdr1, can confer resistance to quinine (136). Interestingly, there seems to be an inverse relationship between chloroquine resistance and quinine and mefloquine resistance in P. falciparum.

The clinical pharmacokinetics of quinine have been reviewed (121, 137). Quinine disposition is linear over the dose range of 250–1000 mg as single oral doses (138). Profiles were found to be similar following either oral administration or intravenous infusion to patients with acute falciparum malaria (139). Intramuscular quinine also gave predictable profiles in a population pharmacokinetic study of children with severe malaria (140). Details of the pharmacokinetic interactions of quinine with other antimalarial agents such as mefloquine, sulfadoxine/pyrimethamine, antibiotics, and primaquine have been reported (137).

The mechanism of action of quinine may involve inhibition of heme degradation, discussed in detail in the section on chloroquine. Quinidine inhibits hemozoin formation with an $IC_{50}$ value of 90 mM, quinine of 300 mM, and epiquinine of >5 mM, the order that corresponds to their antimalarial activity (32). Quinine binds to both serum albumin and $\alpha$1-acid glycoprotein (AAG) in human plasma (141). In malaria patients, the plasma concentrations of AAG are increased and the amount of quinine binding to AAG is consequently higher than in non-infected subjects. In addition, the hepatic disposition of both quinine and quinidine is altered in malaria patients from that seen in uninfected subjects (142, 143).

Efforts have been made in the last decade to clarify issues related to appropriate dosage regimens for quinine. Given that the main therapeutic use of quinine and quinidine in malaria is in severe cases of falciparum malaria, the rapid establishment of effective drug concentrations is essential. The use of a loading dose of quinine or quinidine has become standard practice, even in children where faster recovery from coma and clearance of parasitemia were observed (144). Both drugs have a low therapeutic index and concerns about toxicity intrude. Children younger than 2 years old represent a large percentage of the cases of severe malaria in Africa. Pharmacokinetic studies in children with severe malaria indicate that younger children may be more susceptible to quinine toxicity than older children (145).

The symptom complex called cinchonism is associated with the use of the cinchona alkaloids. Excess quinine or quinidine can induce the symptoms consisting of hearing loss, tinnitus, visual disturbances, rashes, vertigo, nausea, vomiting, and central nervous system changes including headache, confusion, and loss of consciousness. Tinnitus is one of the early signs of quinine toxicity and occurs in nearly all persons whose plasma concentrations of drug are in excess of 5 mg/l. In sensitive individuals, a single dose can cause problems; more commonly, persons experience cinchonism after a prolonged course of treatment. The symptoms resolve when administration of the drug ceases. However, several cases of irreversible sensorineural hearing loss and tinnitus have been reported.

Avoidance of cardiotoxicity is a primary consideration in risk-benefit assessments for using cinchona alkaloids in severe malaria. Quinine may cause myocardial depression, prolongation of QTc interval on the electrocardiogram, and peripheral vasodilatation. It may also cause initial generalized stimulation of the central nervous system leading to fever, delirium, and increased ventilatory rate. Quinine cardiac effects are amplified by concurrent administration of prochlorperazine, a resistance-reversing agent (146).

Quinidine has more potent intrinsic antimalarial properties than quinine. Since 1991, quinidine gluconate has been the only parenteral antimalarial available for use in the United States. It is indicated for the treatment of patients with life-threatening falciparum malaria. As newer anti-arrhythmics have replaced quinidine for many cardiac indications, some health care facilities have dropped quinidine gluconate from their formularies. The limited availability of and delays in obtaining quinidine gluconate have contributed to adverse patient outcomes (147). The pharmacokinetic data available for quinidine have been reviewed (121). Quinidine has more pronounced cardiac effects than quinine and has a greater tendency to induce hypotension. Given the low therapeutic index of quinidine, careful monitoring of cardiac function is required.

Quinine, like many other antimalarial agents, is photosensitizing; these properties have been reviewed (148). Electron paramagnetic resonance (EPR) studies indicated that oxygen and carbon-centered free radicals were formed during photolysis.

Hypoglycemia is another importance complication associated with cinchona use. Both quinine and quinidine stimulate the release of insulin from the pancreas. The degree of hyperinsulinemia varies greatly from patient to patient, and pregnant women are particularly susceptible to hypoglycemia resulting from excess insulin. Blood glucose levels should be monitored during quinine treatment. Children with severe malaria often present with hypoglycemia as well, even without prior quinine administration.

## 2.2 Chloroquine

Chloroquine (**5**), the main drug among the 4-aminoquinoline class, is one of the most successful antimicrobial agents ever produced

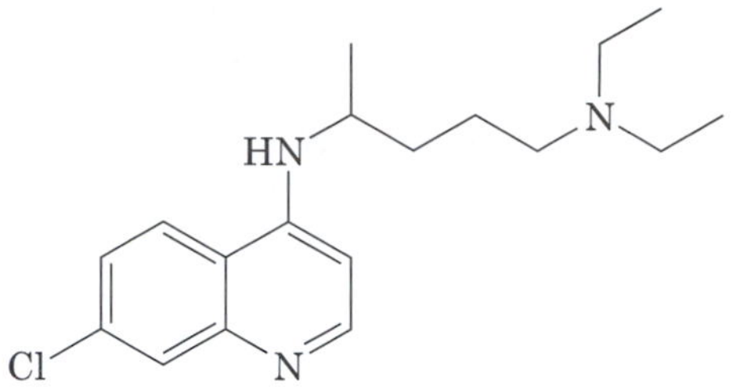

**(5)**
chloroquine

and has had an enormous positive impact on human health. Chloroquine was synthesized in 1934, but at the time it was considered too toxic for human use based on studies of avian malarias and on limited clinical trials. At the outbreak of the Second World War, chloroquine was revisited and its wide-scale use in malaria treatment and prevention began. After decades of use as a first-line therapeutic and prophylactic agent, it still finds use in certain parts of the world more than 65 years after it was first prepared.

Chloroquine remains the favored agent for susceptible malaria. Hydroxychloroquine (**6**) is also active and available, although much

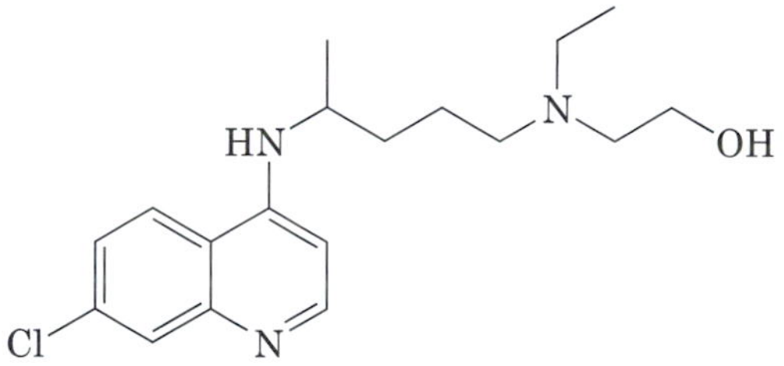

**(6)**
hydroxychloroquine

less widely used, except in the United States. Commercially available chloroquine is racemic. The main metabolite is desethylchloroquine, which is equally active as chloroquine in sensitive-strains of *P. falciparum*. In resistant strains, the metabolite is significantly less active. Chloroquine acts on the intra-erythrocytic stages of the parasite life cycle; at therapeutic concentrations, it has no effect on the sporogonic, exoerythrocytic, and mature sexual stages. Chloroquine was found to increase the gametocytogenesis *in vitro* of *P. falciparum* (149, 150); this observation may explain the relative ineffectiveness of chloroquine therapy to reduce malaria transmission, because there is a positive relationship between gametocyte densities and infectivity to mosquitoes. A review of the pharmacokinetics of 4-aminoquinolines as well as information about pharmacokinetic interactions of chloroquine with other antimalarial agents may be found in Ref. 137. The clinical pharmacokinetics of chloroquine have been reviewed (121, 151), and some reviewers have specifically examined the pharmacokinetics of chloroquine against *P. vivax* (152, 153). The pharmacodynamics of chloroquine in severe malaria has been studied (154). Chloroquine in pregnancy is discussed in Ref. 155. Chloroquine was found to be mutagenic but not clastogenic using mammalian *in vivo* assays (156).

*P. ovale* and *P. malariae* are fully susceptible to chloroquine in all geographic areas at present. For more than 40 years, chloroquine had been nearly 100% effective in treating blood stage infections of *P. vivax*. Although drug resistance of *P. falciparum* is an expected reality of antimalarial therapy, the resistance of *P. vivax* to drugs is comparatively recent and has been documented in Papua New Guinea (157), Myanmar (158), Sumatra (159), Sulawesi (160), Irian Jaya (161, 162), and South America (163). Further studies are needed to find alternatives to chloroquine for treatment and prophylaxis of vivax malaria.

*P. falciparum*, on the other hand, is susceptible to chloroquine in only a few areas, and those are likely to decrease with time (see Section 1.5). As an example of the consequences of increasing resistance, a prospective study of malaria mortality from 1984 to 1995 in Senegal suggested that the emergence of chloroquine resistance was accompanied by an increase in the risk of malaria death, especially among children (164). In sub-Saharan Africa, many countries are faced with increasing levels of chloroquine resistance. Some have already replaced chloroquine as the first-line therapy while others are assessing the current status of drug resistance and making national policy-level decisions. Standardized protocols and processes of making such decisions have been suggested (165).

Chloroquine is one of the most widely available and widely consumed drugs. It has been estimated that self-medication may account for as much as one-half the consumption of antimalarial drugs, most often chloroquine. Self-medication, inadequate dosing, and subtherapeutic blood levels are frequent and are believed to be predominate factors contributing to the development of chloroquine resistance. In Africa, many patients presenting for treatment have evidence of recent intake of chloroquine. A high prevalence of sub-curative chloroquine in the blood of Nigerian children and considerable chloroquine resistance were confirmed (166). One study has shown a beneficial effect on clinical outcome from chloroquine self-medication perhaps because of the inhibitory action of chloroquine on cytokine production, particularly tumor necrosis factor or nitric oxide, both of which are implicated in the pathogenesis of cerebral malaria (167), although other studies have not documented this phenomenon (168).

Even though chloroquine has been used for years as a therapeutic and prophylactic agent, improvements in the dosing regimens are continuously sought. In chemoprophylaxis, lower doses of chloroquine taken daily were shown to provide adequate drug concentrations that were as good as higher doses weekly or twice weekly (169). Compliance may be better with a daily tablet either as a single agent or in a combined formulation with proguanil. A prophylactic regimen of a loading dose followed by weekly chloroquine provided protection in the first week of dosing and better compliance; no post-exposure dosing was needed (170). Use of chloroquine prophylaxis in pregnancy provided a protective effect for the fetus such as a lower risk of neonatal infection, higher birth weights, and fewer perinatal deaths (171). The side effects associated with chloroquine use are blurring of vision, vertigo, ocular toxicity, pruritis, nausea, and headache. Hypotension and cardiovascular toxicity have been seen in children who were treated by intramuscular injection. The drug can induce cardiac conduction disturbances, especially in extended use (172).

Over the years many proposals have been put forward on the mechanism of action of chloroquine and related antimalarials. A summary of these may be found in Refs. 28, 173, and 174. Most researchers in the field have focused attention recently on the inhibition of hemozoin formation by quinoline agents. For a brief discussion of the parasite's pathway for the metabolism of hemoglobin and the production of hemozoin, see Section 1.3.

Considerable evidence supports the proposition that inhibition of hemozoin formation is central to the mechanism of action of quinoline antimalarial agents (174–176). Much of the data on the effect of chloroquine on hemozoin formation was acquired when hemozoin/β-hematin was believed to be a polymer of hematin. Previous literature therefore describes the action of chloroquine as inhibiting heme "polymerization." Now that it has been shown that hemozoin is a crystalline dimer of hematin rather than a polymer (30), it would be more appropriate to discuss chloroquine's action on hemozoin formation, hematin dimerization, or heme sequestration.

The location of action of chloroquine is the parasite food vacuole, an acidic organelle wherein hemoglobin is digested. Chloroquine accumulates in the vacuole with an absolute dependence on the existence of a pH gradient between the acidic vacuoles and the extracellular medium, demonstrated in *P. falciparum*-infected human erythrocytes (177, 178). At pharmacological concentrations of chloroquine ($10^{-9} M$), drug concentrations in the food vacuole can reach millimolar levels at physiological pH. Compounds that inhibit the degradation of hemoglobin were found to be antagonistic to the action of chloroquine, quinine, amodiaquine, mefloquine, and halofantrine (179) supporting the view that this group of antimalarials is acting downstream of hemoglobin digestion in the food vacuole. That site of action appears to be the heme or hematin released by hemoglobin degradation. Chloroquine binds with high affinity to free heme (180), an observation that led to the original proposal that chloroquine action is mediated by binding to heme. Details of the binding of

chloroquine with hematin have been studied using isothermal titration calorimetry (181). In addition, chloroquine accumulates bound to hemozoin and this binding is dependent on the presence of free heme (182). The binding of chloroquine/heme to hemozoin is saturable, specific, and has high affinity (39). A range of quinoline structures can compete for the binding, suggesting a common mode of action involving drug-heme-hemozoin interactions. Saturable chloroquine uptake at equilibrium is caused solely by the binding of chloroquine to hematin rather than to active uptake (183). In a study of chloroquine analogs, the structural features of binding to hematin was explored (184, 185). An electron withdrawing group at the 7-position of the quinoline nucleus, in particular a chloro group, was required for inhibition of both hemozoin formation and parasite growth.

The mere binding of chloroquine to hematin may prevent the sequestration of hematin into hemazoin (36, 186, 187). Hemozoin formation in not mediated by a classical enzymic process and no protein is required (34). The inhibition of both spontaneous and parasite extract-catalyzed β-hematin formation by chloroquine and other 4-aminoquinolines has been confirmed *in vitro* by a number of researchers (34, 187–189). Compounds with little or no antimalarial activity such as 9-epiquinine did not affect β-hematin formation. Heme sequestration activity may be caused, however, by the presence of preformed hemozoin in parasite extracts that act as nucleation sites for further dimerization (34, 190). The antimalarial action of a variety of quinoline agents correlates very well with their ability to inhibit hemozoin formation (174). The antimalarial activity of chloroquine and other quinolines has been shown to be a function of both the accumulation of the drug to appropriate concentrations at the site of action, the food vacuole, and interference with the process of hemozoin formation (191). By delaying the sequestration of heme, chloroquine may allow either hematin or a hematin-chloroquine complex to exert toxic effects (192–196).

However, there is good evidence (38) that heme binding proteins may also play a role, serving as a structural focus for the initiation of heme sequestration. *P. falciparum* histidine-rich protein-2 (Pfhrp-2) is a major heme-binding protein in the parasite (40). There seems to be at least 18 heme binding sites on Pfhrp-2, perhaps associated with a recurrent hexapeptide repeat unit. Pfhrp2 promotes the conversion of heme to β-hematin under conditions that mimic the parasite food vacuole (197). Chloroquine was shown to inhibit the binding of heme to Pfhrp-2, although there was no detectable interaction of chloroquine with the protein in the absence of heme (40). Perhaps binding of heme to Pfhrp-2 is a necessary preliminary step before dimerization to hemozoin.

This may not be the complete picture however. Association of the aminoquinolines with hematin was shown to be necessary but not sufficient for inhibiting formation of β-hematin. Further, inhibition of β-hematin formation was necessary but not sufficient for antiparasitic activity (185). And inhibition of hemozoin formation may not be the only affect that chloroquine has on malaria parasites. There is evidence that chloroquine also interferes with the degradation of heme by peroxidation processes and that this might contribute to the observed lethal effects of the drug (198).

Given chloroquine's long history as a safe, well-tolerated, effective (until relatively recently), and affordable drug, understanding the details of the mechanisms of parasite resistance to chloroquine is critical. In chloroquine-resistant *P. falciparum*, the accumulation of chloroquine inside the parasite food vacuole is reduced (178, 182, 183, 199). This has been attributed to a rapid efflux of drug rather than decreased uptake. Verapamil inhibits multi-drug resistance in mammalian tumor cells and was found partially to reverse chloroquine resistance in malaria parasites as well (200). This action is mediated by P-glycoprotein in the multi-drug resistant tumor cases, a protein coded by the gene *mdr*. Reversal of resistance is specific for genetically resistant parasites and does not result simply from a change in the pH gradient from external medium to vacuole (201). A number of other agents have been shown to have similar resistance-reversing behavior (202–205). A few studies demonstrated benefits from combining chloroquine with a resistance reversal agent in animal models (206) and in the clinic (207). For a review of the reversal of chloroquine resistance, see Ref. 208. Highly-resistant lines of *P. berghei*, however, were associated with a loss of drug potentiation by verapamil (209), suggesting that parasites are able to avoid the activity of the reversing agents.

The parasite version of *mdr*, the *P. falciparum* multi-drug resistance gene *pfmdr1*, which encodes a P-glycoprotein homologue (Pgh1), has been proposed as the determinant of chloroquine resistance, and an association has been reported between resistance to chloroquine and amplification or mutation of *pfmdr1* (210, 211). The chloroquine-resistance phenotype was dissociated from inheritance of the *pfmdr1* gene, located on chromosome 5, in genetic studies, however (212–215), suggesting that other factors were most likely involved in the molecular mechanism of resistance. Some studies found an association between *pfmdr1* mutations and chloroquine resistance in field isolates (216–220) while others did not (221–228). It was shown recently that chloroquine-sensitive *P. falciparum* parasites that acquired *pfmdr1* mutations in transformation experiments did not become resistant to chloroquine (136). A genetic cross allowed the mapping of the locus for chloroquine resistance to a segment of chromosome 7. One gene on chromosome 7, *cg2*, and its polymorphisms were highly associated with chloroquine resistance (229–232), but allelic modification experiments failed to establish *cg2* as playing a role in chloroquine resistance (233).

Recently, a new gene of interest, *pfcrt*, was identified near *cg2* on chromosome 7 (234), a location that had been implicated earlier in chloroquine resistance (235). This gene encodes PfCRT, a transmembrane protein that seems to function as a transporter in the parasite food vacuole. Point mutations in *pfcrt* were completely associated with chloroquine resistance *in vitro* in laboratory lines of *P. falciparum* from Africa, South America, and Asia (234). One mutation, a replacement of lysine with threonine at position 76, was present in all resistant isolates and absent from all sensitive isolates tested *in vitro*. Transformation experiments demonstrated that mutant forms of *pfcrt* were able to confer chloroquine resistance in clones originally sensitive to chloroquine. These studies point to a key role for the *pfcrt* T76 mutation in *in vitro* chloroquine resistance (236).

Many researchers then began to examine the role of *pfcrt* mutations in clinical cases of chloroquine resistance. There was a significant association, often complete, between the T76 mutations of *pfcrt* with *in vitro* and *in vivo* response in field isolates from Mali (237), Sudan (238), Uganda (239), Cameroon (240), Mozambique (241), Nigeria (242), Indonesia (243), Laos (243), Papua New Guinea (244), and non-immune travelers returning to France (245). Where the correlation was less than perfect, researchers suggested this as most likely caused by the presence of multiclonal infections. Mutations in *pfcrt, pfmdr1,* and *pfdhfr* (the dihydrofolate reductase gene) were frequent in Ghana and often occurred together (246). A number of haplotypes of *pfcrt* in the region of residues 72–76 were identified in isolates from Papua New Guinea and all contained threonine at residue 76 (244). The *pfcrt* T76 mutation has been suggested as a molecular marker for chloroquine resistant *P. falciparum* (237). Recent field studies showing the role of mutations in *pfcrt* in chloroquine resistance for *P. falciparum* have been reviewed (247).

Another class of agents does not restore chloroquine accumulation but acts by some other mechanism to reverse chloroquine resistance (248, 249). The prototype of this group is CDRI 87/209 (**7**). It is only weakly anti-plas-

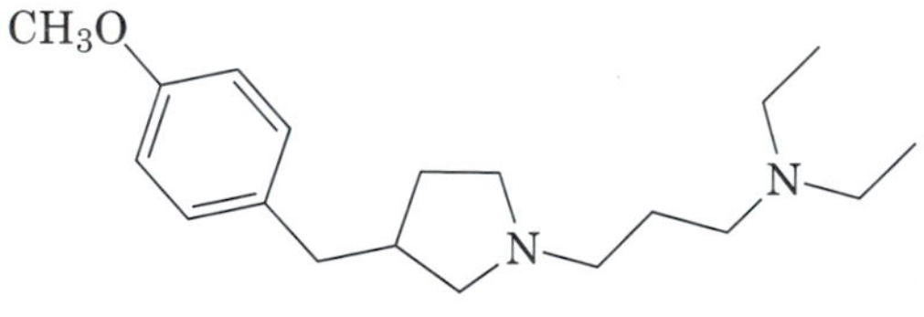

(7)
CDRI 87/209

modicidal alone, but an increase in drug susceptibility was observed only in resistant parasites and not in sensitive strains. Rotationally constrained analogs of CDRI 87/209 were prepared and evaluated (250). A possible mode of action for the most active compound

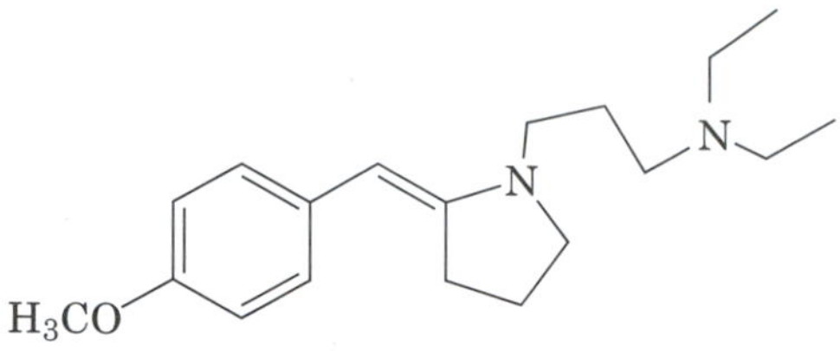

(8)

in the series (**8**) relates to the heme degradation pathway and more specifically to heme oxygenase of the chloroquine-resistant parasite. Drug treatment leads to increased concentration of heme in the food vacuole and to increased levels of a heme-chloroquine complex resulting in toxicity to the parasite. Along similar lines, a pyrrolidino alkaneamine, WR268954 (**9**), is active *in vitro* as a chloro-

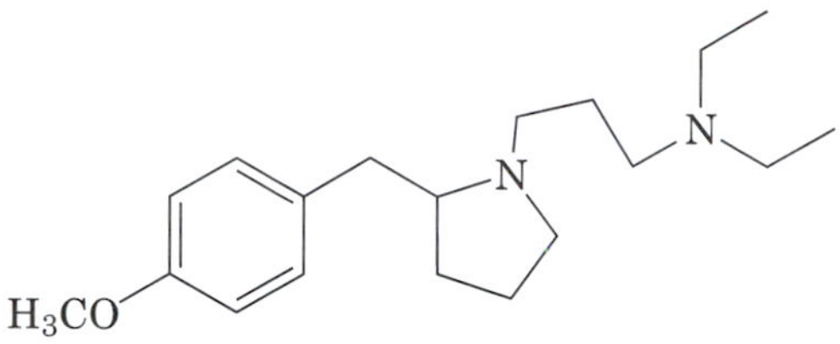

(9)
WR268954

quine resistance modulator (251). Besides increasing the effects of chloroquine in resistant strains, the effects of quinine were also increased; mefloquine resistance was not affected.

Efforts continue to identify new aminoquinolines that would be effective against chloroquine-resistant strains and to clarify the structural features needed for good activity (173, 252, 253). The length of the chain between the aminoquinoline and the terminal tertiary amine was found to be critical. Replacement of the 7-chloro group of chloroquine with a 7-bromo or 7-iodo group and altering the side-chain length yielded aminoquinolines with good activity against resistant strains *in vitro* (254). Another strategy was to link two chloroquine moieties by bisamide units (255). Although the observed *in vitro* activity was not particularly high, enhanced activity against chloroquine-resistant strains was noted. Strategies to prolong the usefulness of chloroquine have included exploring some metal-chloroquine combinations. A complex of gold with chloroquine was found to be active *in vitro* against chloroquine resistant strains of *P. falciparum* and *in vivo* against *P. berghei* in mice (256). Chloroquine analogs incorporating a ferrocene unit have been prepared and evaluated for antimalarial activity (257–259); the metallocene compounds had promising antimalarial activity *in vivo* in mice and *in vitro* against chloroquine-resistant strains of *P. falciparum*. Complexes of chloroquine with either rhodium or ruthenium were synthesized and evaluated for antimalarial activity (260). The complex with ruthenium was more active than chloroquine against *P. berghei* and against chloroquine-resistant *P. falciparum* and had low toxicity.

## 2.3  Mefloquine

In 1963 the Malaria Research Program was re-established at the Walter Reed Army Institute of Research. The stated goal was the identification and development of new antimalarial agents with particular regard for those compounds that would be effective against drug-resistant, especially chloroquine-resistant, strains of *P. falciparum*. Mefloquine (**10**), a 4-quinolinemethanol structurally re-

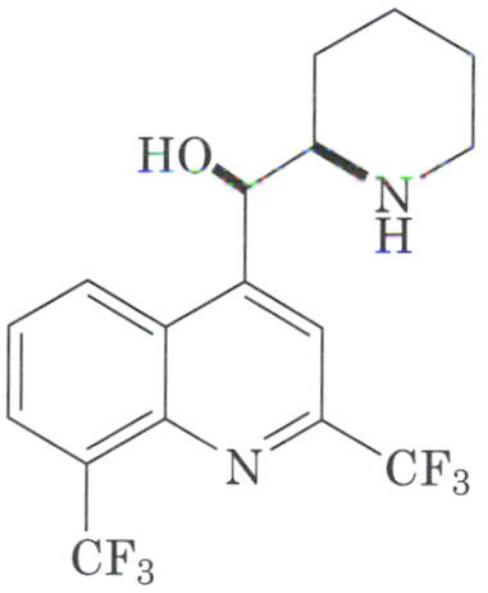

**(10)**
mefloquine

lated to quinine, was a product of that effort and was selected from a group of nearly 300 quinoline methanol agents because of its high activity in animal models. A brief history of the route that led to mefloquine is presented in Ref. 261. Mefloquine was first marketed in

1985 and is perhaps the most studied of all the antimalarial agents. Several excellent and comprehensive reviews of mefloquine have appeared (261–264). A review of physical and pharmaceutical properties includes spectroscopic data (265). The relationship between electronic structure and antimalarial activity for mefloquine and some substituted derivatives has been explored (266, 267).

The synthesis of racemic mefloquine (WR142,490) was first reported in 1971 along with isomers that differed in the position of the trifluoromethyl groups around the quinoline system (268). The 2,8-bis-(trifluoromethyl) arrangement proved to be the most active of the series. Although compounds that possessed an aryl group at the 2-postion of the quinoline ring were found to have augmented antimalarial activity, this structural feature also conferred unacceptable levels of phototoxicity. The replacement in mefloquine of the 2-aryl group with the 2-trifluoromethyl group allowed the compound to retain high activity without photo-toxicity. Hydrogen bonding between the amine and the hydroxyl is critical in the active conformations of mefloquine and related structures. A detailed discussion of structure-activity relationships (SAR) in arylmethanols can be found in Ref. 264. Attempts at improving the synthetic pathway to mefloquine have included the use of a Wittig rearrangement (269), through a cyanohydrin intermediate (270), and by way of a sulfoxide-Grignard (271). Nor-mefloquine, in which the piperidinyl ring has been replaced by a pyrrolidinyl ring, displayed $ED_{50}$ values equivalent to those of mefloquine (272). An interesting analog of mefloquine in which the piperidine ring is replaced by a quinuclidine ring was reported, but no biological data were given (273).

Mefloquine is orally active and effective against the intra-erythrocytic stages of the plasmodium life-cycle. It has no effect on mature gametocytes. Although mefloquine does cause recognizable changes in pre-erythrocytic forms, the effects are insufficient to prevent further development of the tissue parasite. Thus, mefloquine is not a causal prophylactic agent (274). It is a relatively slow-acting drug with greatest activity against the more mature blood stages. Mefloquine approaches the ideal antimalarial agent in that, in areas with low levels of resistance to mefloquine, a single dose of the agent is effective in accomplishing a cure. A decided drawback is that mefloquine costs about 100 times as much as chloroquine.

The mechanism of action of quinoline-containing compounds has been the focus of much research. Despite extensive scrutiny, the exact mode of action of mefloquine is not known. Its effects are confined to the pathogenic blood stages of the parasite. Most recent studies support the proposal that mefloquine interferes with the parasite's ability to detoxify heme. Based on ultra-structural changes observed *in vitro*, the parasite food vacuole is the primary target of mefloquine action and the affinity of mefloquine for free heme is high. Like chloroquine, mefloquine has been demonstrated to inhibit heme sequestration. For a more complete discussion, see Section 2.3.

Mefloquine is metabolized in the liver and excreted in bile and feces, predominately. The main metabolite of mefloquine is carboxymefloquine, a 4-carboxylic acid derivative. This metabolite is present in plasma in concentrations that exceed that of the parent drug by up to threefold. Carboxymefloquine seems to be inactive against *P. falciparum* in culture (275). Quinine and ketoconazole were found to inhibit the metabolism of mefloquine in human liver microsomes (276).

In two studies, the enantiomers of mefloquine [(+)-11$R$,2'$S$ and (−)-11$S$, 2'$R$] were found to be equally active *in vitro* against *P. falciparum* (277) and *P. yoelii* (278). In another, the enantiomers of mefloquine and its threo isomer (WR 177,602) displayed up to a twofold difference in potency (120). The pharmacokinetics of the mefloquine enantiomers in whole blood were observed to be stereoselective both in adults (279) and in children (280), although uptake occurs in a non-stereospecific manner (281). The half-life of the (−)-enantiomer was significantly longer than of the (+)-enantiomer. Brain penetration of (+)-mefloquine is much higher than that of the (−)-enantiomer (282).

Several reviews of mefloquine pharmacokinetics have appeared (137, 262, 283, 284). All the pharmacokinetic studies have been carried out with oral dosage forms because there

is no parenteral formulation available. Generic mefloquine tablets were found not to be bioequivalent to the reference tablet in one study (285, 286). The drug shows marked differences in pharmacokinetics between healthy volunteers and patients and between some ethnic groups. Absorption of mefloquine is generally good and is enhanced by food (287). The elimination half-life of mefloquine is relatively long (14–28 days), and a single day treatment is generally effective. The long half-life also presents the opportunity for resistant strains to arise. The drug does not seem to accumulate with repeated administration.

In animal studies, mefloquine displays low toxicity. In humans, however, the incidence of minor adverse effects seems to be high. A series of non-serious adverse effects have been noted with mefloquine treatment including vomiting, dizziness, nausea, anorexia, dermatological conditions (288), and general malaise (289–291). Early vomiting was correlated with poor treatment outcome, despite re-administration of the drug. Splitting the dose of mefloquine reduced the incidence of adverse effects. Travelers seem to report adverse effects from mefloquine more often than do malaria-experienced populations. The safety of mefloquine in pregnancy has been reviewed (155). In a retrospective study of mefloquine treatment in pregnancy along the western border of Thailand, it was found that women who received mefloquine during, but not before, pregnancy had a significantly greater risk of stillbirth than those exposed to other treatments or those who had no malaria (292). The association remained even after adjusting for all identified confounding factors; mefloquine was not associated with abortion, low birth weight, neurological retardation, nor congenital malformation.

Serious neurological or psychiatric reactions (hallucinations, neurosis, affective disorders, anxiety, and depression) have been associated with mefloquine use (293). Most of the reports relate to its prophylactic use; however, the incidence of serious reaction is higher when mefloquine is used for treatment rather than prophylaxis. The risk of serious neuropsychiatric reaction was seven times higher if mefloquine was given to treat a recrudescent infection that had originally been treated with mefloquine. In one case, neuropsychiatric symptoms were presented after a single dose of mefloquine (294). A combination of mefloquine prophylaxis and ethanol consumption resulted in two episodes of severe psychiatric disturbance in one individual (295). Seizure has also been reported as resulting from mefloquine prophylaxis (296). The drug clearly should not be used in patients with a history of neuropsychiatric disturbances.

To delay the development of resistance to mefloquine, it was recommended that mefloquine be used only in combination. In particular, the combination of mefloquine with pyrimethamine and sulfadoxine (Fansimef) has been widely used in Southeast Asia. Initial clinical trials confirmed the efficacy of the triple combination in preventing and treating drug-resistant falciparum malaria and in delaying the development of mefloquine resistance (297). There seemed to be no pharmacokinetic interaction between the three components (298), and the pharmacokinetics have been reported (299, 300). Unfortunately, the strategy has failed and over a 6-year period, during which mefloquine was used only in the triple combination, resistance to mefloquine increased rapidly. Given the possibility of severe dermatological reaction to the pyrimethamine/sulfadoxine components and the apparent lack of therapeutic advantage at present over mefloquine alone (301) or pyrimethamine/sulfadoxine alone (302), the triple combination is no longer recommended for treatment. Even so, a trial in Nigeria found a low dose regimen of mefloquine/pyrimethamine/sulfadoxine to be significantly more efficacious than chloroquine alone, which is perhaps not surprising given the level of chloroquine resistance in that area (303).

Mefloquine combinations with a variety of artemisinin-derived agents have been the subject of study especially in areas where multi-drug resistant strains of *P. falciparum* are notoriously difficult to treat. The fast-acting artemisinin derivative clears a large percentage of parasites quickly and then the longer circulating mefloquine can act on any remaining parasites. The combination increases cure rates in falciparum patients, reduces transmission, and may slow the development of resistance. Pharmacokinetic studies support the *in vitro* and *in vivo* synergistic antimalarial activity of artemisinin derivatives and mefloquine (304). Artesunate (305, 306), artemether (307), artemisinin (308), and dihydroartemisinin (309) have all been used in combination with mefloquine (310, 311). Two-day courses of treatment have better success than single-day regimens. In many cases, the overall efficacy of the combination was better than with mefloquine alone or than other therapies such as quinine/tetracycline. In studies of the pharmacokinetics of artemisinin derivatives with mefloquine, the bioavailability of mefloquine was reduced in the presence of either artemether or artesunate (312, 313). A sequential schedule of dosing, an artemisinin followed by mefloquine, may be necessary for maximum efficacy. Other mefloquine combinations include those with tetracycline and with pyrimethamine/sulfadoxine/tetracycline (314, 315).

Mefloquine is the only newer antimalarial that is recommended for use as a prophylactic agent. In travelers and malaria-naive persons temporarily exposed to the risk of drug-resistant falciparum malaria, weekly mefloquine seems to be the agent of choice. In a comparison of mefloquine with chloroquine/proguanil for prophylaxis, several studies found both regimens to be well-tolerated and effective. A higher withdrawal rate was noted by some researchers for those taking mefloquine, presumably caused by side effects (316), whereas others detected no significant difference in side effects (317) or in compliance (318) between the groups. None of the subjects developed serious neuro-psychosocial reactions in one report (317), whereas another observed significant differences (319). In an assessment of the risks of mefloquine prophylaxis, one author concluded that mefloquine is the most effective chemoprophylactic drug for much of the tropics, that it is as well-tolerated as less-effective alternatives, and that it rarely causes of life-threatening adverse effects (320). As with mefloquine treatment, mefloquine prophylaxis is clearly contraindicated in persons with a history of psychiatric illness or seizures.

As with all drugs studied, stable resistance to mefloquine could be induced in sensitive strains by applying continuous drug pressure (321). Mefloquine resistance as a clinical problem had been confined to specific areas of Southeast Asia. Up to a 10-fold decrease in sensitivity to mefloquine was observed since 1984 on the Thai-Myanmar border (322). In these multi-drug resistant areas of Thailand, mefloquine prophylaxis was not 100% effective, however, and continued monitoring of the levels of mefloquine sensitivity in such areas will be important (323). Although mefloquine continues to provide a high cure rate in Brazil, cases of emerging resistance have been documented (324). In contrast, mefloquine exhibited 100% growth inhibition against wild isolates from Tanzania in an *in vitro* study (325), and it was effective in areas with high levels of resistance to chloroquine and pyrimethamine/sulfadoxine (326). When resistance emerges, increasing the dose of mefloquine will bring temporary respite but will undoubtedly be associated with increased adverse effects. There is some evidence that resistance to mefloquine can be conferred by mutations in Pgh1, the protein product of *pfmdr1* (136) or by an increase in *pfmdr1* copy number (327).

Selection for mefloquine resistance in *P. falciparum* may be linked to amplification of the *pfmdr1* gene (210, 211, 328) and to cross-resistance to halofantrine and quinine (222, 329, 330), although the data are mixed. In some studies, a clear association is found between increased *pfmdr1* gene copy number (327), *pfmdr1* sequence polymorphism (331), or mutations in Pgh1, the protein product of *pfmdr1*, (136) and mefloquine resistance. But results from other studies suggest that amplification and over-expression of *pfmdr1* are not involved in the increased mefloquine resistance phenotype (332). But, none of these factors is a prerequisite for mefloquine resistance and other mechanisms must also be able to mediate resistance to mefloquine (333). The exact relationship of *pfmdr1* to mefloquine resistance and any clinical significance thereof are not known.

## 2.4 Halofantrine

Phenanthrenemethanols were first discovered to possess antimalarial activity during the drug discovery efforts of World War II. It was

not until 20 years later that further studies were undertaken on this class of compounds. The first two members of the series that were examined in depth were WR33,063 (**11**) and

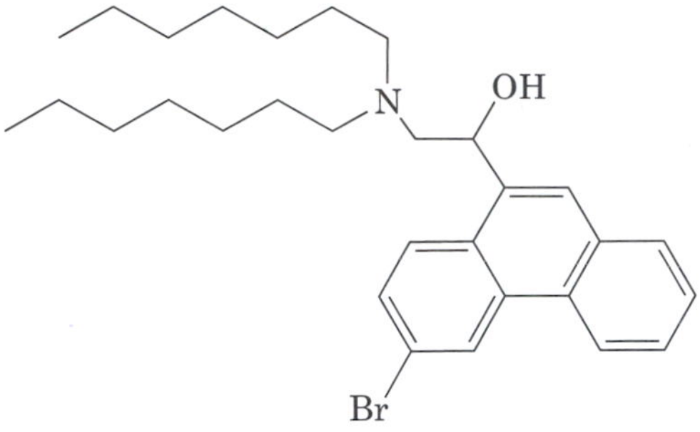

**(11)**
WR 33,063

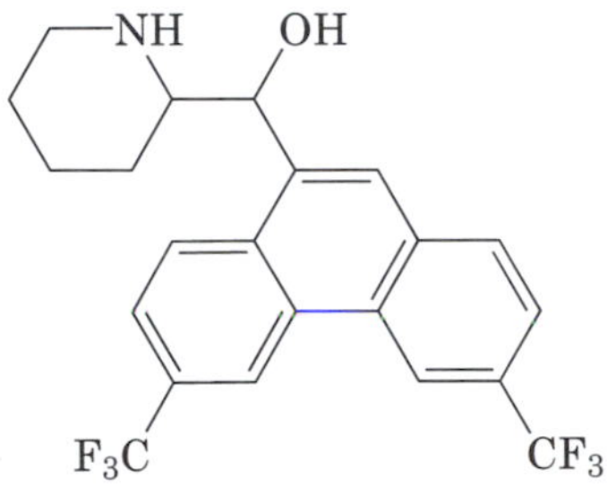

**(12)**
WR 122,455

WR122,455 (**12**), both of which displayed good activity in chloroquine-resistant strains, but high cure rates required multiple doses over several days. Halofantrine (**13**, WR 171,669),

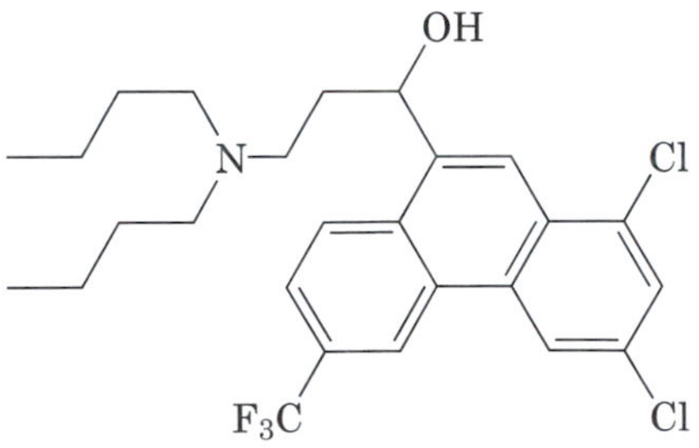

**(13)**
halofantrine

originally synthesized by Colwell et al., (334) was the third synthetic phenanthrenemethanol considered. In structure-activity relationship studies, it was determined that the dibutylaminopropanol side-chain was optimal for the substituted 9-phenanthrene system. Initial human trials confirmed that halofantrine was a promising new agent for the treatment of malaria (335). A study of the molecular electronic properties of carbinolamine antimalarial agents, including halofantrine, indicated that potent compounds share specific molecular electronic properties regardless of the type of aromatic ring and of the nature of the amine side chain, whether cyclic or aliphatic (266). Reviews of the antimalarial activity, pharmacokinetic properties, and therapeutic potential of halofantrine have appeared (283, 336–340).

Halofantrine was placed onto the market in 1988 by Smith, Kline & French, with its first introduction being in Ivory Coast. Halofantrine is marketed as the racemate; (−)-halofantrine is the S isomer as determined by X-ray (341). The separated enantiomers display similar antimalarial activity against P. berghei in mice and in vitro (277, 342). Metabolic studies have shown that on incubation of (±)-halofantrine with either rat liver homogenate (343) or in humans (344), (−)-halofantrine was metabolized preferentially. The primary metabolite is N-desbutylhalofantrine, either enantiomer of which is pharmacologically active in vitro (345, 346). Significant differences were noted in the plasma concentrations of the enantiomers of halofantrine; (+)-halofantrine and its N-monodesbutyl metabolite had higher values than (−)-halofantrine and the corresponding metabolite (347). Minor metabolites have also been identified (348). Halofantrine is metabolized primarily by CYP3A4 but CYP3A5 is also involved; inhibitors or other CYP3A4 substrates, such as ketoconazole, might serve to increase halofantrine concentrations, and consequently, the associated cardiotoxicity (349). Quinine and quinidine, when incubated with halofantrine in human hepatic microsomes, inhibited its metabolism (350). One the other hand, halofantrine is an inhibitor of CYP2D6 activity in healthy Zambians reinforcing a concern over concomitant dosing with agents that are metabolized significantly by CYP2D6 (351).

Halofantrine is schizontocidal with selective activity against the erythrocytic stages of Plasmodia, especially the more mature parasite stages. The drug accumulates in parasitized red blood cells with concentrations up to 60% higher than in uninfected cells; erythrocytes with mature parasites showed the highest accumulations (352). There is no observed action on gametocytes (353) nor hypnozoites.

Clinical trials with halofantrine have confirmed the efficacy of the drug against multidrug resistant strains of P. falciparum and against P. vivax; in the few cases of P. ovale and P. malariae in which halofantrine has been used, good results were obtained with no recrudescence (338). The relative efficacy of halofantrine compared with mefloquine varies with the test system but the effective doses in humans in sensitive strains were similar. The precise dosage regimen seems to be of critical importance. A standard 1-day regimen of halofantrine, even with larger doses, was inadequate therapy for falciparum malaria in multi-drug resistant areas such as the Thai-Burmese border (354). Single-dose treatments, even with larger doses, were plagued by recrudescence. On the other hand, multidose regimens have been studied with good results. One recommended halofantrine regimen is 24 mg/kg followed by a second treatment 7 days later, but cardiac complications are more common after the second treatment. Lowering the follow-up dose to 250 mg, which may reduce the risk of cardiac side effects, was shown to be effective in non-immune people with mild-to-moderate falciparum malaria (355). Treatments 7 days apart offer problems with compliance, however. In a randomized, comparative dose-finding trial in Colombia, the best cure rates were obtained with three doses of 500 mg of the hydrochloride at 6-h intervals (356); the study was carried out on patients from an area known to contain multi-drug resistant P. falciparum. A combination of halofantrine with primaquine was significantly more effective against falciparum malaria than a chloroquine/primaquine combination and similarly effective against vivax in Irian Jaya, Indonesia (357). Fever cleared much faster and was associated with a more rapid and significant decline in malaria-related physical complaints.

As is the case with all antimalarial drugs, the incidence of resistance and cross-resistance is important (330). In one study, halofantrine displayed activity against lines that were primaquine-, cycloguanil-, pyrimethamine-, and menoctone-resistant (358). But parasites resistant to quinine, chloroquine, mefloquine, and amodiaquine showed resistance to halofantrine as well. An artemisinin-resistant organism has a somewhat reduced response to halofantrine. In an *in vitro* study of *P. falciparum* sensitivity in Burkina Faso, resistance to halofantrine was discovered in 1% of the isolates in 1995 and 9.6% in 1996 (133). A single halofantrine resistant isolate was also resistant to chloroquine in 1996. A fourfold decrease in halofantrine sensitivity was noted in Gabon from 1992 to 1994 in *P. falciparum* isolates (359) while halofantrine remained effective in Nigeria in an area where chloroquine sensitivity continued to drop (360). In one set of experiments, researchers uncovered evidence that mutations in Pgh1, the protein product of *pfmdr1*, can confer resistance to halofantrine (136).

A disturbing trend has been observed, however, in locales with highly multi-drug resistant strains. Some strains seem to have diminished sensitivity to halofantrine without drug pressure, that is, in areas where the drug had not been used before. Resistance to halofantrine in *P. falciparum* has been linked to the amplification of the *pfmdr 1* gene. *Pfmdr1*, is found to have increased gene copies and to be overexpressed in association with halofantrine- and mefloquine-reduced susceptibility (222, 329). In contrast, another study supplied evidence that resistance to halofantrine (and mefloquine) was related to reduced drug accumulation within the parasite that occurred without overexpression of the *pfmdr* protein product (361).

There is some evidence to suggest that halofantrine interacts with ferriprotoporphyrin IX in much the same way as chloroquine, although the data are conflicting (338). One study demonstrated the formation of a complex between halofantrine and FP, whereas another failed to observe it. The morphological changes seen in mouse red blood cells infected with *P. berghei* and treated with halofantrine are similar to those observed with quinine and mefloquine, although halofan-

trine also induces mitochondrial damage. Alternatively, the main action of halofantrine may involve inhibition of a proton pump present at the host-parasite interface; halofantrine was observed to inhibit glucose-dependent proton efflux from *P. berghei*–infected erythrocytes.

Halofantrine is a potent and selective inhibitor of the catalytic subunit of cyclic AMP-dependent protein kinase. No inhibition was observed with other protein kinases and halofantrine; antimalarial agents such as chloroquine, pyrimethamine, and artemisinin did not display the effects of halofantrine (362). It is not clear whether the kinase activity is related to antimalarial action, to cardiac side effects, or to neither.

One drawback of halofantrine, a highly lipophilic molecule, is its poor and variable absorption. Drug absorption is improved by food intake (363–365), and the aqueous solubility of halofantrine increases with addition of caffeine or nicotinamide (366). Even so, the practicality and advisability of administering food to very ill persons is questionable. The drug is extensively bound to lipoproteins in human serum (367) and it perturbs lipid bilayers (368). The distribution of halofantrine between lipoprotein fractions, both pre-prandial and post-prandial, has been explored (369, 370). The $IC_{50}$ value of halofantrine was shown to be increased by increasing concentrations of triglycerides (371). The data suggest that altered plasma lipoprotein profiles could influence the pharmacodynamic profile of halofantrine. The authors also argue for care in the interpretation of drug sensitivity data, especially when monitoring the development of drug resistance based on unknown lipoprotein profiles.

When treatment failures have been observed, several reasons have been cited although complete assessment has not been made. One factor associated with treatment failure is low blood concentrations of the parent compound and of the active metabolite, desbutylhalofantrine caused by poor absorption. The possibility of resistance or decreased susceptibility cannot be ruled out, however. Some cross-resistance between halofantrine and mefloquine has been observed; in cases of recrudescence, re-treatment with halofantrine was less effective after failure of mefloquine. Even in these cases, halofantrine treatment was more effective than re-treatment with mefloquine.

Halofantrine was found to be generally well tolerated with less central nervous system or upper gastrointestinal side effects than mefloquine. The most commonly reported adverse effects were abdominal pain, pruritis, vomiting, diarrhea, headache, and rash. Nevertheless, serious concerns have been raised about cardiac effects with halofantrine treatment (372, 373). In a group of children being treated with halofantrine, serious cardiac side effects were noted (374, 375). Because the observed electrocardiographic effects are dose-dependent (376), extra caution is needed at the higher dose used in conjunction with multidrug resistant strains and after mefloquine failure. Mefloquine caused significant changes in the clearance of high dose halofantrine (377), necessitating caution when treating persons who may have been on mefloquine prophylaxis before treatment with halofantrine. Given the long biological half-life of mefloquine, the effect may be exacerbated by drug interaction between halofantrine and mefloquine. Some have suggested that electrocardiography be performed on patients before initiating halofantrine treatment to identify those persons at the greatest risk (378). Cardiotoxicity of the (+)-isomer of halofantrine is greater than for the (−)-compound. This may reflect a more general pattern seen in the greater cardiotoxicity of quinidine over quinine. Mapping of the one chiral center in halofantrine with the secondary carbinol of quinine/quinidine suggests a similar spatial organization (341).

No evidence of mutagenic, teratogenic, or genotoxic effects or of effects on fertility was observed in animal toxicity studies with halofantrine (155, 379). Embryotoxicity could only be demonstrated at doses that caused maternal toxicity. No cases of adverse outcomes of human pregnancy on accidental exposure to halofantrine have been reported.

Halofantrine is currently recommended only for the treatment of malaria. It is not recommended for use as a suppressive prophylactic to minimize the selection of drug resistance and because of inadequate long-term toxicity data. In addition, halofantrine has no action on tissues stages and is not recommended as a causal prophylactic. Given the emerging picture of potentially serious complications, the appropriate use of and prescription guidelines for halofantrine have appeared (380). It has been recommended that halofantrine be restricted to use in cases of suspected chloroquine-, quinine-, or pyrimethamine/sulfadoxine-resistant malaria and that it should not be employed for general use (381). As with other newer antimalarial agents, warnings have been sounded about the use of halofantrine for monotherapy. The cost of halofantrine is about 100 times that of chloroquine.

## 2.5 Primaquine

*P. vivax* and *P. ovale* are the true relapsing malarias in humans, *P. vivax* being by far the more common. Dormant hypnozoites can reside in the liver for months and years after an initial infection, providing a reservoir of parasites that are the cause of malaria relapses. Radical cure of a vivax or ovale infection must involve not only removal of erythrocytic forms of the parasite but of the remaining liver forms as well. Agents that act on liver forms are termed tissue schizontocides and only one drug, primaquine (**14**), an 8-aminoquinoline,

(**14**)
primaquine

is currently available to effect such a radical cure.

The prototype 8-aminoquinoline was pamaquine (**15**, plasmochin), introduced in 1925. The action of pamaquine was clearly different from that of quinine, especially in its ability to prevent relapses. At the beginning of the Second World War, pamaquine was the only anti-relapse drug available. Therapy for vivax malaria involved a combination of pamaquine with quinine or quinacrine. Because of human

**(15)**
pamaquine

toxicity, the use of pamaquine declined, and other 8-aminoquinolines were synthesized. Of these, primaquine (WR 2975, **14**) has been used extensively as the anti-relapse drug of choice. d-, l-, and dl-Primaquine are essentially identical in curative properties against P. cynomolgi (382). Quinocide (**16**), an isomer

**(16)**
quinocide

of primaquine, has been used extensively in Eastern Europe and the former Soviet Union.

Primaquine has a multiplicity of antimalarial activities (339). To clear hypnozoites from the liver and to effect a radical cure of relapsing malarias, primaquine is given daily for 14 days in a usual dose of 15 mg. The standard treatment often involves a combined regimen of chloroquine with primaquine. Chloroquine is given to eradicate the asexual erythrocytic stages of the parasite. In addition, primaquine is active against the primary exoerythrocytic forms and could be used as a causal prophylactic, although until recently it was believed that the drug was too toxic for such prolonged use. A single dose of primaquine (35–45 mg base) is effective against gametocytes of P. falciparum, blocking infection of the vector. For this purpose, it is often given with either mefloquine or quinine to further interrupt transmission of the infection. Finally, primaquine is active (as metabolites) in

inhibiting the development of sporozoites in already infected mosquitoes.

Given that primaquine has been in use for many decades, it comes as no surprise that parasites resistant to it have been reported (68, 383). For example, between 1977 and 1997, the efficacy of a 5-day primaquine regimen in India declined from ~99% to 87%, although a 14-day regimen provided complete clinical cure with no relapses in a 6-month follow-up period (384). Unlike the situation with chloroquine and the 4-aminoquinolines, defining resistance to primaquine is problematic (70). Different strains of P. vivax display widely differing degrees of susceptibility to primaquine, a phenomenon of long-standing, not believed to be a function of varying drug pressure. Strains from New Guinea (the Chesson strain) are the least sensitive to primaquine, whereas those from Vietnam occupy an intermediate position, and the Korean strains are most sensitive. The interval between primary infection and relapse also varies with differing strains of P. vivax. The Chesson strain relapses rapidly, at intervals of 28 days; strains from southern China may relapse on an approximately annual basis. Recent studies have suggested that different populations of hypnozoites may have different susceptibilities to primaquine (385). So the incidence of primaquine-refractive strains may reflect increasing resistance or it may be the natural heterogeneity of P. vivax. Furthermore, primaquine in general is not as effective against erythrocytic parasites as chloroquine and others. When symptoms reappear, it may be a result of recrudescence or re-infection as well as primaquine resistance, and separating these possibilities is often not possible. Unlike the situation with P. falciparum, molecular events involved in resistance of P. vivax to chloroquine do not seem to be associated with mutations in pfcrt (236, 386). For a review on drug-resistant P. vivax, see Ref. 67.

Although primaquine has activity against the blood forms of the parasite, the dose required for monotherapy was considered too toxic for widespread use. However, much of the evidence for this toxicity resulted from data acquired with other, less active, and more toxic 8-aminoquinolines, pamaquine and pentaquine. The intrinsic activity of primaquine against *P. vivax* is certainly lower than that of all other commonly used agents except pyrimethamine/sulfadoxine when evaluated based on parasite clearance times (387). Similar results were observed with primaquine against *P. falciparum* (388). Therefore, for therapy, primaquine is combined with a faster acting agent, often chloroquine. With declining sensitivity of vivax to chloroquine, other agents have been studied as partners with primaquine including Fansidar and artesunate (389).

The causal prophylactic activity of primaquine is well established. Even so, because of problems with toxicity, real or perceived, routine use of primaquine for chemoprophylaxis had not been promoted. The diminishing efficacy of chloroquine and other agents is prompting a re-evaluation of the prophylactic use of primaquine (90). Primaquine prophylaxis has been shown to be safe and effective against both *P. falciparum* and *P. vivax* in Colombian soldiers (390) and in non-immune travelers to Africa (391). It is more effective and better tolerated than prophylactic chloroquine (392, 393). In some cases the rates of protection were better for primaquine than for either mefloquine or doxycycline. The lack of effect of chloroquine is perhaps not surprising given the prevalence of resistance in the areas where these studies were carried out. Indeed, in a study of a prophylactic combination, results showed that the addition of chloroquine did not increase the prophylactic efficacy of primaquine (394). More surprising was the lack of drug-related adverse side effects. Incidence of methemoglobinemia after 50 weeks of daily primaquine prophylaxis were comparable with that seen in a standard 14-day course (395). Administration of the primaquine doses with food may have limited the adverse side effects that had been reported with extended primaquine use. If daily primaquine taken during the period of exposure completely prevents the establishment of liver forms of the parasite, continuation of suppressive treatment after leaving a malarious area would not be required. This would represent an advance in prophylaxis caused by poor compliance with the usual post-exposure regimen.

Primaquine is rapidly absorbed from the gastrointestinal tract reaching maximal concentrations 1-3 h after administration. The distribution of primaquine in blood involves extensive binding to glycoprotein in plasma (396). For reviews, see refs. 137, 339, and 397. The pharmacokinetics of primaquine have been studied in both healthy volunteers (398) and in patients. The clearance of primaquine is reduced significantly during *P. falciparum* infection (399).

Metabolism of primaquine occurs rapidly in the liver, and the elimination half-life is 6 h. In rat liver homogenate, P450 and MAO enzyme systems contributed almost equally to primaquine metabolism (400). The metabolic profile seems to be balanced between activation to various hydroxylated derivatives (401), which may be responsible for the observed pharmacological and toxicological properties of the drug and conversion to the inactive carboxyprimaquine metabolite. Carboxyprimaquine, resulting from oxidative deamination of the side-chain, is the only metabolite of primaquine formed in human liver *in vitro* (402). The half-life of carboxyprimaquine is longer than that of primaquine, and its concentrations are much higher than primaquine itself in plasma. Carboxyprimaquine accumulates on daily dosing with a 14-day course of treatment. A number of hydroxylated metabolites of primaquine have been identified in animal models, although their isolation from humans remains to be accomplished. It has long been recognized that primaquine metabolites have significant effects on the oxidation state of the infected erythrocyte. Studies on 5-hydroxyprimaquine, 5-hydroxy-6-demethylprimaquine, and 5,6-dihydro-8-aminoquinoline have demonstrated the ready oxidation of these metabolites under physiological conditions (403, 404). A proposed mechanism of action for the 8-aminoquinolines involves the oxidation of such metabolites to a quinone-imine derivative, which may mimic ubiquinone. Analogs have been designed in which the metabolic pathway is blocked by a fluoro group at the 5 position (405). Metabolism of primaquine by human liver microsomes is inhibited by ketoconazole, a known inhibitor of cytochrome P450 isozymes. Quinine, artemether, artesunate, halofantrine, and chloroquine did not alter the metabolism of primaquine; only mefloquine had a slight inhibitory effect on metabolite formation (402). Primaquine and some of its newer analogs induce changes in metabolic activity, particularly in CYP1A1 gene expression (406).

The most serious adverse effect associated with primaquine is intravascular hemolysis. Primaquine, when incubated with red blood cells, induces extensive cell lysis and oxidation of oxyhemoglobin to methemoglobin. The two toxic effects however are not causally related; in the presence of a stable nitroxide radical, cell lysis was prevented even though hemoglobin oxidation was enhanced (407). Other studies have also examined the formation of and biological results from radicals generated by primaquine and its metabolites (408–411). These results support the hypothesis that free radicals and redox-active species are involved in the cellular damaging processes induced by primaquine. Hydrogen peroxide has been suggested as the potential toxic product formed from oxidation and redox cycling of primaquine metabolites (404, 412).

Persons who are deficient in the enzyme glucose-6-phosphate dehydrogenase (G6PD) are particularly at risk of hemolysis from primaquine treatment. G6PD plays an important role in the redox processes of the erythrocyte. The conversion of glucose-6-phosphate to 6-phosphonoglucon-D-lactone by G6PD produces NADPH from NADP+. NADPH is used by the erythrocyte to reduce the oxidized form of glutathione back to the reduced form. The reduced glutathione acts as a "sulfhydryl buffer" to maintain cysteine residues of hemoglobin and other proteins in the reduced state. Insufficient levels of the G6PD result in inadequate amounts of NADPH, making erythrocytes especially susceptible to damage from oxidants. As noted above, primaquine generates a number of toxic reactive oxygen species. Clearly, in patients with severe G6PD deficiency, primaquine should be avoided for both treatment and prophylaxis (413).

Methemoglobin formation is another predictable dose-related adverse effect. The usual 14-day curative regimen of primaquine has been reported to lead to a number of other symptoms including anorexia, nausea, cyanosis, epigastric distress, abdominal pain and cramps, malaise, dark urine, vomiting, and vague chest pains. Abdominal cramps are

common when primaquine is taken on an empty stomach, but the discomfort is lessened markedly when it is given with food. Even standard doses of primaquine are to be avoided in pregnancy to minimize the hemolytic effects; the fetus is relatively G6PD-deficient (155). Primaquine was found to induce mutations in a standard assay (414).

Two excellent reviews cover the general structure-activity relationships in the 8-aminoquinoline group (415, 416). A number of studies have been carried out with the aim of identifying drugs to replace primaquine as the agent of choice for radical cure of relapsing malarias; for a summary of the structural types of interest, see Ref. 417. In particular, a reduction in the toxicity associated with primaquine has been a goal. Early on, 4-methylprimaquine was shown to have superior activity to primaquine with less toxicity. Similarly, 4-ethyl and 4-vinyl primaquine retained the radical curative activity of primaquine and had better therapeutic indices (418). Further modification by the introduction of an aryloxy group at the 5-position of the quinoline enhanced antimalarial activity (419–421). The best of this series were fluorine-containing compounds that were more active than primaquine. The U.S. Army has been involved in research on a number of anti-relapse agents. Among these are WR225,448 (17), WR242,511 (18), and WR238,605 (19). All are modified 8-aminoquinolines clearly related to the primaquine structure. Although WR225,448 had good activity against the exoerythrocytic parasite and effective doses much lower

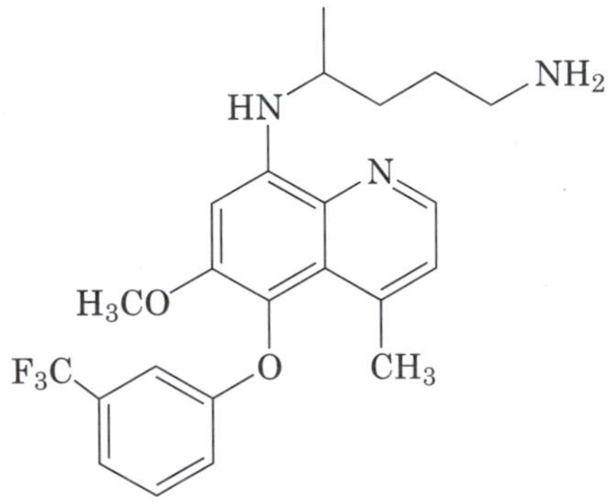

(17)
WR225448

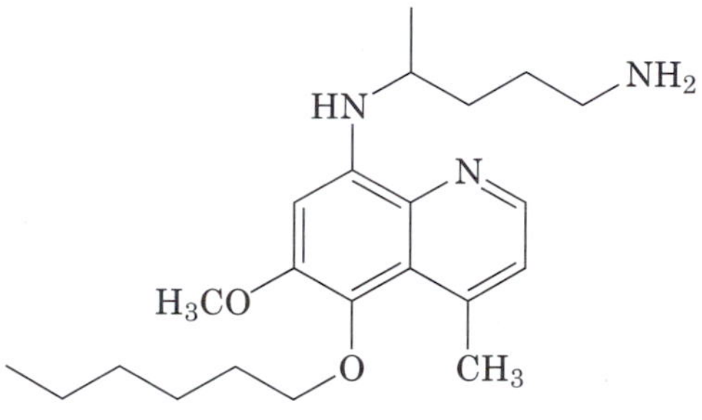

**(18)**
WR242511

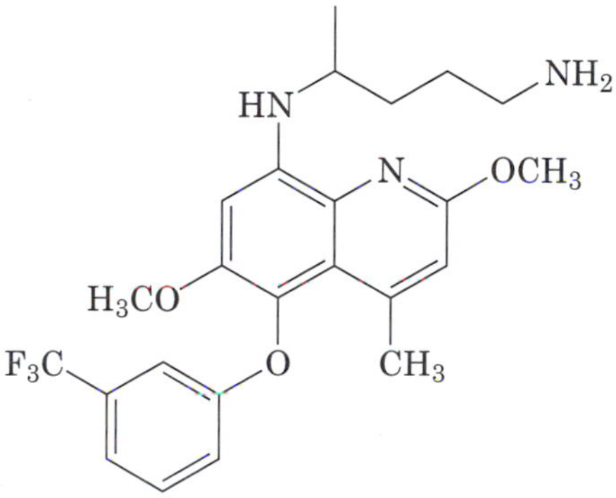

**(19)**
WR 238,605
tafenoquine

of 14 days (427). These workers concluded that the safety, efficacy, and pharmacokinetic properties make the drug an excellent candidate for use as a prophylactic, radical curative, and terminal eradication drug. Tafenoquine is now in phase II clinical studies in semi- and non-immune people.

Another strategy for developing agents with high activity and lower toxicity than primaquine has involved connecting amino acids or peptides to the free amine group of primaquine or primaquine analogs. 4-Methylprimaquines that had an alanine or a lysine conjugated to the free amine were more active and less toxic than primaquine (428). A peptide derivative of primaquine was shown to have equivalent antimalarial activity as the parent compound and less toxicity (429). Dipeptide derivatives of primaquine in which the oxidative deamination of primaquine was blocked were found to be gametocytocidal to *P. berghei* (430). The most active member of this group was gly-gly-PQ. 5-Trifluoroacetylprimaquine (M8506, **20**) was shown to be a causal prophy-

than those of primaquine, the liver toxicity was high (422). The pharmacokinetics of WR242,511 have been studied, and a significant lag was observed between in the appearance of drug in the plasma and the onset of methemoglobinemia. Even so, methemoglobin formation remained a major side effect (423). More promising is WR238,605, now called tafenoquine (424, 425). It is effective against pre-erythrocytic stages, drug-sensitive and multi-resistant erythrocytic stages, gametocytes, and hypnozoites (*P. cynomolgi*), with activity and safety profiles both significantly better than primaquine. Studies of tafenoquine prophylaxis have demonstrated that it was effective and well tolerated, and the duration of protection afforded by the drug is long (426). In a pharmacokinetic study, tafenoquine was administered in single oral doses; it was found to be well tolerated, had a long absorption phase, and was slowly metabolized with a $t_{max}$ of 12 h and an elimination half-life

**(20)**
M8506

lactic agent, equivalent in activity to primaquine in *P. yoelii*–infected mice (431). As a tissue schizonticide, it is more effective than primaquine in *P. cynomolgi*–infected rhesus monkeys and better tolerated than primaquine in mice, rats, and dogs.

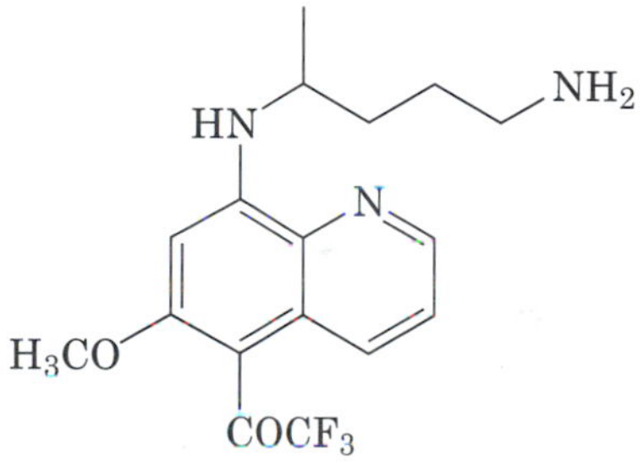

A prodrug of primaquine, CDRI compound 80/53 (**21**, bulaquine), has been shown to possess radical curative and causal prophylactic action against *P. cynomolgi* in rhesus monkeys (432). In addition, this compound caused less methemoglobinemia than did primaquine (433, 434), had very low toxicity (435), and had lessened effects on hepatic enzyme levels

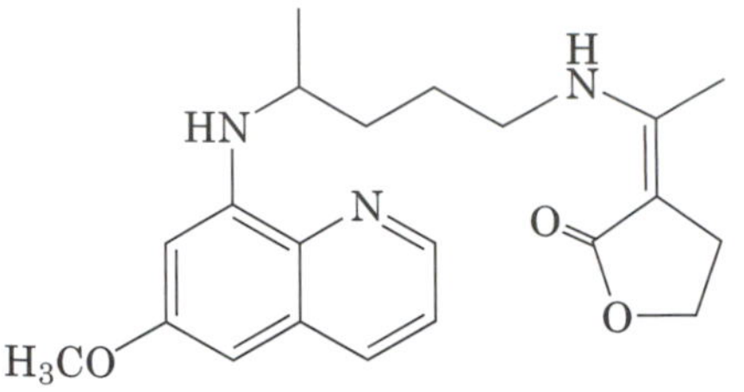

**(21)**
CDRI 80/53, bulaquine

when compared with primaquine (436). Clinical trials have shown that bulaquine was an effective alternative to primaquine in the treatment of vivax malaria (437). Bulaquine has been approved for use in India.

As a result of extensive screening efforts, a number guanylhydrazones were found to have causal prophylactic activity in rodents. One example of this group, WR182,393 (**22**) has

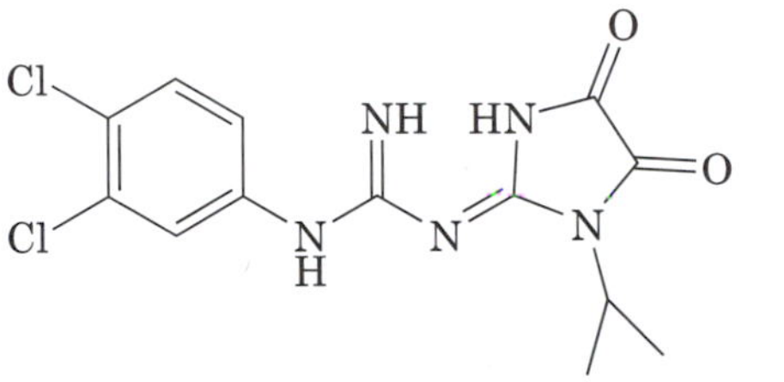

**(22)**
WR182393

been shown to have both causal prophylactic and radical curative properties against a primate relapsing malaria (438). It is the first non-8-aminoquinoline to display both properties.

## 2.6 Antifolates

A number of agents have been discovered that affect antimalarial activity through inhibition of the parasite dihydrofolate pathways. These agents are divided into two types: inhibitors of dihydrofolate reductase (DHFR) and those that inhibit dihydropteroate synthetase (DHPS).

The pharmacokinetics of antifolates have been reviewed (137); this reference also has information about pharmacokinetic interactions of antifolates with other antimalarial agents.

### 2.6.1 Proguanil and Chlorproguanil.

Studies in the synthesis of pyrimidine derivatives led British researchers at ICI to examine the antimalarial activity of biguanides. Proguanil (chloroguanide, Paludrine, **23**) was the most

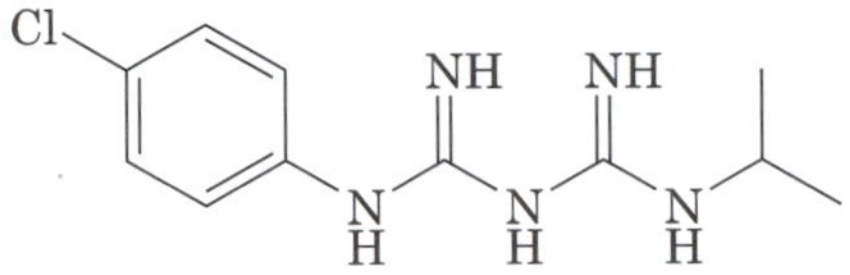

**(23)**
proguanil

active of a series of these compounds screened in the 1940s. Proguanil proved to be an excellent causal prophylactic in falciparum malaria as well as a satisfactory suppressive of vivax malaria, although the drug is not active against hypnozoites (439). A related compound, chlorproguanil (**24**, Lapudrine) has

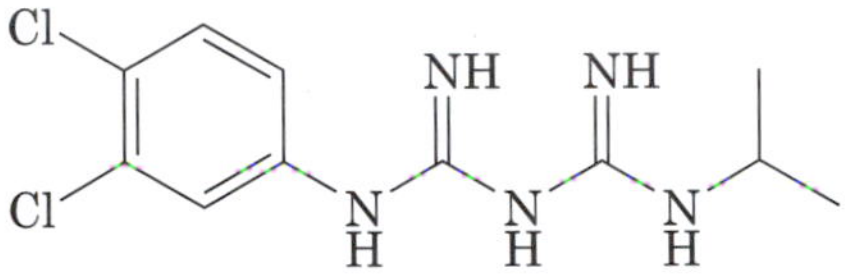

**(24)**
chlorproguanil

been used occasionally, and recent studies suggest that it may be a good alternative to proguanil in areas of drug resistance (440). Proguanil is a tissue schizonticide, active against pre-erythrocytic liver stages, and a sporontocide. Because of their slow action, the availability of faster-acting agents, and the fact that resistance seemed to develop quickly in some strains, these biguanides have not been used for monotherapy. During the 1980s, proguanil was temporarily removed from the list of recommended prophylactic agents because of what seemed to be wide-spread resistance in *P. falciparum*. A combination of proguanil with chloroquine was found to be an effective prophylactic regimen (441, 442).

It has long been thought that the antimalarial activity of proguanil resided exclusively in a cyclized metabolite, cycloguanil (**25**); proguanil would then be a prodrug form of cycloguanil (443). A small percentage of

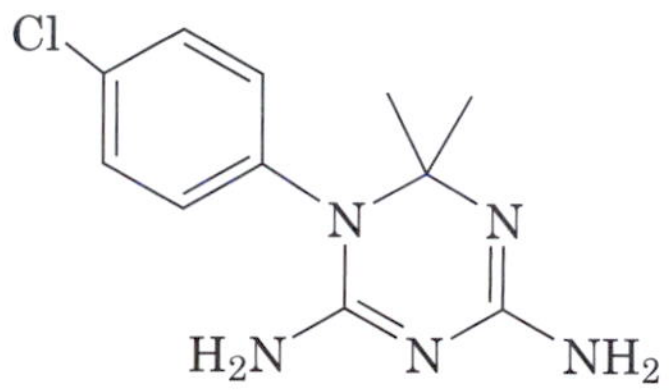

**(25)**
cycloguanil

proguanil is converted into an *N*-dealkylated product, 4-chlorophenylbiguanide, which is not active against malaria. Some confusion has arisen over the question of whether the antimalarial activity associated with proguanil is caused entirely by the metabolite cycloguanil or if the parent compound possesses some inherent activity itself. Some studies have reaffirmed that the activity associated with proguanil administration is completely accounted for by cycloguanil (444), whereas others find intrinsic efficacy against falciparum and vivax independent of cycloguanil (445). The activity of cycloguanil is not affected by physiological concentrations of folic acid or folinic acid in human serum, suggesting that the drug may have additional or different sites of action from inhibition of folate pathways (446). Clinical pharmacokinetic studies of proguanil and cycloguanil have been reported (283, 447, 448). Proguanil lacks serious side effects, is not mutagenic (449), and there is no evidence of any toxic effects of prophylaxis with proguanil during pregnancy from over 40 years of use (155). Cycloguanil acts specifically on *P. falciparum* DHFR and seems to have no other significant target; the action of proguanil may include sites of action other than on DHFR (450, 451).

The enzyme that effects the conversion of proguanil to cycloguanil is a cytochrome P450, specifically CYP2C19. It has been shown that the ability to carry out the metabolic conversion varies widely among individuals and among ethnic groups (452, 453). Between 2% and 3% of Caucasian populations have a relative inability to convert proguanil to cycloguanil; the numbers may be as high as 20% in Oriental subjects and 25% in Kenyans. An association between CYP2C19 mutants and poor metabolism of proguanil was demonstrated

(454). Compounds that inhibit CYP2C19, such as omeprazole and fluvoxamine, were shown to inhibit proguanil metabolism (455, 456). In Asia and Africa, the variable metabolism of proguanil may have decided clinical importance. There was a significant correlation between breakthrough parasitemias and the proguanil/cycloguanil ratios in subjects who were taking proguanil prophylactically (457). A possible solution would be the development of cycloguanil itself as an antimalarial agent.

Proguanil, or more precisely cycloguanil, acts as a mimic of dihydrofolate. The drug competitively inhibits the parasite enzyme dihydrofolate reductase (DHFR, tetrahydrofolate dehydrogenase). Plasmodia require a tetrahydrofolate cofactor to synthesize pyrimidines *de novo* because they have no salvage pathways for pyrimidines. The binding of the drug to the plasmodial enzyme is several hundred times that of binding to the mammalian enzyme. In strains resistant to proguanil/cycloguanil, a mutated DHFR protein is observed that displays a reduced affinity for the normal substrate and for the inhibitor. These strains are hypersensitive to the sulfonamides and sulfones (inhibitors of DHPS), which also reduce the availability of dihydrofolate. Thus, a synergistic effect is observed between inhibitors of DHPS and inhibitors of DHFR; combinations have proven to be particularly useful in chemoprophylaxis. See Section 2.7.2 for information on mutants of plasmodial DHFR and resistance to antifolates.

Based on a three-dimensional homology model of plasmodial DHFR, analogs of cycloguanil were designed and evaluated against both wild-type DHFR and a mutant strain (458). A lead compound (26) was identified that was equally active with cycloguanil in inhibition of wild-type DHFR and 120 times

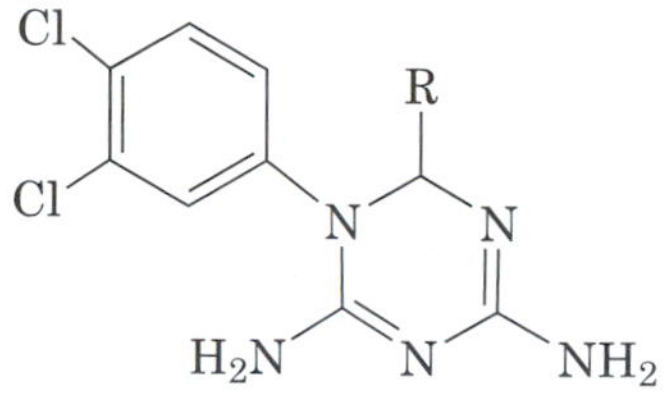

**(26)**
R = CH$_3$, Ph

more effective than cycloguanil against mutated enzyme. Several of the compounds were active both in enzyme assays and against *P. falciparum* clones *in vitro*.

Proguanil and chlorproguanil have been use in combination with a number of drugs. In particular, the combination of proguanil with atovaquone (Malarone) has been extensively studied (see Section 2.8.1).

**2.6.2 Pyrimethamine.** Pyrimethamine (Daraprim, **27**) was developed by Burroughs Well-

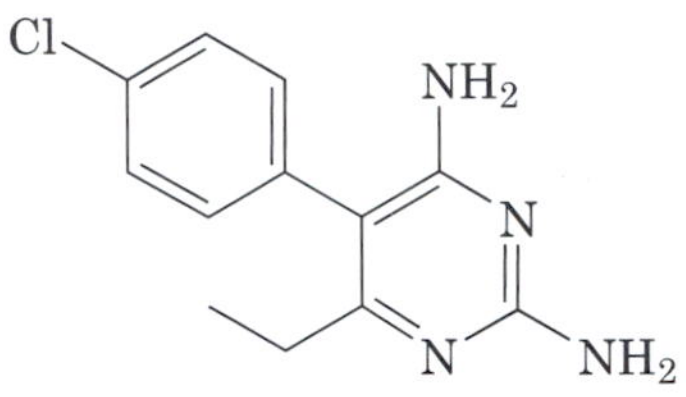

(**27**)
pyrimethamine

come in 1950. The researchers there were investigating folic acid antagonists for use as anticancer agents. A rather tenuous structural similarity between these antifolates and proguanil led them to examine pyrimidines for antimalarial activity. As was later shown, the structural resemblance between cycloguanil, the active metabolite of proguanil, and pyrimethamine is striking. Pyrimethamine is eliminated primarily by oxidation to pyrimethamine 3-*N*-oxide; the P450 isozyme responsible for the oxidation is not known. Despite its wide margin of safety, pyrimethamine is no longer used as a single agent because of the ease with which resistance develops to the drug when administered alone. In addition to combinations with sulfadoxine, pyrimethamine has been studied with artemether. In a 3-day regimen, the combination was more effective than artemether alone, even in an area of pyrimethamine resistance (459). When given during pregnancy, there was no increase in the observed rate of stillbirth, neonatal death, or malformations with pyrimethamine prophylaxis (155), and the drug is not mutagenic (449). A related agent, both structurally and in action, is tri-

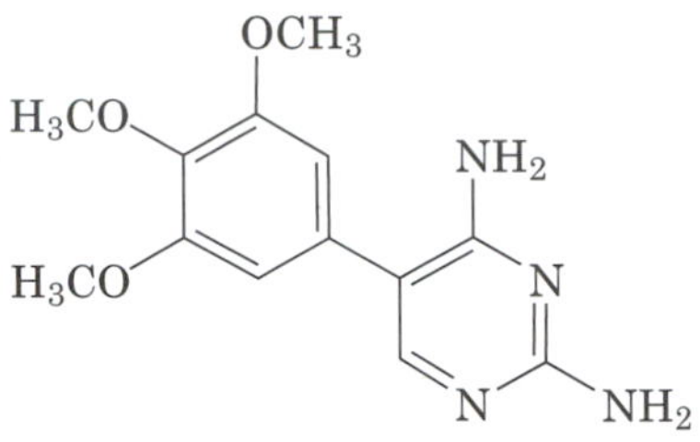

(**28**)
trimethoprim

methoprim (**28**). It is used much less often in the treatment of malaria and always in combination.

Although proguanil and pyrimethamine both act through inhibition of the parasite DHFR, it is clear that some strains resistant to one of these agents can retain sensitivity to the other. In attempting to understand the reasons for the rapid development of resistance to pyrimethamine and the occasional lack of cross-resistance, researchers began to examine the DHFR protein in detail. It was recognized that the DHFR activity was linked on a single protein to the thymidylate synthetase (TS) enzyme (460, 461). This observation led ultimately to the cloning of plasmodial DHFR. Analysis of the DHFR proteins from resistant strains of *P. falciparum* has revealed a structural basis for resistance of antifolate agents (462–465). Mutations in the DHFR gene of *P. falciparum* correlate with antifolate resistance in lab and field isolates.

Cycloguanil resistance involves two point mutations: alanine to valine mutation at position 16 and serine to threonine at position 108. When position 108 is taken by asparagine rather than either serine or threonine, pyrimethamine resistance is conferred. Those strains with the highest levels of pyrimethamine resistance also incorporate a cysteine to arginine mutation at amino acid 59. Significant cross-resistance to both drugs is seen in parasites with a DHFR protein that has asparagine at 108 as well as leucine replacing isoleucine at 164. Using this information, polymerase chain reaction assays can be used in surveillance of drug resistance in the field (466, 467). Such surveillance has led to the acquisition of a sizable quantity of data on DHFR mutants and patterns of resistance

(466, 468–473). Mutations at 108 are the most common; only one example has been reported of a mutant DHFR that was wild-type at 108 (474). Other changes that have been reported occur at amino acids 16, 50, 51, 59, 140, and 164. Although most studies of antifolate resistance have focused on *P. falciparum*, work has also been done on DHFR-TS from *P. vivax*. Point mutations in the vivax enzyme at positions analogous to those seen for *P. falciparum* were a major determinant of antifolate resistance in *P. vivax* (475). Studies have been carried out of the conformations of antifolate drugs and possible modes of binding to DHFR (476).

Two analogs of pyrimethamine were designed and tested for activity against mutant recombinant plasmodial DHFR. CC83 (**29**) and SO3 (**30**) were active against wild-type

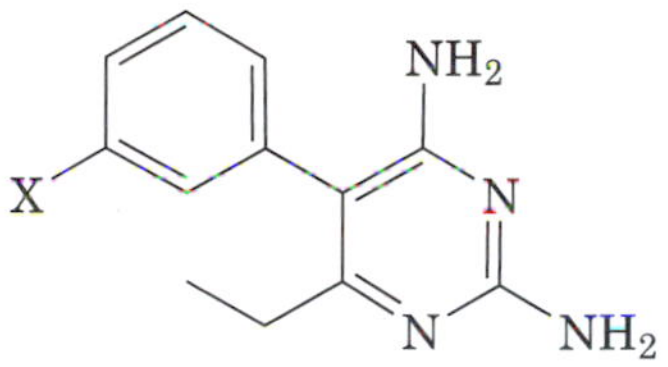

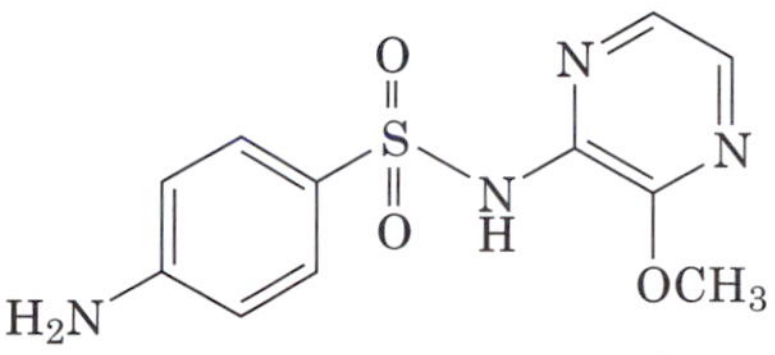

**(32)**
sulfalene

**(33)**
dapsone

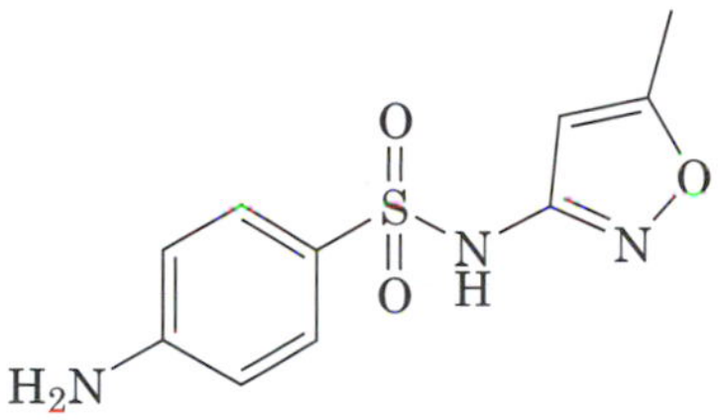

**(29)**
CC83, X = OCH$_3$
**(30)**
SO3, X = Cl

DHFR as well as showing good inhibition of a double mutant (477). The inhibitors retained activity *in vivo* against *P. berghei* in infected mice.

**2.6.3 Sulfonamides and Sulfones.** On discovery that sulfanilamide had appreciable antimalarial activity, a number of related sulfonamides were tested for activity. Two longer acting agents, sulfadoxine (**31**) and sulfalene

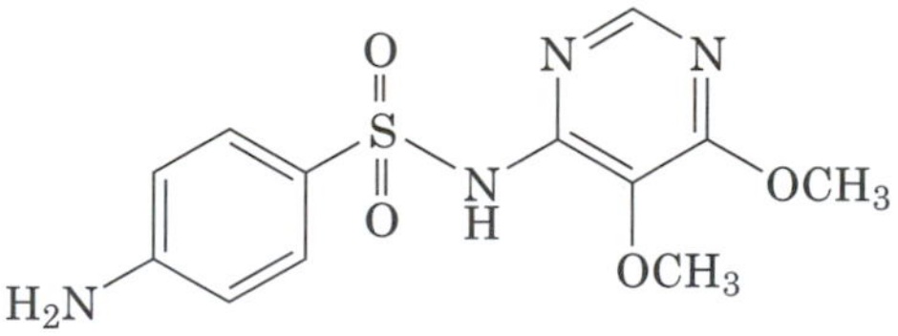

**(34)**
sulfamethoxazole

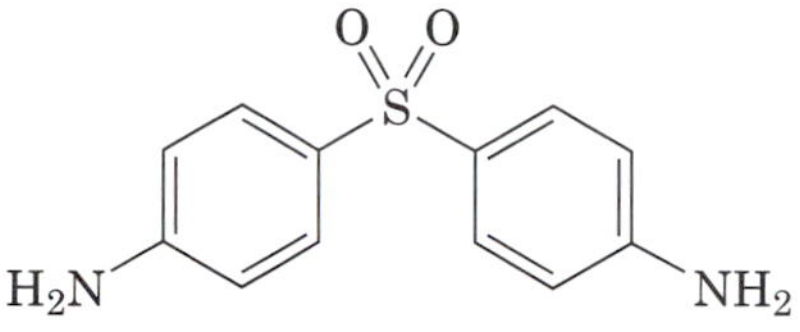

(**32**), were identified as the most promising members of the group. These studies led to the assessment of dapsone (**33**), a sulfone drug, used for leprosy. Sulfamethoxazole (**34**) and sulfisoxazole (**35**) are other examples of this class of agents.

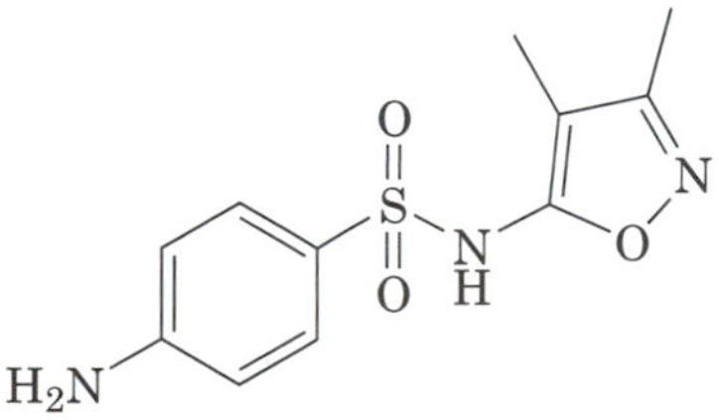

**(35)**
sulfisoxazole

Sulfonamides and sulfones are selectively toxic to malaria parasites because they inhibit a plasmodial enzyme, dihydropteroate synthetase (DHPS), that is not present in mam-

**(31)**
sulfadoxine

mals. DHPS catalyzes the combination of PABA with a substituted pterin. The product, dihydropteroate, is ultimately converted into dihydrofolate. The folate co-factors are necessary in the biosynthesis of pyrimidines. Sulfa drugs compete with PABA and are converted into non-metabolizable adducts; the result is a depletion of the folate cofactor pool. A recent study has noted that several sulfa drugs exert toxic effects on the parasite *in vitro* at concentrations that are two to three orders of magnitude lower than the concentrations needed to inhibit the isolated enzyme (479). The results suggest that (1) sulfa drugs are concentrated by the parasite, (2) partial inhibition of DHPS may be lethal, or (3) sulfa drugs actually act by some other mechanism than the inhibition of DHPS. The general paradigm that holds that plasmodia are unable to use exogenous folate may well be a simplification; folic acid is known to antagonize the antimalarial effects of sulfonamide drugs *in vitro* (480).

The DHPS coding sequence has been determined in *P. falciparum* (481). It forms part of a longer sequence that also specifies the preceding enzyme in the plasmodial folate pathway, 6-hydroxymethyl-7,8-dihydropterin pyrophosphokinase. Parasite lines resistant to sulfadoxine were found to possess mutations in the DHPS sequence. Those with high-level resistance carried either a double mutation at both Ser-436 and Ala-613 or a single mutation at Ala-581. Polymerase chain reaction systems have been developed to further document mutations in DHPS and to track clinical failures as a result of sulfadoxine resistance (482).

Sulfadoxine is not mutagenic nor genotoxic (449). Neither the long-acting sulfonamides (sulfalene, sulfamethoxine, sulfadoxine) nor dapsone in low doses cause teratogenicity (155). Dapsone is excreted in breast milk in sufficient quantity to cause toxicity, however; breastfeeding is not recommended during dapsone therapy. Sulfonamides as a class have been associated with cutaneous reactions and the longer-acting agents seem to be more likely to generate severe reactions. When sulfadoxine alone has been used at dosages higher than that for weekly malaria chemoprophylaxis, the incidence of severe reaction approached 1 in 10,000 (483).

**2.6.4 Combinations.** A number of combinations involving both DHFR and DHPS inhibitors have been studied: sulfadoxine/pyrimethamine (SP, Fansidar), pyrimethamine/dapsone (Maloprim), dapsone/chlorproguanil (484), sulfisoxazole/proguanil (485), sulfamethoxazole/trimethoprim, and sulfalene/trimethoprim. Inhibition at both points in the parasite's folate pathway provides a potentiation of the antimalarial action.

The most widely used combination has been sulfadoxine/pyrimethamine. The actions of the two drugs have been shown to be synergistic (480). The clinical pharmacokinetics of the combination have been reported (283, 339, 486). In many places in Africa, SP has begun to replace chloroquine as the drug of choice in falciparum malaria unless the patient displays sensitivity to sulfonamides. In a number of areas where chloroquine resistance is prevalent, SP has proven to be effective such as in India (487), Côte d'Ivoire (488), Malawi (489), Pakistan (490), and Colombia (491). Increasing SP resistance has been documented in Kenya (492) and Tanzania (493), however. Perhaps not surprisingly, a high proportion of ineffectiveness has been reported with SP in *P. vivax* in patients infected in the Thailand-Myanmar border region, an area of intense multi-drug resistance (389). Given the problem of resistance in Africa, studies have looked at combining SP with chloroquine in Gambian children. The triple combination was more effective than SP alone in controlling symptoms in the first few days after treatment (494). SP has also been combined with artesunate in Gambian children, leading to faster resolution of fever and parasitemia than with SP alone (495, 496).

SP has been used both in a prophylactic regimen as well as for treatment. But severe and even fatal skin reactions have been observed when SP was used for prophylaxis (483, 497). Cases were reported of American travelers using SP for malaria prophylaxis suffering toxic epidermal necrolysis, erythema multiforme, and Stevens-Johnson syndrome. No other risk factors could be identified. Based on these outcomes, the routine use of SP as prophylaxis is not recommended unless persons are at highest risk of acquiring chloroquine-resistant *P. falciparum*. In therapy, however,

SP is still recommended as presumptive self-treatment for travelers to areas of where the presence of chloroquine-resistant *P. falciparum* is documented. Of course, no one with a history of sulfa drug allergy should take SP.

Despite recommendations against using SP during pregnancy (155), studies have shown a beneficial outcome. Treatment of pregnant women with SP, either on a monthly basis or with two prenatal doses, was efficacious in preventing placental malaria in Africa (498). No significant differences were seen between HIV positive and HIV negative women in terms of drug tolerability and adverse side effects (499). Birth weights were higher and fetal growth was improved (500). The risk of severe anemia to the mothers was decreased (501).

The combination of proguanil with dapsone has been used in malaria prophylaxis against chloroquine-resistant falciparum malaria for over 20 years (502). The pharmacokinetic effects of proguanil on dapsone and dapsone on proguanil have been examined. Proguanil was found not to alter the plasma levels of dapsone and its metabolites in healthy individuals (503). The elimination half-life of proguanil was found to be slightly longer in the presence of dapsone; other pharmacokinetic parameters were not affected (504). In a multi-drug resistant area (Thai/Myanmar-Cambodia border), proguanil combined with dapsone offered no significant advantage as a chemoprophylactic agent over pyrimethamine/dapsone against falciparum malaria (505). Proguanil/dapsone was much more effective in preventing cases of vivax malaria, however. The combination of chlorproguanil with dapsone has also received favorable report (506).

The combination of pyrimethamine with dapsone has not been as widely used. It was shown to be more efficacious than chloroquine in Tanzanian children both for treatment (507) and for prophylaxis (508), although dermatological side effects were noted when it was used prophylactically (509). Some have argued that the dapsone/proguanil combination is superior to other antifolate pairs based on pharmacokinetic profiles (504). The elimination half-lives of dapsone (23 h) and proguanil (15 h) are sufficiently similar so that on

discontinuation of administration, suboptimal concentrations of the drug combination in the blood are possible only briefly. In contrast, dapsone (27 h) with pyrimethamine (83 h) or sulfadoxine (204 h) with pyrimethamine (112 h) have markedly different half-lives. This leads to one drug remaining in the blood for long periods without the potentiating effects of the combination. Selective resistance to the longer-lived agent may well result.

Chlorproguanil (Lapudrine) with dapsone is a relatively new combination of antifolate drugs (510) and has been termed "lapdap." For a summary of clinical trials using lapdap, see Ref. 511. The combination is inexpensive, comparable to SP, safe, and effective in uncomplicated falciparum malaria in certain areas (512). In other locations, however, such as in Thailand where multi-drug resistant parasites are prevalent, short-course treatments with lapdap were not effective either for therapy (513) or as prophylaxis (514).

In areas where both malaria and acute respiratory infection in children are endemic, the combination of clotrimazole (**36**) may be

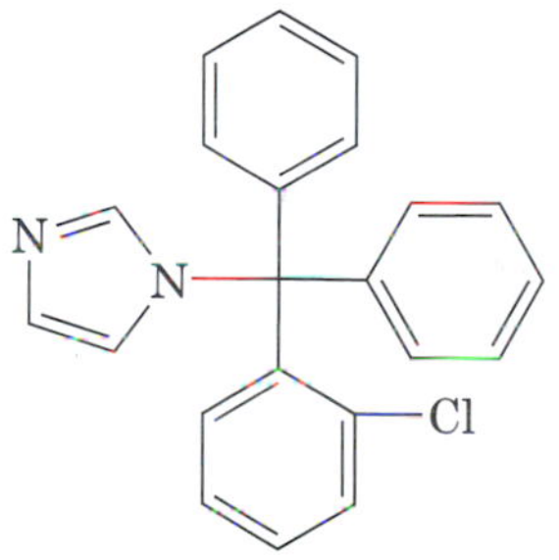

**(36)**
clotrimazole

indicated. Clotrimazole has $IC_{50}$ values higher than SP but is active at doses commonly used in treating bacterial infections (478, 515, 516). There is need for further *in vivo* and clinical studies to assess the usefulness of the combination. Cotrifazid, a fixed combination of rifampicin, isoniazid, sulfamethoxazole, and pyrimethamine, was shown to be effective against malaria in humans (517). The combination was effective either as a first-line treatment or after drug failures of SP, quinine, or

SP and quinine in small studies in Malawi (517) and Kenya (518).

**2.6.5 Related Compounds.** A number of substituted triazines have been prepared and evaluated with some enthusiasm. WR99210 (**37**), a triazine analog of proguanil, was syn-

**(37)**
WR99210

thesized and studied in the 1970s and early 1980s (519). WR99210 is remarkably active against resistant and sensitive *P. falciparum* clones and isolates and lacks cross-resistance with cycloguanil and pyrimethamine, suggesting that the mechanism of WR99210 differs from the other antifolates (520). WR99210 seems to act exclusively on the parasite DHFR displaying no inhibitor effect on the human enzyme (450). Clinical trials uncovered severe gastrointestinal symptoms, however, and development of the compound was dropped. A related compound, clociguanil (**38**, BRL

**(38)**
clociguanil

50216), had a promising activity profile, but its biological half-life was short in human trials and there was no demonstrable benefit over proguanil or pyrimethamine (521). In 1993, the biguanide version of WR99210, PS-15 (**39**, WR250417), was synthesized and studied (522, 523). Like proguanil, PS-15 is an inhibitor of DHFR and is metabolized *in vivo* to WR99210. *In vitro* tests against drug-resistant clones of *P. falciparum* demonstrated that PS-15 was more active than proguanil, and the

**(39)**
PS-15

metabolite WR99210 was more active than cycloguanil. PS-15 was more active than either proguanil or WR99210 in monkeys infected with multi-drug-resistant *P. falciparum*. When administered orally to mice with *P. berghei*, PS-15 was also less toxic than proguanil. In studies designed to assess the chemoprophylactic efficacy of PS-15, neither the parent compound nor the cyclized metabolite prevented primary infection in monkeys (524). PS-15 does not seem to be cross-resistant with other DHFR inhibitors. Studies of combinations of PS-15 with atovaquone, sulfamethoxazole, or dapsone in monkeys demonstrated that the combinations were effective. The combination with atovaquone was particularly promising (525). Additional compounds in this series such as PS-26 (**40**) and

**(40)**
PS-26, X = Cl
**(41)**
PS-33, X = H

PS-33 (**41**) were prepared to identify compounds that retained good activity without the problematic 2,4,5-trichlorophenoxy moiety of PS-15 (526).

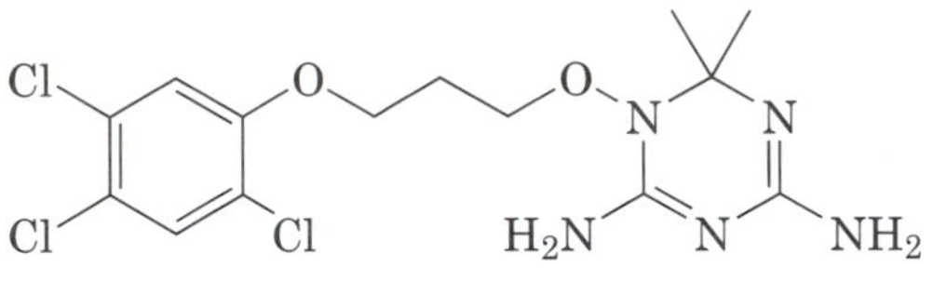

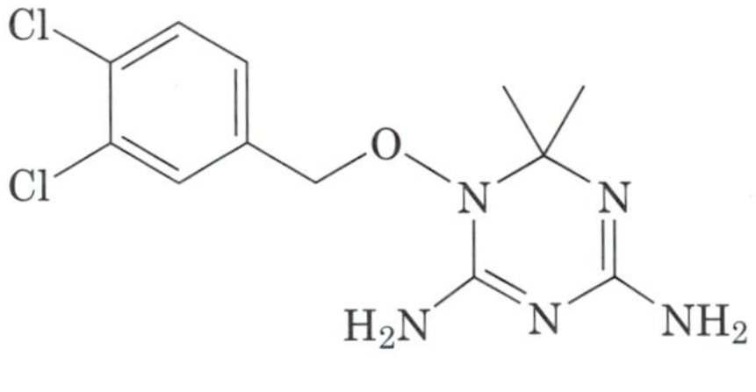

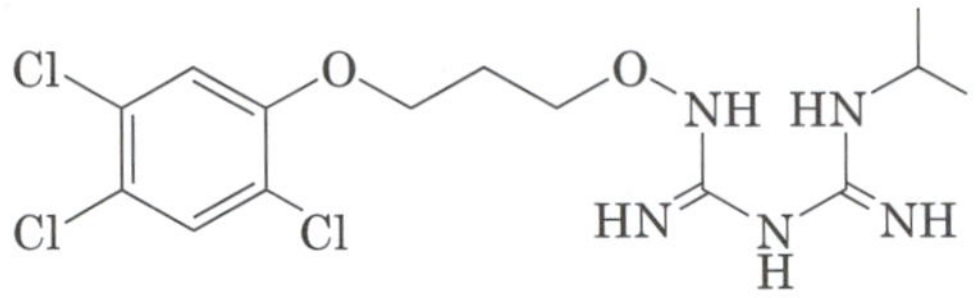

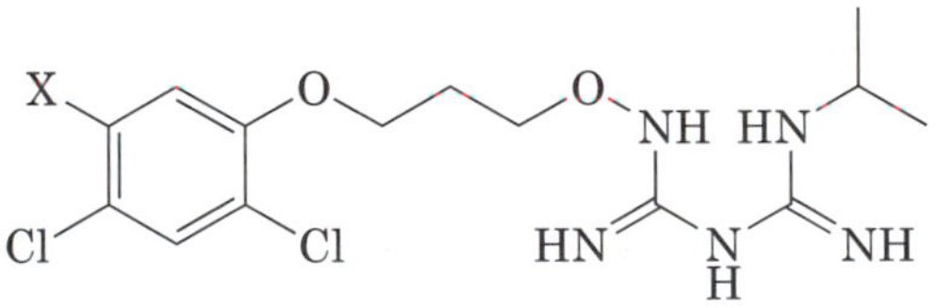

**2.7  Artemisinin and Its Derivatives**

**2.7.1 Background, Isolation, and Chemistry.** The weed *Artemisia annua* (sweet wormwood, sweet annie) has been used for many centuries in Chinese herbal medicine as a treatment for fever and malaria (527). In 1971, Chinese chemists isolated from the leafy portions of the plant the substance responsible for its reputed medicinal action (528–530). The compound, artemisinin (**42**, qinghaosu, artean-

**(42)**
artemisinin

nuin), was determined by X-ray crystallography to be a sesquiterpene lactone bearing an endoperoxide (531). A number of very fine reviews have appeared on artemisinin and its derivatives (527–529, 532–539).

Early results from China that several thousand malaria patients, including those with chloroquine-resistant strains of *Plasmodium falciparum*, were successfully treated with artemisinin (539–541) sparked world-wide interest in the study and development of artemisinin-related compounds. By this writing, the number of persons treated with artemisinin will number in the hundreds of thousands. The practical value of artemisinin itself as a therapeutic agent has been limited by several factors, however. Artemisinin has a high rate of recrudescence, poor oral activity with high doses required (542), a short half-life in plasma (543), and limited solubility in both water and oil (544). In an effort to circumvent these therapeutic and pharmaceutical problems, chemical modifications of artemisinin as well as the synthesis of large numbers of analogs have been extensively pursued.

*A. annua* is being grown commercially and processed for its artemisinin (**42**) and artemisinic acid (**43**) in both China and Vietnam (545). The botany, horticulture, and agricultural production of artemisinin have been reviewed (546, 547). Artemisinin is extracted at low temperature from the aerial portions of the plant. The yield of artemisinin is dependent on the plant strain, the stage of development, soil conditions, and the growing environment. The maximum yield observed is

~1% of the dry plant weight, but values of 0.01–0.5% are more common, especially in plants collected from the wild. Values will also vary depending on whether the plant material assayed was predominately leaves or included stems, which have a lower artemisinin content. Attempts at the improved production of artemisinin by plant cell or tissue culture have not been particularly successful (548). *A. annua* is not the only natural source of artemisinin; small amounts have also been isolated from *A. apiacea* (549) and *A. lancea* (550).

Sesquiterpenoids are known to be synthesized from farnesyl diphosphate (FDP) mediated by a sesquiterpene cyclase (or synthase). Scheme 18.1 outlines a proposed biosynthetic sequence for artemisinin. Amorpha-4,11-diene has recently been isolated from *A. annua* extracts and is suggested as an intermediate in the biosynthesis (551). A key biosynthetic enzyme, amorpha-4,11-diene synthase, has been cloned, expressed, and characterized from *A. annua* (552). A protein accomplishing the transformation of arteannuin B (**44**), the most abundant cadinane from *A. annua*, to artemisinin has been purified to homogeneity from *A. annua* (553). From these studies, it is clear that at least some artemisinin is biosynthesized by an enzyme-mediated peroxidation. Although the conversion of amorpha-4,11-diene to artemisinic acid is, at present, undocumented, the remaining steps through dihydroartemisinic acid (**45**) have been replicated under conditions that mimic those of the plant (554, 555). The steps from dihydroartemisinic acid to artemisinin may be non-enzymatic in the plant (555).

Artemisinin has been prepared by total synthesis in several laboratories (537, 539, 556). Schmid and Hofheinz, of the Hoffmann-LaRoche Company, prepared artemisinin starting from (−)-isopulegol (557). The key step was a photo-oxygenation of a methyl vinyl ether followed by acidic ring closure to give the endoperoxide linkage. The synthesis established the absolute stereochemistry of artemisinin. Avery and co-workers used (*R*)-(+)-pulegone to set the necessary stereochemistry and incorporated the peroxide group by ozonolysis of a vinylsilane in their total synthesis (558, 559). Using similar methods, Avery's group also prepared (+)-9-desmethylartemisi-

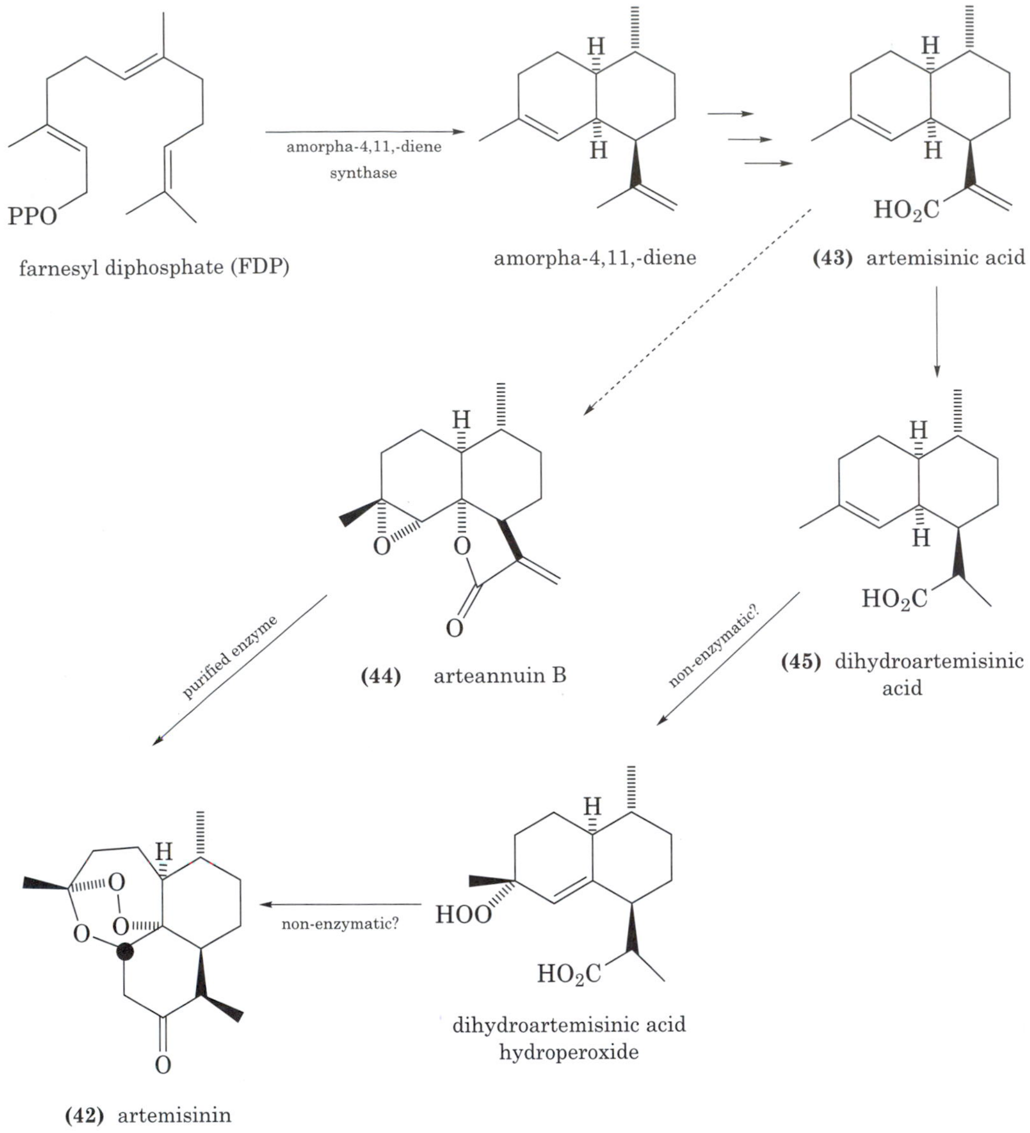

**Scheme 18.1.** Proposed biosynthetic pathways for artemisinin.

nin (559) and (+)-6,9-didesmethylartemisinin (560). Zhou et al., of the Shanghai Institute of Organic Chemistry, began their synthesis from natural artemisinic acid (561). Later, they synthesized artemisinic acid from 10$R$ (+)-citronellal (562), thus bridging the gap for the totally synthetic material (563–565). A group at the University of Alberta reported a total synthesis of (+)-artemisinin starting from (−)-$\beta$-pinene (566). Syntheses of specifically deuterated (567) and $^{14}$C-labeled arte-

misinin has been reported (568, 569). Other noteworthy syntheses have been provided (570–574). Syntheses of artemisinin and its derivatives and analogs have been reviewed (534, 535, 575, 576).

A number of constituents of *A. annua* other than artemisinin have been exploited in developing semi-syntheses of artemisinin and its related compounds. Among these, artemisinic acid (**43**) is 8-10 times more abundant than is artemisinin (547, 577) and can be isolated from the plant source without chromatography (578). Partial syntheses of artemisinin from artemisinic acid or dihydroartemisinic acid (**45**) have been reported by Roth and Acton (579, 580), Ye and Wu (574), Haynes and Vonwiller (581), and Lansbury and Nowak (582, 583), who also used arteannuin B (**44**) as a precursor for artemisinin. Although some very interesting chemistry has resulted from these approaches, the most economical source for artemisinin, both for use itself and for conversion into arteether, artemether, and artesunate, will continue to be *A. annua*.

A fine summary of the chemistry, stability, and reactions of artemisinin is found in (529). The molecule itself is surprisingly robust. It is stable at temperatures above its melting point (156–157°C) (584) and at reflux in neutral solvents (bp < 100°C) for several days. Selective reduction of artemisinin is possible: catalytic hydrogenation or reaction with triphenylphosphine produces deoxyartemisinin (**46**) and borohydride reduction provides dihydroartemisinin (**47**) (585). Although the struc-

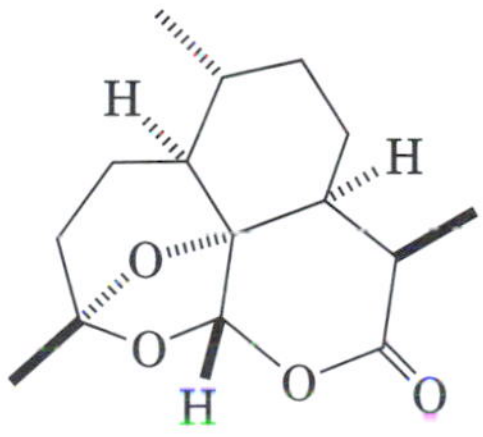

(**46**)
deoxyartemisinin

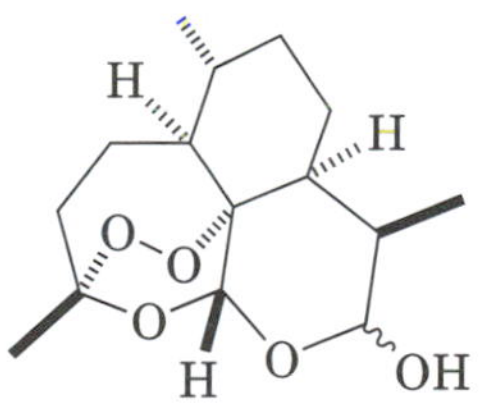

(**47**)
dihydroartemisinin

ture of artemisinin was originally determined by X-ray analysis (531), a more recent X-ray study has provided improved atomic coordinates and a comparison of conformations between artemisinin, dihydroartemisinin, and

artemether (**586**). $^{13}C$ data for a number of sesquiterpenoids and analogs related to artemisinin have been reported (587). Guidelines for assigning stereochemistry in synthetic trioxanes based on 1D and 2D NMR data have been compiled (588).

**2.7.2 Structure-Activity Relationships.** The artemisinins, although fabulously successful, do have drawbacks. Artemisinin has limited availability from the plant, relatively low potency, poor oral activity, poor oil and water solubility, and a short half-life. These issues and concern over neurotoxicity in some derivatives have resulted in a continuing search for improved compounds. A plethora of analogs and derivatives of artemisinin have been prepared and evaluated for antimalarial effect. Activity data on these compounds have been compiled in several excellent review articles (534, 535, 539, 575, 589). The study of these compounds has led to the development of some general structure-activity relationships (534, 575, 590).

The endoperoxide is essential for activity; deoxyartemisinin (**46**), in which the dioxy bridge has been reduced to mono-oxy, is completely inactive (528, 529, 540, 591). A number of synthetic peroxides have been screened for activity (592), and some very simple peroxides such as hydrogen peroxide and *tert*-butyl hydroperoxide, although only weakly active, have been found to suppress parasitemias of several Plasmodia *in vivo* (593). The presence of an endoperoxide is not sufficient on its own to confer activity, however (594, 595); ascaridole, a peroxy natural product, is not active at the maximum tolerated dose (596). Indeed, some bicyclic 1,2,4-trioxanes structurally unrelated to artemisinin do not possess appreciable activity (596–598). Clearly the peroxide grouping is necessary but not sufficient for activity.

These results have led researchers to speculate on what additional criteria need to be met in defining the antimalarial pharmacophore. Some have suggested that activity resides in the array of endoperoxide-ketal-acetal-lactone functionalities (599). The additional oxygen substituent at the 5 position of the 1,2,4-trioxane has been identified as a possible critical feature (598). A strong connection between the lipophilicity of a compound and its activity has led researchers to propose that the position of the peroxide bond relative to hydrophobic groups might be important (600). The investigation of 1,2,4-trioxanes as the central pharmacophore has borne promising preliminary results. The structural requirement of a 1,2,4-trioxane is not absolute, however (601, 602). Activity has been demonstrated in bicyclic peroxy compounds far removed from the artemisinin structure.

It is quite clear that modifications of dihydroartemisinin (47) and deoxoartemisinin (48) are well tolerated, and many derivatives

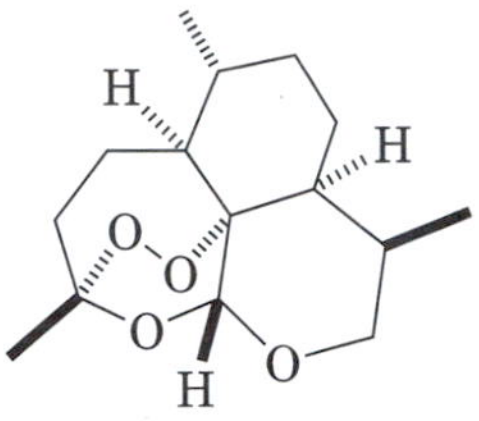

**(48)**
deoxoartemisinin

and analogs of these types have been prepared. Both compounds are more potent than artemisinin itself. Among the agents presently in clinical use, dihydroartemisinin derivatives account for all of them except artemisinin itself [dihydroartemisinin (47), artesunate (49),

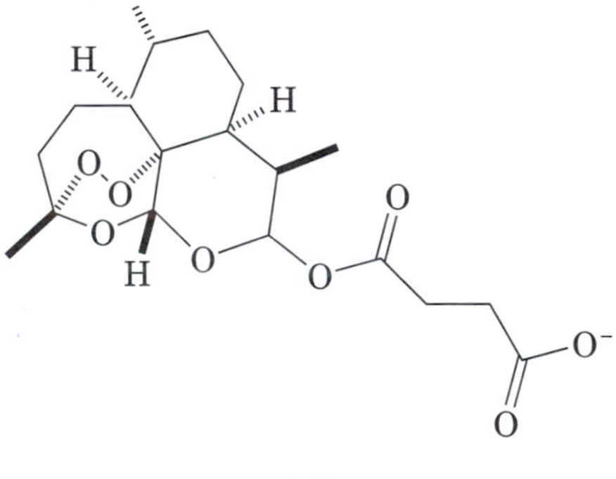

**(49)**
artesunate

artemether (50), arteether (51), and artelinic acid (52)]. It was realized quite early in structural studies that, in general, the ether, ester, and carbonate derivatives of dihydroartemisinin demonstrate appreciable antimalarial ac-

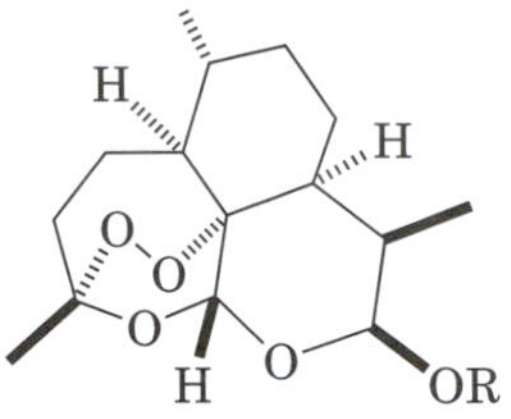

**(50)** R = OCH$_3$
β-artemether
**(51)** R = OEt
β-arteether

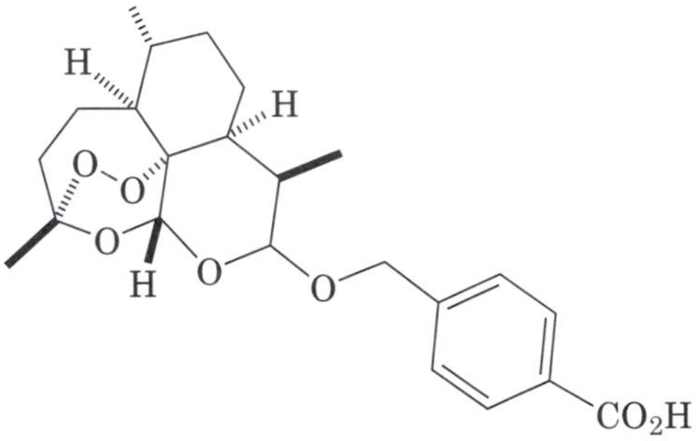

**(52)**
artelinic acid

tivity (603). One conclusion is that increasing polarity and water solubility of an artemisinin-type analog decreases the antimalarial activity. An example is that carboxylate esters have significantly greater antimalarial activity than do the corresponding carboxylic acids in one series of dihydroartemisinin derivatives (604). The lesson from many analog studies seems to be that lipophilicity is an important factor in maintaining and increasing antimalarial activity, even though it might also contribute to the toxicity associated with some analogs. In contradiction of this conclusion, one notes that artesunate (49) and artelinic acid (52), both water-soluble derivatives of dihydroartemisinin, are very active. One particular emphasis has been to identify water-soluble compounds with better stability that artesunate, the only currently available artemisinin drug that can be given intravenously. Several new families of active compounds, including those which are glycosylated (605) or containing amino groups (605–607), have been identified, but none has been of sufficient interest as yet for further development. A

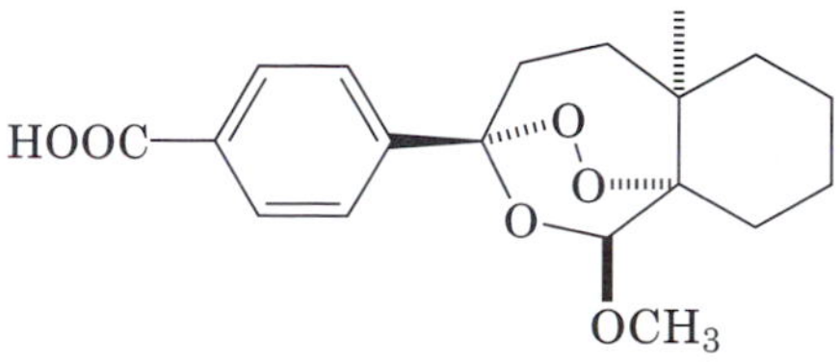

HOOC— OCH₃

**(53)**

**(55)**
isoartemisitene

**(56)**

more promising example, however, is (**53**), which incorporates a phenylcarboxylate at the 3 position (608).

The oxygen functionality on the D ring, either alkoxy or carbonyl, is not required for antimalarial activity. A change in the D ring size, from 6 to 7, results in a decrease in activity perhaps as a result of conformational changes (609). The addition of alkyl groups at position 9 on the artemisinin system provided a number of active analogs of deoxoartemisinin but was not uniformly advantageous to antimalarial activity (610). Introduction of functional groups at this site was similarly shown to be of little value (611). Smaller groups, methyl, ethyl, and propyl, can increase potency, but the effect drops off rapidly. The most lipophilic analogs are not the most active (600). Substitution and stereochemistry at position 9 does impact antimalarial activity. 9-*Epi*-artemisinin (**54**) and isoartemisitene

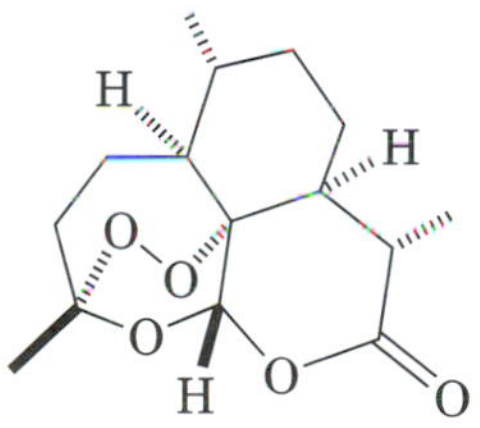

**(54)**
9-*epi*-artemisinin

pounds with an alkoxy group on a bicyclic trioxane tend to be more active than those without. A series of compounds based on alcohol structure (**57**) were synthesized and evaluated

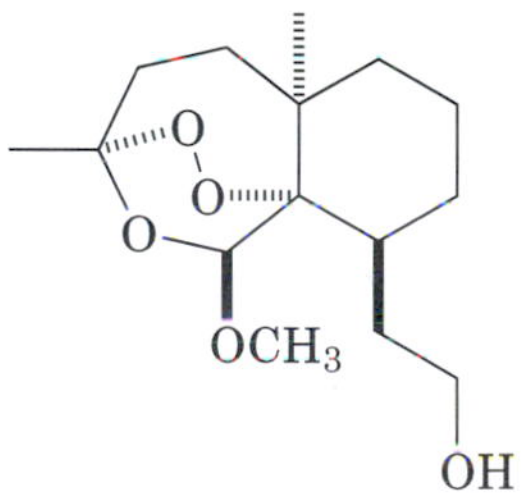

OCH₃

OH

**(57)**

for antimalarial activity (615). Several of the structures approached the activity seen with artemisinin. Most analogs have kept an intact C ring because it provides a convenient scaffold for construction of the bicyclic AB system. There is some suggestion that the 6-methyl contributes to the activity of artemisinin (575). Compounds with a simple dioxane ring instead of the trioxane such as (**58**) are active (601), again demonstrating that the trioxane is not a required component for antimalarial activity. Replacement of the methyl group at position 3 with other alkyl or alkaryl groups yields active compounds (610). In general, incorporation of substituents at position 3 leads

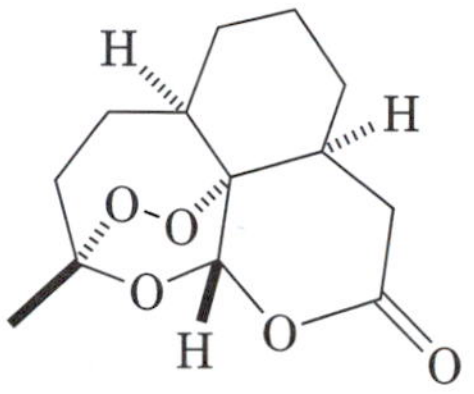

(**55**) are approximately an order of magnitude less active than artemisinin (612, 613). The methyl group in 9-*epi*-artemisinin may be situated such that steric hindrance of the peroxide bond results. 6,9-Didesmethylartemisinin (**56**) was reported to have significant activity *in vitro* (614). So while many changes at position 9 result in active compounds, not all variations lead to good activity. An intact D ring does not seem to be required, but those com-

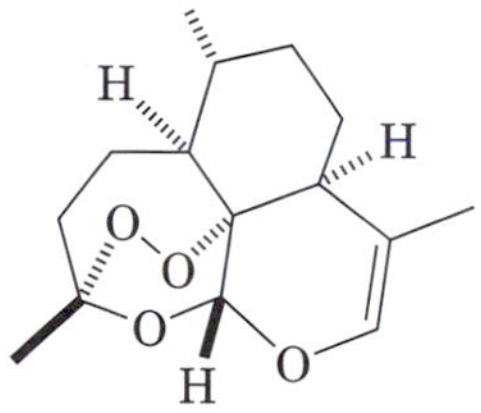

(58)

to compounds with less activity than compounds with substitution at position 9. The presence of an A ring provides compounds with greater activity; this may be because of the need for some degree of rigidity in the AB system.

There is no clear indication, either from the activity of racemic analogs or from the proposed mechanism of action, for chiral preferences in antimalarial behavior. For compounds derived from artemisinin itself, no question of absolute stereochemistry arises. Purely synthetic analogs have most often been prepared in racemic form; these compounds can have significant amounts of activity. Future studies may uncover stereochemical issues in the mechanism of action of the artemisinins with parasite proteins. Differences in relative stereochemistry can produce significant differences in antimalarial activity, however (612, 616). Much might be learned from examining the activity of the enantiomer of naturally occurring artemisinin. None has yet been prepared, as well as can be determined, either by asymmetric synthesis nor by separation of synthetic racemic material.

Acton et al., in the course of describing some artemisinin analogs with substituents at the 9 position, makes useful comments about artemisinin SAR (617). They observed that "lack of activity in a parent compound need not consign derivatives of that compound to the dustbin of chemistry since [structural changes] can convert compounds with little or no discernible activity into derivatives with quite respectable antimalarial behavior." An example of this is found in the work of Posner et al. (618), where a parent tricyclic 1,2,4-trioxane alcohol was relatively inactive while its derivatives showed good activity.

A number of quantitative structure-activity relationship (QSAR) studies have been carried out on artemisinin and on synthetic trioxanes using a number of different approaches and techniques (619–627). For the most part, these models gave satisfactory predictive results. It was suggested that the reactivity of this system is related to the rather dramatic region of negative potential that surrounds the molecule, including all three oxygens of the trioxane. The atoms of the trioxane ring along with the D ring oxygen, the lactol carbon, and the lactol oxygen were located as the important grouping for antimalarial activity. Cyclic voltammetry studies indicate that the irreversible reduction potentials of a number of artemisinins and analogs gave a good correlation with antimalarial activity reconfirming the essential nature of the peroxide moiety (628).

**2.7.3 Mechanism of Action.** The mechanism of action of the artemisinins against malaria has been the subject of fascinating recent work although the situation at present is far from settled. Several reviews are available that address this topic (534, 575, 629, 630). Early on it was determined that artemisinin had a direct parasiticidal action against *P. falciparum* in the erythrocytic stage both *in vitro* and *in vivo* (540, 631, 632). These compounds also show cytotoxicity at considerably higher concentrations than those necessary for antimalarial activity (633). The key morphologic changes observed on artemisinin administration to parasite-infected erythrocytes involves disruption of the food vacuole membranes (634). The breakdown of the vacuole membranes causes the release of the digestive enzymes resulting in harmful effects on the cytoplasm. Artemisinin enhanced the oxidation of lipid membranes in the presence of heme (635). Even so, the antimalarial action of artemisinin is not caused by a direct effect on the lipid structure of the membrane (636). A correlation in time between the morphological changes in the parasite and the biochemical suppression of protein synthesis by artemisinin was noted (637, 638). Tritiated dihydroartemisinin was reversibly and saturably accumulated from low concentrations into erythrocytes infected with *P. falciparum* (639). Uninfected erythrocytes concentrated the drug less than twofold, whereas infected erythrocytes achieved more than 300 times the medium concentration. Artemisinin and its derivatives are transported by a vacuolar network induced by *P. falciparum*; interaction with this membrane network may be implicated in the mechanism of action (640). Interestingly, labeled dihydroartemisinin becomes bound to both proteins and lipids in isolated red cell membrane preparations but does not associate with the same components in intact red cells (641). One explanation is that infected erythrocytes have a altered distribution of membrane lipids which may promote drug binding. Interactions between artemisinin and DNA were not observed (642). Artemisinin has high permeability across biological membranes (643).

The requirement of iron for activity of artemisinin is well known. Whereas hemoglobin and hemozoin react poorly, free iron and heme-iron react readily under a number of experimental conditions (644). The binding affinity of artemisinin derivatives to heme seems to correlate well with antimalarial activity (645). Free radical scavengers and iron chelators are known to antagonize the drug's action (646–648). Interestingly, compounds in which an iron chelating group has been attached to the artemisinin central structure were as active as artemisinin but no more active (649). Iron is not the only metal that can promote the decomposition of artemisinin. When plasmodia infected mice were treated with a combination of artemisinin and a manganese porphyrin complex, which by itself was devoid of antimalarial activity, the effect of artemisinin was increased by 50%. The authors suggest that the activation of artemisinin by a metal porphyrin *in vivo* was confirmed by these results (650).

Although controversy remains about the details of artemisinin's mechanism of action (651, 652), a general scheme based on the radical decomposition of artemisinin by iron has been developed. A number of lines of evidence led Wu and colleagues to propose a "unified" mechanism (Scheme 18.2) (653), developed from the pioneering work of Meshnick (654), Posner (630, 655), and Jefford (656) and their co-workers. The pathway commences with cleavage of the artemisinin peroxy bond by single electron transfer (SET) from Fe(II) of

heme. Two possible oxygen-centered radical anions could be generated by this process, with structures **A** and **B**. These intermediates may be interconvertible, explaining the different ratios of product obtained under various conditions. Further transformation of **A** by a 1,5-hydrogen shift yields a secondary carbon-centered radical **C** (655). Several routes lead from **C** to provide the known products of chemical and biological reactions (**59**), (**60**), (**61**), and (**62**). Alternatively, scission of the C3—C4 bond of **B** leads to primary carbon-centered radical **D** which is proposed to react intramolecularly to form a tetrahydrofuran product (**62**). Other scission pathways are also possible.

In support of this scheme, primary and secondary carbon radicals were detected in EPR experiments using artemisinin with iron in the presence of spin-traps in a ratio of 1:4, consistent with intermediates **D** and **C** (657). Subsequent transformations provided the products observed from a number of studies of the iron-mediated decomposition of artemisinin. Further evidence for the existence of a primary radical such as **D** was obtained from studies of various metalloporphyrins with artemisinin, artemether or Fenozan-50F (658–660). Covalent adducts between C4 of the drug and heme were fully characterized. Reaction of artemisinin with free iron in the presence of a sulfhydryl compound yielded a product that could only have been formed from **D** (661). Additional EPR data also exists in support of a secondary radical generated from arteflene (662).

Structures and stabilities of O-centered and C-centered radicals derived from a model trioxane (2,3,5-trioxabicyclo[2.2.3]nonane) using density functional theory found that the C-centered free radicals were predicted to be stable. The 1° C-radical (**D**) was more stable than the 2° C-radical (**C**) (663). Further calculational studies identified transition state structures for the radical transformations in the model compound and found a low activation energy for the 1,5-hydrogen shift process (664). From the calculated lifetimes of the O-centered radicals, the authors predict that these intermediates should be detectable at low temperature.

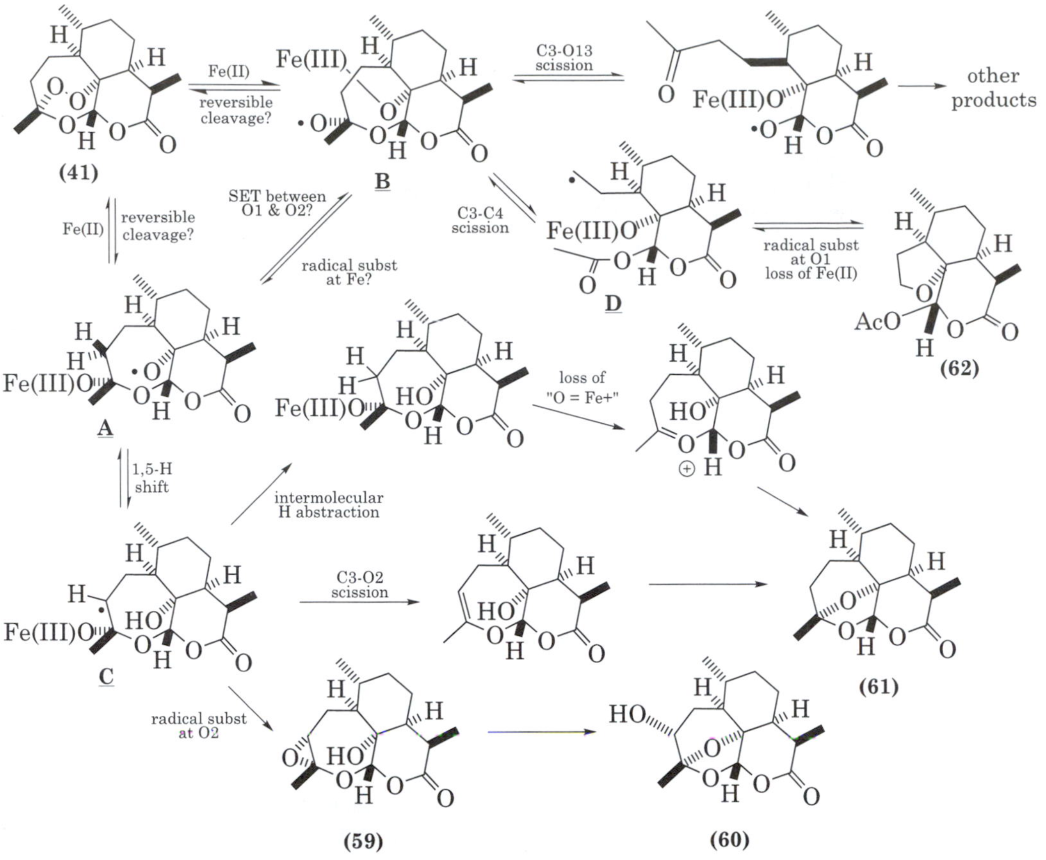

**Scheme 18.2.** Proposed radical decomposition of artemisinin.

Several workers have investigated the Fe(II)-mediated degradation of artemisinin derivatives and synthetic endoperoxides under biomimetic conditions (665–667). The major products obtained from these studies have been shown to be a tetrahydrofuran (**62**), supporting the intermediacy of a primary C-centered radical such as **D** and a hydroxy deoxo derivative (**60**) suggesting the involvement of a secondary C-centered radical intermediate such as **C**.

In additional studies, the possibility of the involvement of a carbon-centered radical (**A**) in the mechanism of action was explored (665). A series of analogs, some with abstractable 4α-hydrogen (**63**) and some without (**64**), were prepared and tested for antimalarial activity. Those compounds with an available 4α-hydrogen showed activity at 12–200 times that of the corresponding 4β-epimers

(668). These results are interpreted to suggest the requirement of an abstractable H at the 4α-position for good antimalarial activity. As further proof of the proposed mechanism of action, Fe(IV)=O has been detected in reactions of artemisinin with ferrous ion (669, 670). In Fourier transform infrared and resonance Raman studies with artemisinin and

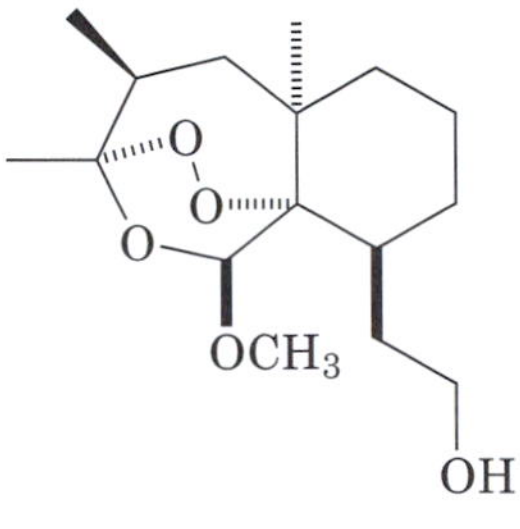

(**63**)

(64)

hemin dimer, researchers report identification of vibrational modes consistent with a heme ferryl-oxo (Fe=O) stretch (671). They believe this provides direct evidence for presence of this species in the interaction of artemisinin with heme. It is suggested that the high-valent iron-oxo species and an electrophilic (and cytotoxic) epoxide (59) derived from artemisinin are formed concurrently.

To test the requirements for the carbon-centered radical, Avery's group prepared an analog designed to stabilize such an intermediate (672). The analog lacked the non-peroxidic oxygen of the trioxane system and did not display significant antimalarial potency when compared with artemisinin. Some analogs with 4β-substituents that would be expected to provide stabilization for a radical were less active than the parent (unsubstituted) system (668). These results indicate that stabilization of a C-4 radical does not necessarily improve antimalarial activity. The exact role of a C-4 radical in the mechanism of action of artemisinin remains under discussion.

The selectivity of artemisinin's toxicity for infected erythrocytes, where the necessary heme-iron is available, is thereby accounted for as is the necessity for the peroxide functionality. Membrane damage is one result of the interaction between hemin and artemisinin (673). Molecular modeling studies of the interaction of artemisinin with heme suggest that a stable docking arrangement brings the endoperoxide bridge in close proximity to the heme iron (674).

It remains unclear how this pathway contributes to the overall destruction of the parasite. Several of the intermediates in the pathway are candidates for a reactive entity that confers cellular damage. The radical species themselves may act to alkylate proteins or heme. And the epoxide (59) has been suggested as a potential site for further reaction.

An alternative viewpoint suggests that whereas radicals may indeed be formed through interaction with iron sources in vivo, such radicals must be short-lived, evidenced by the trapping with porphyrins. The radical intermediates would not/do not have a sufficient lifetime to diffuse to other sites to accomplish parasite damage. Also, although formation of carbon-centered radicals is common in the reductive cleavage of endoperoxides, not all endoperoxides are active antimalarials. Rather, these authors envision a different pathway for artemisinin degradation and action (Scheme 18.3) (675–677). The peroxide may complex with Fe(II) or become protonated which then activates the ring-opening steps to generate an electrophilic cation E. Elimination of a proton yields an open hydroperoxide or a metal peroxide F. This intermediate then may be responsible for the remaining events. The biological significance of hydroperoxides in relation to hydroxylations and autoxidation is well established. It was observed that all active peroxidic antimalarials possessed groups that were capable of stabilizing the positive charge of E induced by heterolytic ring opening.

The next step in the proposed mechanism of action is the alkylation of proteins by artemisinin (678). A clear-cut correlation between antimalarial potency and the alkylative property of synthetic tricyclic trioxanes was reported (679). In the presence of hemin, artemisinin residues become covalently linked to human albumin, probably through thiol and amino groups (680). Artemisinin becomes bound to certain hemoproteins in vitro but not to heme-free globin and not to DNA (681). Further studies with artemisinin and artemisinin derivatives have demonstrated that the endoperoxides react with specific parasitic proteins to form covalently linked species (682). In P. falciparum, one of the main alkylation targets is the translationally controlled tumor protein (TCTP) homolog (683). In resistant parasites, expression of TCTP was increased over sensitive strains, although it is not clear whether this observation is connected with the mechanisms of resistance (684). Artemisinin also forms ad-

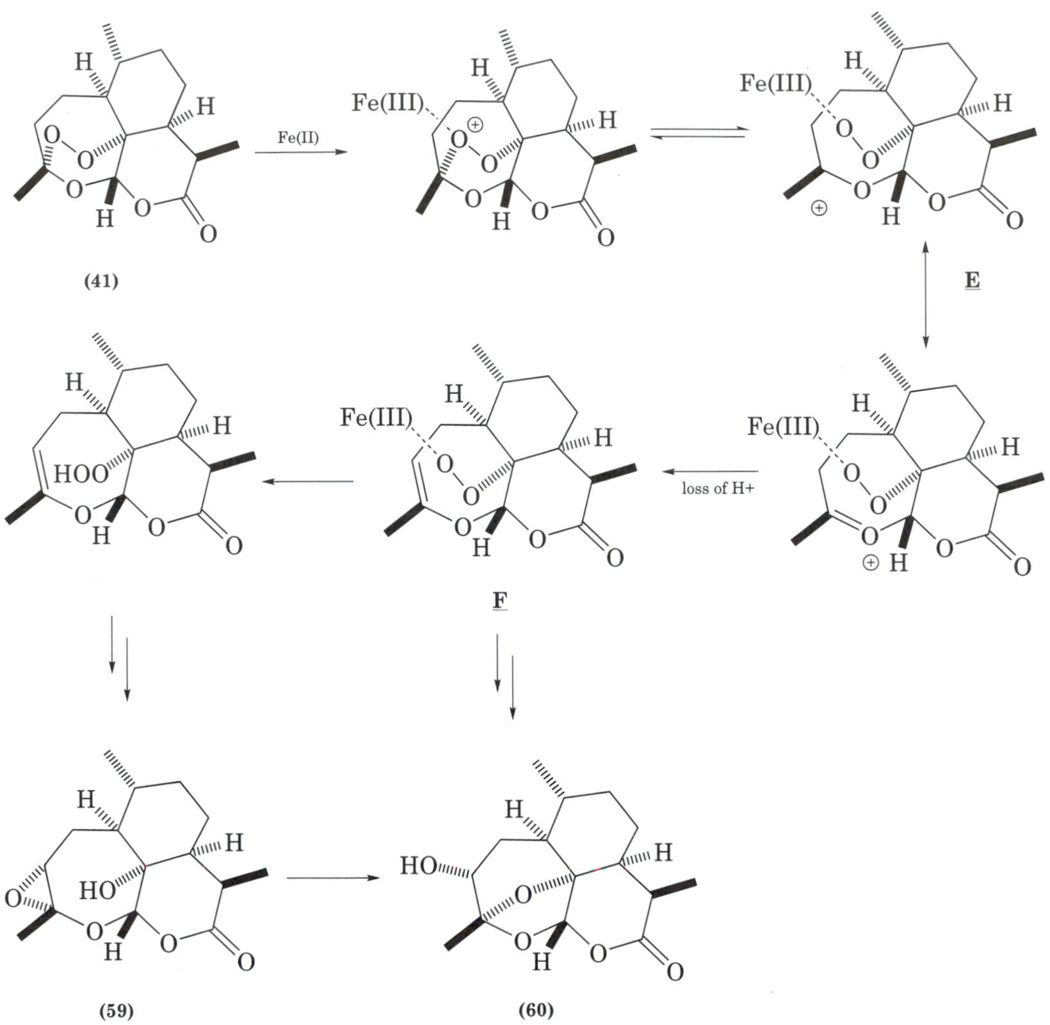

**Scheme 18.3.** Proposed heterolytic decomposition of artemisinin.

ducts with heme; a covalent alkylation product of heme with artemisinin was isolated and fully characterized (658). It is not clear, however, what role these drug-heme adducts may play in the mechanism of action of artemisinin. The heme-artemisinin adduct was not toxic to *in vitro* cultures of *P. falciparum* when added to the media (654).

Artemisinin also forms an adduct with glutathione, and it has been suggested that the reduction in the amount of available glutathione or the inhibition of glutathione reductase by the adduct may contribute to antimalarial activity (685). There are conflicting reports on the effect of artemisinin on hemozoin forma-

tion; one group states that they observed little effect from artemisinin (686) while others noted a great reduction in hemozoin production (687, 688) and potent inhibition of both hemoglobin breakdown and heme sequestration (689).

**2.7.4 First Generation Artemisinins.** More than 2 million patients have been treated with artemisinin agents. Clinical trials comparing various artemisinins in various routes of administration found no particular compound obviously superior to the others, nor was a particular route more efficacious; all treatments with all the compounds tested were safe, well tolerated, and highly effective (690–692). Several formulations of artemisinin and its derivatives have been developed for clinical use. Artemisinin and artemether can be dissolved or suspended in oil for intramuscular injection. An aqueous suspension of artemisinin has been used intramuscularly. Artesunate solutions are available for either intramuscular or intravenous injection. Tablets of artemisinin or artesunate and capsules of artemether are available for oral dosing. Suppositories of artemisinin and of artesunate are increasingly used. When any of the artemisinin drugs are used in monotherapy for uncomplicated malaria, whether given orally, by intramuscular injection, by intravenous injection, or by suppository, the mean times to clearance of parasites and fever are several hours shorter than with other antimalarials (693, 694). In cases of cerebral malaria, shorter coma resolution is observed. The drugs are well tolerated and without evident toxicity, although questions have been raised concerning the neurotoxicity of artemether and arteether. In addition, the artemisinins have the broadest stage specificity of action of any antimalarial currently in use. Treatment with artemisinin derivatives was found to greatly reduce the subsequent carriage of gametocytes providing the potential for interrupting transmission and preventing the spread of multi-drug resistant parasite (695). Given the ease of use and the efficacy of these drugs, it is expected that they will make a strong positive impact in community-based treatment (696).

The percent of recrudescence is high with the artemisinins, with an average value of 24% for over 2000 patients in various trials. Because of the difficulty in separating the incidence of recrudescence from re-infection, this frequency may be overestimated in practice. To achieve good 28-day cure rates with artemisinin drugs alone, treatment over 5–7 days is required. Given the realities of human nature, when symptoms improve rapidly, patients are less likely to continue drug treatment for longer than 3–4 days. The necessity for a prolonged course of therapy has been attributed to the short half-lives of the parent drugs and their active metabolites in plasma. Combinations of artemisinin drugs with other antimalarials are particularly attractive in addressing the problem of recrudescence. Careful choice of the adjunct agent is important because of the observed antagonism of artemisinin with antifolates, chloroquine, and pyrimethamine (697). Synergy has been noted for the combinations of artemisinin with either mefloquine (698) or tetracycline (697). A shorter course of treatment can be adopted when the drugs are used in combination, and thus patients are more likely to complete the prescribed therapy. A wide variety of combination regimens have been studied and several are very effective. The efficacy, pharmacokinetics, pharmacodynamics, clinical properties, and tolerability of the artemisinins have been reviewed (137, 155, 262, 532, 533, 699, 700).

***2.7.4.1 Artemisinin (42).*** Artemisinin has rapid action on all erythrocytic stages of the parasite (701). There is no sporontocidal action and no action against liver stages, but gametocytes are targets of the drug (702). Artemisinin kinetics demonstrated a decrease in plasma concentrations during the course of treatment, a time-dependent phenomenon observed in both healthy adults (703) and in malaria patients (704) that may partially explain the recrudescence commonly seen with these agents. A single dose was poorly absorbed and rapidly cleared with large inter-individual variance (705), and the results do not differ between healthy volunteers and patients with uncomplicated falciparum malaria (706). Food intake had no effect on artemisinin pharmacokinetics (707). In a study of artemisinin suppositories, a formulation that has been demonstrated to be effective and well tolerated, therapeutic concentrations were reached (708). Artemisinin apparently induces its own elimination because it was shown in a study in rats that changes in transport associated with efflux by P-glycoprotein are not involved (709). Artemisinin metabolism is mediated primarily by CYP2B6 in human liver microsomes (710); in addition, artemisinin induces CYP2C19 as well as at least one other enzyme (711). This may have implications for drug metabolism in combination therapy. One of the elimination products of artemisinin is the glucuronide of dihydroartemisinin. The human metabolite has been unambiguously identified as the 12$\alpha$-epimer (712).

The clinical efficacy and pharmacokinetics of artemisinin monotherapy and in combination with mefloquine have been studied (713). The combination produced a more rapid clearance of parasitemia than artemisinin alone and accomplished radical cures. The dose and duration of artemisinin treatment was reduced when used in conjunction with mefloquine. Single-dose combination treatments were equally effective in acute uncomplicated malaria whether the drugs were given together or sequentially (714). The high cost of mefloquine and its association with psychiatric disorders may limit the usefulness of this particular combination, however. A study in Vietnam showed that in uncomplicated falciparum malaria, a single dose of artemisinin along with 5 days of quinine was as effective as the standard 7-day quinine regimen (715). Combinations with tetracycline have also been examined (716).

As with all other known antimalarials, resistance to artemisinin can be induced (717), although this resistance is often unstable (684). In a study of *P. berghei* and *P. yoelii* strains, resistance to artemisinins and synthetic endoperoxides was induced by multiple passages of the parasite under drug pressure. When compared with resistance to amodiaquine, mefloquine, or atovaquone, resistance to the endoperoxides was at a low level and resistant parasites regained sensitivity once drug selection pressure was withdrawn (718). The possibility of cross-resistance with mefloquine, quinine, and halofantrine *in vitro* has been seen in some experiments (719). To provide baseline data for tracking the rise of resistance to the artemisinins, a study was carried out on drug susceptibility in Vietnam shortly after the introduction of artemisinins into nationwide use. All isolates were found to be sensitive to artemisinin, artesunate, dihydroartemisinin, and quinine. All were resistant to chloroquine and mefloquine (720).

WHO has issued guidelines for the use of artemisinin, suggesting that it be used as an alternative to quinine for the treatment of severe malaria in areas where sensitivity to quinine is reduced, that it be restricted to use in cases of multi-drug resistant infections, that it should always be used in combination with another effective drug or, in cases where combi-nations cannot be used, that a 7-day course of treatment is recommended with monitoring for compliance, and that it not be used as a prophylactic (721).

**2.7.4.2 Dihydroartemisinin (47).** Dihydro-artemisinin is more active *in vitro* and *in vivo* than artemisinin or artesunate (722, 723) and more active *in vitro* than mefloquine or halofantrine (724). All currently available artemisinin derivatives are metabolized rapidly to dihydroartemisinin, the main human metabolite. Although clinical trials have been carried out in China (725), the use of dihydroartemisinin has been limited by its low solubility and supposed instability. Treatment is effective and well tolerated in acute uncomplicated falciparum malaria (726). As yet, no evidence of neurotoxicity has been observed. Sequential combinations of rectal dihydroartemisinin with mefloquine was found to be a useful alternative to parenteral drugs in severely ill patients (727). As with all the artemisinins, recrudescence is a problem. In a dosing study, researchers in China determined that cure rates of >97% were achieved with a 7-day regimen of dihydroartemisinin; a 5-day regimen provided cure rates of 80% (728). Pharmacokinetic studies in human volunteers have shown that dihydroartemisinin is more bioavailable than artemisinin (729, 730). The $^1$H NMR spectrum, the X-ray crystal structure (731), and chemical stability studies (732) have been reported.

**2.7.4.3 Artesunate (49).** Artesunic acid is most commonly available as the sodium salt, sodium artesunate. *In vivo*, artesunate has higher activity than artemisinin (723) and significantly lower parasite clearance and fever clearance times (733). Artesunate suffers from intrinsic instability. Aqueous solutions are unstable at neutral pH and must be prepared from the powder by dissolution immediately before either intramuscular or intravenous injection. This is the only artemisinin derivative that can be given intravenously, a route that might allow improved pharmacokinetic parameters over bolus administration (734). The pharmacokinetics and pharmacodynamics of oral, rectal, and intravenous artesunate in healthy volunteers (735) and in uncomplicated falciparum malaria patients has been studied (736–738). The drug is rapidly absorbed, distributed, and eliminated. Oral and intravenous routes were comparable, but the rectal route may offer some advantages. Pharmacokinetic studies of artesunate in *P. vivax* confirms that the drug is rapidly effective either orally or intravenously. The properties were comparable with those reported for uncomplicated falciparum malaria (739). Both artesunate and dihydroartemisinin, its major metabolite, are rapidly cleared leading to concerns that a once daily dosing regimen allows periods of nonsuppression (740). Recent studies have confirmed, however, that once-daily dosing gives results identical with those obtained from more frequent administration, suggesting that pharmacodynamic behavior may be more significant than pharmacokinetic properties in these drugs (741). Artesunate was found to inhibit cytoadherence, the pathological process that contributes to microvascular obstruction in severe falciparum malaria (742). Studies of artesunate with hemin using electrochemical methods indicated that artesunate possesses an identical mechanism of action as does artemisinin (743).

Clinical trials have been carried out in China (725), Vietnam (744), and Thailand (745). Both 5- and 7-day regimens of oral artesunate were effective and safe in treating uncomplicated falciparum malaria in Thailand (746). The completed clinical trials of rectal artesunate have been summarized (511). Short-term monotherapy with artesunate suppositories affected a rapid reduction in parasitemia, especially useful with patients who cannot take medication by mouth, who are unable to travel to health care facilities, or who live in areas where administration of parenteral drugs is problematic (747). In a small study, artesunate was administered to pregnant women and was well tolerated with no adverse drug effects noted; normal neurological development was observed in all the neonates that were able to be followed (748).

A goal in the design of therapeutic regimens with artesunate has been the development of a treatment of short duration that maintains efficacy. In comparison with a 7-day course of quinine/tetracycline, a 5-day course of artesunate monotherapy (749) or a 7-day combination of artesunate and tetracycline (750) produced lower rates of recrudescence,

faster parasite and fever clearance times, fewer side effects, and equally high cure rates. Oral artesunate was superior to an intravenous quinine loading dose in patients at risk of developing life-threatening complications but who were able to take oral medications (751). Obviously, oral dosing is much more convenient especially in community treatment approaches and avoids risks of using needles.

A number of studies have shown the benefits of combining artesunate, administered either orally or as suppositories, with mefloquine in the treatment of multi-drug–resistant falciparum malaria (752–755). A sequential administration of artesunate followed by mefloquine was more effective than either drug alone, and it was well tolerated (756, 757). Artesunate rapidly reduces parasitemia, and mefloquine has a long-term effect in clearing residual parasites. This combination reduced the incidence of falciparum malaria and seemed to halt the progression of mefloquine resistance in a study in Thailand (758). Artesunate is synergistic with both quinine and mefloquine *in vitro* against *P. falciparum* (759).

The WHO has developed artesunate as an emergency treatment for malaria (760). The population most at risk of death are infants and children in Africa. In this group, symptoms can rapidly progress so that they are soon too ill to take an oral medication; the artesunate suppositories provide an effective alternative.

**2.7.4.4 Artemether (50).** Artemether (Paluther) is the methyl acetal derivative of dihydroartemisinin. The details of its structure have been elucidated by $^1$H NMR and X-ray crystallography (731). This oil-soluble derivative has high blood schizonticidal activity comparable with that of arteether *in vitro* and *in vivo* (761). As with artesunate, artemether inhibits cytoadherence (742). The main metabolite of artemether is dihydroartemisinin, for which artemether may be considered a prodrug. A number of hydroxylated metabolites were formed by microbial transformation (762) and as biliary metabolites in rats (763). Pharmacokinetic studies have been carried out in healthy subjects using oral (764, 765) and intramuscular administration (765). Studies of the clinical pharmacokinetics, safety, and tolerance of artemether in healthy volunteers and in patients with acute uncomplicated falciparum malaria have also been reported (339, 766, 767). Pharmacokinetics in patients from Thailand (768) and China (769) using multi-dose regimens were studied as well. The bioavailability of artemether in children with cerebral malaria was found to be highly variable, particularly when respiratory distress was also present (770).

Clinical trials have been carried out in China (725), Vietnam (744), Thailand (745), and India (771). There is a quick resolution of coma in children with cerebral malaria (772) and more rapid clearance of parasites than is seen with chloroquine (773) or quinine (774). In African children with severe falciparum malaria, artemether was highly effective especially against chloroquine and pyrimethamine resistant infections (775–777). Similar results were obtained in studies with Thai adults (778). Artemether compares favorably with, and may be superior to, quinine in treating severe falciparum malaria in children or adults (779–781). Of particular interest has been the identification of an agent to replace intravenous quinine for severe malaria in areas where quinine resistance is increasing. Intramuscular artemether was found to be as efficacious as intravenous quinine in Papua New Guinea (782) and in Nigerian children (783). Intramuscular artemether was completely effective in treating children with recrudescent P. falciparum, those having experienced treatment failures with other regimens (784). A meta-analysis of randomized clinical trials that compared artemether with quinine for severe malaria concluded that artemether was as effective as quinine (785). Substitution of artemether for quinine also avoids any cardiotoxic or hypoglycemic effects associated with quinine use.

A combination of artemether with mefloquine was found to be safe and effective in treating complicated falciparum malaria in Myanmar (786) and in treating drug-resistant malaria during the second and third trimesters of pregnancy (787). Very little recrudescence was observed with the combination. In a study of fresh isolates from Senegal (788) or Gabon (789) there were indications of in vitro cross-resistance of artemether with standard antimalarial agents. This is not necessarily an indication of in vivo cross-resistance but reinforces the need to use new drugs, particularly the artemisinins, in combination only.

### 2.7.4.5 Arteether (51).

Arteether (Artemotil) is the ethyl acetal of dihydroartemisinin. The drug is approved for restricted use as alternative treatment for multi-drug resistant P. falciparum in India (790). Arteether was synthesized in 1988 (541) and chosen for development because of its greater lipophilicity than artemether and the advantage of less toxic metabolites relative to artemether (ethanol versus methanol). Other synthetic methods to the arteethers have been reported (791, 792), and a stereoselective synthesis of α-arteether from artemisinin has appeared (793). Synthesis of the specifically labeled arteethers 11-[$^3$H]-arteether (794) and 14-[$^2$H]-arteether (795) have been accomplished. The β-isomer is crystalline and therefore easily purified; the α-isomer is not and both arteethers are low melting solids. NMR data, both $^1$H and $^{13}$C, are available (796, 797) as is an X-ray structure determination (731). The acid decomposition reactions of arteether have been studied (798, 799); decomposition in simulated stomach acid yielded compounds which retained antimalarial activity (800). A number of metabolites of arteether have been identified: dihydroartemisinin, deoxydihydroartemisinin, 3α-hydroxydeoxydihydroartemisinin, 3α-hydroxyarteether, 9α-hydroxyarteether, 9α-hydroxydihydroartemisinin, 3α-hydroxydeoxyarteether, and tetrahydrofuran derivatives (801–804). A different metabolite, 1α-hydroxyarteether was isolated as a result of microbial metabolism (805). Glucuronides of these hydroxylated metabolites were synthesized and all were significantly less active than arteether (806). β-Arteether is metabolized to dihydroartemisinin primarily by CYP3A4 in human liver microsomes (807).

Arteether is comparable in activity with artemether in vitro and in vivo (761). A strain of P. yoelii that was highly resistant to chloroquine, mefloquine, and quinine was completely susceptible to arteether (808). Although the drug is not active against sporozoites, hypnozoites, nor exo-erythrocytic stages of P. cynomolgi, it is several times more active than artemisinin in curing P. cynomolgi infections (809). Rings and young trophozoites are the most susceptible stages of the parasite life-cycle to arteether (810). The use of $\alpha,\beta$-arteether in patients with severe falciparum malaria in India, given intramuscularly once daily for 3 days, gave good results with no evidence of toxicity. As with artemether, arteether may serve as a good alternative to intravenous quinine in severe cases far from primary health care (811). The pharmacology of $\alpha/\beta$-arteether as a 30:70 mixture has been studied (812). In comparing the isomers, $\beta$ and $\alpha/\beta$ (30:70) are equally effective against *P. cynomolgi* in monkeys; $\alpha$-arteether is slightly less active (813). Pharmacokinetic studies in dog (814) and in humans (766) have shown the drug to be well tolerated with a long elimination half-life of 25–72 h. At higher doses, toxic effects in animals are seen on heart, brain, bone marrow, kidney, and liver (815). Arteether binds human plasma proteins *in vitro* (816), binding to $\alpha$1-glycoprotein more tightly than to albumin; concentrations of $\alpha$1-glycoprotein are markedly increased during malaria infection. Synergism has been observed between arteether and mefloquine or quinine *in vitro* (817).

Both arteether and artemether have been shown to cause neurotoxicity *in vitro* (818, 819) and in animal models (820–822), but the effects only occurred at high doses or after prolonged exposure (823). Arteether was demonstrated to produce irreversible neural injury in monkeys after 14 days of intramuscular administration. It remains unclear whether a 5-day course of treatment that would constitute a typical therapeutic regimen would produce similar damage (824). A single high dose of arteether produced brainstem neuropathology in rats, although behavioral indicators of toxicity were negative (825); this observation points out the difficulty in detecting the onset of arteether-induced toxicity. Arteether was shown to accumulate in plasma of rats after repeated intramuscular dosing of a oil solution. While this may contribute to efficacy, it also contributes to toxicity (826). There was a significant difference in toxicity depending on route of administration and length and amount or dose. Intramuscular administration or oral administration in the presence of peanut oil increased the toxic effects in mice

(827) and dogs (828). On the other hand, once daily oral administration, especially under the usual 3- to 7-day course, was relatively safe.

Brainstem auditory pathways may be particularly vulnerable (829); in rats, there were significant decreases in auditory accuracy at high doses of arteether (830). An auditory evaluation of patients who had been treated with at least two courses of oral artemether or artesunate therapy were found not to differ significantly from controls. In this study, there was no evidence of significant neurotoxicity (831).

Since the original report of neurotoxicity in animals, clinicians have been looking for signs of clinical neurotoxicity. A study of adverse effects in thousands of falciparum malaria patients in Thailand found that artemether or artesunate treatment had fewer side effects than when therapy involved the combination of an artemisinin with mefloquine (832). One case was reported in which a patient suffered an extended period of tremors after two courses of artemether therapy (833).

The results of the *in vitro* studies suggested a specific but as yet unidentified neuronal target. Compared with arteether and artemether, derivatives such as artesunate showed less injury, whereas dihydroartemisinin showed more by parenteral administration (834). *In vitro* studies have shown that the neurotoxic effects of artemisinins were enhanced by the presence of hemin (835). The presence of the endoperoxide is required for toxicity. Higher lipophilicity was associated with greater neurotoxicity in a study of the stereoelectronic properties of the artemisinins (836). Dihydroartemisinin produced changes in differentiating NB2a neuroblastoma cells that may be related to the neurotoxicity in animal models (837). Further, the neurotoxic effects seem to be mediated by drug binding to cellular proteins, binding that is increased in the presence of hemin (838). Antioxidants (ascorbic acid or glutathione) completely protected against the drug induced toxicity *in vitro* (839). Again, the interaction of artemether with hemin was suggested as a possible toxic mechanism of toxicity.

***2.7.4.6 Artelinic Acid (52).*** Artelinic acid was developed as a water-soluble alternative to artesunate, and it has better stability in aqueous solution (840). A stereoselective synthesis of α-artelinic acid has been reported (841). The drug possesses superior activity against *P. berghei in vitro* compared with either artemisinin or artesunate (840). The pharmacokinetics and pharmacodynamics of artelinic acid in rodents have been studied (842) along with its clinical pharmacology (339). Sodium β-artelinate was demonstrated to be an active gametocytocide against *P. cynomolgi* by oral and intravenous administration (843). When compared with other artemisinin derivatives, artelinic acid had the highest plasma concentration, the highest oral bioavailability, the longest half-life, the lowest metabolism rate, and the lowest toxicity at equivalent doses (844), most likely because of its lower extent of conversion to dihydroartemisinin and slower elimination (845). The drug has not yet found wide use.

## 2.8 Other Agents

### 2.8.1 Atovaquone and Malarone.
The potential for anti-protozoal activity of hydroxynaphthoquinones has been known for many years; hydrolapachol (**65**) was found to

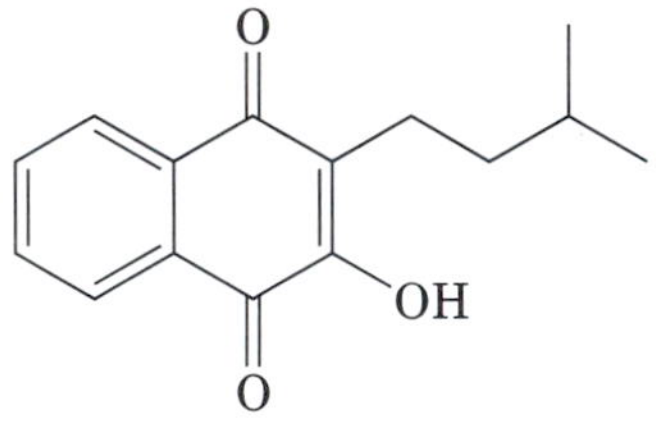

**(65)**
hydrolapachol

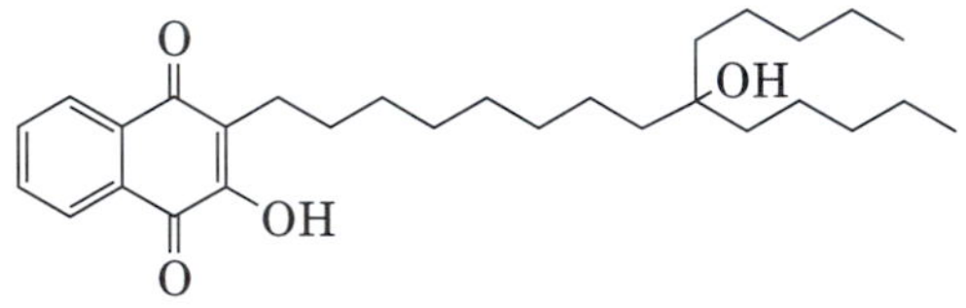

**(66)**
lapinone

be active against an avian malaria. Further studies identified lapinone (**66**) for clinical tests. Although effective against *P. vivax*, parenteral administration was required and in-

terest lagged. Attempts to parlay these observations into clinically active compounds led to the development of, first, menoctone (**67**), and

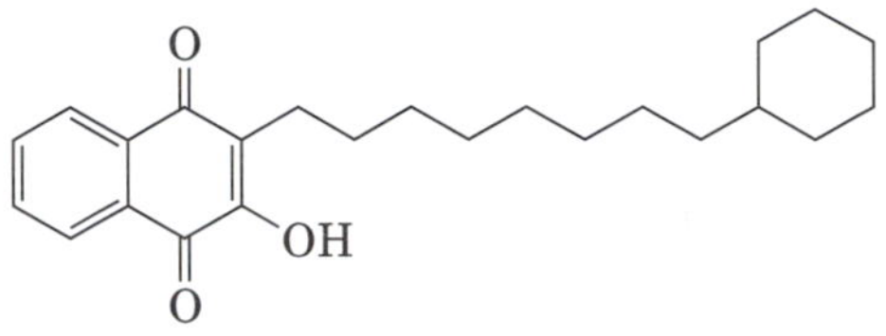

**(67)**
menoctone

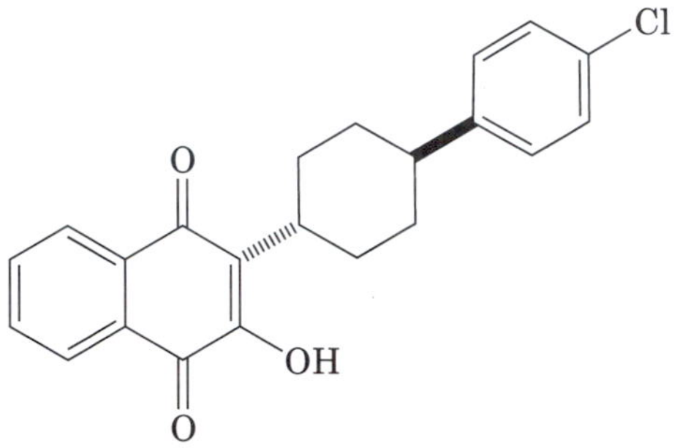

**(68)**
atovaquone

more recently, atovaquone (**68**) (846, 847). A short synthesis of atovaquone has been communicated (848). Menoctone, although active in mice, demonstrated no activity in humans infected with *P. falciparum*. Poor bioavailability was the cause of that disappointing result. Also known as BW 566C80, atovaquone was selected for development from a series of synthetic hydroxynaphthoquinones. Metabolic studies had shown that many of the compounds in the series were unsuitable for clinical use because of extensive human metabolism, even though good activity had been reported in animal models (849). Atovaquone, in contrast, was inert to human liver microsomes, displayed a plasma elimination half-life of 70 h, and was well tolerated. Absorption of the drug is slow and irregular, however. An increase in the plasma concentrations of atovaquone have been observed when dosing while giving with fatty foods (850).

Malaria parasites are unable to salvage pyrimidine bases and nucleotides; acquisition must be through *de novo* synthesis. Dihydroorotate dehydrogenase, a central enzyme in the *de novo* pyrimidine biosynthetic pathway, is moderately inhibited by atovaquone, which results in the accumulation of dihydroorotate and carbamoyl-aspartate, intermediates in pyrimidine biosynthesis (851, 852). Although growth of *P. falciparum* was inhibited by atovaquone, only moderate decreases in UTP, CTP, and dTTP levels were observed. Atovaquone at 15 $\mu M$ decreased the activity of human dihydroorotate dehydrogenase by 50% (853). Kinetic data demonstrated that atovaquone was a competitive inhibitor with respect to the quinone co-substrate and an uncompetitive inhibitor with respect to the substrate dihydroorotate.

A second activity of this naphthoquinone is likely more directly related to antimalarial effect. Atovaquone is thought to act at the cytochrome *bc1* (Complex III) site of the mitochondrial electron transport chain by indirectly inhibiting several metabolic enzymes linked through ubiquinone (847, 854, 855). This results in the collapse of the plasmodial mitochondrial membrane potential. Parasite respiration was inhibited within minutes after drug treatment (856). The inhibitory effects were 1000 times higher in Plasmodia mitochondria than in isolated rat liver mitochondria. And purified ubiquinol-cytochrome *c* reductase from *P. falciparum* was more sensitive to atovaquone than was the mammalian enzyme (857). Intentionally developed resistance to atovaquone was accompanied by mutations in the cytochrome *b* gene of *P. berghei* (858) or *P. yoelii* (859). Recent evidence indicates that *P. falciparum* uses both a cytochrome chain and an alternative oxidase pathway for respiration (860). The action of atovaquone is potentiated by compounds that inhibit this alternative oxidase pathway, for example propyl gallate. Atovaquone inhibits up to 73% of *P. falciparum* oxygen consumption; this is consistent with known activity on the cytochrome *bc1* complex. Taken together, these findings strongly support the interaction of atovaquone with cytochrome *bc1* as a critical component of its mechanism of action.

Trials *in vitro* showed good response and no evidence of cross-resistance (861). Human trials with oral atovaquone indicated that initial responses were good, with prompt clinical and parasitologic response, but that rates of recrudescence were high regardless of length of treatment (862, 863). Re-treatment of the patients that suffered recrudescence with atovaquone were not successful, suggesting the development of drug resistance. Atovaquone demonstrated causal prophylactic activity in non-immune subjects who were challenged with mosquito transmitted *P. falciparum* (864). A number of agents were found to be synergistic with atovaquone including PS-15 (865), doxycycline (866), proguanil, tetracycline, or 5-fluoroorotate *in vitro* (867, 868). The elimination route for atovaquone in humans is almost exclusively in the feces, with no evidence of metabolites (869). Population pharmacokinetics in black, Oriental. and Malay groups have been reported (870).

Against *P. berghei*, atovaquone is highly potent against mosquito stages (871, 872), perhaps because of its ability to block the recruitment of asexual parasites and/or young (stage 1) gametocytes into the population of morphologically identifiable gametocytes (stages 2–5) of *P. falciparum in vitro* (873). Liver stage parasites are also inhibited when cultured *in vitro* (874).

The combination of atovaquone and proguanil has been approved and is now marketed by Glaxo Wellcome as Malarone. The clinical development of Malarone has been reviewed (875). In randomized, controlled clinical trials, treatment with Malarone was significantly more effective than mefloquine (Thailand), amodiaquine (Gabon), chloroquine (Peru and Philippines), pyrimethamine/sulfadoxine (Zambia) (876), and chloroquine plus pyrimethamine/sulfadoxine (Philippines) (877). Overall cure rates exceeded 98% in more than 500 patients with falciparum malaria, even in strains that were resistant to proguanil or had been shown to be refractory to conventional therapy (878, 879). The pharmacokinetics of atovaquone and proguanil and its metabolite cycloguanil were the same whether administered alone or in combination (880). The recommended regimen is 1000 mg atovaquone and 400 mg proguanil once a day for 3 days (863). It is effective in non-immune patients with uncomplicated falciparum malaria (881), safe and effective in children (882), effective when used as the suppressive agent against *P. vivax* infections (883), and effective against *P. ovale* and *P. malariae* infections (884). In studies of Malarone for prophylaxis in various parts of Africa, the drug was found to provide excellent efficacy along with being well tolerated (885–888). The CDC includes Malarone among the agents recommended as prophylaxis for persons traveling to areas known to have chloroquine-resistance (92) and for presumptive self-treatment for travelers.

The two drugs are synergistic. Proguanil itself has no effect on electron transport nor mitochondrial membrane potentials (889). But it significantly enhanced the ability of atovaquone to collapse mitochondrial membrane potentials when used in combination.

Atovaquone is a potent, selective inhibitor of a variety of medically important protozoa as well as Plasmodium species. It is now used to treat a number of AIDS-related opportunistic infections including *Pneumocystis carinii* pneumonia, toxoplasmosis, and babesiosis. The drug is expensive to produce and will not be affordable in poorer areas, especially Africa. Glaxo Wellcome has announced its intention to provide the Malarone through a controlled donation program (890), part of which involves giving away 1,000,000 treatments each year to countries in Africa. The offer has met with skepticism by some governments, but pilot projects using Malarone are now underway (891). Extensive product information for Malarone is available on the web (892).

A triple combination of atovaquone with dapsone and either proguanil or PS-15 holds promise for treating multi-drug resistant parasites. It was more active than the combination of atovaquone with proguanil perhaps because of two pairs of synergistic interactions: proguanil with atovaquone and cycloguanil (the metabolite of proguanil) with dapsone (893, 894).

**2.8.2 Lumefantrine.** Lumefantrine (benflumetol, **69**) was originally synthesized by the Academy of Military Medical Sciences in Beijing and underwent preliminary clinical trials in China. It belongs to the quinoline alcohol group that also includes quinine and mefloquine. A synthesis has been reported (895) as has a crystal structure (896). The compound is

**(69)**
lumefantrine

used in racemic form, and no substantial differences were found between the racemate or each enantiomer separately (897). A metabolite of lumefantrine, desbutyl-lumenfantrine is active in its own right (898). The accepted view of the mechanism of action involves the interaction of the drug with heme (899). Lumefantrine shares several of the pharmacokinetic properties of other antimalarials, notably mefloquine and halofantrine, in that it has a highly variable oral bioavailability between doses and between patients and is eliminated slowly. Pharmacokinetics and pharmacodynamics of lumefantrine in acute *P. falciparum* infections were reported (900).

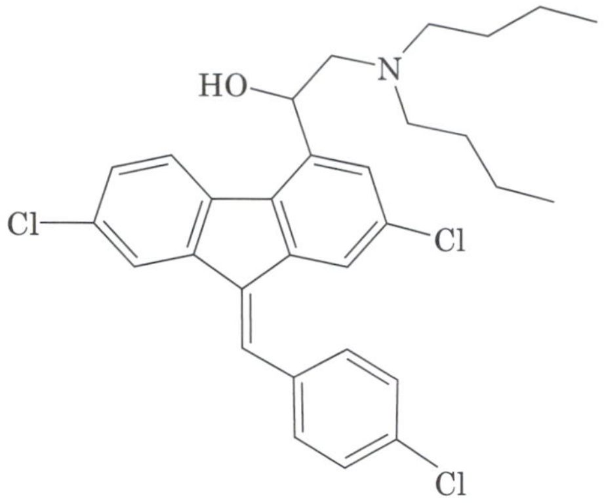

The activity of lumefantrine did not differ between chloroquine-sensitive and chloroquine-resistant strains. Its *in vitro* activity against Cameroonian and Senegalese isolates was similar to that of mefloquine and slightly lower than that of artemisinin or pyronaridine (901, 902).

A fixed combination of lumefantrine with artemether (CGP 56697, Coartem, Riamet) has proven to be particularly promising. Phase II clinical trials of the combination were carried out in China. Parasite reduction at 24 h was 99.4%, the cure rate was 96%, and recrudescence was low (903). No adverse effects were encountered. Subsequent development has been carried out by Novartis Pharmaceuticals, and the combination was approved in Switzerland in 1999 (904). The clinical efficacy of the combination has been confirmed in studies carried out in Thailand

(905, 906), India (907), The Gambia (908), and Tanzania (909). In some cases, a relatively high level of recrudescence was noted (910). In others, the cure rates were lower than that seen with mefloquine (911). A higher dose regimen was suggested to improve these slight drawbacks. For a summary of completed clinical trials of lumefantrine/artemether, see Ref. 511.

Each tablet is 20 mg artemether and 120 mg lumefantrine, and a therapeutic regimen for adults is four doses of four tablets each at 0, 8, 24, and 48 h. In cases of multi-drug resistant infections, six doses of four tablets would be extended over 3 days. The fixed combination achieves its effect through an initial large reduction in parasitemia by the rapidly absorbed artemether and the subsequent removal of any remaining viable parasites by the intrinsically less active but more slowly eliminated lumefantrine (904). The synergism between the two drugs was documented even in a multi-drug resistant strain (912), and reviews of clinical pharmacokinetics (904, 913) and pharmacodynamics (913) of the combination are available.

WHO and Novartis are providing developing countries with Coartem at a cost of approximately US$0.10 per tablet, amounting to less than US$2.50 per treatment for adults and even less for children.

### 2.8.3 Pyronaridine.

The drug pyronaridine (**70**) (or malaridine), first synthesized in 1970,

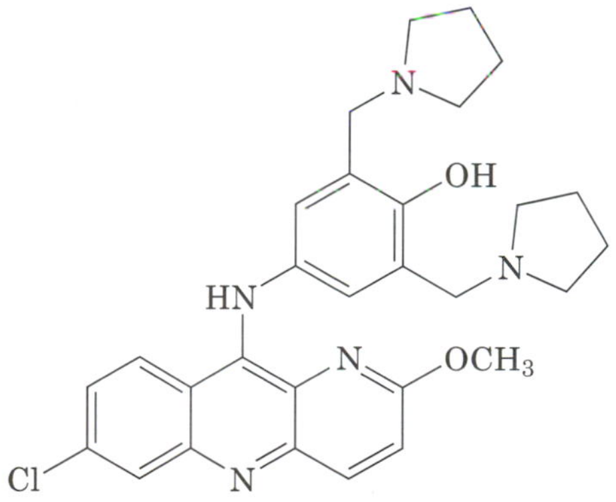

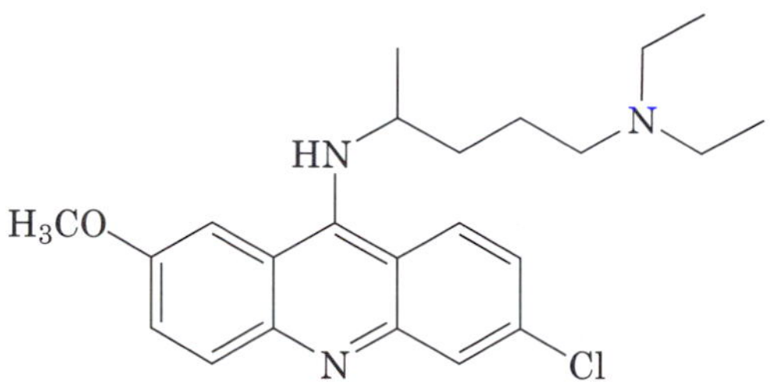

(71)
quinacrine

marily in China. Quinacrine was selected as the lead compound because of the activity of its derivatives against chloroquine-resistant strains of Plasmodium but was not pursued as an antimalarial agent itself because of toxicity. The side-chain of pyronaridine is similar to those of amodiaquine (**72**) and amopyro-

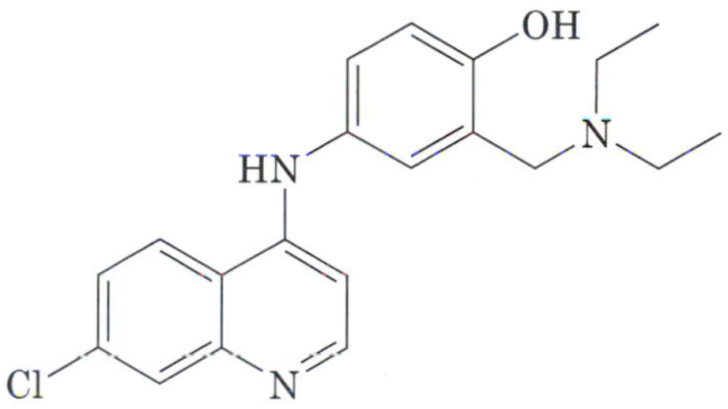

(72)
amodiaquine

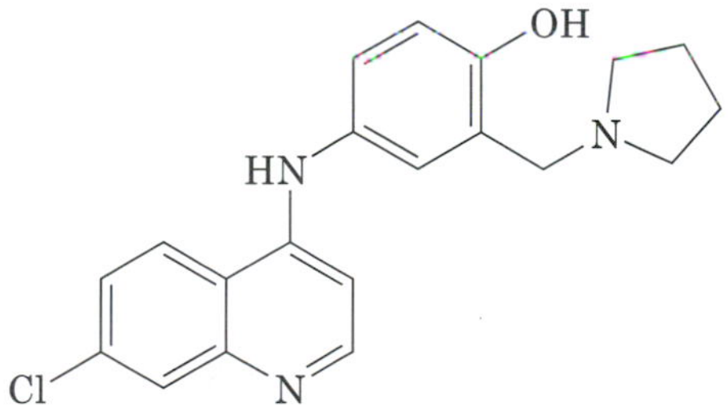

(70)
pyronaridine

is the product of many years of research that began with the quinacrine (mepacrine) nucleus (**71**), work that has been carried out pri-

(73)
amopyroquine

quine (**73**). Issues leading to the design of pyronaridine and its synthesis have been reviewed (914). Reviews has also covered the activity *in vitro* and *in vivo*, toxicity, pharmacokinetics, and clinical studies (914–917). Pyronaridine is active against the erythrocytic stages of malaria, although its mode of action is not well understood. It is not active against gametocytes (918). Early reports suggested that inhibition of the *P. falciparum* topoisomerase II enzyme was associated with the mechanism of action of the drug (919). Pyronaridine inhibited parasite growth in the low nanomolar range but inhibited topoisomerase II only in the micromolar range, with no evidence of drug concentration in the parasite nucleus. In more recent studies, pyronaridine failed to show topoisomerase II activity against *P. falciparum* enzyme *in situ* (920). It has been observed that similar morphologic changes are seen in both pyronaridine- and chloroquine-treated parasites (921) and that the parasitic digestive system is affected (922). Pyronaridine is converted into reactive intermediates by oxidation of its *p*-aminophenol moiety (923).

Preliminary clinical studies in China demonstrated the efficacy of pyronaridine against *P. falciparum* and *P. vivax* (916, 917). Similar results were obtained when the drug was used to treat *P. ovale* and *P. malariae* infections (924). Field testing in China has included several thousand patients, and the drug is well tolerated, with few major adverse effects being reported. Testing in other areas (Thailand, Cameroon) has confirmed the Chinese results (925–927). For a summary of completed clinical trials, see Ref. 511.

Pyronaridine is available as its phosphate salt and can be given orally, intramuscularly, and intravenously. Recrudescence rates with pyronaridine alone are ~12%, which is good but not exceptional (925). Given the propensity for resistance to pyronaridine to develop rapidly in *in vitro* studies (928, 929), most researchers are advocating the development of combinations with pyronaridine rather using the drug in monotherapy. In particular, a triple combination of pyronaridine, sulfadoxine, and pyrimethamine has been studied with promising results; a single optimized dose gave a 100% cure rate in patients with acute falciparum malaria (930). In addition, a 5-year study of the triple combination, during which it was given as the only antimalarial drug at the study site in the Hainan Province, China, demonstrated that its effectiveness remained at the 100% level (931). No evidence of resistance was obtained. In an effort to evaluate alternatives for radical cure of vivax malaria, a combination of pyronaridine with primaquine was shown to have lower toxicity than chloroquine/primaquine in mice and rats (932). The effects of pyronaridine with primaquine were reported to be synergistic (933). Potentiation of antimalarial action was observed when pyronaridine and artemisinin were used in combination with either an artemisinin-resistant or a pyronaridine-resistant strain of *P. yoelii* (929).

Although cross-resistance with chloroquine might be expected based on their structural similarities, pyronaridine is effective against chloroquine-sensitive, chloroquine-resistant, and multi-drug–resistant parasite strains, comparing well with other common antimalarials *in vitro* and in rodent models (934–939). Paradoxically, a pyronaridine-resistant strain was shown to be resistant to chloroquine. Interesting patterns of cross-resistance among pyronaridine and several structural relatives in rodent malarias were noted; resistance to one does not necessarily imply resistance to the whole group (940, 941).

Estimates of the economics of pyronaridine use suggested that in Cameroon, pyronaridine treatment costs about three times as much as chloroquine and about twice as much as amodiaquine or sulfadoxine/pyrimethamine. It is less than one-half as expensive as a full course of oral quinine, halofantrine, or sulfadoxine/pyrimethamine/mefloquine. Given the increasing resistance to chloroquine in Africa and the toxicity of amodiaquine, pyronaridine has been suggested as a cost-effective alternative treatment (926). A note of caution has been sounded, however, about the rapid deployment of this agent in endemic areas such as West Africa (942). Additional detailed pharmacokinetic data are required as well as long-term toxicity studies, especially given the structural similarities between the sidechains of pyronaridine and amodiaquine.

A review of the development of pyronaridine is available (943). The drug has been developed and commercialized in China but is not routinely available elsewhere. At present no commercial partner has seemed interested in working toward international registration.

**2.8.4 Amodiaquine.** Amodiaquine (**72**, camoquine) is a 4-aminoquinoline and is structurally related to chloroquine (944). The crystal structure has been reported (945). The use of amodiaquine in the treatment of uncomplicated malaria has been reviewed (946). It was found that amodiaquine was decidedly more effective than chloroquine, despite the similarities in chemical structure, with times to parasite clearance being significantly shorter and fever clearance marginally faster. As might be expected, amodiaquine was less potent against chloroquine-resistant than against chloroquine-sensitive strains, but more potent than chloroquine against the resistant strains *in vitro* (947). Its toxicity seemed comparable with that of chloroquine when administered at doses up to 35 mg/kg over 3 days. Even though amodiaquine has not been as widely used in the treatment of malaria as chloroquine, good results have been obtained in comparison with other agents (948–950).

A combination of amodiaquine with pyrimethamine/sulfadoxine gave better control of clinical symptoms than did the antifolates alone with no evidence of serious side-effects (951). Resistance to amodiaquine is not prevalent (879) but has been observed (950, 952, 953). An association between resistance to amodiaquine and resistance to chloroquine has been demonstrated in *P. falciparum* isolates, although the resistance mechanism is much less effective for amodiaquine than for chloroquine (954).

Amodiaquine was thought to be well tolerated until it began to be used widely for prophylaxis. At that point, evidence for its long-term toxicity emerged; consequently, amodiaquine is no longer used prophylactically. Severe adverse effects associated with amodiaquine are agranulocytosis (955–957) and liver toxicity (957, 958). It has been suggested that the toxicity of amodiaquine is related to reactive derivatives formed by oxidation of its phenolic side-chain, especially to the formation of a quinone-imine metabolite by biological oxidation (923, 959–962). Much like chloroquine, amodiaquine has been shown to interact with ferriprotoporphyrin IX (963).

The pharmacokinetics of amodiaquine have been studied (121, 964–966). The drug disappears rapidly from the blood with a terminal half-life of about 2 h after intravenous injection. An increased level of amodiaquine accumulation in comparison with chloroquine may be caused by enhanced affinity for a parasite binding site and may also explain the greater inherent activity against *P. falciparum* (967). Amodiaquine could be considered a prodrug; its main metabolite, monodesethyl-amodiaquine, is found in high concentrations in the blood after dosing and is highly active (966, 968, 969). Other metabolites that have been identified are bi-desethylamodiaquine and hydroxydesethylamodiaquine, both of which have negligible activity (970, 971). Amodiaquine was shown to be very weakly mutagenic in an Ames mutagenicity assay (972).

SAR work on amodiaquine had previously shown that wide variations in the side-chain are accommodated with retention of antimalarial activity.

Blocking of bioactivation pathways either through removal of the 5′ phenol or introducing a non-reactive substituents has been the main strategy (973, 974). Replacement of the 3-diethylamino group with cyclic amines or with a tert-butyl group provided analogs (**74**) that were more active than amodiaquine (975–

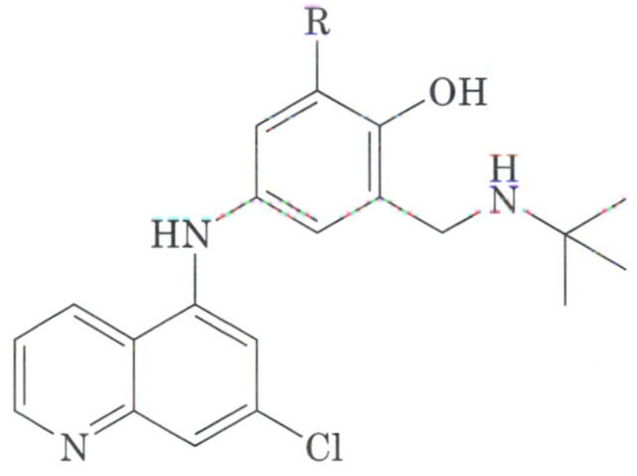

**(74)**
R = propyl, isopropyl

977). Reducing bioactivation also seems to result in compounds with slower elimination and increased tissue accumulation (978). The suggestion is made that a specific binding site for amodiaquine-like compounds may exist and that characterization of such a site would facilitate further drug design. A conformational analysis on structures related to amodiaquine revealed that the interatomic distance between the ring and side-chain nitrogen atoms is an important determinant of antimalarial activity (979). Although amodiaquine does not form a complex with heme, it does inhibit hemozoin formation ($IC_{50}$, 250 m$M$) (32).

One analog of amodiaquine in particular was considered quite promising at one time. Tebuquine (**75**) was found to be more potent

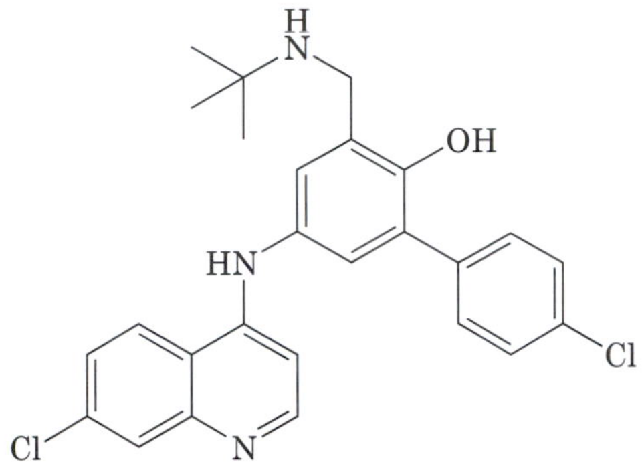

**(75)**
tebuquine

than amodiaquine or chloroquine, have excellent activity against resistant parasite strains, was curative in primate models, and had an extended duration of action (980). Additional tebuquine analogs were prepared and studied, although none was superior to the original (981).

A difference of opinion has arisen about the future of amodiaquine in malaria management. The drug is clearly not appropriate for suppressive prophylaxis. However, there may be more leeway in regard to therapeutic use. At present, amodiaquine is not recommended for treatment or prophylaxis by the World Health Organization because of hepatic and hematological toxicity. Even so, some researchers are advocating an increased use of amodiaquine in treatment because it has higher activity than chloroquine, it has efficacy in areas where chloroquine resistance is reported, and it is often produced locally, making it an inexpensive drug (948, 982). At a time when many countries in Africa are faced with identifying a first-line drug to replace chloroquine, amodiaquine may be a viable alterna-
despite its adverse effects (983–985), al-
h caution is certainly warranted (986).

**2.8.5 Antibiotics.** Early studies revealed that chlortetracycline was active against avian malarias, although as a group the antibiotics were slow-acting, not truly curative, and not causally prophylactic. As such, no benefit from antibiotic use was envisioned. With the increase in drug-resistant strains of Plasmodia, the practicality and use of antibiotics has been re-examined. The most common antibiotics used against malaria are tetracycline (**76**) and doxycycline

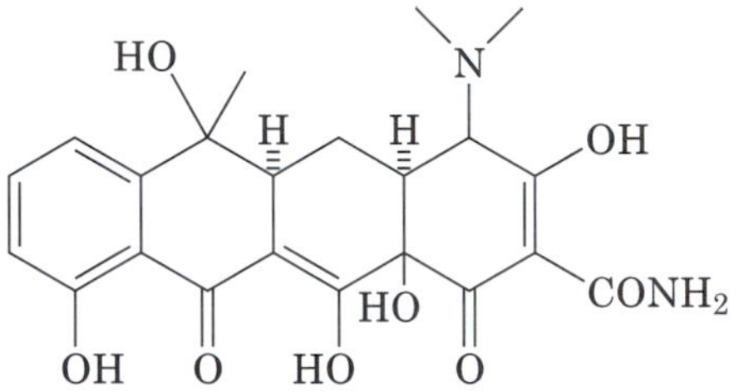

**(76)**
tetracycline

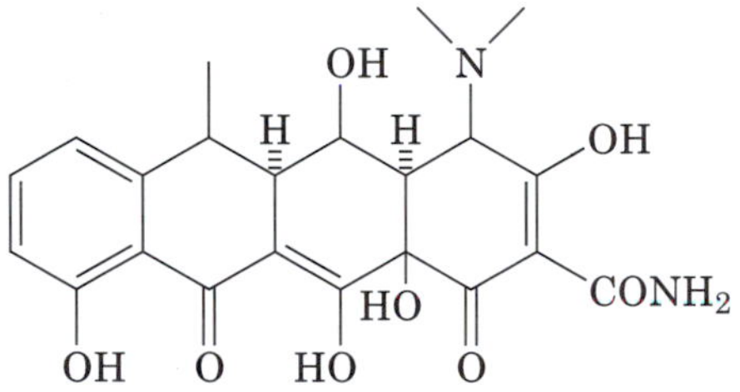

**(77)**
doxycycline

(**77**). Doxycycline is a structural isomer of tetracycline with improved oral absorption; thus, a smaller therapeutic dose is required.

Because of the slow-acting nature of these drugs, they are almost always used in combination with a fast-acting agent such as chloroquine or quinine in treating acute malarial attacks. In particular, a 7-day course of a tetracycline-quinine combination has been found to be useful in treating chloroquine- and multi-drug–resistant strains (987, 988). Tetracycline serves to increase the plasma quinine levels above those seen with equivalent dosing of quinine alone (131).

Although tetracycline displays activity against primary tissue schizonts of chloroquine-resistant strains of *P. falciparum*, its long-term use as a prophylactic agent is not advised. On the other hand, doxycycline has been suggested for short-term prophylaxis by non-immune travelers (505, 989). It was highly efficacious and well tolerated as a prophylactic agent in Indonesian soldiers (990) and in semi-immune volunteers in Kenya (991). Another study with soldiers in hyperendemic areas demonstrated that doxycycline could not be relied on for causal prophylaxis, however, even when combined with primaquine (69). Because of their adverse effects on bones and teeth, tetracyclines should not be given to pregnant women or children less than 8 years old (155), precisely the populations most vulnerable to malaria in endemic areas. The tetracyclines act against the blood stages of Plasmodia. In contrast with what is known about the action of tetracycline on prokaryotes, its mode of action on Plasmodia is not completely understood. The site of action of tetracyclines is the parasite mitochondrion, with inhibition of mitochondrial protein synthesis the result (992). This would account for slow onset of action of the tetracyclines, because mitochondrial replication may be restricted to a limited part of the cell cycle.

Azithromycin (**78**), an erythromycin analog, is used to treat bacterial and chlamydial

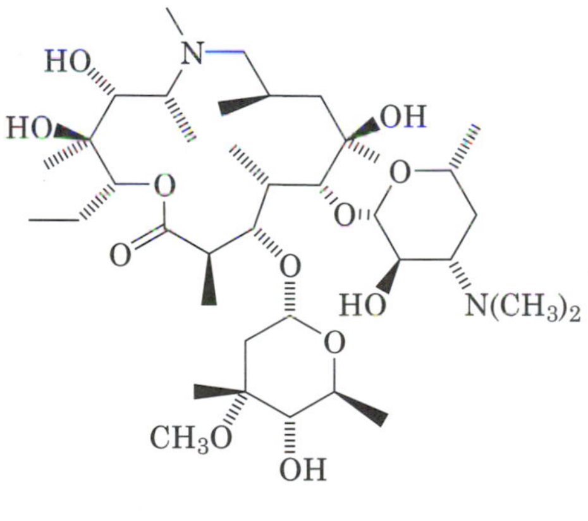

half-life over doxycycline (994). Daily azithromycin provided excellent prophylactic results against *P. vivax*, comparable to those seen with doxycycline, but was less effective against *P. falciparum* in a study in Indonesia (995). A further advantage in substituting azithromycin for tetracycline in combination therapies is that there is no contraindication to administration of azithromycin to pregnant women or young children (996).

Fluoroquinolone antibiotics such as ciprofloxacin (**79**), pefloxacin (**80**), norfloxacin (**81**),

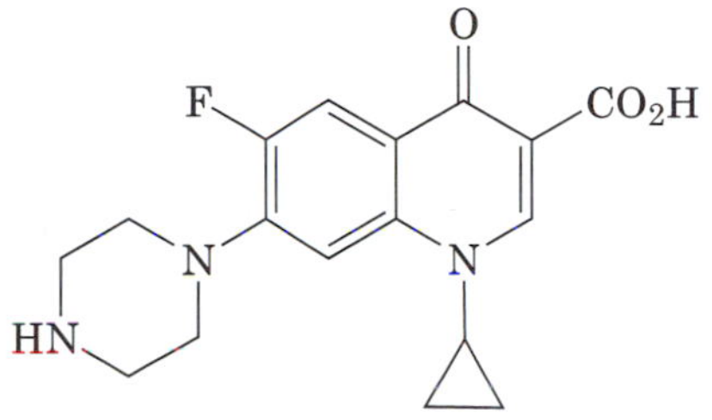

**(79)**
ciprofloxacin

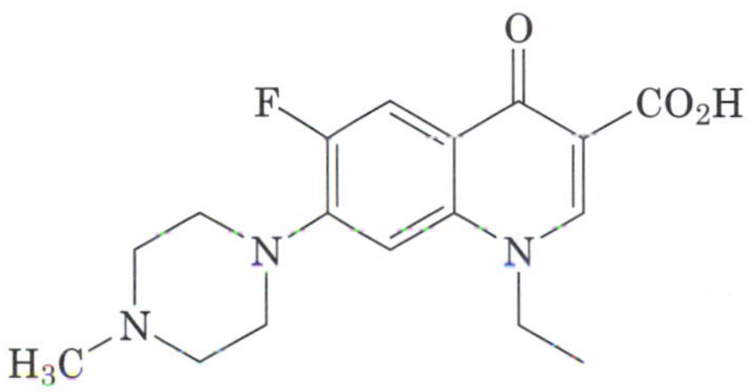

**(80)**
pefloxacin

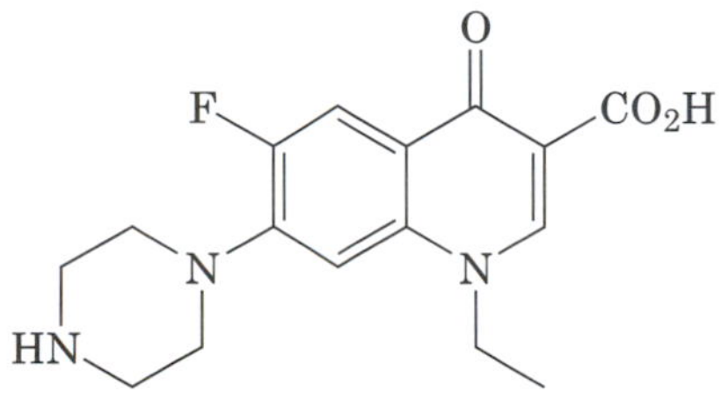

**(78)**
azithromycin

**(81)**
norfloxacin

infections. When combined with fast-acting agents such as quinine, halofantrine, chloroquine, and artemisinin, the effects were additive (993). Azithromycin may also have promise as a prophylactic agent, given its increased

trovafloxacin (**82**) have been shown to be active against *P. falciparum in vitro* (997, 998) and

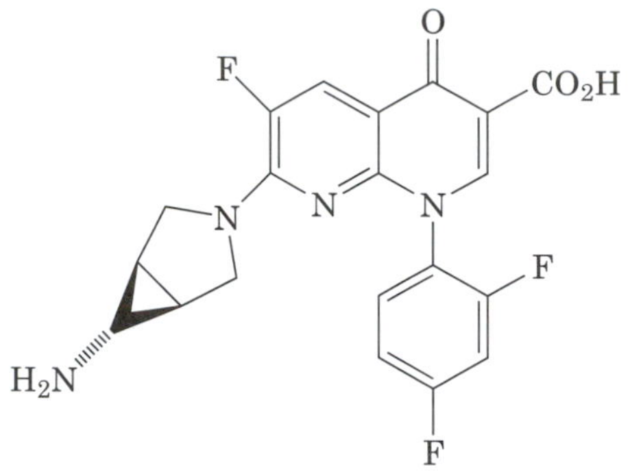

**(82)**
trovafloxacin

against chloroquine-resistant *P. yoelii* in mice *in vivo* (999). Preliminary studies with drugs of this class suggested clinical antimalarial efficacy but more detailed investigations failed to confirm the earlier observations (1000). In combination with quinine, pefloxacin did not potentiate the activity of quinine in terms of speed of parasite or fever clearance (1001). Norfloxacin was less effective than chloroquine in clinical studies (1002). The future role of the fluoroquinolines is not likely to include first-line therapy nor single-agent regimens.

An unusual organelle, the apicomplexan plastid or apicoplast, found in certain parasites including Plasmodium, may be the site of action of ciprofloxacin. This antibiotic inhibits replication of the apicoplast, blocking parasite replication (50). The specific action may involve inhibition of a topoisomerase II activity associated with the apicoplast (1003).

Clindamycin (**83**), a lincosamide antibiotic that can be given to children, has activity against Plasmodia. The addition of clindamy-

cin to a standard quinine treatment significantly improved and shortened chemotherapy in children and adults in Africa and Thailand (1004–1006). A 3-day clindamycin-quinine regimen was well tolerated by children and had an efficacy rate of 97% by day 20 (1007). The mode of action of clindamycin and related compounds against malaria parasites may involve effects on apicoplast replication (50). It was noted that the antibiotic effect of clindamycin may be beneficial in treating the concomitant bacterial infections that are often found in cases of severe malaria.

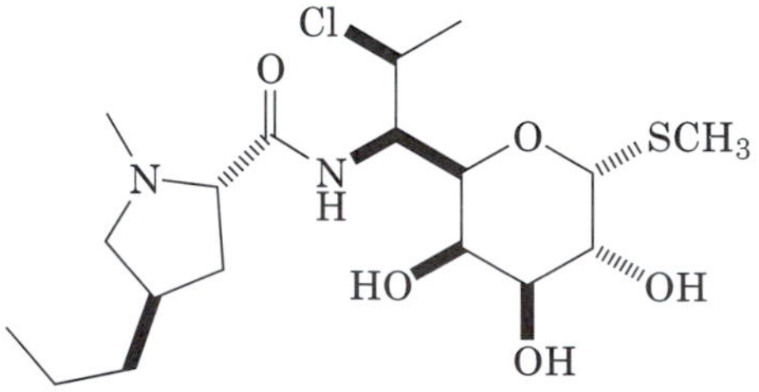

**2.8.6 Yingzhaosu A, Yingzhaosu C, and Arteflene (Ro 42-1611).** Given the success of the peroxide natural product artemisinin in drug development efforts, it was obvious for researchers to explore the antimalarial properties of other naturally occurring peroxides. Two cyclic compounds containing a bisabolene skeleton have been isolated from the roots of the plant yingzhao (*Artabotrys uncinatus*), which has been used as a folk treatment for malaria. Yingzhaosu A (**84**) was first isolated,

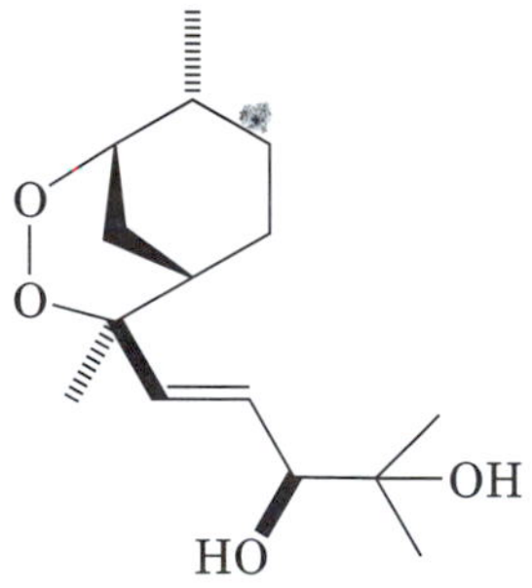

**(84)**
yingzhaosu A

**(83)**
clindamycin

and the structure was proposed in 1979 (1008). The structure and stereochemistry of yingzhaosu C (**85**) were determined by spectroscopic methods, derivatization, and by conversion to a known compound (1009). Both compounds have been synthesized (565, 1010–1012) and both are active antimalarials (592, 1013, 1014). Analogs have been prepared, and some demonstrate significant antimalarial activity *in vitro* (667, 1015).

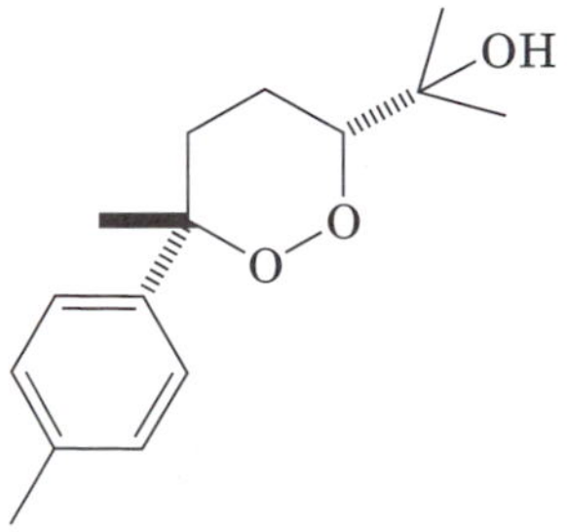

**(85)**
yingzhaosu C

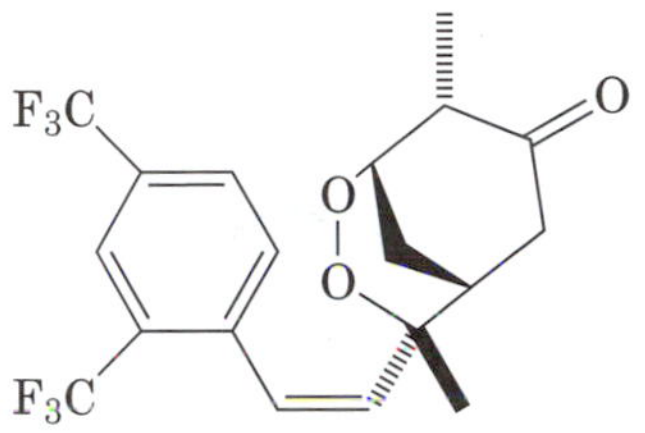

**(86)**
arteflene

# 3 ANTIMALARIAL AGENTS FOR CHEMOTHERAPY AND PROPHYLAXIS: EXPERIMENTAL AGENTS

## 3.1 Synthetic Experimental Antimalarial Agents

A number of new structural lead compounds have been discovered or prepared in the quest for new antimalarial agents. In addition, the analysis of the biochemistry of the parasite has led to the identification of novel molecular targets. The more active and promising frontier areas are discussed below.

Although many trioxane compounds related to artemisinin have been prepared and studied, Fenozan-50F (**87**) is the first trioxane

**(87)**
Fenozan-50F

A particular example is arteflene (**86**), a synthetic antimalarial agent developed as an analog of yingzhaosu A (1016). The central core ring structure (2,3-dioxabicyclo[3.3.1] nonane) was shown to have inherent antimalarial activity. A flexible synthetic strategy was devised that afforded a variety of analogs, with the more lipophilic compounds displaying higher activity. The most active of these was arteflene. It is as active as artemisinin, and has better chemical stability. In a mouse model, arteflene was fast-acting and a single dose lasts significantly longer and results in much lower recrudescence than with artemisinin (1017). The pharmacokinetics and metabolism of arteflene have been studied in animal models (1018) and in humans (1019). Like artemisinin, the compound has low human toxicity and seems to interact with specific parasite proteins (682). In preliminary clinical trials, arteflene is effective in reducing clinical symptoms and is well tolerated (1020–1022). Arteflene may find use both in treatment and as a prophylactic agent, although interest in the drug seems to have waned recently.

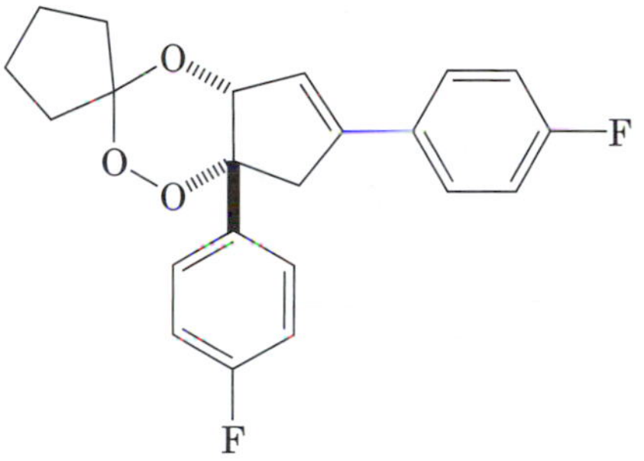

structurally unlike artemisinin to possess the possibility of therapeutic use. Stemming from general studies on the trioxane system, a large number of cis-fused cyclopentenotrioxanes were synthesized (1023). From these, Fenozan-50F was identified as the most active agent (1024, 1025). The compound is active against blood stages of P. berghei and P. yoelii, even against a wide spectrum of drug-resistant parasite strains, and is a potent gametocytocide (1026). There are significant differences between Fenozan-50F and the artemisinins in their blood schizontocidal actions (1026). Fenozan-50F is about one-half as active as arteether in P. berghei infected mice but three times as active as artesunate, and its toxicity seems to be low. The enantiomers of Fenozan-50F are equally active with the racemic mixture (1027). Fenozan-50F exerts complex effects when combined with other antimalarial drugs against mouse parasites (1028). With meflo-quine, for example, observed effects were additive, potentiating, or antagonistic depending on the strain of *Plasmodium* used. Studies on its mechanism of action demonstrated that the trioxane reacted with ferrous ion to produce a number of product esters; oxygen transfer from the peroxide did not occur (1029). Resistance to Fenozan-50F could be established in *P. yoelii*, but the resistance was not stable once drug pressure was removed (718).

Even very simple bicyclic endoperoxides such as (88) were found to have approximately

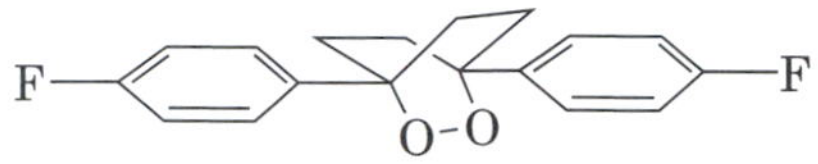

(88)

15% of the activity of artemisinin (1030). The mechanism of action of these compounds may in fact differ from that of the artemisinins; alkylating agents such as epoxy ketones and/or ethylene oxide may be generated *in situ* and may account for the observed parasiticidal activity. Other peroxide systems such as the readily available tetraoxane (89, WR148999)

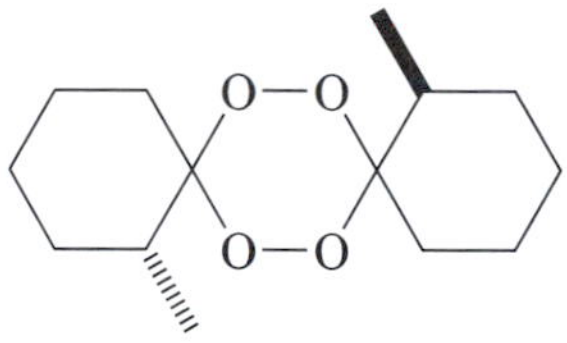

(89)
WR148999

possess IC$_{50}$ values in the nanomolar range (1031), although the oral activity was poor.

Attempts to design compounds with improved oral activity led to the identification of a tetra-methylated compound with *in vivo* activity better than that of artemisinin but not as good as arteether (1032). A number of other structural types of tetraoxacycloalkanes have been produced that have antimalarial activity (1033–1036). Cyclic peroxy ketals, reminiscent of certain marine natural products, have been prepared and tested as antimalarial agents (90) (1037). Some of the compounds

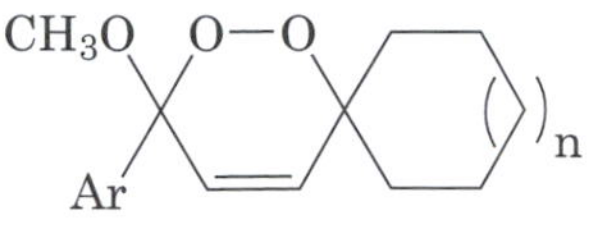

(90)
n = 1, 2

had IC$_{50}$ values in the range of 30–80 nM against *P. falciparum in vitro*. Generalizations about SAR in this series were presented. More complex tetraoxanes could be prepared from steroidal ketones; one compound derived from a cholic acid amide (91) was almost as active as artemisinin and displayed low cyto-toxicity (1038). A hydroperoxide, 15-hydroper-oxyeicosatetraeneoic acid (HPETE), induced a 96% inhibition of *P. falciparum* growth at 40 $\mu M$ (1039).

Compounds incorporating a new antimalarial pharmacophore, a phenyl $\beta$-methoxyacrylate exemplified by (92), have been reported to interfere with mitochondrial electron transport (1040). These compounds were superior to chloroquine both *in vitro* and *in vivo* and may provide a lower cost alternative to atovaquone.

Given the promising introduction of the hydroxynaphthoquinone atovaquone, a series of

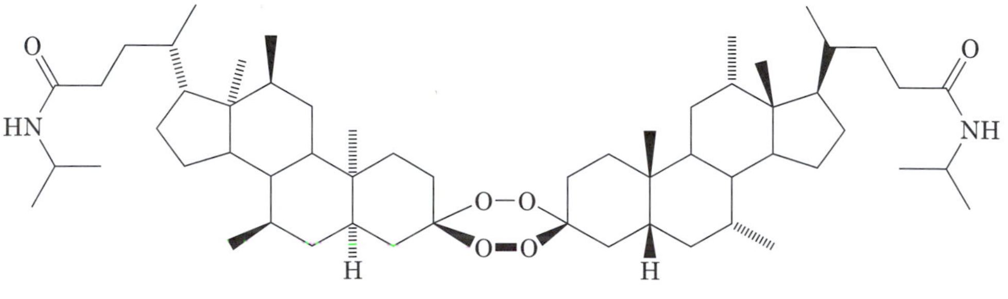

(91)

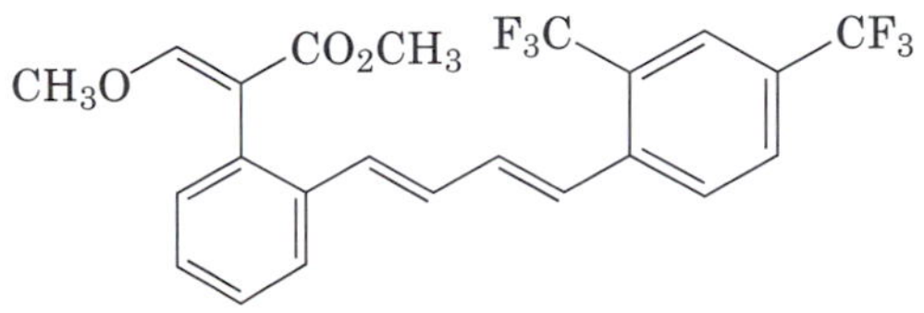

**(92)**

hydroxy- and polyhydroxyanthraquinones were synthesized and screened for antimalarial activity (1041). Rufigallol (**93**), easily pre-

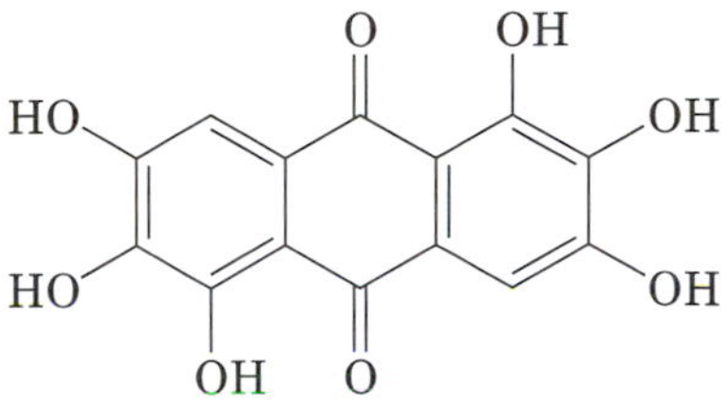

**(93)**
rufigallol

pared by dehydration of gallic acid, displayed an $IC_{50}$ of 35 n$M$ against chloroquine-sensitive or resistant *P. falciparum in vitro* and was the most active compound in the group. When combined with exifone (**94**), a strong potenti-

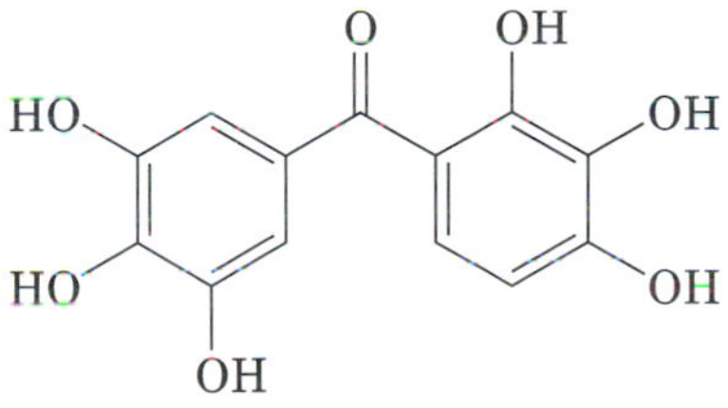

**(94)**
exifone

ation of antimalarial effect was noted (1042). Exifone is also synergistic with other oxidants such as ascorbic acid; the authors suggested

that exifone may undergoes oxidation to a toxic xanthone *in vivo* in presence of rufigallol or ascorbic acid (1043). Hydroxyxanthones such as X5 (**95**) inhibit *P. falciparum* growth

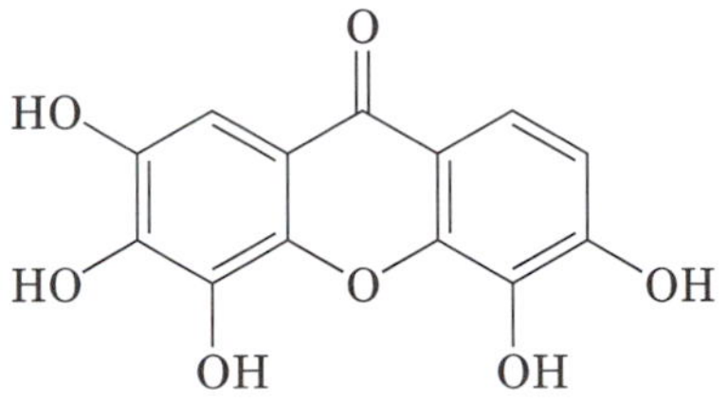

**(95)**
X5

*in vitro*, and they seem to act at the point of heme sequestration (1044). The antimalarial effect occurred during the second half of erythrocytic cycle, when free heme production peaks (1045).

*dl*-α-Difluoromethylornithine and *dl*-α-monomethyldehydroornithine methyl ester, inhibitors of ornithine decarboxylase, part of the polyamine biosynthesis pathway, block exoerythrocytic schizogony (1046). The immunosuppressive agent deoxyspergualin (**96**) has shown activity *in vitro* and *in vivo*, most likely through inhibition of polyamine biosynthesis. Although the drug acted slowly, complete eradication of parasitemia was accomplished in mice (1047). Clinical studies are underway. Another group of ornithine decarboxylase inhibitors includes hydroxylamino compounds. A number of structures including canaline (**97**) and CGP51905A (**98**) were prepared and found to have sub-micromolar inhibition of *P. falciparum in vitro* (1048).

5-Fluoroorotate, a pyrimidine analog, inhibits parasite proliferation at low concentrations with little toxicity to mammalian cells (868, 1049). Inhibition of the plasmodial thymidylate synthase, a key enzyme in the path-

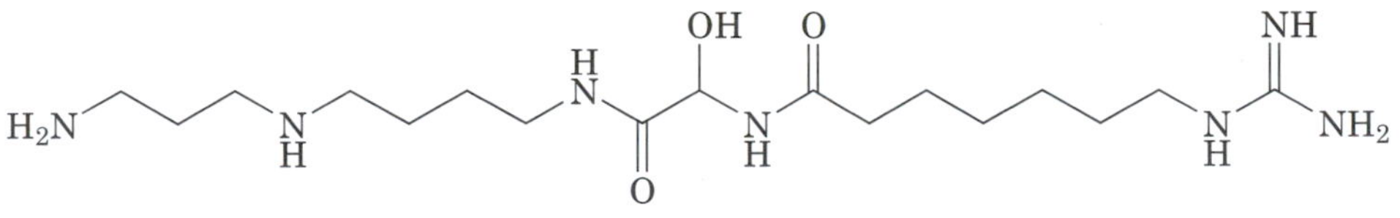

**(96)**
deoxyspergualin

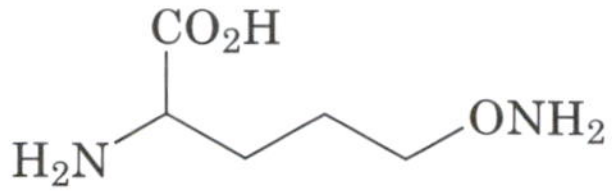

**(97)**
canaline

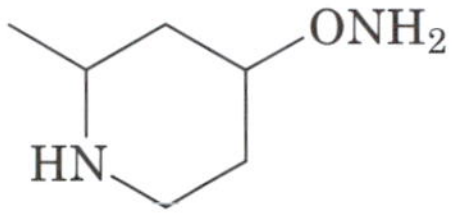

**(98)**
CGP1905A

way of *de novo* pyrimidine synthesis, seems to be the mode of action (1050).

One of the earliest known antimalarial agents after quinine was the synthetic thiazine dye, methylene blue (**99**). The *in vitro*

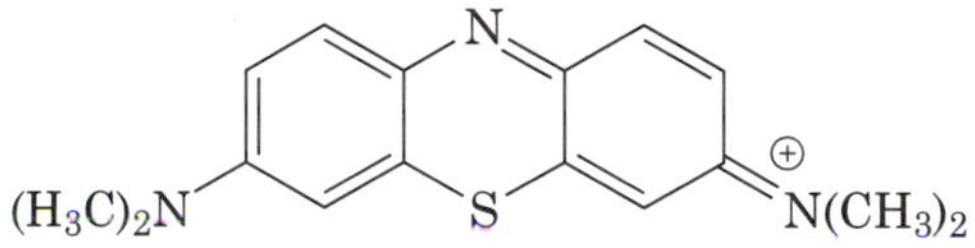

**(99)**
methylene blue

activities of methylene blue and several other dye compounds were reexamined recently (1051). $IC_{50}$ values in the nanomolar range were obtained for the thiazine dyes, and some were more active than chloroquine in sensitive strains. No cross-resistance with chloroquine was observed. The mechanism of action of methylene blue has been studied (1052) and may involve inhibition of plasmodial glutathione reductase (1053). Although the potential toxicity of these compounds will likely prohibit their clinical use as antimalarial agents, the phenothiazine structure may provide a useful lead for further drug design efforts.

Because the end product of hemoglobin digestion is hemozoin, a sequestration product of hematin Fe(III), Fe(II) in hemoglobin is oxidized to Fe(III). It seems that this oxidation occurs at the level of hemoglobin to methemoglobin, because it has been demonstrated that *P. falciparum* increases the levels of methemoglobin in infected erythrocytes and the increase is restricted to ingested hemoglobin

(1054). Riboflavin was active *in vitro* against *P. falciparum* at levels commonly used clinically to treat methemoglobinemia without adverse effects. Both riboflavin and methylene blue may act by reduction of methemoglobin to hemoglobin *in vivo*.

A novel arylene bis(methylketone), compound CNI-H0294 (**100**) was shown to have

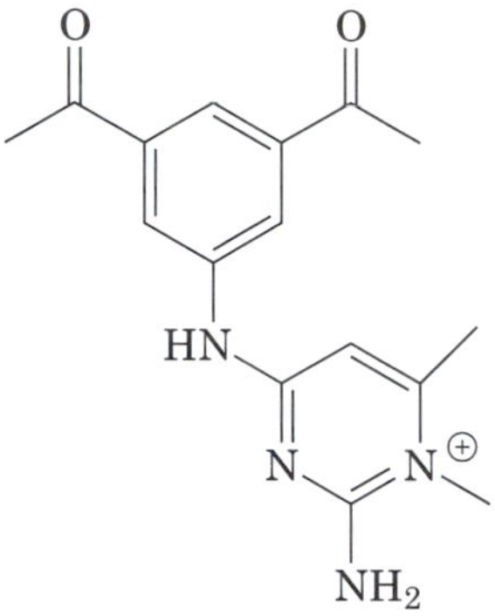

**(100)**
CNI-H0294

good activity *in vitro* against chloroquine and pyrimethamine resistant *P. falciparum in vitro* and against *P. berghei in vivo*. The drug displayed activity against the plasmodial dihydrofolate reductase at achievable concentrations, and this represents a possible mechanism of action for the drug (1055, 1056).

Another target for the design of new antimalarial agents has been the phospholipid metabolism of *P. falciparum*. Phosphatidylcholine, the major phospholipid of infected erythrocytes, is synthesized from choline, drawn mainly from plasma using parasitic enzymes. Interference with the *de novo* biosynthesis of phosphatidylcholine is lethal to the parasite. The transport of choline is impaired by quaternary ammonium compounds (1057, 1058). Several inhibitors showed excellent *in vitro* action against drug-resistant strains of *P. falciparum*. One compound in particular (**101**) had an $IC_{50}$ value in the sub-nanomolar range (1059). From the data obtained for this series, the authors propose a model for binding of these inhibitors to the choline carrier.

Specific inhibitors of the shikimate pathway include the herbicide glyphosate (**102**) (46) and fluorinated derivatives of shikimic

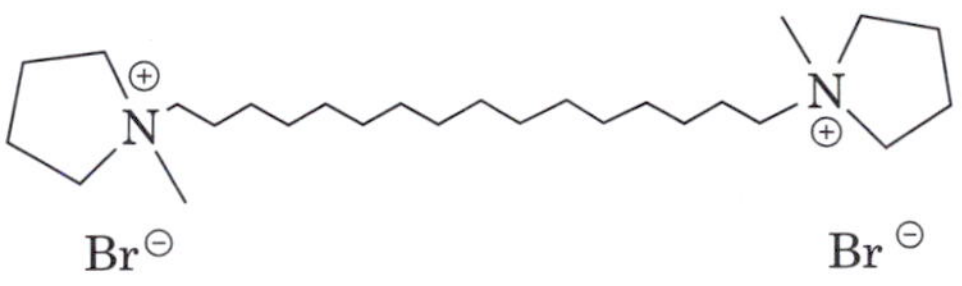

(101)

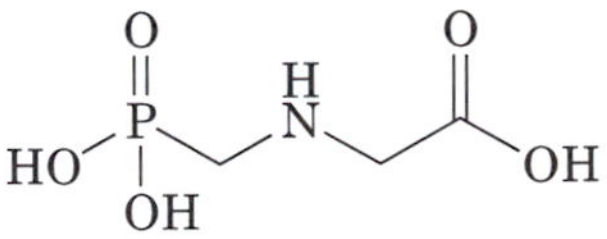

(102)
glyphosate

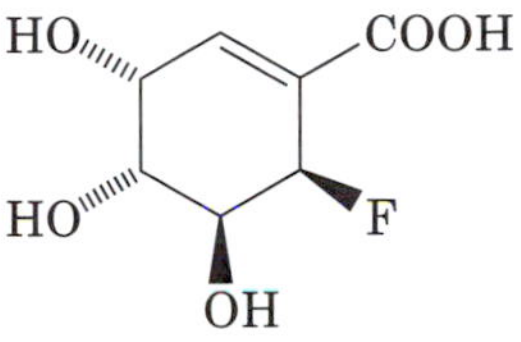

(103)

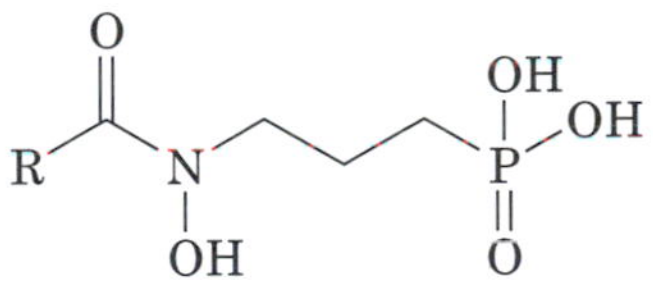

(104)
fosmidomyin, R = H
(105)
FR-900098, R = CH$_3$

dA-dT-rich DNA sequences that make up a large percentage of the *P. falciparum* genome (82%). In contrast with distamycin itself, active analogs such as (106), were much less toxic (1062).

In a study of the antimalarial action of benzimidazole compounds, omeprazole (107),

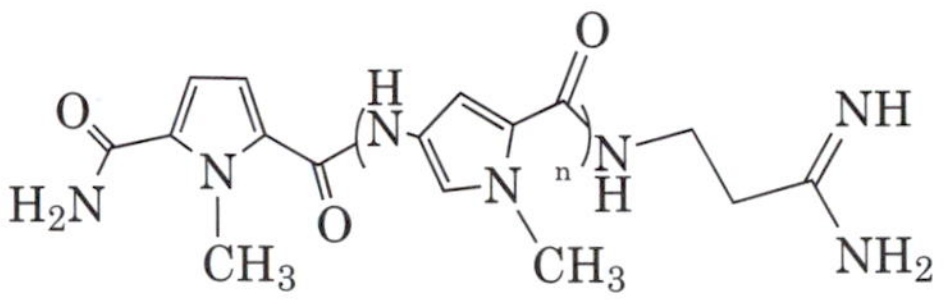

(106)
n = 4

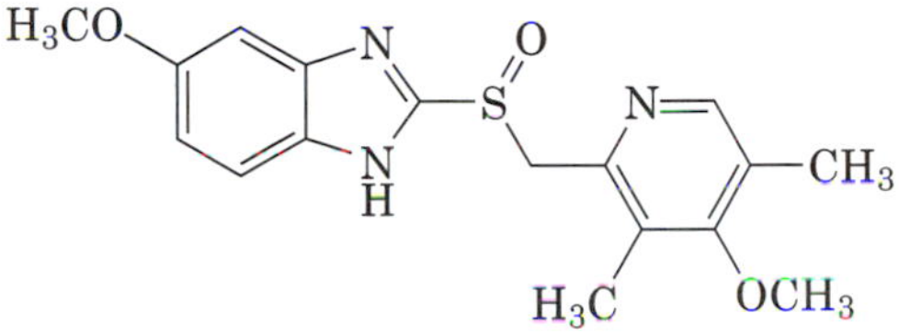

(107)
omeprazole

a proton pump inhibitor, was found to be the most active *in vitro*, primarily against the later stages of parasite development (1063). Its action is synergistic with that of quinine but antagonistic with chloroquine (1064).

Vitamin A (retinol) is central to normal immune function and has been shown to possess antimalarial activity on its own. Supplementation therapy with high dose vitamin A was beneficial in young children in Papua New Guinea (1065), and *in vitro* studies showed that retinol potentiated the actions of quinine and omeprazole (1066).

δ-Carbolines were prepared and tested for antimalarial activity (1067). These compounds (108, 109) were active *in vitro* against

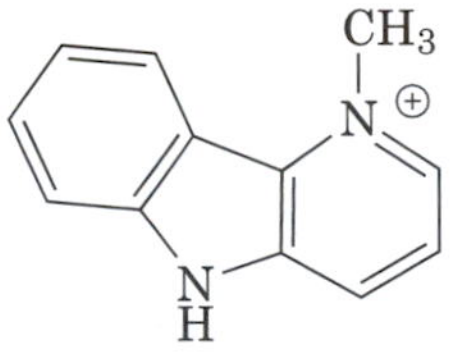

acid such as (103) (1060). Inhibitors of DOXP reductoisomerase were already known, including fosmidomycin (104) and FR900098 (105). These two agents were shown to inhibit *P. falciparum* in culture (~300 and ~140 n*M*, respectively) and active *in vivo* against *P. vinckei* in mice. Both drugs have very low toxicity but also had very short half-lives and may provide leads for new drug development. Indeed, modification of the structure of FR900098 as phosphodiaryl ester resulted in compounds with improved oral activity (1061).

Analogs of distamycin, a compound that binds in the minor groove of duplex DNA, were shown to have antimalarial activity in the sub-micromolar range. Distamycins bind

(108)

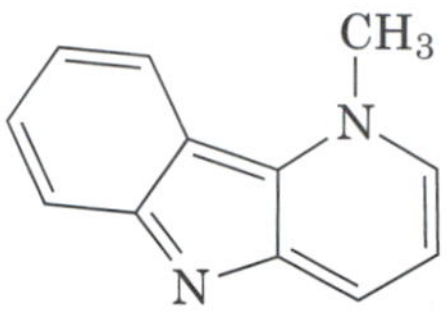

(109)

P. falciparum in the low micromolar range and were weakly cytotoxic. These agents specifically accumulate in the intracellular parasite and may interact with parasite DNA.

A series of 9-anilinoacridines (110) has been synthesized and evaluated for their anti-

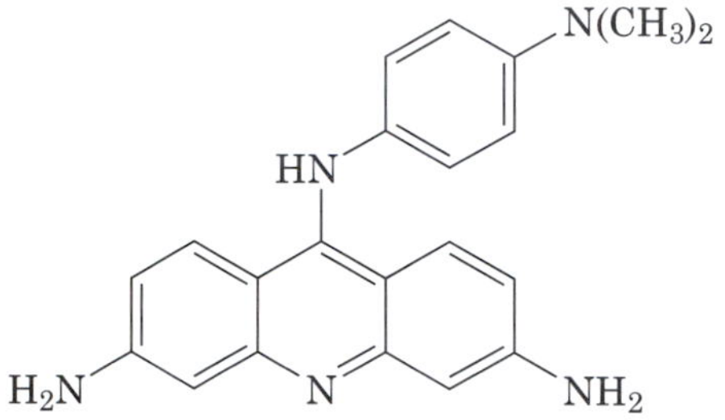

(111)

(110)

malarial activity as topoisomerase inhibitors (919, 1068). It was found that a 3,6-diamino substitution pattern on the acridine conferred high potency against P. falciparum in erythrocyte culture tests. Inhibitors that acted against the parasite topoisomerase II enzyme both in vitro and in vivo were identified based on the ability of the candidate compounds to generate cleavable complexes in situ (920).

Docetaxel (taxotere, 111), a taxol-type drug, was found to inhibit P. falciparum development in vitro with an IC$_{50}$ value in the nanomolar range (1069), although the inherent cytotoxicity of microtubule inhibitors currently available limits the likelihood of their being used in malaria therapy (1070).

Earlier efforts in improving the chloroquine structure involved the development of bisquinoline compounds linked by a covalent bridge, including compounds such as piperaquine (112) (1071). Several of the compounds with simple alkyl bridges exemplified by 113 displayed good activity and were in fact more active against a chloroquine-resistant strain than against a sensitive one (1071).

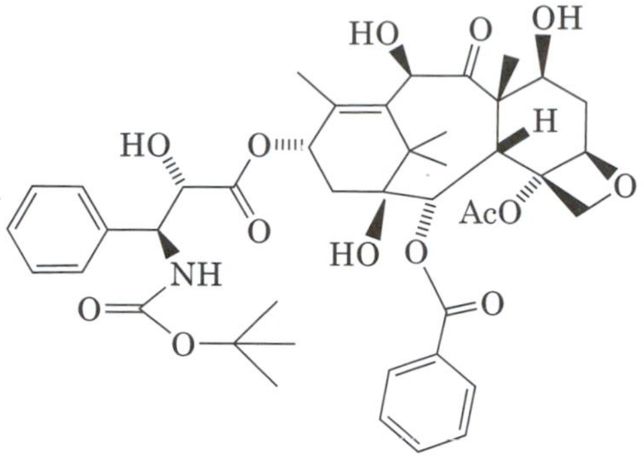

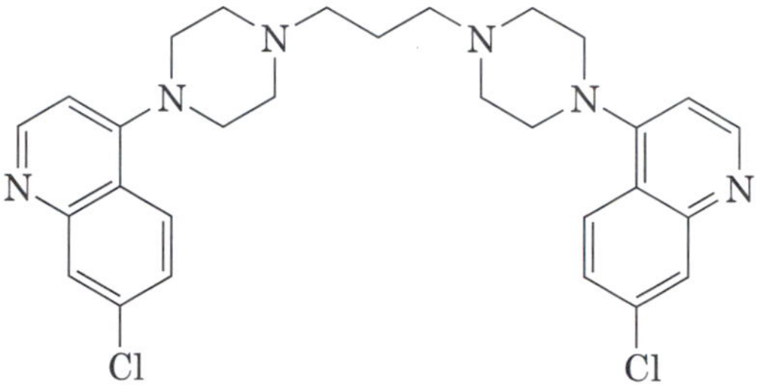

(112)
piperaquine

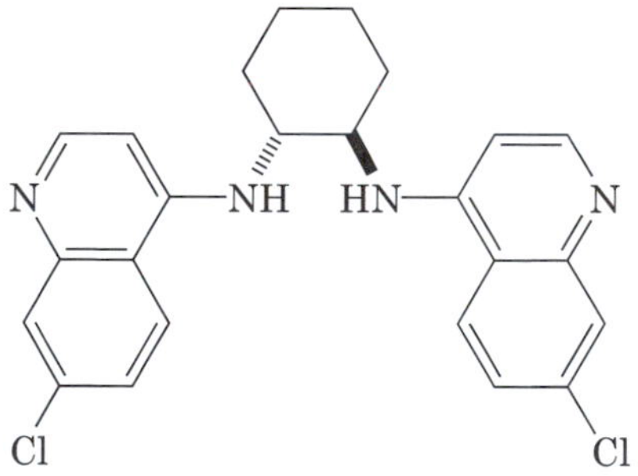

(113)

During parasite development in the host red blood cell, Na$^+$ and K$^+$ levels within the cell are disrupted, and the parasite creates a gradient between its own cytosol and the erythrocytic cytosol. Because ionophores are known to disturb ionic gradients, derivatives of the ionophore monensin were synthesized and evaluated for antimalarial activity (1072). The activity of the ionophores correlated with the transport efficiency for total Na$^+$ and K$^+$ rather than with ion selectivity. Several of the compounds exhibited $IC_{50}$ values in the nanomolar range against *P. falciparum in vitro*. Nigericin (**114**), an ionophore selective for

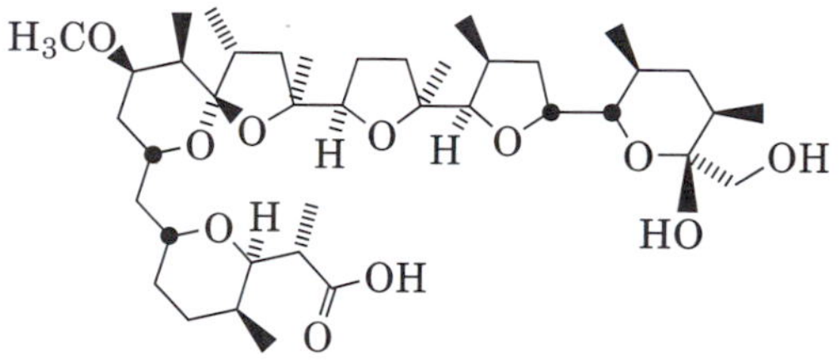

**(114)**
nigericin

$K^+/H^+$ exchange also, had good *in vitro* activity, and its action was synergistic with that of monensin (1073).

Malaria parasites are very susceptible to the action of iron chelators and a number of studies have been carried out to evaluate the clinical potential of agents of this type (1074, 1075). Iron chelators may act by one of two possible mechanisms against malaria parasites. They may act either by depriving the rapidly multiplying parasites of essential iron, or they may form complexes with iron that are ultimately toxic to the parasite. Desferrioxamine (DFO, **115**), a known iron chelator, has

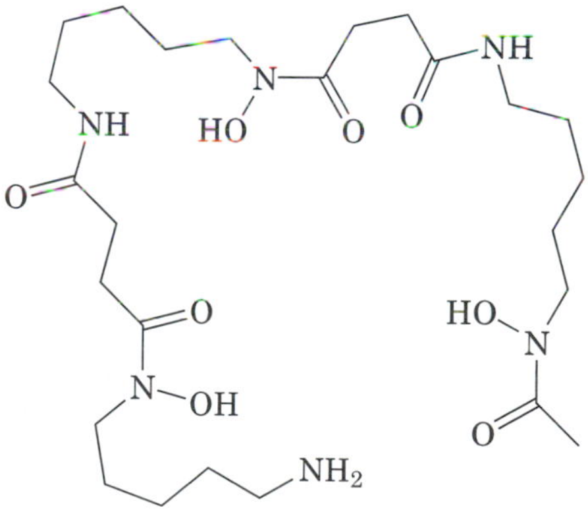

**(115)**
desferrioxamine

been examined as a therapeutic agent with mixed results (1076). DFO inhibits *in vitro* growth of *P. falciparum* in humans. Clinical cases are resolved faster with co-administration of DFO and a traditional antimalarial

(339). For example, parasite clearance and recovery from coma were faster when DFO was used in combination with quinine than when quinine was used alone (1077). Recrudescence occurred in most subjects, however. Clinical efficacy of DFO required continuous parenteral administration, and its use in therapy must therefore be somewhat limited. In addition to effects on iron levels, DFO may protect against iron-mediated oxidant damage (1077). DFO antagonizes the action of chloroquine (1078), apparently by increasing the concentration of soluble forms of hematin and enhancing hematin sequestration (1079). DFO does not seem to extract iron from intact enzyme but may prohibit the initial iron-enzyme interaction necessary to form the active species. Derivatives of DFO demonstrated significant differences in activity against parasites vs. mammalian cells, the parasites being more susceptible to interference from the chelator (1080). Other iron chelators have been studied as well, especially a group of compounds known as reversed siderophores, synthetic analogs of the natural siderophore ferrichrome (1076, 1081, 1082). The most lipophilic of a series was the most potent antimalarial agent and the most efficient in extracting iron from the infected cells (1082). A novel siderophore (**116**) was shown to have sub-micromolar ac-

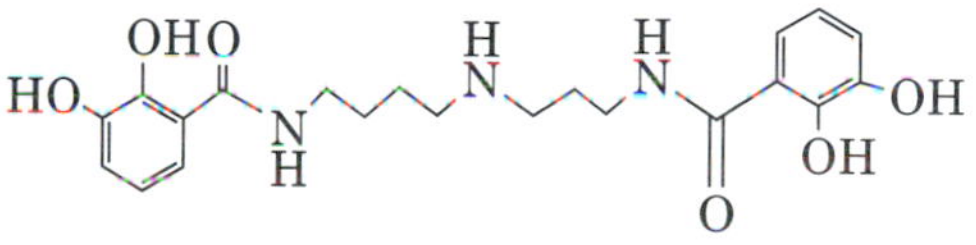

**(116)**

tivity against *P. falciparum* clones, presumably because of iron deprivation (1083).

Compounds that act on the plasmodial proteases that degrade hemoglobin are receiving attention as potential antimalarial agents. Non-peptide inhibitors of plasmepsin II, an aspartyl protease, were identified using combinatorial chemistry and structure-based design. One inhibitor (**117**) had a $K_i$ of 4.3 n$M$ and had good selectivity for the plasmodial enzyme over cathepsin D (1084). Phenothiazines were studied as inhibitors of falcipain, a cysteine protease of *P. falciparum* (1085). Certain

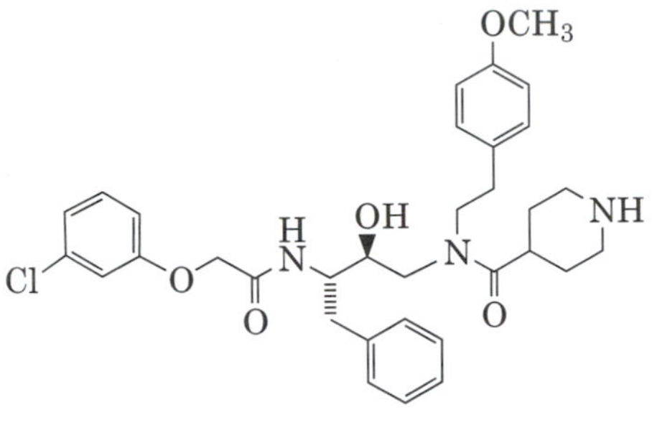

(117)

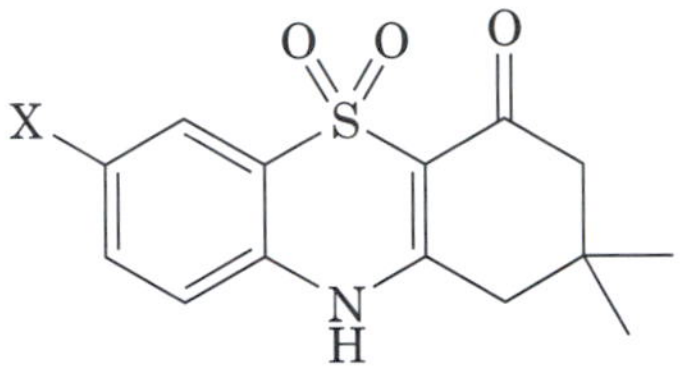

(118)
X = Cl, F

(120)
M = Fe(III), Ga(III)

structures (118) were shown to inhibit parasite development, but the effect required relatively high concentration. Some chalcones (119) have been synthesized that also act on

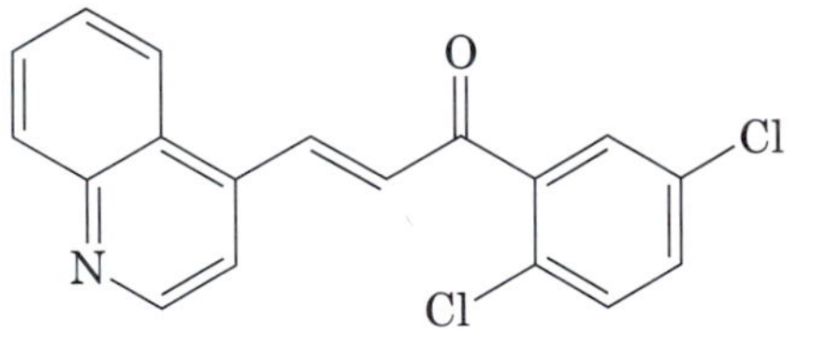

micromolar range (1091, 1092). Both the Fe(III) and Ga(III) complexes seem to inhibit hemozoin aggregation. It has been suggested that the inhibition occurs through formation of a salt complex between the propionate of the heme and the cationic complex (1093). Compounds linking a 4-aminoquinoline to a ferrocenyl unit such as (121) were active

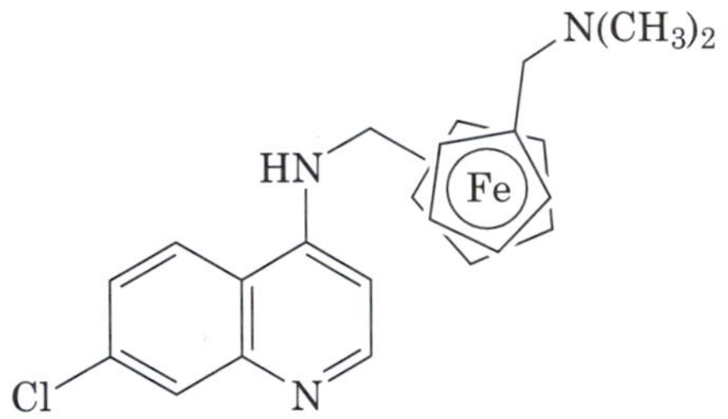

(119)

(121)
ferrochloroquine

falcipain (1086–1088). The compound has an $IC_{50}$ of 200 n$M$ against either chloroquine-sensitive or chloroquine-resistant strains of *P. falciparum*. A combination of agents that inhibit both falcipain and plasmepsin were shown to behave synergistically (1089).

The hemolytic peptide dermaseptin S4 was shown to exert antimalarial activity through lysis of infected cells (1090). Further inquiry into the basis for this action may lead to the design of a new class of antimalarial agents.

A new class of antimalarial agents, metal complexes such as (120), are active in the low

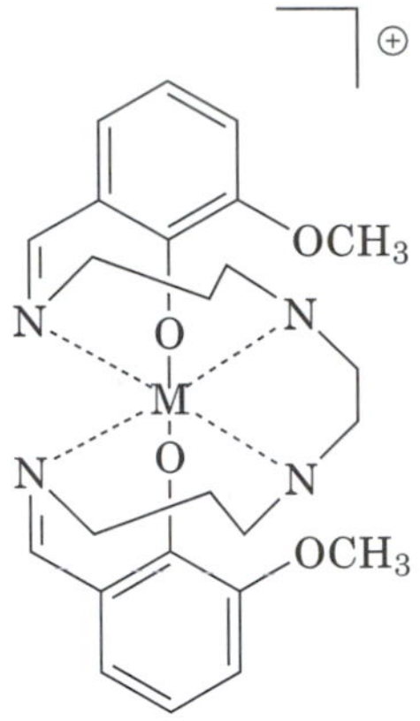

against a chloroquine-resistant strain of *P. falciparum in vitro* and *in vivo* against *P. berghei* and *P. yoelii* (257). Using the metabolites of chloroquine as the model, possible metabolites of ferrochloroquine were synthesized and found to have activity *in vitro* (1094). Amine analogs of ferrochloroquine (122) were more active than chloroquine against both sensitive and resistant strains (1095). Ferrocene-chloroquine analogs were active in clinical isolates, 95% of which were resistant strains (1096). $IC_{50}$ values were in the low

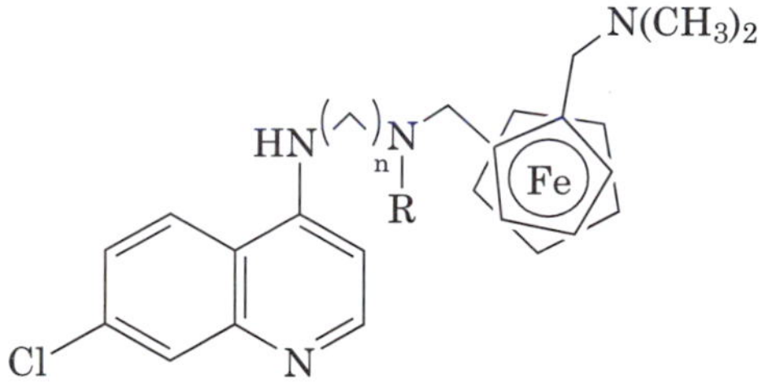

**(122)**

nanomolar range. The use of metal complexes in malaria therapy has been reviewed (1097).

## 3.2 New Targets for Antimalarial Chemotherapy

Many initial efforts towards new and experimental targets have been presented above. A number of writers have compiled lists of plasmodial targets with potential in antimalarial chemotherapy (1098–1100). Of particular interest is the inhibition of the proteases that specifically degrade hemoglobin in the parasite food vacuole, the plasmepsins, falcipain, and perhaps falcilysin (1101). Also very promising are targets associated with pathways specific to the apicoplast, such as the shikimate pathway and the 2-deoxy-D-xylulose-5-phosphate (DOXP) pathway (50, 1102). Cellular and biochemical targets in the mosquito vector are worthy of consideration as well (1103). One group of workers has demonstrated the viability of using a molecular connectivity QSAR approach to identifying new active antimalarial agents (1104).

## 3.3 Antimalarial Natural Products

Given the increasing problem of drug resistance, researchers are turning with renewed interest to natural sources for the discovery of novel structural types for antimalarial investigations. Many natural products have been shown to have antimalarial activity in a variety of screens. Most, however, are only mildly active or have toxicity that precludes any serious interest in pursuing the new structures as lead compounds for drug studies. A very few examples are provided below of compounds providing interesting possibilities or structures for continuing research.

Reviews in the area of antimalarial natural products include a review of traditional medicinal plants as sources of antimalarial agents (1105), a review of plants more generally as sources of antimalarial compounds (1106, 1107), and the selective screening and testing of natural products for antimalarial activity (1108). A number of marine organisms have been screened for natural products with selective antimalarial activity, and the structures of active components have been reported (1109, 1110).

The initial report of isonitrile compounds, isolated from marine organisms, that had significant antimalarial activity was that of axisonitrile-3 (**123**) from the sponge *Acanthella*

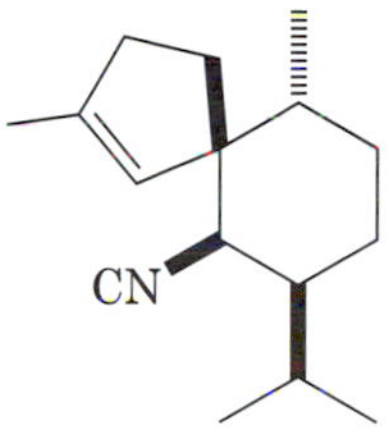

**(123)**
axisonitrile

*klethra* (1111). Further examination of marine sources led to the isolation of a series of diterpene isonitriles and isocyanates from the sponge *Cymbastela hopperi* (1112, 1113). Two of the compounds, (**124**), have $IC_{50}$ values (ng/

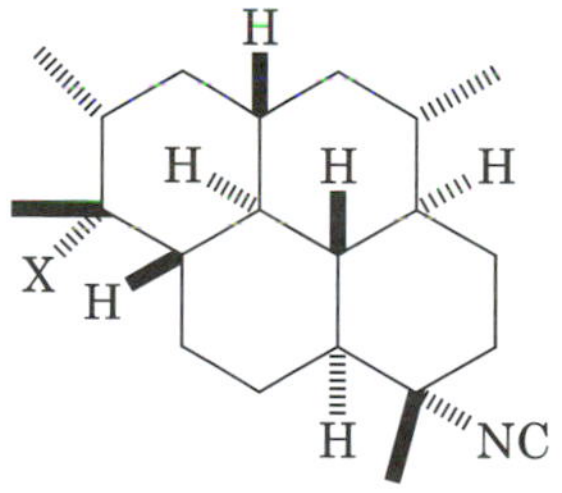

**(124)**
X = NC, NCO

mL) in the 2.5–4.7 range for both chloroquine-sensitive and chloroquine-resistant clones of *P. falciparum*. A marine sponge *Acanthella* sp. provided kalihinol A (**125**) (1114). As with the other isonitrile natural products, this one has activity in the nanomolar range *in vitro* and

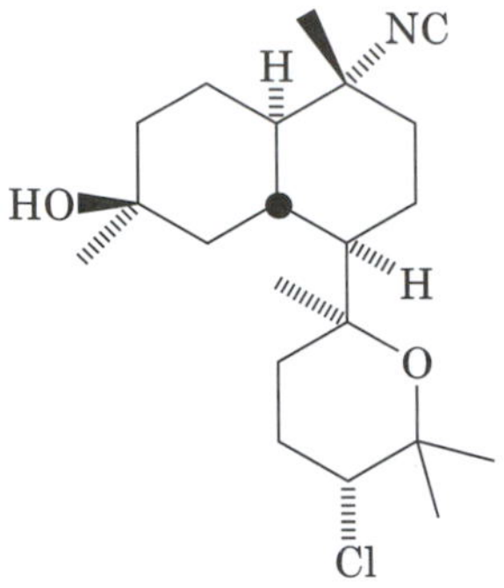

**(125)**
kalihinol A

**(127)**
cycloprodigiosin

was quite selective. Molecular modeling studies employing 3D-QSAR were able to derive a pharmacophore schema consistent with the biological activity (1115). The isonitrile compounds seem to act on heme detoxification processes.

Manzamine A (**126**), an unusual β-carboline isolated from a number of marine

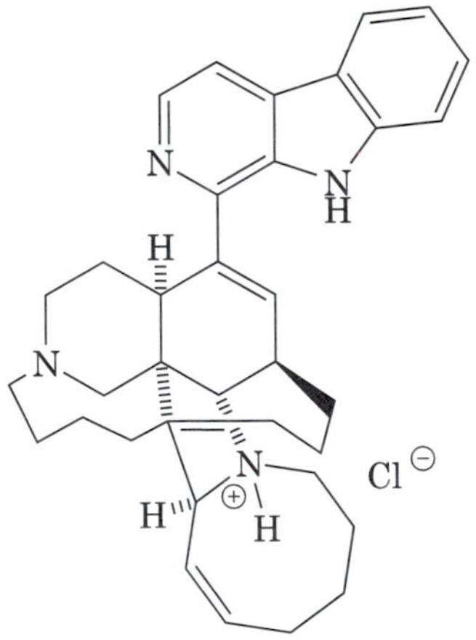

**(126)**
manzamine A

sponges, was an active antimalarial *in vitro* and *in vivo* (1116). Despite a narrow therapeutic index, a recent total synthesis should enable SAR studies to be carried out (1117).

A marine bacterium, *Pseudoalteromonas denitrificans*, was the source of cycloprodigiosin (**127**) (1118). The *in vitro* activity of 11 n*M* was particularly promising given the low tox-

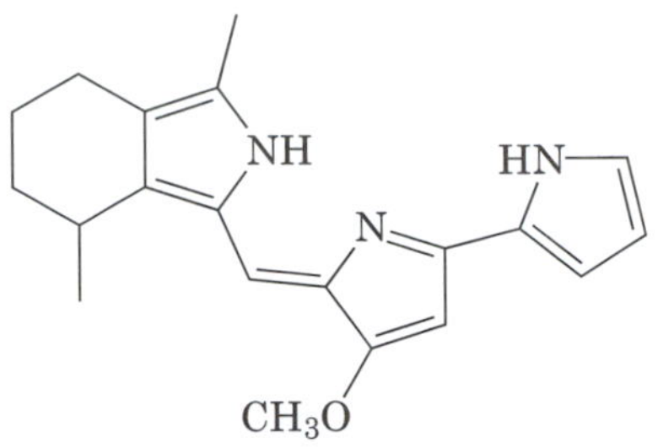

icity. Cycloprodigiosin has been studied for a wide range of biological activities beyond antiprotozoal effects.

Apicidin (**128**) is a tetrapeptide produced by *Fusarium pallidoroseum*. The compound is

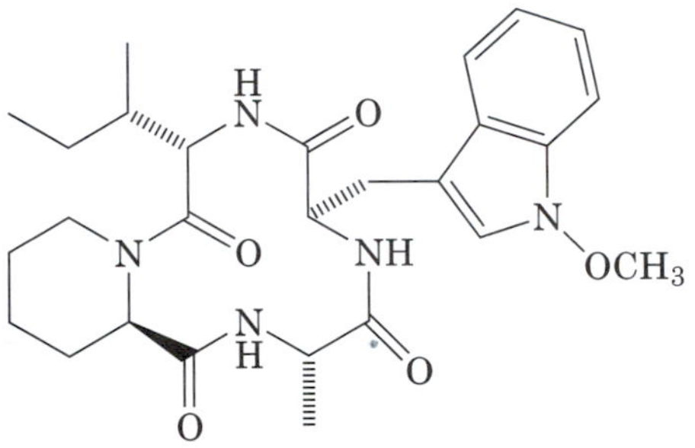

**(128)**
apicidin

orally active against *P. berghei* in mice at less than 10 mg/kg and seems to act by inhibition of the apicomplexan histone deacetylase (1119).

The naphthylisoquinoline alkaloids of *Triphyophyllum peltatum*, the most widespread species of Dioncophyllaceae, have been reviewed (1120). This group includes dioncophylline C (**129**), dioncophylline B (**130**), dioncopeltine A (**131**), dioncolactone A (**132**), and 5'-*O*-demethyldioncophylline A (**133**) (1121). A related compound is ancistrocladine (**134**) from *Ancistrocladus abbreviatus*. All of these display good activity *in vitro*, and some SAR studies of dioncophylline have been carried out (1122). Dioncophylline C and dioncopeltine A were also effective as oral agents against chloroquine-resistant *P. berghei* in mice (1123). Dioncophyllines A and C and an-

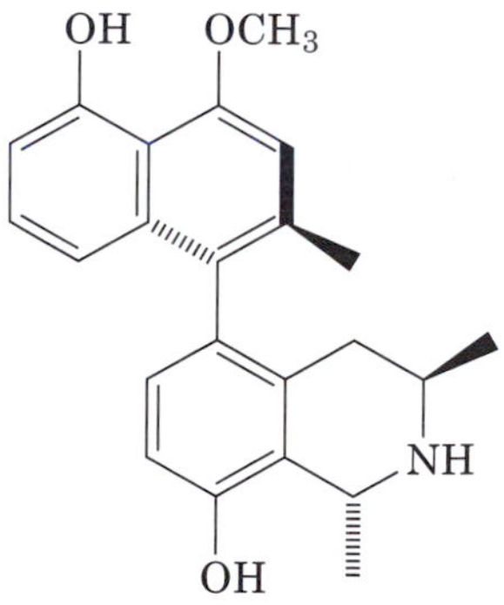

(129)
dionophylline C

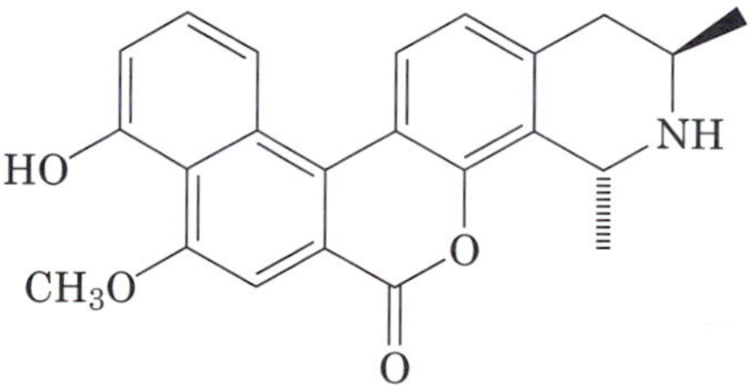

(132)
dioncolactone A

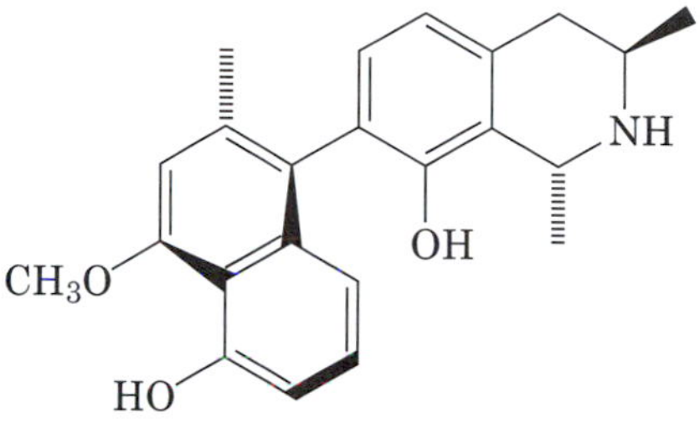

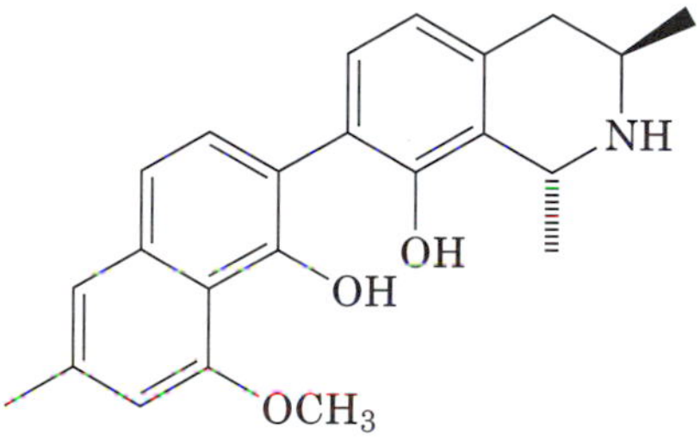

(130)
dioncophylline B

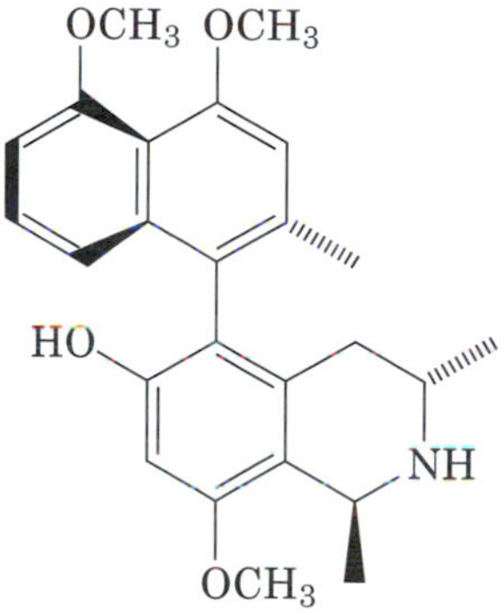

(133)
5'-O-demethyldionophylline A

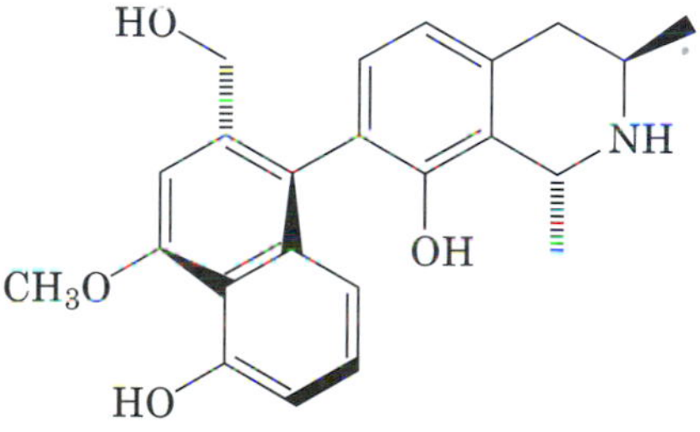

(131)
dioncopeltine A

(134)
ancistrocladine

cistrocladine were shown to be active not only against erythrocytic forms of the parasite but also against exoerythrocytic forms (1124). *Ancistrocladus korupensis*, a tropical liana, has been another rich source of naphthylquinoline alkaloids. Of particular interest are the korupensamines A-E (**135–139**) (1125, 1126). The compounds, being very similar in structure to the alkaloids from *T. peltatum*, are active both *in vitro* and *in vivo* (1127). The naphthyliso-

quinoline alkaloids have inspired a number of total syntheses directed at *T. peltatum* compounds (1128–1130) and those from *A. korupensis* (1131–1133). *A. korupensis* has provided, in addition to the agents mentioned above, a heterodimeric naphthylisoquinoline, korundamine A (**140**) (1134). This is the most potent *in vitro* of the members of this class yet identified.

Bisbenzylisoquinolines make up a large group of natural products, several of which

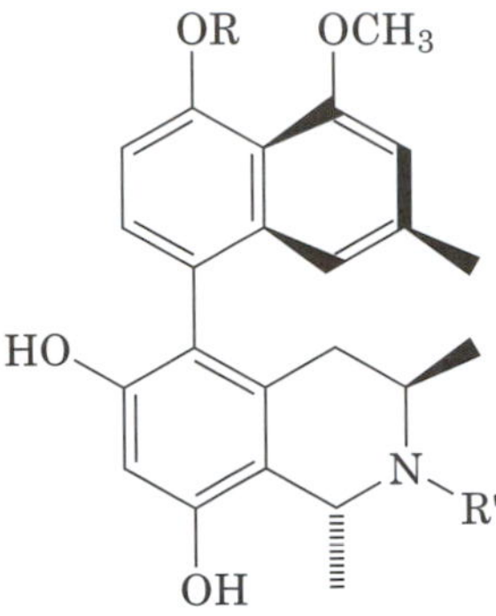

**(135)**
korupensamine A, R = R′ = H
**(137)**
korupensamine C, R = CH₃, R′ = H
**(138)**
korupensamine D, R = H, R′ = CH₃

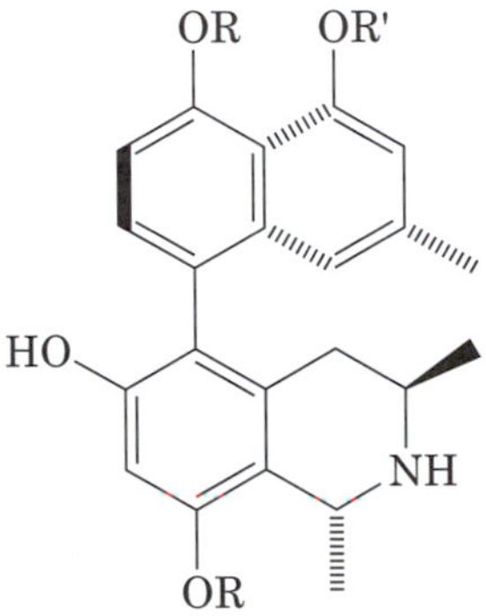

**(140)**
korundamine A

**(136)**
korupensamine B, R = H, R′ = CH₃
**(139)**
korupensamine E, R = CH₃, R′ = H

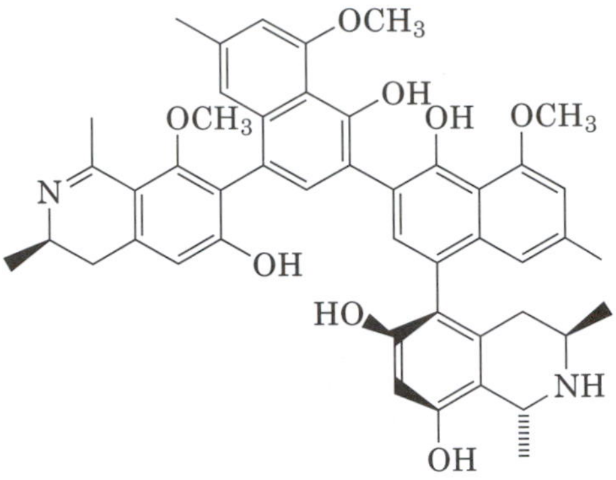

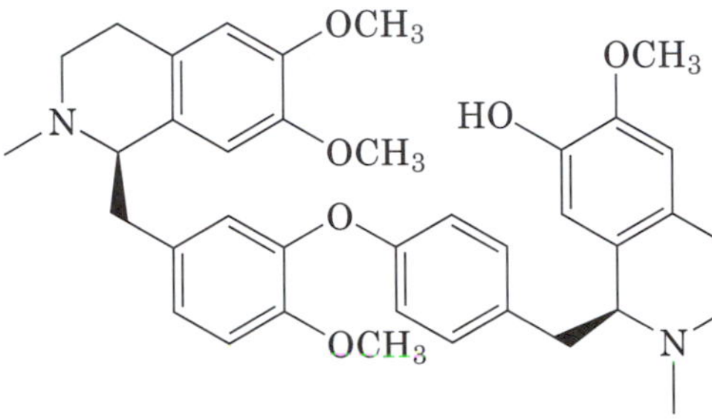

**(141)**
(+)-temuconine

have been shown to possess antimalarial activity (1135). An evaluation of over 50 structures led to the identification of six compounds that not only had appreciable activity against chloroquine-sensitive and resistant strains of *P. falciparum*, but were also somewhat selective.

The roots of *Dichroa febrifuga* have been used for centuries in China to treat malarial fevers. Febrifugine (**147**) and isofebrifugine (**148**) were isolated from *D. febrifuga* (1136–1138) and have attracted considerable attention as antimalarial agents. In studies in mice, febrifugine significantly reduced mortality. The drug seems to potentiate the production of nitric oxide in acute immune responses

(1139). Both compounds have been prepared by asymmetric synthesis (1140).

Folk medicine has often offered leads in locating biologically active compounds. Neem (*Azadirachta indica*) is a popular source of traditional remedies in Africa. Gedunin (**149**) is the main antimalarial component with good activity against both chloroquine sensitive and resistant strains of *P. falciparum* (1141, 1142). The lack of *in vivo* activity against *P. berghei* was surprising, however, given the widespread use of *A. indica* in herbal preparations. Other limonids from neem were isolated and found to have some antimalarial activity (1143). The stem bark of *Tabebuia ochracea* spp. *neochrysantha* has been used by Indians of the Colombian Amazon as an antimalarial. Extracts provided an inseparable mixture of furanonaphthoquinones (**150**) (1144). The mixture demonstrated significant activity *in vitro* against *P. falciparum*. *Vismia guineësis*

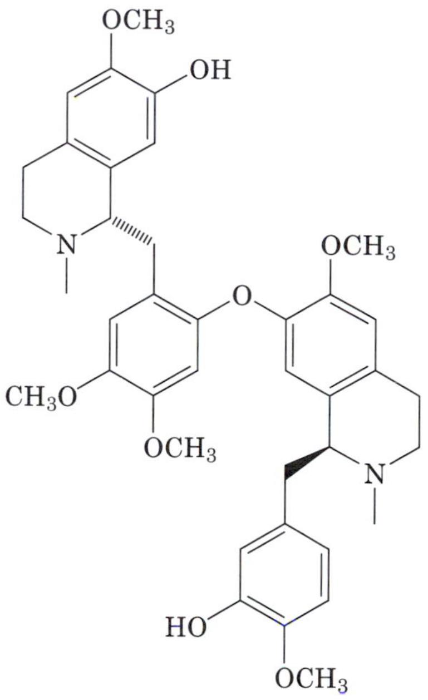

**(142)**
(+)-malekulatine

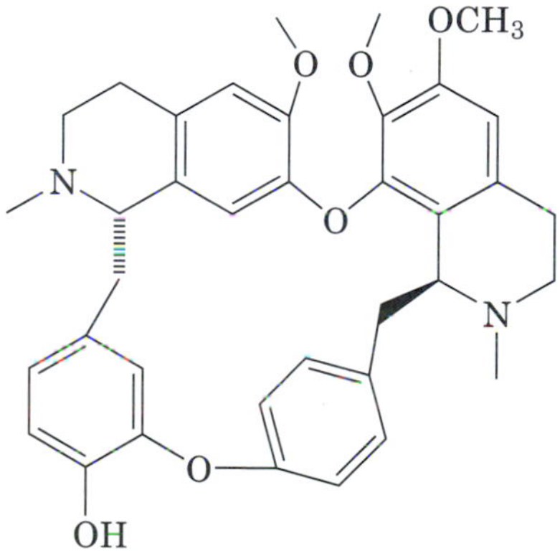

**(143)**
(−)-repandine

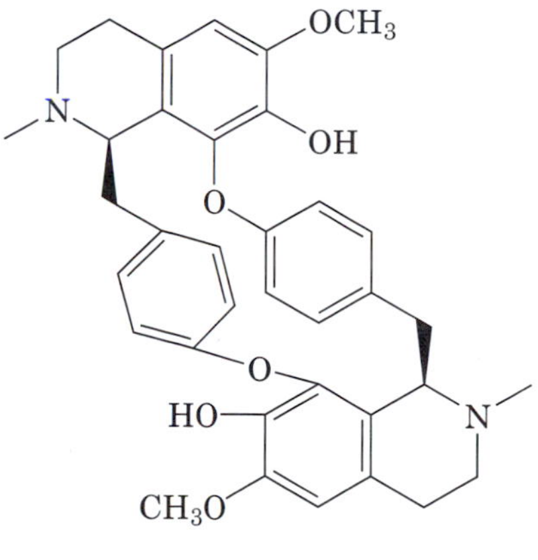

**(144)**
(−)-cycleanine

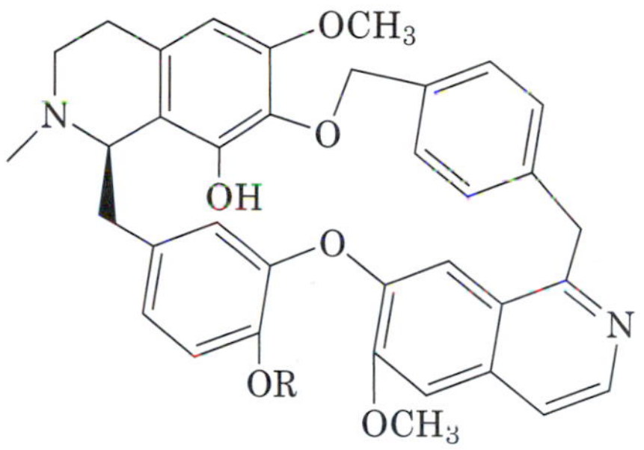

**(145)**
R = CH₃, (+)-cyclatjehenine
**(146)**
R = H, (+)-cycleatjehine

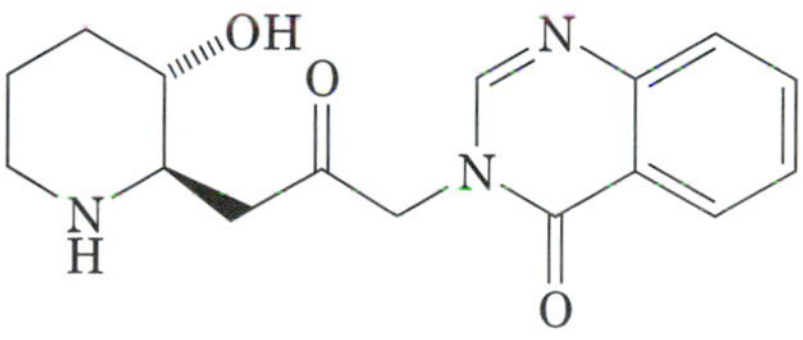

**(147)**
febrifugine

has been used to treat malaria in West Africa; it is the source of vismione H (**151**) (1145). This prenylated preanthraquinone was the active component, and several analogs were studied and found to be active as well. An aporphine, (−)-roemrefidine (**152**), was isolated from *Sparattanthelium amazonum*, a vine of the subtropical rain forests of South American, where it is used by the native com-

munity (1146). *In vivo* against *P. berghei* in mice, the compound was not only effective but displayed little toxicity. Tubulosine (**153**) from *Pogonopus tubulosus*, another source of traditional medicines from the South American subtropical rain forest, was more active than chloroquine in sensitive strains of *P. falciparum* and significantly more active in resis-

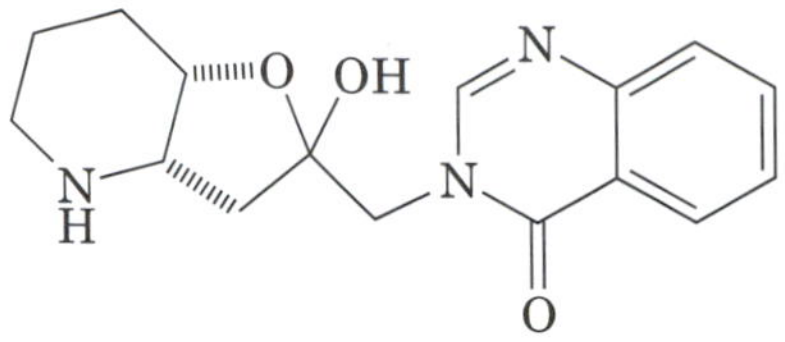

**(148)**
isofebrifugine

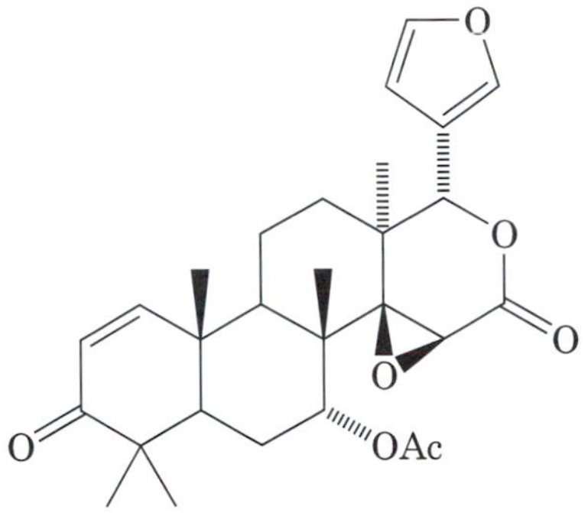

**(152)**
(–)-roemrefidine

**(149)**
gedunin

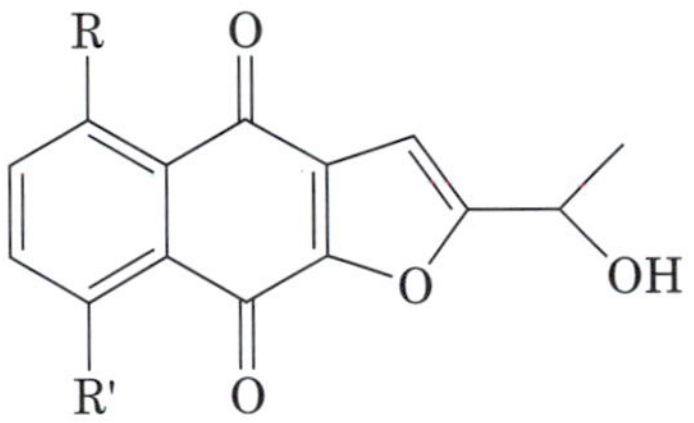

**(153)**
tubulosine

**(150)**
R = H, R' = OH
R = OH, R' = H

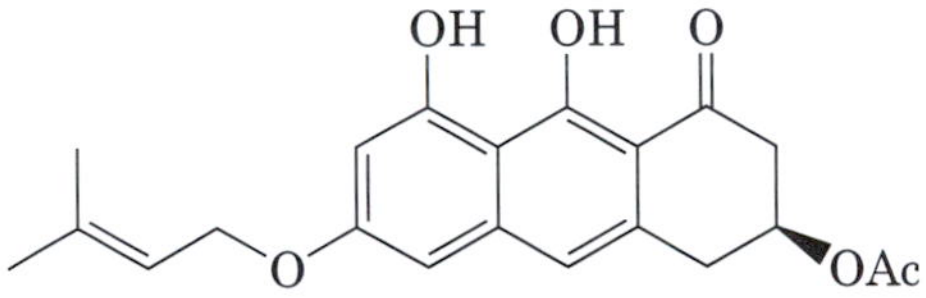

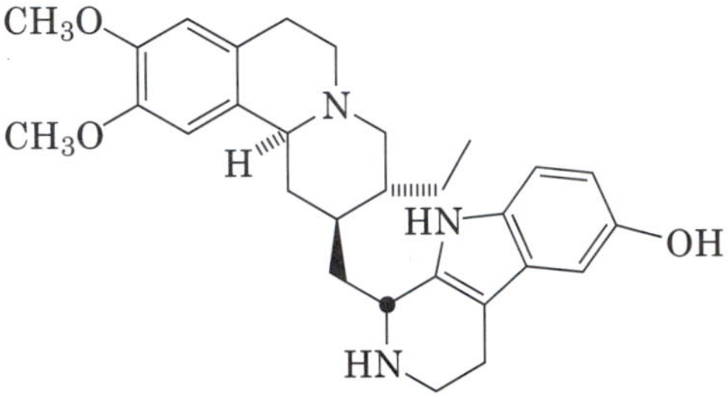

**(154)**
brusatol

**(151)**
vismione H

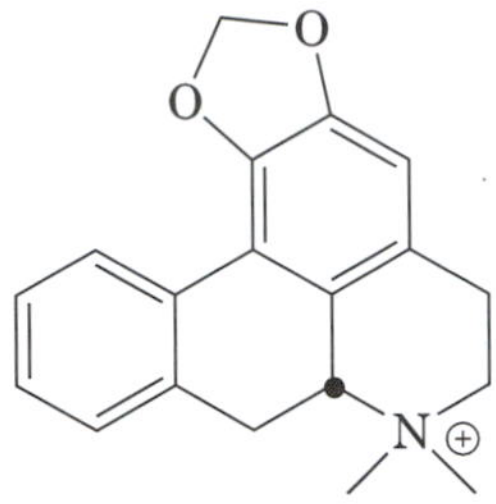

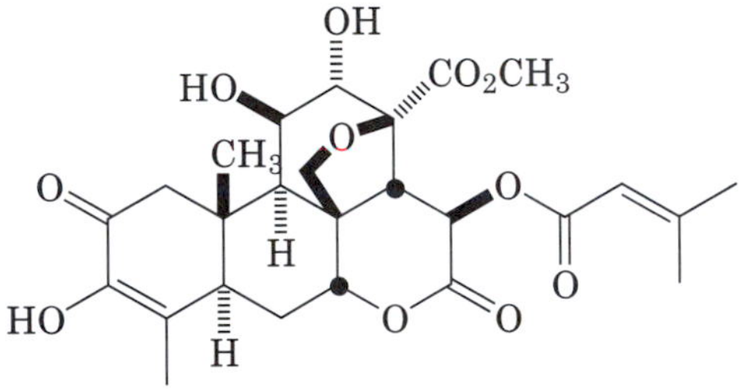

lated derivatives led to the discovery of compounds active *in vivo* with selective toxicities (1149, 1150). Cryptolepine (**155**) was isolated from *Cryptolepis sanguinolenta*, an African climbing liana (1151). Analogs of the natural

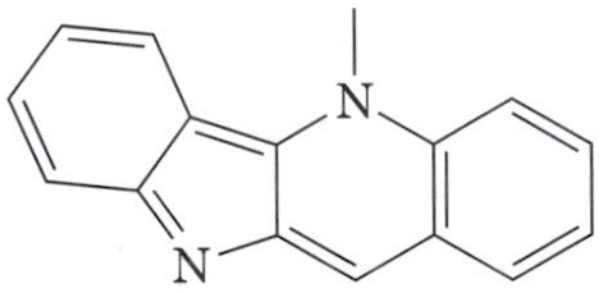

tant ones *in vitro* (1147). Brusatol (**154**), from *Brucea javanica*, a plant used in Chinese herbal remedies for malaria, was shown to be active against a chloroquine-resistant strain of *P. falciparum* (1148). Investigation of acety-

**(155)**
cryptolepine

product were shown to be effective against *P. berghei* in infected mice (1152). The synthetic compounds (**156**) were active in both chlo-

(156)

roquine-sensitive and chloroquine-resistant strains and showed no evidence of cross-resistance with chloroquine. Licochalcone A (**157**) was isolated from Chinese licorice root, a tra-

(157)
licochalcone

ditional treatment for a number of disorders. It acts against chloroquine-sensitive and -resistant strains of *P. falciparum* (1153) and *in vivo* against *P. yoelii* in infected mice (1154). An analog of licochalcone A, 2,4-dimethoxy-4'-butoxychalcone (**158**), was active in

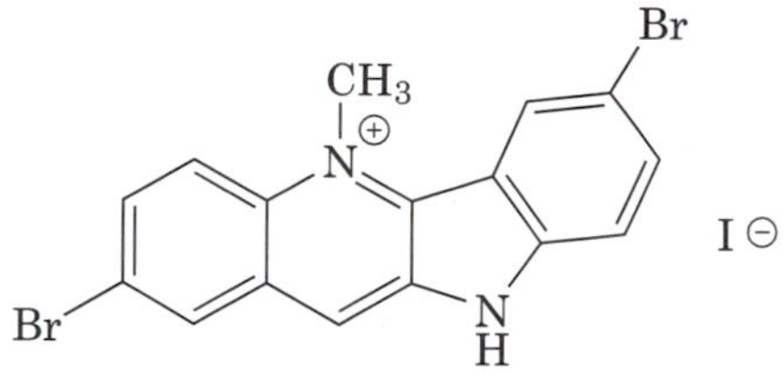

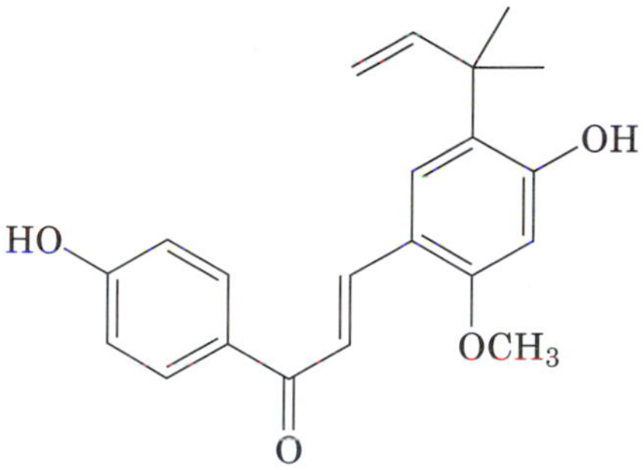

(158)

the micromolar range in rodent models, was orally active, and was much less toxic than the natural product (1155). An aminosteroid, sarachine (**159**), was isolated from *Saracha*

(159)
sarachine

*punctata*, a Bolivian shrub (1156). The compound had good activity both *in vitro* and *in vivo* (*P. vinckei*), although there was some cytotoxicity.

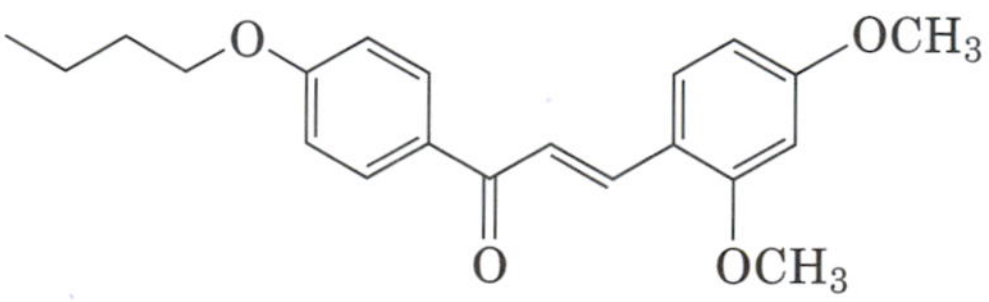

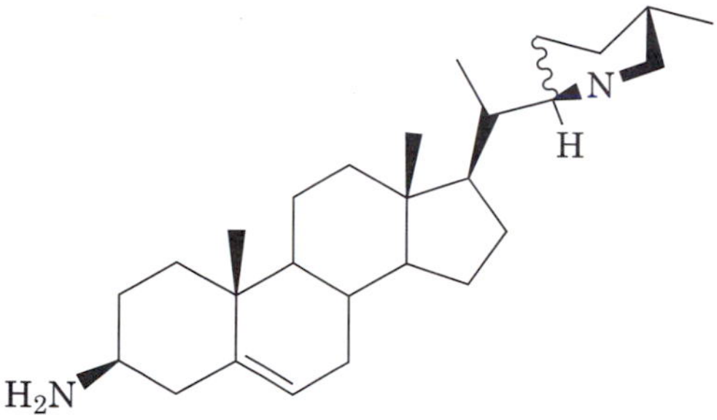

## 4 RESOURCES

General information about malaria can be found on the World Wide Web at sites maintained by the United States Centers for Disease Control (1157) and by the World Health Organization (1158).

## REFERENCES

1. Sir D. J. Macfarlane Burnet, as quoted in Wyler, *N. Engl. J. Med.*, **308**, 875 (1983).

2. W. Shakespeare, *in Henry V, Act II, Scene 1.*

3. E. Bottius, A. Guanzirolli, J.-F. Trape, C. Rogier, L. Konate, and P. Druilhe, *Trans. R. Soc. Trop. Med. Hygiene*, **90**, 15–19 (1996).

4. S. K. Volkman, A. E. Barry, E. J. Lyons, K. M. Nielsen, S. M. Thomas, M. Choi, S. S. Thakore, K. P. Day, D. F. Wirth, and D. L. Hartl, *Science*, **293**, 482–484 (2001).

5. D. J. Conway, C. Fanello, J. M. Lloyd, B. M. A. S. Al-Joubori, A. Baloch, S. D. Somanath, C. Roper, A. M. J. Oduola, B. Mulder, M. M. Povoa, B. Singh, and A. W. Thomas, *Mol. Biochem. Parasitol.*, **111**, 163–171 (2000).

6. S. A. Tishkoff, R. Varkonyi, N. Cahinhinan, S. Abbes, G. Argyropoulos, G. Destro-Bisol, A. Drousiotou, B. Dangerfield, G. Lefranc, J. Loiselet, A. Piro, M. Stoneking, A. Tagarelli, G. Tagarelli, E. H. Touma, S. M. Williams, and A. G. Clark, *Science*, **293**, 455–462 (2001).

7. S. H. Qari, Y.-P. Shhi, I. F. Goldman, V. Udhayakumar, M. P. Alpers, et. al., *Lancet*, **341**, 780–783 (1993).

8. D. D. Despommier and J. W. Karapelou, *Parasite Life Cycles*, Springer-Verlag, New York, 1987.

9. R. Horuk, C. E. Chitnis, W. C. Darbonne, T. J. Colby, A. Rybicki, T. J. Hadley, and L. H. Miller, *Science*, **261**, 1182–1184 (1993).

10. T. J. Hadley and S. C. Peiper, *Blood*, **89**, 3077–3091 (1997).

11. B. L. Pasloske and R. J. Howard, *Ann. Rev. Med.*, **45**, 283 (1994).

12. R. E. Desjardins, C. J. Canfield, D. E. Haynes, and J. D. Chulay, *Antimicrob. Agents Chemother.*, **16**, 710–718 (1979).

13. W. K. Milhous, N. F. Weatherley, J. H. Bowdre, and R. E. Desjardins, *Antimicrob. Agents Chemother.*, **27**, 525–530 (1985).

14. Z. G. Ye, K. Van Dyke, and M. Wimmer, *Exp. Parasitol.*, **64**, 418–423 (1987).

15. A. U. Orjih, J. S. Ryerse, and C. D. Fitch, *Experientia*, **50**, 34–9 (1994).

16. S. E. Francis, D. J. Sullivan, and D. E. Goldberg, *Ann. Rev. Biochem.*, **195**, 97–123 (1997).

17. J. H. McKerrow, E. Sun, P. J. Rosenthal, and J. Bouvier, *Ann. Rev. Microbiol.*, **47**, 821–853 (1993).

18. I. Y. Gluzman, S. E. Francis, A. Oksman, C. E. Smith, K. L. Duffin, and D. E. Goldberg, *J. Clin. Invest.*, **93**, 1602–1608 (1994).

19. S. E. Francis, I. Y. Gluzman, A. Oksman, A. Knickerbocker, R. Mueller, M. L. Bryant, D. R. Sherman, D. G. Russell, and D. E. Goldberg, *EMBO J.*, **13**, 306–317 (1994).

20. R. P. Moon, L. Tyas, U. Certa, K. Rupp, D. Bur, C. Jacquet, H. Matile, H. Loetscher, F. GrueningerLeitch, J. Kay, B. M. Dunn, C. Berry, and R. G. Ridley, *Eur. J. Biochem.*, **244**, 552–560 (1997).

21. S. E. Francis, R. Banerjee, and D. E. Goldberg, *J. Biol. Chem.*, **272**, 14961–14968 (1997).

22. A. M. Silva, A. Y. Lee, S. V. Gulnik, P. Majer, J. Collins, T. N. Bhat, P. J. Collins, R. E. Cachau, K. E. Luker, I. Y. Gluzman, S. E. Francis, A. Oksman, D. E. Goldberg, and J. W. Erickson, *Proc. Natl. Acad. Sci. USA*, **93**, 10034–10040 (1996).

23. A. M. Silva, A. Y. Lee, S. V. Gulnik, P. Majer, J. Collins, T. N. Bhat, P. J. Collins, R. E. Cachau, K. E. Luker, I. Y. Gluzman, S. E. Francis, A. Oksman, D. E. Goldberg, and J. W. Erickson, *Proc. Natl. Acad. Sci. USA*, **93**, 10034–10039 (1996).

24. J. Westling, C. A. Yowell, P. Majer, J. W. Erickson, J. B. Dame, and B. M. Dunn, *Exp. Parasitol.*, **87**, 185–193 (1997).

25. L. Tyas, R. P. Moon, H. Loetscher, B. M. Dunn, J. Kay, R. G. Ridley, and C. Berry, *Adv. Exp. Med. Biol.*, **436**, 407–411 (1998).

26. K. Kolakovich Eggleson, K. L. Duffin, and D. E. Goldberg, *J. Biol. Chem.*, **274**, 32411–32417 (1999).

27. N. D. G. de Domínguez and P. J. Rosenthal, *Blood*, 4448–4454 (1996).

28. M. Foley and L. Tilley, *Pharmacol. Ther.*, **79**, 55–87 (1998).

29. D. A. Bohle, R. E. Dinnebier, S. K. Madsen, and P. W. Stephens, *J. Biol. Chem.*, **272**, 713–716 (1997).

30. S. Pagola, P. W. Stephens, D. S. Bohle, A. D. Kosar, and S. K. Madsen, *Nature*, **404**, 307–310 (2000).

31. C. D. Fitch and P. Kanjananggulpan, *J. Biol. Chem.*, **262**, 15552–15555 (1987).

32. A. F. G. Slater and A. Cerami, *Nature*, **355**, 167–169 (1992).

33. A. C. Chou and C. D. Fitch, *Life Sciences*, **51**, 2073–2078 (1992).

34. A. Dorn, R. Stoffel, H. Matile, A. Bubendorf, and R. G. Ridley, *Nature.*, **374**, 269–271 (1995).

35. C. D. Fitch, *Trans. Am. Clin. Climatolog. Assoc.*, **109**, 97–106 (1998).

36. A. Dorn, S. R. Vippagunta, H. Matile, C. Jaquet, J. L. Vennerstrom, and R. G. Ridley, *Biochem. Pharmacol.*, **55**, 727–736 (1998).

37. C. D. Fitch, G. Cai, Y.-F. Chen, and J. D. Shoemaker, *Biochim. Biophys. Acta*, **1454**, 31–37 (1999).

38. D. J. Sullivan Jr., I. Y. Gluzman, and D. E. Goldberg, *Science.*, **271**, 219–222 (1996).

39. D. J. Sullivan Jr., H. Matile, R. G. Ridley, and D. E. Goldberg, *J. Biol. Chem.*, **273**, 31103–31107 (1998).

40. A. V. Pandey, H. Bisht, V. K. Babbarwal, J. Srivastava, K. C. Pandey, and V. S. Chauhan, *Biochem. J.*, **355**, 333–338 (2001).

41. T. J. Egan, W. W. Mavuso, and K. K. Ncokazi, *Biochemistry*, **40**, 204–213 (2001).

42. D. I. Baruch, B. L. Pasloske, H. B. Singh, X. Bi, X. C. Ma, M. Feldman, T. F. Taraschi, and R. J. Howard, *Cell*, **82**, 77–87 (1995).

43. D. S. Peterson, L. H. Miller, and T. E. Wellems, *Proc. Natl. Acad. Sci. USA*, **92**, 7100–7104 (1995).

44. J. D. Smith, C. E. Chitnis, A. G. Craig, D. J. Roberts, D. E. Hudson-Taylor, D. S. Peterson, R. Pinches, C. I. Newbold, and L. H. Miller, *Cell*, **81**, 101–110 (1995).

45. X. Su, V. M. Heatwole, S. P. Wertheimer, F. Guinet, J. A. Herrfeldt, D. S. Peterson, J. A. Ravetch, and T. E. Wellems, *Cell*, **82**, 89–100 (1995).

46. F. Roberts, C. W. Roberts, J. J. Johnson, D. E. Kyle, T. Krell, J. R. Coggins, G. H. Coombs, W. K. Milhous, S. Tzipori, D. J. Ferguson, D. Chakrabarti, and R. McLeod, *Nature*, **393**, 801–805 (1998).

47. R. J. M. Wilson, P. W. Denny, P. R. Preiser, K. Rangachari, K. Roberts, A. Roy, A. Whyte, M. Strath, D. J. Moore, P. W. Moore, and D. H. Williamson, *J. Mol. Biol.*, **261**, 155–172 (1996).

48. G. I. McFadden, M. E. Reith, J. Munholland, and N. LangUnnasch, *Nature*, **381**, 482 (1996).

49. S. Kohler, C. F. Delwiche, P. W. Denny, L. G. Tilney, P. Webster, R. J. M. Wilson, J. D. Palmer, and D. S. Roos, *Science*, **275**, 1485–1489 (1997).

50. M. E. Fichera and D. S. Roos, *Nature*, **390**, 407–409 (1997).

51. H. Jomaa, J. Wiesner, S. Sanderbrand, B. Altincicek, C. Weidemeyer, M. Hintz, I. Türbachova, M. Eberl, J. Zeidler, H. K. Lichtenthaler, D. Soldati, and E. Beck, *Science*, **285**, 1573–1576 (1999).

52. J. L. Vennerstrom and J. W. Eaton, *J. Med. Chem.*, **31**, 1269–1277 (1988).

53. M. J. Gardner, H. Tettelin, D. J. Carucci, L. M. Cummings, L. Aravind, E. V. Koonin, S. Shallom, T. Mason, K. Yu, C. Fujii, J. Pederson, K. Shen, J. Jing, C. Aston, Z. Lai, D. C. Schwartz, M. Pertea, S. Salzberg, L. Zhou, G. G. Sutton, R. Clayton, O. White, H. O. Smith, C. M. Fraser, M. D. Adams, J. C. Venter, and S. L. Hoffman, *Science*, **282**, 1126–1132 (1998).

54. S. Bowman, D. Lawson, D. Basham, D. Brown, T. Chillingworth, C. M. Churcher, A. Craig, R. M. Davies, K. Devlin, T. Feltwell, S. Gentles, R. Gwilliam, N. Hamlin, D. Harris, S. Holroyd, T. Hornsby, P. Horrocks, K. Jagels, B. Jassal, S. Kyes, J. McLean, S. Moule, K. Mungall, L. Murphy, K. Oliver, M. A. Quail, M. A. Rajandream, S. Rutter, J. Skelton, R. Squares, S. Squares, J. E. Sulston, S. Whitehead, J. R. Woodward, C. Newbold, and B. G. Barrell, *Nature*, **400**, 532–538 (1999).

55. X. Su, M. T. Ferdig, Y. Huang, C. Huynh, A. Liu, J. You, J. C. Wootton, and T. E. Wellems, *Science*, **286**, 1351–1352 (1999).

56. Z. W. Lai, J. P. Jing, X. Aston, V. Clarke, J. Apodaca, E. T. Dimalanta, D. J. Carucci, M. J. Gardner, B. Mishra, T. A. Ananthataman, S. Paxia, S. L. Hoffman, J. C. Venter, E. J. Huff, and D. C. Schwartz, *Nat. Genet.*, **23**, 309–313 (1999).

57. WHO, http://www.wehi.edu.au/MalDB-www/who.html, World Health Organization and Monash University, Australia.

58. WHO, *Wkly. Epidemiological Rep.*, **74**, 265–272 (1999).

59. WHO, http://www.who.int/ith/english/country.htm.

60. WHO, *Wkly. Epidemiological Rep.*, **74**, 265–272 (1999).

61. WHO, WHOSIS, Fact Sheet: Malaria, http://www.who.int/inf-fs/en/fact094.html.

62. J. R. MacArthur, A. R. Levin, M. Mungai, J. Roberts, A. M. Barber, P. B. Bloland, S. P. Kachur, R. D. Newman, R. W. Steketee, and M. E. Parise, *MMWR, Morb. Mortal. Wkly. Rep.*, **50**, 25–44 (2001).

63. W. H. Wernsdorfer and D. Payne, *Pharmacol. Ther.*, **50**, 95–121 (1991).

64. U. D'Alessandro and H. Buttiens, *Trop. Med. Int. Health.*, **6**, 845–848 (2001).

65. W. Peters, *Pharmacol. Ther.*, **47**, 499–508 (1990).

66. B. H. Kean, *J. Am. Med. Assoc.*, **241**, 395 (1979).

67. M. Whitby, *J. Antimicrob. Chemother.*, **40**, 749–752 (1997).

68. T. Jelinek, H. D. Nothdurft, F. Von Sonnenburg, and T. Loscher, *Am. J. Trop. Med. Hygiene*, **52**, 322–324 (1995).

69. G. D. Shanks, A. Barnett, M. D. Edstein, and K. H. Rieckmann, *Med. J. Aust.*, **162**, 306–310 (1995).

70. W. E. Collins and G. M. Jeffery, *Am. J. Trop. Med. Hyg.*, **55**, 243–249 (1996).

71. W. H. Wernsdorfer, T. Chongsuphajaisiddhi, and N. P. Salazar, *Southeast Asian J. Trop. Med. Public Health*, **25**, 11–18 (1994).

72. N. White, *Philos. Trans. R. Soc. London, Ser. B.*, **354**, 739–749 (1999).

73. W. Peters, *J. R. Soc. Med.*, **82**, 14–17 (1989).

74. N. J. White, *Drug Resist. Updates.*, **1**, 3–9 (1998).

75. I. M. Hastings and M. J. Mackinnon, *Parasitology*, **117**, 411–417 (1998).

76. C. V. Plowe and T. E. Wellems, *Adv. Exp. Med. Biol.*, **390**, 197–209 (1995).

77. T. E. Wellems, *Parasitol. Today*, **7**, 110–112 (1991).

78. R. W. Steketee, B. L. Nahlen, M. E. Parise, and C. Menendez, *Am. J. Trop. Med. Hyg.*, **64**, 28–35 (2001).

79. S. Gupta, K. Trenhome, R. M. Anderson, and K. P. Day, *Science*, **263**, 961–963 (1994).

80. M. Fried and P. E. Duffy, *Science*, **272**, 1502–1504 (1996).

81. C. S. Okereke, *Clin. Ther.*, **21**, 1456–1496 (1999).

82. C. Ruwende, S. C. Khoo, R. W. Snow, S. N. R. Yates, D. Kwiatkowski, S. Gupta, P. Warn, C. E. M. Allsopp, S. C. Gilbert, N. Peschu, C. I. Newbold, B. M. Greenwood, K. Marsh, and A. V. S. Hill, *Nature*, **376**, 246–249 (1995).

83. M. Cappadoro, G. Giribaldi, E. O'Brien, F. Turrii, F. Mannu, D. Ulliers, G. Simula, L. Luzzato, and P. Arese, *Blood*, **92**, 2527–2534 (1998).

84. T. R. Jones, *Parasitol. Today*, **13**, 107–111 (1997).

85. J. A. Martiney, A. Cerami, and A. F. G. Slater, *Mol. Med.*, **2**, 236–246 (1996).

86. T. N. Williams, K. S. B. Maitland, M. Ganczakowski, T. E. A. Peto, C. I. Mewbold, D. K. Bowdens, D. J. Weathrall, and J. B. Clegg, *Nature*, **383**, 522–525 (1996).

87. G. Destro-Bisol, E. D'Aloja, G. Spedini, R. Scatena, B. Giardina, and V. Pascali, *Am. J. Phys. Anthropol.*, **109**, 269–273 (1999).

88. S. J. Allen, A. O'Donnell, N. D. E. Alexander, C. S. Mgone, T. E. A. Peto, J. B. Clegg, M. P. Alpers, and D. J. Weatherall, *Am. J. Trop. Med. Hyg.*, **60**, 1056–1060 (1999).

89. K. C. Kain, G. D. Shanks, and J. S. Keystone, *Clin. Infect. Dis.*, **33**, 226–234 (2001).

90. G. D. Shanks, K. C. Kain, and J. S. Keystone, *Clin. Infect. Dis.*, **33**, 381–385 (2001).

91. WHO, http://www.who.int/ith/english/protect.htm.

92. Centers for Disease Control and Prevention, http://www.cdc.gov/travel/yellowbook.pdf.

93. Centers for Disease Control and Prevention, http://www.cdc.gov/travel/.

94. G. Harrison, *Mosquitoes, Malaria & Man: A History of the Hostilities Since 1880*, E. P. Dutton, New York, 1978.

95. F. H. Collins and S. M. Paskewitz, *Annu. Rev. Entomol.*, **40**, 195–219 (1995).

96. N. K. Kere, A. Arabola, B. Bakote'e, O. Qalo, T. R. Burkot, R. H. Webber, and B. A. Southgate, *Med. Vet. Entomol.*, **10**, 145–148 (1996).

97. J. Richards, O. Frank, R. E. Klein, R. Z. Flores, S. Weller, M. Gatica, R. Zeissig, and J. Sexton, *Am. J. Trop. Med. Hyg.*, **49**, 410–418 (1993).

98. R. F. Beach, T. K. Ruebush II, J. D. Sexton, P. L. Bright, A. W. Hightower, J. G. Breman, D. L. Mount, and A. J. Oloo, *Am. J. Trop. Med. Hyg.*, **49**, 290–300 (1993).

99. Z. Premji, P. Lubega, Y. Manisi, E. Mchopa, J. Minjas, W. Checkley, and C. Shiff, *Trop. Med. Parasitol.*, **46**, 147–153 (1995).

100. U. D'Alessandro, *Lancet*, **345**, 479–483 (1995).

101. F. H. Collins and N. J. Besansky, *Science*, **264**, 1874–1875 (1994).

102. M. Balter, *Science*, **285**, 508–509 (1999).

103. D. J. Rogers and S. E. Randolph, *Science*, **289**, 1763–1766 (2000).

104. W. K. Asenso-Okyere and J. A. Dzator, *Soc. Sci. Med.*, **45**, 659–667 (1997).

105. J. L. Gallup and J. D. Sachs, *Am. J. Trop. Med. Hyg.*, **64**, 85–96 (2001).

106. WHO, *Wkly. Epidemiological Rec.*, **75**, 318–319 (2000).

107. J. E. Rosenblatt, *Mayo Clin. Proc.*, **74**, 1161–1175 (1999).

108. WHO, http://mosquito.who.int/.

109. WHO, *Wkly. Epidemiological Rec.*, **76**, 78–79 (2001).

110. WHO, *Wkly. Epidemiological Rec.*, **76**, 89–91 (2001).

111. WHO, www.who.int/inf-fs/en/factXXX.html.

112. D. Greenwood, *J. Antimicrob. Chemother.*, **30**, 417–427 (1992).

113. S. Jarcho, *Quinine's Predecessor Francesco Torti and the Early Histroy of Cinchona*, Johns Hopkins University Press, Baltimore, MD, 1993.

114. R. B. Woodward and W. E. Doering, *J. Am. Chem. Soc.*, **67**, 860–874 (1945).

115. R. B. Woodward and W. E. Doering, *J. Am. Chem. Soc.*, **66**, 849 (1944).

116. G. Stork, D. Niu, A. Fujimoto, E. R. Koft, J. M. Balkovec, J. R. Tata, and G. R. Dake, *J. Am. Chem. Soc.*, **123**, 3239–3242 (2001).

117. J. M. Karle and A. K. Bhattacharjee, *Bioorg. Med. Chem.*, **7**, 1769–1774 (1999).

118. J. M. Karle, I. L. Karle, L. Gerena, and W. K. Milhous, *Antimicrob. Agents Chemother.*, **36**, 1538–1544 (1992).

119. D. L. Wesche and J. Black, *J. Trop. Med. Hyg.*, **93**, 153–159 (1990).

120. J. M. Karle, R. Olmeda, L. Gerena, and W. K. Milhous, *Exp. Parasitol.*, **76**, 345–51 (1993).

121. S. Krishna and N. J. White, *Clin. Pharmacokinet.*, **30**, 263–299 (1996).

122. X.-J. Zhao and T. Ishizaki, *Br. J. Clin. Pharmacol.*, **44**, 505–511 (1997).

123. X. J. Zhao and T. Ishizaki, *Eur. J. Drug Metab. Pharmacokinet.*, **24**, 272–278 (1999).

124. D. Bunnag, J. Karbwang, K. Na-Bangchang, A. Thanavibul, S. Chittamas, and T. Harinasuta, *Southeast Asian J. Trop. Med. Public Health.*, **27**, 15–18 (1996).

125. P. S. A. Sarma, A. K. Mandal, and H. J. Khamis, *Am. J. Trop. Med. Hyg.*, **58**, 454–457 (1998).

126. T. Skinner-Adams and T. M. E. Davis, *Antimicrobial Agents Chemother.*, **43**, 1304–1306 (1999).

127. H. Barennes, E. Pussard, A. M. Sani, F. Clavier, F. Kahiatani, G. Granic, D. Henzel, L. Ravinet, and F. Verdier, *Br. J. Clin. Pharmacol.*, **41**, 389–395 (1996).

128. H. Barennes, J. Munjakazi, F. Verdier, F. Clavier, and E. Pussard, *Trans. R. Soc. Trop. Med. Hyg.*, **92**, 437–440 (1998).

129. P.-E. Kofoed, E. Mapaba, F. Lopes, F. Pussick, P. Aaby, and L. Rombo, *Trans. R. Soc. Trop. Med. Hyg.*, **91**, 462–464 (1997).

130. H. Barennes, F. Verdier, F. Clavier, and E. Pusard, *Clin. Drug Invest.*, **17**, 287–291 (1999).

131. J. Karbwang, P. Molunto, D. Bunnag, and T. Harinasuta, *Southeast Asian J. Trop. Med. Public Health.*, **22**, 72–76 (1991).

132. P. Brasseur, J. Lousmouo, O. Brandicourt, R. Moyou-Somo, and P. Druilhe, *Am. J. Trop. Med. Hyg.*, **39**, 166–172 (1988).

133. J. B. Ouédraogo, Y. Dutheil, H. Tinto, B. Traore, H. Zampa, F. Tall, S. O. Coulibaly, and T. R. Guiguemde, *Trop. Med. Int. Health.*, **3**, 381–384 (1998).

134. T. Jelinek, P. Schelbert, T. Löscher, and D. Eichenlaub, *Trop. Med. Parasitol.*, **46**, 38–40 (1995).

135. M. G. Zalis, L. Pang, M. S. Silveira, W. K. Milhous, and D. F. Wirth, *Am. J. Trop. Med. Hyg.*, **58**, 630–637 (1998).

136. M. B. Reed, K. J. Saliba, S. R. Caruana, K. Kirk, and A. F. Cowman, *Nature*, **403**, 906–909 (2000).

137. P. T. Giao and P. J. de Vries, *Clin. Pharmacokinet.*, **40**, 343–373 (2001).

138. C. P. Babalola, O. O. Bolaji, F. A. Ogunbona, A. Sowunmi, and O. Walker, *Eur. J. Pharm. Biopharm.*, **44**, 143–147 (1997).

139. C. P. Babalola, O. O. Bolaji, F. A. Ogunbona, A. Wowunmi, and O. Walker, *Pharm. World Sci.*, **20**, 118–122 (1998).

140. S. Krishna, N. V. Nagaraja, T. Planche, T. Agbenyega, G. Bedo-Addo, C. Ansong, A. Owusu-Ofori, A. L. Shroads, G. Henderson, A. Hutson, H. Derendorf, and P. W. Stacpoole, *Antimicrol. Agents Chemother.*, **45**, 1803–1809 (2001).

141. S. Wanwimolruk and J. R. Denton, *J. Pharm. Pharmacol.*, **44**, 806–811 (1992).

142. S. M. Mansor, S. A. Ward, G. Edwards, P. E. Hoaksey, and A. M. Breckenridge, *J. Pharm. Pharmacol.*, **42**, 428–432 (1990).

143. S. Pukrittayakamee, S. Looareesuwan, D. Keeratithakul, T. M. E. Davis, P. Teja-Isavadharm, B. Nagachinta, A. Weber, A. L. Smith, D. Kyle, and N. J. White, *Eur. J. Clin. Pharmacol.*, **52**, 487–493 (1997).

144. M. Van Der Torn, P. E. Thuma, G. F. Mabeza, G. Biemba, V. M. Moyo, C. E. McLaren, G. M. Brittenham, and V. R. Gordeuk, *Trans. R. Soc. Trop. Med. Hyg.*, **92**, 325–331 (1998).

145. M. B. van Hensbroek, D. Kwiatkowski, B. van den Berg, F. J. Hoek, C. J. van Boxtel, and P. A. Kager, *Am. J. Trop. Med. Hyg.*, **54**, 237–242 (1996).

146. G. Watt, A. Na-Nakorn, D. N. Bateman, N. Plubha, P. Mothanaprakoon, M. Edstein, and H. K. Webster, *Am. J. Trop. Med. Hyg.*, **49**, 645–649 (1993).

147. Centers for Disease Control and Prevention, *MMWR Morbid. Mortal. Wkly. Rep.*, **49**, 1138–1140 (2000).

148. J. D. Spikes, *J. Photochem. Photobiol., B.*, **42**, 1–11 (1998).

149. B. Hogh, A. Gamage-Mendis, G. A. Butcher, R. Thompson, K. Begtrup, C. Mendis, S. M. Enosse, M. Dgedge, J. Barreto, W. Eling, and R. E. Sinden, *Am. J. Trop. Med. Hyg.*, **58**, 176–182 (1998).

150. A. Buckling, L. C. Ranford-Cartwright, A. Miles, and A. F. Read, *Parasitology*, **118**, 339–346 (1999).

151. J. Ducharme and R. Farinotti, *Clin. Pharmacokinet.*, **31**, 257–274 (1996).

152. G. Edwards, S. Looareesuwan, A. J. Davies, Y. Wattanagoon, R. E. Phillips, and D. A. Warrell, *Br. J. Clin. Pharmacol.*, **25**, 477–485 (1988).

153. K. Na Bangchang, L. Limpaibul, A. Thanavibul, P. Tan-Ariya, and J. Karbwang, *Brit. J. Clin. Pharmacol.*, **38**, 278–281 (1994).

154. F. ter Kuile, N. J. White, P. Holloway, G. Pasvol, and S. Krishna, *Exp. Parasitol.*, **76**, 85–95 (1993).

155. P. A. Phillips-Howard and D. Wood, *Drug Safety*, **14**, 131–145 (1996).

156. A. A. O. Mosuro, *Biokemistrie*, **7**, 9–18 (1997).

157. K. H. Rieckmann, D. R. Davis, and D. C. Hutton, *Lancet*, 1183–1184 (1989).

158. Marlar-Than, Myat-Phone-Kyaw, Aye-Yu-Soe, Khaing-Khaing-Gyi, Ma-Sabai, and Myint-Oo, *Trans. R. Soc. Trop. Med. Hyg.*, **89**, 307–308 (1995).

159. J. K. Baird, M. F. S. Nalim, H. Basri, S. Masbar, B. Leksana, E. Tjitra, R. M. Dewi, D. Khairani, and F. S. Wignall, *Trans. R. Soc. Trop. Med. Hyg.*, **90**, 409–411 (1996).

160. D. J. Fryauff, Soekartono, S. Tuti, B. Leksana, Suradi, S. Tandayu, and J. K. Baird, *Trans. R. Soc. Trop. Med. Hyg.*, **92**, 82–83 (1998).

161. J. K. Baird, I. Wiady, D. J. Fryauff, M. A. Sutanihardja, B. Leksana, H. Widjaya, Kysdarmanto, and B. Subianto, *Am. J. Trop. Med. Hyg.*, **56**, 627–631 (1997).

162. J. K. Baird, B. Leksana, S. Masbar, S. M. A. Sutanihardja, D. J. Fryauff, and B. Subianto, *Am. J. Trop. Med. Hyg.*, **56**, 618–620 (1997).

163. E. J. Phillips, J. S. Keystone, and K. C. Kain, *Clin. Infect. Dis.*, **23**, 1171–1173 (1996).

164. J.-F. Trape, G. Pison, M.-P. Preziosi, C. Enel, A. Desgrées du Loû, V. Delaunay, B. Samb, E. Lagarde, J.-F. Molez, and F. Simondon, *C. R. Acad. Sci Paris, Sciences de la Vie.*, **321**, 689–697 (1998).

165. P. B. Bloland, P. N. Kazembe, A. J. Oloo, B. Himonga, L. M. Barat, and T. K. Ruebush, *Trop. Med. Int. Health.*, **3**, 543–552 (1998).

166. F. P. Mockenhaupt, J. May, Y. Bergovist, O. G. Ademowo, P. E. Olumese, A. G. Falusi, L. Grossterlinden, C. G. Meyer, and U. Bienzle, *Antimicrob. Agents Chemother.*, **44**, 835–839 (2000).

167. S. Picot, A. Nkwelle Akede, J. F. Chaulet, E. Tetanye, P. Ringwald, J. M. Prevosto, C. Boudin, and P. Ambroise-Thomas, *Clin. Exp. Immunol.*, **108**, 279–283 (1997).

168. S. Vanijanota, A. Chantra, N. Phophak, D. Chindanond, R. Clemens, and S. Pukrittayakamee, *Ann. Trop. Med. Parasitol.*, **90**, 269–275 (1996).

169. J. C. F. M. Wetsteyn, P. J. De Vries, B. Oosterhuis, and C. J. Van Boxtel, *Brit. J. Clin. Pharmacol.*, **39**, 696–699 (1995).

170. M. Frisk and G. Gunnert, *Eur. J. Clin. Pharmacol.*, **44**, 271–274 (1993).

171. P. Nyirjesy, T. Kavasya, P. Axelrod, and P. R. Fischer, *Clin. Infect. Dis.*, **16**, 127–132 (1993).

172. M. D. G. Bustos, F. Gay, D. Diquet, P. Thomare, and D. Warot, *Trop. Med. Parasitol.*, **45**, 83–86 (1994).

173. P. M. O'Neill, P. G. Bray, S. R. Hawley, S. A. Ward, and B. K. Park, *Pharmacol. Ther.*, **77**, 29–58 (1998).

174. T. J. Egan and H. M. Marques, *Coord. Chem. Rev.*, **190–192**, 493–517 (1999).

175. M. Foley and L. Tilley, *Int. J. Parasitol.*, **27**, 231–240 (1997).

176. H. Ginsburg and M. Krugliak, *Drug Resist. Updates*, **2**, 180–187 (1999).

177. A. Yayon, Z. I. Cabantchik, and H. Ginsburg, *Proc. Natl. Acad. Sci. USA*, **82**, 2784–2788 (1985).

178. A. Yayon, Z. I. Cabantchik, and H. Ginsburg, *EMBO J.*, **3**, 2695–2700 (1984).

179. M. Mungthin, P. G. Bray, R. G. Ridley, and S. A. Ward, *Antimicrob. Agents Chemother.*, **42**, 2973–2977 (1998).

180. A. C. Chou, R. Chevli, and C. D. Fitch, *Biochemistry.*, **19**, 1543–1549 (1980).

181. S. R. Vippagunta, A. Dorn, R. G. Ridley, and J. L. Vennerstrom, *Biochim. Biophys. Acta*, **1475**, 133–140 (2000).

182. D. J. Sullivan Jr., I. Y. Gluzman, D. G. Russell, and D. E. Goldberg, *Proc. Natl. Acad. Sci., USA*, **93**, 11865–11870 (1996).

183. P. G. Bray, M. Mungthin, R. G. Ridley, and S. A. Ward, *Mol. Pharmacol.*, **54**, 170–179 (1998).

184. S. R. Vippagunta, A. Dorn, H. Matile, A. K. Bhattacharjee, J. M. Karle, W. Y. Ellis, R. G. Ridley, and J. L. Vennerstrom, *J. Med. Chem.*, **42**, 4630–4639 (1999).

185. T. J. Egan, R. Hunter, C. H. Kaschula, H. M. Marques, A. Misplon, and J. Walden, *J. Med. Chem.*, **43**, 283–291 (2000).

186. A. Dorn, S. R. Vippagunta, H. Matile, A. Bubendorf, J. L. Vennerstrom, and R. G. Ridley, *Biochem. Pharmacol.*, **55**, 737–747 (1998).

187. T. J. Egan, D. C. Ross, and P. A. Adams, *FEBS Lett.*, **352**, 54–57 (1994).

188. A. C. Chou and C. D. Fitch, *Biochem. Biophys. Res. Commun.*, **195**, 422–427 (1993).

189. K. Raynes, M. Foley, L. Tilley, and L. W. Deadly, *Biochem. Pharmacol.*, **52**, 551–559 (1996).

190. R. G. Ridley, *Trends Microbiol.*, **4**, 253–254 (1996).

191. S. R. Hawley, P. G. Bray, M. Mungthin, J. D. Atkinson, P. M. O'Neill, and S. A. Ward, *Antimicrob. Agents Chemother.*, **42**, 682–686 (1998).

192. A. C. Chou and C. D. Fitch, *J. Clin. Invest.*, **68**, 672–677 (1981).

193. A. U. Orjih, H. S. Banyal, R. Chevli, and C. D. Fitch, *Science*, **214**, 667–669 (1981).

194. H. S. Banyal and C. D. Fitch, *Life Sciences*, **31**, 1141–1144 (1982).

195. C. D. Fitch, *Parasitol. Today*, **1**, 333–334 (1986).

196. Y. Sugioka and M. Suzuki, *Biochim. Biophys. Acta*, **1074**, 19–24 (1991).

197. V. Papalexis, M. A. Siomos, N. Campanale, X. G. Guo, G. Kocak, M. Foley, and L. Tilley, *Mol. Biochem. Parasitol.*, **115**, 77–86 (2001).

198. P. Loria, S. Miller, M. Foley, and L. Tilley, *Biochem. J.*, **339**, 363–370 (1999).

199. D. J. Krogstad, P. H. Schlesinger, and I. Y. Gluzman, *J. Cell Biol.*, **101**, 2302–2309 (1985).

200. S. K. Martin, A. M. J. Oduola, and W. K. Milhous, *Science*, **235**, 899–901 (1987).

201. J. A. Martiney, A. Cerami, and A. F. G. Slater, *J. Biol. Chem.*, **270**, 22393–22398 (1995).

202. P. Rasoanaivo, S. Ratsimamanga-Urverg, and F. Frappier, *Curr. Med. Chem.*, **3**, 1–10 (1996).

203. A. Sowunmi, A. M. J. Oduola, O. A. T. Ogundahunsi, C. O. Falade, G. O. Gbotosho, and L. A. Salako, *Trans. R. Soc. Trop. Med. Hyg.*, **91**, 63–67 (1997).

204. A. M. J. Oduola, A. Sowunmi, W. K. Milhous, T. G. Brewer, D. E. Kyle, L. Gerena, R. N. Rossan, L. A. Salako, and B. G. Schuster, *Am. J. Trop. Med. Hyg.*, **58**, 625–629 (1998).

205. J. Adovelande, J. Deleze, and J. Schrevel, *Biochem. Pharmacol.*, **55**, 433–440 (1998).

206. S. G. Evans, N. Butkow, C. Stilwell, M. Berk, N. Kirchmann, and I. Havlik, *J. Pharmacol. Exp. Ther.*, **286**, 172–174 (1998).

207. A. Sowunmi, A. M. J. Oduola, O. A. T. Ogundahunsi, and L. A. Salako, *Trop. Med. Int. Health.*, **3**, 177–183 (1998).

208. S. Batra and A. P. Bhaduri, *Adv. Drug Res.*, **30**, 202–232 (1997).

209. D. F. Platel, F. Mangou, and J. Tribouley-Duret, *Int. J. Parasitol.*, **28**, 641–651 (1998).

210. S. J. Foote, J. K. Thompson, A. F. Cowman, and D. J. Kemp, *Cell*, **57**, 921–930 (1989).

211. C. M. Wilson, A. E. Serrano, A. Wasley, M. P. Bogenschutz, A. H. Shakar, and D. F. Wirth, *Science*, **244**, 1184–1186 (1989).

212. T. E. Wellems, L. J. Panton, I. Y. Gluzman, R. V. do Rosario, R. E. Gwadz, J. A. Walker, and D. J. Korgsstad, *Nature*, **345**, 253–255 (1990).

213. L. von Seidlein, M. T. Duraisingh, C. J. Drakeley, R. Bailey, B. M. Greenwood, and M. Pinder, *Trans. R. Soc. Trop. Med. Hyg.*, **91**, 450–453 (1997).

214. I. S. Adagu, W. N. Ogala, D. J. Carucci, M. T. Duraisingh, and D. C. Warhurst, *Ann. Trop. Med. Parasitol.*, **91**, S107–S111 (1997).

215. M. P. Grobusch, I. S. Adagu, P. G. Kremsner, and D. C. Warhurst, *Parasitology*, **116**, 211–217 (1998).

216. J. Cox-Singh, B. Singh, A. Alias, and M. S. Abdullah, *Trans. R. Soc. Trop. Med. Hyg.*, **89**, 436–437 (1995).

217. I. S. Adagu, D. C. Warhurst, D. J. Carucci, and M. T. Duraisingh, *Trans. R. Soc. Trop. Med. Hyg.*, **89**, 132 (1995).

218. L. K. Basco, J. Le Bras, Z. Rhoades, and C. M. Wilson, *Mol. Biochem. Parasitol.*, **74**, 157–166 (1995).

219. I. S. Adagu, F. Dias, L. Pinheiro, L. Rombo, V. do Rosario, and D. C. Warhurst, *Trans. R. Soc. Trop. Med. Hyg.*, **90**, 90–91 (1996).

220. M. T. Duraisingh, C. J. Drakeley, O. Muller, R. Bailey, G. Snounou, G. A. T. Targett, B. M. Greenwood, and D. C. Warhurst, *Parasitology*, **114**, 205–211 (1997).

221. F. M. Awad El Kariem, M. A. Miles, and D. C. Warhurst, *Trans. R. Soc. Trop. Med. Hyg.*, **86**, 587–589 (1992).

222. C. M. Wilson, S. K. Volkman, S. Thaithong, R. K. Martin, D. E. Kyle, W. K. Milhous, and D. F. Wirth, *Mol. Biochem. Parasitol.*, **57**, 151–60 (1993).

223. K. Haruki, P. G. Bray, S. A. Ward, H. Hommel, and G. Y. Ritchie, *Trans. R. Soc. Trop. Med. Hyg.*, **88**, 694 (1994).

224. L. K. Basco, P. E. de Pecoulas, J. Le Bras, and C. M. Wilson, *Exp. Parasitol.*, **82**, 97–103 (1996).

225. L. K. Basco and P. Ringwald, *Am. J. Trop. Med. Hyg.*, **59**, 577–581 (1998).

226. P. R. Bhattacharya and S. K. Biswas, L., *Trans. R. Soc. Trop. Med. Hyg.*, **91**, 454–455 (1997).

227. K. R. McCutcheon, J. A. Freese, J. A. Frean, B. L. Sharp, and M. B. Markus, *Trans. R. Soc. Trop. Med. Hyg.*, **93**, 300–302 (1999).

228. M. M. Póvoa, I. S. Adagu, S. G. Oliveira, R. L. D. Machado, M. A. Miles, and D. C. Warhurst, *Exp. Parasitol.*, **88**, 64–68 (1998).

229. X.-z. Su, L. A. Kirkman, and T. E. Wellems, *Cell*, **91**, 593–603 (1997).

230. L. K. Basco and P. Ringwald, *Am. J. Trop. Med. Hyg.*, **61**, 807–813 (1999).

231. L. K. Basco and P. Ringwald, *J. Infect. Dis.*, **180**, 1979–1986 (1999).

232. I. S. Adagu and D. C. Warhurst, *Parasitology*, **119**, 343–348 (1999).

233. D. A. Fidock, T. Nomura, A. Copper, X. Su, A. K. Talley, and T. E. Wellems, *Mol. Biochem. Parasitol.*, **110**, 1–10 (2000).

234. D. A. Fidock, T. Nomura, A. K. Talley, R. A. Cooper, S. M. Dzekunov, M. T. Ferdig, L. M. B. Ursos, A. B. S. Sidhu, B. Naude, K. W. Deitsch, X. Z. Su, J. C. Wootton, P. D. Roepe, and T. E. Wellems, *Mol. Cell*, **6**, 861–871 (2000).

235. T. E. Wellems, A. Walker-Jonah, and L. J. Panton, *Proc. Natl. Acad. Sci. USA*, **88**, 3382–3386 (1991).

236. J. M. R. Carlton, D. A. Fidock, A. Djimbe, C. V. Plowe, and T. E. Wellems, *Curr. Opin. Microbiol.*, **4**, 415–420 (2001).

237. A. Djimde, O. K. Doumbo, J. F. Cortese, K. Kayentao, S. Doumbo, Y. Diourte, A. Dicko, X. Z. Su, T. Nomura, D. A. Fidock, T. E. Wellems, C. V. Plowe, and D. Coulibaly, *N. Engl. J. Med.*, **344**, 257–263 (2001).

238. H. A. Babiker, S. J. Pringle, A. Abdel-Muhsin, M. Mackinnon, P. Hunt, and D. Walliker, *J. Infect. Dis.*, **183**, 1535–1538 (2001).

239. G. Dorsey, M. R. Kamy, A. Singh, and P. J. Rosenthal, *J. Infect. Dis.*, **183**, 1417–1420 (2001).

240. L. K. Basco and P. Ringwald, *J. Infect. Dis.*, **183**, 1828–1831 (2001).

241. A. G. Mayor, X. Gomez-Olive, J. J. Aponte, S. Casimiro, S. Mabunda, M. Dgedge, A. Barreto, and P. L. Alonso, *J. Infect. Dis.*, **183**, 1413–1416 (2001).

242. I. S. Adagu and D. C. Warhurst, *Parasitol.*, **123**, 219–224 (2001).

243. J. D. Maguire, A. I. Susanti, Krisin, P. Sismadi, D. J. Fryauff, and J. K. Baird, *Ann. Trop. Med. Parasitol.*, **95**, 559–572 (2001).

244. R. K. Mehlotra, H. Fujioka, P. D. Roepe, O. Janneh, L. M. B. Ursos, V. Jacobs-Lorena, D. T. McNamara, M. J. Bockarie, J. W. Kazura, D. E. Kyle, D. A. Fidock, and P. A. Zimmerman, *Proc. Natl. Acad. Sci. USA*, **98**, 12689–12694 (2001).

245. R. Durand, S. Jafari, J. Vauzelle, J. F. Delabre, Z. Jesic, and J. Le Bras, *Mol. Biochem. Parasit.*, **114**, 95–102 (2001).

246. F. P. Mockenhaupt, T. A. Eggelte, H. Till, and U. Bienzle, *Trop. Med. Int. Health.*, **6**, 749–755 (2001).

247. T. E. Wellems and C. V. Plowe, *J. Infect. Dis.*, **184**, 770–776 (2001).

248. R. D. Walter, M. Seth, and A. P. Bhaduri, *Trop. Med. Parasitol.*, **44**, 5–8 (1993).

249. P. Srivastava, V. C. Pandey, and A. P. Bhaduri, *Trop. Med. Parasitol.*, **46**, 83–87 (1995).

250. S. Batra, P. Srivastava, K. Roy, V. C. Pandey, and A. P. Bhaduri, *J. Med. Chem.*, **43**, 3428–3433 (2000).

251. D. De, A. P. Bhaduri, and W. K. Milhous, *Am. J. Trop. Med. Hyg.*, **49**, 113–120 (1993).

252. D. De, F. M. Krogstad, F. B. Cogswell, and D. J. Krogstad, *Am. J. Trop. Med. Hyg.*, **55**, 579–583 (1996).

253. R. G. Ridley, W. Hofheinz, H. Matile, C. Jaquet, A. Dorn, R. Masciadri, S. Jolidon, W. F. Richter, A. Guenzi, et al., *Antimicrob. Agents Chemother.*, **40**, 1846–1854 (1996).

254. D. De, F. M. Krogstad, L. D. Byers, and D. J. Krogstad, *J. Med. Chem.*, **41**, 4918–4926 (1998).

255. K. Raynes, D. Galatis, A. F. Cowman, L. Tilley, and L. W. Deady, *J. Med. Chem.*, **38**, 204–206 (1995).

256. M. Navarro, H. Perez, and R. A. Sanchez-Delgado, *J. Med. Chem.*, **40**, 1937–1939 (1997).

257. C. Biot, G. Glorian, L. A. Maciejewski, J. S. Brocard, O. Domarle, G. Blampain, P. Millet, A. J. Georges, H. Abessolo, D. Dive, and J. Lebibi, *J. Med. Chem.*, **40**, 3715–3718 (1997).

258. C. Biot, L. Delhaes, H. Abessolo, O. Domarle, L. A. Maciejewski, M. Mortuaire, P. Delcourt, P. Deloron, D. Camus, D. Dive, and J. S. Brocard, *J. Organomet. Chem.*, **589**, 59–65 (1999).

259. K. Chibale, J. R. Moss, M. Blackie, D. van Schalkwyk, and P. J. Smith, *Tetrahedron Lett.*, **41**, 6231–6235 (2000).

260. R. A. Sànchez-Delgado, M. Navarro, H. Pérez, and J. A. Urbina, *J. Med. Chem.*, **39**, 1095–1099 (1996).

261. A. D. Wolfe, *Antibiotics*, **6**, 109–120 (1983).

262. F. Nosten and R. N. Price, *Drug Safety*, **12**, 264–273 (1995).

263. K. J. Palmer, S. M. Holliday, and R. N. Brogden, *Drugs*, **45**, 430–435 (1993).

264. T. R. Sweeney, *Med. Res. Rev.*, **1**, 281–301 (1981).

265. P. Lim, *Anal. Profiles Drug Subst.*, **14**, 157–80 (1985).

266. A. K. Bhattacharjee and J. M. Karle, *Bioorg. Med. Chem.*, **6**, 1927–1933 (1998).

267. V. Nguyen-Cong and B. M. Rode, *J. Chem. Inf. Comput. Sci.*, **36**, 114–117 (1996).

268. C. J. Ohnmacht, A. R. Patel, and R. E. Lutz, *J. Med. Chem.*, **14**, 926–928 (1971).

269. S. Adam, *Tetrahedron*, **45**, 1409–1414 (1989).

270. S. Adam, *Tetrahedron*, **47**, 7609–7614 (1991).

271. M. S. Kumar, Y. V. D. Nageshwar, and H. M. Meshram, *Synth. Commun.*, **26**, 1913–1919 (1996).

272. S. Adam, *Bioorg. Med. Chem. Lett.*, **2**, 53–58 (1992).

273. S. Adam, *Tetrahedron*, **50**, 3327–3332 (1994).

274. Y. Boulard, I. Landau, F. Miltgen, W. Peters, and D. S. Ellis, *Ann. Trop. Med. Parasitol.*, **80**, 577–880 (1986).

275. A. Håkanson, A. Landberg-Lindgren, and A. Bjoerk, *Trans. R. Soc. Trop. Med. Hyg.*, **84**, 503–504 (1990).

276. K. Na Bangchang, J. Karbwang, and D. J. Back, *Biochem. Pharmacol.*, **43**, 1957–1961 (1992).

277. L. K. Basco, C. Gillotin, F. Gimenez, R. Farinotti, and J. Le Bras, *Br. J. Clin. Pharmacol.*, **33**, 517–520 (1992).

278. W. Peters, B. L. Robinson, M.-L. Mittelholzer, C. Crevoisier, and D. Stürchler, *Ann. Trop. Med. Parasitol.*, **89**, 465–468 (1995).

279. C. Martin, F. Gimenez, K. Na Bangchang, J. Karbwang, I. W. Wainer, and R. Farinotti, *Eur. J. Clin. Pharmacol.*, **47**, 85–87 (1994).

280. A. Bourahla, C. Martin, F. Gimenez, V. Singhasivanon, P. Attanath, A. Sabchearon, T. Chongsuphajaisiddhi, and R. Farinotti, *Eur. J. Clin. Pharmacol.*, **50**, 241–244 (1996).

281. S. Vidrequin, F. Gimenez, L. K. Basco, C. Martin, J. LeBras, and R. Farinotti, *Drug Met. Disp.*, **24**, 689–691 (1996).

282. Y. T. Pham, F. Nosten, R. Farinotti, N. J. White, and F. Gimenez, *Int. J. Clin. Pharmacol. Ther.*, **37**, 58–61 (1999).

283. G. Edwards, P. A. Winstanley, and S. A. Ward, *Clin. Pharmacokinet.*, **27**, 150–165 (1994).

284. J. Karbwang and K. Na-Bangchang, *Fund. Clin. Pharmacol.*, **8**, 491–502 (1994).

285. E. Weidekamm, G. Rusing, H. Caplain, F. Sorgel, and C. Crevoisier, *Eur. J. Clin. Pharmacol.*, **54**, 615–619 (1998).

286. K. Na-Bangchang, J. Karbwang, P. A. C. Palacios, R. Ubalee, S. Saengtertsilapachai, and W. H. Wernsdorfer, *Eur. J. Clin. Pharmacol.*, **55**, 743–748 (2000).

287. C. Crevoisier, J. Handschin, J. Barre, D. Roumenov, and C. Kleinbloesem, *Eur. J. Clin. Pharmacol.*, **53**, 135–139 (1997).

288. H. R. Smith, A. M. Croft, and M. M. Black, *Clin. Exp. Dermatol.*, **24**, 249–254 (1999).

289. F. O. ter Kuile, F. Nosten, C. Luxemburger, D. Kyle, P. Teja-Isavatharm, L. Phaipun, R. Price, T. Congsuphajaisiddhi, and N. J. White, *Bull. World Health Org.*, **73**, 631–642 (1995).

290. F. O. ter Kuile, C. Luxemburger, and N. J. White, *Trans. R. Soc. Trop. Med. Hyg.*, **89**, 660–664 (1995).

291. A. L. Fontanet and A. M. Walker, *Am. J. Trop. Med. Hyg.*, **49**, 465–472 (1993).

292. F. Nosten, M. Vincenti, J. Simpson, P. Yei, K. L. Thwai, A. De Vries, T. Chongsuphajaisiddhi, and N. J. White, *Clin. Infect. Dis.*, **28**, 808–815 (1999).

293. F. Nosten and M. Van Vugt, *CNS Drugs*, **11**, 1–8 (1999).

294. D. Grupp, A. Rauber, and W. Fröscher, *Akt. Neurol.*, **21**, 134–136 (1994).

295. R. C. Wittes and R. Saginur, *Can. Med. Assoc. J.*, **152**, 515–517 (1995).

296. T. A. Ruff, S. J. Sherwen, and G. A. Donnan, *Med. J. Aust.*, **161**, 453 (1994).

297. P. Tan-Ariya, C. R. Brockelman, and C. Menabandhu, *Southeast Asian J. Trop. Med. Public Health.*, **15**, 531–535 (1984).

298. E. Weidekamm, D. E. Schwarts, U. C. Dubach, and B. Weber, *Exp. Chemother.*, **33**, 259–265 (1987).

299. S. M. Mansor, V. Navaratnam, M. Mohamad, S. Hussein, A. Kumar, A. Jamaludin, and W. H. Wernsdorfer, *Br. J. Clin. Pharmacol.*, **27**, 381–386 (1989).

300. J. Karbwang, D. J. Black, D. Bunnag, and A. M. Breckenridge, *Bull. World Health Org.*, **68**, 633–638 (1990).

301. A. Sowunmi and A. M. J. Oduola, *Trans. R. Soc. Trop. Med. Hyg.*, **89**, 303–305 (1995).

302. B. Lell, L. G. Lehrman, J. R. Schmidt-Ott, D. Sturchler, J. Handschin, and B. G. Schuster, *Am. J. Trop. Med. Hyg.*, **58**, 619–624 (1998).

303. E. N. U. Ezedinachi, O. J. Ekanem, C. M. Chukwuani, M. M. Meremikwu, E. A. Ojar,

A. A. A. Alaribe, A. B. Umotong, and L. Haller, *Am. J. Trop. Med. Hyg.*, **61**, 114–119 (1999).

304. K. Na-Bangchang, P. Tippawangkosol, A. Thanavibul, R. Ubalee, and J. Karbwang, *Int. J. Clin. Pharmacol. Res.*, **19**, 9–17 (1999).

305. R. Price, J. A. Simpson, P. Teja-Isavatharm, M. M. Than, C. Luxemburger, D. G. Heppner, T. Chongsuphajaisiddhi, F. Nosten, and N. J. White, *Antimicrob. Agents Chemother.*, **43**, 341–346 (1999).

306. R. Price, C. Luxemburger, M. Van Vugt, F. Nosten, A. Kham, J. Simpson, S. Looareesuwan, T. Chongsuphajaisiddhi, and N. J. White, *Trans. R. Soc. Trop. Med. Hyg.*, **92**, 207–211 (1998).

307. K. Na-Bangchang, K. Congpuong, J. Sirichaisinthop, K. Suprakorb, and J. Karbwang, *Br. J. Clin. Pharmacol.*, **43**, 639–642 (1997).

308. L. N. Hung, P. J. De Vries, T. D. Le Thuy, B. Lien, H. P. Long, T. N. Hung, N. Van Nam, T. K. Anh, and P. A. Kager, *Trans. R. Soc. Trop. Med. Hyg.*, **91**, 191–194 (1997).

309. K. Na-Bangchang, P. Tippanangkosol, R. Ubalee, S. Chaovanakawee, S. Saenglertsilapachai, and J. Karbwang, *Trop. Med. Int. Health.*, **4**, 602–610 (1999).

310. D. Bunnag, T. Kanda, J. Karbwang, K. Thimasarn, S. Pungpak, and T. Harinasuta, *Trans. R. Soc. Trop. Med. Hyg.*, **90**, 415–417 (1996).

311. D. Bunnag, T. Kanda, J. Karbwang, K. Thimasarn, S. Pungpak, and T. Harinasuta, *Southeast Asian J. Trop. Med. Public Health.*, **28**, 727–730 (1997).

312. K. Na Bangchang, J. Karbwang, P. Molunto, V. Banmairuroi, and A. Thanavibul, *Fund. Clin. Pharmacol.*, **9**, 576–582 (1995).

313. J. Karbwang, K. Na Bangchang, A. Thanavibul, D. J. Black, D. Bunnag, and T. Harinasuta, *Bull. World Health Org.*, **72**, 83–87 (1994).

314. J. Karbwang, K. Na Bangchang, D. J. Back, D. Bunnag, and W. Rooney, *Eur. J. Clin. Pharmacol.*, **43**, 567–569 (1992).

315. D. Bunnag, J. Karbwang, C. Viravan, S. Chitamas, and T. Harinasuta, *Southeast Asian J. Trop. Med. Public Health.*, **23**, 377–382 (1992).

316. A. Croft and P. Garner, *Br. Med. J.*, **315**, 1412–1416 (1997).

317. A. M. J. Croft, T. C. Clayton, and M. J. World, *Trans. R. Soc. Trop. Med. Hyg.*, **91**, 199–203 (1997).

318. M. S. Peragallo, G. Sabatinelli, and G. Sarnicola, *Trans. R. Soc. Trop. Med. Hyg.*, **93**, 73–77 (1999).

319. P. J. Barrett, P. D. Emmins, P. D. Clarke, and D. J. Bradley, *Br. Med. J.*, **313**, 525–528 (1996).

320. P. Winstanley, *Br. J. Clin. Pharmacol.*, **42**, 411–413 (1996).

321. A. M. J. Oduola, W. K. Milhous, N. F. Weatherly, J. H. Bowdre, and R. E. Desjardins, *Exp. Parasitol.*, **67**, 354–360 (1988).

322. E. M. Fevre, G. Barnish, P. Yamokgul, and W. Rooney, *Trans. R. Soc. Trop. Med. Hyg.*, **93**, 180–184 (1999).

323. A. P. C. C. Hopperus Buma, P. P. A. M. van Thiel, H. O. Lobel, C. Ohrt, E. J. C. van Ameijden, R. L. Veltink, D. C. H. Tendeloo, T. van Gool, M. D. Green, G. D. Todd, D. E. Kyle, and P. A. Kager, *J. Infect. Dis.*, **173**, 1506–1509 (1996).

324. C. Cerutti Jr., R. R. Durlacher, F. E. C. De Alencar, A. A. C. Segurado, and L. W. Pang, *J. Infect. Dis.*, **180**, 2077–2080 (1999).

325. M. Hassan Alin, *Parasitology*, **114**, 503–506 (1997).

326. F. M. Smithuis, F. Monti, M. Grundl, A. Z. Oo, T. T. Kyaw, O. Phe, and N. J. White, *Trans. R. Soc. Trop. Med. Hyg.*, **91**, 468–472 (1997).

327. R. N. Price, C. Cassar, A. Brockman, M. Duraisingh, M. Van Vugt, N. J. White, F. Nosten, and S. Krishna, *Antimicrob. Agents Chemother.*, **43**, 2943–2949 (1999).

328. A. F. Cowman, D. Galatis, and J. K. Thompson, *Proc. Natl. Acad. Sci. USA*, **91**, 1143–1147 (1994).

329. S. A. Peel, P. Bright, B. Yount, J. Handy, and R. S. Baric, *Am. J. Trop. Med. Hyg.*, **51**, 648–658 (1994).

330. M. Nateghpour, S. A. Ward, and R. E. Howells, *Antimicrob. Agents Chemother.*, **37**, 2337–2343 (1993).

331. M. T. Duraisingh, C. Roper, D. Walliker, and D. C. Warhurst, *Mol. Microbiol.*, **36**, 955–961 (2000).

332. A. S. Y. Lim, D. Galatis, and A. F. Cowman, *Exp. Parasitol.*, **83**, 295–303 (1996).

333. S. C. Chaiyaroj, A. Buranakiti, P. Angkasekwinai, S. Looareesuwan, and A. F. Cowman, *Am. J. Trop. Med. Hyg.*, **61**, 780–783 (1999).

334. W. T. Colwell, V. Brown, P. Christie, J. Lange, C. Reese, K. Yamamoto, and D. W. Henry, *J. Med. Chem.*, **15**, 771–775 (1972).

335. T. M. Cosgriff, E. F. Boudreau, C. L. Pamplin, E. B. Doberstyn, R. E. Desjardins, and C. J. Canfield, *Am. J. Trop. Med. Hyg.*, **31**, 1075–1079 (1982).

336. J. Karbwang and K. Na Bangchang, *Clin. Pharmacokinet.*, **27**, 104–119 (1994).

337. H. D. Nothdurft, R. Clemens, H. L. Bock, and T. Loescher, *Clin. Invest.*, **71**, 69–73 (1993).

338. H. M. Bryson and K. L. Goa, *Drug Evaluation.*, **43**, 236–258 (1992).

339. J. Karbwang and T. Harinasuta, *Southeast Asian J. Trop. Med. Public Health.*, **23**, 95–109 (1992).

340. R. Thesen, *Pharm. Ztg.*, **6**, 39 (1992).

341. J. M. Karle, *Antimicrob. Agents Chemother.*, **41**, 791–794 (1997).

342. L. K. Basco, C. Gillotin, F. Gimenez, R. Farinotti, and J. Le Bras, *Trans. R. Soc. Trop. Med. Hyg.*, **87**, 78–79 (1993).

343. H. Terefe and G. Blaschke, *J. Chromatogr., Biomed. Appl.*, **615**, 347–351 (1993).

344. H. Terefe and G. Blaschke, *J. Chromatogr., B: Biomed. Appl.*, **657**, 238–242 (1994).

345. L. K. Basco, F. Peytavin, F. Gimenez, B. Genissel, R. Farinotti, and J. LeBras, *Trop. Med. Parasitol.*, **45**, 45–46 (1994).

346. L. K. Basco, C. Gillotin, F. Gimenez, R. Farinotti, and J. Le Bras, *Trans. R. Soc. Trop. Med. Hyg.*, **86**, 12–13 (1992).

347. F. Gimenez, C. Gillotin, L. K. Basco, O. Bouchard, A.-F. Aubry, I. W. Wainer, J. Le Bras, and R. Farinotti, *Eur. J. Clim. Pharmacol.*, **46**, 561–562 (1994).

348. K. N. Cheng, L. F. Elsom, and D. R. Hawkins, *J. Chromatogr.*, **581**, 203–211 (1992).

349. B. Baune, J. P. Flinois, V. Furlan, F. Gimenez, A. M. Taburet, L. Becquemont, and R. Farinotti, *J. Pharm. Pharmacol.*, **51**, 419–426 (1999).

350. B. Baune, V. Furlan, A. M. Taburet, and R. Farinotti, *Drug Metab. Dispos.*, **27**, 565–568 (1999).

351. O. O. Simooya, G. Sijumbil, M. S. Lennard, and G. T. Tucker, *Br. J. Clin. Pharmacol.*, **45**, 315–317 (1998).

352. B. Cenni and B. Betschart, *Chemotherapy (Basel).*, **41**, 153–158 (1995).

353. M. Chutmongkonkul, W. A. Maier, and H. M. Seitz, *Ann. Trop. Med. Parasitol.*, **86**, 103–110 (1992).

354. F. O. ter Kuile, G. Dolan, F. Nosten, M. D. Edstein, C. Luxemburger, L. Phaipun, T. Chongsuphajaisiddhi, H. K. Webster, and N. J. White, *Lancet*, **341**, 1044–1049 (1993).

355. J. E. Touze, *Lancet*, **349**, 255–256 (1997).

356. M. Restrepo, D. Botero, R. E. Marquez, E. F. Boudrea, and V. Navaratnam, *Bull. World Health Org.*, **74**, 591–597 (1996).

357. D. J. Fryauff, J. K. Barid, H. Basri, I. Wiady, Purnomo, M. J. Bangs, B. Subianto, S. Harjosuwarno, E. Tjitra, T. L. Richie, and S. L. Hoffman, *Ann. Trop. Med. Parasitol.*, **91**, 7–16 (1997).

358. W. Peters, B. L. Robinson, and D. S. Ellis, *Ann. Trop. Med. Parasitol.*, **81**, 639–646 (1987).

359. J. Philipps, P. D. Radloff, W. Wernsdorfer, and P. G. Kremsner, *Am. J. Trop. Med. Hyg.*, **58**, 612–618 (1998).

360. C. O. Falade, L. A. Salako, A. Sowunmi, A. M. J. Oduola, and P. Larcier, *Trans. R. Soc. Trop. Med. Hyg.*, **91**, 58–62 (1997).

361. G. Y. Ritchie, M. Mungthin, J. E. Green, P. G. Bray, S. R. Hawley, and S. A. Ward, *Mol. Biochem. Parasitol.*, **83**, 35–46 (1996).

362. B. H. Wang, B. Ternai, and G. M. Polya, *Biol. Chem. Hoppe-Seyler.*, **375**, 527–535 (1994).

363. A. J. Humberstone, C. J. H. Porter, and W. N. Charman, *J. Pharm. Sci.*, **85**, 525–529 (1996).

364. G. D. Shanks, G. Watt, M. D. Edstein, H. K. Webster, L. Loesurriviboon, and S. Wechgritaya, *Trans. Roy. Soc. Trop. Med. Hyg.*, **86**, 233–234 (1992).

365. K. A. Milton, G. Edwards, S. A. Ward, M. L. Orme, and A. M. Breckenridge, *Br. J. Clin. Pharmacol.*, **28**, 71–77 (1989).

366. L. Y. Lim and M. L. Go, *Eur. J. Pharmaceut. Sci.*, **10**, 17–28 (2000).

367. B. Cenni, J. Meyer, R. Brandt, and B. Betschart, *Br. J. Clin. Pharmacol.*, **39**, 519–526 (1995).

368. L. Y. Lim and M. L. Go, *Chem. Pharm. Bull.*, **47**, 732–737 (1999).

369. M. P. McIntosh, C. J. H. Porter, K. M. Wasan, M. Ramaswamy, and W. N. Charman, *J. Pharm. Sci.*, **88**, 378–384 (1999).

370. K. M. Wasan, M. Ramaswamy, M. P. McIntosh, C. J. H. Porter, and W. N. Charman, *J. Pharm. Sci.*, **88**, 185–190 (1999).

371. A. J. Humberstone, A. F. Cowman, J. Horton, and W. N. Charman, *J. Pharm. Sci.*, **87**, 256–258 (1998).

372. P. A. Matson, S. P. Luby, S. C. Redd, H. R. Rolka, and R. A. Meriwether, *Am. J. Trop. Med. Hyg.*, **54**, 229–231 (1996).

373. F. Nosten, F. O. ter Kuile, C. Luxemberger, C. Woodrow, D. E. Kyle, T. Chongsuphajaisiddhi, and N. J. White, *Lancet*, **341**, 1054–1056 (1993).

374. A. Sowunmi, C. O. Falade, A. M. J. Oduola, O. A. T. Ogundahunsi, F. A. Fehintola, G. O. Gbotosho, P. Larcier, and L. A. Salako, *Trans. R. Soc. Trop. Med. Hyg.*, **92**, 446–448 (1998).

375. A. Sowunmi, F. A. Fehintola, O. A. T. Ogundahunsi, A. B. Ofi, T. C. Happi, and A. M. J. Oduola, *Trans. R. Soc. Trop. Med. Hyg.*, **93**, 78–83 (1999).

376. J. E. Touze, J. Bernard, A. Keundjian, P. Imbert, A. Viguier, H. Chaudet, and J. C. Doury, *Am. J. Trop. Med. Hyg.*, **54**, 225–228 (1996).

377. K. U. Leo, D. L. Wesche, M. T. Marino, and T. G. Brewer, *J. Pharm. Pharmacol.*, **48**, 723–728 (1996).

378. E. Monlun, P. Le Metayer, S. Szwandt, D. Neau, M. Longy-Boursier, J. Horton, and M. Le Bras, *Trans. R. Soc. Trop. Med. Hyg.*, **89**, 430–433 (1995).

379. A. A. Mosuro, O. A. Olukoju, and A. O. Olaore, *Biokemistri*, **8**, 101–108 (1998).

380. J. E. Touze, L. Fourcade, F. Peyron, P. Heno, and J. C. Deharo, *Ann. Trop. Med. Parasitol.*, **91**, 867–873 (1997).

381. L. K. Basco and J. Le Bras, *Am. J. Trop. Med. Hyg.*, **47**, 521–527 (1992).

382. L. H. Schmidt, S. Alexander, L. Allen, and J. Rasco, *Antimicrob. Agents Chemother.*, **12**, 51–60 (1977).

383. S. Looareesuwan, K. Buchachart, P. Wilairatana, K. Chalermrut, Y. Rattanapong, S. Amradee, S. Siripiphat, S. Chullawichit, K. Thimasan, M. Ittiverakul, A. Triampon, and D. S. Walsh, *Ann. Trop. Med. Parasitol.*, **91**, 939–943 (1997).

384. N. J. Gogtay, S. Desai, K. D. Kamtekar, V. S. Kadam, S. S. Dalvi, and N. A. Kshirsagar, *Ann. Trop. Med. Parasitol.*, **93**, 809–812 (1999).

385. T. Adak, N. Valecha, and V. P. Sharma, *Clin. Diag. Lab. Immunol.*, **8**, 891–894 (2001).

386. T. Nomura, J. M.-R. Carlton, J. K. Baird, H. A. del Portillo, D. J. Fryauff, D. Rathore, D. A. Fidock, X.-z. Su, W. E. Collins, T. F. McCutchan, J. C. Wootton, and T. E. Wellems, *J. Infect. Dis.*, **183**, 1653–1661 (2001).

387. S. Pukrittayakamee, A. Chantra, J. A. Simpson, S. Vanijanonta, R. Clemens, S. Looareesuwan, and N. J. White, *Antimicrob. Agents Chemother.*, **44**, 1680–1685 (2000).

388. L. K. Basco, J. Bickii, and P. Ringwald, *Ann. Trop. Med. Parasitol.*, **93**, 179–182 (1999).

389. P. Wilairatana, U. Silachamroon, S. Krudsood, P. Singhasivanon, S. Treeprasertsuk, V. Bussaratid, W. Phumratanaprapin, S. Srivilirit, and S. Looareesuwan, *Am. J. Trop. Med. Hyg.*, **61**, 973–977 (1999).

390. J. Soto, J. Toledo, M. Rodriquez, J. Sanchez, R. Herrera, J. Padilla, and J. Berman, *Ann. Intern. Med.*, **129**, 241–244 (1998).

391. E. Schwartz and G. Regev-Yochay, *Clin. Infect. Dis.*, **29**, 1502–1506 (1999).

392. J. K. Baird, D. J. Fryauff, H. Basri, M. J. Bangs, B. Subianto, I. Wiady, Purnomo, B. Leksana, S. Masbar, T. L. Richie, T. R. Jones, E. Tjitra, R. S. Wignall, and S. L. Hoffman, *Am. J. Trop. Med. Hyg.*, **52**, 479–484 (1995).

393. W. R. Weiss, A. J. Oloo, A. Johnson, D. Koech, and S. L. Hoffman, *J. Infect. Dis.*, **171**, 1569–1575 (1995).

394. J. Soto, J. Toledo, M. Rodriquez, J. Sanchez, R. Herrera, J. Padilla, and J. Berman, *Clin. Infect. Dis.*, **29**, 199–201 (1999).

395. D. J. Fryauff, *Lancet.*, **346**, 1190–1193 (1995).

396. E. Kennedy and H. Frischer, *J. Lab. Clin. Med.*, **116**, 871–878 (1990).

397. W. H. Wernsdorfer, P. I. Trigg, Primaquine: pharmacokinetics, metabolism, toxicity and activity, Conference Proceedings 1984, Scientific Working Group on the Chemotherapy of Malaria, UNDP/WORLD BANK/WHO Special Programme for Research and Training in Tropical Diseases, Geneva, Switzerland, 1987.

398. V. Singhasivanon, A. Sabcharoen, P. Attanath, T. Chongsuphajaisiddhi, B. Diquet, and P. Turk, *Southeast Asian J. Trop. Med. Public Health.*, **22**, 527–533 (1991).

399. G. Edwards, C. S. McGrath, S. A. Ward, W. Supanaranond, S. Pukrittayakamee, T. M. E. Davis, and N. J. White, *Brit. J. Clin. Pharmacol.*, **35**, 193–198 (1993).

400. L. Constantino, P. Paixao, R. Moreira, M. J. Portela, V. E. Do Rosario, and J. Iley, *Exp. Toxicol. Pathol.*, **51**, 299–303 (1999).

401. O. R. Idowu, J. O. Peggins, and T. G. Brewer, *Drug Metab. Disp.*, **23**, 18–27 (1995).

402. K. Na Bangchang, J. Karbwang, and D. J. Back, *Biochem. Pharmacol.*, **44**, 587–590 (1992).

403. K. A. Fletcher, P. F. Barton, and J. A. Kelly, *Biochem. Pharmacol.*, **37**, 2683–2690 (1988).

404. J. Vasquez-Vivar and O. Augusto, *J. Biol. Chem.*, **267**, 6848–6854 (1992).

405. P. M. O'Neill, R. C. Storr, and B. K. Park, *Tetrahedron*, **54**, 4615–4622 (1998).

406. F. Fontaine, C. Delescluse, G. De Sousa, P. Lesca, and R. Rahmani, *Biochem. Pharmacol.*, **57**, 255–262 (1999).

407. L. N. Grinberg and A. Samuni, *Biochim. Biophys. Acta*, **1201**, 284–288 (1994).

408. R. Deslauriers, K. Butler, and I. C. P. Smith, *Biochim. Biophys. Acta*, **931**, 267–275 (1987).

409. M. D. Bates, S. R. Meshnick, C. I. Sigler, P. Leland, and M. R. Hollingdale, *Am. J. Trop. Med. Hyg.*, **42**, 532–537 (1990).

410. L. N. Grinberg, O. Shalev, A. Goldfarb, and E. A. Rachmilewitz, *Biochim. Biophys. Acta*, **1139**, 248–250 (1992).

411. J. Vásquez-Vivar and O. Augusto, *Biochem. Pharmacol.*, **47**, 309–316 (1994).

412. Y. Hong, H. Pan, M. D. Scott, and S. R. Meshnick, *Free Radical Biol. Med.*, **12**, 213–218 (1992).

413. D. G. Heppner Jr., R. A. Gasser Jr., and J. Berman, *Ann. Int. Med.*, **130**, 536 (1999).

414. T. Ono, M. Norimatsu, and H. Yoshimura, *Mutat. Res.*, **325**, 7–10 (1994).

415. B. K. Bhat, M. Seth, and A. P. Bhaduri, *Prog. Drug Res.*, **28**, 197–231 (1984).

416. E. A. Nodiff, S. Chatterjee, and H. A. Musallam, *Prog. Med. Chem.*, **28**, 1–40 (1991).

417. D. E. Davidson Jr., A. L. Ager, J. L. Brown, F. E. Chapple, R. E. Whitmire, and R. N. Rossan, *Bull. World Health Org.*, **59**, 463–479 (1981).

418. F. I. Carroll, B. D. Berrang, and C. P. Linn, *J. Med. Chem.*, **22**, 1363 (1979).

419. E. H. Chen, A. J. Saggiomo, K. Tanabe, B. L. Verma, and E. A. Nodiff, *J. Med. Chem.*, **20**, 1107–1109 (1977).

420. E. A. Nodiff, K. Tanabe, E. H. Chen, and A. J. Saggiomo, *J. Med. Chem.*, **25**, 1097–1101 (1982).

421. F. I. Carroll, B. Berrand, and C. P. Pinn, *J. Med. Chem.*, **28**, 1564–1567 (1985).

422. W. Peters and B. L. Robinson, *Ann. Trop. Med. Parasitol.*, **78**, 561–565 (1984).

423. M. T. Marino, J. O. Peggins, L. D. Brown, M. R. Urquhart, and T. G. Brewer, *Drug Metab. Dispos.*, **22**, 358–366 (1994).

424. M. P. LaMontagne, P. Blumbergs, and R. E. Strube, *J. Med. Chem.*, **25**, 1094–1097 (1982).

425. W. H. Wernsdorfer, *Curr. Opin. Anti Infect. Invest. Drugs*, **2**, 88–98 (2000).

426. B. Lell, J.-F. Faucher, M. A. Missinou, S. Bormann, O. Dangelmaier, J. Norton, and P. G. Kremsner, *Lancet*, **355**, 2041–2045 (2000).

427. R. P. Brueckner, K. C. Lasseter, E. T. Lin, and B. G. Schuster, *Am. J. Trop. Med. Hyg.*, **58**, 645–649 (1998).

428. R. Jain, S. Jain, R. C. Gupta, N. Anand, G. P. Dutta, and S. K. Puri, *Indian J. Chem., Sect. B.*, **33B**, 251–254 (1994).

429. A. Philip, J. A. Kepler, B. H. Johnson, and F. I. Carroll, *J. Med. Chem.*, **31**, 870–874 (1988).

430. M. J. Portela, R. Moreira, E. Valente, L. Constantino, J. Iley, J. Pinto, R. Rosa, P. Cravo, and V. E. Do Rosário, *Pharm. Res.*, **16**, 949–955 (1999).

431. B.-R. Shao and X.-Y. Ye, *Southeast Asian J. Trop. Med. Public Health*, **22**, 81–83 (1991).

432. J. K. Paliwal, R. C. Gupta, and P. K. Grover, *J. Chromatogr., Biomed. Appl.*, **616**, 155–160 (1993).

433. P. Srivastava, S. K. Puri, G. P. Dutta, and V. C. Pandey, *Biochem. Pharmacol.*, **46**, 1859–1860 (1993).

434. S. K. Puri, R. Srivastava, V. C. Pandey, N. Sethi, and G. P. Dutta, *Am. J. Trop. Med. Hyg.*, **41**, 638–642 (1989).

435. N. Sethi, S. Srivastava, U. K. Singh, and S. K. Puri, *Indian J. Parasitol.*, **17**, 15–26 (1993).

436. V. C. Pandey, S. K. Puri, S. K. Sahni, P. Srivastava, and G. P. Dutta, *Pharmacol. Res.*, **22**, 701–707 (1990).

437. N. Valecha, T. Adak, A. K. Bagga, O. P. Asthana, J. S. Srivastava, H. Joshi, and V. P. Sharma, *Curr. Sci.*, **80**, 561–563 (2001).

438. K. S. Corcoran, P. Hansukjariya, J. Sattabongkot, M. Ngampochjana, M. D. Edstein, C. D. Smith, G. D. Shanks, and W. K. Milhous, *Am. J. Trop. Med. Hyg.*, **49**, 473–477 (1993).

439. J. B. Jiang, R. S. Bray, W. A. Krotowski, E. U. Canning, D. S. Liang, J. C. Huang, J. Y. Liao, D. S. Li, Z. R. Lun, and I. Landau, *Trans. R. Soc. Trop. Med. Hyg.*, **82**, 56–58 (1988).

440. C. G. Nevill, J. D. Lury, M. K. Mosobo, H. M. Watkins, and W. M. Watkins, *Trans. R. Soc. Trop. Med. Hyg.*, **88**, 319–320 (1994).

441. D. Gozal, C. Hengy, and G. Fadat, *Antimicrob. Agents Chemother.*, **35**, 373–376 (1991).

442. J. E. Touze, A. Keundjian, T. Fusai, and J. C. Doury, *Trop. Med. Parasitol.*, **46**, 158–160 (1995).

443. W. M. Watkins, D. G. Sixsmith, and J. D. Chulay, *Ann. Trop. Med. Parasitol.*, **78**, 273–278 (1984).

444. A. E. T. Yeo, M. D. Edstein, G. D. Shanks, and K. H. Rieckmann, *Ann. Trop. Med. Parasitol.*, **88**, 587–594 (1994).

445. A. Kaneko, Y. Bergqvist, M. Takechi, M. Kalkoa, O. Kaneko, T. Kobayakawa, T. Ishizaki, and A. Bjorkman, *J. Infect. Dis.*, **179**, 974–979 (1999).

446. A. E. T. Yeo, K. K. Seymour, K. H. Riechmann, and R. I. Christopherson, *Ann. Trop. Med. Parasitol.*, **91**, 17–23 (1997).

447. A. Jamaludin, M. Mohamad, V. Navaratnam, P. Y. Yeoh, and W. H. Wernsdorfer, *Trop. Med. Parasitol.*, **41**, 268–272 (1990).

448. Y. Wattanagoon, R. B. Taylor, R. R. Moody, N. A. Ochekpe, S. Looareesuwan, and N. J. White, *Br. J. Clin. Pharmacol.*, **24**, 775–780 (1987).

449. A. A. O. Mosuro, *Biokemistrie*, **7**, 19–23 (1997).

450. D. A. Fidock and T. E. Wellems, *Proc. Natl. Acad. Sci. USA*, **94**, 10931–10936 (1997).

451. D. A. Fidock, T. Nomura, and T. E. Willems, *Mol. Pharmacol.*, **54**, 1140–1147 (1998).

452. N. A. Helsby, S. A. Ward, G. Edwards, R. E. Howells, and A. M. Breckenridge, *Br. J. Clin. Pharmacol.*, **30**, 593–598 (1990).

453. N. A. Helsby, G. Edwards, A. M. Breckenridge, and S. A. Ward, *Br. J. Clin. Pharmacol.*, **35**, 653–656 (1993).

454. A. Kaneko, Y. Bergqvist, G. Taleo, T. Kobayakawa, T. Ishizaki, and A. Bjorkman, *Pharmacogenetics*, **9**, 317–326 (1999).

455. C. Funck-Brentano, L. Becquemont, A. Leneveu, A. Roux, P. Jaillon, and P. Beaune, *J. Pharmacol. Exp. Ther.*, **280**, 730–738 (1997).

456. B. Buur Rasmussen, T. L. Nielsen, and K. Brosen, *Eur. J. Clin. Pharmacol.*, **54**, 735–740 (1998).

457. E. Skjelbo, T. K. Mutabingwa, I. Bygbjerg, K. K. Nielsen, L. F. Gram, and K. Brosen, *Clin. Pharmacol. Therapeut.*, **59**, 304–311 (1996).

458. Y. Yuthavong, T. Vilaivan, N. Chareonsethakul, S. Kachonwongpaisan, W. Sirawaraporn, R. Quarrell, and G. Lowe, *J. Med. Chem.*, **43**, 2738–2744 (2000).

459. K. Na-Bangchang, P. Tipwangso, A. Thanavibul, P. Tan-Ariya, K. Suprakob, T. Kanda, and J. Karbwang, *Southeast Asian J. Trop. Med. Public Health.*, **27**, 19–23 (1996).

460. C. E. Garrett, J. A. Coderre, T. D. Meek, E. P. Garvey, D. M. Claman, S. M. Beverly, and D. V. Santi, *Mol. Biochem. Parasitol.*, **11**, 257–265 (1984).

461. J. E. Hyde, *Pharmacol. Ther.*, **48**, 45–59 (1990).

462. S. J. Foote, D. Galatis, and A. F. Cowman, *Proc. Natl. Acad. Sci. USA*, **87**, 3014–3017 (1990).

463. D. S. Peterson, W. K. Milhous, and T. E. Wellems, *Proc. Natl. Acad. Sci. USA*, **87**, 3018–3022 (1990).

464. T. Triglia and A. F. Cowman, *Drug Resist. Updates*, **2**, 15–19 (1999).

465. P. Wang, R. K. B. Brobey, T. Horii, P. F. G. Sims, and J. E. Hyde, *Mol. Microbiol*, **32**, 1254–1262 (1999).

466. D. S. Peterson, S. M. Di Santi, M. Povoa, V. S. Calvosa, V. Do Rosario, and T. E. Wellems, *Am. J. Trop. Med. Hyg.*, **45**, 492–497 (1991).

467. C. V. Plowe, A. Djimde, M. Bouare, O. Doumbo, and T. E. Wellems, *Am. J. Trop. Med. Hyg.*, **52**, 565–568 (1995).

468. L. E. Giraldo, M. C. Acosta, L. A. Labrada, A. Praba, S. Montenegro-James, N. G. Saravia, and D. J. Krogstad, *Am. J. Trop. Med. Hyg.*, **59**, 124–128 (1998).

469. L. K. Basco and P. Ringwald, *Am. J. Trop. Med. Hyg.*, **58**, 369–373 (1998).

470. B. Khan, S. Omar, J. N. Kanyara, M. Warren-Perry, J. Nyalwidhe, D. S. Peterson, T. Wellems, S. Kaniaru, J. Gitonga, F. J. Mulaa, and D. K. Koech, *Trans. R. Soc. Trop. Med. Hyg.*, **91**, 456–460 (1997).

471. D. Parzy, C. Doerig, B. Pradines, A. Rico, T. Fusai, and J.-C. Doury, *Am. J. Trop. Med. Hyg.*, **57**, 646–650 (1997).

472. W. Sirawaraporn, *Drug Resist. Updates*, **1**, 397–406 (1998).

473. A. M. Nzila, E. K. Mberu, J. Sulo, H. Dayo, P. A. Winstanley, C. H. Sibley, and W. M. Watkins, *Antimicrob. Agents Chemother.*, **44**, 991–996 (2000).

474. P. Wang, C.-S. Lee, R. Bayoumi, A. Djimde, O. Doumbo, G. O. Swedberg, L. D. Dao, H. Mshinda, M. Tanner, W. M. Watkins, P. F. G. Sims, and J. E. Hyde, *Mol. Biochem. Parasitol.*, **89**, 161–177 (1997).

475. P. Eldin de Pécoulas, R. Tahar, T. Ouatas, A. Mazabraud, and L. K. Basco, *Mol. Biochem. Parasitol.*, **92**, 265–273 (1998).

476. D. C. Warhurst, *Drug Discovery Today*, **3**, 538–546 (1998).

477. J. H. McKie, K. T. Douglas, C. Chan, S. A. Roser, R. Yates, M. Read, J. E. Hyde, M. J. Dascombe, Y. Yuthavong, and W. Sirawaraporn, *J. Med. Chem.*, **41**, 1367–1370 (1998).

478. E. Petersen, *Trans. R. Soc. Trop. Med. Hyg.*, **81**, 238–241 (1987).

479. Y. Zhang and S. R. Meshnick, *Antimicrob. Agents Chemother.*, **35**, 267–271 (1991).

480. J. D. Chulay, W. M. Watkins, and D. G. Sixsmith, *Am. J. Trop. Med. Hyg.*, **33**, 325–330 (1984).

481. D. R. Brooks, P. Wang, M. Read, W. M. Watkins, P. F. G. Sims, and J. E. Hyde, *Eur. J. Biochem.*, **224**, 397–405 (1994).

482. P. Wang, D. R. Brooks, P. F. G. Sims, and J. E. Hyde, *Mol. Biochem. Parasitol.*, **71**, 115–25 (1995).

483. K. D. Miller, H. O. Lobel, R. F. Satriale, J. N. Kuritcky, R. Stern, and C. C. Campbell, *Am. J. Trop. Med. Hyg.*, **35**, 451–458 (1986).

484. W. M. Watkins, M. Percy, J. M. Crampton, S. Ward, D. Koech, and R. E. Howells, *Trans. R. Soc. Trop. Med. Hyg.*, **82**, 21–26 (1988).

485. L. W. Pang, N. Limsomwong, P. Singharaj, and C. J. Canfield, *Bull. World Health Org.*, **67**, 51–58 (1989).

486. M. D. Edstein, *Chemotherapy*, **33**, 229–233 (1987).

487. M. R. Garg, N. J. Gogtay, and N. A. Kshirsagar, *JAPI*, **44**, 683–685 (1996).

488. M. C. Henry, T. A. Eggelte, P. Watson, B. D. Van Leeuwen, D. A. Bakker, and J. Kluin, *Trop. Med. Int. Health.*, **1**, 610–615 (1996).

489. F. H. Verhoeff, B. J. Brabin, P. Masache, B. Kachale, P. Kazembe, and H. J. Van Der Kaay, *Ann. Trop. Med. Parasitol.*, **91**, 133–140 (1997).

490. M. Rowland, N. Durrani, S. Hewitt, and E. Sondorp, *Trop. Med. Int. Health*, **2**, 1049–1056 (1997).

491. L. E. Osorio, L. E. Giraldo, L. F. Grajales, A. L. Arriaga, A. L. S. S. Andrade, T. K. Ruebush II, and L. M. Barat, *Am. J. Trop. Med. Hyg.*, **61**, 968–972 (1999).

492. B. R. Ogutu, B. L. Smoak, R. W. Nduati, D. A. Mbori-Ngacha, F. Mwathe, and G. D. Shanks, *Trans. R. Soc. Trop. Med. Hyg.*, **94**, 83–84 (2000).

493. M. Warsame, V. A. E. B. Kilimali, W. H. Wernsdorfer, M. Lebbad, A. S. Rutta, and O. Ericsson, *Trans. R. Soc. Trop. Med. Hyg.*, **93**, 312–313 (1999).

494. K. A. Bojang, G. Schneider, S. Forck, S. K. Obaro, S. Jaffar, M. Pinder, J. Rowley, and B. M. Greenwood, *Trans. R. Soc. Trop. Med. Hyg.*, **92**, 73–76 (1998).

495. J. F. Doherty, A. D. Sadiq, L. Bayo, A. Alloueche, P. Olliaro, P. Milligan, L. Von Seidlein, and M. Pinder, *Trans. R. Soc. Trop. Med. Hyg.*, **93**, 543–546 (1999).

496. L. Von Seidlein, P. Milligan, M. Pinder, K. Bojang, C. Anyalebechi, R. Gosling, R. Coleman, J. I. Ude, A. Sadiq, M. Duraisingh, D. Warhurst, A. Alloueche, G. Targett, K. McAdam, B. Greenwood, G. Walraven, P. Olliaro, and T. Doherty, *Lancet*, **355**, 352–357 (2000).

497. R. Steffen, R. Heusser, R. Mächler, et al., *Bull. World Health Org.*, **68**, 313–322 (1990).

498. L. J. Schultz, R. W. Steketee, L. Chitsulo, A. Macheso, P. Kazembe, and J. J. Wirima, *Am J. Trop. Med. Hyg.*, **55**, 87–94 (1996).

499. M. E. Parise, J. G. Ayisi, B. L. Nahlen, L. J. Schultz, J. M. Roberts, A. Misore, R. Muga, A. J. Oloo, and R. W. Steketee, *Am. J. Trop. Med. Hyg.*, **59**, 813–822 (1998).

500. B. F. H. Verhoeff, B. J. Brabin, L. Chimsuku, P. Kazembe, W. B. Russell, and R. L. Broadhead, *Ann. Trop. Med. Parasitol.*, **92**, 141–150 (1998).

501. C. E. Shulman, E. K. Dorman, F. Cutts, K. Kawuondo, J. N. Bulmer, N. Peshu, and K. Marsh, *Lancet.*, **353**, 632–636 (1999).

502. R. H. Black, *Med. J. Aust.*, **1**, 1265–1270 (1973).

503. M. D. Edstein and K. H. Rieckmann, *Chemotherapy*, **39**, 235–241 (1993).

504. M. D. Edstein, J. R. Veenendaal, and K. H. Rieckmann, *Chemotherapy*, **36**, 169–176 (1990).

505. G. D. Shanks, M. D. Edstein, V. Suriyamongkol, S. Timsaad, and H. K. Webster, *Am. J. Trop. Med. Hyg.*, **46**, 643–648 (1992).

506. W. M. Watkins, A. D. Brandling-Bennett, C. G. Nevill, J. Y. Carter, D. A. Boriga, R. E. Howells, and D. K. Koech, *Trans. R. Soc. Trop. Med. Hyg.*, **82**, 398–403 (1988).

507. H. Mshinda, F. Font, R. Hirt, M. Mashaka, C. Ascaso, and C. Menendez, *Trop. Med. Int. Health.*, **1**, 797–801 (1996).

508. M. M. Lemnge, H. A. Msangeni, A. M. Ronn, F. M. Salum, P. H. Jakobsen, J. I. Mhina, J. A. Akida, and I. C. Bygbjerg, *Trans. R. Soc. Trop. Med. Hyg.*, **91**, 68–73 (1997).

509. K. P. David, N. T. Marbiah, P. Lovgren, B. M. Greenwood, and E. Petersen, *Trans. R. Soc. Trop. Med. Hyg.*, **91**, 204–208 (1997).

510. J. Curtis, *Curr. Opin. Anti Infect. Invest. Drugs*, **2**, 99–102 (2000).

511. P. Winstanley and P. Olliaro, *Expert Opin. Invest. Drugs*, **7**, 261–271 (1998).

512. E. Amukoye, P. A. Winstanley, W. M. Watkins, R. W. Snow, J. Hatcher, M. Mosobo, E. Ngumbao, B. Lowe, M. Ton, G. Minyiri, and K. Marsh, *Antimicrob. Agents Chemother.*, **41**, 2261–2264 (1997).

513. P. Wilairatana, D. E. Kyle, S. Looareesuwan, K. Chinwongprom, S. Amradee, N. J. White, and W. M. Watkins, *Ann. Trop. Med. Parasitol.*, **91**, 125–132 (1997).

514. M. D. Edstein, A. E. T. Yeo, G. D. Shanks, and K. H. Rieckmann, *Acta Tropica.*, **66**, 127–135 (1997).

515. K. J. Saliba and K. Kirk, *Trans. R. Soc. Trop. Med. Hyg.*, **92**, 666–667 (1998).

516. T. Tiffert, H. Ginsburg, M. Krugliak, B. C. Elford, and V. L. Lew, *Proc. Natl. Acad. Sci. USA*, **97**, 331–336 (2000).

517. E. Freerksen, E. W. Kanthumkumva, and A. R. K. Kholowa, *Chemotherapy*, **42**, 391–401 (1996).

518. H. Goerg, S. A. Ochola, and R. Georg, *Chemotherapy*, **45**, 68–76 (1999).

519. D. J. Knight, P. Mamalis, and W. Peters, *Amm. Trop. Med. Parasitol.*, **76**, 1–7 (1982).

520. M. D. Edstein, S. Bahr, B. Kotecka, G. D. Shanks, and K. H. Rieckmann, *Antimicrob. Agents Chemother.*, **41**, 2300–2301 (1997).

521. D. J. Knight and W. Peters, *Ann. Trop. Med. Parasitol.*, **74**, 393–404 (1980).

522. C. J. Canfield, W. K. Mihous, A. L. Ager, R. N. Rossan, T. R. Sweeney, N. J. Lewis, and D. P. Jacobus, *Am. J. Trop. Med. Hyg.*, **49**, 121–126 (1993).

523. P. M. S. Chauhan, *Curr. Opin. Anti Infect. Invest. Drugs*, **2**, 103–108 (2000).

524. M. D. Edstein, K. D. Corcoran, G. D. Shanks, M. Ngampochjana, P. Hansukjariya, J. Sattabongkot, H. K. Webster, and K. H. Rieckmann, *Am. J. Trop. Med. Hyg.*, **50**, 181–186 (1994).

525. A. E. T. Yeo, *J. Parasitol.*, **83**, 515–518 (1997).

526. N. P. Jensen, A. L. Ager, R. A. Bliss, C. J. Canfield, B. M. Kotecka, K. H. Rieckmann, J. Terpinski, and D. P. Jacobus, *J. Med. Chem.*, **44**, 3925–3931 (2001).

527. R. K. Haynes and S. C. Vonwiller, *Acc. Chem. Res.*, **30**, 73–79 (1997).

528. D. L. Klayman, *Science*, **228**, 1049–1055 (1985).

529. X. D. Luo and C. C. Shen, *Med. Res. Rev.*, **7**, 29–52 (1987).

530. Coordinating Group for Research on the Structure of Qing Hau Sau, *K'o Hsueh T'ung Pao*, **22**, 142 (1977).

531. Qinghaosu Research Group, *Sci. Sin. (Engl. Ed.)*, **23**, 380–396 (1980).

532. V. Dhingra, K. V. Rao, and M. L. Narasu, *Life Sci.*, **66**, 279–300 (2000).

533. M. A. Van Agtmael, T. A. Eggelte, and C. J. Van Boxtel, *Trends Pharmacol. Sci.*, **20**, 199–205 (1999).

534. J. A. Vroman, M. Alvim-Gaston, and M. A. Avery, *Curr. Pharamceut. Design.*, **5**, 101–138 (1999).

535. H. Ziffer, R. J. Highet, and D. L. Klayman, *Prog. Chem. Org. Nat. Prod.*, **72**, 121–214 (1997).

536. S. R. Meshnick, T. E. Taylor, and S. Kamchonwongpaisan, *Microbiological Rev.*, **60**, 301–315 (1996).

537. M. Jung, *Curr. Med. Chem.*, **1**, 35–49 (1994).

538. T. T. Hien and N. J. White, *Lancet*, **341**, 603–608 (1993).

539. A. R. Butler and Y.-L. Wu, *Chem. Soc. Rev.*, **21**, 85–90 (1992).

540. Qinghaosu Antimalaria Coordinating Research Group, *Chin. Med. J.*, **92**, 811–816 (1979).

541. A. Brossi, B. Venugopalan, L. Domingues Gerpe, H. J. C. Yeh, J. L. Flippen-Anderson, P. Buchs, and X. D. Luo, *J. Med. Chem.*, **31**, 645–650 (1988).

542. China Cooperative Research Group on Qiughaosu and its Derivatives as Antimalarials, *J. Tradit. Chin. Med.*, **2**, 45–50 (1982).

543. I. S. Lee and C. D. Hufford, *Pharmacol. Ther.*, **48**, 345–355 (1990).

544. China Cooperative Research Group on Qiughaosu and its Derivatives as Antimalarials, *J. Tradit. Chin. Med.*, **2**, 9–16 (1982).

545. R. K. Haynes and S. C. Vonwiller, *Trans. R. Soc. Trop. Me. Hyg.*, **88**, S23–S26 (1994).

546. J. F. S. Ferreira, J. E. Simon, and J. Janick, *Hortic. Rev.*, **19**, 319–371 (1997).

547. J. C. Laughlin, *Trans. R. Soc. Trop. Med. Hyg.*, **88**, S21–S22 (1994).

548. H. J. Woerdenbag, N. Pras, W. van Uden, T. E. Wallaart, A. C. Beekman, and C. B. Lugt, *Pharm. World Science*, **16**, 169–180 (1994).

549. R. Liersch, H. Soicke, C. Stehr, and H.-U. Tüllner, *Planta Med.*, 387–390 (1986).

550. L. Shide, N. Bingmei, H. Wangyu, and X. Jinlun, *J. Nat. Prod.*, **54**, 573–575 (1991).

551. H. J. Bouwmeester, T. E. Wallaart, M. H. A. Janssen, B. Van Loo, B. J. M. Jansen, M. A. Posthumus, C. O. Schmidt, J.-W. De Kraker,

W. A. Konig, and M. C. R. Franssen, *Phytochemistry*, **52**, 843–854 (1999).

552. P. Mercke, M. Bengtsson, H. J. Bouwmeester, M. A. Posthumus, and P. E. Brodelius, *Arch. Biochem. Biophys.*, **381**, 173–180 (2000).

553. V. Dhingra and M. L. Narasu, *Biochem. Biophys. Res. Commun.*, **281**, 558–561 (2001).

554. T. E. Wallaart, W. Van Uden, H. G. M. Lubberink, H. J. Woerdenbag, N. Pras, and W. J. Quax, *J. Nat. Prod.*, **62**, 430–433 (1999).

555. T. E. Wallaart, N. Pras, and W. J. Quax, *J. Nat. Prod.*, **62**, 1160–1162 (1999).

556. H. K. Webster and E. K. Lehnert, *Trans. R. Soc. Trop. Med. Hyg.*, **88** (Suppl 1): 27–29, 1994.

557. G. Schmid and W. Hofheinz, *J. Am. Chem. Soc.*, **105**, 624–625 (1983).

558. M. A. Avery, W. K. M. Chong, and C. Jennings-White, *J. Am. Chem. Soc.*, **114**, 974–979 (1992).

559. M. A. Avery, C. Jennings-White, and W. K. M. Chong, *Tetrahedron Lett.*, **28**, 4629–4632 (1987).

560. M. A. Avery, C. Jennings-White, and W. K. M. Chong, *J. Org. Chem.*, **54**, 1792–1795 (1989).

561. X. Xu, J. Zhu, D. Huang, and W. Zhou, *Huaxue Xuebao*, **41**, 574–576 (1983).

562. X. Xu, J. Zhu, D. Huang, and W. Zhou, *Huaxue Xuebao*, **42**, 940–942 (1984).

563. W. Zhou, *Pure Appl. Chem.*, **58**, 817–824 (1986).

564. X. Xu, J. Zhu, D. Huang, and W. Zhou, *Tetrahedron*, **42**, 819–828 (1986).

565. W.-S. Zhou and X.-X. Xu, *Acc. Chem. Res.*, **27**, 211–216 (1994).

566. H.-J. Liu, W.-L. Yeh, and S. Y. Chew, *Tetrahedron Lett.*, **34**, 4435–4438 (1993).

567. M. A. Avery, J. D. Bonk, and S. Mehrotra, *J. Labelled Compd. Radiopharm.*, **38**, 249–254 (1996).

568. Y. Li, Z. X. Yang, Y. X. Chen, and X. Zhang, *Yaoxue Xuebao*, **29**, 713–716 (1994).

569. M. A. Avery, J. D. Bonk, and J. Bupp, *J. Labelled Compd. Radiopharm.*, **38**, 263–267 (1996).

570. T. Ravindranathan, M. A. Kumar, R. B. Menon, and S. V. Hiremath, *Tetrahedron Lett.*, **31**, 755–758 (1990).

571. T. Ravindranathan, *Curr. Sci.*, **66**, 35–41 (1994).

572. J. B. Bhonsle, B. Pandey, V. H. Deshpande, and T. Ravindranathan, *Tetrahedron Lett.*, **35**, 5489–5492 (1994).

573. M. G. Constantino, M. Beltrame Jr., G. V. J. da Silva, and J. Zukerman-Schpector, *Synthetic Commun*, **26**, 321–329 (1996).

574. B. Ye and Y.-l. Wu, *J. Chem. Soc., Chem. Commun.*, 726 (1990).

575. M. A. Avery, M. Alvim-Gaston, and J. R. Woolfrey, *Adv. Med. Chem.*, **4**, 125–217 (1999).

576. A. K. Bhattacharya and R. P. Sharma, *Heterocycles*, **51**, 1681–1745 (1999).

577. M. Jung, H. N. ElSohly, E. M. Croom, A. T. McPhail, and D. R. McPhail, *J. Org. Chem.*, **51**, 5417–5419 (1986).

578. R. J. Roth and N. Acton, *J. Chem. Ed.*, **66**, 349–350 (1989).

579. R. J. Roth and N. Acton, *J. Nat. Prod.*, **52**, 1183–1185 (1989).

580. R. J. Roth and N. Acton, *J. Chem. Educ.*, **68**, 612–613 (1991).

581. R. K. Haynes and S. C. Vonwiller, *Chemistry Aust.*, **58**, 64–67 (1991).

582. P. T. Lansbury and D. M. Nowak, *Tetrahedron Lett.*, **33**, 1029–1032 (1992).

583. D. M. Nowak and P. T. Lansbury, *Tetrahedron*, **54**, 319–336 (1998).

584. A. J. Lin, D. L. Klayman, J. M. Hoch, J. V. Silverton, and C. F. George, *J. Org. Chem.*, **50**, 4504–4508 (1985).

585. H.-M. Gu, B.-F. Lu, and Z.-X. Qu, *Acta Pharmacol. Sinica (Chung-kuo Yao Li Hsueh Pao)*, **1**, 48–50 (1980).

586. J. N. Lisgarten, B. S. Potter, C. Bantuzeko, and R. A. Palmer, *J. Chem. Crystallogr.*, **28**, 539–543 (1998).

587. P. K. Agrawal and V. Bishnoi, *J. Sci. Ind. Res.*, **55**, 17–26 (1996).

588. C. H. Oh, D. Wang, J. N. Cumming, and G. H. Posner, *Spectrosc. Lett.*, **30**, 241–255 (1997).

589. S. S. Zaman and R. P. Sharma, *Heterocycles*, **32**, 1593–1638 (1991).

590. M. A. Avery, G. McLean, G. Edwards, and A. Ager, Structure-Activity Relationships of Peroxide-Based Artemisinin Antimalarials, Conference Proceedings, 2000, pp. 121–132.

591. D. Y. Zhu, *Zhongguo Yaoli Xuebao*, **4**, 194 (1983).

592. J. L. Vennerstrom, N. Acton, A. J. Lin, and D. L. Klayman, *Drug Des. Del.*, **4**, 45–54 (1989).

593. R. Docampo and S. N. J. Moreno, in W. A. Pryor, Ed., *Free Radicals in Biology*, Academic Press, New York, 1984, pp. 243–288.

594. S. Tani, N. Fukamiya, H. Kiyokawa, H. A. Musallam, R. O. Pick, and K.-H. Lee, *J. Med. Chem.*, **28**, 1743–1744 (1985).

595. J. A. Kepler, A. Philip, Y. W. Lee, H. A. Musallam, and F. I. Carroll, *J. Med. Chem.*, **30**, 1505–1509 (1987).

596. C. W. Jefford, E. C. McGoran, J. Boukouvalas, G. Ricahrdson, B. L. Robinson, and W. Peters, *Helv. Chim. Acta.*, **71**, 1805–1812 (1988).

597. C. W. Jefford, J. Currie, G. D. Richardson, and J.-C. Rossier, *Helv. Chim. Acta.*, **74**, 1239–1246 (1991).

598. J. A. Kepler, A. Philip, Y. W. Lee, M. C. Morey, and F. L. Carroll, *J. Med. Chem.*, **31**, 713–716 (1988).

599. C. C. Shen and X. D. Luo, *Report: Scientific Working Group on the Chemotherapy of Malaria. The Development of Qinghaosu and its Derivatives*, WHO, Geneva, 1986.

600. M. A. Avery, F. Gao, W. K. M. Chong, S. Mehrotra, and W. K. Milhous, *J. Med. Chem.*, **36**, 4264–4275 (1993).

601. M. A. Avery, P. Fan, J. M. Karle, J. D. Bonk, R. Miller, W. Milhous, and D. K. Goins, *J. Med. Chem.*, **39**, 1885–1897 (1996).

602. G. H. Posner, D. Wang, L. González, X. Tao, J. N. Cumming, D. Klinedinst, and T. A. Shapiro, *Tetrahedron Lett.*, **37**, 815–818 (1996).

603. Y. Li, P. L. Yu, Y. X. Chen, L. Q. Li, Y. Z. Gai, D. S. Wang, and Y. P. Zheng, *Yaozue Xuebao.*, **16**, 429–439 (1981).

604. A. J. Lin and R. E. Miller, *J. Med. Chem.*, **38**, 764–770 (1995).

605. M. Jung and J. Bae, *Heterocycles*, **53**, 261–264 (2000).

606. Y. Li, Y.-M. Zhu, H.-J. Jiang, J.-P. Pan, G.-S. Wu, J.-M. Wu, Y.-L. Shi, J.-D. Yang, and B.-A. Wu, *J. Med. Chem.*, **43**, 1635–1640 (2000).

607. P. M. O'Neill, L. P. Bishop, R. C. Storr, S. R. Hawley, J. L. Maggs, S. A. Ward, and B. K. Park, *J. Med. Chem.*, **39**, 4511–4514 (1996).

608. G. H. Posner, H. B. Jeon, M. H. Parker, M. Krasavin, I.-H. Paik, and T. A. Shapiro, *J. Med. Chem.*, **44**, 3054–3058 (2001).

609. M. Jung, D. Yu, D. Bustos, H. N. ElSohly, and J. D. McChesney, *Bioorg. Med. Chem. Lett.*, **1**, 741–744 (1991).

610. M. A. Avery, S. Mehrotra, T. L. Johnson, J. D. Bonk, J. A. Vroman, and R. Miller, *J. Med. Chem.*, **39**, 4149–4155 (1996).

611. J. Han, J.-G. Lee, S.-S. Min, S.-H. Park, C. K. Angerhofer, G. A. Cordell, and S.-U. Kim, *J. Nat. Prod.*, **64**, 1201–1205 (2001).

612. N. Acton and D. L. Klayman, *Planta Med.*, **4**, 266–268 (1987).

613. C. W. Jefford, U. Burger, P. Millasson-Schmidt, and G. Bernardinelli, *Helv. Chim. Acta.*, **83**, 1239–1246 (2000).

614. M. A. Avery, C. Jennings-White, and W. K. M. Chong, *J. Org. Chem.*, **54**, 1792–1795 (1989).

615. J. N. Cumming, D. Wang, S. B. Park, T. A. Shapiro, and G. H. Posner, *J. Med. Chem.*, **41**, 952–964 (1998).

616. A. J. Lin, M. Lee, and D. L. Klayman, *J. Med. Chem.*, **32**, 1249–1252 (1989).

617. N. Acton, J. M. Karle, and R. E. Miller, *J. Med. Chem.*, **36**, 2552–2557 (1993).

618. G. H. Posner, C. H. Oh, L. Gerena, and W. K. Milhous, *J. Med. Chem.*, **35**, 2459–2467 (1992).

619. C. Thomson, M. Cory, and M. Zerner, *Int. J. Quant. Chem.*, **18**, 213–245 (1991).

620. Y. Tang, H. L. Jiang, K. X. Chen, and R. Y. Ji, *Indian J. Chem., Sect. B.*, **35B**, 325–332 (1996).

621. V. Nguyen-Cong, G. Van Dang, and B. M. Rode, *Eur. J. Med. Chem.*, **31**, 797–803 (1996).

622. M. Grigorov, J. Weber, J. M. J. Tronchet, C. W. Jefford, W. K. Milhous, and D. Maric, *J. Chem. Info. Comp. Sci.*, **37**, 124–130 (1997).

623. J. R. Woolfrey, M. A. Avery, and A. M. Doweyko, *J. Comput.-Aided Mol. Des.*, **12**, 165–181 (1998).

624. H.-l. Jiang, K.-X. Chen, H.-W. Wang, Y. Tang, J.-Z. Chen, and R.-Y. Ji, *Acta Pharmacologica Sinica (Zhongguo Yaoli Xuebao)*, **15**, 481–487 (1994).

625. S. Tonmunphean, S. Kokpol, V. Parasuk, P. Wolschann, R. H. Winger, K. R. Liedl, and B. M. Rode, *J. Comput.-Aided Mol. Des.*, **12**, 397–409 (1998).

626. X. Gironés, A. Gallegos, and R. Carbó-Dorca, *J. Chem. Inf. Comput. Sci.*, **40**, 1400–1407 (2000).

627. S. Tonmunphean, V. Parasuk, and S. Kokpol, *Quant. Struct.-Act. Rel.*, **19**, 475–483 (2000).

628. H. L. Jiang, K. X. Chen, Y. Tang, J. Z. Chen, Y. Li, Q. M. Wang, and R. Y. Ji, *Indian J. Chem., Sect. B.*, **36B**, 154–160 (1997).

629. P. Olliaro, *Pharmacol. Therapeut.*, **89**, 207–219 (2001).

630. J. N. Cumming, P. Ploypradith, and G. H. Posner, *Adv. Pharmacol.*, **37**, 253–297 (1997).

631. Anonymous, *J. Trad. Chin. Med.*, **2**, 17 (1982).

632. Z. Ye, Z. Li, M. Gao, X. Fu, and H. Lia, *Yaoxue Tongbao*, **17**, 4247–4248 (1982).

633. H. J. Woerdenbag, T. A. Moskal, N. Pras, T. M. Malingre, F. S. el-Feraly, H. H. Kampinga, and A. W. Konings, *J. Nat. Prod.*, **56**, 849–856 (1993).

634. Y. Maeno, T. Toyoshima, H. Fujioka, Y. Ito, S. R. Meshnick, A. Benakis, W. K. Milhous, and M. Aikawa, *Am. J. Trop. Med. Hyg.*, **49**, 485–491 (1993).

635. P. A. Berman and P. A. Adams, *Free Radical Biol. Med.*, **22**, 1283–1288 (1997).

636. P. M. Browning and R. H. Bisby, *Mol. Biochem. Parasitol.*, **32**, 57–60 (1989).

637. D. S. Ellis, Z. L. Li, H. M. Gu, W. Peters, B. L. Robinson, G. Tovey, and D. C. Warhurst, *Ann. Trop. Med. Parasitol.*, **79**, 367–374 (1985).

638. H. M. Gu, D. C. Warhurst, and W. Peters, *Biochem. Pharmacol.*, **32**, 2463–2466 (1983).

639. H. M. Gu, D. C. Warhurst, and W. Peters, *Trans. R. Soc. Trop. Med. Hyg.*, **78**, 265–270 (1984).

640. T. Akompong, J. VanWye, N. Ghori, and K. Haldar, *Mol. Biochem. Parasitol.*, **101**, 71–79 (1999).

641. W. Asawamahasakda, A. Benakis, and S. R. Meshnick, *J. Lab. Clin. Med.*, **123**, 757–762 (1994).

642. H. Y. Aboul-Enein, *Drug Design Del.*, **4**, 129–133 (1989).

643. P. Augustijns, A. D'Hulst, J. Van Daele, and R. Kinget, *J. Pharm. Sci.*, **85**, 577–579 (1996).

644. S. Kamchonwongpaisan and S. R. Meshnick, *Gen. Pharmacol.*, **27**, 587–592 (1996).

645. S. Paitayatat, B. Tarnchompoo, Y. Thebtaranonth, and Y. Yuthavong, *J. Med. Chem.*, **40**, 633–638 (1997).

646. S. R. Krungkrai and Y. Yuthavong, *Trans. R. Soc. Trop. Med. Hyg.*, **81**, 710–714 (1987).

647. O. A. Levander, A. L. Ager, V. C. Morris, and R. G. May, *Am. J. Clin. Nutr.*, **50**, 346–352 (1989).

648. S. R. Meshnick, T. W. Tsang, F. B. Lin, H. Z. Pan, C. N. Chang, F. Kuypers, D. Chni, and B. Lubin, *Prog. Clin. Biol. Res.*, **313**, 95–104 (1989).

649. S. Kamchonwongpaisan, S. Paitayatat, Y. Thebtaranonth, P. Wilairat, and Y. Yuthavong, *J. Med. Chem.*, **38**, 2311–2316 (1995).

650. F. Benoit-Vical, A. Robert, and B. Meunier, *Antimicrob. Agents Chemother.*, **44**, 2836–2841 (2000).

651. P. L. Olliaro, R. K. Haynes, B. Meunier, and Y. Yuthavong, *Trends Parasitol.*, **17**, 122–126 (2001).

652. G. H. Posner and S. R. Meshnick, *Trends Parasitol.*, **17**, 266–267 (2001).

653. W.-M. Wu, Y. Wu, Y.-L. Wu, Z.-J. Yao, C.-M. Zhou, Y. Li, and F. Shan, *J. Am. Chem. Soc.*, **120**, 3316–3325 (1998).

654. S. R. Meshnick, A. Thomas, A. Ranz, C. M. Xu, and H. Pan, *Mol. Biochem. Parasitol.*, **49**, 181–189 (1991).

655. G. H. Posner and C. H. Oh, *J. Am. Chem. Soc.*, **114**, 8328–8329 (1992).

656. C. W. Jefford, M. G. H. Vicente, Y. Jacquier, J. Mareda, P. Millasson-Schmidt, G. Brunner, and U. Burger, *Helv. Chim. Acta.*, **79**, 1475–1487 (1996).

657. A. R. Butler, B. C. Gilbert, P. Hulme, L. R. Irvine, L. Renton, and A. C. Whitwood, *Free Radical Res.*, **28**, 471–476 (1998).

658. A. Robert, J. Cazelles, and B. Meunier, *Angew. Chem. IEE*, **40**, 1954–1957 (2001).

659. A. Robert and B. Meunier, *Chem. Eur. J.*, **4**, 1287–1296 (1998).

660. D. Y. Wang, Y. L. Wu, Y. K. Wu, J. Liang, and Y. Li, *J. Chem. Soc., Perkin Trans.*, **1**, 605–609 (2001).

661. Y. Wu, Z.-Y. Yue, and Y.-L. Wu, *Angew. Chem., Int. Ed.*, **38**, 2580–2582 (1999).

662. P. M. O'Neill, L. P. D. Bishop, N. L. Searle, J. L. Maggs, R. C. Storr, S. A. Ward, B. K. Park, and F. Mabbs, *J. Org. Chem.*, **65**, 1578–1582 (2000).

663. J. Gu, K. Chen, H. Jiang, and J. Leszczynski, *J. Mol. Struct. (Theochem).*, **491**, 57–66 (1999).

664. J. Gu, K. Chen, H. Jiang, and R. Ji, *Theochemistry*, **459**, 103–111 (1999).

665. G. H. Posner, C. O. Oh, D. Wang, L. Gerena, W. K. Milhous, S. R. Meshnick, and W. Asawamahasadka, *J. Med. Chem.*, **37**, 1256–1258 (1994).

666. P. M. O'Neill, L. P. Bishop, N. L. Searle, J. L. Maggs, S. A. Ward, P. G. Bray, R. C. Storr, and B. K. Park, *Tetrahedron Lett.*, **38**, 4263–4266 (1997).

667. P. M. O'Neill, N. L. Searle, K. J. Raynes, J. L. Maggs, S. A. Ward, R. C. Storr, B. K. Park, and G. H. Posner, *Tetrahedron Lett.*, **39**, 6065–6068 (1998).

668. G. H. Posner, D. Wang, J. N. Cumming, C. H. Oh, A. N. French, A. L. Bodley, and T. A. Shapiro, *J. Med. Chem.*, **38**, 2273–2275 (1995).

669. G. H. Posner, S. B. Park, L. Gonsález, D. Wang, J. N. Cumming, D. Klinedinst, T. A. Shapiro, and M. D. Bachi, *J. Am. Chem. Soc.*, **118**, 3537–3538 (1996).

670. G. H. Posner, J. N. Cumming, P. Ploypradith, and C. H. Oh, *J. Am. Chem. Soc.*, **117**, 5885–5886 (1995).

671. S. Kapetanaki and C. Varotsis, *FEBS Lett.*, **474**, 238–241 (2000).

672. M. A. Avery, P. Fan, J. M. Karle, R. Miller, and K. Goins, *Tetrahedron Lett.*, **36**, 3965–3968 (1995).

673. N. Wei and S. M. H. Sadrzadeh, *Biochem. Pharmacol.*, **48**, 737–741 (1994).

674. K. L. Shukla, T. M. Gund, and S. R. Meshnick, *Journal of Molecular Graphics.*, **13**, 215–222 (1995).

675. R. K. Haynes and S. C. Vonwiller, *Tetrahedron Lett.*, **37**, 253–256 (1996).

676. R. K. Haynes and S. C. Vonwiller, *Tetrahedron Lett.*, **37**, 257–260 (1996).

677. R. K. Haynes, H. Hung-On Pai, and A. Voerste, *Tetrahedron Lett.*, **40**, 4715–4718 (1999).

678. S. R. Meshnick, *Trans. R. Soc. Trop. Med. Hyg.*, **88** (Suppl. 1), 31–32, 1994.

679. O. Provot, B. Camuzat-Dedenis, M. Hamzaoui, H. Moskowitz, J. Mayrargue, A. Robert, J. Cazelles, B. Meunier, F. Zouhiri, D. Desmaele, J. D'Angelo, J. Mahuteau, F. Gay, and L. Ciceron, *Eur. J. Org. Chem.*, 1935–1938 (1999).

680. Y. Z. Yang, W. Asawamahasadka, and S. R. Mechnick, *Biochem. Pahrmacol.*, **46**, 336–339 (1993).

681. Y. Z. Yang, B. Little, and S. R. Meshnick, *Biochem. Pharmacol.*, **48**, 569–573 (1994).

682. W. Asawamahasakda, I. Ittarat, Y. M. Pu, H. Ziffer, and S. R. Meshnick, *Antimicrob. Agents Chemother.*, **38**, 1854–1858 (1994).

683. J. Bhisutthibhan, X.-Q. Pan, P. A. Hossler, D. J. Walker, C. A. Yowell, J. Carlton, J. B. Dame, and S. R. Meshnick, *J. Biol. Chem.*, **273**, 16192–16198 (1998).

684. D. J. Walker, J. L. Pitsch, M. M. Peng, B. L. Robinson, W. Peters, J. Bhisutthibhan, and S. R. Meshnick, *Antimicrob. Agents Chemother.*, **44**, 344–347 (2000).

685. D.-y. Wang and Y.-l. Wu, *J. Chem. Soc., Chem. Commun.*, 2193–2194 (2000).

686. W. Asawamahasakda, I. Ittarat, C.-C. Chang, P. McElroy, and S. R. Meshnick, *Mol. Biochem. Parasitol.*, **67**, 183–191 (1994).

687. A. U. Orjih, *Brit. J. Haematol.*, **92**, 324–328 (1996).

688. N. Basilico, D. Monti, P. Olliaro, and D. Taramelli, *FEBS Lett.*, **409**, 297–299 (1997).

689. A. V. Pandey, B. L. Tekwani, R. L. Singh, and V. S. Chauhan, *J. Biol. Chem.*, **274**, 19383–19388 (1999).

690. R. Price, M. Van Vugt, F. Nosten, C. Luxemburger, A. Brockman, L. Phaipun, T. Chongsuphajaisiddhi, and N. White, *Am. J. Trop. Med. Hyg.*, **59**, 883–888 (1998).

691. H. Vinh, N. N. Huong, T. T. B. Ha, B. M. Cuong, N. H. Phu, T. T. H. Chau, P. T. Quoi, K. Arnold, and T. T. Hien, *Trans. R. Soc. Trop. Med. Hyg.*, **91**, 465–467 (1997).

692. C. X. T. Phuong, D. B. Bethell, P. T. Phuong, T. T. T. Mai, T. T. N. Thuy, N. T. T. Ha, P. T. T. Thuy, N. T. T. Anh, N. P. J. Day, and N. J. White, *Trans. R. Soc. Trop. Med. Hyg.*, **91**, 335–342 (1997).

693. N. J. White, *Trans. R. Soc. Trop. Med. Hyg.*, **88**, S3–S4 (1994).

694. F. Nosten, *Trans. Roy. Soc. Trop. Med. Hyg.*, **88** (Suppl 1), 45–46 (1994).

695. R. N. Price, F. Nosten, C. Luxemburger, F. O. T. Kuile, L. Paiphun, T. Chongsuphajaisiddhi, and N. J. White, *Lancet*, **347**, 1654–1658 (1996).

696. K. Arnold, *Trans. R. Soc. Trop. Med. Hyg.*, **88**, S47–S49 (1994).

697. A. N. Chawira and D. C. Warhurst, *J. Trop. Med. Hyg.*, **90**, 1–8 (1987).

698. N. J. White and F. Nosten, *Curr. Opin. Infect. Dis.*, **6**, 323–330 (1993).

699. P. J. De Vries and T. K. Dien, *Drugs*, **52**, 818–836 (1996).

700. N. J. White, *Trans. R. Soc. Trop. Med. Hyg.*, **88**, 41–43 (1994).

701. M. H. Alin and A. Bjorkman, *Am. J. Trop. Med. Hyg.*, **50**, 771–776 (1994).

702. G. P. Dutta, A. Mohan, and R. Tripath, *J. Parasitol.*, **76**, 849–852 (1990).

703. M. Ashton, T. N. Hai, N. D. Sy, D. X. Huong, N. V. Huong, N. T. Nipu, and L. D. Cong, *Drug Metab. Dispos.*, **26**, 25–27 (1998).

704. M. Ashton, N. Duy Sy, N. Van Huong, T. Gordi, T. N. Hai, D. X. Huong, N. T. Nieu, and L. D. Cong, *Clin. Pharmacol. Ther.*, **63**, 482–493 (1998).

705. M. Ashton, T. Gordi, T. Ngoc Hai, N. Van Huong, N. Duy Sy, N. Thi Nieu, D. Xuan Huong, M. Johansson, and L. Dinh Cong, *Biopharm. Drug Dispos.*, **19**, 245–250 (1998).

706. P. J. De Vries, T. K. Dien, N. X. Khanh, L. N. Binh, P. T. Yen, D. D. Duc, C. J. Van Boxtel, and P. A. Kager, *Am. J. Trop. Med. Hyg.*, **56**, 503–507 (1997).

707. T. K. Dien, P. de Vries, N. X. Khanh, R. Koopmans, L. N. Binh, D. D. Duc, P. A. Kager, and C. J. van Boxtel, *Antimicrob. Agents Chemother.*, **41**, 1069–1072 (1997).

708. R. Koopmans, D. D. Duc, P. A. Kager, N. X. Khanh, T. K. Dien, P. J. De Vries, and C. J. Van Boxtel, *Trans. R. Soc. Trop. Med. Hyg.*, **92**, 434–436 (1998).

709. U. S. H. Svensson, R. Sandstrom, O. Carlborg, H. Lennernas, and M. Ashton, *Drug Metab. Dispos.*, **27**, 227–232 (1999).

710. U. S. H. Svensson and M. Ashton, *Br. J. Clin. Pharmacol.*, **48**, 528–535 (1999).

711. K. Mihara, U. S. H. Svensson, G. Tybring, T. Ngoc Hai, L. Bertilsson, and M. Ashton, *Fundam. Clin. Pharmacol.*, **13**, 671–675 (1999).

712. P. O'Neill, F. Scheinmann, A. V. Stachulski, J. L. Maggs, and B. K. Park, *J. Med. Chem.*, **44**, 1467–1470 (2001).

713. M. H. Alin, M. Ashton, C. M. Kihamia, G. J. Mtey, and A. Björkman, *Br. J. Clin. Pharmacol.*, **41**, 587–592 (1996).

714. T. T. Hien, K. Arnold, T. H. Nguyen, P. L. Pham, T. D. Nguyen, M. C. Bui, M. T. Le, Q. P. Mach, H. V. A. Le, and P. M. Pham, *Trans. R. Soc. Trop. Med. Hyg.*, **88**, 688–691 (1994).

715. P. J. de Vries, N. N. Bich, H. V. Thien, L. N. Hung, T. K. Anh, P. A. Kager, and S. H. Heisterkamp, *Antimicrob. Agents Chemother.*, **44**, 1302–1308 (2000).

716. D. S. Nguyen, B. H. Dao, P. D. Nguyen, V. H. Nguyen, N. B. Le, V. S. Mai, and S. R. Meshnick, *Am. J. Trop. Med. Hyg.*, **48**, 398–402 (1993).

717. A. N. Chawira, D. C. Warhurst, and W. Peters, *Trans. R. Soc. Trop. Med. Hyg.*, **80**, 477–480 (1986).

718. W. Peters and B. L. Robinson, *Ann. Trop. Med. Parasitol.*, **93**, 325–339 (1999).

719. J. C. Doury, P. Ringwald, J. Guelain, and J. Le Bras, *Trop. Med. Parasitol.*, **43**, 197–198 (1992).

720. C. Wongsrichanalai, N. T. Dung, N. T. Trieu, T. Wimonwattrawatee, P. Sookto, D. G. Heppner, and F. Kawamoto, *Acta Trop.*, **63**, 151–158 (1997).

721. WHO, *WHO Drug Info.*, **13**, 14–16 (1999).

722. W. Peters, Z.-l. Li, B. L. Robinson, and D. C. Warhurst, *Ann. Top. Med. Parasitol.*, **80**, 483–489 (1986).

723. C. J. Janse, A. P. Waters, J. Kos, and C. B. Lugi, *Int. J. Parasitol.*, **24**, 589–594 (1994).

724. P. Ringwald, J. Bickii, and L. K. Basco, *Am. J. Trop. Med. Hyg.*, **61**, 187–192 (1999).

725. G. Li, X. B. Guo, L. C. Fu, H. X. Jian, and X. H. Wang, *Trans. R. Soc. Trop. Med. Hyg.*, Suppl., **1**, S5–S6 (1994).

726. S. Looareesuwan, P. Wilairatana, S. Vanijanonta, P. Pitisuttithum, and K. Kraisintu, *Ann. Trop. Med. Parasitol.*, **90**, 21–28 (1996).

727. P. Wilairatana, S. Krudsood, U. Silachamroon, P. Singhasivanon, S. Vannuphan, S. Faithong, M. Klabprasit, S. Ns Bangchang, P. Olliaro, and S. Looareesuwan, *Am. J. Trop. Med. Hyg.*, **63**, 290–294 (2000).

728. G. Q. Li, X. H. Wang, X. B. Guo, L. C. Fu, H. X. Jian, P. Q. Chen, and G. Q. Li, *Southeast Asian J. Trop. Med.*, **30**, 17–19 (1999).

729. K. C. Zhao and Z. Y. Song, *Yao Hsueh Hsueh Pao.*, **28**, 342–346 (1993).

730. K. Na-Bangchang, K. Congpoung, R. Ubalee, A. Thanavibul, P. Tan-Anya, and J. Karbwang, *Southeast Asian J. Trop. Med. Public Health.*, **28**, 731–735 (1997).

731. X. D. Luo, H. J. C. Yeh, A. Brossi, J. L. Flippen-Anderson, and R. Gilardi, *Helv. Chim. Acta.*, **67**, 1515–1522 (1984).

732. A. J. Lin, A. D. Theoharides, and D. L. Klayman, *Tetrahedron*, **42**, 2181–2184 (1986).

733. H. Alin, M. Ashton, C. M. Kihamia, G. J. B. Mtey, and A. Bjorkman, *Trans. R. Soc. Trop. Med. Hyg.*, **90**, 61–65 (1996).

734. K. T. Batty, K. F. Ilett, T. Davis, and M. E. Davis, *J. Pharm. Pharmacol.*, **48**, 22–26 (1996).

735. V. Navaratnam, S. M. Mansor, M. N. Mordi, A. Akbar, and M. N. Abdullah, *Eur. J. Clin. Pharmacol.*, **54**, 411–414 (1998).

736. K. T. Batty, L. T. A. Thu, T. M. E. Davis, K. F. Ilett, T. X. Mai, N. C. Hung, N. P. Tien, S. M. Powell, H. Van Thien, T. Q. Binh, and N. Van Kim, *Br. J. Clin. Pharmacol.*, **45**, 123–129 (1998).

737. P. Newton, Y. Suputtamongkol, P. Teja-Isavadharm, S. Pukrittayakamee, V. Navaratnam, I. Bates, and N. White, *Antimicrob. Agents Chemother.*, **44**, 972–977 (2000).

738. T. M. E. Davis, H. P. Phuong, K. F. Ilett, N. C. Hung, K. T. Batty, V. D. B. Phuong, S. M. Powell, H. V. Thien, and T. Q. Binh, *Antimicrob. Agents Chemother.*, **45**, 181–186 (2001).

739. K. T. Batty, L. T. A. Thu, K. F. Ilett, N. P. Tien, S. M. Powell, N. C. Hung, T. X. Mai, V. V. Chon, H. V. Thien, T. Q. Binh, N. V.

Kim, and T. M. E. Davis, *Am. J. Trop. Med. Hyg.*, **59**, 823–827 (1998).

740. J. Karbwang, K. Na-Bangchang, K. Congpoung, A. Thanavibul, and T. Harinasuta, *Clin. Drg. Invest.*, **15**, 37–43 (1998).

741. D. B. Bethell, P. Teja-Isavadharm, C. X. T. Phuong, P. T. T. Thuy, T. T. T. Mai, T. T. N. Thuy, N. T. T. Ha, P. T. Phuong, D. Kyle, et al., *Trans. R. Soc. Trop. Med. Hyg.*, **91**, 195–198 (1997).

742. R. Udomsangpetch, B. Pipitaporn, S. Krishna, B. Angus, S. Pukrittyakamee, I. Bates, Y. Supputtamongkol, D. E. Kyle, and N. J. White, *J. Infect. Dis.*, **173**, 691–698 (1996).

743. Y. Chen, S.-M. Zhu, H.-Y. Chen, and Y. Li, *Bioelectrochem. Bioenerget.*, **44**, 295–300 (1998).

744. T. T. Hien, *Trans. R. Soc. Trop. Med. Hyg.*, **88**, S7–S8 (1994).

745. S. Looareesuwan, *Trans. R. Soc. Trop. Med. Hyg.*, **88**, S9–S11 (1994).

746. S. Looareesuwan, P. Wilairatana, S. Vanijanonta, P. Pitisuttithum, Y. Ratanapong, and M. Andrial, *Acta Trop.*, **67**, 197–205 (1997).

747. P. Wilairatana, P. Viriyavejakul, S. Looareesuwan, and T. Chongsuphajaisiddhi, *Ann. Trop. Med. Parasitol.*, **91**, 891–896 (1997).

748. R. McGready, T. Cho, J. J. Cho, J. A. Simpson, C. Luxemburger, L. Dubowitz, S. Looareesuwan, N. J. White, and F. Nosten, *Trans. R. Soc. Trop. Med. Hyg.*, **92**, 430–433 (1998).

749. J. Karbwang, K. Na-Bangchang, A. Thanavibul, D. Bunnag, T. Chongsuphajaisiddhi, and T. Harinasuta, *Bull. World Health Org.*, **72**, 233–238 (1994).

750. E. C. Duarte, C. J. F. Fontes, T. W. Gyorkos, and M. Abrahamowicz, *Am. J. Trop. Med. Hyg.*, **54**, 197–202 (1996).

751. C. Luxemburger, F. Nosten, S. D. Raimond, T. Chongsuphajaisiddhi, and N. J. White, *Am. J. Trop. Med. Hyg.*, **53**, 522–525 (1995).

752. S. Looareesuwan, P. Wilairatana, and M. Andrial, *Am. J. Trop. Med. Hyg.*, **57**, 348–353 (1997).

753. K. Thimasarn, J. Sirichaisinthop, P. Chanyakhun, C. Palananth, and W. Rooney, *Southeast Asian J. Trop. Med. Public Health.*, **28**, 465–471 (1997).

754. A. Sabchareon, P. Attanath, P. Chanthavanich, P. Phanuaksook, V. Prarinyanupharb, Y. Poonpanich, D. Mookmanee, P. Teja-Isavadharm, D. G. Heppner, T. G. Brewer, and T.

Chongsuphajaisiddhi, *Am. J. Trop. Med. Hyg.*, **58**, 11–16 (1998).

755. M. Pudney, Antimalarials: From Quinine to Atovaquone, Conference Proceedings: Fifty Years of Antimicrobials: Past Perspectives and Future Trends, Society for General Microbiology Symposium, University of Bath, 1995, pp. 229–247.

756. S. Looareesuwan, C. Viravan, S. Vanijanonta, P. Wilairatana, P. Pitisuttithum, and M. Andrial, *Am. J. Trop. Med. Hyg.*, **54**, 210–213 (1996).

757. S. Looareesuwan, C. Viravan, S. Vanijanonta, P. Wilairatana, P. Suntharasamai, P. Charoenlarp, K. Arnold, D. Kyle, C. Canfield, and K. Webster, *Lancet*, **339**, 821–824 (1992).

758. F. Nosten, M. van Vugt, R. Price, C. Luxemburger, K. L. Thway, R. McGready, F. ter Kuile, and S. Looareesuwan, *Lancet*, **356**, 297–302 (2000).

759. Q. L. Fivelman, J. C. Walden, P. J. Smith, P. I. Folb, and K. I. Barnes, *Trans. R. Soc. Trop. Med. Hyg.*, **93**, 429–432 (1999).

760. *WHO Drug Info.*, **13**, 9–11 (1999).

761. M. J. Shmuklarsky, D. L. Klayman, W. K. Milhous, D. E. Kyle, R. N. Rossan, J. Arba L. Ager, D. B. Tang, M. H. Heiffer, C. J. Canfield, and B. G. Schuster, *Am. J. Trop. Med. Hyg.*, **48**, 377–384 (1993).

762. E. A. Abourashed and C. D. Hufford, *J. Nat. Prod.*, **59**, 251–253 (1996).

763. J. L. Maggs, L. P. D. Bishop, G. Edwards, P. M. O'Neill, S. A. Ward, P. A. Winstanley, and B. K. Park, *Drug Metab. Dispos.*, **28**, 209–217 (2000).

764. M. N. Mordi, S. M. Mansor, V. Navaratnam, and W. H. Wernsdorfer, *Br. J. Clin. Pharmacol.*, **43**, 363–365 (1997).

765. J. Karbwang, K. Na-Bangchang, K. Congpuong, P. Molunto, and A. Thanavibul, *Eur. J. Clin. Pharmacol.*, **52**, 307–310 (1997).

766. P. A. Kager, M. J. Schults, E. E. Zijlstra, B. van den Berg, and C. J. van Boxtel, *Trans. R. Soc. Trop. Med. Hyg.*, **88**, 53–54 (1994).

767. K. Na Bangchang, J. Karbwang, C. G. Thomas, A. Thanavibul, K. Sukontason, S. A. Ward, and G. Edwards, *Br. J. Clin. Pharmacol.*, **37**, 249–253 (1994).

768. J. Karbwang, K. Na-Bangchang, A. Thanavibul, and P. Molunto, *Ann. Trop. Med. Parasitol.*, **92**, 31–36 (1998).

769. M. A. Van Agtmael, S. Cheng-Qi, J. X. Qing, R. Mull, and C. J. Van Boxtel, *Int. J. Antimicrob. Agents*, **12**, 151–158 (1999).

770. S. A. Murphy, E. Mberu, D. Muhia, M. English, J. Crawley, C. Waruiru, B. Lowe, C. R. J. Newton, P. Winstanley, K. Marsh, and W. M. Watkins, *Trans. R. Soc. Trop. Med. Hyg.*, **91**, 331–334 (1997).

771. P. C. Bhattacharya, A. J. Pai-Dhungat, and K. Patel, *Southeast Asian J. Trop. Med. Public Health.*, **28**, 736–740 (1997).

772. T. E. Taylor, B. A. Wills, P. Kazembe, M. Chisale, J. J. Wirima, E. Y. Ratsma, and M. E. Molyneux, *Lancet*, **341**, 661–662 (1993).

773. O. Simooya, S. Mutetwa, S. Chandiwana, P. Neill, S. Mharakurwa, and M. Stein, *Central African J. Med.*, **38**, 257–263 (1992).

774. T. E. Taylor, B. A. Wills, J. M. Courval, and M. E. Molyneux, *Trop. Med. Int. Health*, **3**, 3–8 (1998).

775. A. Sowunmi and A. M. J. Oduola, *Acta Trop.*, **61**, 57–63 (1996).

776. A. Sowunmi, A. M. J. Oduola, and L. A. Salako, *Trans. R. Soc. Trop. Med. Hyg.*, **89**, 435–436 (1995).

777. L. A. Salako, O. Walker, A. Sowunmi, S. J. Omokhodion, R. Adio, and A. M. J. Oduola, *Trans. R. Soc. Trop. Med. Hyg.*, **88**, 13–15 (1994).

778. J. Karbwang, T. Tin, W. Rimchala, K. Sukontason, V. Namsiripongpun, A. Thanavibul, K. Na-Bangchang, P. Laothavorn, D. Bunnag, and T. Harinasuta, *Trans. R. Soc. Trop Med. Hyg.*, **89**, 668–671 (1995).

779. T. T. Hien, N. P. j. Day, N. H. Phu, N. T. Hoang, T. T. Hong, P. P. Loc, D. X. Sinh, L. V. Chuong, H. Vinh, D. Waller, T. E. A. Peto, and N. J. White, *N. Engl. J. Med.*, **335**, 76–83 (1996).

780. M. B. Hensbroek, E. Onyiorah, S. Jaffar, G. Schneider, A. Palmer, J. Frenkel, G. Enwere, S. Forck, A. Nusmeijer, S. Bennett, B. Greenwood, and D. Kwiatkowski, *N. Engl. J. Med.*, **335**, 69–75 (1996).

781. S. Murphy, W. M. Watkins, P. G. Bray, B. Lowe, P. A. Winstanley, N. Peshu, and K. Marsh, *Am. J. Trop. Med. Hyg.*, **53**, 303–305 (1995).

782. R. A. Seaton, A. J. Trevett, J. P. Wembri, N. Nwokolo, S. Naraqi, J. Black, I. F. Laurenson, I. Kevau, A. Saweri, D. G. Lalloo, and D. A. Warrell, *Ann. Trop. Med. Parasitol.*, **92**, 133–139 (1998).

783. P. E. Olumese, A. Bjorkman, R. A. Gbadegesin, A. A. Adeyemo, and O. Walker, *Acta Trop.*, **73**, 231–236 (1999).

784. A. Sowunmi and A. M. J. Odulola, *Trop. Med. Int. Health*, **2**, 631–634 (1997).

785. M. H. Pittler and E. Ernst, *Clin. Infect. Dis.*, **28**, 597–601 (1999).

786. T. Shwe and K. K. Hla, *Southeast Asian J. Trop. Med. Public Health*, **23**, 117–122 (1992).

787. A. Sowunmi, A. M. J. Oduola, O. A. T. Ogundahunsi, F. A. Fehintola, O. A. Ilesanmi, O. O. Akinyinka, and A. O. Arowojolu, *J. Obstet. Gynacol.*, **18**, 322–327 (1998).

788. B. Pradines, C. Rogier, T. Fusai, A. Tall, J. F. Trape, and J. C. Doury, *Am. J. Trop. Med. Hyg.*, **58**, 354–357 (1998).

789. B. Pradines, M. M. Mamfoumbi, D. Parzy, M. O. Medang, C. Lebeau, J. R. M. Mbina, J. C. Doury, and M. Kombila, *Parasitology*, **117**, 541–545 (1998).

790. V. Dev, N. C. K. M. M. Nayak, B. Choudhury, S. Phookan, J. S. Srivastava, O. P. Asthana, and V. P. Sharma, *Current Sci.*, **75**, 758–759 (1998).

791. F. S. El-Feraly, M. A. Al-Yahya, K. Y. Orabi, D. R. McPhail, and A. T. McPhail, *J. Nat. Prod.*, **55**, 878–883 (1992).

792. Y. M. Pu and H. Ziffer, *Heterocycles*, **39**, 649–656 (1994).

793. R. A. Vishwakarma, *J. Nat. Prod.*, **53**, 216–217 (1990).

794. Y. M. Pu and H. Ziffer, *J. Labelled Compd. Radiopharm.*, **33**, 1013–1018 (1993).

795. Y. Hu and H. Ziffer, *J. Labelled Compd. Radiopharm.*, **29**, 1293–1299 (1991).

796. A. R. Butler, L. Conforti, P. Hulme, L. M. Renton, and T. J. Rutherford, *J. Chem. Soc. Perkin Trans.*, **2**, 2089–2092 (1999).

797. L. N. Misra, A. Ahmad, R. S. Thakur, and J. Jakupovic, *Phytochemistry*, **33**, 1461–1464 (1993).

798. N. Acton and R. J. Roth, *Heterocycles*, **41**, 95–102 (1995).

799. J. K. Baker and H. T. Chi, *Heterocycles*, **38**, 1497–1506 (1994).

800. J. K. Baker, J. D. McChesney, and H. T. Chi, *Pharm. Res.*, **10**, 662–666 (1993).

801. H. T. Chi, K. Ramu, J. K. Baker, C. D. Hufford, I. S. Lee, Y. L. Zeng, and J. D. McChesney, *Biol. Mass Spectrom.*, **20**, 609–628 (1991).

802. V. Melendez, J. O. Peggins, T. G. Brewer, and A. D. Theoharides, *J. Pharm. Sci.*, **80**, 132–138 (1991).

803. C. D. Hufford, I. S. Lee, H. N. ElSohly, H. T. Chi, and J. K. Baker, *Pharm. Res.*, **7**, 923–927 (1990).

804. J. K. Baker, R. H. Yarber, C. D. Hufford, I. S. Lee, H. N. ElSohly, and J. D. McChesney, *Biomed. Environ. Mass Spectrom.*, **18**, 337–351 (1989).

805. C. D. Hufford, S. I. Khalifa, K. Y. Orabi, F. T. Wiggers, R. Kumar, R. D. Rogers, and C. F. Campana, *J. Nat. Prod.*, **58**, 751–755 (1995).

806. K. Ramu and J. K. Baker, *J. Med. Chem.*, **38**, 1911–1921 (1995).

807. J. M. Grace, A. J. Aguilar, K. M. Trotman, and T. G. Brewer, *Drug Metab. Dispos.*, **26**, 313–317 (1998).

808. G. P. Dutta, R. Bajpai, and R. A. Vishwakarma, *Pharmacol. Res.*, **21**, 415–419 (1989).

809. G. P. Dutta, R. Bajpai, and R. A. Vishwakarma, *Trans. R. Soc. Trop. Med. Hyg.*, **83**, 56–57 (1989).

810. V. Caillard, A. Beaute-Lafitte, A. G. Chabaud, and I. Landau, *Exp. Parasitol.*, **75**, 449–456 (1992).

811. S. Mohanty, S. K. Mishra, S. K. Satpathy, S. Dash, and J. Patnaik, *Trans. R. Soc. Trop. Med. Hyg.*, **91**, 328–330 (1997).

812. K. Kar, A. Nath, R. Bajpai, G. P. Dutta, and R. A. Vishwakarma, *J. Ethnopharmacol.*, **27**, 297–305 (1989).

813. R. Tripathi, G. P. Dutta, and R. A. Vishwakarma, *Am. J. Trop. Med. Hyg.*, **44**, 560–563 (1991).

814. A. Benakis, C. Schopfer, M. Paris, C. T. Plessas, P. E. Karayannakos, I. Dondas, D. Kotsarelis, S. T. Plessas, and G. Skalkeas, *Eur. J. Drug. Metab. Pharmacokinet.*, **16**, 325–328 (1991).

815. D. E. Davidson Jr., *Trans. R. Soc. Trop. Med. Hyg.*, **88**, 51–52 (1994).

816. S. Wanwimolruk, G. Edwards, S. A. Ward, and A. M. Breckenridge, *J. Pharm. Pharmacol.*, **44**, 940–942 (1992).

817. R. Ekong and D. C. Warhurst, *Trans. R. Soc. Trop. Med. Hyg.*, **84**, 757–758 (1990).

818. J. Fishwick, W. G. McLean, G. Edwards, and S. A. Ward, *Chem. Biol. Interact.*, **96**, 263–271 (1995).

819. D. L. Wesche, M. A. DeCoster, F. C. Tortella, and T. G. Brewer, *Antimicrob. Agents Chemother.*, **38**, 1813–1819 (1994).

820. T. G. Brewer, S. J. Grate, J. O. Peggins, P. J. Weina, J. M. Petras, B. S. Levin, M. H. Heiffer, and B. G. Schuster, *Am. J. Trop. Med. Hyg.*, **51**, 251–259 (1994).

821. T. G. Brewer, J. O. Peggins, S. J. Grate, J. M. Petras, B. S. Levin, P. J. Weina, J. Swearengen, M. H. Heiffer, and B. G. Schuster, *Trans. R. Soc. Trop. Med. Hyg.*, **88**, S33–S36 (1994).

822. T. G. Brewer, J. M. Petras, J. O. Peggins, Q. Li, A. J. Lin, M. Sperry, L. Figueroa, A. Aguilar, and B. G. Schuster, *Am. J. Trop. Med. Hyg.*Suppl., **49**, 292, abstract 413 (1993).

823. S. Kamchonwongpaisan, P. McKeever, P. Hossler, H. Ziffer, and S. R. Meshnick, *Am. J. Trop. Med. Hyg.*, **56**, 7–12 (1997).

824. J. M. Petras, D. E. Kyle, M. Gettayacamin, G. D. Young, R. A. Bauman, H. K. Webster, K. D. Corcoran, J. O. Peggins, M. A. Vane, and T. G. Brewer, *Am. J. Trop. Med. Hyg.*, **56**, 390–396 (1997).

825. R. F. Genovese, D. B. Newman, K. A. Gordon, and T. G. Brewer, *Neurotoxicology*, **20**, 851–860 (1999).

826. Q. G. Li, R. P. Brueckner, J. O. Peggins, K. M. Trotman, and T. G. Brewer, *Eur. J. Drug Metab. Pharmacokinet.*, **24**, 213–223 (1999).

827. A. Nonprasert, S. Pukrittayakamee, M. Nosten-Bertrand, S. Vanijanonta, and N. J. White, *Am. J. Trop. Med. Hyg.*, **62**, 409–412 (2000).

828. W. Classen, B. Altmann, P. Gretener, C. Souppart, P. Skelton-Stroud, and G. Krinke, *Exp. Toxicol. Pathol.*, **51**, 507–516 (1999).

829. R. F. Genovese, D. B. Newman, Q. Li, J. O. Peggins, and T. G. Brewer, *Brain Res. Bull.*, **45**, 199–202 (1998).

830. R. F. Genovese, D. B. Newman, J. M. Petras, and T. G. Brewer, *Pharmacol., Biochem. Behav.*, **60**, 449–458 (1998).

831. M. van Vugt, B. J. Angus, R. N. Price, C. Mann, J. A. Simpson, C. Poletto, S. E. Htoo, S. Looareesuwan, and F. Nosten, *Am. J. Trop. Med. Hyg.*, **62**, 65–69 (2000).

832. R. Price, M. Van Vugt, L. Phaipun, C. Luxemburger, J. Simpson, R. McGready, F. T. Kuile, A. Kham, T. Chongsuphajaisiddhi, N. J. White, and F. Nosten, *Am. J. Trop. Med. Hyg.*, **60**, 547–555 (1999).

833. Z. Elias, E. Bonnet, B. Marchou, and P. Massip, *Clin. Infect. Dis.*, **28**, 1330–1331 (1999).

834. A. Nontprasert, M. Nosten-Bertrand, S. Pukrittayakamee, S. Vanijanonta, B. J. Angus, and N. J. White, *Am. J. Trop. Med. Hyg.*, **59**, 519–522 (1998).

835. S. L. Smith, J. Fishwick, W. G. McLean, G. Edwards, and S. A. Ward, *Biochem. Pharmacol.*, **53**, 5–10 (1996).

836. A. K. Bhattacharjee and J. M. Karle, *Chem. Res. Toxicol.*, **12**, 422–428 (1999).

837. J. Fishwick, G. Edwards, S. A. Ward, and W. G. McLean, *Neurotoxicology*, **19**, 393–403 (1998).

838. J. Fishwick, G. Edwards, S. A. Ward, and W. G. McLean, *Neurotoxicology*, **19**, 405–412 (1998).

839. S. L. Smith, J. L. Maags, G. Edwards, S. A. Ward, B. K. Park, and W. G. McLean, *Neurotoxicology*, **19**, 557–559 (1998).

840. A. J. Lin, D. L. Klayman, and W. K. Milhous, *J. Med. Chem.*, **30**, 2147–2150 (1987).

841. R. A. Vishwakarma, R. Mehrotra, R. Tripathi, and G. P. Dutta, *J. Nat. Prod.*, **55**, 1142–1144 (1992).

842. H. A. Titulaer, W. M. Eling, and J. Zuidema, *J. Pharm. Pharmacol.*, **45**, 830–835 (1993).

843. R. Tripathi, S. K. Puri, and G. P. Dutta, *Exp. Parasitol.*, **82**, 251–254 (1996).

844. Q. G. Li, J. O. Peggins, A. J. Lin, K. J. Masonic, K. M. Trotman, and T. G. Brewer, *Trans. R. Soc. Trop. Med. Hyg.*, **92**, 332–340 (1998).

845. Q.-G. Li, J. O. Peggins, L. L. Fleckenstein, K. Masonic, M. H. Heiffer, and T. G. Brewer, *J. Pharm. Pharmacol.*, **50**, 173–182 (1998).

846. L. G. Haile and J. F. Flaherty, *Ann. Pharmacother.*, **27**, 1488–1494 (1993).

847. A. T. Hudson, *Parasitol. Today*, **9**, 66–68 (1993).

848. D. R. Williams and M. P. Clark, *Tetrahedron Lett.*, **39**, 7629–7632 (1998).

849. A. T. Hudson, M. Dickins, C. D. Ginger, W. E. Gutteridge, T. Holdich, D. B. A. Hutchinson, M. Pudney, A. W. Randall, and V. S. Latter, *Drugs Exp. Clin. Res.*, **17**, 427–435 (1991).

850. P. E. Rolan, A. J. Mercer, B. C. Weatherley, T. Holdich, H. Meire, R. W. Peck, G. Ridout, and J. Posner, *Br. J. Clin. Pharmacol.*, **37**, 13–20 (1994).

851. I. Ittarat, W. Asawamahasakda, and S. R. Meshnick, *Exp. Parasitol.*, **79**, 50–56 (1994).

852. K. K. Seymour, S. D. Lyons, L. Phillips, K. H. Rieckmann, and R. I. Christopherson, *Biochemistry*, **33**, 5268–5274 (1994).

853. W. Knecht, J. Henseling, and M. Loffler, *Chem. Biol. Interact.*, **124**, 61–76 (2000).

854. M. Fry and M. Pudney, *Biochem. Pharmacol.*, **43**, 1545–1553 (1992).

855. K. Lopez-Shirley, F. Zhang, D. Gosser, M. Scott, and S. R. Meshnick, *J. Lab Clin. Med.*, **123**, 126–130 (1994).

856. I. K. Srivastava, H. Rottenberg, and A. B. Vaidya, *J. Biol. Chem.*, **272**, 3961–3966 (1997).

857. J. Krungkrai, S. R. Krungkrai, N. Suraveratum, P. Prapunwattana, K. K. Seymour, A. E. T. Yeo, K. H. Rieckmann, and R. I. Christopherson, *Biochem. Mol. Biol. Int.*, **42**, 1007–1014 (1997).

858. D. Syafruddin, J. E. Siregar, and S. Marzuki, *Mol. Biochem. Parasitol.*, **104**, 185–194 (1999).

859. I. K. Srivastava, J. M. Morrisey, E. Darrouzet, F. Daldal, and A. B. Vaidya, *Mol. Microbiol.*, **33**, 704–711 (1999).

860. A. D. Murphy and N. Lang-Unnasch, *Antimicrob. Agents Chemother.*, **43**, 651–654 (1999).

861. L. K. Basco, O. Ramiliarisoa, and J. L. Bras, *Am. J. Trop. Med. Hyg.*, **53**, 388–391 (1995).

862. P. L. Chiodini, C. P. Conlon, D. B. A. Hutchinson, J. A. Farquhar, A. P. Hall, T. E. A. Peto, H. Birley, and D. A. Warrell, *J. Antimicrob. Chemother.*, **36**, 1073–1078 (1995).

863. S. Looareesuwan, C. Viravan, H. K. Webster, D. E. Kyle, D. B. Hutchinson, and C. J. Canfield, *Am. J. Trop. Med. Hyg.*, **54**, 62–66 (1996).

864. T. A. Shapiro, C. D. Ranasinha, N. Kumar, and P. Barditch-Crovo, *Am. J. Trop. Med. Hyg.*, **60**, 831–836 (1999).

865. A. E. T. Yeo, K. K. Seymour, K. H. Rieckmann, and R. I. Christopherson, *Biochem. Pharmacol.*, **53**, 943–950 (1997).

866. A. E. T. Yeo, M. D. Edstein, G. D. Shanks, and K. H. Rieckmann, *Parasitol. Res.*, **83**, 489–491 (1997).

867. C. J. Canfield, M. Pudney, and W. E. Gutteridge, *Exp. Parasitol.*, **80**, 373–381 (1995).

868. S. Gassis and P. K. Rathod, *Antimicrob. Agents. Chemother.*, **40**, 914–919 (1996).

869. P. E. Rolan, A. J. Mercer, E. Tate, I. Benjamin, and J. Posner, *Antimicrob. Agents Chemother.*, **41**, 1319–1321 (1997).

870. Z. Hussein, J. Eaves, and C. J. Canfield, *Clin. Pharmacol. Therapeut.*, **61**, 518–530 (1997).

871. R. E. Fowler, R. E. Sinden, and M. Pudney, *J. Parasitol.*, **81**, 452–458 (1995).

872. R. E. Fowler, P. F. Billingsley, M. Pudney, and R. E. Sinden, *Parasitology.*, **108**, 383–388 (1994).

873. S. L. Fleck, M. Pudney, and R. E. Sinden, *Trans. R. Soc. Trop. Med. Hyg.*, **90**, 309–312 (1996).

874. C. S. Davies, M. Pudney, J. C. Nicholas, and R. E. Sinden, *Parasitology*, **106**, 1–6 (1993).

875. S. Looareesuwan, J. D. Chulay, C. J. Canfield, and D. B. A. Hutchinson, *Am. J. Trop. Med. Hyg.*, **60**, 533–541 (1999) .

876. M. Mulenga, T. Y. Sukwa, C. J. Canfield, and D. B. A. Hutchinson, *Clin. Ther.*, **21**, 841–852 (1999).

877. D. G. Bustos, C. J. Canfield, E. Canete-Miguel, and D. B. A. Hutchinson, *J. Infect. Dis.*, **179**, 1587–1590 (1999).

878. T. J. Blanchard, D. C. W. Mabey, A. Hunt-Cooke, G. Edwards, D. B. A. Hutchinson, S. Benjamin, and P. L. Chiodini, *Trans. R. Soc. Trop. Med. Hyg.*, **88**, 693 (1994).

879. P. D. Radloff, J. Philipps, M. Nkeyi, D. Hutchinson, and P. G. Kremsner, *Lancet*, **347**, 1511–1514 (1996).

880. C. Gillotin, J. P. Mamet, and L. Veronese, *Eur. J. Clin. Pharmacol.*, **55**, 311–315 (1999).

881. O. Bouchaud, E. Monlun, K. Musanza, A. Fontanet, T. Scott, A. Goestschel, J. D. Chulay, J. Le Bras, M. Danis, M. Le Bras, J. P. Coulaud, and M. Gentilini, *Am. J. Trop. Med. Hyg.*, **63**, 274–279 (2000).

882. G. Anabwani, C. J. Canfield, and D. B. A. Hutchinson, *Pediatr. Infect. Dis. J.*, **18**, 456–461 (1999).

883. S. Looareesuwan, P. Wilairatana, R. Glanarongran, K. A. Indravijit, L. Supeeranontha, S. Chinnapha, T. R. Scott, and J. D. Chulay, *Trans. R. Soc. Trop. Med. Hyg.*, **93**, 637–640 (1999).

884. P. D. Radloff, J. Philipps, D. Hutchinson, and P. G. Kremsner, *Trans. R. Soc. Trop. Med. Hyg.*, **90**, 682 (1996).

885. T. Y. Sukwa, M. Mulenga, N. Chisdaka, N. S. Roskell, and T. R. Scott, *Am. J. Trop. Med. Hyg.*, **60**, 521–525 (1999).

886. G. D. Shanks, D. M. Gordon, F. W. Klotz, G. M. Aleman, A. J. Oloo, D. Sadie, and T. R. Scott, *Clin. Infect. Dis.*, **27**, 494–499 (1998).

887. J. D. Van der Berg, C. S. J. Duvenage, N. S. Roskell, and T. R. Scott, *Clin. Ther.*, **21**, 741–749 (1999).

888. B. Lell, D. Luckner, M. Ndjave, T. Scott, and P. G. Kremsner, *Lancet*, **351**, 709–713 (1998).

889. I. K. Srivastava and A. B. Vaidya, *Antimicrob. Agents Chemother.*, **43**, 1334–1339 (1999).

890. *Who Drug Info.*, **12**, 74–75 (1998).

891. A. Rake, *N. African*, **381**, 17 (2000).

892. GlaxoWellcome, http://www.malarone.com and http://www.glaxowelcome.com/pi/malarone.pdf.

893. A. E. T. Yeo, M. D. Edstein, and K. H. Rieckmann, *Acta Trop.*, **67**, 207–214 (1997).

894. A. E. T. Yeo, and K. H. Rieckmann, *Ann. Trop. Med. Parasitol.*, **91**, 247–251 (1997).

895. R. Deng, J. Zhong, D. Zhao, and J. Wang, *Yaoxue Xuebao*, **35**, 22–25, (2000).

896. J. Zhong, R. Deng, J. Wang, Q. Zheng, and K. Jiao, *Yaoxue Xuebao*, **32**, 824–829 (1997).

897. W. H. Wernsdorfer, B. Landgraf, V. A. E. B. Kilimali, and G. Wernsdorfer, *Acta Trop.*, **70**, 9–15 (1998).

898. H. Noedl, T. Allmendinger, S. Prajakwong, G. Wernsodrfer, and W. H. Wernsdorfer, *Antimicrob. Agents Chemother.*, **45**, 2106–2109 (2001).

899. S. A. Ward, An Overview of Preclinical Investigations with CGP 56697, Conference Proceedings: Conference des Médecines des Voyages, Geneva, Switzerland, 1997.

900. F. Ezzet, M. Van Vugt, F. Nosten, S. Looareesuwan, and N. J. White, *Antimicrob. Agents Chemother.*, **44**, 697–704 (2000).

901. L. K. Basco, J. Bickii, and P. Ringwald, *Antimicrob. Agents Chemother.*, **42**, 2347–2351 (1998).

902. B. Pradines, A. Tall, T. Fusal, A. Spiegel, R. Hienne, C. Rogier, J. F. Trape, J. Le Bras, and D. Parzy, *Antimicrob. Agents Chemother.*, **43**, 418–420 (1999).

903. X. Jiao, G.-Y. Liu, C.-Q. Shan, X. Zhao, X. W. Li, I. Gathmann, and C. Royce, *Southeast Asian J. Trop. Med. Public Health*, **28**, 476–481 (1997).

904. G. Lefèvre and M. S. Thomsen, *Clin. Drug Invest.*, **18**, 467–480 (1999).

905. F. Ezzet, R. Mull, and J. Karbwang, *Br. J. Clin. Pharmacol.*, **46**, 553–561 (1998).

906. M. Van Vugt, A. Brockman, B. Gemperli, C. Luxemburger, I. Gathmann, C. Royce, T. Slight, S. Looareesuwan, N. J. White, and F. Nosten, *Antimicrob. Agents Chemother.*, **42**, 135–139 (1998).

907. N. A. Kshirsagar, N. J. Gogtay, N. S. Moorthy, M. R. Garg, S. S. Dalvi, A. R. Chogle, J. S. Sorabjee, S. N. Marathe, G. H. Tilve, S. Bhatt, S. P. Sane, R. Mull, and I. Gathman, *Am. J. Trop. Med. Hyg.*, **62**, 402–408 (2000).

908. L. Von Seidlein, S. Jaffar, M. Pinder, M. Haywood, G. Snounou, B. Gemperli, I. Gathmann, C. Royce, and B. Greenwood, *J. Infect. Dis.*, **176**, 1113–1116 (1997).

909. C. Hatz, S. Abdulla, R. Mull, D. Schellenberg, I. Gathmann, P. Kibatala, H.-P. Beck, M. Tanner, and C. Royce, *Trop. Med. Int. Health*, **3**, 498–504 (1998).

910. M. van Agtmael, O. Bouchaud, D. Malvy, J. Delmont, M. Danis, S. Barette, C. Gras, J. Bernard, J.-E. Touze, I. Gathmann, R. Mull,

P. J. De Vries, P. A. Kager, T. Van Gool, C. Van Boxtel, T. A. Eggelte, J. P. Coulaud, F. Bricaire, K. Muanza, M. Le Bras, and A. Flechaire, *Int. J. Antimicrob. Agents.*, **12**, 159–169 (1999).

911. S. Looareesuwan, P. Wilairatana, W. Choke-jindachai, K. Chalermrut, W. Wernsdorfer, B. Gemperli, I. Gathmann, and C. Royce, *Am. J. Trop. Med. Hyg.*, **60**, 238–243 (1999).

912. M. Hassan Alin, A. Björkman, and W. H. Wernsdorfer, *Am. J. Trop. Med. Hyg.*, **61**, 439–445 (1999).

913. N. J. White, M. Van Vugt, and F. Ezzet, *Clin. Pharmacokinet.*, **37**, 105–125 (1999).

914. C. Chen and X. Zheng, *Biomed. Environ. Sci.*, **5**, 149–160 (1992).

915. B. R. Shao, *Chin. Med. J.*, **103**, 428–434 (1990).

916. S. Fu and S.-H. Xiao, *Parasitol. Today*, **7**, 310–313 (1991).

917. C. Chen, L.-H. Tang, and J. Chun, *Trans. R. Soc. Trop. Med. Hyg.*, **86**, 7–10 (1992).

918. P. Ringwald, F. Meche, and L. K. Basco, *Am. J. Trop. Med. Hyg.*, **61**, 446–448 (1999).

919. P. Chavalitshewinkoon, P. Wilairat, S. Gamage, W. Denny, D. Figgitt, and R. Ralph, *Antimicrob. Agents Chemother.*, **37**, 403–406 (1993).

920. S. Auparakkitanon and P. Wilairat, *Biochem. Biophys. Res. Commun.*, **269**, 406–409 (2000).

921. L. J. Wu, J. R. Rabbege, H. Nagasawa, G. Jacobs, and M. Aikawa, *Am. J. Trop. Med. Hyg.*, **38**, 30–36 (1988).

922. S. Kawai, S. Kano, C. Chang, and M. Suzuki, *Am. J. Trop. Med. Hyg.*, **55**, 223–229 (1996).

923. D. J. Naisbitt, D. P. Williams, P. M. O'Neill, J. L. Maggs, D. J. Willock, M. Pirmohamed, and B. K. Park, *Chem. Res. Toxicol.*, **11**, 1586–1595 (1998).

924. P. Ringwald, J. Bickii, A. Same-Ekobo, and L. K. Basco, *Antimicrob. Agents Chemother.*, **41**, 2317–2319 (1997).

925. S. Looareesuwan, D. E. Kyle, C. Viravna, S. Vanijanonta, P. Wilairatana, and W. H. Wernsdorfer, *Am. J. Trop. Med. Hyg.*, **54**, 205–209 (1996).

926. P. Ringwald, J. Bickii, and L. Basco, *Lancet*, **347**, 24–28 (1996).

927. P. Ringwald, J. Bickii, and L. K. Basco, *Clin. Infect. Dis.*, **26**, 946–953 (1998).

928. W. Peters, *Lancet*, **347**, 625 (1996).

929. W. Peters and B. L. Robinson, *Ann. Trop. Med. Parasitol.*, **91**, 141–145 (1997).

930. B. R. Shao, Z. S. Huang, X. H. Shi, C. Q. Zhan, F. Meng, X. Y. Ye, J. Huang, and S. H. Ha, *Southeast Asian J. Trop. Med. Public Health*, **20**, 257–263 (1989).

931. B. R. Shao, Z. S. Huang, X. H. Shi, and F. Meng, *Southeast Asian J. Trop. Med. Public Health*, **22**, 65–67 (1991).

932. B. R. Shao, C. Q. Zhan, K. Y. Chen, X. Y. Ye, B. Y. Lin, S. H. Ha, and J. X. Zhang, *Chin. Med. J.*, **103**, 1024–1026 (1990).

933. P. Ringwald, E. C. M. Eboumbou, J. Bickii, and L. K. Basco, *Antimicrob. Agents Chemother.*, **43**, 1525–1527 (1999).

934. R. Tripathi, A. Umesh, M. Mishra, S. K. Puri, and G. P. Dutta, *Exp. Parasitol.*, **94**, 190–193 (2000).

935. B. Pradines, M. M. Mamfoumbi, D. Parzy, M. O. Medang, C. Lebeau, J. R. M. Mbina, J. C. Doury, and M. Kombila, *Am. J. Trop. Med. Hyg.*, **60**, 105–108 (1999).

936. B. R. Shao, X. Y. Ye, and Y. H. Chu, *Southeast Asian J. Trop. Med. Public Health*, **23**, 59–63 (1992).

937. M. H. Alin, A. Bjorkman, and M. Ashton, *Trans. R. Soc. Trop. Med. Hyg.*, **84**, 635–637 (1990).

938. L. K. Basco and J. Le Bras, *Ann. Trop. Med. Parasitol.*, **86**, 447–454 (1992).

939. L. K. Basco and J. Le Bras, *Ann. Trop. Med. Parasitol.*, **88**, 137–144 (1994).

940. E. I. Elueze, S. L. Croft, and D. C. Warhurst, *J. Antimicrob. Chemother.*, **37**, 511–518 (1996).

941. W. Peters and B. L. Robinson, *Ann. Trop. Med. Parasitol.*, **86**, 455–464 (1992).

942. P. Winstanley, *Lancet* **347**, 2–3 (1996).

943. P. Olliaro, *Curr. Opin. Anti-Infect. Invest. Drugs*, **2**, 71–75 (2000).

944. I. Ahmad, T. Ahmad, and K. Usmanghani, *Anal. Profiles Drug Subst. Excipients*, **21**, 43–73 (1992).

945. H. P. Yennawar and M. A. Viswamitra, *Curr. Sci.*, **61**, 39–43 (1991).

946. P. Olliaro, C. Nevill, J. LeBras, P. Ringwald, P. Mussano, P. Garner, and P. Brasseur, *Lancet*, **348**, 1196–1201 (1996).

947. B. Pradines, A. Tall, D. Parzy, A. Spiegel, T. Fusai, R. Hienne, J. F. Trape, and J. C. Doury, *J. Antimicrob. Chemother.*, **42**, 333–339 (1998).

948. L. K. Penali, L. Assi-Coulibaly, B. Kaptue, D. Konan, and A. Ehouman, *Bull. Soc. Pathol. Exot.*, **87**, 244–247 (1994).

949. P. Brasseur, R. Guiguemde, S. Diallo, V. Guiyedi, M. Kombila, P. Ringwald, and P. Olliaro, *Trans. R. Soc. Trop. Med. Hyg.*, **93**, 645–650 (1999).

950. J. van Dillen, M. Custers, A. Wensink, B. Wouters, T. van Voorthuizen, W. Voorn, B. Khan, L. Muller, and C. Nevill, *Trans. R. Soc. Trop. Med. Hyg.*, **93**, 185–188 (1999).

951. H. M. McIntosh and B. M. Greenwood, *Ann. Trop. Med. Parasitol.*, **92**, 265–270 (1998).

952. K. R. Trenhome, D. E. Kum, A. K. Raiko, N. Gibson, A. Narara, and M. P. Alpers, *Trans. R. Soc. Trop. Med. Hyg.*, **87**, 464–466 (1993).

953. L. K. Basco and J. Le Bras, *Am. J. Trop. Med. Hyg.*, **48**, 120–125 (1993).

954. P. G. Bray, S. R. Hawley, and S. A. Ward, *Mol. Pharmacol.*, **50**, 1551–1558 (1996).

955. C. S. R. Hatton, T. E. A. Peto, C. Bunch, O. Pasvol, S. J. Russel, C. R. J. Singer, C. Edwards, and P. Winstanley, *Lancet*, 411–414 (1986).

956. E. G. H. Rhodes, J. Ball, and I. M. Franklin, *Br. Med. J.*, **292**, 717–718 (1986).

957. K. A. Neftel, W. Woodtly, M. Schmid, P. Frick, and J. Fehr, *Br. Med. J.*, **292**, 721–723 (1986).

958. D. Larrey, A. Castot, D. Pessayre, P. Merigot, J. P. Machayekhy, G. Feldman, A. Lenoir, B. Rueff, and J. P. Benhamou, *Ann. Intern. Med.*, **104**, 801–803 (1986).

959. H. Jewell, J. L. Maggs, A. C. Harrison, P. M. O'Neill, J. E. Ruscoe, and B. K. Park, *Xenobiotica*, **25**, 199–217 (1995).

960. R. H. Bisby, *Biochem. Pharmacol.*, **39**, 2051–2055 (1990).

961. P. A. Winstanley, J. W. Coleman, J. L. Maggs, A. M. Breckenridge, and B. K. Park, *Br. J. Clin. Pharmacol.*, **29**, 479–485 (1990).

962. A. C. Harrison, N. R. Ketteringham, J. B. Clarke, and B. K. Park, *Biochem. Pharmacol.*, **43**, 1421–1430 (1992).

963. G. Blauer, M. Akkawi, and E. R. Bauminger, *Biochem. Pharmacol.*, **46**, 1573–1576 (1993).

964. N. J. White, S. Looareesuwan, G. Edwards, R. E. Phillips, J. Karbwang, D. D. Nicholl, C. Bunch, and D. A. Warrell, *Br. J. Clin. Pharmacol.*, **23**, 127–135 (1987).

965. P. Winstanley, G. Edwards, M. Orme, and Breckenridge, *Br. J. Clin. Pharmacol.*, **23**, 1–7 (1987).

966. E. Pussard, F. Verdier, F. Faurisson, J. M. Scherrmann, J. Le Bras, and M. C. Blayo, *Eur. J. Pharmacol.*, **33**, 409–414 (1987).

967. S. R. Hawley, P. G. Bray, B. K. Park, and S. A. Ward, *Mol. Biochem. Parasitol.*, **80**, 15–25 (1996).

968. F. C. Churchill, L. C. Patchen, C. C. Campbell, I. K. Schwartz, P. Nguyen-Dinh, and C. M. Dickinson, *Life Sci.*, **36**, 53–62 (1985).

969. L. A. Salako and O. R. Idowu, *Br. J. Clin. Pharmacol.*, **20**, 307–311 (1985).

970. F. C. Churchill, D. L. Mount, and L. C. Patchen, *J. Chromatogr.*, **377**, 307–318 (1986).

971. E. Pussard, F. Verdier, M. C. Blayo, and J. J. Pocidalo, *C. R. Acad. Sci. Paris, Ser.* **301**, 383–385 (1985).

972. T. Chatterjee, A. Muhkopadhyay, K. A. Khan, and A. K. Giri, *Mutagenesis*, **13**, 619–624 (1998).

973. J. E. Ruscoe, M. D. Tingle, P. M. O'Neill, S. A. Ward, and B. K. Park, *Antimicrob. Agents Chemother.*, **42**, 2410–2416 (1998).

974. D. J. Naisbitt, J. E. Ruscoe, D. Williams, P. M. O'Neill, M. Pirmohamed, and B. K. Park, *J. Pharmacol. Exp. Therap.*, **280**, 884–893 (1997).

975. K. J. Raynes, P. A. Stocks, P. M. O'Neill, B. K. Park, and S. A. Ward, *J. Med. Chem.*, **42**, 2747–2751 (1999).

976. B. M. Kotecka, G. B. Barlin, M. D. Edstein, and K. H. Rieckmann, *Antimicrob. Agents Chemother.*, **41**, 1369–1374 (1997).

977. S. Delarue, S. Girault, L. Maes, M.-A. Debreu-Fontaine, M. Labaeïd, P. Grellier, and C. Sergheraert, *J. Med. Chem.*, **44**, 2827–2833 (2001).

978. S. R. Hawley, P. G. Bray, P. M. O'Neill, B. K. Park, and S. A. Ward, *Biochem. Pharmacol.*, **52**, 723–733 (1996).

979. H. L. Koh, M. L. Go, T. L. Ngiam, and J. W. Mak, *Eur. J. Med. Chem.*, **29**, 107–113 (1994).

980. L. M. Werbel, P. D. Cook, E. F. Elslager, J. H. Hung, J. L. Johnson, S. J. Kesten, D. J. McNamara, D. F. Ortwine, and D. F. Worth, *J. Med. Chem.*, **29**, 924–939 (1986).

981. P. M. O'Neill, D. J. Willock, S. R. Hawley, P. G. Bray, R. C. Storr, S. A. Ward, and B. K. Park, *J. Med. Chem.*, **40**, 437–448 (1997).

982. O. Müller, M. B. Van Hensbroek, S. Jaffar, C. Drakeley, C. Okorie, D. Joof, M. Pinder, and B. Greenwood, *Trop. Med. Int. Health*, **1**, 124–132 (1996).

983. P. Ringwald, J. Bickii, and L. K. Basco, *Trans. R. Soc. Trop. Med. Hyg.*, **92**, 212–213 (1998).

984. J. Y. Le Hesran, C. Boudin, M. Cot, P. Personne, R. Chambon, V. Foumane, J. P. Ver-

have, and C. De Vries, *Ann. Trop. Med. Parasitol.*, **91**, 661–664 (1997).

985. P. Brasseur, P. Agnamey, A. S. Ekobo, G. Samba, L. Favennec, and J. Kouamouo, *Trans. R. Soc. Trop. Med. Hyg.*, **89**, 528–530 (1995).

986. *WHO Drug Info.*, **10**, 178–179 (1996).

987. M. Giboda and M. B. Denis, *J. Trop. Med. Hyg.*, **91**, 205–211 (1988).

988. G. Watt, L. Loesuttivibool, G. D. Shanks, E. F. Boudreau, A. E. Brown, K. Pavanand, H. K. Webster, and S. Wechgritaya, *Am. J. Trop. Med. Hyg.*, **47**, 108–111 (1992).

989. K. H. Rieckmann, *Lancet*, 507–508 (1987).

990. C. Ohrt, T. L. Richie, H. Widjaja, G. D. Shanks, J. Fitriadi, D. J. Fryauff, J. Handschin, D. Tang, B. Sandjaja, E. Tjitra, L. Hadiarso, G. Watt, and F. S. Wignall, *Ann. Intern. Med.*, **126**, 963–972 (1997).

991. S. L. Andersen, A. J. Oloo, D. M. Gordon, O. B. Ragama, G. M. Aleman, J. D. Berman, D. B. Tang, M. W. Dunne, and G. D. Shanks, *Clin. Infect. Dis.*, **26**, 146–150 (1998).

992. R. Kiatfuengfoo, T. Suthiphongchai, P. Prapunwattana, and Y. Yuthavong, *Mol. Biochem. Parasitol.*, **34**, 109–115 (1989).

993. B. A. Gingras and J. B. Jensen, *Am. J. Trop. Med. Hyg.*, **49**, 101–105 (1993).

994. R. A. Kuschner, *Lancet*, **343**, 1396–1397 (1994).

995. W. R. J. Taylor, T. L. Richie, D. J. Fryauff, H. Picarima, C. Ohrt, D. Tang, D. Braitman, G. S. Murphy, H. Widjaja, E. Tjitra, A. Ganjar, T. R. Jones, H. Basri, and J. Berman, *Clin. Infect. Dis.*, **28**, 74–81 (1999).

996. S. T. Sadiq, *Lancet*, **346**, 881–882 (1995).

997. A. A. Divo, A. C. Sartorelli, C. Patton, and F. J. Land Blia, *Antimicrob. Agents Chemother.*, **32**, 1182–1186 (1988).

998. J. Hamzah, T. Skinner-Adams, and T. M. E. Davis, *Acta Trop.*, **74**, 39–42 (2000).

999. N. Singh and S. K. Puri, *J. Parasit. Dis.*, **20**, 45–48 (1996).

1000. P. Deloron, J. P. Lepers, L. Raharimalala, C. Dubois, P. Coulanges, and J. J. Pocidalo, *Ann. Intern. Med.*, **224**, 874–875 (1991).

1001. P. Deloron, P. Aubry, A. Ndayirabije, F. Clavier, and F. Verdier, *Am. J. Trop. Med. Hyg.*, **53**, 646–647 (1995).

1002. K. L. McLean, D. Hitchman, and S. D. Shafran, *J. Infect. Dis.*, **165**, 904–907 (1992).

1003. V. Weissig, T. S. Vetro-Widenhouse, and T. C. Rowe, *DNA Cell Biol.*, **16**, 1483–1492 (1997).

1004. W. Metzger, B. Mordmüller, W. Graninger, U. Bienzle, and P. G. Kremsner, *Antimicrob. Agents Chemother.*, **39**, 245–246 (1995).

1005. P. G. Kremsner, P. Radloff, W. Metzger, E. Wildling, B. Mordmüller, J. Philipps, L. Jenne, M. Nkeyi, J. Prada, U. Bienzle, and W. Graninger, *Antimicrob. Agents Chemother.*, **39**, 1603–1605 (1995).

1006. S. Pukrittayakamee, A. Chantra, S. Vanijanonta, R. Clemens, S. Looareesuwan, and N. J. White, *Antimicrob. Agents Chemother.*, **44**, 2395–2398 (2000).

1007. M. Vaillant, P. Millet, A. Luty, P. Tshopamba, F. Lekoulou, J. Mayombo, A. J. Georges, and P. Deloron, *Trop. Med. Int. Health*, **2**, 917–919 (1997).

1008. X.-T. Liang, D.-Q. Yu, W.-L. Wu, and H.-C. Deng, *Acta Chem. Sinica*, **I**, 215–230 (1979).

1009. L. Zhang, W.-S. Zhou, and X.-X. Xu, *JCSCC*, 523–524 (1988).

1010. X.-X. Xu, J. Zhu, D.-Z. Huang, and W.-S. Zhou, *Tetrahedron Lett.*, **32**, 5785–5788 (1991).

1011. X.-X. Xu and H.-Q. Dong, *J. Org. Chem.*, **60**, 3039–3044 (1995).

1012. X.-X. Xu and H.-Q. Dong, *Tetrahedron Lett.*, **35**, 9429–9432 (1994).

1013. P.-G. Xiao and S.-L. Fu, *Parasitol. Today*, **2**, 353–355 (1986).

1014. C.-c. Shen and L.-g. Zhuang, *Med. Res. Rev.*, **4**, 47–86 (1984).

1015. T. Tokuyasu, A. Masuyama, M. Nojima, H.-S. Kim, and Y. Wataya, *Tetrahedron Lett.*, **41**, 3145–3148 (2000).

1016. W. Hofheinz, H. Bürgin, E. Gocke, C. Jaquet, R. Masciadri, G. Schmid, H. Stohler, and H. Urwyler, *Trop. Med. Parasitol.*, **45**, 261–265 (1994).

1017. C. Jaquet, H. R. Stohler, J. Chollet, and W. Peters, *Trop. Med. Parasitol.*, **45**, 266–271 (1994).

1018. Girometta, R. Jauch, C. Ponelle, A. Guenzi, and R. C. Wiegand-Chou, *Trop. Med. Parasitol.*, **45**, 272–277 (1994).

1019. E. Weidekamm, E. Dumont, and C. Jaquet, *Trop. Med. Parasitol.*, **45**, 278–283 (1994).

1020. R. Somo-Moyou, M.-L. Mittelholzer, F. Sorenson, L. Haller, and D. Stürchler, *Trop. Med. Parasitol.*, **45**, 288–291 (1994).

1021. L. A. Salako, R. Guiguemde, M.-L. Mittelhozer, L. Haller, F. Sorenson, and D. Stürchler, *Trop. Med. Parasitol.*, **45**, 284–287 (1994).

1022. R. d. S. U. da Silva, *Rev. Soc. Brasil. Med. Trop.*, **31**, 323–324 (1998).

1023. W. Peters, B. L. Robinson, J. C. Rossier, and C. J. Jefford, *Ann. Trop. Med. Parasitol.*, **87**, 1–7 (1993).

1024. W. Peters, B. L. Robinson, J. C. Rossiter, D. Misra, and C. W. Jefford, *Ann. Trop. Med. Parasitol.*, **87**, 9–16 (1993).

1025. W. Peters, B. L. Robinson, G. Tovey, J. C. Rossier, and C. W. Jefford, *Ann. Trop. Med. Parasitol.*, **87**, 111–123 (1993).

1026. S. L. Fleck, B. L. Robinson, W. Peters, F. Thevin, Y. Boulard, C. Glenat, V. Caillard, and I. Landau, *Ann. Trop. Med. Parasitol.*, **91**, 25–32 (1997).

1027. C. W. Jefford, S. Kohmoto, D. Jaggi, G. Timári, J.-C. Rossier, M. Rudaz, O. Barbuzzi, D. Gérard, U. Burger, P. Kamalaprija, J. Mareda, and G. Bernardinelli, *Helv. Chim. Acta.*, **78**, 647–662 (1995).

1028. S. L. Fleck, B. L. Robinson, and W. Peters, *Ann. Trop. Med. Parasitol.*, **91**, 33–39 (1997).

1029. C. W. Jefford, F. FaVarger, M. G. H. Vincente, and Y. Jacquier, *Helv. Chim. Acta.*, **78**, 452–458 (1995).

1030. G. H. Posner, X. Tao, J. N. Cumming, D. Klinedinst, and T. A. Shapiro, *Tetrahedron Letters.*, **37**, 7225–7228 (1996).

1031. J. L. Vennerstrom, H.-N. Fu, W. Y. Ellis, A. L. Ager, J. K. Wood, S. L. Andersen, L. Gerena, and W. K. Milhous, *J. Med. Chem.*, **35**, 3023–3027 (1992).

1032. J. L. Vennerstrom, Y. Dong, S. L. Andersen, J. Arba L. Ager, H.-n. Fu, R. E. Miller, D. L. Wesche, D. E. Kyle, L. Gerena, S. M. Walters, J. K. Wood, G. Edwards, A. D. Home, W. G. McLean, and W. K. Milhous, *J. Med. Chem.*, **43**, 2753–2758 (2000).

1033. H.-S. Kim, Y. Shibata, Y. Wataya, K. Tsuchiya, A. Masuyama, and M. Nojima, *J. Med. Chem.*, **42**, 2604–2609 (1999).

1034. K. J. McCullough, Y. Nonami, A. Masuyama, M. Nojima, H.-S. Kim, and Y. Wataya, *Tetrahedron Lett.*, **40**, 9151–9155 (1999).

1035. Y. Nonami, T. Tokuyasu, A. Mauyama, M. Nojima, K. J. McCullough, H.-S. Kim, and Y. Wataya, *Tetrahedron Lett.*, **41**, 4681–4684 (2000).

1036. H.-S. Kim, Y. Nagai, K. Ono, K. Begum, Y. Wataya, Y. Hamada, K. Tsuchiya, A. Masuyama, M. Nojimo, and K. J. McCullough, *J. Med. Chem.*, **44**, 2357–2361 (2001).

1037. G. H. Posner, H. O'Dowd, P. Ploypradith, J. N. Cumming, S. Xie, and T. A. Shapiro, *J. Med. Chem.*, **41**, 2164–2167 (1998).

1038. D. Opsenica, G. Pocsfalvi, Z. Jurani'c, B. Tinant, J.-P. Declercq, D. E. Kyle, W. K. Milhous, and B. A. Solaja, *J. Med. Chem.*, **43**, 3274–3282 (2000).

1039. M. J. Pitt, C. J. Easton, T. A. Robertson, L. M. Kumaratilake, A. Ferrante, A. Poulos, and D. A. Rathjen, *Tetrahedron Lett.*, **39**, 4401–4404 (1998).

1040. J. Alzeer, J. Chollet, I. Keinze-Kruass, C. Hubschwerlen, H. Matile, and R. G. Ridley, *J. Med. Chem.*, **43**, 560–568 (2000).

1041. R. W. Winter, K. A. Cornell, L. L. Johnson, L. M. Isabelle, D. J. Hinrichs, and M. K. Riscoe, *Bioorg. Med. Chem. Lett.*, **5**, 1927–1932 (1995).

1042. R. W. Winter, K. A. Cornell, L. L. Johnson, M. Ignatushchenko, D. J. Hinrichs, and M. K. Riscon, *Antimicrob. Agents Chemother.*, **40**, 1408–1411 (1996).

1043. R. W. Winter, M. Ignatushchenko, O. A. T. Ogundahunsi, K. A. Cornell, A. M. J. Oduola, D. J. Hinrichs, and M. K. Riscoe, *Antimicrob. Agents Chemother.*, **41**, 1449–1454 (1997).

1044. M. V. Ignatushchenko, R. W. Winter, H. P. Bächinger, D. J. Hinrichs, and M. K. Riscoe, *FEBS Lett.*, **409**, 67–73 (1997).

1045. M. V. Ignatushchenko, R. W. Winter, and M. Roscoe, *Am. J. Trop. Med. Hyg.*, **62**, 77–81 (2000).

1046. M. R. Hollingdale, P. McCann, and A. Sjoerdsma, *Exp. Parasitol.*, **60**, 111–117 (1985).

1047. Y. Midorikawa, Q. M. Haque, and S. Nakazawa, *Chemotherapy*, **44**, 409–413 (1998).

1048. B. J. Berger, *Antimicrob. Agents Chemother.*, **44**, 2540–2542 (2000).

1049. J. Krungkai, S. R. Krungkai, and K. Phakanont, *Biochem. Pharmacol.*, **43**, 1295–1301 (1992).

1050. P. K. Rathod, *J. Pharm. Pharmacol.*, **49**, 65–69 (1997).

1051. J. L. Vennerstrom, M. T. Makler, C. K. Angerhofer, and J. A. Williams, *Antimicrob. Agents Chemother.*, **39**, 2671–2677 (1995).

1052. H. Atamna, M. Krugliak, G. Shalmiev, E. Deharo, G. Pescarmona, and H. Ginsberg, *Biochem. Pharmacol.*, **51**, 693–700 (1996).

1053. P. M. Färber, L. D. Arscott, C. H. Williams Jr., K. Becker, and R. H. Schirmer, *FEBS Lett.*, **422**, 311–314 (1998).

1054. T. Akompong, N. Ghori, and K. Haldar, *Antimicrob. Agents Chemother.*, **44**, 88–96 (2000).

1055. B. J. Berger, A. Paciorkowski, M. Suskin, W. W. Dai, A. Cerami, and P. Ulrich, *J. Infect. Dis.*, **174**, 659–662 (1996).

1056. B. J. Berger, W. W. Dai, A. Cerami, and P. Urlrich, *Biochem. Pharmacol.*, **54**, 739–742 (1997).

1057. M. L. Ancelin and H. J. Vial, *Antimicrob. Agents Chemother.*, **29**, 814–820 (1986).

1058. H. J. Vial, M. J. Thuet, M. L. Ancelin, J. R. Philippot, and C. Chavis, *Biochem. Pharmacol.*, **33**, 2761–2770 (1984).

1059. M. Calas, M. L. Ancelin, G. Cordina, P. Portefaix, G. Piquet, V. Vidal-Sailhan, and H. Vial, *J. Med. Chem.*, **43**, 505–516 (2000).

1060. G. A. McConkey, *Antimicrob. Agents Chemother.*, **43**, 175–177 (1999).

1061. A. Reichenberg, J. Wiesner, C. Weidemeyer, E. Dreiseidler, S. Sanderbrand, B. Altincicek, E. Beck, M. Schlitzer, and H. Jomaa, *Bioorg. Med. Chem. Lett.*, **11**, 833–835 (2001).

1062. P. Lombardi and A. Crisanti, *Pharmacol. Ther.*, **76**, 125–133 (1997).

1063. T. S. Skinner-Adams, T. M. E. Davis, L. S. Manning, and W. A. Johnston, *Trans. R. Soc. Trop. Med. Hyg.*, **91**, 580–584 (1997).

1064. T. Skinner-Adams and T. M. E. Davis, *Antimicrob. Agents Chemother.*, **43**, 1304–1306 (1999).

1065. A. H. Shankar, B. Genton, R. D. Semba, M. Baisor, M. Paino, S. Tamja, T. Adiguma, L. Wu, L. Rare, J. M. Tielsch, M. P. Alpers, and K. P. West, *Lancet*, **354**, 230–209 (1999).

1066. T. Skinner-Adams, H. Barrett, and T. M. E. Davis, *Trans. R. Soc. Trop. Med. Hyg.*, **93**, 550–551 (1999).

1067. E. Arzel, P. Rocca, P. Grellier, M. Labaeïd, F. Frappier, F. Guéritte, C. Gaspard, F. Marsia, A. Godard, and G. Quéguiner, *J. Med. Chem.*, **44**, 949–960 (2001).

1068. S. A. Gamage, N. Tepsiri, P. Wilairat, S. J. Wojcik, D. P. Figgitt, R. K. Ralph, and W. A. Denny, *J. Med. Chem.*, **37**, 1486–1494 (1994).

1069. J. Schrével, V. Sinou, P. Grellier, F. Frappier, D. Guénard, and P. Potier, *Proc. Natl. Acad. Sci. USA*, **91**, 8472–8476 (1994).

1070. A. Bell, *Parasitol. Today*, **14**, 234–240 (1998).

1071. J. L. Vennerstrom, W. Y. Elliz, J. Arba L. Ager, S. L. Andersen, L. Gerena, and W. K. Milhous, *J. Med. Chem.*, **35**, 2129–2134 (1992).

1072. M. Rochdi, A.-M. Delort, J. Guyot, M. Sancelme, S. Gibot, J.-G. Gourcy, G. Dauphin, C. Gumila, H. Vial, and G. Jeminet, *J. Med. Chem.*, **39**, 588–595 (1996).

1073. J. Adovelande and J. Schrevel, *Life Sci.*, **59**, 309–315 (1996).

1074. G. F. Mabeza, M. Loyevsky, V. R. Gordeuk, and G. Weiss, *Pharmacol. Ther.*, **81**, 53–75 (1999).

1075. Z. I. Cabantchik, *Parasitol. Today.*, **11**, 74 (1995).

1076. S. D. Lytton, M. Loyevsky, J. Libman, B. Mester, A. Shanzer, and Z. I. Cabantchik, *Adv. Exp. Med. Biol.*, **356**, 385–397 (1994).

1077. G. F. Mabeza, G. Biemba, and V. R. Gordeuk, *Acta Haematologica*, **95**, 78–86 (1996).

1078. R. Jambou, N. A. Ghogomu, D. Kouka-Bemba, and C. Hengy, *Trans. R. Soc. Trop. Med. Hyg.*, **86**, 11 (1992).

1079. S. R. Vippagunta, A. Dorn, A. Bubendorf, R. G. Ridley, and J. L. Vennerstrom, *Biochem. Pharmacol.*, **58**, 817–824 (1999).

1080. H. Glickstein, W. Breuer, M. Loyevsky, A. M. Konijn, J. Libman, A. Shanzer, and Z. I. Cabantchik, *Blood*, **87**, 4871–4878 (1996).

1081. J. Golenser, A. Tsafack, Y. Amichai, J. Libman, A. Shanzer, and Z. I. Cabantchik, *Antimicrob. Agents Chemother.*, **39**, 61–65 (1995).

1082. A. Tsafack, J. Libman, A. Shanzer, and Z. I. Cabantchik, *Antimicrob. Agents Chemother.*, **40**, 2160–2166 (1996).

1083. B. Pradines, F. Ramiandrasoa, L. K. Basco, L. Bricard, G. Kunesch, and J. Le Bras, *Antimicrob. Agents Chemother.*, **40**, 2094–2098 (1996).

1084. T. S. Haque, A. G. Skillman, C. E. Lee, H. Habashita, I. Y. Gluzman, T. J. A. Ewing, D. E. Goldberg, I. D. Wuntz, and J. A. Ellman, *J. Med. Chem.*, **42**, 1428–1440 (1999).

1085. J. N. Domínguez, S. López, J. Charris, L. Iarruso, G. Lobo, A. Semenov, J. E. Olson, and P. J. Rosenthal, *J. Med. Chem.*, **40**, 2726–2732 (1997).

1086. R. Li, G. L. Kenyon, F. E. Cohen, X. Chen, B. Gong, J. N. Dominguez, E. Davidson, G. Kurzban, R. E. Miller, E. O. Nuzum, P. J. Rosenthal, and J. H. McKerrow, *J. Med. Chem.*, **38**, 5031–5037 (1995).

1087. R. Li, X. Chen, B. Gong, P. M. Selzer, Z. Li, E. Davidson, G. Kurzban, R. E. Miller, E. O. Nuzum, et al., *Bioorg. Med. Chem.*, **4**, 1421–1427 (1996).

1088. Y. Zhang, X. Guo, E. T. Lin, and L. Z. Benet, *Pharmacology*, **58**, 147–159 (1999).

1089. A. Semenov, J. E. Olson, and P. J. Rosenthal, *Antimicrob. Agents Chemother.*, **42**, 2254–2258 (1998).

1090. M. Krugliak, R. Feder, V. Y. Zolotarev, L. Gaidukov, A. Dagan, H. Ginsburg, and A. Mor, *Antimicrob. Agents Chemother.*, **44**, 2442–2451 (2000).

1091. V. Sharma, A. Beatty, D. E. Goldberg, and D. Piwnica-Worms, *J. Chem. Soc., Chem. Commun.*, 2223–2224 (1997).

1092. D. E. Goldberg, V. Sharma, A. Oksman, I. Y. Gluzman, T. E. Wellems, and D. Piwnica Worms, *J. Biol. Chem.*, **272**, 6567–6572 (1997).

1093. J. Ziegler, T. Schuerle, L. Pasierb, C. Kelly, A. Elamin, K. A. Cole, and D. W. Wright, *Inorg. Chem.*, **39**, 3731–3733 (2000).

1094. C. Biot, L. Delhaes, C. M. N'Diaye, L. A. Maciejewski, D. Camus, D. Dive, and J. S. Brocard, *Bioorg. Med. Chem.*, **7**, 2843–2847 (1999).

1095. K. Chibale, J. R. Moss, M. Blackie, D. van Schalkwyk, and P. J. Smith, *Tetrahedron Lett.*, **41**, 6231–6235 (2000).

1096. B. Pradines, T. Fusai, W. Daries, V. Laloge, C. Rogier, P. Millet, E. Panconi, M. Kombila, and D. Parzy, *J. Antimicrob. Chemother.*, **48**, 179–184 (2001).

1097. V. Sharma and D. Piwnica-Worms, *Chem. Rev.*, **99**, 2545–2560 (1999).

1098. K. A. Werbovetz, *Curr. Med. Chem.*, **7**, 835–860 (2000).

1099. P. L. Olliaro and T. Yuthavong, *Pharmacol. Ther.*, **81**, 91–110 (1999).

1100. P. Olliaro and D. Wirth, *J. Pharm. Pharmacol.*, **49**, 29–33 (1997).

1101. P. J. Rosenthal, *Emerging Infect. Dis.*, **4**, 49–57 (1998).

1102. S. A. Ralph, M. C. D'Ombrain, and G. I. McFadden, *Drug Resist. Updates*, **4**, 145–151 (2001).

1103. M. Shahabuddin, S. Cociancich, and H. Zieler, *Parasitol. Today*, **14**, 493–497 (1998).

1104. R. Gozalbes, J. Gálvez, A. Moreno, and R. García-Domenech, *J. Pharm. Pharmacol.*, **51**, 111–117 (1999).

1105. J. D. Phillipson and C. W. Wright, *J. Ethnopharmacol.*, **32**, 155–165 (1991).

1106. M. J. O'Neil and J. D. Phillipson, *Rev. Latinoam. Quim.*, **20**, 111–118 (1989).

1107. M. Hamburger, A. Marston, and K. Hostettmann, B. Testa, Ed., *Advances in Drug Research*, Academic Press, New York, 1989.

1108. C. K. Angerhofer, G. M. Konig, A. D. Wright, O. Sticher, W. K. Milhous, G. A. Cordell, N. R. Farnsworth, and J. M. Pezzuto, *Advances in Natural Product Chemistry, Proceedings of the 5th International Symposium and Pakistan-US Binational Workshop on Natural Product Chemistry*, Harwood Academic Publishers, Chur, Switzerland, 1992.

1109. K. A. El Sayed, D. C. Dunbar, and M. T. Hamann, *J. Nat. Toxins*, **5**, 261–285 (1996).

1110. G. M. König, A. D. Wright, O. Sticher, C. K. Angerhofer, and J. M. Pezzuto, *Planta Med.*, **60**, 532–537 (1994).

1111. C. K. Angerhofer, J. M. Pezzuto, G. M. König, A. D. Wright, and O. Sticher, *J. Nat. Prod.*, **55**, 1787 (1992).

1112. A. D. Wright, G. M. König, C. K. Angerhofer, P. Greenidge, A. Linden, and R. Desqueyroux-Faúndez, *J. Nat. Prod.*, **59**, 710–716 (1996).

1113. G. M. König, A. D. Wright, and C. K. Angerholfer, *J. Org. Chem.*, **61**, 3259–3267 (1996).

1114. H. Miyaoka, M. Shimomura, H. Kimura, and Y. Wataya, *Tetrahedron*, **54**, 13467–13474 (1998).

1115. A. D. Wright, H. Wang, M. Gurrath, G. M. König, G. Kocak, G. Neumann, P. Loria, M. Foley, and L. Tilley, *J. Med. Chem.*, **44**, 873–885 (2001).

1116. K. K. H. Ang, M. J. Holmes, T. Higa, M. T. Hamann, and U. A. K. Kara, *Antimicrobiol. Agents Chemother.*, **44**, 1645–1649 (2000).

1117. J. D. Winkler and J. M. Axten, *J. Am. Chem. Soc.*, **120**, 6425–6426 (1998).

1118. H.-S. Kim, M. Hayashi, Y. Shibata, Y. Wataya, T. Mitamura, T. Horii, K. Kawauchi, H. Hirata, S. Tsuboi, and Y. Moriyama, *Biol. Pharm. Bull.*, **22**, 532–534 (1999).

1119. S. J. Darkin-Rattray, A. M. Gurnett, R. W. Myers, P. M. Dulski, T. M. Crumley, J. J. Allocco, C. Cannova, P. T. Meinke, S. L. Colletti, M. A. Bednarek, S. B. Singh, M. A. Goetz, A. W. Dombrowski, J. D. Polishook, and D. M. Schmatz, *Proc. Natl. Acad. Sci. USA*, **93**, 13143–13147 (1996).

1120. G. Bringmann, G. François, L. A. Assi, and J. Schlauer, *Chimia*, **52**, 18–28 (1998).

1121. G. Bringmann, W. Saeb, R. God, M. Schaffer, G. Francois, K. Peters, E.-M. Peters, P. Proksch, K. Hostettmann, and L. A. Assi, *Phytochemistry*, **49**, 1667–1673 (1998).

1122. G. François, G. Timperman, J. Holenz, L. A. Assi, T. Geuder, L. Maes, J. Dubois, M. Hanocq, and G. Bringmann, *Ann. Trop. Med. Parasitol.*, **90**, 115–123 (1996).

1123. G. François, G. Timperman, W. Eling, l. A. Assi, J. Holenz, and G. Bringmann, *Antimicrob. Agents Chemother.*, **41**, 2533–2539 (1997).

1124. G. François, G. Timperman, T. Steenackers, L. A. Assi, J. Holenz, and G. Bringmann, *Parasitol. Res.*, **83**, 673–679 (1997).

1125. Y. F. Hallock, K. P. Manfredi, J. W. Blunt, J. H. Cardellina II, M. Schaeffer, K.-P. Gulden, G. Bringmann, A. Y. Lee, J. Clardy, G. François, and M. R. Boyd, *J. Org. Chem.*, **59**, 6349–6355 (1994).

1126. Y. F. Hallock, K. P. Manfredi, J.-R. Dai, J. H. Cardellina II, R. J. Gulakowski, J. B. McMahon, M. Schäffer, M. Stahl, K.-P. Gulden, G. Bringmann, G. François, and M. R. Boyd, *J. Nat. Prod.*, **60**, 677–683 (1997).

1127. G. Bringmann and D. Feineis, *Act. Chim. Thérapeut.*, **26**, 151 (2000).

1128. G. Bringmann, J. Holenz, R. Weirich, M. Rubenacker, C. Funke, M. R. Boyd, R. J. Gulakowski, and G. Francois, *Tetrahedron*, **54**, 497–512 (1998).

1129. G. Bringmann, W. Saeb, and M. Rubenacker, *Tetrahedron*, **55**, 423–432 (1999).

1130. G. Bringmann and C. Guenther, *Synlett.*, **2**, 216–218 (1999).

1131. G. Bringmann, R. Goetz, S. Harmsen, J. Holenz, and R. Walter, *Liebigs Ann.*, **12**, 2045–2058 (1996).

1132. T. R. Hoye, M. Z. Chen, B. Hoang, L. Mi, and O. P. Priest, *J. Org. Chem.*, **64**, 7184–7201 (1999).

1133. G. Bringmann, M. Ochse, and R. Götz, *J. Org. Chem.*, **65**, 2069–2077 (2000).

1134. Y. F. Hallock, J. H. Cardellina II, M. Schaffer, G. Bringmann, G. Francois, and M. R. Boyd, *Bioorg. Med. Chem. Lett.*, **8**, 1729–1734 (1998).

1135. C. K. Angerhofer, H. Guinaudeau, V. Wongpanich, J. M. Pezzuto, and G. A. Cordell, *J. Nat. Prod.*, **62**, 59–66 (1999).

1136. J. B. Koepfli, J. F. Mead, and J. A. Brockman, *J. Am. Chem. Soc.*, **69**, 1837 (1947).

1137. F. A. Kuehl Jr., C. F. Spencer, and K. Folkers, *J. Am. Chem. Soc.*, **70**, 2091–2093 (1948).

1138. J. B. Koepfli, J. F. Mead, and J. A. Brockman Jr., *J. Am. Chem. Soc.*, **71**, 1048–1054 (1949).

1139. K. Murata, F. Takano, S. Fushiya, and Y. Oshima, *Biochem. Pharmacol.*, **58**, 1593–1601 (1999).

1140. S. Kobayashi, M. Uneo, R. Suzuki, H. Ishitani, H.-S. Kim, and Y. Wataya, *J. Org. Chem.*, **64**, 6833–6841 (1999).

1141. S. MacKinnon, T. Durst, J. T. Arnason, C. Angerhofer, J. Pezzuto, P. E. Sanchez-Vindas, L. J. Poveda, and M. Gbeassor, *J. Nat. Prod.*, **60**, 336–341 (1997).

1142. R. Dhar, K. Zhang, G. P. Talwar, S. Garg, and N. Kumar, *J. Ethnopharmacol.*, **61**, 31–39 (1998).

1143. S. P. Joshi, S. R. Rojatkar, and B. A. Nagasampagi, *J. Med. Aromat. Plant Sci.*, **20**, 1000–1002 (1998).

1144. H. Pérez, F. Díaz, and J. D. Medina, *Int. J. Pharmacogn.*, **35**, 227–231 (1997).

1145. G. François, T. Steenackers, L. A. Assi, W. Steglich, K. Lamottke, J. Holenz, and G. Bringmann, *Parasitol. Res.*, **85**, 582–588 (1999).

1146. V. Muñoz, M. Sauvain, P. Mollinedo, J. Callapa, I. Rojas, A. Gimemez, A. Valentin, and M. Mallié, *Planta Med.*, **65**, 448–449 (1999).

1147. M. Sauvain, C. Moretti, J.-A. Bravo, J. Callapa, V. Munoz, E. Ruiz, B. Richard, and L. L. Men-Olivier, *Phytother. Res.*, **10**, 198–201 (1996).

1148. K.-H. Lee, S. Tani, and Y. Imakura, *J. Nat. Prod.*, **50**, 847–851 (1987).

1149. N. Murakami, T. M. Umezome, M. Sugimoto, M. Kobayashi, Y. Wataya, and H. S. Kim, *Bioorg. Med. Chem. Lett.*, **8**, 459–462 (1998).

1150. H. S. Kim, Y. Shibata, N. Ko, N. Ikemoto, Y. Ishizuka, N. Murakami, M. Sugimoto, M. Kobayashi, and Y. Wataya, *Parasitol. Int.*, **48**, 271–274 (2000).

1151. K. Cimanga, T. De Bruyne, L. Pieters, A. J. Vlietinck, and C. A. Turger, *J. Nat. Prod.*, **60**, 688–691 (1997).

1152. C. W. Wright, J. Addae-Kyereme, A. G. Breen, J. E. Brown, M. F. Cox, S. L. Croft, Y. Gökçek, H. Kendrick, R. M. Phillips, and P. L. Pollet, *J. Med. Chem.*, **44**, 3187–3194 (2001).

1153. M. Chen, T. G. Theander, S. B. Christensen, L. Hviid, L. Zhai, and A. Kharazmi, *Antimicrob. Agents Chemother.*, **38**, 1470–1475 (1994).

1154. A. Kharazmi, M. Chen, T. Theander, and S. B. Christensen, *Ann. Trop. Med. Parasitol.*, **91**, S91–S95 (1997).

1155. M. Chen, S. B. Christensen, L. Zhai, M. H. Rasmussen, T. G. Theander, S. Frokjaer, B. Steffansen, J. Davidsen, and A. Kharazmi, *J. Infect. Dis.*, **176**, 1327–1333 (1997).

1156. C. Moretti, M. Sauvain, C. Lavaud, G. Massiot, J. A. Bravo, and V. Munoz, *J. Nat. Prod.*, **61**, 1390–1393 (1998).

1157. Centers for Disease Control and Prevention, http://www.cdc.gov/ncidod/diseases/submenus/sub_malaria.htm.

1158. WHO, http://www.who.int/home-page/.

# CHAPTER NINETEEN

# Antiprotozoal Agents

DAVID S. FRIES
TJ Long School of Pharmacy
University of the Pacific
Stockton, California

ALAN H. FAIRLAMB
Division of Biological Chemistry and Molecular Microbiology
The Wellcome Trust Biocentre
University of Dundee
Dundee, United Kingdom

## Contents

1 Introduction, 1034
2 Kinetoplastid Protozoan Infections, 1035
  2.1 Human African Trypanosomiasis
    (African Sleeping Sickness), 1035
    2.1.1 Mechanisms of Action of Drugs Used to
      Treat HAT, 1037
      2.1.1.1 Suramin Sodium, 1037
      2.1.1.2 Pentamidine, 1039
      2.1.1.3 Melarsoprol, 1041
      2.1.1.4 Eflornithine, 1043
  2.2 American Trypanosomiasis
    (Chagas' Disease), 1045
    2.2.1 Mechanisms of Action of Drugs Used to
      Treat American Trypanosomiasis, 1046
      2.2.1.1 Nifurtimox, 1046
      2.2.1.2 Benznidazole, 1047
  2.3 Leishmaniasis, 1047
    2.3.1 Mechanisms of Action of Drugs Used to
      Treat Leishmaniasis, 1049
      2.3.1.1 Sodium Stibogluconate and
        Meglumine Antimonate, 1049
      2.3.1.2 Paromomycin, 1051
      2.3.1.3 Amphotericin B-Lipid or
        Deoxycholate, 1051
      2.3.1.4 Pentamidine, 1051
      2.3.1.5 Miltefosine, 1051
3 Perspectives for New Drug Therapies for
  Kinetoplastid Parasitic Diseases, 1052
  3.1 Glycosomal and Other Carbohydrate
    Metabolism, 1052
    3.1.1 Hexose Transport as a Drug Target,
      1055
    3.1.2 Inhibition of Hexose Kinase, 1055

*Burger's Medicinal Chemistry and Drug Discovery*
Sixth Edition, Volume 5: Chemotherapeutic Agents
Edited by Donald J. Abraham
ISBN 0-471-37031-2    © 2003 John Wiley & Sons, Inc.

3.1.3 Inhibition of Glucose-6-Phosphate
         Isomerase, 1055
3.1.4 Inhibition of Phosphofructokinase, 1055
3.1.5 Inhibition of Fructose-1,6-
         Bisphosphate Aldolase, 1056
3.1.6 Inhibition of Triose-Phosphate
         Isomerase, 1056
3.1.7 Inhibition of Glyceraldehyde-3-
         Phosphate Dehydrogenase, 1056
3.1.8 Inhibition of Glycerol-3-Phosphate
         Dehydrogenase, 1056
3.1.9 Inhibition of the Dihydroxyacetone
         Phosphate/Glycero-3-phosphate
         Shuttle, 1057
3.1.10 Inhibition of Glycerol Kinase, 1057
3.1.11 Inhibition of Phosphoglycerate
         Kinase, 1058
3.1.12 Inhibition of Phosphoglycerate
         Mutase and Pyruvate Kinase, 1058
3.1.13 Additional Glycosomal Enzymes as
         Potential Drug Targets, 1058
3.2 Glycosomal Protein Import
     as a Drug Target, 1059
3.3 Inhibitors of the Pentose-Phosphate Shunt,
     1060
3.4 Polyamines and Trypanothione, 1060
     3.4.1 Inhibitors of Ornithine Decarboxylase,
         1060

3.4.2 Inhibitors of $S$-Adenosylmethionine
         Decarboxylase and Related Enzymes,
         1062
3.4.3 Inhibition of the Biosynthesis of
         Trypanothione and of Trypanothione
         Reductase, 1063
         3.4.3.1 Inhibitors of
                 Glutathionylspermidine
                 Synthetase, 1064
         3.4.3.2 Inhibitors of Trypanothione
                 Reductase, 1065
         3.4.3.3 Subversive Substrates, 1067
3.5 Inhibitors of Lipid and Glycolipid
     Metabolism, 1069
     3.5.1 Inhibitors of 14$\alpha$-Demethylase and
         Other Enzymes Related to Sterol
         Biosynthesis, 1069
     3.5.2 Inhibitiors of Prenylation and Protein
         Anchoring, 1071
3.6 Cysteine Protease Inhibitors, 1071
3.7 Inhibitors of Dihydrofolate
     Reductase/Pteridine Reductase, 1072
3.8 Inhibition of Purine Interconversions, 1073
3.9 Miscellaneous Active Compounds, 1075
4 Conclusions, 1076
5 Websites of Interest, 1076

# 1   INTRODUCTION

The topics covered in this chapter are the kinetoplastid protozoan infections, the African and American trypanosomiases, and the leishmaniases. Malaria, the remaining major parasitic protozoan infection, is covered in chapter 18 of this volume. These diseases share the common characteristic of predominately affecting people living in tropical regions of the world. The countries where the diseases are endemic are, for the most part, developing or undeveloped, and they lack the economic resources to combat the diseases successfully. Companies are unwilling to target these diseases for drug development because of the lack of economic incentive. Of the 1393 new drugs introduced between 1975 and 1999, only 16 were indicated for the treatment of tropical diseases that the World Health Organization, Tropical Drug Research (TDR) division, has placed on its list of targeted diseases (Table 19.1) (1, 2). Yet, this group of diseases ac-

counts for an estimated 772 million cases and 2.9 million deaths per year. The statistics accentuate the need for the development of new drugs for these diseases.

Most of the drugs used to treat the diseases reviewed in this chapter were developed more than 50 years ago and their origins can be traced back to the pioneering research of Paul Ehrlich and others on dyes (e.g., suramin), organic arsenicals (e.g., melarsoprol), and antimonials (e.g., Pentostam). Resistance has developed, to varying degrees, to most of the drugs. Many of the existing agents are either too expensive to manufacture, too toxic, or too difficult to administer for use in broad-scale disease treatment or prevention programs in Third World countries.

Because of similarities in kinetoplastid biochemistry and molecular biology, some drug families display selective toxicity across all of the organisms. Conversely, species that are very closely related [e.g., *Trypanosoma brucei gambiense* (*T. b. gambiense*) and *T. brucei rho-*

**Table 19.1   WHO Tropical Disease Research (TDR) List of Target Diseases**

| Disease | Infective Organism | Disease Burden[a] (DALYs) | Cases[b] (worldwide) | Deaths/Year[b] (worldwide) |
|---|---|---|---|---|
| African sleeping sickness | *Trypanosoma brucei* | 1.8 million | 300,000–500,000 | 41,000 |
| Chagas' disease | *Trypanosoma cruzi* | 680,000 | 16–18 million | 21,000 |
| Dengue fever | *Flaviviridae* viruses | 433,000 | 50 million | 12,000 |
| Leishmaniasis | *Leishmania donovani* | 1.7 million | 500,000 | 60,000 |
| Leprosy | *Mycobacterium leprae* | 141,000 | 740,000 | 2,000 |
| Lymphatic filariasis | *Wuchereria bancrofti* *Brugia timori*, and others | 5.5 million | 120 million | 0 |
| Malaria | *Plasmodium* spp. | 40.2 million | 300 million | 1,080,000 |
| Onchocerciasis | *Onchocerca volvulus* | 951,000 | 20 million | 0 |
| Schistosomiasis | *Schistosoma mansoni*, and additional *S.* spp. | 1.6 million | 200 million | 50,000 |
| Tuberculosis | *Mycobacterium tuberculosis* | 35.8 million | 62 million | 1,660,000 |

[a]DALYs, disability adjusted life years = the number of healthy years of life lost because of premature death and disability.
[b]Estimates extracted from WHO Fact Sheets.

*desiense (T. b. rhodesiense)*] may not always respond to the same drug treatments. Nonetheless, the parasites have a number of biochemical pathways that differ significantly from their mammalian hosts and therefore offer opportunities for drug development. However, the potential for the development of selective new drugs has not been aggressively pursued and provides the medicinal chemist with numerous and often validated targets for the design and synthesis of new drug candidates.

## 2   KINETOPLASTID PROTOZOAN INFECTIONS

Trypanosomes and leishmania parasites are single-cell eukaryotes belonging to the order Kinetoplastida (3, 4). This order takes its name from the kinetoplast, a round or oval body situated near the base of the flagellum. As the name suggests, its original function was ascribed to motility of the flagellum. Subsequently, the kinetoplast was identified as a specialized region of the single mitochondrion, constituting a mass of catenated small (minicircles) and large (maxicircles) circular DNA molecules that form the mitochondrial genome of these parasites (5). Unlike other organisms, maxicircle RNA transcripts undergo a remarkable process of RNA editing in which small "guide" RNAs direct the insertion

and deletion of uridine residues to produce functional mRNA (6, 7). The genera *Trypanosoma* and *Leishmania* belong to the suborder Trypanosomatina, family Trypanosomatidae, and are consequently frequently referred to as "trypanosomatids."

## 2.1   Human African Trypanosomiasis (African Sleeping Sickness)

Human African trypanosomiasis (HAT) is also known as sleeping sickness. In cattle, the disease is known as nagana. The parasite is transmitted between vertebrate hosts by the tsetse fly of the genus *Glossina*. The infection can also be spread transplacentally and by accidental contact with the blood of an infected person or animal. There are three subspecies of *Trypanosoma brucei*, two of which cause disease in humans. *T. b. gambiense*, found in West and Central Africa, causes Western African sleeping sickness. *T. b. rhodesiense* occurs in Eastern and Southern Africa, where it causes Eastern African sleeping sickness. The third subspecies, *Trypanosoma brucei brucei*, which is morphologically and biochemically indistinguishable from *T. b. rhodesiense*, does not infect humans because of a lytic factor in the high density lipoprotein fraction of human serum (8, 9). Along with the two other major species *T. congolense* and *T. vivax*, *T. b. brucei* causes nagana in cattle, sheep, and goats. Together, these species constitute the main obstacle to the cattle industry in Africa (6, 10).

*T. b. rhodesiense* is more virulent, but much less common in occurrence than is *T. b. gambiense*. *T. b. rhodesiense* causes acute infection that emerges within a few weeks of the fly's bite. As such, it is much more easy to detect than *T. b. gambiense*, which may not show symptoms for months or years. Once *T. b. gambiense* does emerge, it is already in an advanced stage and difficult to treat. HAT threatens over 60 million people in 36 countries of sub-Saharan Africa. Almost 45,000 cases of HAT were reported in 1999 and the World Health Organization (WHO) estimates that the actual number of cases is between 300,000 and 500,000 (11). *T. b. gambiense* is epidemic (20–50% infection rate in some villages) in Angola, the Democratic Republic of Congo, and southern Sudan and it is highly endemic in Cameroon, Central African Republic, Chad, Congo, Côte d'Ivoire, Guinea, Mozambique, Uganda, and the United Republic of Tanzania.

When an infected (female) tsetse fly bites an uninfected human to obtain a blood meal, trypanosomes are injected with the salivary secretions. The metacyclic trypanosomes multiply at the site of the bite, which can develop into a painful indurated swelling (trypanosomal chancre). The slender trypomastigote forms then migrate through the lymphatic system to reach the bloodstream. Here they multiply by asexual binary fission once every 5 to 10 h (12) and spread to the intercellular spaces of other tissues. In contrast to American trypanosomiasis and leishmaniasis, there are no intracellular stages. In the advanced stage of infection, some trypanosomes pass through the choroid plexus to invade the central nervous system (CNS), causing lesions that lead to the classical symptoms of sleeping sickness. In a chronic relapsing infection, the bloodstream and tissue forms display considerable variation in length and shape (pleomorphism), ranging from "long-slender," through "intermediate" to "short-stumpy." The latter forms are nondividing and are thought to be preadapted for survival when taken up into the tsetse fly midgut, where they transform into procyclic forms to perpetuate the life cycle. The parasites are remarkably well adapted

to the major circulating energy source of their hosts (glucose in mammalian blood and proline in the tsetse hemolymph). In the mammalian stages, they completely lack cytochromes and a citric acid cycle and use glucose as the sole source of energy, whereas in the insect midgut they switch on a functional cytochrome-dependent electron transport system and citric acid cycle to metabolize proline and other citric acid cycle intermediates (13, 14).

The development of host immunity is compromised by the ability of trypanosomes to vary their outer glycoprotein coat (15). Each metacyclic and bloodstream form trypanosome is coated with a monomolecular layer of glycoprotein, known as the variant surface glycoprotein (VSG). Trypanosomes possess a virtually limitless repertoire of immunological distinct forms of VSG. By periodic switching between different antigenic types, the parasite effectively evades the host's antibody response. The ability of the trypanosomes to undergo antigenic variation keeps them one step ahead of the host's immune defense system and renders development of a vaccine unlikely. Thus, chemotherapy remains the best available treatment for HAT.

HAT has a high chance of cure, if it is diagnosed early (16). Early phase symptoms are fever, headaches, joint pains, and pruritus. The second phase is the neurological phase that occurs after the parasite has invaded the CNS. The signs and symptoms of the neurological phase are confusion, sensory disturbances, loss of coordination, disturbance of the sleep cycle (from which the name of the disease originates), coma, and death. Neurological damage can be irreversible, even if late-stage treatment is successful in clearing the parasites from the patients. Left untreated, HAT is always fatal.

Drugs used to treat HAT are listed in Table 19.2 and structures are given in Fig. 19.1. Suramin (**1**) and pentamidine (**2**) are only useful for early-phase disease, whereas melarsoprol (**3**) and eflornithine (**4**) are effective against both phases. The drugs are difficult to administer and all must be given by injection over a relatively long period. In many parts of Africa, medical facilities and staff for long-term treatment and follow-up of patients do not exist. The success rate of the drugs in

**Table 19.2  Drugs Used to Treat Human African Trypanosomiasis**

| Drug | Use | Adult Dose | Pediatric Dose |
|---|---|---|---|
| Suramin sodium | Initial phase, *T. b. rhodesiense* | 100–200 mg (test dose) i.v.; then up to 1 g i.v. on days 1, 3, 7, 14, 21 | 20 mg/kg, days 1, 3, 7, 14, 21 |
| Pentamidine | Initial phase, *T. b. gambiense* | 4 mg kg$^{-1}$ day$^{-1}$; i.m.; 10 days | Same as for adults |
| Melarsoprol | Advanced phase, both subspecies | 2.0–3.6 mg kg$^{-1}$ day$^{-1}$; i.v.; 3 days–week 1<br>3.6 mg kg$^{-1}$ day$^{-1}$; i.v.; 3 day,–week 2<br>Repeat after 10–21 days | 18–25 mg/kg total in 1 month<br>0.36 mg/kg; i.v.; every 1–5 days; total 9–10 doses |
| Eflornithine | Advanced phase, *T. b. gambiense* | 400 mg kg$^{-1}$ day$^{-1}$, i.v.; in four doses × 14 days | |

treating patients with neurological phase HAT is marginal, under the best circumstances. Side effects of the drugs are severe. Supplies of the existing drugs cannot always be guaranteed, and drug companies periodically abandon production because of lack of profitability and environmental issues.

Homidium (ethidium bromide; **5**), diminazene (**6**), and isometamidium (**7**) are available to treat *T. brucei* infections in animals (17) (Fig. 19.1B). These same drugs are used to treat *Trypanosoma evansi* in camels and the *Trypanosoma equiperdum* infection in horses (17). In addition, quinapyramine (**8**) and melarsenoxide cysteamine (**9**; Mel Cy, Cymelarsan) are available for treatment of *T. evansi* infections (18).

### 2.1.1 Mechanisms of Action of Drugs Used to Treat HAT

**2.1.1.1 Suramin Sodium.** Suramin sodium (**1**) is also known as Bayer 205 and Germanin. It is a sulfonated naphthylamine polyanionic dyestuff (MW 1429), chemically related to Trypan Red and Trypan Blue, which, as their names suggest, also possess antitrypanosomal activity. Suramin was introduced in the early 1920s and it remains the drug of choice for treating the early stages of *T. b. rhodesiense* infections. The drug is highly water soluble and must be given by intravenous injection. Suramin is unstable in solution exposed to air and must be dissolved for injection at the time of use. The ineffectiveness of suramin against the neurological stages of the disease is predictable because the highly polar drug does not cross the blood-brain barrier to any signif-

icant extent. The selective toxicity of suramin is explained by the ability of trypanosomes to accumulate the drug. Structure modifications of suramin that result in decreased uptake generally cause loss of antitrypanosomal activity. Fairlamb and Bowman (19, 20) showed that in the presence of serum proteins, trypanosomes take up suramin by a receptor-mediated endocytosis at a rate that is 18-fold higher than could be explained by fluid endocytosis alone. There is evidence that low density lipoproteins are the most important binding protein for suramin endocytosis (21). However, a recent study suggests that uptake of suramin is not mediated through an LDL receptor, but by another as yet unidentified receptor mechanism (22). Fairlamb and Bowman (23) found that in the suramin-treated infected rats, where the plasma suramin reaches a level of 100 $\mu M$, the suramin taken up by *T. brucei* amounts to about 0.5 nmol/mg protein. Assuming a cellular protein concentration of about 200 mg/mL in *T. brucei* (24), the average intracellular suramin concentration is calculated to be about 100 $\mu M$, which is equivalent to the exogenous concentration. If suramin were to be retained within the endocytic compartments, then the local concentration would be much higher.

Once suramin is "inside" trypanosomes (it is debatable whether such a highly charged polysulfonated molecule could exit from the endocytic system and enter the cytosol), the mechanism of its trypanocidal action remains uncertain. Suramin has an inhibitory activity against a number of trypanosomal enzymes and multiple mechanisms are probably in-

Part A.

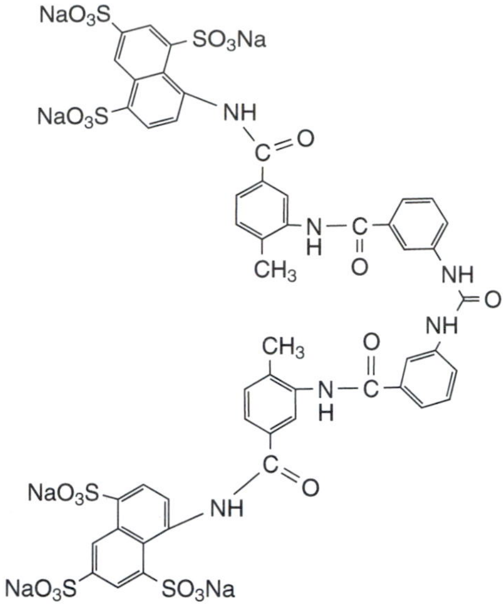

Suramin Sodium (**1**)

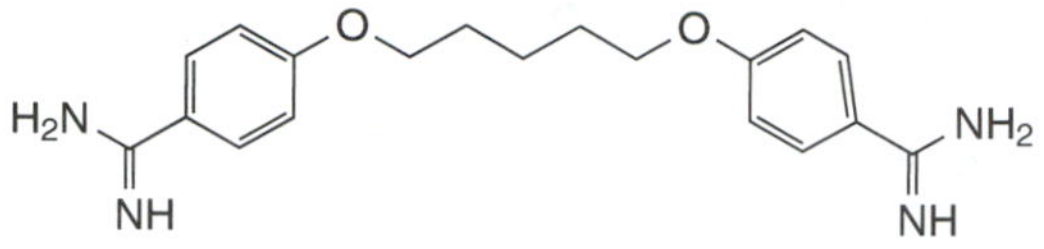

Pentamidine (**2**)

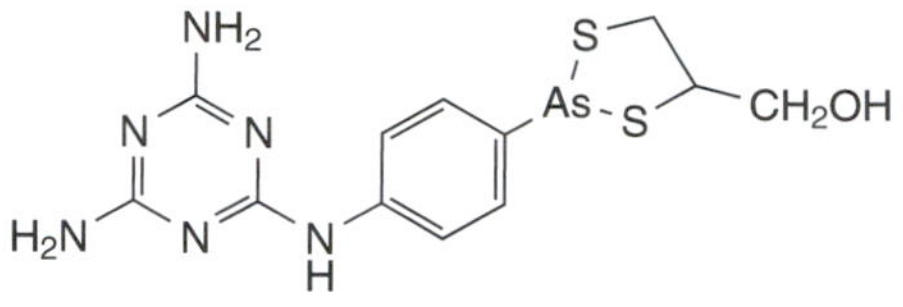

Melarsoprol (**3**)

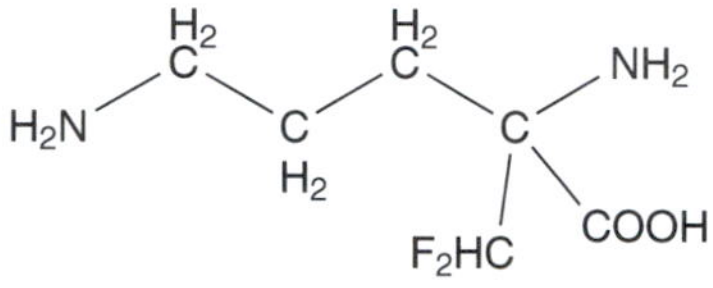

Eflornithine (**4**)

Part B.

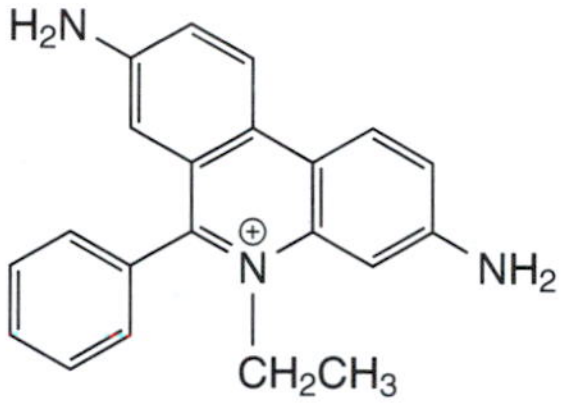

Homidium (**5**)

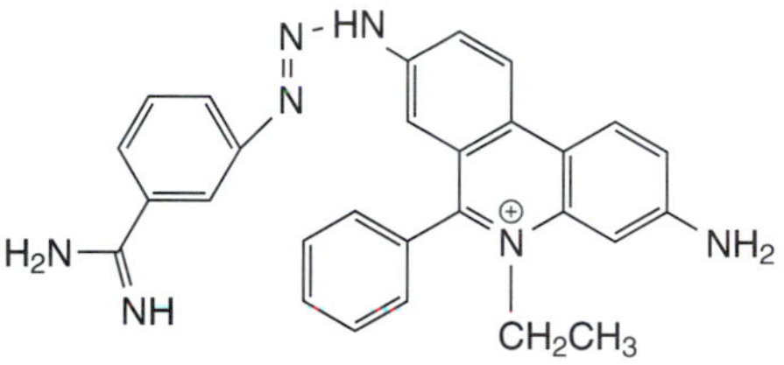

Isometamidium (**6**)

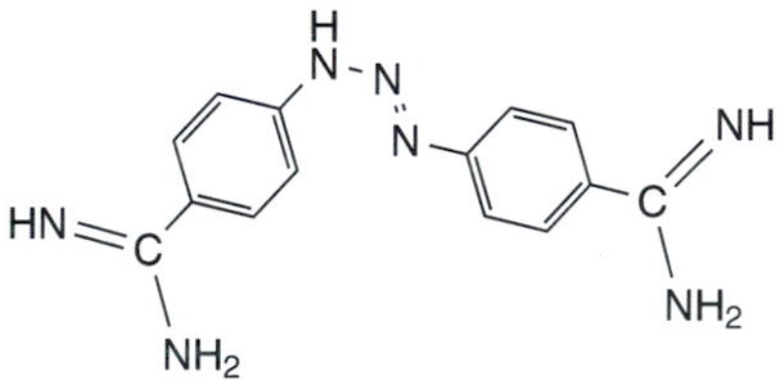

Diminazene (**7**)

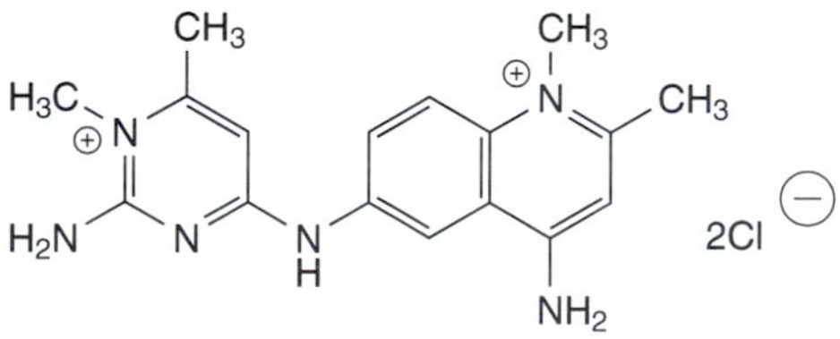

Quinapyramine (**8**)

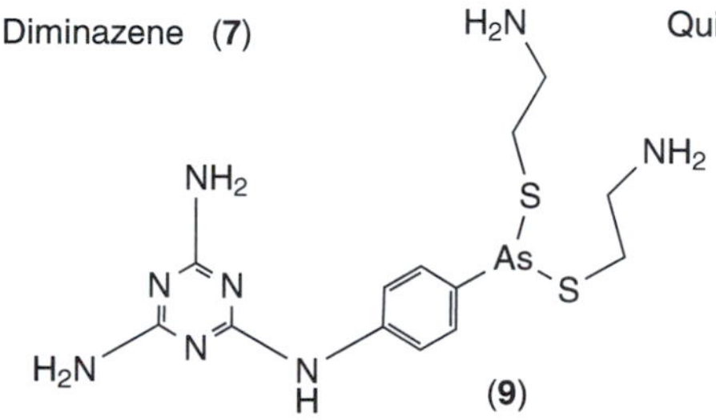

(**9**)

Melarsenoxide cystemine,
Cymelarsen

**Figure 19.1.** Agents used to treat African trypanosomiasis. (Part A) Agents for human disease. (Part B) Agents for livestock.

volved in its therapeutic effect. In *T. brucei*, suramin has been found to inhibit dihydrofolate reductase (25) and thymidine kinase (24). Morty et al. (26) implicated the inhibition of trypanosomal cytosolic serine oligopeptidase in the activity of suramin. It is also a potent inhibitor of the glycolytic enzymes in *T. brucei* (27). However, the glycolytic enzymes are contained within a membrane-bound organelle called the glycosome and it is unlikely that the highly polar suramin can penetrate the organelle's membrane to reach these enzymes (28). Inhibition of the glycolytic enzymes would result in a rapid death and lysis of the trypanosomes because they are totally dependent on glycolysis for energy production; however, it is found that trypanosomes exposed to suramin die slowly over a period of several days (20, 29). The pronounced synergism between difluoromethylornithine (eflornithine) and suramin might implicate polyamine metabolism in its mode of action (30, 31). However, no satisfactory biochemical explanation has been advanced to explain this effect. The fact that no significant resistance to suramin has been reported after 80 years of use is consistent with the drug having multiple sites and mechanisms of action.

Suramin has been found to be a potent inhibitor of HIV reverse transcriptase *in vitro*, but it lacks antiviral activity *in vivo*. It has also shown antiproliferative activity at high doses and has been tested against a number of metastatic tumors (32). Suramin's only other proven clinical use is as an antifilarial agent in treating the disease onchocerciasis. Its action against the adult worm is not understood.

Suramin is 99.7% protein bound after a 1 g intravenous dose. Tight protein binding is believed to be the cause of its long terminal half-life of 90 days (33). Suramin undergoes very little metabolism and about 80% of the drug is eliminated by slow renal clearance. Suramin's long half-life makes it an excellent prophylactic agent for trypanosomiasis. Major side effects frequently induced by suramin include vomiting, pruritus, urticaria, paresthesias, hyperesthesia of hands and feet, photophobia, and peripheral neuropathy. Occasional side effects are kidney damage, blood dyscrasias, shock, and optic atrophy.

**2.1.1.2 Pentamidine.** Pentamidine (**2**), an aromatic diamidine, was first synthesized and tested as a hypoglycemic agent in 1937. It was soon found to have antiprotozoal activity and was introduced to treat trypanosomiasis in 1941. Pentamidine is currently marketed as the di-isethionate (Pentam 300, Nebupent) and dimesylate (Lomidine) salts. It is effective against early-phase *T. b. gambiense* and is used as well against both antimony-resistant *Leishmania donovani* (34) and *Pneumocystis carinii* in patients intolerant to sulfamethoxazole-trimethoprim (35). Pentamidine as the di-isethionate salt is highly water soluble and relatively unstable in solution. It must be given by intravenous or intramuscular dosage and it should not be dissolved for injection until just before use. Diminazene (Berenil, **7**) is a related and more toxic diamidine that is approved for veterinary use (36). Although diminazene has been used for the treatment of HAT, its safety and efficacy for human use has not been properly evaluated (36). The diamidines are nearly fully protonated at both amidine groups at physiological pH and are thus not well absorbed after oral dosing, nor do they readily cross the blood-brain barrier. Their extremely poor penetration into the CNS makes them useless against the neurological phase of trypanosomiasis.

The activity of pentamidine and other diamidines is dependent on active uptake by the parasites (37, 38). One route of entry for pentamidine has been identified to be the high affinity purine 2 (P2) transporter that carries adenine and adenosine into cells (38). Melarsoprol (**3**) and the investigational trypanocide megazol use this same transporter. de Koning and Jarvis (39) identified at least one additional transporter for pentamidine in *T. b. brucei*. Molecules imported by the P2 transporter have a common structural unit (Fig. 19.2) that is recognized by the transporter. Cross-resistance between these agents is known to occur through downregulation of the P2 transporter (40, 41). Some *T. brucei* strains resistant to pentamidine have decreased ability to import diamidines (42), whereas others show no such defect (43), indicating that multiple resistance mechanisms may be involved.

The mechanism by which pentamidine causes a trypanocidal action remains undeter-

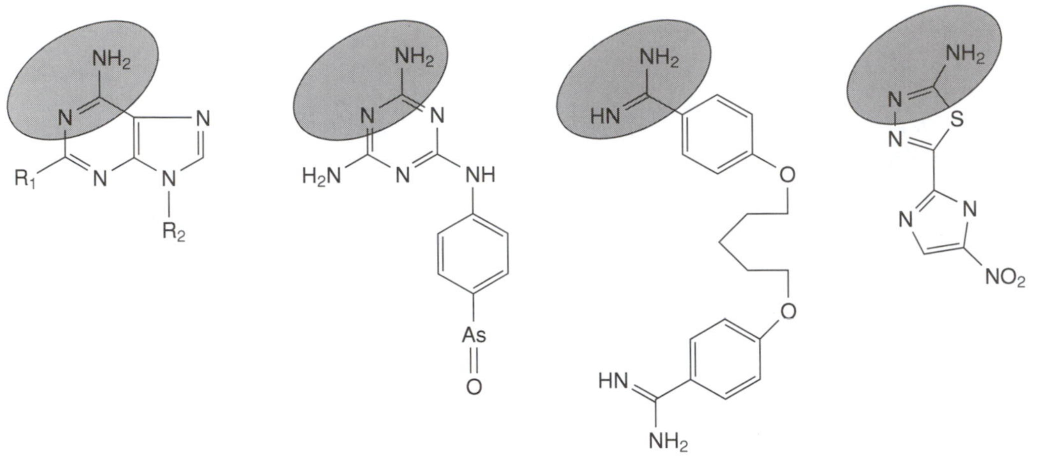

**Figure 19.2.** Substrates for active uptake of drugs into trypanosomes by the P2 transporter. The shaded area of each structure represents the common structural feature thought to be important for transporter recognition.

mined. A number of effects on trypanosome biochemistry have been shown *in vitro*, but none has been confirmed as the major cause of parasite death. Pentamidine reaches millimolar concentrations in cells and has been shown to bind to a number of negatively charged cellular components, including DNA, RNA, phospholipids, and a number of enzymes. Kapusnik and Mills (44) and Bailly et al. (45) reported the binding of pentamidine to nucleic acids and Edwards et al. (46) cocrystallized pentamidine bound to the dodecanucleotide d(CGCGAATFCGCG)2. Pentamidine is found bound, in a crosslinking manner, to the N3 positions of adenines in the 5′-AATT minor groove region of the duplex. Pentamidine is known to bind preferentially to the minor grooves of the kinetoplast DNA in *T. brucei*. It disrupts the kinetoplast DNA (47) and generates dyskinetoplastic cells that retain mitochondrial membranes but lack detectable kinetoplast DNA (48). Shapiro and Englund (49) reported that pentamidine, at 5 $\mu M$, promotes cleavage of the kinetoplast circular DNA to generate linearized DNA in a manner similar to that of a topoisomerase II inhibitor. However, the fact that trypanosomes lacking functional kinetoplasts can survive in the vertebrate host makes the importance of this mechanism uncertain.

Pentamidine has additional actions on trypanosomes. Berger et al. treated rats with pentamidine then infected the rats with *T. b. brucei* (50). After 4 h, they found a 13-fold increase in lysine content and a 2.5-fold increase in arginine in the trypanosomes. The reason for this drug effect was not determined. Benaim et al. (51) reported that pentamidine inhibits a high affinity ( $Ca^{2+}$, $Mg^{2+}$)-ATPase. Although Bitonti et al. (52) reported the *in vitro* inhibition by pentamidine of S-adenosyl-L-methionine decarboxylase (AdoMetDC), a key enzyme in the biosynthesis of polyamines, no perturbation of polyamine metabolism in intact *T. b. brucei* was noted by Berger et al. (50). Moreover, null mutants and overproducers of AdoMetDC in *L. donovani* showed no alterations in sensitivity to pentamidine, berenil, or methylglyoxal bis(guanylhydrazone), eliminating AdoMetDC as the major target for these drugs (53). Pentamidine was also found to inhibit the *in vitro* splicing of a group I intron in the transcripts of ribosomal RNA genes from *P. carinii* (54). It is possible that RNA editing in trypanosomes could also be affected by this drug.

Frequently observed side effects of pentamidine include hypotension, hypoglycemia that may lead to diabetes mellitus, vomiting, blood dyscrasias, renal damage, pain at the injection site, and gastrointestinal disturbances. Occasional side effects are shock, hypocalcemia, hepatotoxicity, cardiotoxicity, delirium, and rash. Rare side effects from pentamidine are Herxheimer-type reactions that include anaphylaxis, acute pancreatitis, hyperkalemia, and ventricular arrhythmias.

*2.1.1.3 Melarsoprol.* Melarsoprol (Mel B; Arsobal; **3**) is the 2,3-dimercaptopropanol adduct of melarsen oxide. It is a 3:1 mixture of two diastereomers (55). Melarsoprol was introduced as an antitrypanosomal agent in 1949. It is water insoluble and is formulated as a 3.6% (w/v) solution in propylene glycol for intravenous dosage through use of a glass syringe. When introduced, it was the only drug effective against late-stage HAT caused by either *T. b. gambiense* or *T. b. rhodesiense*. Resistance to melarsoprol in both subspecies has developed and eflornithine (difluoromethylornithine) is now used for late-stage *T. b. gambiense* infection. The drug crosses the blood-brain barrier in sufficient amounts to kill parasites in the CNS.

Melarsoprol is a prodrug. It is converted in patients to melarsen oxide with a half-life of 30 min. Melarsen oxide disappears relatively rapidly from the serum ($t_{1/2}$ = 3.8 h) and no other free arsenic or organoarsenic compound can be detected by atomic absorption spectroscopy or high pressure liquid chromatography analysis after 24 h (56). In rats, significant biliary excretion of melarsoprol as the glucuronide or as the metabolite melarsen-diglutathione conjugate has been reported (57). Melarsen oxide binds rapidly and reversibly to serum proteins (apparently as unidentified protein-S-As or protein-N-As complexes), which serve as a reservoir from which the melaminophenylarsenical is released with a half-life of 35 h in serum and 120 h in cerebrospinal fluid. Trypanocidal levels of the melaminophenylarsenical do remain in the serum and other tissues and are detectable by bioassay. Selective concentration of trace amounts of free melarsoprol or melarsen oxide by the trypanosome P2-purine transporter may be important in selective toxicity (58). Figure 19.3 summarizes the mechanism of action of melarsoprol.

The controversy about melarsoprol's mechanism of action concerns the identification of the protein-arsenical complex in trypano-somes that is the most important for trypanocidal activity. Historically, the dominant hypothesis for melarsoprol's mechanism of action has been the blockage of enzymes essential for glycolysis in bloodstream forms of the African trypanosome (59). This was inferred to result from inhibition of pyruvate kinase, given that phosphoenolpyruvate accumulated in treated cell suspensions. In fact, this inhibition occurs indirectly because of depletion of fructose-2,6-bisphosphate [Fru(2,6)P$_2$], a potent activator of pyruvate kinase (60). Melarsen oxide was found to inhibit trypanosomal 6-phosphofructo-2-kinase (PFK 2) ($K_i$ < 1 $\mu M$) more than fructose-2,6-bisphosphatase ($K_i$ = 2 $\mu M$), leading to a depletion of the activator. However, the observed lytic effects of the drug preceded depletion of Fru(2,6)P$_2$, and the authors concluded that inhibition of glycolysis is not the cause, but rather the consequence, of lysis (60).

More recently, Fairlamb et al. (61) published that melarsen oxide or melarsoprol can form a stable adduct with trypanothione (**11**; Mel T), a bis(glutathionyl)spermidine adduct essential for redox homeostasis in trypanosomes (Fig. 19.3). Mel T inhibits *T. b. brucei* trypanothione reductase ($K_i$ = 17.2 $\mu M$), a key enzyme in regulating the thiol-disulfide state of trypanothione (62, 63). The combination of depletion of trypanothione and inhibition of trypanothione reductase may be sufficient to kill trypanosomes.

Melarsen oxide readily forms coordination complexes with a variety of dithiol-containing enzymes and lipoic acid (64). In addition to the arsenic-protein complexes described earlier, melarsen forms stable complexes with dihydrolipoamide dehydrogenase and a number of other proteins in which cysteine residues are positioned close together. The drug may be a nonspecific inhibitor of many different enzymes, which may explain the many toxic side effects.

Melarsoprol causes reactive encephalopathy in 5 to 10% of the patients treated and has a fatal outcome in 10 to 50% of those patients (65). There is some controversy about the cause of the encephalopathy. One theory places the cause on covalent binding of the drug or its metabolites to proteins that then trigger immune reactions (66, 67). Another

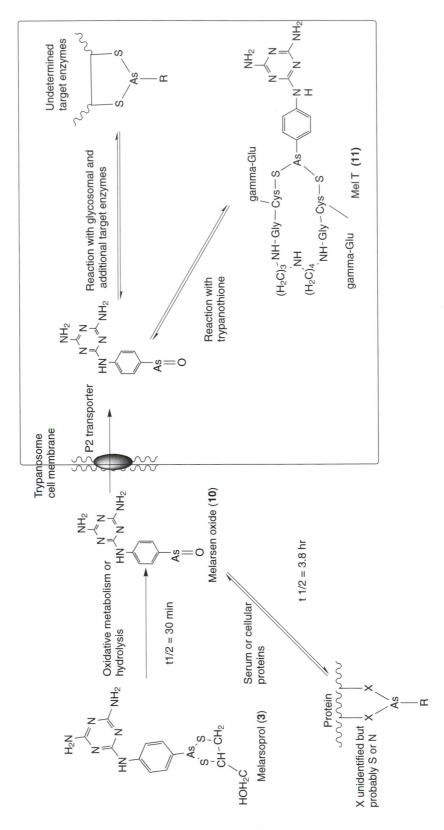

**Figure 19.3.** Mechanism of action of melarsoprol.

theory implicates the release of large concentrations of antigens in the CNS from trypanosomes dying from melarsoprol therapy as the cause of the encephalopathy (68, 69). Interestingly, the occurrence of melarsoprol-induced encephalopathy has been shown to be unrelated to the stage of the disease (peripheral or neurological) or the dosing regimen of the drug (70). Encephalopathy is most likely to occur on the second injection of the drug. These observations support a covalent binding mechanism for induction of encephalopathy. Attempts to reduce the incidence of encephalopathy with a new therapeutic dosing schedule proved unsuccessful (71). Other frequent side effects reported with melarsoprol therapy are peripheral neuropathy, hypertension, myocardial damage, albuminuria, hepatotoxicity, vomiting, and abdominal colic (72).

**2.1.1.4 Eflornithine.** Eflornithine [Ornidyl; DFMO; $(R,S)$-$\alpha$-difluoromethylornithine, **4**] was reported by Metcalf et al. (73) as a potential mechanism-based inhibitor of ornithine decarboxylase (ODC). ODC is an essential and rate-limiting enzyme in the pathway leading to the biosynthesis of polyamines. Initial studies with eflornithine in cell culture showed extensive depletion of the polyamines putrescine and spermidine and eflornithine exhibited a strong cytostatic action. Eflornithine failed in early *in vivo* anticancer screens, probably because of the ability of mammalian cells to induce rapid ODC biosynthesis and to induce transporters for the uptake of polyamines from extracellular fluids (74). Interestingly, eflornithine has been found to be effective in the treatment and chemoprevention of colorectal cancers (75). Bacchi et al. (76) reported eflornithine to have antitrypanosomal activity in 1980, and it was approved in the United States in 1990 and Europe in 1991 for treatment of early- and late-stage *T. b. gambiense* infections (77). Because of its expense and the need for continuous intravenous infusion, it is used only in late-stage disease. Eflornithine lacks effectiveness against *T. b. rhodesiense* and all other *Trypanosoma* spp. on which it has been tested.

Poulin et al. (78) confirmed the proposed mechanism-based enzyme inhibitor action of eflornithine by isolating a covalent adduct between the drug and residue cysteine 360 in

mouse ODC (Fig. 19.4). The adduct is consistent with the predicted mechanism of eflornithine as a suicide inhibitor, where the cysteine 360 acts as an attacking neucleophile in the active pocket of mouse ODC. Human ODC shares 99% identity with mouse ODC (79), and it is presumed to be inhibited by eflornithine in the same manner.

The gene encoding *T. b. brucei* ornithine decarboxylase has been cloned and sequenced, and the recombinant protein expressed in transformed *Escherichia coli* (80, 81). The protein is a homodimer, with an estimated subunit molecular mass of 45 kDa compared to the 53-kDa subunits in the mouse ODC homodimer (82). There is 61.5% sequence identity and 90% similarity between *T. b. brucei* and mouse ODC. The crystal structure of *T. b. brucei* ODC in complex with eflornithine indicates that the drug forms a Schiff base with the pyridoxal phosphate cofactor and is covalently attached to cysteine 360 (83), as predicted for the mammalian enzyme (78).

The reason for the selective toxicity of eflornithine is not fully understood. The susceptibility of ODC from *T. b. brucei* and mouse to inhibition by eflornithine is remarkably similar ($K_i$ values are 220 and 39 $\mu M$ and $k_{inact}$ values are $4.3 \times 10^{-3}$ s$^{-1}$ and $3.7 \times 10^{-3}$ s$^{-1}$ for trypanosome and mouse ODC, respectively) (73, 80). The major difference between the two enzymes is an extra 36 amino acid peptide at the C-terminus of the mouse enzyme (81). This extension contains a PEST sequence, which is found in many eukaryotic proteins that are known to turn over rapidly *in vivo* (84, 85). Because the *T. b. brucei* ODC is highly stable and does not turn over at a detectable rate, it has been proposed that differential toxicity is attributable to the differences in turnover between host and parasite ODCs (81). Expression of *T. b. brucei* ODC in CHO cells demonstrated that the intracellular stability is an intrinsic feature of the trypanosomal enzyme rather than the cellular environment (86). In addition, when full-length mouse ODC was expressed in *T. b. brucei*, it was found to be stable, suggesting that the pathway for degradation was lacking or inactive in *T. b. brucei* (87, 88). Another possibility, which would account for the failure of eflornithine as an anticancer agent, is that mammalian cells are able to bypass inhibition of ODC by taking up putrescine and spermidine from the extracellular

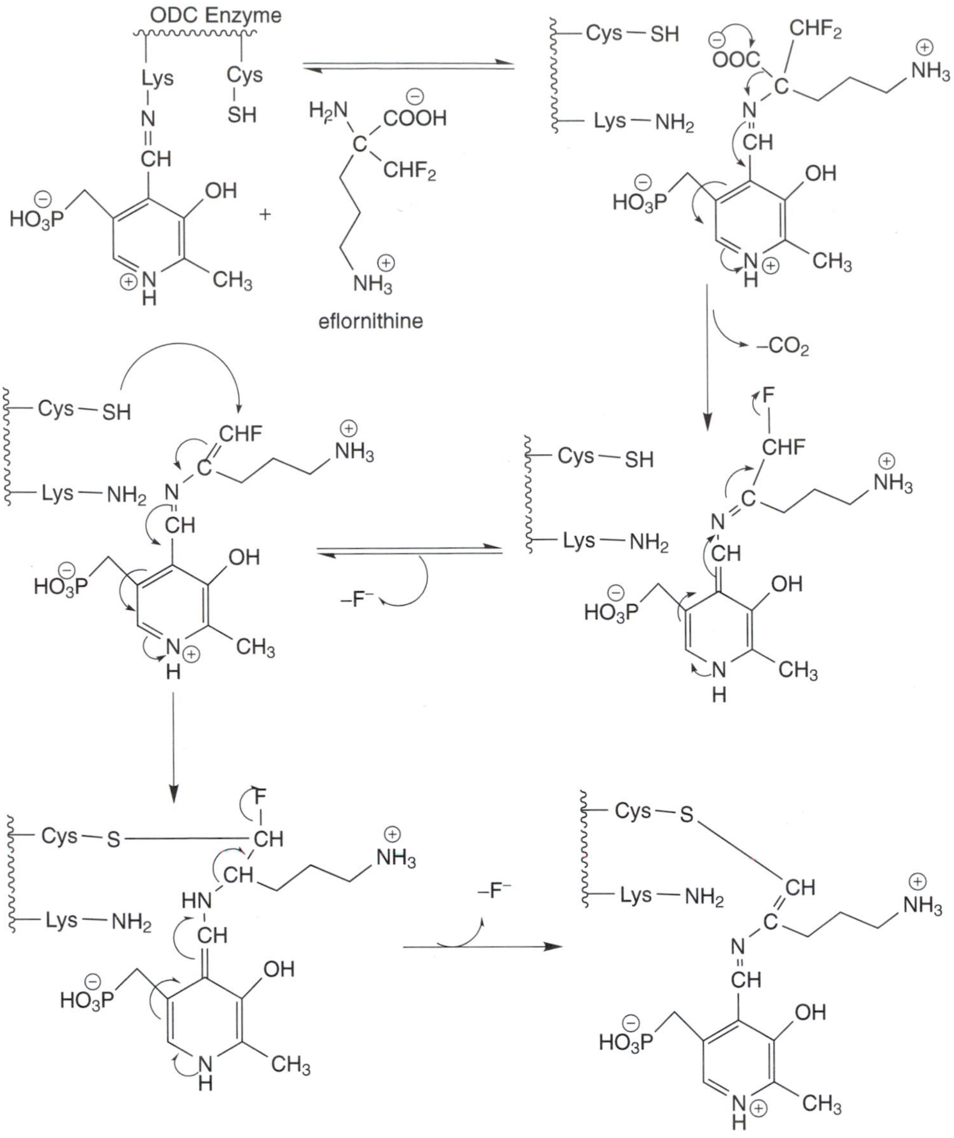

**Figure 19.4.** Proposed mechanism of action for eflornithine: mechanism-based inhibition of ornithine decarboxylase.

medium. Serum concentrations of polyamines are low (89) and *T. b. brucei* has an extremely low capacity for polyamine transport compared to that of *T. cruzi*. Thus, failure to import polyamines could contribute to the selective toxicity of eflornithine toward *T. gambiense*.

The reason for the clinical ineffectiveness of eflornithine against *T. b. rhodesiense* is not fully understood. Drug-susceptible *T. b. gam-*

*biense* and refractory *T. b. rhodesiense* strains showed no difference in uptake of eflornithine or in uptake of physiologically relevant concentrations of putrescine (90). ODC inhibition by eflornithine was not significantly different between the two species. However, ODC activity was threefold higher in the eflornithine-refractory *T. b. rhodesiense* and appeared to have a higher turnover rate than that of *T. b.*

gambiense (90). Because ODC from *T. b. bru-cei* and *T. b. gambiense* are identical at the amino acid level, other cellular factors must be involved in the difference in turnover rate. Bacchi and coworkers also compared drug-susceptible and refractory clinical isolates of *T. b. rhodesiense* (91). No clear consensus emerged from their studies; however, the data suggested that alterations in AdoMet metabolism may be responsible for resistance.

Eflornithine depletes putrescine and spermidine from *T. b. brucei* both *in vitro* and *in vivo* (92). The trypanosomes stop growing and transform to a short, stumpy form that apparently lacks the ability to alter its VSG and is eventually killed by the host's immune system (93). The drug effects can be reversed by putrescine *in vivo* (92, 94) and *in vitro* (92, 95) but not by L-ornithine *in vitro* (95). Depletion of polyamines in eflornithine-treated trypanosomes causes a 40 to 60% decrease in the content of trypanothione and monoglutathionyl-spermidine, respectively (96). It is possible that reduction in trypanothione contributes to eflornithine's mechanism of action as well as accounting for the synergistic effect with arsenical drugs *in vivo*.

## 2.2 American Trypanosomiasis (Chagas' Disease)

American trypanosomiasis, also known as Chagas' disease, is caused by the hemoflagellate *Trypanosoma cruzi*. Carlos Chagas, a Brazilian physician, first described the disease in 1909 (97). The vectors are the bloodsucking triatomine (cone-nosed) bugs of several genera (*Triatoma, Rhodnius, Panstrongylus*), commonly known as reduviid bugs, kissing bugs, and assassin bugs. The type of vector in any particular case will depend on the living conditions (i.e., mud-walled vs. thatched vs. wood huts, etc.) and the alternate animal host involved. Wild and domestic animals of all sorts (cats, dogs, opossums, armadillos, rodents, etc.) are reservoirs for the parasite, which can be transmitted to humans through contaminative inoculation following an insect bite. The disease can also be transmitted by blood transfusions, and this has been a major cause of the spread of the disease in some countries. Transplacental infection (98) and posttransplantation infection (99) have also been documented.

Chagas' is a disease of the Americas. It has a geographic range from Argentina in South America to parts of southern and western United States. It is considered a tropical disease because most of the persons infected with the disease live in poor areas of tropical South and Central America. It is estimated that 24 million people within the geographical area are infected with the parasite and five to six million have developed incurable chronic disease. In the United States between 50,000 to 300,000 people (mostly immigrants) are estimated to be carriers of the disease (98). Most of the persons were infected with the disease before they immigrated to the United States from Central or South America. Spread of the disease in the United States is minimal at this time because of the paucity of vectors and generally higher public health standards than are found in endemic countries (98).

Excellent reviews on the life cycle of *T. cruzi* can be found in Kirchhoff (98) and Tyler and Engman (100). In brief, the parasite, in its metacyclic trypomastigote life form, is transferred to the human, or alternate host animal, from the excreta of the triatomine bug through infection of the bug's bite wound or through the mucous membrane of the host. In humans, a local swelling or skin nodule called a chagoma forms at the site of the insect bite. Here the metacyclic trypomastigotes invade tissue cells and transform into amastigotes. The amastigote stage resides in the host cytoplasm, where it rapidly multiplies to fill the host cell to the bursting point. Amastigotes then transform into actively motile trypomastigotes, which rupture the host cell and escape into the bloodstream. The nondividing extracellular trypomastigotes are then spread throughout the body, after which they invade smooth muscle tissue and ganglia of the heart, esophagus, and colon where they transform back into the intracellular proliferative amastigote stage. Infected tissue can lead to chronic disease manifested as dysrhythmias, cardiomyopathy, megaesophagus, megacolon, and occasionally meningoencephalitis. The initial infective phase of the disease lasts for weeks to months and causes fever, local swelling, skin

**Table 19.3   Drugs Used to Treat American Trypanosomiasis (Chagas' Disease)**

| Drug | Adult Dose | Pediatric Dose |
|------|-----------|----------------|
| Nifurtimox | 8–10 mg kg$^{-1}$ day$^{-1}$; p.o. | 1–10 years old; 15–20 mg kg$^{-1}$ day$^{-1}$; p.o.<br>11–16 years old; 12.5–15 mg kg$^{-1}$ day$^{-1}$ p.o.<br>In four divided doses for 90–120 days for all |
| Benznidazole | 5–7 mg kg$^{-1}$ day$^{-1}$; p.o. | 1–12 years old; 10 mg kg$^{-1}$ day$^{-1}$; p.o.<br>In two divided doses for 30–90 days for all |

rashes, myocarditis, and hepatosplenomegaly. About 10% die in the acute phase of infection, and those who survive enter an asymptomatic phase of chronic infection that can last for more than a decade. Up to 70% of infected individuals never show signs of chronic disease. The remainder develop cardiac and/or gastrointestinal symptoms that often end in death. Sudden death of young adults (age 35–45 years), caused by cardiac arrhythmias induced by *T. cruzi* infection, is a common occurrence in many parts of South America. Chagas' disease is extremely virulent and difficult to treat in patients who also have HIV/AIDS or are immunosuppressed for other reasons.

Vector eradication has proved extremely effective in controlling Chagas' disease in specific geographic areas. Programs that use insecticides to eradicate vectors in Argentina, Brazil, and other "Southern Cone" countries are currently in progress. These programs, coupled with blood bank screening, have greatly reduced the incidence of Chagas' disease in specific geographic areas of South America and provide hope for elimination of the disease (101).

Dosage regimens for nifurtimox (**12**) and benznidazole (**13**), the only two drugs approved to treat *T. cruzi* (American trypanosomiasis), are given in Table 19.3 and structures are given in Fig. 19.5. Both of these drugs give modest cure rates of 50% or less in patients with chronic disease and 70% in the acute stage of infection. Both have a number of toxic side effects. Another disadvantage of the drugs is the requirement for prolonged treatment periods. Improved drugs to treat American trypanosomiasis are desperately needed.

### 2.2.1 Mechanisms of Action of Drugs Used to Treat American Trypanosomiasis

**2.2.1.1 Nifurtimox.** Nifurtimox (Lampit; Bayer 2502, **12**) is a nitrofuran derivative in-

troduced to treat American trypanosomiasis in 1976 (102). The drug is no longer marketed and it appears unlikely that it will continue to be used. It remains the only drug approved in the United States for the treatment of American trypanosomiasis and is available from the Center for Disease Control in Atlanta. In addition to its action against *T. cruzi*, nifurtimox has activity against *T. brucei* and is being tested in combination with other agents in the treatment of drug-resistant strains of *T. b. rhodesiense*.

The exact mechanism of action for nifurtimox remains uncertain. It is clear that nifurtimox requires one electron reduction to form the nitro ion radical (Fig. 19.6). Both NADH and NADPH can serve as electron donors, but the enzyme(s) catalyzing this reaction are not known. The nitro ion radical is thought to reduce molecular oxygen to form superoxide anion and regenerate the parent nitro compound through redox cycling. Overproduction of superoxide anion swamps the cell's capacity to remove it and other reactive oxygen species

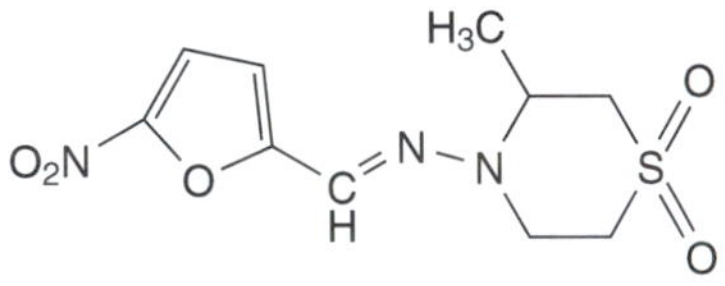

Nifurtimox (**12**)

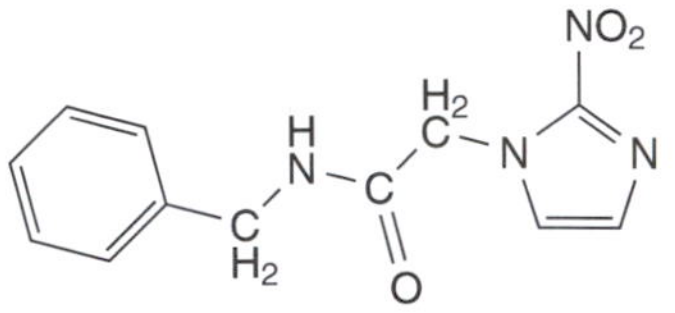

Benznidazole (**13**)

**Figure 19.5.** Agents used to treat American trypanosomiasis (Chagas' disease).

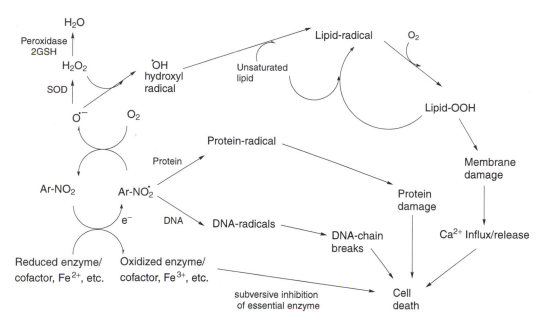

**Figure 19.6.** Mechanism of action for nitroimidazoles and nitrofurans ( $Ar-NO_2$): multiple pathways to cellular toxicity through free-radical generation and induction of oxidative stress. SOD, superoxide dismutase.

(e.g., $H_2O_2$ and $OH^-$) are formed, resulting in lipid peroxidation and damage to membranes, proteins, and DNA. Involvement of reactive oxygen in the mechanism of action of this organism is consistent with the aerobic environment in which they live. *T. cruzi* is reported to be relatively susceptible to oxidative damage (103). Nifurtimox weakly inhibits trypanothione reductase, an essential enzyme for protection of trypanosomes from oxygen free-radical damage (104).

Nifurtimox undergoes extensive metabolism; however, neither the structures of the metabolites nor their antitrypanosomal activity have been determined. The drug has a 3-h elimination half-life with only 0.5% of the drug being excreted unchanged in the urine. Gastrointestinal side effects include nausea, vomiting, abdominal pain, anorexia, and weight loss. Neurological side effects are restlessness, insomnia, paresthesias, polyneuritis, and seizures. Allergic reactions, including dermatitis, fever, icterus, pulmonary infiltrates, and anaphylaxis, may occur. The drug is generally better tolerated in children than in adults.

**2.2.1.2 Benznidazole.** Benznidazole (Rochagan, Roche 7–1051, **13**) is a nitroimidazole analog introduced in 1979 (105). Benznidazole is recognized as the drug of choice for treatment of American trypanosomiasis. Cure rates with benznidazole treatment are nearly identical to those obtained with nifurtimox. Toxic side effects from benznidazole, although still severe, are slightly less than with nifurtimox.

The mechanism of action of benznidazole is similar to that of nifurtimox. Activation of benznidazole, by one-electron transfer from cellular components, is required. The resultant nitro anion radicals cause undetermined cellular damage that leads to, or facilitates, trypanosome eradication. Many isolates of *T. cruzi* are inherently resistant to both nifurtimox and benznidazole. The mechanism of resistance is not known.

### 2.3 Leishmaniasis

More than 20 species of *Leishmania*, distributed worldwide in tropical and subtropical regions, are known to be pathogenic to humans. The World Health Organization estimates that 350 million people are at risk of the disease (106). The severity of the disease depends on the *Leishmania* species causing the disease and the immune response that can be mounted by the host. The most severe infection, known as visceral leishmaniasis or kala-azar disease in India, is caused by *Leishmania donovani* (East India, Bangladesh, Sudan, Northeast Africa). Kala-azar is Hindi for black fever, the name coming from the fever and hyperpigmentation of the skin that often occur as symptoms of the disease. *Leishmania chagasi* (South America) and *Leishmania infantum* (Mediterranean and Middle East) can also develop into visceral infection. Worldwide cases of visceral leishmaniasis are estimated to be 500,000 per year. Cutaneous and mucosal forms of leishmaniasis are more common, but cause less severe pathology than that of visceral forms, but can still be highly disfiguring. Important genera responsible for cutaneous and mucosal leishmaniasis are *Leishmania major* and *Leishmania mexicana*. *Leishmania braziliensis* and *Leishmania panamensis* are known to have a high risk for development into mucocutaneous disease. Cutaneous leishmaniasis will most often heal without treatment over a period of months to years. The disease is usually treated to reduce scarring. Mucocutaneous leishmaniasis will seldom heal spontaneously and requires treatment (107).

The vectors for leishmaniasis are female sandflies of over 30 species from the genus *Lutzomyia* (Americas) or *Phlebotomus* (Europe, Asia, and Africa). The sandfly ingests the parasite in the amastigote form during a blood meal from an infected animal. Alternate animal hosts to humans are most often rodents, opossums, canines, sloths, and other small animals. Amastigotes transform to divide as promastigotes in the fly's gut, before migrating to the mouthparts. Promastigotes are spread to other animals, including humans, by subsequent bites of the fly. In the host, the promastigotes are ingested by macrophages at the bite site. Promastigotes transform to amastigotes that multiply in the phagolysosomes of the mononuclear cell until it fills and bursts. Amastigotes can live and reproduce only in macrophage. Released amastigotes infect new cells and spread the infection. Sandflies then become infected with the amastigotes on taking a blood meal from the host, thus completing the infectious cycle of the parasite. Cutaneous leishmaniasis will present itself as a skin ulcer with raised borders. Mucosal forms of the disease can cause disfiguring erosions of the skin around areas of the mouth, nose, palate, cheeks, and pharynx. Visceral leishmaniasis causes hepatosplenomegaly, fever, weight loss, thrombocytopenia, and hypergammaglobulinemia. Visceral leishmaniasis is fatal if left untreated (108).

Leishmanial infections are most often resolved by the host's immune system. It is not clear why some individuals develop advanced forms of the disease. Certainly, the immunocompetence of the host and the many species of *Leishmania* that can cause disease are important factors. Leishmaniasis is becoming established as a major opportunistic disease in immunocompromised persons. Visceral leishmaniasis in particular is exacerbated in patients coinfected with HIV-associated immunodeficiency (109).

Vector control can sometimes be achieved by the use of insecticides in or near homes and in populated areas. However, much of the disease is sylvan in nature and is difficult to eradicate the vectors in these settings. Use of insect repellents can reduce incidence of infection. To date, no effective vaccine has been developed.

Dosing schedules for the major drugs used to treat leishmaniasis are given in Table 19.4 and structures of the drugs are in Fig. 19.7. Pentavalent antimony has been the agent of choice for treating leishmaniasis. Considerable resistance to the antimony-based drugs has developed and amphotericin B has become the main replacement drug. All of the major drugs used until this time have been relatively toxic, have required parenteral administration, and have relatively long durations of treatment. Amphotericin B-lipid has limited availability and is very expensive. Nonlipid dosage forms of amphotericin B can be effective, but cause a greater incidence of side effects. Interferon-gamma is sometimes used in combination with the antimonial agents and does increase cure rates, but this agent is too expensive for use in undeveloped nations. Miltefosine is potentially a major advancement in the therapy of leishmaniasis. Clinical

**Table 19.4  Drugs Used to Treat Leishmaniasis**

| Drug | Adult Dose | Pediatric Dose |
|---|---|---|
| Amphotericin B-Lipid | 3 mg kg$^{-1}$ day$^{-1}$; i.v.; days 1–5, 14, 21 Immunosuppressed patients: 4 mg kg$^{-1}$ day$^{-1}$; i.v.; days 1–5, 10, 17, 24, 31, 38 | 3 mg kg$^{-1}$ day$^{-1}$; i.v.; days 1–5, 14, 21 |
| Amphotericin B deoxycholate | 0.5–1.0 mg kg$^{-1}$ day$^{-1}$ or alternate day; i.v.; for 8 weeks | 0.5–1.0 mg kg$^{-1}$ day$^{-1}$ or alternate day; i.v.; for 8 weeks |
| Sodium stibogluconate | 20 mg Sb kg$^{-1}$ day$^{-1}$; i.v. or i.m.; for 20–28 days | 20 mg Sb kg$^{-1}$ day$^{-1}$; i.v. or i.m.; for 20–28 days |
| Meglumine antimonate | 20 mg Sb kg$^{-1}$ day$^{-1}$; i.v. or i.m.; for 20–28 days | 20 mg Sb kg$^{-1}$ day$^{-1}$; i.v. or i.m.; for 20–28 days |
| Paromomycin | 25–35 mg kg$^{-1}$ bid$^{-1}$; i.v.; × 15 days | 25–35 mg kg$^{-1}$ bid$^{-1}$; i.v.; × 15 days |
| Pentamidine | 2–4 mg kg$^{-1}$ day$^{-1}$ or alternate day; i.v.; × 15 days | 2–4 mg kg$^{-1}$ day$^{-1}$ or alternate day; i.v.; × 15 days |
| Miltefosine | 100–150 mg/day; p.o.; × 28 days | Undetermined |

trials with the agent have shown that the drug can be dosed orally and gives 95% cure rates in visceral leishmaniasis.

### 2.3.1 Mechanisms of Action of Drugs Used to Treat Leishmaniasis

#### 2.3.1.1 Sodium Stibogluconate and Meglumine Antimonate.
These two pentavalent derivatives of antimony have been the agents of choice for the treatment of leishmaniasis for the past 50 years. Sodium stibogluconate (**14**; Pentostam) is used in most of the world, whereas meglumine antimonate (**15**; Glucantime) is used predominately in French-speaking countries. There is no discernable difference in treatment outcomes with the two drugs. The drugs are prepared by reacting gluconic acid (sodium stibogluconate) or meglumine (N-methyl-D-glucamine; meglumine antimonate) with pentavalent antimony. The reaction mixture is allowed to age and a complex mixture of antimony-sugar polymeric compounds is isolated. It is important to note that structures (**14**) and (**15**) represent the main components of the polymeric mixtures and not the actual chemical composition of the formulated drug. The drugs are assayed for antimony content and aqueous solutions prepared containing 100 mg Sb$^{5+}$/mL. The polymeric composition of meglumine antimonate in solution has been shown to change in time, as is evidenced by a steady increase in osmolarity of solution over 8 days. There is some concern that different batches of these agents give different therapeutic outcomes because of

different polymeric mixtures in the formulations. However, Roberts and Rainey (110) separated sodium stibogluconate into 10 fractions by ion-exchange chromatography and found all fractions to have the same efficacy (based on Sb$^{5+}$ content) against *Leishmania* amastigotes *in vitro*. Some of the confusion concerning the antileishmanial action of Pentostam against the promastigote stage can be attributed to m-chlorocresol added as a stabilizer to the drug preparation (110). Perhaps the variability in treatment outcome may result from differential degrees of resistance of the treated organisms.

Sodium stibogluconate and meglumine antimonate are prodrugs, requiring reduction of Sb$^{5+}$ to Sb$^{3+}$ for activity on leishmanial enzymes (111–113). Activation occurs selectively in the amastigote stage of parasite life cycle (114), indicating that there may be some enzyme or factor expressed only in the amastigote that is responsible for the activation (115). It has been suggested that nonenzymatic reduction by thiols such as glutathione could contribute to activation (116), but this is favored only at low pH. The mode of action of the antimonial drugs is not understood. High concentrations of trivalent antimony have been reported to inhibit glucose catabolism and fatty acid oxidation in amastigotes (117). In schistosomes, trivalent antimony is thought to inhibit glycolysis by selective inhibition of phosphofructokinase (118). However, other studies on phosphofructokinase and other glycolytic enzymes from *L. mexicana* revealed no

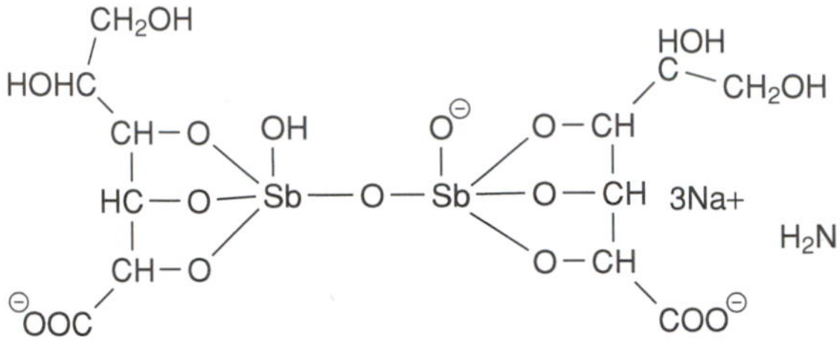

Sodium Stibogluconate
(**14**)

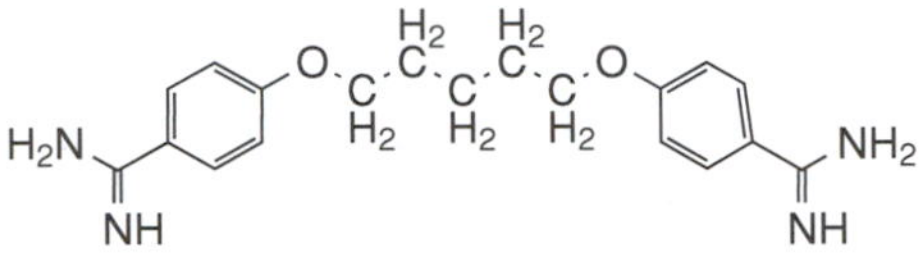

Pentamidine
(**2**)

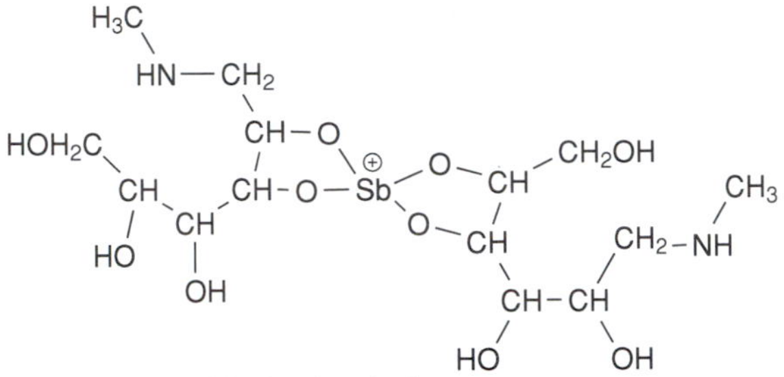

Meglumine Antimonate
(**15**)

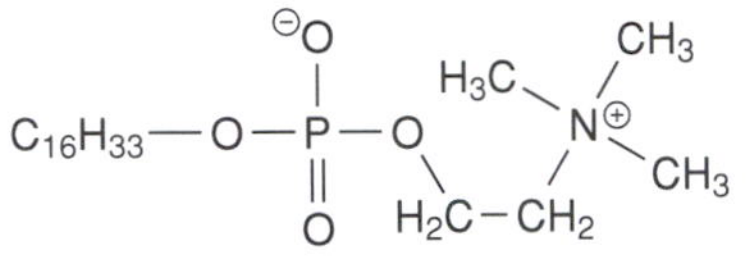

Miltefosine
(**17**)

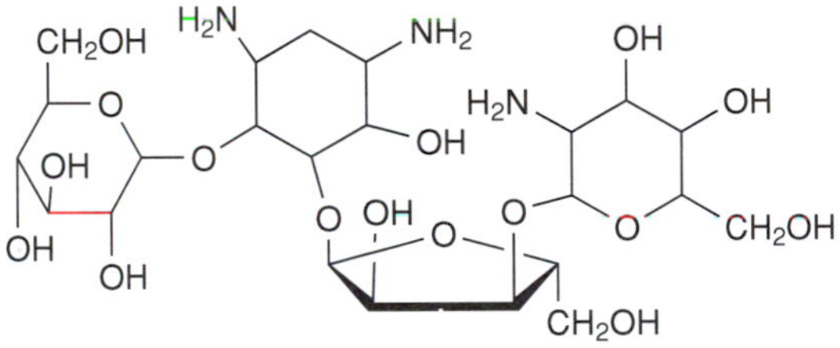

Paromomycin
(**16**)

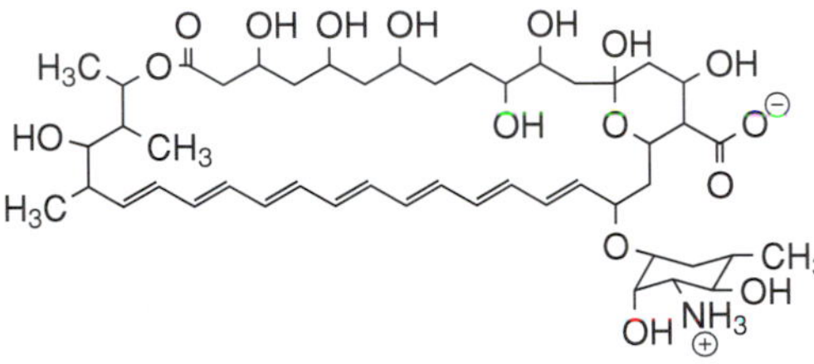

Amphotericin B
(**18**)

**Figure 19.7.** Agents used to treat leishmaniasis.

inhibitory activity and it was concluded that glycolysis is not the target for antimonial drugs (119). Trivalent antimony might be expected to bind to and inhibit a number of sulfhydryl-containing enzymes in the parasite. To date, only trypanothione reductase has been reported to be sensitive to inhibition by trivalent antimony (120). However, the significance of this finding is not clear because glutathione reductase is inhibited to a similar extent (120). Antimonials also induce DNA fragmentation in *L. infantum* (121), but this may be a late consequence of cell killing by another mechanism.

Resistance to the antimonial agents is known and represents an increasing clinical problem, especially in Bihar, India and in parts of Bangladesh. Cure rates in these areas have dropped to 50% or less compared to the normal 90% cure rates of the antimonial agents. Various mechanisms for resistance have been postulated (122), including decreased activation, increased trypanothione levels (123), and increased export of the drug (124, 125) through a pump that actively transports thiol-metal conjugates (126).

Frequent and often severe side effects associated with therapy with the pentavalent antimonials include T-wave flattening or inversions (sometimes leading to fatal arrhythmias), transaminase elevations, muscle and joint pain, fatigue, nausea, and pancreatitis.

*2.3.1.2 Paromomycin.* Paromomycin sulfate (Humatin; aminosidine, **16**) is an effective alternate to the trivalent antimonials for the treatment of all forms of leishmaniasis (127). Paromomycin is an aminoglycoside antibiotic closely related to neomycin. The agent is also effective in combination with the antimonials (128). Aminoglycosides exert a cytocidal action by binding to polysomes, inhibiting protein synthesis, and causing misreading and premature termination of translation of mRNA (129). Paromomycin is effective against antimony-resistant leishmaniasis, but offers no advantage to amphotericin B therapy. Paromomycin is licensed in Europe for parenteral use; however, it is not commercially available at this time. Paromomycin is also useful as an ointment preparation for the topical treatment of cutaneous leishmaniasis (130).

The major toxic effect of paromomycin is nephrotoxicity and damage to the eighth cranial nerve, resulting in hearing loss. To minimize these toxicities care must be taken to avoid exceeding the recommended dosage and the concomitant use of other agents with similar toxicities (130).

*2.3.1.3 Amphotericin B-Lipid or Deoxycholate.* Amphotericin B (**18**) is a polyene antibiotic used extensively against fungal infections. Amphotericin B-lipid complex (AmBisome, Amphocet, Abelcet, and others) and the older Amphotericin B deoxycholate (Fungizone) are highly effective against antimony-resistant strains of *Leishmania* (131, 132). The advantage of the lipid complex dosage forms is the ability to give higher doses with lower side effects. Amphotericin B gives cure rates, on all forms of leishmaniasis, that approach 100%. The problems with the drug are its high cost (especially the lipid dosage form), the necessity for parenteral administration in a hospital setting, and the long treatment periods required.

Amphotericin B works by complexing with ergosterol in the cell membrane of *Leishmania*. Pores are formed in the membrane and ions are allowed to pass into the cell, which leads to cell death (133). Resistance to amphotericin B is rare. When resistance does occur, it is generally the result of decreased ergosterol concentrations in the membrane or the substitution of a structurally altered steroid for ergosterol to which amphotericin B does not bind.

The major adverse side effect of amphotericin B is nephrotoxicity. Acute side effects of fever, chills, muscle spasms, vomiting, headache, hypotension, and anaphylaxis occur more often with amphotericin B deoxycholate preparations than with amphotericin B-lipid complex preparations.

*2.3.1.4 Pentamidine.* Pentamidine (**2**) has activity against leishmaniasis. However, pentamidine has greater overall toxicity, and resistance to it has developed in parts of India; thus, it is considered back-up therapy to the pentavalent antimonials and to amphotericin B-lipid complex. The mechanism of pentamidine is covered earlier in this chapter.

*2.3.1.5 Miltefosine.* Miltefosine (Impavido, **17**) promises to be a breakthrough drug for the treatment of leishmaniasis. The drug has produced impressive results in phase I (134) and Phase II (135) clinical trials against Indian leishmaniasis and the drug was licensed in India in June 2002. A significant advantage of miltefosine is that it is active by oral (p.o.) administration, unlike the other antileishmanial drugs. The effective dose range of oral miltefosine is 50 to 200 mg/day, but the optimal dose is 100–125 mg/day. In combined studies (134, 135), 50 of 51 patients with mild visceral leishmaniasis treated with 100 mg/day miltefosine, p.o., for 28 days, were parasite free 6 months after the end of therapy. Some of the patients had received therapy with antimony agents before the drug trial. The 6-month cure rate with this drug regimen in India was 98%. Trials of miltefosine have been extended to American leishmaniasis. Soto et al. (136) reported a 94% cure rate of American cutaneous leishmaniasis with oral miltefosine after 3- to 4-week treatment with 133–150 mg/day.

Miltefosine (**17**) is the simple ether for ester bioisostere of phosphatidylcholine (lecithin). The agent is a member of the lysophospholipid or ether-lipid class of drugs that were synthesized and developed as potential anti-cancer agents. Early clinical trials as systemic antitumor agents failed, but the compounds have topical antitumor activity. The compound subsequently was found to have *in vitro* activity against leishmaniasis. These findings led to the clinical trials of oral miltefosine described earlier.

The exact mechanism of action of miltefosine is not known. In mammalian cell culture, miltefosine has been shown to have effects on signal transduction (137, 138), lipid metabolism (139, 140), and calcium homeostasis (141). Miltefosine inhibits phosphatidylcholine biosynthesis, resulting from the inhibition of phosphatidylethanolamine-phosphatidylcholine-*N*-methyltransferase in *T. cruzi* (142). Perturbation of ether-lipid remodeling is implicated as a target in *Leishmania* (143).

Side effects from oral miltefosine are significant but generally tolerable. Acute effects are primarily gastrointestinal distress, consisting mainly of vomiting and diarrhea. Gastrointestinal episodes occurred frequently in 62% of the patients and did not dissipate with continuing therapy; however, no patients left the study because of the gastrointestinal side effects, and overall, the patients experienced a slight weight gain over the 4-week treatment period. A more serious side effect was elevated serum aspartate aminotransferase levels. Twelve of 80 patients displayed elevated serum aspartate aminotransferase levels in the initial days of therapy, but these decreased during subsequent days of treatment. One patient had to discontinue therapy because of liver toxicity. Creatinine levels rose in several patients. All abnormal laboratory values returned to normal shortly after discontinuation of therapy.

## 3  PERSPECTIVES FOR NEW DRUG THERAPIES FOR KINETOPLASTID PARASITIC DISEASES

The kinetoplastid parasitic diseases are endemic, if not epidemic, in most of the tropical and subtropical regions of the world. Agents available to treat these diseases are, for the most part, inadequate because of poor efficacy, toxicity, high cost, and route of administration. Many of the older drugs do not meet current standards for regulatory approval for clinical use. Leading medical journals publish editorials on a regular basis calling for the development of new drugs for tropical diseases (144–147). Despite knowledge of a number of enzymes or enzyme systems that could be targeted for development of selectively toxic chemotherapeutic agents, little development of new drugs for these diseases has occurred. The following section explores some of the opportunities for the development of drugs to treat kinetoplastid diseases.

### 3.1  Glycosomal and Other Carbohydrate Metabolism

Bloodstream forms of *T. brucei* lack energy stores and essentially make all of their ATP from glycolysis. Two features of this pathway have attracted considerable interest for drug design. First, these parasites lack cytochromes and respiration is through a plantlike, cyanide-insensitive, glycerophosphate oxidase mitochondrial process. Second, as first described Opperdoes and Borst (148), the first nine enzymes of glucose and glycerol catabolism are contained within a peroxisome-like organelle known as the glycosome. Compartmentalization of the enzymes in the glycosome may make it possible for *T. brucei* to carry out glycolysis at a rate approximately 50 times higher than that of mammalian cells (149), although this has been disputed (150). The high rate of aerobic glycolysis is necessary to compensate for the poor yield of only two ATP molecules per glucose molecule that is processed (151) (Fig. 19.8). Trypanosomes require a large amount of energy to replicate every 6 to 8 h in mammalian blood (152), and to change their surface glycoproteins (VSGs)

**Figure 19.8.** Glycolysis and glycolytic enzymes associated with the glycosomes of bloodstream *Trypanosoma brucei*. The enzymes that catalyze the reactions are: (1) hexose kinase; (2) glucose-6-phospate isomerase; (3) phosphofructokinase; (4) fructose-1,6-bisphosphate aldolase; (5) triose-phosphate isomerase; (6) glycerol-3-phosphate dehydrogenase; (7) glyceraldehyde-3-phosphate dehydrogenase; (8) phosphoglycerate kinase; (9) phosphoglycerate mutase; (10) enolase; (11) pyruvate kinase; (12) glycerol kinase; (13) mitochondrial glycerol-3-phosphate dehydrogenase; (14) ubiquinol oxidase, (13 + 14 = glycerophosphate oxidase); (15) fructose-6-phosphate 2-phosphokinase; (16) fructose-2,6-bisphosphate 2-phosphatase; $\oplus$ = activation of pyruvate kinase.

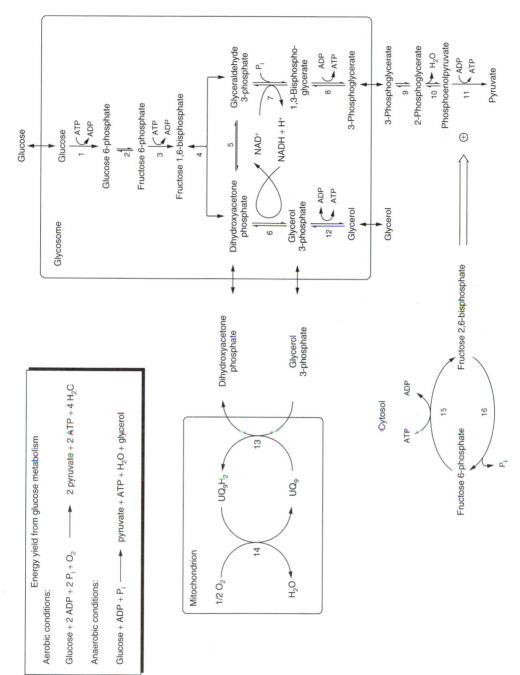

**Figure 19.8.**

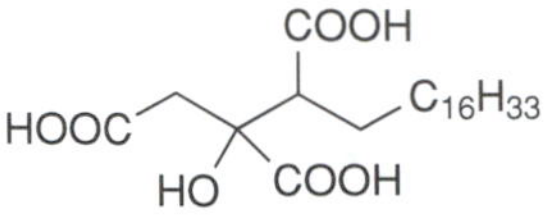

Agaricic acid
(19)

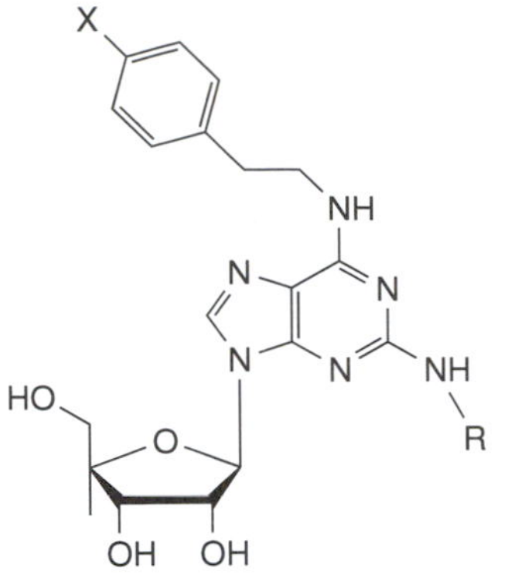

(20a): X = OH, R = H
(20b): X = H, R = $CH_2CH_2C_6H_5$

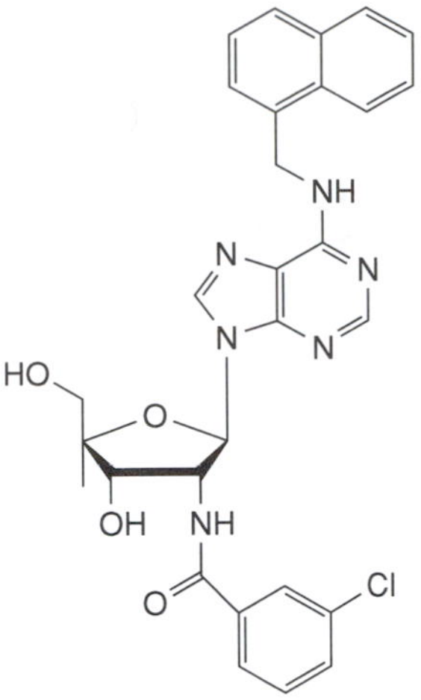

(21)

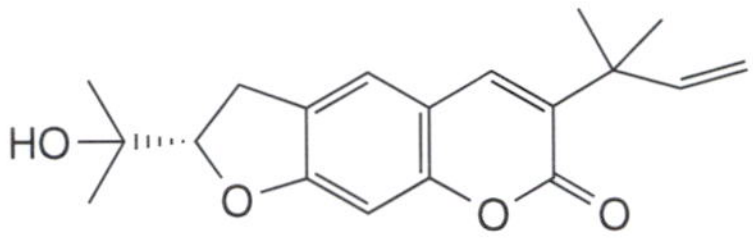

Chalepin
(22)

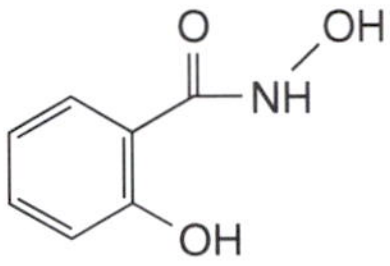

Sham
(23)

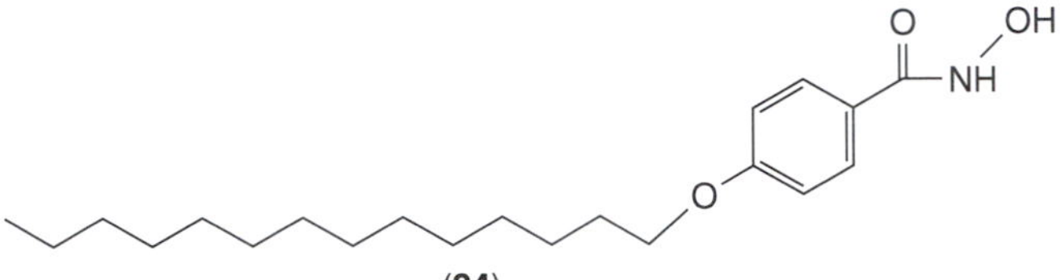

(24)

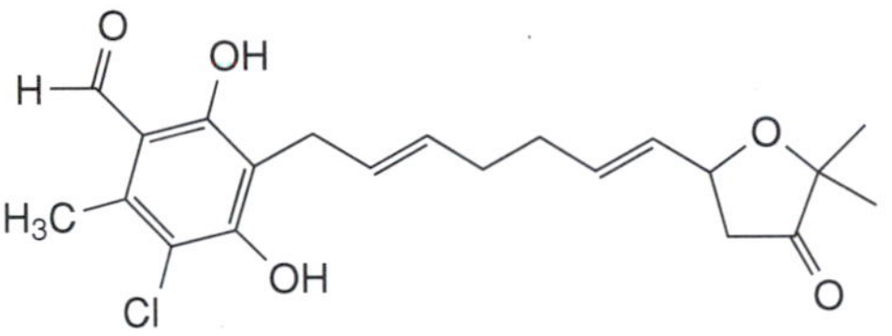

Ascofuranone
(25)

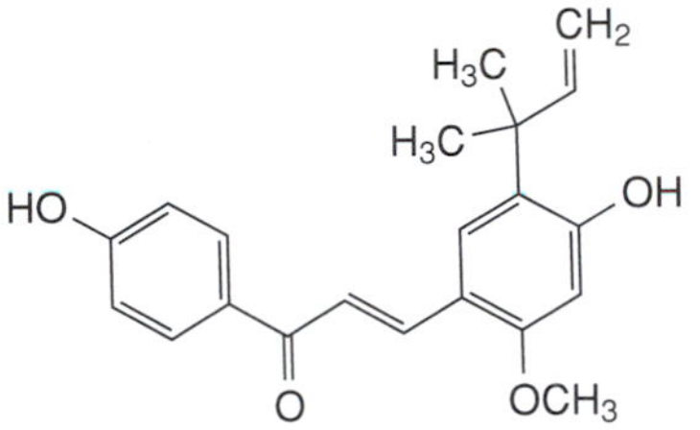

Licochalcone A
(26)

**Figure 19.9.** Inhibitors of glycosomal enzymes and additional enzymes associated with glycolysis in the trypanosome.

at a rate high enough to evade host immune response (15). Other species of the order Kinetoplastida are not as heavily dependent on glycolysis for energy supply as are the African trypanosomes (*T. brucei* spp.); however, many of the enzymes from kinetoplastids have similar structure and function, thus potent inhibitors to key enzymes may have selective toxicity toward many or all of the species (153). Excellent reviews on glycosomes and the enzymes they contain are given by Clayton and Michels (154), Wang (155), and Opperdoes and Michels (156). Structures of compounds known to inhibit glycolysis and respiration are shown in Fig. 19.9.

Glycosomes and the glycosomal enzymes are essential for trypanosome survival and growth. Although the glycosomal enzymes differ significantly in amino acid sequence and structure from the corresponding mammalian enzymes catalyzing glycolysis, this is less evident at their active sites and may present a significant challenge for drug design. Many of the glycosomal enzymes have been sequenced, cloned, and the three-dimensional structures determined. As discussed earlier, the mechanisms of action of the arsenic- and antimony-containing trypanocides and leishmanicides have been attributed to inhibitory activity on enzymes of the glycolysis pathway, but this is far from certain. Because of the uniqueness of these enzymes to trypanosomes and kinetoplastids in general, they are considered by many authorities as excellent targets for structure-based drug design.

Bakker et al. (150, 157, 158) performed mathematical modeling experiments on the enzymes of the glycosomal pathway to determine the enzymes most important for control of "glycolytic flux." The modeling should identify the enzymes most susceptible to control by enzyme inhibitors. Interestingly, transport of glucose into the glycosome was calculated to be rate controlling on the glycolytic flux. Other highly important enzymes are calculated to be aldolase, glyceraldehyde-3-phosphate dehydrogenase, glycerol-3-phosphate dehydrogenase, and phosphoglycerate kinase (158). However, it should be noted that other modeling studies have concluded that there is little prospect of killing trypanosomes by depressing glycolytic flux except by the use of specific irreversible inhibitors (159) or by inhibition of pyruvate export.

### 3.1.1 Hexose Transport as a Drug Target.
Bloodstream *T. brucei* import glucose by facilitated diffusion (160), and the uptake of glucose accounts for greater than 50% control of the glycolytic flux (158, 161). The glucose transporter will also accumulate fructose and mannose. Trypanosomes contain two genes encoding glucose transporters, THT1 and THT2 (162). THT1 glucose transporters are predominantly expressed in the bloodstream form and have low affinity for glucose. Cytochalasin B is a moderately potent inhibitor of THT1 (163). THT2 has high affinity for glucose and is expressed in the procyclic life form of the trypanosomes (164). Both THT1 and THT2 could be considered as potential targets for antitrypanosomal chemotherapy (165), although an opposing view has been published (159). It is also theoretically possible that the

glucose transporter could be used as a carrier to target chemotherapeutic drugs in trypanosomes (166).

### 3.1.2 Inhibition of Hexose Kinase.
Hexose kinase (HK) is the first enzyme in the glycosomal glycolysis pathway (Fig. 19.8, enzyme 1). The enzyme has been sequenced (156) and found to have 36% identity to human HK. The enzyme has low substrate selectivity for glucose, phosphorylating a number of hexoses (167). The enzyme also has low selectivity for ATP, and can even use UTP or CTP in its place (156). The enzyme is not regulated by glucose-6-phosphate or glucose-1,6-bisphosphate as is the mammalian enzyme. Analogs of glucose have been found to inhibit the enzyme *in vitro* (168). Because HK is predicted to have a low degree of control over the glycolytic flux in the glycosome, it is not the best potential target for drug design.

### 3.1.3 Inhibition of Glucose-6-Phosphate Isomerase.
Glucose-6-phosphate isomerase (PGI), the second enzyme in the glycosomal pathway (Fig. 19.8), has been sequenced, cloned, and characterized (169). PGI has 54–56% sequence homology with the equivalent enzymes from yeast and mammalian sources. The enzyme from trypanosomes, but not the enzyme from rabbit muscle, is reversibly inhibited by suramin ($K_i$ = 0.29 m$M$) and irreversibly by agaricic acid (**19**) (169). Hardre et al. (170) synthesized a transition-state analog inhibitor of PGI ($K_i$ = 50 n$M$); however, the compound has poor selectivity for the trypanosomal enzyme and, because of its ionic character, is not likely to have *in vivo* activity. PGI is predicted to have little control on the glycolytic pathway (158); thus its potential as a drug target remains to be validated.

### 3.1.4 Inhibition of Phosphofructokinase.
The next enzyme in the glycolytic pathway of *T. brucei*, phosphofructokinase (PFK; Fig. 19.8, enzyme 3), has been sequenced and cloned (171). PFK from *T. brucei* has the structure of a typical pyrophosphate-dependent kinase, but it will not function with pyrophosphate as a substrate. Rather, it requires ATP for activity. The differences in structure of *T. brucei* PFK, compared to that of the

mammalian enzymes having the equivalent function, make the enzyme attractive as a drug design target. However, the enzyme is not predicted to be important in controlling the glycolytic flux (158) and no inhibitors to the enzyme have been reported.

### 3.1.5 Inhibition of Fructose-1,6-Bisphosphate Aldolase.

This enzyme (Fig. 19.8, enzyme 4) has been cloned (172) and expressed (173), and the structure of the recombinant protein was determined to 1.9-Å resolution (174). Because of the structural similarities in the active sites of the *T. brucei*, *L. mexicana*, and the three human aldolase isoenzymes, selective inhibition of the parasite enzymes is a remote possibility (174). However, the parasite enzymes both contain a type II peroxisomal targeting sequence at their *N*-termini, which might be exploited to inhibit aldolase import into the glycosome (see section 3.2).

### 3.1.6 Inhibition of Triose-Phosphate Isomerase.

Triose-phosphate isomerase (TIM) is an essential enzyme for the conversion of dihydroxyacetone phosphate to glyceraldehyde-3-phosphate. TIM from *T. brucei* has been cloned (175) and the three-dimensional structure determined (176, 177). Helfert et al. (178) used knockout constructs to demonstrate that TIM activity is essential for trypanosomes to survive. Surprisingly, neither the glycolytic scheme in Figure 19.8 (reaction 5) nor the computer model of Bakker et al. (158) predicts this enzyme would be essential for ATP synthesis through glycolysis, even though the gene knockout experiments validate the enzyme as essential. However, although essential, the structural similarities with mammalian TIM do not make it a good target for inhibition.

### 3.1.7 Inhibition of Glyceraldehyde-3-Phosphate Dehydrogenase.

Glyceraldehyde-3-phosphate dehydrogenase (GAPDH) is the most studied enzyme of the glycosomal enzymes (Fig. 19.8, enzyme 7). Crystal structures of GAPDH from *T. brucei* (179), *T. cruzi* (180), and *L. mexicana* (181) are all known and have been compared to the human enzyme. A major difference between the parasitic and mammalian enzymes lies in the region of the enzyme that binds $NAD^+$. Aronov et al. (182, 183) performed molecular modeling on the $NAD^+$ binding site of GAPDH from *L. mexicana* for the design of adenosine analogs as tight binding inhibitors. Adenosine analogs (**20a**), (**20b**), and (**21**) were among the most active compounds reported by this group. Interestingly, compound (**21**) had good activity against cultured *L. mexicana*, *T. brucei*, and *T. cruzi* with $IC_{50}$ values of 10, 30, and 14 $\mu M$, respectively, and showed three- to 15-fold selectivity (based on $ED_{50}$ values on cultured cells) for parasites over mammalian cells. Increasing the potency, selectivity, and bioavailability of the lead compounds represents a significant challenge. Nonetheless, the lethal activity of (**21**) on all three species demonstrates the potential for antiparasitic activity against all kinetoplastid species. Ladame et al. (184) reported the inhibition of GAPDH by phosphorylated epoxides and alpha-enones. These agents are apparently affinity labels for GAPDH and may show some *in vivo* selectivity for the trypanosomal enzyme. Pavao et al. (185) identified chalepin (**22**), a natural product coumarin-related analog, as an inhibitor ($IC_{50}$ = 64 $\mu M$) of GAPDH from *T. cruzi* and have obtained a crystal structure with (**22**) bound to the enzyme. Chalepin binds in the $NAD^+$ binding pocket of the enzyme by a unique mode. The X-ray structure provides a basis for structure-based drug design to maximize activity of (**22**).

### 3.1.8 Inhibition of Glycerol-3-Phosphate Dehydrogenase.

$NAD^+$-dependent glycerol-3-phosphate dehydrogenase (G3PDH) is predicted by mathematical modeling of the glycosomal enzymes to be highly important for the control of the glycolytic flux (Fig. 19.8, enzyme 6). G3PDH has been cloned from several species (186, 187) and three-dimensional structure from *L. mexicana* was reported by Suresh et al. (188). G3PDH from *L. mexicana* is very similar to G3PDH from *T. brucei* (63% sequence homology), but has only 29% amino acid sequence homology with the mammalian enzyme. Agaricic acid, suramin, melarsen oxide, and cymelarsan inhibit trypanosomal G3PDH (188–190).

Denise et al. (191) applied glycosomal protein extracts to affinity chromatography columns containing bound cymelarsan (**9**). Two proteins of molecular weight 36 and 40 kDa bound to the column and could be eluted by high concentrations of cymelarsan. The 36-kDa protein was identified as G3PDH and the 40-kDa protein as fructose-1,6-bisphosphate aldolase. These experiments implicate these two essential enzymes as targets for drug development.

**3.1.9 Inhibition of the Dihydroxyacetone Phosphate/Glycero-3-phosphate Shuttle.** In *T. brucei*, the glycosomal NAD$^+$ used in glycosomal metabolism must be regenerated from NADH through a dihydroxyacetone phosphate (DHAP) for glycerol-3-phosphate shuttle. The shuttle consists of the glycosomal NAD$^+$-dependent glycerol-3-phosphate dehydrogenase and a mitochondrial glycerol-3-phosphate oxidase system (192). The system constitutes a mitochondrial FAD-dependent glycerol-3-phosphate dehydrogenase (Fig. 19.8, enzyme 13) (193) that transfers electrons from glycerol-3-phosphate to ubiquinone, which is then oxidized through a cytochrome-dependent ubiquinol oxidase (Fig. 19.8, enzyme 14), also known as the trypanosome alternative oxidase (194, 195). During anaerobic conditions the oxidase cannot function and glycerol-3-phosphate accumulates in the cell to millimolar concentrations (192). However, net ATP synthesis occurs because of the mass action effect of glycerol-3-phosphate driving the glycerol kinase reaction (Fig. 19.8, enzyme 12) toward the formation of glycerol and ATP (196). As long as glycerol concentrations remain low, the cells can survive considerable periods of anaerobiosis *in vitro*. A similar effect can be demonstrated both *in vivo* and *in vitro* by use of salicylhydroxamic acid (SHAM, **23**), an inhibitor of the glycerol-3-phosphate oxidase. However, addition of glycerol neutralizes this mass action effect and trypanosomes rapidly lose motility and lyse within minutes *in vitro* (197) and *in vivo* (259). Indeed, trypanosome-infected mice can be cured by a combined treatment with SHAM and glycerol (198), but the curative dose of SHAM in the combination caused unacceptable animal mortality rates (199).

Additional attempts have been made to find less toxic agents to apply to this chemothera-peutic approach. Grady et al. (200) made *p-n*-tetradecyloxybenzhydroxamic acid (**24**) and found it to be 70 times more potent as an inhibitor of glycerol-3-phosphate oxidase and 10- to 20-fold more effective as an antitrypanosomal agent *in vitro* than SHAM. However, no therapeutic improvement was found *in vivo*. Ascofuranone (**25**), an antibiotic isolated from *Ascochyta visiae* (201), has been found to potently inhibit the glycerol-3-phosphate-dependent mitochondrial O$_2$ consumption by bloodstream forms of *T. b. brucei in vitro* (202). The mechanism of action of (**25**) is attributed to its binding at the coenzyme Q site of the ubiquinol oxidase, an enzyme that has no counterpart in mammalian metabolism. This finding led to a trial of orally and intraperitoneally administered (**25**) combined with glycerol for the treatment of *T. b. brucei*-infected mice. Either agent alone was ineffective in the dose range tested; however, (**25**) at 100 mg/kg combined with glycerol at 3 g/kg given orally, or (**25**) at 25 mg/kg combined with glycerol at 3 g/kg given intraperitoneally gave a 100% cure rate in five of five mice in each test group. In each experiment, (**25**) was given as a single dose and the glycerol was dosed at 30-min intervals at 1 g/kg dose. Ascofuranone was found to be highly nontoxic to mice, having an LD$_{50}$ value of over 5 g/kg. If the combination of ascofuranone/glycerol is able to cure *T. brucei* in humans and/or livestock by single oral treatment, it would afford a major advancement in the management of African trypanosomiasis. However, glycerol does not readily cross the blood-brain barrier, so the infection would not be eliminated from the CNS. An alternative strategy, therefore, would be to simultaneously inhibit glycerol kinase and the alternative oxidase.

**3.1.10 Inhibition of Glycerol Kinase.** Kralova et al. (203) and Steinborn et al. (204) independently reported the cloning and molecular characterization of glycosomal glycerol kinase (GK) from *T. brucei*. A unique feature, compared to the mammalian enzyme, is the replacement of Ser137 with Ala. The Ala137 substitution results in lower affinity of the substrates for the enzyme compared to those of other GKs. The Ala137 substitution is found in other trypanosome species that can convert glucose directly to glycerol and pyruvate under anaerobic conditions. This unique feature of *T. brucei* GK offers a potential for the design of selectively toxic drugs for use in combination with inhibitors of the alternative oxidase.

### 3.1.11 Inhibition of Phosphoglycerate Kinase.

Phosphoglycerate kinase (PGK, Fig. 19.8, enzyme 8) exists as three isoenzymes in most *Trypanosomatidae*. Mathematical modeling of the glycosomal pathway places PGK as highly important to the overall glycolytic flux in the trypanosome (157, 158). Two of the forms are glycosomal (PGKA and PGKC) and the third (PGKB) is cytosolic in both *T. brucei* and *T. cruzi*. PGKA is expressed at low levels throughout the life cycle, whereas cytosolic PGKB is strongly expressed in the insect procyclic stage, but hardly at all in the mammalian bloodstream form, whereas the reverse is true for glycosomal PGKC. Overexpression of cytosolic PGKB in bloodstream forms is lethal (205), implying that compartmentation plays a vital role in maintaining their energy supply. Cloning and characterization of PGK from multiple species has been accomplished (206) and the structure has been determined (207). PGK from *T. brucei* has a high preference for ATP binding, whereas enzymes from yeast and rabbit muscle will readily accept GTP and ITP as cosubstrates. Suramin (1) selectively inhibits the glycosomal form of PGK from *T. brucei* (208).

### 3.1.12 Inhibition of Phosphoglycerate Mutase and Pyruvate Kinase.

Phosphoglycerate mutase (Fig. 19.8, enzyme 9) in *T. brucei* has been found to be a cofactor-independent enzyme, unlike its mammalian counterpart enzymes that are all cofactor dependent. This major difference in structure of trypanosomal phosphoglycerate mutase is proposed to be an excellent target for drug development (209).

Pyruvate kinases from trypanosomal species possess unique properties that may be exploitable for the purpose of selective drug development. The most notable difference from the host is that the parasite enzyme has a regulatory site on the enzyme that binds fructose-2,6-bisphosphate (210). It is proposed that a tight binding inhibitor to the fructose-2,6-bisphosphate regulatory site would be an effective trypanocide. However, modeling studies on glycolysis predict that this enzyme is likely to be a mediocre drug target (158).

### 3.1.13 Additional Glycosomal Enzymes as Potential Drug Targets.

Several additional enzymes are involved in carbohydrate metabolism of kinetoplastids. Phosphoenolpyruvate carboxykinase (PEPCK) is one such enzyme. In insect stages of *T. brucei*, *T. cruzi*, and *Leishmania*, glucose consumption is associated with $CO_2$ fixation and excretion of significant amounts of succinate. The glycosomes of these life-cycle stages have low amounts of phosphoglycerate kinase and highly elevated levels of PEPCK and malate dehydrogenase (MDH). The combined action of these two enzymes converts phosphoenolpyruvate to malate, with the concomitant oxidation of NADH to $NAD^+$ and conversion of ADP to ATP, thereby maintaining zero net internal balance of ATP and $NAD^+$ in the glycosome (211). Consequently, PEPCK is purported to be a prime candidate for inhibitor development (212, 213). The recently published crystal structure of the *T. cruzi* enzyme (214) will greatly aid efforts at structure-based drug design.

Pyruvate phosphate dikinase (PPDK) is another glycosomal enzyme that is expressed in *T. cruzi*, *Leishmania*, and the procyclic stage of *T. brucei* (but not bloodstream forms) and does not exist in higher eukaryotes. The enzyme has been cloned and its structure determined from several sources and the three-dimensional structure determined (215–217). The functional role and importance of this enzyme in glycosomal metabolism has not been determined. Compartmentalization of this enzyme in the glycosome is important since a futile cycle, resulting in net hydrolysis of ADP, would occur if PPDK were in the cytosol (Scheme 1).

In *T. cruzi*, PDDK could play a role in formation of phosphoenolpyruvate for the synthesis of phosphonopyruvate (catalyzed by phosphoenolpyruvate mutase) and its subsequent conversion to aminoethylphosphonate (AEP). Phosphonopyruvate mutase has been

$$\text{Pyruvate} \: + \: P_i \: + \: \text{ATP} \: \rightarrow \: \text{PEP} \: + \: \text{AMP} \: + \: PP_i \quad (1)$$

$$\text{PEP} \: + \: \text{ADP} \: \rightarrow \: \text{Pyruvate} \: + \: \text{ATP} \quad\quad (2)$$

$$PP_i \: + \: H_2O \: \rightarrow \: 2P_i \quad\quad\quad\quad\quad (3)$$

---

$$\text{ADP} \: + \: H_2O \: \rightarrow \: \text{AMP} \: + \: 2P_i. \quad\quad\quad \text{Sum}$$

**Scheme 1.** Biochemical reactions expected if PPDK were in the cytosol. Where 1 = PPDK, 2 = PK, and 3 = pyrophosphatase.

cloned and expressed from *T. cruzi* (Sarkar and Fairlamb, unpublished), suggesting that the pathway is present. In contrast to *T. brucei* and *Leishmania* spp., AEP is an invariant component of the glycan core of the parasites' surface glycosylinositylphospholipids (GIPLs) and mucins, suggesting an important role in survival. However, further work is required to establish whether this pathway is a viable drug target.

An NADH-dependent fumarate reductase is present in the promastigote stage of *Leishmania* parasites and *T. cruzi*. Mammalian cells do not possess soluble fumarate reductase activity. Chalcones, components of Chinese licorice, have been found to inhibit fumarate reductase and to be lethal to *L. major* and *L. donovani* (218). Nielsen et al. (219) synthesized and performed three-dimensional quantitative structure-activity relationship (QSAR) analysis on a large series of chalcones as potential antileishmanial agents. Licochalcone A (**26**), a lead compound in these studies, has an antileishmanial $IC_{50}$ value of 13 $\mu M$ and is one of the more potent inhibitors of fumarate reductase found to date. In *T. cruzi*, soluble fumarate reductase activity is associated with dihydroorotate dehydrogenase, implicating fumarate reductase in *de novo* pyrimidine biosynthesis (220). It remains to be determined whether this is also the case in *Leishmania*.

## 3.2 Glycosomal Protein Import as a Drug Target

Glycosomal enzymes are synthesized in the cytosol and thus have to be imported into the organelle for glycolysis to function correctly in a coordinated manner. As mentioned earlier, inappropriate expression of PGK in the cytosol disrupts normal glycolytic function in bloodstream *T. brucei* (205). Thus, inhibition of the glycosomal import process is predicted to be lethal. The initial hypothesis that "hot spots" of positive charges spread 40 Å apart on some glycosomal proteins were topogenic signals for import (221) was soon abandoned as details emerged concerning import of proteins into peroxisomes. Given that the requirements for glycosomal import are apparently more relaxed than those for peroxisomal import, selective inhibition of this process could represent a therapeutic strategy (222). Glycosomal import is accomplished by two mechanisms. The first relies on a specific C-terminal peptide sequence similar to the peroxisomal targeting signal (PTS1) that is used by peroxisomes to import proteins (223). Whereas peroxisomes have a rather strict requirement for a serine-lysine-leucine (SKL) carboxy-terminal sequence to undergo import by the PTS1 mechanism, kinetoplastids have a much more degenerate C-terminal target signal for the glycosomal import (224). With the SKL sequence as a starting point, S can be any other polar amino acid; K can be replaced by amino acids capable of forming H bonds; and L is replaceable by other hydrophobic amino acids. Blattner et al. found that a C-terminal SSL, which is not a targeting signal for peroxisomal import (225), is sufficient to direct the import of the cytosolic protein β-glucuronidase into glycosomes (226). Thus, it is possible to attain selective targeting of proteins to the glycosomes of trypanosomes vs. the peroxisomes of mammalian species. The second method for

importing proteins into peroxisomes and glycosomes uses the peroxisomal targeting signal
(PTS2). The PTS2 targets an *N*-terminal or
internal nonapeptide sequence of general
structure H₂N-X-X-R/K-L/V/I-Q-X5-H/Q-L/A
for glycosomal import.

Among the 11 *T. b. brucei* glycosomal proteins whose primary structures have been determined, seven have the C-terminal tripeptide sequences apparently capable of targeting
them for glycosomal import through the
PTS1: GAPDH [AKL; (227)], PGI [SHL;
(169)], PGK [SSL; (226)], PEPCK [SRL;
(228)], G3PDH [SKM; (186)], and PPDK
[AKL; (216)]. The other four proteins, HK
(156), TIM (175), ALDO (172), and 56-kDa
PGK (229), possess C-terminal tripeptide sequences incapable of acting as targeting signals for glycosomal protein import. These proteins have *N*-terminal or internal peptide(s)
PTS2 sequences as glycosomal targeting signals. The *N*-terminal 10 amino acid peptide in
mammalian 3-ketoacyl-CoA thiolase that
serves as a targeting signal for peroxisomal
import (230) does not function in glycosomal
protein import (228), demonstrating that selective targeting or inhibition of the transporter is possible. The recently solved structure of tetrameric aldolases from *T. b. brucei*
and *Leishmania* reveal that two PTS2 non-
apeptides interact to form a dimeric structure
that could serve as a basis for the design of
glycosome import inhibitors (174). A peptidomimetic based on the structure of the C-terminal tripeptide glycosomal targeting signal
could function as an inhibitor of glycosomal
peptide import and result in a new class of
antitrypanosomal agents (231).

The PTS1 or PTS2 sequences of the glycosomal enzymes bind to membrane importer
proteins (mPTS) that translocate them into or
through the glycosomal membrane. Little is
known about the receptor sites on the transporter proteins or the mechanism by which
they internalize the glycosomal enzymes (223,
232).

### 3.3 Inhibitors of the Pentose-Phosphate Shunt

Inhibitors of the pentose-phosphate pathway
(PPP) provide another potential target for
the development of antitrypanosomal agents

(233). The PPP provides NADPH that is essential for many biosynthetic reactions, including the regeneration of reduced trypanothione that is required for protection of the
organisms from oxidative stress. A proportion
of the first two enzymes of the PPP (glucose-
6-phosphate dehydrogenase and 6-phospho-
gluconate dehydrogenase) are associated with
the glycosome (234, 235), where they provide
NADPH for alkoxyphospholipid biosynthesis
(143, 236, 237).

### 3.4 Polyamines and Trypanothione

The biosynthetic routes to polyamines and
trypanothione are outlined in Figure 19.10.
There are at least five validated or highly plausible targets for drug development in the polyamine and trypanothione synthesis pathway.
In *T. brucei* and *Leishmania* ornithine decarboxylase (ODC) and *S*-adenosylmethionine
decarboxylase (AdoMetDC) are co-rate-limit-
ing enzymes in the synthesis of spermidine. *T.
cruzi* lacks ODC activity (238) and obtains putrescine and spermidine from the host by
means of inducible high affinity transporters
(239). With the exception of *T. cruzi*, spermine
is not synthesized in trypanosomes and spermidine is the major polyamine that is essential
for fulfilling the diverse functions of polyamines in the trypanosome cell (240). In addition, spermidine has the unique function in
the Kinetoplastida of serving as a substrate in
the biosynthesis of trypanothione, an essential metabolite for the survival of these organisms. Thus, inhibitors of polyamine biosynthesis or polyamine utilization are potentially
toxic to the Kinetoplastida.

### 3.4.1 Inhibitors of Ornithine Decarbox-
ylase. As discussed previously in the section
on the mechanism of action of eflornithine (4),
the blood form of *T. b. gambiense* appears to be
uniquely susceptible to inhibition of polyamine synthesis. The susceptibility of this one
species may be a result of the slow turnover of
ornithine decarboxylase (ODC) in *T. b. gambiense*, coupled with the low concentrations of
polyamines in the blood, thus negating reple-
tion of polyamines by active transport mechanisms (89). The therapeutic use of eflornithine
is limited by its short serum half-life and its

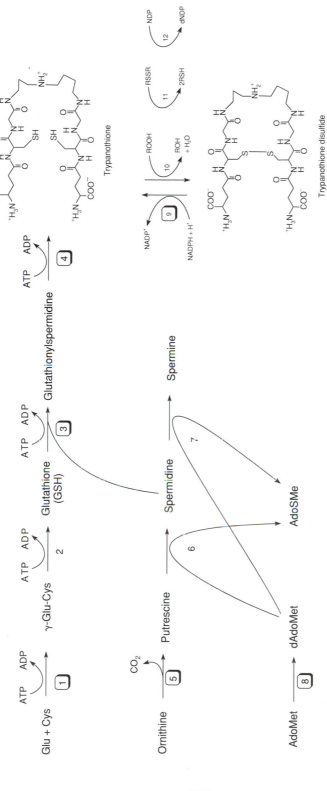

**Figure 19.10.** The biosynthetic pathways to polyamine and trypanothione in *T. brucei*. GSH, glutathione. Enzymes in boxes are validated targets for drug development: (1) γ-glutamylcysteine synthetase; (2) glutathione synthetase; (3) glutathionylspermidine synthetase; (4) trypanothione synthetase; (5) ornithine decarboxylase; (6) spermidine synthetase; (7) spermine synthetase; (8) AdoMet decarboxylase; (9) trypanothione reductase; (10) tryparedoxin peroxidase; (11) dithiol reductase; (12) ribonucleotide reductase.

poor oral bioavailability. A potent ODC inhibitor with a longer serum half-life and good oral bioavailability would certainly simplify treatment of late-stage *T. b. gambiense* infection and might be effective against *Leishmania* species in conjunction with inhibition of the polyamine transporters (241).

### 3.4.2 Inhibitors of *S*-Adenosylmethionine Decarboxylase and Related Enzymes. AdoMetDC and spermidine synthase in trypanosomes are crucial for the synthesis of spermidine (19) and qualify as potential targets for antitrypanosomal chemotherapy. Trypanosomal AdoMetDC does not cross-react with a human AdoMetDC antiserum, indicating a considerable difference in the structures of the two enzymes (242). A number of adenosine analogs have been prepared that inhibit AdoMetDC, and have shown activity against the Kinetoplastida. MDL 73811, 5'-[(*Z*-4-amino-2-butenyl)methylamino]-5'-deoxyadenosine (27) (Fig. 19.11), is a time-dependent irreversible inhibitor to the AdoMetDC in *T. b. brucei*, with a $K_i$ value of 1.5 $\mu M$ and $k_{inact} = 0.039$ s$^{-1}$ (243). Intracellular concentrations of putrescine and AdoMet were increased and the level of spermidine was decreased in trypanosomes treated with (27) (244). Compound (27) also proved to be active against trypanosome infections in animal models (245). The drug is efficiently imported into trypanosome to a high intracellular concentration of 1.9 m$M$ by a parasite-specific adenosine transporter not present in mammalian cells (58, 246). The combination of selective import into trypanosomes and irreversible inhibition of AdoMetDC makes (27) worthy of further investigation.

Adenosine analog (28) inhibits AdoMetDC irreversibly, is a substrate for the purine transporter, and has an IC$_{50}$ value of 0.9 $\mu M$ against *T. brucei in vitro* (247). Nucleotide analogs (29) (hydroxyethylthioadenosine; HETA) and (30) have not been shown to inhibit AdoMetDC. These two agents strongly inhibit protein methylation enzymes (mostly carboxyl methylation) that use AdoMet as a cosubstrate. However, these two agents are transported into trypanosomes by the P1 system and both have submicromolar trypanocidal activity against *T. b. brucei* and strains of trypanosomes resistant to melarsoprol (248, 249).

Methylglyoxal bis(guanylhydrazone) (MGBG, 31) is a potent, but nonselective AdoMetDC inhibitor with additional antimitochondrial actions that has been used in cancer chemotherapy (250). In the early 1990s, scientists at Ciba-Geigy undertook the synthesis and biological evaluation of a large series of MGBG analogs in the hope of discovering new cancer chemotherapeutic AdoMetDC inhibitors (251–253). None of the compounds has been successfully developed into marketable anticancer agents. Tests for antitrypanosomal activity of the MGBG derivatives revealed a number of compounds that had trypanocidal activity below 0.5 $\mu M$ *in vitro* (254), and that the activity was maintained *in vivo* (255). The best of the compounds was CGP 40215A (compound 32; Fig. 19.11), which cured acute infections of *T. brucei* in mice at a dose ≤ 25 mg/kg (255). The structural similarity of (32) and the older trypanocides pentamidine (2) and diminazene (7) is significant. All three of these compounds (32, 2, and 7), as well as AdoMet, use the P2 transporter in trypanosomes for active uptake (38, 256a,b). Once in trypanosomes, the agents all exert a trypanocidal action, although it is far from certain that this action is attributable to inhibition of AdoMetDC (50, 53). Compound (32) remains under development as a potential clinical antitrypanosomal agent. Two additional compounds with bis-guanidine-related structures are in development as antitrypanosomal agents. DB289 (33b), the bis(methyl carbamate) prodrug of furamidine (33a), is orally active against *T. brucei* and is one of two drugs being developed by the WHO on a grant from the Bill Gates Foundation (257). The drug is currently in Phase II clinical trials. Furamidine binds tightly to AT sequences in the minor groove of DNA (258a,b); however, its mechanism of antitrypanosomal activity remains undetermined. Trybizine (SIPI 1029, 34) was found to have IC$_{50}$ values of 0.15 to 2.15 n$M$ in four strains of cultured African trypanosomes (259–261). In the mouse *in vivo* screen, (34) cured 12 of 13 trypanosome isolates at a dose of ≤10 mg/kg i.p. and it showed

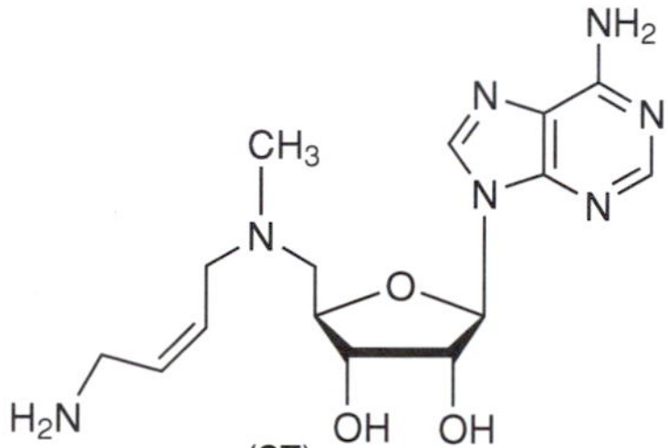

(**27**)

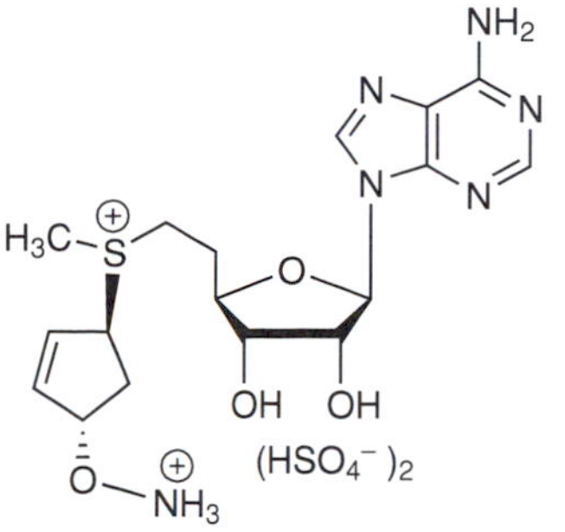

(**28**)

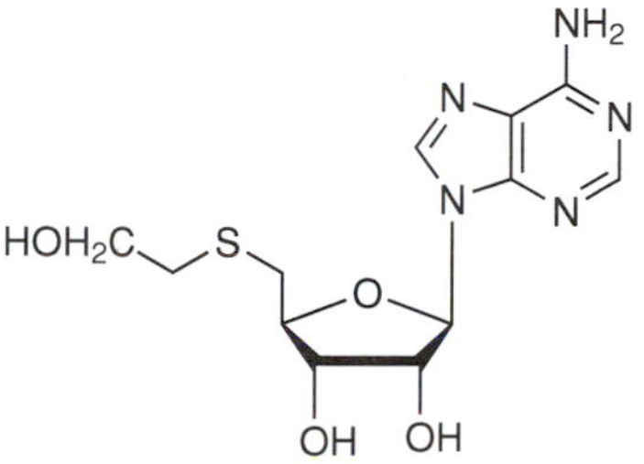

(**29**)

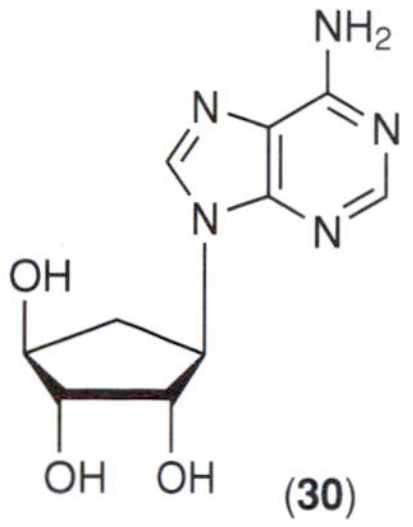

(**30**)

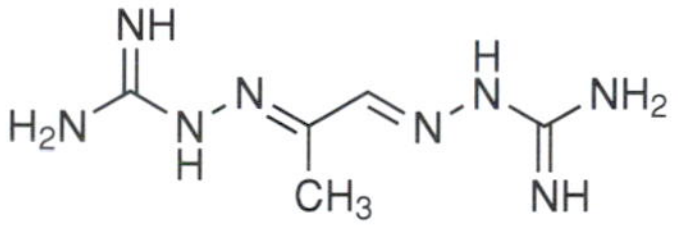

MGBG (**31**)

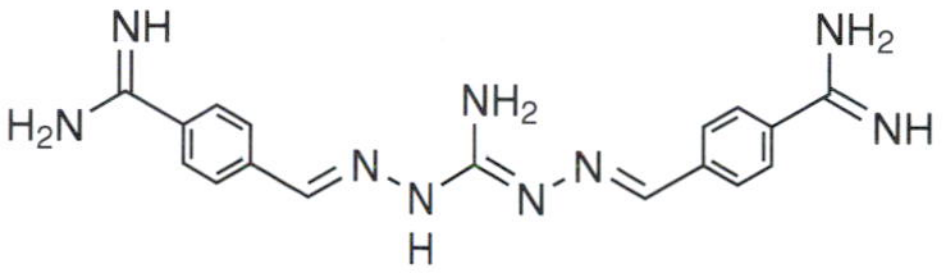

CGP 40215A (**32**)

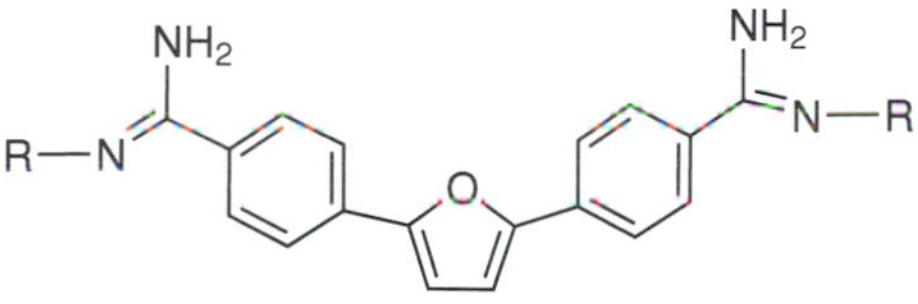

R,R = H; Furamidine (**33a**)
R,R = -COOMe; DB289 (**33b**)

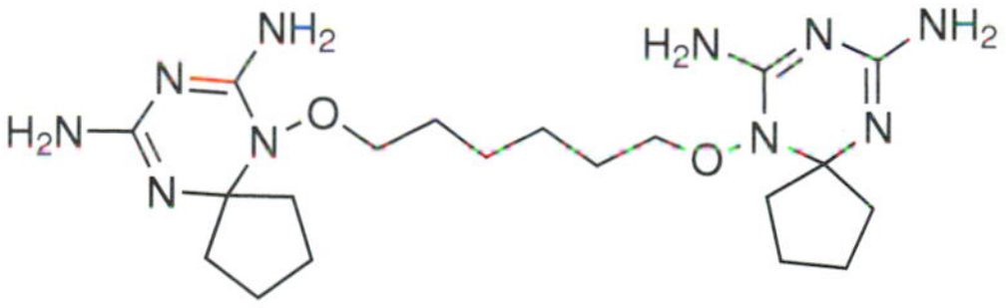

Trybizine (**34**)

**Figure 19.11.** Antitrypanosomal compounds that inhibit AdoMetDC. Several of the compounds have not been shown to be AdoMetDC inhibitors but are included because of their structural similarity to the known AdoMetDC inhibitors in the figure.

oral activity at higher doses. Compound (**34**) inhibits AdoMetDC, with an $IC_{50}$ value of 38 $\mu M$ (261).

**3.4.3 Inhibition of the Biosynthesis of Trypanothione and of Trypanothione Reductase.** Kinetoplastids biosynthesize spermidine and glutathione by the same process as that of mammalian species, yet they do not use glutathione as such. Instead, they conjugate glutathione and spermidine to make a unique substance named trypanothione (262, 263). The enzymes for trypanothione synthesis, glutathionylspermidine synthetase (GspS) and trypanothione synthetase (TryS), and trypanothione reductase (TryR) are unique to these

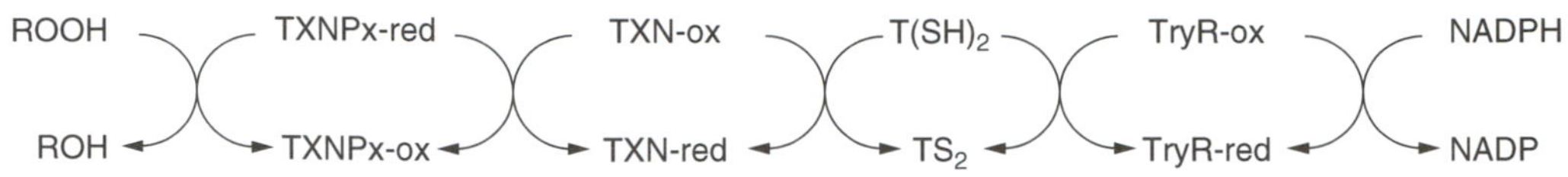

**Figure 19.12.** The trypanothione peroxidase system for detoxifying peroxides. Enzymes trypanothione reductase (TryR) and tryparedoxin peroxidase (TryP) are boxed. T(SH)$_2$ and T(S)$_2$ are trypanothione and trypanothione disulfide, respectively. TryX[SH]$_2$ and TryX[S]$_2$ are the thioredoxin-like proteins tryparedoxin and tryparedoxin disulfide, respectively. ROOH can be hydrogen peroxide or an alkyl or aryl hydroperoxide.

organisms (for reviews see refs. 264–266). Crystal structures of each protein component of the antioxidant pathway in Figure 19.12 have been determined (267–271). Kinetoplastids use trypanothione for most reactions where glutathione is used in mammalian cells. A major function of trypanothione is the protection of the organisms from oxidative stress (Fig. 19.12). Kinetoplastida lack catalase and are highly dependent on the redox cascade described in Figure 19.12 for neutralization of H$_2$O$_2$ and lipid peroxides (272). Tryparedoxin peroxidase and tryparedoxin are members of the thioredoxin family and they may be exploitable as drug targets (see Refs. 273–276 for reviews). The advantage of the biological substitution of trypanothione for glutathione in the kinetoplastids remains unclear. Trypanothione and glutathione have similar redox potentials and thiol-disulfide exchange reactions between them will occur nonenzymatically (265), although trypanothione is slightly more electronegative (−0.242 V for glutathione versus −0.230 V for trypanothione). An additional difference is that the thiols of trypanothione are more acidic than those of glutathione; thus, trypanothione exists more in the thiolate form (15% versus 1% for glutathione) and is more nucleophilic at pH 7.4. Trypanothione remains an effective nucleophile at pH 5.5, whereas glutathione does not (277).

Trypanothione forms stable 1:1 complexes with heavy metal ions such as trivalent arsenicals (61, 62). The stability of the complex is much weaker than that of 2,3-dimercaptopropanol or lipoic acid (61, 62), but probably stronger than with glutathione. The complex between trypanothione and the arsenic of melarsen oxide is believed not to be the major mechanism of action of this drug (63), but may be involved in resistance through export of the trypanothione-arsenical complex (278). Although trypanothione-antimony complexes cannot be separated by HPLC, NMR studies have demonstrated formation of a stable 1:1 complex at pH 7.4, in which trivalent antimony binds to two sulfur atoms and to the oxygen atom of a water molecule (279).

**3.4.3.1 Inhibitors of Glutathionylspermidine Synthetase.** The trypanothione biosynthetic enzymes GspS and TryS have been generally accepted as targets for antitrypanosomal chemotherapy. Both enzymes have been cloned and sequenced (280) from *Crithidia fasciculata*, a kinetoplastid that infects insects. Research on these enzymes has been sparse because of a lack of a ready source of enzymes for study. Oza et al. (281) recently reported the soluble expression of GspS from *C. fasciculata* in *E. coli*. GspS will function in the reverse direction, hydrolyzing $N^1$-glutathionylspermidine to glutathione and spermidine. The amidase activity of GspS lies in the $N$-terminus region of the enzyme and is dependent on a single sulfhydryl group of Cys79. The dual action of the enzyme supports the idea that GspS functions in part as a way to store and retrieve spermidine in the cell, as well as to supply glutathionylspermidine for the synthesis of trypanothione.

The biosynthesis of trypanothione in *T. cruzi* differs from *C. fasciculata* in that a single enzyme with the greatest similarity to TryS from *C. fasciculata* catalyzes formation of trypanothione from glutathione and spermidine (Oza, Tetaud, and Fairlamb, unpublished results). The enzyme shows less specificity toward the polyamine substrate and can also synthesize the homotrypanothione analog (238) from aminopropylcadaverine and glutathione. Like the *Crithidia*, GspS, the *T.*

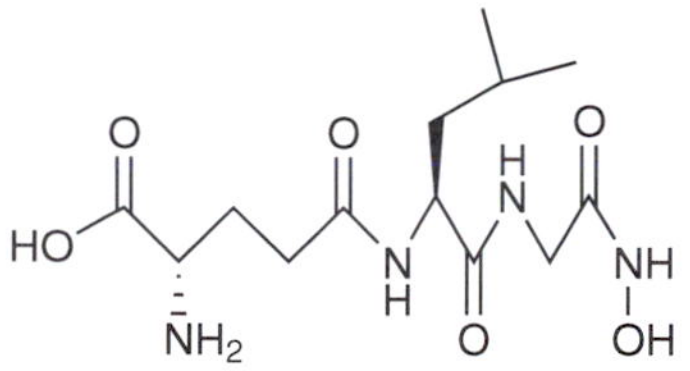

(35)

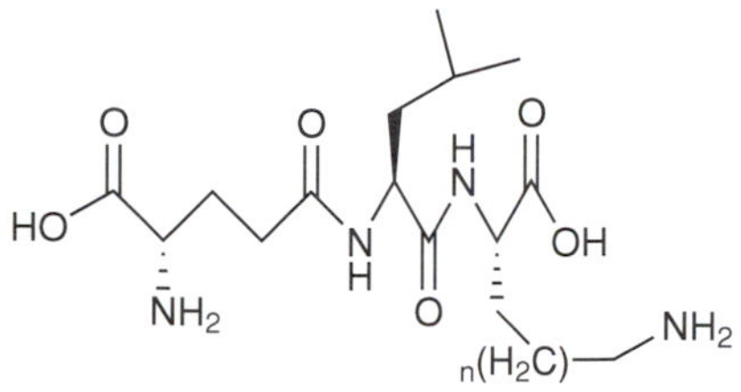

(36a) n = 0
(36b) n = 1

**Figure 19.13.** Inhibitors of glutathionylspermidine synthetase.

*cruzi* TryS, displays weak amidase activity with glutathionylspermidine, trypanothione, and homotrypanothione.

Amssoms et al. (282, 283) reported a series of glutathione-like tripeptides ($\gamma$-Glu-Leu-Gly-X) as potential inhibitors of GspS. The hydroxamic acid derivatization of the glycine carboxylate (**35**, Fig. 19.13) has a $K_i$ value of 2.5 $\mu M$ and substitution of the glycine by diaminopropropanoic acid (**36a**) or lysine (**36b**) gave inhibitors with $K_i$ values of 7.2 and 6.4 $\mu M$, respectively.

### 3.4.3.2 Inhibitors of Trypanothione Reductase.
Trypanothione is an essential metabolite for mediating the redox balance in trypanosomes and trypanothione reductase (TryR) is required to return trypanothione disulfide to its reduced state. TryR uses FAD as a cofactor and NADPH as an electron donor (120). The enzyme was originally purified from *C. fasciculata* (284) and subsequently from other species (120, 285, 286). Gene sequences are available for *C. fasciculata* and medically and veterinary-relevant species (287–291). Detailed X-ray characterizations of the three-dimensional structure were completed on the recombinant TryR from *C. fasciculata* (292). The crystal structure, refined

to 2.6-Å resolution (293), revealed significant differences from that of glutathione reductase, which accounts for the pronounced discriminatory properties of the enzymes for their disulfide substrates (120, 284, 285). Subsequently, crystal structures of TryR (268, 294) and TryR complexed with mepacrine (295) from *T. cruzi* have been reported. The binding site for trypanothione is more open because of the rotations of two helical domains that form part of the active site. In addition, the highly positive charged and hydrophilic region of glutathione reductase, where the glycine carboxylate of glutathione interacts with the enzyme, is replaced by a hydrophobic and negatively charged pocket in trypanothione reductase (292). TryR has a glutamic acid at position 17 that could either form a hydrogen bond with one of the amide linkages between the spermidine and glycine carboxylate (267) or interact with the positively charged secondary amine of spermidine. The marked differences between glutathione and trypanothione reductases in their disulfide binding sites provide an excellent basis for the discovery of specific TryR inhibitors by structure-based drug design methods (296). Validation of TryR as a drug target has been provided by a variety of genetic experiments (297–299), including conditional gene knockouts in *T. brucei* (300). Agents that are proposed to act as inhibitors of TryR are shown in Fig. 19.14.

A number of tricyclic analogs were selected from the computer model of TryR and docking experiments with phenothiazine-based structures (301). Compounds from this lead were obtained or synthesized and evaluated for inhibitory activity on TryR. The first strong lead for an inhibitor was chlorpromazine (**37**), which has an IC$_{50}$ value of 35 $\mu M$ against recombinant trypanothione reductase from *T. cruzi* (302). Open-ring compound (**38**) was made in an attempt to separate the neuroleptic properties of (**37**) from its TryR inhibition activity and resulted in a slight increase in TryR inhibitory activity (46% inhibition of TryR from *T. cruzi* versus 16% for **37**) (303). A series of tricyclic agents related to the antidepressant drug clomipramine (**39**) was prepared and evaluated for TryR inhibitory activity. Clomipramine (IC$_{50}$ = 6.5 $\mu M$) was the most active compound against TryR from *T.*

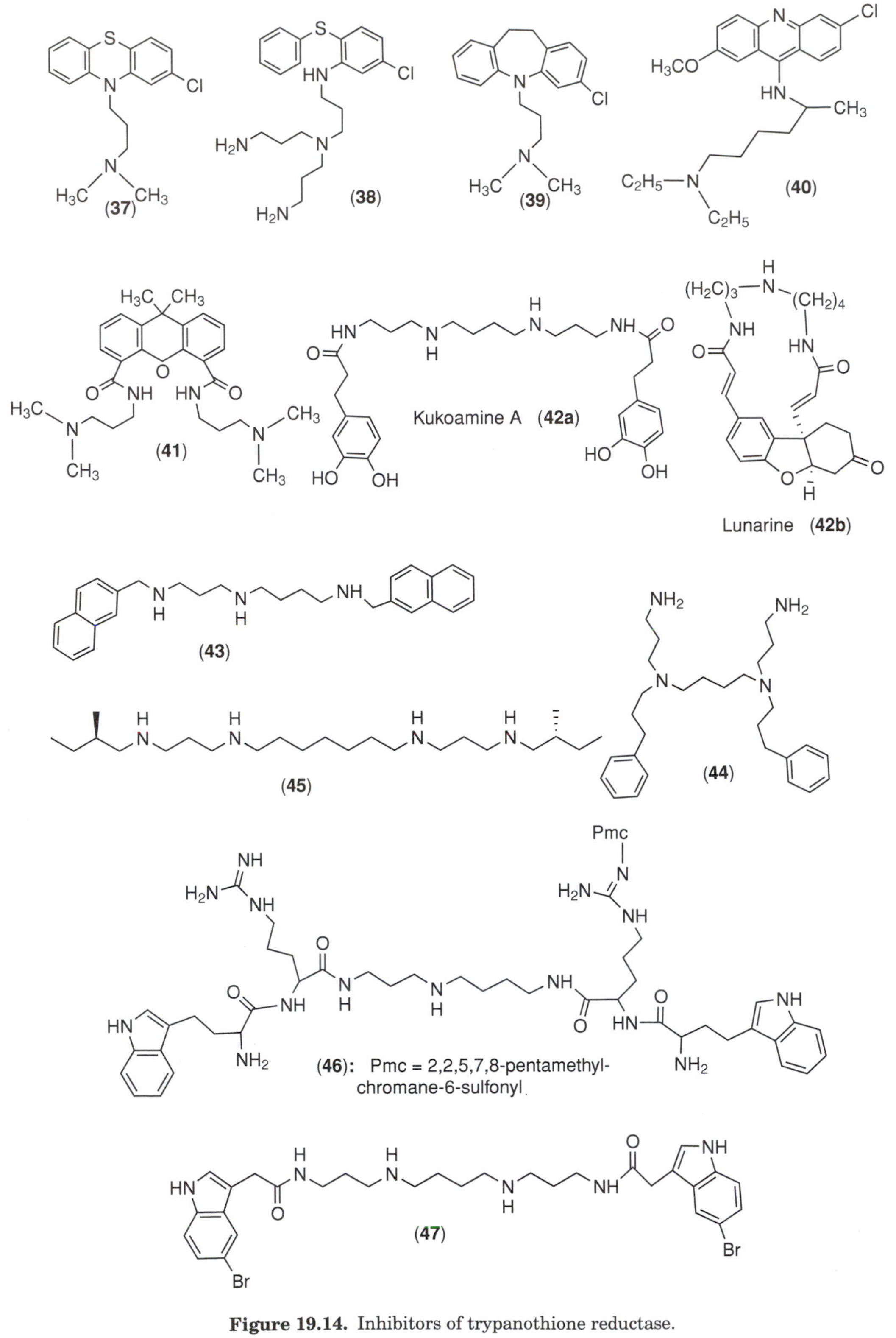

**Figure 19.14.** Inhibitors of trypanothione reductase.

*cruzi* (304). Tricyclic acridine derivatives related to known antiparasitic drugs were investigated for TryR inhibitory activity. The most active compound found was the mepacrine homolog (**40**), which has a $K_i$ value of 5.5 $\mu M$, although simple acridines lacking the entire aminoalkyl substituent from the 9 amino group were nearly as potent (305). The 9-aminoacridines related to (**38**) were competitive inhibitors of TryR, with more than one molecule of inhibitor able to bind to the enzyme. Two 9-thioacridine compounds were prepared and gave $K_i$ values similar to those of the 9-aminoacridines but they exhibited mixed-type kinetics. These findings demonstrate the difficulty in drug design through the use of computer-aided structure fitting or in the development of QSAR for TR inhibitors (305). Chibale et al. (306) prepared 9,9-dimethylxanthines as potential TryR inhibitors. The most active compound in the series was (**41**) ($K_i$ = 13.8 $\mu M$).

A popular approach to the preparation of TryR inhibitors has been the substitution of various groups onto a polyamine scaffold. Natural product lead compounds kukoamine A (**42a**) and lunarine (**42b**) are analogs of spermine and have $K_i$ values of 1.8 $\mu M$ (307) and 144 $\mu M$ ($k_{inact}$ = 0.116 s$^{-1}$) against TryR from *T. cruzi*, respectively (267). O'Sullivan et al. (308, 309) reported several series of aryl alkyl spermidine derivatives. Of the compounds prepared, (**43**) ($K_i$ = 9.5 $\mu M$) and (**44**) ($K_i$ = 3.5 $\mu M$) were the most active. A series of bis-alkyl and aryl-alkyl terminal N-substituted spermine-related compounds, having a 3–7–3 pattern of hydrocarbon chain lengths between the nitrogen atoms, has been prepared. Compound (**45**) is one of the most potent of the compounds, having IC$_{50}$ values of 40 and 165 n$M$ against the K243 and resistant K243-As-10–3 strains of trypanosomes, respectively (310). The $N^1,N^{17}$-bis(benzyl) analog of (**45**) was nearly as potent. Smith and Bradley (311) prepared three 576-compound libraries of spermidine-amino acid conjugates and screened the compounds for their ability to inhibit TryR. Compound (**46**) was the most potent, having a $K_i$ value of 100 n$M$. This work was followed by additional solid-phase synthesis attempting to optimize the activity of the lead compounds. The $N^1,N^{12}$-bis-(5-bromo-3-indole acetic acid) amide derivative of spermine (**47**) had a $K_i$ value of 76 n$M$ when tested against TryR from *T. cruzi* (312).

For compounds in the studies described in the preceding paragraph, it should be noted that there is not a good correlation between their ability to inhibit TryR and their toxicity against cultured trypanosomes. Thus, the mechanism by which these compounds exert a trypanocidal action remains in doubt and multiple mechanisms of activity may be involved. In addition, all of the compounds that have been tested *in vivo* for their ability to cure mice acutely infected with trypanosomes have yielded disappointing results. It is not known whether the lack of *in vivo* potency is attributed to poor uptake by trypanosomes, to rapid metabolism of the compounds by the host, or to other unidentified factors. It should be borne in mind that strictly competitive inhibitors of low nanomolar potency may be required to maintain high levels of inhibition in the face of possibly millimolar levels of trypanothione disulfide accumulating intracellularly as a result of effective inhibition of TryR. Irreversible tight-binding inhibitions may be required. Certainly, the inability of medicinal chemists to find a drug to attack what appears to be a prime target for selective toxicity is frustrating.

***3.4.3.3 Subversive Substrates.*** Henderson et al. attempted the design of subversive substrates for TryR (104). The toxic action of bleomycin on cancer cells might be considered an analogy to the subversive substrate approach to drug design (313). The objective of this strategy is to find a compound that will bind to the active site of the target enzyme (TryR in this case) and subvert electrons from their normal flow from NADPH by way of FAD to trypanothione disulfide (Fig. 19.12). The compound should accept electrons from the FAD cofactor of TryR and divert them into a redox cycling-based mechanism of cell toxicity, as described in Fig. 19.6. The resulting oxygen-derived free radicals would damage the enzyme, or important structures in the surrounding cell, and result in a trypanocidal action. In addition, the normal function of the redox cascade from NADPH to peroxides is disrupted, which in itself might be sufficient for a trypanocidal action. This method of at-

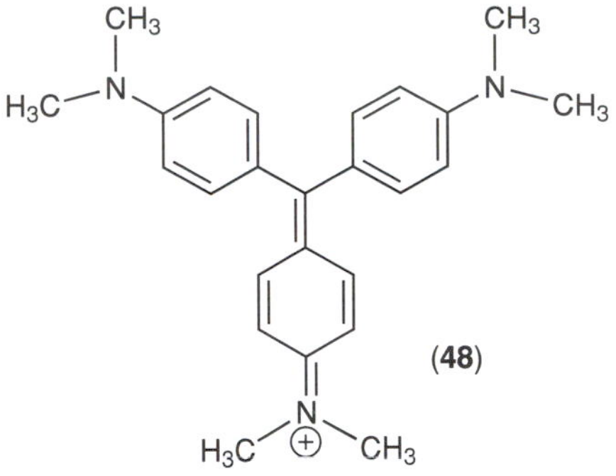

(48)

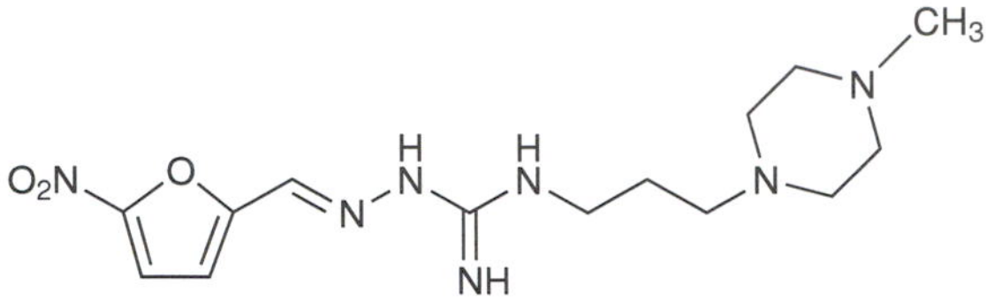

(49)

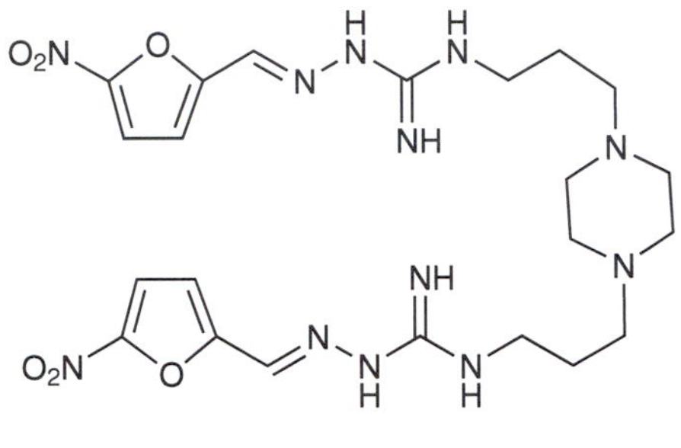

(50)

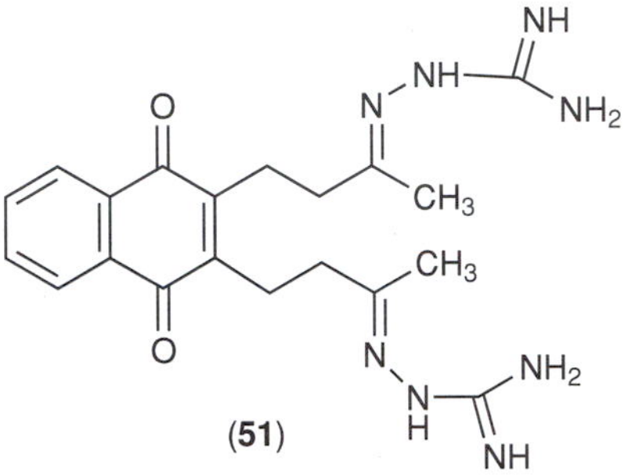

(51)

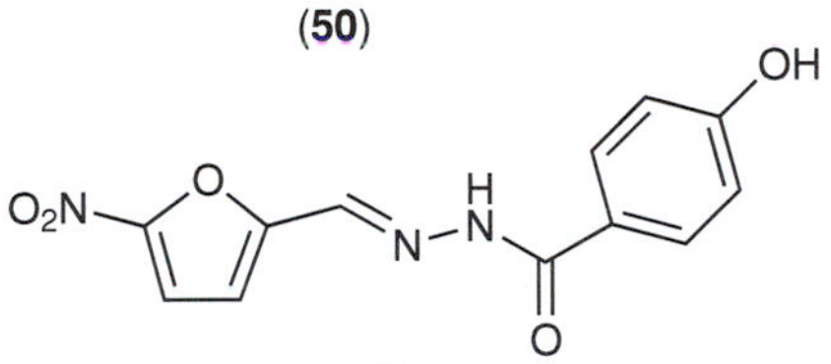

(52)

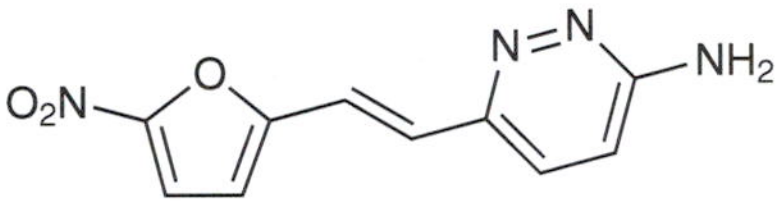

(53)

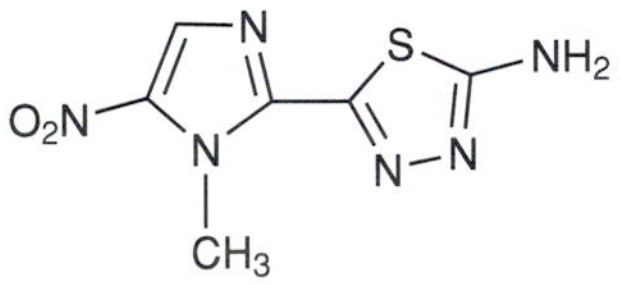

Megazol (54)

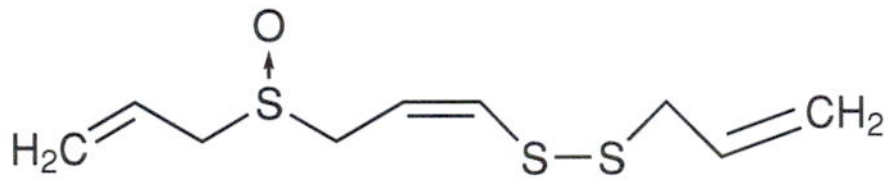

(55)

**Figure 19.15.** Subversive substrates.

tack seemed reasonable on the basis of the known antitrypanosomal activity of compounds like nifurtimox (**12**) and benznidazole (**13**). Structures of other agents that may have a mechanism of action by serving as subversive substrates are given in Fig. 19.15. Interestingly, crystal violet (**48**) is a potent inducer of oxidative stress and has long been used in blood banking to kill *T. cruzi* trypomastigotes and thus prevent the spread of Chagas' disease through blood transfusions (314).

Attempts to find an effective trypanocidal subversive substrate for TryR have been only partially successful. In the initial work, two nitrofuran derivatives (**49** and **50**) and the naphthoquinone (**51**) were identified as trypanocidal subversive substrates. All three of these compounds were effective at a concentration range of 1–5 $\mu M$ in preventing *T. cruzi* trypomastigotes from infecting human saphenous vein smooth muscle cells in culture (104). None of the compounds is very active in *in vivo* screens for trypanocidal activity. Other nitrofuran derivatives such as nifuroxazide (**52**) and nifurprazine (**53**) are effective subversive substrates for TryR and possess trypanocidal activity *in vitro* (315). Cenas et al. (316) published on the mechanism of reduction of quinones by TryR from *T. congolense*. A library of 1360 1,4-naphthquinone derivatives was synthesized and tested for activity against TryR from *T. cruzi*. The most potent compounds from this series had $IC_{50}$ values of 0.3–1.1 $\mu M$ and were highly selective for TryR vs. glutathione reductase (317).

Megazol (**54**), structurally related to the nitrofurans and other subversive substrates, is an important compound that is in development as an antitrypanosomal agent. The mechanism of the drug is not known, but it probably is associated with its ability to undergo cyclical one-electron reduction, thereby generating reactive oxygen species *in vivo*. Megazol has not been shown to inhibit any specific trypanosomal enzyme. Megazol is effective in curing *T. b. brucei* infections in mice (318) and is effective in combination with melarsoprol (319) and suramin (320). Megazol owes part of its effectiveness to its active uptake into trypanosomes by the P2 transporter (321). Studies on the pharmacokinetics, metabolism, and excretion of megazol have been published (322, 323).

Ajoene (**55**), a component of garlic, has been found to be an effective irreversible inhibitor of both TryR and glutathione reductase. Ajoene forms a mixed disulfide bond with Cys58 of TryR (324). In addition, ajoene is a substrate for TryR and leads to the accumulation of reactive oxygen species.

## 3.5 Inhibitors of Lipid and Glycolipid Metabolism

With the exception of bloodstream *T. brucei*, which obtains cholesterol from host LDL, the sterol biosynthesis in other kinetoplastids resembles that of fungi much more closely than that of mammalian species. Ergosterol is the major sterol used by this order in membrane stabilization. Thus, agents that have proved to be effective in treating systemic fungal infections are worthy of investigation for activity against *T. cruzi* and *Leishmania* spp. Amphotericin B (**18**) has proved to be highly effective for the treatment of visceral leishmaniasis. Structures of agents acting as inhibitors of sterol and lipid biosynthesis are given in Figure 19.16.

**3.5.1 Inhibitors of 14$\alpha$-Demethylase and Other Enzymes Related to Sterol Biosynthesis.** The azole class of inhibitors of 14$\alpha$-demethylase, blocking the conversion of lanosterol to 4,4-dimethylcholesta-8,14,24-triol, in the biosynthetic pathway to ergosterol, has yielded varied effectiveness in the treatment of leishmaniasis and Chagas' disease. The gene for 14$\alpha$-demethylase from *T. brucei* has been cloned and characterized (325). In most cases, treatment with an azole agent slows progression of the disease but rarely results in a long-term cure (see Refs. 326–328 and references cited therein). Recent results with newer agents and combination therapy are more encouraging. Alrajhi et al. (328) reported the results of a clinical trial of a 6-week oral fluconazole (**56**) treatment of cutaneous infections caused by *L. major*. The median time to healing in the fluconazole treated group was 8.5 weeks compared to 11.2 weeks in the placebo group ($P < 0.001$). It was concluded that the fluconazole therapy was safe and useful. Ketoconazole (**57**) has been shown to have a synergistic effect in combination with lysophospholipid analogs related to miltefosine (**17**) against the epimastigote form of *T. cruzi* grown in culture (142). A combination of benznidazole (**13**) and (**57**) was found to be effective in curing resistant *T. cruzi* infections in mouse model experiments (329). Third-generation triazole antifungal agents have a more potent trypanocidal action on *T. cruzi* than the older triazoles. Urbina et al. (330) found that D0870 (**58**) is able to cure acute and chronic *T. cruzi* infections in mice and were the first compounds ever shown to have this activity. Resistant strains of *T. cruzi* were cured in mice by (**58**) incorporated into polyethyleneglycol-polylactide nanospheres (331). Unfortunately, the clinical development of (**58**) has been dis-

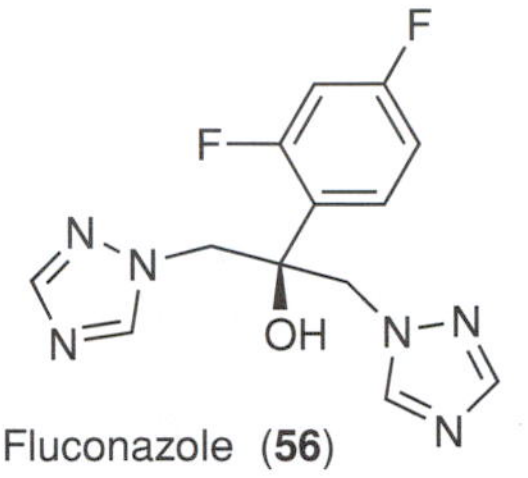

Fluconazole (**56**)

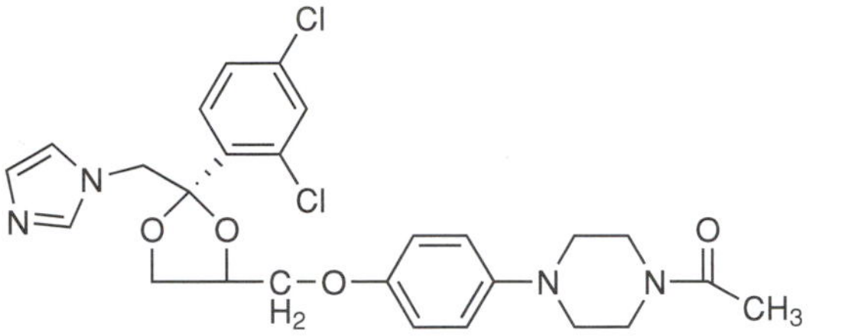

Ketoconazole (**57**)

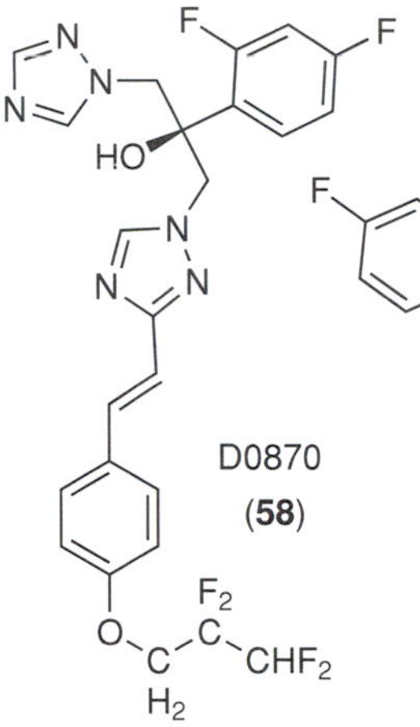

D0870 (**58**)

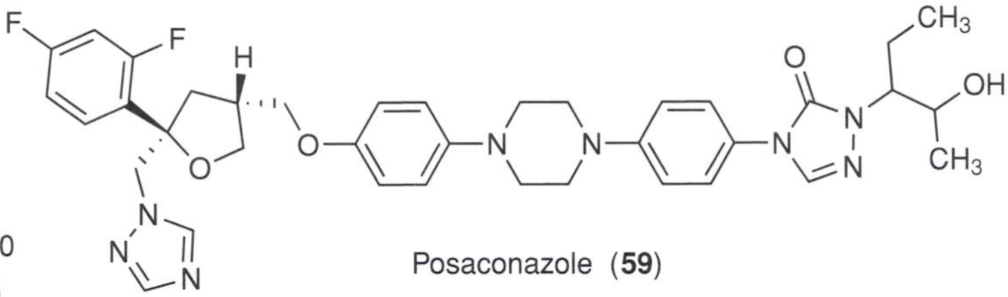

Posaconazole (**59**)

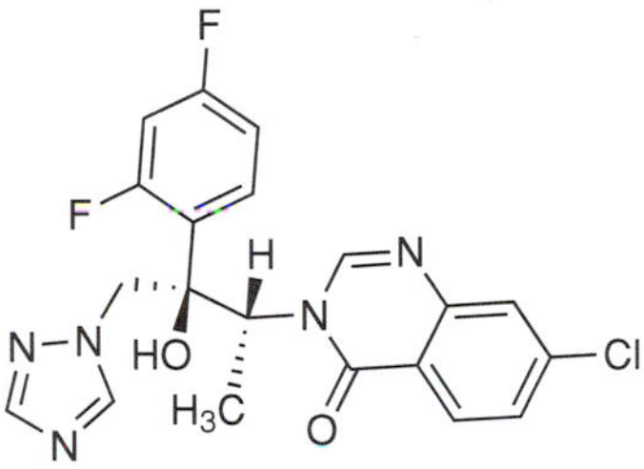

UR-9825 (**60**)

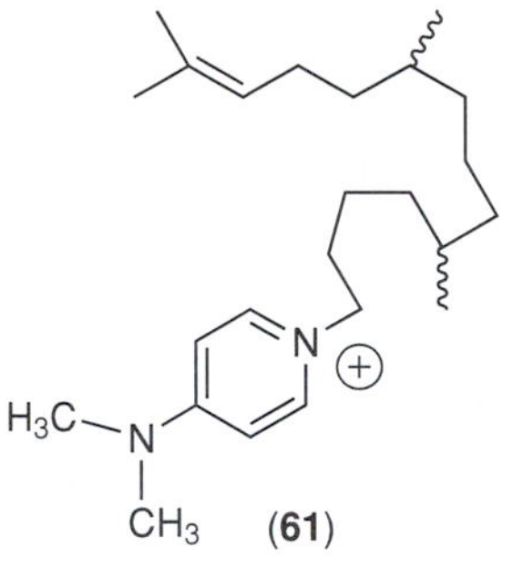

(**61**)

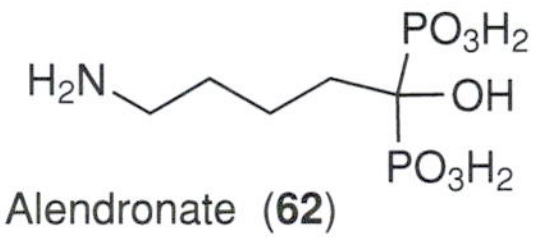

Alendronate (**62**)

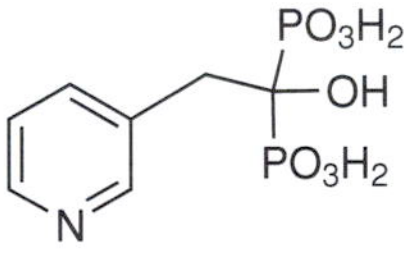

Risedronate (**63**)

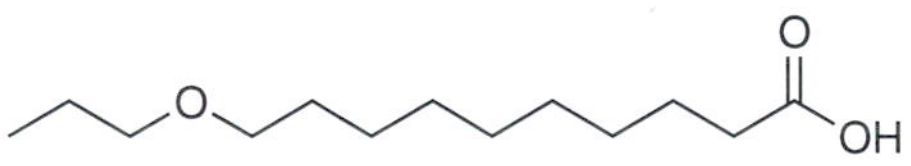

(**64**)

**Figure 19.16.** Investigational agents affecting lipid metabolism.

continued by Zeneca because of adverse reactions. Posaconazole (**59**), another third-generation triazole, has shown activity similar to that of (**58**) and clinical trials with this agent are planned (332, 333). The investigational triazole agent UR-9825 (**60**) also has been found to be extremely potent, with MIC values of 30 and 10 n*M* against *T. cruzi* epimastigotes and amastigotes in cell culture, respectively (334).

Other enzymes in the sterol pathway may offer targets for antitrypanosomal chemotherapeutic agents. Urbina et al. (335) and Ro- drigues et al. (336) reported that a 22,26-aza- sterol analog has potent inhibitory activity against delta$^{24(25)}$ sterol methyl transferase from *T. cruzi*. Sterol biosynthesis is disrupted and parasites die when exposed to 100 n$M$ con- centration of inhibitor in the culture medium. Compound (**61**), an inhibitor of oxidosqualene cyclase, the enzyme that converts 2,3-oxido- squalene to lanosterol, had ED$_{50}$ values of 2–6 n$M$ against four strains of *T. cruzi* in cell cul- ture, whereas the ED$_{50}$ value for inhibition of murine 373 fibroblasts was 880 n$M$ (337).

### 3.5.2 Inhibitiors of Prenylation and Protein Anchoring.

Bisphosphonates, used medicinally as inhibitors of bone resorption, have been found to be active both *in vitro* and *in vivo* against kinetoplastids at concentrations caus- ing little toxicity to host cells (338). The mode of action of the bisphosphonates is thought to involve the inhibition of prenyldiphosphate synthetase and the prenylation of proteins (339–342). The selectivity of the bisphospho- nates may be attributed in part to their accu- mulation in acidocalcisomes, organelles stor- ing high concentrations of pyrophosphate that are specific to trypanosomatids and apicom- plexan parasites (338, 343). Alendronate (**62**) and risedronate (**63**) were among the most ac- tive bisphosphonates tested for inhibition of *T. cruzi* farnesyl pyrophosphate synthetase, having IC$_{50}$ values of 0.77 and 0.037 $\mu M$, re- spectively (342).

The variant surface glycoprotein (VSG) coat of bloodstream African trypanosomes is anchored to the cell surface by a glycosyl phos- phatidylinositol (GPI) that contains myristate as its only fatty acid component (344, 345). *T. brucei* has several hundred VSG genes, only a fraction of which have been cloned (346). In a process termed fatty acid remodeling, two fatty acids with longer carbon chains than those of myristate on the GPI precursor are sequentially replaced by two myristate mole- cules (347). Alternatively, myristoylation can occur by an acyl exchange reaction. Antigenic variations enable the trypanosomes to evade the host immune response and are essential for parasite survival. The process of GPI my- ristate exchange or fatty acid remodeling does not occur in humans. Thus, a drug that would

block the process of fatty acid remodeling, acyl exchange, or replace the myristate with a dys- functional fatty acid might be an effective trypanocidal agent. Doering et al. (348) stud- ied a set of 244 myristate analogs, on the basis of 11-oxatetradecanoic acid (**64**) as a lead com- pound, for their ability to incorporate into the GPI both in a cell-free system and in intact trypanosomes. Twenty compounds that were toxic to trypanosomes at 10 $\mu M$ were discov- ered from the series. Compound (**64**) had an LD$_{50}$ value for the cultured *T. brucei* of 1.0 $\mu M$ and some of the compounds in the series were even more toxic to trypanosome growth. None was subsequently found to be active in mice. The trypanocidal activity of the compounds correlated roughly with their degree of incor- poration into the VSG; however, the mecha- nism of action for the compounds could not be determined with certainty.

In a study related to the VSG, Roper et al. (349) cloned the gene UDP-glucose-4'-epimer- ase, which is responsible for the conversion of UDP-glucose to UDP-galactose in *T. brucei*. A UDP-glucose-4'-epimerase knockout in *T. brucei* was also constructed. The hexose trans- porter of *T. brucei* will not carry galactose and it was believed that this enzyme might be the only source of galactose in the trypanosome. Galactose is an essential component of the VSG coat. Deletion of UDP-glucose-4'-epimer- ase was lethal to the trypanosomes, thus vali- dating the enzyme as a target for drug design.

## 3.6 Cysteine Protease Inhibitors

The major proteases of the order Kinetoplas- tida are of the cathepsin B-like and cathepsin L-like subfamilies of the papain family of cys- teine proteases. Sajid and McKerrow (350) published an excellent review of this topic. *T. cruzi* expresses a large amount of cysteine pro- tease most closely related to cathepsin L. The enzyme has been named cruzain and is synon- ymous with cruzipain. Cruzain is expressed in varying amounts in all life forms to *T. cruzi* (351, 352). Chagasin, a 12-kDa protein iso- lated from *T. cruzi* epimastigote lysates, is a tight-binding inhibitor of cruzain and appar- ently functions as a natural regulator of cruzain's activity (353). Cruzain has been cloned (354) and the three-dimensional struc- ture determined (355, 356). Cruzain has been validated as a target for drug development in studies by Engel et al. (357). The catalytic mechanism of the cysteine proteases is well understood from extensive studies on papain and related enzymes. Resistance to cysteine protease inhibitors can be induced in *T. cruzi* through the mechanism of enhancement of a protein secretory pathway (358).

*T. brucei* expresses a cysteine protease, of the cathepsin L subfamily, similar to cruzain. The enzyme, called brucipain (or trypanopain-Tb), has been cloned and expressed in *E. coli* (359). Brucipain and cruzain have a 59% sequence homology and a approximately 45% homology with mammalian cathepsin L enzymes. Brucipain has been validated as a target for drug development (360); however, selectivity for parasite vs. host enzymes remains a problem.

*Leishmania* spp. express three genes for cysteine proteases: the cathepsin L-like and cathepsin B-like proteases plus calpain-like protease. Inhibitors effective on cruzain and brucipain are effective on the enzyme(s) from *Leishmania* spp. (361), but usually less active on this species because of the multiple forms of the proteases present.

A notable difference between mammalian and Kinetoplastid cathepsin enzymes lies in the carboxy-terminus, where the kinetoplastid enzymes have an extension of about a 100 amino acids, made up of polythreonine residues in cruzain and polyproline residues in brucipain. The function of the additional amino acids is not known, but they are not necessary for the catalytic activity of the enzymes. An additional difference of the parasitic enzymes is their pH-activity profile. Human cathepsin L resides predominantly in lysosomes and has a pH maximum at 4.5, whereas cruzain functions well at higher pH and has a pH maximum near 8.0 (350).

Structures of investigational agents that work through a mechanism of cysteine protease inhibition are given in Figure 19.17. The first cysteine protease inhibitors shown to have activity against trypanosomes were the dipeptide fluoromethylketone derivative (**65a**) and the related diazomethylketone (**65b**) (360, 362). Compound (**65b**) was the more active compound ($ED_{50} = 0.18$ $\mu M$) against *T. brucei* cultures. These compounds

were able to cure parasite infections in mice but produced unacceptable toxicities at high doses (350). A series of dipeptidyl-epoxide derivatives were investigated for their trypanocidal activity. Epoxide (**66**) had an $IC_{50}$ value of 10 n$M$ against cruzain, but no activity of the compound toward trypanosomes has been reported (363). The most promising trypanocidal cysteine protease inhibitor reported to date is the vinyl sulfone (**67**) (364a,b). This drug is highly effective in curing acute and chronic *T. cruzi* in mice. Compound (**67**) showed little or no toxicity to the animals at the treatment doses employed. It is projected that (**67**) will enter clinical trials for the treatment of Chagas' disease in the near future (350). Huang et al. (365) reported a series of mercaptomethyl ketones as potent and selective inhibitors of cruzain. Compound (**68**) had $K_i$ values of 1.1, 1700, and 144 n$M$ for recombinant cruzain, cathepsin B, and cathepsin L, respectively. Compound (**68**) showed time-dependent inhibition and tight binding to cruzain.

Compounds (**65–68**) are all irreversible inhibitors of cysteine proteases. In an attempt to discover competitive inhibitors for the enzyme, an *in silico* docking search of a chemical databank for fit to the cruzain active site was performed. Lead compounds have been prepared and tested, and potency optimization is being pursued. The most potent compounds reported from these studies to date are the hydrazides (**69**) and (**70**), which have $ED_{50}$ values against bloodstream *T. brucei in vitro* of 0.3 and 0.6 $\mu M$, respectively (366).

## 3.7 Inhibitors of Dihydrofolate Reductase/Pteridine Reductase

It has been determined that *Leishmania* spp. are resistant to dihydrofolate reductase (DHFR) inhibitors because they amplify the production of pteridine reductase (PTR), which can reduce folates and bypass the inhibition of DHFR (367). An attempt to find compounds capable of inhibiting these enzymes simultaneously led to the discovery of structures that inhibit both DHFR and PTR (368). Compound (**71**) (Fig. 19.18) was one of several compounds that showed submicromolar $IC_{50}$ values against PTR, DHFR, and cultured *L. major*. Gilbert's group used the Cambridge

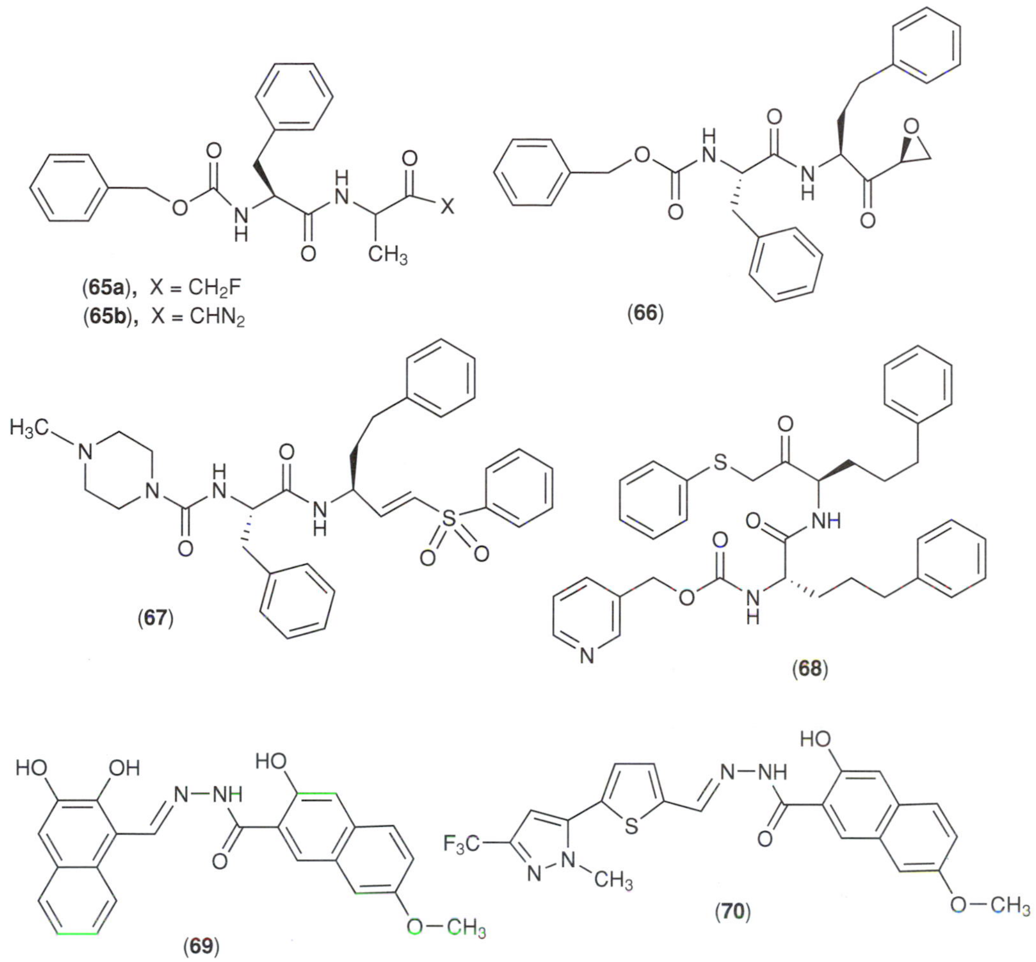

**Figure 19.17.** Cysteine protease inhibitors effective as trypanocides.

Structural Database and the computer program Dock 3.5 to perform a virtual screen on a homology model of *T. cruzi* DHFR. Leads from the virtual screen were tested for activity against *T. cruzi* DHFR and the most potent compounds were assayed against cultured *T. b. rhodesiense*. Structures (**72**) and (**73**) have been identified as two of the more active compounds, each having activity in the low micromolar range (369, 370).

### 3.8  Inhibition of Purine Interconversions

African trypanosomes lack the capability of *de novo* synthesis of purine nucleotides (371). Carter and Fairlamb (58) characterized P1 and P2 adenosine transporters in *T. b. brucei* that function to scavenger purines from the serum of the host. Melarsen oxide, pentamidine, megazol, and other useful antitrypanosomal agents use the P1 or P2 transporters to concentrate their toxic effects in the trypanosomal cells. In addition to the use of the transporters for drug targeting, it has been proposed that potent inhibitors to the transporters might yield trypanocidal drugs. However, the multiplicity of nucleoside and nucleobase transporters make this an extremely challenging proposition.

Once inside the cell, the purines must be converted to useful nucleotides by the purine salvage enzymes in trypanosomes. Thus, the purine salvage enzymes are credible targets

(71)

(72)

(73)

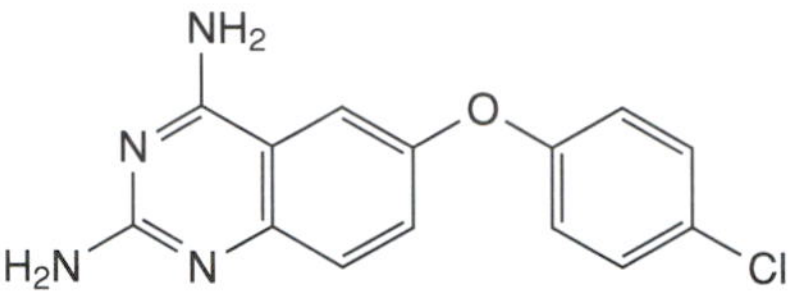

**Figure 19.18.** Inhibitors of trypanosomal dihydrofolate reductase.

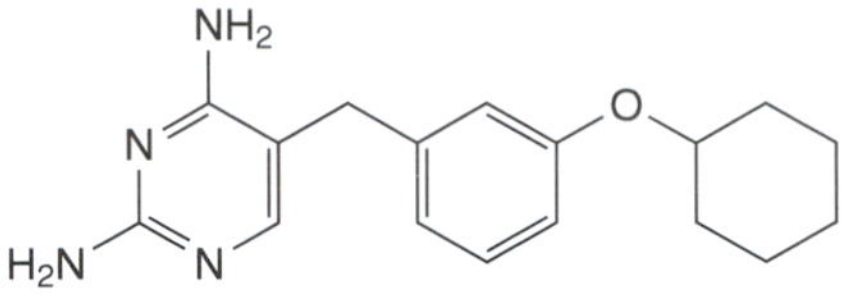

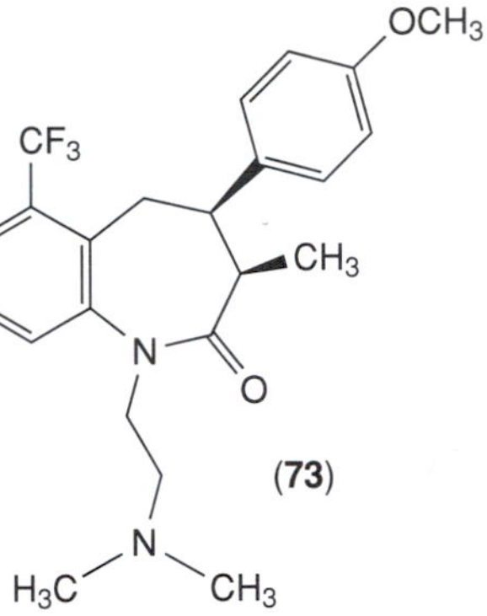

for antitrypanosomal chemotherapy. Again, the considerable redundancy in the salvage and interconversion pathways presents a major challenge for drug design. To date, the most successful purine analogs are prodrugs that are metabolically activated through several enzyme-catalyzed steps. The structures of several drugs that affect purine or pyrimidine interconversions are shown in Fig. 19.19. The most successful example of this approach was the identification of antileishmanial activity in allopurinol (**74**) (372). Allopurinol is a sub-

Allopurinol
(74)

(75)

Acivicin
(76)

DON (77)

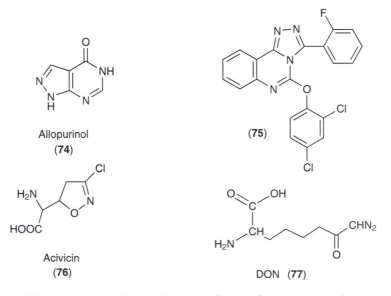

**Figure 19.19.** Inhibitors of purine and pyrimidine interconversion.

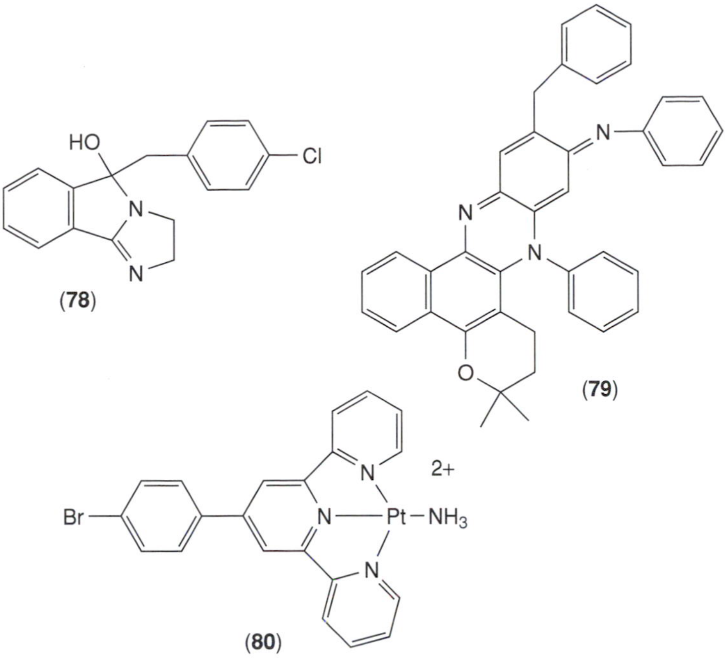

**Figure 19.20.** Miscellaneous new leads for antitrypanosomal drugs.

strate for hypoxanthine phosphoribosyltrans-ferase (HPRT) from *L. donovani*, forming allopurinol-ribose-5'-phosphate (allo-5'-RP). Allo-5'-RP is a substrate for the rest of the parasite enzymes involved in converting IMP to ATP and incorporation of ATP into RNA (204). Allo-5'-RP is an inhibitor of GMP reductase and IMP dehydrogenase, and the allo-5'-RP compounds cause an increase in the breakdown of RNA in *L. donovani* (373). The antileishmanial activity of allopurinol is probably a sum of all of these toxic effects on purine interconversion. The clinical efficacy of allopurinol by itself is weak (374); however, the activity of allopurinol improves when it is combined with fluconazole (375).

The crystal structure of HPRT has been determined from multiple sources (see refs. 376–378 and references cited therein). A computer model of the active site of HPRT from *T. cruzi* was used in a docking search against a database of 34 million compounds (379). Twenty compounds were selected for best fit and screened for inhibitory activity against HPRT from *T. cruzi*. Sixteen of the compounds had IC$_{50}$ values in the range 0.5 to 17 $\mu M$. Four compounds from the 16 were se-

lected for screening against cultured *T. cruzi* in irradiated mouse macrophages. At 10 $\mu M$, two of the compounds showed potent trypanocidal activity. Structure (**75**) is one of the best leads from this study (379). Hofer et al. (380) investigated CTP synthetase in *T. brucei* as a potential target for drug development. They found that CTP pools are very low in *T. brucei* and that this parasite compensates for inhibition of CTP synthetase by the salvage of cytidine. Two inhibitors of CTP synthetase, acivicin (**76**) and DON (**77**), were tested against *T. brucei* in cultured cells and in *T. brucei*-infected mice. Both compounds were effective as trypanocidal agents in these tests (380).

### 3.9 Miscellaneous Active Compounds

Structures of several miscellaneous drugs affecting trypanosomal growth are given in Fig. 19.20. Compound (**78**), synthesized on the basis of a stilbenoid lead, had IC$_{50}$ values of 10 and 7.9 $\mu M$ against cultured *T. cruzi* epimastigotes and trypomastigotes, respectively (381, 382). The mechanism of action of these compounds is undetermined. Neves-Pinto et al. (383) reported derivatives of $\beta$-lapachone, of which the most active compound was (**79**)

($ED_{50}$ = 61 $\mu M$ against *T. cruzi* trypomastigotes). The activity of (**79**) may be related to its structural resemblance to 5-methylphenazinium, a known inducer of free radicals. The (terpyridine) platinum (II) complex (**80**) showed potent trypanocidal action against *L. donovani* (100% at 1 $\mu M$), *T. cruzi* (65% at 1 $\mu M$), and *T. brucei* (100% at 30 n$M$). The mechanism of action of (**80**) is thought to be platination of parasitic DNA (384, 385).

## 4   CONCLUSIONS

Despite their devastating impact on world health, few new drugs to treat the Kinetoplastida diseases have been introduced in the last 25 years. Several drugs are currently in clinical trials. Miltefosine offers hope as the first effective oral treatment to cure visceral leishmaniasis and other members of the lysophospholipid class of agents show promising trypanocidal activity. Megazol has shown good oral activity against African trypanosomiasis and is currently in clinical trials. The third-generation triazole antifungal agents posaconazole and UR-9825 have shown potent *in vivo* activity against *T. cruzi* infections in mice. Members of the existing bisphosphonates class of bone resorption inhibitors have potent trypanocidal activity *in vivo* and may provide new chemotherapeutic drugs for the Kinetoplastida diseases. A potent irreversible vinyl sulfonyl peptidomimetic inhibitor of cysteine proteases is scheduled to begin clinical trials for treatment of Chagas' disease (*T. cruzi*). DB289 and CGP 40215A are two additional agents that have shown potent action against trypanosomatids and are in various stages of drug development.

Active programs to map the genetic code of all the major parasitic pathogens are in progress (see references to web sites). A plethora of hopeful and often validated targets, based on differences in the biochemistry of the parasites vs. the human hosts, have been identified. However, although essentiality of a target can often be demonstrated by genetic methods, insufficient consideration is frequently given to the issue of selectivity between host and parasite enzyme. Modern methods of drug discovery and development (i.e., high throughput screening methods, *in silico* screening, structure-based drug design, parallel synthesis techniques, the preparation of combinatorial libraries, etc.) are being applied to the Kinetoplastida diseases. There is good reason to be hopeful that major advancements in the treatment of these parasitic diseases will occur in the next 10 years, providing there is the political will and economic resources to move some excellent scientific discoveries on to the clinic.

## 5   WEBSITES OF INTEREST

- Information on parasite genome projects, http://parsun1.path.cam.ac.uk/
- World Health Organization, www.who.int
- Leishmania Genome Network, www.ebi.ac.uk
- Centers for Disease Control and Prevention, Emerging Infectious Diseases, www.cdc.gov/ncidod/eid/index.htm
- Parasitological Research Groups worldwide, www.uniduesseldorf.de/WWW/MathNat/Parasitology/paen_ags.htm
- Sanger Center maintains data on nucleotide sequences for many microorganisms, www.sanger.ac.uk./Projects/
- Institute for Genomic Research, http://www.tigr.org
- Website of C. Laveissiere and L. Penchenier on African sleeping sickness, http://www.sleeping-sickness.com

## REFERENCES

1. WHO, *The World Health Report 2001*, World Health Organization, Geneva, Switzerland, 2001.May be accessed at http://www.who.int/tdr/diseases/default.htm
2. P. Trouiller, P. Olliaro, E. Torreele, J. Orbinski, R. Laing, and N. Ford, *Lancet*, **359**, 2188–2194 (2002).
3. J. R. Stevens and W. Gibson, *Parasitol. Today*, **15**, 432–437 (1999).
4. C. A. Hoare, *The Trypanosomes of Mammals, A Zoological Monograph*, Blackwell, Oxford, UK, 1972.
5. T. A. Shapiro and P. T. Englund, *Annu. Rev. Microbiol.*, **49**, 117–143 (1995).
6. K. Stuart, T. E. Allen, M. L. Kable, and S. Lawson, *Curr. Opin. Chem. Biol.*, **1**, 340–346 (1997).

7. A. M. Estevez and L. Simpson, *Gene*, **240**, 247–260 (1999).

8. M. R. Rifkin, *Proc. Natl. Acad. Sci. USA*, **75**, 3450–3454 (1978).

9. S. L. Hajduk, K. M. Hager, and J. D. Esko, *Annu. Rev. Microbiol.*, **48**, 139–162 (1994).

10. A. M. Jordan, *Trypanosomiasis Control and African Rural Development*, Longman, New York, 1986, 45 pp.

11. WHO, *The World Health Report 1997: Conquering Suffering, Enriching Humanity*, World Health Organization, Geneva, Switzerland, 1998.

12. K. Vickerman, *Br. Med. Bull.*, **41**, 105–114 (1985).

13. I. B. R. Bowman and I. W. Flynn in W. H. R. Lumsden and D. A. Evans, Eds., *Biology of the Kinetoplastida*, 1976, pp. 435–476.

14. A. H. Fairlamb and F. R. Opperdoes in M. J. Morgan, Ed., *Carbohydrate Metabolism in Cultured Cells*, Plenum, New York, 1986, pp. 183–224.

15. J. E. Donelson and A. C. Rice-Ficht, *Microbiol. Rev.*, **49**, 107–125 (1985).

16. J. Pepin and J. E. Donelson in R. L. Guerrant, D. H. Walker, and P. F. Weller, Eds., *Essentials of Tropical Infectious Disease*, Churchill-Livingston, Philadelphia, 2001, pp. 360–365.

17. L. D. B. Kinabo, *Acta Trop.*, **54**, 169–183 (1993).

18. E. Zweygarth, J. Ngeranwa, and R. Kaminsky, *Trop. Med. Parasitol.*, **43**, 226–228 (1992).

19. A. H. Fairlamb and I. B. R. Bowman, *Exp. Parasitol.*, **49**, 366–380 (1980).

20. A. H. Fairlamb and I. B. Bowman, *Mol. Biochem. Parasitol.*, **1**, 315–333 (1980).

21. E. L. M. Vansterkenburg, I. Coppens, J. Wilting, O. J. M. Bos, M. J. E. Fischer, L. H. M. Janssen, and F. R. Opperdoes, *Acta Trop.*, **54**, 237–250 (1993).

22. A. Pal, B. S. Hall, and M. C. Field, *Mol. Biochem. Parasitol.*, **122**, 217–221 (2002).

23. A. H. Fairlamb and I. B. R. Bowman, *Exp. Parasitol.*, **43**, 353–361 (1977).

24. P. L. Chello and J. J. Jaffe, *Comp. Biochem. Physiol.*, **43B**, 543–562 (1972).

25. J. J. Jaffe, J. J. McCormack, and E. Meymarian, *Biochem. Pharmacol.*, **21**, 719–731 (1972).

26. R. E. Morty, L. Troeberg, R. N. Pike, R. Jones, P. Nickel, J. D. Lonsdale-Eccles, and T. H. Coetzer, *FEBS Lett.*, **433**, 251–256 (1998).

27. M. Willson, M. Callens, D. A. Kuntz, J. Perie, and F. R. Opperdoes, *Mol. Biochem. Parasitol.*, **59**, 201–210 (1993).

28. F. R. Opperdoes, P. Baudhuin, I. Coppens, C. de Roe, S. W. Edwards, P. J. Weijers, and O. Misset, *J. Cell Biol.*, **98**, 1178–1184 (1984).

29. A. B. Clarkson and F. H. Brohn, *Science*, **194**, 204–206 (1976).

30. C. J. Bacchi, H. C. Nathan, N. Yarlett, B. Goldberg, P. P. McCann, A. Sjoerdsma, M. Saric, and A. B. Clarkson Jr., *Antimicrob. Agents Chemother.*, **38**, 563–569 (1994).

31. A. B. Clarkson Jr., E. J. Bienen, C. J. Bacchi, P. P. McCann, H. C. Nathan, S. H. Hutner, and A. Sjoerdsma, *Am. J. Trop. Med. Hyg.*, **33**, 1073–1077 (1984).

32. S. V. Barrett and M. F. Barrett, *Parasitol. Today*, **16**, 7–9 (2000).

33. J. Pépin and F. Milord, *Adv. Parasitol.*, **33**, 1–47 (1994).

34. R. A. Neal in W. Peters and R. Killick-Kendrick, Eds., The Leishmaniases in Biology and Medicine: Clinical Aspects and Control, Vol. **II**, Academic Press, London, 1987, pp. 793–845.

35. D. Damper and C. L. Patton, *J. Protozool.*, **23**, 349–356 (1976).

36. A. S. Peregrine and M. Mamman, *Acta Trop.*, **54**, 185–203 (1993).

37. M. Sands, M. A. Kron, and R. B. Brown, *Rev. Infect. Dis.*, **7**, 625–634 (1985).

38. N. S. Carter, B. J. Berger, and A. H. Fairlamb, *J. Biol. Chem.*, **270**, 28153–28157 (1995).

39. H. P. De Koning and S. M. Jarvis, *Acta Trop.*, **80**, 245–250 (2001).

40. M. P. Barrett and A. H. Fairlamb, *Parasitol. Today*, **15**, 136–140 (1999).

41. P. Maser, C. Sutterlin, A. Kralli, and R. Kaminsky, *Science*, **285**, 242–244 (1999).

42. H. P. De Koning, *Int. J. Parasitol.*, **31**, 512–522 (2001).

43. B. J. Berger, N. S. Carter, and A. H. Fairlamb, *Mol. Biochem. Parasitol.*, **69**, 289–298 (1995).

44. J. E. Kupusnik and J. Mills in P. K. Peterson and J. Verhoef, Eds., *Antimicrobial Agents Annual*, Elsevier Science Publishers BV, Amsterdam, 1987, pp. 271–281.

45. C. Bailly, I. O. Donkor, D. Gentle, M. Thornalley, and M. J. Waring, *Mol. Pharmacol.*, **46**, 313–322 (1994).

46. K. J. Edwards, T. C. Jenkins, and S. Neidle, *Biochemistry*, **31**, 7104–7109 (1992).

47. B. A. Newton, *Trypanosomiasis and Leishmaniasis*, Elsevier/Excerpta Medica/North Holland, Amsterdam, 1974, pp. 285–307.

48. E. Detain, C. Brack, G. Riou, and B. Festy, *J. Ultrastruct. Res.*, **37**, 200–218 (1971).

49. T. A. Shapiro and P. T. Englund, *Proc. Natl. Acad. Sci. USA*, **87**, 950–954 (1990).

50. B. J. Berger, N. S. Carter, and A. H. Fairlamb, *Acta Trop.*, **54**, 215–224 (1993).

51. G. Benaim, C. Lopez-Estrano, R. Docampo, and S. N. J. Moreno, *Biochem. J.*, **296**, 759–763 (1993).

52. A. J. Bitonti, J. A. Dumont, and P. P. McCann, *Biochem. J.*, **237**, 685–689 (1986).

53. S. C. Roberts, J. Scott, J. E. Gasteier, Y. Q. Jiang, B. Brooks, A. Jardim, N. S. Carter, O. Heby, and B. Ullman, *J. Biol. Chem.*, **277**, 5902–5909 (2002).

54. Y. Liu, R. R. Tidwell, and M. J. Leibowitz, *J. Eukaryot. Microbiol.*, **41**, 31–38 (1994).

55. O. Ericsson, E. K. Schweda, U. Bonner, L. Rombo, M. Friden, and L. L. Gustafsson, *J. Chromatogr. B Biomed. Sci. Appl.*, **690**, 242–251 (1997).

56. J. Keiser, O. Ericsson, and C. Burri, *Clin. Pharmacol. Ther.*, **67**, 478–488 (2000).

57. Z. Gregus and A. Gyurasics, *Biochem. Pharmacol.*, **59**, 1375–1385 (2000).

58. N. S. Carter and A. H. Fairlamb, *Nature*, **361**, 173–176 (1993).

59. I. W. Flynn and I. B. R. Bowman, *Comp. Biochem. Physiol.*, **48B**, 261–273 (1974).

60. E. Van Schaftingen, F. R. Opperdoes, and H. Hers, *Eur. J. Biochem.*, **166**, 653–661 (1987).

61. A. H. Fairlamb, G. B. Henderson, and A. Cerami, *Proc. Natl. Acad. Sci. USA*, **86**, 2607–2611 (1989).

62. P. M. Loiseau, P. Lubert, and J. G. Wolf, *Antimicrob. Agents Chemother.*, **44**, 2954–2961 (2000).

63. M. L. Cunningham, M. J. J. M. Zvelebil, and A. H. Fairlamb, *Eur. J. Biochem.*, **221**, 285–295 (1994).

64. R. M. Johnstone in R. M. Hochster and J. H. Quastel, Eds., *Metabolic Inhibitors*, Academic Press, New York, 1963, pp. 99–118.

65. J. Blum, S. Nkunku, and C. Burri, *Trop. Med. Int. Health*, **6**, 390–400 (2001).

66. D. A. Hess and M. J. Rieder, *Ann. Pharmacother.*, **31**, 1378–1386 (1997).

67. B. K. Park and N. R. Kitteringham, *Drug Metab. Rev.*, **22**, 87–114 (1990).

68. J. Pépin, F. Milord, A. N. Khonde, T. Niyonsenga, L. Loko, B. Mpia, and P. De Wals, *Trans. R. Soc. Trop. Med. Hyg.*, **89**, 92–97 (1995).

69. J. Pépin and F. Milord, *Trans. R. Soc. Trop. Med. Hyg.*, **85**, 222–224 (1991).

70. L. Haller, H. Adams, F. Merouze, and A. Dago, *Am. J. Trop. Med. Hyg.*, **35**, 94–99 (1986).

71. C. Burri, S. Nkunku, A. Merolle, T. Smith, J. Blum, and R. Brun, *Lancet*, **355**, 1419–1425 (2000).

72. R. L. Guerrant, P. F. Walker, and D. L. Weller, Eds., *Essentials of Tropical Infectious Disease*, Churchill-Livingstone, Philadelphia, 2001.

73. B. W. Metcalf, P. Bey, C. Danzin, M. J. Jung, P. Casara, and J. P. Vevert, *J. Am. Chem. Soc.*, **100**, 2551–2553 (1978).

74. V. Quemener, Y. Blanchard, L. Chamaillard, R. Cipolla, and J. P. Moulinoux, *Anticancer Res.*, **14**, 443–448 (1994).

75. F. L. Myskens and E. W. Gerner, *Clin. Cancer Res.*, **5**, 951 (1997).

76. C. J. Bacchi, H. C. Nathan, S. H. Hutner, P. P. McCann, and A. Sjoerdsma, *Science*, **210**, 332–334 (1980).

77. F. A. S. Kuzoe, *Acta Trop.*, **54**, 153–162 (1993).

78. R. Poulin, L. Lu, B. Ackermann, P. Bey, and A. E. Pegg, *J. Biol. Chem.*, **267**, 150–158 (1992).

79. N. J. Hickok, P. Seppanen, G. Gunsalus, and O. A. Janne, *DNA*, **6**, 179–187 (1987).

80. M. A. Phillips, P. Coffino, and C. C. Wang, *J. Biol. Chem.*, **262**, 8721–8727 (1987).

81. M. A. Phillips, P. Coffino, and C. C. Wang, *J. Biol. Chem.*, **263**, 17933–17941 (1988).

82. M. Gupta and P. Coffino, *J. Biol. Chem.*, **260**, 2941–2944 (1985).

83. N. V. Grishin, A. L. Osterman, H. B. Brooks, M. A. Phillips, and E. J. Goldsmith, *Biochemistry*, **38**, 15174–15184 (1999).

84. S. Rogers, R. Wells, and M. Rechsteiner, *Science*, **234**, 364–368 (1986).

85. M. Rechsteiner and S. W. Rogers, *Trends Biochem. Sci.*, **21**, 267–271 (1996).

86. L. Ghoda, M. A. Phillips, K. E. Bass, C. C. Wang, and P. Coffino, *J. Biol. Chem.*, **265**, 11823–11826 (1990).

87. K. E. Bass, J. M. Sommers, Q. L. Cheng, and C. C. Wang, *J. Biol. Chem.*, **267**, 11034–11037 (1992).

88. S. B. Hua, X. Li, P. Coffino, and C. C. Wang, *J. Biol. Chem.*, **270**, 10264–10271 (1995).

89. W. A. Gahl and H. C. Pitot, *In Vitro*, **15**, 252–257 (1979).

90. M. Iten, H. Mett, A. Evans, J. C. K. Enyaru, R. Brun, and R. Kaminsky, *Antimicrob. Agents Chemother.*, **41**, 1922–1925 (1997).

91. C. J. Bacchi, J. Garofalo, M. Ciminelli, D. Rattendi, B. Goldberg, P. P. McCann, and N. Yarlett, *Biochem. Pharmacol.*, **46**, 471–481 (1993).

92. B. F. Giffin, P. P. McCann, A. J. Bitonti, and C. J. Bacchi, *J. Protozool.*, **33**, 238–243 (1986).

93. A. J. Bitonti, P. P. McCann, and A. Sjoerdsma, *Biochem. Pharmacol.*, **35**, 331–334 (1986).

94. H. C. Nathan, C. J. Bacchi, S. H. Hunter, D. Rescigno, P. P. McCann, and A. Sjoerdsma, *Biochem. Pharmacol.*, **30**, 3010–3013 1981.

95. M. A. Phillips and C. C. Wang, *Mol. Biochem. Parasitol.*, **22**, 9–17 (1987).

96. A. H. Fairlamb, G. B. Henderson, C. J. Bacchi, and A. Cerami, *Mol. Biochem. Parasitol.*, **24**, 185–191 (1987).

97. C. Chagas, *Mem. Inst. Oswaldo Cruz.*, **1**, 159–218 (1909).

98. L. V. Kirchhoff in R. L. Guerrant, D. H. Walker, and P. F. Weller, Eds., *Essentials of Tropical Infectious Disease*, Churchill-Livingstone, Philadelphia, 2001, pp. 365–372.

99. *MMWR Morb. Mortal. Wkly. Rep.*, **51**, 210–212 (2002).

100. K. M. Tyler and D. M. Engman, *Int. J. Parasitol.*, **31**, 472–481 (2001).

101. WHO, *Tropical Disease Research: Progress 1995–1996: Thirteenth Programme Report of the UNDP/World Bank/WHO Special Programme for Research and Training in Tropical Diseases*, World Health Organization, Geneva, Switzerland, 1997, 141 pp.

102. Z. Brener, *Pharmacol. Ther.*, **7**, 71–90 (1979).

103. R. Docampo, *Curr. Pharm. Des.*, **7**, 1157–1164 (2001).

104. G. B. Henderson, P. Ulrich, A. H. Fairlamb, I. Rosenberg, M. Pereira, M. Sela, and A. Cerami, *Proc. Natl. Acad. Sci. USA*, **85**, 5374–5378 (1988).

105. J. Reynolds, *J. Pharm. Pharmacol.*, **31** (Suppl.), 29P (1979).

106. WHO, *The World Health Report 2002: Leishmaniasis: The disease and its impact*, World Health Organization, Geneva, Switzerland. May be accessed at http://www.who.int/emc/disease/leish/leisdis1. html

107. R. D. Pearson, S. M. B. Jeronimo, and A. De Q Sousa in R. L. Guerrant, D. H. Walker, and D. L. Weller, Eds., *Essentials of Tropical Infectious Disease*, Churchill-Livingstone, Philadelphia, 2001, pp. 372–379.

108. H. W. Murray, *Int. J. Infect. Dis.*, **4**, 158–177 (2000).

109. H. W. Murray, *AIDS Patient Care STDS*, **13**, 459–465 (1999).

110. W. L. Roberts and P. M. Rainey, *Antimicrob. Agents Chemother.*, **37**, 1842–1846 (1993).

111. J. D. Berman, *Rev. Infect. Dis.*, **10**, 560–586 (1988).

112. W. L. Roberts, W. J. McMurray, and P. M. Rainey, *Antimicrob. Agents Chemother.*, **42**, 1076–1082 (1998).

113. D. Sereno, M. Cavaleyra, K. Zemzoumi, S. Maquaire, A. Ouaissi, and J. L. Lemesre, *Antimicrob. Agents Chemother.*, **42**, 3097–3102 (1998).

114. M. Ephros, A. Bitnun, P. Shaked, E. Waldman, and D. Zilberstein, *Antimicrob. Agents Chemother.*, **43**, 278–282 (1999).

115. P. Shaked-Mishan, N. Ulrich, M. Ephros, and D. Zilberstein, *J. Biol. Chem.*, **276**, 3971–3976 (2001).

116. F. Frezard, C. Demicheli, C. S. Ferreira, and M. A. P. Costa, *Antimicrob. Agents Chemother.*, **45**, 913–916 (2001).

117. J. D. Berman, J. V. Gallalee, and J. M. Best, *Biochem. Pharmacol.*, **36**, 197–201 (1987).

118. J. G. J. Su, J. M. Mansour, and T. E. Mansour, *Mol. Biochem. Parasitol.*, **81**, 171–178 (1996).

119. J. C. Mottram and G. H. Coombs, *Exp. Parasitol.*, **59**, 151–160 (1985).

120. M. L. Cunningham and A. H. Fairlamb, *Eur. J. Biochem.*, **230**, 460–468 (1995).

121. D. Sereno, P. Holzmuller, I. Mangot, G. Cuny, A. Ouaissi, and J. L. Lemesre, *Antimicrob. Agents Chemother.*, **45**, 2064–2069 (2001).

122. P. Borst and M. Ouellette, *Annu. Rev. Microbiol.*, **49**, 427–460 (1995).

123. R. Mukhopadhyay, S. Dey, N. Xu, D. Gage, J. Lightbody, M. Ouellette, and B. P. Rosen, *Proc. Natl. Acad. Sci. USA*, **93**, 10383–10387 (1996).

124. S. Dey, M. Ouellette, J. Lightbody, B. Papadopoulou, and B. P. Rosen, *Proc. Natl. Acad. Sci. USA*, **93**, 2192–2197 (1996).

125. A. Haimeur, C. Brochu, P. A. Genest, B. Papadopoulou, and M. Ouellette, *Mol. Biochem. Parasitol.*, **108**, 131–135 (2000).

126. D. Legare, D. Richard, R. Mukhopadhyay, Y. D. Stierhof, B. P. Rosen, A. Haimeur, B. Papadopoulou, and M. Ouellette, *J. Biol. Chem.*, **276**, 26301–26307 (2001).

127. P. L. Olliaro and A. D. M. Bryceson, *Parasitol. Today*, **9**, 323–328 (1993).

128. C. P. Thakur, T. P. Kanyok, A. K. Pandey, G. P. Sinha, A. E. Zaniewski, H. H. Houlihan, and P. Olliaro, *Trans. R. Soc. Trop. Med. Hyg.*, **94**, 429–431 (2000).

129. B. B. Davis, *J. Antimicrob. Chemother.*, **22**, 1–3 (1988).

130. J. D. Berman, *Clin. Infect. Dis.*, **24**, 684–703 (1997).

131. S. Sundar, A. K. Goyal, D. K. More, M. K. Singh, and H. W. Murray, *Ann. Trop. Med. Parasitol.*, **92**, 755–764 (1998).

132. J. D. Berman, *Clin. Infect. Dis.*, **28**, 49–51 (1999).

133. A. K. Saha, T. Mukherjee, and A. Bhaduri, *Mol. Biochem. Parasitol.*, **19**, 195–200 (1986).

134. S. Sundar, F. Rosenkaimer, E. N. Makhina, et al., *Lancet*, **352**, 1821–1823 (1998).

135. S. Sundar, L. B. Gupta, L. B. Makharia, M. K. Singh, A. Voss, F. Rosenkaimer, J. Engel, and H. W. Murray, *Ann. Trop. Med. Parasitol.*, **93**, 589–597 (1999).

136. J. Soto, J. Toledo, P. Gutierrez, R. S. Nicholls, J. Padilla, J. Engel, C. Fisher, A. Voss, and J. Berman, *Clin. Infect. Dis.*, **33**, E57–E61 (2001).

137. H. Brachwitz and C. Vollgraf, *Pharmacol. Ther.*, **66**, 39–82 (1995).

138. G. Arthur and R. Bittman, *Biochem. Biophys. Acta*, **1391**, 85–102 (1998).

139. T. Wieder, A. Haase, C. C. Geilen, and C. E. Orfonos, *Lipids*, **30**, 389–393 (1995).

140. W. R. Vogler, M. Shoji, D. J. Hayzer, Y. P. Xie, and M. Renshaw, *Leukemia Res.*, **20**, 947–951 (1996).

141. J. Bergmann, I. Junghahn, H. Brachwitz, and P. Langen, *Anticancer Res.*, **14** (Suppl. 4A), 1549–1556 (1994).

142. R. Lira, L. M. Contreras, R. M. S. Rita, and J. A. Urbina, *J. Antimicrob. Chemother.*, **47**, 537–546 (2001).

143. H. Lux, N. Heise, T. Klenner, D. Hart, and F. R. Opperdoes, *Mol. Biochem. Parasitol.*, **111**, 1–14 (2000).

144. C. Sansom, *Lancet Infect. Dis.*, **2**, 134 (2002).

145. M. Schull, *Br. Med. J.*, **321**, 179 (2000).

146. K. Nelson, *Lancet*, **359**, 1042 (2002).

147. J. Stephenson, *J. Am. Med. Assoc.*, **287**, 2061–2063 (2002).

148. F. R. Opperdoes and P. Borst, *FEBS Lett.*, **80**, 360–364 (1977).

149. F. H. Brohn and A. B. Clarkson Jr., *Mol. Biochem. Parasitol.*, **1**, 291–305 (1980).

150. B. M. Bakker, F. I. C. Mensonides, B. Teusink, P. van Hoek, P. A. M. Michels, and H. V. Westerhoff, *Proc. Natl. Acad. Sci. USA*, **97**, 2087–2092 (2000).

151. P. T. Grant and J. R. Sargent, *Biochem. J.*, **76**, 229–237 (1960).

152. K. Vickerman, L. Tetley, K. A. K. Hendry, and C. M. R. Turner, *Biol. Cell.*, **64**, 109–119 (1988).

153. J. M. Sommer and C. C. Wang, *Ann. Rev. Pharmacol.*, **48**, 105–138 (1994).

154. C. E. Clayton and P. Michels, *Parasitol. Today*, **12**, 465–471 (1996).

155. C. C. Wang in M. E. Wolff, Ed., Berger's Medicinal Chemistry and Drug Discovery, Vol. **4**, John Wiley & Sons, New York, 1997, pp. 459–486.

156. F. R. Opperdoes and P. A. M. Michels, *Int. J. Parasitol.*, **31**, 482–490 (2001).

157. B. M. Bakker, M. C. Walsh, B. H. Ter Kuile, F. I. C. Mensonides, P. A. M. Michels, F. R. Opperdoes, and H. V. Westerhoff, *Proc. Natl. Acad. Sci. USA*, **96**, 10098–10103 (1999).

158. B. M. Bakker, P. A. M. Michels, F. R. Opperdoes, and H. V. Westerhoff, *J. Biol. Chem.*, **274**, 14551–14559 (1999).

159. R. Eisenthal and A. Cornish-Bowden, *J. Biol. Chem.*, **273**, 5500–5505 (1998).

160. R. Eisenthal, S. Game, and G. D. Holman, *Biochim. Biophys. Acta*, **985**, 81–89 (1989).

161. B. H. Ter Kuile and H. V. Westerhoff, *FEBS Lett.*, **500**, 169–171 (2001).

162. F. Bringaud and T. Baltz, *Mol. Cell. Biol.*, **13**, 1146–1154 (1993).

163. A. Seyfang and M. Duszenko, *Eur. J. Biochem.*, **214**, 593–597 (1993).

164. M. P. Barrett, E. Tetaud, A. Seyfang, F. Bringaud, and T. Baltz, *Biochem. J.*, **312**, 687–691 (1995).

165. E. Tetaud, M. P. Barrett, F. Bringaud, and T. Baltz, *Biochem. J.*, **325**, 569–580 (1997).

166. A. Walmsley, A. J. Barrett, F. Bringaud, and G. Gould, *Trends Biol. Sci.*, **23**, 476–481 (1998).

167. T. Hara, H. Kanbara, M. Nakao, and T. Fukuma, *Kurume Med. J.*, **44**, 105–113 (1997).

168. M. Trinquier, J. Perie, M. Callens, F. Opperdoes, and M. Willson, *Bioorg. Med. Chem.*, **3**, 1423–1427 (1995).

169. M. Marchand, U. Kooystra, R. K. Wierenga, A. M. Lambeir, J. Van Beeumen, F. R. Opperdoes, and P. A. Michels, *Eur. J. Biochem.*, **455**–464 (1989).

170. R. Hardre, L. Salmon, and F. R. Opperdoes, *J. Enz. Inhib.*, **15**, 509–515 (2000).

171. P. A. M. Michels, N. Chevalier, F. R. Opperdoes, M. H. Rider, and D. J. Rigden, *Eur. J. Biochem.*, **250**, 698–704 (1997).

172. C. E. Clayton, *EMBO J.*, **4**, 2997–3003 (1985).

173. N. Chevalier, M. Callens, and P. A. M. Michels, *Protein Expr. Purif.*, **6**, 39–44 (1995).

174. D. M. Chudzik, P. A. Michels, S. de Walque, and W. G. J. Hol, *J. Mol. Biol.*, **300**, 697–707 (2000).

175. B. W. Swinkels, W. C. Gibson, K. A. Osinga, R. Kramer, G. H. Veeneman, J. H. van Boom, and P. Borst, *EMBO J.*, **5**, 1291–1298 (1986).

176. R. K. Wierenga, K. H. Kalk, and W. G. Hol, *J. Mol. Biol.*, **198**, 109–121 (1987).

177. F. M. D. Vellieux, J. Hajdu, C. L. M. J. Verlinde, H. Groendijk, R. J. Read, T. J. Greenhough, J. W. Campbell, K. H. Kalk, J. A. Littlechild, H. C. Watson, et al., *Proc. Natl. Acad. Sci. USA*, **90**, 2355–2359 (1993).

178. S. Helfert, A. M. Estevez, B. Bakker, P. Michels, and C. Clayton, *Biochem. J.*, **357**, 117–125 (2001).

179. F. M. D. Vellieux, J. Hajdu, and W. G. J. Hol, *Acta Crystallogr. D. Biol. Crystallogr.*, **51**, 575–589 (1995).

180. D. H. F. Souza, R. C. Garratt, A. P. U. Araujo, B. G. Guimaraes, W. D. P. Jesus, P. A. M. Michels, V. Hannaert, and G. Oliva, *FEBS Lett.*, **424**, 131–135 (1998).

181. H. Kim, I. Feil, C. L. M. Verlinde, P. H. Petra, and W. G. J. Hol, *Biochemistry*, **34**, 14975–14986 (1995).

182. A. M. Aronov, S. Suresh, F. S. Buckner, W. C. VanVoorhis, C. L. M. J. Verlinde, F. R. Opperdoes, W. G. J. Hol, and M. H. Gelb, *Proc. Natl. Acad. Sci. USA*, **96**, 4273–4278 (1999).

183. A. M. Aronov, S. Suresh, F. S. Buckner, W. C. Van Voorhis, C. L. M. Verlinde, F. Opperdoes, W. G. Hol, and M. H. Gelb, *J. Med. Chem.*, **41**, 4799 (1998).

184. S. Ladame, M. Bardet, J. Perie, and M. Willson, *Bioorg. Med. Chem.*, **9**, 773–783 (2001).

185. S. Suresh, J. C. Bressi, C. L. Kennedy, C. L. M. Verlinde, M. H. Gelb, and W. G. Hol, *J. Mol. Biol.*, **309**, 423–435 (2001).

186. L. Kohl, T. Drmota, C. D. D. Thi, M. Callens, J. Van Beeumen, F. R. Opperdoes, and P. A. M. Michels, *Mol. Biochem. Parasitol.*, **76**, 159–173 (1996).

187. C. E. Stebeck, U. Frevert, T. P. Mommsen, E. Vassella, I. Roditi, and T. W. Pearson, *Mol. Biochem. Parasitol.*, **76**, 145–158 (1996).

188. S. Suresh, S. Turley, F. R. Opperdoes, P. A. M. Michels, and W. G. J. Hol, *Struct. Fold. Des.*, **8**, 541–552 (2000).

189. D. Kuntz, U. Kooystra, F. Opperdoes, and P. A. M. Michels, *Arch. Int. Phys. Biochim.*, **96**, B99 (1988).

190. F. Pavao, M. S. Castilho, M. T. Pupo, R. L. Dias, A. G. Correa, J. B. Fernandes, M. F. da Ailva, J. Mafezole, P. C. Vieira, and G. Oliva, *FEBS Lett.*, **520**, 13–17 (2002).

191. H. Denise, C. Giroud, M. P. Barrett, and T. Baltz, *Eur. J. Biochem.*, **259**, 339–346 (1999).

192. N. Visser and F. R. Opperdoes, *Eur. J. Biochem.*, **103**, 623–632 (1980).

193. A. H. Fairlamb and I. B. R. Bowman, *Int. J. Biochem.*, **8**, 659–668 (1977).

194. D. A. Berthold, M. E. Andersson, and P. Nordlund, *Biochim. Biophys. Acta*, **1460**, 241–254 (2000).

195. K. Kita, K. Konishi, and Y. Anraku, *Methods Enzymol.*, **126**, 94–113 (1986).

196. D. J. Hammond and I. B. R. Bowman, *Mol. Biochem. Parasitol.*, **2**, 63–75 (1980).

197. A. H. Fairlamb, F. R. Opperdoes, and P. Borst, *Nature*, **265**, 270–271 (1977).

198. D. A. Evans, C. J. Brightman, and M. F. Holland, *Lancet*, **2**, 769 (1977).

199. C. van der Meer and J. A. M. Versluijs-Broers, *Exp. Parasitol.*, **48**, 126–134 (1979).

200. R. W. Grady, E. J. Bienen, and A. B. Clarkson Jr., *Mol. Biochem. Parasitol.*, **19**, 231–240 (1986).

201. H. Sasaki, T. Hosokawa, M. Swada, and K. Ando, *J. Antibiot.*, **26**, 676–680 (1973).

202. N. Minagawa, Y. Yabu, K. Kita, K. Nagai, N. Ohta, K. Meguro, S. Sakajo, and A. Yoshimato, *Mol. Biochem. Parasitol.*, **81**, 127–136 (1996).

203. I. Kralova, D. J. Rigden, F. R. Opperdoes, and P. A. M. Michels, *Eur. J. Biochem.*, **267**, 2323–2333 (2000).

204. K. Steinborn, A. Szallies, D. Mecke, and M. Duszenko, *Biol. Chem.*, **381**, 1071–1077 (2000).

205. J. Blattner, S. Helfert, P. Michels, and C. Clayton, *Proc. Natl. Acad. Sci. USA*, **95**, 11596–11600 (1998).

206. J. L. Concepcion, C. A. Adje, W. Quinones, N. Chevalier, M. Dubourdieu, and P. A. Michels, *Mol. Biochem. Parasitol.*, **118**, 111–121 (2001).

207. B. E. Bernstein, P. A. M. Michels, and W. G. J. Hol, *Nature*, **385**, 275–278 (1997).

208. O. Misset and F. R. Opperdoes, *Eur. J. Biochem.*, **162**, 493–500 (1987).

209. N. Chevalier, D. J. Rigdon, J. Van Roy, F. R. Opperdoes, and P. A. M. Michels, *Eur. J. Biochem.*, **267**, 1464–1472 (2000).

210. I. Ernest, M. Callens, A. D. Uttaro, N. Chevalier, F. R. Opperdoes, H. Muirhead, and P. A. Michels, *Protein Expr. Purif.*, **13**, 373–382 (1998).

211. F. R. Opperdoes, *Annu. Rev. Microbiol.*, **41**, 127–151 (1987).

212. J. A. Urbina, *Parasitol. Today*, **10**, 107–110 (1994).

213. J. J. Cazzulo, *J. Bioenerg. Biomembr.*, **26**, 157–165 (1994).

214. S. Trapani, J. Linss, S. Goldenberg, H. Fischer, A. F. Craievich, and G. Oliva, *J. Mol. Biol.*, **313**, 1059–1072 (2001).

215. F. Bringaud, D. Baltz, and T. Baltz, *Proc. Natl. Acad. Sci. USA*, **95**, 7963–7968 (1998).

216. R. A. Maldonado and A. H. Fairlamb, *Mol. Biochem. Parasitol.*, **112**, 183–191 (2001).

217. L. W. Cosenza, F. Bringaud, T. Baltz, and F. M. Vellieux, *J. Mol. Biol.*, **318**, 1417–1432 (2002).

218. M. Chen, L. Zhai, S. B. Christensen, T. G. Theander, and A. Kharazmi, *Antimicrob. Agents Chemother.*, **45**, 2023–2029 (2001).

219. S. F. Nielsen, S. B. Christensen, G. Cruciani, A. Kharazmi, and T. Liljefors, *J. Med. Chem.*, **41**, 4819–4832 (1998).

220. E. Takashima, D. K. Inaoka, A. Osanai, T. Nara, M. Odaka, T. Aoki, K. Inaka, S. Harada, and K. Kita, *Mol. Biochem. Parasitol.*, **122**, 189–200 (2002).

221. F. R. Opperdoes, R. K. Wierenga, M. E. M. Noble, W. G. J. Hol, M. Willson, D. A. Kuntz, M. Callens, and J. Perié in N. Agabian and A. Cerami, Eds., *Parasites: Molecular Biology, Drug, and Vaccine Design*, Wiley-Liss, New York, 1989, pp. 233–246.

222. M. Persons, T. Furuya, S. Pal, and P. Kesster, *Mol. Biochem. Parasitol.*, **115**, 19–28 (2001).

223. S. Subramani, *Physiol. Rev.*, **78**, 171–188 (1998).

224. J. M. Sommer, Q. L. Cheng, G. A. Keller, and C. C. Wang, *Mol. Biol. Cell*, **3**, 749–759 (1992).

225. J. Blattner, B. Swinkels, H. Dorsam, T. Prospero, S. Subramani, and C. Clayton, *J. Cell Biol.*, **119**, 1129–1136 (1992).

226. J. M. Sommer, G. Peterson, G. A. Keller, M. Parsons, and C. C. Wang, *FEBS Lett.*, **316**, 53–58 (1993).

227. P. A. M. Michels, A. Poliszczak, K. A. Osinga, O. Misset, J. Van Beeuman, R. K. Wierenga, P. Borst, and F. R. Opperdoes, *EMBO J.*, **5**, 1049–1056 (1986).

228. J. M. Sommer, T. T. Nguyen, and C. C. Wang, *FEBS Lett.*, **350**, 125–129 (1994).

229. K. A. Osinga, B. W. Swinkels, W. C. Gibson, and G. H. Veeneman, *EMBO J.*, **4**, 3811–3817 (1985).

230. B. W. Swinkels, S. J. Gould, A. G. Bodnar, R. A. Rachubinski, and S. Subramani, *EMBO J.*, **10**, 3255–3262 (1991).

231. R. J. Simon, R. S. Kania, R. N. Zuckermann, V. D. Huebner, D. A. Jewell, S. Banville, S. Ng, L. Wang, S. Rosenberg, C. K. Marlowe, et al., *Proc. Natl. Acad. Sci. USA*, **89**, 9367–9371 (1992).

232. X. Wang, M. J. Unruh, and J. M. Goodman, *J. Biol. Chem.*, **276**, 10897–10905 (2001).

233. M. P. Barrett, *Parasitol. Today*, **13**, 11–16 (1997).

234. N. Heise and F. R. Opperdoes, *Mol. Biochem. Parasitol.*, **99**, 21–32 (1999).

235. F. Duffieux, J. Van Roy, P. K. M. Michels, and F. R. Opperdoes, *J. Biol. Chem.*, **275**, 27559–27565 (2000).

236. F. R. Opperdoes, *FEBS Lett.*, **169**, 35–39 (1984).

237. N. Heise and F. R. Opperdoes, *Mol. Biochem. Parasitol.*, **89**, 61–72 (1997).

238. K. J. Hunter, S. A. Le Quesne, and A. H. Fairlamb, *Eur. J. Biochem.*, **226**, 1019–1027 (1994).

239. S. A. Le Quesne and A. H. Fairlamb, *Biochem. J.*, **316**, 481–486 (1996).

240. A. H. Fairlamb and S. A. Le Quesne in G. Hide, J. C. Mottram, G. H. Coombs, and P. H. Holmes, Eds., *Trypanosomiasis and Leishmaniasis: Biology and Control*, CAB International, Oxford, UK, 1997, pp. 149–161.

241. M. Basselin, G. H. Coombs, and M. P. Barrett, *Mol. Biochem. Parasitol.*, **109**, 37–46 (2000).

242. B. L. Tekwani, C. J. Bacchi, and A. E. Pegg, *Mol. Cell. Biochem.*, **117**, 53–61 (1992).

243. A. J. Bitonti, T. L. Byers, T. L. Bush, P. J. Casara, C. J. Bacchi, A. B. Clarkson Jr., P. P. McCann, and A. Sjoerdsma, *Antimicrob. Agents Chemother.*, **34**, 1485–1490 (1990).

244. T. L. Byers, T. L. Bush, P. P. McCann, and A. J. Bitonti, *Biochem. J.*, **274**, 527–533 (1991).

245. C. J. Bacchi, H. C. Nathan, N. Yarlett, B. Goldberg, P. P. McCann, A. J. Bitonti, and A. Sjoerdsma, *Antimicrob. Agents Chemother.*, **36**, 2736–2740 (1992).

246. T. L. Byers, P. Casara, and A. J. Bitonti, *Biochem. J.*, **283**, 755–758 (1992).

247. J. Guo, Y. Q. Wu, D. Rattendi, C. J. Bacchi, and P. M. Woster, *J. Med. Chem.*, **38**, 1770–1777 (1995).

248. C. J. Bacchi, R. Brun, S. L. Croft, K. Alicea, and Y. Buhler, *Antimicrob. Agents Chemother.*, **40**, 1448–1453 (1996).

249. K. L. Seley, S. W. Schneller, D. Rattendi, and C. J. Bacchi, *J. Med. Chem.*, **40**, 622–624 (1997).

250. D. D. Von Hoff, *Ann. Oncol.*, **5**, 487–493 (1994).

251. J. Stanek, G. Caravatti, H. G. Capraro, P. Furet, H. Mett, P. Schneider, and U. Regenass, *J. Med. Chem.*, **36**, 46–54 (1993).

252. U. Regenass, G. Caravatti, H. Mett, J. Stanek, P. Schneider, M. Muller, A. Matter, P. Vertino, and C. W. Porter, *Cancer Res.*, **52**, 4712–4718 (1992).

253. U. Regenass, H. Mett, J. Stanek, M. Mueller, D. Kramer, and C. W. Porter, *Cancer Res.*, **54**, 3210–3217 (1994).

254. R. Brun, Y. Buhler, U. Sandmeier, R. Kaminsky, C. J. Bacchi, D. Rattendi, S. Lane, S. L. Croft, D. Snowdon, V. Yardley, G. Caravatti, J. Frei, J. Stanek, and H. Mett, *Antimicrob. Agents Chemother.*, **40**, 1442–1447 (1996).

255. C. J. Bacchi, R. Brun, S. L. Croft, K. Alicea, and Y. Buhler, *Antimicrob. Agents Chemother.*, **40**, 1448–1453 (1996).

256a. B. Goldberg, D. Rattendi, D. Lloyd, J. R. Sufrin, and C. J. Bacchi, *Biochem. Pharmacol.*, **56**, 95–103 (1998).

256b. R. Brun, C. Burri, and C. W. Gichuki, *Trop. Med. Int. Health*, **6**, 362–368 (2001).

257. S. M. Rahmathullah, J. E. Hall, B. C. Bender, D. R. McCurdy, R. R. Tidwell, and D. W. Boykin, *J. Med. Chem.*, **42**, 3994–4000 (1999).

258. C. Bailly, L. Dassonneville, C. Carrasco, D. Lucas, A. Kumar, D. W. Boykin, and W. D. Wilson, *Anticancer Drug Des.*, **14**, 47–60 (1999).

259. W. C. Zhou, Z. M. Xin, X. P. Zhang, J. Shen, and Q. P. Qiu, *Yao Xue. Xue. Bao.*, **31**, 823–830 (1996).

260. R. Kaminsky and R. Brun, *Antimicrob. Agents Chemother.*, **42**, 2858–2862 (1998).

261. C. J. Bacchi, M. Vargas, D. Rattendi, B. Goldberg, and W. Zhou, *Antimicrob. Agents Chemother.*, **42**, 2718–2721 (1998).

262. A. H. Fairlamb, P. Blackburn, P. Ulrich, B. T. Chait, and A. Cerami, *Science*, **227**, 1485–1487 (1985).

263. A. H. Fairlamb and A. Cerami, *Mol. Biochem. Parasitol.*, **14**, 187–198 (1985).

264. A. H. Fairlamb and A. Cerami, *Annu. Rev. Microbiol.*, **46**, 695–729 (1992).

265. K. Augustyns, K. Amssoms, A. Yamani, P. K. Rajan, and A. Haemers, *Curr. Pharm. Des.*, **7**, 1117–1141 (2001).

266. K. A. Werbovetz, *Curr. Med. Chem.*, **7**, 835–860 (2000).

267. C. S. Bond, Y. H. Zhang, M. Berriman, M. L. Cunningham, A. H. Fairlamb, and W. N. Hunter, *Structure*, **7**, 81–89 (1999).

268. Y. Zhang, C. S. Bond, S. Bailey, M. L. Cunningham, A. H. Fairlamb, and W. N. Hunter, *Protein Sci.*, **5**, 52–61 (1996).

269. M. S. Alphey, G. A. Leonard, D. G. Gourley, E. Tetaud, A. H. Fairlamb, and W. N. Hunter, *J. Biol. Chem.*, **274**, 25613–25622 (1999).

270. B. Hofmann, H. Budde, K. Bruns, S. A. Guerrero, H. M. Kalisz, U. Menge, M. Montemartini, E. Nogoceke, P. Steinert, J. B. Wissing, L. Flohé, and H. J. Hecht, *Biol. Chem.*, **382**, 459–471 (2001).

271. M. S. Alphey, C. S. Bond, E. Tetaud, A. H. Fairlamb, and W. N. Hunter, *J. Mol. Biol.*, **300**, 903–916 (2000).

272. L. Flohé, H. J. Hecht, and P. Steinert, *Free Radic. Biol. Med.*, **27**, 966–984 (1999).

273. G. Powis and W. R. Montfort, *Annu. Rev. Pharmacol. Toxicol.*, **41**, 261–295 (2001).

274. C. H. Williams Jr., *FASEB J.*, **9**, 1267–1276 (1995).

275. C. H. Williams, L. D. Arscott, S. Muller, B. W. Lennon, M. L. Ludwig, P. F. Wang, D. M. Veine, K. Becker, and R. H. Schirmer, *Eur. J. Biochem.*, **267**, 6110–6117 (2000).

276. K. Becker, S. Gromer, R. H. Schirmer, and S. Muller, *Eur. J. Biochem.*, **267**, 6118–6125 (2000).

277. M. Moutiez, D. Meziane-Cherif, M. Aumercier, C. Sergheraert, and A. Tartar, *Chem. Pharm. Bull.*, **42**, 2641–2644 (1994).

278. S. K. Shahi, R. L. Krauth-Siegel, and C. E. Clayton, *Mol. Microbiol.*, **43**, 1129–1138 (2002).

279. S. C. Yan, K. Y. Ding, L. Zhang, and H. Z. Sun, *Angew. Chem. Int. Ed. Engl.*, **39**, 4260 (2000).

280. E. Tetaud, F. Manai, M. P. Barrett, K. Nadeau, C. T. Walsh, and A. H. Fairlamb, *J. Biol. Chem.*, **273**, 19383–19390 (1998).

281. S. L. Oza, M. R. Ariyanayagam, and A. H. Fairlamb, *Biochem. J.*, **364**, 679–686 (2002).

282. K. Amssoms, S. L. Oza, A. Ravaschino, A. Yamani, A. M. Lambeir, P. Rajan, G. Bal, J. B. Rodriguez, A. H. Fairlamb, K. Augustyns, and A. Haemers, *Bioorg. Med. Chem. Lett.*, **12** 2553–2556 (2002).

283. K. Amssoms, S. L. Oza, K. Augustyns, S. De Craecker, A. Yamani, A. M. Lambeir, A. H. Fairlamb, and A. Haemers, *Bioorg. Med. Chem. Lett.*, **12**, 2703–2705 (2002).

284. S. L. Shames, A. H. Fairlamb, A. Cerami, and C. T. Walsh, *Biochemistry*, **25**, 3519–3526 (1986).

285. R. L. Krauth-Siegel, B. Enders, G. B. Henderson, A. H. Fairlamb, and R. H. Schirmer, *Eur. J. Biochem.*, **164**, 123–128 (1987).

286. F. X. Sullivan, S. L. Shames, and C. T. Walsh, *Biochemistry*, **28**, 4986–4992 (1989).

287. T. Aboagye-Kwarteng, K. Smith, and A. H. Fairlamb, *Mol. Microbiol.*, **6**, 3089–3099 (1992).

288. M. C. Taylor, *The trypanothione reductase gene of Leishmania donovani*, Ph.D. Thesis, University of London, London, 1992.

289. S. L. Shames, B. E. Kimmel, O. P. Peoples, N. Agabian, and C. T. Walsh, *Biochemistry*, **27**, 5014–5019 (1988).

290. A. Borges, M. L. Cunningham, J. Tovar, and A. H. Fairlamb, *Eur. J. Biochem.*, **228**, 745–752 (1995).

291. F. X. Sullivan and C. T. Walsh, *Mol. Biochem. Parasitol.*, **44**, 145–148 (1991).

292. J. Kuriyan, X.-P. Kong, T. S. R. Krishna, R. M. Sweet, N. J. Murgolo, H. Field, A. Cerami, and G. B. Henderson, *Proc. Natl. Acad. Sci. USA*, **88**, 8764–8767 (1991).

293. S. Bailey, K. Smith, A. H. Fairlamb, and W. N. Hunter, *Eur. J. Biochem.*, **213**, 67–75 (1993).

294. C. B. Lantwin, I. Schlichting, W. Kabsch, E. F. Pai, and R. L. Krauth-Siegel, *Proteins: Struct. Funct. Genet.*, **18**, 161–173 (1994).

295. E. M. Jacoby, I. Schlichting, C. B. Lantwin, W. Kabsch, and R. L. Krauth-Siegel, *Proteins: Struct. Funct. Genet.*, **24**, 73–80 (1996).

296. W. N. Hunter, S. Bailey, J. Habash, S. J. Harrop, J. R. Helliwell, T. Aboagye-Kwarteng, K. Smith, and A. H. Fairlamb, *J. Mol. Biol.*, **227**, 322–333 (1992).

297. C. Dumas, M. Ouellette, J. Tovar, M. L. Cunningham, A. H. Fairlamb, S. Tamar, M. Olivier, and B. Papadopoulou, *EMBO J.*, **16**, 2590–2598 (1997).

298. J. Tovar, M. L. Cunningham, A. C. Smith, S. L. Croft, and A. H. Fairlamb, *Proc. Natl. Acad. Sci. USA*, **95**, 5311–5316 (1998).

299. J. Tovar, S. Wilkinson, J. C. Mottram, and A. H. Fairlamb, *Mol. Microbiol.*, **29**, 653–660 (1998).

300. S. Krieger, W. Schwarz, M. R. Ariyanayagam, A. H. Fairlamb, R. L. Krauth-Siegel, and C. Clayton, *Mol. Microbiol.*, **35**, 542–552 (2000).

301. T. J. Benson, J. H. McKie, J. Garforth, A. Borges, A. H. Fairlamb, and K. T. Douglas, *Biochem. J.*, **286**, 9–11 (1992).

302. M. O. F. Khan, S. E. Austin, C. Chan, H. Yin, D. Marks, S. N. Vaghjiani, H. Kendrick, V. Yardley, S. L. Croft, and K. T. Douglas, *J. Med. Chem.*, **43**, 3148–3156 (2000).

303. S. Baillet, E. Buisine, D. Horvath, L. Maes, B. Bonnet, and C. Sergheraert, *Bioorg. Med. Chem.*, **4**, 891–899 (1996).

304. J. Garforth, H. Yin, J. H. McKie, K. T. Douglas, and A. H. Fairlamb, *J. Enzyme Inhib.*, **12**, 161–173 (1997).

305. S. Bonse, C. Santelli-Rouvier, J. Barbe, and R. L. Krauth-Siegel, *J. Med. Chem.*, **42**, 5448–5454 (1999).

306. K. Chibale, M. Visser, V. Yardley, S. L. Croft, and A. H. Fairlamb, *Bioorg. Med. Chem. Lett.*, **10**, 1147–1150 (2000).

307. J. A. Ponasik, C. Strickland, C. Faerman, S. Savvides, P. A. Karplus, and B. Ganem, *Biochem. J.*, **311**, 371–375 (1995).

308. M. C. O'Sullivan, D. Dalrymple, and Q. Zhou, *J. Enzyme Inhib.*, **11**, 97–114 (1996).

309. M. C. O'Sullivan, Q. B. Zhou, Z. L. Li, T. B. Durham, D. Rattendi, S. Lane, and C. J. Bacchi, *Bioorg. Med. Chem.*, **5**, 2145–2155 (1997).

310. Y. Zou, Z. Wu, N. Sirisoma, P. M. Woster, R. A. Casero Jr., L. M. Weiss, D. Rattendi, S. Lane, and C. J. Bacchi, *Bioorg. Med. Chem. Lett.*, **11**, 1613–1617 (2001).

311. H. K. Smith and M. Bradley, *J. Combin. Chem.*, **1**, 326–332 (1999).

312. B. Chitkul and M. Bradley, *Bioorg. Med. Chem. Lett.*, **10**, 2367–2369 (2000).

313. H. Kappus, I. Mahmutoglu, J. Kostrucha, and M. E. Scheulen, *Free Radic. Res. Commun.*, **2**, 271–277 (1987).

314. R. Docampo and S. N. J. Moreno, *Rev. Infect. Dis.*, **6**, 223–238 (1984).

315. K. Blumenstiel, R. Schoneck, V. Yardley, S. L. Croft, and R. L. Krauth-Siegel, *Biochem. Pharmacol.*, **58**, 1791–1799 (1999).

316. N. K. Cenas, D. Arscott, C. H. Williams Jr., and J. S. Blanchard, *Biochemistry*, **33**, 2509–2515 (1994).

317. L. Salmon-Chemin, A. Lemaire, S. De Freitas, B. Deprez, C. Sergheraert, and E. Davioud-Charvet, *Bioorg. Med. Chem. Lett.*, **10**, 631–635 (2000).

318. B. Bouteille, A. Marie-Daragon, G. Chauviere, C. De Albuquerque, B. Enanga, M. L. Darde, J. M. Vallat, J. Perie, and M. Dumas, *Acta Trop.*, **60**, 73–80 (1995).

319. F. W. Jennings, G. Chauviere, C. Viode, and M. Murray, *Trop. Med. Int. Health*, **1**, 363–366 (1996).

320. B. Enanga, M. Keita, G. Chauviere, M. Dumas, and B. Bouteille, *Trop. Med. Int. Health*, **3**, 736–741 (1998).

321. M. P. Barrett, A. H. Fairlamb, B. Rousseau, G. Chauviere, and J. Perie, *Biochem. Pharmacol.*, **59**, 615–620 (2000).

322. B. Enanga, J. M. Ndong, H. Boudra, L. Debrauwer, G. Dubreuil, B. Bouteille, G. Chauviere, C. Labat, M. Dumas, J. Perie, and G. Houin, *Arzneim.-Forsch.*, **50**, 158–162 (2000).

323. B. Enanga, H. Boudra, G. Chauviere, C. Labat, B. Bouteille, M. Dumas, and G. Houin, *Arzneim.-Forsch.*, **49**, 441–447 (1999).

324. H. Gallwitz, S. Bonse, A. MartinezCruz, I. Schlichting, K. Schumacher, and R. L. Krauth-Siegel, *J. Med. Chem.*, **42**, 364–372 (1999).

325. B. M. Joubert, L. N. Nguyen, S. P. T. Matsuda, and F. S. Buckner, *Mol. Biochem. Parasitol.*, **117**, 115–117 (2001).

326. A. Z. Momeni, T. Jalayer, M. Emamjomeh, N. Bashardost, R. L. Ghassemi, M. Meghdadi, A. Javadi, and M. Aminjavaheri, *Arch. Dermatol.*, **132**, 784–786 (1996).

327. S. Sundar, V. P. Singh, N. K. Agrawal, D. L. Gibbs, and H. W. Murray, *Lancet*, **348**, 614 (1996).

328. A. A. Alrajhi, E. A. Ibrahim, E. B. De Vol, M. Khairat, R. M. Faris, and J. H. Maguire, *N. Engl. J. Med.*, **346**, 891–895 (2002).

329. M. S. Araujo, O. A. Martins-Filho, M. E. Pereira, and Z. Brener, *J. Antimicrob. Chemother.*, **45**, 819–824 (2000).

330. J. A. Urbina, G. Payares, J. Molina, C. Sanoja, A. Liendo, K. Lazardi, M. M. Piras, R. Piras, N. Perez, P. Wincker, and J. F. Ryley, *Science*, **273**, 969–971 (1996).

331. J. Molina, J. Urbina, R. Gref, Z. Brener, and J. M. Rodrigues Jr., *J. Antimicrob. Chemother.*, **47**, 101–104 (2001).

332. J. A. Urbina, G. Payares, L. M. Contreras, A. Liendo, C. Sanoja, J. Molina, M. Piras, R. Piras, N. Perez, P. Wincker, and D. Loebenberg, *Antimicrob. Agents Chemother.*, **42**, 1771–1777 (1998).

333. J. Molina, O. Martins-Filho, Z. Brener, A. J. Romanha, D. Loebenberg, and J. A. Urbina, *Antimicrob. Agents Chemother.*, **44**, 150–155 (2000).

334. J. A. Urbina, R. Lira, G. Visbal, and J. Bartroli, *Antimicrob. Agents Chemother.*, **44**, 2498–2502 (2000).

335. J. A. Urbina, J. Vivas, K. Lazardi, J. Molina, G. Payares, M. M. Piras, and R. Piras, *Chemotherapy*, **42**, 294–307 (1996).

336. J. C. Rodrigues, M. Attias, C. Rodriguez, J. A. Urbina, and W. Souza, *Antimicrob. Agents Chemother.*, **46**, 487–499 (2002).

337. F. S. Buckner, J. H. Griffin, A. J. Wilson, and W. C. Van Voorhis, *Antimicrob. Agents Chemother.*, **45**, 1210–1215 (2001).

338. J. A. Urbina, B. Moreno, S. Vierkotter, E. Oldfield, G. Payares, C. Sanoja, B. N. Bailey, W. Yan, D. A. Scott, S. N. Moreno, and R. Docampo, *J. Biol. Chem.*, **274**, 33609–33615 (1999).

339. M. B. Martin, J. S. Grimley, J. C. Lewis, H. T. Heath, B. N. Bailey, H. Kendrick, V. Yardley, A. Caldera, R. Lira, J. A. Urbina, S. N. J. Moreno, R. Docampo, S. L. Croft, and E. Oldfield, *J. Med. Chem.*, **44**, 909–916 (2001).

340. C. M. Szabo, Y. Matsumura, S. Fukura, M. B. Martin, J. M. Sanders, S. Sengupta, J. A. Cieslak, T. C. Loftus, C. R. Lea, H. J. Lee, A. Koohang, R. M. Coates, H. Sagami, and E. Oldfield, *J. Med. Chem.*, **45**, 2185–2196 (2002).

341. A. Montalvetti, B. N. Bailey, M. B. Martin, G. W. Severin, E. Oldfield, and R. Docampo, *J. Biol. Chem.*, **276**, 33930–33937 (2001).

342. F. S. Buckner, K. Yokoyama, L. Nguyen, A. Grewal, H. Erdjument-Bromage, P. Tempst, C. L. Strickland, L. Xiao, W. C. Van Voorhis, and M. H. Gelb, *J. Biol. Chem.*, **275**, 21870–21876 (2000).

343. R. Docampo and S. N. J. Moreno, *Parasitol. Today*, **15**, 443–448 (1999).

344. M. A. Ferguson and G. A. Cross, *J. Biol. Chem.*, **259**, 3011–3015 (1984).

345. R. M. de Lederkremer and L. E. Bertello, *Curr. Pharm. Des.*, **7**, 1165–1179 (2001).

346. M. Berriman, N. Hall, K. Sheader, F. Bringaud, B. Tiwari, T. Isobe, S. Bowman, C. Corton, L. Clark, G. A. Cross, M. Hoek, T. Zanders, M. Berberof, P. Borst, and G. Rudenko, *Mol. Biochem. Parasitol.*, **122**, 131–140 (2002).

347. W. J. Masterson, J. Raper, T. L. Doering, G. W. Hart, and P. T. Englund, *Cell*, **62**, 73–80 (1990).

348. T. L. Doering, T. Lu, K. A. Werbovetz, G. W. Gokel, G. W. Hart, J. I. Gordon, and P. T. Englund, *Proc. Natl. Acad. Sci. USA*, **91**, 9735–9739 (1994).

349. J. R. Roper, M. L. Guther, K. G. Milne, and M. A. Ferguson, *Proc. Natl. Acad. Sci. USA*, **99**, 5884–5889 (2002).

350. M. Sajid and J. H. McKerrow, *Mol. Biochem. Parasitol.*, **120**, 1–21 (2002).

351. J. J. Cazzulo, R. Couso, A. Raimondi, C. Wernstedt, and U. Hellman, *Mol. Biochem. Parasitol.*, **33**, 33–41 (1989).

352. A. E. Eakin, A. A. Mills, G. Harth, J. H. McKerrow, and C. S. Craik, *J. Biol. Chem.*, **267**, 7411–7420 (1992).

353. A. C. Monteiro, M. Abrahamson, A. P. Lima, M. A. Vannier-Santos, and J. Scharfstein, *J. Cell Sci.*, **114**, 3933–3942 (2001).

354. A. E. Eakin, M. E. McGrath, J. H. McKerrow, R. J. Fletterick, and C. S. Craik, *J. Biol. Chem.*, **268**, 6115–6118 (1993).

355. S. A. Gillmor, C. S. Clark, and R. J. Fletterick, *Protein Sci.*, **6**, 1603–1611 (1997).

356. M. E. McGrath, A. E. Eakin, J. C. Engel, J. H. McKerrow, C. S. Craik, and R. J. Fletterick, *J. Mol. Biol.*, **247**, 251–259 (1995).

357. J. C. Engel, P. S. Doyle, I. Hsieh, and J. H. McKerrow, *J. Exp. Med.*, **188**, 725–734 (1998).

358. J. C. Engel, C. Torres, I. Hsieh, P. S. Doyle, J. H. McKerrow, and C. T. Garcia, *J. Cell Sci.*, **113**, 1345–1354 (2000).

359. E. G. Pamer, C. E. Davis, and M. So, *Infect. Immun.*, **59**, 1074–1078 (1991).

360. S. Scory, C. R. Caffrey, Y. D. Stierhof, A. Ruppel, and D. Steverding, *Exp. Parasitol.*, **91**, 327–333 (1999).

361. P. M. Selzer, S. Pingel, I. Hsieh, B. Ugele, V. J. Chan, J. C. Engel, M. Bogyo, D. G. Russell, J. A. Sakanari, and J. H. McKerrow, *Proc. Natl. Acad. Sci. USA*, **96**, 11015–11022 (1999).

362. K. A. Scheidt, W. R. Roush, J. H. McKerrow, P. M. Selzer, E. Hansell, and P. J. Rosenthal, *Bioorg. Med. Chem.*, **6**, 2477–2494 (1998).

363. W. R. Roush, F. V. Gonzalez, J. H. McKerrow, and E. Hansell, *Bioorg. Med. Chem. Lett.*, **8**, 2809–2812 (1998).

364a. J. T. Palmer, D. Rasnick, J. L. Klaus, and D. Bromme, *J. Med. Chem.*, **38**, 3193–3196 (1995).

364b. W. R. Roush, J. Cheng, B. Knapp-Reed, A. Alverez-Hernandez, J. H. McKerrow, E. Hansell, and J. C. Engel, *Bioorg. Med. Chem. Lett.*, **11**, 2759–2762 (2001).

365. L. Huang, A. Lee, and J. A. Ellman, *J. Med. Chem.*, **45**, 676–684 (2002).

366. C. R. Caffrey, M. Schanz, J. Nkemgu-Njinkeng, M. Brush, E. Hansell, F. E. Cohen, T. M. Flaherty, J. H. McKerrow, and D. Steverding, *Int. J. Antimicrob. Agents*, **19**, 227–231 (2002).

367. B. Nare, J. Luba, L. W. Hardy, and S. M. Beverley, *Parasitology*, **114**, S101–S110 (1997).

368. L. W. Hardy, W. Matthews, B. Nare, and S. M. Beverley, *Exp. Parasitol.*, **87**, 158–170 (1997).

369. S. F. Chowdhury, V. B. Villamor, R. H. Guerrero, I. Leal, R. Brun, S. L. Croft, J. M. Goodman, L. Maes, L. M. Ruiz-Perez, D. G. Pacanowska, and I. H. Gilbert, *J. Med. Chem.*, **42**, 4300–4312 (1999).

370. F. Zuccotto, M. Zvelebil, R. Brun, S. F. Chowdhury, R. Di Lucrezia, I. Leal, L. Maes, L. M. Ruiz-Perez, P. D. Gonzalez, and I. H. Gilbert, *Eur. J. Med. Chem.*, **36**, 395–405 (2001).

371. D. J. Hammond and W. E. Gutteridge, *Mol. Biochem. Parasitol.*, **13**, 243–261 (1984).

372. J. J. Marr and R. L. Berens, *Mol. Biochem. Parasitol.*, **7**, 339–356 (1983).

373. D. L. Looker, J. J. Marr, and R. L. Berens, *J. Biol. Chem.*, **261**, 9412–9415 (1986).

374. P. A. Kager, P. H. Rees, B. T. Wellde, W. T. Hockmeyer, and W. H. Lyerly, *Trans. R. Soc. Trop. Med. Hyg.*, **75**, 556–559 (1981).

375. D. Torrus, V. Boix, B. Massa, J. Portilla, and M. Perez-Mateo, *J. Antimicrob. Chemother.*, **37**, 1042–1043 (1996).

376. J. C. Eads, G. Scapin, Y. Xu, C. Grubmeyer, and J. C. Sacchettini, *Cell*, **78**, 325–334 (1994).

377. J. R. Somoza, M. S. Chin, P. J. Focia, C. C. Wang, and R. J. Fletterick, *Biochemistry*, **35**, 7032–7040 (1996).

378. P. J. Focia, S. P. Craig III, and A. E. Eakin, *Biochemistry*, **37**, 17120–17127 (1998).

379. D. M. Freymann, M. A. Wenck, J. C. Engel, J. Feng, P. J. Focia, A. E. Eakin, and S. P. Craig, *Chem. Biol.*, **7**, 957–968 (2000).

380. A. Hofer, D. Steverding, A. Chabes, R. Brun, and L. Thelander, *Proc. Natl. Acad. Sci. USA*, **98**, 6412–6416 (2001).

381. E. del Olmo, M. G. Armas, J. L. Lopez-Perez, G. Ruiz, F. Vargas, A. Gimenez, E. Deharo, and A. San Feliciano, *Bioorg. Med. Chem. Lett.*, **11**, 2755–2757 (2001).

382. E. del Olmo, M. G. Armas, J. L. Lopez-Perez, V. Munoz, E. Deharo, and A. San Feliciano, *Bioorg. Med. Chem. Lett.*, **11**, 2123–2126 (2001).

383. C. Neves-Pinto, V. R. Malta, M. C. Pinto, R. H. Santos, S. L. de Castro, and A. V. Pinto, *J. Med. Chem.*, **45**, 2112–2115 (2002).

384. G. Lowe, A. S. Droz, T. Vilaivan, G. W. Weaver, L. Tweedale, J. M. Pratt, P. Rock, V. Yardley, and S. L. Croft, *J. Med. Chem.*, **42**, 999–1006 (1999).

385. G. Lowe, A. S. Droz, T. Vilaivan, G. W. Weaver, J. J. Park, J. M. Pratt, L. Tweedale, and L. R. Kelland, *J. Med. Chem.*, **42**, 3167–3174 (1999).

## BIBLIOGRAPHY AND FURTHER READINGS

Aksoy, S., I. Maudlin, C. Dale, A. S. Robinson, and S. L. O'Neill, Prospects for control of African trypanosomiasis by tsetse vector manipulation, *Trends Parasitol.*, **17**, 29–35 (2001).

Caffrey, C. R., S. Scory, and D. Steverding, Cysteine proteinases of trypanosome parasites: novel targets for chemotherapy, *Curr. Drug Targets*, **1**, 155–162 (2000).

Cazzulo, J. J., V. Stoka, and V. Turk, The major cysteine proteinase of Trypanosoma cruzi: a valid target for chemotherapy of Chagas disease, *Curr. Pharm. Des.*, **7**, 1143–1156 (2001).

Choi, C. M. and E. A. Lerner, Leishmaniasis: recognition and management with a focus on the immunocompromised patient, *Am. J. Clin. Dermatol.*, **3**, 91–105 (2002).

Coombs, G. H. and J. C. Mottram, Parasite proteinases and amino acid metabolism: possibilities for chemotherapeutic exploitation. *Parasitology*, **114**, S61–S80 (1997).

Croft, S. L., Pharmacological approaches to antitrypanosomal chemotherapy, *Mem. Inst. Oswaldo Cruz.*, **94**, 215–220 (1999).

Croft, S. L. and V. Yardley, Chemotherapy of leishmaniasis, *Curr. Pharm. Des.*, **8**, 319–342 (2002).

Gull, K., The biology of kinetoplastid parasites: insights and challenges from genomic and post-genomics, *Int. J. Parasitol.*, **31**, 443–452 (2001).

Hunter, W. N., A structure-based approach to drug discovery; crystallography and implications for the development of antiparasitic drugs, *Parasitology*, **114**, S17–S29 (1997).

Keiser, J., A. Stich, and C. Burri. New drugs for the treatment of human African trypanosomiasis: research and development, *Trends Parasitol.*, **17**, 42–49 (2001).

Krauth-Siegel, R. L. and G. H. Coombs, Enzymes of parasite thiol metabolism as drug targets, *Parasitol. Today*, **15**, 404–409 (1999).

Lakhdar-Ghazal, F., C. Blonski, M. Willson, P. Michels, and J. Perie, Glycolysis and proteases as targets for the design of new anti-trypanosome drugs, *Curr. Top. Med. Chem.*, **2**, 439–456 (2002).

Michels, P. A. M., V. Hannaert, and F. Bringard, Metabolic aspects of glycosomal in Trypanosomatidae—new data and views, *Parasitol. Today*, **16**, 482–489 (2000).

Murray, H. W., Clinical and experimental advances in the treatment of visceral leishmaniasis, *Antimicrob. Agents Chemother.*, **45**, 2185–2197 (2001).

Murray, H. W., J. Pepin, T. B. Nutman, S. L. Hoffman, and A. A. F. Mahmoud, Tropical medicine, *Br. Med. J.*, **320**, 490–494 (2000).

Pepin, J., N. Khonde, F. Maiso, F. Doua, S. Jaffar, S. Ngampo, B. Mpia, D. Mbulamberi, and F. Kuzoe, Short-course eflornithine in Gambian trypanosomiasis: a multicentre randomized controlled trial, *Bull. World Health Org.*, **78**, 1284–1295 (2000).

Roberts, L. J., E. Handman, and S. J. Fotte. Leishmaniasis, *Br. Med. J.*, **321**, 801–804 (2000).

Rodriguez, J. B., Specific molecular targets to control tropical diseases, *Curr. Pharm. Des.*, **7**, 1105–1116 (2001).

Seed, J. R., African trypanosomiasis research: 100 years of progress, but questions and problems still remain, *Int. J. Parasitol.*, **31**, 434–442 (2001).

Schmidt, A. and R. L. Krauth-Siegel, Enzymes of the trypanothione metabolism as targets for antitrypanosomal drug development, *Curr. Top. Med. Chem.*, **2**, 1239–1259 (2002).

Tracy, J. W. and L. T. Webster, Drugs used in the chemotherapy of protozoal infections, *Goodman and Gilman's Pharmacological Basis of Therapeutics*, 10th ed., McGraw-Hill Professional, New York, NY, 2002, pp. 1097–1120.

Wang, C. C., Targets for antiparasite chemotherapy, *Parasitology*, **114**, S31–S44 (1997).

Woster, P. M., New therapies for parasitic infections, *Ann. Rep. Med. Chem.*, **36**, 99–108 (2001).

# CHAPTER TWENTY

# Anthelmintics

ROBERT K. GRIFFITH
School of Pharmacy
West Virginia University
Morgantown, West Virginia

## Contents

1 Introduction, 1090
2 Clinical Use of Agents, 1090
   2.1 Benzimidazoles, 1090
   2.2 Praziquantel, 1093
   2.3 Ivermectin, 1094
   2.4 Pyrantel, 1094
   2.5 Oxamniquine, 1095
   2.6 Piperazine, 1095
   2.7 Bithionol, Diethylcarbamazine,
      and Suramin, 1095
3 History, 1096
4 Future Developments, 1096

*Burger's Medicinal Chemistry and Drug Discovery*
Sixth Edition, Volume 5: Chemotherapeutic Agents
Edited by Donald J. Abraham
ISBN 0-471-37031-2   © 2003 John Wiley & Sons, Inc.

# 1  INTRODUCTION

Helminths are parasitic worms, which infect an estimated 2 billion people worldwide, nearly all in poor developing tropical or semi-tropical countries. Schistosomiasis and other helmintic infections account for more than 40% of all tropical disease excluding malaria (1). Although fatal helminth infections are relatively rare, many persons may be simultaneously infected with more than one parasite and these diseases inflict an enormous morbidity burden on some of the poorest of the world's countries, a burden that falls particularly heavily on children. Helminth infections contribute to malnutrition, anemia, stunted growth, cognitive impairment, and increased susceptibility to other diseases. In addition to the human burden, domestic animals are also very susceptible to helminth infections, which adds to the economic burden of developing countries and is also a problem for agriculture in many developed countries. A number of anthelmintics are used in both veterinary practice and human chemotherapy.

Helminths are metazoans broadly classified into the classes of nematodes [roundworms and two types of flatworms (trematodes and flukes)] and cestodes (tapeworms). These various organisms have enormously variable life cycles, routes of infection, and susceptibility to chemotherapy. Most require an intermediate host to complete their life cycles and invade human hosts through the skin or ingestion. *Strongyloides* and *Echinococcus* are exceptions; they can live their entire life cycle within a human host. Filaria, from the Latin word for thread, filum, are a particular group of thread-like nematodes of the family Onchocercidae, which live as adults in human tissue or fluids but require an intermediate host of blood-sucking arthropods. These organisms cause a variety of diseases known collectively as filariasis. References 2 and 3 offer excellent overviews of helmintic parasites and the diseases they cause.

There are only about a dozen or so safe and effective drugs for treating helminth infections, but fortunately many of these are very effective and inexpensive. However, there has been no significant new anthelmintic introduced into clinical use since ivermectin entered clinical trials over 20 years ago.

# 2  CLINICAL USE OF AGENTS

The anthelmintics currently available in the United States are listed in Table 20.1. Of the drugs listed in Table 20.1, only pyrantel is available without a prescription in the United States, and three of them, bithionol, diethylcarbamazine, and suramin, are only available to physicians through the Centers for Disease Control under a compassionate use permit. A summary of the drugs effective against some of the more common helminth infections are summarized in Table 20.2.

## 2.1  Benzimidazoles

There are three benzimidazoles available in the United States, mebendazole (**1**), albendazole (**2**), and thiabendazole (**3**). The discovery in 1961 by Merck that thiabendazole (**3**) had broad spectrum activity against a variety of

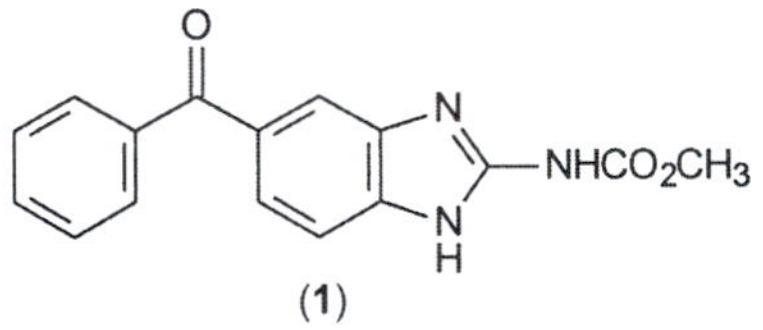

(1)

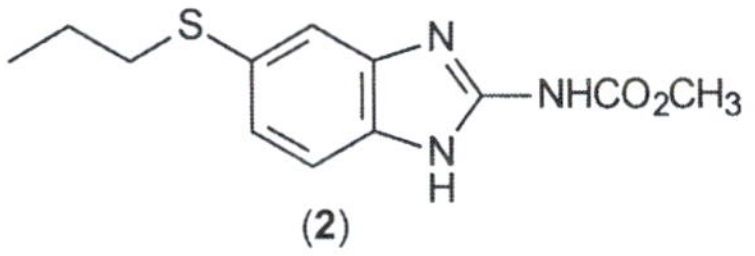

(2)

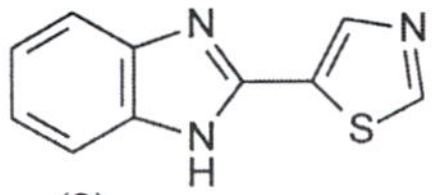

(3)

nematodes infecting the gastrointestinal tract was a major advance in anthelmintic therapy and was followed by the discovery and introduction of the benzimidazole carbamates, me-

# 2 Clinical Use of Agents

**Table 20.1 Anthelmintic Drugs**

| Generic Name | Trade Name | Originator | Chemical Class | Dose[a] |
|---|---|---|---|---|
| Mebendazole (1) | Vermox | Janssen | Benzimidazole | 100 mg, 2× daily |
| Albendazole (2) | Albenza | SmithKlein | Benzimidazole | 400 mg, 2× daily |
| Thiabendazole (3) | Mintezol | Merck | Benzimidazole | 22 mg/kg, 2× daily |
| Praziquante (14) | Biltricide | Bayer | Pyrazinoisoquinoline | 20–25 mg/kg, 3×, one day |
| Ivermectin (19) | Stromectol | Merck | Macrocyclic lactone | 150–200 $\mu$g/kg, once |
| Pyrantel (20) | Antiminth | Pfizer | Tetrahydropyrimidine | 11 mg/kg, 1× daily |
| Oxamniquine (21) | Vansil, Mansil | Pfizer | Tetrahydroquinoline | 12–15 mg/kg, 1× daily |
| Piperazine (24) | Multifuge | | Piperazine | 3.5 g/day for 2 days |
| Bithionol (25) (from CDC) | Lorothidol | Tanabe | Diphenylsulfide | 40–66 mg/kg, for 1 or 2 days |
| Diethylcarbamazine (26) (from CDC) | Hetrazan | Lederle | Piperidine | 13 mg/kg, 1× daily for filariasis; 2 mg/kg, 3× daily for ascariasis |
| Suramin (11) (from CDC, not recommended) | Bayer 205; Antrypol | Bayer | Bisnaphthalene sulfonic acid | 1 g/week for 5–10 weeks |

[a]Adult dose.

bendazole, albendazole, and others. Thiabendazole use has declined substantially since it was first introduced because more recently developed agents are equally or more effective and less toxic. It remains useful however for topical applications in the treatment of cutaneous larva migrans (creeping cruption). Mebendazole and the more recently developed albendazole are widely effective against both intestinal nematodes and most tissue nematodes.

The benzimidazole mechanism of action is inhibition of formation of microtubules through binding to $\beta$-tubulin monomers and preventing polymerization (4). Selectivity for parasite toxicity over host toxicity is caused by tighter binding to the parasite tubulin. There is evidence that susceptibility of a given organism to benzimidazoles is proportional to benzimidazole binding to the organism's tubulin. Further evidence has accumulated recently that benzimidazole resistance is caused by mutations in tubulin (5–9).

When taken orally, thiabendazole is rapidly absorbed and extensively metabolized by the liver (10). Thiabendazole was 12–15% excreted unchanged, and the predominant metabolite is 5-hydroxythiabendazole (4) found as both its free form (22–24%) and glucuronide and sulfate conjugates, 28–29% and 30–

(3) →

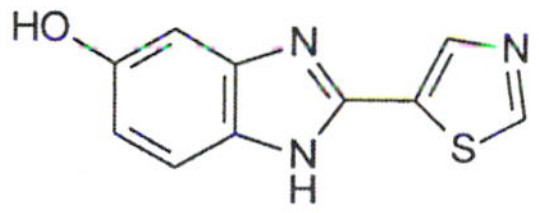

+

5-OH-conjugates

(4)

(5)

31%, respectively. Trace amounts of $N$-methylthiabendazole (5) were also found. All of the metabolites are inactive.

Mebendazole, on the other hand, has poor bioavailability (22%) after oral administration because of irregular absorption and extensive first-pass metabolism (11). The metabolites are inactive (12). The major metabolites are the product of ketone reduction (6) and carbamate hydrolysis (7), isolated both free and as conjugates. A minor metabolite is (8), the product of both reduction and hydrolysis.

Albendazole (2), the most recently developed benzimidazole anthelmintic, is also erratically absorbed and extensively metabo-

**Table 20.2  Major Parasitic Infections and Drugs of Choice**[a]

| Infection (Common Name) | Organism | Drug(s) of Choice |
|---|---|---|
| **Intestinal nematodes** | | |
| Ascariasis (roundworm) | *Ascaris lumbricoides* | Mebendazole, thiabendazole, pyrantel pamoate, or diethylcarbamazine |
| Uncinariasis (hookworm) | *Ancylostoma duodenale, Necator americanus* | Mebendazole or pyrantel[b] |
| Strongyloidiasis (threadworm) | *Strongyloides stercoralis* | Thiabendazole, ivermectin |
| Trichuriasis (whipworm) | *Trichuris trichiura* | Mebendazole |
| Enterobiasis (pinworm) | *Enterobius vermicularis* | Mebendazole, albendazole, thiabendazole, or pyrantel |
| Capillariasis | *Capillaria philippinensis* | Mebendazole, thiabendazole, or albendazole |
| **Tissue nematodes** | | |
| Trichinosis | *Trichinela spiralis* | Albendazole, or mebendazole,[b] plus steroids for severe symptoms |
| Cutaneous larva migrans (Creeping eruption) | *Ancylostoma braziliense* and others | Thiabendazole, albendazole, or ivermectin |
| Onchoceriasis (River blindness) | *Onchocerca volvulus* | Ivermectin, diethylcarbamazine, or suramin[c] |
| Dracontiasis (Guinea worm) | *Dracunculus medinensis* | Thiabendazole or mebendazole |
| Angiostrongyliasis (rat lungworm) | *Angiostrongylus cantonensis* | Thiabendazole or mebendazole |
| Loiasis | *Loa loa* | Diethylcarbamazine |
| **Cestodes** | | |
| Taeniasis (beef tapeworm) | *Taenia saginata* | Praziquantel[b] |
| Taeniasis (pork tapeworm) | *Taenia solium* | Praziquantel[b] or albendazole |
| Diphyllobothriasis (fish tapeworm) | *Diphyllobothrium latum* | Praziquantel[b] |
| Dog tapeworm | *Dipylidium caninum* | Praziquantel[b] |
| Hymenolepiasis (dwarf tapeworm) | *Hymenolepsis nana* | Praziquantel[b] |
| Hydatid cysts | *Echinococcus granulosus* | Albendazole or praziquantel[b] |
| **Trematodes** | | |
| Schistosomiasis | *Schistosoma mansoni* | Praziquantel or oxamniquine |
| | *Schistosoma japonicum* | Praziquantel |
| | *Schistosoma haematobium* | Praziquantel |
| | *Schistosoma mekongi* | Praziquantel |
| | | |
| **Hermaphroditic flukes** | | |
| Fasciolopsiasis (intestinal fluke) | *Fasciolopsis buski* | Praziquantel |
| | *Heterophyes heterophyes* | Praziquantel |
| | *Metagonimus yokogawai* | Praziquantel |
| Clonorchiasis (Chinese liver fluke) | *Clonorchis sinensis* | Praziquantel |
| Fascioliasis (sheep liver fluke) | *Fasciola hepatica* | Praziquantel or bithionol[d] |
| Opisthorchiasis (liver fluke) | *Opisthorchis viverrini* | Praziquantel |
| Paragonimiasis (lung fluke) | *Paragonimus westermani* | Praziquantel or bithionol[d] (alternate) |

[a]Table modified from Drug Facts and Comparisons 2000 [2002, **vol. 110**] with permission.
[b]Unlabeled use.
[c]Available from the Center for Disease Control, although generally not recommended.
[d]Available from the Center for Disease Control.

(1) 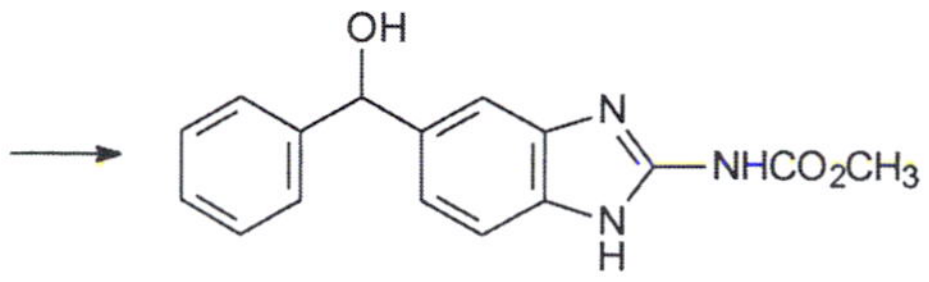 ⟶

(6)

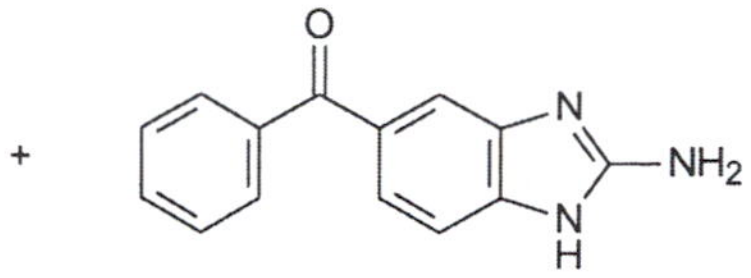

+

(7)

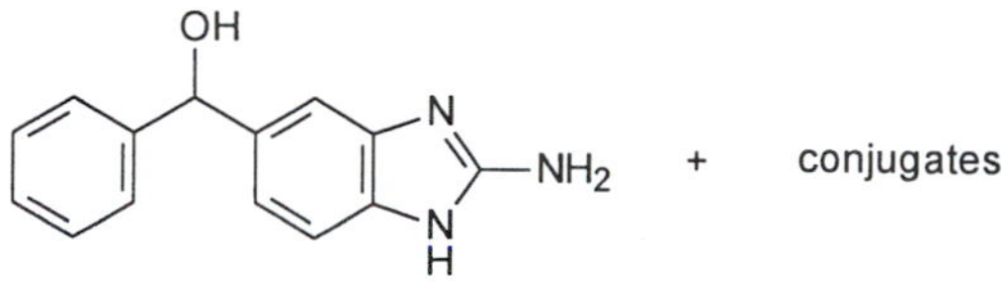

(8)  +  conjugates

lized as shown (12). However the primary metabolite, albendazole sulfoxide (9) is active, and because parent albendazole is barely detectable in plasma, albendazole sulfoxide probably accounts for all the anthelmintic ac-

tivity of the drug (13, 14). Formation of albendazole sulfoxide is carried out by both liver CYP450 and by microsomal flavin-linked monooxygenase. Sulfoxides are chiral, and the flavin enzyme favors formation of the (+)isomer and P450 favors formation of the (−)isomer (15). Both sulfoxide enantiomers are further oxidized to the inactive sulfone (**10**) and on to other minor inactive metabolites (**11–13**).

## 2.2 Praziquantel

Praziquantel (**14**) is a widely employed anthelmintic initially found to be active against schistosomiasis (16) and subsequently found to widely effective against a variety of trematodes and cestodes (17). Both safe and effective, praziquantel is the drug of choice for treating all forms of schistosomiasis. The mechanism of action of praziquantel is complex and as yet not clearly understood (18). Praziquantel rapidly penetrates the tegument of trematodes and cestodes but does not seem to be able to cross the thicker tegument of nematodes, which may account for its selectivity. Once inside, praziquantel alters calcium

(2) 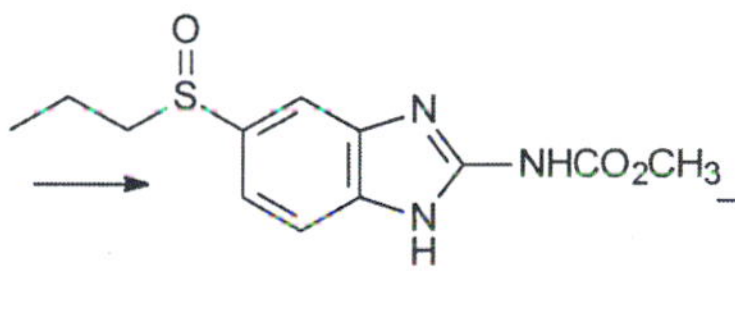  ⟶

(9)

(active)

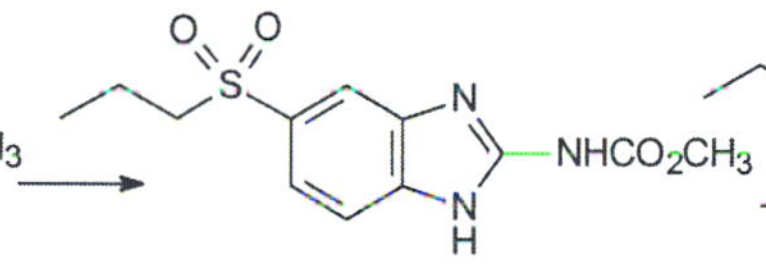

(10)

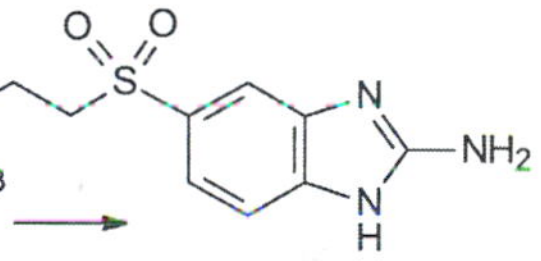

(11)

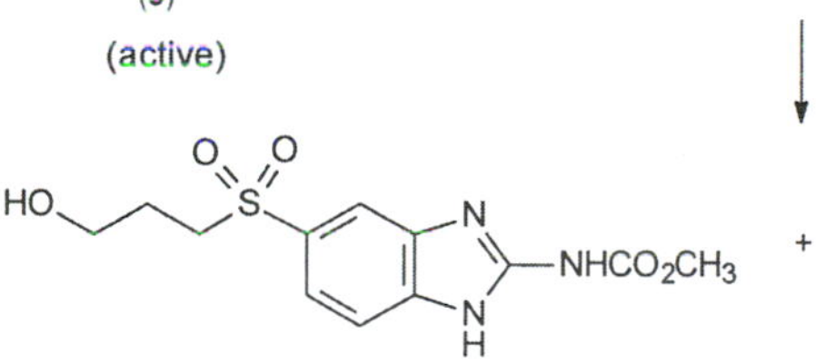

(12) 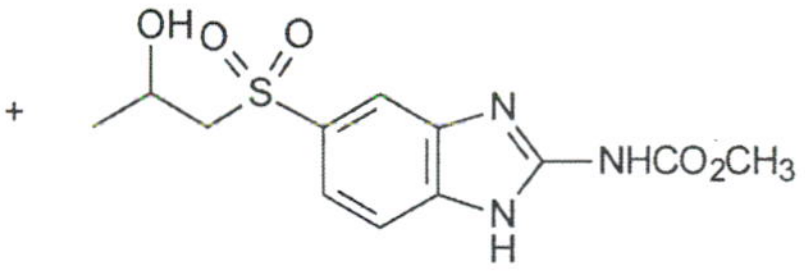  +  (13)

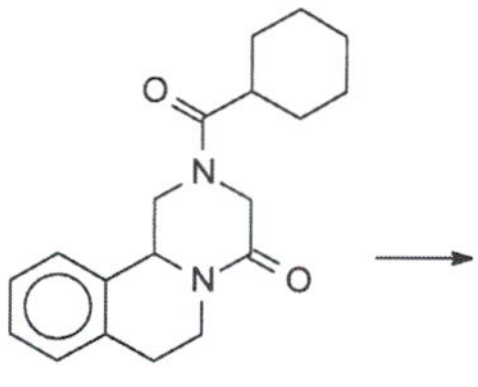

(14) 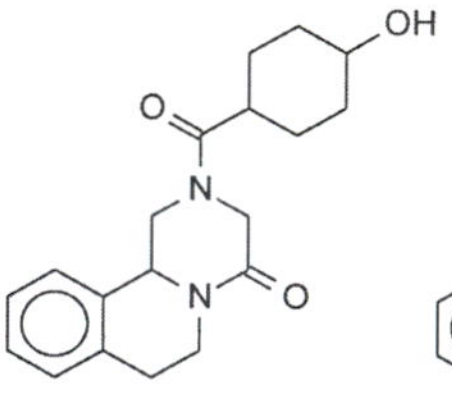  ⟶  (15) 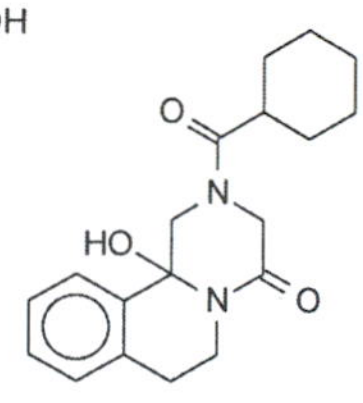  (16) 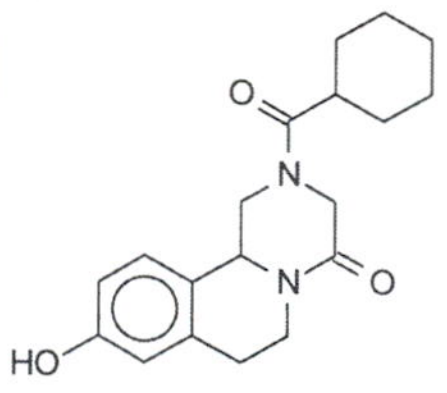  (17) 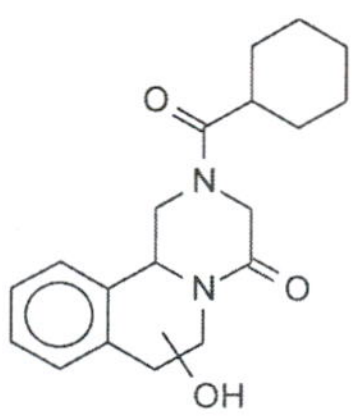  (18)

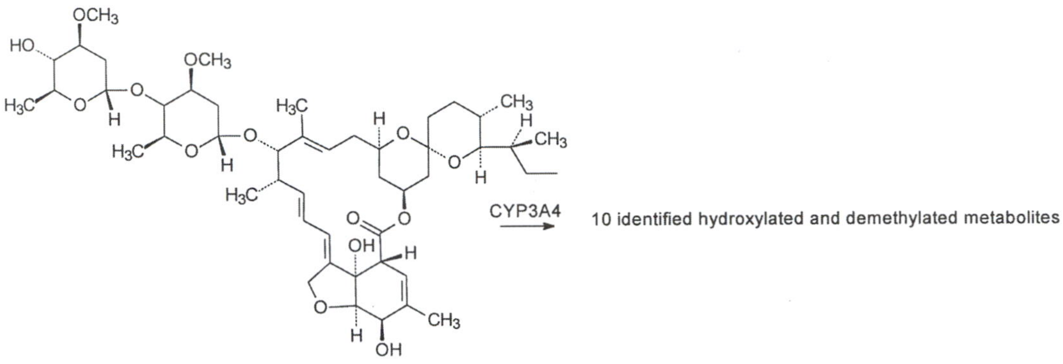

CYP3A4 →  10 identified hydroxylated and demethylated metabolites

(19)

ion flux in the parasite, leading to muscle contractions, and spastic paralysis, leading to detachment of the parasite from the host. Praziquantel also causes damage to the tegument permitting host defenses to attack the weakened parasite. Parasites dislodged from their anchors to host tissue are carried to the liver and attacked by granulocytes (18).

Praziquantel is extensively metabolized by the liver to hydroxylated products (19). The principal product (35%) is the cyclohexane-4-hydroxyl derivative (**15**) of which 80% is the *cis*-isomer and 20% the *trans*-isomer (**20**). Inhibitor studies indicate the metabolism is likely carried out predominantly by isozymes CYP2B1 and CYP3A (21). Other identified but minor hydroxylated metabolites (and their conjugates) are (**16–18**). All the metabolites are inactive.

### 2.3  Ivermectin

Ivermectin, (**19**), produced from avermectin by reduction of one double bond, is another breakthrough drug in anthelmintic chemotherapy. Active against most nematodes and introduced in the mid-1980s, ivermectin was the first safe and effective agent against *Onchocerca volvulus*, the causative agent of River Blindness (22). It is currently the drug of choice for treating onchocerciasis and strongyloides (23). Recent studies have shown ivermectin efficacious in treating lymphatic filarial infections such as those associated with both Wuchereria bancrofti and Brugia malayi infections of humans that cause elephantiasis (24). Ivermectin is replacing the far more toxic

diethylcarbamazine in the treatment of many nematode infections. The antiparasitic mechanism of ivermectin involves disruption of the function of glutamate-gated chloride channels to induce a tonic paralysis of the nematode musculature (25–28). Cestodes and trematodes lack high affinity ivermectin-binding proteins, which may explain why these helminth groups are not sensitive to ivermectin (29). Ivermectin is well absorbed orally and metabolized by the liver CP450 system to at least 10 hydroxylated and demethylated products (30). The principal enzyme involved is CYP3A4.

### 2.4  Pyrantel

Pyrantel (**20**) is the only anthelmintic available without a prescription in the United States. Introduced in the mid-1960s, pyrantel

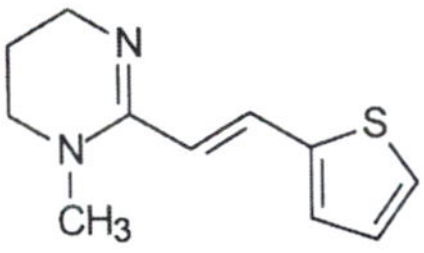

(20)

quickly found use in treating intestinal nematodes (31) and is currently used for treating pinworm, (enterobiasis), roundworm (ascariasis), and hookworm (uncinariasis). The drug acts on nicotinic acetylcholine receptors to cause spastic paralysis of the worms that detach from the host and are swept out in feces (32). The drug is very poorly absorbed, which probably accounts for it's safety for human

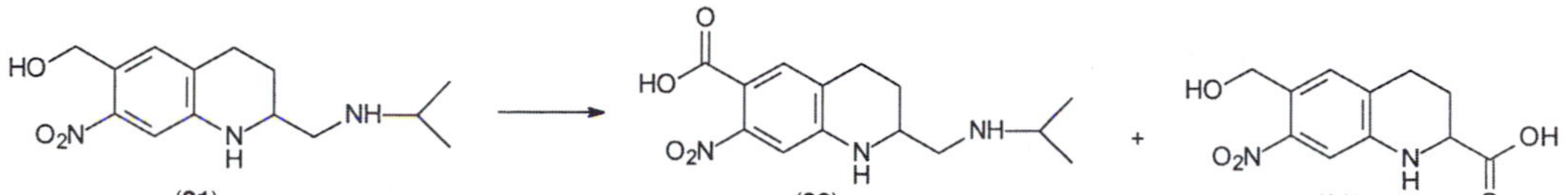

(21)        →        (22)        +        (23)

use without a prescription. Less than 15% is excreted in urine as parent drug or metabolites. Most is excreted unchanged in feces.

## 2.5   Oxamniquine

Oxamniquine (**21**) is an old drug with an intriguingly narrow spectrum of activity. Oxamniquine is an antischistosomal active only against *S. mansoni* (33) and has no effect on the other schistosomes at therapeutic doses. For reasons unknown, male *S. mansoni* concentrate the drug and die, leaving surviving females unable to lay eggs. The mechanism of action is unknown, although the drug exhibits some anticholinergic activity. Oxamniquine is readily absorbed and metabolized to the 6-carboxylic acid (34) (**22**) and 2-carboxylic acid (**23**).

## 2.6   Piperazine

Piperazine (**24**) is useful as an alternative to mebendazole or pyrantel for treating combined ascariasis and Enterobius infections (35). Piperazine acts as a gamma-amino-bu-

tyric acid (GABA)-agonist, increasing membrane permeability to chloride ion and inducing a flaccid paralysis in the parasites (36). The drug is well absorbed, and about 20% is excreted unchanged in the urine (37).

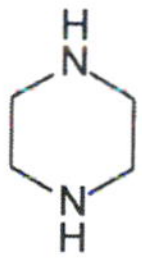

## 2.7   Bithionol, Diethylcarbamazine, and Suramin

Diethylcarbamazine (**25**), bithionol (**26**), and suramin (**27**) are available in the United States only through the government's Centers

(25)

(26)

for Disease Control. These drugs are all very old, have limited efficacy, and have high toxicity. Diethylcarbamazine was at one time the drug of choice for filariasis infections, but has been replaced in most applications by ivermec-

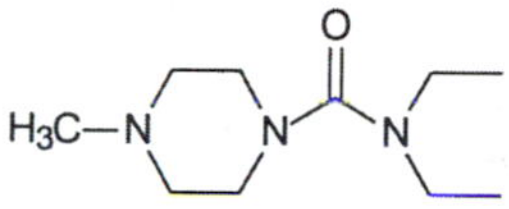

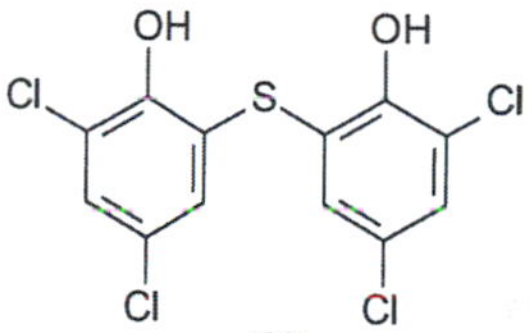

(24)

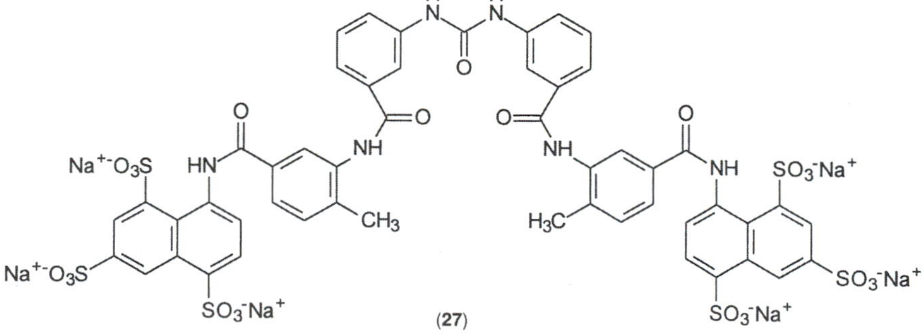

(27)

tin. Suramin was at one time the drug of choice for onchocerciasis but has also been replaced by ivermectin or combinations of ivermectin with albendazole. These agents are available as drugs of last resort when other therapeutic regimens fail.

## 3  HISTORY

An exhaustive history of the development of anthelmintic drugs through the late 1970s is available in a previous edition of this work (38). Ivermectin is the only drug of significance to enter therapeutics in the last 20 years.

## 4  FUTURE DEVELOPMENTS

There is a paucity of research into new anthelmintics for several reasons, most notably that there is very little profit to be made from drugs in this class, and there are few viable molecular targets for application of modern techniques of drug design and high-throughput screening.

## REFERENCES

1. WHO. *Wkly. Epidemiol. Rec.*, **76**, 74–76 (2001).
2. D. J. Wyler, Ed., *Modern Parasite Biology*, W. H. Freeman, New York, 1990.
3. G. L. Mandell, J. E. Bennett, and R. Dolin, Eds., *Mandell, Douglas, and Bennett's Principles and Practice of Infectious Diseases*, 5th ed, Churchill Livingstone, Philadelphia, PA, 2000.
4. E. Lacey, *Parasitol. Today*, **6**, 112–115 (1990).
5. M. H. Roos, *Parasitol. Today*, **6**, 125–127 (1990).
6. M. S. Kwa, et al., *Biochem. Biophys. Res. Commun.*, **191**, 413–419 (1993).
7. M. S. Kwa, J. G. Veenstra, and M. H. Roos, *Mol. Biochem. Parasitol.*, **63**, 299–303 (1994).
8. M. H. Roos, et al., *Pharmacol. Ther.*, **60**, 331–336 (1993).
9. M. S. Kwa, J. G. Veenstra, and M. H. Roos, *Mol. Biochem. Parasitol.*, **60**, 133–143 (1993).
10. T. Tsuchiya, et al., *Chem. Pharm. Bull. (Tokyo)*, **35**, 2985–2993 (1987).
11. P. A. Braithwaite, et al., *Eur. J. Clin. Pharmacol.*, **22**, 161–169 (1982).
12. D. W. Gottschall, V. J. Theodorides, and R. Wang, *Parasitol. Today*, **6**, 115–124 (1990).
13. S. E. Marriner, et al., *Eur. J. Clin. Pharmacol.*, **30**, 705–708 (1986).
14. P. Moroni, et al., *Drug Metab. Dispos.*, **23**, 160–165 (1995).
15. P. Delatour, et al., *Xenobiotica*, **21**, 217–221 (1991).
16. J. E. McMahon and N. Kolstrup, *Br. Med. J.*, **2**, 1396–1398 (1979).
17. P. Andrews, et al., *Med. Res. Rev.*, **3**, 147–200 (1983).
18. W. Harnett, *Parasitol. Today*, **4**, 144–146 (1988).
19. K. U. Buhring, et al., *Eur. J. Drug Metab. Pharmacokinet.*, **3**, 179–190 (1978).
20. A. Hogemann, et al., *Arzneimittelforschung*, **40**, 1159–1162 (1990).
21. C. M. Masimirembwa and J. A. Hasler, *Biochem. Pharmacol.*, **48**, 1779–1783 (1994).
22. J. P. Coulaud, et al., *Lancet*, **2**, 526–527 (1984).
23. O. Zaha, et al., *Intern. Med.*, **39**, 695–700 (2000).
24. K. R. Brown, F. M. Ricci, and E. A. Ottesen, *Parasitology*, **121**, S133–S146 (2000).
25. P. T. Meinke, et al., *J. Med. Chem.*, **35**, 3879–3884 (1992).
26. D. F. Cully, et al., *Parasitology*, **113**(Suppl), S191–S200 (1996).
27. D. J. Brownlee, L. Holden-Dye, and R. J. Walker, *Parasitology*, **115**, 553–561 1997.
28. M. V. Hejmadi, et al., *Parasitology*, **120**, 535–545 (2000).
29. W. L. Shoop, et al., *Int. J. Parasitol.*, **25**, 923–927 (1995).
30. Z. Zeng, et al., *Xenobiotica*, **28**, 313–321 (1998).
31. J. W. McFarland, et al., *J. Med. Chem.*, **12**, 1066–1079 (1969).
32. D. Rayes, et al., *Neuropharmacology*, **41**, 238–245 (2001).
33. L. T. J. Webster in A. G. Gilman, et al., Eds., *Goodman and Gilman's: The Pharmacological Basis of Therapeutics*, Pergamon Press, New York, 1990, pp. 959–977.
34. B. Kaye and N. M. Woolhouse, *Ann. Trop. Med. Parasitol.*, **70**, 323–328 (1976).
35. J. W. Tracy and L. T. J. Webster in J. G. Hardman and L. E. Limbird, Eds., *Goodman and Gilman's:The Pharmacological Basis of Therapeutics*, McGraw-Hill, New York, 2001, pp. 1121–1140.
36. R. J. Martin, *Br. J. Pharmacol.*, **84**, 445–461 (1985).
37. K. A. Fletcher, D. A. Evans, and J. A. Kelly, *Ann. Trop. Med. Parasitol.*, **76**, 77–82 (1982).
38. P. J. Islip in M. E. Wolff, Ed., *Burger's Medicinal Chemistry*, John Wiley & Sons, New York, 1979, pp. 481–530.

# Index

Terms that begin with numbers are indexed as if the number were spelled out; e.g., "3D models" is located as if it were spelled "ThreeD models."

A-87380
  for influenza, 376
A-192558
  for influenza, 376
Aablaquine, 934
Abacavir
  anti-HIV drug, 471, 472
  selective toxicity, 267
ABELCET, 890, 892
*Absidia*, 883
Absorption
  of radiation energy, 153
ABT-675
  for influenza, 376–377
Acedapsone, 830
ACE inhibitors
  as antagonists, 251–252
Achromycin, 740, 935
Acids
  topical antibacterials, 551
*Actinetobacter baumannii*
  penicillin-binding protein, 614, 616
Actinocin
  antitumor natural product, 111
Actinomycin C
  antitumor natural product, 111
Actinomycin D
  antitumor natural product, 111
  PLD repair inhibitor, 192
Actinomycins
  antitumor natural products, 111–115
Activator protein-1, 25
Actonel
  selective toxicity, 273
Acyclovir
  for hepatitis B virus, 331
  for herpes virus, 305–326
  selective toxicity, 254, 267
Acylides, 781–782

Adaptive immunity, 224–226
Adefovir
  anti-HIV drug, 471, 472
  for hepatitis B virus, 332–333, 334, 335
  for papillomavirus, 339
Adeno-associated viruses, 304
  vectors for cancer gene therapy, 42
Adenoma-carcinoma, 4
*S*-Adenosyl-L-homocysteine hydrolase inhibitors, 432–434
Adenoviruses, 303–304
  antiviral agents, 340
  dendritic cell transduction by, 242
  vectors for cancer gene therapy, 41–42
ADEPT (antibody-directed enzyme-prodrug therapy), 83, 87–90
Adozelesin
  alkylating agent, 54, 61–64
Adriamycin
  antitumor natural product, 128
Adrucil
  antimetabolite, 76
African trypanosomiasis (African sleeping sickness)
  antiprotozoal agents for, 1035–1045
Aftate, 891
AG3340
  antiangiogenic agent, 217, 218
AG7088
  picornavirus antiviral, 404–405
Agathisflavone
  influenza antiviral agent, 389
Agonists, 250–251
AIDS, 458–459, 712. *See also* Anti-HIV drugs; HIV protease inhibitors; HIV virus
  associated fungal infections, 884
  and malaria, 930
  and tuberculosis, 823, 824
AK-2123
  radiosensitizer, 179

D-Alanyl-D-alanine transpeptidase, 610
  inhibition by $\beta$-lactams, 611–612
Albendazole
  antihelmintic, 1090–1091, 1091
Alcohols
  topical antibacterials, 547–548
Aldehydes
  topical antibacterials, 551
Aldenosulfone, 830
Alemtuzumab, 229, 231–233
Alendronate
  selective toxicity, 273
Alkeran
  alkylating agent, 54
Alkylating agents
  cyclopropylindoles, 54, 61–64
  mustards, 52, 53–59
  nitrosureas, 54, 64–65
  platinum complexes, 54, 59–61
  triazenes, 54, 65–67
Allegra
  selective toxicity, 275
Allylamine antifungal agents, 890, 891, 899–900
Alvosulfon, 810
AM-1155, 853
Amantadine, 361
  anti-HIV drug, 469
  selective toxicity, 266, 267
Ambisone, 890, 893
AMD3100
  anti-HIV drug, 467
Amentoflavone
  influenza antiviral agent, 389
American trypanosomiasis (Chagas' disease)
  antiprotozoal agents for, 1045–1047
Ametrone
  topoisomerase II inhibitor, 69
Amidines
  topical antibacterials, 554–555
Amikacin, 748
  antimycobacterial application, 810, 844
  efficacy and dosage, 810
  indications, 625

Amikin, 748, 810

3-Aminobenzamide
PLD repair inhibitor, 191–192

9-Aminocamptothecin
antitumor natural product, 133

7-Amino-cephalosporanic acid, 638–639

Aminoglycoside antibiotics. *See also* Gentamicin; Neomycin; Streptomycin
biosynthesis, 751–753
clinical use and currently used drugs, 747–748
drug resistance, 753–757
pharmacology and mode of action, 749–751
recent developments, 757–758
selective toxicity, 260–261
side effects, toxicity, and contraindications, 748–749

8-Aminolaevulinic acid
radiosensitizer, 175

3-Amino nocardicinic acid derivatives, 660–663

6-Aminopenicillanic acid (6-APA)
compounds derived from, 609, 630–637
isolation and characterization, 628–629
structure-activity relationships, 630–637

Aminopenicillins, 632
use recommendations, 626

*p*-Aminopropiophenone
radioprotective agent, 165

Aminopterin
antimetabolite, 76, 78

*p*-Aminosalicylic acid, 830–831
efficacy and dosage, 810, 863
side effects, 831–832

3-Amino-1,2,4-triazole
radioprotective agent, 162

2-Amino-4,6,6-trimethyl-1,3-thiazine
radioprotective agent, 164–165

Amithiozone, 832

Amodiaquine, 979
antimalarial, 981–982
dosage forms, 935

Amopyroquine, 979

Amorolfine, 903
formulations and indications, 891

Amoxicillin, 609

Amoxycillin, 632
antimycobacterial application, 858
*in vitro* antibacterial activity, 635

Amphocil, 890, 893

Amphotec, 890, 893

Amphotericin B, 889–893
with flucytosine, 905
formulations and indications, 890, 891
for leishmaniasis, 1051
selectivity, 886

Amphotericins
selective toxicity, 268

Amphoteric surfactants
topical antibacterials, 557

Ampicillin, 609, 632
antimycobacterial application, 858
pharmacokinetics, 624
use recommendations, 626

Amprenavir
anti-HIV drug, 475–476
selective toxicity, 267

Amsacrine
with etoposide, 71
topoisomerase II inhibitor, 68, 70

Amsidyl
topoisomerase II inhibitor, 68

Anchorage-dependent growth, 12–14, 21

Ancistrocladine
antimalarial, 994–995

Ancobon, 890

Angiogenesis, 216
and cancer, 20–21
compounds inhibiting, 216–220
conditions related to, 220

Angiogenic switch, 20

Angiostatin, 45

Anhydrolides, 781–782

Anidulafungin, 888, 901, 902–903

Anionic surfactants
topical antibacterials, 556–557

Anisoactinomycins
antitumor natural products, 111

Anoxia
radioprotective agents utilizing, 167–168

Antagonists, 250–252

Anthracyclines
alkylating agents, 52
antitumor natural products, 110, 126–128

Antiangiogenic agents, 38–39, 216–220
gene therapy approaches, 45

Antiasthma drugs
selective toxicity, 252

Antibacterial agents, 538–539
broad-spectrum β-lactams, 705–709
drug resistance, 543, 572–573, 586, 590
selective toxicity, 539–543
systemic synthetic, 557–596
topical synthetic, 539–557

Antibiotic resistance, *See* Drug resistance

Antibiotics, 738–739. *See also* Aminoglycoside antibiotics; Erythromycins; Macrolide antibiotics; specific types of Antibiotics, such as Tetracyclines
as antimalarials, 982–983
antimycobacterial agents, 840–853
radioprotective agents, 166
selective toxicity, 259–263

Antibodies, 224–226
general structure, 227

Antibody-dependent cytotoxicty, 225–226

Antibody-directed enzyme-prodrug therapy (ADEPT), 83, 87–90

Antibody-directed immunotherapy, 226–235

Antibody-toxin conjugates (armed antibodies), 83, 92–94
selective toxicity (paclitaxel), 258

Anticancer drugs. *See also* Antitumor drugs; Cancer; Tumor-activated prodrugs
antiangiogenic agents, 215–220
drug resistance in chemotherapy, 281–290
electron-affinic drugs in, 180–183
market for, 109
radioprotective agents with, 172–173
selective toxicity, 257–259

Anticancer gene therapy
  approaches for, 43–46
  prospects for, 39–43
Anticoagulants
  for cancer treatment, 38
Antifolates, 76–79
  antimalarial, 954–960
Antifungal agents
  allylamines, 890, 891, 899–900
  aureobasidins, 906–907
  azoles, 890, 891, 893–899
  benanomycins, 908
  candins, 890, 900–903
  epidemiology of resistance,
    884–885
  fungal efflux pump inhibitors,
    909
  miscellaneous compounds,
    903–910
  morpholines, 903
  N-myristoyl transferase inhib-
    itors, 908–909, 910
  nikkomycins, 906
  polyenes, 889–893, 890, 891
  polyoxins, 906
  pradimicins, 908
  selective toxicity, 254,
    266–268, 886
  sordarins, 907–908
  thiocarbamates, 903
  for treatment of superficial
    cutaneous mycoses, 891
  for treatment of systemic my-
    coses, 890
  trends in, 910–911
Antigen-presenting cells,
    224–225, 235
  and organ transplant rejec-
    tion, 487
Antihelmintics, 1090–1096
Antihistamines
  selective toxicity, 275–276
Anti-HIV drugs. See also HIV pro-
    tease inhibitors; HIV reverse
    transcriptase inhibitors
  combination therapies, 479
  drug resistance, 478–479
  drugs in clinical use, 477–478
  HIV vaccines, 465
  inhibitors of Gp120 binding to
    T-cell receptors, 467–468
  inhibitors of HIV gene expres-
    sion, 473–474
  inhibitors of HIV integrase,
    465, 473
  inhibitors of HIV-1 protease,
    465, 475–477

  inhibitors of HIV reverse tran-
    scription, 465, 469–473
  inhibitors of HIV ribonucle-
    ase, 473
  inhibitors of viral entry,
    465–468
  inhibitors of viral fusion,
    468–469
  inhibitors of viral maturation,
    475–477
  inhibitors of viral uncoating,
    469
  inhibitors of virion assembly,
    474–475
  need for new, 478–479
  selective toxicity, 267
  viral adsorption inhibitors,
    465–467
  virus "knock-out" strategy,
    478–479
Antiinfectives
  selective toxicity, 259–263
Anti-inflammatory therapeutic
    antibodies, 233–235
Antimalarial agents
  currently used drugs, 933–985
  experimental agents, 985–993
  natural products, 993–999
  new targets, 993
  resistance, 921, 928–930
  selective toxicity, 254, 268
Antimetabolites, 75–76
  antifolates, 76–79
  purine analogs, 76, 81–82
  pyrimidine analogs, 76, 79–81
Antimycobacterial agents,
    809–813
  antibiotics, 840–853
  currently used drugs,
    829–840
  drugs under investigation,
    853–860
  future developments, 866–867
  recent developments and
    present status of chemo-
    therapy, 860–867
  screening and evaluation,
    826–829
  selective toxicity, 266
Antioxidants, 33
Antiparasitic drugs. See also An-
    timalarial agents
  antihelmintics, 1090–1096
  selective toxicity, 254, 268
Antiprotozoal agents, 1034–1035
  for African trypanosomiasis,
    1035–1045

  for American trypanosomiasis,
    1045–1047
  for Leishmaniasis, 1047–1052
  new drug therapy perspec-
    tives, 1052–1076
Antiradiation testing, 154
Antisense oligonucleotides
  HBV virus antiviral agents,
    337
  hepatitis C virus antiviral
    agents, 415–417
  influenza antiviral agents,
    384–385
Antisense technology
  organ transplant rejection ap-
    plications, 524–525
Antiseptics, 539
  testing, 544–545
Antitumor drugs. See also Anti-
    cancer drugs; Cancer; Tu-
    mor-activated prodrugs
  drug resistance in chemother-
    apy, 281–290
  macrolides, 784–785
  radioprotective agents with,
    172–173
  selective toxicity, 257–259
Antitumor natural products,
    109–111
  drugs attacking DNA, 110,
    111–124
  drugs inhibiting DNA process-
    ing enzymes, 110, 124–136
  drugs interfering with tubulin
    polymerization/depolymer-
    ization, 110, 136–143
Antiviral agents. See also Anti-
    HIV drugs
  selective toxicity, 254, 266
Antiviral agents, DNA, 305
  adenoviruses, 340
  for hepatitis B virus, 326–337
  for herpesviruses, 305–326
  papillomaviruses, 337–339
  parvoviruses, 341–342
  polyomaviruses, 339–340
  poxviruses, 340–341
Antiviral agents, RNA, 360–361
  arenaviruses, 430
  broad spectrum, 432–435
  flaviviruses, 427–430
  hepatitis C virus, 361–364,
    415–427
  influenza A and B viruses,
    364–390
  measles virus, 390–396

Antiviral agents, RNA (*Continued*)
  orthopoxviruses, 435–440
  parainfluenza virus, 390–398
  paramyxovirus fusion proteins, 390–396
  picornaviruses, 398–415
  respiratory syncytial virus, 390–398
  rotaviruses, 430–431
  rubella virus, 431–432
Apalcillin, 633
APC-6336
  hepatitis C antiviral, 422–433
APC (adenomatous polyposis coli) gene, 5–6, 27, 29–30
  gene therapy target, 44–45
  hypermethylation, 9
Apicidin
  antimalarial, 994
Apoptosis
  and adaptive immunity, 225
  and carcinogenesis, 16–19
  and drug resistance in cancer chemotherapy, 287–289
Apoptotic tumor cells, 239
AQ4N
  radiosensitizer, 182
  tumor-activated prodrug, 83, 84, 85
Aralen, 934
Arenaviruses
  antiviral agents, 430
Armed antibodies, 83, 92–94
Arsumax, 935
Arteannuin, 961
Arteether, 964
  antimalarial, 974–975
  dosage forms, 935
Arteflene
  antimalarial, 984–985
Artelinic acid, 964, 974–975
Artemether, 964
  antimalarial, 973–974
  dosage forms, 935
Artemisinin, 960–963
  dosage forms, 935
  first generation, 970–976
  mechanism of action, 966–970
  structure-activity relationship, 963–966
9-*epi*-Artemisinin, 965
Artemotil, 935
Arteries, 216
Artesunate, 964, 972–973
  dosage forms, 935

Artificial organs, 526
Asepsis, 539
Asparenomycin A, 670, 671
  *in vitro* activity, 692
Aspergillosis, 883
*Aspergillus,* 883, 884
*Aspergillus fumigatus,* 882
Aspirin
  selective toxicity, 274
Asulacrine
  topoisomerase II inhibitor, 68–69, 71
Ataxia telangiectasia, 32
Atazanavir
  anti-HIV drug, 476–477
Atovaquone
  antimalarial, 976–978
  dosage forms, 935
ATP-binding cassette genetic superfamily
  and drug resistance, 285
Augmentin, 695
Aureobasidins, 906–907
Aureomycin, 740
Autolysins, 610–613
Avloclor, 934
Axin, 29–30
Axisonitrile
  antimalarial, 993
Azadicarbonamide
  anti-HIV drug, 469, 470
Azapeptides
  picornavirus antiviral, 408–409
Azathioprine, 489, 490
  clinical use for organ transplants, 513–516
  side effects, 516
6-Azauridine
  flavivirus antiviral, 427–428
Azide
  radioprotective agent, 162
Azidocillin, 609
Azithromycin, 763
  antimalarial, 935, 983
  antimycobacterial application, 857–858
  selective toxicity, 261
Azlocillin, 633
  indications, 625
  pharmacokinetics, 624
Azolastone
  alkylating agent, 54
Azole antifungal agents, 890, 891, 893–899
AZT, *See* Zidovudine

Aztreonam, 609, 666
  activity, 667
  indications, 625
  β-lactamase classification based on activity, 620
  pharmacokinetics, 624
  side effects and interactions, 625
  use recommendations, 626

B-283, 838
B-663, 838
Bacampicillin, 609, 632
Bacterial infections, 738
Bacterial radiosensitizers, 197
Bacterial resistance, *See* Drug resistance
Bacterial ribosome, 738–739
Bactrim, 934
Bad protein, 24
Bafilomycin A1
  flavivirus antiviral, 429
Bak protein, 19
BANA compounds
  antiviral application, 372–374
Basal lamina, 216
Basiliximab, 490, 510–513
Batimastat, 217, 218
Bax protein, 19
Bay 12–9566
  antiangiogenic agent, 217
BBNH
  anti-HIV drug, 473
BBR 2778
  topoisomerase II inhibitor, 68, 71
BBR 3464
  alkylating agent, 54, 61
B-cell receptors, 224
B-cells
  and adaptive immunity, 224–225
  exosome secretion, 241
Bcl-2 protein, 22, 24, 26
  and apoptosis, 19
  and COX-2, 34
  and drug resistance, 287, 288–289
Bcl-x$_L$ protein, 19
BCNU
  alkylating agent, 54, 64–65
Bcr-abl tyrosine kinase, 22
BCX-1812
  for influenza, 364, 374–376
Benadryl
  selective toxicity, 275
Benanomycins, 908

Benflumetol, 978
Benicillin, 632
Benzanthrone derivatives
    RSV antiviral agents, 395
Benzimidazole antihelmintics,
    1090–1093
Benznidazole
    for American trypanosomiasis
        (Chagas' disease),
        1046–1047
Benzylpenicillin. *See also* Peni-
    cillin G
    β-lactamase classification
        based on activity, 619
    structure, 628
Benzyl penicillin
    *in vitro* antibacterial activity,
        667
Betamipron, 693
Biapenem, 693
    *in vitro* activity, 692
Bibenzimidazoles
    radioprotective agents, 172
Bicyclam
    anti-HIV drug, 467
Bid protein, 19
Biguanides
    topical antibacterials, 554–555
Bik protein, 19
Biochemical shock
    radioprotective agents, 169
Bioreductives (hypoxia-activated
    prodrugs), 83–87
    radiosensitizers, 181–182
Bisindolylmaleimides
    influenza antiviral agents,
        388–389
Bisphosphonates
    selective toxicity, 273–274
Bistriazoloacridone
    anti-HIV drug, 474
Bithionol
    antihelmintic, 1091,
        1095–1096
BL-1743
    influenza antiviral agent, 388
*Blastomyces dermatitidis,* 883
Blastomycosis, 883
Blenoxane
    antitumor natural product,
        115–120
Bleomycin
    antitumor natural product,
        110, 115–120
    with tirapazamine, 85
    with vinblastine, 140
Bleomycinic acid, 116–117, 118

Blood vessels
    angiogenesis, 216
    angiogenesis inhibiting com-
        pounds, 216–220
Bloom's syndrome, 32
B-lymphocytes, *See* B-cells
BMS-40383, 680
BMS-40591, 680
BMS-45047, 681
BMS-45742, 680
BMS-181139, 680
BMS-181184, 908, 909
BMS-182880, 681
BMS-200475
    for hepatitis B virus, 331–332
BMS-275291
    antiangiogenic agent, 217
BMY-26270
    influenza antiviral agent, 387
BMY-27709
    influenza antiviral agent,
        377–378
BMY-27946, 681
BMY-40732, 681
BMY-40886, 681
BMY-183021
    influenza antiviral agent, 387
BMY-183355
    influenza antiviral agent, 387
Bone marrow transplantation
    and organ transplant rejec-
        tion, 524
Boron neutron capture therapy,
    173–174
BRCA-1 gene, 8, 10, 27
BRCA-2 gene, 27
Breast cancer resistance protein,
    285, 286
Brefeldin
    rotavirus antiviral, 430
Brequinar, 489
    clinical use for organ trans-
        plants, 516–518
    with leflunomide, 523
    side effects, 518
Brivudin
    for herpes virus, 308
BRL-36650, 636–637
BRL-41897, 651–652
BRL-42715, 698–699
    amoxycillin MIC, 695
    β-lactamase inhibitory activ-
        ity, 695
BRL-44154, 634
    *in vitro* antibacterial activity,
        635
BRL-57342, 643

Broad spectrum antiviral agents,
    432–435
Brobactam, 700
Bromodeoxyuridine
    radiosensitizer, 194–195
Brusatol
    antimalarial, 998
BTA-188
    picornavirus antiviral, 402
Bulaquine, 953–954
    dosage forms, 934
Bupivacaine
    selective toxicity, 273
Bush classification, of β-lactam-
    ases, 621
Butenafine, 899–900
    formulations and indications,
        891
Butoconazole
    formulations and indications,
        891
B19 virus, 304
B-W 322, 490

C-1311
    topoisomerase II inhibitor, 72
C22
    influenza antiviral agent, 379
Cactinomycin-3, 111
Cadherins
    and carcinogenesis, 8, 10
    hypermethylation, 9
    targets of antiangiogenic
        agents, 220
Caffeine
    radiosensitizer, 193
Calcium
    and cancer prevention, 33
Calmette-Guerin, 819
Campath, 229, 233
*Camptotheca acuminata,* 130
Camptothecins
    antitumor natural products,
        110, 130–134
Canaline
    antimalarial, 987, 988
Cancer. *See also* Anticancer
        drugs; Antitumor drugs;
        Carcinogenesis; Oncogenes
    drug resistance in chemother-
        apy, 281–290
    electron-affinic drugs in che-
        motherapy, 180–183
    epigenetic changes, 7–10
    genes related to, 21–32
    and genetic instability, 19–20
    and genetic variability, 5–7

Cancer (*Continued*)
  heritable syndromes, 32
  impact of, 2
  interventions, 32–46
  molecular basis of phenotypes,
    10–21
  prevention strategies, 32–33
  radioprotective agents in ther-
    apy, 172–173
  selective toxicity in chemo-
    therapy, 257–259
  tumorigenesis, 2–3, 5–7
Cancidas, 890, 900–901
*Candida*, 882, 883
  diagnostic and microbiological
    issues, 885
  instrinsic resistance to anti-
    fungals, 885
*Candida albicans*, 882, 883
  biosynthesis and function of
    ergosterol, 886–888
*Candida glabrata*, 882
*Candida krusei*, 882
*Candida parapsilosis*, 882
*Candida tropicalis*, 882
Candidiasis, 882, 883
Candin antifungal agents, 890,
  900–903
Capreomycin
  antimycobacterial application,
    811, 844–846
  efficacy and dosage, 811, 863
  side effects, 845–846
Capstat, 811
Carbacephalosporins, 654
Carbacephems
  brief description, 609
Carbapenams
  brief description, 609
Carbapenems, 668–671
  anti-gram-positive, 703–705
  biological properties, 688–694
  brief description, 609
  β-lactamase inhibitors,
    699–700
  orally active, 709–710
  pharmacokinetics, 623–624
  polycyclic, 683–688
  synthesis, 671–675
  use recommendations, 626
Carbecillin, 609
Carbenicillin, 632
  indications, 625
  β-lactamase classification
    based on activity, 619, 620,
    621
  pharmacokinetics, 624

Carbonic anhydrase inhibitors
  selective toxicity, 264
Carboplatin
  alkylating agent, 54, 59
  with RSR13, 190
  selective toxicity, 257–258
Carboxylic 3-deazaadenosine
  activity against orthopoxvi-
    ruses, 439, 440
Carcinogenesis, 3–5
  and apoptosis, 16–19
  and diet, 7
  and DNA repair, 4, 5
  epigenetic changes, 7–10
  and ionizing radiation, 4, 33
  and transforming growth fac-
    tor β, 30–32
Caretakers, 26
Carfecillin, 609
Carmustine
  alkylating agent, 54, 64–65
Carpetimycin A, 670, 671
Carumonam, 666
  *in vitro* activity, 667
Carzelesin
  alkylating agent, 54, 61–64
Caspase-3
  and apoptosis, 18
Caspase-6
  and apoptosis, 18
Caspase-7
  and apoptosis, 18
Caspase-8
  and apoptosis, 17–18
Caspase-9
  and apoptosis, 18, 19
Caspases, 17–19
Caspofungin, 888, 900–901
  formulations and indications,
    890
*Catharanthus roseus*, vinca alka-
  loids from, 139
Cationic dyes
  radiosensitizers, 175
Cationic surfactants
  topical antibacterials, 555–556
CB-1954
  tumor-activated prodrug, 83,
    91–92
CB-3717
  with permetrexed, 77
CC-83
  antimalarial, 957
CC-1065
  alkylating agent, 61
  for arming antibodies, 94

CCNU
  alkylating agent, 54, 64–65
CDK-4 gene, 27
cDNA microarray chips
  cancer studies using, 35–37
CD8-positive cytotoxic T-lym-
    phocytes, 224–225
CD4-positive T-helper cells, 224
CDRI 80/53, 953–954
CDRI 87/209, 942
Cefaclor, 609, 639, 640
  activity, 644
  oral absorption, 623
  synthesis, 646–649
Cefadroxil
  oral absorption, 623
Cefamandole, 609, 639–640
  activity, 644
  pharmacokinetics, 624
Cefazolin, 639
  activity, 644
  pharmacokinetics, 624
Cefepime
  activity, 641–642, 643, 644
  pharmacokinetics, 624
  use recommendations, 627
Cefixime
  activity, 640, 643, 644
Cefmetazole, 651
  synthesis, 650
Cefonicid
  pharmacokinetics, 624
Cefoperazone
  activity, 640, 641
  biliary excretion, 623
  indications, 625
  pharmacokinetics, 624
  use recommendations, 627
Cefotaxime
  activity, 640, 641, 644
  β-lactamase classification
    based on activity, 620
  pharmacokinetics, 624
Cefotetan, 651
  activity, 640, 641
  use recommendations, 627
Cefoxitin, 609, 651
  activity, 640
  use recommendations, 627
Cefpirome
  activity, 641–642, 643, 644
Cefprozil
  selective toxicity, 269
Cefratrizine
  oral absorption, 623
Cefsulodin
  activity, 640, 642

Ceftazidime
  activity, 640, 642, 644, 667
  indications, 625
  pharmacokinetics, 624
  use recommendations, 627
Ceftizoxime
  pharmacokinetics, 624
Ceftriaxone
  activity, 640, 641, 644
  biliary excretion, 623
  indications, 625
  pharmacokinetics, 624
  use recommendations, 627
Cefuroxime, 609, 639-640
  activity, 644
  pharmacokinetics, 624
  use recommendations, 627
Cefuroxime axetil, 639, 640
Celebrex
  selective toxicity, 274
Celecoxib
  selective toxicity, 274
CellCept, 519, 520
Cell cycle
  changes in carcinogenesis,
    14–16
Cell proliferation, 4
Cell suicide, *See* Apoptosis
Central nervous system (CNS)
    drugs, *See* CNS drugs
Cephabacins, 638
  synthesis, 649–650
Cephalexin, 639
  activity, 644
  oral absorption, 623
  synthesis, 646–649
Cephaloglycine
  oral absorption, 623
Cephalolexin, 609
Cephaloridine
  β-lactamase classification
    based on activity, 620, 621
Cephalosporin C, 609
  derivatives, 638–639
  discovery, 637
  total synthesis, 645–649
Cephalosporins, 637–639
  anti-gram-positive, 703
  antimycobacterial application,
    858
  carbacephalosporins, 654
  C7-α-formamido derivatives,
    651–652
  C7-α-methoxylated deriva-
    tives, 650–651
  oxacephalosporins, 652–654
  pharmacokinetics, 623

  selective toxicity, 253–254,
    260
  side effects and interactions,
    624–625
  structure-activity relationship,
    639–645
  use recommendations, 627
*Cephalosporium acremonium,*
    609, 637
Cephalothin, 609, 639
  activity, 644
  indications, 625
  pharmacokinetics, 624
Cephamycins, 609, 634, 638
  synthesis, 649
Cephazolin, 609
Cephems
  brief description, 609
Cephradine
  oral absorption, 623
  pharmacokinetics, 624
Cerubidine
  antitumor natural product,
    126–128
CGP 37697, 660
CGP 51905A
  antimalarial, 987, 988
CGS 27023A
  antiangiogenic agent, 217
Chagas' disease (American
    trypanosomiasis)
  antiprotozoal agents for, 1035,
    1045–1047
Chalicheamicin-type DNA cleav-
    ing agents, 92
Chemokines
  production in T-cell activation
    by dendritic cells, 235
Chemotherapy, 538
L-Chicoric acid
  anti-HIV drug, 473, 474
Chimeric antibodies, 226
Chinchonism, 937
Chirality
  and selective toxicity, 268–273
Chloproguanil
  antimalarial, 954–956
Chlopromazine
  radioprotective agent, 165
Chloprothixene
  radioprotective agent, 165
Chlorambucil
  alkylating agent, 53, 54
Chloramphenicol, 789–792
Chlorine topical antibacterials,
    545–547
Chlormidazole, 893–897

Chlorocardicin, 661
2-Chloro-2'-deoxyadenosine
  antimetabolite, 76
Chloroquine
  antimalarial, 938–943
  dosage forms, 934
  resistance to, 921, 928–930
Chlorothiazide
  selective toxicity, 264
Chlorpheniramine
  selective toxicity, 275
Chlorproguanil
  with dapsone, 959
  dosage forms, 934
Chlorpromazine
  arenavirus antiviral, 430
Chlorpropamide
  selective toxicity, 264
Chlortetracycline, 740, 742
Chlor-Trimetron
  selective toxicity, 275
Chromatin remodeling, 8
Chromomycosis, 883
Chromosomes
  acquisition of extra, and carci-
    nogenesis, 19
Chromosome X inactivation
  and carcinogenesis, 8
Chrysophanic acid
  picornavirus antiviral, 414
Chrysosplenol C
  picornavirus antiviral, 413
Cialis
  selective toxicity, 277
Ciclopirox, 906
  formulations and indications,
    891
Ciclosporine, 490, 491
Cidofovir
  for adenoviruses, 340
  for herpes virus, 308, 317
  for orthopoxviruses, 439
  for papillomavirus, 338–339
  for polyomaviruses, 339
  for poxviruses, 341
  selective toxicity, 267
  for smallpox, 437–440
Cilastin
  coadministration with imi-
    penem, 623, 693
Cilofungin, 901
Cimetidine
  selective toxicity, 275
Cinchona bark, quinine from,
    933
Cinchonidine, 933
Cinchonine, 933

Ciprofloxacin, 853–855
  antimalarial, 935, 983
Cisplatin
  alkylating agent, 54, 59
  with gemcitabine, 80, 81
  with ionizing radiation, 196
  with nolatrexed, 79
  with permetrexed, 78
  selective toxicity, 257–258
  structure-activity relationship,
    60–61
  with tirapazamine, 86
  with vinblastine, 140
Cisplatin rescue, 172
CL-61917
  influenza antiviral agent, 378
CL-191121, 709
CL-253824
  for herpes virus, 314
CL-309623
  RSV antiviral agent, 390–393
CL-387626
  RSV antiviral agent, 390–393
*Cladosporium*, 883
Cladribine
  antimetabolite, 76, 81–82
Clarithromycin, 763
  antimycobacterial application,
    856–857
  selective toxicity, 261
Claritin
  selective toxicity, 275
Clavulanic acid, 609
  amoxycillin MIC, 695
  antimycobacterial application,
    858
  discovery, 694–697
  indications, 625
  $\beta$-lactamase inhibitory activ-
    ity, 695
  use recommendations, 626
Cleland's reagent
  radioprotective agent,
    156–157
Cleocin, 935
Cleomycins
  antitumor natural products,
    115, 117
Clevudine
  for hepatitis B virus, 327,
    329–330
Clindamycin, 788–789
  antimalarial, 935, 984
Clinical reaction modifiers, 152.
    *See also* Radioprotective
    agents; Radiosensitizers
Clociguanil, 960

Clofazimine, 837–839
  efficacy and dosage, 810
  side effects, 839
Clomiphene
  selective toxicity, 268
Clotrimazole, 893–897, 959
  formulations and indications,
    891
  selective toxicity, 268
Cloxacillin, 631
  *in vitro* antibacterial activity,
    635
CMDA
  tumor-activated prodrug,
    87–88
C-myc protein, 25–26
  and APC gene, 30
  cDNA microarray studies, 37
CNI-H0294
  antimalaria8, 987
CNS drugs
  radioprotective agents, 165
Coartem, 935
*Coccidioides immitis*, 883
Coccidioidomycosis, 883
Colchicine
  radioprotective agent, 165
Cold sores, 298
Collagenases, 13
Collagen I, 216
Collagen III, 216
Collagen IV, 12, 216
Collagen VII, 216
Combined modality therapy, 173
Combretastatin
  tumor-activated prodrug, 87
Combretastatin A4, 219, 220
Complementarity determining
  regions, 227
Conjugative transposons, 614,
  616
Copper complexes
  radioprotective agents, 163
Cord-factor, 817
Cordycepin
  PLD repair inhibitor, 192
Cosalane
  anti-HIV drug, 466
Cosmegen
  antitumor natural product,
    111–115
Costimulatry signals, 487
  agents capable of inhibiting,
    525–526
Cotrifazid, 959
Cotrim, 934

Cotrimoxasole
  systemic antibacterial, 596
COX-1
  and cancer, 34
COX-2
  and cancer, 33–35
COX-1 inhibitors
  selective toxicity, 252,
    274–275
COX-2 inhibitors. *See also* Leu-
    kotrienes
  selective toxicity, 274–275
CP 70429, 660
CpG island methylated pheno-
    type (CIMP), 289
CPT-11
  antitumor natural product,
    131
Cruex, 891
Cryptococcal meningitis, 884
Cryptococcosis, 883
*Cryptococcus neoformans*, 883,
    884
  diagnostic and microbiological
    issues, 885
Cryptolepine
  antimalarial, 998–999
CS-834, 709–710
CTLA4-Ig, 525–526
Cyanide
  radioprotective agent, 162
Cyclacillin, 632
Cyclin dependent kinases
  in carcinogenesis, 14–16
  and organ transplant rejec-
    tion, 487
Cyclins
  in carcinogenesis, 14–16
  and organ transplant rejec-
    tion, 487
Cycloguanil, 955, 956
Cyclooxygenase 1/2 inhibitors,
  *See* COX-1 inhibitors;
    COX-2 inhibitors
Cyclophosphamide
  alkylating agent, 53, 54, 55–56
  with AQ4N, 85
  with dactinomycin, 111
  with daunorubicin, 126
  with MEA as radioprotective
    agent, 172
  as prodrug, 83
  with tirapazamine, 85
  tumor-activated prodrug, 91
Cycloprodigiosin
  antimalarial, 994

Cyclopropylindoles
   alkylating agents, 54, 61–64
Cyclopropylindolines
   alkylating agents, 52
Cycloserine
   antimycobacterial application,
      811, 846
   efficacy and dosage, 811, 863
   side effects, 846
Cyclosporine
   clinical use for organ trans-
      plants, 489–494
   with leflunomide, 523
   pharmacokinetics, 491–492
   pharmacology, 492–494
   side effects, 494
   structure-activity relationship,
      494
Cyclosporine A
   multidrug resistant protein
      inhibition, 290
Cysteamine
   radioprotective agent, 152,
      155–160
Cysteine
   radioprotective agent, 152, 155
Cytallene
   for hepatitis B virus, 330
Cytochrome c
   and apoptosis, 17
Cytokines
   in cancer, 43
   production in T-cell activation
      by dendritic cells, 235
   radioprotective agents, 171
   radiosensitizers, 194
   tumor secreted, 226
Cytomegalovirus, 298
Cytomegalovirus antiviral drugs
   selective toxicity, 267
Cytoplasmic proteins, 26
Cytosar
   antimetabolite, 76
Cytosine arabinoside
   antimetabolite, 75, 76, 79–82
Cytosine deaminase tumor-acti-
      vated prodrugs, 90–91
   for cancer therapy, 44
Cytotoxic drugs, 52, 82, 109. See
      also Antitumor drugs; Tu-
      mor-activated prodrugs
Cytotoxic T-lymphocytes,
      224–225
Cytovene
   tumor-activated prodrug, 83
Cytoxan
   alkylating agent, 54

DA-1131, 708
DACA
   dual topoisomerase I/II inhibi-
      tor, 68, 73
DACA N-oxide, 85
Dacarbazine
   alkylating agent, 54, 65–66
Daclizumab, 490, 510–513
   with mycophenolate mofetil,
      520
Dactinomycin
   antitumor natural product,
      110, 111–115
Dapsone, 936, 957
   antimycobacterial application,
      829–830
   with chlorproguanil, 959
   dosage forms, 934
   efficacy and dosage, 810
   with proguanil, 959
   with pyrimethamine, 959
Daraprim, 934
Darvon
   selective toxicity, 269–270
Daunomycin
   antitumor natural product,
      126–128, 130
Daunorubicin
   antitumor natural product,
      110, 124, 126–128
Death domain, 17
Death effector domain, 17
Death-inducing signaling com-
      plex, 17
Death receptors, 17
3-Deazaneplanocin A
   activity against orthopoxvi-
      ruses, 439, 440
   for Ebola virus, 433
Declomycin, 740
Decontamination, 539
Delavirdine
   anti-HIV drug, 472
   selective toxicity, 267
Deltaprim, 934
Demethylchlortetracycline, 740
5′-O-Demethyldioncophylline A
   antimalarial, 994–995
Dendritic cell immunotherapy,
      235–243
Dendritic cells
   and adaptive immunity, 224
   physiology, 235–236
   T-cell activation by, 235–236
   transduction by viral vectors,
      241–243
   tumor cell fusion, 240–241

Dengue fever, 1035
Dengue virus, 427
Deoxoartemisinin, 964
Deoxyartemisinin, 963
2′-Deoxycoformycin
   antimetabolite, 76
15-Deoxyspergualin, 489, 490
   antimalarial, 987
   clinical use for organ trans-
      plants, 504–506
   with leflunomide, 523
   side effects, 506
Dermatophytoses, 883
Desciclovir
   for herpes virus, 318
Desenex, 891
Desferrioxamine B (DFO)
   antimalarial, 991
Desmethylmisonidazole
   radiosensitizer, 178
Dexamisole
   selective toxicity, 273
Dexrazoxane
   radioprotective agent, 172
Dextrans
   radioprotective agent, 166
Dextromethorphan
   selective toxicity, 255
Dextropropoxyphene
   selective toxicity, 269–270
DFO, See Desferrioxamine B
DHFR, See Dihydrofolate reduc-
      tase
Diaminopyrimidines
   selective toxicity, 263–264
Diarylureas
   topical antibacterials, 554–555
Diazenes
   radiosensitizers, 192–193
Dibenzofurans
   picornavirus antiviral agents,
      403
Dibenzosuberanes
   picornavirus antiviral agents,
      403
Dicloxacillin, 631
Didanosine
   anti-HIV drug, 471, 472
   selective toxicity, 267
Diet
   for cancer prevention, 33
   and carcinogenesis, 7
Diethylcarbamazine
   antihelmintic, 1091,
      1095–1096
Diethylmaleate
   radiosensitizer, 190–191

Difluoromethylornithine, 35
Digoxin
  with bleomycin, 116
Dihydroartemisinin, 963,
    971–972
  dosage forms, 935
Dihydrofolate reductase
  antimalarial target, 925
  increased expression leading
    to drug resistance in cancer
    chemotherapy, 283–284
Dihydrofolate reductase inhibi-
    tors
  antimetabolites, 76–78
  selective toxicity, 264
Dihydroorotate dehydrogenase
  antimalarial target, 925
Dihydropteroate synthetase in-
    hibitors
  antimalarials, 925
  selective toxicity, 264
Dihydrostreptomycin, 841
Diiodofluorescein
  influenza antiviral agent, 379
2,6-Diketopiperazine derivatives
  influenza antiviral agents,
    386–387
Dimeric vinca alkaloids
  antitumor natural product,
    139–143
Dimethyl fumarate
  radiosensitizer, 193
Dimethyl sulfoxide (DMSO)
  radiosensitizer, 176
Dioncolactone A
  antimalarial, 994–995
Dioncopeltine A
  antimalarial, 994–995
Dioncophylline B
  antimalarial, 994–995
Dioncophylline C
  antimalarial, 994–995
2,4-Dioxobutanoic acid deriva-
    tives
  influenza antiviral agents, 386
Diphenhydramine
  selective toxicity, 275
2,3-Diphosphoglycerate
    (2,3-DPG)
  radiosensitizer, 183–184
Diphtheria toxin
  selective toxicity, 259
Dirithromycin, 763
Disinfectants, 539
  testing, 544
Disoxaril
  picornavirus antiviral, 400

Ditercalinium
  dual topoisomerase I/II inhibi-
    tor, 75
Dithiothreitol
  radioprotective agent,
    156–157
DK-35C, 706, 708
DMP 840
  dual topoisomerase I/II inhibi-
    tor, 68, 75
DMSO, See Dimethyl sulfoxide
DMXAA
  with tirapazamine, 85
DNA-binding drugs. See also
    Alkylating agents
  antitumor natural products,
    111–124
  chemotherapeutic agents and
    related tumor-activated
    drugs, 51–94
DNA gyrase inhibitors
  selective toxicity, 263
DNA hypermethylation
  and drug resistance, 289
  and epigenetic changes, 8–10
DNA hypomethylation, 8
DNA-intercalating topoisomer-
    ase inhibitors, See Topo-
    isomerase II inhibitors; To-
    poisomerase I inhibitors
DNA lesions, 4, 176
DNA methylation
  accumulation, 3
  and drug resistance, 289
  and epigenetic changes, 8–10
DNA methyltransferases, 8
DNA polymerases, 295–296
DNA radicals, 176–180
DNA repair
  and carcinogenesis, 4, 5
  and drug resistance in cancer
    chemotherapy, 287
  enhancement by radioprotec-
    tive agents, 169–170
  inhibition by radiosensitizers,
    191–193
DNA repair-deficiency diseases,
    32
DNA topoisomerase 1, See Topo-
    isomerase I
DNA topoisomerase 2, See Topo-
    isomerase II
DNA viruses, 294–295. See also
    Antiviral agents, DNA
DNAzymes
  influenza antiviral agents,
    385–386

Docetaxel
  antimalarial, 990
  antitumor natural product,
    110, 137, 138–139
Dolastatin 10, 140
Donor bone marrow
  organ transplant rejection ap-
    plications, 524
L-Dopa
  selective toxicity, 273
Dopaminergic receptors
  and selective toxicity, 259
Dose reduction factor, radiopro-
    tective agents, 154
Double minute chromosomes, 20
Doxcyclin, 935
Doxepin
  selective toxicity, 268–269
DOX-GA3, 90
Doxorubicin
  with amsacrine, 68
  antibody conjugates, 92
  antitumor natural product,
    83, 110, 124, 128, 129
  resistance to, 288
  tumor-activated prodrug, 88,
    89–90
Doxycin, 740
Doxycycline, 740, 742
  antimalarial, 935, 982
Dramamine
  selective toxicity, 275
Drug detoxification, 286–287
Drug resistance
  aminoglycoside antibiotics,
    753–757
  antibacterial agents, 543,
    572–573, 586, 590
  antifungal agents, 884–885
  anti-HIV drugs, 478–479
  antimalarial agents, 921,
    928–930
  antimycobacterial agents,
    860–861
  in cancer chemotherapy,
    281–290
  HBV virus antiviral agents,
    333–334
  herpes virus antiviral agents,
    309–310
  to β-lactams, 613–623
  linezolid, 590
  macrolide antibiotics, 774–776
  quinolones, 586
  sulfonamides and sulfones,
    572–573
  tetracyclines, 743–746

DTIC
   alkylating agent, 54, 65
DU-86
   alkylating agent, 61
DU-257
   tumor-activated prodrug, 83, 94
Duocarmycin
   for arming antibodies, 94
DuP-721
   antimycobacterial, 859
DX-8739, 682
DX-8951f
   antitumor natural product, 133
Dyes
   topical antibacterials, 553–554

Ebola virus, 432–434
E-cadherin, 8, 10, 14
   hypermethylation, 9
Echinocandins
   antifungals, 900–903
Econazole, 893–897
   formulations and indications, 891
Edatrexate
   antimetabolite, 76, 78
EDTA, *See* Ethylenediaminetet-
      raacetic acid
Efaproxiral
   radiosensitizer, 184–190
Efavirenz
   anti-HIV drug, 472
   selective toxicity, 267
Eflornithine
   for African trypanosomiasis,
      1043–1045
EICAR
   flavivirus antiviral, 428
Electron-affinic drugs, 180–183
Ellence
   antitumor natural product, 128
Elutherobin
   antitumor activity, 139
Embryonic stem cells, 524
EMD121974
   antiangiogenic agent, 218–219
Emerging pathogens, 713
Emivirine
   anti-HIV drug, 472
Emtricitabine
   for hepatitis B virus, 327, 329
Enamelysin, 13
Enbrel, 229, 234
Endostatin, 45
Endothelial cells, 216
   cytoskeleton disruptors,
      219–220
   inhibitors of migration, 219

Entecavir
   for hepatitis B virus, 331–332
*Enterococcus*
   bridge in peptidoglycans, 612
   methicillin-resistant, 618
   penicillin-binding protein, 615
   porin loss in, 621
Enviradene
   picornavirus antiviral, 411
Enviroxime
   picornavirus antiviral agent,
      411–413
Epicillin, 632
Epidermal growth factor
   decreased dependence on in
      carcinogenesis, 11
   radiosensitizer, 194
Epidermal growth factor receptor, 11, 23
Epigenetic changes
   and carcinogenesis, 7–10
   and drug resistance in cancer
      chemotherapy, 289
Epipodophylloxins
   alkylating agents, 52
Epirubicin
   antitumor natural product,
      110, 124, 128
   tumor-activated prodrug, 89
Epothilones
   alkylating agents, 52
   antitumor activity, 139
Epoxides
   topical antibacterials, 550–551
Epstein Barr virus, 296–297,
   299
Erb-B1 growth factor receptor,
   22, 23
Erb-B2/HER2/neu growth factor
   receptor, 22, 23
Ergosterol
   biosynthesis by fungal patho-
      gens, 886–888
Erythrocytes
   exosome secretion, 241
   in malaria, 920–926
Erythromycins, 763
   antimycobacterials, 855–858
   selective toxicity, 261–263
Estrogen receptors
   hypermethylation and carcino-
      genesis, 9
Estrogens
   radioprotective agents, 165
Etanercept, 229, 234–235
Etanidazole
   radiosensitizer, 178

Ethambutol, 839–840
   drug interactions, 864
   efficacy and dosage, 810, 863
   side effects, 840
Ethionamide, 835–836
   efficacy and dosage, 810, 863
   side effects, 836
Ethylenediaminetetraacetic acid
   (EDTA)
   radioprotective agent, 163
*N*-Ethylmaleimide
   radiosensitizer, 177
Etopophos, 135
Etoposide
   with amsacrine, 71
   antitumor natural product,
      110, 134–135
   resistance to, 288
   with tirapazamine, 85
   tumor-activated prodrug, 87
Evernimycin, 793
Evista
   selective toxicity, 276
Exelderm, 891
Exifone
   antimalarial, 987
Exosomes, 241
Extended-spectrum β-lactamases, 619
External guide sequences
   influenza antiviral agents,
      385–386

Famciclovir
   for hepatitis B virus, 330–331
   for herpes virus, 309, 310
   selective toxicity, 267
Fanconi's anemia, 32
Fansidar, 934
Fansimef, 934
Faropenem, 659, 660
Fas-associated death domain, 17
Fas ligand
   and adaptive immunity, 225
Fattiviracin A1
   influenza antiviral agent, 390
FCE 22101, 659, 660
FCE 22891, 659
FCE 29464, 660
Fd4C
   for hepatitis B virus, 327, 329,
      333
Febrifugine
   antimalarial, 996, 997
Femstat, 891
Fenozan-50F
   antimalarial, 985–986

Fenpropimorph, 903
Ferricenium salts
    radiosensitizers, 196
Ferrochloroquine
    antimalarial, 992
Fetal stem cells, 524
Fexofenadine
    selective toxicity, 275
Fibroblast growth factor,
    217–218
Fibroblast growth factor inhibi-
    tors
    antiangiogenic agents,
        218–219
Fibronectin, 12, 216, 219
Fiflucan, 890
FK 463, 888, 902, 903
FK 506
    clinical use for organ trans-
        plants, 494–500
    pharmacokinetics, 495
    pharmacology, 495–499
    side effects, 500
    structure-activity relationship,
        499–500
FK-506 binding protein, 496
Flaviviruses, 427
    antiviral agents, 427–430
Flomoxef, 652–653
Flucloxacillin, 609, 631
Fluconazole, 893–897
    formulations and indications,
        890
    intrinsic resistance to, 885
    selective toxicity, 268
Flucytosine, 903–905
    formulations and indications,
        890
Fludara
    antimetabolite, 76
Fludarabine
    antimetabolite, 76, 81–82
5-Fluoroorotate
    antimalarial, 987, 988
Fluoroquinoline K-12
    anti-HIV drug, 474
Fluoroquinolones
    antimycobacterials, 853–855
    selective toxicity, 263
    side effects, 855
5-Fluorouracil
    antimetabolite, 75, 76, 77,
        79–81
    with ionizing radiation, 195
    with irinotecan, 131
    with oxaliplatin, 60
    for papillomavirus, 338

resistance to, 284, 288
    tumor-activated prodrug, 44,
        90–91
Fluosol
    radiosensitizer, 183
Flurithromycin, 763
Flutimide
    influenza antiviral agent,
        386–387
Fms growth factor receptor, 22
FNQ3
    flavivirus antiviral, 429–430
Folate analog antimetabolites,
    76–79
Fomivirsen
    for herpes virus, 309
*Fonsecaea,* 883
Formadicins, 664
Formidacillin, 636–637
    *in vitro* antibacterial activity, 635
Fosamax
    selective toxicity, 273
Foscarnet
    for herpes virus, 308
    selective toxicity, 267
Fosmidomycin
    antimalarial, 989
Fos protein, 22, 25
FR191512
    influenza antiviral agent, 390
FR900098
    antimalarial, 989
FR901379, 902
Free radicals
    inhibition by radioprotective
        agents, 168
    production by ionizing radia-
        tion, 153
FTY 720, 489
    clinical use for organ trans-
        plants, 506–509
    side effects, 509
Fullerene C$_{60}$
    influenza antiviral agent, 390
Fumigation, 539
Fungal cell wall, 888–889
Fungal diseases, 882–884. *See
    also* Antifungal agents
    diagnostic and microbiological
        issues, 885–886
    trends in incidence, 884
Fungal efflux pump inhibitors, 909
Fungal pathogens, 882–884
    biosynthesis and function of
        ergosterol, 886–888
    opportunistic filamentous, 884
Fungi, 882

Fungizone, 890, 891
Furosemide
    selective toxicity, 264
Fusidic acid, 792

Gamma rays
    antiradiation testing, 154
Ganciclovir
    for hepatitis B virus, 330–331
    for herpes virus, 307–308
    with HSV-based suicidal gene
        therapy, 44
    tumor-activated prodrug, 83, 90
Gantanol, 934
Gantrisin, 934
Gatekeepers, 26
GDEPT (gene-directed enzyme-
    prodrug therapy), 83, 90–92
Gedunin
    antimalarial, 996, 998
Gelatinases, 13
Gemcitabine
    antimetabolite, 76, 80–81
    with permetrexed, 78
    radiosensitizer, 195
Gemtuzumab, 228, 231–233
Gemtuzumab ozogamycin
    tumor-activated prodrug, 83,
        92
Gemzar
    antimetabolite, 76, 80
GeneChips, 35
Gene-directed enzyme-prodrug
    therapy (GDEPT), 83,
        90–92
Gene silencing
    and carcinogenesis, 8
Gentamicin, 748
Gentamycin, 748
    selective toxicity, 260–261
Glabranine
    flavivirus antiviral, 429
Glaucine fumarate
    picornavirus antiviral, 414
Gleevec, 110, 284
    selective toxicity, 257
Glimepiride
    selective toxicity, 264
Gliotoxin
    hepatitis C antiviral, 426
Glipizide
    selective toxicity, 264
Gloximonam, 666, 668
Glucocorticoid response ele-
    ments
    and oncogenes, 25

Glucose 6-phosphate dehydroge-
   nase
   and malaria, 931
Glucosulfone sodium, 829
Glucuronidase tumor-activated
   prodrugs, 89–90
Glutathione
   radiosensitization by depletion
      of, 190–191
Glyburide
   selective toxicity, 264
Glyphosate
   antimalarial, 988–989
GM 237354, 907–908
Gp100, 237–238
Gp120-CCR5 binding, 462
   inhibitors, 468
Gp120 coreceptors, 467
Gp120-CXCR4 binding, 462
   inhibitors, 467–468
G proteins, 23–24
GR69153, 643, 645
Gram-negative bacteria
   action of β-lactams against,
      611–613
   cell wall, 610–611
   β-lactam indications, 625
   penicillin-binding protein, 614
Gram-positive bacteria
   action of β-lactams against,
      611–613
   cell wall, 610–611
   β-lactam indications, 625
   penicillin-binding protein, 615
Granule enzymes, 225
Granulocytes
   and adaptive immunity, 225
   and innate immunity, 224
Griseofulvin, 905–906
   formulations and indications,
      891
GS-4071
   for influenza, 370, 371, 473
GS-4104
   for influenza, 368
GTPase, 24
GTPase-activating protein, 25
Guanine nucleotide exchange
   factors, 22, 24
Gusperimus, 490
GV104326, 684, 685
GV129606X, 687, 689, 694, 708
GV143253X, 687, 690, 694, 705
Gymnochrome D
   flavivirus antiviral, 429

H-961
   for herpes virus, 321
   for poxviruses, 341
Haemophilus influenzae
   penicillin-binding protein,
      614, 616, 617
Halfan, 934
Halofantrine
   antimalarial, 946–949
   dosage forms, 934
Halogenated topical antibacteri-
   als, 545–547
Haloprogin, 906
   formulations and indications,
      891
Halotex, 891
Ha-ras gene, 4
Hayflick limit, 11
HBB
   picornavirus antiviral, 413
HCMV (cytomegalovirus), 298
Heart transplantation, 487
Heat shock proteins, 819
Heavy metals
   topical antibacterials, 552–553
Helicobacter Pylori, 740
   metronidazole-resistant, 712
Helminths, 1090
   antihelmintics for, 1090–1096
Heparin sulfate proteoglycans,
   21, 216
Hepatitis B virus, 297, 299–301
Hepatitis B virus antiviral
   agents, 326
   antisense oligonucleotides,
      337
   combination treatments,
      334–335
   purines, 330–333
   pyrimidines, 326–330
   resistance, 333–334
   targeting drugs to liver,
      335–337
Hepatitis C virus, 361–362
   and HBV infection, 300
   model systems to study repli-
      cation, 427
Hepatitis C virus antiviral
   agents
   inhibitors of HCV 5′ untrans-
      lated region and core gene,
      415–417
   inhibitors of NS5B RNA-de-
      pendent RNA polymerase,
      425–426
   inhibitors of NS3 helicase, 425

   inhibitors of NS3 protease,
      418–425
   interferon-α applications,
      361–364
Hepatitis C virus NS3 protein,
   417–418
Herceptin, 228, 230
   selective toxicity, 257
Herpes simplex virus
   in suicidal gene cancer ther-
      apy, 44
Herpes simplex virus-1, 297, 298
Herpes simplex virus-2, 298
Herpes virus antiviral agents, 305
   compounds under develop-
      ment, 310–326
   currently approved drugs,
      305–310
   immune modulators, 325–326
   nonnucleoside inhibitors and
      other targets, 321–324
   prodrugs increasing oral avail-
      ability, 309
   prodrugs under development,
      318–321
   resistance to, 309–310
   selective toxicity, 267
Herpes viruses, 294–295,
   297–299
α-Herpesviruses, 298
β-Herpesviruses, 298
γ-Herpesviruses, 298, 299
Herpes virus protease inhibitors,
   324–325
Herpes zoster, 298
Hesperidin
   rotavirus antiviral, 431
Heterocyclic compounds
   radioprotective agents,
      163–165
6-Heterocyclylmethylene pen-
   ems, 698–699
HHV-6, 298–299
HHV-7, 298–299
HHV-8, 299
Highly active antiretroviral
   therapy (HAART), 884
HIV/AIDS, 479
Histone deacetylases, 8
Histoplasma capsulatum, 883
Histoplasmosis, 883
HIV-DNA polymerase (reverse
   transcriptase), 462–463
HIV integrase inhibitors, 464
HIV protease, 464

HIV protease inhibitors,
    475–477
  selective toxicity, 267
HIV reverse transcriptase,
    462–463
HIV reverse transcriptase en-
    zyme, 469
HIV reverse transcriptase inhib-
    itors, 465, 469–470
  nonnucleoside reverse tran-
    scriptase inhibitors,
    472–473
  nucleoside reverse transcrip-
    tase inhibitors, 470–472
HIV ribonuclease inhibitors,
    465, 473
HIV vaccines, 465
HIV virion, 460
  inhibitors of assembly,
    474–475
HIV virus, 458–459
  assembly of viral proteins:
    virion formation, 464
  binding to host cell mem-
    brane, 462
  cellular picture of infection,
    459–460
  clinical picture of infection,
    460
  DNA entry to host cell nu-
    cleus, 463–464
  DNA formation by reverse
    transcriptase, 462–463
  fusion with host cell cyto-
    plasm and its uncoating,
    462
  genome, 461
  integration of viral DNA into
    host genome, 464
  life cycle, 461–464
  maturation, 464
  migration of viral mRNA to
    cytoplasm, 464
  mode of viral transmission,
    459
  splicing of viral mRNA into
    host genome, 464
  structure and molecular biol-
    ogy, 460–461
  targets for drug design,
    464–477
  viral budding out of host cell,
    464
HIV-1 virus, 459
  and malaria, 930
  vector for cancer gene ther-
    apy, 41

HIV-2 virus, 459
hMLH1, 27, 32
  hypermethylation, 9
HMPA
  for poxviruses, 341
hMSH2, 27, 32
Hormone replacement therapies,
    250–251
HPMPA
  for adenoviruses, 340
  for polyomaviruses, 340
hPMS1 gene, 27
hPMS2 gene, 27
H-ras protein, 22, 24
  and COX-2, 34
5-HT, See Serotonin
HTLV I (Human T Leukemia
    Virus I), 459
HTLV II (Human T Leukemia
    Virus II), 459
HTLV III (Human T Leukemia
    Virus III: HIV virus), 459.
    See also HIV virus
Human herpesvirus-8 virus, 296
Humanized antibodies, 226
  for organ transplant rejection,
    509
  selective toxicity, 257
Human leukocyte antigens,
    236–237
Hybridomas, 226
Hydrolapachol
  antimalarial, 976
Hydroxychloroquine, 938
  dosage forms, 934
Hydroxylamine
  radioprotective agent, 162
Hydroxyl-containing compounds
  radioprotective agents, 163
8-Hydroxyquinoline
  radioprotective agent, 163
Hydroxystreptomycin, 840
5-Hydroxytryptophan
  radioprotective agent, 165
Hydroxyurea
  for herpes, 322
  radiosensitizer, 195
Hypoxanthine aminopterin-thy-
    midine, 226
Hypoxanthine phosphoribosyl
    transferase, 226
Hypoxia
  nitroimidazole binding for de-
    tection, 180
  radioprotective agents utiliz-
    ing, 154, 167–168

Hypoxia-activated prodrugs,
    83–87
Hypoxia inducible factor-1$\alpha$
  and carcinogenesis, 20

IAZGP
  radiosensitizer, 180
Idarubicin
  antitumor natural product,
    110, 124, 128
Ifex
  alkylating agent, 54
IFN-$\alpha$, See Interferon-$\alpha$
Ifosfamide
  alkylating agent, 53, 54, 55
Ile-2-AM (actinomycin C), 111
Imatinib
  selective toxicity, 257
Imidacloprid
  selective toxicity, 278
Imipenem, 671
  antimycobacterial application,
    858
  coadministration of cilastin,
    623–624
  discovery, 668–669
  indications, 625
  $\beta$-lactamase classification
    based on activity, 620, 621
  MIC 90, 685
  pharmacokinetics, 624
  side effects and interactions,
    625
  use recommendations, 626
  in vitro activity, 692, 693
  in vivo efficacy in systemic
    murine infections, 685
Imipramine
  radioprotective agent, 165
Immortality, 10–11
Immune system, 224–226
Immunity, 224–226
Immunodeficiency, 294
Immunoglobulins
  and adaptive immunity, 225,
    227
  RSV antiviral agents, 395
Immunomodulation, 43–44
  herpes virus antiviral agents,
    325–326
Immunosuppressive drugs, 486.
    See also Cyclosporine; FK
    506; Organ transplant drugs
Immunotherapy
  antibody-directed, 226–235
  anti-inflammatory therapeutic
    antibodies, 233–235

chimeric antibodies, 226
dendritic cell, 235–243
humanized antibodies, 226
Immunotoxins, 92
Imprinting
and carcinogenesis, 8
Indinavir
anti-HIV drug, 475–476
Indocin
selective toxicity, 274
Indomethacin
selective toxicity, 274
Infliximab, 229, 233–234
Influenza A virus, 364
Influenza B virus, 364
Influenza hemagglutinin inhibitors, 377–384
Influenza M2 protein inhibitors, 388
Influenza neuraminidase inhibitors, 364–377
Influenza RNA transcriptase inhibitors, 384–388
Influenza virus antiviral agents, 364
inhibitors of influenza hemagglutinin, 377–384
inhibitors of influenza M2 protein, 388
inhibitors of influenza neuraminidase, 364–377
inhibitors of RNA transcriptase, 384–388
other inhibitors, 388–390
selective toxicity, 267
Initiation, in carcinogenesis, 3–5
Innate immunity, 224
Insecticides
selective toxicity, 277–279
Insulin
radiosensitizer, 194
Insulin and hypoglycemic agents, 250–251
Insulin-like growth factor 1
radiosensitizer, 194
Integrase, 616
Integrins, 13
Integrons
and $\beta$-lactam resistance, 614–616
Intercalation, 67
Interferon-$\alpha$
for hepatitis B virus, 326, 334
for hepatitis C virus, 361, 362–364
for papillomavirus, 337–338
radioprotective agents, 171

Interferon $\gamma$
and adaptive immunity, 225
Interleukin-1
radioprotective agent, 171
Interleukin-2
and adaptive immunity, 225
Interleukin-4
and adaptive immunity, 225
radioprotective agent, 171
Interleukin-5
and adaptive immunity, 225
Interleukin-6
and adaptive immunity, 225
radioprotective agent, 171
Interleukin-10
and adaptive immunity, 225, 226
Internal ribosome entry signal, 41
Interstitial collagenase, 13
Intestine transplantation, 487
Int-2 growth factor, 22
Intoplicine
dual topoisomerase I/II inhibitor, 68, 73, 74
Intraepithelial neoplasia, 35
Intratransguanylation, 158
Invadopodia, 13
Iodine topical antibacterials, 547
5-Iododeoxyuridine
radiosensitizer, 194
Ionizing radiation. See also Radioprotective agents
and carcinogenesis, 4, 33
damage from, 153–154
p53 activation, 28
Iproniazid, 834
Iressa, 12
Irinotecan
antitumor natural product, 110, 131
Isoartemisitene, 965
Isobombycol
radioprotective agent, 167
Isofebrifugine
antimalarial, 998
Isoniazid, 833–834
activity and mechanism of action, 834–835
drug interactions, 862, 864
efficacy and dosage, 810, 863
selective toxicity, 266
side effects, 835
Isopodophyllotoxins
antitumor natural products, 110, 134–136
Isosulfazecin, 664

Isothiazoles
picornavirus antiviral agents, 403
Isoxyl, 810
Itraconazole, 893–897
formulations and indications, 890, 891
selective toxicity, 268
Ivermectin
antihelmintic, 1091, 1093–1094

J-111,225, 682, 684, 708
in vivo efficacy in systemic murine infections, 685
J-111,347, 682, 683, 684
J-114,870, 682, 684, 685
J-114,871, 683, 684, 685
Japanese encephalitis virus, 427, 429–430
JM-83
alkylating agent, 54
JM-216
alkylating agent, 54, 60
JM-473
alkylating agent, 60
Josamycin, 763
Jun protein, 22, 25

Kalihinol A
antimalarial, 993–994
Kanamycin
antimycobacterial application, 811, 842–844
efficacy and dosage, 811, 863
selective toxicity, 260–261
side effects, 844
Kanamycin A, 842–843
Kanamycin B, 842–843
Kanamycin C, 842–843
Kantrex, 811
Kaposi's sarcoma, 299
Ketoconazole, 893–897
formulations and indications, 890, 891
selective toxicity, 268
Ketolides, 776–781
Kidney transplantation, 487, 489
1-year survival rate, 488
KIN-802
radiosensitizer, 179
KIN-804
radiosensitizer, 179
Kinase tumor-activated prodrugs, 90

Kinetoplastid parasitic diseases, 1035
  antiprotozoal agents for, 1035–1076
KM231-DU257
  tumor-activated prodrug, 83, 94
Korundamine A
  antimalarial, 995, 996
Korupensamines
  antimalarial, 995, 996
KP-736, 643, 646
K-ras protein, 22, 24
  and COX-2, 34
KRM-1648, 847, 852
KU-2285
  radiosensitizer, 180
KW-2189
  alkylating agent, 54, 61, 64

L-696,256, 703, 706
L-708,906
  anti-HIV drug, 473, 474
L-731,988
  anti-HIV drug, 473, 474
L-741,462, 703, 706
L-742,728, 703, 706, 707
L-786,392, 703, 705, 706, 707
L-793,422, 908
β-Lactam antibiotics, 608–610.
  See also specific antibiotics and classes of antibiotics, such as Cephalosporins or Penicillins
  antibacterial broad-spectrum, 705–709
  antimycobacterials, 858
  clinical application, 623–628
  combination therapies, 625, 626
  current trends driving industry, 711–714
  discovery, 628–637
  hydrolysis, 609, 619–621
  mechanism of action, 610–613
  mechanism of resistance to, 613–623
  mosiac gene formation, 616–618
  recent developments, 701–711
  selective toxicity, 254, 260
  side effects and interactions, 624–625
β-Lactamase inhibitors
  antimycobacterials, 858
  classification, 694–701

development of, 609
  recent developments, 710–711
β-Lactamases, 608–609
  classification and relevance in clinics, 618–621
  evolution, 618
  extended-spectrum, 619
  molecular and phenotypic classification, 620
β-Lactamase tumor-activated prodrugs, 88
Lamiazid, 810
Laminin, 12, 216
Lamisil, 890, 891
Lamivudine
  anti-HIV drug, 471, 472
  drug resistance, 333–334
  for hepatitis B virus, 327–329, 335–336
  selective toxicity, 267
Lamprene, 810
Langerhans cells, 224
Lanvis
  antimetabolite, 76
Lapinone
  antimalarial, 976
Lapudrine, 959
Lariam, 934
Latamoxef, 652
Lck tyrosine kinase, 22
Leflunomide, 489
  clinical use for organ transplants, 522–523
  side effects, 523
Leishmaniasis
  antiprotozoal agents for, 1035, 1047–1052
Lenampicillin, 632
Lentiviruses
  vectors for cancer gene therapy, 41
Leprosy, 809–813, 1035
  pathogenesis and epidemiology, 825–826
  recent developments and present status of chemotherapy, 865–866
Leucovorin
  with irinotecan, 131
Leukeran
  alkylating agent, 54
Leukotrienes
  radioprotective agent, 171–172
Levamisole
  selective toxicity, 273

Levobupivacaine
  selective toxicity, 273
Levofloxacin, 853
Levopropoxyphene
  selective toxicity, 269–270
Levothyroxine, 250–251
L-FMAU
  for hepatitis B virus, 327, 329–330
Licochalcone A
  antimalarial, 999
Lincomycin, 787–788
Lincosamides, 787–789
Linezolid, 792–793
  systemic antibacterial, 588–590
Lipophilic antifolates, 79
Lipoplex
  for nonviral gene delivery, 42–43
Liposomes
  with influenza antivirals, 382
  with vincristine, 142
Liver transplantation, 487
LJC-084, 709–710
LJC-11036, 709
Lobucivar
  for hepatitis B virus, 333
Loceryl, 891
Lomefloxacin, 853
Lomustine
  alkylating agent, 54, 64–65
Lopinavir
  anti-HIV drug, 476–477
  selective toxicity, 267
Loprox, 891
Loracarbef, 654
Loratidine
  selective toxicity, 275
Losoxantrone
  topoisomerase II inhibitor, 68, 72
L1210 screening, for anitumor agents, 109
Lumefantrine
  antimalarial, 978–979
  dosage forms, 935
Lung transplantation, 487
Lurotecan
  antitumor natural product, 133
LY-303366, 902
LY-338387
  picornavirus antiviral, 407
LY-343814
  picornavirus antiviral, 409–410

LY-353349
  picornavirus antiviral, 409
LY-353352
  picornavirus antiviral, 409
LY-466700
  hepatitis C antiviral, 416
Lymphadenopathy associated
  virus (LAV/HIV-2), 459
Lymphatic filariasis, 1035
Lysozyme, 224

M-8506, 953
M-14659, 643
Macrolide antibiotics. See also
  Erythromycins
  antimycobacterials, 855–858
  biosynthesis, 765–774
  4"-carbamate, 782
  clinical use and currently used
    drugs, 758–761
  drug resistance, 774–776
  future developments, 784–785
  nonbacterial activity, 782–784
  pharmacology and mode of
    action, 762–765
  recent developments, 776–782
  selective toxicity, 261–263
  side effects, toxicity, and con-
    traindications, 761–762
  structure-activity relation-
    ships, 765–774
Macrophages
  and adaptive immunity, 225,
    227
  and innate immunity, 224
Major histocompatibility com-
    plex proteins, 224–225, 235
  and organ transplant rejec-
    tion, 487
  and peptide immunogenicity,
    238–239
Malaria, 920–921, 1035. See also
    Antimalarial agents
  economic and political issues,
    932–933
  global incidence, 926–928
  immunity and prophylaxis,
    930–932
Malaria parasites, 921–924
  biochemistry and genetics,
    924–926
  vector control, 931–932
Malaridine, 979
Malarone, 935
  antimalarial, 976–978
(+)-Malekulatine
  antimalarial, 997

Maloprim, 934
L-Mannaric acid
  anti-HIV drug, 476–477
Mannosidostreptomycin, 840
Manzamine A
  antimalarial, 994
Mappia foetida, 130
Marboran
  smallpox antiviral agent, 435,
    437
Marcellomycin
  antitumor natural product,
    124, 126
Marimastat, 217, 218
Matrilysin, 13, 14
Matrix metalloproteinase inhibi-
    tors
  antiangiogenic agents,
    216–217
  for cancer, 45–46
  for cancer treatment, 37–38,
    45
Matrix metalloproteinases
  and angiogenesis, 216
  and carcinogenesis, 12–14
Maximum tolerated dose (MTD),
    154
May apple, drugs derived from,
    134
MC-02331, 703, 704
MC-02363, 703, 704
MC-02479, 703, 704
MC-510,027, 909, 910
Mdm2 gene
  and p53 gene, 29
Measles virus, 390
  antiviral agents, 390–396
Mebendazole
  antihelmintic, 1090–1091,
    1091
Mechlorethamine
  alkylating agent, 53, 54, 55
Mecillinam, 633–634
  in vitro antibacterial activity,
    635
Mefloquine
  antimalarial, 943–946
  dosage forms, 934
Meglumine
  for leishmaniasis, 1049–1051
Melarsoprol
  for African trypanosomiasis,
    1041–1043
Mellein
  hepatitis C antiviral, 422, 424
Melphalan
  alkylating agent, 53, 54

MEN1 gene, 27
Menoctone
  antimalarial, 976
Mentax, 891
Mephaquine, 934
Mepron, 935
2-Mercaptoethylamine (MEA)
  radioprotective agent, 152,
    155–160, 169–170, 172
2-Mercaptoethylguanidine
    (MEG)
  radioprotective agent,
    158–160, 161
2-Mercaptopropylguanidine
    (MPG)
  radioprotective agent, 158, 161
6-Mercaptopurine, 513
  antimetabolite, 75, 76, 81
  resistance to, 290
2-Mercaptothiolazine
  radioprotective agent, 160
Mercury compounds
  topical antibacterials, 552–553
Meropenem
  antimycobacterial application,
    858
  discovery, 668–669, 680–681
  pharmacokinetics, 623–624
  side effects and interactions,
    625
  use recommendations, 626
  in vitro activity, 692
Merozoites, 920, 922
Mesna
  radioprotective agent, 172
Messenger RNA, See mRNA
Metabolic inhibitors
  radioprotective agents,
    162–163
  radiosensitizers, 193–196
Metal ion complexes
  radioprotective agents, 163
  radiosensitizers, 196
Metalloelastase, 13
Metastasis, 12
Methenamine
  systemic antibacterial, 596
Methicillin, 631
  in vitro antibacterial activity,
    635
Methicillin-resistant staphylo-
    cocci, 617–618
Methisazone
  activity against orthopoxvi-
    ruses, 439
  smallpox antiviral agent, 435,
    437

Methotrexate
  antimetabolite, 75, 76, 77,
    78–79
  flavivirus antiviral, 428
  with infliximab, 234
  selective toxicity, 264
Methylating agents, 54
Methylation, See DNA hyper-
    methylation; DNA methyl-
    ation
1β-Methylcarbapenems,
    675–683
Methyl-CpG-binding domain-
    containing proteins, 8
Methylene blue
  antimalarial, 988
1-Methyl-3-nitro-1-nitrosourea
  alkylating agent, 64
2-Methylpiperazinedithiofor-
    mate
  radioprotective agent,
    160–161
Meticillin, 609
Metronidzole
  radiosensitizer, 178
Mezlocillin, 633
  indications, 625
  pharmacokinetics, 624
Micafungin, 888, 901, 903
Micatin, 891
Miconazole, 893–897
  formulations and indications,
    891
  selective toxicity, 268
Microtubules, 136
Midecamycin, 763
Miltefosine
  for leishmaniasis, 1051–1052
Minocin, 740
Minocycline, 740, 742
Miokamycin, 763
Mismatch repair genes, 8
  and cancer heritable syn-
    dromes, 32
Misonidazole, 84
  radiosensitizer, 177
Mithracin
  antitumor natural product,
    122–124
Mithramycin
  antitumor natural product,
    110, 122–124
Mitogen-activated protein ki-
    nases
  and carcinogenesis, 24

Mitomycin A
  antitumor natural product,
    120, 121
Mitomycin C
  antitumor natural product,
    120–122
  hypoxic selectivity, 84–85
  radiosensitizer, 182
Mitomycins
  antitumor natural product,
    110, 120–122
  tumor-activated prodrug, 87
Mitosis promoting factor, 193
Mitoxantrone, 85
  topoisomerase II inhibitor, 68,
    69, 70–71
Mitozolomide
  alkylating agent, 54, 65,
    66–67
Mixed disulfide hypothesis, of
    radioprotective agents,
    168–169
Mizoribine, 489
  clinical use for organ trans-
    plants, 521–522
  side effects, 522
MK-0787, 668–669
MK-826, 682, 684, 708
MM-4550, 670
  activity, 690
MM-13902, 670
  activity, 690
  in vitro activity, 692
MM-17880, 670
MM-22380, 670
MM-22381, 670, 675
MM-22382, 670
MM-22383, 670, 674
mob genes, 614
Monistat, 891
Monobactams, 664–668
  brief description, 609
  use recommendations, 626
Monoclonal antibodies. See also
    Humanized antibodies
  anti-HIV drug, 475–476
  clinical use for organ trans-
    plants, 489, 509–513,
    525–526
  pharmacokinetics, 510
  pharmacology, 510–513
  selective toxicity, 257
  side effects, 513
Montelukast sodium (MK-0476)
  selective toxicity, 252
Morphazinamide, 836
Morpholine antifungal agents, 903

Mort1, 17
Mosaic genes, 616–618
Mos protein kinase, 22
Moxalactam, 652–653
Mozenavir
  anti-HIV drug, 476–477
MRL-1237
  picornavirus antiviral, 413
mRNA
  bacterial, 738
Mucor, 883
Mucormycosis, 883
Multidrug resistance proteins,
    285–286
Multidrug resistance-related
    protein, 285–286
Murine leukemia virus
  vectors for cancer gene ther-
    apy, 41
Muromonab CD3, 490
Musettamycin
  antitumor natural product,
    124, 126
Mustards, See Nitrogen mus-
    tards
Mustargen
  alkylating agent, 54
Mutamycin
  antitumor natural product,
    120–122
Mutated in multiple advanced
    cancers (MMAC) tumor sup-
    pressor gene, 30
Mutation
  ionizing radiation, 153–154
  and tumorigenesis, 2
Mutation hot spots, 5
Myambutol, 810
Mycobacteria, 813–821. See also
    Antimycobacterial agents
  pathogenesis and epidemiol-
    ogy, 821–826
Mycobacteria other than tuber-
    culosis, 814, 824
Mycobacterium, 813–814
Mycobacterium abscessus, 815
  regimen for treatment, 867
Mycobacterium africanum, 814
Mycobacterium avium, 809, 814,
    815
  regimen for treatment, 866
Mycobacterium bovis, 813, 814,
    815, 820–821, 823
Mycobacterium chelonae, 815
Mycobacterium fortuitum, 815
  regimen for treatment, 867

*Mycobacterium haemophilum,* 815
*Mycobacterium kansasii,* 815
  regimen for treatment, 866
*Mycobacterium leprae,* 809, 825–826
  susceptibility testing, 827
*Mycobacterium malmoense,* 815
  regimen for treatment, 866
*Mycobacterium marinum,* 815
  regimen for treatment, 866
*Mycobacterium microti,* 814
*Mycobacterium scrofulaceum,* 815
  regimen for treatment, 866
*Mycobacterium simiae*
  regimen for treatment, 866
*Mycobacterium tuberculosis,* 809, 813–825
  susceptibility testing, 826–829
*Mycobacterium ulcerans,* 815
*Mycobacterium xenopi,* 815
Mycobactin M, 818
Mycobactin P, 818
Mycobactin S, 818
Mycobutin, 811
Mycolic acid, 816
Mycophenolate mofetil, 489, 490
  clinical use for organ transplants, 518–521
  for herpes, 321–322
  with leflunomide, 523
  side effects, 521
  structure-activity relationship, 520–521
Mycophenolic acid, 518
  flavivirus antiviral, 428
  for herpes virus, 321
Myc protein, 22, 25–26
Mylotarg, 228, 231
  tumor-activated prodrug, 83, 92
*N*-Myristoyl transferase inhibitors, 908–909, 910
M1 zinc finger peptides
  influenza antiviral agents, 387

Nafcillin, 631
  biliary excretion, 623
  pharmacokinetics, 624
Naftifine, 899–900
  formulations and indications, 891
Naftin, 891
Naked plasmid DNA
  for cancer therapy, 39

National Cancer Institute database
  antitumor natural products, 110
Natural killer cells, 225
Natural products. *See also* Antitumor natural products
  antimalarial agents, 993–999
Navelbine
  antitumor natural product, 142–143
Neem, 996
Neflinavir
  anti-HIV drug, 475–476
*Neisseria*
  penicillin-binding protein, 614, 616, 617
Nelfinavir
  selective toxicity, 267
Neohesperidin
  rotavirus antiviral, 431
Neomycin, 748
  selective toxicity, 260–261
Neonicotinoids
  selective toxicity, 278
Neoral, 491, 492
Neosporin, 748
Netilmicin, 748
Neutral red
  hepatitis C antiviral, 417
NeuTrexin
  antimetabolite, 76
Neutrophils
  and adaptive immunity, 225, 227
  and innate immunity, 224
Nevirapine
  anti-HIV drug, 472
  selective toxicity, 267
NF1 gene, 27
Nicotinamide
  radiosensitizer, 184
Niftifine
  selective toxicity, 268
Nifurtimox
  for American trypanosomiasis (Chagas' disease), 1046–1047
Nigericin
  antimalarial, 991
Nikkomycins, 906
Nimorazole
  radiosensitizer, 178–179
Nitidine
  dual topoisomerase I/II inhibitor, 73

Nitracine
  radiosensitizer, 179–180
Nitracine *N*-oxide, 84, 85–86
Nitriles
  radioprotective agent, 162–163
Nitrilotriacetate
  radioprotective agent, 163
Nitrobenzamide
  anti-HIV drug, 469, 470
9-Nitrocamptothecin
  antitumor natural product, 133
Nitrofurans
  systemic antibacterial, 594–596
Nitrogen mustards
  alkylating agents, 52, 53–57
  minor groove targeting, 57–59
Nitroimidazoles
  binding to hypoxic cells, 180–182
  radiosensitizers, 177
Nitroimidazopyrans
  antimycobacterial application, 858–859
*N*-Nitroso-*N*-phenylhydroxylamine
  radioprotective agent, 163
Nitrosureas
  alkylating agents, 54, 64–65
Nitroxides
  radioprotective agent, 171
Nivaquin, 934
Nizoral, 890, 891
NK 109
  dual topoisomerase I/II inhibitor, 73
NLP-1
  radiosensitizer, 179
NMSO3, 394
Nocardicin A, 660, 661, 663–664
Nocardicin B, 661, 663
Nocardicin C, 660, 661, 663
Nocardicin D, 661, 663
Nocardicin E, 661, 663
Nocardicin F, 661, 663
Nocardicin G, 663
Nocardicins
  biological activity, 663–664
  synthesis, 660–663
Nolatrexed
  antimetabolite, 76, 79
Nolvadex
  selective toxicity, 276
Nonnucleoside HIV reverse transcriptase inhibitors, 472–473

Nonreceptor tyrosine kinases, 24–25
Nonsteroidal anti-inflammatory drugs
and decreased cancer risk, 34
Nontuberculosis mycobacteria, 814, 824
Norfloxacin
antimalarial, 936, 983
Novantrone
topoisomerase II inhibitor, 68
Novrad
selective toxicity, 269–270
NP-27, 891
N-ras protein, 22, 24
NSC 356894, 490
Nucleoside HIV reverse transcriptase inhibitors, 470–472
Nydrazid, 810
Nystatin, 889, 893
formulations and indications, 891

OCA-983, 709
Ofloxacin, 853–855
Old tuberculin, 818–819
Oligonucleotide therapeutics
RSV antiviral agents, 396–397
Omaciclovir
for herpes virus, 310
Omeprazole
antimalarial, 989
with quinine, 936
Onchocerciasis, 1035
Oncogenes, 3, 21–26
targeting loss of function, 44–45
transcription factors as, 25–26
types of, 22
Oncovin
antitumor natural product, 142
OPTCOL method, 381
Organ transplantation, 486–487
recent developments, 523–526
trends in management of, 487–489
Organ transplant drugs
agents blocking nucleotide synthesis, 513–523
agents blocking T-cell function, 489–513
clinical use, 489
recent developments, 523–526

Organ transplant rejection, 486–487
t-cell's role in, 487–488
Ormaplatin
alkylating agent, 54, 59
Ornithine decarboxylase
genetic variability in expression, 7
Orthopoxviruses, 435–437
antiviral agents, 435–440
Oseltamivir, 361
for influenza, 368–372
selective toxicity, 266, 267
Osteoporosis, 273
Oxacephalosporins, 652–654
1-Ox-acephalothin, 652
Oxacephems, 652–654
brief description, 609
Oxacillin, 631
biliary excretion, 623
β-lactamase classification based on activity, 619, 620, 621
pharmacokinetics, 624
1-Oxadethiacephems, 652–654
Oxaliplatin
alkylating agent, 54, 59–60, 61
Oxamniquine
antihelmintic, 1091, 1095
Oxapenams
brief description, 609
Oxazolidinones, 792–793
antimycobacterial application, 859
systemic antibacterials, 587–594
Oxiconazole, 904
formulations and indications, 891
Oxidase tumor-activated prodrugs, 91
Oxidizing agents
topical antibacterials, 551–552
Oximonam, 666, 668
in vitro activity, 667
Oxistat, 891
Oxytetracycline, 740, 742

p53 gene, 27–29
and drug resistance, 287–289
gene therapy target, 44–45
role in cell cycle, 16
PA-505
antimycobacterial, 859
PA-824
antimycobacterial, 859

Pacific yew, paclitaxel from, 136
Paclitaxel
antitumor natural product, 110, 136–138
with RSR13, 190
selective toxicity of antibody conjugate, 258
with tirapazamine, 85
Palivizumab, 227, 395
Paludrine, 934
Paluther, 935
Pamaquine, 950
Pancreas transplantation, 487
1-year survival rate, 488
Panipenem, 677, 693
Papillomaviruses, 297, 301–302
antiviral agents, 337–339
Paracoccidioides brasiliensis, 883
Paracoccidioidomycosis, 883
Parainfluenza virus, 390
antiviral agents, 390–398
Paramyxovirus fusion proteins
antiviral agents, 390–396
Paraplatin
alkylating agent, 54
Parasitic infections, 1035, 1090.
See also Malaria
Paromomycin
for leishmaniasis, 1051
Parvoviruses, 304–305
antiviral agents, 341–342
Paser, 810
Pefloxacin
antimalarial, 983
Pegasys, 363
Pegylated agents, See Polyethylene glycol (PEG)
Penams, 609
β-lactamase inhibitors, 698–699
Penam sulfones
β-lactamase inhibitors, 698–699
Penciclovir
for hepatitis B virus, 330–331, 334
for herpes virus, 307–308
Penems, 609
β-lactamase inhibitors, 698–699
synthesis, 654–659
Penicillinase-resistant penicillins
use recommendations, 626
Penicillin-binding proteins, 613
evolution, 618

modification of native,
  616–618
and resistance to β-lactams,
  613–616
Penicillin dihydro F, 629
Penicillin F, 629
Penicillin G, 608, 609, 631. *See
  also* Benzylpenicillin
  pharmacokinetics, 624
  structure, 628, 629
  use recommendations, 626
  *in vitro* antibacterial activity,
  635, 667
Penicillin K, 629
Penicillin N, 629
Penicillin *iso*-N, 629
Penicillin O, 629
Penicillins
  antimycobacterial application,
  858
  brief description, 609
  discovery of, 628–637
  β-lactamase classification
    based on activity, 620, 621
  pharmacokinetics, 623
  selective toxicity, 253–254,
  260
  side effects and interactions,
  624–625
  synthesis, 629–630
Penicillin sulfoxide-cephalosporin conversion, 646–649
Penicillin T, 629
Penicillin V, 608, 609, 629, 631
Penicillin X, 629
*Penicillium,* 608
*Penicillium chrysogenum,* 609,
  628
*Penicillium notatum,* 628
Pentafuside
  anti-HIV drug, 468–469
Pentamidine
  for African trypanosomiasis,
  1037–1039
  for leishmaniasis, 1051–1052
Pentostatin
  antimetabolite, 76, 81
Pentoxifylline
  radiosensitizer, 184
Pepleomycins
  antitumor natural products,
  115
Peptidase tumor-activated pro-
  drugs, 87–88
Peptides
  RSV antiviral agents, 395–396
Peptidoglycan, 610–612

Peptidoglycan transglycosylases,
  610
Peramivir, 374–376
Pericytes, 216
Permetrexed
  antimetabolite, 76, 78
Pestivirus, 427
Phenethicillin, 631
Phenols
  topical antibacterials, 548–550
Phenytoin
  with bleomycin, 116
*Phialophora,* 883
Philadelphia chromosome, 19
Phleomycins
  antitumor natural products,
  115, 117
Phosphatase and tensin homo-
  logue (PTEN) tumor sup-
  pressor gene, 30
Phosphatase tumor-activated
  prodrugs, 87
Phosphodiesterase inhibitors
  selective toxicity, 277
Phosphoinositol 3-kinase, 24
Photodynamic therapy,
  174–175
Photofrin, 174–175
Phthalocyanines
  radiosensitizers, 175
Phthisis, 809–810
Phthivazid, 834
Picornaviral protein 2C inhibitors, 413
Picornaviruse protease inhibitors
  nonpeptidic, 409–411
  peptidic, 404–409
Picornaviruses, 398
  antiviral agents, 398–415
Picovir
  picornavirus antiviral,
  399–400
Picropodophyllotoxin
  antitumor natural product,
  134
Pimonidazole
  radiosensitizer, 179, 180
Piperacillin, 609, 633
  antimycobacterial application,
  858
  indications, 625
  pharmacokinetics, 624
  use recommendations, 626
  *in vitro* antibacterial activity,
  635

Piperaquine
  antimalarial, 990
Piperazine
  antihelmintic, 1091, 1095
Piritrexim
  antimetabolite, 76, 79
Pirodavir
  picornavirus antiviral, 400–401
Piroxantrone
  topoisomerase II inhibitor, 68,
  72
Pivampicillin, 632
Plaquenil, 934
Plasmid DNA-mediated gene
  therapy
  for cancer, 39–40
Plasmids
  and β-lactam resistance,
  614–616
Plasminogen activator inhibi-
  tor-1
  for angiogenesis control, 45
*Plasmodium berghei,* 923
*Plasmodium brasilianum,* 921
*Plasmodium cynomolgi,* 923
*Plasmodium falciparum,* 920
  biochemistry and genetics,
  924–926
  drug resistance, 928–930
  global distribution, 927
  life cycle, 921–924
*Plasmodium fragile,* 921
*Plasmodium knowlesi,* 923
*Plasmodium malariae,* 920
  life cycle, 921–924
*Plasmodium ovale,* 920
  life cycle, 921–924
*Plasmodium simiovale,* 921
*Plasmodium vinckei,* 923
*Plasmodium vivax,* 920
  biochemistry and genetics,
  924
  drug resistance, 928–930
  global distribution, 927
  life cycle, 921–924
*Plasmodium yoelii,* 923
Plasmotrim, 935
Platelet-derived growth factor,
  217–218
Platelet-derived growth factor
  inhibitors
  antiangiogenic agents,
  218–219
Platelets
  exosome secretion, 241

Platinum complex alkylating
    agents, 54, 59–61. *See also*
    Cisplatin
Platomycins
    antitumor natural products,
    117
Pleconaril
    picornavirus antiviral,
    399–400
Plicamycin
    antitumor natural product,
    110, 122–124
Pluracidomycin A, 670, 671
    *in vitro* activity, 692
Plymecillin, 609
*Pneumocystis carinii*, 460, 883
Pneumocystosis, 883
Podophyllin
    for papillomavirus, 338
Podophyllotoxins
    alkylating agents, 52
    antitumor natural product,
    134
*Podophyllum peltatum* (May apple), 134
Polycyclic carbapenems,
    683–688
Polyene antifungal agents,
    889–893, 890, 891
Polyethylene glycol (PEG)
    with interferon-α for hepatitis
    C, 363–364
Polymerase chain reaction
    with mycobacteria, 826
Polymeric radioprotective
    agents, 166–167
Polyomaviruses, 302–303
    antiviral agents, 339–340
Polyoxins, 906
Polyoxometalates
    broad spectrum antiviral
    agents, 435
Polyplex
    for nonviral gene delivery, 42
Polyvinyl alcohol sulfate
    anti-HIV drug, 466
Porfimer sodium, 174–175
Porfiromycin
    tumor-activated prodrug, 83,
    85, 86
Porphiromycin
    antitumor natural product,
    120
Posaconazole, 898–899
Potentially lethal damage repair,
    191–192
Pott's disease, 809

Poxviruses, 305
    antiviral agents, 340–341
    dendritic cell transduction by,
    242–243
Pradimicins, 908
Praziquantel
    antihelmintic, 1091,
    1093–1094
Preincubation effect, radiosensitizers, 180
Primaquine
    antimalarial, 949–954
    dosage forms, 934
Primaxin, 693
PRO440
    anti-HIV drug, 468
Probenecid
    with cephalosporins, 623
Procaine
    radioprotective agent, 166
Procaspase-8
    and apoptosis, 17
Procaspase-9
    and apoptosis, 18–19
Proflavine, 67
Prograf, 495
Progression, in carcinogenesis,
    3–5
Progressive multifocal leukoencephalopathy, 302–303
Proguanil
    antimalarial, 936, 954–956
    with dapsone, 959
    dosage forms, 934
Promotion, in carcinogenesis,
    3–5
Propicillin, 609, 631
Propoxyphene
    selective toxicity, 269–270
Prostacyclin
    for cancer treatment, 38
Prostaglandin H synthase 1, *See*
    COX-1
Prostaglandin H synthase 2, *See*
    COX-2
Prostaglandins
    broad spectrum antiviral
    agents, 432–435
Protein swapping, 259
Proteoglycans, 12
Proteolysis
    inhibitors of, 216–217
Prothionamide, 836
    efficacy and dosage, 810, 863
Protooncogenes, 45
Provirus, 463

PS-5, 670, 671, 675
    *in vitro* activity, 692
PS-6, 670, 671
PS-15
    antimalarial, 960
PS-26
    antimalarial, 960
PS-33
    antimalarial, 960
P388 screening, for anitumor
    agents, 109
*Pseudoallescheria boydii*, 883
Pseudoallescheriasis, 883
*Pseudomonas aeruginosa*
    intrinsic resistance to β-lactams, 621–623
    β-lactams active against,
    632–633
Psilocybine, 165
Purine analog antimetabolites,
    76, 81–82
Purine derivatives
    HBV virus antiviral agents,
    330–333
    RSV antiviral agents, 389
Purinethol
    antimetabolite, 76
Purpurins
    radiosensitizers, 175
Pyrantel
    antihelmintic, 1091,
    1094–1095
Pyrazinamide, 836–837
    drug interactions, 864
    efficacy and dosage, 810, 863
    side effects, 837
Pyrimethamine
    antimalarial, 936, 956–957
    with dapsone, 959
    dosage forms, 934
    selective toxicity, 254, 264
    with sulfadoxine, 958–959
    useful lifetime before resistance develops, 930
Pyrimidine analog antimetabolites, 76, 79–81
Pyrimidine derivatives
    HBV virus antiviral agents,
    326–330
    influenza antiviral agents, 389
    RSV antiviral agents, 389
Pyrimidines
    halogenated, as radiosensitizers, 194–195
Pyronaridine
    antimalarial, 979–981
    dosage forms, 935

Pyrrolidine-based influenza inhibitors, 376–377
Pyrromycin
  antitumor natural product, 124, 126, 130

Qinghaosu (artemisinin), 961
Quinacrine, 979
Quinamm, 934
Quinidine
  antimalarial, 933–938
  dosage forms, 934
Quinimax, 933, 936–937
Quinine
  antimalarial, 933–938
  dosage forms, 934
Quinocide, 950
Quinolones
  selective toxicity, 263
  systemic antibacterials, 582–587
Quinones
  radiosensitizers, 177

R-95867, 709, 710
R-170591, 393–394
Radiation damage, 153–154
Radiation therapy. See also Radioprotective agents
  radioprotective agents in, 172–173
Radioprotective agents, 151–153
  antiradiation testing, 154
  biochemical shock, 169
  DNA breakdown control, 169–170
  free-radical process inhibition, 168
  heterocyclic compounds, 163–165
  hydroxyl-containing compounds, 163
  metabolic effects, 170–172
  metabolic inhibitors, 162–163
  metabolites and naturally occurring compounds, 166
  metal ion containing agents, 163
  miscellaneous substances, 166–167
  mixed disulfide hypothesis, 168–169
  physiologically active substances, 165–166
  polymeric substances, 166–167

protection by anoxia/hypoxia, 154, 167–168
  sulfur-containing compounds, 160–162
  thiols and thiol derivatives, 152, 155–160
  use in radiotherapy and cancer chemotherapy, 172–173
Radiosensitizers, 173
  alteration of energy absorption, 173–175
  alteration of oxygen delivery, 183–190
  alteration of primary radiolytic products, 175–176
  bacterial sensitizers, 197
  for boron neutron capture therapy, 173–174
  depletion of endogenous protectors, 190–191
  electron-affinic drugs in cancer therapy, 180–183
  and hypoxia-activated prodrugs, 83
  inhibition of DNA repair, 191–193
  metal ion complexes, 196
  perturbation of cellular metabolism, 193–196
  for photodynamic therapy, 174–175
  reaction with DNA radicals, 176–180
  thiols, 196–197
  unknown mechanism agents, 196–197
Raf protein kinase, 22, 24
Raloxifene
  selective toxicity, 276
Raltitrexed
  antimetabolite, 76, 77–78
RANTES ligand
  and CCR5, 467
Rapamune, 500, 503
Rapamycin, 490
  clinical use for organ transplants, 500–504
RASI-1, 13
Ras signaling pathways, 23–24
Ravuconazole, 898–899
RB-6145
  radiosensitizer, 182
RB-90740
  radiosensitizer, 181
Rb gene, 27
  gene therapy target, 44–45

RD2-4039
  hepatitis C antiviral, 421, 423
RD3-0028, 394–395
RD3-4082
  hepatitis C antiviral, 421, 423
RD4-6205
  hepatitis C antiviral, 422, 423
Reactive oxygen species, 33
Receptor interacting protein, 17
Receptor tyrosine kinases
  and angiogenesis, 217–219
  and carcinogenesis, 11–12, 22–23
Recrudescence, 922–923
Reduced folate carrier, 285
Reductase tumor-activated prodrugs, 91–92
Relenza, 364
Remicade, 229, 233
(-)-Repandine
  antimalarial, 997
Reserpine
  radioprotective agent, 165
Resiquimod
  for herpes, 325
Resochin, 934
RespiGam, 361, 395
Respiratory syncytial virus
  antiviral agents, 390–398
Restoration (chemical radioprotection), 153
Restriction fragment length polymorphisms
  and carcinogenesis, 6
Retinoblastoma, 27, 28
Retroviruses. See also HIV virus
  defined, 459
  vectors for cancer gene therapy, 40–42
RFI-641, 390–393
Rhinacanthins
  influenza antiviral agents, 389
Rhizopus, 883
Rhodacyanine dye analog, cationic
  selective toxicity, 258
Rhodococcus
  penicillin-binding protein, 615
Rhodomycins
  antitumor natural products, 124, 126
Riamet, 935
Ribavirin
  activity against orthopoxviruses, 439, 440
  flavivirus antiviral, 428–429
  for herpes virus, 321

Ribavirin (*Continued*)
  with interferon-α for hepatitis C, 363–364
  for poxviruses, 341
Riboflavin
  radiosensitizer, 196
Ribozymes
  influenza antiviral agents, 385–386
Richmond-Sykes classification, of β-lactamases, 619–620
Rifabutin
  antimycobacterial application, 811, 847, 850, 852
  drug interactions, 864
  efficacy and dosage, 811, 863
Rifadin, 811
Rifalazil, 847, 850, 852–853
Rifampicin
  antimycobacterial application, 811, 850–852
  drug interactions, 862, 864
  efficacy and dosage, 811, 863
Rifampin, 850
  drug interactions, 267
  efficacy and dosage, 811
  selective toxicity, 266
Rifamycin B, 847, 848
Rifamycin O, 847
Rifamycin S, 847
Rifamycins
  antimycobacterial application, 811, 846–853
Rifamycin SV, 847, 848
  efficacy and dosage, 811
Rifamycin T9, 847, 850, 853
Rifapentine
  antimycobacterial application, 811, 847, 850, 852
  efficacy and dosage, 811
Rifazalil
  antimycobacterial application, 859
Rifocin, 811
Rimactane, 811
Rimantadine, 361
  selective toxicity, 266, 267
Ringworm/tinea, 883
Risedronate
  selective toxicity, 273
Ritipenem acoxil, 659, 660
Ritonavir
  anti-HIV drug, 475–476
  selective toxicity, 267
Rituxan, 228, 230
Rituximab, 228, 230–231
RNA polymerases, 295

RNA viruses, 360–361. *See also* Antiviral agents, RNA
Ro-09-3143, 907
Ro-19-8928, 678, 680
Ro-42-1611
  antimalarial, 984–985
Ro-48-1220, 710, 711
Ro-63-9141, 703, 704
Robustaflavone
  influenza antiviral agent, 389
(-)-Roemrefidine
  antimalarial, 997, 998
Rofecoxib
  selective toxicity, 274
Rokitamycin, 763
Ros growth factor receptor, 22
Rotaviruses, 430
  antiviral agents, 430–431
Roxithromycin, 763
RPR103611
  anti-HIV drug, 469
rRNA
  bacterial, 738
RS 61443, 490, 518, 519
RSHZ19
  RSV antiviral agent, 395
RSR-13
  radiosensitizer, 184–190
RSU 1069
  radiosensitizer, 181–182
RU 44790, 668, 669
Rubella virus, 431
  antiviral agents, 431–432
Rubex
  antitumor natural product, 128
Rubidomycin
  antitumor natural product, 126–128
Rufigallol
  antimalarial, 987
RWJ-54428, 703, 704
RWJ-270201, 364, 374–376
RWJ-333441, 703, 704

S-2242
  for adenoviruses, 340
  for herpes virus, 317–318, 321
  for poxviruses, 341
S-4661, 708
Saintopin
  dual topoisomerase I/II inhibitor, 73
Sandimmune, 491, 492
Sanfetrinem, 684, 685, 686, 693–694
  *in vitro* activity, 692

Sanitizers, 539
Saquinavir
  anti-HIV drug, 475–476
  selective toxicity, 267
Sarachine
  antimalarial, 999
Satraplatin
  alkylating agent, 54, 60
SB-206999, 699
SB-209763
  RSV antiviral agent, 395
SB-408075
  tumor-activated prodrug, 83, 94
SCH 29482, 659, 660, 698
SCH 34343, 660
SCH 48973
  picornavirus antiviral, 400–401
SCH 68631
  hepatitis C antiviral, 422, 424
SCH 351633
  hepatitis C antiviral, 422, 424
Schistosomiasis, 1035
SDZ 880–061
  picornavirus antiviral, 401–402
Selective estrogen receptor modulators
  selective toxicity, 276–277
Selective toxicity, 250–252
  aminoglycosides, 260–261
  antibacterial agents, 539–543
  antifungal drugs, 254, 266–268, 886
  antihistamines, 275–276
  antiinfectives, 259–263
  antimycobacterial drugs, 266
  antiparasite drugs, 254, 268
  antiviral drugs, 254, 266
  bisphosphonates, 273–274
  cancer chemotherapy examples, 257–259
  cationic rhodacyanine dye analog, 258
  cisplatin/carboplatin, 257–258
  comparative biochemistry, 253–254
  comparative cytology, 254
  comparative distribution, 252–253
  comparative stereochemistry, 254–257
  COX inhibitors, 274–275
  diphtheria toxin, 259
  and dopaminergic receptors, 259

and drug chirality, 268–273
estrogen receptor modulators, 276–277
imatinib, 257
insect growth regulators, 278
insecticides, 277–279
β-lactams, 254, 260
macrolides, 261–263
monoclonal antibodies, 257
neonicotinoids, 278
paclitaxel antibody conjugate, 258
phosphodiesterase inhibitors, 277
quinolones/fluoroquinolones, 263
spinosyns, 278
sulfonamides/sulfanilamides, 263–264
target organism examples, 264–268
tetracyclines, 261
tricyclic thiophene, 258–259
Serial analysis of gene expression (SAGE)
cancer studies using, 36–37
Serine kinases, 24
Seromycin, 811
Serotonin
radioprotective agent, 165, 167
Serum complement, 224
SGN-15
tumor-activated prodrug, 83, 92
Shingles, 298
Siamycin
anti-HIV drug, 469
Sickle trait
and malaria, 931
Signal peptidase enzymes, 712
Silatecans
antitumor natural products, 133
Sildenafil
selective toxicity, 277
Silver salts
topical antibacterials, 553
Simulcet, 510, 513
Sindbis virus, 431
Single nucleotide polymorphisms (SNPs)
and carcinogenesis, 6
Sirolimus, 489, 490
clinical use for organ transplants, 500–504
pharmacokinetics, 500

pharmacology, 500–503
side effects, 503–504
structure-activity relationship, 503
Sis growth factor, 22
Skin-derived dendritic (Langerhans) cells, 224
SM 7338, 668–669
Smallpox, 435–440
SN-24771
radiosensitizer, 196
SNPs, See Single nucleotide polymorphisms
SO3
antimalarial, 957
Sodium cysteinethiosulfate
radioprotective agent, 161
Sodium 2,3-dimercaptopropane sulfonate
radioprotective agent, 157
Sodium selenate
radioprotective agent, 162
Sodium stibogluconate
for leishmaniasis, 1051
Solasulfone, 830
Soluble intercellular adhesion molecule-1
picornavirus antiviral agent, 403–404
Sordarins, 907–908
Southeast Asian ovalocytosis
and malaria, 931
Sparfloxacin, 855
Spectazole, 891
Spergualin, 504
Spinosyns
selective toxicity, 278
Spiramycin, 763
Sporanox, 890, 891
Sporothrix schenkii, 883
Sporotrichosis, 883
SQ 26180, 664, 665
SQ 83360, 668
SR-4233
radiosensitizer, 181
SR-4482
radiosensitizer, 181
SR-3727A
for herpes virus, 316
Src-homology domains, 23, 25
Src tyrosine kinase, 22
SRR-SB3
anti-HIV drug, 469, 470
Stachyflin
influenza antiviral agent, 379–380

Staphylococcus
bridge in peptidoglycans, 612
penicillin-binding protein, 615
Staphylococcus aureus
antibiotics active against, 631–632
methicillin-resistant, 617–618
Staurosporine
radiosensitizer, 193
Stavudine
selective toxicity, 267
Stem cells
organ transplant rejection applications, 523–524
Sterilization, 539
STI571, 284
Streptococcus
bridge in peptidoglycans, 612
penicillin-binding protein, 615, 617
Streptococcus pneumonia
β-lactam resistance, 613, 712
Streptogramins, 785–787
Streptomyces, 111, 609
Streptomyces argillaceus, 122, 123
Streptomyces caespitosus, 120
Streptomyces peucetius, 124
Streptomyces plicatus, 122
Streptomyces tanashiensis, 123
Streptomyces verticillus, 115
Streptomycin, 748
antimycobacterial application, 811, 840–842
drug interactions, 864
efficacy and dosage, 811, 863
mechanism of action, 841–842
selective toxicity, 260–261
structure-activity relationship, 840–841
Streptozotocin
alkylating agent, 54, 65
Stress proteins, 819
Stromelysins, 13
SU5416
antiangiogenic agent, 218–219
SU6668
antiangiogenic agent, 218–219
Suicide gene therapy, 90
for cancer, 39, 44
Sulbactam, 697
amoxycillin MIC, 695
antimycobacterial application, 858
β-lactamase inhibitory activity, 695
use recommendations, 626

Sulbenicillin, 632
Sulconazole, 904
    formulations and indications,
        891
Sulfadoxine, 957
    with pyrimethamine, 958–959
    useful lifetime before resis-
        tance develops, 930
Sulfalene, 957
    dosage forms, 935
Sulfamethoxazole, 957
    dosage forms, 934
Sulfanilamide
    selective toxicity, 263–264
Sulfazecin, 609, 664, 665
Sulfisoxazole, 957
    dosage forms, 934
Sulfonamides
    antimalarial, 957–958
    selective toxicity, 263–264
    systemic antibacterials,
        557–582
Sulfones
    antimalarial, 957–958
    antimycobacterial, 829–830
    systemic antibacterials,
        557–582
Sulfonylureas
    selective toxicity, 264
Sulfur-containing compounds
    radioprotective agents, 152,
        155–162
Sultamicillin, 697
SUN 5555, 659, 660
Sunamed, 763, 935
Suramin
    for African trypanosomiasis,
        1037–1039
    antihelmintic, 1091,
        1095–1096
Surfactants
    topical antibacterials, 555–557
Synagis, 227
    RSV antiviral agent, 395
Synercid, 793
Synergis, 361
Synkovit
    radiosensitizer, 196
Synthetic antibacterials
    systemic, 557–594
    topical, 539–557

T-705
    influenza antiviral agent, 390
T-5575, 637
T-5578, 637

T-157602
    for herpes, 323
Tacrolimus, 489, 490. See also
    FK 506
    clinical use for organ trans-
        plants, 494–500
    with leflunomide, 523
Tafenoquine, 953
Tagamet
    selective toxicity, 275
TAK-779
    anti-HIV drug, 468
Talampicillin, 632
Tallimustine
    alkylating agent, 57–58
Tallysomycins
    antitumor natural products,
        115, 117
Talompicillin, 609
Tamiflu, 364
Tamoxifen
    selective toxicity, 276
Targeted mustards, 57
TAS 103
    dual topoisomerase I/II inhibi-
        tor, 68, 73, 74
Taxol
    antitumor natural product,
        110, 136–138
Taxotere
    antimalarial, 990
    antitumor natural product,
        138–139
Taxus diterpenes
    antitumor natural product,
        110, 136–139
Tazobactam, 697–698
    amoxycillin MIC, 695
    antimycobacterial application,
        858
    β-lactamase inhibitory activ-
        ity, 695
    use recommendations, 626
T-cell receptors, 224
T-cells
    activation by dendritic cells,
        235–236
    and adaptive immunity,
        224–226
    exosome secretion, 241
    HIV infection, 458–464
    organ transplant drugs block-
        ing function, 489–513
    role in organ rejection,
        487–488
Tebrazid, 810
Tebuquine, 982

Telithromycin, 763
Telomerase, 11
Telomeres, 11
Teloxantrone
    topoisomerase II inhibitor, 72
Temacrazine
    anti-HIV drug, 474
Temocillin, 609, 636
    in vitro antibacterial activity,
        635
Temodar
    alkylating agent, 54
Temozolomide
    alkylating agent, 54, 65, 67
Tempol
    radioprotective agent,
        163–164, 171
(+)-Temuconine
    antimalarial, 996
Teniposide
    antitumor natural product,
        110, 135–136
Tenofovir
    anti-HIV drug, 471, 472
Terazole, 891
Terbinafine, 899–900
    formulations and indications,
        890, 891
    selective toxicity, 268
Terconazole, 904
    formulations and indications,
        891
    selective toxicity, 268
Tetracyclines
    antimalarials, 935, 982–983
    biosynthesis, 742–743
    clinical use and currently used
        drugs, 739–741
    drug resistance, 743–746
    pharmacology and mode of
        action, 741–742
    radioprotective agents, 166
    recent developments, 746–747
    selective toxicity, 261
    side effects, toxicity, and con-
        traindications, 741
Tetraoxane
    antimalarial, 986
Tetraplatin
    alkylating agent, 54, 59, 61
T-helper lymphocytes, 224–225
Therapeutic gain, radiosensitiz-
    ers, 173
Thiabendazole
    antihelmintic, 1090–1091,
        1091

Thiamine diphosphate
   radiosensitizer, 196
Thiazide diuretics
   selective toxicity, 264
Thienamycin, 609, 669, 670
   activity, 688–693
   synthesis, 672–674
Thioacetazone, 832
   efficacy and dosage, 810, 863
   side effects, 832–833
Thiocarbamate antifungal
      agents, 903
Thiocarlide
   efficacy and dosage, 810
Thioglycolic acid
   radioprotective agent, 161
6-Thioguanine
   antimetabolite, 76, 81
Thiolactomycins
   antimycobacterial application,
      859–860
Thiols
   radioprotective agents, 152,
      155–160
   radiosensitizers, 192–193,
      196–197
Thiourea
   radioprotective agent, 152,
      170
Threonine kinases, 24
Thrush, 882
Thymidylate synthase inhibitors
   antimetabolites, 76–77
Thymitaq
   antimetabolite, 76, 79
Ticarcillin, 632, 633
   antimycobacterial application,
      858
   indications, 625
   pharmacokinetics, 624
   use recommendations, 626
   in vitro antibacterial activity,
      635
Tigemonam, 666
   in vitro activity, 667
Timentin, 695
   use recommendations, 626
Timoxicillin, 633
Tinactin, 891
Ting, 891
Tioconazole, 904
   formulations and indications,
      891
Tipranavir
   anti-HIV drug, 476–477
Tirapazamine
   radiosensitizer, 181

tumor-activated prodrug, 83,
      84, 85, 86
Tirazone
   tumor-activated prodrug, 83
Tissue inhibitor of matrix metal-
      loproteinases-1, 12
Tissue transplantation, 486
T-lymphocytes, See T-cells
TNFR-associated death domain,
      17
TOC-39, 703, 705
TOC-50, 703, 705
Tolnaftate, 903
   formulations and indications,
      891
   selective toxicity, 268
Tomudex
   antimetabolite, 76, 77–78
Topoisomerase I, 67–68
Topoisomerase II, 67–68, 135
Topoisomerase I inhibitors
   camptothecins, 131, 132
   dual I/II inhibitors, 72–75
Topoisomerase II inhibitors
   dual I/II inhibitors, 72–75
   etoposide, 135
   resistance to, 284
   synthetic, 68–72
   tenoposide, 136
Topotecan
   antitumor natural product,
      110, 131
Toxicity. See also Selective toxic-
      ity; specific drugs and
      classes of drugs
   aminoglycosides, 748–749
tra genes, 614
Transforming growth factor β,
      226
   and carcinogenesis, 30–32
Transglycosylases, 610
   inhibition of bacterial by
      β-lactams, 611–613
Transpeptidases, 610
   inhibition of bacterial by
      β-lactams, 611–613
Transplant rejection, 486–487
Trastuzumab, 228, 230
   selective toxicity, 257
Trecator, 810
Trevintix, 810
Triacetyloleandomycin, 763
Triazenes
   alkylating agents, 54, 65–67
Triciribine
   for herpes, 322–323

Tricyclic thiophene
   selective toxicity, 258–259
Trifluorperazine
   arenavirus antiviral, 430
Trimethoprim, 956
   dosage forms, 934
   selective toxicity, 264
Trimetrexate
   antimetabolite, 76, 79
Trinems, 609, 683–688
   anti-gram-positive, 703–705
   biological properties, 688–694
   orally active, 709–710
Trk growth factor, 22
tRNA
   bacterial, 738
Tropical Disease Research List
      (WHO), 1035
Trovafloxacin
   antimalarial, 983–984
Tryptanthrin
   antimycobacterial application,
      859
T-tropic HIV, 468
Tuberculin, 818–819
Tuberculosis, 809–813, 1035
   pathogenesis and epidemiol-
      ogy, 821–825
   recent developments and
      present status of chemo-
      therapy, 860–864
Tubulin
   antitumor natural products
      interfering with, 110,
      136–143
α-Tubulin, 136
β-Tubulin, 136
Tubulosine
   antimalarial, 997, 998
Tumor-activated prodrugs, 44,
      82–83
   for antibody-directed enzyme-
      prodrug therapy (ADEPT),
      83, 87–90
   antibody-toxin conjugates
      (armed antibodies), 83,
      92–94
   for gene-directed enzyme-pro-
      drug therapy (GDEPT), 83,
      90–92
   hypoxia-activated prodrugs,
      83–87
Tumorigenesis, 2–3
   and genetic variability, 5–7
Tumor-infiltrating lymphocytes,
      226
Tumor lysates, 239–240

Tumor necrosis factor-$\alpha$
  and mycobacterial infection, 822
  radioprotective agent, 171
  radiosensitizer, 194
Tumor necrosis factors, 17
Tumor promoter TPA-responsive element, 25
Tumors, 225–226
Tumor suppressor genes, 3, 26–32
  methylation and drug resistance, 289
  targeting loss of function, 44–45

U-100480
  antimycobacterial, 859
UK-109, 496, 897
Ultraviolet radiation
  and carcinogenesis, 4
Unasyn
  use recommendations, 626
Undecylenate, 906
  formulations and indications, 891
Unithiol
  radioprotective agent, 157
Urethane
  radioprotective agent, 165

V-002, 902
Vaccinia immune globulin
  for smallpox, 435
Vaccinia virus
  vector for cancer gene therapy, 41
Vagistat, 891
Val-2-AM (dactinomycin), 111
Valciclovir
  for herpes virus, 309, 310
Valrubicin
  antitumor natural product, 110, 124, 128–130
Vancomycin
  with aztreonam, 625
  in vivo efficacy in systemic murine infections, 685
Vardenafil
  selective toxicity, 277
Varicella zoster virus, 298
Vascular endothelial growth factor, 217–218
  and angiogenesis, 20–21
Vascular endothelial growth factor inhibitors
  antiangiogenic agents, 218–219

Veins, 216
Velban
  antitumor natural product, 140–142
Verapamil
  selective toxicity, 270–271
Verazide, 834
Vfend, 890, 897–898
Viagra
  selective toxicity, 277
Vibramycin, 740, 935
Victomycins
  antitumor natural products, 117
Vidarabine
  for herpes virus, 309, 315
  for papillomavirus, 338
Vinblastine
  antitumor natural product, 110, 140–142
Vinca alkaloids, 52
  antitumor natural product, 110, 139–143
Vincasar PFS
  antitumor natural product, 142
Vincristine
  antitumor natural product, 110, 142
  with dactinomycin, 111
Vindesine
  antitumor natural product, 142
Vinflunine
  antitumor natural product, 142–143
Vinorelbine
  antitumor natural product, 110, 142–143
Viomycin
  antimycobacterial application, 811, 844–846
  efficacy and dosage, 811, 863
  side effects, 845–846
Vioxx
  selective toxicity, 274
Viruses, 294. See also Antiviral agents, DNA; Antiviral agents, RNA
  classes of, 294–305
  progency production, 295–296
Vismione H
  antimalarial, 997, 998
Vitamin A family
  antimalarial, 989
  and cancer risk, 33

Vitamin B$_6$ family
  radioprotective derivatives, 166
Vitamin replacement therapy, 251
Vitamins
  and cancer prevention, 33
Von Hippel-Lindau gene, 8, 10
Voriconazole, 897–898
  formulations and indications, 890
VP-14637, 393
V-sis, 22
VX-497
  hepatitis C virus antiviral agents, 426–427
  for herpes, 322

Warfarin
  selective toxicity, 271
Warts, 301
Wax D, 817
WAY-150138
  for herpes virus, 314–315
WAY-154641, 390–393
Wellvone, 935
West Nile virus, 427, 428
WIN 5723, 853
WIN 51711
  picornavirus antiviral, 400
WIN 54954
  picornavirus antiviral, 399
WIN 59075
  radiosensitizer, 181
WIN 61893
  picornavirus antiviral, 399
WMC-26
  dual topoisomerase I/II inhibitor, 75
WR1065
  radioprotective agent, 167–168
WR2721
  radioprotective agent, 157, 167, 172
WR33,063
  antimalarial, 946
WR99,210
  antimalarial, 960
WR122,455
  antimalarial, 946
WR148,999
  antimalarial, 986
WR182,393
  antimalarial, 954
WR225,448
  antimalarial, 952–953

WR238,605
   antimalarial, 952–953
WR242,511
   antimalarial, 952–953
WR250,417
   antimalarial, 960
WR268,954
   antimalarial, 942
WT-1 gene, 27

Xanthates
   radioprotective agent,
      161–162
Xenotransplantation, 489
   future developments, 526
Xeroderma pigmentosum, 32
XR-5000
   dual topoisomerase I/II inhibi-
      tor, 68, 73
XR-5944
   dual topoisomerase I/II inhibi-
      tor, 75

X-rays. *See also* Radioprotective
      agents
   antiradiation testing, 154

Yellow fever virus, 427
Yingzhaosu A
   antimalarial, 984–985
Yingzhaosu C
   antimalarial, 984–985

Zafirlukast
   selective toxicity, 252
Zalcitabine
   anti-HIV drug, 471, 472
   selective toxicity, 267
Zanamivir, 361
   for influenza, 365–368
   selective toxicity, 266, 267
Zanosar
   alkylating agent, 54
ZD-0473
   alkylating agent, 54, 60, 61
ZD-1839, 12

ZD-2767
   tumor-activated prodrug, 83, 88
ZD-4190
   antiangiogenic agent, 218–219
Zenepax, 510
Zidovudine
   anti-HIV drug, 471–472
   selective toxicity, 267
Zileuton
   selective toxicity, 252
Zinamide, 810
Zinc aspartate
   radioprotective agent, 163
Zithromax, 763, 935
Zorbamycins
   antitumor natural products,
      115, 117
Zorbonamycins
   antitumor natural products,
      117
Zosyn
   use recommendations, 626
Zymosterol, 887